Biographical Encyclopedia
of Scientists

Third Edition

Biographical Encyclopedia
of Scientists

Third Edition

Edited by
John Daintith

CRC Press
Taylor & Francis Group
Boca Raton   London   New York

CRC Press is an imprint of the
Taylor & Francis Group, an **informa** business

A TAYLOR & FRANCIS BOOK

CRC Press
Taylor & Francis Group
6000 Broken Sound Parkway NW, Suite 300
Boca Raton, FL 33487-2742

© 2009 by Market House Books Ltd.
CRC Press is an imprint of Taylor & Francis Group, an Informa business

International Standard Book Number-13: 978-1-4200-7271-6 (Hardcover)

Library of Congress Cataloging-in-Publication Data

Biographical encyclopedia of scientists / editor, John Daintith. -- 3rd ed.
    p. cm.
  Includes bibliographical references and index.
  ISBN 978-1-4200-7271-6 (alk. paper)
  1. Scientists--Biography--Dictionaries. I. Daintith, John. II. Title.
  Q141.B53 2008

509.2'2--dc22
[B]   2008019577

Visit the Taylor & Francis Web site at
http://www.taylorandfrancis.com

and the CRC Press Web site at
http://www.crcpress.com

Preface

This reference book presents biographical entries on important scientists – from the earliest times to the present day. It is an updated version of *A Biographical Encyclopedia of Scientists*, which was published by the Institute of Physics in 1993. This in turn was based o n earlier editions.

In this new edition, we have updated all the entries and added over 200 new biographies. We have also included a simple pronunciation guide in entries where this is appropriate. The respelling system used is shown on page vi. Another added feature of this edition is the inclusion of a selection of quotations by or about certain people. There are also cross references to other entries in the book – these are indicated by SMALL CAPITALS. In the appendix to the book there is a chronology of key events in science, a list of useful web sites, and a subject index.

In compiling such a book there are several difficult decisions to make on the selection of material. The most general one is that of the scope of the book – what areas of knowledge it should cover. This work concentrates on what might be called the 'traditional' pure sciences – physics, chemistry, biology, astronomy, and the earth sciences. It also covers medicine and mathematics, and includes a selection of people who have made important contributions to engineering and technology. A few of the entries cover workers in such fields as anthropology and psychology, and a small number of philosophers have also been allowed in.

Our intention has been to produce a book as much about science itself as about scientists, and this has governed our approach to selecting information: the entries contain basic biographical data – place and date of birth, posts held, etc. – but do not give exhaustive personal details about the subject's family, prizes, honorary degrees, etc. Most of the space has been given to their main scientific achievements and the nature and importance of these achievements. This has not always been easy; in particular, it has not always been possible to explain in relatively simple terms work in the higher reaches of abstract mathematics or modern theoretical physics.

Perhaps the most difficult problem was compiling the entry list. We have attempted to include people who have produced major advances in theory or have made influential or well-known discoveries. A particular difficulty has been the selection of contemporary scientists, in view of the fact that of all scientists who have ever lived, the vast majority are still alive. In this we have been guided by lists of prizes and awards made by scientific societies. It should also be said that the compilers and editors have used their own judgment in choosing what is important or useful. In some cases entries have been added simply because we found them interesting. We hope that the reader will find all the entries useful and share our interest in them.

JD 2008

Pronunciation Guide

A guide to pronunciation is given for foreign names and names of foreign origin; it appears in brackets after the first mention of the name in the main text of the article.

a *as in* bat
ah *as in* palm (pahm)
air *as in* dare (dair), pear (pair)
ar *as in* tar
aw *as in* jaw, ball (bawl)
ay *as in* gray, ale (ayl)
ch *as in* chin
e *as in* red
ee *as in* see, me (mee)
eer *as in* ear (eer)
er *as in* fern, layer
f *as in* fat, phase (fayz)
g *as in* gag
i *as in* pit
I *as in* mile (mIl), by (bI)
j *as in* jaw, age (ayj), gem (jem)
k *as in* keep, cactus (**kak**-tus), quite (kwIt)
ks *as in* ox (oks)
ng *as in* hang, rank (rangk)
o *as in* pot

oh *as in* home (hohm), post (pohst)
oi *as in* boil, toy (toi)
oo *as in* food, fluke (flook)
or *as in* organ, quarter (**kwor**-ter)
ow *as in* powder, loud (lowd)
s *as in* skin, cell (sel)
sh *as in* shall
th *as in* bath
th as in feather (**feth**-er)
ts *as in* quartz (kworts)
u *as in* buck (buk), blood (blud), one (wun)
u(r) *as in* urn (but without sounding the "r")
uu *as in* book (buuk)
v *as in* van, of (ov)
y *as in* yet, menu (**men**-yoo), onion (**un**-yon)
z *as in* zoo, lose (looz)
zh *as in* treasure (**tre**-zher)

Names of two or more syllables are broken up into small units, each of one syllable, separated by hyphens. The stressed syllable in a word of two or more syllables is shown in **bold** type.

We have used a simple pronunciation system based on the phonetic respelling of names, which avoids the use of unfamiliar symbols. The sounds represented are as follows (the phonetic respelling is given in parentheses after the example word, if this is not pronounced as it is spelled):

The consonants b, d, h, l, m, n, p, r, t, and w have their normal sounds and are not listed in the table.

In our pronunciation guide a consonant is occasionally doubled to avoid confusing the syllable with a familiar word, for example, -iss rather than -is (which is normally pronounced -iz); -off rather than -of (which is normally pronounced -ov).

Contents

Acknowledgments

General Editor
John Daintith BSc, PhD

Contributors and Editors
Eve Daintith BSc
Rosalind Dunning BA
Derek Gjertsen BA
Garry Hammond BSc
Robert Hine BSc, MSc
Valerie Illingworth BSc, MPhil
Elizabeth Martin MA
Sarah Mitchell BA
Susan O'Neill BSc
W. J. Palmer MSc
Roger F. Picken BSc, PhD
Richard Rennie BSc, PhD
Carol Russell BSc
W. J. Sherratt BSc, MSc, PhD
Jackie Smith BA
B. D. Sorsby BSc, PhD
Elizabeth Tootill BSc, MSc
P. Welch DPhil
Anthony Wootton

Computer Systems
Anne Stibbs BA

Artwork
Lynn Williams
Nick Hain
O. Levenok

A

Abbe, Cleveland (1838–1916) *American meteorologist*

Abbe (**ab**-ee) was born in New York City and educated there at City College; he later taught at the University of Michigan. He then spent two years (1864–66) in Russia at the Pulkovo Observatory under Otto STRUVE. On his return to America he worked as director of the Cincinnati Observatory (1868–70).

Abbe was the first official weather forecaster in America. He was appointed, in 1871, chief meteorologist with the weather service, which was later formed into the U.S. Weather Bureau (1891), and remained in this organization for the rest of his life. He was one of the first scientists to see the revolutionary role the telegraph had to play in weather forecasting and used reports conveyed to him from all over the country.

Abbe published over 300 papers on meteorology and from 1893 he was in charge of the journals published by the U.S. Weather Bureau. He was also responsible for the division of America into time zones in 1883.

Abbe, Ernst (1840–1905) *German physicist*

Abbe (**ab**-ee), who was born in Eisenach in Germany, came from poor parents but managed to become a lecturer at the University of Jena, where in 1886 he collaborated with Carl Zeiss, a supplier of optical instruments to the university, to improve the quality of microscope production. Up till then, this had been an empirical art without rigorous theory to aid design. Abbe's contribution was his knowledge of optical theory. He is known for the *Abbe sine condition* – a necessary condition for the elimination of spherical aberration in an optical system; such a system he described as aplanatic. He also invented the apochromatic lens system (1886), which eliminated both primary and secondary color distortions, and the *Abbe condenser* (1872) – a combination of lenses for converging light onto the specimen in microscopes.

The partnership between Abbe and Zeiss was a productive combination of Zeiss's practical knowledge and Abbe's mathematical and theoretical ability. After Zeiss's death, Abbe became the sole owner of the Zeiss company.

Abegg, Richard (1869–1910) *German physical chemist*

Abegg (**ah**-beg) was born in the German port of Danzig (now Gdańsk in Poland); he studied chemistry at Kiel, Tübingen, and Berlin. He graduated in 1891 as a pupil of Wilhelm Hofmann. Initially an organic chemist, he was attracted by the advances being made in physical chemistry, and in 1894 moved to Göttingen. Here, he worked on electrochemical and related problems and with G. Bodländer produced an important paper on valence, *Die Elektronaffinität* (1869; Electron Affinity). He is remembered for *Abegg's rule* (partially anticipated by Dmitri Mendeleev), which states that each element has two valences: a normal valence and a contravalence, the sum of which is eight. In 1899 he became a professor at Breslau (now Wrocław in Poland) and was about to become the director of the Physico-Chemical Institute there when he was killed in a ballooning accident.

Abel, Sir Frederick Augustus (1827–1902) *British chemist*

> The investigator must associate himself with those who have labored in fields where molecules and atoms rather than multicellular tissues or even unicellular organisms are the units of study.
> —Describing chemical research

Abel was born in London, the son of a well-known musician and the grandson of a court painter to the grand duke of Mecklenberg-Schwerin. Despite this artistic background, Abel developed an early interest in science after visiting his uncle A. J. Abel, a mineralogist and pupil of BERZELIUS. In 1845 he was one of the first of the pupils to study at the Royal College of Chemistry, remaining there until 1851. After a brief appointment as chemical demonstrator at St. Bartholomew's Hospital, London, he succeeded Michael FARADAY in 1852 as a lecturer in chemistry at the Royal Military Academy at Woolwich. In 1854 he became ordnance chemist and chemist to the war department.

Abel's career was thus devoted exclusively to the chemistry of explosives. New and powerful explosives, including guncotton and nitroglycerin, had recently been invented but were unsafe to use. Abel's first achievement was to show

how guncotton could be rendered stable and safe. His method was to remove all traces of the sulfuric and nitric acids used in its manufacture by mincing, washing in soda until all the acid had been removed, and drying. In 1888 he was appointed president of a government committee to find new high explosives. The two existing propellants, Poudre B and ballistite, had various defects, most important of which was a tendency to deteriorate during storage. Together with Sir James DEWAR, Abel introduced the new explosive, cordite, in 1889. This was a mixture of guncotton and nitroglycerin with camphor and petroleum added as stabilizers and preservatives.

Abel was honored for his services by being made a knight in 1891 and a baronet in 1893.

Abel, John Jacob (1857–1938) *American biochemist*

Abel was born in Cleveland, Ohio, the son of a farmer. He was educated at the University of Michigan and Johns Hopkins University. He spent the years 1884–90 in Europe studying at Leipzig, Heidelberg, Würzburg, Vienna, Bern, and Strasbourg, where he gained an MD in 1888. On his return to America he worked briefly at the University of Michigan before being appointed in 1893 to the first chair of pharmacology at Johns Hopkins, a post he retained until his retirement in 1932.

Abel approached biology with a first-rate training in chemistry and with the conviction that the study of molecules and atoms was as important as the observation of multicellular tissues under the microscope. He thus began by working on the chemical composition of various bodily tissues and fluids and, in 1897, succeeded in isolating a physiologically active substance from the adrenal glands, named by him epinephrine, also known as adrenalin. This extract was actually the monobenzoyl derivative of the hormone. It was left to Jokichi TAKAMINE to purify it in 1900.

As early as 1912 Abel clearly formulated the idea of an artificial kidney and in 1914 isolated for the first time amino acids from the blood. He was less successful with his search (1917–24) for the pituitary hormone, being unaware that he was dealing with not one but several hormones. His announcement in 1926 that he had crystallized insulin met with considerable skepticism, especially regarding its protein nature. This work was not generally accepted until the mid 1930s.

After his retirement Abel devoted himself to a study of the tetanus toxin.

Abel, Niels Henrik (1802–1829) *Norwegian mathematician*

By studying the masters and not their pupils.
—In reply to a question about how he gained his expertise

Abel (**ah**-bel) was born in Froland, the son of a poor pastor; he was educated in mathematics at the University of Christiania (Oslo). After the death of his father, Abel had to support a large family; he earned what he could by private teaching and was also helped out by his teacher. He was eventually given a grant by the Norwegian government to make a trip to France and Germany to visit mathematicians. In Germany he met the engineer and mathematician August Crelle, who was to be of great assistance to him. Crelle published Abel's work and exerted what influence he could to obtain him a post in Germany. Tragically Abel died just when Crelle had succeeded in getting him the chair in mathematics at Berlin.

With Evariste GALOIS (whom he never met), Abel founded the theory of groups (commutative groups are known as *Abelian groups* in his honor), and his early death ranks as one of the great tragedies of 19th-century mathematics. One of Abel's first achievements was to solve the longstanding problem of whether the general quintic (of the fifth degree) equation was solvable by algebraic methods. He showed that the general quintic is not solvable algebraically and sent this proof to Karl GAUSS, but unfortunately Gauss threw it away unread, having assumed that it was yet another unsuccessful attempt to solve the quintic.

Abel's greatest work was in the theory of elliptic and transcendental functions. Mathematicians had previously focused their attention on problems associated with elliptic integrals. Abel showed that these problems could be immensely simplified by considering the inverse functions of these integrals – the so-called "elliptic functions." He also proved a fundamental theorem, *Abel's theorem*, on transcendental functions, which he submitted to Augustin CAUCHY (and unfortunately fared no better than with Gauss). The study of elliptic functions inaugurated by Abel was to occupy many of the best mathematicians for the remainder of the 19th century. He also made very important contributions to the theory of infinite series.

Abelson, Philip Hauge (1913–2004) *American physical chemist*

Part of the strength of science is that it has tended to attract individuals who love knowledge and the creation of it … Thus, it is the communication process which is at the core of the vitality and integrity of science.
—*The Roots of Scientific Integrity* (1963)

Abelson, who was born in Tacoma, was educated at Washington State College and at the University of California at Berkeley, where he obtained his PhD in 1939. Apart from the war years at the Naval Research Laboratory in Washington, he spent most of his career at the Carnegie Institution, Washington, serving as the director of the geophysics laboratory from

1953, and as president from 1971 to 1978. He subsequently became the editor of a number of scientific journals including the important periodical *Science*, which he edited from 1962 to 1985.

In 1940 he assisted Edwin MCMILLAN in creating the first transuranic element, neptunium, by bombardment of uranium with neutrons in the Berkeley cyclotron. Abelson next worked on separating the isotopes of uranium. It was clear that a nuclear explosion was possible only if sufficient quantities of the rare isotope uranium-235 (only 7 out of every 1,000 uranium atoms) could be obtained. The method Abelson chose was that of thermal diffusion. This involved circulating uranium hexafluoride vapor in a narrow space between a hot and a cold pipe; the lighter isotope tended to accumulate nearer the hot surface. Collecting sufficient uranium-235 involved Abelson in one of those massive research and engineering projects only possible in war time. In the Philadelphia Navy Yard, he constructed a hundred or so 48-foot (15-meter) precision-engineered pipes through which steam was pumped. From this Abelson was able to obtain uranium enriched to 14 U-235 atoms per 1,000.

Although this was still too weak a mixture for a bomb, it was sufficiently enriched to use in other separation processes. Consequently a bigger plant, consisting of over 2,000 towers, was constructed at Oak Ridge, Tennessee, and provided enriched material for the separation process from which came the fuel for the first atom bomb.

After the war Abelson extended the important work of Stanley MILLER on the origin of vital biological molecules. He found that amino acids could be produced from a variety of gases if carbon, nitrogen, hydrogen, and oxygen were present. He was also able to show (1955) the great stability of amino acids by identifying them in 300-million-year-old fossils and later (1956) identified the presence of fatty acids in rocks.

Abrikosov, Alexei A. (1928–) *Soviet-born American physicist*

Abrikosov was born in Moscow and in 1951 received his PhD from the Kapitsa Institute for Physical Problems, Moscow, for a thesis on the theory of thermal diffusion in plasmas. Four years later the same Institute made him a Doctor of Physical and Mathematical Sciences for his work on quantum electrodynamics at high energies. He worked at several universities and scientific institutions in the Soviet Union and in 1991 he joined the Materials Science Division of the Argonne National Laboratory, Illinois. He is also (since 1965) a professor at Moscow State University.

Most of Abrikosov's work has concerned the theory of solids, including metals, semimetals, semiconductors, and superconductors. For example, he discovered type-II superconductors (which, unlike type-I, remain superconducting in powerful magnetic fields) and explained their magnetic properties, which involves the so-called *Abrikosov vortex lattice*. He continued his research in Argonne, discovering quantum magnetoresistance in certain silver compounds and studying high-temperature superconductors. Such materials have practical applications in medical magnetic resonance imaging (MRI) equipment and particle accelerators. During his long career in science he has received many honors and awards. In 2003 he shared the Nobel Prize for physics with Vitaly GINZBURG and Anthony LEGGETT for their pioneering work on superconductors and superfluids.

Adams, John Couch (1819–1892) *British astronomer*

Adams was born in the small Cornish town of Launceston in the west of England, where his father was a tenant farmer. He developed an early interest in astronomy, constructing his own sundial and observing solar altitudes, and pursuing his astronomical studies in the local Mechanics Institute. He graduated brilliantly from Cambridge University in 1843, and became Lowndean Professor of Astronomy and Geometry in 1858; in 1860 he was appointed director of the Cambridge Observatory.

His fame rests largely on the dramatic events surrounding the discovery of the planet Neptune in 1846. Astronomers had detected a discrepancy between the observed and predicted positions of Uranus and thus it appeared that either NEWTON's theory of gravitation was not as universal as had been supposed, or there was an as yet undetected body exerting a significant gravitational influence over the orbit of Uranus. There is evidence that Adams had decided to work on this problem as early as 1841. He had a general solution to the problem by 1843 and a complete solution by September 1845. It was then that he visited George AIRY, the Astronomer Royal, with a prediction of the exact position of the new planet.

Airy gave little attention to it and was moved to action only when, in June 1846, the French astronomer, Urbain LE VERRIER, also announced the position of a new planet. It was within one degree of the position predicted by Adams the previous year. Airy asked James Challis, director of the Cambridge Observatory, to start looking for the new planet with his large 25-inch (63.5-cm) refractor. Unfortunately Challis decided to cover a much wider area of the sky than was necessary and also lacked up-to-date and complete charts of the area. His start was soon lost and Johann GALLE in Berlin had no difficulty in discovering the planet on his first night of observation. All the fame, prizes, and honors initially went to Le Verrier.

When it was publicly pointed out, by Challis and John HERSCHEL, that Adams's work had priority over Le Verrier's, the shy Adams wanted no part of the controversy that followed. In fact he seemed genuinely uninterested in honors. He declined both a knighthood and the post of Astronomer Royal, which was offered him after Airy's retirement in 1881. He later worked on the planetary perturbations (1866), and on the secular variation of the mean motion of the Moon (1852), both difficult questions of mathematical astronomy. His scientific papers were published by his brother in two volumes, in 1876 and 1901.

Adams, Roger (1889–1971) *American organic chemist*

Born in Boston, Massachusetts, Adams studied chemistry at Harvard, where he obtained his PhD in 1912, and at the University of Berlin. After working briefly at Harvard he joined the staff of the University of Illinois in 1916 and later served as professor of organic chemistry from 1919 until his retirement in 1957.

Adams developed a simple but effective method for catalyzing the hydrogenation of unsaturated organic compounds using finely divided platinum or palladium dioxides, which were reduced to the metal. He also worked out the structure of a number of naturally occurring physiologically active compounds, such as chaulmoogra oil (which was used to treat leprosy) and gossypol (a poisonous constituent of animal feedstuffs derived from cotton). In the late 1930s Adams was asked by the American Narcotics Bureau to examine the chemistry of the marijuana alkaloids and he succeeded in isolating and identifying the active ingredient, tetrahydrocannabinol.

Adams, Walter Sydney (1876–1956) *American astronomer*

> Professor Adams has ... killed two birds with one stone. He has carried out a new test of Einstein's general theory of relativity, and he has shown that matter at least 2,000 times denser than platinum is not only possible, but actually exists in the stellar universe.
> —Sir Arthur Eddington, *Stars and Atoms* (1927)

Adams was born in Antioch (now in Turkey). He was the son of missionaries working in Syria, then part of the Ottoman Empire, who returned to America in 1885. Adams graduated from Dartmouth College in 1898 and obtained his MA from the University of Chicago in 1900. After a year in Munich he began his career in astronomy as assistant to George HALE in 1901 at the Yerkes Observatory. He moved with Hale to the newly established Mount Wilson Observatory in 1904 where he served as assistant director, 1913–23, and then as director from 1923 until his retirement in 1946.

At Mount Wilson Adams was able to use first the 60-inch (1.5-m) and from 1917 the 100-inch (2.5-m) reflecting telescopes in whose design and construction he had been closely associated. His early work was mainly concerned with solar spectroscopy, when he studied sunspots and solar rotation, but he gradually turned to stellar spectroscopy. In 1914 he showed how it was possible to distinguish between a dwarf and a giant star merely from their spectra. He also demonstrated that it was possible to determine the luminosity, i.e., intrinsic brightness, of a star from its spectrum. This led to Adams introducing the method of spectroscopic parallax whereby the luminosity deduced from a star's spectrum could be used to estimate its distance. The distances of many thousands of stars have been calculated by this method.

He is however better known for his work on the orbiting companion of Sirius, named Sirius B. Friedrich BESSEL had first shown in 1844 that Sirius must have a companion and had worked out its mass as about the same as our Sun. The faint star was first observed telescopically by Alvan CLARK in 1862. Adams succeeded in obtaining the spectrum of Sirius B in 1915 and found the star to be considerably hotter than the Sun. Adams realized that such a hot body, just eight light-years distant, could only remain invisible to the naked eye if it was very much smaller than the Sun, no bigger in fact than the Earth. In that case it must have an extremely high density, exceeding 100,000 times the density of water. Adams had thus discovered the first "white dwarf" – a star that has collapsed into a highly compressed object after its nuclear fuel is exhausted.

If such an interpretation was correct then Sirius B should possess a very strong gravitational field. According to Einstein's general theory of relativity, this strong field should shift the wavelengths of light waves emitted by it toward the red end of the spectrum. In 1924 Adams succeeded in making the difficult spectroscopic observations and did in fact detect the predicted red shift, which confirmed his own account of Sirius B and provided strong evidence for general relativity.

Addison, Thomas (1793–1860) *British physician*

Addison, who was born in Longbenton, near Newcastle upon Tyne, England, graduated in medicine from Edinburgh University in 1815 and soon afterwards moved to practice in London. He entered Guy's Hospital in 1817, was appointed assistant physician in 1824, and later became physician and joint lecturer with Richard BRIGHT, an eminent contemporary.

Addison was the first to describe, in 1849, the disease caused by pathological changes in the adrenal (suprarenal) glands, which is now known as *Addison's disease*. He described the characteristic anemia, bronzed skin, and other symptoms in his famous paper, *On the Constitutional and Local Effects of Disease of the Supra-Renal Capsules* (1855), and distinguished it from another form of anemia, now called pernicious anemia, which results from other causes entirely. This is sometimes called *Addisonian anemia*.

Addison also wrote papers concerning tuberculosis, skin diseases, the anatomy of the lung, and other topics, which were published in a collected edition in 1868. He collaborated with Bright in writing *Elements of the Practice of Medicine* (1839) of which only the first volume was completed.

An eloquent if rather aloof lecturer, Addison's fame as a physician came largely after his death.

Adhemar, Alphonse Joseph (1797–1862) *French mathematician*

Adhemar (ad-ay-**mar**), who was born and died in Paris, France, was a private mathematics tutor who also produced a number of popular mathematical textbooks.

His most important scientific work was his *Les Revolutions de la mer* (1842; The Upheavals of the Sea) in which he was the first to propose a plausible mechanism by which astronomical events could produce ice ages on Earth. It had been known for some time that while the Earth moved in an elliptical orbit around the Sun it also rotated about an axis that was tilted to its orbital plane. Because the orbit is elliptical and the Sun is at one focus, the Earth is closer to the Sun at certain times of year. As a result, the southern hemisphere has a slightly longer winter than its northern counterpart. Adhemar saw this as a possible cause of the great Antarctic icesheet for, as this received about 170 hours less solar radiation per year than the Arctic, this could just be sufficient to keep temperatures cold enough to permit the ice to build up.

Adhemar was also aware that the Earth's axis does not always point in the same direction but itself moves around a small circular orbit every 26,000 years. Thus he postulated a 26,000-year cycle developing in the occurrence of glacial periods, but his views received little support.

Adler, Alfred (1870–1937) *Austrian psychologist*

> It is one of the triumphs of human wit...to conquer by humility and submissiveness...to cause pain to others by one's own suffering,...to make oneself small in order to appear great...such are often the expedients of the neurotic.
> —*The Neurotic Constitution* (1912)

Adler (**ad**-ler or **ahd**-ler) was born in Penzing, Austria, the son of a corn merchant, and was educated at the University of Vienna, where he obtained his MD in 1895. After two years at the Vienna General Hospital he set up in private practice in 1898.

In about 1900 Adler began investigating psychopathology and in 1902 he became an original member of Sigmund FREUD's circle, which met to discuss psychoanalytical matters. His disagreements with Freud began as early as 1907 – he dismissed Freud's view that sexual conflicts in early childhood cause mental illness – and he finally broke away from the psychoanalytic movement in 1911 to form his own school of individual psychology. Adler tended to minimize the role of the unconscious and sexual repression and instead to see the neurotic as overcompensating for his or her "inferiority complex," a term he himself introduced. His system was fully expounded in his *Practice and Theory of Individual Psychology* (1927). In 1921 Adler founded his first child-guidance clinic in Vienna, which was to be followed by over 30 more before the Nazi regime in Vienna forced their closure in 1932. From 1926 onward he began to spend more and more time in America, finally settling there permanently in 1932 and taking a professorship of psychiatry at the Long Island College of Medicine, New York, a post he retained until his death from a heart attack while lecturing in Aberdeen in Scotland.

Adrian, Edgar Douglas, Baron Adrian of Cambridge (1889–1977) *British neurophysiologist*

Adrian, a lawyer's son, was born in London and studied at Cambridge University and St. Bartholomew's Hospital, London, where he obtained his MD in 1915. He returned to Cambridge in 1919, was appointed professor of physiology in 1937, and became the master of Trinity College, Cambridge, in 1951, an office he retained until his retirement in 1965. He was raised to the British peerage in 1955.

Adrian's greatest contribution to neurophysiology was his work on the nerve impulse. When he began it was known that nerves transmit nerve impulses as signals, but knowledge of the frequency and control of such impulses was minimal. The first insight into this process came from Adrian's colleague Keith LUCAS, who demonstrated in 1905 that the impulse obeyed the "all-or-none" law. This asserted that below a certain threshold of stimulation a nerve does not respond. However, once the threshold is reached the nerve continues to respond by a fixed amount however much the stimulation increases. Thus increased stimulation, although it stimulates more fibers, does not affect the magnitude of the signal itself.

It was not until 1925 that Adrian advanced beyond this position. By painstaking surgical techniques he succeeded in separating individual nerve fibers and amplifying and recording the small action potentials in these fibers. By studying the effect of stretching the sternocutaneous muscle of the frog, Adrian demonstrated how the nerve, even though it transmits an impulse of fixed strength, can still convey a complex message. He found that as the extension increased so did the frequency of the nerve impulse, rising from 10 to 50 impulses per second. Thus he concluded that the message is conveyed by changes in the frequency of the discharge. For this work Adrian shared the 1932 Nobel Prize for physiology or medicine with Charles SHERRINGTON.

Aganice (Aglaonike or Aglaonice) (*fl.* ?3rd–1st century BC) *Egyptian sorceress*

Aganice was a natural philosopher and sorceress who is said to have originated from Thessaly in Greece and been associated with the court of the Pharaoh Sesostris III (ruled 1878–1843 BC) in ancient Egypt. However, there is no hard evidence for this, and some modern authors assert that she may have lived much earlier, or simply belong to the canon of ancient mythology. Her status stemmed from a supposed ability to predict lunar and solar eclipses, and to convince gullible audiences that she had the power to make the Moon disappear. According to the Greek biographer Plutarch, she was the daughter of Hegetor and "thoroughly conversant with the periods of the Full Moon when it is subject to eclipse, and knowing beforehand when the Moon was due to be overtaken by the Earth's shadow, imposed upon audiences of women and made them all believe that she drew down the Moon." Some modern astronomers have taken this as evidence for a series of dark lunar eclipses in the first century BC, whereas others dispute that such eclipses were common, casting doubt on the credibility of Aganice's reputation. In any case, her name is lent to a Greek expression for a braggart: "Yes, as the Moon obeys Aganice."

Agassiz, Jean Louis Rodolphe (1807–1873) *Swiss–American biologist*

> The world has arisen in some way or another. How it originated is the great question, and Darwin's theory, like all other attempts to explain the origin of life, is thus far merely conjectural. I believe he has not even made the best conjecture possible in the present state of our knowledge.
> —*Evolution and Permanence of Type* (1874)

Generally considered the foremost naturalist of 19th-century America, Agassiz (**ag**-a-see) was born in Motier-en-Vuly, Switzerland. He was educated at the universities of Zurich, Heidelberg, and Munich, where he studied under the embryologist Ignaz DÖLLINGER. At the instigation of Georges CUVIER, he cataloged and de-scribed the fishes brought back from Brazil by C. F. P. von Martius and J. B. von Spix (*Fishes of Brazil*, 1829), following this with his *History of the Freshwater Fishes of Central Europe* (1839–42) and an extensive pioneering work on fossil fishes, which eventually ran to five volumes: *Recherches sur les poissons fossiles* (1833–43; Researches on Fossil Fishes). These works, completed while Agassiz was professor of natural history at Neuchâtel (1832–46), established his reputation as the greatest ichthyologist of his day. Agassiz's best-known discovery, however, was that of the Ice Ages. Extensive field studies in the Swiss Alps, and later in America and Britain, led him to postulate glacier movements and the former advance and retreat of ice sheets; his findings were published in *Etudes sur les glaciers* (1840; Studies on Glaciers).

A successful series of lectures given at Boston, Massachusetts, in 1846 led to his permanent settlement in America. In 1847 he was appointed professor of zoology and geology at Harvard, where he also established the Museum of Comparative Zoology (1859). Agassiz's subsequent teachings introduced a departure from established practice in emphasizing the importance of first-hand investigation of natural phenomena, thus helping to transform academic study in America. His embryological studies led to a recognition of the similarity between the developing stages of living animals and complete but more primitive species in the fossil record. Agassiz did not, however, share Darwin's view of a gradual evolution of species, but, like Cuvier, considered that there had been repeated separate creations and extinctions of species – thus explaining changes and the appearance of new forms. Unfortunately, one of Agassiz's most influential pronouncements was that there were several species, as distinct from races, of man: an argument used by slavers to justify their subjugation of Black people as an inferior species. His ambitious *Contributions to the Natural History of the United States* (4 vols., 1857–62) remained uncompleted at his death.

Agre, Peter (1949–) *American cell biologist*

Agre was born in Northfield, Minnesota, and in 1970 gained his BA in chemistry from Augsburg College, Minneapolis. He then switched to medicine and attended Johns Hopkins University School of Medicine in Baltimore, where he qualified as MD in 1974. He spent his medical residency at Case Western University and gained a clinical fellowship at the University of North Carolina at Chapel Hill. He then returned to Johns Hopkins for a research fellowship in the cell biology department, becoming a faculty member in the department of medicine in 1984. In 1993 he became professor of biological chemistry and professor of medicine at Johns Hopkins School of Medicine, specializing in blood disorders.

For body cells to function, there has to be a way in which substances such as water and various ions can pass through the wall of the cell to maintain an even pressure within it – the cell's outer membrane must contain pores. From the mid-1980s Agre's main area of research was to discover the nature of these pores and thus how water molecules cross the cell wall. In 1988 he isolated a protein from the outer membrane of red blood cells, and within two years he demonstrated that this protein forms a channel through which water molecules can travel. Agre named the protein *aquaporin*. The 1991 discovery is now known as aquaporin-1 and various other types of aquaporins have since been found in a wide range of other cells of bacteria, plants, and animals. At last there was an explanation of, for example, how osmosis takes place in plant cells and how blood cells in the kidney reabsorb water from primary urine. The discovery was important in the understanding and possible treatment of kidney disease. For his discovery of water channels in cells, Agre shared the 2003 Nobel Prize for chemistry with the American cell biologist Roderick MacKinnon, who had earlier discovered ion channels.

Agricola, Georgius (1494–1555) *German metallurgist*

Agricola (a-**grik**-o-la) was born Georg Bauer (**bow**-er) but, as was the custom of the day, he Latinized his name (Agricola and Bauer both mean "farmer"). Beyond his place of birth – Glauchau in Germany – little is known about him until his entry into the University of Leipzig in 1514. He later pursued his studies of philosophy and medicine in Italy at Bologna, Padua, and Venice (1523–27). In 1527 he was engaged as physician to the Bohemian city of Joachimsthal – the center of a rich mining area – moving in 1534 to another celebrated mining town, Chemnitz, near his birthplace. Here he became burgomaster in 1545. He wrote seven books on geological subjects but these were also so illuminating of other subjects that he was known in his lifetime as "the Saxon Pliny."

His most famous work, *De re metallica* (1556; On the Subject of Metals), concentrates on mining and metallurgy with a wealth of information on the conditions of the time, such as management of the mines, the machinery used (e.g., pumps, windmills, and water power), and the processes employed. The book is still in print having the unique distinction of being translated and edited (1912) by a president of the United States, Herbert Hoover, with Lou Henry Hoover (his wife).

Agricola is often regarded as the father of modern mineralogy. In the Middle Ages, the subject was based on the accumulated lore from the Orient, the Arabs, and antiquity. Stones were believed to come in male and female form, to have digestive organs, and to possess medicinal and supernatural powers. Agricola began to reject these theories and to provide the basis for a new discipline. Thus in his *De ortu et causis subterraneorum* (1546; On the Origin and Cause of Subterranean Things) he introduced the idea of a lapidifying juice (or *succus lapidescens*) from which stones condensed as a result of heat. This fluid was supposedly subterranean water mixed with rain, which collects earthy material when percolating through the ground.

Agricola also, in *De natura fossilium* (1546; On the Nature of Fossils), introduced a new basis for the classification of minerals (called "fossils" at the time). Although far from modern, it was an enormous improvement on earlier works. Agricola based his system on the physical properties of minerals, which he listed as color, weight, transparency, taste, odor, texture, solubility, combustibility, and so on. In this way he tried to distinguish between earths, stones, gems, marbles, metals, building stone, and mineral solutions, carefully describing his terms, which should not be assumed to be synonymous with today's terms, in each case.

Aiken, Howard Hathaway (1900–1973) *American computer pioneer*

Born in Hoboken, New Jersey, Aiken graduated from the University of Wisconsin and began work with the Madison Gas and Engineering Company in 1923. He moved to Westinghouse Electric, Chicago, in 1927 but decided ten years later to pursue an academic career. He was awarded his Harvard PhD in 1939 and was immediately appointed to the faculty, becoming professor of applied mathematics and director of the computation laboratory in 1946, positions he held until his retirement in 1961.

In 1937 Aiken began to look for support to build an electromechanical calculator to reduce the time and effort spent in the numerical solution of differential equations. He obtained backing from IBM in 1939. The result, known as Harvard Mark 1 or ASCC (Automatic Sequence-Controlled Calculator), was completed in 1943. It was 8 feet tall, 50 feet long, weighed 5 tons, and contained 750,000 parts. The Mark 1, which operated decimally, was programmed by punched tape. It was driven by a long metal shaft, which was in turn operated by the main sequence-control mechanism. Consequently, it could calculate no faster than the speed of its mechanical parts. Although essentially obsolete at the time of its appearance, it continued in use. It was employed for calculating mathematical tables for a further sixteen years.

Aiken went on to build a larger Mark II version for the U.S. Navy in 1948. Two further machines were built by Aiken. Both of these were operated by electronic circuits, rather than mechanically: the Mark III in 1949, again for the Navy, and the Mark IV in 1952 for the Air Force.

Airy, Sir George Biddell (1801–1892)
British astronomer

> I had made considerable advance ... in calculations on my favourite numerical lunar theory, when I discovered that, under the heavy pressure of unusual matters (two transits of Venus and some eclipses) I had committed a grievous error in the first stage of giving numerical value to my theory. My spirit in the work was broken, and I have never heartily proceeded with it since.
> —Describing his theoretical work on the orbital motion of the Moon

Airy, the son of a tax collector, was born in Alnwick in the north-east of England. He attended school in Colchester before going to Cambridge University in 1819. He met with early success, producing a mathematical textbook in 1826 and numerous papers on optics. He became Lucasian Professor of Mathematics at Cambridge in 1826 and two years later was made Plumian Professor of Astronomy and director of the Cambridge Observatory. In 1835 he was appointed Astronomer Royal, a post he held for 46 years.

Airy was a very energetic, innovative, and successful Astronomer Royal. He reequipped the observatory, installing an altazimuth for lunar observation in 1847, a new transit circle and zenith tube in 1851, and a 13-inch (33-cm) equatorial telescope in 1859. He created a magnetic and meteorological department in 1838, began spectroscopic investigations in 1868, and started keeping a daily record of sunspots with the Kew Observatory heliograph in 1873. In optics he investigated the use of cylindrical lenses to correct astigmatism (Airy was astigmatic) and examined the disklike image in the diffraction pattern of a point source of light (in an optical device with a central aperture) now called the *Airy disk*. Also named for him is his hypothesis of isostasy: the theory that mountain ranges must have root structures of lower density, proportional to their height, in order to maintain isostatic equilibrium.

Despite his many successes he is now mainly, and unfairly, remembered for his lapses. When John ADAMS came to him in September 1845, with news of the position of a new planet, Airy unwisely ignored him, leaving it to others to win fame as the discoverers of Neptune. He also dismissed Michael FARADAY's new field theory.

Aitken, Robert Grant (1864–1951) *American astronomer*

Born in Jackson, California, Aitken obtained his BA in 1887 and his MA in 1892 from Williams College, Massachusetts. He began his career at the University of the Pacific, then in San Jose, as professor of mathematics from 1891 until 1895 when he joined the staff of Lick Observatory, Mount Hamilton, California. He remained at Lick for his entire career, serving as its director from 1930 until his retirement in 1935.

Aitken did much to advance knowledge of binary stars, i.e., pairs of stars orbiting about the same point under their mutual gravitational attraction. He described over 3,000 binary systems and published in 1932 the comprehensive work *New General Catalogue of Double Stars Within 120° of the North Pole*. He also produced the standard work *The Binary Stars* (1918).

Akers, Sir Wallace Allen (1888–1954)
British industrial chemist

Akers, the son of a London accountant, was educated at Oxford University. He first worked for the chemical company Brunner Mond from 1911 to 1924 and, after four years in Borneo with an oil company, returned in 1928. By then Brunner Mond had become part of ICI. From 1931 he was in charge of the Billingham Research Laboratory and from 1944 was the company director responsible for all ICI research.

In World War II Akers worked under Sir John Anderson, the government minister responsible for work on the atom bomb. He was put in charge of Tube Alloys, the Ministry of Supply's front for secret nuclear work. It was Akers who led the mission of British scientists in 1943 to America to work out details of collaboration, although he proved unacceptable to the Americans and was replaced by James CHADWICK. Akers returned to head Tube Alloys in the UK.

After the war one of his main tasks was the setting up of the Central Research Laboratory for ICI at Welwyn near London, later named the Akers Research Laboratory. He was knighted in 1946.

Al-Battani (*or* Albategnius) (*c.* 858–929)
Arab astronomer

> Nobody is known in Islam who reached similar perfection in observing the stars and in scrutinizing their motions.
> —Ibn al-Qifti (13th century)

Al-Battani (al-ba-**tah**-nee) was the son of a maker of astronomical instruments in Harran (now in Turkey). He worked mainly in Raqqah on the Euphrates (now ar-Raqqah in Syria) and was basically a follower of PTOLEMY, devoting himself to refining and perfecting the work of his master. He improved Ptolemy's measurement of the obliquity of the ecliptic (the angle between the Earth's orbital and equatorial planes), the determination of the equinoxes, and the length of the year. He also corrected Ptolemy in various matters, in particular in his discovery of the movement of the solar perigee (the Sun's nearest point to the Earth) relative to the equinoxes. His work was widely known in the medieval period, having been translated by PLATO of Tivoli in about 1120 as *De motu stellarum* (On Stellar Motion), which was finally published in Nuremberg in 1537.

Albertus Magnus, St. (c. 1200–1280) *German scholastic philosopher*

Albert Magnus was the son of a German lord, his real name being Count von Bollstädt. Born in the south German town of Lauingen, he studied at the University of Padua and then became, in 1223, a member of the Dominican order against his family's wishes. He continued his studies throughout Europe and then taught theology. From about 1245 he lectured in Paris, where the philosopher and theologian Thomas Aquinas became one of his pupils, and in 1248 he was sent to Cologne to establish a Dominican study center, returning to Paris in 1254.

Albertus Magnus's voluminous writings, including treatises on theology, physics, and natural history, were generally Aristotelian in spirit, though he stressed the importance of direct observation of nature rather than strict adherence to textual authority. He also conducted alchemical experiments and in his *De mineralibus* (1569; On Minerals) he claimed to have tested alchemical gold "and found after six or seven ignitions that it was converted to powder."

He retired, in 1270, to his convent at Cologne. He was made a doctor of the Church and canonized in 1931.

Albinus, Bernhard Siegfried (1697–1770) *German anatomist*

Albinus (al-**bI**-nus) was born in Frankfurt an der Oder in Germany and was educated at the University of Leiden, where he subsequently held professorships in anatomy and surgery and later medicine. A popular lecturer on his subject, Albinus also carried out his own studies of human anatomy, and is chiefly known for his detailed classification of contemporary and traditional knowledge in this field. He published, with his former teacher Hermann BOERHAAVE, the complete works of Andreas VESALIUS, and also edited a new edition of the works of Hieronymus FABRICIUS. His own work, in which he emphasized the importance of illustrating the "anatomical norm," was published in 1747 in *Tabulae sceleti et musculorum corporis humani* (Plates of the Skeleton and Muscles of the Human Body), which contains numerous excellent drawings.

Albright, Arthur (1811–1900) *British chemist and industrialist*

Albright came from a Quaker family in the village of Charlbury, near Oxford, England. He was first apprenticed as an apothecary in Bristol and became a partner in the Birmingham firm of John and Edward Sturge, manufacturing chemists. In 1844 the partners established a new plant at Selly Oak to produce white phosphorus for the match industry. This was manufactured from calcium phosphate, derived from imported bones, and sulfuric acid.

White phosphorus was a difficult material to handle, being toxic and spontaneously flammable. Another allotropic form – red phosphorus – was discovered in 1845 by Anton SCHRÖTTER. Albright bought the patents and improved the process for producing red phosphorus. A larger factory was opened at Oldbury. J. E. Lundström of Sweden showed, in 1855, how Albright's phosphorus could be used in the production of safety matches. In this system the chlorate was confined to the tip of the match and the phosphorus was used only on the box.

In 1855 Albright's partnership with the Sturges was dissolved and he joined with J. W. Wilson in the following year, eventually forming the chemical company Albright and Wilson.

Alcmaeon (about 450 BC) *Greek philosopher and physician*

Alcmaeon (alk-**mee**-on) was born in Croton (now Crotone in Italy). Details of his work come from the surviving fragments of his book and through references by later authors, including ARISTOTLE. He was probably influenced by the school of thought founded by PYTHAGORAS in Croton and originated the notion that health was dependent on maintaining a balance between all the pairs of opposite qualities in the body, i.e., wet and dry, hot and cold, etc. Imbalance of these qualities resulted in illness. This theory was later developed by HIPPOCRATES and his followers.

Alcmaeon performed dissections of animals and possibly of human cadavers also. He demonstrated various anatomical features of the eye and ear, including their connections with the brain, and correctly asserted that the brain was the control center of bodily functions and the seat of intelligence.

Alder, Kurt (1902–1958) *German organic chemist*

Alder (**ahl**-der), who was born in Königshütte, Germany (now Chorzów in Poland), studied chemistry in Berlin and Kiel, receiving his doctorate in 1926 under Otto DIELS. In 1928 Diels and Alder discovered an important type of organic chemical reaction in which a compound containing two double carbon–carbon bonds separated by a single carbon–carbon bond adds to a compound containing a double carbon–carbon bond. The resulting molecule contains an aromatic ring structure with alternating double and single bonds. Alder was professor of chemistry at Kiel (1934), chemist with I. G. Farben at Leverkusen (1936), and director of the Chemical Institute at the University of Cologne (1940). In 1950 Diels and Alder jointly received the Nobel Prize for chemistry for their discovery of what is now known as the Diels–Alder reaction.

Aldrovandi, Ulisse (1522–1605) *Italian naturalist*

Aldrovandi (al-droh-**van**-dee) was educated at Bologna (where he was born and died) and Padua, where he obtained a medical degree in 1553. He was professor of botany and natural history at Bologna and in 1567 founded the Bologna Botanic Gardens. Under the patronage of popes Gregory XIII and Sixtus V, Aldrovandi traveled over a great part of Europe observing natural phenomena and collecting specimens, which he planned to incorporate in a multivolume natural history. Only four volumes appeared during his lifetime, *Ornithologiae* (3 vols., 1599; Ornithology), and *De animalibus insectis* (1602; On Insects). Other volumes appeared posthumously and include works on fish and whales (1613), monsters (1640), serpents and dragons (1640), and quadrupeds (1613; 1616; 1621; 1637).

Aldrovandi was in no sense a modern zoologist. The many-headed hydra and the basilisk appear on his pages along with cows and goats. Further, while he could write at great length about creatures, 31 pages for example on the peacock, little of this would be recognized today as relevant to zoology. Much of the material is linguistic, covering the names of, for example, the peacock in various languages together with synonyms for the bird and its use in proverbs. Much also is concerned with the emblematic appearance and significance as it is found on coins, medals, and coats of arms.

Alembert, Jean Le Rond d' *See* D'ALEMBERT, JEAN LE ROND.

Alferov, Zhores Ivanovich (1930–) *Russian physicist*

Alferov (**al**-fe-rof) was born in Vitebsk, Belorussia (then part of the Soviet Union), and educated at the Lenin Electrochemical Institute in St. Petersburg (then Leningrad). Since 1953 he has been a staff member of the A. F. Ioffe Physico-Technical Institute in St. Petersburg, being director from 1987 to the present time.

Since 1995 Alferov has also been a member of the Russian parliament (the Duma), being reelected in 2007 as a member of the Communist Party.

Since the early 1960s Alferov has been one of the leading figures in the development of the physics and technology of layered semiconductor structures known as semiconductor heterostructures, including applications to lasers, solar cells, and light-emitting diodes (LED). This work has had a major impact on information technology because it has enabled information systems based on semiconductor heterostructures to be built that have the desirable features of being both fast and very small. It is necessary for information systems to be fast because large amounts of information have to be transferred in a short time. It is very useful for information systems to be small so that they can then be used in homes, offices, or even briefcases or pockets.

Fast small transistors based on semiconductor heterostructures have many technological applications, including radio-link satellites and the base stations of mobile telephones. The same type of technology is used in the laser diodes that drive the flow of information in the fiber-optical cables used in the Internet. Other technological applications of semiconductor heterostructures include CD players, bar-code readers, and light-emitting diodes.

Alferov has been awarded many prizes for his pioneering work in semiconductor heterostructures, culminating in the Nobel Prize for physics in 2000, which he shared with Herbert KROEMER and Jack KILBY for their related work.

Alfvén, Hannes Olof Gösta (1908–1995) *Swedish physicist*

Alfvén (**al**-ven), who was born in Norrkoeping, Sweden, was educated at the University of Uppsala where he received his PhD in 1934. He subsequently worked at the Royal Institute of Technology, Stockholm, where he served as professor of the theory of electricity (1940–45), professor of electronics (1945–63), and professor of plasma physics (1963–73).

Alfvén is noted for his pioneering theoretical research in the field of magnetohydrodynamics – the study of conducting fluids and their interaction with magnetic fields. This work, for which he shared the 1970 Nobel Prize for physics with Louis NEEL, was mainly concerned with plasmas, i.e., ionized gases containing positive and negative particles. He investigated the interactions of electrical and magnetic fields and showed theoretically that the magnetic field, under certain circumstances, can move with the plasma. In 1942 he postulated the existence of waves in plasmas; these *Alfvén waves* were later observed in both liquid metals and ionized plasmas.

Alfvén also applied his theories to the motion of particles in the Earth's magnetic field and to the properties of plasmas in stars. In 1942, and later in the 1950s, he developed a theory of the origin of the solar system. This he assumed to have formed from a magnetic plasma, which condensed into small particles that clustered together into larger bodies. His work is also applicable to the properties of plasmas in experimental nuclear fusion reactors. Alfvén's books include *Cosmical Electrodynamics* (1950), which collects his early work, *On the Origin of the Solar System* (1954), and *On the Evolution of the Solar System* (1976, with G. Arrhenius).

In his later years Alfvén argued against the current orthodoxy of the big-bang theory of the origin of the universe. Space, he argued, is full of immensely long plasma filaments. The electromagnetic forces produced have caused the plasma to condense into galaxies. As for the expansion of the universe, he attributed this to the energy released by the collision of matter and antimatter. Whereas Alfvén's critics charged him with vagueness, he responded by arguing that cosmologists derive their theories more from mathematical considerations than from laboratory experiments.

Alhazen (*or* **Abu Ali Al-Hassan Ibn Al Haytham**) (*c.* 965–1038) *Arabian scientist*

> Now the known things are of five kinds: the known in number, the known in magnitude, the known in ratio, the known in position, and the known in species … In the examples of analysis we give in the present treatise we shall prove the known things used, whether or not we have found them in other works.
> —*Opticae Thesaurus* (1572; The Treasury of Optics)

Born in Basra (now in Iraq), Alhazen (al-ha-**zen**) was one of the most original scientists of his time. About a hundred works are attributed to him; the main one was translated into Latin in the 12th century and finally published in 1572 as *Opticae Thesaurus* (The Treasury of Optics). This was widely studied and extremely influential. It was the first authoritative work to reject the curious Greek view that the eye sends out rays to the object looked at. Alhazen also made detailed measurements of angles of incidence and refraction. He studied spherical and parabolic mirrors, the camera obscura, and the role of the lens in vision. While the Greeks had had a good understanding of the formation of an image in a plane mirror, Alhazen tackled the much more difficult problem of the formation of images in spherical and parabolic mirrors and offered geometrical solutions. It is difficult to think of any other writer who had surpassed the Greeks in any branch of the exact sciences by the 14th let alone the 11th century. He was, however, unfortunate in his relationship with the deranged caliph al-Hakim.

Having rashly claimed that he could regulate the flooding of the Nile, he was forced to simulate madness to escape execution until the caliph died in 1021.

Al-Khwarizmi, Abu Ja'far Muhammad Ibn Musa (*c.* 800–*c.* 847) *Arab mathematician, astronomer, and geographer*

> With my two algorithms, one can solve all problems – without error, if God wills it!
> —*Algebra*

Al-Khwarizmi (al-**kwah**-riz-mee) takes his name from his birthplace, Khwarizm (now Khiva in Uzbekistan). His importance lies chiefly in the knowledge he transmitted to others. Very little is known about his life except that he was a member of the academy of sciences in Baghdad, which flourished during the rule (813–33) of caliph al-Ma'mun. Al-Khwarizmi's main astronomical treatise and his chief mathematical work, the *Algebra*, are dedicated to the caliph. The *Algebra* enlarged upon the work of DIOPHANTUS and is largely concerned with methods for solving practical computational problems rather than algebra as the term is now understood. Insofar as he did discuss algebra, al-Khwarizmi confined his discussion to equations of the first and second degrees.

His astronomical work, *Zij al-sindhind*, is also based largely on the work of other scientists. As with the *Algebra*, its chief interest is as the earliest Arab work on the subject still in existence.

Al-Khwarizmi's other main surviving works are a treatise on the Hindu system of numerals and a treatise on geography. The Hindu number system, with its epoch-making innovations, for example the incorporation of a symbol for zero, was introduced to Europe via a Latin translation (*De numero indorum*; On the Hindu Art of Reckoning) of al-Khwarizmi's work. Only the Latin translation remains but it seems certain that al-Khwarizmi was the first Arab mathematician to expound the new number system systematically. The term "algorithm" (a rule of calculation) is a corrupted form of his name. His geographical treatise marked a considerable improvement over earlier work, notably in correcting some of the influential errors and misconceptions that had gained currency owing to PTOLEMY's *Geography*.

Allbutt, Sir Thomas Clifford (1836–1925) *British physician and medical historian*

Allbutt was born in Dewsbury in Yorkshire, England, and studied natural sciences at Cambridge University, graduating in 1860. He then studied medicine at St. George's Hospital, London, gaining his MB in 1861. He was appointed physician to Leeds General Infirmary in 1864 where, in 1866, he invented the short clinical thermometer, which was a great advance on previous highly cumbersome instruments. His major interest was cardiology and he was the first to describe the effects of syphilis on the cerebral arteries. His *System of Medicine* (8 vols. 1896–99) became an important medical text. In 1892 he was appointed professor of medicine at Cambridge, where he lectured and wrote several notable works on the history of medicine, including *Greek Medicine in Rome* (1921).

Allen, Edgar (1892–1943) *American endocrinologist*

Allen, the son of a physician, was born in Canon City, Colorado, and educated at Brown University. After war service he worked at Washington University, St. Louis, before being appointed (1923) to the chair of anatomy at the University of Missouri. In 1933 he moved to a similar post at Yale and remained there until his death.

In 1923 Allen, working with Edward DOISY, began the modern study of the sex hormones. It was widely thought that the female reproductive cycle was under the control of some substance found in the corpus luteum, the body formed in the ovary after ovulation. Allen thought rather that the active ingredient was probably in the follicles surrounding the ovum. To test this he made an extract of the follicular fluid and found that on injection it induced the physiological changes normally found only in the estrous cycle. Allen had in fact discovered estrogen although it was only identified some six years later by Adolf BUTENANDT.

Allen, James Alfred Van *See* VAN ALLEN, JAMES ALFRED.

Alpher, Ralph Asher (1921–2007) *American physicist*

Alpher was born in Washington, DC, and educated there at George Washington University; he obtained his BS in 1943 and his PhD in 1948. He spent the war as a physicist with the U.S. Navy's Naval Ordnance Laboratory in Washington followed by a period (1944–55) with the applied physics laboratory of Johns Hopkins University, Baltimore. In 1955 Alpher joined the staff of the General Electric Research and Development Center, Schenectady, New York, remaining with them until his retirement in 1986.

At George Washington, Alpher came under the influence of the physicist George GAMOW with whom he collaborated in a number of papers. They produced, with Hans BETHE, a major paper on the origin of the chemical elements, sometimes called the *Alpher–Bethe–Gamow theory*, which was incorporated into Gamow's modern form of the big-bang theory of the ori-

gin of the universe, published in 1948. It is said that Bethe's name was added to make the title sound like "alpha-beta-gamma," the first three letters of the Greek alphabet. In the theory it was supposed that the universe was initially very hot and dense and was composed entirely of neutrons. The neutrons decayed to protons (hydrogen nuclei), which could then capture neutrons to form deuterium nuclei. A further series of reactions produced helium. It was also proposed that a further succession of reactions, mainly the capture of neutrons, could produce other elements. Calculations on the abundance of the elements as predicted by the theory were performed by Alpher and Robert C. Herman. Although it is accepted that hydrogen and helium were indeed formed in the primitive universe by such a process, heavier elements are now known to be synthesized in the stars.

The same year, 1948, saw the publication of a remarkable paper by Alpher and Herman in which they predicted that the big bang should have produced intense radiation that gradually lost energy as the universe expanded and by now would be characteristic of a temperature of about 5 kelvins (–268°C). Unlike its later and independent formulation by Robert DICKE in 1964, the 1948 paper had surprisingly little impact. Alpher did approach a number of radar experts but was informed that it was then impossible to detect such radiation. When it was discovered by Arno PENZIAS and Robert WILSON in 1964–65 its effect was to initiate a major revolution in cosmology and astrophysics.

In 1953 Alpher and James Follin performed calculations that took account of the change in the relative numbers of protons and neutrons in the early Universe. This has been described as one of the first modern attempts to analyze the early history of the Universe. In 2001 Alpher and Herman published a book, *Genesis of the Big Bang*, on their work together and with Gamow on this topic.

Alpini, Prospero (1553–1617) *Italian botanist and physician*

Alpini (al-**pee**-nee), who was born in Venice, Italy, studied medicine at Padua University, receiving his MD in 1578. He became physician to Giorgio Emo, the Venetian consul in Egypt, and between 1580 and 1583 traveled widely in Egypt and the Greek Islands. This enabled him to make an extensive study of the Egyptian and Mediterranean floras, and he was the first European to describe the coffee and banana plants and a genus of the ginger family, later named *Alpinia*. He also studied the date palm and was the first to fertilize dates artificially, having realized that this palm has separate male and female trees. In 1593 Alpini was appointed professor of botany at Padua and he also became director of the Botanic Gardens there, introducing many Egyptian plants.

Alter, David (1807–1881) *American inventor and physicist*

Alter was born in Westmoreland County, Pennsylvania. Largely self-taught, he graduated from the Reformed Medical College in New York in 1831. During his lifetime Alter experimented at home using apparatus constructed by himself. He worked on a wide range of subjects but was little recognized for his achievements. His best work was on spectrum analysis, suggesting that each element had its own characteristic spectrum that would enable qualitative analysis to be carried out. This was later proved correct by Gustav KIRCHHOFF and Robert BUNSEN. Alter's inventions included a new method for purifying bromine, a process for obtaining oil from coal, an electric clock, and an electric telegraph that would spell out words with the aid of a pointer.

Altman, Sidney (1939–   ) *American chemist*

Born in Montreal, Canada, Altman was educated at the University of Colorado, Boulder, where he obtained his PhD in 1967. He moved to Yale in 1971, becoming professor of biology in 1980 and a naturalized U.S. citizen in 1984.

In 1982 Thomas CECH at Colorado had shown that RNA sometimes served as a biocatalyst – a role previously thought to be exclusive to protein enzymes. Cech's work was on a reaction in which the RNA was a self-catalyst. Altman set out to investigate other catalytic activity of RNA.

He worked with ribonuclease-P, an enzyme composed of both RNA and a protein, which catalyzes the processing of transfer RNA (tRNA). For the enzyme to work at the cellular level, it was thought that both protein and RNA were needed. It could, however, be possible that the RNA was merely a kind of structural support for the protein enzyme. Altman found that, *in vitro*, ribonuclease-P alone could splice the tRNA molecule at the correct place; the unaccompanied protein displayed no such activity.

Final proof came when a recombinant DNA template was used to produce only the RNA part of the ribonuclease-P. The artificial RNA still catalyzed the appropriate activity without any associated protein whatsoever. Altman had thus helped to break down the previously unquestioned dogma that molecules could either carry information, like RNA, or catalyze chemical reactions, like proteins, but they could not do both. The discovery could also throw light on the puzzle that if proteins are needed to assemble RNA, and RNA to assemble proteins, then how did the process ever get started? The answer could lie in the catalytic activity of RNA itself.

For his work on ribonuclease-P Altman shared the 1989 Nobel Prize for chemistry with Thomas Cech.

Alvarez, Luis Walter (1911–1988) *American physicist*

> There is no democracy in physics. We can't say that some second-rate guy has as much right to opinion as Fermi.
> —Quoted by D. S. Greenberg in *The Politics of Pure Science* (1967)

Alvarez, the son of a research physiologist, was born in San Francisco and educated at the University of Chicago where he gained his PhD in 1936. He moved soon after to the University of California, Berkeley. Apart from wartime work on radar at the Massachusetts Institute of Technology Radiation Laboratory (1940–43) and on the Manhattan Project at Los Alamos (1943–45), Alvarez spent his entire career at Berkeley, serving as professor of physics from 1945 until his retirement in 1978.

In 1938 Alvarez reported his first major discovery, namely the phenomenon of orbital electron capture. In 1936 Hans BETHE had argued that an excited nucleus could decay by capturing one of its own orbiting electrons, a process known as K-capture as the electron is taken from the innermost (K) electron shell. Alvarez succeeded in detecting the process experimentally by identifying the characteristic x-rays emitted during K capture as a result of electrons moving from outer orbits into the vacant K orbit.

Alvarez followed this by making (1939) the first measurement, with Felix BLOCH, of the neutron's magnetic moment. He also demonstrated that hydrogen-3 (tritium) was radioactive, work which proved to be of significance in the later development of the hydrogen bomb.

While working on radar during the war Alvarez had what he later described as one of his most valuable ideas. If radar could be used to track approaching aircraft then, he argued, the same information should be adequate to guide a pilot to a safe landing in bad weather. There were many obstacles to be overcome before GCA (Ground Controlled Approach) could be adopted. By early 1943, however, Alvarez was able to talk down a distant plane he could follow only on radar.

Soon after he moved to Los Alamos where he worked on the problem of detonating the bomb. It was necessary for 32 detonators to fire simultaneously. Alvarez was an observer in a follow-up plane of the Hiroshima bomb.

After the war Alvarez remained as creative as ever. His most important work was in the field of particle physics. By the early 1950s experimentalists had begun to find it difficult to track particles. Cloud chambers took too long to operate, emulsions could only pick up charged particles, and consequently much was being missed. In April 1953 Alvarez was introduced by Donald GLASER to the idea that particles passing through a small glass bulb containing diethyl ether would produce bubble tracks. The chamber operated by suddenly reducing the pressure causing the liquid to "boil" and leave a bubble track where a particle had passed.

Alvarez immediately began to design a much larger bubble chamber using liquid hydrogen as a fluid. After a few test runs with some small chambers Alvarez proposed to build a 72-inch model at a cost of $2.5 million. It first came into operation in March 1959 and was used to discover a large number of elementary particles. For his work in this field Alvarez was awarded the 1968 Nobel Prize for physics.

In 1962 Alvarez first saw the Giza pyramids in Egypt. Ever capable of seeing a problem calling for scientific analysis, he wondered if there were any hidden chambers in the Cephren pyramid as there were in the other Giza structures. If the pyramid was radiated with muons, more muons would pass through the less dense plane of a hidden chamber and would show up clearly on a scatter plot. After several years' persuasion of politicians and officials Alvarez's team made the appropriate measurement in 1968. At first it seemed that the pyramid contained a massive chamber and Alvarez excitedly made plans to tunnel into the tomb. Calculations quickly showed, however, that no such chamber could exist as it would have caused a collapse long ago. Eventually a computer error was detected. After several years' work Alvarez conceded that the Cephren pyramid was quite solid.

Pyramid archaeology was not the only excursion taken by Alvarez outside conventional physics. In 1977 his son Walt, a geologist, showed him a rock from Gubbio in the Italian Apennines. It was aged 65 million years and consisted of two layers of limestone, one from the Cretaceous, the other from the Tertiary, separated by a thin clay strip. During the rock's formation the dinosaurs had flourished and passed into extinction.

Alvarez was intrigued by the presence in the clay of unusually high concentrations of iridium. No more than about 0.03 parts per billion are normally to be found in the Earth's crust. The geologists, however, reported that there was 300 times as much iridium in the clay layer than in the surrounding limestone samples (an example of what is now known as the "iridium anomaly"). The clay, it was calculated, had formed over a mere 1,000 years, and was located in time at the KT boundary (K = *Kreide*, German for Cretaceous, T = Tertiary). Could the thin strip of clay and its iridium content throw any light on the mass extinctions that were taking place during its formation?

He first suggested that the iridium could have come from a nearby supernova explosion. This was soon rejected after a fruitless search in the clay for traces of plutonium–244, another supernova byproduct. Alvarez began to consider another possibility, namely, a collision with a large asteroid. It would certainly bring along with it the observed iridium, but it was not immediately apparent how the asteroid could produce a global extinction. Further reflection suggested that an asteroid 10 kilometers in diameter would throw sufficient dust into the atmosphere to darken the sky for several years. This in turn would prevent photosynthesis, destroy plant life and, along the way, all other dependent creatures.

Alvarez published his theory in 1980 and spent much of the remaining decade of his life explaining and defending his views. Some geologists objected that dinosaurs had become extinct some 20,000 years before the iridium layer was deposited. Others claimed that prolonged darkness would have been as damaging to marine as to terrestrial life, whereas marine life suffered no comparable mass extinction. Despite these and other objections Alvarez's impact theory survived the 1980s as the most favored account of the death of the dinosaurs.

Alvarez left a vivid account of his life in his *Alvarez, Adventures of a Physicist* (1987).

Alzheimer, Alois (1864–1915) *German psychiatrist*

Alzheimer (**ahlts**-hI-mer or **awlts**-hI-mer) was born in Markbreit in Germany and studied medicine at the universities of Würzburg and Berlin. After working in hospitals in Frankfurt and Heidelberg, he joined the Munich Psychiatric Clinic of Emil Kraepelin (1856–1926) as head of the anatomy department. He worked in Munich from 1904 until 1912 when he was appointed professor of psychiatry and neurology at the University of Breslau (now Wrocław in Poland).

In 1907 Alzheimer treated a 51-year-old woman with a growing memory loss. Her condition rapidly deteriorated into severe dementia. On autopsy, he identified a number of pathological conditions including shrinking of the cortex and the presence of neurofibrillary tangles and neuritic plaques. The plaques and tangles were distinctive enough to warrant a diagnosis of senile dementia or, as it later became known, *Alzheimer's disease*.

Amagat, Emile Hilaire (1841–1915) *French physicist*

Born at Saint-Satur, Amagat (am-a-**ga**) obtained his doctorate in 1872 from Paris and became a professor of physics at the Faculté Libre des Sciences at Lyons and eventually a full member of the French Academy of Sciences.

He is noted for his work on the behavior of gases. He started work plotting isotherms of carbon dioxide at high pressures, expanding the results of Thomas Andrews; this research was published in 1872 as his doctoral thesis. In 1877 followed a publication on the coefficient of compressibility of fluids, showing conclusively that this decreased with an increase in pressure, a result contradicting the results of other scientists. Between 1879 and 1882 Amagat investigated a number of gases, publishing data on isotherms and reaching the limit of pressures obtainable using glass apparatus – about 400 atmospheres. To get yet further Amagat invented a hydraulic manometer that could produce and measure up to 3200 atmospheres. (This manometer was later used in firearms factories for testing purposes.)

Amaldi, Edoardo (1908–1989) *Italian physicist*

Amaldi (am-**al**-dee) was born in Carpaneto in Italy; in 1929 he graduated from the University of Rome, where he had studied under Enrico FERMI. Together with Fermi and others he discovered that fast-moving neutrons are slowed down (moderated) by substances containing hydrogen, and can thus be brought to energies at which they more easily interact with nuclei. He also contributed to the study of tau mesons, hyperons, and antiprotons.

In 1937 he was made professor of general physics at the University of Rome, and from 1952 until 1954 he was secretary-general of the European Organization for Nuclear Research. He also served as president of the International Union of Pure and Applied Physics (1957–60) and president of the Istituto Nazionale di Fisica Nucleare (1960–65).

Amaldi played a key role in the creation and development of CERN (Conseil européen pour la Recherche nucléaire; European Laboratory for Particle Physics). On a visit to Europe in 1948 Luis ALVAREZ had asked Amaldi why the Europeans did not "do physics jointly" and build, as several American universities had done, a large accelerator. Amaldi began to lobby for such a project and persuaded a reluctant UK to enter, having overcome, in a stormy meeting, the strong objections of Lord Cherwell. When, in 1952, eleven governments set up CERN Amaldi was offered the post of director general. He refused, accepting instead the position of vice director in charge of administration, returned to his university post in 1954, and concentrated on research into the detection of gravitational waves.

Ambartsumian (*or* Ambartsumyan), Viktor Amazaspovich (1908–1996) *Armenian astrophysicist*

Ambartsumian (am-bart-**soo**-mee-an) was born in Tbilisi (now in Georgia), the son of a distinguished Armenian philologist. He graduated from the University of Leningrad in 1928 and did graduate work at Pulkovo Observatory, near Leningrad, from 1928 to 1931. He was professor of astrophysics from 1934 to 1946 at Leningrad and held the same post at the State

University at Yerevan in Armenia from 1947 until his death. In 1946 he organized the construction, near Yerevan, of the Byurakan Astronomical Observatory, having been appointed its director in 1944. He remained as director until 1988.

Ambartsumian's work was mainly concerned with the evolution of stellar systems, both galaxies and smaller clusters of stars, and the processes taking place during the evolution of stars. The idea of a stellar "association" was introduced into astronomy by Ambartsumian in 1947. Associations are loose clusters of hot stars that lie in or near the disk-shaped plane of our galaxy. They must be young, no more than a few million years old, as the gravitational field of the galaxy will tend to disperse them. This must mean that star formation is still going on in the galaxy.

He also argued in 1955 that the idea of colliding galaxies proposed by Rudolph MINKOWSKI and Walter BAADE to explain such radio sources as Cygnus A would not produce the required energy. Instead, he proposed that the source of energy was gigantic explosions occurring in the dense central regions of galaxies and these would be adequate to provide the 10^{55} joules emitted by the most energetic radio sources.

Amdahl, Gene Myron (1922–   ) *American computer engineer*

The son of a South Dakota farmer, Amdahl was educated at South Dakota State College and, following war service, the University of Wisconsin, where he received his doctorate in 1952. Shortly after, he joined IBM to work on the design of their new mainframe computers, the 701, 704, and 709. Amdahl left IBM in 1955 in a disagreement over the minor role he had been assigned in the development of their new model, the 7030. He returned to IBM in 1960, after working for the Ford subsidiary Aeronautics on systems design, and began work on the IBM 360 series. This was one of the first models to incorporate integrated circuits and proved to be one of the most successful mainframes ever produced.

In 1965 Amdahl was made an IBM Fellow, a five-year appointment allowing him to work on any project he chose. During this period he founded the IBM Advanced Computing Systems Lab in Menlo Park, California.

Before the expiry of his fellowship Amdahl resigned once more. This time the disagreement arose over IBM's traditional policy of pricing mainframes on the basis of their computing power rather than their cost. Such an approach, Amdahl argued, gave IBM no incentive to build more powerful machines. In 1970 he founded the Amdahl Corporation with the aim of designing IBM-compatible mainframes. Large-scale integration chips were just coming into use and, while previously designers had struggled to link 25–30 circuits on a chip, Amdahl aimed to have 100. Within three years Amdahl's new computer, the Amdahl 470, three times more powerful than the IBM 360s but no more expensive at $3.5 million, had created sales of $320 million. IBM responded with price cuts and Amdahl consequently sold his company in 1979 to Fujitsu.

He immediately started his second company, Trilogy, using the new technology of wafer-scale integration which, in theory, should allow 2,000 chips to be replaced by 20 wafers. Amdahl was predicting sales of $1 billion within two years. The technology, however, was too new and the models proved to be less powerful than expected. They also had a tendency to short-circuit and came onto the market three years behind schedule. Amdahl left the mainframe business and conceded the market to IBM. Trilogy merged with Elxsi in 1985.

Amdahl subsequently founded other companies: Amdor International (in 1987) and Commercial Data Servers (in 1995). Since 2004 he has been on the board of advisors of Massively Parallel Technologies.

Amici, Giovanni Battista (1786–1863)
Italian astronomer and instrument maker

Amici (a-**mee**-chee) was professor of mathematics at the University of Modena and in 1835 became the director of the observatory at the Royal Museum in Florence. He made great improvements in the design of parabolic mirrors for reflecting telescopes, and constructed and designed prismatic spectroscopes. In 1840 he made two achromatic objective lenses with diameters of 9.5 and 11 inches (24 and 28 cm), which were used by Giovanni DONATI. He also made advances in microscopy, improving the compound microscope and using it to study plant reproduction.

Amontons, Guillaume (1663–1705)
French physicist

Amontons (a-mon-**ton**), a Parisian, who had been deaf since childhood, invented and perfected various scientific instruments. In 1687 he made a hygrometer (an instrument for measuring moisture in the air); in 1695 he produced an improved barometer; and in 1702–03 a constant-volume air thermometer. In 1699 he published the results of his studies on the effects of change in temperature on the volume and pressure of air. He noticed that equal drops in temperature resulted in equal drops in pressure and realized that at a low enough temperature the volume and pressure of the air would become zero – an early recognition of the idea of absolute zero. These results lay largely unnoticed and the relationship between temperature and pressure of gases was not reexamined

until the next century (by scientists such as Jacques CHARLES).

Amontons also published in 1699 the results of his studies on friction, which he considered to be proportional to load.

Ampère, André Marie (1775–1836)
French physicist and mathematician

Ampère (ahm-**pair**) was born in Lyons, France, where his father was a wealthy merchant. He was privately tutored, and to a large extent self-taught. His genius was evident at an early age. He was particularly proficient at mathematics and, following his marriage in 1799 he was able to make a modest living as a mathematics teacher in Lyons. In 1802 he moved first to Bourg-en-Bresse to take up an appointment, then to Paris as professor of physics and chemistry at the Ecole Centrale.

His first publication was on the statistics of games of chance *Considérations sur la théorie mathématique de jeu* (1802; Considerations on the Mathematical Theory of Games) and his work at Bourg led to his appointment as professor of mathematics at the Lyceum of Lyons, and then in 1809 as professor of analysis at the Ecole Polytechnique in Paris. His talents were recognized by Napoleon, who in 1808 appointed him inspector general of the newly formed university system – a post Ampère held until his death.

Ampère's most famous scientific work was in establishing a mathematical basis for electromagnetism. The Danish physicist Hans Christian OERSTED had made the important experimental discovery that a current passing through a wire could cause the movement of a magnetic compass needle. Ampère witnessed a demonstration of electromagnetism by François ARAGO at the Academy of Science on 11 September 1820. He set to work immediately on his own investigations, and within seven days was able to report the results of his experiments.

In a succession of presentations to the academy in the next four months, he developed a mathematical theory to explain the interaction between electricity and magnetism, to which he gave the name "electrodynamics" (now more commonly: electromagnetism) to distinguish it from the study of stationary electric forces, which he christened "electrostatics."

Having recognized that electric currents in wires caused the motion of magnets, and that a magnet can affect another magnet, he looked for evidence that electric currents could influence other electric currents. The simplest example of this interaction is found by arranging for currents to flow through two parallel wires. Ampère discovered that if the currents passed in the same direction the wires were attracted to each other, but if they passed in opposite directions the wires were repelled. From this he went on to consider more complex configurations of closed loops, helices, and other geometrical figures, and was able to provide a mathematical analysis that allowed quantitative predictions.

In 1825 he had been able to deduce an empirical law of forces (*Ampère's law*) between two current-carrying elements, which showed an inverse-square law (the force decreases as the square of the distance between the two elements, and is proportional to the product of the two currents). By 1827 he was able to give a precise mathematical formulation of the law, and it was in this year that his most famous work *Mémoirs sur la théorie mathématique des phénomènes electrodynamiques uniquement déducte de l'expérience* (Notes on the Mathematical Theory of Electrodynamic Phenomena, Solely Deduced from Experiment) was published.

Besides explaining the macroscopic effects of electromagnetism, he attempted to construct a microscopic theory that would fit the phenomenon, and postulated an electrodynamic molecule in which electric-fluid currents circulated, giving each molecule a magnetic field.

In his honor, the unit of electric current is named for him, and in fact the ampere is defined in terms of the force between two parallel current-carrying wires.

Anaxagoras of Clazomenae (about 500 BC–428 BC) *Greek philosopher*

Anaxagoras (an-ak-**sag**-o-rus) left his birthplace in Asia Minor (now Turkey) in about 480 BC and taught in Athens during its most brilliant period under Pericles, who was himself one of Anaxagoras's pupils. In about 450 BC he was exiled to Lampsacus after being prosecuted for impiety by the enemies of Pericles.

Although he wrote a book, *On Nature*, only fragments of his writings survive; his work is known through later writers, notably ARISTOTLE and Simplicius, and is open to contradictory interpretations. The difficulty consists in reconciling his principle of homoemereity, which states that matter is infinitely divisible and retains its character on division, with his statement "there is a portion of everything in everything." His work can be seen as a criticism of the Eleatic school of PARMENIDES and ZENO of Elea, who had argued against plurality and even motion.

Anaxagoras's astronomy was more rational than that of his predecessors; he stated that the Sun and stars were incandescent stones, that the Moon derived its light from the Sun, and he gave the modern explanation for eclipses of the Sun and Moon.

Anaximander of Miletus (c. 611 BC–c. 547 BC) *Greek philosopher*

Anaximander...said that a certain infinite nature is first principle of the things that exist. From it come the heavens and the worlds in them.
—Hippolytus, *Refutation of All Heresies* (2nd century AD)

Anaximander (a-nak-si-**man**-der), who was born and died in Miletus (now in Turkey), belonged to the first school of natural philosophy and was the pupil of THALES. He wrote one of the earliest treatises but none of his writings survive and his work is known only through later writers, notably ARISTOTLE and THEOPHRASTUS.

Anaximander criticized Thales's idea that water was the basic element of the universe by pointing out that no one element gains the upper hand and that "they make retribution and pay the penalty to one another ... according to the ordering of time." From this he deduced that the primal matter was what he called the *apeiron* or the indefinite. This idea was later developed by the atomists. He was the first to realize that the Earth did not have to float on water or be supported in any way; he stated that it was in equilibrium with the other bodies in the universe.

Anaximander was the first philosopher to speculate on the origin of man. He is also credited with the first determinations of the solstices and equinoxes and the production of the first map of the world as he knew it. He was the first to recognize that the Earth's surface is curved but believed it was curved only in the north–south direction and consequently represented the Earth as a cylinder.

Anaximenes of Miletus (about 546 BC) *Greek philosopher*

Anaximenes (an-ak-**sim**-e-neez) was the last of the great Milesian philosophers. He was probably a pupil of ANAXIMANDER of Miletus and, like THALES before him, he identified one of the tangible elements as the primal substance. For Anaximenes this was air, which by processes of condensation and rarefaction could produce every other kind of matter. He used the rather mystical argument that since air is the breath of life for man it must also be the main principle of the universe.

Anderson, Carl David (1905–1991) *American physicist*

Anderson, the son of Swedish immigrants, was born in New York City and educated at the California Institute of Technology where he obtained his PhD in 1930 and where he remained for his entire career, serving as professor of physics from 1939 until his retirement in 1978.

Anderson was deeply involved in the discovery of two new elementary particles. In 1930 he began to study cosmic rays by photographing their tracks in a cloud chamber and noted that particles of positive charge occurred as abundantly as those of negative charge. The negative particles were clearly electrons but those of positive charge could not be protons (the only positive particles known at the time) as they did not produce sufficient ionization in the chamber. Eventually Anderson concluded that such results "could logically be interpreted only in terms of particles of a positive charge and a mass of the same order of magnitude as that normally possessed by a free negative electron." It was in fact the positron or positive electron, whose existence he announced in September 1932. In the following year his results were confirmed by Patrick BLACKETT and Giuseppe OCCHIALINI and won for Anderson the 1936 Nobel Prize for physics.

In the same year Anderson noted some further unusual cosmic-ray tracks. As they appeared to be made by a particle more massive than an electron but lighter than a proton it was at first thought to be the particle predicted by Hideki YUKAWA that was thought to carry the strong nuclear force and hold the nucleus together. The particle was initially named the "mesotron" or "yukon." However, this identification proved to be premature, as its interaction with nucleons was found to be so infrequent that it could not possibly perform the role described by Yukawa. From 1938 the particle became known as the meson, and the confusion was partly dispelled in 1947 when Cecil POWELL discovered another and more active meson, to be known as the pi-meson or pion to distinguish it from Anderson's mu-meson or muon. While the role of the pion is readily explained, that of Anderson's muon is still far from clear.

Anderson, Philip Warren (1923–) American physicist

Anderson was born in Indianapolis and obtained his BS (1943), MS (1947), and PhD (1949) at Harvard University, doing his doctoral thesis under John VAN VLECK. The period 1943–45 was spent at the Naval Research Laboratory working on antenna engineering. Upon receiving his doctorate, Anderson joined the Bell Telephone Laboratories at Murray Hill, New Jersey, where he worked until 1984. He is currently Joseph Henry Professor of Physics at Princeton.

Anderson's main research has been in the physics of the solid state, incorporating such topics as spectral-line broadening, exchange interactions in insulators, the JOSEPHSON effect, quantum coherence, superconductors, and nuclear theory. Under Van Vleck he worked initially on elucidating the phenomenon of pressure broadening of lines in microwave, infrared, and optical spectroscopy. In 1959 he developed a theory to explain "superexchange" – the coupling of spins of two magnetic atoms in a crystal through their interaction with a nonmagnetic atom located between them. He went on to develop the theoretical treatments of antiferromagnetics, ferroelectrics, and superconductors.

In 1961 Anderson conceived a theoretical model to describe what happens where an impurity atom is present in a metal – now widely known and used as the *Anderson model*. Also named for him is the phenomenon of *Anderson localization*, describing the migration of impurities within a crystal. In the 1960s Anderson concentrated particularly on superconductivity and superfluidity, predicting the existence of resistance in superconductors and (with Pierre Morel) pointing out the nature of the possible superfluid states of ^{3}He. In 1971 he returned to disordered media, working on low-temperature properties of glass and later studying spin glasses and high-temperature superconductivity.

Along with his Harvard tutor Van Vleck and the British physicist Nevill MOTT, Anderson shared the 1977 Nobel Prize for physics "for their fundamental theoretical investigation of the electronic structure of magnetic and disordered systems."

In the late 1980s Anderson became a controversial figure in the physics community by arguing before Congress that the proposed SSC (Superconducting Super Collider) to be built in Texas at a cost of $8 billion would yield neither practical benefits nor any fundamental truths that could not be gained elsewhere and more cheaply. When Congress killed the plan in 1993 Anderson commented that he was only sorry that Congress had allowed the project to go on so long.

Anderson, Thomas (1819–1874) British organic chemist

Born in the port of Leith in Scotland, where his father was a physician, Anderson also studied medicine, gaining his MD from Edinburgh in 1841. His main interest, however, lay in chemistry which he pursued in 1842 with BERZELIUS in Sweden and with LIEBIG in 1843 in his famous Giessen Laboratory. Anderson was appointed Regius Professor of Chemistry at the University of Glasgow in 1852.

Anderson's most important work was concerned with the chemistry of pyridine (C_5H_5N), a heterocyclic compound, which he extracted in 1851 from bone oil. He had earlier (1849) extracted the related substances, lutidine (C_5H_3N), picoline (C_5H_4N), and collidine (C_5H_2N). Anderson went on to suggest that he had discovered an homologous series in which the radical C_5H_n was being substituted for H_3 in the ammonia molecule. The pyridines are used as powerful and versatile organic solvents capable of dissolving fats, mineral oils, and rubber.

Anderson also did much to introduce the "agricultural chemistry" of Liebig into Scotland. He worked from 1848 as a chemist with the Highland and Agricultural Society, for which he determined the chemical composition of such staples as turnip, wheat, and beans. Anderson wrote one of the earliest texts on the subject in his *Elements of Agricultural Chemistry* (1860).

Andrade, Edward Neville da Costa
(1887–1971) *British physicist*

Born in London, Andrade was educated at University College, London, and at Heidelberg where he obtained his PhD in 1911. He then worked with Ernest RUTHERFORD at Manchester before joining the Royal Artillery in 1914. After the war he was appointed professor of physics at the Artillery College, Woolwich, and moved to a similar chair at University College, London, in 1928. Andrade resigned in 1950 to take up the directorship of the Royal Institution, a post he held until his retirement in 1952.

Andrade worked mainly on the physics of metals and the viscosity of liquids. On the former subject he made the first serious scientific study of creep in metals while on the latter subject he investigated the effect of an electric field on viscosity. In addition to writing a number of popular works on science, Andrade was also widely known as a student of 17th-century physics. He was an expert on Robert HOOKE and as chairman in 1947 of the Royal Society Newton letters committee he played an important role in beginning the monumental task of the publication of Newton's letters, a task requiring 40 years for its completion (from 1938).

Andreessen, Marc (1971–   ) *American computer scientist*

Andreesen (**an**-dree-sen) was born in New Lisbon, Wisconsin, and was educated at the University of Illinois at Urbana.

In his senior year at Illinois he worked at the National Center for Supercomputer Applications (NCSA), a high-tech think tank, where he came across the Internet, then in its early days. He immediately realized that a browser that could be easily used would be very useful. In a few weeks Andreessen, together with Eric BINA and some colleagues from NCSA, produced an operational browser called *Mosaic*.

In 1994 Andreessen and Jim Clark founded a company called Mosaic Corp. with the intention of making Internet browsers. They soon changed the name to Netscape Communications after legal objections to the previous name by the University of Illinois. Andreessen and his team, including Bina, produced an improved version of Mosaic, which they called *Navigator*.

Navigator was initially very successful but was eclipsed when Microsoft introduced its own browser *Internet Explorer*. In 1998 Andreessen became technology officer of America Online (AOL), which had bought out Netscape. He left AOL in 1999 with the intention of playing a more direct role in the development of the Internet. He is currently involved in a number of projects including Ning, a platform for creating social websites and networks.

Andrews, Roy Chapman (1884–1960)
American naturalist and paleontologist

Andrews was born in Beloit, Wisconsin, and was educated there at Beloit College. After graduating, he took up a post at the American Museum of Natural History, New York. His early interest lay in whales and other aquatic mammals, and these he collected assiduously on a number of museum-sponsored expeditions to Alaska, North Korea, and the Dutch East Indies (Indonesia) between 1908 and 1913. It was largely through Andrews's efforts that the collection of cetaceans at the American Museum of Natural History became one of the most complete in the world.

Andrews is best known for his discovery of previously unknown Asiatic fossils. Most of his findings were made on three expeditions to Asia, which he led as chief of the Asiatic Exploration Division of the American Museum of Natural History. The first of these was to Tibet, southwestern China, and Burma (1916–17); he then visited northern China and Outer Mongolia (1919), and central Asia (1921–22 and 1925). The third Asian expedition produced major finds of fossil reptiles and mammals, including remains of the largest known land mammal, the Paraceratherium (formerly called Baluchitherium), an Oligocene relative of the modern rhinoceros, which stood some 17–18 feet (5.5 m) at the shoulder. In Mongolia, Andrews discovered the first known fossil dinosaur eggs. He was also able to trace previously unknown geological strata, and unearthed evidence of primitive human life on the central Asian plateau.

Andrews was appointed director of the American Museum of Natural History in 1935, but resigned in 1942 in order to devote himself entirely to writing about his travels and discoveries.

Andrews, Thomas (1813–1885) *Irish physical chemist*

> There exists for every liquid a temperature at which no amount of pressure is sufficient to retain it in the liquid form.
> —Quoted by W. A. Miller in *Elements of Chemistry* (1863)

The son of a linen merchant from Belfast (now in Northern Ireland), Andrews studied chemistry under Thomas Thomson at Glasgow, under Jean DUMAS in Paris, and under Justus von LIEBIG at Giessen. He also studied medicine at Edinburgh and obtained his MD in 1835. He practiced medicine in Belfast before becoming vice-president of Queen's College, Belfast, in 1845 and professor of chemistry in 1849.

Andrews made experimental studies on the heat evolved in chemical reactions and also showed that ozone is an allotrope of oxygen. He was a brilliant experimentalist and his work on the liquefaction of gases brought order to a confused subject. Andrews performed a famous series of experiments on the variation of the volume of carbon dioxide gas with pressure. He studied the behavior of the gas at different temperatures, and showed that there was a certain temperature – the critical temperature – above which the gas could not be liquefied by pressure alone. This work, which was published as *On the Continuity of the Liquid and Gaseous States of Matter* (1869) led to the liquefaction of those gases previously held to be "permanent" gases.

Anfinsen, Christian Boehmer (1916–1995) *American biochemist*

> Anfinsen was a true pioneer in the field of protein structure and protein folding.
> —Daniel Nathans

Born in Monessen, Pennsylvania, Anfinsen was educated at Swarthmore College, the University of Pennsylvania, and Harvard, where he obtained his PhD in 1943. He taught at Harvard Medical School from 1943 to 1950, when he moved to the National Heart Institute at Bethesda, Maryland, where from 1952 to 1962 he served as head of the laboratory of cellular physiology. In 1963 Anfinsen joined the National Institute of Arthritis and Metabolic Diseases at Bethesda, where he was appointed head of the laboratory of chemical biology. In 1982 he became professor of biology at Johns Hopkins University.

By 1960 Stanford MOORE and William STEIN had fully determined the sequence of the 124 amino acids in ribonuclease, the first enzyme to be so analyzed. Anfinsen, however, was more concerned with the shape and structure of the enzyme and the forces that permit it always to adopt the same unique configuration. The molecule of ribonuclease – a globular protein – consists of one chain twisted into a ball and held together by four disulfide bridges. By chemical means, the sulfur bridges can be separated so that the enzyme becomes a simple polypeptide chain with no power to hydrolyze ribonucleic acid, i.e., it becomes denatured. Once the bridges are broken they can be reunited in any one of 105 different ways. Anfinsen found that the minimum of chemical intervention – merely putting the enzyme into a favorable environment – was sufficient to induce the ribonucle-

ase to adopt the one configuration that restores enzymatic activity.

The important conclusion Anfinsen drew from this observation was that all the information for the assembly of the three-dimensional protein must be contained in the protein's sequence of amino acids – its primary structure. He went on to show similar behavior in other proteins. For this work Anfinsen shared the 1972 Nobel Prize for physiology or medicine with Moore and Stein.

Ångström, Anders Jonas (1814–1874) *Swedish physicist and astronomer*

Ångström (**ang**-strom or **awng**-stru(r)m) was born the son of a chaplain in Lögdö, Sweden. He studied and taught physics and astronomy at the University of Uppsala, where he obtained his doctorate (1839) and later became professor of physics (1858), a position he held up to his death.

Ångström was one of the pioneers of spectroscopy. His most important work was *Optiska Undersökningar* (1853; Optical Investigations), in which he published measurements on atomic spectra, particularly of electric sparks. He noted spectral lines that were characteristic of both the gas and the electrodes used. Ångström applied EULER's theory of resonance to his measurements and deduced that a hot gas emits light at precisely the same wavelength at which it absorbs light when it is cool. In this he anticipated the experimental proof of Gustav KIRCHHOFF. He was also able to show the composite nature of the spectra of alloys.

Having established the principles of spectroscopy in the laboratory, Ångström turned his attention to the Sun's spectrum, publishing *Recherches sur le spectre solaire* (1868; Researches on the Solar Spectrum) in which he made the inference that hydrogen was present in the Sun. In this work he also reported the wavelengths of some 1,000 FRAUNHOFER lines measured to six significant figures in units of 10^{-8} centimeters. Since 1905 his name has been officially honored as a unit of length used by spectroscopists and microscopists; 1 angstrom (Å) = 10^{-8} centimeters. His map of the *Normal Solar Spectrum* (1869) became a standard reference for some 20 years. Ångström was also the first to examine the spectrum of the aurora borealis and to measure the characteristic bright yellow–green light sometimes named for him.

Anning, Mary (1799–1847) *British fossil hunter*

Anning was born in Lyme Regis, Dorset, on the south coast of England, the daughter of a cabinetmaker. Her father died in 1810, leaving his widow and two surviving children in debt. To make a meager living, the family turned to hunting for fossils along the cliffs bordering the small seaside town, and selling them to collectors and museums. From this modest beginning, and in spite of her poverty and low social class, Mary Anning defied the conventions of her age to become well known and respected among the male-dominated scientific community. In 1817 Lieutenant-Colonel Thomas Birch, an affluent amateur collector of fossils, became the family's patron, and he arranged to sell his own collection for the benefit of the Anning family. During this time, Mary was effectively running the family fossil business, acquiring a wealth of anatomical knowledge about the specimens she gathered and prepared for sale, and becoming a prominent figure in Lyme and beyond. She is credited with several major discoveries of prehistoric marine reptiles, including the first fossil ichthyosaur (in about 1809–1811; now in the Natural History Museum, London), and also the first plesiosaur (in 1821). Although many of Anning's finds were sold on with little or no credit given to their discoverer, her reputation grew, and with it came scientific recognition and reward. From 1838 she received an annuity from the British Association for the Advancement of Science, and shortly before her death she was named as the first honorary member of the newly founded Dorset County Museum.

Antoniadi, Eugène Michael (1870–1944) *Greek–French astronomer*

Antoniadi (an-ton-**yah**-dee) was born in Constantinople (now Istanbul, Turkey). He established quite early a reputation as a brilliant observer and in 1893 was invited by Camille FLAMMARION to work at his observatory at Juvisy near Paris. From 1909 he worked mainly with the 33-inch (84-cm) refracting telescope at the observatory at Meudon. He became a French citizen in 1928.

In his two works *La Planète Mars* (1930; The Planet Mars) and *La Planète Mercure* (1934; The Planet Mercury), Antoniadi published the results of many years' observations and presented the best maps of Mars and Mercury to appear until the space probes of recent times. With regard to Mars he took the strong line: "Nobody has ever seen a genuine canal on Mars," attributing the "completely illusory canals," "seen" by astronomers such as Percival LOWELL and Flammarion, to irregular natural features of the Martian surface. Antoniadi also observed the great Martian storms of 1909, 1911, and 1924 noting, after the last one, that the planet had become covered with yellow clouds and presented a color similar to Jupiter.

On Mercury his observations made between 1914 and 1929 seemed to confirm Giovanni SCHIAPARELLI's rotation period of 88 days, identical with the planet's period of revolution

around the Sun. The effect of this would be for Mercury always to turn the same face to the Sun, in the same way as the Moon always turns the same face to the Earth. Antoniadi cited nearly 300 observations of identifiable features always in the same position, as required by the 88-day rotation period.

However, radar studies of Mercury in 1965 revealed a 59-day rotation period for Mercury. This time is however very close to half the synodic period of Mercury (116 days) so that when the planet returns to the same favorable viewing position in the sky, at intervals of 116 days, it does present the same face to observers.

Antoniadi also wrote on the history of astronomy, publishing *L'Astronomie Egyptienne* (Egyptian Astronomy) in 1934.

Apker, Leroy (1915–1978) *American physicist*

Apker was born in Rochester, New York, and educated there at the University of Rochester; he obtained his PhD in 1941. He later joined the staff of General Electric where he worked as a research associate.

Apker worked on the photoelectric effect applied to semiconductors and ionic crystals. A particular aspect of his work was photoemission from the alkali-metal halides, such as potassium iodide. Crystals of these compounds can contain a type of defect involving a missing negative ion replaced by an electron – called a "color center" (because such defects color the crystal). Apker studied the photoemission from such crystals, and the interaction of excitons with color centers causing the ejection of electrons.

Apollonius of Perga (*c.* 262 BC–*c.* 190 BC) *Greek mathematician and geometer*

Apollonius (ap-o-**loh**-nee-us) moved from his birthplace Perga (now in Turkey) to study in the Egyptian city of Alexandria, possibly under pupils of EUCLID. Later he taught in Alexandria himself. One of the great Greek geometers, Apollonius's major work was in the study of conic sections and the only one of his many works to have survived is his eight-book work on this subject, the *Conics*. Apollonius's work on conics makes full use of the work of his predecessors, notably Euclid and CONON of Samos, but it is a great advance in terms of its thoroughness and systematic treatment. The *Conics* also contains a large number of important new theorems that are entirely Apollonius's creation. He was the first to define the parabola, hyperbola, and ellipse. In addition, he considered the general problem of finding normals from a given point to a given curve (i.e., lines at right angles to a tangent at a point on the curve).

Apart from the geometrical work that has

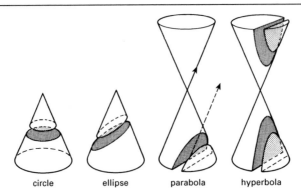

circle   ellipse   parabola   hyperbola

*CONIC SECTIONS   Apollonius showed that the circle, ellipse, parabola, and hyperbola could be
generated by different sections of a conical surface.*

survived, Apollonius is known to have contributed to optics – in particular to the study of the properties of mirrors of various shapes. This work, however, is now lost.

Appel, Kenneth (1932–) *American mathematician*

Appel, who was born in Brooklyn, New York City, was educated at the University of Michigan, where he completed his PhD in 1959. After working for two years at the Institute for Defense Analysis at Princeton, he joined the faculty of the University of Illinois, Urbana, where he served as professor of mathematics from 1991 to 1993. He then took up the chairmanship of the mathematics department at the University of New Hampshire until 2002.

In 1976, in collaboration with Wolfgang HAKEN, Appel announced the solution to one of mathematics long-standing unsolved problems, the four-color map problem. In 1852 Francis Guthrie had noticed that it seemed to be possible to color any map, assuming countries with common borders were colored differently, with no more than four colors. Guthrie was sufficiently intrigued by the point to raise it with the mathematician de Morgan and ask for a proof of the conjecture. De Morgan found the problem unexpectedly difficult, as did succeeding generations of mathematicians.

Appel and Hagen used a variation of a method first tried by Arthur Kempe in 1879. It depends on the fact that maps must contain certain unavoidable configurations – Appel and Hagen recognized 1,482 of these. They then used a computer to show that all of these could be reduced to four-color configurations. They began work in 1972, but it was not until 1976 that they were satisfied with their analysis and their program. It took more than 1,200 hours of computer time to prove the theorem.

Appert, Nicolas-François (1749–1841) *French inventor*

Appert (a-**pair**), who was born in Chalôns-sur-Marne, France, was a chef, confectioner, and distiller who invented the canning of food. In 1795 he started to experiment with sealed containers, using corked glass sealed with wax. He succeeded in preserving fruits, soups, marmalades, etc., for several years. To claim a 12,000-franc award in 1810 he published his findings in *L'art de conserver, pendant plusiers années, toutes les substances animal et végétales* (The Art of Preserving All Kinds of Animal and Vegetable Substances for Several Years). He used the money to set up the first commercial cannery in the world, the House of Appert, at Massy, which was open from 1812 to 1833.

Appleton, Sir Edward Victor (1892–1965) *British physicist*

Appleton was born in Bradford, England, and studied physics at Cambridge University from 1910 to 1913. During World War I, while he was serving in the Royal Engineers, he developed the interest in radio that was to influence his later research. After the war he returned to Cambridge and worked in the Cavendish Laboratory from 1920. In 1924 he was appointed Wheatstone Professor of Experimental Physics at King's College, London.

Here, in his first year, he used a BBC transmitter to conduct a famous experiment, which established beyond doubt the presence of a layer of ionized gas in the upper atmosphere capable of reflecting radio waves. The existence of such a layer had been postulated by Oliver HEAVISIDE and Arthur KENNELLY to explain MARCONI's transatlantic radio transmissions. By varying the frequency of a transmitter in Bournemouth and detecting the signal some 140 miles (225 km) away in Cambridge, he showed that interference occurred between direct (ground) waves and waves reflected off the layer (sky waves). Furthermore, the experiment measured the height of the layer, which he estimated at 60 miles (96 km). He proceeded to do theoretical work on the reflection or transmission of radio waves by an ionized layer and found, using further measurements, a second layer above the Heaviside–Kennelly layer. The *Appleton layer* undergoes daily fluctuations in ionization and he established a link between these variations and the occurrence of sunspots.

In 1936 he became the Jacksonian Professor of Natural Philosophy at Cambridge, and during the war years until 1949 he was secretary of the department of scientific and industrial research, in which period he led research into radar and the atomic bomb.

For his great achievements in ionospheric physics he was knighted in 1941 and in 1947 won the Nobel Prize for physics. From 1949 until his death he was principal of Edinburgh University.

Arago, Dominique François Jean (1786–1853) *French physicist*

> To get to know, to discover, to publish – this is the destiny of a scientist.
> —Quoted by A. L. MacKay in *A Harvest of a Quiet Eye* (1977)

Born in Estagel, France, Arago (a-ra-**goh** or **a**-ra-goh) was educated at the Ecole Polytechnique in Paris and then spent some years in Spain, where he accompanied Jean Baptiste Biot on a measurement of an arc of meridian. On his return to Paris in 1809 he was elected to the Académie des Sciences and received the chair of analytical geometry at the Ecole Polytechnique. In 1830 he succeeded J. B. J. Fourier as the permanent secretary of the Ecole Polytechnique. Arago worked in a number of branches of physics.

His first investigations concerned the polarization of light and in 1811 he discovered chromatic polarization. He was a vigorous defender of A. J. Fresnel's wave theory of light against the criticisms of Laplace and Biot, who both supported the corpuscular theory. In 1838 he described an experiment to decide the issue by comparing the speed of light in air with that in a denser medium. Shortly before Arago's death, Léon Foucault and Armand Fizeau proved that the experiment supported the wave theory.

Arago also worked on electromagnetism, showing that a coil of wire carrying a current could act as a magnet. He also found that a rotating copper disk could deflect a magnetic needle suspended above it. (This arrangement, known as *Arago's disk,* depends on magnetic induction.)

In astronomy, Arago discovered the Sun's chromosphere. He also played a part in the discovery of Neptune by Urbain Le Verrier.

Arago was a fierce republican and, from 1830 onward, he was involved in political life as deputy for the Pyrenées Orientales. In 1848 he became a government minister and, among other measures, abolished slavery in the French colonies.

Arber, Werner (1929–) *Swiss microbiologist*

Arber (**ar**-ber), who was born in Gränichen, Switzerland, graduated from the Swiss Federal Institute of Technology in 1953 and gained his PhD from the University of Geneva in 1958. He spent a year at the University of Southern California before returning to Geneva where he became professor of molecular genetics in 1965. In 1971 Arber moved to Basel to take the chair of molecular biology.

In the early 1950s Giuseppe Bertani reported a phenomenon he described as "host-controlled variation" in which phages (the viruses that infect bacteria) successfully growing on one host found it difficult to establish themselves on a different bacterium. In 1962, he proposed that bacteria possess highly specific enzymes capable of destroying invading phages by cutting up their DNA. The existence of such "restriction enzymes," as they came to be called, was later established by Hamilton Smith.

It turned out that, as Arber had proposed, the enzymes attack the invading DNA at a specific site, always cutting them at exactly the same place. It was this property that endowed restriction enzymes with such interest for if strands of DNA could be so manipulated to be cut at particular known points, it only needed the power to join such strands together in desired combinations for genetic engineering to be a reality. As restriction enzymes were found to leave DNA strands "sticky" and ready to combine with certain other "sticky" strands it was soon apparent to molecular biologists that genetic engineering was at last a practical proposition.

For his work on restriction enzymes Arber shared the 1978 Nobel Prize for physiology or medicine with Smith and Daniel Nathans.

Archimedes (287 BC–212 BC) *Greek mathematician*

> Give me a lever long enough, and a firm place to stand on, and with my own weight I will move the Earth.
> —On the Lever

The father of Archimedes (ar-ki-**mee**-deez) was an astronomer and he himself inherited an interest in the subject. He was educated in Alexandria and spent most of the rest of his life in his birthplace, Syracuse, under the patronage of King Hieron. Archimedes was without question the greatest mathematician and scientist that classical Greek civilization produced and is usually considered to be one of the greatest mathematicians of all time. He was held in very high regard even by his contemporaries, and Karl Friedrich GAUSS thought that only Isaac NEWTON was Archimedes's equal as a mathematician. Archimedes was as much an applied mathematician as a pure mathematician. He was very much interested in putting his theoretical discoveries to practical use and is known to have been skilled in making his own equipment and carrying out his own experiments. It is no exaggeration to describe Archimedes as the creator of the science of mechanics. Naturally before his time many isolated facts had been discovered, but it was only with him that mechanics became a unified body of theory capable of yielding new and unexpected practical applications.

Archimedes was able to find methods for determining the center of gravity of a variety of bodies. He also gave the first general theory of levers, and organized a practical demonstration to show how, with a suitable series of levers, a very small force is capable of moving a very large weight. He amazed his contemporaries by arranging for the king of Syracuse to move a large ship simply by pressing a small lever. In connection with his work on levers Archimedes made one of his famous statements, "Give me a firm place to stand on and I will move the Earth." Archimedes also had a practical interest in optics, although no writings of his on the subject have come down to us. He put all this newfound theoretical knowledge to deadly effect when Syracuse was besieged by the Romans, by designing and building a variety of war machines. Among these were enormous mirrors to focus the Sun's rays and set fire to the Roman ships, and a variety of catapults.

Archimedes also successfully applied his scientific discoveries in hydrostatics. He designed all sorts of pumps, and the Archimedean water-screw is still widely used. But his most famous practical success was in solving a problem presented to him by King Hieron. Hieron wished to know whether a newly made crown, which was supposed to be of pure gold was, as he suspected, partly silver. Archimedes solved the problem by grasping the concept of relative density. By immersing successively the crown itself and pieces of gold and silver of equal weight in full containers of water and observing the amount of water each displaced, Archimedes was able to show that the crown was indeed not made of pure gold. One of the famous stories associated with Archimedes tells how this occurred to him when he was getting into his bath and observed how the more of his body was immersed the more water overflowed from the bath. He saw instantly how to solve his problem, leaped from the bath, and rushed through the streets, stark naked, shouting "Eureka!" (I have found it).

Archimedes's work in applied mathematics and science ensured his great contemporary fame, but some of his greatest work was probably in his more esoteric researches in pure mathematics. Like all Greek mathematicians, his interest was primarily concentrated on geometry. Arithmetic was greatly hampered by a very cumbersome system of notation. Although Archimedes himself invented a much improved system for notation of very large numbers, algebra had yet to be invented, in Europe at least. Archimedes's most profound achievement was to perfect the "method of exhaustion" for calculating the areas and volumes of curved figures. The method involves successively approximating the figure concerned by inscribed and circumscribed polygons. This method essentially used the concept of limit – a concept that took some time for later European mathematicians to grasp. Archimedes used this method to determine an approximate value for π, which was not to be improved on for many centuries.

Archimedes was put to death by a Roman soldier when the Romans, under general Marcellus, finally successfully besieged Syracuse. The killing was against the orders of Marcellus who respected Archimedes and wished for him to be protected. Archimedes was apparently drawing mathematical symbols in the sand when killed.

Argelander, Friedrich Wilhelm August
(1799–1875) *German astronomer*

Born in the Baltic port of Memel (now Klaipeda in Lithuania), Argelander (ar-ge-**lahn**-der) was the son of a wealthy Finnish merchant and a German mother. He was educated at Königsberg, where his interest in astronomy was aroused by the lectures of Friedrich BESSEL. Argelander began his career in 1820 as an assistant in Bessel's Königsberg Observatory. In 1823 he moved to the Åbo Observatory in Finland, then part of Russia. The observatory burned down in 1827 and Argelander began the design and construction of a new observatory in Helsinki, which was completed in 1832. In 1836 he was appointed professor of astronomy at Bonn. Here Friedrich Wilhelm IV built for Argelander an impressive new observatory. They were in fact old friends. In 1806, following Prussia's defeat by Napoleon, Friedrich Wilhelm, then the crown prince, had sought refuge in the Argelander home in Memel, East Prussia.

Argelander's name continues to be known by astronomers for his compilation of the *Bonner Dorchmusterung* (1859–63; 3 vols; Bonn Survey), still the largest and most comprehensive of prephotographic catalogs. Under Bessel he had begun a survey of the sky from 15°S to 45°N. This was extended at Bonn to an area from 90°N to 2°S and when complete listed the positions of 324,198 stars down to the ninth magnitude. His work was continued by his successor, E. Schonfeld, who in the *Southern Bonner Dorchmusterung* (1886; Southern Bonn Survey) added a further 133,659 stars taken from the southern skies (2°S–23°S).

Aristarchus of Samos (*c.* 320 BC–*c.* 250 BC) *Greek astronomer*

Little is known of the life of Aristarchus (a-ristar-kus), but ARCHIMEDES reported that Aristarchus had proposed that, while the Sun and the fixed stars are motionless, the Earth moves around the Sun on the circumference of a circle. Just what led Aristarchus to this view and how firmly he held it is not known. It received no support until the late medieval period.

One short work of Aristarchus has survived – *On the Sizes and Distances of the Sun and Moon*. In this work he calculated that the Earth is about 18 times further away from the Sun than from the Moon. His method was to use the fact that when the Moon is exactly in the second quarter it will form a right-angled triangle with the Earth and the Sun, and the relative lengths of the sides of the triangle can be determined by angular measurement. Aristarchus's method is correct, but his measurement was inaccurate (the Sun is roughly 400 times further away). Despite the size of the error it was nevertheless the first attempt to come to grips with astronomical distances by something more sophisticated than revelation or guesswork.

Aristotle (384 BC–322 BC) *Greek philosopher, logician, and scientist*

> Every science and every inquiry, and similarly every activity and pursuit, is thought to aim at some good.
> —*The Nicomachean Ethics*, Bk I

Aristotle (**a**-ri-stot-el), the son of Nicomachus, physician at the court of Mayntas II of Macedon, was born in Chalcis and moved to Athens in 367 BC, where he was a member of the academy until PLATO's death in 347. For the next 12 years he worked in Assos in Asia Minor, Mytilene on Lesbos, and, from 342 until 335, in Macedon as the tutor of the young Alexander the Great. Unfortunately little is known of this legendary relationship apart from the fact that Alexander took with him on his campaigns a copy of Homer's *Iliad* annotated by Aristotle. Also, Plutarch quotes a letter from Alexander rebuking his former tutor for publishing his *Metaphysics* and revealing to all what had been privately and, he assumed, exclusively taught to him. Following Alexander's accession to the throne of Macedon in 335 Aristotle returned to Athens to found his own school, the Lyceum. When, however, Athens, with little cause to love the power of Macedon, heard of the death of Alexander (323) they turned against Aristotle

and accused him, as they had SOCRATES earlier in the century, of impiety. To prevent Athens from "sinning twice against philosophy" he moved to Chalcis where he died the following year.

Aristotle not only developed an original and systematic philosophy but applied it in a deliberate manner to most areas of the knowledge of his day. The resulting treatises on such subjects as physics, cosmology, embryology, and mineralogy acquired a considerable authority, becoming for medieval scholars if not the last word on any subject then invariably the first. Aristotelian science was not overthrown until the great scientific revolution of the 16th and 17th centuries.

In cosmology Aristotle basically accepted the scheme in which the Earth was at the center of the universe with the planets and fixed stars moving around it with uniform speed in perfectly circular orbits. (He also believed, on empirical grounds, that the Earth was round.) But Aristotle was not content simply to construct models of the universe and faced the problem of how to account for the various forms of motion. He began by accepting that matter was composed of the four elements of EMPEDOCLES – earth, water, fire, and air. Left to themselves the elements would either fall freely, like earth and water, or rise naturally like air and fire. This for Aristotle was natural motion, self-explanatory and consisting simply of bodies freely falling or rising to their natural place in the universe. For a stone to fall to the ground no one had to push or pull it but merely to remove all constraints for it to fall in a straight line to the Earth.

But the heavenly bodies do not move up or down in straight lines. Therefore, Aristotle concluded, they must consist of a fifth element, aether (or *quinta essentia* to the medieval schoolmen), whose natural motion was circular. Thus in the Aristotelian universe different bodies obey different laws; celestial and terrestrial bodies move differently because the laws of motion are different in the heavens from those operating below the Moon. Nor was this the only distinction. For Aristotle the heavens were, with their supposed regularity, incorruptible, without change or decay; such processes were only too apparent on the Earth.

Aristotle also produced a number of volumes on biological problems. In particular his *De partibus animalium* (On the Parts of Animals) and his *De generatione animalium* (On the Generation of Animals) show a detailed knowledge of the fauna of the Mediterranean world and a concern to understand their anatomy and physiology. Over 500 species of animal are referred to by Aristotle. He was also a keen observer and had obviously made empirical investigations on the development of the chick embryo for example, noting the appearance of its heart on the fourth day. In fact some of his observations were only confirmed by zoologists in the

19th century and had for long been thought to be as erroneous as his physics.

In embryology he was also able to refute by dissection the view that the sex of the embryo is determined by its site in the womb. He also argued against the doctrine of pangenesis, that the seed comes from the whole of the body, as he also did against the classical version of preformationism, that the embryo contains all parts already preformed. His physiology, which could not be obtained so readily from simple dissection, was less acute. Respiration was thought to cool the body, an exercise unnecessary for fish who could cool themselves merely by drawing water through their gills.

He, further, produced a rudimentary taxonomy that went to some length to show that divisions based on number of limbs turned out to be obviously arbitrary. Instead, he proposed that mode of reproduction be used. This gave him the basic division between viviparous (exclusively mammalian) and the oviparous, subdivided into birds and reptiles laying proper eggs and the fishes laying "imperfect eggs." He added the insects, who lay no eggs at all but simply produce larvae.

If Aristotle had produced only his *Organon* – works on logic – he would have been considered a prolific and powerful thinker. His style of logic lasted unchallenged even longer than his physics for it was not until 1847 that George BOOLE laid the foundations of a more modern logic and it was not until the present century that non-Aristotelian logics were systematically developed.

Arkwright, Sir Richard (1732–1792) *British inventor*

> In the evening I walked to Cromford and saw the children coming from their work out of one of Mr. Arkwright's manufactories. I was glad to see them look in general very healthy and many with fine, rosy complexions.
> —Joseph Farington, *Diary of Joseph Farington* (ed. James Greig 1922)

Arkwright was born in the town of Preston in northern England. Apprenticed to a barber at the age of 18, he became a wigmaker. Through travel and self-education he developed an interest in spinning machinery and in 1769 he patented a water-powered machine that, unlike previous machines, produced a cotton yarn strong enough for use as warp. This machine was used in the horse-driven mill he established at Nottingham (1768) to produce machine-spun yarn. In 1771 he set up a water-powered mill in Cromford, Derbyshire, and in the following years established a number of factories employing machinery for all processes of textile manufacture from carding to spinning. This established the cotton industry as the main industry in the north of Eng-

land. He was knighted in 1786 and died a wealthy man.

Armstrong, Henry Edward (1848–1937) *British chemist and teacher*

The son of a London provisions merchant, Armstrong studied under FRANKLAND at the Royal College of Chemistry and completed his doctorate in 1870 under Adolph KOLBE at Leipzig. A man of great energy and wide interests, he was a pioneer of British technical education, a prolific researcher, and leader of a major school of chemical research. After various teaching posts in London, he was appointed (1884–1911) professor at the Central Technical College in London (forerunner of the Imperial College of Science and Technology). He served on many committees and was secretary (1883–93) and later president (1893–95) of the Chemical Society.

Armstrong's research work covered many fields, including aromatic substitution, crystallography, stereochemistry, terpenes, and enzymes. He proposed the quinone theory of color and the work carried out by his school on orientation and isomeric change in naphthalene derivatives, although little regarded in Britain, was of fundamental importance to the German dyestuffs industry.

Arnald of Villanova (*c.* 1235–1313) *Spanish alchemist*

Arnald (**ar**-nald), who was born in Valencia, Spain, was educated in Paris and Montpellier, and studied medicine at Naples. He became a famous physician much in demand by popes and monarchs. In 1285 he became a professor at the University at Montpellier, but came into conflict with the Church in 1299 and was charged with heresy in Paris. He was imprisoned but finally released about 1303 and died at sea between Naples and Genoa in about 1313.

Arnald was one of the first scholars to mention alcohol. In medicine he used it to extract the "virtues of herbs," which became known as tinctures. Some of his other medical ideas were less progressive: he wrote at length on the efficacy of seals and amulets claiming to be able to provide one that would defend its wearer from witchcraft, storms, quinsy, inflammation of the brain, and financial difficulties.

He produced many works on medicine, most notably *Medicinalium introductionum speculum* (An Introductory View of Medicine), and also works on theology and chemistry. The many alchemical works, including *Rosarium philosophorum* (A Rose Garden of Philosophers), that were attributed to him and that had considerable influence in the following centuries are now thought not to be his work.

Arrhenius, Svante August (1859–1927)
Swedish physical chemist

> These new theories [of Arrhenius] ... suffered
> from the misfortune that nobody really knew
> where to place them. Chemists would not
> recognize them as chemistry; nor physicists as
> physics. They have in fact built a bridge
> between the two.
> —Per Theodor Cleve (1903)

Arrhenius (ar-**ray**-ni-us) was born in Wijk, near
Uppsala, Sweden. He originally went to Uppsala University to study chemistry, changing
later to physics. Finding the standard mediocre,
he transferred to Stockholm in 1881 to do research under the physicist Erik Edlund, working initially on electrical polarization and then
on the conductivity of solutions (electrolytes).

At the time it was known that solutions of
certain compounds conduct electricity and that
chemical reactions could occur when a current
was passed. It was thought that the current
decomposed the substance. In 1883 Arrhenius
proposed a theory that substances were partly
converted into an active form when dissolved.
The active part was responsible for conductivity. In the case of acids and bases, he correlated
the strength with the degree of decomposition
on solution. This work was published as
*Recherches sur la conductibilité galvanique des
electrolytes* (1884; Researches on the Electrical
Conductivity of Electrolytes) and submitted as
his doctoral dissertation. The paper's great
merit was not recognized by the Swedish authorities and he was awarded only a fourth-
class doctorate. Arrhenius sent his work to
several leading physical chemists, including Jacobus VAN'T HOFF, Friedrich OSTWALD, and
Rudolf CLAUSIUS, who were immediately impressed. This led to a period of travel and work
in various European laboratories in the period
1885–91.

In 1887 van't Hoff showed that although the
gas law ($pV = RT$) could be applied to the osmotic pressure of solutions, certain solutions
behaved as if there were more molecules than
expected. Arrhenius immediately realized that
this was due to dissociation – a conclusion confirmed by further experimental work and published in the classic paper *Über die Dissociation
der in Wasser gelösten Stoffe* (1887; On the Dissociation of Substances in Water). The idea that
electrolytes were dissociated even without a
current being passed proved difficult for many
chemists but the theory has stood the test of
time.

This work won Arrhenius a high international reputation but only limited acclaim in
Sweden. Despite this he returned to Stockholm
in 1891 as lecturer at the Technical Institute
and in 1895 became professor there. In 1903 he
was awarded the Nobel Prize for chemistry, and
in 1905 he became the director of the Nobel Institute, a post he held until shortly before his
death.

Arrhenius is also remembered for the *Arrhenius equation*, which relates the rate of a
chemical reaction to the temperature. He was a
man of wide-ranging intellect and besides developing his work on solutions, he worked on
cosmogony and on serum therapy, being especially interested in the relation between toxins
and antitoxins. He also investigated the greenhouse effect by which carbon dioxide regulates
atmospheric temperature and calculated the
changes that would have been necessary to
have produced the Ice Ages.

Aschoff, Karl Albert Ludwig (1866–1942) *German pathologist*

Educated at Bonn, Berlin (his birthplace), and
Strasbourg, Aschoff (**ah**-shoff) was later professor of pathological anatomy, firstly at Marburg (1903–06) and then at Freiburg, where he
remained for the rest of his career. He carried
out investigations of a number of human pathological conditions, including jaundice, appendicitis, cholecystitis, tuberculosis, and
thrombosis. In 1904 he described the inflammatory nodules located in the muscle of the
heart and associated with rheumatism
(*Aschoff's bodies*). He recognized the bacteria-
engulfing activity of the phagocytes in various
tissues and named them the reticuloendothelial
system. The pathological institute that Aschoff
built up at Freiburg-im-Breisgau was attended
by students from all over the world.

Aselli, Gaspare (c. 1581–1625) *Italian anatomist*

Aselli (ah-**sel**-ee), who was born into a prosperous family in Cremona, Italy, was educated
at the University of Pavia, where he later
served as professor of anatomy and surgery.

In 1622 while dissecting a recently fed dog he
noticed various white vessels spread throughout the intestines. As they exuded a milky fluid
when pricked he called them the "lacteals" or
the "albas venas." He claimed to trace them to
the liver and not unnaturally assumed them to
be the vessels transporting the chyle, broken
down food products, to the liver to be changed
into blood – a process demanded by the current
physiology of GALEN. Aselli's observations were
fully described in the posthumously published
work, *De lactibus* (1627; On the Lacteals), a
work that also contained the first colored
anatomical illustrations.

It was not until 1651 that Jean PECQUET
showed that lacteals did not go to the liver.

Aspect, Alain (1947–) *French physicist*

Alain Aspect was educated at the École Normale Supérieure de Cachan. In 1981–82 at
Orsay in Paris, Aspect and his coworkers performed a series of experiments involving coincidence measurements on polarized photons,

designed to check the inequalities put forward by John BELL in 1964. Bell's inequalities allow for an experimental test of whether there are local hidden variables in quantum mechanics. It is generally believed by physicists that the *Aspect experiment* provided proof that hidden variables are not involved. Aspect is currently working on Bose–Einstein condensates.

Astbury, William Thomas (1889–1961)
British x-ray crystallographer and molecular biologist

> [Molecular biology] is concerned particularly with the forms of biological molecules and with the evolution, exploitation, and ramification of these forms in the ascent to higher and higher levels of organization.
> —Harvey Lectures, 1950–1951

William Astbury was born in Longton, England, were his father was a potter. In 1916 he won a scholarship to Cambridge University, to study chemistry, physics, and mathematics, and graduated in 1921 after spending two years of the war doing x-ray work for the army. He then joined William Henry BRAGG's brilliant group of crystallographers, first at University College, London, and from 1923 at the Royal Institution. In 1945 Astbury was appointed to the new chair of biomolecular structure at Leeds.

Astbury's early structural studies were carried out on tartaric acid but in 1926 Bragg asked him to prepare some x-ray photographs of fibers for his lectures. The results stimulated an interest in biological macromolecules that Astbury retained for the rest of his life. In 1928 he moved to the University of Leeds as lecturer in textile physics and by 1930 had produced an explanation of the extensibility of wool in terms of two keratin structures: α-keratin in which the polypeptides were hexagonally folded (unextended wool) and β-keratin in which the chain was drawn out in zigzag fashion. A popular account of this work was given in *Fundamentals of Fibre Structure* (1933).

The keratin structure established his reputation, and he quickly extended his studies to other fibers and proteins. He showed that the globular proteins consisted of three-dimensionally folded chains that could be denatured and drawn out into protein fibers. This work laid the foundation for the x-ray structural investigations of hemoglobin and myoglobin. The hexagonal α-keratin structure dominated British crystallographic protein studies until 1951, when it was shown to be incorrect by Linus PAULING who demonstrated the α-helical structure of polypeptide chains.

In 1935 Astbury began to study nucleic acids by x-ray crystallography, and in 1938 he and his research student Florence Bell produced the first hypothetical structure of DNA.

Aston, Francis William (1877–1945)
British chemist and physicist

Aston was born in Harborne, England, the son of a metal merchant. He was educated at Mason College, the forerunner of Birmingham University, where he studied chemistry. From 1898 until 1900 he did research on optical rotation. He left Birmingham in 1900 to work in a Wolverhampton brewery for three years. During this time he continued with scientific research in a home laboratory, where he worked on the production of vacua for x-ray discharge tubes. This work came to the notice of J. H. POYNTING of the University of Birmingham who invited Aston to work with him. He remained at Birmingham until 1910 when he moved to Cambridge as research assistant to J. J. THOMSON. He became a research fellow at Cambridge in 1920 and stayed there for the rest of his life, apart from the war years spent at the Royal Aircraft Establishment, Farnborough. Aston's main work, for which he received the Nobel Prize for chemistry in 1922, was on the design and use of the mass spectrograph, which was used to clear up several outstanding problems and became one of the basic tools of the new atomic physics.

Thomson had invented an earlier form of spectrograph in which a beam of positive rays from a discharge tube passed through a magnetic and an electric field, which deflected the beam both horizontally and vertically. All particles (ions) with the same mass fell onto a fluorescent screen in a parabola. Aston improved the design by using a suitable magnetic field, so that ions of the same mass were focused in a straight line rather than a parabola. Different ions were deflected by different amounts, and the spectrograph produced a photographic record of a series of lines, each corresponding to one type of ion. The deflections allowed accurate calculation of the mass of the ions.

Aston's first spectrograph was ready in 1919 and with it he was soon able to throw light on one outstanding problem about the nature of the elements. In 1816 William PROUT had put forward his hypothesis that all elements are built up from the hydrogen atom and that their atomic weights are integral multiples of that of hydrogen. Although receiving considerable support it was eventually rejected when it was found that many elements have nonintegral weights (e.g., chlorine: 35.453). Frederick SODDY in 1913 had introduced the idea of isotopes, that is, the same chemical element in different forms having differing weights. Aston established that isotopes are not restricted to radioactive elements but are common throughout the periodic table. He also saw that they could explain Prout's hypothesis. Thus he found that neon was made from the two isotopes ^{20}Ne and ^{22}Ne in the proportion of 10 to 1. This will give a weighted average of 20.2 for a large number of neon atoms. The value of 35.453 for chlorine can be similarly explained. The whole-number rule is his principle that atoms have a mass that is equivalent to a whole number of hydrogen atoms.

Aston then went on to determine as many atomic weights as accurately as his instruments would allow. His first spectrograph was only suitable for gases but by 1927 he had introduced a new model capable of dealing with solids. From 1927 to 1935 he resurveyed the atomic weights of the elements with his new instrument.

In the course of this activity he found some minor discrepancies with the whole-number rule. Thus the atomic weight of hydrogen is given not as 1 but 1.008, of oxygen–16 as 15.9949 and of oxygen–17 as 16.99913. Aston attempted to show why these values are so tantalizingly close to the integral values of Prout – why the isotopes of oxygen are not simple 16 and 17 times as massive as the hydrogen atom. He argued that the missing mass is in fact, by the mass–energy equivalence of EINSTEIN, not really missing but present as the binding energy of the nucleus. By dividing the missing mass by the mass number and multiplying by 10,000, Aston went on to calculate what was later called the "packing fraction" and is a measure of the stability of the atom and the amount of energy required to break up or transform the nucleus.

Thus contained in Aston's work were the implications of atomic energy and destruction and he believed in the possibility of using nuclear energy – he also warned of the dangers. He lived just long enough to see the dropping of the first atomic bomb in August 1945.

Atanasoff, John Vincent (1904–1995)
American physicist and computer pioneer

Atanasoff was born in Hamilton, New York, and educated at the universities of Florida, Iowa, and Wisconsin, where he gained his PhD in 1930. He taught at the Iowa State University from 1930 until 1942, when he moved to the Naval Ordnance Laboratory at White Oak, Maryland. After World War II, Atanasoff worked for various technical companies, eventually serving as president of Cybernetics Inc. from 1961 until 1982.

The son of a Bulgarian immigrant who was an electrical engineer, Atanasoff was introduced to calculation at the age of nine when his father gave him a slide rule. This was of little use when, in 1930, he was trying to complete his thesis on the electrical properties of helium. Not even a desk calculator could significantly lighten the extensive computations. He began to think about how things could be improved. By 1937 he had opted for a machine that operates digitally, uses capacitors to store binary numbers, and calculates by logic circuits. Work-

ing with his assistant, Clifford Berry, Atanasoff built a prototype in 1939 of the suitably named ABC (Atanasoff–Berry Computer). This was good enough to raise sufficient funds to build an operating machine, which was completed in 1942.

Although the ABC was the first device to incorporate a number of key notions, it was unsatisfactory as a working machine. It was slow, could not be programmed, had to be controlled at all times, and suffered from a number of systematic errors. Clearly, it could be improved but the outbreak of war in 1942 took Atanasoff away to other duties. By the time he was free to work on the ABC other workers had seized the initiative. Atanasoff's work long lay forgotten.

This was corrected in a 1973 court case involving two American companies. Sperry Rand had bought the patent to ENIAC and was seeking to charge royalties to other computer manufacturers. Honeywell Inc. resisted, claiming that ENIAC was derived from the ABC and from information passed to ENIAC's designer, John MAUCHLY, by Atanasoff in the early 1940s. Atanasoff gave evidence and the judge found that ENIAC was not the first "automatic electronic digital computer," and that it was "derived from one Dr. John Vincent Atanasoff."

Audubon, John James (1785–1851)
American ornithologist and naturalist

The illegitimate son of a French sea captain and his Creole mistress, Audubon (**aw**-de-bon) was born in Les Cayes on the Caribbean island of Haiti and was brought up in Nantes in his father's family. He studied painting in Paris, spending six months in the studio of Jacques Louis David. In 1803, to escape conscription into Napoleon's army, Audubon was sent to Pennsylvania where his father owned a farm. Neither the farm nor any of Audubon's other business interests flourished and he was declared bankrupt in 1819 and imprisoned.

No doubt one cause of Audubon's commercial failure was the time spent hunting and observing birds and other animals in the wild. The first hint that his skills as an artist and naturalist could be combined to make money came in 1810 when Alexander Wilson passed through Louisville, Louisiana, where Audubon was operating a general store. Wilson was looking for subscribers to his lavishly illustrated *American Ornithology* (9 vols; 1808–14).

By 1820 Audubon had decided to publish his own collection of animals and birds. He spent a further four years traveling through Louisiana and Mississippi shooting specimens. As no American publisher appeared to be interested in his work, Audubon took his paintings to Britain in 1826. He eventually found a printer in Edinburgh willing to work on his "double elephant size" engravings (39″ × 29″). Sets of five plates were sold to subscribers for about

$10 to finance the next set. In this way 200 full sets of *Birds of America* (1827–38) were published in Britain in 87 parts with 435 plates. Full sets are rarely available for sale – when auctioned they are unlikely to raise less than a million dollars.

Audubon returned to America in 1839, where he bought an estate on the Hudson and began to prepare his *Viviparous Quadrupeds of North America* (3 vols; 1845–48).

Auenbrugger von Auenbrugg, Joseph Leopold (1722–1809) *Austrian physician*

Born in Graz in Austria, Auenbrugger (**ow**-en-broo-ger) learned medicine at the University of Vienna and subsequently worked (1751–62) at the Spanish Hospital, Vienna, where he became chief physician. In the course of his work, he noticed how the note made by tapping the chest altered between healthy and diseased patients and described how this technique of chest percussion could be used to diagnose congestion of the lungs and other conditions. Auenbrugger published his findings in *Inventum novum* (1761; A New Discovery) but their value was slow in gaining recognition. Only after a French translation by Jean-Nicholas CORVISART in 1808 did chest percussion achieve widespread application. It is still used today.

Auenbrugger became one of Vienna's most celebrated physicians and was ennobled by Emperor Joseph II in 1784.

Auer, Karl, Baron von Welsbach (1858–1929) *Austrian chemist*

Auer (**ow**-er) was born in Vienna, the son of the director of the Imperial Printing Press. He was educated at the Vienna Polytechnic and at Heidelberg University, where he was a pupil of Robert BUNSEN.

In 1885 he made a major contribution to knowledge of the lanthanoid (rare-earth) elements. In 1840, Carl MOSANDER had isolated a new "element" called didymium. Auer showed (1885) that this contained, on fractionation, green and rose-red portions. He named them *praseodymia* ("green twin") and *neodymia* ("new twin").

Auer was also one of the first to find some use for the rare-earth elements. Gas had been in use as an illuminant since the beginning of the century and, although an improvement on the early oil lamps, it had many disadvantages of its own. It was expensive, hot, smoky, and smelly. Auer realized that it would be better to use the gas to heat a solid that would itself provide light, rather than use the luminosity of the flame. He used a mantle over the flame, impregnated with thorium oxide and a small amount of cerium. The *Welsbach mantle*, patented in 1885, delayed the end of gas lighting for a few years. Unfortunately for Auer, his invention was too late for, in 1879, EDISON had managed to burn an electric bulb for 40 continuous hours.

Later, in 1898, Auer tried to improve the electric lamp by replacing its carbon filament by metallic osmium, which has a melting point of 2,700°C. Once more he failed, for the future lay with tungsten, which has a higher melting point of 3,410°C.

He was more successful with the so-called *Auer metal* – an alloy consisting mainly of cerium with other lanthanoid elements. It is also called Mischmetal (German: mixed metal) and is used for flints in cigarette lighters.

Auger, Pierre Victor (1899–1993) *French physicist*

Auger (**ow**-zher or **ow**-ger) was born in Paris and educated there at the Ecole Normale Supérieure, where he obtained his doctorate in 1926. He was later appointed to the staff of the University of Paris and after serving there as professor of physics from 1937 became director of higher education for France in 1945. From 1948 until 1960 he was director of the science department of UNESCO; he left UNESCO to become president of the French Space Commission but in 1964 he took the post of director-general of the European Space and Research Organization, a post he retained until his retirement in 1967.

Auger worked mainly on nuclear physics and cosmic rays. In 1925 he discovered the *Auger effect* in which an excited atom emits an electron (rather than a photon) in reverting to a lower energy state. In 1938 Auger made a careful study of "air showers," a cascade of particles produced by a cosmic ray entering the atmosphere and later known as an *Auger shower*. Auger had an interest in the popularization of science. He also published volumes of poetry.

Averroës (1126–1198) *Spanish–Muslim physician and philosopher*

> Knowledge is the conformity of the object and the intellect.
> —*Destructio destructionum* (Refuting Refutations)

Averroës (a-**ve**-roh-eez), also known simply as The Commentator to the Latin West, or Ibn Rushd (from his full name, Abu Al-Walid Muhammad Ibn Ahmad Ibn Rushd), came from a family of jurists and was born in Cordoba in Moorish Spain. He himself trained in law and medicine and later served as *qadi* or judge in Seville and Cordoba. In 1182 he was appointed physician to the court of caliph Abu Ya'qub Yusuf in Marrakesh and to his son, Abu Yusuf Ya'qub, in 1195 but was recalled shortly before his death.

In the field of medicine Averroës produced his *Kulliyat fi al tib* (General Medicine) between 1162 and 1169. He is however better known for his great commentaries on ARISTOTLE but, above all, for his *Tahafut al-Tahafut* (The Incoherence of the Incoherence), a strong attack on the Muslim philosopher al-Ghazzali's *Tahafut al-Falasifah* (The Incoherence of the Philosophers). The work was more influential in the Latin Christian West than in the Muslim East, and its contents paved the way for the medieval separation of faith and reason.

Avery, Oswald Theodore (1877–1955)
American bacteriologist

Avery was born in Halifax, the capital of Nova Scotia, Canada. Educated at Colgate University, he received his BA in 1900 and his medical degree in 1904. After a time at the Hoagland Laboratory, New York, as a lecturer and researcher in bacteriology, he joined the Rockefeller Institute Hospital (1913–48). While investigating the pneumococcus bacteria responsible for causing lobar pneumonia, Avery found that the bacteria produced soluble substances, derived from the cell wall and identified as polysaccharides, that were specific in their chemical composition for each different type of pneumococcus. This work provided a basis for establishing the immunologic identity of a cell in biochemical terms.

In 1932 Avery started work on the phenomenon of transformation in bacteria. It had already been shown that heat-killed cells of a virulent pneumococcus strain could transform a living avirulent strain into the virulent form. In 1944 Avery and his colleagues Maclyn MC-CARTY and Colin MACLEOD extracted and purified the transforming substance and showed it to be deoxyribonucleic acid (DNA). Previously it had been thought that protein was the hereditary material and thus Avery's work was an important step toward the eventual discovery, made nine years later by James WATSON and Francis CRICK, of the chemical basis of heredity.

Avicenna (*or* Abu-'Ali Al-Husayn Ibn-Sina) (980–1037) *Persian physician and philosopher*

> Writing about erotics is a perfectly respectable function of medicine, and about the way to make the woman enjoy sex; these are an important part of reproductive physiology.
> —*Canon*

Avicenna (av-i-**sen**-a), whose works span the entire spectrum of arts and sciences, is one of the most famous figures of Persian culture. Born in Bokhara (now in Uzbekistan), he was a child prodigy, learning and practicing medicine in his teens and gaining the position of court physician to Prince Nuh ibn Mansur when still only 18. This gave Avicenna access to the library of the Samanid court, of which he took full advantage. However, the Samanid

rulers were overthrown by Turkish forces and Avicenna was forced to flee. After a period spent traveling and in several short-lived posts, he became physician to Prince Shams ad-Dawlah in Hamadan. Here he started on a massive medical textbook, the *Canon* (*Al Qanun*). In this, Avicenna collated virtually all preceding medical knowledge and set down his own ideas. Comprising five books, the *Canon* deals with anatomy, physiology, etiology, diagnosis, obstetrics, drugs, and so on, and runs to over one million words. It subsequently became a standard text throughout Europe and the Middle East until the 17th century, being translated into Hebrew and Latin. During this period, he also wrote his *Book of Healing*, a comprehensive encyclopedia covering mathematics, logic, natural sciences, and metaphysics and based largely on the ideas of ARISTOTLE and other Greek philosophers.

In 1022, the death of prince ad-Dawlah led Avicenna to leave Hamadan. He found refuge at the court of Prince 'Ala ad-Dawlah, where his immense output of writings continued. This included his account of the paths to spiritual enlightenment, the *Book of Directives and Remarks*. He also composed some notable works of poetry and wrote on many other topics, including astronomy, physics, and chemistry.

Avicenna was frequently involved in the political turmoil of warring Persian states and on several occasions had to escape possible capture and imprisonment. He was fond of wine, women, and, doubtless, song too and prescribed himself a glass of wine to combat fatigue. He died while accompanying the prince on a campaign, evidently of colic and exhaustion.

Avogadro, Amedeo (1776–1856) *Italian physicist and chemist*

Avogadro (av-oh-**gah**-droh) was born in Turin in northern Italy and came from a long line of lawyers. He too was trained in law and practiced for some years before taking up the study of mathematics and physics in 1800. His early work was carried out in the field of electricity, and in 1809 he became professor of physics at the Royal College at Vercelli. He was professor of mathematical physics at Turin from 1820 until 1822 and from 1834 to 1850.

His fame rests on his paper *Essai d'une manière de determiner les masses relatives des molecules des corps et les proportions selon lesquelles entrent dans cet combinaisons* (1811; On a Way of Finding the Relative Masses of Molecules and the Proportions in Which They Combine), published in the *Journal de Physique*. This states the famous hypothesis that equal volumes of gases at the same temperature and pressure contain equal numbers of molecules. It follows from the hypothesis that relative molecular weights can be obtained from vapor densities and that the proportion by volume in which gases combine reflects the combining ratio of the molecules. Using this theory, Avogadro showed that simple gases such as hydrogen and oxygen are diatomic (H_2, O_2) and assigned the formula H_2O to water, whereas John DALTON had arbitrarily assumed that the simplest compound of two elements would have the formula HO.

Avogadro's work provided the essential link between Gay Lussac's law of combining volumes and Dalton's atomic theory. This was not, however, realized at the time and, as a consequence, the determination of a self-consistent set of atomic weights was delayed for 50 years. The French physicist André AMPÈRE was one of the few who accepted the theory and for many years it was taken to be Ampère's own.

Avogadro's contribution to chemisty was not appreciated in his own lifetime. The importance and truth of the theory was unrecognized until 1860 when his fellow Italian, Stanislao CANNIZZARO, forcefully restated it at the Karlsruhe Conference and demonstrated that it was the key needed to unlock the problem of atomic and molecular weights. The number of particles in one mole of a substance was named *Avogadro's constant* in his honor. It is equal to $6.022\,52 \times 10^{23}$.

Axel, Richard (1946–   ) *American neuroscientist*

Born in New York, Axel attended Columbia University, graduating AB in 1967, and then studied medicine at Johns Hopkins University School of Medicine, obtaining his MD in 1970. He returned to Columbia University to study pathology, and in 1978 was appointed professor in the pathology and biochemistry department. He became investigator in the Howard Hughes Medical Institute in 1984, and university professor in 1999. He was professor at Harvard Medical School (2001–02) and in 2003 he became affiliate professor in the Department of Physiology and Biophysics at the University of Washington, Seattle.

Axel is best known for his work on the sense of smell, or olfaction. In 1991 he published, jointly with Linda BUCK, a landmark paper that described a large family of genes for receptors of odor molecules. The scientists discovered about a thousand such genes in the mouse, each encoding a receptor protein with unique binding properties, and thus capable of responding to just one or a few odorant molecules. The receptors are anchored in the cell membrane of olfactory receptor cells within the nose. Axel and Buck showed that binding of an odor molecule to its corresponding receptor causes the receptor to activate a G protein inside the cell, which in turn stimulates the formation of cyclic AMP (cAMP). This acts as a second messenger, causing the opening of ion channels in the cell membrane and triggering electrical signals from the cell to the brain via long slender cellular processes. Hence, this work established that odor receptors belong to the class known as G protein-coupled receptors.

Axel and Buck subsequently pursued independent but often complementary lines of enquiry into olfactory reception. They showed that each olfactory receptor cell bears only one type of odor receptor molecule on its surface. However, odor molecules typically can bind to several different receptor types. Thus a single odor will elicit a characteristic pattern of signals from the olfactory receptor cells. This is why an animal such as the mouse can discriminate between as many as 10,000 different smells, even with only 1,000 or so types of receptor cells.

Subsequently, Axel's team at Columbia used genetically engineered mice to reveal that the processes of all olfactory sensory cells carrying the same receptor type converge on the same relay station, or glomerulus, in the olfactory bulb in the brain. Hence, any given odor will trigger signals to a particular set of glomeruli, creating a corresponding spatial "map" in the brain; different odors trigger overlapping but nonidentical combinations of glomeruli. Signals from the glomeruli are transmitted to higher regions of the brain, where the information is processed to produce a conscious experience, and memories of smells are stored.

Axel's attention then turned to the fruit fly *Drosophila* as a simple experimental model. The team confirmed that the fly brain follows the same principles of spatial patterning in olfaction as seen in mice, and that this has both functional and behavioral significance. For example, Axel's team could abolish the normal aversive response to an alarm substance (CO_2) by engineering flies in which the specific glomerulus for CO_2 was inhibited.

For his work on "odorant receptors and the organization of the olfactory system," Axel was awarded the 2004 Nobel Prize for physiology or medicine, jointly with his erstwhile colleague, Linda Buck.

Axelrod, Julius (1912–2005) *American neuropharmacologist*

Axelrod was born in New York City, becoming a student at New York University in 1929, and, after a year, transferring to the City College, where he took a degree in chemistry and biology. He wanted to study medicine but was turned down by several medical schools, despite his qualifications. Axelrod felt that his rejection was influenced by anti-Semitism. From 1933 to 1935 he was a laboratory assistant at the Department of Bacteriology, New York University Medical School. From 1935 to 1945 he had a job as a technician in a laboratory of industrial hygiene that had just been set up in New York. But still with an ambition for a career in scientific research, and after some years at the Goldwater Memorial Hospital and the National Heart Institute, he took a year off in 1955, obtained a PhD from George Washington University, and moved to the National Institute of Mental Health (NIMH) as chief of the pharmacology section. He held this post until his retirement in 1984, while also continuing to work in the cell biology laboratory at the NIMH.

Axelrod's research involved the mechanisms underlying the transmission between nerve cells, in particular, the action of the catecholamines, the neurotransmitters of the sympathetic nervous system. The most important of these is norepinephrine, first identified as a neurotransmitter by Ulf VON EULER in 1946. Axelrod realized that once the molecule had interacted with its target cell some mechanism must come into action to switch it off. Later he was able to describe the role of two enzymes, catechol-o-methyltransferase (COMT) and monoamine oxidase (MAO), which degrade the catecholamines.

However, studies with radioactive norepinephrine showed its persistence in the sympathetic nerves for some hours. This led Axelrod to propose that norepinephrine is taken up into, as well as released from, sympathetic nerves. This recapture inactivates the neurotransmitter. In general, Axelrod's research led to the development of drugs designed to affect neurotransmitters, in particular antidepressants that inhibit the uptake of serotonin. Axelrod also worked on a number of other topics, including the metabolism of the hallucinogenic drug LSD and the working of the pineal gland.

For work on the catecholamines Axelrod shared the 1970 Nobel Prize for physiology or medicine with von Euler and Bernard KATZ.

Ayala, Francisco José (1934–) *Spanish–American biologist*

Ayala (ah-**yah**-la), who was born in Madrid, Spain, began his higher education there at the University of Madrid, moved to America in 1961, and obtained his PhD from Columbia in 1964. He worked initially at Rockefeller before joining the Davis campus of the University of California in 1971, where he was later appointed professor of genetics in 1974. He was professor of biology at the University of California, Irving, from 1987 to 1989, when he became Donald Bren Professor of Biological Sciences.

Ayala has worked extensively in the field of molecular evolution. He has also sought to measure genetic variation in natural populations, rates of evolution, and the amount of genetic change needed to produce new species. Many of his results were published in his *Molecular Evolution* (1976) and in a work he coauthored in 1977 entitled *Evolution*. He has written a number of other books including *Molecular Genetics* (1984). Ayala has campaigned against government restrictions on stem-cell research.

Ayrton, William Edward (1847–1908) *British physicist*

Ayrton was born in London, where his father was a lawyer. After attending University College, London, and Glasgow University, he worked for the Indian Telegraph Company, and in 1873 was appointed to teach natural philosophy and telegraphy at the Imperial Engineering College, Tokyo, Japan. He returned to London in 1879 and became a professor at the City and Guilds College and later at Finsbury Technical College (1881). In 1884 he became a professor at the Central College.

B

Baade, Wilhelm Heinrich Walter (1893–1960) *German–American astronomer*

> Almost every one of Baade's papers turned out to have far-reaching consequences.
> —Sir Fred Hoyle

Baade (**bah**-de), born the son of a schoolteacher in Schröttinghausen, Germany, was educated at the universities of Münster and Göttingen, where he obtained his PhD in 1919. He worked at the University of Hamburg's Bergedorf Observatory from 1919 to 1931, when he moved to America. He spent the rest of his career at the Mount Wilson and Palomar Observatories, retiring in 1958.

In 1920 Baade discovered the minor planet Hidalgo, whose immense orbit extends to that of Saturn. He was also, in 1949, to detect the minor planet Icarus, whose orbit, which lies within that of Mercury, can bring it very close to Earth. In the 1930s he did important work with Fritz ZWICKY on supernovae, with Edwin HUBBLE on galactic distances, and with his old Hamburg colleague, Rudolph MINKOWSKI, on the optical identification of radio sources.

Baade's most significant work however began in 1942. As he was of German origin he was precluded from the general induction of scientists into military research, being allowed to spend the war observing the heavens. In early 1943 he was blessed with ideal viewing conditions. Los Angeles was blacked out because of wartime restrictions and, for a short while, the air was calm and the temperature constant. Under these near-perfect conditions Baade took some famous photographs with the 100-inch (2.5-m) reflecting telescope of the central region of the Andromeda galaxy. To his great excitement he was able to resolve stars in the inner region where Hubble before him had found only a blur of light.

These observations allowed Baade to introduce a fundamental distinction between types of stars. The first type, Population I stars, he found in the spiral arms of the Andromeda galaxy. They were young hot blue stars as opposed to the Population II stars of the central part of the galaxy, which were older and redder with a lower metal content. This distinction, now much expanded, has played a crucial role in theories of galactic evolution.

Some of the stars that Baade observed in the Andromeda galaxy were Cepheid variables, stars that vary regularly in brightness. His realization that there were two kinds of Cepheids had an immediate impact. The relationship between period and luminosity of Cepheids had been discovered by Henrietta LEAVITT in 1912 and put into a quantified form by Harlow SHAPLEY so that it could be used in the determination of stellar distances of great magnitude. In the 1920s Hubble had found Cepheids in the outer part of the Andromeda galaxy, and, using the period–luminosity rule, had calculated its distance as 800,000 light-years. Since then the relationship had been used by many astronomers.

Baade, by 1952, was able to show that the original period–luminosity relationship was valid only for Population II Cepheids whereas Hubble's calculation involved Population I Cepheids. Baade worked out a new period-luminosity relationship for these Cepheids and found that the Andromeda galaxy was two million light-years distant.

The distance to the Andromeda galaxy had been used by Hubble to estimate the age of the universe as two billion years. Baade's revised figure gave the age as five billion years. This result was greeted with considerable relief by astronomers as Hubble's figure conflicted with the three to four billion years that the geologists were demanding for the age of the Earth. Further, with Baade's revision of the distance of the Andromeda galaxy without any change in its luminosity, it was now clear that its size must also be increased together with the size of all the other galaxies for which it had been a yardstick. Baade was thus able to establish that while our galaxy was somewhat bigger than normal it was not the largest, as Hubble's work had implied.

Babbage, Charles (1792–1871) *British mathematician*

> The whole of the development and operations of analysis are now capable of being executed by machinery.... As soon as an Analytical Engine exists, it will necessarily guide the future course of science.
> —*Passages from the Life of a Philosopher* (1864)

> From church we went, by his special invitation, to see Babbage's calculating

machine; and I must say, that during an explanation which lasted between two and three hours, given by himself with great spirit, the wonder at its incomprehensible powers grew upon us every moment.
—George Ticknor, *Life, Letters, and Journals of George Ticknor* (1876)

Babbage, whose father was a banker, was born in Teignmouth in southwest England and studied at Cambridge. He played a major role in ending the isolationist attitudes prevalent in British mathematical circles in the early 19th century. In 1815 he helped to found the Analytical Society, which aimed to make the work of Continental mathematicians better known in Britain. Babbage's interest in stimulating British scientific activity was by no means confined to mathematics. In 1820 he was a founder of the Royal Astronomical Society and in 1834 of the Statistical Society, and he continued to attack the British public for their lack of interest in science. Among his inventions were a speedometer and the locomotive "cowcatcher." Babbage also did mathematical work that contributed to the setting up of the British postal system in 1840. From 1828 to 1839 he was Lucasian Professor of Mathematics at Cambridge University.

Babbage is best known for his work in designing and attempting to build three mechanical computers. He had been struck by the discrepancies found in mathematical tables and the persistence of error. "I wish to God these calculations had been executed by steam," he lamented in 1821. Mechanical execution, he argued, would eliminate error. Consequently he began work in 1823 on the machine later known as his Difference Engine No. 1. It operated by the method of finite differences and thus allowed values of functions to be obtained by addition rather than by multiplication. The engine was an analog decimal machine in which numbers were represented by the rotation of various wheels. After a decade of work the project was abandoned when Babbage's credit ran out. It had cost the equivalent of about $85,000, was 8-feet high, and was made from 25,000 parts.

He later designed a simpler version, Difference Engine No. 2, with only a third of the number of parts. Plans were drawn up in 1847 and offered to an uninterested government in 1852. Without financial support Babbage never saw the project develop beyond the design stage.

The more ambitious analytical engine, first described in 1834, was similarly unsuccessful. Unlike the Difference Engine, this was to be a general computing machine in the manner of a modern computer, and was intended to be programmed with punched cards. One of Babbage's more enthusiastic supporters in this work was Ada, Countess of Lovelace.

In 1985, Doron Swade and his colleagues at the Science Museum in London set out to build a full-size Babbage computer based upon his original designs. They chose to work on No. 2 and hoped to have it ready for Babbage's bicentenary in 1992. The construction was carried out in full public view on the floor of the museum. It was completed in May 1991, cost $500,000 to build, and has worked satisfactorily ever since.

Babbage was influential in a number of other areas. His *Reflections on the Decline of Science in England* (1830) began the move to the professionalization of British science. In his *On the Economy of Machinery and Manufactures* (1832), a work closely studied by Marx, Babbage argued that industry could only flourish by adopting a scientific approach to both technical and commercial matters. He also campaigned against street noises and was largely responsible for "Babbage's Bill" of 1864, restricting the rights of street musicians. The subject was sufficiently important to him to form a chapter in his revealing *Passages from the Life of a Philosopher* (1864).

Babcock, Harold Delos (1882–1968)
American astronomer

Babcock, who was born in Edgerton, Wisconsin, was educated at the University of California, Berkeley, where he graduated in 1907. In 1908 he joined the staff of the Mount Wilson Observatory where he remained until his retirement in 1948. After his formal retirement he continued to work for many years with his son Horace Babcock.

When he first joined the observatory George HALE had just discovered the presence of strong magnetic fields in sunspots by noting the splitting of their spectral lines, the so-called "Zeeman effect" first described by Pieter ZEEMAN in 1896. Babcock's first task was to supply the basic laboratory data on the effects of strong magnetic fields on various chemical elements.

Many years later, in collaboration with his son, he used their joint invention, the magnetograph, to detect the presence of weak and more generalized magnetic fields on the Sun. They also, in 1948, revealed the existence of strong magnetic fields in certain stars.

Babcock, Horace Welcome (1912–2003)
American astronomer

Babcock was born in Pasadena, California, the son of Harold Delos Babcock, a distinguished American astronomer who spent a lifetime observing at the Mount Wilson Observatory. Horace Babcock graduated in 1934 from the California Institute of Technology (Cal Tech) and obtained his PhD in 1938 from the University of California. He worked initially at Lick Observatory from 1938 to 1939 and at the Yerkes and McDonald observatories from 1939 to 1941. He then engaged in war work at the radiation laboratory at the Massachusetts Institute of Technology (1941–42) and at Cal Tech (1942–45). In 1946 Babcock returned to astronomy and joined his father at Mount Wilson where they began an enormously profitable collaboration. Babcock later served from 1964 until his retirement in 1978 as director of the Mount Wilson and Palomar Observatories, which became known in 1969 as the Hale Observatories.

In 1908 George Hale had detected splitting of the spectral lines in the light from sunspots. Such an effect results from the presence of a magnetic field, an effect first described by Pieter ZEEMAN in 1896. The fields observed by Hale were of considerable strength, ranging up to some 4,000 gauss. The field of the Earth by contrast is less than one gauss. The question then arose as to whether the Sun itself possessed a general magnetic field distinct from fields associated with sunspots. The problem facing early investigators was how to detect weak fields and this was not overcome until 1948 when the Babcocks successfully developed their magnetograph, permitting them to measure and record the Zeeman effect continuously and automatically. By the late 1940s they were able to report the presence of weak magnetic fields on the Sun, about one gauss in strength and restricted to latitudes greater than 55°. Further unexpected features were changes in polarity discovered in the 1950s: when examined in 1955 the north solar pole possessed positive polarity, the south negative polarity; by 1958 the situation was completely reversed.

In 1948 the Babcocks announced the further major discovery of stellar magnetic fields. By 1958 they had established the presence of magnetic fields in some 89 stars. The fields tended to be strong, of the order of several thousand gauss, and seemed to belong mainly to stars of spectral types O and B. Attempts to explain the presence of such fields were made considerably more difficult by the realization that some stars were "magnetic variables": the field of the brighter component of the binary star Alpha Canes Venatici was found to vary, with reversing polarity, from +5,000 to –4,000 gauss in 5.5 days. Such studies have done much to stimulate work on magnetohydrodynamics.

Babcock, Stephen Moulton (1843–1931)
American agricultural chemist

A farmer's son from New York State, Babcock gained his BA degree from Tufts College, Massachusetts, in 1866 and after a period of farming became a chemistry assistant and (from 1875) instructor at Cornell University. In 1879 he gained his doctorate under Hans Hübner at Göttingen, Germany. After a further spell at Cornell on his return, he became chemist at the New York Agricultural Station in 1882, where he worked on the analysis of milk.

In 1888 Babcock became professor of agricultural chemistry at the University of Wisconsin. Here, in 1890, he devised an efficient test (the *Babcock test*), which quickly became standard, for measuring the butterfat content of milk. Studies followed on rennet, fermentation, metabolic water, and animal nutrition. In 1907 Babcock's associates began studies in which cattle were fed balanced diets derived from a single source – corn, wheat, or oats. The results obtained provided further evidence for the existence of accessory food factors and Babcock's school played an important part in the vitamin studies that followed.

Babinet, Jacques (1794–1872) *French physicist*

Babinet (ba-bee-**nay**), who was born in Lusignan, France, studied in Paris at the Ecole Polytechnique and from 1820 he was a professor at the Collège Louis le Grand. He was elected to the Académie des Sciences in 1840.

His major work was devoted to the diffraction of light; he used diffraction to measure wavelengths more accurately than before, and did theoretical work on general diffraction systems. The *Babinet theorem* states that there is an approximate equivalence between the diffraction pattern of a large system and that of the complementary system, which is opaque where the original system is transparent and vice versa.

Furthermore he showed an interest in the optical properties of minerals, developing new instruments for the measurement of angles and polarizations. He also studied meteorological phenomena, especially those of an optical nature, investigating rainbows and the polarization of skylight. Babinet was the first to suggest (1829) that the wavelength of a given spectral line could be used as a fundamental standard of length. The idea was adopted in 1960, when the meter was defined as 1,650,763.73 wavelengths of the radiation emitted by an atom of krypton–86 in a specified transition. (This definition was changed in 1983 to the distance traveled by light in a certain fraction of a second.)

Babo, Lambert Heinrich Clemens von (1818–1899) *German chemist*

Babo (**bah**-boh) was born in Ladenberg, Germany, and studied at Giessen under Justus von LIEBIG. He was appointed as an assistant professor at Freiburg University in 1845, later being appointed professor of chemistry in 1859.

In 1847 he showed that the vapor pressure of a liquid can be lowered by dissolving substances in it. He also succeeded in demonstrating that the degree of depression of the vapor pressure is, in general, proportional to the concentration of the solution (known as *Babo's law*). François RAOULT was able to use Babo's discovery in 1886 to establish some general rules and to determine molecular weights.

Bache, Alexander Dallas (1806–1867)
American geophysicist

Bache (baych), the great-grandson of Benjamin FRANKLIN, was born in Philadelphia and graduated from West Point in 1825. After two years in the army he became professor of natural science and chemistry (1828–41) at the University of Pennsylvania. He spent the period 1836–38 studying the European educational system, publishing his findings in *Education in Europe* (1839).

In 1840 Bache founded the first American magnetic observatory at Girard College. He became, in 1843, superintendent of the U.S. Coast Survey, which he built up into a major institution. He had the entire coastline surveyed during his lifetime, his own particular research being into the Gulf Stream, an area also studied by his great-grandfather.

Dissatisfied with the American Association for the Advancement of Science he gathered around himself a group of scientists known as the "Lazzorconi," or beggars. He was successful in persuading Congress to create, in 1863, the National Academy of Sciences "to investigate, examine, experiment, and report upon any question of science and art." Bache was made its first president.

Backus, John (1924–2007) *American computer scientist*

Backus was born in Philadelphia. After graduating from Columbia University, New York, he joined the staff of IBM in 1952 and remained with them until his retirement in 1991. From 1959 until 1963 he worked at the IBM Research Center, Yorktown Heights, New York, and thereafter as an IBM Fellow at the IBM Research Laboratory, San Jose, California.

Backus has reported on the state of programming when he joined IBM. It was, he noted, "a black art, a private arcane matter." All programming was done using machine or assembly language. There were no compilers, no index registers, and the programmer spent most of his time debugging the program and feeding it into the computer. The programmers actually cost more than the computer. Backus commented, "They dismissed as foolish plans to make programming accessible to a larger population"; it was inconceivable "that any mechanical process could possibly perform the mysterious feats of invention required to write an efficient program."

In 1954 Backus led an IBM team determined to free computer programming from the professional élite. As the speed of computers increased it made no sense to have them standing idle while a programmer struggled to operate them. The problem was made more pressing by the development of the new and more powerful IBM 704. By late 1954 some of the main details of the high-level language Fortran (from *Formula Translation*) had been established. Backus defined his aim as "to design a language which would make it possible for engineers and scientists to write programs for the 704." The language itself was available in 1957 and soon became the most widely used programming language of the time.

Bacon, Francis (1561–1626) *British philosopher*

> They are ill discoverers that think there is no land, when they can see nothing but sea.
> —*The Advancement of Learning* (1605)

> I have taken all knowledge to be my province.
> —Letter to Lord Burleigh, 1592

Francis Bacon, Baron Verulam and Viscount St. Albans, was born in London into the Elizabethan ruling class; his father was Sir Nicholas Bacon, lord keeper of the Seal, and his uncle was Lord Burleigh. He entered Cambridge University in 1573 to study law, qualified in 1582, and entered parliament in 1584. His political career was noted for his ability both to attach himself to the side of royal favorites (and thereby rise) and to make sure that he was on the opposing side when they fell from favor (and thereby rise further). He held many state offices including those of attorney general (1613) and lord chancellor (1618).

Bacon's first work of philosophy was *The Advancement of Learning* (1605), a review of the current state of knowledge. He planned an encyclopedia of all knowledge, the *Instauratio Magna* (Great Renewal), but this was never completed, the most substantial fragment being the *Novum organum* (1620; New Organum alluding to the *Organum* of ARISTOTLE concerning logic). Bacon rejected completely deference to the authority of the ancients, in particular the deductive logic of Aristotle, in dealing with science and the investigation of the world. He asserted that nature could be understood and even controlled by man. Bacon's method to accomplish this was induction, by which he understood a method of proceeding from the particular to the general by a process of exclusions, generalizing from particular experiments and investigations. Bacon had a great influence on the first generation of British experimental scientists. Despite his urge toward completeness, Bacon's knowledge of the science of his day was inadequate. In one field, however, he was highly prescient: by induction he concluded that heat was a form of motion.

In 1621 Bacon was accused of bribery, imprisoned for a few days, fined, and banished from parliament and the Court. He retired to his estate in Hertfordshire and in this last period of his life he wrote a revision of *The Advancement of Learning* known by its Latin title *De augmentis scientiarum* (1623). He also produced some fragments of the *Instauratio* and *The New Atlantis*, a utopia that foreshadowed the scientific societies founded later in the century.

Bacon's death was brought on by his last scientific experiment. He had the idea that snow might preserve flesh and to test this he stuffed a chicken with snow, which he himself collected. It is said that as a result he caught a chill and died soon after.

Bacon, Roger (1220–1292) *English philosopher and alchemist*

> All science requires mathematics.
> —*Opus majus* (1733; Major Work)

> Without experience nothing can be known sufficiently.
> —As above

Bacon, who was born in Ilchester, southwest England, studied at Oxford and then at Paris (1234–50) under Petrus Peregrinus. In about 1257 he joined the Franciscan Order and from about 1250 until 1277 he was at Oxford, where he studied under Grosseteste. He is supposed to have been confined in Paris from 1277.

Bacon was often in disgrace with the authorities but he had a considerable reputation as a philosopher and alchemist, being called *Doctor Mirabilis* (Miraculous Teacher). From about 1247 until 1257 he concentrated on research in mathematics, optics, alchemy, and astronomy and during this developed a magnifying glass, defined reflection and refraction, and also mentioned gunpowder in his writings. Bacon distinguished two kinds of alchemy, speculative and operative, and he believed firmly in the practical benefits of science. His writings, the *Opus majus* (Major Work, published in 1733), *Opus secundus* (Second Work), and *Opus tertius* (Third Work), include a number of predictions, of powered cars, aircraft, and ships, and of machines for extending the powers of man. He was committed to the belief that the Earth was round, and suggested that it could be circumnavigated. He stressed that experimentation was essential to the progress of science.

Baekeland, Leo Hendrik (1863–1944) *Belgian–American industrial chemist*

Baekeland (**bayk**-land) was born in Ghent and educated at the university there, graduating in 1884. He was professor of physics and chemistry at Bruges in 1887 and returned to Ghent the next year as assistant professor of chemistry. But Baekeland grew impatient with academic life and in 1889 a honeymoon tour took him to America where he settled.

Baekeland worked at first as a photographic chemist and in 1891 he opened his own consulting laboratory. In 1893 he began to manufacture a photographic paper, which he called Velox, and six years later his company was bought out by the Kodak Corporation for one

BAKELITE   The polymeric structure of bakelite.

million dollars. Now financially independent, Baekeland returned to Europe to study at the Technical Institute at Charlottenburg.

On his return to America, Baekeland began to investigate, as a synthetic substitute for shellac, the phenol–formaldehyde resins discovered by BAEYER in 1871. Since nothing remotely like shellac emerged, he began to look for other uses for this material. By choosing suitable reaction conditions he produced a hard amberlike resin, which could be cast and machined and which had excellent durability and electrical properties. Bakelite was finally unveiled in 1909, when Baekeland set up the General Bakelite Corporation.

In 1922 Baekeland's company merged with two rivals and in 1939 it became a subsidiary of the Union Carbide and Carbon Corporation. Baekeland continued to produce scientific papers throughout this period. He received many honors and held many professional posts, including that of president of the American Chemical Society.

Baer, Karl Ernst von (1792–1876) German–Estonian biologist, comparative anatomist, and embryologist

Baer (bair) is generally considered the father of modern embryology. He was born on his family's estate in Piep, Estonia, and received private tutoring and schooling before entering Dorpat University to study medicine. He graduated in 1814 and then studied comparative anatomy at the University of Würzburg, where he was introduced to embryology by Ignaz DÖLLINGER. In 1817 Baer became professor of zoology at Königsberg and in 1834 was appointed academician and librarian of the Academy of Sciences at St. Petersburg.

It was prior to his move to St. Petersburg that Baer did most of his pioneering work in laying the foundation of comparative embryology as a separate discipline. In distinguishing the mammalian ovum within the Graafian follicle he established that all mammals, including man, develop from eggs. He also traced the development of the fertilized egg and the order in which the organs of the body appear and develop, showing that similar (homologous) organs arise from the same germ layers in different animals, thus extending the work of Kaspar WOLFF and the German anatomist Christian Pander. His expounding of the "biogenetic law," demonstrating the increasing similarity and lack of specialization in the embryos of different animals as one investigates younger and younger embryos, provided DARWIN with basic arguments for his evolutionary theory. Baer was, however, opposed to the idea of there being a common ancestor for all animal life, although he conceded that some animals and some races of man might have had common ancestry. His other notable discoveries included

the mammalian notochord and the neural folds as the precursors of the nervous system. Baer intended his embryological work to be, at least partly, a means of improving animal classification by demonstrating vertebrate affinities. Indeed modern zoological classification is now based partly on biogenetic principles. His great work on the mammalian egg, *De ovi mammalium et hominis genesi* (1827; On the Origin of the Mammalian and Human Ovum) was followed (1828–37) by *Über Entwickelungsgeschichte der Tiere* (On the Development of Animals), in which he surveyed all existing knowledge of vertebrate development.

A man of wide interests, Baer did much work in other scientific disciplines. He was instrumental in founding the German Anthropological Society and helped to found the Russian Geographical and Entomological Societies.

Baeyer, Johann Friedrich Adolph von (1835–1917) German organic chemist

Baeyer (**bay**-er) was the son of a member of the Prussian General Staff and his mother was the daughter of a celebrated jurist and literary historian. Born in Berlin, Baeyer went to Heidelberg in 1856 to study chemistry with Robert BUNSEN. Here he met August KEKULE, who had a profound influence on his development as a chemist and gave him the theoretical foundation for his work. After obtaining his PhD (1858) Baeyer took up a teaching position in 1860 at a small technical school, the Gewerbe-Institut, in Berlin. In 1872 he was appointed professor of chemistry at Strasbourg and in 1875 succeeded LIEBIG as professor of chemistry at Munich, where he remained for the rest of his life.

In 1864, continuing the work of WÖHLER, Liebig, and Schlieper on uric acid, Baeyer characterized a related series of derivatives including alloxan, parabanic acid, hydantoin, and barbituric acid. In 1871 he discovered the phthalein dyes, phenolphthalein and fluorescein, by heating phenols with phthalic anhydride. In the course of this work he discovered the phenol–formaldehyde resins, which were later developed commercially by BAEKELAND. The centerpiece of Baeyer's prolific researches, however, was his work on indigo, which started in 1865 and lasted for 20 years.

The first step consisted of the reduction of indigo to its parent substance, indole, which Baeyer accomplished by the new method of heating with zinc dust. The first synthesis was a lengthy one, starting from phenylacetic acid. This was soon followed by shorter methods starting from *o*-nitrocinnamic acid and *o*-nitrophenylpropionic acid. In 1883 he gave a structure of indigo that was correct except for the stereochemical arrangement of the double bond, which was later shown to be *trans* by x-ray crystallography (1928). Baeyer's syntheses proved too costly for commercial manufacture and he took no part in the industrial development of indigo, terminating his work in 1885. Commercial synthetic indigo was eventually produced in 1890. Baeyer's work also led to the production of many other new dyes.

From indigo Baeyer turned to the polyacetylenes, compounds whose explosive properties led him to consider the stability of carbon–carbon bonds in unsaturated and ring compounds. He formulated the *Baeyer strain theory*, stating that compounds are less stable the more their bond angles depart from the ideal tetrahedral arrangement. Baeyer's other researches included work on oxonium compounds; on the reduction of aromatic compounds, in which he observed a loss of aromaticity on reduction; and on terpenes, including the first synthesis of a terpene in 1888.

The strain theory was one of Baeyer's few theoretical contributions; he was a virtuoso of test-tube chemistry at a time when this could produce extraordinary results. In 1905 he received the Nobel Prize for chemistry for his work on indigo and aromatic compounds.

Bahcall, John Noris (1934–2005) *American physicist*

Bahcall was born in Shreveport, Louisiana, and educated at the universities of California, Chicago, and Harvard, where he obtained his PhD in 1961. He immediately moved to the faculty of the California Institute of Technology, and remained there until 1971 when he was appointed to the Institute of Advanced Studies, Princeton.

In the 1960s Bahcall began to consider the emission of neutrinos from the Sun. One of the apparent early triumphs of nuclear physics was the light it threw on the internal workings of the Sun. Theorists such as Hans BETHE had proposed the existence of a number of cyclic fusion reactions producing vast amounts of energy, heavier elements, and a certain number of neutrinos. As neutrinos have a low probability of interacting with other particles, some solar neutrinos should be received at the Earth's surface. Bahcall calculated that there occurs one event per second for every 10^{36} target atoms, for which one solar neutrino unit (SNU) should be detectable. The matter was put to the test by Raymond DAVIS, who used a detector consisting of a tank of 100,000 gallons of cleaning fluid in a one-mile-deep mine.

Bahcall predicted that Davis would observe a flux of about 8 SNU. In fact, from 1967 onwards, Davis recorded a flux of about 2 SNU. The burden was placed upon the theorists to account for the anomalous results, or to revise the theory in ways that would make the results acceptable. This is the *solar neutrino problem*.

A number of options were considered by Bahcall. Perhaps, the Sun was passing through a quiet phase and over long periods of time neutrino output would agree with theory. It was also thought that the issue would eventually be resolved when the abundance of radioactive technetium, produced by interactions deep in the Earth between neutrinos and molybdenum, has been accurately measured.

Or, perhaps, our solar model is wrong. This, Bahcall pointed out, leads nowhere. Alternative solar models agree with the standard model in the rate of neutrino production. Other theorists have challenged generally accepted physical principles. Bahcall considered the possibility that *neutrino oscillations* occurred, i.e., the transition from one type of neutrino to another type. This process is only possible if neutrinos have nonzero masses. This is necessarily the case in most grand unified theories. Since there is evidence from other sources that neutrinos do have nonzero masses, the current consensus is that these neutrino oscillations explain the solar neutrino problem, although this has not been conclusively demonstrated.

As a number of new and more sensitive detectors are being built, theorists seem inclined to await fresh data before judging between competing theories. Until then, as Bahcall noted, physicists will dismiss the problem as a matter of astronomy while astronomers will attribute the anomaly to the failings of physics.

Baillie, Matthew (1761–1823) *British physician*

Baillie, who was born in Scotland, studied classics, mathematics, and philosophy at Glasgow University and then arts and medicine at Oxford University. He graduated MD in 1787 and was appointed physician to St. George's Hospital, London. A nephew of William and John HUNTER, he inherited William Hunter's house and medical school at Windmill Street in 1783. Baillie's major work was the *Morbid Anatomy of Some of the Most Important Parts of the Human Body* (1793). This was a pioneering work, illustrated by a series of engravings, that helped establish pathology as a separate subject. From 1810 onward he was a physician to the royal family, attending George III during his final illness.

His sister was the poet and dramatist, Joanna Baillie.

Baily, Francis (1774–1844) *British astronomer*

Born in Newbury in southern England, Baily was a prosperous stockbroker who, on retirement, devoted himself to astronomy. During the total eclipse of the Sun in 1836 he noted that immediately before and after totality a number of bright points of light appear around the edge of the Moon. This effect, known as

Baily's beads, is caused by light from the Sun shining through the lunar valleys.

Baird, John Logie (1888–1946) *British inventor*

Baird, who was born in Helensburgh, Scotland, studied electrical engineering at the Royal Technical College in Glasgow and then went to Glasgow University. His poor health prevented him from active service during World War I and from completing various business enterprises in the years following the war.

After a breakdown in 1922 he retired to Hastings and engaged in amateur experiments on the transmission of pictures. Using primitive equipment he succeeded in transmitting an image over a distance of a couple of feet, and in 1926 he demonstrated his apparatus before a group of scientists. Recognition followed, and the next year he transmitted pictures by telephone wire between London and Glasgow. In the same year he set up the Baird Television Development Company. He continued to work on improvements and on 30 September 1929 gave the first experimental BBC broadcast. Synchronization of sound and vision was achieved a few months later. In 1937, however, the Baird system of mechanical scanning was ousted by the all-electronic system put forward by Marconi–EMI. Baird was at the forefront of virtually all developments in television and continued research into color, stereoscopic, and big-screen television until his death.

Baird, Spencer Fullerton (1823–1887) *American biologist*

Baird was born in Reading, Pennsylvania, and educated at Dickinson College. From 1845 to 1850 he was professor of natural sciences at the college, making immense collections of North American fauna from several expeditions. In 1874 he was appointed United States Commissioner of Fish and Fisheries. He was assistant secretary and then full secretary (1878–87) of the Smithsonian Institution and in 1857 was one of the founders of the Institution's National Museum. Baird was also one of the principal founders of the Woods Hole Marine Laboratory. A colleague of Louis AGASSIZ and John J. Audubon, Baird did much to introduce field study in botany and zoology in the United States, in ornithology in particular, stressing the importance of extreme accuracy in descriptions. From 1853 to 1884 he published several accounts of North American reptiles, birds, and mammals.

Bakewell, Robert (1725–1795) *British stock breeder*

Little is known about Bakewell's early life except that he helped his father on their 440-acre rented farm at Dishley in Leicestershire in central England, which he took over in 1760 on the death of his father. Bakewell's aim was to "get beasts to weigh where you want them to weigh." His most impressive achievement was the production of his Leicester breed of sheep. Within 50 years the breed had spread throughout the world and Bakewell apparently succeeded in producing two pounds of mutton where there had only been one before. He also introduced the custom of letting out his rams for breeding. He was, however, less successful with his Leicester long horn cattle, which, though good meat-producers, did not yield much milk.

Bakker, Robert (1945–) *American paleontologist*

> Let dinosaurs be dinosaurs. Let the Dinosauria stand proudly alone, a class by itself. They merit it. And let us squarely face the dinosaurness of birds and the birdness of the Dinosauria.
> —*The Dinosaur Heresies* (1986)

The son of an electrical engineer, Bakker was born in Ridgewood, New Jersey, and educated at Yale and at Harvard, where he completed his PhD in 1976. After teaching for eight years at Johns Hopkins University, Baltimore, he moved to Boulder, Colorado, in 1984 to work as an independent paleontologist.

In the early 1970s, while still a graduate student, Bakker argued that traditional views on the nature of dinosaurs were misguided. Dinosaurs, he claimed, were warm-blooded, like mammals, and not cold-blooded like reptiles. In support Bakker offered three main arguments derived from comparative anatomy, latitudinal zonation, and ecology.

Anatomically the bones of endotherms (warm-blooded creatures) are rich in blood vessels and show no growth rings. Precisely these features are found in mammals, birds, and, Bakker noted, dinosaurs as well. They are lacking in cold-blooded reptiles (ectotherms).

Further, endotherms can cope with most temperature variations and can be found in temperate, arctic, and equatorial zones. Large reptiles, however, cannot survive cool winters. Yet, during the Cretaceous, dinosaurs could have been found in the far north of Canada, well within the Arctic Circle.

Finally Bakker points to predator–prey ratios. The Komodo dragon, the largest living lizard, consumes its own weight every 60 days, a lion in only eight days. Such are the demands of endothermy. As a consequence a community can support fewer warm-blooded predators than cold-blooded predators. The predator–prey ratio, Bakker argued, is a constant characteristic of the predator's metabolism. Calculations revealed that a given biomass can support a warm-blooded predator biomass of 1–3% and a cold-blooded predator biomass of up to 40%. Given that some fossil deposits yield thousands of individuals, their predator– prey ratio should be measurable. And Bakker did find that among the reptiles of the Permian (285–225 million years ago) the ratio was very high (35–60%), while the dinosaurs of the Triassic (225–195 million years) had a ratio of only 1–3%.

Bakker's work called for a revision of vertebrate systematics. In traditional classifications birds and dinosaurs are seen as collateral descendants of thecodonts, that is, animals with teeth set in sockets. Bakker, however, has proposed that birds are descended from dinosaurs. Such views, vigorously and frequently expressed, have made Bakker a controversial and well-known figure. It is often said that Bakker was the inspiration for the main character in the popular film *Jurassic Park*.

Bakker has sought to reach an even wider public with his *Raptor Red* (1995), a popular novel dealing with a year in the life of a large predatory dinosaur. He now lives in Wyoming, where he conducts guided digs.

Balard, Antoine-Jérôme (1802–1876)
French chemist

Balard (ba-**lar**) was born in Montpellier in southern France, and studied there at the School of Pharmacy. After graduating in 1826, he remained at Montpellier as a demonstrator in chemistry. In 1825, while investigating the salts contained in seawater, he discovered a dark red liquid, which he proved was an element with properties similar to chlorine and iodine. Balard proposed the name "muride" but the editors of the journal *Annales de chimie* (Annals of Chemistry) preferred "brome" (because of the element's strong odor, from the Greek for "stink") and the element came to be called bromine. Balard also (1834) discovered dichlorine oxide (Cl_2O) and chloric(I) acid (HClO).

In 1833 he became professor at Montpellier and in 1843 succeeded Louis Thenard at the Sorbonne as professor of chemistry. In 1854 he was appointed professor of general chemistry at the Collège de France, where he remained until his death.

Balfour, Francis Maitland (1851–1882)
British zoologist

The younger brother of the British statesman Earl Balfour, Francis Balfour's career was cut short when, while convalescing from typhoid fever in Switzerland, he died attempting an ascent of the Aiguille Blanche, Mont Blanc. Balfour, who was born in Edinburgh, Scotland, held the position of animal morphologist at the Naples Zoological Station and in 1882 was appointed to the specially created post of professor of animal morphology at Cambridge University. Much influenced by the work of

Michael FOSTER, with whom he wrote *Elements of Embryology* (1883), Balfour showed the evolutionary connection between vertebrates and certain invertebrates, both of which have a notochord (a flexible rod of cells extending the length of the body) in their embryonic stages. Similar research was being conducted at that time by Aleksandr KOVALEVSKI. Balfour proposed the term Chordata for all animals possessing a notochord at some stage in their development, the Vertebrata (backboned animals) being a subphylum of the Chordata. He was an early exponent of recapitulation – the theory that ancestral forms are repeated in successive embryonic stages undergone by modern species. Balfour also did pioneer work on the development of the kidneys and related organs, as well as the spinal nervous system. His other important publications include *On the Development of Elasmobranch Fishes* (1878) and *Comparative Embryology* (1880–81) published in two volumes (invertebrates and vertebrates), the latter forming the basis of modern embryological study.

Balmer, Johann Jakob (1825–1898)
Swiss mathematician

Born in Lausanne, Switzerland, Balmer (**bahl**-mer) was not a professional scientist but worked as a school teacher in Basel from 1859. In 1885 he discovered that there was a simple mathematical formula that gave the wavelengths of the spectral lines of hydrogen – the *Balmer series*. This formula proved to be of great importance in atomic spectroscopy and in developing the atomic theory. Balmer arrived at his result purely from empirical evidence and was unable to explain why it yielded correct answers. Not until the further development of the atomic theory by Niels BOHR and others was this possible.

Baltimore, David (1938–) *American molecular biologist*

Baltimore was born in New York City and studied chemistry at Swarthmore College. He continued with postgraduate work at the Massachusetts Institute of Technology, and at Rockefeller University, where he obtained his PhD in 1964. After three years at the Salk Institute in California, he returned to MIT in 1968 where, in 1972, he became professor of biology.

Francis CRICK had formulated what came to be known as the Central Dogma of molecular biology, namely, that information could flow from DNA to RNA to protein but could not flow backward from protein to either DNA or RNA. Although he had not actually excluded the passage of information from RNA to DNA it became widely assumed that such a flow was equally forbidden. In June 1970 Baltimore and, quite independently, Howard TEMIN announced the discovery of an enzyme later to be known as reverse transcriptase, which is capable of transcribing RNA into DNA. Apparently certain viruses, like the RNA tumor viruses used by Baltimore, could produce DNA from an RNA template. For this work Baltimore shared the 1975 Nobel Prize for physiology or medicine with Temin and Renato DULBECCO. A few years later their work took on an added significance when GALLO and MONTAGNIER identified a retrovirus as the cause of AIDS.

Earlier (1968) Baltimore had done important work on the replication of the polio virus. He revealed that the RNA of the virus first constructed a "polyprotein" (or giant protein molecule), which then split into a number of smaller protein molecules. Two of these polymerized further RNA while the remainder formed the protein coat of the new viral particles.

In 1982 Baltimore became founding director of the Whitehead Institute, Cambridge, Massachusetts, a research biomedical foundation backed by the industrialist E. C. Whitehead. While at Whitehead, in collaboration with D. Schatz, he identified two antibody genes, RAG-1 and RAG-2. In 1990 Baltimore was appointed president of Rockefeller University; it was not to prove a fruitful or happy time. Many staff opposed the appointment and Baltimore became involved in a bitter controversy. It had been claimed that a paper coauthored by Baltimore and published in 1986 in the journal *Cell* was based on falsified data. Although Baltimore withdrew his name from the paper, the public controversy persisted in Congressional hearings and the correspondence columns of the British scientific journal *Nature*. Baltimore resigned the presidency in 1992 and returned to MIT in 1994 as professor of molecular biology. He was appointed president of the California Institute of Technology in 1997, resigning in 2006. Baltimore is currently president of the American Association for the Advancement of Science.

Bamberger, Eugen (1857–1932) *German chemist*

Bamberger (**bahm**-ber-ger) studied at Breslau, Heidelberg, and in his native city of Berlin, where he graduated. After working as an assistant, first to Karl Rammelsberg in Berlin (1882) and then to Adolf von BAEYER in Munich (1883), he became professor of chemistry at the Federal Institute of Technology in Zurich in 1893.

Bamberger worked on a number of topics in organic chemistry, including the synthesis of nitroso compounds and quinols, the conversion of sulfonic acids into sulfanilic acids, and the production of diazonium anhydrides. He also extended Baeyer's ideas on benzene structure

to naphthalene and first proposed the term "alicyclic" to describe unsaturated organic ring compounds.

Banach, Stefan (1892–1945) *Polish mathematician*

Banach (**bah**-nak) was born into a poor family in Krakow, Poland, and was largely self-educated. Although he attended the Lwow Institute of Technology he did not graduate. He taught mathematics privately for some years before being appointed in 1922 to the University of Lwow, where he was elected professor of mathematics in 1927. Lwow was occupied by the Germans in 1941 and although Banach managed to survive the war years he died from lung cancer in 1945.

Banach is one of the few mathematicians, along with EUCLID and HILBERT, who have given their name to a space. A *Banach space* is a type of vector space more general than a Hilbert space.

In 1924, Banach, with A. TARSKI, proved an apparently nonsensical theorem. The *Banach–Tarski paradox* claims that it is possible to dissect a sphere into a finite number of pieces which can be reassembled to form two spheres the same size as the original. It was later shown that the sphere must be dissected into at least five pieces.

Banks, Sir Joseph (1743–1820) *British botanist*

> He [Banks] stopped and, looking round, involuntarily exclaimed, "How beautiful!" After some reflection he said to himself, "It is surely more natural that I should be taught to know all these productions of Nature, in preference to Greek and Latin; but the latter is my father's command and it is my duty to obey him; I will however make myself acquainted with all these different plants for my own pleasure and gratification."
> —Sir Everard Home, *Hunterian Oration* (1822)

The son of William Banks of Revesby Abbey, Lincolnshire, Joseph Banks inherited a large fortune when he came of age, and later used this money to finance his scientific expeditions. Born in London, he studied botany at Oxford, graduating in 1763, and three years later traveled abroad for the first time as naturalist on a fishery-protection vessel heading for Labrador and Newfoundland. On the voyage he was able to collect many new species of plants and insects and, on his return, was elected a fellow of the Royal Society.

In London Banks learned that the Royal Society was organizing a voyage to the South Pacific to observe the transit of Venus across the Sun. In 1768 James COOK set sail in the *Endeavour* and Banks, together with a team of artists and the botanist Daniel Solander, accompanied him. Cook landed in Australia, a continent with a flora and fauna different from any found elsewhere. Banks found that most of the Australian mammals were marsupials, which are more primitive, in evolutionary terms, than the placental mammals of other continents.

After three years with the *Endeavour* Banks returned, with a large collection of unique specimens, to find himself famous. George III, interested in hearing a first-hand account of Banks's travels, invited him to Windsor. This visit was the start of a long friendship with the king, which helped Banks establish many influential contacts – possibly a factor in his election as president of the Royal Society in 1778, a post that he held until his death.

Throughout his life Banks retained his interest in natural history and in the specimens collected on the many expeditions mounted during that period. As honorary director of Kew Gardens he played a major part in establishing living representatives of as many species as possible at Kew and in providing a center for advice on the practical use of plants. He initiated many successful projects, including the introduction of the tea plant to India from its native China and the transport of the breadfruit from Tahiti to the West Indies. By George III's request, he also played an active role in importing merino sheep into Britain from Spain.

The British Museum (Natural History) now houses Banks's library and herbarium, both regarded as major collections.

Banting, Sir Frederick Grant (1891–1941) *Canadian physiologist*

Banting, a farmer's son from Alliston, Ontario, began studying to be a medical missionary at Victoria College, Toronto, in 1910. During his studies he concentrated increasingly on medicine and graduated MD in 1916, whereupon he immediately joined the Canadian Army Medical Corps. In 1918 he was awarded the Military Cross for gallantry in action and was invalided out of the army.

Banting then returned to Toronto and worked for a time studying children's diseases before setting up practice in London, Ontario, in 1920. He also began work at the London Medical School, specializing in studies on the pancreas, particularly the small patches of pancreatic cells known as the islets of Langerhans. Earlier work had shown a connection between the pancreas and diabetes and Banting wondered if a hormone was produced in the islets of Langerhans that regulated glucose metabolism. In 1921 he approached John Macleod, professor of physiology at Toronto University, who was initially skeptical. Feeling that Banting needed help in physiological and biochemical methods, Macleod suggested the assistance of a young research student, Charles BEST, and eventually merely granted Banting and Best

some laboratory space during the vacation, while he went abroad.

Over the next six months Banting and Best devised a series of elegant experiments. They tied off the pancreatic ducts of dogs and made extracts of the islets of Langerhans free from other pancreatic substances. These extracts, called "isletin," were found to have some effect against diabetes in dogs. Prior to trials on humans, Macleod asked a biochemist, James Collip, to purify the extracts and the purification method for what was now known as insulin was patented by Banting, Best, and Collip in 1923. They allowed manufacturers freedom to produce the hormone but required a small royalty to be paid to finance future medical research.

The pharmaceutical firm Eli Lilly began industrial production of insulin in 1923 and in the same year Banting was awarded the chair of medical research at Toronto University and a government annuity of $7,000. The Nobel Prize for physiology or medicine was awarded jointly to Banting and Macleod in 1923; Banting was furious that Best had not been included in the award and shared his part of the prize money with him. Macleod shared his portion with Collip. In 1930 the Banting and Best Department was opened at the University of Toronto and Banting became its director. Under Best's guidance, it became the center of medical research in Canada. His own later researches were into cancer and also the function of the adrenal cortex.

Banting was knighted in 1934. When war broke out in 1939 he joined an army medical unit and worked on many committees linking Canadian and British wartime medical research. His bravery was much in evidence at this time, particularly his personal involvement in research into mustard gas and blackout problems experienced by airmen. In 1941 on a flight from Gander, Newfoundland, to Britain his plane crashed and he died in the snow.

Bárány, Robert (1876–1936) *Austro-Hungarian physician*

Bárány (**bah**-rah-nee) was born in Vienna and educated at the university there, graduating in medicine in 1900. After studying at various German clinics, he returned to Vienna to become an assistant at the university's ear clinic. In 1909 he was appointed lecturer in otology. Through his work at the clinic he devised a test, now called the *Bárány test*, for diagnosing disease of the semicircular canals of the inner ear by syringing the ear with either hot or cold water. For this he was awarded the 1914 Nobel Prize in physiology or medicine. At this time he was being held as a prisoner of war in Siberia, but through the offices of the Swedish Red Cross he was released for the presentation.

In 1917 Bárány was appointed professor at

Uppsala University, where he continued his investigations on the inner ear and the role of the cerebellum in the brain in controlling body movement. *Bárány's pointing test* is used to test for brain lesions.

Barcroft, Sir Joseph (1872–1947) *Irish physiologist*

Barcroft was born in Glen Newry in Northern Ireland. As professor of physiology at Cambridge University (1926–37) and director of animal physiology for the Agricultural Research Council (1941–47), he carried out extensive research into human embryology, physiology, and histology. He investigated the oxygen-carrying role of hemoglobin, and devised (1908) an apparatus for the analysis of blood gases.

Barcroft led three high-altitude expeditions – to Tenerife (1910); Monte Rosa, Italy (1911); and the Peruvian Andes (1922) – in order to study acclimatization and the effects of rarefied atmospheres on respiration. He found that, at low oxygen pressures, the human lung is not able to secrete oxygen into the blood at a higher pressure than the pressure of the inhaled air.

During World War I, Barcroft was chief physiologist at the Experimental Station at Porton, Wiltshire, where he studied the effects of poisonous gases. Elected a fellow of the Royal Society in 1910 and knighted in 1935, Barcroft wrote the famous text *Respiratory Function of the Blood* (1914), as well as publications on the brain and its environment and on prenatal conditions.

Bardeen, John (1908–1991) *American physicist*

Bardeen, the son of a professor of anatomy, was born in Madison, Wisconsin, and studied electrical engineering at the University of Wisconsin. He obtained his PhD in mathematical physics at Princeton in 1936. Bardeen began work as a geophysicist with Gulf Research and Development Corporation, Pittsburgh, in 1931 but in 1935 entered academic life as a junior fellow at Harvard, moving to the University of Minnesota in 1938. He spent the war years at the Naval Ordnance Laboratory, followed by six creative years from 1945 until 1951 at the Bell Telephone Laboratory, after which he was appointed professor of physics and electrical engineering at the University of Illinois, a post he held until 1975.

Bardeen is remarkable as a recipient of two Nobel Prizes for physics. The first, awarded in 1956, he shared with Walter BRATTAIN and William SHOCKLEY for their development of the point-contact transistor (1947), thus preparing the way for the development of the more efficient junction transistor by Shockley.

Bardeen's second prize was awarded in 1972 for his formulation, in collaboration with Leon COOPER and John SCHRIEFFER, of the first satisfactory theory of superconductivity – the so-called BCS theory. In 1911 Heike KAMERLINGH-ONNES had discovered that mercury lost all electrical resistance when its temperature was lowered to 4.2 kelvins. Superconductivity was also shown to be a property of many other metals, yet despite much effort to understand the phenomenon, a full explanation was not given until 1957. The basic innovation of the BCS theory was that the current in a superconductor is carried not by individual electrons but by bound pairs of them, later known as *Cooper pairs*. The pairs form as a result of interactions between the electrons and vibrations of the atoms in the crystal. The scattering of one electron by a lattice atom does not change the total momentum of the pair, and the flow of electrons continues indefinitely.

The success of the BCS theory led to an enormous revival of interest in both the theory of superconductors and their practical application. Beginning in the 1970s, there began to emerge a new industry capable of exploiting superconducting materials, especially in devices based on the effects discovered by Brian JOSEPHSON.

Barger, George (1878–1939) *British organic chemist*

Barger, of Anglo-Dutch parentage, was born in Manchester, England, attended school in Holland, and studied natural sciences at Cambridge University, where he graduated with equal distinction in chemistry and botany.

While a demonstrator in botany at the University of Brussels (1901–03) Barger discovered a method of determining the molecular weight of small samples by vapor-pressure measurements. From 1903 to 1909 he was a researcher at the Wellcome Physiological Research Laboratories, where he worked mainly on ergot, isolating ergotoxine in collaboration with Francis Carr (1906).

In the course of the work on ergot Barger and Henry DALE isolated a series of related amines, derived from tyramine, which had sympathomimetic activity. This work led to a better understanding of the nervous system and to the development of new drugs. The work on active amines was collected in *The Simpler Natural Bases* (1914) and the work on ergot culminated in the definitive monograph on every facet of the subject, *Ergot and Ergotism* (1931).

From 1909 to 1919 Barger worked in London, as head of chemistry at Goldsmiths' College, professor of chemistry at Royal Holloway College, and from 1914 as chemist with the National Institute of Medical Research. In 1919 he was elected a fellow of the Royal Society and appointed to the new chair of chemistry in relation to medicine at Edinburgh. The most notable researches of this period were Charles Harington's structural elucidation and synthesis of thyroxine, in which Barger collaborated, and Barger's synthesis of methionine. His school did important work on physostigmine, cholinesterase, and vitamin B_1. In 1938 Barger became professor of chemistry at the University of Glasgow. He was a pioneer of medicinal chemistry, an excellent linguist, and a tireless ambassador for science.

Barkhausen, Heinrich Georg (1881–1956) *German physicist*

After attending the gymnasium and engineering college in his native city of Bremen, Barkhausen (**bark**-how-zen) gained his PhD in Göttingen and in 1911 became professor of electrical engineering in Dresden. Here he formulated the basic equations governing the coefficients of the amplifier valve.

In 1919 he discovered the *Barkhausen effect*, observing that a slow continuous increase in the magnetic field applied to a ferromagnetic material gave rise to discontinuous leaps in magnetization, which could be heard as distinct clicking sounds through a loudspeaker. This effect is caused by domains of elementary magnets changing direction or size as the field increases.

In 1920, with K. Kurz he developed an ultrahighfrequency oscillator, which became the forerunner of microwave-technology developments. After World War II he returned to Dresden to aid the reconstruction of his Institute of High-Frequency Electron-Tube Technology, which had been destroyed by bombing, and remained there until his death.

Barkla, Charles Glover (1877–1944) *British x-ray physicist*

Barkla was born in Widnes in the northwest of England. After taking his master's degree in 1899 at Liverpool, Barkla went to Trinity College, Cambridge but, because of his passion for singing, he transferred to King's College to sing in the choir. At King's College he started his important research on x-rays. In 1902 he returned to Liverpool as Oliver Lodge Fellow and in 1909 became Wheatstone Professor at King's College, London. From 1913 onward he was professor of natural philosophy at Edinburgh University.

His scientific work, for which he received the 1917 Nobel Prize for physics, concerned the properties of x-rays – in particular, the way in which they are scattered by various materials. He showed in 1903 that the scattering of x-rays by gases depends on the molecular weight of the gas. In 1904 he observed the polarization of x-rays – a result that indicated that x-rays are a form of electromagnetic radiation like light. Further confirmation of this was obtained

in 1907 when he performed certain experiments on the direction of scattering of a beam of x-rays as evidence to resolve a controversy with William Henry BRAGG who argued, at the time, that x-rays were particles.

Barkla also demonstrated x-ray fluorescence, in which primary x-rays are absorbed and the excited atoms then emit characteristic secondary x-rays. The frequencies of the characteristic x-rays depend on the atomic number of the element, as shown by Henry Mosely, who could well have shared Barkla's Nobel Prize but for his untimely death.

From about 1916, Barkla became isolated from modern physics with an increasingly dogmatic attitude, a tendency to cite only his own papers, and a concentration on untenable theories.

Barnard, Christiaan Neethling (1922–2001) *South African surgeon*

> The prime goal is to alleviate suffering and not to prolong life. And if your treatment does not alleviate suffering but only prolongs life, that treatment should be stopped.

Barnard, who was born in Beaufort West, South Africa, was awarded his MD from the University of Cape Town in 1953 and joined the Medical Faculty as a research fellow in surgery. After three years at the University of Minnesota (1955–58), where he studied heart surgery, he returned to Cape Town as director of surgical research. He concentrated on improving techniques for artificially sustaining bodily functions during surgery and for keeping organs alive outside the body. On 2 December 1967 Barnard performed the first heart-transplant operation on a human patient, a 54-year-old grocer named Louis Washkansky. He received the heart of Denise Duvall, a traffic accident victim. The heart functioned but the recipient died of pneumonia 18 days after the operation. The body's immune system had broken down following the administration of drugs to suppress rejection of the new heart as foreign tissue.

Barnard subsequently performed further similar operations with improved postoperative treatment giving much greater success. In 1974 he performed the first successful double heart-transplant operation. His pioneering work generated worldwide publicity for heart-transplant surgery and it is now widely practiced. In 1984 he became professor emeritus at the University of Cape Town. He wrote a number of books, including his interesting autobiography *One Life* (1970).

Barnard, Edward Emerson (1857–1923) *American astronomer*

Although Barnard was born into a poor family in Nashville, Tennessee, and received little formal education, he developed a great interest in astronomy and also became familiar with photographic techniques from his work in a portrait studio. He managed both to study and instruct at Vanderbilt University from 1883 to 1887. From 1888 he worked at the Lick Observatory until in 1895 he became professor of astronomy at Chicago and was thus able to work at the newly established Yerkes Observatory.

Barnard was a keen observer and had detected more than ten comets by 1887 and several more in subsequent years. In 1892 he became the first astronomer after GALILEO to discover a new satellite of Jupiter, subsequently named Amalthea, which lay inside the orbits of the four Galilean satellites and was much smaller and fainter. In 1916 he discovered a nearby red star with a very pronounced proper motion of 10.3 seconds of arc per year: in 180 years it will appear to us to have moved a very considerable distance, equal to the diameter of the Moon. The star is now called *Barnard's star*.

Barnard's other discoveries included various novae, variable stars, and binary stars. He was also one of the first to appreciate that dark nebulae were not areas of the sky containing no stars at all (as William HERSCHEL had thought) but, as Barnard and Max WOLF demonstrated, were enormous clouds of dust and gas that shielded the stars behind them from our view. By 1919 he had discovered nearly 200 such nebulae.

Barnard, Joseph Edwin (1870–1949)
British physicist

While working in his father's business, Barnard used his spare time to study at the Lister Institute, King's College, London, where he developed an interest in microscopy, especially photomicrography, which led to his receiving a chair at the Charing Cross Medical School. He was a fellow and three times president of the Royal Microscopical Society and in 1920 became honorary director of the applied-optics department at the National Institute for Medical Research.

His research and experience in photomicrography led him to write *Practical Photomicrography* (1911), which became a standard work. Later he developed a technique for using ultraviolet radiation, which is of shorter wavelength than visible light, and therefore gives greater resolution. With W. E. Gye he used this method to identify several ultramicroscopic organisms connected with malignant growths that were too small to see using standard microscopy.

Barr, Murray Llewellyn (1908–1995)
Canadian geneticist and anatomist

Barr was born in Belmont, Ontario, and was educated at the University of Western Ontario, gaining his BA in 1930, MD in 1933, and MSc in 1938. His association with Western Ontario was continued with his appointment as an instructor in 1936. He subsequently became professor of microscopical anatomy (1952), professor of anatomy and head of the anatomy department (1964), and emeritus professor (1979).

Barr is best known for his discovery, made in 1949 in conjunction with Ewart Bertram, of the densely staining nuclear bodies present in the somatic cells of female humans and other female mammals. These are called sex chromatin or *Barr bodies*. Later studies by Barr and others revealed that the single Barr body in normal cells is one of the two X-chromosomes in a highly condensed and genetically inactive state. The other X-chromosome is in the diffuse state and is genetically active.

Their discovery enabled Barr and his coworkers to devise a relatively simple diagnostic test for certain genetic abnormalities, in which cells rubbed from the lining of the mouth cavity (a buccal smear) were stained and examined microscopically. For instance, individuals suffering from Turner's syndrome, which usually affects females, have only one X-chromosome and lack Barr bodies. In contrast, males affected by Klinefelter's syndrome possess an extra X-chromosome and exhibit Barr bodies in their cells.

Besides his work in cytogenetics and inherited human disorders, Barr is also noted for his descriptions of nervous-system anatomy. His publications included *The Human Nervous System: an Anatomical Viewpoint* (1972; 5th edition, with J. A. Kiernan, 1988).

Barringer, Daniel Moreau (1860–1929)
American mining engineer and geologist

Barringer, who was born in Raleigh, North Carolina, graduated from Princeton in 1879 and then studied law at Pennsylvania, geology at Harvard, and chemistry and mineralogy at the University of Virginia. In 1890 he established himself as a consulting mining engineer and geologist; he was the author of the standard work *Law of Mines and Mining in the U.S.* (1907).

Barringer is remembered for his investigation of the massive Diabolo Crater in Arizona, which is nearly 600 feet (200 m) deep and over 4,000 feet (1,200 m) in diameter. The cause of such a gigantic hole was a matter of speculation, most considering it to be of volcanic origin. Barringer, finding numerous nickel–iron rocks in the area, became convinced that the remains of an enormous meteorite lay buried at the center of the crater. He began drilling in 1902 but failed to find anything of significance. He later concluded that a meteorite would have been unlikely to enter vertically; after experimenting with projectiles he established that it probably entered at an angle of 45° and would therefore be embedded to one side. In 1922 drilling began again but rapid flooding of the shafts caused Barringer to abandon the search. After his death the crater became known as the *Barringer Meteor Crater*.

Barrow, Isaac (1630–1677) *British mathematician*

Born the son of a prosperous London linen draper, Barrow was educated at Cambridge University. Because of his royalist sympathies he was rejected, on Cromwell's instructions, as a candidate for the professorship of Greek. Consequently he began in 1655 an extensive tour of Europe. With the restoration of Charles II, he returned to Britain in 1660 and was finally elected professor of Greek at Cambridge. In 1663 he accepted the newly created Lucasian Professorship of Mathematics, a post he resigned from in 1669.

The claim has often been made that Barrow resigned his chair in favor of his pupil, Isaac NEWTON. In reality Newton was not the pupil of Barrow and, while he appreciated Newton's mathematical genius and saw to it that Newton succeeded him, Barrow was more interested in advancing his own career. He was appointed chaplain to Charles II in 1669 and in 1673 returned to Cambridge as Master of Trinity College, an office in the gift of the king. He died soon after in 1677 from, according to John Aubrey, an overdose of opium, an addiction that he had acquired in Turkey.

Barrow is best known for his *Lectiones opticae* (1669; Lectures on Optics) and *Lectiones geometricae* (1670; Lectures on Geometry), both edited by Newton, and the posthumously published *Lectiones mathematicae* (1683; Lectures on Mathematics). Unfortunately for Barrow's reputation, his work in both optics and mathematics was soon overshadowed by Newton's own publications.

Bartholin, Erasmus (1625–1698) *Danish mathematician*

Bartholin (bar-**too**-lin), the son of Caspar and brother of Thomas Bartholin, who were both distinguished anatomists, was born in Roskilde, Denmark, and educated in Leiden and Padua, where he obtained his MD in 1654. After further travel in France and England he returned to Denmark in 1656 and held chairs in mathematics and medicine at the University of Copenhagen from 1657 until his death.

Bartholin worked on the theory of equations and with Ole RØMER made an unsuccessful attempt to calculate the orbits of the comets prominent in the late 1660s. He is however best remembered for his discovery of double refraction announced in his *Experimenta crystalli Islandici disdiaclastici* (1669; Experiments on Icelandic Double–Refracting Crystal). In it he described how Icelandic feldspar (calcite) produces a double image of objects observed through it. This discovery greatly puzzled scientists and was much discussed by NEWTON and Christiaan HUYGENS, who tried unsuccessfully to incorporate the strange phenomenon into their respective theories of light.

Double refraction proved remarkably recalcitrant to all proposed explanations for well over a century and it was only with the work of Etienne MALUS on polarized light in 1808, and that of Augustin FRESNEL in 1817, that Bartholin's observations could at last be understood.

Bartlett, Neil (1932–) *British–American chemist*

Bartlett was born in Newcastle upon Tyne in northeast England and educated at the University of Durham, where he obtained his PhD in 1957. He taught at the University of British Columbia, Canada, and at Princeton before being appointed to a chemistry professorship in 1969 at the University of California, Berkeley. Bartlett retired in 1993. He became a naturalized American citizen in 2000.

Bartlett was studying metal fluorides and found that the compound platinum hexafluoride (PtF_6) is extremely active. In fact it reacted with molecular oxygen to form the novel compound $O_2{}^+PtF_6{}^-$. This was the first example of a compound containing the oxygen cation. At the time it was an unquestioned assumption of chemistry that the noble gases – helium, neon, argon, krypton, and xenon – were completely inert, incapable of forming any compounds whatsoever. Further, there was a solid body of valence theory that provided good reasons why this should be so. So struck was Bartlett with the ability of PtF_6 to react with other substances that he tried, in 1962, to form a compound between it and xenon. He knew that the ionization potential of xenon was not too much greater than the ionization potential of the oxygen molecule. To his and other chemists' surprise xenon fluoroplatinate ($XePtF_6$) was produced – the first compound of a noble gas. Once the first compound had been detected xenon was soon shown to form other compounds, such as xenon fluoride (XeF_4) and oxyfluoride ($XeOF_4$). Krypton and radon were also found to form compounds although the lighter inert gases have so far remained inactive.

Bartlett, Paul Doughty (1907–1998) *American chemist*

Bartlett was born in Ann Arbor, Michigan, and educated at Amherst College and Harvard, where he obtained his PhD in 1931. After teaching briefly at the Rockefeller Institute and the University of Minnesota he returned to Harvard in 1934 and served there as professor of chemistry from 1948 until his retirement in 1975.

Bartlett worked mainly on the mechanisms involved in organic reactions; for example, the behavior of free radicals and the kinetics of polymerization reactions. He also investigated the chemistry of elemental sulfur and the terpenes (a family of hydrocarbons found in the essential oils of plants).

Barton, Sir Derek Harold Richard
(1918–1998) *British chemist*

Barton was born in the port of Gravesend in southeast England, and was educated at Imperial College, London, where he obtained his PhD in 1942. After doing some industrial research he spent a year as visiting lecturer at Harvard before being appointed reader (1950) and then professor (1953) in organic chemistry at Birkbeck College, London. Barton moved to a similar chair at Glasgow University in 1955 but returned to Imperial College in 1957 and held the chair of chemistry until 1978, when he became director of the Institute for the Chemistry of Natural Substances at Gif-sur-Yvette in France. In 1986 he became a distinguished professor at Texas Agricultural and Mechanical University.

In 1950 Barton published a fundamental paper on conformational analysis in which he proposed that the orientations in space of functional groups affect the rates of reaction in isomers. Barton discussed six-membered organic rings, particularly, following the earlier work of Odd HASSELL, the "chair" conformation of cyclohexane, and explained its distinctive stability.

This was done in terms of the distinction between equatorial conformations, in which the hydrogen atoms lie in the same plane as the carbon ring, and axial conformations, where they are perpendicular to the ring. He confirmed these notions with further work on the stability and reactivity of steroids and terpenes.

It was for this work that he shared the 1969 Nobel Prize for chemistry with Hassell. Barton's later work on oxyradicals and his predictions about their behavior in reactions helped in the development of a simple method for synthesizing the hormone aldosterone.

Basov, Nikolai Gennediyevich (1922–1998) *Russian physicist*

Basov (**bah**-sof), who was born in Voronezh in western Russia, served in the Soviet army in World War II, following which he graduated from the Moscow Institute of Engineering Physics (1950). He studied at the Lebedev Institute of Physics of the Soviet Academy of Sciences in Moscow, gaining his doctoral degree in 1956 and going on to become deputy director (1958) and later director (1973). In 1989 he became director of the quantum radiophysics division.

Basov's major contribution was in the development of the maser (*m*icrowave *a*mplification by *s*timulated *e*mission of *r*adiation), the forerunner of the laser. From 1952 he had been researching the possibility of amplifying electromagnetic radiation using excited atoms or molecules. His colleague at the Lebedev Institute, Aleksandr PROKHOROV, was involved in the microwave spectroscopy of gases, with the aim of creating a precise frequency standard, for use in very accurate clocks and navigational systems. Their work led to theories and experiments designed to produce a state of "population inversion" in molecular beams, through which amplification of radiofrequency radiation became possible.

Together Basov and Prokhorov in 1955 developed a generator using a beam of excited ammonia molecules. This was the maser, developed simultaneously but independently in America by Charles TOWNES. Basov, Prokhorov, and Townes received the 1964 Nobel Prize for physics for this work.

The first masers used a method of selecting the more excited molecules from a beam, but a more efficient method was proposed by Basov and Prokhorov in 1955, the so-called "three-level" method of producing population inversion by "pumping" with a powerful auxiliary source of radiation. The next year the method was applied by Nicolaas BLOEMBERGEN in America in a quantum amplifier.

Basov went on to develop the laser principle, and in 1958 introduced the idea of using semiconductors to achieve laser action. In the years 1960–65 he realized many of his ideas in practical systems. He subsequently did considerable theoretical work on pulsed ruby and neodymium-glass lasers, which are now in common use, and on the interaction of radiation with matter. In particular, he worked on the production of short powerful pulses of coherent light.

Bassi, Agostino (1773–1856) *Italian microbiologist*

Bassi (**bah**-see), who was born in the northern Italian town of Lodi and educated at the University of Pavia, conducted valuable research into animal diseases. Anticipating PASTEUR, he suggested that certain diseases are caused by minute animal or plant parasites. Some of his most important studies were concerned with cholera and with pellagra, a deficiency disease of the skin. In 1835 he was able to show that the disease of silkworms known as muscardine was caused by a fungus, subsequently named *Botrytis bassiana* in his honor. He demonstrated that muscardine is contagious and formulated methods to prevent and eliminate the disease. Bassi also published accounts of work on potato cultivation, cheese-making, and vinification.

Bates, Henry Walter (1825–1892) *British naturalist and explorer*

The son of a stocking-factory owner in the central English town of Leicester, Bates left school at 13 and was apprenticed to a hosiery manufacturer, but still found time for indulging his hobby of beetle collecting. In 1844 he met Alfred WALLACE and stimulated the latter's interest in entomology. This led, three years later, to Wallace suggesting they should travel together to the tropics to collect specimens and data that might throw light on the evolution of species.

In May 1848 they arrived at Pará, Brazil, near the mouth of the Amazon. After two years collecting together they split up, and Bates spent a further nine years in the Amazon basin. By the time he returned to England in 1859, he estimated he had collected 14,712 species, 8,000 of which were new to science.

While collecting Bates had noted startling similarities between certain butterfly species – a phenomenon later to be termed *Batesian mimicry*. He attributed this to natural selection, since palatable butterflies that closely resembled noxious species would be left alone by predators and thus tend to increase. His paper on this, *Contributions to an Insect Fauna of the Amazon Valley, Lepidoptera: Heliconidae* (1861) provided strong supportive evidence for the Darwin–Wallace evolutionary theory published three years earlier.

DARWIN persuaded Bates to write a book on his travels, which resulted in the appearance of *The Naturalist on the River Amazon* (1863), an objective account of the animals, humans, and natural phenomena Bates encountered. Although one of the best and most popular books of its kind, Bates was to comment that he would rather spend a further 11 years on the Amazon than write another book. He became assistant secretary of the Royal Geographic Society in 1864.

Bates, Leslie Fleetwood (1897–1978) *British physicist*

Bates was educated at the university in his native city of Bristol and at Cambridge University, where he obtained his PhD in 1922. After teaching first at University College, London, he moved to Nottingham, where he served as professor of physics from 1936, apart from wartime duties on the degaussing of ships, until his retirement in 1964.

Most of Bates's work was on the magnetic properties of materials. He was the author of a widely used textbook, which went through many editions, *Modern Magnetism* (1938).

Bateson, William (1861–1926) *British geneticist*

Born in the coastal town of Whitby in northeast England, Bateson graduated in natural sciences from Cambridge University in 1883, having specialized in zoology. He then traveled to America, where he studied the embryology of the wormlike marine creature *Balanoglossus*. He discovered that, although its larval stage resembles that of the echinoderms (e.g., starfish), it also has gill slits, the beginnings of a notochord, and a dorsal nerve cord, proving it to be a primitive chordate. This was the first evidence that the chordates have affinities with the echinoderms.

Back at Cambridge Bateson began studying variation within populations and soon found instances of discontinuous variation that could not simply be related to environmental conditions. He believed this to be of evolutionary importance, and began breeding experiments to investigate the phenomenon more fully. These prepared him to accept MENDEL's work when it was rediscovered in 1900, although other British scientists were largely skeptical of the work. Bateson translated Mendel's paper into English and set up a research group at Grantchester to investigate heredity in plants and animals.

Through his study of the inheritance of comb shape in poultry, Bateson demonstrated that Mendelian ratios are found in animal crosses (as well as plants). He turned up various deviations from the normal dihybrid ratio (9:3:3:1), which he rightly attributed to gene interaction. He also found that certain traits are governed by two or more genes, and in his sweet-pea crosses showed that some characters are not inherited independently. This was the first hint that genes are linked on chromosomes, but Bateson never accepted T. H. MORGAN's explanation of linkage or the chromosome theory of inheritance.

In 1908 Bateson became the first professor of the subject he himself named – genetics. However, he left Cambridge only a year later and in 1910 became director of the newly formed John Innes Horticultural Institution at Merton, Surrey, where he remained until his death. He was the leading proponent of Mendelian genetics in Britain and became involved in a heated controversy with supporters of biometrical genetics such as Karl PEARSON. The views of both sides were later reconciled by the work of Ronald FISHER. Bateson wrote a number of books, including the controversial *Materials for the Study of Variation* (1894) and *Mendelian Heredity – A Defence* (1902); he also founded, with R. C. Punnett, the *Journal of Genetics* in 1910.

Bauer, Georg *See* AGRICOLA, GEORGIUS.

Bawden, Sir Frederick Charles (1908–1972) *British plant pathologist*

Bawden was born in North Tawton, England, and was educated at Cambridge University, receiving his MA in 1933. From 1936 to 1940 he worked in the virus physiology department at Rothamsted Experimental Station, becoming the head of the plant pathology department in 1940. He was director of the station from 1958 until his death.

In 1937 Bawden discovered that the tobacco mosaic virus (TMV) contains ribonucleic acid, this being the first demonstration that nucleic acids occur in viruses. With Norman PIRIE, Bawden isolated TMV in crystalline form and made important contributions to elucidating the structure of viruses and the ways in which they multiply. Bawden's work also helped in revealing the mechanisms of protein formation.

Bayer, Johann (1572–1625) *German astronomer*

Born in Rhain in Germany, Bayer (**bI**-er) was an advocate (lawyer) by profession. In 1603 he published *Uranometria* (Measurement of the Heavens), the most complete catalog of pretelescopic astronomy. To Tycho BRAHE's catalog of 1602, he added nearly a thousand new stars and twelve new southern constellations. The catalog's main importance, however, rests on Bayer's innovation of naming stars by letters of the Greek alphabet. Before Bayer, prominent stars were given proper names, mainly Arabic ones such as Altair and Rigel. If not individually named, they would be referred to by their position in the constellation. Bayer introduced the scheme, which is still used, of referring to the brightest star of a constellation by "alpha," the second brightest by "beta," and so on. Thus Altair, which is the brightest star in the constellation Aquila, is systematically named Alpha Aquilae. If there were more stars than letters of the Greek alphabet, the dimmer ones could be denoted by letters of the Roman alphabet and, if necessary, numbers.

Bayer's other proposed innovation – to name constellations after characters in the Bible – was less successful.

Bayliss, Sir William Maddock (1860–1924) *British physiologist*

Bayliss was the son of a wealthy iron manufacturer in the industrial town of Wolverhampton in central England. In 1881 he entered University College, London, as a medical student but when he failed his second MB exam in anatomy he gave up medicine to concentrate on physiology. He graduated from Oxford University in 1888, then returned to University College, London, where he worked for the rest of his life, holding the chair of general physiology from 1912. Bayliss was elected a fellow of the Royal Society in 1903 and was knighted in 1922.

He was chiefly interested in the physiology of the nervous, digestive, and vascular systems, on which he worked in association with his brother-in-law, Ernest STARLING. Their most important work, published in 1902, was the discovery of the action of a hormone (secretin) in controlling digestion. They showed that in normal digestion the acidic contents of the stomach stimulate production of the hormone secretin when they reach the duodenum. Secretin is transported in the bloodstream to initiate secretion of digestive juices by the pancreas. In 1915 Bayliss produced a standard textbook on physiology, *Principles of General Physiology*, which treated the subject from a physicochemical point of view.

Beadle, George Wells (1903–1989) *American geneticist*

Beadle was born in Wahoo, Nebraska, and graduated from the University of Nebraska in 1926; he gained his PhD from Cornell University in 1931. He then spent two years doing research in genetics under T. H. MORGAN at the California Institute of Technology. Beadle was a professor at the California Institute of Technology from 1946 until 1961 and was president of the University of Chicago from 1961 until 1968. In 1937 Beadle went to Stanford University, where in 1940 he began working with Edward TATUM on the mold *Neurospora*. They used nutritional mutants, which were unable to synthesize certain essential dietary compounds, to determine the sequence of various metabolic pathways. Substances similar to the missing compound were added to the mutant mold cultures to find whether or not they could substitute for the lacking chemical. If the culture survived then it could be assumed that the mold could convert the substance into the chemical it needed, showing that the nutrient was likely to be a precursor of the missing chemical.

From this and similar work Beadle and Tatum concluded that the function of a gene was to control the production of a particular enzyme and that a mutation in any one gene would cause the formation of an abnormal enzyme that would be unable to catalyze a certain step in a chain of reactions. This reasoning led to the formulation of the one gene–one enzyme hypothesis, for which Beadle and Tatum received the 1958 Nobel Prize for physiology or medicine, sharing the prize with Joshua LEDERBERG, who had worked with Tatum on bacterial genetics.

Beaufort, Sir Francis (1774–1857) *British hydrographer*

Beaufort was born in Navan in Ireland; his father was a cleric of Huguenot origin who took an active interest in geography and topography, publishing in 1792 one of the earliest detailed maps of Ireland. Beaufort joined the East India Company in 1789 and enlisted in the Royal Navy the following year, remaining on active service until 1812.

He proposed, in 1806, the wind scale named for him. This was an objective scale ranging from calm (0) up to storm (13) in which wind strength was correlated with the amount of sail a full-rigged ship would carry appropriate to the wind conditions. It was first used officially by Robert FITZROY in 1831 and adopted by the British Admiralty in 1838. When sail gave way to steam the scale was modified by defining levels on it in terms of the state of the sea or, following George SIMPSON, wind speed.

In 1812 Beaufort surveyed and charted the Turkish coast, later writing his account of the expedition, *Karamania* (1817). He was appointed hydrographer to the Royal Navy in 1829. In this office Beaufort commissioned voyages to survey and chart areas of the world, such as those of the *Beagle* with Charles DARWIN and the *Erebus* with Joseph HOOKER. The sea north of Alaska was named for him.

Beaumont, Elie de (1798–1874) *French geologist*

Beaumont (**boh**-mon), who was born and died in Canon, France, was educated at the Ecole Polytechnique and the School of Mines, Paris, and taught at the School of Mines from 1827, later becoming professor of geology there (1835). He is remembered chiefly for his theory on the origin of mountains. He published his views in 1830 in his *Revolutions de la surface du globe*, in which he argued that mountain ranges came into existence suddenly and were the result of distortions produced by the cooling crust of the Earth. Such a view fitted in well with the catastrophism of such zoologists as Georges CUVIER. Beaumont summarized his theories in his *Notice sur les systèmes des montagnes* (1852; On Mountain Systems).

Beaumont served as engineer-in-chief of mines for the period 1833–47. He also collaborated with Ours Pierre Dufrénoy in compiling the great geological map of France, published in 1840.

Beaumont, William (1785–1853) *American physician*

Born in Lebanon, Connecticut, Beaumont started out as a farmhand and then a schoolteacher before becoming a doctor's apprentice. He received his license to practice medicine in 1812 and joined the army as a surgeon's mate. While serving as post surgeon at Fort Mackinac, Michigan, he treated a Canadian trapper, Alexis St. Martin, for wounds caused by a close-range shotgun blast. His patient recovered but

was left with an opening (fistula) in his stomach wall. Beaumont took this opportunity to conduct pioneering experiments on the process of digestion in the human stomach, sampling its contents through the fistula and studying the rates of digestion of various foods and the effects of emotional changes on digestion. His *Experiments and Observations on the Gastric Juice and the Physiology of Digestion* (1833) helped trigger further research in this field.

Beche, Sir Henry Thomas de la *See* DE LA BECHE, SIR HENRY THOMAS.

Becher, Johann Joachim (1635–1682) *German chemist and physician*

Becher (**bek**-er), a pastor's son from Speyer in southwest Germany, was largely self-taught. After traveling throughout Europe he gained his MD from the University of Mainz in 1661 and became professor of medicine there in 1663. Short spells as court physician in Mainz (1663–64) and physician to the elector of Bavaria (1664) were followed by a period in Vienna from 1665, where he attempted to realize several economic projects, including a Rhine–Danube canal. He fell from favor and was forced to flee via Holland to England, where he died.

Becher is a transitional figure between alchemy and modern chemistry. He claimed to have a process for producing gold from silver and sand and his chemical system was an extension of the Paracelsian *Tria prima* of sulfur, salt, and mercury. He stated that all inorganic bodies were a mixture of water and three earthy principles: vitreous earth (*terra fusilis*), combustible earth (*terra pinguis*), and mercurial earth (*terra fluida*). *Terra pinguis* was identified as the cause of combustion and this became the phlogiston of Georg STAHL's later theory. Becher's most influential work was *Physicae subterraneae* (1669; Subterranean Physics), which was republished by Stahl in 1703. This work contained his theories on the nature of minerals and experiments.

Beckmann, Ernst Otto (1853–1923) *German organic and physical chemist*

Beckmann (**bek**-man) was born in the German industrial city of Solingen in the Ruhr. His father was an industrial chemist who independently discovered the pigment Paris green. Ernst served an apprenticeship as a pharmacist before studying chemistry at the University of Leipzig, where he graduated in 1878. In 1886, while *Privatdozent* at Leipzig, he discovered the *Beckmann rearrangement* whereby ketoximes are converted into amides. This reaction was soon to be used by Arthur HANTZSCH to determine the stereochemistry of the oximes.

Needing an efficient method for finding molecular weights, Beckmann devised the *Beckmann thermometer*, suitable for measuring changes in temperature over a small range. It is used in finding the elevation in boiling point or depression of freezing point for a solution – which can then be used to calculate the molecular weight of the solute. His other research included work on terpenes and the chlorides of sulfur and selenium. Beckmann was professor at Giessen (1891), Erlangen (1892), and Leipzig (1897). In 1912 he became the first director of the Kaiser Wilhelm Institute at Dahlem, a post with opportunities that never materialized owing to the demands of World War I.

Becquerel, Antoine Henri (1852–1908)
French physicist

Becquerel (bek-er-**el**) was born in Paris, France; his early scientific and engineering training was at the Ecole Polytechnique and the School of Bridges and Highways, and in 1876 he started teaching at the Polytechnique. From 1875 he researched into various aspects of optics and obtained his doctorate in 1888. In 1899 he was elected to the French Academy of Sciences, continuing the family tradition as his father and grandfather, both renowned physicists, had also been members. He held chairs at the Ecole Polytechnique, the Museum of Natural History, and the National Conservatory of Arts and Crafts, and became chief engineer in the department of bridges and highways.

Becquerel is remembered as the discoverer of radioactivity in 1896. Following Wilhelm RÖNTGEN's discovery of x-rays the previous year, Becquerel began to look for x-rays in the fluorescence observed when certain salts absorb ultraviolet radiation. His method was to take crystals of potassium uranyl sulfate and place them in sunlight next to a piece of photographic film wrapped in black paper. The reasoning was that the sunlight induced fluorescence in the crystals and any x-rays present would penetrate the black paper and darken the film.

The experiments appeared to work and his first conclusion was that x-rays were present in the fluorescence. The true explanation of the darkened plate was discovered by chance. He left a plate in black paper next to some crystals in a drawer and some time later developed the plate. He found that this too was fogged, even though the crystals were not fluorescing. Becquerel investigated further and discovered that the salt gave off a penetrating radiation independently, without ultraviolet radiation. He deduced that the radiation came from the uranium in the salt.

Becquerel went on to study the properties of this radiation; in 1899 he showed that part of it could be deflected by a magnetic field and thus consisted of charged particles. In 1903 he shared the Nobel Prize for physics with Pierre and Marie CURIE.

Beddoes, Thomas (1760–1808) *British chemist and physician*

Beddoes, who was born in Shifnal, England, studied classics at Oxford and medicine and chemistry at Edinburgh, where he was the pupil of Joseph BLACK. After obtaining his MD in 1787 he returned to Oxford as reader in chemistry.

When, in 1792, Beddoes lost his Oxford post because of his sympathy for the French Revolution, he approached the Lunar Society of Birmingham, which included Josiah Wedgwood, Joseph PRIESTLEY, Matthew BOULTON, and James WATT, with the idea of forming an institute to investigate the medicinal uses of the gases discovered in the previous 20 years. As a result one of the first specialized research institutes, the Pneumatic Medical Institute, was opened at Clifton, Bristol, in 1799 with Humphry DAVY, then 19 years old, as Beddoes's assistant. The work proved disappointing except in the case of dinitrogen oxide (nitrous oxide, N_2O), the euphoriant properties of which provided Davy with his first serious chemical paper.

Beddoes's work on gases was published as *Considerations on the Medicinal Use of Factitious Airs* (1794–96). Beddoes later moved to London where he built up a successful medical practice.

Bednorz, Johannes Georg (1950–   )
German physicist

Bednorz (**bed**-norts) was educated at the Federal Institute of Technology, Zurich, where he gained his PhD in 1982. He immediately joined the staff of the IBM Research Center in Zurich, where he now leads the high-temperature superconductivity research group.

Here he was invited by his senior colleague, Alex MULLER, to collaborate in a search for superconductors with higher critical temperatures. Little progress had been made in this area for a decade and, as a young unknown scientist, Bednorz's decision to work in such an unpromising field appeared to many to be somewhat rash. Success, however, came relatively quickly and in 1986 Bednorz and Muller found a mixed lanthanum, barium, and copper oxide that had a critical temperature of 35 K (−238°C), which was significantly higher than that of any other superconductor known at the time. Their work was quickly recognized and in 1987 Muller and Bednorz were awarded the Nobel Prize for physics.

Beebe, Charles William (1887–1962)
American naturalist

Beebe (**bee**-bee) was born in New York City and graduated from Columbia University in 1898. The following year, he began organizing and building up the bird collection of the New

York Zoological Park. After serving as a fighter pilot in World War I he became, in 1919, director of the Department of Tropical Research of the New York Zoological Society.

Beebe is noted as one of the pioneers of deep-sea exploration. His first observation capsule was a cylinder; later collaboration with the geologist and engineer Otis Barton resulted in the design of a spherical capsule (the bathysphere). Various dives were made and in August 1934 Beebe and Barton were lowered to a (then) record depth of 3,028 feet (923 m) near Bermuda. Beebe made many interesting observations, such as the absence of light at 2,000 feet (610 m), phosphorescent organisms, and an apparently unknown animal estimated to be some 20 feet (6.1 m) long. He abandoned deep-sea exploration after making thirty dives. Descents to even greater depths were subsequently made by Auguste PICCARD and others.

Beer, Wilhelm (1797–1850) *German astronomer*

Beer (bayr) was born in Berlin, where he spent his whole adult life working as a banker. Like his friend Johann Heinrich von Mädler, Beer was a very competent amateur observer of Mars and the Moon. In 1830 they published the first reasonable map of Mars. It did not show the canals that Giovanni SCHIAPARELLI and others were later to "observe." In 1836 they published a large map of the Moon that was the most comprehensive of its time. Beer also measured the heights of the larger lunar mountains.

Béguyer de Chancourtois, Alexandre Emile (1820–1886) *French mineralogist*

The Parisian Béguyer de Chancourtois (bay-gee-**ay** de shahng-kor-**twah**), who was inspector-general of mines in France, is remembered for his work on the classification of the elements. In 1862 he proposed his system, known as the "telluric screw," in which the elements were arranged in order of atomic weight and then plotted on a line descending at an angle of 45° from the top of a cylinder. It was found that elements on the same vertical line resembled each other.

The proposal received little attention, largely because it was published without the explanatory diagram, which made the article virtually impossible to understand. As with John NEWLANDS's system of classifying the elements, Béguyer de Chancourtois's proposal was overshadowed by that of Dmitri MENDELÉEV, first published in 1869.

Behring, Emil Adolf von (1854–1917) *German immunologist*

> For hundreds of thousands of years, the wisest physicians and scientists have studied the properties of blood and its relation to health and illness, without ever suspecting the specific antibodies appearing in the blood as a result of an infectious disease, which are capable of rendering infectious toxins harmless.
> —Describing his work on immunization

Behring (**bay**-ring) was born in Hansdorf in Germany. He graduated in medicine at Berlin University and entered the Army Medical Corps before becoming (in 1888) a lecturer in the Army Medical College, Berlin. In 1889 he moved to Robert KOCH's Institute of Hygiene and transferred to the Institute of Infectious Diseases in 1891, when Koch was appointed its chief.

In 1890, working with Shibasaburo KITASATO, Behring showed that injections of blood serum from an animal suffering from tetanus could confer immunity to the disease in other animals. Behring found that the same was true for diphtheria and this led to the development of a diphtheria antitoxin for human patients, in collaboration with Paul EHRLICH. This treatment was first used in 1891 and subsequently caused a dramatic fall in mortality due to diphtheria.

Behring's success brought him many prizes, including the first Nobel Prize in physiology or medicine, awarded in 1901. He was appointed professor of hygiene at Halle University in 1894 and one year later moved to a similar post at Marburg. In 1913 he introduced toxin–antitoxin mixtures to immunize against diphtheria, a refinement of the immunization technique already in use. He also devised a vaccine for the immunization of calves against tuberculosis.

Beilby, Sir George Thomas (1850–1924) *British industrial chemist*

Beilby (**beel**-bee), the son of a clergyman, was born in the Scottish capital of Edinburgh and educated at the university there. He began work with an oil-shale company as a chemist in 1869 and increased the yield of paraffin and ammonia from oil shales by improving the process of their distillation. He also worked on cyanides, patenting, in 1890, a process for the synthesis of potassium cyanide in which ammonia was passed over a heated mixture of charcoal and potassium carbonate. This had wide use in the gold-extracting industry.

From 1907 to 1923 Beilby was chairman of the Royal Technical College, Glasgow, later the University of Strathclyde. He became interested in the economic use of fuel and smoke prevention, submitting evidence to the Royal Commission on Coal Supplies in 1902. In 1917 he was appointed as the first chairman of the Fuel Research Board. He was knighted in 1916.

Beilstein, Friedrich Konrad (1838–1906) *Russian organic chemist*

Born to German parents in the Russian city of

St. Petersburg, Beilstein (**bIl**-shtIn) studied chemistry in Germany under BUNSEN, LIEBIG, and WURTZ and gained his PhD under WÖHLER at Göttingen (1858). He was lecturer at Göttingen (1860–66) and from 1866 professor of chemistry at the Technological Institute at St. Petersburg.

Beilstein's many researches in organic chemistry included work on isomeric benzene derivatives. He is better remembered, however, for his monumental *Handbuch der organischen Chemie* (1880–82; Handbook of Organic Chemistry), in which he set out to record systematically all that was known of every organic compound. He produced the second (1886) and third (1900) editions, after which the work was assigned to the Deutsch Chemische Gesellschaft, who have published it ever since.

Békésy, Georg von (1899–1972) *Hungarian–American physicist*

Békésy (**bay**-ke-shee), the son of a diplomat in the Hungarian capital, Budapest, studied chemistry at the University of Bern and physics at Budapest University, where he obtained his PhD in 1923. He immediately joined the research staff of the Hungarian Telephone Laboratory where he remained until 1946 while simultaneously holding the chair of experimental physics at Budapest University from 1939. He left Hungary in 1947, via the Swedish Karolinska Institute, for America, where he served first as a senior fellow in psychophysics at Harvard from 1949 to 1966 and finally as professor of sensory science at the University of Hawaii from 1966 until his death.

Békésy first worked on problems of long-distance telephone communication before moving to the study of the physical mechanisms of the cochlea within the inner ear. When he began this study it was generally thought, following the work of Hermann von HELMHOLTZ, that sound waves entering the ear selectively stimulated a particular fiber of the basilar membrane; this in turn stimulated hairs of the organ of Corti resting on it, which transferred the signal to the auditory nerve.

Using the techniques of microsurgery, Békésy was able to show that a different mechanism is involved. He found that when sound enters the cochlea, a traveling wave sweeps along the basilar membrane. The wave amplitude increases to a maximum, falling sharply thereafter; it is this maximum point to which the organ of Corti is sensitive. For this insight into the mechanism of hearing, Békésy was awarded the 1961 Nobel Prize for physiology or medicine.

Bell, Alexander Graham (1847–1922) *British inventor*

> Mr Watson come here; I want you.
> —Speaking to his assistant Thomas Watson in

the world's first telephone conversation, 10 March 1876

Bell's family were practitioners in elocution and speech correction and he himself trained in this. Born in Edinburgh, as a child he was taught mainly at home. For a short time he attended Edinburgh University and University College, London, after which he taught music and elocution at a school in Elgin, Scotland. It was in Elgin that he carried out his first studies on sound.

From 1868 Bell worked in London as his father's assistant, but after the death of his two brothers from tuberculosis, the family moved to Canada, where Alexander, who had also become ill, recovered. In 1871 he went to Boston where he gave lectures on his father's method of "visible speech" – a system of phonetic symbols for teaching the deaf to speak. A year later he opened a school for teachers of the deaf. In 1873 he became professor of vocal physiology at Boston University.

With financial help from two of his deaf students, Bell experimented with the transmission of sound by electricity, aided by Thomas Watson, his technician. His multiple telegraph was patented in 1875 and, in 1876, the patent for the telephone was also granted. Bell's wife Mabel Hubbard, whom he married in 1877, was deaf. Later she founded the Aerial Experiment Association.

In 1880 he received the Volta Prize from France and the money was used to fund the laboratories in which an improved form of the gramophone was invented by Thomas EDISON. Although best known as the inventor of the telephone, Bell investigated a wide range of related technical subjects, including sonar and various equipment for the deaf. In 1885 he bought land and established laboratories and a summer home on Cape Breton Island.

Bell, Sir Charles (1774–1842) *British physician*

Bell, who was born in Edinburgh, Scotland, studied under his elder brother John, a surgeon, and attended lectures at Edinburgh University. In 1804, Charles moved to London, where he started his own surgical practice and began his investigations into the nervous system. In *A New Idea of the Anatomy of the Brain* (1811), Bell described how nerves are not single structures but bundles of many nerve fibers. He also showed that the anterior and posterior roots of each spinal nerve carry different types of nerve fibers, which he later clarified as being respectively excitatory (motor) and sensory in function. The French physiologist, François MAGENDIE, is also credited with this discovery and the functional differentiation of the spinal nerve roots is sometimes called the *Bell–Magendie law*. In *The Nervous System of the Human Body* (1830), Bell amplified his findings, establishing that a nerve fiber can carry impulses in one direction only and that each muscle must be supplied with both excitatory and sensory fibers.

Bell also contributed to knowledge of facial paralysis and to other surgical fields besides lecturing at the Great Windmill Street School of Anatomy and the Middlesex Hospital. In addition, he wrote *Essays on the Anatomy of Expression in Painting* (1806). He received the first medal awarded by the Royal Society in 1829 and was knighted two years later.

Bell, John Stuart (1928–1990) *British physicist*

> I think that when we have solved the problem of the interface between the classical and the quantum-mechanical worlds, there will be something different in the theory.
> —Quoted by J. Bernstein in *Quantum Profiles*

Born into a poor family in the Northern Irish capital of Belfast, Bell was encouraged by his mother to continue his education after leaving school at sixteen. Consequently, after working for a year as a laboratory assistant in the physics department of Queen's University, Belfast, he enrolled as a student and graduated in 1949. Rather than pursue a PhD and burden his family further, Bell began work immediately at the Atomic Energy Research Establishment at Harwell. He worked initially on the design of the first accelerator at CERN (Conseil Européen pour la Recherche Nucléaire; European Organization for Nuclear Research), the Proton Synchrotron. He was also given a year's leave of absence to work on a doctorate at Birmingham University. On his return to Harwell he turned to the theoretical study of elementary particles. Bell moved to CERN in Geneva in 1960 where he remained for the rest of his life. He was accompanied by his wife, Mary Bell, also a physicist, who worked at CERN on accelerator design.

In 1964 Bell published what for many has become the single most important theoretical paper in physics to appear since 1945; it was entitled *On the Einstein Podolsky Rosen Paradox*. The title referred to a thought experiment proposed by Einstein and others in 1935 sharply challenging the basis of quantum theory. He proposed a principle of reality stating that: "If, without in any way disturbing a system we can predict with certainty … the value of a physical quantity then there exists an element of physical reality corresponding to this physical quantity." For example, electrons have a spin that can take one of two values, conveniently classed as positive or negative. Spin, like angular momentum, is conserved. Consequently, if a particle with zero spin decays into an electron/positron (e^-/p^+) pair, the two particles must have equal and opposite spins. Knowing, for example, that the

electron has a negative spin, it can be inferred that the positron must have a positive spin.

But this, according to Einstein, gives us a way to measure the spin of a particle without disturbance. If the p^+ spin is measured and found to be positive, the measurement may well disturb the p^+, but on this basis the spin of the e^- can be concluded to be negative without in any way disturbing the e^-. It follows from Einstein's reality principle that the negative spin of e^- is a real property of the electron. This view, however, conflicts with the usual interpretation of quantum mechanics, which sees the spin of the electron as a superposition of both spin states, a condition only resolved when the electron is observed and the wave function collapses. Nor can it be said that the state of the electron is in any way influenced by the outcome of the observation of the positron's spin for, as no signal can travel faster than the speed of light, instantaneous communication between separated particles is impossible.

The theoretical physicist is therefore presented with an uncomfortable choice. He or she can accept that electrons have intrinsic spin, in accordance with the reality principle and against quantum mechanics, or adopt what Einstein scornfully termed a "spooky action at distance." One weekend in 1964 Bell saw a way in which the matter could be resolved.

The spin of a particle is complicated in that it can be independently measured along three coordinates x, y, and z at right angles to each other. Further, a measurement of the electron's spin in the x direction will influence the spin of the positron in the x direction also; it will, however, have no effect on measurements along the y and z directions. Similar rules apply to measurements along the y and z axes. Bell argued that, if the reality principle is correct, then one would expect to find for a large number of observations:

$$x^+y^+ < (x^+z^+ + y^+z^+)$$

That is, the number of particles with a positive spin along the x and y axes is smaller than the number found on both the x^+z^+ and y^+z^+ axes. The result is known variously as *Bell's inequality* and *Bell's theorem*. Although it proved impossible to test Bell's inequality in terms of the reactions described in the 1964 paper, later workers have produced equivalent formulations that are testable. The most convincing of these, the *Aspect experiment* performed by Alain Aspect of the Institute of Optics at the University of Paris in 1982, using correlations between polarized photons, established that the inequality did not hold. The conclusion seemed to be that nature preferred to act "spookily" at a distance rather than using Einstein's reality principle.

At first Bell's five-page paper was ignored. Only when experimentalists such as John Clauser at Berkeley in 1969 took his work up did Bell's argument become widely known.

Bell's views on his own work, more tentative and less extreme than those of many of his followers and popularizers, were collected in his *Speakable and Unspeakable in Quantum Mechanics* (1987).

Bell Burnell, Dame Susan Jocelyn
(1943–   ) *British astronomer*

> A bit of scruff.
> —Referring to the pen-recorder trace of the first recorded pulsar

Born in Belfast, Northern Ireland, the daughter of the architect who designed the Armagh planetarium, Jocelyn Bell developed an early interest in radioastronomy. She was advised by Bernard LOVELL to study physics first and consequently found herself the only woman in a class of 50 physics students at Glasgow University. After graduating from Glasgow she moved to Cambridge, where she completed her PhD in 1969.

As part of her duties she visited the Mullard Radio Astronomy Laboratory each day, filled the inkwells and monitored the 100-foot length of paper chart produced daily by the 4.5-acre telescope. The sky was scanned every four days and, with little computer power available at that time, the data had to be analyzed by hand. One day she identified a strange signal. As it meant nothing to her, Bell simply put a question mark next to it. When she noticed it again she drew it to the attention of her supervisor, Antony HEWISH. But nothing more was seen for a month. Perhaps, Hewish suggested, it had been a one-time event, a flare that had been and gone.

But after a month it reappeared. On examination Bell found the signals were equally spaced out at intervals of about 1.3 seconds. It was soon established that the signal was genuine and not an instrumental malfunction; nor was it produced by satellites or any terrestrial interference. The source was shown to lie outside the solar system and to have a sidereal motion. A more detailed examination of the signals revealed a regular sequence of pulses at intervals of 1.33730113 seconds with an accuracy better than one part per hundred million. Bell discovered a second signal coming from a different part of the sky in December, and a third and fourth the following month.

Once it had been decided that no LGMs, or "Little Green Men," were involved in the signals, Hewish made the discovery public in 1968. The name "pulsar" was soon coined and soon after Thomas GOLD proposed that the signals were emitted by a small, rapidly rotating neutron star. In an earlier age the first pulsar would have been known as *Bell's Star*; today it carries number CP 1919. Bell's work helped Hewish to gain the 1974 Nobel Prize for physics.

Bell herself married and became Bell Burnell and was appointed in 1968 to a research fellowship at the University of Southampton, where she worked on gamma rays. In 1973 she moved to the Mullard Space Science Laboratory in London to work on x-ray astronomy. Bell moved again in 1982 to head the James Clerk Maxwell Telescope project at the Royal Observatory, Edinburgh. In 1991 she was appointed professor of physics at the Open University, Milton Keynes, a position she held for ten years. In 2001 she became Dean of Science at the University of Bath, retiring in 2004. At present she is Visiting Professor of Astrophysics at Oxford University. She became a Dame in 2001.

Belon, Pierre (1517–1564) *French naturalist*

Belon (be-**lon**) was born in Le Mans, France, and studied medicine in Paris. In 1540 he went to Germany to study botany, becoming a leading figure in the 16th-century revival of natural history that followed the great voyages, the invention of printing, and the new artistic realism of the Renaissance.

Between 1546 and 1549 Belon traveled in the eastern Mediterranean countries, comparing the animals and plants he observed with their descriptions by classical authors. The results were published as *Les Observations des plusieurs singularitez et choses mémorables trouvées en Grèce, Asie, Judée, Egypte, Arabie et autre pays éstranges* (1553; Observations of Many Singularities and Memorable Items in Greece, Asia, Judea, Egypt, Arabia, and Other Foreign Countries). On his travels, Belon was in the habit of investigating the birds and fishes that came to market, and in England he met the Venetian Daniel Barbaro, who had made many drawings of Adriatic fishes. From these sources Belon produced two books on fishes: *L'Histoire naturelle des éstranges poissons marins* (1551; The Natural History of Foreign Sea Fish) and *De aquatilibus* (1553; On Water Creatures). The first is notable for its dissertation on the dolphin, in which he identified the common Atlantic species with the dolphin of the ancients and distinguished it from the porpoise.

Belon's principal achievement is a history of birds, *L'Histoire de la nature des oyseaux* (1555; The Natural History of Birds). An illustrated book of the kind inspired by the drawings of Albrecht Dürer and Leonardo da Vinci, it describes about 200 birds, mostly of European origin. He drew attention to the correspondence between the skeletons of birds and man, an early hint of the discipline of comparative anatomy.

Belon was also interested in geology and botany and is reputed to have introduced the cedar of Lebanon into western Europe. He also established two botanical gardens in France and suggested that many exotic plants might be acclimatized and grown in temperate regions. In many ways a typical figure of the Renaissance, Belon's end was all too typical of that time, for he was murdered in the Bois de Boulogne in 1564.

Belousov, Vladimir Vladimirovich (1907–1990) *Russian geologist and geophysicist*

Belousov (byel-**oo**-sof), a Muscovite by birth, became head of the department of geodynamics at the Soviet Academy of Sciences, Moscow, in 1942 and was later (1953) made professor of geophysics at Moscow University. His main work concentrated on the structure and development of the Earth's crust. In 1942 he put forward his theory on Earth movements, in which he proposed that the Earth's material has gradually separated according to its density and this is responsible for movements in the crust. He at first rejected theories on continental drift.

Belousov became chairman of the Soviet Joint Geophysical Committee in 1961. His works include *Principles of Geotectonics* (1975).

Benacerraf, Baruj (1920–) *American immunologist*

Benacerraf (bay-nah-**se**-raf), who was born in the Venezuelan capital of Caracas, was brought up in France but moved to America in 1940, becoming naturalized in 1943. He studied at Columbia and the University of Virginia where he obtained his MD in 1945. He worked first at the Columbia Medical School before spending the period 1950–56 at the Hospital Broussais in Paris. He returned to America in 1956 to the New York Medical School where he served from 1960 to 1968 as professor of pathology. After a short period at the National Institute of Allergy and Infectious Diseases at Bethesda, Maryland, Benacerraf accepted the chair of comparative pathology at Harvard in 1970. From 1980 until 1992 he was president of the Dana-Farber Cancer Institute, Boston.

In the 1960s, working with guinea pigs, Benacerraf began to reveal some of the complex activity of the H-2 system, described by George SNELL. In particular he identified the Ir (immune response) genes of the H-2 segment as playing a crucial role in the immune system. This was achieved by injecting simple, synthetic, and controllable "antigens" into his experimental animals and noting that some strains responded immunologically while others were quite tolerant. Such differential responses have so far indicated there are over 30 Ir genes in the H-2 complex.

Later work began to show how virtually all responses of the immune system, whether to grafts, tumor cells, bacteria, or viruses, are under the control of the H-2 region. Benacerraf and his colleagues continued to explore its genetic and immunologic properties and also to extend their work to the analogous HLA system in humans. This work may well be important in the study of certain diseases, such as multiple sclerosis and ankylosing spondylitis, which have been shown to entail defective immune responses.

In 1980 Benacerraf was awarded for this work, together with George Snell and Jean DAUSSET, the Nobel Prize for physiology or medicine.

Beneden, Edouard van (1846–1910) *Belgian cytologist and embryologist*

Beneden (be-**nay**-den) was born in Louvain, Belgium, the son of the zoologist Pierre-Joseph van Beneden. Edouard followed his father's footsteps, taking charge of zoology teaching at Liège University in 1870. Here he extended Walther FLEMMING's work on cell division. Working with the horse intestinal worm, *Ascaris megalocephala*, Beneden had demonstrated by 1887 that the chromosome number in all the body cells is constant, but that this number is halved in the germ cells. This halving is achieved because the two successive divisions preceding ova and spermatozoa formation are accompanied by only one chromosome doubling. Such a reduction division (meiosis) is necessary to prevent chromosome numbers from doubling on fertilization.

Beneden also made important contributions to embryology from his studies on the cleavage and gastrulation of the fertilized egg.

Bennet, Abraham (1750–1799) *British physicist*

Bennet's scientific work was in electrostatics, the study of stationary electric charges and their effects, and he contributed many experiments and observations to the early development of this field. He was the inventor of the gold-leaf electrometer (an instrument for detecting and measuring electric charges) and did various experiments on electrostatic induction (the effect by which one body induces a charge on a nearby uncharged body). He also tried, though unsuccessfully, to demonstrate that light has momentum by focusing light rays on a sheet of paper hanging free in a vacuum.

Bentham, George (1800–1884) *British botanist*

Bentham, son of the naval architect Samuel Bentham, was born in the southwestern English county of Devon; he first became interested in botany at the age of 17, while living in France with his parents. There he read Augustin Pyrame de CANDOLLE's revision of J. B. LAMARCK's *Flore Française* (French Flora) and was much impressed with its analytical keys for plant identification. Thus began his consuming interest in plant taxonomy, on which he consistently worked during his leisure time.

From 1826 to 1832 he was secretary to his uncle, the famous jurist and philosopher, Jeremy Bentham, and studied for the bar at Lincoln's Inn. However, in 1833 he abandoned law for his growing botanical collection and library, which he generously presented to the Royal Botanic Gardens, Kew, in 1854. He then worked at Kew for the rest of his life.

His first botanical work, *Catalogue des plantes indigènes des Pyrénées et du bas Languedoc* (Catalog of the Indigenous Plants of the Pyrenees and Lower Languedoc), was published in Paris in November 1826. On his return to England he published *Outlines of a New System of Logic* (1827). Then in the early 1830s, Bentham turned his attention more to botany and his first important work in this field, *Labiatarum Genera et Species* (Genera and Species of the Labiatae), appeared between 1832 and 1836. While at Kew he published his popular *Handbook of the British Flora* (1858) and contributed to the Kew series of colonial floras with his *Flora Hongkongensis* (1861; Flora of Hong Kong) and the seven-volume *Flora Australiensis* (1863–78; Flora of Australia). In collaboration with Joseph HOOKER he produced his greatest work, the *Genera Plantarum* (1862–83; Plant Genera), which remains a standard in plant classification.

Benz, Karl Friedrich (1844–1929) *German engineer*

Benz was born in the city of Karlsruhe in southwest Germany, where his father was a railway engineer. After training in mechanical engineering (1853–64), he opened his own machine-tools works in Mannheim and in 1877 began experimenting with a two-cycle engine. In 1885 he ran his first Benz car, a three-wheeled vehicle now preserved in Munich, which was the first practical automobile powered by an internal-combustion engine. The vehicle was not patented until January 1886. The company founded by Benz to manufacture the vehicle, Benz & Cie., produced its first four-wheeled automobile in 1893 and its first series of racing cars in 1899. Benz left the company in about 1906; it later merged to form Daimler–Benz (1926), the makers of Mercedes-Benz automobiles.

Benzer, Seymour (1921–2007) *American geneticist*

Benzer was born in New York City, graduated from Brooklyn College in 1942, and gained his PhD in physics from Purdue University in 1947. He spent the years 1948–52 at various research institutes to familiarize himself with biological techniques, returning to Purdue as assistant professor in biophysics.

Benzer hoped to disprove the theory that the gene is an indivisible unit by demonstrating that recombination can occur within genes. However, such recombinations would be expected to occur so rarely that huge numbers of organisms would need to be studied to find one. In 1954 Benzer found a suitable organism to work on – the virus T4, a bacteriophage that infects the bacterium *Escherichia coli*. It multiplies approximately 100 fold in 20 minutes, giving millions of phage in a relatively short period.

Benzer found various mutants of T4, termed rII mutants, that had lost the ability to multiply on a specific strain of *E. coli*. He mixed together mutants that had mutated in different parts of the same gene and placed the mixture on a dish of *E. coli*. Any recombination within the gene that restored its function (and thus its ability to multiply in and destroy the bacteria) could be easily identified as clear areas in the *E. coli* dish. Such areas were indeed found, which proved that the gene can be split into recombining elements and verified James WATSON's and Francis CRICK's model of the gene as consisting of many nucleotide pairs.

Benzer went on to show that the number of distinguishable mutation sites corresponds with the estimated number of nucleotide pairs for that gene and that the sites are arranged linearly. He identified functionally independent units within the gene, naming these "cistrons."

Benzer also did important work on chemical mutagens and on the degeneracy of the genetic code. More recently he concentrated on the genetic control of behavior and the application of molecular biology to brain function. He moved from Purdue University to the California Institute of Technology in 1965, becoming professor of biology in 1967 and Boswell Professor of Neuroscience in 1975.

Berg, Paul (1926–) *American molecular biologist*

Berg was born in New York City and educated at Pennsylvania State University and Case Western Reserve University, where he obtained his PhD in 1952. He taught first at the School of Medicine at Washington University, St. Louis, moving to Stanford University in California in 1959, where he was professor of biochemistry from 1959 to 1970 and Willson Professor of Biochemistry from 1970. From 1985 until 2000 he was director of the Center for Molecular and Genetic Medicine. Berg is professor emeritus at Stanford. He ceased his research work in 2000.

In 1955 Francis CRICK proposed his adaptor hypothesis, in which he argued that amino acids did not interact directly with the RNA template but were brought together by an adaptor molecule. Crick offered little information on the nature of such molecules, merely arguing that they were unlikely to be large protein molecules and suggesting that there might well be a specific adaptor for each of the 20 amino acids. In 1956 Berg successfully identified such an adaptor, later known as transfer RNA, even though he was then unaware of Crick's hypothesis. He found a small RNA molecule that appeared to be quite specific to the amino acid methionine.

Berg's name later became known to a much wider public with the publication in *Science* (24 July 1974) of the "Berg letter," written with the backing of many leading molecular biologists, in which he gave clear warning of the dangers inherent in the uncontrolled practice of recombinant DNA experiments. It had become possible, Berg stated, to excise portions of DNA from one organism, using specialized enzymes, and to insert them into the DNA of another organism. For example, the harmless microorganism *Escherichia coli*, found in all laboratories, could be implanted with active DNA from the tumor-causing virus SV 40 and perhaps allowed to spread throughout a human population with quite unpredictable results. Berg consequently proposed an absolute voluntary moratorium on certain types of experiments and strict control on a large number of others. An international conference was held in Asilomar, California, followed by the publication of strict guidelines by the National Institutes of Health in 1976. That such agreement could be reached and maintained, it has been claimed, was largely a result of the integrity and authority of Berg. Ironically Berg was awarded the Nobel Prize for chemistry in 1980 for the large part he played in developing the splicing techniques that made recombinant DNA techniques possible in the first place.

Berger, Hans (1873–1941) *German psychiatrist*

Berger (**bair**-ger) was born in Neuses, Germany, and studied medicine at the University of Jena; having joined the university psychiatric clinic in 1897 as an assistant, he eventually served as its director and professor of psychiatry (1919–38). In his early work, he attempted to correlate physical factors in the brain, such as blood flow and temperature, with brain function. Disappointing results in this area made Berger turn to investigating the electrical activity of the brain. In 1924 he made the first human electroencephalogram by recording, as a trace, the minute changes in electrical potential measured between two electrodes placed on the surface of the head. Berger subsequently characterized the resultant wave patterns, including alpha and beta waves, and published his findings in 1929. The technique of electroencephalography is now used to diagnose such diseases as brain tumors and

epilepsy. It is also used in psychiatric research and in diagnosing brain death.

Bergeron, Tor Harold Percival (1891–1977) *Swedish meteorologist*

Bergeron (**bair**-ger-on), who was born in Stockholm, studied at the universities of Stockholm and Leipzig. During the period 1925–28 he worked at the famous Geophysical Institute at Bergen before taking a teaching appointment at Oslo University (1929–35). He held various appointments in the Swedish Meteorological Institute and was elected to the chair of meteorology at Uppsala in 1947.

Bergeron is best known for his work on cloud formation and in 1935 published the fundamental paper *On the Physics of Clouds and Precipitation*. Clouds consist of minute drops of water, but these drops will only fall as rain when they coalesce to form sufficiently large drops. Bergeron considered various processes, such as electric attraction and collisions caused by turbulence, but dismissed these as being too slow and inefficient. He therefore proposed a mechanism in which both ice crystals and water droplets are present in clouds. The water droplets tend to evaporate and the vapor then condenses onto the crystals. These fall, melt, and produce rain. Thus all rain, according to Bergeron, begins as snow and without the presence of ice crystals in the upper reaches of clouds there can be no rain. This theory was supported by the experimental and observational work of Walter Findeisen in 1939 and became known as the *Bergeron–Findeisen theory*. It does not explain precipitation from tropical clouds where temperatures are above freezing point.

Bergeron also produced important work on weather fronts, methods of weather forecasting, and the growth of ice sheets.

Bergius, Friedrich Karl Rudolph (1884–1949) *German industrial chemist*

The son of a chemicals industrialist, Bergius (**bair**-gee-uus), who was born in Goldschmieden, Poland, obtained his doctorate at Leipzig (1907) and worked with Hermann NERNST at Berlin and Fritz HABER at Karlsruhe, where he became interested in high-pressure chemical reactions. He was a professor at the Technical University at Hannover (1909–14) and then worked for the Goldschmidt Organization until 1945.

He is noted for his development of the *Bergius process* – a method of treating coal or heavy oil with hydrogen in the presence of catalysts, so as to produce lower-molecular-weight hydrocarbons. The process was important as a German source of gasoline in World War II. After the war Bergius lived in Austria and Spain before settling in Argentina as a technical adviser to the government, working on the production of sugar, alcohol, and cattle feed from wood. He shared the Nobel Prize for chemistry with Carl BOSCH in 1931.

Bergman, Torbern Olaf (1735–1785) *Swedish chemist*

Bergman (**bairg**-man), who was born in Katrineberg, Sweden, studied at the University of Uppsala, at first reading law and theology before turning to science and mathematics. He was a prolific scientist, working in physics, mathematics, and physical geography as well as chemistry. After graduating with a master's degree in 1758, he became professor of mathematics at Uppsala in 1761 and later professor of chemistry and pharmacy in 1767.

Bergman carried out many quantitative analyses, especially of minerals, and he extended the chemical classification of minerals devised by Axel CRONSTEDT. He remained an adherent of the phlogiston theory and although he firmly supported the doctrine of constant composition his analyses were not as solidly based as those of his later compatriot Jöns BERZELIUS. His most influential work was probably *Disquisitio de Attractionibus Electivis* (1785; A Dissertation on Elective Attractions). He compiled extensive tables listing relative chemical affinities of acids and bases. Bergman gave early encouragement to Karl SCHEELE, some of whose work he published.

Bergmann, Max (1886–1944) *German organic chemist and biochemist*

Bergmann (**bairg**-mahn), who was born in Fuerth in Germany, studied in Munich and Berlin and gained his PhD under Emil FISCHER in 1911. He worked as Fischer's assistant in Berlin until the latter's death in 1919. From 1921 to 1934 he was director of the Kaiser Wilhelm Institute for Leather Research, Dresden, from which he resigned on Hitler's coming to power. He then emigrated to America where he worked as a member of the Rockefeller Institute for Medical Research.

Bergmann's research interests were those of his teacher, Fischer: carbohydrates and amino acids. In 1932 he discovered the carbobenzoxy method of peptide synthesis, the greatest advance in this field since Fischer's first peptide synthesis in 1901. In this method the amino group of amino acids is "protected" by the carbobenzoxy group during condensation to form the peptide linkage and later freed by hydrolysis. Following Bergmann's work, many other protective groups have been used in peptide syntheses.

In America Bergmann investigated the specificity of proteinase enzymes and discovered (1937) that enzymes like papain were capable of splitting quite small peptides at precise link-ages. The last three years of his life were devoted mainly to problems connected with the war.

Bergström, Sune (1916–2004) *Swedish biochemist*

Bergström (**bairg**-stru(r)m) was born in Stockholm and educated at the Karolinska Institute there, where he obtained his MD in 1943. In 1947 he was appointed to the chair of biochemistry at Lund. In 1958 he moved to a comparable position at the Karolinska Institute, which he left in 1981.

In the 1930s Ulf VON EULER found an active substance in human semen capable of lowering blood pressure and causing muscle tissue to contract. He named it *prostaglandin* on the assumption that it came from the prostate gland. It soon became clear that there was not one such substance but a good many closely related ones with a variety of important physiological roles, but as they were produced in small quantities and rapidly broken down by enzymatic action, they proved to be very difficult to isolate and analyze. From 100 kilograms of rams' seminal vesicles, Bergström was able to extract a minute dose. To his surprise, however, he found the prostaglandin "extraordinarily active in virtually nonexistent doses."

In the 1950s Bergström succeeded in extracting the prostaglandins referred to as PGD_2, PGE_2, and PGF_2. He went on to demonstrate that they were derived from arachidonic acid ($C_{20}H_{36}O_2$), a fatty acid present in the adrenal gland, liver, and brain. Bergström's discovery opened up the study of prostaglandins by allowing them to be produced in the laboratory. For his pioneering work in this field he shared the 1982 Nobel Prize for physiology or medicine with John VANE and Bengt SAMUELSSON.

Beringer, Johann Bartholomaeus Adam (1667–1740) *German geologist*

Beringer (**bay**-ring-er) was a native of Würzburg, Germany, where his father was the dean of medicine at the university. He obtained his doctorate from Würzburg in 1693 and was appointed professor in 1694 – a position he held for the rest of his life.

He is largely remembered today for his extreme gullibility. Some colleagues, knowing him to be a keen fossil collector, decided to see if he could recognize artificial "fossils." They therefore prepared and scattered on the local hillside "fossils" of an increasingly unlikely nature. Nothing seemed to alert the suspicions of Beringer and indeed, the more bizarre the figure, the more excited he became. He published an account of them in 1726 with full illustrations despite warnings that he was dealing with fakes. Finally, the story goes although there is no documentary evidence for this, he found a stone with his own name on it and spent the rest of his life trying to buy up copies of his book.

Bernal, John Desmond (1901–1971)
British crystallographer

> Life is a partial, continuous, progressive, multiform and conditionally interactive self-realization of the potentialities of atomic electron states.
> —*The Origin of Life* (1967)

> The full area of ignorance is not mapped: we are at present only exploring its fringes.
> —Quoted by Sagittarius and George in *The Perpetual Pessimist*

Bernal's family were farmers in Nenagh, now in the Republic of Ireland; his mother was an American journalist. He was educated at Cambridge University, where his first work on crystallography was done as an undergraduate on the mathematical theory of crystal symmetry. William BRAGG offered him a post at the Royal Institution, which he joined in 1922.

Bernal was one of the most influential scientists of his generation. He had decided early in his career that x-ray crystallography would turn out to be the most likely tool to reveal details of the structure of matter. In addition to his intellectual mastery of the subject, he also possessed the ability to transmit his own enthusiasm to others and to attract around him a large number of highly talented and ambitious colleagues. To this group he was always known as "Sage."

His first success came in 1924 when he worked out the structure of graphite. He also began to work on bronze. In 1927 Bernal moved to Cambridge to a newly created lectureship in structural crystallography. While at Cambridge he worked on the structure of vitamin B_1 (1933), pepsin (1934), vitamin D_2 (1935), the sterols (1936), and the tobacco mosaic virus (1937).

Much of this research came not from Bernal alone; in most of his Cambridge studies he collaborated closely with Dorothy HODGKIN and many others came to work with Bernal, including Max PERUTZ, Francis CRICK, and Rosalind FRANKLIN.

In 1937 he was appointed professor of physics at Birkbeck College, London. With the outbreak of war in 1939 he joined the Ministry of Home Security and carried out with Solly ZUCKERMAN an important analysis of the effects of enemy bombing. Later in the war he served as scientific adviser to Lord Mountbatten, the Chief of Combined Operations. Bernal's main duties were connected with the planned Normandy landings. He spent much time establishing the physical condition of the beaches the Allies would land on in 1944. Maps, he soon discovered, were inaccurate. "Do you realize," he would tell his staff, "no one knows where France is?" He was one of the first to land on the Normandy beaches on D-day.

Bernal's duties were performed despite the fact that he was one of Britain's best known communists, having joined the party in 1924. While many of his friends abandoned the party at some stage of their life, some because of the Stalinist purges, others because of the Molotov pact, and most of those remaining because of the Hungarian uprising, Bernal remained with the party throughout his life. He traveled frequently in Eastern Europe, Russia, and China, and he was probably the only significant Western scientist to give permanent support to the work of LYSENKO.

In 1963 Bernal suffered the first of several serious strokes. He became progressively less mobile and in the last two years of his life, unable to speak, he was confined to a wheelchair.

Bernard, Claude (1813–1878) *French physiologist*

> Science allows no exceptions; without this there would be no determinism in science, or rather, there would be no science at all.
> —*Leçons de la pathologie expérimentale* (1871; Lessons on Experimental Pathology)

Bernard (bair-**nar**), the son of a poor wine grower from St. Julien, began writing plays to earn money but turned to medicine on the advice of a literary critic. His first experiences of medicine were discouraging but, following his appointment as assistant to François MAGENDIE at the Collège de France, he began a period of extremely productive research. He drew attention to the importance of the pancreas in producing secretions for breaking down fat molecules into fatty acid and glycerine and showed that the main processes of digestion occur in the small intestine and not, as was previously thought, in the stomach. In 1856 he discovered glycogen, the starchlike substance in the liver, whose role is to build up a reserve of carbohydrate, which can be broken down to sugars as required; normally the sugar content of the blood remains steady as a result of this interaction. The digestive system, he found, is not just catabolic (breaking down complex molecules into simple ones), but anabolic, producing complex molecules (such as glycogen) from simple ones (such as sugars).

Bernard also did valuable work on the vasomotor system, demonstrating that certain nerves control the dilation and constriction of blood vessels; in hot weather blood vessels of the skin expand, releasing surplus heat, contracting during cold to conserve heat. The body is thus able to maintain a constant environment separate from outside influences. Apart from elucidating the role of the red blood corpuscles in transporting oxygen, Bernard's investigation of the action of carbon monoxide on the blood proved that the gas combines with hemoglobin, the effect being to cause oxygen starvation. He also carried out important work on the actions of drugs, such as the opium alkaloids and curare (curarine), on the sympathetic nerves.

Bernard's health deteriorated from 1860 and he spent less time in the laboratory. He thus turned to the philosophy of science and in 1865 published the famous *Introduction à la médecine expérimentale* (An Introduction to the Study of Experimental Medicine). The book discusses the importance of the constancy of the internal environment, refutes the notion of the "vital force" to explain life, and emphasizes the need in planning experiments for a clear hypothesis to be stated, which may then be either proved or disproved. On the strength of this work he was elected to the French Academy in 1869.

Berners-Lee, Sir Timothy John (1955–) *British computer scientist*

Berners-Lee was born in London and educated at Oxford University, graduating in 1976. When he was at Oxford, he built his first computer. After graduating, he worked for two years with Plessey Telecommunications and in 1978 he joined D. G. Nash, where he wrote software. During a spell of a year and a half that he spent as an independent consultant he was at CERN for six months as a consultant software engineer. Here he wrote a program called *Enquire*, which was never published, to store information. Between 1981 and 1984 he was at John Poole's Image Computer Systems Ltd., working on technical design.

In 1984 Berners-Lee started a fellowship at CERN, where he worked on systems for the acquisition of scientific data. In 1989 he built on his previous work on *Enquire* by proposing a system now known as the World Wide Web, which would enable people to combine information in a web of hypertext documents. This vision was realized with the release of the World Wide Web program, initially internally at CERN in 1990 and then on the Internet in 1991. Berners-Lee continued developing the Web between 1991 and 1993.

He joined the Laboratory for Computer Science at the Massachusetts Institute of Technology (MIT) in 1994 and was a founder of the World Wide Web Consortium, the organization that coordinates the development of the Web. In 2004 he was appointed to the chair of computer science at Southampton University. Berners-Lee is currently working on the Semantic Web – an extension of the World Wide Web in which information content is expressed in both natural language and in a form that can be understood by software. Berners-Lee was knighted in 2004.

Bernoulli, Daniel (1700–1782) *Swiss mathematician*

Daniel was a son of Johann I Bernoulli (bernoo-lee). Of all the Bernoulli family he was probably the most outstanding mathematician and certainly the one with the widest scientific interests. Daniel, who was born in Groningen in the Netherlands, studied at the universities of Basel, Strasbourg, and Heidelberg. His studies, which reflected his already wide interests, included logic, philosophy, and medicine in addition to mathematics.

In 1724 Daniel produced his first important piece of mathematical research – a work on differential equations, which sufficiently impressed the European scientific community to earn him an invitation to the St. Petersburg Academy of Sciences as a professor of mathematics. Once installed in Russia he continued to pursue his varied interests and obtained a post at the academy for his friend Leonhard EULER. In 1733 he left Russia to return to Switzerland to take up a chair in mathematics at Basel. Bernoulli's wide interests continued to occupy him and during his time at Basel he also held posts in botany, anatomy, physiology, and physics.

In Switzerland Daniel did the work for which he is best known, namely his virtual founding of the modern science of hydrodynamics using Isaac NEWTON's laws of force. He published these ideas in his *Hydrodynamica* (1738; Hydrodynamics). Apart from his work in fluid dynamics Daniel made distinguished contributions to probability theory and differential equations in mathematics, and to electrostatics in physics.

He also laid the basis for the kinetic theory of gases. Like his uncle, Jakob I Bernoulli, Daniel corresponded voluminously with many scholars throughout Europe, thus extensively disseminating his new ideas.

Bernoulli, Jakob I (1654–1705) *Swiss mathematician*

Jakob I (or Jacques) was the first of the Bernoulli family of scientists to achieve fame as a mathematician. As with the two other particularly outstanding Bernoullis – his brother, Johann I, and nephew, Daniel – Jakob I played an important role in the development and popularization of the then recently invented integral and differential calculus of Isaac NEWTON and Gottfried LEIBNIZ. His particular contribution to the calculus consisted in showing how it could be applied to a wide variety of fields of applied mathematics.

Jakob I, who was born in Basel, Switzerland, began studying theology and in 1676 traveled through Europe where he met many of the important scientists of the day, such as Robert BOYLE in England. He returned to Basel in 1682 where he began lecturing on mechanics and held a chair in mathematics at Basel University from 1687 until his death. Apart from his mathematical work he was an influential figure in the European scientific community through his voluminous correspondence.

His most important contributions to mathematics were in the fields of probability and in the calculus of variations. His work on probability is contained in his treatise the *Ars conjectandi* (1713; The Art of Conjecturing) in which he made numerous important contributions to the subject, among which was his discovery of what is now known as the "law of large numbers." The law has a number of forms. In effect it says that for an event of probability P in a large number of trials n the number of actual events approaches n P as n increases. *Ars conjectandi* also contains Bernoulli's work on permutations and combinations.

The Bernoulli family were always prone to rivalry and Jakob I and his younger brother, Johann I, became involved in a controversy over the problem of finding the shortest path between two points of a particle moving solely under the influence of gravity. The result of this vigorous dispute was the creation of the calculus of variations, a field that Leonhard EULER was later to develop. In addition to this Jakob I did important and useful work in the study of the catenary, which he applied to the design of bridges.

Bernoulli, Johann I (1667–1748) *Swiss mathematician*

Johann I (or Jean) was the brother of Jakob I Bernoulli and was born in Basel, Switzerland. As in the case of several of the Bernoulli family Johann I's father did not encourage him to make a career of mathematics and he graduated in medicine in 1694.

Once he had abandoned medicine for mathematics he became chiefly interested in applying the calculus to physical problems. He played an important role as a propagandist for the calculus in general and in particular as a champion of Gottfried LEIBNIZ's priority over Isaac NEWTON. Johann I held a chair in mathematics at Groningen, Holland, from 1695 and returned to Switzerland to take up a chair in mathematics at Basel on the death of his brother in 1705. Johann I's interests ranged over many fields outside mathematics including physics, chemistry, and astronomy. His mathematical work also included particularly important contributions to optics, to the theory of differential equations, and to the mathematics of ship sails.

Bert, Paul (1833–1886) *French physiologist and politician*

Bert (bair), who was born in Auxerre in eastern France, initially studied engineering and law but turned to medicine, becoming a pupil of the eminent physiologist, Claude Bernard, at the Sorbonne, Paris. From his studies of the effects of low and high pressures on the human body, Bert showed how deep-sea divers and others working at high external pressure could avoid the condition known as the bends by returning gradually to normal pressure conditions. In this way, the nitrogen gas that dissolves in blood at high pressure is removed slowly instead of forming bubbles in the blood and causing agonizing and possibly fatal cramps. Bert wrote *La Pression barométrique* (1878; Barometric Pressure), which was translated into English in 1943 for the benefit of aircrews flying at high altitudes during World War II.

Bert entered politics in 1876 as deputy for Yonne and served briefly as minister of education and welfare (1881–82). He was staunchly anticlerical and left-wing in his views. In 1886 he was appointed governor-general to the Annan and Tonkin provinces in Indochina but died of dysentery in the same year.

Berthelot, Pierre Eugène Marcellin (1827–1907) *French chemist*

The Parisian-born son of a doctor, Berthelot (bair-te-**loh**) studied medicine at the Collège de France but became interested in chemistry, becoming assistant to Antoine-Jérôme BALARD in 1851. He was professor of organic chemistry at the Ecole Supérieure de Pharmacie (1859–76) and professor of chemistry at the Collège de France (1864–1907).

Alcohols were Berthelot's early research interest and he introduced the terms mono-, di-, and polyatomic alcohols. He showed that glycerin was a triatomic alcohol and in 1854 he synthesized fats from glycerin and fatty acids. He carried out a great deal of work on sugars, which he recognized as being both polyhydric alcohols and aldehydes. Berthelot was one of the pioneers of organic synthesis. Before his time, organic chemists had mainly been concerned with degradations of natural products but Berthelot, in keeping with his logical systematic nature, began with the simplest molecules; his syntheses included methane, methanol, formic acid, ethanol, acetylene, benzene, naphthalene, and anthracene. His favored techniques were reduction using red-hot copper and the silent electric discharge. His methods were somewhat crude and the yields were low. Berthelot's work on organic synthesis was published as *Chimie organique fondée sur la synthèse* (1860; Organic Chemistry Based on Synthesis).

Arising from his interest in esterification, Berthelot studied the kinetics of reversible reactions. In 1862, working with Péan de Saint Gilles, he produced an equation for the reaction velocity. This was incorrect but it inspired Cato GULDBERG and Peter WAAGE to enunciate the law of mass action (1864).

In 1864 Berthelot turned to thermochemistry. In his book *Mecanique chimique* (1879; Chemical Mechanics) he introduced the terms "endothermic" and "exothermic" to describe reactions that respectively absorb and release heat. He also introduced the bomb calorimeter for the determination of heats of reaction and investigated the kinetics of explosions.

Berthelot's interest in agricultural chemistry was stimulated by his discovery of nitrogen uptake by plants in the presence of an electrical discharge. In 1883 he established an agricultural station at Meudon, where fundamental work on the nitrogen cycle was carried out. He looked forward to the day when poverty and squalor would be eradicated by the application of synthetic chemistry and new sources of energy.

Berthelot was a pioneer of historical studies in chemistry. In this he was influenced by his friend, the scholar Renan. In later life he became increasingly involved in affairs of state, mostly concerned with education, and in 1895–96 he served as foreign minister.

Berthollet, Comte Claude-Louis (1748–1822) *French chemist*

Born in Talloires, France, Berthollet (bair-to-**lay**) studied medicine at Turin and gained his MD in 1768. He went to Paris in 1772 where he began publishing chemical researches in 1776 and was elected a member of the Académie Française in 1780. His Italian medical degree was not recognized in France so he obtained a Parisian degree in 1778.

When Berthollet published his important paper on chlorine (which came from sea water), *Mémoire sur l'acide marin déphlogistiqué* (1785; Memoir on a Marine Dephlogistonated Acid), he was the first French chemist to accept Antoine LAVOISIER's new system. Unfortunately, he also accepted Lavoisier's erroneous idea that chlorine contains oxygen. In 1784 Berthollet became inspector of a dyeworks and he discovered and developed the use of chlorine as a bleach. He published a standard text on dyeing *Eléments de l'art de la teinture* (1791; Elements of the Art of Dyeing).

Berthollet was neither a great manipulator nor a persuasive lecturer, but he did original work in many fields. He analyzed ammonia (1785), prussic acid (hydrogen cyanide, 1787), hydrogen sulfide (1798), and discovered potassium chlorate (1787). Although a convert, he remained skeptical about Lavoisier's oxygen theory of acidity: his analyses showed no oxygen in prussic acid or hydrogen sulfide, despite their undoubted acidity. Berthollet attempted to use his newly discovered potassium chlorate in gunpowder but it proved too unstable, destroying a powder mill at Essones in 1788. More productive were his analyses of iron and steel, which resulted in better quality steel.

After the French Revolution of 1789 Berthollet was a member of various commissions and in 1795 he became a director of the national mint. In 1798 he was entrusted by Napoleon with the organization of scientific work on the expedition to Egypt and he established an Institute of Egypt. On his return to Paris in 1799 Berthollet bought a large house at Arcueil in the suburbs of Paris, where he set up a laboratory and subsequently founded the Société d'Arcueil, which included Pierre de LAPLACE, Alexander von HUMBOLDT, Jean BIOT, Louis Thenard, and Joseph GAY-LUSSAC. At Arcueil, Berthollet produced his magnum opus, the *Essai de statique chimique* (1803; Essay on Chemical Statics), in which he propounded a theory of indefinite proportions. By 1808, following the work of John DALTON, Jöns BERZELIUS, and Gay-Lussac, indefinite proportions was decisively rejected, but Berthollet's idea that mass influences the course of chemical reactions was eventually vindicated in the law of mass action of Cato GULDBERG and Peter WAAGE (1864).

Berthollet was made a senator in 1804 and in his later years was regarded as the elder statesman of French science.

Berzelius, Jöns Jacob (1779–1848) *Swedish chemist*

> I ... have seldom experienced a moment of such pure and deep happiness as when the glowing stick which was thrust into it [oxygen] lighted up and illuminated with unaccustomed brilliancy my windowless laboratory.
> —*Autobiographical Notes*

Born in Väversunda, Sweden, Berzelius (ber-**zee**-lee-us) struggled to obtain a satisfactory education. In 1796 he entered the University of Uppsala but his studies were interrupted because of lack of funds. He began his chemical experiments without any official encouragement and from 1799 he worked during the summers as a physician at Medevi Springs where he analyzed the waters. He finally obtained his MD degree in 1802 with a dissertation on the medical uses of the voltaic pile.

After graduating Berzelius moved to Stockholm where he did research with Wilhelm HISINGER, a mining chemist. Their first success came in 1803 with the isolation of cerium but they were anticipated in this by Martin KLAPROTH. Berzelius later discovered selenium (1817), thorium (1828), and his coworkers discovered lithium (1818) and vanadium (1830). In 1807 Berzelius was appointed professor at the School of Surgery in Stockholm (later the Karolinska Institute), and he was soon able to abandon medicine and to concentrate on chemistry.

Berzelius was a meticulous experimenter and systemizer of chemistry. His early work was on electrochemistry and he formed a "dualistic" view of compounds, in which they were composed of positive and negative parts. He was an ardent supporter of John DALTON's atomic theory, but, like LAVOISIER, believed in the importance of oxygen – thus he argued for many years that chlorine contained oxygen.

In 1810 Berzelius began a long series of studies on combining proportions that established Dalton's atomic theory on a quantitative basis. This work led to tables of atomic weights that were generally very accurate, but he never accepted Amedeo AVOGADRO's hypothesis and this led to some confusion. He was a prolific author with about 250 papers to his credit. His *Lärbok i kemien* (1808–1818; Textbook of Chemistry) subsequently passed through many editions and was translated into most languages except English. Pupils who came to study with him included Friedrich WÖHLER, Leopold GMELIN, and Eilhardt MITSCHERLICH. His ideas on chemical proportions and electrochemistry are set out in *Essai sur la théorie des proportions chimiques et sur l'influence chimique de l'électricité* (1819; Essay on the Theory of Chemical Proportions and on the Chemical Effects of Electricity).

Berzelius's work in organic chemistry was less fruitful than the rest of his work but he improved organic analysis by introducing a tube of calcium chloride for the collection of water and the use of copper(II) oxide as an oxidizing agent. From 1835 Berzelius's rigid adherence to the dualistic theory proved obstructive to progress in organic chemistry, although it was given a certain plausibility by Wöhler and Justus von LIEBIG's discovery of the benzoyl radical (1832).

Berzelius introduced much of the familiar chemical apparatus, including rubber tubing and filter paper, and the modern chemical symbols, although these were little used in his lifetime. He had a knack of coining words for phenomena and substances – "catalysis," "protein," and "isomerism" were all introduced by him.

Besicovitch, Abram Samoilovich (1891–1970) *Ukrainian–British mathematician*

Born in the Ukrainian town of Berdjansk, Besicovitch (be-**sik**-o-vich) graduated in 1912 from the University of St. Petersburg, having studied under A. A. MARKOV. In 1917 he became professor of mathematics at Perm (later Molotov University) and then taught at the Leningrad Pedagogical Institute and Leningrad University. He left the Soviet Union in 1924 to work briefly in Copenhagen before moving to England. G. H. HARDY was sufficiently impressed by Besicovitch's analytical powers to secure him a lectureship at Liverpool University. After a year he became a lecturer at Cambridge and in 1950

became Rouse Ball Professor of Mathematics, a post that he held for eight years.

Besicovitch was primarily an analyst. The subject with which he is most associated is that of almost periodic functions, an interest stemming from his collaboration with Bohr. 1932 saw the publication of his book *Almost Periodic Functions*. He also did important work on the "Karkeya problem," general real analysis, complex analysis, and various geometric problems. His other main work was on geometric measure theory.

Bessel, Friedrich Wilhelm (1784–1846) *German astronomer*

Bessel (**bes**-el) was born into a poor family in Minden, Germany, and started work as a clerk. His interest in and aptitude for astronomy brought him to the attention of Heinrich OLBERS, who obtained a position for him in the observatory at Lilienthal. Four years later he was entrusted with the construction of the observatory at Königsberg and appointed its director.

Bessel made many advances in astronomy. He cataloged the position of 50,000 stars down to the ninth magnitude between 15°S and 45°N and, using James BRADLEY's results, achieved new levels of accuracy. He also made careful observations of 61 Cygni and was able to detect a parallax of 30 arc seconds and to calculate the star's distance – the first such determination – as 10.3 light years. (The distance is now known to be 11.2 light-years.) Although Bessel was the first to announce the detection of parallax (1838), Thomas HENDERSON had in fact measured it in 1832 in his observations of Alpha Centauri.

Bessel's other great discovery came after observing a slight displacement in the proper motion of Sirius, which he explained as the effect of an orbit around an unseen star, and announced in 1844 that Sirius was a double star system having a dark companion. Sirius B was detected optically by Alvan CLARK in 1862. Bessel made a similar claim for Procyon whose companion was discovered optically in 1895. He also noted irregularities in the motion of Uranus and suggested that they were caused by an unknown planet, but died just before the discovery of Neptune.

In mathematics Bessel worked on the theory of the functions, named for him, that he introduced to determine motions of bodies under mutual gravitation and planetary perturbations. They still have a wide application in modern physics.

Bessemer, Sir Henry (1813–1898) *British inventor and engineer*

Bessemer was the son of a mechanical engineer who had fled from the French Revolution. After leaving the village school in Charlton,

England, where he was born, he worked as a type-caster, until the family moved to London in 1830. At the age of 17 he set up his own business to produce metal alloys and bronze powder. In 1843 he had an idea that made his fortune. On purchasing some "gold" paint (made of brass) for his sister he was horrified at its high price. He designed an automatic plant to manufacture the paint and made sufficient money to pursue a career as a professional inventor.

During the Crimean War (1853–56) Bessemer invented a new type of gun with a rifled barrel. To manufacture the gun he needed a strong metal that could be run into a mold in a fluid state. At that time cast iron (pig iron) contained carbon and silicon impurities, which made it brittle. Wrought iron, which was relatively pure, was made by a laborious process of refining pig iron. The temperature of the furnace, while sufficient to melt the pig iron, was not sufficient to keep the purer iron molten. The refined metal was extracted in lumps after which it was "wrought." Bessemer proposed burning away the impurities by blowing air through the molten metal. The *Bessemer converter* that he invented is a cylindrical vessel mounted in such a way that it can be tilted to receive a charge of molten metal from the blast furnace. It is then brought upright for the "blow" to take place. Air is blown in through a series of nozzles at the base and the carbon impurities are oxidized and carried away by the stream of air.

Bessemer announced his discovery in 1856. At first his idea was accepted enthusiastically and within weeks he obtained the equivalent of about $135,000 in license fees. However, though the process had worked for him, elsewhere it failed dismally because of excess oxygen trapped in the metal, and because of the presence of phosphorus in the ores. (By chance Bessemer's ore had been phosphorus-free.) His invention was dropped and Bessemer found himself the subject of much ridicule and criticism. Bessemer established his own steelworks in Sheffield (1859) using imported phosphorus-free iron ore.

Robert Mushet (about 1856) solved the problem of the excess oxygen by the addition of an alloy of iron, manganese, and carbon to the melt. Bessemer's process then worked provided nonphosphoric ores were used, but it took much time and determination to convince ironworkers after the initial failure. The invention eventually reduced the price of steel to a fifth of its former cost, made it possible to produce it in large quantities, and made possible its use in a variety of new products. The problem of dealing with the phosphorus impurities was solved in 1878 by Sydney Gilchrist THOMAS and Percy Carlyle GILCHRIST. Bessemer retired a rich man in 1873.

Best, Charles Herbert (1899–1978)
American–Canadian physiologist

Best, who was born in West Pembroke, Maine, graduated in physiology and biochemistry from the University of Toronto in 1921. In the summer of that year he gave up a lucrative holiday playing professional football and baseball to begin work with Frederick BANTING. Together they isolated the hormone insulin, and showed its use in the treatment of diabetes. Banting was furious when Best was not awarded a share in the 1923 Nobel Prize for physiology or medicine for the discovery of insulin.

Best remained at the University of Toronto and gained his MB in 1925. He was made head of the physiology department in 1929 and became director of the Banting and Best Department of Medical Research when Banting was killed in 1941. He continued the work on insulin throughout these years and in an important paper published in 1936 he suggested the administration of zinc along with insulin to reduce its rate of absorption and make it more effective over a longer time. He also studied cardiovascular disease and established the clinical use of heparin as an anticoagulant for blood in the treatment of thrombosis. He discovered the vitamin choline, which prevents liver damage, and the important enzyme histaminase, which takes part in local inflammation reactions, breaking down histamine.

Bethe, Hans Albrecht (1906–2005) *German–American physicist*

Bethe (**bay**-te), the son of an academic, was born in Strasbourg in France and educated at the universities of Frankfurt and Munich, where he obtained his doctorate in 1928. He taught in Germany until 1933 when he moved first to England and, in 1935, to America. In America he held the chair of physics at Cornell from 1937 until his retirement in 1975, serving in the war as director of theoretical physics at Los Alamos from 1943 to 1946.

Bethe soon established a reputation for his impressive knowledge of nuclear reactions. This was, in part, based on three review articles on nuclear physics he published in 1936 and 1937, which have been described as the first presentation of this field as a branch of science and which are sometimes known as "Bethe's Bible."

In 1938 Bethe was invited by Edward TELLER to contribute a paper on astrophysics for a conference he was organizing. Bethe at first pleaded ignorance of the subject but, under pressure from Teller, he finally agreed to search for a relevant topic. He noted that most astrophysicists seemed to be puzzling over the origin of the chemical elements. He therefore decided to consider another issue, namely, the sources of stellar energy. He managed to find, as he reported in his 1939 paper *Energy Production in Stars,* " ... the only nuclear reaction which gives the correct rate of stellar energy production within the limits of the theory."

Bethe was referring to the carbon cycle (or CNO cycle). The cycle begins with a hydrogen nucleus or proton (^{1}H) and a carbon-12 atom; it has six stages:

$$^{12}C + {}^1H \rightarrow {}^{13}N + \gamma$$
$$^{13}N \rightarrow {}^{13}C + e^+ + \nu$$
$$^{13}C + {}^1H \rightarrow {}^{14}N + \gamma$$
$$^{14}N + {}^1H \rightarrow {}^{15}O + \gamma$$
$$^{15}O \rightarrow {}^{15}N + e^+ + \nu$$
$$^{15}N + {}^1H \rightarrow {}^{12}C + {}^4He$$

Here, γ is a gamma ray, e^+ a positron, and ν a neutrino. The net result of the cycle is to convert four protons ($4\,{}^1$H) into a helium nucleus (He^4), while the carbon-12-atom remains available after step 6 to repeat the cycle once more. In the process 27 MeV (million electron volts) are released.

Although the amount of energy produced per cycle is modest, the large amount of stellar matter involved is sufficient to generate the enormous energies met with inside stars. Bethe's CNO cycle, however, gives no indication of the origin of the carbon-12 that starts the cycle. It was left to Fred HOYLE and his colleagues to resolve this issue in the 1950s. Bethe did, however, contribute, with Ralph ALPHER, to George GAMOW's famous 1948 alpha-beta-gamma paper on the origin of the elements and the big bang. Unfortunately their paper advanced no further than the isotopes of hydrogen and helium.

Bethe also played a significant part in public affairs. He quarreled in the 1940s with Teller on the need to build the hydrogen bomb, and in the 1980s over the viability of the U.S. Strategic Defense Initiative (known as "Star Wars"). In 1958 he served as a delegate to the Geneva Conference that negotiated the first test-ban treaty. He also worked for the 1963 ban on atmospheric testing. More recently, in 1992, Bethe called upon the United States and Russia to reduce their nuclear arsenals to a thousand warheads each.

Bethe also worked in other branches of physics. In 1929 he published a classic paper initiating crystal-field theory, and in 1949 he made a pragmatic calculation of the LAMB shift. He also continued to work at physics in his retirement. He collaborated with John BAHCALL on several papers dealing with the solar-neutrino problem. He also began to tackle the problem of explaining how stars explode. Modern computer models, he complained, lead to moderate explosions when compared with the massive eruptions of a supernova.

For his earlier work on the theory of nuclear reactions, and for his contributions to astrophysics, Bethe was awarded the 1967 Nobel Prize for physics.

Bevan, Edward John (1856–1921) British industrial chemist

Bevan, who was born at Birkenhead in Lancashire, northwest England, was educated privately and in 1873 began employment with the Runcorn Soap and Alkali Company. From 1877 to 1879 he studied chemistry at Owens College, Manchester, where he met Charles CROSS with whom he went into partnership in 1885 as consulting chemists.

Their main interest was cellulose and in 1892 they patented the viscose process of rayon manufacture in which cellulose (woodpulp) is dissolved in carbon disulfide and alkali. The cellulose is regenerated by acid, either in the form of yarn (rayon) or film (cellophane). He wrote several books, including *Cellulose* (1895), which he coauthored with Cross.

Bichat, Marie François Xavier (1771–1802) French pathologist

> Life is the ensemble of functions that resist death.
> —*Recherches physiologiques sur la vie et la mort* (1800; Physiological Researches on Life and Death)

Bichat (bee-**sha**), the son of a physician, was born at Thoirette, France, and studied humanities at Montpellier and philosophy at Lyons. He switched to studies in surgery and anatomy and moved in 1793 to Paris, where he worked for some years under the patronage of Pierre Desault, the leading French surgeon. He was appointed physician at the Hôtel-Dieu in 1800 shortly before his death from tuberculous meningitis.

Despite his short life Bichat produced two highly influential works, *Traité des membranes* (1800; Treatise on Membranes) and *Anatomie générale* (1801; General Anatomy). He revealed the inadequacies of Giovanni MORGAGNI's claim that disease resided in the organs of the body, and instead stressed the role of the component tissues in determining health and disease. He consequently went on to make a detailed investigation of the tissues of the body, distinguishing 21 different types. He argued that in an organ made up of a number of tissues it is often found that while one tissue type is diseased the remainder are healthy.

This work was significant in bridging the gap between the organ pathology of Morgagni and the cell pathology of Rudolf VIRCHOW.

Biela, Wilhelm von (1784–1856) Austrian astronomer

Born in Rossia, Austria, Biela (**bee**-la) was an army officer and an amateur astronomer. In 1826 he observed a comet with a short period of 6.6 years. When *Biela's comet* reappeared in 1845 it had split into two parts. It was last seen in 1852 when it presumably broke up, for in November 1872 what was probably its remains was seen as a fantastic meteor shower, which caused a worldwide sensation. The shower can still be seen faintly in Andromeda in late November. The meteors are sometimes called the Bielids or, alternatively, the Andromedids.

Biffen, Sir Rowland Harry (1894–1949)
British geneticist and plant breeder

Biffen was born in Cheltenham in Gloucestershire, England; after graduating in natural sciences from Cambridge in 1896, he joined a team investigating rubber production in Mexico, the West Indies, and Brazil. On his return he was appointed lecturer in botany at Cambridge and patented a method for handling rubber latex.

Biffen was inclined more toward applied than pure botany and joined the Cambridge School of Agriculture shortly after its foundation in 1899. He began conducting cereal trials in order to select improved types, and when Gregor MENDEL's laws of inheritance were rediscovered in 1900, he realized immediately that they could be applied to improving plant-breeding methods. Biffen speculated that physiological as well as morphological traits would prove to be inherited in Mendelian ratios, and in 1905 demonstrated that this was true for resistance to yellow rust, a fungal disease of wheat.

Little Joss and Yeoman, two wheat varieties bred by Biffen, were unequaled for many years. In 1912 Biffen became director of the Plant Breeding Institute at Trumpington, a newly formed research center established by the government to promote Biffen's work and the application of scientific principles to plant breeding. Biffen was also professor of agricultural botany at the university from 1908 to 1931 and was instrumental in setting up the National Institute of Agricultural Botany at Cambridge. He was knighted for his services to agriculture in 1925.

Billroth, Christian Albert Theodor (1829–1894) *Austrian surgeon*

Billroth (**bil**-roht), who was born in Bergen, Germany, first studied natural sciences and later medicine at the University of Göttingen and subsequently in Berlin, becoming qualified to practice in 1852. In the following year he was appointed to the surgical clinic at Berlin University and in 1860 became director of a similar clinic at the University of Zurich, where he wrote his *Allgemeine chirurgische Pathologie und Therapie* (1863; General Surgical Pathology and Therapy). However, it was as professor of surgery at the University of Vienna (1867–94) that Billroth made his greatest contributions to surgery. Here, he introduced the new antiseptic methods in abdominal surgery, thus enabling hitherto impossibly dangerous operations. Billroth was the first to perform the removal of a section of the esophagus (1872), to undertake a complete laryngectomy (1873), and to excise the lower half of the stomach, rejoining the upper half to the duodenum. This is sometimes called *Billroth's operation*.

Apart from being an eminent surgeon of international repute, Billroth was also a gifted musician and like Theodor ENGELMANN a friend of Johannes Brahms.

Bina, Eric (1964–   ) *American computer scientist*

Bina (**bee**-na) worked at the National Center for Supercomputer Applications (NCSA) at the University of Illinois at Urbana in the early 1990s. While there, he collaborated with Marc ANDREESSEN and some other colleagues in producing a prototype Internet browser called *Mosaic*.

After Andreessen founded Mosaic Corp. with Jim Clark in 1994, Bina joined the new company, subsequently named Netscape Communications, and helped to develop an improved version of *Mosaic*, which was called *Navigator*. This invention was a key development in the very rapid growth of the Internet.

Binnig, Gerd (1947–   ) *German physicist*

Binnig was born in Frankfurt, Germany, and educated at the Goethe University there, where he obtained his PhD in 1978. He immediately joined the staff of the IBM Research Laboratory at Roschliken, Zurich.

One limitation of the conventional electron microscope developed in the 1930s by E. RUSKA was its failure to reveal much information about the surface of a material. Overall surface structure can be distinguished by the techniques of low-energy electron diffraction (LEED) and some indication of the detailed structure can be gained from the field-emission and field-ion microscopes developed by Erwin MUELLER. Binnig, in collaboration with Heinrich ROHRER, began work on a high-resolution *scanning tunneling microscope* (STM) in 1978. By 1981 they were able to resolve details on the surface of some calcium–iridium–tin crystals.

The STM employs a small conducting probe which is held close to the surface and slowly scanned across it. Electrons tunnel between the surface and the probe, and the probe is raised or lowered so as to keep the signal constant. A computer-generated map of the surface is produced. Like the similar *atomic force microscope* (ATM), the device can, under suitable conditions, resolve individual atoms or molecules on the surface.

For their work in this field Binnig and Rohrer shared the 1986 Nobel Prize for physics.

OPTICAL ACTIVITY *Biot showed that compounds such as tartaric acid could rotate polarized light in solution. The explanation was later given by Le Bel and van't Hoff, who suggested that carbon molecules have a three-dimensional tetrahedral structure. Tartaric acid (above) has two isomers that are mirror images of each other. One rotates the light to the left (levo); the other to the right (dextro).*

Biot, Jean Baptiste (1774–1862) *French physicist*

Biot (**bI**-oh), a Parisian by birth, grew up during the French Revolution and at the age of 18 he joined the army as a gunner. He left a year later to study mathematics at the Ecole Polytechnique in Paris. On leaving he taught at a school in Beauvais but soon returned to Paris to become a professor of physics at the Collège de France.

In 1804 Biot made an ascent in a balloon with Joseph GAY-LUSSAC. They reached a height of three miles and made many observations, including the fact that the Earth's magnetism was not measurably weaker at that height.

For the next few years Biot collaborated with François ARAGO in many fields of research and they traveled to Spain together to measure the length of an arc of meridian, in order to calibrate a standard unit of length. Biot later went on a number of other important expeditions.

His most famous work was on optical activity, for which, in 1840, he was awarded the Rumford medal of the Royal Society. He was the first to show that certain liquids and solutions, as well as solids, can rotate the plane of polarized light passing through them. Biot suggested that this is due to asymmetry in the molecules. From this idea grew the technique of polarimetry as a method of measuring the concentration of solutions.

Birkeland, Kristian Olaf Bernhard (1867–1917) *Norwegian physicist and chemist*

Birkeland (**beer**-ke-lan) was born in the Norwegian capital, Christiania (now Oslo), and studied in Paris, Geneva, and Bonn where he was a pupil of Robert BUNSEN. In 1898 he was appointed to the chair of physics in Oslo University. He is remembered today for his discovery of a means for the fixation of nitrogen (the *Birkeland–Eyde process*).

Energy crises are not monopolies of the 1970s. In 1898 William CROOKES in his presidential address to the British Association had pointed out that, given the world demand for nitrogeneous fertilizers, the deposits of nitrates would rapidly be exhausted. As there is a virtually unlimited supply of nitrogen in the atmosphere the obvious solution was to find some way in which it could be used. Birkeland, in collaboration with Samuel EYDE, solved the problem in 1903 by passing air through an electric arc to form oxides of nitrogen, which could then be absorbed in water to give nitric acid. This was mixed with lime to give calcium nitrate. The process is particularly useful in regions (as in Scandinavia) where there is a plentiful supply of hydroelectric power, although the Haber process is now the main industrial method of fixing nitrogen.

Birkeland also spent much time studying the aurora borealis, making several expeditions and establishing a geophysical laboratory as far north as 70°. In 1896 he was the first to suggest the correct explanation that the aurora borealis could be the result of charged rays emitted by the Sun and trapped in the Earth's magnetic field near the poles. He derived this idea from the resemblance between the newly discovered cathode rays and the aurora.

Birkhoff, George David (1884–1944) *American mathematician*

Birkhoff, who was born in Overisel, Michigan, studied at the Lewis Institute (now the Illinois Institute of Technology) from 1896 to 1902, and subsequently at the University of Chicago and at Harvard. In 1907 he obtained his PhD from Chicago and took up a teaching post at the University of Wisconsin, moving to Princeton in 1909. In 1912 he became assistant professor at Harvard and, in 1919, professor there, a post he held until 1939.

Birkhoff's mathematical interests were wide, and among the many areas to which he made notable contributions were differential equations, celestial mechanics, difference equations, and the three-body problem. His main field of research was mathematical analysis, especially applied to dynamics. In the course of his work on dynamical systems Birkhoff obtained a famous proof of a conjecture made by Henri POIN-CARÉ in topology, usually known as Poincaré's last geometric theorem. In 1923 he proved a result, now known as *Birkhoff's theorem*, for spherically symmetric systems in General Relativity Theory. This has turned out to be important in the theory of black holes and cosmology. The ergodic theorem, a result concerned with the formal mathematics of probability theory that Birkhoff proved in 1931, is another of his outstanding achievements. Modern dynamics received an enormous impetus from Birkhoff's work, and he also worked on the foundations of relativity and quantum mechanics.

Bíró, Ladislao José (1899–1985) *Argentinian inventor*

Throughout the 1930s Bíró (**bI**-roh) worked as a journalist and artist in Budapest, the capital of his native Hungary. While thus engaged he noticed how quickly printer's ink dried and began to think of a new type of pen. Before Bíró the alternatives were expensive and unreliable fountain pens, or steel nibs with bottles of ink. To meet the demand Birmingham was producing nearly 200 million nibs each year. Bíró saw that a ball rotating in an ink supply would write when pushed along a sheet of paper. The ink bottle and leaking fountain pen could be dispensed with. In 1938, with his brother Georg, a chemist, Bíró applied for a patent.

Europe, however, was no place to develop a new product in 1938. Bíró had happened to meet president Justo of Argentina while on holiday in Yugoslavia. Intrigued by the new pen, the president invited Bíró to come and work in Argentina. Bíró accepted the invitation and soon found backers in return for a substantial share in any future profits. In 1944 the North American rights were sold for $2 million. But Bíró had little taste for business and resigned from the company in 1947 to devote more time to art. He also continued to invent and produced, among other items, a heat-proof tile and a pick-proof lock.

The durability of Bíró's work can be seen partly in the billions of ballpoint pens that are made and thrown away each year, and partly in the fact that in most of the world's languages a ballpoint pen is also known as a "biro."

Bishop, John Michael (1936–   ) *American immunologist and microbiologist*

Bishop attended Gettysburg College and studied medicine at Harvard University. In 1962 he secured an internship at Massachusetts General Hospital in Boston, and in 1964 he moved to the National Institutes of Health, Washington, DC, as research associate in vi-

rology, later becoming senior investigator (1966) and assistant professor (1968). He was appointed professor of microbiology and immunology at the University of California Medical Center, San Francisco, in 1972, and in 1981 he became director of the G. W. Hooper Research Foundation. He became chancellor of the University of California in 1998.

Bishop, working in collaboration with Harold VARMUS, demonstrated for the first time that cancer-causing genes (*oncogenes*) carried by certain viruses are derived from normal genes present in the cells of their host, known as *proto-oncogenes*. This work by the team at the University of California, published in 1976, led to the discovery of many more such cellular genes, and represented a major advance in cancer research. In recognition of this, Bishop and Varmus were jointly awarded the 1989 Nobel Prize for physiology or medicine.

Bittner, John Joseph (1904–1961) *American experimental biologist*

Bittner, who was born in Meadville, Pennsylvania, gained his doctorate at the University of Michigan and spent the greater part of his academic life involved in cancer research. He was George Chase Christian Professor of Cancer Research at the University of Michigan and director of cancer biology of the University of Minnesota's medical school (1942–57), and later professor of experimental biology.

While working at Ben Harbor Research Station, Maine (1936), Bittner found that some strains of mice were highly resistant to cancer, while others were very prone to it. If the young of cancer-resistant mice were transferred to cancer-prone mothers they became cancerous, apparently via the mothers' milk, whereas cancer-resistant parents induced resistance in cancer-prone young. Bittner's discovery of viruslike organisms in the milk of cancer-prone parents suggested that these organisms are the cause of the cancer. Bittner's findings followed, and may be linked with, those of Francis ROUS, who made the controversial finding that other viruslike organisms are, perhaps, the cause of sarcomas (tumors originating in connective tissue) in chickens. Such work does not, of course, suggest that all cancers are caused by viruses or viruslike organisms, merely that some forms may be.

Bjerknes, Jacob Aall Bonnevie (1897–1975) *Norwegian meteorologist*

Bjerknes (**byairk**-nays), the son of Vilhelm Bjerknes, followed the example of his father in studying meteorology. Born in the Swedish capital, Stockholm, he was educated at the University of Oslo, where he obtained his PhD in 1924, and worked at the Geophysical Institute at Bergen with his father from 1917, remaining there when Vilhelm moved to Oslo in 1926.

During World War I Bjerknes worked with his father in establishing a series of weather observation stations throughout Norway. From the data collected, and working with other notable meteorologists, including Tor BERGERON, they developed their theory of polar fronts, also known as the *Bergen theory* or the *frontal theory*. They had established that the atmosphere is composed of distinct air masses possessing different characteristics and applied the term "front" to the boundary between two air masses. The polar front theory showed how cyclones (low-pressure centers) originated from atmospheric fronts over the Atlantic Ocean where a warm air mass met a cold air mass.

In 1939 Bjerknes moved to America and, unable to return to occupied Norway, became professor of meteorology at the University of California where he continued to study atmospheric circulation. In 1952 he became one of the first to use space techniques for meteorological research when he used photographs of cloud cover taken by research rockets for weather analysis.

Bjerknes, Vilhelm Friman Koren (1862–1951) *Norwegian meteorologist*

Bjerknes was born in the Norwegian capital, Christiania (now Oslo), where his father, Carl, was professor of mathematics at the university. In 1890 Vilhelm traveled to Germany and became assistant to Heinrich HERTZ. He was made professor of applied mechanics and mathematical physics at the University of Stockholm in 1895. He returned to Norway in 1907 and in 1912 moved to the University of Leipzig to become professor of geophysics.

Bjerknes's important contributions to meteorology and weather forecasting include his mathematical models (1897) of atmospheric and oceanic motions, which later led to the theory of air masses. In 1904 he produced a program for weather prediction based on physical principles. He returned to Norway in 1907 and in 1912 moved to the University of Leipzig to become professor of geophysics.

During World War I he founded the famous Bergen Geophysical Institute, gathering there a group of notable meteorologists, including his son, Jacob Bjerknes, Tor BERGERON, and Carl-Gustaf ROSSBY. In 1921 he produced an important work, *On the Dynamics of the Circular Vortex with Applications to the Atmospheric Vortex and Wave Motion*.

Bjerrum, Niels (1879–1958) *Danish physical chemist*

The son of a professor of ophthalmology, Bjerrum (**byair**-um), who was born in the Danish capital, Copenhagen, received his master's degree in 1902 and doctorate in 1908, becoming professor at the Royal Veterinary and Agricultural College (1914–49).

His first notable work was an extensive study of chromium complexes by means of conductivity, equilibrium constant, and absorption spectrum measurements. In 1909 he proposed (contrary to Svante ARRHENIUS's original dissociation theory) that strong electrolytes are completely dissociated. In 1911, working with Hermann NERNST, he applied the quantum theory of a harmonic oscillator to the temperature dependence of the specific heats of gases. This led to work on molecular rotation and vibration and hence to pioneer work on the infrared spectra of polyatomic molecules.

Black, Sir James Whyte (1924–) *British pharmacologist*

Black graduated from St. Andrews University, Scotland, in 1946, and, after a number of academic posts, joined ICI as a pharmacologist (1958–64). After working for Smith, Kline, and French he became professor of pharmacology at University College, London (1973–77), before joining Wellcome as Director of Therapeutic Research (1978–84). From 1984 until 1992 he was professor of analytical pharmacology at King's College Hospital, London. From 1992 to 2006 he was chancellor of the University of Dundee. Black was knighted in 1991.

Black has been associated with two important advances in pharmacology. In the 1950s he isolated the first beta blockers. These are compounds that prevent the stimulation of certain nerve endings (beta receptors) in the sympathetic nervous system, thus reducing heart activity. Beta blockers are widely used to treat hypertension and angina. His subsequent work has been concerned with the control of gastric ulcers and his discovery of the drug cimetidine, which reduces acid secretion in the stomach and is used to treat ulcers in the stomach and duodenum. For this work and his work on beta blockers he was awarded the 1988 Nobel Prize for physiology or medicine.

Black, Joseph (1728–1799) *British physician and chemist*

> Upon the whole, Chymistry is as yet but an opening science, closely connected with the usefull and ornamental Arts, and worthy the attention of a liberal mind ... While our knowledge is imperfect, it is apt to run into error: but Experiment is the thread that will lead us out of the labyrinth.

Black, the son of a wine merchant, was born in Bordeaux, France, and studied languages and natural philosophy, and later, medicine and chemistry at Glasgow University (1746–50). He moved to Edinburgh in 1751, where he presented his thesis in 1754. Black published very little and the thesis, expanded and published as

Experiments upon Magnesia Alba, Quicklime, and some other Alcaline Substances (1756), contained his most influential work. The paper in fact marked the beginning of modern chemistry. Black investigated quantitatively the cycle of reactions:

limestone→quicklime→slaked lime→
limestone

and showed that the gas evolved ("fixed air" or carbon dioxide) is distinct from and a constituent of atmospheric air, and is the cause of the effervescence of limestone with acids. He proved that mild alkalis will become more alkaline when they lose carbon dioxide and they are converted back to mild alkalis through reabsorption of the gas.

Black's other great discovery was that of latent heat (the heat required to produce a change of state). The concept of latent heat came to him in 1757 and the experimental determination of the latent heat of fusion of ice was made in 1761. The next year he determined the latent heat of formation of steam. Black also distinguished the difference between heat and temperature and conceived the idea of specific heat.

Black was professor of medicine and lecturer in chemistry at Glasgow (1756–66) and then professor of chemistry at Edinburgh for the rest of his life. Black's lectures, which he gave for over 30 years, were immensely popular and were published in 1803.

Blackett, Patrick Maynard Stuart (1897–1974) *British physicist*

Blackett, the son of a London stockbroker, attended the Royal Naval College at Dartmouth. After serving with the navy in World War I, during which he fought at the Battle of the Falklands and Jutland, he entered Cambridge University, resigned his commission, and decided to become a scientist. He worked in the 1920s with Ernest RUTHERFORD at the Cavendish Laboratory and, in 1933, was appointed professor of physics at London University. In 1937 he moved to Manchester, returning to London in 1953 to take the chair at Imperial College where he remained until his retirement in 1963. During World War II he worked on numerous advisory bodies and from 1942 to 1945 was director of operational research at the admiralty.

Just as Blackett was beginning his research career Rutherford had announced his discovery of the atomic transmutation of nitrogen into oxygen by bombardment with alpha particles. Blackett, using a cloud chamber, took some 23,000 photographs containing some 400,000 alpha particle tracks in nitrogen and found in 1925 just eight branched tracks in which the ejected proton was clearly separated from the newly formed oxygen isotope.

Blackett continued with the Wilson cloud chamber and began, in collaboration with Giuseppe OCCHIALINI, to use it to detect cosmic rays. As the appearance of cosmic rays is unpredictable it was standard practice to set up the chamber to take a photograph every 15 seconds, producing a vast amount of worthless material for analysis. To avoid this Blackett introduced in 1932 the counter-controlled chamber. Geiger counters were so arranged above and below the chamber that when a cosmic ray passed through both, it activated the expansion of the chamber and photographed the ion tracks produced by the ray. Using this device they confirmed in 1933 Carl ANDERSON's discovery of the positron. They also suggested that the positron was produced by the interaction of gamma rays with matter, such that a photon is converted into an electron–positron pair. The phenomenon is known as pair production.

After the war Blackett's research interests moved from cosmic rays to terrestrial magnetism. Using new sensitive magnetometers his group began a major survey of the magnetic history of the Earth. By 1960 they could report that there had been considerable change in the relative positions of the continents over the past 500 million years, thus providing further support for the doctrine of continental drift.

Blackett was also active in public affairs and a noted opponent of nuclear weapons. In 1948 he was awarded the Nobel Prize for physics for "his development of the Wilson cloud chamber and his discoveries therewith in the field of nuclear physics and cosmic radiation." He was raised to the British peerage as Baron Blackett in 1969.

Blackman, Frederick Frost (1866–1947) *British plant physiologist*

Blackman, born in London the son of a doctor, studied medicine at St. Bartholomew's Hospital there and natural sciences at Cambridge University. He remained in Cambridge for the whole of his career where he served as head of plant physiology until his retirement in 1936.

Blackman is mainly remembered for his classic 1905 paper, *Optima and Limiting Factors*, in which he demonstrated that where a process depends on a number of independent factors, the rate at which it can take place is limited by the rate of the slowest factor. This paper was stimulated by the research of one of his students, who showed that raising the temperature only increased the rate of photosynthesis if the level of illumination was high. Increased temperatures had no effect at low light intensities.

He had earlier, in 1895, provided convincing experimental support for the long held view that gaseous exchange between the leaves and the atmosphere takes place through the stomata, the pores on the leaf's surface.

Blackwell, Elizabeth (1821–1910)
British–American physician

Blackwell is best remembered for being the first woman to receive a medical degree in the United States, in 1849. However, she also did much in promoting women's health and training for women physicians. She was born in Bristol, England, and emigrated to America as a child. The early death of her father forced her into teaching to support the family, but she soon determined on a career in medicine. Yet formidable obstacles faced her; apart from lack of money to finance her preliminary studies, no woman had ever been admitted to a medical school. While continuing to teach she arranged to lodge in the house of a physician, Samuel Henry Dickson of Charleston, where she could acquire some basic medical knowledge. She was turned down by all the major medical colleges, and was accepted by just one of the minor medical schools, Geneva Medical College, New York State. She overcame the initial skepticism and embarrassment of staff at the college, and the hostility of the townspeople, to complete her course and graduate first in her class. To continue her medical training she then went to Europe, initially to La maternité lying-in hospital in Paris, where she contracted an eye disease that left her blind in one eye. Thereafter she studied at St. Bart's Hospital in London, where she was generally welcomed by the teaching staff, an exception being the professor of midwifery: according to Blackwell, "[he told me that] his neglecting to give me aid was owing to no disrespect for me as a lady, but to his condemnation of my object!" Blackwell returned to America in 1851 and settled in New York, where she opened a free part-time dispensary for poor women and children. In 1857 she founded the New York Infirmary for Indigent Women and Children. Apart from medical care, an important aspect was the training for women medical and nursing students. In 1859, while on a lecture tour of Britain, Blackwell become the first woman to be included in the United Kingdom's Medical Register. Back in America, in the face of continuing hostility to women from the established medical schools, Blackwell opened a Women's Medical College in 1868, with an initial intake of 15 students and nine faculty members. The following year she returned to live in England, leaving her sister Emily in charge of the College. Elizabeth went on to promote medical education in England, helping to set up the National Health Society and founding the London School of Medicine for Women. In 1875 she was appointed professor of gynecology at the London School of Medicine for Children.

Blaeu, Willem Janszoon (1521–1638)
Dutch cartographer

Blaeu (blow), who was born in Alkmaar, Holland, began work as a carpenter. In 1595 he spent a year with Tycho Brahe at his observatory at Hven. He opened his cartographic shop in Amsterdam soon after his return to Holland and specialized in producing globes. He published his first world map in 1605 and his first atlas in 1633.

After his death the business was carried on by his sons, Jan and Cornelis. In 1648 they produced their world map, which contained much that was new including the coastline of north and west Australia and parts of New Zealand and the Antarctic. It also abandoned the great southern continent previously believed to exist.

Blagden, Sir Charles (1748–1820) *English physician and chemist*

Blagden, born in Wotton under Edge in England, studied medicine at Edinburgh, where one of his professors was Joseph Black, and graduated in 1768. He became a medical officer in the British army in the same year and theoretically remained in that post until 1814. From 1782 to 1789 Blagden was assistant to Henry Cavendish, a post that involved him in the so-called "water controversy," a dispute between James Watt, Cavendish, and Antoine Lavoisier concerning the priority of the discovery of the synthesis of water from its elements. Blagden was friendly with the great French scientists of the day, especially Claude Berthollet, and on a visit to Paris in 1783 he told Lavoisier of Cavendish's synthesis, an experiment that Lavoisier repeated in Blagden's presence. Blagden became secretary of the Royal Society soon afterward, in which capacity he published Watt's papers on the same subject. The dispute was largely artificial because the three men drew different conclusions from their work.

Blagden's own scientific work was concerned with the freezing of mercury, the supercooling of water, and the freezing of salt solutions. He discovered, in 1788, that the lowering of the freezing point of a solution is proportional to the concentration of the solute present. This became known as *Blagden's law*. Blagden was knighted in 1792.

Blakeslee, Albert Francis (1874–1954)
American botanist and geneticist

Blakeslee was born at Geneseo in New York State and educated at Wesleyan University, Connecticut, graduating in 1896. He taught science for four years before entering Harvard to do postgraduate research, gaining his PhD in 1904. In this year he discovered that the bread molds (*Mucorales*) exhibit heterothallism (self-sterility) and spent the next two years in Germany making further investigations on the fungi.

From 1907 to 1914 Blakeslee was professor of botany at Connecticut Agricultural College. In 1915 he moved to the department of genetics at the Carnegie Institution, where he remained until 1941. In 1924 he began work on the alkaloid colchicine, which is found in the autumn crocus, and 13 years later he discovered that plants soaked in this alkaloid had multiple sets of chromosomes in their cells. Such plants, termed polyploids, often exhibit gigantism and this discovery proved of immediate use in the horticultural industry in producing giant varieties of popular ornamentals. More importantly, however, colchicine often converts sterile hybrids into fertile polyploids and is therefore an invaluable tool in crop-breeding research.

Other contributions made by Blakeslee to plant genetics include his study of inheritance in the jimson weed and his research on embryo culture as a method of growing hybrid embryos that would abort if left on the parent plant.

Blane, Sir Gilbert (1749–1834) *British physician*

Blane was born in Blanefield, Scotland, and studied medicine at Edinburgh and Glasgow, receiving his MD in 1778. In the following year he sailed with the fleet as personal physician to Admiral Lord Rodney and later as physician to the fleet. While in the West Indies, he made intelligent use of James LIND's results and introduced the provision of lime juice and other citrus fruits to combat scurvy among the seamen. He also generally improved the standards of hygiene on board ship.

He returned to London in 1783 and became a physician at St. Thomas's Hospital, London, and later attended both George IV and William IV as physician-in-ordinary. He was instrumental in enforcing his health regulations throughout the Royal Navy and helped draft the Quarantine Act of 1799. He was made a baronet in 1812.

His rather grave manner earned him the nickname of "chilblaine" among his colleagues.

Blau, Marietta (1894–1970) *Austrian–American physicist*

Noted for her pioneering work in the photographic detection of subatomic particles, Blau was born in Vienna, Austria, and attended the university there, receiving a PhD in 1919. After working briefly in Berlin and at the University-Institute of Frankfurt-am-Main, she returned to Vienna to work unpaid at the Radiuminstitut and the Second Physical Institute, from 1923 until 1938. During this time she relied on the financial support of her family, apart from a grant (1933–34) from the Aus-

trian Association of University Women, which permitted her brief spells working in Gottingen and at the Curie Institute, Paris. In the face of growing anti-Jewish feeling, Blau left Vienna just before the Nazis annexed Austria; initially she went to Oslo, then, on the recommendation of Albert EINSTEIN, to the Technical University, Mexico City (1939–44). She subsequently worked in the metallurgical industry (1944–48) in North America, and as a physicist at Columbia University (1948–50) and the Brookhaven National Laboratory (1950–55) before being appointed associate professor at the University of Miami (1955–60).

As early as 1925 in Vienna, Blau created photographic emulsions that could detect the tracks of high-energy protons, and in collaboration with Hertha WAMBACHER was the first (in 1932) to expose emulsions to neutron beams to determine the spectrum of neutrons arising from reactions in the emulsion. In 1937 the two women showed how cosmic rays cause the disintegration of heavy nuclei in photographic emulsions, producing so-called Blau–Wambacher stars. In North America, after World War II, Blau was involved in the early development of photomultiplier tubes, and continued her researches into elementary particles. In 1962 she was awarded the Erwin Schrödinger Prize, shared (posthumously) with Wambacher.

Bliss, Nathaniel (1700–1764) *British astronomer*

Bliss was born in Bisley, Gloucestershire, in western England and was educated at Oxford. In 1736 he was appointed rector of St. Ebbe's church, Oxford, and succeeded HALLEY in 1742 as Savillian Professor of Geometry, also in Oxford. He left Oxford in 1762 when he was appointed fourth Astronomer Royal in succession to James BRADLEY.

The tenure of his position at Greenwich was too short for Bliss to leave much of a mark on the Royal Observatory. He did, however, observe in 1761 that rare astronomical event, a transit of Venus, enabling him to calculate the horizontal parallax of the Sun as 10.3″ (compared with the modern figure of 8.8″).

Blobel, Günter (1936–) *German-born American cell biologist*

Blobel (**blo**-bel) was born in Waltersdorf, Silesia, Germany (now in Poland). He graduated in medicine from the University of Tübingen in 1960, and in 1962 secured a fellowship at the University of Wisconsin, Madison, to study oncology. After gaining his PhD in 1967, Blobel moved to the Cell Biology Laboratory at Rockefeller University, New York. He became assistant professor in 1969, associate professor in 1973, and professor in 1976.

Blobel's work has focused on how the many

different proteins made by living cells are sorted and distributed to their correct destinations. He followed on from pioneering studies of cell structure and protein secretion undertaken by his Rockefeller colleague and mentor George PALADE, helping to unravel some of the fundamental mechanisms of cell biology.

In 1971, Blobel proposed the "signal hypothesis," which postulated that proteins destined for transport out of the cell are given a special tag, rather like a baggage label, that directs the protein to its destination. His subsequent work described how this tag consists of a short sequence of amino acids – the signal peptide – that transiently forms part of the protein.

Proteins are assembled from amino acids by particles called ribosomes, which are associated with an intracellular network of membranous chambers called the endoplasmic reticulum (ER). The signal peptide instructs the ribosome to attach to the ER membrane over a membrane channel. The signal peptide leads the growing chain of amino acids (polypeptide) through the channel, until eventually the signal peptide is removed and the complete polypeptide is released into the lumen of the ER. From here it is ultimately transported out of the cell.

Blobel's team, as well as other research groups, went on to show that similar signal peptides are employed in targeting proteins to organelles (miniorgans) within the cell, such as mitochondria and chloroplasts. In 1980 Blobel proposed the general principle that proteins contain specific amino acid sequences (topogenic sequences) that determine not only whether they are exported from the cell or retained, but also their position and orientation in a particular target membrane. Subsequent work by various groups has revealed a variety of topogenic sequences, and established their universal importance to protein targeting in all cell types.

Blobel's work has provided insights into how certain diseases are caused by proteins having faulty address tags, which means that they fail to reach their proper destinations. It also paves the way for the development of drugs that can be targeted more precisely to particular cell locations.

For his discovery that proteins have intrinsic signals that govern their transport and localization in the cell, Blobel was awarded the 1999 Nobel Prize for physiology or medicine.

Bloch, Felix (1905–1983) *Swiss–American physicist*

Bloch (blok) was born in Zürich, Switzerland, and educated at the Federal Institute of Technology there and at the University of Leipzig, where he obtained his PhD in 1928. He taught briefly in Germany and in 1933 moved to America, via various institutions in Italy, Denmark, and Holland. In 1934 he joined the Stanford staff, remaining there until his retirement in 1971 and serving from 1936 onward as professor of physics. He also served briefly as first director of the international laboratory for high-energy physics in Geneva, known as CERN (1954–55).

In 1946, Bloch and Edward PURCELL independently introduced the technique of nuclear magnetic resonance (NMR). This utilizes the magnetic property of a nucleus, which will interact with an applied magnetic field such that it takes certain orientations in the field (a quantum mechanical effect known as space quantization). The different orientations have slightly different energies and a nucleus can change from one state to another by absorbing a photon of electromagnetic radiation (in the radiofrequency region of the spectrum). The technique was used initially to determine the magnetic moment (i.e., the torque felt by a magnet in a magnetic field at right angles to it) of the proton and of the neutron. It has since, however, been developed into a powerful tool for the analysis of the more complex molecules of organic chemistry. The energy states of the nucleus are affected slightly by the surrounding electrons, and the precise frequency at which a nucleus absorbs depends on its position in the molecule. In 1952 Bloch shared the Nobel Prize for physics with Purcell for this work on NMR.

Bloch worked extensively in the field of solid-state physics developing a detailed theory of the behavior of electrons in crystals and revealing much about the properties of ferromagnetic domains.

Bloch, Konrad Emil (1912–2000) *German–American biochemist*

Born in Neisse (now Nysa in Poland), Bloch (blok) was educated at the Technical University, Munich, and – after his emigration to America in 1936 – at Columbia University, New York, where he obtained his PhD in 1938. He then taught at Columbia until 1946, when he moved to the University of Chicago, becoming professor of biochemistry there in 1950. In 1954 Bloch accepted the position of Higgins Professor of Chemistry at Harvard, a post he retained until his retirement in 1978.

In 1940 the important radioisotope carbon-14 was discovered by Martin KAMEN and Samuel Ruben. Bloch was quick to see that it could be used to determine the biosynthesis of such complex molecules as cholesterol, a basic constituent of animal tissues characterized by four rings of carbon atoms. Thus in 1942, in collaboration with David Rittenberg, Bloch was able to confirm the earlier supposition that cholesterol was partly derived from the two-carbon acetate molecule.

The many steps through which acetate develops into the 27-carbon cholesterol took years of analysis to establish. The breakthrough came in 1953, when Bloch and R. Langdon identified squalene as an intermediate in cholesterol synthesis. Squalene, a terpene with an open chain of 30 carbon atoms, initiates the folding necessary to produce the four rings of cholesterol. For this work Bloch shared the 1964 Nobel Prize for physiology or medicine with Feodor LYNEN.

Bloembergen, Nicolaas (1920–)
Dutch–American physicist

Bloembergen (**bloom**-ber-gen) was born in Dordrecht in the Netherlands and was educated at the universities of Utrecht and Leiden, where he obtained his PhD in 1948. He moved to America soon afterward, joined the Harvard staff in 1949, and served from 1957 as Gordon McKay Professor of Applied Physics; from 1974 to 1980 he was also Rumford Professor of Physics. He became Gerhard Gade university professor in 1980, a post he held until his retirement in 1990.

In the mid 1950s Bloembergen introduced a simple yet effective modification to the design of the maser. First built by Charles TOWNES in 1953, the early maser could only work intermittently: once the electrons in the higher energy level had been stimulated they would fall down to the lower energy level and nothing further could happen until they had been raised to the higher level once more. Bloembergen developed the three-level and multilevel masers, which were also worked on by Nikolai BASOV and Aleksandr PROKHOROV in the Soviet Union. In the three-level maser, electrons are pumped to the highest level and stimulated. They consequently emit microwave radiation and fall down to the middle level where they can once more be stimulated and emit energy of a lower frequency. At the same time more electrons are being pumped from the lowest to the highest level making the process continuous. Bloembergen has worked extensively on nonlinear optics – i.e., on effects produced by high intensities of radiation. He has particularly investigated the use of lasers to excite or break particular bonds in a chemical reaction. For his work he shared the 1981 Nobel Prize for physics with Arthur SCHAWLOW (and Kai Siegbahn).

Bloembergen wrote *Nuclear Magnetic Relaxations* (1948) and *Nonlinear Optics* (1965).

Blumberg, Baruch Samuel (1925–)
American physician

Blumberg was born in New York City and studied physics and mathematics at Union College, Schenectady, and at Columbia, where, after a year, he changed to medical studies. He received his MD from Columbia in 1951 and his PhD in biochemistry from Oxford University in 1957. After working at the National Institutes of Health in Bethesda from 1957 until 1964 Blumberg was appointed professor of medicine at the University of Pennsylvania, a position he held until his retirement in 1994.

In 1963, while examining literally thousands of blood samples in a study of the variation in serum proteins in different populations, Blumberg made the important discovery of what soon became known as the "Australian antigen." He found in the blood of an Australian aborigine an antigen that reacted with an antibody in the serum of an American thalassemia patient. It turned out that the antigen was found frequently in the serum of those suffering from viral hepatitis, hepatitis B, and was in fact a hepatitis B antigen.

It was hoped that from this discovery techniques for the control of the virus would develop. It certainly made it easier to screen blood for transfusion for the presence of the hepatitis virus; it also permitted the development of a vaccine from the serum of those with the Australian antigen. Blumberg has also suggested that the virus is involved in primary liver cancer.

For his work on the Australian antigen Blumberg shared the 1976 Nobel Prize for physiology or medicine with Carleton GAJDUSEK.

Blumenbach, Johann Friedrich (1752–1840) *German anthropologist*

Blumenbach (**bloo**-men-bahk), the son of an assistant headmaster of the local gymnasium, was born in Gotha, Germany, and educated at the universities of Jena and Göttingen, where he obtained his MD in 1775. He remained at Göttingen for the whole of his career, serving as professor of medicine from 1778 onward.

In his *De generis humani variatate nativa* (1776; Concerning the Natural Diversity of the Human Race) Blumenbach took up the problem posed by LINNAEUS in 1747, which challenged scientists to find a generic character that would distinguish between men and apes. Blumenbach's solution was to attribute two hands to humans and four hands to monkeys. This enabled him to form a separate order, Bimanus, for man alone while including apes, monkeys, and lemurs in the Quadrumana.

In Bimanus he included five races – Caucasian, Ethiopian, Mongolian, Malayan, and American Indian – insisting that they were "all related and only differing from each other in degree." Nonetheless he favored a monogenetic view of mankind, which led him to see the Caucasian race as primary, with the other four races existing as "degenerate" forms. Blumenbach's version of this common 18th-century view was relatively mild. He argued that the degeneration was an acquired trait, the result of climatic and dietary influences. The rampant racism found in some of his contemporaries is on the whole absent from the works of Blumenbach. To show that "degenerate" forms were not the same as inferior forms he produced in 1799 a paper in which he provided biographies of those African poets, philosophers, and jurists who had established themselves in European society.

Bode, Johann Elert (1747–1826) *German astronomer*

Born in Hamburg, Germany, Bode (**boh**-de) was the director of the Berlin Observatory and popularized a discovery made earlier in 1772 by Johann Titius of Wittenberg. This was a simple but inexplicable numerical rule governing the distance of the planets from the Sun measured in astronomical units (the mean distance of the Earth from the Sun). The rule, known as *Bode's law*, is to take the series 0, 3, 6, 12, 24 ... , add 4 to each member, and divide by 10. The result is the distance in astronomical units of the planets from the Sun. The law and its application can be tabulated and, provided that the asteroids are counted as a single planet, quite an impressive fit can be achieved. It breaks down for Neptune and is hopelessly wrong for Pluto. It played a role in the discovery of Neptune by Urbain LE VERRIER in 1846. It is not known whether the law is simply a pure coincidence, or whether it is a consequence of the way in which the solar system formed.

Bodenstein, Max Ernst August (1871–1942) *German physical chemist*

Bodenstein (**boh**-den-shtIn), who was born in Magdeburg, Germany, gained his doctorate at Heidelberg (1893). He subsequently worked with Wilhelm OSTWALD at Leipzig before becoming a professor at Hannover (1908–23) and at the Institute for Physical Chemistry, Berlin (1923–36). He made a series of classic studies on the equilibria of gaseous reactions, especially that of hydrogen and iodine (1897). His technique was to mix hydrogen and iodine in a sealed tube, which he placed in a thermostat and held at a constant high temperature. The reaction eventually reached an equilibrium, at which the rate of formation of hydrogen iodide (HI) was equal to the rate of decomposition to the original reactants:

$$H_2 + I_2 \rightleftharpoons 2HI$$

The equilibrium mixture of H_2, I_2, and HI was "frozen" by rapid cooling, and the amount of hydrogen iodide present could be analyzed. Using different amounts of initial reactants, Bodenstein could vary the amounts present at equilibrium and verify the law of chemical equilibrium proposed in 1863 by Cato GULDBERG and Peter WAAGE.

Bodenstein also worked in photochemistry and was the first to show how the large yield per quantum for the reaction of hydrogen and chlorine could be explained by a chain reaction.

Boerhaave, Hermann (1668–1738) *Dutch physician and chemist*

> It will now perhaps be universally granted that our professor [Boerhaave] has indeed supplied us with the best system from an unparallel'd fund of medical learning happily digested.
> —W. Burton, *An Account of the Life and Writings of Hermann Boerhaave* (1743)

Boerhaave (**bor**-hah-ve) was born at Voorhout, near Leiden, in the Netherlands, and studied philosophy at the University of Leiden. He was intended to enter the Church, but an accusation of impiety turned him from theology to medicine and he obtained his MD in 1693 at Harderwijk. Boerhaave returned to Leiden where he began to teach medicine and chemistry and practiced medicine privately. Boerhaave was a polymath, learned in medicine, botany, and chemistry. His lectures were immensely successful and he occupied the chairs of medicine and botany (1709), practical medicine (1714), and chemistry (1718). Boerhaave was celebrated during his life for his medical works, *Institutiones medicae* (1708; Institutions of Medicine), and *Aphorismi de cognoscendis et curandis morbis* (1709; The Book of Aphorisms) but posthumously he exerted great influence as a chemist through his excellent work *Elementia chemiae* (1723; Elements of Chemistry). Boerhaave made no great discoveries but he had unerring judgment in selecting that which was most valuable in the work of others. He tried to apply the mechanistic philosophy of NEWTON and Robert BOYLE to both medicine and chemistry. His early work on calorimetry was the foundation on which Joseph BLACK subsequently built. He was elected a fellow of the Royal Society and during his lifetime was as famous as Newton.

Boethius, Anicius Manlius Severinus (c. 480–524) *Roman philosopher*

Boethius (boh-**ee**-thee-us), a member of the noble Anicii family, was born in a Rome conquered and ruled by the Ostrogoths. He flourished under the Ostrogothic king Theodoric, being appointed consul in 510 and *magister officiorum*, the head of the Imperial civil administration, in about 520. For reasons that are not clear Boethius was imprisoned in Pavia, tortured, and finally put to death.

As a scholar Boethius had expressed a wish to translate "the whole work of ARISTOTLE into the Roman idiom" but only succeeded in dealing with the logical works. He also produced works on arithmetic, geometry, astronomy, and music, later to form the basis for the teaching of the "quadrivium" (the course of studies in medieval universities), a term he may well have introduced. Such works were based closely on earlier Greek works of Nichomacus and EuCLID. Boethius is historically important for keeping the tradition of Greek thought and science alive in Europe until the works of the great Arab scholars were translated some six centuries later.

Bogoliubov, Nikolai Nikolaevich (1909–1992) *Russian mathematician and physicist*

Born in Nizhny Novgorod in Russia, Bogoliubov (bu-gu-**lyoo**-bof) was accepted for graduate work at the Academy of Sciences of the Ukrainian SSR, and subsequently worked there and at the Soviet Academy of Sciences.

His main contribution was in the application of mathematical techniques to theoretical physics. He developed a method of distribution-function for nonequilibrium processes, and also worked on superfluidity, quantum field theory, and superconductivity. His work was partly paralleled by the work of John BARDEEN, Leon COOPER, and John Robert SCHRIEFFER.

He was also active in founding scientific schools in nonlinear mechanics, statistical physics, and quantum-field theory. Bogoliubov was a prolific author, with many of his books on these topics translated into English. From 1963 he was academician-secretary of the Soviet Academy of Sciences and was made director of the Joint Institute for Nuclear Research in Dubna in 1965. The next year he was also made a deputy to the Supreme Soviet. He was honored by scientific societies throughout the world, and in his own country received the State Prize three times (1947, 1953, and 1984) and the Lenin Prize (1958).

Bohm, David Joseph (1917–1992) *American physicist*

> There are no things, only processes.
> —Quoted by C. H. Waddington in *The Evolution of an Evolutionist* (1975)

Bohm's father, an Austrian immigrant, ran a successful furniture business. Born in Wilkes-Barre, Pennsylvania, Bohm attended The Pennsylvania State University before moving to the University of California, Berkeley, where he gained his PhD in 1943. After a period working on the development of the atomic bomb under the supervision of OPPENHEIMER at the Radiation Laboratory, Berkeley, Bohm joined the Princeton physics faculty in 1947. Here he ran into political trouble; called to testify before the House Un-American Activities Committee he pleaded the Fifth Amendment and refused to give evidence against his colleagues. He was cited for contempt and threatened with prison. When his Princeton position expired in 1951, Bohm found himself unemployable in the U.S. Oppenheimer advised him to leave the country and he worked in Brazil at São Paulo Univer

sity (1951–55) and in Israel at Haifa (1955–57). He then settled in Britain, working first at Bristol University as a research fellow before being appointed professor of theoretical physics at Birkbeck College, London (1961), a post he held until his retirement in 1983.

In 1951 Bohm published a much respected textbook, *Quantum Theory*. He was, however, unhappy with the traditional account of quantum theory. His concern lay not with its lack of determinism, but with the fact that "it had no place in it for an adequate notion of an independent actuality." Electrons could be both wave and particle and, because of HEISENBERG's uncertainty principle, we can never simultaneously know an electron's position and momentum. One way around this difficulty is to suppose that quantum theory presents an imperfect view of nature, with a more complete and deterministic underlying reality. Such approaches, it was widely believed, had been shown by VON NEUMANN in 1932 to be incompatible with quantum theory. Against this Bohm argued that von Neumann's proof was entirely mathematical and consequently based on axioms and presuppositions that were always open to question. In the 1950s he began to seek for the "hidden variables," which would allow him better to understand quantum theory.

He proposed that the electron is a real particle with a definite position and momentum, but it is also connected with a "pilot wave." Bohm regarded this new wave as real and known only by its effects on the electron. Electrons, of course, in Bohm's theory still display wave–particle duality because of the effect of the pilot wave. As the wave reacts with both the electron and the environment, the wave–particle complex responds accordingly to a particular type of measurement. It is a necessary consequence of this view that signals can be conveyed instantaneously from the pilot wave to the electron.

In a later work, *Wholeness and the Implicate Order* (1980), Bohm sought to establish a more general position. Bohm was inspired by an analogy with a device that he saw at a science exhibition. Two concentric glass cylinders have a layer of glycerine between them. If a localized spot of ink is placed in the glycerine and one cylinder is rotated with respect to the other, the ink is smeared out and the spot of ink disappears (in Bohm's terminology, it is "enfolded" or "implicated"). Turning the cylinder in the opposite direction brings the ink back into a spot (it is "unfolded" or "explicate"). Thus while order may appear to have been lost in a system, it may in fact be enfolded in the system, and could be unfolded under the right conditions.

Bohm appears to have been the first person to suggest in a general way that holography might be a key feature of fundamental physics. He also made important contributions to the theory of collective effects of ions in plasmas and electrons in metals (in collaboration with David Pines) and, together with Yakir Aharonov, predicted an effect, known as the *Aharonov–Bohm effect*, which demonstrated the physical reality of vector potentials in electrodynamics (and more generally gauge theories).

Bohn, René (1862–1922) *German organic chemist*

Born in Dornach, France, Bohn (bohn) studied chemistry at the Federal Institute of Technology at Zürich and obtained his doctorate from the University of Zürich in 1883. From 1884 he was a chemist with the Badische Anilin und Soda Fabrik at Ludwigshafen, becoming a director in 1906. He worked mainly on the synthesis of anthraquinone dyes and his most notable synthesis was that of the blue vat dye, indanthrone (1901), the first nonindigo vat dye. He also did important work on the red alizarin dyes.

Bohr, Aage Niels (1922–   ) *Danish physicist*

Bohr (bor), the son of Niels Bohr, was born in the Danish capital, Copenhagen, and educated at the university there. After postgraduate work at the University of London from 1942 to 1945 he returned to Copenhagen to the Institute of Theoretical Physics, where he served as professor of physics from 1958 to 1981.

When Bohr began his research career the shell model of the nucleus of Maria GOEPPERT-MAYER and Hans JENSEN had just been proposed in 1949. Almost immediately Leo James RAINWATER produced experimental results at odds with the predictions derived from a spherical shell model, and proposed that some nuclei were distorted rather than perfectly spherical.

Bohr, in collaboration with Benjamin MOTTELSON, followed Rainwater's work by proposing their collective model of nuclear structure (1952), so called because it was argued that the distorted nuclear shape was produced by the participation of many nucleons. For this work Bohr shared the 1975 Nobel Prize for physics with Rainwater and Mottelson.

Bohr, Niels Hendrik David (1885–1962) *Danish physicist*

> When it comes to atoms, language can be used only as in poetry. The poet too is not nearly so concerned with describing facts as with creating images.
> —Quoted by J. Bronowski in *The Ascent of Man* (1975)

Niels Bohr came from a very distinguished scientific family in Copenhagen, Denmark. His father, Christian, was professor of physiology at Copenhagen and his brother Harald was a mathematician of great distinction. (His own

son, Aage, was later to win the 1975 Nobel Prize for physics.) Bohr was educated at the University of Copenhagen where he obtained his PhD in 1911. After four productive years with Ernest RUTHERFORD in Manchester, Bohr returned to Denmark becoming in 1918 director of the newly created Institute of Theoretical Physics.

Under Bohr (who after Albert EINSTEIN was probably the most respected theoretical physicist of the century) the institute became one of the most exciting research centers in the world. A generation of physicists from around the world were to pass through it and eventually it was to bestow on the orthodox account of quantum theory the apt description of the "Copenhagen interpretation."

In 1913 Bohr published a classic paper, *On the Constitution of Atoms and Molecules*, in which he used the quantum of energy introduced into physics by Max PLANCK in 1900, to rescue Rutherford's account of atomic structure from a vital objection and also to account for the line spectrum of hydrogen. The first problem Bohr faced was to explain the stability of the atom. Rutherford's 1911 model of the atom with electrons orbiting a central nucleus (the so-called planetary model) was theoretically unstable. This was because, unlike planets orbiting the Sun, electrons are charged particles, which, according to classical physics, should radiate energy and consequently spiral in toward the nucleus.

Bohr began by assuming that there were "stationary" orbits for the electrons in which the electron did not radiate energy. He further assumed that such orbits occurred when the electron had definite values of angular momentum, specifically values $h/2\pi$, $2h/2\pi$, $3h/2\pi$, etc., where h is Planck's constant. Using this idea he was able to calculate energies E_1, E_2, E_3, etc., for possible orbits of the electron. He further postulated that emission of light occurred when an electron moved from one orbit to a lower-energy orbit; absorption was accompanied by a change to a higher-energy orbit. In each case the energy difference produced radiation of energy $h\nu$, where ν is the frequency. In 1913 he realized that, using this idea, he could obtain a theoretical formula similar to the empirical formula of Johann BALMER for a series of lines in the hydrogen spectrum. Bohr received the Nobel Prize for physics for this work in 1922. The Bohr theory was developed further by Arnold SOMMERFELD.

Bohr also made other major contributions to this early development of quantum theory. The "correspondence principle" (1916) is his principle that the quantum-theory description of the atom corresponds to classical physics at large magnitudes.

In 1927 Bohr publicly formulated the "complementarity principle." This argued against continuing attempts to eliminate such supposed difficulties as the wave–particle duality of light

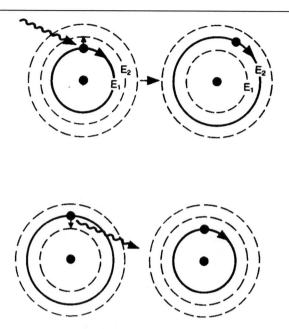

BOHR THEORY *In the Bohr theory of the atom, initially the atom has an electron in its inner orbit and a vacant outer orbit (top). In absorbing a photon, the electron moves to an outer orbit and the atom is in an excited state. Emission (bottom) is the opposite process. The photon energy ν is given by hν = E₂ – E₁.*

and many other atomic phenomena. His starting point was the impossibility to distinguish satisfactorily between the actual behavior of atomic objects and their interaction with the measuring instruments that serve to define the conditions under which the phenomena appear. Examine light with one instrument, the argument went, and it undulates like a wave; select another and it scatters like a particle. His conclusion was that evidence obtained under different experimental conditions cannot be comprehended within a single picture, but must be regarded as complementary in the sense that only the totality of the phenomenon exhausts the possible information about the objects. It was a principle Bohr remained faithful to, even representing it on his coat of arms in 1947 with the motto *Contraria sunt complementa* (Opposites are complements) above the Yin/Yang symbols. Together with the indeterminancy principle of Werner HEISENBERG and the probability waves of Max BORN, this principle emerged from the 1930 Solvay conference (the last one Einstein attended) as the most authoritative and widely accepted theory to describe atomic phenomena.

Bohr also made major contributions to the work on radioactivity that led to the discovery and exploitation of nuclear fission. Bohr's liquid-drop model of the nucleus, which was published in 1936, provided the basis for the first theoretical account of fission worked out in col-laboration with John WHEELER in 1939. It was also Bohr who, in 1939, made the crucial suggestion that fission was more likely to occur with the rarer isotope uranium–235 than the more common variety uranium–238.

In 1943 Bohr, who had a Jewish mother, felt it necessary to escape from occupied Denmark and eventually made his way to Los Alamos in America where he served as a consultant on the atomic bomb project. He was quick to appreciate the consequences of using such weapons and in 1944 made an early approach to Roosevelt and Churchill proposing that such obvious danger could perhaps be used to bring about a rapprochement between Russia and the West. Scientists were in a unique position, he argued, in having the Soviet contacts and the knowledge to make the first approach. Much of Bohr's time after the war was spent working, among scientists, for adequate controls of nuclear weapons and in 1955 he organized the first Atoms for Peace conference in Geneva.

Bok, Bart Jan (1906–1983) *Dutch–American astronomer*

Bok (bok), who was born at Hoorn in the Netherlands, studied at the universities of Leiden (1924–27) and Groningen (1927–29) and obtained his PhD from Groningen in 1932. He had moved to America in 1929, becoming naturalized in 1938, and served at Harvard from 1929 to 1957 with the appointment of professor of astronomy from 1947 onward. Bok spent the period 1957–66 in Australia as director of the Mount Stromlo Observatory, Canberra, and professor of astronomy at the Australian National University. He returned to America in 1966 to become director of the Steward Observatory, Arizona, until 1970 and professor of astronomy (from 1974 emeritus professor) at the University of Arizona, Tucson.

Bok's major interest was the structure of our galaxy, the Milky Way. With his wife, Priscilla, he published a survey of the subject: *The Milky Way* (1941). Although it had been long assumed that the Milky Way had a spiral structure it was not until Walter BAADE identified in the 1940s the hot young O and B stars of the Andromeda galaxy as spiral markers that such a conjecture could be confirmed. The actual structure was first worked out in some detail by William MORGAN. The existence of 21-centimeter radio signals from clouds of neutral hydrogen in the galaxy was predicted by Hendrik VAN DE HULST and their discovery in 1951 provided a second tracer. It was clear to Bok by the late 1950s that the radio data, which were expected to support the optical picture, instead contradicted it. He consequently attempted to harmonize the two structures by modifying Morgan's somewhat elliptical arms, making them much more spherical, and giving more emphasis to the Carina–Centaurus arm.

Bok's name is also associated with his discovery in 1947 of small dark circular clouds visible against a background of stars or luminous gas and since known as *Bok globules*. Since they are thought to be precursors of stars, as Bok himself conjectured, they have received considerable attention in recent years.

Boltwood, Bertram Borden (1870–1927)
American chemist and physicist

Boltwood, the son of a lawyer, was born in Amherst, Massachusetts, and educated at Yale and the University of Munich. Apart from the period 1900–06, when he served as a private consultant, and the year 1909–10, which he spent with Ernest RUTHERFORD at the University of Manchester, England, he devoted the whole of his academic career to Yale. He occupied the chair of physics (1906–10), the chair of radioactivity (1910–18), and the chair of chemistry from 1918 until his death by suicide in 1927.

Boltwood made a number of contributions to the study of radioactivity. The radioactivity of uranium and radium had been discovered in the 1890s by Henri BECQUEREL and Marie CURIE. Starting in 1902 Rutherford and Frederick SODDY had shown that radium, uranium, and other radioactive elements broke down in a quite complicated sequence into other elements. Boltwood worked on the breakdown of uranium into radium, a process that Soddy had not found easy to demonstrate. Soddy had tried to obtain radium directly from uranium in 1904 and failed. Boltwood postulated that this was because uranium did not decay directly into radium but into some intermediate element, and began to search for it. After much effort Boltwood eventually found what he was looking for in "actinium X," which, as it appeared different from anything else, he felt confident enough to claim as a new element and named it "ionium" in 1907. This claim ran into trouble when ionium was found to behave very much like thorium and in 1908 it was shown by B. Keetman that if ionium and thorium are mixed together no chemical technique can separate them. Soddy decided the matter in 1913 when he was able to obtain a spectrograph of ionium and found it to be the same as thorium. Although he was wrong in detail, the general picture Boltwood had developed of the decay of uranium to radium was valid until superseded by Soddy's idea of the isotope.

One important byproduct of Boltwood's work was his demonstration in 1905 that lead was always found in uranium and was probably the final stable product of its decay. He argued that in minerals of the same age the lead–uranium ratio would be constant, and in minerals of different ages the ratio would be different. He calculated some estimates of the ages of several rocks based on the estimates then accepted for decay rates and came up with some good results. This was the beginning of attempts to date rocks and fossils by radiation measurements and other physical techniques, which have revolutionized geology and archeology.

Boltzmann, Ludwig Edward (1844–1906) *Austrian theoretical physicist*

> Who sees the future? Let us have free scope for all directions of research; away with all dogmatism, either atomistic or antiatomistic!
> —*Lectures on the Theory of Gases,* Vol. I (1896)

> In my opinion it would be a great tragedy for science if the theory of gases were temporarily thrown into oblivion because of a momentary hostile attitude toward it, as was for example the wave theory because of Newton's authority.
> —*Lectures on the Theory of Gases*, Vol. II (1898)

Boltzmann (**bohts**-mahn) studied at the university in his native city of Vienna, where he received his doctorate in 1866. He held professorships in physics or mathematics at Graz (1869–73; 1876–79), Vienna (1873–76; 1894–1900; 1902–06), Munich (1889–93), and Leipzig (1900–02).

Boltzmann made important contributions to the kinetic theory of gases. He developed the law of equipartition of energy, which states that the total energy of an atom or molecule is, on average, equally distributed over the motions (degrees of freedom). He also produced an equation showing how the energy of a gas was distributed among the molecules (called the *Maxwell–Boltzmann distribution*).

Boltzmann also worked on thermodynamics, in which he developed the idea that heat, entropy, and other thermodynamic properties were the result of the behavior of large numbers of atoms, and could be treated by mechanics and statistics. In particular, Boltzmann showed that entropy – introduced by Rudolf CLAUSIUS – was a measure of the disorder of a system. *Boltzmann's equation* (1896) relates entropy (S) to probability (p): $S = k\log p + b$. The constant k is known as *Boltzmann's constant* and has the value 1.38054×10^{-23} joule per kelvin. The equation is engraved on his gravestone.

Boltzmann's work in this field was heavily criticized by opponents of atomism, particularly Wilhelm OSTWALD. It did however lead to the science of statistical mechanics developed later by Josiah Willard GIBBS and others. Boltzmann also worked on electromagnetism. He is further noted for a theoretical derivation of the law of radiation discovered by Josef STEFAN.

Toward the end of his life he suffered from illness and depression, and committed suicide in 1906.

Bolyai, Janos (1802–1860) *Hungarian mathematician*

> I have created a new and different world out of nothing.
> —Describing his discovery of non-Euclidean geometry in a letter to his father, Farkas Bolyai (1823)

Bolyai (**boh**-lyoi), who was born in Koloszvár (now Cluj) in Romania, was the son of Farkas Bolyai, a distinguished mathematician who had an obsession with the status of EUCLID's famous parallel postulate and devoted his life to trying to prove it. Despite his father's warnings that it would ruin his health, peace of mind, and happiness, Janos too started working on this axiom until, in about 1820, he came to the conclusion that it could not be proved. He went on to develop a consistent geometry in which the parallel postulate is not used, thus establishing the independence of this axiom from the others. His discovery was published in an appendix to a treatise by his father, Farkas Bolyai, dated 1829 but appearing in 1832. Although his discovery had been anticipated by Nikolai LOBACHEVSKY and Karl GAUSS he was unaware of their work.

The discovery of the possibility of non-Euclidean geometries had a tremendous impact on both mathematics and philosophy. In mathematics it opened the way for a far more general and abstract approach to geometry than had previously been pursued, and in philosophy it settled once and for all the arguments about the supposed privileged status of Euclid's geometry. Bolyai also did valuable work in the theory of complex numbers.

Bond, George Phillips (1825–1865) *American astronomer*

> There is nothing, then, so extravagant in predicting a future application of photography on a most magnificent scale ... What more admirable method can be imagined for the study of the orbits of the fixed stars and for resolving the problem of their annual parallax.
> —Diary entry, 1857

George Bond, who was born in Dorchester, Massachusetts, spent most of his early life assisting his father, William Bond, whom he succeeded as director of the Harvard Observatory. He therefore contributed to most of his father's observational and photographic work. Apart from the joint discovery of Hyperion with his father he is best known for showing how stellar magnitude could be calculated from photographs. In 1857 he noted that the size of the image is relative to the brightness of the star and the length of the exposure. It is this basic fact that has been used by the compilers of the Astrographic Catalog to record measurements of stellar magnitudes. He was also the first to photograph a double star, Mizar, in 1857.

Bond, William Cranch (1789–1859) *American astronomer*

Bond, who was born in Portland, Maine, was probably the first American astronomer to make a European reputation for himself. He was a watchmaker whose passion for astronomy was awakened by an eclipse, witnessed in 1806. He built his own observatory, which soon became one of the best in America. In 1847 the self-educated craftsman was invited to be director of the Harvard Observatory.

Bond's achievements were mainly observational and photographic. He made the first photograph of a star (Vega) in 1850, and was also the first to photograph the Moon. In 1848 with his son George Bond he discovered the eighth satellite of Saturn, Hyperion, and in 1850 he detected the third ring of Saturn between the two brighter ones – the so-called "crepe" ring. He was succeeded by his son as director of the observatory.

Bondi, Sir Hermann (1919–2005) *British–Austrian mathematician and cosmologist*

> Science is above all a cooperative enterprise.
> —Nature (1977)

Bondi (**bon**-dee) was born in Vienna, Austria, and studied at Cambridge University, where he later taught. In 1954 he moved to London to take up the chair in mathematics at King's College. Bondi was always actively interested in the wider implications of science and the scientific outlook; he was a member of the British Humanist Association and the Science Policy Foundation. He served as chief scientific adviser to the Ministry of Defence (1971–77), chief scientist at the Department of Energy (1977–80), and chairman of the Natural Environment Research Council (1980–84). He was knighted in 1973 and served as master of Churchill College, Cambridge, from 1983 to 1990.

Bondi's most important work was in applied mathematics and especially in cosmology. In collaboration with Thomas GOLD, he propounded, in 1948, a new version of the steady-state theory of the universe. This, among other topics, forms the substance of Bondi's book *Cosmology* (1952). The idea of a steady-state theory had first been suggested by Fred HOYLE, the purpose being to devise a model of the universe that could accommodate both the fact that the universe is the same throughout, and yet is expanding. Bondi and Gold's model was innovatory in postulating that there is continuous creation of matter in order to maintain the universe's homogeneity despite its expansion. Although it enjoyed considerable popularity, Bondi and Gold's steady-state model is now considered to have been decisively refuted by observational evidence and the big-bang theory is favored.

Bonner, James Frederick (1910–1996)
American biologist

Born at Ansley, Nebraska, Bonner graduated in chemistry from the University of Utah in 1931 but turned to biology under the influence of Theodosius DOBZHANSKY. He received his PhD from the California Institute of Technology (Cal Tech) in 1934, which was then becoming known as the main center for molecular biology. Here he became interested in developmental biology and the question of why only some genes of the chromosome complement of an organism are expressed in any one cell. He discovered that histone, a protein that is found associated with the chromosomes, is responsible for shutting off gene activity, and that if the histone is removed then the repressed genes become functional again. He also discovered that certain hormones act by repressing and derepressing genes.

Bonner in addition conducted research on the artificial synthesis of ribonucleic acid and studied ribosomes and mitochondria. From 1946 to 1981 he was professor of biology at Cal Tech, becoming Professor Emeritus in 1981. He wrote many books, including *The Nucleohistones* (1964) and *The Molecular Biology of Development* (1965).

Bonnet, Charles (1720–1793) *Swiss naturalist*

> Insects held my attention for some years. The strenuousness with which I worked on this study strained my eyes to such an extent that I was forced to interrupt it. Deprived thus of what had so far been my greatest pleasure, I tried to console myself by changing subjects. I then turned toward the physics of plants – a matter less animated, less fertile in discoveries, but of a more generally recognized usefulness.
> —*Recherches sur l'usage des feuilles dans les plantes* (1754; Researches on the Use of Leaves in Plants)

Born in Geneva, Switzerland, Bonnet (bo-**nay**) studied law, gaining his doctorate in 1743. In the same year he was elected a fellow of the Royal Society for his work on regeneration in lower animals and his demonstration of breathing pores (stigmata or spiracles) in caterpillars and butterflies. He is chiefly remembered however for discovering parthenogenesis (reproduction without fertilization) in the spindle-tree aphid and for the ideas on evolution he proposed following this observation.

Bonnet believed all organisms are preformed and that the germs of every subsequent generation are contained within the female. Such thinking implied that species remain constant, leaving Bonnet to explain how species become extinct as evidenced by fossil remains. He argued that the Earth had experienced periodic catastrophes, each destroying many life forms, but the remaining species all evolved to some degree. (Bonnet was the first to use the term "evolution" in a biological context.) Thus after the next catastrophe apes progress to men, and men become angels. The catastrophism theory was adopted by Georges CUVIER, and strongly influenced geological thinking until the 1820s.

Boole, George (1815–1864) *British mathematician*

> Pure mathematics was discovered by Boole in a work which he called "The Laws of Thought"... His work was concerned with formal logic and this is the same thing as mathematics.
> —Bertrand Russell

Boole came from a poor background in the English city of Lincoln and was virtually self-taught in mathematics. He discovered for himself the theory of invariants. Before he obtained an academic post Boole spent several years as a school teacher, first in Yorkshire and later at a school he opened himself. In 1849 he became professor of mathematics at Queen's College, Cork, Ireland.

Boole's main work was in showing how mathematical techniques could be applied to the study of logic. His book *The Laws of Thought* (1854) is a landmark in the study of logic. Boole laid the foundations for an axiomatic treatment of logic that proved essential for the further fundamental developments soon to be made in the subject by such workers as Gottlob FREGE and Bertrand Russell.

Boole's own logical algebra is essentially an algebra of classes, being based on such concepts as complement and union of classes. His work was an important advance in considering algebraic operations abstractly – that is, studying the formal properties of operations and their combinations without reference to their interpretation or "meaning." Fundamental formal properties like commutativity and associativity were first studied in purely abstract terms by Boole.

Boole's work led to the recognition of a new and fundamental algebraic structure, the *Boolean algebra*, alongside such structures as the field, ring, and group. The study of Boolean algebras both in themselves and their application to other areas of mathematics has been an important concern of modern mathematics. Boolean algebras find important applications in such diverse fields as topology, measure theory, probability and statistics, and computing.

Boot, Henry Albert Howard (1917–1983) *British physicist*

Born in the industrial city of Birmingham, England, the son of an electrical engineer, Boot was educated at the university there, completing his PhD in 1941. He was immediately recruited by his head of department, Marcus OLIPHANT, to work on the generation of microwaves for use in radar sets.

Radar operates by reflecting radio waves off objects, but only a tiny part of the radiation transmitted was reflected back, and only a tiny part of this was picked up by the receiving antenna. For the system to work, high power must be transmitted. Moreover, to ensure that the direction of the target can be precisely identified a narrow beam must be used. Some early transmitters used a wavelength as long as 15 meters and spread the signal over a 100° sector. The demand was for "centrimetric radar"; that is, a high-intensity radar beam with a wavelength of a centimeter or two.

The solution came from Boot and his colleague John RANDALL in November 1939 with the *cavity magnetron*, later to be described as the most important invention of the war. The device consists of a block of copper with accurately machined cavities, such that electrons can circulate in a magnetic field through the cavities and thereby generate electromagnetic radiation. When tested, the magnetron delivered 400 watts on a wavelength of 9 centimeters. It was passed to the research department of General Electric who increased its power to 10 kW. Magnetron radar was used first by Bomber Command but soon became available to fighters and convoy escorts.

Boot remained in the Scientific Civil Service after the war. He worked mainly on microwaves

at SERL, the Services Electronics Research Laboratory at Baldock, Hertfordshire, until his retirement in 1977.

Bopp, Thomas (1949–) *American astronomer*

Thomas Bopp was born in Denver, Colorado. His interest in astronomy led him to join a local informal group of amateur astronomers in the Phoenix area. Here, on July 22, 1995, he was the first to see a new comet, now known as *Hale–Bopp*. The comet was independently found by the astronomer Alan HALE.

Borda, Jean-Charles de (1733–1791) *French mathematician and nautical astronomer*

Born in Dax, France, Borda (bor-**da**) was educated at La Flèche and entered the army as a military engineer. He entered the navy in 1767 and participated in a number of scientific voyages. In 1782 he was captured by the English while participating in the American War of Independence.

His main scientific work was in such practical areas of science as fluid mechanics and in the design of instruments for geodesy and navigation. His work in fluid mechanics had a strongly experimental slant and he applied his discoveries to such practical military problems as the design of artillery, ships, and pumps. He is notable for having shown that the theory of fluid resistance put forward by NEWTON was erroneous. Borda was also involved in instituting the metric system of weights and measures and introduced the term "meter." He made contributions to the calculus of variations and was notable for his hostility to religion.

Bordet, Jules Jean Baptiste Vincent (1870–1961) *Belgian immunologist*

Bordet (bor-**day**) was born in Soignies, Belgium, and graduated in medicine from Brussels University in 1892. In 1894 joined the Pasteur Institute, Paris, where he worked under the bacteriologist Elie METCHNIKOFF. In collaboration with Octave Gengou, Bordet discovered that in an immunized animal the antibodies produced by the immune response work in conjunction with another component of blood (which Bordet termed "alexin" but which is now called "complement") to destroy foreign cells that invade the body. This component, Bordet found, was present in both immunized and non-immunized animals and was destroyed by heating to over 55°C. This work formed the basis of the *complement-fixation test*, a particularly sensitive means of detecting the presence of any specific type of cell or its specific antibody. A notable application of this was the test to detect syphilis devised by August von Wasserman.

In 1901 Bordet left Paris to found and direct

the Pasteur Institute in Brussels and in 1907 he was appointed professor of pathology and bacteriology at Brussels University. In 1906 Bordet isolated the bacterium responsible for whooping cough, which is named after him: *Bordetella (Haemophilus) pertussis.* For his discovery of complement and other contributions to medicine, he was awarded the 1919 Nobel Prize in physiology or medicine.

Borelli, Giovanni Alfonso (1608–1679)
Italian mathematician and physiologist

Borelli (bo-**rel**-ee) was born in Naples. His mathematical training – he was professor at Messina and Pisa – led him to apply mathematical and mechanical laws to his two main interests, astronomy and animal physiology. He rightly explained muscular action and the movements of bones in terms of levers, and also carried out detailed studies of the flight mechanism of birds. However, his extension of such principles to internal organs, such as the heart, stomach, and lungs, overlooked the essential chemical actions that take place in these organs. Borelli's *De motu animalium* (1680; On the Movement of Animals), which includes his theory of blood circulation, is thus in part erroneous.

In astronomy, in his *Theoricae mediceorum planetarum* (Theory of the Medicean Planets; 1666), Borelli presented a novel and influential account of the motions of the Medicean satellites around their parent plant Jupiter. He accounted for their elliptical orbits in terms of two distinct motions. The first, "perpetual and uniform," whereby the satellite is attracted rectilinearly to Jupiter as iron moves in a straight line to a magnet; the second, and "directly contrary ... continually decreasing," arises from the manner in which the satellite "is driven out from the sun by the force of its circular motion." NEWTON was aware of Borelli's work and appreciated the originality of his approach in the use of elliptical orbits and also his appreciation that orbital motion was complex.

Borelli was also one of the first astronomers, in his *Del movimento della cometa di Decembre 1664* (1665; The Motion of the Comet of December 1664), to propose, on the basis of observations and calculations, that comets also move in elliptical orbits. Earlier workers, including KEPLER, had taken comets to be transient visitors passing through the solar system in a straight line. As the church opposed such views, Borelli chose to publish under the pseudonym Pier Maria Mutoli.

Borlaug, Norman Ernest (1914–)
American agronomist and plant breeder

> Dr Borlaug has saved more lives than any person who has ever lived....
> —Congressional Tribute to Dr Norman E.

Borlaug Act, passed unanimously by the US Senate in 2006.

Borlaug (**bor**-lawg) was born in Cresco, Iowa, and graduated in forestry from Minnesota University in 1937, gaining his doctorate in plant pathology in 1941. He then spent three years with the Du Pont Chemical Company, testing the effects of chemicals on plants and plant diseases. In 1944 he joined the newly formed International Maize and Wheat Improvement Center in Mexico and began the breeding work that was to produce the highly adaptable dwarf wheats that played so large a part in the "Green Revolution" of the late 1960s and early 1970s.

Borlaug's high-yielding cereals increased agricultural production in the developing countries to the extent that many became self-sufficient for grain. For his major role in temporarily alleviating world famine, Borlaug was awarded the Nobel Peace Prize in 1970. He is still actively campaigning for agricultural improvement in the developing world, despite some criticisms from environmentalists. In 2007 he was presented with the US Congressional Gold Medal.

Born, Max (1882–1970) *German physicist*

> I am now convinced that theoretical physics is actual philosophy.
> —*Autobiography*

Born (born) was the son of an embryologist, a professor of anatomy at the University of Breslau (now Wrocław) in Poland. He was educated at the university in his native city of Breslau, and at the universities of Heidelberg, Zurich, and Göttingen, where he obtained his PhD in 1907. From 1909 until 1933 he taught at Göttingen, being appointed professor of physics in 1921. With the rise of Hitler he moved to Britain, and from 1936 served as professor of natural philosophy at the University of Edinburgh, returning to Germany on his retirement in 1953.

Born's early work was on crystals, particularly the vibrations of atoms in crystal lattices. The *Born–Haber cycle* is a theoretical cycle of reactions and changes by which it is possible to calculate the lattice energy of ionic crystals. He is noted for his role in the development of the new quantum theory. Together with Pascual JORDAN, he developed (1925) the matrix mechanics introduced by Werner HEISENBERG. He also showed how to interpret the theoretical results of Louis de BROGLIE and the experiments of such people as Clinton J. DAVISSON, which showed that particles have wavelike behavior.

At the time, it was known that in some circumstances light, electrons, etc., behaved as waves whereas in others they acted like particles. (William BRAGG once suggested using the corpuscular (particle) theory on Monday, Wednesday, and Friday, and the undulatory (wave) theory on Tuesday, Thursday, and Saturday.) Mathematical treatments could be used to predict behavior, but there was a problem in finding some accepted physical picture of how electrons, for instance, could act in this way. Erwin SCHRÖDINGER, who developed wave mechanics, interpreted particles as "wave packets," but this was unsatisfactory because such packets would dissipate in time. Born's interpretation was that the particles exist but are "guided" by a wave. At any point, the amplitude (actually the square of the amplitude) indicates the probability of finding a particle there.

An essential part of this idea of electrons, atoms, etc., is that it depends on probability – there is no predetermined way in which absolute predictions can be made, as in classical physics. A similar result is embodied in the uncertainty principle of Werner Heisenberg. EINSTEIN, among others, could never accept this and Born corresponded with him on the subject (the *Born–Einstein letters* were published in 1971).

After the development of quantum mechanics Born returned to the study of lattice dynamics and explained it fully in quantum mechanical terms. He was a prolific author and produced classic works on atomic physics, lattice dynamics, quantum mechanics, and optics as well as several popular books and an autobiography.

Born shared (with Walter Bothe) the 1954 Nobel Prize for physics. He is buried in Göttingen, where his tombstone displays his fundamental equation of matrix mechanics:

$$pq - qp = h/2\pi i$$

Borodin, Alexsander Porfirevich (1834–1887) *Russian chemist and musician*

Born the illegitimate son of a Russian prince in St. Petersburg, Borodin (**bor**-o-deen) was educated at the Medical Academy there from 1850 to 1856, where he showed a strong preference for chemistry over medicine. He traveled widely in Europe for several years and attended, with MENDELEEV, the famous First International Congress of Chemists at Karlsruhe in 1860 as a member of the Russian delegation. On his return to Russia, he was appointed to the faculty of the St. Petersburg Academy where he became professor of chemistry in 1864.

Although known as a composer, Borodin actually spent most of his time working as an organic chemist. In 1861 he developed a method for the fluorination of organic compounds. He later, in 1876, devised a widely used method for the quantitative determination of urea. His musical works include two symphonies, the tone poem *In the Steppes of Central Asia* (1880), and the opera *Prince Igor*, which was finished after his death by Rimsky-Korsakov and Glazunov.

Bosch, Carl (1874–1940) *German industrial chemist*

Born in Cologne, Germany, Bosch (bosh) was trained as both metallurgist and chemist and gained his doctorate under Johannes WISLICENUS at Leipzig (1898). He joined the large German dyestuffs company, Badische Anilin und Soda Fabrik (BASF), in 1899. Following Fritz HABER's successful small-scale ammonia synthesis in 1909, Bosch began to develop a high-pressure ammonia plant at Oppau for BASF. The plant was opened in 1912 – a successful application of the Haber process on a large scale. Bosch also introduced the use of the water-gas shift reaction as a source of hydrogen for the process:

$$CO + H_2O = CO_2 + H_2$$

After World War I the large-scale ammonia fertilizer industry was established and the high-pressure technique was extended by BASF to the synthesis of methanol from carbon monoxide and hydrogen in 1923. Bosch was chairman of BASF's successor, IG Farben (1935–40) and concurrently director of the Kaiser Wilhelm institutes. He shared the Nobel Prize for chemistry with Friedrich BERGIUS in 1931.

Bose, Sir Jagadis Chandra (1858–1937) *Indian plant physiologist and physicist*

Bose (bohs), who was born in Mymensingh (which is now in Bangladesh), began his studies in London as a medical student. He then won a scholarship to Cambridge University, from where he graduated in natural sciences in 1884. He was appointed professor of physical science at Presidency College, Calcutta, in 1885 and retained this post until 1915. In 1917 he founded and became director of the Bose Research Institute, Calcutta. He was knighted in 1917 and in 1920 became the first Indian to be elected a fellow of the Royal Society.

Bose's early research was on the properties of very short radio waves – work in which he showed their similarity to light. He also designed an improved version of Oliver LODGE's coherer, then used to detect radio waves, and as a result was able to put forward a general theory of the properties of contact-sensitive materials.

His most famous work concerned his investigations into plant physiology and the similarities between the behavioral response of plant and animal tissue. By devising extremely sensitive instruments he was able to demonstrate the minute movements of plants to external stimuli and to measure their rate of growth. While his experimental skill was widely admired, this work did not at the time gain universal acceptance.

Bose, Satyendra Nath (1894–1974) *Indian physicist*

Bose was educated at Presidency College, in his native Calcutta. Among his teachers was the eminent Indian physicist Jagadis Chandra Bose. Bose held the post of lecturer at the Calcutta University College of Science from 1917 until he left in 1921 to become a reader in physics at the new University of Dacca in East Bengal. His work ranged over many aspects of physics, among them statistical mechanics, the electromagnetic properties of the ionosphere, theories of x-ray crystallography, and unified field theory. However it is for his work in quantum statistics that he is best known.

Bose attracted the attention of Albert EINSTEIN and other European physicists by publishing a paper in 1924 in which he was able to derive Max PLANCK's black body radiation law, but without using the classical electrodynamics as Planck himself had done. On the strength of this work Bose was able to get two years' study leave in Europe and during his visit he came into contact with many of the great physicists of the day, such as Louis de BROGLIE, Max BORN, and Einstein. Einstein's generalization of Bose's work led to the system of statistical quantum mechanics now known as *Bose–Einstein statistics* and the prediction of a phase transition at low temperatures to a state known as a *Bose–Einstein condensate*. This state was first observed in the late 20th century. The Bose–Einstein system of statistics contrasts with the rival Fermi–Dirac statistics in that it applies only to particles not limited to single occupancy of the same state, i.e., particles (known as bosons) that do not obey the Pauli exclusion principle.

Boss, Lewis (1846–1912) *American astronomer*

> The standard positions and proper motions of all stars brighter than the seventh magnitude, extending from the north to the south pole, and some thousands of additional fainter stars promising to yield reasonably accurate proper motions.
> —Describing the contents of the *General Catalogue* (1937)

Boss, who was born in Providence, Rhode Island, studied at Dartmouth College where he graduated in 1870. In 1872 he was appointed assistant astronomer to the 49th parallel survey for the American–Canadian boundary, accurately locating stations from which the surveyors could work with confidence. In 1876 Boss was appointed director of the Dudley Observatory, Albany, a post he held until his death in 1912.

While working on the parallel survey Boss became aware of the many errors made in the measurement of stellar positions, which thus caused the current catalogs to be inaccurate. He consequently made his own observations, publishing the positions of 500 stars in 1878. He then undertook the observation and reduction of a zone for the Leipzig *Astronomische Gesellschaft* and also began work on his own catalog. This was published in 1910 as the *Preliminary General Catalogue* containing the position and proper motion of 6,188 of the brighter stars. His work was extended by his son, Benjamin Boss, who published in 1937 the *General Catalogue* containing comparable details of 33,342 stars.

Bothe, Walther Wilhelm Georg Franz (1891–1957) *German atomic physicist*

Bothe (**boh**-te), who was born in Oranienburg, Germany, studied at the University of Berlin under Max PLANCK and received his PhD in 1914. For the next few years, he was a prisoner of war in Russia but, on his return to Germany in 1920, he started teaching at Berlin and worked in Hans GEIGER's radioactivity laboratory.

He devised the "coincidence method" of detecting the emission of electrons by x-rays in which electrons passing through two adjacent Geiger tubes at almost the same time are registered as a coincidental event. He used it to show that momentum and energy are conserved at the atomic level. In 1929 he applied the method to the study of cosmic rays and was able to show that they consisted of massive particles rather than photons. For this research he shared the 1954 Nobel Prize for physics with Max BORN.

By 1930 his reputation was established and he was appointed professor of physics at Giessen. The same year he observed a strange radiation emitted from beryllium when it was exposed to alpha particles. This radiation was later identified by CHADWICK as consisting of neutrons.

While director of the Max Planck Institute in Heidelberg, Bothe supervised the construction of Germany's first cyclotron. This work was finished in 1943 and during World War II he led German scientists in their search for atomic energy. When the war ended he was given the chair of physics at Heidelberg, which he retained until his death.

Boucher de Crevecoeur de Perthes, Jacques (1788–1868) *French archeologist*

Boucher de Perthes (boo-**shay** de pairt) was born in Rethel, France, the son of a customs official, and followed in his father's profession. In 1825 he was appointed director of the Abbeville customs, a post he occupied for his entire career.

His hobby was investigating the fossil-rich beds of the nearby Somme Valley and in *Antiquités celtiques et antédiluviennes* (3 vols. 1847–64; Celtic Antediluvian Antiquities) Boucher de Perthes first revealed the existence of a prehistoric world occupied by man. He reported finding such objects as polished stone axes, which he attributed to the people he called "celtiques," and also older tools linked with the remains of extinct mammalian species. These he claimed must have belonged to "homme antédiluvien."

Inevitably such revolutionary ideas initially received little support in France where Georges CUVIER's views still dominated evolutionary thinking. British scientists were more sympathetic, and in 1859 Boucher de Perthes's views were publicly and authoritatively acknowledged before the Royal Society by two British geologists, Joseph PRESTWICH and Hugh Falconer. Later evidence that some of the finds at Abbeville were forgeries tended to discredit Boucher de Perthes's early work, even though it was never satisfactorily established whether or not he played a part in producing the fakes.

Bouguer, Pierre (1698–1758) *French physicist and mathematician*

Bouguer (boo-**gair**), the son of a hydrographer and mathematician, was born in Le Croisic, France, and followed into his father's profession. He was a child prodigy and obtained a post as professor of hydrography at the remarkably early age of 15. The study of the problems associated with navigation and ship design was his chief interest. Bouguer took part in an extended expedition to Peru led by Charles de la Condamine to determine the length of a degree of the meridian near the equator. While on this expedition Bouguer also did a great deal of other valuable experimental work.

One of Bouguer's most successful inventions was the heliometer to measure the light of the Sun and other luminous bodies. Although it was not his chief interest the research for which Bouguer is now best remembered was on photometry. Here too he did much valuable experimental work and one of his major discoveries was of the law now named for him. This states that in a medium of uniform transparency the intensity of light remaining in a collimated beam (i.e., a beam of light in which all the rays are parallel) decreases exponentially with the length of its path in the medium. The law is sometimes unjustly attributed to Johann LAMBERT. Bouguer's work in optics can be seen as the beginning of the science of atmospheric optics.

Boulton, Matthew (1728–1809) *British engineer*

> I shall never forget Mr. Boulton's expression to me: "I sell here, Sir, what all the world desires to have – POWER." He had about seven hundred people at work. I contemplated him as an iron chieftain, and he seemed to be a father to his tribe.

> —James Boswell, *The Life of Samuel Johnson* (1791)

Boulton, who was born in the industrial city of Birmingham in central England, was the manufacturer who supported and financed James WATT's steam engine. He had founded a small factory at Soho near Birmingham (1762), which produced metal articles, and realized that better power sources were needed. He met James Watt in 1768 and recognized the potential of Watt's engine. They became partners in 1775. In 1786 he designed steam-powered coining machinery, which was patented in 1790, and supplied machinery to the Royal Mint and coins to the East India Company.

Bourbaki, Nicolas *French group of mathematicians*

> Structures are the weapons of the mathematician.

"Bourbaki" (**bor**-ba-kee) is the collective *nom de plume* of a group of some of the most outstanding of contemporary mathematicians. The precise membership of Bourbaki, which naturally has changed over the years, is a closely guarded secret but it is known that most of the members are French.

Since 1939, Bourbaki has been publishing a monumental work, the *Eléments de mathématique* (Elements of Mathematics), of which over thirty volumes have so far appeared. In this Bourbaki attempts to expound and display the architecture of the whole mathematical edifice starting from certain carefully chosen logical and set-theoretic concepts. The emphasis throughout the *Eléments* is on the interrelationships to be found between the various structures present in mathematics, and to a certain extent this means that Bourbaki's exposition cuts across traditional boundaries, such as that between algebra and topology. Indeed for Bourbaki, pure mathematics is to be thought of as nothing other than the study of pure structure.

Since the members of Bourbaki are all working mathematicians, rather than pure logicians, in contrast to other foundational enterprises (e.g., those of Gottlob FREGE, Bertrand RUSSELL, and A. N. Whitehead) the influence of Bourbaki's writings on contemporary mathematicians and their conception of the subject has been immense.

Boussingault, Jean Baptiste (1802–1887) *French agricultural chemist*

Born in Paris, France, Boussingault (boo-sang-**goh**) began his career as a mineralogist after studying at the mining school there. He spent several years in Latin America, where he is reported to have fought in the wars of independence under Bolívar. He returned to France in 1832 and shortly afterward became professor of chemistry at Lyons. In 1839 he moved to the

Sorbonne in Paris. Boussingault entered politics in 1848 but returned to chemistry after the coup of 1851. Between 1860 and 1874 he published a comprehensive eight-volume work on agricultural chemistry.

His most important work was on the role of nitrogen in plants. In 1837 he began a series of experiments proving that leguminous plants (peas, clover, etc.) are capable of using atmospheric nitrogen. By growing a wide variety of plants in nitrogen-deficient soils he was able to show that grains could not support themselves but that the legumes could survive (he was not sure how the nitrogen was taken up by the plants). He further demonstrated that animals could not utilize atmospheric nitrogen when living on nitrogen-free diets.

In 1864 Boussingault made the fundamental discovery that the volume of carbon dioxide absorbed by a plant is equal to the volume of oxygen given out.

Boveri, Theodor Heinrich (1862–1915)
German zoologist

Boveri (boh-**vay**-ree) was born in Bamberg, Germany, and graduated in medicine from Munich in 1885. He remained at Munich to do cytological research until his appointment as professor of zoology and comparative anatomy at Würzburg in 1893. In 1888, he coined the term *centrosome* for the region of the cell that contains the centriole, first discovered by Edouard van BENEDEN. Boveri also proved Beneden's theory that equal numbers of chromosomes are contributed by the egg and the sperm to the zygote. Boveri accurately described the formation of the polar bodies following meiosis in the egg cell, and made pioneering studies of sperm formation (spermatogenesis), introducing a diagrammatic representation of the process (1892), which is still in use today.

Bovet, Daniel (1907–1992) *Swiss–Italian pharmacologist*

Bovet (boh-**vay**) was born in Neuchâtel, Switzerland. In 1929 he gained a DSc in zoology and comparative anatomy from the University of Geneva where his father was professor of pedagogy. He continued research work at the Pasteur Institute in Paris, serving as director from 1936 until 1947 and following up the discoveries of Gerhard DOMAGK on prontosil. He and his coworkers were able to show that the sulfonamide group is responsible for the antibacterial action of prontosil. The drug is only active *in vivo* as the animals metabolize the parent drug into sulfanilamide, which is the antibacterial compound.

Prontosil was a dye, protected by patents and expensive. Sulfanilamide was colorless, freely available, cheap to manufacture, and equally as effective as a bactericide. Many analogs, known as sulfa-drugs, have been made and these are widely used against streptococcal infections such as pneumonia, meningitis, and scarlet fever.

These researches led Bovet to develop the earlier ideas of Paul EHRLICH, Emil FISCHER, and Juda Quastel into a more refined "antimetabolite hypothesis," which is one of the fundamental lines of approach in modern drug research. It is based on the idea that a chemical compound whose properties and molecular shape resemble those of a normal body metabolite may affect the functions of that metabolite. Just as a lock (a metabolic reaction) is opened by just one shape of key (a metabolite) so another slightly different shape of key (an antimetabolite) may jam the lock and prevent the new key from fitting. These ideas led Bovet to develop the antihistamine drug "933F" in 1937 and this gave rise to a series of drugs that are useful for asthma and hay fever.

Later, after a trip to Brazil, Bovet became interested in the Indian nerve poison curare. The structure of curare had already been worked out and in 1946 Bovet began work on analogs, which led to the use of succinylcholine as a muscle relaxant in surgical operations.

In 1947 Bovet became head of pharmacology and chemotherapeutics at the Superior Institute of Health in Rome, where he remained until 1964, when he was appointed professor of pharmacology at the University of Sassari. He became an Italian citizen and in his later years carried out research work on tranquilizers and anesthetics. In 1971 he accepted the chair of psychobiology at the University of Rome, finally retiring in 1982. He was awarded the Nobel Prize for physiology or medicine in 1957 for his work on curare and antihistamines.

Bowen, Edmund John (1898–1980)
British physical chemist

Bowen, who was born in Worcester, England, spent his entire career at Oxford University, beginning in 1915 as a student at Balliol College and continuing after graduation as a fellow of University College from 1922 until his retirement in 1965.

He worked mainly on photochemistry, investigating a large number of photochemical reactions and producing a survey of the subject, *Chemical Aspects of Light* (1942). He also produced, with F. Wokes, the more specialized *The Fluorescence of Solutions* (1953) and edited and contributed to *Luminescence in Chemistry* (1968).

Bowen, Ira Sprague (1898–1973) *American astronomer*

Bowen, who was born in Seneca Falls, New York, graduated from Oberlin College, Chicago in 1919 and gained his PhD from the California Institute of Technology, Pasadena, in 1926. He taught physics at Cal Tech from 1921 to 1945, serving from 1931 as professor. In 1946 he was made director of the Mount Wilson Observatory and in 1948 of the newly opened Palomar Observatory, posts he continued to hold until his retirement in 1964.

In 1928 Bowen tackled the problem of the strange lines first observed by William HUGGINS in the 1860s in the spectrum of planetary nebulae and the Orion nebula. The difficulty was, according to Bowen, that the strong lines had not been reproduced in the laboratory. Spectrographic evidence showed that such lines must be emitted by an element of low atomic weight. Talk of a new element, known as "nebulium," that could produce the observed spectral lines was however dismissed by Bowen as nonsense.

Bowen was able to show that the lines were in fact due to radiation emitted from ionized atoms of oxygen and nitrogen as they decayed into more stable lower-energy levels. Specifically he was able to show that triply and doubly ionized oxygen as well as doubly ionized nitrogen would radiate at the wavelengths attributed to "nebulium" but only in the highly rarefied conditions of nebulae where collisions between atoms are very infrequent. It is this radiation that contributes to the green and red colors observed in emission and planetary nebulae.

Bowman, Sir William (1816–1892)
British physician

Bowman, who was born at Nantwich, Cheshire, in England, studied medicine at King's College, London, and in 1840 was appointed assistant surgeon at the newly founded King's College Hospital. His main interest was the investigation of the microscopic structure of tissues (histology). Although he studied voluntary (striated) muscle and liver, it is for his work on the kidney that Bowman is best known. He showed that the capsule (now called Bowman's capsule) that surrounds the bunch of blood capillaries (glomerulus) in each Malpighian corpuscle is continuous with the renal tubule, which in turn leads to the collecting ducts and ultimately to the ureter. This piece of structural evidence prompted him to propose that fluid passed from the glomerulus into the tubule. He reported his findings in his famous paper, *On the Structure and Use of the Malpighian Bodies of the Kidney* (1842).

In the following years, he increasingly devoted his attentions to eye surgery and in 1846 became assistant surgeon at the Royal London Ophthalmic Hospital (now Moorfields Eye Hospital) and full surgeon in 1851. One of the most celebrated eye surgeons of his day, he introduced the newly invented ophthalmoscope into his practice and described many anatomical features of the eye including the basement membrane of the cornea and the radial fibers of the ciliary muscle, both of which bear his name. He also advanced techniques in eye surgery.

Bowman's *The Physiological Anatomy and Physiology of Man* (5 vols. 1845–56), which he wrote with his teacher, Robert Todd, was one of the first texts in which histology assumed real significance in medicine and testifies to the valuable contribution of its author. A popular lecturer with his students, Bowman was also a methodical and kindly practitioner. In 1880 he founded the Ophthalmological Society and was its first president. He was knighted in 1884.

Boyd, William Clouser (1903–1984)
American biochemist

Boyd was born in Dearborn, Missouri, and educated at Harvard and Boston, where he obtained his PhD in 1930. He later taught in the medical school there, serving as professor of immunochemistry from 1948 until his retirement in 1969.

Karl LANDSTEINER's discovery of blood groups in 1902 and subsequent studies of their global distribution permitted a far more accurate estimate of racial types than had previously been possible. To this end Boyd began the systematic collection and analysis of blood samples from all over the world.

Eventually, in the 1950s, in such works as *Genetics and the Races of Man* (1950) he began to present evidence for the existence of 13 races – early European, northern and eastern European, Lapp, Mediterranean, African, Asian, Dravidian, Amerind, Indonesian, Melanesian, Polynesian, and Australian aborigine.

Boyer, Herbert Wayne (1936–) *American biochemist*

Boyer was born in Pittsburgh, Pennsylvania, and educated at St. Vincent College, Latrobe, and the University of Pittsburgh where he obtained his PhD in 1963. He joined the faculty of the University of California, San Francisco, shortly afterward in 1966 and served as professor of biochemistry from 1976 to 1991.

Much of Boyer's work has been concerned with developing some of the basic techniques of recombinant DNA, known more popularly as genetic engineering. Thus in 1973 he succeeded with Robert Helling, and independently of the work of Stanley COHEN and Annie Chang, in constructing functional DNA from two different sources. Such chimeras, as they became called, were initially engineered by splicing together segments from two different plasmids (extrachromosomal DNA found in some bacteria) from the *Escherichia coli* bacillus. The chimera was then inserted into *E. coli* and was found to replicate and, equally significant, to express traits derived from both plasmids.

Development after 1973 was so rapid that by 1976 it had occurred to Boyer and a number of other workers that recombinant DNA could be used to produce such important proteins as insulin, interferon, and growth hormone in commercial quantities. Consequently in 1976 he joined with financier Robert Swanson to invest $500 each to form the company Genentech, which went public in 1980.

Despite successfully developing techniques for the production of somastatin in 1977, insulin in 1978, and growth hormone in 1979, the position of Genentech was far from secure at the beginning of 1981 with the emergence of competition from a number of rival companies and legal problems concerned with the ownership of genes.

In 2004 Boyer and Stanley Cohen received the Albany Medical Center Prize in Medicine and Biomedical Research for their pioneering work in gene splicing and recombination.

Boyer, Paul D. (1918–   ) *American biochemist*

Boyer was born in Provo, Utah, and educated at Brigham Young University. He obtained his PhD in 1943 from the Unversity of Wisconsin-Madison and worked first at Stanford and later at the University of Minnesota. In 1963 he became professor of chemistry and biochemistry at the University of California, Los Angeles. His main research has been on the elucidation of the enzyme mechanism for the synthesis of adenosine triphosphate. For this research, Boyer shared the 1997 Nobel Prize for chemistry with John WALKER. The prize was also shared with Jens SKOU.

Boyle, Robert (1627–1691) *British chemist and physicist*

> ... to beget a good understanding 'twixt the chymists and the natural philosophers.
> —The stated aim of his *Essay on Nitre* (1664)

The son of the Earl of Cork, and born at Lismore Castle, now in the Republic of Ireland, Boyle was a member of an aristocratic and wealthy family. He spent four years at Eton College and from 1638 studied at Geneva, returning to London in 1644. He then retired to his estate at Stalbridge, Dorset, where he took up the life of a scientific "virtuoso."

In 1654 Boyle moved to Oxford, where he worked on pneumatics. In 1658–59 he had an air pump built for him by Robert HOOKE, after the type invented by Otto von GUERICKE in 1654. Boyle was ably assisted by Hooke in various pioneering experiments in which he showed that air was essential for the transmission of sound, and for respiration and combustion – and that the last two processes exhausted only part of the air.

In Boyle's most famous experiment he took a U-shaped tube with a shorter closed end, and a longer open end into which he poured mercury, thus isolating a given volume of air in the shorter end. When the mercury was level in both "limbs" the air was under atmospheric pressure, and by adding more mercury to the longer limb the pressure could be increased. Boyle found that the volume was halved if the pressure was doubled, reduced to a third if the pressure was tripled, and so on – and that this process was reversible. Boyle's work on the compressibility of air was published in *New Experiments Physico-Mechanicall, Touching the Spring of the Air and its Effects* (1660) but the famous law stating that the pressure and volume of air are inversely proportional was not stated explicitly until the second edition (1662). The law, known as *Boyle's law* in America and Britain but in Europe as *Mariotte's law*, can be expressed (where C is a constant) as $p \times V = C$; that is, the product of the pressure and volume of a gas remains constant if, as Edmé Mariotte noted, the temperature remains constant. This law (together with its companion gas law, that of Jacques CHARLES) is true only for ideal gases, but approximately holds for real gases at very low pressures and at high temperatures.

Boyle developed a mechanical corpuscular philosophy of his own, derived from the Greek tradition and the work of GALILEO and Pierre GASSENDI. In Boyle's conception all physical phenomena could be explained by corpuscles of different shapes, sizes, and motions, this corpuscular matter being capable of infinite transformations (which allowed the possibility of alchemy and excluded the existence of elements).

However, in *The Sceptical Chymist* (1661) Boyle proposed a view of matter that presaged modern views and certainly disposed effectively of the Aristotelian doctrine of the four elements. He supposed that all matter was composed of primary particles, some of which joined together to form semi-indivisible corpuscles and whose organization and motion explained all qualities of matter.

Boyle's main contribution to chemistry was his insistence on experiment, precision, and accurate observation. He devised many analytical tests including the use of vegetable dyes as acid–base indicators and of flame tests to detect metals. The chemist's concern for the purity of his materials began with Boyle. Although he prepared hydrogen by the action of acids on iron and observed the oxidation of mercury and its subsequent regeneration on further heating, the "fixation" of gases in bodies remained unexplained in his work. Likewise, he observed the increase in weight of metals on calcination but attributed this to heat, which he sometimes regarded as material.

Boyle left Oxford for London in 1668 where, despite his scholarly nature and poor health, he was very much at the center of scientific life as a founder member of the Royal Society. He believed, like his hero Francis BACON, that science could be put to practical use.

Boys, Sir Charles Vernon (1855–1944)
British physicist

The son of a clergyman, Boys was born in Wing in the eastern English county of Rutland and educated at the Royal School of Mines, London. He later taught at the Royal College of Science, South Kensington. Boys left the College in 1897 to take up the appointment of Metropolitan gas referee, a post that he held until 1939 when the job was abolished. The post was something of a sinecure, and required Boys to do little more than supervise the work of his assistants and to monitor methods used to test the quality of the gas the Board supplied to its customers. Boys also found time to establish himself as an expert witness, appearing in numerous patent and other technical disputes.

Boys is best known for his determination of the gravitational constant in 1897. The measurement was first made by Henry CAVENDISH in 1797 and was expressed in terms of the Earth's density. NEWTON had proposed a density between five and six in 1687; Cavendish found experimentally a density of 5.448.

Whereas Cavendish had used a six-foot beam in his torsion rod experiment, Boys opted for a mere nine inches. The decrease in size was made possible by using some exceedingly fine quartz fibers in the torsion balance. Boys drew these fine filaments by attaching the end of an arrow to a piece of molten quartz and firing it with a crossbow. As a result uniform temperatures were easier to maintain, and convection current disturbances were minimized. Boys calculated on the basis of his measurements that two 1-gram point masses 1 centimeter apart would attract each other with a force of 6.6576 $\times 10^{-8}$ dyne, and consequently the density of the Earth would be 5.527 g/cm^3, figures which compare well with the modern figure of 6.670 $\times$ 10^{-8} dyne, and 5.517 g/cm^3.

There are a number of other instruments linked to Boys. Among these are an integraph (1881) for mechanical integration, a radiomicrometer (1890) for measuring stellar radiation, a rotating lens camera (1900) for photographing the flight of a bullet, and, in his capacity as gas referee, an improved calorimeter to measure the calorific value of gas (1905).

Brachet, Jean Louis Auguste (1909–1998) *Belgian cell biologist*

Brachet (bra-**shay**) was educated at the Free University in his native city of Brussels where his father, an embryologist, was rector. After gaining his MD in 1934 he joined the faculty as an anatomy instructor and in 1943 was appointed professor of general biology.

Brachet began his career by studying the then poorly understood nucleic acids. It had been thought that plant cells contain RNA and animal cells DNA but, in 1933, Brachet demonstrated that both types of nucleic acid occur in both plant and animal cells. He proved this by developing a cytochemical technique that made it possible to localize the RNA-containing structures in the cell. Brachet also noted that cells rich in RNA tend to be those actively engaged in protein synthesis. On this basis Brachet was led in 1942 to propose the important hypothesis that ribonucleoprotein granules could be the agents of protein synthesis. Such granules, later termed ribosomes, were indeed shown to function in this way by George PALADE in 1956.

Later experiments, in which Brachet removed the nucleus from the cell, showed that although protein synthesis continued for a while the amount of RNA in the cytoplasm decreased until there was none left. This indicated that the production of RNA occurs in the nucleus and that it is then transported from the nucleus to the cytoplasm.

Braconnot, Henri (1781–1855) *French naturalist and biochemist*

Born in Commercy, France, and educated at Strasbourg and at Paris, Braconnot (bra-co-**noh**) became professor of natural history at the Lyceum, Nancy, and director of the botanical gardens. Braconnot was interested in the chemical makeup of plants and discovered (1819) a means of obtaining glucose from various plant products lacking in starch, such as tree bark and sawdust. (Until then, glucose had been obtained by boiling starch with acid.) The nonstarch material was later isolated and named cellulose by Anselme PAYEN. Braconnot also did work on saponification and discovered some of the acids occurring in plants.

Bradley, James (1693–1762) *British astronomer*

Bradley was born in Sherborne in Dorset, England, and educated at Oxford University. He was taught astronomy by his uncle, the Rev. James Pound, who was also an astronomer. In 1721 Bradley became Savilian Professor of Astronomy at Oxford and in 1742 he succeeded Edmond HALLEY as Astronomer Royal. His astronomical career began with a determined effort to detect parallax – the angular displacement of a body when viewed from spatially separate positions (or, more significantly, one position on a *moving* Earth). He fixed a telescope in as vertical a position as possible to minimize the effects of atmospheric refraction and began to observe the star Gamma Draconis. He soon found that the star had apparently moved position but prolonged observation convinced him that it could not be parallax he had measured, for he found the greatest shift in position in September and March and not in December and June as it should have been if he was observing parallax. However, the change in position was so regular (every six months) that it could be due only to the observer being on a moving Earth. It took him until 1729 to find the precise cause of the change in position. He realized that as light has a finite speed it will therefore take some time, however small, to travel down the length of the telescope. While it is traveling from the top to the bottom of the telescope the bottom of the instrument will have been carried by the orbital motion of the Earth. The image of the star will therefore be slightly displaced. Bradley realized that he had at last produced hard observational evidence for the Earth's motion, for the finite speed of light, and for a new aberration that had to be taken into account if truly accurate stellar positions were to be calculated. He worked out the constant of aberration at between 20″ and 20″.5 – a very accurate figure. He also discovered another small displacement, which, because it had the same period as the regression of the nodes of the Moon, he identified as the result of the 5° inclination of the Moon's orbit to the ecliptic. This caused a slight wobble of the Earth's axis, which he called *nutation*. Friedrich BESSEL later used Bradley's observations to construct a catalog of unprecedented accuracy.

Bragg, Sir William Henry (1862–1942) *British physicist*

> Physicists use the wave theory on Mondays, Wednesdays, and Fridays, and the particle theory on Tuesdays, Thursdays, and Saturdays.
> —Quoted by A.L. Mackay in *The Harvest of a Quiet Eye* (1977)

Bragg's father was a merchant seaman turned farmer. William Henry Bragg was born in Westwood in England and educated at a variety of schools before going as a scholar to Cambridge University. He graduated in 1884 and after a year's research under J. J. THOMSON took the chair of mathematics and physics at the University of Adelaide, Australia, in 1886. He returned to England as professor of physics at Leeds University in 1909, moving from there to University College, London, in 1915.

In Australia, Bragg concentrated on lecturing and started original research late in life (in 1904). He first worked on alpha radiation, investigating the range of the particles. Later he turned his attention to x-rays, originally believing (in opposition to Charles Barkla) that they were neutral particles. With the observation of x-ray diffraction by Max VON LAUE, he ac-

cepted that the x-rays were waves and constructed (1915) the first x-ray spectrometer to measure the wavelengths of x-rays. Much of his work was on x-ray crystallography, in collaboration with his son, William Lawrence Bragg. They shared the Nobel Prize for physics in 1915.

During the war Bragg worked on the development of hydrophones for the admiralty. In some ways his most significant work was done at the Royal Institution, London, where he was director from 1923. Under James DEWAR's directorship the research functions of the Royal Institution had virtually disappeared. Bragg recruited several young and brilliant crystallographers who shared with him a commitment to applying the new technique to the analysis of organic compounds. There was no reason to suppose there was much chance of success but as early as the 1920s Bragg was planning to investigate biological molecules with x-rays. His first attempts were made on anthracene and naphthalene in 1921.

Bragg, Sir William Lawrence (1890–1971) *British physicist*

> The important thing in science is not so much to obtain new facts as to discover new ways of thinking about them.
> —Quoted by A. Koestler and J. R. Smithies in *Beyond Reductionism* (1958)

William Lawrence Bragg was the son of William Henry Bragg. Born in Adelaide, Australia, he was educated at the university there and at Cambridge University, where he became a fellow and lecturer. After the war, in 1919, he was appointed professor of physics at Manchester University. He succeeded Ernest RUTHERFORD in 1938, after a short period in 1937 as director of the National Physical Laboratory, as head of the Cavendish Laboratory and Cavendish Professor at Cambridge. Finally, in 1953, he became director of the Royal Institution, London, a post his father had held previously and which he held until his retirement in 1961.

Success came very early to Bragg, who shared the Nobel Prize for physics with his father in 1915. Following Max VON LAUE's discovery of x-ray diffraction by crystals in 1912, Lawrence Bragg in the same year formulated what is now known as the *Bragg law*:

$$n\lambda = 2d \sin \theta$$

which relates the wavelength of x-rays (λ), the angle of incidence on a crystal θ, and the spacing of crystal planes d, for x-ray diffraction. n is an integer (1, 2, 3, etc.).

Bragg collaborated with his father in working out the crystal structures of a number of substances. Early in this work they showed that sodium chloride does not have individual molecules in the solid, but is an array of sodium and chloride ions. In 1915 the Braggs published their book *X-rays and Crystal Structure*.

Lawrence Bragg later worked on silicates and on metallurgy. He was responsible for setting up a program for structure determinations of proteins.

Brahe, Tycho (1546–1601) *Danish astronomer*

> Now it is quite clear to me that there are no solid spheres in the heavens, and those that have been devised by the authors to save the appearances, exist only in the imagination, for the purpose of permitting the mind to conceive the motion which the heavenly bodies trace in their courses.
> —*De mundi aetherei recentoribus phaenomenis* (1588; On the Most Recent Phenomena of the Aetherial World)

Tycho (**tee**-koh), whose father Otto was the governor of Helsingborg castle, was born in Knudstrup, Denmark. Kidnapped by and brought up by his uncle Jörgen, an admiral in the Danish navy, he was sent to Leipzig University in 1562 to study law. However, his interest in astronomy had already been kindled. He witnessed a partial solar eclipse in 1560 in Copenhagen whose predictability so impressed him that he began a serious study of PTOLEMY's *Almagest*. He was allowed to continue with the formal study of astronomy and began a tour of the universities of northern Europe. It was while at Rostock in 1566 that, according to tradition, he became involved in a dispute with another young Danish nobleman over who was the better mathematician. The dispute led to a duel in which Tycho lost part of his nose. This he replaced with a mixture of gold, silver, and wax; the nose is clearly visible in contemporary engravings.

Tycho became aware that the successful solar-eclipse prediction of 1560 was not a typical index of the state of 16th-century astronomy. For instance, a conjunction of Jupiter and Saturn predicted by current tables was wrong by ten days. Tycho therefore began his long apprenticeship, traveling through northern Europe meeting the astronomers, instrument makers, and patrons who would support him later on.

His international reputation was made with the dramatic events that centered upon the nova of 1572 – ever since known as *Tycho's star*. Not since the days of HIPPARCHUS (second century BC) had a new star visible to the naked eye appeared in the sky. In *De nova stella* (1573; On the New Star), Tycho was able to demonstrate that the new star showed no parallax and therefore truly belonged to the sphere of the fixed stars. This was important cosmologically because according to ARISTOTLE no change could take place in the heavens, which were supposed to be eternal and incorruptible; change could take place only in the sublunary

sphere. By demonstrating that the new star of 1572 and the great comet of 1577 were changes in the heavens, Tycho was providing new evidence against the traditional Aristotelian cosmology.

In order to induce him to stay in Denmark, Tycho's monarch, Frederick II, offered him the island of Hven and unlimited funds to build an observatory there at Uraniborg. Tycho moved there in 1577, building an observatory/castle stocked with the best instruments then in existence, and constructing enormous quadrants and sextants. He became the greatest observational astronomer of the pretelescopic age. Before Tycho, astronomers tended to work with observations many centuries old. COPERNICUS would be more likely to use the tables of Ptolemy (second century AD) than to make his own observations. When modern tables, such as the *Prutenic Tables* of Erasmus REINHOLD, were constructed, although based on the Copernican system, they were scarcely more reliable. Tycho changed all this. Twenty years' careful observation using accurate instruments enabled him to determine the positions of 777 stars with unparalleled accuracy. He did not, however, accept the Copernican heliocentric system. Instead he proposed a compromise between that and the Ptolemaic, suggesting that the Earth remains at the center, immobile; the Sun and Moon move round the Earth; and all other bodies move round the Sun. His system received hardly any support.

After the death of Frederick II in 1588, Tycho quarreled with his successor, Christian IV, on his coming of age. The last recorded observation made at Hven was on 15 March 1596. Tycho set off once more on his travels, this time encumbered by his enormous instruments. Eventually he found a new patron, one even stranger than himself, the mad Holy Roman Emperor Rudolph II. Tycho was made Imperial Mathematician in 1599, given yet another castle at Benatek outside Prague, and, more important, given the young Johannes KEPLER as an assistant. Although the relationship was a stormy one, both benefited enormously. Tycho died suddenly in 1601 after a short illness leaving Kepler to publish Tycho's *Rudolphine Tables* posthumously in 1627. His last words were: "Let me not seemed to have lived in vain." That such a fear was groundless is witnessed on the title page of Kepler's great work, *Astronomia nova* (New Astronomy):

"Founded on observations of the noble Tycho Brahe."

Brahmagupta (c. 598–c. 665) *Indian mathematician and astronomer*

Brahmagupta (brah-ma-**guup**-ta) was the director of the observatory at Ujjain, a town in central India. In 628 he wrote the *Brahmasphuta-siddhānta* (The Opening of the Universe) half of which is about mathematics and half about astronomy. He introduced negative numbers into India, gave a satisfactory rule for the solution of quadratic equations, and attempted to apply algebra to astronomy. He discovered the formula for the area of a cyclic quadrilateral:

$$A = \sqrt{[(s-a)(s-b)(s-c)(s-d)]}$$

where s is half the perimeter and a, b, c, and d are the lengths of the sides. One innovation introduced by him was the use of oval epicycles for Mars and Venus. He rejected the view that the stars are stationary and the Earth moves, on the grounds that any such movement would cause high buildings to fall.

Braid, James (1795–1860) *British physician*

Braid was born in Rylaw House, Scotland, and educated at Edinburgh University before becoming apprenticed to a doctor in Leith. On obtaining his medical diploma, he became a surgeon to colliery workers but later moved to Manchester where he established in practice as a surgeon. It was here that he began his scientific study of mesmerism – the induction of a trancelike state in a patient, first made famous by Anton MESMER. Braid discovered that by concentrating on a moving pendulum or similar object, a patient could enter into a state of what he termed "neurohypnotism," which he later shortened to "hypnosis." In his paper, *A Practical Essay on the Curative Agency of Neuro-Hypnotism* (1842), Braid suggested several therapeutic applications for his techniques, including the alleviation of pain and anxiety and the treatment of certain nervous diseases.

Although his work encountered criticism from some quarters, Braid undoubtedly influenced his successors, notably the French school of neuropsychiatry.

Bramah, Joseph (1748–1814) *British engineer and inventor*

A former cabinetmaker who was born in Stainborough, England, Bramah designed and built many of the machine tools that facilitated the expansion of British manufacturing in the 19th century. In order to make a pick-proof lock, he developed very precisely engineered tools, many of which were made by Henry MAUDSLAY, then a young blacksmith. Bramah displayed the lock in his shop window (1784) offering a prize to whoever could pick it. This was not achieved until 67 years later when a mechanic took 51 hours to open it.

Bramah also invented a hydraulic press, an improved toilet, a wood-planing machine, and a machine for numbering banknotes.

Brambell, Francis William Rogers (1901–1970) *British zoologist*

Brambell, who was born in Sandycove, now in the Republic of Ireland, gained both his BA and PhD from Trinity College, Dublin. He then worked successively at University College and King's College, London. In 1930 he became professor of zoology at Bangor University, remaining there until his retirement in 1968.

Brambell's work on prenatal mortality in wild rabbits led to his discovery that protein molecules are transferred from mother to fetus not through the placenta as previously supposed but by the uterine cavity and fetal yolk sac. He further found that this process is selective; for example the gamma globulins are transferred more readily than other serum proteins. Such work is important in studies of resistance to disease in the newborn and of hemolytic diseases of infants. For this work Brambell received the Royal Medal of the Royal Society of London in 1964.

Brand, Hennig (about 1670) *German alchemist*

Brand (brant) is remembered as the discoverer of phosphorus, probably in 1669 in Hamburg, Germany, and as such the first known discoverer of an element. The discovery was of particular importance because few new elements had been discovered since classical times. True to the traditions of alchemy, Brand kept his method secret but samples and hints on the method were obtained by Johann Daniel Kraft and Johann Kunckel. The latter prepared phosphorus independently in 1676 by the distillation of concentrated urine with sand but still the method remained unpublished. Eventually, Robert BOYLE, having obtained a sample from Kraft, repeated the isolation; naming the substance *aerial noctiluca* (substance that shines by night in air), he published his findings in a book of the same title (1680).

Brandt, Georg (1694–1768) *Swedish chemist*

Brandt (brant) was the son of an ironworker and former apothecary in Riddarhyta, Sweden, and from an early age he helped his father with metallurgical experiments. He studied medicine and chemistry at Leiden, and gained his MD at Rheims in 1726. He was later made warden of the Stockholm mint (1730), and professor of chemistry at the University of Uppsala.

In 1733 he systematically investigated arsenic and its compounds. He invented the classification of semimetals (now called metalloids), in which he included arsenic, bismuth, antimony, mercury, and zinc.

In 1735 Brandt postulated that the blue color of the ore known as smalt was due to the presence of an unknown metal or semimetal. He named this "cobalt rex" from the Old Teutonic "kobold," originally meaning "demon," later applied to the "false ores" that did not yield metals under the traditional processes. In 1742 Brandt isolated cobalt, and found it was magnetic and alloyed readily with iron. His results were confirmed in 1780 by Torbern BERGMAN, who first obtained fairly pure cobalt.

Brandt also, in 1748, experimented with the dissolution of gold in hot concentrated acid, and with its precipitation from solution. These experiments clarified some of the alleged transmutations of silver into gold. Indeed, Brandt devoted his later years to exposing fraudulent transmutations of metals into gold, and it was said of him that no chemist did more to combat alchemy.

Brans, Carl Henry (1935–) *American mathematical physicist*

Brans was born in Dallas, Texas, and graduated in 1957 from Loyola University, Louisiana. Having obtained his PhD from Princeton in 1961, he returned to Loyola in 1960 and in 1970 was appointed professor of physics.

Brans has worked mainly in the field of general relativity. He is best known for his production with Robert DICKE in 1961 of a variant of EINSTEIN's theory in which the gravitational constant varies with time. A number of very accurate measurements made in the late 1970s has failed to detect this and some of the other predictions made by the *Brans–Dicke theory*.

Brattain, Walter Houser (1902–1987) *American physicist*

Brattain, who was born in Amoy, China, was brought up on a cattle ranch. He was educated at Whitman College, at the University of Oregon, and at Minnesota, where he obtained his PhD in 1929. He immediately joined the Bell Telephone Company with which he worked as a research physicist until his retirement in 1967. After leaving Bell, Brattain taught at Whitman College doing research there on phospholipids.

Brattain's main field of work was the surface properties of semiconductors. It was known that a junction at a semiconductor would rectify an alternating current and that this effect was a surface property. Brattain was particularly interested in using semiconductors to amplify signals. Working with John BARDEEN, he investigated various arrangements for achieving this – originally studying silicon in contact with electrolytes, but later using germanium in contact with gold. Their first efficient point-contact transistor (1947) consisted of a thin wafer of germanium with two close point contacts on one side and a large normal contact on the other. It had a power amplification of 18. Bardeen and Brattain shared the 1956 Nobel

Prize for physics with William SHOCKLEY for their development of the transistor.

Braun, Karl Ferdinand (1850–1918) *German physicist*

Braun (brown), who was born in Fulda, Germany, studied at Marburg and, in 1872, received a doctorate from the University of Berlin. He taught in various university posts. In 1885 he became professor of experimental physics at Tübingen and in 1895 he became professor of physics at Strasbourg.

In 1874, Braun observed that certain semiconducting crystals could be used as rectifiers to convert alternating to direct currents. At the turn of the century, he used this fact in the invention of crystal diodes, which led to the crystal radio. He also adapted the cathode-ray tube so that the electron beam was deflected by a changing voltage, thus inventing the oscilloscope and providing the basic component of a television receiver. His fame comes mainly from his improvements to MARCONI's wireless communication system and, in 1909, they shared the Nobel Prize for physics. Braun's system, which used magnetically coupled resonant circuits, was the main one used in all receivers and transmitters in the first half of the 20th century.

Braun went to America to testify in litigation about radio patents but, when the United States entered World War I in 1917, he was detained as an alien and died in New York a year later.

Bredt, Konrad Julius (1855–1937) *German organic chemist*

Bredt (bret) was born in Berlin and studied in Leipzig and Frankfurt before gaining his doctorate at Strasbourg in 1880 under Rudolph FITTIG. He became professor at the technical institute at Aachen in 1897. His most important research was the structural elucidation of camphor in 1893. The empirical formula had been determined by Jean DUMAS in 1833 and over 30 formulae were proposed before Bredt deduced the correct structure.

Breit, Gregory (1899–1981) *Russian–American physicist*

Although born in Nicholaev, Russia, Breit (brīt) moved to America in 1915 and became a naturalized citizen in 1918. He studied at Johns Hopkins University, gaining his PhD in 1921. From 1921 until 1924 he worked successively at the universities of Leiden, Harvard, and Minnesota, before joining the Carnegie Institution, Washington (1924–29).

At Carnegie, Breit worked in the department of terrestrial magnetism as a mathematical physicist, and it was there that he conducted, with Merle A. TUVE, some of the earliest experiments to measure the height and density of the ionosphere. Their technique was to transmit short bursts of radio waves and analyze the reflected waves received. Their work is now seen as a significant step in the historical development of radar.

Besides his pioneering work on the ionosphere, Breit researched quantum theory, nuclear physics, and quantum electrodynamics. In particular, he and Eugene WIGNER were able to show that the experimental observations of the interactions of neutrons and protons indicated that the particles differed only in their charge and other electrical properties, and not in their nuclear forces. The *Breit–Wigner formula* is a formula for the energy dependence of the absorption cross-section of a compound nucleus in a nuclear reaction.

Between 1929 and 1973 Breit held professorial posts at the universities of New York, Wisconsin, and Yale, and the State University of New York, Buffalo.

Brenner, Sydney (1927–) *South African–British molecular biologist*

The son of a Lithuanian exile, Brenner was born in Germiston, South Africa, and educated at the universities of Witwatersrand and Oxford, where he obtained his DPhil in 1954. In 1957 he joined the staff of the Medical Research Council's molecular biology laboratory in Cambridge, and later served as its director (1979–86). In 1996 he founded the Molecular Sciences Institute, Berkeley, California. He is presently involved with a number of organizations including the Salk Institute.

Brenner's first major success came in 1957 when he demonstrated that the triplets of nucleotide bases that form the genetic code do not overlap along the genetic material (DNA). The basic idea was that the amino-acid sequence of a protein is determined by the sequence of the four nucleotides – A, T, C, and G – in the DNA, with a specific amino acid being specified by a sequence of three nucleotides. Thus in an *overlapping* code the sequence:

ATTAGTACGTCGA...

would yield the following triplets, ATT, TTA, TAG, AGT, GTA, etc., each of which specified a particular amino acid.

Brenner, however, pointed out that such a code imposed severe restrictions on the permitted order of bases. ATT, for example, in an overlapping code, could be followed by the four base triplets TTA, TTT, TTC, and TTG only. This was relatively easy to test without in any way understanding the true nature of the code, and it was soon shown that such implied restrictions were frequently broken.

A greater triumph followed in 1961 when Brenner, in collaboration with Francis CRICK and others, reported the results of careful experiments with the bacteriophage T4, which clearly showed that the code did consist of base triplets that neither overlapped nor appeared to be separated by "punctuation marks."

The same year also saw Brenner, this time in collaboration with François Jacob and Matthew MESELSON, introducing a new form of RNA, messenger RNA (mRNA). With this came one of the central insights of molecular biology – an explanation of the mechanism of information transfer whereby the protein-synthesizing centers (ribosomes) play the role of nonspecific constituents that can synthesize different proteins, according to specific instructions, which they receive from the genes through mRNA. In 1963 Brenner turned his attention to the tiny nematode (roundworm) *Caenorhabditis elegans*. He realized that this simple multicellular animal, consisting of about 1000 cells, had potential to serve as a model system for studying genetic control systems in complex organisms. By his example, Brenner convinced others that genetic experimentation on *C. elegans* would reveal insights into processes such as development, behavior, aging, and disease in humans and other animals. Among his colleagues in Cambridge was John E. SULSTON, who later discovered the importance of programmed cell death during the development of the worm's embryo. Brenner's work paved the way for the sequencing of the entire *C. elegans* genome during the 1980s and 1990s, and enabled further discoveries about genetic control of developmental processes, notably those made by another of Brenner's collaborators, Robert HORVITZ. In 2002, Brenner was awarded the Nobel Prize in physiology or medicine, jointly with Sulston and Horvitz, for discoveries concerning "genetic regulation of organ development and programmed cell death."

Bretonneau, Pierre Fidèle (1778–1862) *French physician*

Bretonneau (bre-ton-**oh**) was born in St. Georges-sur-Cher, France. A surgeon's son, he went to clinical lectures at the Ecole de Santé in Paris, graduated in 1815, and later practiced in Tours, where he served for some time as the chief physician at the Tours hospital. Bretonneau carefully studied a particular disease, which, beginning in 1818, reached epidemic proportions in Tours. In 1821 he proposed to call it diphtheria, from the Greek word for leather, as the disease was characterized by the formation of a membrane over the mucosa of the throat. He realized it was contagious despite his failure to transmit it to animals. He did however distinguish it from scarlet fever and, in 1825, performed the first successful tracheotomy (incision of the trachea through the neck muscles to prevent asphyxiation) on a diphtheria patient.

He further made an early attempt to describe

typhoid and distinguished it from typhus. His theories on infectious diseases partly anticipated the germ theory of PASTEUR.

Brewster, Sir David (1781–1868) *British physicist*

Brewster, who was born in Jedburgh, Scotland, started by studying for the ministry at Edinburgh University but, after completing the course, he abandoned the Church for science. He earned his living by editing various journals and spent much time popularizing science.

Brewster published almost 300 papers, mainly concerning optical measurements. He was an early worker in spectroscopy, obtaining (1832) spectra of gases and of colored glass. His most famous work was on the polarization of light. In 1813 he discovered *Brewster's law*, which states that if a beam of light is split into a reflected ray and a refracted ray at a glass surface, then they are polarized, and the polarization is complete when the two rays are at right angles. The angle of incidence at which this occurs is called the *Brewster angle*. He is also known for his invention, in 1816, of the kaleidoscope.

Brewster was knighted in 1832. From 1859 he was principal of Edinburgh University.

Bricklin, Daniel (1951–) *American computer engineer*

Bricklin graduated in electrical engineering and computer science from MIT in 1973. Before entering the Harvard Business School in 1977 he spent some years at Digital, where he helped to develop their earliest word-processing system. While working for his Harvard MBA he became aware how frustrating and tedious business planning could be, particularly when any small correction or change in one minor item could call for thousands of necessary recalculations.

Bricklin began to develop what has now become a familiar item, namely, the spreadsheet program. It was written in Basic for an Apple II. Initially it took 20 seconds to recalculate 100 cells and would not scroll. A company, Software Arts, was set up in 1979 with two colleagues, Robert Framkston and Dan Fylstra, to produce the spreadsheet they named VisiCalc.

Bricklin's invention was most opportune. Personal computers (PCs), most notably the Apple II, had already appeared. Initially, however, they appealed only to the hobbyist – there was very little anyone could do with an early PC; it had no applications and no software to run. Bricklin's VisiCalc filled this gap to the extent that some customers would buy an Apple II in order to run VisiCalc.

Bricklin's idea was enormously successful and other companies developed spreadsheet programs of their own. By 1986 over seven million spreadsheet programs had been bought. Software Arts was sold to Lotus in 1985 and Bricklin went on to set up his own mail-order business, Software Gardens Inc., to sell his own software programs on a much smaller scale. He is also involved in the development of a web-based spreadsheet (WikiCalc).

Bridgman, Percy Williams (1882–1961) *American physicist*

Doubtless the most influential single discovery was that of a method of producing high hydrostatic pressure without leak. The discovery of the method had a strong element of accident.
—Reply to the National Academy of Sciences questionnaire "Discoveries Which You Regard as Most Important"

It is the merest truism, evident at once to unsophisticated observation, that mathematics is a human invention.
—*The Logic of Modern Physics* (1927)

Bridgman, the son of a journalist, was born in Cambridge, Massachusetts, and educated at Harvard where he obtained his PhD in 1908. He immediately joined the faculty, leaving only on his retirement in 1954 after serving as professor of physics from 1919 to 1926, professor of mathematics and natural philosophy from 1926 to 1950, and as Huggins Professor from 1950 to 1954.

Most of Bridgman's research was in the field of high-pressure physics. When he began he found it necessary to design and build virtually all his own equipment and instruments. In 1909 he introduced the self-tightening joint and, with the appearance of high-tensile steels, he could aim for pressures well beyond the scope of earlier workers. With his equipment, Bridgman was regularly able to attain pressures of 100,000 kg/cm^2.

Bridgman used such pressures to explore the properties of numerous liquids and solids. In the course of this work he discovered two new forms of ice, freezing at temperatures above 0°C. He also, in 1955, transformed graphite into synthetic diamond. Bridgman was awarded the 1946 Nobel Prize for physics for his work on extremely high pressures.

He was also widely known as a philosopher of science and in his book *The Logic of Modern Physics* (1927) formulated his theory of "operationalism" in which he argued that a concept is simply a set of operations. In his seventies Bridgman developed Paget's disease, which gave him considerable pain and little prospect of relief. He committed suicide in 1961.

Briggs, Henry (1561–1630) *English mathematician*

The mirrour of the age for excellent skill in geometry.
—William Oughtred (17th century)

Born in Warley Wood, England, Briggs became a fellow of Cambridge University in 1588 and was later made a lecturer (1592) and a professor (1596) of geometry at Gresham College, London.

He is remembered chiefly for the modifications he made to John NAPIER's logarithms, which were first published in 1614. Napier had produced these to base e (natural logarithms) but Briggs considerably improved their convenience of use by introducing the base 10 (common logarithms). He also introduced the modern method of long division. Briggs became Savilian Professor of Geometry at Oxford in 1619.

Bright, Richard (1789–1858) *British physician*

Born in Bristol and educated at Edinburgh University and Guy's Hospital, London, Bright received his MD from Edinburgh in 1813. After a period spent traveling in Europe, he was appointed assistant physician to Guy's Hospital in 1820, becoming full physician four years later. One of Bright's major interests was kidney disease. He was able to correlate symptoms in his patients with the results of postmortem examinations, and described how the presence of albumin in urine (albuminuria) and the accumulation of fluid in the body (dropsy) were caused by pathological changes in the kidneys. The term *Bright's disease* is now used for several renal diseases all sharing the above symptoms.

Bright also studied and wrote on subjects such as abdominal tumors, jaundice, and nervous diseases, and collaborated with his famous contemporary at Guy's, Thomas ADDISON, in writing *Elements of the Practice of Medicine* (vol. 1, 1839). An amiable and distinguished consultant, Bright also did much to develop teaching methods at Guy's and helped found the *Guy's Hospital Reports*, in which he published much of his work.

Broca, Pierre Paul (1824–1880) *French physician and anthropologist*

Broca (bro-**kah**), who was born in Sainte-Foula-Grande, France, studied at the University of Paris and received his MD in 1849. In 1853 he was appointed assistant professor in the faculty of medicine. His specialty was the brain and, through surgical work and postmortem examination, he was able to demonstrate that damage to one particular region (the left inferior frontal gyrus, now also known as Broca's convolution) of the cortex was associated with impairment or loss of speech. This was one of the first conclusive demonstrations that control of different bodily functions resides in localized regions of the cerebral cortex.

Broca applied his knowledge of the brain to anthropology. He devised techniques of accurately measuring skulls to enable comparison between the different races of modern man and skulls unearthed at prehistoric sites. Broca's findings supported the then highly contentious theory of Charles DARWIN that man, like other living things, had evolved from primitive ancestors. Broca helped found several notable anthropological institutions, including the Société d'Anthropologie de Paris (1859) and the Ecole d'Anthropologie (1876), thus helping to establish anthropology as a respectable branch of science.

Brockhouse, Bertram Neville (1918–2003) *Canadian physicist*

Brockhouse gained his PhD from the University of Toronto in 1950. He worked initially with the Atomic Energy Commission of Canada at the Chalk River Nuclear Laboratory, Ontario. In 1962 he moved to MacMaster University, Hamilton, Ontario, where he remained until his retirement in 1984.

The construction of nuclear reactors in Canada and the United States in the 1940s allowed physicists, once the war had ended, to use neutron beams to explore atomic structure. Neutrons are more effective probes than protons because they are electrically neutral and consequently do not interact with the orbiting electrons. Because neutrons can behave as waves they produce diffraction patterns as a result of collisions with their target atomic nuclei. The effect is similar to that of x-ray diffraction, in which the crystal lattice acts as a diffraction grating for the particles. Neutron diffraction from crystals can be used to select beams of neutrons with the same energy. These "monochromatic" beams can then be used in neutron-scattering experiments.

Brockhouse chose to study the inelastic scattering of neutrons as they bombarded atoms bound in a crystal lattice. In this procedure neutrons give up or gain energy from the atoms they collide with. Monochromatic neutron beams were directed at a crystal target and the energies of the scattered neutrons measured as they emerged. It was thus possible to determine how much energy had been gained or lost. With these data Brockhouse was able to obtain information about the vibration of atoms in the crystal and such important properties as the ability of the crystal to conduct heat and electricity.

For his work on atomic structure Brockhouse shared the 1994 Nobel Prize for physics with Clifford SHULL.

Brodie, Sir Benjamin Collins (1783–1862) *British physician*

Born in Wintersloy, England, Brodie learned medicine in London, first as apprentice to an apothecary and then at the Windmill Street School of Anatomy. He later studied at St. George's Hospital, where he was subsequently appointed house surgeon (1805) and assistant surgeon (1808). Between 1809 and 1814 he contributed six papers to the Royal Society in which he challenged the prevailing contemporary view that body heat was caused, more or less, by simple chemical combustion. Brodie argued that it was a complex process under the influence of the brain, which controlled the functioning of vital organs, such as the heart. This brought Brodie to prominence and he was elected a fellow of the Royal Society in 1810.

Brodie soon had a large and lucrative practice. He wrote *Diseases of the Joints* (1818), which ran to several editions, and in 1819 he was appointed professor of comparative anatomy and physiology at the Royal College of Surgeons, of which he later (1844) became president. A surgeon to both George IV and William IV, Brodie was made a baronet in 1834. As one of the most prominent members of the medical establishment, he was chosen (1858) to be the first president of the newly formed General Medical Council, which administered the registration of all persons qualified to practice medicine.

Brodie, Bernard Beryl (1909–1989)
American pharmacologist

Born in Liverpool in northwest England, Brodie was educated at McGill University in Canada and at New York University, where he obtained his PhD in 1935. He worked at the Medical School there from 1943 to 1950 when he moved to the National Institutes of Health at Bethesda, Maryland, where he served as chief of the chemical pharmacology laboratory until 1970.

Brodie worked in a wide variety of fields including chemotherapy, anesthesia, drug metabolism, and neuropharmacology. In 1955 Brodie and his colleagues produced some results that once more raised the possibility of a chemical basis of mental disease. Basically they showed that the tranquilizer reserpine – an alkaloid extracted from the roots of *Rauwolfia* – can produce a profound fall in the level of serotonin, a naturally occurring monoamine in the brain. The question then arose as to whether the tranquilizing effect of reserpine is due to its reduction of too high a level of serotonin.

It was further shown that some of the actions of serotonin could be neutralized by the presence of the hallucinogen LSD. As the structures of the two molecules are somewhat similar the possibility arose that LSD could monopolize the enzyme that normally breaks down serotonin and thus permit the accumulation of unusually high levels of serotonin. It is perhaps this action that causes the hallu-

cinogenic state and which, it has been argued, mimics the schizophrenic state.

In reality the speculations arising from Brodie's work have turned out to be surprisingly difficult to confirm or reject.

Broglie, Prince Louis Victor de *See* DE BROGLIE, PRINCE LOUIS VICTOR.

Brongniart, Alexandre (1770–1847)
French geologist and paleontologist

Brongniart (bro-**nyar**) was the son of an architect. He was born in Paris and educated at the School of Mines there, serving in the army as an engineer before being appointed director of the Sèvres porcelain factory (1800–47). He also served as professor of natural history at the Ecole Centrale des Quatre Nations, Paris, from 1797. In 1822 he succeeded René HAÜY to the chair of mineralogy at the Natural History Museum, Paris.

Brongniart's early work included his *Essai d'une classification naturelle des reptiles* (1800; Essay on the Classification of Reptiles) in which he divided the Reptilia into the Chelonia, Sauria, Ophidia, and Batrachia. In 1822 he published the first full-length account of trilobites, the important paleozoic arthropods.

His most significant work was done in collaboration with Georges CUVIER on the geology of the Paris region, published jointly by them in 1811 as *Essai sur la géographie minéralogique des environs de Paris* (Essay on the Mineralogical Geography of the Environs of Paris). In this monograph they were among the first geologists to identify strata within a formation by their fossil content; earlier geologists had tended to rely on the characteristics of the rocks rather than their content. They were able to show a constant order of fossil sequence, mainly of mollusks, over the whole Paris region.

They also produced evidence that counted strongly against the neptunism of Abraham WERNER in their discovery of alternate strata of fresh-and sea-water mollusks. The solution to the problem of how fresh-and sea-water strata could alternate was crucial for the new science of geology and was to lead to the catastrophism of Cuvier and the evolution theory of Charles DARWIN.

Outside science his major achievement was the revival of the Sèvres factory which, when he took over in 1800, was at a particularly low point in its history. He abandoned the production of soft-paste ware and developed a new range of colors, producing works as richly colored as paintings.

His son, Adolph, was a distinguished botanist who, in 1828, published a history of fossil plants.

Bronk, Detlev Wulf (1897–1975) *American physiologist*

Bronk, the son of a Baptist minister, came from Dutch stock: the family name survives in the Bronx district of New York City, where he was born. He was educated at Swarthmore, Pennsylvania, and at Michigan, where he obtained his PhD in 1922. Bronk spent some time in England working with Edgar ADRIAN in Cambridge before accepting, in 1929, the post of director of the Johnson Institute, attached to the University of Pennsylvania, where he was already serving as professor of biophysics and director of the Institute of Neurology. In 1949 Bronk became president of Johns Hopkins University but left there in 1953 to take the presidency of the Rockefeller Institute (later Rockefeller University) in New York, where he remained until his retirement in 1968.

Bronk established an early reputation with his fundamental work with Edgar Adrian on nerve impulses (1928–29). They demonstrated that both motor and sensory nerves transmit their messages by varying the number of impulses sent rather than by using impulses of different intensities. By careful experiments they established that the range of the impulses is between 5 and 150 per second, with greater stimuli producing higher frequencies.

In the 1930s Bronk worked mainly on the autonomic nervous system. He is, however, mainly remembered for his crucial role in organizing the institutional structure of American science.

Brønsted, Johannes Nicolaus (1879–1947) *Danish physical chemist*

Born in Varde, Denmark, Brønsted (**bron**-sted or **bron**-ste*th*) studied at the Polytechnic Institute, Copenhagen, from 1897, obtaining degrees in engineering (1899) and chemistry (1902), and a doctorate (1908). The same year he became professor of chemistry at Copenhagen.

Brønsted worked mainly in thermodynamics, especially in the fields of electrochemistry and reaction kinetics. In 1923 he proposed, concurrently with Thomas LOWRY, a new definition of acids and bases. This, the *Lowry–Brønsted theory*, states that an acid is a substance that tends to lose a proton and a base is a substance that tends to gain a proton. In 1924, working with V. K. La Mer on the activity coefficients of dilute solutions, he confirmed the theoretical expression of Peter DEBYE and Erich HÜCKEL for electrolytes.

Broom, Robert (1866–1951) *British–South African morphologist and paleontologist*

Broom, who was born in Paisley, Scotland, graduated in medicine from Glasgow University in 1889. He traveled to Australia in 1892 and in 1897 settled in South Africa where he practiced medicine, often in remote rural communities, until 1928. He also held posts as professor of geology and zoology (1903–10) at Victoria College, now Stellenbosch University, South Africa, and curator of paleontology at the Transvaal Museum, Pretoria, from 1934 until his death.

Apart from studies of the embryology of Australian marsupials and monotremes, Broom's major contributions to science have been concerned with the evolutionary origins of mammals, including man. He excavated and studied the fossils of the Karroo beds of the Cape, and in the 1940s discovered numbers of Australopithecine skeletons in Pleistocene age quarries at Sterkfontein, Transvaal. These latter have proved of considerable importance in investigations of man's ancestry and Broom's account of their discovery is given in *Finding the Missing Link* (1950).

Brouncker, William, Viscount (1620–1685) *English mathematician and experimental scientist*

Brouncker (**brung**-ker) graduated from Oxford University in 1647 with a degree in medicine. He held a variety of official posts, including serving as member of parliament and president of Gresham's College. He was a friend of the eminent mathematician John WALLIS and his own most notable work was also in mathematics. He was a founder and first president of the Royal Society (1662–77) and as such carried out experimental work. Brouncker usually contented himself with solving problems arising from the work of other mathematicians rather than doing creative work himself but was the first to use continued fractions. He was a friend of Samuel Pepys and frequently figures in Pepys's *Diary*. Apart from science Brouncker had a lively interest in music.

Brouwer, Dirk (1902–1966) *Dutch–American astronomer*

Brouwer (**brow**-er), the son of a civil servant from Rotterdam in the Netherlands, studied at the University of Leiden where he obtained his PhD under Willem de SITTER in 1927. He then moved to America to do postdoctoral research at the University of California at Berkeley. He joined the Yale faculty in 1928, serving from 1941 until his death as professor of astronomy and director of the Yale Observatory.

Brouwer worked mainly in celestial mechanics, particularly in the analysis of observations concerning orbiting bodies and on planetary theory, providing new methods by which the motion of a planet could be determined and by which the very long-term changes in orbits could be calculated. He collaborated with Gerald CLEMENCE and W. J. Eckert in the production in 1951 of the basic paper in which the accurate orbits of the outer planets were given for the years 1653–2060. Brouwer's other major contribution was to astrometry, or positional astronomy, where he introduced new techniques and initiated programs for the measurement of stellar positions, especially in the southern sky.

He was also involved in the decision to adopt a new time scale, known at Brouwer's suggestion as "Ephemeris Time" (ET). This became necessary when the rotation rate of the Earth, on which time measurements had been based, was found to vary very slightly. Ephemeris Time is derived from the orbital motions of the Moon and the Earth and is perfectly uniform. It is only used by astronomers however; more general timekeeping now involves atomic clocks.

Brouwer, Luitzen Egbertus Jan (1881–1966) *Dutch mathematician and philosopher of mathematics*

Born in Overschie in the Netherlands, Brouwer took his first degree and doctorate at the University of Amsterdam, where he became successively *Privatdozent* and professor in the mathematics department. From 1903 to 1909 he did important work in topology, presenting several fundamental results, including the fixed-point theorem. This is the principle that, given a circle (or sphere) and the points inside it, then any transformation of all points to other points in the circle (or sphere) must leave at least one point unchanged. A physical example is stirring a cup of coffee – there will always be at least one particle of liquid that returns to its original position no matter how well the coffee is stirred.

Brouwer's best-known achievement was the creation of the philosophy of mathematics known as *intuitionism*. The central ideas of intuitionism are a rejection of the concept of the completed infinite (and hence of the transfinite set theory of Georg Cantor) and an insistence that acceptable mathematical proofs be constructive. That is, they must not merely show that a certain mathematical entity (e.g., a number or a function) *exists*, but must actually be able to construct it. This view leads to the rejection of large amounts of widely accepted classical mathematics and one of the three fundamental laws of logic, the law of excluded middle (either *p* or not-*p*; a proposition is either true or not true).

Brouwer was able to re-prove many classical results in an intuitively acceptable way, including his own fixed-point theorem.

Brown, Alexander Crum *See* CRUM BROWN, ALEXANDER.

Brown, Herbert Charles (1912–2004) *American chemist*

Brown moved from London, where he was born, to Chicago with his family when he was two years old. His father, originally a cabinet maker, ran a hardware store but Brown had to leave school to help support his mother and three sisters. When he finally did get to college, Crane Junior, it was forced to close in 1933 for lack of funds. He eventually made it to the University of Chicago where he obtained his doctorate in 1938. Brown then worked at Wayne University, Detroit, from 1943 until 1947, when he moved to Purdue University, Indiana, where he served as professor of inorganic chemistry until his retirement in 1978.

Brown was particularly noted for his work on compounds of boron. He discovered a method of making sodium borohydride ($NaBH_4$), a reagent used extensively in organic chemistry for reduction. He also found a simple way of preparing diborane (B_2H_6). By reacting diborane (B_2H_6) with alkenes (unsaturated hydrocarbons containing a double bond) he produced a new class of compounds, organoboranes, which are also useful in organic chemistry. Brown also used addition compounds of amines with boron compounds to investigate the role of steric effects in organic chemistry. He received the 1979 Nobel Prize for chemistry.

Brown, Michael Stuart (1941–) *American medical biochemist and geneticist*

Born in New York, Brown gained his BA from the University of Pennsylvania in 1962, and was awarded his MD in 1966. His first post was as an intern at Massachusetts General Hospital, Boston (1966–68), after which he joined the National Institutes of Health as a clinical associate (1968–71). In 1971 he was appointed assistant professor at the University of Texas Southwestern Medical School, Dallas, and in 1977 he became professor of genetics and director of the Center of Genetic Diseases.

Brown's research interests have included digestive enzymes, particularly their role in the metabolism of cholesterol. However, he is perhaps best known for his studies of lipid receptors on body cells and their importance in removing cholesterol from the blood. This work was done in conjunction with Joseph GOLDSTEIN, with whom Brown has had a long and fruitful scientific partnership dating from 1966 when they met as interns at Massachusetts General Hospital. Most notably, their work on familial hypercholesterolemia, an inherited disorder of cholesterol metabolism, earned them the 1985 Nobel Prize for physiology or medicine.

Brown, Robert (1773–1858) *British botanist*

He [Brown] seemed to me to be chiefly

remarkable for the minuteness of his observations and their perfect accuracy ... His knowledge was extraordinarily great, and much died with him, owing to his excessive fear of never making a mistake.
—Charles Darwin, *Autobiography*

Brown, a clergyman's son from Montrose, Scotland, studied medicine at Edinburgh University. He joined the Fifeshire Regiment of Fencibles in 1795 and served five years in Ireland as a medical officer. During a visit to London in 1798 he was introduced to Sir Joseph BANKS. This led, two years later, to his being recommended by Banks for the post of naturalist on the *Investigator* in an expedition to survey the coast of New Holland (Australia) under the command of Matthew Flinders. Brown accepted the appointment and the *Investigator* set sail for the Cape of Good Hope and Australia in 1801. During his five years with the expedition Brown collected 4,000 plant specimens, and on his return to England spent another five years classifying these. Rather than use LINNAEUS's artificial classification, he followed Antoine de JUSSIEU's more natural system, adding his own modifications and using microscopic characters to help delimit species. By 1810 he had described 2,200 species, over 1,700 of which were new (including 140 new genera). He intended to produce an extensive treatise on Australian plants but the poor sales of the first volume, which appeared in 1810, led him to discontinue publication of the remainder.

In the course of his painstaking work Brown became very familiar with plant morphology, which led him to make many important observations. He found that in conifers and related plants the ovary around the ovule is missing, thus establishing the basic difference between these plants and flowering plants or between the gymnosperms and the angiosperms, as the two groups of seed-bearing plants were later named. He also observed and named the nucleus, recognizing it as an essential part of living cells.

In 1827, while examining a suspension of pollen grains in water, under a microscope, Brown observed that the grains were in continuous erratic motion. Initially he believed that this movement was caused by some life force in the pollen, but when he extended his observations to inanimate particles suspended in water he found the same effect. This phenomenon was named *Brownian motion* and remained unexplained until the kinetic theory was developed.

From 1806 to 1822 Brown was librarian of the Linnean Society; in 1810 he also became librarian and curator at Banks's Soho Square residence. Banks stipulated in his will that on his death Brown should take charge of his house, library, and herbarium. In 1820 Brown duly inherited this responsibility and in 1827

he donated Banks's library and herbarium to the British Museum on the understanding that the trustees established an independent botany department in the museum. Thus a botanical collection became accessible to the general public for the first time in Britain.

Brown, Robert Hanbury (1916–2002)
British radio astronomer

Brown, the son of a British army officer, was born in Aruvankadu, India, and educated at the City and Guilds College, London. From 1936 to 1942 he worked on the development of radar at the Air Ministry Research Station, Bawdsey. This was followed by three years in Washington with the British Air Commission and two years with the Ministry of Supply. Brown published an account of his war work in *Boffin* (1991). By the late 1940s Brown was keen to enter academic research and persuaded Bernard LOVELL to admit him as a research student at the Jodrell Bank radio observatory. Brown later served there as professor of radio astronomy from 1960 to 1963 when he was appointed to a professorship in physics at the University of Sydney, a post he held until his retirement in 1981.

In 1950 Brown plotted the first radio map of an external galaxy, the spiral nebula in Andromeda. This was followed by the identification of emissions from four other extragalactic nebulae. Brown further proposed radical changes in the design of standard interferometers that allowed him to measure in 1952 a radio source in Cygnus with a diameter of only 30 seconds of arc. Further work produced in 1954 a new type of interferometer – the so-called optical intensity interferometer.

Brown-Séquard, Charles-Edouard
(1817–1894) *British–French physiologist and neurologist*

Brown-Séquard (brown-say-**kar**) was born in Port Louis, Mauritius, and studied medicine in Paris, graduating in 1846. He was professor of physiology and pathology at Harvard (1864–68) and in 1878 succeeded Claude Bernard as professor of experimental medicine at the Collège de France. The intervening years were spent in a variety of posts in New York, London, and Paris. He is perhaps best known for his work on the adrenal gland. In his experiments on hormonal secretions, he demonstrated the connection between excision of the adrenal glands and ADDISON's disease.

Continuing the work of GALEN on dissection of the spinal cord, he discovered the *Brown-Séquard syndrome* (crossed hemiplegia), a condition of motor nerve paralysis resulting from the lesion of one side of the spinal cord. This produces an absence of sensation on the opposite side of the body to the nerve paralysis.

Brown-Séquard also investigated the possibility of prolonging human life by the use of extracts prepared from the testes of sheep. The majority of his research findings were published as papers in the journal *Archives de physiologie* (Archives of Physiology), of which he was one of the founders.

Bruce, Sir David (1855–1931) *British bacteriologist*

> We are all children of one Father. The advance of knowledge in the causation and prevention of disease is not for the benefit of any one country, but for all ...
> —Address to the British Association, Toronto (1924)

Bruce was a one-time colleague of Robert KOCH in Berlin but spent the greater part of his career as a military physician. Born in Melbourne, Australia, he was educated at Edinburgh University. He was assistant professor of pathology at the Army Medical School, Netley (1889–94), and then commandant of the Royal Army Medical College, Millbank, where he was also director of research on tetanus and trench fever (1914–18). He undertook royal commissions of enquiry into various diseases of man and domestic animals in Malta and central Africa. In Malta he was able to trace the cause of Malta fever (brucellosis or undulant fever found in the milk of goats) to a bacterium later named for him as *Brucella melitensis*. Bruce also investigated the cause of nagana, a disease of horses and cattle in central and southern Africa, and found it to be transmitted by a trypanosome parasite carried by the tsetse fly. This work was of great help in his later research on sleeping sickness (trypanosomiasis), which he also proved to be transmitted by the tsetse fly. The recipient of many honors for his humanitarian work, Bruce was chairman of the War Office's Pathological Committee during World War I. He was knighted in 1908.

Brunel, Isambard Kingdom (1806–1859) *British engineer*

Brunel's father, Marc Brunel (1769–1849), a French emigré and distinguished engineer, arrived in England in 1799. He sent his son to Paris in 1820 to learn mathematics and engineering. Brunel returned to England in 1822 to work for his father and in 1825 they began the construction of the Rotherhithe–Wapping tunnel underneath the Thames. Here Brunel quickly learned of the unpredictability of great engineering projects. He also learned about their danger when the tunnel flooded in 1828 and Brunel nearly drowned.

While convalescing, he heard that the city of Bristol was considering building a bridge across the River Avon. A competition was to be held with Thomas TELFORD as the judge. Brunel submitted plans for a suspension bridge at Clifton,

a span still standing and, perhaps, his most durable monument, even though it remained uncompleted during his lifetime. Telford rejected Brunel's design and proposed instead that he himself should build something more appropriate. The selection committee, however, preferred Brunel's plans. Although work began in 1831 it was not until 1864, well after Brunel's death, that the bridge was opened.

While in the Bristol region other commissions came his way. In 1833 he was invited to build the Great Western Railway (GWR) to run between London and Bristol. He decided to adopt a 7-foot gauge rather than the 4 foot 81/2-inch gauge introduced by George STEPHENSON at the beginning of the railway age. The broad gauge enabled trains to run faster and more comfortably. It did not, however, allow the GWR to link up easily with the rest of the growing railway network. The line was opened in 1841 and extended to Exeter by 1844. It was insisted, however, in the interest of establishing a unitary railway system, that after 1846 no more broad-gauge track could be laid down. The last of the track was removed in 1896.

Not all Brunel's projects were as successful or as durable – in particular, the atmospheric railway that he built in south Devon between Exeter and Newton Abbot in the 1840s. The idea was to eliminate the locomotive. A continuous pipe was laid between the rails and attached to the carriages by a suspended piston. Air was evacuated from the pipe by pumping engines located along the route. In practice, it proved too difficult to maintain the leather seal along the pipe through which the connecting rod emerged; it was either eaten by rats, or made brittle by the sea air, or it froze in winter. Whatever the reason, once breached, the train was brought to a halt. The line was opened in November 1847 and closed the following year, having incurred enormous losses.

Brunel's vision could not be limited by the Atlantic. Why not, he asked, continue the line over the ocean with an equally comfortable liner. This was, for the time, a bold decision. The first Atlantic steam crossing had been accomplished by the American *Savannah* in 1819 using steam as well as sail. Conventional wisdom held that to cross the Atlantic on steam alone would require so much coal as to leave no room for freight. Brunel calculated otherwise and dispelled this myth forever with his *Great Western* (1837; 2,340 tons), a timber ship driven by paddles. It crossed the Atlantic in 15 days with 200 tons of coal unused in its bunkers.

Brunel went on to build the equally revolutionary *Great Britain* (1843; 3,676 tons) with an iron hull and screw propellor, which continued in service for 30 years. His final work was the incredible *Great Eastern* (1858; 32,000 tons) with its double iron hull, screws, and paddles; it was later used to lay the first transatlantic cable.

The struggle to complete the *Great Eastern* against considerable financial and engineering difficulties seems to have ruined Brunel's health and probably caused the stroke he suffered soon after his great ship had been finally launched. He died soon after.

Brunhes, Jean (1869–1930) *French geographer*

Brunhes (**broo**-ne) came from an academic background in Toulouse, France, with both his father and brother being professors of physics. He was educated at the Ecole Normal Supérieure, Paris, (1889–92) and taught at the University of Fribourg from 1896 until 1912, when he moved to the Collège de France, Paris, as professor of human geography.

His most important work was his three-volume *Géographie humaine* (1910; Human Geography), which was translated into English in 1920. Following his teacher, Vidal de la Blache, he argued against the geographical determinism implicit in the work of Friedrich RATZEL. Instead he was more concerned to reveal the complicated interplay between man, society, and the environment.

Brunner, Sir John Tomlinson (1842–1919) *British industrialist*

Brunner's father had traveled to England from Switzerland in 1832 and settled in Liverpool. Brunner began working in the office of a shipping clerk in 1857 and in 1861 joined a chemical firm in Widnes and became chief cashier. There he met Ludwig MOND, a research chemist who had acquired the rights to Ernest SOLVAY's new ammonia–soda process. The two entered into a partnership in 1875 and established a firm in Cheshire to produce soda. Within a few years they had converted an initial loss into a profit, which continued to grow in the succeeding years. In 1881 the partnership was converted into a limited company with capital assets of £600,000. Brunner, Mond, and Company merged with other companies in 1926 to form Imperial Chemical Industries (ICI).

Brunner was made a baronet in 1895 and served as a member of parliament (1885–1910).

Bruno, Giordano (1548–1600) *Italian philosopher*

> Perhaps your fear in passing judgment on me is greater than mine in receiving it.
> —To his judges on the last day of his trial, 8 February 1600

The son of a soldier from Nola in Italy, Bruno (**broo**-noh) entered the Dominican Order in 1565 but was forced to leave in 1576 for unspecified reasons. The following 15 years were spent traveling in France, England, and Germany before visiting Venice in 1591 where he was arrested and handed over to the Inquisi-

tion (1592). He was extradited by the Roman Inquisition in 1593. As details of the trial have been destroyed it is no longer known which eight heretical propositions he refused to recant. The results of his action are not however in any doubt: he was gagged and burned alive.

The exact role of Bruno in 16th-century intellectual history remains a matter of considerable controversy and he was clearly a man of many parts. He was first an expert on the art of mnemonics (memory), a renaissance "science" long extinct, and he was also involved with a revival of the occult mystical philosophical system of hermeticism. More importantly Bruno was also a keen supporter of the heliocentric system of Nicolaus COPERNICUS and in his *Cena de le Ceneri* (1584; The Ash Wednesday Supper) added to some rather implausible arguments in defence of Copernicus's claims for the infinity of the universe. His championing of the then unorthodox heliocentric theory was certainly considered heretical and his unhappy end may well have influenced GALILEO's actions before the Inquisition.

Buch, Christian Leopold von *See* VON BUCH, CHRISTIAN LEOPOLD.

Buchner, Eduard (1860–1917) *German organic chemist and biochemist*

Buchner (**book**-ner) studied chemistry under Adolf von BAEYER and botany in his native city of Munich, gaining his doctorate in 1888. In 1897, while associate professor of analytical and pharmaceutical chemistry at Tübingen, he observed fermentation of sugar by cell-free extracts of yeast. Following PASTEUR's work (1860), fermentation had been thought to require intact cells, and Buchner's discovery of zymase was the first proof that fermentation was caused by enzymes and did not require the presence of living cells. The name "enzyme" came from the Greek *en* = in and *zyme* = yeast. Buchner also synthesized pyrazole (1889). He was professor of chemistry at the University of Berlin from 1898 and won the Nobel Prize for chemistry in 1907 for his work on fermentation. He was killed in Romania, while serving as a major in World War I.

Buchner, Hans Ernst Angass (1850–1902) *German bacteriologist*

Buchner, the brother of Eduard Buchner, was born in Munich, Germany, and gained his MD from the University of Leipzig in 1874. He later worked at Munich University serving as professor of hygiene from 1894 until his death.

In 1888 George Nuttall had shown that the ability of blood to destroy invading bacteria lay in the serum. Buchner followed up his work and went on to demonstrate that the bacteriolytic power was lost when the serum was heated to 56°C. He therefore concluded that serum possessed a heat labile substance that he proposed to name alexin. This work was soon extended by Jules BORDET and the alexins were later renamed complement by Paul EHRLICH. Buchner also did basic work on gamma globulins and developed techniques to study anaerobic bacteria.

Buck, Linda B. (1947–   ) *American neurobiologist*

Buck initially studied psychology and microbiology in her native Seattle, at the University of Washington, obtaining her BS in 1975. After gaining a PhD at the University of Texas in 1980, she worked at Columbia University, first as a postdoctoral fellow (1980–84) then as an associate. In 1991 she joined Harvard Medical School as assistant professor in the neurobiology department, subsequently becoming associate professor (1996) and professor (2001). Throughout this period she was also affiliated to the Howard Hughes Medical Institute at Columbia University. In 2002 she moved to the Fred Hutchinson Cancer Research Center in Seattle. Since 2003 she has been affiliate professor in the Department of Physiology and Biophysics, University of Washington, Seattle.

While at Columbia in the 1980s, Buck made groundbreaking discoveries about the nature of olfaction, or the sense of smell. Working on the mouse as a model organism, in collaboration with Richard AXEL, Buck identified a family of some one thousand genes that encoded the odor receptor molecules found on sensory cells of the olfactory epithelium inside the nose. She and Axel established that these were G protein-coupled receptors. In subsequent work, independently of Axel, Buck demonstrated that each sensory cell carries on its surface just one type of receptor molecule, and that each type of receptor cell sends its electrical signals to a specific relay station, or glomerulus, in the olfactory bulb of the brain.

Buck's later studies showed how this specificity of information from smell receptors is maintained as signals travel from the glomeruli via mitral cells to higher levels of the brain. Micro-regions of the brain cortex receive impulses from mitral cells and combine these to form patterns that are characteristic for the corresponding odors. Here the information is processed to create the perception of smell and relate it to experiences stored as memory.

Buck has also discovered that pheromones – chemical signals released into the environment from other individuals or other species – are sensed by distinct families of G protein-coupled receptors in separate parts of the olfactory epithelium. Her latest work is concerned with instinctive behaviors, and also the possible influence of the brain on aging and lifespan.

She was awarded the 2004 Nobel Prize for physiology or medicine, jointly with Axel, for discoveries of "odorant receptors and the organization of the olfactory system."

Buckland, Francis Trevelyan (1826–1880) *British surgeon and naturalist*

Buckland was born in Oxford, the son of William (Dean) Buckland, founder of the Oxford University Museum of Geology. Educated at Oxford and at St. George's Hospital, London, Buckland practiced medicine and was an army surgeon from 1852 to 1863. Buckland was an authority on pisciculture, investigating the economic aspects of artificial salmon supply, devising ladders to assist the salmon in reaching their marine spawning beds, and observing the preponderance of male as opposed to female trout. He also carried out research into the homing instinct of salmon, deducing that the fish were able to recognize rivers by chemotactile (taste/smell) means. A popular science writer, Buckland is perhaps best known for his many volumes of *Curiosities of Natural History* (1857–72). Buckland also did much to secure international agreement to prevent the extermination of the North Atlantic fur seal.

Budd, William (1811–1880) *British physician*

> A man of the highest genius, [whose] doctrines are now everywhere victorious, each succeeding discovery furnishing an illustration of his marvellous prescience.
> —John Tyndall (19th century)

Budd, the son of a surgeon from North Tawton, England, was the fifth of nine sons, six of whom became doctors. He studied medicine at the Ecole de Médecine in Paris and at the University of Edinburgh where he obtained his MD in 1838. He then practiced in Bristol where he also served as physician at the Royal Infirmary and lectured in medicine at the medical college.

From observations on an outbreak of typhoid fever in North Tawton in 1839 Budd developed a clearly argued case against the view that such outbreaks were due to a generalized atmospheric miasma. He published a number of studies over the years, collected later in his *Typhoid Fever* (1873). He argued that the causative factor could not be atmospheric and stressed that poor sanitary conditions did not of themselves generate typhoid but were important in spreading it. The cause was related to the movement of people, as could be demonstrated by its spread from village to village, and Budd postulated the existence of infective agents, released in the excreta of typhoid sufferers. He thus emphasized the importance of disinfection and a clean water supply. Such views however had little impact on the leading typhoid expert of the day, Charles Murchison, who in his *Treatise on the Continued Fevers* (1873) continued to insist that disease arose spontaneously from dirt and excrement.

Budd also wrote on cholera and in *Malignant Cholera* (1849) narrowly missed anticipating the claim of John SNOW that the disease was waterborne.

Buffon, Comte Georges Louis Leclerc de (1707–1788) *French naturalist*

> The noblest conquest man has ever made.
> —Of the horse. *L'Histoire des mammifères* (The History of Mammals)

The son of wealthy Burgundian landowners, Buffon (boo-**fon**) was born in Montbard; he studied law at Dijon and medicine at Angers. After traveling in Italy and England, he inherited his mother's estate upon her death in 1732. The estate flourished under his direction, benefiting from Buffon's knowledge of silviculture and the ironworks he installed, thus allowing him to concentrate upon scientific matters.

He began by translating S. HALES's *Vegetable Statics* (1735) and NEWTON's *The Method of Fluxions* (1740) into French. In 1739 he was appointed keeper of the Jardin du Roi, a post he occupied until his death. Buffon restored, extended, and embellished the institution, which was renamed as the Natural History Museum during the Revolution.

Buffon began work on his *Histoire naturelle* (Natural History), a work that would dominate the rest of his life and which would eventually run to 44 volumes. The completed *Histoire* consisted of:

Vols. 1–15. *Quadrupeds*, 1749–67, with the assistance of Louis Daubenton who provided the anatomical details.

Vols. 16–24. *Birds*, 1770–83, with the assistance of the Abbé Bexon and G. de Montbeillard.

Vols. 25–31. *Supplementary Volumes*. These deal mainly with the quadrupeds, but Vol. 5 (1778) contains Buffon's important *Epochs of Nature*.

Vols. 32–36. *Minerals*, 1783–88.

The final 8 volumes, *Reptiles* (2 vols., 1788–89), *Fish* (5 vols., 1798–1803), and *Cetacea* (1804) were prepared by E. de Lacepede.

Vol. 1 contained an influential *Preliminary Discourse*. Nature, Buffon argued, was a continuum, and any attempts to divide it into apparently natural classes such as cats and dogs were misguided. Only individuals existed in nature; the rest, genera, species, classes, and orders, were bogus. In accordance with such views Buffon moved in the *Histoire*, quite artificially, from the familiar to the unfamiliar. He began with Man and familiar domestic animals such as dogs, horses, and cows, before moving on to savage animals. The horse was followed by the dog, not the zebra.

In later volumes of the *Histoire* Buffon modified these initial extreme views. He conceded that "two animals belong to the same species as long as they can perpetuate themselves," and also accepted that there did seem to be, beneath superficial differences, "a single plan of structure" present in all quadrupeds. This did not, however, imply a common descent. If, he argued, the ass was derived from the horse, where were the intermediate forms?

Buffon took a bolder line in his *Epochs of Nature*. He argued against the traditional Biblical chronology of about 6,000 years for the Earth's age, claiming instead a period of 78,000 years between the formation of the solar system and the emergence of Man. The estimates were based upon assumptions concerning the rate at which hot bodies of known size and temperature cooled. His calculations allowed him to go further and predict that temperatures will continue to fall, and when they reach 1/25th of the present temperature after 93,000 years, life on Earth will be extinguished.

Bullard, Sir Edward Crisp (1907–1980) *British geophysicist*

Bullard was born in Norwich and educated at Cambridge University, England. After war service in naval research he returned to Cambridge as a reader in geophysics before accepting a post as head of the physics department of the University of Toronto (1948) and visiting the Scripps Institute of Oceanography, California (1949). After a five-year spell as director of the National Physical Laboratory, he returned to Cambridge as a reader and later, in 1964, professor of geophysics and director of the department of geodesy and geophysics. Here he remained until his retirement in 1974.

Bullard made a number of contributions to the revolution in the Earth sciences that took place in the 1950s and 1960s. He carried out major work on the measurement of the heat flow from the Earth. It had been assumed that as the ocean floor was less rich in radioactive material than the continental crust, it would be measurably cooler. The technical difficulties of actually measuring the temperature of the ocean floor were not overcome until 1950, and in 1954 Bullard was able to announce that there was no significant temperature difference between the continental crust and the ocean floor. This led Bullard to reintroduce the idea of convection currents.

In 1965 Bullard studied continental drift, using a computer to analyze the fit between the Atlantic continents. An excellent fit was found for the South Atlantic at the 500-fathom contour line. However, a reasonable fit could only be made for the North Atlantic if a number of assumptions, such as deformation and sedimentation since the continents drifted apart, were taken into account. Later, when independent evidence for these assumptions was obtained, it gave powerful support for the theory of continental drift.

Bullard was knighted in 1953.

Bullen, Keith Edward (1906–1976) *Australian applied mathematician and geophysicist*

Bullen, the son of Anglo-Irish parents, was born in Auckland, New Zealand; he was educated at the universities of Auckland, Melbourne, and Cambridge, England. He began his career as a teacher in Auckland then lectured in mathematics at Melbourne and Hull, England. In 1946 he became professor of applied mathematics at the University of Sydney.

Bullen made his chief contributions to science from his mathematical studies of earthquake waves and the ellipticity of the Earth. In 1936 he gave values of the density inside the Earth down to a depth of 3,100 miles (5,000 km). He also determined values for the pressure, gravitation intensity, compressibility, and rigidity throughout the interior of the Earth as a result of his mathematical studies. From the results on the Earth's density he inferred that the core was solid and he also applied the results to the internal structure of the planets Mars, Venus, and Mercury and to the origin of the Moon.

Bullen conducted some of his early work in collaboration with Harold JEFFREYS on earthquake travel times. This resulted in the publication of the *Jeffreys–Bullen* (JB) *tables* in 1940.

Bunsen, Robert Wilhelm (1811–1899) *German chemist*

Bunsen (**bun**-sen or **buun**-zen), the son of a professor of linguistics, gained his doctorate at the university in his native city of Göttingen (1830) with a thesis on hygrometers. After an extensive scientific tour in Europe, he became a lecturer at Göttingen in 1834. He was professor of chemistry at Kassel (1836), Marburg (1841), and Heidelberg (1852–89).

Bunsen carried out one great series of researches in organic chemistry, *Studies in the Cacodyl Series* (1837–42), after which he abandoned organic for analytical and inorganic chemistry. During his research on the highly toxic cacodyl compound he lost one eye in an explosion and twice nearly killed himself through arsenic poisoning. He prepared various derivatives of cacodyl (tetramethylarsine, $(CH_3)_2As_2(CH_3)_2$), including the chloride, iodide, fluoride, and cyanide, and his work was eagerly welcomed by Jöns BERZELIUS as confirmation of his theory that organic chemistry mirrored inorganic, the "radical theory."

Bunsen was a great experimentalist, an expert in gas analysis and glass blowing, and a pioneer of photochemistry and spectroscopy. He also worked in electrochemistry, devising an improved version of the Grove cell. At Heidelberg he used his new cell to produce metals by electrodeposition. The classic paper *Chemical Analysis through Observation of the Spectrum* (1860) by Bunsen and Gustav KIRCHHOFF ushered in the era of chemical spectroscopy. The spectroscope was an extremely sensitive analytical instrument. With it Bunsen discovered two new elements: rubidium and cesium.

The famous *Bunsen burner* was introduced by him in 1855, although a similar burner, used by Michael FARADAY, did exist before Bunsen and the regulating collar was a later refinement. He greatly refined gas analysis and wrote a standard treatise on the subject, *Gasometrische Methoden* (1857, Methods in Gas Measurement).

Bunsen was a great teacher and at Heidelberg he became a legend. Chemists who came to study with him included Adolph KOLBE, Edward FRANKLAND, Victor and Lothar MEYER, Friedrich BEILSTEIN, and Johann BAEYER.

Burbank, Luther (1849–1926) *American plant breeder*

Burbank was brought up on a farm in Lancaster, Massachusetts, and received only an elementary education. He began breeding plants in 1870, when he bought a 7-hectare (17.3-acre) plot of land. After about a year he had developed the Burbank potato, which was introduced to Ireland to help combat the blight epidemics. By selling the rights to this potato he made $150, which he used to travel to California, where three of his brothers had already settled.

Burbank established a nursery and experimental farm in Santa Rosa, where the climate was especially conducive to fruit and flower breeding – his occupation for the next 50 years. He worked by making multiple crosses between native and introduced strains, using his remarkable skill to select commercially promising types. These were then grafted onto mature plants to hasten development, so that their value could be rapidly assessed. In this way he produced numerous new cultivated varieties of plums, lilies, and many other ornamentals and fruits.

The works of Charles DARWIN, particularly *The Variation of Animals and Plants under Domestication*, greatly influenced Burbank. However his success in varying plant characters reinforced his belief in the inheritance of acquired characteristics, even though he knew of Gregor MENDEL's research.

Burbidge, Eleanor Margaret (1919–   ) *British astronomer*

Born Margaret Peachey in Davenport, England, Burbidge studied physics at the University of London. After graduation in 1948 she joined the University of London Observatory where she obtained her PhD and served as acting director (1950–51). She then went to America as a research fellow, first at the Yerkes Observatory of the University of Chicago (1951–53) and then at the California Institute of Technology (1955–57). The period 1953–55 was spent in highly productive work at the Cavendish Laboratory in Cambridge, England. She returned to Yerkes in 1957, serving as associate professor of astronomy from 1959 to 1962 and then transferred to the University of California, San Diego, where she was professor of astronomy from 1964 until 1990 and emeritus professor from 1990. She also served (1979–88) as director of the Center for Astrophysics and Space Sciences.

Burbidge returned briefly to England in 1972 on leave of absence to become director of the Royal Greenwich Observatory, now situated at Herstmonceux Castle in Sussex. She declared her aim to be to strengthen optical astronomy in Britain. But as the 98-inch (2.5-m) Isaac Newton telescope at Herstmonceux was only a few hundred feet above sea level and sited above a marsh her opportunities for observation at the Royal Observatory were somewhat limited. A little over a year later, in October 1973, Burbidge resigned amid much speculation declaring simply that she preferred "to return to her own research work rather than devote a major part of her time to administrative matters."

In 1948 she married Geoffrey Burbidge, a theoretical physicist, and began a highly productive partnership. They collaborated with Fred HOYLE and William FOWLER in 1957 in publishing a key paper on the synthesis of the chemical elements in stars. They also produced one of the first comprehensive works on quasars in their *Quasi-Stellar Objects* (1967). She had earlier recorded the spectra of a number of quasars with the 120-inch (3-m) Lick reflector and discovered that their spectral lines displayed different red shifts, probably indicating the ejection of matter at very high speeds.

The first accurate estimates of the masses of galaxies were based on Margaret Burbidge's careful observation of their rotation. She was awarded the Bruce Medal in 1999, with her husband Geoffrey Burbidge.

Burbidge, Geoffrey (1925–) *British astrophysicist*

Born at Chipping Norton in central England, Burbidge graduated in 1946 from the University of Bristol and obtained his PhD in 1951 from the University of London. In the period 1950–58 he held junior university positions at London, Harvard, Chicago, Cambridge (England), and the Mount Wilson and Palomar Observatories in California. He became associate professor at Chicago (1958–62) before being appointed associate professor (1962), then professor of physics (1963) at the University of California, San Diego. Burbidge was director of the Kitt Peak National Observatory, Arizona, from 1978 until 1984, when he moved to the University of California, San Diego, as professor of astronomy.

Burbidge began his research career studying particle physics but after his marriage in 1948 to Margaret Peachey, who was to become one of the world's leading optical astronomers, he turned to astrophysics and began a productive research partnership with his wife. The Burbidges worked on the mysterious quasars, first described by Allan SANDAGE in 1960, and produced in their *Quasi-Stellar Objects* (1967) one of the earliest surveys of the subject. Geoffrey Burbidge was far from convinced that quasars were "cosmologically distant" in accordance with the orthodox interpretation of their massive red shifts. In 1965 he proposed with HOYLE that perhaps they were comparatively small objects ejected at relativistic speeds from highly active radio galaxies such as Centaurus A. The effect of this would be to place the main body of quasars only 3–30 million light years from our Galaxy and not the 3 billion light years or more demanded by the generally accepted view.

He was equally reluctant to accept without reservation that other emerging orthodoxy of the 1960s, the big-bang theory on the origin of the universe. In 1971 he published a paper in which he maintained that we still do not know whether the big-bang occurred and that much more effort must be devoted to cosmological tests. Although such views have found little favor, Burbidge has continued, like Hoyle, to be highly productive, rich in new ideas, and yet to remain outside and somewhat skeptical of prevailing cosmological and astrophysical orthodoxy. He was awarded the Bruce Medal in 1999, with his wife Margaret Burbidge.

Buridan, Jean (1300–1358) *French philosopher and logician*

Buridan (**byoo**-ri-dan *or* boo-ree-**dan**) was born in Béthune, France, and studied at the University of Paris under William of Ockham before being appointed professor of philosophy there. It is likely that he remained there all his life as a secular cleric as he is reported to have served as rector of the university in 1328 and 1340.

In his physical works Buridan developed a devastating critique of ARISTOTLE's account of motion. In its place he proposed an impetus theory in which an enforced motion such as the spinning of a millstone is explained by something being impressed into it by the original motive force. Without any external resistance, according to Buridan, the stone would continue spinning forever. The greater the mass and speed of a body the greater its impetus. With such an explanatory framework Buridan and later theorists could begin to tackle some of the basic problems of motion.

Buridan was also an acute logician with his *Sophismata* (Fallacies) containing argument sufficiently subtle to interest contemporary logicians. The animal known as *Buridan's ass* (which was unable to choose between two equidistant bundles of hay) does not, however, appear in any of his works. A similar example does occur in Buridan's commentary on Aristotle's *De caelo* (On the Heavens) where the ass is in fact a dog.

Burkitt, Denis Parsons (1911–1993)
British surgeon

Born in Enniskillen, now in Northern Ireland, Burkitt attended Dublin University, receiving his BA in 1933 and MB in 1935. Having become a fellow of Edinburgh's Royal College of Surgeons in 1938, Burkitt served in the Royal Army Medical Corps during World War II (1941–46). After the war he worked in Uganda as government surgeon (1946–64), being appointed senior consultant surgeon to the Ugandan Ministry of Health in 1961.

In the late 1950s, Burkitt began studying a form of lymphoma that affected children in his part of Africa. Typically these patients had malignant swellings of the facial bones, although tumors could also be found in the ovaries and abdominal lymph nodes. Burkitt demonstrated that all cases were characterized by infiltration of the affected tissues by lymphocytes, and that the various clinical manifestations were all part of the same cancerous condition, now

BURKITT'S LYMPHOMA   Map showing how the incidence of the disease correlates with the incidence of malaria in Africa.

known as *Burkitt's lymphoma*. With his colleagues Edward Williams and Clifford Nelson, Burkitt undertook a geographical survey of the incidence of the disease and found it to be correlated with the same temperature and rainfall zones as malaria. This suggested that the occurrence of the disease may be linked with the distribution of certain insect carriers, as with malaria. Also, an association has now been established between the disease and Epstein-Barr virus, which is isolated from many cases. Burkitt's lymphoma survey is regarded as one of the pioneering studies of geographical pathology.

In later life, and drawing on his experiences of the contrasting diets of developed and developing countries, Burkitt did much to promulgate the benefits to health of a high-fiber diet, and argued that certain diseases of affluent societies, such as bowel cancer and appendicitis, are attributable to dietary fiber deficiency. His publications include *Burkitt's Lymphoma* (1970), *Refined Carbohydrate Foods and Disease* (1975), and *Western Diseases, Their Emergence and Prevention* (1981).

Burnet, Sir Frank Macfarlane (1899–1985) *Australian virologist*

> There is virtually nothing that has come from molecular biology that can be of any value to human living in the conventional sense of what is good, and quite tremendous possibilities of evil, again in the conventional sense.
> —*Changing Patterns* (1968)

Burnet's father, a bank manager, had emigrated to Australia from Scotland as a young man. Burnet was born in Traralgon and studied medicine at Melbourne University, gaining his MD in 1924. After a period abroad at the Lister Institute, London, where he gained his PhD in 1928, Burnet returned to Melbourne to work at the Walter and Eliza Hall Institute; he remained here until his retirement in 1965, having been its director since 1944.

From 1932 until 1933 Burnet worked with the Medical Research Council virology unit in London on influenza. He continued to work with the flu virus in Melbourne, searching for something more convenient than ferrets in which to cultivate the virus. Following the lead of Ernest GOODPASTURE, Burnet showed (1935) that flu virus could be grown in chick embryo (hen's eggs). While developing this new technique Burnet made an unexpected discovery. Adult hens could be infected with flu and, as was well known, develop antibodies against the virus. Yet the chicks born from the eggs used to grow the flu virus failed to develop any flu antibodies. It appeared that there was a period in development before which an organism was "immunologically illiterate"; it could not distinguish between its own tissue and alien tissue.

In 1949 Burnet drew the immunological conclusions from his work. If an antigen were injected into an animal before birth it should develop an immunological tolerance to that antigen, and consequently fail to produce antibodies if ever exposed later in life. But, Burnet discovered, this did not happen. While a young chick exposed to the antigen as an embryo would fail to develop antibodies, such chicks in adulthood display the usual intolerance and produce antibodies to the appropriate antigen. Burnet had failed to realize that the exposure to the antigen must be continuous for tolerance not only to develop but be maintained. The point was later established by Peter MEDAWAR and his colleagues in 1953. It was for this work that Burnet shared the 1960 Nobel Prize for physiology or medicine with Medawar.

Burnet himself found his work on antibodies more satisfying. How, he asked, are organisms able to respond so quickly and so effectively to antigens never before encountered? In 1957, in a paper entitled *Antibody Production Using the Concept of Clonal Selection*, Burnet argued that antibodies, or more accurately the lymphocytes that produce the antibodies, are so comprehensive in their diversity that there is likely to be an antibody in circulation to match any conceivable antigen. The lymphocytes are specialized cells and can respond to just one kind of antigen by producing the appropriately matching antibodies. Once stimulated, however, the lymphocytes will pump out vast numbers of antibodies indefinitely.

Burnet described his work on his clonal selection theory in his autobiography, *Changing Patterns* (1968).

Burnet, Thomas (1635–1715) *English cleric and geologist*

Burnet was born at Croft in England and was educated at Cambridge University, becoming a fellow of Christ's College in 1657. After a period as tutor to various noblemen he was appointed master of Charterhouse School in 1685. He was also appointed chaplain to William III in 1686 but was later forced to resign (1692) because of his controversial account of the history of the Earth.

In 1681 he published his *Telluris theoria sacra*, which was published in an English version as *The Sacred Theory of the Earth* in 1684 and revised and extended in 1691. In this he tackled the problem that was to face all geologists until the last century – how to write a history of the Earth that was consistent with the account given in Genesis. His aim was to take the facts of scripture and show how they could be used to give a rational account of the development of the Earth.

His theory was that the Earth had once been entirely smooth, trapping beneath its shell a large volume of water. Owing to the action of the Sun this shell – the Earth's crust – cracked and released the flood of water; parts of the shattered crust remaining formed the mountains.

Burnet's attempt to explain the history of the Earth in natural terms met a torrent of opposition, both theological and scientific. The strongest argument against this explanation was the presence of marine fossils in mountains for, if the Earth's crust from which the mountains were formed was created before the flood, how could it have come to contain evidence of marine life?

Burt, Cyril Lodowic (1883–1971) *British psychologist*

The son of a London physician, Burt was educated at Oxford University where he studied classics and philosophy. He was introduced to psychology by Oxford's single psychologist, William McDougall. After a period of study in Germany, Burt was appointed to a lectureship in experimental psychology at Liverpool University in 1908. He returned to London in 1912 and remained there for the rest of his career, first as educational psychologist to the London County Council, and from 1932 until his retirement in 1950 as professor of psychology at University College, London.

Much of Burt's life was devoted to the study of intelligence. The statistician Charles Spearman had claimed in 1905 to be able to measure "g," the factor of general intelligence, objectively. In this he was followed by Burt, who further insisted that intelligence was innate as well as general. The surest way to establish this would be, Burt realized, to measure the intelligence quotients of identical twins separated at birth. If intelligence really was inherited, then the IQ of separated twins should show a high degree of correlation, even though they would have been raised in different homes, and educated in different ways. Consequently, he began to collect data from 1912 onwards. By 1966 Burt had collected a sample of 53 identical separated pairs and the correlation of their IQs was the high figure of 0.771.

Soon after, the journalist Oliver Gillie and the psychologist L. J. Kamin began to raise questions about Burt's work. Not only could research collaborators not be found, but their very existence could not be established. Again, the correlation figure of 0.771 appeared to remain constant over the years, despite changes in sample size, a most unlikely statistical outcome. Further, it turned out that much of Burt's work was based not on measurement, but on estimates of home background and intelligence that he made at a distance. For these and other reasons, Burt's work in this field has been largely discounted.

Butenandt, Adolf Friedrich Johann
(1903–1995) *German organic chemist and biochemist*

Butenandt (**boo**-te-nant), who was born at Bremerhaven-Lehe (now Wesermünde) in Germany, took his first degree in chemistry at the University of Marburg and gained his doctorate in 1927 under Adolf WINDAUS at Göttingen. He remained at Göttingen as *Privatdozent* until 1933. Following the work of Windaus on cholesterol, Butenandt investigated the sex hormones and in 1929 he isolated the first pure sex hormone, estrone, from the urine of pregnant women (the compound was also discovered independently by Edward Doisy). A search for the male sex hormone resulted in the isolation in 1931 of 15 milligrams of androsterone from 3,960 gallons of urine.

In 1933 he became professor of organic chemistry at the Danzig Institute of Technology and here he demonstrated the similarities between the molecular structures of androsterone and cholesterol. His proposed structure for androsterone was confirmed by Leopold RUŽIČKA's synthesis in 1934. The male hormone testosterone was synthesized by Butenandt and Ružička only months after its isolation in 1935. Butenandt and Ružička were jointly awarded the Nobel Prize for chemistry in 1939 but Butenandt was forbidden to accept it by the Nazi government. Butenandt was also the first to crystallize an insect hormone, ecdysone, and found that this too was a derivative of cholesterol. Later he led research on the isolation and synthesis of the pheromones.

From 1936 to 1945 Butenandt was director of the Max Planck Institute for Biochemistry at Tübingen and from 1945 to 1956 professor of physiological chemistry there. He retained these posts when the institute moved to Munich in 1956, and in 1960 he succeeded Otto HAHN as president of the Max Planck Society, becoming honorary president in 1972.

Butlerov, Aleksandr Mikhailovich
(1828–1886) *Russian chemist*

Butlerov (**boot**-lyer-of), who was born in Chistopol in Russia, studied at the University of Kazan, graduating as a master of chemistry in 1851. He received his doctorate in 1854 at Moscow and then worked at Kazan, as lecturer (1852) and professor (1857). Russian chemistry, which had been largely dominated by Germans, was backward and Butlerov spent the period 1857–58 traveling in Europe, where he met Friedrich KEKULÉ, Justus von LIEBIG, Friedrich WÖHLER, and Archibald COUPER, and worked in Charles WURTZ's laboratory in Paris.

Butlerov was mainly a theoretician and he extended Kekulé's concepts of organic structure. He proposed that each organic compound has a unique configuration and he invented the term "chemical structure." In 1864 he predicted the existence of tertiary alcohols and in 1876 first introduced the idea of isomers in chemical equilibrium (tautomerism). He wrote *An Introduction to the Full Study of Organic Compounds* (1864). Butlerov was appointed rector of the University of Kazan in 1860 but amid student unrest and clashes between the Russian and German factions he resigned, was reinstated, and finally resigned in 1863. In 1867 he became professor of chemistry at St. Petersburg, where he became increasingly involved in other activities, such as beekeeping and spiritualism.

Buys Ballot, Christoph Hendrik Diederik (1817–1890) *Dutch meteorologist*

Buys Ballot (bIs ba-**lot**) was the son of a minister from Kloetinge in the Netherlands. He was educated at the University of Utrecht, obtaining his PhD in 1844, and became professor of mathematics in 1847 and professor of physics in 1867. He did much to organize the observation and collection of meteorological data in the Netherlands and founded, in 1854, the Netherlands Meteorological Institute.

He is best remembered for the law on wind direction he formulated in 1857. This states that an observer facing the wind in the northern hemisphere has the lower pressure on his right and the higher pressure on his left; in the southern hemisphere this is reversed. Its justification is clearer if stated in the equivalent form: in the northern hemisphere winds circulate counterclockwise around low-pressure areas and clockwise around high-pressure ones.

Byrd, Richard Evelyn (1888–1957) *American polar explorer*

Born in Winchester, Virginia, Byrd graduated from the U.S. Naval Academy in 1912. Forced in 1916 to resign from the navy with an injured leg, he joined the air force. After the end of World War I he became interested in polar aviation. He joined the MacMillan expedition to the Arctic (1924) and, with Floyd Bennett, reputedly made the first flight over the North Pole on 9 May 1926.

Byrd then became interested in Antarctic exploration, leading his first expedition there in 1928. He established his base camp, Little America, on the Ross Ice Shelf and, in 1929, flew over the South Pole. The Antarctic was the largest remaining unmapped and unexplored area of the world and Byrd contributed greatly to opening up and mapping the continent. A more extensive expedition was undertaken in 1933–35, during which Byrd spent five months alone at an Antarctic weather station in 1934, followed by three more, the last in 1955–56.

A large section of the Antarctic was named Marie Byrd Land, after his wife.

Byron, Augusta Ada, Countess of Lovelace (1815–1852) *British computer pioneer*

Ada Lovelace was the daughter of Annabella Millbanke and the poet Lord Byron. Ada's mother left her husband after a month of marriage and Ada never saw her father. Born in London, she was educated privately, studying mathematics and astronomy in addition to the more traditional topics. She seems to have developed an early ambition to be a famous scientist. Her correspondence, however, with Mary SOMERVILLE and Augustus De MORGAN, two of her informal teachers, shows that her formal skills were very limited. De Morgan saw her as a talented beginner who could have become an original mathematician if given the chance to receive a rigorous formal Cambridge-style training. Such routes were not open to young Victorian women.

In 1834 she heard Charles BABBAGE lecture on his famous "difference engine." She offered Babbage her support and they became good friends. In 1842 she translated from French an account of Babbage's analytical engine by the Italian engineer, L. F. Manabrea. At Babbage's suggestion she added some explanatory notes. They constitute one of the primary sources of his work.

C

Cagniard de la Tour, Charles (1777–1859) *French physicist*

> Ferments ... are composed of very simple organized microscopic bodies ... brewer's yeast is a mass of small globulous bodies capable of reproducing themselves ... it is very probably through some effect of their growth that they release carbon dioxide and ... convert [a sugary solution] into a spirituous liquor.
> —Describing the role of yeast in fermentation

Born in Paris, France, Cagniard de la Tour (ka-**nyar** de la toor) was educated at the Ecole Polytechnique and then spent his time as an amateur inventor. In 1819 he invented the disk siren, in which the sound is produced by air blowing through holes in a rotating disk, the pitch being determined by the speed of rotation. He made his most famous discovery in 1822; when he heated certain liquids in sealed tubes he observed that at a particular temperature and pressure the meniscus dividing liquid from vapor disappeared. Under these conditions – known as the *critical state* – the densities of liquid and vapor become the same and the two are identical.

In the field of biology Cagniard de la Tour discovered, independently of Theodor SCHWANN, the role of yeast in alcoholic fermentation. He also studied the physics of the human larynx and the sounds produced by it and invented a machine for studying bird flight.

Cahours, August André Thomas (1813–1891) *French organic chemist*

> These classic works [Cahours's] have become one of the solid foundations on which we today base the fundamental proposition that as a perfect gas the [gram] molecular weight of most substances occupies the same volume in the vapor state.
> —Armand Gautier (1891)

Cahours (ka-**hor**), who was born in Paris, France, was professor at the Ecole Polytechnique there (1871) and warden of the Paris mint. He made many notable discoveries, including methyl salicylate, amyl alcohol in fusel oil, anisole and phellandrene in fennel oil, and the alkyl compounds of aluminum, arsenic, and beryllium.

Cailletet, Louis Paul (1832–1913) *French physicist*

Born in Chatillon-sur-Seine in France, the son of a metallurgist, Cailletet (ka-ye-**tay**) studied in Paris and then became a manager at his father's foundry.

He is most famous for his work on the liquefaction of gases. Cailletet realized that the failure of others to liquefy the permanent gases, even under enormous pressures, was explained by Thomas Andrew's concept of critical temperature. In 1877 he succeeded in producing liquid oxygen by allowing the cold, compressed gas to expand. This technique, depending on the effect discovered by JOULE and THOMSON, cooled the gas to below its critical temperature. In later experiments he liquefied nitrogen and air. Raoul PICTET, working independently, used a similar technique. In 1884 Cailletet was elected to the Paris Academy for his work. He is also the inventor of the altimeter and the high-pressure manometer.

Callippus (*c.* 370 BC–*c.* 300 BC) *Greek astronomer*

Callippus (ka-**lip**-us), who was born in Cyzicus (now in Turkey), was possibly a pupil of EUDOXUS. He is also known to have had discussions on astronomy with ARISTOTLE in Athens. Although none of his works have survived he is known to have reformed the Eudoxan system. Callippus is also reported as suggesting improvements in the Metonic cycle, the period – of 19 years or 235 synodic (lunar) months – after which the phases of the Moon recur on the same days of the year. He proposed taking a period of 76 years (4 Metonic cycles) made from 940 lunar months, 28 of which would be intercalary (inserted). This would give a solar year of precisely 365 1/4 days. The Greeks never actually adopted any of these cycles but they do give some indication of how accurately their astronomers had worked out the length of the year.

Calvin, Melvin (1911–1997) *American chemist and biochemist*

Born in St. Paul, Minnesota, Calvin studied chemistry at the Michigan College of Mining and Technology and gained his BS degree in

1931. After obtaining his PhD from the University of Minnesota in 1935 he spent two years at the Victoria University of Manchester, England, working with Michael POLANYI. Here he became interested in chlorophyll and its role in the photosynthetic process in plants. Calvin began a long association with the University of California at Berkeley in 1937. From 1941 to 1945 he worked on scientific problems connected with the war, including two years on the Manhattan Project (the atomic bomb).

In 1946 Calvin became director of the Bio-organic Division of the Lawrence Radiation Laboratory at Berkeley, where he used the new analytical techniques developed during the war – ion-exchange chromatography, paper chromatography, and radioisotopes – to investigate the "dark reactions" of photosynthesis, i.e., those reactions that do not need the presence of light. Plant cells were allowed to absorb carbon dioxide labeled with the radioisotope carbon–14, then immersed at varying intervals in boiling alcohol so that the compounds they synthesized could be identified. In this way the cycle of photosynthetic reactions (known as the *Calvin cycle*) was elucidated and shown to be related in part to the familiar cycle of cell respiration. This work, which was collected in *The Path of Carbon in Photosynthesis* (1957), earned Calvin the Nobel Prize for chemistry in 1961.

Calvin remained at Berkeley, as director of the Laboratory of Chemical Biodynamics (1960–63), professor of molecular biology (1963–71), and professor of chemistry (1971). He continued to work on problems of photosynthesis (especially on the role of chlorophyll in quantum conversion) and on the evolution of photosynthesis.

Campbell, Keith (1954–) *British cell biologist*

Campbell initially qualified as a medical technician, and then studied microbiology at Queen Elizabeth College, London, where he obtained a BSc. He worked briefly in the Yemen as a chief medical laboratory technician and, on returning to the UK, embarked on a project to help control Dutch elm disease. In 1983 he joined the University of Sussex, where he was subsequently awarded a DPhil. Following two postdoctoral positions he joined the Roslin Institute, near Edinburgh, in 1990. He left the Institute in 1997 to become head of embryology at a biotech company, PPL Therapeutics, and in 1999 he was appointed professor of animal development at Nottingham University.

From the outset, Campbell's main scientific interest was the cell cycle and its control mechanisms. From the time it is formed to the time it divides, every cell goes through a sequence of events known as the *cell cycle*. For his doctoral thesis at Sussex University, Campbell investigated the control exerted on nuclear division in yeast and *Xenopus* (a toad) by factors in a cell's cytoplasm.

At the Roslin Institute, Campbell collaborated with the embryologist Ian WILMUT, who was refining the technique of nuclear transfer for cloning mammals. This technique, employed by cloning pioneers such as Steen WILLADSEN, involved taking a donor cell from an early embryo and fusing it with a recipient unfertilized egg cell from which the nucleus had been removed. The fused cell was then implanted in a surrogate mother and allowed to develop. Campbell devised a new method of synchronizing the cell cycles of the donor and recipient cells, as a means of reducing the high failure rate. He initially placed the donor cells in a low-nutrient medium, thereby starving them into a state of quiescence in which all but their most essential genes were switched off. This coordinated their cell cycle with that of the recipient egg cells. As a result of this modification, in 1995 Wilmut and Campbell successfully cloned two sheep from embryo cells – Megan and Morag – and in 1996 came their most famous arrival – another sheep named Dolly, the first mammal to be cloned from fully differentiated body cells.

After moving to PPL Therapeutics, a company with close ties to the Roslin Institute, Campbell focused on applying cloning technology to developing transgenic livestock with potential medical applications, for example as a source of stem cells or tissue for human transplants. He continues to research mechanisms of development and differentiation, and the cell cycle.

Camerarius, Rudolph Jacob (1665–1721) *German botanist*

Born the son of a professor of medicine at Tübingen in Germany, Camerarius (kam-er-**ar**-ee-uus) was himself educated there and received his doctorate in 1687. He joined the staff at Tübingen and following his father's death in 1695 was appointed professor of medicine and director of the botanic gardens, posts he occupied until his death from TB in 1721.

In 1694 in *De sexu plantarum* (On the Sex of Plants) Camerarius produced clear experimental evidence for the sexuality of plants first proposed by John RAY and Nehemiah Grew. By isolating pistillate (female) dioecious plants from staminate (male) plants (dioecious referring to plant species where the male and female flowers are borne on separate plants), he was able to show that although the pistillate plants produced fruit, they lacked seeds. With monoecious plants (those that bear separate unisexual male and female flowers on the same plant, e.g., corn) he found that removing the male inflorescence also resulted in sterile fruit.

In his description of plant anatomy Camerarius identified the stamens as the male organ and the style and stigma as the female part. He also described the role of pollination.

Cameron, Sir Gordon Roy (1899–1966)
Australian pathologist

Born at Echuca, Australia, the son of a Methodist minister, Cameron studied medicine at Melbourne University, graduating in 1922. He worked first at the university and then at the Walter and Eliza Hall Institute before leaving for Europe in 1927 to do postgraduate work at Freiburg. Shortly afterward he was appointed to the staff of University College Hospital, London, where he served as professor of morbid anatomy from 1937 until his retirement in 1964.

Cameron worked on a wide range of problems including pulmonary edema, inflammation, and the pathology of the spleen and liver. He also produced a major survey of all aspects of the field with his *Pathology of the Cell* (1952).

Candolle, Augustin Pyrame de (1778–1841) *Swiss botanist*

Candolle (kan-**dol**), who was born in Geneva, Switzerland, studied medicine for two years at the academy there before moving to Paris in 1796 to study both medicine and natural sciences. In Paris he met many distinguished naturalists, including Georges CUVIER and Jean Baptiste LAMARCK, and quickly established his own reputation through the publication of many outstanding monographs on plants. He received his MD from the University of Paris in 1804 and, at the request of the French government, made a botanical and agricultural survey of France between 1806 and 1812.

In 1813 he published his famous *Théorie élémentaire de la botanique* (Elementary Theory of Botany), in which he introduced the term "taxonomy" to mean classification. This work was based on the natural classificatory systems of Cuvier and Antoine JUSSIEU, and in it Candolle maintained that relationships between plants could be established through similarities in the plan of symmetry of their sexual parts. He realized that the symmetry could be disguised by fusion, degeneration, or loss of sexual organs, making structures with a common ancestry appear different. Candolle thus formulated the idea of homologous parts – a concept that lends much weight to the theory of evolution, but surprisingly he continued to believe in the immutability of species. Candolle's classification replaced that of LINNAEUS and was used widely until George BENTHAM and Joseph HOOKER produced their improved system 50 years later.

Candolle also made important contributions to plant geography, realizing that the distribution of vegetation can be profoundly influenced by soil type. The relationships he described between plants and soil were backed up by personal observations from his travels in Brazil, East India, and North China.

From 1808 to 1816 Candolle was professor of botany at Montpelier University, after which he returned to Geneva to take the chair of natural history at the Academy. On his arrival in Geneva he completely reorganized the gardens. Between 1824 and 1839 he published the first seven volumes of his huge *Prodromus Systematis Naturalis Regni Vegetabilis* (Guide to Natural Classification for the Plant Kingdom), an encyclopedia of the plant kingdom. His son, Alphonse de Candolle, saw to the publication of the remaining ten volumes after his father's death and also carried on many of his father's other schemes.

Cannizzaro, Stanislao (1826–1910) *Italian chemist*

Born the son of a magistrate in Palermo, Sicily, Cannizzaro (kan-i-**zah**-roh) studied physiology in his native city and at Naples. He turned to organic chemistry after realizing the importance of chemical processes in neurophysiology, and from 1845 to 1847 worked as a laboratory assistant to R. Piria at Pisa. Cannizzaro was an ardent liberal and in 1847 he returned to Sicily to fight as an officer in the insurrection against the ruling Bourbon regime. Following the abortive revolution of 1848 he went into exile and returned to chemistry, working with Michel Eugène CHEVREUL in Paris (1849–51).

Cannizzaro returned to Italy in 1851 as professor of chemistry and physics at the Collegio Nazionale at Alessandria. In 1853 he discovered the reaction known as *Cannizzaro's reaction*, in which an aromatic aldehyde is simultaneously oxidized and reduced in the presence of concentrated alkali to give an acid and an alcohol.

In 1855 Cannizzaro moved to Genoa as professor of chemistry and here he produced the work for which he is chiefly remembered. His pamphlet *Sunto di un corso di filosofia chimica* (1858; Epitome of a Course of Chemical Philosophy) finally resolved more than 50 years of confusion about atomic weights. In 1860 a conference was held at Karlsruhe, Germany, to discuss the problem. No agreement was reached but Cannizzaro's pamphlet was circulated and soon after was widely accepted. In it Cannizzaro restated the hypothesis first put forward by Amedeo AVOGADRO, clearly defined atoms and molecules, and showed that molecular weights could be determined from vapor-density measurements.

Politics intervened once more in Cannizzaro's life and in the struggle to reunite Italy he returned to Palermo in 1860 to join Garibaldi. He was professor of inorganic and organic chemistry at Palermo until 1870, when he went to Rome to found the Italian Institute of Chemistry. The most notable research of this last period was that on santonin, a compound derived from species of *Artemisia* (wormwoods) that is active against intestinal worms, which Cannizzaro showed to be a derivative of naphthalene. He was widely honored and became a senator in 1871.

Cannon, Annie Jump (1863–1941) *American astronomer*

> Miss Cannon was not given to theorizing; it is probable that she never published a controversial word or a speculative thought. That was the strength of her scientific work – her classification was dispassionate and unbiased.
> —Cecilia Payne-Gaposchkin, "Miss Cannon and Stellar Spectroscopy," *The Telescope*, No. 8 (1941)

Annie Cannon was the daughter of a Delaware state senator and was born in Dover in that state. She was one of the first girls from Delaware to attend university, being a student at Wellesley College from 1880 to 1884. After a decade spent at home, where she became deaf through scarlet fever, she entered Radcliffe College in 1895 to study astronomy. In 1896 she was appointed to the staff of the Harvard College Observatory, as it was the practice of the observatory, under the directorship of Edward PICKERING, to employ young well-educated women to do calculations. She worked there for the rest of her career, serving from 1911 to 1932 as curator of astronomical photographs. In 1938, after nearly half a century of distinguished service, she was appointed William Cranch Bond Astronomer.

One of the main programs of the observatory was the preparation of the *Henry Draper Catalogue* of a quarter of a million stellar spectra. Stars were originally to be classified into one of the 17 spectral types, A to Q, which were ordered alphabetically in terms of the intensity of the hydrogen absorption lines. Cannon saw that a more natural order could be achieved if some classes were omitted, others added, and the total reordered in terms of decreasing surface temperature. This produced the sequence O, B, A, F, G, K, M, R, N, and S. Cannon showed that the great majority of stars can be placed in one of the groups between O and M. Her classification scheme has since only been slightly altered.

Cannon developed a phenomenal skill in cataloging stars and at the height of her power it was claimed that she could classify three stars a minute. Her classification of over 225,000 stars, brighter than 9th or 10th magnitude, and the compilation of the *Catalogue* took many years. It was finally published, between 1918 and 1924, as volumes 91 to 99 of the *Annals of Harvard College Observatory*. She continued the work unabated, later publications including an additional 47,000 classifications in the *Henry Draper Extension* (*Annals*, vol. 100, 1925–36). Even as late as 1936 when she was over 70 she undertook the classification of 10,000 faint stars submitted to her by the Cape of Good Hope Observatory.

Cannon, Walter Bradford (1871–1945) *American physiologist*

> Only the moral degenerate is capable of inflicting the torment that the anti-visectionist imagines. No one who is acquainted with the leaders in medical research who are responsible for the work done in the laboratories can believe for a moment that they are moral degenerates.
> —*Journal of the American Medical Association*, No. 51 (1908)

Cannon, who was born in Prairie du Chien, Wisconsin, graduated from Harvard in 1896 and was professor of physiology there from 1906 to 1942. His early work included studies of the digestive system, in particular the use of x-rays to study stomach disorders. For this he introduced the *bismuth meal*. Most of his working life, however, was spent studying the nervous system, particularly the way in which various body functions are regulated by hormones. As early as 1915 he showed the connection between secretions of the endocrine glands and the emotions. In the 1930s he worked on the role of epinephrine in helping the body to meet "fight or flight" situations. He also studied the way hormonelike substances are involved in transmitting messages along nerves.

Canton, John (1718–1772) *British physicist*

Canton, who was born in Stroud in western England, was a schoolmaster in London and a gifted amateur physicist. He invented a new technique for making artificial magnets in 1749 and was elected to the Royal Society, which gave him a medal for this work in 1751.

He followed Benjamin FRANKLIN's theories of electrostatics and was the first in England to repeat Franklin's experiments with lightning. He also discovered a few more properties of electrostatic induction, including the fact that both glass and clouds could become negatively charged instead of, as was usual, positively. In 1757, he made the first observations of fluctuations in the Earth's magnetic field, which, at the end of the next century, led to the discovery of charged layers in the atmosphere. In 1762 he demonstrated that – in spite of scientific opinion to the contrary – water was compressible.

Cantor, Georg Ferdinand Ludwig Philipp (1845–1918) *German mathematician*

The son of a prosperous merchant of St. Petersburg, at that time the capital of Russia, Cantor (**kan**-tor or **kahn**-tor) was educated at the University of Berlin where he completed his PhD in 1868. In 1870 he joined the faculty of the University of Halle and was appointed professor of mathematics in 1879. He spent his entire career at Halle, although it was a career repeatedly interrupted after 1884 by mental illness; he was a manic depressive and was hospitalized first in 1899 and several times thereafter. After 1897 he made no further contribution to mathematics and died of heart failure in 1918 in a mental institution.

Although Cantor's earliest work was concerned with FOURIER series, his reputation rests upon his contribution to transfinite set theory. He began with the definition of infinite sets proposed by DEDEKIND in 1872: a set is infinite when it is similar to a proper part of itself. Sets with this property, such as the set of natural numbers, are said to be "denumerable" or "countable."

In 1874, Cantor published a remarkable paper in Crelle's *Journal*. Here he first showed that the rational numbers (numbers that can be expressed by dividing one integer by another: 1/2, 1/3, 1/4, 2/3, etc.) are denumerable – they can be put in a one-to-one correspondence with the natural numbers (1, 2, 3, etc.). The usual method of demonstrating this is to set up an array in which the first line contains all the rationals in which the denominator is 1 (1/1, 2/1, 3/1, etc.), the second line has all the rationals with a denominator 2 (1/2, 2/2, 3/2, etc.), and so on. It is then possible to "count" all the fractions in the array by moving diagonally backwards and forwards through the array, and it is clear that every rational number can be put in one-to-one correspondence with an integer. This technique, known as *Cantor's diagonal procedure*, is not the one Cantor used in the 1874 paper, although he did give the diagonal demonstration later. The set of rational numbers and the set of natural numbers are said to have the same "power."

Cantor then went on to show that the set of all real numbers is not denumerable. He did this by a *reductio ad absurdum* method. First he assumed that all the real numbers between 0 and 1 are denumerable and expressed as infinite decimal fractions (e.g., 1/3 = 0.333...). They are arranged in denumerable order:

$$a_1 = 0.a_{11}\,a_{12}\,a_{13}\,a_{14}\,...$$
$$a_2 = 0.a_{21}\,a_{22}\,a_{23}\,a_{24}\,....\ \text{etc.}$$

Here, a_1 is the first real number and a_{11} the first digit, a_{12} the second, etc. The first digit of the second number a_2 is a_{21}, etc. Cantor then showed that it is possible to construct an infinite decimal that is not in the above set by taking the diagonal containing a_{11}, a_{22}, a_{33}, etc., and changing the digit to 9 if it is 1 and changing the digit to 1 for all other digits. This gives

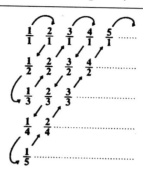

CANTOR'S METHOD   The diagonal procedure for demonstrating that the rational numbers are countable.

a number that is a real number between 0 and 1, yet is not in the above set. In other words, the real numbers are not denumerable – there is a sense in which there are "more" real numbers than rational numbers or natural numbers. The set of real numbers has a higher power than the set of natural numbers.

Cantor designated the set of natural numbers, the smallest transfinite set, with the symbol $\aleph_0$ and the set of real numbers by the letter c, the number of the continuum. $\aleph$ is the first letter of the Hebrew alphabet, called "aleph." Cantor's symbol $\aleph_0$ is referred to as "aleph nul."

Cantor went on to show that there were in fact an infinite number of transfinite sets. The power set of a set S consists of the subsets of S. Thus let $S = (1,2)$, the power set of S, P(S) = [(1), (2), (1,2), ($\varnothing$)], bearing in mind that every set is a member of itself, and that the empty set ($\varnothing$) belongs to every set. In general Cantor demonstrated that if a set S has n members then P(S) will have 2^n members, that P(S) > S, a result since known as *Cantor's theorem*.

The theorem applies to all sets, infinite as well as finite. Thus the power set of $\aleph_0$ will be greater than $\aleph_0$ and the process can be continued with the power set of the power set of $\aleph_0$, and so on indefinitely. Cantor had shown that the set of natural numbers had a cardinality of $\aleph_0$, and the real numbers, c, had a cardinality of $2^{\aleph_0}$. This enabled him to pose in 1897 the hypothesis that $2^{\aleph_0} = \aleph_1$, or, that the continuum (c) is the next highest infinite number after $\aleph_0$. Cantor made little progress with the continuum problem. It remained for Godel and COHEN to illuminate the issue many years later.

Capecci, Mario R. (1937–) *American geneticist*

Born in Verona, Italy, Capecchi had a turbulent childhood. His mother was arrested by the Gestapo in 1941 and deported to Dachau concentration camp; the young Capecchi lived with a local peasant family in the Italian Alps, but after a year was forced to set off alone and live off his wits. Despite numerous hardships, he survived, and after the war was reunited with his mother while being treated for malnutrition in a children's hospital. Mother and son sailed for the USA and Capecchi lived with his aunt and uncle near Philadelphia. Despite his lack of formal education, he made rapid progress at high school and, subsequently, at Antioch College, where he obtained a BS in 1961. He then moved to Harvard University, and the laboratory of James D. WATSON, of Watson and CRICK fame. Following the award of a PhD in 1967 he became assistant professor of biochemistry (1969) and then associate professor (1971). He joined the University of Utah, as professor of biology, in 1973, becoming professor of human genetics in 1989, and distinguished professor of human genetics and biology in 1993. In 1988 he was also appointed investigator for the Howard Hughes Medical Institute.

As early as 1977, Capecchi was developing the techniques that would be crucial for his groundbreaking work in the targeted mutation of genes, using the mouse as an experimental model. He demonstrated that DNA containing functional genes could be injected into the nucleus of a cell using an ultrafine glass needle controlled by micromanipulators, and that the genes would be stably inherited by the progeny cells on subsequent cell division. However, such injected genes were integrated at random into the host chromosome. What was needed was a more precise method of controlling where an introduced gene would be incorporated. This came with the discovery by Capecchi's team that somatic cells (i.e. body cells) were capable of performing homologous recombination, a process hitherto assumed to be confined to germline cells during the production of eggs or sperm. By this process, a mutant gene introduced into a nucleus can, at a low frequency, find and swap places with its homologue, a counterpart gene with the same sequence of bases in its DNA. Hence, appropriate modification of the introduced gene can be used to alter or disable that particular gene in the cell and in all its progeny.

Capecchi and colleagues initially reported their findings in 1982, and by 1984 were confident that the technique offered a viable strategy for gene targeting of cultured mammalian cells. However, was it possible to extend the technique to entire animals, to obtain strains of mice carrying the altered gene? To attempt this Capecchi made use of cultured embryonic stem cells. The British geneticist Martin EVANS had shown that such cells could be genetically modified, and then injected into a developing mouse embryo at the blastocyst stage, ultimately giving rise to various tissues in the adult animal. Such an organism, containing body cells of different genetic makeup, is called a chimera. Capecchi now selected a particular gene, the *Hprt* gene, encoding the enzyme hypoxanthine

phosphoribosyltransferase (HPRT), and disabled it by inserting a gene conferring resistance to the antibiotic neomycin. Embryonic stem (ES) cells transfected with the mutant gene were shown to have lost the ability to make the HPRT enzyme and gained resistance to neomycin, proving that the targeted modification was successfully incorporated. These findings were complemented by the work of Oliver SMITHIES, who had been exploring the potential of gene targeting in attempts to repair defective genes in cultured human tissue cells.

The following year, in 1988, Capecchi and colleagues reported an improved method for selecting ES cells carrying a desired genetic modification. This was the positive-negative selection strategy, which used a vector containing two 'reporter' genes in addition to the 'foreign' gene. Such a strategy enabled researchers to distinguish cells in which the foreign gene had recombined at the target site from cells in which it had integrated at random. This was another significant step toward producing mouse strains with precisely targeted mutations transmitted stably in the germline cells, and ushered in a new era.

The results from Capecchi's team encouraged many others to enter the field. By 1989 the first 'knockout' mouse strains had been bred, with specific genes disabled using the new gene targeting technique. These allowed researchers to identify the effects of targeted mutations on the development or health of the animals, and gave a much clearer insight into the function of the genes concerned. Subsequent developments brought the ability to knockout a gene only in particular tissues or at certain stages of development, and to replace a gene with another, different gene – creating so-called 'knockin' mice.

Capecchi has continued to investigate the effect of targeted mutations on mouse development, and to unravel the roles of various genes, notably Hox genes, in patterning the early mammalian embryo and in adult life. Through such work his team has produced mouse models of certain human genetic disorders, such as synovial sarcoma and alveolar rhabdomyosarcoma, both of which are aggressive childhood cancers in humans.

In 2001 Capecchi was awarded the Albert Lasker Award for basic medical research, jointly with Evans and Smithies. The trio were honoured again in 2007, by receiving the Nobel Prize for physiology or medicine, for discovering principles for introducing specific gene modifications in mice by the use of embryonic stem cells.

Cardano, Girolamo (1501–1576) *Italian mathematician, physician, and astrologer*

Oh, the madness of men to give heed to vanity rather than the fundamental things of life!
—*Ars Magna* (1545; The Great Skill)

The work of Cardano (kar-**dah**-noh) constitutes a landmark in the development of algebra and yet in his own time he was chiefly known as a physician. He studied medicine at the University of Pavia, the city of his birth, and at the University of Padua, receiving his degree in 1526. He spent much of his life as a practicing physician, becoming professor of medicine at Pavia in 1543. One of his notable nonmathematical achievements was to give the first clinical description of typhus fever.

It was however in mathematics that Cardano's real talents lay. His chief work was the *Ars magna* (1545; The Great Skill) in which he gave ways of solving both the general cubic and the general quartic. This was the first important printed treatise on algebra. The solution of the general cubic equation was revealed to him by Niccoló TARTAGLIA in confidence and Cardano's publication aroused a bitter controversy between the two. Cardano's former servant, Lodovico Ferrari, had discovered the solution of the general quartic equation. In his later *Liber de ludo aleae* (Book on Games of Chance) Cardano did some pioneering work in the mathematical theory of probability.

Cardano's interests were not, however, limited to mathematics and medicine. He also indulged in philosophical and astrological speculation and this had the unfortunate consequence that in 1570 he was charged with heresy by the Church. He was briefly jailed but was soon released after the necessary recantation. As a result of this episode Cardano lost his post as a professor at the University of Bologna, which he had held since 1562.

Carlson, Chester Floyd (1906–1968)
American inventor and entrepreneur

The son of a barber from Seattle, Washington, Carlson studied physics at the California Institute of Technology. After graduating in 1930 he began to work for Bell Telephones. Soon after, however, Carlson opted for a change of career, took a law degree, and began to work in the patent department of a New York electronics firm. While there he worked with a refugee German physicist, Otto Kornei.

Carlson was struck by repeated demands for copies of patent applications and the difficulty of meeting this demand rapidly and accurately. Typing could not cope with complex diagrams, and other methods of reproduction were far too slow for a busy legal office. Consequently Carlson and Kornei began to consider what they initially called "electro-photography." The first patent was taken out in 1937 and the first actual electrostatic copy made the following year when Carlson made a clear copy of the date

and site of their achievement, namely, "10-22-38 Astoria."

Carlson's aim was to produce dry, clear, multiple, black-and-white copies of almost any material, rapidly, accurately, and cheaply. Shunning traditional photographic methods as too slow and costly, Carlson began to look at electrostatic forms of reproduction. He began by coating a rotating drum with positively charged selenium, and focusing an image on to the drum. As selenium is photoconductive, exposure to light will remove all charge except in the areas covered by the focused image. The charged image can then be made to pick up the toner, a special black carbon dust, which can in turn be transferred to specially charged paper. After passing through a heating stage to fuse the toner on the paper, and under a brush to remove excess dust, the copy was ready for use.

Once the formidable problem of ensuring the machine's reliability had been overcome, Carlson faced the even greater difficulty of selling his product. After having been turned down by all the leading office-equipment manufacturers, the Batelle Memorial Institute, a research organization in Columbus, Ohio, offered to back Carlson. In 1947 the Haloid Company of Rochester, New York, bought the rights to produce Carlson's new copier. They became Haloid Xerox in 1958, and just plain Xerox in 1961. The term *xerography*, from the Greek word *xeros*, dry, first appeared on 23 October 1948, in the *New York Times*.

With the introduction of the 914 series, the first push-button copier, Xerox was about to conquer the world and turn Carlson into a multimillionaire in the process.

Carlsson, Arvid (1923–   ) *Swedish pharmacologist*

Carlsson was born in Uppsala, Sweden. He graduated MD in 1951 from the University of Lund, where in the same year he was appointed assistant professor, becoming associate professor in 1956. In 1959 he moved to Göteborg University as professor of pharmacology.

Carlsson's main interest has been the nature and role of signaling molecules (neurotransmitters) in the nerve cells of the brain. In the late 1950s he discovered that dopamine is a neurotransmitter in the brain, particularly in parts of the brain called the basal ganglia, which play an important role in controlling muscular activity.

Using experimental animals, Carlsson showed that if dopamine was depleted, the animals developed signs similar to those of Parkinson's disease in humans. These signs could be alleviated, and levels of dopamine could be restored, by administering the dopamine precursor L-dopa. Subsequently it was found that sufferers of Parkinson's disease do indeed have unusually low levels of dopamine in their basal ganglia. As a result, L-dopa was developed as a drug to treat the symptoms of the disease in humans.

Carlsson's findings have also contributed to the development of more effective drugs for the treatment of depresssion, notably the new generation of antidepressive drugs known as selective serotonin uptake blockers.

For his work on signal transduction in the nervous system, Carlsson was awarded the 2000 Nobel Prize for physiology or medicine, jointly with Paul GREENGARD and Eric R. KANDEL.

Carnap, Rudolf (1891–1970) *German philosopher and logician*

Carnap (**kar**-nap) had a rigorous scientific and philosophical education, which was reflected in the style and content of all his later work. Born in Ronsdorf, Germany, he studied mathematics, physics, and philosophy at the universities of Jena and Freiburg (1910–14) and obtained a doctorate from Jena with a thesis on the concept of space in 1921.

In 1926 Carnap was invited to take up a post at the university of Vienna where he became a major figure in the Vienna Circle – a group of philosophers and mathematicians founded by Moritz Schlick. This group had an empiricist outlook, i.e., that all our ideas, concepts, and beliefs about the external world derive from our immediate sensory experience. Out of this evolved the logical empiricist or logical positivist school of thought, which states that the meaningful statements we can make are just those that have logical consequences that are observably verifiable. That is, meaningful statements must be testable by experience; those that are not, like the propositions of metaphysics and religion, are, strictly, meaningless.

Throughout his life Carnap used the tools of symbolic logic to bring a greater precision to philosophical inquiry, including investigations into the philosophy of language and into probability and inductive reasoning. He produced his first major work in 1928, *Der Logisch Aufbau der Welt*, translated into English in 1967 as *The Logical Structure of the World*. In this he developed a version of the empiricist reducibility thesis, holding that scientific theories and theoretical sentences must be reducible to sentences that describe immediate experiences, which are observably verifiable. His other works included *Logishe Syntax der Sprache* (1934; The Logical Syntax of Language) and *Logical Foundations of Probability* (1950).

He emigrated to America in 1936 becoming professor of philosophy at Chicago until 1952 and at UCLA until 1961.

Carnot, Nicolas Leonard Sadi (1796–1832) *French physicist*

Carnot (kar-**noh**) came from a distinguished Parisian political family; his father, Lazare, was a leading politician under Napoleon Bonaparte. He studied at the Ecole Polytechnique, from which he graduated in 1814. For the next few years he worked as a military engineer, but the political climate had changed with the fall of Bonaparte and, in 1819, he transferred to Paris and concentrated on scientific research.

The fruits of this work ripened in 1824 in the form of a book called *Réflexions sur la puissance motrice de feu* (On the Motive Power of Fire). The main theme of this masterpiece was an analysis of the efficiency of engines in converting heat into work. He found a simple formula depending only on the temperature differences in the engine and not on intermediate stages through which the engine passed. He also introduced the concept of reversibility in the form of the ideal *Carnot cycle*. Using these ideas he derived an early form of the second law of thermodynamics, stating that heat always flows from hot to cold. It became an inspiration, many years later, for Rudolf CLAUSIUS's formulations of thermodynamics. Carnot died of cholera at the age of 36.

Caro, Heinrich (1834–1910) *German organic chemist*

Caro (**kah**-roh), who was born in Posen (now Poznan in Poland), studied at the University of Berlin (1852–55) before entering the dyestuffs industry, first in Mülheim then in Manchester, England, where he worked on mauveine, the first synthetic dye. He joined the Badische Anilin und Soda Fabrik in 1866, becoming a director in 1868. In 1869 Caro, GRAEBE, and Liebermann synthesized alizarin, the first natural dye to be synthesized (also discovered simultaneously by William Perkin). Caro made many subsequent discoveries, including methylene blue (1877) and Fast Red A, the first acidic azo dye (1878). He discovered *Caro's acid*, a powerful oxidizing agent (H_2SO_5), in 1898.

Carothers, Wallace Hume (1896–1937) *American industrial chemist*

Carothers, the son of a teacher, was born in Burlington, Iowa, and gained a BS degree from Tarkio College, Missouri (1920), after working his way through college. He gained his PhD in 1924 from the University of Illinois and was an instructor in chemistry at Illinois and Harvard before joining the Du Pont company at Wilmington, Delaware, as head of organic chemistry research in 1928.

Carother's early work was in the application of electronic theory to organic chemistry but at du Pont he worked on polymerization. His first great success was the production of the synthetic rubber neoprene (1931). Working with acetylenes he discovered that the action of hydrochloric acid on monovinylacetylene produced 2-chloro-buta-1,3-diene (chloroprene), which polymerized very readily to give a polymer that was superior in some respects to natural rubber.

In a systematic search for synthetic analogs of silk and cellulose he prepared many condensation polymers, especially polyesters and polyethers. In 1935 one polyamide, produced by condensation of adipic acid and hexamethylenediamine, proved outstanding in its properties and came into full-scale production in 1940 as Nylon 66. But Carothers did not live to see the results of his achievements; despite his brilliant successes he suffered from fits of depression and took his own life at the age of 41.

Carrel, Alexis (1873–1944) *French surgeon*

> Intelligence is almost useless to the person whose only quality it is.
> —*Man, the Unknown* (1935)

Carrel (ka-**rel**) received his medical degree from the university in his native city of Lyons in 1900. In 1902 he started to investigate techniques for joining (suturing) blood vessels end to end. He continued his work at the University of Chicago (1904) and later (1906) at the Rockefeller Institute for Medical Research, New York. Carrel's techniques, which minimized tissue damage and infection and reduced the risk of blood clots, were a major advance in vascular surgery and paved the way for the replacement and transplantation of organs. In recognition of this work, Carrel was awarded the 1912 Nobel Prize for physiology or medicine.

During World War I, Carrel served in the French army. With the chemist Henry Dakin, he formulated the Carrel–Dakin antiseptic for deep wounds. Returning to the Rockefeller Institute after the war, Carrel turned his attention to methods of keeping tissues and organs alive outside the body. He maintained chick embryo heart tissue for many years on artificial nutrient solutions and with the aviator Charles Lindbergh he devised a so-called artificial heart that could pump physiological fluids through large organs, such as the heart or kidneys.

In *Man, the Unknown* (1935), Carrel published his controversial views about the possible role of science in organizing and improving society along rather authoritarian lines. During World War II he founded and directed the Carrel Foundation for the Study of Human Problems under the Vichy government, in Paris. Following the Allied liberation, Carrel faced charges of collaboration but died before a trial was arranged.

Cartan, Elie Joseph (1869–1951) *French mathematician*

Cartan (kar-**tan**) is now recognized as one of the most powerful and original mathematicians of the 20th century, but his work only became widely known toward the end of his life. Cartan, who was born in Dolomieu, France, studied at the Ecole Normale Supérieure in Paris and held teaching posts at the universities of Montpellier, Lyons, Nancy, and, from 1912 to 1940, Paris.

Cartan's most significant work was in developing the concept of analysis on differentiable manifolds, which now occupies a central place in mathematics. He began his research career with a dissertation on LIE groups – a topic that led him on to his pioneering work on differential systems. The most important innovation in his work on Lie groups was his creation of methods for studying their global properties. Similarly his work on differential systems was distinguished by its global approach. One of his most useful inventions was the "calculus of exterior differential forms," which he applied to problems in many fields including differential geometry, Lie groups, analytical dynamics, and general relativity. Cartan's son Henri is also an eminent mathematician.

Carter, Brandon (1942–) *British physicist*

Born in Sydney, Australia, Carter was educated at the University of St. Andrews, Scotland, and at Cambridge University, where he completed his PhD in 1968. He remained in Cambridge as a research fellow at the Institute of Astronomy until 1973. Carter then moved to France to join the Centre National de la Recherche Scientifique (CNRS), working from 1986 as director of research at the Paris-Meudon Observatory.

In 1974 Carter formulated what is known as the *anthropic principle*. The argument began with COPERNICUS whose heliocentric system is often thought to have removed man from any special privileged position in the universe. Carter, however, insisted that it is privileged to the extent that our location in the universe must be compatible with our existence as observers. That is, if the universe had differed significantly in its size, age, and character then intelligent life would not now be present to observe it. If, for example, the strength of the gravitational force differed by just one part in 10^{40} all stars would be either blue giants or red dwarves; with no Sun-like stars to nourish life, the universe would be without observers. The fact that it has observers, therefore, presupposes that the nuclear, gravitational, and electromagnetic forces all fall within some very narrow limits.

This is sometimes known as the "weak form" of the anthropic principle. Carter advanced from this to the strong version with his claim that "The universe must have those properties which allow life to develop within it at some stage in its history." While some physicists have seen in the anthropic principle a profound key to the secrets of nature, others have dismissed it on the grounds that it is immune to falsification, makes no significant predictions, and offers all its explanations after the event.

Cartwright, Edmund (1743–1823) *British inventor*

Cartwright, who was born in Marnham, England, began his career as a clergyman. However, after visiting Richard ARKWRIGHT's cotton-spinning mills in Derbyshire in 1784, he became interested in developing machinery for the textile industry. He built a simple form of power loom, which was patented in 1785, and set up a weaving and spinning factory in Doncaster the same year. In 1789 he patented a wool-carding machine. He was not financially successful – his factory went bankrupt in 1794 – but in 1809 he was awarded £10,000 by the House of Commons in recognition of the benefit to the nation from his invention of the power loom.

Carty, John Joseph (1861–1932) *American telephone engineer*

Carty was born and went to school in Cambridge, Massachusetts, but because of eye trouble was delayed in entering college. During this delay his imagination was captured by the invention of the telephone (1875) by Alexander Graham BELL in neighboring Boston. He consequently joined the Bell System in 1879, eventually becoming chief engineer there. He served as chief engineer of the New York Telephone Company (1889–1907) and of the American Telephone and Telegraph Company (1907–19), and finally became vice president of the latter company until his retirement in 1930.

Carty was responsible for a number of technical innovations in the development of the commercial use of the telephone. These included the introduction of the "common" battery, which, by providing current from a central source to a number of interconnected telephones, allowed the development of a complex urban network. He also directed the project to provide the first transcontinental telephone wire line, completing in 1915 the 3,400-mile (5,400-km) link between New York and San Francisco.

Carver, George Washington (1864–1943) *American agricultural chemist*

Carver was born a slave in Diamond Grove, Missouri. Nevertheless, he managed to acquire some elementary education and went on to study at the Iowa State Agricultural College from which he graduated in 1892. He taught at Iowa until 1896, when he returned to the South to become director of the department of agricultural research at the Tuskegee Institute, Alabama. There he stayed despite lucrative offers to work for such magnates as Henry Ford and Thomas EDISON.

His main achievement was to introduce new crops into the agricultural system of the South, in particular arguing for large-scale plantings of peanuts and sweet potatoes. He saw that such new crops were vital if only to replenish the soil, which had become impoverished by the regular growth of cotton and tobacco.

But he did much more than introduce new crops for he tried to show that they could be used to develop many new products. He showed that peanuts contained several different kinds of oil. So successful was he in this that by the 1930s the South was producing $60 million worth of oil a year. Peanut butter was another of his innovations. In all he is reported to have developed over 300 new products from peanuts and over 100 from sweet potatoes.

Casimir, Hendrik Brugt Gerhard (1909–2000) *Dutch physicist*

Born in the Dutch capital city, The Hague, Casimir (**kaz**-i-meer) studied at the universities of Leiden, Copenhagen, and Zurich, and held various research positions between 1933 and 1942.

He published many papers in the fields of theoretical physics, applied mathematics, and low-temperature physics. His most notable work was in the theory of the superconducting state. Following the work of W. MEISSNER, Casimir and his colleague, Cornelis Gorter, advanced a "two-fluid" model of superconductivity in 1934 in which a fraction of the electrons were regarded as superconducting, while the rest remained "normal" electrons. They were successful in explaining the high degree of interrelationship between the magnetic and thermal properties of superconductors.

Casimir is also known for the *Casimir effect*, which he predicted in 1948. This is the occurrence of a very small attractive force between two parallel metal plates, caused by quantum fluctuations in the vacuum state in quantum electrodynamics. The effect was first observed in 1958. Casimir also did related work on intermolecular forces.

From 1942, Casimir pursued a highly successful career with the Philips company, becoming director of the Philips Research Laboratories in 1946, and a member of the board of management (1957–72). He supervised Philips's research activities in several countries.

Caspersson, Torbjörn Oskar (1910–1997) *Swedish cytochemist*

Caspersson (**kas**-per-son), who was born in Motala, Sweden, gained his MD from Stockholm University in 1936. He then joined the staff of the Nobel Institute serving as professor of medical cellular research and genetics from 1944 to 1977. In 1977 he was appointed professor and head of the medical cell research and genetics department at the Kungliga Karolinska Mediko-Kirurgiska Instituet in Stockholm.

In the late 1930s Caspersson spent a few years working on DNA. In 1936 with the Swiss chemist Rudolf Signer he made fundamental measurements of the molecule that suggested a molecular weight between 500,000 and a million, so showing the nucleic acids to be larger than protein molecules.

Further important data were collected by a photoelectric spectrophotometer developed by Caspersson. This allowed the movement of RNA in the cell to be followed by its characteristic absorption peak in the ultraviolet at 2,600 angstroms and to establish that protein synthesis in the cell was associated with an abundance of RNA. Despite this Caspersson remained committed to the orthodox view that genes were proteins and believed nucleic acids to be a structure-determining supporting substance.

In 1970 Caspersson made a major breakthrough in the study of chromosomes. Before 1970 the only way to identify a chromosome was by its length. Caspersson argued that if genes differed in their concentration of the four bases (guanine, adenine, cytosine, and thymine), then they would be distributed differently in each chromosome. If a dye could be found that bound to one of the bases only, then a characteristic chromosomal pattern would be displayed. Precisely this happened when Caspersson found a quina-acrine mustard with an affinity for guanine. When illuminated with ultraviolet the now familiar pattern of bright and dark bands was displayed with startling clarity. Within a year Caspersson had used distinctive banding patterns to characterize all human chromosomes.

Cassegrain, N. (about 1672) *French telescope designer*

Little is known about the life of Cassegrain (**kas**-e-gran). He was apparently a physician in Chartres and was credited, in 1672, as the inventor of a reflecting telescope, which is named for him. James Short was one of the first to use the design, producing his telescopes in about 1740.

Cassini, Giovanni Domenico (1625–1712) *Italian–French astronomer*

Born in Perinaldo, Italy, Cassini (ka-**see**-nee) was educated in Genoa and at the age of 25 became professor of astronomy at Bologna. He remained there until 1669 when he moved to France in order to take charge of Louis XIV's new Paris Observatory. He became a French citizen in 1673.

While still at Bologna he worked out, fairly accurately, the rotational periods of Jupiter and Mars. In 1668 he constructed a table of the movements of the Medici planets – the satellites of Jupiter discovered by GALILEO. It was this table that allowed Ole RØMER to calculate the speed of light. In Paris, using aerial telescopes up to 150 feet (45.7 m) long, he discovered four new satellites of Saturn – Iapetus in 1671, Rhea in 1672, and Dione and Tethys in 1684. In 1675 he discovered the gap that divides Saturn's rings into two parts and has since been called *Cassini's division*.

Cassini's most important work concerned the size of the solar system. Using data collected by Jean RICHER in Cayenne, together with his own observations in Paris, he was able to work out the parallax of Mars and thus calculate the astronomical unit (AU) – the mean distance between the Earth and the Sun. His figure of 87 million miles (140 million km) may have been 7% too low, but compared with earlier figures of Tycho BRAHE (5 million miles) and Johannes KEPLER (15 million miles) his results gave mankind a realistic picture of the size of the universe for the first time. Cassini also made fundamental measurements on the size and shape of the Earth concluding, erroneously, that it was a prolate spheroid. He became blind in 1710 and was succeeded in the directorship of his observatory by both his son, Jacques Cassini, and his grandson, César François Cassini.

Castner, Hamilton Young (1858–1898) *American chemist*

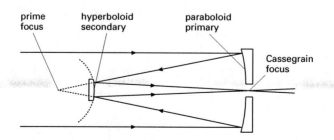

CASSEGRAINIAN TELESCOPE   The ray path in Cassegrain's reflecting telescope.

Born in New York City, Castner studied at Brooklyn Polytechnic and at Columbia University, New York. He started as a chemical consultant in 1879 and moved to Britain in 1886 when he failed to gain any backing in America for his process for the production of sodium.

Henri Sainte-Claire DEVILLE had developed a system in which caustic soda could be reduced to sodium with charcoal at high temperatures. However he ran into a variety of practical difficulties, which were satisfactorily cleared up by Castner. Castner intended to use the sodium for producing aluminum by reduction of aluminum chloride – at the time aluminum was a very expensive metal. A factory was opened in 1888 at Oldbury, Birmingham, England, to manufacture 100,000 pounds of aluminum per annum. But it was too late for, two years earlier, Charles HALL in America and Paul HÉROULT in France had independently discovered a cheap way to produce aluminum by electrolysis. Castner quickly had to invent some uses for his sodium, for which there was little demand at the time. One was the manufacture of sodium peroxide (by burning sodium in air), used as a bleach. By passing ammonia over molten sodium and charcoal he produced sodium cyanide, which was used in the extraction of gold.

By the early 1890s, with the growing demand for his products, his problem was an inability to produce enough sodium. He solved this with a new method of making sodium by the electrolysis of brine using a mercury cathode. The process had been anticipated by Karl KELLNER in Austria; rather than litigate the two chemists cooperated and in 1897 set up the Castner–Kellner Alkali Company in Runcorn, Cheshire, England, where there was a cheap and abundant supply of salt. In the year of his death it was already producing 20 tons of caustic soda a day with a production of 40 tons of bleaching powder daily as a byproduct.

Cauchy, Baron Augustin Louis (1789–1857) *French mathematician*

> Men pass away, but their deeds abide.
> —Attributed

Cauchy (koh-**shee**) showed great mathematical talent at an early age and came to the attention of Joseph LAGRANGE and Pierre LAPLACE, who encouraged him in his studies. Born in Paris, France, he was educated at the Ecole Polytechnique, where he later lectured and became professor of mechanics in 1816, and worked briefly as an engineer in Napoleon's army. He held extreme conservative views in religion and politics, typical of which was the strong allegiance to the Bourbon dynasty that caused him to follow Charles X (who had ennobled Cauchy) into exile in 1830. Cauchy then became professor of mathematics at Turin, but returned to France in 1838 and resumed his post at the Ecole Polytechnique.

Cauchy was an extremely prolific mathematician who made outstanding contributions to many branches of the subject, ranging from pure algebra and analysis to mathematical physics and astronomy. He was also an outstanding teacher. His greatest achievements were in the fields of real and complex analysis in which he was one of the first mathematicians to insist on the high standards of rigor now taken for granted in mathematics. He gave the first fully satisfactory definitions of the fundamentally important concepts of limit and convergence.

Cavalieri, Francesco Bonaventura (1598–1647) *Italian mathematician and geometer*

> Few, if any, since Archimedes, have delved as far and as deep into the science of geometry.
> —Galileo Galilei, letter to Cesare Marsili (1629)

Born in the Italian city of Milan, Cavalieri (ka-va-**lyair**-ee) joined the Jesuits as a boy. He became interested in mathematics while studying EUCLID's works and met GALILEO, whose follower he became.

Cavalieri's fame rests chiefly on his work in geometry in which he paved the way for the development of the integral calculus by Isaac NEWTON and Gottfried LEIBNIZ. In 1629 Cavalieri became professor of mathematics at Bologna, a post he held for the rest of his life. At Bologna he developed his "method of indivisibles," published in his *Geometria indivisibilibus continuorum nova quadam ratione promota* (1635; A Certain Method for the Development of a New Geometry of Continuous Indivisibles), which had much in common with the basic ideas of integral calculus.

Cavalieri also helped to popularize the use of logarithms in Italy through the publication of his *Directorium generale uranometricum* (1632; A General Directory of Uranometry).

Cavalli-Sforza, Luigi Luca (1922–   ) *Italian geneticist*

Cavalli-Sforza (ka-val-ee-**sfort**-za), who was born in Genoa, Italy, was educated at the University of Pavia where he gained his MD in 1944. After working on bacterial genetics at Cambridge (1948–50) and Milan (1950–57) he has held chairs in genetics at Parma (1958–62) and Pavia (1962–70). In 1970 he was appointed professor of genetics at Stanford University, a position he held until his retirement in 1992. He is still actively working as emeritus professor at Stanford.

Cavalli-Sforza has specialized mainly in the genetics of human populations, producing with Walter Bodmer a comprehensive survey of the subject in their *Genetics, Evolution and Man* (1976).

He has also done much to show how genetic data from present human racial groups could be used to reconstruct their past separations. This reconstruction, based on the analysis of 58 genes, yields a bifurcated evolutionary tree with Caucasian and African races in one branch and Orientals, Oceanians, and Amerinds in the other. The main division appeared, according to Cavalli-Sforza, some 35–40,000 years ago.

Cavendish, Henry (1731–1810) *English chemist and physicist*

Cavendish, who was born in Nice in the south of France, was the son of Lord Charles Cavendish, himself a fellow of the Royal Society and administrator of the British Museum. Henry was educated at Cambridge University (1749–53), but left without a degree. Following this he devoted the rest of his life to science. He inherited from his uncle a vast fortune with which he built up a large library and financed his scientific interests. Throughout his life he was an eccentric recluse, appearing only rarely in public and then chiefly at scientific meetings. He communicated with his housekeeper by a system of notes and was such a misogynist that he ordered all his female domestics to keep out of his sight.

Cavendish's first published work was *Three Papers containing Experiments on Factitious Airs* (1766). In these he clearly distinguished hydrogen ("inflammable air") and carbon dioxide ("fixed air") as gases separate from common air. Some of the work on fixed air duplicated that of Joseph BLACK, little of which had been published, but Cavendish was the first to weigh gases accurately.

Much of Cavendish's work remained unpublished in his lifetime and he is now known to have anticipated or come very close to several major discoveries. His electrical studies, which were edited by Clerk MAXWELL in 1879, following the discovery of his notebooks and manuscripts, included the clear distinction between electrical quantity and potential, the measurement of capacitance, and the anticipation of OHM's law (1781). He had the concept of specific heat in 1765 but the work was not published. In 1778, working on the effect of water vapor on the compressibility of air, he arrived at what is essentially the law of partial pressures. One important physical investigation that was published was the determination of the mean density of the Earth (1798) by means of the torsion balance in what became known as the *Cavendish experiment*.

In his chemical work Cavendish came close to the concepts of equivalent weights and multiple proportions but he was not a generalizer and the concepts only became explicit in the works of others. His most illustrious and con-

troversial work was his synthesis of water. The paper *Experiments on Air* (1784) reported his researches on exploding hydrogen with oxygen and air. He concluded that air consisted of a mixture of oxygen and nitrogen in a ratio of 1:4 and that hydrogen and oxygen mixed in proportions of 2:1 yielded their own weight of water. This work was carried out in 1781, and although Cavendish's priority is quite clear a dispute ensued between James WATT, Antoine LAVOISIER, and Cavendish. It was discovered from this work that water is not an element but a compound.

The reason for Cavendish's three-year delay in publishing his work on water was the persistent discovery of nitric acid (then called nitrous acid) in the water after sparking hydrogen and air. In further experiments he accomplished the conversion of nitrogen to nitric acid by sparking over alkali, which then formed potassium nitrate. This synthesis was the basis of the commercial production of nitric acid until 1789. In the course of his work on gases Cavendish refined the eudiometer and his measurements of the oxygen content of air showed it to be the same everywhere.

On his death Cavendish left over a million pounds sterling to his relatives. From this the endowment of the famous Cavendish Laboratory was made to Cambridge University in 1871.

Cavendish, Margaret, Duchess of Newcastle (1623–1673) *British natural philosopher*

The daughter of a wealthy landowner, Margaret was born at St. John's, Colchester, in the eastern English county of Essex and inherited £10,000 on his death. She received a scant education from "an ancient decayed gentlewoman." With the outbreak of the civil war the family lost its estates, two brothers died fighting for the king, and in 1641 Margaret entered the royal court as a maid of honor to Henrietta Maria, wife to Charles I. In 1643 she fled with the queen to Paris and spent the next eighteen years in impoverished exile in Europe.

In 1645 she married William Cavendish, Duke of Newcastle. While in exile in Paris and in Antwerp she moved in circles where the ideas of Hobbes, DESCARTES, and GASSENDI were frequently discussed. From such discussions Margaret was introduced to mechanical philosophy.

While in exile she began to write on topics in natural philosophy, producing her *Philosophical and Physical Opinions* (1655). She continued to write about science after her return from exile with the restored Charles II in 1660, publishing her *Observations upon Natural Philosophy* (1666). She was highly critical of the new science. Her main objection was the familiar one that not all natural phenomena could be explained by "the Dusty motion of Atoms." Consequently, she argued, every atom must be "animated with life and knowledge." She was also critical of the newly invented microscope, claiming that it distorted nature.

Much of her later life was spent in seclusion at Welbeck writing verse and plays as well as concerning herself with natural philosophy. She did, however, pay a well-documented visit to London in 1667 when she visited the Royal Society and witnessed experiments performed by BOYLE and HOOKE. But membership would remain closed to her or any other woman for a further three centuries.

Caventou, Jean Bienaimé (1795–1877) *French pharmacist and organic chemist*

Caventou (ka-vahn-**too**), the son of an army apothecary from St. Omer in France, studied pharmacy in Paris, eventually becoming professor of toxicology at the Ecole de Pharmacie (1835–60); he also had a pharmacy business. He learned the technique of solvent extraction of alkaloids from plants from Joseph PELLETIER, with whom (1817–20) he isolated the alkaloids strychnine, brucine, cinchonine, quinine, veratrine, and colchicine. Caventou and Pelletier are regarded as the founders of alkaloid chemistry. Curiously, until 1823 they believed that alkaloids contained no nitrogen when in fact they are now defined as nitrogenous substances. Caventou's early success was not repeated in later life and after Pelletier's death in 1842 he published nothing.

Cayley, Arthur (1821–1895) *British mathematician*

> As for everything else, so for a mathematical theory: beauty can be perceived but not explained.
> —Quoted by J. R. Newman (editor) in *The World of Mathematics*

Born in Richmond in southern England, Cayley studied mathematics at Cambridge University, but before becoming a professional mathematician spent 14 years working as a barrister. He was forced to do this since he was unwilling to take holy orders – at that time a necessary condition of continuing his mathematical career at Cambridge. When this requirement was dropped, Cayley was able to return to Cambridge and in 1863 became Sadlerian Professor there.

Cayley was an extremely prolific mathematician. His greatest work was the creation of the theory of invariants, in which he worked closely with his friend James Joseph SYLVESTER. Cayley developed this theory as a branch of pure mathematics but it turned out to play a crucial role in the theory of relativity, as it is important in the calculation of space–time relationships in physics. He also developed the theory of matrices and made major contributions to the study of n-dimensional geometry. He went a considerable way toward unifying the study of geometry. Cayley also did important work in the theory of elliptic functions.

One of Cayley's notable nonmathematical achievements was playing a large role in persuading the University of Cambridge to admit women as students.

Cayley, Sir George (1773–1857) *British inventor*

Cayley, a man of independent means, was born in the northern English coastal town of Scarborough and succeeded to the family estates on the death of his father in 1797. He had been educated at schools in York and Nottingham and had learned some science from a nonconformist clergyman who was also a fellow of the Royal Society.

Cayley is recognized as the founder of the science of aerodynamics. As early as 1799, in a design engraved on a small silver disk, he made an important step in the history of aeronautics by separating for the first time the system providing power from that contributing lift. He realized that man would never fly by flapping his arms but only by building a rigid wing to which he could attach an external power source. This was followed by a series of papers produced during the period 1804–55 in which he worked out many of the details implicit in his original idea.

Cayley designed an undercarriage fitted with tension wheels (later to find use on the bicycle) in 1808, streamlining in 1809, and a glider that was capable of lifting him up for a few yards. In the same year he produced his important paper *On aerial navigation*, which introduced the cambered wing, followed by his designs for a hot-air airship in 1816 and a helicopter in 1818. In the 1840s and 1850s he designed a large number of powered gliders, culminating in his construction of the first man-carrying glider in 1853, which was reluctantly tested by his coachman on the first successful manned glider flight in the same year.

In addition to his aeronautical works Cayley produced designs for the caterpillar tractor and for artificial limbs, and was interested in land reclamation and railway engineering. He also founded the Regent Street Polytechnic Institution, London, in 1839.

Cech, Thomas Robert (1947–   ) *American chemist*

Chicago-born Cech (chek) was educated at the University of California, Berkeley, where he gained his PhD in 1975. He then joined the faculty of the University of Colorado, Boulder, where he became professor of chemistry in 1983. In 2000 he became president of the

Howard Hughes Medical Institute in Maryland and also works at the Institute in Boulder, Colorado (the Cech Laboratory).

In 1977 Phillip SHARP discovered long stretches of noncoding DNA, later called "introns." In 1982 Cech began to investigate how these supposedly redundant sequences could be removed from the RNA molecule after it had been copied from the DNA template. For such a complex process to happen so rapidly it must, it seemed, be catalyzed by an enzyme. Since the first enzyme to be synthesized (by James SUMNER in 1926) had turned out to be a protein, and since all of the hundreds of other cellular enzymes had also turned out to be proteins, Cech confidently began to search for the protein enzyme responsible for snipping introns from RNA molecules.

He worked with ribosomal RNA (rRNA) of the protozoan *Tetrahymena thermophilia*. He began with some unspliced rRNA and some protozoa nuclei mixed together *in vitro*. He assumed that the nuclear enzymes would catalyze the splicing. But, although the RNA introns were indeed neatly snipped away, none of the nuclear enzymes appeared to have been used. At first Cech assumed that the RNA itself harbored the protein enzyme responsible. To eliminate this possibility Cech synthesized his own pre-rRNA from a recombinant DNA template. The result was a viable but artificial rRNA, which, never having been in contact with a cell, could not possibly contain any cellular splicing enzymes. Despite this the introns were still removed.

RNA, Cech concluded in 1982, must be self-splicing. It acted like an enzyme in catalyzing a specific reaction, at a greatly accelerated rate, yet, unlike an enzyme, it operated upon itself. To mark the difference Cech proposed the name "ribozyme."

Confirmation of Cech's researches soon came from the work of Sidney ALTMAN on ribonuclease. For this work, Cech and Altman shared the 1989 Nobel Prize for chemistry.

Celsius, Anders (1701–1744) *Swedish astronomer*

Celsius (**sel**-see-us), the son of a mathematician, became professor of astronomy at the university in his native city of Uppsala, where he opened an observatory in 1740. In 1742 he devised a temperature scale in which the temperature of melting ice was taken as 100° and the temperature of boiling water was taken to be 0°. The modern *Celsius* (or *centigrade*) *scale* has the opposite fixed points (ice point 0°; steam point 100°C).

Celsus, Aulus Cornelius (b. *c.* 10 BC) *Roman encyclopedist*

Celsus (**sel**-sus), a member of a noble Roman family, was the author of a comprehensive encyclopedia of knowledge covering many topics, including agriculture, military art, and philosophy. However, only the part dealing with medicine has survived, and it is for this work, *De medicina*, that he is remembered. Comprising eight sections, it covers medical history, diet, symptoms and treatments of various diseases, surgical techniques, and drugs. Celsus's work was rediscovered by Pope Nicholas V and became, in 1478, one of the first medical texts to be printed. Its lucid and elegant Latin prose made it widely acclaimed during the Renaissance period.

Cerenkov, Pavel Alekseyevich *See* CHERENKOV, PAVEL ALEKSEYEVICH.

Cerf, Vinton G. (1943–) *American computer scientist*

Cerf (serf) was educated at Stanford University, where he obtained a BSc in mathematics, and at the University of California at Los Angeles (UCLA), where he obtained MSc and PhD degrees in computer science. It was during his time as a graduate student at UCLA between 1967 and 1972 that he started the work that led to the introduction of the Internet. Cerf is frequently referred to as the "Father of the Internet" because, together with Robert KAHN, he published in 1974 papers on Transmission Control Protocol/Internet Protocol (TCP/IP) for networking computers.

Cerf worked at the US Department of Defense's Research Projects Agency (DARPA) between 1976 and 1982 on Internet-related technology. Between 1982 and 1986 he was vice president of MCI Digital Information. While there he was responsible for the engineering of MCI mail, which was the first commercial email system to be connected to the Internet. Between 1986 and 1994 Cerf was vice president of the Corporation for Research Initiatives (CNRI). He rejoined MCI in 1994.

Cerf was the founding president of the Internet Society between 1992 and 1995 and was chairman of the board from 1998 to 1999. Cerf and Robert Kahn were awarded the US Medal of Technology in 1997 for their role in founding the Internet.

Cesalpino, Andrea (1519–1603) *Italian physician and botanist*

Born at Arezzo in Italy, Cesalpino (chay-zal-**pee**-noh) studied anatomy and medicine at the University of Pisa, where he graduated in 1551; in 1555 he became professor of medicine and director of the botanic garden there. In 1592 he became physician to Pope Clement VIII and professor at the Sapienza University.

He is most famous for his plant classification based on fruit and seed characteristics, which is described in his work, *De plantis* (1583; On Plants). This work also discusses the whole of theoretical botany and had great influence on later botanists.

Cesalpino also wrote a number of anatomical books in which he partly anticipated the theory of the circulation of the blood proposed by William HARVEY.

Chadwick, Sir James (1891–1974)
British physicist

Chadwick was born in Macclesfield in northern England and was educated at the University of Manchester, where he graduated in 1911 and remained as a graduate student under Ernest RUTHERFORD. In 1913 he went to Leipzig to work under Hans GEIGER and found himself interned in 1914 near Spandau as an enemy alien. There he remained for the duration of the war, cold and hungry but permitted, with the help of Walther NERNST, to carry out rudimentary research.

On his return to England in 1919 he was invited by Rutherford to accompany him to Cambridge University where, from 1922 until 1935, he served as assistant director of research at the Cavendish Laboratory. It was during this period that Chadwick, in 1932, made his greatest discovery – the neutron. Before this, physicists had accepted the existence of only two elementary particles: the proton (p) with a positive charge, and the electron (e) with a negative charge. It was however clear to all that these two particles could not account for all the atomic phenomena observed. The helium atom, for example, was thought to consist of four protons; that it only possessed a positive charge of two was due to the nucleus also containing two "internal electrons" which neutralized the charge on two of the protons. The difficulty of such a view was the failure of a disintegrating nucleus to produce the electrons supposedly contained within it.

In 1920 Rutherford had provided an alternative solution by introducing the possibility of "an atom of mass 1 which has zero nuclear charge." Chadwick attempted unsuccessfully to discover such a particle in the 1920s by bombarding aluminum with alpha particles (helium nuclei). More promising, however, was the report in 1930 that the bombardment of beryllium with alpha particles yielded a very penetrating radiation. In 1932 Irène and Frédéric JOLIOT-CURIE found that this radiation could eject protons with considerable velocities from matter containing hydrogen. They thought such radiation consisted of gamma rays – electromagnetic radiation of very short wavelength. Chadwick showed that the gamma rays would not eject protons, but that the result was explained if the particles had nearly the same mass as protons but no charge, i.e., the particles were neutrons. It was for this work that Chadwick was awarded the 1935 Nobel Prize for physics.

By 1936 a certain amount of friction had begun to appear between Chadwick, who wished to build a cyclotron at the Cavendish Laboratory, and Rutherford, who initially was violently opposed to any such project. It was therefore with some relief that Chadwick decided in 1935 to accept the offer of the chair of physics at Liverpool University. There he built Britain's first cyclotron and was on hand at the outbreak of war to support the claims made by Otto FRISCH and Rudolph PEIERLS on the feasibility of the atomic bomb. Chadwick consequently spent most of the war in America as head of the British mission to the Manhattan project.

For this service he was knighted in 1945. He returned to Cambridge in 1958 as Master of Gonville and Caius College, in which office he remained until his retirement in 1958.

Chain, Ernst Boris (1906–1979) *German–British biochemist*

Chain, born the son of a chemist in Berlin, Germany, graduated in 1930 from the Friedrich-Wilhelm University with a degree in chemistry. He left Germany for England in 1933 and, after two years' research at Cambridge University, joined Howard FLOREY at Oxford. Here his brilliance as a biochemist was put to good use in the difficult isolation and purification of penicillin – work that Alexander FLEMING had been unable to carry out. He shared the 1945 Nobel Prize for physiology or medicine with Florey and Fleming for this achievement.

After 1945 he was professor of biochemistry at the Superior Institute of Health in Rome, returning to England in 1961 for the chair of biochemistry at Imperial College, London. During this time he discovered penicillinase – an enzyme that some bacteria can synthesize and so destroy the drug. He also worked on tumor metabolism and the mode of action of insulin in diabetes.

Chamberlain, Owen (1920–2006) *American physicist*

The son of the prominent radiologist Edward Chamberlain, Owen Chamberlain followed his father's interest in physics. Born in San Francisco, he graduated from Dartmouth College in 1941 and gained his doctorate in physics from the University of Chicago in 1949. From 1948 until 1950 he was an instructor in physics at the University of California at Berkeley, becoming associate professor in 1954, professor in 1958, and emeritus professor from 1989.

The onset of America's involvement in World War II interrupted his university studies, and he spent the years 1942–1946 under the leadership of Emilio SEGRÉ working on the Manhattan atom-bomb project at Los Alamos. There he investigated spontaneous fission of the heavy elements and nuclear cross-sections. Later he worked with Enrico FERMI on neutron diffraction by liquids.

At Berkeley, Chamberlain experimented with the bevatron particle accelerator of the Lawrence Radiation Laboratory, and in 1955 (together with Segrè, C. Weigand, and T. Ypsilantis) discovered the antiproton – a particle with the same mass as the proton, but of opposite (negative) charge. For their discovery, Chamberlain and Segrè received the 1959 Nobel Prize for physics. The existence of antiparticles had been predicted by Paul DIRAC's theory of 1926, and the first of these, the positive electron (or positron) had been found by Carl David ANDERSON in cosmic radiation in 1931.

Chamberlain's later research involved working on the development and construction of the Time-Projection-Chamber at the Stanford Linear Accelerator Center.

Chamberlin, Thomas Chrowder (1843–1928) *American geologist*

Chamberlin came from a farming background in Mattoon, Illinois. His discovery of fossils in a local limestone quarry aroused his interest in geology, which he pursued at the University of Michigan. He worked for the Wisconsin Geological Survey from 1873, serving as chief geologist for the period 1876–82. From 1881 until 1904 he was in charge of the glacial division of the U.S. Geological Survey. After a period as president of the University of Wisconsin (1887–92) he became professor of geology at the University of Chicago (1892–1918).

Apart from his work on the geological surveys, Chamberlin's most significant work was in the field of glaciation. Early work on glaciation had assumed that there had been one great ice age but James Geikie, in his *The Great Ice Age* (1874–84), had begun collecting evidence that there had been several ice ages separated by nonglacial epochs. Chamberlin contributed the chapter on North America to Geikie's work. He showed that drift deposits are composed of at least three layers and went on to establish four major ice ages, which were named the Nebraskan, Kansan, Illinoian, and Wisconsin after the states in which they were most easily studied.

Together with the astronomer Forest MOULTON, Chamberlin formulated, in 1906, the planetismal hypothesis on the origin of the planets in the solar system. They supposed that a star had passed close to the Sun causing matter to be pulled out of both. Within the gravitational field of the Sun this gaseous matter would condense into small planetesimals, and eventually into planets. The theory was published in *The Two Solar Families* (1928) but it has little support today as it cannot account for the distribution of angular momentum in the solar system.

Chambers, Robert (1802–1871) *Scottish geologist and writer*

The son of a cotton manufacturer from Peebles in Scotland, Chambers was largely self-educated, drawing much of his learning from the *Encyclopaedia Britannica*. After the family moved to Edinburgh, Chambers set up as a bookseller at the age of 16. Soon after he formed the partnership W. & R. Chambers with his brother William and thus created a publishing house still active in Edinburgh today.

Chambers himself published a number of works on Scottish antiquities and history. None, however, had the impact of his *Vestiges of Creation* (1844), which by 1860 had been through eleven British editions, none of which Chambers ever acknowledged as his own. It was only with the posthumous twelfth edition that the author's name appeared on the title page. The work, dealing with the "development" of the animal kingdom, was highly controversial and rumors about its anonymous author abounded, extending as far as the unlikely Prince Albert.

Chambers began by showing how physics could account for the origin and development of the solar system in terms of the nebular hypothesis of LAPLACE. If the inorganic world developed in accordance with physical law, it could not be absurd to suppose that the organic world followed a similar pattern. Chambers then presented a history of the Earth in which there had been a general progression from lower to higher forms. It was, he argued, no credit to God to suppose that each species must have been created separately. Life, he thought, could arise from a "chemico-electric reaction."

And how is one form transmuted into another? Here he turned to the views of Von BAER and argued that, influenced by external conditions, the embryonic stage could be prolonged until it developed into the next higher stage. For example, a fish that remains an embryonic fish longer than the norm develops into a reptile.

The savage reception the work received must have justified the decision of Chambers to remain anonymous. Asked to defend his decision in later life, he replied to a close friend, referring to his eleven children, that he had eleven good reasons for his secrecy. Nevertheless, unable to tolerate references by the geologist Adam SEDGWICK to "the inner deformity and foulness of the work," Chambers replied in his anonymous *Explanations* (1845).

Chamisso, Adelbert von (1781–1838) *French–German naturalist and poet*

Born in Champagne in northeastern France,

Chamisso (shah-**mis**-oh) studied medicine and botany at Berlin. From 1815 to 1818 he accompanied Otto Kotzebue as naturalist on his Russian scientific expedition in which an attempt was made to find a sea passage through the Arctic. In 1819 he was appointed curator of the Berlin Botanic Gardens, and in the same year studies of tunicates (primitive vertebrates) and mollusks led him to observe the occurrence of both sexual and asexual forms in their life cycles.

Chamisso is however better known as a talented lyricist and wrote a number of well-known stories, ballads, and poems.

Chance, Alexander Macomb (1844–1917) British chemical industrialist

Chance was the son of a Birmingham glassmaker who had branched out into the manufacture of soda to guarantee his supply of raw materials. In 1868 he became manager of an alkali works.

During the latter half of the 19th century the LEBLANC process for producing soda was facing competition from newer methods, such as the ammonia–soda process of Ernest SOLVAY. Chance extended the Leblanc process, making it commercially viable, by finding a way of recovering sulfur from calcium sulfide, which was one of the waste products. His method was to pump carbon dioxide through the calcium sulfide solution, freeing hydrogen sulfide, which was then partially oxidized to sulfur.

Chance, Britton (1913–   ) American biophysicist

Chance was born an engineer's son in Wilkes-Barre, Pennsylvania; he was educated at the University of Pennsylvania, where he obtained his PhD in 1940 and where he served as Eldridge Reeves Johnson Professor of Biophysics from 1949 to 1983. He is still professor emeritus at Pennsylvania.

In 1943 he carried out a spectroscopic analysis that provided firm evidence for the enzyme–substrate complex whose existence had been confidently assumed by biochemists since the beginning of the century. Working with the iron-containing enzyme peroxidase, which strongly absorbs certain wavelengths of light, he found that variations in light absorption could be precisely correlated with rates of production of the enzyme–substrate complex. This was seen as confirming the important work of Leonor MICHAELIS.

Chance has also contributed to one of the great achievements of modern biochemistry, namely the unraveling of the complicated maze through which energy is released at the cellular level. He found that the concentration of ADP (adenosine diphosphate), as well as the oxygen concentration, determined the oxida-

tion and reduction states of the proteins in the respiratory (electron-transport) chain. His studies of changes in ADP concentration led to a better understanding of how glucose is used in the body.

Chandler, Seth Carlo (1846–1913) American astronomer

Chandler, who was born in Boston, Massachusetts, graduated from Harvard in 1861 and then acted as assistant to Benjamin GOULD, an astronomer with the U.S. Coast Survey, from 1861 to 1864. He remained with the Survey until 1870 when he started work as an actuary, returning to scientific work with the Harvard Observatory in 1881. From 1885 he devoted himself to private research.

Chandler is best known for his discovery of the variation in the location of the geographic poles – and, hence, of the variation in latitude of points on the Earth's surface. In 1891 he announced the discovery of a 428-day cycle during which latitude varied by 0.3 second. This variation in the Earth's rotation became known as the *Chandler wobble* and was soon confirmed by the International Latitude Service, established in 1900.

Chandrasekhar, Subrahmanyan (1910–1995) Indian–American astrophysicist

Chandrasekhar (chan-dra-**see**-ker), who was born in Lahore, which is now in Pakistan, studied at the Presidency College, Madras, gaining his MA in 1930. He then went to Cambridge University, England, where in 1933 he both obtained his PhD and was elected to a fellowship. In 1936 he moved to America and worked from 1937 at the University of Chicago and the Yerkes Observatory, serving as the Morton D. Hull Distinguished Service professor of Theoretical Astrophysics from 1952 to 1986, and as professor emeritus from 1986. He became an American citizen in 1953.

Chandrasekhar's major fields of study were stellar evolution and stellar structure and the processes of energy transfer within stars. It was known that stars could end their life either dramatically and explosively as a supernova or as an extremely small dense star of low luminosity known as a white dwarf. But what decided the particular path a star took was answered by Chandrasekhar in his *Introduction to the Study of Stellar Structure* (1939). He showed that when a star has exhausted its nuclear fuel, an inward gravitational collapse will begin. This will eventually be halted in most stars by the outward pressure exerted by a degenerate gas, i.e., a gas that is completely ionized, with the electrons stripped away from the atomic nuclei, and that is very highly compressed. The star will therefore have shrunk into an object composed of material so dense that a matchbox of it would weigh many tons.

Chandrasekhar showed that such a star would have the unusual property that the larger its mass, the smaller its radius. There will therefore be a point at which the mass of a star is too great for it to evolve into a white dwarf. He calculated this mass to be 1.4 times the mass of the Sun. This has since become known as the *Chandrasekhar limit*. A star lying above this limit must either lose mass before it can become a white dwarf or take a different evolutionary path. In support of Chandrasekhar's theoretical work, it has been established that all known white dwarfs fall within the predicted limit.

In the 1970s Chandrasekhar devoted much time to the mathematical theory of black holes, publishing a classic book on the subject in 1983. He also made a detailed study of NEWTON's work and published his results in his *Newton's Principia for the Common Reader*.

For his numerous contributions to astrophysics, Chandrasekhar shared the 1983 Nobel Prize for physics with William FOWLER.

Chang, Min Chueh (1908–1991) *Chinese–American biologist*

Chang, who was born in T'ai-yüan in China, was educated at the Tsinghua University in Peking, and at Cambridge, England, where he obtained his PhD in 1941. He emigrated to America in 1945 and joined the Worcester Foundation in Shrewsbury, Massachusetts, where he subsequently remained. From 1961 he also served as professor of reproductive biology at Boston University.

Chang carried out a number of major research projects from which emerged not only greater understanding of the mechanisms of mammalian fertilization, but also such practical consequences as oral contraceptives and the transplantation of human ova (eggs) fertilized *in vitro* (Latin meaning literally "in glass," or in a test tube) by Robert Edwards and Patrick Steptoe in 1978. In 1951, at the same time as Colin Austin, Chang discovered that a "period of time in the female tract is required for the spermatozoa to acquire their fertilizing capacity," a phenomenon known later as capacitation. He further demonstrated, in 1957, that there is a decapacitation factor in the seminal fluid, which, although it can be removed by centrifugation, has resisted further attempts at identification.

Chang also made the important advance in 1959 of fertilizing rabbit eggs *in vitro* and transplanting them into a recipient doe. This was followed in 1964 by comparable work for the first time with rodents. It was also Chang who provided much of the experimental basis for Gregory PINCUS's 1953 paper showing that injections of progesterone into rabbits could serve as a contraceptive by inhibiting ovulation.

Chang Heng (78–142 AD) *Chinese astronomer, mathematician, and instrument maker*

Chang Heng was the Astronomer Royal at the court of the emperors of the Later Han. Although none of his works have survived there are detailed reports of his achievements, which are, by any standard, numerous and impressive. As a mathematician he is reported to have given 3.1622 or the square root of 10 as the value of π, which was as good as any other attempt of that period, apart from that of ARCHIMEDES. In astronomy he gave a detailed description of the figure of the universe in which the Earth lies at the center of a large sphere like the yolk of an egg, which was an improvement on the earlier conception of a hemispherical heaven standing over the Earth like an umbrella.

His real originality however lay in the introduction and design of scientific instruments. Thus he introduced a complete armilliary sphere at about the same time as his western contemporary, PTOLEMY. This was used to determine positions of celestial bodies and consisted of an interconnected set of such main circles of the celestial sphere as the equator, ecliptic, horizon, and meridian. Chang Heng went much further and constructed one that rotated by the force of flowing water in such a way that its movement coincided with the rising and setting of the stars. What is intriguing about this is whether he had devised some primitive form of clockwork – some early escapement, preparing the way for SU SUNG's water clock of the 11th century – 1,200 years before Giovanni de DONDI was to introduce it into Europe.

Even more impressive is his construction of the world's first seismograph. It is clearly described as consisting of a vessel on the outside of which were eight dragon heads containing a ball. When an earthquake occurred the ball was propelled out of a dragon's mouth and caught by a bronze toad waiting underneath. Inside there was, presumably, some pendulum mechanism, which would release just one ball selectively giving the direction of the shock. It is interesting to note that the first seismograph recorded in the west, depending on the spilling of an overfilled saucer of mercury, was in 1703.

Chapman, Sydney (1888–1970) *British mathematician and geophysicist*

Born in Eccles in northern England, Chapman entered Manchester University in 1904 to study engineering. After graduating in 1907, his interest was diverted into more strictly mathematical areas, and he went to Cambridge to study mathematics, graduating in 1910. His first post was as chief assistant at the Royal Observatory, Greenwich, and his work there sparked off his lasting interest in a number of fields of applied mathematics, notably geomagnetism. In 1914 Chapman returned to Cambridge as a lecturer in mathematics, and in 1919 he moved back to Manchester as professor of mathematics, remaining there for five years. From 1924 to 1946 he was professor of mathematics at Imperial College, London. After working at the War Office during World War II he moved to Oxford to take up the Sedleian Chair in natural philosophy, from which he retired in 1953. However, his retirement meant no lessening in his teaching and research activity, which continued for many years at the Geophysical Institute, Alaska, and at the High Altitude Observatory at Boulder, Colorado.

The two main topics of Chapman's mathematical work were the kinetic theory of gases and geomagnetism. In the 19th century James Clerk MAXWELL and Ludwig BOLTZMANN had put forward ideas about the properties of gases as determined by the motion of the molecules of the gas. Chapman's work, which he began in 1911, was the next major step in the development of a full mathematical treatment of the kinetic theory. The Swedish mathematician Enskog had been working, independently of Chapman, along similar lines, and the resulting theory is now generally known as the *Chapman–Enskog theory of gases*. While working in 1917 on mixtures of gases Chapman predicted the phenomenon of gaseous thermal diffusion. His subsequent work on the upper atmosphere was a practical application of his earlier more theoretical study of gases.

Highlights of Chapman's work on geomagnetism are his work on the variations in the Earth's magnetic field in periods of a lunar day (27.3 days) and its submultiples. This he showed to be the result of a small tidal movement set up in the Earth's atmosphere by the Moon. He also developed, in 1930, in collaboration with one of his students, what has become known as the *Chapman–Ferraro theory* of magnetic storms. In collaboration with Julius Bartels, Chapman wrote *Geomagnetism* (2 vols. 1940), which soon established itself as a standard work.

Chappe, Claude (1763–1805) *French engineer*

Chappe (shap), a former cleric who was born in Brûlon in France, invented the semaphore arm-signaling system that was first used during the French Revolution to signal between Lille and Paris. His brother was a member of the Legislative Assembly and put forward Claude's idea for building a series of towers equipped with telescopes and two-arm semaphores. In 1794 it took less than an hour to semaphore to Paris the news that Condé-sur-l'Escaut had been taken from the Austrians. After others challenged his claim to be the inventor of semaphore, Chappe killed himself in a fit of depression.

Chaptal, Jean Antoine Claude (1756–1832) *French chemist*

> It is possible to dye a beautiful scarlet without being a chemist; but the operations ... of the dyer are not the less founded upon invariable principles, the knowledge of which would be of infinite utility to the artist.
> —*Elémens de chimie* (1790–1803; Elements of Chemistry)

Chaptal (shap-**tal**), the son of an apothecary from Nogaret, France, studied medicine at Montpellier, graduating in 1777. He later switched to chemistry, becoming professor at Montpellier in 1781. During the French Revolution he was arrested but then released to manage the saltpeter works at Grenelle. He also helped to organize the introduction of the metric system and published a textbook, *Elémens de chimie* (1790–1803; Elements of Chemistry).

Chaptal is mainly remembered as an industrial chemist; he was the first to produce sulfuric acid commercially in France at his factory at Montpellier. His early paper on bleaching (1787) was translated and published in England in 1790 by Robert Kerr. In 1800 he proposed a new method of bleaching using vapor from a boiling alkaline liquor, which was soon introduced into England. Chaptal also wrote one of the first books on industrial chemistry, *Chimie appliquée aux arts* (1807; Chemistry Applied to the Arts).

Charcot, Jean-Martin (1825–1893) *French neurologist*

Parisian-born Charcot (shar-**koh**) studied medicine in his native city and received his MD in 1853. His interest in disease of the nervous system led to his appointment, in 1862, to the Salpêtrière Hospital for nervous and mental disorders. This marked the beginning of a long and distinguished association. Charcot described the pathological changes associated with several degenerative conditions of the nervous system, including the disintegration of ligaments and joint surfaces (known as *Charcot's disease*) that occurs in advanced stages of locomotor ataxia. His studies of brain damage in cases of speech loss (aphasia) and epilepsy supported the findings of his contemporary, Paul Broca, that is, different bodily functions are controlled by different regions of the cerebral cortex.

In 1872, Charcot was appointed professor of pathological anatomy at the faculty of medicine and later (1882) became professor of neurology at the Salpêtrière. He was increasingly concerned with the link between mind and body in cases of hysteria and trauma. With his eloquent manner and a dramatic presentation of his lectures on a small stage, he became a widely celebrated teacher. Among many famous students was Sigmund FREUD, who was influenced by Charcot's use of hypnosis on patients.

Charcot's son, Jean, became a famous polar explorer.

Chardonnet, Hilaire Bernigaud, Comte de (1839–1924) *French chemist*

Chardonnet (shar-don-**ay**), who was born at Besançon in eastern France, acted as an assistant to Louis PASTEUR while he was working on silk worms. This stimulated his interest in the chemistry of fibers and led him to search for means to produce a synthetic fiber.

In 1884 Chardonnet took out a patent for a process for producing the world's first artificial fiber, which he made by dissolving nitrocellulose in alcohol and ether, and then forcing the solution through tiny holes leaving thin threads once the solvent had evaporated. Products of the fiber, called "Chardonnet silk" or rayon, were first exhibited in the Paris Exposition of 1889. Despite the fiber's origin in nitrocellulose it was not actually explosive but was highly flammable. Modifications to the process that made rayon less flammable enabled it to be manufactured for a mass market.

Chargaff, Erwin (1905–2002) *Austrian–American biochemist*

Chargaff (**char**-gaf), who was born at Czernowitz (now Chernovtsy in Ukraine), gained his PhD from the University of Vienna in 1928 and then spent two years at Yale University. He returned to Europe, working first in Berlin and then at the Pasteur Institute, Paris, before returning permanently to America in 1935.

Initially Chargaff's work covered a range of biochemical fields, including lipid metabolism and the process of blood coagulation. Later his attention became concentrated on the DNA molecule, following the announcement in 1944 by Oswald AVERY that the factor causing the heritable transformation of bacteria is pure DNA. Chargaff reasoned that, if this were so, there must be many more different types of DNA molecules than people had believed. He examined DNA using the recently developed techniques of paper chromatography and ultraviolet spectroscopy and found the composition of DNA to be constant within a species but to differ widely between species. This led him to conclude that there must be as many different types of DNA as there are different species. However, some interesting and very important consistencies emerged. Firstly the number of purine bases (adenine and guanine) was always equal to the number of pyrimidine bases

(cytosine and thymine), and secondly the number of adenine bases is equal to the number of thymine bases and the number of guanine bases equals the number of cytosine bases. This information, announced by Chargaff in 1950, was of crucial importance in constructing the Watson–Crick model of DNA.

From 1935 Chargaff worked at Columbia University, as professor of biochemistry from 1952 and as emeritus professor from 1974.

Charles, Jacques Alexandre César (1746–1823) *French physicist and physical chemist*

Born in Beaugency, France, Charles (sharl or charlz) was a clerk in the finance ministry who developed an interest in science, especially in the preparation of gases. Eventually he became professor of physics at the Conservatoire des Arts et Métiers in Paris. He constructed the first hydrogen balloons, making an ascent to over 3,000 meters (1.9 mi) in 1783. This feat brought him popular fame and royal patronage.

His name is chiefly remembered, however, for his discovery of *Charles's law*, which states that the volume of a fixed quantity of gas at constant pressure is inversely proportional to its temperature. Hence all gases, at the same pressure, expand equally for the same rise in temperature. Strictly speaking, the law holds only for ideal gases but it is valid for real gases at low pressures and high temperatures. Charles deduced the law in about 1787, working with oxygen, nitrogen, carbon dioxide, and hydrogen, but he did not publish it. He communicated his results to Joseph GAY-LUSSAC, who published his own experimental results in 1802, six months after DALTON had also deduced the law. The priority, as Gay-Lussac himself pointed out, belongs to Charles but Gay-Lussac's figures were more accurate (and thus the law is sometimes referred to as *Gay-Lussac's law*). This law and that formulated by Robert BOYLE comprise the gas laws.

Charpak, Georges (1924–) *French physicist*

Charpak (**shar**-pak or shar-**pak**), who was born in Dabrovica, Poland, was educated at the Ecole des Mines, Paris. He was imprisoned at Dachau from 1943 until 1945. He then worked in France on nuclear research, mainly at the Centre National de la Recherche Scientifique (CNRS). He moved in 1959 to CERN (Conseil Européen pour la Recherche Nucléaire; European Laboratory for Particle Physics), in Geneva, retiring in 1991.

By the time Charpak arrived at CERN nuclear physicists had begun to search for ever-more-elusive particles. To detect their fleeting and rare appearances could require the examination of thousands of particle tracks. Yet the older particle detectors – bubble chamber, cloud chamber, etc. – could handle only a small proportion of the data pouring from the newer and more powerful accelerators.

In 1968 Charpak described his newly designed drift chamber in which charged wires are strung 1.2 millimeters apart, layer on layer, in a gas-filled container. A voltage is applied to the wires in such a way that the central wires are charged positively, and the outer ones negatively. If a charged particle enters the detector, it ionizes atoms of gas, and the ions drift to the central wires, triggering a signal. As the wires criss-cross through the chamber it is possible to reconstruct a three-dimensional picture of the ion's tracks from the signals obtained.

When linked to computers the drift chamber can handle a million nuclear events per second. It played a vital role in the 1983 discovery of the W and Z bosons by Carlo RUBBIA. It also won for Charpak the 1992 Nobel Prize for physics. In more recent times Charpak turned his attention to biochemical reactions. In 2001 he published *Megawatts and Megatons: A Turning Point in the Nuclear Age* (with Richard L. Garvin).

Charpentier, Jean de (1786–1885) *Swiss geologist and glaciologist*

Charpentier (shar-pahn-**tyay**) studied under Abraham WERNER at the Mining Academy in his native city of Freiberg, where his father was also a professor. He worked as an engineer in the Silesian mines before being appointed director of the Bex salt mines in 1813.

He studied the problem of the widely scattered and impressively large erratic boulders and soon rejected the current theories of their origin. The theory that such boulders were meteorites was unlikely for they were identical in composition with other Alpine rocks. The flood theory, supported by Charles LYELL, supposed that they had been distributed by boulder-laden icebergs. However, this raised the problems of where the water had come from and where it had gone to.

Charpentier concluded that the agent responsible was glaciation and first presented his glacial theory publicly in Lucerne in 1835. He gained little support but did attract the attention of Louis AGASSIZ. In 1841 Charpentier published his results in his *Essai sur les glaciers* (Essay on Glaciers) but was anticipated by Agassiz's earlier publication, in 1840, of his *Etudes sur les glaciers* (Studies on Glaciers).

Châtelet, Emilie Le Tonnelier de Breteuil, Marquise du) (1706–1749) *French physicist, mathematician, philosopher, and translator*

Long-overlooked as a scholar in her own right, and caricatured as merely the long-standing mistress of French writer VOLTAIRE, du Châtelet (**shah**-te-lay) is now regarded as one of the great scientific minds of the 18th century. She was born into a wealthy aristocratic family in Paris, where her father held a prominent position at court. Emilie spurned typical girlish interests, and her father arranged for tutors to provide her with a wide-ranging private education. She quickly became a skilled linguist and mathematician, was well versed in science, and listened avidly to scholarly visitors to the court at Versailles. In 1725 she made an arranged marriage to an army officer, and subsequently bore him three children. du Châtelet met Voltaire in 1733, and they fell in love. The following year, with Voltaire in trouble with the authorities for his criticism of the French state, the two sought refuge at Chateau de Cirey, near the German border, a dilapidated mansion belonging to Emilie's husband. Here, they embarked on a fruitful collaboration, with du Châtelet publishing scientific papers and corresponding with leading scientists. However, it is their mutual interest in the works of Isaac NEWTON that would transform science in France and beyond. In 1738 Voltaire published *Element's of Newton's Philosophy*, a clear exposition of Newton's discoveries in astronomy and optics that introduced these ideas to the wider French public. Although cited as sole author, Voltaire emphasized in the preface the crucial role of du Châtelet's scientific expertise in writing the book. Fired with enthusiasm for Newtonianism, du Châtelet then wrote *Institutions de Physique* (Foundations of Physics; 1740), in which she attempted to reconcile conflicts in the astronomical systems of DESCARTES, LEIBNIZ, and Newton. But her greatest achievement was a translation from the original Latin of Newton's *Principia Mathematica*. She augmented the original text with her own interpretations and examples, algebraic commentary, and contemporary experimental confirmations, thus demonstrating to French scholars the full significance of Newtonian mechanics. Published in 1759, this remains the standard French translation of Newton's great work. In 1748 du Châtelet started an affair with a poet and much younger man, the Marquis de Saint-Lambert, and the following year found she was pregnant. In spite of this, Voltaire remained a steadfast friend, and du Châtelet persevered to finish her *Principia* translation. Both she and her newborn daughter died shortly after the birth. Voltaire wryly observed that du Châtelet "was a great man whose only fault was being a woman."

Chatelier, Henri Louis Le *See* LE CHATELIER, HENRI LOUIS.

Cherenkov (*or* **Cerenkov**), **Pavel Alekseyevich** (1904–1990) *Soviet physicist*

Cherenkov (che-**reng**-kof) came from a peasant family in Voronezh, Russia, and was educated at the university there, graduating in 1928. From 1930 he was a member of the Lebedev Institute of Physics in Moscow, serving there from 1953 as professor of experimental physics.

In 1934 Cherenkov was investigating the absorption of radioactive radiation by water when he noticed that the water was emitting an unusual blue light. At first he thought it was due simply to fluorescence but was forced to reject this idea when it became apparent that the blue radiation was independent of the composition of the liquid and depended only on the presence of fast-moving electrons passing through the medium.

It was later shown by Ilya FRANK and Igor TAMM in 1937 that the radiation was caused by electrons traveling through the water with a speed greater than that of light in water (though not of course greater than that of light in a vacuum). This *Cherenkov radiation* can be produced by other charged particles and can be used as a method of detecting elementary particles. Cherenkov, Frank, and Tamm shared the Nobel Prize for physics in 1958.

Chauvin, Yves (1930–   ) *French chemist*

Chauvin is Directeur de recherché honoraire à l'Institut Français du Pétrole, Rueil-Malmaison, France. He is noted for elucidating the mechanism of metathesis, a type of reaction in organic chemistry with many applications.

The word metathesis comes from the Greek word meta, meaning change, and thesis, meaning position. A simple example of metathesis occurs in inorganic chemistry – for example, silver nitrate with potassium chloride gives silver chloride and potassium nitrate. In inorganic chemistry, metathesis is often called "double decomposition." In organic chemistry, metathesis reactions occur in organic compounds with double bonds, which are broken and remade in such a way that groups of atoms change places. A metathesis reaction is frequently compared to a dance in which the couples change partners. In organic chemistry, this type of reaction requires certain complex metal compounds to act as a catalyst.

This type of reaction was discovered in the 1950s. However, the molecular mechanism by which it worked was not known, thus making it impossible to utilize this type of reaction in organic synthesis in any systematic way. In 1971 Chauvin proposed a mechanism for metathesis reactions and with a consequent understanding of what types of catalysts were effective, thus paving the way for systematic investigation into suitable catalysts. Together with his student Jean-Louis Hérrison, Chauvin postulated that the catalyst is a metal carbene (subsequently called metal alkylide), i.e., a compound in which a metal atom is attached to a carbon

atom by a double bond. Unlike previous attempts to explain metathesis reactions, the mechanism which Chauvin proposed was in accord with experimental results and subsequent investigations have confirmed that his proposed mechanism is correct.

The importance of Chauvin's work to organic synthesis, both in academic and industrial research, was rewarded by his share in the 2005 Nobel Prize for chemistry, along with Robert GRUBBS and Richard SCHROCK.

Chevreul, Michel Eugène (1786–1889) *French organic chemist*

> Where the eye sees at the same time two contiguous colours, they will appear as dissimilar as possible, both in their optical composition and in the height of their tone. We have then, *at the same time*, simultaneous contrast of colour properly so called, and contrast of tone.
> —*The Principles of Harmony and Contrast of Colours* (1839)

One of the longest-lived of all chemists, Chevreul (she-**vru(r)l**), who was born at Angers in France, studied at the Collège de France (1803). He was an assistant to Antoine François de FOURCROY (1809), assistant at the Musée d'Histoire Naturelle (1810), then professor of physics at the Lycée Charlemagne (1813–30).

In 1810 Chevreul began a great program of research into fats, which was published in his book *Recherches chimiques sur les corps gras d'origine animale* (1823; Chemical Researches on Animal Fats). By acidification of soaps derived from animal fats and subsequent crystallization from alcohol he was able to identify for the first time various fatty acids: oleic acid, "margaric acid" (a mixture of stearic and palmitic acids), butyric acid, capric and caproic acids, and valeric acid. He recognized that fats are esters (called "ethers" in the nomenclature of the day) of glycerol and fatty acids and that saponification produces salts of the fatty acids (soaps) and glycerol. In 1825 Chevreul and Joseph GAY-LUSSAC patented a process for making candles from crude stearic acid. Other fats investigated by Chevreul were spermaceti, lanolin, and cholesterol.

In 1824 Chevreul became director of the dyeworks for the Gobelins Tapestry, where he did important work on coloring matters, discovering hematoxylin in logwood, quercetin in yellow oak, and preparing the reduced colorless form of indigo. He also investigated the science and art of color with special application to the production of massed color by aggregations of small monochromatic "dots," as in the threads of a tapestry.

Chevreul's later appointments were professor of chemistry at the Musée d'Histoire Naturelle (1830) and director there (1864). His other work included the discovery of creatine (1832) and studies on the history of chemistry.

Chittenden, Russell Henry (1856–1943)
American physiologist and biochemist

As part of his undergraduate course at Yale, Chittenden, who was born in New Haven, Connecticut, was asked to investigate why scallops taste sweeter when reheated from a previous meal than when freshly cooked. This project led to his discovery of glycogen and glycine in the muscle tissue – the first demonstration of the free occurrence of glycine (or glycocoll as it was then known) in nature. The work attracted the attention of Wilhelm KÜHNE at Heidelberg who invited Chittenden to his laboratory. Later collaboration between Chittenden (at New Haven) and Kühne (in Heidelberg) provided a strong foundation for studies in enzymology.

Chittenden also did important work in toxicology and on the protein requirements of man, showing that the so-called Voit standard, which recommended 118 grams of protein per day, was a vast overestimate, and that good health could be maintained on 50 grams a day. He played a major part in the establishment of physiological chemistry (biochemistry) as a science in its own right.

Chladni, Ernst Florens Friedrich (1756–1827) *German physicist*

Born in Wittenberg in Germany, Chladni (**kladnee**) was forced to study law by his father and obtained his degree from Leipzig in 1782. When his father died, Chladni turned to science. He is noted for his work on acoustics, being the first to analyze sound in a rigorous mathematical way. For this he invented the sand-pattern technique, in which thin metal plates covered in sand are made to vibrate. The sand collects in the nodal lines producing symmetrical patterns (called *Chladni's figures*).

Chladni also had a great interest in music and designed two musical instruments: the euphonium and the clavicylinder. He also measured the speed of sound in gases other than air by filling organ pipes with the gas and measuring the change in pitch.

Chladni was one of the first scientists to believe that meteorites fell from the sky but his opinion was treated with disdain until Jean Baptiste BIOT proved him to be correct in 1803.

Chou Kung (about 12th century BC) *Chinese mathematician*

Chou Kung (choh kuung), or the duke of Chou, was the brother of Wu Wang, the founder of the Chou dynasty. He served briefly as regent on his brother's death.

He is remembered for his name in the *Chou li* (Rites of Chou), one of the earliest Chinese mathematical works, in which he supposedly takes part in a dialog with someone called Shang Kao. The dialog was thought to date back to the time of Chou Kung but scholars now think this extravagant. Although they are prepared to accept some parts as going back to the sixth century BC, the bulk of it they assign to the Han dynasty (200 BC – AD 200). The work is important in providing hard evidence for the state of early Chinese mathematics.

The most significant feature of the work is a demonstration of the truth of PYTHAGORAS's theorem for triangles with sides of 3, 4, and 5 units. The "proof," described as "piling up the rectangles," is purely diagrammatic. The work also shows knowledge of the multiplication and division of fractions, the finding of common denominators, and the extraction of square roots. Compared with Greek works of a comparable period, such as ARCHIMEDES, the work is unimpressive.

Christie, Sir William Henry Mahoney (1845–1922) *British astronomer*

Christie, a Londoner, was the grandson of the founder of the firm of London auctioneers and the son of a mathematician. He graduated from Cambridge University in 1868 and immediately joined the staff of the Royal Greenwich Observatory. He later served from 1881 to 1910 as Astronomer Royal and was knighted in 1904.

During Christie's period of office the observatory saw considerable expansion and refurbishment. The Physical Observatory, later known as the South Building, was built at Greenwich between 1891 and 1899 and equipped with new telescopes: a 28-inch (71-cm) refractor for visual use and a 26-inch (66-cm) for photographic purposes were provided in the 1890s.

The Observatory agreed to cooperate in an international project, involving 18 observatories, to produce the first photographic chart of the heavens. The ambitious project was proposed by the director of the Paris Observatory, Admiral E. Mouchez, in 1887 and the chart became known as the "Carte du Ciel." The Observatory was made responsible for the necessary observations and measurements for the large area of the sky from the north celestial pole to 65°N declination. This work was completed and published by 1909, long after the estimated time but before the other participants had finished.

Christie's own research was spent mainly on sunspot activity in collaboration with Edward MAUNDER and on unsuccessful attempts to measure stellar radial velocities.

Chu, Paul Ching-Wu (1941–) *American physicist*

Chu was born in Hunan, China, but his parents were members of the Nationalist Party and the family fled to Taiwan in 1949 for political reasons. After graduating in physics from Chengkung University, Chu moved to America in 1963 and gained his PhD in 1968 from the University of California, San Diego. After spending some time working for the company AT & T, Chu entered academic life, first at Cleveland State University, and since 1979 as professor of physics at the University of Houston. Since 1987 he has been director of the Texas Center for Superconductivity at Houston. He has also served as president of the Hong Kong University of Science and Technology (since 2001).

Much of Chu's work has been in the field of superconductivity. A major breakthrough had been achieved in 1986 when Alex MULLER had discovered some materials that become superconductive below the relatively high critical temperature of 35 K (–238°C). This temperature was still too low to be economic. The vital temperature was 77.4 K (–195.8°C) – the temperature below which nitrogen becomes liquid. The aim was to find materials that could be cooled to a superconducting state using relatively cheap liquid nitrogen, rather than the extremely expensive liquid helium (b.p. –268.9°C). Chu was determined that his Houston laboratory would be the first to find a superconductor with a critical temperature above 77.4 K.

The superconductor found by Muller was a ceramic material composed of barium, lanthanum, and copper oxide (Ba–La–CuO). Chu began by reproducing Muller's work. He next developed new methods of synthesis for this type of compound and began first to vary the ratio of elements in the compound. Initial results obtained by reducing the amount of copper were encouraging, but could not be repeated. However, at high pressures of 10,000 atmospheres it was possible to increase the critical temperature to about 40 K. Changing the proportions of the elements could raise the temperature to 52.5 K, but this was still at high pressures.

The original Muller compound contained three metals in the ratio 2:1:4. Many researchers concentrated on substituting other metals in the same ratio. Copper seemed to play a special bonding role and was judged by Chu to be indispensable. Chu decided to replace the lanthanum with other related lanthanoid elements. One he chose to work with was yttrium (Y). Finally, in January 1987, just a year after Muller's breakthrough, Chu found that the critical temperature of $Y_{1.2}Ba_{0.8}CuO_4$ was 93 K and that the effect was stable and permanent.

Chu, Steven (1948–   ) *American physicist*

Chu was born in St Louis, the second son of Chinese immigrants, and moved to New York in 1950. He attended the University of Rochester, majoring in mathematics and physics and graduating AB and BS in 1970. Graduate studies under Eugene Commins at the University of California, Berkeley, earned Chu his PhD in 1976. After a spell as postdoctoral research fellow at Berkeley, he moved to the Bell Laboratories, Murray Hill (1978–83), and in 1983 became head of the Quantum Electronics Research Department, Holmdel. In 1987 Chu moved to Stanford University as professor of physics and applied physics, and in 2004 was appointed director of the Lawrence Berkeley National Laboratory.

While at Bell Laboratories, in 1982 Chu and his colleague, Allen Mills, achieved the feat of accurately measuring the energy levels of positronium, the bound state of an electron and positron, using laser spectroscopy. Then, in 1985, he led a group that developed a pioneering technique for cooling and trapping atoms using laser beams. This technique can also be adapted as "optical tweezers," to trap microscopic particles, including DNA or other biomolecules, in water. Chu's team arranged three pairs of opposed laser beams at right angles so they trapped sodium atoms, in a vacuum, at the intersection of the six beams. Photons of the appropriate energy pushed back the sodium atoms as they tended to move out of the area, so confining them in the form of a pea-sized cloud consisting of about one million atoms. In this way the atoms cooled to within a few millionths of a degree of absolute zero, a phenomenon known as DOPPLER cooling. In the cloud the atoms move as if in a thick liquid, described as "optical molasses." Chu went on to demonstrate that the sodium atoms could be effectively trapped by the laser beams in combination with magnetic coils, forming a magneto-optical trap (MOT). William D. PHILLIPS subsequently demonstrated Doppler cooling to temperatures lower than had been theoretically predicted; this led to a reappraisal of the entire theory of Doppler cooling. For this work, Chu was awarded the 1997 Nobel Prize for physics, jointly with Phillips and Claude COHEN-TANNOUDJI.

The laser-cooling technique inspired Chu to develop the atomic fountain, in which laser-cooled atoms are sprayed like the jets of a water fountain in a vacuum-filled chamber. Essentially, cooling the atoms decreases their velocity, and improves the accuracy of the device. This now forms the basis of atomic clocks with much greater accuracy than their predecessors. Chu and his Stanford team continue to develop methods of laser cooling and trapping, and to investigate applications for laser-based optical tweezers, for example in studying the behavior of RNA and proteins during translation at ribosomes in biological systems.

Ciechanover, Aaron (1947–   ) *Israeli biochemist*

Born in Haifa, Ciechanover studied medicine at the Hebrew University School of Medicine, Jerusalem, obtaining his MD in 1974. After service in the Israeli Defense Forces (1974–77) he joined the Israel Institute of Technology (Technion) in Haifa, and gained a PhD in medicine in 1981. He moved to the USA as a postdoctoral fellow at Massachusetts Institute of Technology (1982–84) before returning to the Technion, where he was subsequently appointed associate professor (1987) and professor (1992). He is currently director of the Rappaport Family Institute for Research in Medical Sciences at the Technion.

While still a postgraduate student, Ciechanover collaborated with his colleague at the Technion, the Israeli biochemist Avram HERSHKO, in investigating energy-dependent protein degradation mechanisms in living cells. The duo worked with cell-free extracts obtained from immature red blood cells, and discovered that a crucial component of the process was a polypeptide of approximate molecular mass 9000, later named ubiquitin. Ciechanover and Hershko took sabbatical leave to visit the laboratory of US biochemist Irwin ROSE in Philadelphia. Here the three scientists were able to show that ubiquitin molecules bind covalently to various proteins. They went on to describe the enzymes involved in this ubiquitination of proteins, and how this ubiquitin tagging marks proteins for degradation by the cell. Ciechanover subsequently studied various aspects of protein ubiquitination, and its relevance to the cell cycle and other cellular activities.

In 2004 he was awarded the Nobel Prize for chemistry, jointly with Rose and Hershko, "for the discovery of ubiquitin-mediated protein degradation."

Clairaut, Alexis Claude (1713–1765)
French mathematical physicist

> I intended to go back to what might have given rise to geometry; and I attempted to develop its principles by a method natural enough so that one might assume it to be the same as that of geometry's first inventors, attempting only to avoid any false steps that they might have had to take.
> —*Elémens de géométrie* (1741; Elements of Geometry)

The Parisian-born son of a mathematics teacher, Clairaut (kle-**roh**) was introduced to the subject at an early age. By the age of ten he was studying L'Hôpital's work on conic sections and two years later he read a paper to the French Académie des sciences. He was elected to the Académie at the age of 18 following the publication in 1731 of his *Recherches sur les courbes à double courbes* (Researches on the Curves of Double Curves).

Soon after, in 1736, he accompanied MAUPERTUIS on an expedition to Lapland to determine the length of 1° of a meridian within the Arctic circle. The aim of the expedition was to determine the shape of the Earth by measuring its curvature at the places where it differed most – the equator and poles. A similar expedition under the direction of LA CONDAMINE measured the equatorial curvature in the Andes. Behind the expeditions lay a crucial test of Newtonian mechanics that the Earth's rotation should cause it to bulge at the equator and flatten at the poles. Cartesian science predicted that the reverse position should hold. As Clairaut revealed in his *Théorie de la figure de la terre* (1743; Theory of the Shape of the Earth) the Earth had, as NEWTON had claimed, a larger diameter through the equator than through the poles, a shape known to geometers as an "oblate spheroid."

In 1747 Clairaut turned his attention to the Moon and once again the issue was the accuracy of Newtonian mechanics. The motion of the lunar apogee – the point in the lunar orbit furthest away from the Earth – differed from Newton's predicted value by a factor of two. At first Clairaut was tempted to question the validity of Newton's inverse square law, but in 1749 he discovered that no such drastic step need be taken; several factors of the lunar orbit had been ignored. When these were included in Newton's lunar equations the correct value for the motion of the lunar apogee was obtained.

Clairaut was also involved in the dramatic events surrounding the return of HALLEY's comet. Halley had claimed that the comet of 1682 would return in 1758. Clairaut realized that if Newtonian mechanics was to be an exact science it must make a more precise prediction. He informed the Académie in November 1758 that the comet would be at perihelion on 13 April 1759; the actual date was 13 March, just within the allowed-for margins of error.

Clairaut also collaborated with the Marquise du CHÂTELET in her French translation of Newton's *Principia*.

Claisen, Ludwig (1851–1930) *German organic chemist*

Claisen (**klI**-sen), who was born at Cologne in Germany, gained his PhD under Friedrich KEKULÉ at Bonn, later becoming professor at Kiel (1897–1904) and honorary professor at Berlin.

He is best known for the condensation reactions that bear his name. The *Claisen–Schmidt condensation* is a method of synthesizing α,β-unsaturated aldehydes and ketones, by reaction of an aldehyde with another aldehyde or a ketone. This type of reaction was first discovered by J. Gustav Schmidt in Zurich (1880). The *Claisen condensation* (1890) involves condensation between similar or different esters, or esters and ketones, in the presence of sodium ethoxide to give unsaturated esters. The

Claisen flask, used for vacuum distillation, was developed in 1893. Claisen also made an important contribution to the study of tautomerism by isolating (simultaneously with Johannes Wislicenus) the keto-and enol-forms of acetyldibenzoylmethane and dibenzoylacetone.

Clark, Alvan Graham (1832–1897) *American astronomer and instrument-maker*

Clark, the son of the instrument-maker Alvan Clark, was born at Fall River, Massachusetts. He started life as a portrait painter but soon joined his father's firm and became a lens grinder, preparing the mirrors and lenses for some of the best telescopes of the late 19th century. In 1861 he had made a lens for Edward BARNARD at the University of Mississippi. Testing it before parting with it he looked through it at Sirius and to his surprise observed a faint image near the star. It was, in fact, Sirius B, the famous companion predicted by Friedrich BESSEL in 1844. Clark made many more observations, and discovered 16 double stars.

The Clark firm provided Simon NEWCOMB, head of the U.S. Naval Observatory, with a 26-inch (66-cm) refractor. It was with this that the very small satellites of Mars, Phobos and Deimos, were detected by Asaph HALL in 1877. In 1888 Clark built the 36-inch (91-cm) refractor for the Lick Observatory and his final achievement, just before his death, was to install his 40-inch (101-cm) refractor in the Yerkes Observatory. A practical limit is reached in using lenses larger than this and after Clark's death astronomers put their faith in mirrors rather than lenses. For this reason the Yerkes 40-inch and the Lick 36-inch are still the largest and the second largest refractors in the world.

Clarke, Sir Cyril Astley (1907–2000) *British physician*

Clarke, who was born in Leicester in eastern England, was educated at the University of Cambridge and Guy's Hospital, London, where he qualified in 1932. He remained at Guy's until 1936 when he engaged in life-insurance work before spending the war years in the Royal Navy. From 1946 Clarke worked as a consultant physician in Liverpool until 1958, when he joined the staff of the university. Here he later served as professor of medicine from 1965 to 1972 and also, from 1963 to 1972, as director of the Nuffield unit of medical genetics.

Although a consultant physician, Clarke was also a skilled amateur lepidopterist. In 1952 he became interested in the genetics of the wing colors of swallowtail butterflies and began a collaboration with Philip Sheppard, a professional geneticist who later became a colleague at Liverpool University. In particular, they worked on the inheritance of mimicry in the wing patterns of certain swallowtails. They noted that the gene controlling the wing pattern is actually a group of closely linked genes behaving as a single unit – a supergene. They also found that even though the males also carry such supergenes, the patterns only show in the females.

At this point Clarke was struck by certain striking parallels between the inheritance of swallowtail wing patterns and human blood types. Above all it aroused his interest in Rhesus babies. This condition arises when an Rh-negative mother, that is someone whose blood lacks the Rh factor or antigen, and an Rh-positive father produce an Rh-positive child. Occasionally the fetus's blood leaks from the placenta into the mother's blood and stimulates the production of Rh antibodies. This will cause her to destroy unwittingly the red cells of any subsequent Rh-positive babies she may carry.

Clarke and Sheppard puzzled over how to prevent the mother producing the destructive Rh antibodies. The answer eventually came from Clarke's wife who, in an inspired moment, told him to inject the Rh-negative mothers with Rh antibodies. As this is what destroys the blood of the fetus in the first place, the answer initially sounds absurd. However, the Rh antibodies should destroy incompatible Rh-positive cells before the mother's own antibody machinery acted, that is, before the mother could become sensitized to Rh-positive blood.

In 1964 Clarke and his colleagues were able to announce a major breakthrough in preventive medicine. Since then thousands of women have received injections of Rh antibodies with only a few failures.

Claude, Albert (1898–1983) *Belgian–American cell biologist*

Claude (klohd), who was born at Longlier in Belgium, was educated at the University of Liège where he obtained his doctorate in 1928. He joined the staff of the Rockefeller Institute, New York, in 1929 and in 1941 adopted American citizenship. Claude returned to Belgium in 1948 to serve as director of the Jules Bordet Research Institute, a post he retained until his retirement in 1972.

In the 1930s Claude attempted to purify Peyton ROUS's chicken sarcoma virus (RSV) using a centrifuge. He succeeded in producing a fraction with an enhanced sarcogenic power, noting that small granules containing nucleoprotein were present. Suspecting these granules to be the cause of the RSV, he was somewhat surprised to find similar granules present in centrifuged cells taken from uninfected chicken embryo.

Over the next 20 years, using electron microscopes as well as improved centrifuges,

Claude began to chart the constitution of the protoplasm. Although the mitochondria had first been described as early as 1897, Claude was able to distinguish them from what he originally termed "microsomes." Among such microsomes he could make out a "lacelike reticulum" spread throughout the cytoplasm, a structure later named the endoplasmic reticulum. Another member of Claude's laboratory, George PALADE, went on to identify the ribosome.

For his work in opening up the study of cell structures Claude shared the 1974 Nobel Prize for physiology or medicine with Palade and Christian de DUVE.

Claude, Georges (1870–1960) *French chemist*

Born in Paris, France, Claude was educated at the Ecole de Physique et Chimie after which he worked as an engineer in various industries. Claude made a number of important contributions to technology, including the discovery (1886) that acetylene (ethyne) could be handled with safety if dissolved in acetone, and a method of liquefying air (1902), which he used for the large-scale production of nitrogen and oxygen. In 1910 he introduced neon lighting, using neon gas at low pressure excited by an electric discharge to emit a bright red light.

The latter part of his life was, however, less successful. From 1926 onward he worked on new sources of energy. In particular, he tried to show how it could be extracted from the temperature difference between the surface and the bottom of the sea. Although his argument was sound he never overcame the formidable engineering difficulties.

Although an old man of 75 when World War II ended Claude was imprisoned as a Vichy sympathizer.

Clausius, Rudolf Julius Emmanuel (1822–1888) *German physicist*

Clausius (**klow**-zee-uus), who was born in Köslin (now Koszalin in Poland), studied at the University of Berlin and obtained his doctorate from Halle in 1848. He was professor of physics at the Royal Artillery and Engineering School, Berlin (1850–55) and professor of mathematical physics at Zurich (1855–67). He then transferred to the University of Würzburg (1867) and, from there, moved to Bonn (1869).

He is noted for his formulation of what is now known as the second law of thermodynamics. Clausius arrived at this by considering the theorem of Sadi CARNOT on heat engines and attempting to reconcile this with the mechanical theory of heat, which was developing at the time. In 1850 he published a famous paper *Uber die bewegende Kraft der Wärme* (On the Motive Force of Heat), in which he first introduced the principle that "it is impossible by a cyclic process to transfer heat from a colder to a warmer reservoir without net changes in other bodies." An alternative statement of this, the second law, is "heat does not flow spontaneously from a colder to a hotter body." The second law of thermodynamics was independently recognized by Lord KELVIN.

Clausius gave the law a mathematical statement in 1854 and published a number of papers on the topic over the next few years. In 1865 he introduced the term "entropy" as a measure of the availability of heat. The change in entropy of a system is the heat absorbed or lost at a given temperature divided by the temperature. An increase in entropy corresponds to a lower availability of heat for performing work.

The second law of thermodynamics is one of the fundamental principles of physics. It describes the fact that, although the total energy in a system is conserved, the availability of energy for performing work is lost. Clausius showed that in any nonideal (irreversible) process the entropy increased. The first and second laws of thermodynamics are encapsulated in his famous statement, "The energy of the universe is a constant; the entropy of the universe always tends toward a maximum."

Clausius also followed the work of James JOULE on the kinetic theory of gases, introducing the ideas of effective diameter and mean free path (the average distance between collisions). A contribution to electrochemistry was his idea that substances dissociated into ions on solution. In the field of electrodynamics he produced a theoretical expression for the force between two moving electrons – a formula later used by Hendrik LORENTZ.

Clemence, Gerald Maurice (1908–1974) *American astronomer*

Clemence, who was from Smithfield, Rhode Island, studied mathematics at Brown University, Rhode Island. After graduating he joined the staff of the U.S. Naval Observatory in 1930 where he remained until 1963, serving as head astronomer and director of the Nautical Almanac from 1945 to 1958 and science director of the Observatory from 1958. In 1963 he was appointed senior research associate and lecturer at Yale, becoming professor of astronomy in 1966, a post he held until his death.

Clemence's work was primarily concerned with the orbital motions of the Earth, Moon, and planets. In 1951, in collaboration with Dirk BROUWER and W. J. ECKERT, Clemence published the basic paper *Coordinates of the Five Outer Planets 1653–2060*. This was a considerable advance on the tables for the outer planets calculated by Simon NEWCOMB and George W. Hill 50 years earlier. Clemence and his colleagues calculated the precise positions of the outer planets at 40-day intervals over a period of 400 years. It was the first time that the influence of the planets on each other was calculated at each step instead of the prevailing custom of assuming that the paths of all except one were known in advance.

Such an ambitious scheme was only made possible by the emergence of high-speed computers, one of which was made available to them by IBM from 1948. For each step some 800 multiplications and several hundred other arithmetical operations were required and would, Clemence commented, have taken a human computer 80 years if he could have completed the work without committing any errors en route.

Clemence also conceived the idea that Brouwer named "Ephemeris Time," by which time could be determined very accurately from the orbital positions of the Moon and the Earth. This followed the discovery that the Earth's period of rotation was not constant and should not therefore be used in the measurement of time. Ephemeris Time eventually came into use in 1958, although it has been superseded for most purposes by the more convenient and even more accurate atomic time scale.

Cleve, Per Teodor (1840–1905) *Swedish chemist*

Cleve (**klay**-ve), who was born in the Swedish capital Stockholm, became assistant professor of chemistry at the University of Uppsala in 1868 and was later made professor of general and agricultural chemistry there. He is mainly remembered for his work on the rare earth elements.

In 1874 Cleve concluded that didymium was in fact two elements; this was proved in 1885 and the two elements named neodymium and praseodymium. In 1879 he showed that the element scandium, newly discovered by Lars NILSON, was in fact the eka-boron predicted by Dmitri MENDELEEV in his periodic table. In the same year, working with a sample of erbia from which he had removed all traces of scandia and ytterbia, Cleve found two new earths, which he named holmium, after Stockholm, and thulium, after the old name for Scandinavia. Holmium in fact turned out to be a mixture for, in 1886, Lecoq de Boisbaudran discovered that it also contained the new element dysprosium.

Cleve is also remembered as the teacher of Svante ARRHENIUS.

Coblentz, William Weber (1873–1962) *American physicist*

Coblentz, the son of a farmer from North Lima, Ohio, was educated at the Case Institute of Technology and at Cornell, where he obtained his PhD in 1903. In 1904 he joined the National Bureau of Standards in Washington and in the following year founded the radiometry section of the bureau, where he remained until his retirement in 1945.

Coblentz worked mainly on studies of infrared radiation. At the Lick Observatory he began, in 1914, a series of measurements aimed at determining the heat radiated by stars. He was also one of the pioneers of absorption spectroscopy in the infrared region as a technique for identifying compounds.

Cockcroft, Sir John Douglas (1897–1967) *British physicist*

Cockcroft, who was born at Todmorden in northern England, entered Manchester University, England, in 1914 to study mathematics, but left the following year to join the army. After World War I he was apprenticed to the engineering firm Metropolitan Vickers, which sent him to read electrical engineering at the Manchester College of Technology. He later went to Cambridge University, graduated in mathematics, and joined Ernest RUTHERFORD's team at the Cavendish Laboratory.

Cockcroft soon became interested in designing a device for accelerating protons and, with E. T. S. WALTON, constructed a voltage multiplier. Using this, Cockcroft and Walton bombarded nuclei of lithium with protons and, in 1932, brought about the first nuclear transformation by artificial means: $^{7}_{3}\text{Li} + ^{1}_{1}\text{H} \rightarrow ^{4}_{2}\text{He} + ^{4}_{2}\text{He} + 17.2$ MeV.

For this work Cockcroft and Walton received the 1951 Nobel Prize for physics. During World War II Cockcroft played a leading part in the development of radar. In 1940 he visited America as a member of the Tizard mission to negotiate exchanges of military, scientific, and technological information. In 1944 he became director of the Anglo-Canadian Atomic Energy Commission. He returned to Britain in 1946 to direct the new Atomic Energy Research Establishment at Harwell and remained there until 1959, when he was appointed master of Churchill College, Cambridge, a new college devoted especially to science and technology. Cockcroft received a knighthood in 1948.

Cocker, Edward (1631–1675) *English engraver and mathematician*

Cocker was famous for writing and engraving a very influential and popular textbook, the *Arithmetic* (1678). He was also a teacher of arithmetic and writing. Cocker produced notable textbooks on other subjects, including several writing manuals and an English dictionary but none of these rivaled in popularity his book on arithmetic, which went through over 100 editions. Cocker is mentioned by Samuel Pepys who thought his skill as an engraver sufficient to comment favorably on it in his *Diary*.

Cockerell, Sir Christopher Sydney
(1910–1999) *British engineer and inventor*

Cockerell was born in Cambridge and educated at Cambridge University, graduating in engineering in 1931. Initially he joined a small engineering firm, then returned to Cambridge to study electronics, and in 1935 he joined the Marconi Company as an electronics engineer. Here he worked on the development of airborne navigational equipment and on radar.

In 1950 he left Marconi to set up his own boat-hire business on the Norfolk broads. As an amateur yachtsman, Cockerell was interested in the effect of water drag on the hull of a boat and had the idea of raising the boat above the water on a cushion of air. In 1954 he performed a crucial experiment using kitchen scales, tin cans, and a vacuum cleaner to show that a properly directed stream of air could produce the required lift. The next year he built a working model out of balsa wood, powered by a model-aircraft engine.

He was granted a patent on his idea in 1955 and in 1957 the Ministry of Supply commissioned a full-size craft from the company Saunders Roe. The first prototype, SR-N1, weighed 7 tons and was capable of 60 knots. It crossed the English Channel in 1959 (with Cockerell aboard). Hovercraft entered regular cross-channel service in 1968.

Cockerell was a consultant to Hovercraft Development Ltd. until 1979. He was also interested in the development of wave-power generators.

Cohen, Paul Joseph (1934–2007) *American mathematician*

Cohen, who was born at Long Branch, New Jersey, was educated at Brooklyn College and at the University of Chicago, where he obtained his PhD in 1958. He spent a year at the Massachusetts Institute of Technology and two years at the Institute for Advanced Studies, Princeton, before moving to Stanford. He was appointed professor of mathematics at Stanford from 1959 to 2004, when he became an emeritus professor.

Mathematicians had been introduced to transfinite arithmetic by Georg CANTOR from the 1870s onwards. Cantor had identified two distinct infinite sets, namely the set of natural numbers and the set of real numbers, represented by $\aleph_0$ and c respectively. He had also proved that there were an infinite number of infinite numbers, that following $\aleph_0$, there came $\aleph_1$, $\aleph_2, \aleph_3, \ldots$ indefinitely. Where did c fit into this sequence? Cantor answered by proposing that $c = \aleph_1$, a supposition since known as the "continuum hypothesis." It was the first member on David HILBERT's 1900 list of outstanding unsolved mathematical problems.

Little progress was made upon the problem before 1938 when Kurt GÖDEL demonstrated that set theory remains consistent if the continuum hypothesis is added as an axiom. This did not, however, constitute a proof of the hypothesis, for set theory's own absolute consistency has never been proved. Nonetheless, Gödel's work did show that the continuum hypothesis could not be shown to be false within set theory.

In 1963 Cohen proposed to develop a non-Cantorian set theory that contained not the continuum hypothesis but its negation. He showed that no contradiction ensued and it seemed to follow that the continuum hypothesis was quite independent of set theory and that it could be neither proved nor disproved within any standard system of set theory.

Cohen, Seymour Stanley (1917–)
American biochemist

Cohen was educated in his native New York, at the City College and at Columbia, where he obtained his PhD in 1941. He joined the University of Pennsylvania in 1943, serving as professor of biochemistry from 1954 until 1971, when he moved to the University of Denver, Colorado, as professor of microbiology. Cohen returned to New York in 1976 to take the chair of pharmaceutical sciences at the State University, Stony Brook, becoming emeritus professor in 1985.

In 1946 Cohen began a series of studies in molecular biology using the technique of radioactive labeling. The common microorganism *Escherichia coli* could be infected in the laboratory with the bacteriophage known as T2. Within a matter of minutes the bacterial cell would burst releasing several hundred replicas of the invading T2. The problem was to understand the process. It was known that phages were nucleoproteins consisting of a protein coat surrounding a mass of nucleic acid (DNA in the case of T2). But, as Cohen realized, nucleic acid differed from protein in containing measurable amounts of phosphorus. This could in theory be traced through any biochemical reaction by labeling it with the radioactive isotope phosphorus-32.

Cohen used this technique in a number of experiments in the late 1940s that suggested rather than demonstrated the vital role of DNA in heredity. It was not until 1952, when Alfred HERSHEY and Martha Chase used Cohen's labeling technique, that more substantial results were available.

Cohen, Stanley (1922–) *American biochemist*

A native New Yorker, Cohen was educated at Brooklyn and Oberlin colleges and at the University of Michigan, where he was appointed teaching fellow in the department of biochemistry in 1946. He moved to the University of

Colorado School of Medicine in 1948, and in 1952 he took up the post of American Cancer Society postdoctoral fellow at Washington University, St. Louis. His long association with the Vanderbilt University School of Medicine, Nashville, began in 1959, with his appointment as assistant professor of biochemistry. He subsequently became associate professor (1962), professor (1967), and distinguished professor (1986).

Cohen's appointment to Washington University in 1952 marked the start of a fruitful collaboration with the Italian cell biologist, Rita LEVI-MONTALCINI, who had discovered a chemical produced by a culture of mouse tumor cells that influenced the number of nerve cells growing in chick embryos. Cohen set about trying to characterize this growth factor (later termed "nerve growth factor"), and found the same chemical in snake venom and in the salivary glands of adult male mice.

His findings led Cohen to investigate another growth factor that influences the embryological development of such tissues as those of eyes and teeth, which are derived from epidermis. He was able to identify a receptor on the cell membrane that was responsive to this epidermal growth factor. This was of great significance, suggesting a mechanism by which cells are able to interact with chemical messengers such as hormones, which control their growth or normal functions. Such cell-surface receptors are also a crucial element in the abnormal uncontrolled growth of cells in cancer.

For his work on growth factors and membrane receptors, Cohen was awarded the 1986 Nobel Prize for physiology or medicine, jointly with Levi-Montalcini.

Cohen-Tannoudji, Claude (1933–) *Algerian-born French physicist*

Cohen-Tannoudji was educated at the Ecole Normale Supérieure. In 1973 he was appointed professor of atomic and molecular physics at the Collège de France, Paris.

Cohen-Tannoudji, following on from the work of William PHILLIPS and Steven CHU, sought to understand and improve the process of optical cooling of single atoms. He proposed that laser traps operate by a process of what has since been called Sisyphus cooling. The laser beams, he argued, produce a series of standing waves of light polarized in different directions. As the atoms pass through the various fields their energy levels and thereby temperature are lowered.

Early efforts at cooling had found that laser traps unfortunately also tend to agitate atoms causing them to move out of the beam. One solution proposed by Cohen-Tannoudji allowed helium atoms to be cooled to 0.18 microkelvins. The method exploited the fact that atoms could occupy a particular combination of two distinct quantum states with different velocities, in which they remain invisible to any additional photons. Thus, once in this state, their energy cannot be increased by any further photon collisions.

For his work in this field Cohen-Tannoudji shared the 1997 Nobel Prize for physics with Chu and Phillips.

Cohn, Ferdinand Julius (1828–1898) *German botanist and bacteriologist*

> Its perusal makes one feel like passing from ancient history to modern times.
> —On Cohn's 1872 treatise *Untersuchungen über Bacterien* (Researches on Bacteria).
> William Bulloch, *The History of Bacteriology* (1938)

Cohn (kohn), who was born in Breslau (now Wrocław in Poland), was an extremely intelligent child and progressed through school rapidly, being admitted to the philosophy department at Breslau University at the early age of 14. He later developed an interest in botany but was prevented from graduating at Breslau by the university's anti-Semitic regulations. He therefore moved to Berlin, where he received his doctorate in botany in 1847.

Cohn returned to Breslau, becoming professor of botany there in 1872. He had long argued that the state should be responsible for the establishment of botanical research institutes, and as a result of his campaign the world's first institute for plant physiology was set up in Breslau in 1866. Cohn was director of this institute until his death and in 1870 he founded the journal *Beiträge zur Biologie der Pflanzen* (Contributions to Plant Biology), mainly for the purpose of publishing work carried out at Breslau.

Cohn's early research concentrated on the morphology and life histories of the microscopic algae and fungi, which led to his demonstration that the protoplasm of plant and animal cells is essentially similar. Later, stimulated by the work of Louis PASTEUR, he became increasingly interested in bacteria. His classic treatise *Untersuchungen über Bacterien* (Researches on Bacteria), published in his journal in 1872, laid the foundations of modern bacteriology. In it he defined bacteria, used the constancy of their external form to divide them into four groups, and described six genera under these groups. This widely accepted classification was the first systematic attempt to classify bacteria and its fundamental divisions are still used in today's nomenclature.

Although Cohn did not believe in the theory of spontaneous generation he was aware that bacteria could develop in boiled infusions kept in sealed containers. He postulated the existence of a resistant developmental stage and through careful observation was able to demonstrate the formation of heat-resistant spores by *Bacillus subtilis*.

Through his book *Die Pflanze* (1872; The Plant) and the printing of many of his popular lectures, Cohn presented the study of biology to a wide and appreciative public. He was also responsible for the publication of Robert KOCH's important work on the life cycle of the anthrax bacillus.

Cohnheim, Julius (1839–1884) *German pathologist*

Born in Demmin in Germany, Cohnheim (**kohn**-hIm) graduated from the University of Berlin in 1861, remaining there as assistant to the pathologist Rudolf VIRCHOW. He later held chairs of pathological anatomy at Kiel, Breslau, and, from 1878, at Leipzig, until his death from gout complications.

Cohnheim threw considerable light on the process of inflammation. He inflicted relatively minor wounds on the tongue and intestines of frogs and observed the consequences. He noted that at the site of injury large numbers of white blood cells (leukocytes) pass through the walls of the veins to produce the swelling characteristic of inflammation. This disproved Virchow's theory that pus corpuscles originate at the point of wounding and showed them instead to be disintegrated leukocytes.

Cohnheim also did important work on tuberculosis, confirming Jean VILLEMIN's theory that the disease is contagious by injecting tuberculous material into the chamber of a rabbit's eye and watching its development through the transparent cornea.

Cohnheim's technique of freezing material before sectioning is now a standard laboratory practice.

Colombo, Matteo Realdo (c. 1516–1559) *Italian anatomist*

Born in Cremona, Italy, Colombo (ko-**lohm**-boh) became a pupil of the famous anatomist Andreas VESALIUS and succeeded his teacher to the chair of anatomy at Padua University. His book, *De re anatomica* (On Anatomy), published after his death in 1559, contained descriptions of the pleura, peritoneum, and other organs that were more accurate than preceding ones. However, his most important contribution to medicine was to demonstrate that blood from the lungs returns to the heart via the pulmonary vein.

Compton, Arthur Holly (1892–1962) *American physicist*

Compton came from a distinguished intellectual family in Wooster, Ohio. His father, Elias, was a professor of philosophy at Wooster College while his brother, Karl, also a physicist, became president of the Massachusetts Institute of Technology. He was educated at Wooster College and at Princeton, where he obtained his PhD in 1916. He began his career by teaching at the University of Minnesota and, after two years with the Westinghouse Corporation in Pittsburgh, he returned to academic life when in 1920 he was appointed professor of physics at Washington University, St. Louis, Missouri. The main part of his career however was spent at the University of Chicago where he served as professor of physics from 1923 to 1945. Compton then returned to the University of Washington first as chancellor and then (1953–61) as professor of natural philosophy.

Compton is best remembered for the discovery and explanation in 1923 of the effect named for him for which, in 1927, he shared the Nobel Prize for physics with Charles T. R. WILSON. He was investigating the scattering of x-rays by light elements such as carbon, and found that some of the scattered radiation had an increased wavelength, an increase that varied with the angle of scattering. According to classical physics there should be no such change, for it is difficult to see how the scattering of a wave can increase its wavelength, and Compton was led to seek its explanation elsewhere.

He thus assumed that the x-rays also exhibited particle-like behavior. Hence they could collide with an electron, being scattered and losing some of their energy in the process. This would lead to a lowering of the frequency with a corresponding increase in the wavelength. Compton went on to work out the formula that would predict the change of wavelength produced in the secondary x-rays and found that his precise predictions were fully confirmed by measurements made of cloud-chamber tracks by Wilson. Significantly, this was to provide the first hard experimental evidence for the dual nature of electromagnetic radiation; that is, that it could behave both as a wave and a particle. This would be developed much further in the 1920s as one of the cornerstones of the new quantum physics.

In the 1930s Compton concentrated on a major investigation into the nature of cosmic rays. The crucial issue, following the work of Robert MILLIKAN, was whether or not a variation in the distribution of cosmic rays with latitude could be detected. Such an effect would show that the rays were charged particles, deflected by the Earth's magnetic field, and not electromagnetic radiation. As a result of much travel and the organization of a considerable amount of the research and measurements of others Compton was by 1938 able to establish conclusively that there was a clearly marked latitude effect.

During the war Compton was an important figure in the manufacture of the atomic bomb as a member of the committee directing research on the Manhattan project. He also set up at Chicago the Metallurgical Laboratory, which acted as a cover for the construction of the first atomic pile under the direction of Enrico FERMI and took responsibility for the production of plutonium. Compton later wrote a full account of this work in his book *Atomic Quest* (1956).

Comte, Auguste Isidore (1798–1857)
French positivist philosopher

> To understand a science it is necessary to know its history.
> —*Cours de philosophie positive* (1830–1842; Treatise on Positive Philosophy)

The son of a government official from Montpellier in southern France, Comte (kont) started his education at the Ecole Polytechnique, Paris; after only two years he was expelled in 1816 for challenging the authorities. Soon after, he met Henri Saint-Simon (1760–1825), a political sociologist, and became his secretary. However, always unstable, Comte soon quarreled with his mentor and resigned in 1824. There then followed a bout of insanity, a suicide attempt by throwing himself into the Seine, and his marriage to a former prostitute, Caroline Massin.

By this time Comte had started lecturing on his "positive philosophy" and he earned some money from tutoring and examining. Later in his career the English philosopher John Stuart Mill ensured that sufficient private funds were available to meet Comte's needs.

Between 1830 and 1842 Comte published six volumes of his *Cours de philosophie positive* (Treatise on Positive Philosophy), covering the application of positivism to mathematics, astronomy, physics, chemistry, history, and social studies. All subjects, he noted, pass through three stages. In the first theological stage, phenomena are explained by reference to divine beings. The second, metaphysical stage substitutes abstract forces for gods. Finally, in the positive stage, the inadequacy of all underlying causes comes to be recognized, and positivists seek only for relations of constant conjunction between phenomena.

In later life Comte sought to go much further and establish what some saw as "Catholicism without Christianity." He proposed to introduce a Positivist Calendar in which new patron saints such as NEWTON and GALILEO would replace Christian figures. There would be a catechism and a chief priest.

Despite such excesses Comte's work proved to be enormously influential. The rejection of metaphysics and the emphasis on expressing laws of nature in terms of experience found later echoes in the work of Ernst MACH and the early work of EINSTEIN. Comte is also remembered for his presumption in claiming that there are some things, such as the chemistry of the stars, which will always lie outside our experience. Shortly afterwards, in 1859, Gustav Kirchoff demonstrated that the chemical com-position of stars could be inferred from their spectral lines.

Conant, James Bryant (1893–1978)
American chemist

> Science owes more to the steam engine than the steam engine owes to science.
> —*Science and Common Sense* (1961)

Conant, who was born in Dorchester, Massachusetts, studied at Harvard, gaining his PhD in 1916; he was subsequently professor of chemistry there (1919–33), and then president (1933–53). Conant had a multifaceted career as a research chemist, administrator, diplomat, and educationalist. In the research phase he worked on organic reaction mechanisms and he also showed that oxyhemoglobin contains ferrous iron (1923). During World War II he was head of the National Defense Research Committee and deputy head of the Office of Scientific Research and Development; he played an important role in the development of the atomic bomb. On retiring from Harvard he became commissioner and then ambassador to Germany (1953–57) and wrote many controversial books on education.

Connes, Alain (1947–   ) *French mathematician*

Connes was born in Draguignan, France, and educated at the Ecole Normale Supérieure between 1966 and 1970. He was a research fellow at the Centre National de la Recherche Scientifique (CNRS) between 1970 and 1974, followed by a year as a visiting research fellow at Queen's University, Kingston, Canada. He was an associated professor, and subsequently professor at the University of Paris VI between 1976 and 1980 and became the director of research at CNRS in 1981. Since 1979 he has been long-term professor and Léon Motchane professor at the Institut des Hautes Etudes Scientifiques (IHES). Since 1984 he has also been a professor at the Collège de France. In 1982 Connes was awarded the Fields Medal for his work on the abstract algebras associated with sets of operators.

Connes is best known for developing the branch of mathematics known as noncommutative geometry, i.e., geometry in which the fundamental entities are noncommuting objects such as matrices, and applying it to theoretical physics. He summarized the early part of his work on this topic in a lengthy book entitled *Noncommutative Geometry*, the English translation of which was published in 1994. This book included his work of the early 1990s in which he expressed the Standard Model of elementary particle theory in terms of noncommutative geometry. He later extended this work to include gravity. Subsequently he has applied noncommutative geometry to superstring theory and the process of renormalization in quantum field theory.

Conon of Samos (about 245 BC) *Greek mathematician and astronomer*

> He discerned all the lights of the vast universe, and disclosed the risings and settings of the stars, how the fiery brightness of the sun is darkened, and how the stars retreat at fixed times.
> —Catullus (1st century BC)

Conon (**koh**-non) settled in Alexandria and was employed as court astronomer to the Egyptian monarch Ptolemy III. None of Conon's own writings survive and what is known of his work is through secondhand references to him by other Greek mathematicians. For example, Conon's work on conics was made use of by APOLLONIUS of Perga in his famous treatise on conics.

Among Conon's activities as an astronomer was the compilation of tables of the times of the rising and setting of the stars, known as the *parapegma*. He was also responsible for naming a constellation of stars. The consort of Ptolemy III, Berenice II, presented her hair as an offering at the temple of Aphrodite. This disappeared and Conon claimed that the hair now hung as a new constellation of stars, which he named Coma Berenices ("Berenice's Hair").

Conon was known to have been a friend of ARCHIMEDES and it is probable that the "Spiral of Archimedes," a mathematical curve, was in fact Conon's discovery.

Conway, John Horton (1937–) *British mathematician*

Born in the city of Liverpool in the northwest of England, Conway was educated at Cambridge where he obtained his PhD in 1964. He was appointed professor of mathematics at Cambridge in 1983. In 1988 he moved to America to become John von Neumann Professor of Mathematics at Princeton.

Conway's name became familiar to a large number of nonmathematicians through his invention of the "game of life" in 1970. The game is played on an infinite two-dimensional cellular array. Each cell has eight immediate neighbors, and can be in one of two states: on or off, occupied or empty, alive or dead. Two simple rules govern the outcome of any initial state:
1. A live cell will remain alive in the next generation if it has either two or three live neighbors.
2. An empty or dead cell will be occupied or come to life in the next generation if it has exactly three live neighbors.

In all other situations living cells die and dead cells remain dead. The game starts with an initial pattern of live cells and proceeds by a series of discrete changes – at each change all the cells change simultaneously to give a new pattern in the next generation.

Is there a general way, Conway asked, to determine the fate of any pattern? He also offered a $50 prize to anyone who could produce a pattern that grew indefinitely, or demonstrated that no such pattern could exist. The problem so intrigued computer operators that the search for interesting *Life* forms is thought to have cost millions of dollars in unauthorized computer time. The prize was awarded to a group at the Massachusetts Institute of Technology which discovered a pattern known as a "glider," which every thirty generations produced another glider. Conway used gliders to demonstrate in 1982 that there are patterns that behave like self-replicating animals. The snag is that spontaneous creations of such patterns would require a computer screen larger than the solar system.

Conway has also made major contributions to group theory and knot theory. Is it possible to classify all finite simple groups? These are groups which, in the manner of prime numbers, cannot be decomposed into smaller groups. While many groups fitted into clearly defined classes, several others, known as "sporadic groups," fitted into no recognized class. Five such groups were identified by Mathieu in the 1860s. A sixth was discovered in 1965 and Conway identified a further three in 1968. By 1975 twenty-six sporadic groups were known. This completed the classification theorem, also known as the "Enormous Theorem," which clas-

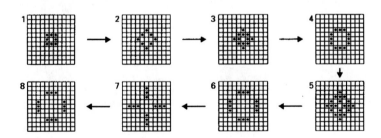

GAME OF LIFE   *Successive generations of a simple starting pattern.*

sifies all finite simple groups and has been estimated to be 15,000 pages long.

One aim of knot theorists is to distinguish between different types of knots. This can be done by calculating the crossing number, that is, the number of points the string crosses itself. Unfortunately, many different knots can have the same crossing number, and the number itself may be difficult to calculate. In 1960 Conway introduced a new and simpler way to find crossing numbers. It allowed him to establish, for example, that there are at least 801 distinct knots with a crossing number no higher than 11.

Conway is also the author, along with E. Burkelamp and R. Guy, of a book on mathematics and games – *Winning Ways* (1982).

Conybeare, William Daniel (1787–1857)
British geologist

Conybeare, the son of a clergyman from St. Botolph, England, was educated at Oxford University. He entered the Church himself and was vicar of Axminster, Devon, before being made dean of Llandaff in Wales in 1844.

He became interested in geology and was an early member of the Geological Society, founded in 1807. His most significant work was *Outlines of the Geology of England and Wales*, which was published in collaboration with William PHILLIPS in 1822. In this influential work fossils were used to date sedimentary strata and the stratigraphy of British rocks was outlined beginning with the 350-million-year-old Carboniferous, which they named from its abundant carbon deposits.

Conybeare also wrote on the geology of the coalfields of the Bristol area with William Buckland and was one of the first to use geological cross-sections. He participated in the dramatic discoveries of the remains of giant reptiles and was instrumental in the naming of Gideon MANTELL's iguanodon from the Tilgate Forest in Sussex and of Buckland's *Megalosaurus*. In 1824 he succeeded in reconstructing the plesiosaur from its remains, claiming it as a link between the ichthyosaur and the crocodiles.

In theory, he largely followed his friend Buckland in attempting to explain the structure of the Earth and the disappearance of species in terms of a series of catastrophes. He thus argued strongly against Charles LYELL.

Cook, James (1728–1778) *British navigator and explorer*

> The great quantity of new plants etc. Mr Banks and Dr Solander collected in this place occasioned my giving it the name of *Botany Bay*. It is situated in the latitude of 34°0′S, longitude 208°37′W; it is capacious safe and commodious, it may be known by the land on the sea-coast which is of a pretty even and moderate height....
> —*The Journals of Captain James Cook* (1770)

Cook, the son of a Scottish farm laborer, was born at Marston in England. He was educated at the local village school and joined the Royal Navy as an able seaman in 1755. He became a ship's master in 1759, spending eight years on survey work before being appointed by the Royal Society to take command of the *Endeavour* in 1768 on its voyage to the islands of Tahiti. He made two further major voyages of discovery in 1772–75 and in 1776.

In many ways Cook's journeys were the first modern voyages. His voyage in 1768 was to be the first of the great scientific expeditions that were to become so common in the following century. One of his main duties was to carry Royal Society observers to Tahiti to watch the transit of Venus across the Sun; such transits of planets were valuable for determining the distance between the Earth and the Sun. The scientists on board included the distinguished naturalists Joseph Banks and his assistant, Daniel Solander, and the expedition also carried artists to maintain a visual record.

The voyage's second main objective was to discover the southern continent, Terra Australis, which was believed to exist. It was assumed that the northern land mass of Eurasia must be symmetrically balanced by a southern land mass. Cook found New Zealand and extensively charted this over a period of six months and then, continuing his voyage, sighted the southeast coast of Australia on 19 April 1770. He continued up the east coast of Australia successfully navigating the treacherous Great Barrier Reef. The *Endeavour* returned to England with a vast collection of scientific observations. Cook also won fame for preventing any of his crew members from dying of scurvy by insisting on a diet that included forms of fresh fruit and vegetables.

Cook led a second expedition (1772–75) to the South Seas in the *Resolution* and the *Adventure* in which he circumnavigated the high latitudes and traveled as far south as latitude 72°. He discovered new lands, including New Caledonia and the South Sandwich Islands, but found no trace of the "great southern continent." It was also on the second voyage that the chronometer was used as a standard issue after its successful testing. Before 1772 navigators determined their longitude either by guesswork or by some very complicated calculations based on the Moon. Now, merely by noting the time and making comparatively simple calculations, it was possible to determine positions east or west of Greenwich. On his return he was made a fellow of the Royal Society and, for his paper on scurvy and its prevention, was awarded the Copley Medal.

Cook's third voyage (1776), again in the *Resolution*, ended in disaster. In trying to recover one of the ship's boats, which had been stolen by Polynesian islanders, Cook was attacked and killed by the natives on the beach of Kealakekua Bay.

Cooke, Sir William Fothergill (1806–1879) *British physicist*

William Cooke was the son of a London surgeon. After a period spent at Durham and Edinburgh Universities he joined the Indian Army and, on his return, started to study medicine at Heidelberg, but his interest was caught by the practical potential of the electric telegraph.

Returning to England in 1837, he entered into partnership with Charles WHEATSTONE and, together, they patented many telegraphic alarm systems for use on the railroads. Cooke received his knighthood in 1869.

Cooper, Leon Neil (1930–) *American physicist*

Cooper, who was born in New York City, was educated at Columbia where he obtained his PhD in 1954. After brief spells at the Institute for Advanced Study, Princeton, the University of Illinois, and Ohio State University, he moved in 1958 to Brown University, Providence, and was later (1962) appointed to a professorship of physics.

Cooper's early work was in nuclear physics. In 1955 he began work with John BARDEEN and John Robert SCHRIEFFER on the theory of superconductivity. In 1956 he showed theoretically that at low temperatures electrons in a conductor could act in bound pairs (now called *Cooper pairs*). Bardeen, Cooper, and Schrieffer showed that such pairs act together with the result that there is no electrical resistance to flow of electrons through the solid. The resulting BCS theory stimulated further theoretical and experimental work on superconductivity and won its three authors the 1972 Nobel Prize for physics.

Cooper has also worked on the superfluid state at low temperatures and, in a different field, on the theory of the central nervous system.

Cope, Edward Drinker (1840–1897) *American vertebrate paleontologist and comparative anatomist*

Educated in his home town of Philadelphia, Cope displayed an interest in natural history from earliest youth. From 1864 to 1867 he was professor of comparative zoology and botany at Haverford College, Pennsylvania. In 1872 Cope joined the U.S. Geological Survey, and was subsequently professor of geology and mineralogy (1889–95) and of zoology and comparative anatomy (1895–97) at the University of Pennsylvania.

Cope's early studies were mainly concerned with living fishes, reptiles, and amphibians, and in 1861 he went to Washington to study the Smithsonian Institution's herpetological collections. Having resigned from his Haverford College professorship in order to be free to study and collect North American animals, Cope's interests gradually began to turn to fossils, and it was during his long association with the U.S. Geological Survey that he made the valuable contributions to paleontological knowledge associated with his name. During explorations of western America, from Texas to Wyoming, Cope discovered a great many new species of extinct (Tertiary) vertebrates (fishes, reptiles, mammals, etc.). Like his rival Othniel MARSH he is credited with the discovery of about 1,000 species of fossils new to science. With Joseph Leidy he described those fossils collected by the Ferdinand Hayden Survey in Wyoming. He also traced the evolutionary history of the horse and other mammals, proposed a theory for the origin and evolution of mammalian teeth (since somewhat modified), and made important contributions to knowledge of the stratigraphy of North America, indicating parallels with European strata.

Cope was a leading paleontological protagonist for the revival of Jean Baptiste LAMARCK's theory of the inheritance of acquired characters, based on his own experience and theory of kinetogenesis, which suggested that the limbs and other moving parts of an animal were altered and modified according to their use or disuse. Cope's Lamarckian views were elaborated in *The Origin of the Fittest* (1886) and *Primary Factors in Organic Evolution* (1896).

In 1878 Cope became editor of the *American Naturalist* but soon resigned owing to differences with Marsh, head of the U.S. Geological Survey.

Copernicus, Nicolaus (1473–1543) *Polish astronomer*

> Mathematics is written for mathematicians.
> —*De Revolutionibus orbium coelestium* (1543; On the Revolution of the Celestial Spheres)

Following the death of his father, a merchant of Torun, Poland, in 1484, Copernicus (koh-**per**-ni-kus) was brought up by his maternal uncle Lucas, the bishop of Ermeland. In 1491 he entered the University of Cracow where he became interested in astronomy. He went to Italy in 1496, studying law and medicine at the universities of Bologna and Padua and finally taking a doctorate in canon law at the University of Ferrara in 1503. By this time he had become, through a literal case of nepotism, a canon of Frauenburg, a post he was to hold until his death. In 1506 Copernicus returned home to serve his uncle as his doctor and secretary at Heinsberg castle. When his uncle died in 1512 Copernicus moved to Frauenburg to take up his modest duties as a canon.

Copernicus's pursuit of his interest in astronomy both brought him a distinguished reputation and led to a dissatisfaction with the prevailing system of astronomy. However his first statement of his revolutionary views, the *Commentariolus* (Commentary), written between about 1510 and 1514, was circulated privately in manuscript form. The system Copernicus was rebelling against goes back to the Greece of PLATO and received its fullest development in the *Almagest* of PTOLEMY in the second century AD. It assumed that the Earth, unmoving, was at the center of the universe around which not only the Moon but the Sun and the other known planets revolved with perfect uniform circular motion. However, in order to fit the complicated movements of the planets into such a simple scheme, all kinds of compromises had to be made and complications brought in. Hence the introduction of such concepts as epicycles, eccentrics, and equants into the basically simple system. The second weakness was its failure to predict at all accurately the movement of the planets. Thus a conjunction of the major planets in 1503 predicted by the almanacs of the day was as much as ten days off. Copernicus, unlike Tycho BRAHE, seemed unworried about the second point and, in all his writings, emphasized the urgency of a return to uniform circular motion. How, he asked himself in the *Commentariolus*, could this be achieved? It is at this point that he came up with his revolutionary hypothesis, "All the spheres revolve about the Sun as their midpoint and therefore the Sun is the center of the universe."

Copernicus then worked his system out in detail. His great work *De revolutionibus orbium coelestium* (On the Revolution of the Celestial Spheres) although finished by the early 1530s was to be published only in the month of his death in 1543. Why Copernicus withheld his masterpiece from the world is a matter for speculation. News of his system seems to have spread quite widely. The popes Leo X and Clement VII refer to it without any obvious hostility although Luther makes an abusive reference to it. The publication of *De revolutionibus* was due to a young professor of mathematics from Wittenberg, RHETICUS. Having heard of Copernicus's system and wishing to study it at first hand he turned up in Frauenburg bearing the typical gifts of the humanist scholar, the first printed editions of EUCLID and Ptolemy. But Copernicus was still reluctant to publish and would only agree to Rheticus writing a description of the new system, which appeared as *Narratio prima* (1539; First Narrative). Copernicus finally agreed, under strong pressure from Rheticus and his friends, to Rheticus's copying and publishing his manuscript. Rheticus went to Nuremberg intending to see the work through the press, but before its completion he had to leave to take up a new appointment at Leipzig. The task of seeing the work through the press was left with Andreas

Osiander, a Nuremburg theologian who added to the work a famous and unauthorized preface asserting that the heliocentric hypothesis was not intended to be a true description of the universe but was merely a useful supposition. He was presumably trying to avoid any church opposition. It was finally published in March 1543. It is recorded by a friend of Copernicus's that "he only saw his completed book at the last moment, on the day of his death." The book did meet opposition from theologians who found that it conflicted with the Bible. The Aristotelians were opposed to it and to many it seemed simply absurd that the Earth could be flying through space. Even professional astronomers like Tycho found it unacceptable. A moving Earth ought to imply apparent movement in the fixed stars, but none could be observed. Acceptance of Copernicus's explanation – that the stars were too far away for parallax to be observed – would involve a radical change in the accepted size of the universe. Moreover, although the heliocentric theory explained the movements of the Moon and the planets in a much more elegant way than the Ptolemaic system, Copernicus's insistence on perfect circular orbits involved nearly as much complexity as was found in Ptolemy.

However, there was no real official opposition to *De revolutionibus* and the system outlined in it until it was placed on the index of those books banned by the Catholic Church in 1616 (from which it was not removed until 1835). But it did find acceptance with many mathematicians and astronomers, so that by the end of the century the issue had switched from whether to accept Ptolemy or Copernicus, to how one should accept Copernicus – as a true description or a useful mathematical trick.

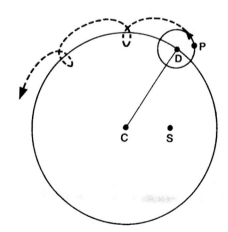

COPERNICAN SYSTEM   *The epicyclic motion of a planet. The center C is the Earth; S is the position of the Sun.*

Corey, Elias James (1928–   ) *American chemist*

Born in Methuen, Massachusetts, Corey was educated at the Massachusetts Institute of Technology, where he originally intended to train as an electrical engineer. He switched to chemistry after attending a lecture course on organic chemistry. He obtained his doctorate in 1950 and, after a period at the University of Illinois, moved to Harvard in 1959 as professor of chemistry.

Corey is a synthetic chemist with over a hundred first syntheses to his credit. These include a number of substances used medicinally, such as ginkgolide B (a compound extracted from the ginkgo tree and used to treat asthma) and the synthetic prostaglandins (hormonelike compounds used to induce labor and to treat infertility).

Yet Corey has done much more than synthesize any number of complex molecules. He has also worked out and described in detail a new and fruitful systematic approach to synthetic chemistry. The difficulty facing the chemist presented with the problem of making a known complex compound is to determine which of several possible routes are worth pursuing.

Corey proposed a systematic scheme known as "retrosynthetic analysis." In this the targeted compound is broken in stages into smaller and smaller subunits, at the same time ensuring that all the steps could be reversed at each stage. The starting point is a catalog of the main features of the compound in terms of chains, rings, branches, etc. Molecular complexity is then reduced by, for example, breaking chains and removing branches to obtain a set of rules that leads from compound to reactants and back to compound again.

Corey has given an account of his method in his book, *The Logic of Chemical Synthesis* (1989). He has also devised a computer program, LHASA (Logic and Heuristics Applied to Synthetic Analysis), to generate synthetic paths. For his work on retrosynthetic analysis Corey was awarded the 1990 Nobel Prize for chemistry.

Cori, Carl Ferdinand (1896–1984) *Czech–American biochemist*

Cori (**kor**-ee), who was born in the Czech capital of Prague, was educated at the gymnasium in Trieste, where his father was director of the Marine Biological Station, and at the University of Prague Medical School. He graduated in 1920, the year he married Gerty Radnitz, a fellow student who was to become his collaborator until her death in 1957. The Coris moved to America in 1922, taking up an appointment at the New York State Institute for the Study of Malignant Diseases in Buffalo. In 1931 they both transferred to the Washington University Medical School, where Cori was successively professor of pharmacology and of biochemistry until his retirement in 1966.

The great French physiologist Claude Bernard had shown as long ago as 1850 that glucose is converted in the body into the complex carbohydrate glycogen. This is stored in the liver and muscle, ready to be converted back into glucose as the body needs a further energy supply. Just what steps are involved in this process was the fundamental problem the Coris began to tackle in the mid 1930s.

The first clue came in 1935, when they discovered an unknown compound in minced frog muscle. This was glucose-1-phosphate, in which the phosphate molecule is joined to the glucose 6-carbon ring at the standard position (1). It was next established that when this new compound, or *Cori ester* as it was soon called, was added to a frog or rabbit muscle extract, it was converted rapidly to glucose-6-phosphate by an enzyme that was named phosphoglucomutase, a process that was reversible. As only glucose itself can enter the cells of the body, glucose-6-phosphate must be converted to glucose by the enzyme phosphatase.

Although the actual pathway of glycolysis is much more detailed and took several years to elucidate, the value of the Coris' work is undeniable. Above all they pointed the way to the crucial role of phosphates in the provision of cellular energy, the details of which were soon to be worked out by Fritz LIPMANN.

For their work the Coris shared the 1947 Nobel Prize for physiology or medicine with Bernardo HOUSSAY.

Cori, Gerty Theresa Radnitz (1896–1957) *Czech–American biochemist*

Gerty Radnitz was born in the Czech capital of Prague and graduated from the Medical School there in 1920, the year in which she married her lifelong collaborator Carl Cori. She moved with him to America, taking a post in 1922 at the New York State Institute for the Study of Malignant Diseases in Buffalo. In 1931 she went with her husband to the Washington University Medical School, where she became professor of biochemistry in 1947.

In 1947 the Coris and Bernardo HOUSSAY shared the Nobel Prize for physiology or medicine for their discovery of how glycogen is broken down and resynthesized in the body.

Coriolis, Gustave-Gaspard (1792–1843) *French physicist*

Coriolis (kor-i-**oh**-lis), a Parisian by birth, studied and taught at the Ecole Polytechnique, becoming assistant professor of analysis and mechanics in 1816. He was the first to give precise definitions of work and kinetic energy in his work *Du calcul de l'effet des machines* (1829; On the Calculation of Mathematical Action) and he particularly studied the apparent effect of a change in the coordinate system on these quantities.

From this latter research grew his most famous discovery. In 1835, while studying rotating coordinate systems, he arrived at the idea of the *Coriolis force*. This is an inertial force which acts on a rotating surface at right angles to its direction of motion causing a body to follow a curved path instead of a straight line. This force is of particular significance to astrophysics, ballistics, and to earth sciences, particularly meteorology and oceanography. It affects terrestrial air and sea currents; currents moving away from the equator will have a greater eastward velocity than the ground underneath them, and so will appear to be deflected. The idea was developed independently by William FERREL in America.

In 1838 Coriolis stopped teaching and became director of studies at the Polytechnique, but his poor health grew worse and he died five years later.

Cormack, Allan Macleod (1924–1998)
South African physicist

Born in Johannesburg in South Africa, Cormack was educated at the University of Cape Town. He became interested in x-ray imaging at the Groote Schuur Hospital in Johannesburg, where he worked as a physicist in the radioisotopes department. In 1956 he moved to America, where he served as professor at Tufts University, Massachusetts, until his retirement in 1994.

Cormack was the first to analyze theoretically the possibilities of developing a radiological cross-section of a biological system. Independently of the British engineer Godfrey HOUNSFIELD, he developed the mathematical basis for the technique of computer-assisted x-ray tomography (CAT), describing this in two papers in 1963 and 1964, and provided the first practical demonstration. X-ray tomography is a process by which a picture of an imaginary slice through an object (or the human body) is built up from information from detectors rotating around the body. The application of this technique to medical x-ray imaging was to lead to diagnostic machines that could provide very accurate pictures of tissue distribution in the human brain and body. Hounsfield was unaware of the work of Cormack when he developed the first commercially successful CAT scanners for EMI in England.

Cormack also pointed out that the reconstruction technique might equally be applied to proton tomography, or to gamma radiation from positron annihilations within a patient, and he investigated these as possible imaging techniques.

Cormack shared the 1979 Nobel Prize for physiology or medicine with Hounsfield for the development of CAT.

Cornell, Eric A. (1961–   ) *American physicist*

Cornell was born in Palo Alto, California. He was educated at Stanford University and gained his PhD in physics at MIT in 1990. He did postdoctoral research at the Joint Institute for Laboratory Astrophysics (JILA), Boulder, from 1990 to 1992. Since 1992, he has been in the physics department at the University of Colorado, Boulder, and at the National Institute of Standards and Technology (NIST), Boulder, and, since 1994, also at JILA.

In 1995 Cornell and Carl WIEMAN succeeded in obtaining a type of matter known as a Bose–Einstein condensate. Wolfgang KETTERLE also obtained a Bose–Einstein condensate in 1995, independently of Cornell and Wieman.

In 1924 Satyendra Nath BOSE discovered Bose–Einstein statistics in the course of calculations in which he put the law of black-body radiation, found by Max PLANCK in 1900, on a firm theoretical footing. He sent his work to EINSTEIN, who arranged for it to be published. Einstein subsequently extended the work of Bose. In a paper published in 1925 Einstein predicted that if a gas of a certain type of atom was cooled to a very low temperature near absolute zero then all the atoms would form a single quantum state together, with the transition to this state, now known as a Bose–Einstein condensate, being analogous to the formation of a liquid from a gas.

In 1995 Cornell and Wieman succeeded in producing a Bose–Einstein condensate in the form of 2000 atoms of rubidium at the extremely low temperature of 20 nK (nanokelvin), which is 0.000 000 02 degrees above the absolute zero temperature.

It may well turn out to be the case that the discovery of Bose–Einstein condensates will have important technological applications in areas such as precision measurement and nanotechnology. Cornell, Ketterle, and Wieman shared the 2001 Nobel Prize for physics for their work.

Corner, Edred John Henry (1906–1996)
British botanist

Corner, a Londoner, graduated from Cambridge University in 1929. He then spent 20 years overseas, first as assistant director of the Gardens Department of the Straits Settlements (Singapore) and from 1947 to 1948 as a field officer for UNESCO in Latin America. He returned to Cambridge in 1949 and later, from 1966 until his retirement in 1973, served as professor of tropical botany.

Although Corner originally began as a mycologist (someone who studies fungi), such was the general level of knowledge of tropical plants that he felt compelled to work in a less specialized field. He produced a large number of books on the subject, of which his *Life of Plants* (1964) and *Natural History of Palms* (1966) have become widely known.

Cornforth, Sir John Warcup (1917–)
Australian chemist

> For him [the scientist], truth is so seldom the sudden light that shows new order and beauty; more often, truth is the uncharted rock that sinks his ship in the dark.
> —Nobel Prize address, 1975

Cornforth was educated at the university in his native city of Sydney and at Oxford University, where he obtained his DPhil in 1941. He spent the war in Oxford working on the structure of penicillin before joining the staff of the Medical Research Council in 1946. In 1962 Cornforth moved to the Shell research center at Sittingbourne in Kent to serve as director of the Milstead Laboratory of Chemical Enzymology. In 1975 he accepted the post of Royal Society Research Professor at Sussex University, where he served until 1982.

In 1951 the American chemist Robert WOOD-WARD had succeeded in synthesizing the important steroid, cholesterol; Cornforth was interested in how the molecule is actually synthesized in the cell. Using labeled isotopes of hydrogen, he traced out in considerable detail the chemical steps used to form the $C_{27}H_{45}OH$ molecule of cholesterol from the initial CH_3COOH of acetic acid. It was for this work that he shared the 1975 Nobel Prize for chemistry with Vladimir PRELOG. He was knighted in 1977. Cornforth has also synthesized alkenes, oxazoles, and the plant hormone abscisic acid.

Correns, Karl Erich (1864–1933) *German botanist and geneticist*

Correns (**kor**-rens), the only child of the painter Erich Correns, was born in Munich, Germany. He studied at Tübingen University, where he began his research on the effect of foreign pollen in changing the visible characters of the endosperm (nutritive tissue surrounding the plant embryo). In some of his crossing experiments Correns used varieties of pea plants, following the ratios of certain characters in the progeny of these. After four generations he had gathered sufficient evidence to formulate the basic laws of inheritance. Not until he searched for relevant literature did he find that Gregor MENDEL had reached the same conclusion a generation earlier. Correns's own work, published in 1900, thus only provided further proof for Mendel's theories. His later research concentrated on establishing how widely Mendel's laws could be applied. Using variegated plants he obtained, in 1909, the first conclusive evidence for cytoplasmic, or non-Mendelian, inheritance, in which certain features of the offspring are determined by the cytoplasm of the egg cell. Other contributions to plant genetics include his proposal that genes must be physically linked to explain why some characters are always inherited together. Correns was also the first to relate Mendelian segregation (the separation of paired genes, or alleles) to meiosis and the first to obtain evidence for differential fertilization between gametes. From 1914 until his death he was director of the Kaiser Wilhelm Institute for Biology in Berlin.

Cort, Henry (1740–1800) *British metallurgist and inventor*

Cort, who was born in Lancaster in northern England, worked as a navy agent in London before taking over an ironworks near Gosport (1775). He experimented with ways of improving wrought iron and in 1783 he patented a method of piling and rolling iron into bars using grooved rollers. The following year he patented his dry-puddling process for converting pig iron into wrought iron. Although his techniques of ironworking were an important contribution to the Industrial Revolution, Cort was financially ruined by the debts of a partnership and lost his patents.

Corvisart, Jean-Nicolas (1755–1821) *French physician*

> I do not believe in medicine, but I do believe in Corvisart.
> —Napoleon Bonaparte

Corvisart (kor-vee-**sar**), the son of an attorney from Dricourt in France, entered medicine against strong opposition from his family, who wished him to pursue a legal career. He nevertheless graduated in medicine in 1785 and later held posts at the Charité and the Collège de France. He was then appointed personal physician to Napoleon.

Corvisart, who emphasized the importance of thorough clinical examinations, did much to introduce the new technique of percussion, first described by Leopold AUENBRUGGER. He not only practiced it himself but also, in 1808, produced a French translation of Auenbrugger's classic work. He was less receptive, though, to the introduction by his pupil René Laennec of the stethoscope, an instrument he seems not to have used.

He also produced *Essais sur les maladies organiques du coeur et des gros vaisseaux* (1806; Essays on the Organic Diseases of the Heart and Large Vessels), one of the earliest works devoted to diseases of the heart.

Coster, Dirk (1889–1950) *Dutch physicist*

Coster (**kos**-ter), who was born in Amsterdam in the Netherlands, was educated at the University of Leiden where he obtained his doctorate in 1922. From 1924 to 1949 he was professor of physics at the University of Groningen. In 1923, in collaboration with Georg von HEVESY, he discovered the element hafnium. The element was named for Hafnia, an old Roman name for Copenhagen. Copenhagen was the home of Niels BOHR, who had suggested to them that the new element would most likely be found in zirconium ores.

Cottrell, Sir Alan Howard (1919–)
British physicist and metallurgist

Cottrell, who was born in the English Midlands city of Birmingham, studied at the university there, gaining his BSc in 1939 and PhD in 1949. After leaving Birmingham in 1955, he took the post of deputy head of the metallurgy division of the Atomic Energy Research Establishment at Harwell until 1958, when he became Goldsmiths Professor of Metallurgy at the University of Cambridge. From 1965 he held a number of posts as a scientific adviser to the British government. In 1974 he became master of Jesus College, Cambridge, and subsequently vice-chancellor of the University of Cambridge (1977–79).

Cottrell's most notable research was in the study of dislocations in crystals. In experiments on the yield points of carbon steels he discovered that, because of the interactions of the carbon atoms with dislocations in the steel, a region enriched with carbon atoms develops around the dislocations (the *Cottrell atmosphere*). This leads to an increase in yield stress compared to the value in the pure material.

Cottrell's analysis of the forces between impurity atoms is basic to dislocation theories of yield point and strain aging. In 1949 Cottrell and B. A. Bilby correctly predicted the time law of strain aging from the rate of diffusion of impurity atoms to dislocations. Cottrell also worked on the effect of radiation on crystal structure – work that was important in design changes to the fuel rods in early nuclear reactors.

Coulomb, Charles Augustin de (1736–1806) *French physicist*

> Coulomb's contributions to the science of friction were exceptionally great. Without exaggeration, one can say that he created this science.
>
> —I. V. Kragelsky and V. S. Schedrov,
> *Development of the Science of Friction – Dry Friction* (1956)

Coulomb (koo-**lom**), who was born in Angoulême, France, was educated in Paris and then joined the army, serving as an engineer. He spent nine years in Martinique designing and building fortifications, returning to France because of ill health. On his return he accepted several public offices but with the coming of

the revolution he withdrew from Paris, spending his time quietly and safely at Blois and devoting himself to science. He returned to public life under Napoleon, serving as an inspector of public instruction from 1802.

He made an early reputation for himself by publishing work on problems of statics and friction. Some of the concepts that he introduced and analyzed are still used in engineering theory, for example, the notion of a thrust line. This describes how a building must be constructed if it is to control the oblique force arising from such items as roof members. Coulomb gave a general solution to the problem.

He is however most widely remembered for his statement of the inverse square laws of electrical and magnetic attraction and repulsion published in 1785. The secret of his work was the invention of a simple but successful torsion balance, which he used with great experimental skill. It was so sensitive that a force equivalent to about 1/100,000 of a gram could be detected. The balance consisted of a silken thread carrying a carefully balanced straw covered with wax. The straw, to which a charged sphere could be fixed, was free to rotate in a large glass tube that was marked in degrees around its circumference. He could now bring another charged ball within various distances of the rotatable straw and measure the amount of twist produced. By varying the distances involved and the nature and amount of the charge, Coulomb was able to deduce a number of laws. He stated his "fundamental law of electricity" as "the repulsive forces between two small spheres charged with the same sort of electricity is in the inverse ratio of the squares of the distances between the centers of the two spheres." That is, two like charged bodies will repel each other and the force of that repulsion will fall off with the square of the distance separating them: if a body moves twice as far away the repulsive force will be four times weaker, if the body moves three times as far away the repulsive force will be nine times weaker, and so on for any distance between them. Coulomb went on to show that the same form of law applies to magnetic as well as electrical attraction and repulsion. What is surprising about Coulomb (and his contemporaries) was an inability to see any relationship between electricity and magnetism. Despite having demonstrated that the two phenomena obey basically the same laws he insisted that they consisted of two distinct fluids.

Coulomb was immortalized by having the unit of electric charge named in his honor: the quantity of electricity carried by a current of one ampere in one second is a *coulomb*.

Coulson, Charles Alfred (1910–1974)
British theoretical chemist, physicist, and mathematician

Coulson was born in Dudley in the English Midlands and his father was principal of a local technical college. He was educated at Cambridge University, where he obtained his PhD in 1935, and afterward taught mathematics at the universities of Dundee and Oxford. Coulson then successively held appointments as professor of theoretical physics at King's College, London (1947–52), Rouse Ball Professor of Mathematics at Oxford (1952–72), and, still at Oxford, professor of theoretical chemistry from 1972 until his death in 1974.

As a physicist Coulson wrote the widely read *Waves* (1941) and *Electricity* (1948). His most creative work however was as a theoretical chemist. In 1933 he did early work on calculating the energy levels in polyatomic molecules and in 1937 he provided a theory of partial bond order (e.g., chemical bonding intermediate between double and single bonding). He also worked on aromatic molecules (benzene, naphthalene, etc.) with Christopher LONGUET-HIGGINS. His book *Valence* (1952) covers the application of quantum mechanics to chemical bonding. Later Coulson turned to theoretical studies of carcinogens, drugs, and other topics of biological interest.

Coulson was also one of the leading Methodists of his generation and, from 1965 to 1971, chairman of Oxfam, the third-world charity. He produced a number of works on the relationship between Christianity and science of which *Science and Christian Belief* (1955) is a typical example.

Couper, Archibald Scott (1831–1892)
British organic chemist

The son of a mill owner in Kirkintilloch, Scotland, Couper was educated at Glasgow and Edinburgh universities, reading humanities, classics, and philosophy. He traveled extensively in Europe and about 1854 began to study chemistry, working with Charles WURTZ in Paris (1854–56).

In 1858 he produced an important paper, *On a New Chemical Theory*, in which he anticipated August KEKULÉ's theory of the structure of carbon compounds. The publication of Couper's paper, which he had entrusted to Wurtz, was delayed and Kekulé thus had priority. Couper returned to Edinburgh in 1858 as assistant to Lyon PLAYFAIR but suffered a nervous breakdown soon after.

Couper did no more work and he was largely forgotten until Richard Anschütz, who was Kekulé's successor at Bonn, discovered some work on salicylic acid in which Couper had used an early form of graphic formulae based on dotted lines for the valence bonds. The paper on chemical theory was also rediscovered and Couper's work achieved full recognition.

Cournand, André Frederic (1895–1988)
French–American physician

Cournand (**kuur**-nand or koor-**nahn**), the son of a Paris physician, was educated at the Sorbonne and, after serving in World War I, at the University of Paris where he finally obtained his MD in 1930. He then went to America for postgraduate work, at Bellevue Hospital, New York, and began working in collaboration with Dickinson RICHARDS. Cournand remained in America, became naturalized in 1941, and continued at Bellevue where he served as professor of medicine from 1951 until his retirement in 1964.

In 1941 Cournand, in collaboration with H. Ranges, continued the earlier work of Werner FORSSMANN and developed cardiac catheterization as a tool of physiological research. He found, contrary to expectation, that the technique did not lead to blood clotting and involved virtually no discomfort.

Cournand spent much time in attempting to determine the pressure drop across the pulmonary system. He investigated the effect of shock on cardiac function and assessed the consequences of various congenital heart defects. He also looked at the action of drugs, notably the digitalin type, on the heart.

In 1945 Cournand introduced an improved catheter with two branches through which simultaneous pressures in two adjacent heart chambers could be recorded. This led to greatly improved diagnoses of anatomical abnormalities, which consequently provided a better guide to treatment.

For his "discoveries concerning heart catheterization" Cournand shared the 1956 Nobel Prize for physiology or medicine with Forssmann and Richards.

Courtois, Bernard (1777–1838) *French chemist*

Courtois (koor-**twah**) was the son of a saltpeter manufacturer from Dijon in France and as a small boy he worked in the plant showing an alert interest. He was later apprenticed to a pharmacist and subsequently studied at the Ecole Polytechnique under Antoine FOURCROY. During his military service as a pharmacist he became the first to isolate morphine in its pure form from opium.

Meanwhile his father's saltpeter business had been running into difficulties because the product could be manufactured more cheaply in India, and Courtois returned to help his father. Saltpeter was obtained from the seaweed washed ashore in Normandy; the ashes (known as "varec") were leached for sodium and potassium salts. Courtois noticed that the copper vats in which the lye was stored were becoming corroded by some unknown substance. By chance, in 1811, during the process of extracting the salts, he added excess concentrated sulfuric acid to the lye (the solution obtained by leaching) and was astonished to see "a vapor of a superb violet color" that condensed on cold surfaces to form brilliant crystalline plates. Courtois suspected that this was a new element but lacked the confidence and the laboratory equipment to establish this and asked Charles Bernard DÉSORMES and Nicolas Clément to continue his researches. His discovery was announced in 1813, and Joseph GAY-LUSSAC and Humphry DAVY soon verified that it was an element, Gay-Lussac naming it "iodine" (from the Greek for "violet").

Cousteau, Jacques Yves (1910–1997)
French oceanographer

> No children ever opened a Christmas present with more excitement than we did when we unpacked the first "aqualung." If it worked, diving could be revolutionized.... With this equipment harnessed to the back, a watertight glass mask over the eyes and nose, and rubber foot fins, we intended to make unencumbered flights in the depths of the sea.
> —*The Silent World* (1956)

Cousteau (koo-**stoh**), who was born at Saint André-de-Cubzac in France, studied at the Ecole Navale in Brest and on graduation entered the French navy. During World War II he served in the French resistance and was awarded the Légion d'Honneur and the Croix de Guerre. During the war he also made his first two underwater films and, despite German occupation of France, designed and tested an aqualung. For this he adapted a valve (hitherto used by Emile Gagnan to enable car engines to work with cooking gas) into an underwater breathing apparatus. Using compressed air, the *Cousteau–Gagnan aqualung* allows long periods of underwater investigation at depths of more than 200 feet (61 m) and frees the diver of the need for a heavy suit and lifeline.

In 1946 Cousteau became head of the French navy's Underwater Research Group and the following year set a world record of 300 feet (91 m) for free-diving. In 1951–52 he traveled to the Red Sea in a converted British minesweeper, the *Calypso*, and made the first underwater color film to be taken at a depth of 150 feet (46 m). Later he helped Auguste PICCARD in the development of the first bathyscaphes. Cousteau also designed and worked on a floating island off the French coast that enabled long-term study of marine life. He was the designer of an underwater diving "saucer" capable of descents to more than 600 feet (183 m) and able to stay submerged for 20 hours. More significantly, he worked on the future exploitation of the sea bed as a living environment for man, conducting various experiments on short-term undersea living.

Cousteau may be said to have made more significant contributions to undersea explo-ration and study than any other individual. His work was brought to a wide audience by means of movies and television, and a number of his films, for example, *The Silent World* (1956), won Academy Awards. His many books, such as *The Living Sea* (1963), and his television documentaries in the series *The Undersea World of Jacques Cousteau* captured the imagination of millions.

Crafts, James Mason (1839–1917) *American chemist*

Crafts, who was born in Boston, Massachusetts, initially studied engineering and mining at Lawrence Scientific School, graduating in 1858. He then went to Europe, where he studied chemistry at Freiburg and worked with Robert BUNSEN and Charles Adolphe WURTZ before returning to America in 1866. He became professor of chemistry at Cornell (1868) and held the same post at the Massachusetts Institute of Technology (1871). He returned to Europe in 1874 for health reasons and worked with Charles FRIEDEL at the mining school in Paris. Here in 1877 they discovered the *Friedel–Crafts reaction*, a versatile synthetic method involving the alkylation or acylation of aromatic hydrocarbons with aluminum chloride as a catalyst. Crafts returned to MIT in 1891, becoming president (1898–1900) before retiring through ill health. His other work included research on silicon derivatives, catalysis, and thermometry.

Craig, Lyman Creighton (1906–1974) *American biochemist*

Born in Palmyra, Iowa, Craig was educated at Iowa State University where he obtained his PhD in 1931. After two years at Johns Hopkins University he moved to Rockefeller University, New York, in 1933 where he was appointed professor of chemistry in 1949.

Craig concentrated on devising and improving techniques for separating the constituents of mixtures. His development of a fractional extraction method named countercurrent distribution (CCD) proved to be particularly good for preparing pure forms of several antibiotics and hormones. The method also established that the molecular weight of insulin is half the weight previously suggested. Craig also used CCD to separate the two protein chains of hemoglobin.

During work on ergot alkaloids Craig, with W. A. Jacobs, isolated an unknown amino acid, which they named lysergic acid. Other workers managed to prepare the dimethyl amide of this acid and found the compound, LSD, to have considerable physiological effects.

Cram, Donald James (1919–2001) *American chemist*

Born in Chester, Vermont, Cram was educated

at the University of Nebraska and at Harvard, where he completed his PhD in 1947. He moved immediately to the University of California, Los Angeles, where he served as professor of chemistry from 1956 until 1995.

In 1963 Charles PEDERSEN announced his discovery of the first of the crown ethers. Cram, it was reported, spent the next 48 hours in his laboratory fiddling with model kits and making a variety of new structures. He soon came to see that crown ethers could be modified in such a way as to distinguish between different forms of chiral molecules, i.e., molecules and their mirror images. By 1973 he had succeeded in devising crown ethers that could identify optically active amino acids. Cram introduced the name "host-guest" chemistry to describe such reactions.

For his work in this new field Cram shared the 1987 Nobel Prize for chemistry with Pedersen and J-M. LEHN.

Crick, Francis Harry Compton (1916–2004) *British molecular biologist*

> I also suspect that many workers in this field [i.e. molecular biology] and related fields have been strongly motivated by the desire, rarely actually expressed, to refute vitalism.
> —*British Medical Bulletin* (1965)

The son of a shoe manufacturer from Northampton in England, Crick was educated at University College, London. After graduating in physics in 1938 he began his research career under E. N. ANDRADE, working on the measurement of the viscosity of water. With the outbreak of war he was posted to the Admiralty to work on the design of acoustic and magnetic mines. Crick found himself at the end of the war at a loss what to do. He was drawn towards pure science and after reading SCHRÖDINGER's book *What Is Life?* (1944), Crick decided that he wanted to work on a "major mystery – the mys-

tery of life and the mystery of consciousness." With backing from the Medical Research Council (MRC) Crick began his odyssey in 1947 at the Strangeways Laboratory, Cambridge, working on tissue culture. Two years later he moved to the newly formed MRC unit at the Cavendish Laboratory, studying the structure of proteins by x-ray diffraction analysis.

In 1951 a young American student, James WATSON, arrived at the unit. Watson suggested to Crick that it was necessary to find the molecular structure of the hereditary material, DNA, before its function could be properly understood. Much was already known about the chemical and physical nature of DNA from the studies of such scientists as Phoebus LEVENE, Erwin CHARGAFF, Alexander TODD, and Linus PAULING. Using this knowledge and the x-ray diffraction data of Maurice WILKINS and Rosalind FRANKLIN, Crick and Watson had built, by 1953, a molecular model incorporating all the known features of DNA. Fundamental to the model was their conception of DNA as a double helix. Despite the significance of Crick's work on DNA he remained officially a graduate student. Consequently he returned to his work on protein structure and completed his PhD in 1953 at the age of 37.

Ten years' intensive research in many laboratories around the world all tended to confirm Crick and Watson's model. For their work, which has been called the most significant discovery of the 20th century, they were awarded, with Wilkins, the 1962 Nobel Prize for physiology or medicine.

Crick, in collaboration with Sydney BRENNER, made important contributions to the understanding of the genetic code and introduced the term "codon" to describe a set of three adjacent bases that together code for one amino acid. He also formulated the adaptor hypothesis in which he suggested that, in protein syn-

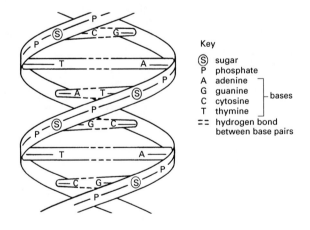

Key

- Ⓢ   sugar
- P   phosphate
- A   adenine ⎤
- G   guanine ⎥
- C   cytosine ⎥ bases
- T   thymine ⎦
- ==   hydrogen bond between base pairs

DNA   *The double-helix structure of DNA according to Crick and Watson.*

thesis, small adaptor molecules act as intermediaries between the messenger RNA template and the amino acids. Such adaptors, or transfer RNAs, were identified independently by Robert HOLLEY and Paul BERG in 1956. Crick is also known for his formulation of the *Central Dogma* of molecular genetics, which assumes that the passage of genetic information is from DNA to RNA to protein. David BALTIMORE was later to show that in certain cases, information can actually go from RNA to DNA.

In 1977 Crick moved to the Salk Institute, San Diego, California. While at Salk he worked on the second of the great mysteries that had inspired him in 1947, namely, the nature of consciousness. At an early stage he rejected computer models of the mind and the "neural Darwinism" of G. EDELMAN. He went on to publish his mature views on the nature of mind in his *The Astonishing Hypothesis* (1994), in which he argued that "your joys and your sorrows, your memories and ambitions, your sense of personal identity and free will, are in fact no more than the behavior of a vast assembly of nerve cells and their associated miracles." He had previously published in 1983, with Graeme Mitchinson, a novel account of dreams. "We dream in order to forget," he claims. Dreams allow the brain to eliminate the unwanted information collected during the day which would otherwise clog up the system. Crick also published his intellectual biography, *What Mad Pursuit* (New York, 1988).

Croll, James (1821–1890) *British geologist*

Croll, the son of a stonemason and crofter from Cargill in Scotland, started work as a millwright. He became caretaker at Anderson's College, Glasgow, in 1859, and was later made resident geologist in the Edinburgh office of the Geological Survey, where he remained until his retirement in 1880.

In 1864 he studied the work of A. J. ADHEMAR and began research into the idea of an astronomical causation of ice ages. He developed the theory that the answer lay in the orbital history of the Earth. Using work done by Urbain LE VERRIER in 1843, he found that the degree of eccentricity had been subjected to substantial change – 100,000 years ago it was highly eccentric while 10,000 years ago its eccentricity was quite small. He concluded that if winter occurred when the Earth was furthest away from the Sun in its precessional cycle and if the orbit of the Earth was at its most eccentric, then the two factors would produce an ice age. He followed Adhemar in seeing this as alternating between the two hemispheres and having a period of about 26,000 years.

This view was generally accepted by other geologists, notably by James Geikie in his pioneering work *The Great Ice Age* (1874–84), but tests made on the theory were too rudimentary to be conclusive.

Croll's work was published in his *Climate and Time* (1875) and *Climate and Cosmology* (1885).

Cronin, James Watson (1931–   ) *American physicist*

Cronin, who was born in Chicago, Illinois, was educated at the Southern Methodist University and at Chicago, where he obtained his PhD in 1955. After a period at the Brookhaven National Laboratory he moved to Princeton in 1958 and later served as professor of physics from 1965 until 1971 when he was appointed to a comparable chair at the University of Chicago.

In 1956 Tsung Dao LEE and Chen Ning YANG made the startling claim that parity (P) was not conserved in weak interactions. To the surprise of physicists their bold conjecture was confirmed in a matter of months. It was however widely assumed that in a reaction the combination of parity and a property called charge conjugation (C) was conserved. Cronin and Val FITCH, together with James Christenson and René Turlay, tested this CP conservation in 1964 by investigating the decay of neutral kaons. It was known that one type of kaon could decay into two pions; the other could not without violating CP conservation. Cronin and his colleagues discovered a small number of decays of the second type into two pions, clearly demonstrating that CP is violated.

The result is of fundamental interest for it is known that the combined properties of charge conjugation (C), parity (P), and time (T) are conserved – so that if CP is violated then the decay of the kaons is not symmetrical with respect to time reversal.

Cronin and Fitch shared the 1980 Nobel Prize for physics for this work.

Cronstedt, Axel Frederic (1722–1765) *Swedish chemist and mineralogist*

Born in Södermanland in Sweden, Cronstedt (**kroon**-stet) studied chemistry at the University of Uppsala and the School of Mines until 1748, becoming a member of the Swedish Academy of Sciences soon after; he held no academic post. Cronstedt used the blowpipe extensively for studying minerals and classified minerals according to chemical principles rather than the physical characteristics that had been used previously. This work was published (at first anonymously) in *An Essay toward a System of Mineralogy* (1758). In 1751 he discovered nickel by reducing a mineral known as devil's copper, or *kupfernickel*, and observed its magnetic properties.

Crookes, Sir William (1832–1919) *British chemist and physicist*

> We have actually touched the border land where matter and force seem to merge into one another...I venture to think that the greatest scientific problems of the future will find their solution in this border land … .
> —*Chemical News* No. 40 (1879)

Crookes studied at the Royal College of Chemistry in his native city of London, under August von Hofmann (1848). After working at the Radcliffe Observatory, Oxford, and the Chester College of Science, he returned to London in 1856, where, having inherited a large fortune, he edited *Chemical News* and spent his time on research.

Following the invention of the spectroscope by Robert BUNSEN and Gustav KIRCHHOFF, Crookes discovered the element thallium (1861) by means of its spectrum. In investigating the properties and molecular weight of thallium, he noticed unusual effects in the vacuum balance that he was using. This led him to investigate effects at low pressure and eventually to invent the instrument known as the *Crookes radiometer* (1875). This device is a small evacuated glass bulb containing an arrangement of four light metal vanes. Alternate sides of the vanes are polished and blackened. When radiant heat falls on the instrument, the vanes rotate. The effect depends on the low pressure of gas in the bulb; molecules leaving the dark (hotter) surfaces have greater momentum than those leaving the bright (cooler) surfaces. Although the instrument had little practical use, it was important evidence for the kinetic theory of gases.

Crookes went on to investigate electrical discharges in gases at low pressure, producing an improved vacuum tube (the *Crookes tube*). He also investigated cathode rays and radioactivity. *Crookes glass* is a type of glass invented to protect the eyes of industrial workers from intense radiation.

From about 1870, Crookes became interested in spiritualism and became one of the leading investigators of psychic phenomena.

Cross, Charles Frederick (1855–1935) *British industrial chemist*

Cross, who was born in Brentford, Middlesex, in England, studied chemistry at King's College, London, in Zurich, and at Owens College, Manchester. There he met Edward BEVAN, whose partner he became and with whom he developed the viscose process of rayon manufacture. Cross was subsequently involved in the industrial development of this process. He wrote several books on cellulose and papermaking.

Crum Brown, Alexander (1838–1922) *British organic chemist*

The son of a Presbyterian minister in the Scottish capital Edinburgh, Crum Brown studied arts and then medicine and chemistry there, gaining an MA degree in 1858 and his MD in 1861. He was also awarded a doctorate from London University (1862) and worked with Robert BUNSEN at Heidelberg and Hermann KOLBE at Marburg before returning to Edinburgh as a lecturer in 1863, becoming professor of chemistry in 1869.

Crum Brown was essentially a theoretician of organic chemistry and his structural formulae, introduced in his MD thesis *On the Theory of Chemical Combination* (1861) and taken up by Edward FRANKLAND in 1866, are essentially the symbols used today. In 1867–68, with T. R. Fraser, he carried out pioneering work in what is now called structure/activity relationships in pharmacology. In 1892 (with J. Gibson) he proposed a rule (*Crum Brown's rule*) concerning the effect of substitution of an organic group into a benzene ring that already contains a group. The rule can be used to predict the position into which the existing group will direct the second group. Other research interests were physiology (the function of the semicircular canals in the ear), phonetics, mathematics, and crystallography.

Crutzen, Paul (1933–   ) *Dutch chemist and meteorologist*

Crutzen (**kruut**-sen) first worked in Sweden, at Stockholm University, moving in 1974 to the National Oceanic and Atmospheric Administration and the National Center for Atmospheric Research in Boulder, Colorado. In 1980 he moved to Germany, where he served as director of the Department of Atmospheric Chemistry at the Max Planck Institute for Chemistry, Mainz, until 2000. Crutzen has also held part-time professorships at Chicago (1987–91), California (1992–2000), and Utrecht (1997–2000).

It began to be suspected in the 1950s that the concentration of ozone in the stratosphere was lower than expected. In 1970 Crutzen argued that nitrous oxide, arising from the use of nitrogen-rich fertilizers and the combustion of fossil fuels, could be responsible. As it was relatively unreactive, nitrous oxide could rise unchanged into the stratosphere, where, under the influence of ultraviolet radiation, it could initiate a series of reactions that would lead to the conversion of ozone into molecular oxygen.

Little notice was taken of Crutzen's argument – if only because it was felt that the amount of nitrous oxide produced was too insignificant to cause any noticeable depletion of the ozone layer. The debate, however, was revived by the growing fear in the early 1970s that a proposed armada of supersonic transport aircraft (SSTs) would emit large quantities of nitrous oxide from their exhausts. Consequently the anti-SST lobby seized upon Crutzen's work.

In fact, the SSTs were never constructed. Soon after, Crutzen's warnings were overshadowed by the greater threat from the chlorofluorocarbons (CFCs), first identified in 1974 by F. Sherwood ROWLAND and Mario MOLINA, with whom Crutzen shared the 1995 Nobel Prize for chemistry. Crutzen was also one of the first scientists to warn of the dangers of a "nuclear winter." In 1982, two years before Carl SAGAN and his colleagues published their famous paper on the subject, Crutzen argued that fires lit by large nuclear explosions would be extensive enough to generate massive amounts of smoke, which, added to the dust produced by the bomb, would profoundly restrict the amount of sunlight reaching the ground.

Cugnot, Nicolas-Joseph (1725–1804)
French engineer

Cugnot (koo-**nyoh**), who was born in Poid, France, served as a military engineer in the Austro-Hungarian army during the Seven Years' War (1756–63). Following this he built what was probably the first fuel-driven vehicle. Independently of James WATT and Thomas NEWCOMEN he designed, in 1770, a two-piston steam boiler, demonstrating that steam traction was feasible. It was intended for pulling heavy artillery and was the first engine to use high-pressure steam, although its application was limited by problems with the water supply.

Culpeper, Nicholas (1616–1654) *English medical writer and astrologer*

Culpeper, a Londoner, studied briefly at Cambridge University, where he learned Latin and Greek, before being apprenticed to an apothecary. In 1640 he established himself as an astrologer and physician in Spitalfields, London. Culpeper invoked the indignation of the College of Physicians by publishing, in 1649, an unauthorized translation of their *Pharmacopeia*, which listed the preparation and uses of all major drugs. In 1652, he published, with Peter Cole, *The English Physician*, in which he attacked the medical establishment. The enlarged version, published the following year, listed many herbal medicines and became very popular, running into several editions and being dubbed "Culpeper's Herbal." His prolific output included works dealing with astrology, medicine, and politics. He was a fervent supporter of the Parliamentary side in the English Civil War and was apparently wounded in battle. This may have contributed to his early death.

Cunitz, Marie (?1604–1664) *Polish astronomer*

Born in Silesia, the daughter of a physician, Cunitz defied the social conventions of her time to become arguably the first woman scientist of the modern era. Yet much remains unknown about her life and career. She was educated chiefly by her father, and became known for her skills in languages, painting, poetry and music. However, her main interest was astronomy, one shared by her husband, the physician Elias von Löwen, whom she married in 1630. Cunitz is best remembered for her sole publication, *Urania propitia* (1650). This contained new simplified planetary tables based on a recalculation of the Rudolphine tables compiled by the astronomer Johannes KEPLER. Cunitz corrected some of the errors of Kepler, but also introduced new ones of her own. The work was privately published, and few copies remain in existence. She corresponded with other eminent scientists of her day, but many of her letters and other papers were lost in a fire in 1656. Nevertheless, Cunitz is credited as being one of the first astronomers to comprehend fully the significance of Kepler's work on the laws of planetary motion.

Curie, Marie Skłodowska (1867–1934)
Polish–French chemist

> In this shed with an asphalt floor, whose glass roof offered us only incomplete protection against the rain, which was like a greenhouse in summer and which an iron stove barely heated in winter, we passed the best and happiest years of our existence, consecrating our entire days to the work...I remember the delight we experienced when we happened to enter our domain at night and saw on all sides the palely luminescent silhouettes of the products of our work.
> —In the Preface to Pierre Curie's *Collected Works*

Marie Curie (koo-**ree**) was born Skłodowska; her father was a physics teacher and her mother the principal of a girls' school in the Polish capital Warsaw. She acquired from her father a positivism and an interest in science although to aid the family finances she was forced, in 1885, to become a governess. She seems to have been on the fringe of nationalist revolutionary politics at a time when Polish language and culture were very much under Russian domination, but her main interest at this time appears to have been science. There was no way in which a girl could receive any form of higher scientific education in Poland in the 1880s, and so in 1891 she followed her elder sister to Paris. Living in poverty and working hard she graduated in physics from the Sorbonne in 1893, taking first place. She received a scholarship from Poland, which enabled her to spend a year studying mathematics; this time she graduated in second place.

In 1894 she met Pierre Curie and they married the following year. He was a physicist of some distinction, having already made several important discoveries, and was working as chief of the laboratory of the school of Industrial Physics and Chemistry. Marie was at this time looking for a topic for research for a higher degree. Her husband was in full sympathy with her desire to continue with research, by no means a common attitude in late 19th-century France. She was also fortunate in her timing and choice of topic – the study of radioactivity. In 1896 Henri BECQUEREL had discovered radioactivity in uranium. Marie Curie had reason to believe that there might be a new element in the samples of uranium ore (pitchblende) that Becquerel had handled, but first she needed a place to work and a supply of the ore. It was agreed that she could work in her husband's laboratory. Her first task was to see if substances other than uranium were radioactive. Her method was to place the substance on one of the plates of Pierre's sensitive electrometer to see if it produced an electric current between the plates. In a short time she found that thorium is also radioactive.

Her next discovery was in many ways the most fundamental. She tried to see whether different compounds of uranium or thorium would have differing amounts of radioactivity. Her conclusion was that it made no difference what she mixed the uranium with, whether it was wet or dry, in powder form or solution; the only factor that counted was the amount of uranium present. This meant that radioactivity must be a property of the uranium itself and not of its interaction with something else. Radioactivity had to be an atomic property; it would soon be recognized as an effect of the nucleus.

One further advance was made by Marie Curie in 1898; she found that two uranium minerals, pitchblende and chalcolite, were more active than uranium itself. She drew the correct conclusion from this, namely that they must contain new radioactive elements. She immediately began the search for them. By the end of the year she had demonstrated the existence of two new elements, radium and polonium, both of which were highly radioactive. No precautions were taken at this time against the levels of radiation, as their harmful effects were not recognized. (Indeed, her notebooks of this period are still too dangerous to handle.)

Her next aim was to produce some pure radium. The difficulty here was that radium is present in pitchblende in such small quantities that vast amounts of the ore were needed. The Curies managed to acquire, quite cheaply, several tons of pitchblende from the Bohemian mines thanks to the intercession of the Austrian government. As there was too much material for her small laboratory she was offered the use of an old dissecting room in the yard of the school. It was freezing in winter and unbearably hot in summer – Wilhelm OSTWALD later described it as a cross between a stable and a potato cellar. The work was heavy and

monotonous. The limitations of her equipment meant that she could only deal with batches of 20 kilograms at a time, which had to be carefully dissolved, filtered, and crystallized. This procedure went on month after month, in all kinds of weather. By early 1902 she had obtained one tenth of a gram of radium chloride. She took it to Eugene DEMARÇAY who had first identified the new elements spectroscopically. He now had enough to determine its atomic weight, which he calculated as 225.93.

The crucial question arising from the discovery of these new elements was, what was the nature of the radiation emitted? It was thought that there were at least two different kinds of rays. One kind could be deflected by a magnetic field while the other was unaffected and would only travel a few centimeters before disappearing. (These were identified as the alpha and beta rays by Ernest RUTHERFORD.) A further question was the nature of the source of the energy. Pierre Curie showed that one gram of radium gave out about a hundred calories per hour. One further mystery at this time was the discovery of induced radioactivity – they had found that metal plates that had been close to, but not in contact with, samples of radium became radioactive themselves and remained so for some time.

The mysteries of radioactivity were explained not by the Curies but by Rutherford and his pupils. Although Marie Curie was no great theorist, she was an industrious experimentalist who with great strength and single-mindedness would pursue important but basically tedious experimental procedures for years. Her thesis was presented in 1903 and she became the first woman to be awarded an advanced scientific research degree in France. In the same year she was awarded the Nobel Prize for physics jointly with her husband and Becquerel for their work on radioactivity.

In 1904, when her husband was given a chair at the Sorbonne, Marie was offered a part-time post as a physics teacher at a girls' Normal School at Sèvres. In the same year her second daughter Eve was born. It is also about this time that she first appears to have suffered from radiation sickness. Given all these distractions it is not surprising that for a few years after the completion of her thesis she had little time for research. In 1906 Pierre Curie died in a tragic accident. The Sorbonne elected her to her husband's chair and the rest of her life was largely spent in organizing the research of others and attempting to raise funds. She made two long trips to America in 1921 and 1929. On her first trip she had been asked what she would most like to have. A gram of radium of her own was her reply, and she returned from America with a gram, valued at $100,000. She also received $50,000 from the Carnegie Institution. In 1912 the Sorbonne founded the Curie laboratory for the study of radioactivity. It was opened in 1914 but its real work could only begin after the war, during which Marie Curie spent most of her time training radiologists. Later her laboratory, with its gram of radium, was to become one of the great research centers of the world.

Her position in France was somewhat odd. As a foreigner and a woman France was never quite sure how to treat her. She was clearly very distinguished for in 1911 she was awarded her second Nobel Prize, this time in chemistry for her discovery of radium and polonium. Her eminence was recognized by the creation of the Curie laboratory, yet at almost the same time she found herself rejected by the Académie des Sciences. She allowed her name to go forward in 1910 as the first serious female contender but was defeated. There is no doubt that this offended her. She refused to allow her name to be submitted for election again and for ten years refused to allow her work to be published in the proceedings of the Académie.

The following year, 1911, worse was to happen and she became the center of a major scandal. The physicist and former pupil of her husband, Paul LANGEVIN was accused of having an affair with her. Langevin had left his wife and four children, but although he was close to Madame Curie it is by no means clear that there were grounds for the accusations. Some of her letters to Langevin were stolen and published in the popular press and doubts were raised about Pierre Curie's death. Most of the attacks seem to have emanated from Gustave Téry, editor of *L'Oeuvre* (The Work) and a former classmate of Langevin. Langevin retaliated by challenging Téry to a duel. Langevin faced Téry late in 1911 at 25 yards with a loaded pistol in his hand. Both refused to fire and shortly afterward the scandal died down.

Her major published work was the massive two-volume *Treatise on Radioactivity* (1910). The Curies' daughter Irène and her husband Frédéric JOLIOT-CURIE continued the pioneering work on radioactivity and themselves received the Nobel Prize for physics.

Curie, Pierre (1859–1906) *French physicist*

> We might still consider that in criminal hands radium might become very dangerous; and here we must ask ourselves if mankind can benefit by knowing the secrets of nature, if man is mature enough to take advantage of them, or if this knowledge will not be harmful to the world.
> —Nobel Prize address, 1903

Pierre Curie was the son of a Paris physician. He was educated at the Sorbonne where he became an assistant in 1878. In 1882 he was made laboratory chief at the School of Industrial Physics and Chemistry where he remained until he was appointed professor of physics at the Sorbonne in 1904. In 1895 he married Marie Skłodowska, with whom he conducted research into the radioactivity of radium and with whom he shared the Nobel Prize for physics in 1903.

His scientific career falls naturally into two periods, the time before the discovery of radioactivity by Henri BECQUEREL, when he worked on magnetism and crystallography, and the time after when he collaborated with his wife Marie Curie on this new phenomenon.

In 1880 with his brother Jacques he had discovered piezoelectricity. "Piezo" comes from the Greek for "to press" and refers to the fact that certain crystals when mechanically deformed will develop opposite charges on opposite faces. The converse will also happen; i.e., an electric charge applied to a crystal will produce a deformation. The brothers used the effect to construct an electrometer to measure small electric currents. Marie Curie later used the instrument to investigate whether radiation from substances other than uranium would cause conductivity in air. Pierre Curie's second major discovery was in the effect of temperature on the magnetic properties of substances, which he was studying for his doctorate. In 1895 he showed that at a certain temperature specific to a substance it will lose its ferromagnetic properties; this critical temperature is now known as the *Curie point*.

Shortly after this discovery he began to work intensively with his wife on the new phenomenon of radioactivity. Two new elements, radium and polonium, were discovered in 1898. The rays these elements produced were investigated and enormous efforts were made to produce a sample of pure radium.

He received little recognition in his own country. He was initially passed over for the chairs of physical chemistry and mineralogy in the Sorbonne and was defeated when he applied for membership of the Académie in 1902. He was however later admitted in 1905. The only reason he seems eventually to have been given a chair (in 1904) was that he had been offered a post in Geneva and was seriously thinking of leaving France. Partly this may have been because his political sympathies were very much to the left and because he was unwilling to participate in the science policies of the Third Republic.

Pierre Curie was possibly one of the first to suffer from radiation sickness. No attempts were made in the early days to restrict the levels of radiation received. He died accidentally in 1906 in rather strange circumstances – he slipped while crossing a Paris street, fell under a passing horse cab, and was kicked to death. The unit of activity of a radioactive substance, the *curie*, was named for him in 1910.

The Curies' daughter Irène JOLIOT-CURIE carried on research in radioactivity and also received the Nobel Prize for work done with her husband Frédéric.

Curl, Robert Floyd Jr. (1933–   ) *American chemist*

Curl was educated at Rice University, Texas, and the University of California, Berkeley, where he gained his PhD in 1957. After working at Harvard he returned to Rice in 1967 as professor of chemistry.

Curl's initial work was on small clusters of atoms of semiconductors, such as germanium and silicon. In 1984, under the influence of Harold KROTO, he became interested in the possibility of producing long-chain carbon molecules, and persuaded his colleague Richard SMALLEY to deploy the resources of his laboratory towards this end. Although they expected on theoretical grounds to discover linear chain clusters with up to 33 carbon atoms, they in fact came across an unexpected molecule with 60 carbon atoms and with a cage-like structure. The discovery of this new allotrope of carbon, later named buckminsterfullerene, opened up a new branch of materials science.

Curl shared the 1996 Nobel Prize for chemistry with Smalley and Kroto.

Curtis, Heber Doust (1872–1942) *American astronomer*

Born in Muskegon, Michigan, Curtis obtained his BA (1892) and MA (1893) from the University of Michigan, where he studied classics. He moved to California in 1894 where he became professor of Latin and Greek at Napa College. There his interest in astronomy was aroused and from 1897 to 1900 he was professor of mathematics and astronomy after the college merged with the University (now the College) of the Pacific. After obtaining his PhD from the University of Virginia in 1902 he joined the staff of the Lick Observatory where he remained until 1920 when he became director of the Allegheny Observatory of the University of Pittsburgh. Finally, in 1930 he was appointed director of the University of Michigan's observatory.

Curtis's early work was concerned with the measurement of the radial velocities of the brighter stars. From 1910 however he was involved in research on the nature of "spiral nebulae" and became convinced that these were isolated independent star systems. In 1917 he argued that the observed brightness of novae found by him and by George RITCHEY on photographs of the nebulae indicated that the nebulae lay well beyond our Galaxy. He also maintained that extremely bright novae (later identified as supernovae) could not be included with the novae as distance indicators. He estimated the Andromeda nebula to be 500,000 light-years away.

Curtis's view was opposed by many, including Harlow SHAPLEY who proposed that our Galaxy was 300,000 light-years in diameter, far larger than previously assumed, and that the spiral nebulae were associated with the Galaxy. In 1920, at a meeting of the National Academy of Sciences, Curtis engaged in a famous debate with Shapley over the size of the Galaxy and the distance of the spiral nebulae. Owing to incomplete and incorrect evidence the matter was not settled until 1924 when Edwin HUBBLE redetermined the distance of the Andromeda nebula and demonstrated that it lay well beyond the Galaxy.

Curtius, Theodor (1857–1928) *German organic chemist*

Born at Duisburg, in Germany, Curtius (**kuur**-tsee-uus) gained his doctorate under Adolph KOLBE at Leipzig and later became professor at Kiel (1889), Bonn (1897), and Heidelberg (1898). Two reactions are named for Curtius; the first, discovered in 1894, is a method for the conversion of an acid into an amine via the azide and urethane, and the second, discovered in 1913, is a method of converting an azide into an isocyanate. Curtius also discovered diazoacetic ester, which was the first known aliphatic diazo compound (1883), hydrazine (1887), and hydrazoic acid (1890).

Cushing, Harvey Williams (1869–1939) *American surgeon*

Born in Cleveland, Ohio, Cushing received his MD from Harvard Medical School in 1895. He then joined Massachusetts General Hospital, Boston, before moving to Johns Hopkins Hospital, where he progressed to associate professor of surgery. In 1912 he was appointed professor of surgery at Harvard. Cushing's specialty was brain surgery and he pioneered several important techniques in this field, especially in the control of blood pressure and bleeding during surgery. From his many case histories, he distinguished several classes of brain tumors and made great improvements in their treatment. Cushing also demonstrated the vital role of the pituitary gland in regulating many bodily functions. He was the first to associate adenoma (tumor) of the basophilic cells of the anterior pituitary with the chronic wasting disease now known as *Cushing's syndrome*. The characteristic symptoms include thin arms and legs, atrophied skin with red lines, obesity of the trunk and face, and high blood pressure. This syndrome is now known to be caused by any of several disorders that result in the increased secretion of corticosteroid hormones by the adrenal glands.

Cushing had a lifelong interest in the history of medicine and wrote the *Life of Sir William Osler* (1925), which won him a Pulitzer Prize.

He donated his large collection of books and papers to the Yale Medical Library.

Cuvier, Baron Georges (1769–1832)
French anatomist and taxonomist

Cuvier (**kyoo**-vee-ay or koo-**vyay**) was born in Montbéliard (now in France) and as a child was greatly influenced by Georges BUFFON's books. In 1795 he became assistant to the professor of comparative anatomy at the Museum of Natural History in Paris – then the world's largest scientific research establishment. During his lifetime he greatly enlarged the comparative anatomy section from a few hundred skeletons to 13,000 specimens. Cuvier extended LINNAEUS's classification, creating another level, the phylum, into which he grouped related classes. He recognized four phyla in the animal kingdom and his work on one of these, the fishes, is recognized as the foundation for modern ichthyology. Together with Achille Valenciennes, Cuvier compiled the lengthy *Histoire des poissons* (History of Fish), nine volumes of which had appeared by his death. The fish families delimited in this work remain as orders or suborders in today's classification. Cuvier was the first to classify fossils and named the pterodactyl. His results from investigations of the Tertiary formations near Paris are published in four volumes as *Recherches sur les ossements fossiles des quadrupèdes* (1812; Researches on the Fossil Bones of Quadrupeds).

In 1799 Cuvier became professor of natural history at the Collège de France and in 1802 was also made professor at the Jardin des Plantes. In his later life he became increasingly involved in educational administration and played a large part in organizing the new Sorbonne.

D

d'Abano, Pietro (*c.* 1250–1316) *Italian physician and philosopher*

D'Abano (**dah**-bah-noh) was born at Abano near Padua in Italy and studied in Greece and Constantinople before learning medicine in Paris. He became professor of medicine at Padua University in 1306 and soon became a celebrated teacher and physician.

His knowledge of Greek enabled him to study the texts of the ancient scholars and physicians and to attempt to reconcile medicine and philosophy in his most famous book known as the *Conciliator*. This rationalist stance together with his interest in astrology brought d'Abano into conflict with the Catholic Church. In 1315 he was accused of being a heretic and was twice brought before the Inquisition. Acquitted the first time, he died before a second trial was completed and his body was hidden by friends. He was nevertheless found guilty and the Inquisition ordered that his effigy be burned instead.

Daguerre, Louis-Jacques-Mandé (1789–1851) *French physicist, inventor, and painter*

Daguerre (da-**gair**), the inventor of the daguerreotype (the first practical photograph), first became interested in the effect of light on films from the artistic point of view. Born in Cormeilles near Paris, France, he worked first as a tax officer, later becoming a painter of opera scenery.

Working with Charles-Marie Bouton he invented the diorama – a display of paintings on semitransparent linen that transmitted and reflected light – and opened a diorama in Paris (1822).

From 1826 Daguerre turned his attention to heliography (photographing the sun) and he was partnered in this by Joseph-Nicéphore NIEPCE until Niepce's death in 1833. Daguerre continued his work and in 1839 presented to the French Academy of Sciences the daguerreotype, which needed only about 25-minutes exposure time to produce an image, compared with over eight hours for Niepce's previous attempts. In the daguerreotype a photographic image was obtained on a copper plate coated with a light-sensitive layer of silver iodide and bromide.

d'Ailly, Pierre (1350–1420) *French geographer, cosmologist, and theologian*

D'Ailly (da-**yee**), who was born in Compiègne, France, studied at the University of Paris, serving as its chancellor from 1389 until 1395. He then became a bishop, first of Le Puy and subsequently of Cambrai. He became a cardinal in 1411.

Primarily concerned with church affairs, in which he advocated reform, d'Ailly was also interested in science and was the leading geographical theorist of his time. He was the author of the influential work *Imago mundi* (Image of the World), which was completed by 1410 and printed in about 1483. This was a summary of the classical and Arabian geographers, typical of the medieval period, and was influenced by Roger BACON. D'Ailly's underestimate of the Earth's circumference and exaggeration of the size of Asia may well have influenced Columbus, whose copy of d'Ailly's work, with his marginal annotations, still survives. D'Ailly's work had been written without the benefit of PTOLEMY's *Geography*, which only became available in a Latin version shortly afterward. He produced, in 1413, the *Compendium cosmographiae* (Compendium of Cosmography). This was basically a summary of Ptolemy, reviving uncritically the full fabric of Ptolemaic geography, which was already being dismantled by such travelers as Marco Polo.

Daimler, Gottlieb Wilhelm (1834–1900) *German engineer and inventor*

Daimler (**dIm**-ler or daym-ler), who was born in Schorndorf, Germany, became a gunsmith's apprentice in 1848. He studied at Stuttgart technical school, worked for a period at an engineering plant, and completed his education at Stuttgart Polytechnic. After traveling in England and France, he worked from 1863 in German engineering companies, beginning work on an internal-combustion engine in 1872.

In 1885 Daimler set up a company with Wilhelm Maybach and in 1883 and 1885 designs for an internal-combustion engine suitable for light vehicles were patented. In 1890 he founded the Daimler-Motoren-Gesellschaft company, which, in 1899, built the first Mercedes car.

Dainton, Frederick Sydney (1914–1997)
British physical chemist and scientific administrator

> Perhaps science will only regain its lost primacy as peoples and governments begin to recognize that sound scientific work is the only secure basis for the construction of policies to ensure the survival of mankind without irreversible damage to Planet Earth.
> —*New Scientist*, 3 March 1990

Born in the English industrial city of Sheffield, Dainton was educated at the universities of Oxford and Cambridge. After World War II he remained in Cambridge until 1950, when he moved to the University of Leeds as professor of physical chemistry. In 1965 Dainton was appointed vice-chancellor of the University of Nottingham. He returned to chemistry in 1970, when he was elected professor of chemistry at Oxford. Soon after, in 1973, he was made chairman of the University Grants Committee, a post he held until his retirement in 1978.

Dainton's early work in physical chemistry was on the kinetics and thermodynamics of polymerization reactions. From about 1945 he turned his attention to studies of radiolysis – i.e., chemical changes produced by high-energy radiation (alpha, beta, or gamma rays). In particular, he studied the properties and reactions of hydrated electrons in liquids.

From 1965 he was a member of the Council for Scientific Policy and was its chairman from 1969 to 1973. While holding this office he was influential in decisions made about the way British academic research is financed. He was raised to the British peerage, as Baron Dainton of Hallam Moors, in 1986.

Dale, Sir Henry Hallett (1875–1968)
British physiologist

Educated at Cambridge University and St. Bartholomew's Hospital in his native city of London, Dale became, in 1904, director of the Wellcome Physiological Research Laboratories. His work there over the next ten years included the isolation (with Arthur Ewins) from ergot fungi of a pharmacologically active extract – acetylcholine – which he found had similar effects to the parasympathetic nervous system on various organs. It was later shown by Otto Loewi that a substance released by electrical stimulation of the vagus nerve was responsible for effecting changes in heartbeat. Following up this work, Dale showed that the substance is in fact acetylcholine, thus establishing that chemical as well as electrical stimuli are involved in nerve action. For this research Dale and Loewi shared the 1936 Nobel Prize for physiology or medicine. Dale also worked on the properties of histamine and related substances, including their actions in allergic and anaphylactic conditions. He was the chairman of an international committee responsible for the standardization of biological preparations, and from 1928 to 1942 was director of the National Institute for Medical Research.

d'Alembert, Jean Le Rond (1717–1783)
French mathematician, encyclopedist, and philosopher

> Push on, and faith will catch up with you.
> —Advice to people unable to accept the methods of the calculus

D'Alembert (da-lahm-**bair**) was the illegitimate son of a Parisian society hostess, Mme de Tenzin, and was abandoned on the steps of a Paris church, from which he was named. He was brought up by a glazier and his wife, and his father, the chevalier Destouches, made sufficient money available to ensure that d'Alembert received a good education although he never acknowledged that d'Alembert was his son. He graduated from Mazarin College in 1735 and was admitted to the Academy of Sciences in 1741.

D'Alembert's mathematical work was chiefly in various fields of applied mathematics, in particular dynamics. In 1743 he published his *Traité de dynamique* (Treatise on Dynamics), in which the famous *d'Alembert principle* is enunciated. This principle is a generalization of NEWTON's third law of motion, and it states that Newton's law holds not only for fixed bodies but also for those that are free to move. D'Alembert wrote numerous other mathematical works on such subjects as fluid dynamics, the theory of winds, and the properties of vibrating strings. His most significant purely mathematical innovation was his invention and development of the theory of partial differential equations. Between 1761 and 1780 he published eight volumes of mathematical studies.

Apart from his mathematical work he is perhaps more widely known for his work on Denis Diderot's *Encyclopédie* (Encyclopedia) as editor of the mathematical and scientific articles, and his association with the philosophes. D'Alembert was a friend of VOLTAIRE's and he had a lively interest in theater and music, which led him to conduct experiments on the properties of sound and to write a number of theoretical treatises on such matters as harmony. He was elected to the French Academy in 1754 and became its permanent secretary in 1772 but he refused the presidency of the Berlin Academy.

Dalén, Nils Gustaf (1869–1937) *Swedish engineer*

Born in Stenstorp, Sweden, Dalén (da-**layn**) graduated in mechanical engineering in 1896 from the Chalmers Institute at Göteborg and then spent a year at the Swiss Federal Institute of Technology at Zurich. For several years he researched and improved hot-air turbines, compressors, and air pumps and from 1900 to 1905 worked with an engineering firm, Dalén and Alsing. He then became plant manager for the Swedish Carbide and Acetylene Company, which in 1909 became the Swedish Gas Accumulator Company with Dalén as managing director.

Dalén is remembered principally for his inventions relating to acetylene lighting for lighthouses and other navigational aids, and in particular an automatic light-controlled valve, for which he received the 1912 Nobel Prize for physics. The valve, known as "Solventil," used the difference in heat-absorbing properties between a dull black surface and a highly polished one to produce differential expansion of gases, and thus to regulate the main gas valve of an acetylene-burning lamp. The lamp could thus be automatically dimmed or extinguished in daylight, and this allowed buoys and lighthouses to be left unattended and less gas to be used. The system soon came into widespread use and is still in use today.

Another invention of Dalén's was a porous filler for acetylene tanks, "Agamassan," that prevented explosions. It was ironic that in 1912 he was himself blinded by an explosion during the course of an experiment. This did not, however, deter him from continuing his experimental work up to his death.

Dalton, John (1766–1844) *British chemist and physicist*

> An enquiry into the relative weights of the ultimate particles of bodies is a subject, as far as I know, entirely new; I have lately been prosecuting this enquiry with remarkable success.
> —*On the Absorption of Gases by Water* (1802)

The son of a hand-loom weaver from Eaglesfield in the northwest of England, Dalton was born into the nonconformist tradition of the region and remained a Quaker all his life. He was educated at the village school until the age of 11, and received tuition from Elihu Robinson, a wealthy Quaker, meteorologist, and instrument maker, who first encouraged Dalton's interest in meteorology. At the age of only 12, Dalton himself was teaching in the village. He then worked on the land for two years before moving to Kendal with his brother to teach (1781). In 1793 he moved to Manchester where he first taught at the Manchester New College, a Presbyterian institute. In 1794 he was elected to the Manchester Literary and Philosophical Society at which most of his papers were read.

From 1787 until his death Dalton maintained a diary of meteorological observations of the Lake District where he lived. His first published work, *Meteorological Observations and Essays* (1793), contained the first of his laws concerning the behavior of compound atmospheres: that the same weight of water vapor is taken up by a given space in air and in a vacuum. Both Dalton and his brother were color blind and he was the first to describe the condition, sometimes known as daltonism, in his work *Extraordinary Facts Relating to the Vision of Colours* (1794).

In 1801 Dalton read four important papers to the Manchester Philosophical Society. *On the Constitution of Mixed Gases* contains what is now known as the law of partial pressures and asserts that air is a mixture, not a compound, in which the various gases exert pressure on the walls of a vessel independently of each other. *On the Force of Steam* includes the first explanation of the dew point and hence the founding of exact hygrometry. It also demonstrates that water vapor behaves like any other gas. The third paper, *On Evaporation*, shows that the quantity of water evaporated is proportional to the vapor pressure. *On the Expansion of Gases by Heat* contains the important conclusion that all gases expand equally by heat. This law had been discovered by Jacques CHARLES in 1787 but Dalton was the first to publish. During this time, Dalton was developing his atomic theory, for which he is best known. A physical clue to the theory was provided by the solubility of gases in water. Dalton expected to find that all gases had the same solubility in water but the fact that they did not helped to confirm his idea that the atoms of different gases had different weights. The first table of atomic weights was appended to the paper *On the Absorption of Gases by Water*, read in 1802 but not printed until 1805. In another paper read in 1802 and printed in 1805 he showed that when nitric oxide is used to absorb oxygen in a eudiometer they combine in two definite ratios depending on the method of mixing. This was the beginning of the law of multiple proportions and led Dalton to much work on the oxides of nitrogen and the hydrocarbons methane and ethylene to confirm the law.

The atomic theory was first explicitly stated by Dalton at a Royal Institution lecture in December 1803 and first appeared in print in Thomas Thomson's *System of Chemistry* (1807). Dalton's own full exposition appeared in *A New System of Chemical Philosophy* (1808), with further volumes in 1810 and 1827. The basic postulates of the theory are that matter consists of atoms; that atoms can neither be created nor destroyed; that all atoms of the same element are identical, and different elements have different types of atoms; that chemical reactions take place by a rearrangement of atoms; and that compounds consist of "compound atoms" formed from atoms of the constituent elements.

Using this theory, Dalton was able to rationalize the various laws of chemical combination (conservation of mass, definite proportions, multiple proportions) and show how they followed from the theory. He did, however, make

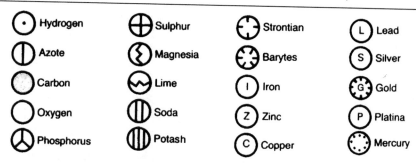

Hydrogen	Sulphur	Strontian	Lead
Azote	Magnesia	Barytes	Silver
Carbon	Lime	Iron	Gold
Oxygen	Soda	Zinc	Platina
Phosphorus	Potash	Copper	Mercury

CHEMICAL ELEMENTS   The symbols for the chemical elements used by John Dalton.

the mistake of assuming "greatest simplicity": i.e., that the simplest compound of two elements must be binary (e.g., water was HO). His system of atomic weights was not very accurate (e.g., he gave oxygen an atomic weight of seven rather than eight). Dalton's theory remained open to dispute until 1858 when Stanislao CANNIZZARO's rediscovery of Amedeo AVOGADRO's work removed the last objections to the theory. Dalton's symbols for atoms and molecules were spherical and he used wooden molecular models similar to the modern version.

Dam, Carl Peter Henrik (1895–1976) *Danish biochemist*

Dam was born in Copenhagen and educated at the polytechnic and the university there, obtaining his doctorate in 1934. He taught at the university from 1923 until 1941, when – although stranded in America because of the war – he was appointed professor of biochemistry at the Copenhagen Polytechnic. From 1956 until 1963 Dam served as director of the Biochemical Division of the Danish Fat Research Institute.

From 1928 to 1930 Dam worked on the problem of cholesterol metabolism in chickens. Cholesterol, first analyzed by Heinrich WIELAND, is a sterol with an important role in mammalian physiology. It was known that many mammals could readily synthesize it, but it was assumed that chickens lacked this ability. To test this assumption Dam began to rear chickens on a cholesterol-free diet enriched with vitamins A and D.

As it turned out he found that chickens could synthesize cholesterol but, more importantly, he also found that if kept on such a diet for two to three weeks the chickens developed hemorrhages under the skin, and blood removed for examination showed delayed coagulation. Supplementing the diet with fat, vitamin C, and cholesterol made no appreciable difference, so Dam concluded that the condition was due to lack of a hitherto unrecognized factor in the diet.

The missing factor, found to be present in green leaves and pig liver, was designated vitamin K by Dam in 1935 (K being the initial letter of "koagulation," the Scandinavian and German form of the word). Using ether, Dam went on to extract the fat-soluble vitamin K from such sources as alfalfa, and in 1939 succeeded, with Paul KARRER, in isolating it. It was for this work that Dam shared the 1943 Nobel Prize for physiology or medicine with Edward DOISY.

Dana, James Dwight (1813–1895) *American geologist, mineralogist, and zoologist*

> We should give a high place in our estimate to all investigation tending to evaluate the variation of permanence of species, their mutability or immutability.
> —*Thoughts on Species* (1856)

Dana was born at Utica in New York State and educated at Yale (1830–33) where he became interested in geology. He worked initially as assistant to Benjamin SILLIMAN and published, in 1837, *A System of Mineralogy*, one of the major textbooks on the subject.

He sailed as geologist and naturalist on the Wilkes expedition (1838–42) visiting the Antarctic and Pacific. On his return, Dana published a series of research reports on the voyage during the period 1844–54, which established his reputation as an important scientist. These included *Zoophytes* (1846), *Geology* (1849), and *Crustacea* (1852).

In 1847 Dana formulated his geosynclinal theory of the origin of mountains. He introduced the term geosyncline to refer to troughs or dips in the Earth's surface that became filled with sediment. These huge deposits of sediment could then, Dana proposed, be compressed and folded into mountain chains.

He was appointed to the chair of natural history at Yale in 1856 and in 1864 to the chair of geology and mineralogy where he remained until his retirement in 1890. He published several important books while at Yale, including his most notable textbook, *Manual of Geology* (1863), and the synthesis of his work on coral reefs in *Corals and Coral Islands* (1872). In agreement with Charles DARWIN's ideas, published in 1842, Dana argued that coral islands are the result of subsidence of the island together with the upward growth of corals.

Daniell, John Frederic (1790–1845) *British chemist and meteorologist*

> His extensive intercourse with men in general added to his natural perspicacity, gave him a clear insight into character, and conferred on him advantages which men of science in general do not possess.
> —Obituary notice of Daniell (1845)

Daniell was the son of a London lawyer. He started work in the sugar-refining factory of a relative and, on the basis of early researches, he was elected to the Royal Society at the age of 23. He was appointed as first professor of chemistry at the newly opened King's College, London, in 1831.

Daniell invented a number of scientific instruments, including a hygrometer (1820) to measure humidity in the atmosphere. His theories on the atmosphere and wind movements were published in *Meteorological Essays and Observations* (1823). He also stressed the importance of moisture in hothouse management.

Daniell is best remembered for his introduction in 1836 of a new type of electric cell. The voltaic cell, introduced by Alessandro VOLTA in 1797, lost power once the current was drawn. This was due to bubbles of hydrogen collecting on the copper plate and producing resistance to the free flow of the circuit (polarization). With the growth of telegraphy there was a real need for a cell that could deliver a constant current over a long period of time. In the *Daniell cell* a zinc rod is immersed in a dilute solution of sulfuric acid contained in a porous pot, which stands in a solution of copper sulfate surrounded by copper. Hydrogen reacts with the copper sulfate. The porous pot prevents the two electrolytes from mixing, and at the positive (copper) electrode, copper is deposited from the copper sulfate. Thus no hydrogen bubbles can form on this electrode.

Daniels, Farrington (1889–1972) *American chemist*

Born in Minneapolis in Minnesota, Daniels was educated at the University of Minnesota and at Harvard, where he obtained his PhD in 1914. He moved to the University of Wisconsin in 1920, spending his whole career there and serving as professor of chemistry from 1928 until his retirement in 1959.

Daniels worked on a wide variety of chemical problems. In addition to a textbook, *Outlines of Physical Chemistry* (1931), he wrote on photochemistry, nitrogen fixation, and thermoluminescence.

He was also interested in the utilization of solar energy, publishing a book on the subject, *Direct Use of the Sun's Energy* (1964), and organizing a symposium on it in 1954, many years before the discussion of solar energy had become fashionable.

Dansgaard, Willi (1922–) *Danish meteorologist*

Dansgaard (**dans**-gord) was born in the Danish capital of Copenhagen and educated at the university there, obtaining his PhD in 1961. He is currently emeritus professor of geophysics.

Dansgaard has studied the applications of environmental isotopes to meteorological, hydrological, and glaciological problems, and in particular to the climate of the last 100,000 years. Oxygen is present in two stable isotopes – the normal oxygen-16 and a much smaller proportion of oxygen-18 with two extra neutrons in its nucleus. In 1947 Harold UREY demonstrated that the variation of the two isotopes in sea water depended on temperature, i.e., the colder the temperature the smaller the oxygen-18 content of the seas. He had further established that a slight change of temperature would produce a measurable alteration in oxygen-18 levels.

In the early 1960s the U.S. army drilled down into the Greenland icecap, producing an ice core 4,600 feet (1,400 m) long and with a 100,000-year history. Dansgaard realized that by making careful measurement of the core's varying oxygen-18 level he should be able to reconstruct the climatic history of the last 100,000 years. The most recent ice age, ending 10,000 years ago, was clearly marked, as was evidence of a weather cycle during the last 1,000 years.

Darby, Abraham (c. 1678–1717) *British metallurgist*

Darby, who was born in Dudley, England, developed the first successful method of smelting iron ore with coke, and in doing so made possible the large-scale production of good quality iron castings. Previously only wood charcoal had been used but supplies of this were depleted and there were insufficient fuel supplies to provide the growing volume of iron. Attempts to use coal were unsuccessful because the sulfur impurities it contained spoiled the iron. By coking the coal the sulfur could be removed. Darby, who had experience in smelting copper with coke, founded a factory at Coalbrookdale to smelt iron using coke, achieving success in 1709. Abundant coal supplies and Darby's process meant that brass could be replaced by iron from large furnaces. Thomas NEWCOMEN's new steam engine (1712) needed hard-wearing metal parts for cylinders, etc. The iron castings introduced by Darby soon became an integral part of the industrial developments of the day.

Darby's grandson, also named Abraham,

built the world's first iron bridge at Coalbrookdale (1779).

Darlington, Cyril Dean (1903–1981)
British geneticist

> Mankind ... will not willingly admit that its destiny can be revealed by the breeding of flies or the counting of chiasmata.
> —Royal Society Tercentenary Lecture, 1960

Born at Chorley in Lancashire, England, Darlington graduated in agriculture from Wye College, London, in 1923 and joined the John Innes Horticultural Institution, which was then under the directorship of William BATESON. He studied nuclear division, comparing mitosis (normal cell division) with meiosis (the reduction division that halves the chromosome number prior to gamete formation). He demonstrated that the chromosomes have already replicated by the first stage of mitosis whereas the chromosomes are still single in the earliest stage of meiosis, a discovery basic to nuclear cytology.

Darlington was also extremely interested in the crossing over of chromosomes that occurs at meiosis. He saw it as a mechanism that not only allows for recombination of genes between chromosomes; it is also able to account for the complete succession of meiotic events.

Darlington became director of the John Innes Institute in 1939 and presided over the move of the Institute from Merton to Bayfordbury in 1949. He became professor of botany at Oxford in 1953 and became emeritus professor in 1971. Darlington wrote a number of books on genetics, including *Evolution of Genetic Systems* (1939). With Ronald FISHER, he founded the journal *Heredity* in 1947.

Dart, Raymond Arthur (1893–1988) *Australian anatomist*

> The loathsome cruelty of mankind to man forms one of his inescapable, characteristic, and differentiative features; it is explicable only in terms of his cannibalistic origins.
> —Quoted by Richard Leakey in *The Making of Mankind* (1981)

Born in Toowong, Australia, Dart was educated at the universities of Queensland and Sydney where he qualified as a physician in 1917. After a short period (1919–22) at University College, London, Dart moved to South Africa to serve as professor of anatomy at the University of the Witwatersrand, Johannesburg, a post he held until his retirement in 1958.

In 1924 Dart was privileged to make one of the great paleontological discoveries of the century, the Taung skull. For this he was indebted to his student Josephine Salmons who brought him in the summer of 1924 a fossil collected from a mine at Taung, Bechuanaland. Dart named it *Australopithecus africanus*, meaning southern African ape, and declared it to be in-

termediate between anthropoids and man. Such a claim was far from acceptable to many scholars at the time who, like Arthur Keith, dismissed the skull as that of a young anthropoid. Other and older australopithecine remains were later discovered by Robert BROOM in South Africa, East Africa, and Asia, making it clear that they were in fact hominid. It is still however a matter of controversy whether *Australopithecus* lies in the direct line of descent to *Homo sapiens* or whether it represents a quite separate and unsuccessful evolutionary sideline.

Darwin, Charles Robert (1809–1882)
British naturalist

> We must, however, acknowledge, as it seems to me, that man with all his noble qualities ... still bears in his bodily frame the indelible stamp of his lowly origin.
> —*The Descent of Man* (1871)

> The preservation of favourable variations and the rejection of injurious variations, I call Natural Selection, or Survival of the Fittest.
> —*On the Origin of Species* (1859)

Darwin, who was born in Shrewsbury in England, began his university education by studying medicine at Edinburgh (1825), but finding he had no taste for the subject he entered Cambridge University to prepare for the Church. At Cambridge his interest in natural history, first stimulated by the geologist Adam SEDGWICK, was encouraged by the professor of botany John Henslow. Their friendship led to Henslow's recommending Darwin to the admiralty for the position of naturalist on HMS *Beagle*, which was preparing to survey the coast of South America and the Pacific.

The *Beagle* sailed in 1831 and Darwin, armed with a copy of Charles LYELL's *Principles of Geology*, initially concerned himself more with the geological aspects of his work. However, his observations of animal species – particularly the way in which they gradually change from region to region – also led him to speculate on the development of life. He was particularly struck by the variation found in the finches of the Galápagos Islands, where he recorded some 14 different species, each thriving in a particular region of the islands. Darwin reasoned that it was highly unlikely that each species was individually created; more probably they had evolved from a parent species of finch on mainland Ecuador. Further considerations, back in England, as to the mechanism that brought this about resulted in probably the most important book in the history of biology.

On returning to England in 1836, Darwin first concerned himself with recording his travels in *A Naturalist's Voyage on the Beagle* (1839), which received the acclaim of Alexander von HUMBOLDT. His interest in geology was reflected in *Structure and Distribution of Coral*

Reefs (1842) and *Geological Observations on Volcanic Islands* (1844). These early works, which established his name in the scientific community and won the respect of Lyell, were fundamental to the development of his theories on evolution.

Early on Darwin had perceived that many questions in animal geography, comparative anatomy, and paleontology could only be answered by disregarding the theory of the immutability of the species (an idea widely held at the time) and accepting that one species evolved from another. The idea was not original but Darwin's contribution was to propose a means by which evolution could have occurred and to present his case clearly, backed up by a wealth of evidence. In 1838 he read Thomas MALTHUS's *An Essay on the Principle of Population* and quickly saw that Malthus's argument could be extended from man to all other forms of life. Thus environmental pressures, particularly the availability of food, act to select better adapted individuals, which survive to pass on their traits to subsequent generations. Valuable characteristics that arise through natural variability are therefore preserved while others with no survival value die out. If environmental conditions change, the population itself will gradually change as it adapts to the new conditions, and with time this will lead to the formation of new species. Darwin spent over 20 years amassing evidence in support of this theory of evolution by natural selection, so as to provide a buffer against the inevitable uproar that would greet his work on publication. In this period the nature of his studies was divulged only to close friends, such as Joseph HOOKER, T. H. HUXLEY, and Charles Lyell.

The stimulus to publish came in June 1858 when Darwin received, quite unexpectedly, a communication from Alfred Russel WALLACE that was effectively a synopsis of his own ideas. The question of priority was resolved through the action of Lyell and Hooker, who arranged for a joint paper to be read to the Linnean Society in July 1858. This consisted of Wallace's essay and a letter, dated 1857, from Darwin to the American botanist Asa GRAY outlining Darwin's theories. Darwin later prepared an "abstract" of his work, published in November 1859 as *On the Origin of Species by Means of Natural Selection*.

As expected the work made him many enemies among orthodox scientists and churchmen since beliefs in the Creation and divine guidance were threatened by Darwin's revelations. Darwin, a retiring man, chose not to defend his views publicly – a task left to (and seemingly immensely enjoyed by) Huxley, "Darwin's bulldog," notably at the famous Oxford debate in 1860. Darwin continued quietly with his work, publishing books that extended and amplified his theories. One of these was The

Descent of Man (1871), in which he applied his theory to the evolution of man from subhuman creatures. Many of his books are seen as pioneering works in various fields of biology, such as ecology and ethology.

Darwin was, however, troubled by one flaw in his theory – if inheritance were blending, i.e., if offspring received an average of the features of their parents (the then-held view of heredity), then how could the variation, so essential for natural selection to act on, come about? This problem was put in a nutshell by Fleeming Jenkin, professor of engineering at University College, London, who wrote a review of the *Origin* in 1867. In this Jenkin pointed out that any individual with a useful trait, assuming it mated with a normal partner, would pass on only 50% of the character to its children, 25% to its grandchildren, 12½% to its greatgrandchildren, and so on until the useful feature disappeared. The logic of this drove Darwin to resort to Lamarckian ideas of inheritance (of acquired characteristics) as elaborated in his theory of pangenesis in the sixth edition of the *Origin*. The question was not resolved until the rediscovery, nearly 20 years after Darwin's death, of Gregor MENDEL's work, which demonstrated the particulate nature of inheritance, i.e., that hereditary characteristics are transmitted from parents to offspring by discrete entities later known as genes.

Darwin was troubled through most of his life by continuous illness, which most probably was due to infection by the trypanosome parasite causing Chagas's disease, contracted during his travels on the *Beagle*. On his death he was buried, despite his agnosticism, in Westminster Abbey.

Darwin, Erasmus (1731–1802) *British physician*

> Soon shall thy arm, unconquer'd steam, afar
> Drag the slow barge, or drive the rapid car;
> Or on wide-waving wings expanded bear
> The flying chariot through the field of air.
> —Predicting the uses of steam power.
> *The Botanic Garden* (1791)

Darwin was born at Elston in England. He studied medicine at the universities of Cambridge and Edinburgh, obtaining his MB from Cambridge in 1755. Darwin set up practice in Lichfield, where he soon established a reputation such that George III asked him to move to London to become his personal physician – an offer Darwin declined. He remained in Lichfield and founded, with friends, the Lunar Society of Birmingham – so called because of the monthly meetings held at members' houses. It included such eminent men as Joseph PRIESTLEY, Josiah Wedgwood, James WATT, and Matthew BOULTON.

Darwin was something of an inventor, but is best remembered for his scientific writings, which often appeared in verse form. These were generally well received until the politician George Canning produced a very damaging parody of his work. This was part of a general campaign by the government against the Lunar Society for its support of the French and American revolutions and its denouncement of slavery.

In his work *Zoonomia* (1794–96), Darwin advanced an evolutionary theory stating that changes in an organism are caused by the direct influence of the environment, a proposal similar to that put forward by Jean Baptiste LAMARCK some 15 years later.

Darwin was the grandfather, by his first wife, of Charles Robert Darwin and, by his second wife, of Francis GALTON.

Darwin, Sir George Howard (1845–1912) *British astronomer and geophysicist*

> I appeal ... for mercy to the applied mathematician ... If our methods are often wanting in elegance and do but little to satisfy that aesthetic sense of which I spoke before, yet they constitute honest attempts to unravel the secrets of the universe in which we live.
> —Speech to the Fifth International Congress of Mathematicians, 1912

Darwin, the second son of the famous biologist Charles Darwin, was born at Down in England. He was educated at Clapham Grammar School, where the astronomer Charles PRITCHARD was headmaster, and Cambridge University. He became a fellow in 1868 and, in 1883, Plumian Professor of Astronomy, a post he held until his death. He was knighted in 1905.

His most significant work was on the evolution of the Earth–Moon system. His basic premise was that the effect of the tides has been to slow the Earth's rotation thus lengthening the day and to cause the Moon to recede from the Earth. He gave a mathematical analysis of the consequences of this, extrapolating into both the future and the past. He argued that some 4.5 billion years ago the Moon and the Earth would have been very close, with a day being less than five hours. Before this time the two bodies would actually have been one, with the Moon residing in what is now the Pacific Ocean. The Moon would have been torn away from the Earth by powerful solar tides that would have deformed the Earth every 2.5 hours.

Darwin's theory, worked out in collaboration with Osmond Fisher in 1879, explains both the low density of the Moon as being a part of the Earth's mantle, and also the absence of a granite layer on the Pacific floor. However, the theory is not widely accepted by astronomers. It runs against the Roche limit, which claims that no satellite can come closer than 2.44 times the planet's radius without breaking up; there are also problems with angular momentum. Astronomers today favor the view that

the Moon has formed by processes of condensation and accretion. Whatever its faults, Darwin's theory is important as being the first real attempt to work out a cosmology on the principles of mathematical physics.

Daubrée, Gabriel Auguste (1814–1896) *French geologist*

Daubrée (doh-**bray**), who was born at Metz in France, was educated at the University of Strasbourg. After investigating the tin mines of England he was employed, in 1838, as an engineer for Bas-Rhin and prepared a geological map of the area (1840–48). He was then appointed professor of geology at the University of Strasbourg, becoming professor of geology at the Museum of Natural History, Paris, in 1861. Finally, in 1862 he became director of the Imperial School of Mines.

Daubrée was a pioneer of experimental geology, publishing his research in his most significant work, *Etudes synthétiques de géologie expérimentale* (1879; Synthesis Studies on Experimental Geology), in which he tried to show that an understanding of geochemical processes can be attained by reproducing them in the laboratory. In particular he worked on the effect that heating water at great depths in the Earth would have on the production of metamorphic rocks.

Daubrée also built up an extensive collection of meteorites and published his findings in *Météorites et la constitution géologique du globe* (1886; Meteorites and the Geological Constitution of the World). He concluded from his study of the composition of meteorites that the Earth has an iron core.

Dausset, Jean (1916–   ) *French physician and immunologist*

Dausset (doh-**say**), the son of a doctor from Toulouse in southern France, gained his MD from the University of Paris in 1945 following wartime service in the blood transfusion unit. He was professor of hematology at the University of Paris from 1958 and professor of immunohematology from 1968. He was professor of experimental medicine at the Collège de France from 1977 to 1987 and is currently professor of medical genetics at the University of Paris.

Dausset's war experience stimulated his interest in transfusion reactions, and in 1951 he showed that the blood of certain universal donors (those of blood group O), which had been assumed safe to use in all transfusions, could nonetheless be dangerous. This was because of the presence of strong immune antibodies in their plasma, which develop following antidiphtheria and antitetanus injections. Donor blood is now systematically tested for such antibodies.

In the 1950s Dausset noticed a peculiar feature in the histories of patients who had received a number of blood transfusions: they developed a low white blood cell (leukocyte) count. He suspected that the blood transfused could well have contained antigens that stimulated the production of antibodies against the leukocytes. With insight and considerable courage Dausset went on to claim that the antigen on the blood cells, soon to be known as the HLA or human lymphocyte antigen, was the equivalent of the mouse H-2 system, described by George SNELL.

The significance of Dausset's work was enormous. It meant that tissues could be typed quickly and cheaply by simple blood agglutination tests as opposed to the complicated and lengthy procedure of seeing if skin grafts would take. Such work made the technically feasible operation of kidney transplantation a practical medical option, for at last the danger of rejection could be minimized by rapid, simple, and accurate tissue typing. Further confirmation of Dausset's work was obtained when the specific regions of the HLA gene complex were later identified by J. van Rood and R. Ceppellini as a single locus on human chromosome 6.

Dausset later shared the 1980 Nobel Prize for physiology or medicine with Snell and Baruj BENACERRAF.

Davaine, Casimir Joseph (1812–1882) *French physician*

> I congratulate myself on having often carried on your clever researches.
> —Louis Pasteur, letter to Davaine (1879)

Born in St. Amand-les-Eaux, France, Davaine (da-**ven**) studied medicine and worked for most of his life in general practice in Paris. Around 1850 he began studying anthrax in cattle and became, arguably, the first scientist to recognize the role of a specific bacteria as the causal agent of an identifiable disease.

The rodlike organisms of anthrax were first described by Franz Pollender in 1849. Pollender found these organisms in the blood of cattle killed by anthrax but was unsure whether they were the cause or consequence of the disease. In 1863 Davaine took the crucial step of demonstrating that the disease could be transmitted to other cattle by inoculating them with the blood of diseased animals. However, if the blood was heated the disease could no longer be transmitted. Further, Davaine found, if the blood was mixed with water and the mixture allowed to stand then fluid taken from the top of the vessel proved harmless but anthrax could still be transmitted with a sample from the bottom. Such was the evidence assembled by Davaine to support the existence of disease transmitting "bactéridies."

His theory, however, did not explain the well-established fact that anthrax could break out in apparently uncontaminated areas. It was only when Robert KOCH was able to show in 1876 that the bacillus formed spores, which could exist unchanged in the soil, that the full force of Davaine's work became clear.

Davenport, Charles Benedict (1866–1944) *American zoologist and geneticist*

Davenport, who was born in Stamford, Connecticut, obtained a zoology doctorate at Harvard in 1892, where he taught until 1899. From 1901 until 1904 he was curator of the Zoological Museum at the University of Chicago, and from 1904 until 1934 was director of the Carnegie Institution's Department of Genetics at Cold Spring Harbor. In 1910 Davenport founded the Eugenics Record Office, directing it until 1934.

Davenport's early studies of animal genetics, using chickens and canaries, were carried out at the turn of the century, and he was among the first to accept Gregor MENDEL's rediscovered theory of heredity. He later turned his attention to man, and in *Heredity in Relation to Eugenics* (1912) offered evidence for the inheritance of particular human traits, suggesting that the application of genetic principles to human breeding might improve the race (eugenics). From 1898 Davenport was assistant editor of the *Journal of Experimental Zoology*; he was also editor of both *Genetics* and the *Journal of Physical Anthropology*.

Davis, Raymond (1914–2006) *American chemist*

Born in Washington, DC, Davis was educated at the universities of Maryland and Yale where he obtained his PhD in 1942. After serving four years in the U.S.A.A.F., Davis took the post of senior chemist at the Brookhaven National Lab, New York, and remained there until his retirement in 1984. He continued to work, however, as a research professor in the astronomy department of the University of Pennsylvania, Philadelphia.

For many years Davis, an experimentalist, worked on the detection of neutrinos emitted by the Sun. In working out the reactions that power the Sun, theorists, such as John BAHCALL, predict that a certain number of neutrinos should be produced, and that a measurable number should be detectable on Earth.

The problem is that neutrinos have a very low probability of interaction with matter. Millions of them pass unimpeded through the Earth every second. The average time lapse for an interaction of a neutrino with an atom is 10^{36} seconds. To increase the probability of detecting a neutrino it was necessary to use a detector containing a large number of atoms. The result was a 100,000-gallon tank of cleaning fluid (tetrachloroethene), containing about 10^{30}

atoms. To exclude confusing interactions with cosmic rays Davis deposited his tank in 1969 at the bottom of the one-mile-deep Homestake Mine at Lead, South Dakota.

Davis was looking for a specific reaction. Neutrinos can react with the isotope chlorine-37 (about a quarter of chlorine atoms) converting it into the radioactive argon-37. The argon atoms could be removed at regular intervals and counted. Theory predicted that Davis should observe 7.9 ± 2.6 solar neutrinos per second, otherwise known as solar neutrino units (SNUs). Actually Davis began by observing about 2 SNUs, and after twenty years of continuous observation he observed no more than 2 SNUs.

Davis sought to eliminate the possibility that the anomalous results were the products of a faulty experimental design. After twenty years spent refining his work, he remained convinced that any errors were unlikely to be traced to the experiment. Further, other workers have produced very similar results. This discrepancy between theory and experiment constitutes the *solar neutrino problem*.

Davis, William Morris (1850–1934) *American physical geographer*

> No one now regards a river and its valley as ready-made features of the earth's surface. All are convinced that rivers have come to be what they are by slow processes of natural development ...
> —*National Geographic Magazine* (1889)

Davis, who was born in Philadelphia, Pennsylvania, was educated at Harvard. He returned to teach there in 1877 after a period as a meteorologist in Argentina and as an assistant with the North Pacific Survey. He became professor of physical geography in 1890 and of geology in 1898.

Davis is acknowledged as the founder of the science of geomorphology, the study of landforms. In his *The Rivers and Valleys of Pennsylvania* (1889) he first introduced what later became known as the *Davisian systems* of landscape analysis. His aim was to provide an explanatory description of how landforms change in an ideal situation and his most important contribution to this was his introduction of the "cycle of erosion" into geographical thought.

He proposed a complete cycle of youth, maturity, and old age to describe the evolution of a landscape. In youth rivers occupy steep V-shaped valleys while in old age the valleys are broad. The end product would be a flat featureless plain he called a "peneplain." This was an ideal cycle but in practice the cycle would invariably be interrupted by Earth movements. It was, nevertheless, strongly attacked by German geographers, who objected to it on the grounds that it neglected such vital factors as weathering and climate in transforming the landscape. They also believed him to be undermining their argument that landforms could only be discovered by local fieldwork and the production of regional monographs.

Davis also produced an influential work, *Elementary Meteorology* (1894), which was used as a textbook for over 30 years, and, in 1928, published *The Coral Reef Problem*.

Davisson, Clinton Joseph (1881–1958) *American physicist*

Davisson, who was born in Bloomington, Illinois, was educated at the University of Chicago and at Princeton, where he obtained his PhD in 1911. After working for a short period at the Carnegie Institute of Technology, Pittsburgh, Davisson joined the Bell Telephone Laboratory (then Western Electric) in 1917 and remained there until his retirement in 1946.

Davisson began his work by investigating the emission of electrons from a platinum oxide surface under bombardment by positive ions. He moved from this to studying the effect of electron bombardment on surfaces, and observed (1925) the angle of reflection could depend on crystal orientation. Following Louis de Broglie's theory of the wave nature of particles, he realized that his results could be due to diffraction of electrons by the pattern of atoms on the crystal surface.

In 1927 he performed a classic experiment with Lester Germer in which a beam of electrons of known momentum (p) was directed at an angle onto a nickel surface. The angles of reflected (diffracted) electrons were measured and the results were in agreement with de Broglie's equation for the electron wavelength ($\lambda = h/p$). In 1937 he shared the Nobel Prize for physics with George Thomson for "their experimental discovery of the diffraction of electrons by crystals."

Davy, Sir Humphry (1778–1829) *British chemist*

> Sir Humphrey Davy
> Abominated gravy.
> He lived in the odium
> Of having discovered sodium.
> —Edmund Clerihew Bentley, *Biography for Beginners* (1905)

> The progression of physical science is much more connected with your prosperity than is usually imagined. You owe to experimental philosophy some of the most important and peculiar of your advantages. It is not by foreign conquests chiefly that you are become great, but by a conquest of nature in your own country.
> —Lecture at the Royal Institution (1809)

The son of a small landowner and wood-carver, Davy went to school in his native town of Penzance and in Truro. At the age of 17 he was apprenticed to an apothecary and surgeon with a view to qualifying in medicine. He was self-reliant and inquisitive from an early age and taught himself chemistry from textbooks. In 1798 he was appointed to Thomas BEDDOES's Pneumatic Institute at Clifton, Bristol, to investigate the medicinal properties of gases. Davy's first papers were published by Beddoes in 1799. In one he concluded, independently of Count RUMFORD, that heat was a form of motion; the other contained some fanciful speculations on oxygen, which he called phosoxygen. Davy soon discovered the inebriating effect of nitrous oxide and his paper *Researches, Chemical and Philosophical; chiefly concerning Nitrous Oxide* (1800), and the subsequent fashion for taking the "airs," made him famous. At Clifton he met many eminent people, including the poets William Wordsworth, Samuel Taylor Coleridge, and Robert Southey (Davy was himself a Romantic poet), and his flirtation with fashionable society began.

In 1801 Davy moved to London, to the Royal Institution, where his lectures were spectacularly successful. At Clifton he had begun to experiment in electrochemistry, following William NICHOLSON's electrolysis of water, and this was to prove his most fruitful field. In the early years at the Royal Institution, however, he did much work of an applied nature, for example on tanning and on agricultural chemistry. In his 1806 Bakerian Lecture to the Royal Society he predicted that electricity would be capable of resolving compounds into their elements and in the following year he was able to announce the isolation of potassium and sodium from potash and soda. This result cast doubts on Antoine LAVOISIER's oxygen theory of acidity. Davy was essentially a speculative and manipulative chemist, not a theorist, and he reasoned incorrectly that ammonia (because of its alkaline properties), and hence nitrogen, might contain oxygen. He remained skeptical about the elementary nature of bodies for many years and tried to show that sulfur and phosphorus contained hydrogen.

Davy's work in the years immediately following the discovery of sodium was hindered by his social success and competition for priority with the French chemists Joseph GAY-LUSSAC and Louis Thenard. He prepared boron, calcium, barium, and strontium by electrolysis but his priority was disputed. In 1810 he published a paper on chlorine, which established that it contained no oxygen – another blow against the oxygen theory of acidity – and was in fact an element. The name "chlorine" was proposed by Davy.

In 1812 Davy was knighted, married a wealthy widow, and published his book *Elements of Chemical Philosophy*. In 1813 he appointed Michael FARADAY as his assistant and the Davys and Faraday visited France. Working in Michel CHEVREUL's laboratory, he established that iodine, discovered two years before by Bernard COURTOIS, was an element similar in many properties to but heavier than chlorine. On his return to England, Davy was commissioned to investigate the problem of firedamp (methane) explosions in mines. In 1816, only six months after beginning the investigation, he produced the famous safety lamp, the *Davy lamp*, in which the flame was surrounded by a wire gauze. Davy became president of the Royal Society in 1820 and the rest of his life was much taken up by traveling on the Continent. Despite his successes there is something incomplete about his life. He never accepted the atomic theory of DALTON, his great contemporary. He had in fact more in common with his Romantic poet friends than he did with Dalton. Jöns BERZELIUS said of him that his work consisted of "brilliant fragments."

Dawes, William Rutter (1799–1868)
British astronomer

Dawes's father was an astronomer, often on colonial service, and later taught mathematics in a London school. Dawes, who was born in London, was originally intended for the Church but finding the views of the established church uncongenial, studied medicine at St. Bartholomew's Hospital, London, instead. He practiced for some time in Berkshire and in 1826 went to Liverpool where he was persuaded to exchange medicine for the post of a Dissenting minister in Ormskirk, Lancashire.

It was in Ormskirk that Dawes took up the serious study of astronomy in 1829. He built his own observatory and equipped it with a Dollond refractor. In 1839 Dawes moved to London to take charge of George Bishop's observatory in Regent's Park. Bishop had made a fortune in the wine trade and hoped to gain recognition by sponsoring some important discovery in his observatory. A second marriage to a wealthy widow in 1842 allowed Dawes to leave the somewhat disagreeable service of Bishop in 1844 and, despite persistent ill health, to devote himself to his own private astronomical researches first in Kent at Cranbrook and Maidstone and after 1857 at Haddenham in Buckinghamshire.

Dawes, the keenest of observers, worked for many years on double stars. During his Ormskirk period he published details on over 200 of them. Basic data on a further 250 were published in 1852 in Bishop's *Astronomical Observations at South Villa*. In November 1850 he narrowly missed an important discovery when he reported that while observing Saturn's rings he noted in the ansa, that part visible at the side of the planet and sticking out like a handle, "a light ... at both ends." Dawes had in fact observed the faint inner "crepe ring" of Saturn, Ring C, but had been narrowly anticipated in this by the American astronomer William C. BOND.

Dawkins, Richard (1941–) *British ethologist*

> We are survival machines – robot vehicles blindly programmed to preserve the selfish molecules known as genes. This is a truth which still fills me with astonishment.
> —*The Selfish Gene* (1976)

Dawkins was educated at Oxford University where he worked for his doctorate under Nicolaas TINBERGEN. He initially taught at the University of California, Berkeley, before returning to Oxford in 1970 as lecturer in animal behavior. He was appointed reader in zoology in 1989. In 1995 he was awarded the Charles Simonyi Chair for the Public Understanding of Science.

In *The Selfish Gene* (1976) Dawkins did much to introduce the work of such scholars as William Hamilton, Robert L. Trivers, and John MAYNARD SMITH to a wider public. He tried to show that such apparently altruistic behavior as birds risking their lives to warn the flock of an approaching predator can be seen as the "selfish" gene ensuring its own survival (by ensuring the survival of the descendants and relatives of the "heroic" bird) – indeed that such behavior is as relentlessly under the control of the selfish gene as the compulsive rutting of the dominant stag. The work was immensely successful, being translated into eleven languages and selling 150,000 copies in English alone. *The Extended Phenotype* (1982) is a further work on evolutionary theory.

In 1986 Dawkins published another successful work, *The Blind Watchmaker*. The title refers to the image used by William Paley in his *Natural Theology* (1802). If anyone were to find a watch he would be able to infer from its mechanism that it had a maker; equally with nature, where the mechanisms of hand, eye, heart, and brain demand the existence of a designer just as strongly. Dawkins accepted the argument but insisted that the watchmaker was merely the operation of natural selection. In case after case he argued that the same effects could be produced by natural selection a good deal more plausibly than by a divine watchmaker.

One of the most original features of Dawkins's work was his demonstration that with few simple recursive rules, and some very simple starting points, various complex life forms or biomorphs were produced on his computer screen. And, he emphasized, the biomorphs were produced not by Dawkins as designer (he was as surprised as anyone else by the outcome) but by the application of simple rules to a large number of apparently random initial positions.

In 1993 in an essay *Viruses of the Mind* he introduced the idea of a "mental virus" analogous to biological and computer viruses. He argues that religious belief is a mental virus.

Dawkins has continued to write on evolutionary theory in *River out of Eden* (1995) and most notably in his *Climbing Mount Improbable* (1996), where he shows how such unlikely candidates as a spider's web and the vertebrate eye can have evolved under the guiding power of natural selection.

Dawkins is a noted atheist and campaigner against religious extremism in television programs, radio programs, lectures, and through his website. In 2006 he published *The God Delusion* openly attacking religion. In 2006 he also founded The Richard Dawkins Foundation for Reason and Science.

Day, David Talbot (1859–1925) *American chemist*

Day was born in East Rockport, Ohio, and educated at Johns Hopkins University, gaining his PhD in 1884. He started his career as a demonstrator in chemistry at the University of Maryland but left to become head of the Mineral Resources Division of the U.S. Geological Survey in 1886.

Day investigated the reasons for the differences found in the composition of various petroleum deposits and concluded that percolation through mineral deposits affected the nature of the underlying petroleum. His experiments, in which he ran crude petroleum through fuller's earth, anticipated the later development of adsorption chromatography by Mikhail TSVET.

Day did much to stimulate the growth of the petroleum industry and in 1922 produced one of its basic texts, *Handbook of the Petroleum Industry*. From 1914 to 1920 he was a consultant chemist with the Bureau of Mines.

Deacon, Henry (1822–1876) *British industrial chemist*

Deacon, a Londoner, started his career as an apprentice with an engineering firm at the age of 14. He later joined the firm of NASMYTH and GASKELL, near Manchester, before joining Pilkington's glassworks at St. Helens. There he introduced, in 1844, a new method for grinding and smoothing glass. Leaving the glassworks in 1851, Deacon went into partnership with the younger William Pilkington in 1853 to manufacture alkali at Widnes.

Deacon made a significant improvement in the LEBLANC process for producing alkali, using one of its by-products, hydrochloric acid, to produce chlorine. William GOSSAGE had introduced his tower to condense and collect the toxic hydrochloric acid fumes in 1836. In 1870 Deacon patented his method where the hydrochloric acid was passed over clay balls soaked in copper chloride, which, in air, oxidized the acid, yielding chlorine. The chlorine was used in making bleaching powder for the textiles industry. The process was supplanted by an alternative method of converting the hydrochloric acid into chlorine introduced by Walter WELDON (1866–69).

de Bary, Heinrich Anton (1831–1888)
German botanist

De Bary (de bah-**ree**), who was born in Frankfurt am Main in Germany, gained his medical degree from Berlin in 1853 and practiced briefly in Frankfurt before devoting all his time to botany. He became *Privatdozent* (nonstipendiary lecturer) in botany at Tübingen and then professor of botany, first at Freiburg im Breisgau (1855), then Halle (1867), and finally Strassburg (1872), where he remained until his death.

When de Bary began working on fungi some people still believed in the spontaneous generation of fungi, and the general ignorance of fungal life cycles severely impeded the development of intelligent control measures against fungal epidemics of crops. De Bary's first mycological publication, *Researches on Fungal Blights* (1853), dealt with the rust and smut diseases of plants and maintained that fungi are the cause, and not the effect, of these diseases. In 1865 he demonstrated that the fungus that causes stem rust of wheat, *Puccinia graminis*, needs two hosts – wheat and barberry – to complete its life cycle. De Bary showed in 1866 that individual lichens consist of a fungus and an alga in intimate association and in 1879 he introduced the term "symbiosis" to describe mutually advantageous partnerships between dissimilar organisms.

De Bary's work was instrumental in encouraging a more developmental approach to mycology and his research on host–parasite interactions greatly helped in the fight against plant diseases.

De Beer, Sir Gavin Rylands (1899–1972)
British zoologist

Born in London, De Beer graduated from Oxford University, where he was a fellow from 1923 to 1938. He served in both World Wars; during World War II he landed in Normandy in 1944, where he was in charge of psychological warfare. He was professor of embryology at University College, London, 1945–50, and then director of the British Museum (Natural History) in London (1950–60).

In an early publication, *Introduction to Experimental Biology* (1926), de Beer finally disproved the germ-layer theory. Embryological investigations had indicated that vertebrate structures such as cartilage and certain bone cells were formed from the ectoderm, or outer layer of the embryo, and not, as was previously thought, from the mesoderm. As this goes against the germ-layer theory, orthodox embryologists argued that the experimental manipulations involved in such work altered the normal course of development. De Beer's contribution was to find a system that does not involve such manipulations so establishing the validity of earlier work.

De Beer also did work to show that adult animals retain some of the juvenile characters of their evolutionary ancestors (*pedomorphosis*), thus refuting Ernst HAECKEL's theory of recapitulation. He suggested that gaps in the evolutionary development of animals may be accounted for by the impermanence of the soft tissues of young ancestors. Studies of *Archeopteryx*, the earliest known bird, led him to propose piecemeal evolutionary changes in such animals, thus explaining the combination of reptilian and avian characters (e.g., teeth and feathers). De Beer also carried out research into the functions of the pituitary gland. In the field of ancient history, de Beer applied scientific methods to various problems, for example, the origin of the Etruscans and Hannibal's journey across the Alps. His other books include *Embryology and Evolution of Chordate Animals* (1962, with Julian Huxley), *The Elements of Experimental Embryology* (1962), and a biography of Charles DARWIN (1961). He was knighted in 1954.

Debierne, André Louis (1874–1949)
French chemist

Born in Paris, France, Debierne (de-**byairn**) was educated at the Ecole de Physique et Chemie. After graduation he worked at the Sorbonne and as an assistant to Pierre and Marie CURIE, finally succeeding the latter as director of the Radium Institute. On his retirement in 1949 he in turn was succeeded by Marie Curie's daughter, Irène JOLIOT-CURIE.

Debierne was principally a radiochemist; his first triumph came in 1900 with the discovery of a new radioactive element, actinium, which he isolated while working with pitchblende. In 1905 he went on to show that actinium, like radium, formed helium. This was of some significance in helping Ernest RUTHERFORD to appreciate that some radioactive elements decay by emitting an alpha particle (or, as it turned out to be, the nucleus of a helium atom). In 1910, in collaboration with Marie Curie, he isolated pure metallic radium.

de Broglie, Prince Louis Victor Pierre Raymond (1892–1987) *French physicist*

De Broglie (de broh-**glee**) was descended from a French family ennobled by Louis XIV. He was born in Dieppe, France, and educated at the Sorbonne. Originally a historian, he became interested in science in World War I when he was posted to the Eiffel Tower as a member of a signals unit. He pursued this interest after the war and finally obtained his doctorate in physics from the Sorbonne in 1924. He taught there from 1926, serving as professor of theoretical physics at the newly founded Henri Poincaré Institute (1928–62).

De Broglie is famous for his theory that particles (matter) can have wavelike properties. At the start of the 20th century physicists explained phenomena in terms of particles (such as the electron or proton) and electromagnetic radiation (light, ultraviolet radiation, etc.). Particles were "matter" – conceived as discrete entities forming atoms and molecules; electromagnetic radiation was a wave motion involving changing electric and magnetic fields.

In 1905 two papers by Albert EINSTEIN began a change in this conventional view of the physical world. His work on the special theory of relativity led to the idea that matter is itself a form of energy. More specifically he explained the photoelectric effect by the concept that electromagnetic radiation (a wave) can also behave as particles (photons). Later, in 1923, Arthur COMPTON produced further evidence for this view in explaining the scattering of x-rays by electrons.

In 1924 de Broglie, influenced by Einstein's work, put forward the converse idea – that just as waves can behave as particles, particles can also behave as waves. He proposed that an electron, for instance, can behave as if it were a wave motion (a *de Broglie wave*) with wavelength h/p, where p is the momentum of the electron and h is PLANCK's constant. This revolutionary theory was put forward in de Broglie's doctoral thesis. Experimental support for it was obtained independently by George THOMSON and by Clinton J. DAVISSON and the wavelike behavior of particles was used by Erwin SCHRÖDINGER in his formulation of wave mechanics.

The fact that particles can behave as waves, and vice versa, is known as wave–particle duality and has caused intense debate as to the "real" nature of particles and electromagnetic radiation. De Broglie took the view that there is a true deterministic physical process underlying quantum mechanics – i.e., that the current indeterminate approach in terms of probability can be replaced by a more fundamental theory. He based his ideas on the concept of particles that are concentrations of energy guided through space by a real wave and exchanging energy with a "subquantum medium."

De Broglie received the 1929 Nobel Prize for physics for his "discovery of the wave nature of the electron."

Debye, Peter Joseph William (1884–1966) *Dutch–American physicist and physical chemist*

> Only experiments can decide ...
> —*Notes on Magnetization at Low Temperatures* (1926)

Born at Maastricht in the Netherlands, Debye (de-**bI**) studied electrical engineering at Aachen and gained his PhD at Munich in 1910. He held chairs of physics at Zurich (1911–12 and 1919–27), Utrecht (1912–14), Göttingen (1914–19), and Leipzig. He was director of the Kaiser Wilhelm Institute for Theoretical Physics (1935–40) before emigrating to America where he was professor of chemistry at Cornell (1940–50).

Debye was essentially a theoretician and most of his work, although varied, had a common theme: the application of physical methods to problems of molecular structure. An early work was the derivation of a relation governing the change of the specific heat capacity of solids with temperature. In 1915 he gave a theoretical treatment of electron diffraction by gases, not realized in practice until 1930. At Göttingen, Debye and P. Scherrer discovered a method of producing x-ray diffraction patterns from powders. This was later extended to the production of diffraction patterns from simple molecules such as CCl_4 (1928).

A major part of Debye's work was devoted to dipole moments, beginning in 1912. He used these to determine the degree of polarity of covalent bonds and to determine bond angles. Together with his x-ray work and results from rotational spectra, this enabled the precise spatial configuration of small molecules to be deduced. For example, the planarity of the benzene molecule was confirmed by dipole moment measurements. Debye is probably better known, however, for the *Debye–Hückel theory* of electrolytes (1923). This was a theory that could be applied to concentrated solutions of ionic compounds, and was a great advance on the theories of the time, which applied only to very dilute solutions. The Debye–Hückel theory takes account of the fact that an ion in solution tends to attract other ions of opposite charge.

Dedekind, (Julius Wilhelm) Richard (1831–1916) *German mathematician*

> For what I have accomplished and what I have become, I have to thank my ... indefatigable working rather than any outstanding talent.
> —Quoted by Hans Zincke in *Richard Dedekind Recalled* (1916)

The son of an academic lawyer from Braunschweig, Germany, Dedekind (**day**-de-kint) was educated at the Caroline College there and at Göttingen, where he gained his doctorate in 1852. After four years spent teaching at Göttingen, he was appointed professor of mathematics at the Zurich Polytechnic. In 1862 he returned to Braunschweig to the Technical High School where he remained until his retirement in 1912.

In 1872 Dedekind published his most important work *Stetigkeit und Irrationale Zahlen* (Continuity and Irrational Numbers) in which he provided a rigorous definition of the irrational numbers. He began by "cutting" or dividing the rational numbers into two nonempty disjoint sets A and B such that if x belongs to A and y to B, then $x < y$. If A has a greatest member A' then A' is a rational number; if B has a smallest number B' then B' will also be a rational number. But if A has no greatest number and B no smallest, then the cut defines an irrational number.

In the same work Dedekind gave the first precise definition of an infinite set. A set is infinite, he argued, when it is "similar to a proper part of itself." Thus the set N of natural numbers can be shown to be "similar," that is, matched or put into a one-to-one correspondence with a proper part, in this case $2N$:

$$N \quad 1 \quad 2 \quad 3 \quad 4 \quad 5 \quad 6 \quad 7 \quad 8 \quad 9 \ldots$$
$$\quad\;\; \downarrow \;\; \downarrow \;\; \downarrow \;\; \downarrow \;\; \downarrow \;\; \downarrow \;\; \downarrow \;\; \downarrow \;\; \downarrow$$
$$2N \quad 2 \quad 4 \quad 6 \quad 8 \quad 10 \quad 12 \quad 14 \quad 16 \quad 18 \ldots$$

Whereas the only thing a finite set can be matched with is the set itself.

In a later work, *Was sind und was sollen die Zahlen?* (What Numbers Are and Should Be, 1888) Dedekind demonstrated how arithmetic could be derived from a set of axioms. A simpler but equivalent version, formulated by PEANO in 1889, is much better known.

de Duve, Christian René (1917–   ) *Belgian biochemist*

De Duve (de **doo**-ve) was born at Thames Ditton in southern England and educated at the Catholic University of Louvain, where he obtained his MD in 1941. After holding brief appointments at the Nobel Institute in Stockholm and at Washington University, he returned to Louvain in 1947 and was appointed professor of biochemistry in 1951. From 1962 to 1988 he held a similar appointment at Rockefeller University in New York.

In 1949 de Duve was working on the metabolism of carbohydrates in the liver of the rat. By using centrifugal fractionation techniques to separate the contents of the cell, he was able to show that the enzyme glucose-6-phosphatase is associated with the microsomes – organelles whose role was only speculative until de Duve began this work. He also noted that the process of homogenization led to the release of the enzyme acid phosphatase, the amount of which seemed to vary with the degree of damage inflicted on the cells. This suggested to de Duve that the enzyme in the cell was normally enclosed by some kind of membrane. If true, the supposition would remove a problem that had long troubled cytologists – namely how it was that such powerful enzymes did not attack the normal molecules of the cell. This question could now be answered by proposing a self-contained organelle, which neatly isolated the digestive enzymes. Confirmation

for this view came in 1955 with the identification of such a body with the aid of the electron microscope. As its role is digestive or lytic, de Duve proposed the name *lysosome*. The peroxisomes (organelles containing hydrogen peroxide in which oxidation reactions take place) were also discovered in de Duve's laboratory.

For such discoveries de Duve shared the 1974 Nobel Prize for physiology or medicine with Albert Claude and George PALADE.

De Forest, Lee (1873–1961) *American physicist and inventor*

> The radio was conceived as a potent instrumentality for culture, fine music, the uplifting of America's mass intelligence. You have debased this child, you have sent him out in the street in rags of ragtime, tatters of jive and boogie-woogie, to collect money from all and sundry.
> —Complaining to radio executives about the quality of broadcasting

De Forest, who was born in Council Bluffs, Iowa, was interested in science from the age of 13. His father, a congregational minister, wanted him to study for the Church, but De Forest refused, going instead, in 1893, to the Sheffield Scientific School at Yale University. His PhD thesis, *Hertzian Waves from the Ends of Parallel Wires* (1899), was probably the first PhD thesis on radio in America, and drew on the work of Heinrich HERTZ and Guglielmo MARCONI. While working for the Western Electric Company in Chicago, he developed an electrolytic detector and an alternating-current transmitter.

In 1907 De Forest patented the Audion tube, a thermionic grid-triode vacuum tube that was a very sensitive receiver of electrical signals. This invention was crucial to the development of telecommunications equipment. In 1912 he had the idea of "cascading" these to amplify high-frequency radio signals, making possible the powerful signals needed for long-distance telephones and for radio broadcasting. His invention formed the basis of radio, radar, telephones, and computers until the advent of solid-state electronics.

Throughout his career De Forest pushed for the acceptance of radio broadcasting. He was not a very good business manager, however, and had to sell many of his patents. Later he worked on a sound film system that was similar to the one eventually adopted. In the 1930s he designed Audion diathermy machines for medical use and during World War II he worked on military research at the Bell Telephone Laboratories.

De Geer, Charles (1720–1778) *Swedish entomologist*

De Geer (de yayr) was born in Finspang, Sweden, and educated in the classics at the Uni-

versity of Utrecht; he then studied under LIN-
NAEUS at Uppsala. His extensive *Mémoires pour
servir à l'histoire des insectes* (7 vols., 1752–78;
Contributory Notes on the History of Insects)
include excellent drawings and probably the
earliest published accounts of the maternal instinct in such nonsocial insects as the earwig
Forficula auricula and the shield bug *Elas-
mucha griseus*. He also initiated a system of insect classification based on the wings and
mouthparts.

Dehmelt, Hans Georg (1922–   ) *American physicist*

Born at Gorlitz in Germany, Dehmelt (**day**-
melt) left the Berlin Gymnasium in 1940 to
join the German army. He was allowed for a
time to study physics at Breslau University
but, in 1945, he was taken prisoner by the
Americans at Bastogne, Belgium. After the war
he continued his education at Göttingen, gaining his PhD there in 1950. He went to America
in 1952 as a postdoctoral student at Duke University, North Carolina, and remained there
until 1955. He then moved to the University of
Washington, Seattle, where he was appointed
professor of physics in 1961, the same year he
became a naturalized American citizen. He retired in 2002.

Dehmelt has worked for many years on the
seemingly impossible task of imprisoning a single electron for an extended period in a suitable
container. In this manner Dehmelt hoped to
measure more accurately the magnetic moment
(*g*) of the electron. Earlier experiments by H. R.
Crane at the University of Michigan had involved passing a beam of electrons through a
magnetic field. But the evidence gathered in
this manner necessarily involves the interactions of other electrons.

In 1955 Dehmelt began work on what later
become known as a *Penning trap*. In 1973 he
succeeded in isolating a single electron and
went on to show (1975) how accuracy could be
further improved by "cooling" the electron (i.e.,
decreasing its kinetic energy). In this way it
proved possible to measure *g* with an accuracy
of 4 parts in a trillion.

The Penning trap operates with a combination of electrical and magnetic fields. An electron in a uniform magnetic field cannot move
across the field lines, but is able to escape by
moving parallel to the field. To avoid this, an
electric field is imposed upon the magnetic
field. This field is produced by three electrodes
– two negatively charged end traps and a positively charged encircling nickel ring.

For his work in this area Dehmelt shared
the 1989 Nobel Prize for physics with Wolfgang
Paul and Norman RAMSEY.

Deisenhofer, Johann (1943–   ) *German chemist*

Deisenhofer (**dee**-zen-hof-er), who was born in
Zusamaltheim, Germany, obtained his PhD
from the Max Planck Institute for Biochemistry, at Martinsried near Munich, in 1974. He
remained at the Institute until 1987 when he
moved to America to work at the Howard
Hughes Medical Institute, Dallas, Texas. He is
professor of biochemistry and also professor of
bimolecular science at the University of Texas.

In 1982 Hartmut MICHEL had succeeded in
crystallizing the membrane proteins of the photosynthetic reaction center. Clearly, to understand how photosynthesis worked at the
molecular level it would be necessary to determine the structure of these proteins, and Michel
invited his colleague Deisenhofer to tackle the
problem. By 1985, using the well-established
techniques of x-ray crystallography, Deisenhofer's group had managed to locate the position of more than 10,000 atoms.

Deisenhofer's analysis revealed complex
protein structures holding a molecular cluster
containing four chorophyll molecules, two pheophytins (molecules resembling chlorophyll), two
quinones (dehydrogenizing agents), and a single iron atom. It has been possible to show how
this center can transform energy from incident
photons. On absorbing a photon, one of the
chlorophyll molecules releases an electron. This
is transferred by the pheophytins and quinones
to the membrane's outer surface. At the same
time, an adjoining cytochrome molecule donates
an electron to one of the chlorophyll molecules
and thus gains a positive charge. In this way
the photon energy has been stored in the charge
separation of the negative electron and the positive cytochrome. And so begins the molecular
process of photosynthesis.

For this work Deisenhofer shared the 1988
Nobel Prize for chemistry with his Institute
colleagues Hartmut Michel and Robert HUBER.

De la Beche, Sir Henry Thomas (1796–1855) *British geologist*

De la Beche (de la besh) entered the army but
at the end of the Napoleonic Wars he chose to
devote himself to geology instead. After traveling extensively in Europe and Jamaica on his
own research work, he became, in 1835, director of the Geological Survey of Great Britain,
which had been recently formed largely on his
initiative. He was also instrumental in setting
up the Royal School of Mines in 1851, of which
he was the first principal.

He wrote extensively on the geology of southwest England and Jamaica, publishing the first
account of the geology of Jamaica in 1827 and
his report on the geology of Devon during the
period 1832–35.

In 1834, while working in Devon, he made
his most significant discovery. He observed that
some rock strata contained fossil plants similar
to those of the Carboniferous system, discov-

ered by William CONYBEARE in 1822, but did not contain any of the fossils of the preceding Silurian system, recently discovered by Roderick MURCHISON. The Silurian was believed to merge directly into the Carboniferous and De la Beche assumed the strata he had discovered came before the Silurian. However, William Lonsdale, librarian of the Geological Society, convincingly argued for a system, later named the Devonian, which overlay the Silurian and underlay the Carboniferous.

De la Beche wrote extensively on geology; his *A Geological Manual* (1831), *How to Observe* (1835), and *Geological Observer* (1851) were in part aimed at satisfying the growing popular interest in geology.

Delambre, Jean Baptiste Joseph
(1749–1822) *French astronomer and mathematician*

Born in Amiens in northern France, Delambre (de-**lahm**-bre) was most unusual for a mathematician and astronomer in that he did not begin the serious study of his subject until he was well over 30 years old. As a student he had been interested in the classics and only turned to the exact sciences when he was 36. He published tables of Jupiter and Saturn in 1789 and of Uranus in 1792. He also measured an arc of the meridian between Dunkirk and Barcelona to establish a basis for the new metric system. He succeeded Joseph de LALANDE as professor of astronomy at the Collège de France in 1795. In his later years he devoted himself to a monumental six-volume *Histoire de l'astronomie* (1817–27; History of Astronomy).

De la Rue, Warren (1815–1889) *British astronomer*

Born in Guernsey, in the Channel Islands, De la Rue was the son of a printer and worked most of his life in his father's business. He was educated in Paris and studied science privately. He was initially interested in chemistry, being a friend of and working with August Hofmann, but later, at the suggestion of James NASMYTH, he took up astronomy, building a small observatory for himself.

De la Rue devoted himself to problems of photographic astronomy. He was the first to apply the collodion process (invented by Frederick Archer in 1851) to photographing the Moon. In 1852 he took some photographs that were sharper than any previously produced and that could be enlarged without blurring. Ten years later he was producing photographs that could show as much as could be seen through any telescope. In 1854 he designed the photoheliograph, a device for taking telescopic photographs of the Sun. In 1860 he used it to take dramatic photographs of prominences during the total eclipse in Spain, proving that they were solar (and not, as had been thought, lunar) in origin. De la Rue gave up active astronomical investigation in 1873, donating his telescope to the observatory at Oxford and devoting the rest of his life to his business and to his chemical researches.

Delbrück, Max (1906–1981) *German-born American physicist and molecular biologist*

> While the artist's communication is linked forever with its original form, that of the scientist is modified, amplified, fused with the ideas and results of others.
> —*The Eighth Day of Creation*

The son of a history professor, Delbrück (**del**-bruuk**) trained as a physicist first in his native city of Berlin, and then at Tübingen, Bonn, and Göttingen, where he completed his doctorate in 1930. After spending the period from 1931 to 1933 in Copenhagen, Delbrück was appointed to the Kaiser Wilhelm Institute for Chemistry, Berlin. He left Germany for America in 1937, working first at the California Institute of Technology and from 1940 until 1947 at Vanderbilt University, Nashville. Delbrück returned to Cal Tech in 1947 and remained there as professor of biology until his retirement in 1976. He became a naturalized American citizen in 1945.

While at Copenhagen, under the influence of Niels BOHR, Delbrück's interest was diverted from atomic physics to questions about the nature of life. In the late 1930s he began to work with bacteriophages, the viruses discovered by D'Hérelle that infect and destroy bacteria. They were relatively simple, reproduced quickly, and were easy to handle; an ideal organism, Delbrück argued, in which to study the mechanisms of replication and development.

In 1939, with E. Ellis, he first demonstrated the phenomenon of "one-step growth." Working with the phage T4 he found that "a virus particle enters a bacterial cell and after a certain period (between 13 and 40 minutes, depending on the virus, on the dot for any particular type), the bacterial cell is lysed and 100 particles are liberated." How can one particle, Delbrück asked, become 100 in a mere 20 minutes?

Soon after he began to collaborate with Salvador LURIA. In 1943 they published a paper, *Mutations of Bacteria from Virus Sensitivity to Virus Resistance*. How, they asked, do bacteria acquire resistance to lethal phage? Is it induced by contact, or does it arise from a fortunate mutation? Luria and Delbrück realized that the dynamics of bacterial growth would differ in each case. The number of resistant strains found in bacterial colonies exposed to phage should fluctuate more than if the resistance was induced. Delbrück worked out the statistics and Luria performed the experiment; the results clearly revealed that bacteria underwent mutations.

Delbrück went on to show in 1945, in collaboration with W. Bailey, that phage can reproduce sexually. They were working with the two viruses T2 and T4r, both of which could be bred in bacterium B. They found that:

T2 formed small colonies and attacked bacterium A.

T4r formed large colonies and attacked bacterium C.

When both T2 and T4r were bred together in B, the parent types produced two new strains:

Strain 1: formed small colonies and attacked bacterium C.

Strain 2: formed large colonies and attacked bacterium A.

Obviously, Delbrück concluded, "the parents had got together and exchanged something."

By this time Delbrück had begun to be recognized as the leader of what became known as "the phage group." From 1945 onwards he ran an annual summer phage course at Cold Spring Harbor Laboratory, New York, which was attended over the years by most of the leading molecular biologists of the following decade. For Delbrück himself, however, the mid-1950s seemed to be a good time to move on. He turned to the study of sensory mechanisms in the fungus *Phycomyces*. It grew towards the light, against gravity, and into the wind. How did it sense these stimuli? What range of light did it respond to? These and other questions were tackled by Delbrück and his coworkers in what was soon called "the Phycomyces group." His last published paper in 1981 was in this field and proposed that the chemical photoreceptor of *Phycomyces* was a flavin and not, as had been supposed, a carotene.

For his earlier work with the phage group Delbrück shared the 1969 Nobel Prize for physiology or medicine with Salvador Luria and Alfred HERSHEY

D'Elhuyar, Don Fausto (1755–1833)
Spanish chemist and mineralogist

> No city on the New Continent, not even in the United States, offers scientific establishments so vast and so solid as does the capital of Mexico. It is enough to cite here the School of Mines, of which the scholar D'Elhuyar is director.
>
> —Alexander von Humboldt, *Political Essay on the Kingdom of New Spain* (1826)

Born in Logroño, Spain, D'Elhuyar (**del**-yoo-ar) studied mineralogy with his brother, Juan José, at the Freiberg Mining Academy under Abraham WERNER. He then studied chemistry in Paris (1772–77). He returned to Spain shortly after and was sent to Mexico in 1788 to supervise mining operations. On his return to Spain in 1821 he was made director general of mines.

The D'Elhuyar brothers working together in 1783 discovered the element tungsten (formerly also known as wolfram). Two very dense minerals were known to chemists in the 18th century: "tungsten" (Swedish meaning "heavy stone") and wolframite. In 1781 Karl SCHEELE had discovered that "tungsten" (now known as scheelite) contained tungstic acid. The brothers proved that the same acid is present in wolframite, from which mineral they succeeded in isolating the element tungsten.

DeLisi, Charles (1941–   ) *American biophysicist*

DeLisi (de-**lee**-see) was educated at City College, in his native New York, and at New York University, where he completed his PhD in 1969. After periods at Yale and Los Alamos, he moved, in 1975, to the National Institutes of Health, Bethesda, Maryland, as head of mathematical biology. In 1985, he moved to Washington to head the Office of Health and Environmental Research (O.H.E.R.), a part of the U.S. Department of Energy. DeLisi returned to academic life in 1988 and since 1990 he has been professor of biomedical engineering at Boston University.

While at O.H.E.R. DeLisi was concerned with the health effects of radiation. Some people, he realized, were genetically disposed to develop cancer when exposed to low levels of radiation. But, before one could spot such tendencies in any individual, it would first be necessary to know more about the human genome. As a physicist DeLisi was used to thinking in terms of large expensive projects and consequently began to consider the possibility of mapping the entire human genome. He soon heard that similar ideas has been canvassed by Robert Sinsheimer, a molecular biologist at the University of California, at a small conference in 1985 at Santa Cruz.

Del Rio, Andrès Manuel (1764–1849)
Spanish mineralogist

Del Rio (del-**ree**-oh) was born in the Spanish capital Madrid and graduated in Spain in 1781 before going on to study in France, England, and Germany, where he was a pupil of Abraham WERNER at the Freiberg Mining Academy. He had been chosen by Charles III to develop and modernize the mining industry in the Spanish empire. Consequently he was sent to Mexico City to become, in 1794, professor of mineralogy at the School of Mines set up by Fausto D'Elhuyar. While in Mexico he published the *Elementos de orictognosia* (1795; Principles of the Science of Mining), possibly the first mineralogical textbook published in the Americas. He was forced into exile in the period 1829–34 after Mexico's war of independence but on his return he tried to reestablish the scientific tradition he had first introduced.

As a scientist he is best remembered for his independent discovery of the element vanadium in 1801. He had found what he took to be a new metal in some lead ore from the Mexican mines and named it "erythronium" (from the Greek *erythros*, red) as its salts turned red when ignited. However, he failed to press his claim, being persuaded by other scientists that it was probably a compound of lead and chromium. Nils Gabriel SEFSTRÖM rediscovered the metal in 1830 and named it vanadium. Its identity with Del Rio's erythronium was demonstrated by Friedrich WÖHLER in 1831.

De Luc, Jean André (1727–1817) *Swiss geologist and meteorologist*

> [De Luc] was probably one of the most accurate observers of nature that ever existed.
> —John Frederic Daniell, *Meteorological Essays and Observations* (1823)

De Luc (de look) came from an Italian family, which had moved to Switzerland from Tuscany in the 15th century; he was born in the Swiss lakeside city of Geneva. He initially concentrated on commercial activities with science as a side line but, in 1773, after the collapse of his business, he moved to England where he devoted himself to science. He was appointed as reader to Queen Charlotte, retaining that post until his death.

In a series of letters *Sur l'histoire de la terre* (On the History of the Earth) addressed to Queen Charlotte in 1779, James HUTTON in 1790, and Johann BLUMENBACH in 1798, De Luc, following in the tradition of Thomas BURNET, tried to write a history of the Earth that took account of the advances in geology yet was still compatible with the Creation as described in Genesis.

De Luc proposed that the Earth itself was old though the flood was recent. The flood was caused by a collapse of the existing lands causing their inundation by the oceans and the emergence of the present continents. As these had been the prediluvial ocean floor it was only reasonable to suppose that they should contain marine fossils. De Luc thus explained one of the puzzles facing early geologists – the presence of marine fossils in the center of continents.

De Luc opposed Hutton's fluvial theory that such major terrestrial features as valleys are the result of the still continuing action of the rivers. He pointed out that many valleys contain no rivers, that rivers far from eroding actually deposit material, and that there seems to be no relation between the size of the river and the valley it is supposed to have created. His main objection was over downstream lakes, for in this case, when the enormous amount of material eroded from the valley is considered, De Luc argued that the lake should have been filled in long before. Hutton's unsatisfactory answer was that such infilling does take place but that the lakes are much younger than the rivers. This issue was not finally resolved until

the crucial role of glaciation was established by Louis AGASSIZ some fifty years later.

De Luc was also a major figure in meteorological research. His two works, *Recherches sur la modification de l'atmosphère* (1772; Studies on Atmospheric Change) and *Idées sur la météorologie* (1786–87; Thoughts on Meteorology), made important suggestions for advances in instrumental design. His most important achievement was his formula, in 1791, for converting barometric readings into height, which provided the first accurate measurements of mountain heights.

Demarçay, Eugene Anatole (1852–1904) *French chemist*

Born in Paris, France, Demarçay (de-mar-**say**) was a research chemist who also maintained his own private laboratory in the city. In 1901, while working with a sample of the newly discovered element samarium (discovered by Paul Lecoq de Boisbaudran in 1879) he found traces of an additional element, europium. More dramatic, however, was his earlier work with the Curies in 1898. They had discovered one new radioactive element, polonium, early in the year. But they found further radioactivity in their sample of pitchblende after the removal of polonium. They took the small sample they had extracted to Demarçay, an expert spectroscopist, who was able to find a new line in the spectrum. This enabled the Curies to announce the existence of a much more strongly radioactive element, radium.

Demerec, Milislav (1895–1966) *Croatian–American geneticist*

Demerec (**dem**-er-ek) was born in Kostajnica in Croatia and graduated from the College of Agriculture in Krizevci in 1916. After a few years' work at the Krizevci Experimental Station, he moved to America. He gained his PhD in genetics from Cornell University in 1923 and then worked at the Carnegie Institution, Cold Spring Harbor, where he remained for most of his career, becoming director in 1943.

Demerec was concerned with gene structure and function, especially the effect of mutations. He found that certain unstable genes are more likely to mutate than others and that the rate of mutation is affected by various biological factors, such as the stage in the life cycle. He also demonstrated that chromosome segments that break away and rejoin in the wrong place may cause suppression of genes near the new region of attachment. This lent additional support to the idea of the "position effect," first demonstrated by Alfred STURTEVANT.

Demerec's work with the bacterium *Salmonella* revealed that genes controlling related functions are grouped together on the chromosome rather than being randomly distributed through the chromosome complement. Such units were later termed "operons." His radiation treatment of the fungus *Penicillium* yielded a mutant strain producing much larger quantities of penicillin – a discovery of great use in World War II. He showed that antibiotics should be administered initially in large doses, so that resistant mutations do not develop, and should be given in combinations, because any bacterium resistant to one is most unlikely to have resistance to both.

Demerec greatly increased the reputation of Cold Spring Harbor while director there and also served on many important committees. He founded the journal *Advances in Genetics* and wrote some 200 scientific articles.

Democritus of Abdera (*c.* 460 BC–*c.* 370 BC) *Greek philosopher*

> Everything existing in the Universe is the fruit of chance and necessity.
> —Quoted by Diogenes Laertius (3rd century)

Democritus (de-**mok**-ri-tus) was reputedly a prolific author but only fragments of his work still exist and little is known about his life. He is believed to have been born a wealthy citizen of Abdera and to have traveled in Egypt and Persia. He wrote on many subjects, and was reputedly the most learned man of his time.

Democritus is best known for his atomic theory. Despite the fact that LEUCIPPUS is generally regarded as the originator of the atomic theory, and the difficulty in separating Democritus's contribution from that of the later EPICURUS and LUCRETIUS, Democritus was acknowledged by ARISTOTLE (the principle source for Democritus's ideas) to be the leading exponent of the theory. In the "classical" atomic theory, coming into being and dissolution were explained by the linking and flying apart of small hard indestructible particles. Conservation of matter was recognized ("nothing is created out of nothing") and the important doctrine of primary and secondary qualities (later taken up by GALILEO and John Locke) was enunciated by Democritus in the memorable aphorism: "Ostensibly there are sweet and bitter, hot and cold, and color; in reality only atoms and the void." The primary qualities were size, shape, and position. Whether or not Democritus attributed weight to atoms is controversial. The theory was deterministic in that the atomic interactions were thought to be ordered by "necessity." Atomism was ignored from the time of Aristotle until the mid-17th century when it was reintroduced by Pierre GASSENDI and Galileo.

De Moivre, Abraham (1667–1754) *French mathematician*

> The manner of his [De Moivre's] death has a certain interest for psychologists. Shortly before it he declared that it was necessary for him to sleep some ten minutes or quarter of an hour longer each day than the preceding one. The day after he had thus reached a total of something over twenty-three hours he slept up to the limit of twenty-four hours, and then died in his sleep.
> —W. W. Rouse Ball, *A Short Account of the History of Mathematics*

Although born at Vitry in France, as a Huguenot De Moivre (de **mwah**-vre) was forced to flee to England to escape the religious persecution that flared up in 1685 after the revocation of the Edict of Nantes. In England he came to know both Isaac NEWTON and Edmond HALLEY, eventually becoming a fellow of the Royal Society of London himself in 1697.

De Moivre made important contributions to mathematics in the fields of probability and trigonometry. His interest in probability was no doubt stimulated by the fact that despite his abilities he was unable to find a permanent post as a mathematician and so was forced to earn his living by, among other things, gambling. De Moivre was the first to define the concept of statistical independence and to introduce analytical techniques into probability. His work on this was published in *The Doctrine of Chances* (1718), later followed by *Miscellanea analytica* (1730; Analytical Miscellany). De Moivre also introduced the use of complex numbers into trigonometry. *De Moivre's theorem* is the relationship $(\cos A + i \sin A)^n = \cos nA + i \sin nA$.

Dempster, Arthur Jeffrey (1886–1950) *Canadian–American physicist*

Dempster was born in Toronto, Ontario, in Canada and educated at the university there. He emigrated to America in 1914, attended the University of Chicago, obtained his PhD in 1916, and began teaching in 1919. In 1927, he was made professor of physics.

He is noted for his early developments of and work with the mass spectrograph (invented by Francis W. Aston). In 1935, he was able to show that uranium did not consist solely of the isotope uranium–238, for seven out of every thousand uranium atoms were in fact uranium–235. It was this isotope, ^{235}U, that was later predicted by Niels BOHR to be capable of sustaining a chain reaction that could release large amounts of atomic fission energy.

Derham, William (1657–1735) *British physicist*

> Let us ransack all the globe, let us with the greatest accuracy inspect every part thereof, search out the innermost secrets of any of the creatures; let us examine them with all our gauges ... pry into them with all our microscopes and most exquisite instruments, till we find them to bear testimony to their infinite workman.
> —*Physico-theology* (1713)

Born at Stoughton in Worcestershire, England, and educated at Trinity College, Oxford, Derham was ordained in 1682. He was appointed to the living of Upminister where he remained for the rest of his life.

Derham is best known for his attempt to measure the speed of sound. Marin MERSENNE in 1640 had claimed a value of 1,038 feet per second while NEWTON, in the first edition of *Principia* (1687), had calculated it to be 968 feet per second. In 1705 Derham observed from the tower of his Upminiser church the flash of cannons being fired 12 miles away across the Thames at Blackheath. By timing the interval between the flash and roar of the cannon he was able to calculate the speed of sound to be 1,142 feet per second, a result in good agreement with the 1,130 feet per second at 20°C given in modern textbooks. In the second edition of his *Principia* (1713), Newton revised his calculation in the light of Derham's published results.

Derham was also the author of two immensely popular works: *Physico-theology* (1713) and *Astro-theology* (1715). Based on his BOYLE lectures, they set out to show that the basic facts of Newtonian mechanics and cosmology were convincing evidence for the "being and attributes of God."

Also known as an editor, Derham published a number of posthumous works of John RAY as well as *The Philosophical Experiments* (1726) of Robert HOOKE.

Desaguliers, John Theophilus (1683–1744) *French–English physicist*

Desaguliers (day-sa-goo-**lyay**) was born into a Huguenot family in La Rochelle, France. The family was forced by religious persecution to flee to England; John began giving popular lectures on science and its applications in Oxford in 1710. In 1713 he moved to London where he continued to lecture and became experimental assistant to Isaac NEWTON until Newton's death in 1727. He is important in spreading Newtonian theory both in England and on the Continent. He was also an experimenter – particularly on the flow of electricity, being the first to use the terms "conductor" and "insulator" – and an inventor, improving the design of Thomas SAVERY's steam engine by adding a safety valve and an internal water jet.

Desargues, Girard (1591–1661) *French mathematician and engineer*

Little is known of the early life of Desargues (day-**zarg**) except that he was born in Lyons in France. He did serve as an engineer at the siege of La Rochelle (1628) and later became a technical adviser to Cardinal de Richelieu and the French government. He is said to have known René DESCARTES.

Around 1630 Desargues joined a group of mathematicians in Paris and concentrated on geometry. In his most famous work, *Brouillon projet d'une atteinte aux événements des rencontres d'une cône avec un plan* (1639; Proposed Draft of an Attempt to Deal with the Events of the Meeting of a Cone with a Plane), he applied projective geometry to conic sections. *Desargues's theorem* states that if the corresponding points of two triangles in nonparallel planes in space are joined by three lines that intersect at a single point, then the pairs of lines that are the extensions of corresponding sides will each intersect on the same line. Blaise PASCAL was greatly influenced by Desargues, whose contribution to projective geometry was not recognized until a handwritten copy of his work was found in 1845. This oversight probably arose because he used obscure botanical symbols instead of the better-known Cartesian symbolism.

Descartes, René du Perron (1596–1650) *French mathematician, philosopher, and scientist*

> Cogito, ergo sum. (I think, therefore I am.)
> —*Discours de la méthode* (1637; Discourse on Method)

> It is contrary to reason to say that there is a vacuum or space in which there is absolutely nothing.
> —*Principia philosophiae* (1644; Principles of Philosophy)

Descartes (day-**kart**) was the son of a counselor of the Britanny *parlement*; his mother, who died shortly after his birth, left him sufficient funds to make him financially independent. Born at La Haye in France, he was educated by the Jesuits of La Flèche (1604–12) and at the University of Poitiers, where he graduated in law in 1616. For the next decade Descartes spent much of his time in travel throughout Europe and in military service, first with the army of the Prince of Orange, Maurice of Nassau, and later with the Duke of Bavaria, Maximilian, with whom he was present at the battle of the White Mountain outside Prague in 1620. In the years 1628–49 Descartes settled in the freer atmosphere of Holland. There, living quietly, he worked on the exposition and development of his system. Somewhat unwisely, he allowed himself to be enticed into the personal service of Queen Christina of Sweden in Stockholm in 1649. Forced to indulge the Queen's passion for philosophy by holding tutorials with her at 5 a.m. on icy Swedish mornings Descartes, who normally loved to lie thinking in a warm bed, died within a year from pneumonia and the copious bleeding inflicted by the enthusiastic Swedish doctors.

Descartes is in many ways, in mathematics, philosophy, and science, the first of the moderns. The moment of modernity can be dated precisely to 10 November 1619, when, as later described in his *Discours de la méthode* (1637; Discourse on Method), he spent the whole day in seclusion in a *poêle* (an overheated room). He began systematically to doubt all supposed knowledge and resolved to accept only "what was presented to my mind so clearly and distinctly as to exclude all ground of doubt."

Descartes thus managed to pose in a single night the problem whose solution would obsess philosophers for the next 300 years. The same night also provided him with one of the basic insights of modern mathematics – that the position of a point can be uniquely defined by coordinates locating its distance from a fixed point in the direction of two or more straight lines. This was revealed in his *La Geométrie* (1637; Geometry), published as an appendix to his *Discourse*, and describing the invention of analytic or coordinate geometry, by which the geometric properties of curves and figures could be written as and investigated by algebraic equations. The system is known as a *Cartesian coordinate* system.

His theories on physics were published in his *Principia philosophiae* (1644; Principles of Philosophy). "Give me matter and motion and I will construct the universe," Descartes had proclaimed. The difficulty for him arose from his account of matter which, on metaphysical grounds, he argued, "does not at all consist in hardness, or gravity or color or that which is sensible in another manner, but alone in length, width, and depth," or, in other words, extension. From this initial handicap Descartes was forced to deny the existence of the void and face such apparently intractable problems as how bodies of the same extension could possess different weights. With such restrictions he was led to describe the universe as a system of vortices. Matter came in three forms – ordinary matter opaque to light, the ether of the heavens transmitting light, and the subtle particles of light itself. With considerable ingenuity and precious little concern for reality Descartes used such a framework within which he was able to deal with the basic phenomena of light, heat, and motion. Despite its initial difficulties it was developed by a generation of Cartesian disciples to pose as a viable alternative to the mechanics worked out later in the century by NEWTON. Unlike many less radical thinkers Descartes did not shrink from applying his mechanical principles to physiology, seeing the human body purely in terms of a physicomechanical system with the mind as a separate entity interacting with the body via the pineal gland – the supposed seat of the soul. The fundamental impact of Descartes's work was basically one of demystification. Apart from the residual enigma of the precise relationship between mind and body, the main areas of physics and physiology had been swept clear of such talk as that of occult powers and hidden forms.

Desch, Cyril Henry (1874–1958) *British metallurgist*

Desch, the son of a London surveyor, was educated at King's College, London. He taught in Glasgow from 1909 until 1920, when he took up an appointment at the University of Sheffield where he served as professor of metallurgy from 1920 to 1931. Desch then moved to the National Physical Laboratory at Teddington, Middlesex, where he was in charge of the metallurgy department until his retirement in 1939.

Desch is mainly known for his publication in 1910 of his *Textbook of Metallography*, a work that served as the standard account of the subject for the first half of the century.

de Sitter, Willem (1872–1934) *Dutch astronomer and mathematician*

De Sitter (de **sit**-er), the son of a judge from Leiden in the Netherlands, studied mathematics and physics at the University of Groningen, his interest in astronomy being aroused by Jacobus KAPTEYN. After serving at the Cape Town Observatory from 1897 to 1899 and, back at Groningen, as assistant to Kapteyn from 1899 to 1908, he was appointed to the chair of astronomy at the University of Leiden. He also served as director of Leiden Observatory from 1919 to 1934.

De Sitter is remembered for his proposal in 1917 of what came to be called the "de Sitter universe" in contrast to the "Einstein universe." EINSTEIN had solved the cosmological equations of his general relativity theory by the introduction of the cosmological constant, which yielded a static universe. But de Sitter, in 1917, showed that there was another solution to the equations that produced a static universe if no matter was present. The contrast was summarized in the statement that Einstein's universe contained matter but no motion while de Sitter's involved motion without matter.

The Russian mathematician Aleksandr FRIEDMANN in 1922 and the Belgian Georges LEMAÎTRE independently in 1927 introduced the idea of an expanding universe that contained moving matter. It was then shown in 1928 that the de Sitter universe could be transformed mathematically into an expanding universe. This model, the "Einstein–de Sitter universe," comprised normal Euclidean space and was a simpler version of the Friedmann–Lemaître models in which space was curved.

De Sitter also spent much time trying to calculate the mass of Jupiter's satellites from the small perturbations in their orbits. The results were published in 1925 in his *New Mathematical Theory of Jupiter's Satellites*.

Desmarest, Nicolas (1725–1815) *French geologist*

Desmarest (day-ma-**ray**) was the son of a school teacher from Soulaines-Dhuys in France. He first came to notice when he won a prize essay set by the Amiens Academy in 1751 on whether England and France had ever been joined together. Working for a while in Paris as an editor of scholarly works, he eventually started work for the department of commerce in 1757 investigating and reporting on various trades and industries. He served as inspector-general of manufactures (1788–91).

In 1763, following the work of Jean GUETTARD, he noticed large basalt deposits and traced these back to ancient volcanic activity in the Auvergne region. He mapped the area and worked out the geology of the volcanoes and their eruptions in great detail, publishing his work in the *Encyclopédie* (Encyclopedia) of 1768. This work disproved the theory that all rocks were sedimentary by revealing basalt's igneous origins. He later produced an influential work, *Géographie physique* (1794; Physical Geography).

Désormes, Charles Bernard (1777–1862) *French chemist*

Désormes (day-**zorm**), who was born at Dijon in France, studied at the Ecole Polytechnique and was an assistant to Guyton de Morveau until 1804. With Jacques and Joseph MONTGOLFIER and his son-in-law, Nicolas Clément, he was coowner of a chemical factory at Verberie.

Clément and Désormes discovered carbon monoxide (1801), investigated the catalytic effect of nitric oxide in the lead-chamber process of sulfuric acid manufacture, and were involved in the early work on iodine, which was discovered by Bernard COURTOIS in 1811. Their most important work, however, was in physical chemistry on the specific heats of gases. In 1819 they published an important paper on the determination of the ratio of the principal specific heats (i.e., the ratio of the specific heat of a gas at constant pressure to that at constant volume).

Deville, Henri Etienne Sainte-Claire (1818–1881) *French chemist*

The son of a wealthy shipowner from the West Indies island of St. Thomas, Deville (de-**veel**) studied medicine in Paris but became interested in chemistry by attending Louis Thenard's lectures. He isolated toluene and methyl benzoate from tolu balsam and investigated other natural products before turning to inorganic chemistry, following his appointment as professor of chemistry at Besançon (1845).

Deville's first major discovery was that of nitrogen pentoxide (1849). Following this success he became professor of chemistry at the Ecole Normale Supérieure (1851) and also lectured at the Sorbonne from 1853. Deville is best known for his work on the large-scale production of aluminum. This had been obtained by Kaspar Wöhler in 1827 but had been produced only in small quantities. Deville developed a commercially successful process involving reduction of aluminum chloride by sodium; the first ingot was produced in 1855. Deville was an expert on the purification of metals and produced (among others) crystalline silicon (1854) and boron (1856), pure magnesium (1857), and pure titanium (1857; with Wöhler). He did much work on the purification of platinum and in 1872 was commissioned to produce the standard kilogram.

After his work on aluminum, Deville's most important researches were those on dissociation. Working with L. J. Troost, he discovered that many molecules were dissociated at high temperature, giving rise to anomalous vapor-density results. Deville's work explained these results and helped to confirm Amedeo AVOGADRO's hypothesis. His other work included the production of artificial gemstones and improved furnaces.

de Vries, Hugo (1848–1935) *Dutch plant physiologist and geneticist*

Born the son of a politician at Haarlem in the Netherlands, de Vries (de vrees) studied botany at Leiden and Heidelberg. He became an expert on the Netherlands flora and later turned his attention from classification to physiology and evolution. He entered Julius von SACHS's laboratory at Würzburg University, where he conducted important experiments on the water of plant cells. He demonstrated that the pressure (turgor) of the cell fluid is responsible for about 10% of extension growth, and introduced the term *plasmolysis* to describe the condition in nonturgid cells in which the cell contents contract away from the cell wall. His work in this field led to Jacobus VAN'T HOFF's theory of osmosis.

During the 1880s, de Vries became interested in heredity. In 1889 he published *Intracellular Pangenesis*, in which he critically reviewed previous research on inheritance and advanced the theory that elements in the nucleus, "pangenes," determine hereditary traits. To investigate his theories, he began breeding plants in 1892 and by 1896 had obtained clear evidence for the segregation of characters in the offspring of crosses in 3:1 ratios. He delayed publishing these results, proposing to include them in a larger book, but in 1900 he came across the work of Gregor MENDEL, published 34 years earlier, and announced his own findings. This stimulated both Karl CORRENS and Erich von Tschermak-Seysenegg to publish their essentially similar observations.

De Vries's work on the evening primrose,

Oenothera lamarckiana, began in 1886 when he noticed distinctly differing types within a colony of the plants. He considered these to be mutants and formulated the idea of evolution proceeding by distinct changes such as those he observed, believing also that new species could arise through a single drastic mutation. He published his observations in *The Mutation Theory* (1901–03). It was later shown that his *Oenothera* "mutants" were in fact triploids or tetraploids (i.e., they had extra sets of chromosomes) and thus gave a misleading impression of the apparent rate and magnitude of mutations. However, the theory is still important for demonstrating how variation, essential for evolution, can occur in a species.

De Vries was professor of botany at Amsterdam from 1878 to 1918 and was elected a fellow of the Royal Society in 1905.

Dewar, Sir James (1842–1923) *British chemist and physicist*

Dewar, the son of a wine merchant, was born at Kincardine-on-Forth in Scotland. He was educated at Edinburgh University where he was a pupil of Lyon PLAYFAIR. In 1869 he was appointed lecturer in chemistry at the Royal Veterinary College, Edinburgh, and from 1873 also held the post of assistant chemist to the Highland and Agricultural Society of Scotland. In 1875 Dewar became Jacksonian Professor of Experimental Philosophy at Cambridge University, England, and from 1877 he was also Fullerian Professor of Chemistry at the Royal Institution, London. He did most of his work in London where the facilities for experimental work were much better.

Dewar conducted his most important work in the field of low temperatures and the liquefaction of gases. In 1878 he demonstrated Louis CAILLETET's apparatus for the liquefaction of oxygen and by 1891 he was able to produce liquid oxygen in quantity. In about 1872 he devised a double-walled flask with a vacuum between its highly reflective walls, the *Dewar flask*, and used this to store liquefied oxygen at extremely low temperatures. This vessel (the thermos flask) has come into everyday use for keeping substances either hot or cold.

Hydrogen had so far resisted liquefaction and Dewar now turned his attention to this. Using the Joule–Thomson effect together with Karl von LINDE's improvements of this, he produced a machine with which he obtained temperatures as low as 14 K and he produced liquid hydrogen in 1898 and solid hydrogen in 1899. Only helium now resisted liquefaction; this was achieved by Heike KAMERLINGH-ONNES in 1908. From about 1891 Dewar also studied explosives and with Frederick ABEL he developed the smokeless powder, cordite. He was knighted in 1904.

Dewar, Michael James Stuart (1918–1997) *British–American chemist*

Born at Ahmednagar in India, Dewar was educated at Oxford University, where he obtained his DPhil in 1942. After research at Oxford he worked in industry as a physical chemist for the Courtauld company from 1945 until his appointment in 1951 as professor of chemistry at Queen Mary College, London. In 1959 Dewar moved to America and served successively as professor of chemistry at the University of Chicago and from 1963 at the University of Texas. In 1990 he was appointed graduate research professor at the University of Florida.

Dewar is noted for his contributions to theoretical chemistry. In his *Electronic Theory of Organic Chemistry* (1949) he argued strongly for the molecular-orbital theory introduced by Robert MULLIKEN. He did much to improve molecular-orbital calculations and by the 1960s he was able to claim that he and his colleagues could rapidly and accurately calculate a number of chemical and physical properties of molecules.

d'Hérelle, Felix (1873–1949) *French–Canadian bacteriologist*

D'Hérelle (day-**rel**), the son of a Canadian father and Dutch mother, was born in Montreal, Quebec, and went to school in Paris, later studying medicine at the University of Montreal. He worked as a bacteriologist in Guatemala and Mexico from 1901 until 1909, when he returned to Europe to take up a position at the Pasteur Institute in Paris. D'Hérelle moved to the University of Leiden in 1921 but after only a short stay resigned to become director of the Egyptian Bacteriological Service (1923). Finally, in 1926, d'Hérelle was appointed to the chair of bacteriology at Yale, a position he held until his retirement in 1933.

D'Hérelle is best known for his discovery of the bacteriophage – a type of virus that destroys bacteria. This work began in 1910 in Yucatan, when he was investigating diarrhea in locusts as a means of locust control.

While developing cultures of the causative agent, a coccobacillus, d'Hérelle found that occasionally there would develop on a culture a clear spot, completely free of any bacteria. The cause of these clean spots became clear to him in 1915, while investigating a more orthodox form of dysentery in a cavalry squadron in Paris. He mixed a filtrate from the clear area with a culture of the dysentery bacilli and incubated the resulting broth overnight. The next morning the culture, which had been very turbid, was perfectly clear: all the bacteria had vanished. He concluded that this was the action of "a filterable virus, but a virus parasitic on bacteria."

A similar discovery of what d'Hérelle termed a "bacteriolytic agent" was announced independently by Frederick TWORT in 1915. D'Hérelle published his own account first in 1917, followed by his monograph *The Bacteriophage, Its Role in Immunity* (1921). He spent the rest of his career attempting to develop bacteriophages as therapeutic agents. Thus he tried to cure cholera in India in 1927 and bubonic plague in Egypt in 1926 by administering to the patients the appropriate phage. D'Hérelle himself claimed good results with his treatment, although in the hands of other workers the effect of phage on such diseases as cholera and plague appeared to be minimal. This conclusion d'Hérelle continued to resist until his death, claiming that no proper test using his methods had ever been carried out.

However, the importance of the bacteriophage as a research tool in molecular biology cannot be disputed. It was the so-called phage group, centered on Max DELBRÜCK, that made many of the early advances in this discipline in the 1940s.

Dicearchus of Messina (about 310 BC)
Greek geographer and philosopher

Dicearchus (dI-see-**ar**-kus) was a pupil of ARISTOTLE and spent most of his life in Sparta. As none of his works have survived it is difficult to be sure of his contributions to geography. He wrote on a large number of topics including the soul, prophecy, and political theory.

His main work in geography was entitled *Periodos ges* (Tour of the Earth) and he also wrote a history of Greek civilization entitled *Bios Hellados* (Life of Greece). As Dicearchus was writing so soon after Alexander the Great's campaigns it is assumed that his works would have contained much new information on the geography of Asia. He is variously reported to have been the first to establish lines of latitude on maps, to have included the heights of mountains, and to have made a reasonable attempt at measuring the size of the Earth using methods that ERATOSTHENES was later to perfect.

Dicke, Robert Henry (1916–1997) *American physicist*

Dicke, who was born in St. Louis, Missouri, graduated in 1939 from Princeton University and obtained his PhD in 1941 from the University of Rochester. He spent the war at the radiation laboratory of the Massachusetts Institute of Technology, joining the Princeton faculty in 1946. In 1957 he was appointed professor of physics and served from 1975 to 1984 as Albert Einstein Professor of Science. In 1984 he was appointed Albert Einstein Emeritus Professor of Science.

In 1964, unaware that he was repeating a line of thought pursued earlier by George GAMOW, Ralph ALPHER, and Robert C. Herman in 1948, Dicke began to think about the consequences of a big-bang origin of the universe. Assuming a cataclysmic explosion some 18 billion years ago with a temperature one minute after of about 10 billion degrees, then intense radiation would have been produced in addition to particles of matter. As the universe expanded this radiation would gradually lose energy. Could there still be any trace left of this "primeval fireball"? It would in fact be detected as black-body radiation, characteristic of the temperature of the black body, which is a perfect emitter of radiation. At Dicke's instigation his colleague P. J. E. Peebles made the necessary calculations and concluded that the remnant radiation should now have a temperature of only about 10 K, later corrected to about 3 K, i.e., −270°C. At this temperature a black body should radiate a weak signal at microwave wavelengths from 0.05 millimeter to 50 centimeters with a peak at about 2 millimeters. Further, the signal should be constant throughout the entire universe.

Dicke began to organize a search for such radiation and had actually begun to install an antenna on his laboratory roof when he heard from Arno PENZIAS and Robert WILSON that they had detected background microwave radiation at a wavelength of 7 centimeters. It was this confluence of theory, calculation, and observation that really established the big-bang theory.

Another major area of study for Dicke is gravitation. In the 1960s he carried out a major evaluation of the experiment originally performed by Roland von EÖTVÖS to confirm that the gravitational mass of a body is equal to its inertial mass. Dicke was able to establish the accuracy of the equivalence to one part in 10^{11}. This equivalence is basic to Einstein's theory of general relativity.

In 1961, following a suggestion of Paul DIRAC in 1937, Dicke and Carl BRANS proposed that the gravitational constant was not in fact a constant, but slowly decreases at a rate of one part in 10^{11} per year. The resulting *Brans–Dicke theory* differs somewhat from Einstein's general relativity at a number of points. Thus while Einstein predicts that a ray of light should be deflected by the Sun's gravitational field 1.75 seconds (″) of arc, the Brans–Dicke theory leads to a figure of 1.62″; such a difference is within the range of observational error and so is not readily detectable. Again the perihelion of Mercury should advance for Einstein by 43″ per century, for Brans–Dicke a mere 39″. A value of 43″ has in fact been measured but Dicke maintains that part of this value, 4″, could be explained by the Sun's nonspherical shape. It has however been claimed that very precise measurements of radio pulses from pulsars appear to favor Einstein. The theory was concurrently and independently developed by Pascual JORDAN, and is thus sometimes known as the

Brans–Dicke–Jordan theory. The idea of a changing gravitational constant was put forward by Paul Dirac.

Diels, Otto Paul Hermann (1876–1954)
German organic chemist

The son of Hermann Diels (deelz), a famous classical scholar, Diels was born in Hamburg, Germany. He gained his doctorate under Emil FISCHER in Berlin (1899), becoming professor there in 1906. From 1916 until his retirement in 1948 he was professor at Kiel. In 1906 he made an extremely unexpected discovery, that of a new oxide of carbon, carbon suboxide (C_3O_2), which he prepared by dehydrating malonic acid with phosphorus pentachloride. Diels's second major discovery was a method of removing hydrogen from steroids by means of selenium. He used this method in research on cholesterol and bile acids, obtaining aromatic hydrocarbons that enabled the structures of the steroids to be deduced.

In 1928 Diels and his assistant Kurt ALDER discovered a synthetic reaction in which a diene (compound containing two double bonds) is added to a compound containing one double bond flanked by carbonyl or carboxyl groups to give a ring structure. The reaction proceeds in the mildest conditions, is of general application, and hence of great utility in synthesis. It has been used in the synthesis of natural products, such as sterols, vitamin K, and cantharides, and of synthetic polymers. For this discovery Diels and Alder were jointly awarded the Nobel Prize for chemistry in 1950.

Diesel, Rudolph Christian Carl (1858–1913)
German engineer and inventor

Diesel (**dee**-zel), the designer of the *diesel engine*, was born to German parents in Paris and brought up there until the age of 12. He was academically talented, but his schooling was interrupted in August 1870, when the Franco-Prussian war broke out and the Diesels were deported to London. His cousin, a teacher in Augsburg, Bavaria, invited Diesel to go there to study and he later won a scholarship to the Munich Institute of Technology.

After graduating, Diesel worked as a mechanic for two years in Switzerland and then worked in Paris as a thermal engineer. He was a devout Lutheran and a dedicated pacifist, believing in international religious liberation. In the laboratory that he set up in 1885, an accident with ammonia gas gave him the idea of using chemical firework-type weapons instead of lethal bombs and bullets on the battlefields. In 1893, he demonstrated his first engine and, although the first few attempts failed, within three years he had developed a pressure-ignited heat engine with an efficiency of 75.6%.

(Equivalent steam engines had an efficiency of 10%.)

By 1898 Diesel was a millionaire but his fortune soon disappeared. He toured the world giving lectures and visited America in 1912. His health was bad, he suffered from gout, and was depressed by the buildup to World War I. On the ferry returning from London in 1913, after dining apparently happily with a friend, he disappeared and was assumed to have drowned in the English Channel.

Diophantus of Alexandria (about 250)
Greek mathematician

> This tomb holds Diophantus. Ah, how great a marvel! The tomb tells scientifically the measure of his life. God granted him to be a boy for the sixth part of his life...five years after his marriage He granted him a son. Alas! late-born wretched child; after attaining the measure of half his father's life, chill Fate took him. After consoling his grief by this science of numbers for four years he ended his life.
> —*The Greek Anthology*

Diophantus (dI-oh-**fan**-tus) was one of the outstanding mathematicians of his era but almost nothing is known of his life and his writings survive only in fragmentary form. His most famous work was in the field of number theory and of the so-called *Diophantine equations* named for him. His major work, the *Arithmetica* (Arithmetic), contained many new methods and results in this field. It originally consisted of 13 books but only 6 survived to be translated by the Arabs. However Diophantus was not solely interested in equations with only integral (whole number) solutions and also considered rational solutions. Diophantus made considerable innovations in the use of symbolism in Greek mathematics – the lack of suitable symbolism had previously hampered work in algebra.

Dioscorides, Pedanius (*c*. 40 AD–*c*. 90 AD)
Greek physician

> If you have not sufficient facility in reading Greek, then you can turn to the herbal of Dioscorides, which describes and draws the herbs of the field with wonderful faithfulness.
> —Cassiodorus (6th century)

Little is known of the life of Dioscorides (dI-os-**kor**-i-deez) except that he was born in Anazarbus (now in Turkey) and became a surgeon to Emperor Nero's armies, having most probably learned his skills at Alexandria and Tarsus. Many writings are attributed to him but the only book for which his authorship is undisputed is *De materia medica* (On Medicine). This pharmacopeia remained the standard medical text until the 17th century, undergoing many revisions and additions and greatly influencing both Western and Islamic cultures. It describes animal derivatives and minerals used therapeutically but is most important for the description of over 600 plants, including notes on their habitat and the methods of preparation and medicinal use of the drugs they contain. Many of the common and scientific plant names in use today originate from Dioscorides and the yam family, Dioscoreaceae, is named for him.

Dirac, Paul Adrien Maurice (1902–1984)
British mathematician and physicist

> It seems that if one is working from the point of view of getting beauty in one's equations, and if one has really a sound insight, one is on a sure line of progress.
> —*Scientific American*, May 1963

Dirac (di-**rak**), whose father was Swiss, was born in Bristol in the west of England. After graduating in 1921 in electrical engineering at Bristol University, Dirac went on to study mathematics at Cambridge University, where he obtained his PhD in 1926. After several years spent lecturing in America, he was appointed (1932) to the Lucasian Professorship of Mathematics at Cambridge, a post he held until his retirement in 1969. In 1971 he became professor of physics at Florida State University.

Dirac is acknowledged as one of the most creative of the theoreticians of the early 20th century. In 1926, slightly later than Max BORN and Pascual JORDAN in Germany, he developed a general formalism for quantum mechanics. In 1928 he produced his relativistic theory to describe the properties of the electron. The wave equations developed by Erwin SCHRÖDINGER to describe the behavior of electrons were nonrelativistic. A significant deficiency in the Schrödinger equation was its failure to account for the electron spin discovered in 1925 by Samuel GOUDSMIT and George UHLENBECK. Dirac's rewriting of the equations to incorporate relativity had considerable value for it not only predicted the correct energy levels of the hydrogen atom but also revealed that some of those levels were no longer single but could be split into two. It is just such a splitting of spectral lines that is characteristic of a spinning electron.

Dirac also predicted from these equations that there must be states of negative energy for the electron. In 1930 he proposed a theory to account for this that was soon to receive dramatic confirmation. He began by taking negative energy states to refer to those energy states below the lowest positive energy state, the ground state. If there were a lower energy state for the electron below the ground state then, the question arises, why do some electrons not fall into it? Dirac's answer was that such states have already been filled with other electrons and he conjured up a picture in which space is not really empty but full of particles of negative energy. If one of these particles were to collide with a sufficiently energetic photon it would acquire positive energy and be observable as a normal electron, apparently appearing from nowhere. But it would not appear alone for it would leave behind an empty hole, which was really an absence of a negatively charged particle or, in other words, the presence of a positively charged particle. Further, if the electron were to fall back into the empty hole it would once more disappear, appearing to be annihilated together with the positively charged particle, or positron as it was later called.

Out of this theory there emerged three predictions. Firstly, that there was a positively charged electron, secondly, that it could only appear in conjunction with a normal electron, and, finally, that a collision between them resulted in their total common annihilation. Such predictions were soon confirmed following the discovery of the positron by Carl ANDERSON in 1932. Dirac had in fact added a new dimension of matter to the universe, namely antimatter. It was soon appreciated that Dirac's argument was sufficiently general to apply to all particles.

In 1937 Dirac published a paper entitled *The Cosmological Constants* in which he considered "large-number coincidences," i.e., certain relationships that appear to exist between the numerical properties of some natural constants. An example is to compare the force of electrostatic attraction between an electron and a proton with the gravitational attraction due to their masses. The ratio of these is about $10^{40}:1$. Similarly, it is also found that the characteristic "radius" of the universe is 10^{40} times as large as the characteristic radius of an electron. Moreover, 10^{40} is approximately the square root of the number of particles in the universe.

These coincidences are remarkable and many physicists have speculated that these apparently unrelated things may be connected in some way. The ratios were first considered in the 1930s by Arthur EDDINGTON, who believed that he could calculate such constants and that they arose from the way in which physics observes and interprets nature. Dirac used the 10^{40} number above in a model of the universe. He argued that there was a connection between the force ratio and the radius ratio. Since the radius of the universe increased with age the gravitational "constant," on which the force ratio depends, may decrease with time. There is modern tentative evidence that this could be the case.

Above all else, however, Dirac was a quantum theorist. In 1930 he published the first edition of his classic work *The Principles of Quantum Mechanics*. In 1933 he shared the Nobel Prize for physics with Schrödinger.

Dirichlet, (Peter Gustav) Lejeune (1805–1859) *German mathematician*

> In mathematics as in other fields, to find

oneself lost in wonder at some manifestation is frequently the half of a new discovery.
—*Works*, Vol. II

Born in Düren (now in Germany), Dirichlet (dee-ri-**klay**) studied mathematics at Göttingen where he was a pupil of Karl GAUSS and Karl JACOBI. He also studied briefly in Paris where he met Joseph FOURIER, who stimulated his interest in trigonometric series. In 1826 he returned to Germany and taught at Breslau and later at the Military Academy in Berlin. He then moved to the University of Berlin, which he only left 27 years later when he returned to Göttingen to fill the chair left vacant by Gauss's death.

Dirichlet's work in number theory was very much inspired by Gauss's great work in that field, and Dirichlet's own book, the *Vorlesungen über Zahlentheorie* (1863; Lectures on Number Theory), is of comparable historical importance to Gauss's *Disquisitiones*. He made many very significant discoveries in the field and his work on a problem connected with primes led him to make the fundamentally important innovation of using analytical techniques to obtain results in number theory.

His stay in Paris had stimulated Dirichlet's interest in Fourier series and in 1829 he was able to solve the outstanding problem of stating the conditions sufficient for a Fourier series to converge. (The other problem of giving necessary conditions is still unsolved.) Fourier also gave the young Dirichlet an interest in mathematical physics, which led him to important work on multiple integrals and the boundary-value problem, now known as the *Dirichlet problem*, concerning the formulation and solution of those partial differential equations occurring in the study of heat flow and electrostatics. These are of great importance in many other areas of physics. The growth of a more rigorous understanding of analysis owes to Dirichlet what is essentially the modern definition of the concept of a function.

Djerassi, Carl (1923–) *American chemist*

Djerassi (jer-**as**-i) was born in the Austrian capital of Vienna, the son of a Bulgarian physician and an Austrian mother. As both parents were Jewish, Djerassi emigrated to America in 1939. He was educated at Kenyon College, Ohio, and at the University of Wisconsin, where he completed his PhD in 1945, the same year in which he became an American citizen. From 1945 to 1949 he worked for the pharmaceutical company CIBA in Summit, New Jersey, as a research chemist. In 1949 Djerassi decided to join a new pharmaceutical company, Syntex, in Mexico City, to work on the extraction of cortisone from plants. At that time it was being produced from cattle bile at a cost of $200 a gram. Despite competition from other leading laboratories, Syntex was the first to extract cortisone ($C_{21}H_{28}O_5$) from a vegetable source, namely diosgenin ($C_{27}H_{42}O_3$), a steroid derived from a variety of wild Mexican yam. Following their initial success Djerassi and his team turned their attention to the steroid hormone progesterone. Known as "nature's contraceptive," the hormone inhibits ovulation. Why, then, could it not be taken as a simple, natural contraceptive? The difficulty was that taken orally it lost most of its activity. Further, as hormones were extracted from such animal sources as human urine, bull's testicles, and sow's ovaries where they occur in small amounts, they tended to be very expensive. The first step was to produce progesterone synthetically. This was achieved at Syntex by Djerassi and others in the early 1950s, the price of progesterone dropped dramatically and it became available in large quantities.

Chemists were inhibited by the belief that steroid hormones were structure-specific; change the structure, the claim went, and the potency is lost. Djerassi was aware, however, that Max Ehrenstein had destroyed this myth a decade earlier and that it was at least conceivable that progesterone could be changed into an oral form without necessarily changing its potency.

Progesterone ($C_{21}H_{30}O_2$) contains four rings of carbon atoms. Following some hints in the literature Djerassi thought that the removal of the methyl group at position 19, thus forming 19-norprogesterone, would increase its potency. His hunch proved to be sound. He was also aware that an acetylene bond introduced into position 17 of the male hormone testosterone increased its oral activity; although known as "ethisterone," it had found no use.

Djerassi's crucial step was to propose that ethisterone's potency could be enhanced, as with progesterone, by removing a methyl group. By October 1951 he had produced testosterone minus a methyl group, but with an added acetylene group. The precise result was 19-nor-17a-ethinyltestosterone, which proved to be a highly active oral progestational hormone. A patent was filed in November 1951. After the appropriate testing it received Federal approval in 1962 under the name Ortho-Novum. Djerassi received one dollar for the patent, a standard payment by a pharmaceutical company to its staff. In 1951 Djerassi left Syntex for Wayne State University, Detroit, where he remained until 1959 when he was appointed professor of chemistry at Stanford. He continued to work for Syntex, as vice-president in charge of research (1957–69) and as president of research (1969–72).

He has also served since 1968 with Zoecon, a company partly owned by Syntex and specializing in pest control by using natural juvenile hormones that prevent insects maturing and breeding. In 1977 Zoecon was taken over by

Occidental Petroleum and was then sold in 1982 to Sandoz, a Swiss pharmaceutical company. Djerassi has remained as chairman of the board.

Djerassi's business interests have had little impact on his productivity with over 600 papers to his credit. He has also written four science-fiction novels (*Cantor's Dilemma, The Bowbaki Gambit, Menachem's Seed,* and *No*) along with a collection of verse, and his autobiography, *The Pill, Pigmy Chimps, and Degas' Horse* (1992). He has also written and collaborated on plays.

Döbereiner, Johann Wolfgang (1780–1849) *German chemist*

> I love science more than money.
> —Explaining his failure to patent his invention of the Döbereiner lamp

Born the son of a coachman in Hof an der Saale, Germany, Döbereiner (**du(r)**-be-rIn-er) had little formal education and worked as an assistant to apothecaries in several places from the age of 14. He was largely self-taught in chemistry and was encouraged by Leopold GMELIN whom he met at Strasbourg. After several failures in business, he was appointed assistant professor of chemistry at Jena (1810).

In 1823 he discovered that hydrogen would ignite spontaneously in air over platinum sponge, and subsequently developed the *Döbereiner lamp* to exploit this phenomenon. Döbereiner was interested in catalysis in general and discovered the catalytic action of manganese dioxide in the decomposition of potassium chlorate. His law of triads (1829), based on his observation of regular increments of atomic weight in elements with similar properties, was an important step on the way to Dmitri MENDELEEV's periodic table. Thus in triads such as calcium, strontium, and barium or chlorine, bromine, and iodine, the middle element has an atomic weight that is approximately the average of the other two. It is also intermediate in chemical properties between the other two elements. Döbereiner also worked in organic chemistry.

Dobzhansky, Theodosius (1900–1975) *Russian–American geneticist*

> Nothing in biology makes sense except in the light of evolution.
> —Title of article in the *American Biology Teacher* (1973)

Dobzhansky (dob-**zhan**-ski), who was born in Nemirov, in Ukraine, graduated in zoology from Kiev University in 1921; he remained there to teach zoology before moving to Leningrad, where he taught genetics. In 1927 he took up a fellowship at Columbia University, New York, where he worked with T. H. MORGAN. Morgan was impressed by Dobzhansky's ability and, when the fellowship was completed, offered him

a teaching post at the California Institute of Technology. Dobzhansky accepted and became an American citizen in 1937.

Dobzhansky studied the fruit fly (*Drosophila*) and demonstrated that the genetic variability within populations was far greater than had been imagined. The high frequency of potentially deleterious genes had previously been overlooked because their effects are masked by corresponding dominant genes. Dobzhansky found that such debilitating genes actually conferred an advantage to the organism when present with the normal type of gene, and therefore they tended to be maintained at a high level in the population. Populations with a high genetic load – i.e., many concealed lethal genes – proved to be more versatile in changing environments. This work profoundly influenced the theories on the mathematics of evolution and natural selection with regard to Mendelism.

In addition, Dobzhansky wrote many influential books, including *Genetics and the Origin of Species* (1937), a milestone in evolutionary genetics.

Doherty, Peter Charles (1940–) *Australian immunologist*

Doherty was educated at Queensland and later at Edinburgh, where he gained his PhD in 1970. After serving at the Wistar Institute in Philadelphia, he was appointed in 1982 professor of experimental pathology at the Curtin Institute, Canberra. In 1988 Doherty moved to St. Jude's Children's Research Hospital, Memphis, as chairman of the Immunology Department. From 2002 he became a professor in the Department of Microbiology and Immunology at the University of Melbourne.

In 1974 Doherty, in collaboration with Rolf ZINKERNAGEL, began to consider the response of the mouse immune system to viral meningitis. It was commonly held at this time that the presence of bacterial or viral invaders were sufficient in themselves to initiate an immune response. Consequently, as expected, the infected cells were attacked by the mouse's own T lymphocytes. Yet, to their surprise, when T cells from one mouse strain were deployed against infected cells from another strain, the lymphocytes failed to respond. Clearly an additional factor was required to trigger the immune response.

They were aware, following the work of George SNELL and others, that major histocompatibility (MHC) antigens played a significant role in controlling the immune response to transplants. They consequently began to examine the role of MHC proteins in the immune response. Working with mice from various strains, they found that T cells from one strain could be provoked to attack infected cells only if the invaders shared at least one MHC antigen.

For his work in this field Doherty shared the 1997 Nobel Prize for physiology or medicine with Rolf Zinkernagel. He is currently working on viral immunity. In 1995 he published an account of science: *The Beginner's Guide to Winning the Nobel Prize*.

Doisy, Edward Adelbert (1893–1986)
American biochemist

Doisy was born in Hume, Illinois, and educated at the University of Illinois and at Harvard, where he obtained his PhD in 1920. From 1919 to 1923 he worked at the Washington University School of Medicine. In 1923 he was appointed to the chair of biochemistry at the St. Louis University Medical School, a position he retained until his retirement in 1965.

Doisy worked initially on ovarian biochemistry. In 1929 he succeeded in isolating the hormone estrone ($C_{18}H_{12}O_2$), and soon after the more potent estradiol ($C_{18}H_{24}O_2$). In 1938 he isolated, and in the following year synthesized, vitamin K, recently discovered by Carl DAM. He discovered that the vitamin existed in two forms, the physiologically active form K_1, extracted from alfalfa, and K_2, differing in a side chain and derivable from rotten fish. For his discovery of the chemical nature of vitamin K, technically a naphthoquinone, Doisy shared the 1943 Nobel Prize for physiology or medicine with Dam.

Dokuchaev, Vasily Vasilievich (1846–1903) *Russian soil scientist*

> Climate, vegetation, and animal life are distributed on the earth's surface, from north to south, in a strictly determined order.
> —*Collected Works,* Vol. VI

Dokuchaev (dok-uu-**cha**-ef) was born in Milyukovo near Smolensk, Russia, the son of the village priest. He too was originally trained for the priesthood but later turned to the study of science at St. Petersburg University where he graduated in 1871. He was immediately appointed to the faculty, initially as curator of the geological collection but he also served as professor of geology until poor health forced him to retire in 1897.

Dokuchaev made the first comprehensive scientific study of the soils of Russia, details of which are to be found in his *Collected Works* (9 vols., 1949–61). He also, in the 1890s, set up at the Kharkov Institute of Agriculture and Forestry, the first department of soil science in Russia.

In the West he is mainly known for his work on the classification of soils, his insistence that soil is a geobiological formation, and his use of soil to define the different geographical zones. It is also owing to Dokuchaev that the Russian term "chernozem," used to describe a black soil rich in humus and carbonates, has entered most languages.

Doll, Sir (William) Richard Shaboe (1912–2005) *British epidemiologist*

Doll was born in Hampton, Greater London, and in 1937 graduated from St. Thomas' Hospital Medical School in London. Following the outbreak of World War II in 1939, Doll served in the Royal Army Medical Corps. In 1946 he started working for the statistical research unit of the Medical Research Council (MRC), becoming its director in 1961. In 1969 he was appointed Regius Professor of Medicine at Oxford University.

Doll was a key figure in the development of modern epidemiology and preventive medicine, most notably through his confirmation of the causal link between smoking and lung cancer. In 1951 he and colleagues at the MRC published strong evidence that smoking is associated with lung cancer: they surveyed 649 such patients and found that only 2 were nonsmokers. So convincing were the results that Doll himself gave up smoking while carrying out the study. Doll's team repeated their survey using the much larger sample of all 40,000 doctors in the UK. The results, published in 1954, prompted a sea change in public attitudes to smoking. Doll's statistical epidemiological approach had produced convincing evidence that people who smoked ran a greatly increased risk of contracting lung cancer.

In subsequent work, Doll turned to numerous other public health issues, including the hazards of oral contraceptives, electrical power lines, radiation, and environmental contaminants. He was no stranger to controversy, often seeming to downplay the cancer risks posed by certain chemicals and forms of radiation, and instead arguing that smoking and other lifestyle factors were preeminent as factors. Nonetheless his work earned him many honors and awards, including the United Nations Award for Cancer Research in 1962 and the European Cancer Society's Gold Medal in 2000. He received a knighthood in 1971.

Dollfus, Audouin Charles (1924–)
French astronomer

Dollfus (**dol**-foos), the son of an aeronaut, studied at the university in his native city of Paris. In 1946 he joined the staff of the astrophysical division of the Paris Observatory at Meudon and is now head of the Laboratory for Physics of the Solar System.

Dollfus has established a reputation as an authority on the solar system and a leading planetary observer. He began a study of Saturn's rings in 1948 and soon noted the presence of occasional brightness ripples. Work in the

1950s showed that they could not be explained by the effects of the known satellites and Dollfus concluded that other forces seemed to be acting, possibly attributable to an unknown satellite very close to the rings. A favorable viewing time, with the rings appearing edge-on to an Earth-bound observer, came in 1966. Although early visual attempts to detect the satellite failed, Dollfus was more successful when he examined photographic plates taken with the 43-inch (1.1-m) reflecting telescope of the Pic du Midi Observatory in December 1966, virtually the last opportunity for favorable viewing before 1980. On these he found the image of an unknown satellite.

Following independent confirmation, the satellite was named Janus and was widely accepted as Saturn's tenth satellite, its period having been calculated by Dollfus as 18 hours. When no such satellite was found by the Pioneer II and Voyager I spacecraft in 1979 and 1980, Dollfus rechecked his measurements. It now appears that he photographed one of the "twin" satellites that appeared in the spacecraft pictures and that recent calculations have shown to move in nearly identical orbits with periods of about 16.7 hours.

Döllinger, Ignaz Christoph von (1770–1841) German biologist

After studying medicine in his native city of Bamberg, then at Würzburg, Vienna, and Pavia (Italy), Döllinger (**du(r)l**-ing-er) gained his doctorate at Bamberg in 1794. In 1796 he was appointed professor of medicine at the University of Bamberg, and in 1803 became professor of physiology and normal and pathological anatomy at Würzburg. From 1823 Döllinger was curator of academic sciences at Munich and then became (1826) professor of anatomy and physiology and director of the anatomical museum of the University of Munich. He was one of the pioneers of the use of the microscope in medical studies and made investigations of blood circulation, the spleen and liver, glandular secretions, and the eye, as well as comparative anatomy studies. His embryological work exercised a considerable influence on Louis AGASSIZ.

Dollond, John (1706–1761) British optician

Dollond was born in London, the son of Huguenot refugees. He started life as a silk weaver but later joined his eldest son, Peter, in making optical instruments, and devoted years of experiment to developing an achromatic lens. The problem confronting lens makers at the time was chromatic aberration – the fringe of colors that surrounds and disturbs images formed by a lens. This put a limit on the power of lenses (and of refracting telescopes), for the stronger the lens, the more chromatically disturbed the images became. Chromatic aberration is caused by the different wavelengths that make up white light being refracted to different extents by the glass, each being focused at a different point.

In 1758 Dollond succeeded in making lenses without this defect by using two different lenses, one of crown glass and one of flint glass (one convex and one concave), so made that the chromatic aberration of one was neutralized by the aberration of the other. In fact he was not the first to make such a lens, since Chester Hall had already done so in 1753, but Dollond managed to patent the idea because he was the first to publicize the possibility.

In 1761 he was appointed optician to George III but died of apoplexy later that year.

Domagk, Gerhard (1895–1964) German biochemist

> If I could start again, I would perhaps become a psychiatrist and search for a causal therapy of mental disease, which is the most terrifying problem of our times.
> —On his scientific career

Domagk (**doh**-mahk), who was born in Lagow, now in Poland, graduated in medicine from the University of Kiel in 1921 and began teaching at the University of Greifswald and later at the University of Münster. At this time he carried out important researches into phagocytes – special cells that attack bacteria in the body. He became interested in chemotherapy and in 1927 he was appointed director of research in experimental pathology and pathological anatomy at the giant chemical factory I. G. Farbenindustrie at Wuppertal-Elberfeld. Pursuing the ideas of Paul EHRLICH, Domagk tested new dyes produced by the Elberfeld chemists for their effect against various infections. In 1935 he reported the effectiveness of an orange-red dye called prontosil in combating streptococcal infections. For the first time a chemical had been found to be active *in vivo* (in a living organism) against a common small bacterium. Earlier dyes used as drugs were active only against infections caused by much larger protozoa.

The work was followed up in research laboratories throughout the world – Alexander FLEMING neglected penicillin to work on prontosil in the early 1930s – but the most significant ramifications were discovered by Daniele BOVET and his coworkers. Prontosil and the sulfa drugs that followed were effective in saving many lives, including those of Franklin D. Roosevelt Jr., Winston Churchill, and Domagk's own daughter. In 1939 Domagk was offered the Nobel Prize for physiology or medicine. The Nazis forced him to withdraw his acceptance because Hitler was annoyed with the Nobel Committee for awarding the 1935 Peace Prize to a German, Carl von Ossietzky, whom Hitler had imprisoned. In 1947 Domagk was finally able to accept the prize. In his later years he undertook drug research into cancer and tuberculosis.

Donati, Giovanni Battista (1826–1873) *Italian astronomer*

After graduating from the university in his native city of Pisa, Donati (do-**nah**-tee) joined the staff of the Florence Observatory in 1852, and was appointed director in 1864. He died from bubonic plague in 1873.

Much of his work was concerned with comets. He discovered six new comets, one of which, first appearing in June 1858 has since been known as *Donati's comet*. He went on in 1864 to make the first observations of a comet's chemical composition. Spectroscopic observation of the 1864 comet produced a line spectrum with three lines named alpha, beta, and gamma by Donati. The three lines were also seen in an 1866 comet by SECCHI. The lines were shown by HUGGINS in 1868 to be due to the presence of carbon.

Dondi, Giovanni de (1318–1389) *Italian astronomer and clockmaker*

Dondi (**don**-dee), born the son of a professor of medicine and clock designer in Chiogga, Italy, taught medicine and astronomy at the universities of Padua and Pavia. He became famous throughout the whole of Europe for his construction of a marvelous astronomical clock that he began in 1348 and that took him 16 years to complete. It was built for the duke of Milan and was put in his library in Pavia. Although the clock has long been destroyed, Dondi wrote a treatise on it, which has survived. It was weight driven, had a verge and foliot escapement, and was made of brass and bronze. It was completely unlike a modern clock being so unconcerned with the time of day that it did not even have hands. It was intended to show the movements of the planets and the time of the movable and fixed ecclesiastical festivals. It involved advanced gear work and showed mechanical skill of a high order. Such was his fame that Dondi was known throughout Europe as "John of the Clock."

Donnan, Frederick George (1879–1956) *British chemist*

Donnan, the son of a Belfast merchant, was born in Colombo, the capital of Ceylon (now Sri Lanka). He was educated at Queen's College, Belfast, and the universities of Leipzig and Berlin where he obtained his PhD in 1896 and some of the German expertise in physical chemistry. On his return to England he worked at University College, London, with Sir William RAMSAY from 1898 until 1904, when he accepted the post of professor of physical chemistry at the University of Liverpool. In 1913 Donnan returned to succeed Ramsay at University College, remaining there until his retirement in 1937.

Donnan is mainly remembered for the *Donnan membrane equilibrium* (1911) – a theory describing the equilibrium which arises in the passage of ions through membranes.

Doodson, Arthur Thomas (1890–1968) *British mathematical physicist*

Doodson, the son of a manager of a cotton mill in Worsley in the north of England, was educated at the University of Liverpool. After working at University College, London, from 1916 to 1918 he joined the Tidal Institute, Liverpool, in 1919 as its secretary. Doodson remained through its re-formation as the Liverpool Observatory and Tidal Institute as assistant director (1929–45) and as director until his retirement in 1960.

Much of Doodson's early work was on the production of mathematical tables and the calculation of trajectories for artillery. Proving himself an ingenious, powerful, and practical mathematician he found an ideal subject for his talents in the complicated behavior of the tides. He made many innovations in their accurate computation and, with H. WARBURG in 1942, produced the *Admiralty Manual of Tides*.

Doppler, Christian Johann (1803–1853) *Austrian physicist*

Christian Doppler, the son of a stonemason from the Austrian city of Salzburg, studied mathematics at the Vienna Polytechnic. In 1835 he started teaching at a school in Prague and six years later was appointed professor of mathematics at the Technical Academy there.

Doppler's fame comes from his discovery in 1842 of the *Doppler effect* – the fact that the observed frequency of a wave depends on the velocity of the source relative to the observer. The effect can be observed with sound waves. If the source is moving toward the observer, the pitch is higher; if it moves away, the pitch is lower. A common example is the fall in frequency of a train's whistle or a vehicle siren as it passes. Doppler's principle was tested experimentally in 1843 by Christoph BUYS BALLOT, who used a train to pull trumpeters at different speeds past musicians who had perfect pitch.

Doppler also tried to apply his principle to light waves, with limited success. It was Armand FIZEAU in 1848 who suggested that at high relative velocities the apparent color of the source would be changed by the motion: an object moving toward the observer would appear bluer; one moving away would appear redder. The shift in the spectra of celestial objects (the *Doppler shift*) is used to measure the rate of recession or approach relative to the Earth.

Dorn, Friedrich Ernst (1848–1916) *German physicist*

Dorn, who was born in Guttstadt (now Dobre Miasto in Poland), studied at Königsberg and in 1873 was made professor of physics at Breslau. In 1886 he transferred to a professorship at Halle and started working with x-rays. He is noted for his discovery, in 1900, that the radioactive element radium gives off a radioactive gas, which Dorn called "radium emanation." The gas was isolated in 1908 by William RAMSAY, who named it "niton." The name radon was adopted in 1923. Dorn's discovery is the first established demonstration of a transmutation of one element into another.

Douglas, Donald Wills (1892–1981) *American aircraft engineer*

Douglas, the son of an assistant bank cashier in New York City, was educated at the U.S. Naval Academy and the Massachusetts Institute of Technology. He gained his first experience of aircraft design working for the Glenn L. Martin Company of California on the development of a heavy bomber.

In 1920 he set up on his own with $600 and the backing of David Davis, a wealthy sportsman willing to invest $40,000 to produce a plane capable of flying nonstop across America. Although the *Cloudster*, the result of their venture, only reached Texas it was in fact the first aircraft in history capable of lifting a useful load exceeding its own weight.

By 1928, on the strength of some profitable navy contracts, Douglas was ready to go public with his new Douglas Aircraft Company. The company had many years of success, with such planes as the DC-3, first flown in 1935, contributing substantially to their profits. In 1967 the company was taken over by the McDonnell Aircraft Company and reformed as the McDonnell–Douglas Corporation.

Douglass, Andrew Ellicott (1867–1962) *American astronomer and dendrochronologist*

Douglass came from a family of academics in Windsor, Vermont, with both his father and grandfather being college presidents. He graduated from Trinity College, Hartford, Connecticut in 1889 and in the same year was appointed to an assistantship at Harvard College Observatory. In 1894 he went with Percival LOWELL to the new Lowell Observatory in Flagstaff, Arizona, moving to the University of Arizona in 1906 as professor of astronomy and physics.

Douglass's first interest was the 11-year sunspot cycle. In trying to trace its history he was led to the examination of tree rings in the hope that he would find some identifiable correlation of sunspot activity with terrestrial climate and vegetation. Soon the tree rings became the center of his studies.

The only previously established method of dating the past, except by inscriptions, was the geological varve-counting technique, which was developed from 1878. But this was of no use if there were no varves (thin seasonally deposited layers of sediment in glacial lakes) to be found. Douglass soon found that he could identify local tree rings with confidence and use them in dating past climatic trends. He thus founded the field of dendrochronology. By the late 1920s he had a sequence of over a thousand tree rings with six thin rings, presumably records of a severe drought, correlated with the end of the 13th century. In 1929 he found some timber that contained the six thin rings and a further 500 in addition. This took him to the eighth century and over the years he managed to get as far as the first century. This was extended still further and by careful analysis scholars have now established a sequence going back almost to 5000 BC.

The dated rings of Arizona and New Mexico were found however not to correlate with sequences from other parts of the world: the tree-ring clock was a purely local one. The search for a more universal clock continued, and the method of radiocarbon dating was developed by Willard LIBBY in 1949.

Drake, Frank Donald (1930–   ) *American astronomer*

Drake, who was born in Chicago, Illinois, graduated in 1952 from Cornell University and obtained his PhD in 1958 from Harvard. He worked initially at the National Radio Astronomy Observatory (NRAO), West Virginia (1958–63) and at the Jet Propulsion Laboratory, California (1963–64) before returning to Cornell and serving as professor of astronomy from 1964. He was appointed professor of astronomy at the University of California in 1984.

Although Drake has made significant contributions to radio astronomy, including radio studies of the planets, he is perhaps best known for his pioneering search for extraterrestrial intelligence. In April 1959 he managed to gain approval from the director at NRAO, Otto STRUVE, to proceed with his search, which was called "Project Ozma." The name was taken from the Oz stories of Frank Baum. Drake began in 1960, using the NRAO 26-meter radio telescope to listen for possible signals from planets of the Sunlike stars Tau Ceti and Epsilon Eridani, both about 11 light-years away. He decided to tune to the frequency of 1,420 megahertz at which radio emission from hydrogen occurs. This would have considerable significance for any civilization capable of building radio transmitters. No signals were received although at one time excitement was generated when signals from a secret military radar establishment were received while the antenna was pointed at Epsilon Eridani. In July 1960 the project was terminated to allow the telescope to fulfill some of its other obligations. Drake revived the project in 1975, in collaboration with Carl SAGAN, when they began using the Arecibo 1,000-foot (305-meter) radio telescope to listen to several nearby galaxies on frequencies of 1,420, 1,653, and 2,380 megahertz. No contact was made nor was it likely, they declared, for "A search of hundreds of thousands of stars in the hope of detecting one message would require remarkable dedication and would probably take several decades." He has published a number of works on this issue including *Is Anyone There? The Search for Extra Terrestrial Intelligence* (1992).

Draper, Henry (1837–1882) *American astronomer*

> I think we are by no means at the end of what can be done. If I can stand 6 hours' exposure in midwinter another step forward will result.
> —Remark made in the last months of his life, describing his work photographing stars and planets

Draper, the son of the distinguished physician and chemist John W. Draper, was born in Prince Edward County, Virginia. He studied at the City University of New York, completing the course in medicine in 1857 before he was old enough to graduate. He obtained his MD in 1858, spending the preceding months in Europe where his interest in astronomy was aroused by a visit to the observatory of the third earl of ROSSE at Parsonstown, Ireland. On his return to New York he joined the Bellevue Hospital and was later appointed professor of natural science at the City University in 1860. Draper later held chairs of physiology (1866–73) and analytical chemistry (1870–82) and in 1882 succeeded his father briefly as professor of chemistry. He retired in 1882 in order to devote himself to astronomical research but died prematurely soon after.

One of the most important events in Draper's life was his marriage in 1867 to Anna Palmer, daughter and heiress to Courtlandt Palmer who had made a fortune in hardware and New York real estate. His wife's money allowed him to purchase a 28-inch (71-cm) reflecting telescope and to begin a 15-year research partnership.

Draper was interested in the application of the new technique of photography to astronomy. He started by making daguerrotypes of the Sun and Moon but in 1872 succeeded for the first time in obtaining a photograph of a stellar spectrum, that of Vega. In 1879 he found that dry photographic plates had been developed and that these were more sensitive and con-

venient than wet collodion. By 1882 he had obtained photographs of over a hundred stellar spectra plus spectra of the Moon, Mars, Jupiter, and the Orion nebula. He also succeeded in directly photographing the Orion nebula, first with a 50-minute exposure in 1880 and then, using a more accurate clock-driven telescope, with a 140-minute exposure. He thus helped to establish photographic astronomy as an important means of studying the heavens.

At the time of his death his widow hoped to continue his work herself, but with prompting from Edward PICKERING at the Harvard College Observatory, she set up the Henry Draper Memorial Fund. It was with the aid of this fund that the famous *Henry Draper Catalogue*, some nine volumes with details of the spectra of 225,000 stars, was published from 1918 to 1924 through the labors of Pickering and Annie CANNON.

Draper, John William (1811–1882) *British–American chemist*

> How is it that the Church produced no geometers in her autocratic reign of twelve hundred years?
> —*The Conflict Between Science and Religion* (1890)

Draper, who was born in St. Helens, Lancashire, in northwestern England, was educated at University College, London, before he emigrated to America in 1833. He qualified in medicine at the University of Pennsylvania in 1836. After a short period teaching in Virginia he moved to New York University (1838) where he taught chemistry and in 1841 helped to start the medical school of which he became president in 1850.

Most of his chemical work was done in the field of photochemistry. He was one of the first scientists to use Louis DAGUERRE's new invention (1837) of photography. He took the first photograph of the Moon in 1840 and in the same year took a photograph of his sister, Dorothy, which is the oldest surviving photographic study of the human face. In 1843 he obtained the first photographic plate of the solar spectrum. He was also one of the first to take photographs of specimens under a microscope. On the theoretical level Draper was one of the earliest to grasp that only those rays that are absorbed produce chemical change and that not all rays are equally powerful in their effect. He also, in a series of papers (1841–45), showed that the amount of chemical change is proportional to the intensity of the absorbed radiation multiplied by the time it has to act. Draper's work was continued and largely confirmed by the work of Robert BUNSEN and Henry ROSCOE in 1857. Draper's work also resulted in the development of actinometers (instruments to measure the intensity of light) which he named "tithonometers." He also wrote on a wide variety of other topics.

Draper's son Henry was an astronomer of note after whom the famous Harvard catalog of stellar spectra was named.

Dreyer, Johann Louis Emil (1852–1926) *Danish astronomer*

Dreyer (**drI**-er), the son of a general of the Danish Army, was born in Copenhagen and studied at the university there, obtaining his MA in 1874 and his PhD in 1882. He began his career in 1874 as an assistant at the observatory of William PARSONS, third earl of ROSSE, in Parsonstown, Ireland, moving in 1878 to another assistantship at the Dunsink Observatory, Dublin. In 1882 Dreyer was appointed director of the Armagh Observatory, in what is now Northern Ireland, a post he occupied until his retirement in 1916 when he settled in Oxford, England.

Dreyer's major contribution to astronomy was his three catalogs. He began by preparing in 1878 a supplement to John HERSCHEL's catalog of 5,000 nebulae. The work naturally led to the production of a totally new work, the *New General Catalogue of Nebulae and Clusters of Stars* (1888) containing details of 7,840 celestial objects and known invariably as the NGC. It listed all the nebulae and clusters that had been discovered up to 1888 and included many galaxies that had not yet been identified as such. This in turn was followed by the two *Index Catalogues* (IC), the first in 1895 containing details of a further 1,529 nebulae and clusters (and galaxies) and the second in 1908 adding a further 3,857. The catalogs with over 13,000 nebulae, galaxies, and clusters were reissued in a single volume in 1953. Many of the objects listed are still referred to by their NGC or IC numbers.

Dreyer is also remembered as a historian of astronomy. In 1890 he published a biography of his countryman, *Tycho Brahe*, and followed this with Brahe's *Omnia opera* (15 vols. 1913–29; Complete Works). He had earlier published his *History of Planetary Systems from Thales to Kepler* (1906), a work, despite its age, still without any competitor in the English language.

Drickamer, Harry George (1918–2002) *American physicist*

Born in Cleveland, Ohio, Drickamer was educated at the University of Michigan, where he obtained his PhD in 1946. He then joined the staff of the University of Illinois, Urbana, serving first as professor of physical chemistry and then as professor of chemical engineering from 1953 until his retirement in 1990. He specialized in the study of the structure of solids by means of high pressures, producing in the course of his researches pressures of the order of some 500,000 atmospheres.

Driesch, Hans Adolf Eduard (1867–1941) *German biologist*

Born in Bad Kreuznach, in southwest Germany, Driesch (dreesh) held professorships at Heidelberg, Cologne, and Leipzig, and was visiting professor to China and America. A student of zoology at Freiburg, Jena, and Munich, he was for some years on the staff of the Naples Zoological Station.

Driesch carried out pioneering work in experimental embryology. He separated the two cells formed by the first division of a sea-urchin embryo and observed that each developed into a complete larva, thus demonstrating the capacity of the cell to form identical copies on division. He was also the first to demonstrate the phenomenon of embryonic induction, whereby the position of and interaction between cells within the embryo determine their subsequent differentiation.

Driesch is perhaps best known for his concept of entelechy – a vitalistic philosophy that postulates the origin of life to lie in some unknown vital force separate from biochemical and physiological influences. This also led him to investigate psychic research and parapsychology.

Dubois, Eugène (1858–1940) *Dutch physician and paleontologist*

Dubois (doo-**bwah**) was born in Eijsden in the Netherlands and studied medicine at the University of Amsterdam. After briefly working there as a lecturer in anatomy, he served as a military surgeon in the Dutch East Indies, now Indonesia, from 1887 to 1895. On his return to Amsterdam he held the chair of geology, paleontology, and mineralogy from 1899 until his retirement in 1928.

The decision to go to the Indies was no accident. Dubois was determined to find the "missing link" and had reasoned that such a creature would have originated in proximity to the apes of Africa or the orang-utang of the Indies. After several years fruitless search in Sumatra, Dubois moved to Java and in 1890 discovered his first humanoid remains (a jaw fragment) at Kedung Brubus. The following year, at Trinil on the Solo river, he found the skullcap, femur, and two teeth of what he was later to name *Pithecanthropus erectus*, more commonly known as Java man. He published these findings in 1894.

Although Dubois's estimate of the cranial capacity of *Pithecanthropus* was, at 850 cubic centimeters (later estimates ranged up to 940 cubic centimeters), on the low side for a hominid, the femur it had been found with indicated to Dubois that it must be a form with a very erect posture. However, many doubted this, stating the usual objections that the remains belonged to different creatures, to apes or (Rudolf VIR-CHOW's view) to deformed humans. So irritated did Dubois become by this reception that he withdrew the fossils from view, keeping them locked up for some 30 years.

When they were once more made available to scholars in 1923 and Peking man was discovered in 1926 it at last became widely agreed that *Pithecanthropus* was, as Dubois had earlier claimed, a link connecting apes and man. By this time however Dubois would have no part of such a consensus and began to insist the bones were those of a giant gibbon, a view he maintained until his death.

Du Bois-Reymond, Emil Heinrich (1818–1896) *German neurophysiologist*

> The more one advances in the knowledge of physiology, the more one will have reason for ceasing to believe that the phenomena of life are essentially different from physical phenomena.

Of Swiss and Huguenot descent, Du Bois-Reymond (doo bwah-ray-**mon**) was born in Berlin and educated at the university there and in Neuchâtel (Switzerland). He is famous as the first to demonstrate how electrical currents in nerve and muscle fibers are generated. He began his studies under the eminent physiologist Johannes MÜLLER at Berlin with work on fish capable of discharging electric currents as an external shock (e.g., eels). Turning his attention to nerve and muscle activity he then showed (1843) that applying a stimulus to the nerve brings about a drop in the electrical potential at the point of stimulus. This reduction in potential is the impulse, which travels along the nerve as waves of "relative negativity." This variation in negativity is the main cause of muscle contraction. Du Bois-Reymond's pioneering research, for which he devised a specially sensitive galvanometer capable of measuring the small amounts of electricity involved, was published as *Untersuchungen über tierische Elektricität* (2 vols. 1848–84; Researches on Animal Electricity): a landmark in electrophysiology, although subject to later elaboration. Du Bois-Reymond's collaboration with fellow physiologists Hermann von HELM-HOLTZ, Carl Ludwig, and Ernst von Brücke was of great significance in linking animal physiology with physical and chemical laws.

Du Bois-Reymond was elected a member of the Berlin Academy of Sciences in 1851 and succeeded Müller as professor of physiology at Berlin in 1858. He was also instrumental in establishing the Berlin Physiological Institute, opened in 1877, then the finest establishment of its kind.

Dubos, René Jules (1901–1982) *French–American microbiologist*

Dubos (doo-**bos**) was born in Saint Brice, France, and graduated in agricultural sciences from the National Agronomy Institute in 1921. After a period with the International Institute of Agriculture in Rome as assistant editor, he emigrated to America in 1924.

Dubos was awarded his PhD in 1927 from Rutgers University for research on soil microorganisms, continuing his work in this field at the Rockefeller Institute for Medical Research. Reports that soil microorganisms produce antibacterial substances particularly interested him and in 1939 he isolated a substance from *Bacillus brevis* that he named tyrothricin. This is effective against many types of bacteria but unfortunately also kills red blood cells and its medical use is therefore limited. However, the discovery stimulated such workers as Selman WAKSMAN and Benjamin DUGGAR to search for useful antibiotics and led to the discovery of the tetracyclines. He won the 1969 Pulitzer Prize for his book *So Human an Animal*.

Duesberg, Peter (1936–) *American molecular biologist*

Born in Germany, Duesberg (**dooz**-berg) was educated at the University of Frankfurt where he obtained his PhD in chemistry in 1963. He immediately moved to the U.S. to work at the University of California, Berkeley, and was appointed professor of molecular biology in 1974.

Duesberg established his reputation in molecular biology by his discovery of cancer-causing genes (oncogenes) in the retrovirus first described by Peyton ROUS in 1910. In 1970 Duesberg and his Berkeley colleagues identified three genes, gag, pol, and env, which encode the proteins of the viral capsid, the enzyme reverse transcriptase, and the proteins of the viral envelope, respectively.

Consequently when GALLO and MONTAGNIER identified the HIV retrovirus in 1983 as the cause of AIDS, Duesberg was well qualified to comment on their judgment. In 1987 he published in *Cancer Research* a paper entitled *Retroviruses as Carcinogens and Pathogens: Expectation and Reality* in which he surveyed the published literature, citing 278 references in the process. The paper turned Duesberg into an international celebrity, winning praise from a few but savage rejection and complaints of irresponsibility from the majority of his colleagues.

Duesberg attacked what he saw as a complacent orthodoxy by denying there was any evidence for the claim that HIV was responsible for AIDS. Being antibody positive would cause him no worry, he insisted, and he would be even prepared to inject himself with pure HIV to establish his point.

In defence of his claim Duesberg offered six main arguments:

1. Many AIDS cases are HIV negative.
2. T-cells, the site of HIV attack, are regenerated more quickly than they are destroyed.
3. Viruses typically cause disease in the absence of antibodies; HIV stimulates the production of antibodies.
4. HIV should cause AIDS on infection, not years later.
5. In general retroviruses sustain rather than destroy cells.
6. No known virus discriminates between men and women.

Duesberg's views have been vigorously defended in a popular book by Jad Adams, *AIDS: the HIV Myth* (London, 1989). They have also been vigorously rejected by Robert Gallo and other virologists, and dismissed by them as more inept than challenging.

Despite a hostile reaction to his work from lay and professional press alike, Duesberg continues to campaign vigorously for his views. So much so that in 1995 he collected his previously published papers on the issue in his *Infectious AIDS: Have We Been Misled?* and went on to produce in 1996 a new comprehensive survey of the whole field in his *Inventing the AIDS Virus*.

Du Fay, Charles François de Cisternay
(1698–1739) *French chemist*

Du Fay (doo fay), a Parisian by birth, started his career in the French army, rising to the rank of captain. He left to become a chemist in the Académie Française and in 1732 became superintendent of the Jardin du Roi. His great achievement was to discover the two kinds of electricity, positive and negative, which he named "vitreous" and "resinous." This was based on his discovery that a piece of gold leaf charged from an electrified glass rod would attract and not repel a piece of electrified amber. This was the "two-fluid theory" of electricity, which was to be opposed by Benjamin FRANKLIN's "one-fluid theory" later in the century.

Duggar, Benjamin Minge (1872–1956)
American plant pathologist

Duggar was born into the farming community of Gallion, Alabama, and soon developed an interest in agriculture. He graduated with honors from the Mississippi Agricultural and Mechanical College, in 1891. He devoted his career to studying plant diseases, and while professor of plant physiology at Cornell University wrote *Fungus Diseases of Plants* (1909), the first publication to deal purely with plant pathology.

Duggar is known for his work on cotton diseases and mushroom culture, but he made his most important discovery after retiring from academic life. In 1945 he became consultant to the American Cyanamid Company, and soon isolated the fungus *Streptomyces aureofaciens*. Three years' work with this organism resulted in Duggar extracting and purifying the compound chlortetracycline (marketed as Aureomycin), the first of the tetracycline antibiotics. This drug was on the market by December 1948, and has proved useful in combating many infectious diseases.

Duggar was one of the foremost coordinators of plant science research in America and was editor of many important publications.

Duhamel du Monceau, Henri-Louis
(1700–1782) *French agriculturalist and technologist*

Duhamel (doo-a-**mel**) first took an interest in science following lectures at the Jardin du Roi in his native city of Paris during the 1720s. His study of the parasitic fungus found to attack saffron bulbs earned him admission to the French Academy of Sciences in 1728. In 1739 Duhamel was appointed inspector-general of the navy, his duties including supervision of the timber used by the French fleet.

During the 1730s he undertook a series of chemical investigations in collaboration with the chemist Jean Grosse. His most important work during this period was *Sur la base du sel marin* (1736; On the Composition of Sea Salt) in which he distinguished between potassium and sodium salts.

His major work was his contribution to agriculture with studies of French and English methods of practice. He published a series of writings entitled *Traité de la culture des terres* (1775; Treatise on Land Cultivation) in which he adapted Jethro TULL's system to France taking into account his own readings, experiments, and case histories.

Duhem, Pierre Maurice Marie (1861–1916) *French physicist, philosopher, and historian*

> A physical theory ... is a system of mathematical propositions, deduced from a small number of principles, which has the object of representing a set of experimental laws as simply, as completely, and as exactly as possible.
> —*The Aim and Structure of Physical Theory* (1906)

The son of a commercial traveler, Duhem (doo-**em**) was born in Paris and educated at the Ecole Normale. In his doctoral thesis he had managed to annoy BERTHELOT, an influential figure in French science and politics, and consequently found himself permanently exiled to the provinces. After teaching at the universities of Lille and Rennes (1887–94), he settled finally at the University of Bordeaux where he remained until his death from a heart attack in

1916. There had been a move shortly before his death to create a chair in the history of science for Duhem in Paris at the Collège de France. Duhem would have none of it, insisting that he was a physicist and that "he would not enter Paris by a side door."

As a historian, his reputation is immense. His most important work, *Le système du monde* (vols. 1–5, 1913–17; vols. 6–10, 1954–59; The Global System), was one of the first attempts to argue for a deep continuity in the history of science. He saw and documented an unbroken tradition linking medieval and modern science, and was able to trace the influence of such medieval writers as BURIDAN and Jordanus upon such moderns as Leonardo and GALILEO.

Duhem's contribution to the philosophy of science has also been influential in the form of the *Duhem-Quine thesis*. Scientific theories, he argued, are seldom falsified by experience; more commonly they are modified by such techniques as redefining terms and introducing new hypotheses. His views were presented in his *The Aim and Structure of Physical Theory* (Paris, 1906; New York, 1954).

As a scientist Duhem published books on thermodynamics (1886), hydrodynamics (1891), chemical mechanics (1897–99), electricity (1902), and elasticity (1906). He took the general position that it was impossible to reduce physics and chemistry to mechanics. For, he argued, mechanics had become "a branch of a more general science. This science embraces not only movement which displaces bodies in space but also every change of qualities, properties, physical state, and chemical constitution. This science is contemporary thermodynamics or, according to the word created by RANKINE, Energetics." Duhem's final attempted synthesis, *Traité d'énergétique* (1911; Treatise on Energetics), shunned atomism and failed to extend the new science to electricity and magnetism.

Dujardin, Félix (1801–1860) *French biologist and cytologist*

Largely self-educated, Dujardin (doo-zhar-**dan**), who was born in Tours, France, studied geology, botany, optics, and crystallography while working variously as a hydraulics engineer, librarian, and teacher of geometry and chemistry at Tours. In 1839 he was elected to the chair of geology and mineralogy at Toulouse, and in the following year was appointed professor of botany and zoology and dean of the Faculty of Sciences at Rennes. As a skilled microscopist, Dujardin carried out extensive studies of the microorganisms (infusoria) occurring in decaying matter. These led him, in 1834, to suggest the separation of a new group of protozoan animals, which he called the rhizopods (i.e., root-feet). He was the first to recognize and appreciate the contractile nature of the proto-

plasm (which he termed the *sarcode*) and also demonstrated the role of the vacuole for evacuating waste matter. Such studies enabled Dujardin to refute the supposition, reintroduced by Christian EHRENBERG, that microorganisms have organs similar to those of the higher animals. Dujardin also investigated the cnidarians (jellyfish, sea anemones, corals, etc.), echinoderms (sea-urchins, starfish, etc.), as well as the platyhelminths, or flatworms, the last mentioned providing the basis for subsequent parasitological investigations.

Dulbecco, Renato (1914–) *Italian–American physician and molecular biologist*

Born in Catanzaro, Italy, Dulbecco (dul-**bek**-oh) obtained his MD from the University of Turin in 1936 and taught there until 1947 when he moved to America. He taught briefly at Indiana before moving to California in 1949, where he served as professor of biology (1952–63) at the California Institute of Technology. Dulbecco then joined the staff of the Salk Institute where, apart from the period 1971–74 at the Imperial Cancer Research Fund in London, he has remained.

Beginning in 1959 Dulbecco introduced the idea of cell transformation into biology. In this process special cells are mixed *in vitro* (Latin, meaning literally "in glass," i.e., in a test tube) with such tumor-producing viruses as the polyoma and SV40 virus. With some cells a "productive infection" results, where the virus multiplies unchecked in the cell and finally kills its host. However, in other cells this unlimited multiplication does not occur and the virus instead induces changes similar to those in cancer cells; that is, the virus alters the cell so that it reproduces without restraint and does not respond to the presence of neighboring cells. A normal cell had in fact been transformed into a "cancer cell" *in vitro*.

The significance of this work was to provide an experimental setup where the processes by which a normal cell becomes cancerous can be studied in a relatively simplified form. It was for this work that Dulbecco was awarded the Nobel Prize for physiology or medicine in 1975, sharing it with Howard TEMIN and David BALTIMORE.

In March 1986 Dulbecco published a widely read paper in *Science*, entitled *A Turning Point in Science*, in which he argued that "if we wish to learn more about cancer, we must now concentrate on the cellular genome." The paper appeared shortly after various groups of scientists had held a meeting at Sante Fe to discuss sequencing the entire human genome. Dulbecco's timely paper publicized the project, gave it some authority, and linked it with a practical purpose. He was still working and publishing in 2007.

Dulong, Pierre-Louis (1785–1838)
French chemist and physicist

Born in Rouen, France, Dulong (doo-**long**) studied chemistry at the Ecole Polytechnique (1801–03) and later studied medicine. He was an assistant to Claude-Louis BERTHOLLET before becoming professor of physics at the Ecole Polytechnique (1820), and later its director (1830).

In 1813 Dulong accidentally discovered the highly explosive nitrogen trichloride, losing an eye and nearly a hand in the process. He is best known for the law of atomic heats (1819), discovered in collaboration with Alexis-Thérèse PETIT.

Dumas, Jean Baptiste André (1800–1884) *French chemist*

> I do not claim to have discovered it, for it does no more than reproduce more precisely and in a more generalized form opinions that could be found in the writings of a large number of chemists.
> —Concerning his substitution theory of chemistry

Dumas (doo-**mah**) was educated in classics at the college in his native city of Alais and intended to serve in the navy. However, after Napoleon's final defeat he changed his mind and became apprenticed to an apothecary. In 1816 he went to Geneva, again to work for an apothecary. His first research was in physiological chemistry, investigating the use of iodine in goiter (1818). He also studied chemistry in Geneva and was encouraged by Friedrich von HUMBOLDT to go to Paris, where he became assistant lecturer to Louis Thenard at the Ecole Polytechnique (1823). He subsequently worked in many of the Parisian institutes, becoming professor at the Ecole Polytechnique (1835) and at the Sorbonne (1841).

Dumas's early work included a method for measuring vapor density (1826), the synthesis of oxamide (1830), and the discoveries of the terpene cymene (1832), anthracene in coal tar (1832), and urethane (1833). In 1834 Dumas and Eugene PELIGOT discovered methyl alcohol (methanol) and Dumas recognized that it differed from ethyl alcohol by one $-CH_2$ group. The subsequent discovery that CHEVREUL's "ethal" was cetyl alcohol (1836) led Dumas to conceive the idea of a series of compounds of the same type (this was formalized into the concept of homologous series by Charles Gerhardt).

Dumas was both a prolific experimentalist and a leading theorist and he took a vigorous part in the many controversies that bedeviled organic chemistry at the time. He was originally an exponent of the "etherin" theory (in which ethyl alcohol (ethanol) and diethyl ether were considered to be compounds of etherin (ethene) with one and two molecules of water, respectively). However, he was converted to the radical theory (an attempt to formulate organic chemistry along the dualistic lines familiar in inorganic chemistry) by Justus von LIEBIG in 1837. He then introduced his own theory – the substitution theory – which was his greatest work. It had been noticed that candles bleached with chlorine gave off fumes of hydrogen chloride when they burned. Dumas discovered that during bleaching the hydrogen in the hydrocarbon oil of turpentine became replaced by chlorine. This seemed to contradict Jöns BERZELIUS's electrochemical theory and the latter was bitterly opposed to the substitution theory. Liebig, too, was hostile at first. Dumas then prepared trichloroacetic acid (1838) and showed that its properties were similar to those of the parent acetic acid. This convinced Liebig but not Berzelius. Further work on this series of acids, combined with the substitution theory, led him to a theory of types (1840), essentially similar to the modern concept of functional groups, although the credit for this theory was disputed between Dumas and Auguste LAURENT.

Dumas also carried out important work on atomic weights. He had been an early supporter of Amedeo AVOGADRO but he never properly distinguished between atoms and molecules and the problems this raised caused him to abandon the theory. He also supported William PROUT's hypothesis that atomic weights were whole-number multiples of that of hydrogen. In 1840, working with Jean STAS, he obtained the figure 12.000 for carbon instead of the figure 12.24 in use at that time.

Following the revolution of 1848 Dumas became involved in administration, becoming minister of agriculture and commerce (1849–51), minister of education, and permanent secretary of the Academy of Sciences (1868).

Dunning, John Ray (1907–1975) *American physicist*

Born in Shelby, Nebraska, Dunning was educated at the Wesleyan University, Nebraska, and at Columbia University, New York, where he obtained his PhD in 1934. He took up an appointment at Columbia in 1933, being made professor of physics in 1950.

Dunning was one of the key figures in the Manhattan project to build the first atomic bomb. It had been shown by Niels BOHR that the isotope uranium–235 would be more likely to sustain a neutron chain reaction than normal uranium. Only 7 out of every 1,000 uranium atoms occurring naturally are uranium–235, which presents difficulties in extraction. Various techniques were tried and Dunning was placed in charge of the process of separation known as gaseous diffusion. This involved turning the uranium into a volatile compound (uranium hexafluoride, UF_6) and passing the vapor through a diffusion "filter." As ^{235}U atoms are slightly less massive than the normal ^{238}U they pass through the filter a little faster and can thus be concentrated. The difference in mass is so small, however, that simply to produce a gas enriched with ^{235}U atoms required its passage through thousands of filters.

As early as 1939 Dunning had shown that the process would work but to produce ^{235}U in the quantities required by the Manhattan project was a daunting prospect and the engineering problems were immense. The gas is extraordinarily corrosive and a single leak in one of the hundreds of thousands of filters would lose the precious ^{235}U. But as the other projects ran into even more formidable difficulties it was largely through gaseous diffusion that sufficient enriched uranium was made available for the bomb to be built.

Durand, William Frederick (1859–1958)
American engineer

Born in Beacon Falls, Connecticut, Durand graduated from the U.S. Naval Academy in 1880 and entered the Engineering Corps of the U.S. Navy (1880–87). He then took a post as professor of mechanical engineering at Michigan State College in 1887. He moved to Cornell in 1891 to the chair of marine engineering and, in 1904, he accepted the professorship of mechanical engineering at Stanford, a position he held until his retirement in 1924.

Durand worked mainly on problems connected with the propeller, both marine and, after 1914, aeronautical. He was general editor of an important standard work, *Aerodynamic Theory*, produced in six volumes (1929–36). He also served on the National Advisory Committee for Civil Aeronautics (NACA) (1915–33) and, in 1941, was recalled to advise the government on the construction of an American jet-propelled airplane.

Du Toit, Alexander Logie (1878–1949)
South African geologist

Du Toit (doo toit), who was born at Rondebosch, near Cape Town in South Africa, studied at the South Africa College (now the University of Cape Town), the Royal Technical College, Glasgow, and the Royal College of Science, London.

After a short period teaching at Glasgow University (1901–03) he returned to South Africa and worked with the Geological Commission of the Cape of Good Hope (1903–20), during which he explored the geology of South Africa. For the next seven years he worked for the Irrigation Department and produced six detailed monographs on South African geology. He served as a consulting geologist to De Beers Consolidated Mines during the period 1927–41.

Following a visit to South America in 1923, du Toit became one of the earliest supporters of Alfred WEGENER's theory of continental drift, publishing his observations in *A Geological Comparison of South America with South Africa* (1927). He noted the similarity between the continents and developed his ideas in *Our Wandering Continents* (1937), in which he argued for the separation of Wegener's Pangaea into the two supercontinents, Laurasia and Gondwanaland.

Dutrochet, René Joachim Henri (1776–1847) *French physiologist*

Born in Néon, France, Dutrochet (doo-tro-**shay**) began medical studies while serving in the army in Paris in 1802. After graduating in 1806 he served as an army surgeon in Spain. However, through illness he resigned his post in 1809 and thereafter devoted his time to natural science.

In 1814 he published his investigations into animal development, suggesting a unity of the main features during the early stages. Later research into plant and animal physiology led to his assertion that respiration is similar in both plants and animals. In 1832, Dutrochet showed that gas exchange in plants was via minute openings (stomata) on the surface of leaves and the deep cavities with which they communicate. He further demonstrated that only cells containing chlorophyll can fix carbon and thus transform light energy into chemical energy. Dutrochet studied osmosis and suggested it may be the cause of ascent and descent of sap in plants. Although sometimes lacking in accuracy, the importance of his work lies mainly in his endeavor to demonstrate that the vital phenomena of life can be explained on the basis of physics and chemistry.

Dutton, Clarence Edward (1841–1912) *American geologist*

Born in Wallingford, Connecticut, Dutton graduated from Yale in 1860 and then entered the Yale Theological Seminary. He joined the army in 1862 during the Civil War and remained in the army although not always on active service. He became interested in geology and joined the Geographical and Geological Survey of the Rocky Mountains and the West in 1875.

The term "isostasy" was introduced into geology by Dutton in 1889. This described a theory propounded by George AIRY in which it is supposed that mountain ranges and continents rest on a much denser base. As mountains are eroded the land rises while the settling sediment will compensate by depressing some other part of the Earth.

After his return to the army in 1890 Dutton turned to the study of earthquakes and volcanoes. His research was published in 1904 in his *Earthquakes in the Light of the New Seismology*.

Duve, Christian René de *See* DE DUVE, CHRISTIAN RENÉ.

Du Vigneaud, Vincent (1901–1978)
American biochemist

Born in Chicago, Illinois, Du Vigneaud (doo **veen**-yoh) graduated from the University of Illinois in 1923; he remained there to take his master's degree before going to the University of Rochester. There he studied the hormone insulin, gaining his PhD in 1927. The research on insulin marked the beginning of his interest in sulfur compounds, particularly the sulfur-containing amino acids – methionine, cystine, and cysteine.

In 1938 Du Vigneaud became head of the biochemistry department of Cornell University Medical College. Two years later he had isolated vitamin H (biotin) and by 1942 had determined its structure. He then went on to examine the hormones secreted by the posterior pituitary gland, especially oxytocin and vasopressin. He found oxytocin to be composed of eight amino acids, worked out the order of these, and in 1954 synthesized artificial oxytocin, which was shown to be as effective as the natural hormone in inducing labor and milk flow. This was the first protein to be synthesized and for this achievement Du Vigneaud received the Nobel Prize for chemistry in 1955. Du Vigneaud's other work included research on penicillin and on methyl groups. He was professor of chemistry at Cornell University from 1967 to 1975 and subsequently professor of biochemistry there.

Dyson, Sir Frank Watson (1868–1939)
British astronomer

Dyson, the son of a Baptist minister from Ashby-de-la-Zouche in Leicestershire, England, graduated in mathematics from Cambridge University in 1889; in 1891 he was elected to a fellowship. After first working as chief assistant at the Royal Observatory at Greenwich from 1894 to 1905, he was Astronomer Royal for Scotland from 1905 to 1910 and then returned to Greenwich as Astronomer Royal, serving from 1910 to 1933. He was knighted in 1915.

Dyson's early observational work was done in collaboration with William G. Thackeray: they measured the positions of over 4,000 circumpolar stars that had first been observed by Stephen Groombridge at the beginning of the

19th century. They were thus able to determine the proper motions of the stars. Dyson could then extend the work of Jacobus KAPTEYN on star streaming to fainter stars.

Dyson observed the total solar eclipses of 1900, 1901, and 1905, obtaining spectra of the atmospheric layers of the Sun. He also organized the detailed observations of the total solar eclipse in 1919, sending expeditions to Principe in the Gulf of Guinea and Sobral in Brazil. The measured positions of stars near the Sun's rim during the eclipse provided evidence for the bending of light in a gravitational field, as predicted by EINSTEIN in his theory of general relativity; this was the first experimental support for the theory.

Dyson, Freeman John (1923–   ) *British–American theoretical physicist*

The son of Sir George Dyson, director of the Royal College of Music, Dyson was born at Crowthorne in England and educated at Cambridge University. During World War II he worked at the headquarters of Bomber Command. In 1947 he went on a Commonwealth Fellowship to Cornell University and in 1953 joined the Institute of Advanced Studies, Princeton, where he served as professor of physics until his retirement in 1994.

Dyson has worked on a number of topics but is best known for his contribution to quantum electrodynamics, i.e., the application of quantum theory to interactions between particles and electromagnetic radiation. The observation in 1946 by Willis LAMB of a small difference between the lowest energy levels of the hydrogen atom was an experimental result against which such theories could be tested. In the period 1946–48 independent formulations of quantum electrodynamics were put forward by Julian SCHWINGER, Sin-Itiro TOMONAGA, and Richard FEYNMAN. Dyson showed that the three methods were all consistent and brought them together into a single general theory.

Dyson later became known to a wider public through his work on the nuclear test ban treaty and for his quite serious considerations of space travel and the "greening of the galaxy." He also reached a wider audience with the publication of his autobiography *Disturbing the Universe* (1980) and his 1985 Gifford Lectures, *Infinite in All Directions* (1988). More recently (2007) he claimed that the global warming problem had been exaggerated. He has long advocated reducing carbon dioxide by planting large areas with trees.

E

Eastman, George (1854–1932) *American inventor*

Eastman, who was born in Waterville, New York, began his career in banking and insurance but turned from this to photography. In 1880 he perfected the dry-plate photographic film and began manufacturing this. He produced a transparent roll film in 1884 and in the same year founded the Eastman Dry Plate and Film Company. In 1888 he introduced the simple hand-held box camera that made popular photography possible. The Kodak camera with a roll of transparent film was cheap enough for all pockets and could be used by a child. It was followed by the Brownie camera, which cost just one dollar.

Eastman gave away a considerable part of his fortune to educational institutions, including the Massachusetts Institute of Technology. He committed suicide in 1932.

Ebashi, Setsuro (1922–2006) *Japanese biochemist*

Ebashi (e-**bash**-ee), who was born in the Japanese capital of Tokyo, received his MD from the university there in 1944 and his PhD in 1954. He became professor of pharmacology there in 1959 and, from 1963, held the chair of biochemistry.

Ebashi was, for many years, one of the leading workers in the field of muscle contraction. His work threw considerable light on the identity and workings of the so-called "relaxing factor." As early as 1952, B. Marsh had isolated a substance from muscle that produces relaxation in muscle fibers, and he noted that its effect could be neutralized by the presence of calcium ions.

While the process of muscle contraction appeared to be initiated by the release of calcium ions from the sarcoplasmic reticulum, Ebashi and his colleagues were able to show, in the 1960s, that such ions were not enough. The presence of two globular proteins, troponin and tropomyosin, is also necessary. Neither protein alone is sufficient as only the complex of both sensitizes muscle to calcium ions.

The globular proteins appear to prevent the interaction of myosin and actin as first described by Hugh HUXLEY. However, once a certain level of calcium is reached such inhibition is prevented and contraction occurs. Later x-ray analysis by many workers seems to have confirmed Ebashi's model.

Eccles, Sir John Carew (1903–1997) *Australian physiologist*

Born in Melbourne, Australia, Eccles was educated at the university there and at Oxford University. In Oxford he worked with Charles SHERRINGTON on muscular reflexes and nervous transmission across the synapses (nerve junctions) from 1927 to 1937. He then worked in Australia at the Institute of Pathology from 1937 to 1943. After a period in New Zealand, as professor of physiology at the University of Otago from 1944 to 1951, Eccles returned to Australia to the Australian National University, Canberra, where he served as professor of physiology from 1951 to 1966. In 1966 Eccles moved to the U.S., working first in Chicago and finally, from 1968 until his retirement in 1975, at the State University of New York, Buffalo.

While at Canberra Eccles carried out work on the chemical changes that take place at synapses, pursuing the findings of Alan HODGKIN and Andrew HUXLEY, with whom he subsequently shared the 1963 Nobel Prize for physiology or medicine. Eccles showed that excitation of different nerve cells causes the synapses to release a substance (probably acetylcholine) that promotes the passage of sodium and potassium ions and effects an alternation in the polarity of the electric charge. It is in this way that nervous impulses are communicated or inhibited by nerve cells. Eccles is the author of *Reflex Activity of the Spinal Cord* (1932) and *The Physiology of Nerve Cells* (1957).

After his retirement Eccles began to publish a number of works on the mind-body problem. Notable among them are *The Self and the Brain* (1977), written in collaboration with Karl POPPER, *The Human Mystery* (1979), and *The Creation of the Self* (1989).

Eckert, John Presper Jr. (1919–1995) *American computer scientist*

Born in Philadelphia, Pennsylvania, Eckert was educated at the Moore School of Electrical Engineering at the University of Pennsylvania in his native city. After graduating in 1941 he immediately joined the faculty. Soon after he began his long and profitable career with his colleague, J. W. MAUCHLY. Together they built the historically important computers, ENIAC, EDSAC, and UNIVAC.

In 1946 Eckert resigned from the Moore School to set up the Electronic Control Co. with his colleague, Mauchly. As they failed to raise any money on Wall Street they began with a $25,000 loan from Eckert's father. Initially things went well but the constraints of a fixed-price contract with the U.S. Census Bureau led them into virtual bankruptcy. Consequently, they were forced to sell out to Remington Rand in 1950. They received $200,000 plus a guaranteed executive position for eight years in a separate UNIVAC Division.

While some of the initial ideas emerged from Mauchly, Eckert has been described as "the mainspring of the whole operation." Thus it was Eckert who solved the problem of the vacuum tubes. How does one ensure that sufficient numbers of the 17,000 tubes of ENIAC work at any one time to keep the computer running? The answer was stringent testing and, less obviously, to run the tubes well below their rated voltage and so extend their working lives.

It was also Eckert who devised the mercury delay lines, used as memory stores in early computers, and successfully deployed by M. V. WILKES in EDSAC.

Eddington, Sir Arthur Stanley (1882–1944) *British astrophysicist and mathematician*

> We used to think that if we knew one, we knew two, because one and one are two. We are finding that we must learn a great deal more about "and".
> —Quoted by A. L. Mackay in *The Harvest of a Quiet Eye* (1977)

Born at Kendal in northwestern England, Eddington moved with his mother and sister to the southwestern county of Somerset after the death of his father in 1884. He was a brilliant scholar, graduating from Owens College (now the University of Manchester) in 1902 and from Cambridge University in 1905. From 1906 to 1913 he was chief assistant to the Astronomer Royal at Greenwich after which he returned to Cambridge as Plumian Professor of Astronomy. He was knighted in 1930. Eddington was a Quaker throughout his life.

Eddington was the major British astronomer of the interwar period. His early work on the motions of stars was followed, from 1916 onward, by his work on the interior of stars, which was published in his first major book, *The Internal Constitution of the Stars* (1926). He introduced "a phenomenon ignored in early investigations, which may have considerable effect on the equilibrium of a star, viz. the pres-

sure of radiation." He showed that for equilibrium to be maintained in a star, the inwardly directed force of gravitation must be balanced by the outwardly directed forces of both gas pressure and radiation pressure. He also proposed that heat energy was transported from the center to the outer regions of a star not by convection, as thought hitherto, but by radiation.

It was in this work that Eddington gave a full account of his mass-luminosity relationship, which was discovered in 1924 and shows that the more massive a star the more luminous it will be. The value of the relation is that it allows the mass of a star to be determined if its intrinsic brightness is known. This is of considerable significance since only the masses of binary stars can be directly calculated. Eddington realized that there was a limit to the size of stars: relatively few would have masses exceeding 10 times the mass of the Sun while any exceeding 50 solar masses would be unstable owing to excessive radiation pressure. Eddington wrote a number of books for both scientists and laymen. His more popular books, including *The Expanding Universe* (1933), were widely read, went through many editions, and opened new worlds to many enquiring minds of the interwar years. It was through Eddington that EINSTEIN's general theory of relativity reached the English-speaking world. He was greatly impressed by the theory and was able to provide experimental evidence for it. He observed the total solar eclipse of 1919 and submitted a report that captured the intellectual imagination of his generation. He reported that a very precise and unexpected prediction made by Einstein in his general theory had been successfully observed; this was the very slight bending of light by the gravitational field of a star – the Sun. Further support came in 1924 when Einstein's prediction of the reddening of starlight by the gravitational field of the star was tested: at Eddington's request Walter ADAMS detected and measured the shift in wavelength of the spectral lines of Sirius B, the dense white-dwarf companion of the star Sirius. Eddington thus did much to establish Einstein's theory on a sound and rigorous foundation and gave a very fine presentation of the subject in his *Mathematical Theory of Relativity* (1923).

Eddington also worked for many years on an obscure but challenging theory, which was only published in his posthumous work, *Fundamental Theory* (1946). Basically, he claimed that the fundamental constants of science, such as the mass of the proton and the mass and charge of the electron, were a "natural and complete specification for constructing a universe" and that their values were not accidental. He then set out to develop a theory from which such values would follow as a consequence, but never completed it.

Edelman, Gerald Maurice (1929–　)
American biochemist

Born in New York City, Edelman was educated at Ursinus College, the University of Pennsylvania, and Rockefeller University, where he obtained his PhD on human immunoglobulins in 1960. He remained at Rockefeller where he was appointed professor of biochemistry in 1966 and Vincent Astor Distinguished Professor in 1974. Edelman left Rockefeller in 1992 to set up and direct the Neuroscience Institute at the Scripps Research Institute, La Jolla, California.

Edelman was interested in determining the structure of human immunoglobulin. The molecule is very large and it was first necessary to break it into smaller portions, which was achieved by reducing and splitting the disulfide bonds. Following this, Edelman proposed that the molecule contained more than one polypeptide chain and, moreover, that two kinds of chain exist, light and heavy. Such studies helped Rodney PORTER propose a structure for the antibody immunoglobulin G (IgG) in 1962.

Edelman was more interested in attempting to work out the complete amino-acid sequence of IgG. As it contained 1,330 amino acids it was by far the largest protein then attempted. By 1969 he was ready to announce the results of his impressive work, the complete sequence, and was able to show that while much of the molecule was unchanging the tips of the Y-like structure were highly variable in their amino-acid sequence. It thus seemed obvious that such an area would be identical with the active antigen binding region in Porter's structure and that such variability represented the ability of IgG to bind many different antigens. It was for this work that Edelman and Porter shared the 1972 Nobel Prize for physiology or medicine.

Edelman has also speculated on antibody formation and the mechanism behind the spurt in production after contact with an antigen. In the former area he argued in 1966 for a major modification of the clonal theory of Macfarlane BURNET. In the latter case he suggested, in 1970, that the signal to the immune system to increase production is set off by the change in shape of the antibody molecule as it combines with its antigen.

Following his biochemical successes Edelman turned to the neurosciences. In such works as *Neural Darwinism* (1987) and *Bright Air, Brilliant Mind* (1993), he produced a distinctive theory of the development and nature of the mind. We are, he claims, at the beginning of a neuroscientific revolution from which we will learn "how the mind works, what governs our nature, and how we know the world." Edelman was struck by a number of similarities between the immune system and the nervous system. Just as a lymphocyte can recognize and respond to a new antigen, the nervous system can respond similarly to novel stimuli. Neural mechanisms are selected, he argued, in the same manner as antibodies. Although the 10^9 cells of the nervous system do not replicate, there is considerable scope for development and variation in the connections that form between the cells. Frequently used connections will be selected, others will decay or be diverted to other uses. There are two kinds of selection: developmental, which takes place before birth, and experiential. There are also innate "values," built in preferences that is, for such features as light and warmth over the dark and the cold. In Edelman's model, higher consciousness, including self-awareness and the ability to create scenes in the mind, have required the emergence during evolution of a new neuronal circuit. To remember a chair or one's grandmother is not to recall a bit of coded data from a specific location; it is rather to create a unity out of scattered mappings, a process called by Edelman a "reentry." Edelman's views have been dismissed by many as obscure; some neurologists, however, consider Edelman to have begun what will eventually turn out to be a major revolution in the neurosciences.

Edison, Thomas Alva (1847–1931) *American physicist and inventor*

> In 1879 the performance of the Edison telephone was illustrated in the theatre of the Royal Institution…. Perhaps the most striking illustration of the pliant power of the instrument was its capability to reproduce a whistled tune. Mr Edison's whistling at the Circus was heard in Albemarle Street almost as distinctly as if it had been produced upon the spot.
> —John Tyndall, *On Sound* (1883)

> My personal desire would be to prohibit entirely the use of alternating currents. They are unnecessary as they are dangerous … I can therefore see no justification for the introduction of a system which has no element of permanency and every element of danger to life and property.
> —Quoted by R. L. Weber in *A Random Walk in Science*

> Genius is one percent inspiration and ninety-nine percent perspiration.
> —Quoted in *Life* (1932)

Edison was born in Milan, Ohio, and was taught at home by his mother – he had been expelled from school as "retarded," perhaps because of his deafness. From the age of seven he lived in Port Huron, Michigan, and when he was twelve years old began to spend much of his time on the railroad between Port Huron and Detroit, selling candy and newspapers to make money. However, he was also fascinated by the telegraph system, designing his own experiments and training himself in telegraphy. He became a casual worker on telegraphy (1862–68), reading and experimenting as he traveled. At the age of 21 he bought a copy of FARADAY's *Experimental Researches in Electricity* and was inspired to undertake serious systematic experimental work.

While Edison was living in a Wall Street basement (1869) he was called in to carry out an emergency repair on a new telegraphic gold-price indicator in the Gold Exchange. He was so successful that he was taken on as a supervisor. Later he remodeled the equipment and, soon after being commissioned to improve other equipment, his skill became legendary.

For a while Edison had a well-paid job with the Western Union Telegraph Company, but he gave it up to set up a laboratory of his own at Menlo Park, New Jersey. This he furbished with a wide range of scientific equipment, costing $40,000, and an extensive library. He employed 20 technicians and later a mathematical physicist. The laboratory was the first organized research center outside a university and produced many inventions. In 1877 Edison became known internationally after the phonograph was invented. His original instrument used a cylinder coated with tinfoil to record sounds, and was not commercially practical. In 1878, after seeing an exhibition of glaring electric arc lights, he declared that he would invent a milder cheap alternative that could replace the gas lamp. Because of his past successes, he managed to raise the capital to do this and the Edison Electric Light Company was set up. It took 14 months to find a filament material but by October 1879, Edison was able to demonstrate 30 incandescent electric lamps connected in parallel with separate switches. Three years later a power station was opened in New York and this was the start of modern large-scale electricity generation. Edison later merged his electric-light company with that of Joseph SWAN who developed the carbon-filament light independently. Also in his work on incandescent filaments, Edison discovered that a current flows in one direction only between the filament and a nearby electrode. The use of this *Edison effect* in the thermionic valve was independently achieved by J. A. FLEMING. In 1887 Edison's laboratory moved to larger premises in West Orange, now a national monument. In his lifetime he took out over 1,000 patents covering a variety of applications, including telephone transmission, cinematography, office machinery, cement manufacture, and storage batteries. No other inventor has been so productive.

Edlén, Bengt (1906–1993) *Swedish physicist*

Born at Gusum in Sweden, Edlén (ed-**lyen**) studied at the University of Uppsala, where he gained his PhD in 1934 and also served on the faculty from 1928 until 1944. He was then appointed professor of physics at the University of

Lund, a post he retained until his retirement in 1973.

Edlén is recognized for his research on atomic spectra and its applications to astrophysics. In the early 1940s he carried out important work on the emission lines, first described in 1870, in the Sun's corona, i.e., the outermost layer of the solar atmosphere. The problem with the 20 "well-measured lines" was that none had ever been observed in a laboratory light source. At one time they were thought to indicate the presence of an unknown element, conveniently described as "coronium," but it had become apparent when Edlén began his work that the periodic table no longer contained any suitable gaps.

Edlén succeeded in showing in 1941 that the coronal lines were mainly caused by iron, nickel, calcium, and argon atoms deprived of 10–15 electrons; i.e., by highly charged positive ions. The implications of such extreme ionization were not lost on Edlén, who was quick to point out that it must indicate temperatures of over a quarter of a million degrees in the solar corona.

Edsall, John Tileston (1902–2002) *American biochemist*

Edsall was born in Philadelphia, Pennsylvania, and educated at Harvard and at Cambridge University, England. He joined the Harvard faculty in 1928 and from 1951 to 1973 served there as professor of biochemistry.

Edsall was basically a protein chemist. He spent much time establishing basic data on the constitution and properties of numerous proteins – information that has since been reproduced in innumerable textbooks. With Edwin Cohn he was the author of the authoritative work *Proteins, Amino Acids and Peptides* (1943).

In later years Edsall turned his attention to the history of biochemistry, his books in this field including *Blood and Hemoglobin* (1952).

Egas Moniz, Antonio (1874–1955) *Portuguese neurologist*

> Frontal leukotomy ... one of the most important discoveries ever made in psychiatric therapy...a great number of suffering people and total invalids have recovered and have been socially rehabilitated.
> —Herbert Olivecrona, presentation speech at award of the 1949 Nobel Prize

Egas Moniz (eg-**ash** mon-**ish**) was born at Avanca in Portugal and educated at the University of Coimbra, where he gained his MD in 1899. After postgraduate work in Paris and Bordeaux he returned to Coimbra, becoming a professor in the medical faculty in 1902. He moved to Lisbon in 1911 to a newly created chair of neurology, a post he retained until his retirement in 1944. At the same time he was pursuing a successful political career, being elected to the National Assembly in 1900. He served as ambassador to Spain in 1917 and in the following year became foreign minister, leading his country's delegation to the Paris Peace Conference.

Egas Moniz achieved his first major success in the 1920s in the field of angiography (the study of the cardiovascular system using dyes that are opaque to x-rays). In collaboration with Almeida Lima, he injected such radiopaque dyes into the arteries, enabling the blood vessels of the brain to be photographed. In 1927 he was able to show that displacements in the cerebral circulation could be used to infer the presence and location of brain tumors, publishing a detailed account of his technique in 1931.

Egas Moniz is better known for his introduction in 1935 of the operation of prefrontal leukotomy. It was for this work, described by the Nobel authorities as "one of the most important discoveries ever made in psychiatric therapy," that they awarded him the 1949 Nobel Prize for physiology or medicine.

The operation consisted of inserting a sharp knife into the prefrontal lobe of the brain, roughly the area above and between the eyes; it required the minimum of equipment and lasted less than five minutes. The technique was suggested to Egas Moniz on hearing an account (by John Fulton and Carlyle Jacobsen in 1935) of a refractory chimpanzee that became less aggressive after its frontal lobes had been excised. Egas Moniz believed that a similar surgical operation would relieve severe emotional tension in psychiatric patients. He claimed that 14 of the first 20 patients operated upon were either cured or improved. The operation generated much controversy, since the extent of the improvement in the patients' symptoms was not easy to judge and the procedure often produced severe side-effects. Today a more refined version of the operation, in which selective incisions are made in smaller areas of the brain, is still quite widely practiced.

Ehrenberg, Christian Gottfried (1795–1876) *German biologist and microscopist*

> Until now my favorite pursuit has been neither naked systematizing nor unsystematic observation, and whenever time and circumstances, together with my ability, allow it, I prefer getting down to the grass-roots level.
> —Letter to Nees von Esenbeck (1821)

Ehrenberg (**air**-en-berg) was born in Delitzsch, which is now in Germany, and took an MD degree at the University of Berlin in 1818. In the same year he was elected a member of the Leopoldine German Academy of Researchers in Natural Sciences. Two years later he took

part in a scientific expedition to Egypt, Libya, the Sudan, and the Red Sea, sponsored by the Prussian Academy of Sciences and the University of Berlin. During these travels (1820–25) Ehrenberg collected and classified some 75,000 plant and animal specimens, both terrestrial and marine, including microorganisms. On a further expedition in 1829 to Central Asia and Siberia, sponsored by Czar Nicholas I, he was accompanied by Alexander von HUMBOLDT. In 1827 Ehrenberg was appointed assistant professor of zoology at Berlin and was elected a member of the Berlin Academy of Sciences. He became professor of natural science at Berlin in 1839.

Ehrenberg's studies in natural science were primarily concerned with microorganisms, especially the protozoans. His paleontological investigations led him to demonstrate the presence of single-celled fossils in certain rock layers of various geological formations, and he was also able to show that marine phosphorescence (strictly, bioluminescence) was due to the activity of species of animal plankton. He described fungal development from spores, as well as their sexual reproduction, and also carried out detailed studies of corals. His most important thesis lay in the belief that microorganisms were complete in the sense of sharing the same organs as higher animals, and that social behavior provided the basis for a new approach to animal classification. Ehrenberg's theory was demolished, with the help of experimental evidence, by Félix DUJARDIN. Ehrenberg's publications include *Travels in Egypt, Libya, Nubia and Dongola* (1828) and *The Infusoria as Complete Animals* (1838). His descriptions and classification of fossil protozoans were published as *Microgeology* (1854).

Ehrlich, Paul (1854–1915) *German physician, bacteriologist, and chemist*

> Success in research needs four Gs: *Glück, Geduld, Geschick* und *Geld* [luck, patience, skill, and money].
> —Quoted by M. Peruz in *Nature* (1988)

Born in Strehlen (now Strzelin in Poland), Ehrlich (**air**-lik) studied medicine at the universities of Breslau, Strasbourg, and Freiburg, gaining a physician's degree at Breslau in 1878. For the next nine years he worked at the Charité Hospital, Berlin, on many topics including typhoid fever, tuberculosis, and pernicious anemia. He was awarded the title of professor by the Prussian Ministry of Education in 1884 for his impressive work in these fields. In 1887 he became a teacher at the University of Berlin but was not paid because of the antisemitic feeling at the time – Ehrlich would not renounce his Jewish upbringing. As a result of his laboratory work he contracted tuberculosis and was not restored to health until 1890, when he set up his own small research laboratory at Steglitz on the outskirts of Berlin.

In 1890 Robert KOCH announced the discovery of tuberculin and suggested its use in preventing and curing tuberculosis. He asked Ehrlich to work on it with him at the Moabit Hospital in Berlin. Ehrlich accepted and for six years studied TB and cholera. In 1896 he accepted the post of director of the new Institute for Serum Research and Serum Investigation at Steglitz and in 1899 moved to the Institute of Experimental Therapy in Frankfurt. Here he investigated African sleeping sickness and syphilis along with his other studies. In 1908 he was awarded the Nobel Prize for physiology or medicine for his work on immunity and serum therapy.

Two years later he announced his most famous discovery, Salvarsan – a synthetic chemical that was effective against syphilis – and until the end of his life he worked on the problems associated with the treatment of patients using this compound of arsenic.

Ehrlich is considered to be the founder of modern chemotherapy because he developed systematic scientific techniques to search for new synthetic chemicals that could specifically attack disease-causing microorganisms. Ehrlich sought for these "magic bullets" by carefully altering the chemical structure of dye molecules that selectively stained the microorganisms observable in his microscope but did not stain cells in the host. He was persevering and optimistic – Salvarsan (compound number 606) was not "rediscovered" until almost 1,000 compounds had been synthesized and tried. He made and tested about 3,000 compounds based on the structure of Salvarsan in an attempt to make a drug that was bacteriocidal to streptococci.

Eichler, August Wilhelm (1839–1887) *German botanist*

Eichler (**Ik**-ler) was born in Neukirchen (which is now in Germany) and studied natural science and mathematics at the University of Marburg, gaining his PhD in 1861. Eichler then became assistant to Karl von Martius at Munich. With Martius he began editing the 15-volume *Flora of Brazil*, continuing this work single-handed after Martius's death in 1868. In 1878 he was appointed professor of systematic and morphologic botany at Berlin and also became director of the university's herbarium. The same year he published the second volume of his two-volume *Diagrams of Flowers*, describing the comparative structure of flowers.

In 1886, the year before his death, Eichler developed a plant classification system in which the plant kingdom is split into four divisions: Thallophyta (algae, fungi), Bryophyta (mosses, liverworts), Pteridophyta (ferns, horsetails), and Spermatophyta (seed plants). He further subdivided the Spermatophyta into the Gymnospermae (conifers, cycads, ginkgos) and Angiospermae (flowering plants). This system was later adopted by nearly all botanists, but because of his early death from leukemia Eichler did not live to see the general acceptance of his work.

Eigen, Manfred (1927–) *German physical chemist*

> A theory has only the alternative of being right or wrong. A model has a third possibility: it may be right, but irrelevant.
> —*The Physicist's Conception of Nature* (1973)

Eigen (**I**-gen), the son of a musician, was born at Bochum in Germany and educated at the University of Göttingen where he obtained his PhD in 1951. He joined the staff of the Max Planck Institute for Physical Chemistry at Göttingen in 1953 and served as its director from 1964 until his retirement in 1995.

In 1954 Eigen introduced the so-called relaxation techniques for the study of extremely fast chemical reactions (those taking less than a millisecond). Eigen's general method was to take a solution in equilibrium for a given temperature and pressure. If a short disturbance was applied to the solution the equilibrium would be very briefly destroyed and a new equilibrium quickly reached. Eigen studied exactly what happened in this very short time by means of absorption spectroscopy. He applied disturbances to the equilibrium by a variety of methods, such as pulses of electric current, sudden changes in temperature or pressure, or changes in electric field.

The first reaction he investigated was the apparently simple formation of a water molecule from the hydrogen ion, H^+, and the hydroxide ion, OH^-. Calculations of reaction rates made it clear that they could not be produced by the collision of the simple ions H^+ and OH^-. Eigen went on to show that the reacting ions are the unexpectedly large $H_9O_4^+$ and $H_7O_4^-$, a proton hydrated with four water molecules and a hydroxyl ion with three water molecules. For this work Eigen shared the 1967 Nobel Prize for chemistry with George PORTER and Ronald NORRISH.

Eigen later applied his relaxation techniques to complex biochemical reactions. He has also become interested in the origin of nucleic acids and proteins; with his colleague R. WINKLER he has proposed a possible mechanism to explain their formation. Much of this and subsequent work was described by Eigen in his *Laws of the Game: How Principles of Nature Govern Chance* (1982).

Eijkman, Christiaan (1858–1930) *Dutch physician*

Eijkman (**Ik**-man) was born at Nijkerk in the

Netherlands and qualified as a physician from the University of Amsterdam in 1883. He served as an army medical officer in the Dutch East Indies from 1883 to 1885, when he was forced to return to the Netherlands to recuperate from a severe attack of malaria. In 1886 he returned to the East Indies as a member of an official government committee to investigate beriberi. After the completion of the committee's work, Eijkman remained in Batavia (now Djakarta) as director of a newly established bacteriological laboratory. In 1896 he took up the post of professor of public health at the University of Utrecht.

Eijkman was responsible for the first real understanding of the nature and possible cure of beriberi. For this work he shared the 1929 Nobel Prize for physiology or medicine with Frederick Gowland HOPKINS. Beriberi is a disorder caused by dietary deficiency, producing fatal lesions in the nervous and cardiovascular systems. Physicians of the late 19th century, however, were not trained to recognize its cause: with the clear success of the germ theory recently demonstrated by Robert KOCH it was difficult to realize that symptoms could be produced by the absence of something rather than by the more obvious presence of a visible pathogen. Eijkman's discovery was prompted by the outbreak of a disease very similar to human beriberi among the laboratory chickens. Despite the most thorough search no causative microorganisms could be identified, and then, for no obvious reason, the disease disappeared.

On investigation, Eijkman discovered that the symptoms of the disease had developed during a period of five months in which the chickens' diet was changed to hulled and polished rice. With a return to their normal diet of commercial chicken feed the symptoms disappeared. Eijkman subsequently found that he could induce the disease with a diet of hulled and polished rice and cure it with one of whole rice. However, he failed to conclude that beriberi was a deficiency disease. He argued that the endosperm of the rice produced a toxin that was neutralized by the outer hull: by eating polished rice the toxin would be released in its unneutralized form. Thus although Eijkman had clearly demonstrated how to cure and prevent beriberi it was left to Hopkins to identify its cause as a vitamin deficiency. It was not until the early 1930s that Robert WILLIAMS identified the vitamin as vitamin B_1 (thiamine).

Einstein, Albert (1879–1955) *German– Swiss–American theoretical physicist*

> Everything should be made as simple as possible, but not simpler.
> —Quoted in *Reader's Digest*, October 1977

> Imagination is more important than knowledge.
> —*On Science*

Einstein was born at Ulm in Germany where his father was a manufacturer of electrical equipment. Business failure led his father to move the family first to Munich, where Einstein entered the local gymnasium in 1889, and later to Milan. There were no early indications of Einstein's later achievements for he did not begin to talk until the age of three, nor was he fluent at the age of nine, causing his parents to fear that he might even be backward. It appears that in 1894 he was expelled from his Munich gymnasium on the official grounds that his presence was "disruptive." At this point he did something rather remarkable for a fifteen-year-old boy. He had developed such a hatred for things German that he could no longer bear to be a German citizen. He persuaded his father to apply for a revocation of his son's citizenship, a request the authorities granted in 1896. Until 1901, when he obtained Swiss citizenship, he was in fact stateless.

After completing his secondary education at Aarao in Switzerland he passed the entrance examination, at the second attempt, to the Swiss Federal Institute of Technology, Zurich, in 1896. He did not appear to be a particularly exceptional student, finding the process of working for examinations repellent. Disappointed not to be offered an academic post, he survived as a private tutor until 1902, when he obtained the post of technical expert, third class, in the Swiss Patent Office in Bern. Here he continued to think about and work on physical problems. In 1905 he published four papers in the journal *Annalen der Physik* (Annals of Physics) – works that were to direct the progress of physics during the 20th century. The first, and most straightforward, was on Brownian motion – first described by Robert BROWN in 1828. Einstein derived a formula for the average displacement of particles in suspension, based on the idea that the motion is caused by bombardment of the particles by molecules of the liquid. The formula was confirmed by Jean PERRIN in 1908 – it represented the first direct evidence for the existence of atoms and molecules of a definite size. The paper was entitled *Über der von molekularkinetischen Theorie der Wärme geförderte Bewegung von in ruhenden Flüssigkeiten suspendierten Teilchen* (On the Motion of Small Particles Suspended in a Stationary Liquid According to the Molecular Kinetic Theory of Heat).

His second paper of 1905 was *Über einen die Erzeugung und Verwandlung des Lichtes betreffenden heuristischen Gesichtspunkt* (On a Heuristic Point of View about the Creation and Conversion of Light). In this Einstein was concerned with the nature of electromagnetic radiation, which at the time was regarded as a wave propagated throughout space according to

Clerk MAXWELL's equations. Einstein was concerned with the difference between this wave picture and the theoretical picture physicists had of matter. His particular concern in this paper was the difficulty in explaining the photoelectric effect, investigated in 1902 by Philipp LENARD. It was found that ultraviolet radiation of low frequency could eject electrons from a solid surface. The number of electrons depended on the intensity of the radiation and the energy of the electrons depended on the frequency. This dependence on frequency was difficult to explain using classical theory.

Einstein resolved this by suggesting that electromagnetic radiation is a flow of discrete particles – quanta (or photons as they are now known). The intensity of the radiation is the flux of these quanta. The energy per quantum, he proposed, was $h\nu$, where ν is the frequency of the radiation and h is the constant introduced in 1900 by Max PLANCK. In this way Einstein was able to account for the observed photoelectric behavior. The work was one of the early results introducing the quantum theory into physics and it won for Einstein the 1921 Nobel Prize for physics.

The third of his 1905 papers is the one that is the most famous: *Zur Elektrodynamik bewegter Körper* (On the Electrodynamics of Moving Bodies). It is this paper that first introduced the special theory of relativity to science. The term "special" denotes that the theory is restricted to certain special circumstances – namely for bodies at rest or moving with uniform relative velocities.

The theory was developed to account for a major problem in physics at the time. Traditionally in mechanics, there was a simple procedure for treating relative velocities. A simple example is of a car moving along a road at 40 mph with a second car moving toward it at 60 mph. A stationary observer would say that the second car was moving at 60 mph relative to him. The driver of the first car would say that, relative to him, the second car was approaching at 100 mph. This common-sense method of dealing with relative motion was well established. The mathematical equations involved are called the Galilean transformations – they are simple equations for changing velocities in one frame of reference to another frame of reference. The problem was that the method did not appear to work for electromagnetic radiation, which was thought of as a wave motion through the ether, described by the equations derived by Maxwell. In these, the speed of light is independent of the motion of the source or the observer. At the time, Albert MICHELSON and Edward MORLEY had performed a series of experiments to attempt to detect the Earth's motion through the ether, with negative results. Hendrik LORENTZ proposed that this result could be explained by a change of size of moving bodies (the Lorentz–Fitzgerald contraction).

Although Einstein was unaware of the Michelson–Morley experiment, he did appreciate the incompatibility of classical mechanics and classical electrodynamics. His solution was a quite radical one. He proposed that the speed of light *is* a constant for all frames of reference that are moving uniformly relative to each other. He also put forward his "relativity principle" that the laws of nature are the same in all frames of reference moving uniformly relative to each other. To reconcile the two principles he abandoned the Galilean transformations – the simple method of adding and subtracting velocities for bodies in relative motion. He arrived at this rejection by arguments about the idea of simultaneity – showing that the time between two events depends on the motions of the bodies involved. In his special theory of relativity, Einstein rejected the ideas of absolute space and absolute time. Later it was developed in terms of events specified by three spatial coordinates and one coordinate of time – a space–time continuum. The theory had a number of unusual consequences. Thus the length of a body along its direction of motion decreases with increasing velocity. The mass increases as the velocity increases, becoming infinitely large in theory at the speed of light. Time slows down for a moving body – a phenomenon known as time dilation. These effects apply to all bodies but only become significant at velocities close to the speed of light – under normal conditions the effects are so small that classical laws appear to be obeyed. However, the predictions of the special theory – unusual as they may seem – have been verified experimentally. Thus increase in mass is observed for particles accelerated in a synchrocyclotron. Similarly, the lifetimes of unstable particles are increased at high velocities.

In that same year of 1905 Einstein had one more fundamental paper to contribute: *Ist die Trägheit eines Körpers von seinem Energieinhalt abhängig?* (Does the Inertia of a Body Depend on Its Energy Content?). It was in this two-page paper that he concluded that if a body gives off energy E in the form of radiation, its mass diminishes by E/c^2 (where c is the speed of light) obtaining the celebrated equation $E = mc^2$ relating mass and energy.

Within a short time Einstein's work on relativity was widely recognized to be original and profound. In 1908 he obtained an academic post at the University of Bern. Over the next three years he held major posts at Zurich (1909), Prague (1911), and the Zurich Federal Institute of Technology (1912) before taking a post in Berlin in 1914. This was probably due in part to the respect in which he held the Berlin physicists, Max Planck and Walther NERNST.

By 1907 Einstein was ready to remove the restrictions imposed on the special theory showing that, on certain assumptions, accelerated motion could be incorporated into his new, general theory of relativity. The theory begins with the fact that the mass of a body can be defined in two ways. The inertial mass depends on the way it resists change in motion, as in NEWTON's second law. The gravitational mass depends on forces of gravitational attraction between masses. The two concepts – inertia and gravity – seem dissimilar yet the inertial and gravitational masses of a body are always the same. Einstein considered that this was unlikely to be a coincidence and it became the basis of his principle of equivalence.

The principle states that it is impossible to distinguish between an inertial force (that is, an accelerating force) and a gravitational one; the two are, in fact, equivalent. The point can be demonstrated with a thought experiment. Consider an observer in an enclosed box somewhere in space far removed from gravitational forces. Suppose that the box is suddenly accelerated upward, followed by the observer releasing two balls of different weights. Subject to an inertial force they will both fall to the floor at the same rate. But this is exactly how they would behave if the box was in a gravitational field and the observer could conclude that the balls fall under the influence of gravity. It was on the basis of this equivalence that Einstein made his dramatic prediction that rays of light in a gravitational field move in a curved path. For if a ray of light enters the box at one side and exits at the other then, with the upward acceleration of the box, it will appear to exit at a point lower down than its entrance. But if we take the equivalence principle seriously we must expect to find the same effect in a gravitational field. In 1911 he predicted that starlight just grazing the Sun should be deflected by 0.83 seconds of arc, later increased to 1″.7, which, though small, should be detectable in a total eclipse by the apparent displacement of the star from its usual position. In 1919 such an eclipse took place; it was observed by Arthur EDDINGTON at Principe in West Africa, who reported a displacement of 1″.61, well within the limits of experimental error. It was from this moment that Einstein became known to a wider public, for this dramatic confirmation of an unexpected phenomenon seemed to capture the popular imagination. Even the London *Times* was moved to comment in an editorial, as if to a recalcitrant government, that "the scientific conception of the fabric of the universe must be changed." In 1916 Einstein was ready to publish the final and authoritative form of his general theory: *Die Grundlage der allgemeinen Relativitätstheorie* (The Foundation of the General Theory of Relativity). It is this work that gained for Einstein the reputation for producing theories that were comprehensible to the very few. Eddington on being informed that there were only three people capable of understanding the theory is reported to have replied, "Who's the third?" It is true that Einstein introduced into gravitational

theory a type of mathematics that was then unfamiliar to most physicists, thus presenting an initial impression of incomprehension. In his theory Einstein used the space–time continuum introduced by Hermann MINKOWSKI in 1907, the non-Euclidean geometry developed by Bernhard RIEMANN in 1854, and the tensor calculus published by Gregorio Ricci in 1887. He was assisted in the mathematics by his friend Grossmann. The theory of gravitation produced is one that depends on the geometry of space–time. In simple terms, the idea is that a body "warps" the space around it so that another body moves in a curved path (hence the notion that space is curved). Einstein and Grossmann in 1915 succeeded in deriving a good theoretical value for the small (and hitherto anomalous) advance in the perihelion of Mercury. The theory was put to an early test. Because of perturbations in the orbit of Mercury produced by the gravitational attractions of other planets, its perihelion (point in the orbit closest to the Sun) actually precesses by a small amount (9′ 34″ per century). When these perturbation effects were calculated on the basis of Newtonian mechanics, they could only account for a precession rate of 8′ 51″ per century, a figure 43″ too small. In 1915 Einstein, while completing his 1916 paper on General Relativity, calculated Mercury's perihelion precession on the basis of his own theory and found that, without making any extra assumptions, the missing 43″ were accounted for. The discovery, Einstein later reported, gave him palpitations and "for a few days I was beside myself with joyous excitement."

The theory also predicted (1907) that electromagnetic radiation in a strong gravitation field would be shifted to longer wavelengths – the *Einstein shift*. This was used by Walter ADAMS in 1925 to explain the spectrum of Sirius B. In 1959 Robert Pound and Glen Rebka demonstrated it on Earth using the Mossbauer effect. They found that at a height of 75 feet (23 m) above the ground gamma rays from a radioactive source had a longer wavelength than at ground level. Physicists have been less successful, however, with the prediction in 1916 of the existence of gravitational waves. Despite an intensive search from 1964 onward by Joseph WEBER and others, they have yet to be detected.

Einstein was less successful in applying his theory to the construction of a cosmological model of the universe, which he assumed to be uniform in density, static, and lacking infinite distances. He found himself forced to complicate his equations with a cosmological constant, λ. It was left to Aleksandr FRIEDMANN in 1922 to show that the term could be dropped and a solution found that yielded an expanding universe, a solution that Einstein eventually adopted. He later described the cosmological constant as "my greatest mistake." By the early

1920s Einstein's great work was virtually complete. He wrote in 1921 that: "Discovery in the grand manner is for young people ... and hence for me a thing of the past." From the early 1920s he rejected quantum theory – the theory he had done much to establish himself. His basic objection was to the later formulation that included the probability interpretation. "God does not play dice," he said, and, "He may be subtle, but he is not malicious." He felt, like Louis de BROGLIE, that although the new quantum mechanics was clearly a powerful and successful theory it was an imperfect one, with an underlying undiscovered deterministic basis. For the last 30 years of his life he also pursued a quest for a unified field theory – a single theory to explain both electromagnetic and gravitational fields. He published several attempts at such a theory but all were inadequate. This work was carried out right up to his death.

Also, from about 1925 onward, Einstein engaged in a debate with Niels BOHR on the soundness of quantum theory. He would present Bohr with a series of thought experiments, which seemed undeniable even though they were clearly incompatible with quantum mechanics. The best known of these was presented in a paper written with Boris Podolsky and Nathan Rosen and entitled: *Can Quantum Mechanical Description of Physical Reality Be Considered Complete?* (1935). The *EPR experiment*, as it soon became known, assumed that, after interacting, two particles become widely separated. Quantum theory allows the total momentum of the pair (A, B) to be measured accurately. Thus if the momentum of B is also measured accurately, it is a simple matter, as momentum is conserved, to calculate the momentum of particle A. We can then measure the position of A with as much precision as is practically possible. It would therefore seem to follow that, without violating any laws of physics, both the position and momentum of particle A have been accurately determined. But, according to the uncertainty principle of HEISENBERG, we are prevented from ever knowing accurately a particle's position and momentum simultaneously.

The paper troubled Bohr. He spent six weeks going through the text word by word and analyzing every possibility. Eventually he saw that the measurements of A's position and momentum are separate and distinct. The uncertainty principle insisted that no *single* measurement could determine a particle's precise position and momentum, and this central claim remained unchallenged by the EPR experiment. Einstein's correspondence with Bohr about quantum mechanics is published as the *Bohr–Einstein letters*.

Einstein was also involved in a considerable amount of political activity. When Hitler came to power in 1933 Einstein made his permanent home in America where he worked at Princeton. In 1939 he was persuaded to write to President Roosevelt warning him about the possibility of an atomic bomb and urging American research. He was, in later years, a convinced campaigning pacifist. He was also a strong supporter of Zionist causes and, on the death of Chaim Weizmann in 1952, was asked to become president of Israel, but declined.

Einthoven, Willem (1860–1927) *Dutch physiologist*

Einthoven (*Int*-hoh-ven), the son of a physician, was born at Semarang on the Indonesian island of Java and educated at the University of Utrecht in the Netherlands, where he gained his MD in 1885. In the following year he moved to Leiden as professor of physiology.

As early as 1887 the English physiologist Augustus Waller had recorded electric currents generated by the heart. He had used the capillary electrometer invented by Gabriel LIPPMANN in 1873, which – although sensitive to changes of a millivolt – turned out to be too complicated and inaccurate for general use. In 1901 Einthoven first described a recording system using a string galvanometer, which he claimed would overcome the inadequacies of Waller's device.

A string galvanometer consists of a fine wire thread stretched between the poles of a magnet. When carrying a current it is displaced at right angles to the directions of the magnetic lines of force to an extent proportional to the strength of the current. By linking this up to an optical system the movement of the wire can be magnified and photographically recorded. As the differences in potential developed in the heart are conducted to different parts of the body it was possible to lead the current from the hands and feet to the recording instrument to obtain a curve that was later called an *electrocardiogram* (ECG).

Having demonstrated the potentiality of such a machine, two further problems needed solution. Einthoven first had to standardize his ECG so that different machines or two recordings of the same machine would produce comparable readings. It was therefore later established that a 1 millivolt potential would deflect a recording stylus 1 centimeter on standardized paper. The second problem was how to interpret such a curve in order to distinguish normal readings from recordings of diseased hearts. By 1913 Einthoven had worked out the interpretation of the normal tracing and, by correlating abnormal readings with specific cardiac defects identified at post mortem, was able to use the ECG as a diagnostic tool.

For his development of the electrocardiogram Einthoven was awarded the 1924 Nobel Prize for physiology or medicine.

Ekeberg, Anders Gustaf (1767–1813)
Swedish chemist

Ekeberg (**ay**-ke-berg), who was born in the Swedish capital of Stockholm, graduated from the University of Uppsala in 1788 and, after traveling in Europe, began teaching chemistry at Uppsala in 1794. He was an early convert to the system of Antoine LAVOISIER and introduced this new chemistry to Sweden. He was partially deaf from a childhood illness but the further loss of an eye (1801) caused by an exploding flask did not impede his work.

Ekeberg is remembered chiefly for his discovery of the element tantalum. In 1802, while analyzing minerals from Ytterby quarry, Sweden, he isolated the new metal. The name supposedly comes from its failure to dissolve in acid, looking like Tantalus in the waters of Hell. It was a long time before it was recognized as a separate element as it was difficult to distinguish from niobium, isolated by Charles HATCHETT in 1801. WOLLASTON failed to distinguish between them and it was as late as 1865 that Jean MARIGNAC conclusively demonstrated the distinctness of the two new metals.

Ekman, Vagn Walfrid (1874–1954)
Swedish oceanographer

Ekman (**ayk**-man), the son of an oceanographer, was born in the Swedish capital of Stockholm and educated at the University of Uppsala, graduating in 1902. He worked at the International Laboratory for Oceanographic Research in Oslo (1902–08) before he moved to Lund, Sweden, as a lecturer in mathematical physics, being made a professor in 1910.

In 1905 Ekman published a fundamental paper, *On the Influence of the Earth's Rotation on Ocean Currents*. This work originated from an observation made by the explorer Fridtjof NANSEN that in the Arctic drift ice did not follow wind direction but deviated to the right. He showed that the motion, since known as the *Ekman spiral*, is produced as a complex interaction between the force of the wind on the water surface, the deflecting force due to the Earth's rotation (Coriolis force), and the frictional forces within the water layers.

Ekman also studied the phenomenon of dead water, a thin layer of fresh water from melting ice spreading over the sea, which could halt slow-moving ships. This, he established, resulted from the waves formed between water layers of different densities. The *Ekman current meter*, invented by him, is still in use.

Elhuyar, Don Fausto D' *See* D'ELHUYAR, DON FAUSTO.

Elion, Gertrude Belle (1918–1999) *American biochemist*

It was watching her beloved grandfather die slowly and painfully from stomach cancer that prompted the young Gertrude Elion, in 1933, to decide on a career in biochemistry. Thus began a vocation that would result in the discovery of many new drugs to combat cancer, AIDS, and other diseases, and bring Elion a Nobel Prize. She was born in New York and attended Hunter College, graduating BA in chemistry in 1937. She began graduate studies at New York University while working as a high-school teacher, and gained her MS degree in 1941. However, prejudice against women in scientific research meant that she had to take a series of humdrum jobs, and opportunity came only with the entry of the United States into World War II. Now, women were in demand to fill jobs vacated by servicemen, and in 1944 Elion joined the Wellcome Research Laboratories, Tuckahoe, NY, where she began a long and fruitful collaboration with George HITCHINGS. Guided by Hitchings, Elion worked on identifying compounds that would interfere with the synthesis of DNA by cancer cells. By 1951 she had synthesized a purine analogue, 6-mercaptopurine (6MP), that halted DNA synthesis in leukemia cells and subsequently proved to be an effective treatment for certain types of human cancer. The drug was also shown to suppress the immune system in dogs, which motivated Elion and Hitchings to search for other immunosuppressive drugs. This led to the discovery of the 6MP derivative, azathioprine, which is still used to prevent rejection of human kidney transplants by the immune system. Several other notable discoveries were made by Elion and Hitchings during the 1950s and 1960s, including allopurinol (used to treat gout), pyrimethamine (an antimalaria drug), and trimethoprim (an antibacterial drug). Another breakthrough, this time against viral disease, came in 1969, when one of Elion's compounds yielded the forerunner of acyclovir, a drug that combats herpesviruses responsible for shingles, chickenpox, and other infections. The first anti-HIV drug, azidothymidine (AZT), was developed in Elion's lab shortly before her retirement in 1983. Elion received many honours and awards, including the 1988 Nobel Prize, shared with her long-time colleague, George Hitchings, and with James BLACK.

Ellet, Charles (1810–1862) *American civil engineer*

Born at Penns Manor in Pennsylvania, Ellet began work as a surveyor and assistant engineer, and then traveled to Europe to study engineering, returning to America in 1832. He concentrated on designing suspension bridges and built his first in 1842 over the Schuylkill River in Pennsylvania, with a span of 358 feet (109 m). He went on to build the world's first long-span wire-cable suspension bridge, with a central span of 1,010 feet (308 m), over the

Ohio River (1846–49). The bridge failed in 1854 because of aerodynamic instability.

Ellet also devised a steam-powered ram that helped to win the Mississippi River for the Union in the Battle of Memphis on 6 June 1862. Ellet was mortally wounded in this battle.

Elliott, Thomas Renton (1877–1961)
British physician

Elliott was the son of a retailer from the northeastern English city of Durham. He was educated at Cambridge University and University College Hospital, London, where he later served as professor of clinical medicine from the end of World War I until his retirement in 1939.

It was as a research student under John LANGLEY at Cambridge that Elliott made his greatest discovery. In 1901 Langley had injected animals with a crude extract from the adrenal gland and noted that the extract stimulated the action of the sympathetic nerves. Adrenaline (epinephrine) had earlier (1898) been isolated by John ABEL at Johns Hopkins University. Elliott therefore decided to inject adrenaline into animals to see if he got the same response as Langley had with the adrenal gland extract. He did indeed achieve increases in heart beat, blood pressure, etc., characteristic of stimulation of the sympathetic nervous system. Elliott is remembered for his subsequent suggestion that adrenaline may be released from sympathetic nerve endings – the first hint of neurotransmitters. Langley discouraged such speculation but later work by Henry DALE and Otto LOEWI on acetylcholine supported Elliott's early work.

Elsasser, Walter Maurice (1904–1991)
German–American geophysicist

Elsasser was born in the German city of Mannheim and educated at the University of Göttingen, where he obtained his doctorate in 1927. He worked at the University of Frankfurt before leaving Germany in 1933 following Hitler's rise to power. He taught at the Sorbonne, Paris, before emigrating to America (1936) where he joined the staff of the California Institute of Technology. He became professor of physics at the University of Pennsylvania (1947–50) and at the University of Utah (1950–58). In 1962 he became professor of geophysics at Princeton and he was appointed research professor at the University of Maryland from 1968 until his retirement in 1974.

Elsasser made fundamental proposals on the question of the origin of the Earth's magnetic field. It had been known for some time that this could not be due to the Earth's iron core for its temperature is too high for it to serve as a simple magnet. Instead he proposed that the molten liquid core contains eddies set up by the Earth's rotation. These eddies produce an electric current that causes the familiar terrestrial magnetic field. Elsasser also made predictions of electron diffraction (1925) and neutron diffraction (1936). His works include *The Physical Foundation of Biology* (1958) and *Atom and Organism* (1966).

Elton, Charles Sutherland (1900–1991)
British ecologist

Born in the English city of Liverpool, Elton graduated in zoology from Oxford University in 1922. He was assistant to Julian HUXLEY on the Oxford University expedition to Spitzbergen (1921), where Elton carried out ecological studies of the region's animal life. Further Arctic expeditions were made in 1923, 1924, and 1930. Such experience prompted his appointment as biological consultant to the Hudson's Bay Company, for which he carried out investigations into variations in the numbers of fur-bearing animals, using trapper's records dating back to 1736. In 1932 Elton helped establish the Bureau of Animal Population at Oxford – an institution that subsequently became an international center for information on and research into animal numbers and their ecology. In the same year he became editor of the new *Journal of Animal Ecology*, launched by the British Ecological Society, and in 1936 was appointed reader in animal ecology as well as a senior research fellow by Oxford University.

Elton was one of the first biologists to study animals in relation to their environment and other animals and plants. His demonstration of the nature of food chains and cycles, as well as such topics as the reasons for differences in animal numbers, were discussed in *Animal Ecology* (1927). In 1930 *Animal Ecology and Evolution* was published in which he advanced the notion that animals were not invariably at the mercy of their environment but commonly, perhaps through migration, practiced environmental selection by changing their habitats. Work on the rodent population of Britain, and how it is affected by a changing environment, was turned to eminently practical account at the outbreak of World War II when Elton conducted intensive research into methods of controlling rats and mice and thus conserving food for the war effort. *Voles, Mice and Lemmings: Problems in Population Dynamics* was published in 1942, and *The Control of Rats and Mice* in 1954, the latter becoming accepted as the standard work on the subject.

Elvehjem, Conrad Arnold (1901–1962)
American biochemist

Elvehjem (**el**-ve-yem), the son of a farmer from McFarland, Wisconsin, graduated from and spent his whole career at the University of Wisconsin. He obtained his PhD in 1927, and served as professor of biochemistry from 1936 until 1958, when he became president of the university, a position held until his retirement in 1962.

In 1937, following discoveries by Casimir FUNK and Joseph GOLDBERGER, Elvehjem succeeded in producing a new treatment for pellagra. In the 1920s Goldberger had postulated that this disease was caused by a deficiency of "P-P" (pellagra preventive) factor present in milk. In 1913 Funk, while searching for a cure for beriberi, came across nicotinic acid in rice husks. Although it was of little use against beriberi, Elvehjem found that even in minute doses it would dramatically remove the symptoms of blacktongue, the canine equivalent of pellagra. Tests on humans revealed the same remarkable effects on pellagra.

Elvehjem, a prolific author with over 800 papers to his credit, also worked on the role of trace elements in nutrition, showing the essential role played by such minerals as copper, zinc, and cobalt.

Embden, Gustav George (1874–1933) *German physiologist*

Embden, the son of a lawyer from Hamburg in Germany, was educated at the universities of Freiburg, Munich, Berlin, and Strasbourg. From 1904 he was director of the chemical laboratory in the medical clinic of the Frankfurt hospital, becoming in 1907 director of the Physiological Institute (which evolved from the medical clinic) and in 1914 director of the Institute for Vegetative Physiology (which in its turn evolved from the Physiological Institute).

In 1918 Otto MEYERHOF threw considerable light on the process of cellular metabolism by showing that it involved the breakdown of glucose to lactic acid. Embden spent much time in working out the precise steps involved in such a breakdown, as did many other chemists and physiologists. By the time of his death the details of the metabolic sequence from glycogen to lactic acid, later known as the *Embden–Meyerhof pathway*, had been worked out. Embden's earlier work concentrated on the metabolic processes carried out by the liver. In his experiments he used a new perfusion technique to maintain the condition of the dissected livers. In this way he discovered the breakdown of amino acids by oxidative deamination, realized that abnormal sugar metabolism can lead to the formation of acetone and acetoacetic acid, and showed that sugar is synthesized from lactic acid.

Emeleus, Harry Julius (1903–1993) *British inorganic chemist*

Emeleus was educated at Imperial College, in his native city of London, and at Karlsruhe and Princeton. He returned to Imperial College in 1931 and taught there until 1945 when he moved to Cambridge University, where he served as professor of inorganic chemistry until his retirement in 1970.

In 1938, in collaboration with John Anderson, he published the well-known work, *Modern Aspects of Inorganic Chemistry*. He also worked on fluorine, publishing a monograph on the subject in 1969: *The Chemistry of Fluorine and its Compounds*.

Emiliani, Cesare (1922–1995) *Italian–American geologist*

Emiliani (e-meel-**yah**-ni) was born in the Italian city of Bologna and educated at the university there. He moved to America in 1948, obtaining his PhD from the University of Chicago in 1950. After teaching at Chicago (1950–56) he moved to the University of Miami, where he served as professor of geology at the Institute of Marine Science from 1963 until his retirement in 1993.

Emiliani specialized in using oxygen isotopic analysis of pelagic microfossils from ocean sediments. Albrecht PENCK and Eduard Brückner had, in 1909, established the long-held orthodox view that four ice ages had occurred during the Pleistocene. In his fundamental paper *Pleistocene Temperatures* (1955), Emiliani produced evidence that there had been more. Using the principle established by Harold UREY that the climate of past ages can be estimated by the ratio of oxygen-16 to oxygen-18 present in water (i.e., the less oxygen-18 present the colder the climate must have been), he examined the oxygen-18 content of fossils brought up from the mud of the Caribbean. By choosing fossils that he knew had lived near the surface he could reconstruct the climatic history, and consequently identified seven complete glacial cycles.

Empedocles of Acragas (*c.* 490 BC–*c.* 430 BC) *Greek philosopher*

Empedocles (em-**ped**-oh-kleez) was a poet and a physician as well as a philosopher. Born at Acragas in Greece, he was probably a pupil of PARMENIDES. Much legend surrounds what is known of his life. Styling himself as a god, he reputedly brought about his own death in an attempt to persuade his followers of his divinity by throwing himself into the volcanic crater of Mount Etna. Fragments of two poems by Empedocles survive: *On Nature* and *Purifications*. There is some difficulty in reconciling the two because the first is purely physical while the second deals with the progress of the soul from fall to redemption.

Encke, Johann Franz (1791–1865) *German astronomer*

The son of a Lutheran pastor from Hamburg in Germany, Encke (**eng**-ke) was educated at Göttingen where he impressed Karl GAUSS. In 1816 he was appointed to the staff of the Seeberg Observatory, Gotha, where he remained until 1825 when he moved to the Berlin Observatory as its director.

A faint comet had first been observed by P. Mechain in 1786. Over the years other reports were made in 1795, 1805, and 1818 of a series of faint comets. Although several astronomers suspected there was only one comet involved, it was Encke who provided in 1819 the necessary computations. He showed that *Encke's comet*, as it became known, had a period of 3.3 years and predicted that it would be at perihelion again on 24 May 1822. His prediction was accurate to within a few hours and Encke thus became, after HALLEY, only the second man to predict the return of a periodic comet successfully.

Enders, John Franklin (1897–1985)
American microbiologist

Enders, the son of a wealthy banker from West Hartford in Connecticut, was educated at Yale and Harvard where he obtained his PhD in 1930. His career was somewhat delayed by the war, in which he served as a flying instructor, and also by his initial intention to study Germanic and Celtic languages. This was upset by the influence of the bacteriologist Hans Zinsser who "seduced" Enders into science in the late 1920s.

In 1946 Enders set up an Infectious Diseases Laboratory at the Boston Children's Hospital; it was here that he did the work to be later described as opening up a "new epoch in the history of virus research." This referred to his success, in collaboration with Thomas WELLER and Frederick ROBBINS, in 1949 in cultivating polio virus in test tube cultures of human tissue for the first time. They further demonstrated that the virus could be grown on a wide variety of tissue and not just nerve cells.

This at last allowed the polio virus to be studied, typed, and produced in quantity. Without such an advance the triumphs of Albert SABIN and Jonas SALK in developing a vaccine against polio in the 1950s would have been impossible. In 1954 Enders, Weller, and Robbins were awarded the Nobel Prize for physiology or medicine.

By this time Enders had already begun to work on the cultivation of the measles virus. This time, working with T. Peebles, they followed up their success in cultivating the virus with, in 1957, the production of the first measles vaccine.

Engelbart, Douglas C. (1925–   ) *American computer scientist*

Engelbart was born in Oregon and educated at Oregon State University, where he studied electrical engineering. His university education was disrupted by World War II. Having joined the navy he spent two years in the Philippines as a radar technician. Here he read a visionary and influential article by Vannevar Bush entitled *As We May Think*, in which Bush envisaged the possibility of a machine that could augment human memory by letting the user of the machine store and retrieve documents that are linked by associations.

After the war Engelbart returned to university and obtained his degree in 1948. After graduating he worked for the NACA Ames Laboratory, the predecessor of NASA. Engelbart then became a graduate student in electrical engineering at the University of California at Berkeley, obtaining his PhD in 1955. He remained at Berkeley for a short time as an assistant professor before obtaining a research position at the Stanford Research Institute (SRI).

In 1962 he wrote the definitive exposition of his views in a paper entitled *Augmenting Human Intellect: A Conceptual Framework*, in which he discussed how technology could be used to help to find solutions to problems. Engelbart set up his own research laboratory, which he called the Augmentation Research Center, in 1963. In the 1960s and 1970s this laboratory developed a system called oNLine System (NLS), which made the creation of digital libraries and the storing and retrieving of documents easier. This was done using *hypertext*, the first time the idea had been successfully implemented. The NLS also improved the computer interaction with Engelbart's invention of the mouse, although the device was not commonly used until the 1980s. The NLS produced some other technological innovations such as on-screen video teleconferencing.

Engelbart was a central figure in the ARPANET project, which pioneered computer networking. He founded the Bootstrap Institute in 1989 with the purpose of encouraging collaboration in the development of technology. Engelbart's work has been widely recognized. In 2000 he was awarded the US National Medal of Technology and in 2001 he was awarded the British Computer Society's Lovelace Medal. In 2005 he received a grant to fund his Hyperscope Project.

Engelmann, George (1809–1884) *American botanist*

Born the son of a schoolmaster at Frankfurt in Germany, Engelmann was educated at the universities of Heidelberg, Berlin, and Würzburg where he obtained his MD in 1831. In the following year he visited America to invest in some land for a wealthy uncle and decided in 1835 to settle and practice medicine in St. Louis.

Engelmann was not only a plant collector of some importance; he also did much to initiate and organize major collecting expeditions of the flora of the West. It was thus through Engelmann that many of the newly collected specimens passed on their way to eastern scholars as Asa GRAY at Harvard. Engelmann's role became more official with the setting up of the Missouri Botanical Garden in 1859 with the backing of the St. Louis businessman Henry Shaw.

He is also remembered for his demonstration that some stocks of American vine were resistant to the pest *Phylloxera*, which had begun to devastate the vineyards of Europe from 1863 onward.

Engelmann, Theodor Wilhelm (1843–1909) *German physiologist*

Engelmann, the son of a publisher, was educated at Jena, Heidelberg, and Göttingen, before obtaining his PhD from the university in his native city of Leipzig in 1867. He immediately joined the faculty of the University of Utrecht, serving there as professor of physiology from 1888 until 1897 when he returned to Germany to a similar chair at the University of Berlin, where he remained until his retirement in 1908.

Between 1873 and 1895 Engelmann published a number of papers on muscle contraction. By this time, following the work of such physiologists as William BOWMAN, the main anatomical details of striated muscle had been established. However an explanation was needed as to why the anisotropic or A bands refract polarized light quite differently to the isotropic or I bands. Engelmann had noted that in contraction the A bands increased in volume while the I bands decreased. He consequently proposed his "imbibition" theory in which the contraction of striped muscle is attributed to a flow of fluid from the I to the A bands.

Engelmann also worked on the nature and mechanism of the heartbeat and in 1875 devised an experiment that proved the heartbeat is myogenic; that is, the contraction originates in the heart muscle and not from an external nerve stimulus. In 1881 he discovered the chemotactic response of certain bacteria to oxygen, and he also demonstrated that red and blue light are far more effective in stimulating plant chloroplasts during photosynthesis than other parts of the spectrum.

Engler, Heinrich Gustav Adolf (1844–1930) *German botanist*

Engler was born in Sagan (which is now in Poland) and studied botany at Breslau University, gaining his PhD in 1866 for his thesis on the genus *Saxifraga*. After teaching natural history he became custodian of the Munich botanical collection and then professor of botany at Kiel University; in 1884 he returned to Breslau to succeed his former teacher in the chair of botany. At Breslau Engler replanned the botanic garden, ordering the plants according to their geographical distribution. In 1887 he took up the important chair of botany at Berlin and successfully reestablished the garden in Dahlem.

Between 1878 and his retirement from Berlin in 1921, Engler contributed greatly to the development of plant taxonomy with his classifications, presented in such books as *The Natural Plant Families* (1887–1911) and *The Plant Kingdom* (1900–37). Much of his work was drawn from first-hand observation gained during travels through Africa, Europe, India, China, Japan, and America.

Eötvös, Baron Roland von (1848–1919) *Hungarian physicist*

Born in Budapest, Eötvös (**u(r)t**-vu(r)sh) studied at the University of Königsberg and at Heidelberg where he obtained his PhD in 1870 for a thesis concerning a method of detecting motion through the ether by measuring light intensity. At Königsberg in 1886 he introduced the *Eötvös law* – an equation approximately relating surface tension, temperature, density, and relative molecular mass of a liquid.

He then started teaching at Budapest University, where he was appointed professor in 1872. His work from then on centered on gravitation. In 1888 he developed the *Eötvös torsion balance*, consisting of a bar with two attached weights, the bar being suspended by a torsion fiber. He argued that if the two weights were made from different materials, and if the inertial and gravitational forces were not equivalent, there would be a discernible twisting force, which would cause a slight rotation of the bar about a vertical axis. Observations were made with copper, aluminum, asbestos, platinum, and other materials. No torque was found and Eötvös concluded that the masses of different materials were equivalent to a few parts per billion. His experiments were repeated in the 1960s by DICKE and in 1970 by Braginsky, with results affirming the equivalence to 1 part per 100 billion and 1 part per trillion respectively. The experiment became one of the foundation stones of general relativity since, by failing to distinguish between inertial and gravitational mass experimentally, it supported EINSTEIN's principle of equivalence.

Eötvös spent much of his time trying to improve the Hungarian education system and for a short time was minister of instruction. He was also an excellent mountain climber and a peak in the Dolomites is named for him.

Epicurus (*c.* 341 BC–270 BC) *Greek philosopher*

The atoms come together in different order

and position, like the letters, which, though they are few, yet, by being placed together in different ways, produce innumerable words.
—Quoted by Max Muller in *Science of Language* (1871)

Epicurus, who was born on the Greek island of Samos, traveled to Athens when he was about 18 years old, and received military training. He then taught at Mytilene and Lampsacus before returning to Athens (305 BC) where he founded a school of philosophy and attracted a substantial following.

Epicurus revived Democritean atomism and was little influenced by his predecessors, PLATO and ARISTOTLE. His work is known through substantial fragments in the writings of Diogenes Laërtius and especially through the long poem, *De rerum natura* (On the Nature of Things), by his Roman disciple LUCRETIUS. The Epicurean philosophy aimed at the attainment of a happy, though simple, life and used the atomic theory to sanction the banishment of the old fears and superstitions. Epicurus also made important additions to the atomic theory, asserting the primacy of sense-perception where DEMOCRITUS had distrusted the senses, and he introduced the concept of random atomic "swerve" to preserve free will in an otherwise deterministic system.

Epstein, Sir Michael Anthony (1921–)
British virologist

Epstein, a Londoner, was educated at Cambridge University and at the Middlesex Hospital Medical School in his native city. After serving in the Royal Army Medical Corps (1945–47), he returned to the Middlesex Hospital as an assistant pathologist. He left the Middlesex in 1965 and in 1968 he was appointed professor of pathology at Bristol University, a position he held until his retirement in 1985. He has continued to work in the Department of Clinical Medicine at Oxford.

In 1961 Epstein heard Denis BURKITT describe the distribution of a particularly savage lymphoma throughout Africa. Epstein saw that "anything which has geographical factors such as climate affecting distribution must have some kind of biological cause." That biological cause, Epstein suspected, for no very good reason, was a virus. Although Epstein received tumor samples from Burkitt, he found them impossible to culture and saw no trace of any virus. After struggling unsuccessfully for two years, Epstein and his assistant Yvonne Barr developed a new approach. Instead of working with small tumor lumps they divided the pieces into single cells. The technique proved successful and for the first time ever human lymphocytes were being grown in a continuous culture. Yet Epstein initially found no virus until he examined some cells under an electron microscope. The virus was named the *Epstein–Barr*

virus and proved to be a member of the herpes family. The virus turned out to have a worldwide distribution and was identified as the cause of mononeucleosis. Clearly, its presence alone is insufficient to cause lymphoma. For if most of us have the virus, why is lymphoma not distributed more widely? Epstein and Burkitt argued that only in cases in which malaria or some other chronic condition has suppressed a child's immature immune system could the virus provoke lymphoid cells into malignant growth.

Erasistratus of Chios (*c.* 304 BC–*c.* 250 BC) *Greek anatomist and physician*

Erasistratus (er-a-**sis**-stra-tus), who was born on the Greek island of Chios, came from a distinctly medical background and studied in Athens, Cos, and Alexandria. Following HEROPHILUS he became the leading figure in the Alexandrian School of Anatomy.

It is possible with Erasistratus, unlike his contemporaries, to make out at least the outline of his physiological system. Every organ and part of the body was served by a "threefold network" of vein, artery, and nerve. Indeed, he believed the body tissues were a plaiting of such vessels, which at their extremities became so fine as to be invisible. The veins carried blood and the nerves and arteries transported nervous and animal spirits respectively.

As an atomist he rejected all attractive and occult forces seeking instead to explain everything in terms of atoms and the void. He thus accounted for the bleeding of severed arteries by assuming the escaped pneuma left a vacuum that was filled by blood from adjoining veins.

One of the most interesting aspects of his thought was his unusual rejection of the humoral theory of disease which, formulated by the Hippocratics and authorized by GALEN, became the sterile orthodoxy of Western medicine for 2,000 years. Instead he seems to have argued for a more mechanical concept of disease, attributing it to a "plethora" of blood, vital spirit, or food, which produces a blocking and inflammation of the various vessels.

His objection to the humoral theory found little support and with the passing of Erasistratus the great innovative period of Alexandrian medicine came to an end.

Eratosthenes of Cyrene (*c.* 276 BC–*c.* 194 BC) *Greek astronomer*

Eratosthenes (er-a-**tos**-the-neez) was born in Cyrene, now in Libya, and educated at Athens. He then taught in Alexandria where he became tutor to the son of Ptolemy III and librarian. He was prominent in history, poetry, mathematics, and astronomy and was known by the nickname "beta" because, some say, he was the second PLATO.

In number theory he introduced the procedure named for him to collect the prime numbers by filtering out all the composites. The method, called the *sieve of Eratosthenes*, was to write down a list of ordered numbers and to strike out every second number after 2, every third number after 3, every fourth number after 4, and so on. The numbers remaining are primes.

Eratosthenes achieved his greatest fame by using a most ingenious and simple method to measure the circumference of the Earth. He was aware that on a certain day the Sun at Syene (now Aswan) was exactly at its zenith (it was known to shine directly down a deep well on that day). He found that on the same day at Alexandria it was south of its zenith by an angle corresponding to 1/50 of a circle (7° 12'). He also knew that the distance between Syene and Alexandria was 5,000 stadia – a distance that he estimated from the time it took a camel train to make the journey. Therefore, 5,000 stadia must be 1/50 of the circumference of the Earth; that is, 250,000 stadia. (Since the exact length of a stade is not known it is impossible to work out exactly how accurate his measurement was but it has been thought to be within 50 miles of the presently accepted value.) Eratosthenes also established an improved figure for the obliquity of the ecliptic (the tilt of the Earth's axis) of 23°51'20". Finally, he produced the first map of the world, as he knew it, based on meridians of longitude and parallels of latitude.

Ercker, Lazarus (*c.* 1530–1594) *Bohemian metallurgist*

Ercker was born in Annaberg, now in Germany, and studied at the University of Wittenberg (1547–48). In 1554 he was made assayer in Dresden, Saxony, and later he became control tester of coins in a town near Prague.

Ercker's main contribution to metallurgy was to write the first systematic account of analytical and metallurgical chemistry. This manual, *Beschreibung allerfürnemisten mineralischen Ertzt und Berckwercksarten* (1574; Description of Leading Ore Processing and Mining Methods), described the testing of alloys, minerals, and compounds containing silver, gold, copper, antimony, mercury, bismuth, and lead.

Erdös, Paul (1913–1996) *Hungarian mathematician*

Paul Erdös has the theory that God has a book containing all the theorems of mathematics with their absolutely most beautiful proofs, and when he wants to express particular appreciation of a proof he exclaims, "This is one from the book!"
—Ross Honsberger, *Mathematical Morsels*

The son of two mathematics teachers from the Hungarian capital of Budapest, Erdös (erdu(r)sh) devoted his life to mathematics to an unequalled degree. He was educated at the University of Budapest where he obtained his PhD in 1934. Sensing difficult political times ahead, Erdös moved to Manchester, England, on a four-year fellowship. Shortly before the outbreak of war he left for America. At the height of the McCarthy era he was denied a reentry permit after attending a conference in Amsterdam. Erdös then settled in Israel for several years until, in the 1960s, he was allowed a U.S. visa once more.

With no official job and no family, Erdös lived out of a suitcase and traveled from one mathematical center to another. His life was spent working up to 20 hours a day on mathematical problems, usually with colleagues. In the process he produced over 1,000 papers. He lived on lecture fees, prizes, grants and the hospitality of his wide circle of collaborators. He wrote up to 1,500 letters a year and had more than 250 published collaborators.

Erlanger, Joseph (1874–1965) *American neurophysiologist*

Erlanger, the son of a German immigrant drawn to California in the gold rush, was born in San Francisco. He was educated at the University of California and at Johns Hopkins University in Baltimore, where in 1899 he obtained his MD. After working on the staff for a few years Erlanger moved to the University of Wisconsin (1906) to accept the chair of physiology. In 1910 he moved to Washington University, St. Louis, where he held the chair of physiology in the Medical School until his retirement in 1944.

Between 1921 and 1931 Erlanger carried out some fundamental research on the functions of nerve fibers with his former pupil and colleague, Herbert GASSER. They investigated the transmission of a nerve impulse along a frog nerve kept in a moist chamber at constant temperature. Their innovation was to study the transmission with the cathode-ray oscillograph, invented by Ferdinand BRAUN in 1897, which enabled them to picture the changes the impulse underwent as it traveled along the nerve.

Erlanger and Gasser found that on stimulating a nerve, the resulting electrical activity indicating the passage of an impulse was composed of three waves, as observed on the oscillograph. They explained this by proposing that the one stimulus activated three different groups of nerve fibers, each of which had its own conduction rate. They went on to measure these rates, concluding that the fastest fibers (the A-fibers) conduct with a speed of up to 100 meters per second (mps) while the slowest (the C-fibers) could manage speeds of no more than 2 mps. The intermediate B-fibers conducted in the range 2–14 mps. Erlanger and Gasser were able to relate this variation to the thickness of the different nerve fibers, A-fibers being the largest.

It was a short step from this to the theory of differentiated function, in which it was proposed that the slender C-fibers carry pain impulses whereas the thicker A-fibers transmit motor impulses. But it was soon demonstrated that while such propositions may be broadly true the detailed picture is more complex.

Erlanger and Gasser produced an account of their collaboration in *Electrical Signs of Nervous Activity* (1937); they were awarded the 1944 Nobel Prize for physiology or medicine for their work.

Erlenmeyer, Richard August Carl Emil
(1825–1909) *German chemist*

Born near Wiesbaden in Germany, Erlenmeyer (**er**-len-mI-er) studied at Giessen and practiced at first as a pharmacist. In 1855 he became a private pupil of August KEKULÉ at Heidelberg and later was appointed professor at the Munich Polytechnic (1868–83). He synthesized guanidine and was the first to give its correct formula (1868). He also synthesized tyrosine and formulated the *Erlenmeyer rule*, which states the impossibility of two hydroxy groups occurring on the same carbon atom or of a hydroxy group occurring adjacent to a carbon–carbon double bond (chloral hydrate is an exception to this rule). His son F. G. C. E. Erlenmeyer introduced the Erlenmeyer synthesis of amino acids and synthesized cystine, serine, and phenylalanine.

Ernst, Richard Robert (1933–   ) *Swiss chemist*

Born at Winterthur in Switzerland, Ernst (ernst) was educated at the Federal Institute of Technology, Zurich, where he obtained his PhD in 1962. He spent the period from 1963 until 1968 working as a research chemist for Varian Associates, Palo Alto, California, before returning to the Federal Institute where he was appointed professor of physical chemistry in 1976. He became emeritus professor in 1988.

The technique of nuclear magnetic resonance (NMR) described by I. I. RABI in 1944, and developed by Felix BLOCH and Edward PURCELL in the late 1940s, quickly became a recognized tool for the exploration of atomic nuclei. As nuclei possess a magnetic moment they will tend to align themselves with any strong magnetic field. If, however, nuclei are subjected to radiowaves of the appropriate frequency, they will be raised to a higher energy level, and align themselves in a different direction with respect to the field. With the removal of the radio signal, the nuclei will revert to their original energy state by emitting radiation of a characteristic frequency. The frequency of the radiation emitted allows nuclei to be identified, and the structure of certain molecules determined. But, the process was time-consuming because, in order to find which radiofrequency a sample responded to, it was necessary to sweep the applied frequency through a range of frequencies. Ernst developed a technique in which the sample was subjected to a single high-energy radio pulse. In this way numerous nuclei would respond and emit an apparently jumbled signal. But Ernst showed that, with the aid of FOURIER analysis and a computer, the signal could be unraveled into its separate components. Ernst's procedure considerably increased the sensitivity of NMR.

In 1970 Ernst made a further advance. He found that if he subjected his samples to a sequence of high-energy pulses instead of to a single pulse, it enabled him to use NMR techniques to study much larger molecules. Ernst's "two-dimensional analysis," as it became known, opened the way to investigate complex biological molecules such as proteins. His work also laid the foundation for the development by Peter MANSFIELD and others of MRI (magnetic resonance imaging).

For his work on NMR Ernst was awarded the 1991 Nobel Prize for chemistry.

Ertl, Gerhard (1936–   ) *German chemist and physicist*

Gerhard Ertl was born in Bad Cannstadt in Germany. He studied at the Technical University of Stuttgart where he received a Diploma in physics in 1961. In 1965 he received his PhD from the Technical University of Munich. Subsequently, Ertl held positions at the Technical University of Munich (1965–1968), the Technical University of Hannover (1968–1973), and the Ludwig Maximilians University of Munich (1973–1986). Since 1986 he has been at the Fritz Haber Institut der Max Planck Gesellschaft in Germany, being the Director of the Institut from 1986 until his retirement in 2004.

Ertl's most significant research has been his experimental work on the molecular process involved in heterogeneous catalysis. Ertl has used a number of experimental techniques in surface studies, including low-energy electron diffraction (LEED), ultraviolet photoelectron spectroscopy (UPS), ultraviolet photoelectron microscopy (UPM), and scanning electron microscopy (SEM). In particular, he investigated the catalytic synthesis of ammonia (the Haber process) and the catalytic oxidation of carbon monoxide in great detail and was able to elucidate the mechanisms involved. This work on heterogeneous reactions has many industrial applications. The importance of Ertl's work has been recognized by the award of many prizes, culminating in the Nobel Prize for chemistry in 2007.

Esaki, Leo (1925–   ) *Japanese physicist*

Born in the Japanese city of Osaka, Esaki (e-**sah**-ki) graduated in physics at the University of Tokyo in 1947, gaining his doctorate there in 1959. His doctoral work was on the physics of semiconductors, and in 1958 he reported an effect known as "tunneling," which he had observed in narrow p–n junctions of germanium that were heavily doped with impurities. The phenomenon of tunneling is a quantum-mechanical effect in which an electron can penetrate a potential barrier through a narrow region of solid, where classical theory predicts it could not pass.

Esaki was quick to see the possibility of applying the tunnel effect, and in 1960 reported the construction of a device with diodelike properties – the tunnel (or *Esaki*) diode. With negative bias potential, the diode acts as a short circuit, while under certain conditions of forward bias it can have effectively negative resistance (the current decreasing with increasing voltage). Important characteristics of the tunnel diode are its very fast speed of operation, its small physical size, and its low power consumption. It has found application in many fields of electronics, principally in computers, microwave devices, and where low electronic noise is required. Esaki shared the Nobel Prize for physics in 1973 with Brian JOSEPHSON and Ivar GIAEVER.

Esaki worked for the computer firm International Business Machines at the Thomas J. Watson Research Center, Yorktown Heights, New York, until 1992, when he returned to Japan to become president of Tsukuba University, Ibaraki. He is also president of the Yokahama College of Pharmacy.

Eschenmoser, Albert (1925–   ) *Swiss chemist*

Born at Erstfeld in Switzerland, Eschenmoser (**esh**-en-moh-ser) was educated at the Federal Institute of Technology, Zurich, where he has taught since 1956 and where, in 1960, he was appointed professor of organic chemistry.

He is best known for his work in synthesizing a number of complex organic compounds. His first success came with colchicine – an alkaloid found in the autumn crocus – which has important applications in genetical research. He also collaborated with Robert WOODWARD on the synthesis of vitamin B_{12} (cyanocobalamin), which had first been isolated and crystallized by Karl FOLKERS in 1948. Its empirical formula was soon established and in 1956 Dorothy HODGKIN established its structure. It took many years with samples passing between Zurich and Harvard before Eschenmoser and Woodward were finally able to announce its synthesis in 1965.

Eskola, Pentti Elias (1883–1964) *Finnish geologist*

The son of a farmer from Lellainen in Finland, Eskola (**es**-ko-la) was educated at the universities of Helsinki and Freiburg, Germany, where he obtained his PhD in 1915. He joined the faculty of Helsinki University in 1916, and became professor of geology in 1924, a position he held until his retirement in 1953.

Early in his career Eskola introduced the important notion of metamorphic facies. He pointed out (1915) that in any rock of metamorphic formation that has arrived at chemical equilibrium under conditions of constant temperature and pressure, the mineral composition is controlled only by its chemical composition. Initially he recognized five types of facies, namely, sanidine, hornfels, greenschist, amphibolite, and eclogite. By 1939 he had raised the number to eight while later workers have continued to add further types.

Espy, James Pollard (1785–1860) *American meteorologist*

Espy was born in Washington County, Pennsylvania, and graduated from Transylvania University in Lexington, Kentucky, in 1808. He became principal of the Academy in Cumberland, Maryland (1812–17), and, after teaching mathematics and physics in Philadelphia, he became a full-time meteorologist in 1835. He was made state meteorologist in 1839 and from 1842 took up a national appointment, being attached at various times to the War Department, the Navy, and, after 1852, the Smithsonian, which eventually took over his network of observers and out of which the Weather Bureau was later to grow.

In addition to collecting and issuing basic data, Espy also wrote on questions of theoretical meteorology. He was the first to argue for the convectional theory of precipitation in which the ascent and cooling of moist air with the release of latent heat leads to condensation and the formation of clouds. Much of the evidence for this came from his "nephelescope," a device whereby he could simulate cloud behavior and measure cooling rates. Espy published, in 1841, his *Philosophy of Storms* and then became involved in a prolonged and bitter controversy with William REDFIELD on the cause and nature of storms. Here he argued that storms consisted of radially convergent winds with the air escaping up the middle. Although his views on storms were influential in Europe, Redfield and Elias LOOMIS were able to produce convincing contrary evidence.

Euclid (*c.* 330 BC–*c.* 260 BC) *Greek mathematician*

> A line is length without breadth.
> —*Elements*

Euclid (**yoo**-klid) is one of the best known and most influential of classical Greek mathematicians but almost nothing is known about his life. He was a founder and member of the academy in Alexandria, and may have been a pupil of PLATO in Athens. Despite his great fame Euclid was not one of the greatest of Greek mathematicians and not of the same caliber as ARCHIMEDES.

Euclid's most celebrated work is the *Elements*, which is primarily a treatise on geometry contained in 13 books. The influence of this work not only on the future development of geometry, mathematics, and science, but on the whole of Western thought is hard to exaggerate. Some idea of the importance that has been attached to the *Elements* is gained from the fact that there have probably been more commentaries written on it than on the Bible. The *Elements* systematized and organized the work of many previous Greek geometers, such as Theaetetus and EUDOXUS, as well as containing many new discoveries that Euclid had made himself. Although mainly concerned with geometry it also deals with such topics as number theory and the theory of irrational quantities. One of the most celebrated number theoretic results is Euclid's proof that there are an infinite number of primes. The *Elements* is in many ways a synthesis and culmination of Greek mathematics. Euclid and APOLLONIUS of Perga were the last Greek mathematicians of any distinction, and after their time Greek civilization as a whole soon became decadent and sterile.

Euclid's *Elements* owed its enormously high status to a number of reasons. The most influential single feature was Euclid's use of the axiomatic method whereby all the theorems were laid out as deductions from certain self-evident basic propositions or axioms in such a way that in each successive proof only propositions already proved or axioms were used. This became accepted as the paradigmatically rigorous way of setting out any body of knowledge, and attempts were made to apply it not just to mathematics, but to natural science, theology, and even philosophy and ethics.

However, despite being revered as an almost perfect example of rigorous thinking for almost 2,000 years there are considerable defects in Euclid's reasoning. A number of his proofs were found to contain mistakes, the status of the initial axioms themselves was increasingly considered to be problematic, and the definitions of such basic terms as "line" and "point" were found to be unsatisfactory. The most celebrated case is that of the parallel axiom, which states that there is only one straight line passing through a given point and parallel to a given straight line. The status of this axiom was long recognized as problematic, and many unsuccessful attempts were made to deduce it from the remaining axioms. The question was only settled in the 19th century when Janos BOLYAI

and Nicolai Lobachevski showed that it was perfectly possible to construct a consistent geometry in which Euclid's other axioms were true but in which the parallel axiom was false. This epoch-making discovery displaced Euclidean geometry from the privileged position it had occupied. The question of the relation of Euclid's geometry to the properties of physical space had to wait until the early 20th century for a full answer. Until then it was believed that Euclid's geometry gave a fully accurate description of physical space. No less a thinker than Immanuel KANT had thought that it was logically impossible for space to obey any other geometry. However when Albert EINSTEIN developed his theory of relativity he found that the appropriate geometry for space was not Euclid's but that developed by Bernhard RIEMANN. It was subsequently experimentally verified that the geometry of space is indeed non-Euclidean.

In mathematical terms too, the discovery of non-Euclidean geometries was of great importance, since it led to a broadening of the conception of geometry and the development by such mathematicians as Felix KLEIN of many new geometries very different from Euclid's. It also made mathematicians scrutinize the logical structure of Euclid's geometry far more closely and in 1899 David HILBERT at last gave a definitively rigorous axiomatic treatment of geometry and made an exhaustive investigation of the relations of dependence and independence between the axioms, and of the consistency of the various possible geometries so produced. Euclid wrote a number of other works besides the *Elements*, although many of them are now lost and known only through references to them by other classical authors. Those that do survive include *Data*, containing 94 propositions, *On Divisions*, and the *Optics*. One of his sayings has come down to us. When asked by Ptolemy I Soter, the reigning king of Egypt, if there was any quicker way to master geometry than by studying the *Elements* Euclid replied, "There is no royal road to geometry."

Eudoxus of Cnidus (*c.* 400 BC–*c.* 350 BC)
Greek astronomer and mathematician

Born in Cnidus, which is now in Turkey, Eudoxus (yoo-**dok**-sus) is reported as having tudied mathematics under Archytas, a Pythagorean. He also studied under PLATO and in Egypt. Although none of his works have survived they are quoted extensively by HIPPARCHUS. Eudoxus was the first astronomer who had a complete understanding of the celestial sphere. It is only this understanding that reveals the irregularities of the movements of the planets that must be taken into account in giving an accurate description of the heavens. For Eudoxus the Earth was at rest and around this center 27 concentric spheres rotated. The outermost sphere carried the fixed stars, each of the planets required four spheres, and the Sun and the Moon three each. All these spheres were necessary to account for the daily and annual relative motions of the heavenly bodies. He also described the constellations and the changes in the rising and setting of the fixed stars in the course of a year.

In mathematics, Eudoxus is thought to have contributed the theory of proportion to be found in Book V of EUCLID – the importance of this being its applicability to irrational as well as rational numbers. The method of exhaustion in Book XII is also attributed to Eudoxus. This tackled in a mathematical way for the first time the difficult problem of calculating an area bounded by a curve.

Euler, Leonhard (1707–1783) *Swiss mathematician*

> Mathematicians have tried in vain to this day to discover some order in the sequence of prime numbers, and we have reason to believe that it is a mystery into which the human mind will never penetrate.
> —Quoted by G. Simmons in *Calculus Gems*

Euler (**oi**-ler) was one of the outstanding figures of 18th-century mathematics, and also the most prolific. He was born at Basel in Switzerland and studied at the university there, where he came to the notice of the BERNOULLI family; he became, in particular, a friend of Daniel Bernoulli. Having completed his studies Euler applied unsuccessfuly for a post at Basel University. He was encouraged by his friend Bernoulli to join him at the St. Petersburg Academy of Science. Bernoulli obtained for Euler a post in the medical section of the academy and in 1727 Euler left for Russia to find on arriving that the empress had just died and that the future of the academy was doubtful. Fortunately it survived intact and he managed to find his way into the mathematical section. In 1733 Bernoulli left Russia to return to Switzerland and Euler was appointed to replace him as head of the mathematical section of the academy. Euler was not particularly happy in the highly repressive political climate and devoted himself almost exclusively to his mathematics. In 1740 he eventually left Russia to join Frederick the Great's Berlin Academy. He returned to Russia in 1766 at the invitation of Catherine the Great, remaining there for the rest of his life. He had lost the sight of his right eye through observing the Sun during his first stay in Russia and shortly after returning there his blindness became total. However, such was his facility for mental calculation and the power of his memory that this did not affect his mathematical creativity.

Euler contributed to almost all areas of mathematics, and did equally important work in both pure and applied mathematics. One of his most significant contributions was to the development of analysis. Although he was working before the development of modern standards of rigor by such mathematicians as Karl Friedrich GAUSS and Augustin CAUCHY and thus lacked a rigorous treatment of such key topics as convergence, nonetheless his work in analysis constitutes a major advance over previous work in the area. He wrote three treatises on different aspects of analysis, which together collect, systematize, and develop what the mathematicians of the 17th century had achieved. These treatises became important and influential textbooks. It is worth noting that unlike some great mathematicians, such as Gauss, Euler was a highly successful and effective teacher. It is a measure of his intuitive insight into mathematics that even though he did lack truly rigorous analytical techniques he was still able to arrive at so many novel and important results. His phenomenal ability for calculation came to his aid here, for he frequently arrived at results, for example about infinite series, by induction from a great many calculations and gave only a highly dubious proof, leaving future mathematicians to give properly rigorous proofs of results that were indeed quite correct.

Outside analysis Euler made extremely important contributions both to the calculus of variations and mechanics. Mechanics was transformed by his treatment of it in his treatise of 1736. In essence he transformed the subject into one to which the full resources of analytical techniques could be applied. In doing so he paved the way for the work of mathematicians such as Joseph LAGRANGE.

Euler put his expertise in mechanics to practical use in his work on the three-body problem. He was interested in this problem because he wished to investigate the motion of the Moon. He published his first lunar-motion theory in 1753 and then a second theory in 1772. Euler was able to invent methods of approximating solutions that were accurate enough to be of practical use in navigation. Here his prodigious ability as a sheer calculator came into its own.

In addition to these pieces of work Euler made notable contributions to number theory and geometry. Numerous theorems and methods are named for him.

Euler, Ulf Svante von *See* VON EULER, ULF SVANTE.

Euler-Chelpin, Hans Karl August Simon von (1873–1964) *German–Swedish biochemist*

Euler-Chelpin (oi-ler-**kel**-pin) was born at Augsburg in Germany and educated at the universities of Berlin, Strasbourg, and Göttingen and at the Pasteur Institute. In 1898 he moved to Sweden, being appointed to the staff of the University of Stockholm, where in 1906 he became professor of general and inorganic chemistry. In 1929 he also became director of the Institute of Biochemistry where he remained until his retirement in 1941. Although he became a Swedish citizen in 1902 he served Germany in both world wars.

In 1904 important work by Arthur HARDEN had shown that enzymes contain an easily removable nonprotein part, a coenzyme. In 1923 Euler-Chelpin worked out the structure of the yeast coenzyme. He showed that the molecule is made up from a nucleotide similar to that found in nucleic acid. It was named diphosphopyridine nucleotide (now known as NAD). Euler-Chelpin shared the 1929 Nobel Prize for chemistry with Harden for this work. His son, Ulf VON EULER, was also a Nobel prizewinner.

Eustachio, Bartolommeo (c. 1524–1574)
Italian anatomist

The son of a physician from San Severino in Italy, Eustachio (ay-oo-**stah**-kyoh) received a sound humanist education acquiring a good knowledge of Latin, Greek, and Arabic. In the early 1540s he was appointed physician to the duke of Urbino and, from 1547, served the duke's brother, Cardinal Rovere, with whom he moved to Rome in 1549. While there he also served from 1562 as professor of anatomy at the Sapienza or Papal College.

In 1564 he collected some of his earlier work in his *Opuscula anatomica* (Anatomical Works). It contains his *De auditus organis* (On the Auditory Organs), which describes the eustachian tubes connecting the middle ear to the pharynx and named in his honor (although these were first described by Alcmaeon). It also contained the *De renum structura* (On the Structure of the Kidney), in which he published the first description of the adrenal glands.

Eustachio was not in fact as influential an anatomist as he might have been for his anatomical plates, prepared for a text never completed, were lost and only rediscovered in the early 18th century. They were finally published as the *Tabulae anatomica* in 1714 and republished with full notes by Bernhard ALBINUS in 1744. It has been claimed that the plates manifest an originality equaled only by Andreas VESALIUS and Leonardo da Vinci and that Eustachio's plates of the sympathetic nervous system are the best yet produced.

Evans, Martin J. (1941–) *British developmental biologist*

Evans graduated from Cambridge University in 1963 with a BA in biochemistry, and began postgraduate research at University College, London, in the department of anatomy and embryology, gaining his PhD in 1969. He stayed on as a lecturer until returning to Cambridge in 1978 to join the department of genetics. In 1999 he was appointed professor of mammalian genetics at Cardiff University, and director of the school of biosciences.

From his early postgraduate work at University College, Evans' prime interest was the genetic control of vertebrate development, and he sought a means of studying this using cultured cells that could be genetically manipulated. A promising avenue involved undifferentiated cells derived from a mouse testicular tumor, termed a teratocarcinoma, that were capable of differentiating into a range of tissue cells, in ways that mimicked embryonic development. Evans devised a method of maintaining these embryonal carcinoma (EC) cells in culture, whereby their ability to differentiate was retained indefinitely. So, for example, such cells could give rise to skin, nerve, and even beating heart muscle.

Evans and a collaborator, Richard Gardner, saw the potential for creating mice containing cells derived from genetically modified EC cells. They injected EC cells into early mouse embryos (blastocysts) and reimplanted the embryos in surrogate mothers to complete development. The resulting offspring were genetic chimeras, comprising a mix of normal cells from the natural parents, and EC cells, but they developed multiple tumors and failed to establish breeding lines of mice. Another way had to be found to achieve this.

In 1981 Evans and colleague Matt Kaufmann reported their discovery in normal mouse embryos of cells that had properties similar to the EC cells. Laboriously, they devised ways of isolating such cells from mouse embryos, and maintaining them in culture. Evans' team subsequently showed that when these embryonic stem (ES) cells were transfected with foreign DNA and injected into blastocysts, the embryos completed their development to form chimeric mice incorporating cells containing the foreign DNA. Moreover, unlike the EC cells, the foreign DNA could be transmitted through the germline to descendants of the chimeric adults. In 1987 Evans and coworkers confirmed the strength of this approach by inserting into ES cells several copies of a mutant gene for the enzyme hypoxanthine phosphoribosyltransferase (HPRT), and producing mouse chimeras that contained the mutation in their cells. The mutation was inherited through the germline, and affected progeny could be identified by virtue of the defective HPRT enzyme.

In the meantime, other researchers, notably Mario Capecchi and Oliver SMITHIES, had independently developed a method of introducing targeted genetic modifications to a genome using the phenomenon of homologous recombination. Applying this process to the ES cells discovered by Evans represented a breakthrough that heralded a new era in developmental biology and experimental medicine.

Within a few years, numerous mouse lines had been bred to contain specific stably inherited genetic alterations; these were often 'knockout' mice, with a particular gene disabled, or mice containing other, more subtle, genetic changes.

Evans remained at the forefront of this emerging field, using the new technique to engineer mouse models that mimicked human diseases, such as cystic fibrosis and breast cancer. Besides providing insights into the genetic mechanisms underlying such disease, such models also suggest novel approaches to treatment, including gene therapy.

In 2001 Evans was awarded the Albert Lasker Award for basic medical research, jointly with Capecchi and Smithies. The trio were honoured again in 2007, by receiving the Nobel Prize for physiology or medicine, for discovering principles for introducing specific gene modifications in mice by the use of embryonic stem cells.

Evans, Robley Dunglison (1907–1996)
American physicist

Born in University Place, Nebraska, Evans was educated at the California Institute of Technology, where he obtained his PhD in 1932. He went to the Massachusetts Institute of Technology in 1934 and was appointed professor of physics there in 1945.

In 1940 Evans suggested that radioactive potassium-40 could be of use in geologic dating. It is widespread in the Earth's crust and has an exceptionally long halflife of over a thousand million years. It decays to the stable isotope argon-40, and determination of the ratio of ^{40}K to ^{40}Ar allows estimations of the age of potassium-bearing rocks ranging from 100,000 to about 10 million years. The technique proved to be particularly valuable as it permitted accurate dating beyond the limits of Willard LIBBY's carbon-14 technique.

Everett III, Hugh (1930–1982) *American physicist*

Everett was a doctoral pupil of John WHEELER in the 1950s at Princeton. In 1957 he published a famous paper on the foundations of quantum mechanics describing what has become known as the "many worlds" interpretation. The paper was entitled *Relative State Formulation of Quantum Mechanics*.

The traditional Copenhagen interpretation of quantum mechanics applied only to the submicroscopic world. Everett broke away from this tradition and attempted to apply quantum mechanics to the universe. He established a universal wave function that could be applied to both microscopic entities and macroscopic observers. As a consequence, there is no collapse of the wave function and quantum paradoxes, such as Schrödinger's cat, are avoided.

This approach, however, is not without paradoxical conclusions of its own. In Everett's formulation, the result of a measurement is to split the universe into as many ways as to allow all possible outcomes of the measurement. Thus if an observer were to check the outcome of a die throw, the universe would split into six copies with each one containing one of the six possible outcomes of the throw. Everett proposed that each outcome is realized in a number of parallel universes between which there is no communication.

While Everett's work has inevitably been taken up by many science fiction writers, it has also been taken seriously by other scientists. GELL-MANN, for example, has tried to develop a version of quantum theory that eliminates the role of the observer, in the manner of Everett, but reduces the idea of "many worlds" to one of possible histories of the universe to which a probability value can be assigned.

Ewing, Sir James Alfred (1855–1935)
British physicist

The son of a minister of the Free Church of Scotland, Ewing was educated at the University of Edinburgh where he studied engineering. He served as professor of engineering at the Imperial University, Tokyo, from 1878 until 1883 when he returned to Scotland to a similar post at the University of Dundee. In 1890 he was appointed professor of applied mechanics at Cambridge University, but in 1903 moved into higher levels of administration, first as director of naval education and from 1916 until his retirement in 1929 as principal and vice-chancellor of Edinburgh University.

In Japan he worked on problems in seismology and in 1883 published *Treatise on Earthquake Measurement*. However, his most notable achievement as a physicist was his work on hysteresis, first described by him in 1881. Hysteresis is an effect in which there are two properties, M and N, such that cyclic changes of N cause cyclic variations of M. If the changes of M lag behind those of N, there is hysteresis in the relation of M to N. Ewing came across the phenomena when working on the effects of stress on the thermoelectric properties of a wire. Hysteresis effects were later shown to apply to many aspects of the behavior of materials, in particular in magnetization.

Ewing was put in charge of the cryptologists at the Admiralty from 1914 to 1916. He described his work there in his book *The Man in Room 40* (1939).

Ewing, William Maurice (1906–1974)
American oceanographer

Ewing was born at Lockney in Texas and educated at the Rice Institute, Houston, obtaining his PhD in 1931. He taught at Lehigh University, Pennsylvania from 1934 until moving in 1944 to Columbia University, New York, where he organized the new Lamont Geological Observatory into one of the most important research institutions in the world.

Ewing pioneered seismic techniques to obtain basic data on the ocean floors. He was able to establish that the Earth's crust below the oceans is only about 3–5 miles (5–8 km) thick while the corresponding continental crust averages 25 miles (40 km).

Although the Mid-Atlantic Ridge had been discovered when cables were laid across the Atlantic, its dimensions were unsuspected. In 1956 Ewing and his colleagues were able to show that the ridge constituted a mountain range extending throughout the oceans of the world and was some 40,000 miles (64,000 km) long. In 1957, working with Marie Tharp and Bruce HEEZEN, he revealed that the ridge was divided by a central rift, which was in places twice as deep and wide as the Grand Canyon.

His group found that the oceanic sediment, expected to be about 10,000 feet (3,000 m) thick, was nonexistent on or within about 30 miles (50 km) of the ridge. Beyond this it had a thickness of about 130 feet (40 m) – much less than the depth of the corresponding continental sediment. All this seemed to be consistent with the new sea-floor spreading hypothesis of Harry H. HESS. Ewing was however reluctant to support it until Frederick VINE and Drummond H. MATTHEWS showed how the magnetic reversals discovered by B. Brunhes in 1909 could be used to test the theory.

Ewing also proposed, with William Donn, a mechanism to explain the periodic ice ages. If the Arctic waters were icefree and open to warm currents this source of water vapor would produce greater accumulations of snowfall. This would increase the Earth's reflectivity and reduce the amount of solar radiation absorbed. Temperatures would fall and glaciers move south, but with the freezing of the Arctic seas the supply of water vapor would be cut off and the ice sheets would retreat. This would cause an increase in solar radiation absorbed and the cycle would begin again. No hard evidence has yet been found in support of the theory.

Ewins, Arthur James (1882–1957) *British pharmaceutical chemist*

Ewins was born in Norwood on the outskirts of London and went straight from school to the brilliant team of researchers at the Wellcome Physiological Research Laboratories at Beckenham in Kent. Here he worked on alkaloids with George BARGER and in 1914, with Henry DALE, isolated the important neurotransmitter acetylcholine from ergot. Following wartime experience in manufacturing arsenicals with the Medical Research Council, he became head of research with the pharmaceutical manufac-

turers May and Baker, where he remained until his retirement in 1952. Under Ewins in 1939, the team produced sulfapyridine, one of the most important of the new sulfonamide drugs. An important later discovery was the antiprotozoal drug pentamidine (1948).

Eyde, Samuel (1866–1940) *Norwegian engineer and industrialist*

Born in Arendal in Norway, Eyde (Id) trained as a civil engineer in Berlin. Up until 1900 he worked in Germany and also on harbor and railroad-station projects in Scandinavia. Here he became interested in industrial electrochemical processes – a subject of some potential in Scandinavia because of the availability of cheap hydroelectric power.

In 1901 Eyde met Kristian BIRKELAND with whom he developed a process (1903) for reacting atmospheric nitrogen with oxygen in an electric arc. Processes in which nitrogen in the atmosphere is "fixed" in nitrogen compounds are of immense importance in the production of nitrogenous fertilizers for agriculture. The main method is the catalytic reaction with hydrogen developed by Fritz HABER. The *Birkeland–Eyde process*, which gives nitrogen oxides, needed plentiful and cheap supplies of electricity, leading to a significant growth in the production of hydroelectric power. In 1900 Norway had an output of little more than 100,000 kilowatts; by 1905 production had jumped to 850,000 kilowatts. In the same year Birkeland started the company Norsk Hydro-Elektrisk.

Kvaelstof, with the help of French capital, began to produce fertilizers by the Birkeland–Eyde process. As a result of this, Norway's export of chemicals was to treble before the start of World War I. Eyde retired from the firm in 1917. As well as his industrial interests, he was also a member of the Norwegian parliament.

Eyring, Henry (1901–1981) *Mexican–American physical and theoretical chemist*

Eyring, a grandson of American missionaries who had become Mexican citizens, was born at Colonia Juarez in Mexico. He thus first came to America in 1912 as a Mexican citizen and did not take American citizenship until 1935. He was educated at the University of Arizona and the University of California, where he obtained his PhD in 1927. He then held a number of junior appointments before joining the Princeton faculty in 1931, becoming professor of chemistry there in 1938. Eyring moved to a similar chair at the University of Utah, holding the post until his retirement in 1966.

Eyring, the author of 9 books and over 600 papers, was as creative a chemist as he was productive. His main work was probably in the field of chemical kinetics with his transition-state theory. Since the time of Svante ARRHE-

NIUS it had been appreciated that the rate constant of a chemical reaction depended on temperature according to an equation of the form:

$$k = Ae^{-E/RT}$$

The constant A is the frequency factor of the reaction; E_A is the activation energy. The values of A and E_A can be found experimentally for given reactions. Eyring's contribution to the field was to develop a theory capable of predicting reaction rates.

In a reaction, the atoms move – i.e., molecules break and new molecules form. If the potential energy of a set of atoms is plotted against the distances between atoms for chosen arrangements, the result is a surface. Positions of low energy on the surface correspond to molecules; a reaction can be thought of as a change from a low-energy point, over a higher energy barrier, to another low-energy position.

A. Marcelin, in 1915, had shown that reactions could be represented in this way, and in 1928 Fritz London pointed out that it was possible to calculate potential surfaces using quantum mechanics. Eyring, with Michael Polyani,

first calculated such a surface (1929–30) for three hydrogen atoms and Eyring later went on to calculate the potential surfaces for a number of reactions. The activation energy of the reaction is the energy barrier that the system must surmount.

Eyring later (1935) showed how to calculate the frequency factor (A). He assumed that the configuration of atoms at the top of the energy barrier – the "activated complex" – could be treated as a normal molecule except for a vibrational motion in the direction of the reaction path. Assuming that the activated complex was in equilibrium with the reactants and applying statistical mechanics, Eyring derived a general expression for reaction rate. Eyring's theory, called absolute-rate theory, is described in his book (with Samuel Glasstone and Keith J. Laidler) *The Theory of Rate Processes* (1941). It can also be applied to other processes, such as viscosity and diffusion.

Eyring also worked on the theory of the liquid state and made contributions in molecular biology.

F

Fabre, Jean Henri (1823–1915) *French entomologist*

> History ... records the names of royal bastards, but cannot tell us the origin of wheat.
> —*Souvenirs entomologiques* (1879–1907; Entomological Recollections)

> If there is one vegetable which is God-given, it is the haricot bean.
> —As above

Although world famous for his detailed studies of insect habits and life histories, Fabre (**fah**-bre) did not attain stature as a field entomologist until late in his life, his earlier years being spent under great difficulty and comparative poverty. Described by DARWIN, with whom he corresponded, as "an inimitable observer," Fabre's best-known entomological observations were largely made in his native Provence and the Rhône valley. His enthusiasm for his subject had been stimulated by reading an essay on the habits of the *Cerceris* wasp, which prompted Fabre to make his own detailed observations of these and other parasitic wasps, as well as many other insect groups. His descriptions of their development and behavior, written in a clear simple style, still stand as models of accurate observation. Fabre's earliest entomological observations appeared in *Annales des sciences naturelles* (Annals of the Natural Sciences), but his major publication is the classic *Souvenirs entomologiques* (10 vols. 1879–1907; Entomological Recollections).

Fabricius, David (1564–1617) *German astronomer*

Fabricius (fah-**bree**-tsee-uus), who was born in Essen, now in Germany, was a clergyman and amateur astronomer. He engaged in a prolonged correspondence with Johannes KEPLER but remained unconvinced by Kepler's Copernican arguments. In 1596 he noticed the variability of the star Omicron Ceti and called it "Mira" (the marvelous). Fabricius was murdered by one of his parishioners. His son Johannes Fabricius was also an astronomer and was one of the discoverers of sunspots.

Fabricius ab Aquapendente, Hieronymus (1537–1619) *Italian anatomist and embryologist*

Fabricius (fa-**brish**-ee-us) was born at Aquapendente in Italy and educated at the University of Padua where he studied under Gabriel FALLOPIUS, succeeding him, in 1565, as professor of anatomy.

As an anatomist his most significant work was his *De venarum ostiolis* (1603; On the Valves of the Veins), which contains a clear and detailed description of the venous system and which exercised a considerable influence on his most famous pupil, William HARVEY. Fabricius himself entertained no such idea as the circulation of the blood, explaining the role of the valves as retarding the blood flow, thus allowing the tissues to absorb necessary nutriment.

He spent much time observing the development of the chick embryo and published two works *De formato foetu* (1600; On the Formation of the Fetus) and *De formatione ovi et pulli* (1612; On the Development of the Egg and the Chick). These were hailed as elevating embryology into an independent science but they still contain many incorrect assumptions.

Thus for Fabricius semen did not enter the egg but rather initiated the process of generation from a distance in some mysterious way. He also made a now totally unfamiliar distinction between what nourishes and what produces the embryo. Thus he believed both the yolk and albumen merely nourished the embryo. Having eliminated the sperm, yolk, and albumen, Fabricius claimed that the chalaza – the spiral threads holding the yolk in position – produces the chick.

It was while engaged upon this work that he discovered and described the bursa of Fabricius. This is a small pouch in the oviduct of the hen, which Fabricius thought to be a store for semen. In the 1950s however the young research student B. Glick showed that this obscure organ plays a key role in the immune system of chickens, and by implication of humans who must possess a comparable system.

Fahrenheit, (Gabriel) Daniel (1686–1736) *German physicist*

Possibly owing to a business failure, Fahrenheit (**fah**-ren-hIt) emigrated to Amsterdam from his native Danzig (now Gdańsk in Poland) to become a glass blower and instrument maker. He specialized in the making of meteorological instruments, and proceeded to develop a reliable and accurate thermometer. GALILEO had invented the thermometer in about 1600, using changes in air volume as an indicator. Since the volume of air also varied considerably with changes in atmospheric pressure liquids of various kinds were quickly substituted. Fahrenheit was the first to use mercury in 1714. He fixed his zero point by using the freezing point of a mixture of ice and salt as this gave him the lowest temperature he could reach. His other fixed point was taken from the temperature of the human body, which he put at 96°. Given these two fixed points the freezing and boiling points of water then work out at the familiar 32° and 212°. One advantage of the system is that, for most ordinary purposes, negative degrees are rarely needed.

Using his thermometer, Fahrenheit measured the boiling point of various liquids and found that each had a characteristic boiling point, which changed with changes in atmospheric pressure.

Fairbank, William (1917–1989) *American physicist*

Fairbank was born at Minneapolis in Minnesota and educated at Whitman College, Walla Walla, Washington, and at the University of Washington. He gained his PhD at Yale in 1948. He spent the war years at the Radiation Laboratory of the Massachusetts Institute of Technology. After working at Amherst, Maryland (1947–52) and Duke University, North Carolina (1952–59), Fairbank was appointed professor of physics at Stanford, a post he held until his death from a heart attack in 1989.

In 1977, Fairbank, in collaboration with George Larue, claimed to have experimental evidence for the existence of a quark. The concept of quarks, with an electric charge −1/3 or +2/3 the electron charge, had been proposed by Murray GELL-MANN in 1963 to explain the behavior of hadrons. It was known that it would be unlikely that quarks could be produced at the energies available in particle accelerators. However, it was possible that some might be created in the atmosphere as a result of high-energy cosmic rays. A number of physicists set up ingenious sensitive experiments to "hunt the quark."

Fairbank's technique was a much more sensitive and sophisticated version of Robert MILLIKAN's oil-drop experiment for measuring the charge of the electron. A small sphere (0.25 millimeter diameter) of niobium was suspended between metal plates at a temperature close to absolute zero. The charge on the sphere could be measured by the electric field between the plates.

When Fairbank examined his results he found that in the case of one ball there was "a nonzero residual change of magnitude −0.37 ±

0.03." At first, Fairbank warned, the results did not necessarily imply the presence of a quark as there could well be spurious charge forces present. Consequently, Fairbank spent a good deal of time eliminating these and numerous other possible distortions from his experimental setup.

Theorists were suspicious of Fairbank's work because free quarks are thought to be impossible to produce – the doctrine of "quark confinement." Despite this, Fairbank announced in 1979 that, using modified apparatus, he had detected a second particle with a fractional charge.

While no one has managed to reproduce Fairbank's experiments, he was sufficiently respected as a careful and skillful experimentalist for his work to be taken seriously. Consequently, for some particle physicists at least, the issue of quark confinement remains an open question.

Fajans, Kasimir (1887–1975) *Polish–American physical chemist*

Fajans (**fah**-yahns) was born in the Polish capital of Warsaw and gained his doctorate at Heidelberg (1909). He was professor of physical chemistry at Munich (1925–35) before emigrating to America in 1936. He was a member of the Faculty of the University of Michigan (1936–57) and emeritus professor from 1957.

He is best known for *Fajan's laws* (1913), which state that elements emitting alpha particles decrease in atomic number by two while those undergoing beta emission gain in atomic number by one. These were independently discovered by Frederick SODDY. In 1917 Fajans discovered, with Otto Gohring, uranium X_2 – a form of the element protactinium ^{234}Pr.

Fallopius, Gabriel (1523–1562) *Italian anatomist*

Fallopius (fa-**loh**-pee-us) was born in Modena, Italy. He originally intended to enter the Church and served as a canon in his native city for some time before turning to the study of medicine at Ferrara and Padua where he was a pupil of the great Andreas VESALIUS. He held the chair of anatomy at the University of Pisa from 1548 until 1551 when he moved to a similar post at the University of Padua, remaining there until his early death at the age of 39.

On the strength of his masterpiece *Observationes anatomicae* (1561; Anatomical Observations) he has been described as a better and more accurate anatomist than his teacher, Vesalius. His main innovations were in the anatomy of the skull and the generative system. In the former area he introduced the terms cochlea and labyrinth, going on to give a clear account of the auditory system.

In his account of the reproductive system he described the clitoris and introduced the term "vagina" into anatomy. He is mainly remembered today, however, for his description of the eponymous fallopian tubes. The tubes, connecting the uterus and the ovaries, were described by him as resembling at their extremity a "brass trumpet" (tuba). This somehow became mistranslated into English as tube. It was many years before their function was understood.

Faraday, Michael (1791–1867) *British physicist and chemist*

> Mr. Faraday is not only a man of profound chemical and physical science (which all Europe knows), but a very respectable lecturer. He speaks with ease and freedom, but not with a gossiping, unequal tone, alternately inaudible and bawling, as some very learned professors do; he delivers himself with clearness, precision, and ability.
> —Friedrich Ludwig Georg von Raumer in *Cyclopaedia of Literary and Scientific Anecdote* (19th century)

Faraday's father was a blacksmith who suffered from poor health and could only work irregularly. Faraday, who was born in Newington, England, knew real poverty as a child and his education was limited for he left school at the age of 13. He began work for a bookseller and binder in 1804 and was apprenticed the following year. His interest in science seems to have been aroused by his reading the 127-page entry on electricity in an Encyclopaedia Britannica he was binding and this stimulated him to buy the ingredients to make a Leyden jar and to perform some simple experiments. He joined the City Philosophical Society, which he attended regularly, broadening his intellectual background still further. The turning point in his life came when he attended some lectures by Humphry DAVY at the Royal Institution in 1812. He took very full notes of these lectures, which he bound himself.

By now he was no longer satisfied with his amateur experiments and evening lectures and wanted desperately to have a full-time career in science. He wrote to the President of the Royal Society, Joseph BANKS, asking for his help in obtaining any post but received no reply. Faraday now had a little luck. Davy had had an accident and needed some temporary assistance. Faraday's name was mentioned and proved acceptable. While working with Davy he showed him the lecture notes he had taken and bound. When a little later, in 1813, a vacancy for a laboratory assistant arose, Davy remembered the serious young man and hired him at a salary of a guinea a week (less than Faraday had been earning as a bookbinder).

Faraday was to spend the rest of his working life at the Royal Institution, from which he finally resigned in 1861. In 1815 he was promoted to the post of assistant and superintendent of the apparatus of the laboratory and meteorological collection. In 1825 he was made director of the laboratory and, in 1833, he was elected to the newly endowed Fullerian Professorship of Chemistry at the Royal Institution. He had earlier turned down the offer of the chair of chemistry at University College, London, in 1827.

The paucity of the salary paid him was made up by Faraday with consultancy fees and a part-time lectureship he held at the Royal Military Academy, Woolwich. These extra sources took up his time and in 1831, when he was working as hard as he could on his electrical experiments, he gave up all his consultancies. This left him in some financial difficulties and moves were made to arrange for a government pension. He called on the prime minister of the day, Lord Melbourne, who made some sneering remark about such pensions being, in his view, a "gross humbug." This was enough to make Faraday refuse the pension. In fact, Faraday was one of nature's great refusers. Apart from the pension and the chair at University College, he also refused a knighthood and, what must surely be a record, the presidency of the Royal Society, not once, but twice. Faraday also had strong views on awards – "I have always felt that there is something degrading in offering rewards for intellectual exertion, and that societies or academies, or even kings and emperors, should mingle in the matter does not remove the degradation." He had become a fellow of the Royal Society in 1824 but not without some friction between himself and the president, Davy. He was asked to withdraw his application by Davy. Just why Davy behaved in this way is not clear. Some have seen it as jealousy by Davy of someone whose talents so clearly surpassed his own. There is no evidence of this but it is reasonably clear that when Faraday insisted on going ahead with his application Davy voted against him.

Faraday's financial problems were solved when, in 1835, Melbourne apologized, enabling him to accept the pension. After his labors of the 1830s he suffered some kind of breakdown in 1841 and went into the country to rest. Just what was wrong is not known; he wrote in 1842 that he could see no visitors because of "ill health connected with my head." For two years he did no work at all until in 1844 he seemed to be able to resume his experiments. Faraday continued to work but by the 1850s his creativity was in decline. He gave his last childrens' lectures at the Royal Institution in 1860 and resigned from it the following year, taking up residence in a house at Hampton Court made available to him by Prince Albert in 1858.

Faraday's first real successes were made in chemistry. In 1823 he unwittingly liquefied chlorine. He was simply heating a chlorine compound in a sealed tube and noticed the formation of some droplets at the cold end. He realized that this was the result of both temperature and pressure and on and off over the years applied the method to other gases. In 1825 he discovered benzene (C_6H_6) when asked to examine the residue collecting in cylinders of illuminating gas; he called the new compound "bicarburet of hydrogen" because he took its formula to be C_2H. As a working chemist Faraday was one of the best analysts of his day. All his working life he was working and publishing as a chemist but in 1820 he also turned to a new field that was to dominate his life.

Faraday had begun by accepting the view that electricity is composed of two fluids. It was common in the 18th century to see such phenomena as light, heat, magnetism, and electricity to be the result of weightless fluids. In 1820 Hans Christian OERSTED made a most surprising discovery: he had found that a wire carrying a current is capable of deflecting a compass needle; the direction in which the needle turned depended on whether the wire was under or over the needle and the direction in which the current was flowing. André Marie AMPÈRE found that two parallel wires attract each other if the current in each is traveling in the same direction but repel each other if the currents are moving in opposite directions. Finally François ARAGO discovered that a copper disk rotating freely on its own axis would produce rotation in a compass needle suspended over it.

These phenomena were difficult to fit into fluid theories of electricity and magnetism. They enabled Faraday to make his first important discovery in 1821, that of electromagnetic rotation. A magnet was placed upright in a tube of mercury and secured firmly at the bottom with the pole of the magnet above the surface. A wire dipping into the mercury but free to rotate was suspended over the pole. When a current was passed through the mercury and through the wire, the wire rotated around the magnet. If the wire was secured and the magnet allowed to move, then the current caused the magnet to rotate. The first electric motor had been constructed.

When Faraday published his results they were to cause him much distress. William WOLLASTON had spoken of the possibility of such rotation and many concluded that Faraday had stolen his ideas. Faraday was only too aware of the stories about him but found there was little he could do about them. It may well have been this that Davy thought disqualified him from membership of the Royal Society.

In any case it was not really electromagnetic rotation that interested Faraday. All the new results involved the production of a magnetic force by an electric current and Faraday, with many others, was sure that it should also be possible to induce an electric current by magnetic action. He tried intermittently for ten years without success until in 1832 he hit upon an apparatus in which an iron ring was wound with two quite separate coils of wire. One was connected to a voltaic cell; the other to a simple galvanometer. He showed that on making and breaking the current in the cell circuit, the galvanometer momentarily registered the presence of a current in its circuit. The following few months were some of the most active of his life. He showed that the same results can be obtained without a battery: a magnet moved in and out of a coil of wire produced a current. A steady current could be produced by rotating a copper disk between the poles of a powerful magnet. His results were published in his *Experimental Researches in Electricity, first series* (1831).

Faraday found this deeply satisfying for it reinforced one of his strongest convictions about nature "that the various forms under which the forces of matter are made manifest have one common origin." That electricity and magnetism could interact made this view more plausible. At the time it was by no means clear that the various types of electricity – static, voltaic, animal, magnetic, and thermoelectric – were the same and FARADAY spent the period 1833–34 on this problem, publishing his results in the third series of his *Experimental Researches*.

Faraday had also continued the work of Davy on electrolysis – i.e., on the chemical reaction produced by passing an electric current through a liquid. He applied his ideas on the quantity of electricity to this chemical effect and produced what are now known as *Faraday's laws of electrolysis*. By careful analysis he showed that the chemical action of a current is constant for a constant quantity of electricity. This was his first law, that equal amounts of electricity produce equal amounts of decomposition. In the second law he found that the quantities of different substances deposited on the electrode by the passage of the same quantity of electricity were proportional to their equivalent weights.

In his explanations of magnetic and electrical phenomena Faraday did not use the fluid theories of the time. Instead he introduced the concept of lines of force (or tension) through a body or through space. (A similar earlier idea had been put forward by R. J. Boscovich with his picture of point atoms surrounded by shells of force.) Thus Faraday saw the connection between electrical and magnetic effects as vibrations of electrical lines communicated to magnetic lines. His experiments on induction were described in terms of the cutting of magnetic lines of force, which induces the electrical current. He explained electrical induction in dielectrics by the strain in "tubes of induction" – and electrolysis was complete breakdown under such strain.

Faraday was no mathematician, relying instead on his wonderful experimental skill and his imagination. His lines of force were taken up by others more skillful mathematically. In the latter half of the century Clerk MAXWELL developed Faraday's ideas into a rigorous and powerful theory, creating an orthodoxy in physics that lasted until the time of EINSTEIN. Faraday's greatness rests in his courage and insight in rejecting the traditional physics and creating an entirely new one. Few can compete with Faraday at the level of originality.

One further effect discovered by Faraday lay in optics. His discovery of *Faraday rotation* in 1845 was one that gave him pleasure for it seemed to be further evidence for the unity of nature by showing that "magnetic force and light were proved to have a relation to each other." Here, he showed that if polarized light is passed through a transparent medium in a magnetic field its plane of polarization will be rotated. Not the least of Faraday's achievements was as a lecturer and popularizer of science. In 1826 he started the famous Christmas lectures to children at the Royal Institution in London and gave 19 of these lecture courses. For most only the notes exist but a couple of lectures were taken down in shorthand and later published: *The Chemical History of a Candle* and *Lectures on Various Forces of Matter*. The children's Christmas lectures still continue to be given every year by eminent scientists.

Farman, Joe (1930–) *British atmospheric chemist*

Farman has been engaged in the study of the Antarctic atmosphere since 1957, working in recent years for the British Antarctic Survey based in Cambridge, England. The Survey operates five Antarctic stations.

One of the tasks of Farman's team has been to measure Antarctic ozone levels. This is normally done with a spectrophotometer recording the amount of ultraviolet radiation (UV) reaching the Earth's surface. As ozone absorbs UV, the less ozone in the atmosphere, the greater the amount of UV reaching the Earth. The ozone level was measured in Dobson units (DU) and normally had a value of about 400 DUs (after the designer of the spectrophotometer, G. M. Dobson). In 1982 Farman first noted a very low reading of 130 DUs. Initially he merely assumed that his instruments were breaking down. Further, ozone levels were known to fluctuate widely and shortlived, unusual levels, both high and low, had been known before. And as nothing unusual had been detected by the Nimbus satellite, Farman suspected his readings were, in some unknown way, unreliable. Nonetheless new equipment was installed and when similar low levels were recorded in the following season Farman began to take the matter seriously. An examination of past records revealed that there had been a decline in ozone levels since 1977. He further checked

the readings by organizing measurements at another research station 1,000 miles away.

By the end of the 1984 season Farman was convinced that he had detected a persistent seasonal fall in Antarctic ozone levels of about 40%. Soon after the southern winter's end in October, the polar vortex collapses and ozone levels rise as air is drawn in from elsewhere. A check on the Nimbus satellite revealed that it was programmed to ignore readings that varied from the average by more than a third.

Farman was aware of the work of ROWLAND and began to suspect that ozone depletion was connected with atmospheric CFCs. He published his results in *Nature* in May 1985. All subsequent work, whether by Farman or others, has confirmed his initial measurements of 1982.

Fechner, Gustav Theodor (1801–1887) *German physicist and psychologist*

Fechner (**fek**-ner), who was born at Gross-Särchen in Germany, studied medicine at the University of Leipzig; after graduating, however, his interest turned toward physics. He did some research on the galvanic battery and, on the strength of this, was appointed professor of physics at Leipzig in 1834. However, partial blindness, caused by overlong study of the Sun, and mental illness forced him to resign five years later.

It was after his recovery that he started trying to put psychology on a scientific basis. He studied the relationship between the intensities of external stimulae and subjective sensations, and elaborated Ernst WEBER's work to develop the *Weber–Fechner law* governing this connection. This work, published in 1860, founded the science of psychophysics.

Fechner also became interested in spiritualism, like many scientists of his day, and wrote much on mystical and philosophical subjects. He also wrote satirical poems under the pseudonym of Dr. Mises.

Fehling, Hermann Christian von (1812–1885) *German organic chemist*

Fehling (**fay**-ling), who was born in Lübeck, now in Germany, gained his doctorate at Heidelberg in 1837 and studied with Justus von LIEBIG at Giessen. He was professor of chemistry at the polytechnic school at Stuttgart from 1839 to 1882.

Fehling did much work in organic chemistry, including the preparation and hydrolysis of benzonitrile and the discovery of succinosuccinic ester. However, he is best known for his invention of *Fehling's solution*, used as a test for aldehydes and reducing sugars (e.g., glucose, fructose, lactose). It consists of a solution of copper(II) sulfate (Fehling's I) and an alkaline solution of a tartaric acid salt (Fehling's II).

The solutions are boiled with the test material and a positive result is a brick-red precipitate (insoluble copper(I) oxide), formed by the reduction of copper(II) sulfate.

Feigenbaum, Edward Albert (1936–)
American computer scientist

Born at Weehawken, New Jersey, Feigenbaum (**fI**-gen-bowm) first studied electrical engineering at Carnegie Institute of Technology, Pittsburgh; there, under the influence of Herbert SIMON, his interests were diverted into the newly emerging field of artificial intelligence (AI). Consequently after completing his PhD in 1960 at Carnegie in cognitive psychology, Feigenbaum spent the period from 1960 to 1964 at the University of California, Berkeley, before being appointed professor of computer science at Stanford University in 1965.

Feigenbaum was an early pioneer in the expert-systems approach to AI. Workers like Allan NEWELL had tried to provide computers with a general understanding that would enable them to tackle a wide variety of problems. Their ambitions, however, remained unfulfilled. Feigenbaum opted for a more specialized alternative. "We found," he commented, "it was better to be knowledgeable than smart"; to build computers, that is, capable of playing high-level chess, but unable to understand the rules of tick-tack-toe.

Against this background Feigenbaum set out to program a computer to be able, on the basis of data from mass spectrometers and elsewhere, to identify organic compounds. By 1971 he had developed DENDRAL which was capable of formulating hypotheses about molecular structure, and then testing the hypothesis against further data. It finally offered a list of compounds ranked in terms of decreasing likelihood. DENDRAL has proved to be remarkably successful and, allowing for later improvements and refinements, is recognized to be almost as proficient as a human chemist.

Fenn, John B. (1917–) *American chemist*

Fenn was born in New York City and educated at Berea College, where he obtained a BA in chemistry in 1937, and at Yale University, where he obtained a PhD in 1940. He then worked for about a dozen years at Monsanto and Sharples Chemical in Michigan and subsequently for seven years at Richmond, working for a small company. Between 1959 and 1967 he was the director of Project SQUID, a program of research in jet propulsion run by Princeton University. Between 1967 and 1980 he was a professor of applied science and chemistry at Yale University, and between 1980 and 1987 he was a professor of chemical engineering at Yale. He remained at Yale as an emeritus professor until 1994, when he moved to Virginia Commonwealth University. He celebrated his 90th birthday in 2007.

In 1988 Fenn published papers on a method of mass spectrometry known as electrospray ionization (ESI). In this method charged droplets of protein solution are produced. As the water evaporates, the droplets shrink, thus increasing their charge density, and after some time protein ions remain. The masses of these protein ions are determined by getting them to move over some fixed distance and measuring their time of flight. The technique was an important advance in extending the use of mass spectrometry to large molecules that could not be vaporized. Fenn shared the 2002 Nobel Prize for chemistry with Koichi TANAKA and Kurt WÜTHRICH for this work.

Fermat, Pierre de (1601–1665) *French mathematician and physicist*

> It cannot be denied that he [Fermat] has had many exceptional ideas, and that he is a highly intelligent man. For my part, however, I have always been taught to take a broad overview of things, in order to be able to deduce from them general rules, which might be applicable elsewhere.
> —René Descartes

Fermat (fer-**mah**) was one of the leading mathematicians of the early 17th century although not a professional mathematician. Born at Beaumont-de-Lomagne in France, he studied law and spent his working life as a magistrate in the small provincial town of Castres. Although mathematics was only a spare-time activity, Fermat was an extremely creative and original mathematician who opened up whole new fields of enquiry.

Fermat's work in algebra built on and greatly developed the then new theory of equations, which had been largely founded by François VIÈTE. With PASCAL, Fermat stands as one of the founders of the mathematical theory of probability. In his work on methods of finding tangents to curves and their maxima and minima he anticipated some of the central concepts of Isaac NEWTON's and Gottfried LEIBNIZ's differential calculus. Another area of mathematics that Fermat played a major role in founding, independently of René Descartes, was analytical geometry. This work led to violent controversies over questions of priority with Descartes. Nor were Fermat's disagreements with Descartes limited to mathematics. Descartes had produced a major treatise on optics – the *Dioptrics* – which Fermat greatly disliked. He particularly objected to Descartes's attempt to reach conclusions about the physical sciences by purely *a priori* rationalistic reasoning without due regard for empirical observation. By contrast Fermat's view of science was grounded in a thoroughly empirical and observational approach, and to demonstrate the errors of

Descartes's ways he set about experimental work in optics himself. Among the important contributions that Fermat made to optics are his discovery that light travels more slowly in a denser medium, and his formulation of the principle that light always takes the quickest path.

Fermat is probably best known for his work in number theory, and he made numerous important discoveries in this field. But he also left one of the famous problems of mathematics – *Fermat's last theorem*. This theorem states that the algebraic analog of PYTHAGORAS's theorem has no whole number solution for a power greater than 2, i.e., the equation:

$$a^n + b^n = c^n$$

has no solutions for n greater than 2, if a, b, and c are all integers. In the margin of a copy of a book *Arithmetica of Diophantos*, an early treatise on equations, he wrote: "To resolve a cube into the sum of two cubes, a fourth power into two fourth powers, or in general any power higher than the second into two of the same kind, is impossible; of which I have found a remarkable proof. The margin is too small to contain it." Fermat never wrote down his "remarkable proof" and the equation was the subject of much investigation for over 350 years. In June 1993 the British-born mathematician Andrew WILES presented a proof to a conference at Cambridge in a lecture entitled "Modular forms, elliptic curves, and GALOIS representations." His proof ran to 1,000 pages – rather more space than Fermat's margin – and it is generally believed that Fermat, given the mathematical techniques available at the time, must have been mistaken in believing that he had a proof of the conjecture.

Fermi, Enrico (1901–1954) *Italian–American physicist*

> Whatever Nature has in store for mankind, unpleasant as it may be, men must accept, for ignorance is never better than knowledge.
> —Quoted by Laura Fermi in *Atoms in the Family*

Fermi (**fer**-mee) was without doubt the greatest Italian scientist since GALILEO and in the period 1925–50 was one of the most creative physicists in the world. Unusually in an age of ever-growing specialization he excelled as both an experimentalist and a theoretician.

He was born in Rome and brought up in the prosperous home of his father who, beginning as a railroad official, progressed to a senior position in government service. Fermi's intelligence and quickness of mind were apparent from an early age and he had little difficulty in gaining admission in 1918 to the Scuola Normale in Pisa, a school for the intellectual élite of Italy. He later completed his education at the University of Pisa where he gained his PhD

in 1924. After spending some time abroad in Göttingen and Leiden, Fermi returned to Italy where, after some initial setbacks, he was appointed to a professorship of physics at the University of Rome. This in itself was a considerable achievement for one so young, considering the traditional and bureaucratic nature of Italian universities. It was no doubt due to the reputation he had already established with the publication of some 30 substantial papers, and the support of O. M. Corbino, the most distinguished Italian physicist at the time and also a senator. Corbino was determined to modernize Italian physics and had the good sense to see that Fermi, despite his youth, was the ideal man to advance his cause.

Fermi began by publishing the first Italian text on modern physics, *Introduzione alla Fisica Atomica* (1928; Introduction to Nuclear Physics). Soon his reputation attracted around him the brightest of the younger Italian physicists. But the growth of fascism in Italy led to the dispersal of its scientific talent. By 1938 Fermi, with a Jewish wife, was sufficiently alarmed by the growing anti-Semitism of the government to join the general exodus and move to America.

However, before his departure, his period in Rome turned out to be remarkably productive, with major advances being made in both the theoretical and the experimental field. His experimental work arose out of attempts to advance the efforts of Irène and Frédéric JOLIOT-CURIE who had announced in 1934 the production of artificial radioactive isotopes by the bombardment of boron and aluminum with helium nuclei (alpha particles). Fermi realized that the neutron, discovered by James CHADWICK in 1932, was perhaps an even better tool for creating new isotopes. Although less massive than an alpha particle, the neutron's charge neutrality allowed it to overcome the positive charge of a target nucleus without dissipating its energy.

Fermi reported that in 1934 he had impulsively and for no apparent reason interposed paraffin between the neutron source and the target. "It was with no advance warning, no conscious prior reasoning ... I took some odd piece of paraffin" and placed it in front of the incident neutrons. The effect was to increase the activation intensity by a factor that ranged from a few tens to a few hundreds. Fermi had stumbled on the phenomenon of slow neutrons. What was happening was that the neutrons were slowing down as the result of collisions with the light hydrocarbon molecules. This in turn meant that they remained in the vicinity of the target nucleus sufficiently long to increase their chance of absorption.

The production of slow neutrons was later to have a profound impact in the field of nuclear energy, both civil and military. However, Fermi's immediate task was to use them to irradiate as many of the elements as possible and to produce and investigate the properties of a large number of newly created radioactive isotopes. It was for this work, for "the discovery of new radioactive substances ... and for the discovery of the selective power of slow neutrons" that Fermi was awarded the 1938 Nobel Prize for physics.

He did however miss one significant phenomenon. In the course of their systematic irradiation of the elements Fermi and his colleagues naturally bombarded uranium with slow neutrons. This would inevitably lead to nuclear fission, but Fermi thought that transuranic elements were being produced and in his Nobel address actually referred to his production of elements 93 and 94, which he named "ausonium" and "hesperium." In 1938 Otto FRISCH and Lise MEITNER first realized that nuclear fission was taking place in such reactions.

On the theoretical level Fermi's major achievement while at Rome was his theory of beta decay. This is the process in unstable nuclei whereby a neutron is converted into a proton with the emission of an electron and an antineutrino ($n \rightarrow p + e^- + \nu$). Fermi gave a detailed analysis which introduced a new force into science, the so-called "weak" force. An account was published in Italian in 1933 as an original English version was rejected by the journal *Nature* as being too speculative.

In America Fermi soon found himself caught up in the attempt to create a controlled nuclear chain reaction. In 1942 he succeeded in building the first atomic pile in the stadium of the University of Chicago at Stagg Field. Using pure graphite as a moderator to slow the neutrons, and enriched uranium as the fissile material, Fermi and his colleagues began the construction of the pile. It consisted of some 40,000 graphite blocks, specially produced to exclude impurities, in which some 22,000 holes were drilled to permit the insertion of several tons of uranium. At 2:20 p.m. on 2 December 1942, the atomic age began as Fermi's pile went critical, supporting a self-supporting chain reaction for 28 minutes. In a historic telephone call afterwards Arthur COMPTON informed the managing committee that "the Italian navigator has just landed in the new world," and that the natives were friendly.

Fermi continued to work on the project and was in fact present in July 1945 when the first test bomb was exploded in the New Mexico desert. He is reported to have dropped scraps of paper as the blast reached him and, from their displacement, to have calculated the force as corresponding to 10,000 tons of TNT.

After the war Fermi accepted an appointment as professor of physics at the University of Chicago where he remained until his untimely death from cancer. His name has been commemorated in physics in various ways. El-

ement 100, *fermium*, and the unit of length of 10^{-13} centimeter, the *fermi*, were named for him, as was the National Accelerator Laboratory, Fermilab, at Batavia, near Chicago.

Fernel, Jean François (1497–1558)
French physician

> From his [Fernel's] school there went forth skilled physicians more numerous than soldiers from the Trojan horse, and spread over all regions and quarters of Europe.
> —Guillaume Plancy, *Life of Fernel* (1577)

Fernel (fer-**nel**), the son of an innkeeper from Clermont in France, was educated at the Collège de Sainte Barbe in Paris where he graduated in 1519. After some years devoted to such subjects as philosophy and cosmology, Fernel took up the study of medicine, qualifying in 1530 and being appointed professor of medicine in 1534 at the University of Paris. He was also appointed as physician to Henri II after successfully treating Henri's mistress Diane de Poitiers, although his failure to cure Henri's father, Francis I, of syphilis appears not to have been held against him.

In 1554 Fernel published his *Medicina*, which went through some 30 editions and was one of the standard texts of the late 16th century. It is here that he introduced the terms "physiology" and "pathology" into medicine. The work was however mainly traditional, describing the physiology of GALEN with but a few modifications. Fernel also wrote an interesting account of the status of medicine in his *On the Hidden Causes of Things* (1548) in which he began to question long accepted magical and astrological accounts of diseases. He fell back on the standard medical objection to therapies offered by competitors, namely that any relief obtained would be only superficial and temporary.

Before turning to medicine Fernel had published his *Cosmotheoria* (1528), a work that contained one of the earliest measurements of a meridian, measured with an odometer between Paris and Amiens.

Ferrel, William (1817–1891) *American meteorologist*

> [Ferrel gave] to the science of meteorology a foundation in mechanics as solid as that which Newton laid for astronomy.
> —Cleveland Abbe, *Biographical Memoirs* (1895)

Born in Fulton County, Pennsylvania, Ferrel moved with his family to farm in West Virginia in 1829. Since he received only the most rudimentary education, his early scientific knowledge was entirely self acquired. Despite this he developed an interest in mathematical physics and, after graduating from Bethany College in West Virginia in 1844, began to study the *Principia* of Isaac Newton and the *Mé-*

canique céleste (Celestial Mechanics) of Pierre Simon de LAPLACE. He earned his living as a school teacher from 1844 until 1857 when, having established his scientific reputation, he was appointed to the staff of the American Ephemeris and Nautical Almanac. He worked there until 1867 when he joined the U.S. Coast and Geodetic Survey.

In 1856 he published his most significant work, *Essay on the Winds and Currents of the Oceans*. He showed that all atmospheric motion, as well as ocean currents, are deflected by the Earth's rotation. He went on in 1858 to formulate his law, which states that if a mass of air is moving in any direction there is a force arising from the Earth's rotation that always deflects it to the right in the northern hemisphere and to the left in the southern hemisphere. The air tends to move in a circle whose radius depends on its velocity and distance from the equator. Ferrel went on to show how this law could be used to explain storms and the pattern of winds and currents. He was in some ways anticipated by Gustave-Gaspard CORIOLIS whose name is much better known.

Ferrel also did fundamental work on the solar system. He was able to correct Laplace and show that the tidal action of the Sun and Moon on the Earth is slowly retarding the Earth's rotation. In 1864 he provided the first mathematical treatment of tidal friction. His other works included his three-volume *Meteorological Researches* (1877–82). In 1880 he invented a machine to predict tidal maxima and minima.

Ferrier, Sir David (1843–1928) *British neurologist*

Born in Aberdeen, Scotland, Ferrier graduated in philosophy from the university there in 1863 and then spent some time at Heidelberg studying psychology. He returned to Scotland to study medicine at Edinburgh University and graduated in 1868. Initially he worked in private practice until his appointment as professor of forensic medicine at King's College Hospital, London (1872–89). The chair of neuropathology was created for him in 1889, a post he retained until his retirement in 1908.

Following the fundamental paper of Gustav FRITSCH and Eduard HITZIG on cerebral localization, Ferrier began, in 1873, a series of experiments that both confirmed and extended their work. Whereas they had worked only with dogs, Ferrier used a wide variety of mammals, including primates. He was thus able to identify many more different areas in the cerebral hemispheres capable of eliciting movement. This work was soon shown to have practical implications. The surgeon William MACEWEN saw that the technique could be reversed and that disturbances of movement in a patient

could be used to indicate the site of a possible brain tumor.

Ferrier was an important figure in the newly emerging discipline of neurophysiology. He was one of the first editors of the influential journal *Brain* (founded 1878) and was also a founder member of the Physiological Society.

Fert, Albert (1938–) *French physicist*

Albert Fert was born in Carcassonne, France. He graduated in physics in 1962 from the Ecole Normale Supérieure in Paris. Fert obtained his masters degree in 1963 at the University of Paris and was awarded his PhD from the Université de Paris-Sud in 1970, where he has been a professor since 1976.

In 1988 he discovered the phenomenon of giant magnetoresistance in layers of iron and chromium. Ordinary magnetoresistance is the small change (about 5%) in the electrical resistance of a metal when it is placed in a magnetic field. In certain materials, such as thin films in which there are alternating layers of ferromagnetic metals and nonferromagnetic metals, there is a large change in the electrical resistance of the non-magnetic layer in the presence of a magnetic field. This phenomenon, known as giant magnetoresistance, is a subtle quantum mechanical effect involving the spins of the electrons in the metals. There are many important applications of the effect. This is because it is ideally suited for reading data from hard disks and has led to vastly increased storage capacities. The technology resulting from giant magnetoresistance is regarded as the first example of spintronics, i.e., electronic devices that exploit the spins of electrons.

Giant magnetoresistance was also discovered independently by Peter GRÜNBERG in 1988. Fert and Grünberg shared several major prizes for their discovery, notably the Wolf Prize in 2006 and the Nobel Prize for physics in 2007.

Fessenden, Reginald Aubrey (1866–1932) *Canadian electrical engineer*

Fessenden was born at Milton in Quebec, Canada. After an education in Canada, he worked for EDISON as an engineer and later as head chemist (1886–90) and as an engineer (1890–91) for WESTINGHOUSE, Edison's great rival. He was professor of electrical engineering at Purdue University, Indiana (1882–83), and the Western University of Pennsylvania (now the University of Pittsburgh) from 1893 to 1900. He was then appointed special agent for the U.S. Weather Bureau, adapting the technique of radio telegraphy, newly developed by Guglielmo MARCONI, to weather forecasting and storm warning. In 1902 he became general manager of the National Electric Signaling Company, a company formed by two Pittsburgh financiers to exploit his ideas. From 1910 he was consultant engineer at the Submarine Signal Company.

Fessenden's inventions were prolific and varied: at the time of his death he held over 500 patents. In 1900 he developed an electrolytic detector that was sufficiently sensitive to make radio telephony feasible. His most significant invention was the technique of amplitude modulation. This involved the use of "carrier waves" to transmit audio signals; he varied the amplitude of a steady high-frequency radio signal so that it corresponded to variations in the sound waves and thus "carried" the audio information. Using this principle he transmitted, on Christmas Eve 1906, what was probably the first program of music and speech broadcast in America. The program was heard by ships' radio operators up to a distance of several hundred miles.

Fessenden had a choleric temperament and a fear of being outwitted by businessmen. He was involved in various lawsuits, one of which was against his financial backers in the National Electric Signaling Company: Fessenden won a judgment of $406,000 and sent the company into bankruptcy.

Feyerabend, Paul Karl (1924–1994) *Austrian philosopher of science*

> Unanimity of opinion may be fitting for a church, for the frightened or greedy victims of some (ancient or modern) myth, or for the weak and willing followers of some tyrant. Variety of opinion is necessary for objective knowledge.
> —*Against Method* (1975)

Feyerabend (**fī**-er-ah-bent), who was born in the Austrian capital of Vienna, enlisted in the German army immediately after leaving school in 1942. He was seriously wounded on the Russian front in 1945 and was subsequently awarded the Iron Cross. His education after the war was varied and unusual. He was a student of the Vienna Music Academy, but he also studied history, physics, and astronomy at the University of Vienna. In 1952 he moved to London and attended the seminars given by Karl POPPER at the London School of Economics. Though he had never studied philosophy formally, he was appointed in 1955 to a philosophy lectureship at Bristol University. In 1959 Feyerabend moved to the U.S. and in 1962 was appointed professor of philosophy at the University of California, Berkeley, a position he held, along with a similar appointment from 1979 at the Zurich Federal Institute of Technology, until his retirement in 1990.

Feyerabend had arrived at Berkeley at an exciting time. Political radicalism, the free speech movement, and increasing racial toleration were to be found in Berkeley in an exceptionally intense form. Here he began to reexamine the claims of western science to offer the only valid account of nature. Why, he asked, should modern science be preferred to Hopi cosmology or Aristotelian metaphysics? On examination, he argued, the superiority of western science was no more than an assumption. It assumed the achievements of science were gained through the exercise of a distinctive scientific method. Yet, an examination of the works of COPERNICUS, GALILEO, and others revealed that they had succeeded by adopting extrascientific assumptions. Further, early prescientific societies had made such major advances as metalworking and agriculture without recourse to science. Nor, he pointed out, did science give its rivals a fair hearing. Astrology was dismissed out of hand; its principles were untested, and astrologers were denied access to scientific periodicals.

Science had thus, for Feyerabend, flourished by dubious rhetorical devices and by imposing a general consensus by totalitarian methods. As an alternative, in his *Against Method* (London, 1975), a work translated into 16 languages, he argued for an "anarchistic theory of knowledge," guided only by the principle "Anything goes." In *Science and a Free Society* (1978) Feyerabend argued that the democratic process itself is threatened by the authoritarian nature of modern science. His views have been vigorously contested and he has been described in the pages of *Nature* as "currently the worst enemy of science." Shortly before he died Feyerabend completed his autobiography *Killing Time* (1995), in which he revealed that in 1993 he had been diagnosed as having an inoperable brain tumor.

Feynman, Richard Phillips (1918–1988) *American theoretical physicist*

> For a successful technology, reality must take precedence over public relations, for Nature cannot be fooled.
> —Final report of the inquiry into the *Challenger* space shuttle disaster of 1986

The father of Feynman (**fīn**-man) had been brought with his immigrant parents from Minsk, Byelorussia, in 1895. Feynman himself was born in New York and educated at the Massachusetts Institute of Technology and at Princeton, where he completed his PhD in 1942 under the supervision of John WHEELER. In 1943 Feynman moved to Los Alamos to work on the Manhattan Project in the theoretical division under Hans BETHE. He was soon recognized to be, in the words of Robert OPPENHEIMER, "the most brilliant young physicist here." Feynman's own writings about the period deal less with the bomb than with his wife, Arline, who was dying of TB. He had married her in 1942 against much family opposition. She moved into a sanatorium in nearby Albuquerque and died in June 1945, a month before the first atomic bomb was tested.

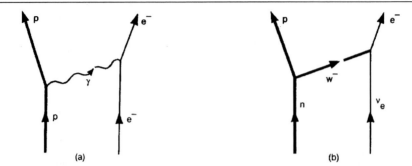

*FEYNMAN DIAGRAM Diagrams for (a) electromagnetic scattering of an electron by a proton;
(b) scattering of a neutrino by a neutron.*

In 1945 Feynman moved to Cornell as professor of physics, a post he held until 1950, when he was appointed to a similar position at the California Institute of Technology, where he remained for the rest of his career. While at Cornell he began to consider anew some of the outstanding problems in quantum electrodynamics (QED) – an area of physics dealing with the interactions between electrons and photons. The electron was seen as a point charge. As the strength of a charged body diminishes with distance in accordance with the inverse square law, it will vary as $1/r^2$. But what about the strength of the charge at the electron itself where $r = 0$? At this point the charge (and for a point, density) of the electron must be infinite. To handle this and other similar absurdities physicists developed a number of artificial mathematical techniques which would allow them to "renormalize" their equations so as to remove the infinite terms. Yet the charge on the electron is finite and can be measured accurately. Theoretical calculations, it was felt, should reach the same value without requiring artificial manipulation.

Freeman DYSON has described Feynman at this time as claiming that "he couldn't understand the official version of quantum mechanics," and that he had to "reinvent quantum mechanics" in a form he could understand. Feynman first presented his new approach in a paper, turned down by the *Physical Review*, entitled *Space-Time Approach to Non-Relativistic Quantum Mechanics* (1948), in which he introduced the notion of path integrals, also referred to as "sum over histories."

In Feynman's approach, the probability of an event that can happen in a number of different ways, such as finding an electron at a certain place, was the sum of the probabilities of all the possible ways the event could happen. When all the probabilities were added, Feynman noted, the result was SCHRÖDINGER's wave function.

In a 1949 paper, *Space-Time Approach to Quantum Electrodynamics*, Feynman showed how to calculate these path integrals using simple sketches, which have since become widely known as *Feynman diagrams*. It was for work in this field that Feynman shared the 1965 Nobel Prize for physics with Julian SCHWINGER and Sin-Itiro TOMONAGA. His first reaction had been to decline what he termed "Alfred Nobel's other mistake" on the grounds that he would thereafter become a celebrity and not just someone who wanted to talk about physics. Warned that a refusal would mark him in the media as an even bigger celebrity, Feynman agreed to accept the award.

Feynman also worked on problems connected with superconductivity and with particle physics. In 1955 he proposed a new model to represent the structure of liquid helium. In 1958 he published a paper with Murray GELL-MANN on the theory of weak interactions, which extended FERMI's earlier theory to take parity violations into account. During a visit to Stanford in 1968 Feynman began to work on the strong nuclear interaction. He attributed to the proton a set of constituents he named "partons," which were pointlike and did not interact with each other. Their value lay in their ability to explain the inelastic scattering results emerging from the Stanford Linear Accelerator (SLAC).

With the publication in 1963 of the *Feynman Lectures in Physics*, Feynman began to be known outside the small community of theoretical physicists. However, he gained international celebrity in the 1980s following a number of TV programs and the publication of *Surely You're Joking Mr. Feynman* (1985). The picture of a man of rare seriousness and honesty, with little time for honors, institutions, and formality, yet who clearly enjoyed the life of the flesh as well as the mind, had rarely been presented with such clarity. Despite his reluctance to accept the Nobel award, Feynman seems to have enjoyed his fame, and even to accentuate his unconventional character.

When he was asked to serve in 1986 on the presidential commission to investigate the explosion of the *Challenger* space shuttle, he assumed that he was being asked to contribute to a genuine scientific investigation. Pointed in the right direction by sympathetic colleagues, Feynman soon realized that the immediate cause of the explosion had been the O-ring seals used in the booster rocket. On the morning of the shuttle's launch the temperature was below freezing and the seals failed to retain their elasticity at low temperatures. Feynman also found out that the NASA officials had been warned about this potential failure. Feynman demonstrated the point, unannounced, at a televised meeting of the commission. He placed the O-rings in a glass of iced water for a few minutes and showed that, for several seconds after their removal, the seals had lost their resilience.

Feynman wrote up his findings as a separate appendix. He was aware that, while the commission would not actually suppress any evidence, much that was critical of NASA could be scattered throughout the report and consequently picked up by only the most careful of readers. Before the commission agreed to publish his report in a form acceptable to Feynman, he first found it necessary to threaten to resign and issue the report elsewhere. FEYNMAN found his Washington experiences genuinely distressing. Much of his life had been spent trying to understand various natural phenomena. The work had been hard, demanding many hours of intense intellectual concentration. He went to Washington intending to put the same effort into the service of the commission. Yet he found himself working with people who, though not crooks, liars, or lazy, were only marginally interested in the truth. It was more important for them to find a story acceptable to a community consisting of the Washington establishment, NASA, and "the American people," than to set out to establish what happened.

By this time, however, Feynman was a seriously ill man. In 1978 a malignant growth had been removed from his abdomen. A second cancer, involving bone marrow, was diagnosed in 1986. The abdominal cancer returned in late 1987 and soon after Feynman died from kidney failure. His influential lectures on gravitation and computation were published after his death.

Fibiger, Johannes Andreas Grib (1867–1928) *Danish physician*

Fibiger (**fee**-bi-ger), born the son of a physician at Silkeborg in Denmark, was educated at the University of Copenhagen, completing his medical studies in 1890. After some hospital work and further study in Berlin under Robert KOCH and Emil von BEHRING, Fibiger joined the Institute of Pathological Anatomy at the University of Copenhagen in 1897, serving there as its director from 1900.

It was realized that cancers could be chemically induced by factors in the environment but all attempts to induce such cancers artificially

had failed. Fibiger thought he could change this when, in 1907, he observed extensive papillomatous tumors virtually filling the stomachs of three wild rats. Microscopic examination showed the presence in the stomachs of formations similar to nematode worms, and Fibiger naturally concluded that these parasites were the cause of the tumors. A search of a further 1,200 wild rats, however, produced no additional cases of cancer. This suggested to him that the nematodes were transmitted by an intermediate host, and a report published in 1878 confirmed that such nematodes had been found as parasites of a common kind of cockroach. Before long Fibiger found rats from a sugar refinery that fed regularly on the cockroaches there: examination of 61 of these rats showed that 40 had nematodes in their stomachs and 7 of these 40 had the earlier identified tumor. By 1913 Fibiger was able to claim that he could induce such malignancies in rats by feeding them with cockroaches infested with nematode larvae, noting a proportional relationship between the number of parasites and the degree of anatomic change in the stomach. It was for this work, described somewhat extravagantly by the Nobel Committee as the "greatest contribution to experimental medicine in our generation," that Fibiger was awarded the 1926 Nobel Prize for physiology or medicine.

Although no one disputed that Fibiger had induced cancer it was never completely accepted that such growths were caused by the nematodes. In any case Fibiger's work had little impact on experimental cancer research: simpler methods of carcinogenesis were almost universally preferred.

Fibonacci, Leonardo (c. 1170– c. 1250) *Italian mathematician*

Fibonacci (fee-boh-**nah**-chee) lived in Pisa and is often referred to as Leonardo of Pisa. Although he was probably the most outstanding mathematician of the Middle Ages virtually nothing is known of his life. The modern system of numerals, which originated in India and had first been introduced to the West by al-Khwarizmi, first became widely used in Europe owing to Fibonacci's popularization of it. His father served as a consul in North Africa and it is known that Fibonacci studied with an Arabian mathematician in his youth, from whom he probably learned the decimal system of notation.

Fibonacci's main work was his *Liber abaci* (1202; Book of the Abacus) in which he expounded the virtues of the new system of numerals and showed how they could be used to simplify highly complex calculations. Fibonacci also worked extensively on the theory of proportion and on techniques for determining the roots of equations, and included a treatment of these subjects in the *Liber abaci*. In addition it contains contributions to geometry and Fibonacci later published his *Practica geometriae* (1220; Practice of Geometry), a shorter work that was devoted entirely to the subject.

Fibonacci was fortunate in being able to gain the patronage of the Holy Roman Emperor, Frederick II, and a later work, the *Liber quadratorum* (1225; Book of Square Numbers) was dedicated to his patron. This book, whch is generally considered Fibonacci's greatest achievement, deals with second order Diophantine equations. It contains the most advanced contributions to number theory since the work of DIOPHANTUS, which were not to be equaled until the work of FERMAT. He discovered the "Fibonacci sequence" of integers in whch each number is equal to the sum of the preceding two (1, 1, 2, 3, 5, 8, ...).

Finch, George Ingle (1888–1970) *Australian physical chemist*

Finch was born at Orange in Australia and educated at Wolaroi College in his native country and the Ecole de Médecine in Paris. Finding the study of medicine unappealing, Finch moved to Switzerland where he studied physics and chemistry, first at the Federal Institute of Technology in Zurich and afterward at the University of Geneva. On moving to Britain in 1912 he worked briefly as a research chemist at the Royal Arsenal, Woolwich, joining the staff of Imperial College, London, in 1913. Finch remained there until his retirement in 1952, having been appointed professor of applied physical chemistry in 1936. After his retirement Finch spent the period 1952–57 in India as director of the National Chemical Laboratory.

Finch worked mainly on the properties of solid surfaces. In the 1930s he developed the technique of low-energy electron diffraction, using the wavelike properties of electrons, demonstrated by George THOMSON and Clinton J. DAVISSON in 1927, to investigate the structure of surfaces. X-rays are neutral and too penetrating to provide much information about surfaces or thin films; electrons, however, are charged and are deflected after penetrating no more than a few atoms below the surface. With this new and powerful tool Finch began the study of lubricants and the BEILBY layer – a thin surface layer produced by polishing, and differing in its properties from the underlying material. Finch had earlier worked on the mechanism of combustion in gases as initiated by an electric discharge.

Finch was also widely known as a mountaineer. As one of the leading climbers of his generation he was a member of the 1922 Everest expedition, climbing to the then unequaled height of 27,300 feet (8,321 m).

Finlay, Carlos Juan (1833–1915) *Cuban physician*

Finlay's father was a Scottish physician who had fought with Simon Bolivar. Carlos, who was born at Puerto Principe, now Camagüey, in Cuba, was educated in Paris and at the Jefferson Medical College, Philadelphia, where he graduated in 1855. He then returned to Cuba where he spent his life in general practice.

In 1881 Finlay published a prophetic paper *The Mosquito Hypothetically Considered as the Agent of the Transmission of Yellow Fever*, naming the species *Culex fasciatus* (now *Aëdes aegypti*) as the vector. He had been struck by the presence of *Aëdes* in houses during epidemics and noted that the yellow fever and mosquito seasons seemed to coincide. Although he campaigned vigorously for his theory he only succeeded in turning himself into a figure of mild ridicule.

There were a number of complicating factors that undermined Finlay's work. Firstly yellow fever is caused by a virus, a microorganism only discovered in 1898 and far too small to be detected by the microscopic techniques of the late 19th century. Further, with the newly established germ theory of Robert KOCH, scientists were demanding visual evidence for the existence of the supposed pathogen. As Finlay could not isolate the causative organism he attempted to demonstrate its existence by such clinical techniques as transmitting the disease from a sick patient via *Aëdes* to a healthy individual. But here too, although he repeated such experiments many times, he failed to produce any coherent results. As it turned out, it was only the female mosquito, and then only one who had bitten a victim in the first three days of infection who could, two weeks later, transmit the disease. Consequently many of Finlay's failures could be explained by using the wrong type of mosquito or the right type at the wrong time. But to discover such facts required more ambitious resources than Finlay commanded.

He did however live to see such resources come to Cuba with the arrival of the 4th U.S. Yellow Fever Commission in 1900 under the command of Walter REED. Finlay interested Reed in his views, and Reed, in a classic series of trials, completely vindicated them.

Finsen, Niels Ryberg (1860–1904) *Danish physician*

Finsen (**fin**-sen), the son of a leading civil servant, was born at Thorshavn on the Faeroe Islands, which are part of Denmark; he was educated in Reykjavik and at the University of Copenhagen, where he qualified as a physician in 1890. After teaching anatomy for some time Finsen founded (1895) the Institute of Phototherapy, which he directed until his early death at the age of 43.

In the 1890s, following up some earlier work

suggesting that light had the ability to kill bacteria, Finsen began a systematic appraisal of its therapeutic effects. Arguing that it was light, acting slowly and weakly, rather than heat that was effective, he devised various filters and lenses to separate and concentrate the different components of sunlight. He found that it was the short ultraviolet rays, either natural or artificial, that turned out to have the greatest bactericidal power.

Finsen found phototherapy to be of most use against lupus vulgaris, a skin infection produced by the tubercle bacillus. He claimed that on exposure to ultraviolet rays the skin regained its normal color and the ulcerations began to heal. For this Finsen received the third Nobel Prize for physiology or medicine in 1903.

It was, however, an avenue that few physicians were willing to explore. The use of ultraviolet radiation was mainly restricted to the treatment of lupus vulgaris and even this was superseded by x-rays and, more importantly, by such drugs as cortisone when they became available in the 1950s.

Fire, Andrew Zachary (1959–   ) *American molecular geneticist*

Born in Palo Alto, California, Fire was awarded a BA in mathematics by the University of California, Berkeley, when still just 19. He then joined the Massachusetts Institute of Technology to study biology, receiving a PhD in 1983. After a three-year spell as a postdoctoral fellow in Cambridge, UK, Fire joined the staff of the Carnegie Institution's Department of Embryology in Baltimore. In 1989 he became adjunct professor of biology at Johns Hopkins University, and in 2003 he joined Stanford University School of Medicine as professor of pathology and genetics.

Fire is lauded for his discovery, in collaboration with Craig MELLO and others, of the phenomenon known as RNA interference (RNAi). In 1998 Fire and Mello published a paper in *Nature* that broke new ground in our understanding of how cells control the flow of genetic information. It was already established that information is encoded by the DNA (deoxyribonucleic acid) in the genes within a cell's chromosomes (located in the nucleus of the cell). Also, when a gene is "switched on," the sequence of chemical bases in the DNA, which contains the genetic code, is copied over (transcribed) into specific messenger molecules, consisting of another type of nucleic acid termed messenger ribonucleic acid (mRNA). The mRNA molecules are exported from the nucleus to the surrounding cytoplasm, where they relay the genetic information to the cell's protein-manufacturing machinery, in a process called translation. Here the machinery "reads" the instructions contained in each mRNA molecule, and assembles each protein accordingly. Hence,

different mRNAs give rise to all the diverse proteins required by the cell, including enzymes, cell receptors, and cell scaffolding.

It was known that RNA could in certain circumstances inhibit the activity of genes. This latter phenomenon was caused by RNA molecules whose base sequence matched that of the gene in question, but it was poorly understood. Then came Fire and Mello. Their experiments were conducted on a tiny nematode worm called *Caenorhabditis elegans*, and focused on a particular gene that affected muscle action. Into the worm they injected RNA whose base sequence matched that of the target gene. When "silencing" of the target gene occurred, it caused a muscle defect in the worm and consequent twitching movements.

The researchers found that their target gene was effectively silenced only by double-stranded RNA molecules (dsRNAs). These comprise two parallel strands of linked bases, one strand (the "antisense" strand) with a base sequence complementary to that of the target gene's mRNA, and hence capable of binding to it by base pairing; the other strand (the "sense strand") with a sequence exactly matching the target mRNA, but unable to bind to it.

Fire and Mello confirmed that the silencing was specific, affecting only the target gene mRNAs, and that it involved fully formed, or "mature" mRNAs, and hence probably took place in the cytoplasm. Also, it could be achieved using only a few molecules of dsRNA, indicating that the dsRNA acted not as a reagent but most likely as some form of catalyst. Moreover, the target mRNA disappeared from the cell, meaning that it was degraded. These findings explained hitherto puzzling aspects of RNA-induced gene inhibition, and pointed to a possible new mechanism for control of gene expression in living cells.

In subsequent studies, Fire and Mello discovered more details about RNAi in living cells. For example, they established that the dsRNA molecules are cut into short double-stranded segments, and that the antisense strand then binds to its complementary sequence on the mRNA molecule, which triggers degradation of the mRNA. Later, the identities of other key components of the RNAi molecular machinery were determined, notably the large protein complex called RISC (RNA-induced silencing complex), which links the antisense RNA and the mRNA, and the endonuclease called Dicer, which cuts the dsRNA into short fragments.

RNAi is now known to be significant in several ways. Firstly, it helps to protect cells against attack by certain viruses, by targeting viral RNA for destruction. Second, it is believed to silence potentially mobile segments of the genome called transposons. These are short segments of DNA derived from viral nucleic acid that have become integrated within the organism's DNA over evolutionary time, and replicate by making RNA copies of their DNA sequence. Without RNAi to keep them in check, transposons might replicate excessively and overwhelm the genome. Thirdly, RNAi-type cellular machinery in a wide variety of organisms is now known to produce a class of small RNAs called micro-RNAs (miRNAs). These regulate gene expression by binding to mRNAs, causing them to be degraded, or suppressing translation into protein. This form of gene regulation is widespread; for example, in mammals about 30% of all genes are controlled by this mechanism.

Another aspect of RNAi is its enormous importance as an experimental tool, and potential as a therapeutic strategy to treat genetic diseases. As Fire and Mello showed with their worm, dsRNA can be tailor-made to target specific genes. This enables extremely precise and efficient "knockout" of genes to investigate their impact on the organism. It could also lead to new forms of targeted gene therapy, by preventing the expression of faulty genes and so restoring normal cell function.

For his work on RNAi, Fire was awarded the 2006 Nobel Prize in physiology or medicine, jointly with Craig Mello.

Fischer, Debra *American astronomer*

Fischer studied physics, gaining a BS from the University of Iowa in 1975. In 1986 she embarked on graduate studies in the department of physics and astronomy at San Francisco State University (SFSU), becoming a lecturer (1990–92) and gaining an MS in physics (1992). In 1994 she joined the University of California, Santa Cruz, as research assistant, obtaining her PhD in 1998. After a spell in the astronomy department at the University of California, Berkeley (1999–2003), she returned to CFSU as assistant professor.

While a graduate student at SFSU, Fischer joined Geoff Marcy as a member of the newly founded SFU Planet Search project, working at the Lick Observatory, near San Jose. Thus began her career, which has focused on finding planets in other solar systems. Fischer discovered her first planet in orbit around the star HD 217107, using the DOPPLER technique, which detects fluctuations in a star's velocity resulting from small gravitational forces exerted by orbiting planets. Then in 1998 she analyzed data from the star Upsilon Andromeda, and found that they were best explained by three orbiting planets, dubbed Dinky, Twopiter (twice the mass of Jupiter) and Fourpiter. This was the first discovery of multiple extrasolar planets around a star. Subsequently, the SFU project, now the California and Carnegie Planet Search Project, has discovered many of the over 200 extrasolar planets so far identified. Fischer selects stars with a high abundance of metallic elements as candidates for finding planets. Some of these planets are massive "hot Jupiters" that

orbit close to their host stars; a wealth of information can be gleaned as such planets transit in front of a bright host star. In 2002 Fischer received the Carl Sagan Award of the American Astronautical Society.

Fischer, Edmond H. (1920–) *American biochemist*

Born in Shanghai, China, Fischer was educated at the University of Geneva, where, after graduation, he worked as assistant in the organic chemistry laboratories (1946–47). After spending two years as a research fellow with the Swiss National Foundation (1948–50) he moved to the University of Washington, Seattle, as a Rockefeller Foundation research fellow (1950–53). After a brief spell at the California Institute of Technology in 1953, he was appointed assistant professor of biochemistry at the University of Washington, subsequently becoming associate professor (1956–61) and professor (1961–90).

Fischer's most acclaimed work was done at Seattle in the 1950s and 1960s in collaboration with the biochemist Edwin KREBS. In 1955–56 the pair discovered how the enzyme (glycogen phosphorylase) that catalyzes the release of glucose from glycogen in the body is "switched on." The enzyme receives a phosphate group from ATP (adenosine triphosphate – the body's major energy carrier) in a transfer reaction catalyzed by a second enzyme, which Fischer and Krebs termed a "protein kinase." They went on to show that glycogen phosphorylase is then switched off by removal of the phosphate group by a further enzyme (called a protein phosphatase). The addition and removal of the phosphate group reversibly changes the shape of the enzyme molecule, thereby switching it between the inactive and active forms.

These findings opened the way to a major new field of research suggesting how enzymes might function in various physiological processes, such as hormone regulatory mechanisms, gene expression, and fertilization of the egg. Their work also had implications for the understanding of certain diseases; for instance, abnormally phosphorylated proteins have been identified in muscular dystrophy and diabetes, while protein kinases may play a significant role in certain cancers and in airway constriction in asthma.

In recognition of his contributions to our understanding of enzymes Fischer was awarded the 1992 Nobel Prize for physiology or medicine, which he shared with his long-time coworker, Krebs.

Fischer, Emil Hermann (1852–1919) *German organic chemist and biochemist*

The son of a successful businessman from Euskirchen, now in Germany, Fischer joined his father's firm on leaving school (1869) but left in 1871 to study chemistry with August KEKULÉ at Bonn. He was not happy with the chemistry instruction there and came close to abandoning chemistry for physics. In 1872, however, he moved to Strasbourg to study with Adolf von BAEYER. Here, he gained his doctorate in 1874 for work on phthaleins. The same year he made the vital discovery of phenylhydrazine, a compound that was later to prove the vital key for unlocking the structures of the sugars.

Fischer became Baeyer's assistant and together they moved to Munich (1875). At Munich, working with his cousin, Otto Fischer, he proved that the natural rosaniline dyes are derivatives of triphenylmethane. In 1879 Fischer became assistant professor of analytical chemistry and soon after became financially independent. He was then professor at Erlangen (1882), Würzburg (1885), and Berlin (1892). Fischer has some claim to be called the father of biochemistry. He carried out extremely comprehensive work in three main fields: purines, sugars, and peptides, the last two effectively founding biochemistry on a firm basis of organic chemistry. The work on purines, begun in 1882, resulted in the synthesis of many important compounds, including the alkaloids caffeine and theobromine, and purine itself (1898). Fischer's early structures were incorrect but from 1897 the correct structures were used.

In 1884 Fischer discovered that phenylhydrazine produces well-defined crystalline compounds with sugars, thus affording a reliable means of identification. In 1887 he synthesized first fructose (from acrolein dibromide) and later mannose and glucose. By 1891 he was able to deduce the configurations of the 16 possible aldohexoses, which he represented in the form of the famous *Fischer projection formulae.*

In 1899 Fischer turned to amino acids and peptides and devised a peptide synthesis that eventually produced a polypeptide containing 18 amino acids (1907). Fischer's other work included the first synthesis of a nucleotide (1914), the "lock-and-key" hypothesis of enzyme action,

work on tannins, and attempts to prepare very high-molecular-weight compounds. He was awarded the 1902 Nobel Prize for chemistry for his work on purines and sugars.

Fischer, Ernst Otto (1918–1994) *German inorganic chemist*

Fischer, the son of a physics professor, was educated in his native city at the Munich Institute of Technology, where he obtained his PhD in 1952. He taught at the University of Munich, serving as professor of inorganic chemistry from 1957 to 1964, when he became the director of the Institute for Inorganic Chemistry at the Institute of Technology.

Fischer is noted for his work on inorganic complexes. In 1951 two chemists, T. Kealy and P. Pauson, were attempting to join two five-carbon (cyclopentadiene) rings together and discovered a compound, $C_5H_5FeC_5H_5$, which they proposed had an iron atom joined to a carbon atom on each ring.

Fischer, on reflection, considered such a structure inadequate for he failed to see how it could provide sufficient stability with its carbon–iron–carbon bonds. The British chemist Geoffrey WILKINSON suggested a more novel structure in which the iron atom was sandwiched between two parallel rings and thus formed bonds with the electrons in the rings, rather than with individual carbon atoms. Compounds of this type are called "sandwich compounds." By careful x-ray analysis Fischer confirmed the proposed structure of ferrocene, as the compound was called, and for this work shared the Nobel Prize for chemistry with Wilkinson in 1973. Fischer went on to do further work on transition-metal complexes with organic compounds and was one of the leading workers in the field of organometallic chemistry.

Fischer, Hans (1881–1945) *German organic chemist*

Fischer, the son of a chemicals industrialist

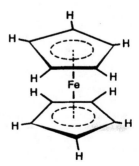

FERROCENE   Originally the structure was thought to be that on the left. E. O. Fischer showed it to be a sandwich compound with the iron atom positioned between the two rings.

from Höchst-am-Main in Germany, gained his doctorate in chemistry at the University of Marburg in 1904. He also studied medicine at the University of Munich, gaining his MD in 1908. He was assistant to Emil Fischer before occupying chairs of medical chemistry at Innsbruck (1916) and Vienna (1918). In 1921 he succeeded Heinrich WIELAND as professor at the Technical Institute in Munich.

Fischer's life work was the study of the immensely important biological molecules hemoglobin, chlorophyll, and the bile pigments, especially bilirubin. He showed that hemin – the nonprotein, iron-containing portion of the hemoglobin molecule – consists of a system of four pyrrole rings, linked by bridges, with iron in the center. He synthesized hemin in 1929 and extensively investigated similar molecules – the porphyrins. He was awarded the Nobel Prize for chemistry for this work in 1930. He then turned to the chlorophylls and showed that they are substituted porphins with magnesium rather than iron in the center. The bile acids were shown by Fischer to be degraded porphins, and he synthesized bilirubin in 1944. Fischer took his own life at the end of World War II, after his laboratories had been destroyed in the bombing of Munich.

Fischer, Otto Philipp (1852–1932) *German organic chemist*

Otto Fischer was the cousin of Emil Fischer, with whom he collaborated in some of his work. He was born at Euskirchen, now in Germany, gained his doctorate at Strasbourg (1874), and was professor of chemistry at Erlangen from 1885 to 1925. His most notable work was with dyestuffs: he discovered malachite green, a triphenylmethane dye (1877), and determined the structures of pararosaniline (1880; with Emil) and mauveine (1890).

Fisher, Sir Ronald Aylmer (1890–1962) *British statistician and geneticist*

> Natural selection is a mechanism for generating an exceedingly high degree of improbability.
> —Quoted by A. L. Mackay in *The Harvest of a Quiet Eye* (1977)

Fisher, a Londoner by birth, studied mathematics and physics at Cambridge University, graduating in 1912. In the years before joining Rothamsted Experimental Station in 1919 he undertook a variety of jobs, including farm work in Canada, employment with an investment company, and teaching in various private schools. In this period he also produced two important papers marking his interest in both statistics and genetics. The first, published in 1915, described a solution for the exact distribution of the correlation coefficient, a problem that had been perplexing other statisticians.

The second paper was *The Correlation between Relatives on the Supposition of Mendelian Inheritance* (1918). This demonstrated that the inheritance of continuous variation, which had been thought of as non-Mendelian, is in fact governed by many additive genes, each of small effect and each inherited in a Mendelian manner. Thus continuous variation may be analyzed following Mendelian rules. This work later led to the development of the science of biometric genetics. At Rothamsted, Fisher was appointed to sort out the accumulation of over 60 years' data on field trials. He modified the significance test, enabling more confident conclusions to be drawn from small samples of data, and developed the analysis of variance technique. He emphasized the need for random rather than systematic experimental design so that error due to environmental variation could be analyzed quantitatively. His book *Statistical Methods for Research Workers* (1925) is one of the most influential works in statistics.

Fisher's major researches in genetics at Rothamsted were brought together in *The Genetical Theory of Natural Selection* (1930). In this book he argued that Mendelism, far from contradicting Darwinism as some people believed, actually provides the missing link in the theory of evolution by natural selection by showing that inheritance is by means of particulate entities (genes) rather than by physical blending of parental characteristics. (In 1936 Fisher published a paper arguing that probabilistically MENDEL's famous results were "too good to be true.") The book also summarizes his views on eugenics and on genes controlling dominant characteristics. He believed that dominance develops gradually by selection, showing selection rather than mutation to be the driving force in evolution. *The Genetical Theory of Natural Selection* led to Fisher's appointment as Galton Professor of Genetics at University College, London, in 1933. Here he did important work clarifying the genetics of the Rhesus blood groups. He accepted the chair of genetics at Cambridge University in 1943, remaining there until 1959 although he retired officially in 1957. He spent the last three years of his life working for the Commonwealth Scientific and Industrial Research Organization (CSIRO) in Adelaide. Fisher was knighted in 1952.

Fitch, Val Logsdon (1923–) *American physicist*

Born in Merriman, Nebraska, Fitch was educated at McGill and Columbia universities and obtained his PhD from Columbia in 1954. He then joined the Princeton staff, being appointed professor of physics there in 1960.

Working with Leo James RAINWATER, Fitch was the first to observe radiation from muonic atoms, i.e., from species in which a muon is orbiting a nucleus rather than an electron. This work indicated that the sizes of atomic nuclei were smaller than had been supposed. He went on to study kaons and in 1964 collaborated with James CRONIN, James Christenson, and René Turley in an experiment that disproved charge–parity conservation. In 1980 Fitch and Cronin shared the Nobel Prize for physics for this fundamental work.

Fittig, Rudolph (1835–1910) *German organic chemist*

Fittig (**fit**-ik) was born in Hamburg, now in Germany, and gained his doctorate at Göttingen in 1858, becoming a professor at Tübingen (1870) and at Strasbourg (1876–1902). He was a prolific experimentalist with many discoveries and syntheses to his credit, including pinacol, diphenyl, mesitylene, cymene, coumarone, and phenanthrene in coal tar. He also did extensive work on lactones and unsaturated acids. His name is remembered in the *Wurtz–Fittig reaction*, a variation of the WURTZ reaction for synthesizing alkylaryl hydrocarbons. An example is the reaction to form methylbenzene (toluene):

$$CH_3Cl + C_6H_5Cl + 2Na \rightarrow C_2H_5CH_3 + 2NaCl$$

Fitzgerald, George Francis (1851–1901) *Irish physicist*

> I am not in the very least sensitive to having made mistakes. I rush out with all sorts of crude notions in hope that they may set others thinking and lead to some advance.
> —Letter to Oliver Heaviside, 4 February 1889

Fitzgerald graduated from Trinity College in his native Dublin in 1871 and then joined the staff there, rising to the position of professor in 1881. His work was mainly concerned with the development of electromagnetism from James Clerk MAXWELL's equations, in which he became interested after Heinrich HERTZ had demonstrated the existence of the radio waves predicted by theory. Like Hendrik LORENTZ, he constructed an electromagnetic theory of the reflection and refraction of light.

His name is best known for the *Lorentz–Fitzgerald contraction* – an effect suggested to explain the negative result of the Michelson–Morley experiment. The suggestion was that a body moving relative to an observer contracted slightly in the dimension parallel to the direction of motion, the amount of the contraction being dependent on the velocity. Thus, light emitted by a moving body would have a different speed, but would travel over a different path length. The contraction in size was supposed to result from the effect of the ether on the electromagnetic forces holding the atoms together. The Lorentz–Fitzgerald contraction received an alternative explanation in EINSTEIN's theory of relativity.

Fitzroy, Robert (1805–1865) *British hydrographer and meteorologist*

Fitzroy, who was born near Bury St. Edmunds in the eastern English county of Suffolk, was educated at the Royal Naval College, Portsmouth, and in 1828 took command of the *Beagle*, on a survey of the South American coast. In 1831 he made a second voyage in the *Beagle* to complete his survey, taking with him as naturalist the then unknown Charles DARWIN. After his return in 1836 he devoted himself to the publication, in 1839, of his *Narrative*, a three-volume account of the voyage. The third volume, known as *The Voyage of the Beagle*, was the work of Darwin.

Fitzroy served as a member of Parliament (1841–43) and was then appointed governor of New Zealand until his dismissal in 1845. He retired from active duty in 1850 but later (1854) took up the newly created post of meteorological statist and remained there until his death. He organized a number of observation stations, designed the barometer named for him, and began the publication of storm warnings that evolved into daily weather forecasts. One of the earliest textbooks on meteorology, *The Weather Book* (1863), was published by him.

He had a reputation for having a quick temper and, while suffering from depression, committed suicide.

Fizeau, Armand Hippolyte Louis (1819–1896) *French physicist*

Fizeau (fee-**zoh**), a Parisian by birth, started by studying medicine but his interest turned to optics before he finished the course. In collaboration with Léon FOUCAULT he first tried to improve the newly developed process of photography and, in 1845, they took the first clear pictures of the Sun.

In 1849 he obtained a value for the speed of light in air, using an ingenious toothed-wheel apparatus. Light was directed through a gap between two teeth and reflected back between the teeth from a distant mirror. The wheel was rotated, the rate of rotation being changed until the reflected flashes were blocked by the tooth of the wheel. The speed of light could then be calculated from the rate of rotation of the wheel. Fizeau's experiment was performed using a path of 8 kilometers (5 mi) between Suresnes and Montmartre.

The next year both he and Foucault simultaneously proved that light traveled faster in air than in water, thus giving experimental support to the wave theory of light. Fizeau is also known for analyzing the DOPPLER effect for light waves. The change in wavelength with relative speed is sometimes called the *Doppler–Fizeau shift*.

Fizeau was elected a member of the Paris Academy in 1860 and was awarded the Royal Society's Rumford medal in 1875.

Flammarion, Nicolas Camille (1842–1925) *French astronomer*

Owing to his family's poverty, Flammarion (fla-ma-**ryon**), who was born at Montigny-le-Roi, France, was forced to leave school and abandon his intention of going into the priesthood. Instead he got a post in the Paris Observatory in 1858 but eventually was dismissed by Urbain LE VERRIER who accused him of being a poet rather than a scientist. Flammarion made serious contributions to astronomy, including founding the French Astronomical Society in 1887, and built himself a private observatory at Juvisy. However, he is best known as a popularizer of astronomy. His general attitude to astronomy was made clear when he published at the age of 19 *Plurality of Inhabited Worlds* (1862). This was translated into 12 languages, including Chinese and Arabic, and became a best seller in France, making Flammarion famous. His *Popular Astronomy* (1879) was one of the most popular books of the late 19th century. Giovanni SCHIAPARELLI's discovery of "canals" on Mars in 1877 received enthusiastic support from Flammarion. He went even further, claiming that he could discern changes in the lunar craters that could be cultivated fields.

Flamsteed, John (1646–1719) *English astronomer*

> In carrying out views of practical utility, with a scrupulous attention to accuracy in the most minute details, in fortitude of resolution under adverse circumstances, and persevering adherence to continuity and regularity of observation throughout a long career, he had few rivals in any age or country.
> —Robert Grant, *History of Physical Astronomy* (1852)

Flamsteed was born in Denby, England. Because of ill health, which was to dog his career, he was forced to leave school early and was therefore largely self-educated. He started his scientific career under the patronage of William BROUNCKER, the first president of the Royal Society, having impressed him by computing an almanac of celestial events for 1670.

A major problem of the time – one tackled at some time by all major astronomers of the 17th century – was the determination of longitude at sea. A suggestion had been made that the motion of the Moon against the stellar background could be used to determine standard time. Flamsteed, asked by Brouncker to comment on this proposal, pointed out that the scheme was impractical because of the inaccuracy of contemporary tables. Charles II subsequently commanded that accurate tables should be constructed, appointing Flamsteed as first Astronomer Royal with this responsibility in 1675, and building the Royal Greenwich Observatory for him, which was opened in 1676. The limited nature of the royal patronage is indicated by the fact that Flamsteed was paid a salary of £100 a year but was expected to provide his own instruments and staff. He eventually managed to put together two small telescopes and then began his decades of observation, made more difficult by his lack of staff and the crippling headaches from which he suffered. In order to make ends meet he was forced to become a clergyman at Burstow in Surrey from 1684 until his death.

The results of his labors were eventually published posthumously in 1725 as the *Historia coelestis Britannica* (British Celestial Record). It contains the position of over 3,000 stars calculated to an accuracy of ten seconds of arc. It was the first great modern comprehensive telescopic catalog and established Greenwich as one of the leading observatories of the world. The publication of the work was not without its difficulties. It involved Flamsteed in a long and bitter dispute with NEWTON. Flamsteed was reluctant to rush into print with his catalog, claiming, it seemed to Newton, far too much time for the checking of his numerous observations. The dispute lasted from Newton's assumption of the presidency of the Royal Society in 1703 until Flamsteed's death. It involved the virtual seizure of Flamsteed's papers by Newton, the editing and partial publication by Edmond HALLEY, and their total rejection by Flamsteed who even went so far as to acquire 300 of the 400 printed copies of his own work and burn them. He managed, however, to revise the first volume to his satisfaction before his death in 1719.

Fleischmann, Martin (1927–) *British chemist*

> A cold fusion kit has been marketed by a subsidiary of Japan's *Nippon Telegraph and Telephone* for $565,000.
> —*Wall St. Journal*, 27 November 1992

Born in Karlsbad, Czechoslovakia, Fleischmann (**flIsh**-man) and his family fled to Britain in 1939. He was educated at Imperial College, London, where he gained his PhD in 1951. He taught chemistry at the University of Durham (1952–57) and at Newcastle University (1957–67). In 1967 he was appointed professor of electrochemistry at the University of Southampton.

By 1988 Fleischmann had an international reputation as a productive and innovative electrochemist. Between 1985 and 1988 he coauthored 29 papers with his former student, Stanley PONS. Quite unexpectedly, in March 1989, they announced that they had achieved nuclear fusion by an electrolytic method under laboratory conditions at room temperature.

Nuclear fusion, the fusion of two light atomic nuclei to produce a heavier nucleus, is a process accompanied by large amounts of released energy. For fusion to occur the nuclei have to be brought close together, and this involves overcoming a high energy barrier caused by the mutual repulsion of the nuclei.

In practical cases this is achieved by high temperatures, as in the Sun or in thermonuclear weapons. Experimental thermonuclear reactors such as the tokomak also use high temperatures (about 300 million degrees) to initiate fusion. There is, however, considerable interest in methods of initiating fusion at low temperatures – so-called "cold fusion." One approach to this has been *inertial confinement*, in which a sample of material is compressed by intense laser beams or particle beams.

Fleischmann and Pons also thought that pressure might be a way of initiating cold fusion. Palladium metal has a high affinity for hydrogen and, under the right conditions, can absorb large quantities of it. They electrolyzed water containing the deuterium isotope using a palladium cathode and reasoned that the palladium might absorb so much deuterium that the effective deuterium pressure within the electrode would be high enough to cause nuclear fusion. If this occurred, there would be a large increase in temperature, over and above that produced by the heating effect of the current. Fleischmann and Pons reported just such an effect.

Fleischmann was reluctant to reveal too many details of their work lest it prejudice their patent application. He did, however, collaborate with Harwell, the British Atomic Energy Research Establishment, who were best suited to replicate cold fusion. They found no evidence of fusion, no excess heat. Whereas positive results had been obtained by several leading institutions, they were subsequently withdrawn when errors were detected. Despite such setbacks Fleischmann has remained convinced of the essential soundness of his work with Pons.

Since 1990 Fleischmann's work was mainly supported by Minora Toyoda, a Japanese businessman. He worked at Sophia Antipolis, a research center established by Toyoda outside Nice, France, concerned with the development of future technology. In 1995 he moved to Southampton, where he stayed until 1999. In 2006 he joined the San Francisco company D2 Fusion.

Fleming, Sir Alexander (1881–1955)
British bacteriologist

"Pain in the mind" was not the spur that drove him to do research ... but rather an urge to do a job better than the next man. Competition was the breath of life to him.

—Leonard Colebrook in *Biographical Memoirs of Fellows of the Royal Society*

Fleming was born at Lochfield in Scotland. After his early education at Kilmarnock Academy and the London Polytechnic Institute, he began his career at the age of 16 as a shipping clerk in a London office. With encouragement from his brother, who was a doctor, he became a medical student at St. Mary's Hospital Medical School in 1902 and graduated from the University of London in 1908. He worked at St. Mary's all his life apart from 1914–18, when he served in the Royal Army Medical Corps. During this time he became interested in the control of wound infections and was a vigorous supporter of the physiological treatment of wounds rather than treatment using harsh chemicals, such as carbolic acid. In the 1920s he studied various body secretions and their effects on bacteria. Thus he discovered lysozyme, a bacteriolytic enzyme that is present in serum, saliva, and tears, publishing his findings in 1922.

In 1928 Fleming was appointed professor of bacteriology and in the same year he made his most important discovery. After accidentally leaving a dish of staphylococcus bacteria uncovered, Fleming noticed certain clear areas in the culture. He found these areas were due to contamination by a mold he identified as *Penicillium notatum*, which produced a substance that killed the bacteria. Fleming named this substance "penicillin" and tested the bactericidal effect of the mold on various different bacteria, observing that it killed some but not others. He appreciated the potential of his discovery but was unable to isolate and identify the compound. It was not until World War II, with the urgent need for new antibacterial drugs, that penicillin – the first antibiotic – was finally isolated by Howard FLOREY and Ernst CHAIN.

Fleming was awarded the 1945 Nobel Prize for physiology or medicine jointly with Florey and Chain for his discovery, which initiated a whole new range of lifesaving antibiotics. He received a knighthood in 1944 and many other honors.

Fleming, Sir John Ambrose (1849–1945)
British physicist and electrical engineer

This war [World War I] is a war quite as much of chemists and engineers as of soldiers and sailors.
—*Nature*

Fleming, who was born at Lancaster in northwestern England, studied for a short time at University College, London, but left before graduating. However, he continued his work for a science degree in his leisure hours while employed first in a shipwright's drawing office and later as a stockbroker's clerk. Between 1871 and 1880 he had alternate periods of school science teaching and further study, including working under James Clerk MAXWELL from 1877 at the new Cavendish Laboratory in Cambridge. In 1881 he was appointed professor of mathematics and physics at University College, Nottingham. From 1882 to 1885 he worked as consultant to the Edison Electric Light Company in London. He was then appointed professor of electrical technology at University College, London, a post he held for 41 years.

At University College Fleming gave special courses and experimented on wireless telegraphy, cooperating a great deal with Guglielmo MARCONI. One of Fleming's outstandingly important inventions was the thermionic vacuum tube, a rectifying device based on an effect discovered by Thomas Edison. Fleming's diode consisted of a glass bulb containing two electrodes. One, a metal filament, was heated to incandescence by an electric current, so that it emitted electrons by thermionic emission. The second electrode (the anode) could collect electrons if held at a positive potential with respect to the filament (the cathode) and a current would flow. Current could not flow in the opposite direction – hence the name "valve" for such devices. Lee DE FOREST developed the device into the triode for amplifying current.

Other scientific contributions by Fleming included investigations into the property of materials, transformer design, electrical measurements, and photometry. He was an outstanding teacher and highly successful as a popular lecturer. *Fleming's left-hand rule* and *right-hand rule* are mnemonics for relating the direction of motion, magnetic field, and electric current in electric motors and generators respectively.

Fleming, Williamina (1857–1911) *Scottish–American astronomer*

> Sparkling and friendly though she was, her reputation as a strict disciplinarian lived after her, and as late as the 1930s elderly ladies who had worked with her in their youth still regarded her with awe.
> —Dorrit Hofleit, *Notable American Women* (1971)

Williamina Paton, as she was born, came from Dundee in Scotland and worked for several years as a schoolteacher. In 1877 she married James Fleming and emigrated with him to Boston, Massachusetts, in 1878. Her marriage broke up and, forced to support her young son, she worked for Edward PICKERING, director of the Harvard College Observatory, as a maid. As it was his policy to employ young women at the observatory as computers, Pickering, who quickly recognized her intelligence, offered her temporary employment as a copyist and computer in 1879. She was given a permanent post in 1881. She remained at the observatory for

the rest of her life, serving as curator of astronomical photographs from 1899 until her death.

She worked with Pickering on the basic classification of stars into spectral types and was thus involved in the introduction of the original 17 classes arranged alphabetically from A to Q in terms of the intensity of the hydrogen spectral lines. This system was later modified and improved by her colleagues Annie CANNON and Antonia MAURY.

Fleming was largely responsible for the classification of over 10,000 stars, published in 1890 in the *Draper Catalogue of Stellar Spectra*. In the course of her work she discovered 10 novae and over 200 variable stars, and estimated that by 1910 she had examined nearly 200,000 photographic plates.

Flemming, Walther (1843–1905) *German cytologist*

Flemming (**flem**-ing) was born at Sachsenberg in Germany, and graduated in medicine from the University of Rostock in 1868. However, after a short period working in a hospital, he turned to physiology and became assistant to Willy Kuhne at the Institute of Physiology in Amsterdam. After serving as a physician in the Franco-Prussian War he held professional posts at Prague (1873) and Kiel (1876).

By making use of the newly synthesized aniline dyes Flemming was able to discern the threadlike structures in the cell nucleus, which Heinrich Waldeyer was later to term chromosomes. The new staining techniques made it possible for Flemming to follow in far greater detail the process of cell division, which he named "mitosis" from the Greek for thread. Most importantly, Flemming detailed the fundamental process of mitosis, that is, the splitting of the chromosomes along their lengths into two identical halves. These results were published in the seminal book *Zell-substanz, Kern und Zelltheilung* (1882; Cytoplasm, Nucleus and Cell Division). It was another 20 years before the significance of Flemming's work was truly realized with the rediscovery of Gregor MENDEL's rules of heredity.

Flerov, Georgii Nikolaevich (1913–1974) *Russian nuclear physicist*

Born at Rostov-on-Don, now in Russia, and educated at the Leningrad Industrial Institute of Science, Flerov (**flyair**-of) started his career at the Leningrad Institute of Physics and Technology in 1938. He later became chief of the laboratory of multicharged ions at the Kurchatov Institute of Atomic Energy, Moscow.

Throughout his life, Flerov was involved in the search for new elements and isotopes through synthesis and discovery. In many ways his work paralleled that of Glenn Theodore SEABORG and his research team in America.

Flerov and his coworkers synthesized and analyzed isotopes of elements 102, 103, 104, 105, 106, and 107 (members of the actinide group and transactinides) by bombarding nuclei of heavy elements with heavy ions in a cyclotron. In particular, they have a claim to the first discovery or identification of transactinide elements 104 (1964) and 107 (1968). The correct attribution of these discoveries is still in dispute. Besides his work on the transuranic elements, Flerov was also involved in the search, both by synthesis and discovery in nature (possibly in cosmic rays), of the postulated superheavy elements. Many theorists believe that although elements beyond the actinides in the periodic table would be highly unstable, there may be "islands of stability" at higher atomic numbers, if they can only be reached.

In 1960 Flerov became director of the nuclear radiation laboratories of the Joint Institute for Nuclear Research, Dubna, near Moscow.

Florey, Howard Walter (1898–1968) *Australian experimental pathologist*

> If the Professor suggests an experiment always do it, for it is sure to work.
> —Colleague of Florey, quoted in obituary notice, *The Times*, 23 February 1968

Florey was born and educated in Adelaide and graduated in medicine from the university there in 1921. Early in 1922 he arrived in Oxford, England, on a Rhodes scholarship and studied physiology for two years under Charles SHERRINGTON. He then moved to Cambridge University where he studied the various roles and behavior of cells and their constituents for his PhD degree.

In 1931 he was appointed professor of pathology at Sheffield University and for four years studied mucus secretions and the role of the cell in inflammation. He was especially interested in the chemical action of lysozyme (an enzyme discovered in 1921 by Alexander Fleming), which is an antibacterial agent that catalyzes the destruction of the cell walls of certain bacteria. In 1935 Florey became head of the Sir William Dunn School of Pathology at Oxford. Here, along with the biochemist Ernst CHAIN, he took up FLEMING's neglected studies on *Penicillium* mold and in 1939 they succeeded in extracting an impure form of the highly reactive compound penicillin. Florey's work on penicillin was a natural extension of his earlier antibacterial work, inspired by the necessity for efficient antibiotics in wartime. Work continued over the next few years on the purification of the drug. The main problem was that vast quantities of mold needed to be grown for just a few milligrams of penicillin. In wartime Britain the necessary financial backing for these innovative biochemical engineering developments could not be obtained, and permis-

sion was given to use companies in the U.S. for the manufacture of the drug. The considerable problems of large-scale production were overcome and from 1943 onward sufficient penicillin was available to treat war casualties as well as cases of pneumonia, meningitis, syphilis, and diphtheria.

Florey is important as a scientist who took Fleming's discovery and made it into a workable treatment for disease – 15 years after the original discovery. He shared the Nobel Prize for physiology or medicine with Chain and Fleming in 1945. He had been knighted in 1944, and in 1965 he was raised to the British peerage.

Flory, Paul John (1910–1985) *American polymer chemist*

Flory was born at Sterling, Illinois, and educated at Ohio State University, where he obtained his PhD in 1934. His career was divided between industry and university. He worked with Du Pont from 1934 until 1938 on synthetic polymers and then spent the next two years at the University of Cincinnati. After working for Standard Oil from 1940 until 1943, Flory served as Director of Fundamental Research for the Goodyear Tire Company in Akron, Ohio, until 1948. He was then appointed to the chair of chemistry at Cornell. He left Cornell in 1957 to become director of research of the Mellon Institute in Pittsburgh and, finally, in 1961, accepted the chair of chemistry at Stanford University, California.

Flory was one of the people who, in the 1930s, began working on the properties of polymers. A particular problem at the time was that polymer molecules do not have a definite size and structure; a given polymeric material consists of a large number of macromolecules with different chain lengths. Flory approached this problem using statistical methods, obtaining expressions for the distribution of chain lengths.

In further work he developed a theory of nonlinear polymers, which involve cross linkages between molecular chains. He showed how such extended structures can form from a solution of linear polymers. A particular innovation was the concept of *Flory temperature* – a temperature for a given solution at which meaningful measurements can be made of the properties of the polymer.

In later work Flory considered the elasticity of rubbers and similar polymeric materials. He published two authoritative books: *Principles of Polymer Chemistry* (1953) and *Statistical Mechanics of Chain Molecules* (1969). For his major contribution in the field Flory was awarded the Nobel Prize for chemistry in 1974.

Flourens, Jean Pierre Marie (1794–1867) *French physician and anatomist*

Flourens (floo-**rahn**), who was born at Maureilhan in France, studied medicine at the University of Montpellier, graduating in 1813. Moving to Paris he was fortunate enough to be taken in hand by the powerful Georges CUVIER, serving as his deputy at the Collège de France from 1828. After Cuvier's death in 1832, Flourens succeeded him as professor of anatomy and secretary of the Académie des Sciences.

In 1824 Flourens published his highly influential *Recherches expérimentales sur les propriétés et les fonctions du système nerveux dans les animaux vertébrés* (Experimental Researches on the Properties and Functions of the Nervous System in Vertebrates) in which he demonstrated the main roles of different parts of the central nervous system. Extending the work of the Italian anatomist Luigi Rolando on the nervous system, Flourens removed various parts of the brain and carefully observed the resulting changes. Thus he found that removal of the cerebral hemispheres of a pigeon destroyed the sense of perception. Removal of the cerebellum destroyed coordination and equilibrium and excision of the medulla oblongata caused respiration to cease. He also exposed the spinal cord of a dog from head to tail and found that while stimulation lower down would produce movement there came a point higher up where no muscular reaction could be elicited. Flourens is also known for important work on the semicircular canals in the ear, demonstrating their function in balance.

Although Flourens assigned different roles to different anatomical parts of the brain he was not prepared to go further and localize different roles and powers within each part. It was not until 1870 that Gustav FRITSCH and Eduard HITZIG were able to break this unitary picture and establish cerebral localization experimentally.

Flourens is also remembered for his attack on DARWIN in his *Examen du livre de M. Darwin* (1864; Examination of Mr. Darwin's Book) in which he poured scorn on Darwin's "childish and out of date personifications."

Floyer, Sir John (1649–1734) *English physician*

Born at Hintes in England, Floyer graduated MD from Oxford University in 1680, after which he spent most of his life in Lichfield in general practice. In *The Physician's Pulse Watch* (2 vols., 1707–10) he described measuring the pulse with a special pulse watch made to run for exactly one minute. The instrument was a considerable improvement over the "string pendulum" of SANCTORIUS.

He also published a number of works on the therapeutic powers of cold baths, a topic on which he became somewhat obsessive. His work on asthma was important in recognizing the lung condition emphysema in connection with one of the forms of asthma.

Fock, Vladimir Alexandrovich (1898–1974) *Soviet theoretical physicist*

Fock (fok) was educated at the University of Petrograd (now St. Petersburg) where he graduated in 1922. He later worked at the Leningrad Institute of Physics (1924–36) and in 1939 joined the Academy's Institute of Physics. He was also appointed professor of physics at Leningrad University in 1961.

Although Erwin SCHRÖDINGER had published a solution of the wave equation for the hydrogen atom in 1926 it was by no means clear how it could be applied to atoms with more than one electron. The problem was solved between 1927 and 1932 by Fock and, independently, R. Hartree, who developed the method known as the *Hartree–Fock approximation*.

Fock also worked extensively on problems in general relativity.

Folkers, Karl August (1906–1997) *American organic chemist*

Folkers was born in Decatur, Illinois, graduated in chemistry from the University of Illinois in 1928, and gained his PhD from the University of Wisconsin in 1931. After postdoctoral work at Yale he joined the pharmaceutical manufacturers Merck and Company in 1934, becoming director of organic and biochemical research in 1945. He was president of the Stanford Research Institute from 1963 to 1968, and then director of the Institute for Biomedical Research at the University of Texas.

In 1948 Folkers's team isolated the antipernicious anemia factor, vitamin B_{12} (cyanocobalamin) and they played a major role in the lengthy process of determining the structure of this molecule. Folkers has been involved in many investigations of biologically active compounds, especially antibiotics, and the structure of streptomycin was largely determined by his group in 1948.

Forbes, Edward (1815–1854) *British naturalist*

Forbes was born in Douglas on the Isle of Man. Although his education was primarily in medicine at Edinburgh University, his interests included natural history, and in 1833 he visited Norway, collecting mollusks and plants. In 1841–42 he went as naturalist on the HMS *Beacon* to investigate the botany and geology of the Mediterranean and Asia Minor. After lecturing for the Geological Society, Forbes became its curator, and in 1843 was elected to the chair of botany at King's College, London. From 1844 he was paleontologist to the Geological Survey, and subsequently became professor of natural history both at the Royal School of Mines and Edinburgh University.

Forbes traveled widely over Europe, North Africa, and the Middle East, collecting plants and animals (especially mollusks) and studying their relationships. His observations of the littoral zones produced evidence of oceanic sedimentation, while he also investigated molluskan migration, and speculated on the origin and distribution of animal and vegetable life – stimulating later research along these lines. Some of Forbes's most important work was concerned with the starfishes, the British species of which he was the first to classify systematically (*History of British Starfishes*; 1842). His dredging up of a starfish from a depth of a quarter of a mile in the Mediterranean confounded previously held views that life was largely confined to the upper layers of the sea. Forbes also wrote *A History of British Mollusca* (1848–52) and *Naked-eyed Medusae* (1848). His geological investigations of the Purbeck Beds (1849) proved them to be of the Oolitic series.

Ford, Edmund Brisco (1901–1988)
British geneticist

> One of the most far-reaching results of recent work on ecological genetics is the discovery that unexpectedly great selective forces are normally operating to maintain or to adjust the adaptations of organisms in natural conditions.
> —*Ecological Genetics* (1964)

Ford, who was born at Papcastle in England, studied at Oxford University and began research there, in collaboration with Julian HUXLEY, on the genetic control of growth. Together they demonstrated that one type of gene action is to affect the rate rather than nature of chemical reactions. During the same period (1923–26) Ford was also working with Ronald FISHER and was stimulated by Fisher's work on genetic dominance to expand his own work on adaptation and evolution in natural populations.

Having gained his MA in 1927, Ford developed techniques to identify the conditions that promote rapid adaptation and evolution, and then subjected such conditions to ecological investigation. He termed these methods "ecological genetics" and the application of his methods was later to show that selection for beneficial characters is some 30 to 40 times greater than had been believed – an important consideration in evolutionary studies.

Most of Ford's work was conducted on moths and butterflies. He showed that in a given moth population the amount of genetic diversity corresponds to the population size, proportionally greater variation occurring when numbers are higher. He studied the existence of distinct forms in populations of the butterfly *Maniola jartina* and formulated a definition of such ge-

netic polymorphisms. He speculated that the different human blood groups, which are an example of polymorphism, are maintained by the association of certain blood groups with specific diseases, a theory that was proved correct in 1953.

Ford became professor of ecological genetics at Oxford in 1963 and emeritus professor in 1969. His publications include *Ecological Genetics* (1964) – the culmination of over 30 years' research – and *Genetic Polymorphism* (1965).

Forest, Lee de *See* DE FOREST, LEE.

Forrester, Jay (1918–   ) *American computer engineer*

Forrester, born on a cattle ranch in Nebraska, attended a small country school before studying electrical engineering at the University of Nebraska. He went on to do graduate work at MIT on servomechanisms.

This led him, in 1945, to begin work on the design of a flight simulator for the U.S. Navy. He soon discovered that, without high-speed servomechanisms, realistic systems could not be developed. At this point he was directed, in 1946, toward the possibility that digital computers could be used. Forrester set up a laboratory to tackle what became known as the "Whirlwind Project." The Whirlwind, the largest computer of the time, became operational in the early 1950s. Problems, however, soon emerged. With several thousand vacuum tubes, each with a life of about 500 hours, regular breakdowns occurred.

Forrester's first advance was to increase the life of the tubes using new materials and a checking system. However, the main problem was with the machine's memory, which consisted of electrostatic storage tubes. These were expensive and unreliable, with each tube lasting no more than 1 month and costing $1,000 to replace. Consequently, Forrester began to think about magnetic systems of data storage. He used magnetic ferrite rings on a grid of wires in a three-dimensional array. Each ring could be magnetized in one of two directions to represent the binary digits 1 or 0. The method was first used in 1953, and gave an access time twice as fast as that using storage tubes.

The improvements were opportune. Following the political crises of the early cold-war years, the SAGE (Semi Automatic Ground Environment) project was initiated by the U.S. Navy under the supervision of Forrester. Very reliable and very fast computers were needed to analyze air traffic, identify any likely threat, and guide interceptors to hostile planes or missiles. SAGE proved remarkably effective, coming into full operation in 1958 and remaining active until 1984. Forrester left the project in 1956, moving to the MIT Sloan Management School as professor of management with the aim of developing computer systems capable of simulating economic and social systems. He has explained his approach in a number of works, including *Industrial Dynamics* (1961), *Principles of Systems* (1968), and *World Dynamics* (1971). More recently he has worked on a model of the U.S. economy.

Forssmann, Werner Theodor Otto (1904–1979) *German surgeon and urologist*

Forssmann (**fors**-mahn) was educated at the university in his native city of Berlin where he qualified as a physician in 1929. He then worked in the 1930s as a surgeon in various German hospitals. After the war he practiced as a urologist at Bad Kreuznach from 1950 until 1958 when he moved to Düsseldorf as head of surgery at the Evangelical Hospital.

In 1929 Forssmann introduced the procedure of cardiac catheterization into medicine. He was struck by the danger inherent in the direct injection of drugs into the heart frequently demanded in an emergency. The alternative that he proposed sounded no less alarming – introducing a catheter through the venous system from a vein in the elbow directly into the right atrium of the heart. Drugs could then be introduced through this.

After practice on cadavers and an unsuccessful attempt on himself made with the aid of a nervous colleague, Forssmann decided to do the whole thing himself. He consequently introduced a 65-centimeter (25.6-in) catheter for its entire length, walked up several flights of stairs to the x-ray department and calmly confirmed that the tip of the catheter had in fact reached his heart. There had been no pain or discomfort.

Unfortunately further development was inhibited by criticism from the medical profession, which assumed the method must be dangerous. Consequently it was left to André COURNAND and Dickinson RICHARDS to develop the technique into a routine clinical tool in the 1940s; for this work they shared the Nobel Prize for physiology or medicine with Forssmann in 1956.

Foster, Sir Michael (1836–1907) *British physiologist*

Foster, the son of a surgeon from Huntingdon in eastern England, was educated at University College, London, where he obtained his MD in 1859; after several years in private practice with his father, he returned in 1867 as a lecturer in physiology. He moved to Cambridge University in 1870 where he later served as the first professor of physiology from 1883 until his retirement in 1903.

Foster's greatest achievement was undoubtedly the Cambridge school of physiology. Just as the late 19th century saw the emergence in Cambridge of a major physics laboratory, the Cavendish, so too did it witness the creation of a major center for physiological research. Foster did much to establish physiology as a major scientific discipline in Cambridge and, indeed, in Britain. He pressed for the construction of new laboratories, produced the *Textbook of Physiology* (1877), and founded the Physiological Society (1875) and the *Journal of Physiology* (1878). From Foster's laboratory there emerged, as clear proof of his success, such scholars as Charles SHERRINGTON, Henry DALE, John LANGLEY, and Walter GASKELL.

Much of Foster's work as a research physiologist was devoted to the problem of the genesis of the heartbeat, namely whether it was under nervous control. The role of the vagus nerve in inhibiting the heartbeat had been known since the 1840s, as had the presence of nervous ganglia in the heart itself. Against such impressive evidence Foster pointed out, in 1859, that in the snail's heart neither the inhibitory nerve nor ganglia could be found, yet small pieces taken from it would continue to contract rhythmically for some time. He concluded from this that the cause of the heartbeat was a "peculiar property of general cardiac tissue." Several years later this work was extended and confirmed by Gaskell.

After 1880 Foster did little original work, being more concerned with the demands of his growing organization. When he began there were 20 students studying physiology at Cambridge; when he retired there were over 300.

Foucault, Jean Bernard Léon (1819–1868) *French physicist*

Foucault (foo-**koh**), the son of a Parisian bookseller, originally intended to study medicine, but transferred his interest to physical science. In 1855 he became a physicist at the Paris Observatory.

His main work was on measurements of the speed of light. He helped Armand FIZEAU in his toothed-wheel experiment and, in 1850, took over D. F. J. ARAGO's experiments on comparing the speed of light in air with that in water. The experiment was important for distinguishing between the wave and particle theories of light: the wave theory predicted that light should travel faster in air than in water; the particle theory predicted the opposite. In 1850, Foucault showed that the wave-theory prediction was correct. In 1862 he obtained the first accurate value for the speed of light using a rotating-mirror apparatus.

Foucault also worked on other topics. Thus he noted (1849) that a bright yellow line in the spectrum of sodium corresponded to a dark line in the FRAUNHOFER spectrum, although he failed to follow this up.

His most famous experiments began in 1850 and involved pendulums. While trying to construct an accurate timing device for his work on light, he noticed that a pendulum remained swinging in the same plane when he rotated the apparatus. He then used a pendulum to demonstrate the rotation of the Earth. Over a long period of time the plane in which a pendulum is swinging will appear to rotate. In fact the pendulum swings in a fixed plane relative to the fixed stars, and the Earth rotates "underneath" it.

At the Earth's poles, the plane of the pendulum will make one full rotation every 24 hours; this period increases as the equator is approached. Foucault derived an equation relating the time of rotation to the latitude. He also gave public exhibitions of the effect, including one in which he suspended an iron ball of 28 kilograms (62 lbs) by steel wire 67 meters (222 ft) long from the dome of the Panthéon in Paris. Foucault also invented the gyroscope.

Fourcroy, Antoine François de (1755–1809) *French chemist*

Fourcroy (foor-**krwah**) was born and grew up in poverty in Paris and began to study medicine in 1773. Graduating in 1780, he became professor of chemistry at the Jardin du Roi in 1784. He was an excellent lecturer, a prolific writer, and played a leading role in the scientific life of France. He helped to found the Ecole de Santé (1795) and the Ecole Polytechnique (1795) and was minister of public instruction under Napoleon (1802–08). With VAUQUELIN he carried out many researches, especially on natural products, and they were the first to prepare a relatively pure form of urea and to name this compound. Fourcroy was an important advocate of the system of Antoine LAVOISIER, which he taught from 1786. Among his many writings was the massive *Système de connaissance chimique* (1801–02; System of Chemical Knowledge).

Fourier, Baron (Jean Baptiste) Joseph (1768–1830) *French mathematician*

> There cannot be a language more universal and more simple, more free from errors and obscurities ... more worthy to express the invariable relations of natural things [than mathematics].
> —*Théorie analytique de la chaleur* (1822; The Analytical Theory of Heat)

Fourier (**foor**-ee-ay), the son of a tailor from Auxerre in France, was educated at the local military school and later at the Ecole Normale in Paris. He held posts at both the Ecole Normale and the Ecole Polytechnique where he was a very effective and influential teacher. In 1798 he accompanied Napoleon on the invasion of Egypt and later contributed to and oversaw the publication of the *Description de l'Egypte* (1808–25; Description of Egypt), a massive compilation of the cultural and scientific materials brought back from the expedition.

Fourier's most important mathematical work is contained in his *Théorie analytique de la chaleur* (1822; The Analytical Theory of Heat), a pioneering analysis of the conduction of heat in solid bodies in terms of infinite trigonometric series, now known as *Fourier series*. Fourier was led to consider these series when attempting to solve certain boundary-value problems in physics and his interest was always in the physical applications of mathematics rather than in its development for its own sake. His work continues to be extremely important in many areas of mathematical physics, but it has also been developed and generalized to yield a whole new branch of mathematical analysis, namely, the theory of harmonic analysis.

Fourneau, Ernest François Auguste (1872–1949) *French medicinal chemist and pharmacologist*

Fourneau (foor-**noh**) was born in Biarritz, France. After graduating from the Lyceum of Bayonne, he worked in a Paris hospital pharmacy (1892–96) before becoming director of the research laboratories at the pharmaceutical manufacturers Poulenc Frères (1900–1912). From 1912 to 1942 he was head of the therapeutic chemistry laboratories at the Pasteur Institute. In 1924 he synthesized suramin, an effective drug against African sleeping sickness, which had first been discovered by the German BAYER company in 1920 but had not been published by them. This was followed by acetarsol (1926), an amebicide, and plasmocid (1930), an antimalarial agent. In 1935, working with the newly discovered antibacterial dye, Prontosil, he showed that its metabolite, now known as sulfanilamide, was the active part of the molecule.

Fourneyron, Benoît (1802–1837) *French engineer and inventor*

Fourneyron (foor-nay-**ron**), who was born the son of a mathematician in Saint-Etienne, France, studied at the then newly opened engineering school there until 1816. His professor, Claude Burdin, had written a paper describing a new kind of waterwheel called a "turbine." This design principle is now basic to electricity generators and many propulsion engines, but Burdin's paper was rejected by the French Academy of Sciences and by the Society for the Encouragement of Industry. Fourneyron, while he was working in an ironworks at Le Creuset, studied the idea and, in 1827, actually built a six-horsepower unit, in which the force of water flowing outward from a central source onto angled blades turned a rotor.

Ten years later, Fourneyron built a turbine with a wheel one foot (30 cm) in diameter, which could turn at 2,300 revolutions per minute and had an efficiency of 80%. Weighing only 40 pounds (18 kg), it was very much smaller than an equivalent waterwheel and could be mounted horizontally with a vertical shaft. Fourneyron thought of building steam turbines but was unable to do so because, at that time, no suitable materials and manufacturing techniques were available.

Fowler, Alfred (1868–1940) British astrophysicist

Although born into a poor family in the industrial city of Bradford, England, Fowler gained a scholarship and in 1882 went to the Normal School of Science (later to become the Royal College of Science and now part of Imperial College, London). After graduating with a diploma in mechanics, he became assistant to Norman LOCKYER at the Solar Physics Observatory in South Kensington, London. He remained there after Lockyer's retirement in 1901, being made professor of astrophysics in 1915. Finally, from 1923 to 1934 he served as Yarrow Research Professor of the Royal Society. Fowler was one of the leading figures behind the founding of the International Astronomical Union in 1919, serving as its first general secretary until 1925.

Not surprisingly Fowler worked very much in the Lockyer tradition of solar and stellar spectroscopy. He became particularly skilled in identifying difficult spectra, using his experience in producing different spectra in the laboratory. He thus detected magnesium hydride in sunspots and carbon monoxide in the tails of comets, and showed that the band spectra of cool M-type stars were due to titanium oxide. In addition, following the announcement in 1913 of the BOHR theory of the atom, Fowler was outstanding in analyzing the structure of atoms from their special characteristics.

Fowler, William Alfred (1911–1995) American physicist

Fowler was born in Pittsburgh, Pennsylvania, graduated in 1933 from Ohio State University, and obtained his PhD in 1936 from the California Institute of Technology. He was immediately appointed to the staff, serving as professor of physics there from 1946 to 1970; he was Institute Professor from 1970 and professor emeritus from 1982.

Fowler worked mainly in nuclear physics, especially on the nuclear reactions that occur in stars and by which energy is produced and the elements synthesized, on nuclear forces, and on nuclear spectroscopy. In 1957 Margaret and Geoffrey BURBIDGE, Fred HOYLE, and Fowler published a key paper dealing with the problem of the creation of the chemical elements in the interiors of stars. They were aware that the hot big bang proposed by George GAMOW could produce nothing heavier than helium. Clearly the elements were produced later. They therefore had to identify nuclear reactions that could occur at the immense temperatures of stellar cores; the type of process changed, and hence changed the elements being produced, as the temperature increased and conditions altered inside the stars. A later and fuller version was published by Fowler and Hoyle in their *Nucleosynthesis in Massive Stars and Supernovae* (1965).

Fowler subsequently worked on such fundamental questions as the amount of helium and deuterium in the universe, the answers to such questions having profound implications for knowledge of the age and future development of the universe.

For his work on nuclear astrophysics, Fowler shared the 1983 Nobel Prize for physics with Subrahmanyan CHANDRASEKHAR.

Fox, Harold Munro (1889–1967) British zoologist

Fox (originally Fuchs), the son of an officer in the Prussian army, was born in London and educated at Cambridge University; after war service, he served as a fellow from 1920 to 1928. He then moved to Birmingham as professor of zoology, a post he held until 1941 when he accepted a similar chair at Bedford College, London, where he remained until his retirement in 1954.

Fox, a zoologist of wide interests, is best known for his work on invertebrate blood pigments. In 1871 Ray LANKESTER had noted that the red blood of the water flea *Daphnia*, a crustacean, was due to the presence of the pigment hemoglobin. Other crustaceans, such as lobsters, were blue-blooded with the pigment hemocyanin in their blood.

It had been observed that the transparent *Daphnia* could become redder or paler by synthesizing or breaking down blood hemoglobin. This was done at a rate far in excess of that noted in any other creature. Fox showed by laboratory experiments in the 1940s that the response was controlled by the level of dissolved oxygen in the water. If this were low, hemoglobin, with its affinity for oxygen, was synthesized; if high, *Daphnia* lose their hemoglobin and become colorless.

Earlier, in 1923, Fox had repeated the controversial experiments of Paul KAMMERER on the supposed elongations produced in the siphons of the sea squirt *Ciona intestinalis*. He reported that in none of the operated animals was there any further growth of the siphons once the original length had been attained, and went on to suggest that Kammerer's results could have been produced by keeping the animals in a highly nutritious solution, an explanation Kammerer was quick to dismiss.

Fox, Sidney Walter (1912–1998) *American biochemist*

Born in Los Angeles, Fox was educated at the University of California in his native city and at the California Institute of Technology, where he obtained his PhD in 1940. He taught briefly at Berkeley and the University of Michigan before moving to Iowa State University in 1943, serving as professor of biochemistry from 1947 until 1954. Fox then moved to Florida State University as professor of chemistry, joining the University of Miami in 1964 to take up an appointment as director of the Institute of Molecular and Cellular Evolution. He retired in 1989.

Biochemists dealing with the origin of life at some time or other have to face the problem of which came first, proteins or nucleic acids? In this classic dispute Fox placed himself firmly in favor of the protein-first hypothesis.

He demonstrated in 1958 the existence of what he termed "proteinoids," polymers of amino acids produced by the application of heat alone. Furthermore he found that on cooling, the proteinoids produce microspheres – tiny spheres, a micrometer or two in diameter, apparently resembling bacteria and supposed by Fox to be protocells. Since they appeared to develop some kind of membrane, produced buds, and sometimes divided, the comparison was far from outrageous. One further significant consideration is that the proteinoids are not just a random grouping of amino acids; a selection principle of some kind operates to give preference to some amino acids over others. There is, however, a major difficulty implicit in Fox's work. To construct the proteinoids he was forced to use an above-average concentration of the amino acids lysine, glutamic acid, and aspartic acid, and it is unlikely that the early terrestrial environment consisted of such a specialized distribution of amino acids.

Fracastoro, Girolamo (1478–1553) *Italian physician*

Fracastoro (frah-kah-**stor**-oh) was born in Verona, Italy, and educated at the University of Padua where he graduated in 1502. He taught logic there until 1508, when he returned to the family estates at Incaffi, near Verona. He thereafter seems to have divided his time between practicing medicine in Verona and attending to the family estates.

Fracastoro was the author of two important medical works. The first, written as the poem *Syphilis sive morbus Gallicus* (1530; Syphilis or the French Disease), introduced the term "syphilis" into most European languages.

Syphilis seems to have appeared first in Europe at the end of the 15th century with the siege of Naples by the French army of Charles VIII. Called the "Mal de Naples" by the French and the "Mal Francese" by the Neapolitans, the disease spread rapidly throughout Europe. It was because of this rapidity that Fracastoro rejected the even then popular view attributing the infection to the New World. His descriptions of the symptoms suggest syphilis was then more virulent than it is now.

The second work of Fracastoro, *De contagione et contagiosis morbis* (1546; On Contagion and Contagious Diseases), is claimed to be a precursor of the germ theory of disease. He distinguished three types of infection: by direct contact, by indirect contact through an intermediary, and those, like some fevers, that appear to operate at a distance. He spoke of "seeds of contagion," imperceptible and passing from person to person, which he believed could generate more identical germs. Having made such a theoretical advance in introducing his "seminaria contagium" it was difficult to see what further steps could be taken at that time. Consequently his work remained an isolated curiosity rather than a part of the main tradition of medicine.

Fracastoro also, in 1538, published an astronomical work, *Homocentrica sive de stellis liber* (Homocentricity or the Book of Stars), which attempted to eliminate the epicycles and eccentrics from the geocentric system of PTOLEMY and replace them with the original concentric spheres of EUDOXUS. This work influenced COPERNICUS, a contemporary of Fracastoro at Padua.

Fraenkel-Conrat, Heinz L. (1910–1999) *German–American biochemist*

Fraenkel-Conrat (freng-kel-**kon**-rat), son of the noted gynecologist Ludwig Fraenkel, was born in Breslau (which was then in Germany and is now Wrocław in Poland). He left Germany for Britain after graduating MD from the University of Breslau in 1934. Having gained his PhD for work on ergot alkaloids and thiamine from the University of Edinburgh in 1936, he moved on to America, where he settled and became an American citizen in 1941. He joined the faculty of the University of California at Berkeley in 1951, becoming professor of virology in 1955 and professor emeritus in 1981.

Fraenkel-Conrat, working with the tobacco mosaic virus (TMV), an RNA virus, provided evidence that RNA, like DNA, can act as the genetic material. This he did by separating the RNA and protein portions of the virus, and then reassembling them to make a fully infective virus. Moreover Fraenkel-Conrat demonstrated that, while the isolated protein was quite dead, the isolated RNA showed slight signs of infectivity. This work was reported in a paper by Fraenkel-Conrat and Robley C. WILLIAMS entitled *Reconstitution of Tobacco Mosaic Virus from Its Inactive Protein and Nucleic Acid Components* (1955). In later work with Wendell Stanley the complete amino-acid sequence, consisting of 158 amino acids, of the TMV protein was established.

Fraenkel-Conrat, in collaboration with R. R. Wagner, edited one of the basic texts of modern virology, *Comprehensive Virology* (19 vols.; 1974–84).

Francis, Thomas Jr. (1900–1969) *American virologist*

Francis, the son of a methodist clergyman from Gas City, Indiana, was educated at Allegheny College and Yale where he obtained his MD in 1925. He worked with the Rockefeller Institute from 1925 to 1938 and, after serving as professor and chairman of bacteriology at the New York University College of Medicine, moved to the University of Michigan in 1941 as professor of epidemiology, a post he retained until his death.

Francis became known to a wide public when, in 1954, he reported on the SALK polio vaccine trial. Before this however he had worked for over 20 years on the epidemiology of the influenza virus. The first such virus, the A-type, had been detected by Christopher Andrewes and his colleagues in 1933. In the following year Francis found a further strain of the A-type, the PR 8, present in the Puerto Rican epidemic of 1934. In 1940 he went on to detect a completely distinct type, B, with no immunological relationship to the A-type.

The U.S. Army, fearful of a repeat of the 1918 flu epidemic, set up in 1941 a commission to develop a vaccine and asked Francis to be its chairman. By 1942 he was ready to vaccinate 8,000 soldiers with his vaccine but, perversely, flu was scarce that year. It was not until 1943 that he was able to report that those vaccinated were 70% less likely to be hospitalized compared with the control group. This encouraged the army to vaccinate some 1,250,000 troops in 1947 but this time it disconcertingly seemed to offer no protection at all.

It soon became clear to Francis why the vaccine had failed – the arrival of a new strain of A-type virus, known as A^1. Francis was thus able to present the dilemma facing flu epidemiologists, namely that while it was certainly possible to develop a vaccine against flu it was more than likely that it would end up as a vaccine against yesterday's flu.

Franck, James (1882–1964) *German–American physicist*

Franck, the son of a banker from Hamburg in Germany, was educated at Heidelberg and Berlin where he obtained his doctorate in 1906. After distinguished war service, in which he won two iron crosses, he was appointed to the chair of experimental physics at Göttingen. Although exempt from the 1933 Nazi law that excluded Jews from public office because of his

military service, he insisted on publicly resigning. After spending a year in Copenhagen, he emigrated to America in 1935 where he served as professor of physical chemistry at the University of Chicago from 1938 to 1949.

In collaboration with Gustav HERTZ he produced experimental evidence of the quantized nature of energy transfer, work that won them the 1925 Nobel Prize for physics. Their experiment, conducted in 1914, consisted of bombarding mercury atoms with electrons. Most of the electrons simply bounced off, losing no energy in the process. When the velocity of electrons was increased it was found that on collision with mercury atoms they lost precisely 4.9 electronvolts (eV) of energy. If an electron possessed less energy than 4.9 eV it lost none at all on collision; if it had more than 4.9 eV it made no difference – only 4.9 eV was absorbed by the mercury atoms. Franck and Hertz had thus succeeded in showing that energy can only be absorbed in quite definite and precise amounts. For mercury the minimum amount was 4.9 eV. Their results were quickly confirmed and shown to hold for other atoms.

In America Franck worked mainly on the physical chemistry involved in photosynthesis although he is better known as the author of the *Franck Report* published in 1946. This report, actually produced by a number of distinguished scientists of whom Leo SZILARD was probably the most important, was sent to the Secretary of State for War in June 1945. It argued that it was not necessary to drop the recently produced atomic bomb on Japan as its explosion on a barren island would be sufficient to force the Japanese into submission.

Frank, Ilya Mikhailovich (1908–1990) *Russian physicist*

Frank was born at St. Petersburg in Russia and educated at Moscow University. After working at the State Optical Institute from 1930 to 1934, he was made an associate of the Physics Institute of the Soviet Academy of Sciences. He was professor of physics at Moscow University from 1944 to 1990.

Frank and Igor TAMM first gave an explanation of the radiation discovered in 1934 by Pavel CHERENKOV. This occurs when a charged particle is traveling through a medium faster than the speed of light in that medium. The particle displaces electrons in the atoms of the medium and an electromagnetic wave is produced. The effect is similar to the sonic boom produced when a body moves faster than sound in air.

Frank, Cherenkov, and Tamm shared the Nobel Prize for physics in 1958.

Frankland, Sir Edward (1825–1899) *British organic chemist*

Born at Churchtown, near Lancaster in northwestern England, Frankland was first apprenticed to a pharmacist in Lancaster; he was later encouraged to go to London to study chemistry under Lyon PLAYFAIR at the Royal College of Engineers (1845). He became Playfair's assistant in 1847 and studied extensively in Europe with Robert BUNSEN and Justus von LIEBIG. He succeeded Playfair as professor of chemistry in 1850, holding the same position at Owens College, Manchester (1851–57), and in London at St. Bartholomew's Hospital (1857), the Royal Institution (1863), and the Royal School of Mines (1865), later the Royal College of Science.

Frankland's first important work was on methylalkyls (1849). He prepared zinc methyl and zinc ethyl but, being at that time an adherent of the radical theory, was led into the error of believing that "methyl hydride" and "ethyl hydride" (actually both ethane) were different compounds.

Frankland is generally credited as the originator of the theory of valence – this being the number of chemical bonds that a given atom or group can make with other atoms or groups in forming a compound. In 1852 he noticed that coordination with an alkyl group could change the combining power of a metal. He showed that the concept of valence could reconcile the radical and type theories and in 1866 he elaborated the concept of a maximum valence for each element. In 1864, working with B. F. Duppa, Frankland pointed out that the carboxyl group (–COOH, which he called "oxatyl") is a constant feature of the series of organic acids. He was also interested in applied chemistry: he investigated the luminosity of flames and his later work was in the field of coal-gas supply and water purification. He was knighted in 1897.

Franklin, Benjamin (1706–1790) *American scientist, statesman, diplomat, printer, and inventor*

> He snatched lightning from the heavens and sceptres from kings.
> —Turgot, epitaph on Franklin

Franklin's father left England in 1682 and the following year settled in Boston, Massachusetts, where he worked as a candle maker and soap boiler. Although originally intended for the clergy, Franklin, who was born in Boston, was forced to leave school at the age of ten for financial reasons; after helping his father for some time he was apprenticed to his brother, a printer, in 1718.

He continued his trade as a printer in London (1724–26), and thereafter in Philadelphia where he published, from 1729, the *Pennsylvania Gazette* and, from 1733, the hugely successful *Poor Richard's Almanac*. Shortly afterward Franklin began his life in public affairs serving as clerk of the State Assembly (1736–

51) and as deputy postmaster representing the colonies (1753–74), during which he saw further service in London (1757–62; 1764–75). On his return to America Franklin played an active role in the revolution and was one of the five who drafted the Declaration of Independence in 1776. He was sent to France in 1776 to seek military and financial aid for the colonies and largely through his own popularity succeeded in achieving an alliance in 1778. Returning to America in 1785, he performed his last public duty as a member of the Constitution Convention in 1787 before retiring from public life in 1788.

Despite such an active political and public life Franklin also made important contributions to 18th-century physical theory in the period 1743–52. By conversation with scholars in London, reading, and correspondence with friends, his interest in the newly discovered phenomena of electricity had been aroused.

Franklin would have known of the work of Stephen GRAY and Charles Dufay and the basic distinction established between electrics (such as glass and amber), which could be electrified by rubbing, and nonelectrics (such as metals), which resisted such treatment. Electrics were further divided into vitreous substances, such as glass, and resinous substances, such as amber. Dufay had concluded that there were two distinct electric fluids – the vitreous and the resinous – in his two-fluid theory.

Franklin agreed with Dufay that electricity was a fluid. More significant properties of electricity emerged out of Franklin's experiments from 1747 onward. From these experiments, including those on the Leyden jar, Franklin devised his one-fluid theory of electricity. He also introduced the terminology of "positive" and "negative" into the science.

Practical gains emerged from Franklin's discovery that lightning is an electric charge. He knew that the electric fluid was attracted by points but wondered if lightning would also be attracted. In 1752 he performed his famous, yet hazardous, experiment with a kite during a thunderstorm and established the identity of lightning with electricity. Following this he suggested the use of lightning rods on tall buildings to conduct electricity away from the building and direct to ground.

Franklin also published works on the problems of light, heat, and dynamics. Outside of physics Franklin's most important scientific work was in oceanography with his study of the Gulf Stream. He measured its temperature at different places and depths, estimated the current's velocity, and analyzed its effects on the weather. From reports supplied to him by Nantucket sea captains he also constructed the first printed chart of the Gulf Stream. Franklin is also remembered for his large number of inventions that included (in addition to his lightning rod) bifocal spectacles, the rocking chair, and an efficient stove.

Franklin, Rosalind (1920–1958) *British x-ray crystallographer*

Franklin was a Londoner by birth. After graduating from Cambridge University, she joined the staff of the British Coal Utilisation Research Association in 1942, moving in 1947 to the Laboratoire Centrale des Services Chimique de L'Etat in Paris. She returned to England in 1950 and held research appointments at London University, initially at King's College from 1951 to 1953 and thereafter at Birkbeck College until her untimely death from cancer at the age of 37.

Franklin played a major part in the discovery of the structure of DNA by James WATSON and Francis CRICK. With the unflattering and distorted picture presented by Watson in his *The Double Helix* (1968) her role in this has become somewhat controversial. At King's, she had been recruited to work on biological molecules and her director, John RANDALL, had specifically instructed her to work on the structure of DNA. When she later learned that Maurice WILKINS, a colleague at King's, also intended to work on DNA, she felt unable to cooperate with him. Nor did she feel much respect for the early attempts of Watson and Crick in Cambridge to establish the structure.

The causes of friction were various, ranging from simple personality clashes to, it has been said, male hostility to the invasion of their private club by a woman. Despite this unsatisfactory background Franklin did obtain results without which the structure established by Watson and Crick would have been at the least delayed. The most important of these was her x-ray photograph of hydrated DNA, the so-called B form, the most revealing such photograph then available. Watson first saw it in 1952 at a seminar given by Franklin, and recognized that it clearly indicated a helix. Franklin also appreciated, unlike Watson and Crick, that in the DNA molecule the phosphate groups lie on the outside rather than inside the helix. Despite such insights it was Watson and Crick who first realized that DNA has a double helix. By March 1953 Franklin had overcome her earlier opposition to helical structures and was in fact producing a draft paper on 17 March 1953, in which she proposed a double-chain helical structure for DNA. It did not, however, contain the crucial idea of base pairing, nor did she realize that the two chains must run in opposite directions. She first heard of the Watson–Crick model on the following day.

Frasch, Herman (1851–1914) *German–American industrial chemist*

The son of a wealthy apothecary from Gaildorf,

in Germany, Frasch (frahsh) was educated in Halle and was apprenticed to an apothecary before emigrating to America in 1868. He worked in the field of petroleum chemistry and in 1876 he joined the Standard Oil Company in Cleveland.

In 1882 Frasch developed a process for reducing the sulfur content of poor-quality crude oil through the use of metallic oxides, thus making it commercially acceptable, and in 1884 he bought such a field at London, Ontario. When, in 1886, Standard Oil discovered high-sulfur oil in Ohio Frasch rejoined them as a part-time consultant to solve the problem. From sulfur as an impurity he turned to the problems of extracting sulfur as a product in its own right. In 1894 he was the first to raise sulfur from deep underground by means of superheated water. In the *Frasch process* three concentric pipes are sunk down to the sulfur deposit. Superheated water is pumped down the outside pipe to liquefy the sulfur, which is forced up the middle pipe by compressed air passed down the center pipe. The process became a commercial proposition in 1900 thanks largely to the availability of cheap local oil to heat the water. Frasch had a 50% share in the company, Union Sulfur, and the Sicilian sulfur monopoly was broken. Eventually other companies used the process, Frasch lost his patent-infringement actions, and his company faded away.

Fraunhofer, Josef von (1787–1826) *German physicist and optician*

Fraunhofer (**frown**-hoh-fer), whose family was in the optical trade, was born in Straubing, Germany; he was apprenticed to an optician in Munich after his parents died. He subsequently moved to the Utzschneider optical institute near Munich.

His great ambition was to perfect the achromatic lens, which John DOLLOND had developed a century earlier, and his scientific discoveries came as by-products of this work. The major difficulty was to measure the refractive indices of the different types of glass used in these lenses. In 1814, while testing prisms in order to determine these constants, he observed that the Sun's spectrum was covered with fine dark lines. He also noticed that these *Fraunhofer lines* occurred in the spectra of bright stars, but that their positions were different. The lines had been observed earlier by William WOLLASTON, but Fraunhofer studied them in detail, measuring the positions of 576 of them and giving the main ones letters A–G. He also found the lines to be present in spectra produced by reflection from a grating (1821–22), thus proving them to be a characteristic of the light, not the glass of the prism.

Fraunhofer had in his grasp the key to finding the composition of the stars, but this step was taken half a century later by Gustav

KIRCHHOFF, who showed that lines in the solar spectrum resulted from characteristic absorption by elements in the atmosphere of the Sun.

Fredholm, Erik Ivar (1866–1927) *Swedish mathematician*

Fredholm (**fred**-holm), who was born in the Swedish capital of Stockholm, was a pioneer in the theory of integral equations. He studied at the Stockholm Polytechnic Institute, the University of Stockholm, and the University of Uppsala, from which he received his doctorate in 1898. The same year he was appointed lecturer in mathematical physics at the University of Stockholm, becoming professor there in 1906.

In the field of mathematical physics Fredholm's work on the deformation of anisotropic media (such as crystals) led him to the study of partial differential equations. His most important work was on integral equations, establishing the field in which David HILBERT was to do some of his greatest work. A number of results and concepts are named for him, including the *Fredholm integral equations* and the *Fredholm operator.*

Frege, Gottlob (1848–1925) *German philosopher and mathematician*

Born at Wismar in Germany, Frege (**fray**-ge) studied at the universities of Jena and Göttingen, where he obtained his PhD in 1873. He then returned to Jena as a lecturer, where he remained for the rest of his working life, rising to the position of professor in 1896. In a series of seminal works Frege laid the foundations of modern mathematical logic, transforming logic with an understanding of and notation for the problem of multiple generality – propositions containing predicates, quantifiers, and variables – and showing how the basic concepts and operations of mathematics could be formalized. He also revolutionized modern philosophy through his influence on the philosophy of language. However, Frege's work was almost completely ignored, misunderstood, or treated with hostility by his contemporaries – notable exceptions were Bertrand RUSSELL and Giuseppe PEANO.

In his first major work, *Begriffsschrift* (Concept Writing, 1879), he provided a new formalism containing an adequate symbolism and an axiomatic base for the rigorous derivation of both propositional and predicate logic. While a few workers, such as the American C. S. Pierce, had been moving in this direction, their work was completely overshadowed by the comprehensive nature of Frege's work.

In *Die Grundlagen der Arithmetik* (1884; The Foundations of Arithmetic) Frege gave a formal definition of cardinal number and showed how basic properties of numbers could be logically derived from it. In *Grundgesetze der Arithmetik* (1893 and 1903; Basic Laws of Arithmetic) he went further in attempting to derive arithmetic from formal logic. The *Grundgesetze* is still regarded as a massive achievement, but his main aim was doomed to failure. On the eve of the publication of the second volume Russell wrote to Frege pointing out a contradiction – Russell's paradox – that could be derived from his system. This, as Frege acknowledged, vitiated his whole project.

Fresnel, Augustin Jean (1788–1827) *French physicist*

> I must confess that I am greatly tempted to attribute the transmission of light and heat to the vibrations of a special fluid...One day the cause of electrical phenomena may well be sought in the disequilibrium of this fluid also.
> —On ether. Letter, 5 July 1814

Fresnel (fray-**nel**) was born in Broglie, France, and grew up in the time of the French Revolution; by the time he was 26, Napoleon had been exiled and Louis XVIII was on the throne. At this time Fresnel was a qualified engineer but, when Napoleon returned from Elba, Fresnel supported the royalists and lost his job as a result.

Fresnel started studying optics in 1814 and was one of the major supporters of the wave theory of light. He worked on interference, at first being unaware of the work of Thomas YOUNG, and produced a number of devices for giving interference effects. *Fresnel's biprism* is a single prism formed of two identical narrow-angled prisms base-to-base. Placed in front of a single source it splits the beam into two parts, which can produce interference fringes. Initially, Fresnel believed that light was a longitudinal wave motion, but he later decided that it must be transverse to account for the phenomenon of polarization.

Another important part of Fresnel's work was his development of optical systems for lighthouses. He invented the *Fresnel lens* – a lens with a stepped surface – to replace the heavy metal mirrors that were in use at the time.

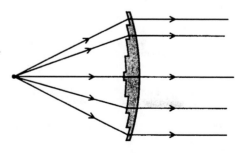

FRESNEL LENS   *The stepped structure of a Fresnel lens.*

Fresnel became a member of the French Academy of Sciences in 1823 and four years later, shortly before he died, the Royal Society awarded him the Rumford medal.

Freud, Sigmund (1856–1939) *Austrian psychoanalyst*

> My life and work has been aimed at one goal only: to infer or guess how the mental apparatus is constructed and what forces interplay and counteract in it.
> —Quoted by E. Jones in *Life and Work of Sigmund Freud* (1983)

The son of a wool merchant from Freiberg (now Příbor in the Czech Republic), Freud (froid) graduated from the University of Vienna with an MD in 1881 having also spent much time in the study of physiology. He worked at the Vienna General Hospital until 1885 and, after a further period of study in Paris under the neurologist Jean-Martin CHARCOT, set up in private practice in Vienna in the same year. He took the post of part-time head of the neurological outpatients clinic at the Children's Hospital and also held the position of *Privatdozent* in neuropathology at the University of Vienna.

Before Freud worked out the basic principles of psychoanalysis in the 1890s he had produced a substantial body of research in more orthodox fields. In addition to early work on comparative neuroanatomy, he discovered the euphoric effects of cocaine in 1884 and produced two sizable monographs – one on aphasia (1891) and the other on paralysis in children (1893).

By this time he had developed a more ambitious research program, clearly stated in his unpublished *Project for a Scientific Psychology* (1895). In this unfinished work he aimed to explain "the theory of mental functioning" in terms of quantitative physical concepts that would apply to both normal and abnormal psychology. To this end he went into considerable detail, even supposing the existence of three types of neurones with different physiological properties. Freud however drew back from such a neurological approach. In 1893 he collaborated with Josef Breuer on *The Psychical Mechanism of Hysterical Phenomena*, later expanded into *Studien über Hysterie* (1895; Studies in Hysteria, 1955), a work that marked the beginning of psychoanalysis. During the period 1892–95 Freud evolved his psychoanalytical method using the technique of free association. Following this he developed his theory that neuroses were rooted in suppressed sexual desires.

Freud's major work, *Die Traumdeutung* (1899; The Interpretation of Dreams, 1953), is regarded as his most original. In this he analyzed dreams in terms of unconscious desires and experiences. His other works included *Zur Psychopathologie des Alltagslebens* (1904; Psychopathology of Everyday Life, 1960), *Totem und Tabu* (1913; Totem and Taboo, 1955), *Jenseits des Lustprinzips* (1920; Beyond the Pleasure Principle, 1955), and *Das Ich und das Es* (1923; The Ego and the Id, 1961).

In 1902 Freud established a circle of colleagues who met to discuss psychoanalytical matters once a week at his house. The group's original members were Alfred ADLER, Max Kahane, Rudolf Reitler, and Wilhelm Stekel. This grew, and later became the Vienna Psycho-Analytical Society (1908), and finally, the International Psycho-Analytical Association (1910). Freud, now becoming famous in Europe, made a tour of America in 1909 where he was well received. By 1911 the International Psycho-Analytical Association had begun to break up through differences of opinion, Carl JUNG and Alfred Adler being among the most significant to leave. However, by the 1920s Freud had become one of the most famous thinkers of the century.

In 1923 Freud was diagnosed as having cancer of the jaw. During the next 16 years he was to suffer more than 30 operations and be compelled to live with a prosthesis which, by substituting for his excised jaw and palate, allowed him to eat, drink, smoke, and talk.

In 1938 he was forced to leave Vienna by the Nazis for exile in London. He continued to see patients and to work on his last book, *Der Mann Moses und die monotheistische Religion* (1939; Moses and Monotheism, 1960), but within a matter of months it was clear to him that he could continue to work no more. It was then that he reminded his doctor: "My dear Schur...you promised you would help me when I could no longer carry on." Schur honored his pledge with morphine ensuring a peaceful death.

Freundlich, Herbert Max Finlay (1880–1941) *German–American physical chemist*

Freundlich (**froint**-lik) was born at Charlottenburg, now a district of Berlin, Germany. He studied physical chemistry at Munich and at Leipzig where he became assistant to Wilhelm OSTWALD. From 1911 to 1917 he was professor at the Institute of Technology in Brunswick. He was professor at the Kaiser Wilhelm Institute, Berlin-Dahlem, from 1917 to 1934 when he left Germany because of the Nazi regime. He traveled first to Britain and worked at University College, London, and from 1938 to his death was professor of colloid chemistry at the University of Minnesota.

Freundlich's life work was in colloid chemistry. He showed how colloid stability was changed by the addition of electrolytes and he is well known for the *Freundlich adsorption isotherm*, a theoretical formula for the amount of adsorption on a surface at constant temperature.

Frey-Wyssling, Albert Friedrich (1900–1988) *Swiss botanist*

Born in Küsnacht, Switzerland, Frey-Wyssling (frI-**voos**-ling) graduated in biology from the Federal Institute of Technology, Zurich, in 1923 and gained his PhD the following year. He then studied optics at the University of Jena, followed by plant physiology at the Sorbonne. From 1928 to 1932 he was plant physiologist at the Rubber Experimental Station in Medan, Sumatra, where he investigated latex flow. On his return to Switzerland he took up a lectureship at the Federal Institute of Technology, becoming professor of botany in 1938, and finally serving as rector from 1957 to 1967.

Frey-Wyssling was concerned about the rift in biology between histologists and biochemists and decided to help unite the two camps by promoting research in the intermediate area of submicroscopic morphology. To this end he reintroduced research with the polarizing microscope, using it to study, indirectly, the structure of macromolecules, which are beyond the resolving power of the light microscope. Many of his results were corroborated by the work of x-ray crystallographers, such as William ASTBURY, and later verified directly when the electron microscope was introduced in 1940. His work on "macromolecular chemistry" helped lay the foundations of ultrastructural research and molecular biology.

Frey-Wyssling published a number of books, including (with K. Mühlethaler) *Ultrastructural Plant Cytology* (1959) and *Plant Cell Walls* (1976).

Friedel, Charles (1832–1899) *French chemist*

After studying in his native city of Strasbourg with Louis PASTEUR and at the Sorbonne with Charles Adolphe WURTZ, Friedel (free-**del**) became curator of the mineral collections at the mining school in Paris (1856). Although his most important work was in organic chemistry, he also did much work on the synthesis of minerals. In 1862 he discovered secondary propyl alcohol, thus verifying Hermann KOLBE's prediction of its existence. His most notable work was that carried out with James CRAFTS on the alkylation and acylation of aromatic hydrocarbons – the *Friedel-Crafts reaction* (1877). This was a method of synthesizing hydrocarbons or ketones from aromatic hydrocarbons using aluminum chloride as a catalyst. Friedel and Crafts also did much work on the synthesis of organosilicon compounds. Friedel was professor of mineralogy at the mining school in Paris (1876–84) and professor of organic chemistry at the Sorbonne (1884–99).

Friedländer, Paul (1857–1923) *German organic chemist*

Friedländer (**freed**-lent-er) was assistant to Adolf von BAEYER at Munich during the latter's great researches on indigo, and much of Friedländer's work was concerned with indigo derivatives. He synthesized many derivatives and showed that the ancient dyestuff Tyrian purple is dibromoindigo (1909). He also developed a useful quinoline synthesis (1882). He was the holder of 235 patents concerned with dyestuffs.

Friedman, Herbert (1916–2000) *American physicist and astronomer*

Born in New York City, Friedman graduated from Brooklyn College in 1936 and obtained his PhD in 1940 from Johns Hopkins University, Baltimore. Shortly afterwards he joined the U.S. Naval Research Laboratory in Washington, where he spent his whole career, being appointed in 1958 as superintendent of the atmosphere and astrophysics division, and in 1963 superintendent of the space science division. Also in 1963 he became chief scientist at the E. O. Hulbert Center for Space Research. In addition he served as adjunct professor at the universities of Maryland and Pennsylvania.

Friedman was a pioneer in both rocket astronomy and in the study of the x-ray sky. The two went hand in hand in the early days of x-ray astronomy for without rockets it would have been impossible to detect any significant x-ray activity in space since x-rays are absorbed by the Earth's atmosphere. Solar x-rays were detected as early as 1948 by T. R. Burnright and were systematically investigated from 1949 by Friedman and his colleagues, who observed x-ray activity throughout a full solar cycle of 11 years. Friedman also studied ultraviolet radiation from the Sun, and in 1960 produced the first x-ray and ultraviolet photographs of the Sun.

X-ray astronomy really came of age in 1962 when nonsolar x-ray activity was first discovered by Bruno ROSSI. The x-rays came from a source in Scorpio, since named Sco X–1. A second source was discovered in 1963 in Taurus and named Tau X–1. Friedman made the first attempt to locate accurately an x-ray source two years later: when the Moon passed in front of the Crab nebula, a luminous supernova remnant in Taurus, the x-ray activity of Tau X–1 was found to fade out gradually. Tau X–1 was therefore identified with the Crab nebula and seemed to be a source about a light-year across lying in the center of the nebula.

Since then satellites carrying x-ray equipment have been launched, including Uhuru in 1970 and the Einstein Observatory in 1978. These have enormously extended the scope of x-ray astronomy and shown its value in the search for neutron stars and black holes.

Friedman, Jerome Isaac (1930–)
American physicist

Chicago-born Friedman was educated at the university in his native city and gained his PhD there in 1956. After spending three years in California at Stanford, Friedman moved to the Massachusetts Institute of Technology in 1961, and was later appointed to a chair of physics in 1967. He is currently Emeritus Institute Professor of Physics at MIT.

Working with his MIT colleague Henry KENDALL and with Richard TAYLOR from Stanford, Friedman began to study the internal structure of the proton. They worked with the 3-kilometer linear accelerator recently opened at Stanford (SLAC). Electrons were accelerated to an energy of 20,000 million electronvolts and directed against a target of liquid hydrogen. In a manner reminiscent of the 1911 experiments of Ernest RUTHERFORD, they analyzed the angles and energies of the electrons and protons of the hydrogen nuclei as they scattered after collision. Similar experiments had been performed by Robert HOFSTADTER in the 1950s and he had found protons not to be mere points, but fuzzy blobs spread out over an area of about 10^{-15} meter. In 1967, however, higher energies were available to Friedman and his colleagues, which led them to hope that they might see into the proton with a little more precision.

In cases of elastic scattering, where beam and target particles retain their identity, the deflections were minor and occurred as expected. When, however, the scattering was inelastic and the protons were struck with sufficient energy to produce new particles, such as pions, the electrons were deflected through much wider angles than expected.

These latter scattering results proved difficult to explain. A possible answer was proposed by Richard Feynmann in 1968 on a visit to SLAC. Protons, he suggested, could be composed of a number of pointlike particles, which he called "partons." From such charged points, electrons could be scattered through large angles. Further, it followed from the angular distribution of the scattered electrons that the partons must have a spin of one half.

As these were the properties calculated for the hypothetical quarks proposed by Murray GELL-MANN, the SLAC experiment was soon taken to be the first experimental evidence for the existence of quarks. It was for this work that Friedman shared the 1990 Nobel Prize for physics with his collaborators Kendall and Taylor.

Friedmann, Aleksandr Alexandrovich (1888–1925) *Russian astronomer*

Born the son of a composer in St. Petersburg, Russia, Friedmann (**freed**-man) was educated at the university there. He began his scientific career in 1913 at the Pavlovsk Observatory in St. Petersburg and, after war service, was appointed professor of theoretical mechanics at Perm University in 1918. In 1920 he returned to the St. Petersburg Observatory where he became director shortly before his early death from typhoid at the age of 37.

Friedmann established an early reputation for his work on atmospheric and meteorological physics. He is, however, better known for his 1922 paper on the expanding universe. This arose from work of EINSTEIN in 1917 in which he attempted to apply his equations of general relativity to cosmology. Friedmann developed a theoretical model of the universe using Einstein's theory, in which the average mass density is constant and space has a constant curvature. Different cosmological models are possible depending on whether the curvature is zero, negative, or positive. Such models are called *Friedmann universes*.

Friedmann, Herbert (1900–1987) *American biologist and ornithologist*

Born in New York City, Friedmann gained a PhD in ornithology from Cornell University in 1923; he then held a fellowship at Harvard for three years during which he spent much time in Argentina and Africa. After a series of academic posts he became, in 1929, curator of birds at the Smithsonian Institution, Washington, and was its head curator of zoology from 1957 to 1961.

Friedmann was especially interested in reproductive parasitism, particularly of birds, studying for example the African cuckoos and parasitic weaverbirds. He also investigated the parasitic honey guides, one species of which was observed to guide people to wild bees' nests. It had been thought that such birds fed on the honey of bee larvae and eggs, but Friedmann demonstrated the birds were actually after the beeswax. This discovery was particularly interesting as it had been assumed wax was indigestible. Friedmann went on to find that a wax-breaking bacterium, *Micrococcus cerolyticus*, was present in the birds' intestine.

Friedmann also investigated the systematics of birds of Africa, North America, and South America. His other work included studies of animal and plant symbolism in European medieval and Renaissance art.

Fries, Elias Magnus (1794–1878) *Swedish mycologist*

Fries (frees), who was born in Femsjö, Sweden, grew up in a region particularly abundant in fungi; this – together with botanical instruction he received from his father – resulted in his lifelong interest in mycology (the study of fungi). Fries entered Lund University in 1811, gradu-

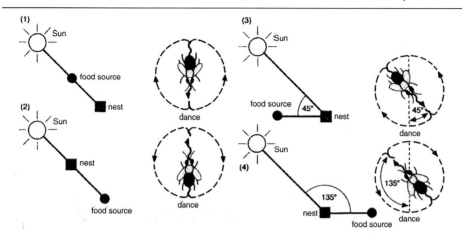

THE TAIL-WAGGING DANCE OF THE BEE   The relationship between the angle of the straight run of the tail-wagging dance on the vertical surface of the comb and the angle between Sun, nest, and food source.

ating in 1814 with a dissertation on the flora of Sweden. He continued at Lund until 1835, producing the three-volume *Systema Mycologicum* (1921–32; A System of Mycology), which classifies the fungi by their developmental stages and morphological relationships. This work remains the foundation of mycological taxonomy today for all but three groups of fungi.

Fries also produced an important work on the European lichens and issued floras spanning all of Scandinavia. In 1835 he took up the chair of botany at Uppsala University, becoming a member of the Linnean Society in the same year, and in 1875 he was elected a fellow of the Royal Society.

Frisch, Karl von (1886–1982) *Austrian zoologist, entomologist, and ethologist*

Frisch (frish) was born in Vienna, Austria, and educated at home and at convent school. He began his academic career by studying medicine at the University of Vienna, but gave this up to study zoology at the Zoological Institute in Munich, an internationally recognized center for research in experimental zoology. He also studied marine biology at the Trieste Biological Institute for Marine Research, taking his PhD for work on the color adaptation and light perception of minnows (1910). Frisch taught at the Munich Zoological Institute and in 1919 became assistant professor of zoology. He then held academic posts at the universities of Rostock and Breslau before returning to Munich in 1925 as director of the Zoological Institute where he continued to work until the end of World War II. In 1946 he assumed the chair of zoology at Graz but returned to Munich where he remained until his retirement in 1958.

Interested in animals from childhood, Frisch devoted 40 years to an intense study of the senses, communication, and social organization of honey (or hive) bees. By means of ingenious experiments he showed that bees can find their way back to the hive, even when the sun is obscured by cloud, by using polarized or ultraviolet light, and that they are able to communicate discovery of a new food source by means of a special "dance." The bee performs its dance on the vertical surface of the comb. Depending on the distance of the food supply, the bee may perform either the round dance – food within about 80 feet (24 m) – or the tail-wagging dance – food beyond about 325 feet (99 m). At intermediate distances various transitional dance forms between these are seen. In the tail-wagging dance the bee makes a straight run over a short distance, wagging its abdomen rapidly, and then makes a semicircle back to the starting point. This movement is repeated, making a semicircle in the opposite direction. The dance gives information on the direction of the food supply because the angle that the straight run makes with the vertical surface of the comb is the same as the angle between the direction of the food and the Sun at the hive or nest.

Frisch also showed that the bees are unable to distinguish between certain shapes, that they have a limited range of color perception, but can see light of longer wavelength than man. They do not, for example, distinguish red but can see ultraviolet, which is reflected by many flowers. Red poppies are seen as wholly ultraviolet, while many yellow flowers are seen either as yellow or in varying shades of ultraviolet. Frisch's discoveries have proved of practical benefit to beekeepers in that if hives are painted certain colors – a yellow hive next to a blue one for example – this aids the bees' homing. His major contribution to ethology was recognized in 1973 when he was awarded the Nobel Prize for physiology or medicine, jointly with Konrad LORENZ and Nicolaas TINBERGEN. His books include *The Dancing Bees* (1927, trans. 1954) and *Animal Architecture* (1974).

Frisch, Otto Robert (1904–1979) *Austrian–British physicist*

> A really good scientist is one who knows how to draw correct conclusions from incorrect assumptions.
> —*What Little I Remember* (1979)

Frisch, the son of a Viennese printer and publisher, was educated at the University of Vienna where he obtained his doctorate in 1926. He was employed in Berlin (1927–30) at the German national physical laboratory, the Physikalisch Technische Reichsanstalt, and moved to the University of Hamburg in 1930. However, with the introduction of Hitler's racial laws, he was fired in 1933 and consequently traveled via Copenhagen to England. After working at the universities of Birmingham and Liverpool (1939–43) he moved to America and spent the period 1943–45 at Los Alamos, working on the development of the atom bomb. With the end of the war Frisch worked briefly at the Atomic Energy Research Establishment at Harwell, England, leaving in 1947 to take up the Jackson Chair of Physics at Cambridge, a post he held until his retirement in 1972.

In 1939 Frisch, with his aunt, Lise MEITNER, was closely involved in the crucial discovery of nuclear fission. He spent Christmas in Sweden visiting Meitner, who reported to him some strange results obtained by her former colleague Otto HAHN. Hahn found that when uranium was bombarded with neutrons, one of its decay products was the much lighter element barium. Frisch said that his first reaction was that Hahn had made a mistake, but Meitner was more inclined to trust Hahn's qualities as a good chemist. After some thought and calculation they concluded that this must in fact be what was later called nuclear fission. Frisch rushed back to Copenhagen to inform Niels BOHR who was able to confirm Hahn's experiments. But in all this excitement the most important point had been missed – the mechanism of the neutron chain reaction. However, the thought did occur independently to many others.

Frisch did further work on fission while at Birmingham, collaborating with Rudolph PEIERLS in confirming Bohr's suggestion that a chain reaction would be more likely to result with uranium–235 rather than with the more common isotope, uranium–238. After much work Frisch came to the basic and frightening conclusion that an "explosive chain reaction" could be produced with a pound or two of uranium–235 rather than the tons of it which he first thought would be necessary. Frisch and Peierls were therefore probably the first two people in the world to be aware not just of the possibility of a nuclear bomb but of its practicality. They immediately wrote a report that was sent to Henry TIZARD, a scientific adviser to the British government, which Frisch claimed was decisive in getting the British Government to take the atomic bomb seriously.

In 1979 Frisch produced his fascinating and witty memoirs, *What Little I Remember.*

Fritsch, Felix Eugen (1879–1954) *British algologist*

Fritsch (frich), the son of a London headmaster, was educated at the University of London and at Munich, where he obtained his DPhil in 1899. He taught at University College, London, from 1902 until 1911, when he moved to Queen Mary College, London, where he served as professor of botany from 1924 until his retirement in 1948.

Before Fritsch there was no comprehensive work on algae. He remedied this with his classic work *The Structure and Reproduction of the Algae* (2 vols. 1935–45). He was also instrumental in the foundation of the Freshwater Biological Association in 1929.

Fritsch, Gustav Theodor (1838–1927) *German ethnographer, anatomist, and neurologist*

Fritsch, the son of a building inspector from Cottbus, near Berlin in Germany, was educated at the gymnasium in Breslau and the University of Berlin where he obtained his MD in 1862. From 1867 to 1900 Fritsch worked at the Institute of Anatomy in Berlin, also becoming head of the histological section at the Physiological Institute.

Fritsch was a much traveled ethnographer. From 1863 to 1866 he was in South Africa where he made a careful study of the anatomy and culture of the Bushmen. He also made research expeditions to Egypt, Syria, and Persia. He is however best known for his collaboration with Eduard HITZIG in 1870 in which they clearly demonstrated the phenomenon of cerebral localization.

Frontinus, Sextus Julius (*c*. 35 AD–*c*. 104 AD) *Roman engineer*

Frontinus (fron-**tIn**-us), praetor in 70 and consul in 74, was governor of Britain from 74 until 78, during which time he subdued the Silures. In 97 he was appointed *curator aquarum* and went on to write an account of the Roman water supply in his *De aquis urbis Romae* (On the Aquaducts of Rome), the main source of information on the water-supply systems of his time.

He devized a method for measuring water

flow based upon a standard pipe called a *quinaria*, made from bronze and about 2 centimeters in diameter and 22 centimeters long. Using figures derived from Frontinus it has been calculated that a Roman citizen used as much water daily as a modern European.

Fuchs, Klaus Emil Julius (1911–1988)
German physicist

Born the son of a Lutheran pastor in Russellheim, Germany, Fuchs (fuuks) was educated at the universities of Kiel and Leipzig. Originally a Social Democrat, Fuchs joined the Communist Party in 1932 as an expression of his strong opposition to the Nazis. Following the Reichstag fire in 1932, and the occasional beating up, Fuchs fled to Britain in 1933. He soon found a job as assistant to Nevill MOTT in Bristol where he began to work for his PhD. He remained there until 1937 when he moved to Edinburgh to a research post under Max BORN.

On the outbreak of war Fuchs was interred as an enemy alien, first on the Isle of Man, and soon after in Canada. The detention camp contained one future Nobel laureate, Max PERUTZ, two founders of the Steady State Theory, Thomas GOLD and Hermann BONDI, and, surprisingly, three rabbis. The authorities, under pressure from Born, released Fuchs in late 1941.

In 1940 Otto FRISCH and Rudolph PEIERLS had calculated that only a relatively small amount of uranium-235 would be necessary to produce an atomic bomb. They were asked by the government to continue their work and Peierls invited Fuchs to join him in Birmingham to work on the project. The British intelligence agency, MI5, although familiar with Fuchs's background, raised no objections to his appointment.

Following the German invasion of the Soviet Union in June 1941 Fuchs made his first contact with the Soviet Embassy and began to pass information to them later in the year. Soon after, in 1942, Fuchs became a British citizen. By this time the center of research had moved to America and, in 1943, Fuchs followed it, first to New York, and eventually to Los Alamos where he worked in the Theoretical Physics Division under Hans BETHE on the actual design and construction of the bomb. Once more he made contact with Soviet agents.

By this time Fuchs knew as much about the bomb as almost anyone else on the project. He learned that two bombs were being developed, namely, the original uranium bomb and a plutonium device. He also gathered details about complicated ignition devices needed to explode the bombs. Fuchs further heard of plans to build the "Super," a fusion (hydrogen) bomb. All information that came his way was passed to his connection. He was offered, but refused, $1,500 for his efforts.

Following the successful explosions of both atomic bombs in 1945, and the end of the war, Fuchs returned to Britain and in the following year was appointed Head of the Theoretical Division at the Atomic Energy Research Establishment, Harwell. It was at Harwell that Britain's own atomic weapons would be made; it was also from Harwell that Fuchs informed his Soviet contact of Britain's plans.

But new forces were at work in Fuchs's life. He was becoming increasingly discontented with his own duplicity. He seems to have felt a genuine personal loyalty to his Harwell colleagues, and he was looking for ways to withdraw from his espionage activities. However, by this time, the security forces had become aware that a high-level scientific spy was passing atom secrets to the Soviet Union. Suspicion fell on Fuchs and after some initial questioning by MI5 he confessed everything. He was arrested, tried, and in 1950 was sentenced to fourteen years' imprisonment and the loss of his British citizenship.

He served eight years and, on his release in 1959, Fuchs headed straight for East Germany. He was granted citizenship and allowed to join the Communist Party. Until his retirement in 1979, the much honored and bemedalled Fuchs served as Deputy Director of the Institute for Nuclear Research, Rossendorf. He died suddenly in 1988. Had he lived a little longer Fuchs would have seen the collapse of the communist system that he had served so faithfully and so long.

Fuchs, Leonhard (1501–1566) *German botanist and physician*

Fuchs was born in Wemding, Germany, and was professor of medicine at Tübingen University from 1535 until his death, his main interest being the medicinal properties of plants. He is remembered for his herbal manual *De historia stirpium* (1542; On the History of Roots), in which about four hundred native and a hundred foreign plants are arranged alphabetically, with original descriptions of the form, habitat, and best season for collection of each plant. The herbal is notable for its glossary of botanical terms, accurate woodcut illustrations, and detailed descriptions.

A genus of ornamental shrubs, *Fuchsia*, is named for him.

Fuchs, Sir Vivian Ernest (1908–1999)
British explorer and geologist

Fuchs, the son of a farmer of German origin, was born at Freshwater in England. He was educated at Cambridge University where he obtained his MA in geology in 1929 and his PhD (1935). From 1929 he traveled on a series of expeditions as geologist, the first being the Cambridge East Greenland Expedition (1929). In

1947 he became a member of the Falkland Islands Dependencies Survey, and from 1951 to 1960 its director. He also served as director of the British Antarctic Survey from 1958 to 1973.

As a result of his involvement with the survey, Fuchs was made leader of an expedition to Graham Land in Antarctica (1948) to reassert the British claim to the territory, which was being challenged by Argentina. He was stranded on Storington Island by bad weather until 1950.

Fuchs is best known for the Commonwealth Trans-Antarctic Expedition he jointly led with Edmund Hillary in the International Geophysical Year (1957–58). Leaving Shackleton Base on the Weddell Sea on 24 November 1957, he reached the South Pole and met Hillary on 19 January 1958, then continued to reach Scott Base, Victoria Land, on 2 March. From the scientific studies made during this it was established that a single continent exists below the polar ice cap.

Fuchs was knighted in 1958.

Fukui, Kenichi (1918–1998) *Japanese theoretical and physical chemist*

Born in Nara, Japan, Fukui (foo-**koo**-ee) was a lecturer in chemistry at Kyoto Imperial University, becoming professor of physical chemistry from 1951 to 1982. He is noted for his theoretical work on the change in molecular orbitals during reactions, especially reactions of methyl radicals. He also investigated the reaction of molecular nitrogen with transition-metal complexes – a topic potentially important for the fixation of nitrogen.

For his work on frontier orbital theory, Fukui shared the 1981 Nobel Prize for chemistry with Roald HOFFMANN.

Funk, Casimir (1884–1967) *Polish–American biochemist*

Funk (fuungk), born the son of a dermatologist in the Polish capital of Warsaw, obtained his doctorate from the University of Bern in 1904. He worked at the Pasteur Institute in Paris, the University of Berlin, and the Lister Institute in London, before emigrating to America in 1915. Although he became naturalized in 1920, Funk returned to Poland in 1923 as director of the Warsaw Institute of Hygiene, but finding the political conditions unattractive moved to Paris in 1927, where he acted as a consultant to a drug company and founded a private research institute, the Casa Biochemica. In 1939 Funk returned to America, where he served as consultant to the U.S. Vitamin Corporation and as president of the Funk Foundation for medical research.

It was while working at the Lister Institute in 1912 that Funk first clearly formulated his crucial idea that certain diseases are caused by food deficiencies. He was working on the antiberiberi factor, which he succeeded in extracting from rice husks. He went on to postulate that there were comparable ingredients whose absence from a regular diet would produce scurvy, rickets, and pellagra.

Noting that the antiberiberi factor contained an amine ($-NH_2$) group, Funk proposed to call such ingredients "vital amines," or "vitamines." When it became clear that the amine group was not present in all "vitamines" the term "vitamin" came to be preferred.

Furchgott, Robert F. (1916–   ) *American pharmacologist*

Furchgott was born in Charleston, South Carolina, and graduated with a BS in chemistry from the University of North Carolina in 1937. Three years later he gained his PhD in biochemistry at Northwestern University. From 1956 to 1988 he worked in the department of pharmacology at the State University of New York, and since 1988 has been Distinguished Professor, State University of New York Health Science Center. In 1998 he shared the Nobel Prize for physiology or medicine with Louis IGNARRO and Ferid MURAD for their discovery that molecules of the gas nitrogen monoxide (nitric oxide, NO) can transmit signals in the cardiovascular system.

Nitrogen monoxide, produced by one cell, acts by penetrating membranes and regulating the function of another cell. Nerves and hormones are well known as signal carriers, but this discovery was a totally new signaling principle in a biological system.

Furchgott began by researching the actions of drugs on blood vessels, but often observed contradictory results. Sometimes a drug acted to contract a vessel, but at other times it dilated the vessel. In 1980 he discovered that the neurotransmitter acetylcholine did not act to dilate blood vessels if the endothelium (surface cells) of the receiving vessel were damaged. He deduced that the cells of the endothelium produce another hitherto unknown signal substance that makes the smooth muscle cells of the vascular system relax. He called the substance endothelium-derived relaxing factor (EDRF). By 1986 he had shown that the substance was in fact the simple compound nitric oxide (NO). Furchgott was awarded the 1998 Nobel Prize for physiology or medicine for this discovery. The prize was shared with Louis Ignarro and Ferid Murad.

G

Gabor, Dennis (1900–1979) *Hungarian–British physicist*

> The most important and urgent problems of the technology of today are no longer the satisfactions of the primary needs or of archetypal wishes, but the reparation of the evils and damages wrought by the technology of yesterday.
> —*Innovations: Scientific, Technological and Social* (1970)

> Short of a compulsory humanistic indoctrination of all scientists and engineers, with a "Hippocratic oath" of never using their brains to kill people, I believe that the best makeshift solution at present is to give the alpha minuses alternative outlets for their dangerous brain-power, and this may well be provided by space-research.
> —*Inventing the Future* (1964)

Gabor (**gah**-bor or ga-**bor**), the son of a businessman, was born in the Hungarian capital of Budapest; he was educated at the technological university there and in Berlin, where he obtained a doctorate in engineering in 1927. He worked initially as a research engineer for Siemens and Halske from 1927 until 1933 when, with the rise of Hitler, he decided to leave Germany and took a post with the British Thomson–Houston Company, Rugby. In 1948 he joined the staff of Imperial College, London, later serving as professor of applied electron physics from 1958 until his retirement in 1967.

Gabor is credited with the invention of the technique of holography – a method of photographically recording and reproducing three-dimensional images. The modern technique uses lasers to form such images, but the invention came out of work by Gabor on improving the resolution of the electron microscope – work done in 1948, twelve years before the introduction of the laser.

The electron microscope has theoretically much higher resolution than the optical microscope because of the shorter wavelength of electrons (resolution is limited by diffraction effects). One method of improving the resolution of the electron microscope is to improve the electron lenses used to deflect and focus the beam of electrons. Gabor was interested in increasing the resolution to the point at which atoms in a lattice could be "seen." Rather than work on the electron optics of the system, he had the idea of extracting more information from the electron micrographs produced by existing instruments, and to do this he proposed forming a diffraction pattern between the incident beam of electrons and a background beam that was coherent with it (i.e., one with the same wavelength and phase). The principle was that the image produced would have information on the phase of the electrons as well as the intensity and that it would be possible to reconstruct a true image from the resulting electron micrograph.

Gabor began experiments with light to investigate the technique (which he named holography from the Greek *holos* meaning whole – the record contained the whole information about the specimen). He used a mercury lamp and pinhole to form the first, imperfect, holograms. Subsequently, in 1961, E. Leith and J. Upatnieks produced holograms using laser light. The technique is to illuminate a specimen with light from a laser and form an interference pattern between light reflected from the specimen and direct light from the source, the pattern being recorded on a photographic plate. If the photographic record is then illuminated with the laser light, a three-dimensional image of the specimen is generated.

Gabor received the Nobel Prize for physics for his work in 1971.

Gabriel, Siegmund (1851–1924) *German chemist*

Gabriel (gab-ri-**el**) studied under August von Hofmann in his native Berlin and under Robert BUNSEN in Heidelberg. He returned to Berlin in 1874, being appointed to the chair of chemistry there in 1886.

In 1880 Zdenko SKRAUP had succeeded in synthesizing the heterocyclic compound quinoline. Gabriel helped work out the chemistry of such structures by synthesizing isoquinoline in 1885. Isoquinoline is the parent substance of several opium alkaloids. In 1887, he also worked out methods for the synthesis of primary amines from haloalkanes and potassium phthalimide with subsequent hydrolysis.

Gadolin, Johan (1760–1852) *Finnish chemist*

Gadolin (**gad**-o-lin) was born the son of an astronomer and physicist in Åbo, now Turku in Finland. He studied under Torbern BERGMAN at Uppsala and taught at Åbo from 1785, becoming professor of chemistry from 1797 until 1822.

In 1794 Gadolin examined a black mineral from Ytterby, a quarry in Sweden. The rocks from this quarry were found to contain a dozen or so new elements. Gadolin isolated the first lanthanoid element from it in the form of its oxide and named it yttria. The element was named gadolinium after him in 1886 by Lecoq de Boisbaudran. Gadolin also worked on specific heat and published a set of standard tables.

Gaffky, Georg Theodor August (1850–1918) *German bacteriologist*

Gaffky (**gahf**-kee) was born in Hannover, Germany, and educated at the University of Berlin where, after service in the Franco–Prussian War, he obtained his MD in 1873. He served as Robert KOCH's assistant from 1880 to 1885 and, after holding the chair of hygiene at the University of Giessen (1888–1904), succeeded Koch in Berlin as director of the Institute for Infectious Diseases, renamed the Koch Institute in 1912.

In 1880 Carl Eberth had first described the microorganism responsible for typhoid fever. As it was easily confused with several other intestinal organisms, it was only when Gaffky succeeded in obtaining a pure culture in 1884 that Eberth's work could be confirmed.

Gaffky also served on a number of important field investigations. He was with Koch on his trips to Egypt and India in 1883–84, during which the vibrio responsible for cholera was discovered, and wrote the official report of the investigation. Gaffky made a further trip to Egypt in 1897 to work on the bubonic plague.

Gahn, Johan Gottlieb (1745–1818) *Swedish chemist and mineralogist*

> A number of entirely new minerals were discovered ... and analysed at the time in Gahn's excellently equipped laboratory.
> —Jöns Jacob Berzelius, *Autobiographical Notes* (1816)

Born at Voxna in Sweden, Gahn (gahn) graduated from the University of Uppsala in 1770 and then taught at the College of Mines. Working with Karl SCHEELE, he showed that phosphorus is present in bones (1770). He also first isolated the element manganese, suspected by Scheele to be present in pyrolusite – an oxide known to the ancients and used chiefly to color glass. Gahn's extraction involved igniting a paste of pyrolusite (manganese(IV) oxide) and charcoal. Gahn also contributed to the discovery of selenium by Jöns Berzelius in 1818.

He became superintendent of mines in 1782

and later became a mine owner himself. The mineral ghanite (zinc spinel) is named for him.

Gajdusek, Daniel Carleton (1923–) *American virologist*

Gajdusek (**gI**-duu-shek) was born at Yonkers in New York and educated at the University of Rochester and at Harvard, where he obtained his MD in 1946. He specialized in pediatrics, working at Harvard, the Pasteur Institute in Teheran, and the Walter and Eliza Hall Institute for Medical Research in Melbourne, before joining the National Institutes of Health in Bethesda, Maryland, in 1958. Since 1970 he has been with the National Institute of Neurological Diseases.

In 1963 he made an intriguing discovery that could well have profound consequences for the control of a number of serious but little-understood diseases. In the 1950s he began studying the Fore people of New Guinea, a supposedly cannibalistic tribe who suffered from a very localized neurological complaint they called "kuru." With the aid of the district medical officer, who first drew his attention to the disease, Gajdusek spent much of the next ten years among the Fore looking for the cause of kuru. He suspected the disease was transmitted by the Fore custom of ritually eating parts of the brain of their deceased relatives, so he collected samples from the brains of several kuru victims.

Failing to detect any obvious signs of an organism in the brain tissue, he injected filtered extracts into the brains of chimpanzees and waited. After about 12 months the disease at last appeared. This was the first of the so-called "slow virus infections" to be observed in humans. By 1968 Gajdusek and his colleagues had shown that kuru was not unique and that the rare neurological complaint Creutzfeldt-Jakob disease, a presenile dementia, is transmitted after a comparable delay. This immediately opened the still unresolved question as to whether such complaints as Parkinson's disease and multiple sclerosis are also due to slow viruses.

For his work on kuru Gajdusek shared the 1976 Nobel Prize for physiology or medicine with Baruch BLUMBERG. The prize itself he used to set up a trust for the education of the Fore people.

In his work in New Guinea Gajdusek had brought back more than 50 children to America, where he cared for them and paid for their education. In 1996 he was accused of molesting one teenager whom he had brought back some years before. Gajdusek pleaded guilty and spent a year in prison.

Galen (*c.* 130–*c.* 200) *Greek physician*

> That physician will hardly be thought very

careful of the health of others who neglects his own.
—*Of Protecting the Health*, Bk. V

Galen (**gay**-len) was educated as a doctor supposedly because his father had a dream in which Asklepios, the god of medicine, appeared to him. After an initial period of training in his native city of Pergamum (now Bergama in Turkey), Galen spent the years 148–57 traveling and studying at Corinth, Smyrna, and Alexandria. He then returned home and took the post of surgeon to the Pergamum gladiators. Eventually Galen left for Rome where he made a considerable reputation for himself serving the emperors Marcus Aurelius and Commodus.

In addition to his imperial duties Galen wrote extensively and more than 130 of his texts have survived. Even though some are undoubtedly spurious, there still remains an impressive opus on virtually every aspect of the medicine of his times.

It is this opus that acquired an unprecedented authority, which persisted, unopposed, for another 1,500 years. It was not until William HARVEY proposed his new theory of the circulation of the blood (1628) that anatomists were presented with a viable alternative to the traditional system of Galen.

Galen made few advances in pathology, believing as he did in HIPPOCRATES's humoral theory – that disease is caused by an imbalance of the four body humors; phlegm, black bile, yellow bile, and blood. It was as an anatomist that Galen's true originality lay. He stressed the importance of dissecting personally and frequently, and combined his anatomical studies with a number of neat and conclusive experiments. Thus he clearly demonstrated the falsity of ERASISTRATUS's view that the arteries carry air not blood by placing ligatures both above and below the point of incision and noting the absence of any air escaping before the discharge of blood. Also, by similar experiments he was able to show that urine passes from the kidneys to the bladder irreversibly down the ureters.

Galen however was less successful in describing the complete system and operation of the body. He thought that blood was made in the liver from the food brought by the portal vein from the stomach. This was then transported by the venous system to nourish all parts of the body. Some of the blood however passed along the vena cava to the right side of the heart where it passed to the left side by some supposed perforations in the dividing wall or septum. Belief in such perforations persisted well into the 16th century. In the left ventricle the blood was mixed with air brought from the lungs and distributed round the body via the arterial system carrying the vital spirit or innate heat. Some of this blood was carried by the arteries to the head where in the *rete mirabile* (a vascular network not actually found in man) it was mixed with animal spirit and distributed to the senses and muscles by the supposedly hollow nerves. It was this spirit or pneuma that produced consciousness.

Thus instead of a single basic circulation Galen has a tripartite system in which the liver, heart, and brain each inject into the body three different spirits – natural, vital, and animal, which travel through the body via the venous, arterial, and nervous channels respectively. Although such a scheme is now recognized as totally misguided it possessed sufficient plausibility and experimental support to persist into the 17th century.

Galileo (1564–1642) *Italian astronomer and physicist*

> In questions of science the authority of a thousand is not worth the humble reasoning of a single individual.
> —Quoted in *Smithsonian Report* (1874)

> I wish, my dear Kepler, that we could have a good laugh together at the extraordinary stupidity of the mob. What do you think of the foremost philosophers of this University? In spite of my oft-repeated efforts and invitations, they have refused, with the obstinacy of a glutted adder, to look at the planets or the moon or my glass [telescope].
> —*Works*

Galileo (gal-i-**lay**-oh or gal-i-**lee**-oh), whose full name was Galileo Galilei (gal-i-**lay**-ee), made major contributions to most branches of physics (especially mechanics), invented and so deployed the telescope to change our view of the nature of the universe completely, and became engaged in a highly dramatic confrontation with the Church.

Born in Pisa, Italy, Galileo was the son of a scholar and musician of some distinction. He entered the University of Pisa in 1581 to study medicine – a subject in which he showed little interest – and failed to complete the course, developing instead a passion for mathematics. He is thought to have made his first important observation in 1583, two years before he left the university. While in Pisa cathedral he noticed that the lamps swinging in the wind took the same time for their swing whatever its amplitude. He timed the swing against his pulse. In 1586 he invented a hydrostatic balance for the determination of relative densities. In 1589 he was appointed to the chair of mathematics at Pisa and later moved to the chair in Padua, in 1592.

It was while in Padua in 1610 that he designed and constructed a simple refracting telescope. He may not have been the first to do so, and there are many other claimants, but he was certainly the first to use the instrument constructively. His initial reaction was sheer amazement at the number of stars in the sky, "So numerous as to be almost beyond belief," he asserted. Merely looking at the Moon immediately revealed that it is not the smooth unchangeable object of Aristotelian theory. He also discovered sunspots. His most exciting moment came in January 1610, when he observed Jupiter for the first time telescopically. To his astonishment he found that the planet has four satellites. Contrary to received opinion as to what was possible, these were new bodies, unmentioned in ARISTOTLE, and certainly not circling the Earth. He published his observations immediately, in 1610, in *Sidereus nuncius* (Starry Messenger), dedicating them to his patron and former pupil Cosimo II of Tuscany, in whose honor he named the satellites *Sidera Medicea* (Medici Stars). Within weeks he had received an invitation to a well-paid research chair in Pisa and returned there at the end of 1610.

In a series of works he was also tackling the problem of motion. His mature views were expressed in his *Discorsi...a due nuove scienze ...* (1638; Discourse on Two New Sciences). In a series of brilliant experiments rolling balls down inclined planes (and not, as is commonly thought, by dropping weights from the leaning tower of Pisa) he showed that the speed with which bodies fall is independent of their weight and correctly formulated the law $s = 1/2at^2$ (where s is speed, a acceleration, and t time). He lacked the concept of inertia and could only accept circular motion as being natural since, for Galileo, a body without any force acting on it would move in perfect circular motion. This was one aspect of his medieval heritage from which he was unable to break away.

After his return to Pisa, Galileo was beginning to meet opposition. Some of this was merely personal; in his criticism of others he wrote with a wit and savagery that many found wounding and impossible to forgive. He also found himself in dispute with many over his numerous discoveries. In these he was never inclined to take a charitable view over the claims of others, and there were many waiting for a chance to humiliate him on his return from the safety of Padua (then part of the independent Venetian republic). Their chance was his open support for the Copernican system. That the Earth moves round the Sun was so contrary to Scripture for many churchmen that those who enthusiastically campaigned for such a view were seen as heretics if not atheists. Galileo made no secret of his views in his writings, talk, and lectures. He openly ridiculed the Aristotelian scholars and supporters of the Ptolemaic (geocentric) system, many of whom were, or were to become, high officers of the Church. To stop the squabbling, Pope Paul V, on the advice of Cardinal Bellarmine, placed COPERNICUS on the Index (a list of books banned by the Catholic Church) in 1616, summoned Galileo to Rome, and informed him that he could no longer support Copernicus publicly. At this point Galileo had no choice; continued support for Copernicus would be to call the authority of

the pope into question and to send him to the stake as it had sent Giordano BRUNO in 1600.

In 1623 an old friend of Galileo's, Cardinal Barberini, was elected pope as Urban VIII. Galileo wasted no time in dedicating his new book *Il saggiatore* (1623; The Assayer) to him. It was a work that was savagely critical of Aristotle's account of comets. Although Urban was not prepared to go back on the decision of 1616 publicly, Galileo seems to have thought that he would be safe unless he supported Copernicus specifically. One thing that should have made him more prudent was the death of his patron, Cosimo, and the succession of a powerless minor to the duchy of Tuscany.

Galileo thought he could avoid the problem by writing a dialogue between a (Ptolemaic-) Aristotelian and a Copernican without ostensibly committing himself to either side, and the Church gave him permission to do so. Thus he wrote the *Dialogue Concerning the Two Chief World Systems* (1632). However, the form of the dialogue fooled no one. The Aristotelian, Simplicius, is no match for the brilliance of the Copernican, Salviati. It was even suggested to the pope that the bumbling Simplicius was a portrait of Urban himself. Galileo was once more summoned to Rome, threatened with torture, and forced to renounce Copernicus in the most abject terms. There is a tradition that after his renunciation he whispered, "Eppur si muove" ("Yet it moves"), but this is unlikely. Galileo was truly frightened, and it must be remembered that he was nearly 70 years old and facing a powerful body, which might have tortured and burned him at the stake. In the end he was treated reasonably well. He was allowed to return to his villa at Arcetri in isolation. Later, after he became blind and after the death of his daughter, his disciples Vincenzo VIVIANI and Evangelista Toricelli were allowed to stay with him.

Gall, Franz Joseph (1758–1828) *German physician and anatomist*

> The author of the true anatomy of the brain.
> —Jean Pierre Marie Flourens, *On Phrenology and Genuine Studies of the Brain* (1863)

Born in Tiefenbronn, Germany, Gall (gahl) studied medicine at Strasbourg and then Vienna, where he obtained his MD in 1785 and set up practice. In 1805, seeking a less restrictive scientific community, he left Vienna, where his lectures had been forbidden by the chancellor and after a European tour settled in Paris in 1807.

Although Gall is largely remembered as the founder of phrenology, now a totally discredited system, his crucial place in the development of cerebral localization is little known. Gall began as a conventional anatomist and established that the white matter of the brain consists of nerve fibers. However, he was distracted from such work by a couple of chance observations.

He noted that the two acquaintances of his with the best memory both had prominent staring eyes. This impressed him to search for cases of distinct traits or skills being linked with a distinctive physiognomy. This path led to him linking specific parts of the brain with specific behavior, a theory that could have been a major advance in anatomy, if properly handled. Unfortunately Gall went too far, recognizing 27 faculties. His collaborator Johann Spurzheim, who worked closely with him for several years and actually coined the term phrenology, added a further eight faculties. The evidence for these was largely anecdotal and uncritically accepted. Thus paintings of the great depicting the appropriate bump would confirm the theory; if, however, the bump was missing this would merely show the painter was working according to standards of beauty not truth.

Operations to test the theory were later performed by Pierre Flourens and tended not to support Gall. It was not in fact until 1861 that Paul Broca obtained any hard surgical evidence for the localization of cerebral faculties while the full proof of the hypothesis had to wait for the work of Gustav Fritsch and Eduard Hitzig in 1870.

Galle, Johann Gottfried (1812–1910)
German astronomer

> Returning with the chart to the telescope I discovered a star of the eighth magnitude – not at first glance, to tell the truth, but after several comparisons. Its absence from the chart was so obvious that we had to try to observe it.
> —On the discovery of Neptune. *Astronomical News* (1877)

Born at Pabsthaus, in Germany, Galle (**gahl**-e) was chief assistant to Johann Encke at the Berlin Observatory at a crucial moment in the history of planetary astronomy. Urbain Le Verrier had worked out what he considered to be the position of an as yet undiscovered planet. Having had some contact with Galle, Le Verrier wrote to him on 18 September 1846, asking him to try to check his prediction. Galle started observing on 23 September. He was favored by having an unpublished copy of a new star chart covering the right part of the sky and, aided by Louis D'Arrest, he found a star that was not on the chart. A wait of 24 hours showed that it had moved against the background of the fixed stars and so was a planet – it was the planet Neptune.

Galle also made an important contribution to determine the mean distance of the Sun from the Earth (the astronomical unit or AU). Conventional means of determining the AU had leaned heavily on the two transits of Venus in each century. In practice it turned out to be dif-

ficult to measure accurately the moment of first contact. Galle proposed instead, in 1872, that measuring the parallax of the planetoids would give a more reliable figure. (Harold Spencer Jones followed this procedure in 1931 when the planetoid Eros came within 16 million miles of the Earth. He was able to calculate the AU to within 10,000 miles.) In 1851 Galle became director of the Breslau Observatory. He lived long enough to receive the congratulations of the astronomical world on the 50th anniversary of the discovery of Neptune in 1896.

Gallo, Robert Charles (1937–) *American physician*

The son of a metallurgist, Gallo was born at Waterbury in Connecticut and educated at Providence College, Rhode Island, and Jefferson Medical College, Philadelphia, where he received his MD in 1963. He served his internship at the University of Chicago. In 1965, Gallo joined the staff of the National Cancer Institute at the National Institutes of Health, Bethesda, Maryland, and since 1972 he has been head of the Institute's Tumor Cell Biology Laboratory.

Gallo is noted as one of the people who first identified the virus responsible for AIDS. The discovery came out of work in his laboratory on leukemia. The first success by Gallo's team was the identification in 1976 of interleukin-2, a factor that stimulates growth in the lymphocytes known as T-cells. This was followed in 1979 by the crucial discovery of the first human retrovirus, HTLV-1 (human T-cell lymphotropic/leukemia virus). Retroviruses were first described in 1970 by Temin and Baltimore and, unlike other viruses, their genetic material is encoded in DNA rather than RNA. HTLV-1 and another virus discovered by Gallo in 1982, HTLV-2, both cause rare forms of leukemia.

In the early 1980s concern was growing about the emergence and spread of AIDS, a disease characterized by suppression of the patient's immune system. Gallo was aware that HTL viruses acted by attacking the immune system of leukemia patients. He made the bold conjecture that AIDS was caused by yet another retrovirus recently discovered in his laboratory, namely, HTLV-3. Similar conclusions were reached by Luc Montagnier at the Pasteur Institute in Paris, who had succeeded in isolating a retrovirus from an AIDS patient. He named the virus LAV (lymphadenopathy associated virus) and sent Gallo a sample in September 1983.

Gallo began work on finding a test for the AIDS virus. At first the virus proved impossible to grow in sufficient quantities, until Mika Popovich in Gallo's laboratory found a particular strain of T-cell (HUT-78 H9) in which the virus replicated without killing the cell. It was consequently a relatively simple matter, once large amounts of virus were available, to test for antibodies in AIDS sufferers' blood.

In April 1984, before Gallo had published his results, the U.S. Department of Health announced that he had found the cause of AIDS, HTLV-3, and took out patents on Gallo's blood test for AIDS antibodies. Gallo published his results in May 1984. He pointed out that he had identified the virus in 48 out of 167 cases from a risk group and that no evidence of its presence was found in the blood of "115 healthy heterosexuals."

The award in May 1985 of an exclusive patent for Gallo's test provoked a strong response from the Pasteur Institute. They argued that Gallo's blood test was based upon a virus substantially identical to the LAV strain first isolated by Montagnier. The Institute sued the U.S. Government, and heads of state became involved in the dispute. The issue was finally resolved in 1986 at a meeting in Frankfurt between Gallo and Montagnier. An agreed chronology about the discovery was established and it was also decided that 80% of royalties from the test should go to a new AIDS research foundation. The names of both Gallo and Montagnier would appear on the patent. The issue of the name of the AIDS virus was resolved in 1986 when the International Committee on the Taxonomy of Viruses diplomatically ignored both LAV and HTLV-3 and proposed the name HIV (human immunodeficiency virus).

Gallo's work made him the most cited scientist of the decade with the 418 papers he published between 1981 and 1990 gaining 36,789 citations. He has published a full and popular account of his work in *Virus Hunting: AIDS, Cancer and the Human Retrovirus* (1991).

Galois, Evariste (1811–1832) *French mathematician*

> A mind that had the power to perceive at once the totality of mathematical truths – not just those known to us, but all the truths possible – would be able to deduce them regularly and, as it were, mechanically ... but it does not happen like that.
> —*Writings and Mathematical Memoirs of Evariste Galois*

> Science progresses by a series of combinations in which chance does not play the smallest role; its life is unreasoning and planless and resembles that of minerals that grow by juxtaposition.
> —As above

Galois (ga-**lwah**) was born at Bourg-la-Reine, near Paris, France, during the rule of Napoleon. He entered the Collège Royale de Louis-le-Grand in Paris in 1823 and it was here that his precocious mathematical genius first emerged. He published several papers while still a student and at the age of about 16 embarked upon his noted work on algebraic equations. But his career was marred by lack of advancement, as-

sociated with political bitterness. Twice, in 1827 and 1829, he was rejected by the Paris Ecole Polytechnique, and three papers submitted to the Academy of Sciences were rejected or lost. In 1830 he entered the Ecole Normale Supérieure to train as a teacher. That year revolution in Paris caused the abdication of Charles X, who was succeeded by Louis Philippe. Galois – fiercely republican – was expelled for writing an antiroyalist newspaper letter. In 1831 he was arrested twice: once for a speech against the king and the second time for wearing an illegal uniform and carrying arms – for this he received six-months' imprisonment. In the spring of 1832 he died in a duel; the details are uncertain but it may have been provoked by political opponents.

Galois seems to have anticipated that he was to die, for the night before was spent desperately recording his mathematical ideas in a letter to his former schoolmaster, Auguste Chevalier. Here he outlined his work on elliptic integrals and set out a theory of the roots (solutions) of equations, in which he considered the properties of permutations of the roots. Admissible permutations – ones in which the roots obey the same relations after permutation – form what is now known as a *Galois group*, having properties that throw light on the solvability of the equations. The manuscripts were published in 1846 and his work recognized. With the equally tragic Norwegian, Niels Henrik ABEL, he is regarded as the founder of modern group theory.

Galton, Sir Francis (1822–1911) *British anthropologist and explorer*

> I always think of you in the same way as converts from barbarism think of the teacher who first relieved them from the intolerable burden of their superstitions.... Consequently the appearance of your *Origin of Species* formed a real crisis in my life; your book drove away the constraint of my old superstition as if it had been a nightmare, and was the first to give me freedom of thought.
> —Letter to Charles Darwin, 24 December 1869

Even though Galton's exceptional intelligence was apparent at an early age, his higher education was unremarkable. Born in the English Midlands city of Birmingham, he studied mathematics at Cambridge University and studied medicine in London but abandoned his studies on inheriting his father's fortune, which enabled him to indulge his passion for travel. Following consultations with the Royal Geographical Society, Galton set out to cover various uncharted regions of Africa, and became known as an intrepid explorer. He collected much valuable information and was elected first a fellow of the Royal Geographical Society and three years later, in 1856, a fellow of the Royal Society.

Galton made important contributions to the

science of meteorology, identifying and naming anticyclones and developing the present techniques of weather mapping. This work was published as *Meteorographica* (1863; Weather Mapping). He was also instrumental in establishing the Meteorological Office and the National Physical Laboratory, but he is remembered chiefly for his researches on human heredity, which were stimulated by the publication of *The Origin of Species* by his cousin, Charles Darwin. This led Galton to speculate that the human race could be improved by controlled breeding and he later gave the name *eugenics* to the study of means by which this might be achieved.

Galton studied the histories of notable families to determine whether intelligence is inherited, and concluded that it is. This aroused much controversy among those who believed environment is all important. Galton was the first to use identical twins to try to assess environmental influences. His work was characterized by its quantitative approach and he was also the first to stress the importance to biology of statistical analysis, introducing regression and correlation into statistics.

At a time when most scientists believed in blending inheritance, Galton deviated from contemporary thought and, in a letter to Darwin, outlined a theory of particulate inheritance, which anticipated Gregor MENDEL's work, then still undiscovered. Galton also discussed a concept similar to the phenotypes and genotypes of Wilhelm JOHANNSEN, under the terms patent and latent characteristics.

Galton was knighted in 1909. In his will he left a large sum of money to endow a chair of eugenics at University College, London, which was first held by Karl PEARSON, an energetic advocate of Galton's ideas on eugenics.

Galvani, Luigi (1737–1798) *Italian anatomist and physiologist*

> It is easy in experimenting to deceive ourselves, and to imagine we see things we hope to see.
> —*De viribus electricitatis in motu musculari commentarius* (1791; Commentary on the Effect of Electricity on Muscular Motion)

Galvani (gal-**vah**-nee) studied medicine at the university in his native city of Bologna, gaining his MD in 1762 for his thesis on the structure and development of bones. He stayed at Bologna to teach anatomy and in the late 1770s began his experiments in electrophysiology. He observed that the muscles of a dissected frog twitched when touched by a spark from an electric machine or condenser, such as a Leyden jar. Similar responses could be obtained when such muscles were laid out on metal during a thunderstorm, or even by simple contact with two different metals, without the deliberate application of an electric current. Galvani concluded that the source of the electricity therefore lay in the living tissue, and did not derive from outside. His finding was later disproved by Alessandro VOLTA. However, Galvani is celebrated for his discovery of Galvanic electricity (the metallic arc), as well as for applications of his principle to the galvanization of iron and steel and the invention of the galvanometer, named in Galvani's honor by André AMPÈRE. Galvani's animal electricity theory was published as *De viribus electricitatis in motu musculari commentarius* (1791; Commentary on the Effect of Electricity on Muscular Motion).

Gamble, Josias Christopher (1776–1848) *British industrial chemist*

Gamble came from a Presbyterian background in Enniskillen, now in Northern Ireland, and was educated at Glasgow University, graduating in 1797. He first became a minister, going to Belfast in that capacity in 1804, but soon entered the bleaching industry. He had picked up enough knowledge of Charles Tennant's St. Rollox process to open a factory in Dublin to manufacture chlorine bleach powder. He also produced sulfuric acid, potash, and alum. In 1829 he returned to England and partnered James MUSPRATT to open the first chemical plant in St. Helens, a LEBLANC soda plant, which became a highly profitable business.

Gamow, George (1904–1968) *Ukrainian–American physicist*

> Perhaps the last example of amateurism in scientific work on a grand scale.
> —Stanislaw Ulam on Gamow's work

Gamow (**gam**-of) was born the son of a teacher at Odessa, now in Ukraine. He was educated at the University of Leningrad where he obtained his doctorate in 1928 and later served as professor of physics (1931–34). Before his move to America in 1934 he spent long periods at Göttingen, Copenhagen, and Cambridge, England, the major centers of the revolution then taking place in physics. In America he spent his career as professor of physics at George Washington University (1934–55) and then at the University of Colorado (1956–68).

Gamow made many contributions to nuclear and atomic physics, but is mainly noted for his work on interesting problems in cosmology and molecular biology.

In cosmology he revised and extended the big-bang theory of the creation of the universe (first formulated by Georges Lemaître). This postulates that the universe expanded from a single point in space and time. It was first announced in Gamow's famous "alpha beta gamma" paper in 1948, which he wrote in collaboration with Ralph ALPHER and Hans BETHE. A fuller account was later published by Gamow in his *Creation of the Universe* (1952). Gamow dated the expansion to about 17 billion years ago, probably the result of an earlier contraction. The difficulty with any such theory was in accounting for the formation of the chemical elements. He supposed the primeval atom to consist of "Ylem," an old word used by Gamow to refer to a mixture of protons, electrons, and neutrons. Using the conditions of temperature and density prevailing in the first half hour of the universe's history, he tried to work out ways in which the elements could be formed by nuclear aggregation. There was no difficulty in showing that ^1H, ^2H, ^3H, ^3He, and ^4He would be formed but at that point he could see no way to advance the chain further, for there is no stable element with an atomic weight of 5. Add either a proton or a neutron to the nucleus of ^4He and either ^5Li or ^5He will be formed, both of which are unstable and decay in less than 10^{-20} second back to the original ^4He.

The only solution was to suppose that more than one particle collided with the ^4He nucleus simultaneously but, as Gamow realized, the universe by this time would be insufficiently dense and hot enough to permit such collisions to occur with the required frequency. He was therefore forced to conclude in 1956 that most of the heavy elements have been formed later in the hot interior of stars. One prediction that did emerge from his work and was to have important consequences for cosmology was his claim that the original explosion would produce a uniform radiation background; the discovery of such radiation in 1964 by Arno PENZIAS and Robert WILSON did more than anything else to stimulate interest once more in Gamow's theory.

Gamow later moved from showing how the universe began to the no less interesting question of how life began. He was quick to see the significance of the DNA model proposed by James WATSON and Francis CRICK in 1953. The problem was to show how the sequences of the four nucleic acid bases that constitute the DNA chain could control the construction of proteins, which may be made from 20 or more amino acids. Gamow had the insight to see that the bases must contain a code for the construction of amino acids. But the question of how this worked still remained. It could not be one base to one amino acid for then there would be only four amino acids. Nor would two bases be sufficient for they could produce only $4 \times 4 = 16$ amino acids. It would therefore need a sequence of three bases to produce one amino acid, a language with a capacity of $4 \times 4 \times 4 = 64$ words, which was more than adequate for the construction of all proteins. Gamow also produced convincing arguments to show that the code is not overlapping.

The work on DNA allowed Gamow to indulge his passion for science fantasy. He founded the RNA tie club for which he actually designed a tie. It was restricted to 20 members, one for each amino acid. Each member took the name of one of the acids – Gamow was "phe" (the usual abbreviation for phenylalanine) while Crick was "tyr" (tyrosine). Meetings were held, information was exchanged, and considerable progress was made.

Gamow was also known as one of the most successful popular science writers of his day. He wrote many books, most of which are still in print, which convey much of the excitement of the revolution in physics that he lived through.

Garrod, Sir Archibald Edward (1857–1936) *British physician*

Garrod's father was also a physician, a specialist in joint diseases who in 1848 demonstrated the presence of uric acid in the joints of his patients with gout. His son was born in London, graduated in natural science from Oxford University in 1880, and qualified in medicine from St. Bartholomew's Hospital, London. Garrod remained at St. Bartholomew's until 1920 when he returned to Oxford to succeed William OSLER as professor of medicine, a post he retained until his retirement in 1927.

In a lecture he gave to the Royal College of Physicians in 1908, *Inborn Errors of Metabolism*, Garrod introduced to the medical world an entirely new and unsuspected class of disease. An early interest in the pathological aspects of urinary pigments led Garrod to investigate alkaptonuria, a complaint in which the urine turns black on exposure to air. This was known to be due to the presence of large amounts of alkapton or homogentisic acid, a breakdown product of the amino acids tyrosine and phenylalanine. It was at first thought that some intestinal microbe was responsible for the disorder but Garrod's initial researches quickly disposed of this supposition when he found a tendency for siblings to be affected while their parents remained quite normal. He therefore argued that the disease was genetically determined and inherited as a recessive Mendelian trait.

He went on to propose a mechanism whereby in a normal individual the homogentisic acid derived from the breakdown of amino acids was itself further reduced to its harmless ingredients. In some individuals, however, there was a "metabolic block." They lacked the specific enzyme that breaks down homogentisic acid into harmless carbon dioxide and water, so allowing the acid to accumulate and overflow into the urine.

Nor was alkaptonuria the only such metabolic disorder; others identified by Garrod included cystinuria, pentosuria and, the cause of George III's madness, porphyria.

Garrod's daughter Dorothy, a distinguished archeologist, was appointed to the Disney Professorship at Cambridge University in 1939, thus becoming the first woman to hold a Cambridge chair. Of his three sons, two died in action in World War I and the third died of pneumonia.

Gaskell, Walter Holbrook (1847–1914)
British physiologist

Gaskell, a lawyer's son, was born at Naples in Italy and graduated in mathematics from Cambridge University in 1869. He then studied medicine at University College Hospital, London, but returned to Cambridge to serve as lecturer in physiology from 1883 until his death. His first studies investigated whether the heartbeat is under external nervous control or is an inherent property of the cardiac musculature (myogenic). Skillful work with tortoises and crocodiles showed that the heart's rhythm is indeed myogenic.

Gaskell also greatly increased knowledge of the structure of the autonomic (involuntary) nervous system. In 1886 he noted three major "outflows" of nerves from the spinal cord and lower part of the brain: the cervico-cranial, thoracic, and sacral. On leaving the central nervous system each nerve passes through a ganglion, or relay station, sited alongside the spine. Gaskell discovered two key properties of the system. He first noted that the nerves of all three groups are enclosed in a white sheath of myelin before entering their adjacent ganglion; on leaving the ganglion however the nerves of the thoracic outflow have lost their sheath in contrast to the still myelinated nerves of the other two outflows. He had thus succeeded in finding a simple anatomical distinction between the myelinated nerves, sacral and cervico-cranial, of the parasympathetic system, and the unmyelinated nerves, thoracic, of the sympathetic nervous system.

He also noted that most parts of the body receive nerves of both types and that their actions seem to be antagonistic. That is, while the myelinated nerves of the parasympathetic system inhibited the action of involuntary muscle, those of the sympathetic system seemed to increase its activity. Although much of Gaskell's work was done on reptiles he realized it had wider implications and boldly predicted that it would apply also to mammals, a prediction soon confirmed. His work was published posthumously in *The Involuntary Nervous System* (1916).

Much of Gaskell's later life was spent studying mammalian evolution. He tried to show how mammals could have evolved from arthropods rather than echinoderms (the orthodox view). His ideas were published in *The Origin of the Vertebrates* (1908), which contains the results of 20 years' work, but his theories have been largely ignored.

Gassendi, Pierre (1592–1655) *French physicist and philosopher*

> In his Epicurean philosophy, M. Gassendi strongly refutes everything that stands in opposition to Christianity and in so doing, as you rightly remark, he is taking precautions.
> —Marin Mersenne, letter to Rivet, 8 February 1642

Gassendi (ga-sahn-**dee**) was born at Champtercier in France. After being educated in Aix and Paris, he gained a doctorate in theology from Avignon in 1616, was ordained in 1617, and in the same year was appointed to the chair of philosophy at Aix. In 1624 Gassendi moved to Digne where he served as provost of the cathedral until 1645 when he was elected to the professorship of mathematics at the Collège Royale in Paris, resigning because of illness in 1648.

As a practicing astronomer Gassendi made a large number of observations of comets, eclipses, and such celestial phenomena as the aurora borealis – a term he introduced himself. His most significant observation was of the 1631 transit of Mercury, the first transit to be observed, which he recorded in his *Mercurius in sole visus* (1632; Mercury in the Face of the Sun) as support for the new astronomy of Johannes KEPLER.

In physics Gassendi attempted to measure the speed of sound and obtained the (too high) figure of 1,473 feet per second. He also, in 1640, performed the much contemplated experiment of releasing a ball from the mast of a moving ship; as he expected, it fell to the foot of the mast in a straight line.

Gassendi's importance to science rests with his role as a propagandist and philosopher rather than as an experimentalist. Even though the Paris parliament declared in 1624 that on penalty of death "no person should either hold or teach any doctrine opposed to ARISTOTLE," Gassendi published in the same year his *Excertitationes ... adversus Aristoteleos* (Dissertations ... against Aristotle), the first of his many works attacking both medieval Scholasticism and Aristotelianism. Nor did Gassendi find much attraction in the then emerging system of René DESCARTES. Instead he sought in his influential *Animadversiones in decimum librum Diogenes Laertii* (1649; Observations on the Tenth Book of Diogenes Laertius) to revive the classical atomism of EPICURUS, suitably modified to ensure its compatibility with 17th-century Christianity. Unlike Epicurus he insisted that the atoms were created by God who also bestowed on man an immaterial soul; against Descartes he admitted the existence of the void within which his atoms could interact.

Gassendi's works were well known in England and exercised considerable influence on such leading scientists as Robert BOYLE.

Gasser, Herbert Spencer (1888–1963)
American physiologist

Gasser, the son of a country doctor from Platteville, Wisconsin, was educated at the University of Wisconsin and at Johns Hopkins University. Having qualified as a physician in 1915, he moved to Washington University, St. Louis, to take up an appointment as professor of pharmacology. Here he joined his old teacher, Joseph ERLANGER, in a famous collaboration that resulted in their sharing the 1944 Nobel Prize for physiology or medicine for work on the differentiated function of nerve fibers. In 1931 Gasser was appointed to the chair of physiology at Cornell Medical School. Finally, in 1935, he was made director of the Rockefeller Institute in New York, a post he retained until his retirement in 1953.

Gates, William Henry (Bill) (1955–   ) *American software designer and entrepreneur*

The son of a lawyer, Gates developed an early interest in computers and wrote his first piece of software – a tic-tac-toe program – when only 13. He entered Harvard in 1973 but soon dropped out along with an old school friend, Paul Allen, to write and sell software. Gates later noted that, even at this early time, he feared that if he remained at Harvard to complete his education "the revolution would happen without us."

Gates and Allen had already worked out a version of BASIC to run on the newly available Intel 8008 chip. In 1975, the Altair 8800 was produced in kit form by MITS (Micro Instrumentation and Telemetry Systems). Gates realized that to broaden the PC's appeal it had to be programed in BASIC rather than in machine code. Gates and Allen moved to Albuquerque, New Mexico, the home of MITS, established Microsoft, and began to write software.

There were copyright disputes with MITS, litigation followed, and in 1978 Gates and Allen moved Microsoft to their home town of Seattle. The key moment in their fortune came in 1980 when IBM came to Seattle shopping for an operating system for their newly planned PC. Gates passed them over to Gary KILDALL, owner of the operating system CP/M. IBM failed to reach an agreeement with Kildall and consequently returned to Microsoft with an invitation to develop an operating system.

Gates and Allen realized that they needed an operating system in a hurry. They knew that a local company, Seattle Computer, owned by Rod Brock, had such a system known as 86-DOS. Gates bought the system outright for $50,000. He had in fact purchased one of the most commercially successful properties of the century. Renamed MS-DOS (Microsoft Disk Operating System), over 100 million copies have been sold.

Subsequently, Microsoft went on to develop a graphical user interface (GUI), in which the operator uses a mouse and icons. The program was introduced to compete with the successful system used on the Apple Macintosh computer. Microsoft's system, called "Windows," was first introduced in 1983; Windows 3 followed in 1990, Windows 95 in 1995, with other later versons. Gates's success has made him one of the richest men in the world. In 2006 he cut down his work at Microsoft in order to devote more time to philanthropic work. The Bill and Melinda Gates Foundation has supported charities and humanitarian research in many countries. He has described both his plans and his vision of the future in his book *The Road Ahead* (1995).

Gattermann, Ludwig (1860–1920) *German organic chemist*

Gattermann (**gah**-ter-man) was born at Goslar in Germany, and studied chemistry at Göttingen University, where he became Victor MEYER's assistant, moving with Meyer to Heidelberg in 1889. In 1900 Gattermann was appointed professor of chemistry at Freiburg University.

Gatterman is noted for his work on derivatives of benzene and a number of reactions are named for him. The *Gattermann–Koch synthesis* (with J. C. Koch, 1897) is a method of introducing the formyl group (CHO) onto a benzene ring using a mixture of carbon monoxide and hydrogen chloride with a metal-chloride catalyst. The *Gattermann synthesis* (1907) uses liquid hydrogen cyanide in place of carbon monoxide to achieve the same result. The *Gattermann–Sandmeyer reaction* (1890) is a method of promoting the conversion of diazonium compounds into other benzene derivatives using freshly precipitated copper powder.

Gauss, Karl Friedrich (1777–1855) *German mathematician*

> I have had my results for a long time: but I do not yet know how I am to arrive at them.
> —Quoted by A. Arber in *The Mind and the Eye* (1954)

> We must admit with humility that, while number is purely a product of our minds, space has a reality outside our minds, so that we cannot completely prescribe its properties *a priori*.
> —Letter to Friedrich Wilhelm Bessel (1830)

Gauss (gows) came of a peasant background in Brunswick, Germany, and his extraordinary talent for mathematics showed itself at a very early age. By the age of three, he had discovered for himself enough arithmetic to be able to correct his father's calculations when he heard him working out the wages for his laborers. Gauss retained a staggering ability for mental calculation and memorizing throughout his life. At the age of ten he astonished his schoolteacher by discovering for himself the formula for the sum of an arithmetical progression. As a result of such precocity the young Gauss obtained the generous patronage of the duke of Brunwick. The duke paid for Gauss to attend the Caroline College in Brunswick and the University of Göttingen, and continued to support him until his death in 1806. Gauss then accepted an offer of the directorship of the observatory at Göttingen. This post probably suited him better than a more usual university appointment since he had little enthusiasm for teaching. Working at the observatory no doubt also stimulated his interest in applied mathematics and astronomy.

Gauss's life was uneventful. He remained director of the observatory for the rest of his life and indeed only rarely left Göttingen. Apart from mathematics he had a very keen interest in languages and at one stage hesitated between a career in mathematics and one in philology. His linguistic ability was evidently very great for he was able to teach himself fluent Russian in under two years. He also had a lively interest in world affairs, although in politics as in literature his views were somewhat conservative.

Gauss's contributions to mathematics were profound and they have affected almost every area of mathematics and mathematical physics. In addition to being a brilliant and original theoretician he was a practical experimentalist and a very accurate observer. His influence was naturally very great, but it would have been very much greater had he published all his discoveries. Many of his major results had to be rediscovered by some of the best mathematicians of the 19th century, although the extent to which this was the case was only revealed after Gauss's death. To give but two of many examples – Janos BOLYAI and Nikolai LOBACHEVSKY are both known as the creators of non-Euclidean geometry, but their work had been anticipated by Gauss 30 years earlier. CAUCHY's great pioneering work in complex analysis is justly famous, yet Gauss had proved but not published the fundamental Cauchy theorem years before Cauchy reached it. The reason for Gauss's extreme reluctance to publish seems to have been the very high standard he set himself and he was unwilling to publish any work in a field unless he could present a complete and finished treatment of it.

Gauss received his doctorate in 1799 from the University of Helmstedt for a proof of the fundamental theorem of algebra, i.e., the theorem that every equation of degree n with complex coefficients has at least one root that is a complex number. This was the first genuine proof to be given; all the supposed previous proofs had contained errors, and it is this standard of rigor that really marks Gauss's work out from that of his predecessors. (Mathematicians of the 18th century and earlier had often possessed an intuitive ability to conjecture

mathematical theorems that were in fact true, but their ideas of rigorous mathematical proof fell short of modern standards.)

Gauss's first publication is generally accepted as his finest single achievement. This is the *Disquisitiones Arithmeticae* (Examinations of Arithmetic) of 1801. Appropriately it was dedicated to Gauss's patron the duke of Brunswick. The *Disquisitiones* is devoted to the area of mathematics that Gauss always considered to be the most beautiful, namely the theory of numbers or "higher arithmetic." Gauss's prodigious ability for mental calculation enabled him to arrive at many of his theorems by generalizing from large numbers of examples. Among many other striking results Gauss was able to prove in the *Disquisitiones* the impossibility of constructing a regular heptagon with straight edge and compass – a problem that had baffled geometers since antiquity.

Gauss's interest was not confined to pure mathematics and he made contributions to many areas of applied mathematics and mathematical physics. Thus he discovered the *Gaussian error curve* and also the method of least squares, which he used in his work on geodesy. In his work on electromagnetism he collaborated with Wilhelm WEBER on studies that led to the invention of the electric telegraph. The invention of the bifilar magnetometer for his own experimental work was another practical consequence of Gauss's interest in electromagnetism. His interest in mathematical astronomy resulted in many valuable innovations; he obtained a formula for calculating parallax in 1799 and in 1808 he published a work on planetary motion. When in 1801 the asteroid Ceres was first observed and then "lost" by Giuseppe PIAZZI, Gauss was able to predict correctly where it would reappear. He also made improvements in the design of the astronomical instruments in use at his observatory. Gauss's work transformed mathematics and he is generally considered to be, with NEWTON and ARCHIMEDES, one of the greatest mathematicians of all time. The cgs unit of magnetic flux density is named in his honor.

Gay-Lussac, Joseph-Louis (1778–1850)
French chemist and physicist

> There is a general rule that in every case where the same elements can form compounds of different stability (but capable of existing simultaneously under the same given conditions), the first to be formed is the least stable.
> —*Annals of Chemistry* (1842)

Gay-Lussac (gay-loo-**sak**) was the son of a judge who was later imprisoned during the French Revolution. Born at St. Léonard in France, he entered the recently founded Ecole Polytechnique in 1797 and graduated in 1800. His career was thereafter one of steady promotion. Originally studying engineering, in 1801 he attracted the attention of the chemist Claude-Louis BERTHOLLET who made him his assistant at Arcueil, near Paris. The science of chemistry was then in its infancy. Few chemists were actively engaged in research, and the equipment used was primitive. Chemical symbols had just been introduced, and no chemical formulae were known with certainty. During his career Gay-Lussac contributed to the advancement of all branches of chemistry by his discoveries, and greatly improved and developed experimental techniques.

In 1802, following the researches of the chemist Jacques CHARLES, Gay-Lussac formulated the law now alternatively attributed to himself and Charles – that gases expand equally with the same change of temperature, provided the pressure remains constant. Using superior experimental techniques, Gay-Lussac largely eliminated the errors of his predecessors in this field, in particular by developing a method of drying the gases. He measured the coefficient of expansion of gases between 0°C and 100°C, thus forming the basis for the idea of the absolute zero of temperature. His law was received with satisfaction as complementary to BOYLE's law. It was later shown that Gay-Lussac's and Boyle's laws applied exactly only to a hypothetical "ideal gas"; real gases obey the law approximately.

Gay-Lussac made his first daring balloon ascent in 1804 with Jean BIOT, during which they made scientific observations and established that there was no change in either the composition of the air or in the Earth's magnetic force at the heights they reached. Gay-Lussac made a second ascent alone, reaching a height of 23,018 feet.

In 1805, by exploding together given volumes of hydrogen and oxygen, Gay-Lussac discovered that one volume of oxygen combined with two volumes of hydrogen to form water. In 1808, after researches using other gases, he formulated his famous law of combining volumes – that when gases combine their relative volumes bear a simple numerical relation to each other (e.g., 1:1, 2:1) and to the volumes of their gaseous product, provided pressure and temperature remain constant. The English chemist John DALTON was immediately interested in Gay-Lussac's discovery, but when, on investigation, the law appeared to conflict with his own theory of the indivisibility of atoms, Dalton rejected the law and sought to discredit Gay-Lussac's experimental methods. The reason for the apparent conflict was that the difference between an atom and a molecule was not clearly understood, and it was left to the Italian chemist Amedeo AVOGADRO to formulate a theory reconciling the two laws, thus laying the basis of modern molecular theory.

From 1808 Gay-Lussac worked with the chemist Louis Thenard. Following Humphry DAVY's isolation of minute amounts of sodium and potassium, the two chemists in 1808 prepared these metals in reasonable quantities. It was during his experiments with potassium as a reagent that Gay-Lussac blew up his laboratory, temporarily blinding himself. In collaboration with Thenard he isolated and named the element boron. Simultaneously with Davy, Gay-Lussac investigated in 1813 a substance first isolated by Bernard COURTOIS and established that it was an element similar to chlorine. He named it iodine from the Greek "iode" meaning "violet." In 1815 he prepared cyanogen and described it as a compound radical. He proved that prussic acid (hydrogen cyanide) was made up of this radical and hydrogen, completing the overthrow of LAVOISIER's theory that all acids must contain oxygen. His recognition of compound radicals laid the basis of modern organic chemistry.

Gay-Lussac also investigated fermentation, the phenomenon of supercooling, the growth of alum crystals in solution, the compounds of sulfur, and the various stages of oxidation of nitrogen. With the young student Justus von LIEBIG he investigated the fulminates. In his later years he improved on experimental techniques, and laid the basis of modern volumetric analysis.

In 1827 he devised the *Gay-Lussac tower*. Oxides of nitrogen arising from the preparation of sulfuric acid by the lead-chamber process, which formerly escaped into the atmosphere, are absorbed by passing them up a chimney packed with coke, over which concentrated sulfuric acid is trickled. This tower and its modifications are used in many chemically based industries today.

Gay-Lussac was a chemist of brilliance and determination. Although said to be cold and reserved as a man, as a researcher he was bold and energetic. Shortly before his death he expressed regret at the experiments that he would never be able to perform.

Geber (about 14th century) *Spanish alchemist*

The name Geber (jee-ber), the Latinized form of Jabir, was adopted by an anonymous medieval writer, probably because of the reputation of the great Arabian alchemist, Jabir ibn Hayyan, who is also known better as Geber.

Four of Geber's works are known, the longest being the *Summi perfectionis magisterii*, which was translated into English as *The Sum of Perfection* or *The Perfect Magistery*. The other works are *De investigatione perfectionis* (The Investigation of Perfection), *De inventione veritatis* (The Invention of Verity), and *Liber fornacum* (Book of Furnaces). The four manuscripts were translated into English in 1678.

Geber's major contribution was to spread Arabian alchemical theories through Europe where they had considerable influence.

Geber (c. 721–c. 815) *Arabian alchemist*

Geber (whose Arabic name was Jabir ibn Hayyan) seems to have spent his life among the political uncertainties of the decline of the Umayyad dynasty. His father was executed for his part in a plot to oust the caliph and Geber, who was born in Tus (now in Iran), was sent to southern Arabia. He became a courtier to Harun al-Rashid (of *Arabian Nights* fame) but fell out of favor in 803 and left Baghdad for Kufah, where he probably remained for the rest of his life. He was a Sufi as well as being connected with the Isma'ilite sect.

A large number of works carry his name, which is now usually taken to refer to a corpus as a whole without implying actual authorship by one individual. The most important works are *The 112 Books, The 70 Books, The 10 Books of Rectification*, and *The Books of the Balances*. Geber believed that everything is composed of a combination of earth, water, fire, and air. These elements combined to form mercury and sulfur from which he believed all metals are formed, a view that continued until Robert BOYLE. He further held that if the right proportions of each were combined they would produce gold. Geber's theory was of considerable influence on alchemy and the early development of chemistry.

Geber is regarded as the father of Arabian chemistry. In the 14th century his name was adopted by an anonymous Spanish alchemist to add authority to his work.

Geer, Charles de *See* DE GEER, CHARLES.

Geer, Gerard Jacob de (1858–1943) *Swedish geologist*

Geer (yayr) came from a noble Swedish family in Stockholm and both his father and brother served as prime minister of Sweden. He graduated from Uppsala University in 1879 and worked initially with the Geological Survey on the problem of raised beaches before taking up an appointment as professor of geology at Uppsala in 1897. In 1924 he became the first director of the Stockholm Geochronological Institute.

Geer originated the varve-counting method for dating the geological past in years, a system that gave unprecedented accuracy in age determinations. In 1878 he had begun a study of the Quaternary Period in Sweden and soon became aware of the layered deposits, known as varves, laid down in glacial lakes. Seasonal differences in the material deposited enabled individual years to be identified, the summer layer consisting of light-colored coarse-grained material and the winter layer of dark-colored fine material, and Geer noticed the analogy of the varves to tree rings.

He tried to see if the sequence of varves from one region would correlate in any way with those of other areas and found that this could be done for most parts of Sweden. However, this would only allow him to say that two samples came from the same time without being able to say whether that time was a century or a millenium in the past. He was able eventually to establish a base year at 6839 BC from which point individual years could be counted in either direction.

Geer's work was a major breakthrough although it was soon to be overshadowed by radioactive dating and was limited to certain glaciated areas. In his later years, Geer tried to apply his techniques and to establish correlations with other areas of the world, but with varying success.

Gegenbaur, Karl (1826–1903) *German comparative anatomist*

> The human organism is not isolated in nature, but is only one member of an endless series within which knowledge of the whole will illuminate that of the individual.
> —*Manual of Human Anatomy* (1883)

Educated at the university in his native city of Würzburg, Gegenbaur (**gay**-gen-bowr) was professor of zoology and comparative anatomy at the University of Jena from 1855 to 1873. He then held a similar post at Heidelberg until 1901.

Gegenbaur's work was noteworthy in emphasizing the importance of comparative anatomy to the concept of evolution. One of the leading champions of Darwinism in Europe, he may be said to have laid the foundations of modern comparative anatomy with his embryological investigations, which led to his demonstration (1861) that all vertebrate eggs and sperm are unicellular, thus developing an earlier supposition of Theodor SCHWANN. Like T. H. HUXLEY, Gegenbaur denied that the vertebrate skull derived from expanded vertebrae, basing his opinion on studies of cartilaginous fishes. He also investigated the disappearance of the gill clefts in mammalian development and evolution. In his standard textbook on evolutionary morphology, *Grundriss der vergleichenden Anatomie* (1859; Elements of Comparative Anatomy), Gegenbaur expounded his view that the most reliable clues to animal evolutionary relationships lay in homology, e.g., the arm of a man as compared with the foreleg of a horse or the wing of a bird. Gegenbaur was editor of the *Morphologisches Jahrbuch* (Morphology Yearbook) from 1875, and published his autobiography in 1901.

Geiger, Hans Wilhelm (1882–1945) *German physicist*

Geiger (**gI**-ger) was born at Neustadt in Germany and studied physics at the universities of

Munich and Erlangen, obtaining his doctorate (1906) for work on electrical discharges in gases. He then took up a position at the University of Manchester, England, where he worked with Ernest RUTHERFORD from 1907 to 1912. In 1912 he returned to Germany, from then until his death holding a series of important university positions, including director of the German physical laboratory, the Physikalisch Technische Reichsanstalt, in Berlin (1912), and professor of physics at Kiel University (1925).

Geiger, a pioneer in nuclear physics, developed a variety of instruments and techniques for detecting and counting individual charged particles. In 1908 Rutherford and Geiger, investigating the charge and nature of alpha particles, devised an instrument to detect and count these particles. The instrument consisted of a tube containing gas with a wire at high voltage along the axis. A particle passing through the gas caused ionization, and initiated a brief discharge in the gas, and the resulting pulse of current could be detected on a meter. This was the prototype, which Geiger subsequently improved and made more sensitive; in 1928 he produced, with W. Müller, a design of counter that is now widely used (and known as the *Geiger–Muller counter*). With their primitive counter Rutherford and Geiger established that alpha particles are doubly charged helium atoms.

Other important work of Geiger was his investigation with E. Marsden in 1909, of the scattering of alpha particles by gold leaf; this led Rutherford to propose a nuclear theory for the atom.

Geikie, Sir Archibald (1835–1924)
British geologist

Geikie (**gee**-kee) was born in Edinburgh, the son of a musician. Although intended to be a banker, he was allowed to pursue his interests in natural history and fossils by attending Edinburgh University. With the help of Hugh MILLER he joined the staff of the Geological Survey of Great Britain in 1855 under Roderick MURCHISON. In 1867 he became director of the Scottish branch of the Geological Survey, a position he continued to hold after becoming professor of geology at Edinburgh in 1871. Geikie moved to London to become head of the Geological Survey in 1882, a post he occupied until his retirement in 1901.

Geikie was especially concerned with erosion processes and believed that rivers were a major factor in soil erosion. His studies were confined chiefly to the Scottish landscape and he was an early supporter of the role of glaciation in its formation.

A prolific writer, Geikie published works on the geology of Edinburgh (1861) and of Fife (1900) for the Geological Survey. His other works include his *Textbook of Geology* (1882), and a fine historical survey, *Founders of Geology* (1897). Perhaps his most original geological work was contained in his *Ancient Volcanoes of Great Britain* (1897).

From 1908 until 1913 Geikie served as president of the Royal Society. He was knighted in 1907.

Gelfand, Israil (or Israel) Moiseevich
(1913–) *Russian mathematician and biologist*

Born in Krasnye Orny, now in Moldova, Gelfand (**gyel**-fand) showed mathematical ability at an early age. Although he had not completed the usual university education course his mathematical expertise was sufficient for him to be admitted to do postgraduate work at the Moscow State University at the age of 19. He began teaching there in 1932 and was appointed an assistant professor in 1935. From 1939 Gelfand worked at the V. A. Steklov Institute of Mathematics of the Soviet Academy of Sciences, and from 1940 to 1991 he was professor of mathematics at the Moscow State University. In 1989–90 he taught at Harvard and in 1990 he also taught at MIT. The same year he emigrated to the U.S., becoming Distinguished Visiting Professor of Mathematics at Rutgers University, New Jersey.

Gelfand is one of the most fertile and brilliant Russian mathematicians of his generation, and his work has been equally influential in both pure and applied mathematics. His first major original contribution was made in his doctoral dissertation, presented in 1940, in which he made advances of great importance in the theory of commutative normed rings. Other outstanding achievements of Gelfand's include his contributions to group theory, in particular his work on infinite dimensional representations of continuous groups, and on the harmonic analysis of noncompact groups.

Gelfand carried out very important studies on the theory of generalized functions. This work focused on the application of such functions to be found in dealing with differential equations. As a result of this research the whole theory of integral transformations was placed on a geometrical basis, and thus became amenable to a whole new range of techniques. In applied mathematics Gelfand's work has provided key mathematical tools needed in developing a theory of symmetry for elementary particles.

From about 1958 Gelfand became interested in biology and physiology and he made studies of the nervous system and of cell biology. He applied mathematical techniques to these studies, for example, devising mathematical models of neurophysiological systems.

Geller, Margaret Joan (1947–   ) *American astronomer*

The daughter of a crystallographer from Ithaca, New York, Geller was encouraged as a small child to study science and mathematics. She was educated at the University of California, Berkeley, and at Princeton where she obtained her PhD in 1975. After a period at the Institute of Astronomy, Cambridge, England, Geller moved to Harvard in 1980 and was appointed professor of astronomy in 1988. She is also a staff member of the Smithsonian Astrophysical Observatory.

Since the early 1980s Geller in collaboration with John HUCHRA has been carrying out for the Center for Astrophysics (CfA) a red-shift survey of some 15,000 galaxies. The intention is to map all galaxies above a certain brightness, out to about 650 million light years, in a particular sector of the heavens. They were aware that to some observers the sky lacked the uniformity predicted by the big-bang theory. In 1981, for example, a 100-million-light-year gap had been discovered in the constellation Bootes. Geller considered the possibility that this was a local phenomenon, and that the predicted homogeneity would become more apparent on a much larger scale. Further investigations were expected to show a uniform distribution.

But when they came to plot the distribution of galaxies they saw neither a uniform spread, nor a random scattering of galaxies, but large-scale clusters grouped into enormous structures. The largest of these, dubbed the *Great Wall*, stretches for more than 500 million light years. It was difficult to see how anything as massive could have been formed within the context of current cosmological theory; when Geller reported the initial results of the CfA survey in 1989 she noted, "Something fundamental is missing in our models." Geller has been active in promoting public education in science, with many lectures and TV and radio interviews.

Gellibrand, Henry (1597–1636) *English mathematician and astronomer*

Gellibrand, a Londoner, was educated at Oxford University. He was described by the diarist John Aubrey as "good for little a great while, till at last it happened accidentally, that he heard a Geometrie lecture. He was so taken with it, that immediately he fell to studying it, and quickly made great progress in it." In 1626 he became Gresham Professor of Astronomy at Oxford. He did important work on evidence for the variation of the Earth's magnetic field, publishing, in 1635, his observation of the 7° shift in direction of the compass needle over the previous 50 years.

Gell-Mann, Murray (1929–   ) *American theoretical physicist*

> I employed the sound "quork" for several weeks in 1963 before noticing "quark" in *Finnegans Wake*...The allusion to three quarks seemed perfect....
> —Quoted in *A Supplement to the Oxford English Dictionary* (1982)

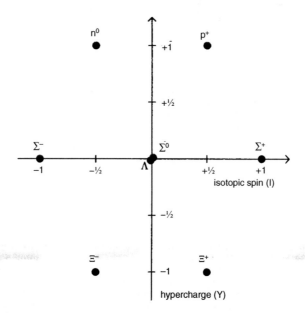

UNITARY SYMMETRY   An example of an octet consisting of the proton, neutron, sigma, lambda, and xi particles.

Gell-Mann was born in New York City, the son of Austrian immigrants. Having entered Yale at the age of fifteen, he went on to the Massachusetts Institute of Technology, where he completed his PhD in 1951. After spending a year at the Institute for Advanced Study, Princeton, and four years at the University of Chicago, Gell-Mann was appointed professor of physics in 1955 at the California Institute of Technology. In 1967 he was elected R. A. Millikan Professor of Theoretical Physics, a post he held until his retirement in 1993. He is a Distinguished Fellow at the Santa Fe Institute.

Gell-Mann has been mainly concerned with the study of elementary particles. In the early 1950s physicists were puzzled by the "strange" behavior of some apparently strongly interacting mesons. The kaons, as they were called, should have a lifetime of only 10^{-23} second; they actually survived for some 10^{-10} second. Gell-Mann suspected that some unknown property was conserved and that this explained the rapid decay of the kaons via the strong force. He named the new property "strangeness" (S) and assigned the kaons a value $S = +1$. S was later defined as $2Q - B$, or twice the charge minus the baryon number. Thus the proton with a charge of $+1/2$ and a baryon number of $+1$, will take the value $S = 0$, and this value will be conserved in all strong interactions. Similar ideas were proposed in 1954 by K. Nishijima in Japan.

New problems emerged. Physicists had become particularly worried by the discovery of large numbers of supposedly elementary particles. In the late 1950s it was possible to list and classify some thirty subatomic particles. Within five years another 70 "elementary" particles had been discovered. The theorist's first priority was to bring some kind of order to this unwelcome abundance. In 1964, using group theory, and a number of conservation principles, Gell-Mann demonstrated that hadrons – i.e., particles such as baryons and mesons that interact strongly – could be classified into multiplets of 1, 8, 10, or 27 members. One such multiplet constructed with the group SU(3), or symmetry unity theory of dimension 3, yielded the octet. A similar proposal was made at the same time by the Israeli physicist Y. Ne'eman. The new synthesis was described in *The Eightfold Way* (1964), a work edited by Gell-Mann and Ne'eman. That there was more to their theory than arbitrary classifications was soon shown when the Ω^- (omega minus) particle, whose existence was previously unsuspected, was discovered in 1964 at Brookhaven with precisely the properties demanded for it by the SU(3) theory. The name itself alludes to the Buddhist path to enlightenment and to the eight quantum numbers required by the theory.

Following the discovery of the Ω^- Gell-Mann began to consider why the theory should be so successful in explaining hadron behavior. In his 1964 paper *A Schematic Model of Baryons and Mesons*, he proposed the existence of a more fundamental level of reality, that hadrons were composite, built out of more basic entities he proposed to call "quarks." The name itself was taken from a phrase in James Joyce's *Finnegans Wake* – "three quarks for muster mark." In Gell-Mann's original formulation he worked with three quarks, up (u), down (d), and strange (s), and three antiquarks ($\bar{u}$, $\bar{d}$, and $\bar{s}$). As their most distinctive feature Gell-Mann assigned them fractional units of charge: s and d with a charge of $-\frac{1}{3}$, and u with $+\frac{2}{3}$. It was a simple matter to show that all baryons could be constructed with three quarks, and all mesons from a quark and an antiquark. Similar views were expressed in 1964 by G. Zweig.

The original model soon required some refashioning. A fourth quark property, "charm," was proposed by Sheldon GLASHOW in 1970 and detected in 1974. This required extending the hadron classification scheme from a SU(3) group to a SU(4) group. Two further quarks have been added, the t and b quarks, known variously as top and bottom, or "beauty" and "truth." The fifth b quark was identified by L. LEDERMAN in 1977. The t quark was discovered in the last few years of the 20th century. A further complication of quark theory emerged with NAMBU's proposal in 1965 that they came with three varieties of "color." In 1972 Gell-Mann and his colleagues put forward quantum chromodynamics, a quantum field theory that describes the strong interactions of quarks with great success.

In the 1980s Gell-Mann turned to the study of complexity. He was a cofounder in 1984 of the Santa Fe Institute, which works on adaptive complex systems. Reportedly fluent in fifteen languages himself, Gell-Mann has also worked in the field of historical linguistics.

For his numerous contributions to the study of elementary particles Gell-Mann was awarded the 1969 Nobel Prize for physics.

Gennes, Pierre Gilles de (1932–)
French physicist

Parisian-born Gennes (zhen) was educated at the Ecole Normale in his native city, completing his PhD there in 1955. He was appointed professor of solid-state physics at the University of Paris in 1961; since 1971 he has served as professor of physics at the Collège de France and from 1976 as director of the College of Physics and Chemistry, Paris. He retired in 2002.

Some areas of science have long been thought to be too unstructured and messy to be conducive to traditional physical analysis. Two such areas, liquid crystals and polymers, were consequently largely ignored by physicists. Gennes, however, saw that they behave in many ways just like other better understood physical processes.

Liquid crystals consist of rodlike molecules in a liquid state. They undergo, like magnets and superconductors, phase changes. Thus in what is known as the smetic A phase, the molecules are oriented with their axes perpendicular to the layers; at lower temperatures they adopt the C phase with a parallel orientation. Using concepts derived from the study of phase changes in other fields Gennes was able to throw light on changes in liquid crystals. The results of his work were described in his *The Physics of Liquid Crystals* (1974). Gennes adopted a similar approach to his study of polymers which he described fully in his *Scaling Concepts of Polymer Physics* (1979).

For his work in these fields Gennes was awarded the 1991 Nobel Prize for physics. His more recent work was on granular materials and also the science of the brain.

Geoffroy Saint-Hilaire, Etienne (1772–1844) *French biologist*

> It seems to me that nature ... has formed all living things upon one plan, varying in minor details only.
> —*Magasin encyclopédique*, Vol. VII (1796; Encyclopedic Storehouse)

The youngest of fourteen children in a poor family from Etampes in France, Geoffroy (zho-**frwah**) was supported by the local clergy, who recognized his precocious intelligence. He took a scholarship to the Collège de Navarre, Paris, and was soon appointed as a demonstrator at the Jardin des Plantes, a precursor of the Muséum National d'Histoire Naturelle, as his predecessor, Bernard Lacépède, fled the Revolution. In 1793 Geoffroy became professor of vertebrate zoology at the museum; the comparable chair of invertebrate zoology was held by LAMARCK.

In 1798 Geoffroy accompanied Napoleon on his conquest of Egypt and contributed to the celebrated 24 volumes of the *Description de l'Egypte* (1809–28; Description of Egypt). He traveled as far down the Nile as Aswan and while in Egypt he examined a number of mummified cats taken from ancient tombs. They were, he noted, identical to the animals of his day. Did this mean that species were fixed? If they were fixed why were there so many similarities between different forms? Why, for example, despite differences in external form, do the skeletons of bats, whales, and dogs resemble each other so closely?

Geoffroy derived his answer from German *Naturphilosophie* (nature philosophy), which claimed to see beneath an apparent diversity of form, mere variations on a single plan. There was a vertebrate type which could be identified in all vertebrates. Thus he demonstrated in 1807 that pectoral fins in fish and the bones of the front limbs of other vertebrates were morphologically and functionally similar.

But CUVIER had identified in the operculum, a bony flap covering gill slits in fishes, an apparently unique structure. It took Geoffroy a decade's investigation before he could explain it away as equivalent to the auditory bones in mammals. He was thus able in his *Philosophie anatomique* (1818; Anatomical Philosophy) to announce the principle of anatomical connection claiming that the same anatomical structural plan could be identified in all vertebrates.

By 1830 Geoffroy had begun to argue that there was a universal "unity of composition," quoting in evidence work claiming to have detected a unity in crustacea, fish, and mollusks. Such views brought a savage onslaught from Cuvier who insisted that there were distinct forms in nature, and that parts were formed to meet functional needs.

But, once having accepted a unity of composition, it becomes possible to see how one species can be transformed into another. If birds and reptiles are built to the same plan, then "an accident that befell one of the reptiles ... could develop in every part of the body the conditions of the ornithological type." Geoffroy was thus moving late in his career to some form of evolutionary theory. A stroke in 1840 which left him blind and paralyzed brought such work to an end.

Geoffroy was succeeded at the museum in 1841 by his son, Isidore (1805–61), also a distinguished biologist and best known for his three-volume work on teratology, *Histoire ... des anomalies de l'organisation chez l'homme et les animaux* (1833–37; Account ... of Irregularities in the Structure of Man and the Animals).

Gerard, John (1545–1612) *British herbalist*

> Upon my conscience, I do not thinke for the knowledge of plants, that he [Gerard] is inferior to any.
> —George Baker, surgeon to Queen Elizabeth I (1597)

Gerard was born at Nantwich in Cheshire, England, and educated nearby in Willaston, taking up an apprenticeship as a barber-surgeon in London at the age of 16. Thereafter he traveled, probably as ship's surgeon, aboard a merchant ship trading in Scandinavia, Russia, and the Baltic. He then settled in London and began his study of plants, paying particular attention to those with medicinal properties. By 1577 he had become superintendent of the gardens of Lord Burghley, a post he held for over 20 years. His knowledge and practical experience led to his appointment as curator of the gardens of the College of Physicians between 1586 and 1603 or 1604, and in 1597 his election as warden of the Barber-Surgeon's Company.

Gerard had cultivated his own extensive garden in London and in 1596 issued a catalog listing over a thousand different plants in his possession. In 1597, he published his most famous work, the *Herball*, a survey of plants then available in western Europe. Although probably an adaptation of an earlier work by Rembertus Dodaens and of questionable accuracy, its descriptive style and over 1800 woodcut illustrations make it the most famous of English herbals.

Gerhardt, Charles Frédéric (1816–1856)
French chemist

> Never before has science been so much a plaything of the imagination as it is today.
> —Quoted by René Taton in *Science in the Nineteenth Century* (1965)

Gerhardt (zhay-**rar**) was the son of an Alsatian chemical manufacturer from Strasbourg. He was educated at the universities of Karlsruhe, Leipzig, and Giessen where he studied under Justus von LIEBIG. From 1838 he worked in Paris as assistant to Jean DUMAS before becoming professor of chemistry at Montpellier (1844). He returned to Paris in 1848 and worked with Auguste LAURENT in their private laboratory until he was appointed to the chair at Strasbourg in 1855. He published two original works: *Précis de chimie organique* (1844–45; Summary of Organic Chemistry) and his *Introduction à l'étude de la chimie par le système unitaire* (1848; Introduction to the Study of Chemistry by Means of the Unified System).

Gerhardt is best known for his attempts to rationalize organic chemistry. Like most chemists he was aware that the dualistic system of Jöns BERZELIUS was unsatisfactory and tried to create an alternative. He adopted what became known as "type theory" in which he saw all organic compounds with reference to four "types" – hydrogen, hydrogen chloride, ammonia, and water. Organic compounds were referred to these types by replacing a hydrogen atom in one of these compounds by a radical (i.e., by a group of atoms).

Germain, Sophie Marie (1776–1831)
French mathematician

The daughter of a prosperous Parisian merchant, Germain (zher-**man**) showed an early interest in mathematics and from the age of thirteen read whatever texts she could obtain. Although the main higher education institutions were closed to her, she managed to acquire the lecture notes of the mathematician J. L. LAGRANGE, which he had delivered at the newly founded Ecole Polytechnique.

She also began to correspond with prominent mathematicians using the pseudonym Le Blanc and allowing them to assume that she was a man. She had been working on number theory and had begun to tackle the celebrated last theorem of FERMAT: that there are no integers x, y, z, n, where $n > 2$ such that:

$$x^n + y^n = z^n.$$

Germain made a major contribution to showing that the equation does not hold for the case in which n is equal to 5. She informed GAUSS of the result but, typically, he failed to reply.

In 1809 Germain began to work on the theory behind the appearance of curious patterns formed by sand placed on vibrating plates. The phenomenon had first been described by E. F. CHLADNI who had demonstrated them to Napoleon in 1808. The emperor had been so intrigued that he had offered a one-kilogram gold medal to the first person to explain what are now known as *Chladni's figures*.

Germain submitted a solution in 1811 based on EULER's theory of elasticity. She was the only entrant but her work contained a number of errors. It did, however, provoke Lagrange to produce a corrected equation to derive the patterns theoretically. The competition was extended, and after two further attempts, Germain was finally awarded the prize in 1815. She published her work privately in 1821 as *Recherches sur la théorie des surfaces élastiques* (Researches on the Theory of Elastic Surfaces).

Sophie Germain developed breast cancer in 1829 and died two years later.

Germer, Lester Halbert (1896–1971)
American physicist

Born in Chicago, Illinois, Germer began research at the Western Electric and the Bell Telephone Laboratories. Later he moved to Cornell University, New York. In 1927 he assisted Clinton J. DAVISSON in the celebrated *Davisson–Germer experiment*, which first demonstrated the wavelike behavior of electrons. Germer's other interests were in thermionics, the erosion of metals, contact physics, and the plating of molybdenum and tungsten, by thermal decomposition of metal carbonyl compounds.

Gesner, Conrad (1516–1565) *Swiss naturalist, encyclopedist, and physician*

Although Gesner (**ges**-ner), who was born in Zurich, Switzerland, graduated in medicine from the University of Basel (1541), his main interest was natural history. In the comprehensive survey *Historiae animalium* (1551–87; Description of the Animals) he attempted, in five volumes, to present the facts rather than the myths then known about the animal kingdom. A comparable work that he planned on plants was never written owing to his untimely death from the plague. He nevertheless collected more than 500 plant species not described in ancient texts, mainly alpine species found while indulging his hobby of mountain climbing. The notes and engravings made in preparation for his botanical encyclopedia were used by botanists for the following 200 years.

Gesner also wrote a major bibliographical work on Latin, Greek, and Hebrew writers, a work on comparative philology, and a book on fossils, containing the first illustrations of these "stony concretions," as Gesner regarded them.

Giacconi, Riccardo (1931–) *Italian-born American astronomer*

Giacconi (**jee**-a-koh-ni) was born in Italy and educated at the University of Milan, where he gained a PhD in cosmic-ray physics. He began working on x-ray astronomy when he joined American Science and Engineering, a research corporation based at Cambridge, Massachusetts. In 1973 he joined the Harvard-Smithsonian Center for Astrophysics. Between 1981 and 1993 he was the director of the Space Telescope Science Institute, based at the Johns Hopkins University, Baltimore, Maryland. He was director general of the European Southern Observatory from 1993 to 1999. Since 1999 he has been president of Associated Universities, Inc. based at Washington, DC. Between 1982 and 1997 he also held the position of professor of physics at Johns Hopkins University and since 1998 has held the position of research professor there.

Giacconi is frequently called the father of x-ray astronomy because of his pioneering work on this topic. Since cosmic x-rays are absorbed in the atmosphere of the Earth it is necessary to place instruments to investigate them into space. Giacconi and his colleagues constructed several x-ray telescopes and observatories, work that led to the discovery of numerous x-ray stars. It is thought that in many of these stars the emission of x-rays is due to matter falling into neutron stars and black holes. His pioneering work in the field led to him sharing the 2002 Nobel Prize for physics with Raymond DAVIS and Masatoshi KOSHIBA.

Giaever, Ivar (1929–) *Norwegian–American physicist*

Born in Bergen, Norway, Giaever (**yay**-ver) studied electrical engineering at the Norwegian Institute of Technology. He did service with the Norwegian Army (1952–53) and worked as a patent examiner in the Norwegian Patent Office (1953–54). In 1954 he emigrated to Canada to take up the post of mechanical engineer with the Canadian General Electric Company, transferring to General Electric's Research and Development Center in Schenectady, New York, in 1956. He gained his doctorate in 1964 from the New York Rensselear Polytechnical Institute, where he became professor of physics in 1988.

At General Electric, Giaever worked on tunneling effects in superconductors, a phenomenon explored by Leo ESAKI. In 1960 he performed experiments with metals separated by a thin insulating film through which electrons tunneled, and found that if one of the metals was in the superconducting state, the current–voltage characteristics of such junctions were highly nonlinear and revealed much about the superconducting state. This laid the foundation for Brian JOSEPHSON's important discovery of the Josephson effect.

Giaever, Josephson, and Esaki shared the 1973 Nobel Prize for physics for their various contributions to knowledge of the phenomenon of tunneling and superconductivity. Their work has had important application in microelectronics and in the precise measurement of electromotive force.

Subsequently, Giaever has also published work in the field of visual observation of the antibody-antigen reaction.

Giauque, William Francis (1895–1982) *Canadian–American physical chemist*

Born at Niagara Falls in Canada, Giauque (jee-**ohk**) spent his whole academic life at the University of California. He began as a student, obtaining his PhD in 1922, and was immediately appointed to the staff at Berkeley, becoming professor of chemistry in 1934.

Giauque was one of the pioneer workers in low-temperature phenomena. His early work in the 1920s concerned the experimental measurement of entropies at very low temperature – work that depended on the use of the third law of thermodynamics introduced in 1906 by Walther NERNST (the Nernst heat theorem). At the same time, Giauque used statistics to calculate the absolute entropies using the energy levels of molecules obtained from spectroscopy. This method, developed by Josiah Willard GIBBS and others, is known as statistical mechanics. Giauque's work provided support for the validity of both statistical thermodynamics and the third law.

Moreover, it led him to a method of attaining very low temperatures, close to absolute zero. The lowest temperature achieved at that time was 0.8 K, reached by Heike KAMERLINGH-ONNES in 1910 by pumping away the vapor of liquid helium and causing it to evaporate under reduced pressure. Giauque, and independently Peter DEBYE, proposed in 1925 a completely different method known as adiabatic demagnetization.

The basic idea is to take a paramagnetic substance surrounded by a coil of wire in a gas-filled container. The sample can be cooled by surrounding the container by liquid helium and magnetized by a current through the coil. It is thus possible to produce a magnetized specimen at liquid-helium temperature, and then to isolate it in a vacuum by removing the gas from the container. In the magnetized specimen the "molecular magnets" are all aligned. If the magnetic field on the specimen is reduced to zero the sample is demagnetized, and in this process the molecular magnets become random again. The entropy increases and work is done against the decreasing external field, causing a decrease in the temperature of the specimen.

There were considerable problems in putting this theory into practice, not least in measuring the temperatures produced. In 1933 Giauque had a working apparatus that improved on Kamerlingh-Onnes's in achieving a temperature of 0.1 K. Giauque received the 1949 Nobel Prize for chemistry for his work on low-temperature phenomena.

He also worked on isotopes, showing in 1929 (with H. L. Johnson) that oxygen was a mixture of ^{16}O, ^{17}O, and ^{18}O.

Gibbs, Josiah Willard (1839–1903) *American mathematician and theoretical physicist*

> Mathematics *is* a language.
> —Remark at a Yale faculty discussion concerning the difference between languages and mathematics

Gibbs came from an academic family in New Haven, Connecticut. He entered Yale in 1854, graduated in 1858, and in 1863 received a PhD for research on the design of gears. The same year he traveled to Europe, returning in 1869 to Yale where he remained until his death. In 1871 he was appointed professor of mathematical physics.

His initial work on the theory of James WATT's steam-engine governor led him into a study of the thermodynamics of chemical systems. In a series of long papers published between 1873 and 1876 he developed, and indeed virtually completed, the theory of chemical thermodynamics. Gibbs's most famous paper, *On the Equilibrium of Heterogeneous Substances* (1876), contains the celebrated *Gibbs phase rule*, describing the equilibrium of heterogeneous systems. His name is also associated with the *Gibbs free energy* – a function that determines the conditions in which a chemical reaction will occur – and with several other equations in thermodynamics.

Gibbs was also active in mathematics and physics. He worked on the theory of William Hamilton's quaternions and introduced the simpler, widely used, vector notation. His book *Vector Analysis* (1901) was very influential in promoting the use of vectors by physicists and engineers. Between 1882 and 1889 he published a series of papers on the electromagnetic theory of light. He also made important contributions to statistical mechanics, introducing the fundamental concept of *Gibbsian ensembles* – collections of large numbers of macroscopic systems with the same thermodynamic properties, used in relating thermodynamic properties to statistical properties.

Gibbs, who never married, lived a quiet retiring life at Yale; he was a poor teacher but a brilliant and productive theorist. His work, carried out far from the European mainstream of science, was largely published in the obscure *Transactions of the Connecticut Academy of Sciences*. However, James Clerk MAXWELL understood the importance of his ideas as early as 1875 and in later life Gibbs was widely recognized. Many regard him as the greatest native-born American scientist.

Gibbs, William Francis (1886–1967) *American naval architect and marine engineer*

Born in Philadelphia, Pennsylvania, Gibbs was educated at Harvard (1906–10) and Columbia (1910–13) where he studied law. His interest in ship design was aroused by the rapid sinking of the *Empress of Ireland* following a collision in 1914 with the loss of 1,000 lives. He argued that simple compartmentalization and properly designed bulkheads could have prevented such a tragedy. Gibbs consequently joined the International Mercantile Marine Company with whom he stayed until 1922 when, with his brother, Frederick, he founded the company Gibbs Brothers, later (1929) Gibbs and Cox.

Beginning in 1933, Gibbs did much to modernize the U.S. Navy, and persuaded it to adopt high-pressure and high-temperature steam turbines into the design of its destroyers. It was largely due to plans laid down by Gibbs that large numbers of Liberty ships could be rapidly built when America entered World War II. During the war Gibbs served as controller of shipbuilding.

Gibbs is probably best-known as the designer of the SS *United States*, a passenger ship launched in 1952, which set new speed records across the Atlantic, achieving an average speed of 35.05 knots (40.3 statute miles per hour).

Gilbert, Walter (1932–) *American molecular biologist*

Born in Boston, Massachusetts, Gilbert was educated at Harvard and at Cambridge University, England, where he obtained his PhD in physics in 1957. He returned to America to take up an appointment in theoretical physics at Harvard. He changed to molecular biology in 1960 under the influence of James WATSON and in 1968 became professor of molecular biology at Harvard. He was elected chairman of the department of cellular and developmental biology in 1987. He is currently an emeritus professor.

In 1961 Jacques MONOD and François Jacob proposed a theoretical answer to one of the most pressing problems of molecular biology, that of genetic control. If the common bacteria *Escherichia coli* is grown in the presence of milk sugar (lactose) it will produce an enzyme, beta-galactosidase, to split it into its component sugars. However, if grown in the absence of lactose, the enzyme will not be produced. There must therefore presumably be a mechanism whereby the gene controlling the production of the enzyme can be switched on and off. Monod and Jacob proposed a detailed account of such a mechanism, part of which involves the existence of a repressor molecule, which could bind itself to the gene and switch it off in the absence of lactose. The lac repressor, as it was called, would be inactivated, thus switching the gene on, by an inducer molecule produced by the lactose itself.

Plausible and powerful though the Monod–Jacob model appeared, it was still only a model until the basic confirmation provided by the isolation and identification of the lac repressor was achieved. Gilbert began such a search in 1965. This was a formidable task as the repressor was known to exist in small quantities only; nor was its chemical nature known. Gilbert himself likened the task to isolating the neutrino.

By 1966, in collaboration with Benno Muller-Hill, Gilbert had devised an ingenious experimental procedure, known as equilibrium dialysis. They used a specially active inducer, isopropyl thiogalactoside (IPTG), discovered by Melvin CALVIN. Cells of *E. coli* were ground up and placed in a bag with a cellular membrane, allowing the passage of water and IPTG molecules but excluding such larger molecules as proteins. The bag was then placed in water containing radioactive IPTG.

As IPTG can pass through the bag an equal concentration of the inducer should be achieved. But if IPTG should bind itself to the lac repressor inside the bag then it will be too large to pass freely through the bag membrane. Consequently the concentration of the IPTG bound to the repressor should start to build up inside the bag and, being radioactive, should be readily detectable. Eventually they were able to report a concentration of IPTG 4% greater inside the bag than out. This was enough to encourage Gilbert and Muller-Hill to proceed to the next stage of fractionating, purifying, and isolating the repressor. This proved more difficult than they had expected but in late 1966 they were able to report the existence of a large protein molecule, the lac repressor. The following year their Harvard colleague M. Ptashne obtained a similar result with the lambda phage repressor.

Gilbert has also developed techniques for determining the sequence of bases in DNA, which though similar to Frederick SANGER's method differs in that it can be applied to single as well as double-stranded DNA. It was for this work that he shared the 1980 Nobel Prize for chemistry with Sanger and Paul BERG.

Gilbert, William (1544–1603) *English physicist and physician*

> Look for knowledge not in books but in things themselves.
> —*De magnete, magnetisque corporibus, et de magno magnete tellure* (1600; On the Magnet, Magnetic Bodies, and the Great Magnet Earth)

Gilbert, who was born at Colchester in Essex, England, was educated at Cambridge University, where he took his degree in 1569 and later became a fellow. He moved to London in 1573, became a member of the Royal College of Physicians, and served as physician to Queen Elizabeth I and briefly to James I. In 1600 he published the first great English scientific work *De magnete, magnetisque corporibus, et de magno magnete tellure* (On the Magnet, Magnetic Bodies, and the Great Magnet Earth) in which he presented his investigations into magnetic bodies and electrical attractions. It is a remarkably "modern" work – rigorously experimental, emphasizing observation, and rejecting as unproved many popular beliefs about magnetism, such as the supposed ability of diamond to magnetize iron. He showed that a compass needle was subject to magnetic dip (pointing downward) and, reasoning from experiments with a spherical lodestone, explained this by concluding that the Earth acts as a bar magnet. He also introduced the term "magnetic pole." The book was widely available on the Continent, there being five editions in Germany and Holland alone before 1628, and was very influential in the creation of the new mechanical view of science.

Gilchrist, Percy Carlyle (1851–1935) *British industrial chemist*

Gilchrist, who was born at Lyme Regis in Dorset, England, was educated at the Royal School of Mines in London. In 1875 he was appointed chemist at the Cwm Avon works in South Wales, and here in his spare time he assisted his cousin Sidney Gilchrist THOMAS in trials of a new process of smelting iron ore. In the *Gilchrist–Thomas process* the furnace was lined with a basic material (such as magnesium oxide), which combined with phosphorus impurities in the iron to form a "basic slag" of phosphates. Gilchrist later made other improvements to the process, which was extremely important at the time in allowing exploitation of phosphoric ores.

Gill, Sir David (1843–1914) *Scottish astronomer*

> By his individual achievements and by his leadership he has exerted an incalculable influence on the progress of all that pertains to precision of observation.
> —Sir Arthur Eddington, *Monthly Notices of the Royal Astronomical Society* (1915)

Born in Aberdeen, Scotland, Gill was educated at Marischal College and Aberdeen University. He was in charge of the Earl of Crawford's private observatory at Dunecht before becoming royal astronomer at the Cape of Good Hope, where he remained until 1907. He was knighted in 1900.

Gill spent much time and thought on improving the accuracy of the astronomical unit (AU – the mean distance between the Earth and the Sun, one of the basic measurements of astronomy), then determined from measurements of the distances of Venus and Mars. In 1874 he went to Mauritius to observe the transit of Venus. The difficulty is that Venus, on magnification, presents a disk whose edges are not absolutely sharp, thus making it difficult to estimate the moment of first contact. In 1877 Gill went to Ascension Island to measure the distance of Mars using the distance from Greenwich as a base line. Although he obtained reasonable results he realized (as had Johann Galle) that a more accurate figure could be obtained if the planetoids were used instead for they came closer to the Earth and on magnification presented a starlike appearance. (This idea was taken up with great success later by Harold SPENCER JONES.) In 1897, with the cooperation of astronomers in Leipzig and New Haven, Gill made a very accurate determination of the solar parallax.

His other main research was extending Friedrich ARGELANDER's catalog to the southern skies. This began in 1882 when he photographed a comet and was impressed with the clarity of the stars visible in the background. Consequently he started photographing the southern skies, collaborating with the Dutch astronomer Jacobus KAPTEYN. In 1904 the *Cape Photographic Dorchmusterung* was published cataloging over 450,000 stars to within 19° of the southern celestial pole.

Gilman, Alfred Goodman (1941–) *American pharmacologist*

The son of Alfred Gilman Snr, a noted pharmacologist, Gilman received his PhD from Case Western Reserve University, Cleveland, Ohio, in 1969. After working at the University of Virginia Medical School from 1971 until 1981, Gilman moved to the University of Texas Southwestern Medical Center, Dallas, where he became professor of pharmacology.

Gilman's work has been on the processes by which hormones, neurotransmitters, and other stimuli – the so-called "first messengers" – influence cellular activity. It had been shown by Earl SUTHERLAND in 1971 that hormones do not actually enter cells. They seem to bind to receptor sites on the cell's surface and then produce a "second messenger," cAMP (cyclic adenosine monophosphate), which initiates the appropriate cellular response.

It was further shown by Martin RODBELL that other factors, namely an amplifier and a transducer, were required in the process. Rodbell identified the enzyme adenylate cyclase (AC) as the amplifier and demonstrated that transducers would only work in the presence of the energy-rich molecule guanine triphosphate (GTP).

Gilman set out to elucidate the process further. He established in the late 1970s that the transducers were in fact proteins. They were initially named "guanine nucleotide binding proteins," a term quickly shortened to G proteins. Gilman went on to outline the main steps in cellular signalling as:

1. A hormone, neurotransmitter, etc., binds to a cell receptor;
2. The receptor binds to and activates a G protein;
3. The activated G protein binds to GTP;
4. The activated GTP stimulates AC to produce cAMP;
5. cAMP produces an appropriate cellular response.

G proteins have been shown to play a number of important physiological roles. In cholera, for example, a toxin is produced that freezes G proteins into their GTP-bound activated state, producing in the body a massive fluid loss with consequent dehydration. G proteins are also thought to be involved in some aspects of diabetes and some types of cancer.

Gilman shared the 1994 Nobel Prize for physiology or medicine with Martin Rodbell for their work on G proteins.

Gilman, Henry (1893–1986) *American chemist*

Gilman was born in Boston, Massachusetts, and educated at Harvard, where he obtained his PhD in 1918. After working briefly at the University of Illinois he moved to the Iowa State College of Agriculture and Mechanical Arts in 1919, spending the rest of his career there and serving as professor of organic chemistry from 1923 until his retirement in 1962. He worked extensively in the field of organometallic chemistry and produced one of the standard texts in the field, *Organic Chemistry, an Advanced Treatise* (1938, 2 vols.; revised and extended 1953, 4 vols.).

Ginzburg, Vitaly L. (1916–) *Russian physicist*

Ginzburg was born in Moscow, where he gained his doctor's degree at the local university. He is former head of the Theory Group at the P. N. Lebedev Physical Institute, Moscow. A renowned 20th-century physicist and member of several National Academies of Sciences, he was in residence at the Department of Physics and Astronomy at the University of Kentucky, Lexington, USA in early 1997.

Ginzburg was among the first people to study superconductors. In particular, in the 1950s he, Lev LANDAU, and their co-workers explained the functioning of type-I semiconductors, which (unlike type-II superconductors) lose their superconductivity in strong magnetic fields. This became known as the *Ginzburg–Landau* theory. For this seminal work, Ginzburg shared the 2003 Nobel Prize for physics with Alexei ABRIKOSOV and Anthony LEGGETT, which was awarded for pioneering contributions to the theory of superconductors and superfluids. (Lev Landau received the 1962 Nobel Prize for physics for different research on condensed matter.)

Glaser, Donald Arthur (1926–) *American physicist*

Glaser was born in Cleveland, Ohio, and took his degree in physics and mathematics at the Case Institute of Technology there. After graduating in 1946, he went on to gain his doctorate for cosmic-ray research from the California Institute of Technology in 1950. From 1949 to 1959, Glaser worked in the physics department of the University of Michigan, becoming professor in 1957. In 1959 he moved to the University of California at Berkeley as a professor of physics and subsequently (1964) as a professor of physics and biology.

While at the University of Michigan, Glaser became interested in techniques for the visualization and recording of elementary particles. The WILSON cloud chamber, using supersaturated vapor, had been in use since the 1920s, but was unsuited to the detection of the highly energetic particles emerging from the new accelerators of the 1950s.

Glaser considered other unstable systems that could be used, and experimented with superheated liquids, in which ionizing particles would leave a trail of vapor bubbles. In 1952 he produced the first radiation-sensitive bubble chamber, in which he used diethyl ether under pressure and controlled temperature. A sudden brief reduction in pressure was used and trails of bubbles forming along the tracks of particles could be captured by high-speed photography before the bulk of the liquid boiled. For this invention, and its subsequent development into a useful research tool, Glaser received the 1960 Nobel Prize for physics.

The bubble chamber, using liquid hydrogen at low temperature, is now a basic component of almost all high-energy physics experiments, and has been the instrument of detection of many strange new particles and phenomena. Present-day bubble chambers are much bigger (and more expensive) than Glaser's original, which was only three cubic centimeters in volume.

In 1962 he turned to molecular biology and was involved in research on bacterial evolution, regulation of cell growth, and the cause of genetic mutation. He later moved into the field of neurobiology.

Glashow, Sheldon Lee (1932–) *American physicist*

Glashow was born in New York City and graduated from the Bronx High School there in 1950. He went on to Cornell University, where he gained his bachelor's degree in 1954. His MA (1955) and PhD in physics (1959) were gained at Harvard University, and his postdoctoral research took him to the Bohr Institute, the European Organization for Nuclear Research (CERN) in Geneva, and the California Institute of Technology. After a year at Stanford he joined the faculty of the University of California at Berkeley (1961–66). In 1967 he returned to Harvard as a professor of physics and remained there until 2000, when he unexpectedly moved to become Metcalf Professor of Mathematics and Physics at Boston University. Glashow is a notable opponent of string theory, a fact that some say explains why he left Harvard.

The award of the 1979 Nobel Prize, shared with Abdus SALAM and Steven WEINBERG, was for the explanation of the forces that bind together elementary particles of matter. The citation was "for their contribution to the theory of the unified weak and electromagnetic interaction between elementary particles, including *inter alia* the prediction of the weak neutral current."

The Weinberg–Salam theory was a major step in unifying two of the four fundamental forces of physics: the electromagnetic interaction and the weak interaction. The theory was originally applied only to the class of particles known as leptons (electrons and neutrinos). Glashow extended the theory to other elementary particles (including the baryons and mesons) by introducing a new property that he called "charm." The quark theory of Murray GELL-MANN could be extended by the introduction of a fourth quark – the "charmed quark" – and combinations of the four types of quark could lead to a group of particles with symmetry SU4. The idea of charm can be used to explain the properties of the J/psi particle, discovered in 1974 by Burton RICHTER and Samuel TING.

Other extensions of the quark theory have since been made involving "colored quarks" – the theory is known as "quantum chromodynamics."

Glauber, Johann Rudolf (1604–1668) *German chemist*

Glauber (**glow**-ber) was the son of a barber in

Karlstadt, now in Germany. He spent the first half of his life traveling and working in Germany, and later settled in Amsterdam. There he made his living chiefly through selling chemicals and medicinals. He made important contributions to chemistry and his fame is largely based on his discovery of a salt called *sal mirabile* (sodium sulfate). He had found that the action of sulfuric acid on common salt produced a new acid, hydrochloric, and a salt, sodium sulfate. This became known as *Glauber's salt* and was originally sold to treat everything from typhus to constipation. It is still widely used, together with Epsom salt (magnesium sulfate), as a laxative.

Glauber's chief work, *Philosophical Furnaces* (1646–48), has been described as the first comprehensive treatise on industrial chemistry outside metallurgy. He seems to have been something of a consultant developing new products and advising on chemical processes in industry. He also speculated on the nature and rules of chemical affinity. His technical works were published as *Opera omnia chymica* (1658; Complete Works on Chemistry) in seven volumes. His *Prosperity of Germany*, written after the Thirty Years' War, looked forward to a peaceful Europe ruled by a Germany whose power arose from a science-based military technology.

Glauber, Roy J. (1925–) *American physicist*

Glauber was born in New York and educated at Harvard University, where he received his PhD in 1949. Most of his subsequent career has been spent as a professor of physics at Harvard University. He was one of the founders of the subject of *quantum optics* – i.e., an extension of the theory of optics in which the electromagnetic field associated with light is analyzed using quantum field theory rather than the classical electrodynamical theory described by MAXWELL's equations.

Glauber first put forward his fundamental work on quantum optics in 1963. This described a quantum theory of optical coherence. In his work, he emphasized that analyzing experiments that detect photons in terms of quantum electrodynamics in a consistent way necessarily involves the idea that, after a photon has been absorbed, the quantum state of the electromagnetic field itself has been changed. Glauber was able to describe interference effects in optical experiments in terms of his theory. In particular, his formalism could give a clear description of the differences between light fom a laser, which is characterized by coherence, and light from a source such as a light bulb, in which there is no coherence. He was also able to describe the circumstances under which it is reasonable to regard the quantum effects on classical light as small fluctuations.

Glauber's work of 1963 provided the foundations for a vast amount of subsequent theoretical and experimental work on quantum optics. Quantum optics has many practical applications and has also been used to investigate the fundamentals of quantum mechanics. For his work in this area Glauber shared the 2005 Nobel Prize for physics with John HALL and Theodor HÄNSCH.

Glazebrook, Sir Richard Tetley (1854–1935) *British physicist*

Glazebrook, a Liverpudlian by birth, graduated from Cambridge University in 1876 and at once joined the Cavendish Laboratory, Cambridge, under the directorship of James Clerk MAXWELL. He became lecturer in physics and mathematics and in 1891 was appointed assistant director of the Cavendish. He left to become first principal of University College, Liverpool (1898–99) and then director of the newly established National Physical Laboratory at Teddington (1900–19). He was knighted in 1917.

Glazebrook conducted a considerable amount of research on electrical standards involving very precise electrical measurements. He also made contributions to the fields of optics and thermometry. From 1909 the National Physical Laboratory undertook research into aeronautics. This work included wind-tunnel experiments for aircraft development and the adoption of scientific apparatus for military and aeronautical engineering purposes.

After his retirement in 1919 Glazebrook took on new teaching and administrative responsibilities, serving for many years as chairman of the Aeronautical Research Committee. He was the author of several books on physics and science in general and also edited the *Dictionary of Applied Physics*.

Glisson, Francis (1597–1677) *English physician*

Born at Rampisham in southwest England, Glisson was educated at Cambridge University where he obtained his MD in 1634. He was appointed professor of physics at Cambridge in 1636 and retained the post until his death. However most of his time was spent in private practice in London, so an assistant was employed to fulfill his Cambridge teaching obligations.

Glisson was a member of the group that, beginning in 1645, met regularly in London and out of which the Royal Society was later to emerge. From this "Invisible College," as it was later known, came one of the earliest examples of cooperative research. A committee of nine was set up in 1645 to investigate rickets but, as

Glisson's contribution far exceeded that of any other contributor, it was agreed that he should publish the report *De rachitide* (1650; On Rickets) under his own name. Although the nature of rickets could only begin to be comprehended with the discovery of vitamins by Casimir FUNK in 1912, Glisson must be credited for his clear description of the disease.

He was more original and influential in his account of irritability, first formulated in his work on the liver, *Anatomia hepatis* (1654; Anatomy of the Liver). He argued that muscular irritability, that is, their tendency to respond to stimuli, was independent of any external input, nervous or otherwise. This was a considerable improvement over the orthodox position adopted by the followers of René DESCARTES who believed that muscle could only respond by being pumped up like a tire, with a subtle nervous spirit rather than air. Glisson later reported a simple experiment where he placed his arm in a tube filled with water and noted that when his muscles contracted the level of water actually fell. This showed quite clearly, he claimed, that there had been no flow of anything into the limb.

It was this idea of irritability which, picked up by Albrecht von HALLER in the following century, was to find a permanent place in physiology.

Gmelin, Leopold (1788–1853) *German chemist*

Gmelin (**gmay**-leen), whose father and grandfather were botanists, was born at Göttingen (in Germany) and studied at the universities of Tübingen, Göttingen, and Vienna. In 1817 he was appointed to the first chair of chemistry at Heidelberg, where he remained until 1851. In 1817 he published the first edition of what was to become the major chemical textbook of the first half of the 19th century, *Handbuch der Chemie* (Handbook of Chemistry), in three volumes. By 1843 the book was in its fourth edition and had been expanded to nine volumes. In this edition Gmelin adopted the atomic theory and devoted much more space to the growing discipline of organic chemistry. The terms *ester* and *ketone* were introduced by him. His book was translated into English in 1848.

He also worked on the chemistry of digestion, discovering several of the constituents of bile, and introduced *Gmelin's test* for bile pigments. In 1822 he discovered potassium ferrocyanide.

Goddard, Robert Hutchings (1882–1945) *American physicist*

> God pity a one-dream man.
> —Quoted by Carl Sagan in *Broca's Brain* (1980)

Goddard was educated in his hometown of Worcester, Massachusetts, gaining his BSc in 1908 from the Polytechnic Institute and his PhD in 1911 from Clark University. He did postgraduate work at Princeton, 1912–13. He returned to Clark in 1914 where he was made professor of physics in 1919 and where he remained until his retirement in 1943.

Goddard is remembered as the designer of the first successful liquid-fuel rocket. He had been interested in space travel as a boy and his early views were first publicly revealed in his famous pamphlet *A Method of Reaching Extreme Altitudes*, published in 1919. He had most of the basic ideas of rocket design and travel clear in his mind in his 1919 paper. He realized that a reaction engine would be necessary and that sufficient thrust could only be developed with liquid fuels. In 1923 he therefore started to design rockets powered by liquid oxygen.

Unlike his Russian contemporary Konstantin TSIOLKOVSKY, he was fortunate to attract almost immediate and generous backing. The Smithsonian Institution supported him until 1929 when, through the influence of Charles Lindbergh, the Guggenheim family backed him with $15,000 a year up to the mid-1940s. His first successful rocket flight was made in 1926 when a four-foot rocket flew for two seconds and reached an altitude of 40 feet (12 m). This was followed by bigger, faster, and inevitably noisier and smellier rockets. To escape the complaints and threatened litigation he set up a research and testing station at Roswell, New Mexico, in 1929, financed by the Guggenheim Foundation.

He continued to make improvements in his design. In 1931 he introduced the now familiar automatic launch sequential system and in 1932 gyroscopic steering. In 1935 his liquid-fuel rockets flew at supersonic speeds. The greatest altitude reached was 12½ miles (20 km).

Goddard's New Mexico station closed down during the war even though he offered his work to the military. He lived just long enough to inspect the captured German V–2 rockets, which, though much bigger than his, were based on the same principles. He was also prudent enough to take out about 100 patents on rocket design. His widow was later to receive $1,000,000 for their use from the Department of Defense.

Gödel, Kurt (1906–1978) *Austrian–American mathematician*

> It is impossible to demonstrate the noncontradictoriness of a logical mathematic system using only the means offered by the system itself.
> —Gödel's statement of his "incompleteness" theorem for arithmetic

Born in Brünn (now Brno in the Czech Republic), Gödel (**gu(r)**-del) initially studied physics at the University of Vienna, but his interest soon turned to mathematics and mathematical logic. He obtained his PhD in 1930 and the same year joined the faculty at Vienna. He became a member of the Institute for Advanced Study, Princeton, in 1938 and in 1940 emigrated to America. He was a professor at the Institute from 1953 to 1976, and received many scientific honors and awards, including the National Medal of Science in 1975. He became a naturalized American citizen in 1948.

In 1930 Gödel published his doctoral dissertation, the proof that first-order logic is complete – that is to say that every sentence of the language of first-order logic is provable or its negation is provable. The completeness of logical systems was then a concept of central importance owing to the various attempts that had been made to reveal a logical axiomatic basis for mathematics. Completeness can be thought of as ensuring that all logically valid statements that a formal (logical) system can produce can be proved from the axioms of the system, and that every invalid statement is disprovable.

In 1931 Gödel presented his famous incompleteness proof for arithmetic. He showed that in any consistent formal system complicated enough to describe simple arithmetic there are propositions or statements that can neither be proved nor disproved on the basis of the axioms of the system – intuitively speaking, there are logical truths that cannot be proved within the system. Moreover, as a corollary Gödel showed (what is known as his second incompleteness theorem) that the *consistency* of any formal system including arithmetic cannot be proved by methods formalizable within that system; consistency can only be proved by using a stronger system – whose own consistency has to be assumed.

Gödel's second great result concerned two important postulates of set theory, whose consistency mathematicians had been trying to prove since the turn of the century. Between 1938 and 1940 he showed that if the axioms of (restricted) set theory are consistent then they remain so upon the addition of the axiom of choice and the continuum hypothesis, and that these postulates cannot, therefore, be disproved by restricted set theory. (In 1963 Paul COHEN showed that they were independent of set theory.)

Gödel also worked on the construction of alternative universes that are models of the general theory of relativity, and produced a rotating-universe model.

Gödel apparently suffered from depression throughout much of his life. In 1936–37 he spent some time in an Austrian sanatorium being treated for the condition. He was also something of a hypochondriac. He retired from the institute in 1976 and when, soon after, his wife underwent major surgery, he seems to have stopped eating. Apparently he was convinced that he was being poisoned. In late 1977 he was admitted to hospital dehydrated and undernourished. He refused to eat and two weeks later died from "malnutrition and inanition caused by personality disturbance."

Godwin, Sir Harry (1901–1985) *British botanist*

Godwin was born at Rotherham in Yorkshire, England, and educated at Cambridge University. He remained at Cambridge for his whole career, serving as demonstrator, lecturer, and reader before becoming director of the subdepartment of Quaternary research in 1948. He held the chair of botany from 1960 until his retirement in 1968.

Godwin's early investigations concerned the development of fen and bog vegetation using the technique of pollen analysis. During these studies he compiled a card index of higher plants with information about their Quaternary history. This led to his *History of the British Flora* (1956), which has become a standard reference for archeological, climatic, geological, and botanical investigations of the Quaternary period in Europe.

Goeppert-Mayer, Maria (1906–1972) *German–American physicist*

Maria Goeppert (**gu(r)**-pert) was born at Kattowitz in Poland and educated at the University of Göttingen where she obtained her PhD in 1930. (She changed her name on marrying the physical chemist, Joseph Mayer.) Emigrating to America in 1931 she was employed at Johns Hopkins University, Baltimore (1931–39), Columbia University, New York (1939–46), and the Argonne National Laboratory (1946–60). Finally, in 1960 she took a post at the University of California, San Diego, at La Jolla.

In 1963 she was awarded the Nobel Prize for physics together with Hans JENSEN and Eugene P. WIGNER for their work on nuclear shell theory. The shell theory of the nucleus is analogous to the shell model of the atom. The theory could help explain why some nuclei were particularly stable and possessed an unusual number of stable isotopes. In particular, in 1948, she argued that the so called "magic numbers" – 2, 8, 20, 50, 82, and 126 – which are the numbers of either protons or neutrons in particularly stable nuclei, can be explained in this way. She supposed that the protons and neutrons are arranged in the nucleus in a series of nucleon shells. The magic numbers thus describe those nuclei in which certain key shells are complete. In this way helium (with 2 protons and 2 neutrons), oxygen (8 of each), calcium (20 of each), and the ten stable isotopes of tin with 50 protons all fit neatly into this pattern. Also significant was the fact that, in general, the more complex a nucleus becomes the less likely it is to be stable (although there are two complex stable nuclei, lead 208 and bismuth 209, both of which have the magic number of 126 neutrons).

Goethe, Johann Wolfgang von (1749–1832) *German poet, novelist, dramatist, and natural philosopher*

> As for what I may have done as a poet, I take no pride in it whatever.... But that in my century I am the only person who knows the truth in the difficult science of colors – of that, I say, I am not a little proud, and here I have a consciousness of superiority to many.
> —Quoted by Johann Peter Eckermann in *Conversations with Goethe* (1837)

> Both the way in which the parts of an animal relate to one another and the particular individual characteristics of those parts are determined by the creature's everyday needs. This explains the distinctive, yet strictly delimited, lifestyles of the various animal species …
> —*First Draft of a General Introduction to Comparative Anatomy* (1795)

Born the son of a lawyer at Frankfurt in Germany, Goethe (**gu(r)**-te) was educated privately until 1765 when, under pressure from his father, he agreed to study law at Leipzig University. In 1776 he joined the court of the Duke of Weimar which remained as his main base for the rest of his life. By this time he had established a European reputation following the publication in 1774 of his semi-autobiographical novel *Die Leiden des Jungen Werthers* (The Sorrows of Young Werther). Goethe was a prolific writer, producing about 133 published works. His greatest work *Faust* (Part I, 1808; Part II, 1832) was completed in the last years of his life.

Throughout his life Goethe was interested in science and philosophy, as well as astrology and the occult. He founded the science that he called *morphology* (a term he introduced around 1817). To Goethe this was a general discipline involving the systematic study of all types of formation and transformation – whether of rocks, clouds, colors, living things, or structures of society. To some extent, this integrated view of natural science is echoed in quite modern ideas, such as those of catastrophe theory, chaos theory, and complexity theory.

In particular, Goethe had ideas about the morphology of plants and animals. These are most clearly enunciated in two didactic poems, *Die Metamorphose der Pflanzen* (1798; The Metamorphosis of Plants) and *Metamorphose der Tiere* (1806; Metamorphosis of Animals). There were, he argued, "Urpflanze" and "Urtier," ideal archetypes of the plant and animal kingdom. Thus the reason many species resembled each other had nothing to do with their history, evolutionary or otherwise, but was because they were built to the same basic design. Organisms were not only variations on a single type, but each type was composed of morphologically identical segments. Thus for Goethe, the different parts of a plant – stamens, pistils, etc. – are all variations of the leaf.

In Goethe's animal morphology, the archetype for a vertebrate was the vertebra, with the skull seen as a vertebral variation. As a consequence of his assumptions Goethe was forced to argue that any structure present in one mammal would have to be found in all others. As ruminants have an intermaxillary bone in their upper jaw, he was compelled to find a similar structure in man. Even though it is present only in human embryos (the premaxilla), Goethe claimed in 1784 to have found the bone in adult males.

Goethe's poems on plant and animal morphology approach the central issue from opposite directions: whereas the earlier work celebrates the diversity of plant life, the later poem emphasizes that even the most unusual evolutionary feature emanates from the archetype. Thus the very organization of these literary works as complementary contrasts is indicative of Goethe's holistic thinking.

Goethe's other main area of scientific interest was represented in his *Zur Farbenlehre* (On the Science of Colors, 1810), a work he perhaps valued more than any of his poetry. It was deeply (and naively) critical of NEWTON and sought to replace reliance on such abstract notions as "rays" of light with a more direct appeal to observed phenomena. There was much more to vision, he insisted, than the measurements of lines and angles. Ultimately, however, Goethe's attempt to ascribe affective qualities to particular colors abandons the realm of empirical science. Significantly, the *Farbenlehre* finds later resonance in the Expressionist painter Wassily Kandinsky's highly subjective color theory, rather than in the modern chromatics of such people as CHEVREUL.

Contemporary chemical theory formed the basis of his novel *Elective Affinities* (Die Wahlverwandtschaften, 1809). The 18th-century chemist spoke of a "double affinity" in which two compounds AB and CD join to exchange parts and create two new bodies, AC and BD. (Chemists now call this type of reaction "double decomposition.") In a similar way, in the novel, a married couple Eduard and Charlotte form a double affinity with their two friends, the Captain and Ottilie.

The mineral form of iron oxide, *goethite*, was named in Goethe's honor by LENZ in 1806.

Gold, Thomas (1920–2004) *Austrian–American astronomer*

> Things are as they are because they were as they were.
> —Quoted by Misner, Thorpe, and Wheeler in *Gravitation*

Born in Vienna, Gold became a refugee from the Austrian Anschluss and gained his BA in 1942 from Cambridge University, England. He lectured there in physics from 1948 to 1952 before he joined the Royal Greenwich Observatory as chief assistant to the Astronomer Royal. He moved to America in 1956, becoming director of the Center for Radiophysics and Space Research at Cornell from 1959 to 1981, and professor of astronomy from 1971 to 1986.

Gold is best known for his contribution to cosmology, the study of the origin, evolution, and large-scale structure of the universe. In the 1940s the prevailing cosmological model was the big-bang theory originally proposed by Georges LEMAÎTRE. Since this theory postulated a "beginning of time" when the incredibly compact universe exploded into being, it was regarded with suspicion and alarm by many astronomers. In 1948 Gold published, with Hermann BONDI, *The Steady-State Theory of the Expanding Universe*. At the heart of this paper was the adoption of what became known as the "perfect cosmological principle." This was an extension of the cosmological principle, which states that the universe looks basically the same from whichever point one observes it; Gold and Bondi added to this that the time of observation was as irrelevant as the place. Thus the universe, on a large scale, is unchanging in time and space. It had no beginning, will never end, and a constant density of matter throughout space will always be maintained. This needed to be reconciled with the work of Edwin HUBBLE, which Gold and Bondi accepted and which showed that the galaxies are receding and the universe is expanding. To maintain the steady state of their universe, Gold and Bondi had to introduce an original and startling proposition, namely, that there must be continuous creation of new matter from nothing. They calculated the amount needed as about one hydrogen atom per cubic kilometer of space every ten years, an amount too small to be detected. Although this proposition conflicted with such deep physical assumptions as the conservation of matter and the laws of thermodynamics they found that it was compatible with all astronomical data.

The steady-state theory proved attractive to a number of cosmologists and crucial evidence only emerged against it in the 1960s. Then Arno PENZIAS and Robert WILSON discovered the background microwave radiation in 1965, and Maarten SCHMIDT produced a survey of the distribution of quasars that seemed to support the evolving universe of the big-bang theory.

In 1968 news of a new type of star, a "pulsar," was published by Jocelyn BELL-BURNELL and Antony HEWISH. The distinguishing features of the pulsar were its high-frequency radio signals, which had a periodicity of the order of a second or less. Gold quickly proposed a structure capable of producing such an effect: rapidly rotating neutron stars. The same theory was proposed independently by Franco Pacini. Neutron stars are extraordinarily dense stars that have undergone such extreme gravitational collapse following exhaustion of their nuclear fuel that their constituent protons and electrons have combined to form neutrons. These stars would be small and dense enough to rotate with a period equivalent to that of the radio pulses. It had also been shown that they would radiate energy in a narrow beam. If the Earth happened to be in the direction of the beam, the beam would be picked up as a source of pulses, much as the beam of a lighthouse is observed as a series of flashes. The theory of Gold and Pacini was eventually accepted once pulsars rotating even faster than the original one were detected in the Crab and Vela nebulae.

Gold was able to make a prediction that has since been confirmed. He argued that pulsars should be slowing down by a small but measurable amount, because of the loss of energy. Following careful observation of the pulsar in the Crab nebula it was found to be slowing down and its period increasing by 3.46×10^{-10} seconds per day.

Goldberger, Joseph (1874–1929) *American physician*

Goldberger, the son of Jewish immigrants, had been born at Giralt in Hungary and was brought to America at the age of six. He was educated at the College of the City of New York and at Bellevue Hospital Medical School. After a brief period in private practice Goldberger joined the U.S. Public Health Service in 1899, remaining there for the rest of his life.

Goldberger worked as a field officer for many years, making contributions to the understanding and control of such diseases as yellow fever, typhus, and dengue. He is, however, mainly remembered for his authoritative investigation of the nature, causation, and treatment of pellagra. This disease, which became widely known in America after the Civil War, is typified by chronic diarrhea, roughening of the skin, a sore tongue, and involvement of the nervous system. Death from secondary infection or general emaciation was not uncommon.

When Goldberger began his work in 1913 it was thought that the disease was caused by an unknown toxin produced by bacterial fermentation during storage of grain. But stimulated by the work of Frederick Gowland HOPKINS and Casimir FUNK, Goldberger directed his attention to deficiency diseases. He began a classic investigation into the connection between pellagra and diet in various asylums and orphanages of the southern states. He was immediately struck by the fact that the staff of such institutions – with a diet containing milk, eggs, cheese, and meat – remained free of the disease while the inmates, subsisting virtually on cereals alone, frequently suffered from epidemics of pellagra.

It was a relatively simple matter to show that the disease could be eliminated by supplementing the inmates' diet with milk. He was further able to trade the offer of a pardon with 11 inmates of a Mississippi prison for their adoption of a diet of corn, rice, sugar, pork fat, potatoes, and turnips. Within a few months 7 of the 11 were showing early symptoms of pellagra. Attempts to transmit the disease by contact with the clothes, excreta, and vomit of the patients ended in failure. All this led Goldberger to propose the existence of a P-P, or pellagra-preventive, factor. Whatever such a factor might be, he was able to show by 1920 that sufficient of it was contained in a daily dose of 15–30 grams of yeast, and by this means alone Goldberger was able to prevent the 10,000 deaths a year attributable to pellagra in the United States.

The active ingredient involved was shown in 1937 by Conrad ELVEHJEM to be nicotinic acid (niacin), part of the vitamin B complex.

Goldhaber, Maurice (1911–) *Austrian–American physicist*

Goldhaber (**gohld**-hahber), who was born at Lemberg (now Lvov in Ukraine), was educated at the universities of Berlin and Cambridge, where he obtained his PhD in 1936. He emigrated to America in 1938 where he first taught at the University of Illinois, becoming professor there in 1945. He moved to the Brookhaven National Laboratory in 1950, serving as its director from 1961 until 1973.

In 1934, while at the Cavendish Laboratory of Cambridge University, Goldhaber codiscovered the nuclear photoelectric effect with James CHADWICK. This is the disintegration of a nucleus by high-energy x-rays or gamma rays. From this it was later established that the neutron is slightly heavier than the proton. Following Enrico FERMI's discovery of slow neutrons, Chadwick and Goldhaber also discovered (1934–35) the neutron disintegration reactions for lithium, boron, and nitrogen. The nitrogen reaction is the major source of radioactive carbon–14 on Earth.

At the University of Illinois (1938) Goldhaber and his wife, Gertrude Scharff-Goldhaber, demonstrated that electrons and beta particles are the same. In 1940 he discovered that beryllium is a good moderator, i.e., it slows down fast neutrons so they more readily split uranium atoms.

He has also proposed a cosmological theory in which an initial "universon" broke up into a "cosmon" (matter) and an "anticosmon" (antimatter), with the anticosmon forming a second universe made of antimatter. After his retirement from Brookhaven, Goldhaber worked on the Irvine–Michigan–Brookhaven (IMB) underground detector studying proton decay and neutrino oscillations.

Goldschmidt, Johann (Hans) Wilhelm (1861–1923) *German chemist*

Goldschmidt (**gohlt**-shmit), the son of an Essen industrialist, was born at Berlin in Germany. After studying at Berlin and Heidelberg he joined the family business in 1888.

In 1905 Goldschmidt introduced cheaper methods for the reduction of metallic oxides to metals (the "Thermit process"). It had been common for the reduction process to use highly reactive but expensive metals, such as potassium and sodium. Goldschmidt showed that if a mixture of the metal oxide and aluminum powder is ignited with magnesium in contact with barium peroxide, great heat is generated and the pure metal produced. The technique was important in the production of certain metals used in alloy steels. It was also used for localized welding of steel (using iron oxide and aluminum).

Goldschmidt, Victor Moritz (1888–1947) *Swiss–Norwegian chemist*

> This poison is for professors of chemistry only. You, as a professor of mechanics, will have to use the rope.
> —Explaining his possession of a cyanide capsule to a university colleague during the Nazi occupation of Norway

Goldschmidt, the son of H. J. Goldschmidt, a physical chemist, was born at Zurich in Switzerland. He attended Christiania (now Oslo) University where he obtained his PhD in 1911, remaining in Norway as director of the Mineralogical Institute until 1929 when he moved to the University of Göttingen in Germany. Being a Jew he returned to Norway in 1935, following the rise of anti-Semitism and the Nazi party. He was later sent to a concentration camp but was released by the Norwegian authorities on the grounds of ill health and escaped to England (1942). His time in England was spent first at the Macaulay Institute for Soil Research near Aberdeen, and later at the Rothamsted Experimental Station, Harpenden. He returned to Oslo after the war.

Goldschmidt is acknowledged as the founder of modern geochemistry. Following the work of Max VON LAUE and W. H. and W. L. BRAGG, he laid the foundation for his work by working out the crystal structure of over 200 compounds. His interest was directed to more practical work when, as a result of the naval blockade in the war, he was called upon to investigate Norway's mineral resources.

By the mid-1920s the atomic radii of elements in various stages of ionization had been established. Using this information, together with his detailed knowledge of crystal structure, Goldschmidt began predicting in which minerals and rocks various elements could or could not be found. His results were published over the years in his eight-volume *Geochemische Verteilungsgesetze der Elemente* (1923–38; The Geochemical Laws of the Distribution of the Elements). His book *Geochemistry* was published posthumously in 1954.

Goldstein, Eugen (1850–1930) *German physicist*

Goldstein (**gohlt**-shtIn), who was born at Gleiwitz (now Gliwice in Poland), studied for a year at the University of Breslau (1869–70) then worked with Hermann von HELMHOLTZ at the University of Berlin. He was appointed physicist at the Berlin Observatory in 1878, took his doctorate in 1881, and later established his own laboratory. In 1927 he became head of the astrophysical section of the Potsdam Observatory.

Goldstein's best-remembered scientific work is his studies of electrical discharges in gases at low pressures. He gave the name "cathode rays" to the invisible emanations coming from the cathode of an evacuated discharge tube, showed that the rays could cast sharp shadows, and demonstrated that they were emitted perpendicular to the cathode surface. He later showed that they could be deflected by magnetic fields. Like most German scientists Goldstein believed that cathode rays were waves like light. J. J. THOMSON later identified cathode rays as a stream of electrons. In 1886 Goldstein announced his discovery of *kanalstrahlen* (canal rays), rays that emerged from channels in the anode in a low-pressure discharge tube. These were identified as positively charged particles by Jean Baptiste PERRIN in 1895, and later exploited in mass spectroscopy.

Goldstein, Joseph Leonard (1940–) *American medical geneticist*

Born at Sumpter in South Carolina, Goldstein attended Washington and Lee University, Virginia, and the University of Texas Southwestern Medical School, where he gained his MD in 1966. For two years he worked at Massachusetts General Hospital, Boston. In 1968 he joined the National Institutes of Health as a clinical associate (1968–70). After a stint of research at the University of Washington, Seattle (1970–72), he joined the University of Texas Southwestern Medical Center in Dallas. In 1977 Goldstein was appointed professor of medicine and chairman of the Department of Molecular Genetics, and in 1985 he was made regental professor.

Goldstein's work has centered on the metabolism of cholesterol, fats, and other lipids in the body; much of it has been done in collaboration with his fellow biochemist and geneticist, Michael BROWN, whom Goldstein met when both were interns at Massachusetts General Hospital in 1966. Starting in the early 1970s, the pair began by studying how cells obtain their cholesterol from blood. Most of the blood's cholesterol is present in the form of low-density lipoproteins (LDLs) – minute particles comprising proteins, cholesterol, and other lipids. Working with cultures of skin cells, Goldstein and Brown discovered receptors on the cell surface that recognize the LDLs and bind them to the cell membrane. The LDL is subsequently enfolded by the cell membrane and taken into the cell where its contents are metabolized (cholesterol, for instance, is a vital component of cell membranes).

They went on to show that there is a deficiency of LDL membrane receptors in individuals suffering from the inherited disorder known as familial hypercholesterolemia. Such persons have abnormally high levels of cholesterol in their blood and run a much greater risk of developing atherosclerosis – the narrowing of the arteries due to a buildup of fatty plaques on their inner surface. This in turn makes them much more prone to heart attacks and strokes. Goldstein and Brown were able to show that in this disorder the gene encoding the LDL receptors is defective, hence the number of such receptors is small and the sufferer's body cells are unable to remove LDLs from the bloodstream. The consequent high blood-cholesterol levels prompt scavenger white cells to remove the cholesterol, turning them into the plaque-forming cells responsible for atherosclerosis.

The work of the Dallas-based duo has covered many other aspects of cholesterol metabolism, particularly how the cholesterol absorbed from the gut into the bloodstream is processed and repackaged by the liver, with the formation of high-density, very-low-density, and intermediate-density lipoproteins. They have not only revealed fundamental features of cellular metabolism but have shown ways in which people with elevated blood cholesterol may be treated, for instance, by increasing the number of LDL receptors on their cells. This, in turn, may reduce their risk of heart attacks and strokes.

For his work on familial hypercholesterolemia and LDL receptors, Goldstein was awarded the 1985 Nobel Prize for physiology or medicine, which he shared with his long-time colleague, Brown.

Golgi, Camillo (1843–1926) *Italian cytologist and histologist*

Born at Corteno near Brescia (now in Italy), Golgi (**gol**-jee) studied medicine at Pavia University and thereafter mainly concerned himself with research on cells and tissues. In 1873, while serving as physician at the home for incurables, Abbiategrasso, he devised a method of staining cells by means of silver salts. This allowed the fine processes of nerve cells to be distinguished in greater detail than before and enabled Golgi to confirm Wilhelm von Waldeyer's view that nerve cells do not touch but are separated by gaps called synapses. Golgi also found a specialized type of nerve cell, later called the Golgi cell, which, by means of fingerlike projections (dendrites), serves to connect many other nerve cells. This discovery led to the formulation (by Waldeyer) and establishment (by Santiago Ramón y Cajal) of the neuron theory – a theory that Golgi was nevertheless strongly opposed to.

Golgi was also the first to draw attention to the Golgi bodies: flattened cavities parallel to the cell's nuclear membrane whose function appears to be packaging and exporting various materials from the cell. Apart from work on the sense organs, muscles, and glands, Golgi studied varying forms of malaria. He found that different species of the protozoan parasite *Plasmodium* are responsible for the two types of intermittent fever – the tertian and quartan. He also established that the onset of fever coincides with the release into the blood of the parasitic spores from the red blood cells.

Golgi served as professor of histology (1876) and then of general pathology (1881) at Pavia University. In 1906 he shared with Ramón y Cajal the Nobel Prize for physiology or medicine for his work on the structure of the human nervous system.

Gomberg, Moses (1866–1947) *Ukrainian–American chemist*

> Gifted with a remarkable memory, he presented his lectures with the full use of a wealth of historical material and so vividly that they left an indelible imprint on his students.
> —C. S. Schoepfle and W. E. Bachmann, *Journal of the American Chemical Society* (1947)

Gomberg's father, an estate owner in Elizavetgrad (now Kirovograd in Ukraine), fled with his family from Russia in 1884 when accused of plotting against the czar. Gomberg was educated at the University of Michigan where he obtained his doctorate in 1894. After a period abroad at Munich and Heidelberg he returned to Michigan where he spent his whole career serving as professor of organic chemistry from 1904 until his retirement in 1936.

Gomberg is noted for the first preparation of a stable free radical – i.e., a group of atoms with an unpaired electron. In 1900, he was trying to make hexaphenylethane, which is simply an ethane molecule (C_2H_6) in which all the hydrogen atoms have been replaced by phenyl groups (C_6H_5). Gomberg found to his surprise, and everyone else's disbelief, that he was obtaining the free radical, triphenylmethyl – (C_6H_5)$_3$C – which clearly has a carbon atom with only three phenyl groups attached to it;

that is, with a forbidden valence of three. In general free radicals are highly reactive short-lived entities. Gomberg's compound, a colorless crystalline substance, was stabilized by the three benzene rings.

Gomberg later discovered a suitable antifreeze for automobile radiators, ethylene glycol.

Good, Robert Alan (1922–2003) *American pathologist and immunologist*

Good was the son of a high-school principal who died of cancer when Good was five. Born in Crosby, Minnesota, he was educated at the University of Minnesota where he simultaneously obtained an MD and PhD in 1947. After this triumph he joined the Minnesota staff and served as professor of pediatrics from 1954 until 1973, when he moved to New York as director of the Sloan-Kettering Institute for Cancer Research. In 1982 Good moved to the University of Oklahoma as professor of microbiology, a post held until 1985, when he was appointed to a similar position in the University of South Florida, Tampa.

One of the great achievements of modern immunology has been the demonstration that the immunological system is not a simple unity but rather a complex interrelationship of a number of different units. The unraveling of this particular tangle was not the work of any one man or, indeed, any one group; Good's contribution was, however, as great as any other.

In the 1940s he showed a link between plasma cells, cells found in lymphoid tissue, and antibodies. Later he noted a simple tendency to recurrent infection among his patients suffering from myeloma (a tumor of bone-marrow cells) despite an abundance of plasma cells. This suggested to Good that there must be more to the immune system than simply the ability to make antibodies. This was reinforced when examining patients with agammaglobulinemia, who had no plasma cells at all, yet who were immunologically active enough to reject foreign skin grafts.

In the mid 1950s Good realized that there are two parts to the immune system: one dealing with defenses against typical bacterial infections; the other more concerned with clearing up "foreign" or unusual cells. By 1961, independently of Jacques MILLER, Good was beginning to suspect that the thymus gland was deeply implicated in providing the latter type of immunity.

Work on chickens' defense mechanisms against bacterial infection had demonstrated that if the bursa (a gland found in the chicken's alimentary canal) was removed, the creature lost the ability to make antibodies in any real quantity. They were in fact just like Good's agammaglobulinemic patients.

Good therefore postulated that there must be

two types of immunity – one related to the thymus and the other related to the human equivalent of the chicken bursa producing antibodies. The details of the two systems and their evolution and interrelationship called for major, and as yet far from complete, research programs by immunologists.

Good became a leading proponent of the view that cancer is somehow the result of an immunological defect, a failure of the system to recognize and destroy the cancerous cell before it has begun to proliferate.

Goodall, Jane (1934–) *British primatologist*

> Just as he is overshadowed by us, so the chimpanzee overshadows all other animals. He has the ability to use and make tools for a variety of purposes, his social structure and methods of communication with his fellows are elaborate, and he shows the beginnings of self-awareness.
> —*In the Shadow of Man* (1971)

In 1957 Goodall, a Londoner, approached Louis LEAKEY for a job of some kind as she "wanted to get closer to animals." Leakey employed her, initially as a secretary, and took her with him to Olduvai. He told her that he had long been searching for someone sufficiently interested in animals to be prepared "to forego the amenities of civilization for long periods of time without difficulty." More precisely he wanted someone to observe the 160 chimps of the Gombe Stream Reserve on the eastern shores of Lake Tanganyika at close quarters over several years. After some initial training at the Royal Free Hospital and the London Zoo, Goodall was installed at Gombe in 1970. She has remained there ever since; under her direction it has become a world-famous and much-respected research center.

In her first full account of her work, *In the Shadow of Man* (1971), Goodall presented what now seems to be a somewhat idealized picture of chimpanzee society. They were seen as mainly vegetarian, living in a relatively peaceful community and spending the bulk of their lives socializing with each other. They were also shown as toolmakers and users, adept at fashioning blades of grass into probes to be inserted into mounds to extract termites.

Her later work, however, presented in *Through a Window* (1990), revealed a darker side of chimpanzee society. The Gombe Reserve was home for three communities of about 50 chimpanzees each. Males will routinely attack and attempt to kill adult females of another group. In 1974, Goodall witnessed the outbreak of war within a single community. At that time the band split into two groups, which she called the Kahama and the Kasakela. Over a period of four years, Goodall noted that the Kasakela systematically and deliberately killed the entire Kahama group, males, females, and infants, presumably to take over their territory. It took several years for Goodall to come to terms with this picture. Goodall also rejected her earlier account of vegetarian chimpanzee bands. She found that they hunted monkeys, baboon infants, bushpig, bushbuck, and other small mammals. Hunting is undertaken by males and always in groups. Cannibalism took place on a number of occasions and the meat, as at other times, was shared within the group.

Because of her prolonged observation Goodall has been able to document the social development of the individual in the community as well as the histories of a number of families. Males establish a dominance hierarchy and protect the group, while females remain with their mothers until they reach sexual maturity at about the age of ten. Young males leave a few years earlier to establish their place in the male hierarchy. Sibling and maternal ties, however, remain strong.

In more recent years Goodall has campaigned vigorously for the conservation of chimpanzees in the wild and for a less barbarous confinement of the many held in research institutions and zoos. To this end she has set up the Jane Goodall Institute for Research Education and Conservation with centers in the U.S., Canada, and Britain. She has reported on her life's work at Gombe in a number of popular and often moving books, most recently in *The Chimpanzee* (1992). She has been the subject of a number of documentaries and a film *Jane Goodall's Wild Chimpanzees* (2002).

Goodman, Henry Nelson (1906–1998) *American philosopher*

> If we lack the means for interpreting counterfactual conditions, we can hardly claim to have any adequate philosophy of science.
> —*Fact, Fiction, and Forecast* (1955)

Goodman was born in Somerville, Massachusetts, and educated at Harvard where he obtained his PhD in 1941. Before this he had worked as a Boston art dealer from 1929 until 1941. After serving in the U.S. Army during the war he taught at Tufts University (1945–51), the University of Pennsylvania (1951–64), Brandeis (1964–67), and finally at Harvard from 1968 until his retirement in 1977.

In his *Fact, Fiction, and Forecast* (London, 1955) Goodman posed his "new riddle of induction." The traditional problem of induction was to justify inferences from the particular to the general; for example, from the observation that "some crows are black" to the conclusion that "all crows are black." Goodman, however, shifted the question from "Why does a positive instance of a hypothesis confirm the hypothesis?" to "What hypotheses are confirmed by positive instances?"

Thus a green emerald might be thought to

support the hypothesis "All emeralds are green." But define the predicate "grue" to refer to objects which are green before the year 2000 or blue thereafter, and a green emerald offers just as much support for the hypothesis that "All emeralds are grue." Why should we, to use Goodman's term, "project" the property green onto emeralds, but feel reluctant to project grueness? Goodman has a response to the objection that "grue" is not a proper predicate, requiring the terms "blue" and "green" to define it. If one defines "bleen" to refer to objects that are blue before the year 2000 or green thereafter, the color green is definable as being grue before 2000 or bleen thereafter.

Goodman's new riddle has been extensively discussed. His own suggestion is that we prefer hypotheses that are "entrenched," that is, the degree with which predicates have been projected in the past.

Goodpasture, Ernest William (1886–1960) *American pathologist*

Goodpasture, the son of a lawyer, was born in Montgomery County, Tennessee. He was educated at Vanderbilt University, Nashville, and at Johns Hopkins University, where he gained his MD in 1912. After working as a pathologist for some years at Johns Hopkins and at Harvard, Goodpasture returned to Vanderbilt in 1924 as professor of pathology, a post he retained until his retirement in 1955.

In 1931 Goodpasture devised a method of virus culture that provided an enormous stimulation to virology. Before this, as viruses will grow only in living tissue, they could be studied experimentally either in a living host or, after the work of Alexis CARREL in 1911, *in vitro* (in a laboratory, the Latin *in vitro* literally means "in glass") in a tissue culture. The first method was expensive and difficult to control while the second, before the advent of antibiotics, was susceptible to contamination by bacteria.

Goodpasture, in collaboration with Alice Woodruff, avoided such difficulties by providing a cheap living environment for viral growth – a fertile egg. Their first success was with fowl pox but within a year they had also grown both cowpox and coldsore viruses. Goodpasture went on in 1933 to show that attenuated cowpox vaccine could be produced in a purer and cheaper form in eggs than by the customary method of production in calf lymph.

Within a few years Goodpasture's technique had made possible the production of vaccines against yellow fever by Max THEILER and influenza by Thomas Francis. Thereafter eggs became as standard a part of the virologist's laboratory as the test tube.

Goodrich, Edwin Stephen (1868–1946) *British zoologist*

Goodrich's father, a clergyman from Weston-super-Mare in the west of England, died when his son was only two weeks old. He was brought up in France by his mother and, as he intended to be an artist, attended the Slade School of Art at University College, London. His interest in zoology was aroused by Ray LANKESTER with whom he moved to Oxford as an assistant in 1892. Goodrich remained at Oxford for the rest of his career, serving as professor of comparative anatomy from 1921 to 1945.

Goodrich produced two important works, *Evolution of Living Organisms* (1912) and *Studies on the Structure and Development of Vertebrates* (1930). His main achievement in the latter work was to distinguish clearly, for the first time, between the nephridium and the coelomoduct, or the primitive kidney and primitive reproductive tract.

More generally his labors are comparable to those of such other paleontologists and comparative anatomists as David Watson and Alfred ROMER who, building on the efforts of previous workers, managed to sort out the muddled story of vertebrate evolution.

Goodricke, John (1764–1786) *Dutch–British astronomer*

> If it were perhaps not too early to hazard even a conjecture on the cause of its [Algol's] variation, I should imagine it could hardly be accounted for otherwise than ... by the interposition of a large body revolving around Algol.
> —*Philosophical Transactions of the Royal Society* (1783)

Born at Groningen in the Netherlands, Goodricke was a deaf mute who, although he died when he was 21, had already done work of such importance as to receive the Copley medal of the Royal Society three years before he died. Variable stars had been discovered by FABRICIUS nearly 200 years before but Goodricke was the first scientist to offer a plausible explanation. Noticing the rapid variation in magnitude of Algol he proposed, in 1782, that it was being regularly eclipsed by a dark companion that passed between it and the Earth. His suggestion was confirmed a century later.

Gordan, Paul Albert (1837–1912) *German mathematician*

Gordan studied in his native Breslau, at Königsberg, and at Berlin before becoming professor of mathematics at the University of Erlangen. For most of his mathematical career his research was concentrated on a single field, the study of indeterminates. The central problem in the field, which Gordan eventually solved, was to prove the existence of a finite basis for binary forms of any given degree. His result was subsequently refined and extended by many workers including Gordan himself.

Gordan's proof was long and complicated and the result was re-proved in 1888 by David HILBERT using newer and far simpler methods. In collaboration with Rudolf Clebsch, Gordan also wrote a book on Abelian functions that included the central theorem now known as the *Clebsch–Gordan theorem*. This work was influential in giving a new direction to algebraic geometry.

Gorer, Peter Alfred (1907–1961) *British immunologist*

Gorer, the son of a wealthy Londoner who died on the Lusitania, was educated at Guy's Hospital, London, graduating in 1929. Ater studying genetics under J. B. S. HALDANE at University College, London, from 1933 to 1934 Gorer worked at the Lister Institute until 1940 when he returned to Guy's as morbid histologist and hematologist. In 1948 he became reader in experimental pathology.

As early as 1936 Gorer tried to see if red cells of mice could be divided into antigenic groups similar to the blood groups of humans. Using his own blood serum he distinguished three kinds of mouse red cell on the basis of their ability to agglutinate his serum. He further found that such a property was inherited by the mice in a Mendelian dominant manner. Such work was supported by comparable results obtained in 1937 with the transplantation of a spontaneously appearing tumor among the various distinguished genetic strains of mice.

Gorer had in fact discovered the histocompatibility antigens and established their control at the genetic level, an outstanding result little appreciated in his lifetime but later to be recognized as of fundamental importance in immunology, genetics, and transplantation surgery. One who did recognize the significance of Gorer's work was George SNELL who worked with him in 1948. For his later work Snell was to receive the 1980 Nobel Prize for physiology or medicine, a prize Gorer would have undoubtedly shared with him if he had not died some 19 years before from lung cancer.

Gorgas, William Crawford (1854–1920) *American physician*

Gorgas was the son of an army ordnance officer from Toulminville, Alabama, who later served as a general in the Confederate Army. He was educated at the University of the South and at Bellevue Hospital, New York, where he obtained his MD in 1879. In the following year he joined the Army Medical Corps, serving in a number of frontier posts before being appointed (1898) chief sanitary officer of Havana.

Gorgas was fortunate to be in Cuba at the time when Walter REED and his colleagues identified the mosquito *Aedes aegypti* as the vector of yellow fever. Given this information Gorgas was able to introduce measures to so control the vector that the disease was virtually eliminated in Cuba. The strategy basically involved reducing mosquito breeding grounds by either draining or covering all standing sources of water, attacking the adult mosquitoes by fumigation, and isolating patients with the disease under netting. Such simple procedures, energetically and conscientiously carried out, worked remarkably quickly.

In 1899 the U.S. Government had obtained the shares of the Panama Railroad Company for $40 million, and when in 1903 they also obtained permission from the newly established state of Panama to attempt the construction of a canal, they realized the enormous medical problems facing them in such a task. The previous attempt (1881–89) by Ferdinand de Lesseps had failed, largely owing to yellow fever killing over 20,000 of his labor force in eight years. On the strength of his success in Cuba, the government appointed Gorgas chief sanitary officer of the Canal Zone in 1904. Unfortunately the chairman of the Canal Commission, Admiral John WALKER, refused Gorgas's request for screening material and sulfur for fumigation on the grounds of economy; privately he tried to get Gorgas removed, proclaiming that "the whole idea of mosquitoes carrying fever is balderdash." Before long, however, Walker resigned. Under his successor, Gorgas was allowed to introduce the measures he had pioneered in Cuba. The extent of Gorgas's achievement is best seen in the mortality figures: when he arrived the death rate from yellow fever was running at about 10%, but after 1906 no further cases were reported.

Gorgas remained in Panama until 1913. In the following year he was appointed surgeon-general of the U.S. Army and was much in demand to advise on public health programs. It was in fact on his way to West Africa to advise on yellow fever control that Gorgas died, of a stroke, in London.

Gossage, William (1799–1877) *British chemist*

Born in Burgh-in-the Marsh, England, Gossage was apprenticed to his uncle, an apothecary in Chesterfield, and later set himself up in business in Leamington selling the spa salt. He moved to Stoke Prior to manufacture salt and alkali by the LEBLANC process.

In 1836 Gossage introduced a basic improvement into the manufacture of alkali. One of the disadvantages of the Leblanc process was the production of large quantities of hydrochloric acid fumes as a waste product. This was released into the atmosphere and as a result polluted the surrounding area. Gossage constructed towers in which the gas was passed up through layers of coke, and then condensed by cold water flowing down. This fluid was fed into streams and rivers where it still acted as a pollutant until Henry DEACON discovered a method for producing chlorine from the solution. Gossage's tower made possible the introduction of the Alkali Act in 1865, which made it a legal obligation for manufacturers to absorb 95% of their waste hydrochloric acid.

Gossage also tried to invent a process to recover the sulfur from calcium sulfide, a waste product of the Leblanc process. He introduced a technique in 1837 that recovered some by partial oxidation. The thiosulfate resulting could be sold to paper mills. In 1850 Gossage moved to Widnes to open a soap and alkali factory and in 1854 he patented a process for introducing sodium silicate into soap, which produced a soap with improved detergent powers.

Goudsmit, Samuel Abraham (1902–1978) *Dutch–American physicist*

Born in The Hague in the Netherlands, Goudsmit (**gowd**-smit) was educated at the universities of Amsterdam and Leiden, where he obtained his PhD in 1927. He emigrated to America shortly afterward, serving as professor of physics at the University of Michigan (1932–46) and Northwestern University (1946–48). He then moved to the Brookhaven National Laboratory on Long Island, New York, where he remained until his retirement in 1970.

In 1925 Goudsmit, in collaboration with George UHLENBECK, put forward the proposal of electron spin. They suggested that electrons rotate about an axis and, as they are charged, set up a magnetic field. This model was successful in clearing up a number of anomalies that were becoming apparent in the fine structure of atomic spectra. A theory of spin was later given by Paul DIRAC.

During World War II Goudsmit worked on radar and then became head of a top secret mission codenamed *Alsos* in 1944. The mission was for Goudsmit to follow the front-line Allied troops in Europe, and even in some cases to precede them, looking for any evidence of German progress in the manufacture of an atomic bomb. He found that the German scientists had, in fact, made little progress and it was clear that Hitler would not be presented with such a weapon before the end of the war. For this war service Goudsmit was awarded the Medal of Freedom from the U.S. Department of Defense and he published his account of the mission in his book *Alsos* (1947).

Gould, Benjamin Apthorp (1824–1896) *American astronomer*

Gould, the son of a merchant and teacher from Boston, Massachusetts, graduated from Harvard in 1844. Having studied for a year at Berlin, he obtained his PhD from Göttingen University in 1848 under the great Karl Friedrich GAUSS. On his return to America he served as head of the longitude department of the U.S. Coast Survey from 1852 to 1867, pioneering the use of the telegraph in measuring longitude. At the same time Gould founded the *Astronomical Journal* in 1849 and edited it until 1861 when its publication was halted by the Civil War. He was also connected with the Dudley Observatory, Albany, from 1855 and served as its director briefly in 1858 before being forced to get out of town in the following year. After his traumatic expulsion from Albany he handled his father's business for some time. He set up a private observatory in Cambridge, financed by his wife, and in 1862 produced a star catalog that brought together measurements made at various observatories. He left for Argentina in 1870.

The 15 years spent in Cordoba were by far the most productive of Gould's career. He established the Argentine National Observatory there and began the first major survey of the southern skies. The Observatory's first survey of naked-eye stars, i.e., down to 7th magnitude, was published as the *Uranometria Argentina* (1879; Argentinian Survey of the Heavens). This was followed by the fuller recording, published in 1884, of 73,160 stars from 23°S to 80°S and in 1886 by the publication of the *Catàlago General* (General Catalogue) containing the more accurate recording of 32,448 stellar coordinates. This important work was continued by Gould's successor, Juan Thomé. An extended band of young stars, cloud, and dust that forms a spur off one of the spiral arms of our Galaxy and was revealed by the southern surveys was subsequently named *Gould's Belt*.

In 1885 Gould returned to Massachusetts where he restarted the *Astronomical Journal* in 1886 and worked on the 1,000 photographic plates of star clusters he brought back with him from Cordoba.

Gould, Stephen Jay (1941–2002) *American biologist*

> A man does not attain the status of Galileo merely because he is persecuted; he must also be right.
> —*Ever Since Darwin* (1977)

The grandson of a Hungarian immigrant and the son of a court stenographer, Gould is reported to have developed his interest in biology as a five-year-old when he first saw *Tyrannosaurus rex* at the American Museum of Natural History. Born in New York City, he was educated at Antioch University, Pennsylvania, and at Columbia, where he completed his PhD in 1967. He immediately moved to Harvard where he served as professor of geology and curator of the Harvard Museum of Comparative Zoology from 1973.

Gould is widely known for the volumes of essays on natural history that he began publishing in 1978. The articles are usually about some aspect of evolution and are rooted firmly in history, carry a detailed argument, and are relevant to some contemporary issue. Throughout over 300 essays Gould maintained an exceptionally high standard and dealt with a wide variety of topics. These include the panda's thumb, the flamingo's smile, Adam's navel, male nipples, Mickey Mouse's development, and changes in the average size of chocolate bars.

Gould also published a number of influential monographs. In *Ontogeny and Phylogeny* (1977) he examined the notion of recapitulation – the view that individual development (ontogeny) is a rerun of evolutionary history (phylogeny). *The Mismeasure of Man* (1984) sought to demonstrate that attempts to measure man's intelligence were often designed to serve political rather than scientific ends. In a further monograph, *Wonderful Life* (1990), Gould surveyed the fossils of the Burgess Shale, first described by C. D. WALCOTT. He used the fossils to illustrate a familiar theme of his work that evolution is not "a ladder of predictable progress," it is rather "a copiously branching bush, continually pruned by the grim reaper of extinction."

In the fields of paleontology and ecology Gould worked for many years on the West Indian land snail, *Cerion*. As an evolutionary theorist he is best known for proposing in 1972, along with Nils Eldredge, the punctuated equilibrium hypothesis, which views evolution as episodic rather than continuous. Relatively short periods of branching speciation, they argued, are followed by much longer periods of stasis.

In 1981 Gould was very much in the news as one of the biologists called as an expert witness in the so-called Scopes II trial in Arkansas. Fundamentalists had claimed as equal a right to teach creationism in the Arkansas public schools as biology teachers had long claimed for Darwinism. Judge William Overton ruled in 1985 that creationism was a religious doctrine and it would therefore be a violation of the constitution if it were to be taught in public schools.

Shortly before his death Gould published a large book, *The Structure of Evolutionary Theory*, which summarized his life's work on this topic.

Graaf, Regnier de (1641–1673) *Dutch anatomist*

Graaf (grahf), who was born at Schoonhoven in the Netherlands, studied at the University of Leiden and later obtained a degree in medicine at the University of Angers, France (1665). He was the first to collect and study the secretions of the pancreas and gall bladder but is best known for his work on the mammalian

sex organs. In 1668 he described the structure of the testicles and in 1673 described the minute follicles of the ovary, which have been called *Graafian follicles* since HALLER coined the term in the mid-18th century. Graaf died of the plague aged 32.

Graaff, Robert Jemison van de *See* VAN DE GRAAFF, ROBERT JEMISON.

Graebe, Karl (1841–1927) *German chemist*

Born at Frankfurt in Germany, Graebe (**greb**-e) graduated from Heidelberg in 1862. He worked as an assistant to BUNSEN and BAEYER before accepting an appointment as professor of chemistry at Königsberg in 1870. In 1878 he moved to the University of Geneva.

With C. T. Liebermann, Graebe made a major contribution to the chemistry of synthetic dyes. While working as assistant to Baeyer he studied alizarin, the coloring matter taken from the madder plant. Graebe and Liebermann vaporized the pigment and passed it over zinc dust and to their surprise found that anthracene was produced. As they had earlier been working on the structure of anthracene, a coal-tar derivative, they had little difficulty in seeing how to synthesize alizarin. They applied for a patent on 15 June 1869, only a day before Sir William Henry PERKIN, the discoverer of mauveine, applied for a patent for the same product.

Graham, Thomas (1805–1869) *Scottish chemist*

Graham was the son of a prosperous Glaswegian manufacturer. He entered Glasgow University at the age of 14 and attended the classes of the chemist Thomas Thomson. Graham's father was determined that he should enter the ministry and on Graham's persistence with his scientific studies his father withdrew his financial support. To continue in chemistry Graham made his living through teaching and writing. In 1829 he became a lecturer at the Mechanics Institution and in 1830 he was elected to the chair of chemistry at Glasgow University. In 1837 he was appointed professor in the recently founded University College, London. He was the first president of the Chemical Society of London, and of the Cavendish Society, which he founded. In 1854 he was made master of the mint.

In 1829 Graham published a paper on the diffusion of gases. Observations on this subject had been made by Joseph PRIESTLEY and Johann DÖBEREINER, but it was Graham who formulated the law of diffusion. He compared the rates at which various gases diffused through porous pots, and also the rate of effusion through a small aperture, and concluded that the rate of diffusion (or effusion) of a gas at constant pressure and temperature is inversely proportional to the square root of its density.

In 1860 Graham examined liquids. He noticed that a colored solution of sugar placed at the bottom of a glass of water gradually extends its color upwards. He called this spontaneous process *diffusion*. He also noticed that substances such as glue, gelatin, albumen, and starch diffuse very slowly. He classified substances into two types: colloids (from Greek *kolla*, glue), which diffuse slowly, and crystalloids, which diffuse quickly. He also found that substances of the two types differ markedly in their ability to pass through a membrane, such as parchment, and he developed the method of dialysis to separate them. Graham is regarded as the father of modern colloid science, and many terms that he invented, such as sol, gel, peptization, and syneresis, are still in use.

In 1833 Graham published the results of his research on phosphates. The composition of the compound that was then called phosphoric acid was expressed by the formula PO_5. Graham proved the existence of three "acid hydrates" to which he gave compositions PO_53HO, PO_52HO, and PO_5HO, thus laying the foundations for Justus von LIEBIG's theory of the polybasicity of acids. He carried out similar studies with the arsenic acids and the arsenates. Other work done by Graham includes research into the water of crystallization in hydrated salts and investigations into the absorption of hydrogen by palladium. Graham was an excellent and successful teacher.

Gram, Hans Christian Joachim (1853–1938) *Danish bacteriologist*

Gram graduated in medicine from the university in his native city of Copenhagen in 1878 and from 1883 to 1885 traveled in Europe, studying pharmacology and bacteriology. While in Berlin (1884) he discovered the method of staining bacteria with which his name has become associated. He followed the method of Paul EHRLICH, using aniline-water and gentian violet solution. After further treatment with Lugol's solution (iodine in aqueous potassium iodide) and ethanol he found that some bacteria (such as pneumococcus) retained the stain (*Gram positive*) while others did not (*Gram negative*). This discovery is of great use in the identification and classification of bacteria. It is also useful in deciding the treatment of bacterial diseases, since penicillin is active only against Gram-positive bacteria; the cell walls of Gram-negative bacteria will not take up either penicillin or Gram's stain.

In 1891 Gram became professor of pharmacology at the University of Copenhagen, where he showed a keen interest in the clinical education of the students. During this time he had a large medical practice in the city. He was chairman of the Pharmacopoeia Commission from 1901 to 1921 and director of the medical department of Frederick's Hospital, Copenhagen, until he retired in 1923.

Granit, Ragnar Arthur (1900–1991) *Finnish neurophysiologist*

Born in the Finnish capital of Helsinki, Granit (**gran**-it) qualified as a physician from the university there in 1927. He taught at the university from 1927 until 1940, serving as professor of physiology from 1935. In 1940 he moved to the Karolinska Institute, Stockholm, becoming professor of neurophysiology at the newly founded Medical Nobel Institute in 1946.

In a long career Granit was a prolific writer on all aspects of the neurophysiology of vision. He demonstrated that light not only stimulates but can also inhibit impulses along the optic nerve. By attaching microelectrodes to individual cells in the retina he showed that color vision does not simply depend on three different types of receptor (cone) cells sensitive to different parts of the spectrum. Rather, some of the eye's nerve fibers are sensitive to the whole spectrum while others respond to a much narrower band and so are color specific.

Granit described his work in *Sensory Mechanisms of the Retina* (1947) and *The Visual Pathway* (1962); for such research he shared the 1967 Nobel Prize for physiology or medicine with George WALD and Haldan HARTLINE. Granit also did important work on the control of muscle spindles by the gamma fibers.

Grassi, Giovanni Battista (1854–1925) *Italian zoologist*

The son of a municipal official from Rovellasca in Italy, Grassi (**grah**-see) was educated at the University of Pavia, where he obtained his MD in 1878. He went on to study zoology at the universities of Heidelberg and Würzburg and on his return to Italy was appointed to the chair of zoology in 1883 at the University of Catania. He later moved to the University of Rome, where he served as professor of comparative anatomy from 1895 until his death.

Although Grassi made a number of contributions to zoology working on bees, termites, eels, and other species, he is mainly remembered for his work on malaria. Unfortunately his own contribution in this field is not precisely known, since not only was much of his work done in collaboration with a large number of colleagues but also virtually all his published results were claimed by Ronald ROSS, as his own.

Despite the accusations by Ross of "brigandage" by Grassi and his colleagues it would seem that they were undisputably involved in three major areas of the complex malaria story. Firstly it was the Italians Angelo Celli and Ettore Marchiafava who, together with Grassi, made the important point in 1889–90 that there were a number of protozoa producing malaria. *Plasmodium falciparum* had been discovered in 1889 and *P. vivax* in 1890: this was enough to allow them to suggest that the different varieties of malaria, the so-called quartan and tertian fevers, could be understood as due to infection with different species of *Plasmodium*.

Secondly it was Grassi, in 1898, who solved one of the major problems in implicating mosquitoes in the production of malaria. For centuries the association between marshes, malaria, and mosquitoes had been known but no one could confidently make the obvious connection between mosquitoes and malaria as there were numerous areas in Italy that were free of malaria despite being plagued with mosquitoes. In 1897 Ross had shown that avian malaria was transmitted by the *Proteosoma* mosquito; in the following year, quite independently of this work, the Italians demonstrated that it was only the *Anopheles* mosquito that could transmit the disease to humans. Thus the connection was at last established between *Anopheles* mosquitoes and malaria.

Grassi's final major contribution was to collaborate with Patrick MANSON in 1900 in providing the crucial experimental confirmation of their claim. Mosquitoes of the species *Anopheles maculipennis* that had bitten malaria patients were sent by Grassi to London, where they were allowed to feed on Manson's son; 15 days later the boy developed a clear case of tertian fever.

In 1900 Grassi published an account of his work in *Studi di uno zoologo sulla malaria* (Studies of a Zoologist on Malaria).

Gray, Asa (1810–1888) *American botanist*

Born in Sauquoit, New York, Gray studied at Fairfield Medical School, obtaining his MD in 1831. He spent the next five years developing his interest in botany, supporting himself by teaching and library work. During this period he met, and made a favorable impression on, John TORREY, a prominent American botanist with whom he later collaborated to produce the two-volume *Flora of North America* (1838–43).

In 1838 Gray made the first of many visits to Europe, in order to examine type specimens of American plants in various European herbaria so that the taxonomy of his flora could be based on original species descriptions. For the next four years Gray worked full-time on the flora, but with his appointment as professor of natural history at Harvard in 1842 further progress was slow due to the pressure of other commitments. Nevertheless the work was hailed in its time as being second only to A. P. de CANDOLLE's *Prodromus* and its publication marked the

adoption of natural systematics in American classification.

Gray helped promote specialization in natural history by accepting the professorship at Harvard only on the condition that he might limit his work to botany. He remained at Harvard until 1873, during which time he developed the library and botanical gardens from almost nothing. He donated his own vast collection of books and plants to the university in 1865 on the understanding that they would be housed in their own building.

During another visit to Europe in 1851, Gray met Charles DARWIN. The two men later corresponded on matters of plant geography, and in 1857 Darwin wrote to Gray concerning his ideas on the origin of species. When Darwin's theory was published, Gray became his main advocate in America and wrote many essays reconciling Darwinism with Christian doctrine. These articles were collected together and published as *Darwinia* in 1876.

Gray's most widely used book is his *Manual of the Botany of the Northern United States* (1848) but he established his worldwide reputation with the publication in 1859 of a monograph on the flora of Japan and its relation to floras of other north temperate zones.

Gray was president of the American Association for the Advancement of Science from 1863 to 1873 and in 1900 was commemorated in the newly founded Hall of Fame for Great Americans.

Gray, Harry Barkus (1935–) *American inorganic chemist*

Gray, who was born at Woodburn in Kentucky, was educated at the University of West Kentucky and at Northwestern University, where he obtained his PhD in 1960. After a year at the University of Copenhagen, he joined the Columbia faculty in 1961 but moved to the California Institute of Technology in 1966 and has since served there as professor of chemistry.

As an inorganic chemist Gray has worked on the electronic structure of metal complexes and on inorganic reaction mechanisms. He has also studied the chemistry of biochemical compounds containing metal ions (e.g., proteins containing iron and copper). Later work on the photochemistry of inorganic compounds led to an interest in using complexes as a method of absorbing and storing solar energy. He was awarded the Wolf Prize in 2004.

Gray, Stephen (c. 1670–1736) *English physicist*

Gray made many studies of electrical phenomena and discovered that electricity could be transmitted from one substance to another. In 1729 he electrified a glass tube (by friction) and found that corks in the end of the tube became electrically charged. Further experiments showed that many other materials could conduct electricity, and that electric charges could be transferred over long distances. The results of Gray's investigations were published in the *Philosophical Transactions* of the Royal Society in the years 1731–32 and 1735–36.

Green, George (1793–1841) *British mathematician*

> Had his life been prolonged, he might have stood eminently high as a mathematician.
> —Obituary notice of Green (1841)

Green's father was a prosperous baker from Nottingham in England. Green worked for his father from the age of nine until his father's death in 1829. His father left him a mill which still stands; it has been restored and was opened to the public in 1979.

Green must have been a largely self-taught mathematician. It is known that he joined the Nottingham Subscription Library in 1823 and that the library had copies of such advanced works as the *Mécanique céleste* (Structure of the Heavens) of LAPLACE. The translator, John Toplis, was head of the local grammar school and may well have influenced Green. In 1828 Green published *An Essay on the Application of Mathematics to Electricity and Magnetism*. It was made available to 51 subscribers and few seem to have been aware of its appearance. The work only became widely known when Lord KELVIN came across a copy in 1845 and was so impressed that he arranged for it to be reissued in *Crelle's Journal*. The *Essay* introduced into science *Green's theorem*, *Green's function*, and the notion of the electric potential.

Following the death of his father, Green felt free to follow his scientific interests. He was encouraged in this by Sir Edward Bromhead, a local landowner, who offered to arrange Green's admission to Cambridge. At first Green was reluctant, doubting that it would be suitable "for a person of my age and Imperfect Classical attainments." Nevertheless he arrived in Cambridge at the age of 40, finished as fourth wrangler, and was elected to a fellowship of his college in 1839. He died two years later after contracting flu, leaving seven illegitimate children.

Greengard, Paul (1925–) *American neuroscientist*

Greengard was born in New York. He graduated with a PhD in biophysics from the Johns Hopkins University, Baltimore, in 1953. Postdoctoral studies in England (London University, Cambridge University and the National Institute for Medical Research) were followed by a stint at the National Institutes of Health, Bethesda. From 1959 to 1967 he was director of the department of biochemistry at Geigy Research Laboratories. He also held academic positions at the Albert Einstein College of Medicine, New York (1961–70), and Yale University School of Medicine (1968–83), before being appointed professor and head of the Laboratory of Molecular and Cellular Neuroscience at Rockefeller University, New York, in 1983.

Greengard has made significant advances in our understanding of how the chemicals (neurotransmitters) that relay signals between nerve cells in the central nervous system exert their effects. These relays occur at special junctions (synapses), where one nerve cell ending lies extremely close to the outer membrane (plasma membrane) of another nerve cell. Neurotransmitters such as dopamine, noradrenaline, and serotonin are responsible for what is called slow synaptic transmission, which alters the basal responsiveness of the recipient nerve cell, and is important in determining levels of alertness and mood, for example.

Greengard investigated slow synaptic transmission, and unraveled the chain of molecular events that follow when dopamine stimulates a receptor on a recipient cell membrane. The first step is a rise in levels of a so-called second messenger, cyclic AMP (cAMP), inside the cell. This activates protein kinases, enzymes that add phosphate groups to (phosphorylate) various proteins in the cell, thereby altering the shape and function of the proteins. For example, phosphorylation of proteins that form ion channels through the plasma membrane affects the excitability of the nerve cell, and hence how many nerve impulses it can transmit.

Building on this work, Greengard showed that in certain nerve cells neurotransmitters cause a cascade of phosphorylations and dephosphorylations (addition and removal of phosphate groups), and that certain proteins play a key regulatory role in governing the overall changes in excitability. His work has direct relevance to understanding how certain drugs work in relation to nerve cell function.

For his discoveries relating to signal transduction in the nervous system, Greengard was awarded the 2000 Nobel Prize for physiology or medicine, jointly with Arvid CARLSSON and Eric R. KANDEL.

Greenstein, Jesse Leonard (1909–2002) *American astronomer*

Born in New York, Greenstein graduated from Harvard in 1929 and then, as he puts it, "rode out four depression years as an operator in real estate and investments." He returned to Harvard and obtained his PhD in 1937. He worked at the Yerkes Observatory from 1937 until 1948 when he moved to the California Institute of Technology. In 1949 he became professor of astrophysics and a staff member at the Mount Wilson and Palomar observatories, where he remained until his retirement in 1981.

Greenstein worked on the constitution of stars, and on how and why the constitution can vary from one star to another, and also made extensive studies of quasars. In 1963, following Maarten SCHMIDT, he interpreted the spectrum of the quasar 3C 48 and was able to show that its peculiarities resulted from a red shift over twice that obtained by Schmidt for the quasar 3C 273.

Gregor, William (1761–1817) *British mineralogist*

Born in Trewarthenick in England, Gregor was educated at Cambridge University. Although elected a fellow of his college he decided instead to pursue a career in the Church and became rector of Creed, Cornwall, in 1793.

In 1791 he found a strange black sand in Manaccan (then spelled Menacchan), Cornwall. This contained iron and manganese plus an additional substance that Gregor could not identify. He called it menacchanine and succeeded in extracting its reddish-brown oxide, which, when dissolved in acid, formed a yellow solution. Martin KLAPROTH isolated the same oxide from a different source in 1795 and demonstrated that it was a new element, naming it titanium.

Gregory, James (1638–1675) *Scottish mathematician and astronomer*

> My salary was ... kept back from me and scholars of most eminent rank were violently kept from me ... the masters persuading them that their brains were not able to endure mathematics.
> —On being persecuted at St. Andrews for his radical scientific ideas (1674)

Gregory was one of the many 17th-century mathematicians who made important contributions to the development of the calculus, although some of his best work remained virtually unknown until long after his death.

Born at Drumoak, Scotland, he studied mathematics at the University of Padua in about 1665 and produced *Vera circuli et hyperbolae quadratura* (1667; The True Areas of Circles and Hyperbolas). He was particularly interested in expressing functions as series,

and he sketched the beginnings of a general theory. It was Gregory who first found series expressions for the trigonometric functions. He introduced the terms "convergent" and "divergent" for series, and was one of the first mathematicians to begin to grasp the difference between the two kinds. Gregory also gave the first proof of the fundamental theorem of calculus.

In addition to his mathematical work Gregory's interests in astronomy led him to do some valuable practical work in optics. He anticipated NEWTON by recommending a reflecting telescope in his *Optica promota* (1663; The Advance of Optics). He realized that refracting telescopes would always be limited by aberrations of various kinds. His solution was to use a concave mirror that reflected (rather than a lens that refracted) to minimize these effects. He solved the problem of the observer by having a hole in the primary mirror through which the light could pass to the observer. However, he was unable to find anyone skilled enough actually to construct the telescope.

Gregory held chairs in mathematics at the University of St. Andrews (1669–74) and the University of Edinburgh (1674–75). He died at the age of 37 shortly after going blind.

Griess, Johann Peter (1829–1888) *German chemist*

Griess (grees), who came from a farming family in Kirklosbach near Kassel in Germany, was educated at the universities of Jena and Marburg. In 1858 he became August Wilhelm Hofmann's assistant at the Royal College of Chemistry in London. In 1862 he took up a post in industry with Allsopp's brewery in Burton-on-Trent, where he stayed for the rest of his career.

In 1862 he made an important discovery concerned with a new class of dyes. He discovered that aromatic amines (such as aniline, $C_6H_5NH_2$) react with nitrous acid at low temperature to form diazonium salts (of the type $C_6H_5N_2^+X^-$). These would react with phenols to produce larger, colored molecules – the azo dyestuffs. Griess himself had no interest in these products and, in fact, received no reward

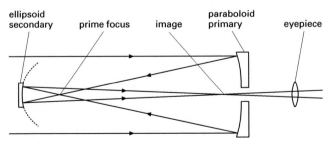

GREGORIAN TELESCOPE   *The path of light rays in Gregory's reflecting telescope.*

from them. The first azo dye, Manchester or Bismarck brown, was discovered within a year.

Griffin, Donald Redfield (1915–2003)
American zoologist

> I always found small mammals enough like ourselves to feel that I could understand what their lives would be like, and yet different enough to make it a sort of adventure and exploration to see what they were doing.
> —*Echoes of Bats and Men* (1959)

Born in Southampton, New York, Griffin was educated at Harvard where he obtained his PhD in 1942. He spent the war applying physiological principles to the design of such military equipment as cold-weather clothing and headphones. After the war he worked initially at Cornell but returned to Harvard in 1953 as professor of biology, a position he held until his retirement in 1986.

Bat navigation had been first studied by Lazzaro SPALLANZANI in 1793 when he noted that blinded bats were as efficient in catching insects on the wing as sighted bats. He also noted that impairment of their hearing produced disorientation. How, Griffin asked in 1938, could ears replace eyes in flight guidance? He was fortunate in that he had a colleague in the physics laboratory, G. W. Pierce, who had an interest in high-frequency sound and was willing to use his specialized equipment on bats. In this way they soon found that bats were issuing sounds, in the case of the horseshoe bat, with frequencies between 60 and 120 kilohertz; the limit of human hearing is between 15 and 20 kilohertz.

That the bats were operating by echolocation was demonstrated when either their mouths or ears were covered. In either case they would collide in a dark room with anything in their path, even the room's walls. But, when operating freely, it was found that the small brown bat, *Myotis lucifugus*, could detect in the dark and fly through a screen with wires no more than 24 centimeters apart. Only when the wires were reduced to a diameter of less than 0.07 millimeter, about the size of a human hair, did their detection system break down. Griffin went on to explore in greater detail the nature of bat sonar systems, describing his work in his *Listening in the Dark* (New Haven, 1958), and the more popular account *Echoes of Bats and Men* (1959).

Griffin also worked extensively on the problem of bird navigation. Early work inevitably consisted of homing experiments. But from these no more can be established than the percentage of birds arriving safely, and the time taken. It was, however, established that the number of returns decreased with distance in a way that suggested that birds navigate by identifying local landmarks. To discover more about the process Griffin taught himself to fly and in the late 1940s spent many hours in a Piper Cub observing the flight paths of gannets and gulls. His observations tended to support the idea that homing from unfamiliar territory was not accurately directed, but succeeded eventually because of exploratory flights.

While this was perhaps a reasonable account of the behavior of gannets and gulls, later work by others showed that in the case of homing pigeons, shearwaters, and starlings there did exist a highly developed ability to select a particular direction. Griffin described his own work and the work of others in his *Bird Migration* (London, 1965).

Griffith, Fred (1881–1941) *British microbiologist*

Griffith, who was born at Hale in southern England, has been described variously as a "virtual recluse" or "quiet and retiring." He worked as a bacteriologist at the Ministry of Health's pathology laboratory in London and was killed working in his laboratory during an air-raid.

Despite the general obscurity of his background Griffith has acquired long after his death an almost legendary role as one of the founding fathers of molecular biology by his discovery in 1928 of bacterial transformation in pneumococci. He had first succeeded in distinguishing two types of pneumococci, the nonvirulent R (rough) of serological type I and the virulent S (smooth) of type III.

He inoculated mice with both live nonvirulent R and heat-killed S pneumococci. Although when either were inoculated separately no infection resulted, together they produced in the mice lethal cases of pneumonia. Further, he recovered from the infected mice living, virulent S pneumococci of type III.

It was this awkward result which later led Oswald AVERY and his colleagues in 1944 to carry out the experiments that succeeded in explaining Griffith's results by suggesting that the power to transform bacteria lay with the nucleic acid of the cell rather than its proteins or sugars.

Grignard, François Auguste Victor (1871–1935) *French chemist*

Born at Cherbourg in northern France, Grignard (gree-**nyar**) first studied mathematics at the University of Lyons before he switched to chemistry. He was a lecturer at the universities of Besançon, Nancy, and Lyons before he was appointed professor of chemistry at Nancy in 1910. In 1919 he moved to the chair of chemistry at Lyons.

In 1901 he discovered an important class of organic reagents now known as *Grignard reagents*. For this work he shared the Nobel Prize for chemistry with Paul SABATIER in 1912. He was searching for a catalyst for a methylation reaction he was trying to induce; chemists had earlier tried to use zinc in combination with various organic compounds and found it moderately successful. Grignard used magnesium mixed with organic halides in ether solution and obtained compounds of the type RMgX, where X is a halogen (Cl, Br, I) and R an organic group. These Grignard reagents are very versatile and permit the synthesis of a large number of different classes of compounds, particularly secondary and tertiary alcohols, hydrocarbons, and carboxylic acids.

In 1935 he began the publication of his *Traité de chimie organique* (Treatise on Organic Chemistry), which was continued after his death and is now a massive multivolume work.

Grimaldi, Francesco Maria (1618–1663)
Italian physicist

Grimaldi (gri-**mahl**-dee) was born at Bologna, Italy, and became a Jesuit. In 1648 he became professor of mathematics at his order's college in his native city, where he acted as assistant to Giovanni RICCIOLI. His discovery of the phenomenon that he named the diffraction of light was reported in his posthumous work *Physicomathesis de lumine, coloribus, et iride* (1665; Physicomathematical Studies of Light, Colors, and the Rainbow). He showed that when a beam of light passed through two successive narrow apertures, the pattern of light produced was a little bigger than it should have been if the light had traveled in an absolutely straight line. Grimaldi considered that the beam had bent outward very slightly, indicating that light must have a wave nature. The result presented difficulties to all 17th-century corpuscular theories of light.

Grisebach, August Heinrich Rudolph (1814–1879) *German plant taxonomist and phytogeographer*

Grisebach was born in Hanover. His uncle was a professor of botany at Göttingen, and Grisebach originally studied there, later moving to Berlin.

His first taxonomic work was on the family Gentianaceae and in 1838 he published the substantial work *Genera et Species Gentianearum*. The next year, inspired by the scientific expeditions of Alexander HUMBOLDT, he set out on a trip to the Balkan peninsula, where he studied the vegetation. This was the start of his work on phytogeography, which he continued for several years and which resulted in his major work, *Vegetation de Erde*, published in 1872. Grisebach also worked on the botany of the Caribbean and South America.

Gross, David J. (1941–) *American theoretical physicist*

Gross was born in Washington, DC. He obtained his PhD in physics from the University of Cal-

ifornia, Berkeley, in 1966, and became a junior fellow at Harvard, where he stayed until 1969, when he moved to Princeton University. In 1997 he moved to the Institute for Theoretical Physics at the University of California, Santa Barbara.

In 1973 Gross together with Frank WILCZEK, and independently David POLITZER, discovered the property known as asymptotic freedom in quantum chromodynamics (QCD). QCD is a theory of the strong nuclear interactions put forward by Murray GELL-MANN and his colleagues in 1973, in which the fractionally charged constituents of hadrons (e.g. protons and neutrons), called quarks, interact with each other via massless spin-1 bosons, called gluons. This is analogous to the way in which electrically charged particles such as electrons interact with each other via photons in quantum electrodynamics (QED). In QED the strength of the electromagnetic interactions between electrically charged particles (e.g. electrons) becomes weaker as the distance between the particles increases. By contrast, Gross, Wilczek, and Politzer discovered that in QCD the strength of the strong nuclear interaction between quarks gets stronger as the distance between the quarks becomes larger, and goes to zero as the distance between them goes to zero. This property is known as asymptotic freedom. The theory explained the results of certain high-energy scattering experiments, which suggested that protons and neutrons have pointlike objects moving around inside them. In addition, the discovery of asymptotic freedom led to the development of perturbative QCD, i.e., the description of high-energy processes in QCD using perturbation theory. Gross, Wilczek, and Politzer were awarded the 2004 Nobel Prize for physics for their discovery of asymptotic freedom. Gross has also made other contributions to quantum field theory and has played an important role in the development of superstring theory.

Grove, Sir William Robert (1811–1896) *British physicist*

> For my part, I must say that science to me generally ceases to be interesting as it becomes useful.
> —Address to the Jubilee meeting of the Chemical Society (1891)

Born at Swansea in South Wales, William Grove was educated at Oxford and, after graduating in 1835, became a barrister. However, ill health turned him away from law to science.

His research concentrated on the newly formed science of electrochemistry and in 1839 he produced an improved version of the voltaic battery, which came to be known as the *Grove cell*. He also constructed the first fuel cell and, in 1845, invented a type of electric bulb, which he hoped would be useful in coal mines.

Grove was elected to the Royal Society in 1840 and, the next year, was appointed professor of physics at the London Institution. However, soon after this he returned to the legal profession and rose to the position of a high court judge. He continued to publish books on science, mainly supporting the idea of energy conservation.

Grubbs, Robert H. (1942–) *American chemist*

Grubbs was born in Calvert City, Kentucky, and educated at Columbia University, New York, receiving his PhD in chemistry in 1968. He subsequently became Victor and Elisabeth Atkins Professor of Chemistry at the California Institute of Technology.

The focus of his research has been metathesis reactions in organic synthesis, particularly in finding convenient catalysts for this type of reaction. After Yves CHAUVIN proposed his mechanism for metathesis reactions in 1971, Grubbs was one of the chemists who demonstrated that this mechanism is correct. In the 1980s Grubbs started to investigate the use of ruthenium compounds as catalysts in organic synthesis. In 1992 Grubbs and his colleagues announced their discovery of a well-defined ruthenium compound that was stable in air and could readily be used. Grubbs has continued to investigate ways of improving ruthenium-based catalysts for metathesis. One of these catalysts is usually referred to as *Grubbs' catalyst*.

The work of Grubbs in finding stable, readily applicable catalysts for metathesis reactions has led to their widespread use in organic synthesis in both academic and industrial research. Grubbs himself has worked on the synthesis of polymers with particular properties. Other examples include the synthesis of natural products that are difficult to produce by any other means. Metathesis reactions using the types of catalyst discovered by Grubbs have shorter routes for their synthesis and have higher yields of product than other synthesis techniques. This means that the chemical production processes are cleaner and "greener," i.e., better for the environment. Drugs and herbicides are among the substances which have been produced in this way.

Grubbs shared the 2005 Nobel Prize for chemistry with Yves Chauvin and Richard SCHROCK for the development of the metathesis method in organic synthesis.

Grünberg, Peter (1939–) *German physicist*

Peter Grünberg was born in Pilsen, now in the Czech Republic. He obtained a Diploma in Physics in 1966 and his PhD in 1969, both from the Darmstadt University of Technology, Germany. Between 1969 and 1972 Grünberg did post-doctoral research in Canada at Carleton University, Ottawa. He subsequently joined the Institute for Solid State Physics at the Jülich Research Centre, where he remained until he retired in 2004.

Independently of Albert FERT, he discovered the phenomenon of giant magnetoresistance in 1988. The phenomenon created much interest as a novel and unexpected phenomenon in solid-state physics, but also because of its many important technological applications, especially in data storage. Fert and Grünberg have shared several prizes for their discovery, notably the Wolf Prize in 2006 and the Nobel Prize in physics in 2007.

Guericke, Otto von (1602–1686) *German physicist and engineer*

Guericke (**gay**-ri-ke), who was born in Magdeburg, Germany, trained in law and mathematics before becoming an engineer in the army of Gustavus Adolphus of Sweden. After the Thirty Years' War he returned to Magdeburg as mayor, there carrying out numerous dramatic experiments on vacuums and the power of the atmosphere.

In 1650 Guericke constructed the first air pump, which he used to create a vacuum in various containers. He showed that sound would not travel in a vacuum, and furthermore that a vacuum would not support combustion or animal life. In 1654 Guericke gave an impressive demonstration in front of the emperor Ferdinand III, of the force of atmospheric pressure. Two identical copper hemispheres 12 feet (3.66 m) in diameter were joined together. When the air was pumped out, 16 horses could not pull them apart although when the air reentered the hemispheres they fell apart by themselves. He also showed that 20 men could not hold a piston in a cylinder once the air had been evacuated from one end of it. The results of these and other experiments were published in his *Experimenta nova Magdeburgica de vacuo spatio* (1672; New Magdeburg Experiments Concerning Empty Space). In 1663 he built the first electrical friction machine by rotating a sulfur globe against a cloth.

Guettard, Jean Etienne (1715–1786) *French geologist*

Guettard (ge-**tar**) was born at Etampes in France. After a training in medicine and chemistry, he worked under the royal patronage of the duc d'Orléans from 1747 as keeper of his natural-history collection. Following the duc's death (1752) he continued this work under the patronage of his son.

In 1751 he made a crucial observation that upset the neptunism theories of Abraham WERNER and his followers. While traveling through the Auvergne region he noticed an abundance of hexagonal basalt rocks and, exploring the region, identified the surrounding mountain peaks as the cones of extinct volcanoes (which would explain the presence of basalt). However, Werner's theory stated that all volcanic activity is recent, only occurring after the land has completely emerged from the oceans. Therefore, according to Wernerian theory, no volcanoes as ancient as the Auvergne ones should exist. Guettard published his findings in 1752 in his memoir, *On Certain Mountains in France which once have been Volcanoes*. He later changed his mind, distinguishing basalt from lava as it was not to be found among the recent volcanoes. He also observed the lack of vitrification found in basalt, then taken to be a sure sign of volcanic origin, and explained its formation by crystallization from an aqueous fluid.

Guettard was the first to map France geologically, publishing in 1780 his *Atlas et description minéralogiques de la France* (Mineralogical Atlas and Description of France). In the preparation of this he discovered (1765) a source of kaolin in Alençon, which made possible the production of the celebrated Sèvres porcelain.

Guillaume, Charles Edouard (1861–1938) *Swiss metrologist*

As a child Guillaume (gee-**yohm**) learned a good deal of science from his father, a clockmaker with a considerable scientific knowledge. Born in Fleurier, Switzerland, in 1878 he entered the Zurich Federal Institute of Technology, gaining his doctorate in 1882. In 1883 Guillaume became an assistant at the newly established International Bureau of Weights and Measures at Sèvres, near Paris. He was appointed director in 1915 and held this post until his retirement in 1936.

Guillaume's early work at the Bureau was concerned with thermometry; his treatise of 1889 on this subject became a standard text for metrologists. He was also involved in developing the international standards for the meter, kilogram, and liter. His research on thermal expansion of possible standards materials led him from 1890 to investigate various alloys. After a methodical study of nickel–steel alloys he devised an alloy that showed a very small expansion with temperature rise. Guillaume's new material ("invar") found immediate practical applications, particularly in clocks, watches, and other precise instruments. He also produced a nickel–chromium–steel alloy, known as "elinvar," with an elasticity that remains nearly constant over a wide range of temperatures. It became widely used, for example, for the hairsprings of watches.

In 1920 Guillaume received the Nobel Prize

for physics for his researches into nickel–steel alloys.

Guillemin, Roger (1924–   ) *French–American physiologist*

Guillemin (gee-e-**man**) was educated at the universities of Dijon (his native city), Lyons, and Montreal, where he gained his PhD in physiology and experimental medicine in 1953. The same year he moved to America to join the staff of the Baylor University Medical School, Houston. In 1970 Guillemin joined the staff of the Salk Institute in La Jolla, California, where he remained until 1989, when he moved to the Whittier Institute for Diabetes and Endocrinology, La Jolla, becoming its director in 1993.

Early in his career Guillemin decided to work on the hypothesis of Geoffrey HARRIS that the pituitary gland is under the control of hormones produced by the hypothalamus. As the anterior pituitary secretes a number of hormones it was far from clear which to begin with. He eventually decided to search for the hypothalamic factor that controls the release of the adrenocorticotrophic hormone (ACTH) from the pituitary – it is known as the corticotrophic releasing factor (CRF). As it turned out, this was an unfortunate choice for after seven years Guillemin had nothing to show for his not inconsiderable efforts. Guillemin then worked for a further six years fruitlessly searching for the thyrotropin releasing factor (TRF), exposing him to skepticism from many other workers in the endocrine field. The main difficulty was that such hormones were present in very small quantities. When Guillemin finally did succeed in 1968 in isolating one milligram of TRF it had come from 5 million sheep's hypothalami. It turned out to be a small, relatively simple tripeptide, easy to synthesize. The development of the radioimmunoassay method for the detection of minute quantities by Rosalyn YALOW was also of considerable help. Other successes quickly followed. Andrew SCHALLY isolated the luteinizing-hormone releasing factor in 1971 and Guillemin in 1972 succeeded with somatostatin, which controls the release of the growth hormone.

In 1977 Guillemin shared the Nobel Prize for physiology or medicine with Schally and Yalow.

Guldberg, Cato Maximilian (1836–1902) *Norwegian chemist*

Guldberg (**guul**-bairg) was educated at the university in his native city of Christiania (now Oslo) and started his career teaching at the Royal Military School there in 1860. He was appointed to the chair of applied mathematics at the university in 1869.

Guldberg's main work was on chemical thermodynamics. In 1863 he formulated the law of mass action in collaboration with his brother-in-law, Peter WAAGE. The law states that the rate of a chemical change depends on the concentrations of the reactants. Thus for a reaction: A + B → C the rate of reaction is proportional to [A][B], where [A] and [B] are concentrations. Guldberg and Waage also investigated the effects of temperature. They did not gain full credit for their work at the time, partly due to their first publishing the law in Norwegian. However, even when published in French (1867) the law received little attention until it was rediscovered by William Esson and Vernon HARCOURT working at Oxford University.

In 1870 Guldberg investigated the way in which the freezing point and vapor pressure of a pure liquid are lowered by a dissolved component. In 1890 he formulated Guldberg's law. This relates boiling point and critical temperature (the point above which a gas cannot be liquefied by pressure alone) on the absolute scale. The law was discovered independently by Phillippe-Auguste Guye.

Gullstrand, Allvar (1862–1930) *Swedish ophthalmologist*

Gullstrand (**gul**-strand), a physician's son from Landskrona, Sweden, was educated at the universities of Uppsala, Vienna, and Stockholm, where he obtained his PhD in 1890. After working briefly at the Karolinska Institute in Stockholm Gullstrand moved to the University of Uppsala, where he served as professor of ophthalmology from 1894 until his retirement in 1927.

In 1911 Gullstrand was awarded the Nobel Prize for physiology or medicine for his work on the dioptrics of the eye. Hermann von HELMHOLTZ had earlier shown that the eye solves the problem of accommodation (how to focus on both near and distant objects) by changing the surface curvature of the lens – the nearer the object, the more convex the lens becomes; the further the object, the more concave the lens. Gullstrand showed that this could in fact account for only two thirds of the accommodation a normal eye could achieve. The remaining third was produced by what Gullstrand termed the "intracapsular mechanism" and depended on the fact that the eye was not a homogeneous medium.

Gutenberg, Beno (1889–1960) *German–American geologist*

Gutenberg (**goo**-ten-berg) was educated at the Technical University in his native city of Darmstadt and at the University of Göttingen, where he obtained his PhD in 1911. He then taught at the University of Freiburg becoming professor of geophysics in 1926. He emigrated to America in 1930, taking a post at the California Institute of Technology, and later served as director of the seismological laboratory (1947–58).

In 1913 Gutenberg suggested a structure of the Earth that would explain the data on earthquake waves. It was known that there were two main types of waves: primary (P) waves, which are longitudinal compression waves, and secondary (S) waves, which are transverse shear waves. On the opposite side of the Earth from an earthquake, in an area known as the shadow zone, no S waves are recorded and the P waves, although they do appear, are of smaller amplitudes and occur later than would be expected. Gutenberg proposed that the Earth's core, first identified by Richard OLDHAM in 1906, is liquid, which would explain the absence of S waves as, being transverse, they cannot be transmitted through liquids. Making detailed calculations he was able to show that the core ends at a depth of about 1,800 miles (2,900 km) below the Earth's surface where it forms a marked discontinuity, now known as the *Gutenberg discontinuity*, with the overlying mantle. Its existence has been confirmed by later work, including precise measurements made after underground nuclear explosions.

In collaboration with Charles RICHTER, Gutenberg produced a major study, *On Seismic Waves* (1934–39), in which, using large quantities of seismic data, they were able to calculate average velocity distributions for the whole of the Earth.

Guth, Alan Harvey (1947–) *American physicist and cosmologist*

> They say that there is no such thing as a free lunch – but the universe is the ultimate free lunch!
> —Speaking about the inflationary model of the universe

Born in New Brunswick, New Jersey, Guth was educated at the Massachusetts Institute of Technology, where he obtained his PhD in 1969. After holding postdoctoral appointments at Princeton, Columbia, Cornell, and Stanford, Guth returned to MIT in 1980, becoming professor of physics in 1986.

Initially Guth worked as a theorist in elementary-particle physics, but, stimulated by the work of Steven WEINBERG, he began to consider some of the outstanding problems of cosmology. These included a number of difficulties raised against the standard interpretation of the big-bang account of the origin of the universe. Many had found the apparent isotropy of the universe puzzling, while James PEEBLES and Charles MISNER had discovered the flatness and horizon problems. The big-bang was clearly in need of revision.

Consequently in 1980 Guth first proposed the *inflationary universe model*. Guth's theory agrees with the standard model after the first

10^{-30} second. Within the initial brief period, he proposed that the universe had undergone an extraordinarily rapid inflation. During this period the diameter of the universe would have increased by a factor of about 10^{50}. An increase of this size would expand a centimeter to 10^{32} light years.

Thus as the region from which the present universe emerged was so small, thermal equilibrium could have been achieved before the inflation. An observer today would, therefore, detect an isotropic cosmic-background radiation. He would also detect a flat universe – an observer on a globe that had increased 10^{50} times in diameter would have seen his universe flattened.

Guth has conceded that even if the inflationary model is correct "it will be difficult for anyone to ever discover observable consequences of the conditions existing before the inflationary phase transition." He has also pointed out that the grand unified theories of physics allow that the observed universe could have evolved from nothing. The inflationary model comes close to this by providing a mechanism by which the observable universe could have evolved from an infinitesimal region.

In the 1990s evidence emerged to support the idea that quantum fluctuations in the early Universe, expanded by inflation, were responsible for the large-scale structure in the Universe, such as galaxies.

Guthrie, Samuel (1782–1848) *American chemist*

Guthrie, who was born at Brimfield, Massachusetts, was a student at the University of Pennsylvania, where he qualified as a doctor. In 1831 he discovered chloroform, which was used as an anesthetic by James SIMPSON in 1847. It was, however, discovered independently by Soubeiran in the same year and by Justus von LIEBIG in 1832. Guthrie also discovered percussion powder, a substance that explodes on impact.

Guyot, Arnold Henry (1807–1884) *Swiss–American geologist and geographer*

Intending to enter the Church, Guyot (**gee**-yoh), who was born at Boudevilliers in Switzer-land, studied at the universities of Neuchâtel, Strasbourg, and Berlin, where his interests in science began to absorb him. After teaching in Paris (1835–40) he was appointed professor of history and physical geography at Neuchâtel in 1839 where he remained until 1848, when he emigrated to America. He taught first at the Lowell Technological Institute in Boston before he was appointed, in 1854, to the chair of geology and physical geography at Princeton University.

While in Switzerland he had studied the structure and movement of glaciers, spending much time testing the new theories of Louis AGASSIZ. In America, under the auspices of the Smithsonian Institution, he began to develop, organize, and equip a number of East Coast meteorological stations. He also surveyed and constructed topographical maps of the Appalachian and Catskill mountains. In 1849 he published his influential work *The Earth and Man*.

Guyton de Morveau, Baron Louis Bernard (1737–1816) *French chemist*

Born at Dijon in France, Guyton (gee-**ton**) began his career as a lawyer; as a member of the Burgundy parliament (1755–82), however, he met the great Georges BUFFON who encouraged his interest in science. In 1782 he gave up law to devote himself to science and he collaborated with Antoine LAVOISIER. During the revolutionary period he reentered politics. He was a founder of and teacher at the Ecole Polytechnique (1795–1805) and in 1800 became master of the mint until his retirement in 1814.

In the period 1776–77 Guyton published his three-volume *Eléments de chimie théorique et pratique* (Rudiments of Theoretical and Practical Chemistry), which was a major attempt to quantify chemical affinities. Guyton was a passionate Newtonian and tried to apply Newtonian laws to chemistry. He tried to do this by floating disks of various metals on mercury and measuring the force necessary to remove them. Thus he obtained figures such as gold needs a force of 446 grains to remove it, lead 397, zinc 204, iron 115, and cobalt 8. He attempted to correlate his figures with the chemical affinities of the elements.

H

Haber, Fritz (1868–1934) *German physical chemist*

> For more than forty years I have selected my collaborators on the basis of their intelligence and their character and not on the basis of their grandmothers, and I am not willing for the rest of my life to change this method which I have found so good.
> —Letter of resignation over the Nazis' demand that he dismiss his Jewish colleagues, 30 April 1933

Haber (**hah**-ber), the son of a merchant, was born at Breslau, now Wrocław in Poland. He was educated at Berlin, Heidelberg, Charlottenburg, and Jena, and in 1894 he became an assistant in physical chemistry at the Technical Institute, Karlsruhe, where he remained until 1911, being promoted to a professorship in 1906. He moved to Berlin in 1911 to become director of the Kaiser Wilhelm Institute of Physical Chemistry. Though an intensely patriotic German he was also a Jew and with the rise of anti-Semitism he resigned his post in 1933 and went into exile in England, where he worked at the Cavendish Laboratory, Cambridge. He died in Basel en route to Italy.

Haber is noted for his discovery of the industrial process for synthesizing ammonia from nitrogen and hydrogen. The need at the time was for nitrogen compounds for use as fertilizers – most plants cannot utilize free nitrogen from the air, and need "fixed" nitrogen. The main source was deposits of nitrate salts in Chile, but these would have a limited life.

Haber, in an attempt to solve this problem, began investigating the reaction:

$$N_2 + 3H_2 = 2NH_3$$

Under normal conditions the yield is very low. Haber (1907–09) showed that practical yields could be achieved at high temperatures (250°C) and pressures (250 atmospheres) using a catalyst (iron is the catalyst now used). The process was developed industrially by Carl Bosch around 1913 and is still the main method for the fixation of nitrogen. Haber received the Nobel Prize for chemistry for this work in 1918.

During World War I, Haber turned his efforts to helping Germany's war effort. In particular he directed the use of poisonous gas. After the war he tried, unsuccessfully, to repay the indemnities imposed on Germany by a process for extracting gold from seawater.

Hadamard, Jacques Salomon (1865–1963) *French mathematician*

> Some intervention of intuition issuing from the unconscious is necessary at least to initiate the logical work.
> —The Psychology of Invention in the Mathematical Field

The son of a Latin teacher, Hadamard (a-da-**mar**) was born at Versailles in France and educated at the Ecole Normale Supérieure in Paris. He taught first at the University of Bordeaux from 1893 until 1897, before returning to Paris to the Sorbonne. In 1909 he took up the chair of mathematics at the Collège de France where he remained until his retirement in 1937.

In his long life Hadamard worked in many areas of mathematics, but remains best known for his proof in 1896 of the *prime number theorem*.

Mathematicians have long been interested in prime numbers. There is no simple formula for determining primes, but it is possible to say something about the distribution of prime numbers. If P_n is the nth prime number, $\pi(n)$ is used to denote the number of primes between 1 and n. Both Gauss and Legendre used the formula

$$\pi(n) = n/\log_e n.$$

This does not work at small values of n but Gauss and Legendre suspected that the ratio of $\log_e n$ to $\pi(n)/n$ would approach 1 as n approaches infinity. They were, however, unable to prove it. Hadamard and, independently, Charles de la Vallée-Poussin produced proofs in 1896 using the Riemann zeta function.

Hadfield, Sir Robert Abbott (1858–1940) *British metallurgist*

Hadfield's father was a steel manufacturer who had opened a factory for the production of steel castings in Sheffield in 1872. Hadfield was educated locally and started work in the laboratory in his father's works, inheriting the business on his father's death (1880). It had been known for some years that if manganese was added to iron the result was hard but was too brittle to be commercially useful. Hadfield discovered that if a large quantity of manganese was added (about 12–14%) and the steel was heated and quenched in water, the resulting alloy was extremely hard and strong. He patented his discovery in 1883. He also worked on the development of other steel alloys and in 1899 was able to show that silicon steels have a high electrical resistance. This made them suitable for use in transformers as they could substantially reduce bulk.

Hadfield was also interested in the history of metallurgy, forming a fine collection, and out of this interest came his two books: *Faraday and his Metallurgical Researches* (1931) and *Metallurgy and its Influence on Modern Progress: With a Survey of Education and Research* (1925). He was knighted in 1908 and made a baronet in 1917.

Hadley, George (1685–1768) *English meteorologist*

The younger brother of the inventor John Hadley, George Hadley was born in London and educated at Oxford University. Although called to the bar in 1709, he became more interested in physics and was made responsible for producing the Royal Society of London's meteorological observations.

In 1686 Edmond HALLEY had offered a partial explanation of the trade winds, pointing out that heated equatorial air will rise and thus cause colder air to move in from the tropics, but could not explain why the winds blew from the northeast in the northern hemisphere and the southeast in the southern.

Hadley put forward the explanation, in his paper *Concerning the Cause of the General Trade Winds* (1735), that the airflow toward the equator was deflected by the Earth's rotation from west to east. This circulation is now known as the *Hadley cell*.

Hadley, John (1682–1744) *English mathematician and inventor*

Little is known of Hadley's life. He is chiefly remembered for developing the reflecting telescope, producing his first in 1721. He was an extremely skilled craftsman and his reflectors were among the first to be useful in astronomy.

Hadley also invented, in 1730, the reflecting quadrant with which a ship's position at sea could be determined by measurements of the Sun or a star above the horizon. This instrument later developed into the sextant.

Haeckel, Ernst Heinrich (1834–1919) *German biologist*

Man is separated from the other animals only by quantitative, not qualitative, differences.
—Expounding his idea that man was part of nature

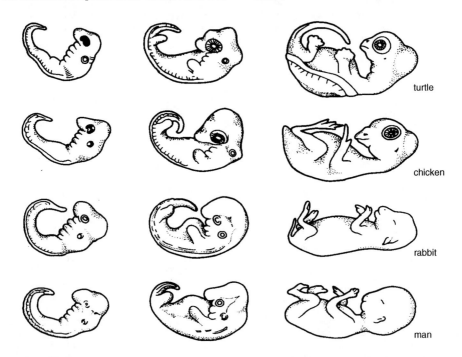

turtle

chicken

rabbit

man

STAGES OF DEVELOPMENT IN VERTEBRATE EMBRYOS These show the similarities, particularly in the early stages, that led Haeckel to suggest his theory of recapitulation.

The son of a government lawyer, Haeckel (**hek-el**) was born at Potsdam in Germany and educated at the universities of Warburg, Vienna, and Berlin, where he qualified as a physician in 1858. His main interests lay elsewhere and, after a brief period in practice, he moved to Jena to study zoology. In 1862 he was appointed professor of zoology and comparative anatomy at Jena, a position he held until his retirement in 1909.

Haeckel's contributions to zoological science were a mixture of sound research and speculation often with insufficient evidence. An advocate of monism, which postulated a totally materialistic view of life as a unity, he based his evolutionary ideas on the embryological laws expounded by Karl von BAER. Expanding the idea of his mentor, Johannes MÜLLER, Haeckel argued that the embryological stages of an animal were a recapitulation of its evolutionary history, and indeed that there had once been complete animals resembling the embryonic stages of higher animal forms living today. He formulated a scheme of evolution for the whole animal kingdom, from inorganic matter upward. His studies, with Müller, of marine life, particularly the crystalline radiolarians, encouraged him to compare the symmetry of crystals with the simplest animals, and led him to postulate an inanimate origin for animal life. In 1866 Haeckel anticipated later proof of the fact that the key to inheritance factors lies in the cell nucleus, outlining this theme in his *Die Perigenesis der Plastidule* (1876; The Generation of Waves in the Small Vital Particles).

Haeckel also proposed the idea that all multicellular animals derived from a hypothetical two-layered (ectoderm and endoderm) animal, the *Gastraea* – a theory that provoked much discussion. He engaged in much valuable research on marine invertebrates, such as the radiolarians, jellyfish, calcareous sponges, and medusae, and wrote a series of monographs on these groups based largely on specimens brought back by the Challenger Expedition. He was also the first to divide the animal kingdom into unicellular (protozoan) and multicellular (metazoan) animals. An ardent Darwinist, Haeckel made several zoological expeditions and founded the Phyletic Museum at Jena and the Ernst Haeckel Haus, which contains his books, archives, and other effects.

In 1906 the Monist League was formed at Jena with Haeckel as its president. The League held a strong commitment to social Darwinism. Man was seen as part of nature and in no way qualitatively distinct from any other organic form. Human society was as much a creation of natural selection as the bird's wing. To such views Haeckel added a strong anticatholicism, a contempt for politicians, and a forecast of impending doom. After the chaos of World War I, Haeckel's views were taken up by eugenicists, the Volk movement and, more significantly, the National Socialists.

Hahn, Otto (1879–1968) *German chemist*

> If Hitler gets an atomic bomb, I shall kill myself.
> —Quoted in Hahn's obituary notice in *The Times*

Hahn's father, a successful merchant in the German city of Frankfurt, was eager for his son to train as an architect and it was against much family opposition that Hahn was finally allowed to study chemistry at the University of Marburg in 1897. After obtaining his doctorate in 1901 he studied abroad, first with William RAMSAY in London and then at McGill University, Canada, with Ernest RUTHERFORD. Hahn returned to Germany in 1907, where he took up an appointment at the University of Berlin, being made professor of chemistry in 1910. Two years later he joined the Kaiser Wilhelm Institute of Chemistry where he served as director from 1928 to 1945.

Hahn had trained as an organic chemist and had really gone to London to learn English in order to prepare himself for an industrial career. Ramsay had however asked him to separate radium from some radioactive material he had recently acquired from Ceylon. In so doing Hahn found a new material, a highly active form of thorium which he named "radiothorium." So impressed was Ramsay with this work that he wrote to Emil FISCHER in Berlin suggesting that he employ Hahn after he had acquired more experience of radioactivity with Rutherford at McGill.

Hahn was thus diverted into an academic career, most of which was spent in research on radioactivity and much of it in collaboration with Lise MEITNER. With her he discovered a new element, protactinium, in 1917. He went on to define, in 1921, the phenomenon of nuclear isomerism. This arises when nuclei with different radioactive properties turn out to be identical in atomic number and mass.

Hahn's most important work, however, was done in the 1930s when, with Meitner and Fritz STRASSMANN, he made one of the most important discoveries of the century, namely nuclear fission. One of the strange features about Hahn's work was that he was repeating experiments already done and formulating hypotheses already rejected as nonsense or due to some contamination of the materials used. Chemists at this time felt that they understood the process of nuclear transformation. After all it was some twenty years since Rutherford had first detected the transformation of nitrogen into oxygen, and a newer form of the same "alchemy" had been described by Irène and Frédéric JOLIOT-CURIE in 1934. Two basic rules were involved in this understanding. First, that nuclear transmutations always involved the emission of either an alpha particle (helium nucleus) or a beta particle (electron); and secondly, that the change could take place only between elements separated by no more than two places in the periodic table. If more substantial transformations appeared to occur, as in the transformation of uranium into lead, this was explained as the result of a series of such intermediate steps, each one taking place with the emission of the appropriate particle.

Thus when in 1938 Hahn bombarded uranium with slow neutrons and detected some strange new half-lives, he assumed that the uranium had changed into radium, a close neighbor, with some undetected alpha particles. But when he tried to remove the radium all he could find was barium. This Hahn simply could not understand, for barium was far too low in the periodic table to be produced by the transmutation of uranium; and if the transformation *was* taking place it should be accompanied by the emission of a prodigious number of alpha particles, which Hahn could not have failed to detect. The thought that the heavy uranium nucleus could split into two lighter ones was too outrageous for him to consider seriously. He could not dismiss it entirely for he asserted at the time that "we must really state that we are not dealing with radium but with barium." But to suppose the barium arose from what he then called nuclear "bursting" he felt would be "in contradiction to all previous experience in nuclear physics." He did realize that something of importance was going on and quickly sent off for publication a joint paper with Strassmann even though, as he recalled twenty years later, "After the manuscript had been mailed, the whole thing once more seemed so improbable to me that I wished I could get the documents back out of the mail." Appropriately enough it was his old collaborator Meitner, in exile from the Nazis in Sweden, and her nephew Otto FRISCH, who made the necessary calculations and announced fission to the world early in 1939. Hahn received the Nobel Prize for chemistry in 1944.

Haken, Wolfgang (c. 1927–) *German-born American mathematician*

Haken (**hah**-ken) was born in Berlin, Germany, and educated at the University of Kiel, where he obtained his PhD in 1953. After working for the Siemens Corporation for a number of years, he moved to America in 1961 to take up a visiting professorship at the University of Illinois, Urbana. He spent the period 1963–65 at the Institute for Advanced Study, Princeton, but returned to Urbana in 1966 as professor of mathematics.

Haken's main work has been in the field of topology. He first established his reputation in 1962 with a proposal for calculating the genus of any knot. The genus is a topological invariant that allows it to be decided whether two given knot diagrams represent the same knot.

He is best known, however, for his solution in 1977, in collaboration with Kenneth APPEL, of the four-color map problem. One unusual feature of the proof was its use of a computer. Some mathematicians objected that unless all parts of a proof can be surveyed by hand and eye, the proof cannot be considered valid. This objection has lost some of its force as mathematicians have offered more and more computer-assisted proofs. Haken has denied that there has been any change in the basic concept of mathematical proof. "What has changed," he claims, "is not the theory but the practice of mathematics."

Haldane, John Burdon Sanderson
(1892–1964) *British geneticist*

> My own suspicion is that the universe is not only queerer than we suppose, but queerer than we can suppose.
> —*Possible Worlds and Other Essays* (1927)

> Religion is still parasitic in the interstices of our knowledge which have not yet been filled. Like bed-bugs in the cracks of walls and furniture, miracles lurk in the lacunae of science. The scientist plasters up these cracks in our knowledge; the more militant Rationalist swats the bugs in the open. Both have their proper sphere and they should realise that they are allies.
> —*The Rationalist Annual* (1934)

> Physical science is … no more than a superficial aspect of ultimate philosophical truth.
> —*Respiration* (1935)

Haldane, who was born at Oxford in England, became involved in scientific research at an early age through helping in the laboratory of his father, the physiologist John Scott Haldane. His interest in genetics was first stimulated as early as 1901, when he heard a lecture on MENDEL's work, and he later applied this by studying inheritance in his sister's (the writer Naomi Mitchison) 300 guinea pigs. On leaving school he studied first mathematics and then the humanities at Oxford University. He served in World War I with the Black Watch Regiment and was wounded at Loos and in Mesopotamia. Some work on gas masks, following the first German gas attacks, marked the beginning of his physiological studies.

In 1919 Haldane took up a fellowship at Oxford, where he continued research on respiration, investigating how the levels of carbon dioxide in the blood affect the muscles regulating breathing. He was next offered a readership in biochemistry at Cambridge, where he conducted some important work on enzymes. These experiments, and later work on conditions in submarines, aroused considerable public interest because he frequently used himself as a guinea pig.

In 1933 Haldane became professor of genetics at University College, London, a position he exchanged in 1937 for the chair of biometry. While at London he prepared a provisional map of the X sex chromosome and showed the genetic linkage between hemophilia and color blindness. He also produced the first estimate of mutation rates in humans from studies of the pedigrees of hemophiliacs, and described the effect of recurring deleterious mutations on a population. With the outbreak of the Spanish Civil War, Haldane joined the Communist Party and advised the republican government on gas precautions. In the 1950s he left the party as a result of Soviet acceptance and promotion of Trofim LYSENKO. In protest at the Anglo-French invasion of Suez, Haldane emigrated to India in 1957, becoming an Indian citizen in 1961. He was director of the laboratory of genetics and biometry at Bhubaneswar from 1962 until his death.

Haldane's books include *Enzymes* (1930), *The Causes of Evolution* (1932), and *The Biochemistry of Genetics* (1954); he also wrote a number of books popularizing science.

Haldane, John Scott (1860–1936) *British physiologist*

Haldane, the son of a lawyer, was educated at the university in his native city of Edinburgh, where he obtained his MD in 1884. He worked first at the University of Dundee but moved to Oxford to assist his uncle, John Burdon-Sanderson, the professor of physiology, in 1887. Haldane was made reader in physiology in 1907 but resigned in 1913 to become director of the Mining Research Laboratory, initially sited in Doncaster but transferred to Birmingham in 1921.

From the beginning of his career Haldane sought to apply the results of physiological research to the solution of practical social and industrial problems. He was much concerned with problems of ventilation in mines and in 1896 published an important report, *Causes of Death in Colliery Explosions*. He was struck by the fact that in a serious explosion in 1896 only 4 out of 57 miners died from the blast and its effects, the vast majority succumbing to carbon monoxide poisoning. Haldane recommended the simple and effective safety procedure of taking mice down the pit: with their higher metabolic rate they would show the effects of carbon monoxide poisoning long before it reached levels dangerous to man.

He also worked for the admiralty on the problems faced by their divers at high pressures. It had been known for some time that rapid decompression produced the liberation of nitrogen bubbles into the bloodstream, with crippling and often lethal effects. Haldane showed how such effects could be minimized by pointing out that however long a diver had been exposed to compressed air it was always safe to halve the pressure; that is, it is just as safe to ascend from six to three atmospheres as from two to one. Using his technique over £5,000,000 of gold was recovered from the wreck of the *Lusitania* between 1917 and 1924.

Haldane also investigated the response of the human body to high temperatures. Haldane's main work as a pure physiologist, however, was on the mechanism of respiration. In 1906 he published his most significant paper, in collaboration with John Priestley, which demonstrated the key role of carbon dioxide in the regulation of breathing. They showed that it was not a deficiency in oxygen that triggers the respiratory reflex but an excess of carbon dioxide in the arterial blood, acting on the respiratory center in the midbrain. Their work was published in full in *Respiration* (1935; 2nd edition).

In much of his work Haldane used for an experimental subject his precocious son J. B. S. Haldane, later to become one of the leading biologists of the 20th century.

Hale, Alan (1958–) *American astronomer*

Alan Hale was raised on the outskirts of Alamogordo, New Mexico. His interest in astronomy developed at school and early in 1970 his father bought him his first telescope – a 4½-inch reflector. After graduating from high school he attended the US Naval Academy, Annapolis, Maryland, where he studied physics. He was then stationed at various naval bases. Throughout this period he continued his astronomical observations, in particular the observation of comets.

In 1983 he left the navy and spent 2½ years working for the Deep Space Network at the Jet Propulsion Laboratory, Pasadena. He then went to New Mexico State University, Las Cruces, where he obtained his PhD in 1992. In 1993 he formed an independent research and education organization, now called the Earthrise Institute.

The organization moved in 1995 to a mountain village of Cloudcroft, New Mexico. It was here on the night of 22–23 July that Alan Hale first observed the comet *Hale–Bopp*. It proved to be perhaps the most prominent comet of the century. The comet was discovered independently by an amateur astronomer, Thomas BOPP.

Hale, George Ellery (1868–1938) *American astrophysicist*

> Hale may be said to be the father of modern solar observational astronomy.
> —Robert Howard

Hale's father, William Hale, was a wealthy manufacturer of elevators in Chicago, Illinois, who stimulated in his son an early interest in designing and making his own instruments. This interest was directed to astronomy by Sherburne W. Burnham, a neighbor and passionate observer of double stars, and increased during his four years at the Massachusetts Institute of Technology, where he studied physics. He built a solar observatory, financed by his father, at Kenwood, Chicago, and after graduating in 1890 became its director. In 1892 he was appointed assistant professor and later professor of astrophysics at the new University of Chicago and from 1895 to 1905 he was director of the university's Yerkes Observatory. From 1904 to 1923 he was director of the newly established Mount Wilson Observatory in California. The last 15 years of his life were spent organizing the equipping and building of the Palomar Observatory in California and in the pursuit of his solar researches in his private observatory in Pasadena.

Hale was undoubtedly one of the key figures in 20th-century astronomy. He saw very clearly and very early that astronomy could only develop if much more powerful telescopes were constructed. Thus with great vision and enormous persistence and energy he spent 40 years acting as midwife to a series of bigger and bigger telescopes. His insight was clearly justified for it was with his telescopes that Harlow SHAPLEY, Edwin HUBBLE, and many others made their observations.

His first triumph came when he persuaded Charles T. Yerkes, a Chicago trolley-car magnate, to provide $349,000 to build a 40-inch (1-m) refracting telescope for the University of Chicago. This was and still is the largest refractor ever built. It was first used in 1897. He was soon anxious, however, to build a large reflecting telescope. In 1896 his father acquired a 60-inch (1.5-m) mirror but the University of Chicago was unable to fund its mounting. Hale once more started raising money. This time he interested the Carnegie Institution of Washington in financing the Mount Wilson Observatory. The observatory was founded in 1904 and the 60-inch reflector eventually went into use in 1908. In 1918 this superb instrument was surpassed by the 100-inch (2.5-m) Hooker telescope, largely financed by a Los Angeles business man, John D. Hooker. For 30 years this was the world's largest telescope and it revolutionized astronomy.

Hale had resigned from his directorship of Mount Wilson Observatory in 1923 on the grounds of ill health but lost little time in seeking to interest the Rockefeller Foundation in building a reflecting telescope that would be the ultimate in size, 200 inches (5 m) across, for Earth-based instruments. In 1929 it was finally agreed that $6 million would be donated for this purpose to an educational institute, the California Institute of Technology, rather than the Carnegie Institution. Hale became chairman of the group directing the planning, construction, and operation of the instrument that was to become his masterpiece.

Thus there began an epic struggle to complete the 200-inch telescope, which was to take nearly 20 years. The first mirror made from fused quartz proved to be a $600,000 failure. Hale next tried Pyrex and the first experimental 200-inch disk cast proved satisfactory. The actual casting was made in December 1934 when the 65 tons of molten Pyrex began its carefully controlled 10 months' cooling. The mirror managed to survive the flooding of the factory, which necessitated shutting down the temperature control for three days, and its long journey in the spring of 1936 from Corning, New York, to California at a maximum speed of 25 mph. The grinding of the mirror was interrupted by the war and took so long that Hale had been dead for nine years when the instrument was finally commissioned as the Hale telescope in 1948. It was set up at the specially constructed Palomar Observatory, which together with the Mount Wilson Observatory was jointly operated by the California Institute of Technology and the Carnegie Institution. The two observatories were renamed the Hale Observatories in 1969. The Hale telescope was the world's largest telescope until the Soviet 6-meter (236-in) reflector went into operation in 1977 but is still considered by many to be the world's finest.

Hale was not just a highly successful scientific entrepreneur for he made major advances in the field of solar spectroscopy. As early as 1889 he had conceived of his spectroheliograph, an instrument that allowed the Sun to be photographed at a particular wavelength. He also designed an appropriate telescope to which it could be attached. In 1908 Hale made his most significant observation. He found that some of the lines in the spectra of sunspots were double. He realized that this demonstrated the presence of strong magnetic fields in sunspots, being due to the effect discovered by Pieter ZEEMAN in 1896, and was the first indication of an extraterrestrial magnetic field.

Hales, Stephen (1677–1761) *English plant physiologist and chemist*

> Since we are assured that the all-wise Creator has observed the most exact proportions, of number, weight and measure, in the make of all things, the most likely way therefore, to get any insight into the nature of those parts of the creation, which come within our observation, must in all reason be to number, weigh and measure.
> —*Vegetable Staticks* (1727)

Born at Bekesbourne in Kent, southeast England, Hales entered Cambridge University in 1696 to study theology. He was ordained in 1703 and appointed curate at Teddington, near London, in 1708 (or 1709). During his time at Cambridge, he studied science and was influenced by Isaac NEWTON's ideas, which still dominated scientific thought at the university and probably accounted for Hales's consistent use of the quantitative method in his biological researches.

Hales was elected a fellow of the Royal Society in 1718 but his first work, *Vegetable Staticks*, was not published until 1727. In this book, which included his most important observations in plant physiology, Hales demonstrated that plant leaves absorb air and that a portion of air is used in plant nutrition. In addition, he realized that light is necessary for growth and investigated growth rates by marking plants at regular intervals. He measured the rate of water loss (transpiration) in plants, finding that it occurred through the leaves and was responsible for an upward flow of sap in plants. From additional measurements of sap flow he concluded that there was no circular movement of sap in plants analogous to blood circulation in animals.

Hales also made important contributions to the understanding of blood circulation by measuring such properties as blood pressure, output per minute from the heart, rate of flow and resistance to flow in vessels. The results were published in *Haemastaticks* (1733; Blood Statics).

Other notable discoveries include the development of methods for collecting gases over water, distilling fresh water from sea water, and preserving foodstuffs with sulfur dioxide. He also invented a ventilator for introducing fresh air into prisons, ships, and granaries.

Hall, Asaph (1829–1907) *American astronomer*

> The chance of finding a satellite appeared to be very slight, so that I might have abandoned the search had it not been for the encouragement of my wife.
> —On discovering the satellites of Mars, August 1877

Born at Goshen in Connecticut, Hall had to leave school at the age of 13 and support his family as a carpenter, following the death of his father. He educated himself, and his interest in astronomy was strong enough for George BOND to employ him as his assistant at Harvard in 1857. In 1863 Hall became professor of mathematics at the Naval Observatory in Washington. He returned to Harvard as professor of astronomy in 1895.

In 1877 Mars was in opposition to the Sun at a distance of about 30 million miles from the Earth. Hall decided to search for Martian satellites using the 26-inch (66-cm) refractor that the Clark firm had provided for the Naval Observatory. On 11 August he discovered a tiny satellite (the smaller moon) but was then compelled to wait a further six nights for the persistent cloud to clear before he could confirm his sighting and discover a further satellite. Both were very small, having diameters of 17 miles (27 km) and 9 miles (15 km) only. He named the larger "Phobos" and the smaller "Deimos" (Fear and Terror), after the sons of Mars. One curious feature of the two satellites was that Jonathan Swift had spoken of Martian satellites in *Gulliver's Travels* (1726). Not only did Swift get their number correct but also spoke accurately of their size and orbital period.

In 1876, by noticing a white spot on the surface of Saturn, Hall was able to work out correctly the rotation period as 10.75 hours, which compares well with today's figure of 10 hours 14 minutes (for its equatorial region).

Hall, Charles Martin (1863–1914) *American chemist*

Hall was born in Thompson, Ohio, and educated at Oberlin College, graduating in 1885. He became interested in the costly process of manufacturing aluminum – until the late 19th century aluminum was a precious metal costing about $5.50 an ounce. Napoleon III would have the majority of his guests served from gold plate; he and the chosen few he wished to impress were served from aluminum plates. Hall was stimulated by a remark of his teacher that anyone who could find a cheap way to make aluminum would win great wealth and fame.

Although the ore itself (bauxite, aluminum oxide) was cheap and plentiful, the metal could only be extracted by electrolysis of the molten ore, and aluminum oxide has a very high melting point. Hall tried various added compounds and, in 1886, found that adding 10–15% of cryolite (sodium aluminum fluoride) reduced the melting point to a little over 1,000°C. Hall produced his first sample in the form of buttons, which soon became known as the "aluminum crown jewels." Paul HÉROULT, working in France, discovered the process independently at about the same time. Hall helped to found the Pittsburgh Reduction Company (later the Aluminum Company of America) of which he became vice-president in 1890.

Hall, Edwin Herbert (1855–1938) *American physicist*

Hall was born in Great Falls, Maine, and educated at Johns Hopkins University, Baltimore, where he received his PhD in 1880. After a year in Europe he joined the Harvard faculty and was appointed professor of physics in 1895, a post he held until his retirement in 1921.

While working for his thesis, Hall began to consider a problem first posed by MAXWELL concerning the force on a conductor carrying a current in a magnetic field. Does the force act on the conductor or the current? Hall argued that if the current was affected by the magnetic field then there should be "a state of stress ... the electricity passing toward one side of the wire." Hall used a thin gold foil and in 1879 detected for the first time an electric potential acting perpendicularly to both the current and the magnetic field. The effect has since been known as the *Hall effect*. A simple interpretation is that the charge carriers moving along the conductor experience a transverse force and tend to drift to one side. The sign of the Hall voltage gives information on whether the charge carriers are positive or negative.

Other so-called galvanomagnetic effects were later discovered by Walther NERNST and others. Hall spent much of his later life attempting to measure the various effects as exactly as possible.

More than a century after its discovery Klaus von KLITZING was awarded the 1985 Nobel Prize for physics for his work on the *Hall effect*.

Hall, James (1811–1898) *American geologist*

> A very great man, not honoured in his family, not well understood in his own community, not always courteously entreated and appreciated by his scientific contemporaries; but on the other hand winning the admiration and acclaim of those great-minded enough to understand his inflexible purpose and the magnitude of his achievement.
> —John M. Clarke, *James Hall of Albany* (1921)

Hall, who came from a poor background in Hingham, Massachusetts, was educated at Rensselaer Polytechnic Institute, New York, where he became assistant professor in 1832. In 1837 he started work as a geologist on a survey of New York State, publishing an important report of these studies, *Geology of New York* (1843). In 1843 he became state paleontologist and was later, in 1865, made the curator and, in 1871, director of the New York State Museum in Albany.

Hall's major work was his massive 13-volume *The Palaeontology of New York* (1847–94). This important work became the standard for much of the later geological exploration of America in the 19th century. He was also active in the organization of the geological surveys of the Far and Midwest. He thus became a figure in 1888 as the first president of the Geological Society of the United States.

As a theorist, Hall is remembered for his account of mountain building published in 1859. This theory, known as the geosynclinal theory, was developed in a more comprehensive form by James DANA.

Hall, Sir James (1761–1832) *British geologist*

Born in Dunglass, Scotland, Hall succeeded to his father's baronetcy and fortune in 1776. He studied at Cambridge University (1777–79) but left without taking a degree. He spent two years traveling in Europe before returning to Edinburgh where he attended the university (1781–83). Following this he once more traveled extensively throughout Europe, meeting most of the scientists and scholars of his day.

Hall performed fundamental experiments to establish the plausibility of his friend James HUTTON's uniformitarian and plutonist theories. The neptunist Abraham WERNER had argued in criticism of James Hutton that great heat cannot have been a major factor in the formation of the Earth. He pointed out that when basalt cools it becomes glassy, not stony, and limestone, when subjected to heat, decomposes. The plutonist theories could therefore not account for the abundance of limestone and stony basalt on the Earth.

In 1798 Hall conducted his first experiment and succeeded in showing that if igneous rocks are allowed to cool slowly they form crystalline rather than glassy rocks. In 1805 he managed to refute Werner's second objection by showing that limestone when heated under pressure does not decompose on cooling but becomes marble.

Although Hall's experiments provided a good deal of support for Hutton's views, Hall himself was only a moderate supporter. In his work *On the Revolutions of the Earth's Surface* (1812) he argued strongly, using his Alpine tours as evidence, for the need to assume the existence of enormous tidal waves and floods in the past to explain the present features of the Earth's surface.

Hall, John L. (1934–) *American physicist*

John Hall was born in Denver, Colorado, and educated at the Carnegie Institute of Technology, Pittsburgh, where he received his PhD in 1961. He became Senior Scientist at the National Institute of Standards and Technology and Fellow, JILA, University of Colorado, Boulder, Colorado, retiring in 2004.

Hall has been a leading figure in the development of precision spectroscopy using laser sources; in particular, he was a pioneer in a technique known as the *optical frequency comb technique*. His research has involved very accurate experimental determinations of the speed of light, and this led to the definition of the unit of length, the meter, in terms of the distance traveled by light in a defined time period (i.e., a second). A practical difficulty with defining the meter in this way is that the frequency has to be measured very precisely. In the 1980s, Hall and his colleagues were able to attain the desired precision by stabilizing the laser frequency using feed-back mechanisms. Together with the work of Theodor HÄNSCH, this approach has led to the possibility of measuring the frequency of electromagnetic radiation with very high precision; in fact, to one part in 10^{18}.

There are many potential technological applications. For example, it could be used practically in satellite navigation systems, such as the Global Positioning System, to improve accuracy. But there are also applications in basic physics. For example, it could be used to determine whether the spectrum of an anti-hydrogen atom (with a positron orbiting an anti-proton) is exactly analogous to the spectrum of normal hydrogen (an electron orbiting a proton). It might also be possible to use the technique to investigate the basic issue of whether fundamental constants, such as the fine-structure constant, can change with time. There is no conclusive evidence for this at present but increased precision might enable this interesting issue to be definitively resolved.

Hall shared the 2005 Nobel Prize for physics with Roy GLAUBER and Theodor Hänsch for his work on laser-based precision spectroscopy.

Hall, Marshall (1790–1857) *British physician*

Hall, born the son of a cotton manufacturer at Basford in England, obtained his MD from the University of Edinburgh in 1812. After a period of further study in Europe, Hall returned to England and set up in private practice, first in Nottingham in 1816 and, after 1826, in London where he became one of the most successful and prosperous physicians of his day. His son, also named Marshall, was a famous lawyer who played a leading role in some of the most celebrated cases of his time.

In various publications from 1832 onward Hall described his investigations into reflex actions. He correctly believed that reflexes are controlled by the spinal cord, but this idea only met with ridicule from the Royal Society, who refused to publish his work. His claims were greeted more favorably by European scientists and were later used by such giants as Charles SHERRINGTON to illuminate the workings of the nervous system.

Hall also worked on the function of the blood capillaries, denounced the practice of bloodletting, and introduced a form of artificial respiration for use in accidents involving drowning.

Haller, Albrecht von (1708–1777) *Swiss physiologist*

> My friend, the artery ceases to beat.
> —On death. Quoted by Alan L. Mackay in *The Harvest of a Quiet Eye* (1977)

Born at Bern in Switzerland, Haller (**hahl**-er) studied under Hermann BOERHAAVE at Leiden, gaining his MD in 1727. He was later appointed professor of anatomy, botany, and medicine (1736–53) at the newly established University of Göttingen. He then retired to Bern to spend more time on his research and writing.

Between 1757 and 1766 Haller published in eight massive volumes his *Elementa Physiologiae Corporis Humani* (Physiological Elements of the Human Body). The work described the advances in physiology made since the time of William HARVEY, enriched with Haller's own experimental researches.

Before Haller, physiology followed the views of René DESCARTES – that bodily systems are essentially mechanical but require some vital principle to overcome their initial inertness. Haller, anticipated somewhat by Francis GLISSON, broke radically with this tradition. When stimulated, muscles contract; such "irritability," according to Haller, is inherent in the fiber and not caused by external factors.

The implications of this work were not immediately apparent to Haller. It was left to the philosophers of the Enlightenment to hammer home the conclusion that if such an inherent force resided in muscles then there no longer remained a need for the assumption of vital principles to imbue them with activity.

Haller also made important contributions to embryology and was a noted botanist, publishing a major work on the Swiss flora. However, his attempt to construct an alternative classification scheme to that of LINNAEUS, based on fruits rather than sexual organs, received little support despite being a more logical system.

Halley, Edmond (1656–1742) *British astronomer and physicist*

> Joyfully seize the gifts of the present hour.
> —Manuscript in the Bodleian Library, Oxford

Edmond (or Edmund) Halley was the son of a wealthy London merchant. He was educated at St. Paul's School, London, and at Oxford University. He left Oxford without a degree in 1676, but having already published his first scientific paper in the *Philosophical Transactions* of the Royal Society on the theory of planetary orbits. Halley's scientific work and his life covered an enormous range. He started his active scientific career by spending two years on St. Helena mapping the southern skies. In 1679 he published *Catalogus stellarum australium* (Catalog of the Southern Stars), the first catalog of telescopically determined star positions. On his return he traveled extensively in Europe meeting such leading astronomers as Johannes HEVELIUS and Giovanni CASSINI.

Halley now began his enormous contribution to just about all branches of physics and astronomy. He prepared extensive maps showing magnetic variation, winds, and tides. In atmospheric physics he formulated the mathematical law relating height and pressure (1686), making many advances in barometric design. He carried out important studies on evaporation and the salinity of lakes (1687–1694), which allowed him to draw conclusions about the age of the Earth. He used

Newtonian mathematical techniques to improve and augment DESCARTES's work on the optics of the rainbow (1697–1721). He almost incidentally constructed mortality tables, estimated the acreage of England and the size of the atom, improved the design of the diving bell, and published numerous articles on natural history and classical studies.

These were sidelines compared to his work in astronomy and to the help he provided NEWTON. It is owing to Halley that Newton's *Principia* was published in the complete form we know it today. He pressed Newton to publish it, paid for the cost himself, saw it through the press, and even contributed some Latin verses in honor of the author. In 1695 he proposed the secular acceleration of the Moon, in 1718 he discovered the proper motion of the stars, but above all in his *Astronomiae cometicae synopsis* (1705; A Synopsis of the Astronomy of Comets) he laid the foundations of modern cometary study. His grasp of the geometry of cometary orbits allowed him to identify the comet (now known as Halley's comet) of 1531 with those of 1607 and 1682, and confidently to predict its return in 1758 – long after his death.

He held an equally varied and bewildering set of appointments. From 1696 until 1698 he was deputy controller of the mint at Chester. From 1698 to 1700 he actually commanded a Royal Navy man-of-war, the *Paramour*, making prolonged and eventful ocean voyages. In 1702 and in 1703 he made two diplomatic missions to Vienna. In 1703 he was elected to the Savillian Chair of Geometry at Oxford, and in 1720 he succeeded John FLAMSTEED as Astronomer Royal. He held this post until his death, making observations of nearly 1,500 lunar meriodional transits and the full 18-year period of the Moon.

Halsted, William Stewart (1852–1922)
American surgeon

Halsted, who was born in New York City, was educated at private schools, Yale, and the College of Physicians and Surgeons, New York. After graduating in 1877 he spent two years as a postgraduate student at the universities of Vienna, Leipzig, and Würzburg. Returning to America he worked as a surgeon in a number of New York hospitals before a growing addiction to cocaine forced him to leave in search of a cure in 1886. He eventually settled in Baltimore, where in 1892 he became professor of surgery at Johns Hopkins University.

The work that led to his addiction was probably his demonstration in 1885 of the localized anesthesia produced by the injection of cocaine into the appropriate nerve. The drug, introduced into medicine by Sigmund FREUD in 1884, quickly became a valuable anesthetic for minor regional surgery.

As a surgeon Halsted pioneered two common operations. In 1889, independently of Eduardo Bassini in Padua, he devised a permanent surgical cure for inguinal hernia. Previous operative techniques involved such a high relapse rate – 40% in four years – that the operation had fallen into disrepute. He also introduced the operation of radical mastectomy, known in America as *Halsted's operation*. This involved treating breast cancer by the excision not only of the breast but also much of the underlying musculature and surrounding lymphatic tissue. He claimed a recurrence rate of 6%, as opposed to 50% produced by more conventional surgery.

One further important innovation by Halsted was his introduction, in 1889, of thin rubber gloves in operating theaters. This came about when he arranged for the Goodyear Rubber Company to make some gloves for a theater nurse, his future wife Caroline Hampton, whose hands were allergic to the antiseptic used. Over the next few decades rubber gloves gradually came to be used by all theater staff.

Hamilton, William Donald (1936–2000)
British theoretical biologist

Hamilton was educated at the universities of Cambridge and London. He served as a lecturer in genetics at Imperial College, London, from 1964 until 1977 when he moved to America to take up an appointment as professor of evolutionary biology at the University of Michigan. He returned to England in 1984 to serve as a Royal Society Research Professor at Oxford.

In the *Origin of Species* (1859) DARWIN raised a "special difficulty," which he at first considered insurmountable. How could natural selection ever lead to the evolution of neuter or sterile insects? Darwin's answer was that selection may be applied to the family, as well as the individual. In a series of papers, beginning in 1964 with *The Genetical Theory of Social Behaviour*, Hamilton pursued these implications and opened the way for the emergence of sociobiology. The key concept deployed by Hamilton is that of inclusive fitness, which covers not only an individual's fitness to survive but also the effects of that individual's behavior on the fitness of his or her kin.

Hamilton, Sir William Rowan (1805–1865) *Irish mathematician*

> Here, as he walked by on the 16th October, 1843, Sir W. R. Hamilton in a flash of genius discovered the fundamental formula for quaternion algebra $[i^2 = j^2 = k^2 = ijk = -1]$ and cut it on a stone of this bridge.
> —Inscription on Brougham Bridge, Dublin

Hamilton was a child prodigy, and not just in mathematics; he also managed to learn an extraordinary number of languages, some of them very obscure. In 1823 he entered Trinity College in his native city of Dublin, and four years later at the age of 22 was appointed professor of astronomy and Astronomer Royal for Ireland; these posts were given to him in order that he could continue to research unhampered by teaching commitments.

In 1827 he produced his first original work, in the theory of optics, expounded in his paper *A Theory of Systems of Rays*. In 1832 he did further theoretical work on rays, and predicted conical refraction under certain conditions in biaxial crystals. This was soon confirmed experimentally. In dynamics he introduced *Hamilton's equations* – a set of equations (similar to equations of Joseph Lagrange) describing the positions and momenta of a collection of particles. The equations involve the *Hamiltonian function*, which is used extensively in quantum mechanics. *Hamilton's principle* is the principle that the integral with respect to time of the kinetic energy minus the potential energy of a system is a minimum.

One of Hamilton's most famous discoveries was that of *quaternions*. These are a generalization of complex numbers with the property that the commutative law does not hold for them (i.e., $A \times B$ does not equal $B \times A$). Hamilton's discovery of such an algebraic system was important for the development of abstract algebra; for instance, the introduction of matrices. Hamilton spent the last 20 years of his life trying to apply quaternions to problems in applied mathematics, although the more limited theory of vector analysis of Josiah Willard GIBBS was eventually preferred. Toward the end of his life Hamilton drank increasingly, eventually dying of gout.

Hämmerling, Joachim August Wilhelm (1901–1980) *German biologist*

Born in Berlin, Germany, Hämmerling (**hem-er-ling**) was educated at the universities of Berlin and Marburg. After graduating he worked at the Kaiser Wilhelm Institute of Biology from 1922 until 1948 when he moved to Wilhelmshaven to serve as director of the Max Planck Institute for Marine Biology until his retirement in 1970.

In 1953 Hämmerling carried out a series of classic experiments on the unicellular alga, *Acetabularia*. It is shaped something like a mushroom with the nucleus included in the stem part, thus making its removal a relatively simple and harmless matter. This Hämmerling did. The aim of the experiment was to see if the nucleus itself produced the proteins necessary for the growth and development of an organism, or whether it merely produced the "machinery" with which protein synthesis could take place outside the nucleus. Hämmerling's work seemed to confirm the latter alternative, for the enucleated alga continued to grow and develop, even proving capable in certain circumstances of regenerating a new cap.

On Hämmerling's death a note was found among his papers disclosing that he had suffered from severe depression for much of his life. He estimated that he must have spent a third of his working life in this way.

Hammond, George Simms (1921–2005) *American chemist*

Hammond, the son of a farmer from Auburn, Maine, was educated at Bates College and at Harvard where he obtained his PhD in 1947. He immediately afterward joined the faculty at Iowa State University, serving as professor of chemistry from 1956 until 1958. He then moved to California, being appointed first to the chair of organic chemistry at the California Institute of Technology and in 1972 to the chemistry chair at Santa Cruz. In 1978 he left academic life to become associate director for corporate research with the Allied Chemical Corporation, a post he held until his retirement in 1988.

Coauthor with Donald CRAM of a widely used textbook, *Organic Chemistry* (1959), Hammond worked mainly on the mechanism of photochemical reactions, particularly the behavior of energy-rich molecules.

Hänsch, Theodor W. (1941–) *German physicist*

Hänsch was born in Heidelberg, Germany, and educated at the University of Heidelberg, where he received his PhD in 1969. He subsequently became the director of the Max-Planck-Institut für Quantenoptik, Garching, and professor of physics at the Ludwig-Maximilians-Universität, Munich, Germany, starting in 1986.

Hänsch's research work has mostly been concerned with laser-based precision spectroscopy. He made crucial contributions to the development of the *optical frequency comb technique*, so-called because it is based on using a range of evenly distributed frequencies, which look like the teeth of a comb when they are plotted on a graph. It is possible to determine an unknown frequency by comparing it with one of the frequencies of the comb. Hänsch made use of lasers in which the whole of the visible part of the electromagnetic spectrum is covered by the frequency comb. This research, some of which involved collaboration with John HALL and his colleagues, was performed in the last decades of the twentieth century and the early years of the twenty-first century. It has led to the development of instruments in which the frequency of an unknown laser can be determined by measuring the beat between the frequency and one of the teeth in the frequency comb.

In the early years of the twenty-first century frequency-comb techniques were extended to very short wavelength ultraviolet radiation. This gives rise to the possibility of even more accurate clocks.

Hänsch shared the 2005 Nobel Prize for physics with Roy GLAUBER and John Hall for his work on laser-based precision spectroscopy.

Hansen, Gerhard Henrik Armauer (1841–1912) *Norwegian bacteriologist*

Hansen (**hahn**-sen) was born the son of a merchant at Bergen in Norway and graduated in medicine from the University of Christiania (now Oslo) in 1866. He began work in the Bergen leprosy hospital in 1868, an institution under the control of Daniel Danielssen, the leading European authority on leprosy and the future father-in-law of Hansen. Leprosy was thought to be a hereditary affliction, but Hansen concluded from epidemiological studies that it was infectious. He thus took the opportunity in 1870 to travel to Bonn and Vienna to extend his knowledge of bacteriology.

Back in Bergen he observed, in 1873, the rod-shaped bacilli in specimen tissues from leprosy patients, since known variously as Hansen's bacillus or *Mycobacterium leprae*. He later proposed this to be the cause of leprosy but his claim was not appreciated for many years. Hansen however never managed to fulfil the postulates of Robert KOCH and transmit the disease via the bacilli to animals or men, a difficulty also met with by all later workers. He was forced to resign from the leprosy hospital in 1880 for injudiciously injecting live leprosy bacilli into a patient without first obtaining her permission; he did, however, continue to advise the Norwegian government on their policy to leprosy and also carried on with his own research.

He succeeded, by a policy of limited isolation, in reducing the Norwegian incidence of leprosy from 2,833 cases in 1850 to 140 in 1923. Hansen was also an ardent proponent of Darwinism and publicized DARWIN's work in Norway.

Hantzsch, Arthur Rudolf (1857–1935) *German chemist*

Hantzsch (hantsh) studied at the polytechnic in his native city of Dresden and at the University of Würzburg, where he took his doctorate in 1880. He taught in Leipzig (1880), Zurich (1885), Würzburg (1893), and finally, in 1903, Leipzig. He retired to Dresden in 1927.

Hantzsch's first success came in 1882 when he announced his method for synthesizing substituted pyridines from aldehyde ammonia compounds and keto esters. In 1887 he synthesized thiazole and later prepared imidazole, oxazole, and selenazole. In 1890 he published with his pupil Alfred WERNER an account of the stereochemistry of the organic nitrogen compounds, oximes. Hantzsch later tried to extend this work to the diazo compounds, which were being investigated by Eugen BAMBERGER.

Hantzsch was a prolific writer publishing over 450 papers. He wrote extensively on the theory of acids and bases, the absorption of light by different compounds, and on nitrophenols.

Harcourt, Sir William Venables Vernon (1789–1871) *British chemist*

Harcourt, whose father was the archbishop of York, was born at Sudbury in England. Before entering Oxford University in 1807, Harcourt received a private education and spent five years in the Royal Navy. On graduation (1811) he moved to Yorkshire as a clergyman and in 1861, on the death of his elder brother, he succeeded to the family estates and retired to Nuneham in Oxfordshire.

Although he set up his own chemical laboratory he was basically an amateur scientist whose importance lies rather with those he stimulated and influenced than in the work he produced himself. He was a friend of John KIDD, William WOLLASTON, and Humphry DAVY, among other early-19th-century chemists. He played a crucial role in the establishment of the British Association for the Advancement of Science in 1830, serving as its first secretary, being responsible for drawing up its laws and constitution, and becoming its president in 1839.

Harden, Sir Arthur (1865–1940) *British biochemist*

Harden was born in Manchester and educated at Owens College there (where he subsequently taught) and at the University of Erlangen, Germany. He was professor of biochemistry at the Jenner (later Lister) Institute of Preventive Medicine, where he began research into alcoholic fermentation, continuing the work of Eduard BUCHNER who had discovered that such reactions can take place in the absence of living cells.

Harden demonstrated that the activity of yeast enzymes was lost following dialysis (the separation of large from small molecules by diffusion of the smaller molecules through a semipermeable membrane). He went on to show that the small molecules are necessary for the successful action of the yeast enzyme and that, whereas the activity of the large molecules was lost on boiling, the activity of the small molecules remained after boiling. This suggested that the large molecules were proteins but the small molecules were probably nonprotein. This was the first evidence for the existence of *coenzymes* – nonprotein molecules that are essential for the activity of enzymes. Harden also discovered that yeast enzymes are not broken down and lost with time, but that the gradual loss of activity with time can be reversed by the addition of phosphates. He found that sugar phosphates are formed during fermentation as intermediates – phosphates are now known to play a vital part in biochemical reactions. Knighted in 1936, Harden shared the 1929 Nobel Prize for chemistry with Hans von EULER-CHELPIN for his work on alcoholic fermentation and enzymes.

Hardy, Godfrey Harold (1877–1947)
British mathematician

> There is no scorn more profound, or on the whole more justifiable, than that of the men who make for the men who explain. Exposition, criticism, appreciation, is work for second-rate minds.
> —*A Mathematician's Apology* (1940)

> A science is said to be useful if its development tends to accentuate the existing inequalities in the distribution of wealth, or more directly promotes the destruction of human life.
> —As above

Born at Cranleigh in southern England, Hardy had his mathematical education at Cambridge University and remained there as a fellow of Trinity College until 1919, when he became Savilian Professor of Geometry at Oxford. From 1931 to 1942 he was back in Cambridge as Sadleirian Professor of Pure Mathematics.

His central field of interest was in analysis and such related areas as convergence and number theory. *Hardy classes* of complex functions are named for him. For 35 years, starting in 1911, Hardy collaborated with J. E. LITTLEWOOD and together they wrote nearly a hundred papers. The principal areas they covered were Diophantine approximations, the theory of numbers, inequalities, series and definite integrals, and the Riemann zeta-function.

Although primarily a pure mathematician Hardy made one lasting contribution to applied mathematics; the *Hardy–Weinberg law* was discovered independently by Hardy and the physician Wilhelm Weinberg in 1908 and proved to be fundamental to the science of population genetics. It gives a mathematical description of the genetic equilibrium in a large random-mating population and explains the surprising fact that, unless there are outside changing forces, the proportion of dominant to recessive genes tends *not* to vary from generation to generation. The law offered strong confirmation for the Darwinian theory of natural selection.

Hardy was one of the outstanding British mathematicians of his day, an excellent teacher, and one of the first to introduce modern work on the rigorous presentation of analysis into Britain. His *Course of Pure Mathematics* (1908) was influential on the teaching of mathematics in British universities. One of his achievements was his discovery of the young Indian mathematician Srinivasa RAMANUJAN. Partly through Hardy's efforts Trinity College made funds available for Ramanujan to go to Cambridge to pursue his mathematical researches under Hardy.

Hardy was a passionate devotee of cricket and an equally passionate enemy of the Christian religion. During World War I he was a staunch supporter of Bertrand RUSSELL when Trinity set about depriving Russell of his position on account of his pacifist activities. Hardy wrote a lively autobiographical sketch, *A Mathematician's Apology* (1940).

Hardy, Sir William Bate (1864–1934)
British biologist and chemist

> You know, this applied science is just as interesting as pure science, and what's more it's a damn sight more difficult.
> —Quoted by Sir Henry Tizard in the Haldane Memorial Lecture, 1955

Born at Erdington in England, Hardy was educated at Cambridge University and remained there for the whole of his life, being appointed a lecturer in physiology in 1913 and superintendent of the Low Temperature Research Station in 1922. In addition Hardy served on a number of advisory bodies of which the most significant were his chairmanship of the Advisory Committee on Fisheries Development (1919–31) and his directorship of the Food Investigation Board (1917–34).

He began his research career as a histologist, publishing a number of papers in the 1890s on the morphology and behavior of the leukocytes (white blood cells). He began, however, to have considerable doubts about the value of staining and fixing living tissue suspecting that certain structures seen in cells after fixation may simply be artefacts caused by the fixing reagents themselves. He emphasized the point by the production of reticulated and fibrillar structures in his laboratory by fixing and staining albumin.

This led Hardy to work on colloid chemistry and the properties of proteins in solution. He later studied molecular films and lubrication. He reported on this work in his Croonian lecture *On Globulins* (1905) and his Bakerian lecture of 1925, *Boundary Lubrication*.

Hare, Robert (1781–1858) *American chemist*

Born in Philadelphia, Pennsylvania, Hare was the son of the owner of a brewery. He is best known for his invention of the oxyhydrogen blowpipe, which he demonstrated to Joseph PRIESTLEY in 1801. This became of great value in the welding process for it is the ancestor of the later welding torches.

Hargreaves, James (d. 1778) *British inventor*

Hargreaves was a poor and uneducated spinner and weaver when, in 1764, he had the idea of a hand-powered spinning machine that could spin many threads at once. The story is that he saw his daughter, Jenny, knock over a spinning wheel and noticed that it kept turning in an upright position. He built a machine, naming this the "spinning jenny," and started to sell his invention. However, hand weavers who were afraid of unemployment broke into his workshop and destroyed his machines. As a result of this he moved to Nottingham and together with a partner, Thomas James, set up a mill in 1768 in which he used the jennies to spin yarn for making hosiery. His machine was patented in 1770. Thereafter he was moderately successful and worked at his mill until his death.

Hariot, Thomas (1560–1621) *English mathematician, astronomer, and physicist*

> As to Hariot, he was so learned, saith Dr. Pell, that had he published all he knew in algebra, he would have left little of the chief mysteries of that art unhandled.
> —John Collins quoted by J. W. Shirley in *Thomas Harriot: a Biography*

Hariot (or Harriot) is best known as a pioneer figure in the British school of algebra, although his interests and activities were very wide ranging. Born in Oxford, England, he was an associate of Sir Walter Raleigh, whom he accompanied on a voyage to Virginia (1585–86) in the capacity of navigator and cartographer. He later wrote a book about this journey – *A Briefe and True Report of the New Found Land of Virginia* (1588). Among his many innovations in algebra Hariot introduced a number of greatly simplified notations. His central mathematical achievements were in the theory of equations, where he discovered important relationships between the coefficients of equations and their roots. This work was published in his *Artis analyticae praxis ad aequationes algebraicas resolvendas* (1631; The Analytical Arts Applied to Solving Algebraic Equations).

Outside mathematics Hariot's achievements as a practical astronomer were noteworthy. He designed and constructed telescopes and made detailed studies of comets and sunspots. Independently of GALILEO he discovered the moons of Jupiter. Hariot also discovered the law governing the refraction of light and was the first to interpret refraction in terms of the scattering of light by atoms. He was granted a pension by the earl of Northumberland and was briefly imprisoned along with the earl during the Gunpowder Plot of 1605. Hariot conducted numerous experiments in a variety of fields including optics, ballistics, and meteorology. However, he published few of his discoveries and it was only after his death that their extent was realized from his voluminous notes and papers.

Harkins, William Draper (1873–1951) *American physical chemist*

Harkins, whose father was a pioneer in the Pennsylvania oil fields, was born in Titusville, Pennsylvania, and educated at Stanford University where he obtained his PhD in 1907. After studying abroad under Fritz HABER at Karlsruhe, he taught briefly at the University of Montana before moving to Chicago in 1912, where he spent the rest of his career and was made professor of physical chemistry in 1917.

Harkins was one of the first Americans to establish an international reputation in the field of nuclear studies, although many of his results were independently established by his European counterparts. Thus in 1915 he proposed the "whole number rule" at the same time as Francis ASTON was publishing his theory of isotopes. In 1920 he predicted the existence of the neutron and an isotope of hydrogen, deuterium, which were also predicted by Ernest RUTHERFORD in the same year.

In 1915 he demonstrated that the fusion of four atoms of hydrogen to form helium involved a mass excess that was available for conversion to energy in accordance with the famous formula $E = mc^2$ of Albert EINSTEIN. He calculated that four grams of hydrogen could form four grams of helium with a release of energy of 10^{12} calories. Such facts of nuclear fusion were taken up by Arthur EDDINGTON in 1920 and proposed as the source of energy in stars.

Harkins also worked on problems of surface tension and, in 1952, published a standard work on the topic, *The Physical Chemistry of Surface Tension*.

Harris, Geoffrey Wingfield (1913–1971) *British endocrinologist*

Harris, the son of a physicist, was born in London and educated at Cambridge University and St. Mary's Hospital, London. After qualifying in 1939, he worked at Cambridge as an anatomy lecturer until 1952 when he moved to the Maudsley Hospital, London, to direct the Laboratory of Experimental Neuroendocrinology. In 1962 Harris was appointed Dr. Lee's Professor of Anatomy at Oxford where he remained until his sudden death in 1971.

In the period 1950–52 Harris published several papers that provided evidence for the important theory that the release of pituitary hormones is controlled by the hypothalamus. Such a mechanism had long been suspected but was only really made plausible when Harris showed that while nervous connection between the two glands could be severed without major effect, cutting the connecting blood supply severely restricted the production of pituitary hormones. (Solly ZUCKERMAN announced conflicting results but Harris was able to show that in the experimental animal used, the ferret, the severed vessels tended to regenerate and restore the humoral connection.)

This led to a massive and prolonged search

for the hypothalamic hormones or "releasing factors." As Harris was unwilling to devote himself exclusively to such a demanding and basically tedious exercise, the first hypothalamic hormones were isolated, purified, and synthesized by Roger GUILLEMIN and Andrew SCHALLY in the late 1960s.

Harrison, John (1693–1776) *English horologist*

John Harrison was born at Foulby near Pontefract in Yorkshire, the son of a carpenter, a trade he originally took up himself.

In the early 18th century exploration and colonial expansion was an important concern of European governments. A number of naval disasters caused by navigation errors prompted the British Government to set up a Board of Longitude in 1714. The board offered a prize of £20,000 to anyone who discovered a practical method of measuring longitude accurately. The key to determining the position of a ship out of sight of land was accurate timekeeping. The board's conditions effectively meant the construction of a chronometer that kept time to an accuracy of 3 seconds per day.

Harrison became interested in the problem in 1728 and in 1735 completed his first instrument, H1. He improved his design over a period of many years and in 1762 his perfected chronometer, H4, was found to be in error by only 5 seconds (corresponding to 1.25' of longitude) after a voyage to Jamaica.

Harrison's chronometers all met the conditions set up by the Board of Longitude but he had problems obtaining the prize money. In 1763 he was given £5000 and it was not until 1773, after the intervention of King George III, that he received the full amount less expenses. His chronometer was used in 1776 by James COOK on his voyage to Australia and New Zealand.

Harrison, Ross Granville (1870–1959) *American biologist and embryologist*

> The reference of developmental processes to the cell was the most important step ever taken in embryology.
> —*Embryology and its Relations* (1937)

Born in Germantown, Philadelphia, Harrison graduated from Johns Hopkins University in 1889 and continued studying experimental embryology for the next ten years at Bryn Mawr College, Johns Hopkins University, and in Germany at the University of Bonn. He had an excellent ear for languages and spoke German fluently. In 1899 he gained his MD degree from the University of Bonn and returned to America to become associate professor of anatomy at Johns Hopkins. From 1907 to 1938 he worked at Yale, first as professor of anatomy then from 1927 as professor of biology.

Harrison's work in experimental embryology formed a bridge between the morphological studies of the 19th century and the new molecular biology of the 20th century based on cell function and structure. In his most influential work (1910) he demonstrated the outgrowth of nerve fibers from ganglion cells in embryonic tissues by devising techniques so that the event could actually be observed. His early attempts used frog-embryo cells hanging in a nutrient medium from the underside of a special microscope slide. The method was gradually refined to give the important new technique of tissue culture. Although Harrison himself did not pursue tissue culture to any great extent the method has proved immensely useful in testing new drugs and in the production of vaccines.

Harrison founded the influential *Journal of Experimental Zoology* in 1906.

Hartline, Haldan Keffer (1903–1983) *American physiologist*

Hartline, who was born in Bloomsburg, Pennsylvania, was educated at Lafayette College, Easton, Pennsylvania, and Johns Hopkins University, Baltimore, where he was professor of biophysics from 1949 to 1953. In 1953 he became professor of physiology at the Rockefeller University, New York City. His work was specially concerned with sense receptors and in particular with the neurophysiology of vision. Using minute electrodes to separate and study individual eye fibers of arthropod and vertebrate eyes, notably horseshoe crabs and frogs, he was able to elucidate the fine working of individual cells in the retina and to show how the eye distinguishes between different shapes. His work in this field led to his sharing the Nobel Prize for physiology or medicine with George WALD and Ragnar GRANIT (1967).

Hartmann, Johannes Franz (1865–1936) *German astronomer*

Hartmann was born the son of a merchant in Erfurt, in Germany. He was educated at the universities of Tübingen, Berlin, and Leipzig where he obtained his PhD in 1891. He worked first at the Leipzig and Potsdam observatories before being appointed in 1909 as professor of astronomy and director of the Göttingen University Observatory. He remained there until 1921 when he became director of the La Plata Observatory in Argentina, only returning to Göttingen in 1935 a few months before his death.

Hartmann was responsible for the important observation in 1904 that provided the first clear evidence for the existence of interstellar gas. He noted that in the spectrum of the star Delta Orion, a binary system, the calcium lines failed to exhibit any periodic DOPPLER effect arising from the orbital motion of the stars: when a star moves in its orbit toward the Earth the wavelength of lines in its spectrum are shifted toward the blue, while as it moves away from the Earth its spectral lines are shifted toward the red. That there were what Hartmann described as "stationary lines" of calcium in the spectrum could only mean that the calcium was not part of the atmosphere of Delta Orion and therefore was not participating in the orbital motion. It must occur somewhere between binary system and observer. The existence of interstellar matter and its significance in the estimation of stellar distances was finally demonstrated by Robert TRUMPLER in 1930.

Hartwell, Leland H. (1939–) *American cell biologist*

Born in Los Angeles, Hartwell was awarded a BS in 1961 by the California Institute of Technology, and a PhD in 1964 by the Massachusetts Institute of Technology. After postdoctoral studies at the Salk Institute (1964–65), he joined the University of Washington, Seattle, in 1968, becoming professor of genetics in 1973. In 1996 he also joined the Fred Hutchinson Cancer Research Center, Seattle, and in 1997 was appointed its president and director.

Since the late 1960s, Hartwell has been one of the foremost investigators of the cell cycle – the sequence of events that mark the growth of a living cell, and prepare it for division into two daughter cells. In particular, he has used baker's yeast, *Saccharomyces cerevisiae*, as a model to determine the genes responsible for controlling the cell cycle, the so-called cell division cycle (CDC) genes.

During 1970–71 Hartwell isolated yeast mutants with defects in various aspects of the cell cycle, and subsequently identified over 100 CDC genes. Outstanding among these was a particular gene designated *CDC28* by Hartwell. This was found to control the so-called START point in the cell cycle, after which the yeast cell is committed to progress through the growth phase (G1) and embark on the "synthesis" phase (S). The S phase is when the chromosomes are duplicated in readiness for division of the cell's nucleus prior to cell division.

From his later yeast studies, Hartwell introduced the concept of checkpoints in the cell cycle. These allow a pause in the cell cycle to enable any damage to the cell's genetic material (DNA) to be repaired. This ensures that errors in the DNA are not transmitted to the daughter cells. Checkpoints also are a means of verifying that all necessary steps have been completed before the cell enters the next phase of the cycle.

Hartwell pointed out the importance of this concept in understanding the behavior of cancer cells. Cancer cells bypass these checkpoints, so transmitting mutations to their progeny cells, and hence the capability to evolve into malignant cells.

For his discoveries of key regulators of the cell cycle, Hartwell received the 2001 Nobel Prize for physiology or medicine, awarded jointly with R. Timothy HUNT and Paul M. NURSE.

Harvey, William (1578–1657) *English physician*

> Everything is from an egg.
> —*Exercitationes de generatione animalium* (1651; Exercises in Animal Reproduction)

Harvey was born in the English coastal town of Folkestone and educated at King's School, Canterbury, and Cambridge University. In 1599 he made the then customary visit to Italy where he studied medicine at the University of Padua under the anatomist FABRICIUS ab Aquapendente, obtaining his MD in 1602. He was appointed physician at St. Bartholomew's Hospital, London, in 1609 and in 1618 began working at the court as physician extraordinary to James I. He also served Charles I, accompanying him on his various travels and campaigns throughout the English Civil War. He was rewarded briefly with the office of warden of Merton College, Oxford, in 1645 but with the surrender of Oxford to the Puritans in 1646 Harvey, suffering much from gout, took the opportunity to retire into private life.

In 1628 Harvey published *De motu cordis et sanguinis in animalibus* (On the Motion of the Heart and Blood in Animals). This announced the single most important discovery of the modern period in anatomy and physiology, namely, the circulation of the blood. The orthodox view, going back to GALEN, saw blood originating in the liver and from there being distributed throughout the body. There was no circulation for Galen and he believed that the arteries and veins carried different substances. Harvey made a simple calculation that revealed the prodigious amounts of blood that would have to be produced if there was no circulation. Harvey also could not understand why the valves in the veins were placed so that they allowed free movement of blood to the heart but not away from it. It did however make sense if blood was pumped to the limbs through the arteries and returned through the veins.

Harvey then set about demonstrating his supposition. He examined the action of the heart of such cold-blooded creatures as frogs, snakes, and fishes as their slower heart rate allows clearer observations to be made. This enabled him to establish that blood passes from the right to the left side of the heart not through the wall, or septum, which was solid, but via the lungs. It was also clear that blood is pumped from the heart into the arteries, for he observed that they begin to fill at the moment when the heart contracts. Between contractions, the heart fills with blood from the veins. The heart is thus, Harvey declared, nothing more than a pump. To show that blood passes from the arteries to the veins rather than vice versa, Harvey resorted to a number of simple and compelling experiments with ligatures.

Harvey's 72-page masterpiece received considerable but by no means universal support. There was, as he was well aware, one weak link in his argument, namely the precise connection, or anastomoses, between the arterial and venous system. He thus had to accept, without observation, that the hair-thin capillaries of the two systems did in fact link up. Harvey only had a magnifying glass at his disposal and it was left to Marcello MALPIGHI to observe the implied anastomoses through his microscope in 1661. The importance of Harvey's discovery lay in providing an alternative to the Galenic theory, thus encouraging other scientists to question the authority of ancient texts.

Harvey also worked in embryology. He argued that all life arose from the egg, thus denying spontaneous generation. To describe the process of generation he thought he had observed in chickens and deer, Harvey coined the term *epigenesis*. By this he meant the female egg possessed an independent existence and was capable of completing its development through the activity of its own vital principle. It did not join with the semen, nor was it fertilized by it. Harvey believed the semen acted by initiating the self-contained development of the egg through touch alone. However with the use of the microscope and the earlier identification of anatomical features within the egg by Malpighi in 1673 the alternative preformationist view began to gain ground.

Harvey became a figure of much influence within the College of Physicians. He served as treasurer in 1628 and although offered the presidency in 1654 felt compelled to decline it on grounds of health.

Hassell, Odd (1897–1981) *Norwegian chemist*

Hassell (**has**-el), who was born in the Norwegian capital of Christiania (now Oslo), was educated at the university there and in Berlin where he obtained his doctorate in 1924. He immediately returned to the University of Oslo and served there as professor of chemistry from 1934 until his retirement in 1964.

Early in his career, following studies on how organic dyes photosensitize silver halides, Hassell discovered adsorption indicators. In 1943 he published an important conformational analysis of cyclohexane but, as he refused to use the language of the German conquerors and published it in Norwegian, its influence was considerably reduced. The molecule exists in two main forms, the so-called boat and chair con-

formations; Hassell had little difficulty in showing the chair form to be the most stable. It was for his work on conformation that he shared the 1969 Nobel Prize for chemistry with Derek BARTON.

Hatchett, Charles (1765–1847) *British chemist*

Hatchett was the son of a London coachbuilder. He is noted as the discoverer of the element niobium, which he found in 1801 in a mineral from Connecticut. The following year, Anders EKEBERG found the element tantalum in Sweden and, for some time, it was unclear whether the elements were the same. It was not until 1865 that Jean MARIGNAC showed that the two were different elements. Hatchett's original name of columbium was replaced by niobium (for Niobe, daughter of Tantalus in mythology). Hatchett later served (1823) with Humphry DAVY on a committee to investigate the corrosion of copper plates on ships. On his father's death he retired from research to take over the family coachbuilding business.

Hauksbee, Francis (c. 1670–c. 1713) *English physicist*

> A beautiful Phaenomenon, viz. a fine purple light, and vivid to that degree, that all the included Apparatus was easily and distinctly discernible by the help of it.
> —On producing light by friction in a vacuum

Hauksbee, a student of Robert BOYLE, conducted numerous experiments on a wide range of topics. They are fully described in his *Physico-Mechanical Experiments* (1709). He worked as demonstrator at the Royal Society and became a fellow in 1705. Under the supervision of NEWTON he conducted a series of experiments on capillary action (the movement of water through pores, caused by surface tension) using tubes and glass plates. He also made improvements to the air pump and made a thorough investigation of static electricity, showing that friction could produce luminous effects in a vacuum.

Hauptman, Herb Aaron (1917–) *American mathematical physicist*

Born in New York, Hauptman graduated from Columbia in 1939. After two years spent with the U.S. Census Bureau he was drafted into the U.S. Air Force where he spent much of the war as a radar instructor. In 1947 he joined the staff of the Naval Research Laboratory, Washington, DC. At the same time he pursued his doctorate (awarded 1952) at the University of Maryland. In 1970 Hauptman moved to the Medical Foundation, Buffalo, as its research director, becoming president in 1988.

In collaboration with his Washington colleague Jerome KARLE, Hauptman has made a significant advance in the use of x-ray crystallography. The classic x-ray photograph of a crystal has a diffraction pattern of dots and it was the task of the crystallographer to find a molecular structure that would give this pattern. A likely structure would be assumed and tested against the x-ray data.

Further progress was made when crystallographers worked out how to apply FOURIER analysis to the scattered x-rays. The difficulty was that, although the amplitudes of the waves could be deduced, their phase could not be determined in this way. In 1953, Hauptman and Karle showed mathematically how the phase problem could be overcome. Their method, known as the *direct method*, was mainly statistical and allowed structures to be determined much more quickly. Whereas previously a 15-atom molecule might require many months' work, the structure of larger molecules, following the work of Hauptman and Karle, could be worked out in a matter of days.

For his work in this area Hauptman shared the 1985 Nobel Prize for chemistry with Karle.

Haüy, René Just (1743–1822) *French mineralogist*

Haüy (a-oo-**ee**), whose father was a poor clothworker, was born in St. Just in France. His interest in church music attracted the attention of the prior of the abbey, who soon recognized Haüy's intelligence and arranged for him to receive a sound education. While in Paris, his interest in mineralogy was awakened by the lectures of Louis Daubenton. He became professor of mineralogy at the Natural History Museum in Paris in 1802. His *Traité de minéralogie* (Treatise on Mineralogy) was published in five volumes in 1801 and *Traité de cristallographie* (Treatise on Crystallography) in three volumes in 1822.

Haüy is regarded as the founder of the science of crystallography through his discovery of the geometrical law of crystallization. In 1781 he accidentally dropped some calcite crystals onto the floor, one of which broke, and found, to his surprise, that the broken pieces were rhombohedral in form. Deliberately breaking other and diverse forms of calcite, he found that it always revealed the same form whatever its source. He concluded that all the molecules of calcite have the same form and it is only how they are joined together that produces different gross structures. Following on from this he suggested that other minerals should show different basic forms. He thought that there were, in fact, six different primitive forms from which all crystals could be derived by being linked in different ways. Using his theory he was able to predict in many cases the correct angles of the crystal face. The work aroused much controversy and was attacked by Eilhard MITSCHERLICH in 1819 when he discovered isomorphism

in which two substances of different composition can have the same crystalline form. Haüy rejected Mitscherlich's arguments.

Haüy also conducted work in pyroelectricity. The mineral *haüyne* was named for him.

Hawking, Stephen William (1942–   ) *British theoretical physicist and cosmologist*

> Even as he [Hawking] sits helpless in his wheelchair, his mind seems to soar even more brilliantly across the vastness of space and time to unlock the secrets of the universe.
> —*Time* (1988)

> Why does the universe go to all the bother of existing? Is the unified theory so compelling that it brings about its own existence? Or does it need a creator, and, if so, does he have any other effect on the universe? And who created him?
> —*A Brief History of Time* (1988)

Hawking, who was born in Oxford, England, graduated from the university there and obtained his PhD from Cambridge University. After holding various Cambridge posts, he became professor of gravitational physics in 1977 and, in 1979, was appointed Lucasian Professor of Mathematics.

Hawking has worked mainly in the field of cosmology, in particular the theory of black holes. In 1965 Roger PENROSE had shown that a star collapsing to form a black hole would ultimately form a singularity – a point at which the density of matter is infinite and at which there is an infinite curvature of space–time. Hawking realized that by reversing the time in Penrose's theory he could show that the big bang originating the Universe must also have come from a singularity. Similarly, if the Universe were to stop expanding and start contracting it would eventually end at a singularity – the so-called "big-crunch." In a classic paper published in 1970 Penrose and Hawking proved that these results hold in a very complete way. The results of Hawking's work using general relativity were summarized in a book with George Ellis, *The Large Scale Structure of Spacetime* (1973).

In fact, at a singularity, with infinite curvature of space–time, the general theory of relativity breaks down and consequently it cannot be applied to the origin of the Universe. This led Hawking to the application of quantum theory to the gravitational interaction. Of the four fundamental interactions, the strong, weak, electromagnetic, and gravitational interactions, the gravitational interaction is the only one not described by quantum theory. Quantum theory has been successfully applied to the other interactions. Gravitational interactions between masses over long distances are important in cosmology and can be described by the nonquantum theory of relativity. However, at the vanishingly small distances necessarily occurring just after the big bang (or just before a total collapse), quantum effects would be important. Hawking and others turned their attention to "quantum gravity."

So far, the general application of quantum mechanics to gravitational interactions has had limited success. One notable discovery has been Hawking's theory showing that black holes are not in fact "black" – they effectively emit energy as if they were a hot body.

The basis of the mechanism behind "hot" black holes is the HEISENBERG uncertainty principle. According to this, free space cannot be empty because a point in space would then have zero energy at a fixed time and this would contradict the principle. In space, pairs of virtual particles and antiparticles are constantly forming and annihilating. One member of the pair has a positive energy and the other has a negative energy. Under normal conditions a virtual particle does not exist in isolation and is not detected. However, Hawking has shown that in the vicinity of a black hole it is possible for the particles to separate. The negative-energy particle can fall into a black hole and its partner may escape to infinity, appearing as emitted energy.

The theory resolves a problem concerning the thermodynamics of black holes. If matter of high entropy falls into a black hole then there has been a net entropy loss unless the black hole itself gains entropy. One interpretation of the entropy of a black hole is its area, which increases whenever matter falls into the hole. However, if a black hole has an entropy it must also have a temperature. Hawking showed that the emission of energy from a black hole was distributed as if it were radiated from a black body at the appropriate temperature.

A black hole produced by a collapsing star would have a temperature within a few millionths of a degree above absolute zero. Hawking has speculated on the existence of "mini black holes" weighing a billion tons but having a size no bigger than a proton (about 10^{-15} meter). These could be produced during the early stages of the big bang and are consequently known as "primordial black holes." Because of their small size they would radiate gigawatts of energy in the x-ray and gamma-ray regions of the electromagnetic spectrum. So far there is no experimental evidence for their existence.

Hawking has innovatively applied quantum gravity to the question of the origin of the Universe, making various modifications to the inflationary theory first proposed by Alan GUTH. He has also put forward an original proposal for the origin and evolution of the Universe applying the "sum over histories" formalism of quantum mechanics of Richard Feynmann. Hawking's model of the Universe is conceptually difficult. It involves a Euclidean space–time in which time is an imaginary quantity (in the mathematical sense). There are no singularities at which the laws of physics break down; space–time is finite but closed, having no boundaries and no beginning or end.

Hawking is generally regarded as one of the foremost theoretical physicists of the last hundred years despite a severe physical handicap. In the early 1960s he developed motor neuron disease and for many years he has been confined to a wheelchair. Most of his communication is through a computer speech synthesizer. Of this, Hawking has said that he "was fortunate in that I chose theoretical physics, because that is all in the mind." In 1988 Hawking published a popular account of cosmology, *A Brief History of Time*. The book and its author captured the public imagination and made Hawking an international celebrity, even to noncosmologists. In it he looks forward to a time when a complete theory could be found understandable to everyone. "If we find the answer to that, it would be the ultimate triumph of human reason – for then we would know the mind of God." Other popular books by Hawking include *Black Holes and Baby Universes and Other Essays* (1993), *The Universe in a Nutshell* (2001), and *A Briefer History of Time* (2005, with Leonard Mlodino).

Haworth, Sir (Walter) Norman (1883–1950) *British chemist*

Haworth, who was born in Chorley, in northwest England, began work in a linoleum factory managed by his father. This required some knowledge of dyes, which led Haworth to chemistry. Despite his family's objections he persisted in private study until he was sufficiently qualified to gain admission to Manchester University in 1903, where he studied under and later worked with William PERKIN, Jr. on terpenes. Haworth did his postgraduate studies at Göttingen where, in 1910, he gained his PhD. In 1912 he joined the staff of St. Andrews University and worked with Thomas Purdie and James IRVINE on carbohydrates. He remained there until 1920 when, after five years at the University of Durham, he was appointed Mason Professor of Chemistry at Birmingham, where he remained until his retirement in 1948.

Emil FISCHER had dominated late 19th-century organic chemistry and, beginning in 1887, had synthesized a number of sugars, taking them to be open-chain structures, most of which were built on a framework of six carbon atoms. Haworth however succeeded in showing that the carbon atoms in sugars are linked by oxygen into rings: either there are five carbon atoms and one oxygen atom, giving a pyranose ring, or there are four carbon atoms and one oxygen atom, giving a furannose ring. When the appropriate oxygen and hydrogen atoms are added to these rings the result is a sugar. He went on to represent the carbohydrate ring by what he called a "perspective formula," today known as a *Haworth formula*.

With Edmund HIRST he went on to establish the point of closure of the ring using the technique of Irvine and Purdie of converting the sugar into its methyl ester. He later investigated the chain structure of various polysaccharides. In 1929 he published his views in *The Constitution of the Sugars*.

In 1933 Haworth and his colleagues achieved a further triumph. Albert Szent-Györgyi had earlier isolated a substance from the adrenal cortex and from orange juice which he named hexuranic acid. It was in fact vitamin C and Haworth, again in collaboration with Hirst, succeeded in synthesizing it. He called it ascorbic acid.

For this work, the first synthesis of a vitamin, Haworth shared the 1937 Nobel Prize for chemistry with Paul KARRER.

Hays, James Douglas (1933–) *American geologist*

Hays was born in Johnstown, New York, and educated at Harvard, Ohio State, and Columbia universities, obtaining his PhD from the last in 1964. He joined Columbia's Lamont–Doherty Geological Observatory, New York, in 1967 as director of the deep-sea sediments core laboratory and in 1975 was appointed professor of geology.

In 1971 Hays reported that from his study of 28 deep-sea piston cores from high and low latitudes it was shown that during the last 2.5 million years eight species of radiolaria had become extinct. Prior to extinction these species were widely distributed and their sudden extinction, in six out of eight cases, was in close proximity to a magnetic reversal, a change in the Earth's magnetic polarity. Hays concluded that the magnetic reversals influenced the radiolarians' extinction.

Heaviside, Oliver (1850–1925) *British electronic engineer and physicist*

> Should I refuse a good dinner simply because I do not understand the process of digestion?
> —On being criticized for using formal mathematical manipulations, without understanding how they worked

Heaviside, a Londoner, was a nephew of Charles WHEATSTONE. Being very deaf, he was hampered in school, and was largely self-taught. He was interested in the transmission of electrical signals and used MAXWELL's equations to develop a practical theory of cable telegraphy, introducing the concepts of self-inductance, impedance, and conductance. However, his early results were not recognized, possibly because the papers were written using his own notation. Heaviside was one of the inventors of vector analysis. He used vector analysis extensively in his writings on electromagnetic theory.

After radio waves had been transmitted across the Atlantic in 1901, he suggested (1902) the existence of a charged atmospheric layer that reflected the waves. The same year Arthur KENNELLY independently suggested the same explanation. The *Heaviside layer* (sometimes called the Kennelly–Heaviside layer) was detected experimentally in 1924 by Edward APPLETON.

Later in life his fame grew and he was awarded an honorary doctorate at Göttingen and was elected a fellow of the Royal Society in 1891.

Hebraea, Maria (Mary the Jewess; Miriam the Prophetess) (?1st century AD) *Egyptian alchemist*

Little is known about the life of Mary Hebraea, but her alchemical theories and practical inventions have featured in numerous texts, albeit often in fragmented or corrupted form, and been quoted by other famous alchemists. She was a central figure in a flourishing alchemical tradition in Alexandria, then a key city of the Roman Empire. In keeping with the secretive nature of her profession, she was known by various names, including Miriam the Prophetess, Sister of Moses, and was heavily influenced by the religious beliefs of gnosticism. The introduction of several key items of alchemical equipment is attributed to her, including a water bath, or *balneum mariae* ("Maria's bath"), which she used to maintain a constant temperature (hence the modern French name, *bain-marie*). Maria also described a three-armed still, known as a *tribikos*. To investigate the effects of arsenic, mercury, and sulfur vapors on metals, she devised an apparatus called *kerotakis*, which comprised essentially a heating chamber, or retort, with a hemispherical cover. Sulfur, for example, was heated in the chamber over a fire, and the vapor allowed to condense on the underside of the cover and flow onto metal held in a tray beneath the cover. This formed a sulfide deposit known as "Mary's black," and was believed to be a first step in the transmutation of base metal into gold. Maria is credited with establishing certain principles of laboratory technique that remained little changed until the modern scientific era.

Hecataeus of Miletus (*c.* 550 BC–*c.* 476 BC) *Greek geographer*

Hecataeus (hek-a-**tee**-us), who was born in Miletus (now in Turkey), flourished during the time of the Persian invasion of Ionia, and was one of the ambassadors sent to Persia. One of the earliest geographical works, the *Periegesis* (Tour Round the World), is attributed to him but only fragments of this now exist. It reportedly contained a map showing the world as Hecataeus believed it to be – a flat disk surrounded by ocean. The work was used by the ancients, notably by the Greek historian Herodotus (who also ridiculed it). Even fewer fragments remain of Hecataeus's other surviving work, *Historiai* (Histories), which gave an account of the traditions and mythology of the Greeks.

Hecht, Selig (1892–1947) *American physiologist*

Hecht (hekt) was born in the Austrian town of Glogow and brought to America in 1898. He was educated at the City College, New York, and at Harvard where he obtained his PhD in 1917. After several junior posts and a prolonged traveling fellowship, Hecht was appointed professor of biophysics at Columbia in 1926, a post he retained until his death.

Hecht is best remembered for his photochemical theory of visual adaptation formulated in the mid-1920s. That the eye can readily adapt to changes in brightness is a familiar experience but the exact mechanism behind this response is far from clear. Hecht proposed that in bright light the visual pigment rhodopsin is somewhat bleached while regeneration takes place in the dark. Under steady illumination the amount of rhodopsin bleached would be balanced by that regenerated. Adaptation is thus simply equated with the amount of rhodopsin in the retinal rods.

Heeger, Alan J. (1936–   ) *American physicist*

Heeger (**hee**-ger) was born in Sioux City, Iowa, and educated at the University of Nebraska, gaining a BS in 1957, and at the University of California, Berkeley, gaining a PhD in 1961. In 1962 he joined the University of Pennsylvania as an assistant professor, becoming an associate professor in 1964 and a full professor in 1967. In 1982 he became a professor of physics at the University of California, Santa Barbara. He retired in 2007.

Heeger was one of the pioneers in the investigation of polymers that conduct electricity. Normally, one thinks of plastics as materials that do not conduct electricity. Indeed, plastic is used for electrical insulation around copper wires in electric cables. However, the work of Heeger, Alan MacDiarmid, and Hideki Shirakawa at the end of the 1970s showed that, when modified in certain ways, some plastics can become electrical conductors.

Plastics are polymers. It is necessary that a polymer has alternate single and double bonds between the carbon atoms of its chain for it to become an electrical conductor. It is also necessary for the polymer to be doped, either by removing electrons by oxidation or introducing them by reduction. The polymer then conducts because of the motion of the holes or extra electrons along the molecule.

There are many important practical applications of these materials. For example, electrically conducting plastics can be used as shields against electromagnetic radiation for computer screens and for "smart" windows to exclude light from the Sun. Research on such polymers is closely associated with rapid progress being made in the subject of molecular electronics. It may well be the case that, in the future, it will be possible to make transistors and other components that consist of single molecules. This would mean that the size of computers would be dramatically reduced while their speed would be dramatically increased. Heeger and his colleagues at UCSB have investigated semiconducting polymers and their applications in light-emitting diodes, light-emitting electrochemical cells, and lasers.

Heeger has won many awards for his pioneering work in this field, culminating in sharing the 2000 Nobel Prize for chemistry with MacDiarmid and Shirakawa.

Heezen, Bruce Charles (1924–1977) *American oceanographer*

Born in Vinton, Iowa, Heezen was educated at Iowa State University, graduating in 1948, and at Columbia, New York, where he received his PhD in 1957. He worked at the Lamont Geological Observatory at Columbia from 1948.

Heezen's work has contributed significantly to knowledge of the ocean floor and the processes that operate within the oceans. In 1952 he produced convincing evidence for the existence of turbidity currents, i.e., currents caused by a mass of water full of suspended sediment. Their existence had been suggested by Reginald Daly in 1936 and proposed as the cause of submarine canyons. Heezen used precise records available from the 1929 Grand Bank earthquake to study these currents. As the area off the Grand Bank was rich with communication cables, exact records of the disturbance caused by the earthquake had been obtained. He was able to reconstruct the movement down the bank of about 25 cubic miles (100 cubic km) of sediment moving with speeds approaching 50 miles per hour (85 km per hour).

In 1957, in collaboration with William Ewing and Marie Tharp, the existence of the worldwide ocean rift was demonstrated and its connection with seismic activity postulated. In 1960 Heezen argued for an expanding Earth in which new material is emerging from the rift, increasing the oceans' width and pushing the continents further apart. Such a view, based on the grounds that the gravitational constant decreases slowly with time, had been suggested earlier by Paul Dirac, but received little support in the early 1960s, particularly when a more plausible mechanism was suggested by Harry H. HESS in 1962.

Heidelberger, Michael (1888–1991)
American immunologist

Heidelberger (**hI**-del-ber-ger) was born in New York City and educated at Columbia where he obtained his PhD in 1911. He first worked at the Rockefeller Institute from 1912 until 1927 when he moved to Columbia, where he served as professor of immunochemistry from 1948 until his retirement in 1956. He then took up the position of adjunct professor of pathology at the New York University Medical School, where he continued to work in his lab when over 100.

Heidelberger in his long career worked on many immunological problems. Between 1928 and 1950 he did much to reveal the chemical structure of antibody and complement, two of the key parts of the immune system.

He also collaborated with a colleague at Rockefeller, Oswald AVERY, and in a famous experiment (1923) demonstrated that the specific antigenic properties of pneumococci are due to certain polysaccharides in their capsules.

Heisenberg, Werner Karl (1901–1976)
German physicist

Heisenberg (**hI**-zen-berg), whose father was a professor of Greek at the University of Munich, was born in Würzburg, Germany. He was educated at the universities of Munich and Göttingen, where in 1923 he obtained his doctorate. After spending the period 1924–26 in Copenhagen working with Niels BOHR, he returned to Germany to take up the professorship of theoretical physics at the University of Leipzig. After the war, Heisenberg returned to Göttingen, where he reestablished the Kaiser Wilhelm Institute for Physics. This was renamed the Max Planck Institute and in 1958 it moved to Munich with Heisenberg as its director, a post he occupied until 1970 when he resigned on the grounds of ill health.

In 1925 Heisenberg formulated a version of quantum theory that became known as matrix mechanics. It was for this work, which was later shown to be formally equivalent to the wave mechanics of Erwin SCHRÖDINGER, that Heisenberg was awarded the 1932 Nobel Prize for physics. Heisenberg began in a very radical way, much influenced by Ernst MACH. Considering the various bizarre results emerging in quantum theory, such as the apparent wave–particle duality of the electron, his first answer was that it is simply a mistake to think of the atom in visual terms at all. What we really know of the atom is what we can observe of it, namely, the light it emits, its frequency, and its intensity. The need therefore was to be able to write a set of equations that would permit the correct prediction of such atomic phenomena.

Heisenberg succeeded in establishing a mathematical formalism that permitted accurate predictions to be made. The method was also developed by Max BORN and Pascual JORDAN. As they used the then relatively unfamiliar matrix mathematics to develop this system, it is not surprising that physicists preferred the more usual language of wave equations used in the equivalent system of Schrödinger.

In 1927 Heisenberg went on to explore a deeper level of physical understanding when he formulated his fundamental "uncertainty principle": that it is impossible to determine exactly both the position and momentum of such particles as the electron. He demonstrated this by simple "thought experiments" of the following type: if we try to locate the exact position of an electron we must use rays with very short wavelengths such as gamma rays. But by so illuminating it the electron's momentum will be changed by its interaction with the energetic gamma rays. Alternatively a lower-energy wave can be used that will not disturb the momentum of the electron so much but, as lower energy implies longer wavelength, such radiation will lack the precision to provide the exact location of the electron. There seems to be no way out of such an impasse and Heisenberg went on to express the limits of the uncertainty mathematically:

$$\Delta x . \Delta p \geq h/2\pi$$

where Δx is the uncertainty in ascertaining the position in a given direction, Δp is the uncertainty in ascertaining the momentum in that direction, and h is the Planck constant. What the equation tells us is that the product of the uncertainties must always be about as great as the Planck constant and can never disappear completely. Further, any attempt made to reduce one element of uncertainty to the minimum can only be done at the expense of increasing the other. The consequence of this failure to know the *exact* position and momentum is an inability to predict accurately the future position of an electron. Thus, like Max BORN, Heisenberg had found it necessary to introduce a basic indeterminacy into physics.

After his great achievements in quantum theory in the 1920s Heisenberg later turned his attention to the theory of elementary particles. Thus in 1932, shortly after the discovery of the neutron by James CHADWICK, Heisenberg proposed that the nucleus consists of both neutrons and protons. He went further, arguing that they were in fact two states of the same basic entity – the "nucleon." As the strong nuclear force does not distinguish between them he proposed that they were "isotopes" with nearly the same mass, distinguished instead by a property he called "isotopic spin." He later attempted the ambitious task of constructing a unified field theory of elementary particles. Although he published a monograph on the topic in 1966 it generated little support.

Unlike many other German scientists Heisenberg remained in Germany throughout the war and the whole Nazi era. He was certainly no Nazi himself but he thought it essential to remain in Germany to preserve traditional scientific values for the next generation. At one time he came under attack from the Nazis for his refusal to compromise his support for the physics of EINSTEIN in any way. Thus when, in 1935, he wished to move to the University of Munich to succeed Arnold SOMMERFELD he was violently attacked by the party press and, eventually, the post went to the little-known W. MÜLLER.

With the outbreak of war in 1939 Heisenberg was soon called upon to come to Berlin to direct the program to construct an atom bomb. His exact role in the program has become a matter of controversy. He has claimed that he never had any real intention of making such a bomb, let alone giving it to Hitler. As long as he played a key role he was, he later claimed, in a position to sabotage the program if it ever looked like being a success. He even went so far as to convey such thoughts to Niels Bohr in 1941 when he met him in Copenhagen, hinting that the Allies' physicists should pursue a similar policy. Bohr later reported that if such comments had been made to him they were done so too cryptically for him to grasp; he was rather under the impression that Heisenberg was trying to find out the progress made by the Allies.

New information on the role of Heisenberg and other senior German scientists was released in 1992. The Allies had gathered the scientists in a bugged house, Farm Hall, near Cambridge and recorded their conversation for six months. When the possibility of microphones was put to Heisenberg, he casually dismissed the suggestion: "Microphones installed? (Laughing) Oh no, they're not as cute as all that...they're a bit old fashioned in that respect."

Heisenberg learned of the Hiroshima bomb on 6 August 1945. His first reaction was of disbelief. He insisted that the announcement could refer only to high explosives. During further discussion he declared: "I never thought we would make a bomb." He felt that as a bomb could not have been completed before the war's end, he lacked the urgency to argue the case strongly enough before the military and politicians. He was also arrogant enough to believe that the Allies would do no better. The question of having to make a moral choice, of deliberately sabotaging a German nuclear program simply never arose.

Helmholtz, Hermann Ludwig von
(1821–1894) *German physiologist and physicist*

> Whoever, in the pursuit of science, seeks after immediate practical utility may rest assured that he seeks in vain.
> —*Academic Discourses* (1862)

> The first discovery of a new law ... is of the same quality as the highest performances of artistic perception in the discovery of new types of expression.
> —"On Thought in Medicine" in *Popular Lectures on Scientific Subjects* (1884)

Born in Potsdam, Germany, Helmholtz (**helm**-holts) studied medicine at the Friedrich Wilhelm Institute in Berlin and obtained his MD in 1842. He returned to Potsdam to become an army surgeon, but returned to civilian life in 1848 and was appointed assistant at the Anatomical Museum in Berlin. He then held a succession of chairs at Königsberg (1849–55), Bonn (1855–58), Heidelberg (1858–71), and Berlin (1871–77) and later became director of the Physico-Technical Institute at Berlin Charlottenburg.

Helmholtz made major contributions to two areas of science: physiology and physics. In physiology he invented (1851) the ophthalmoscope for inspecting the interior of the eye and the ophthalmometer for measuring the eye's curvature. He investigated accommodation, color vision, and color blindness. His book *Handbuch der physiologische Optik* (Handbook of Physiological Optics) was published in 1867. Helmholtz also worked on hearing, showing how the cochlea in the inner ear resonates for different frequencies and analyzes complex sounds into harmonic components. In 1863 he published *Die Lehre von den Tönemfindungen als physiologische Grundlage für die Theorie der Musik* (The Sensation of Tone as a Physiological Basis for the Theory of Music). Another achievement was his measurement of the speed of nerve impulses (1850).

One of Helmholtz's interests had been muscle action and animal heat and this, inspired by his distaste for vitalism, led him to his best-known discovery – the law of conservation of energy. This was developed independently of the work of James JOULE and Julius von MAYER and published as *Über die Erhaltung der Kraft* (1847; On the Conservation of Force). He showed that the total energy of a collection of interacting particles is constant, and later applied this idea to other systems.

Helmholtz also worked in thermodynamics, where he introduced the concept of free energy (energy available to perform work). In electrodynamics he attempted to produce a general unified theory. Heinrich HERTZ, who discovered radio waves in 1888, was Helmholtz's pupil.

Helmont, Jan Baptista van (1579–1644)
Flemish chemist and physician

> A citizen being by a Peer openly disgraced and injured; unto whom he might not answer a word without fear of his utmost ruine; in

silence dissembles and bears the reproach: but straightway after, an Asthma arises.
—*Ortus medicinae* (1648; Origin of Medicine)

Van Helmont (**hel**-mont), who came from a noble Brussels family, was educated at the Catholic University of Louvain in medicine, mysticism, and chemistry, but declined a degree from them. Rejecting all offers of employment, he devoted himself to private research at his home. In 1621 he was involved in a controversy with the Church over the belief that it was possible to heal a wound caused by a weapon by treating the weapon rather than the wound. Van Helmont did not reject this common belief but insisted that it was a natural phenomenon containing no supernatural elements. He was arrested, eventually allowed to remain under house arrest, and forbidden to publish without the prior consent of the Church. He wrote extensively and after his death his collected papers were published by his son as the *Ortus medicinae* (1648; Origin of Medicine).

Van Helmont rejected the works of the ancients, although he did believe in the philosopher's stone. He carried out careful observations and measurements, which led him to discover the elementary nature of water. He regarded water as the chief constituent of matter. He pointed out that fish were nourished by water and that substantial bodies could be reduced to water by dissolving them in acid. To demonstrate his theory he performed a famous experiment in which he grew a willow tree over a period of five years in a measured quantity of earth. The tree increased its weight by 164 pounds despite the fact that only water was added to it. The soil had decreased by only a few ounces.

Van Helmont also introduced the term "gas" into the language, deriving it from the Greek for chaos. When a substance is burned it is reduced to its formative agent and its gas and van Helmont believed that when 62 pounds of wood is burned to an ash weighing 1 pound, 61 pounds have escaped as water or gas. Different substances give off different gases when consumed and van Helmont identified four gases, which he named gas carbonum, two kinds of gas sylvester, and gas pingue. These we would now call carbon dioxide, carbon monoxide, nitrous oxide, and methane.

Hempel, Carl Gustav (1905–1997) *American philosopher of science*

Hempel was born in Oranienburg, Germany, and educated at Göttingen, Heidelberg, Vienna, and Berlin, where he completed his doctorate in 1934. Alarmed by political conditions in Germany, he moved to Holland and finally emigrated to America in 1938. He became a naturalized citizen in 1944. Hempel then taught at the University of Chicago, City College of New York, and Yale before finally settling at Princeton as professor of philosophy, a position he held until his retirement in 1973.

Hempel established his reputation with a paper, *Studies in the Logic of Confirmation* (1945), which challenged the foundations of inductive logic. He pointed out that if we accept the equivalence condition, namely, that whatever confirms either of two equivalent propositions confirms the other, and the rule that positive instances confirm, then a simple paradox follows. For example, the generalization:

(1) All ravens are black

is confirmed by finding anything that is a raven and black. But (1) is equivalent to:

(2) All nonblack things are nonravens.

By the equivalence principle whatever confirms (2) must also confirm (1). But a green pencil, as something that is a nonblack nonraven, confirms (2) and therefore it must also confirm (1). Thus at least according to the rules of inductive logic, it seems that the fact that a pencil is green has confirmed the generalization that all ravens are black. The point can be pushed further to the absurd conclusion that whatever is neither black nor a raven, such as red flags and blue skies, confirms the proposition that all ravens are black.

Hempel went on, in a paper written in collaboration with Paul Oppenheim, *Studies in the Logic of Explanation* (1948), to propose a model of scientific explanation, the Hypothetico-Deductive method (HD), which dominated discussion for a generation. Scientific explanations are seen as deductive consequences from a set of laws and initial conditions. For example, from the initial conditions describing the positions of the Sun, Moon, and Earth together with Newton's laws of motion, the date of a future eclipse can be deduced. The laws and the initial conditions constitute an explanation of the eclipse.

For the reasoning to be valid, Hempel stressed, four conditions must be satisfied, namely, the conclusion must follow logically from the hypotheses, the hypotheses must contain a statement of a law of nature, they must have empirical content, and they must be true. Yet, later critics have argued, even if all conditions are satisfied, the HD method can still lead to error. Further, many explanations in science lie outside the HD framework. Finally, Hempel seems to ignore the crucial question of how the hypotheses are established in the first place. His work is more an account of how scientific discoveries are reported or rationalized; not of how they are arrived at or justified.

Hench, Philip Showalter (1896–1965) *American biochemist*

Born in Pittsburgh, Pennsylvania, Hench was educated at Lafayette College and the University of Pittsburgh, where he obtained his MD in 1920. He spent most of his career working at the Mayo Clinic, becoming head of the section for rheumatic diseases in 1926. Hench was also connected with the Mayo Foundation and the University of Minnesota, where he became professor of medicine in 1947.

For many years Hench had been seeking a method of treating the crippling and painful complaint of rheumatoid arthritis. He suspected that it was not a conventional microbial infection since, among other features, it was relieved by pregnancy and jaundice. Hench therefore felt it was more likely to result from a biochemical disturbance that is transiently corrected by some incidental biological change. The search, he argued, must concentrate on something patients with jaundice had in common with pregnant women. At length he was led to suppose that the antirheumatic substance might be an adrenal hormone, since temporary remissions are often induced by procedures that stimulate the adrenal cortex. Thus in 1948 he was ready to try the newly prepared "compound E," later known as cortisone, of Edward KENDALL on 14 patients. All showed remarkable improvement, which was reversed on withdrawing the drug.

For this development of the first steroid drug Hench shared the 1950 Nobel Prize for physiology or medicine with Kendall and Tadeus REICHSTEIN.

Henderson, Thomas (1798–1844) *British astronomer*

Born in Dundee, Scotland, Henderson started as an attorney's clerk who made a reputation as an amateur astronomer. In 1831 he accepted an appointment as director of a new observatory at the Cape of Good Hope in South Africa. While observing Alpha Centauri he found that it had a considerable proper motion. He realized that this probably meant that the star was comparatively close and a good candidate for the measurement of parallax – the apparent change in position of a (celestial) body when viewed from spatially separate points, or from one point on a moving Earth. All major observational astronomers had tried to detect this small angular measurement and failed. Henderson at last succeeded in 1832 and found that Alpha Centauri had a parallax of just less than one second of arc. The crucial importance of this was that once parallax was known, the distance of the stars could be measured successfully for the first time. Alpha Centauri turned out to be over four light years away. Unfortunately (for Henderson), he delayed publication of his result until it had been thoroughly checked and rechecked. By this time Friedrich BESSEL had already observed and published, in 1839, the parallax of 61 Cygni.

In 1834 Henderson became the first Astronomer Royal of Scotland.

Henle, Friedrich Gustav Jacob (1809–1885) *German physician, anatomist, and pathologist*

> A rational system of histology must employ the transformations of the cells as a principle of classification, so that groups of tissue can be formed according to whether, for example, the cells remain discrete or join lengthwise in rows, or expand into star shapes, or split into fibers, and so forth.
> —*Allgemeine Anatomie* (1841; General Anatomy)

Henle (**hen**-lee), a merchant's son from Fürth, in Germany, was educated at the universities of Heidelberg and Bonn where he obtained his MD in 1832. He began his career as assistant to Johannes MÜLLER in Berlin and, despite various political troubles (he was tried for treason in Berlin), served as professor of anatomy and physiology at the universities of Zurich (1840–44) and Heidelberg (1844–52) before moving to a similar post at Göttingen where he remained until his death.

By the beginning of the 19th century the humoral theory of disease had been finally expelled from orthodox medicine. It was far from clear, however, what to put in its place. A cogent and comprehensive theory, as developed by Louis PASTEUR and Henle's own pupil, Robert KOCH, would not be available for a further 40 years. Henle however took some preliminary steps, notably his declaration that contagious substances are not only organic but indeed are living organisms. He distinguished between miasmas, which arise from the environment, and contagions, which spread from person to person. Such theorizing had little immediate impact on medicine, largely because of the difficulty in providing experimental support. His work consequently was largely ignored as speculative.

Henle also produced two standard and highly influential textbooks: *Allgemeine Anatomie* (1841; General Anatomy) and *Handbuch der rationelle Pathologie* (2 vols., 1846–53; Handbook of Rational Pathology). In them he first described and emphasized the microscopic structure of the epithelium, the cells that cover the internal and external surface of the human body. He has thus frequently been referred to as the founder of modern histology.

As an anatomist Henle's name has been preserved in the *loop of Henle*, a part of the nephron, or urine-secreting tubules, in the kidney.

Henry, Joseph (1797–1878) *American physicist*

> One great object of science is to ameliorate our present condition, by adding to those

advantages we naturally possess ... by a combination of theoretical knowledge with practical skill, machines have been constructed no less useful in their productions than astonishing in their operations.
—*Albany Argus and City Gazette* (1826)

One of the first great American scientists, Henry came from a poor background in Albany, New York, and had to work his way through college. He was educated at the Albany Academy, New York, where he first studied medicine, changing to engineering in 1825. A year later he was appointed a professor of mathematics and physics at Albany. In 1832 he became professor of natural philosophy at Princeton (then the College of New Jersey) where he taught physics, chemistry, mathematics, and geological sciences, and later astronomy and architecture.

Henry is noted for his work on electricity. In 1829 he developed a greatly improved form of the electromagnet by insulating the wire that was to be wrapped around the iron core, thus allowing many more coils, closer together, and greatly increasing the magnet's power. Through this work he discovered, in 1830, the principle of electromagnetic induction. Soon after, and quite independently, Michael FARADAY made the same observation and published first. Faraday is thus credited with the discovery but Henry has the unit of inductance (the *henry*) named for him. However, Henry did publish in 1832 – prior to Faraday and Heinrich LENZ – his discovery of self-induction (in which the magnetic field from a changing electric current induces an electromotive force opposing the current). Earlier (in 1829) he had invented and constructed the first practical electric motor. In 1835 he developed the electric relay in order to overcome the problem of resistance that built up in long wires. This device had an immediate social impact for it was the key step in the invention of the long-distance telegraph, which played a large part in the opening up of the North American continent. In 1846 he became the first secretary of the Smithsonian Institution, which he formed into an extremely efficient body for liaison between scientists and government support of their research. He also did work on solar radiation and on sunspots.

Henry, William (1774–1836) *British physician and chemist*

Henry's father, Thomas Henry, was a manufacturing chemist in Manchester, England, and an analytical chemist of some repute. Initially qualifying as a physician from Edinburgh University, Henry practiced for five years in the Manchester Infirmary. Later he took over the running of the chemical works established by his father.

In 1801 he formulated the law now known as *Henry's law*, which states that the solubility of a gas in water at a given temperature is proportional to its pressure. His close friend John DALTON was encouraged by this finding, seeing it as a confirmation of his own theory of mixed gases, and the two men discussed the methods of experimentation in detail.

Henry also researched into the hydrocarbon gases, following Dalton in clearly distinguishing methane from ethylene (ethene). He determined the molecular formula of ammonia by exploding it with oxygen. He also described the preparation, purification, and analysis of coal gas, and developed a method of analyzing gas mixtures by fractional combustion. His textbook, *Elements of Experimental Chemistry* (1799), went through 11 editions in 30 years.

Hensen, Viktor (1835–1924) *German physiologist and oceanographer*

Hensen (**hen**-zen) was born in Schleswig, Germany, and studied science and medicine at the universities of Würzburg, Berlin, and Kiel, graduating from the latter in 1858. He remained at Kiel to work in the physiology department and later served as professor of physiology (1871–1911).

Hensen worked on comparative studies of vision and hearing but also discovered, independently of Claude Bernard, the compound glycogen. He is better remembered however for his work on plankton. He introduced the term plankton in 1887 to describe the minute drifting animals and plants in the oceans. Moreover he advanced beyond the descriptive stage and introduced numerical methods into marine biology, notably in constructing the *Hensen net*, a simple loop net designed to filter a square meter of water. This enabled the number of plankton in a known area of water to be counted. Hensen tested his equipment in the North Sea and the Baltic in 1885.

Satisfied with his techniques, he made a more ambitious trip in 1889 covering more than 15,000 miles of the Atlantic. One of his more surprising results was the greater concentration of plankton in temperate than in tropical waters.

Heracleides of Pontus (*c.* 390 BC–*c.* 322 BC) *Greek astronomer*

Heracleides of Pontus, when he drew the circle of Venus, and also of the Sun, and assigned a single center to both circles, showed that Venus is sometimes above, sometimes below the Sun.
—Calcidius (5th century AD)

Heracleides (her-a-**klI**-deez), who was born in Heraclea, now in Turkey, was an associate and possibly a pupil of PLATO. Although none of his writings have survived, two views that were unusual for the time have been attributed to him. The philosopher Simplicius of Cilicia, a usually reliable source, reports that "Heracleides supposed that the Earth is in the center and rotates while the heaven is at rest." If this is accurate he must have been the first to state that the Earth rotates, a view that found as little support in antiquity as it did in the medieval period. The second doctrine attributed to him is that Mercury and Venus move around the Sun, which moves around the Earth – a view adopted later by Tycho BRAHE in the 16th century.

Heraclitus of Ephesus (about 500 BC)
Greek natural philosopher

> If you do not expect the unexpected, you will not find it; for it is hard to be sought out, and difficult.
> —Quoted in C. H. Khan, *The Art and Thought of Heraclitus* (1979)

Virtually nothing is known of the life of Heraclitus (her-a-**kll**-tus), and of his book *On Nature* only a few rather obscure fragments survive. His doctrines contrast with those of his near contemporary PARMENIDES for whom, on purely logical grounds, change of any kind was totally impossible. For Heraclitus, everything is continually in a state of change, hence his characteristic aphorism: "We cannot step twice into the same river," and his selection of fire as the fundamental form of matter. The mechanism behind such unremitting change was the constant tension or "strife" between contraries or opposites.

Hérelle, Felix d' *See* D'HÉRELLE, FELIX.

Hermite, Charles (1822–1901) *French mathematician*

Hermite (air-**meet**) was born in Dieuze, France. His mathematical career was almost thwarted in his student days, since he was incapable of passing exams. Fortunately his talent had already been recognized and his examiners eventually let him scrape through. Hermite obtained a post at the Sorbonne where he was an influential teacher.

Hermite began his mathematical career with pioneering work on the theory of Abelian and transcendental functions, and he later used the theory of elliptic functions to give a solution of the general equations of the fifth degree – the quintic. One long-standing problem solved was proving that the number "*e*" is transcendental (i.e., not a solution of a polynomial equation). He also introduced the techniques of analysis into number theory. His most famous work is in algebra, in the theory of *Hermite polynomials*. Although Hermite himself had little interest in applied mathematics this work turned out to be of great use in quantum mechanics.

Hero of Alexandria (about 62 AD) *Greek mathematician and inventor*

> Hero is no scientist, but a practical technician and surveyor.
> —J. L. Heiberg, *History of Mathematics and Science in Classical Antiquity* (1925)

Hero (**heer**-oh) produced several written works on geometry, giving formulas for the areas and volumes of polygons and conics. His formula for the area of a triangle was contained in *Metrica* (Measurement), a work that was lost until 1896. This book also describes a method for finding the square root of a number, a method now used in computers, but known to the Babylonians in 2000 BC. In another of Hero's books, *Pneumatica* (Pneumatics), he wrote on siphons, a coin-operated machine, and the aeolipile – a prototype steam-powered engine that he had built. The engine consisted of a globe with two nozzles positioned so that steam jets from the inside made it turn on its axis. Hero also wrote on land-surveying and he designed war engines based on the ideas of Ctesibius. Yet another of his works, *Mechanica* (Mechanics), was quoted by PAPPUS of Alexandria.

Herophilus of Chalcedon (about 300 BC)
Greek anatomist and physician

Herophilus (her-**off**-i-lus), a pupil of PRAXAGORAS of Cos, was one of the founders of the Alexandrian medical school set up at the end of the 4th century BC under the patronage of Ptolemy I Soter. Although none of his works have survived, GALEN lists some eight titles of which the *Anatomica* (Anatomy) was probably the most significant.

Herophilus is widely, even notoriously, remembered as the result of a famous passage in CELSUS reporting that, with ERASISTRATUS, he practiced vivisection on criminals. The passage has been regarded as suspect by many scholars on the grounds that no such reference occurs in any extant, earlier Greek text. It is however certain that from the results attributed to him he must have undertaken both human and animal dissection. For example, he described a passage from the stomach to the intestines as being "12 finger widths" (*dodekadaktylon*) or in its Latin form, the duodenum; he also named the retina and the prostate and did much work on the brain.

It has been claimed that Herophilus was the first to distinguish between sensory and motor nerves. Nerves, or neura, for Herophilus were simply channels that carried the pneuma or vital air to different parts of the body. Thus while he probably identified sensory nerves it is unlikely that he was able to distinguish between motor nerves and tendons.

Herophilus was reported to have advanced Praxagoras's work on the pulse by counting its frequency against a water clock. Also, according to Galen, he made the important observation that the arteries carried blood as well as pneuma.

Héroult, Paul Louis Toussaint (1863–1914) *French chemist*

Héroult (ay-**roo**), who was born at Thury-Harcourt in France, was a student at the St. Barge Institute in Paris and later worked at the Paris School of Mines. In 1886 he discovered a process for extracting aluminum by electrolysis of molten aluminum oxide, with cryolite (sodium aluminum fluoride) added to lower the melting point. Charles HALL developed a similar process independently in America at about the same time.

Herring, William Conyers (1914–) *American physicist*

Herring was born in Scotia, New York, and educated at the University of Kansas where he obtained a BA in astronomy in 1933. He did research in mathematical physics at Princeton and obtained his PhD at Princeton in 1937. During World War II he worked on underwater warfare and then joined the Bell Telephone Laboratories in 1946. Herring remained with Bell as a research physicist until 1978 when he was appointed professor of applied physics at Stanford University, California, a post he held until his retirement in 1981.

Herring has worked on a wide range of theoretical problems in solid-state physics, including electrical conduction, surface tension of solids, anisotropic effects in superconducting materials, and the magnetic properties of solids.

Herschbach, Dudley Robert (1932–) *American chemist*

Herschbach (**hersh**-bahk), who was born in San José, California, was educated at Stanford University and at Harvard University, where he gained his PhD in 1958. After teaching at the University of California, Berkeley, for four years, he returned to Harvard in 1963 as professor of chemistry.

Herschbach has worked on the details of chemical reactions; for example, a simple reaction in which potassium atoms and iodomethane molecules form potassium iodide and methyl radicals as products, that is:

$$K + CH_3I \rightarrow KI + CH_3\bullet$$

He decided to use molecular beams to examine the nature of the reaction. The reagent molecules were formed into two collimated beams at a sufficiently low pressure to make collisions within the beams a negligible event. The beams were allowed to collide and the direction and velocity of the product molecules measured.

Herschbach was able to draw some conclusions about the reaction. He demonstrated, for example, that the reagents would only react if the incoming potassium atoms struck the iodomethane molecules at the iodide end.

As techniques were refined and extended Herschbach demonstrated that the study of molecular beams could throw considerable light on reaction dynamics. For his work in this field he shared the 1986 Nobel Prize for chemistry with Yuan LEE and John POLANYI.

Herschel, Caroline Lucretia (1750–1848) *German–British astronomer*

Caroline Herschel (**her**-shel), who was born in Hannover, Germany, was the sister and colleague of William HERSCHEL. Having joined her brother as his housekeeper in Bath in 1772, she rapidly graduated to being his assistant and then to original astronomical research of her own. In 1786 she observed her first comet and before 1797 had detected seven more. She also discovered many new nebulae. Her devotion to her brother and his work must have been completely unconditional judging by the many hundreds of nights spent observing. There is a story that she once slipped and fell on a hook attached to the telescope but made no cry lest she disturb her brother's observations. After his death in 1822 she returned to Hannover where she prepared a catalog of about 2,500 nebulae and star clusters. Although it was never published she received the Gold medal of the Royal Astronomical Society in 1828 for it. Caroline Herschel was awarded money by the government to allow her to continue work. She was the first woman in England to hold a government post.

Herschel, Sir (Frederick) William (1738–1822) *German–British astronomer*

> I saw, with the greatest pleasure, that most of the nebulae … yielded to the force of my light and power, and were resolved into stars.
> —On the great light-gathering power of his large telescopes

Herschel, who was born in Hannover, Germany, started life in the same occupation as his father – an oboist with the band of the Hannoverian footguards. He moved permanently to England in 1757, where he worked as a freelance itinerant musician until 1767 he was appointed as organist of a church in Bath. His sister Caroline Herschel joined him in Bath in 1772. He was led by his interest in musical theory to a study of mathematics and ultimately astronomy. Herschel made his own telescopes and his early observations were significant enough to be drawn to the attention of George III in 1782. The king, who had a passionate interest in astronomy and clockwork, was sufficiently impressed with Herschel to employ him as his private astronomer at an initial salary of £200 a year and to finance the construction of very large telescopes. At first Herschel settled at Datchet, near Windsor, but in 1786 he moved to

Slough where he remained for the rest of his life.

Herschel's contributions to astronomy were enormous. He was fortunate to live at a time when prolonged viewing with a large reflector could not but be fruitful and he took full advantage of his fortune. He made his early reputation by his discovery in 1781 of the first new planet since ancient times. He wished to name it after his patron as "Georgium Sidus" (George's Star) but Johann BODE's suggestion of "Uranus" was adopted. Herschel's work is notable for the unbelievable comprehensiveness with which he extended the observations of others. Thus he extended Charles MESSIER's catalog of just over 100 nebulae by a series of publications listing over 2,000 nebulae. He not only began the study of double stars but cataloged 800 of them. He also discovered two satellites of Uranus – Titania and Oberon (1787) – and two of Saturn – Mimas and Enceladus (1789–90). He built a large number of telescopes of various sizes, culminating in his enormous 40-foot (12-m) reflector. This cost George III £4,000 plus £200 a year for its upkeep. The eyepiece was attached to the open end, thus eliminating the loss of light caused by the secondary mirror used in the Newtonian and Gregorian reflectors. The disadvantage was the danger of climbing up to the open end of the 40-foot instrument in the dark. One eminent astronomer, Giuseppe PIAZZI, failing to master this skill, fell and broke his arm. It was finally dismantled in 1839 while William's son John conducted his family in a special requiem he had composed for the occasion.

Herschel produced not only observational work but theoretical contributions on the structure of the universe. He established the motion of the Sun in the direction of Hercules and tried to calculate its speed (1806). But, above all, he was the first to begin to see the structure of our Galaxy. Conducting a large number of star counts, he established that stars are much more numerous in the Milky Way and the plane of the celestial equator, becoming progressively fewer toward the celestial poles. He explained this by supposing that the Galaxy is shaped like a grindstone. If we look through its short axis we see few stars and much dark space; through its long axis we see a stellar multitude. Herschel was supported in his astronomical life by his sister Caroline. His son John also became an astronomer of note.

Herschel, Sir John Frederick William (1792–1871) *British astronomer*

John Herschel, who was born at Slough in the southeast of England, read mathematics at Cambridge University and then began to study law. Although he was the son of the astronomer William HERSCHEL he did not take up astronomy seriously until 1816 when he began, some-

what reluctantly, to assist his father with his observations.

John Herschel went to South Africa in 1834 to make a comprehensive survey of the skies of the southern hemisphere, and succeeded Thomas HENDERSON as director of the Cape of Good Hope Observatory, doing for the southern skies what his father had done for the northern. He discovered and described some 2,000 nebulae and some 2,000 double stars, publishing the results of his surveys in 1847. He seems to have given up astronomical observation on his return from South Africa in 1838, instead becoming interested in photography (introducing the terms "positive" and "negative") and pioneering the use of photographic techniques in astronomy. He also experimented on the spectral lines discovered by Joseph von FRAUNHOFER because he began to see a connection between the absorption and emission lines. He was also a major figure in the regeneration and reorganization of British science in the first half of the 19th century. He was one of the founder members of the Royal Astronomical Society in 1830, and took on many public duties, becoming, like Newton, master of the mint from 1850 to 1855. This, however, proved too taxing for him and he suffered a nervous breakdown, which led to his retirement from public life. His study of scientific method *Discourse on the Study of Natural Philosophy* (1830) influenced the philosopher John Stuart Mill and Charles DARWIN.

Hershey, Alfred Day (1908–1997) *American biologist*

Hershey was born in Owosso, Michigan, and graduated from Michigan State College in 1930, remaining there to do his PhD thesis on the chemistry of *Brucella* bacteria. Having received his doctorate in 1934, he taught at Washington University, St. Louis, until 1950, when he moved to the Genetics Research Unit of the Carnegie Institute, Washington. In 1962 he became director of the unit, a position he retained until his retirement in 1974.

Hershey, along with Salvador LURIA and Max DELBRÜCK, was one of the founders in the early 1940s of the so-called "phage group." In 1945, independently of Luria, he demonstrated that spontaneous mutations must occur in bacterial viruses (phage). In the following year he established, at the same time as Delbrück, that genetic recombination takes place between phages present in the same cell.

Hershey is best known, however, for the experiment conducted with Martha Chase and reported in their 1952 paper, *Independent Functions of Viral Proteins and Nucleic Acid in Growth of Bacteriophage*. At the time it was still uncertain whether genes were composed of protein, nucleic acid, or some complex mixture of the two. They utilized the fact that DNA contains phosphorus but no sulfur, while phage protein has some sulfur in its structure but no phosphorus. Phage with a protein coat labeled with radioactive sulfur and DNA with radioactive phosphorus was allowed to infect bacteria. After infection the bacteria were spun in a blender. The labeled protein was stripped off the bacteria while the radioactive DNA remained inside the bacterial cell. When allowed to incubate, the bacteria proved capable of producing a new phage crop. The experiment would seem to show that DNA was more involved than protein in the process of replication. Many think that Martha Chase was given insufficient credit for her contribution to this work.

For his fundamental contributions to molecular biology, Hershey received the 1958 Albert Lasker Award and the 1965 Kimber Genetics Award. However, it was not until 1969 that Hershey, together with Delbrück and Luria, was awarded the Nobel Prize for physiology or medicine.

Hershko, Avram (1937–   ) *Israeli biochemist*

Hershko was born in Karcag, Hungary, but his family emigrated to Israel in 1950. He studied medicine at the Hebrew University, Jerusalem, gaining his MD in 1965. Following a stint as a physician in the Israel Defense Force (1965–67), he obtained his PhD in 1969. After working as a postdoctoral fellow at the University of California (1969–72), he joined the Israel Institute of Technology (Technion) in Haifa as associate professor (1972–80), then professor. He is currently distinguished professor at the Rappaport Family Institute for Research in Medical Sciences at the Technion.

In the late 1970s and early 1980s Hershko worked closely with his colleague at the Technion, Aaron CIECHANOVER, to investigate mechanisms of energy-dependent protein degradation in cells. The two, along with the American biochemist Irwin ROSE, discovered the crucial role played by the polypeptide ubiquitin, which covalently attaches to proteins as a label, marking out the protein as a target for cutting up by protein complexes called proteasomes. This work has since proved fundamentally important as a control mechanism for many cell activities, and of relevance to the development of cancers and other diseases. In recognition of this, Hershko was awarded the 2004 Nobel Prize for chemistry, jointly with Ciechanover and Rose.

Hertz, Gustav (1887–1975) *German physicist*

A nephew of the distinguished physicist Heinrich HERTZ (herts), Gustav Hertz was born in Hamburg, Germany, and educated at the universities of Munich and Berlin. He taught in

Berlin and Halle before his appointment in 1928 to the professorship of experimental physics at the Technical University, Berlin. Hertz, as a Jew, was dismissed from his post in 1935. He worked for the Siemens company from 1935 until 1945, somehow managing to survive the war, when he was captured by the Russians. He reemerged in 1955 to become director of the Physics Institute in Leipzig, then in East Germany.

In 1925 Hertz was awarded the Nobel Prize for physics for his work with James FRANCK on the quantized nature of energy transfer.

Hertz, Heinrich Rudolf (1857–1894) *German physicist*

> One cannot escape the feeling that these mathematical formulae have an independent existence and an intelligence of their own, that they are wiser than we are, wiser even than their discoverers, that we get more out of them than was originally put into them.
> —Quoted by E. T. Bell in *Men of Mathematics*

Hertz came from a prosperous and cultured Hamburg family. In 1875 he went to Frankfurt to gain practical experience in engineering and after a year of military service (1876–77) spent a year at the University of Munich. He had decided on an academic and scientific career rather than one in engineering, and in 1878 chose to continue his studies at the University of Berlin under Hermann von HELMHOLTZ. Hertz obtained his PhD in 1880 and continued as Helmholtz's assistant for a further three years. He then went to work at the University of Kiel. In 1885 he was appointed professor of physics at Karlsruhe Technical College and in 1889 became professor of physics at the University of Bonn. His tragic early death from blood poisoning occurred after several years of poor health and cut short a brilliant career.

Hertz's early work at Berlin was diverse but included several pieces of research into electrical phenomena and equipment. With no laboratory facilities at Kiel he had considered more theoretical aspects of physics and had become more interested in the recent work of James Clerk MAXWELL on electromagnetic theory. Helmholtz had suggested an experimental investigation of the theory to Hertz in 1879 but it was not until 1885 in Karlsruhe that Hertz found the equipment needed for what became his most famous experiments. In 1888 he succeeded in producing electromagnetic waves using an electric circuit; the circuit contained a metal rod that had a small gap at its midpoint, and when sparks crossed this gap violent oscillations of high frequency were set up in the rod. Hertz proved that these waves were transmitted through air by detecting them with another similar circuit some distance away. He also showed that like light waves they were reflected and refracted and, most important, that they traveled at the same speed as light but had a much longer wavelength. These waves, originally called *Hertzian waves* but now known as radio waves, conclusively confirmed Maxwell's prediction on the existence of electromagnetic waves, both in the form of light and radio waves.

Once at Bonn Hertz continued his analysis of Maxwell's theory, publishing two papers in 1890. His experimental and theoretical work put the field of electrodynamics on a much firmer footing. It should also be noted that in 1887 he inadvertently discovered the photoelectric effect whereby ultraviolet radiation releases electrons from the surface of a metal. Although realizing its significance, he left others to investigate it.

Hertz's results produced enormous activity among scientists but he died before seeing Guglielmo MARCONI make his discovery of radio waves a practical means of communication. In his honor the unit of frequency is now called the hertz.

Hertzsprung, Ejnar (1873–1967) *Danish astronomer*

Hertzsprung (**hairt**-spruung) was born in Frederiksberg, Denmark, the son of a senior civil servant who had a deep interest in mathematics and astronomy but who was anxious to see that his son received a more practical education. Consequently Hertzsprung was trained as a chemical engineer at the Copenhagen Polytechnic, graduating in 1898. He worked as a chemist in St. Petersburg and then studied photochemistry under Wilhelm OSTWALD in Leipzig before returning to Denmark in 1902. His first

professional appointment as an astronomer was in 1909 at the Potsdam Observatory. The bulk of his career, from 1919 to 1944, was spent at the University of Leiden where from 1935 he served as director of the observatory. After his retirement in 1944 he returned to Denmark where he continued his studies for a further 20 years.

Hertzsprung's name is linked with that of Henry RUSSELL as independent innovators of the Hertzsprung–Russell (H–R) diagram. In the late 19th and early 20th centuries, techniques used in photographic spectroscopy were being greatly improved. With his background in photochemistry, Hertzsprung was able to devise methods by which he could determine the intrinsic brightness, i.e., luminosity, of stars. He showed that the luminosity of most of the stars he studied decreased as their color changed from white through yellow to red, i.e., as their temperature decreased. He also found that a few stars were very much brighter than those of the same color. Hertzsprung thus discovered the two main groupings of stars: the highly luminous giant and supergiant stars and the more numerous but fainter dwarf or main-sequence stars. Hertzsprung published his results, although not in diagrammatic form, in 1905 and 1907 in an obscure photographic journal. His work therefore did not become generally known and credit initially went to Russell who published the eponymous diagram in 1913. It would be difficult to exaggerate the importance or usefulness of the H–R diagram, which has been the starting point for discussions of stellar evolution ever since.

Much of Hertzsprung's work concerned open clusters of stars. In 1911 he published the first color-magnitude diagrams of the Pleiades and

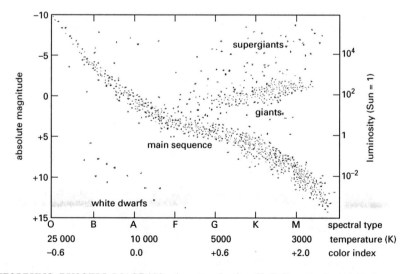

HERTZSPRUNG–RUSSELL DIAGRAM   An example of an H–R diagram for bright stars.

Hyades clusters, showing how the color of member stars varied with observed brightness. He also measured the proper motions of stars, i.e., their angular motions in a direction perpendicular to the observer's line of sight, and used the results to establish membership of clusters.

One other major achievement of Hertzsprung was the development of a method for the determination of stellar and galactic distances. In the 19th century Friedrich BESSEL and Friedrich Georg STRUVE had been the first to use measurements of annual parallax to calculate stellar distances but this was only accurate up to distances of about a hundred light-years. In 1913, when Hertzsprung announced his results, astronomers had made little progress in measuring distances. The work of Henrietta LEAVITT in 1912 had shown that the period of light variation of a group of stars known as Cepheid variables was related to their observed mean brightness. These Cepheids lay in the Magellanic Clouds. Hertzsprung assumed that at the great distance of the Clouds all member stars could be considered to have approximately the same distance. Since observed and intrinsic brightness of a star are directly linked by its distance, the periods of light variation of Cepheids in the Clouds were thus also related to their intrinsic brightness. By extrapolation, Cepheids could thus be an invaluable means of measuring the distance of any group of stars containing a Cepheid by observing the period and apparent brightness of the Cepheid.

The work of establishing the period-luminosity relation on a numerical basis was begun by Hertzsprung and continued by Harlow SHAPLEY. Hertzsprung determined the distances of several nearby Cepheids from measurements of their proper motions. Using his results and Leavitt's values for the periods and apparent brightness of Cepheids in the Small Magellanic Cloud (SMC) he was then able to calculate the distance to the SMC. Although somewhat smaller than today's value, this was the first measurement of an extragalactic distance.

Herzberg, Gerhard (1904–1999) *Canadian spectroscopist*

Born in Hamburg, Germany, Herzberg (**herts**berg) was educated at the universities of Göttingen and Berlin. He taught at the Darmstadt Institute of Technology from 1930 until 1935 when, with the rise to power of the Nazis, he emigrated to Canada, where he was research professor of physics at the University of Saskatchewan from 1935 until 1945. He returned to Canada in 1948 after spending three years as professor of spectroscopy at the Yerkes Observatory, Wisconsin. From 1949 until his retirement in 1969 he was director of the division of pure physics for the National Research Council in Ottawa.

Herzberg is noted for his extensive work on the technique and interpretation of the spectra of molecules. He elucidated the properties of many molecules, ions, and radicals and also contributed to the use of spectroscopy in astronomy (e.g., in detecting hydrogen in space). His work included the first measurements of the LAMB shifts (important in quantum electrodynamics) in deuterium, helium, and the positive lithium ion.

Herzberg also produced a number of books, notably the two classic surveys *Atomic Spectra and Atomic Structure* (1937) and *Molecular Spectra and Molecular Structure* (4 vols., 1939–79). He received the Nobel Prize for chemistry in 1971 for his "contributions to the knowledge of electronic structure and geometry of molecules, particularly free radicals."

Hess, Germain Henri (1802–1850) *Swiss–Russian chemist*

Hess, who was born at Geneva in Switzerland, was taken to Russia as a child by his parents. He studied medicine at the University of Dorpat (1822–25) and started his career by practicing medicine in Irkutsk. In 1830 he moved to St. Petersburg, becoming professor of chemistry at the Technological Institute of the university. While there he wrote a chemistry textbook in Russian, which became a standard work.

Hess worked on minerals and on sugars, but his main work was on the theory of heat. By carefully measuring the heat given off in various chemical changes, he was able to conclude in 1840 that in any chemical reaction, regardless of how many stages there are, the amount of heat developed in the overall reaction is constant. *Hess's law*, also called the law of constant heat summation, is in fact a special case of the law of conservation of energy.

Hess, Harry Hammond (1906–1969) *American geologist*

Hess was born in New York City and educated at Yale, graduating in 1927, and Princeton where he gained his PhD in 1932. He worked first as a field geologist in Northern Rhodesia (now Zambia) in the period 1928–29. After a year at Rutgers in 1932 he moved to Princeton in 1934, becoming professor of geology in 1948.

Hess was a key figure in the postwar revolution in the Earth sciences. He was the first to draw up theories using the considerable discoveries on the nature of the ocean floor that were made in the postwar period. Hess himself discovered about 160 flat-topped summits on the ocean bed, which he named guyots for an earlier Princeton geologist, Arnold GUYOT. As they failed to produce atolls he dated them to the Precambrian, 600 million years ago, before the appearance of corals. But in 1956 Cretaceous fossils, from only 100 million years ago,

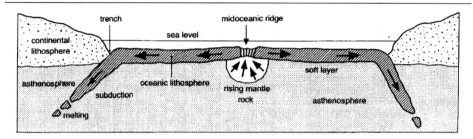

SEA-FLOOR SPREADING   Magma rises from the Earth's mantle to the surface along the midoceanic ridge and cools to form the new oceanic crust. H. H. Hess estimated the oceanic crust to be spreading apart along the midoceanic ridge at the rate of about 1 to 2 inches (2.5–5 cm) a year.

were found in Pacific guyots. The whole of the ocean floor was discovered to be surprisingly young, dating only as far back as the Mesozoic, while the continental rocks were much older.

In 1962 Hess published his important paper, *History of Ocean Basins*. The ocean floors were young, he argued, as they were constantly being renewed by magma flowing from the mantle up through the oceanic rifts, discovered by William Morris EWING, and spreading out laterally. This became known as the sea-floor spreading hypothesis and was a development of the convection-currents theory proposed by Arthur HOLMES in 1929. The hypothesis has been modified since its proposal, notably through the work of Drummond Hoyle MATTHEWS and Frederick VINE on magnetic anomalies, but remains largely accepted.

Hess, Victor Francis (1883–1964) *Austrian–American physicist*

Hess, the son of a forester, was born at Waldstein in Austria and educated at the University of Graz where he obtained his doctorate in 1906. He worked at the Institute for Radium Research, Vienna, from 1910 to 1920 and then took up an appointment at the University of Graz where he became professor in 1925. In 1931 he set up a cosmic-ray observatory near Innsbruck but in 1938 he was dismissed from all his official positions as he was a Roman Catholic. Leaving Nazi Austria, he emigrated to America where he served as professor of physics at Fordham University, New York, from 1938 to 1956.

In 1911–12 Hess made the fundamental discovery of cosmic rays, as they were later called by Robert MILLIKAN in 1925. For this work he shared the Nobel Prize for physics with Carl Andersen in 1936. The work stemmed from an attempt to explain why gases are always slightly ionized; thus a gold-leaf electroscope, however well insulated it might be, will discharge itself over a period of time. Radiation was clearly coming from somewhere and the most likely source was the Earth itself. To test this, attempts were made to see if the rate of

discharge decreased with altitude. But both T. Wulf, who took an electroscope to the top of the Eiffel Tower in 1910, and A. Gockel, who took one up in a balloon in 1912, failed to obtain any clear results.

However when Hess ascended in a balloon to a height of 16,000 feet (4,880 m) he found that although the electroscope's rate of discharge decreased initially up to about 2,000 feet (610 m), thereafter it increased considerably, being four times faster at 16,000 feet than at sea level. He concluded that his results were best explained by the assumption that a radiation of very great penetrating power enters our atmosphere from above.

He was able to eliminate the Sun as the sole cause for he found that the effect was produced both by day and at night. Further, in 1912, he made a balloon ascent during a total eclipse of the Sun and found that during the period when the Sun was completely obscured there was no significant effect on the rate of discharge. Hess however failed to convince everyone that cosmic rays came from outside the Earth's atmosphere as it could still be argued that the source of the radiation was such atmospheric disturbances as thunderstorms. It was left to Millikan in 1925 finally to refute this objection.

Hess, Walter Rudolf (1881–1973) *Swiss neurophysiologist*

The son of a physics teacher from Frauenfel in Switzerland, Hess was educated at the universities of Lausanne, Bern, Berlin, Kiel, and Zurich where he obtained his MD in 1906. Although he actually began as an ophthalmologist, building up a prosperous practice, he decided in 1912 to abandon it for a career in physiology. After junior posts in Zurich and Bonn he was appointed in 1917 to the directorship of the physiology department at the University of Zurich, where he remained until his retirement in 1951.

In the early 1920s Hess began an important investigation of the interbrain and hypothalamus. To do this he inserted fine electrodes into the brains of cats, and used these to stimulate specific groups of cells. His most startling discovery was that when electrodes in the posterior interbrain were switched on this would instantaneously turn a friendly cat into an aggressive spitting creature – a transformation instantly reversed by a further press of the switch. Other areas found by Hess would induce flight, sleep, or defecation.

Less dramatic perhaps but no less significant were the two main areas identified by Hess in the hypothalamus. Stimulation of the posterior region prepared the animal for action but stimulation of the anterior region tended to cause relaxation. Hess had discovered the control center for the sympathetic and parasympathetic systems.

Hess's work was enormously influential and led to a detailed mapping of the interbrain and hypothalamus by many different workers in various centers over a number of years. For his discovery of "the functional organization of the interbrain" Hess was awarded the 1949 Nobel Prize for physiology or medicine, sharing it with Antonio EGAS MONIZ.

Hevelius, Johannes (1611–1687) *German astronomer*

> I prefer the unaided eye.

Hevelius (hay-**vay**-lee-uus) was the son of a prosperous brewer from Danzig (now Gdańsk in Poland). He followed his father in the family business as well as devoting himself to civic duties. After studying in Leiden, he established his own observatory on the rooftops of several houses overlooking the Vistula, an observatory which soon gained him an international reputation.

He published several major works of observational astronomy. Four years' telescopic study of the Moon, using telescopes of long focal power, led to his *Selenographia* (1647; Pictures of the Moon). Making his own engravings of the Moon's surface he assigned names to the lunar mountains, craters, and plains taken from the Earth, placing, with what the writer Sir Thomas Browne called "witty congruity," " … the Mediterranean Sea, Mauritania, Sicily, and Asia Minor in the Moon." This system of naming, apart from the Alps, did not survive long, Giovanni RICCIOLI's alternative system of scientific eponomy being preferred. His star catalog *Prodromus astronomiae* (Guide to Astronomy) was published posthumously in 1690.

Hevelius is today best remembered for his "aerial" telescopes of enormous focal length and his rejection of telescopic sights for stellar observation and positional measurement. He was widely criticized for the latter eccentricity and in 1679 was paid a famous visit by Edmond HALLEY who had been instructed by Robert HOOKE and John FLAMSTEED to persuade him of the advantages of the new telescopic sights. Hevelius claimed he could do as well with his quadrant and alidade. Halley tested him thoroughly, finding to his surprise that Hevelius could measure both consistently and accurately. He is therefore the last astronomer to do major observational work without a telescope.

Hevesy, George Charles von (1885–1966) *Hungarian–Swedish chemist*

Hevesy (**he**-ve-shee) came from a family of wealthy industrialists in Budapest, the Hungarian capital. He was educated in Budapest and at the University of Freiburg where he obtained his doctorate in 1908. He then worked in Zurich, Karlsruhe, Manchester, and Copenhagen, before his appointment to the chair of physical chemistry in 1926 at Freiburg. In 1935 he left Germany for Denmark, fleeing from the Nazis who caught up with him once more in 1942, when he sought refuge in Sweden at the University of Stockholm.

In 1923 Hevesy discovered the new element hafnium in collaboration with Dirk Coster. His most important work, however, began in 1911 in the Manchester laboratory of Ernest RUTHERFORD, where he worked on the separation of "radium D" from a sample of lead. In fact radium D was a radioactive isotope of lead (lead–210) and could not be separated by chemical means. Hevesy was quick to see the significance of this and began exploring the use of radioactive isotopes as tracers. In 1913, with Friedrich Adolph PANETH, he used radioactive salts of lead and bismuth to determine their solubilities. In 1923 Hevesy made the first application of a radioactive tracer – Pb–212 – to a biological system. The Pb–212 was used to label a lead salt that plants took up in solution. At various time intervals plants were burned and the amount of lead taken up could be determined by simple measurements of the amount of radioactivity present. The drawback of this technique was the high toxicity of lead to most biological systems and it was only with the discovery of artificial radioactivity by Irène and Frédéric JOLIOT-CURIE in 1934 that Hevesy's radioactive tracers developed into one of the most widely used and powerful techniques for the investigation of living and of complex systems. For his work in the development of radioactive tracers Hevesy was awarded the 1943 Nobel Prize for chemistry.

Hewish, Antony (1924–) *British radio astronomer*

Hewish was born at Fowey in Cornwall, England, and studied at Cambridge University. He obtained his BA in 1948 and his PhD in 1952 after wartime work with the Royal Aircraft Establishment, Farnborough. He lectured in physics at Cambridge until in 1969 he was made reader and in 1971 professor of radio astronomy, becoming professor emeritus in 1989.

He continues to work part-time at Cambridge. In 1974 he was awarded the Nobel Prize for physics jointly with Martin RYLE.

One of Hewish's research projects was the study of radio scintillation using the 4.5-acre telescope, which consisted of a regular array of 2,048 dipoles operating at a wavelength of 3.7 meters. Radio scintillation is a phenomenon, similar to the twinkling of visible stars, arising from random deflections of radio waves by ionized gas. The three types of scintillation are caused by ionized gas in the interstellar medium, in the interplanetary medium, and in the Earth's atmosphere. All three types were discovered at Cambridge and Hewish was involved in their investigation. In 1967 a research student, Jocelyn BELL-BURNELL, noticed a rapidly fluctuating but unusually regular radio signal that turned out to have a periodicity of 1.33730113 seconds. She had discovered the first pulsar.

To determine the nature of the signal, Hewish's first job was to eliminate such man-made sources as satellites, radar echoes, and the like. Measurements indicated that it must be well beyond the solar system. It seemed possible that it had been transmitted by an alien intelligence and the LGM (Little Green Men) hypothesis, as it became known, was seriously considered at Cambridge, but with the rapid discovery of three more pulsars it was soon dropped.

Hewish did however manage to establish some of the main properties of the pulsar from a careful analysis of its radio signal. Apart from its striking regularity (it was later shown to be slowing down very slightly) it was extremely small, no more than a few thousand kilometers, and was situated in our Galaxy.

By the end of February 1968 Hewish was ready to publish. His account received wide publicity in the popular press and stimulated much thought among astronomers as to the possible mechanism. The proposal made by Thomas GOLD and others that pulsars were rapidly rotating neutron stars has since won acceptance.

Heymans, Corneille Jean François
(1892–1968) *Belgian physiologist and pharmacologist*

Heymans (**hI**-mahns) was educated at the university in his native city of Ghent, where his father was professor of pharmacology, obtaining his MD in 1920. He began as a pharmacology lecturer there in 1923 and in 1930 succeeded his father, holding the chair until his retirement in 1963.

In 1924 Heymans began a series of important cross-circulation experiments. The relationship between respiration and blood pressure had been known for some time – high arterial pressure (hypertension) inhibited respiration while low pressure (hypotension) stimulated it – but the mechanism of such a response was far from clear. Heymans's basic experiment consisted of separating the head of a dog from its body in such a way that its only remaining contact with the body was the nervous supply to the heart. The body of the dog could be made to respire artificially while its head could be linked up to the blood supply of a second dog. Even in such circumstances, hypotension produced an increase in the rate of respiration while hypertension inhibited it. This suggested to Heymans that the process was due not to direct action of the blood pressure on the respiratory center but to nervous control.

Heymans went on to show the important role played in the regulation of heart rate and blood pressure by the carotid sinus, an enlargement of the carotid artery in the neck. By severing the sinus from its own blood supply while maintaining its nervous connection and linking it up to the blood supply of another animal he was able to show that changes of pressure initiated nervous reflexes that automatically reversed the process. The sinus was in fact a sensitive pressure receptor. He also demonstrated that a nearby glandlike structure, the glomus caroticum, was a chemoreceptor, responding to changes in the oxygen/carbon dioxide ratio in the blood.

For his work on the regulation of respiration Heymans was awarded the 1938 Nobel Prize for physiology or medicine.

Heyrovský, Jaroslav (1890–1967) *Czech physical chemist*

> The reason why I keep some 38 years to ... the dropping mercury electrode is its exquisite property as electrode material.
> —Nobel Prize address, 1959

Heyrovský (**hay**-rawf-skee), the son of an academic lawyer, was educated at Charles University in his native Prague and at University College, London. He joined the staff of Charles University in 1919 where he served as professor of physics from 1919 until 1954. From 1950 he was also head of the Central Polarographic Institute, which, since 1964, has borne his name.

Heyrovský is best known for his discovery and development of polarography, which he described in 1922. This is one of the most versatile of all analytical techniques. It depends on the fact that in electrolysis the ions are discharged at an electrode and, if the electrode is small, the current may be limited by the rate of movement of ions to the electrode surface. In polarography the cathode is a small drop of mercury (constantly forming and dropping to keep the surface clean). The voltage is increased slowly and the current plotted against voltage. The current increases in steps, each corresponding to a particular type of positive ion in the solution. The height of the steps indicates the concentration of the ion. For his work, Heyrovský was awarded the Nobel Prize for chemistry in 1959.

Higgins, William (1763–1825) *Irish chemist*

Born at Colooney in Ireland, Higgins worked in London as a young man with his uncle, Bryan Higgins, the chemist. He studied at Oxford University from 1786 and on his return to Ireland became, in 1791, chemist to the Apothecaries Company of Ireland. He moved to Dublin in 1795 to become chemist and librarian to the Royal Dublin Society, this post being made into a professorship in 1800. From 1795 to 1822 he was chemist to the Irish Linen Board.

Higgins is remembered for his contributions to the new atomic theory and for his claim to have anticipated John DALTON. His claim is based on his work *A Comparative View of the Phlogistic and Antiphlogistic Theories with Inductions* (1789), which was written as a reply to Richard KIRWAN's work.

He introduced a clearer symbolism system than that of Dalton but did not follow up his work on atomism until he published a strong attack on Dalton's work in his eight-volume work *Experiments and Observations on the Atomic Theory and Electrical Phenomena* (1814).

Higgins spent the intervening years between these publications trying to introduce new chemical technology into Ireland. In 1799 he published an *Essay on the Theory and Practice of Bleaching*, a work written specifically for the bleachers themselves.

Higgs, Peter Ware (1929–) *British theoretical physicist*

Higgs was born at Bristol in the west of England and educated at Kings College, London, where he completed his PhD in 1955. He worked initially at the University of London but moved to Edinburgh University in 1960 and was elected professor of theoretical physics there in 1980. He is currently emeritus professor. In 2004 he was awarded the Wolf Prize for physics.

Along with many other physicists in the 1960s, Higgs worked on proposals to unify the weak and the electromagnetic forces into a single electroweak theory. At very high temperatures the two forces and their carriers, photons for the electromagnetic force and the W and Z bosons for the weak force, would be indistinguishable. But, at lower temperatures, the symmetry breaks down and massless photons are obviously distinct from the massive W and Z bosons. In 1964 Higgs worked out a mechanism for the breakdown in symmetry, since known as

the *Higgs field*, which would endow the bosons with mass. At the same time, he noted, the mechanism would also produce another massive particle, the *Higgs boson*.

The existence of this particle, it has been claimed, will be the ultimate test of the correctness of the electroweak theory, and of the standard model of particle physics itself. Yet no sign of the particle has so far been detected. The failure is normally explained away by pointing out that the Higgs boson probably has a mass in excess of 1 TeV (10^9 electron volts), well beyond the capacity of any current accelerator. It will therefore be the first task of the Superconducting Proton Synchotron, with an expected capacity of 20 TeV, if ever completed, to search for the Higgs boson.

Hilbert, David (1862–1943) *German mathematician*

> Physics is much too hard for physicists.
> —Quoted by Constance Reid in *Hilbert* (1970)

> One hears a good deal nowadays of the hostility between science and technology. I don't think that is true, gentlemen. I am quite sure that it isn't true, gentlemen. It almost certainly isn't true. It really can't be true. *Sie haben ja gar nichts einander zu tun* [They have nothing whatever to do with one another].
> —Quoted by J. R. Oppenheimer in *Physics in the Contemporary World*

Hilbert (**hil**-bairt) studied at the university in his native city of Königsberg (now Kaliningrad in Russia) and at Heidelberg; he also spent brief periods in Paris and Leipzig. He took his PhD in 1885, the next year became *Privatdozent* at Königsberg, and by 1892 had become professor there. In 1895 he moved to Göttingen to take up the chair that he occupied until his official retirement in 1930.

Hilbert's mathematical work was very wide ranging and during his long life there were few fields to which he did not make some contribution and many he completely transformed. His attention was first turned to the newly created theory of invariants and in the period 1885–88 he virtually completed the subject by solving all the central problems. However, his work on invariants was very fruitful as he created entirely new methods for tackling problems in the context of a much wider general theory. The fruit of this work consisted of many new and fundamental theorems in algebra and in particular in the theory of polynomial rings. Much of his work on invariants turned out later to have important application in the new subject of homological algebra.

Hilbert now turned to algebraic number theory where he did what is probably his finest research. Hilbert and MINKOWSKI had been asked to prepare a report surveying the current state of number theory but Minkowski soon dropped out, leaving Hilbert to produce not only a masterly account but also a substantial body of original and fundamental new discoveries. The work was presented in the *Zahlbericht* (1897; Report on Numbers) with an elegance and lucidity of exposition that has rarely been equaled.

Hilbert then moved to another area of mathematics and wrote the *Grundlagen der Geometrie* (1899; Foundations of Geometry), giving an account of geometry as it had developed through the 19th century. Here his interest lay chiefly in expounding and illuminating the work of others in a systematic way rather than in making new developments of the subject. He devised an abstract axiomatic system that could admit many different geometries – Euclidean and non-Euclidean – as models and by this means go much further than had previously been done in obtaining consistency and independence proofs for various sets of geometrical axioms. Apart from its importance for pure geometry his work led to the development of a number of new algebraic concepts and was particularly important to Hilbert himself because his experience with the axiomatic method and his interest in consistency proofs shaped his approach to mathematical logic and the foundations of mathematics.

In mathematical logic and the philosophy of mathematics Hilbert is a key figure, being one of the major proponents of the formalist view, which he expounded with much greater precision than had his 19th-century precursors. This philosophical view of mathematics had a formative impact on the development of mathematical logic because of the central role it gave to the formalization of mathematics into axiomatic systems and the study of their properties by metamathematical means. Hilbert aimed at formalizing as much of mathematics as possible and finding consistency proofs for the resulting formal systems. It was soon shown by Kurt GÖDEL that *Hilbert's program*, as this proposal is called, could not be carried out, at least in its original form, but it is nonetheless true that Gödel's own revolutionary metamathematical work would have been inconceivable without Hilbert. Hilbert's contribution to mathematical logic was important, especially to the development of proof theory, as further developed by such mathematicians as Gerhard Gentzen.

Hilbert also made notable contributions to analysis, to the calculus of variations, and to mathematical physics. His work on operators and on *Hilbert space* (a type of infinite-dimensional space) was of crucial importance to quantum mechanics. His considerable influence on mathematical physics was also exerted through his colleagues at Göttingen who included Minkowski, Hermann WEYL, Erwin SCHRÖDINGER, and Werner HEISENBERG.

In 1900 Hilbert presented a list of 23 outstanding unsolved mathematical problems to the International Congress of Mathematicians in Paris. A number of these problems still remain unsolved and the mathematics that has been created in solving the others has fully vindicated his deep insight into his subject. Hilbert was an excellent teacher and during his time at Göttingen continued the tradition begun in the 19th century and built the university into an outstanding center of mathematical research, which it remained until the dispersal of the intellectual community by the Nazis in 1933. Hilbert is generally considered one of the greatest mathematicians of the 20th century and indeed of all time.

Hildebrand, Joel Henry (1881–1983)
American chemist

> A child of the new generation
> Refused to learn multiplication.
> He said, "Don't conclude
> That I'm stupid or rude;
> I am simply without motivation."
> —*Perspectives in Biology and Medicine* (1970)

Hildebrand was born in Camden, New Jersey, and educated at the University of Pennsylvania and at the University of Berlin, where he obtained his PhD in 1906. He returned to Pennsylvania in 1907, but moved to the University of California, Berkeley, in 1913 and served there as professor of chemistry from 1918 until his retirement in 1952.

Hildebrand, the author with R. Powell of a widely read textbook, *Principles of Chemistry* (1964; 7th edition), was engaged in chemical research for virtually the whole of the century. He worked on fluorine chemistry, intermolecular forces, and, above all, on the theory of solubility. Even in his nineties Hildebrand continued with his researches, producing in 1976 a substantial monograph on the subject of viscosity and diffusion.

Hilditch, Thomas Percy (1886–1965)
British chemist

A Londoner, Hilditch was educated at the University of London, where he was awarded a DSc in 1911. After further studies in Jena and Geneva he joined the research laboratories of Joseph Crosfield and Sons Ltd., a soap and chemical manufacturer (1911–25). There he began his studies of fat chemistry for which he is best known. He published over 300 papers in this field and produced two standard works, *The Chemical Constitution of Natural Fats* (1940) and *Industrial Chemistry of Fats and Waxes* (1927).

In 1926 Hilditch became the first John Campbell Brown Professor of Industrial Chemistry at Liverpool University, where he remained until his retirement in 1951. He established a large research school of fat chemistry at Liverpool, where with his coworkers he established the fatty acid and glyceride composition of a large number of fats and oils – over 1,450 such substances were dealt with in the fourth edition of his 1940 work published in 1964.

Hill, Archibald Vivian (1886–1977)
British physiologist and biochemist

Born at Bristol in the west of England, Hill was professor of physiology at Manchester University (1920–23) and then Jodrell Professor at University College, London, from 1923 to 1925 (and honorary professor from 1926 to 1951). He was Foulerton Research Professor of the Royal Society (1926–51), of which he was also for some years both secretary and foreign secretary. From 1940 until 1946 he was the Independent Conservative member of Parliament for Cambridge University and a member of the War Cabinet Scientific Advisory Committee.

Hill's major research was directed toward accurately recording the minute quantities of heat produced during muscle action. For this he used thermocouples, which recorded the smallest variations in heat generated after the muscle had completed its movement. He was able to show that oxygen was only consumed *after* muscular contraction, and not during it, indicating that molecular oxygen is required only for muscle recovery. In 1922 he shared the Nobel Prize for physiology or medicine (with Otto Meyerhof) for this work on the physiology of muscular contraction.

Hill, James Peter (1873–1954) *British embryologist*

Hill, the son of a farmer from Kennoway in Scotland, was educated at the University of Edinburgh and the Royal College of Science, London. He taught in Australia at the University of Sydney from 1892 until 1906 when he returned to England as Jodrell Professor of Zoology at University College, London. In 1921 he became professor of embryology and histology, a post he retained until his retirement in 1938.

Beginning during his time in Australia, Hill spent most of his life working on monotreme and marsupial embryology, a virtually unexplored area. Hill totally dominated the field and in his numerous monographs on the subject revealed the evolutionary relationships between the primitive mammals.

Hillier, James (1915–2005) *Canadian–American physicist*

Born in Brantford, Ontario, Hillier was educated at the University of Toronto, where he gained successively his BA (1937), MA (1938), and PhD in physics (1941). He went to live and work in America in 1940 and became a naturalized citizen in 1945. From 1940 to 1953 he worked for the Radio Corporation of America (RCA) Laboratories as a research physicist, primarily on the development of the electron microscope.

Many efforts were being made around the world to develop a commercial electron microscope that could offer higher resolution than optical microscopes. It had been known since the 1920s that a shaped magnetic field could act as a "lens" for electrons, and in the 1930s the first electron micrographs had been taken. Hillier and his colleagues at RCA designed and built the first successful high-resolution electron microscope in America in 1940; they had in fact been anticipated by Ernst RUSKA and Max Knoll who had produced a similar machine for the Siemens and Halske Company in Germany in 1938. The outbreak of war prevented commercial development and exploitation of the German machine.

Hillier made many instrumental advances to the electron microscope. By 1946 he had achieved resolutions (magnifications) approaching close to the theoretical limits. He also involved himself in the development of techniques for the preparation of viral and bacteriological samples for examination.

Hillier's career at RCA continued with only a short break to his position of executive vice-president for research engineering. He was principally concerned with research management and served on various American governmental, research, and engineering committees.

Hinshelwood, Sir Cyril Norman (1897–1967) *British chemist*

Hinshelwood, a Londoner, was educated at Oxford University, where he was elected to a fellowship in 1920 and obtained his doctorate in 1924. In 1937 he became Dr. Lee's Professor of Chemistry at Oxford. He retired in 1964 when he moved to Imperial College, London, as a senior research fellow.

Hinshelwood worked mainly in the field of chemical reaction kinetics. He produced a major text on the subject, *The Kinetics of Chemical Change in Gaseous Systems* (1926) and, in 1956, shared the Nobel Prize for chemistry with Nicolay SEMENOV for his work. He later applied his work to a relatively new field in his book, *The Chemical Kinetics of the Bacterial Cell* (1954).

In some papers published earlier, in 1950, Hinshelwood came very close to the true meaning of DNA, established by James WATSON and Francis CRICK three years later. He declared that in the synthesis of protein the nucleic acid guides the order in which the various amino acids are laid down. Little attention was paid to Hinshelwood's proposal at the time although Crick later declared it to be the first serious suggestion of how DNA might work.

Hinshelwood was a linguist and classical scholar as well as a scientist; he had the unique distinction of serving as president of both the

Royal Society (1955–60) and the Classical Association. He was knighted in 1948.

Hipparchus (c. 170 BC–c. 120 BC) Greek astronomer and geographer

Born at Nicaea, which is now in Turkey, Hipparchus (hi-**par**-kus) worked in Rhodes, where he built an observatory, and in Alexandria. None of his works have survived but many of them were recorded by PTOLEMY. In 134 BC he observed a new star in the constellation of Scorpio. This led him to construct a catalog of about 850 stars. By comparing the position of the stars of his day with those given 150 years earlier he found that Spica, which was then 6° from the autumn equinox, had previously been 8°. He used this observation to deduce not the movement of Spica but the east to west precession (motion) of the equinoctial point. He calculated the rate of the precession as about 45 seconds of arc a year – a value close to the 50.27 seconds now accepted. He also introduced the practice of dividing the stars into different classes of magnitude based on their apparent brightness. The brightest stars he classed as first magnitude and those just visible to the naked eye he classed as sixth magnitude.

As a theorist Hipparchus worked on the orbits of the Sun and Moon. He established more accurate lengths of both the year and the month and was able to produce more accurate eclipse predictions. One of his lasting achievements was the construction of a table of chords, which virtually began the discipline of trigonometry. The concept of a sine had not yet been developed. Instead, Hipparchus calculated the ratio of the chord to the diameter of its own circle, which was divided into 120 parts. Thus if a chord produced by an angle of 60° is half the length of the radius, it would have, for Hipparchus, 60 parts. He much improved the geography of ERATOSTHENES, fixing the parallels astronomically.

Hippocrates of Cos (c. 460 BC–c. 370 BC) Greek physician

> We must turn to nature itself, to the observations of the body in health and disease to learn the truth.
> —Aphorisms

> Science is the father of knowledge, but opinion breeds ignorance.
> —The Canon Law

Very little is known of the life of Hippocrates (hi-**pok**-ra-teez) except that he was born on the Greek island of Cos. The main source, Soranus, dates from the second century AD and was clearly telling a traditional tale rather than writing a biography. Hippocrates is reported to have studied under his father Heraclides, also a physician, and with the atomist DEMOCRITUS and the sophist Gorgias. He then seems to have

spent most of his life traveling around the Greek world curing the great of obscure diseases and ridding grateful cities of plagues and pestilence.

After the fantasy of his life there is the reality of the *Corpus Hippocraticum* (The Hippocratic Collection). This consists of some 70 works though whether any were actually written by Hippocrates himself will probably always remain a matter of speculation. What is clear, on stylistic and paleographic grounds, is that the corpus was produced by many hands in the second half of the fifth century and the first part of the fourth. Nor do the works represent a single "Hippocratic" point of view but, it has been suggested, probably formed the library of a physician and acquired the name of its first owner or collector.

Of more importance is the character of these remarkable works. They are surprisingly free of any attempt to explain disease in theological, astrological, diabolic, or any other spiritual terms. Diseases in the *Corpus* are natural events, which arise in a normal manner from the food one has eaten or some such factor as the weather. The cause of the disease is for the Hippocratic basically a malfunction of the veins leading to the brain which, though no doubt false, is the same kind of rational, material, and verifiable claim that could be found in any late 20th-century neurological textbook.

Such rationality was not to rule for many years for in the fourth century BC new cults entered Greece and with them the dream, the charm, and other such superstitions entered medicine. More successful in the length of its survival was the actual theory of disease contained in the *Corpus*. This was the view, first formulated by ALCMAEON in the fifth century BC, that health consists of an *isonomia* or equal rule of the bodily elements rather than a *monarchia* or domination by a single element. By the time of Hippocrates it was accepted that there were just four elements, earth, air, fire, and water with their corresponding qualities, coldness, dryness, heat, and wetness. If present in the human body in the right amounts in the right places health resulted, but if equilibrium was destroyed then so too was health.

A new terminology developed to describe such pathological conditions, a terminology still apparent in most western languages. Thus an excess of earth, the cold/dry element, produced an excess of black bile, or in Greek melancholic, in the body; too much water, the cold/moist element, made one phlegmatic.

One striking contrast between Hippocratic and later medicine is the curious yet impressive reluctance of the former to attempt cures for various disorders: the emphasis is rather on prognosis. For example, the *Epidemics* describes the course, but not treatment, of various complaints. At least knowing the expected course and outcome of an illness helped the practitioner to inform his patient what to expect, information that could be useful and reassuring. Further, if it is known which conditions lead to a disease such conditions could sometimes be avoided.

The works *Regimen in Acute Diseases* and *Regimen in Health*, which deal specifically with therapy, tend to restrict themselves to diet, exercise, bathing, and emetics. Thus the Hippocratic doctor may not have cured many of his patients but he was certainly less likely than his 18th-century counterpart to actually kill them.

Hirsch, Sir Peter Bernhard (1925–) *British metallurgist*

Hirsch, who was born in Berlin, Germany, gained his doctorate from Cambridge University in 1951. Continuing at Cambridge he lectured in physics from 1959 and took a readership in 1964. In 1966 he moved to the University of Oxford where he was Isaac Wolfson Professor of Metallurgy from 1966 to 1992. His research interests include the development of the Oxford field-emission scanning transmission electron microscope, the weak-beam technique of electron microscopy, and electron microscopy of the chemical behavior of metal oxides. Hirsch is known principally for his pioneering work in applying the electron microscope to the study of imperfections in the crystalline structure of metals, and in relating these defects to mechanical properties. He showed, for instance, that dislocations (faults) play an important part in theories of work-hardening.

Hirst, Sir Edmund Langley (1898–1975) *British chemist*

Hirst, the son of a baptist minister from Preston in the north of England, was educated at St. Andrews University, where he worked with Norman HAWORTH and obtained his PhD in 1921. He continued working with Haworth at the universities of Durham and Birmingham, where Hirst served as reader in natural products from 1934 until 1936 when he was appointed professor of organic chemistry at Bristol University. He moved to Manchester in 1944 but held the chair of organic chemistry there only until 1947 when he accepted a similar post at the University of Edinburgh. Here he remained until his retirement in 1968.

Hirst worked mainly in the field of carbohydrate chemistry, collaborating with Haworth over a long period in working out the structure and synthesis of various sugars. He succeeded in showing that the ring structure of stable methyl glycosides, for example of xylose, rhamnose, arabinose, and glucose, are actually six-membered and not, as previously assumed, five-membered. That is, they have the pyranose rather than the furanose ring structure.

Hirst is also known for determining the structure of vitamin C and, in collaboration with Haworth, managing to synthesize this compound – the first chemical synthesis of a vitamin.

His, Wilhelm (1831–1904) *Swiss anatomist and physiologist*

His (hiss), a merchant's son from Basel in Switzerland, studied at a number of German and Swiss universities before graduating from the university in his native city in 1854. He served on the staff there as professor of anatomy and physiology from 1857 until 1872 when he moved to a comparable post at the University of Leipzig, which he retained until his death.

His did much to redirect embryological research onto a more fruitful track by arguing powerfully against the so-called biogenetic law as formulated by Ernst HAECKEL. His detailed how, starting from the assumption that the embryonic layers are basically elastic sheets, the principal organs could be constructed by such straightforward processes as cutting, bending, pinching, and folding. He was the first to describe accurately the development of the human embryo. His also worked on the development of the nervous system and was able to show that nerve fibers grow from specialized kinds of cell he termed "neuroblasts."

He had earlier, in 1866, introduced into science that invaluable instrument the microtome. This permitted the cutting of extremely thin slices, a few microns thick, for staining and microscopic examination.

One discovery frequently attributed to His, namely the *bundles of His* – a specialized bundle of fibers connecting the cardiac auricles and ventricles – have nothing to do with him. They were in fact first described by his son, also named Wilhelm, professor of medicine at the University of Berlin.

Hisinger, Wilhelm (1766–1852) *Swedish mineralogist*

Hisinger (**hee**-sing-er) was the son of a wealthy iron foundry owner from Skinnatersberg in Sweden. Following his father's death he ran the family business and also devoted himself to private geological research.

One of his iron mines at Bastnäs produced a mineral with an unexpectedly high density and Hisinger studied this mineral over the years, sending samples to some of the most expert analysts of Europe. Martin KLAPROTH who examined it in 1803 became convinced that it contained a new element. Shortly afterwards both Jöns BERZELIUS and Hisinger isolated a new element from it, which they called cerium, after the new minor planet Ceres, discovered by Giussepe PIAZZI in 1801.

Hisinger also made major contributions to the growth of geological knowledge, publishing a geological map of southern and central Sweden (1832) and an account of the fossils of Sweden (1837–41).

Hitchings, George Herbert (1905–1998)
American pharmacologist

Hitchings, the son of a naval architect, was born in Hoquiam, Washington, and educated at the University of Washington and at Harvard where he obtained his PhD in 1933 and where he taught until 1939. He then moved briefly to Western Reserve University until in 1942 he joined the Wellcome Research Laboratories, where he spent the rest of his career, serving as vice president in charge of research from 1966 until his retirement in 1975.

Hitchings was one of the most productive of modern chemical pharmacologists. He began in 1942 with the study of purines and pyrimidines on the grounds that, as important ingredients in cell metabolism, their manipulation could lead to the control of important diseases at the cellular level. This insight led to the synthesis in 1951 of the purine analog, 6-mercaptopurine (6MP), which, as it inhibited DNA synthesis and thus cellular proliferation, proved valuable in the treatment of cancer, particularly leukemia.

In 1959 6MP was found to inhibit the ability of rabbits to produce antibodies against foreign proteins. A less toxic form, azathioprine or Imuran, was quickly developed by Hitchings and used in 1960 by the surgeon Roy Calne to control rejection of transplanted kidneys.

One further drug was developed from work on 6MP when it was realized that it was broken down in the body by the enzyme xanthine oxidase, the same enzyme that converts purines into uric acid. As gout is caused by an excess of uric acid Hitchings developed allopurinol, which blocks uric acid production by competing for xanthine oxidase.

Other drugs developed by Hitchings include the malarial prophylactic pyrimethamine, or Daraprim, and the antibacterial, trimethoprim. Hitchings shared the 1998 Nobel Prize for physiology or medicine with his long-time collaborator Gertrude ELION and with James BLACK.

Hittorf, Johann Wilhelm (1824–1914)
German chemist and physicist

Born at Bonn in Germany, Hittorf (**hit**-orf) became professor of chemistry and physics at the University of Münster in 1852 and later became director of laboratories there (1879–89).

He carried out fundamental work on electrolytes, publishing, in 1853, his paper *The Migration of Ions during Electrolysis*. Hittorf showed that changes in concentration around electrodes during electrolysis could be understood if it was assumed that not all ions move with the same speed. He showed how the relative speeds of ions could be calculated from changes in concentration. He also introduced the ideas of complex ions and transport numbers (sometimes called the Hittorf numbers). The transport number of a given ion in an electrolyte is the fraction of total current carried by that ion.

Hittorf was also one of the first to experiment on cathode rays, noting as early as 1868 that obstacles put in their way would cast shadows.

Hitzig, Eduard (1838–1907) *German psychiatrist*

> I believe that intelligence, or more accurately, the storage of ideas, is to be looked for in all portions of the cerebral cortex, or rather in all parts of the brain.
> —*Über die Funktionen der Grosshirnrinde* (1874; On the Function of the Cerebral Cortex)

> Incorrigible conceit and vanity complicated by Prussianism.
> —Anon. on Hitzig

Hitzig (**hit**-sik), the son of a Berlin architect, was educated at the university there and obtained his MD in 1862. He was later appointed, in 1875, director of the Berghölzi asylum and professor of psychiatry at the University of Zurich. In 1885 Hitzig moved to similar posts at the University of Halle, posts he retained until his retirement in 1903.

In 1870, in collaboration with Gustav FRITSCH, Hitzig published a fundamental paper, *On the Excitability of the Cerebrum*, which provided the first experimental evidence for cerebral localization. Following the important work of Pierre FLOURENS in 1824 it was widely accepted that, despite the discoveries of Paul BROCA and John Neethlings Jackson, the cerebral hemispheres constituted a unity, the seat of intelligence, sensation, and volition and not the source of movement.

This was shown to be false when Hitzig and Fritsch electrically stimulated the cerebral cortex of a dog and elicited distinct muscular contractions. They identified five localized centers, which produced various movements on the side of the dog opposite to the side of the brain stimulated. Their work was soon confirmed by David FERRIER and opened up a vast research program, still, a century later, unfinished.

Hitzig himself continued with this work and in 1874 tried to define what soon became known as the motor area of the dog and the monkey. He also tried to identify, though less successfully, the site of intelligence, in the sense of abstract ideas, in the frontal lobes.

Hjelm, Peter Jacob (1746–1813) *Swedish chemist and metallurgist*

Hjelm (yelm), who was born at Sunnerbo, Sweden, studied at the University of Uppsala and became assay master of the Royal Mint in Stockholm in 1782. In the same year he discovered the element molybdenum. At the time, the term "molybdaena" (a Latin form of a Greek word for "lead") was used for a number of substances, including the substances now known as graphite and molybdenite. Karl SCHEELE showed that the mineral molybdenite with nitric acid produced sulfuric acid and an insoluble residue, which he suspected contained a new element. Lacking the appropriate equipment to reduce this, he called on the assistance of Hjelm, who obtained the metallic element molybdenum. This was later obtained in pure form by Jöns BERZELIUS in 1817.

Hoagland, Mahlon Bush (1921–) *American biochemist*

Hoagland was born in Boston, Massachusetts, the son of Hudson Hoagland, a distinguished neurophysiologist. Having obtained his MD from Harvard in 1948, he joined the Huntington Laboratories of the Massachusetts General Hospital. He then served in the Harvard Medical School from 1960 until 1967 when he became professor of biochemistry at Dartmouth. From 1970 to 1985 he was scientific director of the Worcester Institute for Experimental Biology, founded by his father and Gregory PINCUS in 1944.

In early 1955 Francis CRICK published his "adaptor" hypothesis to explain protein synthesis by the cell. Unaware of this work, Hoagland, in collaboration with Paul Zamecnick and Mary Stephenson, provided the experimental confirmation in 1956. It had earlier been shown by George PALADE that protein synthesis occurred outside the nucleus in the ribosomes. Hoagland and Zamecnick discovered that before the amino acids reach the ribosomes to be synthesized into protein, they are first activated by forming a bond with the energy-rich adenosine triphosphate (ATP).

What happened in the ribosome was unveiled by forming a cell-free mixture of ATP, the radioactively labeled amino acid leucine, enzymes, and some of the small soluble RNA molecules found in the cytoplasm. At this point they discovered the crucial step, predicted by Crick, in between the activation of the amino acid and its appearance in the protein; the amino acid became tightly bound to the soluble RNA. Shortly afterward the labeled leucine was no longer bound to the RNA but present in the protein. The discovery of transfer RNA (or tRNA as it soon became known) was also made independently by Paul BERG and Robert HOLLEY.

In his autobiography *Toward the Habit of Truth* (1990), Hoagland described his work as a molecular biologist and sketched the history of the Worcester Institute.

Hodge, Sir William Vallance Douglas (1903–1975) *British mathematician*

Hodge studied at the university in his native city of Edinburgh and then at Cambridge. He later taught at Bristol University (from 1926) and in America at Princeton (1931–32). Most of his career was spent in Cambridge, England, where in 1936 he took up the Lowndean Chair in Mathematics.

Hodge's mathematical work belongs almost entirely to algebraic geometry and although no expert on analysis or physics his work had immense impact in both these fields. Hodge's principal contribution was to the theory of harmonic forms. One of his central results was a uniqueness theorem showing that there is a unique harmonic form with prescribed periods. In general Hodge helped to initiate the shift of focus in mathematics from a search for purely local results to the more ambitious global approach now so influential.

Hodgkin, Sir Alan Lloyd (1914–1998) *British physiologist*

Born at Banbury in England, Hodgkin graduated from Cambridge University and became a fellow in 1936. He spent World War II working on radar for the Air Ministry. He then worked at the physiological laboratory at Cambridge, where he served as Foulerton Research Professor from 1952 to 1969 and as professor of biophysics from 1970 until 1981. He also served from 1978 to 1984 as master of Trinity College, Cambridge; he was knighted in 1972.

In 1951, with Andrew HUXLEY and Bernard KATZ, he worked out the sodium theory to explain the difference in action and resting potentials in nerve fibers. Using the single nerve fiber (giant axon) of a squid, they were able to demonstrate that there is an exchange of sodium and potassium ions between the cell and its surroundings during a nervous impulse, which enables the nerve fiber to carry a further impulse. Hodgkin also showed that the nerve fiber's potential for electrical conduction was greater during the actual passage of an impulse than when the fiber is resting. For their work on the "sodium pump" mechanism and the chemical basis of nerve transmission, Hodgkin, Huxley, and John ECCLES shared the Nobel Prize for physiology or medicine in 1963. Hodgkin wrote *Conduction of the Nervous Impulse* (1964) and in 1992 he published his autobiography *Chance and Design: Reminiscences of Science in Peace and War.*

Hodgkin, Dorothy Crowfoot (1910–1994) *British chemist*

I'm really an experimentalist. I used to say "I think with my hands." I just like manipulation.
—Quoted by L. Wolpert and A. Richards in *A Passion for Science* (1988)

Born Dorothy Crowfoot in Cairo, Egypt, she was educated at Somerville College, Oxford. After a brief period as a postgraduate student at Cambridge University, she returned to Oxford in 1934 and spent her entire academic career there. She served as Wolfson Research Professor of the Royal Society from 1960 until 1977.

Hodgkin had the good fortune to fall under the influence of the inspiring and scientifically imaginative physicist J. D. BERNAL at Cambridge. Bernal was eager to use the technique of x-ray diffraction analysis, introduced by Max VON LAUE in 1912, to investigate important complex organic molecules. He gathered around him a group of enthusiastic scientists to work out the appropriate techniques. Of the Bernal group, Hodgkin was probably the most talented; she also possessed a greater single-mindedness than Bernal himself and, despite the demands of three young children and a busy political life, it was her persistence and talent that produced some of the first great successes of x-ray analysis.

Her first major result came in 1949 when, with Charles Bunn, she published the three-dimensional structure of penicillin. This was followed by the structure of vitamin B_{12} (by 1956) and, in 1969, that of insulin. For her work on vitamin B_{12} she was awarded the Nobel Prize for chemistry in 1964.

Hodgkin, Thomas (1798–1866) *British pathologist*

Hodgkin, the son of the grammarian John Hodgkin, was born in London and graduated in medicine from Edinburgh in 1823. After further study abroad and practice in London he was appointed in 1825 as pathologist at Guy's Hospital. He resigned in 1837 to devote himself to his practice and, increasingly, to the affairs of the Aborigines' Protection Society, which he helped to found in 1838.

Hodgkin is widely known for his description of lymphadenoma, first described in his paper *On Some Morbid Appearances of the Absorbent Glands and the Spleen* (1832), and named Hodgkin's disease by Samuel Wilks in 1865. Hodgkin reported six cases, in all of whom he found enlargement of the glands in the neck, armpit, and groin together with, in five of the cases, a diseased spleen. However, later studies by Wilks employing the microscope, which Hodgkin did not use, revealed that some of Hodgkin's cases were actually different conditions.

Hoff, Marcian Edward (1937–) *American computer engineer*

It's like a light bulb. When it's broken, unplug it, throw it away and plug in another.
—When asked how he would repair a chip

Hoff gained his doctorate in 1962 at Stanford, where he worked for a further six years as a research associate. In 1968 he was invited by Robert NOYCE to join his newly formed semiconductor firm, Intel.

Noyce had earlier shown how to assemble a large number of transistors into an integrated circuit (IC). Shortly after joining Intel, Hoff was asked to help some Japanese engineers design a number of IC chips to be used in desktop calculators.

Hoff proposed a calculator that could perform simple hardware instructions but could store complex sequences of these instructions in read-only memory (ROM) on a chip. The result of his idea was the first microprocessor – the Intel 4004 – released in 1971. Despite initial debate about its use and marketability, it became the forerunner of a whole range of advanced microprocessors, leading to a new generation of computers.

Hoff left Intel in 1982 to move to the computer company Atari to investigate new products. When Atari was sold in 1984 Hoff set up as an independent consultant.

Hoffleit, (Ellen) Dorrit (1907–2007) *American astronomer*

The daughter of German immigrants, Hoffleit was born in Florence, Alabama, and raised in Pennsylvania. She attended Radcliffe College and graduated with a BA in mathematics in 1928. Her career began as a research assistant in the astronomy department of Harvard University, and in 1938 she received her PhD. Her chief interest was the spectroscopic analysis of the brightness of stars. During World War II she computed missile trajectories for the Ballistic Research Laboratory. In 1956 she left Harvard to run the star cataloging program at Yale University, and became renowned as the author of *The Bright Star Catalogue*. Now in its fifth edition, this contains basic astronomical data for 9110 stars visible to the naked eye, and has become a standard compendium for astronomers. Hoffleit also became a coauthor of another Yale publication, the *General Catalogue of Trignometric Stellar Parallaxes*, which contains precise distance measurements of the major stars. Hoffleit was noted for her dedicated and painstaking research, and was director (1957–78) of the Maria Mitchell Observatory, Nantucket, where she spent summers teaching and inspiring a new generation of young women astronomers. She wrote hundreds of research articles, and also an autobiography, *Misfortunes as Blessings in Disguise*. The asteroid Dorrit was named after her in 1987 in recognition of her immense contribution to astronomy.

Hoffmann, Friedrich (1660–1742) *German physician*

The son of a physician, Hoffman (**hof**-mahn) was born in Halle, Germany, and studied medicine at the University of Jena where he qualified in 1681. After a period of travel and further study in Holland and England, Hoffmann returned to Germany where he practiced medicine in Minden and Halberstadt. In 1693 Hoffmann was appointed professor of medicine at the University of Halle where he remained for the rest of his life apart from two periods, 1709–12 and 1734, when he served as physician at the Brandenburg court.

Hoffmann belonged to the period of medical history that had come to reject the humoral theory of disease, being one of the new generation of theorists who tried to reconstruct some alternative scheme out of the mechanical philosophy of René DESCARTES. He declared that medicine can only be a science insofar as it uses the four mechanical principles of physics – size, shape, motion, and rest. All natural phenomena and effects may be explained by resort to these principles. However, after this radical start he slipped back, believing the body consisted of an indeterminate number of elements. Various mixtures of these explained temperament, and imbalance accounted for disease.

It is clear that Hoffmann had simply translated ancient medicine into a modern terminology – there had in fact been no radical break with the past.

Hoffmann, Roald (1937–) *Polish–American chemist*

Born in Zloczow, Poland (now Zolochez in Ukraine), Hoffmann was moved at the age of four with his family to a labor camp. His father was executed for trying to escape, but Hoffmann and his mother were smuggled out in 1943 and spent the rest of the war hiding in the attic of a schoolhouse. Hoffmann has noted that only 80 of the 12,000 Jews of Zloczow survived the war. Following the liberation in mid-1944, Hoffmann's mother returned to Poland and emigrated with her son to America in 1949; he became a naturalized citizen in 1955.

Hoffmann was educated at Columbia and at Harvard, where he obtained his PhD in 1962. He moved to Cornell in 1965 and was appointed professor of chemistry in 1974.

In the mid-1960s Hoffmann began a research collaboration with R. B. WOODWARD on molecular orbital theory. Their work led to the formulation in 1965 of what are now known as the *Woodward-Hoffmann rules*. These laid down general conditions under which certain organic reactions can occur. The rules apply to pericyclic reactions. In reactions of this kind bond breaking and formation occur simultaneously without the presence of intermediates, i.e., they are said to be "concerted." The reactions also involve cyclic structures. Woodward and Hoffmann published their work in their *Conservation of Orbital Symmetry* (1969). Hoffmann's collaboration with Woodward won him a share of the 1981 Nobel Prize for chemistry with K. Fukui; Woodward's death in 1979, however, robbed him of his second Nobel Prize.

Hoffmann has also published two volumes of verse, *The Metamict State* and *Gaps and Verges*. He has also written and presented a number of television programs, *The World of Chemistry* and *The Molecular World*, in which he has attempted to introduce chemistry to a wider audience. A similar approach can be seen in his *The Same and Not the Same* (1995), in which he tries to describe for a popular audience how the world behaves at the molecular level.

Hofmann, August Wilhelm von (1818–1892) *German chemist*

Hofmann was born at Giessen and initially studied law at Göttingen. He then changed his interest to chemistry and studied with Justus von LIEBIG at the University of Giessen. In 1845 he moved to London to become director of the Royal College of Chemistry, a post he obtained through the influence of Prince Albert. In 1864 he returned to Germany, becoming a professor at the University of Berlin the following year.

August Hofmann is noted for the pioneering work he did in all areas of organic chemistry. In his investigations of coal tar he discovered aniline and this led eventually to his work on amines and organic bases. The standard reactions, the *Hofmann rearrangement* and the *Hofmann elimination*, are named for him. He had a number of important students including Richard ABEGG, Fritz HABER, and William PERKIN. Hofmann was also the originator of the ball-and-stick models used in chemistry and the colors now used (carbon = black, hydrogen = white, oxygen = red, etc.) are the ones he originally chose. It is said that the colors are the result of his use of croquet balls in the original models.

Hofmann, Johann Wilhelm (1818–1892) *German chemist*

Born at Giessen, Germany, Hofmann started as a law student but turned his attention to chemistry, becoming assistant to Justus von LIEBIG. In 1842 Liebig had visited England, where his impressive chemical knowledge drew attention to the lack of chemical skills and training in the UK. To overcome this problem a group led by Prince Albert opened the Royal College of Chemistry in London in 1845. Inevitably the college had to rely heavily on foreign staff to begin with, and Liebig recommended Hofmann as its director. He remained there until 1865, when he returned to Germany to take the chair at Berlin. His appointment in London was a great success; among his staff and pupils were William Henry PERKIN, Edward FRANKLAND, and William ODLING.

It is partly owing to Hofmann that Perkin was able to develop the aniline dyes, since he was working under Hofmann's direction when he made his famous discovery of mauveine, the first synthetic dye, in 1856. Hofmann himself developed a series of violet dyes based on magenta. In theoretical chemistry he worked on type theory, showing that amines are derivatives of ammonia in which a hydrogen atom is replaced by a compound radical.

Hofmeister, Wilhelm Friedrich Benedict (1824–1877) *German botanist*

He is represented to us as one of the leading botanists in Germany, a man with the talent of a genius, highest diligence, and excellent powers of exposition ...
—Government report quoted by E. Pfitzer in *Wilhelm Hofmeister* (1903)

The father of Hofmeister (**hof**-mI-ster), a music and book publisher from Leipzig, Germany, was also a keen amateur botanist and encouraged his son's interest in botany. Wilhelm left school at 15 and served a two-year apprenticeship in a music shop before entering his father's business in 1841. He soon began to study botany seriously in his spare time and was greatly influenced by the views of Matthias SCHLEIDEN, who believed that botany could advance rapidly if researchers concentrated on studying cell structure and life histories.

Using procedures recommended by Schleiden, Hofmeister's first work was to disprove Schleiden's theory that the plant embryo develops from the tip of the pollen tube. He believed that a preexisting cell in the embryo sac gave rise to the embryo and his paper *The Genesis of the Embryo in Phanerogams* (1849) gained him an honorary doctorate from Rostock University.

Hofmeister's major discovery, however, was to demonstrate the alternation of generations between sporophyte and gametophyte in the lower plants. The work, published in 1851 as *Vergleichende Untersuchungen* (Comparative Investigations), showed the homologies between the higher seed-bearing plants (phanerogams) and the mosses and ferns (cryptogams) and demonstrated the true position of the gymnosperms between the angiosperms and the cryptogams. In 1863 Hofmeister was appointed professor at Heidelberg University and director of the botanic gardens there, and in 1872 moved to Tübingen University to succeed Hugo von MOHL.

Hofstadter, Douglas Richard (1945–) *American computer scientist*

Hofstadter's Law: it always takes longer than you expect, even when you take into account Hofstadter's Law.
—*Godel, Escher, Bach* (1979)

Hofstadter (**hof**-stat-er), the son of Nobel laureate Robert Hofstadter, was born in New York and educated at Stanford and at the University of Oregon, where he completed his PhD (1976) in theoretical physics. He worked initially at the University of Indiana, Bloomington, but moved to the University of Michigan in 1984 to take up an appointment as professor of cognitive and computer science. In 1988 he returned to Bloomington as professor of computer science.

In 1979 Hofstadter published a remarkable book, *Godel, Escher, Bach*, described by him as "a metaphorical fugue on minds and machines." In the following year he was invited to succeed the much respected Martin Gardner as monthly columnist with *Scientific American*. For two and a half years Hofstadter contributed a column, *Metamagical Themas* (an anagram of Gardner's *Mathematical Games*).

Hofstadter, Robert (1915–1990) *American physicist*

A New Yorker by birth, Hofstadter graduated from the College of the City of New York in 1935 and gained his MA and PhD at Princeton University in 1938. From 1939 he held a fellowship at the University of Pennsylvania, and in 1941 returned to the College of the City of New York as an instructor in physics. From 1943 to 1946 Hofstadter worked at the Norden Laboratory Corporation, and from there took on an assistant professorship in physics at Princeton University. In 1950 he moved to Stanford University as an associate professor and was made full professor in 1954.

His early research was in the fields of infrared spectroscopy, the hydrogen bond, and photoconductivity. One of his first notable achievements, in 1948, was the invention of a scintillation counter using sodium iodide activated with thallium. He is noted for his studies of the atomic nucleus, for which he received the 1961 Nobel Prize for physics (shared with Rudolph Mössbauer).

At Stanford, Hofstadter used the linear accelerator to study the scattering effects of high electrons fired at atomic nuclei. In many ways these experiments were similar in concept to RUTHERFORD's original scattering experiments. He found that the distribution of charge density in the nucleus was constant in the core, and then decreased sharply at a peripheral "skin." The radial distribution of charge was found to vary in a mathematical relationship that depended upon the nuclear mass. Further, Hofstadter was able to show that nucleons (protons and neutrons) were not simply point particles, but had definite size and form. Both appeared to be composed of charged mesonic clouds (or shells) with the charges adding together in the proton, but canceling each other out in the neutral neutron. This led him to predict the existence of the rho-meson and omega-meson, which were later detected.

Hofstadter served as director of the high-energy physics laboratory at Stanford from 1967 to 1974.

Hollerith, Herman (1860–1929) *American engineer*

The father of Hollerith (**hol**-e-rith) was a German immigrant who settled in Buffalo, New York, and taught classics. Hollerith was educated at the Columbia School of Mines, New York. After graduating in 1879 he assisted his teacher, W. P. Trowbridge, working with him on the 1880 U.S. census. He also helped the head of vital statistics, John Billings, to prepare his final report.

Like many others, Hollerith and his supervisors were dismayed by the thought that, while it took only a few months to carry out the census, it would take the best part of a decade to tabulate and analyze the data collected. They might not even be finished before the 1890 census was taken. Clearly some form of mechanical aid was required. Hollerith spent the 1880s working first as an instructor at the Massachusetts Institute of Technology (1882–84) and for the rest of the decade at the Patent Office in Washington. During much of this time he worked on the census problem. He also developed a set of electrically controlled air brakes for freight trains, which he patented in 1885. His design, however, was not adopted; the Westinghouse system was preferred.

More successful was the tabulating system he invented for the 1890 census. He began with a system of punched tape, run over a metal drum and under some brushes. Whenever the brushes passed over a hole, current flowed and one bit of data was recorded. But, to retrieve an item, operators soon found, the whole tape had to be scanned. The obvious solution of turning the tape into cards was quickly introduced.

Hollerith's punched cards and related equipment operated so well in 1890 that it reduced the number of working days spent tabulating the data by two-thirds. It was quickly adopted by many other countries. By 1896 Hollerith felt sufficiently confident to set up his own company, Tabulating Machine Co., to manufacture and market machinery. The business thrived as it became apparent that Hollerith's tabulating machines could be used to record and analyze data of almost any kind. In 1911, a victim of high blood pressure, Hollerith sold out to the entrepreneur Charles Flint for $450 a share, a deal which brought Hollerith about $1.25 million. Hollerith's company thereby became the Computing-Tabulating-Recording Co. and eventually, in 1924, IBM.

Holley, Robert William (1922–1993) *American biochemist*

Holley was born in Urbana, Illinois. After graduating in chemistry from Illinois University in 1942, he joined the team at Cornell Medical School that achieved the first artificial synthesis of penicillin. He remained at Cornell to receive his PhD in organic chemistry in 1947.

Two years (1955–56) spent at the California Institute of Technology marked the beginning of Holley's important research on the nucleic acids. He decided that to work out the structure of a nucleic acid he first needed a very pure specimen of the molecule. Back again at Cornell, his research team spent three years isolating one gram of alanine transfer RNA (alanine tRNA) from some 90 kilograms of yeast. In March 1965 he was able to announce that they had worked out the complete sequence of 77 nucleotides in alanine tRNA. For this work Holley received the 1968 Nobel Prize for physiology or medicine, an award he shared with Marshall NIRENBERG and Har Gobind KHORANA.

Holmes, Arthur (1890–1965) *British geologist*

Holmes came from a farming background in Hebburn-on-Tyne in the northeast of England. He graduated from Imperial College, London, in 1910, and went on to work with Lord RAYLEIGH on radioactivity. After an expedition to Mozambique in 1911 he taught at Imperial College until 1920 when he went to Burma as an oil geologist. In 1925 he returned to England to become professor of geology at Durham University, where he remained until 1943 when he moved to Edinburgh University.

Holmes conducted major work on the use of radioactive techniques to determine the age of rocks, leading to his proposal of the first quantitative geological time scale in 1913 and to his estimate of the age of the Earth being about 1.6 billion years. He continued to revise this estimate throughout his life, producing a figure in 1959 some three times larger.

Holmes also made a major contribution to the theory of continental drift proposed by Alfred WEGENER in 1915. One of the early difficulties the theory faced was that geologists could not envisage a force capable of moving the continents in the way described by Wegener. In 1929 Holmes proposed the existence of convection currents in the Earth's mantle. Rocks in the Earth's interior are, according to Holmes, heated by radioactivity, causing them to rise and spread out and, when cold and dense, to sink back to the interior. It was only after World War II that hard evidence for such a view could be produced.

In 1944 Holmes published his *Principles of Physical Geology*, a major work on the subject.

A substantially revised edition of this book was published in 1965, shortly before Holmes's death.

Holmes, Oliver Wendell (1809–1894) *American physician*

> Science is the topography of ignorance.
> —*Medical Essays*

> The truth is, that medicine, professedly founded on observation, is as sensitive to outside influences, political, religious, philosophical, imaginative, as is the barometer to the changes of atmospheric density.
> —As above

Holmes was the son of a congregational minister and the father of the identically named jurist and Supreme Court judge. Born in Cambridge, Massachusetts, he graduated in law from Harvard in 1829 and in medicine in 1836. He was professor of anatomy and physiology at Harvard from 1847 until 1882.

In 1843 Holmes published the classic paper *The Contagiousness of Puerperal Fever* in which he repeated some of the earlier arguments of a number of notable physicians and anticipated some of the work of Ignaz SEMMELWEIS and Joseph LISTER. Puerperal (childbed) fever, a streptococcal infection of maternity wards in the early 19th century, had an average mortality of 5–10%, occasionally rising to as much as 30% in particularly virulent outbreaks. Holmes stated that puerperal fever is frequently carried from patient to patient by physicians and nurses. He advised that no doctor or nurse who had recently participated in a post-mortem should treat a patient in labor, and he also recommended that they should always wash their hands before examining patients. So little attention was paid to Holmes's paper, however, that he felt it necessary to republish it in 1855.

In addition to being a distinguished man of American letters with numerous volumes of essays, verse, and fiction to his credit, Holmes is also remembered for a letter to William MORTON in 1846 concerning the condition induced by Morton's ether inhalation. In this, he wrote, "The state should, I think, be called 'Anaesthesia.' This signifies insensibility." The term had previously been employed in antiquity by PLATO and DIOSCORIDES.

Honda, Kotaro (1870–1954) *Japanese metallurgist*

Born in Aichi, Japan, Honda was educated at the Imperial University, Tokyo, graduating in 1897. After studying abroad at the universities of Göttingen and Berlin, he returned to Japan to take up an appointment at the Tohoku Imperial University. In 1922 a research institute for iron and steel was attached to the university and Honda became its director. He finally, in 1931, became president of the university.

Honda is noted for his research on magnetic alloys. In 1917 he found that an alloy of 57% iron, 35% cobalt, 2% chrome, 5% tungsten, and 1% carbon was the most highly magnetic material then known. Its use in such instruments as magnetos and dynamos permitted a marked decrease in their size.

Honda, Soichiro (1906–1991) *Japanese motor engineer*

Honda was the eldest son of a poor village blacksmith from Tenryu in the Shizuoka Prefecture, Japan. He showed an early interest in engines, although it was not until he was eight that he actually saw an automobile, an old Model-T. In 1922 he began working as an apprentice in a Tokyo car repair shop and six years later opened his own shop in Hamamatsu.

But Honda had ambitions to do more than repair vehicles; he aimed eventually to design and build them himself. After the war, during which he made piston rings, he set up the Honda Motor Company in 1947. Initially he built small engines to be fitted to bikes. Known in Japan as "bata-bata," they were very successful. Honda went on to build real motorcycles; they proved to be so effective that in 1961 his machines took the first five positions in both the 125 cc and 250 cc classes at the Tourist Trophy races on the Isle of Man.

In 1957 Honda entered the car market and within a few years had overtaken all his competitors other than Toyota and Nissan. Honda had the foresight to develop lightweight engines, which proved to be well suited to the energy crises of the 1970s. The Honda Civic of 1972, for example, was the first Japanese car to meet American pollution standards and allowed Honda to sell more cars in America than any other Japanese manufacturer apart from Toyota. In 1982 Honda established his own production lines in the United States. Honda retired in 1973.

Hood, Leroy Edward (1938–) *American biologist*

Born in Missoula, Montana, Hood was educated at the California Institute of Technology, where he obtained his PhD in biochemistry in 1964, and at Johns Hopkins Medical School, Baltimore, where he qualified as an MD in 1964. He immediately joined the staff of the National Institutes of Health, Bethesda, working in the area of immunology. In 1970 Hood returned to the California Institute of Technology and was appointed professor of biology in 1975.

In May 1985 at a meeting in Santa Cruz, California, plans were laid to map the human genome (the Human Genome Project). As the genome consists of 3 billion base pairs of DNA, the ability to sequence the genes rapidly would be a crucial factor.

Fortunately an automatic sequencer was almost at hand in 1985, developed by Hood and his colleague Lloyd Smith. The sequencer operates with fluorescent dyes. Each of the four DNA bases – adenine (A), cytosine (C), guanine (G), and thymine (T) – can be tagged with a different dye. Unsequenced dye-tagged DNA fragments are analyzed by gel electrophoresis, in which they migrate at different rates. The dyes are excited by an argon laser and the light emitted is turned into a digital signal by photomultiplier tubes. The digital signals can be analyzed by a computer and identified as A, T, C, or G.

Hood's automatic sequencer enabled work that once took a week or more to be carried out overnight. Later commercial models of the device can read 12,000 base pairs a day, and operate more accurately than any manual sequencing.

In 1992 Bill GATES of Microsoft presented the University of Washington Medical School, Seattle, with $12 million to establish a department of molecular biotechnology. Hood was persuaded to move to Seattle to head the new department, to work on a faster DNA sequencer, and to analyze the genes controlling the human immune response. That same year Hood, in collaboration with Ronald Cape, a former head of the biotechnology firm Cetus, founded Darwin Molecular in Seattle. Their aim is to develop new drugs utilizing processes comparable to natural selection.

Hooke, Robert (1635–1703) *English physicist*

> The truth is, the science of Nature has been already too long made only a work of the brain and the fancy. It is now high time that it should return to the plainness and soundness of observations on material and obvious things.
> —*Micrographia* (1665; Micrography)

> He is certainly the greatest Mechanick this day in the World.
> —John Aubrey, *Brief Lives* (17th century)

Hooke, whose father was a clergyman from Freshwater on the Isle of Wight, England, was educated at Oxford University. While at Oxford he acted as assistant to Robert BOYLE, constructing the air pump for him. In 1662 Boyle arranged for Hooke to become first curator of experiments to the Royal Society. There he agreed to "furnish the Society every day they meet with three or four considerable experiments." Even though the society only met once a week, the pressure on Hooke was still great and may explain why he never fully developed any of his ideas into a comprehensive treatise. He was also something of an invalid.

Hooke made numerous discoveries, perhaps the best known being his law of elasticity, which states that, within the elastic limit, the strain (fractional change in size) of an elastic material is directly proportional to the stress (force per unit area) producing that strain. He was the first to show that thermal expansion is a general property of matter. He also designed a balance spring for use in watches, built the first Gregorian (reflecting) telescope, and invented a number of scientific instruments, including the compound microscope and the wheel barometer.

In 1665 he published his main work *Micrographia* (Micrography), which was an account – fully and beautifully illustrated – of the investigations he had made with his improved version of the microscope. It also contained theories of color, and of light, which he suggested was wavelike. This led to one of the major controversies – over the nature of light and the priority of theories – that he had with Isaac NEWTON. The other conflict was over the discovery of universal gravitation and the inverse square law. It is true that Hooke had revealed, in a letter to Newton in 1680, that he had an intuitive understanding of the form the inverse square law must take. Newton's reply to Hooke's charge of plagiarism was to distinguish between Hooke's general intuition that may have been well founded, and his own careful mathematical treatment of the law and detailed working out of its main consequences.

Hooke was also a capable architect, having written on the theory of the arch and designed parts of London after the great fire of 1666.

Hooker, Sir Joseph Dalton (1817–1911)
British plant taxonomist and explorer

> From my earliest childhood I nourished and cherished the desire to make a creditable journey in a new country and write such a respectable account of its natural features as should give me a niche among the scientific explorers of the globe I inhabit, and hand my name down as a useful contributor of original matter.
> —Letter to Charles Darwin (1854)

Hooker was born at Halesworth in Suffolk, England, and studied medicine at Glasgow University, where his father William Hooker was professor of botany. After graduating in 1839, he joined the Antarctic expedition on HMS *Erebus* (1839–43), nominally as assistant surgeon but primarily as naturalist. Between 1844 and 1860, using collections made on the expedition, Hooker produced a six-volume flora of the Antarctic Islands, New Zealand, and Tasmania.

When he returned from the Antarctic expedition Hooker was congratulated on his work by Charles Darwin, who had been following his progress, and in 1844 Darwin confided to Hooker his theory of evolution by natural selection. This communication later proved important in establishing Darwin's precedence

when his theory – together with Alfred Russel
WALLACE's essentially identical conclusions –
was presented by Hooker and Lyell at the famous Linnaean Society meeting of July 1858.

Following his unsuccessful application in
1845 for the botany chair at Edinburgh University, Hooker was employed to identify fossils
for a geological survey, but he took time off between 1847 and 1850 to explore the Indian sub-
continent. He visited Sikkim and Assam, Nepal,
and Bengal, introducing the brilliant Sikkim
rhododendrons into cultivation through the
botanical gardens at Kew. Later (1872–97) he
produced a seven-volume flora of British India.

In 1855 Hooker was appointed assistant director at Kew Gardens and in 1865 succeeded
his father as director. In his 20 years as head of
the institute he founded the Jodrell Laboratory and Marianne North Gallery, extended the
herbarium, and developed the rock garden. His
efforts established Kew as an international center for botanical research and in 1872 he successfully fought a move from the commissioner
of works to relegate the gardens to a pleasure
park. With George BENTHAM he produced a
world flora, *Genera Plantarum* (1862–83; Genera of Plants) – a major work describing 7,569
genera and 97,000 species. The Kew herbarium is still arranged according to this classification.

Hooker retired from the directorship of Kew
in 1885 owing to ill health but continued working until his death.

Hooker, William Jackson (1785–1865) *British botanist*

> The great secret of his success was that he
> deemed nothing too small for his notice, if it
> illustrated any fact of science or economy, and
> nothing too difficult to be attempted.
> —William Henry Harvey

The son of a merchant's clerk, Hooker was born
in Norwich in the east of England; he attended
the grammar school there but had little formal
education. An interest in botany led to his first
voyage, in 1809, to Iceland, which was followed
by an extensive study of the English flora. From
1820 until 1842 Hooker held the botany chair
at Glasgow. His main interest was in ferns,
mosses, and fungi but his works also include
some important regional floras and he was a pioneer of economic botany. Hooker's herbarium
was accessible to all scholars, and with his publications (more than 20 major books and numerous articles) and journals he became the
leading British botanist of his day.

In 1841 Hooker was appointed the first director of Kew Gardens, a position he held until
his death. Under his direction Kew became the
world's most important botanical institution
and here he founded the Museum of Economic
Botany in 1847.

Hope, James (1801–1841) *British cardiologist*

Hope, the son of a wealthy merchant from
Stockport near Manchester, England, studied
medicine at Edinburgh University and various
London hospitals. He became assistant physician at St. George's Hospital in 1834 and in
1839, shortly before his early death from consumption (tuberculosis), was appointed full
physician.

In the 1820s use of René Laennec's stethoscope became widespread in Britain. It still remained to link the sounds heard through the
stethoscope and actual events in the heart, a
task begun by Hope and reported in his influential *A Treatise on the Diseases of the Heart
and Great Vessels* (1831), a work that went
through three editions in his short life.

Hope made valuable observations on heart
murmurs, valvular disease, and aneurism, and
thus began the important work of transforming
heart complaints from being merely a set of
symptoms into identifiable and specific lesions
of the heart.

Hope, Thomas Charles (1766–1844) *British chemist*

Hope's father, John Hope, was a professor of
botany at Edinburgh University and founder of
the new Edinburgh botanic gardens. Thomas,
who was born in Edinburgh, studied medicine
there and became professor of chemistry at
Glasgow in 1787. He returned to Edinburgh in
1795 as joint professor of chemistry with Joseph
BLACK, succeeding Black on his death in 1799.
He remained as chemistry professor until 1843.

In 1787 Hope isolated the new element
strontium and named it after the town of Stron-
tian in Scotland where it was discovered. At
first it was thought to be barium carbonate and
was only established as a new metal in 1791.
Martin KLAPROTH made the same discovery independently but a little later.

Hope was also the first to show the expansion
of water on freezing and demonstrated that
water attains a maximum density a few degrees above its freezing point (actually 3.98°C).
He published his results in his paper *Experiment on the Contraction of Water by Heat*
(1805).

Hopkins, Sir Frederick Gowland (1861–1947) *British biochemist*

> Life is a dynamic equilibrium in a polyphasic
> system.
> —Joseph Needham, *Order and Life* (1936)

Hopkins was the son of a bookseller and publisher and a distant cousin of the poet Gerard
Manley Hopkins. He was born at Eastbourne on
the south coast of England and, after attending
the City of London School, was apprenticed as
a chemist in a commercial laboratory, where

for three years he performed routine analyses. An inheritance in 1881 allowed him to study chemistry at the Royal School of Mines and at University College, London. His work there brought him to the attention of Thomas Stevenson, who offered Hopkins the post of assistant in his laboratory at Guy's Hospital. Feeling the need of more formal qualifications he began to work for a medical degree at Guy's in 1889, finally qualifying in 1894. In 1898 Hopkins moved to Cambridge, where he remained for the rest of his long life and not only served as professor of biochemistry (1914–43) but also established one of the great research institutions of the century.

In 1901 Hopkins made a major contribution to protein chemistry when he discovered a new amino acid, tryptophan. He went on to show its essential role in the diet, since mice fed on the protein zein, lacking tryptophan, died within a fortnight; the same diet with the amino acid added was life-supporting. This work initiated vast research programs in biochemical laboratories.

In 1906–07 Hopkins performed a classic series of experiments by which he became convinced that mice could not survive upon a mixture of basic foodstuffs alone. This ran against the prevailing orthodoxy, which supposed that as long as an animal received sufficient calories it would thrive. He began by feeding fat, starch, casein (or milk protein), and essential salts to mice, noting that they eventually ceased to grow. Addition of a small amount of milk, however, was sufficient to restart growth. It took several years of careful experiments before, in 1912, Hopkins was prepared to announce publicly that there was an unknown constituent of normal diets that was not represented in a synthetic diet of protein, pure carbohydrate, fats, and salts. Hopkins had in fact discovered what were soon to be called vitamins, and for this work he shared the 1929 Nobel Prize for physiology or medicine with Christiaan EIJKMAN.

At the same time Hopkins was working with Walter Fletcher on the chemistry of muscle contraction. In 1907 they provided the first clear proof that muscle contraction and the production of lactic acid are, as had long been suspected, causally connected. This discovery formed the basis for much of the later work done in this field. Hopkins later isolated the tripeptide glutathione, which is important as a hydrogen acceptor in a number of biochemical reactions.

In England Hopkins did more than anyone else to establish biochemistry as it is now practiced. He had to fight on many fronts to establish the discipline, since many claimed that the chemistry of life involved complex substances that defied ordinary chemical analysis. Instead he was able to demonstrate that it was a chemistry of simple substances undergoing complex reactions. Hopkins was knighted in 1925.

Hopper, Grace (1906–1992) *American mathematician and computer scientist*

Hopper, born Grace Murray, was educated at Vassar and at Yale, where she gained her PhD in 1934. She taught at Vassar until 1944, when she enlisted in the U.S. Naval Reserve and was immediately assigned to Harvard to work with Howard AIKEN on the Mark I computer, the ASCC (Automatic Sequence Controled Calculator), for which she wrote the manual. Although she hoped to remain in the Navy after the war, her age prevented this and she had to be satisfied with the Naval reserve as a second choice. Consequently she remained at Harvard working on the Mark II and the Mark III computers.

In 1949 Hopper moved to Philadelphia to work with J. P. Eckert and John MAUCHLY on the development of BINAC and remained with the Eckert-Mauchly Computer Corporation, despite several changes of ownership, until 1967. During this period she made a number of basic contributions to computer programming. In 1952 she devised the first compiler, a program that translated a high-level language into machine code, named A-O. She went on to produce a data-processing compiler known as Flow-matic.

It was apparent by this time to Hopper and other programmers that the business world lacked an agreed and adequate computer language. Hopper lobbied for a combined effort from the large computer companies and consequently a committee was established in 1959 under the guidance of the Defense Department to develop a common business language. Although she did not serve on the committee, the language developed by them, COBOL (Common Business Oriented Language), was derived in many respects from Flow-matic. For this reason Hopper has often been referred to as "the mother of Cobol."

Although Hopper was forced through age to resign from the U.S. Naval reserve in 1966 she was recalled a year later to work on their payroll program. She remained in the Naval reserve until 1986, having been promoted to the rank of rear admiral in 1985.

Hoppe-Seyler, (Ernst) Felix Immanuel (1825–1895) *German biochemist*

Hoppe-Seyler (hop-e-**zI**-ler) was born in Freiberg, Germany, and early in his career became assistant to Rudolf VIRCHOW in Berlin. He became professor of physiological chemistry (biochemistry) at Strasbourg, where he established the first exclusively biochemical laboratory, and later founded the first biochemistry journal. In 1871 Hoppe-Seyler discovered the enzyme invertase, which aids the conversion of sucrose into the simpler sugars glucose and fructose. He was also the first to prepare hemoglobin in crystalline form. He isolated the fatlike compound lecithin, one of the class of compounds called phospholipids. Hoppe-Seyler's classification of the proteins (1875) is still accepted today.

Horrocks, Jeremiah (1619–1641) *English astronomer*

Little is known about the early life of Horrocks (or Horrox) other than that he was born into a Puritan family in Toxteth, Liverpool, and was admitted to Cambridge in 1632. Even though he died "in his twenty second year" he had made major contributions to astronomy and several original observations.

Horrocks noted that as the orbits of Venus and Mercury fall between the earth and the Sun, it would seem possible that at certain times the inner planets would appear to an observer on the Earth to cross the face of the Sun. The events, known as transits, are so rare that they are unlikely to be seen by chance. Only five transits of Venus have been observed, those of 1639, 1761, 1769, 1874, 1882, and 2004.

At Cambridge, Horrocks had mastered the new astronomy of KEPLER. From Kepler's recently published *Rudolphine Tables* (1627), he worked out that a transit of Venus was due on 24 November 1639 at 3 p.m. At this time he was probably working as a curate at Hoole near Preston in Lancashire. He prepared for the transit by directing the solar image onto a large sheet of paper in a darkened room. However, a late November afternoon in Lancashire is not the best time to observe the Sun. For Horrocks there was another problem. The predicted day was a Sunday which meant that the puritan curate could well find himself in church at the crucial moment.

Horrocks was successful in observing the transit, however, and left an account of the day in his *Venus in Sole Visa* (Venus in the Face of the Sun), published posthumously in 1662. The day was cloudy but at 3.15, "as if by divine interposition" the clouds dispersed. He noted a spot of unusual magnitude on the solar disc and began to trace its path; but, he added, "she was not visible to me longer than half an hour, on account of the Sun quickly setting."

With the aid of his observations Horrocks could establish the apparent diameter of Venus as 1′ 12″ compared with the Sun's diameter of 30′, a figure much smaller than the 11′ assigned by Kepler. Horrocks also attempted to determine the solar parallax, and derived, although with little confidence, a figure of 15″, compared with a modern value of 8″.8.

Before his death Horrocks was working on an *Astronomia Kepleriana* (Astronomy of Kepler), and essays on comets, tides, and the Moon. Un-

fortunately none of this was published until long after his death. Much of his work had been lost in the chaos of the Civil War. Other material sent to a London bookseller was burned in the Great Fire of 1666. The remainder of his papers were published by John WALLIS as *Opera posthuma* (1678; Posthumous Works).

Horsfall, James Gordon (1905–1995) *American plant pathologist*

Born in Mountain Grove, Missouri, Horsfall graduated in soil science from the University of Arkansas and gained his PhD in plant pathology from Cornell in 1929. He remained at Cornell until 1939 when he became chief of the department of plant pathology (1939–48) and director (1948–71) of the Connecticut Agricultural Experimental Station.

A leading plant pathologist, Horsfall coedited, with A. E. Dimond, the three-volume work *Plant Pathology* (1959–60). More recently, with Ellis Cowling, he coedited *Plant Disease: An Advanced Treatise* (5 vols., 1977–80). He had earlier published extensively on fungicides.

Horvitz, H. Robert (1947–) *American cell biologist*

Horvitz (**hor**-vich) initially studied mathematics and economics at the Massachusetts Institute of Technology (MIT), receiving his BA in 1968. He then turned to biology at Harvard University, graduating with an MA in 1972, and gaining his PhD in 1974. After working as a researcher in the UK at the Medical Research Council's Laboratory of Molecular Biology in Cambridge, he returned to MIT in 1978 as assistant professor of biology, subsequently becoming associate professor (1981) and professor (1986). In 1988 he was appointed Investigator at the Howard Hughes Medical Institute, MIT.

Horvitz has devoted much of his career to studying the development and genetics of the nematode (roundworm) *Caenorhabditis elegans*. During his stint at the MRC Laboratory of Molecular Biology in Cambridge in the mid-1970s, he saw at first hand how Sydney BRENNER and his colleagues were developing the potential of this tiny transparent hermaphrodite worm as a model experimental organism. At this time, one of Brenner's team, John E. SULSTON, was mapping the developmental lineage of each of the adult worm's thousand or so cells. Sulston showed how certain cells were destined to die during development by a process called *programmed cell death* (*apoptosis*), and that this was a normal part of organ formation and intrinsic to the patterning of body systems.

By determining the effects of particular mutations on the worm's development, Horvitz was able to pinpoint certain key genes that control apoptosis. For example, in worms with mutations of the *ced-3* or *ced-4* genes, apoptosis does not occur and all cells survive, whereas a mutation of the *ced-9* gene causes the death of all cells. The *ced-3* gene encodes an enzyme called a *caspase*, which cleaves certain target proteins in the cytoskeleton and nuclear framework, and is instrumental in dismantling the cell during apoptosis. However, the caspase must be activated by CED-4, the protein encoded by the *ced-4* gene. Normally, CED-4 is bound by CED-9 (the protein encoded by *ced-9*). Apoptosis is triggered by the activity of a killer gene, *egl-1*, whose product causes the release of CED-4 from CED-9, and hence the activation of the caspase. Counterparts of these genes and their products occur in humans and other vertebrates, and the apoptosis pathway seems to be highly conserved (i.e., very similar) in all animals.

C. *elegans* has provided Horvitz and his team at MIT with many other valuable insights into the genetic regulation of animal development. In particular, Horvitz has focused on the differentiation and growth of the worm's vulva, which consists of just 22 cells. For example, he has identified a gene, *let-60*, whose protein functions as a switch for vulval induction, and characterized other genes that exert control over vulval development.

Another area illuminated by this simple organism is the relationship between structural features of the nervous system and the animal's behavior. Horvitz's team have discovered that receptors to the neurotransmitter serotonin are important in modulating the worm's rate of locomotion in response to its experience. They have also identified certain genes that control aspects of neurotransmission and muscle contraction.

Through his work, Horvitz has demonstrated that the humble worm C. *elegans* reveals many fundamental aspects of cell biology, and gives us a better understanding of human biology and disease. A complete picture of apoptosis, for instance, will help in developing new treatments for cancer and other diseases in which apoptosis is defective or misregulated.

In 2002, Horvitz was awarded the Nobel Prize in physiology or medicine, jointly with Brenner and Sulston, for their work on the genetic regulation of organ development and on programmed cell death.

Hounsfield, Sir Godfrey Newbold
(1919–2004) *British engineer*

Hounsfield was born at Newark in Nottinghamshire, England, and educated in that county before going on to the City and Guilds College, London, and the Faraday House College of Electrical Engineering in London. Having spent the war years in the RAF, he worked for Electrical and Musical Industries (EMI) from 1951 and led the design effort for Britain's first large solid-state computer. Later he worked on problems of pattern recognition. Although he had no formal university education he was granted an honorary doctorate in medicine by the City University, London (1975).

Hounsfield was awarded the 1979 Nobel Prize for medicine, together with the South-African-born physicist Allan CORMACK, for his pioneering work on the application of computer techniques to x-ray examination of the human body. He was knighted in 1981. Working at the Central Research Laboratories of EMI he developed the first commercially successful machines to use computer-assisted tomography, also known as computerized axial tomography (CAT). In CAT a high-resolution x-ray picture of an imaginary slice through the body (or head) is built up from information taken from detectors rotating around the patient. These "scanners" allow delineation of very small changes in tissue density. Introduced in 1973, early machines were used to overcome obstacles in the diagnosis of diseases of the brain, but the technique has now been extended to the whole body. Although Cormack worked on essentially the same problems of CAT, the two men did not collaborate, or even meet.

Houssay, Bernardo Alberto (1887–1971)
Argentinian physiologist

Born in the Argentinian capital, Houssay (oo-sI) was the founder and director of the Buenos Aires Institute of Biology and Experimental Medicine. He was also professor of physiology at Buenos Aires from 1910 until 1965, apart from the years 1943–55 when he was relieved of his post by the regime of Juan Perón.

Houssay's work centered upon the role of the pituitary gland in regulating the amount of sugar in the blood, as well as its effects in aggravating or inducing diabetes. Working initially with dogs, he found that diabetic sufferers could have their condition eased by extraction of the pituitary gland, since its hormonal effect is to increase the amount of sugar in the blood and thus counter the influence of insulin. Deliberate injection of pituitary extracts actually increases the severity of diabetes or may induce it when the condition did not previously exist. He was also able to isolate at least one of the pituitary's hormones that had the reverse effect to insulin. Houssay's work on hormones led to his award, in 1947, of the Nobel Prize for physiology or medicine, which he shared with Carl and Gerty CORI. He was the author of *Human Physiology* (1951).

Hoyle, Sir Fred (1915–2001) *British astronomer*

> Space isn't remote at all. It's only an hour's drive away if your car could go straight upwards.
> —*The Observer*, 9 September 1979

It is the true nature of mankind to learn from mistakes not from example.
—*Into Deepest Space* (1974)

[We must] recognize ourselves for what we are – the priests of a not very popular religion.
—*Physics Today*, April 1968

The son of a textile merchant from Bingley in Yorkshire, England, Hoyle was educated at Cambridge. After graduating in 1936 he remained at Cambridge as a graduate student before being elected to a fellowship at St. John's College in 1939. Hoyle spent the war working on the development of radar at the Admiralty. After the war he returned to Cambridge and was appointed Plumian Professor of Astronomy in 1958.

Hoyle first came to prominence in 1948 with his formulation of the "steady-state theory" of the universe. He was aware that cosmology at the time was inadequate in that it required a smaller age for the universe than geologists had attributed to the Earth. Hoyle's ideas about the steady-state theory were provoked one night in 1946, when he went to see a ghost film with Hermann BONDI and Thomas GOLD. The film was in four parts but linked the sections together to create a circular plot in which the end of the film became its beginning. Hoyle later noted that it showed him that unchanging situations need not be static. The universe could perhaps be both unchanging and dynamic.

Hoyle worked out some of the detailed implications of this view in his 1948 paper *A New Model for the Expanding Universe*. Matter, he argued, was created continually. It arose from a field generated by the matter that already exists – that is, in the manner of the film, "Matter chases its own tail." Created matter is spread throughout the whole of space and, according to the theory, is being produced at a rate of about one atom per year in a volume equal to that of a large building. It is this creation that drives the expansion of the universe. Matter is distributed evenly through space and therefore new clusters of galaxies are forming as other galaxies are receding into the distance.

Although Hoyle's work was initially treated sympathetically, the steady-state theory failed to cope with new evidence emerging in the 1960s from radio astronomy. Counts of radio sources by Martin RYLE in the 1960s and, in particular, the discovery by Robert WILSON and Arno PENZIAS of the cosmic background radiation in 1964, convinced most scientists that the universe had begun with a big bang. Hoyle defended his theory strongly, objecting to the accuracy of the radio counts by arguing that they were so constructed as to allow every error to count against the theory. "Properly analyzed," Hoyle wrote in 1980, "the disproof of the theory claimed in the 1950s and early 1960s fails completely." He also suggested that there could be alternative explanations for the background radiation.

Hoyle subsequently felt that he was not committed to the details of any cosmological orthodoxy, such as either the big-bang or the steady-state theory of 1948. He spent much time exploring the implications of both theories and, in collaboration with Jayant Narlikar, developing a new theory of gravity. In 1964 they proposed, following some early arguments of Ernst MACH, that the inertia of any piece of matter derives from the rest of the matter in the universe. They also predicted that the gravitational constant changes over time.

Hoyle also worked in the 1950s on the formation of the elements. It was widely believed that carbon could be formed, along with many other elements, in the interior of stars. One reaction proposed required three helium nuclei to fuse into a carbon atom as in:

$$^4\text{He} + {}^4\text{He} + {}^4\text{He} \rightarrow {}^{12}\text{C}$$

Hoyle realized that the reaction would take place too infrequently to account for the abundance of carbon in the universe. Another possibility was a two-stage reaction:

$$^4\text{He} + {}^4\text{He} \rightarrow {}^8\text{Be}$$

$$^8\text{Be} + {}^4\text{He} \rightarrow {}^{12}\text{C}$$

In this, two helium nuclei first form a beryllium nucleus, which fuses in turn with another helium nucleus to form carbon. As the Be has a longer life-time than the collision time of two ^{4}He nuclei, the reaction should make the production of carbon more likely. Something more was needed and in 1954 Hoyle predicted that there must be a resonance channel easing the two reaction steps. Hoyle's prediction was confirmed when it was shown experimentally that there was an energy level of 7.65 million electronvolts (MeV) in the ^{12}C nucleus, just above the energy of the Be + ^{4}He structure of 7.366 MeV.

Further work on the formation of the elements was carried out by Hoyle in collaboration with William FOWLER and Geoffrey and Margaret BURBIDGE. In 1957 their work resulted in a paper, commonly referred to as B^2FH, that is one of the most authoritative and comprehensive works of modern science. It describes precisely how all the naturally occurring elements other than hydrogen and helium are formed in the interior of stars.

Hoyle spent much of the early 1960s working in the U.S. at the Hale Observatories and at Princeton. In 1967 he was appointed director of the newly formed Institute of Theoretical Astronomy at Cambridge. It was not a happy time. There were bitter disputes with Martin Ryle and the radio astronomers, demands for apologies, and threats of legal action. Hoyle had problems with his requests for funds from the research councils. In 1973 he resigned and subsequently held no permanent post.

He did, however, continue to publish on a wide variety of topics. Much of this later work,

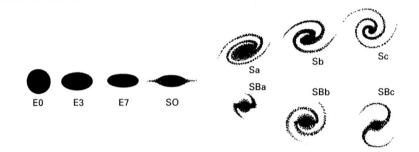

HUBBLE'S CLASSIFICATION OF GALAXIES   *This is known as his "tuning-fork" diagram.*

often in collaboration with Chandra Wickramsingh of Cardiff University, stemmed from his claim that the blind operation of physicochemical laws would have been insufficient to shuffle an assortment of amino acids into an enzyme. The odds against this happening by chance were 1 in $10^{40,000}$, as were the chances that an atom with the properties of carbon could be produced by nature. Such considerations led Hoyle to attack the notion of evolution by natural selection.

Hoyle began his campaign with a frontal attack; he asserted that the fossil *Archaeopteryx* was a fake. Probably the most famous of all fossils, *Archaeopteryx* had been bought by the British Museum in 1862 for 700 pounds and supposedly links reptiles with birds. Hoyle published a paper in 1985 claiming that the skeleton was genuine and of a reptile, but that the feathers had been glued on. Hoyle went on to publish a book on the issue, *Archaeopteryx, the Story of a Fake* (1987), in which he identified Richard Owen as the culprit. Tests of the fossil by the British Museum have failed to detect any glue or cement.

But if enzymes, let alone organisms, could not have evolved on Earth, where did they originate? In *Lifecloud* (1978), *Diseases from Space* (1979), and *Space Travellers* (1981), Hoyle argued that life must have come from space. Hoyle was partly led to this view by a long-standing interest in interstellar grains. They had long been thought to be made of ice, but, as they failed to reveal the appropriate infrared absorption bands, this view had to be ruled out. Hoyle pursued the matter and struggled for twenty years to find a particle with the observed spectral properties of the interstellar grains. In 1980 he decided to compare the grains with bacteria and found, to his great surprise, agreement so close that he was forced to conclude that "hitherto unidentified components of dust clouds were in fact bacterial cells."

HOYLE's new theory allowed him to explain not only the origin of life on Earth but also much about the spread of disease. The abrupt appearance of a new disease, such as syphilis in the 15th century, can be seen as a bacterial seeding from a passing comet. Other epidemiological problems can also be solved in this way. How, for example, Hoyle asks, did a group of Amerindians in Suriname, isolated from all alien contact until recently, become infected with the polio virus? Because, Hoyle believes, both forest and city dwellers were infected with pathogens rained on them from above.

Hoyle's numerous other publications covered such areas as the history of astronomy in his *Copernicus* (1973), an important textbook, *Astronomy and Cosmology* (1975), archeoastronomy in *From Stonehenge to Modern Cosmology* (1972), and a first volume of autobiography in his *The Small World of Fred Hoyle* (1986). He also wrote fourteen science-fiction novels, the first being *The Black Cloud* (1957). In 1994 Hoyle published his long-awaited autobiography, *Home Is Where the Wind Blows: Chapters from a Cosmologist's Life*.

Hubble, Edwin Powell (1889–1953)
American astronomer and cosmologist

> Equipped with his five senses, man explores the universe around him and calls the adventure Science.
> —*Science*

Hubble, who was born in Marshfield, Missouri, was the son of a lawyer. He was educated at the University of Chicago where he was influenced by the astronomer George HALE and, as a good athlete, was once offered the role of "Great White Hope" in a match against the world heavyweight champion, Jack Johnson. Instead he went to England, accepting a Rhodes scholarship to Oxford University where, between 1910 and 1913, he studied jurisprudence, represented Oxford in athletics, and fought the French boxer, Georges Carpentier. On his return to America he practiced law briefly before returning in 1914 to the study of astronomy at the Yerkes Observatory of the University of Chicago. He obtained his PhD in 1917. After being wounded in France in World War I he took up an appointment in 1919 at the Mount Wilson Observatory in California where Hale was director and where he spent the rest of his career.

Hubble's early work involved studies of faint nebulae, which in the telescopes of the day appeared as fuzzy extended images. He considered that while some were members of our Galaxy and were clouds of luminous gas and dust, others, known as spiral nebulae, probably lay beyond the Galaxy. After the powerful 100-inch (2.5-m) telescope went into operation at Mount Wilson he produced some of the most dramatic and significant astronomy of the 20th century. In 1923 he succeeded in resolving the outer region of the Andromeda nebula into "dense swarms of images which in no way differ from those of ordinary stars." To his delight he found that several of them were Cepheids, which allowed him to use Harlow SHAPLEY's calibration of the period-luminosity curve to determine their distance as the unexpectedly large 900,000 light-years. Although this conflicted sharply with the results of Adriaan van Maanen, Hubble continued with his observations. Between 1925 and 1929 he published three major papers showing that the spiral nebulae were at enormous distances, well outside our own Galaxy, and were in fact isolated systems of stars, now called spiral galaxies. This was in agreement with the work of Heber CURTIS. In 1935 van Maanen reexamined his data and, appreciating their unsatisfactory nature, withdrew the final objection to Hubble's results.

In 1929 Hubble went on to make his most significant discovery and announced what came to be known as *Hubble's law*. Using his own determination of the distance of 18 galaxies and the measurements of radial velocities from galactic red shifts carried out by Vesto SLIPHER and Milton HUMASON, he saw that the recessional velocity of the galaxies increased proportionately with their distance from us, i.e., $v = Hd$, where v is the velocity, d the distance, and H is known as *Hubble's constant*. Further measurements made by Hubble in the 1930s seemed to confirm his earlier insight. It was this work that demonstrated to astronomers that the idea of an expanding universe, proposed earlier in the 1920s by Aleksandr FRIEDMANN and Georges LEMAÎTRE, was indeed correct. The expansion of the universe is now fundamental to every cosmological model.

Hubble's law was soon seen as containing the key to the size, age, and future of the universe. Hubble's constant can be found from the mean value of v/d. Hubble himself gave it a value approximately ten times its presently accepted figure. The constant permits a calculation of the observable size of the universe to be made. The limiting value of recession must be the speed of light (c). If we divide this by H we get a "knowable" universe with a radius of about 18 billion light-years. Beyond that no signal transmitted could ever reach us, for to do so it would need to exceed the speed of light.

It is also possible to calculate the time that must have elapsed since the original highly compact state of the universe, i.e., the age of the universe. Hubble's own estimate was 2 billion years but with revisions of his constant, cosmologists now, none too precisely, assign a value of between 12 and 20 billion years.

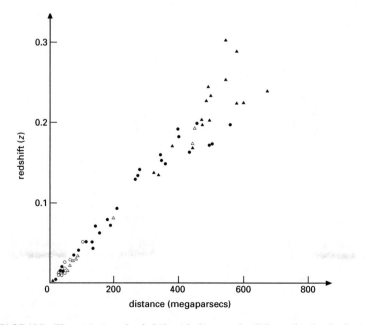

HUBBLE DIAGRAM   *The variation of red shift with distance for different kinds of galaxies.*

Hubble also made a major contribution to the study of galactic evolution by producing the first significant classification of galaxies. William HERSCHEL had simply classified them as bright or faint, large or small, while his son John Herschel introduced five categories in terms of size, brightness, roundness, condensation, and resolvability, each with five subdivisions. Hubble published his scheme in 1926. It involved dividing galaxies into two classes, elliptical and spiral. Ellipticals could be subdivided on the basis of their degree of ellipticity, ranging from the circular form (E0) to the elongated (E7). Spirals could be either barred or normal spirals which were subdivided in terms of their degree of openness. Although anomalous objects were later discovered that failed to fit it, Hubble's scheme is still used as the basis for galactic classification.

Hubel, David Hunter (1926–) *Canadian–American neurophysiologist*

Born in Windsor, Ontario, Hubel was educated at McGill University and then worked at the Montreal Neurological Institute. He moved to America in 1954 and after working at Johns Hopkins joined the Harvard Medical School in 1959 where he was appointed professor of neurobiology from 1968 to 1982.

Beginning in the 1960s, Hubel, in collaboration with Torsten WIESEL, published a number of remarkable papers that explained for the first time the mechanism of visual perception at the cortical level.

Their work was made possible by a number of technical advances. From the early 1950s onward it became possible to use microelectrodes to monitor the activity of a single neuron. Further, the work of Louis SOKOLOFF allowed workers to identify precise areas of neural activity. Using this latter technique it was thus possible to identify the region known as the striate cortex, located at the back of the cortex in the occipital lobes, as one of the key centers of activity during the visual process.

The cells of the striate cortex seemed to be arranged into columns, or "hypercolumns" as they were soon described, that run the length of the cortex (3–4 millimeters) from the outer surface to the underlying white matter. Such hypercolumns were further clearly divided into distinct layers. Hubel and Weisel went on to probe the structure, function, and contents of such columns in great detail.

Above all they succeeded in establishing two crucial points. First that the retinal image was mapped in some way onto the striate cortex. That is, to each point on the retina there corresponded a group of cells in the striate cortex that would respond to a stimulation of that point and of no other.

Furthermore, the response could be evoked only by a relatively precise stimulus. Thus there were cells that would respond to a spot of light but not to a line. Cells that responded to lines would do so only to those lines with a specific tilt and if the angle of tilt was changed by as little as 10°, in either direction, the cells' ability to react would be diminished or even abolished.

As a result of such work the visual cortex has become the best known of all cortical regions. Hubel and Wiesel shared the 1981 Nobel Prize for physiology or medicine with Roger SPERRY.

Huber, Robert (1937–) *German chemist*

Huber (**hoo**-ber) was educated at the Technical University in his native city of Munich where he obtained his PhD in 1972. He moved to the Max Planck Institute for Biochemistry, Munich, where he later became head of the Institute's crystallography group. In 2006 he moved to Cardiff University to work on the development of structural biology.

In 1982 Hartmut MICHEL had succeeded in crystallizing the membrane proteins of the photosynthetic reaction center. His colleague Johann DEISENHOFER had managed by 1985 to work out the molecular structure of the crystals. For this work they shared the 1988 Nobel Prize for chemistry. The prize also went to Huber as head of the unit in which Michel and Deisenhofer carried out their work.

Huchra, John Peter (1948–) *American astronomer*

Huchra, who was born in Jersey City, New Jersey, was educated at the Massachusetts Institute of Technology and the California Institute of Technology, where he obtained his PhD in 1976. He then moved to Harvard, serving as professor of astronomy and as a staff member at the Smithsonian Astrophysical Observatory.

In the early 1980s Huchra worked on the Tully–Fisher relation, which links the intrinsic luminosity of a spiral galaxy with the rotational velocity of its stars. He found that with regard to galaxies in the Coma cluster there was a departure of up to 20% from the supposed correlation. Huchra went on in 1982 with a number of colleagues to apply the relation to the Local Group of galaxies to see if the peculiar motion tentatively identified by Vera RUBIN could be detected. The peculiar velocities of several hundred galaxies in the region of the Virgo cluster were measured. They found that velocities in the direction of Virgo steadily increased, while decreasing on the other side. The result has been interpreted by some as evidence for the existence of "The Great Attractor," a proposed massive concentration of galaxies lying beyond the Hydra-Centaurus supercluster.

In 1986, in collaboration with Margaret GELLER, Huchra began a galactic survey for the Smithsonian Center for Astrophysics (CfA).

They used a 1.5-meter telescope located on Mount Hopkins, Tucson, Arizona, and sought to measure the red shifts of galaxies below a magnitude of 15.5 and falling within a wedge of sky 6° wide, 120° long, and out to a distance of about 300 million light years. By 1989 they had mapped the positions of 10,000 galaxies.

To their surprise, instead of producing a uniform distribution their maps revealed large voids within which huge clusters of galaxies were distributed. The largest structure they observed, dubbed "The Great Wall," stretched 500 million light years without its edge being found.

Hückel, Erich Armand Arthur Joseph (1896–1980) *German chemist*

Born the son of a physician in Berlin, Germany, Hückel (**hoo**-kel) was educated at the University of Göttingen, gaining his PhD in 1921. He worked at a number of institutions, including the Zurich Technische Hochschule and in Copenhagen, Leipzig, and Stuttgart, before taking the chair of theoretical physics at Marburg in 1937.

Initially Hückel worked with DEBYE on electrolyte solutions. From 1930, however, he turned his attention to organic compounds. Since Friedrich KEKULÉ had discovered the structure of benzene (C_6H_6) in 1865, it had continued to puzzle chemists. Kekulé had shown that the six carbon atoms of benzene were formed into a ring joined by alternating single and double bonds. Organic chemists call such molecules as benzene "aromatic," thereby indicating, among other things, the molecule's great stability. Yet, double bonds normally make a molecule reactive. How, then, it was asked, can certain molecules like benzene with double bonds be so stable?

In the 1930s Hückel developed an answer to this problem based upon molecular orbital theory. Molecular orbitals are formed from overlapping atomic orbitals. Hückel proposed that the electrons of the pi-orbitals were delocalized and spread diffusely above and below the plane of the carbon ring. As this configuration was energetically more stable than placing electrons in isolated double bonds, benzene's stability followed directly from the model.

Hückel went on to generalize his model to cover other cyclic molecules containing alternating double and single bonds. Aromatic molecules were planar compounds which had precisely $4n + 2$ pi-electrons, where $n = 0, 1, 2, 3 \ldots$. This is known as the *Hückel rule*. Benzene represents the case where $n = 1$; and $n = 2$ and $n = 3$ represent the 10 and 14 member aromatic rings of naphthalene and anthracene. For $n = 0$, the predicted aromaticity of a 3 member ring was confirmed in 1962 with the discovery of the cyclopropenyl cation.

Huffman, Donald Ray (1935–   ) *American physicist*

Born at Fort Worth in Texas, Huffman was educated at Texas Agricultural and Mechanical College, at Rice University, Houston, and at the University of California, Riverside, where he completed his PhD in 1966. After spending a postdoctoral year at the University of Frankfurt, Huffman moved to the University of Arizona, Tucson, in 1967 and was later appointed professor of physics in 1975. He is currently emeritus professor at Arizona.

In 1985 in the laboratory of Richard SMALLEY a new form of carbon had been discovered: C_{60}, known as "buckminsterfullerene." The C_{60} was produced by vaporizing a graphite target with a pulsed laser beam. The sooty carbon produced in this manner certainly contained a detectable amount of C_{60}, but all efforts to extract the substance from the residue in amounts sufficient to carry out a detailed spectroscopic study failed.

Huffman, in collaboration with Wolfgang Kratschmer of the Max Planck Institute for Physical Chemistry, Heidelberg, was involved in the discovery of the new forms of carbon known as *fullerenes*. For many years they had been interested in the nature of interstellar dust, which they believed to be mainly carbon. The interstellar matter has a characteristic broad absorption spectrum and Huffman and Kratschmer were experimenting with various forms of finely divided carbon produced in electric arcs. During this work, around 1982, they found a form of carbon with a peculiar double hump, which they called "the camel."

When, in 1985, they heard of the discovery of C_{60}, buckminsterfullerene, they suspected that this might be the cause of their camel spectrum. Huffman and Kratschmer reproduced their earlier experimental conditions, in which they had formed a carbon powder by striking an arc between graphite electrodes in a low pressure of helium.

They treated the resulting soot with benzene, from which they crystallized a light-yellow solid, which they named "fullerite." It was later found to contain about 75% of C_{60} together with 25% of another fullerene, C_{70}. The method has allowed the production of fullerenes in large quantities.

Huggins, Charles Brenton (1901–1997) *Canadian–American surgeon*

Huggins, who was born at Halifax in Nova Scotia, was educated at Acadia University and at the Harvard Medical School, where he obtained his MD in 1924. After graduate training at the University of Michigan he moved to the University of Chicago in 1927 where he served as professor of surgery from 1936 and director of the May Laboratory of Cancer Research from 1951 until 1969.

In 1939 Huggins made a very simple inference that led to the development of new forms of cancer therapy. Noting that the prostate gland was under the control of androgens (male sex hormones) he concluded that cancer of the prostate might be treated by preventing the production of androgens. Admittedly his proposed treatment of orchiectomy (castration) might appear somewhat severe but it did lead to remissions in some cases and an alleviation of the condition in others.

Huggins soon appreciated, however, that the same results could probably be achieved by the less drastic procedure of the administration of female sex hormones to neutralize the effect of androgens produced by the testicles. Consequently in 1941 he began to inject his patients with the hormones stilbestrol and hexestrol. He was able to report later that of the first 20 patients so treated 4 were still alive after 12 years. Later workers, inspired by Huggins's work, treated women suffering from cancer of the breast with the male hormone testosterone and claimed improvement in some 20% of the cases.

It was for this work that Huggins shared the 1966 Nobel Prize for physiology or medicine with Peyton ROUS.

Huggins, Sir William (1824–1910) *British astronomer and astrophysicist*

> It is remarkable that the elements most widely diffused through the host of stars are some of those most closely connected with the living organisms of our globe, including hydrogen, sodium, magnesium, and iron. May it not be that, at least, the brighter stars are like our Sun, the upholding and energizing centres of systems of worlds, adapted to be the abode of living beings?
> —Report on his observations (1863)

Huggins, the son of a London silk merchant, attended school for a short period before being educated privately. After a few years in business he retired to devote himself exclusively to the study of science. His first interest was in microscopy but he became absorbed in the work of Gustav KIRCHHOFF and Robert BUNSEN on spectroscopy and the solar spectrum and decided that he would try to do the same with the stars. He equipped himself with the best of instruments, including a superb 8-inch (20-cm) glass from Alvan CLARK. He spent some time making maps of the terrestrial elements before moving to the stars, collaborating with William MILLER, professor of chemistry at King's College, London. He then began the first major intensive spectral investigation of the stars, which lasted until he was 84 years old, when he found that he could no longer see clearly enough. In later life he was also helped by his wife, Margaret, whom he married in 1875.

Huggins's first observations, published in 1863, showed the stars to be composed of known elements occurring on the Earth and in the Sun. His next great discovery came when he obtained the spectra of those nebula that earlier astronomers had failed to resolve into stars. His excitement is apparent in his report: "I looked into the spectroscope. No spectrum such as I expected! A single bright line only!...The riddle of the nebula was solved...Not an aggregation of stars, but a luminous gas." He quickly examined the spectra of over 50 nebulae and found that a third were gaseous. In the same year he obtained the spectra of a comet and found that it contained hydrocarbons. In 1866 he showed that a nova was rich in hydrogen. He also discovered previously unidentified bright emission lines in the spectra of certain nebulae and attributed them to a new element "nebulium." The true explanation for these forbidden lines was not provided until the next century, by Ira BOWEN.

In 1868 Huggins successfully employed a use of spectroscopy that has had a more profound impact on cosmology than anything else. It had been shown by Christian DOPPLER and Armand FIZEAU that the light waves of an object leaving an observer would have a lower frequency, and the frequency of an object approaching an observer should increase. In spectral terms this means that the spectra of the former object should be shifted toward the red and the latter toward the blue. In 1868 Huggins examined the spectrum of Sirius and found a noticeable red shift. As the degree of the shift is proportional to the velocity, Huggins was able to calculate that the speed of recession of Sirius was about 25 miles (40 km) per second. He quickly determined the velocity of many other stars. He and Lady Huggins published their spectral work in its entirety as the *Atlas of Representative Stellar Spectra* in 1899. Huggins had tried to photograph Sirius but was only successful in 1876 by which time the gelatine dry plate had been developed.

Huggins was knighted in 1897, and was president of the Royal Society from 1900 to 1905.

Hughes, John Russell (1928–) *American neurophysiologist*

Born in Du Bois, Pennsylvania, Hughes was educated at Oxford University and at Harvard, where he obtained his PhD in 1954. After some time at the National Institutes of Health and the State University of New York, he moved to Northwestern University where he has served as professor of neurophysiology since 1964.

Among other problems Hughes has worked on the way in which information is transmitted within the central nervous system. In particular he has made a detailed study of the "language" used by the olfactory bulb to inform the brain of the nature of the olfactory medium. His method consisted of implanting electrodes in the olfactory bulb and recording their response to gauze soaked with a certain chemical held at a standard distance away. He found that the message transmitted essentially comprises a mixture of various frequency components.

Huisgen, Rolf (1920–   ) *German chemist*

Huisgen (**hoos**-gen), born the son of a surgeon in Gerolstein, Germany, studied chemistry at the universities of Bonn and Munich where he obtained his PhD in 1943. He initially taught at Tübingen before moving in 1952 to the University of Munich as professor of chemistry and director of the Institute of Physical Chemistry. He retired in 1988.

Huisgen has worked on the mechanisms of organic chemical reactions, and particularly on reactions leading to the formation of ring compounds. Many of his studies have been on reactions to which the principle of orbital symmetry, introduced by Robert Burns WOODWARD and Roald Hoffman, applies.

Hulse, Russell Alan (1950–   ) *American astrophysicist*

In 1974 Hulse was working as a graduate student at the University of Massachusetts, Amherst, under the supervision of Joseph TAYLOR. It was arranged that he would spend the summer in Puerto Rico using the Arecibo Radio Telescope to search for pulsars, a type of star first observed by Jocelyn BELL-BURNELL in 1967. Among several pulsars detected by Hulse one particular example, named 1913+16 in the constellation Aquila, proved to be of special significance.

Hulse initially found that the pulsar had a short period of 0.059 seconds. More detailed examination, however, revealed that the pulse rate was not constant but varied by some 5 microseconds from day to day. At first Hulse suspected a computer fault. But despite writing a new program, the variability remained. Eventually Hulse spotted that the variation was cyclical, repeating itself every 7.75 hours.

Such phenomena, Hulse argued, would arise naturally if the pulsar was a binary, orbiting an undetected companion star. This would produce a DOPPLER effect. That is, when the pulsar travels in its orbit toward the Earth the pulses would be crowded together, giving a greater than average pulse rate; when, however, it traveled away from the Earth the pulses would be more spread out and yield a lower than average frequency.

In collaboration with Taylor, Hulse went on to establish some of the basic properties of the pulsar. It appeared to have a mass equivalent to 2.8 solar masses, was thought to be a neutron star with a diameter no more than 20–30 kilometers, and to have an approaching velocity of 300 kps (kilometers per second) and a receding velocity of 75 kps. For his work in this field Hulse shared the 1993 Nobel Prize for physics with Joseph Taylor.

After completing his work on the pulsar 1913+16 in 1977 Hulse moved to Princeton, abandoned astronomy, and began to work at the Plasma Physics Laboratory.

Hulst, Hendrik Christoffell van de *See* VAN DE HULST, HENDRIK CHRISTOFFELL.

Humason, Milton La Salle (1891–1972) *American astronomer*

Born in Dodge Center, Minnesota, Humason had no formal university training – in fact he began work as a donkey driver moving supplies to the Mount Wilson Observatory in southern California. Here he quickly developed an interest in astronomy and its techniques, an interest that was stimulated by the staff of the observatory. He was taken on as janitor and by 1919 he was competent enough to be appointed assistant astronomer on the staff of the Mount Wilson Observatory and, after 1948, of the Palomar Observatory, where he spent the rest of his career.

In the 1920s Edwin HUBBLE formulated his law that the distance of the galaxies was proportional to their recessional velocity. This work was based on the careful, painstaking, and difficult measurements of galactic red shifts made by Humason and also by Vesto SLIPHER. Humason developed extraordinary skill in this field. By 1936, using long photographic exposures of a day or more, he was able to measure a recessional velocity of 40,000 kilometers per second, which took him to the limits of the 100-inch (2.5-m) reflecting telescope at Mount Wilson.

With the opening of the Palomar Observatory he was able to use the 200-inch (5-m) Hale reflector and by the late 1950s was obtaining velocities of over 100,000 km per second; this corresponded to a distance, according to Hubble's law, of about six billion light-years.

Humboldt, Alexander, Baron von (1769–1859) *German explorer and scientist*

> An object well worthy of research, and which has long fixed my attention, is the small number of simple substances (earthy and metallic) that enter into the composition of animated beings, and which alone appear fitted to maintain what we may call the chemical movement of vitality.
> —*Personal Narrative of Travels to the Equinoctial Regions of America* (1804)

Humboldt (**hum**-bolt or **huum**-bolt) was born in Berlin, Germany. He initially showed little enthusiasm for his studies, but while taking an engineering course in Berlin suddenly became interested in botany, and a year at the University of Göttingen further increased his interest in the sciences. Geology and mineralogy particularly intrigued him and he went on to join the School of Mines in Freiberg, Saxony, staying there for two years. He then worked for the mining department in Ansbach-Bayreuth, reorganizing and supervising the mines in the region.

In 1796 Humboldt inherited enough money to finance himself as a scientific explorer, and he gave up mining to do two years' intensive preparatory studies in geological measuring methods. Initially his expeditionary plans were thwarted by the Napoleonic Wars but, in 1799, he finally managed to sail with a ship bound for Latin America.

The French botanist, Aimé Bonpland, accompanied him on the five-year journey, during which they navigated the Orinoco River and traveled widely through Peru, Venezuela, Ecuador, and Mexico, collecting scientific specimens and data and covering over 6,000 miles. Humboldt studied the Pacific coastal currents, the *Humboldt current* (now the Peru current) being named for him, and he was the first to propose building a canal through Panama. He investigated American volcanoes, noting their tendency to follow geological faults, and concluded that volcanic action had played a major part in the development of the Earth's crust, thus finally disproving the neptunist theory of Abraham WERNER. He climbed the Chimborazo volcano to what was then a world record height of 19,280 feet (5,876 m), and was the first to attribute mountain sickness to oxygen deficiency. He also measured changes in temperature with altitude and noted its effect on vegetation.

On his return to Europe in 1804, Humboldt settled in Paris and began to publish the data gathered on his travels, a task that took 20 years and filled 30 volumes. He introduced isobars and isotherms on his weather maps, so pioneering the subject of comparative climatology, and also helped initiate ecological studies with his discussions on the relationship between a region's geography and its flora and fauna.

By 1827 Humboldt's finances were severely depleted and he returned to Berlin as tutor to the Prussian crown prince. Two years later he was invited by the Russian finance minister to visit Siberia, and Humboldt made use of the trip to take more geological and meteorological measurements. He also organized a series of meteorological and magnetic observatories through Russia, Asia, and the British Empire, to trace the fluctuations in the Earth's magnetic field.

Humboldt spent the last years of his life writing *Kosmos* (The Cosmos), a synthesis of the knowledge about the universe then known, of which four volumes were published during his lifetime.

Hume, David (1711–1776) *British philosopher*

> If we take in our hand any volume, of divinity or school metaphysics, for instance, let us ask, Does it contain any abstract reasoning concerning quantity or number? No. Does it contain any experimental reasoning concerning matter of fact and existence? No. Commit it then to the flames; for it can contain nothing but sophistry and illusion.
> —*An Enquiry Concerning Human Understanding* (1748)

Hume was the younger son of a laird. Born in Edinburgh, Scotland, and educated at the university there, he intended to study law but found himself stricken with an "insurmountable aversion to everything but the pursuit of philosophy and general learning." He became a tutor, served in a variety of diplomatic posts, and spent the period 1752 to 1763 as librarian of the Edinburgh Faculty of Advocates where he wrote his *History of England* (6 vols., 1754–62). After another period of diplomatic service abroad in Paris, Hume retired to Edinburgh in 1769. In the spring of 1775 Hume contracted cancer. Shortly before his death in the summer of 1776 he was visited by James Boswell, eager to report back to Dr. Johnson that the notorious atheist, Hume, was facing death with apprehension. Instead he found Hume placid and cheerful, denying that death was to be feared and declaring survival to be "an unreasonable fancy." Boswell was most disturbed.

In 1738 Hume published his masterpiece, *A Treatise of Human Nature*, in which his aim was to do for the moral sciences, the science of man, what NEWTON had done for the natural sciences. In pursuing this aim Hume was lead into a profound study of causality – the idea of a "necessary connection" between two physical events when one is said to be the cause of the other – work that carried important implications for science. Against Rationalists such as DESCARTES, Hume argued that empirical knowledge can never be deduced a priori, thus aiding the destruction of the Cartesian tradition in science. Hume put forward the basis of modern Empiricism – that all knowledge of matters of fact (and thus scientific knowledge) is based on experience and evidence and is therefore only probable, capable of denial, and can never be logically necessary. Thus in his analysis of causation Hume argued that there is no necessary connection between physical events (e.g., when one is said to be the *cause* of the other); there is no mystical "power" in one event that "brings about" another. We merely observe that one type of event is invariably followed by another, and because of this association of ideas in our minds, when we observe an event of the first type we "predict" an event of the second.

But such a practice will only be sound if the conjunctions we have observed in the past continue to hold in the future. How can, asked

Hume, this latter assumption be justified? Not by pure reason, nor from experience, for we cannot conclude from the fact that nature has behaved uniformly in the past that it will continue to do so without assuming the truth of the very principle we are examining. In this way Hume presented the classical "problem of induction" that has since been puzzled over by generations of philosophers and scientists.

Hume-Rothery, William (1899–1968)
British metallurgist

The son of a lawyer, Hume-Rothery was born at Worcester Park in Surrey, England. He originally intended to pursue a military career and consequently entered the Royal Military Academy, Woolwich, on leaving school. An attack of meningitis which left him totally deaf forced him to leave the army and he turned instead to chemistry. Although refused entry to his father's college, Trinity College, Cambridge, because of his deafness, he was more graciously received by Magdalen College, Oxford. After obtaining his PhD from the Royal School of Mines in 1925, Hume-Rothery returned to Oxford where he remained for the rest of his life, being appointed in 1958 to the university's first chair of metallurgy.

With 178 published papers to his credit Hume-Rothery illuminated many areas of metallurgy. His best-known work was concerned with alloys that are solid solutions, in which atoms of the constituent metals share a common lattice. The *Hume-Rothery rules* give the conditions that have to be satisfied for metallic solid solutions to form. The first concerns the atomic size factor and claims that if the atomic diameter of the solvent differs in size from that of the solute by more than 14%, the chances of solubility are small. Secondly, the more electronegative is one component and the more electropositive the other, the more they are likely to form compounds rather than solutions. And, finally, a metal of lower valency is more likely to dissolve one of higher valency than vice versa. Much of his work in this field was published in his book *The Structure of Metals and Alloys* (1936).

Hund, Freidrich (1896–1997) *German physicist*

Hund (hunt) was born in Karlsruhe, Germany, and educated at the Universities of Marburg and Göttingen, gaining his PhD from Göttingen under the supervision of Max BORN in 1922. He stayed at Göttingen between 1922 and 1927 before moving to the University of Rostock (1927–29) and the University of Leipzig (1929–46). He was a professor of physics at the University of Jena between 1946 and 1951, the University of Frankfurt-am-Main between 1951 and 1956, and the University of Göttingen between 1956 and his retirement in 1964.

Hund worked on molecular orbital theory and interpreted molecular spectra in terms of this theory in the late 1920s. He is perhaps best known to chemists and physicists for *Hund's rules* – a series of rules for determining the ground state of a many-electron atom. Hund derived these rules using a combination of the old quantum theory of Niels BOHR and Arnold SOMMERFELD and the empirical study of atomic spectra. They were subsequently derived from the quantum theory of many-electron systems. He also wrote a number of books and reviews on atomic and molecular theory and produced a history of quantum theory.

Hunsaker, Jerome Clarke (1886–1984) *American aeronautical engineer*

Hunsaker was born in Cheston, Iowa, and educated at the U.S. Naval Academy, graduating in 1908. He served in the navy from 1909 until 1926, reaching the rank of commander. Selected for the Construction Corps, he studied naval architecture at the Massachusetts Institute of Technology and aeronautical engineering in Europe. In 1914 he established the first course in aeronautical engineering at MIT. Recalled by the navy (1916) to put aeronautical engineering into practice, he worked on zeppelins, producing the *Shenandoah* (1923), and flying boats. Hunsaker worked in business from 1926, initially at the Bell Telephone Laboratories and then (1928–33) at Goodyear where he built zeppelins. In 1933 Hunsaker was appointed professor of mechanical engineering at MIT, also serving concurrently as head of the department of aeronautical engineering until his retirement in 1951.

He was the author of *Aeronautics at the Midcentury* (1952), a review of the problems facing the industry and the strategies open to it.

Hunt, Sir Timothy R. (1943–) *British cell biologist*

Hunt gained a BA from Cambridge University in 1964, and a PhD in 1968. In 1981 he was appointed university lecturer in biochemistry at Cambridge, and in 1990 joined the Imperial Cancer Research Fund. Hunt was knighted in 2006.

Hunt is noted for his discovery in the early 1980s of the first cyclin molecule. Cyclins are proteins that periodically rise and fall in concentration during the cell cycle, hence their name. Hunt isolated his first cyclin in the sea urchin *Arbacia*, after finding a protein that underwent periodic degradations in step with phases of the cell cycle – a characteristic feature of cyclins. Hunt went on to find cyclins in other species, and showed that they had very similar molecular structures (i.e. were structurally conserved).

Cyclins are crucial regulators of the cell cycle. They form complexes with another family of proteins, the cyclin-dependent kinases. The cyclin component controls the activity of the complex, determining which proteins the CDK component phosphorylates (adds phosphate groups to) to drive the cycle forward. As the cyclin levels rise and fall, so does the activity of the CDKs. Hence the progress of the cell cycle is controlled.

Hunt's work helps our understanding of cancer. For example, abnormally high or sustained levels of cyclins and CDKs are found in certain human tumors, such as breast cancer and brain tumors, indicating malfunction of the cell-cycle control mechanisms. Inhibitors of these key molecules may have the potential to treat certain cancers.

For his discoveries of key regulators of the cell cycle Hunt was awarded the 2001 Nobel Prize for physiology or medicine, jointly with Leland HARTWELL and Paul NURSE.

Hunter, John (1728–1793) *British surgeon and anatomist*

> … A gifted interpreter of the Divine Power and Wisdom at work in the Laws of Organic Life, and … the founder of Scientific Surgery.
> —Memorial plaque to Hunter in Westminster Abbey (1859)

Hunter, who was born at Long Calderwood in Scotland, joined his elder brother William, the famous obstetrician, in London in 1748. He there assisted his brother and attended surgical classes at Chelsea Hospital. Disputes with William over their research led John to branch out on his own and in 1759 he joined the army to serve as a surgeon in Portugal during the Seven Years' War. On his return to London in 1762 he set up as a private teacher and in 1767 was appointed surgeon at St. George's Hospital, London.

As a surgeon Hunter's major innovation was in the treatment of aneurysm, a bulge appearing at a weak spot in the wall of an artery. Rather than follow the drastic procedure of amputation Hunter instead tied the artery some distance from the diseased part and found that, with the pressure of circulation removed from the aneurysmal sack, the progress of the disease is halted. He also made radical proposals, based on his military experience, for the treatment of gunshot wounds. He wisely argued in his *A Treatise on the Blood, Inflammation and Gunshot Wounds* (1794) that unless the missile in the body was actually endangering life the surgeon should leave it alone and under no circumstances enlarge the wound by opening it.

Hunter also wrote a famous work on venereal disease (1786), inadvertently producing much confusion. In the late 18th century it was still a matter of dispute whether syphilis with its chancre and gonorrhea with its purulent discharge were separate complaints. In 1767 Hunter decided to resolve the issue by inoculating himself with gonorrhea. He developed both gonorrhea and the typically hard chancre of syphilis, concluding therefore that discharge from a gonorrhea produces chancres. It seems not to have occurred to Hunter that his "gonorrhea" was also infected with syphilis.

Hunter's main claim to fame however lay in his superb anatomical collection. He was supposed to have dissected over 500 different species and at his death his collection contained over 13,000 items. Included in his museum in Leicester Square, London, was the skeleton of the Irish giant, C. Byrne, who was so keenly aware of the desire of Hunter for his 7-foot 7-inch frame that he arranged to be secretly buried at sea. Hunter was widely reported to have paid the undertakers the sum of £500 for the corpse.

His collection was purchased by the government after his death and in 1795 was presented to the Royal College of Surgeons in London where, despite some losses from bombing in World War II, it has remained ever since.

Hunter, William (1718–1783) *British obstetrician*

> Some physiologists will have it that the stomach is a mill; – others, that it is a fermenting vat; – others again that it is a stew-pan; – but in my view of the matter, it is neither a mill, a fermenting vat, nor a stew-pan – but a stomach, gentlemen, a stomach.
> —In a lecture to medical students

A brother of John Hunter, the famous surgeon, William was born at Long Calderwood in Scotland and studied medicine at Edinburgh University before moving to London in 1740. Hunter went on to specialize in obstetrics and became an eminent practitioner, attending the royal family. He founded the Great Windmill Street School of Anatomy and his most famous work is *The Anatomy of the Human Gravid Uterus* (1774), a collation of 25 years' work, which contains 34 detailed plates produced from engravings of dissections.

Hunter built up a large and valuable collection of coins, medals, pictures, and books, which he bequeathed to Glasgow University. It is now housed in the Hunterian Museum.

Huntington, George (1851–1916) *American physician*

Huntington was born in East Hampton, New York, the son and grandson of physicians. He was initially trained by his father before following a more formal course of medicine at the College of Physicians and Surgeons, Columbia University. He worked for some years with his father in East Hampton before moving to

Palmyra, Ohio. Huntington returned later to Duchess County, New York.

In 1872 he published a paper, *On Chorea*, in which he described an encounter with two women near East Hampton, a mother and a daughter. They were, he noted, "tall, thin, almost cadaverous, both bowing, twisting, grimacing." The disease was common in the area and he went on to note some of its main features. The condition was invariably fatal and progressed "gradually but surely until the hapless sufferer is but a quivering wreck." He also noted that it was "an heirloom from the past" in that it was confined to a few families – if the parents contracted the chorea then so would one or more of the children. Huntington also reported that it was an adult complaint, unknown in anyone under thirty. Not much more was known about the complaint until a century later when Nancy WEXLER began her search for the responsible gene. The disease, now known as Huntington's chorea, was thought to have been brought to East Hampton by a settler from Suffolk a century earlier.

Hurter, Ferdinand (1844–1898) *Swiss chemist*

Born at Schaffhausen in Switzerland, Hurter (**huur**-ter) was educated at the Federal Institute of Technology, Zurich, and at Heidelberg University, where he was a pupil of Robert BUNSEN. In 1867 he moved to England and became the chief chemist at Holbrook GASKELL and Henry DEACON's alkali factory in Widnes. In 1890, when the United Alkali Company was formed, Hurter was appointed its chief chemist and set up one of the first industrial research laboratories. He collaborated with Lunge in producing *The Alkali Maker's Handbook*, a work describing the LEBLANC process in technical detail. Hurter also worked on photography with V. C. Driffield.

Hutchinson, George Evelyn (1903–1991) *American biologist*

Hutchinson was born in Cambridge, England, and graduated from the university there in 1924. He was senior lecturer at the University of Witwatersrand (1926–28) before emigrating to America where he served as Sterling Professor of Zoology at Yale from 1945 until 1971. He received American citizenship in 1941.

Hutchinson's most important work was concerned with aquatic ecosystems and the physical, chemical, meteorological, and biological conditions of lakes. He made particular studies of the classification and distribution of aquatic bugs (Hemiptera), and investigated water mixing and movement in stratified lakes, proving the circulation of phosphorus. He also studied lake sediments and investigated certain aspects of evolution. His work took him to many different regions, including the lakes of western Transvaal, Tibet, and northeastern North America. Hutchinson published much of his life's work in his *A Treatise on Limnology* (3 vols., 1957–75); a fourth volume was completed shortly before his death and published in 1993.

Hutchinson, John (1884–1972) *British botanist*

Hutchinson, who was educated at the village school in Wark-on-Tyne, England, where he was born, began work in 1900 under his father, the head gardener on a large estate. In 1904 he was appointed to a junior post at the Royal Botanic Gardens, Kew, where he remained for the rest of his career. Starting as an assistant in the herbarium he was in charge of the Africa section from 1919 until 1936 when he became keeper of the Museum of Economic Botany, a post he occupied until his retirement in 1948.

Hutchinson's most significant work was his *Families of Flowering Plants* (2 vols. 1926–34; 2nd edition 1959), which contains details of 342 dicotyledon and 168 monocotyledon families. Hutchinson drew most of the illustrations for this work himself. In it he concentrated on the different plant families that various workers had considered the most primitive. He concluded that bisexual flowers with free petals, sepals, etc., as seen in the magnolia and buttercup families, are more ancient than the generally unisexual catkinlike flowers found in the nettle and beech families, which lack these parts. This conclusion supported the classification of George BENTHAM and Joseph HOOKER and added weight to arguments against the system of Adolf ENGLER. Furthermore Hutchinson stated that families with apparently more simple flowers are in fact more advanced, and have evolved by reduction from more complex structures; that is, the families show retrograde evolution. In this, the now generally accepted view, Hutchinson was developing the earlier ideas of the German botanist, Alexander Braun.

An enormously prolific and industrious worker, Hutchinson also published, with John Dalziel, the standard work, *Flora of West Tropical Africa* (1927–36) and at the time of his death was engaged in a revision of the *Genera Plantarum* (Genera of Plants) of Bentham and Hooker.

Hutton, James (1726–1797) *British geologist*

> There is presently laying at the bottom of the ocean the foundation of a future land, which is to appear after an indefinite space of time.
> —*Concerning the System of the Earth, its Duration, and Stability* (1788)

> It is scarcely more difficult to procure the secrets of science from Nature herself, than to dig them from the writings of this philosopher.
> —Thomas Thomson (1801)

Hutton was born in Edinburgh, Scotland, the son of a merchant who became the city treasurer. He was educated at Edinburgh University, which he left in 1743 to be apprenticed to a lawyer. This did not retain his interest long for, in 1744, he returned to the university to read medicine. He studied in Paris for two years and finally gained his MD from Leiden in 1749. He next devoted several years to agriculture and industry, farming in Berwickshire and commercially producing sal ammoniac. In 1768 he returned to Edinburgh, financially independent, and devoted himself to scientific studies, especially of geology, for the rest of his life.

Hutton's uniformitarian theories were first published as a paper in 1788 and later extended into a two-volume work, *Theory of the Earth* (1795). This work proved difficult to read and it only reached a wide audience when his friend John PLAYFAIR edited and summarized it as *Illustrations of the Huttonian Theory* (1802). It marked a turning point in geology. The prevailing theory of the day, the neptunism of Abraham WERNER, was that rocks had been laid down as mineral deposits in the oceans. However, Hutton maintained that water could not be the only answer for it was mainly erosive. The water could not account for the non-conformities caused by the foldings and intrusions characteristic of the Earth's strata. Hutton showed that the geological processes that had formed the Earth's features could be observed continuing at the present day. The heat of the Earth was the productive power, according to Hutton, that caused sedimentary rocks to fuse into the granites and flints, which could be produced in no other way. It could also produce the upheaval of strata, their folding and twisting, and the creation of mountains.

A long time scale is essential to Hutton's theory of uniformitarianism as the forces of erosion and combustion work, in general, only slowly, as demonstrated by the presence of visible Roman roads. He concluded that on the face of the Earth "we find no vestige of a beginning – no prospect of an end."

Hutton's work was accepted with little delay by most geologists, including the leading Edinburgh neptunist, Robert Jameson. In the 19th century Charles LYELL expanded the theories of uniformitarianism and these were to influence Charles DARWIN in his theory of evolution.

Huxley, Sir Andrew Fielding (1917–)
British physiologist

Huxley, a grandson of T. H. Huxley, was born in London and graduated in 1938 from Cambridge University, receiving his MA there three years later. He is best known for his collaboration with Alan HODGKIN in elucidating the "sodium pump" mechanism by which nerve impulses are transmitted, for which they were awarded,

with John ECCLES, the Nobel Prize for physiology or medicine (1963). He has also done important work on muscular contraction theory and has been involved in the development of the interference microscope and ultramicrotome. Huxley was reader in experimental biophysics at Cambridge (1959–60), and from 1960 to 1969 was Jodrell Professor of Physiology at University College, London. In 1969 he was elected research professor becoming emeritus professor in 1983. In 1980 he succeeded Alexander TODD as president of the Royal Society, a position he held until 1985. He also served as master of Trinity College, Cambridge, from 1984 to 1990 and was knighted in 1974.

Huxley, Hugh Esmor (1924–) *British molecular biologist*

Huxley (no relation to T. H. Huxley or any of his descendants) was born at Birkenhead in northwest England. He studied physics at Cambridge University where he obtained his PhD in 1952 after wartime research on the development of radar. Like many other physicists after the war Huxley was interested in applying physics to biological problems. After two years in America at the Massachusetts Institute of Technology and the period 1956–61 at the biophysics unit of the University of London, he returned to Cambridge to join the staff of the Medical Research Council's molecular biology laboratory, where he remained until 1987. In 1988 he became director and professor of biology at Brandeis University, Boston. He also served from 1988 to 1994 as director of the Rosenstiel Basic Medical Sciences Research Centre at Brandeis. Huxley was awarded the Copley Medal of the Royal Society in 1998.

In 1953, in collaboration with Jean Hanson, Huxley proposed the sliding-filament theory of muscle contraction. This was based on his earlier study of myofibrils, the contractile apparatus of muscle, with the electron microscope. He found that myofibrils are made of two kinds of filament, one type about twice the width of the other. Each filament is aligned with other filaments of the same kind to form a band across the myofibril, and the bands of thick and thin filaments overlap for part of their length. The bands are also linked by an elaborate system of crossbridges. When the muscle changes length the two sets of filaments slide past each other. Further, the two sorts of filaments can be identified with the two chief proteins of muscle, myosin in the thick filament and actin in the thin. This made possible an elegant solution to how muscles contract at the molecular level.

In the areas where both kinds of protein are in contact, Huxley suggested that one, most probably myosin, serves as an enzyme, splitting a phosphate from ATP and so releasing the en-

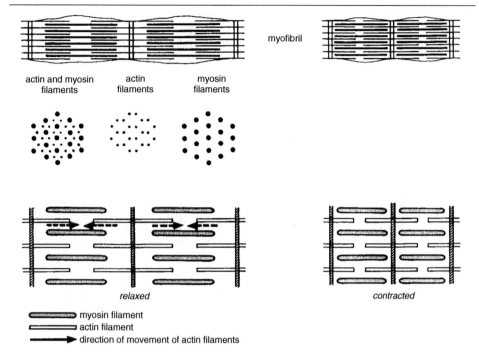

MUSCLE CONTRACTION   *The sliding-filament theory proposed by Hugh Huxley and Jean Hanson.*

ergy required for contraction. He concluded that the evidence of the combination of actin and myosin is seen in the bridges between the two kinds of filaments.

The theory has since been much enlarged and taken to deeper levels of molecular understanding. Despite this, the basic insight of Huxley and Hanson has remained intact.

Huxley, Sir Julian Sorell (1887–1975)
British biologist

> Operationally, God is beginning to resemble not a ruler but the last fading smile of a cosmic Cheshire cat.
> —*Religion without Revelation* (1927)

A grandson of T. H. Huxley, Julian Huxley was born in London and graduated in zoology from Oxford University in 1909. He did research on sponges (Porifera) at the Naples Zoological Station (1909–10) before taking up the post of lecturer in biology at Oxford (1910–12). From 1912 until 1916 he worked at the Rice Institute, Houston, Texas, where he met the famous American geneticist Hermann MULLER. Before returning to Oxford to take up the post of senior demonstrator in zoology (1919–25) he saw war service in Italy. He was next appointed professor of zoology at King's College, London (1925–27), resigning from this post to devote more time to writing and research.

Huxley was a keen ornithologist and published, in 1914, a classic paper on the courtship of the great crested grebe. In the 1930s he was involved in the production of natural-history films, the most notable of which was the highly praised *Private Life of the Gannet* (1934), which he produced with the help of R. M. Lockley. One of the leading popularizers of science of modern times (especially the years before and just after World War II), Huxley spent much of his life explaining advances in natural science to the layman and in advocating the application of science to the benefit of mankind. To many he is best remembered as a most capable and lucid educationalist, but Huxley was also eminent in many other fields.

In 1946 he was appointed the first director-general of UNESCO, a post he held for two years. As an administrator, he also did much to transform the Zoological Society's collections at Regent's Park (London Zoo). Viewing man as "the sole agent of further evolutionary advance on this planet," he caused considerable controversy by advocating the deliberate physical and mental improvement of the human race through eugenics. Huxley's biological research was also extensive, carrying out work on animal hormones, physiology, ecology, and animal (especially bird) behavior as it relates to evolution. He was president of the Institute of Animal Behaviour and the originator of the term "ethology," now in general use to define the science of animal behavior. He also introduced several other scientific terms, such as cline and clade.

Huxley's publications are extensive and include *Evolution: the Modern Synthesis* (1942, 1963). He was knighted in 1958.

Huxley, Thomas Henry (1825–1895) *British biologist*

> The great tragedy of science – the slaying of a beautiful hypothesis by an ugly fact.
> —"Biogenesis and Abiogenesis," *Collected Essays*, Vol. VIII

Huxley, the seventh child of a school teacher, was born in Ealing in southeast England and received only two years' schooling. From the age of 10 he educated himself, doing sufficiently well to be admitted to Charing Cross Hospital to study medicine. He graduated in 1845 and the following year was employed as surgeon on HMS *Rattlesnake*, which was due to survey the Torres Strait between Australia and Papua. During the voyage Huxley studied the marine life of tropical waters and wrote an important paper on the medusae (jellyfish) and related species, naming a new phylum, the Coelenterata, into which these were placed. Recognizing the value of this work, the Royal Society elected Huxley a member in 1851. In 1854 he became lecturer in natural history at the Royal School of Mines (later the Royal College of Science) and while there gave a lecture on "The Theory of the Vertebrate Skull," which disproved the idea that the skull originates from the vertebrae.

Huxley is best remembered as the main advocate of Charles DARWIN's theory of evolution, and in 1860 – the year following the publication of *The Origin of Species* – he took part in the famous debate with the bishop of Oxford, Samuel Wilberforce, at the Oxford meeting of the British Association for the Advancement of Science. During the discussion Wilberforce asked whether Huxley traced his ancestry to the apes on his mother's or father's side of the family. Huxley answered witheringly that given the choice of a miserable ape and a man who could make such a remark at a serious scientific gathering, he would select the ape. The meeting resulted in a triumph for science, and after it Huxley continued to gain the better of many other distinguished theologians in long academic wrangles. He introduced the term "agnosticism" to describe his own view that since knowledge rested on scientific evidence and reasoning (and not blind faith) knowledge of the nature and certainty about the very existence of God was impossible.

Huxley worked hard to better educational standards for the working classes and spoke out against the traditional method of learning by rote. He opened Josiah Mason College (later Birmingham University), Owens College Med-

ical School (later part of Manchester University), and Johns Hopkins University, Baltimore. Huxley was the grandfather of the author Aldous Huxley, the Nobel Prize winner Andrew Huxley, and the biologist Sir Julian Huxley.

Huygens, Christiaan (1629–1695) *Dutch physicist and astronomer*

> A man that is of Copernicus's opinion, that this earth of ours is a Planet, carry'd round and enlighten'd by the Sun, like the rest of them, cannot but sometimes have a fancy ... that the rest of the Planets have their Dress and Furniture, nay and their Inhabitants too ...
> —*New Conjectures Concerning the Planetary Worlds, Their Inhabitants, and Productions* (c. 1690)

Huygens (**hI**-genz or **hoi**-gens), whose father was the famous Renaissance poet Constantin Huygens, was born in The Hague and studied at the University of Leiden and the College of Breda. He worked in Paris as one of the founding members of the French Academy of Sciences from 1666 to 1681 when, as a Protestant, he found the growing religious intolerance threatening, and returned to The Hague. His first work was in mathematics, but his greatest achievements were in physical optics and dynamics, and his importance to 17th-century science is second only to that of NEWTON.

Huygens's first great success was the invention of the pendulum clock. GALILEO had noted in 1581 that a pendulum would keep the same time whatever its amplitude. Many, including Galileo himself, had tried unsuccessfully to use this insight to construct a more reliable clock. Huygens showed that a pendulum that moves in the arc of a circle does not move with an exactly equal swing. To produce an isochronous (equal-timed) swing it would need to move in a curve called a cycloid. It should be emphasized that Huygens worked this out largely from first principles. He also showed how the pendulum could be constructed so to move in a cycloidal path and how to make the connection to the escapement. The first clock was made to his design by Salomon COSTER in 1657 and was described in Huygens's book *Horologium* (1658; The Clock). The pendulum became one of the basic tools of 17th-century scientific investigation.

Huygens also made major contributions to astronomy as a designer of improved telescopes and as an observer of Saturn. He discovered Titan, Saturn's largest satellite, in 1655 and after prolonged observation was able to describe Saturn's rings correctly.

In 1673 Huygens published *Horologium oscillatorium* (The Clock Pendulum), a brilliant mathematical analysis of dynamics, including discussions of the relationship between the length of a pendulum and its period of oscillation, and the laws of centrifugal force. It also included an early formulation of Newton's first law of motion: that without some external force, such as gravity, a body once set in motion would continue in a straight line. His views on gravity were worked out in *Discours de la cause de la pesanteur* (1690; Discourse on the Cause of Gravity). As a Cartesian (a follower of René Descartes) he could not accept Newtonian action at a distance or, in fact, any talk of forces. Instead he would only accept a mechanical explanation, which meant a return to some kind of vortex theory. That is, bodies can only be heavy not because they are attracted by another body but because they are pushed by other bodies.

Huygens's greatest achievement was his development of the wave theory of light, described fully in his *Traité de la lumière* (1690; Treatise on Light). He assumed that space was pervaded by ether formed of particles, the disturbance of which constituted the radiation of light with the disturbance of one particle being passed on to its neighbor and so on. The disturbances can be considered as waves spreading in a regular spherical form from the point of origin – the particles disturbed in phase constituting a wave front. Each point on a wave front may be regarded as a source of new secondary wavelets and a surface tangent joining such wavelets (i.e., the envelope of the secondary wavelets) can be considered as a new wave front. This method of treating light waves is known as the *Huygens construction*. Using it, Huygens dealt with reflection and refraction and predicted – as Newtonian theory did not – that light should travel more slowly in a denser medium. But as Huygens considered the waves to be longitudinal, the theory could not explain polarization.

Newton's *Opticks* (1704) presented a corpuscular (particle) theory of light, and the wave theory lay dormant until it was taken up by Thomas YOUNG and his contemporaries.

Hyman, Libbie Henrietta (1888–1969)
American zoologist

Born in Des Moines, Iowa, and educated at the University of Chicago, Hyman also held a research post there from 1916 until 1931 where, under Charles Manning Child, she worked on the physiology and morphology of the lower invertebrates, particularly the planarians (flatworms). She published a number of works on invertebrate and vertebrate zoology, anatomy, physiology, and embryology, but her major compilation is *The Invertebrates*, a monumental work, six volumes of which had been published (1940–68) at the time of her death.

Hypatia (c. 370–415) *Greek mathematician*

> He who influences the thought of his times, influences all the times that follow. He has made his impress on eternity.
> —Quoted by Elbert Hubbard in *Little Journeys to the Homes of Great Teachers*

Hypatia (hI-**pay**-sha), who was born in Alexandria, Egypt, was the daughter of Theon of Alexandria, the author of a well-known commentary on PTOLEMY. In 400 she was reported to be head of the Neoplatonic school in Alexandria. None of her work has survived, although some information about her comes from the letters of her pupil Synesius of Cyrene. To her have been attributed commentaries on Ptolemy's *Almagest*, DIOPHANTUS's *Arithmetic*, and APOLLONIUS's *Conics*. She also designed several scientific instruments including an astrolabe, a hydrometer, and a still.

Learning and science came to a violent conclusion in Alexandria and in the West, as did Hypatia. In conflict with Cyril, bishop of Alexandria, through her friendship with Orestes, the Roman prefect of the city, she was killed by a Christian mob. The circumstances of her death in March 415 have been described by the fifth-century historian Socrates Scholasticus:

> All men did both reverence and had her in admiration for the singular modesty of her mind. Wherefore she had great spite and envy owed unto her, and because she conferred oft, and had great familiarity with Orestes, the people charged her that she was the cause why the bishop and Orestes were not become friends. To be short, certain heady and rash cockbrains whose guide and captain was Peter, a reader of that Church, watched this woman coming home from some place or other, they pull her out of her chariot: they hail her into the Church called Caesarium: they strip her stark naked: they raze the skin and rend the flesh of her body with sharp shells, until the breath departed out of her body: they quarter her body: they bring her quarters unto a place called Cinaron and burn them to ashes.

The manner of her death and reports of her intellect and beauty have made her a romantic figure. For many centuries, with the possible exception of the alchemist Maria HEBRAEA, she was regarded as the only woman scientist of the Ancient World.

I

Ignarro, Louis J. (1941–   ) *American pharmacologist*

Ignarro, who was born in Brooklyn, New York, graduated with a BA in pharmacy from Columbia University in 1962 and four years later gained his PhD in pharmacology at the University of Minnesota. From 1979 to 1985 he worked in the department of pharmacology at Tulane University School of Medicine, New Orleans. Since 1985 he has been at the UCLA School of Medicine, Los Angeles, California. In 1998 he shared the Nobel Prize for physiology or medicine with Robert FURCHGOTT and Ferid MURAD for their discovery that molecules of the gas nitrogen monoxide (nitric oxide, NO) can transmit signals in the cardiovascular system. The hitherto unknown substance endothelium-derived relaxing factor (EDRF) had been discovered by Furchgott in 1980. Ignarro made a series of brilliant analyses to determine the chemical nature of EDRF and in 1986 concluded, independently of and together with Furchgott, that the mystery substance was indeed nitrogen monoxide.

I-Hsing (*c.* 681–*c.* 727) *Chinese mathematician and astronomer*

I-Hsing (I-*shing*) was a Buddhist monk around whom many legends have grown. Only a small portion of his work has survived so it is difficult to appreciate it in detail. There is, however, no reason to doubt his involvement in two major astronomical achievements. In the period 723–26, in collaboration with the Astronomer Royal, Nankung Yueh, expeditions were organized to measure, astronomically, the length of a meridional line. Over a distance of 1,553 miles (2,500 km) along this line, simultaneous measurements of the Sun's solstitial shadow were made at nine stations. The estimated length of a degree, on the basis of their measurements, was far too large and it must be supposed that some systematic error in the method of observation was taking place. However, when it is appreciated that research expeditions to determine the length of a meridional degree were not organized in Europe until the 17th century, the amazing nature of I-Hsing's work can be appreciated. He also probably anticipated SU SUNG in the use of an escapement in an astronomical clock. It was described in a 13th-century encyclopedia: "Water, flowing into scoops, turned a wheel automatically, rotating it one complete revolution in one day and one night." This turned various rings representing the motion of the celestial bodies. It was soon reported to be corroded, relegated to a museum, and to have fallen into disuse.

Imbrie, John (1925–   ) *American geologist*

Imbrie was born at Penn Yan, New York and educated at Princeton and Yale, where he obtained his PhD in 1951. He joined the Columbia faculty in 1952 and, after serving as professor of geology (1961–66), he moved to Brown University where he was professor of geology until 1975. He was then appointed to the chair of oceanography and is currently an emeritus professor.

In his 1956 paper *Biometrical Methods in the Study of Invertebrate Fossils* Imbrie showed how statistical techniques could be applied to the analysis of variation in fossil assemblies. He has also worked on the paleoecology of the Great Bahamas Bank and in his *Ice Ages* (1979) produced a comprehensive and popular survey of the subject.

Ingenhousz, Jan (1730–1799) *Dutch plant physiologist and physician*

Ingenhousz (**ing**-en-hows), who was born at Breda in the Netherlands, studied medicine, chemistry, and physics at the universities of Louvain and Leiden, receiving his MD from Louvain in 1752. In 1765 he visited London and became expert at administering smallpox inoculations using Edward JENNER's method. News of his expertise spread and he was invited to Vienna in 1768 by the Empress Maria Theresa to inoculate her family and to become court physician.

In 1779 Ingenhousz returned to England and published his work on gaseous exchange in plants. His experiments demonstrated that plants absorb carbon dioxide and give off oxygen (in his words, "purify the air") only in the light, and that the reverse process occurs in the dark. The light process later became known as photosynthesis. Ingenhousz also conducted research on soils and on plant nutrition, improved apparatus for generating static electricity, and studied heat conduction in metals.

Ingold, Sir Christopher Kelk (1893–1970) *British chemist*

> The calculations must have been dreadful … but one structure like this brings more certainty into organic chemistry than generations of activity by us professionals.
> —On the structures of hexamethyl benzene and hexachlor benzene as revealed by x-ray crystallography. *Nature*, 1928

Ingold, a Londoner, was educated at the University of Southampton and at Imperial College, London. After serving as professor of organic chemistry at the University of Leeds from 1924 until 1930, he moved to the chair of chemistry at University College, London, where he remained until his retirement in 1961.

With over 400 papers to his credit and as the author of the classic text *Structure and Mechanism in Organic Chemistry* (1953), Ingold was one of the leading figures in British chemistry. The basic aim running through all his work was to understand the mechanism of organic reactions, particularly the kinetics of elimination and substitution reactions. In 1926 he introduced the idea of mesomerism, fully explained in his paper *Principles of an Electronic Theory of Organic Reactions* (1934). This was similar to the concept of resonance proposed by Linus PAULING in the early 1930s. The basic idea was that if a molecule could exist in two electronic structures then its normal state was neither one nor the other but some "hybrid" form. This theory was substantiated by measuring bond lengths in appropriate molecules.

Ingram, Vernon Martin (1924–2006) *German–British–American biochemist*

Ingram, born Immerwahr (**im**-er-var) in Breslau (now Wrocław in Poland), was brought to Britain as a child refugee from Nazi Germany. He was educated at Birkbeck College, London, where he obtained his PhD in 1949. After working briefly at Rockefeller and Yale he returned to England and joined the staff of the Medical Research Council's molecular biology unit at the Cavendish Laboratory, Cambridge, in 1952. In 1958, however, he moved to the Massachusetts Institute of Technology where he served as professor of biochemistry from 1961 and as John and Dorothy Wilson Professor of Biology from 1988.

By the mid-1950s it was clear to Francis CRICK that it should be possible, and was indeed essential, for molecular biology to be able to show that mutant genes produced changes in the amino acid sequences of proteins. Although such a claim was central to the supposed revolution in molecular biology, there was, as Crick realized in 1955, no direct evidence that proteins are in fact coded by genes.

Consequently Crick and Ingram attempted to reveal such a change in the lysozyme of fowl eggs. However, although they succeeded in distinguishing differences between lysozymes from such different birds as duck and pheasant, they failed to find any difference in lysozymes between two hens of the same species. At this point, however, Max PERUTZ gave Ingram some sickle-cell hemoglobin (hemoglobin S) to work with. (Hemoglobin S, possessed by sufferers of a crippling anemia, had been distinguished from normal hemoglobin A by Linus PAULING and his student Harvey Itano in 1949.) Ingram split the hemoglobin into smaller units by using the enzyme trypsin to break the peptide bonds. He then separated these units by electrophoresis and paper chromatography. This allowed him to show that hemoglobin S differs from normal hemoglobin at just one site, where the amino acid valine replaces the glutamic acid of the normal A form. Although it came as a surprise that the alteration of one amino acid in over 500 could produce such major effects, it also dramatically established that molecular biology was not just an abstract and remote branch of structural chemistry.

Ingram went on to show that this and other point mutations of hemoglobin could be used to trace the evolutionary history of vertebrates, work reported in his *The Hemoglobins in Genetics and Evolution* (1963). In later years he focused on the molecular biology of aging and of ALZHEIMER's disease.

Ipatieff, Vladimir Nikolayevich (1867–1952) *Russian–American chemist*

Ipatieff (i-**pa**-tyef), a Muscovite by birth, became an officer in the Imperial Russian Army in 1887 and was educated at the Mikhail Artillery Academy (1889–92) in St. Petersburg. After further study in Germany and France he returned to the academy in 1898 and became professor of chemistry until 1906.

While in Munich (1897) Ipatieff achieved the synthesis of isoprene, the basic unit of the rubber molecule. On his return to Russia he carried out important work on high-pressure catalytic reactions. The first breakthrough in organic catalysis had been due to Paul SABATIER who had demonstrated the use of finely ground nickel to catalyze hydrogenation of unsaturated hydrocarbons (1897). Ipatieff greatly extended this work. He showed how it could be applied to liquids and demonstrated that the process became much more powerful and adaptable at high pressures. To this end he designed the so-called *Ipatieff bomb* – an autoclave that permitted the heating of substances under pressure to above their boiling point. Thus before World War I Ipatieff had synthesized isooctane, and had polymerized ethylene.

During World War I and after the revolutionary years in Russia Ipatieff held a number of important advisory posts, in addition to continuing with his own research, despite his anti-Communist feelings. In 1930, worried for his own safety, he traveled to America. Despite being 64 when he arrived in America Ipatieff still had much to offer, publishing over 150 papers in this last phase of his career. He was appointed professor of chemistry at Northwestern University, Illinois (1931–35) and also acted as a consultant to the Universal Oil Products Company of Chicago which, in 1938, established at Northwestern University the Ipatieff High Pressure Laboratory, which he directed. With the growth of the petrochemical industry after 1918, Ipatieff's techniques became widely used. Working in America he showed how low-octane gasolines could be converted to high-octane gasoline by "cracking" hydrocarbons at high temperatures.

Irvine, Sir James Colquhoun (1877–1952) *British chemist*

Irvine studied chemistry at the Royal Technical College in his native Glasgow and at Leipzig. His father was a manufacturer of light iron castings and appears to have been a capable mathematician. Irvine's whole career was spent at St. Andrews beginning in 1901 as a lecturer, being appointed professor of chemistry in 1909, and finally, in 1921, becoming vice-chancellor of the university. Under Irvine the tradition of carbohydrate studies established by Thomas Purdie was to continue. Irvine's work involved the application of Purdie's methylation technique to carbohydrates. He realized that the constitution of disaccharides and other compound carbohydrates might be found by methylating them and he isolated the first methylated sugars, trimethyl and tetramethyl glucose.

Irvine's fruitful line of research was to be continued at St. Andrews and later at Birmingham by Norman HAWORTH.

Isaacs, Alick (1921–1967) *British virologist and biologist*

Isaacs, who was born in Glasgow, Scotland, graduated in medicine from the university there in 1944. After three years' work in the department of bacteriology he moved to Sheffield University for a year and then spent two years in Australia at the Hall Institute for Medical Research, Melbourne. During this time he studied influenza, in particular the genetic variation of the various strains of the virus and also the

response of the body to attack by the virus. He continued with this work from 1950 at the National Institute for Medical Research in London, where he was director of the World Influenza Centre.

In 1957, together with the Swiss virologist Jean Lindenmann, Isaacs reported that a specific low-molecular-weight protein, which interfered with the multiplication of viruses, was produced by animal cells when under viral attack. This was interferon, which he studied closely for the rest of his life, investigating problems associated with its production and isolation, its mechanism of action, and its chemical and physical properties. Isaacs's work formed the basis for all present-day research on this potentially important drug, the full effects of which are still being closely studied.

In the early 1960s his health began to deteriorate but he continued work as head of the Laboratory for Research on Interferon at the National Institute.

Ivanovsky, Dmitri Iosifovich (1864–1920) *Russian botanist*

Ivanovsky (i-van-**of**-skee) was born in Gdov, Russia, and studied natural sciences at St. Petersburg University, graduating in 1888. He obtained his master's degree in botany in 1895 and worked (1896–1901) as an instructor in plant anatomy and physiology at the Technological Institute, St. Petersburg. In 1908 he was appointed professor at the University of Warsaw.

In 1892, following his investigations of tobacco mosaic disease in the Crimea, he demonstrated that a filtrate of the sap from infected tobacco plants had the ability to transmit the disease to healthy plants. Ivanovsky showed that minute crystalline particles were present in the filtrate and asserted that they were somehow linked to the disease. However, he wrongly attributed the cause of the disease to minute bacteria. Ivanovsky's work was confirmed in a publication by the Dutch bacteriologist Martinus Beijerinck in 1898. It was Beijerinck who stated that such infective agents are not bacterial and coined the term "virus." This, together with the work of the French bacteriologist Charles Chamberland on rabies, was one of the earliest pieces of evidence for the existence of viruses although it was not until 1935 that Wendell Stanley confirmed this.

J

Jackson, Charles Thomas (1805–1880)
American chemist

No true man of science will ever disgrace himself by asking for a patent; and if he should, might not know what to do with it any more than the man did who drew an elephant at a raffle. He cannot and will not leave his scientific pursuits to turn showman, mechanic, or merchant.
—Address to the American Institute (1851)

Born in Plymouth, Massachusetts, Jackson studied medicine at Harvard and continued his education at the Sorbonne in Paris, working on chemistry and geology. He returned to America and set up a practice in Boston. Jackson's professional career consisted of a series of spectacular claims to the work of others. These started on his homeward voyage and were to persist until he finally became insane in 1873.

While sailing from France to America in 1832 Jackson befriended a fellow American, the portrait painter Samuel MORSE, with whom he discussed the possibilities of electric telegraphy. When Morse exhibited his telegraph to Congress in 1837 he found that he had to establish a right to his own invention against Jackson's claim that Morse had stolen it from him. It took Morse seven years to prove the validity of his claim.

In July 1844 Jackson recommended to William MORTON, a young dentist lodging with him, that he should try treating his patients using ether, which was commonly used by medical students as a joke. Morton took up his suggestion and found it promising. He experimented on himself, gave up his practice to work on dosages and systems of inhalation, and introduced the anesthetic to the medical profession. Nothing was heard from Jackson until it was clear that money and fame were going to be awarded to someone. When Morton went to Congress to ask for compensation for yielding his patent to the U.S. government he found some senators who took him for a thief. When he went to Paris in 1847 to lecture on his discovery he found that Jackson had already lodged a sealed envelope with the Académie claiming a priority going back to 1842. Committees were set up by governments, states, academies, and professional bodies but Jackson managed to so confuse the issue that when Morton collapsed and died in 1868 he was still fighting his claim and still penniless.

Jackson became obsessive about his "discovery," ignored his other work, took to drink, and spent the last seven years of his life in a lunatic asylum. He even wrote a book on the subject, *A Manual of Etherisation* (1861). Curiously, both Morton and Jackson have monuments in the same cemetery, both proudly proclaiming their triumph in alleviating the misery of mankind.

Jackson, John Hughlings (1835–1911)
British neurologist

Jackson was born at Green Hammerton in England and educated at York Hospital and St. Bartholomew's Hospital, London. He received his MD from St. Andrews University in 1860. He served on the staff of the London Hospital as assistant physician (1863) and physician (1874–94) and in 1862 began his long association with the National Hospital for the Paralysed and Epileptic, London. Here he specialized in neurology and ultimately exercised a profound influence on the development of clinical neurology. Through his work with epileptics, he described the condition, now called *Jacksonian seizure* or *Jacksonian epilepsy*, in which part of the leg, arm, or face undergoes spasmodic contraction due to local disease of the cerebral cortex in the brain.

Jackson's work supported the findings of Paul BROCA and others – that different bodily functions are controlled by different regions of the cerebral cortex. Jackson also described a local paralysis of the tongue and throat caused by disease of the corresponding cranial nerves. This is now known as *Jackson's syndrome*.

Jacob, François (1920–   ) *French biologist*

Myths and science fulfill a similar function: they both provide human beings with a representation of the world and of the forces that are supposed to govern it. They both fix limits of what is considered as possible.
—*The Possible and the Actual* (1981)

Born at Nancy in France, Jacob (zha-**kob**) served with the Free French forces during World War II. Although badly wounded, he resumed his medical studies in 1945, obtaining his MD from the University of Paris in 1947. In

1950 he became André LWOFF's assistant at the Pasteur Institute, Paris, and, with Elie Wollman, began working on the bacteria, discovered by Lwoff, that carry a nonvirulent virus incorporated in their genetic material. In 1961 they introduced the term "episomes" for genetic elements that become established in bacterial cells. Jacob and Wollman also studied conjugation in bacteria, the process by which genetic material is transferred from one cell to another. They found that the genes of the donor cell enter the recipient cell in a specific order and by interrupting the process, the position of given genes on the chromosome could be determined.

In 1958 Jacob began collaborating with Jacques MONOD and Arthur PARDEE on the control of bacterial enzyme production, research that culminated in a greatly increased understanding of the regulation of gene activity. In 1960 Jacob and Monod proposed the existence of the operon, consisting of an operator gene and structural genes that code for the enzymes needed in a given biosynthetic pathway. When the enzymes are not required another gene outside the operon, the regulator gene, produces a protein that binds with the operator and renders the operon ineffective. Jacob and Monod received the 1965 Nobel Prize for physiology or medicine for this research, sharing the award with Lwoff.

From 1964 until 1991, Jacob occupied the chair of cellular genetics at the Collège de France; the chair was created in his honor. He became a foreign member of the Royal Society in 1973 and a member of the Academy of Sciences, Paris, in 1977. He has also written on some of the wider implications of biology in *The Possible and the Actual* (1981) and has published an autobiography, *The Statue Within* (1987).

Jacobi, Karl Gustav Jacob (1804–1851) *German mathematician*

> Mathematics is the science of things that are self-evident.
> —Describing the nature of mathematics

Jacobi (yah-**koh**-bee) was born at Potsdam, in Germany. After studying in Berlin, he became a lecturer at Königsberg where he managed to attract the favorable attention of Karl Friedrich GAUSS. He was a superb teacher and had an astonishing manipulative skill with formulae. He made a brief but disastrous foray into politics that resulted in his losing a pension he had been granted by the king of Prussia.

Jacobi's most important contributions to mathematics were in the field of elliptic functions. Niels Henrik ABEL had partially anticipated some of Jacobi's work, but the two were equally important in the creation of this subject. Jacobi also worked on Abelian functions and discovered the hyperelliptic functions. He applied his work in elliptic functions to number theory.

Jacobi worked in many other areas of mathematics as well as the theory of functions. He was a pioneer in the study of determinants and a certain type of determinant arising in connection with partial differential equations is known as the *Jacobian* in his honor. This work was the result of his interest in dynamics, in which field he continued and developed the work of William Hamilton, and produced results that are important in quantum mechanics.

Jansky, Karl Guthe (1905–1950) *American radio engineer*

Born in Norman, Oklahoma, Jansky was educated at the University of Wisconsin and started his career with the Bell Telephone Laboratories in 1928. He was given the task of investigating factors that could interfere with radio waves used for long-distance communication. He designed a linear directional antenna, which, mounted on wheels from a Model T Ford, could scan the sky. He identified all the sources of interference, such as thunderstorms, except for one weak emission. This he found to be unconnected with the Sun and in 1931 he discovered that the radio interference came from the stars.

Jansky published his findings in the *Proceedings of the Institute of Radio Engineers* in December 1932, the date that marks precisely the beginnings of radio astronomy. In his paper Jansky made two astute comments: he suggested that the radio emission was somehow connected with the Milky Way and that it originated not from the stars but from interstellar ionized gas. He did not pursue his suggestions and it was left to Grote REBER, the amateur astronomer, to keep the subject alive until it developed into a major research field after 1945.

The unit of radio-wave emission strength was named the *jansky* in his honor.

Janssen, Pierre Jules César (1824–1907) *French astronomer*

> The 20th century … will see … the terrestrial atmosphere navigated by apparatuses that will take possession of it to make a daily and systematic study of it, or to establish among nations communications … that will take continents, seas, and oceans in their stride.
> —On the future role of aviation. Address to the International Aeronautical Congress, Paris (1889)

Janssen (zhahn-**sen**) studied mathematics and physics at Paris University before becoming professor of general science at the school of architecture. In 1857 he went to Peru to determine the magnetic equator. He observed the transits of Venus of 1874 and 1882 in Japan and Algeria and went on all the major eclipse expeditions. So eager was he to witness the 1870 eclipse in Algeria that he had to escape from the siege of Paris by balloon. While in India in 1868, observing the solar eclipse spectroscopically, he noticed the hydrogen lines visible in the solar prominences and wondered if they could still be detected after the eclipse. The next day he found them still visible. This meant that while photography and observation would still depend on eclipse work the spectroscope could be used almost anywhere anytime. Janssen made one further important discovery on the same trip; he discovered lines in the solar spectrum that he could not identify. He sent his results to Norman LOCKYER who suggested that they were produced by some element found only on the Sun, which Lockyer called "helium." In 1895 William RAMSAY discovered a substance on Earth that matched exactly with Janssen's spectral lines.

In later life Janssen arranged for an observatory to be built on Mont Blanc in order to avoid as much atmospheric interference as possible. Using data from observations made there, he showed that absorption lines in the solar spectrum are caused by elements in the Earth's atmosphere.

Janssen, Zacharias (1580–c. 1638) *Dutch instrument maker*

Zacharias Janssen (**yahn**-sen) was born in Middelburg, now in the Netherlands. Together with his father, Hans, he is believed to have invented the first compound microscope in 1590. He is also credited with having made the first telescope in 1608, although Hans LIPPERSHEY, who also lived in Middelburg, and Jacobus METIUS share claims to this invention.

Jeans, Sir James Hopwood (1877–1946) *British mathematician, physicist, and astronomer*

> Life exists in the universe only because the carbon atom possesses certain exceptional properties.
> —*The Mysterious Universe* (1930)

Jeans, the son of a London journalist, graduated from Cambridge University in 1900 and obtained his MA in 1903. After lecturing at Cambridge (1904–05) he became professor of applied mathematics at Princeton University (1905–09), and, back in England, Stokes Lecturer in applied mathematics at Cambridge (1910–12). There followed a period of writing and research during which his interest turned to astronomy. In 1923 he became a research assistant at the Mount Wilson Observatory, Pasadena, California, where he worked until 1944. Jeans was also professor of astronomy at the Royal Institution in London from 1935 until his death. He was knighted in 1928.

Jeans is best known as an astronomer and as a writer both of popular books on science and of several excellent textbooks. His earlier books were devoted to physics and included *Dynamical Theory of Gases* (1904) and *Mathematical Theory of Electricity and Magnetism* (1908). His serious astronomical works included *Problems of Cosmogony and Stellar Dynamics* (1919) and *Astronomy and Cosmogony* (1928), while among his popular books were *The Universe Around Us* (1929) and *The Mysterious Universe* (1930).

Jeans pioneered various ideas in astronomy and astrophysics. He showed that Pierre Simon LAPLACE's theory of the origin of the solar system, in which the Sun and planets condensed from a contracting cloud of gas and dust, was untenable. In collaboration with Harold JEFFREYS he proposed a new view in its place. According to this "tidal theory" a star had passed close by the newly formed Sun and the planets had been formed from the cigar-shaped filament of material drawn away from this star. Jeans and Jeffreys based their theory on a similar idea proposed earlier by Thomas CHAMBERLIN and Forest Ray MOULTON. The tidal theory was eventually superseded in the 1940s by revamped versions of Laplace's nebular theory.

Jeans also investigated other astronomical phenomena, among them spiral nebulae, binary and multiple star systems, and the source of energy in stars, which he concluded involved radioactivity. In 1902 he published a classic paper on galaxy formation in which he postulated that large-scale structures such as galaxies emerged because fluctuations in the density of matter in the universe caused matter to clump together unevenly because of gravitational collapse. This picture is the basis, when extended to general relativity theory describing an expanding universe, of all subsequent work on the origin of large-scale structures, such as galaxies, in the Universe.

Jeans is also well known for the *Rayleigh–Jeans law* of black-body radiation. He put forward this law in 1905 as a modification of a law found by Lord RAYLEIGH in 1900 to describe black-body radiation using classical statistical mechanics. The limited applicability of the Rayleigh–Jeans law, particularly its breakdown at high frequencies, clearly indicates the necessity of introducing quantum theory.

Jeffreys, Sir Alec John (1950–) *British geneticist*

Jeffreys was born at Luton in Bedfordshire, England, and educated at Oxford, where he completed his PhD in 1975. After spending two years at the University of Amsterdam as a research fellow he joined the genetics department of the University of Leicester. He was appointed professor of genetics in 1987 and knighted in 1994.

Jeffreys is noted as the discoverer of the technique known as "genetic (or DNA) fingerprinting." In 1984 he was working on the gene that codes for the protein, myoglobin. Part of the gene consisted of short sequences repeated a number of times. The number of repeats was found to vary between individuals and became known as VNTRs ("variable number tandem repeats"). Initially Jeffreys saw the VNTRs as no more than useful gene markers of the myoglobin gene. Later he came to the conclusion that they were unique to the individual – they could act like a fingerprint.

The marker sequences can be identified by cleaving the DNA with restriction enzymes and applying a gene probe – a single-strand fragment of DNA or RNA with a base sequence complementary to that of the marker. If the bases are labeled with a radioactive tracer, they can be identified on separation by electrophoresis.

Very small samples of DNA can be used, obtained, for example, from blood, semen, saliva, etc., and the technique has been exploited in forensic science and in investigating paternity and other family relationships.

Jeffreys' team has been investigating the effects of chronic irradiation post-Chernobyl, analyzing human genome instability and recombination processes, and the effects of ionizing radiation on germ-line mutation.

Jeffreys, Sir Harold (1891–1989) *British astronomer and geophysicist*

Jeffreys was born in Birtley in the northeast of England and educated in Newcastle upon Tyne and at Cambridge University. After graduating in 1913 he was made a fellow of his college. He was reader in geophysics from 1931 to 1946 before being elected to the Plumian Professorship of Astronomy and Experimental Philosophy where he remained until his retirement in 1958.

In 1924 Jeffreys produced one of the fundamental works in geophysics of the first half of the 20th century, *The Earth: Its Origin, History, and Physical Constitution*. In this he argued forcibly against Alfred WEGENER's proposed theory of continental drift. He demonstrated that the forces proposed by Wegener were inadequate. This did much to inhibit interest and research into drift theory for a while but much new evidence in its favor has since been uncovered.

Jeffreys was also joint author, with Keith BULLEN, of the *Seismological Tables* (1935). These, more frequently known as the *JB Tables*, were revised in 1940 and are the present standard tables of travel times of earthquake waves. They allow observers to determine from the elapsed time between the arrival of the primary (P) waves and the secondary (S) waves the distance between the observer and the earthquake.

Jeffreys's work in astronomy included studies on the origins of the universe. He developed James JEANS's theory of tidal evolution. He also devised models for the planetary structure of Jupiter, Saturn, Uranus, and Neptune. He was knighted in 1953.

Jemison, Mae Carol (1956–) *American physician and astronaut*

Jemison studied chemical engineering at Stanford and then medicine at Cornell, where she gained her MD in 1981. After serving her internship in Los Angeles, she joined the Peace Corps and spent the period from 1983 until 1985 working as area medical officer for Liberia and Sierra Leone.

Following her return to the United States in 1985 Jemison applied to join NASA as an astronaut. However, following the 1986 *Challenger* disaster, NASA had stopped recruiting at all levels, and it was not until 1987 that Jemison was accepted into the program. She was assigned in 1989 to Mission STS-47 Spacelab J, a joint U.S.–Japanese project during which it was proposed to study space sickness and the effects of weightlessness on the development of several species of animal.

On 12 September 1992 Jemison became the first black woman in space when the *Endeavor* was successfully launched. Accompanying the crew of seven on the *Endeavor*'s eight-day flight were two fish, four frogs, 180 hornets, and 7,600 flies.

Jemison left NASA in 1993 to accept a position on the faculty at Dartmouth College, New Hampshire.

Jenner, Edward (1749–1823) *British physician*

> The scepticism that appeared, even among the most enlightened of medical men when my sentiments on the important subject of the cow-pox were first promulgated, was highly laudable. To have admitted the truth of a doctrine, at once so novel and so unlike anything that ever appeared in the annals of medicine, without the test of the most rigid scrutiny, would have bordered on temerity.
> —*A Continuation of Facts and Observations Relative to the Variolae Vaccinae*

Jenner, born a vicar's son at Berkeley in Gloucestershire, England, was apprenticed to the London surgeon John HUNTER from 1770 to 1772. He then returned to country practice and established a reputation as a field naturalist. In 1787 Jenner observed that the newly hatched cuckoo, rather than the adult cuckoo, was responsible for removing the other eggs from the nest. He was elected a fellow of the Royal Society in 1789 partly on the basis of this work. Jenner's lasting contribution to science, however is his investigations into the disease smallpox.

In 17th-century London some 10% of all deaths were due to smallpox. In response to this the practice of variolation – inoculation with material taken from fresh smallpox sores – was widely adopted. This was first described in England in 1713. However variation suffered from two major defects for, if too virulent a dose was given, a lethal case of smallpox would develop and, secondly, the subject inoculated, unless isolated, was only too likely to start an epidemic among those in contact with him.

Jenner had heard reports that milkmaids once infected with cowpox developed a lifelong immunity to smallpox. On 14 May 1796 he made the crucial experiment and took an eight-year-old boy and injected him with cowpox. He followed this on 1 July with injections taken from smallpox pustules, repeating the procedure several months later. On both occasions the boy did not develop smallpox and the same happy result was later observed with other experimental subjects. Jenner's conclusion that cowpox infection protects people from smallpox infection was first published in *An Inquiry into the Causes and Effects of the Variolae Vaccinae* (1798).

General acceptance of Jenner's work was almost immediate. In 1802 he was awarded £10,000 by a grateful House of Commons and in 1804 he was honored by Napoleon who made vaccination compulsory in the French army. Variolation was made illegal in England in 1840 and in 1853 further legislation made the vaccination of infants compulsory. As a consequence of this deaths from smallpox, running at a rate of 40 per 10,000 at the beginning of the 19th century, fell to 1 in 10,000 by the end.

Jensen, Johannes Hans Daniel (1907–1973) *German physicist*

Jensen (**yen**-zen) was educated at the university in his native city of Hamburg and at Freiburg, where he obtained his doctorate in 1932. He worked at Hamburg and Hannover before his appointment in 1949 as professor of physics at Heidelberg. In 1963 Jensen was awarded the Nobel Prize for physics with Maria GOEPPERT-MAYER for their independent publication of the "shell" theory of the nucleus.

Jerne, Niels Kaj (1911–1994) *Danish immunologist*

London-born Jerne (**yairn**-e) was educated at the University of Copenhagen, where he gained his doctorate in medicine in 1951 while working as a researcher at the Danish State Serum Institute (1943–54). After a period of research at the California Institute of Technology (1954–55), he was appointed chief medical offi-

cer with the World Health Organization in Geneva (1956–62) and also professor of biophysics at the University of Geneva (1960–62). He returned to America in 1962 to become head of the department of microbiology at the University of Pittsburgh. Subsequently he served for three years (1966–69) as director of the Paul Ehrlich Institute, Frankfurt, before leaving to found the Basel Institute for Immunology, where he served as director until 1980.

Jerne is noted for his theories concerning the diversity and production of antibodies. In 1955 he proposed the *clonal selection theory* of antibody formation to account for how the body's white blood cells (lymphocytes) are able, potentially, to manufacture such a huge range of different antibodies. He refuted the idea that antibodies are formed from scratch as and when required. Instead, Jerne proposed that different cells, each capable of producing a particular antibody, are present in the body from birth. When an agent such as a virus or bacteria enters the body, its chemical components (antigens) activate the relevant lymphocytes and cause them to divide repeatedly, thereby producing a clone of cells and enhancing manufacture of the appropriate antibody. The theory has since been shown to be correct.

The immense diversity of antibodies presents the problem of how the genome accommodates all the genetic information. Jerne was one of the first to advance the notion that some form of somatic mutation may be involved, an idea that led to the theory of so-called "jumping genes" and its demonstration in mouse cells by Jerne's colleague, Susumu TONEGAWA.

Jerne also constructed a model of immune-system self-regulation based on the interactions of antibodies. Although a valuable contribution, the model does not anticipate the great complexity of control mechanisms revealed by recent discoveries of numerous chemical modulators of the immune system.

For his work, which helped to inspire a whole generation of immunologists, Jerne received the 1984 Nobel Prize for physiology or medicine. The prize was shared with César MILSTEIN and Georges KÖHLER, another colleague of Jerne's working at Basel.

Jobs, Steven Paul (1955–) *American computer engineer and entrepreneur*

An orphan, Jobs was adopted by a machinist working for a company manufacturing scanners. In 1960 the family moved to Palo Alto, where Jobs showed little interest in school but an early attraction to electronics. After graduating from high school in 1972 and a single term at the liberal arts college Reed, in Portland, Oregon, Jobs began to explore a number of alternative lifestyles before joining the computer-game manufacturer Atari in 1974.

He had already joined the Homebrew Computer Club, a meeting ground for computer en- thusiasts, and had collaborated with Steve Woz- niak to make the so-called "blue boxes" – electronic devices used to make free telephone calls. They began making simple computers and in April 1976 they had the Apple ready to be exhibited at a Homebrew meeting. They sold 150 devices for $666 each and suddenly found investors willing to back them.

Apple II was the first recognizable personal computer by later standards. It had a keyboard, memory, expansion slots, video terminal, and color graphics. When fully assembled it sold at an affordable price of $1,350. A disk drive was added in 1978, and in 1979 the first spread- sheet, VisCalc, could be installed for an extra $100. Sales soared with 2 million models sold by 1984. Apple went public in 1980, and Jobs became a multimillionaire.

The problem was, however, to sustain Apple's success against competition, firstly from IBM, which entered the personal-computer market in 1982, and later from the clones that quickly followed. Apple III could sell no more than 90,000 models. Jobs himself took over the pro- duction of the next model, the Apple Macin- tosh (named after a variety of eating apple), which was notable for its user-friendly graphi- cal user interface (GUI) employing a mouse and icons.

Jobs resigned from the company in 1985 to pursue other interests. He set up a new com- pany, NeXT, financed by the sale of $100 million of his own Apple stock, to develop educational computers. He also acquired Pixar, a computer graphics firm.

Johannsen, Wilhelm Ludwig (1857–1927) *Danish botanist and geneticist*

> It is a pure waste of time to lose oneself in such an author's ... views; they are just not worth a bean.
> —On the vitalist theories of the French philosopher Henri Bergson. *Falske Analogier* (1914; False Analogies)

Johannsen (yoh-**han**-sen) was born in the Dan- ish capital Copenhagen. On leaving school in 1872 he became apprenticed to a pharmacist as his father could not afford university fees. From his work in Danish and German pharmacies, Johannsen taught himself chemistry and de- veloped an interest in botany. In 1881 he began work under Johan KJELDAHL in the chemistry department of the Carlsberg laboratories, in- vestigating dormancy in seeds, tubers, and buds.

In 1892 Johannsen became lecturer at the Copenhagen Agricultural College. On reading Francis GALTON's *Theory of Heredity* he was im- pressed by experiments demonstrating that se- lection is ineffective if applied to the progeny of self-fertilizing plants. Johannsen repeated this work using the Princess bean, but found that

selection did work on the offspring of a mixed population of self-fertilizing beans. It was only when plants were derived from a single parent that selection had no effect. He called the de- scendants of a single parent a "pure line" and argued that individuals in a pure line are ge- netically identical: any variation among them is due to environmental effects, which are not her- itable. In 1905 he coined the terms "genotype" to describe the genetic constitution of an indi- vidual and "phenotype" to describe the visible result of the interaction between genotype and environment.

Johannsen explained his ideas in *On Hered- ity and Variation* (1896), which he revised and lengthened with the rediscovery of Gregor MENDEL's laws and reissued as *Elements of Heredity* in 1905. The enlarged German edi- tion of this work became available in 1909 and proved the most influential book on genetics in Europe. In the same year Johannsen pro- posed the term "genes" to describe Mendel's factors of inheritance. Johannsen's research, with its emphasis on the quantitative varia- tion of characters in populations and the ap- plication of statistical methods, played a major role in the development of modern genetics from 19th-century ideas.

In 1905 Johannsen became professor of plant physiology at Copenhagen University and was made rector of the University in 1917. He spent his later years writing on the history of sci- ence.

Johanson, Donald Carl (1943–) *Amer- ican paleoanthropologist*

> It's clear from these fossils [found at Hadar] that upright walking happened long before brain expansion. Hominid brains don't show any striking signs of getting particularly big until two to two-and-a-half million years ago, and yet these creatures were bipedal at least a million years before that.
> —Quoted by Richard Leakey in *The Making of Mankind* (1981)

Johanson, who was born in Chicago, gained his BA in anthropology from the University of Illi- nois in 1966; he received an MA from the Uni- versity of Chicago in 1970 and a PhD in 1972. Two years later he was appointed professor of anthropology at Case Western Reserve Uni- versity, Cleveland, and was also made associate curator of anthropology at the Cleveland Mu- seum of Natural History. He was later given the position of curator of physical anthropology and director of scientific research at the Museum. In 1981 he moved to California as director of the newly founded Institute of Human Origins at Berkeley, and later became professor of an- thropology at Stanford University (1983–89).

In 1973 Johanson led his first expedition to Hadar about 100 miles northeast of Addis Ababa. Here he found a hominid knee joint. The following year he discovered further remains of a new species of fossil primate that challenged the existing theories of the evolution of modern man (*Homo sapiens*) and other hominids. The remains were reconstructed to form, remarkably, a 40% complete skeleton, revealing a female hominid about three and a half feet tall with a bipedal stance and a relatively small brain. The fossil proved to be some 3 million years old, making it the oldest known fossil member of the human tribe. Johanson named it *Australopithecus afarensis*, after the Afar triangle of northeast Ethiopia where the find was made. The skeleton is popularly called "Lucy," prompted by the Beatles' song "Lucy in the sky with diamonds," which was playing in the camp site of Johanson's team on the evening following their momentous discovery.

During the 1975 season Johanson's team made another dramatic find. Scattered in a single hillside were more than 350 fossil pieces from a group of thirteen men, women, and children, all dating from the same time as Lucy. The "first family," as it was later called, was Johanson's last major find at Hadar. Following the 1976 expedition a series of military coups, civil wars, and famines closed Ethiopia to scientific expeditions.

Johanson's analysis of the "Lucy" skeleton showed it to belong to an upright chimpanzee-like creature with an apelike face, a slightly bow-legged gait, and curved toe and finger bones. According to Johanson, Lucy demonstrated that bipedalism preceded enlarged brain capacity, rather than vice versa, and marked a crucial step toward the evolution of all other antecedents of modern man, as well as the later australopithecines identified by Raymond DART.

The findings of Johanson's team were published in 1979, and sparked controversy among other workers in the field, notably Richard Leakey. He maintained that the genus *Homo* could be traced back to an age comparable with the Lucy skeleton, and was descended not from an australopithecine ancestor, such as Lucy, but from some earlier, hypothetical, hominid, perhaps some 4–5 million years old.

Although the precise relationship of *A. afarensis* to the early human ancestors remains in doubt, the significance of Johanson's discovery is unquestioned. His account of the discovery of Lucy was published as *Lucy: The Beginnings of Humankind* (with Maitland A. Edey; 1981). Johanson and Edey have also written *Blueprints: Solving the Mystery of Evolution* (1989).

Joliot-Curie, Frédéric (1900–1958)
French physicist

I have always attached great importance to the manner in which an experiment is set up and conducted ... the experiment should be set up to open as many windows as possible on the unforeseen.
—Describing research in physics

Frédéric Joliot (zho-**lyoh**), the son of a prosperous Paris tradesman, was educated at the School of Industrial Physics and Chemistry. In 1923 he began his research career at the Radium Institute under Marie CURIE, where he obtained his doctorate in 1930. He was appointed to a new chair of nuclear chemistry at the Collège de France in 1937 and, after World War II, in which he played an important part in the French Resistance, was head of the new Commissariat à l'Energie Atomique (1946–50). In 1956 he became head of the Radium Institute.

In 1926 Joliot married the daughter of Marie Curie, Irène, and changed his name to Joliot-Curie. In 1931 they began research that was to win them the Nobel Prize for chemistry in 1935 for their fundamental discovery of artificial radioactivity (1934). His description of the crucial experiment is as follows: "We bombarded aluminum with alpha rays [the heavy nucleus of a helium atom, made of two protons and two neutrons]...then after a certain period of irradiation, we removed the source of alpha rays. We now observed that the sheet of aluminum continued to emit positive electrons over a period of several minutes." What had happened was that the stable aluminum atom had absorbed an alpha-particle and transmuted into an (until then) unknown isotope of silicon, which was radioactive with a half-life of about 3.5 minutes. The significance of this was that it produced the first clear chemical evidence for transmutation and opened the door to a virtually new discipline. Soon large numbers of radioisotopes were created, and they became an indispensable tool in various branches of science. Dramatic confirmation of the Joliot-Curies' discovery was provided when Frédéric realized that the cyclotron at the laboratory of Ernest Lawrence in California would have been producing artificial elements unwittingly. He cabled them to switch off their cyclotron and listen. To their surprise the Geiger counter continued clicking away, registering for the first time the radioactivity of nitrogen–13.

In 1939 Joliot-Curie was quick to see the significance of the discovery of nuclear fission by Otto HAHN. He confirmed Hahn's work and saw the likelihood of a chain reaction. He further realized that the chain reaction could only be produced in the presence of a moderator to slow the neutrons down. A good moderator was the heavy water that was produced on a large scale only in Norway at Telemark. With considerable foresight Joliot-Curie managed to persuade the French government to obtain this entire stock of heavy water, 185 kilograms in all, and to arrange for its shipment to England out of the reach of the advancing German army.

Joliot-Curie, Irène (1897–1956) *French physicist*

> That one must do some work seriously and must be independent and not merely amuse oneself in life – this our mother has told us always, but never that science was the only career worth following.
> —Recalling the advice of her mother, Marie Curie. Quoted in Mary Margaret McBride, *A Long Way from Missouri*

Irène Curie was born in Paris, the daughter of Pierre and Marie Curie, the discoverers of radium. She received little formal schooling, attending instead informal classes where she was taught physics by her mother, mathematics by Paul LANGEVIN, and chemistry by Jean Baptiste PERRIN. She later attended the Sorbonne although she first served as a radiologist at the front during World War I. In 1921 she began work at her mother's Radium Institute with which she maintained her connection for the rest of her life, becoming its director in 1946. She was also, from 1937, a professor at the Sorbonne.

In 1926 Irène Curie married Frédéric Joliot and took the name Joliot-Curie. As in so many other things she followed her mother in being awarded the Nobel Prize for distinguished work done in collaboration with her husband. Thus in 1935 the Joliot-Curies won the chemistry prize for their discovery in 1934 of artificial radioactivity.

Irène later almost anticipated Otto HAHN's discovery of nuclear fission but like many other physicists at that time found it too difficult to accept the simple hypothesis that heavy elements like uranium could split into lighter elements when bombarded with neutrons. Instead she tried to find heavier elements produced by the decay of uranium.

Like her mother, Irène Joliot-Curie produced a further generation of scientists. Her daughter, Hélène, married the son of Marie Curie's old companion, Paul Langevin, and, together with her brother, Paul, became a distinguished physicist.

Joly, John (1857–1933) *Irish geologist and physicist*

Joly was the son of a clergyman from Hollywood, now in the Republic of Ireland. He entered Trinity College, Dublin, in 1876 where he studied literature and engineering. He taught in the engineering school from 1883 and was appointed professor of geology and mineralogy in 1897, a post he held until his death.

Joly's major geological work was in the field of geochronology. He first tried to estimate the age of the Earth by using Edmond HALLEY's method of measuring the degree of salinity of the oceans, and then by examining the radioactive decay in rocks. In 1898 he assigned an age of 80–90 million years to the Earth, later revising this figure to 100 million years. He published *Radioactivity and Geology* in 1909 in which he demonstrated that the rate of radioactive decay has been more or less constant through time.

Joly also carried out important work on radium extraction (1914) and pioneered its use for the treatment of cancer. His inventions in physics included a constant-volume gas thermometer, a photometer, and a differential steam calorimeter for measuring the specific heat capacity of gases at constant volume.

Jones, Sir Ewart Ray Herbert (1911–2002) *British chemist*

Born in Wrexham, North Wales, Jones was educated at the University College of Wales at Bangor and at Manchester University. He taught at Imperial College, London, from 1938 until 1947, when he returned to Manchester as professor of chemistry. In 1955 he moved to a similar chair at Oxford, in which post he remained until his retirement in 1978.

Jones worked mainly on the structure, synthesis, and biogenesis of natural products, particularly the steroids, terpenes, and vitamins.

Jordan, (Marie-Ennemond) Camille (1838–1922) *French mathematician*

> I shall never forget the astonishment with which I read that remarkable work [Jordan's *Cours d'analyse* (A Course of Analysis)], the first inspiration for so many mathematicians of my generation, and learnt for the first time as I read it what mathematics really meant.
> —Godfrey Harold Hardy, *A Mathematician's Apology* (1940)

Born at Lyons in France, Jordan (zhor-**dahn**) studied in Paris at the Ecole Polytechnique, where he trained as an engineer. Later he taught at both the Ecole Polytechnique and the Collège de France until his retirement in 1912. His interests lay chiefly in pure mathematics, although he made contributions to a wide range of mathematical subjects.

Jordan's most important and enduring work was in group theory and analysis. He was especially interested in groups of permutations and grasped the intimate connection of this subject with questions about the solvability of polynomial equations. This basic insight was one of the fundamental achievements of the seminal work of Evariste GALOIS, and Jordan was the first mathematician to draw attention to Galois's work, which had until then been almost entirely ignored. Jordan played a major role in starting the systematic investigation of the areas of research opened up by Galois. He also introduced the idea of an *infinite* group.

Jordan also passed on his interest in group theory to two of his most outstanding pupils, Felix KLEIN and Sophus LIE, both of whom were to develop the subject in novel and important ways.

Jordan, Ernst Pascual (1902–1980) *German theoretical physicist and mathematician*

Jordan (**yor**-dahn) was educated at the Institute of Technology in his native city of Hannover and at the University of Göttingen, where he obtained his doctorate in 1924. He left Göttingen in 1929 for the University of Rostock and after being appointed professor of physics there in 1935, later held chairs of theoretical physics at Berlin from 1944 to 1952 and at Hamburg from 1951 until his retirement in 1970.

Jordan was one of the founders of the modern quantum theory. In 1925 he collaborated with Max BORN and in 1926 with Werner HEISENBERG in the formulation of quantum mechanics. He also did early work on quantum electrodynamics. He developed a new theory of gravitation at the same time as Carl BRANS and Robert DICKE.

Josephson, Brian David (1940–) *British physicist*

Josephson was born in Cardiff and educated at Cambridge University, where he obtained his PhD in 1964. He remained at Cambridge and in 1974 was appointed to a professorship of physics.

His name is associated with the *Josephson effects* described in 1962 while still a graduate student. The work came out of theoretical speculations on electrons in semiconductors involving the exchange of electrons between two superconducting regions separated by a thin insulating layer (a *Josephson junction*). He showed theoretically that a current can flow across the junction in the absence of an applied voltage. Furthermore, a small direct voltage across the junction produces an alternating current with a frequency that is inversely proportional to the voltage. The effects have been verified experimentally, thus supporting the BCS theory of superconductivity of John BARDEEN and his colleagues. They have been used in making accurate physical measurements and in measuring weak magnetic fields. Josephson junctions can also be used as very fast switching devices in computers. For this work Josephson shared the 1973 Nobel Prize for physics with Leo Eskai and Ivar Giaevar.

More recently, Josephson has turned his attention to the study of the mind and is director of the Mind-Matter Unification Project. This looks at topics outside 'conventional' science, such as homeopathy, cold fusion, and parapsychology. In 1991 controversy was caused by Josephson writing a short piece in a booklet to accompany the Royal Mail's issue of a set of stamps to commemorate the centenary of the Nobel Prizes. Here he said, "Quantum theory is now being fruitfully combined with theories of information and computation. These developments may lead to an explanation of processes still not understood within conventional science such as telepathy, an area where Britain is at the forefront of research."

Joule, James Prescott (1818–1889) *British physicist*

> The phenomena of nature, whether mechanical, chemical, or vital, consist almost entirely in continual conversion of attraction through space, living force, and heat into one another. Thus it is that order is maintained in the universe – nothing is destroyed, nothing ever lost, but the entire machinery, complicated as it is, works smoothly and harmoniously.
> —*Manchester Courier*, 12 May 1847

Joule, the son of a brewer from Salford in England, received little formal education, was never appointed to an academic post, and remained a brewer all his life. He began work in a private laboratory that his father built near the brewery.

His first major research was concerned with determining the quantity of heat produced by an electric current and, in 1840, Joule discovered a simple law connecting the current and resistance with the heat generated. For the next few years he carried out a series of experiments in which he investigated the conversion of electrical and mechanical work into heat. In 1849 he read his paper *On the Mechanical Equivalent of Heat* to the Royal Society. Joule's work (unlike that of Julius Mayer) was instantly recognized.

In 1848 Joule published a paper on the kinetic theory of gases, in which he estimated the speed of gas molecules. From 1852 he worked with William Thomson (later Lord KELVIN) on experiments on thermodynamics. Their best known result is the *Joule–Kelvin effect* – the effect in which an expanding gas, under certain conditions, is cooled by the expansion. The SI unit of work and energy, the *joule*, was named in his honor.

Julian, Percy Lavon (1899–1975) *American organic chemist*

Julian's mother and father were a railway clerk and schoolteacher respectively. His grandfather had been born into slavery and had been punished for daring to learn to write by having two right-hand fingers amputated. Julian was educated at DePauw University, Indiana, and at Harvard, where he gained his MA in 1923. Despite his obvious talents, he found it difficult to find either funding or an appropriate position. Consequently he decided in 1929 to move to the University of Vienna to acquire further experience in the synthesis of complex organic molecules and to work for his PhD, which he finally received in 1931. He returned soon after to the United States and was appointed professor of chemistry at Howard University, Washington, DC.

Julian's first major success came in 1934, when he announced that he had worked out the structure of, and synthesized, physostigmine, an alkaloid derived from the Calabar bean and used in the treatment of glaucoma. At this time Julian had a dispute with the Howard administration. As no suitably senior academic posts were open elsewhere to a black chemist, he turned to industry. In 1936 he was appointed chief chemist at the Glidden Company, Chicago.

While in Vienna Julian had been made aware that the soya bean was a useful source for producing many valuable and biologically active molecules. He began to investigate whether sex hormones, such as progesterone and testosterone, could be derived from soya beans. He also worked out ways to synthesize a cheap and active substitute for cortisone using soya as a starting point. In 1954 Julian decided to leave Glidden and establish his own business, Julian Laboratories Inc., which was based in Chicago and Mexico City and specialized in deriving drugs and hormones from soya beans and wild yam. In 1961 he sold his business to Smith, Kline, and French and went on to establish the Julian Research Institute in Franklin Park, Illinois.

Jung, Carl Gustav (1875–1961) *Swiss psychologist and psychiatrist*

> Show me a sane man and I will cure him for you.
> —*Modern Man in Search of a Soul*

> The separation of psychology from the premises of biology is purely artificial, because the human psyche lives in indissoluble union with the body.
> —*Factors Determining Human Behavior*

Born the son of a pastor in Kesswil, Switzerland, Jung (yuung) studied medicine at the universities of Basel (1895–1900) and Zurich, where he obtained his MD in 1902. From 1902 until 1909 he worked under the direction of Eugen Bleuler at the Burghölzi Psychiatric Clinic, Zurich, while at the same time lecturing in psychiatry at the University of Zurich (1905–13). In 1907 Jung met Sigmund FREUD, whose chief collaborator he became. Following the formation of the International Psycho-Analytical Association (1910) he served as its first president from 1911 until his break with Freud in 1912.

Jung continued to practice in Zurich and to develop his own system of analytical psychology. He became professor of psychology at the Federal Institute of Technology in Zurich (1933–41) and was appointed professor of med-

ical psychology at the University of Basel in 1943 but was forced to resign almost immediately for health reasons. He continued however to write, hold regular seminars, and take patients until well over 80.

Like Alfred ADLER, who had broken away from Freudian orthodoxy earlier, Jung minimized the sexual cause of neuroses but, unlike Adler, he continued to emphasize the role of the unconscious. His final break with Freud followed publication of his *Wandlungen und Symbole de Libido* (1912) translated into English in 1916 as *Psychology of the Unconscious*. To the "personal" unconscious of the Freudian he added the "collective unconscious" stocked with a number of "congenital conditions of intuition" or archetypes. In search of such archetypes Jung spent long periods with the Pueblo of Arizona, and visited Kenya, North Africa, and India, and also sought for them in dreams, folklore, and the literature of alchemy.

Jung also emphasized the importance of personality and in his *Psychologische Typen* (1921; Psychological Types) introduced his distinction between introverts and extroverts.

Jussieu, Antoine-Laurent de (1748–1836) *French plant taxonomist*

Jussieu (zhoo-**syu(r)**) was born into a family of eminent botanists from Lyons in France. His uncles Antoine, Bernard, and Joseph de Jussieu all made important contributions to botany and his son, Adrien, subsequently continued the family tradition.

After graduating from the Jardin du Roi in 1770, Jussieu continued to work there, becoming subdemonstrator of botany in 1778. In his first publication in 1773, which reexamined the taxonomy of the Ranunculaceae, he advanced the idea of relative values of characters; the following year he applied this principle to other plant families.

Jussieu is remembered for introducing a natural classification system that distinguishes relationships between plants by considering a large number of characters, unlike the artificial Linnean system, which relies on only a few. In producing the famous *Genera Plantarum* (1789; Genera of Plants) Jussieu had access to a number of collections, including LINNAEUS's herbarium and some of Joseph BANKS's Australian specimens. He was also able to include many tropical angiosperm families thanks to the collection made by Philibert Commesson. From all this material he distinguished 15 classes and 100 families, and the value of his work can be seen in the fact that 76 of his 100 families remain in botanical nomenclature today. Both Georges CUVIER and Augustin de CANDOLLE built on Jussieu's system.

Jussieu was in charge of the hospital of Paris during the French Revolution and was professor of botany at the National Natural History

Museum (formerly the Jardin du Roi) from 1793 to 1826.

Just, Ernest Everett (1883–1941) *American biologist*

> Few investigators subscribe to the naive but seriously meant comparison that the experimenter on an egg seeks to know its development by wrecking it...The days of experimental embryology as a punitive expedition against the egg, let us hope, have passed.
> —*American Naturalist* (1933)

The grandson of a freed slave and the son of an alcoholic, Just was brought up by his mother, a teacher. He was educated at school in Vermont and at Dartmouth College, where he was the only black student. He graduated in biology in 1907 and took up a post at the leading black college, Howard University, Washington, DC.

Just had ambitions to work as a research scientist. In 1909 he was given a summer assistantship at the Marine Biological Laboratory (MBL), at Woods Hole, Massachusetts. Here he was encouraged to pursue his research interests and work for his doctorate, which he finally gained in 1916 from the University of Chicago. He also continued to work each summer at the MBL and published his results.

In an early paper Just demonstrated that in the eggs of marine invertebrates the initial plane of cleavage was determined simply by the point of fertilization, which could occur anywhere on the egg's surface – a result strongly suggesting that the embryo was not preformed. Just also came to realize that the cell surface did more than enclose an all-powerful nucleus; it too was biologically active. His ideas were put forward in his most important work, *The Biology of the Cell Surface* (1938).

Despite Just's skills as an experimental embryologist, which were widely recognized, his numerous applications for full-time research posts were all rejected. Finally realizing that he would never receive proper recognition at home, Just left the United States in 1929 and settled in Europe. For the next decade he worked in Berlin at the Kaiser Wilhelm Institute for Biology, at the Naples Marine Institute, and at the Sorbonne in Paris. After the fall of France in 1940, Just returned to Howard but died soon after from pancreatic cancer.

K

Kahn, Robert (1938–) *American computer scientist*

Kahn (kan) was educated at Princeton University, where he obtained his PhD. For a time he worked at AT&T Bell Laboratories, before becoming a professor of electrical engineering at the Massachusetts Institute of Technology (MIT).

In the early 1970s Kahn was involved in work on the ARPANET, a pioneering project in the networking of computers. His experience at this time persuaded him that an open-architecture network needed to be developed. In 1973 Vinton CERF got involved in this area and the subsequent collaboration led to the key papers of Cerf and Kahn in 1974 on Transmission Control Protocol/Internet Protocol (TCP/IP) – work that led to the introduction of the Internet.

Since that time Kahn has continued to be involved with the development of the Internet. He is president of the Corporation for National Research Initiatives, an organization that conducts research in the development of network-based information technologies. In 1997, Cerf and Kahn were awarded the US Medal of Technology for their role in founding the Internet.

Kaluza, Theodor Franz Eduard (1885–1954) *German mathematical physicist*

> Your idea of determining electromagnetic fields through a five-dimensional manifold has never occurred to me and is, I believe, thoroughly original. This notion of yours pleases me enormously.
> —Albert Einstein, letter to Kaluza, 21 April 1919

The son of a phonetician, Kaluza (kah-**loot**-sa) was born at Ratibor in Germany and educated at the University of Königsberg where he served (1902–29) as a privatdocent (a largely unpaid teaching assistant). On Einstein's recommendation he was appointed in 1929 to a professorship in physics at the University of Kiel. He remained there until 1935, when he moved to a similar appointment at Göttingen University.

In Einstein's theory of general relativity, space and time are joined together into a four-dimensional space–time. In 1921 Kaluza decided to supplement Einstein's model with a fifth spatial dimension. Within this model it proved possible to derive Einstein's four-dimensional gravitational equations as well as the equations for the electromagnetic field. Thus in a world of five dimensions gravity and electromagnetism were not distinct forces.

There were, however, two major defects in Kaluza's theory. Firstly, he could give no indication of the nature of this fifth dimension. Moreover, his theory assumed that bodies behave classically and quantum-mechanical effects were not considered. An attempt to remedy these defects was made in 1926 by Oskar KLEIN. This revised form, known as the *Kaluza–Klein theory*, has proved to be of considerable interest to string theorists such as Edward WITTEN.

Kamen, Martin David (1913–2002) *American biochemist*

The father of Kamen (**kah**-men) arrived in Canada in 1906 as a Russian political exile with a forged passport. He moved to Chicago in 1911 where he worked as a photographer. Kamen himself was born in Toronto, Canada, and was educated at the University of Chicago. After completing his PhD in 1936 on nuclear chemistry, he moved to the Radiation Laboratory at the University of California, Berkeley.

Kamen had a number of early successes. In 1940, in collaboration with Sam Ruben, he discovered the isotope carbon-14. Soon after, Kamen, working with Ruben, used oxygen-18 to show that the oxygen liberated in the photosynthetic process came from water and not from CO_2.

Soon after Kamen became part of the Manhattan project. He worked at the Radiation Laboratory on uranium-235 separation. In July 1944, however, Kamen was dismissed from the project without notice; the only ground given was that of security. Many colleagues stood by Kamen and in 1945 he was invited to join the Medical School at Washington University, St. Louis.

Kamen retained his productivity. Working with Sol SPIEGELMAN in 1947, he proposed a process in which nucleic genes produced partial copies that transferred genetic information to the cytoplasm; an anticipation, perhaps, of messenger RNA. At about the same time he established that certain nitrogen-fixing bacteria evolve hydrogen in the light. In the 1950s Kamen worked on the respiratory pigment, cytochrome c.

At the same time Kamen was under intense pressure from the security services and the media, so intense that at one point he even attempted suicide. His passport was taken away and colleagues felt it wiser not to be seen with him. In 1947 he was called to Washington to testify before the House Committee on Un-American Activities (HUAC). Leaks from HUAC appeared in the *Chicago Times* linking Kamen with a Soviet "atom-spy ring."

Kamen fought back by suing the *Times* for libel and the Federal Government for withholding his passport. The processes were lengthy, expensive, and involved several severe setbacks. Kamen persisted and in 1955 he finally received his passport and won $7,500 compensatory damages from the *Times*. During the trial the HUAC file on Kamen was produced. It contained no more than the charges that Kamen had discussed "atoms" at the Berkeley Faculty Club, and that, with his sister, he had been a member in the 1930s of the American Student Union. It contained no reference to Soviet agents and left Kamen bitterly wondering how such innocuous data could impose upon him such a monstrous burden. Kamen published a vivid account of both aspects of his life in his *Radiant Science, Dark Politics* (1985).

Following the resolution of Kamen's political troubles he moved to Brandeis University, Massachusetts, in 1957 to establish a new postgraduate school of biochemistry. He undertook a similar task in 1961 when he joined Roger REVELLE to build a new campus of the University of California at La Jolla.

Kamerlingh-Onnes, Heike (1853–1926) *Dutch physicist*

> I should like to write *"Door meten tot weten"* ["Through measuring to knowing"] as a motto above each physics laboratory.
> —Inaugural address on appointment to the chair of physics, University of Leiden (1882)

Born at Groningen in the Netherlands, Kamerlingh-Onnes (kah-mer-ling-**o**-nes) was educated at the university there, obtaining his doctorate in 1879. In 1882 he was appointed professor of physics at Leiden, where he remained for the rest of his career. There he started the study of low-temperature physics, at first in order to gather experimental evidence for the atomic theory of matter. However, his interest turned to the problems involved in reaching extremely low temperatures and, in 1908, he became the first to succeed in liquefying helium. Matter at low temperatures – only a few degrees above absolute zero – has such strange properties that a completely new field of cryogenic physics was opened up. The first of these properties to be studied was superconductivity, which Kamerlingh-Onnes discovered in 1911. This phenomenon involves the total loss of resistance by certain metals at low temperatures.

Kamerlingh-Onnes was elected to the Royal Academy of Sciences in Amsterdam for this research and, in 1913, was awarded the Nobel Prize for physics.

Kamin, Leon (1927–   ) *American psychologist*

Kamin, who was born in Taunton, Massachusetts, has served since 1968 as professor of psychology at Princeton.

Psychologists have devoted much effort to the notion of intelligence. They have introduced a measure of intelligence, namely the intelligence quotient (IQ), and have claimed that they have worked out how to determine it accurately and reliably. In the process a vast amount of material has been gathered about the IQs of peoples of all ages and races and how they relate to occupations, wealth, marriages, health, etc. Although much of this material was controversial, and some of it implausible, it was claimed that IQ research met the highest scientific and statistical standards.

In the 1970s Kamin began to challenge many of these claims. In his *The Science and Politics of I.Q.* he surveyed much of the literature on the subject and found it far from rigorous. Not only did he expose the dubious nature of much of the material collected for separated identical twins by Sir Cyril BURT, but he also found flaws in other twin studies. Repeatedly he claimed to find evidence slanted toward a hereditarian position. When Kamin came to examine the early history of IQ testing he claimed that the evidence shows how it was linked with bogus racist theories and a confused eugenics lobby. This led him to conclude that the IQ test "has served as an instrument of oppression against the poor dressed in the trappings of science, rather than politics."

Kammerer, Paul (1880–1926) *Austrian zoologist*

Kammerer (**kam**-e-rer), the son of a prosperous factory owner, was educated at the university in his native city of Vienna, where he obtained his PhD. He then joined the staff of the university's recently opened Institute of Experimental Biology, where he worked until the end of 1922 and soon established a reputation as a skilled experimentalist. Much of his work appeared to support the unorthodox doctrine of the inheritance of acquired characteristics associated with Jean LAMARCK. The most famous of Kammerer's experiments concerned the breeding behavior of *Alytes obstetricans*, the midwife toad. Unlike most other toads this species mates on land; the male consequently lacks the nuptial pads, blackish swellings on the hand, possessed by water-breeding males in the mating season to enable them to grasp the female during copulation.

Kammerer undertook the experiment of inducing several generations of *Alytes* to copulate in water to see what changes resulted. This involved overcoming the difficult task of rearing the eggs in water and ensuring the developing tadpoles were kept free of fungal infection. After almost ten years following this line he noted that in the F_3 generation (the great grandchildren of the original parents) grayish-black swellings, resembling rudimentary nuptial pads, could be seen on the upper, outer, and palmar sides of the first finger.

In 1923 Kammerer visited Britain in the hope of resolving a controversy that had arisen between himself and the leading Cambridge geneticist William BATESON. As virtually all his animals had been destroyed in the war, he brought with him as evidence one preserved specimen and slides of the nuptial pads from the F_5 generation made some ten years earlier. His lectures at Cambridge and to the Linnean Society were successful and none of the eminent biologists who examined Kammerer's specimen noticed anything suspect.

However, when, early in 1926, G. Noble of the American Museum of Natural History came to examine the specimen in Vienna he found no nuptial pads, only blackened areas caused by the injection of ink. Despite the support of the institute's director, Hans Przibram, several possible explanations of the obvious fraud, and a still-open invitation from Moscow to establish an experimental institute there, Kammerer shot himself some six months after Noble's visit.

Kammerer had in fact carried out a whole series of experiments of which the work with *Alytes* was but a part, and for him not the most important part. In 1909 he claimed to have induced inherited color adaptation in salamander, and by cutting the siphons of the sea squirt *Ciona intestinalis*, to have induced hereditary elongations. The few people who attempted to repeat Kammerer's results were unsuccessful although in certain cases Kammerer was able to claim, with some justification, that his protocols had not been scrupulously followed.

Kamp, Peter van de *See* VAN DE KAMP, PETER.

Kandel, Eric R. (1929–) *American neuroscientist*

Born in Vienna, Austria, Kandel received a medical degree from New York University in 1956. He followed this with postdoctoral training in neurophysiology at the National Institutes of Health, and residency in psychiatry (1960–64) at Harvard Medical School. After serving as associate professor in the Department of Physiology and Psychiatry at New York University (1965–74), he joined Columbia University, New York, becoming director of the Center for Neurobiology and Behavior (1974–83).

Kandel used animal models to investigate the molecular events that underlie basic mechanisms of learning and memory. His most significant discoveries were made using the sea slug *Aplysia*, an invertebrate animal with a relatively simple nervous system. Kandel focused on the animal's gill-withdrawal reflex, and how repeated stimulation leads to strengthening of this reflex – a rudimentary form of learning.

Kandel discovered that this learning depends on changes in the junctions (synapses) between nerve cells. Weaker stimuli elicit short-term learning, lasting minutes to hours. This depends on amplification of the synaptic signal through the nerve cell releasing greater amounts of the chemical (neurotransmitter) that relays nerve impulses across the synapse. This increase in neurotransmitter levels is brought about by the phosphorylation of ion channel proteins by protein kinases.

Stronger stimuli result in long-term memory, lasting for weeks. This involves enlargement of the synapse, and hence a longer-lived enhancement of synaptic function. The strong stimuli cause signals to be sent to the nerve cell's nucleus, bringing about changes in the proteins being manufactured by the cell, and consequent changes to the shape of the synapse.

During the 1990s, Kandel demonstrated that similar mechanisms are responsible for long-term learning in mice, and are also applicable to other mammals, including humans. Thus, he has laid the foundation for an understanding of the complexities of learning and memory.

For his contributions to knowledge of signal transduction in the nervous system, Kandel received the 2000 Nobel Prize for physiology or medicine, awarded jointly with Arvid CARLSSON and Paul GREENGARD.

Kane, Sir Robert John (1809–1890) *Irish chemist and educationalist*

Kane, born the son of a manufacturing chemist in Dublin, studied medicine at Trinity College there and became professor of chemistry in 1831. The following year he founded the *Dublin Journal of Medical Science*. He was president of Queen's College, Cork, from 1845 until 1873 and president of the Royal Irish Academy in 1877. In 1873 he was appointed the commissioner of national education and in 1880 he became vice-chancellor of Queen's University, Belfast. He was knighted in 1846.

In his books Kane did much to try and spread the new chemistry and show its relevance to industrial Ireland. After his early work *Elements of Practical Pharmacy*, he published his *Elements of Practical Chemistry* (1841–43). His most famous work, however, was his *Industrial Resources of Ireland* (1844), which caught the attention of Peel and led to his becoming an adviser to the government on the development of industry and education in Ireland and his sitting on the commission in 1846 to investigate the potato blight.

Kane's main work was that of administering and encouraging institutions rather than that of a creative scientist. His attempts to stimulate Irish industry and science were unfortunately held back by the famine and its consequences.

Kant, Immanuel (1724–1804) *German philosopher*

> Two things fill the mind with ever new and increasing wonder and awe, the more often and more seriously reflection concentrates upon them: the starry heaven above me and the moral law within me.
> —*Critique of Practical Reason* (1788)

The son of a saddle maker and the grandson of a Scottish immigrant, Kant (kant or kahnt) was born in Königsberg (now Kaliningrad in Russia) and educated at the university there. Owing to interruptions necessary to fulfill family obligations it was not until 1755 that Kant, who had studied mathematics and physics, received his doctorate. He remained on the university staff as a *Privatdozent* until 1770 when he was appointed to the chair of logic and metaphysics, a post he occupied until his retirement in 1797.

Apart from his influential philosophical works, Kant's first significant scientific publication was his *The Theory of the Heavens* (1755), which contained the first statement of the nebular hypothesis, an account of the origin of solar systems perhaps better known in the later version of Pierre Simon LAPLACE.

A more pervasive influence was exerted by Kant however in his *Metaphysische Angfangsgründe der Naturwissenschaft* (1786; Metaphysical Foundations of Natural Science). Here he squarely faced the problem of action at a distance arising from Newtonian mechanics. How could gravity act over the vast distances of space once the idea that causes act continuously in space had been rejected? He answered that there were two basic forces, attractive or gravitational and repulsive or elastic. While the latter requires physical contact to operate, the former is "possible without a medium," acts immediately at a distance, and "penetrates space without filling it."

Such ideas, together with his rejection of classical atomism in favor of the infinite divisibility of matter, were not just idle philosophical speculations. They were to exercise much influence over Michael FARADAY in his later development of field theory, one of the great ideas of modern science.

Kapitza, Pyotr Leonidovich (1894–1984) *Russian physicist*

> Theory is a good thing but a good experiment lasts forever.
> —*Experiments, Theory, Practice* (1980)

> The year that Rutherford died (1938) there disappeared forever the happy days of free scientific work, which gave us such delight in our youth. Science has lost her freedom. Science has become a productive force. She has become rich but she has become enslaved and part of her is veiled in secrecy. I do not know whether Rutherford would continue to joke and laugh as he used to.
> —*Science Policy News*, No. 2 (1969)

Pyotr (or Peter) Kapitza (kah-**pyi**-tsa or ka-**pit**-sa) was born in Kronstadt, Russia, and educated (1918–21) at the Polytechnic Institute and the Physical and Technical Institute in Petrograd (now St. Petersburg). He lectured at the Polytechnic Institute from 1919 to 1921. From 1921 to 1924 he was involved in magnetic research at the Cavendish Laboratory of Cambridge University, England, under Ernest Rutherford and gained his PhD there in 1923. He was made director of the Royal Society Mond Laboratory at Cambridge in 1930. In 1934 he paid a visit to his homeland but was detained by the Soviet authorities. The next year Kapitza was made director of a newly founded research institute in Moscow – the Institute for Physical Problems – and was able to continue the line of his Cambridge research through the purchase of his original equipment. He worked there until 1946 when, apparently, he fell into disfavor with Stalin for declining to work on nuclear weapons. He was held under house arrest until 1955, when he was able to resume his work at the Institute. Kapitza had shown similar courage earlier in 1938 when he had intervened on behalf of his colleague Lev LANDAU who had been arrested as a supposed German spy. Without Landau, Kapitza insisted, he would be unable to complete work considered to be important by the authorities. Soon after, Landau was released.

Kapitza's most significant work in low-temperature physics was on the viscosity of the form of liquid helium known as He–II. This he (and, independently, J. F. Allen and A. D. Misener) found to exist in a "superfluid" state – escaping from tightly sealed vessels and exhibiting unusual flow behavior. Kapitza found that He–II is in a macroscopic quantum state with perfect atomic order. In a series of experiments, he found also that a novel form of internal convection occurs in this form of helium.

Besides work on the unusual properties of helium, Kapitza also devised a liquefaction technique for the gas, which is the basis of present-day helium liquefiers, and was able to produce large quantities of liquid hydrogen, helium, and oxygen. The availability of liquid helium has led to the production of electric superconductors and enabled much other work at extremely low temperatures to proceed. Kapitza also created very high magnetic fields for his experiments, and his record of 500 kilogauss in 1924 was not surpassed until 1956. Kapitza's low-temperature work was honored after almost forty years by the award of the 1978 Nobel Prize for physics.

From 1955 Kapitza headed the Soviet Committee for Interplanetary Flight and played an important part in the preparations for the first Soviet satellite launchings. In his career Kapitza collected many awards from scientific institutions of both East and West, including the Order of Lenin on six occasions. In 1965 he was finally allowed to travel outside the Soviet Union. He first visited Copenhagen and in 1966 he spent some time in Cambridge, England, with his colleagues of the 1930s, John Cockroft and Paul DIRAC.

Kapoor, Mitchell David (1950–) *American computer scientist*

Kapoor was educated at Yale and MIT. He worked initially at a variety of odd jobs before setting up as a freelance consultant in the software business in 1978. He worked as product manager with VisiCorp from 1980 until 1982, when he began the Lotus Development Corporation in Cambridge, Massachusetts, serving as its president from 1982 until 1986.

The major product of the company was Lotus 1-2-3, comprising spreadsheet, graphics, and database in one software package. This was a major competitor to VisiCalc – the spreadsheet program invented by Daniel BRICKLIN. Lotus 1-2-3 was a highly successful product designed to run on an IBM PC. Its availability increased the sale of IBM machines significantly.

Kapoor went on to develop further products. These included Symphony, a word processing package to add onto 1-2-3, and Jazz, an integrated package for Apple. Neither was particularly successful. By this time Kapoor was beginning to tire of running a big company. Consequently he resigned from Lotus in 1986. He went on in 1990 to set up the Electronic Frontier Foundation to explore more original forms of software.

Kapteyn, Jacobus Cornelius (1851–1922) *Dutch astronomer*

Born at Barneveld in the Netherlands, Kapteyn (kahp-**tIn**) studied at Utrecht University and became professor of astronomy at the University of Groningen in 1878. He was a very careful stellar observer and using David GILL's photographs of the southern hemisphere skies, he published in 1904 a catalog of over 450,000 stars within 19 degrees of the south celestial pole. He repeated William HERSCHEL's count of the stars by sampling various parts of the heavens and supported Herschel's view that the Galaxy was lens-shaped with the Sun near the center; but his estimate of its size was different from Herschel's – 55,000 light-years long and 11,000 light-years thick. He pioneered new methods for investigating the distribution of stars in space.

Kapteyn also discovered the star, now called *Kapteyn's star*, with the second greatest proper motion – 8.73 seconds annual motion compared to the 10″.3 of BARNARD's star. He found this as part of a wider study of the general distribution of the motions of stars in the sky. To his surprise he found, in 1904, that they could be divided into two clear streams: about $^3/_5$ of all stars seem to be heading in one direction and the other $^2/_5$ in the opposite direction. The first stream is directed toward Orion and the second to Scutum, and a line joining them would be parallel to the Milky Way. Kapteyn was unable to explain this phenomenon; it was left to his pupil Jan OORT to point out that this is a straightforward consequence of galactic rotation.

Karle, Isabella Helen (1921–) *American crystallographer*

Karle was born Isabella Lugoski (loo-**gos**-kee) in Detroit, Michigan, the daughter of Polish immigrants; her father was a house painter and her mother a seamstress. She first heard English spoken only when she began school. She was educated at the University of Michigan, where she obtained her PhD in 1943. Here she met Jerome Karle, a physicist who would win the 1985 Nobel Prize for chemistry for work in x-ray crystallography. They married in 1942 and worked together during the war on the Manhattan Project in Chicago. After the war they moved in 1946 to the Naval Research Laboratory in Washington, DC.

With over 200 papers to her credit Karle has made a number of major contributions to the development of x-ray crystallography. In the 1950s, Jerome Karle and Herb HAUPTMAN had developed new and powerful techniques to enable the phase of diffracted x-rays to be calculated directly. Isabella Karle was one of the first to deploy the new method successfully, and thereby to draw the attention of other workers to its potential.

In her first major success in 1969 she established the structure of venom extracted from South American frogs. This was followed in 1975 with the structure of valinomycin, a polypeptide that transports potassium ions across biological membranes. At the time it was the largest molecule to be worked out directly. In 1979 the structure of another peptide, antamanide, was solved. More recently she has determined the structure of the natural opiate, enkephalin.

Karle, Jerome (1918–) *American physicist*

Born in Brooklyn, New York, Karle was educated at the City College there and at the University of Michigan, where he obtained his PhD in 1943. After working on the Manhattan Project in Chicago, Karle moved in 1946 to the Naval Research Laboratory, Washington, DC, becoming chief scientist in the lab for the structure of matter in 1968.

While in Washington Karle began an important collaboration with Herb HAUPTMAN exploring new ways to determine the structure of crystals using x-ray diffraction techniques. Before their work the structure of anything but the simplest molecule was usually worked out with the so-called "heavy-atom" technique. This involved substituting a heavy atom, such as mercury, in a definite position in the structure. The changes produced in the intensities of the diffraction patterns allowed the phases to be inferred. The method, however, is limited and time consuming.

In 1953 Karle and Hauptman published a monograph, *The Phases and Magnitudes of the Structure Factors*, in which they demonstrated how phase structures could be inferred directly from diffraction patterns. For their work in this field Hauptman and Karle shared the 1985 Nobel Prize for chemistry.

In 1942 Karle had married Isabella Lugoski, also a crystallographer. She was one of the earliest workers to apply the new direct method to a number of important molecules.

Karrer, Paul (1889–1971) *Swiss chemist*

Karrer (**kar**-er), the son of a Moscow dentist, was educated at the University of Zurich where he obtained his PhD. After working in Frankfurt he returned to the University of Zurich in 1918, where he served as professor of chemistry until his retirement in 1959.

He began his research career working on the chemistry of plant pigments. Although Karrer tackled a wide variety of such pigments his most significant result was his determination, by 1930, of the structure of carotene, the yellow pigment found in such vegetables as carrots. By 1931 he had also worked out the structure of vitamin A and synthesized it. The similarity between the two molecules did not escape Karrer's attention and it was later shown that vitamin A is derived from the breakdown of carotene in the liver. Karrer went on to synthesize vitamin B_2 (riboflavin) in 1935 and vitamin E (tocopherol) in 1938.

In 1937 Karrer was awarded, with Norman HAWORTH, the Nobel Prize for chemistry for his work on the "constitution of carotenoids, flavins, and vitamins A and B." Karrer was also the author of a much respected textbook, *Lehrbuch der organischen Chemie* (1927; Textbook of Organic Chemistry).

Kastler, Alfred (1902–1984) *French physicist*

Kastler (kast-**lair**), who was born in Gebweiler (now Guebwiller in France) was educated at the Ecole Normale Supérieure. He then taught at the University of Bordeaux where he became professor of physics in 1938. He moved to the University of Paris in 1941 where he remained until his retirement in 1972.

He worked on double-resonance techniques of spectroscopy, using absorption by both optical and radiofrequency radiation to study energy levels in atoms. He also introduced the technique known as "optical pumping" – a method of exciting atoms to a different energy state. In practical terms Kastler's work led to new frequency standards and new methods for the measurement of weak magnetic fields. Kastler received the 1966 Nobel Prize for physics for his work on double resonance.

Katz, Sir Bernard (1911–2003) *German–British neurophysiologist*

Born at Leipzig in Germany, Katz (kats) received his MD from the university there in 1934 and his PhD, under Archibald Hill, from the University of London in 1938. He spent the war in Australia first working with John ECCLES and later in the Royal Australian Air Force as a radar operator. Katz returned to London in 1946 to University College and in 1952 became professor of biophysics, a post he retained until he retired in 1978.

In 1936 Henry DALE demonstrated that peripheral nerves act by releasing the chemical acetylcholine in response to a nerve impulse. To find how this secretion takes place Katz, working in collaboration with the British biophysicist Paul Fatt, inserted a micropipette at a neuromuscular junction to record the "end-plate potential" or EPP. He noted a random deflection on the oscilloscope with an amplitude of about 0.5 millivolt even in the absence of all stimulation. At first he assumed such a reading to be interference arising from the machine but the application of curare, an acetylcholine antagonist, by abolishing the apparently random EPPs, showed the activity in the nerves to be real.

Consequently Katz proposed his quantum hypothesis. He suggested that nerve endings secrete small amounts of acetylcholine in a random manner in specific amounts (or quanta). When a nerve is stimulated it enormously increases the number of quanta of acetylcholine released. Katz was able to produce a good deal of evidence for this hypothesis, which he later presented in his important work *Nerve, Muscle and Synapse* (1966).

It was mainly for this work that Katz shared the 1970 Nobel Prize for physiology or medicine with Julius AXELROD and Ulf VON EULER.

Keeler, James Edward (1857–1900)
American astronomer

Born in La Salle, Illinois, Keeler graduated from Johns Hopkins University in 1881, becoming an assistant at the Allegheny Observatory. In 1888 he moved to the Lick Observatory at Mount Hamilton, California, for a short period but returned as director to Allegheny in 1891, being appointed professor of astrophysics in the Western University of Pennsylvania in the same year. He became director of Lick in 1898.

Using spectroscopic methods Keeler made several important discoveries. In 1895 he showed that the rings of Saturn do not rotate uniformly but that the inner border is rotating much faster than the outer. The difference is quite striking, as the innermost edge revolves in about 4 hours while the system's outermost edge needs over 14 hours. This is only possible if the rings are not solid but made from numerous small particles. He also worked on nebulae, photographing and taking the spectra of hundreds. He showed that about ¾ of them had a spiral structure and demonstrated that their line-of-sight motion showed that they are receding and advancing like the stars. He further studied the spectra of the Orion Nebula and showed that the bright lines in its spectra correspond to the dark lines in the stellar spectra. He was one of the growing band of astronomers who failed to see the supposed canals on Mars despite using a new 36-inch (91-cm) refractor at Allegheny. He died when only 42.

Keenan, Philip Childs (1908–2000)
American astronomer

Keenan, who was born in Bellevue, Pennsylvania, graduated in 1929 from the University of Arizona and obtained his PhD from the University of Chicago in 1932. He worked initially at Chicago's Yerkes Observatory from 1929 until 1942, when he joined the Bureau of Ordnance. With the return of peace, Keenan was appointed to the staff of the Perkins Observatory of Ohio State University, becoming professor of astronomy in 1956.

Keenan is best known for his work with William MORGAN and Edith Kellman on their *Atlas of Stellar Spectra with an Outline of Spectral Classification* (1943). It was this work that formed the basis for the MKK system of classifying stars by their luminosity in addition to their spectral type.

Keilin, David (1887–1963) *British biologist and entomologist*

Keilin was born in Moscow, Russia, and educated at Cambridge University, England. He was subsequently professor of biology at Cambridge from 1931 until 1952 and also director of the Cambridge Moltena Institute. His most important research was the discovery of the respiratory pigment cytochrome, which, he demonstrated, is present in animal, yeast, and higher plant cells. He also studied the biochemistry of the Diptera (true flies), and investigated the respiratory systems and adaptations of certain dipterous larvae and pupae.

Keir, James (1735–1820) *British chemist and industrialist*

Keir was the youngest of 18 children. He came from a prosperous Edinburgh family and was educated at the high school and the university there, where he started a lifelong friendship with Erasmus DARWIN. He left the university without graduating to join the army and served in the West Indies, reaching the rank of captain before resigning in 1768.

Keir settled near Birmingham and became a leading member of the famous Lunar society, an organization founded to promote interest in science and its applications. At this time he translated Pierre MACQUER's *Dictionary of Chemistry* (1776), which was one of the key volumes whereby chemical knowledge was transmitted to mechanics and engineers. He served as an assistant to Joseph PRIESTLEY during his stay in Birmingham. In 1778 he acted as general manager for the firm of Boulton and Watt.

Together with James WATT and Matthew BOULTON, Keir started a venture to obtain soda from nonvegetable sources. He tried to extract it from potassium and sodium sulfates, which were waste products of the vitriol industry. He passed these waste products slowly through a sludge of lime, producing an insoluble calcium sulfate and a weak solution of alkali. He then used this in soap production. By 1801 Keir's alkali factory, which he founded in 1780, was paying excise duty on the production of £10,000 worth of soap. As a pure producer of alkali the venture was not a success; the future was to lie with the LEBLANC process.

Keith, Sir Arthur (1866–1955) *British anatomist*

Keith, the son of a farmer from Old Machan in Scotland, was educated at the University of Aberdeen, where he qualified as a doctor in 1888. He served as a medical officer in Siam (now Thailand) from 1889 until 1892, when his interest in the comparative anatomy of the primates was first aroused. On his return to Europe he studied anatomy in Leipzig and

London before being appointed (1895) demonstrator in anatomy at the London Hospital. In 1908 Keith moved to the Royal College of Surgeons, where he served as curator of the Hunterian Museum until his retirement in 1933.

On 18 December 1912, Arthur WOODWARD and Charles Dawson announced to the Geological Society the discovery at Piltdown in Sussex of a remarkable skull, which apparently combined the mandible of an ape with the cranium of a man. Here at last, it was felt, was solid evidence for the antiquity of man. Although some at the meeting were skeptical of the find, suggesting that the skull and jaw must have come from two different individuals, Keith was not among them. It thus appeared that a man with a cranial capacity of 1,500 cubic centimeters (as estimated by Keith) and with the jaw of an ape had coexisted with the mastodon. Keith, in the first edition of his *Antiquity of Man* (1915), dated Piltdown man to the beginning of the Pliocene, which was then assumed to be about a million years ago. With the change in geological fashion Keith was forced to halve the date of Piltdown man in the second edition of his book (1925).

In 1915 Keith estimated the actual separation of man from the apes to have taken place in the lower Miocene, then considered to be some 2–4 million years ago. This meant that Keith was unable to accommodate the discovery of the famous Taung skull by Raymond DART in 1924, and consequently he denied that Dart's *Australopithecus* was either man or a link between ape and man, considering it to be a pure ape having affinities with two living apes, the gorilla and the chimpanzee.

Keith lived long enough to witness the exposure of Piltdown man by Kenneth OAKLEY in 1949, using modern fluorine dating techniques. These showed the fossil to date back only as far as the Pleistocene, while later work (1953) revealed its fraudulent nature by assigning markedly different dates to the skull and jaw. When Oakley made a special journey to the 87-year-old Keith to inform him of his results, Keith commented "I think you are probably right, but it will take me some time to adjust myself to the new view."

Kekulé von Stradonitz, Friedrich August (1829–1896) *German chemist*

Kekulé (**kay**-koo-lay) was born in Darmstadt, Germany. As a youth he showed considerable skill in drawing and was consequently encouraged to be an architect. Although he began as a student of architecture at Giessen he soon switched, despite family opposition, to the study of chemistry, which he continued abroad. He first went to Paris in the period 1851–52 where he studied under Jean DUMAS and Charles GERHARDT who influenced him greatly. He worked

in Switzerland for a while before taking a post in England in 1854–55 as a laboratory assistant at St. Bartholomew's Hospital, London. While in London he met and was influenced by Alexander WILLIAMSON and William ODLING. He accepted an unsalaried post at the University of Heidelberg before his appointment to the chair of chemistry at Ghent in 1858. He then moved to the chemistry chair at the University of Bonn in 1867, where he remained for the rest of his life.

Kekulé's main work was done on the structure of the carbon atom and its compounds. It has often been claimed that he had changed his career from the architecture of buildings to the architecture of molecules. Certainly, after Kekulé it was much easier to visualize the form of atoms and their combinations. In 1852 Edward FRANKLAND had pointed out that each kind of atom can combine with only so many other atoms. Thus hydrogen can combine with only one other atom at a time, oxygen could combine with two, nitrogen with three, and carbon with four. Such combining power soon became known as the valency (valence) of an atom. Each atom would be either uni-, bi-, tri-, quadrivalent, or some higher figure.

In 1858 both Kekulé and Archibald COUPER saw how to use this insight of Frankland to revolutionize organic chemistry. They both assumed that carbon was quadrivalent and that one of the four bonds of the carbon atom could be used to join with another carbon atom. The idea came to him, he later claimed, while traveling on a London bus to Clapham Road. He fell into a reverie, "and lo, the atoms were gamboling before my eyes...I saw frequently how two smaller atoms united to form a pair; how a larger one embraced two smaller ones; how still larger ones kept hold of three or even four of the smaller...I saw how the longer ones formed a chain...(and then) the cry of the conductor 'Clapham Road' awakened me from my dreaming; but I spent part of the night in putting on paper at least sketches of these dream forms." He published his results in 1858 in his paper *The Constitution and the Metamorphoses of Chemical Compounds and the Chemical Nature of Carbon* and in the first volume of his *Lehrbuch der organische Chemie* (1859; Textbook of Organic Chemistry).

The diagrams of carbon compounds used today come not from Kekulé but from Alexander CRUM BROWN in 1865. Kekulé's own notation, known as "Kekulé sausages," in which atoms were represented by a cumbersome system of circles, was soon dropped. The gains from such representations were immediate. It can be seen why two molecules could have the same number of atoms of each element and yet differ in properties. Thus C_2H_6O represents both ethanol and dimethyl ether. If the rules of valence are observed these are the only two ways in which two carbon, six hydrogen, and one oxygen atom can be combined and indeed these are the only two compounds of the formula ever observed.

While Kekulé had dramatic success demonstrating how organic compounds could be constructed from carbon chains, one set of compounds, the aromatics, resisted all such treatment. Benzene, discovered by Michael FARADAY in 1825, had the formula C_6H_6, which, on the assumption of a quadrivalent carbon atom, just could not be represented as any kind of chain. The best that could be done with alternating single and double carbon bonds would still violate the valence rules, for at the end of the chain the carbon atoms will both have an unfilled bond. Kekulé once more has left a description of how the solution of the puzzle came to him. In 1890 he recalled that while working on his textbook in 1865, "I dozed off. Again the atoms danced before my eyes. This time the smaller groups remained in the background. My inner eye...now distinguished bigger forms of manifold configurations. Long rows, more densely joined; everything in motion, contorting and turning like snakes. And behold what was that? One of the snakes took hold of its own tail and whirled derisively before my eyes. I woke up as though I had been struck by lightning; again I spent the rest of the night working out the consequences."

The snake with its tail in its mouth is in fact an ancient alchemical symbol and is named Ouroboros but, to Kekulé, it meant a more prosaic image, that of a ring. For if the two ends of the benzene chain are joined to each other then benzene will have been shown to have a ring structure in which the valence rules have all been observed. Again the rewards in understanding were immediate. It was now obvious why substitution for one of benzene's hydrogen atoms always produced the same compound. The mono-substituted derivative C_6H_5X was completely symmetrical whichever H atom it replaced. Each of the hydrogen atoms was replaced by NH_2 and in each case the same compound, aniline $C_6H_5.NH_2$, was obtained.

Such was the revolution in organic chemistry initiated by Kekulé. Together with new methods introduced by Stanislao CANNIZZARO at Karlsruhe in 1860 for the determination of atomic weight, a new age of chemistry was about to dawn in which the conflicts and uncertainties of the first half of the 19th century would be replaced by a unified chemical theory, notation, and practice. After this it comes as something of a shock to discover that Kekulé had no firm belief in the existence of atoms. Whether they exist, he argued in 1867, "has but little significance from a chemical point of view; its discussion belongs rather to metaphysics. In chemistry we have only to decide whether the assumption of atoms is an hypothesis adapted to the explanation of chemical phenomena."

Kellner, Karl (1851–1905) *Austrian chemical engineer*

Kellner (**kel**-ner) worked as an engineer in his native city of Vienna. In 1894 he took out a patent on the manufacture of caustic soda from the electrolysis of brine and founded the Konsortium für Electrochemische Industrie at Salzburg for its exploitation. The same discovery had also been made quite independently by the American, Hamilton CASTNER. To avoid costly litigation the two inventors exchanged patents, and plants using the *Castner–Kellner process* were opened at Niagara Falls in 1896 and in England in 1897 at Runcorn in Cheshire.

Kelvin, William Thomson, Baron (1824–1907) *British theoretical and experimental physicist*

> When thermal energy is spent in conducting heat through a solid, what becomes of the mechanical effect it might produce? Nothing can be lost in the operations of nature – no energy can be destroyed.
> —*Account of Carnot's Theory* (1849)

Born in Belfast, Northern Ireland, William Thomson was an extremely precocious child intellectually and matriculated at Glasgow University at the astonishingly early age of 10. He went on to Cambridge, after which he returned to Glasgow to become professor of natural philosophy. He was to occupy this chair for 53 years. It was in Glasgow that he organized and ran one of Britain's first adequately equipped physical laboratories. In 1892 in recognition of his contributions to science he was raised to the British peerage as Baron Kelvin of Largs. He was also a devout member of the Scottish Free Church.

Kelvin's work on electromagnetism is second only to that of Michael FARADAY and James

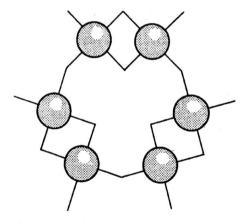

BENZENE   Kekulé's structure of the benzene molecule.

Clerk MAXWELL. Together with Faraday he was responsible for the introduction of the concept of an electromagnetic field. Kelvin was of a much more mathematical turn of mind than Faraday, but it was left to Maxwell to weld the ideas of Faraday and Kelvin together into a powerful, elegant, and succinct mathematical theory. But Maxwell's work would have been greatly hampered without some of the penetrating suggestions made by Kelvin. Particularly important is a fundamental paper of 1847 in which Kelvin drew an analogy between an electrostatic field and an incompressible elastic solid. Kelvin made many other innovations, including the introduction of the use of vectors to represent magnetic induction and magnetic force. He also put his knowledge of electromagnetism to use in many practical inventions of which the transatlantic electric telegraph cable and the mirror galvanometer were among the most important.

Kelvin's other great area of work was thermodynamics. He was one of the first to understand and appreciate the importance of James JOULE's seminal work in the field. In his 1852 paper on the *Dissipation of Mechanical Energy* Kelvin set out the fundamentally important law of conservation of energy that was to be so important in the physics of the second half of the 19th century. In his work on thermodynamics Kelvin assimilated and developed the work of the great pioneers of the subject, Nicolas Carnot and James Joule. He also collaborated with Joule in experimental work. One of the important results of Kelvin's work was his introduction of the concept of *absolute zero* and his recognition of the theoretical importance of the absolute scale of temperature, which is named in his honor. Kelvin was able to calculate the value of absolute zero from theoretical considerations. One of the first formulations of the second law of thermodynamics was given by Kelvin. With Joule he first demonstrated the *Joule–Kelvin effect*. He also made important contributions to the theory of elasticity and some basic contributions to hydrodynamics in which he collaborated with George STOKES.

Kemeny, John George (1926–1992) *Hungarian–American mathematician*

Kemeny was born in Budapest, Hungary. In 1938 his father was so alarmed by the Nazi annexation of Austria that he moved to the United States. The family followed in 1940 and Kemeny entered Princeton in 1943 to study mathematics. A year later he was drafted onto the Manhattan Project and sent to Los Alamos where he operated an IBM calculator. He returned to Princeton in 1946, completed his PhD in 1949, and moved to Dartmouth in 1953, serving as professor of mathematics from 1956 until 1968, as president of the college from 1970 until 1981, and once more, from 1981 until his retirement in 1990, as professor of mathematics.

Between 1963 and 1964 Kemeny, working with a Dartmouth colleague, Thomas Kurtz, developed BASIC (Beginner's All Purpose Symbolic Instruction Code), probably the best known of all computer languages. Previously the large computers could only be approached through specialized computer programmers. BASIC was conceived initially as something for Dartmouth students to use on Dartmouth computers. With a few simple self-evident commands and an equally simple syntax and vocabulary, it proved remarkably easy to use.

As it was meant to be freely available to students, the software was placed in the public domain. Subsequently it became the most widely used computer language of the 1970s and 1980s.

Kemeny himself became something of a public figure. It was during his presidency that Dartmouth became coeducational and he did much to open up the college to minorities. He also campaigned against the Vietnam War. Kemeny was one of the main campaigners for the not altogether successful "new math" introduced into America in the 1970s. In 1979 he was invited by President Carter to chair the committee set up to investigate the nuclear accident at Three Mile Island.

Kendall, Edward Calvin (1886–1972) *American biochemist*

Kendall, a dentist's son from South Norwalk, Connecticut, studied chemistry at Columbia University where he obtained his PhD in 1910. After working briefly at St. Luke's Hospital in New York from 1911 to 1914, Kendall moved to the Mayo Foundation in Rochester, Minnesota, where from 1921 to 1951 he served as professor of physiological chemistry.

In 1914 Kendall achieved an early success by isolating the active constituent of the thyroid gland. The importance of hormones in the physiology of the body had become apparent through the work of William BAYLISS and Ernest STARLING on the pancreas. Kendall was able to demonstrate the presence of a physiologically active compound of the amino acid tyrosine and iodine, which he named thyroxin.

Kendall was led from this to investigate the more complex activity of the adrenal gland. This gland secretes a large number of steroids, many of which he succeeded in isolating. Four compounds, labeled A, B, E, and F, seemed to possess significant physiological activity. They were shown to affect the metabolism of proteins and carbohydrates and in their absence animals seemed to lose the ability to deal with toxic substances. It was therefore hoped that some of these compounds might turn out to be therapeutically useful. After much effort sufficient compound A was obtained but, to

Kendall's surprise and disappointment, it was shown to have little effect on ADDISON's disease, a complaint caused by a deficient secretion from the adrenal cortex. Kendall was more successful with his compound E – later known as cortisone to avoid confusion with vitamin E – when in 1947 a practical method for its production was established. Clinical trials showed it to be effective against rheumatoid arthritis. It was for this work that Kendall shared the 1950 Nobel Prize for physiology or medicine with Tadeus REICHSTEIN and Philip HENCH.

Kendall, Henry Way (1926–1999) *American physicist*

Kendall, who was born in Boston, Massachusetts, was educated at Amherst College and the Massachusetts Institute of Technology, where he gained his PhD in 1955. After a four-year spell in California at Stanford, Kendall returned to MIT in 1961 and was appointed to the chair of physics in 1967.

In a series of experiments at Stanford in 1967, Kendall and his collaborators, Richard TAYLOR and Jerome FRIEDMAN, provided the first experimental evidence that the proton has an inner structure, and that it could indeed contain the quarks first described by Murray GELL-MANN in 1964. The Stanford experiment won for Kendall a share of the 1990 Nobel Prize for physics along with Taylor and Friedman.

Kendrew, Sir John Cowdery (1917–1997) *British biochemist*

Kendrew, who was born at Oxford in England, graduated in natural science from Cambridge University in 1939. He spent the war years working for the Ministry of Aircraft Production, becoming an honorary wing commander in 1944. In 1946 he joined Max PERUTZ at Cambridge and, like Perutz, used x-ray diffraction techniques to study the crystalline structure of proteins, particularly that of the muscle protein myoglobin. X-ray diffraction, or crystallography, involves placing a crystal in front of a photographic plate and rotating the crystal in a beam of x-rays. The pattern of dots that is formed on the plate by the x-rays can be analyzed to find the positions of the atoms in the crystal. The technique had been used successfully to show the structures of small molecules but Kendrew's progress with the much larger myoglobin structure was slow, especially since diffraction patterns yield no information on the phases of the directed x-rays. However, in 1953 Perutz made a breakthrough by incorporating atoms of heavy elements into the protein crystals. Kendrew modified this new method and applied it successfully in his myoglobin studies, so that four years later he had built up a rough model of the three-dimensional structure of myoglobin. By 1959 he had greatly clarified the structure and could pinpoint most of the atoms.

Kendrew and Perutz received the 1962 Nobel Prize for chemistry for their work on protein structure. Kendrew was knighted in 1974 and served as director general of the European Molecular Biology Laboratory in Heidelberg from 1975 to 1982.

Kennelly, Arthur Edwin (1861–1939) *British–American electrical engineer*

Kennelly, the son of an Irish-born employee of the East India Company, was born near Bombay in India and was educated in Europe. He left school at the age of 14 to become office boy to the London Society of Telegraph Engineers. From 1876 to 1886 he worked for the Eastern Telegraph Company, acquiring an engineering education through practice and independent study. He emigrated to America in 1887 and became an assistant to Thomas EDISON and a consulting engineer. In 1894, together with E. J. Houston, he founded his own consulting firm. Kennelly was professor of electrical engineering at Harvard University from 1902 to his retirement in 1930; between 1913 and 1925 he held a second appointment as professor of electrical communication at the Massachusetts Institute of Technology.

Kennelly made many contributions to the theory and practice of electrical engineering. These included the representation of quantities by complex variables, a mathematical treatment that helped in understanding the behavior of electrical circuits. In 1902 he explained the Atlantic transmission of radio waves by suggesting that they were reflected back to Earth by some layer of electrically charged particles in the upper atmosphere (suggested independently by Oliver HEAVISIDE and called the *Kennelly–Heaviside layer*). Kennelly was a great scientific administrator and made contributions to the development of electrical units and standards.

Kepler, Johannes (1571–1630) *German astronomer*

> Oh God, I am thinking Thy thoughts after Thee.
> —On his studies of astronomy

> It may well wait a century for a reader, as God has waited 6,000 years for an observer.
> —On the publication of his astronomical observations. Quoted by David Brewster in *Martyrs of Science* (1841)

The grandfather of Kepler (**kep**-ler) had been the local burgomaster but his father seems to have been a humble soldier away on military service for most of Kepler's early youth. His mother was described by Kepler as "quarrelsome, of a bad disposition." She was later to be accused of witchcraft. Kepler was born at

Würtenburg, in Germany, and was originally intended for the Church; he graduated from the University of Tübingen in 1591 and went on to study in the theological faculty. In 1594 he was offered a teaching post in mathematics in the seminary at Gratz in Styria. It was from his teacher, Mästlin, who was one of the earliest scholars fully to comprehend and accept the work of Nicolaus COPERNICUS, that the young Kepler acquired his early Copernicanism. In addition to his teaching at Gratz and such usual duties as mathematicians were expected to do in those days Kepler published his first book – *Mysterium cosmographicum* (1596; Mystery of Cosmography). The book expresses very clearly the belief in a mathematical harmony underlying the universe, a harmony he was to spend the rest of his life searching for. In this work he tried to show that the universe was structured on the model of PLATO's five regular solids. Although the work verges on the cranky and obsessive it shows that Kepler was already searching for some more general mathematical relationship than could be found in Copernicus.

He married in 1597 shortly before he was forced to leave Gratz when, in 1598, all Lutheran teachers and preachers were ordered to leave the city immediately. Fortunately for Kepler, he had an invitation to work with Tycho BRAHE who had recently become the Imperial Mathematician in Prague. Tycho was the greatest observational astronomer of the century and he had with him the results of his last 20 years' observations. Kepler joined him in 1600 and although their relationship was not an easy one it was certainly profitable. Tycho assigned him the task of working out the orbit of Mars. Somewhat rashly Kepler boasted he would solve it in a week – it took him eight years of unremitting effort. Not only did Kepler lack the computing assistance now taken for granted but he was also working before the invention of logarithms. It was during this period

that he discovered his first two laws and thus, with GALILEO, began to offer an alternative physics to that of ARISTOTLE. The first law asserts that planets describe elliptic orbits with the Sun at one focus while the second law asserts that the line joining the Sun to a planet sweeps out equal areas in equal times. The laws were published in his magnum opus *Astronomia nova* (1610; New Astronomy).

Tycho had died in 1601 leaving Kepler with his post, his observations, and a strong obligation to complete and publish his tables under the patronage of their master, the emperor Rudolph II. This obligation was to prove even more onerous and time consuming than the orbit of Mars. It involved dealing with Tycho's predatory kin, attempting, vainly, to extract money from the emperor to pay for the work, which he ended up financing himself, and trying to find a suitable printer. All this, it must be realized, was done against the background of the Thirty Years' War, marauding soldiery, and numerous epidemics. His work, the *Tabulae Rudolphinae* (Rudolphine Tables), was not completed until 1627 but remained the standard work for the best part of a century.

While serving the emperor in Prague, Kepler had also produced a major work, *Optics* (1604), which included a good approximation of SNELL's law, improved refraction tables, and discussion of the pinhole camera. In the same year he observed only the second new star visible to the naked eye since antiquity. He showed, as Tycho had done with the new star of 1572, that it exhibited no parallax and must therefore be situated far beyond the solar system. He studied and wrote upon the bright comet of 1607 – later to be called HALLEY's comet – and those of 1618 in his *Three Tracts on Comets* (1619). His final work in Prague, the *Dioptrics* (1611), has been called the first work of geometrical optics.

In 1611 Kepler's wife and son died, civil war broke out in Prague, and Rudolph was forced to abdicate. Kepler moved to Linz in the following year to take up a post as a mathematics teacher and surveyor. Here he stayed for 14 years. He married again in 1613. While in Linz he produced a work that, starting from the simple problem of measuring the volume of his wine cask, moved on to more general problems of mensuration – *Nova stereometria* (1615; New Measurements of Volume). One further crisis he had to face was his mother's trial for witchcraft in Würtemburg. The trial dragged on for three years before she was finally freed. His greatest work of this period, *Harmonices mundi* (1619; Harmonics of the World), returns to the search for the underlying mathematical harmony expressed in his first work of 1596. It is here that he stated his third law: the squares of the periods of any two planets are proportional to the cubes of their mean distance from the Sun. After the completion of the Rudolphine

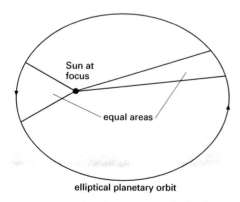

Sun at focus

equal areas

elliptical planetary orbit

KEPLER'S SECOND LAW   The planet moves such that equal areas are swept out in equal times.

tables Kepler took service under a new patron, the Imperial General Wallenstein. He settled at Sagan in Silesia. In return for the horoscopes Wallenstein expected from him, Kepler was provided with a press, a generous salary, and the peace to publish his ephemerides and to prepare his work of science fiction – *A Dream, or Astronomy of the Moon* (1634). He left Sagan in 1630 to see the emperor in Ratisbon, hoping for a payment of the 12,000 florins still owed him. He died there of a fever a few days later.

As a scientist Kepler is of immense importance. Copernicus was in many ways a traditional thinker, still passionately committed to circles. Kepler broke away from this mode of thought and in so doing posed questions of planetary motion that it took a Newton to answer.

Kerr, John (1824–1907) *British physicist*

Born at Ardrossan in Scotland, Kerr studied at Glasgow University and carried out research work under Lord KELVIN. He taught mathematics at a training college in Glasgow. He is remembered for his work on polarized light, in which he discovered that certain substances placed in a strong electric field exhibit birefringence (the *Kerr effect*). Kerr also described the behavior of polarized light when reflected from the polished pole of an electromagnet (the *Faraday effect*).

Kerst, Donald William (1911–1993) *American physicist*

Kerst's name is primarily associated with the development of the betatron, a machine capable of accelerating beta particles (electrons) to speeds approaching that of light. Born at Galena, Illinois, Kerst graduated from the University of Wisconsin in 1937, going on to be assistant professor at the University of Illinois, becoming full professor in 1943. In 1939 he developed the idea of the cyclotron particle accelerator a stage further. He circulated electrons in a doughnut-shaped vacuum tube, guiding them round the circle in a magnetic field similar to that of a toroidal electrical transformer. The electrons are accelerated, but are kept in their circular orbits by increasing the magnetic field. Kerst was responsible for the building of the largest such machine (the betatron), completed at the University of Illinois in 1950, in which electrons attained energies of up to 310 MeV (million electronvolts).

Kerst's work on the betatron, and other nuclear physics work primarily with the Van de Graaff generator, led to his involvement during World War II in the Los Alamos thermonuclear (atom-bomb) project, after which he returned to Illinois. He returned to the University of Wisconsin as professor of physics in 1962.

Ketterle, Wolfgang (1957–) *German physicist*

Ketterle (**ket**-erl) was born in Heidelberg, Germany. He was educated at the Technical University of Munich and gained his PhD in physics from the University of Munich in 1986. He did postdoctoral work at the Max-Planck Institute for Quantum Optics, Garching, the department of physical chemistry, University of Heidelberg, and the department of physics, MIT. He became a member of the physics faculty at MIT in 1993, becoming a full professor in 1997. Since 1998 he has been the John D. MacArthur Professor of Physics at MIT.

In the early 1990s one of his major scientific activities was the cooling of atoms to extremely low temperatures very close to absolute zero temperature. This led to his discovery in 1995 of a Bose–Einstein condensate independently of Eric CORNELL and Carl WIEMAN. Ketterle produced a Bose–Einstein condensate with a large number of sodium atoms at a very low temperature.

Since making this discovery Ketterle and his colleagues have investigated a number of aspects of the physics of Bose–Einstein condensates. For example, two initially separate Bose–Einstein condensates which were then allowed to overlap showed interference effects characteristic of coherent waves. The process of the formation of Bose–Einstein condensates and their optical properties have also been studied. Ketterle has used drops of Bose–Einstein condensate to produce a type of laserlike atomic beam.

Ketterle has won many prizes for his pioneering work in atomic physics at very low temperatures, culminating in sharing the 2001 Nobel Prize for physics with Cornell and Wieman.

Kettlewell, Henry Bernard Davis (1907–1979) *British geneticist and lepidopterist*

Kettlewell was educated at Cambridge University and St. Bartholomew's Hospital, London, where he gained his medical qualification in 1933. He practiced in Cranleigh and then worked as an anesthetist in Surrey. After the war he worked in South Africa at the International Locust Control Centre in Cape Town before returning to Britain in 1952 as research fellow in genetics at Oxford, a post he continued to hold until his retirement in 1974.

Kettlewell is best known for his work on the occurrence of melanism – black pigmentation in the epidermis of animals. In 1953 he set out to explain why, in the mid-19th century, certain moth species had a light coloration, which camouflaged them on such backgrounds as light tree trunks where they sat motionless during the day. However, by the 1950s, of 760 species of larger moths in Britain 70 had changed their light color and markings for dark or even totally black coloration.

Kettlewell suspected that the success of the melanic form was linked with the industrial revolution and the consequent darkening of the trees by the vast amounts of smoke produced by the 19th-century factories. To test his hypothesis he released large numbers of the dark and light forms of the peppered moth, *Biston betularia*, in the polluted woods around Birmingham and in a distant unpolluted forest. As many of the released moths as possible were recaptured and when the results were analyzed it was found that the light form had a clear advantage over the dark in the unpolluted forest but in the polluted Birmingham woods the result was just the opposite. From this Kettlewell concluded that if the environment of a moth changes so that it is conspicuous by day, then the species is ruthlessly hunted by predators until it mutates to a form better suited to its new environment. His work was seen as a convincing and dramatic confirmation of the Darwinian hypothesis of natural selection.

Kety, Seymour Solomon (1915–2000)
American neurophysiologist

Kety was born in Philadelphia and educated at the University of Pennsylvania where he obtained his MD in 1940. He continued working there and was appointed professor of clinical physiology in 1948. In 1951 he also took on the directorship of the National Institutes of Health. From 1967 to 1980 he was professor of psychiatry at Harvard, becoming professor of neuroscience in 1980 and emeritus professor in 1983.

Kety's first major success was his development of a treatment for lead poisoning using the lead-citrate complex. He later concentrated on measuring the blood flow and energy metabolism of the brain. One of the most interesting results arising from this was his demonstration that the brain's energy consumption during sleep is the same as in the conscious state.

Kety later worked mainly on the role of biological mechanisms in mental illness. He pointed out that both schizophrenia and manic depression run in families. This led him to conclude that because genes can only express themselves through biochemical mechanisms, the hereditary nature of mental disorders suggests the involvement of biochemical substrates. However, although he started to examine the blood and urine of schizophrenics as early as 1957 for such biochemical substrates, none has yet been specifically identified.

Khorana, Har Gobind (1922–1993) *Indian–American chemist*

Khorana (koh-**rah**-na), who was born at Raipur

(now in Pakistan), gained his BSc (1943) and MSc (1945) from the University of Punjab. He then traveled to Liverpool University to work for his doctorate. On receiving his PhD in 1948, he did two years postdoctoral research in Switzerland before taking up a Nuffield Fellowship at Cambridge University. There he worked with Alexander TODD, who fired his interest in nucleic acid research – the field in which Khorana later made his name.

Shortly after Khorana joined Wisconsin University in 1960 he became interested in unraveling the genetic code. He synthesized each of the 64 nucleotide triplets that make up the code, and for this work received the Nobel Prize for physiology or medicine in 1968, sharing the award with Marshall NIRENBERG and Robert HOLLEY.

Khorana's next major achievement came in 1970, when he announced the synthesis of the first artificial gene. The same year he moved to the Massachusetts Institute of Technology, where, by 1976, his team had made a second gene, which (unlike the first) was capable of functioning in a living cell.

Kidd, John (1775–1851) *British chemist*

Kidd was the son of a London merchant captain. He was educated at Oxford University, where he graduated in 1797, spending the next four years at Guy's Hospital in London. In 1803 he became the first Aldrichian Professor of Chemistry at Oxford. He stayed in this chair until 1822, when he became professor of physics.

In 1819 Kidd obtained naphthalene from coal tar. This aromatic hydrocarbon, which is used in mothballs, played an important role in the development of aniline dyes by Sir William PERKIN 40 years later. Kidd also published in 1809 *Outlines of Mineralogy*. In 1833 he contributed to the "Bridgewater Treatises" – a series commissioned by the Earl of Bridgewater in his will in which eight scientists selected by the Royal Society would demonstrate "the Power, Wisdom, and Goodness of God, as manifested in the creation." Kidd's treatise was on *The Adaptation of External Nature to the Physical Condition of Man*.

Kiddinu (about 379 BC) *Babylonian astronomer*

Almost nothing is known about Kiddinu (**kid**-i-noo), although he was head of the astronomical school in the Babylonian city of Sippar, and some late classical writers such as PLINY refer to him as Kidenas. There are references to some lunar eclipse tables that Kiddinu had prepared and he is credited with a new method of construction of ephemerides (tables of planetary motion), the so-called "System B." He is also thought to have discovered the precession of the equinoxes – the slow westward motion of the equinoctial points that is caused by the rotation of the Earth's axis.

Kilby, Jack St. Clair (1923–2005) *American inventor*

Kilby was born in Jefferson City, Missouri. He was educated as an electrical engineer at the University of Illinois, gaining a BS in 1947, and at the University of Wisconsin, gaining his MS in 1950. From 1947 to 1958 he worked on circuits for electronic products at the Centralab Division of Global Union Inc. in Milwaukee. From 1958 he was based with Texas Instruments Incorporated in Dallas, Texas. Although he officially retired from Texas Instruments in the 1980s he continued to be involved with the company. From 1978 to 1984 he was a distinguished professor of electrical engineering at Texas A & M University, College Station, Texas.

In 1958, soon after he joined Texas Instruments, he constructed the first electronic circuit in which all the components were in a single small amount of semiconductor, about half the size of a paper clip. This was the first integrated circuit, known as the *microchip*. The invention of the microchip revolutionized electronics and paved the way for many technological applications. For example, the microchip made possible high-speed computers and semiconductors with large-capacity memories. Kilby was a major figure in developing applications of the technology based on microchips. For example, he and his colleagues invented the first computer to incorporate integrated circuits. He also co-invented the hand-held calculator. Kilby also investigated other topics in electronics such as technology based on the element silicon to generate electrical power from light from the Sun. He held more than sixty United States patents for his inventions.

His major pioneering work in the invention of the integrated circuit was recognized by the award of many prestigious honors. In 1970, he received the National Medal of Science at the White House. In 1982, his induction into the National Inventors Hall of Fame took place. In 2000, he shared the Nobel Prize for physics with Zhores ALFEROV and Herbert KROEMER, two other pioneers of the information age.

Kildall, Gary (1942–1995) *American computer scientist*

Kildall was educated at the University of Washington where he gained his PhD in 1972. Rather than be drafted to Vietnam, he opted to teach computer science at the Naval Postgraduate School, Monterey, California.

With the emergence of the personal computer in the 1970s, operating systems were needed in order to run the various machines coming onto the market. One of the first and most successful of these early systems was Kildall's CP/M (Control Program for Microcomputers). It was flexible enough to work on most computers with 16K of memory and an 8080 Intel chip; it would also, unlike any other system at that time, control floppy disks. In 1976 Kildall resigned from the Naval School to found Digital Research to develop and market CP/M. The system sold well in its various versions, with about 200 million copies produced.

The software could have been even more successful for, in 1980, Kildall was approached by IBM, which was looking for a system to operate its new 16-bit PC. In the event, IBM elected to work with Bill GATES of Microsoft and MS-DOS, rather than CP/M, became the industry-standard operating system for personal computers.

Kildall continued to develop software. He produced, for example, one of the earliest multitask operating systems, Concurrent CP/M, in 1983, bringing out a DOS compatible version in 1984. Later work was more concerned with CD ROMs.

Kimura, Doreen (c. 1935–) *Canadian psychologist*

Kimura (ki-**muur**-a) was born at Winnipeg, Manitoba, and educated at McGill University, Montreal, where she obtained her PhD in 1961. After working briefly at the University of California, Los Angeles, and at McMaster University, Hamilton (1964–67), Kimura moved to the University of Western Ontario, London, and was later appointed professor of psychology in 1974.

Kimura's early work was concerned with possible differences in function between the left and right sides of the brain. Different series of numbers were played simultaneously into the left and right ears of her experimental subjects. The subjects could more easily remember numbers that were fed into their right ear – i.e., the ear connected directly with the left side of the brain. When melodies were substituted for numbers, subjects recognized tunes more easily with the left ear and, accordingly, the right side of the brain. Kimura's work thus supported a slowly emerging hypothesis that there was a functional asymmetry to be found in the cerebral hemispheres. More precisely, the left side was thought to process verbal data, while the right side handled such nonverbal data as music and spatial relationships.

In the 1970s Kimura extended her work to cover differences between the sexes. Among other differences, she found women better at matching items and arithmetical calculation, and that they also had a greater verbal fluency. Men seemed to perform better at certain spatial tasks, at mathematical reasoning, and at disembedding tests.

Kimura proposed that the answer may well

lie in the level of sex hormones such as estrogen and testosterone. She found, for example, that women's spatial skills increased when they had low estrogen levels; similarly, men's skills improved when testosterone levels were low.

More recently Kimura has begun to study body asymmetry. She has found that men tend to be larger on the right side and women larger on the left. Much to her surprise Kimura found that body differences might correlate with intellectual skills. Thus mathematical problems are handled better by right-larger individuals, be they male or female.

Kimura, Hisashi (1870–1943) *Japanese astronomer*

Kimura, who was born at Kanazawa in Japan, graduated from Tokyo University in 1892. He worked mainly on latitude variation at the Mizusawa Latitude Observatory, where he served as director from 1899 onward. The International Geodetic Association had organized an International Latitude Service in 1899 provided by six latitude observatories set up on the line of latitude 39°08′ N; Mizusawa was one of the six observatories.

Variations in latitude had been identified first by Leonhard EULER in 1765. A perfectly symmetrical sphere will have a stable spin. The earth's mass, however, is not distributed uniformly about its axis of rotation. Consequently the axis of rotation and the axis of figure, also known as the axis of the moment of inertia, do not coincide. This will lead to a slight wobble with a periodicity of about 14 months. The distances are small and over a fourteen month period the poles will drift no more than 72 feet (22 m). The variation was first detected in 1891 when observations made from Berlin and, 180° away in Waikiki, found that an increase in Berlin's latitude was matched by a decrease in that of Waikiki. Knowledge of the variation is important when very precise and accurate astronomical measurements need to be made.

In 1902 Kimura announced the discovery of a new annual term in the variation of latitude, independent of the components of the pole's motion.

Kimura, Motoo (1924–1994) *Japanese population geneticist*

Kimura was born in Okazaki, Japan, and educated at Kyoto University and the University of Wisconsin, where he gained his PhD in 1956. He joined the research staff of the National Institute of Genetics, Mishima, and served from 1964 as head of the population genetics department.

From 1968 Kimura developed a cogent alternative to the neo-Darwinian synthesis as it emerged in the 1930s in the works of such scholars as J. B. S. HALDANE. He gathered evidence to show that certain mutations can increase in a population without necessarily having any selective advantage. He examined a number of mutant genes whose effects were not apparent in the phenotype and could only be detected by advanced chemical techniques. He found that adaptively they were neither better nor worse than the genes they replaced, concluding that, at the molecular level, most evolutionary changes are the result of "random drift" of selectively equivalent mutant genes.

Kimura allowed that at the level of the phenotype evolution is basically Darwinian but insisted that the laws governing molecular evolution are clearly different. Such views have met with much opposition from Darwinians. They have argued that many of the apparently neutral mutations are, on closer examination, found to be selective; also many cases, such as human hemoglobin, do not seem to show the variants expected from Kimura's theory.

King, Charles Glen (1896–1986) *American biochemist*

Born in Entiat, Washington, King was educated at Washington State University and the University of Pittsburgh, where he obtained his PhD in 1923 and became professor of chemistry in 1930. He later moved to Columbia University in New York, where he held the chair of chemistry from 1946 until his retirement in 1962.

In 1928 Albert Szent-Györgyi isolated from the adrenal gland a substance that he named "hexuronic acid"; he had in fact discovered vitamin C. It was left to King in 1932 to complete the work. He isolated from lemon juice and cabbages a substance, identical to Szent-Györgyi's hexuronic acid, that possessed powerful antiscorbutic properties. It was vitamin C, later called ascorbic acid. In the following year King determined its formula ($C_6H_8O_6$).

Kinsey, Alfred Charles (1894–1956) *American zoologist*

> There were wives and husbands in the older generation, who did not even know that orgasm was possible for a female; or if they knew that it was possible, they did not comprehend that it could be desirable.
> —Alfred Kinsey, *Sexual Behavior in the Human Female* (1953)

Kinsey was born in Hoboken, New Jersey, and educated at Bowdoin College and Harvard. He was professor of zoology (from 1920) and director of the Institute for Sex Research, Indiana University, which he helped found, from 1942 until his death.

Kinsey's researches on human sexual behavior, published as *Sexual Behavior in the Human Male* (1948) and *Sexual Behavior in the Human Female* (1953), have attracted much interest and some controversy. His work demonstrated that there was considerable variation in behavior in all social classes and helped to dispose of certain erroneous ideas, for example, with regard to juvenile sexual activity and homosexuality. Even though based on many (about 18,500) carefully conducted personal interviews, Kinsey's findings have been criticized for sampling limitations and the general unreliability of personal communication in this sphere of human activity.

Kipping, Frederic Stanley (1863–1949)
British chemist

> The prospect of any immediate and important advance in this section of organic chemistry [silicone polymer synthesis] does not seem to be very hopeful.
> —*Proceedings of the Royal Society* (1937)

Kipping, a banker's son from Manchester in England, was educated at Owens College, the forerunner of Manchester University. After a period as chemist at the Manchester gas plant and postgraduate work in Munich, Kipping took up his first academic appointment at Heriot-Watt College in Edinburgh in 1885. From 1890 to 1897 he was chief demonstrator at the City and Guilds College in London and in 1897 was appointed to the chair of chemistry at Nottingham University, where he remained until his retirement in 1936.

Kipping is best known as the author, with William PERKIN, Jr., of *Organic Chemistry* (1894). This was one of the first works to be devoted to organic chemistry alone, and was the basic textbook for organic chemists for over fifty years.

One of the burning issues of the day when Kipping was an undergraduate was that of stereoisomerism, which had earlier been demonstrated in carbon compounds by Jacobus VAN'T HOFF. Working with William POPE, Kipping showed that such isomerism was not exclusive to carbon but was detectable in compounds of nitrogen and other atoms. Thus Kipping went on to show the same effect in silicon, discovering between 1905 and 1907 a number of asymmetric silicon compounds. This was the beginning of Kipping's exhaustive study of the chemistry of silicon, on which he published 51 papers.

Kirchhoff, Gustav Robert (1824–1887)
German physicist

Kirchhoff (**keerk**-hof) studied in his native city of Königsberg (now Kaliningrad in Russia), graduating in 1847. Three years later he was appointed a professor at Breslau. He moved to Heidelberg, where Robert BUNSEN was professor of chemistry, in 1854.

Kirchhoff was one of the foremost physicists of the 19th century and is remembered as one of the founders of the science of spectroscopy. He is also known for *Kirchhoff's laws*, formu-

lated in 1845 while he was still a student, which refer to the currents and electromotive forces in electrical networks.

In 1859 he published an explanation of the dark lines in the solar spectrum discovered by Josef von FRAUNHOFER, in which he suggested that they are due to absorption of certain wavelengths by substances in the Sun's atmosphere. He later formulated *Kirchhoff's law of radiation*, which concerns the emission and absorption of radiation by a hot body. It states that the rate of emission of energy by a body is equal to the rate at which the body absorbs energy (both emission and absorption being in a given direction at a given wavelength). Kirchhoff gave a final proof of this in 1861.

In about 1860 Bunsen was analyzing the colors given off by heating chemicals to incandescence, using colored glass to distinguish between similar shades. Kirchhoff joined this research when he suggested that the observation of spectral lines, by dispersing the light with a prism, would be a more precise way of testing the color of the light. Kirchhoff and Bunsen found that each substance emitted light that had its own unique pattern of spectral lines – a discovery that began the spectroscopic method of chemical analysis. In 1860, a few months after publishing these results, they discovered a new metal, which they called cesium and the next year found rubidium. Kirchhoff and Bunsen also constructed improved forms of the spectroscope for such work and Kirchhoff showed that, if a gas emitted certain wavelengths of light then it would absorb those wavelengths from light passing through it.

Kirchhoff was crippled by an accident in midlife but remained in good spirits and, when his health forced him to stop experimental work in 1875, he was offered the chair of theoretical physics in Berlin. He remained there until his death 12 years later.

Kirkwood, Daniel (1814–1895) *American astronomer*

> Planets and comets have not formed from rings but rings from planets and comets.
> —*The Divisions in Saturn's Rings* (1883)

Born in Harford County, Maryland, Kirkwood became professor of mathematics at the University of Delaware in 1851, moving to the University of Indiana in 1856. In 1857 he noted that the asteroids (planetoids) are not evenly distributed in between the orbits of Mars and Jupiter but that there are areas in which no – or very few – asteroids orbit. He showed that these gaps in the asteroid belt – since known as *Kirkwood gaps* – occur where the period of revolution of an asteroid would have been an exact simple fraction of the Jovian period. Kirkwood explained that any asteroids in these areas would eventually be forced into other orbits by perturbations caused by Jupiter. Similarly he was able to explain gaps in the rings of Saturn (the Cassini division) as being caused by the satellite Mimas. Kirkwood published his findings in *The Asteroids* (1887).

Kirwan, Richard (1733–1812) *Irish chemist and mineralogist*

Born in Galway, now in the Republic of Ireland, Kirwan studied in France to become a Jesuit but returned to Ireland after only a year. There he inherited the family estates following his brother's death in a duel in 1755. He was called to the Irish bar in 1766 but gave up law just two years later to devote himself to science. He was made a fellow of the Royal Society in 1780 and won the Copley medal in 1782 for his work on chemical affinity. In 1787 he settled in Dublin, where he remained until his death. In 1799 he was elected president of the Royal Irish Academy.

Kirwan was at first a staunch supporter of the phlogiston theory and in 1787 he wrote his *Essay on Phlogiston*, which was translated into French. He conceded to Antoine LAVOISIER's criticism of this and Lavoisier's subsequent evidence of oxygen by giving up his support of the theory in 1791.

Kirwan was also an eminent mineralogist and published, in 1784, his *Elements of Mineralogy*, which has been described as the first systematic work on the subject in English. His *Geological Essays* (1799) brought him into conflict with James HUTTON over the chemical composition of rocks. Kirwan was also involved in industrial chemistry. He received news from Karl SCHEELE of the bleaching properties of chlorine and quickly had it tested and marketed in both Lancashire and Ireland.

Kistiakowsky, George Bogdan (1900– 1982) *Russian–American chemist*

Kistiakowsky (kis-ti-a-**kow**-skee) came from a family of academics in Kiev, now in Ukraine. He began his education in his native city but, after fighting against the Bolsheviks, completed it in Berlin. He moved to America in 1926, working first at Princeton before moving to Harvard where he was appointed professor of chemistry in 1937, a post he retained until his retirement in 1971.

His most important work during the war was as head of the Explosives Division at Los Alamos (1944–45). On being told of the project his initial reaction had been: "Dr. Oppenheimer is mad to think this thing will make a bomb." The basic device, proposed by Seth Neddermeyer, consisted of a thin hollow sphere of uranium that would become critical only when "squeezed" together. In theory this was achieved by surrounding the subcritical uranium with conventional explosives whose detonation would compress the radioactive material into a critical mass. To work the process must take place in less than a millionth of a second and with great precision and accuracy. Right to the very end there was considerable doubt as to whether Kistiakowsky could solve the technical problems involved.

After the war Kistiakowsky, very much a figure of the scientific establishment, spent much time advising numerous governmental bodies. From 1959 until 1961 he served as special assistant for science and technology to President Eisenhower, later writing an account of this period in *A Scientist at the White House* (1976). Toward the end of his life he spoke out about the dangers of nuclear weapons.

Kitasato, Baron Shibasaburo (1852– 1931) *Japanese bacteriologist*

Born in Oguni, Japan, Kitasato (kee-tah-**sah**-toh) graduated from the medical school of the University of Tokyo in 1883 and then went to Berlin to study under Robert KOCH. A close and long-lasting friendship developed between the two men.

While in Berlin Kitasato worked with Emil von BEHRING and in 1890 they announced the discovery of antitoxins of diphtheria and tetanus. They showed that if nonimmune animals were injected with increasing sublethal doses of tetanus toxin, the animals became resistant to the disease. Their paper laid the basis for all future treatment with antitoxins and founded a new field in science, that of serology. Kitasato returned to Japan and became director of the Institute of Infectious Diseases in 1892. Two years later there was an outbreak of bubonic plague in Hong Kong and he succeeded in isolating the plague bacillus, *Pasteurella pestis*. In 1898 he isolated the microorganism that causes dysentery.

He founded the Kitasato Institute for Medical Research in 1914 and became dean of the medical school, Keio University, Tokyo. In 1924 he was created a baron. In 1908 Koch visited Japan and Kitasato secretly obtained clippings of the visitor's hair and fingernails. When Koch died in May 1910, Kitasato built a small shrine for the relics in front of his laboratory; when Kitasato died, his remains were placed in the same shrine, next to those of his respected master.

Kittel, Charles (1916–   ) *American physicist*

Kittel is regarded by many as the leading authority on the physics of the solid state. Born in New York City, he graduated from Cambridge University, England, with a BA in 1938 and gained his PhD from the University of Wisconsin in 1941. He was an experimental physicist at the Naval Ordnance Laboratories, Washington (1940–42), and an operations analyst with the U.S. Fleet (1943–45). After a short spell as a physics research associate at the Massachusetts Institute of Technology, he worked as a research physicist with the Bell Telephone Laboratories (1947–50). From there he took up an associate professorship at the University of California at Berkeley, becoming full professor of physics in 1951. He retired in 1978.

During the 1950s Kittel published several important papers on the properties and structure of solids. These were concerned with: antiferroelectric crystals; electron-spin resonance in the study of conduction electrons in metals; the nature of "holes" in the process of electrical conduction; plasma-resonance effects in semiconductor crystals; ferromagnetic resonance and domain theory; and spin-resonance absorption in antiferromagnetic crystals. He is widely known for his textbook *Introduction to Solid State Physics* (1953), successive editions of which have served generations of physics students. He also wrote *Quantum Theory of Solids* (1963).

Kjeldahl, Johan Gustav Christoffer Thorsager (1849–1900) *Danish chemist*

Kjeldahl (**kel**-dahl), the son of a physician, was born at Jagerpris in Denmark and educated at the Roskilde Gymnasium and the Technical University of Denmark, Copenhagen. After working briefly at the Royal Veterinary and Agricultural University he joined the laboratory set up by the brewer Carl Jacobsen in 1876 to introduce scientific methods into his Carlsberg brewery founded the previous year. Kjeldahl directed the chemistry department of the laboratory from 1876 until his fatal heart attack in 1900.

Kjeldahl is still widely known to chemists for the method named for him, first described in 1883, for the estimation of the nitrogen content of compounds. It was much quicker, more accurate, and capable of being operated on a larger scale than the earlier combustion-tube method dating back to Jean DUMAS. It utilized the fact that the nitrogen in a nitrogenous organic compound heated with concentrated sulfuric acid will be converted into ammonium sulfate. The ammonia can then be released by introducing an alkaline solution, and then distilled into a standard acid, its amount being determined by titration.

His name is also remembered with the *Kjeldahl flask*, the round-bottomed long-necked flask used by him in the operation of his method.

Klaproth, Martin Heinrich (1743–1817) *German chemist*

Born in Wernigerode, Germany, Klaproth (**klahp**-roht) was apprenticed as an apothecary. After working in Hannover and Danzig he moved to Berlin where he set up his own business. In 1792 he became lecturer in chemistry at the Berlin Artillery School and in 1810 he became the first professor of chemistry at the University of Berlin.

His main fame as a chemist rests on his discovery of several new elements. In 1789 he discovered zirconium, named from zircon, the mineral from which it was isolated. In the same year he extracted uranium from pitchblende and named it for the newly discovered planet, Uranus. He also rediscovered titanium in 1795, about four years after its original discovery, and discovered chromium in 1798. Klaproth used the Latin *tellus* (earth) in his naming of tellurium (1798), which had been discovered by Muller von Richtenstein in 1782. In 1803 he discovered cerium oxide, named for the newly discovered asteroid, Ceres. He made important improvements to chemical analysis by bringing samples to a constant weight through drying and ignition.

Klaproth's son, Heinrich Julius, became a noted orientalist.

Klein, (Christian) Felix (1849–1925) *German mathematician*

Klein (klIn), one of the great formative influences on the development of modern geometry, was born in Düsseldorf, Germany, and studied at Bonn, Göttingen, and Berlin. He worked with Sophus LIE – a collaboration that was particularly fruitful for both of them and led to the theory of groups of geometrical transformations. This work was later to play a crucial role in Klein's own ideas on geometry.

Klein took up the chair in mathematics at the University of Erlangen in 1872 and his inaugural lecture was the occasion of his formulation of his famous *Erlangen Programm*, a suggestion of a way in which the study of geometry could be both unified and generalized. Throughout the 19th century, with the work of such mathematicians as Karl Friedrich GAUSS, János BOLYAI, Nikolai LOBACHEVSKY, and Bernhard RIEMANN, the idea of what a "geometry" could be had been taken increasingly beyond the conception EUCLID had of it and Klein's ideas helped show how these diverse geometries could all be seen as particular cases of one general concept. Klein's central idea was to think of a geometry as the theory of the invariants of a particular group of transformations. His *Erlangen Programm* was justly influential in guiding the further development of the subject. In particular Klein's ideas led to an even closer connection between geometry and algebra.

Klein also worked on projective geometry, which he generalized beyond three dimensions, and on the wider application of group theory, for example, to the rotational symmetries of regular solids. His name is remembered in topology for the *Klein bottle*, a one-sided closed surface, not constructible in three-dimensional Euclidean space. In 1886 Klein took up a chair at Göttingen and was influential in building Göttingen up into a great center for mathematics.

Klein, Oskar Benjamin (1894–1977)
Swedish physicist

Klein (klIn) was born in Stockholm, Sweden, and educated at the University of Stockholm, gaining his PhD in 1922. He had spells at the University of Michigan (1923–25) and the University of Copenhagen (1926–33) before returning to the University of Stockholm as a professor of physics and director of the Institute of Mechanics in 1933. He retained these positions at Stockholm until he retired in 1968.

In 1926 Klein modified the theory of Theodor KALUZA, which attempted to unify gravitation and electromagnetism in terms of the existence of a fifth space–time dimension by combining it with quantum theory. Klein's modification of Kaluza's theory provided an explanation of why the fifth dimension is not observed by making it possible for this fifth dimension to be rolled up very tightly like a sheet rolled up as a straw. The theory resulting from Klein's modification of Kaluza's theory is called the *Kaluza–Klein theory*. The way in which the fifth dimension becomes very small is called *compactification*. The idea that unified theories of forces and particles are naturally formulated in more than four space–time dimensions, with the higher dimensions being compactified, has become generally accepted. Klein made several other significant contributions to relativistic quantum mechanics and quantum field theory.

Klingenstierna, Samuel (1698–1765)
Swedish mathematician and physicist

Klingenstierna (klin-gen-shair-na) was born at Linköping in Sweden and, before embarking on his mathematical and scientific studies, studied law at Uppsala. He was appointed secretary to the Swedish treasury (1720) but also had interests in philosophy and science and was allowed to continue his studies at Uppsala. In 1727 he was awarded a scholarship, which enabled him to travel in Europe. He traveled to Marburg where he studied with the Leibnizian philosopher Christian Wolff and also to Basel to study mathematics with Johann I BERNOULLI. Klingenstierna became professor of mathematics at Uppsala and later professor of physics there (1750). His last appointment was the highly prestigious one of tutor to the crown prince (1756–64).

Klingenstierna's most notable scientific work was in the field of optics. He was able to show that some of NEWTON's views on the refraction of light were incorrect and made practical use of this discovery in producing designs for lenses free from chromatic and spherical aberration.

Klitzing, Klaus von (1943–) *German physicist*

Von Klitzing (fon **klit**-sing) was born at Poznan but when the town was restored to Poland after 1945 his family moved to West Germany. He was educated at the universities of Brunswick and Würzburg, where he received his PhD in 1972 and where he remained as a teaching fellow until 1980. After serving as professor of physics at the Technical University, Munich, from 1980 to 1984, he was appointed director of solid-state research at the Max Planck Institute, Stuttgart.

In 1980 von Klitzing began work on the Hall effect, first described by Edwin HALL in 1879. Hall noted that when a current flows in a conductor placed in a magnetic field perpendicular to the sample's surface, a potential difference (the *Hall voltage*) is produced acting at right angles to both the current and field directions. It is possible to measure a *Hall resistance*, defined in the normal way, by dividing the Hall voltage by the current it produces.

Von Klitzing set out to make extremely precise measurements of the Hall resistance working with a two-dimensional electron gas. This can be formed by using a special kind of transistor in which electrons can be drawn into a layer between an insulator and a semiconductor. When the layer is thin enough, of the order of 1 nanometer (10^{-9} meter), and the temperature is as low as 1.5 K, the electrons are forced into a two-dimensional plane parallel to the surface of the semiconductor.

Under normal conditions the Hall resistance increases directly with the strength of the magnetic field. In contrast von Klitzing found that under his experimental conditions the resistance became quantized, varying in a series of steps as the magnetic field was changed. Von Klitzing went on to establish that the Hall resistance at each step was a function of Planck's constant h, the fundamental constant of quantum theory, and could be used to measure h very accurately. For his discovery of the *quantized Hall effect* von Klitzing was awarded the 1985 Nobel Prize for physics.

Klug, Sir Aaron (1926–) *South African biophysicist*

> Visiting Americans often think that the tea and coffee breaks are a waste of time but some of them learn better.
> —*The Sunday Times*, 24 October 1982

Klug (kloog) was born in Lithuania of South African parents, and moved to South Africa at the age of three. He studied medicine for a year at Witwatersrand University, Johannesburg, and then changed to study science. He moved to

Cape Town in 1947 where he took a master's degree in crystallography. In 1949 Klug moved to England, where he worked at the Cavendish Laboratory, Cambridge. In 1954 he went to Birkbeck College, London, and was director of the Virus Structure Research Group there from 1958 to 1962, when he returned to Cambridge.

Klug originally developed a technique of improving the results of electron microscopy by illuminating the micrograph with laser light. For micrographs of regular structures, a diffraction pattern is produced, from which extra information on the specimen can be obtained. At Birkbeck, Klug worked with J. D. BERNAL on the polio virus. With Donald Caspar he went on to study small viruses. These are either rod-shaped or spherical, and consist of nucleic acids covered by a protein coat. Klug and Caspar developed a theory of how the coat could be formed by an arrangement of smallish quasi-equivalent protein molecules. At Cambridge, Klug worked on helical viruses showing how the protein units form. He went on to investigate the structure and action of transfer DNA in animal cells. More recently he has worked on chromatin in cells. Klug received the 1982 Nobel Prize for chemistry for his work in molecular biology. He was president of the Royal Society (1995–2000).

Klumpke Roberts, Dorothea (1861–1942) *American astronomer*

Dorothea Klumpke was born in San Francisco, into a German immigrant family. In 1877 she and her four sisters were taken to Europe for their education in various private schools. Dorothea initially studied music, but switched to astronomy, and in 1886 received her BS degree from the Sorbonne, Paris. She secured a post at the Paris Observatory, and began working with astrophotographers Paul and Prosper Henry, photographing the minor planets through a refracting telescope. In 1887 Klumpke became involved in an international project to create an atlas of the stars, the Carte du Ciel, and in 1891 beat 50 male applicants to be appointed director of the Bureau of Measurements, responsible for processing the photographic plates taken for the project by the Paris Observatory. She became the first woman doctor of science and mathematics at the Sorbonne (in 1893), and the first woman to make astronomical observations from a balloon, when in 1899 she lofted over Paris to view the Leonid meteor shower. In 1901 Klumpke moved to England after marrying Dr. Isaac ROBERTS, a retired Welsh entrepreneur and noted astronomer. She assisted her husband at his Sussex home, photographing the 52 HERSCHEL regions of nebulosity. But after her husband's death in 1904, Klumpke Roberts returned to Paris, staying with her sister Anna. She continued to process the results of Isaac's studies,

and in 1929, to coincide with the centenary of his birth, she published *The Isaac Roberts Atlas of 52 Regions*, followed by a supplement in 1932. Two years later, in recognition of her services to French astronomy, she was elected a Chevalier de la Légion d'Honneur. Thereafter she returned to the United States with her sister to spend her retirement in San Francisco, where she remained active in scientific and artistic circles, and endowed various awards to benefit young astronomers.

Knowles, William S. (1917–) *American chemist*

Knowles was educated at Columbia University, gaining his PhD in 1942. He worked at the Monsanto Company, St. Louis, Missouri, retiring in 1986. Knowles was one of the leading figures in the development of what is called *catalytic asymmetric synthesis*. It is common for molecules to exist in two forms that are nonsuperimposable mirror images. These molecules are said to be chiral because the two forms of the molecule are related to each other in a way that is analogous to the way that left and right hands are related to each other. It is often the case that in Nature one of the forms predominates, which usually means that only one of the forms has a useful biological function. Indeed, the other form which does not fit can be biologically harmful. Since pharmaceutical products are frequently chiral molecules it is very important to be able to synthesize the two different chiral forms separately.

Knowles, Ryoji NOYORI, and Barry SHARPLESS developed molecules which act as catalysts in chemical reactions producing only one form of the mirror image molecules. The catalyst molecule is itself a chiral molecule. It is possible for a single catalyst molecule to produce millions of molecules of the required chiral form.

In 1968 Knowles discovered that certain transition-metal compounds act as chiral catalysts in an important type of reaction known as hydrogenation, i.e., the addition of hydrogen to carbon–carbon multiple bonds, with the sought mirror image form of the molecule being formed. This work soon led to an industrial process for producing the drug L-DOPA, used to treat Parkinson's disease. The work of Knowles on chiral catalysts for hydrogenation was extended by Noyori.

The pioneering research by Knowles, Noyori, and Sharpless initiated a large branch of chemistry, which makes it possible to synthesize molecules and materials with novel properties. There are now many applications of their research in the industrial syntheses of pharmaceutical products including antibiotics, anti-inflammatory drugs, and heart medicines. Knowles, Noyori, and Sharpless shared the 2001 Nobel Prize for chemistry for their work in this area.

Knuth, Donald Ervin (1938–   ) *American computer programmer*

Knuth showed an early interest in words and numbers. While in the 8th grade he entered a competition to find as many words as possible from the letters in the phrase "Ziegler's Giant Bar" and came up with 4,500 – some 2,000 more than the judges had compiled. At first he considered devoting himself to music but opted for mathematics and physics, which he studied at Case Institute for Technology, Cleveland, Ohio, and at the California Institute of Technology, where he gained his PhD in 1963. He remained at Cal Tech until 1968 when he moved to Stanford as professor of computer science. He resigned in 1992 to concentrate on writing.

In the 1960s Knuth began compiling what is now widely recognized as the fundamental work on computer science, *The Art of Computer Programming*. It is planned for seven volumes – the first three have already appeared: *Fundamental Algorithms* (1968), *Seminumerical Algorithms* (1969), and *Sorting and Searching* (1973). Having completed the first three volumes, Knuth spent several years exploring typography. He had long been interested in printing and it occurred to him in 1977 that "printing was a computer science problem." The result was the much-studied book *Tex and Metafont* (1979) and the five-volume *Computers and Typesetting* (1986). Metafont allows the user to construct a custom-designed typesetting font. Tex (which Knuth prints as T$_E$X, and which is pronounced "tek") is an automatic typesetting and page makeup program. It is widely available and popular with academic users. Knuth has said that he intends to return to music once he has completed all seven volumes of his computer book – his house is built around a two-story pipe organ that he designed himself. He has also written a remarkable science-fiction novel, *Surreal Numbers: How Two Ex-Students Turned On to Pure Mathematics and Found Total Happiness* (1974), based on a number system invented by the mathematician John CONWAY.

Koch, (Heinrich Hermann) Robert (1843–1910) *German bacteriologist*

> I have undertaken my investigations in the interests of public health and I hope the greatest benefits will accrue therefrom.
>
> The master of us all in bacteriology.
> —Theobald Smith, describing Koch

Born the son of a mining official in Klausthal, Germany, Koch (kok) studied medicine at the University of Göttingen where he was a pupil of Jacob HENLE. After graduating in 1866 and serving in the Franco–Prussian War, Koch was appointed district medical officer in Wollstein. Here, working alone with only the most modest of resources, Koch began the research that was to make him, with PASTEUR, one of the two founders of the new science of bacteriology.

Koch saw more clearly than anyone before what was involved in bacteriological research and achieved his first success with anthrax. Whereas Casimir DAVAINE had succeeded in transmitting anthrax from one cow to another by the injection of blood, Koch saw that the emerging germ theory required something more specific. It needed to be the germ that was injected rather than a fluid that could only be presumed to contain the organism. He thus spent three years devising techniques to isolate the anthrax bacillus from the blood of infected cattle and then to produce pure cultures of the germ. By 1876 Koch was ready to publish the life history of the anthrax bacillus. He had found that while the bacillus in its normal state is somewhat sensitive and unable to survive long outside the body of its host, it forms resting spores, which are particularly hardy. Such spores, persisting in deserted ground, are responsible for the apparently spontaneous outbreaks of anthrax in healthy and isolated herds. Koch followed this with work on septicemia, during which he developed techniques for obtaining pure cultures. With these triumphs behind him Koch at last achieved official recognition, being appointed to the Imperial Health Office in Berlin in 1880. In this office Koch, with the aid of his assistants Friedrich LÖFFLER and Georg GAFFKY, began one of the great periods of medical discovery. Much of this was based upon techniques of staining and culture growth developed by Koch in the obscurity of Wollstein. He developed culture media suitable for bacterial growth, proceeding from liquid media to boiled potato, to the still commonly used agar plates. Agar plates together with the new stains derived from aniline dyes constituted the heart of Koch's technique. With them he made, in 1882, his most famous discovery – the bacillus responsible for TB. In the second half of the 19th century TB, responsible for one in seven of all European deaths, was the most feared of all diseases. The difficulties Koch faced were formidable. The bacillus was only about a third of the size of the anthrax bacillus, grew much more slowly, and in general was more difficult to detect. With great patience Koch managed to culture the thin rod-shaped bacilli, which he used to inoculate four guinea pigs. All four developed TB while two uninoculated controls remained uninfected. The fame won by Koch for this work brought him into open competition with Pasteur, a competition fanned by Franco-German nationalism in the aftermath of the 1870 war. In 1883 both sides met in Egypt to study cholera. The French, under Emile ROUX, seem to have mistakenly confused platelets in the blood with the vibrio responsible for cholera. Koch, noting microorganisms in the small intestines of the victims, took them to be the cause of the disease, an assumption confirmed when he observed the same comma-shaped rods in the intestines of Indian victims. At this point Koch was forced to violate his own rules because, although he failed to infect experimental animals with pure strains of the vibrio, he nonetheless declared the organism to be the cause. In 1885 Koch was appointed professor of hygiene at the University of Berlin and in 1891 became head of the newly formed Institute of Infectious Diseases. The pressure on Koch to "earn" this latter appointment led him to announce in 1890 the discovery of a substance he claimed could prevent the growth of tubercle bacilli in the body. The new drug, a sterile liquid containing dead tubercle bacilli, which he named tuberculin, was consequently in huge demand. However, it had little effect in most cases and probably exacerbated some. It later proved useful however in testing whether patients have experienced tuberculosis infection, by noting their local reaction to an injection of tuberculin. Koch resigned the directorship of the Institute in 1904 to become one of the first of the emerging breed of international experts. Indulging his passion for travel, Koch spent his last years advising South Africa on rinderpest, India on bubonic plague, Java on malaria, and East Africa on sleeping sickness. In 1905 Koch was awarded the Nobel Prize for physiology or medicine for "his discoveries in regard to tuberculosis." Perhaps as important as this, or any of his other specific discoveries, was his formulation of the so-called Koch's postulates. To establish that an organism is the cause of a disease we must, Koch declared, first find it in all cases of the disease examined; secondly it must be prepared and maintained in a pure culture; and finally it must, though several generations away from the original germ, still be capable of producing the original infection. Such postulates, first formulated fully in 1890, rigorously followed, established clinical bacteriology as a scientific practice.

Kocher, Emil Theodor (1841–1917) *Swiss surgeon*

Kocher (**ko**-ker), an engineer's son from Bern in Switzerland, graduated in medicine from the university there in 1865. He later studied surgery in Berlin, Paris, and in London under Joseph LISTER, and in Vienna under Theodor BILLROTH. Kocher served as professor of clinical surgery at the University of Bern from 1872 until his retirement in 1911 although he continued as head of the university surgical clinic until his death.

Using the antiseptic techniques developed by Lister, Kocher, following the initiative of Billroth, played an important role in developing the operation of thyroidectomy for the treatment of goiter, a not uncommon complaint in Switzerland. By 1914 Kocher was able to report

a mortality of only 4.5% from over 2,000 operations. Earlier however, Kocher discovered that while technically successful the operation was responsible for the unnecessary ruin of many lives. In 1883, he found to his horror that something like a third of his patients who had undergone thyroidectomy were suffering from what was politely termed operative myxedema; they had in fact been turned into cretins once the source of the thyroid hormone (thyroxine) had been removed. Kocher showed that such tragedies could be prevented by not removing the whole of the thyroid, for even a small portion possesses sufficient physiological activity to prevent such appalling consequences. For this work Kocher was awarded the 1909 Nobel Prize for physiology or medicine.

Köhler, Georges J. F. (1946–1995) *German immunologist*

Born at Munich in Germany, Köhler (**ku**-ler) was educated at the University of Freiburg, receiving his doctorate in 1974. He then worked in Cambridge at the Medical Research Council Laboratory of Molecular Biology (1974–76) and at the Basel Institute for Immunology (1977–84). In 1985 he was appointed director of the Max Planck Institute for Immunology in Freiburg.

In 1975, Köhler's work at the MRC Laboratory in Cambridge, with the molecular biologist César MILSTEIN, yielded the first monoclonal antibody, i.e., antibody of one specific type manufactured by a culture of genetically identical cells, representing a single clone. This proved to be one of the major scientific breakthroughs of the decade, as anticipated by Köhler and Milstein in their letter to *Nature* briefly describing their achievement, published on 7 August 1975: "Such cells can be grown *in vitro* in massive cultures to provide specific antibody. Such cultures could be valuable for medical and industrial use." The technique of Köhler and Milstein involved fusing a white blood cell (lymphocyte), which manufactures antibody but has a short lifespan, with a cancer cell, which ensures continuous production of the antibody. Specifically, lymphocytes were obtained from the spleen of a mouse that had previously been injected with sheep red blood cells (the antigen) to stimulate antibody production in the mouse cells. The lymphocytes were then mixed with cells from a bone-marrow tumor culture to create a hybridoma – an antibody-secreting cell line. Köhler and Milstein later adapted their technique to produce a range of hybridoma cultures, each manufacturing pure antibody to specific known antigen. Monoclonal antibodies have revolutionized biological and medical diagnostic tests and assays and have found applications in various therapeutic techniques. Their great advantage is that they can be tailor-made to target a particular type of cell or subcellular component, and used as a vehicle for precision delivery of powerful drugs. For their achievement, Köhler and Milstein shared the 1984 Nobel Prize for physiology or medicine with Niels JERNE.

Kohlrausch, Friedrich Wilhelm Georg
(1840–1910) *German physicist*

Kohlrausch (**kohl**-rowsh) was born at Rinteln, in Germany, the son of R. H. A. Kohlrausch, a famous physicist who served as professor at the University of Erlangen. Friedrich studied at Erlangen and then at Göttingen, where he gained his PhD in 1863. He held a series of professorial appointments at Göttingen, Frankfurt, Darmstadt, Würzburg, Strasbourg, and Berlin and was elected to the Academy of Sciences in Berlin in 1895.

Kohlrausch is remembered for his work on the electrical conductivity of solutions. He was able to measure the electrical resistance of electrolytes (substances that, by transferring ions, conduct electricity in solutions) by introducing an alternating current to prevent polarization of the electrodes. In this way he recorded the conductivity of electrolytes at various solute concentrations and discovered that conductivity increases with increased dilution. This finding led to the formulation of *Kohlrausch's law* of the independent migration of ions.

Kohn, Walter (1923–) *Austrian–American physicist*

A native of Vienna, Kohn (kohn) emigrated to England in 1939 and was interned and sent to Canada in 1940. Here he studied at the University of Toronto and later at Harvard, where he gained his PhD in 1948. After holding a junior appointment at Harvard he moved to the Carnegie Institute in 1950, remaining there until 1960 when he became professor of physics at the University of California, San Diego. From 1979 to 1984 Kohn also served as director of the Institute for Theoretical Physics, University of California, San Diego. In 1984 he became professor of physics at the University of California, Santa Barbara.

Kohn has worked mainly in the field of solid-state physics, publishing work on the electronic structure of solids and solid surfaces. Kohn shared the 1998 Nobel Prize for chemistry with John POPLE for his pioneering work on a method of calculating the electronic structure of many-electron systems known as *density functional theory*.

Ko Hung (c. 283–343) *Chinese alchemist*

Ko Hung (koh huung), the son of a provincial governor from Jiangsu Province, China, served in various official posts until he retired to the mountains to devote himself to his chemical researches. He received an early education in Confucian ethics but became interested in Taoism, with its cult of physical immortality, and attempted to combine the two.

In about 317 he finished his classic work the *Pao-piu-tzu* or the *Book of the Preservation-of-Solidarity Master*, which is divided into two parts. The first part contains Ko Hung's alchemical studies and the search for the elixir of immortality. It includes a recipe for an elixir called gold cinnabar. The gold implied in this recipe was not in fact the natural element but the artificial transmuted gold for which there were numerous obscure recipes. Attempts to produce this gold helped alchemists gain a considerable amount of knowledge of and skill with chemical reactions and equipment. Ko Hung, for example, gives the first account of the process for making tin(IV) sulfide, which was widely used in "gold" paints. This process was not described in Europe until the 14th century. One unfortunate result of Ko Hung's work was that it led to numerous cases of elixir poisoning as the artificial "golds" produced by the alchemists tended to be rich in such toxic ingredients as mercury, arsenic, and silver. The second part of Ko Hung's book concentrates on Confucian ethical principles.

Kolbe, Adolph Wilhelm Hermann
(1818–1884) *German chemist*

Kolbe (**kol**-be), the son of a clergyman from Göttingen, in Germany, was the eldest of 15 children. He studied under Friedrich WÖHLER at Göttingen and then, in 1842, went to Marburg as Robert BUNSEN's assistant and learned his method of gas analysis. In 1845 he went to London to work as Lyon PLAYFAIR's assistant on the analysis of mine gases for a commission set up to investigate recent explosions in coal mines. He was professor of chemistry at Marburg from 1851 until he moved to Leipzig to succeed Justus von LIEBIG in 1865.

Kolbe made a number of advances in organic chemistry. He was the first to synthesize acetic acid from inorganic materials (following Wöhler's synthesis of urea). The *Kolbe method* is a technique for making hydrocarbons by electrolysis of solutions of salts of fatty acids. He also produced a *Textbook of Organic Chemistry* (1854–60), which collected together all the methods of preparing organic compounds and in 1854 he edited Liebig and Wöhler's *Dictionary of Chemistry*.

Koller, Carl (1857–1944) *Austrian–American ophthalmologist*

Koller (**kol**-er) was born in Schüttenhoffen (now Susice in the Czech Republic) and educated at the University of Vienna, where he obtained his MD in 1882. After a few years working for the department of ophthalmology as an intern, he emigrated to America in 1888 setting up in private practice in New York.

In 1884 Sigmund FREUD drew the attention of the world to the drug cocaine. Although he failed to appreciate its true value, he did arouse the interest of his colleagues Koller and another ophthalmologist, L. Königstein. While Freud and Königstein spent the summer away from Vienna, Koller performed a number of crucial experiments on animals and himself. He discovered that after bathing his eye in a solution of cocaine he could take a pin and prick the cornea without any awareness of the touch, let alone pain. Without delay Koller published his discovery of local anesthesia on 15 September 1884. By the time Freud and Königstein were ready to begin their own experiments they found that Koller had already gained all the credit. Although all three were later to dispute to whom the real credit for the discovery belonged, it was appropriate that when, in 1885, Freud's father needed an operation for glaucoma, the local anesthetic was administered by Koller, assisted by Freud, with the actual surgery performed by Königstein.

Kölliker, Rudolph Albert von (1817–1905) *Swiss histologist and embryologist*

Born in Zurich, Switzerland, Kölliker (**kul**-i-ker) qualified in medicine at Heidelberg in 1842 and later held professorships at Zurich and Würzburg. Celebrated for his microscopic work on tissues, he provided much evidence to show that cells cannot arise freely, but only from existing cells. He was the first to isolate the cells of smooth muscle (1848), as expounded in *Handbuch der Gewebelehre des Menschen* (1852; Manual of Human Histology): probably the best early text on the subject. He showed that nerve fibers are elongated parts of cells, thus anticipating the neuron theory, and demonstrated the cellular nature of eggs and sperm, showing for example that sperm are formed from the tubular walls of the testis, just as pollen grains are formed from cells of the anthers. Again anticipating modern discoveries, Kölliker believed the cell nucleus carried the key to heredity. His pioneering studies of cellular embryology mark him as one of the founders of the science. His book *Entwicklungsgeschichte des Menschen und der höheren Tiere* (1861; Embryology of Man and Higher Animals) is a classic text in embryology.

Kolmogorov, Andrei Nikolaievich (1903–1987) *Russian mathematician*

Kolmogorov (kol-**mog**-or-of) was born at Tambov in Russia and educated at Moscow State University, graduating in 1925. He became a research associate at the university, later a professor (1931), and in 1933 was appointed director of the Institute of Mathematics there.

He made distinguished contributions to a wide variety of mathematical topics. He is best known for his work on the theoretical foundations of probability, but also made lasting contributions to such diverse subjects as FOURIER analysis, automata theory, and intuitionism.

In 1933 Kolmogorov published his major treatise on probability, translated into English in 1950 as *The Foundations of the Theory of Probability*. The book is a landmark in the development of the theory, for in it Kolmogorov presented the first fully axiomatic treatment of the subject. It also contains the first full realization of the basic and underivable nature of the so-called "additivity assumption" about probability, first put forward by Jakob BERNOULLI. This claims simply that if an event can be realized in any one of an infinite number of mutually exclusive ways, the probability of the event is simply the sum of the probabilities of each of these ways. This assumption is fundamental to the whole measure-theoretic study of probability. Kolmogorov's interest in Luitzen BROUWER's intuitionism led him to prove that intuitionistic arithmetic, as formalized by Brouwer's disciple Arend Heyting, is consistent if and only if classical arithmetic is. In 1936 Kolmogorov settled a key problem in Fourier analysis when he constructed a function that is (Lebesgue) integrable, but whose Fourier series diverges at every point. In 1939, the same year in which he was elected an academician of the Soviet Academy of Sciences, he published a paper on the extrapolation of time series. This was later taken much further by Norbert WIENER and became known as "single-series prediction."

Kolmogorov also made a number of important contributions to theoretical physics. These include a theory of turbulence, which has been the starting point for much subsequent work. He also suggested a theorem on perturbation theory in celestial mechanics, subsequently proved by his student Vladimir Arnold and extended by the German mathematician Jürgen Moser, which shows that the Solar System is probably a stable dynamical system. This theorem is known as the *Kolmogorov–Arnold–Moser (KAM) theorem*. In addition, Kolmogorov and some of his colleagues extended the concept of entropy from thermodynamics to general dynamical systems.

Kopp, Hermann Franz Moritz (1817–1892) *German chemist and historian of chemistry*

Kopp was the son of a physician from Hanau, in Germany. He studied chemistry at the University of Heidelberg and obtained his PhD from Marburg. He was professor of chemistry at Giessen (1852–63) before he moved to the chair at Heidelberg.

Kopp is best remembered for his monumental work *Geschichte der Chemie* (1843–47; History of Chemistry), begun in 1841 while he was an unsalaried lecturer at Giessen. As a chemist he was noted for his extensive and precise work in measuring such properties as boiling points, thermal expansion, molecular volume, and specific gravity (relative density) for many elements and compounds.

Köppen, Wladimir Peter (1846–1940)
Russian–German climatologist

Köppen (**kup**-en) was educated at the university in his native city of St. Petersburg and then at Heidelberg and Leipzig. Although he began his career in 1872 with the Russian meteorological service, he moved to Germany shortly afterward where, in 1875, he was appointed director of the meteorological research department of the German Naval Observatory at Hamburg, a post he retained until the end of World War I when he was succeeded by his son-in-law Alfred WEGENER.

Köppen is mainly remembered today for the mathematical system of climatic classification he first formulated in 1900 and subsequently modified several times before 1936. He began by distinguishing between five broad climatic types – tropical rainy, dry, warm temperate, cold forest, and polar – symbolized by the letters A to E respectively. He further defined three patterns of precipitation: a climate with no dry period (f), with a dry summer period (s), and with a dry winter period (w). Four geographical zones were also introduced – steppe (S), desert (W), tundra (T), and perpetual frost (F). With such a technique some 60 climatic types are theoretically possible, although Köppen argued that only 11 are in fact realized. Köppen further modified his scheme by introducing six temperature categories, which enabled him to make fine adjustments to his initial 11 classes. Though by no means the only such classification in existence, Köppen's system is still widely and conveniently used on climatic maps. In his long career Köppen produced a number of substantial volumes, including a joint work with Wegener, *Die Klimate der geologischen Vorzeit* (1924; The Climate of Geological Prehistory), one of the founding texts of paleoclimatology. He also coedited, with Rudolph Geiger, a five-volume *Handbuch der Klimatologie* (Handbook of Climatology), begun in 1927 and nearing completion on his death in 1940.

Kornberg, Arthur (1918–2007) *American biochemist*

Kornberg was born in Brooklyn, New York, and graduated from the City College of New York in biology and chemistry in 1937. He then studied medicine at Rochester University, gaining his MD in 1941. He joined the National Institutes of Health, Bethesda, where from 1942 to 1953 he directed research on enzymes. During this period he helped elucidate the reactions leading to the formation of two important coenzymes, flavin adenine dinucleotide (FAD) and diphosphopyridine nucleotide (DPN; later renamed nicotinamide adenine dinucleotide – NAD).

From 1953 to 1959 Kornberg was professor of microbiology at Washington University. In 1956, while investigating the synthesis of coenzymes, he discovered an enzyme that catalyzes the formation of polynucleotides from nucleoside triphosphates. This enzyme, which he named DNA polymerase, can be used to synthesize short DNA molecules in a test tube, given the appropriate triphosphate bases and a DNA template. For the discovery and isolation of this enzyme, Kornberg was awarded the 1959 Nobel Prize for physiology or medicine, sharing the award with Severo OCHOA, who discovered the enzyme catalyzing the formation of RNA. Kornberg was chairman of the biochemistry department at Stanford University from 1959 and his work there contributed to the understanding of the synthesis of phospholipids and many reactions of the tricarboxylic acid, or Krebs, cycle.

Kornberg, Roger David (1947–) *American biochemist*

Kornberg was born in St. Louis, Missouri, the son of the eminent biochemist and Nobel Laureate, Arthur Kornberg. He gained his bachelor's degree from Harvard University in 1967, and his PhD from Stanford University in 1972. After postdoctoral research at the Medical Research Council Laboratory of Molecular Biology in Cambridge, UK, he joined Harvard Medical School as assistant professor in the biology department. He moved to Stanford in 1978, where he is now professor of structural biology.

Kornberg has enormously increased our understanding of how the cells of eukaryotic organisms (i.e., organisms other than bacteria, such as plants and animals) use the information in their genes to create proteins, which are the principal tools and building blocks of cells. The chromosomes of a cell contain its genes, which carry information as a genetic code. This code takes the form of the sequence of chemical bases that make up the DNA (deoxyribonucleic acid) of the genes. To access the information, the base sequence of the gene is first "copied over," or transcribed, by the enzyme RNA polymerase into a complementary base sequence carried by a messenger ribonucleic acid (mRNA) molecule. Transcription is crucial to all aspects of cell life, and must be performed accurately and only when the corresponding protein is needed. Kornberg used electron microscopy and x-ray crystallography to produce three-dimensional pictures of the molecules engaged in the process of transcription, providing vivid insights into the molecular mechanisms.

After joining Stanford, Kornberg employed baker's yeast, *Saccharomyces cerevisiae*, as an experimental organism. In the late 1980s and 1990s, his team devised and developed techniques that enabled them to reconstitute the DNA transcription machinery of this single-celled eukaryotic organism outside the cell (i.e., *in vitro*). This process led to the discovery of a key component of transcription, a multiprotein complex called Mediator.

In 2001 Kornberg and co-workers published the crystal structure of a yeast RNA polymerase in action, with a DNA molecule and newly formed RNA chain in place. This and subsequent structures have revealed, for instance, that the DNA occupies a cleft between two subunits of the polymerase, with the cleft bridged by an alpha-helix. This helix is thought to act like a ratchet, moving the DNA–RNA hybrid molecule along as successive nucleotide components are added to the RNA chain at the active site.

Kornberg's work has revealed many other aspects of transcription, such as how the process is initiated, how the newly synthesized RNA strand separates from the DNA template, and how incoming nucleotides are selected for inclusion at the appropriate place in the molecule. Such knowledge is vital for a proper understanding of how defects in transcription contribute to diseases such as cancer, heart disease, and inflammation. It also opens the way for manipulating transcription, for example in directing stem cells to develop into certain cell types.

For his studies of the molecular basis of eukaryotic transcription Kornberg was awarded the 2006 Nobel Prize for chemistry.

Korolev, Sergei Pavlovich (1907–1966) *Ukrainian rocket engineer*

Both the parents of Korolev (ko-ro-**lyef**) were teachers. He was born at Zhitomir, in Ukraine, and educated at the Kiev Polytechnic and the Moscow Higher Technical School from 1926 to 1929. An early interest in gliders and the writings of Konstantin TSIOLKOVSKY aroused in Korolev a fascination with the problems of space flight. He helped to found and direct the Gruppa Izucheniya Reaktivnogo Dvizheniya (GIRD; the Institute for Jet Research) in 1931 in Moscow.

At some later time, for unknown reasons, Korolev and most of his colleagues at GIRD were caught up in the Great Terror and imprisoned in the Arctic. He appears to have spent most of the war in one of the many special prison research centers before being released to cooperate with and direct the captured German rocket experts. The details of Korolev's later work were only released on his death

when he was described in his official obituary as responsible for "the manned space ships in which Man made his first flight to the Cosmos." In his lifetime he was referred to simply as the "chief spacecraft designer." Nevertheless his successes included the first satellite ever launched, on 4 October 1957, the development of the Vostok, Voskhod, and Soyuz manned spacecraft, and the Venus and Mars probes of 1961–62.

Koshiba, Masatoshi (1926–) *Japanese physicist*

Koshiba (**ko**-shee-ba) was born in Toyohashi, Japan, and educated at the University of Tokyo, graduating in 1951, and at the University of Rochester, New York, where he gained his PhD in 1955. He spent much of his career at Tokyo where he became a professor in the physics department and an emeritus professor when he retired in 1987.

Koshiba and his colleagues are noted for their construction of a neutrino detector, which they called *Kamiokande*. It consisted of a very large tank full of water placed underground at the Kamioka mine in Japan. The presence of neutrinos is detected by electrons being released when the neutrinos interact with the nuclei of atoms in the water. The Kamiokande experiment confirmed the results of Raymond DAVIS that there are fewer neutrinos coming from the Sun than theoretically expected. In 1987 the Kamiokande detector detected neutrinos from the supernova explosion 1987A.

In 1996 Koshiba and his colleagues completed the construction of an improved neutrino detector, *Super Kamiokande*. This detector has observed neutrinos produced in the atmosphere of the Earth. These observations indicate the occurrence of *neutrino oscillations*, a phenomenon in which one type of neutrino can change into another type. This phenomenon is of great significance both in elementary particle theory and in cosmology.

The pioneering work of Koshiba and Davis in the subject now known as *neutrino astronomy* led to them sharing the 2002 Nobel Prize for physics with Riccardo GIACCONI.

Kossel, Albrecht (1853–1927) *German biochemist*

Born in Rostock, Germany, Kossel (**kos**-el) was professor of physiology at the universities of Marburg and Heidelberg (1895–1923). He had first studied medicine, but turned his attention to biochemistry under the influence of Felix HOPPE-SEYLER, whose assistant he was at Strasbourg (1877–81). Kossel was also for a time a colleague of Emil DU BOIS-REYMOND. While with Hoppe-Seyler, Kossel continued the latter's investigations of the cell substance called nuclein, demonstrating that it contained both protein and nonprotein (nucleic acid) parts. He was further able to show that the nucleic acids, when broken down, produced nitrogen-bearing compounds (purines and pyrimidines) as well as carbohydrates. Kossel also studied the proteins in spermatozoa, being the first to isolate the amino acid histidine. He was awarded the Nobel Prize for physiology or medicine in 1910 for his work on cells and proteins.

Kosterlitz, Hans Walter (1903–1996) *German–British psychopharmacologist*

Kosterlitz (**kost**-er-lits) was educated at the universities of Heidelberg, Freiburg, and Berlin where he obtained his MD in 1929. He remained there as an assistant until 1933 when, with the rise of the Nazis, he sought safety in Britain. He joined the staff of Aberdeen University in 1934 where he later served as professor of pharmacology and chemistry from 1968 until 1973 when he became director of the university's drug addiction research unit.

For many years Kosterlitz had been working on the effects of morphine on mammalian physiology when in 1975, in collaboration with J. Hughes, he made his most dramatic discovery. They were investigating the effect of morphine in inhibiting electrically induced contractions in the guinea-pig intestine and, to their surprise, discovered that the same effect could be produced by extracts of brain tissue. When it turned out that the effect of the extract could be inhibited by naloxone, a morphia antagonist, it seemed likely to them that they had in fact stumbled on the endogenous opiates, discussed earlier by Solomon SNYDER, and named by them enkephalins. They quickly succeeded in isolating two such enkephalins from pigs' brains and found them to be almost identical peptides consisting of five amino acids differing at one site only and consequently known as methionine and leucine enkephalins. When it was further shown that they possessed analgesic properties hopes were raised once more of the possibility of developing a nonaddictive yet powerful pain killer. Research is still continuing.

Kouwenhoven, William Bennett (1886–1975) *American electrical engineer*

Kouwenhoven (**koh**-en-hoh-ven), who was born in Brooklyn, New York, began his career as an instructor in physics at the Brooklyn Polytechnic in 1906. After a brief period as an instructor in electrical engineering at Washington University from 1913 to 1914 he moved to Johns Hopkins University where he later served as professor of electrical engineering from 1930 until his retirement in 1954.

In the 1930s Kouwenhoven introduced the first practical electrical defibrillator for delivering an alternating current discharge to the heart. It was not, however, until the postwar years that the technique became widely used. The method is used to cure ventricular fibrillation, a heart condition in which the normal rhythmical contractions are modified or cease owing to irregular twitchings of the heart wall. In 1959 Kouwenhoven also introduced the technique of closed chest cardiac massage, a first-aid method that is used to keep alive people whose hearts have stopped beating or who have stopped breathing. The technique can maintain life for up to an hour.

Kovalevski, Aleksandr Onufrievich (1840–1901) *Russian zoologist and embryologist*

> [Kovalevski's work] regards the fossil not as a petrified skeleton, but as having belonged to a moving and feeding animal; every joint and facet has a meaning.
> —H. Osborn, *The Age of Mammals in Europe, Asia, and North America* (1910)

Kovalevski (kov-a-**lef**-skee), who was born at Shustyanka, near Dvinsk (now Daugavpils in Latvia), took a science doctorate at the University of St. Petersburg, where he later taught and became professor (1891–93). He also taught at the universities of Kazan (1868–69), Kiev (1869–74), and Odessa (1874–90) and in 1890 was elected to the Russian Academy of Sciences.

One of Kovalevski's most notable contributions to zoological science and the fuller understanding of evolution lay in his demonstration that all multicellular animals display a common pattern of physiological development. His research into the embryology of primitive chordates, such as *Amphioxus* (the lancelet), *Balanoglossus* (the acorn worm), and the sea squirts, particularly his demonstration of the links between them and the craniates, provided the basis for later studies of the evolutionary history of the vertebrates and led to HAECKEL's theory that all multicellular animals are derived from a hypothetical ancestor with two cell layers. Kovalevski's most important publications are *Development of Amphioxus lanceolatus* (1865) and *Anatomy and Development of Phoronis* (1887).

Kovalevsky, Sonya (1850–1891) *Russian mathematician*

> While Saturn's rings still shine,
> While mortals breathe,
> The world will ever remember your name.
> —Fritz Leffler, obituary poem on Kovalevsky (1891)

The daughter of a wealthy artillery general, Kovalevsky (kov-a-**lef**-skee) was born in Moscow and brought up in the traditional manner on a large country estate, being educated only by an English governess. Her introduction to mathematics was unusual. One of the children's rooms of the family's large country house had temporarily been papered with Ostragradsky's lecture notes on calculus. Kovalevsky studied the notes as a child, and this gave her a grounding in the subject.

In 1867 the family moved to St. Petersburg and Kovalevsky managed to receive some more formal teaching at the Naval Academy. But the only socially acceptable way to continue her education was as a married woman. Consequently, in 1868 she entered into a marriage of convenience with Vladimir Kovalevsky, a young paleontologist and a translator of DARWIN. They moved to Heidelberg in 1869 and two years later to Berlin. Although she could not be admitted to public lectures, Kovalevsky received private classes from Karl WEIERSTRASS, who was so impressed with her work that he persuaded the Göttingen authorities to award her a doctorate in 1874 for a thesis on partial differential equations. Despite this she remained unemployable as a female professional mathematician. In 1878 the Kovalevskys returned to Russia and speculated unwisely in property. With mounting debts Vladimir committed suicide in 1883. By this time Weierstrass had arranged for Kovalevsky to be appointed to a lectureship in mathematics at the University of Stockholm. She died from pneumonia in 1891. Kovalevsky is best known for her work in partial differential equations in which she extended some earlier results of CAUCHY. She also won the Borodin prize of the French Académie des Sciences in 1888 for her memoir *On the Rotation of a Solid Body about a Fixed Point.*

Kozyrev, Nikolay Aleksandrovich (1908–1983) *Russian astronomer*

Kozyrev (ko-**zeer**-ef), who was born in St. Petersburg, studied at the university there, graduating in 1928. In 1931 he joined the Pulkovo Astronomical Observatory near St. Petersburg (or Leningrad, as it was then known); he also worked at the Kharkov and the Crimean astrophysical observatories. From 1936 to 1948 he was imprisoned under the Stalin regime.

Kozyrev's work included planetary studies and research into stellar atmospheres and stellar structure. In 1958 he reported some remarkable observations of the Moon that demonstrated that it was not completely inert. He was observing the central peak of the Alphonsus crater when "it became strongly washed out and of an unusual reddish hue." The disturbance lasted for at least half an hour and was distinguished from all other reports of strange transient lunar phenomena by being supported with a spectrogram. This revealed a marked increase in temperature together with the release of a cloud of carbon particles. Since Kozyrev's report other transient red spots have been seen and an examination of past records has shown that they were noticed as far back as the 18th century. It is thought that Kozyrev was observing some form of volcanic activity.

Kraft, Robert Paul (1927–) *American astronomer*

Born in Seattle, Washington, Kraft graduated in 1947 from the University of Washington and obtained his PhD in 1955 from the University of California, Berkeley. He held brief appointments at Whittier College, the Mount Wilson and Palomar Observatories, and the Universities of Indiana and Chicago before returning to the Mount Wilson and Palomar Observatories in 1960. In 1967 Kraft moved to Lick Observatory as professor of astronomy, a post he held until his retirement in 1992. He is best known for his work on novae and supernovae. He has proposed that most if not all stars on which novae erupt are small dense white dwarfs that are members of binary systems. This led him to construct a dynamical model of novae based on the passage of material from the larger to the smaller white-dwarf member of the binary.

Kramer, Paul Jackson (1904–1995) *American plant physiologist*

Kramer was born in Brookville, Indiana, and graduated in botany from the University of Miami, obtaining his PhD from Ohio State University in 1931. He immediately joined the faculty of Duke University, South Carolina, and spent his entire career there serving as professor of botany from 1945 until his retirement in 1974.

Kramer worked on problems of the absorption of water by plants, surveying the subject in his *Plant and Soil Water Relationships* (1949). He demonstrated that two different mechanisms are involved in water uptake by roots, depending on whether the plants are transpiring quickly or slowly. He also showed the importance of taking plant water stress into account when making correlations between soil moisture and plant growth. In studies using radioactively labeled elements he found that the region of maximum absorption in roots is not the tip but the area several centimeters behind the tip where the xylem-conducting vessels are fully formed. Other researches led Kramer to the conclusion that substantial amounts of minerals enter plant roots passively in the transpiration stream. Kramer also worked on the physiology of trees, publishing with Theodore Kozlowski *The Physiology of Woody Plants* (1979), an update of an earlier 1960 joint work.

Kratzer, Nicolas (1486–1550) *German astronomer, horologist, and instrument maker*

Although born at Munich, in Germany, Kratzer (**kraht**-ser) spent 30 years in England. He worked for part of that time at the court of Henry VIII and played an important part in introducing knowledge of scientific instruments and techniques of instrument making to England. He had his portrait painted by Holbein whose famous picture "The Ambassadors" apparently contains many of Kratzer's instruments.

Kraus, Charles August (1875–1967) *American chemist*

Kraus (krows) was born in Knightsville, Indiana, and brought up on a Kansas farm. He was educated at the University of Kansas, Johns Hopkins University, and the Massachusetts Institute of Technology where he obtained his PhD in 1908. After holding junior appointments at the University of California and MIT Kraus served successively as professor of chemistry at Clark University (1914–24) and at Brown University from 1924 until his retirement in 1946.

A noted experimentalist, Kraus worked on liquid ammonia. He was especially concerned with the solubility of various elements, particularly the alkali metals, in ammonia. He also made an important contribution to industrial chemistry in 1922 when, at the request of the Standard Oil Company of New Jersey, he developed processes for the commercial production of tetraethyl lead. This permitted Thomas MIDGLEY's discovery of its antiknock properties to be fully utilized and thus made possible the high-compression engine. During World War II Kraus worked on the Manhattan Project (the atom-bomb project) on problems connected with the purification of uranium salts.

Krebs, Edwin Gerhard (1918–   ) *American biochemist*

Born in Lansing, Iowa, Krebs was educated at the University of Illinois and Washington University, Seattle, gaining his MD in 1943. For the following two years he worked as an intern and assistant resident at Barnes Hospital, St. Louis, and in 1946 returned to Washington University as a National Institutes of Health research fellow. In 1968, by now professor of biochemistry, Krebs moved to a similar position at the University of California, Davis. In 1977 he again returned to Seattle as professor of pharmacology in the school of medicine. In the same year he joined the Howard Hughes Medical Institute as investigator, becoming senior investigator in 1980 and emeritus professor in 1991.

Krebs is perhaps best known for his work on the regulation of enzyme activity, which he did at Seattle in conjunction with the biochemist, Edmond FISCHER. For this, he and Fischer shared the 1992 Nobel Prize for physiology or medicine.

Krebs, Sir Hans Adolf (1900–1981) *German–British biochemist*

Krebs, the son of an ear, nose, and throat specialist, was born in Hildesheim, Germany, and educated at the universities of Göttingen, Freiburg, Munich, Berlin, and Hamburg, obtaining his MD in 1925. He taught at the Kaiser Wilhelm Institute, Berlin, and the University of Freiburg but in 1933, with the growth of the

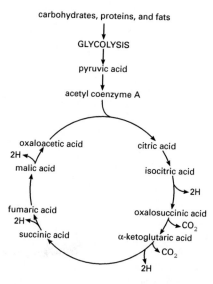

THE TRICARBOXYLIC ACID CYCLE   (Krebs cycle) The purpose of the TCA cycle is to complete the oxidation of glucose begun in glycolysis. In the sequence of reactions from citric acid to oxaloacetic acid, pyruvic acid is oxidized to carbon dioxide and water. During this process considerable amounts of energy are released – 93% of the total energy released in glucose oxidation.

Nazi movement, decided to leave Germany. Consequently he moved to England, where from 1935 to 1954 he served as professor of biochemistry at Sheffield University; after 1945 he was appointed director of the Medical Research Council's Cell Metabolism Unit at Sheffield. In 1954 Krebs moved to Oxford to take the Whitley Chair of Biochemistry, a post he held until his retirement in 1967.

Krebs is best known for his discovery of the *Krebs cycle* (or tricarboxylic acid cycle) in 1937. This is a continuation of the work of Carl and Gerty CORI, who had shown how carbohydrates, such as glycogen, are broken down in the body to lactic acid; Krebs completed the process by working out how the lactic acid is metabolized to carbon dioxide and water. When he began this work little was known apart from the fact that the process involved the consumption of oxygen, which could be increased, according to Albert Szent-Györgyi, by the four-carbon compounds succinic acid, fumaric acid, malic acid, and oxaloacetic acid. Krebs himself demonstrated in 1937 that the six-carbon citric acid is also involved in the cycle. By studying the process in pigeon breast muscle Krebs was able to piece together the clues already collected into a coherent scheme. The three-carbon lactic acid is first broken down to a two-carbon molecule unfamiliar to Krebs; it was in fact later identified by Fritz LIPMANN as coenzyme A. This then combines with the four-carbon oxaloacetic acid to form the six-carbon citric acid. The citric acid then undergoes a cycle of reactions to be converted to oxaloacetic acid once more. During this cycle two molecules of carbon dioxide are given up and hydrogen atoms are released; the hydrogen is then oxidized in the electron transport chain with the production of energy. Much of the detail of this aspect of the cycle was later filled in by Lipmann, with whom Krebs shared the 1953 Nobel Prize for physiology or medicine. Krebs fully appreciated the significance of the cycle, pointing out the important fact that it is the common terminal pathway for the chemical breakdown of *all* foodstuffs. In 1932, with K. Henselheit, Krebs was responsible for the introduction of another cycle. This was the urea cycle, whereby amino acids (the constituents of proteins) eliminate their nitrogen in the form of urea, which is excreted in urine. This left the remainder of the amino acid to give up its potential energy and participate in a variety of metabolic pathways.

Kroemer, Herbert (1928–) *German-born American physicist*

Kroemer (**kroh**-mer) was born in Weimar, Germany, and educated at the University of Göttingen, from where he received a PhD in theoretical physics in 1952 on the topic of the effects of hot electrons on the then recently invented transistor. Kroemer then worked in several research laboratories in Germany and the United States. In 1976 he started a research group into compound semiconductor technology in the Electrical and Computer Engineering Department at the University of California, Santa Barbara (UCSB).

His career has been dedicated to research on the physics and technology of semiconductors and semiconductor devices. Kroemer has worked on problems that are many years ahead of well-established technology. In the mid-1950s he was the first person to point out that layered semiconductor structures called semiconductor heterostructures could lead to greatly improved performances in certain semiconductor devices. In 1963 he put forward the idea of the double-heterostructure laser, which is fundamental in the subject of semiconductor lasers. These ideas did not become part of the mainstream of technology until 1980, and the development of epitaxial growth technology, i.e., the technology associated with the growth of a layer of one substance on a single crystal of another in such a way that the crystal structure of the layer is the same as that of the substrate. In the 1980s heterostructures dominated the field of compound semiconductors, being used for lasers, light-emitting diodes, and integrated circuits.

After joining UCSB, Kroemer and his colleagues investigated properties of novel materials such as GaAs on silicon. Since the mid-1980s he has been interested in the properties and technological applications of materials such as InAs, GaSb, and AlSb. Kroemer shared the 2000 Nobel Prize for physics with Zhores ALFEROV and Jack KILBY, two other leading pioneers of the information age, for his pioneering work on semiconductor heterostructures and their applications.

Krogh, Schack August Steenberg (1874–1949) *Danish physiologist*

> We may fondly imagine that we are impartial seekers after truth, but with a few exceptions, to which I know that I do not belong, we are influenced – and sometimes strongly – by our personal bias; and we give our best thoughts to those ideas which we have to defend.
>
> —Comment made in 1929

Krogh (krog), the son of a brewer from Grenaa in Denmark, was educated at the Aarhus gymnasium and the University of Copenhagen, where he obtained his PhD in 1903. He spent his whole career at the university, serving as professor of animal physiology from 1916 until his retirement in 1945.

Krogh first worked on problems of respiration. He argued, against his teacher Christian Bohr, that the absorption of oxygen in the lungs and the elimination of carbon dioxide took place by diffusion alone. He made precise measurements to show that the oxygen pressure was always higher in the air sacs than in the blood and, consequently, there was no need to assume any kind of nervous control. It was, however, with his studies of the capillary system that Krogh achieved his most dramatic success. The simplest explanation of its action was to assume it was under the direct hydraulic control of the heart and arteries: the stronger the heart beat, the greater the amount of blood flowing through the capillaries. Krogh had little difficulty in showing the inadequacy of such a scheme by demonstrating that even among a group of capillaries fed by the same arteriole some were so narrow that they almost prevented the passage of red cells while others were quite dilated, allowing the free passage of the blood. Not content with this descriptive account Krogh went on to make a more quantitative demonstration. Working with frogs, which he injected with Indian ink shortly before killing, he showed that in sample areas of resting muscle the number of visible (stained) capillaries was about 5 per square millimeter; in stimulated muscle, however, the number was increased to 190 per square millimeter. From this he concluded that there must be a physiological mechanism to control the action of the capillaries in response to the needs of the body. It was for this work, fully described in *The Anatomy and Physiology of Capillaries* (1922), that Krogh was awarded the 1920 Nobel Prize for physiology or medicine.

Kronecker, Leopold (1823–1891) *German mathematician*

Born in Liegnitz, now Legnica in Poland, Kronecker (**kroh**-nek-er) studied mathematics at Berlin but he did not become a professional mathematician until relatively late in life. He worked, highly successfully, as a businessman until he had made enough money to abandon commerce and devote himself fully to mathematics. He taught at Berlin from 1861, and, in 1883, was appointed professor. Outside mathematics Kronecker's interests were wide. He was a highly cultured man who used his wealth to patronize the arts. He also had a deep interest in philosophy and Christian theology, although he was not converted to Christianity until shortly before his death.

Kronecker's mathematical work was almost entirely in the fields of number theory and higher algebra, although he also made some contributions to the theory of elliptic functions. His work on algebraic numbers was inspired by his constructivist outlook, which involved a distrust of nonconstructive proofs in mathematics and a suspicion of the infinite and all kinds of number other than the natural numbers. This attitude led him to rewrite large areas of algebraic number theory in order to avoid reference to such suspect entities as imaginary or irrational numbers. Kronecker's constructivism

is summed up in a famous remark he made during an after-dinner speech: "God made the integers, all else is the work of man." His suspicion of nonconstructive methods led Kronecker into fierce controversy with two of the leading mathematicians of his day, Karl WEIERSTRASS and Georg CANTOR. His outlook anticipates to a considerable extent the views of the Dutch mathematician L. E. J. BROUWER. Kronecker was also one of the first to understand thoroughly and use Evariste GALOIS's work in the theory of equations. The *Kronecker delta function* is named for him.

Kroto, Sir Harold Walter (1939–   ) *British chemist*

Born at Bolton in Lancashire, England, Kroto was educated at the University of Sheffield where he completed his PhD in 1964. After a postdoctoral fellowship with the National Research Council of Canada, and two years with Bell Laboratories, New Jersey, Kroto moved to the University of Sussex (1967) where he was appointed professor of chemistry in 1985. He was knighted in 1996.

Along with his Sussex colleague David Walton, Kroto had a long-standing interest in molecules containing carbon chains linked by alternate triple and single bonds. Such chains with five, seven, and nine carbon atoms had been identified by radioastronomers in space. In 1984 Kroto heard that the American chemist Richard SMALLEY had developed new techniques involving laser bombardment for the production of clusters of atoms. He suspected they might be suitable to produce the chains of carbon atoms which interested him. Consequently in 1985 he visited Smalley in Houston. Kroto persuaded Smalley to direct his laser beam at a graphite target. Clusters of carbon atoms were indeed produced but, more interesting than the small chains he was looking for, Kroto found a mass-spectrum signal for a molecule of exactly 60 carbon atoms. The first suggestion was that it had a sandwichlike graphite structure. However, such a planar fragment would have reactive carbon atoms at the edges, whereas C_{60} appeared to be stable. An alternative structure with the necessary lack of reactivity would be a spherical one in which the 60 carbon atoms are positioned at the vertices of a polyhedron. In Kroto and Smalley's model, which was later shown to be correct, the faces of the polyhedron are pentagons or hexagons arranged like the panels on a modern soccer ball. The framework also resembles the structure of the geodesic dome designed by the architect Buckminster Fuller. Kroto called C_{60} *buckminsterfullerene*, a name subsequently shortened to *fullerene*. Carbon molecules with this type of structure are informally called "bucky balls." Before the structure could be put beyond doubt sufficient C_{60} would have to be

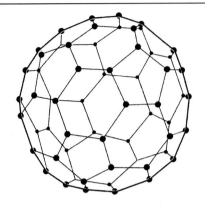

BUCKMINSTERFULLERENE   The ball-shaped carbon structure of the C_{60} molecule.

prepared for detailed spectroscopic analysis. This, however, proved unexpectedly difficult. On his return to Sussex Kroto began to work on the problem. Unfortunately funding was unavailable, even for relatively cheap experiments with graphite electrodes. It was consequently left to Kratschmer and HUFFMAN to announce in 1990 that they had synthesized a new form of carbon. When he read a full description of the synthesis Kroto learned that he had been pursuing the same method two years earlier when he had been forced to abandon the project through lack of funds. He shared the 1996 Nobel Prize for chemistry for his work on fullerene.

Kuffler, Stephen William (1913–1990) *American neurophysiologist*

Kuffler (**kuuf**-ler) was born in Tap, Hungary, and educated at the University of Vienna where he obtained his MD in 1937. He worked in Sydney, Australia, with John ECCLES and Bernard KATZ from 1938 to 1945 when he moved to America. There Kuffler worked first at Johns Hopkins from 1947 until 1959 when he became professor of neurophysiology at Harvard.

In 1953 Kuffler reported the results of certain experiments on the ganglion cells of the retina that did much to stimulate the later important work of David HUBEL and Torsten Weisel. Electrodes were inserted in a cat's brain and recordings made when a small spot of light, about 0.2 millimeter in diameter, was shone onto the cat's receptive field. He found that in one area the spot of light excites a ganglion cell and produces an "on" response, but such an "on" response can be converted to an inhibitory "off" response by simply shifting the spot by 1 millimeter or less across the retinal surface. He thus went on to propose that there are only two basic receptive field types in the cat's retina, the "on" and "off" center. Kuffler also worked on synaptic transmission and collaborated with the American neurophysiologist John Nicholls on the stimulating text, *From Neuron to Brain* (1976). In 1976 he received the Wakeman award for his contributions to the treatment of paraplegia.

Kuhn, Richard (1900–1967) *Austrian–German chemist*

Kuhn (koon) was educated at the university in his native city of Vienna and then at Munich, where he obtained his PhD in 1922. He worked at the Federal Institute of Technology in Zurich from 1926 to 1929, when he moved to the University of Heidelberg to serve as professor of chemistry and, from 1950, as professor of biochemistry.

Like Paul KARRER, Kuhn worked mainly on the chemistry of plant pigments and vitamins, repeating many of Karrer's results. In particular Kuhn, independently of Karrer, worked out the structures of vitamins A and B_2, and, in 1938, he also synthesized vitamin B_6. For his work on carotenoids and vitamins Kuhn was awarded the Nobel Prize for chemistry in 1938, the year following the same award to Karrer. Hitler however objected to the award and Kuhn was forced to wait until the end of the war before he was allowed to receive the prize.

Kuhn, Thomas Samuel (1922–1996) *American philosopher and historian of science*

> Nevertheless, paradigm changes do cause scientists to see the world of their research-engagement differently. In so far as their only recourse to that world is through what they see and do, we may want to say that after a revolution, scientists are responding to a different world.
> —*The Structure of Scientific Revolutions* (1962)

Kuhn was born in Cincinatti, Ohio, and educated at Harvard. After serving as a junior fellow and working as a scientist in the Radio Research Laboratory there, he moved to the History of Science Faculty in 1951. He subsequently held posts at Berkeley (1957–64) and Princeton (1964–79), including a spell at the Institute for Advanced Studies (1972–79). From 1979 to 1990 he worked at the Massachusetts Institute of Technology.

Kuhn's first book, *The Copernican Revolution* (1957), gave some indication of the direction of his thought. He is best known for his second book, *The Structure of Scientific Revolutions* (1962). Not only has it been extremely influential in the field since its publication, but its influence has also spread to related areas and his terminology of "paradigms" has become part of general intellectual discourse. Kuhn began by considering the nature of scientific change. This had long been seen as progressive – the addition of new, more precise truths to an already existing stock of truths. In contrast, Kuhn argued, change was revolutionary and total. The old order was rejected and a new order imposed as when, for example, a Darwinian world of natural selection overthrew a world of design and teleology or when a quantum-mechanical world of uncertainty overthrew the deterministic mechanics of NEWTON and LAPLACE. But science is not in a state of permanent revolution; it undergoes periods of stability as well as change. Kuhn identified such stable periods as times of "normal science." These follow a successful revolution and the broad acceptance of a common view, or paradigm in Kuhn's terminology. Here scientists practice normal science – the aim is not to challenge the paradigm, but to describe the world in its paradigmatic terms. Because changes from one revolutionary period to another – paradigm shifts – are so total, different paradigms cannot conflict; they cannot even be compared; they are "incommensurable." What leads scientists to switch paradigms? This is seldom brought about by a single issue. Rather, it is a growing accumulation of anomalous cases, in which the paradigm seems unable to cope with newer data. At first, changes are made to "save the theory," but eventually a point is reached at which this is no longer tenable. A new paradigm is then introduced and the Kuhnian cycle once more enters a new phase.

Kühne, Wilhelm Friedrich (1837–1900)
German physiologist

Willy Kühne (**koo**-ne), the son of a wealthy Hamburg merchant, was educated at the University of Göttingen where he obtained his PhD on induced diabetes in frogs in 1856. He studied further in Jena, Berlin, Paris (under Claude Bernard), and Vienna before joining Rudolf VIRCHOW's Berlin institute in 1861. Kühne later held chairs of physiology, first at Amsterdam from 1868 and from 1871 until his retirement in 1899 at Heidelberg.

Kühne worked with Russell CHITTENDEN on problems of digestion, and he isolated trypsin from pancreatic juice. In 1859, working with the sartorius muscle, he demonstrated that nerve fibers can conduct impulses both ways, and also showed that chemical and electrical stimuli can be used to excite muscle fibers directly. He also, in the late 1870s, coined the term "rhodopsin" for the substance, also known as visual purple, first discovered in the retinal rods by Franz Boll in 1876. It was soon realized that the pigment was bleached out of the retina by light and resynthesized in the dark. Kühne realized that this could be used to photograph the eye, to take what he termed an "optogram" by the process of "optography." To achieve this he placed a rabbit facing a barred window after having its head covered with cloth to allow the rhodopsin to accumulate. After three minutes it

was decapitated and the retina removed and fixed in alum, clearly revealing a picture of a barred window. Later investigations of rhodopsin by such scholars as George WALD revealed much about the mechanism of vision.

Kuiper, Gerard Peter (1905–1973)
Dutch–American astronomer

Born at Harenkarspel in the Netherlands, Kuiper (**kI**-per or **koip**-er) studied at the University of Leiden, where he obtained his BSc in 1927 and his PhD in 1933. He immediately emigrated to America where he took up an appointment at the Lick Observatory in California and then lectured (1935–36) at Harvard. In 1936 he joined the Yerkes Observatory and in 1939 moved to the McDonald Observatory in Texas, both run by the University of Chicago. He served as their director (1947–49, 1957–60) and was also professor of astronomy (1943–60). From 1960 to 1973 he was head of the Lunar and Planetary Laboratory of the University of Arizona.

Kuiper's main research work was on the solar system. He discovered two new satellites: Miranda, the fifth Uranian satellite, in 1948 and Nereid, the second Neptunian satellite, in 1949. He also investigated planetary atmospheres and succeeded in detecting carbon dioxide in the Martian atmosphere in 1948. Four years earlier he had found evidence of methane in the atmosphere of Saturn's largest satellite, Titan. In 1950 he produced some intriguing data on Pluto. Based on observations made with the 200-inch (5-m) reflector at the Palomar Observatory he estimated the diameter of Pluto as 0.23 seconds of arc, which was equivalent to about 3,600 miles (5,800 km) or half the Earth's diameter. But as its mass was supposed to be roughly the same as that of the Earth it implied the unlikely conclusion that Pluto had a density of about ten times that of the Earth. Recent measurements, however, reveal that Pluto's mass is only 0.2% and its diameter roughly a quarter of Earth's. Kuiper also speculated on the origin of the planets and proposed in 1949 his theory that each planet evolved from its own gaseous cloud that was not initially part of the Sun. This is not generally accepted. With the advent of the space age, Kuiper became closely involved with several space missions, including the Ranger program, 1961–65, in which the first close-up photographs of the Moon were obtained, and the Mariner 10 flight to Venus and Mercury, which was launched in 1973 shortly before his death.

Kundt, August (1839–1894) *German physicist*

Kundt (**kuunt**), who was born at Schwerin, in Germany, studied at Berlin University, where he received his doctorate in 1864. He was made a professor of physics at the Federal Institute of Technology, Zurich, in 1868 and, three years later, obtained a similar post in Strasbourg.

His fame rests largely on his experimental determination of the speed of sound. His apparatus involved a pipe closed at one end and containing a small amount of fine powder. A vibrating disk at the open end was used to set up standing waves in the pipe, causing the dust to settle in regular nodes down the pipe. From the distance between the nodes the speed of sound could be calculated. The apparatus is known as *Kundt's tube*. In 1888 he was appointed to the chair of physics in Berlin and, with his student, Wilhelm RÖNTGEN, he demonstrated the rotation of the plane of polarization for light traveling through a gas in a magnetic field.

Kurchatov, Igor Vasilievich (1903–1960)
Russian physicist

Kurchatov (**kuur**-cha-tof), born the son of a surveyor at Sim in Russia, was educated at the University of the Crimea, from which he graduated in 1923. Shortly afterwards he was appointed to the Leningrad Physico-Technical Institute where, in 1938, he became director of the nuclear physics laboratory. At some time during the war he moved to Moscow to take control of his country's military and industrial atomic research.

Under Kurchatov's direction the Soviet atomic program was remarkably successful. The Soviet Union exploded its first atomic bomb in 1949, its first hydrogen bomb in 1952, and constructed a nuclear power station in 1954. Before 1978, element-104 was known in the Soviet Union as *kurchatovium*. (The American claimants to the discovery named it *rutherfordium*; it now has the systematic name *unnilquadium*.)

Kurti, Nicholas (1908–1998) *Hungarian–British physicist*

Kurti (**kuur**-tee) was educated at the gymnasium in his native city of Budapest and at the universities of Paris and Berlin. After a brief period teaching at Breslau, he moved to England in 1933 and joined the Clarendon Laboratory at Oxford. He remained there, apart from the war years spent on the development of the UK atomic-energy project, and eventually served as professor of physics from 1967 until his retirement in 1975.

Kurti was one of the pioneers in work on low-temperature physics. At the Clarendon Laboratory he worked on methods of achieving low temperatures and studying thermal and magnetic properties at these temperatures. In his work he produced temperatures as low as 10^{-6}K to study the ordering of nuclear spins under such conditions.

Kusch, Polykarp (1911–1993) *American physicist*

Although born in Blankenburg, Germany, Kusch (kuush) spent all but the first year of his life as a resident of the United States, becoming a naturalized citizen in 1922. His early education was in the Midwest of the United States, and his undergraduate studies were at the Case Institute of Technology, Cleveland, Ohio. Starting in chemistry, he made an early switch to physics, and gained his BS in 1931. He followed this with work as a research assistant at the University of Illinois on problems of optical molecular spectroscopy, gaining his MS there in 1933 and his PhD in 1936. This was followed by a short period at the University of Minnesota researching mass spectroscopy (1936–37). From 1937 to 1972 he was associated with the physics department of Columbia University, New York City, apart from interruptions in World War II when he was engaged in special research on the military applications of vacuum tubes and microwave generators at Westinghouse Electric Corporation (1941–42) and Bell Telephone Laboratories (1944–46).

At Columbia, Kusch began work under Isidor RABI on the first radiofrequency spectroscopy experiments using atomic and molecular beams. His research was principally on the fine details of the interactions of the constituent particles of atoms and molecules with each other and with an externally applied magnetic field. In particular, Kusch made very accurate determinations of the magnetic moment of the electron as deduced from the hyperfine structure of the energy levels in certain elements, and in 1947 found a discrepancy of about 0.1% between the observed value and that predicted by theory. Although minute, this anomaly was of great significance to theories of the interactions of electrons and electromagnetic radiation (now known as quantum electrodynamics). It was for his precise work in measuring the electron's magnetic moment that he received the 1955 Nobel Prize for physics, sharing it with Willis LAMB, who performed independent but related experiments at Columbia University on the hyperfine structure of the hydrogen atom. Kusch's career at Columbia University took him to associate professorship (1946), professorship (1949), executive director of the Columbia Radiation Laboratory (1952–60), vice president and dean of the faculty (1969–70), and executive vice president and provost (1970–71). In 1972 he left to become professor of physics, and then Eugene McDermott Professor, at the University of Texas at Dallas.

L

Lacaille, Nicolas Louis de (1713–1762) French astronomer

> The so-called nebulous stars offer to the eyes of the observers a spectacle so varied that their exact and detailed description can occupy astronomers for a long time and give rise to a great number of curious reflections on the part of philosophers.
> —In *Memoirs of the Royal Academy of Sciences* (1755)

Lacaille (la-**kah**-ye), who was born at Rumigny in France, started his career as a theology student, becoming a deacon before taking up astronomy. He became professor of mathematics at Mazarin College. In the late 1730s he worked with Jacques CASSINI on measuring the French meridian by means of triangulation. In 1751 he went to South Africa while Joseph de LALANDE went to Berlin to attempt to determine the lunar parallax, their separation by 85° giving them a sufficient base for an accurate result. While in South Africa he observed 10,000 stars, publishing a catalog of 2,000 of them in *Coelum australe stelliferum* (1763; Catalog of the Southern Sky).

Lack, David Lambert (1910–1973) British ornithologist

Lack, the son of a well-known and prosperous London surgeon, was educated at Cambridge University. From 1933 to 1940 he taught at Dartington Hall, a progressive private school, except for a year's leave in 1938 spent studying the birds of the Galapagos. During World War II he served with the Army Operational Research Group, helping to develop radar. This experience was later applied when he used radar in his studies of bird migration. In 1945 he finally became a professional ornithologist and served as director of the Edward Grey Institute of Field Ornithology, Oxford, until his death.

Lack's first substantial work, the widely read *Life of the Robin* (1943), was followed by the publication of his Galapagos material in *Darwin's Finches* (1947), a fascinating account of the 14 specialized species of finch that have evolved from an original invading flock of ordinary seed-eating finches. Such fieldwork inevitably led Lack to the consideration of more theoretical questions. In particular, he studied the factors controlling numbers in natural populations and concluded that such factors act more severely when numbers are high than when they are low. The irregularities of population fluctuation suggested to Lack that the control mechanisms must be very complex. His data and theories are discussed in such works as *Natural Regulation of Animal Numbers* (1954) and *Population Studies of Birds* (1966). Such data was variously interpreted, for instance by Richard DAWKINS, who claimed that it supported the theory of the "selfish gene"; Lack himself was more inclined to use the data to support the idea of group selection.

La Condamine, Charles Marie de (1710–1774) French geographer and explorer

> The Immortals of the Academy
> Have admitted La Condamine to their ranks today
> He's stone deaf – so much the better for him
> But not mute – so much the worse for them.
> —Alexis Piron, epigram on La Condamine's election to the French Academy

Born in Paris, France, La Condamine (la kon-da-**meen**) began his career in the army. In 1735 he took part, with Pierre BOUGUER, in an expedition to Peru organized by the French Academy. The aim was to measure the length of a degree of the meridian near the equator to determine the true shape of the Earth. This was disputed by Newtonians, who claimed it to be an oblate spheroid (a larger equatorial than polar diameter), and the followers of Jacques CASSINI, who saw it more as a sphere that had been squeezed rather than flattened, with a larger polar diameter. The expedition's results confirmed NEWTON's views.

In 1743 La Condamine traveled by raft down the Amazon. Among the scientific observations he made, he was particularly impressed by a product the Omagua tribe called "Cahoutcha" and used to make unbreakable containers. He took samples back with him to France and, in 1751, read a paper to the Academy on the substance – the first mention in Europe of rubber.

Laënnec, René Théophile Hyacinthe (1781–1826) French physician

Laënnec (lay-**nek**) was born in Quimper, Brittany, the son of an unsuccessful lawyer. After his mother's death he was brought up mainly by his uncle G. F. Laënnec, professor of medicine at the University of Nantes. He was a pupil of Jean CORVISART at the Charité in Paris, qualifying as a doctor in 1804. He worked at a number of hospitals before becoming in 1814 physician-in-chief at the Hôpital Necker, where he remained until 1822. Then he was appointed professor of medicine at the Collège de France. In 1826 he was forced to retire to Brittany where he died of consumption.

In 1819 Laënnec published one of the classic texts of modern medicine, *De l'auscultation médiate* (On Mediate Auscultation). It advanced the work of Leopold AUENBRUGGER in describing sounds detected within the body and the various diseases and anatomical defects they were related to. The work is, however, best known for its description of the situation leading to his invention of the stethoscope. In 1816 he was consulted by a young woman with heart trouble whose age and sex inhibited him from examining her by his usual method, namely, placing his ear on her breasts. Instead, Laënnec tightly rolled a sheaf of paper and placed one end over the heart and the other to his ear. He was surprised and pleased to find that the heartbeat could be heard far more clearly and distinctly than before. The work became widely known and was translated into English in 1821 and only two years later was published in America, where it was vigorously promoted by Austin Flint. The stethoscope itself, in improved flexible versions, rapidly became a standard part of the physician's equipment.

Lagrange, Comte Joseph Louis (1736–1813) *Italian–French mathematician and theoretical physicist*

> The reader will find no diagrams in this work. The methods which I set forth do not require either constructions or geometrical or mechanical reasonings but only algebraic operations, subject to a regular and uniform rule of procedure.
> —*Mécanique analytique* (1788; Analytical Mechanics)

Lagrange (la-**grahnzh**) was born at Turin in Italy of French ancestry. At school he was at first more interested in the classics until he read an essay by the astronomer Edmond HALLEY on the calculus, which converted him to mathematics. By the age of 18 he was teaching at the Artillery School in Turin, eventually becoming professor of mathematics there. He also began a discussion group, which evolved to form the Turin Academy of Sciences in 1758.

By 1760 Lagrange's reputation as one of Europe's greatest mathematicians was established; he received a prize from the Paris Academy of Sciences in 1764 for an essay on the libration of the Moon, i.e., the way in which the Moon apparently oscillates on its axis, so that 59% of its surface can be observed over a 30-year period. On the invitation of Frederick the Great he succeeded Leonhard EULER as mathematical director of the Berlin Academy in 1766, serving there until 1787 when, following Frederick's death, he moved to Paris and a post in the Academy of Sciences. Lagrange produced his greatest work, *Mécanique analytique* (1788; Analytical Mechanics), in Paris. This summarized the research in mechanics since Isaac NEWTON, based on Lagrange's own calculus of variations, and finally placed the mechanical theory of solids and fluids on a rigorous and analytical foundation. The book broke away from Euclidean tradition, and Lagrange commented in it that "the reader will find no diagrams in this work." Lagrange also made many contributions to astronomy and number theory. His work on the theory of equations helped Niels ABEL in his later development of group theory. In astronomy he found a special solution to the three-body problem showing that asteroids will tend to oscillate around a certain point now called the *Lagrangian point*. Much of the credit for the revolutionary introduction of the metric system is also due to him. Despite Lagrange's being technically a foreigner and also a friend of aristocrats, he was regarded with respect during the French Revolution, which broke out in 1789, and was made president of the commission for metrication in 1793. He founded the mathematics department in the Ecole Normale in 1795 and later in the Ecole Polytechnique in 1797. Napoleon honored him, making him a count and a senator.

Lakatos, Imre (1922–1974) *Hungarian philosopher of science*

> Blind commitment to a theory is not an intellectual virtue: it is an intellectual crime.
> —*Philosophical Papers*, Vol. I (1977)

The life of Lakatos (**lok**-o-tohsh) was one of great variety and much tragedy. Born at Debrecen in Hungary, he was educated at the university there and at Budapest, while simultaneously fighting the Nazi invaders as a member of the resistance. Both his mother and grandmother died at Auschwitz. Lakatos had been born Imre Lipschitz but changed his name to Molnar at the beginning of the war, and to Lakatos at the end.

By this time he was a communist and served initially as a secretary in the Hungarian Ministry of Education and spent the year 1949 in Moscow. The authorities, however, began to distrust his reliability, and in 1950 he was arrested as a dissident. During his three years' incarceration, it was later said, one of his inquisitors collapsed under the strain of interrogating him. When released in 1953 he survived by translating works for the Mathematical Institute. Following the Hungarian uprising, he eventually made his way to Cambridge, England, where he gained his PhD in 1958. Soon after, he was appointed to a post at the London School of Economics, where in 1969 he succeeded Karl POPPER as professor of logic and scientific method. Lakatos had earlier made his reputation with the publication (1963–64) of his doctoral thesis as a four-part article. This was later published as *Proofs and Refutations* (Cambridge, England, 1976). The work was concerned with the formula $V - E + F = 2$, first noticed by EULER in 1758 to be true of regular polyhedra; V is the number of vertices, E the number of edges, and F the number of faces. The work, in the form of a dialogue between a teacher and his pupils, aimed to show that mathematical discovery involves more than finding an abstract proof. There are, surprisingly, refutations and counterexamples in the manner of the empirical sciences. Proof in this context becomes more a matter of explanation than a merely formal procedure. Lakatos initially saw science in the same terms as Karl Popper. Later he moved beyond taking falsification as a central feature and developed his notion of a "research program." This consists of a set of rules indicating which research paths to avoid (negative heuristic) and which to pursue (positive heuristic). Further, not all hypotheses of the program have equal status. Some were so central as to be beyond the reach of disproof and constituted the "hard core" of the program; other, more peripheral, parts were clearly dispensable and formed a "protective belt" against the vagaries of experience. In the model proposed by Lakatos, research programs can be either "progressive," leading to new discoveries, or "degenerating" and leading to a possible change of program. Lakatos was still developing his views when he died suddenly aged 51. Much of his later work, however, can be found in the posthumously published *Philosophical Papers* (Cambridge, England, 2 vols., 1977).

Lalande, Joseph de (1732–1807) *French astronomer*

> I am an oilskin for insults and a sponge for praise.
> —On his attitude to publicity

Lalande (la-**lahnd**), who was born at Boug-en-Bresse in France, initially studied law. He later became interested in astronomy, in which field he achieved success at the age of 20 when he and Nicolas de LACAILLE observed lunar parallax and determined the distance of the Moon. He was the professor of astronomy at the Collège de France from 1762. In 1795 he became the director of the Paris Observatory. In 1751 he went to Berlin to make direct observations of the Moon while Lacaille made similar observations 85° away in South Africa. They used these observations to work out the lunar parallax.

Lamarck, Jean Baptiste, Chevalier de (1744–1829) *French biologist*

> The most important ... thoughts and theories which have passed through the mind of a scientific investigator have been crushed in silence and secrecy by his own severe criticism and adverse examination; ... in the most successful instances not a tenth of the suggestions, the hopes, the wishes, the preliminary conclusions have been realized.
> —Quoted by A. L. Mackay in *A Dictionary of Scientific Quotations* (1991)

Lamarck (la-**mark**) was born at Bazantin in France. Although his parents had intended him to enter the Church, when his father died in 1759 Lamarck left his Jesuit school and joined the army, where he was honored for his bravery in the Seven Years' War. He became interested in botany during his postings to camps in the Mediterranean and eastern France and completed his *Flore française* (Flowers of France) in 1778, having left the army some ten years earlier because of poor health.

To simplify the identification of plants, Lamarck presented the flora as a dichotomous key (a systematic list of characteristics in which there are two choices at each step in the classification). This system, which was far easier to use than other classifications of the time, impressed Georges de BUFFON, who arranged for it to be published and also ensured that Lamarck was elected to the French Academy of Sciences. Lamarck was employed as botanist at the Jardin du Roi until the institute was reorganized during the French Revolution. In the newly named Muséum National d'Histoire Naturelle Lamarck was placed in charge of animals without backbones – a group he later named "invertebrates." He differentiated the arachnids, insects, crustaceans, and echinoderms and wrote a seven-volume *Natural History of Invertebrates* (1815–22). Lamarck was also interested in meteorology, geology, chemistry, and paleontology and it is thought that his observations in these fields contributed to the formulation of his evolutionary theory, which he first put forward in *Zoological Philosophy* (1809). Until the late 1790s, he had believed that species remained unchanged, but fossil evidence, and his nonbelief in extinction, combined to change his mind. He saw evolution as a natural tendency to greater complexity and put forward four laws to explain how such complexity is brought about. The second law states that "the production of a new organ in an animal body results from a new need that continues to make itself felt," and the fourth law, for which he is best remembered, states that such acquired characteristics are inherited. A much-quoted example of this view is the neck of the giraffe, which, through stretching for the uppermost leaves, becomes gradually elongated, and this adaptation is passed on to its offspring. Today Lamarckism has largely been rejected in favor of DARWIN's theory of evolution by natural selection, especially in the light of knowledge gained about genetic mutation as a source of the variation on which Darwin's theory is based. However, there have been strong advocates of Lamarck's theory, notably Luther BUR-BANK, Paul KAMMERER, and Trofim LYSENKO. Some support for Lamarck's views comes with the interesting work of the immunologists Edward Steele and R. M. Gorczynski who claim to have shown that acquired immunological tolerance is inherited in mice.

Lamb, Sir Horace (1849–1934) *British mathematician and geophysicist*

Born at Stockport in northern England, Lamb studied at Cambridge University. His earliest interest had been in classics, but soon after arriving in Cambridge he discovered his true vocation lay in mathematics. Among his teachers were George STOKES and the great mathematical physicist James Clerk MAXWELL. Lamb's own interest in mathematics was very much shaped by contact with the ideas of these two men, and was almost entirely in applied mathematics. In 1875 he went to Australia to become the first professor of mathematics at the newly founded University of Adelaide. In 1885 he returned to England to take up the chair of mathematics at Manchester University, and held this post until his retirement in 1920. He was knighted in 1931.

Among the great variety of fields of applied mathematics to which Lamb made important contributions are electricity and magnetism, elasticity, acoustics, vibrations and wave motion, statics and dynamics, seismology, and the theory of tides and terrestrial magnetism. However, it is for his work in fluid mechanics that Lamb is most celebrated. He wrote a book on the subject, *Hydrodynamics* (1895), which immediately became a classic and by 1932 had gone through six editions. Lamb's work in geophysics was also important. He wrote a paper in 1904 on the propagation of waves over the surface of an elastic solid, in which he virtually laid the whole theoretical foundations for modern mathematical seismology.

Lamb, Hubert Horace (1913–1997) *British climatologist*

Lamb graduated in 1935 and, after working at the Meteorological Office, London, where he led the climatic variation research, he started the Climate Research Unit at the University of East Anglia, Norwich. He eventually became honorary professor in the School of Environmental Science at that university.

His main aim was to build up a detailed picture of the climates of the past and to acquire sufficient understanding of them to be able to see how climate might develop. By examining such records as ships' logs he was able to reconstruct much of the climate of the last two to three hundred years; for earlier periods he used such evidence as the fluctuations in tree-ring width. The results of his researches were published in his two-volume *Climate: Present, Past and Future* (1972, 1977). To understand such major climatic changes as the "Little Ice Age" (1550–1850) or the unusually warm weather of the 1900–50 period Lamb searched for possible physical mechanisms. To investigate the effects of volcanic dust, for example, he introduced the Dust Veil Index, which measures the amount of dust released each year on a scale that assigns the arbitrary value 1,000 to the great Krakatoa eruption of 1883. As the present cooling has begun in a period with a low index it cannot, Lamb argued, be the crucial factor.

Lamb, Willis Eugene, Jr. (1913–) *American physicist*

Born in Los Angeles, Lamb was a student at the University of California, Berkeley, graduating in chemistry in 1934 and gaining his PhD in physics in 1938. His thesis research, on the electromagnetic properties of nuclear systems, was directed by J. Robert OPPENHEIMER. In 1938 he became an instructor in physics at Columbia University, New York, becoming a professor in 1948, and from 1943 to 1951 he was also associated with the Columbia Radiation Laboratory. It was at Columbia that he performed the experiments on the fine structure of the hydrogen spectrum that led to his receiving the 1955 Nobel Prize for physics.

Shortly after World War II, Lamb began his work to check the accuracy of the predictions of Paul DIRAC as they related to the energy levels and spectral lines of hydrogen. Dirac's quantum mechanical theory predicted that the hydrogen atom had two possible energy states with equal energies. Lamb's accurate work using radiofrequency resonance techniques, reported in 1947, revealed that there was a minute difference in these energy levels. Small as it was, this *Lamb shift* necessitated a revision of the theory of the interaction of the electron with electromagnetic radiation. For this work Lamb was awarded the Nobel Prize for physics, which he shared with another leader of research at Columbia, Polykarp KUSCH, with whom he had performed wartime research in developing microwave radar. In 1951 Lamb was made a professor of physics at Stanford. There he devised microwave techniques for examining the hyperfine structure of the spectral lines of helium. In 1956 he took a professorship in England at Oxford University, and in 1962 returned to America to a professorship at Yale. From 1974 he was professor of physical and optical sciences at the University of Arizona's department of physics. His publications include *Laser Physics* (1974), written in collaboration with M. Sargent and M. O. Scully.

Lambert, Johann Heinrich (1728–1777)
German mathematician, physicist, astronomer, and philosopher

> I bought some books in order to learn the first principles of philosophy. The first object of my endeavors was the means to become perfect and happy. I understood that the will could not be improved before the mind had been enlightened.

Lambert (**lahm**-bert or **lam**-bert), the son of a poor tailor from Mulhouse (now in France), was largely self-educated. He spent the early part of his life in various occupations, including teaching and bookkeeping, and followed his scientific interests in his spare time. In 1764 he moved to Berlin, where he attracted the notice of Frederick the Great and became a member of the Berlin Academy.

Lambert contributed to numerous branches of science and learning generally. His main mathematical achievement was to prove that "π" is irrational. He did this by use of continued fractions and published the proof in 1768. He also studied the hyperbolic functions and introduced their use into trigonometry. Lambert did some remarkable work in non-Euclidean geometry, but this remained totally unknown until the end of the 19th century when it was published. In addition to mathematics Lambert was an astronomer of note. The suggestion that there might be further galaxies beyond our own was first made by him and this was subsequently confirmed observationally by William HERSCHEL. Lambert was the first to invent an accurate way of measuring light intensities and the lambert, a measurement of light intensity, was named for him. In his philosophical ideas Lambert largely developed the ideas of the great German rationalist philosopher Gottfried LEIBNIZ and his chief philosophical work, *Neues Organon* (New Organon), was published in 1764.

Lamont, Johann von (1805–1879) *Scottish–German astronomer*

Lamont (**lah**-mont or la-**mont**), whose father worked in the customs service, came from an old but impoverished family in Braemar, Scotland. After his father's death in 1816 the young Lamont was sent to the Scottish Benedictine monastery at Ratisbon and given a rigorous mathematical training from the abbot. He was admitted to the Munich Academy of Science in 1827 and became assistant astronomer at the Bogenhausen Observatory in 1828 and director in 1835. In 1852 he became professor of astronomy at Munich University.

Lamont published a major catalog in six volumes (1866–74) of 34,674 small stars but his most important work was on terrestrial magnetism. He began keeping local records in 1836 and performed magnetic surveys of Bavaria, France, Spain, Denmark, and North Germany. He announced the important discovery of the Earth's (approximately) ten-year magnetic cycle in 1850, only a few years after Heinrich Schwabe had announced a similar cycle for sunspots. It was not long before scientists began to speculate on a possible connection.

Lancisi, Giovanni Maria (1654–1720)
Italian physician

Lancisi (lan-**chee**-zi) came from a wealthy Roman family. He was educated at the Sapienza, the Papal College in Rome, graduating in medicine in 1672. After some hospital work he returned in 1684 to the Sapienza as professor of anatomy. From 1688 he also served as papal doctor.

In 1717 Lancisi published a prophetic work, *De noxiis paludum effluviis* (On the Noxious Effluvia of Marshes), in which he drew attention to the interconnections between the swamps, mosquitoes, and malaria of Rome. He even proposed the draining of the swamps. Lancisi also wrote extensively on heart disease, describing for the first time changes in the valves produced by syphilis. It is owing to Lancisi that the lost plates of Bartolommeo EUSTACHIO, rediscovered in the early 18th century, were published in an appropriate form in 1714.

Land, Edwin Herbert (1909–1991) *American inventor*

Born in Bridgeport, Connecticut, Land began his education at Harvard but left to develop a number of commercial ideas.

His first success was a method of producing a relatively cheap filter that would transmit polarized light. The material, sold under the tradename Polaroid, was a plastic containing aligned crystals, which restrict the light vibrations to one plane. Land set up the Polaroid Corporation in 1937 in Cambridge, Massachusetts. With the onset of World War II there was an increased demand for polarizing filters for sunglasses, binoculars, and other optical instruments, and Land's company flourished. Following the end of the war Land sought new areas to exploit. He chose photography and concentrated on designing an instant camera. Chemicals to develop the film were contained in a lead pod pierced as it was squeezed through a pair of rollers. Early instant cameras required a wait of a minute or more before peeling away the protective plastic. Land's SX70 camera, launched in 1972, ejected the print instantly for the image to develop within seconds. Land also tried to develop a new theory of color vision that began by rejecting the old trichromatic theory linked with NEWTON and YOUNG. How, he asked, does the eye cope with an excess of red in a room lit by an incandescent tungsten light? Familiar objects like green apples and yellow lemons do not appear to redden in this artificial light, despite the fact that it does not have the same spectral distribution as sunlight. How do objects retain their "color identity" under a great variety of lighting conditions? It cannot simply be the responses of the retinal photoreceptors to radiant energy; also involved are high-level brain processes. Thus for Land, vision was a retina-and-cortex system, which he called a "retinex system."

Landau, Lev Davidovich (1908–1968)
Azerbaijani theoretical physicist

> I spent a year in prison and it was clear that I would be unable to live for even another half year.
> —On his imprisonment by Stalin. Quoted in Landau's obituary notice in *The Times*, 3 April 1968

Landau (lahn-**dow**), whose father was a petroleum engineer and whose mother was a physician, was born in Baku, the capital of Azerbaijan. He studied at the university in his native city (1922–24) and at Leningrad (now St. Petersburg) (1924–27), graduating in 1927. In 1929 he visited various scientific centers in Europe, including Copenhagen where he developed a long-lasting friendship and working relationship with Niels BOHR. He returned to the Soviet Union and in 1932 went from Leningrad to Kharkov to head the theoretical physics groups at two of the institutes there. He was appointed professor of physics at Kharkov University in 1935. In 1937, at the request of Pyotr KAPITZA, he moved to Moscow as director of theoretical physics at the Institute of Physical Problems and in 1943 became professor of physics at the Moscow State University.

Landau was one of the major theoretical physicists of his day, making numerous contributions to many branches of physics. These included quantum mechanics, atomic and nuclear physics, astrophysics, thermodynamics, particle physics, quantum electrodynamics, and low-temperature physics. In Moscow he collaborated with E. M. Lifshitz on a highly successful series of monographs on theoretical physics, published in 1938. In 1962 he was awarded the Nobel Prize for physics for his work on condensed matter (i.e., matter in the solid or liquid state), especially liquid helium. Liquid helium has such unusual properties when its temperature falls below 2.2 kelvin that physicists describe it as helium II, as opposed to helium I above 2.2 K. To explain the strange superfluidity and superconductivity of helium II, Landau introduced the idea of a "phonon," a quantum of thermal energy, and a "roton," an elementary quantum of vortex motion. The existence of

such entities has since been confirmed experimentally. Under Landau a vigorous school of theoretical physics was created in Moscow. Tragically he was involved in a serious motor accident in 1962; his physical powers never returned to normal and he died six years later.

Landolt, Hans Heinrich (1831–1910)
German physical chemist

Landolt (**lahn**-dolt) studied in his native city of Zurich, then at Breslau, Berlin, and at Heidelberg under Robert BUNSEN. He began teaching at Breslau in 1856 and successively held the chairs of chemistry at the universities of Bonn (1867), Aachen (1869), and Berlin (1891).

The specific refractivity of a compound is given by the formula $(n − 1)/d$ where n is the refractive index and d the density. Landolt studied the specific refractivities of compounds and introduced molar refractivity (specific refractivity divided by molecular weight), which he showed was related to the chemical structure of the compound. In 1883, Landolt published, with R. Börstein, a compendium of physical chemistry, *Physikalisch Chemische Tabellen* (Physical Chemistry Tables).

Landsteiner, Karl (1868–1943) *Austrian–American pathologist*

Landsteiner (**land**-stI-ner or **lahnt**-shtI-ner), the son of a prominent Viennese journalist, was educated at the University of Vienna, where he obtained his MD in 1891. After studying chemistry in Germany under Emil FISCHER and in Switzerland under the German chemist Arthur HANTZSCH, Landsteiner returned to the University of Vienna to work as a pathologist, serving as professor of pathology from 1911 to 1919. He then spent a couple of years in Holland before moving to America, where he took up an appointment with the Rockefeller Institute, New York, in 1922, remaining there until his death.

In 1902 Landsteiner announced one of the major medical discoveries of the century, that of the ABO blood group system. It was already known that the proteins in any animal or plant species were specific to that species and differed from those of other species, but Landsteiner went on to suggest that individuals within a species showed similar though small differences in their proteins. He knew that if serum (blood from which the cells and clotting factors have been removed) of one species is mixed with the erythrocytes (red cells) of another species the resulting mixture will agglutinate (clump together); he therefore decided to see what would happen when serum and erythrocytes from different humans were combined. In many cases there was no agglutination – it was as if the blood cells were mixed with their own serum – but in others he noted that agglutination occurred with just as strong an interaction as that between serum and cells of different species. The pattern of agglutination was such that Landsteiner proposed the existence of four distinct human blood groups, which he named A, B, AB, and O, based on the presence or absence in the blood of one or both of two antigens (substances against which antibodies react), which he named A and B. On this supposition individuals of blood group A (i.e., with antigen A) possess in their serum an antibody to antigen B, while group B individuals possess an antibody against antigen A; type AB individuals possess both antigens A and B (and therefore neither anti-A nor anti-B antibody), while type O individuals possess neither antigen and both antibodies. Not only did Landsteiner's work at last permit successful blood transfusions and save many thousands of lives, it also raised profound questions about the nature of the immunological system – questions still being vigorously pursued. The ABO grouping was the first of many different groups to be discovered; Landsteiner himself, in 1927, discovered the second and third systems, the MN and the P. For his work on blood groups Landsteiner was awarded the 1930 Nobel Prize for physiology or medicine. He also produced major results outside the field of serology, making (in 1908) one of the earliest breakthroughs in the conquest of polio. By taking pieces of the spinal cord of a polio victim and soaking them in liquid, he produced a mixture capable of infecting monkeys. Further work led him to conclude that a virus caused the disease. Landsteiner's approach permitted laboratory investigation and experimentation, which is the initial step in gaining understanding and control of any infective organism. In the field of immunology Landsteiner demonstrated the specificity of antibodies by introducing the concept of the hapten. Haptens are small organic molecules that can stimulate antibody production only when combined with a protein molecule. Landsteiner combined haptens of known structures with such proteins as albumin and showed that small changes in the hapten would radically affect the production of antibodies. Landsteiner was fortunate to be able to continue with creative scientific work virtually to the end of his life: he in fact suffered his fatal heart attack while working at his laboratory bench with a pipette in his hand. He was over 70 when, in 1940, he announced the discovery of the rhesus (Rh) factor, then responsible for the consequent serious illness or death of 1 in 200 white babies. The factor was so named as it was first detected in the blood of rhesus monkeys.

Langevin, Paul (1872–1946) *French physicist*

> Langevin's scientific thought displayed an extraordinary clarity and vivacity combined with a quick and sure intuition for the essential point.
> —Albert Einstein, *Thought* (1947)

Langevin (lahnzh-**van**), a native Parisian, studied at the Cavendish Laboratory, Cambridge, England, under J. J. THOMSON and at the Sorbonne, where he obtained his PhD under Pierre CURIE in 1902. He became physics professor at the Collège de France in 1904 and at the Sorbonne in 1909.

Langevin worked on the application of ultrasonic vibrations, which, following Pierre Curie's discovery of piezoelectricity, could be generated by applying a rapidly changing electric potential to a crystal, making it vibrate and produce sound waves in the ultrasonic region. Because ultrasonic wavelengths are shorter than those in the audible range, they are better reflected and Langevin saw that this might be put to military use in World War I. His development of echo location to detect submarines in fact came too late to be used in the war, but this work was the grounding for the later sonar. Langevin also studied paramagnetism and gave a modern explanation of it incorporating electron theory. In this way he was able to deduce a formula correlating paramagnetism with absolute temperature, which gave a theoretical explanation of the experimental observation that paramagnetic moment changes inversely with temperature. The formula also enabled Langevin to predict the occurrence of paramagnetic saturation – a prediction confirmed experimentally by Heike KAMERLINGH-ONNES. Langevin also studied the properties of ionized gases, and Brownian movement in gases. He publicized in France Einstein's views on the equivalence of energy and mass.

Langley, John Newport (1852–1925) *British physiologist*

Langley, the son of a schoolmaster, was born in Newbury, England, and educated at Exeter Grammar School and Cambridge University. Although he originally intended to join the Indian Civil Service, Langley fell under the influence of Michael FOSTER, professor of physiology at Cambridge, and turned to physiology instead. He was appointed lecturer in histology in 1884 and in 1903 he succeeded Foster as professor of physiology, a post he retained until his death.

Langley began his research career in 1875, when Foster asked him to observe the physiological effects of jaborandi juice, an alkaloid derived from an American shrub. Foster was particularly interested in its effects on the heart but Langley was more impressed by its power to evoke copious secretion, a process he spent the next 15 years studying. His most significant work, however, was done on the autonomic nervous system, a term he himself coined in 1898. In collaboration with William Dickinson, Langley began (in 1889) to explore the sympathetic nervous system in some detail. This was made possible by the discovery that nicotine would selectively block nerve impulses at the sympathetic ganglia, enabling Langley to distinguish those nerves that actually ended at the ganglia from those that merely passed through them. Langley was thus able to show that only one ganglion lay along the sympathetic nerve on its path from the spinal cord to its goal. In 1901 Langley followed this by exploring the problem of how nerves communicate with the muscle cells with which they form junctions. Basically the sympathetic system increases heartbeat, raises blood pressure, and contracts smooth muscle, i.e., it prepares an animal for action. Langley noted that when an extract from the adrenal gland was injected into an animal, these responses were produced. It was left to the British physician Thomas EL-LIOTT, a pupil of Langley, to propose that it was the adrenal extract, later termed adrenaline (epinephrine), that was released from sympathetic nerves to stimulate muscle. Langley later published the details of his work in his *Autonomic Nervous System* (1921).

Langley, Samuel Pierpont (1834–1906) *American astronomer and aviation pioneer*

> Mechanical flight is possible with engines we now possess.
> —Prediction made in 1891

Langley was born in Roxbury, Massachusetts. Although he received no higher education, he became an astronomical assistant at Harvard and later became director of the Allegheny Observatory and professor of physics and astronomy at the University of Pittsburgh. He was appointed secretary of the Smithsonian Institution in 1887.

His most important innovation as an astronomer was his invention of the bolometer, an instrument that allows very small temperature measurements. It is made from a fine blackened platinum wire in a circuit with a galvanometer that is placed parallel to the lines of the solar spectrum and moved along its length. Changes in the energy absorbed by different parts of the spectrum will produce different temperatures and thus produce changes in the electrical resistivity and current, which will show on the galvanometer. The bolometer can thus measure variations of a hundred-thousandth of a degree and enabled Langley to study the far infrared region of the solar spectrum. Langley first became interested in aviation in 1886, but it was not until 1896 with his fifth model that he became the first to achieve sustained unmanned flight. By 1903 he had developed a plane with a span of 48 feet (14.6 m) and powered by a 52-horsepower petrol engine. Its two trials were disasters due to the catapult launching adopted

by Langley, not the aerodynamic stability of the plane. Langley's last trial was on 8 December 1903, just nine days before the successful flight of the WRIGHT brothers.

Langmuir, Irving (1881–1957) *American chemist*

Langmuir, who was born in Brooklyn, New York, studied metallurgical engineering at the Columbia School of Mines, New York. He then went on to do postgraduate work under Walther NERNST at Göttingen, where he obtained his PhD in 1906. On his return to America he taught for a short time at the Stevens Institute of Technology, New Jersey, before joining the research laboratory of the General Electric Company (GEC), Schenectady, New York (1909), where he remained until his retirement in 1950.

Langmuir was an extremely original and productive industrial physical chemist. In 1913 he achieved a major breakthrough in the design of electric light bulbs. The vacuum tubes (tungsten bulbs) then in use contained an incandescent tungsten wire that tended to break and also deposited a black film inside the bulb. Most research to rectify this was concentrating on improving the quality of the vacuum in the bulb. Langmuir saw that the same effect could be obtained more cheaply and efficiently by filling the bulb with an inert gas. After much experimentation he found that a mixture of nitrogen and argon did not attack the tungsten filament and eliminated the oxidation on the bulb. In 1919 Langmuir tried to develop the theory of the electron structure of the atom published by Gilbert LEWIS in 1916. Lewis had only dealt with the first two rows of the periodic table and Langmuir tried to extend it. He proposed that electrons tend to surround the nucleus in successive layers of 2, 8, 8, 18, 18, and 32 electrons respectively. Then, using similar arguments to those of Lewis, he went on to try and explain the basic facts of chemical combination. It was not until after the development of quantum theory in the 1920s that a definitive account could be provided by Linus PAULING. Langmuir also developed a vacuum pump, constructed a hydrogen blowtorch (1927) for welding metals at high temperatures, and worked on the production of artificial rain with Vincent SCHAEFER in 1947. In his research career he conducted a prolonged investigation into the chemistry of surfaces, tackling such problems as how and why certain substances spread on water and how gases interact with metal surfaces. Langmuir introduced the idea of adsorption of a single layer of atoms (a monolayer) on a surface and the theory that surface reactions (as in heterogeneous catalysis) take place between adsorbed molecules or atoms. The *Langmuir isotherm* is an expression relating the amount of adsorption on a surface with the gas pressure (at constant temperature). He was awarded the Nobel Prize for chemistry in 1932 for his research into surface reactions.

Lankester, Sir Edwin Ray (1847–1929)
British zoologist

Lankester was born in London and educated at both Cambridge and Oxford universities. His precocity as a zoologist is indicated by the fact that his first scientific paper (on the fossil fish *Pteraspis*) was published at the age of 15. In 1874 he was appointed to the chair of zoology and comparative anatomy at University College, London, and from 1890 to 1898 was Linacre Professor of Zoology and Comparative Anatomy at Oxford. Founder (1884) and president (1892) of the Marine Biological Association, Lankester was also professor of physiology at the Royal Institution (1898–1900) and from 1898 to 1907 was director of the British Museum (Natural History). He was knighted in 1907.

Lankester was an ardent Darwinist, and his work on invertebrate morphology and embryology (e.g., of the mollusks) did much to strengthen arguments in favor of evolution. He believed that an inherited ability to learn played an important part in the evolution of man, and he expounded this controversial idea in *The Significance of the Increased Size of the Cerebrum in Recent as Compared with Extinct Animals* (1899). Lankester was also one of the first to describe the protozoan parasites in the blood of vertebrate animals, a crucial step in the diagnosis and treatment of diseases such as malaria. His publications include both purely scientific works such as *Comparative Longevity in Man and the Lower Animals* (1870) and popular works such as *Science from an Easy Chair* (1910–12), based on his newspaper column of the same title. He edited the *Treatise on Zoology* (1900–09), which enjoyed considerable popularity, and from 1869 to 1920 was editor of the *Quarterly Journal of Microscopical Science*, the journal begun by his father in 1860.

Laplace, Marquis Pierre Simon de
(1749–1827) *French mathematician, astronomer, and physicist*

All the effects of nature are only mathematical results of a small number of immutable laws.
—*Complete Works*, Vol. VII

I have no need of that hypothesis.
—On being asked by Napoleon why he had made no mention of God in his book about the universe

Laplace (la-**plas**), born the son of a small estate owner in Beaumont-en-Auge, France, was educated at the University of Caen. Jean D'Alembert, impressed by a letter on mechanics sent to him by Laplace, arranged for him to become professor of mathematics in the Ecole Militaire

in Paris. He became a full member of the French Academy of Sciences in 1785. Laplace prospered during the reign of Napoleon who had a genuine interest in mathematics, taking pleasure in rewarding the eminent mathematicians of his day. Laplace served briefly as Napoleon's minister of the interior in 1799 and much longer as one of his senators. He was also made a count by the emperor. Laplace was clearly one of nature's survivors, for with the Bourbon restoration in 1814 he found no embarrassment in signing the decree banishing his patron. He was made a marquis by Louis XVIII.

Laplace's greatest work was the *Traité de mécanique céleste* (Celestial Mechanics) published in five volumes between 1799 and 1825. Although NEWTON had derived the laws that govern the movements of the heavenly bodies it was clear to him that there were certain irregularities or perturbations in the movements of the planets that would lead to the end of the universe if not corrected. Newton was willing to accept God's intervention in the system to prevent its collapse. What Laplace showed was that while it was true that there were irregularities in the system they were periodic and not cumulative. The system was basically stable. When Napoleon pointed out to him that he had not mentioned God in his book Laplace proudly retorted that he had no need of that hypothesis. A popular account, *Exposition du système du monde* (Exposition of the System of the World), was published in 1796. It is in this work that he proposed the nebular hypothesis. Here he argued that the solar system had evolved from a rotating mass of gas that had condensed to form the Sun. From this had been thrown off the planets which in turn threw off their various satellites. The theory had in fact been proposed earlier by Immanuel KANT (in 1755) and was soon to run into trouble with the discovery of retrograde orbits in the system. However, a new form of the nebular hypothesis has been introduced into astronomy by Carl von WEIZSÄCKER. Laplace's second great achievement was the establishment of probability theory on a rigorous basis in his *Théorie analytique des probabilités* (1812; Analytic Theory of Probabilities) and *Essai philosophique sur les probabilités* (1814; Philosophical Essay on Probabilities). His third great achievement was his development of the concept of a "potential" and its description by the Laplace equation:

$$\nabla^2 u = \partial^2 u/\partial x^2 + \partial^2 u/\partial y^2 + \partial^2 u/\partial z^2 = 0$$

For Laplace u was a "velocity potential" but it was soon seen to have an enormous number of applications in all kinds of areas and became an essential means of dealing with motion in any field, be it electromagnetic, gravitational, or hydrodynamic.

Although Laplace has the reputation among

mathematicians of being both a careerist and a plagiarizer he did much to help others and to encourage the growth of French science. His home at Arcueil, Paris, was the center for the greatest concentration of scientific talent of his or any other time. He died there uttering what sound like some well-prepared last words: "What we know is very slight; what we don't know is immense."

Lapworth, Arthur (1872–1941) *British chemist*

Lapworth, the son of the geologist Charles Lapworth, was born at Galashiels in Scotland. Arthur Lapworth was educated at Birmingham University and the City and Guilds College, London. In 1900 he was appointed head of the chemistry department at Goldsmith's College, London, but the same year moved to Manchester as senior lecturer in physical chemistry, becoming professor of organic chemistry in 1913. In 1922 he changed to the chair of physical and inorganic chemistry.

In 1922 Lapworth published an early account of organic chemical reactions based on electron theory. It involved the idea of polar atoms and alternating positive and negative centers at which reactions occurred.

Lapworth, Charles (1842–1920) *British geologist*

Lapworth was born at Faringdon in Oxfordshire, England. In 1864 he became a schoolmaster in Scotland where his interest in geology developed. There he studied the Lower Paleozoic rocks, publishing his results between 1878 and 1882. He used graptolites, the fossil skeletons of colonial animals found in the Lower Paleozoic, to determine the stratigraphy of Scotland and later published a series of monographs on the fossils (1901–18).

In 1879 Lapworth introduced the Ordovician system of geological strata. There had been a bitter dispute between Adam SEDGWICK, who considered this complex series of strata to be Upper Cambrian, and Roderick MURCHISON, who considered it to be Lower Silurian. Lapworth's suggestion that it was in fact a separate system, which he named Ordovician after a Welsh tribe, settled the dispute. Lapworth was appointed to the chair of geology at Birmingham University in 1881. His son, Arthur, was a chemist of some distinction.

Larmor, Sir Joseph (1857–1942) *Irish physicist*

> Larmor made few friends, perhaps; but while he lived, and they lived, he lost none.
> —Sir D'Arcy Wentworth Thompson in *Yearbook of the Royal Society of Edinburgh* (1942)

Born at Magheragall in Ireland, Larmor gained his BA and MA from Queen's University, Belfast; he entered Cambridge University in 1877, gaining a fellowship in 1880. He then became professor of natural philosophy at Queen's College, Galway. In 1885 he returned to Cambridge as a university lecturer in mathematics and in 1903 became Lucasian Professor of Mathematics. He retired from this post in 1932. Apart from his scientific work Larmor served as member of parliament for Cambridge University from 1911 to 1922.

Larmor's central interests were in applied mathematics and physics, specifically in electromagnetic theory, optics, mechanics, and the dynamics of the Earth. Like the work of his contemporary, Hendrik LORENTZ, Larmor's work belongs to the final phase of classical physics that paved the way for the revolutions of relativity and quantum theory. An example of Larmor's basic scientific conservatism was his support of the concept of the ether as the wave-bearing medium thought to pervade all space and his work, published in 1900 as *Aether and Matter*, on the motion of matter through the ether. He believed that matter could only interact with the ether through the effects of electrically charged particles that formed part of the ether. Larmor made two particularly important contributions to electrodynamics. He was the first to predict in 1897 the *Larmor precession*. This is the wobbling motion of the orbital plane of an electron moving in an atom when subjected to a magnetic field. The axis at right angles to the plane of the orbit sweeps out a conical area. He also derived a nonrelativisitic formula that expresses the power radiated by an accelerated electron as being proportional to the square of the product of charge and acceleration.

Lartet, Edouard Armand Isidore Hippolyte (1801–1871) *French paleontologist*

> For fifteen years he made excavations at his own expense and devoted late evenings to the study and classification of this material.
> —Louis Lartet, quoted in *Life and Works of Edouard Lartet* (1872)

After studying law at Toulouse, Lartet (lar-**tay**) – the son of a wealthy landowner – spent the earlier part of his life managing the family estates at St. Guiraud. His interest in paleontology was aroused in the early 1830s when he was shown a locally discovered mastodon tooth. Beginning with his discovery in 1836 of *Pliopithecus*, the ancestor of the gibbon, he made a number of significant fossil discoveries. In 1856 he found remains of *Dryopithecus*, the fossil ancestor of the other apes, and in 1868, at Cro-Magnon in the Dordogne, he came across several adult human skeletons, later to be named Cro-Magnon man, that were found to belong neither to Neanderthal nor to modern man. These are the earliest known fossils of man in

Europe. They have modern teeth and jaw bones and a brain capacity averaging about 1,300 cubic centimeters.

Lartet's most important work, however, helped to solve one of the major problems of 19th-century science, namely the antiquity of man. Early proposals that man's history could be pushed back into glacial times lacked conclusive proof: attempts to link human remains with the bones of extinct animals were always open to the objection that they were accidental assemblies that had come together at some historic time. Nor was the argument that some of the animal bones had been worked by man any more persuasive, since the working of ancient bones could have taken place at any time. In the early 1860s Lartet began a careful study of the caves of the Dordogne and the Pyrenees in collaboration with the British banker Henry Christy. In a cave at La Madelaine in 1863 he found a piece of ivory with a woolly mammoth clearly engraved upon it. Excluding forgery, there seemed no other explanation than that an animal of the ice age and a human witness had coexisted. Lartet went on to propose a sequence for the Paleolithic based on the presence of animals, but scholars preferred alternative classifications based on tools. The results of his researches were published posthumously in *Reliquiae Aquitanicae* (1875; Aquitainian Remains), a work that did much to establish the prime importance of the archeological sites of southern France. In 1869 Lartet was appointed professor of paleontology at the museum of the Jardin des Plantes – a post he held until his death.

Lassell, William (1799–1880) *British astronomer*

Lassell, who was born at Bolton in Lancashire, England, was a brewer by profession who became interested in astronomy. He built an observatory for himself at Starfield near Liverpool and developed an interest in techniques for building very large reflectors, which he used himself. In 1844 he began building a 24-inch (61-cm) reflector with which – just 17 days after the discovery of Neptune in 1846 – he discovered Neptune's largest satellite, Triton. He also discovered Hyperion (a satellite of Saturn) and the crepe ring of Saturn independently of William BOND. In 1851 he discovered two satellites of Uranus – Ariel and Umbriel – from observations made in Malta (where the atmosphere was clearer than in industrial England). In Malta in 1861 he built a 48-inch (122-cm) reflector and observed and cataloged hundreds of new nebulae.

Laue, Max von *See* VON LAUE, MAX THEODOR FELIX.

Laughlin, Robert B. (1950–) *American physicst*

Laughlin was born in Visalia, California, and gained his PhD in physics in 1979 from the Massachusetts Institute of Technology. In 1989 he became professor of physics at Stanford University, where he did research on the fractional quantum Hall effect. For this work he shared the 1998 Nobel Prize for physics with Horst STÖRMER and Daniel TSUI, for explaining their discovery of a new form of quantum fluid with fractionally charged excitations.

Laughlin showed how electrons in a powerful magnetic field can condense to form a so-called "quantum fluid" similar to those that occur in liquid helium and in superconductors. The theory derives ultimately from the Hall effect (the production of a voltage in a current-carrying conductor or semiconductor at right angles to a magnetic field), discovered in 1879 by the American physicist Edwin Hall. It occurs because electrons – the charge carriers – are deflected laterally in the magnetic field. A century later the German physicist Klaus von KLITZING discovered that in a powerful magnetic field at extremely low temperatures the Hall resistance of a semiconductor is quantized in integral steps.

Using even stronger magnetic fields and lower temperatures, Störmer and Tsui discovered more steps, called the *fractional quantum Hall effect*. A year later Laughlin theorized that the low temperature and powerful magnetic field forced the electrons to form a new type of quantum fluid. The addition of a single electron to this superfluid produced a number of fractionally charged quasiparticles, with the correct charges to account for the results of Störmer and Tsui.

Laurent, Auguste (1807–1853) *French chemist*

> Chemistry claims to teach us not only the properties of bodies that do not exist, but even of bodies that cannot possibly exist.
> —Quoted by C. de Milt in *Chymia* (1953)

The son of a mining engineer, Laurent (lo-**rahn**) was born at La Folie in France. He was educated at the mining school in Paris and worked for some time as a mining engineer before becoming an assistant to Jean DUMAS in Paris. He was appointed professor of chemistry at Bordeaux in 1838. He returned to Paris in 1846 but he found it difficult to find employment, largely, it is thought, because of his unpopular chemical views. He worked for a short time as an assayer in the Mint in 1848 and later at Sèvres. Laurent worked closely with Charles GERHARDT, so that it is not always possible to separate their ideas very precisely. His collected papers were published posthumously in his *Methode de Chimie* (1854; Method of Chemistry).

Dumas had formulated his theory of substitution in 1834. According to this theory hydrogen can be actually removed from certain substances and replaced by other substances. Laurent used this to demolish BERZELIUS's electrochemical theory, pointing out that in the synthesis of trichloroacetic acid the electronegative chlorine replaces the electropositive hydrogen. Laurent's powerful arguments against the orthodox chemistry won him little support and less popularity. With Gerhardt, he introduced type theory, in which organic compounds are recognized as having common structural features by which they can be assigned to various types. Thus alcohol, water, ammonia, ether, and numerous other types were distinguished. Type theory, soon to be superseded by the chemistry of Stanislao CANNIZZARO and August KEKULÉ, was valuable in that it provided a tolerable basis of classification and some understanding of the chemistry of organic compounds. It was Laurent who saw that chemists must distinguish clearly between atoms, molecules, and equivalents. He regarded the molecules of hydrogen, oxygen, and others as consisting of two atoms, forming what he called an "homogeneous compound," which, by double decomposition, could form "heterogeneous compounds." This provided a sound base for the accurate determination of atomic weights. Unfortunately, Laurent did not have the funds, facilities, or time to provide the necessary experimental support for his theories, and without the essential laboratory work his views were ignored.

Lauterbur, Paul C. (1929–2007) *American chemist*

Lauterbur was born in Sydney, Ohio, and in 1951 received his BS in chemistry at the Case Institute of Technology in Cleveland. He went on to obtain his PhD at the University of Pittsburgh in 1962, and in 1969 became professor of chemistry and radiology at New York University at Stony Brook. In 1985 he moved to the University of Illinois College of Medicine at Chicago, where he remained until 1990. Also in 1985 he became professor and director of the Biomedical Magnetic Resonance Laboratory, University of Illinois College of Medicine at Urbana-Champaign.

Magnetic resonance imaging (MRI) is a noninvasive technique for producing images of internal organs for use in diagnosis and treatment. The American physicists Felix BLOCH and Edward PURCELL, who were awarded the 1952 Nobel Prize for physics for their work, pioneered the method in the late 1940s. It depends on the fact that, in a strong magnetic field, the nuclei of certain atoms can have quantized energies resulting from the interaction of their spin with the field. The absorption of radio waves of a particular frequency – a type of resonance – increases their energy, and when the nuclei return to their former energy level they emit characteristic radio waves. Hydrogen atoms, in particular, are affected in this way and because water (containing hydrogen) is a major component of the human body, MRI can be used to investigate internal organs.

In the early 1970s Lauterbur discovered that the introduction of gradients in the magnetic field, achieved by adding so-called gradient magnets to the main magnet, enabled him to create two-dimensional images of the body's internal structures. He could determine the origin of the emitted radio waves by analyzing their characteristics. In this way he obtained images that could not be obtained using other methods. The technique was developed further by the British physicist Sir Peter MANSFIELD and in 2003 Lauterbur and Mansfield shared the Nobel Prize for physiology or medicine for their discoveries concerning MRI.

Laveran, Charles Louis Alphonse (1845–1922) *French physician and parasitologist*

The son of a military surgeon, Laveran (lav-e-**rahn**) was born in Paris and studied medicine at the University of Strasbourg, obtaining his MD in 1867. Like his father, he joined the Army Medical Service and served in Algeria (1878–83).

In 1880 Laveran made one of the most important discoveries of 19th-century medicine, namely, the causative agent of malaria. In Algeria he frequently performed autopsies on malaria victims who, he noted, had numerous pigmented bodies in their blood. Although some of these bodies were in the red blood cells he also noted other free bodies, at the edge of which he observed moveable filaments or flagella. The extremely rapid and varied movements of these flagella indicated to Laveran that they must be parasites. He found such parasites in 148 out of 192 cases and thus assumed them to be the cause of malaria. He called the parasite *Oscillaria malariae* but the Italian name *Plasmodium* later won favor. Laveran also speculated that mosquitoes might play a part in transmitting malaria but he failed to follow up this insight. In 1883 he returned to France to become professor of military hygiene and parasitology at the Val-de-Grace School of Military Medicine and in 1897 moved to the Pasteur Institute where he remained until his death. Here he published important works on leishmaniasis and trypanosomiasis. In 1884 Laveran published *Traité des fièvres palustres* (Treatise on Marsh Fevers), which later won him the 1907 Nobel Prize for physiology or medicine for showing the role played by protozoa in causing disease. With the prize money he founded a laboratory of tropical medicine at the Pasteur Institute.

Lavoisier, Antoine Laurent (1743–1794)
French chemist

> Only a moment to cut off that head, and a hundred years may not give us another like it.
> —Joseph Louis Lagrange on the execution of Lavoisier (1794). Quoted by D. McKie in *Antoine Lavoisier*

Lavoisier (la-vwah-**zyay**) is regarded as the founder of modern chemistry. Born in Paris, he studied both law and science, but after graduating concentrated his attention on science. He invested his money in a private tax-collecting company, the Ferme Générale, and thereby became rich enough to build a large and well-equipped laboratory. He then proceeded to study combustion.

During the 18th century combustible matter was thought to contain a substance called phlogiston, which was released when combustion took place. This theory, which was developed by Georg STAHL, had one obvious flaw; substances frequently increase in weight as a result of combustion. Lavoisier performed a number of experiments in which he burned phosphorus, lead, and other elements in closed containers and noted that the weight of the container and its contents did not increase though that of the solid did. In 1772 he recorded his observations that phosphorus and sulfur burn with a gain in weight caused by their combination with air. In 1774 he discovered that when a calx (oxide) was heated with charcoal (carbon) the gas produced was the "fixed air" found by Joseph BLACK. This suggested that the element was combining with air. Soon after this, Lavoisier was visited by Joseph PRIESTLEY in Paris. Priestley had recently discovered that mercury(II) oxide gives off a gas when heated, leaving behind mercury. Priestley called the gas "dephlogisticated air." Lavoisier repeated Priestley's experiments and, by 1778, had convinced himself of the existence in the air of a gas that combines with substances during combustion and is the same gas as that given off by heating mercury(II) oxide. Lavoisier named the gas "oxygine" from the Greek "acid producing" – he held the erroneous belief that all acids contain oxygen. He further recognized the existence of a second, inert, gas in the air – "azote" (named from the Greek "no life" and later renamed nitrogen). In 1783 he explained the formation of water from hydrogen and oxygen, work that led to a controversy involving Henry CAVENDISH. Lavoisier then went on to study respiration and deduced that oxygen is essential to animal life. In 1783 he collaborated with Claude BERTHOLLET, Antoine François de FOURCROY, and L. B. Guyton de Morveau in publishing *Méthode de nomenclature chimique* (System of Chemical Nomenclature), which proposed new names for the elements. In 1789 he published *Traité élémentaire de chimie* (Elementary Treatise on Chemistry) – an influential work summarizing his ideas and stating the law of conservation of mass. The book also contained a list of the known elements, although this included light and heat (caloric). Lavoisier became, in 1775, director of the government powder mills and as such considerably improved the production of gunpowder. He also made contributions to agriculture and demonstrated the advantages of scientific farming on a model farm at Fréchines. He was a member of the commission, appointed in 1790, that eventually led to the adoption of the metric system in France. Tragically Lavoisier's involvement with the tax-gathering company was to prove his downfall. In the 1790s France was in the middle of a protracted revolution and, in 1794, Paris was ruled by a radical group of republicans – the Jacobins – who ruthlessly executed thousands of alleged opponents of the revolution. Among these were included the tax gatherers and Lavoisier, despite his work for the state, was tried as a farmer of taxes, found guilty, and guillotined in Paris.

Lavoisier, Marie Anne Pierrette (1758–1836) *French chemist*

In 1771 Antoine Lavoisier married the fourteen-year-old Marie Paulze. Her father, like Lavoisier himself, was a farmer-general, a member of a private consortium that had bought the right to collect indirect taxes. She studied under the painter David and learned Latin and English. Her skills as a draftswoman are evident in the thirteen pages of copperplate illustration accompanying Lavoisier's *Traité* and signed "Paulze Lavoisier sculpsit" ("engraved by Paulze Lavoisier"). Mme Lavoisier also translated a number of important chemical texts, such as Richard KIRWAN's *Essay on Phlogiston* (1784).

It is indisputable that they worked together in Lavoisier's laboratory and there is a celebrated painting by David from about 1788 showing them so occupied. The nature of the collaboration, however, remains unclear. In 1794 both her husband and her father were guillotined as tax farmers. Mme Lavoisier's last service to her husband was to collect and publish his *Mémoires de chemie* (1803, 2 vols.; Memoirs of Chemistry). Soon after, she met the physicist Count RUMFORD. They were married in 1805, whereupon she began to refer to herself as Mme Lavoisier de Rumford. The marriage was not a happy one and they separated in 1809 (Rumford is said to have suggested that Lavoisier was lucky to have been guillotined).

Lawes, Sir John Bennet (1814–1900)
British agricultural chemist

Lawes was the only son of the lord of the manor of Harpenden in Hertfordshire, England, and inherited his father's estates in 1822. He was educated at Eton and Oxford University, but he left without taking a degree. He developed an interest in science, particularly chemistry, and at the age of 20 he constructed a laboratory for himself at his home.

He turned his attention to the problems of agricultural chemistry when a neighbor pointed out to him that on some local farms bone meal increased turnip production, while on others it seemed to have no effect. This started him on his life's work, the chemistry of fertilizers. After experimentation, Lawes showed that it was necessary to make the phosphate in the bones more readily soluble in the soil for absorption by plants. This he achieved by adding sulfuric acid to the crushed bones. Lawes took out a patent on these "superphosphates" in 1842, opening his first factory for their production in 1843. By the 1870s he was producing 40,000 tons of superphosphates a year using phosphate rock rather than bones. Also in 1843, Lawes was joined by Henry Gilbert, beginning a lifelong collaboration, and he started the Rothamsted Experimental Station, the first agricultural research station in the world. Experiments were conducted on different fertilizers; crops which were normally grown in rotation were grown here year after year on the same plot using a variety of manures and fertilizers. Animal feed was also examined and varied to find the most economical and efficient. Well over 100 papers were produced by Lawes and GILBERT on their Rothamsted work. Lawes established the Lawes Agricultural Trust in 1889 to safeguard the continuation of research following his death. He was created a baronet in 1882.

Lawless, Theodore Kenneth (1893–1971) *American physician*

The son of a minister, Lawless was educated at Talladega College, Alabama, the University of Kansas, and Northwestern University, where he received his MD in 1919. He chose to specialize in dermatology and venereology. To this end he set up in private practice in Chicago in 1924 and was appointed instructor at Northwestern University Medical School.

Despite a fruitful research career, racial considerations made his position at Northwestern initially uncongenial and finally untenable. When, in 1941, he found that he had been assigned no students at all he resigned from Northwestern and devoted himself to private medicine.

Lawless sought to develop new therapies against general paralysis, an until then incurable consequence of late syphilis. It had long been thought that a high fever and syphilis were incompatible. Consequently, in 1917 the Viennese physician and Nobel laureate Wagner von Juaregg decided to induce high fevers in his patients by inoculating them with tertian

malaria, a procedure that could itself be lethal. In 1936 Lawless proposed a safer approach, electropyrexia, in which the patient's body temperature was raised by passing an electric current into tissue. Although initial results appeared promising all such procedures were quickly dropped when penicillin became widely available.

Lawrence, Ernest Orlando (1901–1958) *American physicist*

Lawrence was the son of a superintendent of public schools in Canton, South Dakota. He was educated at the universities of South Dakota, Minnesota, Chicago, and Yale (where he obtained his PhD in 1925). He taught at Yale before moving in 1928 to the University of California at Berkeley. He was appointed professor there in 1930 and director of the radiation laboratory in 1936.

Lawrence was awarded the 1939 Nobel Prize for physics for his invention of the cyclotron. In the 1920s, experiments on bombardment of nuclei relied on low-energy linear accelerators. Lawrence, in 1930, began experiments to construct a cyclic accelerator. In this device charged particles move in spiral paths under the influence of a vertical magnetic field. The particles move inside two hollow D-shaped metal pieces arranged with a small gap between them. A high-frequency electric field applied between the "dees" gives a "kick" to the particle each time it crosses the gap. By early 1931 the first model (4 inches (10.2 cm) in diameter) produced energies of 13,000 electronvolts. Subsequently, Lawrence, and other workers, developed larger machines capable of achieving much higher energies for nuclear research. Lawrence also played an important part in the development of the atomic bomb. He was responsible for developing the radiation laboratory (now named for him) into one of the world's leading centers for high-energy physics. Element 103, lawrencium, is named in his honor.

Lax, Benjamin (1915–) *Hungarian–American physicist*

Lax was born in Miskolz, Hungary, but moved to America in 1926. He studied mechanical engineering at Cooper Union, New York, and was a radar officer during World War II. He then went to the Massachusetts Institute of Technology, where he gained his degree in physics in 1949. Shortly afterward he joined the staff of MIT.

Lax originally worked on the effect of magnetic fields on microwave discharges in gases. Inspired by a suggestion of William SHOCKLEY, he turned from this to studies of solids and in 1953 managed to detect cyclotron resonance in germanium. This phenomenon involves the application of a magnetic field with a simultaneous oscillating electromagnetic field (i.e., electromagnetic radiation). At a certain frequency the electrons move in spiral paths with absorption of the radiation. The technique, applied to semiconductors, gives information on the energies of electrons in the solid. Lax was one of the people instrumental in setting up the Francis Bitter National Magnet Laboratory at MIT for research involving high magnetic fields. He became director of the laboratory in 1960 and professor of physics in 1965.

Lazear, Jesse Williams (1866–1900) *American physician*

Lazear (la-**zeer**) was born in Baltimore, Maryland, and educated at Johns Hopkins University and Columbia University, New York, where he obtained his MD in 1892. After some hospital work and a period of postgraduate study at the Pasteur Institute in Paris, Lazear joined the staff at Johns Hopkins.

By 1900 Lazear was beginning to establish a reputation in the new discipline of bacteriology: he was, for example, the first person in America to isolate *Diplococcus*, the organism causing pneumonia. It was therefore natural for him to be chosen in 1900 to work with Walter REED in investigating the control of yellow fever in Cuba. Tragically Lazear allowed a mosquito, unconnected with any experiment, to bite him as he thought it did not belong to the species *Aedes aegypti*, the vector for yellow fever. He was wrong and died within days.

Leakey, Louis Seymour Bazett (1903–1972) *British anthropologist and archeologist*

Leakey was born at Kabete in Kenya and educated at Cambridge University, where he studied French and Kikuyu. He held various academic posts at British and American universities, and was curator of the Coryndon Memorial Museum at Nairobi (1945–61). Apart from anthropological studies, notably of the Kikuyu people, Leakey is best known for his excavations of fossils of early man, notably in Tanzania's Olduvai Gorge. Here, in 1959, jaw, skull, and huge teeth fragments of a species that Leakey called *Zinjanthropus* (*Australopithecus*) were uncovered by his wife Mary. The following year his son Jonathan discovered remains of the larger-brained *Homo habilis*. Both have been estimated at between 1,750,000 and 2,000,000 years old, but Leakey considered that only *H. habilis* was the true ancestor of modern man, *Zinjanthropus* having died out, a view not shared by other researchers. Leakey also found, in western Kenya, remains of the earliest known hominid *Proconsul africanus*. Leakey's work has not only provided evidence

for the greater age of man but suggests that Africa, and not, as was previously thought, Asia, may have been the original center of human evolution.

Leakey, Mary (1913–1997) *British paleoanthropologist*

The daughter of Erskine Nicol, the landscape painter, Mary Leakey is related on her mother's side to the antiquarian John Frere (1769–1841). Born in London, but widely traveled in her youth and skimpily educated, Mary first made contact with the world of the professional paleontologist through her skill as a draftswoman. She met Louis Leakey in the early 1930s and married him in 1936. Thereafter she spent the next 36 years before his death working with Leakey on his East African field trips and collaborating with him as excavator, author, and paleontologist. From 1960 she was Director of Research of Olduvai Gorge Excavations. She was also the mother of Richard Leakey, another paleontologist.

Many of the more dramatic discoveries associated with the Leakeys were in fact made by Mary and not her better-known husband. Thus it was Mary who discovered in Kenya in 1947 the skull of *Proconsul africanus*, the first fossil ape skull ever to be found. It was also Mary, at Olduvai in 1959, who found the skull of *Zinjanthropus boisei*, a 1.75 million-year-old new species of *Australopithecus*, and claimed by the Leakeys as the true ancestor of man. From 1975 she worked in northern Tanzania in the Laetoli beds near Lake Eyasi. It was there in 1976, in beds older than the lowest Olduvai levels, that she made what she has described as "the most remarkable find" of her whole career. Still preserved in the volcanic ash she had found footprints of hominids, clear evidence that man's ancestors had already adopted an upright posture some 3.75 million years ago. An account of her own researches was included in her autobiography *Disclosing the Past* (1984).

Leakey, Richard Erskine (1944–) *Kenyan anthropologist*

> For me, the search for our ancestors has provided a source of hope. We share our heritage and we share our future. With an unparalleled ability to choose our destiny, I know that global catastrophe at our own hands is not inevitable. The choice is ours.
> —*The Making of Mankind* (1981)

Richard Leakey was born at Nairobi in Kenya, the son of the famous scholars Louis and Mary Leakey. Having left school at sixteen, he first worked as a hunter and animal collector before turning in 1964 to the search for fossil man. His parents had spent much of their lives exploring the Rift Valley and working at Olduvai in Northern Tanzania.

In contrast Leakey undertook his first field

trip to the Omo valley in Ethiopia. In 1965 he shifted his interest to Lake Turkana in northern Kenya, concentrating his work in the Koobi Fora area. At the same time he was appointed to the directorship of the Kenya National Museum, Nairobi. In 1972 he made his first major find at Koobi Fora. This was a skull with a brain capacity of about 800 cc, given the number 1470. Leakey identified 1470 as *Homo* rather than an australopithecine precursor, and took it to be *Homo habilis*. The age of the skull, however, was in dispute, varying from 1.8 to 2.4 million years; the former age was eventually accepted. In 1975 a second skull was found, this time *Homo erectus*, a more advanced form than 1470. By this time Leakey was finding that the demands of administration, producing TV series, and writing popular accounts of his work were limiting his research activities. Moreover, he suffered the onset of kidney failure in 1979. The donation of a kidney by his brother Philip restored Leakey to what he termed in his autobiography *One Life* (1983) the beginning of his "second life." Much of this second life has been devoted to conservation and Leakey has been a leading figure in the fight to preserve the African elephant by banning the trade in ivory. In 1990 he was appointed director and executive chairman of the Kenyan Wildlife Service. However, during his fight against ivory smugglers, Leakey made many enemies. His determination, outspokenness, and ruthlessness alienated him from many leading Kenyan politicians and administrators. Consequently, in 1994, he resigned his post with the Kenyan Wildlife Service and decided to enter politics, despite an airplane crash in 1993 that led to the amputation of both legs. In 1995 Leakey formed a new political party, Safina (Noah's Ark), and announced his intention to challenge the ruling KANU (Kenya African National Union) in the next elections. Throughout his career Leakey has described the development of man in terms similar to those adopted by his father. He has rejected the claims of Donald JOHANSON that "Lucy," *Australopithecus afarensis*, is a joint ancestor of *Homo* as well as the australopithecines first described by Raymond DART. Leakey has continued to claim that it is too simple to present the human evolutionary tree as having only two branches; rather, there were at least three, and it was more than likely that future discoveries would add to the number. Human evolution, for Leakey, seems more like a bush than a tree.

Leavitt, Henrietta Swan (1868–1921)
American astronomer

> Her most important work required greater understanding and even more meticulous care … even though it lacked the glamor and popular appeal of the newly opened field of stellar spectroscopy.
> —Dorrit Hoffleit, *Notable American Women* (1971)

Henrietta Leavitt was born the daughter of a Congregational minister in Lancaster, Massachusetts. Her interest in astronomy was aroused while she was at Radcliffe College (then the Society for the Collegiate Instruction of Women), from which she graduated in 1892. In 1895 she became a volunteer research assistant at the Harvard College Observatory, receiving a permanent post in 1902. She was soon head of the department of photographic photometry. Like her colleague Annie CANNON, she was extremely deaf.

Leavitt's work involved the determination of the photographic magnitudes of stars, i.e., their brightness as recorded on a photographic plate. The photographic magnitude of a star differs somewhat from its visual magnitude since a photographic emulsion is more sensitive to blue light than the eye. The accurate measurement of visual magnitudes had been part of the program of the Harvard College Observatory since the 1870s. In 1907 the director of the observatory, Edward PICKERING, announced plans to redetermine stellar magnitudes by photographic techniques. The photographic magnitudes of a group of stars near the north celestial pole were to act as standards of reference for other stars. Leavitt was selected to measure these magnitudes, known as the "north polar sequence," and the results were published in 1917 in the *Annals of Harvard College Observatory* (vol. 71, no. 3). She also spent many years measuring secondary stellar magnitudes, based on the north polar sequence, which was adopted as an international standard until superseded by photometric measurements of magnitude. Leavitt also did much work on variable stars, discovering about 2,400 – roughly one half of those known in her time. She is best known, however, for her studies of Cepheid variables. At Harvard Observatory's field station at Arequipo, Peru, a series of photographic plates had been taken of the Magellanic Clouds (now known to be small neighboring galaxies). From her analysis of the plates, Leavitt detected nearly 1,800 variable stars, some of which belonged to a class known as Cepheid variables. The variation in brightness of Cepheids is extremely regular and in 1908 Leavitt noted that the brighter Cepheids had the longer periods. By 1912 she was able to show that the apparent magnitude, i.e., observed brightness, of Cepheids decreased linearly with the logarithm of the period. It was this seemingly simple discovery that led to an invaluable means for determining very great distances; previous to this only distances out to a hundred light-years could be estimated. Leavitt's work on the light variation of Cepheids was extended first by Ejnar HERTZSPRUNG and

Harlow SHAPLEY and then by Walter BAADE to give the period–luminosity relation of Cepheids. Using this relation the luminosity, or intrinsic brightness, of a Cepheid can be determined directly from a measure of its period and this in turn allows the distance – of the Cepheid and its surroundings – to be calculated. Distances of galaxies up to ten million light-years away can be determined this way.

Lebedev, Pyotr Nicolayevich (1866–1912) *Russian physicist*

A Muscovite, Lebedev (**lyay**-bye-dyef) originally attended the Highest Technical School in his native city, going on to study physics at Strasbourg University and taking his doctorate in 1891. On his return to Moscow later in 1891 he became a teacher at the university there. In 1900 he was awarded a doctorate and was appointed a professor of physics in 1902. In 1911 he resigned in protest against the actions of the Minister of Education in Moscow and he died soon afterward.

Lebedev's major work was concerned with the effect that light and other waves have on small bodies. A consequence of James Clerk MAXWELL's theory of electromagnetism was that electromagnetic waves, including light, should exert a minute pressure on matter. Lebedev succeeded not only in demonstrating the existence of this light pressure, now known as radiation pressure, but also in measuring it. This result helped to confirm Maxwell's theory. It also showed that for tiny particles of cosmic dust, the pressure exerted by the Sun's radiation could be greater than the gravitational attraction, and Lebedev argued that this was why the tails of comets always pointed away from the Sun. This was accepted until the much greater effect of the solar wind on cometary matter was discovered.

Le Bel, Joseph Achille (1847–1930) *French chemist*

Born in Pechelbron, France, Le Bel (le bel) came from a family with oil interests. He was a student at the Ecole Polytechnique and for a short while was assistant to Charles Würtz at the Ecole de Médecine. He sold his share of the family oil interests and devoted himself to independent scientific research.

He is best known for his account of the asymmetric carbon atom, although his achievement was overshadowed by the almost simultaneous account given by Jacobus VAN'T HOFF. Le Bel's account was published in November 1874, two months after that of van't Hoff. Their work is virtually identical; what difference there is arises from a difference of origin. Le Bel wished to explain the molecular asymmetry of Louis PASTEUR while van't Hoff was more concerned with understanding the quadrivalent carbon atom recently introduced by August KEKULÉ.

Lebesgue, Henri Léon (1875–1941) *French mathematician*

> In my opinion, a mathematician, in so far as he is a mathematician, need not preoccupy himself with philosophy – an opinion, moreover, which has been expressed by many philosophers.
> —Quoted in *Scientific American*, September 1964

Lebesgue (le-**beg**), who was born at Beauvais in northern France, studied at the Ecole Normale Supérieure. He obtained posts at Rennes (1902) and Poitiers (1906) universities, at the Sorbonne (1910), and at the Collège de France (1921).

Lebesgue's extremely important contributions to measure theory and the theory of integration were stimulated by the earlier work of Emille Borel and by Camille JORDAN's famous

CHEMICAL FORMULAS   *The formulas of simple chemical compounds according to Le Bel.*

Cours d'Analyse (Lessons in Analysis). The importance of Lebesgue's work resides in the fact that he was the first mathematician to develop integration in a measure-theoretic context. This allowed natural generalizations of both concepts to be made. Lebesgue's definition of integral is considered, in many respects, smoother and more useful than those that came before. Lebesgue also worked on the theory of point sets, the calculus of variations, and dimension theory. He wrote a very large number of books and papers and had interests in the pedagogy and history of mathematics.

Leblanc, Nicolas (1742–1806) *French physician*

Leblanc (le-**blahng**), who was born at Issoudun in central France, began his career as an apprentice to an apothecary. He later trained as a physician, becoming in 1780 personal surgeon to the duc d'Orleans.

One of the great problems facing the industries of the late 18th century was to find a means of producing soda ash (sodium carbonate) independently of vegetable sources. Soda ash was essential in the soap, bleaching, and glass-making industries but its only sources were wood, seaweed, and barilla (the saltwort). Such sources were either in short supply or expensive to transport and process, and in 1775 the French Academy of Sciences announced a prize for the first person who managed to convert salt into soda ash. Leblanc devised a cheap simple process and qualified for the prize in 1783, although he never received it. In this process salt was dissolved in sulfuric acid. The resulting sodium sulfate (salt cake) was then roasted with coal and limestone, yielding a black ash consisting chiefly of sodium carbonate and calcium sulfide. This was dissolved in water and the sodium carbonate crystallized out. One of the side effects of the process was to stimulate the demand for sulfuric acid. In 1791 Leblanc received a patent for his invention and, financed by the duc d'Orleans, he built a factory at St. Denis on the outskirts of Paris. His patron was guillotined in 1793 and Leblanc's factory confiscated. Although it was returned to him by Napoleon in 1802 he had no funds to develop his discovery and it was left to others to make their fortunes from it. In 1806 he committed suicide.

Le Chatelier, Henri Louis (1850–1936) *French chemist*

> Every system in chemical equilibrium, under the influence of a change of any single one of the factors of equilibrium, undergoes a transformation in such direction that, if this transformation took place alone, it would produce a change in the opposite direction of the factors of equilibrium and temperature, pressure, and electromotive force,

corresponding to the three forms of energy – heat, electricity, and mechanical energy.
> —*Researches on Chemical Equilibrium* (1888)

Le Chatelier (le sha-tel-**yay**) was born in Paris, the son of the inspector-general of mines for France. He himself began studying mining engineering, before becoming professor of chemistry at the School of Mines in 1877. He later became professor of mineral chemistry at the Collège de France and finally took the chemistry chair at the Sorbonne in 1907.

He was particularly interested in metallurgy, cements, ceramics, and glass, and his studies of flames led him to study heat and its measurement. He made a number of contributions to thermometry, the most important of which was his first successful design of a platinum and rhodium thermocouple for measuring high temperatures (1887). This was based on the principle shown by Thomas SEEBECK in 1826 that if a circuit is made from two different metals and heated, a current will flow, and that the current is proportional to the temperature difference between the junctions. It was quickly appreciated that the Seebeck effect could be used in a variety of measuring devices; if one junction was placed on the object to be measured and the other kept at a known constant temperature then the first temperature could be calculated by measuring the current. By using platinum and platinum–rhodium alloy rods, Le Chatelier succeeded where others had failed. His most important discovery, *Le Chatelier's principle*, was made in 1884. This simply states that any change made in a system of equilibrium results in a shift of the equilibrium in the direction that minimizes the change. In his original 1884 version he referred only to pressure but soon generalized the principle to cover any kind of external constraint. Le Chatelier published his principle in 1888 as the *Loi de stabilité de l'equilibre chimique* (Law of Stability of Chemical Equilibrium). The principle is important in studies of chemical equilibrium for predicting the effects of pressure and temperature on an equilibrium reaction. Le Chatelier's principle fitted in well with the law of mass action recently formulated by Cato GULDBERG and Peter WAAGE and the new chemical thermodynamics of Josiah Willard GIBBS, whose work Le Chatelier was responsible for introducing to France. The principle was soon shown to have industrial implications, for Fritz HABER successfully utilized it in his process for the production of ammonia.

Leclanché, Georges (1839–1882) *French engineer and inventor*

Leclanché (le-klahn-**shay**), who was born in Paris, France, is best known for his invention of the electrical battery, now known as the dry cell. This he developed in 1866, six years after completing his formal technical education and starting work as an engineer. The cell, which uses ammonium chloride as the electrolyte and zinc and manganese dioxide as the electrodes, was used extensively in the telegraph system from 1868 onward.

Lecoq de Boisbaudran, Paul-Emile (1838–1912) *French chemist*

Lecoq de Boisbaudran (le-**kok** de bwah-boh-**drahng**) came from a wealthy family of distillers from Cognac in southwestern France. Of independent means and excited by the new spectroscopy of Gustav KIRCHHOFF, he built his own laboratory. In 1859, using spectroscopic techniques, he began searching for new minerals and elements.

In 1874, while examining a sample of zinc ore from the Pyrenees, Lecoq de Boisbaudran noticed some new spectral lines and discovered a new element, which he named "gallium" after the old name of his country. On hearing of the new element in 1875 Dmitri MENDELEEV claimed this to be his long-predicted eka-aluminum, thus providing the first dramatic confirmation of his periodic table. Lecoq de Boisbaudran later discovered two more elements: samarium (1879) and dysprosium (1886).

Lederberg, Joshua (1925–2008) *American geneticist*

Lederberg was born in Montclair, New Jersey, and educated at Columbia and Yale where he gained his PhD in 1948. He later held chairs of genetics at the University of Wisconsin, where he had taught since 1947, and at Stanford from 1959 until his appointment as president of Rockefeller University in 1978, a post he held until 1990.

When Lederberg began work as a graduate student it was widely believed that bacteria had no genes, nuclei, or sex. However, as a result of experiments undertaken with his supervisor Edward TATUM he was able to show in 1946 that bacteria do possess genetic and behavior systems with nuclei, genes, and in certain cases even true sexual mechanisms. The technique used to demonstrate recombination and hence sexuality was a spin-off from those developed by Tatum and George BEADLE in their work on the fungus *Neurospora*. Two distinct mutants of the K 12 strain of the *Escherichia coli* bacillus were used; the first (A) was incapable of synthesizing the essential ingredients methionine and biotin (M^-B^-) while the second (B) was incapable of producing proline and threonine (P^-T^-). As long as they were grown in a medium rich in the essential ingredients both the A and B mutants could flourish. If, however, they were grown in a medium totally deficient in all four essential ingredients then, assuming reproduction by fission without

genetic interaction, both mutants should fail to develop. In such a context Lederberg in fact found that about one in every ten million bacteria did yield visible colonies. That is, from plaque containing only the A-type $M^-B^-P^+T^+$ and the B-type $M^+B^+P^-T^-$ a normal form, $M^+B^+P^+T^+$, had emerged, which could only have arisen from a sexual mating process. Lederberg went on to show that the process of "conjugation," as he called it, was common and could be used to map the bacterial genes. Lederberg, working with Norton ZINDER in 1952, made the equally significant discovery of bacterial transduction. Here they took two strains of *Salmonella typhimurium* lacking in the ability to synthesize different but essential amino acids. They were placed on either side of a U-tube separated by a fine filter positioned in the bend. The nutrient broth, which alone could pass through the filter, lacked the relevant amino acids. Yet despite the isolation of the two *Salmonella* strains, recombinant bacteria appeared and multiplied. Something must therefore have carried genetic information through the barrier. But it could not be the bacteria themselves as the filter was too small nor, they showed, was it DNA for the process continued in the presence of an enzyme which destroyed free DNA. They finally established that it must be a bacteriophage (a bacteria-infecting virus). It was this experiment that gave the first hint that genes could be deliberately inserted in cells or, in a later jargon, that genes could be engineered. For his discoveries concerning genetic recombination and the organization of the genetic apparatus of bacteria Lederberg shared the 1958 Nobel Prize for physiology or medicine with Tatum and Beadle.

Lederman, Leon Max (1922–) *American physicist*

The son of Russian immigrants, Lederman was born in New York and educated at City College there. After three years with the U.S. Signal Corps during the war, he went to Columbia where he gained his PhD in 1951. He was appointed professor of physics in 1958 and remained at Columbia until 1979, when he accepted the directorship of the Fermi National Accelerator Laboratory, Batavia, Illinois, a post he held until his retirement in 1989.

In 1959 T. D. Lee asked his Columbia colleagues Lederman, Melvin SCHWARTZ, and Jack STEINBERGER if it was possible to study the weak fundamental interaction at high energies. While well understood at low energies, Lee noted, theories of weak interactions at high energies led to absurdities. Yet it was difficult to explore the interaction experimentally, for at high energies other forces tended to obscure all other reactions. Lederman and his coworkers began to investigate decay processes that lead to neutrinos. These proceed by a weak interaction, and there are two processes in which they can occur. In one, pions decay to give muons and neutrinos. The other, beta decay, is decay of a neutron to give a proton, an electron, and a neutrino. In what has become known as the two-neutrino experiment, the team investigated the question of whether the two types of neutrino were identical – whether the muon neutrino was the same particle as the electron neutrino. The experiment was difficult since neutrinos have a very low probability of interacting with matter. It required an intense beam of high-energy neutrinos and a large detector to have any chance of yielding a measurable number of events. Using the Alternating Gradient Synchrotron at Brookhaven, a beam of 10^{11} protons per second was directed with an energy of 30 billion electronvolts (30 GeV) at a beryllium target. This produced a large number of pions, which rapidly decay into muons and neutrinos. The muons were filtered out by a steel barrier 44 feet thick built from the plates of an old battleship. The neutrinos passed through untouched into a tenton aluminum detector. The experiment ran for ten days and diverted 10^{14} high-energy neutrinos through the detector. If there was only one type of neutrino it should react in the experiment with neutrons to produce an equal number of muons and electrons; if, however, the experiment produced a unique muon-linked neutrino, only muons should be created. Fifty-one neutrino collisions were recorded by the detector; all produced muons and none an electron. For this work Lederman and his Columbia colleagues Schwartz and Steinberger shared the 1988 Nobel Prize for physics. In 1977 Lederman led another team that made a second fundamental discovery. Working with the Fermilab accelerator they discovered a new particle nine times heavier than the proton. It was named the "upsilon particle" and provided the first evidence of the fifth quark – the so-called "bottom quark." Lederman has given a popular account of particle physics in his 1992 book *The God Particle* (the title refers to the Higgs boson). Lederman was a key figure in the campaign to build a superconducting super collider (SSC), the giant accelerator which would supposedly finally detect the Higgs boson. To further the project Lederman made a ten-minute video for President Reagan to explain what they hoped to achieve. On the strength of the video, Reagan agreed to back the SSC. A later administration, however, decided in 1993 that the planned expenditure of $8 billion could not be supported and killed the project.

Lee, David Morris (1931–) *American physicist*

Lee was born in Rye, a small town just outside New York City. He originally studied physics at

Harvard, graduating in 1952. After a period spent in the army, he joined the University of Connecticut in 1954, and in 1955 enrolled at Yale to work for a PhD in the low-temperature research group.

Superfluidity in the helium isotope, helium-4, had first been detected by Pyotr KAPITZA in 1938 at a temperature of about two degrees above absolute zero (2.17 K). Helium-4 has a nucleus containing two protons and two neutrons and has two orbiting electrons. This means that it has an integral spin and belongs to the class of particles known as bosons. It was recognized that helium-3, with a nucleus consisting of one neutron and two protons, would have a spin of $+\frac{1}{2}$ and therefore must be a fermion. As only bosons could occupy the same quantum state, only bosons, it was thought, could ever become superfluids. However, theoretical considerations proposed by John BARDEEN and his colleagues suggested that under certain conditions fermions could behave like bosons and that helium-3 could possibly display superfluidity.

In 1971, Lee's graduate student Douglas OSHEROFF stumbled on precisely the conditions that would lead to superfluidity. Further work by Lee and his colleagues established that there are three distinct superfluid phases of helium-3 at 0.0027 K, 0.0021 K, and 0.0018 K.

For this work, Lee was awarded the 1996 Nobel prize for physics with Douglas Osheroff and Robert RICHARDSON.

Lee, Tsung-Dao (1926–) *Chinese–American physicist*

Lee was born in Shanghai, China. His early studies at the National Chekiang University in Guizhou province, southern China, were interrupted by the Japanese invasion during World War II. He fled to Kunming, Yunnan, where from 1945 to 1946 he studied at the National Southwest Associated University. In 1946 he received a Chinese government scholarship, which enabled him to study at the University of Chicago in America. In 1950 he gained his PhD there for his astrophysics work on the composition of certain types of stars. In the years 1950–51 he worked as a research associate in astronomy at the Yerkes Astronomical Observatory, Wisconsin, and taught physics at the University of California at Berkeley. The next two years he spent at the Princeton Institute of Advanced Study, leaving to take up an assistant professorship in physics at Columbia University. He was made full professor in 1956.

While at Berkeley and Princeton, Lee worked with a fellow countryman he had known briefly in Kunming – Chen Ning YANG. These two maintained contact while Lee was at Columbia, working on problems of elementary particle physics. In a great insight, the two men challenged one of the fundamental concepts of that time – the conservation of parity. Put simply, it had been assumed that the laws of nature are unchanged in mirror-image transformations. Lee and Yang realized that this assumption had never been explicitly tested, and that it might not be valid in the case of the so-called "weak" interactions between particles. They published a controversial paper in 1956, and within months experiments had been performed (by another Chinese, Chien Shiung Wu) which showed that the "law" of parity is indeed violated in such interactions. In 1957, only a year later, Lee and Yang were jointly honored with the Nobel Prize for physics. Lee went on to consider some of the implications of this discovery, particularly as it affected ideas about the neutrino. He also made contributions in the fields of statistical mechanics, nuclear physics, field theory, and turbulence. With the exception of a three-year break (1960–63) at the Princeton Institute of Advanced Study, he continued his work at the physics department of the University of Columbia.

Lee, Yuan Tseh (1936–) *American chemist*

Born at Hsinchu, Taiwan, Lee was educated at Taiwan University and at the University of California, Berkeley, where he obtained his PhD in 1965. He moved soon after to the University of Chicago, but returned to Berkeley in 1974 as professor of chemistry. The same year he became a naturalized American citizen.

Lee worked initially as an experimentalist. In 1967 he began to help Dudley HERSCHBACH with his colliding-molecular-beam experiments. Lee introduced a newly designed and extremely sensitive mass spectrometer allowing them to work with much greater accuracy and precision. He shared the 1986 Nobel Prize for chemistry with Herschbach and John POLANYI.

From 1986 he continued his research in chemical dynamics and also worked on various photochemical processes. He returned to Taiwan in 1994 to become president of Academia Sinica, actively supporting scientific research and promoting cultural and educational activities.

Leeuwenhoek, Antonij van (1632–1723) *Dutch microscopist*

> What struck me as the most extraordinary thing was that I never came upon a single one of these small creatures [aphids] from which I was unable to extract a number of young in the process of formation ... or which I could possibly consider as being a male.
> —Letter (1695). *The Collected Letters of A. van Leeuwenhoek* (ed. A. Schierbeeck)

Born the son of a basket maker at Delft in the Netherlands, Leeuwenhoek (**lay**-ven-hook) received little formal education and was apprenticed to a linendraper at the age of 16. In about 1654 he set up in business in Delft as a draper. He also served from 1660 as chamberlain of the town's law courts.

In 1673 Henry Oldenburg, secretary of the Royal Society, received a letter from a Dutch correspondent informing him that "a certain most ingenious person here Leeuwenhoek has devised microscopes which far surpass those manufactured by others." A letter from Leeuwenhoek describing his observations of bees, mold, and lice was enclosed. It was published in the *Philosophical Transactions* of the Royal Society in 1673. It was the first of 165 letters reporting Leeuwenhoek's observations which would appear in the Transactions between 1672 and his death in 1723. Leeuwenhoek wrote no books and, lacking Latin, reported his work in Dutch, which was then translated into English or Latin for publication. Among the highlights of the *Letters* are his 1674 observations of his "little animalcules" (protozoa) discovered in rainwater that had stood for a few days. They were, he estimated, some 10,000 times smaller than water fleas. He also gave some idea of the profusion of nature by calculating that "there were upwards of 1 million living creatures in one drop of pepperwater." He was sufficiently detached to examine with his microscope his own feces and note that "when of ordinary thickness" no animalcules were observed but whenever "the stuff was a bit looser than ordinary I have seen animalcules therein." Leeuwenhoek also announced in 1679 his discovery of human spermatazoa. In 1677 a Mr. Ham brought him "the spontaneously discharged semen of a man who had lain with an unclean woman and was suffering from gonorrhea." He observed "animalcula in semine masculino" ("animalcules in human semen"), and noted they had tails and lived for a few hours only. He went on to examine the sperm of birds, frogs, insects, cattle, and several other species. A further important biological observation was Leeuwenhoek's 1684 description of red blood cells, which he estimated to be 25,000 times smaller than a fine sand grain. Leeuwenhoek's instruments were all simple microscopes with a single small lens clamped between two metal plates. The object was placed on a fine pin and its distance from the plates adjusted by turning a screw. On his death he left 247 completed microscopes and 172 mounted lenses. They were auctioned and dispersed in 1747. A further 26 mounted in silver were bequeathed to the Royal Society but disappeared without trace in the mid-19th century. Nine of Leeuwenhoek's original microscopes have survived, with a highest magnification of 266 and resolution of 2 micrometers. Toward the end of his life Leeuwenhoek became something of a European celebrity. Monarchs such as Peter the Great and Queen Mary traveled to Delft to be shown the "little animalcules" by Leeuwenhoek himself.

Leffall, LaSalle (1930–   ) *American surgeon and oncologist*

Leffall was educated at Florida A & M University and at Howard University College of Medicine, Washington, DC, where he gained his MB in 1952. After serving his internship in St. Louis, and his surgical residency in Washington, Leffall spent the period from 1957 until 1959 at the Sloan–Kettering Cancer Center, New York, where he trained in cancer surgery. In 1962 Leffall returned to Howard University and was appointed professor of surgery there in 1970. In 1979 he became the first African-American president of the American College of Surgeons. In 2002 Leffall was appointed chairman of the President's Cancer Panel.

In addition to the heavy demands of surgery and administration, Leffall has also been extremely active in publicizing the major health problems facing minorities in the United States. In particular, in 1973 he published a long study demonstrating the alarming degree to which cancer mortality has continued to climb among the U.S. black population. Much of this increase, he insisted, is due to the simple truth that many black people have been denied access to adequate health care.

Leggett, Anthony J. (1938–   ) *British-born American physicist*

Leggett was born in London and gained his doctor's degree in physics in 1964 from Oxford University. From 1967 to 1983 he carried out research at the University of Sussex. He is currently the John D. and Catherine T. MacArthur Professor at the University of Illinois at Urbana-Champaign. In 2007 he was also appointed scientific advisor at the Institute for Quantum Computing at the University of Waterloo, Ontario, Canada.

At very low temperatures, liquid helium completely loses all viscosity and becomes a superfluid, as discovered by Pyotr Kapitska *et al.* in the late 1930s. The Azerbaijani physicist Lev LANDAU provided the theoretical explanation of this phenomenon. Then in the early 1970s American physicists discovered that the rare isotope helium-3 (or ^{3}He; the common isotope is ^{4}He) is also a superfluid at extremely low temperatures. While working at the University of Sussex, Leggett researched the superfluid state of this isotope, in particular how the helium atoms are ordered – they form pairs – and how they interact. The paired atoms have magnetic properties, and magnetic measurements revealed that the superfluid comprises a mixture of three different phases whose composition depends on the external physical conditions. Leggett's theoretical explanation of this phenomenon has been applied also in cosmology and particle physics. Leggett shared the 2003 Nobel Prize for physics with Alexei ABRIKOSOV and Vitaly GINZBURG for their pioneering contributions to the theory of superconductors and superfluids.

Le Gros Clark, Sir Wilfrid Edward (1895–1971) *British anatomist and anthropologist*

Le Gros Clark (le groh klark), who was born at Hemel Hempstead in southeastern England, qualified in medicine at St. Thomas's Hospital, London, in 1916. He then served with the Royal Army Medical Corps in France for the remainder of World War I. After the war he returned to St. Thomas's and taught anatomy for two years before going out to Sarawak, Borneo, as principal medical officer. In Borneo he studied tarsiers and tree shrews and on his return to England expanded his work on these animals, showing that they should be classified as primates rather than insectivores.

Le Gros Clark also did very useful research on the anatomy of the brain, investigating the relation of the cerebral cortex to the thalamus and mapping the hypothalamus. He held posts in anatomy successively at St. Bartholomew's Hospital, St. Thomas's, and Oxford University, retiring in 1962. In the fifties he was one of the experts to expose the Piltdown Man hoax.

Lehmann, Inge (1888–1993) *Danish seismologist*

It was Lehmann (**lay**-man) who, in 1936, first put forward the view that the Earth's core consisted of two parts – an inner and an outer, separated by a discontinuity. For many years it had been thought that the Earth consisted merely of a core, mantle, and crust that were separated by the discontinuities discovered by Beno GUTENBERG and Andrija MOHOROVIČIĆ. This was partly based on the realization that the primary (P) waves of an earthquake are not detected in a shadow zone between 105° and 145° from the epicenter. The reason for this was their diffraction by the Earth's core.

Lehmann found that P-waves increased their velocity quite sharply within the core. She reasoned from this that there was an outer core and an inner core separated by a further discontinuity about 700 miles (1,200 km) from the center of the Earth and sufficient to bend some of the P-waves into the shadow zone.

Lehn, Jean Marie Pierre (1939–   ) *French chemist*

Born at Rosheim in France, Lehn (len) was educated at Strasbourg, where he obtained his PhD in 1963, and at Harvard. After working in Strasbourg from 1966 to 1970, Lehn returned to Harvard as professor of chemistry. In 1979 he moved to the Collège de France, Paris, where he currently holds the chair of chemistry of molecular interactions.

In 1963 Charles PEDERSEN had discovered

(2,2,2) cryptand   spherical cryptand

CRYPTANDS   Chemical structures of cryptands discovered by Lehn.

the first of a type of compound known as "crown ethers." These have molecules containing large rings of carbon and oxygen atoms. They are able to form strongly bonded complexes with metal ions. The formation is highly selective, depending on the size of the ion that is complexed. Consequently, crown ethers have a number of important uses in separation of mixtures and in analysis. While Pedersen had worked with two-dimensional rings, Lehn sought to extend his work into three dimensions. If two nitrogen atoms replaced the oxygen atoms of the original crown ether, Lehn found, two crowns could be made to combine into a cagelike structure; a "cryptand" in Lehn's terminology. He found that cryptands were capable of binding metal cations more selectively than the crown ethers. Lehn went on to develop cryptands that would bind selectively with other molecules, including important biologically active molecules such as the neurotransmitter acetylcholine. The molecules found in this way are known as "supramolecules" and their discovery has opened up an important new field known as "host–guest chemistry." For his work in this new field Lehn shared the 1987 Nobel Prize for chemistry with Pedersen and Donald CRAM.

Leibniz, Gottfried Wilhelm (1646–1716)
German mathematician, philosopher, historian, and physicist

> The art of discovering the causes of phenomena, or true hypotheses, is like the art of deciphering, in which an ingenious conjecture greatly shortens the road.
> —*New Essays Concerning Human Understanding* (1765)

> It would be difficult to name a man more remarkable for the greatness of his intellectual powers than Leibniz.
> —John Stuart Mill, *A System of Logic* (1843)

Born the son of a Lutheran professor of moral philosophy in Leipzig, Germany, Leibniz (**llb**-nits or **llp**-nits) was educated at the universities of Leipzig, Jena, and Altdorf where he gained his doctorate in 1666. In 1667 he entered the service of the elector of Mainz for whom he spent the period 1673–76 on a diplo-

matic mission to Paris. Through meeting with such scholars as Christiaan HUYGENS in Paris and with members of the Royal Society, including Robert BOYLE, during two trips to London in 1673 and 1676, Leibniz was introduced to the outstanding problems challenging the mathematicians and physicists of Europe. On leaving Paris he joined the staff of John Frederick the duke of Brunswick-Lüneburg, also Elector of Hannover, where he was given the commission to write the history of the House of Brunswick and the position of librarian. For the remaining 40 years of his life Leibniz dissipated his prodigious talents under three electors, including the future George I of Great Britain and Ireland, constructing genealogies of the numerous Brunswick progeny, both legitimate and, even more numerous, illegitimate. He also undertook a variety of administrative and diplomatic duties of which his attempt to unite the Protestant and Catholic churches in 1683 and the founding of the Berlin Academy of Sciences in 1700 are the most noteworthy. It was also to his Brunswick years that most of his philosophical writings belong although many of them remained unpublished until well after his death.

Leibniz's greatest achievement was undoubtedly his discovery of the differential and integral calculus, work which was to involve him in a bitter priority dispute with Isaac NEWTON. Newton's ideas on the calculus were developed first, as early as 1665, but remained unpublished until 1687; Leibniz, however, began work on problems of the calculus during his Paris years and published his results in 1684 in *Nova methodus pro maximis et minimis* (New Method for the Greatest and the Least). It was later suggested in 1699 that Leibniz's original inspiration may well have come from conversations in London in 1673 and in 1676 as well as letters of Newton to Henry Oldenburg shown to Leibniz. From this point the dispute became open to all and was conducted with considerable ferocity and not a little dishonesty. In fact, as became clear later on, the discoveries were made independently; the final triumph lay with Liebniz for it was his notation of differentiation and integration, rather than the fluxions of Newton, that have survived in modern textbooks. In physics, Leibniz's metaphysical principles also led him to deny Newtonian gravity acting at a distance on the grounds that: "A body is never moved naturally, except by another body which touches it and pushes it; after that it continues until it is prevented by another body which touches it. Any other kind of operation on bodies is either miraculous or imaginary." He also rejected Newtonian concepts of absolute space and time, arguing more plausibly that space was simply "the order of bodies among themselves" while time was their order of succession. Leibniz also, with Huygens, developed the concept of kinetic energy. As well as his contributions to metaphysics and philosophy, Leibniz established the foundations of symbolic logic, probability theory, and combinatorial analysis, and was led to design a practical calculating machine. It was actually built and shown to the Royal Society in 1794.

Leishman, Sir William Boog (1865–1926) *British bacteriologist*

Leishman was born in Glasgow, the son of the regius professor of medicine at the university there. He himself was educated at the university, where he obtained his MD in 1886. He immediately joined the Army Medical Service and began his career in India, where he served from 1890 to 1897. On his return to England he took up an appointment at the Army Medical School at Netley as an assistant to Almroth WRIGHT, succeeding him in 1903 as professor of pathology when the school moved to Millbank. In 1913 Leishman transferred to the War Office, where he served in various advisory positions before being appointed director of pathology (1919) and director general of Army Medical Services (1923), a post he held until his death.

Leishman's first major success was his discovery in 1900 of the protozoan parasite (*Leishmania*) responsible for the disease known variously as kala-azar and dumdum fever. As he delayed publication until 1903 he was forced to share his discovery with C. Donovan, who independently repeated his work (the form of the parasite found in man became known as the *Leishman–Donovan body*). The disease caused by the parasite is now known as *leishmaniasis*. In 1900 he went on to develop the widely used *Leishman's stain*. This is a compound of methylene blue and eosin that soon became adopted as the standard stain for the detection of such protozoan parasites as *Plasmodium* (malaria parasite) in the blood. Leishman also made major contributions to the development of various vaccines, particularly those used against typhoid. By 1896 Wright had developed a safe vaccine of killed typhoid bacilli, which he persuaded the Army to test during the Boer War (1899–1902). The extent of the protection provided by the vaccine became a matter of violent controversy between Wright and the English statistician Karl Pearson; the Army Council therefore invited Leishman in 1904 to resolve the dispute. By 1909 Leishman was able to report that those inoculated in India carried a significantly smaller risk of dying from enteric complaints (5 died out of 10,378 vaccinated, compared with 46 out of the 8,936 not vaccinated). It was mainly as a result of this work, together with improvements introduced by Leishman in the actual quality of the vaccine, that a policy of mass vaccination was adopted in 1914. Consequently only 1,191 deaths due to typhoid were reported by the British Army throughout the whole of World War I. Leishman was knighted in 1909.

Leloir, Luis Frederico (1906–1987) *Argentinian biochemist*

Born in Paris, France, Leloir (le-**lwar**) was educated at Buenos Aires University, obtaining his MD there in 1932. He spent a year in Cambridge, England, studying under Gowland HOPKINS, returning to Argentina to work at the Institute of Physiology until 1944, when – in conflict with the president, Juan Peron – he went into exile in America. In 1945 Leloir returned to Argentina, where he worked at the Institute of Biology and Experimental Medicine, set up in Buenos Aires by Bernard Hussey with private funding.

Despite working well away from the main biochemical research centers and using equipment that would have been thrown out of more fashionable laboratories, Leloir and his colleagues managed to surprise the biochemical world and make one of the major discoveries of the postwar years. In the 1930s Carl and Gerty CORI had demonstrated a process by which glycogen is synthesized and broken down. It was assumed that because there were enzymes capable, *in vitro*, of both breaking down glycogen into lactic acid and reversing the whole process, that this is what actually happened in the body. It was therefore a matter of some surprise when Leloir and his colleagues announced in 1957 an alternative mechanism for the synthesis of glycogen. They discovered a new coenzyme, uridine triphosphate (UTP), analogous to adenosine triphosphate (ATP), which combined with glucose-1-phosphate to form a new sugar nucleotide, uridine diphosphate glucose (UDPG). In the presence of a specific enzyme and a primer UDPG will yield uridine diphosphate (UDP) and transfer the glucose to the growing glycogen chain. In the presence of ATP, UDP is converted back into UTP and the reaction can continue. It was soon made clear that this is the actual process of glycogen synthesis taking place in the body; the Cori process is, in contrast, mainly concerned with the degradation of glycogen. It was for this work that Leloir was awarded the 1970 Nobel Prize for chemistry, the first Argentinian to be thus honored.

Lemaître, Abbé Georges Edouard
(1894–1966) *Belgian astronomer and cosmologist*

Lemaître (le-**me**-tre) was born at Charleroi in Belgium. After serving in World War I, he studied at the University of Louvain in Belgium from where he graduated in 1920. He then attended a seminary at Mailines, becoming ordained as a Roman Catholic priest in 1923. Before taking up an appointment at the University of Louvain in 1925, he spent a year at Cambridge, England, where he worked with Arthur EDDINGTON, and a year in America where he worked at the Harvard College Observatory and the Massachusetts Institute of Technology. He remained at Louvain for the whole of his career, being made professor of astronomy in 1927.

Lemaître was one of the propounders of the big-bang theory of the origin of the universe. EINSTEIN's theory of general relativity, announced in 1916, had led to various cosmological models, including Einstein's own model of a static universe. Lemaître in 1927 (and, independently, Aleksandr FRIEDMANN in 1922) discovered a family of solutions to Einstein's field equations of relativity that described not a static but an expanding universe. This idea of an expanding universe was demonstrated experimentally in 1929 by Edwin HUBBLE who was unaware of the work of Lemaître and Friedmann. Lemaître's model of the universe received little notice until Eddington arranged for it to be translated and reprinted in the *Monthly Notices of the Royal Astronomical Society* in 1931. It was not only the idea of an expanding universe which was so important in Lemaître's work, on which others were soon working, but also his attempt to think of the cause and beginning of the expansion. If matter is everywhere receding, it would seem natural to suppose that in the distant past it was closer together. If we go far enough back, argued Lemaître, we reach the "primal atom," a time at which the entire universe was in an extremely compact and compressed state. He spoke of some instability being produced by radioactive decay of the primal atom that was sufficient to cause an immense explosion that initiated the expansion. This big-bang model did not fit too well with the available time scales of the 1930s. Nor did Lemaître provide enough mathematical detail to attract serious cosmologists. Its importance today is due more to the revival and revision it received at the hands of George GAMOW in 1946.

Lémery, Nicolas (1645–1715) *French chemist*

Lémery (**lay**-me-ree or laym-**ree**), who was born at Rouen in northern France, trained with an apothecary before studying chemistry at Montpellier, where he also lectured. After moving to Paris in 1672, he opened a pharmacy and laboratory there and gave public lectures to large audiences. In 1675 he published his popular *Cours de chymie* (Lectures on Chemistry), which went through 31 editions in the next 100 years and was translated into most European languages. This was basically a work on pharmaceutical chemistry, dealing with preparations from vegetables, minerals, salts, and sulfur. Lémery rejected alchemical beliefs and as a theorist followed René DESCARTES, attempting to explain all chemical reactions by the shape and movement of the fundamental particles of substances. Thus acids consist of sharp pointed particles while metals are porous. The acids are able to enter the pores in the metals and tear them to pieces with their sharp points.

Lémery produced a method for obtaining sulfuric acid by burning sulfur and saltpeter in an enclosed container. It is possible that Joshua WARD met Lémery during his exile in France and acquired his process for the commercial manufacture of sulfuric acid, which he set up in Surrey, England.

Lenard, Philipp Eduard Anton (1862–1947) *German physicist*

> No entry to Jews and Members of the German Physics Society.
> —Notice on Lenard's door

Born the son of a wine merchant in Pozsony (now Bratislava in Slovakia), Lenard (**lay**-nart) studied at the universities of Budapest, Vienna, Berlin, and Heidelberg, where he obtained his doctorate. He taught at the universities of Bonn (1893), Breslau (1894), Aachen (1895), and Heidelberg (1896). In 1898 he was appointed professor of experimental physics at Kiel. He returned to Heidelberg in 1907, where he remained until his retirement in 1931.

Lenard's career falls naturally into two distinct periods. Before 1914 he made several major contributions to fundamental physics. In particular he investigated the photoelectric effect. It had been known for some time that light falling on certain metals would cause the emission of electrons. Starting in 1899 Lenard investigated why the effect could only be produced by ultraviolet or shortwave light. In the course of his experiments he established two anomalous results. He found that the speed with which the electron was emitted was a function of the wavelength of the light used – the shorter the wavelength the faster the electron. Increasing the intensity of the light did not affect the speed but did, surprisingly, increase the number of electrons emitted. It was left to Albert EINSTEIN to explain the significance of these results in 1905 by linking them to the new quantum theory of Max PLANCK. Lenard also did important work on cathode rays (electrons) for which he received the Nobel Prize for physics in 1905. He demonstrated how they could be induced to leave the evacuated tube in which they were produced, penetrate thin metal sheets, and travel a short distance in the air, which would become conducting. On the basis of this work he proposed a model of the atom in which it is made from "dynamids," units of positive and negative charge. This was, however, soon superseded by the nuclear atom of Ernest RUTHERFORD. Lenard also seems to have come close to making two other discoveries. He almost discovered x-rays and felt that if he had not moved to Aachen in 1895 he would have been successful. He did in fact help their discoverer Wilhelm RÖNTGEN with equipment – aid which, he argued, was never duly acknowledged. He also felt that J. J. THOMSON had used some of his work without due recognition. His suspicions of other workers were the first signs that Lenard was developing a somewhat idiosyncratic view of physics. The latter half of his career, from 1919, was spent arguing for the establishment of a new physics, a "German" physics untainted with Jewish theories. Although Lenard was a German patriot who was deeply affected by Germany's defeat in 1918, he was not simply an antisemite. He attacked Einstein as a socialist, a pacifist, and, indeed, as a Jew; however, his strongest abuse was directed toward him as a *theoretical* physicist. In 1920 Lenard organized a conference at Bad Neuheim to discuss relativity theory and attacked Einstein for somehow misleading people with a very abstract theory with little experimental support. He was also deeply upset by Einstein's dismissal of theories of the ether. The only course for him was to develop a non-Jewish physics and to this end he produced a curious four-volume work, *Deutsche Physik* (1936–37; German Physics). Faced with the objection that science is international he replied, "It is false. Science like every other human product is racial and conditioned by blood." The atmosphere produced by Lenard did much to cause the general exodus of scientists from Germany and to destroy creative science there for a generation. Just why Lenard was transformed from a talented experimentalist into a bigoted and almost pathological crank is not clear. Germany's losing the war followed by the death of his son and the loss of all his savings in the postwar inflation no doubt contributed, but the ultimate source seems to have been his distaste, as an experimentalist, for the increasing mathematical abstraction introduced into physics by such scientists as Einstein.

Lenat, Douglas (1950–) *American computer scientist*

Born in Philadelphia, Lenat was educated at the University of Pennsylvania and at Stanford, where he obtained his PhD in 1976 and remained until 1984.

Lenat has worked mainly in the field of artificial intelligence (AI). In 1976 he introduced Eurisko, an expert system designed to discover new information. It was given a limited amount of information, but stocked also with hundreds of general heuristic rules. In 1984 Lenat began a new project backed by Microelectronics and Computer Technology (MCC), a consortium of 56 high-tech companies based in Austin, Texas. The project, called Cyc, became Cycorp in 1994.

Lennard-Jones, Sir John Edward (1894–1954) *British theoretical chemist*

Lennard-Jones was born at Leigh in England, the son of a retail furnisher. He was educated at Manchester University and, after service in the Royal Flying Corps, at Cambridge University, where he obtained his PhD in 1924. He moved to Bristol University in 1925, serving as professor of theoretical physics from 1927 until 1932 when he returned to Cambridge as professor of theoretical chemistry. He resigned in 1952, shortly before his death, to become principal of the University of Keele.

Lennard-Jones began his research career in the early 1920s by attempting to produce a formula from which interatomic forces could be calculated. He later moved into the field of theoretical chemistry, doing much to promote the molecular-orbital theory of Robert MULLIKEN.

Lenoir, Jean Joseph Etienne (1822–1900) *Belgian mechanical engineer*

Though self-taught, Lenoir (le-**nwar**) was clearly an imaginative and effective engineer. Born at Mussy-la-Ville in Belgium, he moved to France in 1838 and worked initially in Paris as a metal enameler. He quickly made money from his inventions, developing a new electroplating method and devising a number of improvements to, among other things, railway signaling, telegraph transmission, and the electric motor.

Lenoir is best known for the patent he took out in 1860 for an internal-combustion engine. The basic idea of burning fuel in the cylinder of the engine, and thus eliminating the boiler and firebox at a stroke, was extremely attractive. Lenoir operated with a horizontally mounted water-cooled engine, which generated up to 4 horsepower. A mixture of coal-gas and air was fed into a cylinder and ignited by an electric spark. It drove a piston connected by a crank to a flywheel. The engine was double-acting with ignition on both sides of the piston. As there was no compression in the cylinder, Lenoir's engine closely resembled the double-acting atmospheric steam engines of his day. The crucial step of introducing compression with a four-stroke cycle was taken a decade later by Nicolaus Otto. The engine was bulky, inefficient, and required an external fuel supply. It was clearly unsuitable to power a vehicle. Uses, however, were soon found in such operations as pumping water and driving lathes. More than 500 were built on license, including about 100 in Britain, and a smaller number were produced by the Lenoir Gas Engine Company of New York.

Lenz, Heinrich Friedrich Emil (1804–1865) *Russian physicist*

While a student at the university in his native city of Dorpat (now Tartu in Estonia), Lenz (lents) accompanied a voyage around the world as a geophysicist. Soon after his return he started teaching at the University of St. Petersburg, where he became professor in 1836.

Lenz worked on electrical conduction and electromagnetism. In 1833 he reported investigations into the way electrical resistance changes with temperature, showing that an increase in temperature increases the resistance (for a metal). He is best known for *Lenz's law*, which he discovered in 1834 while investigating magnetic induction. It states that the current induced by a change flows so as to oppose the effect producing the change. Lenz's law is a consequence of the, more general, law of conservation of energy.

Lepaute, Nicole-Reine (1723–1788) *French astronomer*

Lepaute was born in the Luxembourg Palace, Paris, where her father worked, and was noted as a studious child with an interest in science. In 1749 she married Jean-André Lepaute, a royal clockmaker, and in the early 1750s she helped him by calculating a table of oscillations for a pendulum, which was published in his book *Traité d'horlogerie*. During this time she became acquainted with the astronomer Jérôme LALANDE, who was involved in designing one of Lepaute's clocks. Subsequently, Lalande engaged Nicole-Reine to help in calculating the date of the return of HALLEY's comet; this was a mammoth enterprise, in which they collaborated with the mathematician Alexis CLAIRAUT. Much acclaim greeted them when the comet returned just one month earlier than predicted, on 13 March 1759. However, relations soured when Clairaut omitted to credit Lepaute's significant contribution in his book. But Lepaute and Lalande continued to work closely together in compiling astronomical tables. In 1762 Lepaute calculated the timing of a forthcoming solar eclipse, predicted for 1764, and compiled a map showing the visibility of the eclipse across Europe in 15-minute intervals. Her achievements were recognized by France's scientific community, and she was influential in the decision of her husband's nephew, Joseph Lepaute, to pursue a career in astronomy.

Leucippus (about 500 BC–450 BC) *Greek philosopher*

> Not one thing comes to be randomly, but all things from reason and necessity.
> —Quoted by Diels-Kranz in *Fragments of the Presocratic Philosophers*

Very little is known about the life of Leucippus (loo-**sip**-us); he probably came from Miletus in Asia Minor, although Elea, Italy, and Abdera in Thrace have also been suggested.

Our knowledge of Leucippus comes from the writings of ARISTOTLE and THEOPHRASTUS. He is said to have been the teacher of DEMOCRITUS and author of the *Great World System* and *On Mind*. He is also credited with being the originator of atomic theory, although it is difficult to distinguish his contributions from those of his pupil Democritus.

Leuckart, Karl Georg Friedrich Rudolf (1822–1898) *German zoologist*

> It is not possible for man, as a thinking being, to close his mind to the knowledge that he is ruled by the same power as is the animal world.
> —*Die Menschlichen Parasiten*, Vol. II (1876; The Parasites of Man)

Leuckart (**loik**-art) was born in Helmstedt, Germany, and educated at the University of Göttingen, where he also taught. He became professor of zoology at Giessen in 1850 and at Leipzig in 1869.

Leuckart was the first to describe the life histories of such parasites as tapeworms and flukes, and he elucidated the causes of many diseases, finding, for example, that trichinosis is due to the actions of a nematode worm. His wide-ranging investigations in this field were published as *Die Menschlichen Parasiten* (2 vols., 1863–76; The Parasites of Man). Leuckart was also interested in animal classification, notably of the invertebrates; for example, he separated the coelenterates (jellyfish) from the echinoderms (starfish), recognizing that the radial symmetry of the two groups did not indicate close affinity. Leuckart investigated the alternation of generations in the coelenterates and developed this into a much wider field of polymorphism.

Levene, Phoebus Aaron Theodor (1869–1940) *Russian–American biochemist*

Levene was born in Sagor, Russia, and gained his MD from St. Petersburg in 1891. He then emigrated with his family to America where he attended courses in chemistry at Columbia University, New York. He continued his chemical studies in Germany under Emil FISCHER and Albrecht KOSSEL, who introduced him to the study of nucleic acids. In 1905 he joined the newly formed Rockefeller Institute for Medical Research where he remained for the rest of his career.

It was known that nucleic acid exists in two forms, one found in the thymus of animals and the other in yeast. Kossel had shown that thymus nucleic acid contained the four nitrogen compounds adenine, guanine, cytosine, and thymine, whereas yeast nucleic acid differed by containing uracil instead of thymine. Carbohydrate and phosphorus were also known to be present. Virtually nothing, however, was known about its structure and function. The work of Levene allowed some conclusions to be drawn on these issues. In 1909 Levene found that the carbohydrate present in yeast nucleic acid is the pentose sugar ribose; it was not, however, until 1929 that he succeeded in identifying the carbohydrate in thymus nucleic acid. It is also a pentose sugar but lacks one oxygen atom of ribose and was therefore called deoxyribose. These facts enabled Levene to suggest a simple tetranucleotide structure for the inevitably named ribonucleic and deoxyribonucleic acids (RNA and DNA). (A nucleotide is simply one of the four bases plus a sugar and a phosphate group.) According to Levene each of the four bases occurred just once in each DNA and RNA molecule and were joined together by the sugar and phosphate groups. This structure could then be repeated to form a polynucleotide with the bases occurring in the same order throughout. Levene had succeeded in establishing the nucleic acids as genuine molecules existing independently of the proteins but the price he paid for this clarification was to impose on them an absurdly simple and repetitive structure. Consequently, when the search for biological individuality reached the molecular level, the far more complex and varied structure of the proteins was favored over the "monotonous" form of the nucleic acids, and a generation of biochemists mistakenly sought for the structure of the gene among the inexhaustible potential of the amino acids. When Levene was told, shortly before his death, of the classic work of Oswald AVERY, which showed the crucial part played by DNA, he was reported to be skeptical. It took a further 13 years before James WATSON and Francis CRICK came up with their famous double helical structure and completed the revolution begun by Levene and other biochemists earlier in the century.

Le Verrier, Urbain Jean Joseph (1811–1877) *French astronomer*

> The planet whose position you indicated *really* exists.

> —Johann Galle, letter to Le Verrier, 25 September 1846

Born the son of a local government official in St. Lô, northern France, Le Verrier (le ve-**ryay** or le **ve**-ree-ay) was educated at the Ecole Polytechnique and worked afterward on chemical problems with Joseph GAY-LUSSAC. He became a lecturer in astronomy at the Ecole Polytechnique in 1836 and succeeded Dominique ARAGO as director of the Paris Observatory in 1854.

Le Verrier worked on celestial mechanics, and in particular considered the problems associated with the motion of Uranus. In 1821 Alexis Bouvard, of the Paris Observatory, had published a set of tables of the motion of Uranus. Within a few years there was a noticeable discrepancy between the predicted and the observed position of Uranus. Assuming the correctness of Bouvard's work there were only two possibilities: either NEWTON's gravitational theory was not as universal as had been supposed, or there was an undetected body further out than Uranus but exerting a significant gravitational influence over its orbit. After much effort Le Verrier managed to deduce the mass and position that such a body would have to have to cause such disturbances in the orbit. (Le Verrier was unaware – as were most astronomers – that John Couch ADAMS had made these calculations in the previous year.) He asked Johann Galle in Berlin to search for the proposed planet. Galle was immediately successful, sighting Neptune on his first night of observation, 23 September 1846. The new planet was named Neptune and Le Verrier immediately became famous. He went into politics for a short time but wisely returned to astronomy in 1851. Continuing with problems of celestial mechanics, Le Verrier reworked and revised much of the work of Pierre Simon LAPLACE. He discovered the advance of the perihelion (the point of the orbit nearest the Sun) of Mercury and was convinced that this anomaly was caused by an undiscovered planet between Mercury and the Sun. So confident was he of its existence that he named it Vulcan, but despite much searching Vulcan still remained undetected. (The discrepancies in the position of Uranus could be seen as an impressive vindication of Newtonian mechanics, but the true explanation of the anomalous motion of Mercury was to play a vital role in confirming EINSTEIN's general theory of relativity.) Camille FLAMMARION claimed that Le Verrier had never taken the trouble to look through a telescope at Neptune, being satisfied with his equations and the words of others.

Levi-Montalcini, Rita (1909–   ) *Italian cell biologist*

Levi-Montalcini (lay-vee-mont-al-**chee**-nee) was educated at the university in her native city of Turin, graduating from medical school just before the outbreak of World War II. Being of Italian-Jewish descent, she found that posts in Italy's academic establishments were closed to her as a result of growing antisemitism. Undaunted, she converted her bedroom into a makeshift laboratory and proceeded with her studies of the development of chick embryos. In this she was joined by her former professor, Giuseppe Levi, a Jew who had been purged from his job by the Fascists. Between 1941 and 1943, Levi-Montalcini lived in a country cottage in the Piedmont region, then in hiding in Florence. After the Allied liberation of Italy in 1944 she worked as a doctor among refugees in Florence and in 1945 she returned to the University of Turin. Two years later she moved to the Washington University, St. Louis, becoming associate professor (1956) and professor (1958–77). She was appointed director of the Institute of Cell Biology of the Italian National Research Council in Rome in 1969, a post she held until her retirement in 1978. She was a member of the Italian Senate.

After moving to St. Louis in 1947, Levi-Montalcini continued her work on chick embryos under professor Viktor Hamburger. By the early 1950s she had demonstrated that the number of nerve cells produced in these embryos could be influenced by an agent (later termed nerve growth factor) obtained from a mouse tumor-cell culture. In 1952 the Italian embryologist was joined by an American biochemist, Stanley COHEN, who collaborated with her in determining the chemical nature of this growth factor. Cohen went on to investigate another growth factor, epidermal growth factor, which controls the embryological development of tissues such as eyes and teeth. The early studies of Levi-Montalcini represent a key advance in the understanding of mechanisms controlling embryological tissue development. Indeed, in the 1980s it was established that the nerve growth factor discovered by Levi-Montalcini influences the growth of nerves in the brain and spinal cord. The value of her work earned her the 1986 Nobel Prize for physiology or medicine, which she shared with Stanley Cohen.

Levinstein, Ivan (1845–1916) *German–British chemical industrialist*

Born at Charlottenburg near Berlin in Germany, Levinstein (**lev**-in-stIn or **lay**-vin-shtIn) studied at the University of Berlin and the Berlin Institute of Technology. Having emigrated to England in 1864, he set up a factory to manufacture synthetic dyestuffs at Blackley, Manchester.

He manufactured a large number of dyes but he is best remembered for his battle to change the patent law. As the act stood there was nothing to stop a foreign firm taking out a patent in the UK, even though they had no intention of working it. This was sufficient to stop anyone else using it. Some German firms were behaving in this way and shipping their own locally manufactured products to England, which was legally prevented from competing. Levinstein's long battle resulted in the patent law being changed in 1907 with a compulsory working clause added. Levinstein's business eventually became the foundation of the dyestuffs division of ICI in 1926.

Lewis, Edward B. (1918–2004) *American geneticist*

Lewis was educated at the University of Minnesota and at the California Institute of Technology (Cal Tech), gaining his PhD in 1942. In 1946 he joined the Cal Tech faculty, where he served as professor of biology from 1956 until his retirement in 1988.

Lewis worked mainly in the field of developmental biology, concentrating on the manner in which genes control the development of the fruit fly *Drosophila melanogaster*. In 1894, William BATESON described a characteristic set of mutations, named by him "homeotic mutations," in which one body structure is replaced by a different structure. For example, an insect leg may be replaced by an insect wing. In the late 1940s, Lewis began to study a group of genes known as the bithorax complex, which control the manner in which *Drosophila* embryos become segmented as they develop. After decades spent breeding numerous generations of fruit flies Lewis finally published his main results in 1978. *Drosophila* is divided into one head, three thoracic, and eight abdominal segments. The development of the head and first thoracic segment are controlled by the antennapedia complex, the remaining segments by the bithorax complex. Lewis found that a minimum of eight genes, clustered on chromosome 3, were involved in the segmentation of the fly's abdomen and thorax. Lewis demonstrated that the production of the second thoracic segment, which is the first to be controlled by the bithorax complex, was controlled by the fewest homeotic genes. Each later segment required the activation of one or more additional genes. The sequence of genes along the chromosome exactly matched the segments of the insect's body. He also realized that a single mutation could lead to major homeotic transformations even though, for example, hundreds of active genes would be required to create misplaced legs and wings. This could only mean that mutations were also taking place in a master gene of some kind, a gene capable of controlling the activity of many other subordinate genes. For his work on homeotic genes Lewis shared the 1995 Nobel Prize for physiology or medicine with Eric WIESCHAUS and Christiane NÜSSLEIN-VOLHARD.

Lewis, Gilbert Newton (1875–1946)
American physical chemist

Lewis, born the son of a lawyer in Weymouth, Massachusetts, was educated at the University of Nebraska and at Harvard, where he obtained his PhD in 1899. After a period abroad at Göttingen and Leipzig he returned to teach at Harvard and the Massachusetts Institute of Technology until 1912, when he moved to the University of California at Berkeley to take up an appointment as professor of physical chemistry.

In about 1916 he first introduced the notion of a covalent bond, in which the chemical combination between two atoms derives from the sharing of a pair of electrons, with one electron contributed by each atom. This was part of Lewis's more general octet theory and he published his views in *Valence and the Structure of Atoms and Molecules* (1923). Here he proposed that the electrons in an atom are arranged in concentric cubes and that a neutral atom of each element contains one more electron than a neutral atom of the element preceding it. The cube of eight electrons is reached in the atoms of the rare gases. These simple ideas enabled Lewis to explain many of the facts of chemical combination. Thus neon and argon with all vertices of the cube occupied are obviously inert, having no space to interact with other atoms. The tendency is for other atoms to attain the same configuration. Thus sodium with one vertex occupied will react readily with the seven occupied vertices of chlorine to produce a combination with all vertices occupied. And so, with considerable success, Lewis went on to explain the basic combinations of the lighter elements. The theory was extremely influential in the development of ideas about chemical valence. The notion of a stable octet of electrons is now explained by quantum theory. The theory became widely known as the Lewis–Langmuir theory. This was partly due to the failure of Lewis, a shy and reserved man, to publicize his theory and the willingness of Irving LANGMUIR, a brilliant lecturer, to fill the gap. Lewis also carried out significant work in the field of chemical thermodynamics and published, with Merle Randall, *Thermodynamics and the Free Energy of Chemical Substances* (1923), which did much to introduce the basic ideas of Josiah Willard GIBBS. For many years, the study of "Lewis and Randall" was an essential part of the education of generations of chemistry students. Lewis is also known for his ideas about acids and bases. The traditional idea of an acid, due to ARRHENIUS, is that it is a compound that can produce hydrogen ions, H^+, in solution. A base reacts with an acid to give a salt and water. The concept of an acid was extended by BRØNSTED and LOWRY. In their theory, an acid is a proton donor and a base a proton acceptor (a proton is a hydrogen ion, H^+). Lewis made a further extension in 1923. In his theory, a *Lewis acid* is a compound that can accept a pair of electrons and a *Lewis base* can donate a pair of electrons.

Lewis, Julian Herman (1891–1989)
American physiologist

Lewis's father had been born into slavery. After the Civil War, as a freed slave, he was educated at Berea College, Kentucky, and later became, along with his wife, a primary school teacher. It was necessary, however, for the younger Lewis to travel to Illinois to escape discrimination and to complete his education at the University of Illinois, Urbana, and at the University of Chicago. In 1915 he was awarded his PhD in physiology at Chicago. He went on to gain his MD from Rush Medical College, Chicago, in 1917. Soon after, Lewis was appointed as a pathologist at the University of Chicago, where he remained until 1943.

Lewis is best known for his *The Biology of the Negro* (1942), a pioneering and well-documented work in which Lewis sought to analyze the characteristics of a particular race in the same manner and as dispassionately as a biologist would examine the nature of any other life form.

In the 1960s a number of psychologists and sociologists had claimed to demonstrate that people of Afro-Caribbean descent tended to have a lower IQ than Asians and whites. The research was highly contentious and caused much debate about the nature of IQ tests and their possible social bias. In his work, Lewis insisted that, from a purely scientific viewpoint, there are no grounds for concluding that any race is biologically inferior to any other.

Lewis, Timothy Richard (1841–1889)
British physician

> Lewis was, like [Sir Patrick] Manson, a pioneer.
> —C. Dobell, *Parasitology* (1922)

After attending the local grammar school in Llanboidy, Wales, Lewis was apprenticed to the town pharmacist. He later moved to London where he attended classes at University College. He finally completed his medical education at the University of Aberdeen where he graduated in 1867. He entered the Army Medical School and was commissioned assistant surgeon in 1868. For some years Lewis served in India where he investigated cholera and described amoebae in the human intestine. On his return he was appointed to the staff of the Army Medical School at Netley where he served as professor of pathology from 1886 to 1889.

Lewis is mainly remembered for his observation in 1872 of nematode (roundworm) embryos in human blood. He named the organism *Filaria sanguinis hominis*, later renamed *Filaria bancrofti* for Joseph Bancroft, who discovered the adult form of the worm independently in 1876. In a later publication Lewis described the discovery that subsequent development of these embryos occurred in the mosquito. This work intrigued Patrick Manson and led him to the realization that diseases could be transmitted by insect vectors. Lewis was to make further discoveries of parasites in animal bodies, the most important of these being his detection in 1878 of trypanosomes in rats' blood, the first clear discovery of such organisms in mammals.

L'Hôpital, Marquis Guillaume François Antoine de (1661–1704) *French mathematician*

L'Hôpital (loh-pee-**tal**), a Parisian by birth, began his career as a cavalry officer. However, he was forced to resign because of his shortsightedness and devoted the rest of his life to mathematical study and research. To this end he invited the German mathematician Jean BERNOULLI to his chateau in 1691 to teach him the details of his newly worked-out differential calculus. Shortly afterward L'Hôpital published his *Analyse des infiniment petits pour l'intelligence des lignes courbes* (1696; Analysis of Infinitely Small Quantities for the Understanding of Curved Lines), the first calculus textbook ever to appear. L'Hôpital's basic assumption was that "...a quantity, which is increased or decreased only by an infinitely smaller quantity, may be considered as remaining the same." It was in this work that he first formulated the rule for finding the limiting value of a fraction with a numerator and denominator simultaneously tending to zero (0/0), since known to mathematicians as *L'Hôpital's rule*.

Bernoulli appeared none too pleased with L'Hôpital's book, considering it to be largely his own work, a belief supported by the discovery in 1921 of *Die Differentialrechnung* (Differential Calculus), a manuscript of Bernoulli, on which L'Hôpital had clearly based his own text.

Lhwyd, Edward (1660–1709) *British geologist and botanist*

> They who have no other aim than the search of Truth, are no ways concerned for the honour of their opinions: and for my part I have always been ... so much less an admirer of hypotheses, as I have been a lover of Natural History.
> —Letter to John Ray (*c.* 1695)

Lhwyd (**hloo**-wid) was born in Cardiganshire, west Wales, and educated at Oxford University. After he graduated (1682) he moved to the newly opened Ashmolean Museum in Oxford and became head keeper (1690–1709).

Lhwyd emphasized the importance of fieldwork in science, and traveled widely in his specific areas of study, the Celtic parts of Britain. His most important work was his eight-volume

Lithophylaci Britannici ichnographia (1699; A Plan of the British Fossil Collection), a catalog of the fossils of the Ashmolean. In an appendix to the work he tackled the problem – which was to worry geologists for over a century – of how to account for the presence of marine fossils deep within the Earth. His theory was that vapors from the sea could contain the seeds of various forms of marine life. These would be carried by the wind and eventually fall with the rain to percolate through cracks in the ground deep into the Earth's crust, where they would germinate. Lhwyd also played an important role in the revival of interest in the antiquities of Britain. His *Archaeologia Britannica* (1707; Archeology of Britain) contained some of the earliest descriptions of the prehistoric remains and languages of Celtic Britain.

Li, Choh Hao (1913–1987) *Chinese–American biochemist*

Li was born in Canton, China, and studied at the University of Nanjing. After his emigration to America in 1935, he attended the University of California at Berkeley, where in 1938 he obtained his PhD. He continued working at Berkeley and in 1950 was appointed professor of biochemistry and experimental endocrinology.

Li worked mainly on isolating and determining the structure of several important pituitary hormones. His first success was with adrenocorticotropic hormone (ACTH), which stimulates the adrenal gland and thus controls the level of such hormones as cortisone in the body. By 1956 he and his colleagues had determined the number (39) and sequence of the amino acids constituting ACTH. They were further able to show the general function performed by different portions of the amino-acid chain. The main biological activity was found to reside in the amino acids 1–24; 25–33 seemed to indicate the human origin of the hormone, for this stretch varies from species to species. The function of the remaining members (34–39) of the sequence was not clear. By 1966 Li had also succeeded in establishing the structure of growth hormone, or somatotropin. This consists of 256 amino acids and was first synthesized by Li in 1970.

Li, Shih-Chen (1518–1593) *Chinese pharmacist*

Born the son of a physician at the Imperial Medical Academy in Wa-hsiao-pa, China, Li turned to medicine after having failed his civil service entrance examinations. He went on to become, in the words of Joseph NEEDHAM, "probably the greatest naturalist in Chinese history." He worked, though only briefly, for the noble family Chhu, and at the Imperial Medical Academy. The bulk of Li's life was spent reading older texts and compiling his massive *Pen Tshao Kang Mu*, or *The Great Pharmacopoeia*, published posthumously in 1596.

His apparent starting point was the confusion found in earlier texts. In these, he noted, "gems, minerals, waters, and earths were all inextricably confused. Insects were not distinguished from fishes, nor fishes from shellfishes. Indeed, some insects were placed in the section on trees, and some trees in that of herbs." The work was divided into 52 chapters, with 1,895 entries of which 275 were mineral, 446 animal, and 1,094 plant. Li introduced 374 new substances, recorded 11,096 prescriptions, and provided 1,100 illustrations. He adopted the convention that the first recorded name given to plants or animals be adopted as standard, a convention still followed today. His classification system proceeded from the smaller to the larger and along a scale of nature that began with the elements (water, fire, earth, and metal) and proceeded through minerals, herbs, grains, vegetables, fruit, insects, and scaly creatures to birds, animals, and man. As a pharmacist Li was clearly a man of his time, relying on correspondences between the microcosm and macrocosm and the presence of yin or yang to determine a drug's use. Despite these limitations, Needham notes, Li was a careful and critical observer of nature. His most notable observation concerned the extraction of what were probably steroid hormones from large quantities (up to 150 gallons) of human urine. They were used to treat a variety of sexual conditions, including impotence, dysmenorrhea, and hypogonadism. Most later works of the following two centuries were inevitably based upon the *Pen Tshao*. Li's work was introduced into Europe by a copy sent by G. E. Rumpf to Holland in about 1691. Charles DARWIN was aware of the work, citing a "Chinese Encyclopaedia published in 1596" for details of seven breeds of fowls in his *Variation of Animals and Plants* (London, 1868).

Libavius, Andreas (c. 1540–1616) German chemist

> Andreas Libavius, physician of Halle in Saxony, most renowned teacher and most accurate and diligent researcher into natural phenomena, most zealous champion of true chemistry …
> —Joseph Duchesne, *On the Truth of Hermetic Medicine* (1605)

Libavius (li-**bah**-vee-uus) was born at Halle, Germany, the son of a weaver. He studied at the University of Jena where he became professor of history and poetry in 1586. He left after 1591 to become town physician at Rothenburg where he probably lectured on chemistry. In 1605 he moved to Coburg, establishing his own gymnasium there.

He wrote many works of which the most fa-

mous was *Alchymia* (1606; Alchemy). This has been described as the first modern chemistry textbook. Although very critical of PARACELSUS, he accepted most of his doctrines, such as the transmutation of base metals into gold. As a practical chemist, Libavius was successful in preparing for the first time tin(IV) chloride and ammonium sulfate and in analyzing many mineral waters. He also invented many new analytical methods.

Libby, Willard Frank (1908–1980) American chemist

Born in Grand Valley, Colorado, Libby was educated at the University of California at Berkeley where he obtained his PhD in 1933 and began teaching. In 1941 he moved to Columbia University to work on the development of the atom bomb. After the war he was appointed professor of chemistry at the Institute for Nuclear Studies at the University of Chicago before returning to the University of California (1959) as director of the Institute of Geophysics.

Libby was responsible for considerably improving dating techniques. In 1939 Serge Korff discovered the existence of the radioactive isotope carbon-14. This is different from the common stable isotope carbon-12 in that it contains an extra two neutrons in its nucleus. It is absorbed by all carbon users, such as animals and plants, during their lifetimes. It was established that the ratio of carbon-12 to carbon-14 in living organisms was constant and that on death the carbon-14 in the organism began to decay into nitrogen at a constant and measurable rate – carbon-14 has a half-life of 5,730 years. In 1947 Libby and his students at the University of Chicago's Institute for Nuclear Studies developed the radiocarbon dating technique using a highly sensitive geiger counter. He tested the process on objects of known age, such as timbers from Egyptian tombs. The test proved the technique to be reliable for the past 5,000 years and it was assumed from this to be accurate as far back as radiocarbon could be measured, about 50,000 years. A later improvement extended the range to about 70,000 years. The radiocarbon dating technique proved of immense value to the earth sciences, archeology, and anthropology, and for its development Libby was awarded the 1960 Nobel Prize for chemistry. His published works included *Radiocarbon Dating* (1952).

Lie, (Marius) Sophus (1842–1899) Norwegian mathematician

> Sophus Lie, great comparative anatomist of geometric theories.
> —C. J. Keyser, *Lectures on Science, Philosophy, and Art*

Born at Nordfjordeid in Norway, Lie (lee) was a friend of Felix KLEIN, whose ideas influenced him. Among Lie's most important work is his founding of the theory of continuous groups, which are now called *Lie groups* in his honor. Another contribution was his discovery of contact transformations. On both these subjects Lie wrote major treatises. In 1886 he became professor of mathematics at Leipzig and in 1898 he returned to Norway to take up a post that had been instituted for him at the University of Kristiania. By now, however, his health was poor and he died in Kristiania the following year. Lie also did notable work on differential geometry and on the study of differential equations.

Liebig, Justus von (1803–1873) *German chemist*

> God has ordered all his Creation by weight and measure.
> —Inscription over the door of the instruction laboratory at Giessen

Liebig (**lee**-bik), who was born in Darmstadt, Germany, was the son of a dealer in drugs, dyes, and associated chemicals. Aided by his father he developed an early interest in chemistry and was apprenticed to an apothecary. He studied chemistry at Bonn and Erlangen, after which the grand duke of his native Hesse was persuaded to finance Liebig to pursue his chemical studies overseas. Consequently he went to Paris, where he worked in the laboratory of Joseph GAY-LUSSAC. While there he came into contact with Alexander von HUMBOLDT, who exercised his patronage to have Liebig appointed to the chair of chemistry at Giessen in 1825, when he was still only 21. He remained there until 1852, when he moved to the University of Munich.

Liebig did much to establish chemistry as a discipline. At Giessen he set up one of the first laboratories for student instruction through which most of the great chemists of the 19th century passed. He also started the first scholarly chemical periodical. In 1832 he took over the *Annalen der Pharmacie* (Annals of Pharmacy) and renamed it in 1840 the *Annalen der Chemie* (Annals of Chemistry; the periodical still exists). He was constantly looking for ways to spread chemistry into new areas, to assert its dominance in previously autonomous disciplines. Thus in a series of works after 1840, when he moved from pure to applied organic chemistry, he tried to show that such studies as agriculture, physiology, and pathology were only intelligible when based on sound chemical principles. His *Chemistry in its Applications to Agriculture and Physiology* (1840) was one of the great books of the century. By 1848 it had gone through 17 editions and appeared in 8 languages. It was followed two years later by his *Organic Chemistry in its Application to Pathology and Physiology*. Liebig's first significant discovery was made with the aid of

Friedrich WÖHLER, his lifelong collaborator and friend. While working in the laboratory of Gay-Lussac, Liebig had prepared silver fulminate; Wöhler working in Sweden in the laboratory of BERZELIUS had prepared silver cyanate. To their surprise these two different chemicals appeared to have the same formula. They had unwittingly discovered what Berzelius was to call isomerism, that is, the condition in which two different chemical compounds have the same molecular formula. They decided to work together on the growing crisis in organic chemistry: how to deal with the sheer size and complexity of the molecules. (Molecules of inorganic compounds tend to be relatively small and straightforward and thus presented fewer problems.) Together they developed a method of analyzing the amounts of carbon and hydrogen present in organic compounds. Liebig and Wöhler came up with a theory of compound radicals. In 1832 they introduced the benzoyl radical, arguing for the existence of a family of chemicals all made from the same radical with the addition of one or more atoms to differentiate them. Thus to the benzoyl radical, $C_6H_5.CO$, can be added OH to make benzoic acid, H to make oil of bitter almonds (benzaldehyde), Cl for benzoyl chloride, Br for benzoyl bromide, and so on. Unfortunately it was difficult to find another radical as productive and convincing as benzoyl. However, this could not detract from the important fact that they had shown that organic compounds could be dealt with in a rational way. After organic chemistry Liebig's greatest work was carried out in agricultural science. His first achievement was in rejecting the current humus theory – the belief that plants absorb carbon from humus, the organic part of the soil, and turn it into the minerals they need. He demonstrated the falsity of this by showing that some crops left the soil richer in carbon than they found it, claiming that plants obtain carbon from the air. On burning plants he found various minerals present and argued that these must be obtained from the soil. He also thought that nitrogen was obtained from the ammonia in the soil, which ultimately derived from the rain. Thus plant growth could be stimulated with nitrates, manures, and minerals in which the soil was deficient. He experimented on a plot of land from 1845 until 1849 but had very disappointing results. Fearful of his additives being leached away, he was using a fertilizer too insoluble for the plants to absorb. Once this was corrected, he demonstrated the power of minerals and nitrates in increasing crop yield. During his visit to England he was shocked to observe the sewage of Britain being sent out to sea. He delivered a tremendous tirade against the British for their practice of importing bones from Europe instead of using their sewage as a fertilizer. In the field of biochemistry Liebig became involved in a famous dispute with Louis

PASTEUR. As a supporter of Berzelius, he claimed that all chemical changes were brought about by catalysts and that no organisms were involved. In 1869 he argued that there was nothing biological about fermentation. Pasteur, however, managed to demonstrate that vinegar produced by wine souring on contact with air resulted from the action of yeast. In chemical physiology Liebig showed that animal heat could be entirely accounted for by the oxidation of food. Although he misrepresented the role of protein he pioneered attempts to calculate the calorific values of different foods. Liebig was remarkable for the wide range of his work. There were greater chemists in the 19th century but none who worked with such authority over such an enormous field.

Lighthill, Sir Michael James (1924–1998) *British physicist*

Born in Paris, France, Lighthill was educated at Cambridge University, graduating in mathematics in 1943. During World War II he worked at the National Physical Laboratory. He joined the staff of Manchester University in 1946 and later served as professor of applied mathematics from 1950 until 1959 when he was appointed director of the Royal Aircraft Establishment, Farnborough. Lighthill returned to academic life in 1964, serving first as Royal Society Research Professor at Imperial College, London, and then as Lucasian Professor of Applied Mathematics at Cambridge (1969–79). He was provost of University College, London, from 1979 to 1989.

Lighthill was one of the leading aerodynamicists of the postwar years. He worked on the aerodynamics of high-speed aircraft and missiles. He also studied the theory of jet noise, work that much influenced the later design of silencers in jet exhausts. In the field of pure mathematics Lighthill also worked on the generalized theory of FOURIER analysis.

Lilienthal, Otto (1848–1896) *German aviation pioneer*

> To those who from a modest beginning, and with gradually increased extent and elevation of flight, have gained full control of the apparatus [glider], it is not in the least dangerous to cross deep and broad ravines.
> —Quoted in Donald Clark (editor), *Great Inventors and Discoveries* (1978)

Born at Anklam in Prussia, Lilienthal (**lee**-lee-en-tahl or **lil**-ee-en-thawl) trained as an engineer at the Trade School, Potsdam, and at the Berlin Trade Academy. He went on to manage a factory producing steam engines. His main interest from childhood onward, however, was in flight.

His approach to the subject had been stimulated by observing the flight of birds, which he saw as something to be copied in any future air-craft design. His observations and ideas were presented in his *Der Vogelflug als Grundlage der Fliegerkunst* (1889; Bird Flight as the Basis of Aviation). Lilienthal first experimented with flapping birdlike wings (an ornithopter). He soon, however, came to see that such motions were far too complex to produce mechanically. Consequently he turned to gliders, building his first model in 1889. It was a monoplane with rounded wings stiffened with radiating spars. Control was by adjusting the center of gravity by changing the position of the pilot. With his third model in 1891 he achieved a few small flights. Later models had larger wing areas and introduced such novelties as fins, tailplanes, edge flaps, and air brakes. With model number 6 he flew for 700 feet, but his aim to fly in a circle was never attained. Lilienthal also experimented with a model powered by a two-horsepower carbonic-acid engine. In his final flight in 1896 in Berlin the glider stalled and slide-slipped to the ground, breaking Lilienthal's spine. He died the following day. Despite the tragic outcome of his work, Lilienthal's designs and writings influenced other pioneers in Europe and America, including the American inventors Orville and Wilbur WRIGHT.

Lin, Chia-Chiao (1916–) *Chinese–American mathematician*

Born in Fujian, China, Lin graduated from the National Tsing Hua University in Taiwan. He then obtained his MA from the University of Toronto in 1941 and his PhD in 1944 from the California Institute of Technology, Pasadena. After teaching briefly at Brown University, Rhode Island (1945–47), he moved to the Massachusetts Institute of Technology. He was appointed professor of applied mathematics in 1953, becoming Institute Professor in 1966. He retired from MIT in 1987 but is still active as an emeritus institute professor.

Lin has worked on problems of hydrodynamics and turbulent flow in general. At a more particular level he has considered and is widely known for his account of how the spiral structure of spiral galaxies is sustained. Lin's account, known as the "density-wave theory," is based on work by Bertil LINDBLAD and was worked out in collaboration with Frank Shu in 1964. They propose that the spiral structure is a rotating density wave that sweeps through the galaxy. The spiral pattern is always there but the material in the pattern is continuously changing under the influence of gravity. It is further supposed that as the spiral wave moves through a region it compresses the gas there sufficiently to trigger the process of star formation along the lines of compression. Young stars should therefore be found in the arms of a spiral galaxy, as indeed they are. The model has received some additional observational support by predicting that spiral arms must trail and cannot lead as the galaxy rotates. Its universality has, however, been challenged by H. Gerola and P. Seiden who proposed in 1977 an alternative mechanism triggered by supernovae, without needing to assume the presence of an underlying density wave.

Linacre, Thomas (c. 1460–1524) *English physician and humanist*

> Owing to the studies of Thomas Linacre, Galen has begun to be so eloquent and informative [in Latin] that even in his own tongue he may seem to be less so.
> —Desiderius Erasmus, on Linacre's prime role in spreading the teachings of the Greek physician Galen

Linacre was born at Canterbury in southeast England and studied classics at Oxford University (1480–84). He then spent some time in Italy, obtaining an MD from the University of Padua in 1496. On his return to England he was appointed tutor to the heir to the throne, Prince Arthur, son of Henry VII. He later served as physician to Henry VIII and tutor to Mary Tudor (Mary I) for whom he wrote a Latin grammar.

Linacre produced new editions of many classical texts, his most important work being his publication of a number of Latin translations of various works of Galen between 1517 and 1524. His most lasting work in medicine was his creation of the College of Physicians in 1518 of which he was the founding president, an office he held until his death.

Lind, James (1716–1794) *British physician*

> The number of seamen in time of war who died by shipwreck, capture, famine, fire, or sword [was far less than those killed by] ship diseases and the usual maladies of intemperate climates.
> —*Two Papers on Fevers and Infections* (1763)

Born in Edinburgh, the Scottish capital, Lind entered medicine as an apprentice to a surgeon in his native city. In 1739 he joined the navy, serving as a surgeon in West African and Caribbean waters. Lind left the navy in 1748 and, after obtaining his MD from the University of Edinburgh, spent some years in private practice. He then took up an appointment as physician at the Haslar Naval Hospital, Portsmouth, a post he continued to occupy until his death.

In 1754 Lind published his classic work in the field of preventive medicine, *A Treatise on Scurvy*. Lind was personally familiar with the disease, noting that in 1746, on a ten-week voyage, he calculated some 80 out of 350 crew were afflicted. This debilitating disease, which claimed more lives in wartime than active combat, had had a profound impact on naval history and marine exploration when Lind began

his experiments in 1747. He started by noting the failure of all existing treatments. He then took 12 seamen sailing on HMS *Salisbury*, all suffering from scurvy to a comparable degree. All took the common sailor's diet of the day and, in addition, Lind divided them into six pairs who respectively received daily supplements of a quart of cider, 25 drops of vitriol, six spoonfuls of vinegar, a half pint of sea water, a concoction of various dried herbs and, for the final pair, two oranges and one lemon. The orange and lemon eaters were fit for duty after six days. Lind went on to show how the juice of the fruit could be extracted and stored for several years. It was, however, only after his death that the British Admiralty, in 1796, ordered the regular issue of such concentrated fruit juice. Lind also wrote one of the earliest works on tropical medicine, *Essay on Diseases Incidental to Europeans in Hot Climates* (1768).

Lindblad, Bertil (1895–1965) *Swedish astronomer*

Lindblad (**lind**-blahd) was born in Örebro, Sweden, and educated at the University of Uppsala where he obtained his PhD in 1920. After two years in America at the Lick and Mount Wilson observatories in California he returned to Uppsala. He was appointed in 1927 to the directorship of the Stockholm Observatory while serving at the same time as professor of astronomy at Stockholm University. He was followed by his son, Per Olof, who became director at the observatory in 1967.

It was Lindblad who in 1926 put forward the fundamental idea of the rotation of our Galaxy. This was based partly on the discovery by Jacobus KAPTEYN in 1904 of the two main streams of stars that appear to be moving in opposite directions and also on studies of the motions of stars with high radial velocity. Lindblad realized that these and other phenomena only made sense on the assumption of galactic rotation. Confirmation of Lindblad's conjecture was soon provided by Jan OORT. Lindblad also studied the structure of the Galaxy and suggested a mechanism, known as a density wave, whereby its spiral structure is sustained.

Linde, Karl von (1842–1934) *German engineer*

Linde (**lin**-de), who was born in Berndorf (now in Austria), studied engineering under Rudolf CLAUSIUS in Zurich. He taught in Munich at the Polytechnic as assistant professor of machine design from 1868 before moving into the refrigeration business in the 1870s.

His first breakthrough came in 1876 when he produced an ammonia refrigerator. This was a much more efficient cooler than the compression machine introduced by Jacob Perkins in 1834. By 1908 the Linde Company had sold

2,600 machines, of which just over half were bought by breweries. Linde also developed an equally successful domestic version. Linde was also the first industrialist to use the new developments in low-temperature physics. Oxygen had been first liquefied by Raoul PICTET and Louis CAILLETET in 1877. In 1895 Linde set up the first large-scale plant for the manufacture of liquid air using the Joule–Kelvin effect. Within six years he also developed a method for separating liquid oxygen from liquid air on a large scale. New industrial processes needed oxygen, and consequently Linde's process was rapidly taken up.

Lindemann, Carl Louis Ferdinand von
(1875–1939) *German mathematician*

Born at Hannover in Germany, Lindemann (**lin**-de-man) is principally known for solving one particular mathematical problem, namely the question of whether or not the number π is transcendental, i.e., whether or not it can be the root of a certain type of equation. In his paper *Über die Zahl π* (1882; On the Number π) he showed that π is a transcendental number. This was relevant to the ancient problem of squaring the circle, i.e., whether it is possible using only straight edge and compasses to construct a square equal in area to a given circle. Lindemann's result established its impossibility. Lindemann held posts as professor of mathematics at both the universities of Königsberg and Munich. His other mathematical work was chiefly in analysis and geometry.

Lindemann, Frederick, Viscount Cherwell
(1886–1957) *British physicist*

You know the definition of the perfectly designed machine ... one in which all its working parts wear out simultaneously. I am that machine.
—Remark made shortly before his death.
Quoted by Lord Birkenhead in *The Prof in Two Worlds* (1961)

Lindemann's father, a naturalized Briton, was a wealthy Alsatian by birth; his mother was American. Born at Baden-Baden, he attended schools in Scotland and Germany and studied under Walther NERNST in Berlin where, in 1910, he obtained his doctorate. During World War I he worked at the Royal Aircraft Establishment, Farnborough, after which he was appointed in 1919 to the directorship of the Clarendon Laboratory and the chair of experimental philosophy at Oxford. He held these posts until 1956 apart from a period during World War II when he served as scientific adviser to Churchill, and 1951–53 when he was paymaster-general in Churchill's last administration.

Although Lindemann's early scientific reputation was very high, his contributions to science were relatively modest. Apart from some early work with Nernst on specific heats

(1910–12) and the derivation of a formula relating the melting point of crystals to the amplitude of their atomic vibration, he tended to treat science as a hobby. His major contribution to science was the transformation of the Clarendon Laboratory in Oxford from the moribund institution it was in 1919 to one of the world's leading centers of physical research. When he took over it lacked electricity, had an annual budget of £2,000, and had an academic staff of only two, with one technician. Lindemann raised substantial funds, increased the academic staff to 20 by 1939, and attracted several brilliant Jewish physicists fleeing from Hitler's Germany. After 1933 Oxford University became the world's leading institution for research into low-temperature physics. His role in public affairs has been the cause of much controversy since C. P. Snow first publicized his work in 1960 in his lecture *Science and Government*. Snow argued that Lindemann opposed the development of radar and supported an unreasonable emphasis on the mass bombing of civilian populations. Such a view was not shared by Lindemann's wartime colleagues, R. Harrod, R. V. Jones, and Churchill himself, who have written at length praising his war work. His main achievement in government after 1945 was the creation of the UK Atomic Energy Authority in 1954. In 1956 he was made a viscount, taking the title Lord Cherwell.

Linnaeus, Carolus (1707–1778) *Swedish botanist*

Minerals grow; plants grow and live; animals grow and live and feel.
—*Systema Naturae* (1735; The System of Nature)

Nature does not make progress by leaps and bounds.
—*Philosophia Botanica* (1751; Botanical Philosophy)

Linnaeus (li-**nee**-us), born Carl Linné (lin-**ay**), a pastor's son in Råshult, Sweden, began studying medicine at the University of Lund in 1727, transferring to Uppsala University the following year. While at college he investigated the newly proposed theory that plants exhibit sexuality and, by 1730, had begun formulating a taxonomic system based on stamens and pistils. He extended his knowledge of plants on travels through Lapland in 1732, where he discovered a hundred new species, and around Europe from 1733 to 1735.

In 1735 he settled in Holland and published his first major work, *Systema Naturae* (The System of Nature), in which he systematically arranged the animal, plant, and mineral kingdoms. In it, he classified whales and similar creatures as mammals and recognized man's affinity to the apes to the extent of naming the orang-utan *Homo troglodytes*. The flowering plants were divided into classes, depending on the number and arrangement of their stamens, and subdivived into orders, according to the number of their pistils. This system, because it was based simply on sexual characters, only partly showed the natural relationships between plants. It was undoubtedly useful in its time, however, for ordering the many new species that were arriving in Europe from all over the world. Linnaeus's lasting contribution to taxonomy was his introduction, in 1749, of binomial nomenclature, which he applied in *Species Plantarum* (1753; Species of Plants) by giving each plant a generic and a specific name. For example, applying the Linnean system, the Texas bluebonnet is named *Lupinus subcarnosus*, where *Lupinus* is the generic name and *subcarnosus* the specific name. Until then scientific plant names were polynomial – a short Latin description of the distinguishing features. This combination of name and description was unsatisfactory, being too long for the name and too brief for the description. Linnaeus's innovation, separating the two functions, is the basis of modern nomenclature. Linnaeus had returned to Sweden in 1738 and practiced there as a physician until he was appointed professor at Uppsala University in 1741. His botanical teaching stimulated many pupils, such as Daniel Solander, Carl Per THUNBERG, and Anders Dahl, to travel widely collecting specimens. On Linnaeus's death his collection was bought by the English naturalist Sir James Smith. The London-based Linnean Society, founded by Smith in 1788, purchased the books and herbarium specimens from Smith's widow in 1828.

Linstead, Sir Reginald Patrick (1902–1966) *British chemist*

Linstead was born in London, the son of a chemist with the pharmaceutical firm Burroughs Wellcome. Having studied chemistry at Imperial College, London, where he obtained his PhD in 1926, he worked briefly with the Anglo-Persian Oil Company before returning as a lecturer to Imperial College in 1929. In 1938 he became Firth Professor of Chemistry at the University of Sheffield, but the following year moved to Harvard to take the chair of organic chemistry. Linstead returned to England in 1942 to work at the Ministry of Supply. After the war he worked at the National Physical Laboratory before returning once more to Imperial College in 1949, where he served as professor of organic chemistry until 1955, when he became rector of the college, a post he retained until his death.

The first chromophores emerged from Linstead's basic work on the structure of the phthalocyanines. These are nitrogen derivatives of phthalic acid, which yield brilliant blues and greens resistant to chemical attack and to light.

Liouville, Joseph (1809–1882) *French mathematician*

Born in Saint-Omer, France, Liouville (lyooveel) was a professor at the Ecole Polytechnique in Paris and edited an influential mathematical journal, the *Journal de Mathématiques Pures et Appliquées* (Journal of Pure and Applied Mathematics; often known as *Liouville's Journal*). The achievement for which he is best known is his work on transcendental numbers. Not only did he show that such numbers exist but he also devised methods of exhibiting infinitely many of them. Apart from this work on numbers, his other areas included differential equaions and boundary-value problems. His work on boundary-value problems became part of what is now known as *Stürm–Liouville theory* and came to be of considerable importance in physics. Liouville also made contributions to differential geometry, in particular to the theory of conformal transformations, and to complex analysis (named for him), which is of fundamental importance in statistical mechanics and measure theory.

Lipmann, Fritz Albert (1899–1986) *German–American biochemist*

After attending the university in his native city of Königsberg (now Kaliningrad in Russia), Lipmann studied in Berlin, where he obtained his MD in 1922 and his doctorate in 1927. He then worked with Otto MEYERHOF in Heidelberg and taught at the Kaiser Wilhelm Institute in Berlin (1927–31), but with the rise of the Nazis decided to abandon Germany and consequently accepted a position with the Carlsberg Foundation in Copenhagen. In 1939 he moved to America, where he worked at the Cornell Medical School (1939–41), Harvard (1941–49), and the Massachusetts General Hospital in Boston (1949–57), before becoming professor of biochemistry at the Rockefeller Institute for Medical Research in New York, a post he occupied until his retirement in 1970.

It was widely known that the breakdown of such carbohydrates as glucose provides energy for the body's cells, but just how the cell obtains the energy released was a mystery. Lipmann's work, recently described by a historian of molecular biology as "the most magnificent achievement of late-classical biochemistry," began in 1937, when he was working on the breakdown of glucose by a particular bacterium. Quite fortuitously he found that a certain oxidation would not proceed without the addition of some phosphate. This was all he needed to see that the real purpose of metabolism was to deliver energy into the cell. Lipmann sought for the phosphate that delivered the energy and found a molecule, adenosine triphosphate (ATP), which had been identified as the probable source of muscular energy by K. Lohmann in 1929. The molecule consisted of adenosine monophosphate (a nucleotide of the nucleic acid RNA), with the addition of two energy-rich phosphate bonds. When ATP is hydrolyzed to adenosine diphosphate (ADP), some of this energy is released ready for use in the cell. It was not for this work, however, that Lipmann shared the 1953 Nobel Prize for physiology or medicine with Hans KREBS but for his discovery of coenzyme A and its importance for intermediary metabolism in 1947. While working on the role of phosphate in cell metabolism, Lipmann discovered that a heat-stable factor was acting as a carrier of acetyl (CH_3CO-) groups. It could not be replaced by any other known cofactor. Lipmann eventually isolated and identified what he termed "cofactor A," or CoA (the A stands for acetylation), showing it to contain pantothenic acid (vitamin B_2). He also realized that the two-carbon compound in the Krebs cycle that joined with oxaloacetic acid to form citric acid was in fact acetyl CoA. The coenzyme was soon shown to have wider application than the Krebs cycle, when in 1950 Feodor LYNEN found that it played a key role in the metabolism of fats.

Lippershey, Hans (c. 1570–c. 1619) *Dutch spectacle maker*

Lippershey (**lip**-ers-hI) was a maker of eyeglasses in Wesel, Germany. According to tradition, an apprentice playing with a couple of lenses suddenly found that a distant weathercock looked much bigger and nearer. Lippershey realized the significance of this and made the first telescope in 1608. He offered his invention to the Estates of Holland for use in warfare and initially an attempt was made by the Estates to keep the invention secret, but it was much too easy to reconstruct. All GALILEO needed was a report of the "Dutch invention" to allow him to make his own. There are many other claimants to Lippershey's invention, including his compatriot Zacharias JANSSEN.

Lippmann, Gabriel (1845–1921) *French physicist*

> For his method, based on the interference phenomenon, for reproducing color photographically.
> —Citation accompanying the award of the Nobel Prize for physics, 1908

Born at Hollerich in Luxembourg, Lippmann (leep-**man**) was educated at the Ecole Normale in Paris. After conducting research in Germany he became professor of probability and mathematical physics at the Sorbonne in 1883. In 1886 he became director of the laboratories for physical research and professor of physics at the Sorbonne.

In 1873 he invented the *Lippmann capillary electrometer*, an instrument for measuring extremely small voltages. Lippmann is, however,

better known for producing the first color photographic plate (1891). His color-photography process involved placing a coat of mercury behind the emulsion on the photographic plate. It is the only direct method of color photography but requires a long exposure time. For this work Lippmann was awarded the Nobel Prize for physics in 1908. Lippmann's other inventions included a galvanometer, a seismograph, and a coelostat.

Lipscomb, William Nunn (1919–) *American inorganic chemist*

Born in Cleveland, Ohio, Lipscomb was educated at the University of Kentucky (graduating in 1942) and the California Institute of Technology where he obtained his PhD in 1946. He worked at the University of Minnesota from 1946 to 1959, being appointed professor of chemistry in 1954. In 1959 Lipscomb moved to the chair of chemistry at Harvard, where he remained until his retirement in 1990. He is currently emeritus professor of chemistry and chemical biology.

Lipscomb is best known for his work on boranes – hydrides of boron first investigated by Alfred STOCK in the early part of the century. Boranes have such typical formulae as B_2H_6, B_4H_{10}, $B_{10}H_{14}$, and $B_{18}H_{22}$, which immediately appear to the chemist as analogous to the comparable hydrocarbon series, CH_4, C_2H_6, C_4H_{10}, etc. However, as boron has only three electrons in its outer shell, it was difficult to see how the covalent electron-pair bonds could work with boron hydrides. Using low-temperature x-ray diffraction analysis, Lipscomb tackled the problem of investigating the notoriously unstable boranes, producing evidence of some remarkable structures, totally original and completely unsuspected by earlier chemists. The basic concept of a three-center bond was derived from a structure for diborane proposed by H. C. LONGUET-HIGGINS. This differs from the normal covalent bond found in hydrocarbons where adjacent carbon and hydrogen atoms share two electrons. In a three-center bond, a pair of electrons is shared equally by three atoms. Lipscomb's work on boron hydrides involved new techniques that proved to have a wider application in chemistry and produced results that led to the formulation of more general theories. In particular, Lipscomb produced a theory of chemical effects in nuclear magnetic resonance studies of complex molecules. He also worked on the quantum mechanics of large complex molecules. His group has also applied low-temperature x-ray diffraction techniques to other substances, including single crystals of such gases as oxygen and nitrogen, other inorganic compounds, and naturally occurring organic compounds. More recently he has turned to determinations of the structures of proteins, enzymes, and other substances of biochemical interest. Lipscomb received the Nobel Prize for chemistry in 1976.

Lissajous, Jules Antoine (1822–1880)
French physicist

Born at Versailles in France, Lissajous (lee-sa-**zhoo**) graduated from the Ecole Normale Supérieure in 1847 and then taught physics at a school in Paris.

He was interested in finding a way of making sound vibrations visible. CHLADNI had already produced his sand-pattern method but this only determined the nodal lines, where there was no vibration. Lissajous's method was to reflect light off mirrors attached to two tuning forks set at right angles. The superposition of the vibrations formed dynamic patterns on a screen, which are now called *Lissajous figures*. From the form of these curves he could calculate the relative frequencies of the forks and thus provided a precise way of measuring pitch. He also invented the vibrating microscope, which produced Lissajous figures when vibrating objects, such as violin strings, were viewed through it.

Lister, Joseph, Baron (1827–1912)
British physician

> Since the antiseptic treatment has been brought into full operation, and wounds and abscesses no longer poison the atmosphere with putrid exhalations, my wards … have completely changed their character; so that during the last nine months not a single instance of pyaemia, hospital gangrene, or erysipelas has occurred in them.
> —*British Medical Journal* (1867)

Lister, the son of Joseph Jackson Lister, was born at Upton in England and educated at Quaker schools before entering University College, London, in 1843. University College was, at the time, the only English university open to religious dissenters. After graduating in arts, Lister studied medicine, obtaining his MB in 1852. He then served as assistant to the leading Scottish surgeon James Syme at the Royal Infirmary, Edinburgh, from 1854 until 1860 when he was appointed professor of surgery at Glasgow University. Lister returned to Edinburgh in 1869 as professor of clinical surgery but in 1877 became professor of surgery at King's College, London, serving there until his retirement in 1892.

In 1867 Lister published two short but revolutionary papers, which introduced the principles of antiseptic surgery into medicine. In 1846 he had been present when Robert Liston had first successfully used ether as an anesthetic in England. Yet this, apart from making surgery tolerable for both patient and surgeon, had not greatly advanced the profession. The full potential of anesthesia did not develop because of the high mortality produced by the infection that inevitably followed major surgery. The in-

evitable consequence was that a surprisingly small amount of surgery was actually attempted, even in the major centers with ready access to anesthetics. Lister acknowledged the twin sources of his innovations in his 1867 papers. The first and most important were the writings of Louis PASTEUR. These revealed the cause of the widespread surgical sepsis to be the germs present in the air. To control them Lister reported that he had been impressed by an account of the effects produced by carbolic acid on sewage in Carlisle. Carbolic acid (phenol, C_6H_5OH) is a weak acid derived from benzene. Although Lister's first attempt to use it as an antiseptic in March 1865 ended in failure he persisted and in August dressed a compound fracture of the leg, that is, one in which the skin has been broken, with a piece of lint dipped in liquid carbolic acid. The wound healed well. This encouraged Lister to introduce the carbolic acid dressings into his regular surgical procedure. By 1870 he claimed that mortality for amputations had dropped from over 40% to 15%. He later analyzed his figures for his Edinburgh period, reporting that from 1871 to 1877 he performed 725 major operations with a mortality of only 5.1%. Another of Lister's major innovations was his introduction, in 1869, of cat-gut ligatures to replace the traditional silk thread, which was a major source of infection. Lister's experiments showed that cat-gut ligatures were absorbed by the body and if soaked in carbolic acid could be made sterile. He also attempted to maintain an antiseptic atmosphere in the operating theater by introducing a carbolic spray. This, however, made working conditions very unpleasant and the procedure was abandoned. The Listerian system appears to have been accepted with little dissent and remarkable speed for by 1880 it had become the standard mode of surgical procedure virtually everywhere. Lister, an intensely shy and reserved man, achieved considerable fame and received many honors. In 1897 he became the first physician to be made a peer and sit in the House of Lords.

Lister, Joseph Jackson (1786–1869)
British physicist

> The pillar and source of all the microscopy of the age.
> —Obituary notice of Lister, *Monthly Microscopical Journal* (1870)

Lister, a Londoner, was the son of a Quaker wine merchant and left school at the age of 14 to join his father in the wine trade. As a boy he was interested in optics and owned a telescope, but it was not until about 1824 that he turned his attention to improving the microscope. He made public his work of the next six years in a paper *On the Improvement of Compound Microscopes* read before the Royal Society in 1830. In this he announced his discovery of the existence of two aplanatic (free from spherical aberration) foci in a double achromatic object glass, and described how, by using two sets of achromatic lens combinations, distortions could be reduced. This design allowed for dramatically clearer images and removed the trial-and-error procedure hitherto necessary in the making of microscope objective lenses. It was to help turn the microscope into a serious scientific instrument. As a result of this paper Lister was elected a fellow of the Royal Society in 1832.

Lister's other contributions to science included studies on mammalian blood. His second son, Joseph, became famous as the founder of antiseptic surgery.

Littlewood, John Edensor (1885–1977)
British mathematician

> Millions of words must have been written on [what God was doing before the Creation]; but He was doing Pure Mathematics and thought it would be a pleasant change to do some Applied.
> —*Littlewood's Miscellany*

Born at Rochester in southeast England, Littlewood studied at Cambridge University and in 1907 obtained a lectureship at Manchester. By 1910 he had returned to Cambridge where in 1928 he became Rouse Ball Professor of Mathematics. He retired in 1950 but continued active mathematical research until shortly before his death.

Littlewood is primarily known for his work in analysis, but he made contributions to many other fields, including mathematical astronomy, physics, differential equations, and probability theory. One of the most notable features of his career was his 35-year collaboration with G. H. HARDY. Among their most important joint work was the systematic investigation of problems of the convergence and summability of FOURIER series. In 1914 Littlewood proved a famous theorem about the error term in the prime number theorem. In collaboration with R. E. A. C. Paley, Littlewood created and developed a new and specific link between trigonometric series and analytical functions. His work with Mary Cartwright on nonlinear differential equations was also of importance.

Lobachevsky, Nikolai Ivanovich (1793–1856) *Russian mathematician*

> [Non-Euclidean geometry] might find application in the intimate sphere of molecular attraction.
> —*Prediction made in 1826*

Lobachevsky (loh-ba-**chef**-skee) was born at Nizhny Novgorod in Russia. Throughout his life he was associated with the University of Kazan; he was a student there and held various posts, including the chair in mathematics and finally the rectorship. Later he was deprived of his position for political reasons.

Lobachevsky's fame is due to his epoch-making discovery, announced in 1826 and published in 1829, that there could be consistent systems of geometry based on other postulates than those of EUCLID. In particular Lobachevsky constructed and studied a type of geometry in which Euclid's parallel postulate is false (the postulate states that through a point not on a certain line only one line can be drawn not meeting the first line). János BOLYAI had, at the same time though quite independently, come to a similar result and the same discovery had in fact been made decades earlier by Karl Friedrich GAUSS, but he never published his work. For centuries the status of Euclid's geometry and in particular of his parallel postulate had been a matter of controversy. Attempts had been made to show that it followed from the other axioms and it was widely held that Euclid's geometry described the necessary structure of space. By revealing the coherence of a non-Euclidean geometry Lobachevsky showed that Euclidean geometry has no such privileged position, helped mathematicians to break free from undue reliance on intuition, and paved the way for the systematic study of different kinds of non-Euclidean geometry in the work of Bernhard RIEMANN and Felix KLEIN. Although it was not well received at first the value of Lobachevsky's work was fully appreciated once Riemann began his investigations into the fundamental concepts of geometry. Perhaps his fullest vindication came with the advent of EINSTEIN's theory of relativity when it was demonstrated experimentally that the geometry of space is not described by Euclid's geometry. Apart from geometry, Lobachevsky also did important work in the theory of infinite series, algebraic equations, integral calculus, and probability.

Locke, John (1632–1704) *English philosopher*

> Did we know the mechanical affections of the particles of rhubarb, hemlock, opium, and a man, as a watchmaker does those of a watch … we should know why rhubarb will purge, hemlock kill, and opium make a man sleep.
> —*Essay Concerning Human Understanding* (1690)

> It is one thing to show a man that he is in error, and another to put him in possession of truth.
> —As above

Locke, the son of a lawyer, was born at Wrington in the west of England and educated at Oxford University. After qualifying as a physician and teaching Greek and moral philosophy at Oxford for some years, in 1667 Locke entered the service of Lord Ashley (later the 1st earl of Shaftesbury), for whom he was physician, adviser, and tutor for his son. He spent the period

1675–80 in France, where he met most of the leading continental scientists and thinkers.

After returning to England Locke published his *Two Treatises on Civil Government* (1690) and the four letters *Concerning Toleration* (1689–93), works that were to have an enormous impact on the political and constitutional life of Britain as they were later to have on the American and French revolutions. His influence on science was exerted through his *Essay Concerning Human Understanding* (1690) in which he attempted to replace the rationalist foundations of knowledge proposed by René DESCARTES with alternative empirical ones – i.e., those based on experience and experiment. Locke himself was much under the influence, and eager to promote the ideas, of scientists such as Robert BOYLE. The *Essay* was influential in spreading the new corpuscular philosophy, which held that the macroscopic physical properties of matter are explained by the arrangement (and rearrangement) of basic microscopic particles.

Lockyer, Sir Joseph Norman (1836–1920) *British astronomer*

Lockyer, born the son of a surgeon-apothecary at Rugby in the English Midlands, started his career as a civil servant. He turned to astronomy and taught at the Royal College of Science, becoming director of the solar physics observatory and professor of astronomical physics from 1890 to 1901. He was one of the founders and the first editor of the British periodical *Nature*. He made many eclipse trips and played a leading role in attempts to reorganize the structure of British science. He wrote numerous books on popular science and virtually created the new discipline of astroarcheology. Lockyer was knighted in 1897.

The spectroscopic work of Robert BUNSEN and Gustav KIRCHHOFF so stimulated Lockyer that he moved from traditional astronomy to spectral studies. He worked mainly on the Sun, publishing *The Chemistry of the Sun* in 1887. He investigated sunspots and solar prominences discovering, with Pierre JANSSEN in 1868, that they could be observed spectroscopically in daylight without an eclipse. He also successfully identified the spectral line observed by Janssen in the 1868 eclipse as being an unknown element (found, he thought, only in the Sun), which he proposed to name helium (from the Greek for Sun: *helios*). His supposition about the existence of the element was confirmed in 1895 when William RAMSAY isolated it from gases in the atmosphere. In 1873 Lockyer published his theory of dissociation to explain the appearance of further unfamiliar spectral lines. William HUGGINS had found a new bright line in the spectra of nebulae and thought it could be a new substance that he proposed to call "nebulium." Lockyer argued instead that it could be an earthly element that had "dissociated" into simpler substances under conditions of great heat and temperature, producing unrecognizable spectral lines. It was, however, difficult to make much sense of this view until the discovery of the electron some 20 years later and the correct explanation for the new spectral lines was not to be provided until the next century (by Ira Bowen). In 1894 Lockyer published *The Dawn of Astronomy*, the first classic of what has since been called astroarchaeology, and in 1906 he produced *Stonehenge and Other British Monuments Astronomically Considered*. His aim in these works was to establish (without, of course, the benefit of computer and TV camera) that many ancient buildings were astronomically aligned. He did a good deal of field work, paying regular trips to Egypt and Greece as well as to the stones of Britain. Not the least of his achievements was the creation of a new type of scientific periodical with *Nature* in 1869. It was by no means obvious that *Nature* would survive and it owes much to Lockyer's half century of editorship. The virtues of *Nature* were in fact the virtues of Lockyer himself – it relished controversy, was tolerant of a wide range of scientific views, and was quick to publish scientific results.

Lodge, Sir Oliver Joseph (1851–1940) *British physicist*

Born at Penkhull in England, Lodge entered his father's business in 1865. However, at the age of 22 he resumed a formal education, studying at the Royal College of Science (now part of Imperial College) and at University College, both in London; he was awarded a DSc in 1877. After several teaching posts he was appointed the first professor of physics at University College, Liverpool, in 1881. In 1900 he became the first principal of the new Birmingham University, remaining there until his retirement in 1919. He was knighted in 1902.

Lodge's principal scientific contributions were concerned with the transmission of electromagnetic waves, which led to developments in radio broadcasting. His experiments in the field of electricity started in the late 1870s. In 1887–88 he discovered that electromagnetic waves could be produced by electrical means and transmitted along conducting wires. These results were somewhat overshadowed by the work of Heinrich HERTZ who in 1888 succeeded in producing electromagnetic waves, transmitted them through air, and demonstrated their similarities with light waves. In 1894 Lodge made his mark, however, by greatly improving the means of detecting these "Hertzian" waves (now known as radio waves) by developing the coherer. This was an electrical device whose function was based on a discovery made in 1890 by E. Branley: that electrical discharges in certain metallic powders, caused by radio waves, resulted in a drop in electrical resistance. Lodge is also remembered for his work on the ether, which had been postulated as the wave-bearing medium filling all space. In 1893 he devised an experiment that helped to discredit the theory. Other scientific work included investigations on lightning, the source of the electromotive force in the voltaic cell, electrolysis, and the application of electricity to the dispersal of fog and smoke. He played a part in establishing the National Physical Laboratory. From 1900 Lodge increasingly devoted himself to administrative work. He was also interested in the history of science and wrote several scientific memoirs. In his writings he made attempts to reconcile what seemed to him the divergence between science and religion. After 1910 he became deeply involved in psychical research. He believed in the possibility of communicating with the dead, a belief sustained by the hope of somehow communicating with his youngest son Raymond, who was killed in World War I.

Loeb, Jacques (1859–1924) *German–American physiologist*

Loeb (lohb) was born at Mayen in Germany. After studying medicine at Strasbourg, he settled in America (1891) where he held professorships at Bryn Mawr College, Pennsylvania, and the universities of Chicago and California, before becoming a member of the Rockefeller Institute for Medical Research (1910).

Much of Loeb's major research was concerned with plant and animal tropisms (involuntary movements in response to stimuli such as light, water, and gravity); he postulated that these occur not only in primitive animals but also in higher animals, including man (*Forced Movements, Tropisms, and Animal Conduct*, 1918). He also carried out important work on artificial parthenogenesis, showing that unfertilized frogs' eggs could be induced to divide by altering their environment. Another discovery was that sea-urchin eggs are hatched by the osmotic pressure exerted by various substances dissolved in water.

Loewi, Otto (1873–1961) *German–American physiologist*

> The night before Easter Sunday [1921] I awoke, turned on the light, and jotted down a few notes on a tiny slip of thin paper. Then I fell asleep again. It occurred to me at 6 o'clock in the morning that during the night I had written down something most important ... It was the design of an experiment to determine whether or not the hypothesis of chemical transmission that I had uttered seventeen years ago was correct.
> —*An Autobiographical Sketch* (1960)

Loewi (loh-ee) was born at Frankfurt am Main in Germany and qualified in medicine at the University of Strasbourg before taking up pro-

fessorships in physiology and pharmacology at Vienna and Graz Universities. For a time he worked under Ernest STARLING in London, and in 1940 emigrated to America where he became research professor at the New York University College of Medicine.

Loewi's most important work was concerned with nerve action in vertebrate animals, demonstrating, for example, that chemical reactions are involved in nerve impulses. In 1921 he discovered that certain chemical substances are released when the nerves of a frog's heart are electrically stimulated. Loewi's "vagus material" (thus named because it was obtained by stimulation of the vagus nerve) was subsequently shown by Henry DALE to be acetylcholine. It can be used to stimulate the activity of another heart without the need for nervous activity. Loewi and Dale shared the Nobel Prize for physiology or medicine in 1936 for their work in this field.

Löffler, Friedrich August Johannes (1852–1915) *German bacteriologist*

Löffler (luf-ler) was born at Frankfurt in Germany, the son of an army surgeon. He was educated at the University of Würzburg and the Berlin Institute of Military Medicine, where – after serving in the Franco-Prussian War – he obtained his MD in 1874. After various official positions Löffler worked with Georg GAFFKY as an assistant to Robert KOCH from 1884 to 1888. He later served as professor of hygiene at the University of Griefswald (1888–1913), after which he succeeded Gaffky as director of the Koch Institute in Berlin, where he remained until his death.

Löffler's major contribution to the new field of bacteriology was the isolation and cultivation in 1884 of the bacillus responsible for diphtheria, which had first been observed by the German physiologist Theodor Klebs in the throats of diphtheria patients (the organism became known as the *Klebs–Löffler bacillus*). The isolation and cultivation of pure cultures involved a number of formidable technical problems; Löffler found it necessary to develop a new medium, thickened serum, as the conventional gelatin used by Koch required temperatures far too low for the diphtheria pathogen. Earlier (in 1882) Löffler had discovered the organism responsible for glanders (a contagious disease, especially of horses) and in 1898, in collaboration with the pathologist Paul Frosch, he succeeded in demonstrating for the first time that viruses could cause diseases in animals. This was achieved by passing foot-and-mouth disease from one cow to another by inoculation with cell-free filtrates taken from lesions. Löffler's work, together with Dmitri IVANOVSKY's demonstration in 1892 that plants were susceptible to viral infections, constituted the start of modern virology.

Lomonosov, Mikhail Vasilievich (1711–1765) *Russian scientist and scholar*

> I will not attack for their errors people who have served the republic of science; rather, I will try to use their good thoughts for useful work.
> —Research note (1743)

> ...the Russian land can give birth to its own Platos and quick-witted Newtons.
> —*The Delight of Earthly Kings and Kingdoms* (1747)

Lomonosov (lom-o-**naw**-sof) was the son of a fisherman from Deniskova, now Lomonosov, in Russia. He left for Moscow in 1730 to obtain an education and studied science there until 1735. In 1736 he attended the St. Petersburg Academy of Sciences before traveling to Marburg where he studied under Christian Wolff. Following his return to St. Petersburg in 1741 he was put under arrest (1743) and in prison began work on his *276 Notes on Corpuscular Philosophy and Physics*, in which he outlined his scientific ideas. He became professor of chemistry at St. Petersburg in 1745.

Lomonosov campaigned successfully for a laboratory for teaching and research at St. Petersburg and this was opened in 1749. On the basis of the results of experiments conducted in the laboratory he set up a glass factory to produce, in particular, colored-glass mosaics. As an administrator he helped to found Moscow University (1755) with Leonhard EULER. As a chemist Lomonosov was opposed to the phlogiston theory and is reported to have anticipated Antoine LAVOISIER on the conservation of mass, Benjamin RUMFORD on the kinetic theory of heat, and Thomas YOUNG on the wave theory of light. Lomonosov also made equally important contributions to Russian literature. He wrote the grammar that systematized the Russian literary language and was himself a poet. His work, *Ancient History of Russia*, published posthumously in 1766, was the first work on the history of Russia.

London, Fritz (1900–1954) *German–American physicist*

London (**lun**-don) was born in Breslau, now Wrocław in Poland, the son of a mathematics professor at Bonn and the elder brother of the well-known physicist, Heinz London. He originally received a classical education at the Universities of Frankfurt, Munich, and Bonn, where in 1921 he was awarded his doctorate for a philosophical thesis. London taught in secondary schools for some years and then began work as a physicist in 1925 at Munich. After appointments in Stuttgart, Zurich, and Berlin he left Germany in 1933 with the rise of Hitler. He worked first in the Clarendon Laboratory in Oxford, England, and from 1935 at the Institut Poincaré, Paris, before moving to America to become professor of theoretical chemistry at Duke University, North Carolina – a post he retained until his death in 1954.

In 1927 London, in collaboration with Walter Heitler, succeeded in providing an account of the covalent bond in the hydrogen molecule using wave mechanics. In the 1930s he collaborated with his brother in gaining a major insight into the nature of superconductivity. This was followed by his own researches, beginning in 1938, into the nature of superfluidity, a phenomenon first described by Pyotr KAPITZA.

London, Heinz (1907–1970) *German–British physicist*

> Perhaps he might be described as a cross between a theoretical physicist and an inventor.
> —D. Shoenberg, obituary notice of London in *Biographical Memoirs of Fellows of the Royal Society* (1971)

> For the second law of thermodynamics, I would die at the stake.
> —Heinz London, quoted in his obituary notice (as above)

Born at Bonn in Germany, London was the son of a mathematics professor and the younger brother of the distinguished physicist, Fritz London, with whom he collaborated on some of his early work. He was educated at the universities of Bonn, Berlin, Munich, and Breslau, where he obtained his PhD under Francis SIMON in 1933. Abandoning Nazi Germany immediately afterwards, London first joined his brother at the Clarendon Laboratory in Oxford but moved to Bristol in 1936, remaining there until his brief internment as an enemy alien in 1940. On his release he worked on the separation of uranium isotopes for the development of the atomic bomb. With the coming of peace, in 1946 London joined the staff of the Atomic Energy Research Establishment at Harwell where he remained until his death.

In Oxford London continued the work of his thesis in his collaboration with his brother on a number of pressing problems in superconductivity. In particular they explained the discoveries of W. MEISSNER that at the moment a metal becomes superconductive it expels the magnetic field produced by an electric current; if, however, a strong external magnetic field is applied normal resistivity will return. At Harwell after the war London worked until the 1950s on isotope separation. His attention was also drawn to the problem of superfluidity in liquid helium, which led him to develop his "dilution refrigerator." This consisted in mixing the two isotopes of helium, helium-3 and helium-4, at temperatures below 1 K in a dilution of 1:1,000. Some of the helium-3 would pass to and fro across the boundary between the isotopes separated by their different densities and reduce the temperature by a small amount with each passage. The machine was first described in 1951 and a working apparatus was built in 1963. The device has reached temperatures as low as 0.005 K.

Long, Crawford Williamson (1815–1878)
American physician

Long, who was born in Danielsville, Georgia, received his MD from the University of Pennsylvania in 1839. He then practiced in the small Georgian village of Jefferson where he became probably the first physician to perform surgery using ether as an anesthetic. (There is one earlier record of the administration of ether, for a tooth extraction: in January 1842, William Clark gave ether to a patient whose tooth was then removed by Elijah Pope.)

The idea of using ether came to Long after he had engaged in "ether frolics" – wild parties at which ether was inhaled for exhilarative effect. Long noticed that he developed many bruises during such parties but had no recollection of sustaining any injuries. This suggested to him the possibility of using it more constructively to provide surgical anesthesia. Consequently, on 30 March 1842, Long removed a small tumor from the neck of an etherized patient who assured him, when he regained consciousness, that he had not experienced any pain. Long followed this up in July by painlessly amputating the toe of a young etherized boy. However, Long had little chance to use his dramatic discovery in major operations and did not publish details until 1849. By this time William MORTON had already (1846) given a public demonstration of the use of ether as an anesthetic and Long thus received little credit for his discovery.

Longuet-Higgins, Hugh Christopher (1923–2004) *British theoretical chemist*

Longuet-Higgins was born at Lenham in England and educated at Oxford University. After holding junior appointments in America at the University of Chicago and in England at the University of Manchester he was appointed in 1952 to the chair of theoretical physics at King's College, London. He moved to Cambridge University in 1954, becoming professor of chemistry there until 1967, when he was appointed Royal Society Professor, first at Edinburgh and, from 1974 to 1989, at the Centre for Research in Perception and Cognition at Sussex University.

Longuet-Higgins was one of the leading figures in postwar theoretical chemistry. He was the first to suggest, in 1943, that borane (B_2H_6) had a structure containing hydrogen bridges – an idea that he extended to volatile metal borohydrides ($M(BH_4)_n$), to certain alkyl compounds, and to beryllium hydride. Later he was able to explain the hydrogen bridges by molecular orbitals, using the idea of three-center bonds. The concept of multicenter bonds was

used by William LIPSCOMB in his work on boron hydrides. From 1945 to 1947, Longuet-Higgins worked with Charles COULSON on conjugated organic molecules (molecules having alternate double and single carbon–carbon bonds). They developed a general orbital theory of such molecules, including the effects of substituting hydrogen atoms by other groups. Longuet-Higgins also made other contributions to theoretical chemistry, including work on molecular spectra, the statistical mechanics of solutions, polymers, and stereospecific reactions. In more recent years, Longuet-Higgins turned his attention from theoretical chemistry, first to information processing in biological systems, and later to problems of the mind. In this latter area he worked on theories of language acquisition, music perception, and the analysis of speech.

Lonsdale, Dame Kathleen (1903–1971)
British crystallographer

> Any scientist who has ever been in love knows that he may understand everything about sex hormones but the actual experience is something quite different.
> —*Universities Quarterly* No. 17 (1963)

The daughter of a postman, Lonsdale (née Yardley) was born at Newbridge in Ireland and moved to England with her family in 1908. She studied physics at Bedford College, London, graduating in 1922, and spent most of the following 20 years based at the Royal Institution in the research team of William Henry BRAGG. In 1946 she moved to University College, London, where she served as professor of chemistry and head of the department of crystallography from 1948 until her retirement in 1968.

Lonsdale was one of the early pioneers of x-ray crystallography, centered on the Royal Institution and the team headed by Bragg and including such scholars as William ASTBURY, John BERNAL, Dorothy HODGKIN, and John ROBERTSON. It was from this group that most of the concepts and techniques of the new discipline emerged in the 1920s and 1930s. Lonsdale herself was responsible for one of the first demonstrations of the power of the new techniques when, in 1929, she published details of the structure of benzene. Working with a large crystal of hexamethylbenzene she established the hexagonal nature of the ring, that it was planar to within 0.1 angstrom ($1 \text{ Å} = 10^{-10}$ m), and that the carbon–carbon bonds were 1.42 Å. This was followed in 1931 by the equally significant structure of the more difficult hexachlorobenzene, the first investigation of an organic compound in which FOURIER analysis was used. Other crystallographic subjects researched by Lonsdale included the magnetic susceptibility of crystals and the structure of synthetic diamonds and, in the 1960s, that of bladder stones. She edited the first three volumes of the *International Tables for X-ray Crystallography* (1952, 1959, 1962) and also produced a survey of the subject in her *Crystals and X-rays* (1948). As a Quaker and a convinced pacifist Lonsdale refused to register in 1939 for government service or civil defense despite the fact that as a mother of three young children she would have been exempted from any such service. Fined £2 in 1943, she refused to pay and served a month in Holloway prison instead. When, 285 years after its foundation, the Royal Society finally decided to admit women to its fellowship Lonsdale was the first to be elected (1945) and she became the society's vice-president in 1960. She was appointed a Dame of the British Empire in 1956 and also became, in 1968, the first woman to serve as president of the British Association for the Advancement of Science.

Loomis, Elias (1811–1889) *American meteorologist*

Loomis, the son of a parson from Wilmington, Connecticut, graduated from Yale in 1830. After a period of study in Paris he was made professor of natural philosophy at the City University of New York (1844–60) and at Yale (1860–89).

Loomis made a detailed study of the great storm of 1836 in which he showed that the physical processes involved were much more complicated than had been supposed. At each meteorological station the pressure would fall to be followed by a sudden rise, the temperature would rise followed by a sudden fall, and the wind would shift to the north. In order to present this complex data in a comprehensible form he designed and published the first synoptic weather map in 1846.

Lorentz, Hendrik Antoon (1853–1928) *Dutch theoretical physicist*

> The greatest and noblest man of our times.
> —Albert Einstein, oration at Lorentz's funeral (1928)

Lorentz (**lor**-ents), who was born at Arnhem in the Netherlands, studied at the University of Leiden and received his doctorate in 1875. In 1877, aged only 24, he became professor of theoretical physics at Leiden. This was the Netherlands' first chair in the newly independent field of theoretical physics, and one of the first in Europe, and Lorentz did a great deal in shaping and developing the field. On his retirement in 1912 he was appointed director of the Teyler Laboratory in Haarlem, a museum of science and art with a laboratory where he could continue his research. He still retained contact with the world of advanced physics, giving every week at Leiden his famous "Monday morning lectures" on current scientific problems.

Lorentz had wide-ranging interests in physics and mathematics, his linguistic abilities allowing him to follow the scientific trends in Europe. His major work, however, was spent in the development of the electromagnetic theory of James Clerk MAXWELL. He brought this to a point where a need for a radical change in the foundations of physics became noticeable and thus provided the inspiration for Einstein's theory of relativity. Lorentz's early work on this highly complex and confused subject followed from the writings of Hermann von HELMHOLTZ and began in his doctoral thesis. Lorentz refined Maxwell's theory so that for the first time various effects including the reflection and refraction (bending) of light could be fully explained. In a series of articles published between 1892 and 1904 Lorentz put forward his "electron theory": he proposed that the atoms and molecules of matter contained small rigid bodies that carried either a positive or negative charge. By 1899 he was referring to these charged particles as "electrons." It was through the effects of these electrons that many phenomena in science were explained. Lorentz believed that matter and the wave-bearing medium known as the "ether" were distinct entities and that the interaction between them was mediated by electrons. He saw that the interaction of light waves and matter resulted from the presence of electrons in matter and that if set into vibration these charged particles would produce light waves, as predicted by Maxwell's equations. In 1895 he described the force, now known as the *Lorentz force*, on charged particles of matter in an electromagnetic field. In 1900 he identified the negatively charged particles that had been found to constitute cathode rays as the negative electrons of his theory. He also used the theory to explain the effect discovered by Pieter ZEEMAN in 1896 whereby the spectral lines of sodium atoms were split by the action of a magnetic field. Lorentz and Zeeman shared the 1902 Nobel Prize for physics for their investigations of the influence of magnetic fields on radiation. However, other phenomena, such as the photoelectric effect, could not be explained and, in fact, were inconsistent with Lorentz's theory; it was these anomalies that inspired the development of the quantum theory. The other work for which Lorentz is famous is his suggested method of resolving the problems raised by the experiments in the 1880s of Albert MICHELSON and Edward MORLEY on the motion of the Earth through the ether. Lorentz showed that if it were assumed that moving bodies contracted in the direction of motion, then the observed effects would follow. This solution was derived independently by George FITZGERALD and came to be known as the *Lorentz–Fitzgerald contraction*. Lorentz extended his idea, putting it on a firmer mathematical footing, and in 1904 published in final form what became known as the *Lorentz transformations*. These transformations of the space and time coordinates of an event in one frame of reference to those in another frame again figured largely in Einstein's theory of special relativity (1905), in which Einstein could be said to have reinterpreted Lorentz's ideas. Lorentz devoted much time to education and the teaching of science and medicine. In later life he was very active in international science conferences, acting as president of the first Solvay Congress for physics in Brussels and continuing as president until his death. He also played a major role in restoring international scientific relations after World War I.

Lorenz, Edward Norton (1917–   ) *American meteorologist*

> The average person, seeing that we can predict tides pretty well a few months ahead, would say, "Why can't we do the same thing with the atmosphere? It's just a different fluid system, the laws are about as complicated." But I realized that *any* physical system that behaved nonperiodically would be unpredictable.
> —On the genesis of his so-called "chaos theory"

Born in West Hartford, Connecticut, Lorenz was educated at Dartmouth College, New Hampshire, at Harvard, and at the Massachusetts Institute of Technology. He joined the MIT faculty in 1946 and served as professor of meteorology from 1962 until 1987.

Lorenz had trained initially as a mathematician, but after serving in the U.S. Army Air Corps as a weather forecaster, he decided to continue as a meteorologist and work on the more theoretical aspects of the subject. Could forecasting, he asked, be significantly improved? If astronomers could predict the return of a comet decades ahead, it should surely be possible, given enough computing power, to forecast the state of the atmosphere for more than a few hours in advance. The pioneer of this method was the mathematician von NEUMANN, who planned a program to predict, and even control, the weather. One day in 1961 Lorenz discovered that there were more than a few simple, practical obstacles to overcome. In order to follow variations in weather conditions Lorenz set up a system in which he fed a set of initial conditions into the computer and allowed it to run on showing graphically the values taken by a single variable, such as temperature, over a long period of time. On one occasion he wished to examine part of one run in greater detail and fed in the initial conditions taken from an earlier run. To his surprise the computer produced a markedly different sequence from the original printout. Eventually Lorenz traced the source of the discrepancy. The initial conditions of the program, stored in the computer memory, had used the number 0.506127, correct to six decimal places; the printout, however, gave just three decimal places, 0.506.

Lorenz, like everyone else, had assumed that so small a difference could have no significant effect. In fact, a small difference can, over a long period of time, build up to produce a large effect. Moreover, the way the difference affects the outcome is very sensitive to small changes. Technically, this is termed "sensitive dependence on initial conditions." More graphically, it is called the "butterfly effect," from the idea that a single butterfly flapping its wings in China might, weeks later, "cause" a hurricane in New York. The butterfly effect occurs because the weather depends on a number of factors – temperature, humidity, air flow, etc. – and these are to a certain extent interdependent. Thus the way the temperature changes depends on the humidity, but this depends on temperature. Consequently, equations relating these factors are nonlinear – a variable is a function of itself. And it is this nonlinearity that causes the sensitive dependence on initial conditions. The weather is a system that repeats itself, giving periods of dry weather and periods of wet, but it repeats itself in an unpredictable way. There are a number of similar nonlinear systems – for example, population cycles or economic cycles – that depend on nonlinear equations. The study of such systems has come to be known as "chaos theory." Lorenz went on to study behavior of this kind in a more rigorous and abstract manner. He considered some simple nonlinear equations describing fluid flow in a system with three degrees of freedom. He appeared to discover in the process a new kind of "attractor." In broad terms an attractor is what the behavior of a system settles down to, or is attracted to. The simplest kind, a fixed-point attractor, is represented by a pendulum subject to friction. The pendulum, no matter how it starts swinging, will always come to rest in the same position. Its final state is predictable. The *Lorenz attractor*, however, proved to be both chaotic and unpredictable. It was the first example of a "strange attractor," a term introduced by David RUELLE in 1971. Lorenz discovered the attractor by examining the changing relationships between the three variables described by the equations. At any moment in time the vari-

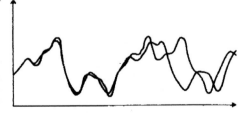

LORENZ CURVES   Two curves of a type produced by Lorenz, showing how small differences in initial conditions can build up to widely different later states.

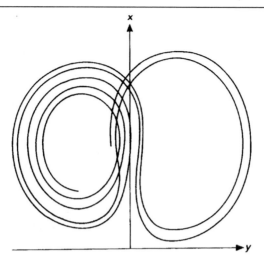

LORENZ ATTRACTOR   *The strange attractor discovered by Lorenz. Note that this is a projection of the curve in one plane. The actual curve in three dimensions does not intersect itself.*

ables will be defined by a point in three-dimensional space. The point can be plotted. Lorenz found that the point never followed the same path, nor did the paths it followed ever intersect; instead it displayed a system of complex loops, something in fact like the wings of a butterfly. Lorenz first published his work in 1963 in a paper entitled *Deterministic Nonperiodic Flow*. Although it received little early attention, it became one of the most cited papers of the 1980s.

Lorenz, Konrad Zacharias (1903–1989)
Austrian ethologist

> It is a good morning exercise for a research scientist to discard a pet hypothesis every day before breakfast. It keeps him young.
> —*On Aggression* (1966)

Alfred Lorenz (**lor**-ents or **loh**-rents), father of Konrad, was a very wealthy Viennese orthopedic surgeon who had developed a new operation for a congenital dislocation of the femur, a common complaint of the period. He was eager for his son to follow him into medicine and consequently, though reluctantly, Lorenz studied medicine at Columbia in New York and at the University of Vienna, where he gained his MD in 1929. He remained at Vienna to complete his PhD in 1933 on the comparative anatomy and evolution of avian wings. Lorenz was appointed lecturer in animal psychology in 1937.

Alfred Lorenz had bought a sizeable estate at Altenberg, a site about twenty miles outside Vienna. It was on this family estate that Lorenz first began his researches on animal behavior. Here he studied a jackdaw colony that had settled on the roof of his father's house. He also began to rear goslings; wild geese had proved to be too difficult to study profitably. Among other

tasks Lorenz systematically classified the signals and behavior patterns of his goslings. Before long he had constructed a "glossary" of their various calls and behavior patterns. They presented a number of problems. What did they mean? How did they originate in the individual? And how could such behavior patterns evolve? In his bird studies Lorenz made good use of the phenomenon of imprinting, first described by Heinroth in 1911. Goslings, as they hatch, tend to take the first object they encounter to be their mother. Lorenz would allow goslings and other birds to imprint themselves upon him and thereby gain easy access to them without actually taming them. It became a common sight at Altenberg to see Lorenz followed by a line of goslings who, if threatened, would scurry to him in alarm. He noted some of the properties of the process and defined it as "a developmental process by which behavior becomes attached to a particular object." No reinforcement is required; mere passive exposure will suffice. It is also irreversible, and is clearly innate. In 1937 Lorenz began to offer an explanatory system – a new theory of instinct – to account for many aspects of animal behavior. Much complex behavior, he noted, came perfectly formed and required no initial learning period. Nor did it necessarily arise from external stimuli. For example, Lorenz noted starlings in midwinter hunting nonexistent flies, presumably responding to some internal drive – a form of behavior he described as "vacuum activities." He went on to characterize instincts in terms of four properties: they were clearly innate; they were species-specific; they involved stereotyped behavior; and instincts also involved what Lorenz termed "action specific energies," which were discharged by the presence of innate releasing mechanisms, also known as "releasers." Thus the sight of a male stickleback's red belly (releaser) in the breeding season induces a stereotyped aggressive response in a rival male. The response is species and action specific. Lorenz likened the process to liquid in a reservoir. Just as water is released by opening a valve, the instinctive behavior innate in the system is discharged when presented with the appropriate releaser. Later ethologists have objected that Lorenz's model underestimates environmental influences. Lorenz's work at Altenberg was interrupted by the onset of World War II. He served as a physician in the German army and was taken prisoner by the Russians in 1944. He was released in 1948 and on his return to Austria he was invited by the Max Planck Institute to establish a Department of Comparative Ethology at Buldern, Westphalia. The department moved in 1961 to the Institute for Behavioral Physiology, Seewiesen, Bavaria. On his retirement in 1973 Lorenz returned, along with his geese, to Grunau in Austria where he established his own research institute with funds provided by the Austrian Academy. By this time Lorenz had become world famous. Two books published in the 1950s, *King Solomon's Ring* (1952) and *Man Meets Dog* (1954), were immensely popular and have remained in print. He assumed a more controversial role in 1966 with his *On Aggression*, in which he argued that aggression was not necessarily an evil as it also served a number of evolutionary purposes. Man, he claimed, actually suffered from "an insufficient discharge of his aggressive drive." Equally controversial was *Man's Eight Deadly Sins* (1974), in which he warned against the genetic deterioration of the human race. For his work on ethology Lorenz shared the 1973 Nobel Prize for physiology or medicine with Nicolaas TINBERGEN and Karl von FRISCH.

Loschmidt, Johann Josef (1821–1895) *Austrian chemist*

Born into a peasant family in Carlsbad, Bohemia, Loschmidt (**loh**-shmit) was educated by the local clergy, who encouraged him to pursue his studies at the Prague Gymnasium and at the German University in Prague. After graduating in 1843 he struggled with various businesses before ending his commercial career in 1854 as a bankrupt. He turned to academic life and in 1856 was appointed to the Vienna Realschule where he remained until 1868 when he moved to the University of Vienna as professor of physical chemistry.

In 1861 Loschmidt published a brief work, *Chemische Studien* (Chemical Studies), in which he listed 368 formulas. Like most chemists of the time, Loschmidt was seeking for ways to express chemical structure and composition accurately and graphically. In his sys-

tem, atoms were represented by circles, with a large circle for carbon and a smaller one for hydrogen. Thus four years before KEKULÉ announced his own results, Loschmidt represented the benzene molecule by a single large ring (the carbon) with six smaller circles (hydrogen) around the rim. Little attention seems to have been paid to his work at the time. He is far better known for a paper published in 1865 entitled *Zur Grosse der Loftmolecule* (On the Magnitude of Air Molecules) in which he made the first accurate estimate of the size of molecules. Using the kinetic theory of gases, Loschmidt derived the equation $s = 8el$, where s is the molecular diameter, l is the mean free path (the distance a molecule moves between successive collisions), and e is the condensation coefficient. This latter factor was derived from changes in volume due to evaporation and condensation. Using published data for e and l, he obtained the value 10^{-7} centimeter. In his honor, the number of particles per unit volume of an ideal gas at standard temperature and pressure is known as *Loschmidt's constant* (or *Loschmidt's number*). It has the value 2.686763×10^{25} m^{-3}.

Louis, Pierre Charles Alexandre (1787–1872) *French physician*

Louis (**loo**-ee or lwee) was born at Champagne in France and educated at Rheims and Paris where he obtained his MD in 1813. He first practiced in Odessa, Russia, but returned to France in 1820. He held appointments at the Charité and the Hôtel-Dieu, large Paris hospitals in which he taught and pursued his researches.

Those researches were of a quite unusual nature for 19th-century medicine in that they involved a considerable use of statistics, which Louis termed his "numerical method." An early example of this is his *Recherches sur la saignée* (1828; Researches on Health) in which he evaluated the effects of blood-letting on various diseases. He collected data on 78 cases, noting how frequently the patients were bled, when in the progress of the disease the bleeding was begun, and the outcome and duration of the disease. In his analysis of the data he concluded that blood-letting had very little influence on the progress of pneumonia, erysipelas of the face, or angina tonsilaris. He suggested that the misguided belief in the efficacy of blood-letting probably arose because it was practiced at an advanced stage of the disease when it had nearly run its course. Despite such careful work only a few physicians, such as Marshall HALL, followed his advice and the leech remained in use for many years afterward. Louis also published substantial works on tuberculosis (1825) and typhoid (1829), both of which made considerable use of his numerical method.

Love, Augustus Edward Hough (1863–1940) *British mathematician and geophysicist*

> My eyes were first opened by Professor Love, who taught me for a few terms and gave me my first serious conception of analysis.
> —Godfrey Harold Hardy, *A Mathematician's Apology* (1940)

Born at Weston-super-Mare in the west of England, Love studied at Cambridge University and was a fellow there from 1886 until 1889. In 1889 he moved to Oxford to take up the Sedleian Chair in Mathematics and remained there for the rest of his career.

Love's most important research was in the theory of deformable media and in theoretical geophysics. He also did important work in the theory of waves and in ballistics. He wrote a major two-volume treatise on elasticity – *A Treatise on the Mathematical Theory of Elasticity* (1892–93) – that went through several subsequent editions. It soon established itself as a classic and is still a standard work of reference. Love's work on geophysics led to a considerable number of practical discoveries. Among his original concepts now much used in geophysics are the so-called *Love numbers* and *Love waves*. Love numbers play a key role in tidal theory. In the formal theory of Love waves, which are surface seismic waves, he was building on and much improving work begun by Siméon-Denis POISSON, Sir George STOKES, and Lord RAYLEIGH. This was perhaps Love's single most significant piece of mathematical work.

Lovell, Sir (Alfred Charles) Bernard (1913–) *British radio astronomer*

> A study of history shows that civilizations that abandon the quest for knowledge are doomed to disintegration.
> —Quoted in *The Observer*, 14 May 1972

Lovell was born at Oldland Common, England, and received his PhD in 1936 from the University of Bristol; in the same year he was appointed as a lecturer in physics at the University of Manchester. In 1945, after war service on the development of radar, he returned to Manchester. He was elected in 1951 to the chair of radio astronomy and the directorship of Jodrell Bank (now the Nuffield Radio Astronomy Laboratories), a post he held until his retirement in 1981. He was knighted in 1961.

Lovell's first research was done in the field of cosmic rays. In the course of his war work he realized that radio waves were a possible tool with which to pursue his studies. Thus in 1945 two trailers of radar equipment that had been used in wartime defense work were parked in a field at Jodrell Bank in Cheshire to begin radio investigation of cosmic rays, meteors, and comets. Lovell soon produced worthwhile results on meteor velocities and other topics and

began to feel that a more permanent and ambitious telescope should be built. Thus Lovell began a heroic ten-year struggle to finance a 250-foot (76-m) steerable radio telescope with a parabolic dish that would be able to receive radio waves as short as 30 centimeters. The main problem was to find sufficient funds to meet the rising costs of the project at times of government cuts. Thus in 1955 the project found itself £250,000 in debt. The Department of Scientific and Industrial Research agreed to find half if Lovell could raise the rest. A public appeal failed to raise more than £65,000 and it required a strong public press campaign to move the Treasury to meet the outstanding costs in 1960, three years after the telescope was first used. The Jodrell Bank telescope came to public notice when it was used to track the first Sputnik in 1957. It was not just an adjunct to the space program, however, but a major tool for astronomical research of which Lovell has given a full account in his *Out of the Zenith* (1973). He there showed the power of the giant telescope to supplement and advance the discoveries of others. Thus it was the Cambridge radio astronomers under Antony HEWISH who discovered pulsars, but they were limited to observing them only for the few minutes each day that the pulsars were on the Cambridge meridian. The steerable Jodrell Bank telescope could observe objects for as long as they were above the horizon and it was no accident that of the 50 pulsars discovered in the northern hemisphere before 1972, 27 were detected at Jodrell Bank. So too with those other mysterious phenomena of the 1960s, quasars. Once more the initial discovery was made elsewhere but Jodrell Bank possessed the instruments to show that some quasars had angular diameters of one second of arc or less, which was surprisingly small for such prodigious sources of energy. Lovell wrote a number of important books recounting the story – political, financial, and scientific – of Jodrell Bank. They include *The Story of Jodrell Bank* (1968), *The Jodrell Bank Telescopes* (1985), and his autobiography *Astronomer by Chance* (1990).

Lovelock, James Ephraim (1919–) *British scientist*

A Londoner, Lovelock was educated at the universities of London and Manchester during the early years of World War II. After graduating in 1941 he joined the staff of the National Institute for Medical Research (NIMR) in London, where he worked on a variety of technical wartime problems including the measurement of blood pressure under water, the freezing of viable cells, and the design of an acoustic anemometer.

After twenty years with the NIMR Lovelock began to feel that his creativity was being stifled by the security of his position as a scientific civil servant. Consequently he resigned and took up a short-term appointment with NASA. He was assigned to work on the first lunar Surveyor mission at the Jet Propulsion Laboratory, California. Lovelock left America in 1964 determined to set up as an "independent" scientist. He claimed that he did not wish to be just one more consultant serving the needs, whatever they might be, of multinational companies – he wished to work at science without constraints, in the manner of a novelist or painter. In this manner Lovelock was able to make a number of important observations. While at NIMR he had developed a sensitive electron-capture detector. In the summer of 1966 he used it to monitor the supposedly clean Atlantic air blowing onto the west coast of Ireland. He detected chlorofluorocarbons (CFCs). Although unable to pursue the matter further through lack of funding, Lovelock managed to travel to the Antarctic in 1971 where, again, he found atmospheric CFCs. It was partly as a result of this work that Sherwood ROWLAND began to ponder their role in the atmosphere. It is, however, as the author of the Gaia hypothesis, first presented in his *Gaia* (London, 1979), and developed further in several sequels, that Lovelock is best known. Gaia has been widely accepted by environmentalists, conservationists, and New-Age thinkers. Lovelock argued in 1979 that the Earth, including its rocks, oceans, and atmosphere, as well as its flora and fauna, was a living organism "maintained and regulated by life on the surface." He referred to it as "Gaia" after the Greek Earth goddess – a suggestion made by the novelist William Golding. The hypothesis has been dismissed by scientists as "crudely anthropomorphic" and "pseudoscientific idiocy." However, some see Gaia as a working hypothesis that can be tested and evaluated in the normal manner.

Lowell, Percival (1855–1916) *American astronomer*

> The broad physical conditions of the planet [Mars] are not antagonistic to some form of life … there turns out to be a network of markings covering the disk precisely counterparting what a system of irrigation would look like.
> —*Mars* (1895)

Born in Boston, Massachusetts, Lowell graduated from Harvard in 1876. His first interest was oriental studies but Giovanni SCHIAPARELLI's report in 1877 of the "canali" (mistranslated as "canals") of Mars had interested him and he finally decided to devote the rest of his life to astronomy. As he was a member of a famous, aristocratic, and wealthy Boston family, he had no difficulty in financing his own observatory. He built it at a height of 7,200 feet (2,200 m) in the clear skies of Arizona, giving him good observing conditions, and began his studies of the planets in 1894. He was ap-

pointed professor of astronomy at the Massachusetts Institute of Technology in 1902.

Lowell spent 15 years observing Mars with an excellent 24-inch (61-cm) refractor built by Alvan CLARK. He had no difficulty in seeing the "canals" of Schiaparelli, also claiming to see oases and clear signs of vegetation. He soon concluded that Mars was inhabited and wrote a series of books on this topic: *Mars* (1895), *Mars and its Canals* (1906), and *Mars as the Abode of Life* (1908). He was by no means alone among professional astronomers in his, what now appears to be extravagant, claim. It should be realized that large telescopes do little to improve the visual appearance of the planets because of the constantly shifting terrestrial atmosphere. It was not until the Martian surface was mapped by the Mariner and Viking spacecraft in the 1960s and 1970s that the idea of artificial canals could be definitely dispelled. Lowell was more successful in his work on a trans-Neptunian planet. Even making full allowances for the disturbing effects of Neptune, the orbit of Uranus still was not free from anomalies. Lowell thought that this could be due to an unknown Planet X still further out in space. He calculated its orbit and position, beginning his search in 1905. He published his negative results in 1914. Fourteen years after his death Clyde TOMBAUGH, observing at the Lowell Observatory in 1930 and using a blink comparator, discovered the new planet. It was named Pluto since like the god it ruled as prince of outer darkness but also because its first two letters stood appropriately for Percival Lowell.

Lower, Richard (1631–1691) *English physician*

> If the blood moves through its own power, why does the Heart need to be so fibrous and so well supplied with Nerves?
> —*Tractatus de corde* (1669; Treatise on the Heart)

Born near Bodmin in Cornwall, Lower was educated at Westminster School and Oxford University. While in Oxford he served as assistant to Thomas WILLIS and was a member of the "Invisible College," an informal network of individuals whose unofficial meetings prepared the way for the formation of the Royal Society in London in 1660. Lower moved to London in 1666 to set up in private practice.

In 1669 he published his *Tractatus de corde* (Treatise on the Heart) in which he made the first step in attempting to explain the reason for respiration. He began with the clear difference in color between the dark venous blood and the bright red arterial blood, which he attributed to the penetration of air into the lungs. He neatly demonstrated that such a "penetration" took place in the lungs by tying the trachea of a dog and noting that the arterial blood remained as dark as the venous. This, Lower noted, could be reversed by blowing into the animal's lungs with a bellows, which resulted in the blood entering the aorta returning to its normal lighter color. Lower thus saw for the first time that the function of respiration was to add something to the blood, which he referred to variously as "nitrous spirits" or "nitrous food-stuffs." More than a century later, in 1775, Joseph PRIESTLEY was able to show that this rather vague "nitrous spirit" was his newly discovered gas, "dephlogisticated air" or oxygen. Lower was also involved in the first attempts at blood transfusion, although the first experiment using a human subject was carried out by the French physician Jean Denys. This practice, however, led to a number of fatalities and it was not until the discovery of blood groups by Karl LANDSTEINER in 1902 that transfusions became widely used in medicine.

Lowry, Thomas Martin (1874–1936)
British chemist

Lowry was the son of a Wesleyan army chaplain from Bradford in England. He was educated at the Central Technical College (later part of Imperial College), London, where from 1896 to 1913 he served as assistant to Henry ARMSTRONG. From 1904 he was also head of chemistry at Westminster Training College until he moved to Guy's Hospital, London, in 1913 to become head of the chemistry department. In 1920 he was appointed as the first professor of physical chemistry at Cambridge University.

As a physical chemist Lowry was largely concerned with the optical activity of certain compounds. In 1898 he first described the phenomenon of mutarotation. He found that the optically active compound nitro-camphor revealed an alteration of rotatory power over time. Later, in the 1920s, he confirmed experimentally that there is a relationship between the optical rotatory power of compounds and the wavelength of light passing through them. He published an account of this aspect of his work in *Optical Rotatory Power* (1935). He is also remembered for his theory of acids and bases which he formulated in 1923 simultaneously with but independently of Johannes Bronsted. They simply defined an acid as any ion or molecule able to produce a proton, while a base is any ion or molecule able to take up a proton.

Lubbock, John, 1st Baron Avebury
(1834–1913) *British biologist, politician, and banker*

Although Lubbock was born in London, his father's estate was situated close to Charles DARWIN's home in Kent and Lubbock benefited greatly from his contact with Darwin from an early age. He left school at 14 to enter his fa-

ther's banking business and his knowledge of natural history was almost entirely self-taught. His first published papers were on specimens collected by Darwin during his voyage on the *Beagle* and he also provided illustrations for Darwin's work.

A convinced evolutionist, Lubbock became interested in primitive man and the origin of civilization. He introduced the terms "Paleolithic" and "Neolithic" for the Old and New Stone Ages respectively, and also found the first fossil remains of the musk ox in Britain, providing evidence for the existence of an ice age. However, Lubbock's best-remembered biological researches are those on insect behavior. He set up an artificial ants' nest bounded by two panes of glass, which enabled him to study the ants without undue disturbance. In this way he was the first to witness the "farming" of aphids by ants. He also studied insect vision and the detection of color by bees, and he tested intelligence by setting up obstacle courses and mazes. His observations are collected in *Ants, Bees and Wasps* (1882). Lubbock also had a very successful political career, sponsoring over 20 bills through the British parliament, including the Wild Birds Protection Act (1880) and the Open Spaces Act (1880).

Lubbock, John William (1803–1865)
British mathematical physicist

Lubbock was the son of a London banker who also had scientific interests. He was educated at Eton and Cambridge University, England, where he graduated in 1825 and then became a partner in his father's bank.

During the period 1831–36 Lubbock published a number of tidal studies. By working on the tidal records of the London docks from 1795 he was able to work out the establishment of port, the time high water falls behind the Moon at a particular place on the Earth. He was much influenced by Pierre Simon de LAPLACE whose innovations in probability theory he used in his theoretical studies of annuities and whose work on the stability of the solar system he advanced. Lubbock was the first vice-chancellor of the University of London (1837–42).

Lucas, Keith (1879–1916) *British neurophysiologist*

> The primary problem of comparative physiology...[is] the question to what extent, and along what lines, the functional capabilities of animal cells have been changed in the course of evolution.
> —*The Evolution of the Animal Kingdom* (1909)

Lucas, the son of an engineer, was born at Greenwich in England and graduated from Cambridge University, where he was elected a fellow in 1904. In 1914 he joined the research department of the Royal Aircraft Establishment, Farnborough, and while carrying out his duties died in a midair collision.

In 1905 and 1909 Lucas published two papers that clearly stated the all-or-none law of nervous stimulation: a stimulus can evoke a maximum possible response or nothing. If stimuli of varying strength appear to elicit responses of increasing contraction this is because, Lucas demonstrated, more nerve fibers are responding rather than that the same number of fibers are reacting with greater vigor. Lucas's work became more widely known through its later development by his youthful collaborator, Edgar ADRIAN.

Lucretius (*c.* 95 BC–*c.* 55 BC) *Roman philosopher*

> Nothing can be created out of nothing.
> —*De rerum natura* (56 BC; On the Nature of Things)
>
> Some races increase, others are reduced, and in a short while the generations of living creatures are changed and like runners relay the torch of life.
> —As above

Little is known of the life of Lucretius (loo-**kree**-shush; full name Titus Lucretius Carus), who was born in Rome, apart from his materialistic (Epicurean) philosophy as set forth in his *De rerum natura* (On the Nature of Things), published in 56 BC. One of his main aims was to demonstrate the natural origin of the universe and its physical, biological, and social development, in which he may be said to have anticipated modern evolutionists. Dismissing ideas of the immortality of man, he denounced beliefs in divine guidance as the one great source of human misery and evil.

Lucretius also believed in the atomic structure of all living things, although he also extended this to include the mind, wind, etc. He recognized the virtual indestructibility of matter, had a notion of gravity and of the nature and speed of light, and demonstrated that the Earth is but one of many worlds in a boundless universe. Lucretius also recognized the incessant struggle for existence in the natural world, and in this again anticipated DARWIN and other evolutionists.

Ludwig, Carl Friedrich Wilhelm (1816–1895) *German physiologist*

Ludwig (**luud**-vig or **loot**-vik), who was born at Witzenhausen in Germany, graduated in medicine from the University of Marburg in 1840. He then held chairs at Marburg, Zurich, and Vienna before moving to Leipzig where he served as professor of physiology from 1865 until his death.

In an influential textbook Ludwig presented the strongest case in the 19th century for a mechanistic physiology. He aimed to reduce the organism to its fundamental constituents and thereafter explain its processes by forces of attraction and repulsion between them. Impressed by the ability of the salivary glands to continue secreting, even after decapitation, when the appropriate nerve is stimulated, Ludwig concluded that there was no role for any "vital" principle in the body. He also introduced some of the most basic physiological tools, for example the kymograph. This consists of the familiar rotating drum on which a pen can trace a continuous record of such things as blood pressure, changes in temperature, or virtually any physiological system. Using this instrument Ludwig showed that blood circulation can be explained by simple mechanical rather than vital forces. He later introduced the mercurial blood pump, which, by passing blood through a vacuum, allows the dissolved gases to be collected and measured. Ludwig's ideas and influence appear in the publications of his many students, for in the latter half of the 19th century students came from all over the world to Ludwig's laboratory.

Lummer, Otto (1860–1925) *German physicist*

> Revelations of a mind in which the divine spark glowed.
> —On Lummer's lectures. Clemens Schaefer, review of Lummer's work on his centenary (1960)

Born at Jena, in Germany, Lummer (**luum**-er) became professor at the University of Breslau in 1905. In 1889 he designed, with Eugen Brodhum, a photometer with an arrangement of prisms, which was an improved version of the grease-spot photometer of Robert BUNSEN.

Lummer's research was chiefly on radiation energy and temperature. Working with Wilhelm WIEN he achieved in practice the blackbody radiator, which had been conceived as a theoretical abstraction in the study of radiant heat. In 1899 he did further work with Ernst PRINGSHEIM on the distribution of energy in black-body radiation – work that eventually led to Max PLANCK's formulation of the quantum theory. Lummer also built, with Leon Arons, a mercury vapor lamp in 1892.

Lundmark, Knut (1889–1958) *Swedish astronomer*

Lundmark (**lund**-mark) was born at Norrbotten in Sweden and educated at the University of Uppsala. After working at the Lick and Mount Wilson observatories in California in the 1920s, he served as professor of astronomy at the University of Lund from 1929 to 1955.

He published a series of papers in the 1920s in which he attacked Adriaan van Maanen's claim to have measured significant amounts of internal rotation in some spiral nebulae and supported the claim of Heber CURTIS that such nebulae were isolated star systems too far away for such rotation to be detected. He repeated the measurements by the same methods and with the same instruments, concluding that "the nebulae are stationary for me and if I got a rotational effect it is very small compared with that of van Maanen." By 1929 Lundmark's views had been backed by the observations of Edwin HUBBLE who demonstrated conclusively the great distances of the spiral nebulae, or spiral galaxies as they are now called. In 1926 Lundmark published a preliminary classification of galaxies, on which Hubble was also working. Hubble, somewhat angrily, accused Lundmark of having "borrowed" his results. Lundmark replied that he had had no access to Hubble's work and in any case the two schemes were different.

Luria, Salvador Edward (1912–1991)
Italian–American biologist

Having studied medicine in his native city of Turin, physics and radiology in Rome, and bacteriophage research techniques in Paris, Luria (**luur**-ee-a) emigrated to America in 1940. There he met Max DELBRÜCK and became associated with the American Phage Group, a body formed to study virus replication. He served as professor of microbiology at the Massachusetts Institute of Technology from 1950 to 1958, and in 1964 became Sedgwick Professor of Biology there.

Luria was interested in the way bacteria acquire resistance to virus infection and investigated whether this is due to an adaptive response or spontaneous mutation. His development of the fluctuation test, and its subsequent mathematical analysis by Delbrück,

showed that resistance is indeed due to spontaneous mutation. The same year (1943) Luria also demonstrated mutation in bacteriophage; he completed his analysis of this phenomenon in 1951. He then worked on lysogeny, transduction, and host-controlled properties of viruses. For his contributions to the genetics of viruses and bacteria Luria shared the 1969 Nobel Prize for physiology or medicine with Delbrück and Alfred HERSHEY.

Lusk, Graham (1866–1932) *American physiologist*

> The work of a man's life is equal to the sum of all the influences he has brought to bear upon the world in which he lives.
> —*Science* No. 65 (1927)

Lusk was the son of an obstetrician from Bridgeport in Connecticut, but because of partial deafness he decided to choose physiology and chemistry rather than medicine as his career. He was educated at the Columbia School of Mines, New York, and in Germany. He studied physiology in Leipzig under Carl Ludwig and chemistry in Munich under Karl von Voit, obtaining his PhD in 1891. On his return to America Lusk held appointments at Yale and Bellevue Hospital, New York, before he moved in 1909 to Cornell as professor of physiology, a post he retained until his retirement in 1932.

Lusk worked almost exclusively on problems concerned with the physiology of food metabolism. Physiologists of this period were interested in tracing the fate of various foods in the body and the energy produced by them. Thus in 1922 Lusk was able to show how the body treats excess intake of protein. After deamination (removal of the amine part of the molecule) protein is converted into glycogen; when, however, the body's ability to store glycogen is

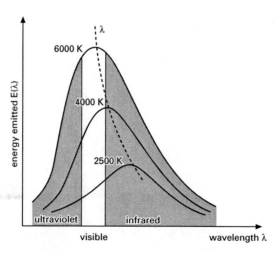

BLACK-BODY RADIATION   The distribution of radiation for a black body shown at three different temperatures as predicted by the work of Lummer and Pringsheim.

exhausted any remaining protein is converted into fat. He also worked for many years on the specific dynamic action (SDA) of various foods. This is a measure of the increase in energy output produced by the consumption of different foods; for example, while fat and carbohydrate have an SDA of about 5%, protein has a markedly higher SDA of 30%. After much effort and many years of doubt Lusk finally agreed with Max RUBNER that the extra energy came from the metabolism of that part of the protein molecule not converted into carbohydrate. In his *Elements of the Science of Nutrition* (1906) Lusk produced one of the standard texts in the field.

Lwoff, André Michael (1902–1994)
French biologist

Born at Ainy-le-Château in France, Lwoff (lwof) graduated in natural sciences in 1921 and joined the Pasteur Institute in the same year. He went on to become head of the laboratory in 1929 and from 1938 was head of the microbiology department. In 1958 Lwoff moved to the Sorbonne, where he served until 1968 as professor of microbiology. During the 1920s he demonstrated that vitamins function as coenzymes and also found that certain characters of protozoans are controlled by genes outside the nucleus. His most notable work, however, was his explanation of the phenomenon of lysogeny in bacteria. Lysogenic bacteria contain the DNA of a virus in their own DNA, the virus duplicating along with the bacterial chromosome and being passed on to subsequent generations. The virus, however, is nonvirulent and rarely destroys its host. When Lwoff began his research the prevailing view of lysogeny was that the phage–host association rendered the bacteria resistant to later viral destruction and that the association was perpetuated by the added presence of exogenous bacteriophage on the surface of the host. Further, it was believed that the increase in phage numbers in a lysogenic bacterial culture was due to the presence of some susceptible bacteria. Lwoff showed firstly that exogenous phage were not necesary to the association, and secondly that the increase of phage numbers in cultures is due to a reversal of the phage state from nonvirulent to virulent, which leads to the multiplication of phage particles in the host and subsequent breakdown or lysis of the host with release of these particles. He named the noninfective structure in lysogenic bacteria the prophage, and showed that ultraviolet light is one agent that can induce the prophage to produce infective viral particles.

In recognition of this work, Lwoff received the 1965 Nobel Prize for physiology or medicine, along with François JACOB and Jacques MONOD.

Lyell, Sir Charles (1797–1875) *British geologist*

> A scientific hypothesis is elegant and exciting insofar as it contradicts common sense.
> —Quoted by S. J. Gould in *Ever Since Darwin* (1978)

Lyell was the son of a notable botanist. He was born at Kinnordy in Scotland and educated at Oxford University, where he developed an interest in geology and attended the lectures of William BUCKLAND. While a student at Oxford he made, in 1819, the first of his many geological trips to the Continent and he met Alexander von HUMBOLDT and Georges CUVIER in France in 1823. He studied law and was called to the bar in 1825, but because of strained eyesight he turned increasingly to geological investigation.

During the following years Lyell traveled extensively on the Continent, his studies culminating in the publication of his three-volume masterpiece *The Principles of Geology* (1830–33). This was to be published in 11 editions in his lifetime and established him as a leading authority on geology. He became professor of geology at King's College, London (1831–33), but gave this up to continue his geological studies, traveling throughout Europe and visiting America in 1841 and 1845. In *Principles* Lyell established the doctrine of uniformitarianism already stated by James HUTTON and John PLAYFAIR and the first volume, published in 1830, was subtitled "Being an Attempt to Explain the Former Changes of the Earth's Surface by Reference to Causes Now in Operation." Lyell explicitly rejected the work of Abraham WERNER, in which some unique deluge is the chief agent producing the Earth's topography. Uniformitarianism also involved the rejection of the catastrophism theory followed by zoologists such as Cuvier to explain dramatic changes in the flora and fauna of the Earth. Instead Lyell saw the crust of the Earth as being shaped by forces operating over unlimited time. Lyell contributed considerable knowledge and analysis to geology. In 1833 he introduced the structure of the Tertiary in which it spread from the Cretaceous to the emergence of man and was subdivided on the basis of the ratio of living to extinct species – Eocene, Miocene, Pliocene, and Pleistocene. His other works included *Elements of Geology* (1838) describing European rocks and fossils from the most recent to the oldest then known. Charles Darwin, in his work *Origin of Species* (1859), drew heavily on Lyell's *Principles*. Lyell did not at first share Darwin's views and it was not until the tenth edition of the *Principles* (1867–68) that he expressed any support for evolutionary theory. Even then in his *The Antiquity of Man* (1863), which was published in four editions before 1873, Lyell denied that the theory could be applied to man. Lyell was knighted in

1848 and created a baron in 1864. He became president of the British Association in 1864.

Lyman, Theodore (1874–1954) *American physicist*

Lyman came from an old and wealthy Boston family and was educated at Harvard where he obtained his PhD in 1900. After a short spell studying abroad at Göttingen, Germany, and Cambridge, England, he returned to Harvard where he served as Hollis Professor of Physics (1921–25) and director of the Jefferson Physical Laboratory (1910–47).

Lyman was a spectroscopist who first developed a technique of investigating spectra in the ultraviolet region. In 1906 he observed the *Lyman series* of lines in the ultraviolet spectrum of hydrogen (similar to the series discovered by Johann BALMER in the visible region).

Lynen, Feodor (1911–1979) *German chemist*

Lynen (**lee**-nen) was educated at the university in his native city of Munich where he obtained his doctorate under Heinrich WIELAND in 1937. That same year he married Wieland's daughter. He was appointed to the faculty at Munich in 1942, being made professor of chemistry in 1947. In 1954 he became director of the Max Planck Institute for Cell Chemistry, Munich.

In 1950 Lynen showed that coenzyme A (described in 1947 by Fritz LIPMANN) plays the central role in the breakdown of fats in the body. Fats are first broken down by the enzyme lipase into a number of free fatty acids. It had been shown in 1904 that these are then degraded two carbon atoms at a time. This is done by coenzyme A combining with the fatty acid and forming, after a number of intermediate steps, acetoacetyl coenzyme A at one end of the chain. This can now react with another molecule of coenzyme A causing a two-carbon fragment of acetyl coenzyme A to split off. The process can now be repeated with the result that a fatty acid chain of n carbon molecules is eventually reduced to half that number of acetyl coenzyme A molecules. For this work on fatty acid metabolism and on cholesterol Lynen shared the 1964 Nobel Prize for physiology or medicine with Konrad BLOCH.

Lyot, Bernard Ferdinand (1897–1952) *French astronomer*

Lyot (lyoh), a Parisian by birth, was the son of a surgeon. He graduated from the Ecole Supérieure d'Electricité in 1917. After working briefly at the Ecole Polytechnique he joined the Paris Observatory at Meudon in 1920, becoming chief astronomer in 1943.

In 1930 Lyot invented the coronagraph. This is an instrument that permits the regular observation of the Sun's corona and chromosphere, which previously could only be briefly studied during a total solar eclipse. An artificial eclipse is produced within the instrument by means of a black disk placed in front of the lens of a telescope. This had been tried many times before with little success but Lyot realized that the slightest imperfection in the lens or the presence of dust in the instrument would scatter the light and thus destroy the image. He therefore succeeded by being particularly careful in these matters and in addition reduced the effects of atmospheric scattering by mounting his coronagraph high in the Pyrenees at the Pic du Midi Observatory. Lyot also developed a set of monochromatic filters to be used in conjunction with the coronagraph. With these he was able to produce images in a very narrow range of wavelengths. Lyot also pioneered the study of polarization of lunar and planetary light. By observing what happens to light when it is reflected from different surfaces in the laboratory he was able to build up standards of comparison for his lunar and planetary results. He thus demonstrated that the Moon's surface could be covered in volcanic ash and that the surface of Mars was dusty. Less convincing was his claim that the atmosphere of Venus contained water vapor: the minute quantities of water vapor now known to exist in the Venusian atmosphere could not have been detected by Lyot's method.

Lysenko, Trofim Denisovich (1898–1976) *Soviet agriculturalist*

> The Party, the Government, and J. V. Stalin personally have taken an unflagging interest in the further development of the [Lysenko's] teaching.
> —Official Report of the Genetics Debate at the Lenin All-Union Academy of Agricultural Sciences, August 1948

A peasant's son from Karlovka in Ukraine, Lysenko (li-**seng**-koh) began work at the Kirovabad Experimental Station in 1925, after completing his studies at the Kiev Agricultural Institute. In 1929 he became a senior specialist in the physiology department at the Institute of Selection and Genetics in Odessa and in the same year he first claimed success using vernalized grain on his father's farm. Vernalization is the cold treatment of soaked grains and it promotes flowering in spring-sown plants that might otherwise take two years to flower. Lysenko was later credited with inventing the method, although it had long been an established agricultural practice.

Lysenko claimed that the effects of vernalization could be inherited and so the treatment need not be repeated each year. This reversion to a belief in the inheritance of acquired characteristics was the hallmark of his career, and he is remembered for his single-minded application of this belief to Soviet biology. The validity of the chromosome theory of inheritance had been generally accepted in the West, especially since the publication of T. H. MORGAN's results. Naturally, many Soviet scientists also followed the Mendel–Morgan theory of heredity, notably Nikolai VAVILOV, who was president of the Lenin All-Union Academy of Agricultural Sciences and head of the Institute of Plant Protection. Vavilov was publicly discredited by Lysenko and in 1940 exiled to Siberia, Lysenko taking over his scientific posts. Other dissenting scientists were brought into line at the genetics debate held at the Lenin All-Union Academy of Agricultural Sciences in 1948. At this meeting Lysenko announced that he had the backing of the central committee of the communist party, and a motion was passed directing all textbooks and courses to be changed in accordance with his views. This dictatorial state of affairs continued until 1964, and Lysenko was finally ousted from his powerful position in 1965.

M

MacArthur, Robert Helmer (1930–1972)
American ecologist

MacArthur was born at Toronto in Canada, but moved to America in 1947. He was educated at Brown University and Yale where he obtained his PhD in 1958. He taught initially at the University of Pennsylvania from 1958 until 1965 when he joined the Princeton faculty as professor of biology from 1968.

One of his first successes was published in his well-known paper *Population Ecology of some Warblers of Northeastern Coniferous Forests* (1958). He investigated how five closely related species of warbler could spend the spring in the spruce forests of New England apparently feeding on the same diet and occupying the same niche. MacArthur managed to establish that the warblers tended to occupy and hunt in different parts of the tree, so preserving the current dogma – that in equilibrium communities no two species occupy the same niche.

At a more theoretical level MacArthur produced the "broken stick" model, which he used to predict the relative abundance of species in a particular ecosystem. Although initially the agreement between his theoretical prediction and census results was impressive, weaknesses eventually emerged in MacArthur's pioneering system. Despite this the possibility of a predictive science of ecology still remained and was reinforced by his collaborative work with Edward Wilson published in their *Theory of Island Biogeography* (1967). They were struck by and sought to explain the apparent stability of the number of species found on islands. They assumed immigration and extinction to be the major forces operating and further assumed that these two processes are in equilibrium. This allowed them to predict that smaller islands should have fewer species than large ones, as too would distant islands over those nearer to the mainland.

In the same 1967 volume MacArthur and Wilson clearly formulated a distinction that was emerging in a variety of forms in the writings of ecologists, that of r and K strategists. r-strategists are opportunistic species living in a variable environment; they typically have high reproductive rates, heavy mortality, short lives, and rapid development. K-strategists on the other hand are larger, develop more slowly, and are in general more stable. In 1962 MacArthur succeeded in showing that the basic theorems of natural selection apply to both r and K species. Although MacArthur met with much initial opposition, being accused of distorting and oversimplifying the complexity of nature, the power and originality of his thought was widely recognized well before his death.

MacDiarmid, Alan G. (1927–2007) *New Zealand-born American chemist*

MacDiarmid was born in New Zealand and educated at the University of New Zealand, the University of Wisconsin, and Cambridge University. He joined the faculty of the University of Pennsylvania in 1955, where he was Blanchard Professor of Chemistry.

In 1975 his interest in organic polymers that conduct electricity started when Hideki SHIRAKAWA introduced him to a new type of polyacetylene. MacDiarmid was the chemist who was responsible for the chemical and electrochemical doping of this compound. The collaboration between MacDiarmid, Shirakawa, and Alan HEEGER led to the discovery of metallic conductivity in organic polymers in 1977 and the subsequent detailed investigation of their properties. This pioneering work initiated a vast amount of research in chemistry, physics, and materials science which is still continuing both from the points of view of the novel chemistry and physics involved and the important technological applications.

MacDiarmid was particularly interested in polyaniline, a substance of great industrial interest because of the combination of its high electrical conductivity and desirable mechanical properties. He won many awards for his pioneering work in the field, the most notable being his share in the 2000 Nobel Prize for chemistry with Heeger and Shirakawa.

Macewen, Sir William (1848–1924) *Scottish surgeon*

Macewen, the son of a marine trader, was born at Rothesay on the Isle of Bute in Scotland and graduated in medicine from the University of Glasgow in 1869. He worked as a surgeon at the Glasgow Royal Infirmary from 1875 and served as professor of surgery at Glasgow University from 1892 until his death.

As an undergraduate Macewen had been a student of Joseph LISTER and was therefore introduced to antiseptic techniques at the beginning of his career. He thus belonged to the generation of surgeons determined to utilize such advances in opening up new areas to surgery.

Macewen was a surgical innovator in two main fields. In the late 1870s he began the surgical treatment of infections and tumors of the brain, claiming cures for 63 of the first 74 cases operated on. In 1893 he published an account of his work in his *Pyogenic Infective Diseases of the Brain and Spinal Cord*.

In 19th-century Glasgow one of the most incapacitating complaints was the brutal bone deformities resulting from the all too prevalent rickets. In the 1870s Macewen showed how such cases could be corrected surgically. He also, in 1880, showed how bone destroyed by disease could be functionally replaced by bone grafts. He was able to claim that by 1884 he had performed 1,800 operations on bones and joints without producing a single case of infection. Knighted in 1902, Macewen was apparently a surgeon in the grand manner – masterful, forceful, dictatorial, unpredictable, and highly successful.

Mach, Ernst (1838–1916) *Austrian physicist*

> Science may be regarded as a minimal problem consisting of the completest presentation of facts with the least possible expenditure of thought.
> —*The Science of Mechanics*

> I have repeatedly observed that I was merely an unprejudiced rambler, endowed with original ideas, in various fields of knowledge.
> —*Die Principien der physikalischen Optik* (1913; Principles of Physical Optics)

Mach (mahk), who was born at Turas (now in the Czech Republic), had a somewhat unorthodox upbringing and education. His father was a man knowledgable in both the classics and the sciences who retired to farm near Vienna where he educated his son academically, but also emphasized the importance of such practical skills as carpentry and farming. He received his higher education from the University of Vienna obtaining his PhD there in 1860. He was appointed professor of mathematics at Graz in 1864 and in 1867 was appointed to the chair of physics at the University of Prague, moving to Vienna in 1895 where he became professor of the history and theory of the inductive sciences. He was forced to retire in 1901 after being partially paralyzed by a stroke.

Mach made a large number of contributions to science in a variety of fields, but it is as a critic of science and as a philosopher that he ex-

ercised such a powerful influence over several generations of scientists. He began his research career in experimental psychology, inspired by the work of such scholars as Gustav FECHNER. For several years he investigated various perceptual processes; in the course of this work he discovered the function of the semicircular canals of the ear.

He is however best known to a wider public for his work on shock waves, which eventually led to the *mach number* being introduced (1929) as a measure of speed (it is the ratio of the speed of an object in a fluid to the speed of sound in the fluid). Mach's interest in shock waves was aroused by a claim made by L. Melsen that the savage wounds of those shot in the Franco-Prussian war were due not to the ammunition used by the French but to the impact of compressed air pushed ahead of the bullet. In 1884 Mach began his attempt to photograph the shock and sound waves produced by projectiles. It was not until 1886 that he was successful and able to conclude that Melsen was wrong.

Mach is also known to cosmologists for his controversial statement of the principle of inertia, called *Mach's principle* by EINSTEIN over whom Mach exerted considerable influence. For NEWTON the inertia of a body was an intrinsic property independent of the presence or absence of any other matter. For Mach the concept of inertia could only acquire quantitative meaning through Newton's laws. But Newton formulated his laws against the background of absolute space and time. This latter point Mach rejected completely, claiming that only relative motion exists.

Thus for Mach, it should make no difference whether we talk of the Earth rotating on its axis relative to the fixed stars or whether we see the Earth as fixed and talk of the stars as moving relative to the Earth. There is, however, the significant difference that only in the first case can the concept of inertia be used to understand such phenomena as the flattening of the Earth and FOUCAULT's pendulum. But Mach was a radical thinker and argued that the inertia is not an intrinsic property of matter but is itself caused by the background of the fixed stars. If these were removed inertia would disappear with them.

Mach's elimination of absolute space was simply part of a more general program in which he hoped to eliminate metaphysics – all those purely "thought-things" which cannot be pointed to in experience – from science. He began by asserting that the world contains nothing but sensations and their connections; scientific laws describe such connections in the simplest possible way; they provide an "economy of thought." His views influenced the important philosophical movement of logical positivism and also had some impact on scientific practice, influencing Einstein in his theory of relativity.

There is no doubt however that his approach could lead him astray. He never, for example, accepted the existence of atoms. They were rather "economical ways of symbolizing experience. But we have as little right to expect from them, as from the symbols of algebra, more than we have put into them." Nor did he ever accept the relativity theory of Einstein.

Macintosh, Charles (1766–1834) *Scottish industrial chemist*

Macintosh's father was the owner of a Glasgow factory producing cudbear, a purple dye obtained from various lichens, which is capable of coloring wool and silk. Macintosh attended local schools then became a clerk, but he also attended the chemical lectures of Joseph BLACK at Glasgow University in his spare time. In 1786 he started in business manufacturing sal ammoniac and sugar of lead (lead acetate) for use as mordants in the dyeing industry. In 1797 he set up the first alum works in Scotland.

Ten years later Macintosh made an important breakthrough in the bleaching industry. He went into partnership with Charles TENNANT in 1797 and together they opened a bleach factory at St. Rollox in Glasgow. In 1799 Macintosh invented a process for making bleaching powder (chloride of lime), a considerable improvement on Claude BERTHOLLET's chlorine gas, and this was patented by Tennant. Production of the bleaching powder enabled more textiles to be produced and stimulated the chemical industry to produce more sulfuric acid, soda, and salt for its manufacture.

In 1823 Macintosh made his second great invention. After his father's death he inherited the cudbear factory. While investigating the potentiality of using waste products from the new Glasgow gasworks, opened in 1818, he noticed that coal-tar naptha dissolved India rubber. He experimented with sandwiching wool fabric together with the substance and produced a flexible waterproof material. From this he produced coats, which came to be named for him. There were initial problems as the rubber cement was affected by the oils in the wool and the coats became stiff in winter and sticky in summer. This was solved when vulcanized rubber was introduced in 1839.

Mackenzie, Sir James (1853–1925) *British cardiologist*

Mackenzie, the son of a farmer from Scone in Scotland, was educated at the University of Edinburgh, where he obtained his MD in 1882. He then spent many years in practice in Burnley before moving to London in 1907 to become a consultant cardiologist.

Despite his relatively modest position in Burnley, Mackenzie was the most creative cardiologist of the period; his two influential works, *Study of the Pulse* (1902) and *Symptoms of the Heart* (1908), did much to turn cardiology from an anatomical study into a medical one. Before Mackenzie it was automatically assumed that all anomalies in the heart's action were inevitably signs of heart disease.

Mackenzie's research began as the result of the death of a pregnant girl from an unexpected heart attack. He suspected he may have missed some warning sign in the heart's action and therefore decided to examine his patients' hearts more closely and also to keep detailed records. To his surprise he found disorders of rate and rhythm, with beats missed and extra ones inserted, quite common in patients who showed no other signs of cardiac disease. He therefore concluded that while some irregularities of the heart were associated with heart disease others did no harm at all.

To distinguish between the different types of arrhythmias Mackenzie designed a polygraph that enabled him to record and correlate simultaneous pulses in veins, arteries, and the heart itself. Using this instrument he was able to undertake a major research program of distinguishing between harmless and pathological cardiac irregularities.

MacKinnon, Roderick (1956–) *American cell biologist*

MacKinnon was raised in Burlington, near Boston, Massachusetts. After initially studying biochemistry and gaining a BA at Brandeis University, Waltham, Massachusetts in 1978, he turned to medicine and in 1982 qualified as a medical doctor (MD) at Tufts Medical School, Boston. For some years he practiced medicine as a resident physician at Beth Israel Hospital at Harvard Medical School and then, in the mid-1980s, began field research, carrying out postdoctoral studies at Brandeis and mastering the techniques of x-ray crystallography to carry out his work. Since 1996 he has been professor of neurobiology and biophysics at the Rockefeller University in New York.

For cells to function, various substances have to enter and leave the cell through its outer membrane. In the 1990s MacKinnon researched the mechanism by which ions, such as potassium and sodium ions, cross the cell membrane of the bacterium *Streptomycetes lividans*. In 1998 he obtained a high-resolution image of a protein channel, a pore in the membrane that allowed the passage of specific ions. He was able to explain why, for example, a particular channel admits potassium ions but does not allow sodium ions to pass. He also showed that there is a molecular sensor that acts as a gate to open and close the ion channel as required. Ion channels are important factors in nerve transmission and disturbances in their function can lead to disorders of the nerves and muscles. With an understanding of how the channels work, pharmaceutical companies can now seek new drugs to treat such disorders. These structural and mechanistic studies of ion channels earned MacKinnon a half-share of the 2003 Nobel Prize for chemistry, along with Peter AGRE, who had discovered water channels in cell membranes.

Maclaurin, Colin (1698–1746) *Scottish mathematician*

> An unlucky accident has happened to the French mathematicians at Peru. It seems that they were shewing some French gallantry to the natives' wives, who have murdered their servants, destroyed their Instruments, and burnt their papers, the gentlemen escaping narrowly themselves. What an ugly article this will make in a journal.
> —Letter to James Stirling (1740)

Maclaurin, who was born at Kilmoden in Scotland, was a child prodigy. He entered Glasgow University at the age of 11 and became professor of mathematics at Marischal College, Aberdeen, at the age of 19.

His chief work was *Geometrica organica; sive descriptio linearum curvarum universalis* (1720; Organic Geometry, with the Description of the Universal Linear Curves) and in this he proposed several theorems that developed along similar lines to those contained in Isaac NEWTON's *Principia*. Maclaurin became a friend of Newton and defended Newton's new theory of calculus against the polemics of the Irish philosopher George Berkeley. Maclaurin was one of the first to treat the theory of maxima and minima properly. He also contributed to the theory of the equilibrium of rotating bodies of fluid. The *Maclaurin expansion*, which is a special case of the Taylor series, was named for him. Maclaurin played an important role in organizing the defense of Edinburgh against the Jacobites in 1745 and when they captured the city he was forced to flee to England.

MacLeod, Colin Munro (1909–1972) *American microbiologist*

MacLeod was born at Port Hastings in Nova Scotia, Canada, and educated at McGill University, Montreal, where he gained his MD in 1932. He moved to America in 1934 and joined the Rockefeller Institute until 1941 when he was appointed professor of microbiology at New York University. He afterward held chairs at the medical school of the University of Pennsylvania from 1956 to 1960 and New York from 1960 until 1966, after which time he served in a large number of administrative and consultative positions.

In 1944 MacLeod collaborated with Oswald AVERY and Maclyn MCCARTY in the seminal paper *Studies on the Chemical Transformation of Pneumococcal Types*. It was this paper that provided the first real evidence that DNA and not protein is responsible for the bacterial transformation described by Fred GRIFFITH in 1928.

MacLeod, John James Rickard (1876–1936) *British physiologist*

> [MacLeod occupied] an outstanding position in the field of carbohydrate metabolism, and it was both appropriate and fortunate that the discovery of insulin should have been made in his laboratory.
> —J. B. Collip, obituary of MacLeod in *Biochemical Journal* (1935)

MacLeod, born the son of a clergyman in Cluny, Scotland, was educated at Aberdeen Grammar School and Aberdeen University, where he gained his MD in 1898. After postgraduate work in Leipzig and hospital work in London, MacLeod moved to America in 1903 as professor of physiology at Western Reserve University, Cleveland, Ohio. In 1918 he accepted a similar post at the University of Toronto but returned to Scotland in 1928 to become professor of physiology at Aberdeen.

Before his move to Toronto, MacLeod worked mainly on problems of carbohydrate metabolism, producing from 1907 a series of papers entitled *Studies in Experimental Glycosuria*. He also produced a standard textbook, *Physiology and Biochemistry in Modern Medicine* (1918), which went through seven editions before his death.

However, his most significant work arose from his association at Toronto with the young surgeon Frederick BANTING. In early 1921 Banting asked if he could attempt to extract the pancreatic hormone believed to control the level of sugar in the blood. MacLeod apparently did his best to discourage Banting, pointing out to him that expert physiologists had made no progress with such a task over many years, but eventually offered him use of the laboratory and the help of a young student, Charles BEST, as a research assistant.

MacLeod left shortly afterward for Scotland, returning in September to find that the two researchers had succeeded in extracting a substance that controlled the level of blood sugar in dogs whose pancreases had been removed. Realizing the importance of obtaining as pure an extract as possible, MacLeod arranged for the laboratory's chemist, James Collip, to work on the problem. By January 1922, Collip had been so successful that they were ready to try the new hormone, named insulin by MacLeod, on a human patient.

In 1923 Banting and MacLeod were awarded the Nobel Prize for physiology or medicine for their discovery of insulin. Banting, distressed that his young colleague Best was apparently to receive no recognition, wished to refuse the prize. He finally decided to accept and shared the money with Best. MacLeod, too, was presumably feeling some discomfort at his award for he shared his prize with Collip.

MacMahon, Percy Alexander (1854–1919) *British mathematician*

MacMahon, who was born on the island of Malta, was a mathematical coach at the Royal Military Academy from 1882 to 1888. Subsequently he was assistant inspector at Woolwich Arsenal (1888–91) and professor of physics at the Artillery College (1891–98). His chief mathematical work was in classical algebra, in particular in combinatorial analysis. He built on the work of Arthur CAYLEY and James Joseph SYLVESTER, and thus belongs squarely in the 19th-century British school of algebraists. MacMahon was particularly interested in symmetric functions. He distilled his ideas on combinatorial analysis into a book that became a classic on the subject.

Macquer, Pierre Joseph (1718–1784) *French chemist and physician*

> Chemistry is a science, the object of which is to recognize the nature and properties of all substances by analyzing them and their compounds.
> —Quoted by A. F. de Fourcroy in *Elémens de chemie* (1791; Elements of Chemistry)

Macquer (ma-**kair**), a Parisian by birth, qualified as a physician in 1742. He became professor of chemistry at the Jardin du Roi in 1771. His greatest influence was exercised by his three-volume textbook *Elémens de chimie* (1741–51; Elements of Chemistry). He also produced, in 1766, the first dictionary of chemistry organized on modern systematic lines. His writings replaced those of Nicholas LÉMERY as the standard work on chemistry until Antoine LAVOISIER's treatise later in the century. Macquer's works show a move away from the orthodox views of Georg STAHL with a willingness to accept some of NEWTON's views on chemistry.

Macquer did some work in industrial chemistry being chief chemist at the state porcelain works at Sèvres where, in 1768, he managed to produce a hard porcelain to rival that of Meissen. He was also important in establishing the supremacy of the French dyeing industry.

Magendie, François (1783–1855) *French physiologist*

> I compare myself to a ragpicker: with my spiked stick in my hand and my basket on my back, I traverse the field of science, and I gather what I find.
> —Quoted by Claude Bernard in *François Magendie* (1856)

Magendie (ma-zhahn-**dee**), who was born at Bordeaux in France, began medical studies in Paris in 1803 and graduated in 1808. From 1813 to 1821 he carried out much research into the effects of various alkaloid arrow poisons. He began by using animals to test the results but once he had established a safe dose, he tried out many of the drugs on himself. He showed the need for direct contact between the chemical compound and the particular organ and by this he laid the foundations of modern experimental pharmacology. In 1822 he published his *Formulaire* (Formulary), which introduced alkaloids such as morphine, strychnine, and quinine into medical practice.

Although initially a vitalist, he nevertheless showed the absorption of liquids through vessel walls to be a physicochemical phenomenon. Later in life he became disenchanted with physiological theories, claiming it was better to stick to the facts rather than any medical system. In 1815, postrevolutionary France was short of food, so Magendie was appointed chairman of a commission to investigate the nutritional value of various food extracts. He showed the need for adequate amounts of the right sort of protein in a diet and thus laid the foundations of the science of nutrition.

In 1822 he carried out a series of experiments to distinguish the separate motor and sensory roots of the spinal nerves. He showed that the anterior roots control movement and the posterior roots, sensations. A bitter dispute over the priority for the discovery then ensued with the distinguished physiologist Charles Bell. Magendie continued working on the nervous system and cerebrospinal fluid and in 1828 he described the canal in the brain that bears his name. In 1842 he published an influential book that helped to reform clinical medicine along physiological lines.

Mahomed, Frederick Henry Horatio Akbar (1849–1884) *British physician*

Mahomed's grandfather, Sake Deen Mahomed, a former surgeon of the East India Company, opened a "hot vapor bath" in the seaside resort of Brighton in 1784, an establishment later managed by his son and much frequented by the Prince Regent. Mahomed, who was born in Brighton, was educated at Cambridge University and Guy's Hospital, London, where he served briefly as an assistant physician from 1881 until his premature death from typhoid fever in 1884.

Using an improved version of the sphygmograph of Etienne MAREY, Mahomed began an important investigation of arterial blood pressure in both sickness and health. He not only established some of the basic parameters of blood pressure but also showed in this way the existence of essential hypertension, a condition of high blood pressure without any underlying pathological basis. He was further able to demonstrate raised blood pressure as an early sign of nephritis, an infection of the kidney.

Maiman, Theodore Harold (1927–   )
American physicist

Maiman, the son of an electrical engineer, was born in Los Angeles, California, and graduated in engineering physics from the University of Colorado in 1949. He gained his PhD from Stanford University in 1955, following which he joined the Hughes Research Laboratories, Miami.

Maiman was especially interested in the maser, which had been developed independently by Charles TOWNES in America and Nikolai BASOV and Aleksandr PROKHOROV in the USSR in 1955. It was realized that the principles by which the maser produced microwaves at selected well-defined wavelengths could be extended to emissions at the shorter visible wavelengths. In 1960, at the Hughes Research Laboratory, Maiman designed and operated the first instrument to achieve this, now known as the laser (*l*ight *a*mplification by *s*timulated *e*mission of *r*adiation). He used a ruby cylinder with mirror-coated ends as a resonant cavity, and succeeded in converting flashes of white (incoherent) light into a pulsed beam of monochromatic coherent light. Such a beam can travel long distances with little dispersion and can concentrate optical energy on a small spot. The first continuous (as opposed to pulsed) laser was made by Ali Javan and his colleagues at the Bell Telephone Laboratories in 1961. The laser has since been developed in many forms as a research and engineering tool.

Maiman founded his own company, Korad Corporation, in 1962, which became the leading developer and manufacturer of high-power lasers. He founded Maiman Associates in 1968, acting as a consultant on lasers and optics, and cofounded the Laser Video Corporation in 1972. In 1977 he joined TRW Electronics of California as assistant for advanced technology.

Maimonides, Moses (1135–1204) *Jewish physician and philosopher*

> One who is ill has not only the right but also the duty to seek medical aid.

Maimonides (mI-**mon**-i-deez) was born in Cordoba, Spain, the son of a judge. With the invasion of Spain in 1148 by the fanatical Almohads, he and his family were compelled to leave their home. They moved to Fez in about 1160 but finally settled permanently in Cairo in 1166 and Maimonides became physician to the famous opponent of the Crusaders, Saladin.

Maimonides is best known for his *Guide for the Perplexed* (1190), a work designed to show how philosophy could justify and elucidate the tenets of revealed religion, attempts to reconcile Aristotelianism with Judaism, and argues

against astrology. He also produced a number of medical monographs, in Arabic, on such topics as asthma, poisons, and hemorrhoids.

Mallet, Robert (1810–1881) *Irish industrialist and seismologist*

After graduating from Trinity College in his native city of Dublin in 1830, Mallet joined his father's foundry business. He remained there until 1861 when, with the completion of the Irish railroad network, there was such a marked decline in his business that he moved to London as a consultant engineer. Mallet was an engineer of some skill; he constructed several bridges, the Fastnet Lighthouse, and erected the 133-ton roof of St. George's in Dublin.

As a geologist he is mainly remembered for pioneering observational and experimental seismology. In 1857 he spent two months in Naples studying the effects of the recent earthquake. He showed that by noting the direction of cracks in walls and the arrangement of fallen masonry, it was possible to determine the epicenter and the depth of the earthquake. His results were published in *The Great Neapolitan Earthquake of 1857: the First Principles of Observational Cosmology* (1862).

In a series of experiments starting in 1850, Mallet attempted to determine the speed of earthquake waves. He did this by setting off small explosives at different depths in different soils and measuring the time taken for the seismic waves to travel varying distances. James Michel, working in the 18th century, had estimated a speed of 20 miles (32 km) per minute but this could not compare with Mallet's precise measurements.

Malpighi, Marcello (1628–1694) *Italian histologist*

> The nature of things, enveloped in shadows, is revealed only by the analogical method ... which enables us to analyze the most complex mechanisms by means of simpler ones that are more easily accessible to the experience of the senses.
> —*Anatomes plantarum idea* (1675; Theory of the Structures of Plants)

Born at Crevalcore in Italy, Malpighi (mahl-**pee**-gee) graduated firstly in philosophy and then medicine from the University of Bologna. He subsequently served as professor of medicine at Pisa, Bologna, and Messina. In later life he became private physician to Pope Innocent XII. Malpighi may be considered the father of microscopy: he was one of the first to reveal the hitherto hidden world of the detailed structure of animals and plants. Some of his most important investigations were concerned with mammalian blood. In 1660 he provided a clue to the mechanism of respiration by his discovery that blood flowed over the lungs by means of a complex system of minute vessels. He also discovered the fine capillaries or blood vessels in the wing membrane of bats by means of the microscope, and demonstrated their linkage with minute arteries and veins, in this way amplifying the blood circulation theories of William HARVEY and Malpighi's contemporary Olof RUDBECK. He was also able to locate the filtering units of the vertebrate kidney, which became known as *Malpighian bodies* or *corpuscles*, and discovered traces of gill-like structures in chick embryos, although the true nature and significance of his discovery was not recognized until later. Using a chick as his model, he was able to describe the formation of the heart from undifferentiated tissue. Studies of the skin resulted in his description of the layer of epidermis (the *Malpighian layer*) next to the dermis in which active cell division takes place.

Malpighi was also interested in insect physiology, showing that the insect respiratory system was based on the simple diffusion of gases in vessels called tracheae. In the first published treatise to deal exclusively with an invertebrate animal, he described the internal organs and anatomy of the silkworm moth. In the world of plants, Malpighi noted the stomata on the undersides of leaves, appreciated the differences in development between monocotyledons and dicotyledons, and described the annual rings of dicotyledon stems.

In 1669, Malpighi became the first Italian to be named an honorary member of the Royal Society, following which much of his research was published in the *Philosophical Transactions* of the society.

Malthus, Thomas Robert (1766–1834)
English economist

> Population, when unchecked, increases in a geometrical ratio. Subsistence only increases in an arithmetical ratio.
> —*An Essay on the Principle of Population* (1798)

Born at Dorking in Surrey, England, Malthus was educated at Cambridge University, graduating in 1788. He grew up at a time when many people believed man was progressing toward an ideal society, and indeed his father was a keen follower of the philosophy of Jean-Jacques Rousseau. Malthus was more realistic and believed that any technological advances would always be accompanied by increases in the human population, and moreover that while the means of subsistence would tend to increase arithmetically, population growth would be geometric. Thus overall the human condition would remain unimproved and numbers would still be controlled by famine, disease, and war. Others such as ARISTOTLE and Benjamin FRANKLIN had already proposed this idea, but Malthus presented his case so well in *An Essay*

on the Principle of Population (1798) that much controversy was aroused. Many people were later to use his arguments as an excuse for neglecting social reform and to justify laissez-faire.

In 1805 Malthus became the first professor of political economy in England at the East India Company's Haileybury College. He continued to enlarge his original work adding evidence collected from travels through Europe, but he never placed it on a sound statistical footing. The population essay later helped Alfred WALLACE and Charles DARWIN reach the theory of evolution by natural selection.

Other contributions Malthus made to economics include formulating the law of diminishing returns, and his advocacy of public and private spending to motivate production. The Malthusian League was founded in the 1860s to promote his research on population, but was suspended in 1927.

Malus, Etienne Louis (1775–1812)
French military engineer and physicist

Malus (ma-**loos**), who was born in Paris, attended the military school in Mezières (1793) and the newly established Ecole Polytechnique (1794–96) where he received his basic scientific education. He was commissioned in 1796 and served as a military engineer in Napoleon's expedition to Egypt and Syria. After his return to France he held various military engineering appointments. He became an examiner in geometry and analysis in 1805 and an examiner in physics in 1806 at the Ecole Polytechnique; these posts brought him into contact with other physicists.

Malus carried out many researches in optics, which was his main scientific interest. He is remembered for his discovery made in 1808 that light rays may be polarized by reflection. He found this while looking through a crystal of Iceland spar at the windows of a house reflecting the rays of the Sun; he noticed that on rotating the crystal the light was extinguished in certain positions. In explaining his observations Malus, a Newtonian and believer in corpuscular theory, argued that light particles have sides or poles and used for the first time the word "polarization" to describe the phenomenon.

Malus's work in optics gave considerable impetus to investigations into polarization and the optical properties of crystals.

Mandelbrot, Benoit (1924–   ) *Polish-born American mathematician*

> Science would be ruined if (like sports) it were to put competition above everything else, and if it were to clarify the rules of competition by withdrawing entirely into narrowly defined specialities. The rare scholars who are

nomads-by-choice are essential to the intellectual welfare of the settled disciplines.
—Comment added to his entry in the U.S. *Who's Who*

The son of a Lithuanian Jewish merchant, Mandelbrot (**man**-del-brot) was born in the Polish capital Warsaw but moved with his parents to Paris in 1936. In 1939 they found it necessary to flee once more and lived in Tulle in southern France for the duration of World War II. Despite an interrupted and irregular education, Mandelbrot gained acceptance at the Ecole Polytechnique after the war even though, he later claimed, he had never learned the alphabet, nor progressed beyond the five-times table. He gained his PhD from the University of Paris in 1952 and spent several years in short-term appointments at the Institute of Advanced Studies, Princeton, and at the University of Geneva and Lille University. In 1958 he moved to the IBM Research Center, Yorktown Heights, New York, where he remained until 1987, when he was appointed professor of mathematics at Yale.

Mandelbrot studied a number of such seemingly unrelated topics as fluctuations in commodity prices, noise in telephone lines, and linguistics. He also considered the seemingly innocent question "How long is the coast of Britain?" Encyclopedias gave lengths differing by as much as 20%. Mandelbrot pointed out that it depended on how the measurement was done. From a distant space craft, many inlets would reveal their own inlets. Mandelbrot dealt with this and other matters in his *The Fractal Geometry of Nature* (1982). "Clouds are not spheres," he declared, "mountains are not cones, coastlines are not circles, and bark is not smooth, nor does lightning travel in a straight line." To understand this structured irregularity of nature Mandelbrot introduced the term "fractal" based on the idea of fractional dimension.

An example of a fractal is the snowflake curve first described by Helge von Koch in 1904. It begins with an equilateral triangle. Each side is divided into three equal parts and the middle section is used as the base of a smaller equilateral triangle, resulting in a six-pointed star. The process can be continued indefinitely and has an infinite perimeter bounding a finite area. It is fractal in the sense that it is self-similar, and also in the sense that it has fractional dimension. Mandelbrot saw that shrinking the unit that a side is measured in by a factor of P increases the number of units along that side by a factor of Q. In the case of the *Koch curve*, shrinking the side by a factor of 3 increases the units by a factor of 4. The fractal dimension A can be defined as:

$$A = \log Q / \log P = \log 4 / \log 3 = 1.2618$$

Mandelbrot went on to determine the fractal dimensions of other similar objects.

Mandelbrot is equally well known for his discovery of the *Mandelbrot set*. The set is constructed from the simple mapping

$$z \rightarrow z^2 + c,$$

where z and c are complex numbers, with z arbitrarily chosen and c fixed. If a fixed value is assigned to c and $z = 0$, the answer is calculated and fed back into the mapping as a new value for z. The process is repeated, substituting each new output for z. Some values for c when plugged back into the mapping rapidly approach infinity; other values remain within a certain boundary. For example, when $c = 1 + 0i$, the sequence begins 0, 1, 2, 5, 26, 677, 458, 330...and is unbounded. But when $c = -1 + 0i$, the sequence is $0, -1, 0, -1, 0, -1$...and is clearly bounded.

The set is constructed by marking a black dot on the complex plane for those points c where the sequence is unbounded, and leaving all other values white. The result, best displayed in color on a computer screen, takes on the distinctive shape described as a warty figure of eight on its side. Yet at higher magnifications borders reveal endless detail and startling images, apparently copies of the original but also displaying small differences.

Mansfield, Peter (1933–   ) *British physicist*

Mansfield was born in London and educated at Queen Mary College there, completing his PhD in 1962. After a two-year period at the University of Illinois as a research associate, he took up a position at Nottingham University in 1964; he was professor of physics there from 1979 to 1994, when he became emeritus professor.

Mansfield began work in the early 1970s on the use of nuclear magnetic resonance (NMR) to investigate conditions in the human body. The NMR technique was investigated in the 1940s by BLOCH and PURCELL. It depends on the fact that the nuclei of certain atoms have a net magnetic moment. In an external magnetic field these nuclei can take up allowed orientations with the field direction, each corresponding to a particular quantized energy state. Electromagnetic radiation in the radiofrequency region of the spectrum can be absorbed at a particular resonance frequency corresponding to a transition from one energy state to a higher one. The nuclei, in reverting to the lower state, emit radiation.

Mansfield and his colleagues used this as a nonintrusive method of producing images of the body by detecting the emitted radiation and forming an image by computer-aided tomography (CAT). The x-ray CAT scanner had been developed earlier by Godfrey HOUNSFIELD. However, light elements such as hydrogen are relatively transparent to x-rays. NMR, on the other hand, is particularly suitable for detecting hydrogen and magnetic resonance imaging (MRI) is especially useful for showing soft tissues. A prototype MRI scanner had been developed by 1980. Initially, it was designed to take cross-sectional images of the brain, but before long whole-body machines were available. In 2003 Mansfield was awarded the Nobel Prize for physiology or medicine. The prize was shared with Paul LAUTERBUR.

Manson, Sir Patrick (1844–1922) *British physician*

Manson was the son of a local laird and farmer in Oldmeldrum, Scotland. He studied medicine at Aberdeen University, obtaining his MD in 1866. He then became a medical officer, firstly in Formosa (1866–71) and then for a further 13 years in Amoy. Manson completed his 23 years' service in the Far East by running a profitable private practice in Hong Kong until 1889. Back in Britain he set up as a consultant in London, served as medical adviser to the Colonial Office from 1897 to 1912, and was the prime mover in the foundation of the London School of Tropical Medicine in 1899. In 1914 Manson finally retired to Ireland, where he spent his last years fishing on Lough Mask in Galway.

Manson was an original and creative scientist. His greatest achievement was to demonstrate conclusively what had long been suspected, namely, that certain diseases are transmitted by insects. His first success, in 1877, was to link the mosquito *Culex fatigans* with the presence of the parasite *Filaria sanguinis hominis* in many of his patients suffering from elephantiasis. Full details of the life cycle of the FSH were published later in 1884. The clinical effects of the parasites, which eventually obstruct the lymphatic system, he published in his monograph, *The Filaria sanguinis hominis, and Certain New Forms of Parasitic Disease* (1883).

With this success behind him Manson was able to throw considerable light on a wide range of further tropical diseases. He thus made numerous suggestions on the mode of transmission of such widespread diseases as sleeping sickness and bilharzia. He also played a crucial role in the working out of the etiology and spread of the biggest killer of all, malaria. It was Manson who suggested to Ronald ROSS in 1894 that mosquitoes carry malaria, who guided Ross throughout his research, and who, in 1900, performed one of the crucial experiments confirming his earlier hypothesis.

Manson used mosquitoes of the species *Anopheles maculipennis*, which were sent him by Giovanni GRASSI from Rome, and allowed them to feed on his son Patrick Manson, then a medical student at Guy's Hospital, London. Within 15 days his son had developed malaria and parasites were clearly visible in his blood.

For such work Manson was elected to the Royal Society in 1900 and knighted in 1903. But although Ross was awarded the 1902 Nobel Prize for physiology or medicine for his work on malaria, Manson was surprisingly ignored.

Mantell, Gideon Algernon (1790–1852) *British geologist*

The son of a shoemaker from Lewes in Sussex, England, Mantell was apprenticed to a local surgeon and set up in practice in Lewes in 1812. As the area around Lewes, the South Downs, is rich in Cretaceous deposits, fossil hunting was something of a local pastime and Mantell began collecting fossils in his schooldays.

His interest grew and in 1822, in the area known as the Tilgate Forest, he made his most important find – a large tooth with a worn, smooth surface. It obviously belonged to a large herbivore and initially reminded Mantell of an elephant's tooth. However, mammals were unknown in the Cretaceous while reptiles, which were common, did not masticate food. Baffled by his find, Mantell sent his tooth to the great Baron CUVIER in Paris for identification. But Cuvier's judgment that it was the upper incisor of a rhinoceros Mantell knew to be nonsense. He continued to search for an answer and eventually in the Museum of the Royal College of Surgeons he found a smaller but identical tooth belonging to the South American iguana. The tooth, he concluded, came from a lizard after all, a giant toothed lizard he named *Iguanadon* (iguana tooth).

In 1834 an *Iguanadon* skeleton was discovered in a Maidstone quarry. Mantell bought the remains for £25 to make, as the joke went, "a famous Mantel-piece." Mantell also discovered, in 1832, the first armored dinosaur, *Hylaeosaurus*.

Mantell described his work in his 67 books and memoirs. The best known of these are the once widely read *The Geology of South East England* (1833) and *The Wonders of Geology* (1838). His private life, however, fared less well. He moved to Brighton in 1835 in search of patronage and a more affluent practice. He failed on both accounts and sold his vast collection of fossils to the British Museum in 1838 for £4,000. Abandoned by his wife and children, he moved to London in 1839, spending most of his time lecturing and writing on geology. He died from an overdose of opium taken to relieve persistent back pain.

Manton, Sidnie Milana (1902–1979) *British zoologist*

Manton was born in London and educated at Cambridge University. She was appointed university demonstrator in comparative anatomy (1927–35) at Cambridge and then director of studies in natural science at Girton College,

Cambridge (1935–42). From 1943 until 1960 she was visiting lecturer, assistant lecturer, and reader in zoology at King's College, London. Until her retirement in 1967 she was research fellow of Queen Mary College, London.

Manton's most significant work was on the structure, physiology (especially locomotion), and evolution of the arthropods. Her revision of the Arthropoda, indicating their different evolutionary lines, into the Chelicerata (Arachnida: spiders, scorpions, mites, ticks, etc.), the Crustacea, and the Uniramia (insects, myriapods, etc.), plus the extinct Trilobita, has now been accepted by most modern phylogenists. She was the author of *The Arthropoda: Habits, Functional Morphology and Evolution* (1977), as well as many papers on the coelenterates and on arthropod evolution, locomotion, mandibular mechanisms, etc. Manton was elected a fellow of the Royal Society in 1948 and received the Linnean Society's Gold medal in 1963. In her retirement she bred new varieties of cats, having published *Colourpoint, Longhair and Himalayan Cats* in 1971.

Manzolini, Anna Morandi (1716–1774)
Italian anatomist

Anna Morandi was a talented artist, born into a Bologna that was home to a rich scientific and artistic tradition. Progress in medical research was exemplified by public dissections of corpses in the Teatro Anatomico of the University. The teaching of anatomy was illustrated with the help of wax models of body organs, in which artistic flair combined with anatomical knowledge. In 1742 the sculptor and painter Ercolle Lelli made the first detailed wax reconstructions of the human skeleton and muscles, with his assistant Giovanni Manzolini. The latter was helped by a childhood friend, Anna Morandi, whom he married in 1740. The couple then set up their own workshop, and became known for their faithful anatomical models, despite Anna's inherent fear of corpses. Giovanni later became a professor of anatomy at the University of Bologna, and when illness prevented him from teaching, Anna took his place. In 1760, five years after her husband's death, she also was appointed professor in the anatomy department, later becoming its chief model maker. Her fame spread throughout Europe, and she traveled to Russia at the invitation of Empress Katherine II to give lectures. The Manzolinis' models set standards for anatomical teaching that remain to this day. Many are still on display at the Anatomical Museum of Bologna University, including a self-portrait by Anna in wax, showing her bejewelled and dressed in silks while preparing to dissect a skull.

Marconi, Marchese Guglielmo (1874–1937) *Italian electrical engineer*

When he came to England and his baggage was examined by customs officials something like panic ensued over the discovery of his apparatus ... [He] was suspected of carrying an infernal machine and notwithstanding all his protests it was promptly dropped into a bucket of water.
—Obituary of Marconi in *The Daily Telegraph*, 21 July 1937

Marconi's most cherished possession was a gold tablet presented to him by 600 survivors of the *Titanic* who had been saved by the fact that the ship's wireless transmitter had been able to call ships from hundreds of miles away to pick up survivors.
—Douglas Walters in *The Daily Herald*, 21 July 1937

Marconi (mar-**koh**-nee), the second son of a prosperous Italian country gentleman and a wealthy Irishwoman, was born in Bologna, Italy, and educated there and in Leghorn (Livorno). He studied physics under several well-known teachers and had the opportunity of learning about the work carried out on electromagnetic radiation by Heinrich HERTZ, Oliver LODGE, Augusto RIGHI, and others.

Marconi became interested in using Hertz's "invisible rays" to signal Morse code and in 1894 began experimenting to this end at his father's estate. Similar work was being done at the time in Russia by Aleksandr POPOV. Although he convinced himself of the importance of this new system and was soon able to transmit radio signals over a distance of more than a mile, he received little encouragement to continue his work in Italy and was advised to go to England.

Shortly after arriving in London in February 1896 Marconi secured the interest of government officials from the war office, the admiralty, and the British postal service. The next five years he spent demonstrating and improving the range and performance of his equipment, and overcoming the prevailing skepticism about the usefulness of this form of transmission. In 1897 he helped to form the Wireless Telegraphic and Signal Co. Ltd., which in 1900 became Marconi's Wireless Telegraph Co. Ltd. He achieved the first international wireless (i.e., radio) transmission, between England and France, in March 1899; this aroused considerable public interest and attracted attention in the world's press. In the same year the British fleet's summer naval maneuvers, for which Marconi equipped several ships with his apparatus, helped to convince the admiralty and mercantile ship owners of the value of radio telegraphy at sea.

In December 1901 Marconi succeeded in transmitting radio signals in Morse code for the first time across the Atlantic, a distance of some 2,000 miles (3,200 km). Already well known, Marconi created a sensation, became world famous overnight, and silenced many of his critics from the scientific world who had believed that because radio waves travel in straight lines they could not follow the curvature of the Earth. This phenomenon was explained by Arthur KENNELLY the following year as being due to the presence of a reflecting layer – the ionosphere – in the Earth's atmosphere. Thus by 1901 radio telegraphy had become a practical system of communication, especially for maritime purposes. Marconi spent the rest of his life improving and extending this form of communication, and managing his companies.

Although a good deal of Marconi's work was based on the ideas and discoveries of others he was granted various patents and was responsible for some notable inventions. These included the first of all patents on radio telegraphy based on the use of waves (1896), the elevated antenna (1894), patent 7777, which enabled several stations to operate on different wavelengths without mutual interference (1900), the magnetic detector (1902), the horizontal directional antenna (1915), and the timed-spark system of generating pseudo-continuous waves (1912).

From about 1916, Marconi began to exploit the use of radio waves of short wavelength, which allowed a more efficient transmission of radiant energy. In 1924 the Marconi company obtained a contract to establish short-wave communication between England and the British Commonwealth countries and by 1927 a worldwide network had been formed. In 1932 Marconi installed a short-wave radio telephone system between the Vatican and Castel Gandolfo, the Pope's summer residence.

Despite having little interest in anything outside radio Marconi several times acted in an official capacity for his government: he was sent as a plenipotentiary delegate at the 1919 Paris peace conference. In 1923 he joined the Fascist Party and became a friend of Mussolini. Marconi received several honorary degrees and many awards, which included the Nobel Prize for physics jointly with Karl BRAUN (1909), being made a marquis (marchese) in 1929, and president of the Royal Italian Academy (1930). At his death he was accorded a state funeral by the Italian government. All Post Office wireless telegraph and wireless telephone services in the British Isles observed a two-minute silence at the hour of his funeral.

Marcus, Rudolph Arthur (1923–   )
Canadian–American chemist

Marcus was born in Montreal and educated at McGill University there. After graduating he taught at the Polytechnic Institute, Brooklyn (1951–64) and at the University of Illinois (1968–78). In 1978 he was appointed professor of chemistry at the California Institute of Technology.

In the 1950s Marcus began to work on electron-transfer reactions. The addition, removal, and transfer of electrons is the driving force behind many basic chemical processes including photosynthesis, respiration, and the production of solar energy. Such reactions are, in principle, very simple, involving the movement of an electron from one ion to form another ion. The rates of such reactions can, however, vary widely. Marcus was able to explain electron-transfer reaction rates in terms of the way in which the solvent molecules, initially configured to solvate the reactant, reorganize to solvate the products.

Marcus was awarded the 1992 Nobel Prize for chemistry for his work in this field.

Marey, Etienne-Jules (1830–1904)
French physiologist

The [human] body is a theater of motion.

Marey (ma-**ray**), the son of a wine merchant from Beaune in France, qualified as a physician in Paris. In 1868 he succeeded Pierre FLOURENS as professor of natural history at the Collège de France, a post he retained until his retirement in 1898.

Marey is mainly remembered today for introducing a number of recording devices into science. His first major success came in 1863 with his invention of the sphygmograph, an ancestor of the modern sphygmometer, to record the strength and duration of the human radial pulse at the wrist. It recorded the pulse by connecting it to a very light lever that registered its movements on a rotating cylinder of blackened metal – the kymograph of Karl Ludwig.

Marey's ingenuity and adeptness with simple recording instruments was shown in 1868 when he established the basic figure-eight motion of insect wings. He simply attached a piece of gold leaf to the wing tip of a fly fixed to a stationary mounting; under a bright light the figure-eight motion clearly stood out. Further details of the wing's motion were revealed by allowing the wing tip to brush against the blackened surface of a revolving cylinder, so as to produce marks on the surface.

He later became absorbed by the problems involved in photographing animal locomotion. Thus in 1888 he first demonstrated his "chambre chronophotographique," a device that had a shutter rotating between lens and film and could record images at the rate of 10–12 a second. He used both high-speed and time-lapse photography to slow down or speed up fast and slow movements. With such an instrument Marey was able to continue his work on insect and bird flight, achieving with it a new depth of detail.

Marggraf, Andreas Sigismund (1709–1782) *German chemist*

Marggraf (**mark**-grahf), who was born in

Berlin, studied at the universities of Strasbourg and Halle, and at the Freiberg School of Mines. His father was court apothecary at Berlin and Marggraf served as his assistant in the period 1735–38. He became a member of the Berlin Academy of Sciences and was made director of its chemical laboratory (1754–60).

Although he was a traditionalist and was the last eminent German scientist still to accept the phlogiston theory, he was also a successful practical chemist and some of his work had profitable industrial implications. Most useful was his discovery of sucrose in beet in 1747. This was utilized by France in the Napoleonic wars when France was cut off from its sources of cane sugar by the continental blockade. The first sugar beet factory was opened in 1801.

Marggraf also worked on phosphorus and showed that potassium and sodium salts were distinguishable by the color they emitted when burned. He worked out the structure of gypsum and established chemical tests for the presence of iron. Marggraf also introduced the microscope into analytical work in chemistry. His works, *Chemische Schriften* (1761–67; Writings on Chemistry), were published in two volumes.

Margulis, Lynn (1938–) *American biologist*

Margulis was educated at the university in her native city of Chicago, at Wisconsin, and at Berkeley where she obtained her PhD in 1965. After working briefly at Brandeis she moved to Boston University in 1966 and served there as professor of biology from 1977 to 1988. She became Distinguished Professor of the University of Massachusetts in 1988.

With the success of modern biochemistry, genetics, and cytology it became apparent that there was a fundamental division in nature between cells with nuclei (eukaryotes) and those without (prokaryotes). In terms of metabolism, chemistry, genetics, and structure, higher organisms differ radically from bacteria and bluegreen algae, the prokaryotes. She studied the question of how eukaryotes evolve and her answer, in terms of hereditary endosymbiosis, was fully formulated in her *Origin of Eukaryotic Cells* (1970). She argued that eukaryotes are basically colonies of prokaryotes and that such features of cells as mitochondria were once freeliving bacteria but have, "over a long period of time, established a hereditary symbiosis with ancestral hosts that ultimately evolved into animal cells." Similarly she proposes that chloroplasts and flagella evolved in the same way.

The actual evolutionary sequence proposed begins with a "fermenting bacterium" entering into a symbiotic relationship with some oxygen-using bacteria, the first mitochondria. Such a complex might join with "a second group of symbionts, flagellumlike bacteria comparable to modern spirochaetes," which, attached to the host's surface, would greatly increase its motility.

As Margulis points out, the proof for such an imaginative model requires that the cell organelles are separated, cultured independently, and then brought back into symbiotic association again. So far no one has managed to grow an organelle outside the cell. In 1982 Margulis published (with Karlene Schwartz) *Five Kingdoms*, "a catalog of the world's living diversity."

Marignac, Jean Charles Galissard de (1817–1894) *Swiss chemist*

Marignac (ma-ree-**nyak**), who was born in Geneva, Switzerland, was educated in Paris at the Ecole Polytechnique and the School of Mines. He worked in Justus von LIEBIG's laboratory in Giessen for a year and at the Sèvres porcelain factory before his appointment to the chair of chemistry at Geneva (1841). He also became professor of mineralogy in 1845 and held the two posts until his retirement in 1878.

He was known as a careful analyst who carried out many accurate determinations of atomic weights. He was an enthusiastic supporter of PROUT's hypothesis that all elements have an atomic weight that is an integral multiple of the hydrogen atom and defended it from the criticism that refined measurements show it to be false (e.g., chlorine has an atomic weight of 35.4) by claiming it to be sufficiently accurate for the practical calculations of chemistry.

Marignac discovered silicotungstic acid in 1862 and was the first to isolate ytterbium (1878). He also codiscovered gadolinium (1880).

Mariotte, Edmé (*c.* 1620–1684) *French physicist*

> The mind of this man was highly capable of all learning, and the works published by him attest to the highest erudition.
> —J.-B. du Hamel, *Regiae scientiarum academiae historia* (1701; History of the Royal Academy of Sciences)

> Mariotte took everything from me … .He should have mentioned me. I told him that one day, and he could not respond.
> —Christiaan Huygens, accusing Mariotte of plagiarism (1690)

Mariotte (ma-ree-**ot**) was born at Dijon in France. A priest, he was prior of Saint-Martin-sous-Beaune and also a founder member of the French Academy of Sciences.

In 1662 Robert BOYLE had stated his law that the pressure and volume of a gas are inversely proportional. In 1676 Mariotte formulated the law quite independently, deriving it in a less inductive manner than Boyle. More importantly, he stated that the law holds only if there is no change in temperature, as he realized that a gas expands with an increase in temperature and contracts with a fall. Mariotte also investigated a large number of other topics connected with pressure. He performed some of the earliest experiments on the physiology of plants showing the high pressure of the sap. In France, Boyle's law is naturally referred to as *Mariotte's law*.

Marius, Simon (1570–1624) *German astronomer*

Born at Guntzenhausen in Germany, Marius (**mah**-ree-uus) was a pupil of Tycho BRAHE. He gained notoriety in his clash with GALILEO in 1614 when he claimed priority in the discovery of the Medicean planets – Jupiter's satellites. A four-year delay in making such a claim gives it little credibility but Marius had one final triumph over Galileo, who had named the new planets after the children of his patron Cosimo, the Grand Duke of Tuscany. Marius suggested the names Io, Europa, Ganymede, and Callisto, names we still use today. He was the first astronomer to observe telescopically the great spiral nebula in Andromeda.

Markov, Andrey Andreyevich (1856–1922) *Russian mathematician*

Born at Ryazan in Russia, Markov (**mar**-kof) studied at the University of St. Petersburg and later held a variety of teaching posts at the same university, eventually becoming a professor in 1893. He was an extremely enthusiastic and effective teacher. His mathematical interests were very wide, ranging over number theory, the theory of continued fractions, and differential equations. It was, however, his work in probability theory that constituted his most profound and enduring contribution to mathematics.

Among Markov's teachers was the eminent Russian mathematician Pafnuti Chebyshev, whose central interest was in probability. One of Markov's first pieces of important research centered on a key theorem of Chebyshev's – "the central limit theorem." He was able to show that Chebyshev's supposed proof of this result was erroneous, and to provide his own, correct, proof of a version of the theorem of much greater generality than that attempted by Chebyshev. In 1900 Markov published his important and influential textbook *Probability Calculus*, and by 1906 he had arrived at the fundamentally new concept of a *Markov chain*. A sequence of random variables is a Markov chain if the two probabilities conditioned on different amounts of information about the early part of the sequence are the same. This aspect of Markov's work gave a major impetus to the subject of stochastic processes.

The great importance of Markov's work was that it enabled probability theory to be applied to a very much wider range of physical phenomena than had previously been possible. As a result of his work a whole range of subjects, among them genetics and such statistical phenomena as the behavior of molecules, became amenable to mathematical probabilistic treatment.

Marsh, James (1794–1846) *British chemist*

Marsh was employed for many years at the Royal Arsenal, Woolwich. In 1829 he became Michael FARADAY's assistant at the Royal Military Academy on a salary of 30 shillings a week, on which he remained until his death. Apart from the invention of electromagnetic apparatus, for which he received the Society of Arts silver medal (1823), he is today best remembered for the test for the detection of arsenic (1836) named for him. Although Marsh was honored by numerous learned bodies for his discovery, he seems not to have gained financially by his work and on his death his wife and children were left in poverty.

Marsh, Othniel Charles (1831–1899) *American paleontologist*

Born in Lockport, New York, Marsh studied at Yale, as well as in Germany. He worked at Yale, from 1866 until his death, as professor of vertebrate paleontology, the first to hold such a post in America.

From 1870 onward Marsh organized and led a number of paleontological expeditions to parts of North America, during which were unearthed a large number of fossils of considerable importance in enlarging the knowledge of extinct North American vertebrates. He was accompanied on such expeditions by William (Buffalo Bill) Cody, who acted as a scout. In 1871 Marsh discovered the first American pterodactyl, as well as remains of Cretaceous toothed birds and ancestors of the horse. These finds were described in a number of monographs, published by the U.S. Government. Marsh's other publications include *Fossil Horses of America* (1874) and *The Dinosaurs of North America* (1896).

Marsh's appointment as head of the U.S. Geological Survey's vertebrate paleontology section in 1882 contributed toward bitter rivalry with his fellow paleontologist Edward COPE.

Marshall, Barry J. (1951–   ) *Australian pathologist*

Born in Kalgoorlie, Western Australia, Marshall studied medicine at the University of Western Australia (1968–74) and obtained his MB BS. He then joined the Royal Perth Hospital as registrar (1977–84) and subsequently was appointed NHMRC Research Fellow in Gastroenterology (1985). In 1986 he moved to the USA as research fellow and professor of medicine at the University of Virginia, becoming professor of research in internal medicine in

1996. The following year he returned to his homeland, as clinical professor of medicine at the University of Western Australia, and in 2003 he was appointed NHMRC Senior Principal Research Fellow.

Marshall is best known for his discoveries concerning the bacterium *Helicobacter pylori*, and his tireless campaigning to convince the medical establishment that it is a major cause of duodenal and stomach ulcers (peptic ulcers) and other stomach disorders. In 1981, while an intern at the Royal Perth Hospital, he started working with the pathologist J. Robin WARREN. In 1979 Warren had reported the presence of a hitherto unknown bacterium in the stomach lining of patients with gastritis (inflammation of the stomach wall). Warren and Marshall studied a large group of patients, and found the bacterium in most cases of stomach ulcer and in all cases of duodenal ulcer. And in 1982, after several months of trying, they managed to culture the spiral-shaped bacterium, which eventually became the first member of a completely new genus, *Helicobacter*.

The duo jointly published their findings in *The Lancet* in 1983, but were met with a welter of scepticism and scorn from the medical community. It had long been thought that no bacteria could survive in the highly acidic conditions inside the stomach. The entrenched position was that stomach ulcers were primarily due to overproduction of acid by glands in the stomach wall, exacerbated by stress or by certain foods. Treatment involved the often lifelong administration of acid suppressant drugs, which were among the best-selling products of the pharmaceutical industry. Marshall became so frustrated with this reception to their work that in 1985 he took the dramatic step of infecting himself with *H. pylori*. One week after swallowing a solution containing the bacteria he developed symptoms of gastritis, including vomiting and pain. He then cured his self-inflicted condition by administering antibiotics. Yet this well-publicized "experiment" was only the beginning of a long road to widespread acceptance by the medical establishment that stomach disorders could be the result of infection instead of being exclusively physiological.

The infection hypothesis was backed up by subsequent work in which Warren and Marshall determined the most effective antibiotic treatments, and demonstrated that elimination of *H. pylori* in patients led to permanent cure of their gastric ulcers. Marshall also discovered a mechanism that enabled the bacteria to survive in the stomach lining. The bacteria synthesized a urease enzyme that converts urea, present in stomach contents, into ammonium and hydrogencarbonate ions, both of which create alkaline conditions in the immediate vicinity of the bacterial cells. Marshall exploited this finding in devising a rapid, convenient, and noninvasive test for the presence of *H. pylori* based on the activity of the urease enzyme. Patients swallow a small amount of radioactively labelled urea; if *H. pylori* is present in an individual, the bacteria break down the urea and radiolabelled carbon dioxide is subsequently detected in the breath of the patient.

Marshall continued to be a vocal champion of the infection hypothesis of gastritis, and in 1994 established the Helicobacter Foundation, to inform people about the diagnosis and treatment of *H. pylori* infection. The findings of Marshall and Warren have transformed medical strategies for dealing with the majority of peptic ulcers, turning a once chronic and often serious condition into a highly treatable one. For his "discovery of the bacterium *Helicobacter pylori* and its role in gastritis and peptic ulcer disease," Marshall received the 2005 Nobel Prize for physiology or medicine, jointly with Robin Warren.

Martin, Archer John Porter (1910–2002)
British chemist

Martin, a Londoner by birth, was educated at Cambridge University, obtaining his PhD in 1936. He worked as a research chemist with the Wool Industries Research Association in Leeds from 1938 to 1946 and with Boots Research Department in Nottingham until 1948, when he joined the Medical Research Council. From 1959 until 1970 Martin was director of Abbotbury Laboratories Ltd.

In 1944 Martin and his colleague Richard SYNGE developed a chromatographic technique that proved indispensable to later workers investigating protein structures. Without this technique the explosive growth of knowledge in biochemistry and molecular biology would have been dampened by prolonged and tedious analyses of complex molecules.

Column chromatography was first invented by Mikhail Tsvett for the analysis of plant pigments in 1906. Martin was trying to isolate vitamin E and developed a new method of separation involving the distribution and separation of molecules between two immiscible solvents flowing in different directions – countercurrent extraction. From this rather cumbersome apparatus evolved the idea of partition chromatography, in which one solvent is stationary and the other moves across it. Martin and Synge tried different substances, such as silica gel and cellulose, to hold the stationary solvent and hit on the idea of using paper. Thus paper chromatography was introduced.

In this process a drop of the mixture to be analyzed is placed at the corner of a piece of absorbent paper, the edge of which is dipped into an organic solvent. This will soak into the paper by capillarity taking with it the components of the mixture to be analyzed to different distances depending on their solubility. In the case of a protein, the identity of the various amino acids can be discovered by comparing positions of the spots with a reference chart. The basic technique is easy to operate, quick, cheap, works on small amounts, and can separate out closely related substances.

For their work Martin and Synge were awarded the 1952 Nobel Prize for chemistry. Martin tended to treat the value of their contribution somewhat dismissively, pointing out that "All the ideas are simple and had peoples' minds been directed that way the method would have flourished perhaps a century earlier."

Marvel, Carl Shipp (1894–1990) *American chemist*

Born in Waynesville, Illinois, Marvel graduated from the Wesleyan University, Illinois, in 1915 and obtained his PhD from the University of Illinois, Urbana, in 1920. Immediately afterward he joined the faculty at Urbana and later served as professor of organic chemistry from 1930 until 1961.

Marvel devoted himself mainly to polymer research, publishing a survey of the subject in his *Introduction to the Organic Chemistry of High Polymers* (1959). He contributed largely to the understanding of vinyl polymers and, with the beginnings of the space program in the mid 1950s, produced many polymers resistant to high temperatures. One of the most successful of these high-temperature polymers was one based on repeating benzimadazole units. During the war Marvel headed the research program into the development of synthetic rubbers. He also worked on the synthesis of organic molecules and in 1930 succeeded in synthesizing the amino acid methionine.

Maskelyne, Nevil (1732–1811) *British astronomer*

Maskelyne, who was born in London, was educated at Westminster School and Cambridge University. He became the fifth Astronomer Royal in 1765 and, from 1782, rector of North Runcton in Norfolk.

Maskelyne spent considerable time trying to solve the problem of determining longitude at sea. His preferred method was by means of lunar observation, since, on a trip to St. Helena in 1761, he had successfully used such a method. To popularize his technique he published his *British Mariner's Guide* in 1763 and started publishing in 1767 the *Nautical Almanac* to provide the necessary information. He was a member of the Board of Longitude, which had been set up in 1714 to decide on the award of the £20,000 prize for a solution to the problem. Perhaps his commitment to his lunar method made him blind to the value of the chronometer invented by John HARRISON, which

he was asked to judge. He refused to recommend it for the award.

Mather, John Cromwell (1946–) *American astrophysicist and cosmologist*

Mather was born at Roanoke, Virginia. He obtained a BA from Swarthmore College, Pennsylvania in 1968 and a PhD from the University of California at Berkeley in 1974. He subsequently became senior astrophysicist at NASA's Goddard Space Flight Center, Greenbelt, Maryland.

After the discovery of the cosmic microwave background radiation in 1964, cosmologists wanted to understand it in quantitative detail, for its fundamental importance in cosmology. They realized that to do so would require a satellite since the atmosphere of the Earth absorbs many of the wavelengths that occur in the spectrum of the radiation, making it impossible to determine whether the spectrum is described by PLANCK's Law of black-body radiation, as predicted by the Big Bang theory.

As a result of an invitation by NASA in 1974 to astronomers and cosmologists to suggest experiments based on satellites in space, the opportunity arose to overcome the difficulties of Earth-based investigations of the cosmic microwave background radiation. This resulted in the COBE (Cosmic Background Explorer) project. John Mather was the leader of this project, which was a collaboration involving more than 1,000 scientists and engineers. Mather also led the specific investigation of whether the radiation is black-body radiation. The COBE satellite was launched on 18 November, 1989. Very soon afterwards it obtained evidence that the background radiation is indeed black-body radiation.

Mather was awarded a share in the 2006 Nobel Prize for physics for his COBE work. He wrote a popular account of this work with John Boslough entitled *The Very First Light*, which was published in 1996.

Mather, Kenneth (1911–1990) *British geneticist*

Born at Nantwich in Cheshire, England, Mather graduated from the University of Manchester in 1931. He then joined the John Innes Horticultural Institution at Merton, Surrey, where the chromosome theory of heredity was then being developed. Here Mather investigated chromosome behavior, especially crossing over, his research being influenced by his association with Cyril DARLINGTON.

Mather gained his PhD in 1933 and then spent a year at the plant breeding institute, Svalöf, Sweden. Experience at Svalöf convinced him that characters that vary continuously through a population are extremely important in breeding work. On his return to England he took up a lectureship at University College, London, under Ronald FISHER, who was developing statistical techniques that could be used to analyze such quantitative variation.

In 1938, after a year with T. H. MORGAN in America, Mather returned to John Innes as head of the genetics department. It was already appreciated that quantitative variation is governed by many genes, each of small effect, and Mather termed such complexes "polygenic systems." He demonstrated that by applying selection to continuously varying characters one could greatly increase the range of variation beyond that found in the normal population. Continuous variation cannot be analyzed satisfactorily by conventional segregation ratios and Mather thus applied statistics to his results, terming this combination "biometrical genetics."

In 1948 Mather became professor of genetics at Birmingham University, where he remained until his appointment as vice-chancellor at Southampton University in 1965. As founder of biometrical genetics he wrote a number of books on the subject, which he greatly developed during his time at Birmingham in collaboration with J. L. Jinks. Mather returned to Birmingham in 1971 as honorary professor of genetics.

Matthews, Drummond Hoyle (1931–1997) *British geologist*

Matthews was educated at Cambridge University, where he obtained his PhD in 1962. After a short period working as a geologist in the Falkland Islands he returned to Cambridge, where he was appointed reader in marine geology from 1971 until 1982 and senior research associate from 1982 until 1990.

In collaboration with Frederick VINE he produced, in 1963, a fundamental paper on magnetic anomalies, *Magnetic Anomalies over Ocean Ridges*, which modified the sea-floor spreading hypothesis of Harry H. HESS.

Matthias, Bernd Teo (1918–1980) *German–American physicist*

> If you see a formula in the *Physical Review* that extends over a quarter of a page, forget it. It's wrong. Nature isn't that complicated.
> —Quoted by A. L. Mackay in *A Dictionary of Scientific Quotations* (1991)

Matthias was born in Frankfurt, Germany, and educated at Rome University and at the Federal Institute of Technology, Zurich, where he obtained his PhD in 1943. He moved to the United States in 1947, became naturalized in 1951, and, after a brief period at the University of Chicago, joined the staff of the Bell Telephone Laboratories. In 1961 he returned to academic life when he was appointed professor of physics at the University of California, San Diego.

Matthias carried out extensive work on superconducting materials. In the early 1950s no existing theory of superconductivity allowed deductions as to which metals were superconductors and at what temperature – their transition point – they became so. Consequently Matthias set out to find such materials by experiment, testing thousands of alloys in the hope that some kind of pattern would emerge. He found that superconductivity depended on the number of outer electrons in the atom; substances with five or seven valence electrons most readily became superconductors and they had transition points furthest above absolute zero. The crystal structure of the solid was another important factor. As a result of these empirical observations, Matthias and his collaborators were able to make new superconducting materials, including a niobium–germanium alloy with a transition temperature of 23 K. Matthias also worked on ferroelectric materials.

Matuyama, Motonori (1884–1958) *Japanese geologist*

Matuyama (ma-too-**yah**-ma), who was born at Uyeda (now Usa) in Japan, was the son of a Zen abbot. He was educated at the University of Hiroshima and the Imperial University in Kyoto, where he was appointed to a lectureship in 1913. After spending the period 1919–21 at the University of Chicago working with Thomas CHAMBERLIN he was made professor of theoretical geology at the Imperial University.

He conducted a gravity survey of Japan during the period 1927–32, extending this to also cover Korea and Manchuria, and studied marine gravity using the Vening–Meinesz pendulum apparatus in a submarine.

Matuyama made a significant discovery of the Earth's magnetic field and announced this in his paper *On the Direction of Magnetization of Basalt* (1929). From studying the remnant magnetization of some rocks he observed that it had appeared that the Earth's magnetic field had changed, even reversing itself in comparatively short times. The period between the late Pliocene and the mid-Pleistocene during which the field appeared to be opposite to present conditions became known as the *Matuyama reversed epoch*. This reversed polarity, particularly as shown by the rocks of the ocean floor, was to prove crucial evidence for the sea-floor spreading hypothesis of Harry H. HESS.

Mauchly, John William (1907–1980) *American computer engineer*

The son of a physicist, Mauchly was born in Cincinnati, Ohio, and educated at Johns Hopkins University, Maryland, where he obtained his PhD in 1932. After teaching physics at Ursinus College, Pennsylvania, from 1933 until 1941, Mauchly joined the staff of the Moore School of Electrical Engineering at Johns Hopkins.

In 1936 Mauchly had become interested in possible connections between sunspots and the weather. Appalled by the amount of statistical data available, Mauchly began to consider whether it could be analyzed automatically. The great breakthrough in Mauchly's ideas came with a visit in 1941 to ATANASOFF, from whom he seems to have learned about arithmetic units made from vacuum tubes operating upon binary numbers. The jump from mechanical devices to electronics was a fundamental step.

At the same time the Ballistic Research Laboratory (BRL) was urgently seeking help to calculate the thousands of trajectories needed for their artillery. They were therefore willing to invest $500,000 in Mauchly's 1942 proposal to build what later became known as ENIAC (Electronic Numerator Integrator and Calculator). It was completed in late 1945 and was used by the BRL until 1955. ENIAC contained over 17,000 vacuum tubes, 70,000 resistors, weighed 30 tons, and consumed 174,000 watts. Fans were needed to dissipate the heat generated. But it was fast and, although the process was lengthy, it was also programmable.

Mauchly had collaborated on the project with ECKERT. On its completion in 1945, they were already thinking of the next major advance, namely, the stored program. Their proposal to build EDVAC (Electronic Digital Variable Computer), a computer with a stored program, was financed by the Ordnance Department. They soon ran into problems with the Moore School and with John von NEUMANN, who competed with them for the patent rights to the stored program. Mauchly and Eckert resigned from the Moore School, which immediately lost its dominant role in American computer science. The EDVAC was finally completed in 1952, although by then the first stored-program computer had already been built by Maurice WILKES in England.

After leaving the Moore School, Mauchly and Eckert set up their own business, the Electronic Control Company, and proposed to build a new computer, UNIVAC (Universal Automatic Computer). They were originally financed by the U.S. Census Bureau for which they rashly contracted to build the first UNIVAC for the fixed sum of $300,000. For a while things went well; another five orders were gained. But the constraints of a fixed-price contract led them to the verge of bankruptcy.

They were rescued by Henry Strauss, Chairman of the American Tote, who had dreams of controlling the totalizators of the racetracks of America through a linked net of UNIVACs. He invested $500,000 in the ailing company. It was insufficient, and when Strauss died in a plane crash in 1949 the Tote withdrew its backing. Another problem emerged when Army Intelligence denied Mauchly security clearance on the grounds that he had been connected with various supposedly Communist "front" organizations. As a result the company lost valuable military orders. Consequently Mauchly and Eckert sold out to Remington Rand in 1950. Long after the damage had been done, Mauchly's security clearance was restored in 1958.

Mauchly and Eckert remained with Remington and managed the production and sale of 46 UNIVACs. It was indeed a UNIVAC that publicly and successfully predicted in 1952 on television the outcome of the U.S. presidential election. But before long, against competition from IBM's 700 range, UNIVAC would become obsolete.

In 1964 Mauchly's patent was finally awarded to him, 17 years after the case had begun. The victory, however, was only temporary. Remington Rand had bought the patent with the company and consequently tried to collect on their new asset. The Honeywell company challenged this and in a case decided in 1972 it was judged that the rights to the invention of the stored program lay with Atanasoff.

Maudslay, Henry (1771–1831) *British engineer and inventor*

Maudslay was the son of a workman from Woolwich, London. He was apprenticed to the locksmith Joseph BRAMAH, from whom he learned about precision metalwork. His skills soon became recognized and he was made Bramah's foreman.

Maudslay later set up his own business and over a period of 30 years he invented many pieces of machinery that were fundamentally important to the Industrial Revolution, including a screw-cutting lathe. Other inventions included machines for printing cloth, desalinating seawater, and making measurements accurate to one ten-thousandth of an inch. He was the first person to realize the importance of accurate plane surfaces for guiding tools.

Maunder, Edward Walter (1851–1928) *British astronomer*

Maunder, who was born in London, took some courses at King's College there but did not obtain a degree. After working briefly in a bank he became photographic and spectroscopic assistant at the Royal Observatory, Greenwich, in 1873. Maunder's appointment allowed Greenwich to branch out from purely positional work, for Maunder began a careful study of the Sun, mainly of sunspots and related phenomena. After 1891 he was assisted by Annie Russell, a Cambridge-trained mathematician, who must have been one of the first women to be so employed. She became his wife in 1895.

It had been known since 1843 that the intensity of sunspot activity went through an 11-year cycle. In 1893 Maunder, while checking the cycle in the past, came across the surprising fact that between 1645 and 1715 there was virtually no sunspot activity at all. For 32 years not a single sunspot was seen on the Sun and in the whole period fewer sunspots were observed than have occurred in an average year since. He wrote papers on his discovery in 1894 and 1922 but they aroused no interest.

More sophisticated techniques developed in recent years have established that Maunder was undoubtedly correct in the detection of the so-called *Maunder minimum*. Also, the realization that the period of the minimum corresponds to a prolonged cold spell suggests that Maunder's discovery is no mere statistical freak. It may throw light on the Sun's part in long-term climatic change and on possible variations in the processes within the Sun that produce the sunspots.

Maupertuis, Pierre-Louis Moreau de
(1698–1759) *French mathematician, physicist, and astronomer*

> These laws [of movement], so beautiful and so simple, are perhaps the only ones which the Creator and Organizer of things has established in matter in order to effect all the phenomena of the visible world.
> —*Works*, Vol. I (1756)

Maupertuis (moh-pair-**twee**), who was born at Saint-Malo in northwest France, joined the army as a youth, leaving in 1723 to teach mathematics at the French Academy of Sciences in Paris. He traveled to England in 1728 where he became an admirer of Isaac NEWTON's work and was made a member of the Royal Society of London. He was responsible for introducing Newton's theories on gravitation into France on his return.

In 1736 Maupertuis led an expedition to Lapland to verify Newton's hypothesis that the Earth is not perfectly spherical by measuring the length of a degree of longitude. This was successful and as a result Maupertuis was invited by Frederick the Great to join his Academy of Sciences in Berlin.

Maupertuis is best known as being, in 1744, one of the first to formulate the principle of least action, which was published in his *Essai de cosmologie* (1750; Essay on Cosmology). A similar principle had previously been formulated by Leonhard EULER as a result of his mathematical work on the calculus of variations, whereas Maupertuis had been led to formulate his version of the principle through his work in optics. In particular Maupertuis's attention was drawn to the need for such a principle by his interest in the work of Willebrord

SNELL and Pierre de FERMAT. Fermat had shown how to explain Snell's law of refraction, which describes the behavior of a ray of light at the boundary of two media of different densities on the assumption that a ray of light takes the least time possible in traveling from the first medium to the second. However, Fermat's explanation implied that light travels more slowly in a denser medium and Maupertuis set out to devise an explanation of Snell's law that did not have this, to him, objectionable consequence. Maupertuis thought of his principle as the fundamental principle of mechanics, and expected that all other mechanical laws ought to be derivable from it. He attempted to derive from his principle a proof of the existence of God.

Maupertuis was not a mathematician of Euler's stature and his version of the principle was not as precisely formulated mathematically. However, his attempts to apply it to a much wider range of problems made it an influential formative principle in 18th- and 19th-century physical thinking. Joseph LAGRANGE, in his work on the calculus of variations, dispensed with the teleological and theological trimmings Maupertuis had given the principle and found wide application for it in mechanics. Subsequently the principle became less influential until it was revived and refined by William Hamilton.

Maupertuis, who had a quarrelsome character, became involved in violent controversy over the principle. Samuel König, another scientist at Frederick's court, claimed that it had been formulated earlier by Gottfried LEIBNIZ. Maupertuis found himself on the receiving end of some of VOLTAIRE's most biting satires, and eventually he was hounded out of Berlin.

Maury, Antonia Caetana de Paiva Pereira (1866–1952) *American astronomer*

> In my opinion the separation by Antonia C. Maury of the c-and ac-stars is the most important advancement in stellar classification since the trials by [Herman Karl] Vogel and Pietro Angelo [Secchi].
> —Ejnar Hertzsprung, letter to William Henry Pickering, 22 July 1908

Maury, who was born at Cold Spring-on-Hudson, New York, came from a family with a distinguished scientific background. She was a cousin of Matthew Maury, the oceanographer, a niece of Henry DRAPER, the physician and astronomer after whom the Harvard star catalog was named, her sister became a paleontologist, while her father, a clergyman, was also a well-known naturalist. She herself was educated at Vassar, graduating in 1887, and in 1889 became an assistant to Edward Pickering at Harvard College Observatory. There she worked alongside an unusually large collection of women astronomers of whom the most eminent were Annie CANNON and Henrietta LEAVITT.

Apart from the years 1899–1908, when she lectured at various eastern colleges, she retained her position at Harvard until her retirement in 1935.

Much of her work was on the classification of stellar spectra for the Harvard catalog. At about the same time that Cannon was revising the system of spectral classification of stars, Maury proposed an additional modification that turned out to be of permanent significance. It was important to notice, she argued, not just the absence or presence of a particular spectral line but also its appearance. Stars with normal lines she marked "a," those with hazy lines "b," and those that were sharp she marked "c"; intermediate cases were marked "ab" or "ac." This has been described as the first step in using spectroscopic criteria for the luminosities of stars. Maury's spectral classifications, including those of 681 bright northern stars, were published in 1896 in the Harvard *Annals*. Ejnar Hertzsprung was quick to see the significance of her classification system and in 1905 pointed out that c-type and ac-type stars were brighter than a- or b-type stars and were in fact giants.

Maury spent many years studying and detecting spectroscopic binary stars and as early as 1889 had determined the period of Mizar, in Ursa Major. This was the first spectroscopic binary to be discovered, identified by Pickering earlier in the year. The two stars in a spectroscopic binary cannot be resolved visually but as they revolve they will each alternately approach and recede from an observer on the Earth. This causes an alternate lengthening and shortening of the emitted light waves and will produce a periodic doubling of the spectral lines. Maury's particular interest was the binary Beta Lyrae, the investigation of which she continued long after her retirement.

Maury, Matthew Fontaine (1806–1873)
American oceanographer

Maury, who was born in Fredericksburg, Virginia, graduated from Harpeth Academy in 1825 and joined the U.S. Navy as a midshipman. A leg injury in 1839 ended his sea career but, having made his reputation by his publication in 1836 of his *Treatise on Navigation*, he was chosen in 1842 to be superintendent of the Depot of Charts and Instruments in Washington. This post carried with it the directorship of the U.S. Naval Observatory and Hydrographic Office. Maury largely ignored astronomical work, emphasizing instead the study of oceanography and meteorology, and consequently aroused the opposition of the scientific establishment centered upon Joseph HENRY and Alexander BACHE.

He resigned his position with the outbreak of the American Civil War (1861) to become a commander in the Confederate Navy. After the war he took on, in 1865, the post of Imperial Commissioner for Immigration to the doomed Emperor Maximilian of Mexico to establish a confederate colony. Following the collapse of the Mexican Empire he spent some time in England writing textbooks before he was permitted to return to America where he became, in 1868, professor of meteorology at the Virginia Military Institute, remaining there until his death.

Maury has often been described as the father of oceanography. He wrote one of the earliest works on the topic, *The Physical Geography of the Sea* (1855), and he demonstrated the rewards to be gained from an increased knowledge of the oceans. From 1847 he began to publish his *Wind and Current* pilot charts of the North Atlantic, which could shorten sailing times dramatically. Claims were made that as much as a month could be saved on the sailing time for the New York–California voyage. This knowledge was acquired by the study of especially prepared logbooks and the collection of data in a systematic way from a growing number of organized observers.

After 1849 Maury had the use of two research vessels and began a study of ocean temperature and a collection of samples of the ocean floor. He was thus able to publish his *Bathymetrical Map of the North Atlantic Basin* (1854) showing a profile of the Atlantic floor between Yucatan and Cape Verde.

Maxwell, James Clerk (1831–1879)
British physicist

> I am very busy with Saturn on top of my regular work. He is all remodelled and recast, but I have more to do to him yet for I wish to redeem the character of mathematicians and make it intelligible.
> —Letter to J. R. Droop (1857). Quoted by L. Campbell and W. Garnett in *The Life of James Clerk Maxwell* (1882)

Maxwell was born in Edinburgh and studied at the university there (1847–50) and at Cambridge (1850–54), becoming a fellow in 1855. He was professor of natural philosophy at Marishal College, Aberdeen, from 1856 until 1860, when he became professor of natural philosophy and astronomy at King's College, London. He resigned in 1865 and worked on his estate in Scotland researching and writing. From 1871 he was professor of experimental physics at Cambridge.

Maxwell is regarded as one of the great physicists of the 19th century. At the age of 15 he produced a paper on methods of drawing oval curves. In 1857 he published a paper on the rings of Saturn, in which he analyzed the dynamics of the rings and proved that they could not be wholly solid or liquid. His own theory was that they were made up of many particles, and he showed that such a system would be stable.

Maxwell is regarded as one of the founders of the kinetic theory of gases – the calculation of the properties of a gas by assuming that it is composed of a large number of atoms (or molecules) in random motion. Maxwell, around 1860, put forward a statistical treatment of gases in *Illustrations of Dynamical Theory of Gases*. Maxwell and Ludwig BOLTZMANN obtained a formula for the way in which the speeds of molecules were distributed over the number of molecules – the *Maxwell–Boltzmann distribution law*. The kinetic theory of gases disposed of the idea of heat as a fluid ("caloric").

One interesting notion coming out of his work on the kinetic theory was the statistical interpretation of thermodynamics. A particular point was the idea of *Maxwell's demon* (1871) – a small hypothetical creature that could open or close a shutter between two compartments in a vessel, separating the fast molecules from the slow ones, and thus causing one part of the gas to become hotter and the other colder. The system would appear to violate the second law of thermodynamics. (In fact it does not; the gas decreases in entropy but there is an increase in entropy in the demon, using the idea that entropy is connected with "information.")

Maxwell's greatest work was his series of papers on the mathematical treatment of the lines of force introduced by Michael FARADAY to visualize electromagnetic phenomena. He showed the connection between magnetism and electricity and demonstrated that oscillating electric charges would produce waves propagated through the electromagnetic field. He showed that the speed of such waves was similar to the experimentally determined speed of light, and concluded that light (and infrared and ultraviolet radiation) was in fact this electromagnetic wave. Maxwell went on to predict the existence of other forms of electromagnetic radiation with frequencies and wavelengths outside the infrared and ultraviolet regions. Heinrich HERTZ first detected radio waves in 1888. Maxwell's theory was developed further by Hendrik LORENTZ.

In *Dynamical Theory of the Electric Field* (1864) Maxwell put forward four famous differential equations (known simply as *Maxwell's equations*) describing the propagation of electromagnetic waves. These equations contain the speed of the waves, c, a value that is independent of the velocity of the source. This was one of the facts that led EINSTEIN to his special theory of relativity. Maxwell also wrote *Treatise on Electricity and Magnetism* (1873).

May, Robert McCredie (1936–) *Australian–American theoretical ecologist*

Not only in research, but in the everyday world of politics and economics we would all be better off if more people realized that simple systems do not necessarily possess simple dynamical properties.
—*Nature* No. 261 (1976)

May was educated at the university in his native city of Sydney, obtaining his PhD in theoretical physics in 1960. He then taught applied mathematics at Harvard and physics at Sydney from 1962 until 1973, when he was appointed professor of zoology at Princeton. In 1988 he was appointed Royal Society research professor of zoology at Oxford University and Imperial College, London. Since 1995 he has also served as chief scientific adviser to the British government. He was knighted in 1996 and made a life peer in 2001. He was president of the Royal Society (2000–2005).

Whereas an earlier generation of theoretical physicists had become interested in biology through their desire to understand the molecules of life, May and his generation were more attracted to the problem of understanding the abundance and distribution of species. Greatly influenced by the pioneering work of Robert MACARTHUR, they could see tempting analogies between the flow of energy in a physical system and the structure and growth of an ecosystem.

May has worked on a number of detailed problems on the population dynamics of various species but is perhaps best known for his influential *Theoretical Ecology* (1976).

Mayer, Julius Robert von (1814–1878) *German physician and physicist*

Mayer (**mI**-er), the son of an apothecary from Heilbronn in Germany, studied medicine at the University of Tübingen, where he seems to have been a mediocre student. He continued his studies abroad in Vienna and Paris before accepting an appointment in 1840 as a ship's physician on a vessel bound for Java. On his return in 1841 he settled in Heilbronn working as a general practitioner.

When Mayer sailed to Java he was familiar with the views of Antoine LAVOISIER that animal heat is produced by slow combustion in the body. Being forced to bleed some of the crew at Surabaya, he found that venous blood was surprisingly bright. Indeed, at first he thought that he had cut an artery by mistake. "This phenomenon riveted my earnest attention," he reported, drawing the correct conclusion that the blood was redder because in the tropics the body does not need to burn as much oxygen to maintain body temperature as it does in temperate regions. The observation led Mayer to speculate about the conversion of food to heat in the body, and also the fact that the body can do work. He came to the view that heat and work are interchangeable – that the same amount of food can be converted to different proportions of heat and work, but the total must be the same.

Moreover, Mayer appreciated that this equivalence should hold universally and tried to apply it to other systems and to make it quantitative. Unfortunately, at the time he was confused about such concepts as force and work and his ideas were presented in an obscure metaphysical style. His first paper on the subject was sent to *Annalen der Physik* (Annals of Physics); the editor, Johann Poggendorf, did not even acknowledge Mayer's letter. The paper was published in 1842 by Justus von LIEBIG in the journal *Annalen der Chemie und Pharmazie* (Annals of Chemistry and Pharmacy). The paper was almost totally ignored and Mayer published, in 1845, a pamphlet at his own expense – *Organic Motion Related to Digestion* – which fared no better than his paper.

In his arguments Mayer used the specific heat capacities of gases, i.e., the heat required to produce unit temperature rise in unit mass of gas. It was known that the specific heat capacity of a gas maintained at constant volume is slightly smaller than that at constant pressure. This difference in heat, for a given quantity of gas, Mayer interpreted as the work done by a gas expanding at constant pressure. He was able to find the amount of work required to produce unit amount of heat – thus obtaining what was later known as the mechanical equivalent of heat (J). He found a weight of 1 gram falling 365 meters corresponds to heating 1 gram of water 1°C. (This is equivalent to a value of J of 3.56 joules per calorie; the modern conversion factor is 4.18 joules per calorie.)

Mayer clearly anticipated James JOULE and Hermann von HELMHOLTZ in the discovery of the law of conservation of energy. The lack of recognition seems to have affected him strongly, for in the early 1850s he attempted suicide. His work was eventually recognized and he received many honors, including the Rumford medal of the Royal Society (1871).

Maynard Smith, John (1920–2004) *British biologist*

> The mixture of intellect and blasphemy was absolutely overwhelming and I've been attracted to that all the rest of my life.
> —On J. B. S. Haldane's *Possible Worlds*. Quoted by L. Wolpert and A. Richards in *A Passion for Science* (1988)

Maynard Smith was educated at Cambridge University, where he qualified as an engineer in 1941. He spent the next six years designing aircraft before deciding they were "noisy and old-fashioned" and moving to University College, London, to study zoology under J. B. S. Haldane. After obtaining his BSc in zoology in 1951 he remained as a lecturer in zoology until 1965 when he was appointed professor of biology at the University of Sussex. He became emeritus professor in 1985.

Maynard Smith, known to a wide public for his lucid *Theory of Evolution* (1958), became one of the leading evolutionary theorists. Much influenced by W. D. Hamilton and Robert MACARTHUR, and using concepts taken from the theory of games formulated by John VON NEUMANN in the 1940s, he introduced in the 1970s the idea of an evolutionarily stable strategy (ESS). Assuming that two animals are in conflict, then an ESS is one that, if adopted by the majority of the population, prevents the invasion of a mutant strategy. Stable strategies by definition thus tend to be mixed strategies.

Much of Maynard Smith's work on ESS was published in his *Mathematical Ideas in Biology* (1968). He also discussed why sexual modes of reproduction predominate over other means in *The Evolution of Sex* (1978). Maynard Smith continued to write on evolutionary theory in such works as *Evolutionary Genetics* (1989) and *The Major Transitions of Evolution* (1995).

Mayow, John (1640–1679) *English physiologist and chemist*

> Some learned and knowing men speak very slightly of the *Tractatus quinque* [Fifth Treatise] of J. M. [John Mayow] ... a particular friend of yours and mine told me yesterday, that as far as he had read him, he would shew to any impartial and considering man more errors than one on every page.
> —Henry Oldenburg. Letter to Robert Boyle, July 1674. Quoted in Robert Boyle, *Works*, Vol. VI (1772)

Mayow was born at Morvah in Cornwall, England, and educated at Oxford University. He became a doctor of law in 1670 but then turned to medicine.

In 1674 he published his *Tractatus quinque* (Fifth Treatise) in which he came close to discovering the composite nature of the atmosphere. He showed that if a mouse is kept in a closed container over water then the air in the container will diminish in quantity, its properties change, and the water will rise up into the container. The same effect, he realized, could be produced by burning a candle. He further pointed out that combustion and respiration stopped before all the air was used up. He preceded Joseph PRIESTLEY and Antoine LAVOISIER by about a hundred years with his discoveries relating to respiration and combustion.

Mayr, Ernst Walter (1904–2005) *German–American zoologist*

Born at Kempten in Germany, Mayr (mIr) was educated at the universities of Griefswald and Berlin, where he obtained his PhD in 1926. He then served as assistant curator of the museum there before moving to America in 1932. After spending many years at the Museum of Natural History, New York, he moved to Harvard in 1953 to serve as Agassiz Professor of Zoology, in which post he remained until his retirement in 1975.

As a field zoologist Mayr worked extensively on the birds of the Pacific. Beginning with his *New Guinea Birds* (1941), he published a number of surveys and monographs on the ornithology of the area.

He was, however, better known for such works as *Systematics and the Origin of Species* (1942), *Animal Species and Evolution* (1963), and *The Growth of Biological Thought* (1982) in which, at the same time as such other scholars as George SIMPSON and Theodosius DOBZHANSKY in America and Julian HUXLEY in Britain, he attempted to establish a neo-Darwinian synthesis. The enterprise has continued to hold together fairly well against the onslaughts of such critics as Motoo KIMURA and has so far absorbed, without major upset, the massive inflow of data from the discipline of molecular biology.

McAdam, John Loudon (1756–1836) *British engineer*

In 1770 McAdam traveled from his native Ayrshire, in Scotland, to New York City, where he became a successful merchant and made his fortune. He then returned to Ayrshire (1783) and bought an estate there, using his own money to experiment with roadbuilding in the surrounding district.

In 1798 McAdam was given a government appointment in Falmouth, Cornwall. He recommended that roads should be raised, so that water could drain away, and should be built with large rocks covered over with smaller ones bound by slag. Bitumen is now used instead of slag, but the basic principle is the same in modern roads. He was appointed surveyor general of the Bristol roads in 1815 and there put his ideas on roadbuilding into practice.

McAdam published several papers and essays on roadbuilding, including *Remarks on the Present System of Road-Making*. In 1823, after a parliamentary enquiry, his recommendations were adopted by the public authorities. The methods spread and improved travel throughout the world. In 1827 he was appointed general surveyor of roads in Britain.

McCarthy, John (1927–　　) *American computer scientist*

Born in Boston, Massachusetts, McCarthy was educated at the California Institute of Technology and at Princeton, where he obtained his PhD in 1951. After holding junior posts successively at Princeton, Stanford, and Dartmouth, McCarthy joined the Massachusetts Institute of Technology in 1957 and immediately set up the Artificial Intelligence (AI) Laboratory. In 1962 he took up the post of professor of computer science at Stanford University. In

the following year he founded the Stanford AI Laboratory. He retired in 2000.

In many ways McCarthy is the founding father of AI. He coined the term in the mid 1950s, organized the 1956 Dartmouth conference, which largely defined the new discipline, and set up the first AI Laboratory in 1957. He was also the inventor (1956–58) of the computer language LISP (List Processing), in which much AI work is pursued. McCarthy was also responsible during the period between 1957 and 1962 for the development of time-sharing in computers.

McCarthy's position on AI has always remained radical and even extreme. Acts of intelligence, he has argued, can be reduced in all cases to propositions expressible in purely logicomathematical terms. Thus mistakes in AI arise not when logic is relied upon too much, but when the program is not rigorous enough. To avoid this it is necessary to formalize such concepts as causation, knowledge, and belief; these tasks have so far eluded McCarthy and his followers.

McCarty, Maclyn (1911–2005) *American microbiologist and geneticist*

McCarty was born at South Bend, Indiana, and educated at Stanford and Johns Hopkins School of Medicine, where he obtained his MD in 1937. After holding a junior appointment there he moved in 1940 to the Rockefeller Institute where he remained until his retirement in 1974, serving as physician-in-chief from 1961 and as vice-president from 1965.

In 1944 McCarty collaborated with Oswald AVERY and Colin MACLEOD in the experiment that first clearly revealed the transforming power of DNA. McCarty went on to work on streptococci and rheumatic fever.

McClintock, Barbara (1902–1992) *American geneticist*

The daughter of a physician, McClintock was born in Hartford, Connecticut, and educated at Cornell's College of Agriculture, where she received her PhD in 1927 for work in botany. She remained at Cornell until 1936 supported by various grants from the National Research Council and the Guggenheim Foundation. But there was no future at Cornell for her as, until 1947, only the department of home economics appointed women professors. Fortunately a new genetics department was being set up in the University of Missouri by Craig Stadler, who knew and admired her work, and she was offered a post as assistant professor there, although it was made clear to her that any further advancement would be unlikely. She left in 1941, and in 1944 was elected to the National Academy of Sciences, becoming only the third woman to be so honored. McClintock then joined the Carnegie Institute's Cold Spring Harbor Laboratory, New York, where she remained until her death.

By the 1920s MORGAN and other geneticists, working mainly with the *Drosophila* fruit fly, had established that gene action was connected with chromosomes and thereby established the new discipline of cytogenetics. *Drosophila* chromosomes, however, before the discovery of the giant salivary chromosomes by T. Painter in 1931, were too small to reveal much detail. McClintock chose to work with a variety of maize that possessed much more visible chromosomes. Further, the development of new staining techniques allowed McClintock to identify, distinguish, and number the ten maize chromosomes.

Morgan and his group had also demonstrated the existence of "linkage groups" in *Drosophila* – groups of genes, such as those for white eyes and maleness, linked together because the genes themselves were sited near each other on a chromosome. In a series of papers published between 1929 and 1931, McClintock established similar linkage groups in maize. Because maize chromosomes were more visible under the microscope than those of *Drosophila*, McClintock was able to identify the chromosomal changes responsible for a change in phenotype and thus confirmed Morgan's work.

McClintock's own Nobel Prize for physiology or medicine, awarded in 1983, was for later work done on the so-called "jumping genes." In the 1940s at Cold Spring Harbor, McClintock planted her maize and began to track a family of mutant genes responsible for changes in pigmentation. She was struck by the fact that mutation rates were variable. After several years' careful breeding, McClintock proposed that in addition to the normal genes responsible for pigmentation there were two other genes involved, which she called "controlling elements."

One controlling element was found fairly close to the pigmentation gene and operated as a switch, activating and turning off the gene. The second element appeared to be located further away on the same chromosome and was a "rate gene," controlling the rate at which the pigment gene was switched on and off. She further discovered that the controlling elements could move along the chromosome to a different site and could even move to different chromosomes where they would control different genes. McClintock gave a full description of the process of "transposition," as it became known, in 1951, in her paper *Chromosome Organization and Genic Expression*. McClintock's work was largely ignored until 1960, when controlling elements were identified in bacteria by MONOD and Jacob.

McCollum, Elmer Verner (1879–1967)
American biochemist

Born in Fort Scott, Kansas, McCollum was educated at the University of Kansas and at Yale, where he obtained his PhD in 1906. He taught at the University of Wisconsin from 1907 until 1917, when he was appointed to the chair of biochemistry at Johns Hopkins University, a post he retained until his retirement in 1944.

McCollum made a number of advances in the study of vitamins. He was the first, in collaboration with M. Davis in 1913, to demonstrate clearly their multiplicity. They found that rats fed with a diet lacking butterfat failed to develop. They assumed, therefore, the existence of a special factor present in butterfat without which the normal growth process could not take place. As it was clearly fat-soluble, it must be distinct from the antiberiberi factor proposed by Casimir FUNK in 1912, which was water-soluble. McCollum named them fat-soluble–A and water-soluble–B, which later became vitamins A and B. In 1920 McCollum was able to extend the alphabet further by naming the antirachitic factor found in cod-liver oil vitamin D (C had already been appropriated to describe the antiscorbutic factor).

McCollum wrote widely on the subject of nutrition, his books including a standard text of the subject, *Newer Knowledge of Nutrition* (1918), which went through many editions, and *A History of Nutrition* (1957).

McConnell, Harden Marsden (1927–   )
American theoretical chemist and biochemist

McConnell, who was born in Richmond, Virginia, graduated from George Washington University in 1947 and received his doctorate from the California Institute of Technology in 1951. He served as professor of chemistry there from 1959 until 1964, when he was appointed to a similar chair at Stanford University. He is now emeritus professor at Stanford.

McConnell's early work was on the movement of electrons in unsaturated hydrocarbons. He found a way of measuring the electron density on carbon atoms in these molecules. He has done extensive work on the theory of nuclear magnetic resonance spectroscopy in organic compounds, and also on electron spin resonance (in which absorption is by unpaired electrons rather than nuclei). Studies of the methyl radical (CH_3) by McConnell and his group have demonstrated that it is almost planar.

His work on magnetic resonance spectroscopy led to the introduction of spin labeling, a method in which biological compounds are labeled by adding a stable organic free radical. This converts normally nonparamagnetic systems to paramagnetic systems, which display resonance. Important facts about the structure of compounds can be obtained by examining this resonance. Thus spin labeling helped McConnell demonstrate that the phospholipids found in biological membranes have the properties of a liquid and so allow rapid transfer of molecules. Work on the double lipid layers of membranes has shown that its physical properties play a large part in the immune response. Such work has since become part of a major research effort based on several research centers.

McKusick, Victor Almon (1921–)
American physician and geneticist

The son of a dairy farmer, McKusick was born in Parkman, Maine, and educated at Tufts University, Medford, Massachusetts. He went on to Johns Hopkins University, Baltimore, where he qualified as an MD in 1946 and where he trained as a cardiologist. McKusick has remained at Johns Hopkins and has served as professor of medicine from 1960 and, from 1985, as professor of medical genetics. He was awarded the 2008 Japan Prize for Medical Genetics and Genomics.

The name of McKusick has become identified with a book, an encyclopedic listing of human gene loci, titled *Mendelian Inheritance in Man*, but more commonly referred to as MIM. In 1911 E. B. Wilson identified the first gene locus – the gene for color blindness on the X-chromosome. Over the years other genes were identified and located. By 1966, when the first edition of MIM appeared, 68 genes had been mapped on the X-chromosome. McKusick went on to describe a total number of 1,487 human genes. By the eighth edition of MIM, published in 1988, improved techniques in cytogenetics allowed the identification of 4,344 gene loci. Two further editions of MIM, in 1990 and 1992, have continued to add to the total.

McLennan, Sir John Cunningham
(1867–1935) *Canadian physicist*

Born the son of a miller and grain dealer from Ingersoll in Ontario, Canada, McLennan was educated at the University of Toronto. After graduating in 1892 he immediately joined the staff and later served as professor of physics from 1907 until his retirement in 1932. He then moved to England where he began, shortly before his death, to work on the use of radium in the treatment of cancer.

He had earlier collaborated with Ernest RUTHERFORD during World War I on work in combating submarines. He is, however, best remembered for his work on the auroral spectrum. The existence of a green line with a wavelength of 5,377 angstroms without any apparent source had long puzzled scientists. In 1924, in collaboration with A. Shrumm, McLennan succeeded in obtaining similar lines in the laboratory from atomic oxygen, thus indicating their source.

It was also under McLennan that Toronto became a leading center for research into low-temperature physics. Helium was first liquefied there in 1923 when they reported the curious behavior of helium as its temperature dropped below 2.2 K. Without being aware of it they had observed the onset of superfluidity at the so-called lambda point where, as was to be reported in 1930 by W. Keesom, the heat conduction of helium suddenly increases enormously.

McMillan, Edwin Mattison (1907–1991)
American physicist

Born in Redondo Beach, California, McMillan was educated at the California Institute of Technology and at Princeton, where he obtained his PhD in 1932. He took up an appointment at the University of California at Berkeley in 1935, being made professor of physics in 1946 and director of the Lawrence Radiation Laboratory in 1958, posts he held until his retirement in 1973.

In 1940 McMillan and Philip ABELSON announced the discovery of the first element heavier than uranium. The new element had a mass number 93 and a relative atomic mass of 239. It was named neptunium after the planet Neptune, just as 150 years earlier Martin KLAPROTH had named uranium after the planet Uranus. McMillan also suspected the existence of element 94 and in the same year was proved right by the discovery of the new element (plutonium) by Glenn SEABORG with whom he was to share the 1951 Nobel Prize for chemistry. The new elements were produced when uranium was bombarded with neutrons and were detected by virtue of their characteristic half-life.

McMillan also made a major advance in the development of Ernest Lawrence's cyclotron, which, in the early 1940s, had run up against a theoretical limit. Lawrence found that as his particles accelerated beyond a certain point their increase in mass, as predicted by EINSTEIN's theory of relativity, was putting them out of phase with the electric impulse they were supposed to receive inside the cyclotron.

In 1945 McMillan proposed a neat solution in the synchrocyclotron (also independently suggested by Vladimir Veksler) in which the fixed frequency of the cyclotron was abandoned. The variable frequency of the synchrocyclotron could thus be adjusted to correspond to the relativistic mass gain of the accelerating particles and once more get into phase with them. In this way accelerators could be built that were forty times more powerful than Lawrence's most advanced cyclotron.

Mechnikov, Ilya Ilych (1845–1916) *Russian microbiologist*

Mechnikov was born in Kharkov (now in Ukraine) and studied natural sciences at Kharkov University. He first worked on a study of marine fauna in Heligoland in the North Sea. After a series of university posts at Giessen, Göttingen, Munich, and St. Petersburg, he was appointed a professor at the University of Odessa.

Mechnikov had an unhappy married life. His first wife died in 1873 after a long battle against tuberculosis and he attempted suicide by taking opium. He again attempted suicide in 1880, when his second wife became very ill with typhoid, this time by injecting himself with relapsing fever. He survived the illness, resigned his post at Odessa, and set up a private laboratory at Messina. Here Mechnikov studied the immune system and made his greatest discovery, that certain white blood cells could destroy harmful bodies, such as bacteria. He made the discovery while working on starfish. Inserting a thorn into a starfish larva, he noticed that some strange cells were gathering around the thorn and eliminating foreign bodies entering the wound. He named these cells "phagocytes."

Mechnikov returned to Odessa and then moved in 1898 to the Pasteur Institute in Paris. Initially his ideas were scorned but eventually they gained general acceptance. In 1908 he was awarded the Nobel Prize for physiology or medicine, which he shared with Paul EHRLICH.

Medawar, Sir Peter Brian (1915–1987) *British immunologist*

> Scientific discovery is a private event, and the delight that accompanies it, or the despair of finding it illusory does not travel.
> —*Hypothesis and Imagination*

The son of an Englishwoman and a Lebanese businessman trading in Brazil, Medawar was born in Rio de Janeiro and brought to Britain at the end of World War I. He was educated at Oxford, graduating in zoology in 1937, and remained there to work under Howard FLOREY. His first researches concerned factors affecting tissue-culture growth but during World War II he turned his attention to medical biology. He subsequently developed a concentrated solution of the blood-clotting protein fibrinogen, which could be used clinically as a biological glue to fix together damaged nerves and keep nerve grafts in position.

The terrible burns of many war casualties led Medawar to study the reasons why skin grafts from donors are rejected. He realized that each individual develops his own immunological system and that the length of time a graft lasts depends on how closely related the recipient and donor are. He found that grafting was successful not only between identical twins but also between nonidentical, or fraternal, twins. It had

already been shown in cattle that tissues, notably the red-cell precursors, are exchanged between twin fetuses. This led to the suggestion by Macfarlane BURNET that the immunological system is not developed at conception but is gradually acquired. Thus if an embryo is injected with the tissues of a future donor, the animal after birth should be tolerant to any grafts from that donor.

Medawar tested this hypothesis by injecting mouse embryos, verifying that they do not have the ability to form antibodies against foreign tissue but do acquire immunologic tolerance to it. For this discovery Medawar and Burnet were awarded the 1960 Nobel Prize for physiology or medicine.

Medawar moved from Oxford in 1947 to the chair of zoology at Birmingham, a post he held until 1951 when he was appointed professor of zoology at University College, London. In 1962 he accepted the important post of director of the National Institute for Medical Research. For some years he tried to combine his research work with a heavy administrative load. But in 1969 Medawar suffered his first stroke. Although he continued as director until 1971, the stroke had seriously restricted his mobility and dexterity. Despite this he continued his research work on cancer at the Clinical Research Centre, London. A second stroke in 1980, and a third in 1984, brought Medawar's research career to an end.

He continued to write, however, and in 1986 published his autobiography, *Memoirs of a Thinking Radish*, and collected most of his early essays in *Pluto's Republic* (1982). It was in one of these essays, first published in 1964, that Medawar characterized science in a much quoted phrase as "the art of the soluble."

Mees, Charles Edward Kenneth (1882–1960) *British-American photochemist*

> The best person to decide what research shall be done is the man who is doing the research. The next best is the head of the department. After that you leave the field of best persons and meet increasingly worse groups. The first of these is the research director, who is probably wrong more than half the time. Then comes a committee which is wrong most of the time. Finally there is a committee of company vice-presidents, which is wrong all the time.
> —On his experiences as research director at Eastman-Kodak in the 1930s. Quoted in *Biographical Memoirs of Fellows of the Royal Society*, Vol. VII (1961)

Mees, the son of a Wesleyan minister at Wellingborough in Northamptonshire, England, was educated at the University of London where he obtained his doctorate in 1906. After working for a company manufacturing photographic plates he emigrated to America in 1912, to become director of the research laboratory of

Eastman-Kodak at Rochester, New York, a post he held until his retirement in 1946.

In England Mees had worked on the development of plates of high sensitivity. This had attracted the attention of astronomers and in 1912 he was invited to join the staff of the Mount Wilson Observatory. Although he joined Eastman instead, he maintained his contacts with Mount Wilson, sending them, in the early 1930s, some of the fastest plates he had ever produced. In 1935 he organized the large-scale production of Kodachrome, the first color film made for the mass market. He also published *The Theory of the Photographic Process* (1942), a comprehensive and authoritative survey of the state of photographic science.

Meissner, Fritz Walther (1882–1974)
German physicist

In 1933 Meissner (**mIs**-ner), in collaboration with R. Oschenfeld, discovered what has since been known as the *Meissner effect*. He was examining the magnetic properties of materials as they became superconductive, a condition met with as the temperature of the element or compound falls below a critical point, i.e., critical temperature (T_c). It was found, quite unexpectedly, that if a solid lead sphere is placed in a magnetic field and the temperature allowed to fall below the T_c of lead, the magnetic field is expelled from the lead sphere, which becomes perfectly diamagnetic. The presence of the Meissner effect is now used as a routine test for superconductivity.

Meitner, Lise (1878–1968) *Austrian–Swedish physicist*

Meitner (**mIt**-ner), the daughter of a lawyer, was born in Vienna and entered the university there in 1901. She studied science under Ludwig BOLTZMANN and obtained her doctorate in 1906. From Vienna she went to Berlin to attend lectures by Max PLANCK on theoretical physics. Here she began to study the new phenomenon of radioactivity in collaboration with Otto HAHN, beginning a partnership that was to last thirty years.

At Berlin she met with remarkable difficulties caused by prejudice against women in academic life. She was forced to work in an old carpentry shop and forbidden, by Emil FISCHER, to enter laboratories in which males were working. In 1914, at the outbreak of World War I, she became a nurse in the Austrian army, continuing work with Hahn during their periods of leave. In 1918 they announced the discovery of the radioactive element protactinium.

After the war Meitner returned to Berlin as head of the department of radiation physics at the Kaiser Wilhelm Institute. Here she investigated the relationship between the gamma and beta rays emitted by radioactive material.

In 1935 she began, with Hahn, work on the transformation of uranium nuclei under neutron bombardment. Confusing results had been obtained earlier by Enrico FERMI.

But by this time she was beginning to fear a different sort of prejudice. Following Hitler's annexation of Austria in 1938 she was no longer safe from persecution and, like many Jewish scientists, left Germany. With the help of Dutch colleagues she found refuge in Sweden, obtaining a post at the Nobel Institute in Stockholm. Hahn, with Fritz Strassman, continued the uranium work and published, in 1939, results showing that nuclei were present that were much lighter than uranium. Shortly afterward Lise Meitner, with Otto FRISCH (her nephew), published an explanation interpreting these results as fission of the uranium nuclei. The nucleus of uranium absorbs a neutron, and the resulting unstable nucleus then breaks into two fragments of roughly equal size. In this induced fission, two or three neutrons are ejected. For this she received a share in the 1966 Enrico Fermi Prize of the Atomic Energy Commission.

Lise Meitner became a Swedish citizen in 1949 and continued work on nuclear physics. In 1960 she retired to Cambridge, England.

Mela, Pomponius (about 44 AD) *Iberian geographer*

> There is no definite decision whether this is the action of the universe through its own heaving breath ... whether there exist some cavernous depressions for the ebb-tides to sink into,...whether the moon is responsible for currents so extensive.
> —On the action of the tides. *De chorographia* (43 AD; Concerning Chorography)

Mela (**mee**-la) probably came from southern Spain. He was the author of *De situ orbis* (A Description of the World), also known as *De chorographia* (Concerning Chorography), which is the first extant Latin geographical work.

The book borrowed largely from earlier Greek works and was simply a descriptive account of the lands surrounding the Mediterranean, with a brief description of the rest of the known world added. Mela retained the surrounding Oceanus of earlier maps and saw the continents of Africa, Asia, and Europe indented by the Caspian, Arabian, Persian, and Mediterranean seas. The Earth was divided into five zones – two arctic, one tropical, and two temperate – only the temperate zones being capable of supporting life. He seems to have been the first to realize that Spain juts out into the Atlantic. According to Strabo its coastline continued down from France in a straight line.

Mellanby, Sir Edward (1884–1955) *British physiologist*

Mellanby, the son of a shipyard manager from West Hartlepool in northeast England, was educated at Cambridge University and at St. Thomas's Hospital, London, where he completed his medical studies. In 1913 he was appointed to the chair of physiology at King's College for Women (now Queen Elizabeth College), London. He remained there until 1920, when he moved to the chair of pharmacology at Sheffield University. However, he resigned this post in 1933 to take up the influential position of secretary to the Medical Research Council (MRC), which he held until his retirement in 1949.

In 1918 Mellanby fed puppies a variety of diets containing all the vitamins then known but found that they developed rickets. If, however, cod-liver oil or butter were added to the diet no symptoms appeared. This led him to suggest that the fat-soluble vitamin A in cod-liver oil and the antirachitic factor were identical.

However in 1922 Elmer McCOLLUM showed that it was not the vitamin A in cod-liver oil that prevented rickets developing: when the oil was heated it would continue to cure xerophthalmia, an eye complaint due to vitamin A deficiency, but lost all antirachitic properties. He therefore proposed the existence of a fourth vitamin, D, later shown to be calciferol.

In fact the etiology of rickets turned out to be more complicated, since vitamin D can be made in the skin from its precursor, 7-hydrocholesterol, in the presence of sunlight. Thus rickets can be cured either by exposure to sunlight or by administration of calciferol in cod-liver oil.

While at the MRC Mellanby's most significant work was his support of Howard FLOREY in the work of isolating penicillin. It was largely owing to Mellanby that Florey was appointed to the Oxford professorship of pathology in 1935. Once there, the research that led to the isolation and extraction of penicillin was partly financed by the MRC on the recommendation of Mellanby.

Mello, Craig Cameron (1960–) *American molecular biochemist*

Mello was born in Worcester, Massachusetts, the son of a paleontologist. He was awarded a BS from Brown University in 1982, and a PhD from Harvard University in 1990. After working as a postdoctoral fellow at the Fred Hutchinson Cancer Research Center, he joined the faculty of the University of Massachusetts Medical School in 1994, where he is currently professor of molecular medicine.

Mello's chief interest is the embryology and development of the tiny nematode worm *Caenorhabditis elegans*. He discovered that it was possible to block the expression of specific genes by injecting custom-made RNA molecules into the worm's cells – the phenomenon of RNA interference (RNAi). He was surprised to find that the blocking effect was transmitted from

cell to cell, and could even be passed from one generation to the next. Mello subsequently collaborated with Andrew FIRE to investigate RNAi using *C. elegans* as an experimental organism, and in 1998 they published a landmark paper in *Nature*, accurately describing the phenomenon. In recognition of this work, Mello was awarded the 2006 Nobel Prize in physiology or medicine.

Melloni, Macedonio (1798–1854) *Italian physicist*

> Light and radiant heat are effects directly produced by two different causes.
> —*Annales de chimie* (1835; Annals of Chemistry)

Melloni (may-**loh**-nee) was professor of physics at the university in his native city of Parma from 1824 to 1831. However, his political activities in the unsuccessful revolutions of 1830 forced him to flee to France. He was later allowed to return to Italy and in 1839 appointed director of a conservatory of physics in Naples. He also directed, until 1848, the Vesuvius Observatory.

Melloni is especially noted for investigations into radiant heat. In 1831, together with Leopoldo NOBILI, he published a description of a sensitive thermopile; further development of this instrument was due almost entirely to him. Melloni emphasized that just as there was variety in light rays so there were different kinds of radiant heat. He made numerous experiments on its absorption by solids and liquids, and introduced the word "diathermancy" (the ability to transmit heat radiations). By 1850 he had shown that infrared rays could be reflected, refracted, polarized, and cause interference effects as could ordinary light. In 1850 he published *La Thermochrôse*, a treatise that embodied his researches on radiant energy and gave details of his childhood love of science.

Melville, Sir Harry Work (1908–2000) *British chemist*

Melville was educated at the university in his native city of Edinburgh, where he obtained his PhD. After a period (1933–40) at Cambridge University, Melville served as professor of chemistry at the University of Aberdeen (1940–48) and at Birmingham University (1948–56).

At this point in his career Melville moved mainly into science administration. He advised such bodies as the Ministry of Power and the Electricity Authority and from 1956 to 1965 served as secretary of the Department of Scientific and Industrial Research. He finally held the post of head of Queen Mary College, London, from 1967 until his retirement in 1976.

As a chemist, Melville is noted for his work on chain reactions involving free radicals, in which he followed up the ideas of Cyril HIN-SHELWOOD and Nikolay SEMENOV and showed experimentally that they were correct. Later he studied the kinetics of polymerization chain reactions. He wrote *Experimental Methods in Gas Reactions* (1938) with Adalbert Farkas.

Melzack, Ronald (1929–) *Canadian psychologist*

Melzack was born at Montreal in Quebec, Canada, and educated at McGill University there, completing his PhD in 1954. He spent the period from 1954 until 1957 as a research fellow at Oregon University Medical School. Soon after, following brief periods at University College, London, and the University of Pisa, Melzack joined the Massachusetts Institute of Technology (1959). He returned to McGill in 1963 as professor of psychology and as director of research at the Pain Clinic at Montreal General Hospital.

Much of Melzack's career has been devoted to the study of the psychology and physiology of pain. He has published a general account of the subject in his *The Puzzle of Pain* (Harmondsworth, 1973) and a massive comprehensive survey, *Textbook of Pain* (London, 1984).

In 1965, in collaboration with Patrick Wall, Melzack proposed the "gate theory" of pain. They saw the central nervous system (CNS) as something more than a channel along which signals passed; the channel contained gates which could be opened or closed to allow or bar the passage of stimuli. Where were the gates located and what controlled them? They were located, Melzack argued, in the dorsal horns of the spinal cord.

The control mechanism was a function of the different types of nerve fiber along which stimuli passed to the spinal cord. The thicker A fibers conduct sensory stimuli both to and from (afferent and efferent) the CNS at a speed of 120 meters per second; in contrast, the afferent fibers for pain are thinner and transmit impulses at no more than 2 meters per second. Melzack proposed that activity in large fibers tends to inhibit transmission (close the gate) while small-fiber activity tends to facilitate transmission (open the gate).

Melzack also noted that the spinal gating mechanism was influenced by nerve impulses descending from the brain. That is, brain activities "subserving attention, emotion, and memories of prior experience" can control sensory input. In this manner, for instance, soldiers wounded in battle may feel no immediate pain.

Melzack has used gate theory to explain such anomalous states as referred pain and phantom-limb pain. He has also sought ways to use his theoretical understanding in the control of pain.

Mendel, Gregor Johann (1822–1884) *Austrian plant geneticist*

> The information technology of the gene is digital. This fact was discovered by Gregor Mendel in the last century, although he wouldn't have put it like that. Mendel showed that we don't blend our inheritance from our parents ... we receive [it] in discrete particles.
> —Richard Dawkins, *The Blind Watchmaker* (1986)

Born in Heinzendorf (now Hynčice in the Czech Republic), Mendel studied at Olmütz University before entering the Augustinian monastery at Brünn (now Brno in the Czech Republic) in 1843. His childhood experience of horticultural work as the son of a peasant farmer had given him an interest in the role of hybrids in evolution, and in 1856 he began plant-breeding experiments. He studied seven characters in pea plants and obtained important results after much laborious recording of character ratios in the progeny of crosses. From his experiments Mendel concluded that each of the characters he studied was determined by two factors of inheritance (one from each parent) and that each gamete (egg or sperm cell) of the organism contained only one factor of each pair. Furthermore he deduced that assortment into gametes of the factors for one character occurred independently of that for the factors of any other pair. Mendel's results are summarized today in his law of segregation and law of independent assortment (*Mendel's laws*).

Mendel's work is now recognized as providing the first mathematical basis to genetics but in its day it stimulated little interest. He read a brief account of his research to the Brünn Natural History Society in 1865 and asked members to extend his methods to other species, but none did. In 1866 he published his work in the society's *Verhandlungen* (Proceedings), a journal distributed to 134 scientific institutions, and sent reprints of the paper to hybridization "experts" of the time. Karl NAEGELI, the Swiss botanist, was skeptical of his results and suggested that he continue work on the hawkweeds (*Hieracium*), a genus now known to show reproductive irregularities and with which Mendel was bound to fail.

Mendel's work with peas, and later with *Matthiola, Zea*, and *Mirabilis*, had shown that characters do not blend on crossing but retain their identity, thus providing an answer to the weakness in Charles DARWIN's theory of natural selection. Mendel read a copy of Darwin's *Origin of Species*, but unfortunately, Darwin never heard of Mendel's work.

Mendel became abbot of the monastery in 1868 and thereafter found less time to devote to his research. It was not until 1900, when Hugo de Vries, Karl CORRENS, and Erich von Tschermak came across his work, that its true value was realized.

Characters of parents	First generation	Numbers in second generation	Ratio in second generation
Stem length			
tall	100%	787	2.84
short	—	277	1
Flower position			
axial	100%	651	3.14
terminal	—	207	1
Pod color (unripe)			
green	100%	428	2.82
yellow	—	152	1
Pod form (ripe)			
smooth	100%	332	2.95
wrinkled	—	299	1
Flower seed coat color			
purple/gray	100%	705	3.15
white/white	—	224	1
Seed form			
round	100%	5474	2.96
wrinkled	—	1850	1
Cotyledon color			
yellow	100%	6022	3.01
green	—	2001	1

MENDEL'S EXPERIMENTS The results of Mendel's pea-crossing experiments, showing the characteristic 3:1 ratio in the second generation.

Mendeleev, Dmitri Ivanovich (1834–1907) *Russian chemist*

There will come a time, when the world will be filled with one science, one truth, one industry, one brotherhood, one friendship with nature...This is my belief, it progresses, it grows stronger, this is worth living for, this is worth waiting for.
—Quoted by Yu A. Urmanster in *The Symmetry of Nature and the Nature of Symmetry* (1974)

Mendeleev (men-de-**lay**-ef) was the youngest child of a large family living in Tobolsk, Siberia. His father was a local school teacher whose career was ended by blindness and to support the family his mother ran a glass factory. Mendeleev learned some science from a political refugee who had married one of his sisters. His father died in 1847, and soon after his mother's factory was destroyed by fire. She left Tobolsk with Mendeleev, determined that her last son should receive a good education, and placed him at the Pedagogic Institute of St. Peters-

burg only ten weeks before her death. He later studied in France under Henri REGNAULT and in Heidelberg with Robert BUNSEN and Gustav KIRCHHOFF.

While abroad Mendeleev attended the famous conference at Karlsruhe in 1860 which did so much to settle the question of atomic weights. He returned to Russia shortly after and in a short time had completed his doctorate, written a textbook, and married. In 1866 he was elected to the chair of chemistry at St. Petersburg University where he remained until his retirement in 1890. His textbook *The Principles of Chemistry* was published between 1868 and 1870.

In 1869 Mendeleev published his classic paper *On the Relation of the Properties to the Atomic Weights of Elements*, which brought order and understanding to this confused subject. His first major proposal was his claim that the only way of classifying the elements is by their atomic weights. Optical, magnetic, and electrical properties vary with the state the

I	II	III	IV	V	VI	VII	VIII		
H									
Li	Be	B	C	N	O	F			
Na	Mg	Al	Si	P	S	Cl			
K — Cu	Ca — Zn	□ — □	□ — Ti	As — V	S — Cr	Br — Mn	Fe	Co	Ni
Rb — Ag	Sr — Cd	In — Y	Sn — Zr	Sb — Nb	Te — Mo	I — □	Ru	Rh	Pd

PERIODIC TABLE   *The table proposed by Mendeleev is shown above. Vertical columns are groups of elements with similar chemical properties. Horizontal rows in the table are called "periods" — across a period there is a gradual change of properties; for example, the period Li to F involves a change from metallic behavior (Li) to nonmetallic behavior (F). In order to group chemically similar elements, Mendeleev left gaps in the table for undiscovered elements.*

body is in at any particular moment; other properties, such as valence, yield conflicting results. When the elements are arranged in order of increasing atomic weight, Mendeleev found that they show a distinct periodicity of their properties. Arranging them in rows of increasing atomic weights produced columns of similar elements.

The table did not at first receive universal acceptance, but its value became apparent during the following 20 years. Through it Mendeleev was able to spot those elements that had been assigned incorrect atomic weights. Thus he suggested that the atomic weights of gold and tellurium must be wrong. There were three missing elements in his table, and he was able to predict their existence, valences, and certain physical properties. The three were eventually discovered – "eka-aluminum" (gallium, Paul Lecoq de Boisbaudran, 1875), "eka-boron" (scandium, Per CLEVE, 1879), and "eka-silicon" (germanium, Clemens WINKLER, 1885).

Mendeleev became the most famous Russian scientist of his day and received numerous medals and prizes although not, surprisingly, the Nobel Prize (in 1906 it was awarded to Ferdinand MOISSAN by one vote). Element 101 was named *mendelevium* in his honor.

Mengoli, Pietro (1625–1686) *Italian mathematician*

Mengoli (men-**goh**-lee) was born in Bologna and studied there with Francesco CAVALIERI. He was ordained a priest and held a chair in mathematics at Bologna. The mathematical work was chiefly on the calculus and infinite series and his most important work was with convergence and with integration. He was one of the first to establish results about the condi-

tions under which infinite series converge. His work constitutes an important stage in the development of the fully fledged calculus of Isaac NEWTON and Gottfried LEIBNIZ from such ideas as Cavalieri's method of indivisibles. Mengoli gave a definition of a definite integral, similar to that given by Augustin CAUCHY.

Menzel, Donald Howard (1901–1976) *American astrophysicist*

Born in Florence, Colorado, Menzel graduated from the University of Denver in 1920 and obtained his PhD from Princeton in 1924. In 1926 he was appointed to the staff of the Lick Observatory in California. He became assistant professor of astronomy at Harvard in 1932, then served as professor of astrophysics (1938–71), and from 1954 to 1966 was director of the Harvard College Observatory.

Menzel worked mainly on the solar system, investigating planetary atmospheres and the constitution of the Sun. He improved on the work of Henry RUSSELL in 1929 on the abundance of the elements in the Sun: he estimated that the Sun contained 81.76% hydrogen and 18.17% helium by volume, leaving 0.07% for the rest of the elements.

Menzel also used spectrographic methods to work out, with J. H. Moore, the rotational period of Neptune, finding it to be 15.8 hours.

Mercator, Gerardus (1512–1594) *Dutch cartographer and geographer*

Mercator (mer-**kay**-ter), originally named Kremer, was born at Rupelmonde, now in Belgium. At the University of Louvain (1530–32) he was a pupil of Gemma Frisius. After learning the basic skills of an instrument maker and engraver, he founded his own studio in Louvain in 1534. Despite accusations of heresy and imprisonment in 1544, he remained in Louvain until 1552, when he moved to Duisburg and opened a cartographic workshop.

Mercator first made his international reputation as a cartographer in 1554 with his map of Europe in which he reduced the size of the Mediterranean from the 62° of PTOLEMY to a more realistic, but still excessive, 52°. He produced his world map in 1569 and his edition of Ptolemy in 1578, while his *Atlas*, begun in 1569, was only published by his son after his death. It was intended to be a whole series of publications describing both the creation of the world and its subsequent history. Mercator was the first to use the term "atlas" for such works, the book having as its frontispiece an illustration of Atlas supporting the world.

The value of Mercator's work lies not just in his skills as an engraver, but also in the introduction of his famous projection in his 1569 map of the world. Navigators wished to be able to sail on what was called a rhumb-line course, or a loxodrome, i.e., to sail between two points on a constant bearing, charting their course with a straight line. On the surface of a globe such lines are curves; to project them onto a plane chart Mercator made the meridians (the lines of longitude) parallel instead of converging at the Poles. This made it straightforward for a navigator to plot his course but it also produced the familiar distortion of the *Mercator projection* – exaggeration of east–west distances and areas in the high latitudes.

The big difference, apart from projection, between Mercator's and classical maps was in the representation of the Americas. He was not the first to use the name America on a map, that distinction belonging to Martin Waldseemüller in 1507, but he was the first to divide the continent into two named parts – *Americae pars septentrionalis* (northern part of America) and *Americae pars meridionalis* (southern part of America).

Mercer, John (1791–1866) *British chemist*

Mercer was the son of a cotton spinner from Dean in England and worked in the industry as a boy until, in 1807, he started a small dyeing business. He spent the rest of his life introducing new dyes and techniques into the industry. He seems to have been largely self-educated, learning his chemistry from one of the available mechanics manuals, Parkinson's *Chemical Pocket Book* (1791).

In 1814 Mercer introduced his first new dye, an orange one produced from sulfide of antimony, followed by a yellow dye from lead chromate and a bronze dye from manganese. He worked mainly in the calico-print side of the industry. His major innovation, the process of *mercerizing*, was patented in 1850. He found

that cloth soaked in caustic soda would become stronger, take up dye more readily, and acquire a lustrous silken sheen. Another contribution of his was a method of printing fabric by a photographic process.

Apart from his industrial chemistry Mercer became interested in more theoretical problems through his friendship with Lyon PLAYFAIR formed in the early 1840s. He suggested an early theory of catalytic action and also contributed papers on the chemical activity of light and on atomic weights.

Merrifield, Robert Bruce (1921–2006) *American biochemist*

Born in Fort Worth, Texas, Merrifield was educated at the University of California, Los Angeles, where he received his PhD in 1949. He began work immediately at Rockefeller University, New York, and was appointed to the chair of biochemistry in 1966, a post he held until his retirement in 1992.

In the 1950s Merrifield began work on solid-phase peptide synthesis (SPPS). Peptides, like proteins, are composed of chains of amino acids, but have shorter and less complicated chains. Naturally occurring ones possess important physiological properties. The ability to synthesize peptides cheaply and quickly would lead to numerous commercial and medical gains. Yet the synthesis of a polypeptide using traditional methods could take many months.

In peptides the amine end ($-NH_2$) of one amino acid reacts with the carboxyl end ($-COOH$) of another. To prepare a pure product of known structure, amino acids have to be coupled in a specific sequence. To achieve this the amine group on one amino acid and carboxyl group on the other must be blocked, so that the other two ends are the ones reacting. And this must be done as each further amino acid is added. In addition, at each stage the product must be isolated and purified. This will involve crystallizing the products. The synthesis of a hundred-unit peptide would involve ninety-nine such procedures. The need for improvement in the technique was painfully clear to peptide chemists.

Merrifield's innovation was to apply an ion-exchange technique by bonding the amino acids, one at a time, to an insoluble solid support. A polystyrene resin was the original choice. As the solid support was insoluble in the various solvents used, all the intermediate products and impurities could be simply washed away by using the appropriate reagent. Much initial work was required in setting up the right kinds of activating agents, blocking agents, and solvents. In 1964 in eight days Merrifield single-handedly synthesized bradykinin, a nine-amino-acid peptide that dilates blood vessels.

One further aspect of Merrifield's process is

that it can be fully automated. To demonstrate the power and potential of his method Merrifield undertook in 1965 the automatic synthesis of insulin. With 51 amino acids and two peptide chains held together by two disulfide bridges, the molecule was a formidable challenge. Although more than 5,000 operations were involved in assembling the chains, most of these were carried out automatically in a few days. The linking of the two chains, however, was achieved by more traditional methods. The resulting insulin was active in the standard biological assay.

For his development of the technique Merrifield was awarded the 1984 Nobel Prize for chemistry.

Merrill, Paul Willard (1887–1961) *American astronomer*

Merrill was born in Minneapolis, Minnesota, and graduated from Stanford University in 1908. He obtained his PhD in 1913 from the University of California. After teaching at the University of Michigan from 1913 to 1916 and serving briefly with the U.S. Bureau of Standards in Washington, he was appointed to the staff of the Mount Wilson Observatory in 1919, remaining there until his retirement in 1952.

It was in his retirement year that he detected the lines of technetium in the spectra of S-type stars, a class of cool red giant stars. This was surprising for the most stable isotope of technetium has a half-life of about 2.6 million years, which is much shorter than the lifetime of a star. Merrill thought it unlikely that there was an unknown stable isotope of technetium found only on S-type stars and argued rather that it was being produced within the stars by some form of nuclear reaction. The technetium lines are now accepted as strong evidence for one of the nuclear processes by which the heavy elements are thought to be created in the interiors of stars.

Mersenne, Marin (1588–1648) *French mathematician, philosopher, and theologian*

> Most men are glad to find work done, but few want to apply themselves to it, and many think that this search is useless or ridiculous.
> —*Les préludes de l'harmonie universelle* (1634; Prelude to Universal Harmony)

> [Mersenne was] a man of simple, innocent, pure heart, without guile ... a man than whom none was more painstaking, inquiring, experienced.
> —Pierre Gassendi, letter to Louis de Valois, 4 September 1648

Mersenne (mair-**sen** or mer-**sen**), who was born at Oize in France, was educated at the famous Jesuit college at La Flèche. After this he studied theology at the Sorbonne in Paris and became a Catholic priest in the Order of Minims in 1611. He spent most of the rest of his life teaching at their convent near Place Royale but he also traveled extensively in Europe.

Despite his commitment to the Catholic faith, Mersenne took a very lively interest in the rapidly developing world of the physical sciences, seeing his task as one of using the new discoveries as a means to defend Catholic orthodoxy. Probably his most important contribution to science was his role as a communication link between scientists. He kept in contact with as many eminent scientists of the day as possible, as his enormous correspondence testifies, and there were few 17th-century men of science of any importance who did not correspond with Father Mersenne. His convent at Place Royale became a regular meeting place for what were in effect conferences of leading scientists and philosophers. Among those who gathered here were Blaise PASCAL, René DESCARTES, Pierre de FERMAT, and the philosopher Thomas Hobbes.

Mersenne himself was particularly interested in problems of scientific methodology. He doubted the range and usefulness of the Euclidean axiomatic method advocated for the sciences, especially by Descartes. Mersenne was a vigorous opponent of the various forms of skepticism that were popular at the time, and was much interested in the viability of the new mechanistic world picture which the new science seemed to bring with it. Mersenne's interests extended to nonscientific aspects of philosophy and he was one of the first to try to popularize the idea of an invented universal language. His interest in science was by no means purely theoretical and he did important practical work in optics and acoustics.

Mersenne proposed, in 1644, his *Mersenne numbers*. These are numbers generated from the formula $2^p - 1$, in which p is a prime. This formula does not represent all primes but it contributed to developments in the theory of numbers.

Merton, Robert (1910–2003) *American sociologist*

> Most institutions demand unqualified faith; but the institution of science makes skepticism a virtue.
> —*Social Theory and Social Structure* (1962)

Merton was educated at Temple University, in his native city of Philadelphia, and then at Harvard, where he obtained his PhD in 1935. He served as tutor at Harvard until 1939, and taught briefly at Tulane, New Orleans, before he moved to Columbia, New York, in 1941. Here he held the post of professor of sociology from 1963 until his retirement in 1979.

Merton's doctoral thesis, *Science, Technology, and Society in Seventeenth-Century England* (1938; 1970 reissue, New York), has been the subject of much discussion ever since its publication. Merton's bold thesis, since known as the

Merton hypothesis, claimed that there were significant links during the period between puritanism and science. He also argued that much of the motivation of early science was practical and linked to such activities as navigation, mining, artillery, and metallurgy. Merton's claims were based upon a careful study of, among other things, the subject matter of articles in the *Philosophical Transactions of the Royal Society* (1665–1702) and an analysis of the religious leanings of members of the Society.

Merton also undertook a study of multiple discoveries in science. A number of discoveries in science appear to have been made independently by two or more people at almost the same time. A notable example is the theory of evolution proposed by DARWIN and by WALLACE in 1859. Merton argued that these were not simple coincidences – they are not uncommon and may even be the norm. By 1961 he had analyzed some 264 cases of which 179 involved two independent discoverers and 8 involved six independent workers.

A further issue illuminated by Merton was the social dynamics of science. It was Merton's work that demonstrated the crucial role that priority of discovery and publication played in science. He also identified, as integral features, communalism, universalism, disinterestedness, and skepticism. Merton's papers in this field, written with wit, learning, and insight, were collected in his *The Sociology of Science* (Chicago, 1979).

Merton, Sir Thomas Ralph (1888–1969)
British physicist and inventor

Merton had an unusual scientific career. He was born in Wimbledon, near London, into a family of considerable wealth that derived from their metal-trading business. After leaving Oxford University in 1910 he moved to London and engaged in spectroscopic research in his private laboratory. He appears to have spent World War I in the Secret Service working on the development of secret inks. He returned to Oxford in 1919 and, although he later held the professorship of spectroscopy from 1923 to 1935, he obviously preferred his private laboratory, for most of his work was performed at his country residence, Winforton House, Herefordshire. From this address he produced a number of important publications on atomic spectra, which appeared until 1927. Thereafter he published nothing for a further twenty years until, in 1947, he produced the first of a number of papers on his newly developed interference microscope.

Some of this time was clearly devoted to perfecting a number of inventions, some of which were of great value in World War II. The most important of these was his two-layer long-persistence screen for cathode-ray tubes used in radar, based on work done in 1905.

Merton built up one of the most important collections of Renaissance paintings in private hands. Many of the important pieces from his collection now grace the National Gallery, London. He was knighted in 1944.

Meselson, Matthew Stanley (1930–   )
American molecular biologist

Meselson, who was born in Denver, Colorado, studied liberal arts at the University of Chicago and physical chemistry at the California Institute of Technology, Pasadena. After he obtained his PhD in 1957 he remained at the Institute until 1960 when he moved to Harvard, where he served from 1964 as professor of biology and from 1976 as Thomas Dudley Cabot professor of natural sciences.

In 1957 Meselson, in collaboration with Franklin STAHL, conducted one of the classic experiments of molecular biology, which clearly revealed the semiconservative nature of DNA replication. It seemed likely that when the double helix of DNA duplicated, each new helix, and hence each daughter cell, would contain one DNA strand from the original helix; in the jargon of the time, replication would be semiconservative. The other possibility was that one daughter molecule would contain both the old strands and the other daughter molecule both the new strands – conservative replication.

Meselson and Stahl grew many generations of the bacterium *Escherichia coli* on a simple culture medium containing ammonium chloride (NH_4Cl), labeled with the heavy isotope of nitrogen, ^{15}N, as the only nitrogen source. They then added normal ^{14}N nitrogen to the medium and removed bacterial cells at intervals, extracting their DNA by ultracentrifugation. The density of the DNA in successive samples could be determined by the method of equilibrium density gradient centrifugation, in which samples of differing density diffuse into discrete bands corresponding to their own effective density. Ultraviolet absorption photographs of these bands allowed the concentration of DNA in each band to be determined. The results showed that (following the introduction of ^{14}N) after one doubling of the *E. coli* bacteria all the DNA molecules contained equal amounts of ^{15}N and ^{14}N, i.e., they were all half labeled.

After two generations there were equal amounts of half-labeled DNA molecules and wholly ^{14}N molecules. This effectively demonstrated that replication is semiconservative. The results were published in the Proceedings of the National Academy of Sciences U.S., as *The Replication of DNA in Escherichia coli* (1958).

Meselson has also worked with Sidney BRENNER and François Jacob on the mechanism of viral infection. In 1961, working with the virus T4, they showed that on invasion of a host bacterial cell the viral DNA releases messenger

RNA, which, when it arrives at the host ribosomes, instructs these to make viral protein rather than bacterial proteins.

Mesmer, Franz Anton (1734–1815) *Austrian physician*

Mesmer (**mez**-mer or **mes**-mer) was born at Iznang (now in Germany). He studied medicine at the University of Vienna, graduating in 1766 with a thesis, clearly indicative of his later work, on *The Influence of the Planets on the Human Body*.

Mesmer's ideas developed from Newtonian physics. The idea that bodies like the Moon and the Sun could influence such terrestrial phenomena as that of the tides was a very powerful one. For, if matter could act over seemingly empty space, then, Mesmer extrapolated, virtually any object could be used to explain the behavior of any other. The clearest statement of Mesmer's theory is contained in his account, in 1779, of the discovery of "animal magnetism." He believed that heavenly bodies, the Earth, and living things affected each other through a universally distributed fluid that could "receive, propagate, and communicate all motion."

He applied his ideas to medical treatment and, using magnets, obtained his first cure in 1774 with a distant relative who appeared to be suffering from such hysterical symptoms as vomiting, convulsions, and paralysis. Later, Mesmer dispensed with applying magnets, replacing them by a variety of devices more likely to increase the patient's amenability to suggestion. However his methods met with hostility in Vienna so he moved in 1778 to Paris, where he won the support of Marie Antoinette. Nevertheless in 1784 a royal commission was set up to investigate his procedures, which included Antoine LAVOISIER, the American minister to France, Benjamin FRANKLIN, and a certain Dr. Guillotin, whose name is now associated with the execution device whose use he advocated during the French Revolution. The report concluded dismissively that "the imagination without magnetism produces convulsions, and that magnetism without imagination produces nothing."

Consequently Mesmer, disappointed by the reception of the French savants, moved on once more. However, in 1843 "mesmerism" received the first sign of scientific recognition when James BRAID wrote that he had separated "animal magnetism" from what he termed "hypnotism," a term he had deliberately introduced to replace "mesmerism."

Messel, Rudolph (1848–1920) *German chemical industrialist*

Messel (**mes**-el) was the son of a barber from Darmstadt in Germany. He was apprenticed to a chemical manufacturer before studying chem-istry at the universities of Zurich, Heidelberg, and Tübingen. In 1870 he traveled to England, where he served initially as assistant to Henry ROSCOE in Manchester.

In 1875 Messel took out a patent for a new process for the production of sulfuric acid. In the contact process developed by Peregrine PHILIPS in the 1830s the use of a catalyst was essential, but one of the drawbacks of the process was that the expensive platinum catalyst soon became contaminated and useless. Messel showed that the contamination of the catalyst could be avoided if the sulfur dioxide used was first carefully purified. With the help of W. S. Squire, he was able to develop a satisfactory industrial version of the process at their Silvertown plant in 1876. The fuming sulfuric acid produced was in great demand by the early years of the 20th century and this demand escalated enormously in World War I.

Messier, Charles (1730–1817) *French astronomer*

> The systematic listing by Messier in 1784 marked an epoch in the recording of observations.
> —Harlow Shapley, *Star Clusters* (1930)

Messier (mes-**yay**), who was born in Badonviller in France, arrived in Paris in 1751 and was taken on as a clerical assistant by J. Delisle at the Naval Observatory sited in the Collège de Cluny. He quickly learned how to use the Observatory instruments and began a lifetime's obsessive search for comets. Dubbed the "comet ferret" by Louis XV, Messier is credited with the discovery of 13 comets between 1759 and 1798. The computation of the cometary orbits, however, was left to his more mathematically sophisticated colleagues.

In 1758 he observed what appeared to be a faint comet in Taurus. Further examination revealed it to be a nebula, an immense cloud of gas. Messier thought it sensible to provide a list of such objects "so that astronomers would not confuse these same nebulae with comets just beginning to shine." He published his first list of 45 nebulae in 1774 under the title *Catalogue des nebeleuses et des amas étoiles* (Catalog of Nebulae and Star Clusters). Two supplements published in 1783 and 1784 increased the number of nebulae to 103.

The nebulae listed in the catalogs were given the identifying letter M and a number; for example, the Andromeda nebula is commonly referred to as M31.

Metchnikoff, Elie (1845–1916) *Russian–French zoologist and cytologist*

> Thus it was in Messina that the great event of my scientific life took place. A zoologist until then, I suddenly became a pathologist.
> —On his observation of phagocytes in starfish larvae

Metchnikoff (mech-ni-**kof** or **myech**-nyi-kof) was born at Ivanovka near Kharkov (now in Ukraine) and educated at Kharkov University. After holding posts under Rudolf LEUCKART at Göttingen and Giessen, and under Karl SIE-BOLD at Munich, he taught zoology at Odessa and St. Petersburg. From 1873 to 1882 he was professor of zoology and comparative anatomy at Odessa.

He spent the years 1882–86 at Messina in Italy where, working on starfish larvae, he first noticed that certain nondigestive cells enclose and engulf foreign particles introduced into the body. These cells he called phagocytes and, extending his studies, he demonstrated that they also occur in man – they are the white blood corpuscles. He realized that they are important in the body's defenses against disease, in engulfing bacteria and other foreign bodies in the blood. These advances were outlined in such publications as *Intra-Cellular Digestion* (1882), *The Comparative Pathology of Inflammation* (1892), and *Immunity in Infectious Diseases* (1905). For his work on phagocytosis, Metchnikoff was awarded, in 1908, the Nobel Prize for physiology or medicine jointly with Paul EHRLICH.

In 1886 Metchnikoff was appointed director of the new bacteriological institute at Odessa; two years later he went to the Pasteur Institute in Paris, which he directed from 1895 to 1916, succeeding Pasteur himself.

In 1903 Metchnikoff, working with Emile ROUX, showed that syphilis could be transferred to apes. He also did research on cholera. His later years were largely concerned with a study of the aging factors in man and means of inducing longevity, discussed in *The Nature of Man* (1904) and *The Prolongation of Human Life* (1910).

Metius, Jacobus (1580–1628) *Dutch instrument maker*

Metius (**may**-tee-u(r)s) was born at Alkmaar in the Netherlands; together with Hans LIP-PERSHEY and Zacharias JANSSEN, he is credited with the invention of the telescope. His father was a cartographer while his brother, Adriaen, was an astronomer and mathematician who had worked with Tycho BRAHE at Hven and had taught the famous mathematician and philosopher René DESCARTES.

Metius put in a claim for a patent for his "perspicilla" (telescope) in October 1608, but he was beaten to it by Lippershey who had put in a similar claim just two weeks earlier.

Meton (about 432 BC) *Greek astronomer*

> Meton the son of Pausanius, who has a reputation in astronomy, set out the so-called nineteen-year period, taking the beginning from the thirteenth of Skirophorion [month in the Athenian civil calendar] at Athens.
> —Diodorus (5th century BC)

Nothing is known about the life and personality of Meton (**mee**-ton). His proposed cycle, called the *Metonic cycle*, which was not accepted by the citizens of Athens, was designed to bring the lunar month and the solar year into some form of acceptable agreement. As the lunar month is about 291/2 days and the solar year is 3651/4 days there is no way a whole number of months can make up a year. An early solution to the problem was the *octaeteris*, in which three intercalary (inserted) months are added to each eight-year cycle. This would lead to an error of alignment of a day and a half each eight years. Meton's suggestion was an improvement on this. He realized that 235 lunar months and 19 solar years are both 6,939 days. To bring the two cycles into phase would need 7 intercalary months spread over the 19 years of the full cycle. This would produce a solar year only 30 minutes too long. The Metonic cycle was eventually adopted by the Greeks and was used until the introduction of the Julian calendar in 46 BC. The Jewish calendar still uses it.

Meyer, Julius Lothar (1830–1895) *German chemist*

Meyer (**mI**-er) was the son of a doctor from Varel in Germany. He qualified in medicine himself in 1854 after studying at Zurich and Würzburg and gained his PhD from the University of Breslau in 1858. At first his interests were physiological but he slowly moved into chemistry. He became professor of chemistry at Karlsruhe in 1868 where he stayed until he moved to the chair at Tübingen (1876–95).

Meyer is best remembered for his early work on the periodic table. He was much impressed by Stanislao CANNIZZARO, expounding his work in his book *Die modernen Theorien der Chemie* (1864; Modern Chemical Theory). In writing his textbook it had occurred to him that the properties of an element seem to depend on its atomic weight. Meyer plotted the values of a certain physical property, atomic volume, against atomic weight. He found clear signs of periodicity, the graph consisting of a series of four sharp peaks. He noticed that elements with similar chemical properties occur at comparable points on the different peaks; e.g., the alkali metals all occur at the tops of the peaks (see illustration overleaf).

Meyer did not publish his table until 1870 so he was preempted by Dmitri MENDELEEV, who had published his periodic table in 1869. Meyer never disputed Mendeleev's priority and later stated that he lacked sufficient courage to have gone on to predict the existence of undiscovered elements.

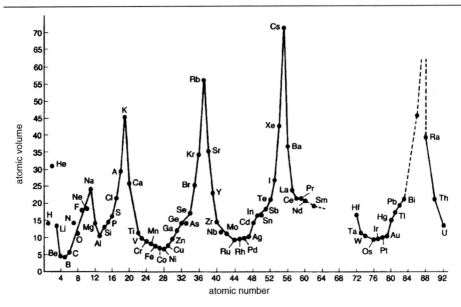

LOTHAR MEYER'S CURVES   *Curves showing the periodicity of the chemical elements.*

Meyer, Karl (1899–1990) *American biochemist*

Meyer gained his MD from the university in his native city of Cologne in 1924 and his PhD from Berlin University in 1928. He moved to the University of California, Berkeley, in 1930, and in 1933 transferred to the College of Physicians and Surgeons at Columbia, where he has spent the rest of his career. He served as professor of biochemistry from 1954 until his retirement in 1967.

Meyer studied the acidic mucopolysaccharides found in connective tissue and isolated two of these, hyaluronic acid and chondroitin sulfate. He also discovered that various bacteria have enzymes – hyaluronidases – that can break down hyaluronic acid. It was later shown that these enzymes are the same as the "spreading factors" isolated from various sources, such as snake venom and leeches. Meyer and his colleagues found that there are three different types of chondroitin sulfate, and in 1953 he isolated a third mucopolysaccharide, keratosulfate, found in the cornea. This was later also found in cartilage. Meyer also investigated the production and distribution of mucopolysaccharides and was able to show that Marfan's syndrome, an inherited disease of connective tissue, is associated with large amounts of keratosulfate in cartilage.

Meyer, Viktor (1848–1897) *German chemist*

Meyer's father was a wealthy Berlin merchant in the textile trade. He studied at the universities of Heidelberg and Berlin, receiving his PhD in 1867. He held chemistry chairs at Stuttgart (1871), Zurich (1872–85), and Göttingen (1885–89) and succeeded Robert BUNSEN at Heidelberg (1889–97).

Meyer synthesized a number of organic compounds. In 1882 he discovered the compound thiophene through the failure of a color test for benzene in one of his lecture demonstrations. He had used a sample of benzene obtained from benzoic acid instead of from petroleum. He also introduced apparatus for the determination of the vapor density of organic substances. It was Meyer who prepared the way for Jacobus VAN'T HOFF by pointing out that isomeric methylene chlorides do not occur, i.e., two different forms of CH_2Cl_2 are never found, and he introduced the term stereoisomerism into chemistry.

Meyerhof, Otto Fritz (1884–1951) *German–American biochemist*

Meyerhof (**mI**-er-hof), who was born at Hannover in Germany, devoted the greater part of his academic life to the study of the biochemistry and metabolism of muscle; he shared the Nobel Prize for physiology or medicine with Archibald Hill in 1922. He held professorships at Kiel and Heidelberg universities, was director of physiology at the Kaiser Wilhelm Institute for Biology, Berlin, and was director of research at the Paris Institute of Biology. In 1940 he emigrated to America, where he joined the medical faculty of the University of Pennsylvania.

Meyerhof demonstrated that the production

of lactic acid in muscle tissue, formed as a result of glycogen breakdown, was effected without the consumption of oxygen (i.e., anaerobically). The lactic acid was reconverted to glycogen through oxidation by molecular oxygen, during muscle rest. This line of research was continued by Gustav EMBDEN and Carl and Gerty CORI who worked out in greater detail the steps by which glycogen is converted to lactic acid – the *Embden–Meyerhof pathway*.

Michaelis, Leonor (1875–1949) *German–American chemist*

Michaelis (mi-kah-**ay**-lis) was educated at the university in his native city of Berlin and at Freiburg. He worked in the laboratory of the Berlin Municipal Hospital from 1906 to 1922, when he took up the post of professor of biochemistry at the Nagoya Medical School, Japan. In 1926 Michaelis emigrated to America and after spending four years at Johns Hopkins moved to the Rockefeller Institute of Medical Research, where he remained until his retirement in 1940.

In 1913 Michaelis, in collaboration with L. M. Menten, formulated one of the earliest precise and quantitative laws applying to biochemical systems. They were trying to picture the relation between an enzyme and its substrate (the substance it catalyzes) and, in particular, how to predict and understand the reaction rate, that is, how much substrate is acted upon by an enzyme per unit time, and the basic factors that stimulate or inhibit this rate. The kind of graph obtained when reaction rate is plotted against substrate concentration showed that additional substrate concentration sharply increases the reaction rate until a certain point is reached when the rate appears to become completely indifferent to the addition of any further amounts of substrate.

Michaelis saw this as indicating that the reaction between enzyme and substrate is a very specific one. In the early phase of the curve there was enzyme lacking substrate; as this was increased more and more enzyme came into play, increasing the reaction rate. Eventually, however, there will come a point when all the enzyme is being used and from that point the addition of any amount of substrate can have no effect on the reaction rate. This variation in rate was subsequently described by the *Michaelis–Menten equation*.

Michaelis's insight into the working of the enzyme–substrate complex was quite remarkable as no hard evidence for its existence was to emerge for a good many years, not in fact until Britton CHANCE was able to produce spectroscopic evidence in 1949.

Michel, Hartmut (1948–) *German chemist*

Michel (**mik**-el) was born at Lüdwigsburg in Germany and educated at the University of Warburg, where he obtained his PhD in 1977. He moved to the Max Planck Institute for Biochemistry at Martinsried, near Munich, and remained there until 1987, when he moved to Frankfurt to head the biophysics division.

By 1970 chemists had succeeded in uncovering the basic chemistry of photosynthesis but little was known about the process at the molecular level. It was established that the process occurred in the photosynthetic reaction centers first identified by Roderick Clayton in the late 1960s. These are to be found embedded in the membranes of the photosynthetic vesicles. Within the reaction centers was a complex protein structure. Before further progress could be made, the structure of the proteins would have to be worked out, but first it would be necessary to crystallize the proteins.

Michel first tackled the problem in 1978. While it was relatively easy to crystallize water-soluble proteins, membrane proteins, which react with both fats and water, were only partially soluble in water. Michel used a molecule in which one end was attracted to water (hydrophilic) while the opposite end was water repellent (hydrophobic). By binding the hydrophobic ends of the organic molecules to the hydrophobic ends of the protein membranes the hydrophilic ends alone would lie exposed. The complex structure could then be dissolved in water and crystallized. By 1982 Michel had succeeded in crystallizing the membrane proteins of the bacterium *Rhodopseudomonas viridis*.

For this work Michel shared the 1988 Nobel Prize for chemistry with Johann DEISENHOFER and Robert HUBER.

Michell, John (1724–1793) *English geologist and astronomer*

Michell studied at Cambridge University and became a fellow. In 1762 he was appointed Woodward Professor of Geology but left academic life to take up a post as rector at Thornhill, Yorkshire, in 1764.

Before his departure from Cambridge he published, in 1760, a fundamental paper, *Conjectures Concerning the Cause, and Observations upon the Phenomena of Earthquakes*. After the great Lisbon earthquake (1755) this was a fashionable subject. Michell assigned the cause of earthquakes to the force generated by high-pressure steam, produced when water suddenly met subterranean fires. He appreciated that such a force would generate waves in the Earth's crust and tried to estimate the velocity of these, giving a not unreasonable figure of 1,200 miles per hour. Finally, Michell showed various means to determine the point of origin of the earthquake.

In 1790 he constructed a torsion balance to

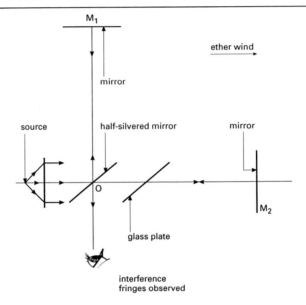

M₁

ether wind

mirror

source half-silvered mirror mirror

O

glass plate

M₂

interference
fringes observed

MICHELSON–MORLEY EXPERIMENT The rays of light are split and recombined in the
telescope to give fringes. The system was mounted on a large bed of stone, which could be rotated
through 90°. It was shown that this rotation did not affect the position of the fringes. Michelson and
Morley did this experiment a large number of times at different times of the day and night and at
different times of the year. On no occasion did they detect any effect.

measure gravitational attraction and thus the mean density of the Earth. Michell was unable to use this before his death, but Henry CAVENDISH carried on his work, deriving a value for the density of the Earth in 1798.

Michell also made contributions to astronomy. In 1767 he published a paper on double stars, pointing out with originality and insight that there are far too many of them to result from a random scattering and therefore they must in many cases constitute a genuine binary system. He also devised a method for calculating the distance of the stars.

Michelson, Albert Abraham (1852–1931)
American physicist

> Physical discoveries in the future are a matter of the sixth decimal place.
> —Quoted by A. L. Mackay in *A Dictionary of Scientific Quotations* (1991)

Michelson, who was born at Strelno (now Strzelno in Poland), came to America with his parents when he was two years old. He graduated from the U.S. Naval Academy in 1873 and remained there to teach physics and chemistry. Some five years later he began his work on measuring the speed of light and to this end he traveled to Europe to study optics at the Collège de France, Heidelberg, and Berlin. When he returned to America he left the navy to become professor of physics at the Case School, Cleveland. In 1882 he estimated the speed of light as

186,320 miles per second. This was the most accurate value then available and remained so for another ten years, when Michelson made an even more accurate measurement.

In the course of this work Michelson developed an interferometer, an instrument that can divide a beam of light in two, send the beams in different directions, and then unite them again. If the two beams traveled the same distance at different speeds (or different distances at the same speed) then, on being brought together again, the waves would be out of step and produce interference fringes on a screen. Michelson used the interferometer to test whether light traveling in the same direction as the Earth moves more slowly than light traveling at right angles to this direction. This was effectively testing the presence of the "ether" – a substance that was supposed to exist in all space. Because the ether was thought to be motionless and the Earth moved through it, then it followed that light traveling in the same direction as the Earth would be impeded. Michelson first conducted this experiment in 1881 in Berlin and got a negative result, that is, there were no interference fringes and thus no evidence that the two beams were traveling at different speeds. He repeated the procedure several times under increasingly elaborate conditions until, in 1887, with Edward MORLEY, the experiment was made under near perfect conditions at the Case School. Again the ether could not be detected. This result questioned much orthodox physical theory, and it remained for EIN-STEIN to develop the special theory of relativity to explain the constancy of the speed of light. Michelson was awarded the 1907 Nobel Prize in physics for this work.

Michelson also applied interference techniques to astronomical measurements and was able to measure the diameters of various heavenly bodies by contrasting the light emitted from both sides. He also continued to make increasingly more accurate estimates of the speed of light and he suggested that the wavelength of light waves should be used as the length standard rather than the platinum–iridium meter in Paris. This suggestion was taken up in 1960 when light waves from the inert gas krypton became the standard measure.

Midgley, Thomas Jr. (1889–1944) *American chemist*

Midgley's father was an inventor and held several patents in wire drawing and for various types of rubber tire. Midgley was born in Beaver Falls, Pennsylvania, and educated at Cornell, graduating in mechanical engineering in 1911. In 1916 he joined the research staff of the Dayton (Ohio) Engineering Laboratories Company.

The first problem Midgley tackled was the sometimes quite violent engine knock produced by the fuel of his day. He began work on this in 1917, shortly after the company had been taken over by General Motors. Midgley was able to show that engine knock was a property of the fuel used and realized that the solution lay in finding an additive to eliminate the knock. Midgley's first breakthrough came when he argued (wrongly) that a dye added to the fuel would allow it to absorb more radiant heat and vaporize sooner. As the only oil-soluble dye available was iodine he tried this and found that it worked. A few simple tests showed that it was the iodine and not the color that was crucial, and Midgley began a long struggle to find a cheaper additive. From his studies of chemistry Midgley was led to tetraethyl lead (1921), which continues to be unsurpassed as an antiknock fuel additive, although it is now banned in many places as an environmental hazard.

The second problem Midgley attempted to resolve for General Motors was to find a nonflammable, cheap, and nontoxic refrigerant gas (several customers having already been poisoned by sulfur dioxide used in refrigerators). Midgley experimented with the fluorides and discovered dichlorodifluoromethane, now commonly known as freon. Within a few years this became the almost universal refrigerant.

In 1940 Midgley contracted polio, from which he died.

Miescher, Johann Friedrich (1844–1895) *Swiss biochemist*

Miescher (**mee**-sher) came from a distinguished scientific family from Basel in Switzerland: both his father, also called Johann Friedrich, and his uncle, Wilhelm His, held the chair of anatomy at the University of Basel. Miescher himself studied medicine at Basel but, feeling that his partial deafness (produced by a severe attack of typhus) would be a drawback for a physician, turned to physiological chemistry. He consequently spent the period from 1868 to 1870 learning organic chemistry under Felix HOPPE-SEYLER at Tübingen and physiology at Leipzig in the laboratory of Carl Ludwig. In 1871 he was appointed professor of physiology at Basel.

It was while working on pus cells at Tübingen in 1869 that Miescher made his fundamental discovery. It was thought that such cells were made largely of protein, but Miescher noted the presence of something that "cannot belong among any of the protein substances known hitherto." In fact he was able to show that it was not protein at all, being unaffected by the protein-digesting enzyme pepsin. He also showed that the new substance was derived from the nucleus of the cell alone and consequently named it "nuclein." Miescher was soon able to show that nuclein could be obtained from many other cells and was unusual in containing phosphorus in addition to the usual ingredients of organic molecules – carbon, oxygen, nitrogen, and hydrogen. It was not until 1871 that Miescher's paper, delayed by Hoppe-Seyler (who wanted to confirm the results), was published. In it he announced the presence of a nonprotein phosphorus-containing molecule in the nuclei of a large number of cells.

Just what precise role the molecule played in the cell was not revealed until the structure of nucleic acid, as it was renamed by Richard Altmann in 1889, was announced by James WATSON and Francis CRICK in 1953. Miescher continued to work on the nuclein extracted from the sperm of the Rhine salmon for the rest of his short life. He spent much time puzzling on the chemistry of fertilization, even speculating in 1874 that "if one wants to assume that a single substance ... is the specific cause of fertilization then one should undoubtedly first of all think of nuclein." Unfortunately Miescher failed to follow up his suggestion, preferring to explore physical models of fertilization. However, his work on nuclein was eagerly taken up by other organic chemists, and by 1893 Albrecht KOSSEL had succeeded in recognizing four nucleic acid bases.

Milankovich, Milutin (1879–1958) *Croatian mathematician*

Milankovich (mi-**lan**-ko-vich), who was born at Dalj in Croatia, was educated at the Institute of Technology in Vienna, where he obtained his PhD in 1904. He then moved to the University of Belgrade, remaining there for the rest of his career except for the period 1914–18, during which he was a prisoner of war, but was allowed to pursue his researches in the library of the Hungarian Academy of Science in Budapest.

Milankovich was the most talented of the scientists who worked in the tradition of James CROLL in trying to explain the development of the Earth's climate by reference to astronomical events. From 1911 to 1941, when he published his *Canon of Insolation and the Ice Age Problem*, he tried obsessively, in numerous works, to reconstruct the past climate of the Earth and the planets.

Milankovich realized that the key to past climates was the amount of solar radiation received by the Earth, which varies at different latitudes and depends upon three basic factors. One is the degree of ellipticity of the Earth's orbit, which varies over 100,000 years from being nearly circular to a noticeable ellipticity and which could reduce the amount of insolation by 30%. Secondly, over about 21,000 years a precessional change occurs, which will determine whether the northern or the southern hemisphere receives the most radiation. Finally, the tilt of the Earth's axis to the plane of its orbit changes over about 40,000 years from 21.8° to 24.4°.

Over a period of 30 years Milankovich constructed radiation curves for the last 650,000 years for the summer northern hemisphere from 5°N to 75°N. At first his results looked most impressive for he identified nine climatic minima, which fitted closely the four ice ages identified by Albrecht PENCK. However, with the advent of more precise and accurate dating techniques his results are now considered doubtful.

Miller, Dayton Clarence (1866–1941) *American physicist*

Born in Strongsville, Ohio, Miller obtained his PhD from Princeton in 1890. He was later appointed to the chair of physics at the Case School of Applied Science, Cleveland, Ohio.

It was at the Case School that the famous experiment to detect the presence of the ether was performed in 1887 by Albert MICHELSON and Edward MORLEY. The existence of an ether would produce different values for the speed of light in different directions. Miller was dissatisfied with the negative result of the experiment and he repeated the experiment with Morley under varying conditions over the period 1897–1908 without coming up with anything conclusive. Morley lost interest but Miller was as determined as he was suspicious.

In 1925 he took his equipment to the 6,000-foot (1,830-m) peak of Mount Wilson in California. At last he thought he had detected a difference of 6 miles per second for light traveling at right angles to the Earth's orbit. Confident of his result he cabled Albert EINSTEIN on Christmas Day 1925, announcing the discovery of ether drift. Einstein suspected, rightly as it turned out, that Miller's results were due to different temperature conditions.

Miller, Hugh (1802–1856) *British geologist*

> Almost every fragment of clay, every splinter of sandstone, every limestone nodule contained its organism – scales, spines, plates, bones, entire fish ... were I to sum up all my happier hours, the hour would not be forgotten in which I sat down on a rounded boulder of granite by the edge of the sea and spread out on the beach before me the spoils of the morning.
> —On his first discovery of fossil fishes in the Old Red Sandstone

Miller was born in Cromarty, northern Scotland, and during his early life apprenticed himself to a stonemason, becoming a journeyman in 1819. The pursuit of his trade led him to travel widely in Scotland and he acquired an unrivaled knowledge of its detailed geology. He also began to write poetry on the political issues of the day. After a period as a bank accountant he was offered the editorship of the newly formed influential Scottish paper, *The Witness*, and became a key figure in the political and theological disputes that led to the 1843 split in the Church of Scotland.

His main geological work was originally published in his newspaper and collected in book form as *The Old Red Sandstone* (1841), which went through 20 editions before 1900. In this he described the fossils he had found in Devonian strata and, as in his later work *Footsteps of the Creator* (1847), he opposed the view that the record of the rocks provided evidence for the evolution of species.

Miller succeeded in showing that the Old Red Sandstone was relatively rich in fossils, particularly fossil fish. He used the evidence provided by the fossil fish to pose a problem to those, such as Robert CHAMBERS and later Charles DARWIN, who wished to think in such terms as the progress, development, or evolution of species. He argued that "if fish could have risen into reptiles, and reptiles into mammalia, we would necessarily expect to find lower orders of fish passing into higher." Miller claimed that in the Old Red Sandstone it is the fish of the higher orders that appear first, thus there was no room for progression.

Miller, Jacques Francis Albert Pierre (1931–) *French–Australian immunologist*

Miller, who was born at Nice in the south of France, was educated at the University of Sydney, Australia, and at University College, London, where he obtained his PhD in 1960. He then held brief appointments at the Chester Beatty Research Institute in London and the National Cancer Institute in Bethesda, Maryland. Miller returned to Australia in 1966 to serve as head of the experimental pathology department at the Hall Institute of Medical Research, Melbourne. He has been semiretired since 1996 but is still working.

In 1961 Miller succeeded in solving an ancient medical mystery. The thymus gland is a large organ placed in the chest beneath the breastbone. Surprisingly, until 1961 scientists lacked any clear idea of the role played by such a prominent body. The normal technique in such a situation is to watch for any changes in the behavior of the subject when the organ has been removed. In this case thymectomy seemed to make no discernible difference to the behavior of any experimental animal.

Working within this tradition Miller performed a surgical operation of great skill, the removal of the thymus from one-day-old mice. As the mice weigh no more than a gram and are no bigger than an inch it is not difficult to see why such an operation had been little attempted before. In this case, however, the excision did lead to dramatic and obvious changes. The mice failed to develop properly and usually died within two to three months of the operation. Just what was wrong with them became clear when Miller went on to test their ability to reject skin grafts, a sure sign of a healthy immune system. Miller's mice could tolerate grafts from unrelated mice and sometimes even from rats. This made it quite clear that the thymus was deeply involved in the body's immune system but just what precise role it played was to occupy immunologists for a decade or more.

Much of Miller's work was performed independently, also in 1961, by a team under the direction of Robert GOOD at Minnesota.

Miller, Stanley Lloyd (1930–2007) *American chemist*

Born at Oakland in California, Miller was educated at the universities of California and Chicago where, in 1954, he was awarded his PhD. From 1960 he taught at the University of California, San Diego, being appointed to a professorship in chemistry in 1968.

In 1953 Miller published a famous paper, *A Production of Amino Acids under Possible Primitive Earth Conditions*, in which he reported the results of an experiment carried out while still a graduate student at Chicago under the direction of Harold UREY.

It was thought that the early atmosphere of the Earth could well have been something like that now existing on Jupiter and Saturn, namely one rich in methane (CH_4) and ammonia (NH_3). Miller mixed water vapor with ammonia, methane, and hydrogen in a closed flask and subjected it to a high-voltage electrical discharge. Sensitive analysis with paper chromatography produced a number of organic molecules. In addition to hydrocyanic acid, formic acid, acetic acid, lactic acid, and urea were two of the simpler amino acids, alanine and glycine.

As it is from the amino acids that the proteins are constructed many scholars saw this as clear evidence for the spontaneous origin of life. It has however been shown that such random processes could not yet have produced a single protein without the assumption of additional operating principles. Such principles have been suggested by Manfred EIGEN and Sol SPIEGELMAN.

Millikan, Robert Andrews (1868–1953) *American physicist*

The son of a Congregational minister from Morrison, Illinois, Millikan was educated at Oberlin, where he studied classics, and Columbia University, where he obtained his PhD in 1895. After a year in Europe, studying under Max PLANCK and Walther NERNST, he took up an appointment in 1896 at the University of Chicago, being promoted to a full professorship in 1910. Millikan moved to the California Institute of Technology in 1921 as director of the Norman Bridge Laboratory, a position he held until his retirement in 1945.

In 1909 Millikan started on a project that was to win for him the 1923 Nobel Prize for physics – the determination of the electric charge of the electron. His apparatus consisted of two horizontal plates that could be made to take opposite charges. Between the plates he introduced a fine spray of oil drops whose mass could be determined by measuring their fall under the influence of gravity and against the resistance of the air. When the air was ionized by x-rays and the plates charged, then an oil drop that had collected a charge would be either repelled from or attracted to the plates depending on whether it had collected a positive or negative charge. By measuring the change in the rate of fall and knowing the intensity of the electric field Millikan was able to calculate the charges on the oil drops. After taking many careful measurements he was able to come to the important conclusion that the charge was always a simple multiple of the same basic unit, which he found to be $4.774 \pm 0.009 \times 10^{-10}$ electrostatic units, a figure whose accuracy was not improved until 1928. Millikan followed this with a prolonged attempt from 1912 to 1916 to demonstrate the validity of the formula introduced by Albert EINSTEIN in 1905 to describe the photoelectric effect, work that was cited in Einstein's Nobel award.

In 1923 he began a major study of cosmic rays, first identified in 1912 by Victor HESS, which was to occupy him for the rest of his career. His first aim was to show that they did not originate in our atmosphere. To do this, he devised an ingenious set of observations made at two lakes in the San Bernadino mountains of southern California. The lakes were many miles apart and differed by 6,700 feet (2,042 m) in altitude. The difference in altitude would have the same effect on intensity of cosmic rays as six feet of water. He found that the intensity of ionization produced by the incoming cosmic rays in the lower lake was the same as the intensity six feet deeper in the higher lake. This showed, he claimed, that the rays do come in definitely from above and that their origin is entirely outside the layer of atmosphere between the levels of the two lakes.

Millikan then went on to theorize about the nature of the cosmic rays. He argued that they were electromagnetic radiation photons, for if they were charged particles they would be influenced by the Earth's magnetic field and therefore more likely to arrive in higher rather than lower latitudes. Millikan had failed to detect any such effect with latitude. In fact Millikan's theories were soon disproved for Arthur COMPTON did detect a latitude effect.

Mills, Bernard Yarnton (1920–) *Australian physicist and radio astronomer*

Mills was born at Manly in New South Wales, Australia. After graduating from the University of Sydney in 1940, he joined the staff of the Commonwealth Scientific and Industrial Research Organization, CSIRO, where he worked initially on the development of radio astronomy. In 1960 he joined the staff of the University of Sydney to form a radio-astronomy group; he subsequently served as professor of physics there from 1965 to 1985.

Mills is best known for his design of a radio interferometer, known as the *Mills cross* radio telescope. In 1954 he completed the construction of the first Mills cross at Fleurs, near Sydney. It consists of two fixed intersecting arrays of dipole antennas running along the ground in an east–west and north–south direction for 1,500 feet (457.2 m). It was with this system that the first radio survey of the southern hemisphere, the MSH, was performed; some 2,000 sources at a wavelength of 11.43 feet (3.5 m) were examined and the results published from 1958 to 1961.

When plans to build an optical telescope at Sydney University fell through, Mills was offered the accumulated funds of 200,000 Australian dollars to construct a "super" cross. The instrument has arms 1 mile long along which cylindrical parabolic reflectors are fixed in position. It is sited at the Molonglo Radio Observatory near Hoskinstown, New South Wales, and was completed in 1960.

Mills, William Hobson (1873–1959) *British chemist*

Mills, the son of a London architect, was educated at Cambridge University, England, and at Tübingen University in Germany. After serving as head of chemistry at the Northern Polytechnic, London (1902–12), Mills returned to Cambridge where he served as reader in stereochemistry from 1931 until his retirement in 1938.

In 1925 Mills, in collaboration with E. WARREN, confirmed the hypothesis of Alfred WERNER concerning the tetrahedral configuration of the ammonium ion. During World War I, he also worked on problems of photography. The then-existing emulsions were sensitive only to the blue and violet part of the spectrum; the military, however, were keen to photograph any changes in the German trench system in the red light of dawn. Working with William POPE, Mills succeeded in remedying the deficiency with the development of the cyanine dyes.

Milne, Edward Arthur (1896–1950) *British mathematician and astrophysicist*

Milne, the son of a headmaster from Hull in eastern England, studied at Cambridge University (1914–16). He returned there in 1919 after working on ballistics during World War I and was appointed as assistant director of the Cambridge Solar Physics Observatory in 1920. He lectured in both mathematics and astrophysics. In 1924 he became professor of applied mathematics at the University of Manchester where he remained until 1928. In 1929 Milne was appointed professor of mathematics at Oxford, a post he held, apart from a period working for the Ordnance Board (1939–44), until his death in 1950.

Milne's wartime work had involved studies of the Earth's atmosphere. From 1920 he extended this to theoretical research on the atmospheres of stars, concentrating on the flow of radiation through these outermost layers of stars and on the ionization of the component atoms. In collaboration with Ralph Fowler, Milne used the ionization theory of Meghnad SAHA to determine and fix the temperature scale for the known sequence of stellar spectra. Thus just by knowing that a star falls into spectral type G permitted its temperature (5,000–6,000 kelvin) to be inferred.

Following three years' research into stellar structure Milne turned in 1932 to the development of his theory of "kinematic relativity." This was an alternative to EINSTEIN's theory of general relativity and contained a new cosmological model of the universe from which he evolved new systems of dynamics and electrodynamics. It was Milne who introduced the "cosmological principle" that simply states that the universe appears essentially the same from wherever it is observed. This is still the basis of much of modern cosmology and became for Milne one of the axioms of an axiomatic cosmology that he hoped to construct in a purely deductive way. "Starting from first principles (like Descartes)" his aim was to "pursue a single path towards the understanding of this unique entity the universe; and it will be a test of the correctness of our path that we should find at no point any bifurcation of possibility. Our path should nowhere provide any alternatives." However, Milne did find alternative universes littering his path and in each case he behaved as DESCARTES had done before him and selected that alternative he favored as being the only one compatible with the rationality of the Creator.

Milne's world model, which is now seen as a misguided curiosity, led nowhere. But for his early death he might well have developed it and brought it more into line with orthodox cosmology.

Milne, John (1850–1913) *British seismologist*

Milne, who was born at Liverpool in England, was educated at King's College, London, and at the Royal School of Mines. After fieldwork in Newfoundland and Labrador (1872–74) he was appointed, in 1875, professor of geology and mining at the Imperial College of Engineering, Tokyo.

Milne developed a passion for seismology and became known as "Earthquake Milne" in Tokyo. He was instrumental in forming the Seismological Society of Japan in 1880. His first priority was to organize the recording, collecting, and distribution of data. He asked the postal authority in each town throughout Japan to return to him a weekly record of the numbers of earthquakes experienced and he also set up over 900 stations for more detailed recording of seismic activity. Milne also invented, in 1880, a seismograph and he spent much time devising simple and hardy seismographs, which could be used by the relatively unskilled in a wide variety of conditions.

Milne returned to England in 1894 and made his home at Shide, on the Isle of Wight. This became the center of an international system for the collection and distribution of seismological data. His publications included *Earthquakes* (1883) and *Seismology* (1898).

Milstein, César (1927–2002) *British molecular biologist*

Milstein was born at Bahia Blanca in Argentina and attended the University of Buenos Aires, receiving his degree in 1952 and his doctorate in 1957. Three years later he was granted a PhD by Cambridge University. Milstein returned to his native Argentina in 1961 to head the Molecular Biology Division of the Instituto Nacional de Microbiología in Buenos Aires. In 1963 he joined the staff of the Medical Research Council's Laboratory of Molecular Biology in Cambridge and in 1983 he was appointed head of the Division of Protein and Nucleic Acid Chemistry, a post in which he remained until his retirement in 1994.

Milstein is best known for producing the first monoclonal antibodies, using a technique developed at the MRC's Laboratory in collaboration with the German immunologist, Georges KÖHLER, and first reported by them in 1975. The pair went on to show how it was possible to manufacture quantities of antibody of any desired specificity employing cultures of so-called "hybridoma" cells. Monoclonal antibodies have found wide-ranging applications in biology, medicine, and industry, especially for diagnostic tests and assays. For his part in developing this revolutionary technology, Milstein was awarded the 1984 Nobel Prize for physiology or medicine, which he shared with Köhler and Niels JERNE.

Minkowski, Hermann (1864–1909) *Russian–German mathematician*

> The views of space and time which I wish to lay before you have sprung from the soil of experimental physics, and therein lies their strength. They are radical. Henceforth space by itself, and time by itself, are doomed to fade away into mere shadows, and only a kind of union of the two will preserve an independent reality.
> —Quoted by L. D. Henderson in *The Fourth Dimension and Non-Euclidean Geometry in Modern Art* (1983)

Minkowski (ming-**kof**-skee) was born at Alexotas in Russia to parents of German origin. In 1872 the family returned to Germany, settling in Königsberg (now Kaliningrad). Minkowski studied alongside David HILBERT at the University of Königsberg, under Adolf Hurwitz, and gained his PhD in 1885. He taught at Bonn (1885–94) and Königsberg (1894–96) and then worked with Hurwitz at the Zurich Federal Institute of Technology (1896–1902). At Hilbert's instigation a new chair of mathematics was created for Minkowski at the University of Göttingen and he worked there (1902–09) until his death.

In 1883, when still 18, Minkowski was awarded the Grand Prix des Sciences Mathématiques of the Paris Academy of Sciences. The award was shared with Henry J. Smith for their work on the theory of quadratic forms. Minkowski remained occupied with the arithmetic of quadratic forms for the rest of his life. In 1896 he gave a detailed account of his "geometry of numbers" in which he developed geometrical methods for the treatment of certain problems in number theory.

During his short period at Göttingen Minkowski worked closely with David Hilbert and decisively influenced Hilbert's interest in mathematical physics. Minkowski's most celebrated work was in developing the mathematics that played a crucial role in EINSTEIN's formulation of the theory of relativity. Einstein knew when he published the special theory of relativity in 1905 that the universe could not be adequately described using normal, or Euclidean, three-dimensional geometry. Minkowski's seminal idea was to view space and time as forming together a single four-dimensional continuum or manifold, known as "space–time," rather than two distinct entities. In normal three-dimensional geometry, any point in space can be identified by three coordinates. The analog of this point in three-dimensional space is an event localized both in space and time in four-dimensional space–time.

Minkowski put forward his concept of space–time, or *Minkowski space* as it is sometimes called, in 1907 in his book *Space and Time*. Einstein himself was very forthright about the extent to which the theory of relativity depended on Minkowski's innovatory work. Space–time was a useful and elegant format for special relativity, and was essential for general relativity, published in 1916, in which space–time is allowed to be curved. It is the curvature of space–time that accounts for the phenomenon of gravitation.

Minkowski, Rudolph Leo (1895–1976)
German–American astronomer

Minkowski was born at Strasbourg (now in France) and educated at the University of Breslau (now Wrocław in Poland), where he obtained his PhD in 1921. He taught at Hamburg University until 1935, when he emigrated to America and took up an appointment at the Mount Wilson Observatory. In 1960 he moved to the University of California at Berkeley where he remained until his retirement in 1965.

In 1951 Minkowski, in collaboration with Walter BAADE, made the first identification of a discrete radio source, Cygnus A. It seemed to be one of the brightest objects in the radio sky and possessed a peculiar spectrum. It also turned out to be an extremely distant object, as determined from the red shift of its spectral lines. It must therefore have an immense output of radio energy. Minkowski allowed himself to be converted to Baade's view that Cygnus A was a pair of colliding galaxies. The fact that its radio map revealed two nuclei seemed to confirm this. Minkowski and Baade made many other optical identifications of radio sources, including 3C 295, which possesses the same distinctive shape and unusual emission lines of Cygnus A. Failure to detect evidence of the appropriate relative galactic motion, which should be apparent if Cygnus A were two colliding galaxies, has led to the rejection of this idea.

Minot, George Richards (1885–1950)
American physician

The son of a physician from Boston, Massachusetts, Minot was educated at Harvard, where he obtained his MD in 1912. After working briefly at Johns Hopkins University he moved back to Harvard in 1915 and served as professor of medicine from 1928 until his retirement in 1948.

In 1926, with his assistant William MURPHY, Minot discovered a cure for pernicious anemia. This was based on earlier work by George WHIPPLE showing that red-cell production could be increased in dogs by adding liver to their diets. In 1924 Minot began feeding some of his patients with small amounts of liver. He was soon able to report that most of their symptoms disappeared within a week and that within 60 days their red cell counts were back to normal. As pernicious anemia was invariably fatal at that time the new therapy was life saving. It was not until 1948 that the vital ingredient in liver, vitamin B_{12}, was actually isolated and purified. For their work on liver therapy, Minot, Whipple, and Murphy shared the 1934 Nobel Prize for physiology or medicine.

Minsky, Marvin Lee (1927–) *American computer scientist*

> Logic doesn't apply to the real world.
> —Quoted by D. R. Hofstadter and D. C. Dennet in *The Mind's I*

The son of a surgeon, Minsky was born in New York and educated at Harvard and at Princeton, where he obtained his PhD in 1954. He taught at Harvard before moving to the Massachusetts Institute of Technology (MIT) in 1957 as professor of mathematics, a post he occupied until 1962 when he became professor of electrical engineering. He also served as director of the Artificial Intelligence (AI) Laboratory (1964–73).

In the summer of 1956 Minsky attended a conference on AI at Dartmouth, New Hampshire. Here, it was generally agreed that powerful modern computers would soon be able to simulate all aspects of human learning and intelligence. Much of Minsky's later career has been spent testing this claim.

Under Minsky's direction a number of AI programs have been developed at MIT. One of the earliest, a program to solve problems in calculus, showed that most problems could be solved by a careful application of about 100 rules. The computer actually received a grade A in an MIT calculus exam. Other programs developed such topics as reasoning by analogy, handling information expressed in English, and how to catch a bouncing ball with a robotic arm.

But Minsky soon became aware that AI had a number of problems to overcome. For example, in one project a computer with a robotic arm and a TV camera was programmed to copy an assembly of bricks. Although it could quickly recognize the bricks and their relationships to each other, it found the stacking more difficult. It tried to stack the blocks from the top down, releasing brick after brick in midair. Computers simply do not have an innate knowledge of gravity or, he has pointed out, many other things we take for granted, such as that chairs painted a different color remain the same chair, or that boxes must be opened before things can be put inside. He also noted problems with "perceptrons," designed by Frank Rosenblatt in 1960 with the supposed ability to respond to and recognize certain patterns with the aid of an array of 400 photocells. In collaboration with Seymour Papert, Minsky published a critical account of this work in *Perceptrons* (1968) showing, in purely formal terms, that the powers of perceptrons were strictly limited. They could not, for example, be relied upon to tell when a figure was connected. A cat's tail protruding from a chair would prevent the perceptron from identifying either the cat or the chair.

In 1974 Minsky introduced the notion of "frames." A frame is a package of knowledge stored in the mind, which allows us to understand many things about a certain topic. For example, the "dog frame" includes what dogs look like, the sorts of things they do, and many other aspects of their nature and behavior. Because we possess numerous such frames we are able to communicate about the world without too much confusion, and to distinguish routinely between the "bark" of a tree and the "bark" of a dog. Only when a computer could be stocked with an enormous number of "frames," some interlocking, others slotted hierarchically in other frames, could it begin to show signs of intelligence.

While speculating about developments in the 1990s Minsky has referred to "societies of the mind." A computer capable of recognizing shadows would be unable to process perspective or parallax. Yet, the untutored human mind can normally handle all three. The aim should therefore be to write a program "that allows each expert system to exploit the body of knowledge that lies buried in the others." It remains to be seen whether Minsky's "society" can be realized. In 2006 Minsky published *The Emotion Machine*.

Misner, Charles William (1932–) *American astronomer*

> Space acts on matter, telling it how to move. In turn, matter reacts back on space telling it how to curve.
> —*Gravitation* (1973)

Born in Jackson, Michigan, Misner was educated at Notre Dame University, Indiana, and at Princeton, where he obtained his PhD in 1957. He remained at Princeton until 1963 when he moved to the University of Maryland, College Park, becoming professor of physics in 1966. He has been emeritus professor since 2000.

In 1969 Misner posed what is known as the "horizon paradox." The universe on a large scale appears to be remarkably homogenous and isotropic, in particular with respect to the cosmic microwave background radiation discovered in 1965. How, given the initial conditions of the big bang, could such uniformity have arisen?

The problem is that no physical process can take place over distances at a speed faster than the speed of light. Consequently there must be, at any particular time, a distance (the "horizon distance") beyond which light or any other process could not have spread since the moment of the big bang itself. Calculations show that sources of the cosmic background radiation would have been separated by 90 times the horizon distance when emitted. How then, Misner asked, could the universe have reached a state of equilibrium and appear so uniform? The standard big-bang model merely assumed the uniformity of the universe as an initial condition.

For a time Misner sought to develop one of the so-called "mixmaster models" in which it was assumed that the universe began in an inhomogenous and anisotropic state and became uniform through the interplay of frictional forces as the universe expanded. This, however, merely trades one set of assumptions for another. A more favored response to the horizon paradox has been presented by the idea of an inflationary model of the big bang proposed in 1980 by GUTH.

Misner is also well known as coauthor with Kip THORNE and John WHEELER of *Gravitation* (1973), one of the seminal textbooks of modern cosmology.

Mitchell, Maria (1818–1889) *American astronomer*

Mitchell was born in Nantucket, Massachusetts, the daughter of William Mitchell, who started life as a cooper and became a school teacher and amateur astronomer of some distinction. Her brother, Henry Mitchell, became the leading American hydrographer. She herself was mainly educated by her father, whom she helped in the checking of chronometers for the local whaling fleet and in determining the longitude of Nantucket during the 1831 eclipse. From 1824 to 1842 she worked as librarian at the Nantucket Athenum and in 1849 she became the first woman to be employed full time by the U.S. Nautical Almanac, with whom she computed the ephemerides of Venus. Finally, in 1865 she was appointed professor of astronomy and director of the observatory at the newly founded Vassar College.

Maria Mitchell was clearly fortunate to come from a highly talented family. She was also helped by coming from Nantucket, an area where women were expected to demonstrate an unusual degree of independence while the local men were absent on their long whaling voyages. It was also an area where it was common for the average person to possess a familiarity with mathematics, astronomy, and navigation. She is mainly remembered today for her discovery, in 1847, of a new comet.

Mitchell, Peter Dennis (1920–1992)
British biochemist

Born at Mitcham in Surrey, England, Mitchell was educated at Cambridge University, where he obtained his PhD in 1950. He remained at Cambridge, teaching in the department of biochemistry, until 1955 when he moved to Edinburgh University as director of the Chemical Biology Unit. In 1964 Mitchell made the unusual decision to set up his own private research institution, the Glynn Research Laboratory, in Bodmin, Cornwall.

It was well known that the cell obtains its energy from the adenosine triphosphate (ATP) molecule; it was also clear that ATP was made by coupling adenosine diphosphate (ADP) to an inorganic phosphate group by the process known to biochemists as oxidative phosphorylation. What was less clear was just how this happened and it was widely assumed that it was controlled by a number of enzymes. Despite considerable effort the proposed enzymes remained surprisingly elusive.

Beginning in 1961 Mitchell proposed a completely different and totally original model, without any obvious precursors and judged to be unorthodox to the point of eccentricity. He suggested a physical mechanism by which an electrochemical gradient is created across the cellular membrane; this, in turn, creates a proton current capable of controlling the phosphorylation.

For his account of such processes Mitchell was awarded the 1978 Nobel Prize for chemistry.

Mitscherlich, Eilhardt (1794–1863) *German chemist*

Mitscherlich (**mich**-er-lik), who was born at Neuende in Germany, studied oriental languages at Heidelberg and Berlin. He then turned to the study of medicine at Göttingen in 1817, where he became interested in crystallography. For two years he worked with Jöns BERZELIUS in Stockholm, returning to Berlin in 1821, where he was appointed to the chair of chemistry.

While working on arsenates and phosphates,

Mitscherlich realized that substances of a similar composition often have the same crystalline form, and from this he formulated, in 1819, his law of isomorphism. This was in opposition to the orthodox view of René HAÜY that each substance has a distinctive crystalline form. Despite Haüy's rejection of the law, Berzelius accepted it and was quick to spot its significance, for if the composition of a substance X is known, and it is also known that X has a similarity of crystalline form with Y, then Y's composition can be derived. Thus knowing the composition of sulfur trioxide as SO_3, and that it has a similar form to "chromic acid," Berzelius was able to give this compound the composition CrO_3. Using this technique Berzelius produced his revised table of atomic weights in 1826.

Mitscherlich also discovered selenic acid (1827), named benzene, and showed, in 1834, that if benzene reacts with nitric acid it forms nitrobenzene.

Möbius, August Ferdinand (1790–1868)
German mathematician

Möbius (**mu(r)**-bee-uus) worked mainly on analytical geometry, topology, and theoretical astronomy. He was born at Schulpforta in Germany and held a chair in theoretical astronomy at Leipzig, making numerous contributions to the field with publications on planetary occultations ("eclipses") and celestial mechanics. His more purely mathematical work centers on geometry and topology.

Möbius is chiefly famed for his discovery of the *Möbius strip*, a one-sided surface formed by giving a rectangular strip a half-twist and then joining the ends together. He introduced the use of homogeneous coordinates into analytical geometry and did significant work in projective geometry, inventing the *Möbius net*, which became of central importance in the future development of the subject.

Mohl, Hugo von (1805–1872) *German botanist*

Mohl (mohl) was born at Stuttgart in Germany and studied medicine at Tübingen, graduating in 1828. He became professor of physiology at Bern in 1832. Optics and botany were his main interests, however, and he followed both by pursuing a career in plant microscopy; in 1835 he accepted the chair of botany at Tübingen, where he remained until his death.

In 1846 Mohl gave the name "protoplasm" to the granular colloidal material around the cell nucleus and in 1851 proposed the now-confirmed view that the secondary cell walls of plants are fibrous in structure. He was the first to suggest that new cells arise through cell division, from observations on the alga *Conferva glomerata*, and also the first to investigate the activity of stomata (pores) in the epidermis.

Mohorovičić, Andrija (1857–1936) *Croatian geologist*

Mohorovičić (moh-ho-**roh**-vi-chich) was born the son of a shipwright in Volosko, Croatia, and educated at the University of Prague. He worked initially as a teacher and at the meteorological station in Bakar before being appointed a professor at the Zagreb Technical School in 1891 and at Zagreb University in 1897.

In 1909 he made his fundamental discovery of the *Mohorovičić discontinuity* (or *Moho*). From data obtained while he was observing a Croatian earthquake in 1909, Mohorovičić noticed that waves penetrating deeper into the Earth arrived sooner than waves traveling along its surface. He deduced from this that the Earth has a layered structure, the crust overlaying a more dense mantle in which earthquake waves could travel more quickly. The abrupt separation between the crust and the mantle Mohorovičić calculated as being about 20 miles (32 km) below the surface of the Earth; this is now called the Mohorovičić discontinuity.

As the crust is much thinner under the ocean beds – in some places only 3 miles thick – a project was set up in the 1960s to drill through the crust to the mantle. Mohole, as it became known, failed, however, largely as a result of the great financial cost involved and the inadequate technological expertise available for such a project.

Mohs, Friedrich (1773–1839) *German mineralogist*

Mohs (mohs or mohz) was born at Gernrode in Saxony and studied at Halle and at the Freiberg Mining Academy under Abraham WERNER. In 1812 he became curator of the mineral collection at the Johanneum in Graz. He succeeded Werner at Freiberg in 1818 and in 1826 he was appointed professor of mineralogy at Vienna.

In 1812 Mohs introduced the scale of mineral hardness – *Mohs scale* – named for him. Ten minerals whose hardness is known are ordered on a scale ranging from 1 (talc) to 10 (diamond), the general rule being that a higher number will scratch all lower numbers. The hardness of a mineral is judged by the ease with which its surface is scratched by these minerals whose values are known, and it can be given a numerical value.

Moissan, Ferdinand Frédéric Henri (1852–1907) *French chemist*

Moissan (mwah-**sahn**) came from a poor background in Paris, France. He was the son of a railroad worker and was apprenticed to a pharmacist before studying chemistry under Edmond Frémy at the Muséum National d'His-toire Naturelle, Paris (1872). From 1880 he worked at the Ecole Supérieure de Pharmacie, being elected to the chair of toxicology in 1886 and the chair of inorganic chemistry in 1889. In the next year he became professor of chemistry at the University of Paris.

Moissan began studying fluorine compounds in 1884 and in 1886 succeeded in isolating fluorine gas by electrolyzing a solution of potassium fluoride in hydrofluoric acid, the whole process being contained in platinum. He received the Nobel Prize for chemistry for this work in 1906.

He also worked on synthetic diamonds. He was impressed by the discovery of tiny diamonds in some meteorites and concluded from this that if the conditions undergone by these in space could be reproduced in the laboratory it would be possible to convert carbon into diamond. He therefore put iron and carbon into a crucible, heated it in an electric furnace, and while white hot cooled it rapidly by plunging it into liquid. In theory, he felt that the cooling should exert sufficient pressure on the carbon to turn it into diamond. He claimed to have succeeded in producing artificial diamonds but there was a suggestion that one of his assistants had smuggled tiny diamonds into the mixture at the beginning of the experiment. Moissan did, however, use his electric furnace for important work in preparing metal nitrides, borides, and carbides, and in extracting a number of less common metallic elements, such as molybdenum, tantalum, and niobium.

Moivre, Abraham De *See* DE MOIVRE, ABRAHAM.

Molina, Mario José (1943–) *Mexican physical chemist*

Molina, the son of a diplomat, studied chemical engineering at the University of Mexico. After further study in Europe at the University of Freiburg and at the Sorbonne, Molina moved to the University of California, Berkeley, where he gained his PhD in 1972. He worked initially as a postdoctoral student at the Irvine campus of the University of California with F. S. ROWLAND. Following a spell at the Jet Propulsion Laboratory he moved to MIT in 1989 as professor of environmental sciences.

Rowland had become interested in the fate of the chlorofluorocarbons (CFCs) used as the propellant in most aerosol cans, and asked his new colleague if he would be interested in working out what happened to them as they rose into the stratosphere. It would be, Molina later confessed, "a nice, interesting, academic exercise."

He quickly worked out that as CFCs were stable they would eventually accumulate in the upper atmosphere. There, he argued, they would be broken up by ultraviolet light and chlorine atoms would be released. Rowland suggested that Molina should analyze how free chlorine atoms would behave. Molina suspected that a chain reaction would be produced, reducing the amount of ozone in the upper atmosphere. Despite this, Molina still thought the effect would be negligible. It was only when he discovered that the amount of CFCs released each year was about 1 million tonnes that he realized that much of the ozone layer could be destroyed. Molina published his results in a joint paper with Rowland in 1974. The National Academy of Sciences issued a report in 1976 confirming the work of Molina and Rowland and in 1978 CFCs used in aerosols were banned in the United States. In 1984 Joe Farman detected a 40% ozone loss over Antarctica.

For his work on CFCs and the ozone layer Molina shared the 1995 Nobel Prize for chemistry with Rowland and Paul CRUTZEN, thus becoming the first Mexican to receive a Nobel Prize for science.

Mond, Ludwig (1839–1909) *German–British industrial chemist*

Mond (mond or mohnt) was the son of a prosperous merchant from Kassel, in Germany. After studying at the local polytechnic he continued his chemical education under Adolph KOLBE at Marburg and Robert BUNSEN at Heidelberg. He ran some small chemical businesses on the Continent before going to Britain in 1862, where he took out a patent for the recovery of sulfur from the waste left by the LEBLANC alkali process. In 1867 he started a factory in Widnes in partnership with John Hutchinson. The process was not really satisfactory and was to be superseded by that of Alexander CHANCE announced in 1882.

The Leblanc process was expensive in waste products and in 1861 Ernest SOLVAY patented a new process that was continuous and theoretically cheaper and cleaner. In 1872 Mond purchased the option to use the new process in Britain. He went into partnership with John BRUNNER at Winnington, Cheshire, in 1873 and they founded the firm of Brunner, Mond, and Company. It was not until 1880 that they made a profit after having twice faced ruin. Mond started his own production plant to guarantee cheap and plentiful supplies of ammonia for the Solvay process. Coal was burnt in an airstream producing ammonia and producer gas, a cheap fuel. He set up a plant to produce and sell the gas to the industrial West Midlands.

The Leblanc process, while not as efficient as the Solvay process in the production of soda, did have the advantage of also producing commercial quantities of bleaching powder. In 1886 Mond invented a process to remedy this in the Solvay process. Ammonium chloride vapor was pumped over nickel oxide, producing nickel chloride. This, when heated, yielded chlorine to be used for the production of bleaching powder and the ammonia could be returned to the start of the operation. Nickel valves used in this process dissolved away in the gas produced and, on investigation, Mond found that when heated to 60°C nickel will combine with carbon monoxide to produce nickel carbonyl. When heated to higher temperatures this breaks down to give pure nickel and reusable carbon monoxide. This discovery led to the formation of the Mond Nickel Company.

Mond became an extremely wealthy man and was the first of a new generation of chemical industrialists. From his private fortune he left £100,000 to equip the Davy–Faraday laboratory at the Royal Institution in London. He was also a connoisseur in art and he left to the National Gallery, London, his superb collection of paintings. Brunner, Mond, and Company became in 1926 part of Imperial Chemical Industries (ICI) through amalgamation.

Mondino de Luzzi (c. 1275–1326) *Italian anatomist*

> Now the bones of the chest are many and are not continuous, in order that it may be expanded and contracted, since it has ever to be in motion.
> —*Anatomica* (1316; Anatomy)

Mondino (mon-**dee**-noh), the son of a Bolognese apothecary, studied medicine first with his father and then at the University of Bologna, where he graduated in 1300. Known as the "Restorer of Anatomy," he completed, in 1316, his *Anatomica* (Anatomy), the text that dominated anatomical teaching for two centuries. The work however made few advances and repeated many ancient errors. Thus many of the classical fictions such as the spherical stomach, the seven-chambered uterus, the five-lobed liver, and the perforations in the septum of the heart are once again carefully described and illustrated in the rather primitive diagrams of his text.

Although Mondino writes of "anatomizing" people it is by no means clear that he actually performed the dissection himself. He may rather have read from the text, as was the custom, while his assistant used the knife and a demonstrator indicated the appropriate parts. Such classes were not intended to be research investigations into human anatomy but rather illustrations of an authoritative text. If there was a disagreement between what the body contained and what the text stated ought to be in it, this would be explained by the state of the body, a diseased organ, normal human variability, or even the incompetence of the dissector.

Mondino's manual owes its success to a growing 14th-century need for such a work. Medical schools were just beginning to insist that students should witness at least one complete dissection and as Mondino's text was available it captured the market.

Monge, Gaspard (1746–1818) *French mathematician*

> Monge knew to an extraordinary degree how to conceive of the most complicated forms in space, to penetrate to their general relation and their most hidden properties with no other help than his hands.
> —Michel Chasles, *Aperçu historique sur l'origine et le développement des méthodes en géométrie* (1837; Historical Note on the Origin and Development of Geometrical Methods)

Monge (monzh), who was born at Beaune in France, was trained as a draftsman at Mézières, where he later became professor of mathematics (1768). During the French Revolution he served on the committee that formulated the metric system (1791), became minister of the navy and the colonies (1792–93), and played a vital part in organizing the defense of France against the counterrevolutionary armies. He contributed significantly to the founding of the Ecole Polytechnique in 1795. Monge met Napoleon in 1796 and saw active service in Napoleon's army during the Egyptian campaign (1798–1801).

Monge's major mathematical achievements were the invention of descriptive geometry and the application of the techniques of analysis to the theory of curvature. The latter ultimately led to the revolutionary work of Bernhard RIEMANN on geometry and curvature.

Following Napoleon's fall from power in 1815, Monge was expelled from the French Academy and deprived of all his honors.

Moniz, Antonio Egas *See* EGAS MONIZ, ANTONIO.

Monod, Jacques Lucien (1910–1976) *French biochemist*

> Language may have created man, rather than man language.
> —Inaugural lecture at the Collège de France, 3 November 1967

> In science, self-satisfaction is death. Personal self-satisfaction is the death of the scientist. Collective self-satisfaction is the death of the research. It is restlessness, anxiety, dissatisfaction, agony of mind that nourish science.
> —*New Scientist*, 17 June 1976

Monod (mo-**noh**) was born in Paris, France, and graduated from the university there in 1931; he became assistant professor of zoology in 1934, having spent the years immediately following his graduation investigating the origin of life. After World War II, in which he served in the Resistance, he joined the Pasteur

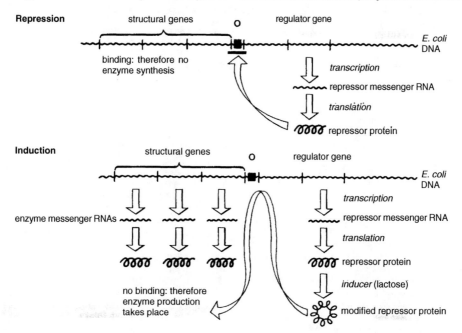

THE OPERON OF MONOD AND JACOB   This operon, which controls lactose metabolism in the bacterium Escherichia coli, consists of stuctural genes and an operator gene (O). Close by is a regulator gene. In the absence of lactose (top), repressor protein inhibits synthesis of the enzyme required for lactose metabolism by binding to the operator. When lactose (the inducer) is present (bottom), it modifies the shape of the repressor protein so that it cannot bind to the operator.

Institute, becoming head of the cellular biochemistry department in 1953.

In 1958 Monod began working with François JACOB and Arthur PARDEE on the regulation of enzyme synthesis in mutant bacteria. This work led to the formulation, by Monod and Jacob, of a theory explaining gene action and particularly how genes are switched on and off as necessary. In 1960 they introduced the term "operon" for a closely linked group of genes, each of which controls a different step in a given biochemical pathway. The following year they postulated the existence of a molecule, messenger RNA, that carries the genetic information necessary for protein synthesis from the operon to the ribosomes, where proteins are made. For this work Monod and Jacob were awarded the 1965 Nobel Prize for physiology or medicine, which they shared with André LWOFF, who was also working on bacterial genetics.

In 1971 Monod became director of the Pasteur Institute and in the same year published the best-selling book *Chance and Necessity*, in which he argued that life arose by chance and progressed to its present level as a necessary consequence of the pressures exerted by natural selection.

Monro, Alexander (Primus) (1697–1767) *Scottish anatomist*

Monro's father, John, was a surgeon serving first in the army of William III before settling in Edinburgh into private practice. Alexander Monro, who was born in London, was educated at the University of Edinburgh and then entered into an apprenticeship with his father. He later pursued his medical studies first in London under William Cheselden and then, in 1718, with Hermann BOERHAAVE in Leiden. He returned to Edinburgh in 1719 and in the following year was appointed to the first chair of anatomy, or indeed of any medical subject, at the university.

This was the beginning of the famous Edinburgh medical school. It was also the beginning of three generations of Monros in charge of the school. As they were all called Alexander they are normally distinguished by their order in the sequence with the founder known as "Primus." Under Monro the school flourished with the number of students admitted quadrupling before his retirement in 1764.

Monro published one original work, *Anatomy of Human Bones* (1726), but his career was mainly devoted to teaching and the establishment of the new medical school. In 1752 he produced one of the earliest figures for mortality arising from amputations. The figures are somewhat puzzling for he claimed a mortality of only 8% while Joseph LISTER in the early days of antiseptic surgery had figures as high as 25%.

He was succeeded by his youngest son, Alexander, or "Secundus."

Monro, Alexander (Secundus) (1733–1817) *Scottish anatomist*

Monro, the son of Alexander Monro "Primus," entered the university in his native city of Edinburgh at the age of 12. He began the study of medicine in 1750 being made conjoint professor of anatomy with his father in 1754, a year before his graduation. He took over completely on his father's retirement in 1764, holding the chair himself until his own retirement in 1807.

Of the three Monros Secundus was the most significant. His most important work, although it involved him in a prolonged priority dispute with William HUNTER, was his *De venis lymphaticis* (1757; On the Lymphatic Veins), which clearly distinguished between the lymphatic and circulatory systems. He also produced one of the earliest works of comparative anatomy in his *Structure and Physiology of Fishes* (1785). His name is preserved by his description in 1764 of the *foramen of Monro*, an opening connecting the lateral and third ventricles in the brain.

Monro continued the family tradition, being succeeded by his son, also called Alexander and referred to as "Tertius."

Montagnier, Luc (1932–) *French virologist*

Montagnier (mon-ta-**nyay**), who was born at Chabris in France, was educated at the universities of Poitiers and Paris. He joined the Viral Oncology Unit of the Pasteur Institute in 1972 and was appointed professor of virology in 1985.

Montagnier's team at the Institute were searching for, among other things, possible links between cancers and retroviruses. The retroviruses had been described in 1970 by TEMIN and Baltimore and were distinguished from other viruses by having RNA rather than DNA genes. In early 1983 they were presented with a blood sample from a patient showing early signs of AIDS. Reverse transcriptase, an enzyme characteristic of retroviruses, was found in the blood. Montagnier sought to identify the virus. It was not HTLV-1, a retrovirus recently discovered by Robert GALLO, as serum from the AIDS patient did not react with samples of HTLV-1 provided by Gallo. The virus was found in T-4 cells, specialized lymphocytes of the immune system, and was therefore named LAV as an acronym for "lymphadenopathy associated virus." Electron micrographs taken of LAV differed from those of HTLV-1.

Montagnier went on to develop a blood test for the presence of LAV. Antibodies to LAV were found in a number of patients with AIDS. As the sensitivity of the test increased, Montagnier was able to identify more and more AIDS patients and by October 1983 he was convinced that LAV was the cause of AIDS. By this time, however, Gallo had isolated a new retrovirus, HTLV-3, which he was equally convinced was the cause of AIDS. It was eventually agreed, despite some considerable initial controversy, that HTLV-3 and LAV were to all intents and purposes the same virus. In 1986 it was officially renamed HIV and the patent for HIV blood tests carried the names of both Gallo and Montagnier.

A further advance was made by Montagnier in late 1985 while examining blood samples from Guinea-Bissau in West Africa. He was puzzled by the fact that some of the samples came from apparently HIV-negative AIDS patients, even though they had been tested with a sensitive new probe. Montagnier resolved the issue by isolating a virus from the samples which differed from electronmicrographs of HIV-1. Montagnier named the virus HIV-2 and demonstrated that antibodies to the new virus were commonly found in blood samples from West African AIDS patients.

Montgolfier, Etienne Jacques de *See* MONTGOLFIER, MICHEL JOSEPH DE.

Montgolfier, Michel Joseph de (1740–1810) *French balloonist*

Michel Montgolfier (mon-gol-**fyay** or mont-**gol**-fee-er) and Etienne Jacques Montgolfier (1745–1799), the sons of a paper manufacturer from Vidalon-les-Annonay, Lyons, engaged themselves in various enterprises. Michel founded his own paper factory in 1771, while Etienne practiced as an architect until 1782 when called upon to run the family factory at Vidalon. In later life Michel abandoned business and was appointed in 1800 to the faculty of the Conservatoire des Arts et Métiers.

Like many before, the brothers had noticed how pieces of paper thrown into the fire would often rise aloft in a column of hot air. They were interested enough to see whether paper bags filled with hot smoke would rise. Satisfied with their small-scale experiments they became convinced that something much larger was viable. On 4 June 1783 they gave the first public demonstration of their work at Annonay. The balloon was made of linen and lined with paper, measured 36 feet across, and weighed 500 pounds. Once inflated over a fire burning chopped straw, the balloon ascended to a height of 6,000 feet before coming down ten minutes later, a mile and a half away. News quickly spread throughout France. Called to Versailles they demonstrated their balloon, this time carrying a sheep, a cock, and a duck, before Louis XVI and Marie Antoinette. The balloon landed two miles away in a wood with the animals none the worse for their journey.

The first manned flight was made by François de Rozier in Paris in October 1783. Of the brothers, only Michel flew in the balloon, making an ascent of 3,000 feet with seven other people in 1784.

Moore, Stanford (1913–1982) *American biochemist*

Moore, who was born in Chicago, Illinois, graduated in chemistry from Vanderbilt University in 1935 and received his PhD from the University of Wisconsin in 1938. He then joined the staff of the Rockefeller Institute, spending his entire career there and serving as professor of biochemistry from 1952 onward.

One of the major achievements of modern science has been the determination by Frederick SANGER in 1955 of the complete amino acid sequence of a protein. Sanger's success with the insulin molecule inspired Moore and his Rockefeller colleague, William STEIN, to tackle the larger molecule of the enzyme ribonuclease. Although their work was lightened by the availability of techniques pioneered by Sanger the labor involved was still immense until eased by their development of the first automatic amino-acid analyzer.

They inserted a small amount of the amino-acid mixture into the top of a five-foot column containing resin. They then washed down the mixture using solutions of varying acidity. The individual amino acids travel down the column at different rates depending on their relative affinity for the solution and for the resin. It is possible to adjust the rates of travel so that the separate amino acids emerge from the bottom of the column at predetermined and well-spaced intervals. The colorless amino acids were then detected with ninhydrin, a reagent that forms a blue color on heating with proteins and amino acids. A continuous plot of the intensity of the blue color gives a series of peaks, each corresponding to a certain amino acid with the area under the peak indicating the amount of each.

By the end of the 1950s Moore and Stein had not only established the sequence of ribonuclease but they were also able to indicate the most likely active site on the single-chained molecule. For this work they shared the 1972 Nobel Prize for chemistry with Christian ANFINSEN.

Mordell, Louis Joel (1888–1972) *American–British mathematician*

Mordell had his first mathematical education at school in his native city of Philadelphia. He decided to try for a scholarship to Cambridge University, England, and having scraped together enough money for a one-way ticket made the crossing, took the exam, and got the scholarship. Mordell taught at Birkbeck College, London, from 1913 to 1920. From 1920 to 1945 he held posts at the Manchester College of Technology and at Manchester University. While in Manchester he got to know Sydney CHAPMAN, and the two remained life-long friends. From 1945 to 1953 he held the Sadleirian Chair in Mathematics at Cambridge.

Mordell's central interest was in the theory of numbers and, in particular, Diophantine equations. He worked on the theory of modular functions and their applications to the theory of numbers. In the 1920s he published his most important single result, the Mordell finite basis theorem. André WEIL later generalized it, and the result, which plays a fundamental role in number theory, is now usually known as the *Mordell–Weil theorem*. Among his other works are results on the estimation of trigometric and character sums, cubic surfaces and hypersurfaces, and the geometry of numbers.

Morgagni, Giovanni Batista (1682–1771) *Italian pathologist*

Morgagni (mor-**gah**-nyee) was born at Forli in Italy and educated at the University of Bologna; after graduating in philosophy and medicine in 1701, he worked as a demonstrator in anatomy and served as assistant to his former teacher, Antonio Valsalva. In 1712 he moved to the University of Padua and served there as professor of anatomy from 1715 until his death.

In 1761 Morgagni published his classic work, *De sedibus et causis morborum per anatomen indagatis* (On the Seats and Causes of Diseases as Investigated by Anatomy) in which he used the reports of 640 autopsies, many carried out personally, to produce this pioneering text in morbid anatomy. There were earlier collections of autopsies, such as the work of Theophilus Bonet whose *Sepulchretum* (1679; Cemetery) reported over 3,000, but they tended to be uncritical compilations with an emphasis on the bizarre.

With a general decline in the plausibility of the humoral theory of disease, scientists began the long search for a viable alternative. Morgagni took one of the first steps by searching for the cause of disease in the lesions present in particular organs. He would begin by presenting a full clinical picture of the disease, which he would then try to link with the presence of some lesion observed during the autopsy. With experience he began to find that he could frequently predict the lesion from the case history alone.

Many were to follow him over the next century. Some, like Marie François BICHAT and Rudolf VIRCHOW, would propose alternative "seats" for disease in the tissues and cells respectively. To others, like Louis PASTEUR, the crucial question raised by the new pathological anatomy of Morgagni was what caused the lesions, the answer to which would ultimately lead to the discovery of pathogens.

Morgan, Thomas Hunt (1866–1945) *American geneticist*

> The participation of a group of scientific men united in a common venture for the advancement of research fires my imagination to the kindling point.
> —Letter to George Ellery Hale, 9 May 1927

Born in Lexington, Kentucky, Morgan studied zoology at the State College of Kentucky, graduating in 1886. He received his PhD from Johns Hopkins University in 1890 and from 1891 to 1904 was associate professor of zoology at Bryn Mawr College. He carried out his most important work between 1904 and 1928, while professor of experimental zoology at Columbia University. Here he became involved in the controversy that followed the rediscovery, in 1900, of Gregor MENDEL's laws of inheritance.

Many scientists had noted that Mendel's segregation ratios fitted in well with the observed pattern of chromosome movement at meiosis. Morgan, however, continued to regard Mendel's laws with skepticism, especially the law of independent assortment, and with good reason. It was known by then that many more characters are determined genetically than there are chromosomes and therefore each chromosome must control many traits. It was also known that chromosomes are inherited as complete units, so various characters must be linked together on a single chromosome and would be expected to be inherited together.

In 1908 Morgan began breeding experiments with the fruit fly *Drosophila melanogaster*, which has four pairs of chromosomes. Morgan's early results with mutant types substantiated Mendel's law of segregation, but he soon found evidence of linkage through his discovery that mutant white-eyed flies are also always male. He thus formulated the only necessary amendment to Mendel's laws – that the law of independent assortment only applies to genes located on different chromosomes.

Morgan found that linkages could be broken when homologous chromosomes paired at meiosis and exchanged material in a process known as "crossing over." Gene linkages are less likely to be broken when the genes are close together on the chromosome, and therefore by recording the frequency of broken linkages, the positions of genes along the chromosome can be mapped. Morgan and his colleagues produced the first chromosome maps in 1911.

For his contributions to genetics, Morgan received the Nobel Prize for physiology or medicine in 1933. A prolific writer, his most influential books – produced with colleagues at Columbia – are *The Mechanism of Mendelian Heredity* (1915) and *The Theory of the Gene* (1926).

Morgan, William Wilson (1906–1994)
American astronomer

Morgan, who was born in Bethesda, Tennessee, studied at the Washington and Lee University and then at the University of Chicago where he obtained his BSc in 1927 and his PhD in 1931. The following year he took up an appointment at the Yerkes Observatory, where he served as professor from 1947 to 1966 and Sunny Distinguished Professor of Astronomy from 1966 to 1974. He was Director of the Yerkes and McDonald observatories from 1960 to 1963.

The standard system of spectral classification of stars, the Henry DRAPER system, assigned the majority of stars to one of the classes, O, B, A, F, G, K, or M, which were each subdivided into ten categories numbered from 0 to 9. In this classification our Sun is assigned the number G2. Useful as the Draper system is, Morgan realized that it had its limitations. He pointed out that the system was based only on the surface temperature of stars and commonly produced cases where two stars, like Procyon in Canis Minor and Mirfak in Perseus, fell into the same spectral class, F5 in this case, yet differed in luminosity by a factor of several hundreds.

Consequently, in collaboration with Philip Childs KEENAN and Edith Kellman, he introduced the *Yerkes system* or *MKK system* (also known as the Morgan–Keenan classification) in 1943 in *An Atlas of Stellar Spectra with an Outline of Spectral Classification*. The new system was two dimensional, containing in addition to the spectral typing a luminosity index. This was used to classify stars in terms of their intrinsic brightness by means of Roman numerals from I to VI, and ranged from supergiants (I), giants (II and III), subgiants (IV), main-sequence stars (V), to subdwarfs (VI). Procyon thus becomes a F5 IV star while Mirfak is a distinguishable F5 I supergiant.

In the 1940s Walter BAADE had shown that hot O and B stars were characteristic members of the spiral arms of a galaxy. Morgan and his colleagues thus began to trace out the structure of our own Galaxy by searching for clouds of hydrogen ionized by O and B stars. By 1953 they claimed to have identified the Perseus, Orion, and Sagittarius arms of the Galaxy, thus providing good evidence for its spiral structure. Morgan also worked on star brightness and discovered so-called "flash" variables – stars that change their luminosity very quickly. He also worked on galaxy classification.

Morley, Edward Williams (1838–1923)
American chemist and physicist

Morley's father was a Congregational minister from Newark, New Jersey. For reasons of health Morley was educated at home and appears to have been a precocious child, starting Latin at the age of 6 and being able to read Greek at 11. He taught himself chemistry by mastering the textbook of Benjamin SILLIMAN when only 14. He received his higher education at Williams College (1857–60) and Andover Theological Seminary (1861–64). As he felt he was not fit enough to follow his father into the ministry he opted instead for a university career and in 1869 was appointed professor of natural history and chemistry at Western Reserve College, Hudson, Ohio.

Morley was a very careful and patient experimentalist. In 1878 he began a study of the variations in oxygen content of the atmosphere. This was followed in the period 1883–94 by a study of the relative weights of oxygen and hydrogen in pure water. After the weighings of literally thousands of samples he produced figures to five decimal places and accurate to one part in 300,000.

He is, however, best remembered for his collaboration with Albert MICHELSON in their famous experiment to detect the Earth's motion through the ether. Morley was not completely convinced by the negative result and did further experiments with Dayton Clarence MILLER. Despite many refinements no conclusive effect was found.

Morse, Samuel (1791–1872) *American inventor and painter*

> What God hath wrought.
> —The first message sent on Morse's telegraph (1844)

Morse, the inventor of telegraphy, was born the son of a Calvinist minister and distinguished geographer in Charlestown, Massachusetts. Although not very keen on study at school, he was sent to Yale. Apart from painting, his main interest, he was stimulated to further study by lectures on electricity. In 1810 he graduated and took a job as a clerk in Boston. He visited England in 1811 to study painting, returning in 1815 to earn his living as a portrait painter. He was a founder and first president (1826–45) of the National Academy of Design at New York. Throughout his life Morse was politically conservative and religiously orthodox, at one point being involved in campaigning against "licentiousness" in the theater.

In 1832, after overhearing a conversation about electromagnets, he had the idea of a design for the electric telegraph. Although the idea itself was not entirely new, his was the first definite design. By 1835 he had built a working model and from 1837 onward he concentrated his attention on developing the telegraph system, helped out by several friends. A year later he devised the *Morse code*, a system of dots and dashes for sending messages as electrical signals. In 1843 Congress financed the first telegraph line, between Washington and Baltimore.

Morton, William Thomas Green (1819–1868) *American dentist*

Morton, who was born the son of a small farmer and village shopkeeper in Charlton City, Massachusetts, is believed to have trained as a dentist at the Baltimore College of Dentistry. After a brief partnership with Horace WELLS, Morton set up in practice in Boston.

To alleviate the pain of tooth extraction Morton experimented with such drugs as opium and alcohol, but only succeeded in making his patients violently sick. The chemist Charles Jackson advised Morton to try ether, an old student standby, as a local anesthetic. This was moderately effective and Morton decided to try ether inhalation to produce general anesthesia. He first used ether to extract a tooth on 30 September 1860. His initial successes left Morton confident enough to offer to demonstrate his technique at the Massachusetts General Hospital. He was successful in using it on a patient who was undergoing a tumor operation. His innovation was well received by the leading surgeon John WARREN and the use of ether quickly gained acceptance in medical practice. The news soon spread to Europe and in December 1846 Robert Liston, the skilled British surgeon, used ether in a painless and successful leg amputation at University College Hospital, London.

Morton subsequently went to a lot of trouble trying to patent his anesthetic and fight off competitors, notably Jackson, who were claiming priority. His wrangling with Jackson, the government, and the law courts achieved little and Morton died virtually penniless while traveling to New York to answer yet another attack on him from Jackson.

Mosander, Carl Gustav (1797–1858) *Swedish chemist*

Born at Kalmar in Sweden, Mosander (moo-san-der) started his career as a physician and became Jöns BERZELIUS's assistant after a time in the army. He became curator of minerals at the Royal Academy of Science in Stockholm before succeeding Berzelius as secretary. In 1832 he became professor of chemistry and mineralogy at the Karolinska Institute, Stockholm.

Mosander worked chiefly on the lanthanoid elements. These had been known since the discovery of yttrium by Johan GADOLIN in 1794 and cerium by Martin KLAPROTH in 1803. He began by examining the earth from which cerium had been isolated, ceria. From this he derived in 1839 the oxide of a new element, which he called lanthanum, from the Greek meaning "to be hidden." In 1843 he announced the discovery of three new rare-earth elements – erbium, terbium, and didymium. As it happened, didymium was not elementary, being shown in 1885 by Karl Auer von Welsbach to consist of two elements – praseodymium and neodymium.

Moseley, Henry Gwyn Jeffreys (1887–1915) *British physicist*

Moseley came from an academic family in Weymouth, southwest England. He graduated in natural sciences from Oxford University in 1910 and then joined Ernest RUTHERFORD at Manchester University to work on radioactivity, although he soon turned his attention to x-ray spectroscopy. He returned to Oxford in 1913 to continue his work under J. S. E. TOWNSEND.

When x-rays are produced by an element a

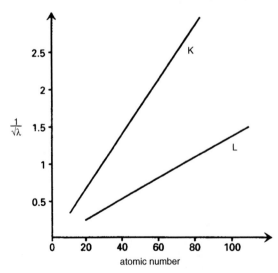

MOSELEY'S LAW   The relationship between characteristic x-rays and atomic number for the chemical elements.

continuous spectrum is emitted together with a more powerful radiation of a few specific wavelengths characteristic of the element. To investigate the positive charge on atomic nuclei Moseley examined these characteristic spectral lines using crystal diffraction. For a number of elements, he discovered a regular shift in the lines with increasing atomic weight. From this he determined for each element an integer approximately proportional to the square root of the frequency of one of its spectral lines. This integer, now called the atomic number (or proton number), equaled the positive charge on the atomic nuclei. Moseley's work led to major improvements in Dmitri MENDELEEV's periodic table and enabled elements to be classified in a new and more satisfactory manner.

At the outbreak of World War I Moseley enlisted in the army and was commissioned in the Royal Engineers. His death, from a sniper's bullet at Gallipoli, cut short what promised to be a most brilliant career in science.

Mössbauer, Rudolph Ludwig (1929–)
German physicist

Born at Munich in Germany, Mössbauer (**mu(r)s**-bow-er or **mos**-bow-er) was educated in Munich-Pasing and, after a year in industrial laboratories, studied physics at the Munich Technical University. There he passed his intermediate degree in 1952, and completed his thesis in 1954. From 1955 to 1957 he did postgraduate research at the Max-Planck Institute for Medical Research in Heidelberg, gaining his doctorate from the Technical University in 1958.

From 1953 he had been studying the absorption of gamma rays in matter, in particular the phenomenon of nuclear resonance absorption. Normally, when an atomic nucleus emits a gamma ray, it will recoil, and this recoil action will influence the wavelength of the gamma ray emitted. Mössbauer discovered that, contrary to classical predictions, at a sufficiently low temperature the nucleus can be locked into position in the crystal lattice, and it is the lattice itself that recoils, with negligible effect on the wavelength. The result is that the wavelength can be defined with extremely high precision (about 1 part in 10^{12}). As with emission, so it is with absorption; a crystal of the same material under similar conditions absorbs gamma rays at the same highly specific wavelength – a resonance phenomenon akin to a well-tuned radio receiver and transmitter. If, however, the conditions are slightly different, the small changes in wavelength can be accurately compensated and thus measured using the DOPPLER effect (by moving the source relative to the receiver).

This phenomenon of recoilless nuclear resonance absorption, now known as the *Mössbauer effect*, has given physicists and chemists a very useful tool through the high precision of measurement it allows. In particular, it allowed the first laboratory testing (and verification in 1960) of the prediction of EINSTEIN's general theory of relativity that the frequency of an electromagnetic radiation (in this case gamma rays) is influenced by gravity. The Mössbauer effect is now commonly employed as a spectroscopic method in chemical and solid-state physics because of its ability to detect differences in the electronic environments surrounding certain nuclei (*Mössbauer spectroscopy*).

In 1960, after finishing his studies at the Technical University, Mössbauer went on to continue his investigations of gamma absorption at the California Institute of Technology, Pasadena, where he was appointed professor of physics the next year. In 1961 he also received the Nobel Prize for physics, sharing the honor with Robert HOFSTADTER who had advanced knowledge of the nucleus by electron-scattering methods. Mössbauer is currently a professor at the Munich Technical University.

Mott, Sir Nevill Francis (1905–1996)
British physicist

Born at Leeds in England, Mott studied at Cambridge University, gaining his bachelor's degree in 1927 and his master's in 1930. He never pursued a doctorate, but from 1930 until 1933 was a lecturer and fellow of Gonville and Caius College, Cambridge. Subsequently he moved to Bristol University as a professor of theoretical physics. In 1948 he became director of Bristol's physics laboratories, but returned later to Cambridge as Cavendish Professor of Experimental Physics, where he served from 1954 until his retirement in 1971.

Mott's work in the early 1930s was on the quantum theory of atomic collisions and scattering. With Harrie Massey he wrote the first of several classic texts, *The Theory of Atomic Collisions* (1933). Other influential texts that followed were on *The Theory of Properties of Metals and Alloys* with H. Jones (1936), *Electronic Processes in Ionic Crystals* with R. W. Gurney (1940), *Electronic Processes in Non-Crystalline Materials* with E. A. Davis (1979), and *Metal–Insulator Transitions* (1990). Each marked a significant phase of active research. Mott began to explore also the defects and surface phenomena involved in the photographic process (explaining latent-image formation), and did significant work on dislocations, defects, and the strength of crystals.

By the mid-1950s, Mott was able to turn his attention to problems of disordered materials, liquid metals, impurity bands in semiconductors, and the glassy semiconductors. His models of the solid state became more and more complex, and included an analysis of electronic processes in metal–insulator transitions, often called *Mott transitions*.

In 1977 Mott shared the Nobel Prize for physics with Philip ANDERSON and John VAN VLECK for their "fundamental theoretical investigations of the electronic structure of magnetic and disordered systems." Mott was knighted in 1962. His autobiography, *A Life in Science*, was published in 1986. In the final years of his life he worked on high-temperature superconductivity.

Mottelson, Benjamin Roy (1926–) *American–Danish physicist*

Mottelson, who was born in Chicago, Illinois, graduated from Purdue University in 1947 and gained his PhD in theoretical physics at Harvard University in 1950.

From Harvard, Mottelson gained a traveling fellowship to the Institute of Theoretical Physics in Copenhagen (now the Neils Bohr Institute). There he worked with Neils Bohr's son, Aage Bohr, on problems of the atomic nucleus. In particular, they considered models of the nucleus and combined the two principal theories current at the time – one based on independent particles regarded as arranged in shells and the other treating the nucleus as a collective entity exhibiting liquid-drop-like behavior – and advanced a unified theory. They worked out the consequences of the interplay between the individual particles and the collective motions, specified the structure of the rotational and vibrational excitations and the coupling between them, and showed how the collective concepts could be applied to the nuclei of various elements. For their work on nuclear structure Mottelson, Bohr, and James RAINWATER (Bohr's earlier collaborator at Columbia University) shared the 1975 Nobel Prize for physics.

Mottelson held a research position in CERN (the European Center for Nuclear Research) from 1953 until 1957, then returned to Copenhagen to take up a professorship at the Nordic Institute for Theoretical Atomic Physics (NORDITA) adjacent to the Neils Bohr Institute. He took Danish nationality in 1973.

Together with Aage Bohr, he published *Nuclear Structure* (2 vols., 1969–75).

Moulton, Forest Ray (1872–1952) *American astronomer*

Born in Osceola County, Michigan, Moulton was educated in frontier schools, Albion College, and the University of Chicago where he obtained his PhD in 1899. He taught there until 1926, being made full professor in 1912. From 1927 to 1936 he worked in business before returning to science as executive secretary of the American Association for the Advancement of Science from 1936 to 1948.

Moulton is still remembered for his formulation of the planetismal theory of the origin of the planets in collaboration with Thomas CHAMBERLIN in 1904. They suggested that a star had passed close to the Sun and that this resulted in the ejection of filaments of matter from both stars. The filaments cooled into tiny solid fragments, "planetesimals." On collision the small particles stuck together (a process known as "accretion"). Thus over a very long period, grains became pebbles, then boulders, then even larger bodies. For larger bodies, the gravitational force of attraction would accelerate. In this way the protoplanets formed. This formation by accretion is still accepted although the stellar origin of the planetesimals has been largely dropped.

Mueller, Erwin Wilhelm (1911–1977) *German–American physicist*

Mueller (**moo**-ler or **myoo**-ler) was born in Berlin and studied engineering at the university there, gaining his PhD in 1935. He worked in Berlin at Siemens and Halske (1935–37) and at Stabilovolt (1937–46). Subsequently he was at the Altenberg Engineering School (1946–47) and the Fritz Haber Institute (1948–52), from where he moved to the Pennsylvania State University. He became a naturalized American in 1962.

He is noted for his fundamental experimental work on solid surfaces. In 1936 he invented the field-emission microscope. In this device, a fine metal point is placed a distance away from a phosphorescent screen in a high vacuum with a very high negative voltage applied to the point. Electrons are emitted from the surface under the influence of the electric field (field emission) and these travel to the screen where they produce a magnified image of the surface of the tip. The instrument is used to study reactions at surfaces.

In 1951 he made a further advance using the principle of field ionization. In the field-ion microscope the tip is at a positive potential in a low pressure of inert gas. Atoms of gas adsorbing on the tip are ionized and the positive ions are repelled from the tip and produce the image. The resolution is much better than in the field-emission microscope; in 1956, by cooling the tip in liquid helium, Mueller was able to resolve individual surface atoms for the first time.

As a further refinement Mueller used a field-ion microscope with a mass spectrometer, so that individual atoms on the surface could be seen, desorbed, and identified (the atom-probe field-ion microscope).

Müller, (Karl) Alex (1927–) *Swiss physicist*

Born at Basel in Switzerland, Müller (**mool**-er)

was educated at the Federal Institute of Technology, Zurich, where he obtained his PhD in 1958. After working for a few years at the Batelle Institute in Geneva, he returned to Zurich (1963) to take up a post at the IBM Research Laboratory at Rüschlikon, where he has remained ever since.

In 1911 Kammerlingh-Onnes discovered the phenomenon of superconductivity. He found that a current passing through mercury at 4 K, that is four degrees above absolute zero, met with no resistance. To utilize this discovery fully the temperature at which materials became superconductive, the critical temperature (T_c), would have to be raised to some more economically accessible level. Yet 75 years' intensive research had raised the critical temperature no higher than 23.3 K for a niobium–germanium alloy. And to cool the alloy to this point requires bathing it in either expensive liquid helium (bp 4.2 K) or the cheaper but flammable liquid hydrogen (bp 20.3 K).

Müller first began to work on the problem in 1983. He ignored the usual candidates for a high critical temperature and turned instead to look at ceramic metal oxides. This was partly because his laboratory had worked with oxides of this kind for many years and had built up a considerable expertise in them. Also, he suspected, their lattice structure was of the right kind to allow superconductivity. In January 1986 Müller, working with his IBM colleague Georg BEDNORZ, found that a mixed lanthanum, barium, and copper oxide showed a change to superconducting behavior below 35 K (–238° C). Once the initial advance had been made, other physicists were quick to follow and to confirm and extend Müller's work.

The significance of Müller's discovery was recognized with unusual speed by the Nobel authorities when, in the following year, they awarded the Nobel Prize for physics jointly to Müller and Bednorz.

Müller, Franz Joseph, Baron von Reichenstein (1740–1825) *Austrian geologist*

Müller (**mool**-er), who was born at Nagyszeben (now in Romania), was the son of a treasury official. He studied law in Vienna and mining at Schemnitz in Hungary. After a period working

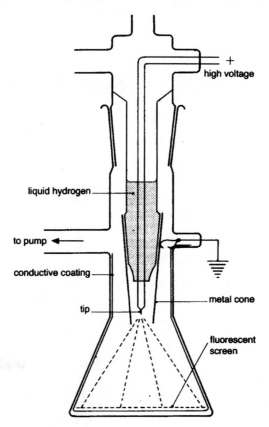

FIELD-ION MICROSCOPE   The field-ion microscope invented by Erwin Mueller.

in the state saltworks in Transylvania, he was appointed successively as director of the state mines of the Tyrol (1775–78), of Transylvania (1778–1802), and of Austria–Hungary (1802–18).

Officials had known for some time that the ores of Transylvania produced less gold and silver than elsewhere and it was assumed that they contained significant amounts of antimony. In 1782 Müller extracted from the local ores the substance responsible for the shortfall – a substance that he thought might be a new element. He sent some samples to Torbern BERGMAN, who confirmed that it was not antimony. Martin KLAPROTH later demonstrated that it was indeed a new element and named it tellurium, from the Latin *tellus* (earth).

Muller, Hermann Joseph (1890–1967)
American geneticist

> Death is an advantage to life ... Its advantage lies chiefly in giving ampler opportunities for the genes of the new generation to have their merits tested out ... by clearing the way for fresh starts.
> —*Science* No. 121 (1955)

Born in New York City, Muller (**mul**-er) was awarded a scholarship to Columbia University in 1907 and specialized in heredity during his undergraduate studies. On graduation he took up a teaching fellowship in physiology at Cornell Medical School, gaining his master's degree in 1912 for research on the transmission of nerve impulses. During this period he continued working at Columbia in his spare time, contributing to the genetic researches on *Drosophila* fruit flies. He was employed officially at Columbia in 1912 and received his PhD in 1916 for his now classic studies on the crossing over of chromosomes. He was also a coauthor of *The Mechanism of Mendelian Heredity* (1915), a fundamental contribution to classical genetics.

In 1915, at the request of Julian HUXLEY, Muller moved to the Rice Institute, Houston, Texas, where he began studying mutation. By 1918 he had found evidence that raising the temperature increases mutation rate. In 1920, after a brief spell back at Columbia, he joined the University of Texas as an associate professor, becoming a professor in 1925. In 1926 he found that x-rays induce mutations, a discovery for which he eventually received the 1946 Nobel Prize for physiology or medicine.

In 1933 Muller spent the first of eight years in Europe at the Institute for Brain Research, Berlin. Hitler's rise to power forced him to leave Germany and he moved to the Academy of Sciences, Leningrad, at the invitation of Nikolai VAVILOV. Muller believed that in a communist state he would be able to develop his own socialist ideas and apply his research to improve the human condition. However the advent of Lysenkoism effectively hampered most genetic research in Russia and Muller left, volunteering to serve in the Spanish Civil War. He then worked at the Institute of Animal Genetics, Edinburgh, returning to America in 1940. He held a position at Amherst College, Massachusetts, from 1942 until 1945, when he became professor of zoology at Indiana University, remaining there for the rest of his life.

Muller made important theoretical contributions to genetics. He visualized the gene as the origin of life, because only genes can replicate themselves, and he believed all selection and therefore evolution acted at the level of the gene. He worried about the increasing number of mutations accumulating in human populations, which can survive because of modern medical technology, and proposed a program of eugenics to overcome the problem. He fully realized the harm to human chromosomes that can result from ionizing radiation and campaigned against excessive use of x-rays in medicine, careless handling of nuclear fuels, and testing of atomic bombs.

Muller is seen by many as the most influential geneticist of the 20th century, mainly through his appreciation of genetic mutation as fundamental to future genetic research. He published over 350 works, the most important paper being *Artificial Transmutation of the Gene* (1927).

Müller, Johannes Peter (1801–1858)
German physiologist

> In intercourse with the external world we continually sense ourselves.
> —*Zur vergleichenden Physiologie des Gesichtssinnes des Menschen und der Tiere* (1826; Comparative Physiology of the Sense of Sight in Humans and Animals)

Müller (**moo**-ler), a shoemaker's son from Koblenz in Germany, graduated in medicine from the University of Bonn in 1822. He worked as a pathologist in Bonn until 1833 when he moved to the University of Berlin as professor of anatomy and physiology, a post he retained until his death.

Müller was the most important figure in 19th-century German physiology. Not only did he number among his pupils such figures as Hermann von HELMHOLTZ, Carl Ludwig, Rudolf VIRCHOW, and Max SCHULTZE but those he did not teach were reached by his influential work, *Handbuch der Physiologie des Menschen* (2 vols., 1834–40; Handbook of Human Physiology).

It was in the field of neurophysiology that Müller made his major contribution to science. In 1831 he neatly and conclusively confirmed the law of Charles Bell and François MAGENDIE, which first clearly distinguished between motor and sensory nerves. Using frogs and dogs, Müller cut through the posterior roots of nerves from a limb as they entered the spinal cord. The limb was insensible but not paralyzed. When however Müller severed the anterior root he found that the limb had become paralyzed but had not lost its sensibility.

He also worked on the cranial nerves and succeeded in showing that the first two branches of the trigeminal nerve are sensory while the third branch, to the jaw, contains motor fibers also. The vagus and the glossopharyngeal were, Müller claimed, mixed nerves.

Müller also formulated, in 1826, the law of specific nervous energies, which claimed that nerves are not merely passive conductors but that each particular type of nerve has its own special qualities. For example, the visual nerves, however they may be stimulated, are only capable of transmitting visual data. More specifically, if such a nerve is stimulated, whether by pressure, electric current, or a flashing light, the result will always be a visual experience.

After the completion of the *Handbuch* in 1840 Müller turned more to problems of anatomy and physiology. He worked with Robert REMAK on embryological problems and was the first to describe what later became known as the *Müllerian duct*. This is a tube found in vertebrate embryos, which develops into the oviduct in females; it is found only vestigially in males. He also spent a large amount of time collecting and classifying zoological specimens.

Müller was much given to fits of depression, frequently feeling that his own creativity was exhausted. Consequently when he was found dead in bed, although no autopsy was ever performed, it was widely assumed that he had died by his own hand.

Müller, Otto Friedrich (1730–1784) *Danish microscopist*

One of the earliest microscopists, Müller was among the first to study and describe microorganisms, such as bacteria, diatoms, and infusoria, of which he discovered many new species. He introduced the terms "bacillum" and "spirillum" for two groups of bacteria, and devised a system of classification of microscopic animals and plants, based on the Linnean system. Born in Copenhagen, Müller was professor of botany at the university there and a proponent of the theory of spontaneous generation. His published works include the continuation of Georg Christian Oeder's *Flora Danica* and other studies of Danish flora and fauna, including works on insects and earthworms.

Müller, Paul Hermann (1899–1965) *Swiss chemist*

Müller, who was born in Olten, Switzerland,

was educated at the University of Basel where he obtained his PhD in 1925. From then until 1961 when he retired Müller worked for the Swiss dye firm of J. R. Geigy as a research chemist.

In 1935 Müller began looking for a potent and persistent insecticide that would nevertheless be harmless to plants and warm-blooded animals. Five years later he took out a patent on a chemical that had first been prepared in 1873. The compound was dichloro-diphenyltrichloroethane which, not surprisingly, was soon abbreviated to DDT. It turned out to be cheap and simple to manufacture, requiring only chlorine, ethanol, benzene, and sulfuric acid, all of which were available in bulk from the heavy chemical industry.

It soon proved its effectiveness as an insecticide during the war. Müller thought it to be toxic only against insects and soon extravagant claims were being made about the elimination of arthropod-borne diseases. Before long, however, the insects appeared to be more resilient than chemists had supposed and DDT more destructive of life and ecosystems than they imagined. Several advanced countries were to ban it.

Müller was awarded the Nobel Prize for physiology or medicine in 1948 for his discovery.

Muller, Richard August (1944–   ) *American physicist*

Born in New York, Muller was educated at Columbia and at the University of California, Berkeley, where he completed his PhD in 1969. He has remained at Berkeley and was appointed professor of physics in 1980.

Muller has made numerous contributions to several areas of physics, including optics, astrophysics, and high-energy physics. In 1983, however, he began to work on a problem which brought him before a much wider audience. His colleague Luis ALVAREZ drew his attention to the claim made by the paleontologist David RAUP that there were increased rates of extinction every 26 million years. What could possibly produce such periodicity?

One possibility was that the extinctions could be caused by collisions with other astronomical objects. Generally this would be random but Muller saw that there could be a way to produce periodic collisions with asteroids. If the Sun had a companion star that approached the earth every 26 million years bringing asteroids in its train, then the periodicity would be explained. Muller later thought the companion star, eventually to be named "Nemesis," brought with it comets from the OORT cloud rather than asteroids.

Muller found further evidence from a study of impact craters. If the craters were caused by the passage of Nemesis then they too would show an appropriate periodicity. Although the data was sketchy Muller found a fair enough fit to maintain his interest in the project. Despite some prolonged searches of the heavens, Nemesis has continued to remain undetected. In his own account of the issue, *Nemesis* (London, 1989), Muller, while still supporting periodic extinctions, sees his own work as "an elegant theory, a marvelous prediction, that needs verification."

Mulliken, Robert Sanderson (1896–1986) *American physicist and chemist*

Mulliken was born in Newburyport, Massachusetts, the son of an organic chemist. He was educated at the Massachusetts Institute of Technology and at the University of Chicago, where he obtained his PhD in 1921. After working briefly at Washington Square College, New York, he was appointed to the staff of the University of Chicago, where he served as professor of physics from 1931 until he retired in 1961. From 1961 he was Distinguished Service Professor of Physics and Chemistry at Chicago and Distinguished Research Professor of Chemical Physics at Florida State University.

It was Mulliken, in 1922, who first suggested a method of isotope separation by evaporative centrifuging. Most of his research career was concerned with the interpretation of molecular spectra and with the application of quantum theory to the electronic states of molecules.

Mulliken, with Freidrich HUND, developed the molecular-orbital theory of chemical bonding, which is based on the idea that electrons in a molecule move in the field produced by all the nuclei. The atomic orbitals of isolated atoms become molecular orbitals, extending over two or more atoms in the molecule. He showed how the relative energies of these orbitals could be obtained from the spectra of the molecule.

Mulliken's approach to finding molecular orbitals was to combine atomic orbitals (LCAO, or linear combination of atomic orbitals). He showed that energies of bonds could be obtained by the amount of overlap of atomic orbitals.

Another of Mulliken's contributions is the application of electronegativity – the ability of a particular atom in a molecule to draw electrons to itself. He showed that this property was given by the formula $1/2(I + E)$, where I is the ionization potential of the atom and E is its electron affinity.

He also made major contributions to the theory and interpretation of molecular spectra. In 1966 he was awarded the Nobel Prize for chemistry for "his fundamental work concerning chemical bonds and the electronic structure of molecules by the molecular-orbital method."

Mullis, Kary Banks (1944–) *American biochemist*

Born in Lenoir, North Carolina, Mullis was educated at Georgia Institute of Technology and at the University of California, Berkeley, where he completed his PhD in 1973. After postdoctoral periods at the University of Kansas Medical School and at the San Francisco campus of the University of California, Mullis joined the Cetus Corporation of Emeryville, California, in 1979.

One Friday night in April 1983 while driving to his weekend cabin, Mullis has recorded, it suddenly struck him that there was a method of producing unlimited copies of DNA fragments simply and *in vitro* (i.e., outside living cells). Previously, fragments could only be produced in limited numbers, in cells, and with much effort. Mullis named his method the "polymerase chain reaction" (PCR). The significance of the reaction can be judged by the price of $300 million placed by Cetus on the PCR patent sold to Hoffman-La Roche in 1991.

The first stage of the process is to heat DNA containing the required genetic segment in order to unravel the helix. Primers can then be added to mark out the target sequence. If, then, the enzyme DNA polymerase together with a number of free bases are added, two copies of the target sequence will be produced. These two copies can then be heated, separated, and once more produce two further copies each. The cycle, lasting no more than a few minutes, can be repeated as long as supplies last, doubling the target sequence each time. With geometric growth of this kind, more than 100 billion copies can be made in a few hours.

Relations between Mullis and Cetus quickly soured. He left the corporation in 1986 to work for a plastics manufacturer. But as the importance of his work began to be recognized Mullis found himself in sufficient demand to warrant his setting up as a consultant. One of his clients was Cetus as they fought off challenges to the PCR patent from DuPont and others. Mullis himself claims to be "tired of PCR" and more interested in "artificial intelligence, tunneling microscopes, science fiction, and surfing lessons."

Munk, Walter Heinrich (1917–) *American geophysicist*

Munk, who was born in the Austrian capital of Vienna, traveled to America in 1932; he was educated at the California Institute of Technology, graduating in 1939, and the University of California where he obtained his PhD in 1947. He taught initially at the Scripps Institution of Oceanography at the University of California before being appointed professor of geophysics there in 1954.

Munk's main fields of study were the irregularities in the rotation of the Earth, tides, ocean waves, currents, and wind stress. Measurements made in the 19th century by such workers as Seth CHANDLER had made it clear that the Earth's rotation was far from uniform either in its rate or on its axis. By 1930 determinations of stellar transits made with pendulum clocks revealed that a January day exceeded a July day by two milliseconds. Such work was confirmed by the more accurate quartz-crystal clocks introduced in the 1950s.

With his colleagues at Scripps, Munk began a major survey of the problem of the Earth's rotation, which culminated in the publication, in collaboration with Gordon MacDonald, of the monograph, *The Rotation of the Earth: A Geophysical Discussion* (1961). Using evidence from ancient sources, from the International Latitude Service, and modern observations of the Moon they attributed variations in the length of the day to seasonal shifts in the terrestrial air masses, ocean tides, the distribution of glaciation, and changes within the Earth's core.

Murad, Ferid (1936–) *American physician and pharmacologist*

Murad was born in Whiting, Indiana, and graduated as an MD from the Western Reserve University School of Medicine, Cleveland, Ohio, in 1965. Three years later he gained his PhD from the department of pharmacology at the same university. He has held many important appointments; since 1988 he has worked at the Northwestern University Medical School, Chicago, Illinois. In 1998 he shared the Nobel Prize for physiology or medicine with Robert FURCHGOTT and Louis IGNARRO for their discovery that molecules of the gas nitrogen monoxide (nitric oxide, NO) can transmit signals in the cardiovascular system.

The hitherto unknown substance endothelium-derived relaxing factor (EDRF) had been discovered by Furchgott in 1980. Murad researched the action of glyceryl trinitrate (nitroglycerin) and related vasodilators, and in 1977 discovered that they release nitrogen monoxide, which relaxes the walls of smooth muscle cells. It thus has the effect of controlling the blood pressure, which is why glyceryl trinitrate is prescribed as a drug to treat the heart condition atherosclerosis (as ironically it once was for Alfred Nobel, who invented the nitroglycerin-containing dynamite). Murad also speculated that hormones and other endogenous factors might act in a similar way, although no evidence of this was forthcoming.

Murchison, Sir Roderick Impey (1792–1871) *British geologist*

Murchison, the son of a Highland landowner from Tarradale in Scotland, was educated in Durham before entering the army at the age of 15. He served in the Napoleonic Wars, leaving the army after Waterloo, and settled in Durham where he intended to devote himself to a life of fox hunting. In 1824 his interest in science was

aroused by his friendship with Humphry Davy; he moved to London and began attending lectures at the Royal Institution.

He became particularly interested in geology, learning field geology from William BUCKLAND, and began exploring the main geological areas of Europe including Scotland, France, and the Alps. He was appointed president of the Geological Society in 1831 and in the same year concentrated his study on the Lower Paleozoic strata of South Wales. The strata he observed consisted chiefly of graywacke; he was soon able to establish that different beds of the graywacke were characterized by different fossils. His findings were outlined in *The Silurian System* (1839).

Together with Adam SEDGWICK, Murchison also worked in southwest England where they identified the Devonian system. He went on an expedition to Russia in 1841 following which he proposed the Permian system based on the stratification of the Perm area of Russia. Noted for his obstinate nature, Murchison quarreled with Sedgwick and Henry De la Bech. He tried unsuccessfully to prevent his Silurian system being divided to form the Silurian, Ordovician, and Cambrian periods.

Murchison became director general of the Geological Survey in 1855. In 1871 he founded the chair of geology and mineralogy at the University of Edinburgh. The *Murchison Falls*, Uganda, were named for him, Murchison being president of the Royal Geographical Society at the time when Richard Burton, John Hanning Speke, and Samuel Baker were searching for the source of the River Nile.

Murphy, William Parry (1892–1987) *American physician*

Murphy, the son of a Congregational minister from Stoughton, Wisconsin, was educated at the University of Oregon and at Harvard where he obtained his MD in 1920. He then served on the staff of both the Harvard Medical School and the Peter Bent Brigham Hospital.

Murphy collaborated with George MINOT in developing liver as a treatment of the then invariably fatal pernicious anemia. In 1934 he was able to report that 42 of the 45 patients originally treated in 1926 had been kept under observation. Of these 31 (75%) were alive and fit after ten years of treatment while the 11 who had died were victims of other complaints.

For this work Murphy shared the 1934 Nobel Prize for physiology or medicine with Minot and George WHIPPLE.

Murray, Sir John (1841–1914) *British marine zoologist and oceanographer*

Murray was born at Cobourg in Canada. A graduate of Edinburgh University, he was one of the naturalists on the *Challenger* expedition of 1872–76, the scientific reports of which he edited, becoming editor-in-chief in 1882. He completed much research on deep-sea deposits as well as observations of marine organisms such as foraminiferas and radiolarians. He was the inventor of a device for sounding and registering the temperature at great depths, conducting a bathymetrical survey of some of the freshwater lochs of Scotland. Murray took part in many other voyages, notably to the North Atlantic, the Faroe Channel, Spitzbergen and the Arctic, and to tropical Oceania. He wrote extensively on marine and freshwater biology, coral reef formation, and oceanography, including *The Ocean: A General Account of the Science of the Sea* (1913).

Murray, Joseph Edward (1919–) *American surgeon*

Murray was born in Milford, Massachusetts. Educated at Holy Cross College and at Harvard University, he embarked on a career in medicine, specializing in plastic surgery. He worked at the Peter Bent Brigham Hospital, Boston, becoming chief plastic surgeon (1964–86), held a similar position at the Children's Hospital Medical Center, Boston (1972–85), and served as professor of surgery at Harvard Medical School from 1970 to 1986.

Murray was a pioneer of kidney transplantation. In December 1954 he performed the first operation to implant a donor kidney into the pelvis of the recipient and attach it via the ureter to the bladder. Earlier attempts had placed the transplanted organ outside the body cavity, at sites such as the groin and armpit. The patient in Murray's operation was Richard Herrick, who received a kidney from his identical twin, Ronald.

The use of an organ from an identical sibling overcame the great obstacle of transplant surgery, namely rejection of the transplanted organ by the recipient's immune system. By receiving an organ of virtually identical tissue type, this first patient survived for eight years. For patients receiving organs from less closely related donors, the outlook was much worse.

Murray endeavored to improve the survival of the transplanted organ by suppressing the recipient's immune responses immediately prior to the operation. He conducted trials of the drug azathioprine, which killed cells of the immune system and so reduced the ability of the patient's own defense mechanism to reject the "foreign" tissue of a transplanted organ. Azathioprine had been developed by the British researcher, Roy Calne, working in collaboration with Murray at Boston. The drug proved to be an effective and much less hazardous alternative to Murray's initial method of using a massive dose of x-rays to suppress the recipient's immune system. (Azathioprine has now been superseded by cyclosporin, also developed by Calne.)

For his work in developing fundamental techniques in transplantation surgery, Murray was awarded the 1990 Nobel Prize for physiology or medicine, jointly with E. Donnall Thomas.

Muspratt, James (1793–1886) *Irish industrial chemist*

Muspratt, the son of a cork-cutter, was born and educated in Dublin before being apprenticed to a druggist. He developed an early interest in chemistry but before going into business he led an adventurous life and fought in Spain in the Peninsular War in both the army and the navy. He returned to Dublin in 1814 where, with the help of a small inheritance received in 1818, he manufactured various chemicals in a small way.

Muspratt moved to Liverpool, where he produced sulfuric acid, in 1822 and was quick to see the importance of the abolition in 1823 of the £30 per ton duty on salt. Cheap salt and the LEBLANC process meant that a plentiful supply of soda could be produced for the large demands of the soap, glass, and dyeing industries. With this financial incentive Muspratt set up on Merseyside the third soda plant in Britain. Close to both the salt mines of Cheshire and the textile industry of Lancashire, he was ideally situated. The need for expansion drove him into partnership with Josias GAMBLE and they founded the alkali industry in St. Helens (1828), but two years later he moved to Newton-le-Willows on his own.

One of the major problems of the Leblanc process was its production of quantities of hydrochloric acid gas as a waste product. This pollution raised protests, such as the letters appearing in a Liverpool paper in 1827 lamenting that the local church could not be seen from a distance of 100 yards (91 m) and was rapidly turning a dark color. The move to St. Helens only delayed the inevitable prosecutions. Muspratt was unwilling to use William GOSSAGE's tower and following litigation (1832–50) he was eventually successfully prosecuted by the neighboring farmers whose land was being destroyed. For this reason he moved his factories to Widnes and Flint in 1850. Following his retirement in 1857 they were run by his sons until in 1890 they became part of the United Alkali Company.

Musschenbroek, Pieter van (1692–1761) *Dutch physicist*

Musschenbroek (**mu(r)s**-en-brook) came from a family of instrument makers in Leiden in the Netherlands. He studied at the University of Leiden, where he gained an MD in 1715 and a PhD in 1719. After holding a chair of medicine at Duisburg (1721–23) and of natural philosophy at Utrecht (1723–40), Musschenbroek returned to Leiden and served as professor of physics until his death.

On 20 April 1746 Musschenbroek reported in a letter to René Reaumur details of a new but dangerous experiment he had carried out. He had suspended, by silk threads, a gun barrel, which received static electricity from a glass globe rapidly turned on its axis and rubbed with the hands. From the other end he suspended a brass wire, which hung into a round glass bottle, partly filled with water. He was in fact trying to "preserve" electricity by storing it in a nonconductor.

When Musschenbroek held the bottle with one hand while trying to draw sparks from the gun-barrel he received a violent electric shock. He had accidentally made the important discovery of the Leyden jar – an early form of electrical capacitor. This event captured both the popular and the scientific imagination and led to efforts by such scientists of the latter half of the 18th century as Benjamin FRANKLIN to understand the nature and behavior of electricity. The German inventor Georg von Kleist independently discovered the Leyden jar in 1745.

Muybridge, Eadweard James (1830–1904) *American photographer*

> He [Muybridge] invented one of the first high-speed shutters, and took a series of photos of a galloping horse by arranging a row of cameras beside the race track, operated by tapes which the horse ran through. This series ... caused consternation, since it showed that all previous ideas of how a horse moves were incorrect ... None of the photos showed the "hobbyhorse attitude," with front legs stretched forwards and hind legs stretched back, traditional in painting.
> —John Carey, *The Faber Book of Science* (1995)

Muybridge (**mI**-brij) was born Edward James Muggeridge at Kingston-on-Thames in Surrey, England. In his early twenties he changed his surname from Muggeridge to Muybridge and forename from Edward to Eadweard (the name of the Saxon kings who were crowned at Kingston in the 10th century). Although Muybridge spent much of his life in America, making his first trip there in 1852, he always retained links with his birthplace. Indeed, following a serious stagecoach accident in 1860 he returned to England to recuperate from his injuries.

By 1867 Muybridge was back in America, working as partner to the San Francisco-based photographer, Carleton E. Watson, and he quickly established a reputation as a skilled exponent of landscapes with a series of prints taken in California's Yosemite Valley. In 1868 he was appointed director of photographic surveys for the U.S. Government, and undertook photographic surveys of several remote regions, including the ports and harbors of newly purchased Alaska.

An interest in high-speed photography can be traced to the year 1872, when Muybridge was commissioned by the wealthy Californian racehorse owner, Leland Stanford, to attempt to settle the contentious issue of whether a trotting or galloping horse lifted all four feet clear of the ground at any point during its stride. Muybridge's attempts to capture this on film were of poor quality and less than convincing.

In October 1874 Muybridge's personal life was shattered when he was arrested for the murder of his wife's lover, whom he suspected was the father of the son born in April that year. Muybridge was held in prison for several months, but after a lengthy trial he was acquitted in February 1875. His wife, who had unsuccessfully sued for divorce, died later that year, leaving Muybridge to support the child.

Following a trip to Central America in 1875, and a dramatic panoramic sequence of pictures taken of San Francisco in 1877, Muybridge returned to his attempts at high-speed photography. He developed a more efficient shutter mechanism for the camera, and by using a battery of 12 cameras he was able to produce 12 sharply defined consecutive images of a galloping horse, all taken within half a second.

It was readily apparent that if such a sequence of pictures were viewed in rapid succession, the motion of the horse or other subject would be reproduced. Muybridge mounted the silhouettes of the horse on a glass disk, which was rotated and projected onto a screen through a device invented by the photographer and called a "zoopraxiscope." This was first demonstrated to the public in 1880, in what some would claim to be the first moving picture.

Muybridge's work was by now attracting considerable scientific interest, and in 1884 he began work at the University of Pennsylvania on what was to prove a celebrated series of high-speed studies of movement in both animals and human subjects. His new multilens camera could take 12 pictures on a single photographic plate in as little as one-fifth of a second. The results of this work were published in 11 volumes as *Animal Locomotion: an electrophotographic investigation of consecutive phases of animal movement* (1887). Included in this were his famous sequences of nude human subjects, often performing bizarre actions such as sweeping with a broom.

The technique used by Muybridge could produce only very short sequences of moving pictures in the zoopraxiscope. However, the American inventor Thomas EDISON was impressed by them, and may have found in them inspiration for his own invention, the cine camera and its perforated roll film. Certainly Muybridge and Edison collaborated on an abortive attempt to match sound to Muybridge's picture sequences.

N

Naegeli, Karl Wilhelm von (1817–1891) *Swiss botanist*

Naegeli (**ne**-ge-lee), the son of a physician from Kilchberg in Switzerland, began medical studies at Zurich but went on to study botany under Alphonse de CANDOLLE at Geneva. After graduating in 1840 he studied philosophy in Berlin but resumed his botanical studies in 1842, when he left for Jena to work with Matthias SCHLEIDEN.

In 1842 Naegeli published an essay on pollen formation in which he accurately described cell division, realizing that the wall formed between two daughter cells is not the cause but the result of cell division. He noted the division of the nucleus and recorded the chromosomes as "transitory cytoblasts." By 1846 these investigations had convinced him that Schleiden's theory of cells budding off the nuclear surface was incorrect.

Naegeli discovered the antherozoids (male gametes) in ferns and archegonia (female sex organs) in *Ricciocarpus* but did not realize the analogy of these to the pollen and ovary of seed plants. In 1845 he began investigating apical growth, which led to his distinguishing between formative (meristematic) and structural tissues in plants. Naegeli's micellar theory, formulated from studies on starch grains, gave information on cell ultrastructure.

In the taxonomic field, Naegeli made a thorough study of the genus *Hieracium* (hawkweeds), investigating crosses in the group. He had strong views on evolution and inheritance, which led him to reject MENDEL's important work on heredity and hybrid ratios.

Nagaoka, Hantaro (1865–1950) *Japanese physicist*

Nagaoka (nah-gah-**oh**-ka) was born in Nagasaki, Japan, and educated at Tokyo University. After graduating in 1887 he worked with a visiting British physicist, C. G. Knott, on magnetism. In 1893 he traveled to Europe, where he continued his education at the universities of Berlin, Munich, and Vienna. He also attended, in 1900, the First International Congress of Physicists in Paris, where he heard Marie CURIE lecture on radioactivity, an event that aroused Nagaoka's interest in atomic physics. Nagaoka returned to Japan in 1901 and served as professor of physics at Tokyo University until 1925.

Physicists in 1900 had just begun to consider the structure of the atom. The recent discovery by J. J. THOMSON of the negatively charged electron implied that a neutral atom must also contain an opposite positive charge. In 1903 Thomson had suggested that the atom was a sphere of uniform positive electrification, with electrons scattered through it like currants in a bun.

Nagaoka rejected Thomson's model on the ground that opposite charges are impenetrable. He proposed an alternative model in which a positively charged center is surrounded by a number of revolving electrons, in the manner of Saturn and its rings. Nagaoka's model was, in fact, unstable and it was left to Ernest RUTHERFORD and Niels BOHR, a decade later, to present a more viable atomic model.

Nambu, Yoichiro (1921–   ) *Japanese physicist*

Nambu (nahm-**boo**) was educated at the university in his native Tokyo, serving (1945–49) as a research assistant there before being appointed professor of physics at Osaka City University. He moved to America in 1952 and, after a two-year spell at the Institute of Advanced Studies, Princeton, he joined the University of Chicago and was appointed professor of physics in 1958, a position he held until his retirement in 1981.

In 1965 Nambu, in collaboration with M. Y. Han, tackled a major problem arising from the supposed nature of quarks. Baryons, that is, particles that interact by the strong force and have half-integer spin, were composed of three quarks. Thus the proton consists of two up and one down quark and consequently has a configuration written uud. But some baryons are composed of three identical quarks. The omega minus (Ω^-) particle, for example, is composed of three strange quarks with an sss configuration. Quarks, however, are fermions and are thus governed by the Pauli exclusion principle – i.e., no two identical particles can be in the same quantum state. As three s quarks will have the same quantum number, and as their spins can be aligned in only two ways, it seemed that at least two of the s quarks of the Ω^- particle occupy the same state.

Nambu proposed that quarks have an extra quantum number, which can take one of three possible values. The quantum number was arbitrarily referred to as "color," and the varieties equally arbitrarily as red, green, and blue. In this manner three up (uuu), down (ddd), or strange (sss) quarks could coexist without violating any quantum rules, as long as they had different colors. Nambu's work has been confirmed experimentally and is part of what is known as the standard model.

Nambu went on to consider the problem of quark confinement. How could it be, he asked, that free quarks were never encountered? When baryons decay they do not break down into quarks, but into different baryons and other particles. In response to this problem Nambu introduced string theory into physics in 1970. Particles were seen not as small spheres, but as massless rotating one-dimensional entities about 10^{-13} centimeter long, with an energy proportional to their length. The quarks are located at the string's ends. In the simplest case, a meson, a quark is located at one end and an antiquark at the other.

The quarks that make up a meson cannot be separated by stretching the string because the energy required rapidly increases with length. Nor would cutting the string suffice, for at the breaking point a newly created quark–antiquark pair would be created, yielding not a free quark but a further meson.

Though Nambu's string theory had its attractions as a theory of elementary particles it soon ran into other difficulties. Nonetheless, it has been revised by such theorists as John SCHWARZ in the form of superstring theory.

Nansen, Fridtjof (1861–1930) *Norwegian explorer and biologist*

> Man wants to know, and when he ceases to do so, he is no longer man.
> —Justifying polar explorations

One of the greatest men in Norway's history, Nansen (**nahn**-sen) is best remembered for his explorations of the Arctic, although he made many contributions to science, humanitarianism, and politics. Born at Store-Froen in Norway, he graduated in zoology from the University of Christiania, now Oslo. Nansen was appointed curator of the Bergen Natural History Museum in 1882, later becoming successively professor of zoology (1896) and professor of oceanography (1908) at the Royal Frederick University, Christiania. He helped found the International Commission for the Study of the Sea and was director of its Central Laboratory from 1901.

In 1888–89, after several preliminary expeditions, Nansen was the first to explore and describe the uncharted Greenland icecap, trekking from east to west and proving that the island is uniformly covered with an ice sheet. While wintering at Godthaab, Nansen spent some time studying the Eskimos, later publishing his observations as *Eskimoliv* (1891; Eskimo Life). Using a specially constructed ship, *Fram* (Forward), designed to withstand ice-pressure, Nansen then (1893–96) proceeded on his epic expedition to the North Pole. Allowing his ship to freeze in the ice, it drifted northwards (thus proving the existence of a warmer current from Siberia to Spitzbergen). Nansen left the ship and continued northward by sled to 86°14′N – only 200 miles (320 km) from the North Pole and further north than anyone had ever been before. Nansen described his Arctic journey in *Farthest North* (2 vols. 1897). He made further oceanographic expeditions to the northeast Atlantic, Spitzbergen, the Barents and Kara Seas, and to the Azores. In addition to explaining the nature of wind-driven sea currents and the formation of deep- and bottom-water, Nansen did much valuable work in improving and designing oceanographic instruments. In a quite different field his paper on the histology of the central nervous system is considered a classic.

In later life Nansen became a dedicated humanitarian. He assisted in famine relief and aid for refugees after World War I, for which he received the Nobel Peace Prize in 1922. As a politician, he influenced the separation of Norway from Sweden (1905), was a member of the Disarmament Committee (1927), and was Norway's first ambassador to Britain (1906–08).

Napier, John (1550–1617) *Scottish mathematician*

> [Napier] hath set my head and hands awork with his new and admirable logarithms. I hope to see him this summer, if it please God, for I never saw book which pleased me better and made me more wonder.
> —Henry Briggs, letter, 10 March 1615

Born in Edinburgh, Napier studied at the University of St. Andrews but left before taking his degree and then traveled extensively throughout Europe. He was a fervent Protestant and wrote a diatribe attacking Catholics and others whose religious views he did not like. Napier was also very active in politics and he designed a number of war-engines of various kinds when it was believed that the Spanish were about to invade Scotland.

Napier devoted his spare time to mathematics, in particular to methods of computation. He introduced the concept of logarithms, publishing his work on this in *Mirifici logarithmorum canonis descriptio* (1614; Description of the Marvelous Canon of Logarithms). Napier's tables used natural logarithms, i.e., to base e, and soon after their publication the tables were slightly modified by Henry Briggs to base 10.

Napier's further work on logarithms was published after his death in *Mirifici logarithmorum canonis constructio* (1619; Construction of the Marvelous Canon of Logarithms). Napier did some other mathematical work, in particular in spherical trigonometry and in perfecting the decimal notation.

Nash, John F. (1928–) *American mathematician and economist*

Nash was born at Bluefield, West Virginia. His father was an electrical engineer and his mother was a teacher. He originally studied chemical engineering at Carnegie Institute of Technology in Pittsburgh but moved courses, first to chemistry and then, encouraged by the mathematics faculty, to mathematics.

Nash then entered Princeton on a fellowship as a graduate student. At Carnegie he had taken a course on international economics and this had led to a paper on what he called "The Bargaining Problem." At Princeton, he developed this further using the ideas of game theory first discussed by VON NEUMANN and Morgenstern. The result was Nash's theory of *noncooperative games*, which he wrote up for his PhD thesis. The theory, which could be applied to any finite number of players, later found applications in economics.

Nash was not certain that this work would be an acceptable topic for a thesis and, during this period, he also made certain discoveries in pure mathematics concerning manifolds. This work was published later when he was an instructor at the Massachusetts Institute of Technology, a post he took up in 1951. At MIT he also worked on problems in differential geometry, which were relevant to general relativity theory.

At this point Nash seemed set for a brilliant mathematical career but, early in 1959, he began to suffer mental problems. In his own words, it was "the time of my change from scientific rationality of thinking into the delusional thinking characteristic of persons who are psychiatrically diagnosed as 'schizophrenic' or 'paranoid schizophrenic'." He resigned his academic post and spent periods in mental hospitals. After some 25 years Nash appears to have recovered and to have started serious mathematical work again. In 1997 he was awarded the Nobel Prize for economics for the work he had done as a young man many years before on noncooperative game theory. His life story was made into a film, *A Beautiful Mind*.

Nasmyth, James (1808–1890) *British engineer*

> I also had the pleasure of showing him [Sir John Herschel] my experiment of cracking a glass globe filled with water and hermetically sealed. The water was then slightly expanded, on which the glass cracked. This was my method of explaining the action which, at some

previous period of the cosmical history of the Moon, had produced those bright radiating lines that diverge from the lunar volcanic craters.
> —*James Nasmyth, Engineer, an Autobiography*

Nasmyth demonstrated his mechanical expertise at an early age, producing model steam engines while only a boy. Born in the Scottish capital Edinburgh, he worked for two years in Henry MAUDSLAY's machine shop in London and in 1836 set up his own foundry to make machine tools. When asked by Isambard Kingdom BRUNEL to make huge paddle wheels for a steamship, the *Great Britain*, he had to design the steam hammer to forge them (1839). Although the paddles were never made because screw propellers were used instead, the steam hammer was a crucial development in forging techniques.

Nasmyth retired from engineering at the age of 48 to study astronomy, especially the surface of the Moon.

Nathans, Daniel (1928–1999) *American molecular biologist*

Born in Wilmington, Delaware, Nathans was educated at the University of Delaware and at Washington University, St. Louis, where he obtained his MD in 1954. After first working at the Presbyterian Hospital and Rockefeller University in New York he moved in 1962 to Johns Hopkins as professor of microbiology.

With the identification of the first restriction enzyme by Hamilton SMITH in 1970 it was clear to many microbiologists that at last a technique was available for the mapping of genes. Nathans immediately began working on the tumor-causing SV40 virus and by 1971 was able to show that it could be cleaved into 11 separate and specific fragments. In the following year he determined the order of such fragments, after which the way was clear for a full mapping. This also helped advance the techniques of DNA recombination.

It was for this work that Nathans shared the 1978 Nobel Prize for physiology or medicine with Smith and Werner ARBER.

Natta, Giulio (1903–1979) *Italian chemist*

Natta (**naht**-ah), who was born at Imperia in Italy, was educated at the Milan Polytechnic Institute where he obtained his doctorate in chemical engineering in 1924. He was professor at the University of Pavia (1933–35), the University of Rome (1935–37), and the University of Turin (1937–38). Natta returned to Milan in 1939 as professor of industrial chemistry. In 1963 he became the first Italian to be awarded the Nobel Prize for chemistry, which he shared with Karl ZIEGLER for their development of *Ziegler–Natta catalysts*.

Natta's early work was on x-ray crystallography and on catalysis. In 1938 he began to organize research in Italy for the production of synthetic rubbers – work that led him on to his discoveries in polymer chemistry. Ziegler in 1953 had introduced catalysts for polymerizing ethene (ethylene) to polyethene (polythene) – these catalysts gave straight-chain polymers producing a superior form of polyethene. Natta applied these catalysts (and later improved catalysts) to propene (CH_3CHCH_2) to form polypropene. In 1954 he showed that polymers could be formed with regular structures with respect to the arrangement of the side groups (CH_3–) along the chain. These so-called stereospecific polymers had useful physical properties (strength, heat resistance, etc.). Natta extended the technique to the polymerization of other molecules.

Naudin, Charles (1815–1899) *French experimental botanist and horticulturist*

> Happy is the professor who enjoys an assured income and whom the government provides with assistance and collaborators.
> —On his own difficulties in obtaining state funding

Born at Autun in eastern France, Naudin (noh-**dan**) received a sparse formal education as a child; nevertheless, his will to learn led to him earning the baccalaureate in science at Montpellier in 1837. He obtained his doctorate in Paris in 1842 and held minor posts until 1846 when he secured a teaching post in botany at the Natural History Museum, Paris. However, two years later he was forced to abandon his public career due to a severe nervous disorder that left him totally deaf. Eventually, in 1869, Naudin established a private experimental garden at Colliaure and in 1878 he became director of a state-owned experimental garden at Antibes.

Naudin's most significant work began in 1854 with experiments in plant hybridization, from which he found that first-generation hybrids display relative uniformity while second-generation hybrids obtained by crossing within the first generation show great diversity of characters. He recognized, in his theory of disjunction, that inheritance is particulate and not blending. However, unlike his contemporary Gregor MENDEL, he failed to recognize the statistical regularity with which different characters appear – a phenomenon now called segregation.

Naudin proposed hybridization to be the prime agent of evolutionary change and not natural selection or environmental action. He held that the present diversity of species is the product of a smaller number of basic forms and might or might not exhibit permanence.

Needham, Dorothy Mary Moyle (1896–1987) *British biochemist*

Needham, a Londoner, was educated at Cambridge University where she spent her whole career as a research worker at the Biochemistry Laboratory from 1920 until her retirement in 1963.

She worked extensively on the biochemistry of muscle and produced in her *Machina Carnis* (1971; Workings of the Flesh) the definitive history of the subject.

In 1948 Dorothy Needham was elected to the Royal Society, thus becoming one of the first female fellows. As her husband, Joseph Needham, had been a fellow since 1941, they were the first husband and wife members since Queen Victoria and Prince Albert.

Needham, John Turberville (1713–1781) *British naturalist*

Born in London and ordained as a Roman Catholic priest in France, Needham was founder and director of the Brussels Academy of Sciences. In 1740 he showed that pollen grains expanded in water, releasing their contents by means of papillae extruded at the pores. Like Lazzaro SPALLANZANI, Needham also demonstrated the revival ability of apparently dead microorganisms (e.g., rotifers and tardigrades) when placed in water. His most significant work, however, was concerned with what he thought to be spontaneous generation of living matter (1748). Having boiled meat broths and sealed them in apparently airtight containers, he later found them teeming with microorganisms. Spallanzani subsequently showed that Needham's experiments (conducted in collaboration with Georges Buffon) were faulty in that the broth had not been boiled long enough to kill the organisms previously present and that the containers had not been properly sealed.

Needham, Joseph (1900–1995) *British biochemist, historian, and sinologist*

> I regarded the nature of biological organization as a purely philosophical question, and excluded it from scientific biology ... I had not seen the full significance of the analogous science of crystallography. I am glad to have an opportunity of cancelling what I then said.
> —*Order and Life* (1936)

The son of a London physician, Needham was educated at Cambridge where he received his doctorate in 1924. He remained in Cambridge and began to work in the field of embryology, publishing in 1931 the comprehensive *Chemical Embryology* (3 vols.). He served as Dunn Reader in Biochemistry from 1933 until 1966, when he was appointed master of Gonville and Caius College.

Needham's life was radically changed in 1937 when three Chinese biochemistry students arrived in Cambridge. One of them, Lu Gwei-

Djen, would become Needham's life-long collaborator and, in 1989, his second wife. At this point Needham became obsessed by Chinese culture and history. He began to learn the language and in 1942 was appointed scientific counsellor at the British embassy in Chungking. During his three-year stay in China he became aware of the enormous amount of material, virtually all unknown to the West, on science in China. He resolved that he would one day publish this material.

Needham had also spotted a major problem. It appeared to him that during much of its history Chinese science was more advanced than Western science. Yet, modern science arose in 17th-century Europe, not in China. The answer Needham suspected lies in the bureaucratic nature of Chinese society, and the rise of capitalism in the West.

After working briefly for UNESCO in Paris after the war, Needham returned to Cambridge and began to plan a multivolume work, *Science and Civilisation in China*. The first part appeared in 1954 and by the time of his death 16 substantial volumes of a planned 25 had been completed. From 1976 until his death Needham served as director of the East Asian History of Science Library, which was renamed the Needham Research Institute in 1986.

Néel, Louis Eugène Félix (1904–2000)
French physicist

Néel (nay-**el**), who was born at Lyons in France, studied at the Ecole Normale Supérieure, later becoming professor of physics at the University of Strasbourg and subsequently at Grenoble. He became director of the Grenoble Polytechnic Institute in 1954 and director of the Center for Nuclear Studies there in 1956. He retired in 1976.

Most of his work was concerned with the magnetic properties of solids. About 1930 he suggested that a new form of magnetic behavior might exist – called antiferromagnetism. This is exhibited by such substances as manganese(II) oxide (MnO), in which the magnetic moments of the Mn atoms and O atoms are equal and parallel but in opposite directions. Above a certain temperature (the *Néel temperature*) this behavior stops. More generally, Néel pointed out (1947) that materials could also exist in which the magnetic moments were unequal – the phenomenon is called ferrimagnetism.

Néel also did considerable work on other magnetic properties, including an explanation of the weak magnetism of certain rocks that has made it possible to study the past history of the Earth's magnetic field. He was awarded the Nobel Prize for physics in 1970.

Ne'eman, Yuval (1925–2006) *Israeli physicist*

Ne'eman (**nay**-man), who was born at Tel Aviv, was educated at the Israel Institute of Technology (Haifa) where he graduated in engineering in 1945. His academic career was interrupted by service in the Israeli army in the post-World War II troubles of Palestine.

In 1948, with the formation of the independent Israeli state, Ne'eman was able to return to his studies, while still serving with the Israeli defense forces. He went to the Ecole de Guerre in Paris and, while serving as a military attaché at the Israeli embassy in London, gained his PhD in physics from the University of London in 1962.

In 1961 Ne'eman and the American Murray GELL-MANN, working independently, developed a mathematical representation for the classification of elementary particles. This was known as the SU(3) theory and it successfully predicted the mass of the omega-minus particle observed for the first time in 1964. The theory was consolidated in a book by the two men with the title *The Eightfold Way* (1964) and later formed the basis of a further significant theoretical development – the "quark" hypothesis.

From 1961 to 1963 Ne'eman was scientific director of the Saraq Research Establishment of the Israeli Atomic Energy Commission, and from 1963 was head of the physics department and an associate professor at Tel Aviv University. In 1964 he became a full professor and vice-rector of the university. He founded the Israel Space Agency in 1983 and served as its chairman until shortly before his death.

Nef, John Ulric (1862–1915) *American chemist*

Nef was born at Herisau in Switzerland. In 1864 his father emigrated to America, where he became a superintendent of a textile factory; four years later Nef, aged six, joined him there. In 1880 he entered Harvard intending to study medicine but became interested in chemistry instead. Awarded a traveling fellowship he spent two years at Munich University where he gained his PhD in 1886. Returning to America, he first taught at Purdue and then Clark before he was finally appointed to the chair of chemistry at the University of Chicago in 1892.

Research on fulminates led Nef to speculate on the valence of carbon, one of the fundamental problems of late 19th-century chemistry. His work supported the theory of Archibald COUPER that the carbon atom may sometimes have a valence of two, and Nef suggested the existence of the methylene radical, :CH_2. Nef's work also popularized Couper's method of representing the structure of carbon compounds, using dashes or dotted lines for carbon bonds.

Ne'eman, Yuval (1925–2006) *Israeli physicist*

Neher, Erwin (1944–) *German biophysicist*

Neher (**nay**-er) was born in Landsberg, Germany. After attending the Technical University of Munich and the University of Wisconsin, he joined the Max Planck Institute of Psychiatry, Munich, in 1970, as a research associate. In 1972 he moved to the Max Planck Institute for Biophysical Chemistry in Göttingen, being appointed research director in 1983. Two periods of research in America took him to Yale University (1975–76) and the California Institute of Technology (1988–89).

Neher is best known for his studies of the minute channels in the membranes of living cells that allow ions to pass in and out of the cell. In the mid-1970s, working in collaboration with Bert SAKMANN, Neher developed the so-called "patch-clamp technique" to detect the tiny electrical currents produced by the passage of ions through the membrane. Detection posed considerable technical challenges, given that the currents associated with each channel are of the order of 10^{-12} ampere, and the channels have a diameter comparable to the diameter of the ions.

The technique involved applying the tip of a saline-containing micropipette to the cell's membrane and applying suction to form a seal around the patch of membrane. The current produced by the ions passing through the ion channel was monitored using a special amplifier. The technique had the great advantage of eliminating electrical noise generated by other parts of the membrane, which hitherto had obscured signals from any one channel.

Using their technique, Neher and Sakmann were able to demonstrate that the ion channels are either "open" or "shut," i.e., producing an "all or nothing" signal. Also, each channel is specific to a particular type of ion.

The patch-clamp technique has proved itself to be both sensitive and elegant, and has found application in many fields of basic and applied research. Ion channels are involved in a range of biological processes, such as the generation of nerve impulses, the fertilization of eggs, and the regulation of the heartbeat. The way in which their behavior is altered by disease or drugs can have far-reaching implications. This crucial development in cellular research techniques earned Neher and Sakmann the 1991 Nobel Prize for physiology or medicine.

Neisser, Albert Ludwig Siegmund
(1855–1916) *German bacteriologist*

Neisser (**nI**-ser) was born in Schweidnitz (now Swidnica in Poland), the son of a physician. He was educated at Munsterberg and at Breslau, where he qualified in medicine (1877) and held the professorship of dermatology (from 1882).

In 1879 Neisser discovered the gonococcus, the causative agent of gonorrhea, which was named for him (*Neisseria*). In the same year he also identified the bacillus *Mycobacterium lep-*

rae as the cause of leprosy, but in this he had been anticipated by Armauer HANSEN. Neisser also worked on syphilis and collaborated with August von Wasserman in 1906 on the development of his diagnostic test.

Nelson, Ted (Theodor Holm) (1937–   )
American computer scientist

Nelson was educated at Swarthmore College in the late 1950s and subsequently at the University of Chicago and Harvard University. His first job was at Miami as a photographer and a film editor. He then taught at Vassar College. Since 2004 he has been a visiting professor at Wadham College, Oxford.

Nelson is credited with originating the concept of *hypertext*, a word he coined in 1963. He envisaged a "docuverse" in which all data were stored once, with all information being linked, and navigable by a set of links. This was a development of earlier visionary ideas of Vannevar Bush published in 1945 in an article entitled *As We May Think*. Many of Nelson's ideas on hypertext were incorporated into the World Wide Web introduced by Timothy BERNERS-LEE.

Nelson has spent his career developing his idea of hypertext and related matters. He has been very concerned with commercial applications of hypertext. In 1983 he created the Xanadu Operating Company, Inc. In 1994 he moved to the Sapporo Hyper Lab., Japan. He became a professor of environmental information at Keio University, Shonan.

Nernst, Walther Hermann (1864–1941)
German physical chemist

> Knowledge is the death of research.
> —Quoted by C. G. Gillespie in *The Dictionary of Scientific Biography* (1981)

Nernst (nairnst), who was born at Briesen in Germany, studied at the universities of Zurich, Berlin, Würzburg, and Graz. After working as assistant to Wilhelm OSTWALD in Leipzig from 1887 he became professor of chemistry at Göttingen in 1890. In 1904 he became professor of physical chemistry at Berlin and later was appointed director of the Institute for Experimental Physics there (1924–33). In 1933, out of favor with the Nazis, he retired to his country estate.

Nernst's early work was in electrochemistry – a field in which he made a number of contributions. Thus in 1889 he introduced the idea of the solubility product, i.e., the product of the concentrations of the different types of ions in a saturated solution. The product is a constant for sparingly soluble compounds (at constant temperature). Nernst also suggested (1903) the use of buffer solutions – mixed solutions of weak acids (or bases) and their salts, which resist changes in pH.

His main work, in 1906, was in thermodynamics. It came out of attempts to predict the course of chemical reactions from measurements of specific heats and heats of reaction. If heat is absorbed during a reaction, the amount absorbed falls with temperature and would become zero at absolute zero. Nernst postulated that the *rate* at which this reduction occurred would also become zero at absolute zero of temperature, and, as a consequence, derived the *Nernst heat theorem*, which states that if a reaction occurs between pure crystalline solids at absolute zero, then there is no change in entropy.

The theorem, stated in a slightly different form, is now known as the third law of thermodynamics. It is equivalent to the statement that absolute zero cannot be attained in a finite number of steps. At the time it allowed the calculation of absolute values of entropy (and then equilibrium constants), rather than changes in entropy. It is now known to be a consequence of the quantum statistics of the particles. For his work in thermodynamics, Nernst received the Nobel Prize for chemistry in 1920.

He also made contributions to photochemistry and, in addition, produced one of the standard texts of the period, *Theoretische Chemie* (1893; Theoretical Chemistry), which went through numerous editions and translations.

He managed to make a large fortune by the turn of the century by selling a form of electric light, which though superior to the EDISON carbon-filament lamp soon became obsolete with the invention of the tungsten-filament lamp.

Neumann, John von *See* VON NEUMANN, JOHN.

Newcomb, Simon (1835–1909) *American astronomer*

The son of an itinerant teacher, Newcomb was born at Wallace in Nova Scotia and had little formal education. He was apprenticed to a herbalist in Nova Scotia but ran away to join his father in the United States. In 1857 he joined the American Nautical Almanac Office, and he graduated from Harvard in 1858. He joined the corps of professors of mathematics in the navy, and became professor of mathematics at the Naval Observatory in Washington in 1861. From 1884 to 1894 he was professor of mathematics and astronomy at Johns Hopkins University. He was also superintendent of the American Nautical Almanac from 1877 to 1897, retiring with the rank of rear admiral. In addition he was the editor of the *American Journal of Mathematics* and the author of over 350 scientific papers and a number of popular works on astronomy.

Newcomb worked for many years on new tables for the planets and the Moon, which were published in 1899. These, together with his or-

ganization of the Nautical Almanac, were his major astronomical work. His tables, the result of detailed observations and sophisticated mathematics, were the most accurate ever made and were in constant use until the middle of the 20th century. Also of major importance was his production and promotion of a new, unified, and more accurate system of astronomical constants, which was adopted worldwide in 1896.

He did much to encourage younger scientists. Hearing the young Albert MICHELSON lecture to the American Association for the Advancement of Science on new methods for accurately determining the speed of light, he went out of his way to raise money for the unknown young scientist to continue with his work.

Newcomen, Thomas (1663–1729) *British engineer*

Newcomen worked as an ironmonger in his home town of Dartmouth, Devon. His familiarity with the Cornish tin mines prompted him to try to improve the horse-driven pumps used to remove water from mines. Together with his assistant, a plumber by the name of John Calley, he built a pump based on the design developed and patented by Thomas SAVERY. In Newcomen's piston-operated steam engine, a counterweighted plunger was pushed to the top of a cylinder by steam pressure. Then, when the steam condensed, the partial vacuum created inside the piston caused the plunger to be pushed back down by atmospheric pressure. A water jet inside the cylinder condensed the steam at each stroke, and the design, unlike Savery's, included automatic valves. Newcomen's engine was the forerunner of James WATT's steam engine.

Newcomen entered into a partnership with Savery and they constructed their first engine near Dudley Castle, Staffordshire, in 1712. Versions of Newcomen's engine were being built until the 1800s and some were still in use at the beginning of the 20th century, both in mines and to power water wheels.

Newell, Allan (1927–1992) *American computer scientist*

Newell was born in San Francisco, the son of a radiologist. He was educated at Princeton and at the Carnegie Institute of Technology, Pittsburgh, where he completed his PhD in 1957. After a period at RAND, Santa Monica, Newell returned to Carnegie in 1961 and was appointed professor of computer science in 1976.

In 1956 Newell helped to organize a seminal conference in Dartmouth, New Hampshire, on the subject of artificial intelligence (AI). Soon after, in collaboration with his RAND colleague Herbert SIMON, Newell produced one of the early successes of the new discipline with Logic

Theorist. This was a program written in IPL (Information Processing Language) designed to prove theorems in RUSSELL and WHITEHEAD's *Principia Mathematica*. They succeeded in deriving 38 of the first 52 theorems of the system. Further, because the proofs were produced in the main within a few minutes, they could not have been derived by any kind of purely random search; some kind of intelligence, it was argued, must have been deployed.

Encouraged by their success with Logic Theorist, Newell and his collaborators attempted to develop a successor, which they called General Problem Solver. It would be sufficiently flexible to play chess, break codes, solve mathematical problems, and deal with other diverse intellectual problems. Although some progress was made the program was eventually abandoned; the ability to solve problems proved to be not as general as was initially supposed. Further, other workers, such as Edward FEIGENBAUM, began to show in the 1970s that more progress could be made by preparing "expert systems."

Newlands, John Alexander Reina (1837–1898) *British chemist*

Newlands, who was born in London, studied under August von Hofmann at the Royal College of Chemistry. In 1860, being of Italian ancestry, he fought with Giuseppe Garibaldi's army in its invasion of Naples. On his return from Naples he set up as a consultant with his brother in 1864 but, after the failure of his business, worked as an industrial chemist in a sugar refinery.

In various papers published in 1864 and 1865 Newlands stated his law of octaves and came close to discovering the periodic table. He claimed that if the elements were listed in the order of their atomic weights a pattern emerged in which properties seemed to repeat themselves after each group of seven elements. He pointed out the analogy of this with the intervals of the musical scale.

Newlands's claim to see a repeating pattern was met with savage ridicule on its announcement. His classification of the elements, he was told, was as arbitrary as putting them in alphabetical order and his paper was rejected for publication by the Chemical Society. It was not until MENDELEEV's periodic table was announced in 1869 that the significance of Newlands's idea was recognized and he was able to publish his paper *On the Discovery of the Periodic Law* (1884).

Newton, Alfred (1829–1907) *British ornithologist*

Newton, the son of the member of Parliament for Ipswich, was born at Geneva in Switzerland and educated at Cambridge University. From 1854 until the early 1860s he traveled widely in Lapland, Iceland, the Caribbean, and America, making one of the first systematic studies of their birds. Thereafter he resided in Cambridge serving as the first occupant of the chair of zoology and comparative anatomy from 1866 until his death.

Newton developed ornithology in Britain into a serious scientific discipline, a cooperative and organized activity. It was, for example, in his Cambridge rooms in 1858 that the decision was made to found the important journal *Ibis*. He was also deeply involved in attempting to initiate bird censuses, the keeping of migration records, and the collecting of skins and eggs.

His own special interest lay with the great auk and dodo. He did, however, publish more general works such as the important *A Dictionary of Birds* (1893–96). Although noted for his reactionary views – he fought against the introduction of women and music into the college chapel with equal passion – he nonetheless had little difficulty in accepting the evolutionary theory of DARWIN and of supporting the experimental physiology of Michael FOSTER.

Newton, Sir Isaac (1642–1727) *English physicist and mathematician*

> I do not know what I may appear to the world, but to myself I seem to have been only like a boy playing on the sea-shore, and diverting myself in now and then finding a smoother pebble or a prettier shell than ordinary, whilst the great ocean of truth lay all undiscovered before me.
> —Quoted by L. T. More in *Isaac Newton*

Newton's father, the owner of the manor of Woolsthorpe in Lincolnshire, England, died three months before Newton was born. The family had land but were neither wealthy nor gentry. Left by his mother in the care of his grandmother, the young Newton is reported to have been quiet, unwilling to play with the village boys, and interested in making things. His mother returned to Woolsthorpe in 1656 after the death of her second husband. By this time Newton was at school in Grantham where he stayed until 1658, lodging with a local apothecary. There is no evidence that he was especially gifted at this time, although he was certainly skillful for he made a water clock, sundials (which still survive at Woolsthorpe), and model furniture for his stepsisters. After two years helping his mother to run the family farm, he went to Cambridge University in 1661, where he stayed for nearly 40 years.

Not much is known of Newton's student life. In 1665 he was forced by the plague to leave Cambridge and return to Woolsthorpe. Here, during his so-called *annus mirabilis* (miraculous year), he began to develop the ideas and insights for which he is so famous. Here he first began to think about gravity, and also devoted time to optics, grinding his own lenses and considering the nature of light. During this period he also worked out his mathematical ideas about "fluxions" (the calculus).

When he returned to Cambridge after the plague had died down he was elected a fellow of his college, Trinity, in 1667, and in 1669 he succeeded BARROW as Lucasian Professor. He served as member of parliament for the university for the periods 1689–90 and 1701–02, although he does not appear to have been politically very active. His public career was pursued through Charles Montague, first earl of Halifax, who was able to introduce Newton into court and society circles. When Montague became chancellor of the exchequer he was able to offer Newton the post of warden of the Mint in 1696. He was made master of the Mint in 1699 and knighted for his services in 1705. From this time he did virtually no new science apart from publishing and revising works already written. He did concern himself with the affairs of the Royal Society, of which he became president in 1703. He resigned his Cambridge post in 1701.

At the Mint, his first task was to supervise "the great recoinage" – the replacement of the old hammered coins with new pieces with milled edges. It was also Newton's business to pursue the counterfeiters and clippers of his day. As ever, he took his duties seriously and could be found regularly visiting suspects in Newgate and other prisons. Between June 1698 and Christmas 1699 he interviewed 200 witnesses on 123 separate occasions. In the same period 27 counterfeiters were executed. Other major tasks undertaken by Newton included the introduction of a union coinage in 1707 following the union of the Kingdoms of England and Scotland, the issue of new copper coins in 1718, the revaluation of the guinea to 21 shillings in 1717, and a general improvement in the assaying of the currency. The Mint made Newton a wealthy man. In addition to a salary of £600 a year he also received a commission on the amount of silver minted, which brought in on average £1,000 a year. At the time of his death Newton had accumulated £30,000 in cash and securities.

Much of Newton's later life was also spent in needless priority disputes. These arose largely through his reluctance to publish his own work. It was not until 1704, when Newton was over 60, that he actually published a mathematical text. Even then his main work on the calculus, *Methodis fluxionum* (Method of Fluxions), composed between 1670 and 1671, was only published posthumously in 1736. At the same time manuscripts of unpublished works were shown to friends and colleagues.

When, therefore, LEIBNIZ published his own work on the differential and integral calculus in 1684, he felt no need to acknowledge any unpublished work of Newton. He had developed his methods and notation largely from his own vast intellectual resources. He had seen some Newtonian manuscripts on a visit to London in 1673, and letters from Newton in 1676 contained further details. None of this, it is now accepted, was sufficient to account for Leibniz's 1684 paper. The dispute began in 1700 when Leibniz objected to the practice of the Newtonians referring to him as the "second inventor" of the calculus. The dispute dragged on until the 1720s, long outlasting the death of Leibniz in 1716. After a decade of bitter and anonymous dispute Leibniz unaccountably applied to the Royal Society in 1712 to conduct an inquiry into the matter. Newton behaved quite shamelessly. He appointed the committee, decided what evidence it should see, and actually drafted the published report himself. Thereupon, in later stages of the dispute, he would appeal to the report, the *Commercium epistolicum* (1713; On the Exchange of Letters), as an independent justification of his position.

Newton is best known for his work on gravitation and mechanics. The most famous story in the history of science has the unusual distinction of being true, at least according to Conduit, who married Newton's niece Catherine. He reported that "In the year 1665, when he retired to his own estate on account of the plague, he first thought of his system of gravity, which he hit upon by observing an apple fall from a tree." His ideas were not published until 1684 when Edmond HALLEY asked Newton to find what force would cause a planet to move in an elliptical orbit. Newton replied that he already had the answer. Finding that he had lost his proof, he worked it out again.

His result was that two bodies – such as the Sun and a planet, or the Earth and the Moon – attract each other with a force that depends on the product of their masses and falls off as the square of their distance apart. Thus the force is proportional to $m_1 m_2 / d^2$, where m_1 and m_2 are the masses and d is their distance apart. Originally he applied this to point masses but in 1685 he proved that a body acted as if its mass were a point mass of the same magnitude acting at the center of the body (for a symmetrical body).

Newton's original work on gravity, in 1665, had been applied to the motion of the Moon. His insight was that the Moon in its motion "falls" to the Earth under the same cause as the apple falls. His calculations at this time used an erroneously low value for the Earth's radius and it was possibly this that made him lay aside his calculations until 1684.

Then in 1684 he took up the subject again and began to write his great work *Philosophiae Naturalis Principia Mathematica* (Mathematical Principles of Natural Philosophy) – known as the *Principia*. The first edition was published in 1687. Here he set out his three laws of motion. His first law states that a body at rest or in uniform motion will continue in that state unless a force is applied. His second law gives a definition of force – that it equals the mass of a body multiplied by the acceleration it produces in the body. His third law puts forward the idea that if a body exerts a force (action) on another there is an equal but opposite force (reaction) on the first body.

What Newton did in his work in mechanics was to establish a unified system: one in which a simple set of basic laws explained a range of diverse phenomena – the motion of the Moon and planets, motion of the Earth, and the tides. Newton did not give any explanation of what gravity actually *is*. How it acts, its mechanism, and its cause were matters that NEWTON claimed we should not frame hypotheses on. That it has a cause Newton was sure of, for the idea that a body may act on another through a vacuum over a long distance "without the mediation of anything else...is to me so great an absurdity that I believe no man can ever fall into it. Gravity must be caused...but whether this agent be material or immaterial I have left to the consideration of my reader." Newton did not always obey his own injunction and in the 1713 edition of the *Principia* speculated about the existence of "certain very subtle spirits that pervade all dense bodies," which might explain light, electricity, sensation, and much besides.

Newton's reluctance to provide a gravitational mechanism was seen as a basic weakness of his system by the Cartesians. Whatever the defects of the physics of René DESCARTES at least he provided mechanisms, in the form of vortices, to explain all movements. To some Newton's gravity seemed a retrograde step in

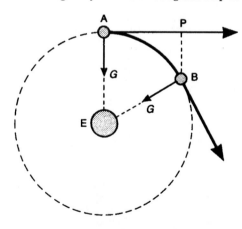

THE FALL OF THE MOON   The Moon at position A experiences a force of gravitational attraction toward the Earth E. In the absence of any force it would continue along a straight line AP. In fact it moves in a circular path AB. Newton saw that this is equivalent to the Moon "falling" toward the Earth – in the diagram the fall is the distance PB in the same direction as AE.

that it was reintroducing into physics the occult, meaningless forces that Descartes had recently eliminated. Nevertheless, Newton's system received great acclaim in England and was to become the model for all succeeding scientific theory.

Newton also worked extensively on light. He began by rejecting the Cartesian account of color. For Descartes white light was natural light; colored light was the modification produced in light by the medium through which it passes. Thus light passing through a prism is spread into a spectrum because the light has been differentially modified by the varying thickness of the prism.

Newton published his own account in 1672 in his first published paper, *New Theory about Light and Colours*. "I procured me a Triangular glass-Prisme to try therewith the celebrated Phenomena of Colours," he began. When he passed a ray of light through a prism he found that it formed an oblong, not a circular image, five times longer than its breadth. He found that, as was well known, light passing through a prism was dispersed and formed a colored spectrum. But when a second prism was taken and colored light rays passed through it, no further change was discernible. Red light remained red, and blue light blue. From this, his famous *experimentum crucis* (cross experiment), he derived two important conclusions: firstly, ordinary white light was composite, a mixture of the various colors of the spectrum; secondly, he concluded, "Light consists of Rays differently Refrangible," and it was this difference in refrangibility that produced the oblong image which had so puzzled Newton.

Newton's views found little favor and over the next few years he was repeatedly called upon to explain and defend his position. By 1676 he had had enough. "I see a man must either resolve to put out nothing new or become a slave to defend it," he wrote to Oldenburg, and published no more on light until 1704.

He had, however, already in 1675 sent a paper to the Royal Society entitled *An Hypothesis Explaining the Properties of Light*. He refused to allow it to be published and it first appeared in 1757, long after Newton's death. The paper contains Newton's analysis of light as "multitudes of unimaginable small and swift corpuscles of various sizes, springing from shining bodies." He dismissed the view that light, like sound, could consist of waves, because light unlike sound could not travel around corners. The paper also contained an account of what have since become known as *Newton's rings*. These consist of a series of concentric colored rings and can be produced by putting a plano-convex lens of large radius of curvature on a flat reflecting surface. They were explained by Newton with some ingenuity and some difficulty in terms of his corpuscular theory of light. New-

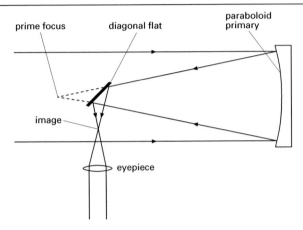

prime focus   diagonal flat   paraboloid primary

image

eyepiece

NEWTONIAN TELESCOPE   The ray path in the reflecting telescope first constructed by Newton.

ton's mature views on these and many other matters were presented in his *Opticks* (1704).

Not all of Newton's work on light was of a theoretical kind. He was an extremely talented experimentalist and in the late 1660s he designed and built the first reflecting telescope. This involved grinding and polishing the mirrors himself. The idea of a reflecting telescope had occurred earlier to James Gregory but his attempts at constructing a model, despite receiving professional help, led nowhere. The advantage of the Newtonian telescope over the refractor of GALILEO is that mirrors do not suffer from chromatic aberration.

But above all else Newton was a mathematician of incomparable power. In 1696 Johann BERNOULLI posed a problem to the mathematicians of Europe, allowing them six months to solve it. Newton solved the problem in a single night and published the result anonymously in the *Transactions* of the Royal Society. Bernoulli was not fooled, claiming to recognize the author or, "the lion by his claw." Again in 1716 Leibniz issued another difficult problem, which Newton solved before going to bed after a day's work at the Mint.

Newton communicated the generalized form of the binomial theorem to Henry Oldenbourg in 1676. It was also in that year that he deposited with Oldenbourg his *epistola prior* (first letter) claiming discovery of his method of fluxions in an anagram. The terminology arose from his considering the path of a continuously moving body as a curve made by a continuously moving point. The moving point he called a fluent and its velocity he called a fluxion. This he symbolized by $\dot{x}$ and its acceleration as $\ddot{x}$. This, independently of Leibniz, was Newton's discovery of the calculus, although Leibniz's notation was the one eventually adopted.

Throughout his life Newton also displayed a deep interest in two other areas: religion and alchemy. At his death over 1,000 manuscript

pages, running to nearly 1.5 million words, and two completed books were discovered, devoted entirely to religious matters. Newton was a unitarian, a matter kept fairly secret during his life as it would have excluded him from his Lucasian chair and his post at the Mint. Much of his life was spent on deep studies of church history, the Bible, and ancient chronology. His aim was to show that the text of the Bible had been corrupted by later trinitarian editors, and that the history of the early church revealed a similar corruption introduced by Athanasius in the fourth century. The matter was dealt with at length in his *Two Notable Corruptions of Scripture* (1754) and in numerous manuscripts.

Equally extensive were Newton's alchemical manuscripts. In his library were 138 books on alchemy, and his manuscripts on the subject exceed 600,000 words. It is less clear, however, whether Newton was a genuine alchemist committed to dreams of transmutation and the philosopher's stone, or whether he was merely using whatever sources he could find to further his chemical interests. His interests in chemistry were sufficient to lead him to establish a laboratory in Trinity College and for a while in the 1680s, it was reported, "the laboratory fire scarcely went out night or day." He published during his life just one brief work on chemistry, *De natura acidorum* (1710; On the Nature of Acids). There are, however, several passages devoted to chemistry scattered among the *Queries* added by Newton to his *Opticks* (1704).

Newton died in 1727 at the age of 85 after a fairly short illness. He managed to preside over a meeting of the Royal Society a fortnight before his death. Shortly after he was diagnosed as having a stone in the bladder and seems to have spent the last days of his life in great pain. He was buried in Westminster Abbey where a most unattractive monument can still be seen. Many words have been written about his greatness as a scientist; the most apposite remain the often quoted words of Alexander Pope, composed for Newton's tomb but, for some reason, rejected:

> Nature and Nature's laws lay hid in night:
> God said, let Newton be! and all was light.

Nicholas of Cusa (1401–1464) *German cardinal, mathematician, and philosopher*

> God has endowed every body with its own nature, orbit, and place, and has set the Earth in the middle of all, decreeing that it be heavy and deviate neither upward nor sideways.
> —*De venatione sapientiae* (1463; On the Pursuit of Knowledge)

The son of a prosperous fisherman from Kues (Cusa), in Germany, Nicholas (**nik**-oh-las or nee-koh-**lah**-uus) studied at Heidelberg and Padua where he obtained an LLD in 1423. He rose through the hierarchy to become a cardinal in 1448 and a figure of some importance in the Church. His most lasting work was his *De docta ignorantia* (1440; On Learned Ignorance) in which he argued against the possibility of definitive knowledge. "Reason stands in the same relation to the truth as the polygon to a circle," he declared in a striking image. "The more vertices it has, the more it resembles a circle...yet never becomes a circle." Using metaphysical rather than astronomical arguments he was thus led to deny the standard picture of the universe as bounded by spheres with the Earth at the center. In its place he proposed a universe in which nothing is fixed, with no center or circumference yet not infinite, with everything in motion and a complexity beyond our understanding. He also proposed that the Earth revolved on its axis and around the Sun, and that the stars were other suns.

Not surprisingly, such an unorthodox construction had no impact on a conventional thinker like Nicolaus COPERNICUS. It did, however, help to overthrow in later minds the traditional idea of a "closed world" and prepare the way for the "infinite universe" of the 17th century.

In other areas of science, Nicholas is also said to have constructed spectacles for the nearsighted (with concave lenses) and to have thought that plants drew nourishment from the air.

Nicholson, Seth Barnes (1891–1963) *American astronomer*

Nicholson was the son of a farmer from Springfield, Illinois. He graduated from Drake University, Des Moines, Iowa, in 1912 and in 1915 obtained his PhD from the University of California. From then until his retirement in 1957 he worked at the Mount Wilson Observatory, California.

While still a graduate student at the Lick Observatory, Nicholson discovered the ninth satellite of Jupiter (Sinope) in 1914, working close to the limits of resolution of the Lick telescope and the photographic plates then available. He went on to discover a further three satellites, Jupiter X (Lysithea) and XI (Carme) in 1938 and Jupiter XII (Ananke) in 1951. All four satellites are very small, about 15–20 kilometers (9–12 mi) in diameter.

Nicholson studied the surface features and spectrum of the Sun and also, in collaboration with Edison Petit, worked on planetary and lunar temperatures. Thus they showed in 1927 that the Moon undergoes enormous temperature changes, for in the course of an eclipse its surface temperature dropped from 160°F to −110°F (71°C to −79°C) in about an hour.

Nicholson, William (1753–1815) *British chemist*

Nicholson was born in London, the son of a lawyer. In 1769, at the age of 16, he joined the East India Company, but left them in 1776 to become an agent in Europe for the pottery manufacturer Josiah Wedgwood. He then started a school of mathematics, as well as translating various scientific works into English. During his career he was also a patent agent, the engineer of the Portsmouth and Gosport waterworks, and the inventor of the hydrometer named for him for measuring the density of liquids.

In 1800 Nicholson heard of the invention of the electric battery by Alessandro VOLTA. Nicholson built the first voltaic pile in England and then discovered that when the leads from the battery were immersed in water the water broke up into bubbles of hydrogen and oxygen; he had discovered electrolysis.

Nicholson's importance also rests on the steady stream of textbooks that he produced, many of which were translations from the French. These included his *Introduction to Natural Philosophy* (1781), *First Principles of Chemistry* (1790), and *Dictionary of Chemistry* (1795). He also started the first independent scientific journal, the *Journal of Natural Philosophy, Chemistry, and the Arts*. Through publications of this kind the new artisans and industrialists learned much of the new chemistry, which they applied to the developing industries.

Nicol, William (1768–1852) *British geologist and physicist*

Little is known about Nicol's early life except that he was born in Scotland. Primarily a geologist, he lectured in natural philosophy at the University of Edinburgh where James Clerk MAXWELL was probably one of his pupils. His first publication came when he was nearly 60.

Nicol is best remembered for his invention,

announced in 1828, of the *Nicol prism*. This device, constructed from a crystal of Iceland spar (a natural form of calcium carbonate), made use of the phenomenon of double refraction discovered by Erasmus BARTHOLIN. The crystal was split along its shorter diagonal and the halves cemented together in their original position by a transparent layer of Canada balsam. The ordinary ray was totally reflected at the layer of Canada balsam while the extraordinary ray, striking the cement at a slightly different angle, was transmitted. Nicol prisms made it easy to produce polarized light. For a long time they became the standard instrument in the study of polarization and played a part in the formation of theories of molecular structure.

Nicol also developed new techniques of preparing thin slices of minerals and fossil wood for microscopic examination. These techniques allowed the samples to be viewed through the microscope by transmitted light rather than by reflected light, which only revealed surface features. His lack of publications resulted in some 40 years elapsing before these techniques were incorporated into studies in petrology.

Nicolle, Charles Jules Henri (1866–1936) *French bacteriologist*

Nicolle (nee-**kol**), the son of a physician, was born at Rouen in France and educated there and in Paris, where he obtained his MD in 1893. He returned to Rouen to join the faculty of the medical school but in 1902 moved to Tunis, where he served as director of the Pasteur Institute until his death.

In 1909 Nicolle revolutionized the study and treatment of typhus. He noticed that typhus patients outside the hospital transmitted the disease to their families, to the doctors who visited them, to the staff admitting them into hospital, to the personnel responsible for taking their clothes and linen, and to hospital laundry staff. But once admitted to the ward the typhus patients did not contaminate any of the other patients, the nurses, or the doctors. Since all newly admitted patients were stripped, washed, and changed, Nicolle concluded that the disease carrier was attached to the patient's skin and clothing and could be removed by washing. The obvious carrier was the louse.

Nicolle lost no time in providing experimental evidence for his reasoning. He transmitted typhus to a monkey by injecting it with blood from an infected chimpanzee. A louse was allowed to feed on the monkey and when transferred to another monkey succeeded in infecting it by its bite alone. It was for this work that Nicolle was awarded the Nobel Prize for physiology or medicine in 1928.

Nicolle actually considered his discovery of "apyretic typhus" the most important of his achievements. He found guinea pigs to be susceptible to typhus but that some of them, with blood capable of infecting other animals, showed no symptoms of the disease at all. He had in fact discovered the carrier state, which was to have significance for the emerging science of immunology.

Nicolle also attempted to develop vaccines against typhus and other infections. He was mildly successful in using serum from patients recovering from typhus and measles to induce a short-lasting passive immunity on those at risk.

Niepce, Joseph-Nicéphore (1765–1833) *French inventor*

> Niepce found that he could fix the camera image using a plate coated in bitumen that hardened on exposure to light. After taking the plate from the camera he washed off the unhardened bitumen in oil of lavender, so developing the picture. A murky view of a farmyard, photographed (with an eight-hour exposure) from his upstairs window, still survives.
> —John Carey, *The Faber Book of Science* (1995)

Niepce (nyeps), who made the first permanent photographic image, came from a wealthy family in Châlon-sur-Saône, eastern France, that fled the French Revolution. He returned to serve with Napoleon Bonaparte's army, but after being dismissed because of ill health, went back to his birthplace (1801) to do scientific research.

With his brother, Niepce built an internal-combustion engine (1807) for boats using carbon and resin for fuel. In 1813 he started the attempt to record images, on paper coated with silver chloride. He produced his first image, a view from his workroom, in 1816, but was only able to fix this partially with nitric acid. In 1822 he produced a photographic copy of an engraving using a glass plate coated with bitumen of Judea. Later (1826) he used a pewter plate to make the first permanent camera photograph. He also devised the first mechanical reproduction process. The main difficulty was the long exposure times needed – over eight hours. Niepce formed a partnership with the Parisian painter Louis-Jacques-Mandé DAGUERRE in 1826 to perfect the process of heliography but he died before seeing the final success of his efforts.

Nieuwland, Julius Arthur (1878–1936) *American chemist*

Nieuwland (**nyoo**-land) was born at Hansbeke in Belgium and brought to America when three years old. He was educated at the University of Notre Dame, Indiana, obtaining his PhD on acetylene chemistry from there in 1904. His whole career was spent at Notre Dame, teaching botany from 1904 to 1918 and organic chemistry from 1918 to 1936. He entered the Roman Catholic priesthood in 1903.

Nieuwland made an important contribution to the chemistry of acetylene when, in 1925, he succeeded in making three molecules join together in a small chain to give divinylacetylene, which, to his surprise, produced a rubberlike solid on polymerization.

Working with chemists at Du Pont, Nieuwland found in 1929 that if monovinylacetylene is treated with acid it forms chloroprene, which can then be polymerized to give neoprene, the first synthetic rubber.

Nilson, Lars Fredrick (1840–1899)
Swedish chemist

Born at Östergötland in Sweden, Nilson (**nil**-son) studied at the University of Uppsala and was appointed professor of chemistry there in 1878. In 1883 he moved to the Agricultural Academy in Stockholm.

In 1879 he isolated a new element, which he named scandium after his native Scandinavia, from the rare earth ytterbia. His colleague, Per CLEVE, showed that its properties were those of Dmitri MENDELEEV's eka-boron, thus providing a second confirmation of the validity of the periodic table.

Nilson also made a significant contribution to agricultural chemistry by showing that the chalk moors of the island of Gotland would grow sugar beet if treated with a potash fertilizer.

Nirenberg, Marshall Warren (1927–)
American biochemist

Nirenberg, who was born in New York City, graduated from the University of Florida in 1948 and gained his PhD in biochemistry from the University of Michigan in 1957. He then joined the National Institutes of Health in Bethesda, Maryland, where he began the work that culminated in the cracking of the genetic code.

When Nirenberg began this research it had already been surmised that different combinations of three nucleotide bases (triplets) each coded for a specific amino acid and that through the operation of this "genetic code" amino acids are aligned in the right order to make proteins. As there are 64 possible triplets (from the combinations of the four bases present in DNA or RNA), the big question was: which triplets code for the 20 amino acids? Severo OCHOA's discovery of the technique to synthesize RNA artificially enabled Nirenberg to make an RNA molecule consisting entirely of uracil nucleotides. Thus the only triplet possible would be a uracil triplet (UUU of the code). Nirenberg found that the protein made by this RNA molecule consisted entirely of the amino acid

phenylalanine, indicating that UUU must code for phenylalanine. With this first important step completed others were quick to unravel the rest of the code.

For this work, Nirenberg shared the 1968 Nobel Prize for physiology or medicine with Har Gobind KHORANA and Robert HOLLEY.

Nobel, Alfred Bernhard (1833–1896)
Swedish chemist, engineer, and inventor

> ... to those who, during the preceding year, shall have conferred the greatest benefit on mankind.
> —Clause in Nobel's will stipulating the beneficiaries of the annual prizes he endowed

Nobel (noh-**bel**) left Stockholm, where he was born, in 1842 to join his father, an engineer, who had moved to St. Petersburg. He was taught chemistry by his tutors and spoke fluently in English, French, German, Swedish, and Russian. In 1850 he went to Paris to study chemistry and then went on to America for four years, before returning to work in his father's factory in St. Petersburg.

In 1859 Nobel moved back to Sweden and set up a factory there (1864) to make nitroglycerin, a liquid explosive. After an explosion at the factory in 1864 in which his brother, Emil, and four others were killed, the Swedish government would not allow the factory to be rebuilt. Nobel then started to experiment to find a more stable explosive. Discovering that nitroglycerin was easily absorbed by a dry organic packing material, he invented dynamite and the detonating cap. These were patented in 1867 (UK) and 1868 (United States). From such work and from oil fields in Russia that he owned, Nobel amassed a vast fortune. He traveled widely and was a committed pacifist. He left the bulk of his money in trust for international awards – the Nobel Prizes for peace, literature, physics, chemistry, and medicine. The Nobel Prize for economics was introduced in his honor in 1969 and financed by the Swedish National Bank.

Nobili, Leopoldo (1784–1835) *Italian physicist*

Nobili (**noh**-bee-lee) was born at Trassilico in Italy and, after a university education, served in the army as an artillery captain. He was later appointed professor of physics at the Florence Museum.

Nobili carried out pioneer work in the newly developing field of electrochemistry, publishing his results in 1828. He is known for his inventions of an astatic galvanometer (1825) and the thermocouple. His astatic galvanometer – a moving-magnet device constructed in such a way as to be independent of the Earth's magnetic field – was far more sensitive than existing instruments. Nobili's thermocouple – a device for measuring temperature – was based on the SEEBECK effect. It consisted of two wires of different metals joined at each end with the junctions maintained at different temperatures; the temperature difference was determined by measuring the current flow. In 1829 Nobili made the first thermopile by joining in series six couples of antimony and bismuth. It was a cumbersome and fairly insensitive instrument. However, after later improvements by Macedonio MELLONI it became fundamental to investigations of radiant heat.

Noddack, Ida Eva Tacke (1896–1979)
German chemist

Noddack (**nod**-ak), who was born at Lackhausen in Germany, was educated at the Technical University, Berlin, where she obtained her doctorate in 1921. After her marriage to the chemist Walter Noddack she worked at the same institutions as her husband.

In 1926 she collaborated with her husband and O. Berg in the discovery of rhenium. More intriguing, however, was her interpretation of a famous experiment of Enrico FERMI in which he had bombarded uranium with slow neutrons in the hope of producing artificial elements. Although their results were not particularly clear Noddack, in 1934, argued that "It is conceivable that in the bombardment of heavy nuclei with neutrons, these nuclei break up into several large fragments which are actually isotopes of known elements, not neighbors of the irradiated element." This is, in fact, the hypothesis of nuclear fission which, when it was published five years later by Otto FRISCH, was immediately seen to be of fundamental importance.

Noddack's contribution seems rather to have passed unnoticed. Fermi was aware of her work as she had sent him a copy, but he remained unconvinced and continued to believe that he had made transuranic elements.

Noddack, Walter (1893–1960) *German chemist*

Noddack was born in Berlin and educated at the university there, obtaining his doctorate in 1920. He worked first at the Physikalische Technische Reichsanstadt, the German national physical laboratory, until 1935, and then held chairs in physical chemistry at Freiburg and Strasbourg until, in 1946, he moved to Bamberg. Noddack taught chemistry at the local Hochschule there before serving as the director of the Bamberg Institute of Geochemistry (1955–60).

In 1926, in collaboration with his wife Ida Tacke, Noddack discovered the element rhenium. They thought that they had found element 43, which they named "masurium." In fact this element was correctly identified in

1937 by Emilio SEGRÈ, who named it technetium.

Noddack is also remembered for arguing for a concept he called *allgegenwartskonzentration* or, literally, omnipresent concentration. This idea, somewhat reminiscent of the early Greek philosopher ANAXAGORAS, assumed that every mineral actually contained every element. The reason they could not all be detected was, of course, because they existed in too small quantities.

Noether, Amalie (Emmy) (1882–1935)
German mathematician

Noether (**no**-ther) was born in Erlangen, Germany, and educated at the University of Erlangen and the University of Göttingen, being awarded a PhD in 1907. At that time women were not allowed to have academic positions in Germany so she became an independent researcher in mathematics, but gave "guest" lectures for her father Max Noether, who was a professor of mathematics at Erlangen, and for David HILBERT at Göttingen. Hilbert attempted to get her a permanent position at Göttingen but was unsuccessful. When the Nazis took over in 1933 Noether, who was Jewish, moved to the United States. She obtained jobs at Bryn Mawr College and at the Institute for Advanced Study, Princeton. She died following an operation.

Noether did fundamental work in pure mathematics on abstract algebra but is best known to physicists for *Noether's theorem*. This result, which was found by Noether in 1918, states that for every continuous symmetry of a mechanical system described by a Lagrangian there is a conserved quantity. For example, Noether's theorem shows that the conservation of linear momentum is associated with translational invariance in space.

Noguchi, (Seisako) Hideyo (1876–1928)
Japanese bacteriologist

Noguchi (no-**goo**-chee) was born at Okinashimamura in Japan. In spite of a humble family background and a physical handicap caused by a childhood accident, he pursued a career in medicine. After considerable perseverance he entered medical school in Tokyo, obtaining his diploma in 1897. Three years later he traveled to America and commenced work at the University of Pennsylvania, studying animal venoms and their antivenins. In 1904, after a year spent in Copenhagen, Noguchi joined the Rockefeller Institute for Medical Research, New York. Here he successfully cultured the spirochete bacterium, *Treponema pallidum*, which causes syphilis. This enabled Noguchi to devise a diagnostic skin test for syphilis using an emulsion of his spirochete culture. He further showed that *T. pallidum* invades the nervous system as the disease progresses. In recognition of his work, Noguchi was awarded the Order of the Rising Sun in his home country in 1915.

Noguchi went on to study the possible causes of other diseases. After investigating Oroya fever in South and Central America, he showed that it was caused by a bacterium, *Bartonella bacilliformis*, which was transmitted to humans by sand flies. Between 1919 and 1922, Noguchi became certain that yellow fever also was caused by a bacterium. However, by 1927 this view had been discredited with the discovery that a virus was responsible.

In the same year, Noguchi went to West Africa and worked doggedly to prove to himself that yellow fever was in fact a virus disease. Within six months he had confirmed this but just before his departure for New York he contracted yellow fever himself and died shortly after.

Nollet, Abbé Jean Antoine (1700–1770) *French physicist*

> No one ever saw him lose his composure or his unfailing consideration; he only became excited when he talked about physics.
> —Grandjean de Fouchy, *Histoire de l'Académie des Sciences* (1770; History of the Academy of Sciences)

Born at Pimprez in France, Nollet (no-**lay**) was one of the great popularizers of the new electrical science in the salons and at the court of 18th-century France. He had collaborated with Charles Dufay in the period 1730–32 and tended to follow him in his electrical theory. Nollet saw electricity as a fluid, subtle enough to penetrate the densest of bodies. In 1746 he first formulated his theory of simultaneous "affluences and effluences" in which he assumed that bodies have two sets of pores in and out of which electrical effluvia might flow. He was later involved with Benjamin FRANKLIN in a dispute over the nature of electricity.

After the discovery of the Leyden jar (a device for storing electrical charge) by Pieter van MUSSCHENBROEK in 1745, Nollet arranged some spectacular demonstrations of its power. He once gave a shock to 180 royal guards and, even more dramatically, joined 700 monks in a circle to a Leyden jar with quite startling results. Nollet also contributed to the theory of sound when he showed in 1743 that sound carried in water (he had taken care to expel the dissolved air from the water first).

Nordenskiöld, Nils Adolf Eric (1832–1901) *Finnish geologist, explorer, and cartographer*

Nordenskiöld (**noor**-den-shu(r)ld) was educated at the university in his native city of Helsinki where be obtained his PhD in 1857. Because of his hostility to the ruling Russians

he emigrated to Sweden in 1858 where he became head of the mineralogy department of the National Museum.

As a geologist he made, between 1858 and 1872, five field trips to Spitsbergen, reaching as far north as latitude 81°. His most significant feat of exploration was his discovery and successful negotiation of the Northeast Passage between the Atlantic and Pacific oceans. He left Tromso, Norway, on board the *Vega* in July 1878, and after being trapped in ice near the Bering Strait throughout the winter entered the Pacific the following July.

Nordenskiöld also produced two works – *Facsimilie Atlas* (1889; Duplicate Atlas) and *Periplus* (1897; Circumnavigation) – which did much to stimulate the modern study of the history of cartography.

Norlund, Niels Erik (1885–1981) *Danish mathematician and geodesist*

Norlund (**nor**-luund), who was born at Stagelse in Denmark, studied at Soro High School and then worked as an assistant at the Copenhagen Observatory. His work there was to influence many of his subsequent scientific interests. After obtaining his doctorate in 1910 from the University of Copenhagen, Norlund took up a post in Sweden as professor of mathematics at Lund University. He remained there until 1922 when he returned to Denmark to become professor of mathematics at Copenhagen University, a post that he held until his retirement in 1956. He held many official posts, among them director of the Royal Geodesic Institute of Denmark (1923–55) and editor of the journal *Acta Mathematica*.

Norlund's interest in geodesy dates back to his time at the Copenhagen Observatory and his publications include works on the mapping of both Denmark and Iceland. He organized a new triangulation of Denmark from 1923, accompanied by gravity measurements and astronomical determinations of longitude, and also carried out a partial triangulation of Greenland. Among his more purely mathematical interests were differential equations and analysis.

Norman, Robert (about 1580) *English navigator and instrument maker*

Norman was described by William GILBERT as a "skilled navigator and ingenious artificer," and by himself as an "unlearned mechanician." By observing the downward tilt of the compass needle, Norman discovered the phenomenon of magnetic dip – the angle made by the Earth's magnetic field with the vertical at a point on the Earth's surface. He wrote a full account of it and published it in *The Newe Attractive* (1581).

Norrish, Ronald George Wreyford
(1897–1978) *British chemist*

Norrish was educated at the university in his native Cambridge. Apart from the war years, he spent his whole career there, serving as professor of physical chemistry from 1937 until 1965.

Norrish made his important contributions to chemistry in the fields of photochemistry and chemical kinetics, being introduced to these by Eric RIDEAL during his PhD work. From 1949 to 1965 he collaborated with his former pupil George PORTER in the development of flash photolysis and kinetic spectroscopy for the investigation of very fast reactions. For their work they shared the 1967 Nobel Prize for chemistry with Manfred EIGEN.

Norrish also made a significant contribution to chemistry when he showed the need to modify DRAPER's law. In the mid-19th century John Draper proposed his law that the amount of photochemical change is proportional to the intensity of the light multiplied by the time for which it acts. Norrish was able to show that the rate should be proportional to the square root of the light intensity.

Northrop, John Howard (1891–1987)
American chemist

The son of biologists, Northrop was born in Yonkers, New York, and educated at Columbia, obtaining his PhD there in 1915. In 1917 he joined the Rockefeller Institute of Medical Research, only leaving on his retirement in 1961.

In the early 1930s Northrop confirmed some earlier results of James SUMNER. Between 1930 and 1935 he and his coworkers succeeded in isolating a number of enzymes, including pepsin, trypsin, chymotrypsin, ribonuclease, and deoxyribonuclease, crystallizing them and unequivocally exhibiting their protein nature. This was sufficient finally to convince chemists that Sumner was correct, and Richard WILLSTÄTTER had been wrong in his assertion that enzymes are nonprotein.

Using Northrop's techniques, Wendell Stanley was able in 1936 to isolate and crystallize the tobacco mosaic virus, and showed it to be composed of nucleoprotein. Subsequently (1938) Northrop isolated a bacteriophage (bacterial virus) and demonstrated that this also consisted of nucleoprotein. For such work on the isolation and crystallization of proteins and viruses, Northrop, Sumner, and Stanley shared the 1946 Nobel Prize for chemistry.

Norton, Thomas (c. 1437–c. 1514) *English alchemist*

It is generally accepted that the anonymous work *The Ordinall of Alchimy* was written by Norton in about 1477. His authorship is revealed by an obvious cipher. The work was first published in a Latin translation by Maier in 1618 and, in its original English, was included in Elias Ashmole's *Theatrum chemicum Brittanicum* (1652; Exhibition of British Chemicals).

Norton claims in his poem to have learned the secrets of alchemy from his master, George Ripley, in 40 days. His objectives included the elixir of life and the philosopher's stone. His work contained much practical information on the craft of alchemy and is one of the classic alchemical works. Norton recognized the importance of color, odor, and taste as guides to chemical analysis.

Noyce, Robert Norton (1927–1990)
American physicist

Noyce was the son of a Congregational minister from Denmark, Iowa. He was educated at Grinnell College, Iowa, and at the Massachusetts Institute of Technology, where he obtained his PhD in 1953. After working briefly for Philco, Philadelphia, Noyce moved to Mountain View, California, to work for William SHOCKLEY, coinventor of the transistor, at his Semiconductor Laboratory. Noyce and a number of colleagues decided to set up in business themselves. Financed by the Fairchild Corporation of New York, they set up Fairchild Semiconductor in the Santa Clara Valley, fifty miles south of San Francisco, a site better known today as "Silicon Valley."

The first major success was the integrated circuit, the foundation of the modern electronics industry. Noyce filed his patent in April 1959, some six weeks after a similar patent had been filed by Jack KILBY at Texas Instruments. Although Noyce's design was more advanced, priority seemed to lie with Kilby. Yet in 1968 the Supreme Court awarded all rights to Noyce and Fairchild on the grounds that Kilby's patent application lacked sufficient clarity.

Whereas Kilby's circuit had used the silicon mesa transistor, Noyce opted for a planar model. Unlike the mesa, Noyce's model had no raised parts to attract contaminants and was more easily protected by a layer of silicon dioxide. Parts were no longer connected by wires but by evaporating the aluminum wires onto the insulating surface. As an extra bonus it also proved much easier to mass-produce planar transistors.

At this point Noyce was able to sell back to Fairchild his initial investment of $500 for $250,000. He went on in 1968 to found Intel (Integrated Electronics). It gained an early success with the production of a one-kilobyte RAM chip. Further improvements were quickly made and, with the 1973 launch of a 4K RAM chip, sales soared above $60 million. Intel's success rapidly made Noyce one of Silicon Valley's first multimillionaires; it continued with the production of the 486 chip in 1989 and the 60 MHz Pentium chip in 1993.

Noyes, William Albert (1857–1941) *American chemist*

Noyes, born the son of a farmer in Independence, Iowa, was educated at Johns Hopkins University where he obtained his PhD in 1882. After working at the University of Tennessee (1883–86) and Rose Polytechnic, Terre Haute, Indiana (1886–1903), he held the post of chief chemist at the Bureau of Standards (1903–06). He then moved to the University of Illinois where he served as professor of chemistry until his retirement in 1926.

Noyes is mainly remembered for his careful and accurate determination of certain crucial atomic weights while at the Bureau of Standards. His measurement of the hydrogen-to-oxygen ratio as 1.00787:16 differs only at the fourth decimal place from currently accepted values. He also, in addition to writing a number of textbooks, founded *Chemical Abstracts* in 1907 and was the first editor of *Chemical Reviews* (1924–26).

Noyori, Ryoji (1938–) *Japanese chemist*

Noyori (**noi**-yori) was born in Kobe, Japan, and educated at Kyoto University, gaining his PhD in 1967. Since 1972 he has been professor of chemistry at Nagoya University, Japan, and since 2000 has been Director of the Research Center for Materials Science, Nagoya University.

Noyori extended the work of William KNOWLES on chirally catalyzed hydrogenation reactions. This research was recognized by the award of the 2001 Nobel Prize for chemistry to Knowles, Noyori, and Barry SHARPLESS.

Nurse, Sir Paul M. (1949–) *British cell biologist*

Nurse was awarded a BSc by the University of Birmingham in 1970 and a PhD by the University of East Anglia in 1973. After postdoctoral research at Edinburgh University (1973–79) and the University of Sussex (1979–84), he served (1984–87) as head of the Cell Cycle Control Laboratory at the Imperial Cancer Research Fund (ICRF), London. From 1987 through 1993 he worked at Oxford University, and in 1993 rejoined the ICRF as director of laboratory research. In 1996 he was appointed director-general of the ICRF. Nurse was knighted in 1999. In 2003 he became president of the Rockefeller Institute, New York.

Nurse's main field of work concerns the mechanisms controlling the cell cycle. During the mid-1970s, Nurse used a type of yeast, *Schizosaccharomyces pombe*, to isolate mutants with defects in cell-cycle control, and hence determine the corresponding genes. A similar approach was used by Leland HARTWELL with baker's yeast. Nurse discovered a gene, designated *cdc2* (cell division cycle 2), with a key role. Mutants lacking the Cdc2 protein (encoded by the *cdc2* gene) failed to enter the mitosis (M) phase of the cell cycle – when the chromosomes separate and segregate to form the nuclei of the new daughter cells. Nurse later found that *cdc2* was functionally analogous to the *CDC28* gene discovered by Hartwell in baker's yeast, and that it controlled other phases of the cell cycle.

In 1987 Nurse isolated the corresponding human gene, now known as *CDK1*. The protein encoded by this gene belongs to a family called cyclin-dependent kinases. These form complexes with another family of proteins, the cyclins, and drive the cell cycle by phosphorylating (adding phosphate groups to) other proteins. These findings have important implications for cancer research, since mutations in CDK and cyclin genes can lead to cancer.

For his discoveries of key regulators of the cell cycle, Nurse received the 2001 Nobel Prize for physiology or medicine, awarded jointly with Hartwell and Timothy HUNT.

Nüsslein-Volhard, Christiane (1942–) *German biologist*

Nüsslein-Volhard (noos-lIn-**fol**-hart) was educated at the University of Tübingen. After a spell at the European Molecular Biology Laboratory at Heidelberg, she moved in 1981 to the Max Planck Institute for Development, Tübingen, where she has served since 1990 as the director of the department of genetics.

From 1978 to 1981 Nüsslein-Volhard collaborated with Eric WIESCHAUS on identifying the genetic factors responsible for the development of the fruit fly *Drosophila melanogaster*. After examining many thousands of specially bred mutant flies they had managed by 1980 to identify the main development sequence in *Drosophila*. Nüsslein-Volhard also succeeded in illuminating the general process of development. It had long been thought that differentiation in early embryos – anterior from posterior, for example, or dorsal from ventral – was caused by varying concentrations of substances along the axes of the egg. The theory of morphological gradients, as it is known, has recently been supported by experimental work carried out by Nüsslein-Volhard and her Tübingen colleagues. Nüsslein-Volhard shared the 1995 Nobel Prize for physiology or medicine with Edward LEWIS and Eric Wieschaus for their work on the development of *Drosophila*.

Oakley, Kenneth Page (1911–1981)
British physical anthropologist

Oakley, the son of a physician, was educated at University College, London, where he obtained his PhD in 1938. After a brief period with the Geological Survey, Oakley joined the British Museum (Natural History); he remained there for his whole career and served as deputy keeper (anthropology) from 1959 until his retirement in 1969.

Oakley, much concerned with the accurate dating of human and animal remains, had developed by the late 1940s a method of dating them by measuring the amount of fluorine present. The technique received considerable publicity in the early 1950s when it was used to discredit Piltdown man, the skull and jaw supposedly discovered by Charles Dawson and Arthur WOODWARD in 1912.

Oberth, Hermann Julius (1894–1989)
German rocket scientist

> We shall make every effort so that the greatest dream of mankind might be fulfilled.
> —On manned space flight. Letter to Konstantin Tsiolkovsky (1929)

The son of a surgeon, Oberth (**oh**-bairt or **oh**-bert) was born in Hermannstadt (then in Austria–Hungary; now Sibiu in Romania). He began by studying medicine in Munich but after the interruption caused by World War I decided to study the exact sciences. This he did at Munich, Göttingen, and Heidelberg, qualifying as a teacher in 1922. He taught first in Romania then in 1938 joined the faculty of the Technical University in Vienna. He became a German citizen in 1940. In 1941 he was transferred to the rocket development center at Peenemünde, on the Baltic Sea, where he worked under Wernher VON BRAUN, and in 1943 was sent to another rocket-research location. After the war he settled for varying periods in Switzerland, Italy, and America before returning to Germany in 1958. He appears to have been deliberately misleading and secretive about his activities in this period.

Like the other space pioneers, Robert GODDARD and Konstantin Tsiolkovsky, the passion for space travel came to him early. He related reading Jules Verne's *From the Earth to the Moon* so frequently as a child that he knew it by heart. He calculated quite early that only with a rocket engine powered by liquid propellants would it be theoretically possible to reach the escape velocity of 25,000 mph (11.2 km/s). Oberth also worked out the equations for practical flight and submitted them to the University of Heidelberg in 1922 for his doctorate. They were rejected but Oberth went ahead and published the thesis in 1923 as *Die Rakete zu den Planetenräumen* (The Rocket into Interplanetary Space). He was unaware of similar calculations by Tsiolkovsky and Goddard but later acknowledged their precedence in the field.

Unlike Goddard and Tsiolkovsky, Oberth's work was greeted enthusiastically by an admittedly small band of devotees. There were certainly enough of them to form a society, the *Verein für Raumschiffahrt*, or Society for Space Travel, by the end of the decade. This partly explains why, when war came in 1939, Germany could so quickly organize an efficient and competent research team. It was made up of people who had worked on and thought of little else other than rockets since the late 1920s.

Oberth almost got a rocket in the air in the 1920s. He was acting as scientific adviser to Fritz Lang who was making a film, *Frau im Mond* (Lady in the Moon), and was persuaded that publicity would flow to the film if rocket flight could be publicly demonstrated just before its opening. Oberth, who was no engineer, confidently accepted Lang's finance, promising a rocket within a year. The flight never materialized, with Oberth disappearing at a crucial moment, exhausted and depressed after nearly having been blinded in an explosion while testing various mixtures of gasoline and liquid air.

Occhialini, Giuseppe Paolo Stanislao (1907–1993) *Italian physicist*

Occhialini (ohk-ya-**lee**-nee) was born at Fossombrone in Italy and educated at the University of Florence, where he also taught from 1932 to 1937. He belonged to a small group of creative physicists, which, centered on Enrico FERMI, emerged in Italy during the interwar period. Like most of his colleagues he fled from Italy with the growth of fascism, and he taught at São Paulo, Brazil, from 1937 to 1944, at Bristol University, England, from 1944 to 1947, and at the University of Brussels from 1948 to 1950. He then returned to Italy where he has held chairs of physics at the University of Genoa (1950–52) and at the University of Milan since 1952.

Occhialini was involved in two major discoveries with his collaborators, both of whom were later awarded Nobel prizes. In 1933 with Patrick BLACKETT he obtained cloud-chamber photographs that showed tracks due to the positive electron (or positron).

In 1947 Occhialini and Cecil POWELL, examining the tracks of cosmic rays in photographic emulsions, noted the track of a particle that was some 300 times more massive than an electron. The particle broke down into the already familiar mu meson (muon); it was in fact the pi meson or pion.

Ochoa, Severo (1905–1993) *Spanish–American biochemist*

Born at Luarca in Spain, Ochoa (oh-**choh**-a) graduated from Málaga University in 1921 and then proceeded to study medicine at Madrid University, receiving his MD in 1929. Having held research positions in Germany, Spain, and England, he became a research associate in medicine at New York University in 1942, taking American citizenship in 1956. He became a full professor in 1976 and in 1985 was appointed honorary director of the center for molecular biology, University of Madrid.

Ochoa was one of the first to demonstrate the role of high-energy phosphates, e.g., adenosine triphosphate, in the storage and release of the body's energy. While investigating the process of oxidative phosphorylation, in which such triphosphates are formed from diphosphates, he discovered the enzyme polynucleotide phosphorylase. This can catalyze the formation of ribonucleic acid (RNA) from appropriate nucleotides and was later used for the synthesis of artificial RNA. Ochoa was awarded the 1959 Nobel Prize for physiology or medicine for this discovery, sharing the prize with Arthur KORNBERG, who synthesized deoxyribonucleic acid (DNA). Ochoa also isolated two enzymes catalyzing certain reactions of the KREBS cycle.

Odling, William (1829–1921) *British chemist*

Odling, the son of a London surgeon, studied medicine at London University before moving into chemistry. He studied in Paris under Charles GERHARDT and in 1863 was appointed professor of chemistry at St. Bartholomew's Hospital, London. In 1867 he succeeded Michael FARADAY as Fullerian Professor at the Royal Institution, London, and in 1872 he moved to Oxford University to take up the Waynflete Chair of Chemistry until his retirement in 1912.

Odling was one of the pioneers of the valence theory first propounded by Edward FRANKLAND in 1852. Although the term "valence" was not in use in 1854 when Odling first wrote on the topic, he had a clear idea of the concept, which he referred to as replaceable or representative value. Odling, like many of his contemporaries, was skeptical of the existence of atoms, and it was not until the 1890s that his misgivings were overcome. From his work on atomic weights he was led to suggest that the atomic weight of oxygen should be 16, not 8. In 1861 he was able to clear up a troublesome problem over oxygen by suggesting that ozone was triatomic; this was later confirmed by J. Soret in 1866. Odling also studied and classified silicates.

Oersted, Hans Christian (1777–1851) *Danish physicist*

> Spirit and nature are one, viewed under two different aspects. Thus we cease to wonder at their harmony.
> —*The Soul in Nature* (1852)

Oersted (**er**-sted) was born at Rudkjöbing in Denmark and studied at Copenhagen University, where he received a PhD in 1799 for a thesis defending Kantian philosophy. To complete his scientific training he then traveled through Europe visiting the numerous physicists working on aspects of electricity. On his return to Denmark he started giving public lectures, which were so successful that, in 1806, he was offered a professorship at Copenhagen. Here he became well known as a great teacher and did much to raise the level of Danish science to that of the rest of Europe.

It was while lecturing that he actually first observed electromagnetism (although for years he had believed in its existence) by showing that a needle was deflected when brought close to a wire through which a current was flowing. By the summer of 1820 he had confirmed the existence of a circular magnetic field around the wire and published his results. They produced an enormous flurry of new activity in the scientific world, which up to that time had accepted COULOMB's opinion that electricity and magnetism were completely independent forces.

Ohm, Georg Simon (1787–1854) *German physicist*

Ohm (ohm), who was born at Erlangen in Germany, seems to have acquired his interest in science from his father, a skilled mechanic. He studied at the University of Erlangen and then taught at the Cologne Polytechnic in 1817. From 1826 to 1833 he taught at the Military Academy in Berlin, moving to the Polytechnic at Nuremburg before finally obtaining a chair in physics at Munich in 1849.

Despite the fact that he published his famous law in 1827 in his *Die galvanische Kette mathematisch bearbeitet* (The Galvanic Circuit Investigated Mathematically) he received no recognition or promotion for more than twenty years. Ohm seems to have been stimulated by the work on heat of Joseph FOURIER. The flow of heat between two points depends on the temperature difference and the conductivity of the medium between them. So too, argued Ohm, with electricity. If this line of thought is pursued it soon leads to the general form of *Ohm's law* that the current is proportional to the voltage. Using wires of different sizes he was able to show that the resistance was proportional to the cross-sectional area of the wire and inversely proportional to its length.

Ohm also worked on sound, suggesting in 1843 that the ear analyzes complex sounds into a combination of pure tones. This result was rediscovered by Hermann von HELMHOLTZ in 1860.

In Ohm's honor the unit of electrical resistance, the ohm, was named for him.

O'Keefe, John Aloysius (1916–2000) *American astronomer and geophysicist*

O'Keefe was born in Lynn, Massachusetts, and studied at Harvard, gaining his BA in 1937, and at the University of Chicago where he obtained his PhD in 1941. He worked as a mathematician in the Army Map Service from 1945 to 1958, and then for the National Aeronautics and Space Administration at their Goddard Space Flight Center in Maryland.

Once with NASA O'Keefe started work on the Vanguard project, which was part of the American program for International Geophysical Year. The first Vanguard satellite was launched in March 1958, two months after the launch of Explorer 1, America's first satellite. Both craft were tracked over a long period – Vanguard 1 transmitted signals for six years – and O'Keefe used the satellite positions to check the size and shape of the Earth. By making careful observations of the orbits adopted by the satellites it is possible to reconstruct the detailed figure of the attracting mass, in this case, the Earth.

The 18th-century surveyors had confirmed the predictions of NEWTON that the Earth was an oblate spheroid, i.e., its equatorial diameter was longer than its polar diameter (by about 26.5 miles or 42.7 km). O'Keefe showed that while this is true in general there are a few modifications that need to be added. In particular he showed that the north pole is about 100 feet (30 m) further from the Earth's center than the south pole (both poles considered to be at sea level) and that the equatorial bulge is not completely symmetrical for its highest points lie a little south of the equator. This means that the Earth is slightly pear-shaped, which has been verified by more recent satellite observations.

Oken, Lorenz (1779–1851) *German nature philosopher*

> In August 1806, while walking in a forest in the Harz Mountains, I saw at my feet a [sheep's] skull bleached by time. This, I cried out, is a vertebral column!
> —On his theory that the skull is composed of a number of vertebrae. Quoted by René Taton in *Science in the Nineteenth Century* (1965)

Oken (**oh**-ken), a farmer's son from Bohlsbach in Germany, was one of the most influential adherents of the system termed "nature philosophy," a school of thought established in Germany at the beginning of the 19th century. Oken's numerous clashes with the authorities over his strongly held political and scientific beliefs meant that he was frequently changing jobs until he finally settled at Zurich University in 1832. He nevertheless managed to found the reputable biological journal *Isis*, and he was also instrumental in organizing regular scientific congresses for biological and medical scientists.

Oken is remembered mainly for the theory proposing that the skull is formed from the fusion of several vertebrae – an idea for which Wolfgang von GOETHE also claimed priority. The theory was later disproved by T. H. HUXLEY, but it served some purpose in preparing people for the evolutionary ideas put forward by Charles DARWIN.

Olah, George Andrew (1927–   ) *Hungarian–American chemist*

Olah gained his PhD from the Technical University, Budapest, in 1949. He moved to Canada in 1956 following the Hungarian uprising and joined the staff of the Dow Chemical Company in Ontario. In 1964 he moved to the U.S. and in 1965 joined the faculty of the Case Western Reserve University, Cleveland, Ohio. In 1977 he moved to the University of Southern California, becoming director of the Loker Hydrocarbon Research Institute in 1991. Olah became a naturalized American citizen in 1970.

In certain chemical reactions involving hydrocarbons, extremely short-lived highly reactive positively charged carbon intermediates are often formed. These have a positive charge on the carbon atom and are known as "carbonium ions" or "carbocations." Because of their short lifetime, little had been established about these intermediates.

Olah, while working at Dow, discovered a way to preserve the intermediates and to allow their properties to be investigated. He found that solutions of a very strong acid, variously described as a "superacid" or a "magic acid," would preserve carbocations for months at a time and thus allow their structure to be determined with such techniques as nuclear magnetic resonance spectroscopy (NMR). Olah's superacids were formed by dissolving compounds such as antimony pentafluoride in water at low temperature. The result was an acid some 10^{18} times stronger than sulfuric acid. The stable carbocations formed in this way proved to be quite unusual, with structures quite unlike the more familiar tetrahedral forms. Olah's work quickly found important applications in industry; it has, for example, been widely used in synthesizing high-octane gasoline.

For his work on carbocations Olah was awarded the 1995 Nobel Prize for chemistry.

Olbers, Heinrich (1758–1840) *German astronomer*

> He [Olbers] was to me the most noble friend. With wise and fatherly counsel he guided my youth.
> —Friedrich Wilhelm Bessel, obituary notice of Olbers (1840)

Olbers (**ohl**-bers), who was born at Arbegen in Germany, was a physician who practiced medicine at Bremen. He became a good amateur astronomer, and converted part of his house into an observatory. He became interested in searching for a planet in the "gap" between Mars and Jupiter, and rediscovered the first minor planet (or asteroid) Ceres after it had been lost by its discoverer Giuseppe PIAZZI. In 1802 he discovered the second asteroid, Pallas, and in 1807 the fourth, Vesta. He also devised a method of calculating comet orbits, called *Olbers's method*, and discovered five comets. The one named for him was last seen in 1956.

Olbers's modern fame, however, rests on his statement of a very simple problem, the solution of which has had a profound impact on modern cosmology. He asked the naive question: why is the sky dark at night? Olbers assumed that the heavens are infinite and unchanging and that the stars are evenly distributed. The amount of light reaching the Earth from very distant stars is very small – in fact, the illumination decreases with the square of the distance. On the other hand, this is compensated for by the increased number of stars – the average number at a given distance increases with the square of the distance. The result is that the whole sky should be about as bright as our Sun. Olbers's solution to this problem was that the light is absorbed by dust in space, but this is an unsatisfactory explanation since the dust would eventually become incandescent and radiate energy. In the 20th century it became clear that the solution lies in the fact that the universe is not uniform, infinite in time, or unchanging; the red shift of the light from distant galaxies results in a reduc-

tion of the energy of the radiation from stars. The paradox had been discussed earlier (1744) by J. P. L. Chesaux.

Oldham, Richard Dixon (1858–1936) *British seismologist and geologist*

Oldham's father, Thomas, was professor of geology at Trinity College, Dublin, and director of the Geological Surveys of India and Ireland. Oldham, who was born in Dublin, was educated at the Royal School of Mines; in 1879 he followed his father in joining the Geological Survey of India, in which he rose to the rank of superintendent. He retired in 1903 and became director of the Indian Museum in Calcutta.

Oldham made two fundamental discoveries. He made a detailed study of the Assam earthquake of 1897 and, in 1900, was the first to identify clearly the primary (P) and secondary (S) seismic waves transmitted through the Earth, which had been predicted by the mathematician Siméon POISSON on theoretical grounds. Secondly, in 1906 he provided the first clear evidence that the Earth had a central core. He found that the arrival of the primary, or compressional, waves was delayed at places opposite to the focal point of an earthquake. He deduced from this that the Earth contains a central core that is less dense and rigid than the rocks of the mantle, and through which compressional waves would travel less fast. A detailed analysis of the arrival and distortion of the P and S waves that had traveled through or near to the center of the Earth later provided much insight into the structure of the Earth.

Oliphant, Marcus Laurence Elwin (1901–2000) *Australian physicist*

Born in Adelaide, Oliphant was educated at the university there and at Cambridge University, England, where he obtained his PhD in 1929. He then worked at the Cavendish Laboratory in Cambridge before being appointed (1937) to the Poynting Professorship of Physics at Birmingham University. Oliphant returned to Australia in 1950 and held research chairs at the Australian National University, Canberra, until his retirement in 1967.

Hydrogen, the simplest of all atoms, normally has a nucleus of a single proton, but in 1932 Harold UREY had discovered a heavier form that he called deuterium, with a nucleus consisting of a proton and a neutron. The enlarged nucleus became known as the deuteron. In 1934 Oliphant and his collaborator Paul Harteck produced an even heavier form of hydrogen by bombarding deuterium with deuterons. This new isotope has a nucleus consisting of one proton and two neutrons (hydrogen-3). They named it "tritium" and called the nucleus a "triton." The isotope is radioactive with a half-life of 12.4 years and for this reason is not found in significant amounts in nature.

During World War II Oliphant did important work on the development of radar. It was in his laboratory that two German refugees, Rudolph PEIERLS and Otto FRISCH, made some of the vital calculations and experiments that revealed the real possibility of an atomic bomb.

Olsen, Kenneth Harry (1926–   ) *American computer engineer and entrepreneur*

Born in Bridgeport, Connecticut, Olsen studied electrical engineering at the Massachusetts Institute of Technology (MIT). He remained at MIT after graduating, joining the Digital Computer Laboratory directed by Jay FORRESTER. While there Olsen worked on the design and construction of the first computers to adopt magnetic-core memory systems.

In 1957 Olsen left MIT to set up Digital Equipment Corporation in partnership with his brother Stan Olsen and a colleague, Harlan Anderson. The company was partly financed by the Boston firm American Research and Development who, for a stake of $70,000, took some 60% of the equity. When they sold their stake in 1972 it was worth $350 million.

The foundation for Digital's success lay with the development of the first minicomputer, the PDP-1 (Programmed Data Processor). Before Olsen the computer market was dominated by mainframe computers built by IBM and Sperry Rand. They cost millions to produce and operated by selling time to their clients. Olsen saw that few organizations needed such computing power, but could readily use a smaller and cheaper model to calculate their payroll, monitor sales, or analyze data. The PDP-1 sold for $120,000; the PDP-8, introduced in 1963 and incorporating magnetic cores and integrated circuits, was about the size of a refrigerator and sold for $18,000. Under Olsen's direction Digital grew into one of the largest computer companies in the world generating by the mid-1980s sales in excess of $4 billion.

Omar Khayyam (*c.* 1048–*c.* 1122) *Persian astronomer, mathematician, and poet*

> The majority of the people who imitate philosophers confuse the true with the false ... they do nothing but deceive and pretend knowledge, and they do not use what they know of the sciences except for base and material purposes.
> —Quoted by H. J. J. Winter and W. 'Arafat in *Journal of the Royal Asiatic Society of Bengal Science* (1950)

Omar Khayyam (oh-**mar** kI-**yahm**), who was born at Nishapur (now in Iran), produced a work on algebra that was used as a textbook in Persia until the last century. He gave a rule for solving quadratic equations, he could solve special cases of the cubic, and – in a last work –

seemed to have some inkling of the binomial theorem. He also worked on the reform of the Persian calendar, which was basically the Egyptian one of 365 days, introducing a sixth epagomenic (extra) day and obtaining an accurate estimate of the tropical year.

Onsager, Lars (1903–1976) *Norwegian–American chemist*

Born at Christiania (now Oslo) in Norway, Onsager (on-**sah**-ger) was educated at the Norwegian Institute of Technology. He moved to America in 1928 and obtained his PhD at Yale in 1935. He spent virtually his whole career at Yale, serving as J. W. Gibbs Professor of Theoretical Chemistry from 1945 until 1972.

Onsager made two important contributions to chemical theory. In 1926 he showed the need to modify the equation established by Peter DEBYE and Erich HÜCKEL in 1923, which described the behavior of ions in a solution, by taking Brownian motion into consideration.

Onsager's main work, however, was in the foundation of the study of nonequilibrium thermodynamics. Here an attempt is made to apply the normal laws of thermodynamics to systems that are not in equilibrium – where there are temperature, pressure, or potential differences of some kind. For his work in this field Onsager was awarded the Nobel Prize for chemistry in 1968. The study of nonequilibrium thermodynamics was further developed by Ilya PRIGOGINE.

Oort, Jan Hendrik (1900–1992) *Dutch astronomer*

The son of a physician from Franeker in the Netherlands, Oort (ohrt) was educated at the University of Gröningen where he worked under Jacobus KAPTEYN and gained his PhD in 1926. After a short period at Yale University in America he was appointed to the staff of the University of Leiden where he was made professor of astronomy in 1935 and from 1945 to 1970 served as director of the Leiden Observatory. He also served as director of the Netherlands Radio Observatory.

Oort's main interest was in the structure and dynamics of our Galaxy. In 1927 he succeeded in confirming the hypothesis of galactic rotation proposed by Bertil LINDBLAD. He argued that just as the outer planets appear to us to be overtaken and passed by the less distant ones in the solar system, so too with the stars if the Galaxy really rotated. It should then be possible to observe distant stars appearing to lag behind and be overtaken by nearer ones. Extensive observation and statistical analysis of the results would thus not only establish the fact of galactic rotation but also allow something of the structure and mass of the Galaxy to be deduced.

Oort was finally able to calculate, on the basis of the various stellar motions, that the Sun was some 30,000 light-years from the center of the Galaxy and took about 225 million years to complete its orbit. He also showed that stars lying in the outer regions of the galactic disk rotated more slowly than those nearer the center. The Galaxy does not therefore rotate as a uniform whole but exhibits what is known as "differential rotation."

Oort was also one of the earliest of the established astronomers to see the potential of the newly emerging discipline of the 1940s, radio astronomy. As one of the few scientists free to do pure research in the war years, he interested Hendrik VAN DE HULST in the work that finally led to the discovery in 1951 of the 21-centimeter radio emission from neutral interstellar hydrogen.

By measuring the distribution of this radiation and thus of the gas clouds Oort and his Leiden colleagues lost little time in tracing the spiral structure of the galactic arms and made substantial improvements to the earlier work of William MORGAN. They were also able to make the first investigation of the central region of the Galaxy: the 21-centimeter radio emission passed unabsorbed through the gas clouds that had hidden the center from optical observation. They found a huge concentration of mass there, later identified as mainly stars, and also discovered that much of the gas in the region was moving rapidly outward away from the center.

Oort made major contributions to two other fields of astronomy. In 1950 he proposed that a huge swarm of comets surrounded the solar system at an immense distance and acted as a cometary reservoir. A comet could be perturbed out of this *Oort cloud* by a star and move into an orbit taking it toward the Sun. In 1956, working with Theodore Walraven, he studied the light emitted from the Crab nebula, a supernova remnant. The light was found to be very strongly polarized and must therefore be synchrotron radiation produced by electrons moving at very great speed in a magnetic field.

Oparin, Aleksandr Ivanovich (1894–1980) *Russian biochemist*

Oparin (o-**pa**-rin), who was born at Uglich near Moscow, studied plant physiology at the Moscow State University, where he later served as professor. He helped found, with the botanist A. N. Bakh, the Bakh Institute of Biochemistry, which the government established in 1935. Oparin became director of the institute in 1946. As early as 1922 Oparin was speculating on how life first originated and made the then controversial suggestion that the first organisms must have been heterotrophic – that is, they could not make their own food from inorganic starting materials, but relied upon organic substances. This questioned the prevailing view that life originated with autotropic organisms, which, like present-day plants, could synthesize their food from simple inorganic materials. Oparin's view has gradually gained acceptance in many circles. Oparin did much to stimulate research on the origin of life and organized the first international meeting to discuss the problem, held in Moscow in 1957.

Opie, Eugene Lindsay (1873–1971) *American pathologist*

Born in Staunton, Virginia, Opie was educated at Johns Hopkins University, where he gained his MD in 1897. He worked at the Rockefeller Institute from 1904 until 1910 when he moved to St. Louis as professor of pathology at Washington University. He later held similar appointments at the University of Pennsylvania (1923–32) and Cornell from 1932 until his retirement in 1941.

It had been shown by Oskar Minkowski in 1889 that removal of the pancreas leads to fatal diabetes. Opie, in 1900, made the further advance of implicating the pancreatic islets of Langerhans in the disease. He pointed out that they showed a characteristic degeneration found only in the pancreas of diabetic patients. It was left to Frederick BANTING and Charles BEST in 1921 to show that it was the failure of the islets to secrete insulin that caused the complaint. Opie also did important work on the epidemiology of tuberculosis and demonstrated that the level of tuberculosis in a population can be accurately assessed from the tuberculin reaction.

Oppenheimer, Julius Robert (1904–1967) *American physicist*

> The physicists have known sin; and this is a knowledge which they cannot lose.
> —*Open Mind* (1955)

> The scientist is not responsible for the laws of nature, but it is a scientist's job to find out how these laws operate. It is the scientist's job to find ways in which these laws can serve the human will. However, it is not the scientist's job to determine whether a hydrogen bomb should be used. This responsibility rests with the American people and their chosen representatives.
> —Quoted by L. Wolpert and A. Richards in *A Passion for Science* (1988)

Oppenheimer came from a wealthy New York City family. He was educated at Harvard, at Cambridge, England, and at Göttingen where he obtained his PhD in 1927. From 1929 to 1942 he was at the University of California, and while there accepted the post of director of the Los Alamos laboratory where he worked on the development of the atom bomb. After the war in 1947 he was appointed director of the Institute for Advanced Studies at Princeton, a post he held until his death. He also served (1947–52) as chairman of the important General Advisory Committee of the Atomic Energy Commission.

He is mainly remembered, however, for his work on the Los Alamos project. It has been argued that only Oppenheimer could have made Los Alamos viable for only he could have commanded the allegiance of the world's best talents in physics, who gathered around him in the New Mexico desert. It was also only Oppenheimer who had sufficient independence and authority to persuade the military and General Groves, his superior, to grant sufficient freedom to the scientists to make the project workable.

Freeman DYSON, who saw Oppenheimer in the early 1950s at Princeton with some of his old colleagues, caught in their talk "a glow of pride and nostalgia. For every one of these people the Los Alamos days had been a great experience, a time of hard work and comradeship and deep happiness." But Oppenheimer stayed on after the war when, by all accounts, there was little comradeship, much divisiveness and, ultimately, tragedy for Oppenheimer and some of his friends. In 1948 he was on the cover of *Time* magazine; four years later he was summarily dismissed from his post with the Atomic Energy Commission.

Oppenheimer had actually been under investigation since 1942, first as a matter of routine and then more rigorously when reports critical of his loyalty began to arrive at the office of Colonel Pash, who was responsible for security at Los Alamos. It should be emphasized that at no time has any evidence been published to suggest that Oppenheimer was disloyal to his country. Suspicions were aroused because some of his friends had been members of the Communist party and because he had moved freely in left-wing circles. Both his wife and brother were well-known left-wing sympathizers, if not communists. Before long the suspicions became more precise: it was felt that a Russian agent had made an approach to Oppenheimer and although he had not responded he was guilty of failing to report the approach to the authorities.

Oppenheimer finally admitted that an approach had been made to him but he refused to disclose any names for he felt the man was no longer involved and in any case had merely been a messenger. In a classic dialogue with his inquisitor he kept insisting that, "I feel that I should not give it. I don't mean that I don't hope that if he's still operating that you will find it ... But I would just bet dollars to doughnuts that he isn't still operating."

Finally, at the end of 1943, the Army lost patience and Groves put it clearly to Oppenheimer that he must either provide names or go. He named Haakon Chevalier, a professor of romance languages at the University of California whom he had known since 1938. Chevalier was of course ruined and, although no charges were ever laid against him, it became impossible for him to find academic employment ever again in America. Whether Oppenheimer had behaved honorably by his own judgment is far from clear as there is too much conflicting evidence about the crucial approach. For some, Chevalier was a totally innocent man maligned by a man consumed by ambition; for others Chevalier was a Russian agent who was lucky not to collect a heavy sentence. Where precisely the truth lies must await the release of further documentation.

Oppenheimer was thus free to develop the bomb and at 5.30 a.m. on 16 July 1945 the first bomb was tested. When Oppenheimer saw the huge cloud rising over the desert, he later reported, a passage from the *Bhagavad Gita* came to him: "I am become Death, the shatterer of worlds." A move by senior scientists led by James FRANCK and Leo SZILARD to arrange for a public demonstration of the bomb's power rather than its military use on a Japanese city was referred to Oppenheimer for comment. He was in favor of using it on a Japanese town.

After the war, when he could reasonably have left Government service and devoted himself to theoretical physics, he – for reasons that are not clear – remained as the leading adviser on nuclear weapons, taking responsibility for the development of the hydrogen bomb. Oppenheimer had made many enemies and when accusations were made that he had in fact obstructed the program to build the fusion bomb they were more than willing to work for his downfall. A commission to investigate his loyalty reported in 1954 that "Dr. Oppenheimer did not show the enthusiastic support for the Super (H-bomb) program which might have been expected of the chief adviser of the Government" and rendered its judgment that he was unfit to serve his country. Although Oppenheimer never regained his security clearance, peace of a sort was made with the authorities when in 1963 he received the FERMI award from President Kennedy. Four years later, after bearing his illness with great courage, he died of cancer of the throat.

Orgel, Leslie Eleazer (1927–2007)
British–American biochemist

An inorganic chemist by training, Orgel became fascinated with the origins and evolution of early life, and was one of the architects of the "RNA world hypothesis." Born in London, he attended Oxford University, receiving a BA in chemistry (1949) and a PhD (1951). Following research fellowships at the California Institute of Technology (1954) and University of Chicago (1955), he became assistant director of the theoretical chemistry department, University of Cambridge. In 1964 he was appointed senior fellow and research professor at the Salk Institute for Biological Studies, La Jolla, California, as head of the chemical evolution laboratory. He later also became a professor in the department of chemistry and biochemistry at the University of California, San Diego.

At Cambridge in the 1950s, Orgel worked on the chemistry of transition metals, and in 1960 published *An Introduction to Transition-Metal Chemistry: Ligand Field Theory.* In the 1960s his focus shifted to the nature of early life forms and the origins of self-replicating genetic molecules such as DNA and RNA. Orgel proposed that the nucleic acid RNA was a more likely genetic blueprint for the earliest organisms, instead of the more complex DNA molecule with its protein scaffolding. Orgel's colleague at the Salk Institute, Francis CRICK, and the microbiologist Carl Woese, espoused similar views. This concept of an early RNA-encoded world subsequently gained wider acceptance, especially with the discovery in the early 1980s of RNA enzymes, or ribozymes. However, Orgel later argued that even RNA was too complex, and sought to synthesize simpler molecules that might have preceded it in the evolution of life. Orgel's team attempted to replicate various chemical reactions that were thought necessary to create the precursors of life, such as the prebiotic synthesis of nucleotides (the building blocks of nucleic acids) and their assembly into polymers. One practical outcome of this work was the discovery of an improved method for synthesizing the nucleotide analogue cytosine arabinoside (cytarabine), an anticancer drug.

The fascination with early life led Orgel to participate in the search for life in the solar system and beyond. For example, in 1975 he helped to design equipment for detecting molecular evidence of life by NASA's Mars Lander program, and later served on NASA's astrobiology oversight committee. Contentiously, he supported the notion that life on earth could have originated from space, perhaps delivered as microorganisms on meteorites. But in 2006 he admitted that, apart from beginning an estimated 4.5 billion years ago, other aspects of the origin of life remained obscure. In addition to over 300 scientific articles, Orgel wrote *The Origins of Life: Molecules and Natural Selection* (1970) and, with Stanley L. MILLER, *The Origins of Life on Earth* (1974).

Ortelius, Abraham (1527–1598) *Flemish cartographer*

Ortelius (or-**tee**-lee-us) was the son of a local merchant from Antwerp, which is now in Belgium. At the age of 20 he began to illustrate maps and, in 1564, issued his first world map. His most famous work, *Theatrum orbis terrarum* (Theater of the Whole World), a collection of 70 maps, was the first modern atlas of the world. Published in 1570, it went through 41 editions and appeared in 7 languages.

The maps in it were quite conventional as Ortelius was neither an explorer nor a scientist. He was prepared to collect and produce the best maps of his day and thus uncritically represented such terrestrial additions as "the great southern continent."

Osborn, Henry Fairfield (1857–1935) *American paleontologist*

Osborn was born in Fairfield, Connecticut, and educated at Princeton. He held professorships firstly in comparative anatomy at Princeton and then in biology followed by zoology at Columbia University. He was curator of the American Museum of Natural History's vertebrate paleontology department, and was also a member of the United States and Canadian geological surveys. Chairman of the executive committee of the New York Zoological Society (1896–1903), Osborn was among the founders of the society's renowned zoological park. At the American Museum of Natural History he built up one of the world's finest collections of vertebrate fossils. Like Ernst MAYR he was a proponent of adaptive radiation – the theory that one species may eventually develop into several different species through individuals dispersing from a common center and subsequently developing different characters in adaptation to new environments. Osborn was a successful popularizer of paleontology and wrote extensively on aspects of evolution.

Osheroff, Douglas (1945–   ) *American physicist*

Osheroff was educated at California Institute of Technology, Pasadena, and at Cornell, where he gained his PhD in 1973. After working at Bell Labs he moved to Stanford in 1987 as professor of physics.

In 1972, while still a Cornell graduate student of David LEE, Osheroff made a crucial observation that led to the discovery of the supposedly impossible superfluidity of helium-3. He was actually studying magnetism in helium-3 but, because the magnetic equipment was required elsewhere, Osheroff was detailed to test a new refrigeration device. He noted an unexpected blip in the cooling rate when the solid–liquid mixture fell to 2.6 millikelvins (2.6 millionth of a degree above absolute zero). He had in fact observed a phase transition. Further work revealed that the transition was from a normal fluid to a superfluid.

For his work on superfluidity Osheroff shared the 1996 Nobel Prize for physics with David Lee and Robert RICHARDSON.

Osler, Sir William (1849–1919) *Canadian physician*

Patients should have rest, food, fresh air, and exercise – the quadrangle of health.
—*Aphorisms*

In science the credit goes to the man who convinces the world, not to the man to whom the idea first occurs.
—*Books and Men*

Born at Bond Head, in Ontario, Osler studied medicine at Toronto Medical School and McGill University, obtaining his degree in 1872. In the following year he discovered the minute bodies in blood, known as platelets, that are involved in the clotting mechanism. After a time spent studying and traveling in Europe, he returned to McGill University and was appointed professor in 1875. Following a spell (1884–88) as professor of clinical medicine at the University of Pennsylvania, Philadelphia, Osler was appointed as the first professor of medicine at the newly founded Johns Hopkins University, Baltimore. Here he pioneered new teaching methods in medicine, including tuition on the wards and an emphasis on laboratory techniques. He also wrote a perennially popular medical textbook, *The Principles and Practice of Medicine* (1892).

In 1905 Osler moved to England as professor of medicine at Oxford University, where he lectured and wrote on medical history. He was made a baronet in 1911.

Ostwald, Friedrich Wilhelm (1853–1932) *German chemist*

If my ideas should prove worthless, they will be put on the shelf here [Britain] more quickly than anywhere else, before they can do harm.
—Faraday Lecture to fellows of the Chemical Society (1904)

Ostwald (**ost**-vahlt) was born of German parents who had settled at Riga, now in Latvia. He was educated at the University of Dorpat and the Riga Polytechnic, where he was professor of physics from 1881 until he left to take the chair of physical chemistry at Leipzig in 1887. He retired from his chair in 1906 and spent the rest of his life mainly in literary, philosophical, and editorial work.

Ostwald probably did more than anyone else to establish the new discipline of physical chemistry. He was a great teacher and built up an important research school at Leipzig through which most of the major chemists passed at some time in their career. He founded in 1887 the *Zeitschrift für physikalische Chemie* (Journal of Physical Chemistry), the first journal in the world devoted to the new discipline, translated the writings of the American physical chemist Josiah Willard GIBBS into German in 1892, and also produced an inspiring two-volume textbook on the subject, *Lehrbuch der allgemeinen Chemie* (1885, 1887; Textbook of General Chemistry). Ostwald's own research was mainly on catalysts, for which he received

the Nobel Prize for chemistry in 1909. He defined catalysis in 1894 as "the acceleration of a chemical reaction, which proceeds slowly, by the presence of a foreign substance." He emphasized that the catalyst for the reaction does not alter the general energy relations or the position of equilibrium. In 1888 he formulated his dilution law, which allows the degree of ionization of a weak electrolyte to be calculated with reasonable accuracy. The *Ostwald process* (patented in 1902) was an industrial process for oxidizing ammonia to nitric acid.

Philosophically Ostwald was a positivist and denied the reality of atoms until well into the 20th century. The chemist, he argued, does not observe atoms but studies the simple and comprehensive laws to which the weight and volume ratios of chemical compounds are subject. He believed that atoms were a hypothetical conception but by 1908 he had been converted to atomism.

Ostwald's son, Wolfgang, also became a chemist of some note.

Otto, Nikolaus (1832–1891) *German mechanical engineer and inventor*

The son of a farmer from Holzhausen in Germany, Otto (**ot**-oh) was very much the self-taught engineer. He left school at sixteen to begin work in a merchant's office, progressing to the position of a traveling salesman. During his work he became aware of the gas engines developed by Jean LENOIR, and Lenoir's attempt to use them to power vehicles. Otto also began to experiment with engines. In 1864 he joined with the industrialist Eugen Langen to design and manufacture engines. A factory was built near Cologne in 1869 and engineers hired, one of whom was Gottlieb DAIMLER.

In 1876 Otto took out a patent on a new engine design. Like the Lenoir engine, it was powered by coal gas but it operated with a four-stroke cycle. Lenoir had used, in the manner of a steam engine, a two-stroke cycle. Otto's four-stroke cycle of intake, compression, expansion, and exhaust had originally seemed a backward step producing only one power stroke for every two revolutions of the crank. Further, it was tied to an external gas supply and very heavy, and clearly unsuitable to power a vehicle. A further problem arose when it was discovered that the four-stroke cycle had been patented in 1862 by A. B. Rochas, thus depriving Otto of his own patent.

Despite this the engine was adaptable enough to be used for lifting and pulling a variety of goods. Some 30,000 were sold before Otto's death in 1891. Thereafter, suitably modified by Daimler to consume gasoline, millions of Otto's engines would eventually be sold each year.

Oughtred, William (1575–1660) *English mathematician*

> The *Clavis* [The Key (to Mathematics) by Oughtred] doth in as little room deliver as much of the fundamental and useful parts of geometry (as well as of arithmetic and algebra) as any book I know.
> —John Wallis, *Algebra* (1695)

Oughtred (**aw**-tred) was born at Eton in England and educated at the famous school there (where his father taught writing) and at Cambridge University. He was ordained a priest in 1603 and eventually became rector of Albury.

Despite his clerical post he found time to work on mathematics and he produced what was to become a very famous book on mathematics, the *Clavis mathematicae* (1631; The Key to Mathematics). This work dealt with arithmetic and algebra, and it is of historical importance because Oughtred managed to put into it more or less everything that was known at that time in those areas of mathematics. It rapidly became an influential and widely used textbook and held in high regard by mathematicians of the stature of Isaac NEWTON and John Wallis, himself a pupil of Oughtred. A number of mathematical symbols that are still used were first introduced by Oughtred. Among these were the sign "×" for multiplication and the "sin" and "cos" notation for trigonometrical functions. Oughtred also invented the earliest form of the slide rule in 1622 but only published this discovery in 1632. As a result he became embroiled in a violent dispute with one of his former students, Richard Delamain, who had made the same invention independently.

Oughtred's religious views were conservative and he was a staunch supporter of the Royalist party, but during the time of Cromwell and the Commonwealth he was able to retain his post as vicar. He lived just long enough to see Charles II installed on the throne.

Owen, Sir Richard (1804–1892) *British anatomist and paleontologist*

> Sir Richard Owen, best and brightest of Victorian anatomists, coined the term "Dinosauria," from Greek roots meaning "terrible lizard," in 1842. When Owen first penned the word "dinosaurs," paleontology was still a brand-new science. Owen invented the term … to describe the huge *land* animals of this age. And his original definition is still good.
> —Robert Bakker, *The Dinosaur Heresies* (1986)

Owen studied medicine in his native city of Lancaster (where he was apprenticed), at Edinburgh University, and at St. Bartholomew's Hospital, London. He then practiced and lectured at St. Bartholomew's under the eminent surgeon John Abernethy. In 1826 Owen was appointed curator of the Hunterian Collection of anatomical specimens of the Royal College of Surgeons. He was Hunterian Professor in 1836, and in the following year became the Royal College's professor of comparative anatomy and physiology and also Fullerian Professor of Comparative Anatomy at the Royal Institution. In 1856 Owen was appointed superintendent of the natural history department of the British Museum, and from then until his retirement in 1883 was responsible for the transfer of specimens to and organization of the new Natural History Museum at South Kensington, London (opened to the public in 1881).

The chief anatomist of his day, Owen gained much of his extensive knowledge of comparative anatomy from studies of the Hunterian Collection, which he described in a series of catalogs published from 1833 to 1840. In 1830–31 he had also studied specimens in the Natural History Museum, Paris, attending lectures by Georges CUVIER. Prompted largely by his interest in the structure and origin of mammalian teeth, he turned his attention to fossil forms, reconstructing a large number of extinct animals, including giant flightless toothed birds. He obtained for the British Museum the first specimen of the earliest known fossil bird, *Archaeopteryx*, describing it in the *Philosophical Transactions of the Royal Society of London* (1863). However reexamination of the specimen in 1954 brought to light errors in Owen's reconstruction. Owen early recognized the difference between homologous (real) and analogous (apparent) structures in different animals, constructing a hypothetical vertebrate ancestor. His thesis that the vertebrate skull derived from modified vertebrae was demolished by T. H. HUXLEY and others.

Owen was a bitter opponent of DARWIN and his theory of natural selection. He published anonymously a damaging and slanted review of *The Origin of Species* in *The Edinburgh Review* (1860), but subsequently changed his views to partial and then total acceptance as support for Darwin grew.

P

Palade, George Emil (1912–) *Romanian–American physiologist and cell biologist*

Palade (pa-**lah**-dee) was born at Iasi in Romania. Educated at Bucharest University, where he was professor of physiology during World War II, he emigrated to America in 1946, becoming a naturalized citizen in 1952. He worked at the Rockefeller Institute for Medical Research, New York, becoming professor of cytology there (1958–72). In 1972 he became director of studies in cell biology at Yale University's medical school. In 1990 he was appointed dean of scientific affairs at the University of California at San Diego, retiring in 2002.

Palade's work was primarily concerned with studies of the fine structure of animal cells, although he also investigated the nature of plant chloroplasts. He showed that minute semisolid structures in cells, known as mitochondria, have an enzymic effect, oxidizing fats and sugars and releasing energy. His discovery of even smaller bodies called microsomes, which function independently of the mitochondria (of which they were previously thought to be part), showed them to be rich in ribonucleic acid (RNA) and therefore the site of protein manufacture. The microsomes were subsequently renamed ribosomes. For his work in cellular biology, Palade received, with Albert Claude and Christian de DUVE, the Nobel Prize for physiology or medicine (1974).

Paley, William (1743–1805) *British natural philosopher*

> I take my stand in human anatomy ... the necessity, in each particular case, of an intelligent designing mind for the contriving and determining of the forms which organized bodies bear.
> —*Natural Theology* (1802)

Born the son of a headmaster in Peterborough, eastern England, Paley graduated from Cambridge in 1763. He taught theology at Cambridge until 1776, after which time he held a number of livings in the Church of England in Westmorland, Carlisle, and Lincoln.

In 1802 Paley published one of the most successful works of the century, *Natural Theology, or Evidence of the Existence and Attributes of the Deity Collected from the Appearances of Nature*; twenty editions were called for before 1820. He began by contrasting the significance between finding a stone and a watch on the heath. The presence of the stone can be explained by saying "it had lain there forever." But the mechanism of the watch is such that we must conclude that there is a maker "who formed it for the purpose which we find it actually to answer."

He went on to argue that if a watch led to a designer then, as "the contrivances of nature surpass the contrivances of art," nature should lead more firmly to a designer. The eye, for example, is more clearly made for vision than a telescope. Throughout the book Paley deploys the facts of anatomy and physiology of a variety of animals and plants to this end. A number of specific mechanisms were identified. By a process of "compensation" the shortness of an elephant's neck is overcome by the length of its trunk. Unnecessary organs such as fetal lungs display "prospective contrivance."

The bulk of Paley's argument derived from the living world. The heavens, he found, were too simple, showing nothing but bright points. He conceded, however, that if they did not lead to the Creator's existence, they did show something of His magnificence.

Paley was not the first to present the argument from design. Nor was he the last. A generation later, eight of the leading British scientists of the day would be commissioned by the Earl of Bridgewater to demonstrate how "the Power, Wisdom, and Goodness of God" can be seen in nature. In more recent times something of Paley's approach can be seen in the writings of Sir Fred HOYLE, Freeman DYSON, and Bernard LOVELL.

The approach has also aroused considerable scientific opposition. Most recently a strong attack on Paley has been launched by Richard DAWKINS in his appropriately titled *The Blind Watchmaker* (1986).

Palmieri, Luigi (1807–1896) *Italian geophysicist and meteorologist*

Palmieri (pahl-**myair**-ee), who was born at Faicchio in Italy and educated at the University of Naples, became professor of physics there in 1847. He later joined the staff of the Mount

Vesuvius Observatory where he was made director in 1854.

He is mainly remembered for his introduction of improved versions of such instruments as the seismograph and the rain gage. He also produced, in 1854, a two-volume work on experimental meteorology.

Paneth, Friedrich Adolf (1887–1958)
Austrian chemist

Paneth (**pah**-nayt), the son of a Viennese physiologist, was educated at Munich, Glasgow, and Vienna where he obtained his PhD in 1910. After working at the Institute for Radium Research in Vienna from 1912 to 1918 he taught in Germany at the universities of Hamburg, Berlin, and Königsberg until he fled to England in 1933. There he worked at Imperial College, London, and in 1939 was appointed to the professorship of chemistry at the University of Durham. In 1953 he returned to Germany to the post of director of the Max Planck Institute.

In 1913 Paneth collaborated with Georg von HEVESY on the possibility of using radioisotopes as tracers. He later, in 1929, advanced the work of Moses GOMBERG when he demonstrated the existence of the free methyl radical ($\cdot CH_3$). He found that by heating lead tetramethyl vapor in a tube, one of the products of the decomposition was a very reactive material that attacked metal films and turned out to be the methyl radical. Unlike the triphenyl methyl of Gomberg, this free radical was short-lived. In the late 1920s Paneth used chemical techniques to determine the age of meteorites. As uranium in its disintegration produces a fixed amount of helium at a constant rate then by establishing the uranium:helium ratio of a sample it should be possible to date it. Paneth's calculations of meteorites eventually yielded a date ranging from 10^9 to 3×10^9 years, which agreed reasonably well with dates established by Arthur HOLMES for the age of the Earth.

Panofsky, Wolfgang Kurt Hermann (1919–2007) *German–American physicist*

Panofsky (pa-**nof**-skee) was born in Berlin, Germany. He was brought to America in 1934 and studied at Princeton University, where he graduated in 1938, and at the California Institute of Technology, where he gained his PhD in 1942. In the same year he became a naturalized American citizen. He helped America's wartime research efforts, working for the Office of Scientific Research and Development (1942–43) and on the Manhattan project in Los Alamos (1943–45).

After World War II, Panofsky worked as a physicist at the Radiation Laboratory of the University of California at Berkeley (1945–46). He then moved to Stanford University, where he rose to become professor of physics in 1951. During the 1940s and 1950s, Panofsky did a series of high-energy physics experiments in which the fundamental nature of the pion particle was uncovered, and another series that probed the connection between pions and the electromagnetic field.

In 1953 he was made director of the High-Energy Physics Laboratory at Stanford and made many contributions to the design of accelerators for fundamental particles and instruments for their detection. In 1961 he was appointed director of the Stanford Linear Accelerator Center, a post he held until his retirement in 1984.

Pantin, Carl Frederick Abel (1899–1967) *British physiologist and marine biologist*

Pantin was born in London. Following a brief period of service with the Royal Engineers in World War I, he entered Cambridge University to study zoology, graduating in 1922. He then worked at the Marine Biological Association's laboratory at Plymouth until 1929, when he returned to Cambridge as a lecturer in zoology. Pantin's research concentrated on marine invertebrates, particularly sea anemones and crabs. He demonstrated that the primitive nerve net found in sea anemones behaves in a way essentially identical to the nervous system of higher animals. His work on crabs, however, emphasized their dissimilarities from other animals.

Papin, Denis (1647–c. 1712) *French physicist and inventor*

Papin (pa-**pan**), who was born at Blois in France, graduated with a medical degree from the University of Angers in 1669. Preferring physics to medicine, he started as an assistant in the laboratory of Christiaan HUYGENS in Paris in 1671. As a Protestant, Papin was attracted to England in 1675 where he worked as an assistant to Robert BOYLE. From 1687 to 1696 he was professor of mathematics at Marburg in Germany. He then worked in Cassel until 1707, when he returned to England.

In 1679 while working with Boyle he invented his steam digester – an early and very efficient pressure cooker or, as he described it, "an Engine for softening bones." It was a container with a tightly fitting lid – the enclosed steam generated a high pressure that raised the boiling point of the water – and a lever-type safety valve, which he also invented.

His most important work however was his description of "a new method of obtaining very great moving power at small cost." This was a simple form of steam engine whereby the condensation of steam under a piston allowed atmospheric pressure to raise a weight of 60 pounds (27 kilograms). This was described in his *Ars nova ad aquam ignis adminiculo efficacissime elevandum* (1707; The New Art of Pumping Water by Using Steam). Papin's design was never developed commercially.

Pappenheimer, John Richard (1915–2007) *American physiologist*

Born in New York City, Pappenheimer was educated at Harvard and in England at Cambridge University, where he obtained his PhD in 1940. After teaching at the College of Physicians and Surgeons, New York, and the University of Pennsylvania he joined the staff of the Harvard Medical School in 1946, serving from 1969 as professor of physiology.

The idea that the brain might secrete daily a chemical that will naturally produce sleep at the appropriate time was first investigated by H. Pieron in 1913. He kept dogs awake for ten days, extracted cerebrospinal fluid from them, and injected it into normal dogs. Pieron claimed that such dogs would sleep for several hours while other dogs used as controls, receiving fluid from regular dogs, would behave just as any other animal.

In 1965 Pappenheimer took up the problem, aiming this time to actually "identify the sleep-promoting factor." After three years' work on 20 sleep-deprived goats Pappenheimer had some six liters of fluid for purification. Eventually he could claim that "a concentrated solution of cerebrospinal fluid...was found to reduce the nocturnal activity of rats to half the normal level for 12 hours." That such power could be destroyed by a proteolytic enzyme indicated that the sleep-factor probably contained an amino acid.

As so little of the factor remains after purification, little more can be said as yet of its chemical constitution. Pappenheimer however claimed to have established that the factor is not species specific.

Pappus of Alexandria (about 320) *Greek mathematician*

> I declare God to be one in form but not in number, the maker of heaven and Earth, as well as the tetrad of the elements and things formed from them, who has further harmonized our rational and intellectual souls with our bodies, and who is borne upon the chariots of the cherubim and hymned by angelic throngs.
> —Oath attributed to Pappus

Pappus (**pap**-us) was the last notable Greek mathematician and is chiefly remembered because his writings contain reports of the work of many earlier Greek mathematicians that would otherwise be lost. His chief work, *Synagogue* (c. 340; Collection), consisted of eight books of which the first and part of the second are now lost. It was intended as a guide to the whole of Greek mathematics and this is what makes it such a significant historical source. Among the mathematicians whose work Pappus expounds are EUCLID, APOLLONIUS of Perga, Aristaeus, and ERATOSTHENES. Pappus did contribute some original work, however, notably in projective geometry.

As with many Greek mathematicians Pappus was as interested in astronomy as in pure mathematics and his other work included comments on PTOLEMY's astronomical system contained in the *Almagest*.

Paracelsus, Philippus Aureolus (1493–1541) *German physician, chemist, and alchemist*

> Medicine is not only a science; it is also an art. It does not consist of compounding pills and plasters; it deals with the very processes of life, which must be understood before they may be guided.
> —*Die grosse Wundarznei* (The Great Surgeons)

Paracelsus (pa-ra-**sel**-sus), born Theophrastus Bombastus von Hohenheim in Einsiedeln, Switzerland, was the son of a physician from whom he received his early training in medicine and alchemy. His assumed name stems from his claim to have surpassed the Roman physician CELSUS. He traveled with his father to Villach in Austria where he worked as an apprentice in the mines and acquired much of his practical knowledge of mineralogy and metallurgy. He left the mines in 1507, attended various German universities, and may have obtained an MD from Ferrara. After practicing medicine in Sweden, Strasbourg, Basel, Nuremburg, and a host of other places, in 1540 Paracelsus settled in Salzburg where he died in the following year.

Paracelsus was the first to reject totally the authority of antiquity, suggesting its replacement by nature and experiment. To show his seriousness he burned the great medieval compilation of medical knowledge, the *Canon* of AVICENNA, before his students. In Basel he was lucky enough to cure the infected limb of an influential publisher Frebenius, for which orthodox physicians had recommended amputation. This led to his appointment as professor of medicine and city physician at Basel.

His contempt for traditional learning and the reason the medical authorities found him so distasteful is best conveyed by his much quoted riposte to them: "Let me tell you this: every little hair on my neck knows more than you and all your scribes, and my shoe-buckles are more learned than your GALEN and Avicenna, and my beard has more experience than all your high colleges." The alternative proposed by Paracelsus contained a number of not particularly coherent strands. There was much in his work that belonged to, or at least overlapped with, the occult tradition but there was an eagerness to embrace new sources of knowledge. He would willingly learn his chemistry from the craftsmen in the mines and claimed to gain knowledge from gypsies, magicians, and elderly country folk.

His greatest influence on 16th-century science arose from his chemical philosophy in which he posed the "tria prima": salt, sulfur, and mercury. These terms were meant to emphasize the principles of solidity, combustibility, and liquidity inherent in any substance. It was by following through the implications of such schemes that later chemists such as Robert BOYLE were led to the corpuscular view of matter.

In medicine Paracelsus took the revolutionary step of introducing chemically prepared drugs rather than persisting exclusively with the herbal medicines or "simples" of antiquity. While he was perhaps not the first to use such new remedies as mercury, sulfur, potassium, and antimony his dramatic use of them, often with supposedly verified cures, brought them sharply before the attention of the public. It is from this work that medicine begins to take on its modern aspect as being concerned with the discovery of specialized drugs providing complete and harmless cures.

Pardee, Arthur Beck (1921–) *American biochemist*

Born in Chicago, Pardee was educated at the University of California, Berkeley, and at the California Institute of Technology where he obtained his PhD in chemistry under Linus PAULING in 1947. He taught at Berkeley until 1961, taking a year off (1957–58) to work with Jacques MONOD at the Pasteur Institute. He was then appointed professor of biochemical science at Princeton and in 1975 joined the Harvard Medical School as professor of pharmacology.

In the 1950s and 1960s Pardee was involved in a number of major advances in the new discipline of molecular biology, for example his demonstration in 1955 with R. Litman of the mutagenic effect on the virus bacteriophage T2 of 5-bromouracil. This was followed in 1958 by his publication with François JACOB and Jacques Monod of their classic paper announcing the results of the so-called PaJaMo experiment. Using a mutant bacterium, which was able to synthesize β-galactosidase without any outside stimulation, they crossed this (by conjugation) with a normal bacterium, which had to be induced to make β-galactosidase, and found that the normal process is dominant to the mutant; that is inducibility is dominant to constitutivity. It was this experiment that led them to formulate the concept of a "repressor molecule" suppressing the production of β-galactosidase by the gene.

Pardee threw considerable light on the control of enzyme synthesis by the cell. The process of end-product feedback by which a metabolic pathway is inhibited by its own end product had been clearly shown to apply to a number of synthetic processes by numerous workers. An elegant development of this mechanism was revealed by Pardee in collaboration with John Gerhart in 1959. They showed that the output of pyrimidine is controlled not only by its own end product but also by the independent purine production line. As both were essential constituents of nucleic acid such a mechanism was presumably designed to keep their production in phase.

Just how the process of feedback inhibition is controlled was proposed by Pardee and Gerhart in 1962; their work was, however, superseded by a more general account of allosteric proteins published by Jacob, Monod, and J.-P. Changeux in 1963.

Paré, Ambroise (1510–1590) *French surgeon*

> I dressed him, God cured him.
> —Motto referring to his surgery

Paré (pa-**ray**), the son of an artisan from Bourg-Hersent in France, received his medical training as an apprentice to a barber-surgeon. He first worked at the Hôtel-Dieu, a large Paris hospital, but beginning with the Italian campaign of 1536, spent much time as a surgeon with the French army. He also served as a surgeon at the courts of Henri II, Charles IX, and Henri III.

Paré's fame as a surgeon rests mainly on his abandonment of the practice of cauterizing gunshot wounds and the introduction of surgical ligatures. He introduced his first innovation during the 1536 campaign to relieve Turin. He had accepted the orthodox view that as gunpowder was poisonous, gunshot wounds should be cauterized with boiling oil. However on one occasion, because of an unusually large number of wounded, Paré found himself running out of boiling oil so instead he dressed the wounds with an ointment. To his surprise the following day revealed a number of such patients with uninflamed wounds while those with cauterized wounds were feverish and in great pain, with the region around their wounds swollen. He thus decided never again to cauterize gunshot wounds, publishing his findings in his *La Méthode de traicter les ployes faites par les arquebuses et autres bastons a feu* (1545; The Method of Treating Wounds Made by Arquebuses and Other Guns).

His second main innovation, the practice of tying rather than cauterizing blood vessels, has been described as the greatest improvement ever in operative surgery but the improvement was only widely adopted after a delay of centuries. The initial advantage in using ligatures was that it permitted more selective surgery for there was a limit to the areas that could be cauterized. The disadvantages were however many. Cauterization was simple, swift, and could be practiced single handed; ligatures were time consuming, requiring many stitches for the amputation of a limb necessitating skilled assistance, and the cooperation of a patient willing to accept long operations. It was only with the invention of the screw tourniquet by Jean Petit in the early 18th century that ligatures became widely used in surgery.

Parkes, Alexander (1813–1890) *British chemist and inventor*

Parkes, the son of a Birmingham lock manufacturer, was apprenticed to a brass founder and started his career in charge of the casting department at Elkington. He took out his first patent, which was for electroplating delicate objects such as works of art, in 1841. Eventually Parkes took out over 50 patents in this field; when the prince consort, Prince Albert, visited Elkington, Parkes presented him with a silver-plated spider's web. In metallurgy, Parkes also invented a process for removing silver from lead by extraction with molten zinc (the *Parkes process*).

He also worked on rubber and plastics. In 1846 he discovered the cold vulcanization process, which was important in the manufacture of thin-walled rubber articles. In 1855 he took out a number of patents on a new product initially called xylonite (or parkesine). Aiming to produce a synthetic form of horn, Parkes found that if the recently discovered nitrocellulose was mixed with camphor and alcohol, a hornlike substance was produced. It was not fully developed by Parkes however and was left to the Hyatt brothers of New Jersey to develop it as celluloid, the first synthetic plastic material.

Parkinson, James (1755–1824) *British physician and paleontologist*

The son of a London surgeon, Parkinson practiced medicine in his native city and engaged in radical politics such as campaigning for universal suffrage. He was the author of a number of pamphlets advocating political reform under the pseudonym "Old Hubert" and, in 1794, was thought by the authorities to be involved in the so-called pop-gun plot to assassinate George III with a poisoned dart. He was called to give evidence before the Privy Council but they failed to implicate him.

In 1817 he published his most significant work, *An Essay on the Shaking Palsy*, which described the disease he referred to as paralysis agitans but is now more commonly known as Parkinson's disease. He also gave, in 1812, the first description in English of appendicitis and established perforation as the cause of death.

Parkinson was also a well-known paleontologist. He was one of the founder members of the London Geological Society in 1807 and author of the influential *Organic Remains of a Former World* (3 vols. 1804–11).

Parmenides of Elea (*c.* 515 BC–*c.* 450 BC) *Greek philosopher*

Virtually nothing is known about the early life of Parmenides (par-**men**-i-deez) except that he was born in Elea (now Velia) in southern Italy. He was the founder of the Eleatic school, one of the leading pre-Socratic forms of Greek thought, and wrote the remarkable poem *On Nature*, 155 lines of which survive. It is generally accepted as the first extended philosophical text in which an idea, however obscure, is argued for instead of the isolated assertions contained in the surviving fragments of earlier thinkers.

The poem argues for an austere Monism, a universe that must necessarily be uncreated, uniform, changeless, complete, and without motion. The dialectic used in the poem awakened the minds of the Greeks to the power of reason as no other work had done; used by another Eleatic, ZENO, it produced a number of deep paradoxes.

Parsons, Sir Charles Algernon (1854–1931) *British engineer*

Parsons, a Londoner, was educated at Cambridge University and in 1877 became an engineer at an engineering plant. In 1884 he invented the multistage steam turbine and in 1889 set up his own engineering plant at Newcastle upon Tyne to produce turbines and dynamos. The dynamo was first used in a steamship in 1897 and revolutionized ship design. It was also crucial to the development of electricity generators. Parsons became a fellow of the Royal Society in 1898, was awarded the Rumford medal in 1902, and was knighted in 1911.

Partington, James Riddick (1886–1965) *British chemist*

Born in Bolton, northwestern England, Partington was educated at Manchester University and in Berlin, where he worked under Walther NERNST. After working during World War I at the Ministry of Munitions, he was appointed, in 1919, to the chair of chemistry at Queen Mary College, London, where he remained until his retirement in 1951.

Partington is remembered chiefly as a historian of chemistry and he was also a prolific writer, compiling over 20 substantial volumes in his lifetime. He produced a number of textbooks of which his five-volume *Advanced Treatise on Physical Chemistry* (1949–54) is perhaps the best known. It is however but one of many, written on all the main branches of chemistry, at all levels, from the introductory to the advanced.

Partington published his first major work on the history of chemistry in 1935, *Origins and Development of Chemistry*. This was followed by *A Short History of Chemistry* (1937), which heralded his vast four-volume *History of Chemistry* (1961), of which one volume remained unfinished on his death. These works have proved invaluable to students and historians of chemistry.

Pascal, Blaise (1623–1662) *French mathematician, physicist, and religious philosopher*

> Contradiction is not a sign of falsity, nor the lack of contradiction a sign of truth.
> —*Pensées* (1670; Thoughts)

> The more intelligence one has the more people one finds original. Commonplace people see no difference between men.
> —As above

Pascal (pas-**kal**) was the son of a respected mathematician and a local administrator in Clermont-Ferrand, France. Early in life Pascal displayed evidence that he was an infant prodigy and apparently discovered EUCLID's first 23 theorems for himself at the age of 11. While only 17 he published an essay on mathematics that René DESCARTES refused to acknowledge as being the work of a youth. Pascal produced (1642–44) a calculating device to aid his father in his local administration; this was in effect the first digital calculator.

Pascal conducted important work in experimental physics, in particular in the study of atmospheric pressure. He tested the theories of Evangelista TORRICELLI (who discovered the principle of the barometer) by using mercury barometers to measure air pressure in Paris and, with the help of his brother-in-law, on the summit of the Puy de Dôme (1646). He found that the height of the column of mercury did indeed fall with increasing altitude. From these studies Pascal invented the hydraulic press and the syringe and formulated his law that pressure applied to a confined liquid is transmitted through the liquid in all directions regardless of the area to which the pressure is applied. He published his work on vacuums in 1647.

Pascal corresponded with a contemporary mathematician, Pierre de FERMAT, and together they founded the mathematical theory of probability. Pascal had been converted to Jansenism in 1646 and religion became increasingly dominant in his life, culminating in the religious revelation he experienced on the night of 23 November 1654. Following this he entered the Jansenist retreat at Port-Royal (1655) and devoted himself to religious studies from then on.

Pasteur, Louis (1822–1895) *French chemist and microbiologist*

> In the field of observation, chance only favors those minds which have been prepared.
> —*Encyclopaedia Britannica*, Vol XX, 11th edition (1911)

> [I have an] invincible belief that Science and Peace will triumph over Ignorance and War, that nations will unite, not to destroy, but to build, and that the future will belong to those who have done most for suffering humanity.
> —Address given at the Sorbonne, Paris, 27 December 1892

Pasteur (pa-**ster**), the son of a tanner, was born at Dôle in France and studied chemistry at the Ecole Normale Supérieure in Paris where he obtained his doctorate for crystallographic studies in 1847. His first appointments were as professor of chemistry firstly at Strasbourg (1849) and then at Lille (1854). In 1857 Pasteur returned to Paris as director of scientific studies at the Ecole Normale but moved to the Sorbonne in 1867 as professor of chemistry. He returned once more to the Ecole Normale in 1874 to direct the physiological chemistry laboratory but spent his last years, from 1888 to 1895, as director of the specially created Pasteur Institute.

Although not a physician, Pasteur was undoubtedly the most important medical scientist working in the 19th century. His work possessed an originality, depth, and precision that was apparent even in his early work, which led

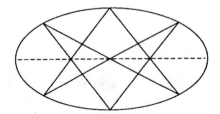

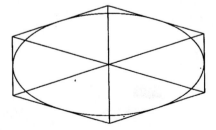

PASCAL'S THEOREM   The opposite sides of a hexagon inscribed in a conic intersect in three points on a line. The dual theorem, that the opposite vertices of a hexagon circumscribed about a conic may be connected by three lines that meet in a point, is Brianchon's theorem.

to the discovery of molecular asymmetry. Tartaric and racemic acid were known to have the same formula, but their crystalline salts possessed different optical properties in solution: tartrate rotates a ray of polarized light to the right, and was accordingly described as dextrorotatory or d-tartrate, while the racemic salts were optically inactive. How, it was asked, could the same compound have such contrasting properties?

Pasteur, in 1848, examined d-tartrate crystals under a magnifying glass and noticed that they all possessed an identical asymmetry, which he assumed to be sufficient to twist a ray of light to the right. The racemate however, while also asymmetrical, appeared to contain crystals divided equally between the d-tartrate form and its mirror image. Pasteur painstakingly separated the crystals into two piles and found that, in solution, one pile behaved exactly as d-tartrate while the other rotated polarized light to the left. Racemic acid was therefore, he concluded, an equal mixture of d-tartrate and its mirror image, the levorotatory l-tartrate, each of which was optically active on its own but together neutralized each other.

Such facts would later have profound consequences for structural chemistry and Pasteur, in 1860, suggested that the effect was a result of the internal arrangement of atoms. Thus he asked "Are the atoms of the dextro acid arranged on a right-hand helix, or positioned at the corners of an irregular tetrahedron, or have they some other asymmetric grouping? We cannot answer these questions. But there is no doubt that an asymmetric arrangement exists that has a nonsuperposable image. It is not less certain that the atoms of the levo acid have exactly the inverse asymmetric arrangement." The idea of an asymmetric tetrahedral carbon atom was put forward in 1874 by Jacobus VAN'T HOFF and Joseph LE BEL. In 1857 Pasteur also noted that a mold accidentally growing on a tartrate solution selected just one of the two racemic forms, the d-form. More generally, he realized that only living organisms could distinguish between such asymmetric forms and even went so far as to argue that this ability marks the only sharply defined difference between the chemistry of dead and living matter.

By 1856 Pasteur had also begun to work on fermentation, beginning with the fermentation of milk into lactic acid. He reported the presence of microorganisms, which continued to bud and multiply; if excluded, fermentation failed to occur but they could be transferred from one ferment to produce another in uncontaminated milk. Further, such organisms were quite specific, as the yeast used to produce beer was incapable of producing lactic acid from milk. With these and many other observations and experiments behind him Pasteur was ready to dispute the chemical theory of Justus von LIEBIG. He declared that all true fermentations are caused by the presence and multiplication of organisms, and are not, as Liebig insisted, purely chemical phenomena. With his germ theory of fermentation Pasteur anticipated much of his later work.

He next turned to the origin of such "organisms" and "ferments" and investigated spontaneous generation. In 1862 Pasteur published his famous paper, *Mémoire sur les corpuscles organisés qui existent dans l'atmosphère* (Note on Organized Corpuscles that Exist in the Atmosphere), which finally brought to an end centuries of earlier debate. Pasteur demonstrated that if sterilized fermentable fluid was placed in a swan-neck flask (a flask with a long curved thin neck that allows air to enter but prevents dust and microorganisms entering) then the fluid remained clear. However if the neck of the flask was broken off, allowing dust to enter, contamination soon resulted.

In 1865 Pasteur was asked to investigate a new disease devastating the silkworms of southern France. Despite his protest that "I have never even seen a silkworm," despite considerable confusion caused by the presence of two quite independent infections, and despite a stroke in 1868, which partially paralyzed his left side, Pasteur still managed to provide a comprehensive analysis of the disease and its prevention.

It was not until 1877 that Pasteur finally turned to human disease and pioneered effective methods of treatment against virulent infections such as anthrax. The breakthrough came in 1880 as a result of an oversight by his assistant, who had inadvertently left a batch of chicken cholera bacilli standing in the laboratory over a long hot summer. On injection into some healthy chickens the culture produced only mild and transient signs of disease. Pasteur then instructed his assistant to prepare a fresh batch of the bacillus and once more inject it into the chickens. They survived unscathed, whereas chickens fresh from the market succumbed rapidly to a similar injection. Pasteur had accidentally discovered an attenuated vaccine for chicken cholera. By May 1882 he had succeeded in deliberately producing a comparable vaccine against anthrax and was ready to test it publicly at Pouilly-le-Fort. Here his success was total with all 24 unvaccinated sheep dying of anthrax while those receiving his vaccine survived.

Events even more dramatic followed in 1885 when Pasteur used a rabies vaccine recently developed by him on a badly bitten nine-year-old boy, Joseph Meister. Against the advice of such colleagues as Emile ROUX he began the course of 14 injections using virus attenuated in the spine of rabbits. Meister survived. He committed suicide 55 years later in 1940 when, as a caretaker at the Pasteur Institute, he preferred to die rather than open the tomb of Pasteur to the invading Nazi forces.

Thus nearly a century after Edward JENNER, Pasteur had introduced only the second vaccine effective against a serious human disease. Others would rapidly follow him so that by the turn of the century several would be in use. Shortly after his triumph Pasteur suffered a second stroke in 1887, one which affected his speech. Although he lived another seven years his long creative period was at an end.

Paul, Wolfgang (1913–1993) *German physicist*

Paul (powl), who was born at Lorenzkirch in Germany, was educated at the universities of Kiel and Berlin, where he obtained his PhD in 1939. After World War II, he taught physics at Göttingen until 1952, when he was appointed professor of physics at the University of Bonn.

During the 1950s he developed the so-called *Paul trap* as a means of confining and studying electrons. The device consists of three electrodes – two end caps and an encircling ring. The ring is connected to an oscillating potential. The direction of the electric field alternates; for half the time the electron is pushed from the caps to the ring and for the other half it is pulled from the ring and pushed towards the caps.

For his work in this field Paul shared the 1989 Nobel Prize for physics with Hans DEHMELT and Norman RAMSEY.

Pauli, Wolfgang (1900–1958) *Austrian–Swiss physicist*

> I don't mind your thinking slowly: I mind your publishing faster than you think.
> —Quoted by A. L. Mackay in *A Dictionary of Scientific Quotations* (1991)

> What God hath put asunder, no man shall ever join.
> —On Einstein's attempts at a unified field theory. Quoted by J. P. S. Uberoi in *Culture and Science*

Born in the Austrian capital of Vienna, Pauli (**pow**-lee or **paw**-lee) was the son of a professor of physical chemistry at the university there and the godson of Ernst MACH. He was educated at the University of Munich, where he obtained his PhD in 1922. After further study in Copenhagen with Niels BOHR and at Göttingen with Max BORN, Pauli taught at Heidelberg before accepting the professorship of physics at the Federal Institute of Technology, Zurich. Apart from the war years, which he spent working in America at the Institute of Advanced Studies, Princeton, he remained there until his early death in 1958.

Pauli was a physicist much respected by his colleagues for his deep insight into the newly emerging quantum theory. His initial reputation was made in relativity theory with his publication in 1921 of his *Relativitätstheorie* (Theory of Relativity). His name is mainly linked with two substantial achievements. The first, formulated in 1925, is known as the *Pauli exclusion principle*. It follows from this that as an electron can spin in only two ways each quantum orbit can hold no more than two electrons. Once both vacancies are full further electrons can fit only into other orbits. With this principle the distribution of orbital electrons at last became clear, that is, they could be explained and predicted in purely quantum terms.

The early model of the atom by Niels Bohr had been extended by Arnold SOMMERFELD in 1915. In the Bohr–Sommerfeld atom, each electron orbiting the nucleus had three quantum numbers: n, l, and m. Pauli introduced a fourth quantum number (s), which could have values of +1/2 or –1/2 and corresponded to possible values of the "spin" of the electron. Pauli's exclusion principle stated that no two electrons in an atom could have the same four quantum numbers (n, l, m, and s). The concept of electron spin was put forward in 1926 by Samuel GOUDSMIT and George UHLENBECK to explain Pauli's fourth quantum number. The exclusion principle explained many aspects of atomic behavior, including a proper understanding of the periodic table. It has also been applied to other particles. It was for his introduction of the exclusion principle that Pauli was awarded the 1945 Nobel Prize for physics.

Pauli's second great insight was in resolving a problem in beta decay – a type of radioactivity in which electrons are emitted by the atomic nucleus. It was found that the energies of the electrons covered a continuous range up to a maximum value. The difficulty was in reconciling this with the law of conservation of energy; specifically, what happened to the "missing" energy when the electrons had lower energies than the maximum? In 1930, in a letter to Lise MEITNER, Pauli suggested that an emitted electron was accompanied by a neutral particle that carried the excess energy. Enrico FERMI suggested the name "neutrino" for this particle, which was first observed in 1953 by Frederick REINES.

Pauli made a number of other important contributions to quantum mechanics and quantum field theory, including the spin–statistics theorem (1940) and the CPT theorem (1955). He also worked out the formulation of non-Abelian gauge theories in 1953, but did not publish this work because he did not see how the observed short-range nuclear forces could emerge from such theories. Pauli wrote several other influential books and reviews in addition to his early work on relativity theory, notably on quantum mechanics and quantum field theory.

Pauling, Linus Carl (1901–1994) *American chemist*

> Science is the search for truth, that is, the effort to understand the world: it involves the

rejection of bias, of dogma, of revelation, but not of morality.

Pauling, a pharmacist's son, was born at Portland, Oregon, and graduated in chemical engineering from Oregon State Agricultural College in 1922. Having gained his PhD in physical chemistry from the California Institute of Technology in 1925, he spent two years in Europe working under such famed scientists as Arnold SOMMERFELD, Niels BOHR, Erwin SCHRÖDINGER, and William Henry BRAGG. He was appointed associate professor of chemistry at Cal Tech in 1927 and full professor in 1931.

Pauling worked on a variety of problems in chemistry and biology. His original work in chemistry was on chemical bonding and molecular structure. He applied physical methods, such as x-ray diffraction, electron diffraction, and magnetic effects, to determining the structure of molecules. He also made significant contributions to applying quantum mechanics to the bonding in chemical compounds. In this field he introduced the idea of hybrid atomic orbitals to account for the shapes of molecules. Another of his innovations was the idea of resonance hybrid – a molecule having a structure intermediate between two different conventional structures. Pauling also worked on the partial ionic character of chemical bonds, using the concept of negativity. Pauling's ideas on chemical bonding were collected in his influential book *The Nature of the Chemical Bond and the Structure of Molecules and Crystals* (1939).

From about 1934 he began to work on more complex biochemical compounds. He studied the properties of hemoglobin using magnetic measurements. This work led to extensive studies of the nature and structure of proteins. With Robert B. COREY, he showed that the amino-acid chain in certain proteins can have a helical structure.

He also made a number of original contributions on other biological topics. In 1940, with Max DELBRÜCK, he introduced a theory of antibody–antigen reactions that depended on molecular shapes. In the 1940s he also studied the genetic disease sickle cell anemia. In 1960 he published a theory of anesthesia and memory. He is noted for his originality and intuition in tackling complex problems and his deep understanding of chemistry. In his book *The Double Helix*, describing the race to determine the structure of DNA, James WATSON describes the concern caused by the knowledge that Pauling was working on the same problem. Pauling was awarded the 1954 Nobel Prize for chemistry. By this time, he had been campaigning for some years against the development of nuclear weapons. He had earlier refused to join the Manhattan project, but had overcome his pacifist principles sufficiently to work on conventional weapons: "Hitler had to be stopped," he noted. By the early 1950s, campaigns against nuclear weapons were being interpreted as "unAmerican" and Pauling's passport was withdrawn making it impossible to travel to Stockholm for the Nobel ceremonies; his passport was returned at the last moment. Pauling continued with his campaign, publishing a book, *No More War* (1958), and organizing a petition of scientists against nuclear testing. He was awarded the Nobel Peace Prize in 1962.

Pauling later began to campaign on another issue, namely the therapeutic value of high doses of vitamin C. In 1971, in his *Vitamin C and the Common Cold*, Pauling claimed that large vitamin C doses, over 10 grams a day, would also reduce the risk of heart disease. Pauling himself took 18 grams of vitamin C daily, a figure 300 times the recommended dose, for the last 27 years of his life. To pursue the matter further Pauling set up in 1973 the Pauling Institute of Science and Medicine in Palo Alto, California.

Pavlov, Ivan Petrovich (1849–1936)
Russian physiologist

> Remember that science demands from a man all his life. If you had two lives that would not be enough for you. Be passionate in your work and in your searching.
> —*Bequest to Academic Youth* (1936)

Born at Ryazan in Russia, Pavlov (**pa**-vlof) studied medicine and general science at the University of St. Petersburg and the Military Medical Academy. He subsequently carried out research in Breslau (now Wrocław, in Poland) and Leipzig (1883–86). Returning to St. Petersburg, he became professor of physiology at the Medical Academy and director of the physiology department of the Institute of Experimental Medicine. Pavlov's early research lay in the physiology of mammalian digestion, showing, for example, that the secretion of digestive juices in the stomach is prompted by the sight of food and nerve stimulation via the brain. For this work Pavlov received the Nobel Prize for physiology or medicine (1904). He then went on to study the way that dogs and other animals may be induced to salivate and show signs of anticipation of food by actions, such as the ringing of a bell or even a powerful electric shock, that they have learned to associate with the appearance of food. Pavlov's work on conditional or acquired reflexes, which he believed to be associated with different areas of the brain cortex, has led to a new psychologically oriented school of physiology and has stimulated ideas as to the probability of many aspects of human behavior being a result of "conditioning."

Pavlov openly criticized communism and the Soviet government. In 1922 he requested and was refused permission to move his laboratory abroad. Following the expulsion of priests' sons from the Medical Academy, Pavlov, himself the son of a priest, resigned from the chair of physiology in protest. Despite such actions his work continued to be supported by state funds and Pavlovian psychology remained popular in the Soviet Union.

Payen, Anselme (1795–1871) *French chemist*

> I never rely on him [Payen] where accuracy is concerned. But when it is a matter of writing pamphlet-fodder for the general public, then he is in his element.
> —Jöns Jacob Berzelius, letter to Friedrich Wöhler, 12 January 1847

Payen (pa-**yahn**), the son of a Parisian industrialist, was educated at the Ecole Polytechnique. At the age of 20 he was put in charge of a borax-refining plant, and in 1835 became professor of chemistry at the Ecole Centrale des Arts et Manufactures.

Payen's achievements were mainly concerned with improving old industrial processes and introducing new ones. He showed how borax, used widely in the glass, ceramics, and soldering industries, and as an antiseptic, could be produced from boric acid. This eliminated French dependence on the Dutch East India monopoly of borax. He then turned to the sugar-beet industry (1820) and in 1822 introduced a decolorization process for beet sugar using charcoal. He also introduced into the sugar industry the enzyme diastase for converting starch into sugar (1833). This was also the first enzyme to be obtained in concentrated form. From his subsequent researches on wood and its components he discovered cellulose.

Payne-Gaposchkin, Cecilia Helena (1900–1979) *British–American astronomer*

One of the 20th century's most renowned astronomers was born Cecilia Payne in Wendover, England. In 1919 she won a scholarship to study natural sciences at Cambridge University, and completed her studies in 1923, at a time when women were still denied degrees. While at Cambridge she attended a lecture on relativity given by the astrophysicist Arthur EDDINGTON, and was inspired to become an astronomer. She then received a fellowship for graduate studies at Harvard University, under Harlow SHAPLEY, the director of the Harvard Observatory. In 1925 she became the first person to receive a doctorate in astronomy at Harvard, for a thesis described by the eminent astronomer Henry RUSSELL as "the best doctoral thesis I ever read." In her dissertation, *Stellar Atmospheres*, Payne used a quantum mechanical understanding of atomic structure to demonstrate that apparent variations in solar spectra were the result of physical, not chemical, differences in stars, and that the stellar universe was essentially homogeneous in

terms of proportions of the different elements. Her analysis also pointed to an extremely high abundance of hydrogen and helium in stellar atmospheres, a conclusion with which Russell initially disagreed. However, in 1929 Russell published a paper that vindicated Payne's earlier finding, and subsequently received the lion's share of the credit for this discovery.

For many years Payne's position at Harvard was ambiguous and poorly paid, and not until 1938 was she made an official faculty member. Meanwhile, she published *Stars of High Luminosity* (1930), and in 1935 she married the exiled Russian astronomer Sergei Gaposchkin. The couple worked together at Harvard, and made pioneering studies of variable stars (i.e., ones of fluctuating luminosity) in the Milky Way and the Magellanic Clouds. They cowrote *Variable Stars* (1938). In 1956 Payne-Gaposchkin was finally made professor and appointed chair of the department of astronomy, the first woman to hold such a position at Harvard. She continued to publish papers and books on stars and stellar evolution, and her undergraduate lectures formed the basis of *Introduction to Astronomy* (1954). In 1977 the minor planet 1974 CA was named for her.

Peano, Giuseppe (1858–1932) *Italian mathematician and logician*

> The purpose of mathematical logic is to analyze the ideas and reasoning that especially figure in the mathematical sciences.
> —Letter to Felix Klein, 19 September 1894

Peano (pay-**ah**-noh), who was born at Spinetta near Cuneo, in Italy, studied at the University of Turin and was an assistant there from 1880. He became extraordinary professor of infinitesimal calculus in 1890 and was full professor from 1895 until his death. He was also professor of the military academy in Turin from 1886 to 1901.

Peano began his mathematical career as an analyst and, like Richard DEDEKIND before him, his interest in philosophical and logical matters was awakened by the lack of rigor in some presentations of the subject. Peano was particularly keen to avoid all illegitimate reliance on intuition in analysis. His discovery in 1890 of a curve that was continuous but filled space went against intuition. A similar discovery was Karl WEIERSTRASS's famous function that was everywhere continuous but nowhere differentiable. As with Weierstrass's function, Peano's curve shows that the concept of a continuous function cannot be identified with that of a graph.

His interest in rigorous and logical presentation of mathematics led Peano naturally to an interest in the mathematical development of logic. In this field he was one of the great pioneers along with George BOOLE, Gottlob FREGE, and Bertrand RUSSELL. Peano's achievement was twofold. First he devised, in his *Notations de logique mathématique* (1894; Notations in Mathematical Logic), a clear and efficient notation for mathematical logic which, as modified by Bertrand Russell, is still widely used. Secondly, he showed how arithmetic can be derived from a purely logical basis. To do this he formulated, in his *Nova methodo exposita* (1889; New Explanation of Method), nine axioms, four dealing with equality, and the remaining five, listed below, characterizing the numbers series:

1 is a number
The successor of any number is a number
No two numbers have the same successor
1 is not the successor of any number
Any property that belongs to 1 and the successor of any number that also has that property, belongs to all numbers (mathematical induction).

Peano's axioms had been proposed, in a more complicated form, by Dedekind a year earlier.

Peano also did notable work in geometry and on the error terms in numerical calculation. Among his extramathematical interests he was a keen propagandist for a proposed international language, Interlingua, which he had developed from Volapük.

Pearson, Karl (1857–1936) *British biometrician*

The right to live does not connote the right of each man to reproduce his kind … As we lessen the stringency of natural selection, and more and more of the weaklings and the unfit survive, we must increase the standard, mental and physical, of parentage.
—*Darwinism, Medical Progress and Parentage* (1912)

Pearson, the son of a London lawyer, studied mathematics at Cambridge University. He then joined University College, London, initially (from 1884) as professor of applied mathematics and mechanics and from 1911 until his retirement in 1933 as Galton Professor of Eugenics.

Pearson's career was spent largely on applying statistics to biology. His interest in this derived ultimately from Francis GALTON and was much reinforced by the work of his colleague Walter WELDON. In 1893 Weldon had argued that variation, heredity, and selection are matters of arithmetic; Pearson started in the 1890s to develop the appropriate "arithmetic" or statistics as it came to be called. Between 1893 and 1906 Pearson published over a hundred papers on statistics in which such now familiar concepts as the standard deviation and the chi-square test for statistical significance were introduced. Later work was published in *Biometrika*, the journal founded by Pearson, Galton, and Weldon in 1901 and edited by Pearson until his death. This he ran with an unashamed partisanship, rejecting outright or correcting without invitation papers expressing views Pearson considered "controversial." It is for this reason that Ronald FISHER, the most creative British statistician of the century, decided after receiving the treatment from Pearson in 1920 to publish elsewhere.

Pearson and Weldon became involved in an important controversy with William BATESON on the nature of evolution and its possible measurement. The biometricians emphasized the importance of continuous variation as the basic material of natural selection and proposed that it be analyzed statistically. Bateson and his supporters, whose views were reinforced by the rediscovery of the works of Gregor MENDEL in 1900, attached more importance to discontinuous variation and argued that breeding studies are the best way to illuminate the mechanism of evolution.

The validity of Mendelism eventually became generally accepted. At the same time, however, the immense value of biometrical techniques in analyzing continuously variable characters like height, which are controlled by many genes, was also recognized. Following Weldon's death in 1906 Pearson spent less time trying to prove the biometricians' case and devoted himself instead to developing statistics as an exact science. He prepared and published many volumes of mathematical tables for statisticians. He also devoted much of his time to the study of eugenics, using Galton's data to issue various volumes of the *Studies in National Deterioration* (1906–24). In 1925 he founded and edited until his death the *Annals of Eugenics*.

To many, Pearson is best known as the author of *Grammar of Science* (1892), a widely read positivist work on the philosophy of science in which he argued, like his earlier teacher Ernst MACH, that science does not explain but rather summarizes our experience in a convenient language.

Pecquet, Jean (1622–1674) *French anatomist*

Born at Dieppe in France, Pecquet (pe-**kay**) was educated in Paris and Montpellier, gaining his medical qualifications at the latter in 1652. He then entered the service of Nicolas Fouquet, Louis XIV's superintendent of finance, and followed him into the Bastille in 1661 with his fall from power. In later life Pecquet was reported to have fallen a victim to the seductive fallacy that brandy could cure all disease.

In 1651 he published the first description of the thoracic duct. The lymphatic system had been discovered by Gaspare ASELLI in 1622 when he first observed the "white vessels" or lacteals of the intestines. It was assumed that they would go to and terminate at the liver. Pecquet was able to show that no such vessels went to the liver. Instead, he went on to trace their path into the "receptaculum chyli," later known as the thoracic duct, which eventually led to and drained into the subclavian vein in the shoulder. The significance of this lay in its rejection of the view of GALEN that food or "chyle" from the intestines is transported to the liver to be turned into blood. This lent support to the work of William HARVEY on the circulation of the blood.

Pedersen, Charles (1904–1989) *American chemist*

The son of Norwegian parents, Pedersen was born in Pusan, Korea, and moved with his family to America in the 1920s. He became a naturalized American citizen in 1953. Pedersen was educated at the Massachusetts Institute of Technology and, for most of his career up to his retirement in 1969, he worked as a research chemist for DuPont.

While working on synthetic rubber, Pedersen noted that one of his materials had been contaminated. He investigated the impurity and found that it had a ring structure of 12 carbon and 6 oxygen atoms, with a pair of carbon atoms between each oxygen. Such structures are known as cyclic polyethers. Normally, organic solvents such as ether and benzene will not dissolve sodium hydroxide. Yet Pedersen found that caustic soda did dissolve in his new compound, with the sodium ions binding loosely to the oxygen atoms of the ether. To accomplish this the polyether formed a nonplanar ring with a crownlike structure, with the sodium ions sitting neatly in the center. For this reason, Pedersen named what turned out to be a new class of compounds "crown ethers." Although he made his first observations in 1964, DuPont delayed publication until 1967.

The implications of Pedersen's work were varied and important. If one crown ether could coordinate sodium ions, it was likely that others of different ring size would be able to bind to other metal ions. Crown ethers could therefore be used as a simple means of gathering specific ions from aqueous solutions.

Other chemists were also quick to see the implications of Pedersen's work and it was with two of these, Jean LEHN and Donald CRAM, that he shared the 1987 Nobel Prize for chemistry.

Peebles, (Philip) James (Edwin) (1935–) *Canadian astronomer*

Born at Winnipeg in Canada, Peebles was educated at the University of Manitoba and at Princeton, where he took his PhD in 1962. He has remained at Princeton ever since, becoming Einstein Professor of Science in 1984.

Peebles has made a number of contributions to modern cosmology. In 1965, in collaboration with Robert DICKE, he made the important prediction that a background radiation should be detectable as a remnant of the big bang. He also calculated that the amount of helium present in the universe as a consequence of the big bang should be about 25–30%, a figure that agrees with current observations. In 1979, again in collaboration with Dicke, Peebles drew the attention of cosmologists to the so-called "flatness problem" and asked how the standard model of the big-bang theory could deal with it. Cosmologists ask if the universe is "open" or "closed." If it is open it will continue to expand forever; if closed, the expansion will cease at some future point and it will begin to contract. To answer the question the value of omega (Ω) must be found.

Omega (Ω) is the ratio of the average density of mass in the universe to the critical mass density. This latter factor is the mass density needed just to halt the universe's expansion. If Ω is less than 1 the universe is open, and if Ω is greater than 1 the universe is closed. If Ω is equal to 1 the universe will be flat, that is, the universe will continue to expand, although at a decreasing rate.

The actual measured value of Ω is close to 1. A bit more matter in the universe and it would have collapsed long ago; a little less matter, and it would have expanded too quickly for galaxies to form. But if Ω is close to 1 now, it must have been close to 1 soon after the big bang. If it had differed significantly, Ω would either be approaching infinity and the universe would have the density of a black hole, or equal to 0 and the universe would be indistinguish-

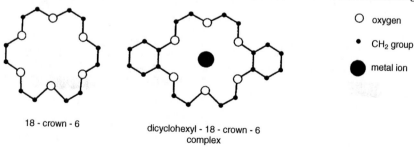

O   oxygen

•   CH$_2$ group

⬤   metal ion

18 - crown - 6

dicyclohexyl - 18 - crown - 6
complex

CROWN ETHERS   *Structures of a simple crown ether and a crown ether complex.*

able from a vacuum. The question thus becomes why so early in the history of the universe was Ω so close to 1? Instead of merely laying this down as an arbitrary initial condition, there should be some way to see why, given the big bang, Ω should have this value. As Peebles's problem seemed to have no solution in the standard model of the big bang, it has been left to astronomers such as GUTH to propose alternative foundations based on an inflationary model.

Peebles is also the author of two important books, *Physical Cosmology* (Princeton, 1971) and *The Large Scale Structure of the Universe* (Princeton, 1980), which have between them done much to define the subject of cosmology for a generation of astronomers.

Peierls, Sir Rudolph Ernst (1907–1995)
German–British theoretical physicist

The son of a businessman, Peierls (perlz) was educated in his native city of Berlin, at Munich, and at Leipzig where he completed his PhD in 1929. He spent the next three years as research assistant to Wolfgang PAULI at the Federal Institute of Technology, Zurich. After short periods at Rome, Copenhagen, and Cambridge, and fearing to return to Germany, Peierls sought a post in England. He worked first at Manchester and in 1937 was appointed professor of physics at Birmingham University. Peierls became a naturalized British citizen in 1940.

In 1939 nuclear fission had been discovered by Otto FRISCH and Lise MEITNER. Immediately the question arose as to whether the process could be harnessed to build a new and powerful bomb. Frisch had arrived in Birmingham soon after announcing his discovery and discussed the matter with Peierls. In 1940 they began to calculate just how much uranium–235 would be needed to sustain a chain reaction. To their astonishment their calculations revealed that an amount as small as one pound would be enough to set off an explosion equivalent to thousands of tons of high explosives.

Peierls was terrified that similar calculations had been made in Germany; consequently Peierls and Frisch quickly prepared a report that eventually reached the appropriate government committee under G. P. Thomson. The committee was just about to disband having convinced itself that there was no immediate prospect for nuclear weapons. Minds were quickly changed and it was recommended that a front organization, Tube Alloys, be set up to start work on the extraction of uranium–235 by means of gaseous diffusion. After some initial confusion because of his German background Peierls was recruited by Tube Alloys and was sent to America to continue his work at Los Alamos.

At the end of the war Peierls returned to his chair at Birmingham. In 1963 he moved to Oxford to become Wykeham Professor of Physics, a position he held until his retirement in 1974. His autobiography, *Bird of Passage*, was published in 1985.

Peirce, Benjamin (1809–1880) *American mathematician*

Peirce (pers) was born in Massachussetts, the son of a state legislator and librarian at Harvard College. Peirce junior studied at Harvard, entering in 1825 and graduating in 1829. After a spell as a school teacher he returned to Harvard, receiving a master's degree in 1833. He remained at Harvard for the rest of his career, becoming professor of mathematics and natural philosophy (1833) and professor of mathematics and astronomy (1842).

In the 1830s and 1840s Peirce published several textbooks on subjects including trigonometry, algebra, and geometry. However, all but the most gifted students found both his books and lecturing style demanding. Peirce's research was no less wide-ranging, encompassing celestial mechanics, linear associative algebra, and number theory. Following the discovery of Neptune in 1846, he helped determine its orbit and calculated the consequent perturbations on the other planets. But perhaps his most lasting achievement was his book *Linear Associative Algebra* (1870). This presented a general theory of linear associative algebra, and introduced the terms "idempotent" and "nilpotent." Peirce presented his results to the National Academy of Sciences, of which he was a founder member (in 1863). A more utilitarian aspect of Peirce's work was his involvement from 1852 with the United States Coastal Survey; he became its director in 1867, supervising the production of a map of the United States. Peirce was a fervent devotee of the science of mathematics, which he described as "the great master-key, which unlocks every door of knowledge."

Peligot, Eugene Melchior (1811–1890) *French chemist*

Peligot (pe-lee-**goh**), a Parisian by birth, studied under Jean DUMAS at the Ecole Polytechnique and became professor of chemistry at the Conservatoire des Arts et Métiers in his native city.

He worked with Dumas on the theory of radicals, but his most practical discovery was made in 1838, with Bouchardat, when they demonstrated that the sweet substance found in the urine of diabetics was glucose. This prepared the way for Hermann von FEHLING to develop his test for the presence and quantity of sugar in 1848. In 1834 Peligot obtained acetone by heating calcium acetate. As acetone is a product of the distillation of wood this was, following Friedrich WÖHLER's work, another example of the synthesis of an organic compound from inorganic materials. He also attempted to explain what happens in the lead chamber in the manufacture of sulfuric acid. He realized that the catalyst nitric oxide forms nitrogen dioxide and then the nitrogen dioxide reacts with the sulfur dioxide present so that nitric oxide is formed again. In 1841 Peligot succeeded in isolating uranium from its oxide.

Pelletier, Pierre Joseph (1788–1842)
French chemist

Pelletier (pel-e-**tyay**) was born the son of a pharmacist in Paris, France. He both studied and taught at the Ecole de Pharmacie until his retirement in 1842. His major work was the investigation of drugs, which he began in 1809. By pioneering the use of mild solvents he successfully isolated numerous important biologically active plant products: working with Bienaimé CAVENTOU he discovered caffeine, strychnine, colchicine, quinine, and veratrine. Their greatest triumph, however, came in 1817, when they discovered chlorophyll – the green pigment in plants that traps light energy necessary for photosynthesis.

Peltier, Jean Charles Athanase (1785–1845) *French physicist*

Born in the Somme department of France, Peltier (pel-**tyay**) was a watchmaker who gave up his profession at the age of 30 to devote himself to experimental physics. In 1821 T. J. SEEBECK had shown that if heat is applied to the junction of a loop of two different conductors a current will be generated. In 1834 Peltier demonstrated the converse effect (the *Peltier effect*). He found that when a current is passed through a circuit of two different conductors a thermal effect will be found at the junctions. There is a rise or fall in temperature at the junction depending on the direction of current flow.

Penck, Albrecht (1858–1945) *German geographer and geologist*

> He used to be liked as much as admired but during the war [World War I] some of his statements have lessened the esteem formerly felt for him.
> —On Penck's outspoken nationalism. William Morris Davis, *Geographical Review* (1920)

Penck (pengk) studied at the university in his native city of Leipzig in Germany. After teaching briefly at the University of Munich, he was appointed professor of physical geography at the University of Vienna in 1885. He moved in 1906 to the chair of geography at Berlin, where he remained until his retirement in 1926.

Penck is remembered chiefly for his collaboration with his assistant Eduard Brückner on the three-volume *Die Alpen im Eiszeitalter* (1901–09; The Alps in the Ice Age). Working in the Bavarian Alps and studying the succession of gravel terraces occurring at different heights above the present river-valley floors, they were able to reconstruct the sequence of past ice ages. They recognized four, which they called Günz, Mindel, Riss, and Würm after the river valleys where they were first identified. For over half a century this scheme provided the framework for the discussion of the European Pleistocene, which they underestimated as lasting for 650,000 years.

Penck produced a fundamental work on geomorphology, a term he is also believed to have introduced, *Morphologie der Erdoberfläche* (1894; Morphology of the Earth's Surface), in which he identified six topographic forms – the plain, scarp, valley, mountain, hollow, and cavern – and discussed their origins.

After World War I he turned more to political geography.

Penney, William George, Baron (1909–1991) *British mathematical physicist*

The son of a sergeant-major in the British army, Penney was born in the British crown colony of Gibraltar. He was educated at Imperial College, London, at the University of Wisconsin, and at Cambridge, where he obtained his PhD in 1935. In the following year he was appointed to a lectureship in mathematics at Imperial College.

With the outbreak of World War II, Penney was recruited by the Admiralty to work on blast waves. In 1944 he was sent to Los Alamos to apply his knowledge to the development of the atomic bomb. He witnessed the Nagasaki bomb and was also present at Bikini in 1946 when the Americans first tested their hydrogen bomb. On this latter occasion Penney was able to calculate the bomb's blast power using his own simple equipment.

Penney remained in government service after the war. When the British government decided to build a nuclear bomb Penney became responsible for its design and production, first at the Ministry of Supply (1946–52) and then at the Atomic Weapons Research Establishment, Harwell (1952–59). The bomb used plutonium made in the Windscale reactor. The task was made harder by the passage in the United States in 1946 of legislation forbidding the release of any information on the design of nuclear weapons. Everything, therefore, had to be dredged from memory or discovered anew. By 1952 Penney was ready to test the first British bomb at Montebello, a small island off the northwest coast of Australia. He went on to direct the production of the first British hydrogen bomb, successfully exploded at Christmas Island in 1957.

The success of the British nuclear program led in 1958 to a bilateral treaty with the United States sanctioning the exchange of information on nuclear weapons. Penney played a significant role in these negotiations, as he also did in the talks that led to the 1963 nuclear test-ban treaty.

In 1959 Penney left Harwell for the UK Atomic Energy Authority (UKAEA) where he served as chairman from 1964 to 1967, when he returned to Imperial College as rector, a post he held until his retirement in 1973. He was raised to the British peerage in 1967, becoming Baron Penney of East Hendred.

Penrose, Sir Roger (1931–   ) *British mathematician and theoretical physicist*

Penrose, the son of the geneticist Lionel Penrose, was born at Colchester in the eastern English county of Essex. He graduated from University College, London, and obtained his PhD in 1957 from Cambridge University. After holding various lecturing and research posts in London, Cambridge, and in America at Princeton, Syracuse, and Texas, Penrose was appointed professor of applied mathematics at Birkbeck College, London, in 1966. In 1973 he was elected Rouse Ball Professor of Mathematics at Oxford. He was knighted in 1994.

Penrose has done much to elucidate the fundamental properties of black holes. These result from the total gravitational collapse of large stars that shrink to such a small volume that not even a light signal can escape from them. There is thus a boundary around a black hole inside which all information about the black hole is trapped; this is known as its "event horizon." With Stephen HAWKING, Penrose proved a theorem of EINSTEIN's general relativity asserting that at the center of a black hole there must evolve a "space–time singularity" of zero

volume and infinite density where the present laws of physics break down. He went on to propose his hypothesis of "cosmic censorship," that such singularities cannot be "naked"; they must possess an event horizon. The effect of this would be to conceal and isolate the singularity with its indifference to the laws of physics.

Despite this Penrose went on in 1969 to describe a mechanism for the extraction of energy from a Kerr black hole, an uncharged rotating body first described by Roy Kerr in 1963. Such bodies are surrounded by an ergosphere within which it is impossible for an object to be at rest. If, Penrose demonstrated, a body fell into this region it would split into two particles; one would fall into the hole and the other would escape with more mass-energy than the initial particle. In this way rotational energy of the black hole is transferred to the particle outside the hole.

From the mid-1960s Penrose worked on the development of a new way of attempting to combine quantum mechanics and general relativity based on a complex geometry. Penrose began with "twistors" – massless objects with both linear and angular momentum in twistor space. From these he attempted to reconstruct the main outlines of modern physics. The matter was pursued not only by Penrose but through a number of "twistor groups" who communicated through a *Twistor Newsletter*. The fullest account of twistor theory is to be found in *Spinors and Space-Time* (2 vols., 1984–86) by Penrose and W. Rindler. In the early 1970s Penrose pioneered the study of *spin networks* as a possible discrete structure for quantum gravity. This approach was revived in the 1990s by Lee Smolin and others.

In 1974 Penrose introduced a novel tiling of the affine plane (*Penrose tiling*). Periodic tilings in which a unit figure is endlessly repeated can be constructed from triangles, squares, and

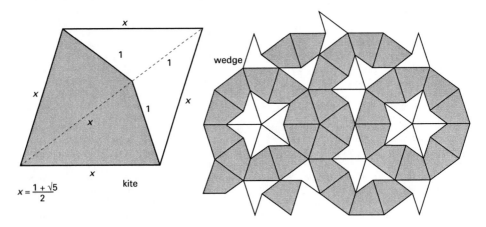

PENROSE TILING

hexagons – figures with three-, four-, or six-fold symmetry. The plane cannot be tiled by pentagons, which have a five-fold symmetry; three pentagons fitted together always leave a crack, known to crystallographers as a "frustration." It was also known that crystal structures could have two-, three-, four-, or six-fold rotational symmetries only. No crystal, that is, could have a five-fold rotational symmetry.

Penrose's method of tiling the plane involved constructing two rhombuses by dividing the diagonal of a regular parallelogram by a golden section. These could be combined according to simple rules so as to cover the plane, even though there was no simple repeated unit cell. The rhombuses can be assembled in such a way as to have an almost five-fold symmetry.

As such they were seen as an interesting oddity, usually discussed in columns devoted to recreational mathematics. However, things changed dramatically in 1984 when Dany Schectman of the National Bureau of Standards and his colleagues found that a rapidly cooled sample of an aluminum–manganese alloy formed crystals that displayed a five-fold symmetry. "Quasicrystals," as they soon became known, developed rapidly into a major new research field and the subject of hundreds of papers.

In addition to continuing his work on twistor theory Penrose also published a widely read book, *The Emperor's New Mind* (1989). The book is an attack on aspects of artificial intelligence. In it he argues that there are aspects of mathematics that cannot be tied to a set of rules. We cannot allow "one universally formal system ... equivalent to all the mathematicians' algorithms for judging mathematical truth." Such a system would violate GÖDEL's theorem. Nor can we accept that algorithms used are so complicated and obscure that their validity can never be known. We do not in fact ascertain

mathematical truth solely through the use of algorithms. "We must see the truth of a mathematical argument to be convinced of its validity," Penrose has insisted. Consequently when we *see* the validity of a theorem, in *seeing* it "we reveal the very nonalgorithmic nature of the 'seeing' process itself."

He further developed his arguments in *Shadows of the Mind* (1994), in which he also answered many of the objections raised against the earlier work. Penrose has also published (in collaboration with Stephen Hawking) *The Nature of Space and Time* (1996), in which the authors develop their own cosmological viewpoints. Thus while Penrose presents his own "twistor view" of the universe, Hawking concentrates on problems connected with "quantum cosmology." Penrose also wrote *The Large, The Small, and The Human Mind* (1997). His views on physics and mathematics are collected in a lengthy mathematical book entitled *The Road to Reality* (2003).

Penzias, Arno Allan (1933–   ) *American astrophysicist*

Penzias (**pent**-see-as), who was born in Munich, Germany, earned his BS at City College, New York, in 1954 after fleeing with his parents as a refugee from the Nazis. He gained his PhD from Columbia University in 1962. In 1961 he joined Bell Laboratories at Holmdel, New Jersey, and was made director of their radio research laboratory in 1976. From 1979 he has been executive director of research in the communications division.

Penzias and his coworker Robert W. WILSON are credited with one of the most important discoveries of modern astrophysics, the cosmic microwave background radiation. This is considered to be the remnant radiation produced in the "big bang" in which the universe was cre-

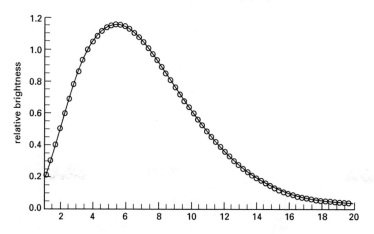

COSMIC BACKGROUND RADIATION   *The distribution of background microwave radiation as found by the COBE space probe.*

ated some billions of years ago. As the universe has expanded, the radiation has lost energy: it has effectively "cooled." Its existence was originally predicted by George GAMOW and Ralph ALPHER in 1948, who calculated that the radiation should now be characteristic of a perfectly emitting body (a black body) at a temperature of about 5 kelvin (–268°C). This radiation should lie in the microwave region of the spectrum. Similar calculations were later made by Robert DICKE and P. J. E. PEEBLES.

The discovery of the remnant radiation was made while Penzias and Wilson were working at the Bell Laboratories. They were using a 20-foot (6-m) directional radio antenna, designed for communication with satellites, and found what appeared to be excessive radio noise in their instrument. They decided to investigate further thinking that it could be due to radio waves from our own Galaxy. In May 1964 they found that there was a background of microwave radiation that came from all directions uniformly and was not accountable simply as instrumental noise. They calculated its effective temperature as about 3.5 kelvin (–269.65°C). An explanation was proposed by Dicke at nearby Princeton University that this was the predicted remnant radiation of the creation of the universe. Subsequent experiments confirmed that it was isotropic and apparently unchanging (on human timescales).

For their discovery Penzias and Wilson were awarded the 1978 Nobel Prize for physics, which they jointly shared with Pyotr L. KAPITZA, who received the award for his (unrelated) developments in low-temperature physics. Penzias and Wilson have continued to collaborate on research into intergalactic hydrogen, galactic radiation, and interstellar abundances of the isotopes. In particular their work has led to the discovery of a large number of interstellar molecules and rare isotopic species.

Perelman, Grigori (Grisha) (1966–) *Russian mathematician*

Grigori Perelman was born in Leningrad in the Soviet Union (now St. Petersburg, Russia). He obtained a Candidate of Science degree (equivalent to a PhD) in the late 1980s at Leningrad State University. Subsequently, he worked at the Steklov Institute of Mathematics. After spending time at various universities in the United States in the early 1990s he returned to the Steklov Institute in 1995.

Perelman is most famous for having proved the POINCARÉ conjecture, which was first advanced by the French mathematician Henri Poincaré in 1904. It concerned the topological classification of three-dimensional surfaces. By the time that Perelman addressed this problem analogues of it has been proved for all other dimensions. Perelman was able to prove the con-

jecture, and generalizations of it, by using novel geometrical techniques. His results were first announced in papers published online in non-peer-reviewed archives in 2002 and 2003. Other mathematicians subsequently published detailed versions of his results. Perelman was offered the prestigious Fields Medal in 2006 but refused to accept it. An eccentric, with no taste for self-promotion, he left the Steklov Institute in 2006 and is reputed to be unemployed and living with his mother. His work may well turn out to be of great importance to physics as well as to mathematics.

Perey, Marguerite Catherine (1909–1975) *French nuclear chemist*

Perey (pe-**ray**), the daughter of an industrialist, was born at Villemomble in France and educated at the Faculté des Sciences de Paris. She began her career in 1929 as an assistant in the Radium Institute in Paris under Marie CURIE. In 1940 she moved to the University of Strasbourg, becoming professor of nuclear chemistry in 1949 and director of the Center for Nuclear Research in 1958.

By the 1930s chemists had discovered all the elements of the periodic table below uranium except for those with atomic numbers 43, 61, 85, and 87. Many claims had been made for the discovery of element 87 with it being variously and prematurely named russium, moldavium, and virginium. In 1939 Perey found in the radioactive decay of actinium–227 the emission of alpha-particles as well as the expected beta-particles. As an alpha-particle is basically a helium nucleus with an atomic mass of 4 this implied that Perey had discovered a nuclide of mass number 223. Further investigation showed it to be one of the missing elements, with an atomic number of 87. She originally called it actinium K but in 1945 named it francium (for France).

Perkin, Sir William Henry (1838–1907) *British chemist*

> Before you began work there was little, almost nothing, known of this subject [Faraday rotation], certainly nothing of practical use to the chemist. You created a new branch of science.
> —W. Bruehl, letter to Perkin (1906)

Perkin was born in London, the youngest son of a builder. His interest in chemistry was aroused early by some experiments shown to him by a young friend and he was fortunate to attend the City of London School, which was one of the few London schools where science was taught. Perkin's teacher there, Thomas Hall, was a pupil of Johann HOFMANN at the Royal College of Chemistry and Hall pleaded with Perkin's father to allow his son to study chemistry and not to force him into a career in architecture. Hall was successful and Perkin entered the college in 1853.

In 1855 he was made Hofmann's assistant and the following year was given the task of synthesizing quinine (despite much effort, this difficult task was not achieved until 1944, by Robert Burns WOODWARD and William von Eggers Doering). Perkin started from the coal-tar derivative allyltoluidine, which has a formula very similar to that of quinine. He thought the conversion could be achieved by removing two hydrogen atoms and adding two of oxygen. Although no quinine was formed by this reaction, it did produce a reddish-brown precipitate. Perkin decided to treat a more simple base in the same manner and tried aniline and potassium dichromate. This time a black precipitate was produced. Addition of alcohol to this precipitate yielded a rich purple color. Perkin soon realized that this coloring matter had the properties of a dye and resisted the action of light very well. He sent some specimens of dyed silk to a dyeing firm in Perth, Scotland, which expressed great interest provided that the cost of the cloth would not be raised unduly. With this behind him Perkin took out the appropriate patents, borrowed his father's life savings, and in 1857 built a dye factory at Greenford Green, near Harrow, for mass production of the first synthetic dye – mauveine.

Initially there were difficulties. Since aniline was not readily available, it had to be produced at the factory from benzene. There was also the conservatism of the dye industry to overcome.

The significance of Perkin's discovery lay in its being the first *synthetic* dye; before this all dyes were derived from such organic sources as insects, plants, and mollusks. Purple had traditionally come from a Mediterranean shellfish and could be produced only at great cost, so that it was used only by royalty. Apart from the difficulty of supply there was also the problem of the quality of the dyes: vegetable and animal dyes were not particularly fast and tended to fade in the light. The market was ripe for anyone who could provide a dye in bulk that was cheap, fast, and did not fade. Perkin quickly made his fortune and stimulated a rush to find other synthetics. Karl GRAEBE and C. T. Liebermann soon synthesized alizarin, the coloring ingredient of madder. Magenta and Bismarck brown were among the other new colors that were soon to flood the market.

In 1874 Perkin sold his factory and retired, a wealthy man, at the age of 35, devoting the rest of his life to research in pure science. He became particularly interested in Faraday rotation and produced over 40 papers on this topic.

Perkin was knighted in 1906. His son and namesake was also a chemist.

Perkin, William Henry, Jr. (1860–1929) *British chemist*

Perkin, who was born at Sudbury near London, was the elder son of the famous chemist who discovered the aniline dyes, also called William Henry Perkin. As a child he assisted his father in his private laboratory. He was educated at the City of London School and then in 1877 followed his father to the Royal College of Chemistry. He then went to Germany where he studied at the universities of Würzburg and Munich. On his return to England he worked at Manchester before being appointed professor of chemistry at Heriot-Watt College, Edinburgh, in 1887. In 1892 he returned to Manchester as professor of organic chemistry and in 1912 he became Waynflete Professor of Chemistry at Oxford.

Perkin was a very practical chemist who, in a long career, achieved many syntheses and analyses. His first success came in his student days in Munich. It had been argued by Victor MEYER in 1876 that no ring with fewer than six carbon atoms could exist. Perkin succeeded in 1884 in preparing rings with four carbon atoms.

Of the many molecules he synthesized are the terpenes, limonene (1904), the oxygenated terpineol (1904), and camphoric acid (1903). He worked on many alkaloids, including strychnine, and on natural coloring compounds like brazilin. Perkin also worked with William POPE, showing that optical activity can be found in compounds in which the carbon atoms are not necessarily asymmetrical.

He produced three chemical works in collaboration with Frederic KIPPING and also did much to stimulate the growth of chemistry at Oxford by campaigning for the new laboratories that were opened there in 1922.

Perl, Martin Lewis (1927–) *American physicist*

Born in New York, Perl first graduated in chemistry in 1948 at the Brooklyn Polytechnic Institute. After working in industry as a chemical engineer for General Electric, Perl became interested in nuclear physics. Consequently he returned to college and in 1955 he was awarded his PhD from Columbia, New York. He immediately moved to the University of Michigan, where he remained until he took up the position of professor of physics at Stanford University, California, in 1963. He is now emeritus professor.

In 1972 physicists were aware of four leptons: the electron, the muon, and their corresponding neutrinos. Further, leptons, unlike hadrons such as the proton, are genuinely pointlike elementary particles, which interact by the weak force. In 1972 Stanford opened its new accelerator, SPEAR (Stanford Positron–Electron Asymmetry Ring). While no new lepton had been found since the discovery of the muon in 1936, Perl decided to use the SPEAR facilities to see whether the lepton family could be extended. Theoretical reasons led him to believe that any new lepton would have a charge of plus or minus 1, have a mass greater than a billion electronvolts (1 GeV), decay in less than a billionth of a second, and respond only to the weak and electromagnetic forces. Like any other particle the new lepton would have to be identified by detecting its characteristic decay products. The particle had been tentatively named the "tau particle" from the initial letter of the Greek word for third, "triton." Perl argued that the tau would decay into either a muon or an electron, plus a neutrino and an antineutrino. In 1974 a sample of 10,000 events yielded twenty four of the predicted kind. Despite some initial skepticism the existence of a heavy lepton with a mass greater than the proton was quickly confirmed.

Perl's discovery had an important theoretical implication. The four previously known leptons were linked with the four known quarks. The discovery of a new lepton suggested the symmetry could only be maintained by the existence of a new quark. The prediction was confirmed in 1977 when Leon LEDERMAN discovered the upsilon particle. For his discovery of the tau particle Perl shared the 1995 Nobel Prize for physics with Frederick REINES.

Perlmutter, Saul (1959–) *American cosmologist*

A leading cosmologist, Perlmutter led the team that made the 1998 breakthrough discovery that the universe is expanding at an accelerating rate. He received a degree in physics at Harvard University (1981) and then started graduate studies under Richard MULLER at the University of California, Berkeley. Here he developed robotic systems for automatically identifying and recording exploding stars, or supernovae, gaining his doctorate in 1986. Continuing to work at Berkeley, Perlmutter collaborated with Carl Pennypacker in starting to look for far-off bright supernovae as a means of measuring the rate of expansion of the universe. These so-called type Ia supernovae would, it was hoped, act as more reliable "standard candles," reference points whose brightness indicated their distance from our solar system. It was not until 1992 that the pair found their first convincing example. But many more of these rare phenomena needed to be recorded to gather statistically meaningful data. This prompted Perlmutter to devise a method for predicting in which galaxies the sought-for high-red-shift supernovae could be found, thus maximizing precious telescope time. This enabled Perlmutter's team, now the Supernova Cosmology Project, to use the world's best telescopes in their quest, and by 1997 they had found 42 suitable supernovae. Their analysis of the results contradicted prevailing cosmological theory, and showed that the universe was expanding at an increasing rate. Perlmutter announced his team's results in 1998, almost coincidentally with matching findings of a team led by Brian SCHMIDT. Other studies have since confirmed the "accelerating universe," which cosmologists claim indicates the existence of EINSTEIN's cosmological constant, or "dark energy," perhaps accounting for three-quarters of all energy in the universe. Perlmutter was appointed professor of physics at UC Berkeley in 2004. He and his team continue to investigate supernovae and dark matter, for example using the Hubble Space Telescope.

Perrault, Claude (1613–1688) *French anatomist, engineer, and architect*

Perrault (pe-**roh**), the son of an advocate, came from a most distinguished Parisian family. Of his two brothers Charles and Pierre, the first was the author in 1697 of one of the most famous of all collections of fairy tales, while the latter was a hydrologist of note. Claude himself was educated at the University of Paris and after graduating in 1639 opened a private practice.

As a founder member of the French Academy of Sciences in 1666 he became a figure of some consequence in French science. He appears to have been put in control of a group of anatomists charged with the dissection of the more exotic species as they died in the Royal Menagerie. His *Descriptions anatomiques de divers animaux* (1682; Anatomical Descriptions of Various Animals) covered the dissection of some 25 mammalian species alone and thus constitutes one of the earliest works on comparative anatomy grounded in reality.

Perrault also published a comprehensive work on physics and was a celebrated architect, taking part in the design of the Louvre. At the age of 75 Perrault is supposed to have died from an infection contracted while dissecting a camel.

Perrin, Jean Baptiste (1870–1942) *French physicist*

> Accustomed as we are, by laboratory work, to clear predictions, we see clearly what the more ignorant still have not realized; and I put into this category of ignorant certain men who are cultivated, but are completely unaware of science and its enormous potential, and who ... think that the future will always have to be like the past and conclude that there will always be wars, poverty, and slavery.
> —*Science and Hope* (1948)

Perrin (pe-**ran**) was born in Lille, France, the son of an army officer. He was educated at the Ecole Normale, where he received his doctorate in 1897. He was appointed to the Sorbonne where he was made professor of physical chemistry in 1910. He remained there until 1941, when he went to America to escape the Nazis.

Perrin's early work was in the developing field of cathode rays and x-rays. In 1895 he established the important result that cathode rays are deflected by a magnetic field and thus carry a negative charge. He began to calculate the ratio of charge to mass for these particles but was anticipated by J. J. THOMSON. In 1901 he produced a work on physical chemistry, *Les Principes* (Principles). His most important work however was on Brownian motion and the molecular hypothesis. In 1828 Robert BROWN had reported that pollen granules immersed in water moved continuously and erratically. However, it was left to Albert EINSTEIN to provide some quantitative explanations for the motion, in 1905. Assuming that the pollen was being moved by water molecules, he showed that the average distance traveled by a particle increased with the square of the elapsed time. Making the necessary corrections for temperature, size of particles, and nature of the liquid involved, Einstein made precise predictions about how far a particle should travel in a given time. In 1908 Perrin finally confirmed Einstein's predictions experimentally. His work was made possible by the development of the ultramicroscope by Richard ZSIGMONDY and Henry Siedentopf in 1903. He worked out from his experimental results and Einstein's formula the size of the water molecule and a precise value for AVOGADRO's number.

The fundamental importance of this work was that it established atomism as something more than a useful hypothesis. It was mainly as a result of Perrin's work that the most eminent skeptic, Wilhelm OSTWALD, at last relented. Perrin was awarded the Nobel Prize for physics in 1926 for his work on Brownian motion and sedimentation.

In 1913 he published *Les Atomes* (Atoms), which collected together not only his own work on molecules but new material from radiochemistry, black-body radiation, and many other fields, to demonstrate the reality of molecules. It was an enormously influential work, going through four editions in its first year and being translated into many languages.

Perutz, Max Ferdinand (1914–2002) *Austrian–British biochemist*

> True science thrives best in glass houses, where everyone can look in. When the windows are blacked out, as in war, the weeds take over; when secrecy muffles criticism, charlatans and cranks flourish.
> —*Is Science Necessary?* (1989)

While studying chemistry at the university in his native Vienna, Perutz (pe-**roots**) became interested in x-ray diffraction techniques; after graduation he went to England to work on the x-ray diffraction of proteins with William L. BRAGG at the Cavendish Laboratory, Cambridge. A meeting in Prague with the biochemist Felix Haurowitz in 1937 turned his attention to the blood protein hemoglobin and he received his PhD in 1940 for work in this field. Soon after, he was arrested as an alien and interned, first on the Isle of Man and then in Canada with Hermann BONDI and Klaus FUCHS. He was released and allowed to return to Britain in 1941. In the following year he joined the staff of Lord Mountbatten, examining various applications of science for the war effort.

After the war Perutz organized the setting up, in 1946, of the molecular biology laboratory in Cambridge, where he was soon joined by John KENDREW. After seven years' hard work Perutz was still far from his objective of working out the three-dimensional structure of hemoglobin, a molecule containing some 12,000 atoms. Then in 1953 he applied the heavy atom or isomorphous replacement technique to his work whereby heavy metal atoms, e.g., mercury or gold, are incorporated into the molecule under study. This alters the diffraction patterns, making it easier to compute the positions of atoms in the molecule. By 1959 he had shown hemoglobin to be composed of four chains, together making a tetrahedral structure, with four heme groups near the molecule's surface.

For this achievement Perutz received the 1962 Nobel Prize in chemistry, sharing it with Kendrew, who had worked out the structure of the muscle protein, myoglobin, using similar methods. In later work Perutz demonstrated that in oxygenated hemoglobin the four subunits are rearranged. This explained the change in structure noted by Haurowitz in 1938. Perutz also investigated the various mutated forms of hemoglobin characteristic of inherited blood diseases.

While indulging his hobby of mountaineering, Perutz made some notable contributions to the understanding of glaciers, particularly by his demonstration that the rate of flow is faster at the glacier surface than at the base.

Perutz continued as head of the Medical Research Council molecular biology unit at Cambridge until his retirement in 1979. He published a brief account of his early life and his views on science in his *Is Science Necessary?* (1989). He also published several books about proteins and nucleic acids.

Peters, Sir Rudolph Albert (1889–1982) *British biochemist*

Peters, the son of a London doctor, was educated at Cambridge University and St. Bartholomew's Hospital, London. After teaching briefly in Cambridge, he accepted the Whitley Chair of Biochemistry at Oxford, which he held from 1923 until his retirement in 1954.

Between 1928 and 1935 Peters and his Oxford colleagues succeeded in showing for the first time the precise activity of a vitamin in the body. Working with vitamin B_1, or thiamine – the antiberiberi factor first described by Christiaan EIJKMAN – they fed pigeons on a diet of polished rice. This was free of thiamine and produced a number of debilitating symptoms in most of the birds. As one of these symptoms was convulsions, Peters suspected that the thiamine deficiency could involve the central nervous system. He consequently began a search of the pigeon's brain for what he termed a "biochemical lesion."

The first hint of the role of thiamine was provided by the failure of minced pigeon brain to take up as much oxygen as the brain of a normally fed bird. The lesion was promptly reversed by the addition of thiamine. Further work showed an accumulation of lactic acid in the pigeon brain. As this is one of the intermediate products in the metabolism of carbohydrates into carbon dioxide and water it seemed clear that thiamine must be an essential ingredient in this metabolic pathway.

Peters's work therefore provided the first proof of the action of any vitamin upon an enzyme system *in vitro*.

Petit, Alexis-Thérèse (1791–1820) *French physicist*

Petit (pe-**tee**), who was born at Vesoul in France, entered the Ecole Polytechnique in 1807. He spent a period teaching physics in Paris, and received a doctorate in 1811. He was one of the professors of physics at the Ecole Polytechnique. Petit did some research with his brother-in-law D. F. J. ARAGO on the variation of refractive index with temperature. However, he is known solely for his work with Pierre DULONG in which they established the law (*Dulong and Petit's law*) that the specific heat of a solid multiplied by the atomic weight is (approximately) a constant for different solids.

Petit, Jean Louis (1674–1750) *French surgeon*

Petit was the leading French surgeon of the first half of the 18th century. Born in Paris, he served as first director of the Royal Academy of Surgery, founded in 1731.

His major surgical innovation was that of the screw tourniquet in 1718. In the 16th century Ambroise PARÉ had introduced the ligature into surgery. It had found little use in such major surgery as the amputation of a limb because of the difficulty involved in arresting the blood flow in order to permit the tying of up to 50 ligatures. Surgeons therefore continued with their customary mode of controlling bleeding, namely cauterization.

Petit's innovation thus allowed surgeons at last to utilize the earlier work of Paré. Petit was also the first to open the mastoid process for the evacuation of pus.

Petri, Julius Richard (1852–1921) *German bacteriologist*

Petri (**pay**-tree or **pet**-ree), who was born at Barmen in Germany, worked initially as an assistant to Robert KOCH before his later appointment to a curatorship of the Hygiene Museum in Berlin. In 1887 Petri introduced a modification into the craft of the bacteriologist, which has persisted virtually unchanged to the present day. Koch's practice was to spread his bacterial sample on a glass slide, which was then placed under a bell jar to avoid contamination. The covered dishes introduced by Petri were less bulky, easier to handle, and could also be conveniently stacked. They consequently gained rapid acceptance throughout the laboratories of the world.

Pettenkofer, Max Joseph von (1818–1901) *German chemist and physician*

> Impurity cleaves longest and most tenaciously to the soil, which suffers no change of place, like air and water.
> —*The Relations of the Air to the Clothes We Wear, the House We Live In, and the Soil We Dwell On* (1872)

Born at Lichtenheim in Germany, Pettenkofer (**pet**-en-koh-fer) qualified as a physician in 1843. He became professor of medical chemistry at Munich in 1847 and later was appointed to the first chair of hygiene. In 1882 he published one of the earliest textbooks on hygiene, *Handbuch der Hygiene* (Handbook of Hygiene). Although he was one of the earliest campaigners for better sanitation and hygiene in general, he was one of the most bitter and articulate opponents of the germ theory of disease, which, through the work of Louis PASTEUR and Robert KOCH, was becoming widely accepted. Pettenkofer supported the miasma theory of disease, which supposed that infection was caused and spread by toxic vapors that tended to concentrate in special areas. The job of the hygienist was to eliminate such concentrations.

In 1883 Koch claimed to have discovered the causative agent of cholera. In 1892 Pettenkofer attempted to disprove Koch's theory by drinking a glass of water swarming with cholera vibrios. Fortunately he survived this dramatic experiment, which supported his belief in the miasma theory until his death.

Earlier he had collaborated with Carl Voit in constructing a calorimeter big enough for a human being, which they used in attempting to calculate the basic metabolic rate of man.

Pfeffer, Wilhelm Friedrich Philipp
(1845–1920) *German botanist*

Born the son of a pharmacist in Grebenstein, Germany, Pfeffer (**pfef**-er) gained his PhD in botany and chemistry from the University of Göttingen in 1865. He continued pharmaceutical studies at Marburg until, realizing his increasing preference for botany, he left in 1867 for Berlin where he studied the germination of *Selaginella*. He continued this research at Würzburg under Julius SACHS who encouraged him to pursue a career in plant physiology. He was professor of botany at Basel, Tubingen, and Leipzig universities.

Pfeffer did much work on irritability in plants, but he is best remembered for his improvement of the semipermeable membrane and its application in the measurement of osmotic pressure. He demonstrated that osmotic pressure is correlated with the concentration of the solution within the membrane and with temperature.

Pfeiffer, Richard Friedrich Johannes
(1858–1945) *German bacteriologist*

Born at Zduny in Poland, Pfeiffer (**pfI**-fer) trained as a military surgeon in Berlin, serving in the German army until 1889. He then worked with Robert KOCH at the Institute of Hygiene before being appointed (1899) professor of hygiene at the University of Königsberg. In 1909 Pfeiffer moved to a similar post at the University of Breslau, where he remained until his retirement in 1926.

In 1892 Pfeiffer discovered the bacillus *Haemophilus influenzae* (known as *Pfeiffer's bacillus*), which he found in the throats of patients in the influenza epidemic of 1889–92 and declared to be responsible for the disease. Influenza was later shown (in 1933) by the British pathologist Christopher Andrewes and his colleagues to be a viral infection, with Pfeiffer's bacillus responsible for many of the complications.

Pfeiffer's most important discovery, however, was that of bacteriolysis in 1894. He had injected cholera germs into a guinea pig already immunized against the infection. When he extracted some of the germs from the guinea pig and examined them under the microscope he observed that they first became motionless and then swelled up and burst. He went on to note that this destruction (lysis) of the bacteria would also take place in an artificial environment but could be stopped by heating to over 60°C.

Pfeiffer was in fact observing and describing for the first time a complicated immune reaction by the body to an invading germ. It was this work that stimulated Jules BORDET to look more closely at the immune system and led to his discovery of complement. It also did much to confirm Emil von BEHRING's theory of antibodies in 1891.

Philips, Peregrine (about 1830) *British chemist*

Philips, who was born in the port of Bristol, western England, became a manufacturer of vinegar. In 1831 he took out a patent for a new way to produce sulfuric acid which, although not immediately taken up, was to account for more than half the world's production later in the 19th century. In the lead chamber process introduced by John Roebuck, sulfur was burned in a lead chamber with water and a potassium nitrate catalyst. The disadvantages of this were that it was slow, produced dilute acid, and required considerable expense in the construction of the lead chambers.

Philips proposed oxidizing the sulfur dioxide into sulfur trioxide by passing it through a tube packed with a catalyst of fine platinum wires. This gas was then dissolved in water or in sulfuric acid. There were difficulties with the process, which prevented its immediate adoption. The platinum soon became contaminated and useless, requiring costly replacement and, although the contact process could make much more concentrated acid there was at the time no real demand for very strong acid. The difficulty was initially overcome by Rudolph MESSEL in 1876, and today the contact process accounts for the production of over 90% of the world's sulfuric acid.

Phillips, David Chilton, Baron (1924–1999) *British biophysicist*

Phillips, who was born at Ellesmere in Shropshire, gained his doctorate from University College, Cardiff. He then moved to the National Research Laboratories in Ottawa, where he worked until his return to Britain in 1955. He worked first at the Royal Institution before being appointed in 1966 to the chair of molecular biophysics at Oxford, a post he held until his retirement in 1990. He was raised to the British peerage in 1994.

In 1965 Phillips achieved the major success of working out the full three-dimensional structure of lysozyme, the first enzyme to be so analyzed and, following the success of John KENDREW with myoglobin and Max PERUTZ with hemoglobin, only the third protein to be so treated. Lysozyme was first identified by Alexander FLEMING in 1922 and shown by him to be capable of dissolving certain bacterial cells. In 1966 Phillips related the structure of the molecule to its bacteriolytic power. He demonstrated that the 129-amino-acid molecule is folded so as to form a cleft that holds the substrate molecule while it is being broken in two. More precisely, he was able to show that hexasaccharides of the bacterial cell wall are split between the fourth and fifth sugar rings by the charged amino-acid residues, aspartic and glutamic acid, which align with them in the enzyme's cleft.

Phillips had thus succeeded in explaining for the first time the catalytic activity of an enzyme in stereochemical terms.

Phillips, William D. (1948–) *American physicist*

Phillips, who was born in Wilkes-Barre, Pennsylvania, was educated at the Massachusetts Institute of Technology, where he gained his PhD in 1976. He then joined the staff of the National Bureau of Standards and Technology, Maryland. He is currently a professor at the University of Maryland.

In the early 1980s Phillips and his colleagues showed how beams of sodium atoms could be cooled down to temperatures of about 0.1 kelvin, just a tenth of a degree above absolute zero. The method involved exposing the sodium atoms to a laser beam tuned to a frequency lower than the resonant frequency of the atoms. Some of the atoms moving toward the laser light, because of an appropriate DOPPLER shift, will absorb and emit light photons. The process, repeated endlessly, will slow the atoms down. Atoms moving in the direction of the laser beam, however, because the frequency will be shifted lower, are less likely to absorb radiation and their motion is little affected.

Soon after this, Steven CHU developed the technique so that it could be applied to a gas rather than a beam of atoms, and isolated and cooled individual atoms to a temperature of 240 microkelvins. Unfortunately, atoms could be contained in Chu's optical laser trap for no more than a second. In 1988 Phillips sought to improve the trap by using a varying magnetic field placed above and below the area in which the laser beams intersected. This reduced the temperature of the confined atoms to 40 microkelvins. With this technique Phillips found that he need only reduce the atom's speed to less than 3.5 meters per second, corresponding to a temperature of 17 millikelvins, for the atoms to be held in the laser trap for several seconds.

For his work in this field Phillips shared the 1997 Nobel Prize for physics with Steven Chu and Claude COHEN-TANNOUDJI.

Philolaus of Croton (about 475 BC) *Greek philosopher*

Little of any reliability is known of Philolaus (fil-oh-**lay**-us). Apart from such stories as that PLATO derived his *Timaeus* from him and that he settled in Tarentum in southern Italy, he is generally only mentioned by classical writers as the author of the Pythagorean cosmology. He proposed, for unknown reasons, that the center of the universe was occupied not by the Earth or even the Sun but by a previously unsuspected fire. The reason this had never been seen was simply that the Earth in its orbit always turned its inhabited face away from it. There was also the counter-Earth or *antichthon*, again out of our view for always being between the Earth and the "central hearth." All ARISTOTLE, the main source of Pythagorean cosmology, could suppose was that it was introduced to make the number of heavenly bodies in orbit – the fixed stars, five planets, Earth, Moon, Sun, and antichthon – equal ten, the Pythagorean sacred number.

The Pythagorean model with its apparently arbitrary features lay well outside the mainstream of Greek cosmology. However, it did propose that the Earth moved in an orbit – the first recorded instance of such a speculation.

Philon of Byzantium (about 100 BC) *Greek scientist*

Philon (**fi**-lon) studied the flow of liquids and gases. He designed an instrument for showing the expansion of air, which may have been one of the earliest thermometers.

Piazzi, Giuseppe (1746–1826) *Italian astronomer*

> I congratulate you on your splendid discovery of this new star. I do not think that others have noticed it, and because of its smallness, it is unlikely that many astronomers will see it.
> —On the discovery of Ceres. Barnaba Oriani, letter to Piazzi, January 1801

Born at Ponte in Valtellina, Italy, Piazzi (**pyaht**-see or pee-**aht**-see) was a monk who originally taught philosophy but later in life developed an interest in astronomy. He became professor of mathematics at Palermo, Sicily, setting up an observatory there in 1787. He was a careful observer, publishing a catalog of 7,646 stars in 1814. In his work he had found that proper motion was not the rarity assumed by some but the property of most stars and he found that 61 Cygni had a very large proper motion of 5.2″. His most dramatic discovery was that of the minor planet Ceres in 1801. He named it after the goddess of agriculture, once widely worshipped in Sicily. Although Piazzi lost the planet, its position was precisely predicted by Karl GAUSS after a staggering feat of calculation based on three observations of Piazzi. Three more similar bodies were quickly found. He had a dispute with William HERSCHEL over their nature. Piazzi proposed that they shoud be called "planetoids" but Herschel's alternative suggestion of "asteroid" has proved more acceptable until quite recently.

Picard, Jean (1620–1682) *French astronomer*

Born at La Flèche in northwestern France, Picard (pee-**kar**) succeeded Pierre GASSENDI as professor of astronomy at the Collège de France in 1655. He helped to found the Paris Observatory and conducted fundamental researches into the size of the Earth. Using new instruments such as William Gascoigne's micrometer he established an accurate baseline and by a series of 17 triangles between Malvoisin and Amiens calculated one degree to be 69.1 miles (111.2 km). This result proved to be extremely valuable to NEWTON in his calculations on the attractive force of the Moon.

Picard also determined accurately the position of Tycho BRAHE's observatory at Uraniborg (this information was necessary in order to analyze and interpret Tycho's observations). He further noted, but was not able to explain, an annual periodic motion of Polaris (approximately 40″). James BRADLEY later explained this as the aberration of light.

Piccard, Auguste (1884–1962) *Swiss physicist*

Piccard (pee-**kar**) was born at Basel in Switzerland, where his father was a professor of chemistry at the university, and was educated at the Federal Institute of Technology, Zurich. Although he taught in America at Chicago and Minnesota and in Switzerland at Lausanne he spent the main part of his career in Belgium where, from 1922 to 1954, he was professor of physics at the Brussels Polytechnic.

Piccard is remembered for his explorations of both the atmosphere and the ocean floor. Dissatisfied with sending up unmanned instruments, he designed a pressurized cabin to be attached to a balloon and with his brother, Jean Felix, ascended to a height of about 11 miles (18 km) in 1931. Later balloonists using his techniques extended this to 20 miles (32 km).

He was equally dissatisfied with attempts to study the ocean floor and introduced (1947) a craft he called a bathyscape, which was maneuverable and on its first test in 1948 reached a depth of nearly a mile. He then built a second craft, the *Trieste*, which descended 21/2 miles (4 km) into the Mediterranean. This was later sold to the U.S. Navy and in 1960 Piccard's son Jacques, together with a naval officer, descended 7 miles (11 km) in it to the bottom of the Marianas trench in the Pacific.

Pickering, Edward Charles (1846–1919) *American astronomer*

Born in Boston, Massachusetts, Pickering graduated from Harvard in 1865. He taught physics at the Massachusetts Institute of Technology before becoming professor of astronomy and director of the observatory at Harvard in 1876, remaining there until his death in 1919.

Pickering made innovations in spectrography. Instead of placing a small prism at the focus to capture the light of a single star, he put a large prism in front of the objective, obtaining at the same time a spectrogram of all the stars in the field sufficiently bright to affect the emulsion. This made possible the massive surveys he wanted to organize and enabled the publication in 1918 of the *Henry Draper Catalogue*, compiled by Annie CANNON, giving the spectral types of 225,300 stars. The other innovation in instruments due to him was the meridian photometer introduced in 1880. In this, images of stars near the meridian would be reflected at the same time as the image of Polaris. The brightness could then be equalized and as the brightness of Polaris was known, that of the meridian stars could easily be calculated. More than a million observations with such instruments permitted the compilation of the Harvard catalog giving the magnitude of some 50,000 stars. He was able to include stars of the southern hemisphere in this catalog, for in 1891 he had established an observatory in Arequipa, Peru, with the help of his brother William Henry Pickering.

One further improvement due to Edward Pickering was his introduction, around 1900, of the alphabetic system of spectral classes.

Pickering, William Henry (1858–1938) *American astronomer*

Pickering, the younger brother of Edward Pickering, was also an astronomer. Born in Boston, Massachusetts, he studied at the Massachusetts Institute of Technology where he worked after graduating in 1879. In 1887 he moved to the Harvard College Observatory where his brother was director. He set up a number of observing stations for Harvard including that at Arequipa, Peru, in 1891 and Mandeville, Jamaica, in 1900. He took charge of the latter in 1911, converting it into his own private observatory following his retirement in 1924.

He also helped Percival LOWELL set up his private observatory in Flagstaff, Arizona, and, also like Lowell, concerned himself with the trans-Neptunian planet. In 1919, on the basis of past records, he predicted that a new planet would be found near the constellation of Gemini but photographic surveys failed to confirm his prediction. When the planet was finally detected in 1930 by Clyde TOMBAUGH, Pickering made a somewhat exaggerated claim to be its discoverer.

He made extensive observations of Mars and claimed, like Lowell, that he saw signs of life on the planet by observing what he took to be oases in 1892. He went further than Lowell, however, when in 1903 he claimed to observe signs of life on the Moon. By comparing descriptions of the Moon from Giovanni RICCIOLI's 1651 chart onward, he thought he had detected changes that could have been due to the growth and decay of vegetation.

He was more successful in 1899 when he discovered Phoebe, the ninth satellite of Saturn. This was the first planetary satellite with retrograde motion to be detected, i.e., with orbital motion directed in an opposite sense to that of the planets. His 1905 report of a tenth satellite, which he confidently named Themis, was not substantiated.

Pictet, Ame (1857–1937) *Swiss chemist*

Pictet (peek-**tay**), a banker's son from Geneva in Switzerland, became professor of chemistry there. He synthesized a number of alkaloids including nicotine (1903), confirming the formula that had been proposed in 1891. He also synthesized laudanosine and papaverine (1909).

Pictet, Raoul Pierre (1846–1929) *Swiss chemist and physicist*

Born in Geneva, Pictet was professor of physics at the university there from 1879 and at the University of Berlin from 1886. He later moved to Paris.

Pictet was first interested in the production of artificial ice and then turned his attention to the study of extremely low temperatures and the liquefaction of gases. On 22 December 1877 he was involved in one of those strange simultaneous discoveries that sometimes occur in science. He announced on that day, by telegram to the French Academy, that he had liquefied oxygen. Just two days later the French physicist Louis CAILLETET made a similar announcement.

Both Pictet and Cailletet had recognized that both cooling and compression were necessary to liquefy oxygen but they had achieved this using different techniques. Pictet had used his cascade method, in which he evaporated liquid sulfur dioxide to liquefy carbon dioxide, which in turn was allowed to evaporate and to cool oxygen to below its critical temperature. The oxygen could then be liquefied by pressure. The advantage over Cailletet's method was that it produced the liquid gas in greater quantity and was easier to apply to other gases.

Pierce, John Robinson (1910–2002) *American communications engineer*

Born in Des Moines, Iowa, Pierce was educated at the California Institute of Technology, graduating in 1933 and obtaining his PhD in 1936. He then worked for the Bell Telephone Laboratories, New York, from 1936 until 1971 when he became professor of engineering at the California Institute of Technology, retiring in 1980.

At the Bell Laboratories Pierce improved the design of traveling wave tubes for microwave equipment. He also invented the electrostati-cally focused electron-multiplier tube and the electron gun, the basis of television sets and other visual display equipment. During World War II he helped to develop the equipment used in the U.S. radar system. In 1952 he became the director of electronics research at the Bell Laboratories. His efforts, ignored at first, later helped to establish the NASA communications satellites, the first attempt being a simple radio-reflector balloon, Echo 1, which was launched in 1960.

After retiring from the Bell Laboratories Pierce became a professor of engineering at the California Institute of Technology. He also wrote science fiction throughout his career using the pseudonym J. J. Coupling.

Pilbeam, David Roger (1940–   ) *British physical anthropologist*

Pilbeam was born in the English coastal town of Brighton and educated at Cambridge University, where he completed his PhD in 1967. He moved to Yale in 1968 and was appointed professor of anthropology in 1974, a post he held until 1981 when he accepted a similar position at Harvard. He is currently Henry Ford Professor of the Social Sciences.

Anthropologists distinguish between hominids, which are human species and their extinct ancestors, and the hominoids, which comprise the superfamily containing humans and apes. Pilbeam has sought to identify the time in the Miocene (between 25 million and 5 million years ago) which saw the separation of the hominids from the hominoids. In 1932 in the Siwalik Hills in India G. E. Lewis had discovered the fossil remains of a creature he named *Ramapithecus* (Rama's ape) in rocks about 15 million years old. In 1961 Elwyn SIMON, later to be a Yale colleague of Pilbeam, was struck by the small canines and the shape of the dental arch of *Ramapithecus*, seeing in them hominid characteristics.

Pilbeam began a general review of Miocene hominoid fossils. In *Ramapithecus* he thought he could see signs of bipedality. With its reduced canines, he argued, food must have been prepared in some way, implying that the hands must have been free, possibly to use tools. The creature could well have been both bipedal and terrestrial. In 1968 Pilbeam argued that about 20 million years ago there had been three species of *Dryopithecus* (tree ape) and that these had been ancestral to the chimpanzee, gorilla, and orangutan (pongids). This left the 15 million year old *Ramapithecus* as a hominid and the possibility that hominid and pongid lines could have been separated for 30 million years and more.

Pilbeam's account received wide support and appeared to become increasingly confirmed with the discovery of further fossils. Yet by the late 1960s evidence of another kind that would seriously question the position of *Ramapithecus* was beginning to appear. Biologists such as Vincent SARICH had begun to use variations in protein structure to measure evolutionary divergence. Their evidence suggested that humans and African apes had been separated for no longer than five million years. Clearly, *Ramapithecus* could not have been a Miocene hominid.

Following Sarich's revelations and further fieldwork, Pilbeam has proposed a new evolutionary sequence. In 1980 in Turkey he discovered a partial skull of *Sivapithecus*, a hominoid fossil very similar to *Ramapithecus*. It also closely resembled the sole surviving Asian great ape, the orangutan, but showed little resemblance to any of the australopithecines. Thus the 15 million year old *Ramapithecus* represents not the hominid–hominoid split, but the hominoid divergence between the African and Asian great apes.

Pilbeam has since offered the opinion that he would "never again cling so firmly to one particular evolutionary scheme," and that "fossils themselves can solve only part of the puzzle, albeit an important part."

Pincus, Gregory Goodwin (1903–1967)
American physiologist

Born in Woodbine, New Jersey, and educated at Cornell and Harvard, Pincus was research director at the Worcester Foundation for Experimental Biology, Shrewsbury, Massachusetts. His most significant work was in reproductive physiology, notably his investigations of human birth control, which led to his developing, with Min Chueh CHANG and John Rock, the now famous "pill." This form of oral contraception is based upon the use of synthetic hormones that have an inhibitory effect on the female reproductive system, preventing fertilization but still allowing sexual freedom. Pincus discovered that the steroid hormone progesterone, which is found in greater concentrations during pregnancy, is responsible for the prevention of ovulation in pregnancy. With the development, in the fifties, of synthetic hormones, similar in action to progesterone, Pincus saw the possibility of using such synthetics as oral contraceptives. The first clinical trials were conducted in 1954 and proved extremely successful.

In 1963, Pincus became the first chairman of the Oral Advisory Group of the International Planned Parenthood Federation.

Pinel, Philippe (1745–1826) *French physician and psychiatrist*

Pinel (pee-**nel**), who was born at Saint-André in France, obtained his MD in 1773 from the University of Toulouse. He then worked as a translator and teacher, visiting mentally disturbed people in his spare time. The articles published as a result of these visits led to his appointment at the Hospice de Bicêtre, the Paris asylum for men, in 1793. Two years later Pinel moved to the Saltpêtrière, the hospital for the poor, aged, and insane, where he served as chief physician until his death in 1826.

Pinel is mainly remembered for his dramatic act in unchaining the insane of Paris, believing "air and liberty" to be a far more effective treatment. His general approach seems to have been to bring the patient to an awareness of his position and a recognition of his surroundings.

Pinel published a full account of his new techniques in his *Traité medico-philosophique sur l'aliénation mentale ou la manie* (1801; Medico-Philosophical Treatise on Mental Alienation or Mania). His influence led to the general adoption of a far more enlightened approach to the treatment of mental patients. In 1798 Pinel published a more conventional work, *Nosographie philosophique* (Philosophical Classification of Diseases), an attempt to classify diseases in the way LINNAEUS had earlier classified animals.

Pinker, Steven (Arthur) (1954–)
Canadian-born American psychologist

Pinker was born in Montreal, Canada, and educated at McGill University, where he obtained a BA in 1976, and Harvard University, where he obtained his PhD in 1979. He did postdoctoral research at the Massachusetts Institute of Technology (MIT) in 1979–80. He was an assistant professor at Harvard University (1981–82) and MIT (1982–89), before becoming a full professor at MIT in 1989.

Pinker's main interest has been the operation of the mind, particularly with regard to the nature of language and the concept of instinct. In this area Pinker has combined ideas from the theory of computation and DARWIN's theory of evolution to obtain insights into the workings of the human mind. These ideas have been expounded in a number of popular books including *Language Learnability and Language Development* (1984), *Learnability and Cognition* (1989), *The Language Instinct* (1994), *How the Mind Works* (1997), *Words and Rules* (1999), and *The Blank Slate* (2002). Pinker has expressed skepticism toward utopian schemes and certain types of social engineering that ignore, or deny the existence of, human nature. In *How the Mind Works* he observed: "In our society, the best predictor of a man's wealth is his wife's looks, and the best predictor of a woman's looks is her husband's wealth." In 2007 he published *The Stuff of Thought*.

Pippard, Sir Alfred Brian (1920–)
British physicist

> The value of a formalism lies not only in the range of problems to which it can be successfully applied, but equally in the degree

to which it encourages physical intuition in guessing the solution of intractable problems.
—*Physics Bulletin* No. 20 (1969)

Pippard, the son of a professor of engineering, was born at Leeds in the north of England and educated at the University of Cambridge. After wartime research on the development of radar, he obtained his PhD from Cambridge in 1947. He remained at the university for the rest of his career, serving as Plummer Professor of Physics from 1960 until 1971 and Cavendish Professor until 1982.

After World War II Pippard began to use microwaves to study superconductors, in particular the conduction in a thin layer at the surface of the material. He introduced the idea of "coherence" in superconductors – the way in which electrons "act together" so that an effect at one point influences electrons a certain distance away. Pippard's ideas were explained by the BCS theory of John BARDEEN and his colleagues. Pippard has also worked on microwave absorption at metal surfaces as a method of investigating the conduction electrons. His book *Dynamics of Conduction Electrons* (1964) deals with metallic conductivity. His most recent work is *Magnetoresistance in Metals* (1989).

Pirie, Norman Wingate (1907–1997)
British biochemist

> My teacher, [Frederick Gowland] Hopkins, often commented on the craving for certainty that led so many physicists into mysticism or into the Church and similar organisations ... Faith seems to be an occupational hazard for physicists.
> —*Penguin New Biology* (1954)

Pirie, son of the painter Sir George Pirie, was educated at Cambridge University where he studied under Frederick Gowland Hopkins. He remained at Cambridge as a demonstrator in the Biochemical Laboratory from 1932 until 1940. Pirie then moved to the Agricultural Research Station at Rothamsted, where he worked first as a virus physiologist (1940–46) and then as head of the biochemistry department (1947–73).

In 1935 Wendell Stanley succeeded in growing crystals of tobacco mosaic virus (TMV), claiming them to be protein. However, when, in collaboration with Frederick BAWDEN, Pirie repeated Stanley's work in 1936 he came across a small amount (about 0.5%) of phosphorus in the virus. As no amino acids contain phosphorus as an ingredient, this could only mean that TMV was not a pure protein. Pirie and Bawden concluded that TMV was in fact a nucleoprotein and went on to show that the nucleic acid present was the same as that derived from yeast, namely, RNA. This finding was of fundamental importance, providing an impetus for work on molecular biology in the following decade.

Pirie, a prolific author, wrote on such topics as contraception, the origin of life, and the organization of science. In later life he turned to the technical problem of extracting edible protein in large quantities from plants, and in such popular books as *Food Resources* (1969) he gave publicity to the nutritional problems of the world.

Pitzer, Kenneth Sanborn (1914–1997)
American theoretical chemist

Born in Pomona, California, Pitzer was educated at the California Institute of Technology and the University of California, Berkeley, where he obtained his PhD in 1937. He immediately joined the staff at Berkeley and served there as professor of chemistry from 1945 until 1961. Pitzer then moved to Rice University, Houston, where he served both as president and professor of chemistry until his return to California in 1968. After a brief period at Stanford, he was once more appointed professor of chemistry at Berkeley.

Pitzer worked extensively on problems of molecular structure and in particular on a conformational analysis of cyclic and polycyclic paraffins. This he linked with detailed studies of the thermodynamic properties of hydrocarbons. His work here began in the late 1930s when it became apparent that supposedly straightforward calculations of the thermodynamic properties of such a simple molecule as ethane, C_2H_6, disagreed with experimental results. Pitzer and his colleagues were able to show that the anomaly was due to an unexpected restriction of the free rotation of the methyl groups, CH_3, around the carbon–carbon bonds. This in turn was shown to be due to the hydrogen atoms of the methyl group adopting different conformations with different potential energies as they rotated around the carbon–carbon bond. Later work by Pitzer concerned more complex molecules and motions and also extended his calculations to substances in the liquid, solid, and dissolved states. He also produced a widely read textbook, *Quantum Chemistry* (1953), and was a co-author of the standard textbook *Thermodynamics* with L. Brewer.

Planck, Max Karl Ernst Ludwig (1858–1947) *German physicist*

> An important scientific innovation rarely makes its way by gradually winning over and converting its opponents: it rarely happens that Saul becomes Paul. What does happen is that its opponents gradually die out, and that the growing generation is familiarized with the ideas from the beginning.
> —*Scientific Autobiography* (1949)

Planck (plahngk or plangk) was born at Kiel in Germany, where his father was a professor of civil law at the university. He was educated at the universities of Berlin and Munich where he obtained his doctorate in 1880. He began his teaching career at the University of Kiel, moving to Berlin in 1889 and being appointed (1892) professor of theoretical physics, a post he held until his retirement in 1928.

Although Planck's early work was in thermodynamics, in 1900 he published a paper, *Zur Theorie der Gesetzes der Energieverteilung im Normal-Spektrum* (On The Theory of the Law of Energy Distribution in the Continuous Spectrum), which ranks him with Albert EINSTEIN as one of the two founders of 20th-century physics. It is from this paper that quantum theory originated.

A major problem in physics at the end of the 19th century lay in explaining the radiation given off by a hot body. It was known that the intensity of such radiation increased with wavelength up to a maximum value and then fell off with increasing wavelength. It was also known that the radiation was produced by vibrations of the atoms in the body. For a perfect emitter (a so-called black body, which emits and absorbs at all wavelengths) it should have been possible to use thermodynamics to give a theoretical expression for black-body radiation. Various "radiation laws" were derived. Thus Wilhelm WIEN in 1896 derived a law that applied only at short wavelengths. Lord RAYLEIGH and James JEANS produced a law applying at long wavelengths, but predicting that the body should have a massive emission of short-wavelength energy – the so-called "ultraviolet catastrophe."

Planck's problem was initially a technical one; he was simply searching for an equation that would allow the emission of radiation of all wavelengths by a hot body to be correctly described. He hit upon the idea of correlating the entropy of the oscillator with its energy. Following his intuition he found himself able to obtain a new radiation formula, which was in close agreement with actual measurements under all conditions.

There was, however, something unusual about the Planck formula. He had found that in seeking a relationship between the energy emitted or absorbed by a body and the frequency of radiation he had to introduce a constant of proportionality, which could only take integral multiples of a certain quantity. Expressed mathematically, $E = nh\nu$, where E is the energy, h is the constant of proportionality, ν is the frequency, and $n = 0, 1, 2, 3, 4$, etc. It follows from this that nature was being selective in the amounts of energy it would allow a body to accept and to emit, allowing only those amounts that were multiples of $h\nu$. The value of h is very small, so that radiation of energy at the macroscopic level where n is very large is likely to *seem* to be emitted continuously.

Planck's introduction of what he called the "elementary quantum of action" was a revolutionary idea – a radical break with classical physics. Soon other workers began to apply the concept that "jumps" in energy could occur. Einstein's explanation of the photoelectric effect (1905), Niels BOHR's theory of the hydrogen atom (1913), and Arthur COMPTON's investigations of x-ray scattering (1923) were early successes of the quantum theory. In 1918 Planck was awarded the Nobel Prize for physics. The constant h (6.626196×10^{-34} joule second) is known as the *Planck constant* – the value "$h = 6.62 \times 10^{-27}$ erg.sec" is engraved on his tombstone in Göttingen.

By the time of his retirement Planck had become the leading figure in German science and was therefore to play a crucial role in its relations with the Nazis. His attitude was that of prudent cooperation with the overriding aim of retaining the integrity of German science and preventing it from falling into international ridicule. Although he did not publicly protest against the harassment of Jewish scientists, considering such barbarisms a temporary madness, he did, in 1933, raise the issue with Hitler himself. He argued that the racial laws of 1933, barring Jews from government positions, would endanger the preeminence of German science. Hitler is reported to have expressed a willingness to do without science for a few years. Nor did Planck succeed in protecting the institutions of German science for in 1939 the presidency of the academy went to a party member, T. Vahlen, who lost no time in turning it virtually into an organ of the party.

Planck's later years, despite the honors that came his way, were indeed bitter ones. "My sorrow cannot be expressed in words," he lamented at one point. During World War I his elder son Karl died from wounds suffered in action, and his twin daughters, Grete and Emma, died during childbirth in 1917 and 1919, respectively. In World War II he was forced to witness the destruction of his country and of German science and its institutions. His own home, with all his possessions, was totally destroyed by allied bombing in 1944. Worst of all, his one surviving child, Erwin, was executed in 1945 for complicity in the 1944 attempt to assassinate Hitler.

Planté, Gaston (1834–1889) *French physicist*

Born at Orthez in France, Planté (plahn-**tay**) worked on electric cells and was the first to design and construct a storage battery (i.e., a cell that could be recharged with electricity). His design used lead electrodes in sulfuric acid, similar to the type used today in automobile batteries.

The *Planté battery* was invented in 1859. It was soon in use to illuminate railroad carriages and drive automobiles. Its main disadvantage was its weight. In 1888 a 666 ampere-hour battery could weigh as much as 2.8 hundredweights (127 kg).

Plaskett, John Stanley (1865–1941)
Canadian astronomer

Plaskett, who was born at Woodstock, Ontario, was initially trained as a mechanic and began work as such for the physics department of the University of Toronto. He eventually graduated from there in physics and mathematics in 1899. When he moved, in 1903, to the Dominion Observatory in Ottawa it was as mechanical superintendent and not as an astronomer. He gradually moved into astronomy, however, and in 1918 became director of the newly established Dominion Astrophysical Observatory at Victoria, British Columbia, for which he had organized the design, construction, and installation of a new 72-inch (1.8-m) reflecting telescope. He retired in 1935.

Plaskett's field of research was spectroscopy, in particular the measurement of radial velocities of celestial bodies, i.e., their velocities along the line of sight, from the shift in their spectral lines. Using the 72-inch reflector and a highly sensitive spectrograph, many spectroscopic binary systems were discovered. In 1922 Plaskett identified an extremely massive star as a binary, now known as *Plaskett's star*. In 1927 Plaskett provided confirmatory evidence for the theory of galactic rotation put forward by Bertil LINDBLAD and Jan OORT.

By 1928 Plaskett, in collaboration with J. A. Pearce, had obtained evidence for the hypothesis formulated by Arthur EDDINGTON in 1926 that interstellar matter was widely distributed throughout the Galaxy; their results showed that interstellar absorption lines, mainly of calcium, took part in the galactic rotation and so the interstellar matter was not confined to separate star clusters. Although this result was first announced by Otto STRUVE in 1929, Plaskett felt he had priority and was convinced that Struve had obtained his results from him.

Plato (c. 428 BC–347 BC) *Greek philosopher*

> Mind is ever the ruler of the universe.
> —*Philebus*

Little is known of Plato's early life. He apparently came from an established Athenian family active in politics. With the execution of SOCRATES in 399 BC he left Athens for some years and visited, among other places, Sicily in about 389 where he made contact with the Pythagoreans and much impressed Dion, the brother-in-law of Dionysius I, the tyrant of Syracuse. On his return to Athens shortly afterward he founded in about 387 the most famous of all institutions of learning, the Academy, which in one form or another remained viable until its closure by the emperor Justinian in 529 AD. On the death of Dionysius I in 367 BC Plato returned to Sicily at the invitation of Dion to try to educate Dionysius II as the new philosopher-king, attempting once more in 361. The visits were disastrous and ended with Plato dismissing Sicily as a place where "happiness was held to consist in filling oneself full twice a day and never sleeping alone at night."

It is virtually impossible to overestimate the impact of Plato on Western thought. His views, preserved and transmitted through the distorting medium of neo-Platonists and Christian fathers alike, came to influence theology, politics, ethics, education, and aesthetics just as much as they have (and still do) metaphysics and logic. Nor were his contributions to the development of science negligible. It was Plato who posed to the astronomers of his day, such as EUDOXUS, the question: "By the assumption of what uniform and orderly motions can the apparent motions of the planets be accounted for?" The request that there should be but one explanation applying to the seemingly disparate observed motions of each planet did much to shape the development of Greek astronomy and to add to it a characteristic dimension of model building lacking, for example, in Babylonian astronomy.

He also, in the *Timaeus*, under the influence of the Pythagoreans of Sicily, introduced an alternative form of atomism to that of DEMOCRITUS. He began with the result of Theaetetus that there can be only five regular solids, the tetrahedron, cube, octahedron, dodecahedron, and the icosahedron, and went on to assign to each of the four elements of EMPEDOCLES a characteristic shape of one of the regular solids. The cube as the most stable is assigned to the least mobile element, earth; the pyramid is assigned to fire; the octahedron to air; and the icosahedron to water. To the remaining figure, the dodecahedron, most closely approaching the sphere, Plato associated the "spherical heaven."

The main significance of Plato's thought for science was thus to establish the vital tradition, originating with the Pythagoreans and finding ready echoes in the work of GALILEO and KEPLER, of the mathematical analysis of nature. It is said that an inscription over the vestibule of the Academy read: "Let no one enter here who is ignorant of Geometry."

Playfair, John (1748–1819) *Scottish mathematician and geologist*

> How different would geological literature be today if men had tried to think and write like Playfair.
> —Archibald Geikie, *The Founders of Geology* (1905)

Born at Benvie in Scotland, Playfair studied at St. Andrews University before becoming minister of Liff and Benvie in 1773. He was made a professor of mathematics at Edinburgh

University in 1785 and professor of natural philosophy in 1805.

Playfair was a friend of the geologist James HUTTON and in his *Illustrations of the Huttonian Theory of the Earth* (1802) he amplified and explained Hutton's uniformitarian ideas. Hutton's own work had been notoriously hard to follow and Playfair brought uniformitarianism to a considerably larger public. He also pioneered the idea that a river carves out its own valley.

Although he is better known as a geologist Playfair did make contributions of note to mathematics, in particular to geometry. In 1795 he published his *Elements of Geometry* in which he set out an alternative version of EUCLID's fifth postulate, which, given the truth of the other postulates, is equivalent to Euclid's original formulation. This postulate is consequently now known as "Playfair's axiom" and asserts that for any line (L) and point (P) not on L there is one and only one line, L', through P parallel to L.

Playfair, Lyon, 1st Baron (1818–1898)
British chemist and politician

Playfair was born at Chumar in India while his father was inspector-general of hospitals in the Bengal area. He traveled to England to be educated and attended St. Andrews University. In 1835 he began studying medicine at the Andersonian Institute, Glasgow, but abandoning medicine because of ill health he later took up the study of chemistry at University College, London, and with Justus von LIEBIG at Giessen. Playfair had an extremely varied career. As an academic he became professor of chemistry at the Royal Institute, Manchester, in 1842 and he later held the chemistry chair at Edinburgh University (1858–69). In 1868 he was elected a member of parliament for the universities of Edinburgh and St. Andrews and served until 1892. As a politician he was reasonably successful, serving as postmaster general in 1873 and as deputy speaker of the House of Commons (1880–83). He was raised to the British peerage in 1892, becoming 1st Baron Playfair of St. Andrews.

For a short period (1841–43) he was manager of a textile-printing factory in Clitheroe. His main work, however, was as a civil servant and a propagandist for science. In 1853 a new government department of science and art was created with Playfair as joint secretary. He was also a member of Prince Albert's circle and organized the Great Exhibition of 1851.

Playfair's own chemical work included a study of the atomic volume and specific gravity (relative density) of hydrated salts, conducted with James JOULE.

Pliny the Elder (23 AD–79 AD) *Roman philosopher*

Why it [quartz] is formed with hexagonal faces cannot be easily explained; and any explanation is complicated by the fact that, on the one hand, its terminal points are not symmetrical and that, on the other hand, its faces are so perfectly smooth that no craftsmanship could achieve the same effect.
—*Natural History*

PLINY (full name Gaius Plinius Secundus) was born at Novum Comum, now Como in Italy. His monumental *Natural History* (published in 77 AD), the only survival of his writings, incorporates a summary of the knowledge of his time, the greater part of it based on the works of earlier writers. In many respects Pliny's work is highly uncritical, including, for example, descriptions of monsters, which were in turn copied by subsequent writers; in others he may be said to have been before his time, recognizing, for example, the spherical shape of the Earth. One of the great merits of the *Natural History* is the general theme of wonder it generates about the natural world. The most learned man of his age, Pliny died in true scientific tradition, succumbing to gases while trying to get as close as possible to the great eruption of Vesuvius (79 AD), which destroyed Pompeii and Herculaneum.

Plücker, Julius (1801–1868) *German mathematician and physicist*

Born at Elberfeld in Germany, Plücker (**plooker**) studied in Heidelberg, Berlin, and Paris and became a professor of mathematics at the universities of Halle and Bonn. He became professor of physics at Bonn in 1847.

He did important and pioneering work in analytic geometry in which he suggested taking straight lines rather than points as the basic geometrical concept. This idea led him to the celebrated principle of duality, stating the equal validity of certain equivalent theorems. He used this in a geometrical context but it now has far wider applications.

In about 1847 Plücker turned from mathematics to physics, studying the discharge of electricity through gases. He was the first to find that cathode rays can be deflected by a magnetic field, thus indicating their electric charge.

Later Plücker returned to his mathematical interests and did further work on geometry.

Poggendorff, Johann Christian (1796–1877) *German physicist, biographer, and bibliographer*

Poggendorff (**poh**-gen-dorf) was born at Hamburg in Germany and, after leaving school, worked for a pharmaceutical chemist, devoting his leisure hours to the study of science. In 1820 he entered the University of Berlin to study physics. In 1824 he became editor of *Annalen der Physik und Chemie* (Annals of

Physics and Chemistry), a well-established periodical. He was appointed extraordinary professor at the University of Berlin in 1834 and was named a member of the Prussian Academy of Sciences in 1839.

Poggendorff was esteemed by his contemporaries for his work in the newly developing sciences of electricity and chemistry. He devised various measuring instruments, including a galvanometer (1821) and a magnetometer (1827), and carried out important experiments. With Leibig he introduced subscript figures into chemical formulas.

Poggendorff's lasting influence lies perhaps in his work as a biographer and bibliographer. As editor of *Annalen* he secured many excellent contributions, placed stress on articles with an experimental basis, presented translations of important foreign papers, and gave the journal a prominent position among scientific periodicals. He also wrote on the history of physics and lectured on that subject at the University of Berlin. In 1863 he produced the first two volumes of *Biographisch-Literarische Handworterbuch zur Geschichte der exakten Wissenschaften* (A Concise Biographical Literary Dictionary on the History of the Exact Sciences; now called "Poggendorff"); these included the dates and bibliographical references for over 8,000 researchers in the exact sciences of all periods and many countries up to 1858. This immensely important reference work is still published and now numbers over twenty volumes.

Pogson, Norman Robert (1829–1891)
British astronomer

Pogson, who was born at Nottingham in England, started his career in 1852 as an assistant at the Radcliffe Observatory in Oxford. While there he discovered four new asteroids: Amphitrite in 1854, Isis in 1856, and Ariadne and Hestia in 1857. He was to discover nine in all, including the first to be discovered on the continent of Asia and consequently called Asia (1891).

In 1860 he was appointed government astronomer at Madras. He remained in India for the rest of his life, conscious of the enormous amount of observational work that could be done there. He constructed star catalogs and a variable star atlas while there.

His most lasting achievement was the introduction of *Pogson's ratio*. It had been realized that the average first-magnitude star is about a hundred times brighter than stars of the sixth magnitude. He therefore proposed that this interval should be represented by five equal magnitudes, that is, one magnitude would equal $\sqrt[5]{100}$, which equals 2.512. This means that stars of increasing magnitude are roughly 2.5 times brighter. The system has survived in the form proposed by Pogson more than a century ago.

Poincaré, (Jules) Henri (1854–1912)
French mathematician and philosopher of science

> The mind uses its faculty for creating only when experience forces it to do so.
> —*Science and Hypothesis* (1905)

> Mathematical discoveries, small or great ... are never born of spontaneous generation. They always presuppose a soil seeded with preliminary knowledge and well prepared by labor, both conscious and subconscious.
> —Quoted by A. L. Mackay in *A Dictionary of Scientific Quotations* (1991)

Poincaré (pwan-ka-**ray**) was born at Nancy in eastern France and studied at the Ecole Polytechnique and the School of Mines. At first he had intended to become an engineer, but fortunately his mathematical interests prevailed and he took his doctorate in 1879 and then taught at the University of Caen. He was professor at the University of Paris from 1881 until his death.

As Poincaré is commonly referred to as the great universalist – the last mathematician to command the whole of the subject – an account of his work would have to cover the whole of mathematics. In pure mathematics he worked on probability theory, differential equations, the theory of numbers, and in his *Analysis situs* (1895; Site Analysis) virtually created the subject of topology. He was, however, hostile to the work on the foundations of mathematics carried out by Bertrand RUSSELL and Gottlob FREGE. The discovery of contradictions in their systems, disasters to Frege and Russell, was happily welcomed by Poincaré: "I see that their work is not as sterile as I supposed; it breeds contradictions."

He also deployed the powerful weapons of modern mathematics against a number of problems in mathematical physics and cosmology. In 1887 Oscar II of Sweden offered a prize of 2,000 krona for a solution to the question of whether or not the solar system is stable. Will the planets continue indefinitely in their present orbits? Or will some bodies move out of the system altogether, or collide catastrophically with each other? Poincaré published his answer in the monograph *Sur les trois corps et les equations de la dynamique* (1889; On the Three Bodies and Equations of Kinetics). The title refers to what is now known as the "three-body problem": given three point masses with known initial positions and velocities, to work out their positions and velocities at any future time. The three-body problem had resisted all previous attempts to find a general solution. Poincaré also failed to find an analytical general solution, but he was awarded the prize for making significant advances in the ways of finding approximate solutions. This work included the ideas behind what is now known as "chaos theory." Poincaré expanded this work into a three-volume treatise on celestial mechanics in the 1890s.

Poincaré also formulated a famous conjecture, which despite considerable effort and many false alarms remains unsolved. To a topologist an ordinary sphere is a two-dimensional manifold (a 2-sphere) – two-dimensional because, although it looks like a three-dimensional solid, only its surface is significant. A loop placed on its surface can be shrunk to a point, or, in the language of topology, the 2-sphere is "simply connected." This is seen as a defining property of a sphere. A torus, on the other hand, is not a sphere because not all loops placed upon it can be shrunk to points.

What about an n-sphere, the surface of an $n+1$-dimensional body? Poincaré's conjecture is that the n-sphere is the only simply connected manifold in higher dimensions, as the 2-sphere is the only simply connected 2-manifold. Stephen Smale proved in 1969 that the conjecture would hold for all dimensions $n > 4$, and in 1984 Michael Freedman added the case $n = 4$. Grigori PERELMAN proved the general case in 2002 and 2003.

Poincaré, in such later books as *Science and Hypothesis* (1905), developed a radical conventionalism. The high-level laws of science, he argued, are conventions, adopted for ease and simplicity and not for "truth." What would happen, he asked, if we found a very large triangle defined by light rays with angles unequal to 180°? As Euclidean geometry is so useful in countless other ways we would more likely sacrifice our physics to preserve our geometry and conclude that light rays do not travel in straight lines.

Poincaré's interest in electrodynamics led him to a point where he was close to discovering special relativity theory. He discovered much of the mathematical structure of the theory, such as the LORENTZ group, but did not appreciate its physical significance.

Poisson, Siméon-Denis (1781–1840) *French mathematician and mathematical physicist*

> At the very end of his [i.e., Poisson's] life, when it had become painful for him to speak, I saw him almost weep, ... for he had become convinced that our young teachers were concerned solely with obtaining a post and possessed no love for science at all.
> —Antoine Augustin Cournot, *Souvenirs* (1913; Recollections)

Poisson (pwah-**son**), born the son of a local government administrator at Pithiviers in France, studied at the Ecole Polytechnique, where his teachers included Pierre Simon LAPLACE and Joseph LAGRANGE. He himself later held various teaching posts at the Ecole. His important mathematical work was largely in mathematical physics and he also did a considerable amount of experimental work on heat and sound. In thermodynamics he played an important role in making the whole subject amenable to mathematical treatment by showing how to quantify heat precisely. He is also one of the principal founders of the mathematical theory of elasticity.

Poisson is possibly best known for his work on probability, and he was something of a pioneer in applying the techniques of mathematical probability to the social sciences, something that was extremely controversial at the time. The term "law of large numbers" was introduced by Poisson in his seminal work, *Recherches sur la probabilité des jugements* (1837; Researches on the Probability of Opinions), in which he put forward his discovery of the *Poisson distribution*. This is the distribution that is a special case of the binomial distribution obtained when the probability of success in a given trial is some constant divided by the number of trials.

Although chiefly an applied mathematician Poisson also made some significant contributions to pure mathematics, in particular to complex analysis. It was Poisson who first thought of integrating complex functions along a path in the complex plane.

Polanyi, John Charles (1929–   ) *Canadian chemist*

John Polanyi (pol-**yah**-nee) is the son of the distinguished physical chemist Michael Polanyi. Born in Berlin, Germany, he was educated at Manchester University and at Princeton, where he obtained his PhD in 1952. He moved soon after to Toronto University, being appointed professor of chemistry in 1962.

Beginning in the 1950s Polanyi has sought to throw light on the nature of chemical reactions. What actually happens, he asked, during the reaction $H + Cl_2 \rightarrow HCl + Cl$? The reaction was known to be strongly exothermic; it was not known, however, how this released energy was distributed in the various degrees of freedom of the reaction products. D. Herschbach had begun detailed investigations of reaction mechanics by measuring the velocities and angular distribution of the reaction products using molecular beams. In contrast Polanyi described his own method as one in which "the molecules formed in chemical reaction do the work by signaling to us their state of excitation ... through infrared emission."

Initially Polanyi and his coworkers had to work with a detector "only slightly more sensitive than the palms of our hands ... a thermocouple." They were soon able to replace this with semiconductor infrared detectors. By analyzing the infrared emission, Polanyi was able to measure how much of the reaction energy was stored as molecular vibration and rotation. In this way he was able to show that in the example cited above two distinct states of the molecule HCl were formed: one with high vibrational and rotational excitation, but low translational energy; and the less common state with low vibrational and rotational energy but high translational energy. Polanyi continued to work in the field of reaction dynamics and developed many new techniques and derived numerous insights into the subject. For his contributions he shared the 1986 Nobel Prize for chemistry with Herschbach and Y. T. LEE.

Polanyi's work in infrared chemical luminescence led to the development of chemical lasers by G. Pimental and J. Kaspar in 1960.

Polanyi, Michael (1891–1976) *Hungarian–British physical chemist and philosopher*

> The pursuit of science can be organized ... in no other manner than by granting complete independence to all mature scientists. They will then distribute themselves over the whole field of possible discoveries, each applying his own special ability to the task that appears most profitable to him. The function of public authorities is not to plan research, but only to provide opportunities for its pursuit.
> —*The Logic of Liberty* (1951)

Polanyi, the son of a civil engineer, was educated at the university in his native Budapest, where he qualified in medicine in 1913, and at the Institute of Technology, Karlsruhe. He remained in Germany, working at the Institute for Fiber Chemistry from 1920 to 1923 and in Fritz HABER's institute. In 1933, with the rise of the Nazis, he moved to Manchester University, England, as professor of physical chemistry. From 1948 until his retirement in 1958 Polanyi, whose interests had become more philosophical, occupied a personal chair of social studies at Manchester.

His earliest success came from his work as an x-ray crystallographer on the structure of natural fibers in 1921. He also, in 1923, made the first use of the rotating-crystal apparatus to determine crystal structure. With Haber he worked extensively on the kinetics of chemical reactions, beginning there an important collaboration with Henry EYRING. In 1931 they constructed the first potential energy surface.

Later he became interested in the philosophy of science, exploring in his widely read book *Personal Knowledge* (1958) just how science can reconcile its claim to objective knowledge with the intensely personal and fallible manner in which it operates.

Politzer, H. David (1949–) *American theoretical physicist*

Politzer was born in the United States and ed-

ucated at Harvard University, where he obtained his PhD in physics in 1974. He subsequently moved to the Department of Physics, California Institute of Technology (Cal Tech). In 1973 he discovered asymptotic freedom, independently of David GROSS and Frank WILCZEK. In 2004 Politzer, Gross, and Wilczek were awarded the Nobel Prize for physics for this discovery.

Poncelet, Jean-Victor (1788–1867) *French mathematician*

Poncelet (pawn-se-**lay**) was born at Metz in eastern France. A military engineer in Napoleon's Russian campaign, he was taken prisoner in 1812 during the retreat from Moscow, having been left for dead by the army. During his two years in Russia as a prisoner of war he set about reconstructing as much as he could remember of the mathematics he had learned as a student. He was able to go beyond merely reconstructing what he had been taught, to do original work.

Poncelet's important contribution to mathematics was in the field of projective geometry. He was one of the first to formulate and make extensive use of the principle of duality – the equivalence of certain related geometric concepts and theorems. He was also the first mathematician to introduce imaginary points into projective geometry.

Pond, John (1767–1836) *British astronomer*

Pond, the son of a London businessman, studied at Cambridge University until ill health forced him to leave. In 1798 he began a series of astronomical observations at Westbury in Wiltshire. These and earlier observations showed him that the instruments at the Royal Observatory at Greenwich had become defective. In 1811, on the recommendation of Nevil MASKELYNE, he was appointed Astronomer Royal, an office he continued to hold until his resignation in 1835.

Pond is remembered for his major reform of the Royal Observatory, in terms of new instruments, new methods of observation, and a larger staff. He installed a six-foot mural circle in 1812 and a new transit instrument in 1816, both built by Edward Troughton. With these he compiled an accurate catalog of 1,113 stars, published by him in 1833. At a less fundamental level Pond introduced in the same year the Greenwich custom of "timeballs" by which, "every day from the top of a pole from the Easter Turret of the Royal Observatory at Greenwich, at the moment of one o'clock P.M." a ball was, and still is, dropped. This has been described as the world's "first public time signal."

Pond's term of office was far from harmonious. He was forced to suffer a series of attacks from Stephen Lee, assistant secretary of the Royal Society, from 1825, being accused of publishing contradictory and inaccurate data and of making insufficient observations. Although Pond was acquitted of "culpable" negligence, two of his assistants were reprimanded by a Royal Society Committee set up to investigate the observatory.

Ponnamperuma, Cyril Andrew (1923–1994) *Singhalese–American chemist*

Ponnamperuma (pon-am-pe-**roo**-ma), who was born at Galle (now in Sri Lanka), was educated at universities spanning three continents – namely Madras in India, Birkbeck College, London, and Berkeley in California where, in 1962, he obtained his PhD. After working in the exobiology division of NASA from 1963 he was appointed, in 1971, professor of chemistry and director of the Laboratory of Chemical Evolution at the University of Maryland.

In 1952 Stanley MILLER illuminated discussion on the origin of life by showing that simple amino acids could be produced in a laboratory test tube under minimal conditions that might well have existed in the distant past. After Miller chemists were eager to see what other organic molecules could be produced in a variety of supposed primitive terrestrial environments.

In this field Ponnamperuma had a number of significant successes. In 1963, with Ruth Mariner and Carl SAGAN, he irradiated adenine, ribose sugar, a nucleic acid, and a phosphate source and detected the formation of ATP (adenosine triphosphate), the basic cellular energy source. Ponnamperuma also managed, with Mariner and Melvin CALVIN, to synthesize adenine by exposing a mixture of methane, ammonia, and water to beta particles and, in 1965, he derived various sugars by exposing formaldehyde to ultraviolet light. Such experimental results led Ponnamperuma to conclude, in 1972, that the basic units that constitute nucleic acids and proteins could have been synthesized by the various forms of energy in the early atmosphere.

However, this is but the first of the three necessary stages – construction of molecules from atoms, formation of molecules into polymers, and formation of polymers into organisms. Some simple dipeptides have been made, but beyond this there are major difficulties. Ponnamperuma's views are expressed fully in his *The Origins of Life* (1972).

Pons, Jean Louis (1761–1831) *French astronomer*

Born at Peyres in France, Pons (pawns) started his astronomical work in Marseilles before moving to Florence as director of the observatory in

1819. He was an assiduous comet hunter, discovering 37. He is known as the codiscoverer of the *Pons–Winnecke* and the *Pons–Brooks* comets. True fame seems to have missed him, however, for though he did in fact discover the comet with the shortest period of 3.3 years it was named for ENCKE, who first worked out its orbital period.

Pons, Stanley (1943–) *American chemist*

Born the son of a textile-mill owner in Valdek, North Carolina, Pons was educated at Wake Forest University, North Carolina, and at Ann Arbor, Michigan. Instead of completing his PhD he was induced in 1966 to enter the family business. His interest in chemistry remained too strong and, consequently, he decided in 1975 to return to graduate school. He opted for the University of Southampton in England where he worked under Martin Fleischman, completing his doctorate in 1978. After holding junior posts at Oakland University, Michigan, and Alberta University, Canada, Pons was appointed in 1980 to a professorship in chemistry at the University of Utah, Salt Lake City.

In March 1989 Pons and his former supervisor Fleischman announced that they had achieved fusion of deuterium nuclei in a test tube at room temperature by using an electrolytic process. Their announcement made headline news around the world. However, subsequent work at other laboratories failed to reproduce the results, although Pons and Fleischman continued to insist on the validity of their claim.

Pontecorvo, Guido (1907–1999) *Italian–British geneticist*

Pontecorvo (pon-tay-**kor**-voh) was born in Pisa, Italy. Having graduated in agricultural sciences from the university in his native city in 1928, he spent the following nine years in Florence, supervising the Tuscany cattle-breeding program. Political conditions caused him to leave Italy in 1938, his intention being to continue with similar work in Peru. However, he first accepted an invitation to the Institute of Animal Genetics in Edinburgh, where he met the famous American geneticist Hermann MULLER, another visitor at the institute. Under Muller's influence Pontecorvo became increasingly interested in pure genetics and together with Muller he devised an elegant method for investigating the genetic differences between species that produce sterile hybrids on crossing.

Pontecorvo remained in Edinburgh working on fruit flies (*Drosophila* species) and gained his PhD in 1941. Two factors then prompted him to change from *Drosophila* genetics to fungal genetics, firstly the dire need for penicillin during World War II and secondly his interest in the structure and function of the gene, a topic more easily investigated in the fungi.

Pontecorvo's work on the fungus *Aspergillus nidulans* led to the discovery, with Joseph Roper, of the parasexual cycle in 1950. This cycle gives rise to genetic reassortment by means other than sexual reproduction and its discovery provided a method of genetically analyzing asexual fungi. Pontecorvo also put forward the idea of the gene as a unit of function, a theory substantiated by Seymour BENZER in 1955. Pontecorvo occupied the first chair of genetics at Glasgow University from 1955 until 1961, when he moved to the Imperial Cancer Research Fund. He retired in 1975.

Pope, Sir William Jackson (1870–1939) *British chemist*

Pope, born the son of a city merchant in London, was educated at Finsbury Technical College and the Central Institution, London (later to become part of Imperial College). After acting as an assistant to Henry ARMSTRONG, he served as head of the chemistry department at Goldsmith's College, London, from 1897 to 1901. He then occupied the chair of chemistry at Manchester until 1908 when he moved to a similar chair at Cambridge.

Pope's main work was in the field of stereochemistry and the optical activity of chemical compounds. Jacobus VAN'T HOFF had shown in 1874 that the tetrahedral structure of the carbon atom would account for the molecular asymmetry of certain carbon compounds. Between 1898 and 1902 Pope was able to show that there were compounds where the asymmetry did not occur at carbon atoms; he showed, for example, that nitrogen, sulfur, and selenium could also act as asymmetric centers. Pope further demonstrated that compounds containing no asymmetric atoms could still be optically active because of overall asymmetry of the molecule.

During World War I Pope concentrated on developing war gases, particularly mustard gas. He was knighted in 1919.

Pople, John A. (1925–2004) *British theoretical chemist*

Pople was born in Burnham-on-Sea, Somerset, and gained his PhD in mathematics in 1951 at Cambridge. In 1964 he became professor of chemical physics at Carnegie-Mellon University, Pittsburgh, and in 1986 moved to Northwestern University as professor of chemistry. In 1998 he shared the Nobel Prize for chemistry with Walter KOHN. Pople received his award for the development of computational methods in quantum chemistry.

Pople developed the whole quantum-chemical methodology currently used in various areas of chemistry. This makes it possible to study the configurations and properties of molecules and how they interact, using the basic principles of quantum mechanics. Details of a molecule or a reaction are fed to a computer, which outputs a list of properties of the molecule or describes how the reaction will proceed. Pople's main contribution to the process was the development in 1970 of a user-friendly program known as GAUSSIAN-70, now used by theoretical research chemists throughout the world. He continued to refine the methodology, accumulating a well-documented model chemistry, which by the early 1990s was able to incorporate Kohn's density-function theory.

Popov, Aleksandr Stepanovich (1859–1906) *Russian physicist and electrical engineer*

Popov (po-**pof**), the son of a priest from Bogoslavsky in Russia, was educated in a seminary to prepare him for a clerical profession. His interest however turned to physics and mathematics, which he studied at the University of St. Petersburg between 1877 and 1882. While still a student he worked in 1881 at the Electrotekhinik artel, which operated Russia's first electric power stations. He taught for a short time at the university and then in 1883 joined the staff of the Torpedo School at Kronstadt, where naval specialists were trained in all branches of electrical engineering and where he was able to conduct his own research. He subsequently became head of the physics department and remained there until 1900. Popov returned to St. Petersburg as professor at the Institute of Electrical Engineering in 1901 and was later appointed its director in 1905. Later that year Popov's health was undermined by his refusal to take severe action against the political disturbances among his students and he died shortly after.

In 1888 Heinrich HERTZ had produced and transmitted electromagnetic waves, arousing the interest of many scientists. Popov began experiments on the transmission and reception of the so-called Hertzian waves (radio waves) somewhat earlier than MARCONI. He modified the coherer developed by Oliver LODGE for detecting these waves, making the first continuously operating detector. Connecting his coherer to a wire antenna, he was able in 1895 to receive and detect the waves produced by an oscillator circuit. His interests at this time, however, seemed more toward the investigation of atmospheric phenomena such as thunderstorms and lightning; he used his coherer connected to lightning conductors for this purpose. Stimulated by the 1896 patent awarded to Marconi, Popov again turned his attention to radio transmission and enlisted the help of the Russian navy. In 1897 he was able to transmit from ship to shore over a distance of 3 miles (5 km) and managed to persuade the naval authorities to begin installing radio equipment in its vessels. By the end of 1899 he had increased the distance of his ship to shore transmissions to 30 miles (48 km). He received little encouragement or support from the Russian government and did not commercialize his discoveries.

The Russian claim that Popov invented radio communication is not widely accepted, although he did publish in January 1896 a description of his receiving apparatus that coincides very closely with that described in Marconi's patent claim of June 1896. Popov is credited however with being the first to use an antenna in the transmission and reception of radio waves.

Popper, Sir Karl Raimund (1902–1994)
Austrian–British philosopher

> Science may be described as the art of systematic oversimplification.
> —*The Observer* (August 1982)

> Our knowledge can only be finite, while our ignorance must necessarily be infinite.
> —*Conjectures and Refutations* (1963)

Popper was born in the Austrian capital Vienna, where his father was a lawyer with an interest in literature and philosophy. After obtaining his PhD from the University of Vienna in 1928, he taught in a secondary school for some years and then lectured at various universities in England in 1935 and 1936. In 1937 he was appointed to a lectureship in philosophy at the University of New Zealand, Christchurch. After the war Popper returned to England and joined the London School of Economics, where he served as professor of logic and scientific method from 1949 until his retirement in 1969.

Popper's view of science, first fully formulated in his *Logik der Forschung* (1934; The Logic of Scientific Discovery, 1959), has found considerable support among working scientists and rather less from philosophers of science. The logic of science is not an inductive one, Popper claimed. Science does not begin by attempting to formulate laws and theories on the basis of carefully collected observations. Science, in fact, begins not with observations but with problems.

The problems are dealt with by constructing theories, laws, or hypotheses, which for Popper function as conjectures or guesses. But there may well be a multitude of conjectures, all proposing plausible solutions. Here Popper insists that the aim of science is not to attempt to select from the competing hypotheses and theories the one that is true, for such constructs can never be shown to be true. No matter how many observations confirm a theory, it is not possible to say that it is correct. We can, however, frequently show that theories are undoubtedly false.

Popper used this insight to insist that the basic procedure of science consists of strenuously attempting to falsify such conjectures and accepting those that have survived the most severe attempts at falsification. This acceptance does not confer truth on the conjecture. That a hypothesis has so far resisted attempts to falsify it is no guarantee that it will continue to pass future tests.

With such an intellectual framework Popper could easily solve the demarcation problem. Scientific theories can conceive of and describe facts that could falsify them, while what he termed the "pseudo sciences," such as Marxism and psychoanalysis, are able to interpret any event within their theory.

This, for Popper, was a point of more than academic significance. In his *The Poverty of Historicism* (1944–45) and *The Open Society and its Enemies* (1945) he argued strongly against inexorable laws of historical destiny. "There can be no scientific theory of historical development serving as a basis for historical prediction," he concluded.

In Popper's later work, as seen in his *Objective Knowledge* (1972), an attempt was made to develop an evolutionary account of the growth of knowledge. He also introduced his notion of "three worlds," arguing that in addition to the familiar worlds of physical and mental states, there exists also a "third world" populated by "theories ... their logical relations ... arguments ... and problems."

Porter, George, Baron (1920–2002)
British chemist

> Should we force science down the throats of those that have no taste for it? Is it our duty to drag them kicking and screaming into the 21st century? I am afraid that it is.
> —Speech, September 1986

Born at Stainforth in Yorkshire, Porter was educated at the universities of Leeds and Cambridge, where he obtained his PhD. After working on radar during World War II, he returned to Cambridge until, in 1955, he was appointed professor of chemistry at Sheffield University. From 1966 until 1985 he held leading positions in British science, namely, the directorship of the Royal Institution and the Fullerian Professorship of Chemistry. In 1987 he was appointed professor (from 1990, chairman) of the Centre for Photomolecular Science at Imperial College, London.

In collaboration with his Cambridge teacher, Ronald NORRISH, Porter developed from 1949 onward the new technique of flash photolysis. There were good reasons for thinking that the course of a chemical reaction was partly determined by a number of intermediate species too short-lived to be detected, let alone investigated. Porter therefore set out to study what he called the spectroscopy of transient substances.

The apparatus used involved a long glass or

quartz tube containing the gas under investigation. This was subjected to a very brief pulse of intense light from flash tubes, causing photochemical reactions in the gas. The free radicals and excited molecules produced have only a transient existence, but could be detected by a second flash of light, directed along the axis of the reaction tube, used to record photographically an absorption spectrum of the reaction mixture. In this way the spectra of many free radicals could be detected.

In addition, it was possible to direct a continuous beam of light down the reaction tube and focus on one particular absorption line of a species known to be present. The change of this line with time allowed kinetic measurements of the rates of very fast gas reactions to be made.

The methods of flash photolysis have since been extended to liquids and solutions, to gas kinetics, and to the study of complex biological molecules such as hemoglobin and chlorophyll. Porter shared the Nobel Prize for chemistry in 1967 with Norrish and with Manfred EIGEN for "their studies of extremely fast reactions effected by disturbing the equilibrium by means of very short pulses of energy."

In 1990 Porter was raised to the British peerage as Baron Porter of Luddenham.

Porter, Keith Roberts (1912–1977) *Canadian biologist*

Porter, who was born at Yarmouth in Nova Scotia, Canada, studied biology at Acadia University and Harvard, receiving his PhD in 1938. After working at the Rockefeller Institute (1939–61) he held chairs of biology, first at Harvard (1961–70) and thereafter at the University of Colorado.

While working with Albert Claude at the Rockefeller Institute, Porter studied the endoplasmic reticulum, a network of membranes within cells. More significant was his study of its equivalent form in muscle fibers, the sarcoplasmic reticulum. Although this had first been discussed by E. Veratti in 1902, it required the development of the electron microscope to permit Porter to describe, in the 1950s, its pervasive character as a network of extremely fine channels enclosing each myofibril. He went on to propose that it served to coordinate and harmonize the complex response of the contractions of millions of fibers.

The actual mechanism of contraction was initiated by the release of calcium ions into the fluid surrounding the muscle fibers. The source of such ions was shown to be the sarcoplasmic reticulum, to which they were quickly returned and stored by what became known as a "calcium pump."

Porter, Rodney Robert (1917–1985) *British biochemist*

Porter was educated at the university in his native city of Liverpool, England, and at Cambridge University where he was a pupil of Frederick SANGER. After working at the National Institute of Medical Research from 1949 to 1960 he moved to the chair of immunology at St. Mary's Hospital, London. Porter remained there until 1967 when he became professor of biochemistry at Oxford.

In 1962 Porter proposed a structure for the important antibody gamma globulin (IgG). Ordinary techniques of protein chemistry revealed that the molecule is built up of four polypeptide chains paired so that the molecule consists of two identical halves, each consisting of one long (or heavy) chain and one short (or light) chain.

Further evidence was obtained by splitting the molecule with the enzyme papain. This split IgG into three large fragments, two similar to each other known as F_{ab} (fragment antigen binding) and still capable of combining with antigen, and a crystalline fragment known as F_c (fragment crystalline) without any activity. This immediately suggested to Porter that, because crystals only form easily from identical molecules, the halves of the heavy chain that make up the F_c fragment are probably the same in all molecules. Thus the complexity is mainly in the F_c fragments where the combining sites are found.

Linking such insights with results obtained by Gerald EDELMAN and data derived from electron microscopy allowed Porter to propose the familiar Y-shaped molecule built from four chains joined by disulfide bridges with the variable combining part at the tips of the arms of the Y.

In 1972 Porter shared with Edelman the Nobel Prize for physiology or medicine for their work in determining the structure of an antibody.

Poseidonius (c. 135 BC–c. 51 BC) *Greek philosopher, historian, and astronomer*

Poseidonius (pos-I-**doh**-nee-us) was born at Apameia in Syria and studied under the Greek Stoic philosopher Panaetius in Athens. Sometime after 100 BC he became head of the Stoic school of philosophy at Rhodes. Although none of his works have survived it is known that he completed a history of the world. He taught Cicero, who appears to have used some of his ideas in his *Dream of Scipio*.

As an astronomer Poseidonius made an ingenious attempt to measure the circumference of the Earth. Assuming Rhodes and Alexandria to be on the same meridian he noted that while the star Canopus just touches the horizon at Rhodes, it has a meridian altitude of 7 degrees 30 minutes or 1/48th of the circumference of a circle at Alexandria. As the distance between the two was 5,000 stadia, he concluded that the Earth's circumference was 48 × 5,000, which equals 240,000 stadia or, assuming 8.75 stadia to the mile (this value is uncertain), about 27,000 miles. The result was reasonably accurate and in broad agreement with that established by ERATOSTHENES by a different method. At some stage, however, Poseidonius reduced the figure to the much too low 180,000 stadia, presumably the result of revising the Rhodes–Alexandria distance. It was this figure, transmitted in the *Geography* of PTOLEMY, that Columbus used more than 1,000 years later, when considering the feasibility of reaching Asia by sailing westward.

Poseidonius made less successful attempts to measure the size of the Sun and Moon. He also constructed a revolving celestial sphere exhibiting the daily motions of the Sun, Moon, and planets.

Postel, Jonathan (1943–1998) *American computer scientist*

Postel worked on the ARPANET, a forerunner of the Internet. He was particularly concerned with computer communication protocols and applications of multi-machine internets. His main contribution to the development of the Internet was establishing and directing the Internet Assigned Numbers Authority (IANA), a nonprofit-making company that assigns unique names and numbering for names and addresses on the Internet. The work of Postel was a key ingredient in the rapid growth of the Internet.

Postel was the director of the Computer Networks division, the Information Services Institute, the University of Southern California. He died following heart surgery.

Poulton, Sir Edward Bagnall (1856–1943) *British zoologist*

Poulton was born at Reading in southern England, the son of an architect. He spent his entire career at the University of Oxford, first as a student at Jesus College and later (1893–1933) as Hope Professor of Zoology.

Influenced by the writings of Alfred Russel WALLACE, Poulton made an intensive and detailed study of the adaptive importance of protective coloring and mimicry in nature. His results, published in *The Colours of Animals* (1890), were strictly Darwinian with an emphasis on the inheritance of factors arising from the continuous variation found in a population.

Poulton maintained this belief despite the rediscovery of the works of Gregor MENDEL and the consequent insistence on the role of mutation in evolution. In such works as *Essays on Evolution* (1908) Poulton argued against the growing Mendelian orthodoxy, insisting that adaptations as complex as mimicry could not have been brought about by mutation.

Powell, Cecil Frank (1903–1969) *British physicist*

The son of a gunsmith, Powell was born at Tonbridge, Kent, and educated at Cambridge University, where he obtained his PhD in 1927. He spent virtually his entire career at Bristol University where he became Wills Professor of Physics in 1948 and director of the Wills Physics Laboratory in 1964.

Under Powell Bristol became a leading center for the study of nuclear particles by means of photographic emulsions. In this technique an ionizing particle crossing a sensitive plate coated with grains of silver bromide leaves clear tracks of its passage. From the size and path of the track much information about the nature of the particle can be inferred. In 1947 Powell, in collaboration with Giuseppe OCCHIALINI, published a standard work on the subject, *Nuclear Physics in Photographs*. It was this technique that allowed Powell to discover the pi-meson (or pion) in 1947 in the plates of cosmic rays. The existence of such a particle had been predicted in 1935 by Hideki YUKAWA, and Powell's discovery thus went some way to establish a coherent picture of nuclear phenomena.

For his discovery Powell was awarded the 1950 Nobel Prize for physics.

Poynting, John Henry (1852–1914) *British physicist*

Poynting was born at Monton, near Manchester in England, and educated at the universities of Manchester and Cambridge (1872–76). He served as professor of physics at Mason Science College (later the University of Birmingham) from 1880 until his death in 1914.

He wrote on electrical phenomena and radiation and is best known for *Poynting's vector*, introduced in his paper *On the Transfer of Energy in the Electromagnetic Field* (1884). In this he showed that the flow of energy at a point can be expressed by a simple formula in terms of the electric and magnetic forces at that point.

In 1891 he determined the mean density of the Earth and made a determination of the gravitational constant in 1893 through the accurate use of torsion balances. His results were published in *The Mean Density of the Earth* (1894) and *The Earth; Its Shape, Size, Weight and Spin* (1913). Poynting was also the first to suggest, in 1903, the existence of the effect of radiation from the Sun that causes smaller particles in orbit about the Sun to spiral close and eventually plunge in. This was developed by the American physicist Howard Robertson and was related to the theory of relativity in 1937, becoming known as the *Poynting–Robertson effect*.

Pratt, John Henry (1809–1871) *British geophysicist*

Pratt was the son of a secretary to the Church Missionary Society; after graduating from Cambridge University in 1833 he went to India as a chaplain with the East India Company. In 1850 he became archdeacon of Calcutta.

An amateur scientist, Pratt became interested in geophysics and his most important work was published when he formulated the theory of isostasy in 1854. While conducting his triangulation of India the surveyor, George Everest, found a discrepancy in the astrogeodetic and triangulation measurements between two stations – Kaliana and Kalianpur – near the Himalayas. From this Pratt surmised that mountain ranges failed to exert the gravitational pull expected of them and thus distorted measurements made with pendulums. He saw the Himalayas as having a lesser density than the crust below, and generalized that the higher the mountain range, the lower is its density. He compared the raising of mountains to fermenting dough in which the density decreases as the dough rises.

Some of the same ideas were present in a paper submitted just six weeks after Pratt's by George AIRY although Airy preferred the image of an iceberg to that of rising dough.

Pratt wrote *Mathematical Principles of Mechanical Philosophy* (1836), a work that was expanded to *On Attractions, Laplace's Functions, and The Figure of the Earth* (1860, 1861, 1865).

Prausnitz, John Michael (1928–) German–American chemist

Prausnitz (**prows**-nits) emigrated to America from his native Berlin in 1937 and became naturalized in 1944. He was educated at the universities of Cornell, Rochester, and Princeton, where he obtained his PhD in 1955. He then joined the Berkeley faculty of the University of California, where he has served since 1963 as professor of chemical engineering. Now emeritus professor, he was awarded the National Medal of Science in 2005.

Prausnitz has worked mainly on the extension of physical chemistry into applications in chemical engineering process design. To this end he has worked on phase equilibria and on molecular thermodynamics. His views are discussed in his book *Molecular Thermodynamics of Fluid Phase Equilibria* (1969).

Praxagoras of Cos (about 350 BC) Greek physician

Little is known of the life of Praxagoras (prak-**sag**-o-ras), except that he was born on the Greek island of Cos, and not much more of his writings other than their titles as recorded by GALEN. He was the teacher of HEROPHILUS and wrote works on *Physics, Anatomy, Diseases,* and *Symptoms.*

He developed the humoral theory of HIPPOCRATES, apparently proposing that the four traditional humors be supplemented with the sweet, acid, salty, and bitter humors among several others. As the bulk of them are suggestive of tastes they could well have been derived from consideration of diet.

Praxagoras is also the first reported to have restricted the pulse to a specific set of vessels, the arteries, and to use it in diagnosis. However he remained committed to the traditional Greek view that the arteries carried not blood but "pneuma" or air. This came from venous blood via the heart and the lungs, continued along the arteries until they joined the neura (meaning for Praxagoras tendons rather than nerves), and ultimately produced muscular action. It was left to his pupil Herophilus to appreciate more fully the nature of the nerves.

Pregl, Fritz (1869–1930) Austrian chemist

Pregl (**pray**-gel), who was born at Laibach (now Ljubljana in Slovenia), was the son of a bank official. He graduated in medicine from Graz (1893) where he became an assistant in physiological chemistry in 1899. In 1910 he became head of the chemistry department at Innsbruck, remaining there until 1913 when he returned to Graz to become director of the Medico-Chemical Institute.

Pregl began research on bile acids in about 1904 but soon found that he could only obtain tiny amounts. This led him to pioneer techniques of microanalysis. Justus von LIEBIG had needed about 1 gram of a substance before he could make an accurate analysis; through his new techniques Pregl was capable of working with 2.5 milligrams. This was achieved by the careful scaling down of his analytic equipment and the design of a new balance, which was produced in collaboration with the instrument maker W. Kuhlmann of Hamburg. With this he was capable of weighing 20 grams to an accuracy of 0.001 milligram.

The techniques developed by Pregl are of immense importance in organic chemistry and he was awarded the Nobel Prize for chemistry in 1923 for this work.

Prelog, Vladimir (1906–1998) Swiss chemist

Born in Sarajevo (now in Bosnia-Hercegovina), Prelog (**prel**-og) studied chemistry at the Prague Institute of Technology where he received his doctorate in 1929. He then worked in Prague as an industrial chemist until 1935 when he moved to the University of Zagreb. With the German invasion of Yugoslavia in 1941 Prelog joined the staff of the Federal Institute of Technology in Zurich, serving there as professor of chemistry from 1950 until his retirement in 1976.

Prelog's early work was with the alkaloids. His research resulted in the solution of the configuration of *Cinchona* alkaloids (antimalarial compounds), the correction of the formulas for *Strychnos* alkaloids, and the elucidation of many other indole, steroid, and aromatic alkaloid configurations. He later investigated the metabolites of certain microorganisms and in so doing discovered many new natural substances including the first natural compound found to contain boron, boromycin.

Prelog intensively studied the relationship between conformation and chemical activity in medium-sized (8–11 ring members) ring structures. This brought to light a new type of reaction that can occur in such compounds. Prelog next showed that conformation affects the outcome of syntheses where different-sized atoms or groups are being substituted into a compound. The regular way in which this occurs allowed the configurations of many important compounds to be worked out. Applying such work to the reactions between enzymes, coenzymes, and substrates gave interesting results about the stereospecificity of microorganisms.

With Christopher INGOLD, Prelog introduced the so-called R–S system into organic chemistry, which allowed, for the first time, enantiomers, or mirror images, to be described unambiguously.

For such wide-ranging work on the "stereochemistry of organic molecules and reactions" Prelog was awarded the 1975 Nobel Prize for chemistry, which he shared with John CORNFORTH.

Prestwich, Sir Joseph (1812–1896) *British geologist*

Prestwich, the son of a London wine merchant, was educated at University College, London. He then spent 40 years in the family wine business before accepting, after his retirement from business in 1874, the chair of geology at Oxford University, a post he held until his final retirement in 1888.

In his spare time Prestwich produced six books and well over 100 papers. He worked on the Quaternary Period of England, Belgium, and France and the Tertiary Period of southeast England, where he established the stratigraphy of the clays of the London basin. These results and other technical work were fully described in his *Geology – Chemical, Physical and Stratigraphical* (2 vols. 1886–88).

In 1859 Prestwich played a more public role than that of technical geologist when he was persuaded to visit BOUCHER DE PERTHES's site at Abbeville and to report on the authenticity of the antiquities found there. He concluded that the flint implements were the work of man and that they were associated with the remains of unknown animals. By doing so he was committing himself to the then revolutionary belief in the antiquity of man. Prestwich was knighted in 1896.

Prévost, Pierre (1751–1839) *Swiss physicist*

Born in Geneva, Switzerland, Prévost (pray-**voh**) was professor of physics at Berlin and then at the university in his native city. In 1792 he published his *Sur l'equilibre du feu* (On the Equilibrium of Heat), which did much to clarify the nature of heat.

If, as was widely believed at the time, heat was a fluid, called caloric, which flowed from hot bodies to colder ones, then it was reasonable to suppose that cold was also a fluid, "frigoric," which flowed from cold bodies to warmer ones. In favor of the existence of frigoric was a body of experimental work that dated back to the 17th century. Thus it was known that if a piece of ice was placed near a thermometer in a room of constant temperature then the temperature of the thermometer would fall. More impressively, if two concave mirrors are arranged so that they face each other and a piece of ice is placed at one focus and a thermometer at the other, then the indicated temperature will fall. Experiments like this readily lent themselves to the interpretation that the fluid frigoric can be emitted and reflected. Prévost argued in 1791 that there is but a single fluid involved. Snow melting in the hand was a case of heat flowing from the hand to the snow rather than conversely. He introduced the idea of dynamic equilibrium in which all bodies are radiating and absorbing heat. When one body is colder than another it absorbs more than it radiates. Its temperature will rise until it is in equilibrium with its surroundings. At this point, it does not stop radiating heat but absorbs just as much as it loses to remain in equilibrium. The idea is known as the *Prévost theory of exchanges*. Although Prévost was a supporter of the caloric theory of heat, his views influenced a later generation of physicists who introduced the kinetic theory of heat on a quantitative basis toward the end of the 19th century.

Priestley, Joseph (1733–1804) *British chemist and Presbyterian minister*

> It was ill policy in Leo the Tenth to patronize polite literature. He was cherishing an enemy in disguise. And the English [church] hierarchy ... has equal reason to tremble even at an air pump or an electrical machine.
> —*Experiments and Observations on Different Kinds of Air* (1775–86)

Priestley was the greatest British chemist of the 18th century and also one of the century's greatest men. Born in the English city of Leeds, his father was a cloth dresser and a Congregationalist. Priestley was educated at a nonconformist seminary, later becoming a minister at Needham Market in 1755. After a few years in

Nantwich in a similar post he went to teach at Warrington Academy in 1761.

On visits to London he met Benjamin FRANKLIN, who aided him in his *History of Electricity* (1767). He moved to Leeds in 1767 and, being near a brewery, "began to make experiments in the fixed air that was continually produced in it." It was around this time that he invented soda water with the ample supply of carbon dioxide ("fixed air") from the brewery. In 1772 he became Lord Shelburne's librarian, which involved only nominal duties and allowed him to do some of his most important work. He left in 1780 to become a minister in Birmingham, where he mixed with such members of the Lunar Society as Erasmus DARWIN, James WATT, Josiah Wedgwood, and Matthew BOULTON. As a dissenting radical he preached against the discrimination suffered by non-Anglicans and, in reply to Edmund Burke's *Reflections on the French Revolution*, wrote in favor of the principles of the French Revolution. This led a Birmingham mob to break into his home (1791) and burn it, destroying all his books, papers, and instruments. He moved to London for a short while but finding no security there moved to America in 1794 to join his sons who had emigrated there earlier. In Pennsylvania he continued with his scientific and theological work until his death.

Priestley attempted to understand the facts of combustion and respiration. His first insight came from the realization that, since even a small candle uses an enormous amount of pure air, there must be a provision in nature to replace it. After trying various techniques to purify the foul air left after combustion he eventually found that a sprig of mint would revive the air so it could support combustion once more (1771). In the next few years, using a variety of new techniques, he isolated various gases – nitrous oxide, hydrogen chloride, sulfur dioxide – and, in 1774, he produced oxygen. Using a powerful magnifying glass to focus the rays of the Sun, he heated oxides of mercury and lead confined in glass tubes over mercury. He found that they gave off large amounts of a gas in which a candle would burn with an enlarged flame. At first he identified the gas as nitrous oxide but found that, unlike that gas, it was barely soluble in water. He next thought it might simply be ordinary air but on putting mice in it he found that they lived longer than in a similar volume of normal air. Being an ardent believer in the phlogiston theory, he named this gas "dephlogisticated air." Antoine LAVOISIER realized the crucial value of this discovery in explaining combustion and named the gas "oxygine."

Priestley continued to discover more compounds. He determined the relative densities of various gases by weighing balloons filled with them and also investigated gaseous diffusion,

conductivity of heat in gas, and the effect of electrical discharge on gases at low pressure.

Prigogine, Ilya (1917–2003) *Belgian chemist*

Prigogine (pree-goh-**zheen**) was born in Moscow and educated at the Free University of Belgium where he served as professor of chemistry from 1947 to 1987. He was appointed director of the Statistical Mechanics and Thermodynamics Center of the University of Texas, Austin, in 1967.

In 1955 Prigogine produced a seminal and revolutionary work, *Thermodynamics of Irreversible Processes*. In this book he pointed out a serious limitation in classical thermodynamics of being restricted to reversible processes and equilibrium states. He argued that a true thermodynamic equilibrium is rarely attained; a more common state is met with in the cell, which continuously exchanges with its surroundings, or in the solar system with the steady flow of energy from the Sun preventing the atmosphere of the Earth from reaching thermodynamic equilibrium.

A beginning had been made by Lars ONSAGER to cover nonequilibrium states but this applied only to states not too far away from equilibrium. Prigogine, in a quite radical way, developed machinery to deal with states far from equilibrium. These he called "dissipative structures." He went on to suggest that, "On a broader scale, it is difficult to avoid the feeling that such instabilities related to dissipative processes should play an extensive role in biological processes." Such a possibility Prigogine began to explore in his *Membranes, Dissipative Structures and Evolution* (1975). Other works by Prigogine include *Order Out of Chaos* (1979) and *From Being to Becoming* (1980).

Prigogine was awarded the 1977 Nobel Prize for chemistry for his work on "nonequilibrium thermodynamics particularly his theory of dissipative structures."

Pringsheim, Ernst (1859–1917) *German physicist*

Pringsheim (**prings**-hIm) was educated at Berlin University, where he gained his PhD in 1882. He later took up an appointment at Breslau as professor of theoretical physics.

Pringsheim did important work on the distribution of radiation from a hot body, and with a colleague at Breslau, Otto LUMMER, showed that the formulas of Max PLANCK and Wilhelm WIEN, accounting for black-body radiation, led to certain inconsistencies. This observation was instrumental in the development by Planck of his quantum theory.

Pringsheim, Nathanael (1823–1894) *German botanist*

Pringsheim, who was born at Wziesko (now in Poland), studied medicine at the universities of Breslau and Leipzig. However, his interest turned to natural science when he moved to the University of Berlin; he gained his PhD in 1848 with a thesis on the growth and thickness of plant cell walls. In 1864 he was appointed professor of botany at the University of Jena but resigned the post in 1868 to conduct private research in a laboratory attached to his home in Berlin.

Pringsheim was one of the leaders in the botanical revival of the 19th century with his contribution to studies of cell development and life history, particularly in the algae and fungi. He was among the first to demonstrate sexual reproduction in algae and observe alternation of generations between the two sexually differentiated motile zoospores and the resting undifferentiated spore that results from their fusion. He further showed that sexual reproduction involves fusion of material of the two sex cells.

From studies (1873) on the complex morphological differentiation in a family of marine algae, the Sphacelariaceae, Pringsheim opposed the Darwinian theory of evolution by natural selection. Like the Swiss botanist, Karl NAEGELI, he believed the increase in structural complexity to be a spontaneous morphological phenomenon, conferring no survival value.

Pringsheim's studies of the origin of plant cells contributed evidence for the theory that cells are only produced by the division of existing cells. With Julius von SACHS, Pringsheim also described the plastids, organelles unique to plant cells. In later years he concentrated more on physiology than morphology but his contributions to this field were not acknowledged or developed by other workers.

He was founder of the *Jahrbücher für Wissenschaftliche Botanik* (1858; Annals of Scientific Botany) and the German Botanical Society (1882). He wrote memoirs on *Vaucheria* (1855), *Oedogonium* and *Coleochaete* (1856–58), *Hydrodictyon* (1861), and *Pandorina* (1869).

Pritchard, Charles (1808–1893) *British astronomer*

Pritchard, who was born at Alderbury in southern England, was the son of an unsuccessful manufacturer. With the help of family friends he was sent to school and to Cambridge University. He became a fellow of his college in 1832 and headmaster of Clapham Grammar School in 1834 where he remained until 1862 – his school became quite famous and was attended by the sons of John HERSCHEL, George AIRY, and Charles DARWIN. In 1870 he became Savilian Professor of Astronomy at Oxford where he built and equipped a new observatory. Although he had a small observatory at Clapham, Pritchard is most unusual for a sci-

entist in that most of his work was done in his seventies. In 1881 he introduced a new kind of photometer – the wedge photometer – which he used in estimating the brightness of stars. He observed and calculated the relative magnitude of 2,784 stars from the celestial pole to 10°S, publishing his results in 1885. He also pioneered attempts to determine stellar parallax by photography. In 1886 he exposed 200 plates of 61 Cygni, establishing a parallax of 0.438 seconds of arc. Between 1888 and 1892, when he was 84, he determined the parallax of a further 28 stars by photographic means.

Proclus (*c*. 410–485 AD) *Greek philosopher*

> It is well known that the man who first made public the theory of irrationals perished in a shipwreck in order that the inexpressible and unimaginable should ever remain veiled. And so the guilty man, who fortuitously touched on and revealed this aspect of living things, was taken to the place where he began and there is for ever beaten by the waves.
> —Commentary on Euclid's *Elements*

Proclus (**proh**-klus) was one of the last significant Greek philosophers. He was born at Constantinople (now Istanbul in Turkey) and studied in Athens under Plutarch and Syriacus. He later became head of the Academy, which PLATO had founded in Athens centuries earlier. Proclus was largely responsible for the wide dissemination of Neoplatonic thought throughout the Byzantine, Roman, and Islamic worlds. Neoplatonism stemmed from the work of Plotinus, a third-century Roman philosopher, founded on a modified system of Platonism with the addition of mysticism.

Proclus's ideas influenced Arabic thinkers and in the Middle Ages his works became the principal source of the then popular Neoplatonism. He wrote a number of commentaries on Plato's dialogues. He was a distinguished mathematician and wrote an important commentary on the first book of Euclid's *Elements*, which survived through the ages.

Proctor, Richard Anthony (1837–1888) *British–American astronomer*

Proctor, the son of a London lawyer, worked as a clerk in a bank before going to Cambridge University in 1856 to study law. In 1866 the failure of a bank he had interests in led him to make his living from writing and lecturing. He had become interested in mathematics and astronomy, and most of his books were straightforward accounts of recent advances in astronomy. He also made lecture tours through America and Australia and in 1881 he settled in America where he founded the scientific magazine *Knowledge*.

His most important work as an astronomer was with Mars. In 1867 he published a chart of Mars that introduced the current, if misleading, nomenclature of seas, oceans, islands, and straits. He worked out the rotational period of Mars very accurately as a little more than 24 hours 37 minutes. He also charted the stars in Friedrich ARGELANDER's catalog, and was one of the first to suggest that meteoric bombardment, and not volcanic activity, was the cause of lunar craters.

Prokhorov, Aleksandr Mikhaylovich
(1916–2002) *Russian physicist*

Prokhorov (**proh**-ko-rof) graduated in 1939 from the faculty of physics of the Leningrad State University, where he later became a doctor of physics, mathematics, and science (1946). During World War II he served in the Russian army.

Subsequently, working at the Physics Institute of the Soviet Academy of Sciences with Nikolai BASOV, Prokhorov performed fundamental work in microwave spectroscopy, which led to the development of the maser in 1955, and later the laser. Basov and Prokhorov, together with the American physicist Charles TOWNES, received the 1964 Nobel Prize for physics for their development of the maser principle, and were pioneers of the new science of quantum radio physics (now referred to by the broader term, quantum electronics).

Proudman, Joseph (1888–1975) *British mathematician and oceanographer*

> The quickest way to get a lot of things done is to do one thing at a time.
> —Maxim

Proudman, who was born at Unsworth, near Manchester in England, studied mathematics and physics at Liverpool University. After graduating in 1910 he continued his studies at Cambridge University. Subsequently Proudman returned to Liverpool (1913) in the capacity of lecturer in mathematics and remained associated with this university for the rest of his career. In 1919 he became the first professor of applied mathematics and in 1933 a chair in oceanography was created specifically for him. He remained at Liverpool, serving also as provice-chancellor (1940–46), until his retirement in 1954.

Proudman's abiding scientific interest was in the study of the tides and other aspects of the sea. He made fundamentally important contributions of both a practical and theoretical nature to oceanography. In 1916 Horace LAMB was asked to prepare a report on tidal research and Proudman contributed to this a paper on the harmonic analysis of tidal observations. In 1919 he founded an institute in Liverpool devoted to research on tidal phenomena, which, after several changes of name, eventually became known as the Bidston Laboratory of the Institute of Oceanographic Sciences. He worked closely on tidal phenomena with another member of the institute, Arthur DOODSON. Proudman's work pioneered the mathematical study of the tides, and only as a result of his work did a unified mathematical approach to oceanography become possible. His works included *Dynamical Oceanography* (1953).

Proust, Joseph Louis (1754–1826) *French chemist*

Proust (proost) was born the son of an apothecary at Angers in northwest France. He studied in Paris and became chief apothecary at the Saltpêtrière Hospital. In 1789 he went to Madrid to become director of the Royal Laboratory under the patronage of Charles IV. After the invasion of Spain by Napoleon, the fall of his patron, and the destruction of his laboratory by the invading army, he returned to France in 1808. He lived in poverty for some years before being awarded a pension by Louis XVIII.

In 1799 Proust formulated his law of definite proportions. He pointed out that copper carbonate must always be made from the same fixed proportions of copper, carbon, and oxygen. From this he generalized that all compounds contained elements in certain definite proportions. Proust's law was not immediately accepted by all chemists; in particular, his proposal led to a long and famous controversy with Claude-Louis BERTHOLLET who argued that elements could combine in a whole range of different proportions. It is now clear that Proust was talking about compounds whereas Berthollet was thinking of solutions or mixtures. Berthollet eventually admitted his error.

The strength of Proust's law was seen a few years later when John DALTON published his atomic theory. The law and the theory fitted exactly – Proust's definite proportions being in fact a definite number of atoms joining together to form molecules.

Prout, William (1785–1850) *British chemist and physiologist*

Prout was born at Horton in England and studied medicine at Edinburgh, graduating in 1811. He established himself as a physician in London and became a pioneer of physiological chemistry, in which he lectured. He wrote on the stomach and urinary diseases and on the chemistry of the blood, urine, and kidney stones. In 1818 he prepared urea for the first time and in 1824 he identified hydrochloric acid in stomach secretions. He was also one of the first to divide food components into the groups of fats, carbohydrates, and protein.

Prout's fame also rests on a paper he published anonymously in 1815, *On the Relation between the Specific Gravities of Bodies in Their Gaseous State and the Weight of Their Atoms*. In this he formulated what has since been called

Prout's hypothesis: the atomic weight of all atoms is an exact multiple of the atomic weight of hydrogen. Determination of atomic weights had made this view plausible. At the time there was considerable interest in the hypothesis as it implied that elements were themselves "compounds" of hydrogen, and Prout suggested that hydrogen was the *prima materia* (basic substance) of the ancients. However, more accurate determinations of atomic weight, particularly by Jean STAS, showed that many were not whole numbers. Stas described the hypothesis as "only an illusion" although he also remarked that there was "something at the bottom of it." Interest was revived with the publication of Dmitri MENDELEEV's periodic table, although Mendeleev described the idea of a *prima materia* as "a torment of classical thought." The discovery of isotopes in the 20th century resolved the position.

Prusiner, Stanley Ben (1942–) *American biochemist*

Prusiner was born in Des Moines, Iowa, and qualified as an MD at the University of Pennsylvania in 1968. In 1972 he began a residency at the University of California School of Medicine, San Francisco, where he was appointed professor of neurology in 1980 and professor of biochemistry in 1988.

Early in his career in 1972, Prusiner, following the death of a patient from Creutzfeldt–Jakob disease (CJD), discovered that virtually nothing was known about the condition. It was thought to be one of a class of diseases known as transmissible spongiform encephalopathies (TSEs), which include scrapie in sheep, the more familiar BSE in cattle, and kuru and CJD in man. Following the work of Carleton GAJ-DUSEK, it was proposed that CJD and the other TSEs were probably caused by slow viruses and could be transmitted by injecting extracts from diseased brains into healthy animal brains.

Prusiner began to suspect that something unusual was going on when he read reports suggesting that the agent responsible for scrapie could lack both DNA and RNA. Working with infected hamster brains, he established that procedures known to damage nucleic acids failed to lessen the infection whereas steps that denatured proteins did reduce infectivity. In 1982 he introduced the term *prion* (standing for *proteinaceous infectious particle*) to refer to an agent that is distinct from viruses, bacteria, fungi, and all other known causes of disease and is responsible for scrapie and other TSEs.

He went on to claim that scrapie prions contained a single protein, PrP (prion protein), which was shown to consist of some 15 amino acids. A gene for PrP was found in the chromosomes of hamsters, mice, humans, and all other examined mammals. Why then did not all mammals suffer from prion diseases? Because, Prusiner argued, PrP could be found in two forms: PrP^c, normal cellular protein, and PrP^{Sc}, the abnormal disease-causing form.

In 2004 Prusiner and colleagues showed that genetically engineered prion protein, made in bacteria and then folded into amyloid fibrils, could induce neurological disease when injected into the brains of mice. This study thus provided strong evidence for the infective nature of prions. It is now believed that the different forms of TSE are caused by different strains of infective prions, which produce PrP^{Sc} proteins with different physical and chemical properties. Prusiner was awarded the 1997 Nobel Prize for physiology or medicine for "finding a new biological principle of infection."

Ptolemy (about 2nd century AD) *Egyptian astronomer*

Virtually nothing is known about the life of Ptolemy (**tol**-e-mee; full name Claudius Ptolemaeus). He was probably a Hellenized Egyptian working in the library at Alexandria. He produced four major works, the *Almagest*, the *Geography*, the *Tetrabiblos* (Four Books), and the *Optics*. The first work – the culmination of five hundred years of Greek astronomical and cosmological thinking – was to dominate science for 13 centuries. Ptolemy naturally relied on his predecessors, especially HIPPARCHUS, as in classical times borrowing the work of others was normal practice. A work of such staggering intellectual power and complexity could never be created by one person alone. The basic problem he faced was to try to explain the movements of the heavens on the assumption that the universe is geocentric and all bodies revolve in perfectly circular orbits moving with uniform velocity. As the heavenly bodies move in elliptical orbits with variable velocity around a center other than the Earth some quite sophisticated geometry is called for to preserve the basic fiction. Ptolemy used three complications of the original scheme: epicycles, eccentrics, and equants. These devices worked reasonably well except that they did not lead to particularly accurate predictions. Nor did they permit Ptolemy to develop a system of the universe as a whole. He could give a reasonable account of the orbit of Mars, and of Venus, and of Mercury, and so on, taken separately, but if they were put together into one scheme then the dimensions and the periods would start to conflict. Whatever its faults the system remained intact for 1,300 years until it was overthrown by COPERNICUS in the 15th century.

In the *Geography* Ptolemy explains fully how lines of latitude and longitude can be mathematically determined. However no longitudes were astronomically determined and only a few latitudes had been so calculated. Positions of places were located on this dubious grid by re-

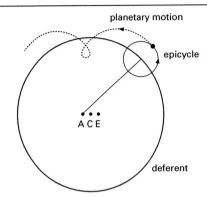

planetary motion

epicycle

A C E

deferent

PTOLEMAIC SYSTEM   The motion of planets as described by Ptolemy.

ducing distances measured on land to degrees. Distances over seas were simply guessed at. As he had put the Canaries 7° east of their true position his whole grid was thrown out of alignment. The *Geography* had almost as great (and as enduring) an influence on the western world-view as the *Almagest*. Columbus might never have sailed without Ptolemy's erroneous view that Asia was closer (westward) than it really is, a view endorsed by the map-makers contemporaneous with Columbus.

The only book of Ptolemy's that is readily available today and still widely read is the *Tetrabiblos*, which is a work on astrology. The work is long and comprehensive and is probably as well argued as the case for astrology can be. It is naturalistic in that he supposes that there might be some form of physical radiation from the heavens that affects mankind. Most of the concepts and arguments of modern astrology can be traced back to this Ptolemaic work. The final major work of Ptolemy, the *Optics*, in which he sets out and demonstrates various elementary principles, is in many ways the most successful of all his works. Although he understood the principles of reflection reasonably well his understanding of refraction seems to be purely empirical. He gives tables he has worked out for the refraction of a ray of light passing from light into water for various angles of incidence.

His main work was known in Greek as the *Syntaxis*; it was the Arabs who named it the *Almagest* from the Arabic definite article "al" and their own pronunciation of the Greek word for "great." Such was the tribute posterity has paid to Ptolemy.

Puck, Theodore Thomas (1916–2005) *American biologist*

Born in Chicago, Illinois, Puck was educated at the university there, obtaining his PhD in 1940. He moved to the University of Colorado in 1948

as professor of biophysics, remaining there until his retirement in 1986.

In 1956 Puck and his students at Colorado extended the technique of cloning (culturing colonies of identical cells) to mammalian cells. The technique has had a profound impact on many areas of biological activity, both theoretical and practical.

In the 1950s, Puck and J. H. Tijo proved that the human genetic material consisted of 46 chromosomes arranged in 23 pairs. He also helped create the Denver system used to classify chromosomes, and was involved in research on ALZHEIMER's disease and Down's syndrome.

Purbach, Georg von (1423–1461) *Austrian astronomer and mathematician*

Purbach (or Peurbach; **poor**-bahk or **poir**-bahk) took his name from his birthplace in Austria. He had traveled in Italy and studied under NICHOLAS of Cusa before becoming professor of mathematics and astronomy at the University of Vienna in about 1450. Purbach's main aim as a scholar was to produce an accurate text of PTOLEMY's *Almagest*. The most common available text was that of Gerard of Cremona, which was a Latin translation of an Arabic translation and was nearly 300 years old. Purbach began by writing a general introduction to Ptolemy that described accurately and briefly the constructions of the *Almagest*. Unfortunately he died before he could embark on the translation. His place was taken by his pupil, REGIOMONTANUS, who completed a textbook begun by Purbach but failed to produce the edition and translation of Ptolemy so much wanted by Purbach.

One of his most significant works was a very thorough table of lunar eclipses, which he published in 1459. Purbach wrote a textbook, *Theoricae novae planetarum* (New Theories about Planets), which became an influential exposition of the Ptolemaic theory of the solar system, a theory whose influence lasted until Tycho BRAHE finally disproved the existence of the solid spheres postulated by Ptolemy. Such was the accuracy of Purbach's set tables that they were still in use almost two hundred years later. He also compiled a table of sines, using Arabic numerals, and was one of the first to popularize their use instead of chords in trigonometry.

Purcell, Edward Mills (1912–1997) *American physicist*

Purcell was born at Taylorville, Illinois. He gained his BSc degree in electrical engineering at Purdue University, Illinois (1933) and his masters degree and PhD from Harvard (1938), having also spent a year in Germany at the Technische Hochschule, Karlsruhe. At Harvard his career advanced from instructor in physics

(1938), to associate professor (1946), full professor (1949), and professor emeritus (1980). During the war years 1941–45 he was a group leader at the Massachusetts Institute of Technology Radiation Laboratories.

Purcell's research spanned nuclear magnetism, radio astronomy, radar, astrophysics, and biophysics. In the field of nuclear magnetism, he was awarded the 1952 Nobel Prize for physics (shared with Felix Bloch) for his work in developing the nuclear magnetic resonance (NMR) method of measuring the magnetic fields of the nuclei of atoms. As a result of these experiments, measurements of nuclear magnetic moment could now be performed on solids and liquids, whereas previously they had been confined to molecular beams of gases. Nuclear magnetic resonance is now commonly applied in chemical analysis.

Purcell's major contribution to astronomy was the first detection of microwave emission from neutral hydrogen in interstellar space at the wavelength of 21 centimeters (1420 Hz). The phenomenon had been predicted theoretically by Hendrik VAN DE HULST and others and was first observed in 1951 by three independent groups of radio astronomers – American, Dutch, and Australian – the American group being the first to report their findings. This, and the subsequent observation of the corresponding absorption line by the Dutch group in 1954, has made possible the mapping of a large part of our own galaxy, and allowed the calculation of the excitation temperature in interstellar space.

Purkinje, Johannes Evangelista (1787–1869) *Czech physiologist*

> With boundless eagerness I investigated within the shortest time all areas of plant and animal histology, and concluded that this new field was inexhaustible.
> —Letter to Rudolph Wagner (1841)

Born at Libochovice (now in the Czech Republic), Purkinje (**puur**-kin-yay or pur-**kin**-jee) began studying to be a priest but changed to medicine and graduated MD from Charles University, Prague, in 1819. He became professor of physiology and pathology at the University of Breslau in 1823 but returned to Charles University in 1850 to take the chair of physiology, which he held until his death. Purkinje's most celebrated research was concerned with the eye, although he also did valuable work on the brain, muscles, sweat glands, digestion, animal and plant cells, and embryology. He explored various aspects of vision, drawing attention, for example, to the fact that in subdued light blue objects appear brighter to the eye than red objects – the *Purkinje effect*. He located *Purkinje cells* in the middle layer of the brain's cerebellar cortex and was the first to apply the term "protoplasm" to the living embryonic material contained in the egg. He also discovered, in the inner walls of the ventricles of the heart, the *Purkinje fibers*, which transmit the pacemaker stimulus. His comparative studies of cellular structure in plants and animals were continued by Matthias SCHLEIDEN and Theodor SCHWANN and led to subsequent increased knowledge of the factors involved in inheritance. Purkinje was among the first to use a microtome for preparing thin slices of tissue for microscopic examination, and may have been the first to teach microscopy and microscopical technique as part of his college courses.

Pythagoras (c. 580 BC–c. 500 BC) *Greek mathematician and philosopher*

> Number is the ruler of forms and ideas, and the cause of gods and demons.
> —Quoted by Robert Graves in *The White Goddess* (1948)

All that is known of the life of Pythagoras (pI-**thag**-or-as) with any certainty is that he left his birthplace, Samos, in about 520 BC to settle in Croton (now Crotone) in southern Italy and, as a result of political trouble, made a final move to Metapontum in about 500.

In Croton Pythagoras established his academy and became a cult leader. His community was governed by a large number of rules, some dietary, such as those commanding abstinence from meat and from beans, and others of obscure origin, such as the commands not to let a swallow nest under the roof or not to sit on a quart measure.

The movement was united by the belief that "all is number." While the exact meaning of this may be none too clear, that it led to one of the great periods of mathematics is beyond doubt. Not only were the properties of numbers explored in a totally new way and important theorems discovered, of which the familiar theorem of Pythagoras is the best example, but there also emerged what is arguably the first really deep mathematical truth – the discovery of irrational numbers with the realization of the incommensurability of the square root of 2.

Pytheas of Massalia (about 4th century BC) *Greek navigator and astronomer*

> Neither land nor sea nor air but a mixture of all, like a sea-lung, in which sea and land and everything swing and which may be the bounds of all, impassable by foot or boat.
> —Description of the hostile sea conditions around Thule (probably Iceland)

Pytheas (**pith**-ee-as) is famous for providing the first account of the seas of northern Europe. He sailed to the tin mines of Cornwall, circumnavigated Britain (without mentioning Ireland), and sailed further north to a land he called Thule, which may have been Iceland. He may also have sailed to the mouth of the Vistula River and visited Heligoland.

Q

Quetelet, Adolphe (1796–1874) *Flemish astronomer, mathematician, and sociologist*

Quetelet (kay-te-**lay**), born the son of a municipal official in Ghent (now in Belgium), was educated at the lycée there. When only 19 he was appointed as an instructor in mathematics at the city's Royal College. In 1819 he moved to Brussels to take the chair of mathematics at the Athenaeum, a post he held until his appointment in 1828 as director of the newly established Royal Observatory.

It is, however, as a statistician and not as an astronomer that Quetelet is remembered. In this field he appeared to have an obsession with the collection and analysis of the variations found in natural phenomena. Thus in 1825 he began by preparing a table of births and deaths in Brussels that was later extended to cover the whole of Belgium. He soon branched out to cover the statistics of crime and tried to show their relationship to such variables as sex, age, climate, and education.

In various influential works Quetelet argued for the use of statistics in the establishment of a social science and for the discovery of social laws. This idea, together with his concept of the "average man" (*l'homme moyen*) caused much controversy.

R

Rabi, Isidor Isaac (1898–1988) *Austrian–American physicist*

> There isn't a scientific community. It is a culture. It is a very undisciplined organization.
> —Quoted by D. S. Greenberg in *The Politics of Pure Science* (1967)

Rabi (**rah**-bee) and his parents emigrated to America from Rymanow in Poland, where he was born, while he was still young. He subsequently grew up in a Yiddish-speaking community in New York, where his father ran a grocery store. He was educated at Cornell, graduating in 1919, and Columbia, where he obtained his PhD in 1927. After two years in Europe he returned to Columbia where he spent his whole career until his retirement in 1967, being appointed professor of physics in 1937 and the first University Professor (a position with no departmental duties) there in 1964.

While in Germany (1927) Rabi had worked under Otto STERN and was impressed with the experiment Stern had performed with Walter Gerlach in which the use of molecular beams led to the discovery of space quantization (1922). Consequently Rabi began a research program at Columbia where he invented the atomic- and molecular-beam magnetic-resonance method of observing atomic spectra, a precise means of determining the magnetic moments of fundamental particles. Using his techniques after World War II, experimentalists were able to measure the magnetic moment of the electron to nine significant figures, thus providing a powerful tool for the testing of theories in quantum electrodynamics. The method had wide applications to the atomic clock, to nuclear magnetic resonance, and to the maser and laser. For this work Rabi was awarded the Nobel Prize for physics in 1944.

During the war Rabi worked on the development of microwave radar. In the postwar years he was a member of the General Advisory Committee of the Atomic Energy Commission, serving as its chairman (1952–56) following the resignation of J. Robert OPPENHEIMER. As a member of the American delegation to UNESCO he originated the movement that led to the foundation of the international laboratory for high-energy physics in Geneva known as CERN.

Rainwater, (Leo) James (1917–1986) *American physicist*

Born in Council, Idaho, and educated at the California Institute of Technology, Rainwater went on to gain his BS, MA, and PhD from Columbia University. At Columbia he progressed through physics assistant (1939–42), to instructor (1946), assistant professor (1947), and associate professor (1949), to become full professor of physics in 1952. In the intervening war years he worked for the Office of Scientific Research and Development and on the Manhattan (atom bomb) project.

Rainwater's principal academic achievement was in explaining the structure and behavior of the atomic nucleus. At the time, two independent models existed, each explaining some of the properties of the atom – the "shell" model of independent particles, and the "liquid-drop" model of collective motion. Rainwater, in collaboration with Aage Bohr, showed how these theories could be unified (1950).

Rainwater, Bohr, and Benjamin MOTTELSON (Bohr's principal collaborator in Denmark) are credited with developing a unified theory that reconciled the individual motions of the nuclear particles with the collective behavior of the nucleus. For this the three men shared the 1975 Nobel Prize for physics.

From 1951 until 1953 and again in the period 1956–61, Rainwater was director of the Nevis Cyclotron Laboratory. From 1965 he spent much of his time supervising the conversion of the synchrocyclotron there.

Raman, Sir Chandrasekhara Venkata (1888–1970) *Indian physicist*

Raman (**rah**-man) was born at Trichinopoly (now Tiruchirappalli) in India and educated at the University of Madras. However, although he revealed considerable talent, he was unable to pursue his education overseas because of ill health. Instead, he chose to enter the civil service where he worked as an auditor for ten years while continuing with his own private research. In 1917 he took up an appointment as professor of physics at the University of Calcutta. In 1933 he moved to Bangalore where he first headed the physics department at the Indian Institute of Science and later, in 1948, became founding director of the Raman Institute.

In 1928 he discovered a spectral effect for which, in 1930, he was awarded the Nobel Prize for physics, thus becoming not only the first Indian but the first Asian to be so honored. The *Raman effect* (as it is now known) occurs when visible radiation is scattered by the molecules in the medium. Not only will the original frequency of the incident light be found but in addition specific new-frequency lines will be detected as a result of the interaction of photons with the molecules. From these new lines in the spectrum (Raman lines) information can be deduced about the molecular structure. The effect is similar to that found by Arthur COMPTON for x-rays and had in fact been predicted by Werner HEISENBERG some years earlier.

Ramanujan, Srinivasa Iyengar (1887–1920) *Indian mathematician*

Ramanujan (rah-**mah**-nuu-jan), the son of a clerk, was born into a poor Brahmin family in Erode near Madras, India. Sometime in 1903, while a student at Kumbakonam High School, he acquired a copy of G. S. Carr's *Synopsis of Elementary Results in Pure Mathematics*. Carr is an unusual work, normally of use as a reference work for a professional mathematician: it consists of about 6,000 theorems presented without comment, explanation, or proof. Ramanujan set himself the task of demonstrating all the formulas, a task only a natural-born mathematician would contemplate, let alone pursue. Indifferent to other subjects, Ramanujan failed every exam he entered. For a time he was supported by Ramachandra Rao, a senior civil servant and secretary of the Indian Mathematical Society (IMS). In 1912 he took a clerical position with the Madras Port Trust. At the same time it was suggested that he should seek the advice of a number of British mathematicians about his work and career.

In January 1913 Ramanujan sent a letter to a number of British mathematicians containing a number of formulas. The only one to respond was the Cambridge mathematician G. H. HARDY. Hardy noted that, while some of the formulas were familiar, others "seemed scarcely possible to believe." Some he thought he could, with difficulty, prove himself; others, he had never seen anything like before, and they defeated Hardy completely. Despite this, it was obvious to Hardy that the formulas must be true and could only come from a mathematician of the very highest class. With Hardy's backing, Ramanujan was awarded a scholarship by the University of Madras and invited to visit Cambridge.

There were, however, religious problems facing the devout Ramanujan but these were resolved when the goddess Namagiri appeared in a dream to Ramanujan's mother absolving him from his traditional obligations. By June 1913 Ramanujan was in Cambridge working with Hardy. They collaborated on five important papers. Ramanujan was elected to the Royal Society in 1918, the first Indian to be honored in this way, and was made a fellow of Trinity College, Cambridge, in 1919. By this time his health had begun to fail. He returned to India in 1919 and died soon after from TB.

Some of Ramanujan's most distinctive work in collaboration with Hardy was on partitions. The partition of a number n, $p(n)$, is the number of ways it can be expressed as a sum of smaller numbers. Thus as the number 4 can be expressed as 5 summands: 4, 3+1, 2+2, 2+1+1, and 1+1+1+1, so p(4)=5. As $p(n)$ grows rapidly as n increases (p(50) for example is 204,266), some way other than counting is needed to determine its value. In 1918 Hardy and Ramanujan established a complicated formula that produced for p(100) the figure 190,569,291.996 while the true figure is 190,569,292. A similar accuracy applied to p(200).

Part of Ramanujan's mathematical ability came from his ability to do mental calculations extremely quickly. It is said that he was traveling in a cab with Hardy when Hardy observed that the number of the cab in front, 1729, was a dull number. "No," replied Ramanujan, "it is a very interesting number; it is the smallest number expressible as a sum of two cubes in two different ways." ($1,729 = 1^3 + 12^3$ and $9^3 + 10^3$.)

Ramón y Cajal, Santiago (1852–1934) *Spanish histologist*

The son of a country doctor from Petilla in Spain, Ramón y Cajal (rah-**mon** ee kah-**hahl**) embarked on medical studies only after being apprenticed first to a barber and then to a shoemaker. He obtained his license to practice in 1873, and after a year's service in the army in Cuba returned to Madrid and graduated as a doctor of medicine in 1877.

He is remembered for his research into the fine structure of nervous tissue. Before the development of the nerve-specific silver nitrate stain by the Italian cytologist Camillo GOLGI in 1873, it was difficult (in neurohistological preparations) to distinguish true nervous elements from the surrounding supporting tissue (neuroglia). Ramón y Cajal refined Golgi's staining technique and subsequently used it to show the intricacy of the structure and connections of cells in the gray matter of the brain and spinal cord. He also used the stain to elucidate the fine structure of the retina of the eye, and the stain has proved useful in the diagnosis of brain tumors.

In 1906 Ramón y Cajal (together with Golgi) was awarded the Nobel Prize for physiology or medicine for establishing the neuron as the fundamental unit of the nervous system – a finding basic to the present understanding of the nerve impulse. He also advanced the neuron theory, which states that the nervous system is made up of numerous discrete cells and is not a system of fused cells.

Between 1884 and 1922 Ramón y Cajal held professorships successively at the universities of Valencia, Barcelona, and Madrid and in 1900 became director of the newly established Instituto Nacional de Higiene. In 1920 the Institute Cajal was commissioned by King Alfonso XIII of Spain and here Ramón y Cajal worked until his death.

Ramsay, Sir William (1852–1916) *British chemist*

> [Ramsay was] a great general who wanted an able chief-of-staff. All his best work was done with a colleague.
> —Morris William Travers, letter to Frederick Soddy, 10 October 1949

Ramsay came from a scientific background in Glasgow, Scotland, his father being an engineer and one of his uncles a professor of geology. He studied at Glasgow University (1866–69) and returned there as an assistant in 1872 after postgraduate work in chemistry at Tübingen, where he studied under Robert BUNSEN. He was appointed professor of chemistry at University College, Bristol, in 1880 and moved to a similar post at University College, London (1887–1912).

Ramsay's early research was mainly in the field of organic chemistry but in 1892 he came across some work of Lord RAYLEIGH that dramatically changed the direction of his work. Rayleigh reported that nitrogen obtained from the atmosphere appeared to be denser than nitrogen derived from chemical compounds. Rayleigh's original view of this was that the synthetic nitrogen was probably contaminated with a lighter gas. Ramsay, however, predicted that the atmosphere contained some unknown denser gas. He favored this view for he remembered some experiments performed by Henry CAVENDISH in 1785 in which he showed that present in the air, after removal of all its oxygen and nitrogen, there remained an unabsorbable 1/20th part of the original. Ramsay experimented with methods of totally removing the oxygen and nitrogen from samples of air and found (1894) that a bubble of gas remained. The gas was identified as a new element by Sir William CROOKES from the lines in its spectrum. Ramsay and Rayleigh announced the discovery of the element in 1898, naming it argon from the Greek "inert."

In the following year Ramsay heard that in America a strange gas had been discovered by heating uranium ores. Ramsay obtained some of the gas from the mineral cleveite and Crookes was able to establish spectroscopically that this was in fact helium, the element whose spectrum had first been observed in a solar eclipse by Pierre JANSSEN in 1868.

From the positions of argon and helium in the periodic table of elements it appeared that three more gases should exist. In 1898 Ramsay began the search for these, assisted by Morris Travers. They liquefied argon and by its fractional distillation were able to collect three new gases, which they named neon, krypton, and xenon, from the Greek for "the new," "the hidden," and "the strange." Ramsay completed his work on the inert gases when, in 1904, with R. Whytlaw-Gray he discovered niton (now known as radon), the radioactive member of the series.

In 1903, with Frederick Soddy, Ramsay demonstrated that helium is continually produced during the radioactive decay of radium. The significance of this was not apparent for some time but its explanation by Ernest RUTHERFORD was to lead to the foundation of the new discipline of nuclear physics.

Ramsay was knighted in 1902 and in 1904 was awarded the Nobel Prize for chemistry. His published works included *The Gases of the Atmosphere* (1896) and his two-volume *Modern Chemistry* (1900).

Ramsden, Jesse (1735–1800) *British instrument maker*

The son of an innkeeper, Ramsden was born at Halifax in England and spent the years 1751–55 as apprentice to a clockmaker. In 1755 he moved to London, where he served a further apprenticeship before setting up his own business in 1762 in the Haymarket, and from 1775 at 199 Piccadilly.

Ramsden became the leading maker of astronomical instruments in Europe in the latter half of the 18th century and at the height of his fame customers were prepared to wait three years for the delivery of one of his sextants, theodolites, or micrometers. His most significant innovation was described in his *Description of an Instrument for Dividing Mathematical Instruments* (1777) for which he received £615 from the Longitude Board. This improved the accuracy of instruments by calibrating both linear and circular scales mechanically – a process previously done with compass and dividers.

Ramsey, Norman Foster (1915–   ) *American physicist*

Born in Washington, DC, Ramsey was educated at Columbia and at Harvard, where he obtained his PhD. He has served as professor of physics at Harvard since 1947.

Ramsey was a student of Isidor RABI and worked with him on the rotational magnetic moments of molecules, showing how these depend on the mass of the nuclei. During World

War II he worked first on radar, and later at the Los Alamos Laboratory. His subsequent work was on both high-energy particle scattering and on low-energy magnetic resonance.

In 1947 he began work on a new, more accurate, molecular-beam resonance technique using two separate radiofrequency fields. This was used to measure nuclear magnetic moments and nuclear quadrupole moments and also to investigate the magnetic interactions within simple molecules. Ramsey used the idea of magnetic shielding to interpret the chemical shifts found in nuclear magnetic resonance spectra.

Ramsey, along with D. Kleppner and H. M. Goldenberg, also developed the hydrogen maser. This used a molecular beam of hydrogen atoms and depended for its action on the hyperfine splitting of energy levels (splitting caused by interaction of electron energy levels with the nuclear magnetic moment). It was a highly accurate device, capable of measuring frequency to 1 part in 10^{12}.

Ramsey has also worked on the thermodynamics and statistical mechanics of systems at low temperatures, pointing out that there is a possibility of certain nuclear spin systems having a negative thermodynamic temperature (i.e., a temperature below absolute zero).

For his work on molecular beams Ramsey was awarded the 1989 Nobel Prize for physics, which he shared with H. Dehmelt and W. Paul. He is also the author of a standard work on the subject, *Molecular Beams* (1956).

Randall, Sir John Turton (1905–1984)
British physicist

The son of a nurseryman, Randall was born at Newton-le-Willows in Lancashire, now in Merseyside, and educated at the University of Manchester. He began work with General Electric at its Wembley laboratory, originally seeking to develop luminescent powders. He was subsequently appointed in 1937 to a Royal Society Fellowship at Birmingham University to conduct research into luminescence.

But the coming of the war stopped all such nonmilitary work and Randall, along with his Birmingham colleagues, began work on improving current radar techniques. By 1943, in collaboration with Boot, he had developed the cavity magnetron, one of the most vital inventions of World War II.

With the end of the war in sight Randall was appointed professor of natural philosophy in 1944 at St. Andrews in Scotland. In 1946 he accepted the post of head of a new department of biophysics set up by the Medical Research Council at King's College, London. Under him were Maurice Wilkins and Rosalind Franklin, two scientists intimately connected with the elucidation in 1953 of the structure of DNA. Randall himself worked on the structure of

collagen, an important protein giving the skin its elasticity; it was found to consist of three helices coiled into a superhelix. On his retirement to Edinburgh in 1970 Randall continued to work with an informal research group on problems in biophysics.

Rankine, William John Macquorn (1820–1872) *British engineer and physicist*

Rankine was the son of an engineer from the Scottish capital Edinburgh and he himself trained as a civil engineer. He spent four years in Ireland working on surveys and in 1855 became professor of civil engineering and mechanics at Glasgow University.

He is best known for his work in thermodynamics, especially his formulation of the *Rankine cycle*, the theoretically ideal process for the operation of turbines and steam engines, in which a condensing vapor is the working fluid. This was contained in his *Manual of the Steam Engine and Other Prime Movers* (1859), and was the first systematic theory for steam engines. He also published studies of soil mechanics and of metal fatigue.

Rankine was the first president of the Institute of Engineers in Scotland and was elected a fellow of the Royal Society in 1853.

Ranvier, Louis-Antoine (1835–1922)
French histologist

Ranvier (rahn-**vyay**), the son of a businessman, studied medicine in his native city of Lyons and in Paris, where he qualified as a physician in 1865. He worked initially as an assistant to Claude Bernard at the Collège de France where he served as professor of general anatomy from 1875 until his retirement in 1900.

Ranvier wrote *Traité technique d'histologie* (1875–82; Technical Treatise on Histology), which served as the leading histological textbook for a generation. He is, however, better known for his description in 1871 of the so-called *nodes of Ranvier*. These are regular gaps in the myelin sheath covering medullated nerves.

He clearly had some insight into the transmission of nerve impulses for in 1878 he went on to compare the function of the myelin sheath with that of the insulation surrounding an underwater telegraph cable. It was, however, left to Ralph Lillie in 1925 to suggest that the nerve impulse was transmitted in a saltatory fashion from node to node; this was later confirmed by I. Tasaki, among others, in the 1930s.

Raoult, François-Marie (1830–1901)
French chemist

Raoult (rah-**ool**) came from a poor background in Fournes-en-Weppes, France. He obtained his PhD in 1863 from the University of Paris and was 37 years old when he took up his first academic appointment at the University of Grenoble, where he was made professor of chemistry in 1870.

Raoult is noted for his work on the properties of solutions, in particular the effect of a dissolved substance in the lowering of freezing points. In 1882 he showed (*Raoult's law*) that the depression in the freezing point of a given solvent was proportional to the mass of substance dissolved divided by the substance's molecular weight. He later showed a similar effect for the vapor pressure of solutions. Measurement of freezing-point depression became an important technique for determining molecular weights.

Raoult's work was also important in validating Jacobus VAN'T HOFF's theory of solutions. Also of significance in his work was his observation that the depression of the freezing point of water caused by an inorganic salt was double that caused by an organic solute (given the same molecular weight). This was one of the anomalies whose explanation led Sven ARRHENIUS to formulate his theory of ionic dissociation.

Ratcliffe, John Ashworth (1902–1987) *British physicist*

Born at Bacup in northern England, Ratcliffe was educated at Cambridge University, where he later lectured in physics (1927–60) and held the title of reader (from 1947). During World War II he was with the Telecommunications Research Establishment, Malvern. In 1960 he was appointed director of the Radio and Space Research Station at Slough, a position he retained until his retirement in 1966.

Ratcliffe worked extensively in the field of atmospheric physics, beginning with his research in the 1920s with Edward APPLETON. They established that radio waves are reflected from the HEAVISIDE layer by electrons and not ions. Details of his researches were published in *Magneto-Ionic Theory and Its Applications to the Ionosphere* (1959) and in his more general work, *Sun, Earth, and Radio* (1970).

Not the least of his achievements lay in his role as one of the founding fathers of radioastronomy in Cambridge. Not only did he maintain the tradition of radio research in the Cavendish Laboratory but encouraged many other scientists in the field, including the young Martin RYLE, who worked on solar radio waves. From such beginnings later emerged the Mullard Radio Astronomy Observatory.

Ratzel, Friedrich (1844–1904) *German geographer and ethnographer*

Ratzel (**raht**-sel) was the son of an official at the court of the grand duke of Bavaria. He was born in Karlsruhe in Germany and educated at the universities of Heidelberg, Jena, and Munich. After a tour of America (1874–75) he taught geography at the Technical University of Munich (1875–86) before being appointed in 1886 to the chair of geography at Leipzig, where he remained until his death.

In his two-volume *Anthropogeographie* (1882; 1891; Human Geography) he laid the foundations for the modern study of human geography. He attempted to describe the distribution of human populations and its relation to migration, and to show the relationship between man and his environment.

Ratzel introduced in his *Politische Geographie* (1897; Political Geography) the concept of *lebensraum*, or "living space," which was to have a powerful attraction for the German Nazi regime. The book was written under the influence of Herbert SPENCER, and in it Ratzel developed the idea of *lebensraum* as part of his theory of the state and society as an organism. The state, he argued, was a spatial entity, which sought to grow and develop to its natural limits.

Ratzel was a prolific author; in addition to a short but much reprinted work on the geography of Germany he also produced a comprehensive statement of his views in his *Die Erde und das Leben: Eine vergleichende Erdkunde* (1901–02; Earth and Life: A Comparative Geography).

Raunkiaer, Christen (1860–1938) *Danish botanist*

Raunkiaer was born in Varde. He graduated in botany in 1885 and was a scientific assistant at the Botanical Garden and Botanical Museum in Copenhagen from 1893 to 1911. From 1912 until 1923 he was professor of botany at the University of Copenhagen.

Raunkiaer is best known for his method of classifying life forms based on the position of the perennating buds of plants and the degree of protection that these give in cold conditions or in drought. In general, the closer these buds are to the ground, the greater the protection. *Raunkiaer's classification* is useful for dividing areas of vegetation into groups and for investigating the occurrence of these groups under different climatic conditions.

Raup, David Malcolm (1933–) *American paleontologist*

Born in Boston, Massachusetts, Raup was educated at the University of Chicago and at Harvard, where he obtained his PhD in 1957. After a brief spell at the California Institute of Technology, Raup spent the period from 1957 to 1965 at Johns Hopkins, Baltimore, and from 1966 to 1978 as professor of geology at the University of Rochester, New York. In 1980 Raup returned to the University of Chicago as professor of geophysics.

Much of Raup's work has been devoted to the problem of extinction. In 1983, in collaboration with John Sepkoski, Raup proposed the controversial thesis that extinction rates were cyclical, peaking periodically every 26 million years. The evidence for the hypothesis was derived from a large body of data collected by Sepkoski.

Initially Raup offered no explanation for such periodicity, other than to suggest that an extraterrestrial cause was more likely than a terrestrial one. Physicists were quick to take up the challenge with Richard MULLER proposing the existence of a companion star of the sun, later named Nemesis, with a 26-million-year orbit, bringing with it periodic asteroid showers. Raup described the controversy which developed in his *The Nemesis Affair* (New York, 1986). But, writing later in his *Extinction* (Oxford, 1993), he has noted that most astronomers have rejected the Nemesis and similar hypotheses. As to the 26-million-year periodicity, expert opinion is apparently evenly divided.

Raup has also published, with S. M. Stanley, a widely used textbook, *Principles of Paleontology* (San Francisco, 1978).

Ray, John (1627–1705) *English naturalist and taxonomist*

> Diseases are the tax on pleasures.
> —*English Proverbs*

Ray, a blacksmith's son from Black Notley, Essex, attended Braintree Grammar School, where he benefited from a trust established to finance needy scholars at Cambridge University. He graduated in 1648 and became a fellow the following year, but his university career ended with the Restoration; as a Puritan, he refused to take the oath required by the Act of Uniformity and he lost his fellowship in 1662.

His activities as a naturalist were funded thereafter by friends from Cambridge, in particular by Francis Willughby, who helped him with the ambitious project of describing all known living things. From 1663 to 1666 Ray and Willughby traveled through Europe, widening their knowledge of the flora and fauna. On their return Ray moved into Willughby's house so that they could collaborate in writing up the work. In 1667 Ray published a catalog of British plants. In the same year he was elected a fellow of the Royal Society.

Willughby died in 1672 but left money in his will to enable Ray to continue their project. Between 1686 and 1704 Ray published *Historia plantarum* (History of Plants), a three-volume encyclopedia of plants describing 18,600 species. In it he emphasized the importance in classification of distinguishing between the monocotyledons and the dicotyledons but more importantly he fixed the species as the basic unit in the taxonomic hierarchy.

Ray also attempted to classify the animal kingdom. In 1693 he published a system based on a number of structural characters, including internal anatomy, which provided a more natural classification than those being produced by his contemporaries.

Ray is also remembered for his theological writings, in which he used the homologies he had perceived in nature as evidence for the necessity of an omniscient creator.

Rayleigh, John William Strutt, 3rd Baron (1842–1919) *British physicist*

> Some proofs command assent. Others woo and charm the intellect. They evoke delight and an overpowering desire to say "Amen, Amen."
> —Quoted by H. E. Hunter in *The Divine Proportion* (1970)

Rayleigh, born at Witham in England, succeeded to his father's title in 1873. He graduated in mathematics from Cambridge University in 1865 and remained at Cambridge until his marriage, in 1871, to Evelyn Balfour, sister of the statesman Lord Balfour. In the following year poor health, which had also disrupted his schooling as a child, necessitated a break from academic life and recuperation in a warmer climate. During this convalescence, which was spent traveling up the Nile in a houseboat, Rayleigh wrote *The Theory of Sound*, which remains a classic in writings on acoustics.

On his return to England, Rayleigh built a laboratory next to his family home. Apart from the period 1879–84, when he succeeded James Clerk MAXWELL as Cavendish Professor of Experimental Physics at Cambridge, Rayleigh carried out most of his work in this private laboratory. Of his early work the best known is his equation to account for the blue color of the sky, which (confirming John TYNDALL's theory) concerned light scattering by small particles in the atmosphere. The amount of scattering depends on the wavelength of the light, and this causes the blue color. From this theory came the scattering law, an important concept in studies of wave propagation. Rayleigh also did a vast amount of work on other problems in physics, particularly in optics and acoustics.

While serving as Cavendish Professor, Rayleigh concerned himself with the precise measuring of electrical standards. He invented the *Rayleigh potentiometer* for precise measurement of potential difference. He extended this precision to the determination of the density of gases, and made the seemingly strange observation that nitrogen from air is always slightly denser than nitrogen obtained from a chemical compound. This led to his collaboration with William RAMSAY that resulted in the discovery of argon. Rayleigh received the Nobel

Prize for physics for this work in 1904; in the same year Ramsay was awarded the chemistry prize.

Read, Herbert Harold (1889–1970)
British geologist

Read, the son of a dairy farmer from Whitstable in Kent, was educated at Imperial College, London, graduating in 1912. After service in World War I he worked with the British Geological Survey of Scotland until 1931 when he was appointed to the chair of geology at Liverpool University. In 1939 he returned as professor of geology to Imperial College, serving there until his retirement in 1955.

Read is best known for his research on the origins of granite. His chief work on this was *The Granite Controversy* (1957), a collection of related papers published during the period 1939–54, in which he argued, in his famous phrase of 1948, "there are granites and granites," or in other words, many different processes had led to a uniform final product. From his fieldwork in the Scottish Highlands, the Shetland Islands, and in Ireland he grouped together the metamorphic, migmatitic, and granitic rocks as plutonic.

Read also proposed a granite series beginning with the early deep-seated autochthonous granites (granites occurring in the same region in which they were formed) and followed by the para-autochthonous, intruded magmatic, and plutonic granites. His work destroyed the simplistic view, dating back to James HUTTON, that granites are simple igneous rocks.

Réaumur, René Antoine Ferchault de (1683–1757) *French entomologist, physicist, and metallurgist*

> The crocodile is certainly a fierce insect, but I am not in the least disturbed about calling it one.
> —On his broad use of the term "insect"

> From the time of Aristotle to the present day I know of but one man who has shown himself Mr. [Charles] Darwin's equal in one field of research – and that is Réaumur.
> —Thomas Henry Huxley. Quoted by Leonard Huxley in *Life and Letters of Thomas Henry Huxley* (1901)

Born at La Rochelle in western France, Réaumur (ray-oh-**muur**) traveled to Paris in 1703 and was admitted to the French Academy of Sciences in 1708. He was commissioned by Louis XIV (1710) to compile a report on the industry and arts of France, published as the *Description des arts et métiers* (Description of the Arts and Skilled Trades).

Réaumur made contributions to many branches of science and industry. He developed improved methods for producing iron and steel; the cupola furnace for melting gray iron was first built by him (1720). In 1740 he produced an opaque form of porcelain, still known as *Réaumur porcelain*. Réaumur also devised a thermometer (1731), using a mixture of alcohol and water, with its freezing point of water at 0° and its boiling point at 80° (the *Réaumur temperature scale*). He also investigated digestion and established that this was a chemical rather than a mechanical process and he isolated gastric juice in 1752.

Perhaps his greatest individual achievement was the six-volume *Memoires pour servir à l'histoire des insectes* (1734–42; Memoirs Serving as a Natural History of Insects), the first serious and comprehensive entomological work.

Reber, Grote (1911–2002) *American radio astronomer*

Reber, who was born in Wheaton, Illinois, studied at the Illinois Institute of Technology and became a radio engineer. His work in radio astronomy took him to many places including Washington, DC, in the late 1940s, where he was chief of the Experimental Microwave Research Section, Hawaii in 1951, and Tasmania in 1954, where he joined the Commonwealth Scientific and Industrial Research Organization. From 1957 to 1961 he worked at the National Radio Astronomy Observatory in Virginia and then returned to Tasmania to complete the studies he had started there.

Reber built the first antenna to be used specifically for extraterrestrial radio observations and was largely responsible for the early developments in radio astronomy. For many years he was probably the world's only radio astronomer. His interest was aroused in 1933 by the work of Karl JANSKY. In 1937 he built, in his own backyard, a 9.4-meter steerable parabolic bowl-shaped radio reflector with an antenna at its focus. Working at a shorter wavelength than Jansky, 60 centimeters instead of 15 meters, he began to spot emission peaks in the Milky Way. These were the intense radio sources in the constellations Cygnus, Taurus, and Cassiopeia. He published his results from 1940 onward and these came to the attention of many astronomers who, although unable to follow him immediately owing to the war, recognized the value of his work. Over the years Reber constructed several telescopes so that he could map the radio sky at different wavelengths. His Hawaiian instrument operated at 5.5–14 meters while in Tasmania he used radio waves of 144 meters.

It was reading Reber's results that stimulated Jan OORT to pose the problem that led to Hendrik VAN DE HULST's discovery of the 21-centimeter hydrogen emission.

Redfield, William C. (1789–1857) *American meteorologist*

Redfield, who was born in Middletown, Connecticut, worked initially as a saddle and harness maker, studying science in his spare time. He then moved to New York in the 1820s to develop steam transport on the Hudson River.

Redfield's most important contribution to meteorology was contained in his 1831 paper *Remarks on the Prevailing Storms of the Atlantic Coast*. By carefully examining ships' logs in addition to his own direct observations of the effects of storms, he concluded that they blow counterclockwise around a center that moves in the direction of the prevailing winds. This conflicted with the view of James ESPY that storms are systems of radially convergent winds and caused a bitter controversy between the two men.

He also worked on, and published an account of, the fossil fish of the American sandstone. In 1848 he was made the first president of the American Association for the Advancement of Science.

Redi, Francesco (1626–1697) *Italian biologist, physician, and poet*

Redi (**ray**-dee), who was born at Arezzo in Italy, studied medicine and philosophy at the University of Pisa, graduating in 1647. He was employed as personal physician to Ferdinand II and Cosimo III, both grand dukes of Tuscany. Intellectually, Redi displayed a variety of talents, being a noted poet, linguist, literary scholar, and student of dialect. On the scientific side, he laid the foundations of helminthology (the study of parasitic worms) and also investigated insect reproduction.

As a biologist he is best known for his experiments to test the theory of spontaneous generation. These were planned to explore the idea, put forward by William HARVEY, that flies and similar vermin do not arise spontaneously but develop from eggs too small to be seen. Redi prepared eight flasks of various meats, with half left open to the air and half sealed. Maggots were found only in the unsealed flasks where flies had been able to enter and lay their eggs. That this effect was not due to the presence or absence of fresh air was shown by a second experiment in which half the flasks were covered with fine gauze. Again, no maggots developed in these. This was one of the earliest examples of a biological experiment planned with proper controls. Redi still believed, however, that spontaneous generation occurred in such animals as intestinal worms and gall flies, and it was not until the time of Louis PASTEUR that the theory of spontaneous generation was finally discredited.

Reed, Walter (1851–1902) *American physician*

> Walter Reed, medical graduate of Virginia, the Army Surgeon who planned and directed in Cuba the experiments that have given man control over that fearful scourge, yellow fever.
> —Citation on the conferment of an honorary degree from Harvard University (1902)

Born in Belroi, Virginia, Reed trained as a doctor at the University of Virginia and at Bellevue Hospital, New York, graduating in 1870. He joined the Army Medical Corps in 1874, spending many years in various frontier posts before being appointed professor of bacteriology at the Army Medical Center, Washington, in 1893.

In the 1890s Reed achieved some success in the investigation of epidemic diseases. For example he was able to show in 1896 that the outbreak of malaria in Washington was not due to bad water and in 1899 he proved that *Bacillus icteroides*, proposed by the Italian bacteriologist Giuseppe Sanarelli as the pathogen of yellow fever, in fact caused hog cholera. Accordingly, in 1900, Reed was appointed to direct a commission to investigate yellow fever in Cuba. The commission began by testing the assumption that yellow fever was spread by infected clothing and bedding. Three nonimmune American volunteers spent 20 nights in a building with tightly screened windows, effectively excluding the entry of insects, sleeping on bedding used by patients with yellow fever. None of the volunteers contracted the disease. They next tested the hypothesis of Carlos FINLAY, that the disease was transmitted by the mosquito *Aedes aegypti*. For the initial trial healthy nonimmune volunteers were bitten by infected mosquitoes; five of the six contracted the disease, while the sixth turned out to have been bitten by a mosquito incapable of transmitting yellow fever. Equally significant was the fact that none of the seven nonimmune members of the group who had not been bitten contracted the disease. Reed thus concluded that an attack of yellow fever could be readily induced in healthy subjects by the bite of *Aedes aegypti* mosquitoes that had previously been fed with the blood of yellow fever patients. (However, the trials were not carried out without tragedy, for in the course of them the entomologist of the commission, Jesse LAZEAR, died of the disease.) In 1901 Reed, in collaboration with James Carroll, the bacteriologist of the expedition, succeeded in establishing the nature of the actual yellow fever pathogen. By showing it to be a microorganism similar to that first discovered by Martinus Beijerinck in 1898, they were the first to implicate a virus in human disease.

Shortly after his return to America Reed died of appendicitis.

Regiomontanus (1436–1476) *German astronomer and mathematician*

> The motion of the stars must vary a tiny bit on account of the motion of the earth.
> —Remark in a letter anticipating Copernicus's idea of a moving earth (1472)

Regiomontanus (ree-jee-oh-mon-**tay**-nus), as befitted a Renaissance humanist, changed his name, Johann Müller, for a Latin version of the name of his home town Königsberg (now Kaliningrad in Russia). His father was a miller. He was educated at Leipzig and Vienna where he was a pupil of Georg von PURBACH. One of the ambitions of Purbach had been to produce a good text of PTOLEMY's *Almagest* based on the original Greek rather than translations from Arabic at third or fourth hand. He had intended to go to Italy in quest of manuscripts of Ptolemy and other ancient scientists with the great Greek scholar Cardinal Bessarion. The death of Purbach in 1461 allowed Regiomontanus to take his place and spend six years in Italy searching for, translating, and editing manuscripts. After his return he settled in Nuremberg where his wealthy benefactor, Bernard Walther, built him an observatory and provided him with instruments. In 1475 he was called to Rome by the pope, Sixtus IV, to help in the reform of the calendar but died of the plague (or, possibly, poison) in 1476.

Regiomontanus was one of the key figures of 15th-century science. In the 1460s he wrote *De triangulis* (On Triangles), a work not printed until 1533 but that was, together with *Tabulae directionum* (1475; Tables of Direction), the main channel for the introduction of modern trigonometry into Europe. In the latter work he broke away from the ancient tradition of chords and instead gave tables of sines for every minute and tangents for every degree.

Regnault, Henri Victor (1800–1878) *French physicist and chemist*

Regnault (re-**nyoh**) came from a poor background in Aachen (now in Germany) and started work as a draper's assistant. He entered the Ecole Polytechnique in 1830 and later worked under Justus von LIEBIG at Giessen. He was successively professor of chemistry at the University of Lyons and the Ecole Polytechnique (1840). He became professor of physics at the Collège de France (1841) and finally director of the Sèvres porcelain factory in 1854.

Regnault's main work was in physics on the properties of gases and in particular the more accurate determination of many physical and chemical effects. Through his meticulous studies he showed, for example, that the law of Pierre DULONG and Alexis PETIT was only approximately true when pure samples were taken and temperatures carefully measured. He also worked on the properties of gases – Joseph GAY-LUSSAC had claimed that a gas will increase by 1/266 of its volume for each increase of temperature of 1°C but Regnault showed that the true increase was 1/273. In addition he made accurate measurements of specific and latent heats and reliable determi-

nations of atomic weights. Regnault is credited with the invention of the air thermometer.

In chemistry, Regnault discovered various organic chlorides that have since become important industrially, including vinyl chloride and carbon tetrachloride. He also took samples of air from different parts of the world and demonstrated that wherever it comes from it contains about 21% oxygen.

Reich, Ferdinand (1799–1882) *German mineralogist*

Reich (rIk), who was born at Bernburg in Germany, studied at the University of Göttingen and taught at the Freiberg Mining Academy. In 1863 he obtained a yellow precipitate from some local zinc ores. Convinced that it contained a new element he asked his assistant Hieronymus RICHTER to examine it spectroscopically (he himself was color blind). Richter found a new line in the dark blue region that confirmed Reich's original conviction; the new element was named "indium" after the bright indigo line characteristic of its spectrum.

Reichenbach, Hans (1891–1953) *American philosopher*

> His impact on his students was that of a blast of fresh, invigorating air; he did all he could to bridge the wide gap of inaccessibility and superiority that typically separated the German professor from his students.
> —Carl Gustav Hempel

The son of a prosperous Hamburg wholesaler, Reichenbach (**rI**-ken-bahk) trained originally as an engineer at the Technical High School, Stuttgart. Further study took him to the universities of Berlin, Göttingen, Munich, and Erlangen, where he gained his PhD for a thesis on probability theory. Soon after he joined the Signals Corps, serving with them on the Russian front. He worked briefly in the radio industry after the war but found time to attend EINSTEIN's seminar on relativity given in Berlin. The two became good friends. In 1920 Reichenbach returned to Stuttgart where he taught radio, surveying, and philosophy. In 1926 he applied for the philosophy of science chair in Berlin. Many, however, were opposed to him because of his well-known indifference to metaphysics. It took the intervention of Einstein, with the accusation, "And what would you have done if the young Schiller had applied?", to win the post for Reichenbach. But in 1933, following the rise of Hitler, Reichenbach was dismissed. He had in fact already fled Germany for Turkey where he served as professor of philosophy at Istanbul University from 1930 to 1938, when he moved to a similar position at the University of California, Los Angeles.

Reichenbach was a logical empiricist. Thus he argued in his *Theory of Probability* (Los Angeles, 1949) that probability statements are about measurable frequencies. That is, to say the probability of a die showing a six is 1/6 means that, in the long run, it will show a six one-sixth of the time. The problem facing the frequency theorist is to explain precisely what is meant by "in the long run." And how it might be applied to such claims as "there is probably life on Mars."

Reichenbach also tackled the problem of induction, favoring the pragmatic theory, with the claim that while we could not show that inductive arguments are probable, we can show that no other method is better.

Toward the end of his life Reichenbach's interest turned to the nature of space and time and his ideas are to be found in two posthumously published works, *The Philosophy of Space and Time* (New York, 1953) and *The Direction of Time* (Los Angeles, 1956).

Reichenbach and CARNAP founded the important journal *Erkenntnis* (Perception) in 1930, the house organ of the Vienna circle, which somehow managed to survive in Nazi Germany until 1938.

Reichstein, Tadeus (1897–1996) *Polish–Swiss biochemist*

Reichstein (**rIk**-shtIn or **rIk**-stIn), the son of an engineer, was born in Wloclawek, Poland, and educated at the Federal Institute of Technology, Zurich, where he obtained his PhD in 1922. After some years in industry, Reichstein returned to work on the staff of the Institute in 1929. In 1938 he moved to the University of Basel, becoming in 1946 head of the Institute of Organic Chemistry, a position he held until his retirement in 1967.

In the 1930s Reichstein began to investigate the chemical role of the adrenal cortex. These are small glands, found on the kidneys, whose removal is invariably fatal. Beginning with a ton of beef adrenals he managed to reduce it to a mere ten grams of biologically active material.

From such samples he had, by 1946, isolated 29 different steroids, six of which he found would prolong the life of an animal with its adrenal gland removed. Of these, aldosterone, corticosterone, and hydrocortisone later proved the most active. Reichstein managed a partial synthesis of desoxycorticosterone, which for many years was the only corticoid that lent itself to large-scale production. At that time it was also the most effective treatment for AD-DISON's disease.

Similar work was being done in America by Edward KENDALL who shared with Reichstein the 1950 Nobel Prize for physiology or medicine together with Philip HENCH.

In 1933 Reichstein succeeded in synthesizing ascorbic acid, vitamin C, at about the same time as Norman HAWORTH in England. He found a better technique for making the vitamin later that year, and this method is still used in commercial production.

Reid, Harry Fielding (1859–1944) *American geologist*

Reid, who was born in Baltimore, Maryland, was educated at Johns Hopkins University there, gaining his PhD in 1885. After a brief period at Chicago, he returned to Johns Hopkins in 1894 and remained there until his retirement in 1930, having served as professor of geology since 1900.

Reid is best known for his work on earthquakes and for his formulation of the *elastic rebound theory*. Rocks, he noted, are elastic and can store energy in the same way as a compressed spring. Elastic strains are caused by movements of the Earth's crust. If these are greater than the rock can bear, it ruptures, and the rock rebounds to relieve the elastic stresses. Reid formulated his theory in his *The California Earthquake of April 19, 1906* (1910).

Reines, Frederick (1918–1988) *American physicist*

Born at Paterson in New Jersey, Reines was educated at the Stevens Institute of Technology and gained his PhD in theoretical physics at New York University in 1944. From 1944 to 1959 he was a group leader at the Los Alamos Scientific Laboratory, concerned with the physics and effects of nuclear explosions.

He was also concerned with investigations into the neutrino, and with Clyde Cowan performed the first experiments that confirmed its existence in the intense radiations from nuclear reactors. The first tentative observation was in 1953, but more definitive experiments were carried out at the Savannah River nuclear reactors in 1956. Detection of the neutrino is difficult because it can travel very long distances through matter before it interacts. Reines subsequently refined the techniques of detection and measurement.

Reines later turned his attention to looking for the relatively small numbers of natural neutrinos originating in cosmic radiation, and to this end constructed underground detectors looking for signs of interactions in huge vats of perchloroethylene. In the course of this work he devised a method of distinguishing cosmic-ray neutrinos from the muons they produce in traveling through the atmosphere.

From 1959 Reines was head of the physics department at the Case Institute of Technology, Cleveland, going on to become professor of physics and dean of physical sciences at the University of California at Irvine in 1966, positions he held until his retirement in 1988. For his codiscovery of the neutrino Reines shared the 1995 Nobel Prize for physics with Martin PERL.

Reinhold, Erasmus (1511–1553) *German astronomer and mathematician*

> Erasmus Reinhold, my teacher ... a man well-versed not only in mathematics but in universal philosophy, and very careful besides.
> —Kaspar Peucer

Little is known about the life of Reinhold (**rIn**-hohlt) other than that he was born at Saalfeld in Germany and became a student at Wittenberg, where he was appointed professor of astronomy in 1536 and rector of the university in 1549. He died a few years later from bubonic plague.

In 1542 Reinhold published a commentary on the *Theoricae novae planetarum* (New Theories about Planets) of Georg von PURBACH, a traditional Ptolemaic text dating from about 1454. He is best known, however, for his *Tabulae Prutenicae* (Prussian Tables, 1551), the first work to provide astronomical tables based upon the new heliocentric system of COPERNICUS. He referred to Copernicus as "a second Atlas, a second Ptolemy," but went on to complain that the computations in Copernicus did not agree with Copernicus's own observations. Comparison between the two based upon recalculations with a computer have established that the accuracy of Reinhold's calculations systematically exceeds that of Copernicus. Although Reinhold's work did much to extend Copernican views, Reinhold made no reference to heliocentric assumptions in his tables.

Remak, Robert (1815–1865) *Polish–German embryologist and anatomist*

Remak (**ray**-mahk), born the son of a shopkeeper in Posen (now in Poland), obtained his MD from the University of Berlin in 1838. Although he spent most of his career there and despite his considerable scientific achievements Remak was denied appropriate promotion and a teaching position because he was a Jew.

In 1838 Remak finally disposed of the ancient myth, probably dating back to ALCMAEON of Croton, that nerves were hollow tubes. In the long history of medicine they had been authoritatively described by centuries of observant anatomists as carrying various spirits, fluids, and airs. Even the introduction of the microscope in the 17th century made no difference. It was left to Remak to point out that the nerve fiber is not hollow, but solid and flat.

In 1844 Remak discovered ganglion cells in the heart, thus showing that it could maintain a rhythmic beat independently of the central nervous system. He further noted that certain fibers of the nervous system, the sympathetic fibers, have a distinctly gray color rather than the more common white. They in fact lack the myelin sheath enclosing other nerve fibers.

In the mid 1840s, in collaboration with Johannes MÜLLER, Remak made a major revision to the orthodox embryology of Karl von BAER. They reduced the four germ layers of von Baer to three by taking the two middle layers as only one. They also at this point introduced the modern terminology of endoderm, mesoderm, and ectoderm.

It was also Remak who, in 1841, first fully described the process of cell division. He went on to insist that the nucleus was a permanent feature of the cell even though it did become less noticeable after cell division. By 1855 Remak was ready to assert the general conclusion implicit in much of the early cell theory: that the production of nuclei or cells is really only division of preexisting nuclei or cells.

Remsen, Ira (1846–1927) *American chemist*

Born in New York City, Remsen was educated at the Free Academy and at the College of Physicians and Surgeons, where he qualified as a physician in 1867. He then studied chemistry in Germany, obtaining his PhD from Göttingen in 1870, and spent the period 1870–72 as assistant to Rudolph FITTIG at Tübingen. He returned to America in 1872 and after teaching joined Johns Hopkins University as the first professor of chemistry in 1876. He spent the rest of his career there, serving as president from 1901 to 1913.

Remsen did much to introduce into America the new, rigorous, and professional chemistry that he had learned in Germany. In 1873 he translated Friedrich WÖHLER's *Outline of Organic Chemistry* and followed this with his own *Theoretical Chemistry* (1877) and numerous other books, popular and educational as well as advanced. He also founded the *American Chemical Journal* in 1879.

His numerous editorial, writing, teaching, and administrative duties made it impossible for him to pursue a very active research life and he is today best remembered as initiating the discovery of saccharin. In 1879 he gave Constantin Fahlberg, one of his research students, the task of oxidizing the sulfamide of toluene with potassium permanganate. The resulting compound was 500 times sweeter than cane sugar. Fahlberg patented the process, naming the compound saccharin; although Remsen had a part in the discovery he never contested Fahlberg's patent.

Retzius, Anders Adolf (1796–1860) *Swedish anatomist*

Retzius (**ret**-si-us) was born in Lund, Sweden, where his father was a professor of natural history at the university; he himself was educated there before he went to the University of Copenhagen. On his return to Sweden he held the chair of anatomy at the Stockholm Karolinska Institute from 1824 until his death.

As a comparative anatomist Retzius did important work on *Amphioxus lanceolatus*, the primitive chordate and key link between the invertebrates and vertebrates. He is, however, best known for his work on human races, which arose from dissatisfaction with the system developed by Johann BLUMENBACH. In 1842 he introduced the idea of a cephalic index, defined as the ratio between a skull's length and width. Using this he classified skulls into brachycephalic and dolichocephalic, or long and short, types defined precisely in terms of the cephalic index. Retzius has thus been claimed as the father of both craniometry and physical anthropology.

Revelle, Roger (1909–1991) *American oceanographer*

Born in Seattle, Washington, Revelle was educated at Pomona College, California, and at the University of California, where he obtained his PhD in 1936. At the same time he joined the Scripps Institute of Oceanography, La Jolla, California, serving as director from 1951 until 1964. He then moved to Harvard as director of the Center for Population Studies, a post he held until 1976.

Revelle was responsible for directing much of the work at Scripps that eventually led to the discovery of sea-floor spreading and magnetic reversals. He also turned in the 1950s to what was then the far from fashionable topic of global warming.

The issue had first been raised by ARRHENIUS in 1895. He had calculated that a doubling of atmospheric carbon dioxide would raise the temperature by about 10°C. This would, in theory, occur because energy from the Sun arrives at the Earth's surface in the form of light and ultraviolet radiation, which are not absorbed by carbon dioxide. The energy is radiated by the Earth as infrared radiation, which is absorbed by carbon dioxide. The atmosphere thus acts in a similar way to the glass in a greenhouse, and the consequent warming is known as the "greenhouse effect." Revelle lobbied for real measurements and as a result gas recorders were set up in 1957 at Mauna Loa, Hawaii, and at the South Pole. By 1990 the carbon dioxide concentration had risen to 350 parts per million, an increase of 11%.

Revelle was also largely responsible for the foundation of the San Diego campus of the University of California in 1959, and was appointed to be its first dean of science. He returned to San Diego in 1976 as professor of science and public policy.

Reynolds, Osborne (1842–1912) *Irish engineer and physicist*

> In my boyhood, I had the advantage of the constant guidance of my father, also a lover of mechanics and a man of no mean attainments in mathematics and their application to physics.
> —Quoted by H. Lamb in his obituary notice of Reynolds in *Proceedings of the Royal Society* (1912–13)

Reynolds was born in Belfast, now in Northern Ireland; after gaining experience in workshop engineering, he went to Cambridge University, graduating in 1867. In 1868 he became the first professor of engineering at Owens College, Manchester. Reynolds made valuable contributions to hydrodynamics and hydraulics. He is best known for the *Reynolds number*, which he introduced (1883–84) as a dimensionless parameter that determines whether fluid flow is smooth or turbulent. His studies of condensation and heat transfer led to a radical revision of boiler and condenser design. He formulated theories of lubrication and of turbulence, helped to develop the idea of group velocity of waves, explained how radiometers work, and determined the mechanical equivalent of heat.

Reynolds became a fellow of the Royal Society in 1877 and a Royal Society medalist in 1888.

Rhazes (*c.* 865 AD–*c.* 925 AD) *Islamic physician*

> Truth in medicine is an unattainable goal, and the art as described in books is far beneath the knowledge of an experienced and thoughtful physician.
> —Quoted by Max Neuburger in *History of Medicine*

Little is known about the life of Rhazes (**ray**-zeez), although there are numerous traditional stories about him. Born at Rayy (now in Iran), he was chief physician at his local hospital and also, at some time, physician at the new hospital in Baghdad. Several of his works on medicine were translated into Latin and he had considerable influence on medical science in the middle ages.

The *Fihrist*, a tenth-century source, lists 113 of his works, some of which survived and were available to the medieval West. His *Kitab al-Mansuri*, known as the *Liber almansuris* (The Book by al-Mansuri) in the West, is mainly a compilation of earlier writings of such authorities as HIPPOCRATES and GALEN and as such was important in making such texts more widely available. Another encyclopedic work, *Kitab al-hawi*, or *Liber continens* (The Complete Book), went through five editions. His most original work, though, was his *Treatise on the Small Pox and Measles*, which contains what is thought to be the first description of smallpox.

Rheticus (1514–1576) *Austrian astronomer and mathematician*

> The planets show again and again all the phenomena which God desired to be seen from the Earth.
> —Quoted by O. Neugebauer in *The Exact Sciences in Antiquity* (1957)

Rheticus (or Rhäticus; ray-tee-kuus) was born Georg Joachim von Lauchen in Feldkirch, Austria, but, in the manner of the time, he adopted a professional name from the Latinized form of his birth district, Rhaetia. After traveling in Italy and after attending several German universities, Rheticus was appointed professor of mathematics at Wittenberg in 1536; the chair of astronomy at the time was held by REINHOLD. Both were Copernicans and both knew that the doctrine was opposed by the authorities, Protestant and Catholic alike.

Nevertheless the Protestant Rheticus traveled to Catholic Poland in 1539 to see COPERNICUS. As it happened Copernicus had completed the manuscript of his *De revolutionibus* (On Revolutions) many years before but, for a number of reasons, was unwilling to publish the work. Rheticus was a man of some charm and consequently persuaded Copernicus to allow a brief summary of his work to appear. The result was the *Narratio Prima of Rheticus* (Danzig, 1539; Basel, 1541; The First Narrative of Rheticus). It caused no major reactions and consequently Copernicus released the manuscript of the *De revolutionibus* to Rheticus.

A brief stay of a few weeks was extended to two years as Rheticus first copied the manuscript and then prepared the text for publication. By May 1542 Rheticus was ready to take the manuscript to the Nuremberg printers. When the work finally appeared in 1543 Copernicus failed to acknowledge the help of Rheticus in any way at all.

By this time Rheticus, reportedly because of his homosexuality, was no longer acceptable at Wittenberg and moved accordingly to the Leipzig chair of mathematics in 1542. But in 1550 he was once more forced into flight charged with indulging in "Italian perversions." Thereafter Rheticus seems to have practiced medicine and is occasionally met with in the service of some noble house.

During the latter part of his life Rheticus had, with the assistance of several calculators, prepared a massive set of trigonometrical tables. Whereas previous workers had expressed the functions in terms of arcs of circles, Rheticus introduced the modern practice of defining them as ratios between the sides of right triangles. Ten-place tables were computed for all six trigonometric functions for every 10″ of arc. The work was incomplete at his death and Rheticus needed his own disciple to publish his manuscript. It finally appeared in 1596 edited by Valentin Otho as the *Opus palatinum de triangulis* (The Imperial Work on Triangles).

Ricci, Matteo (1552–1610) *Italian astronomer and mathematician*

Ricci (**reet**-chee), a Jesuit, was born the son of a pharmacist in Macerata, Italy. He received a rigorous education from the Jesuits in Rome, including classes in mathematics and astronomy from Christopher Clavius. After a few years' missionary work in India he arrived in Macao in 1582 to wait his chance to gain admission to the centralized xenophobic Ming China. The policy of the Celestial Empire incorporated a generalized contempt for all foreigners as uncultured barbarians together with the not unreasonable claim that the least offensive place for such creatures was their own country. If foreigners themselves were despised, foreigners who presumed to possess a superior religion and culture were beyond comprehension. Thus Ricci's options were very limited, forcing him to gain entry in the only practical way, that is, by becoming an unofficial Chinese. To gain respect and influence he had to go further and become a Chinese scholar. It took 20 years for Ricci to reach Peking, for first he had to master Chinese culture, language, and literature.

Once accepted as a scholar he found many in the Mandarinate who were keen to learn western mathematics and astronomy. Thus he translated the first six books of EUCLID into Chinese in 1607, together with works of his teacher Clavius. The Chinese scholars appreciated the many fields in which Ricci was their superior. This had the effect that Chinese science after 1600 ceased to exist as an independent tradition, for although the door could be closed against the Bible and Christianity, it was impossible to shut one's mind to Euclid. Thus Chinese science, because of Ricci, at last became part of universal science. One ironical aspect of this was that Ricci introduced the astronomy of PTOLEMY into China with all its medieval accretions just as it was being rejected in Europe.

Ricci had one further advantage that gained for him his ten years' residence in Peking. He had brought with him a magnificent clock with "self-sounding bells" for the emperor Wan LI. As the Chinese scholars were largely ignorant of their own horological tradition, Ricci and the Jesuits were charged with its installation, its upkeep, and the training of eunuchs to care for it. Such was the respect the emperor felt for Ricci, though they never met, that on Ricci's death, land was made available on which his tomb could be built.

Riccioli, Giovanni Battista (1598–1671) *Italian astronomer*

Born at Ferrara in Italy, Riccioli (reet-**choh**-

lee) was a Jesuit priest who spent most of his life at Bologna where he was professor of astronomy. In 1651 he produced his famous work *Almagestum novum* (The New Almagest). It is in this work that the system of naming craters and mountains on the Moon after famous astronomers was introduced. Although the work is not Copernican – Riccioli presents no less than 77 arguments against COPERNICUS – it is not, despite the title, Ptolemaic either. Riccioli was a follower of Tycho BRAHE, naming the largest lunar crater after him. As an observational astronomer he found that Mizar was a double star. He was also a skilled and patient experimenter who attempted to work out the acceleration due to gravity or *g*. He first tested GALILEO's claim for the isochronicity of the pendulum and the relationship between the period and the square of the length. To measure the time a falling body takes he needed a pendulum that would beat once a second or 86,400 times per sidereal day. This led to the farce of using a team of Jesuits day after day to count the beats of his pendulum but the magic figure of 86,400 escaped them. Eventually the fathers could no longer tolerate staying up night after night counting pendulum beats and he was left with his pupil Francesco GRIMALDI having to accept a less than perfect pendulum. He then performed with Grimaldi the type of experiment Galileo is supposed to have done from the leaning tower of Pisa. He dropped balls of various sizes, shapes, and weights from the 300-foot (92-m) Torre dei Asinelli in Bologna. He succeeded in confirming Galileo's results and establishing a figure for *g* of 30 feet (9.144 m) per second per second, which is close to the value of 9.80665 meters per second per second accepted today.

Richard of Wallingford (c. 1291–1336) *English astronomer and mathematician*

After the death of his father, a blacksmith of Wallingford, Oxfordshire, Richard was adopted by the prior of Wallingford. He was at Oxford University as a student from 1308 to 1314 and taught there from 1317 to 1326 before becoming the abbot of St. Albans. He is thought to have contracted leprosy in early life and there is a manuscript illustration of him in the British Museum that shows him with a spotty or scarred face.

Oxford at this time had gone through a minor renaissance. There were a number of scholars including Richard who were profoundly aware of the limitations imposed by traditional mathematical methods in dealing with virtually any problem of physics. It was Richard who introduced trigonometry into England in its modern form and in a series of manuscripts he produced the basic texts that could have initiated a mathematical revolution. (He was, however, two centuries too soon. The po-

litical troubles of the next 200 years and the Black Death were sufficient to smother any premature intellectual birth.) He was not just a theoretical mathematician for he designed and made his own instruments and, above all, he designed a marvelous clock for his abbey. It has been suggested that he introduced the word "clock" into the English language, from the Latin "clocca" for bell. His clock, the plans for which survive, probably predated that of Giovanni de DONDI in the use of an escapement. It showed the position of the Sun, the Moon, the stars, the state of the tide – in fact, it seemed, like most of the medieval clocks, to do just about everything except tell the time.

Richards, Dickinson Woodruff (1895–1973) *American physician*

Richards, who was born in Orange, New Jersey, was educated at Yale and Columbia, where he obtained his MD in 1922. After a period of postgraduate work in London, Richards worked at various New York hospitals. From 1945 until his retirement in 1961 he served as professor of medicine at Columbia University.

In 1931 Richards began an important collaboration with André COURNAND, which led to the successful development of cardiac catheterization, first used by Werner FORSSMANN in 1929. In the 1940s Richards demonstrated the use of catheterization as both a research and a diagnostic tool. He showed how it could detect congenital heart defects, measure the effects of drugs acting on the heart, and be used to assess the results of heart surgery.

On a more fundamental level he studied (with Cournand) the blood flow to the lungs and began a major study of changes produced in the circulatory system by traumatic shock.

For his discoveries concerning heart catheterization and pathological changes in the circulatory system Richards shared the 1956 Nobel Prize for physiology or medicine with Cournand and Forssmann.

Richards, Theodore William (1868–1928) *American chemist*

Richards came from an artistic background in Germantown, Pennsylvania: his father was a well-known painter and his mother a writer. He originally had ambitions in astronomy but his poor eyesight and the influence of his professor, Josiah P. Cooke, turned him to chemistry. After obtaining his doctorate from Harvard (1888) he continued his studies in Germany before returning to Harvard to take up a professorship in chemistry (1894).

In his doctoral work Richards made an accurate measurement of the ratio of the atomic weight of oxygen to that of hydrogen. His career continued to be devoted almost exclusively to the more accurate determination of atomic weights. He obtained the atomic weights of approximately 60 elements, improving considerably on those achieved by Jean STAS in the 1860s. His determination of the atomic weight of silver, for example, lowered this from Stas's 107.93 to 107.88. In 1913 his team showed that lead present in uranium had a lower atomic weight than normal specimens of lead, thus supporting the idea that it was formed by radioactive decay. In 1914 Richards was awarded the Nobel Prize for chemistry for his work on atomic weights.

In the latter half of his life he became interested in more theoretical problems. In 1902 he published an article which seemed to anticipate some of the ideas of the heat theorem of Nernst; he also worked on the compressibility of the elements.

Richardson, Lewis Fry (1881–1953) *British meteorologist*

> Big(ger) whirls have little whirls
> That feed on their velocity,
> And little whirls have lesser whirls,
> And so on to viscosity.
> —Summarizing his paper *The Supply of Energy from and to Atmospheric Eddies* (1920)

Richardson, the son of a farmer, was born in Newcastle-upon-Tyne in northeast England and educated at Durham College and Cambridge University, graduating in 1903. In 1913 he became superintendent of the meteorological observatory at Eskdalemuir, Scotland. His work was interrupted by World War I and he resigned from the Meteorological Office in 1920. He was head of the physics department at Westminster Training College, London, until in 1929 he became principal of Paisley Technical College, Scotland, where he remained until his retirement in 1940.

Richardson was the first to try to apply mathematical techniques to weather prediction, publishing his ideas in *Weather Prediction by Numerical Process* (1922). In this he argued that the state of the atmosphere is defined by its temperature, pressure, and velocity. Once these were known, he believed that equations could be used to predict future weather conditions. The main problem with implementing his program was the time taken for computation, and it also suffered from a shortage of information. This was partially resolved with the advent of electronic computers following World War II.

The *Richardson number*, a value involving the gradients of temperature and wind velocity, is named for him.

Richardson also attempted to apply a mathematical framework to the study of the causes of war, publishing his work in *Generalized Foreign Politics* (1939), *Arms and Insecurity* (1949), and *Statistics of Deadly Quarrels* (1950).

Richardson, Sir Owen Willans (1879–1959) *British physicist*

Richardson, the son of a woollen manufacturer from Dewsbury in the north of England, was educated at Cambridge University and at London University, where he became a DSc in 1904. He taught at Princeton in America from 1906 to 1913, when he returned to England to become Wheatstone Professor of Physics at King's College, London, where he remained until his retirement in 1944.

Richardson is noted for his work on the emission of electrons from hot surfaces – the phenomenon first observed by Thomas EDISON and used by Edison, John FLEMING, Lee de FOREST, and others in electron tubes. Richardson proposed an explanation of what he named "thermionic emission," suggesting that the electrons came from within the solid and were able to escape provided that they had enough kinetic energy to overcome an energy barrier at the surface – the work function of the solid. Thus thermionic emission of electrons is analogous to evaporation from a liquid. The *Richardson law* (1901) relates the electron current to the temperature, and shows that it increases exponentially with increasing emitter temperature.

Richardson published an account of his extensive work on thermionic emission in his book *The Emission of Electricity from Hot Bodies* (1910). His work was important for the development of electron tubes used in electronic devices, and he was awarded the 1928 Nobel Prize for physics for this work. During World War II he worked on radar.

Richardson, Robert Coleman (1937–   ) *American physicist*

Richardson was educated at Virginia State University and at Duke University, North Carolina, where he obtained his PhD in 1966. He moved immediately to Cornell and was appointed professor of physics in 1975.

In the early 1970s work with Douglas OSHEROFF and David LEE revealed that, contrary to expectations, helium-3 became a superfluid at a temperature of 0.0027 degrees above absolute zero. Richardson shared the 1996 Nobel Prize for physics with Osheroff and Lee for his work in this field.

Richer, Jean (1630–1696) *French astronomer*

> Now several astronomers, sent into remote countries to make astronomical observations, have found that pendulum clocks do accordingly move slower near the equator than in our climates ... In the year 1672, M. Richer took notice of it in the island of Cayenne.
> —Isaac Newton, *Principia* (1687)

Richer (ree-**shay**) is a rather anonymous figure

who is known only through his work with others. In 1671 he went on a scientific mission to Cayenne, where it was intended that he should observe meridian transits of the Sun and also measure the distance of Mars from any nearby stars while Giovanni CASSINI performed similar observations in Paris. Apart from the obvious advantage of having two observers making the same measurements with a long base line, there was also the hope that Cayenne, 5° from the equator, would provide better viewing and that, because of its equatorial position, meridian sightings would be subject to less atmospheric refraction than in Paris.

But more important discoveries were to be made. The observations of Mars were made successfully, allowing Cassini to use the parallax obtained to give a distance of the Sun from the Earth of 86 million miles (138 million km). This was not particularly accurate – being 7% out – but at least it was properly determined and could easily be improved. What, however, really surprised Richer was that his pendulum timings were all slightly different from what they would have been in Paris; the pendulum was running slow, even at sea level. Cassini, at first, seems to have doubted the competence of Richer's observations but was eventually convinced. It was this work that provided Newton with the essential information in working out the size and shape of the Earth. If the pendulum slowed down then there must be a smaller gravitational force operating on it. The only way this could reasonably happen was if Cayenne was further from the center of the Earth than Paris; that is, if the Earth bulged at the equator, or to be precise, is an oblate spheroid.

Richet, Charles Robert (1850–1935) *French physiologist*

> I possess every good quality, but the one that distinguishes me above all is modesty.
> —*The Natural History of a Savant*

Richet (ree-**shay**), the son of a distinguished Parisian surgeon, studied medicine at the University of Paris. After graduation in 1877 he worked at the Collège de France before he returned to the University of Paris where he served as professor of physiology from 1887 until his retirement in 1927.

Richet worked on a wide variety of problems, which ranged from heat regulation in mammals to the unsuccessful development of an anti-TB serum. His most important work began, however, with his investigation of how dogs react to the poison of a sea anemone. He found that he could induce a most violent reaction in dogs that had survived an original injection without any distress. If 22 days later he gave them a second injection of the same amount then they immediately became extremely ill and died in 25 minutes. Richet had discovered the important reaction of anaphylaxis, a term he coined in 1902 to mean the opposite of phylaxis or protection.

By 1903 he was able to show that the same effect could be produced by any protein whether toxic or not as long as there was a crucial interval of three to four weeks between injections. His work was to have profound implications for the newly emerging science of immunology and won for Richet the 1913 Nobel Prize for physiology or medicine.

Richter, Burton (1931–) *American physicist*

Richter (**rik**-ter) was joint winner of the 1976 Nobel Prize for physics. A New Yorker by birth, he studied first at the Massachusetts Institute of Technology, gaining his BS in 1952 and his PhD in physics in 1956. His interest in the physics of elementary particles took him subsequently to Stanford University's high-energy physics laboratory where he became a member of the group building the first pair of electron-storage rings. In this machine, intense beams of particles were made to collide with each other in order to study the validity of quantum electrodynamic theory.

In the 1960s, Richter designed the Stanford Positron Electron Accelerating Ring (SPEAR), which was capable of engineering collisions of much more energetic particles. It was on this machine that, in November 1974, Richter and his collaborators created and detected a new kind of heavy elementary particle, which they labeled psi (ψ). The discovery was announced in a 35-author paper (typical of today's high-energy research teams) in the journal *Physical Review Letters*. The particle is a hadron with a lifetime about one thousand times greater than could be expected from its observed mass. Its discovery was important as its properties are consistent with the idea that it is formed from a fourth type of quark, thus supporting Sheldon GLASHOW's concept of "charm."

Almost simultaneously, another group led by Samuel TING 2,000 miles away at the Brookhaven Laboratory, Long Island, made the same discovery independently in a very different experiment. Richter and Ting met to discuss their findings, and confirmation came quickly from other laboratories when they knew the energy of the new particle and were able to tune their own machines accordingly. Ting called the new particle J; it is now usually referred to as the J/psi in recognition of the simultaneity of its discovery. The discovery led to the finding of many other similar particles as a "family" and has stimulated new attempts to rationalize the underlying structure of matter. Within only two years, Richter and Ting were to be the recipients of the Nobel Prize for physics.

Richter has been a full professor at Stanford University since 1967, taking a sabbatical year at the European Organization for Nuclear Research (CERN) in Geneva (1975–76). He became director of the Linear Accelerator Center in 1984.

Richter, Charles Francis (1900–1985)
American seismologist

Born in Hamilton, Ohio, Richter was educated at the University of Southern California, Stanford, and the California Institute of Technology, where he obtained his PhD in 1928. He worked for the Carnegie Institute (1927–36) before being appointed to the staff of the California Institute of Technology. He became professor of seismology in 1952.

Richter developed his scale to measure the strength of earthquakes in 1935. Earlier scales had been developed by de Rossi in the 1880s and by Giuseppe Mercalli in 1902 but both used a descriptive scale defined in terms of damage to buildings and the behavior and response of the population. This restricted their use to the measurement of earthquakes in populated areas and made the scales relative to the type of building techniques and materials used.

Richter's scale is an absolute one, based on the amplitude of the waves produced by the earthquake. He defined the magnitude of an earthquake as the $\log_{10}$ of the maximum amplitude, measured in microns. This means that waves whose amplitudes differ by a factor of 100 will differ by 2 points on the Richter scale. With Beno GUTENBERG he tried to convert the points on his scale into energy released. In 1956 they showed that magnitude 0 corresponds to about 10^{11} ergs (10^4 joules), while magnitude 9 equals 10^{24} ergs (10^{17} joules). A one unit increase will mean about 30 times more energy being released. The strongest earthquake so far recorded had a Richter-scale value of 8.6. In 1954 Richter and Gutenberg produced one of the basic textbooks on seismology, *Seismicity of the Earth*.

Richter, Hieronymous Theodor (1824–1898) *German chemist*

Born at Dresden in Germany, Richter was assistant to Ferdinand REICH at the Freiberg School of Mines. In 1863 they noticed a brilliant indigo line in the spectrum of some zinc-ore samples they were examining. This led to the discovery of a new element, which they named "indium." Richter later succeeded Reich to become director of the School of Mines at Freiberg (1875).

Ricketts, Howard Taylor (1871–1910)
American pathologist

> Ricketts brought facts to light with brilliance and accuracy and indicated by the methods he used most of the major lines of development subsequently employed in the study of rickettsial diseases.
> —Obituary notice of Ricketts (1910)

Rickets was born at Findlay, Ohio, and graduated in medicine from the Northwestern University Medical School in 1897. Five years later he became associate professor in pathology at the University of Chicago. Shortly before his death he became professor of pathology at the University of Pennsylvania.

His research was concerned with the transmission of disease by insects, and in 1906 he showed that the bite of the wood tick was responsible for the transmission of Rocky Mountain spotted fever. After a further three years' study he was able to describe the microorganism that caused the disease. This is an unusual microorganism because it is smaller than a bacterium and resembles a virus in that it can grow only inside living cells. Ricketts went on to study the related disease typhus and showed that it was transmitted by the body louse. Unfortunately he contracted the disease, but after his death the microorganisms causing typhus and related diseases were given the name *rickettsia* in his honor.

Riddle, Oscar (1877–1968) *American biologist*

Born in Cincinnati, Ohio, Riddle was educated at Indiana University and at Chicago, where he obtained his PhD in 1907. After serving on the Chicago faculty from 1904 to 1911, he joined the research staff of the Carnegie Institution at their Station for Experimental Evolution, Cold Spring Harbor, New York, where he remained until his retirement in 1945.

In 1928 it was found that an extract from the anterior pituitary would stimulate milk secretion in rabbits. Riddle and his colleagues soon succeeded in isolating the hormone, which he named prolactin in 1932, and began a prolonged study of its physiological effects. He discovered that it would stimulate the growth of the crop sac in pigeons and inhibit gonadal growth in a number of animals.

His most dramatic and controversial finding, however, was that he could induce maternal behavior in hens by the injection of prolactin. Rats too were shown to adopt such normal maternal behavior as licking and retrieving despite their virgin state.

Rideal, Sir Eric (Keightley) (1890–1974) *British chemist*

Rideal, born the son of a consultant chemist in London, was educated at Cambridge University, England, and at the University of Bonn where he obtained his doctorate in 1912. After active service in World War I he taught at the University of London (1919–20) before returning to Cambridge. He became professor of colloid science there (1930–46). Rideal moved to the Royal Institution, London, in 1946 as Fullerian Professor of Chemistry but returned to university chemistry in 1950 when he accepted the chair of physical chemistry at King's College, London, where he remained until his retirement in 1955.

Rideal worked on catalysis, producing with Sir Hugh TAYLOR one of the first comprehensive works on the subject, *Catalysis in Theory and Practice* (1919). He later worked on other problems in surface chemistry, particularly in its relation to biology.

Riess, Adam Guy (1969–   ) *American astrophysicist*

Riess was born in Washington, DC, and after high school attended the Massachusetts Institute of Technology, receiving a BS in physics in 1992. Having decided on a career in astrophysics, he entered Harvard University, and began working on exploding stars, or supernovae. He received his PhD in 1996, and moved to the University of California, Berkeley (1996–99) and then to the Space Telescope Institute, Baltimore, headquarters of the Hubble Space Telescope. In 2006 Riess was appointed professor of physics and astronomy at Johns Hopkins University.

In 1994 Riess became a founder member of the High-Z SN Search team, an international project initiated by the astrophysicist Brian SCHMIDT, based in Australia, to look for a class of bright supernovae called type Ia. At Berkeley, Riess analyzed the data from the 15 distant supernovae that the team had identified, and compared them with data amassed during his doctorate studies on near supernovae at Harvard. This enabled him to measure the recent and past rates of expansion of the universe. To his surprise, the results indicated that the universe was expanding at an accelerating rate, rather than slowing down under the force of gravity, as was generally supposed at the time. This indicated that a force was acting to oppose gravity. Riess, Schmidt, and the team invoked EINSTEIN's notion of the cosmological constant, or "dark energy," to explain their findings, and published their startling conclusion in 1998. Almost simultaneously, a competing team led by Saul Perlmutter at UC Berkeley published similar conclusions. Riess remained wary, wondering if the relative dimming of the distant supernovae was due to some other cause. In 2002, he formed the Hubble Higher-z Team to look for even more distant supernovae, and by 2004 had identified a dozen. These were relatively brighter, supporting the model of a matter-dominated universe, in which the universe had only recently begun accelerating after an initial period of deceleration following the formation of cosmic structures. This was consistent with the existence of dark matter. Riess continues to study supernovae and the light they shed on dark matter and the new cosmological model of the universe introduced by his breakthrough discoveries. Among many honors in recognition of his work are the Shaw Prize (2006) and the Gruber Prize in cosmology (2007).

Riemann, (Georg Friedrich) Bernhard (1826–1866) *German mathematician*

> Therefore, either the reality on which our space is based must form a discrete manifold, or else the reason for the metric relationships must be sought for, externally, in the binding forces acting upon it.
> —Lecture on the foundations of geometry. Quoted in *Nature*, 22 March 1990

Riemann (**ree**-man) was born at Breselenz in Germany and, before studying mathematics in earnest, studied theology in preparation for the priesthood at his father's request. Fortunately he was able to persuade his father, a Lutheran pastor, that his real talents lay elsewhere than in theology. He attended the University of Göttingen and his mathematical abilities were such that his doctoral thesis won the rarely given praise of Karl Friedrich GAUSS. After gaining his doctorate Riemann worked on the inaugural lecture necessary in order to gain the post of *Privatdozent* at Göttingen and this too gained Gauss's praise. Eventually Riemann succeeded his friend, Lejeune DIRICHLET, as professor of mathematics at Göttingen in 1859 but by then his health had begun to decline and he died of tuberculosis while on holiday in Italy.

Riemann's work ranges from pure mathematics to mathematical physics and he made influential contributions to both. His work in analysis was profoundly important. The *Riemann integral* is a definite integral formally defined in terms of the limit of a summation of elements as the number of elements tends to infinity and their size becomes infinitesimally small.

One of Riemann's most famous pieces of work was in geometry. This was initiated in his inaugural lecture of 1854 that so impressed Gauss, entitled "Concerning the Hypotheses that Underlie Geometry." What Riemann did was to consider the whole question of what a geometry was from a much more general perspective than anyone had previously done. Riemann asked questions, such as how could concepts like curvature and distance be defined, in such a way as to be applicable to geometries that were not Euclidean. Jnós BOLYAI and Nikolai LOBACHEVSKY (and, at the time unknown to everyone, Gauss) had developed particular non-Euclidean geometries, but Riemann went further and opened up the possibility of a range of geometries different from EUCLID's. This work had far-reaching consequences, not just in pure mathematics but also in the theory of relativity.

Riemann was also interested in applied mathematics and physics and was a coworker of Wilhelm WEBER.

Righi, Augusto (1850–1920) *Italian physicist*

Righi (**ree**-gee) studied in his native city of Bologna, firstly at the technical school and later at the university. For seven years, from 1873, he taught physics at the Bologna Technical School. In 1880 he was appointed professor of experimental physics at the University of Palermo and in 1885 professor of physics at the University of Padua. He returned to Bologna in 1889 as professor at the Institute of Physics of the university, remaining there until his death.

Righi's early work included investigations on the action and efficiency of disk-type electrostatic machines and analysis of the composition of vibrational motion as described by Jules Antoine LISSAJOUS. In 1880 he described magnetic hysteresis – the lagging of magnetic induction behind magnetizing field in ferromagnetic substances. He studied the photoelectric effect and in 1888 demonstrated that two electrodes exposed to ultraviolet radiation act like a voltaic couple. However, Righi is best remembered for his studies on electrical oscillations. He made improvements to HERTZ's vibrator and showed that at least the shorter Hertzian waves displayed the phenomena of reflection, refraction, polarization, and interference. This demonstrated that radio waves differed from light in wavelength rather than in nature and helped to establish the existence of the electromagnetic spectrum.

Ritchey, George Willis (1864–1945) *American astronomer*

Ritchey, who was born in Tupper's Plains, Ohio, was the son of an amateur astronomer and instrument maker. He completed his studies at the University of Cincinnati in 1887, taught at the Chicago Manual Training School from 1888 to 1896, and for a further three years was an optician. He joined the Yerkes Observatory in 1901 as chief optician. In 1906 he moved with George HALE to Mount Wilson as head of instrument construction, working on both the 60-inch (1.5-m) and 100-inch (2.5-m) reflecting telescopes. Shortly after the war Ritchey and the observatory became involved in a serious dispute and he was actually dismissed for supposedly exceeding his authority. He thus spent the period 1923–30 working in Paris for the National Observatory but returned to America to become director of photographic and telescopic research at the U.S. Naval Observatory where he remained until his retirement in 1936.

While in Paris Ritchey worked with Henri Chrétien on a new design for the optics of a reflecting telescope. *Ritchey–Chrétien optics*, first used by Ritchey in 1930 and again in 1936, have since become one of the standard optical configurations of reflectors.

Apart from his strictly instrumental work Ritchey is also remembered for his astronomical observations, especially for his photographs in 1917 of novae in spiral nebulae. In the hands of Heber CURTIS these were to become the basis of a new method for determining the distance of nebulae.

Ritter, Johann Wilhelm (1776–1810) *German scientist*

Ritter (**rit**-er), who was born at Samnitz (now in Poland), left school at the age of fourteen and became an apothecary's apprentice. In 1795 he received a modest inheritance that allowed him to enter the University of Jena the following year. Between 1796 and 1804 he studied and taught at Jena and at Gotha. In 1804 he became a member of the Bavarian Academy of Science and moved to Munich.

Ritter was one of the first investigators to collect hydrogen and oxygen separately from the newly discovered electrolysis of water (1800). In 1802 he built the first dry voltaic pile. He believed in the electrical nature of chemical combination, and explained galvanic and voltaic effects in chemical terms. He was the first to present an electrochemical series (a series of the relative chemical activity of elements based on electrochemical properties). Ritter also examined the effect of light on chemical reactions. His identification of ultraviolet radiation (1801) came after investigating its darkening effects on silver chloride.

Robbins, Frederick Chapman (1916–2003) *American virologist and pediatrician*

Robbins was born in Auburn, Alabama, the son of plant physiologist William Robbins. He obtained his MD from Harvard Medical School in 1940 and from 1942 to 1946 headed the virus and rickettsial section of the U.S. army's 15th medical general laboratory. Here he worked on the isolation of the parasitic microorganisms causing Q fever, which are also responsible for certain kinds of typhus.

After the war Robbins became assistant resident at the Children's Hospital, Boston. In 1948 he became a National Research Fellow in virus diseases, working with John ENDERS and Thomas WELLER. By 1952 Robbins and his coworkers had managed to propagate the poliomyelitis virus in tissue cultures. They established that the polio virus can multiply outside nerve tissue and, in fact, exists in the extraneural tissue of the body, only later attacking the lower section of the brain and parts of the spinal cord.

This research enabled the production of polio

vaccines, the development of sophisticated diagnostic methods, and the isolation of new viruses. In recognition of this work, Robbins, together with Enders and Weller, received the Nobel Prize for physiology or medicine in 1954.

Robbins was director of the pediatrics and contagious diseases department at Cleveland Metropolitan General Hospital, and professor of pediatrics at the Case Western Reserve University, from 1952 until his retirement in 1980. He was married to Alice Havemeyer Northrop, daughter of the Nobel Prize winner John Northrop.

Robert of Chester (c. 1110–c. 1160) *English translator of scientific works*

Robert of Chester was an important figure in the development of medieval science in general and mathematics in particular, not because of any significant discoveries of his own but because he was the first to translate into Latin many important Arab scientific works. Of the scientific texts that thus became available to European scholars for the first time, one of the most important was the *Algebra* of al-Khwarizmi. Robert of Chester was also the first European to translate the Koran from Arabic. He is known to have spent some time in Spain, where he acquired his knowledge of Arabic.

Roberts, John D. (1918–) *American chemist*

Born in Los Angeles, California, Roberts was educated at the university there, obtaining his PhD in 1944. Having taught at the Massachusetts Institute of Technology from 1946 until 1953, he moved to the California Institute of Technology as professor of organic chemistry, a post he held until his retirement in 1988.

Roberts worked on a number of problems in organic chemistry, including the use of carbon–14 tracers in studying the mechanism of reactions, the effect of substituted groups on organic acids, the molecular-orbital theory of organic molecules, and the mechanisms of cycloaddition reactions.

His most important research was on the application of nuclear magnetic resonance to various problems in organic chemistry using absorption of radiofrequency radiation by hydrogen nuclei. He also extended the NMR technique to resonance absorption by naturally occurring isotopes present in molecules, such as nitrogen–15 and carbon–13, which have a net magnetic moment.

Roberts's books include *Nuclear Magnetic Resonance* (1958) and *At the Right Place at the Right Time* (1990).

Roberts, Richard (1943–) *British molecular biologist*

Born in Derby, England, Roberts was educated at the University of Sheffield where he gained his PhD in 1968. He moved soon after to America and, after spending a year at Harvard, he moved to the Cold Spring Harbor Laboratory, New York, in 1971. He is currently serving as research director at New England Biolabs, Beverly, Massachusetts.

By the late 1970s it had become clear that the cells of some organisms seemed to have far too much DNA. Prokaryotic cells, i.e., cells without a nucleus, such as the bacteria *Escherichia coli*, have a single chromosome consisting of about 3 million DNA bases. A protein of about 300 amino acids will require 900 base pairs. Consequently a prokaryotic cell should be able to produce about 3,000 proteins, a figure in reasonable agreement with experience. Eukaryotic cells, however, i.e., cells with a nucleus, as in mammals, have a genome of 3–4 billion base pairs, capable of producing some 3 million proteins, a number far in excess of the 150,000 or so proteins found in mammals. The disparity was solved in 1977 when Roberts, working with adenoviruses, stumbled upon the phenomenon of split genes. While all the DNA of prokaryotic cells was transcribed into messenger RNA (mRNA), which was then used as a template upon which amino acids could be assembled into proteins, something quite different seemed to be happening in the nuclei of eukaryotic cells. Only a part of the DNA, sometimes as little as 10%, was actually transcribed into mRNA. DNA appeared to be composed of several stretches, termed "introns," serving no known purpose, but which separated the active DNA sequences, soon to be called "exons." In the process of transcription the introns were neatly excised and the exons consequently spliced together to form the mature mRNA responsible for the production of protein.

The work of Roberts was independently confirmed by Phillip SHARP, with whom he shared the 1993 Nobel Prize for physiology or medicine.

Robertson, John Monteath (1900–1989) *British x-ray crystallographer*

Robertson was born in Auchterarder, Scotland, and educated at Glasgow University where he obtained his PhD in 1926. From 1928 until 1930 he studied at the University of Michigan and then worked at the Royal Institution throughout the 1930s. After brief periods at the University of Sheffield and with Bomber Command of the Royal Air Force, he returned to Glasgow in 1942 and served as professor of chemistry until his retirement in 1970.

Robertson was one of the key figures who, centered on the Braggs and the Royal Institution, developed x-ray crystallography in the interwar period into one of the basic tools of both the physical and life sciences. He established structures for a large number of molecules, including accurate measurements of bond length in naphthalene, anthracene, and similar hydrocarbons. He also worked on the structure of the important pigment phthalocyanine (1935), durene (1933), pyrene (1941), and copper salts (1951). A notable contribution to the technique was his development of the heavy-atom substitution method, which he used in his investigation of phthalocyanine. This involves substituting a heavy atom into the molecule investigated. The change in intensity of diffracted radiation gives essential information on the phases of scattered waves.

In 1953 Robertson published a full account of his work in his *Organic Crystals and Molecules* in which he demonstrated the growing success in applying the new techniques of x-ray crystallography to complex organic molecules.

Robertson, Sir Robert (1869–1949)
British chemist

Robertson, born the son of a dental surgeon in Cupar, Scotland, was educated at St. Andrews University. After graduating in 1890 he served briefly from 1890 to 1892 in the City Analyst's Office, Glasgow, before entering government service on the staff of the Royal Gunpowder Factory, Waltham Abbey. In 1900 he became chemist in charge, but moved to the Royal Arsenal, Woolwich, in 1907 to serve as superintendent chemist of the research department. In 1921 he became government chemist in charge of the Government Chemical Laboratory in the Strand, London. Robertson remained there until his retirement in 1936 but returned to public service during World War II, which he spent working on explosives at the University of Swansea.

Robertson made a number of advances in the chemistry and technology of explosives. He carried out early work on the decomposition of gun cotton and also improved the process of TNT manufacture. More important was his introduction in 1915 of amatol, a mixture of up to 80% ammonium nitrate to 20% TNT, an explosive more efficient and much cheaper than conventionally produced TNT. It was in fact said of amatol by the director of artillery that it "won the war."

As a pure chemist Robertson was one of the first to see the value of infrared spectroscopy for determining molecular structure. He consequently used it to explore ammonia and arsine (AsH_3).

Robinson, Sir Robert (1886–1975)
British chemist

Robinson's father was a manufacturer of surgical dressings and one of the inventors of cotton wool. Robinson, who was born at Chesterfield in Derbyshire, England, was educated at the University of Manchester where he

obtained a DSc in 1910. From 1912 to 1930 Robinson held chairs in organic chemistry successively at Sydney (1912–15), Liverpool (1915–20), St. Andrews (1921–22), Manchester (1922–28), and University College, London (1928–30). In 1930 he was appointed to the chair of chemistry at Oxford, a post he occupied until his retirement in 1955.

Early in his career, while working with William PERKIN Jr. at Manchester, Robinson became interested in the natural dyes brazilin and hematoxylin. Important advances were achieved in understanding the chemistry of these compounds and their derivatives, which eventually led to his syntheses of anthocyanins and flavones, important plant pigments. Robinson also worked on the physiologically active alkaloids and established the structure of morphine (1925) and strychnine (1946). For his "investigations of plant products of biological importance, especially the alkaloids" Robinson was awarded the 1947 Nobel Prize for chemistry.

From 1945 to 1950 Robinson was president of the Royal Society.

Roche, Edouard Albert (1820–1883)
French mathematician

Roche (rohsh) studied at the university in his native city of Montpellier, obtaining his doctorate there in 1844. After further study in Paris he returned to Montpellier in 1849 and served as professor of pure mathematics from 1852 until his retirement in 1881.

Roche's name is still remembered by astronomers for his proposal in 1850 of the limiting distance since named for him. He calculated that if a satellite and the planet it orbited were of equal density then the satellite could not lie within 2.44 radii, the *Roche limit*, of the larger body without breaking up under the effect of gravity. As the radius of Saturn's outermost ring is 2.3 times that of Saturn it was naturally felt that the rings could well consist of broken-down fragments of a former satellite that had transgressed the forbidden limit. It is now thought, however, that the Roche limit has prevented the fragments from aggregating into a satellite.

Roche later worked on the nebular hypothesis of Pierre Simon de LAPLACE, submitting it to a rigorous mathematical analysis and concluding in 1873 that a rapidly rotating lens-shaped body was in fact unstable. He also published work on the structure and density of the Earth and produced a generalization of TAYLOR's theorem, much used in mathematics.

Rodbell, Martin (1925–1998) *American biochemist*

Rodbell was educated at the University of Washington where he gained his PhD in 1954.

He first worked at the National Institutes of Health, Bethesda, Maryland. In 1985 he was appointed Scientific Director of the National Institute for Environmental Health Sciences, North Carolina.

Rodbell sought to show at the molecular level how cells respond to such chemical signals as hormones and neurotransmitters. It had been shown by Earl SUTHERLAND in 1957 that hormones, also known as "first messengers," do not actually penetrate into the cell but rather stimulate the production of a so-called second messenger, cAMP (cyclic adenosine monophosphate). But, Rodbell asked, how does the binding of a hormone to its receptor stimulate the production of cAMP?

The process proved to be quite complex. In the late 1960s Rodbell found that at least two other factors, referred to as an amplifier and a transducer, were essential. The first extra factor, the amplifier, was needed to initiate the production of cAMP. It was identified as the enzyme AC (adenylate cyclase), which converted the energy-rich molecule ATP (adenosine triphosphate) into the second messenger cAMP. For the second extra factor, the transducer, Rodbell found that no reaction would occur without the presence of a complex energy-rich molecule GTP (guanine triphosphate). The GTP reacted in some way with the AC to initiate cAMP production. If either were absent the cell simply would not respond to such external stimuli as insulin or adrenaline.

For his work on what later became known as G proteins, Rodbell shared the 1994 Nobel Prize for physiology or medicine with Alfred GILMAN.

Roget, Peter Mark (1779–1869) *British physician and encyclopedist*

Roget (roh-**zhay** or ro-**zhay**) was born in London, where his father, a Swiss pastor of the French protestant church in Soho, died when Roget was only four. He was brought up by his mother and uncle, Sir Samuel Romilly (1757–1818), a well-known politician and law reformer. He was educated at the University of Edinburgh where he became a qualified MD in 1798.

Roget began to practice medicine in 1804 at the Manchester Public Infirmary. Before this he had spent the period 1789–99 in Bristol where he learned something of the new "pneumatic medicine" from the famous quack Thomas BEDDOES and from Humphry DAVY. He moved to London in 1800 and worked with Jeremy Bentham on a plan to build a "frigidarium" or ice house for the preservation of food. But Bentham lost interest after a while and in 1802 Roget set off on a European tour as tutor-companion to the sons of John Phillips, a Manchester mill owner. It was an exciting, although dangerous, time to tour Europe. The Treaty of Amiens, signed in 1802, seemed to offer peace-ful times, but was soon ignored and in 1803 Napoleon ordered that all British of military age should be immediately interned. Roget managed to delay the authorities by claiming Swiss, and thereby French, citizenship. While his claim was being investigated he succeeded in escaping over the border to Germany.

On his return to England Roget began to practice medicine, first in Manchester, and from 1809 in London. This, however, was only part of his life. Roget also wrote much, including contributions to the *Encyclopedia Britannica Supplement* (6 vols., London, 1816–24). In one entry, "Cranioscopy", he attacked the then fashionable phrenology, arguing that it lacked direct proof and relied excessively on dubious analogies. Another entry, on BICHAT, was still being used in the 1967 14th edition. Other works of Roget include *Animal and Vegetable Physiology* (1834) and *Outline of Physiology* (1839).

Roget also devoted much of his time to the administration of British science, most notably as secretary of the Royal Society (1827–49). It was a difficult time in the Society's history. Reformers like Charles BABBAGE and Michael FARADAY argued that the Society had fallen into the hands of nonscientists and did little to promote science. Although Roget weathered this storm, a later dispute in 1845 about the award of the Royal Medal to Thomas Beck for his work on nerves of the uterus led to his resignation in 1849.

Although nearly seventy, Roget's release from the Royal Society at last gave him the freedom to work on the book with which he has become famous. Roget had actually begun to compile words for his *Thesaurus* in 1805 and continued intermittently until he began to construct the final version in 1848. It was published in 1852 and went through a further 25 editions before Roget's death in 1869; thereafter it has continued to sell in great quantities – by 1990 well over 20 million copies had been sold. In his classification of words into categories, Roget was influenced by the success of LINNAEUS's classification of plants and animals. He hoped that his classification of language might prove a useful tool for the study of language. Roget also displayed an interest in scientific inventions. He spent much time trying to develop a calculating machine. Though unsuccessful in this field he had in 1814 invented the log-log scale for use on the slide rule. It was for this innovation that Roget was elected to the Royal Society.

Rohrer, Heinrich (1933–) *Swiss physicist*

Born at Buchs in Switzerland, Rohrer (**rohr**-er) was educated at the Federal Institute of Technology, Zurich, where he obtained his PhD in 1960. After two years' postdoctoral work at Rutgers, New Jersey, he returned to Zurich in

1963 to join the staff of the IBM Research Laboratory, Roschliken. Rohrer left IBM in 1997 and moved to Japan where he is doing research in nanotechnology.

The conventional electron microscope developed by RUSKA in the 1930s could present a two-dimensional image only. Further, while atoms have a diameter of 1–2 angstroms (1 angstrom = 10^{-10} meter), electron microscopes could not resolve images below 5 angstroms. Consequently, the surface structure at the atomic level was beyond the range of any existing microscope. To overcome this limitation Rohrer, in collaboration with his IBM colleague, Gerd BINNIG, began work in 1978 on a scanning tunneling microscope (STM).

In the STM a fine probe passes within a few angstroms of the surface of the sample. If a positive voltage is applied to the probe, electrons can move from the sample to the probe by the tunnel effect, and a current can be detected. This current is sensitive to distance from the surface; a slight change in distance will produce a significant change in current. Consequently, in theory at least, a feedback mechanism should be able to keep the probe at a constant distance from the surface, or, in other words, trace the surface's contours. If the tip is allowed to scan the surface by sweeping through a path of parallel lines, a three-dimensional image of the surface can be constructed.

Inevitably practice proved less straightforward than theory. A major difficulty was to eliminate vibration. As the magnification required was of the order of 100 million, any interference would grossly distort the image produced. The microscope was suspended on springs and placed in a vacuum, and further vibrations were dampened by resting the microscope on copper plates positioned between magnets. If the copper plates begin to move an eddy current will be induced by the magnetic field and the interaction between current and field will, in turn, damp the motion of the plates. Vibration was so reduced as to allow a vertical resolution of 0.1 angstrom; the lateral resolution, depending upon the sharpness of the probe, was initially no better than 6 angstroms.

The STM has proved useful in the study of the surfaces of semiconductors and metals. It is also hoped that with increased lateral resolution it will be applicable to biological samples such as viruses. For their work in this field Rohrer and Binnig shared the 1986 Nobel Prize for physics with Ruska.

Rokitansky, Karl (1804–1878) *Austrian pathologist*

Rokitansky (roh-kee-**tahn**-skee) was born at Königgrätz (now Hradec Králové in the Czech Republic), the son of a local government official. He studied philosophy at the University of Prague, later moving to Vienna to study medicine. After graduating in 1828 he accepted a post at the Pathological Institute where he later served as professor of pathological anatomy from 1844 until his retirement in 1875.

He was thus a key figure in the revival of Viennese medicine, making the General Hospital one of the leading centers for medical research in the world. In his *Handbuch der pathologischen Anatomie* (3 vols. 1842–46; Handbook of Pathological Anatomy), based on many thousands of autopsies, he attempted to introduce new standards and criteria of pathology.

Despite the more than 30,000 autopsies personally performed by him, Rokitansky still presented an admittedly modified humoral theory of disease in his Handbook, which led to criticism from Rudolf VIRCHOW. Rokitansky saw sickness as fundamentally due to an imbalance of various substances in the blood, mainly serum proteins such as albumin. Such ideas were omitted from the second edition.

Rokitansky nonetheless distinguished between lobar and lobular pneumonia, described acute yellow fever of the liver, and made fundamental studies of spondylolisthesis, endocarditis, and gastric ulcers.

Ironically, while the eminent pathologist was struggling to modernize the theories of traditional medicine, there worked in his hospital a lowly Hungarian, Ignaz SEMMELWEIS, who by the late 1840s had already achieved considerable success with what was later to develop into the germ theory of disease. Rokitansky was one of the very few physicians to support Semmelweis's work.

Romer, Alfred Sherwood (1894–1973) *American paleontologist*

Romer, who was born at White Plains, New York, and educated at Amherst College and Columbia University, established his reputation with a PhD on comparative myology (musculature), which remains a classic in its field. The impetus for subsequent paleontological fieldwork and research came with his appointment as associate professor in the University of Chicago's department of geology and paleontology, where he was able to study the collections of late Paleozoic fishes, amphibians, and reptiles. Professor of biology at Harvard from 1934, Romer then became Harvard's director of biological laboratories (1945) and director of the Museum of Comparative Zoology (1946).

One of the major figures in paleontology since the 1930s, Romer spent the greater part of his career researching the evolution of vertebrates, based on evidence from comparative anatomy, embryology, and paleontology, and his work has had considerable influence on evolutionary thinking, especially with regard to the lower vertebrates. He paid particular attention to the relationship between animal form and physical function and environment, tracing, for example, the physical changes that occurred during the evolutionary transition of fishes to primitive terrestrial vertebrates. He made extensive collections of fossils of fishes, amphibians, and reptiles from South Africa and Argentina and from the Permian deposits in Texas. His best-known publication is *Man and the Vertebrates* (1933), subsequently revised as *The Vertebrate Story* (1959).

Rømer, Ole Christensen (1644–1710)
Danish astronomer

Born at Aarhus in Denmark, Ole (or Olaus) Rømer (**ru(r)**-mer) was professor of astronomy at the University of Copenhagen when Jean PICARD visited Denmark to inspect Tycho BRAHE's observatory at Uraniborg. Picard recruited him and Rømer joined the Paris Observatory in 1671. In Paris, working on Giovanni CASSINI's table of movements of the satellites of Jupiter, he noticed that whether the eclipses happened earlier or later than Cassini had predicted depended on whether the Earth was moving toward or away from Jupiter. Rømer realized that this anomaly could be explained by assuming that the light from the satellite had a longer (Earth moving away from Jupiter) or shorter (Earth moving toward Jupiter) distance to travel. As Cassini had recently established the distance between the Earth and Jupiter, Rømer realized that he had all the information needed to calculate one of the fundamental constants of nature – the speed of light. In 1676 he announced to the French Academy of Sciences that the speed of light was, in modern figures, 140,000 miles (225,000 km) per second. This value is too small but was an excellent first approximation. In 1681 Rømer was made Astronomer Royal and returned to Copenhagen where he designed and developed the transit telescope.

Röntgen, Wilhelm Conrad (1845–1923)
German physicist

> Pride in one's profession is demanded, but not professional conceit, snobbery or academic arrogance, all of which grow from false egotism.
> —Address on becoming rector of Würzburg University (1894)

Röntgen (**ru(r)nt**-gen or **rent**-gen), who was born at Lennep in Germany, received his early education in the Netherlands; he later studied at the Federal Institute of Technology, Zurich. After receiving his doctorate in 1869 for a thesis on *States of Gases*, he held various important university posts including professor of physics at Würzburg (1888) and professor of physics at Munich (1900). Röntgen researched into many branches of physics including elasticity, capillarity, the specific heat of gases, piezoelectricity, and polarized light. He is

chiefly remembered, however, for his discovery of x-rays made at Würzburg on 8 November 1895.

In 1894 Röntgen had turned his attention to cathode rays and by late 1895 he was investigating the fluorescence caused by these rays using a CROOKES tube. In order to direct a pencil of rays onto a screen, he covered a discharge tube with black cardboard and operated it in a darkened room. Röntgen noticed by chance a weak light on a nearby bench and found that another screen, coated with barium platinocyanide, was fluorescing during the experiment. He had already established that cathode rays could not travel more than a few centimeters in air, and as the screen was about a meter from the discharge tube he realized that he had discovered a new phenomenon. During the succeeding six weeks he devoted himself, feverishly and exclusively, to investigating the properties of the new emanations, which, because of their unknown nature, he called "x-rays". On 28 December 1895 he announced his discovery and gave an accurate description of many of the basic properties of the rays: they were produced by cathode rays (electrons) at the walls of the discharge tube; they traveled in straight lines and could cause shadows; all bodies were to some degree transparent to them; they caused various substances to fluoresce and affected photographic plates; they could not be deflected by magnetic fields. Röntgen concluded that x-rays were quite different from cathode rays but seemed to have some relationship to light rays. He conjectured that they were longitudinal vibrations in the ether (light was known to consist of transverse vibrations). Their true nature was finally established in 1912.

Röntgen's discovery immediately created tremendous interest. It did not solve the contemporary wave–particle controversy on the nature of radiation but it stimulated further investigations that led, among other things, to the discovery of radioactivity; it also provided a valuable tool for research into crystal structures and atomic structure, and x-rays were soon applied to medical diagnosis. Unfortunately their danger to health only became understood very much later; both Röntgen and his technician suffered from x-ray poisoning.

Although Röntgen was subjected to some bitter attacks and attempts to belittle his achievements, his discovery of x-rays earned him several honors, including the first ever Nobel Prize for physics (1901).

Roozeboom, Hendrik Willem Bakhuis
(1856–1907) *Dutch chemist*

Roozeboom (**roh**-ze-bohm) was born at Alkmaar in the Netherlands. Having worked for some time in a butter factory, he became assistant to Jakob van Bemmelen, professor of chemistry at the University of Leiden, from which he graduated in 1884. He succeeded Jacobus VAN'T HOFF as professor of chemistry in the University of Amsterdam in 1886.

Roozeboom's great achievement was the dissemination of Josiah Willard Gibb's phase rule. GIBBS had published his results in the *Transactions of the Connecticut Academy of Sciences* (a journal not read by many European scientists) in the period 1876–78. Roozeboom heard of the work from Johannes VAN DER WAALS and saw it as a major breakthrough in chemical understanding. Not only did he bring it to the attention of Europe, but he also demonstrated its validity and showed its applicability and usefulness, areas in which Gibbs was weak. He showed, for instance, that it had practical applications in the chemistry of alloys and that it led to the discovery of many new ones.

Roscoe, Sir Henry Enfield (1833–1915)
British chemist

> The best of the thing is that vanadium will turn out to be a most valuable substance for calico-printers and dyers – as by its means an aniline-black can be prepared which is far superior to that obtained with copper salts.
> —On the industrial application of vanadium.
> Letter to Robert Bunsen (1876)

The son of a London barrister, Roscoe was educated at University College, London, and at Heidelberg where he studied under Bunsen. On his return to England he was appointed in 1857 to the professorship of chemistry at Owens College, Manchester, the precursor of Manchester University.

Roscoe's main chemical researches were devoted to the study of vanadium and its compounds. Earlier workers, he demonstrated, had used interstitial oxides and nitrides of vanadium rather than the pure metal. Roscoe succeeded for the first time in isolating the pure metal by passing hydrogen over the dichloride. He went on to show that vanadium was a member of Group V of the periodic table.

Not the least of his achievements was the cultivation of science in Victorian Manchester. Owens College was in a state of decline when he arrived in 1857, with few students and little local support. When only 19 students enrolled, Roscoe took his case to local manufacturers and local politicians. To the local politicians he stressed the role of chemistry in sanitary matters, and to the industrialists he spoke of the economies possible from a correct understanding of the chemistry involved in many manufacturing processes. Before long, the Manchester community came to see the need to offer their citizens a scientific education and training. The enrollment at Owens began to grow and long before Roscoe's retirement the college had become a thriving center for teaching and research.

Roscoe made other contributions to public affairs. He was Liberal member of parliament for South Manchester from 1885 to 1889 and served on a large number of Royal and other commissions. He was also the author of the frequently revised and reprinted *Lessons in Elementary Chemistry* (1869).

Rose, Irwin (1926–   ) *American biochemist*

Rose was born in Brooklyn, New York City, and enrolled at Washington State University before his studies were interrupted by World War II, in which he served as a radio technician in the US Navy. After the war he completed his degree at the University of Chicago, and stayed on to obtain a PhD in biochemistry in 1952. After postdoctoral fellowships at Case Western University, Cleveland, and New York University, he joined the faculty of Yale Medical School in 1954. In 1963 he moved to work as a senior member at the Fox Chase Cancer Center in Philadelphia. Two years after his retirement in 1995, he became emeritus researcher at the University of California, Irvine.

The outstanding achievement of Rose's scientific career has been to unravel the way in which living cells break down defective or unwanted proteins in a controlled and energy-dependent manner. He conducted pioneering work on the mechanism of this protein degradation pathway in the late 1970s and early 1980s, in collaboration with two visiting Israeli biochemists, Aaron CIECHANOVER and Avram HERSHKO. Using cell-free extracts obtained from immature red blood cells, the three researchers showed that the crucial step in targeting proteins for degradation was the attachment of a specific protein, subsequently called ubiquitin because it occurs in numerous tissues. Rose and his co-workers published two landmark papers in 1980, demonstrating that the ubiquitin is bound covalently to its target, and that multiple ubiquitin molecules are attached, a process termed polyubiquitination. The ubiquitin acts like a label, identifying the protein as destined for irreversible destruction by complex protein structures within the cell called proteasomes.

In the early 1980s, Rose and his colleagues discovered the three basic enzyme-mediated stages leading to polyubiquitination – their so-called "multistep ubiquitin-tagging hypothesis." The first stage, mediated by E1 enzymes, involves the ATP-dependent activation of ubiquitin molecules. In the second stage the activated ubiquitin transfers to an E2 enzyme. Finally, and crucially, an E3 enzyme identifies the target protein within the cell and causes the ubiquitin to transfer from E2 to the target. The ubiquitin-tagged protein is allowed to enter a barrel-shaped proteasome, inside which it is chopped into small peptides for recycling by the cell.

Rose's work on ubiquitin-labelling and protein disposal opened up a whole new area of cellular research, revealing how cells use the controlled degradation of specific proteins to regulate various cellular activities, including crucial aspects of chromosome pairing and the cell cycle. Subsequently the ubiquitin system has proved to play a role in various forms of cancer and other diseases, and in inflammation and certain immune responses.

Rose was awarded the 2004 Nobel Prize for chemistry "for the discovery of ubiquitin-mediated protein degradation," shared with Ciechanover and Hershko.

Rose, William Cumming (1887–1984) *American biochemist*

Born in Greenville, South Carolina, Rose was educated at Davidson College, North Carolina, and at Yale, where he obtained his PhD in 1911. He taught at the University of Texas from 1913 to 1922, when he moved to the University of Illinois as professor of physiological chemistry; from 1936 until his retirement in 1955 he was professor of biochemistry there.

In the late 1930s Rose was responsible for a beautifully precise set of experiments that introduced the idea of an essential amino acid into nutrition, demonstrating its effect on both human and rodent diet. It had been known for a long time to nutritionists that rats fed on a diet in which the only protein was zein (found in corn), despite enrichment with vitamins, would inevitably die. Rose worked with the constituent amino acids rather than proteins; he still found, however, that whatever combination of amino acids he tried the rats died. However, if the milk protein, casein, was added to their diet the ailing rats recovered.

It was obvious from this that casein must contain an amino acid, not present in zein and then unknown, that was essential for life. Rose began a long series of experiments extracting and testing various fragments of casein until at last he found, in 1936, threonine, the essential amino acid that provided a satisfactory rodent diet when added to the other amino acids. Rose argued that if there was one essential amino acid there could well be others. Over several years he therefore continued to manipulate the rodent diet and finally established the primary importance of ten amino acids: lysine, tryptophan, histidine, phenylalanine, leucine, isoleucine, methionine, valine, and arginine, in addition to the newly discovered threonine. With these in adequate quantities the rats were capable of synthesizing any of the other amino acids if and when they were needed.

In 1942 Rose began a ten-year research project on human diet. By persuading students to restrict their diet in various ways Rose eventually established that there are eight essential amino acids for humans: unlike rats we can survive without arginine and histidine. Since then, however, it has been suggested that these two amino acids are probably required to sustain growth in infants.

Ross, Sir Ronald (1857–1932) *British physician*

> I find Thy Cunning seeds,
> O million-murdering Death.
> I know this little thing
> A myriad men will save
> O Death where is thy sting?
> Thy victory, O Grave?
> —Describing his discovery in 1897 of the life-cycle of the malaria parasite

The son of an Indian Army officer, Ross was born in Almora, India. He originally wished to be an artist but his father was determined that he should join the Indian Medical Service. Consequently, after a medical education at St. Bartholomew's Hospital, London, Ross entered the Indian Medical Service in 1881.

Much of Ross's early career was spent in literary pursuits, writing poetry and verse dramas; he published some 15 literary works between 1883 and 1920. It was also during this first period in India that Ross developed his passion for mathematics. This was a lifelong interest and he published some seven titles between 1901 and 1921; in his *Algebra of Space* (1901) he claimed to have anticipated some of the work of A. N. WHITEHEAD.

On leave in England in 1889, Ross took a diploma in public health and attended courses in the newly established discipline of bacteriology. He became interested in malaria and in 1894 approached Patrick MANSON with a request to be shown how to detect the causative parasite of malaria, first described by Charles LAVERAN in 1880. With his guidance and encouragement, Manson turned out to be the major influence in Ross's scientific career. It was Manson who suggested to Ross that mosquitoes might be the vectors of malaria, and when Ross returned to India he spent the next four years researching this theory.

His first strategy, to try and demonstrate the transmission of the disease from mosquitoes to man, met with little success: attempts to infect a colleague with bites from a mosquito fed on malaria patients failed, possibly because the species he used was not a carrier of the disease. He therefore decided to study the natural history of the mosquito in more detail and by 1897 had succeeded in identifying malaria parasites (plasmodia) in the bodies of *Anopheles* mosquitoes fed on blood from infected patients. Ross then attempted to show what happened to the parasite in the mosquito and how it reached a new human victim. He decided to work with avian malaria and its vector *Culex fatigans*, giving him a control over his experimental subjects impossible to attain with man. By 1898 he had succeeded in identifying the *Proteosoma* parasite responsible for avian malaria in the salivary glands of the mosquito, thus proving that the parasite was transmitted to its avian host by the bite of the mosquito. Manson was able to report Ross's work to the meeting of the British Medical Association in Edinburgh and by the end of the year Italian workers under Giovanni GRASSI had been able to show similar results in the *Anopheles* mosquito, the vector of human malaria.

In 1899 Ross resigned from the Indian Medical Service and accepted a post at the Liverpool School of Tropical Medicine, remaining there until 1912, when he moved to London to become a consultant. During this period he spent much time on the problem of mosquito control, advising many tropical countries on appropriate strategies.

For his work on malaria Ross was awarded the 1902 Nobel Prize for physiology or medicine.

Rossby, Carl-Gustaf Arvid (1898–1957) *Swedish–American meteorologist*

Rossby was born the son of an engineer in Stockholm and educated at the university there. In 1919 he joined the Geophysical Institute at Bergen, which at the time, under Vilhelm BJERKNES, was the world's main center for meteorological research. In 1926 he emigrated to America and was appointed professor of the first meteorology department in America at the Massachusetts Institute of Technology in 1928. After two years as assistant head of the Weather Bureau he became professor of meteorology at the University of Chicago in 1941.

Rossby carried out fundamental work on the upper atmosphere, showing how it affects the long-term weather conditions of the lower air masses. Measurements recorded with instrumented balloons had demonstrated that in high latitudes in the upper atmosphere there is a circumpolar westerly wind, which overlays the system of cyclones and anticyclones lower down. In 1940 Rossby demonstrated that long sinusoidal waves of large amplitude, now known as *Rossby waves*, would be generated by perturbations caused in the westerlies by variations in velocity with latitude. Rossby also showed the importance of the strength of the circumpolar westerlies in determining global weather. When these are weak, cold polar air will sweep south, but when they are strong, the normal sequence of cyclones and anticyclones will develop.

Rossby is credited with having discovered the jet stream. He also devised mathematical models to predict the weather which were simpler than those of Lewis F. RICHARDSON. His school provided the "dynamic meteorology" that allowed, with the coming of computers and weather satellites, the long-term prediction of weather.

Rosse, William Parsons, 3rd Earl of (1800–1867) *Irish astronomer and telescope builder*

The eldest son of the 2nd Earl of Rosse, William Parsons was born at York in England. He was educated at Trinity College, Dublin, and Oxford University, where he graduated in 1822. He was a member of parliament from 1822 until 1834, when he resigned to devote himself to science.

Rosse's main aim was to build a telescope at least as large as those of William HERSCHEL. As Herschel had left no details of how to grind large mirrors, Rosse had to rediscover all this for himself. It was not until 1839 that he had made a 3-inch (8-cm) mirror; this was followed by mirrors of 15 inches (38 cm), 24 inches (61 cm), and 36 inches (91 cm) until, in 1842, he felt confident enough to start work on his 72-inch (183-cm) masterpiece. He was only successful on the fifth casting. It weighed 8,960 pounds (4,064 kg), cost £12,000, and became known as the "Leviathan of Corkstown." Its tube was over 50 feet (15 m) long and because of winds it had to be protected by two masonry piers 50 feet (15 m) high and 23 feet (7 m) apart in which it was supported by an elaborate system of platforms, chains, and pulleys.

The giant reflector suffered, despite the cost and time, from two major defects. The climate of central Ireland is such that very few nights of viewing are possible during the year. Also, viewing (when possible), was restricted by the piers to a few degrees of the north–south meridian. Despite this Rosse made a couple of discoveries. He was the first to identify a spiral nebula and went on to discover 15 of them. He also named and studied the Crab nebula, which has been so important to contemporary astronomy. The telescope was finally dismantled in 1908. More than the individual discoveries made by Rosse, the Leviathan was important in the warnings it gave telescope builders. Good big mirrors were needed but they were by no means sufficient; in addition a good site and an adequate mounting were necessary.

Rossi, Bruno Benedetti (1905–1994) *Italian–American physicist*

Rossi (**ros**-ee), born the son of an electrical engineer in Venice, Italy, was educated at the universities of Padua and Bologna. He first taught at the universities of Florence and Padua before emigrating to America in 1938. There he worked at Chicago and Cornell universities and in 1943 moved to Los Alamos to work on the development of the atom bomb. After World War II he was appointed, in 1946, to the chair of physics at the Massachusetts Institute of Technology where he remained until his retirement in 1970.

Rossi's main work was in the field of cosmic rays. These had first been detected by Victor

HESS in 1911 but, by 1930, there was little agreement on their real nature; it was not even certain whether or not they were charged particles. To answer this question most physicists had been inconclusively searching for a variation in intensity with latitude.

Instead, in 1930 Rossi proposed an experimental arrangement that would search for any east–west asymmetry. Charged particles coming from outer space would be deflected by the Earth's magnetic field eastward if positively charged and westward if negatively charged. To detect them Rossi suggested that two or more Geiger counters be arranged pointing eastward with their centers arranged in a straight line. A similar arrangement should be set up pointing westward. Thus only particles coming from the direction along the axis of the counters would register simultaneously on both or all of them. In 1934 Rossi set up his counters in the mountains of Eritrea and found a 26% excess of particles traveling eastward, thus showing that the majority of cosmic-ray particles are positively charged.

Rossi was the author of several books, including *Cosmic Rays* (1964), which has been used by generations of physics students.

Rossini, Frederick Dominic (1899–1994)
American physical chemist

Rossini, who was born at Monongahela, Pennsylvania, was educated at the Carnegie Institute of Technology (now the Carnegie–Mellon University), Pittsburgh, and the University of California where he obtained his PhD in 1928. He worked at the National Bureau of Standards, Washington, DC, from 1928 to 1950, serving as head of the Thermochemistry and Hydrocarbon section from 1936. In 1950, however, he returned to Carnegie as professor of chemistry and director of the Petroleum Research Laboratory, moving later to chairs of chemistry at Notre Dame in 1960 and Rice University in 1971, in which latter position he remained until his retirement in 1978.

Rossini worked mainly in the fields of thermodynamics and the chemistry of hydrocarbons. On the former subject he published a textbook, *Chemical Thermodynamics* (1950); on the latter he edited and contributed to a standard work, *Hydrocarbons from Petrol* (1953), which described the fractionation, analysis, isolation, purification, and properties of petroleum hydrocarbons.

Rotblat, Sir Joseph (1908–2005) *Polish–British physicist*

Rotblat was educated at the University of Warsaw. In 1939 he was appointed to a research fellowship at the University of Liverpool to work on neutron fission in the laboratory of James CHADWICK. Like many other Europeans, he worked during World War II at Los Alamos on the development of the atom bomb. But unlike most other scientists as soon as it was clear in early 1945 that Germany was defeated and would be unable to produce a nuclear weapon in time, Rotblat felt unable to justify working any longer on the development of such weapons. He was also alarmed by the attitude of some of his senior colleagues. General Groves, for example, the head of the Manhattan Project, was overheard by Rotblat insisting that the real purpose of the bomb was to subdue Soviet Russia.

Consequently he resigned from Los Alamos, arousing the deep suspicion of the security officers, who suspected him of being a Soviet agent, and he returned to Liverpool in early 1945. He remained at Liverpool until 1950 when he was appointed professor of physics at Bart's Hospital Medical College, a position he held until his retirement in 1976.

Rotblat's scientific work was mainly concerned with radiation medicine. He did, however, concern himself with other matters. After the explosion of the Bikini H-bomb on Bikini atoll, the Atomic Energy Commission reported that the fallout from this and all other explosions was no more damaging to the individual than the exposure received from a single x-ray. The announcement did nothing to reassure Rotblat. The Bikini bomb, he worked out independently, would have been much more dangerous than the authorities admitted. Further, he reasoned, chest x-rays screen only chests; nuclear bombs radiated the whole body including wombs, ovaries, and testicles.

Rotblat was determined not to let the matter rest. He drafted an appeal for peace, backed by Bertrand RUSSELL and Albert EINSTEIN among others, to be delivered to world leaders. He realized, however, that something more permanent and constructive was needed. Consequently he set about raising money to hold a series of conferences in which technical matters could be debated, authoritative proposals made, and contacts established between scientists from different disciplines and countries.

The American millionaire Cyrus Eaton offered to finance the first conference as long as it was held in the Nova Scotian fishing village of Pugwash, Eaton's birthplace. The first conference was held in 1957 and was attended by 22 scientists. Known as Pugwash conferences ever since, they have continued to be held around the world. Rotblat served as the Secretary-General of Pugwash from its inception until 1973, and from 1988 he held the office of Pugwash President.

Rotblat was awarded the 1995 Nobel Peace Prize, an award he shared with the organization he helped to found, the Pugwash Conferences on Science and World Affairs. The award was made, the Nobel committee announced, to mark the 50th anniversary of Hiroshima and

Nagasaki and to protest against the French nuclear tests in the Pacific. Rotblat was knighted in 1998. He wrote a number of books about the dangers of nuclear weapons.

Rouelle, Guillaume François (1703–1770) *French chemist*

> His [Rouelle's] eloquence was not a matter of words; he presented his ideas the way nature does her productions, in a disorder which was always pleasing and with an abundance which was never wearisome.
> —Vicq d'Azyr

Rouelle (roo-**el**) came from a farming background in Mathieu, France. He studied at the University of Caen and in Paris, becoming an apothecary in 1725. During the period 1742–68 he was chemical demonstrator at the Jardin du Roi and earned a reputation for being an enthusiastic lecturer. He was a follower of Georg STAHL and was one of the first to teach phlogiston theories in France. His students included most of the French chemists of the late 18th century and Antoine LAVOISIER was probably his most famous pupil. Rouelle's own work included a classification of salts.

Rouelle's brother, Hilaire Martin Rouelle, acted first as his assistant and then succeeded him in 1768.

Rous, Francis Peyton (1879–1970) *American pathologist*

Rous was born in Baltimore, Maryland, and educated at Johns Hopkins University, obtaining his MD in 1905. After working as a pathologist at the University of Michigan, he moved in 1908 to the Rockefeller Institute of Medical Research in New York, remaining there until his official retirement in 1945. Unofficially, Rous continued to work in his laboratory until his death at the age of 90.

In 1909 Rous began to investigate a particular malignant tumor of connective tissues in chickens – later to be known as *Rous chicken-sarcoma*. He ground up the tumor and passed it through a fine filter, extracting what would normally be accepted as a cell-free filtrate. On injection of this filtrate into other chickens, identical tumors developed. In 1911 Rous published his results in a paper with the significant title *Transmission of a Malignant New Growth by means of a Cell-free Filtrate*, significant because nowhere in the title (or even in the paper) does the expected term "virus" occur.

It was well known by 1911 that only a virus could be present in such a filtrate, but Rous was unwilling to use the term for fear of offending his more senior colleagues. Scientists were reluctant to accept that cancer could be caused by viruses since the epidemiology of the disease was obviously different from that of such viral infections as influenza. Rous persisted with his work, however, and by the late 1930s it was widely accepted that a number of animal cancers were caused by viruses. In 1966 Rous was awarded the Nobel Prize for physiology or medicine for the discovery he had announced some 55 years earlier.

Rous also worked on the development of a number of culture techniques for both viruses and cells, techniques that have since become standard laboratory practice.

Roux, Paul Emile (1853–1933) *French microbiologist*

Roux (roo), who was born at Conforens in France, was educated at the universities of Clermont-Ferrand and Paris, where he obtained his MD in 1881. Before this he had worked as an assistant in the laboratory of Louis PASTEUR. In 1888 Roux joined the newly created Pasteur Institute and in 1904 became the director, a post he retained until his death.

Roux worked with Pasteur on most of the latter's major medical discoveries; he assisted him at Pouilly-la-Fort in 1882 with the testing of the anthrax vaccine. He also did much of the early work on the development of a rabies vaccine, but later came to disagree with Pasteur on the speed with which the vaccine was applied to humans and consequently withdrew from the project. His most important work, however, was done quite independently of Pasteur and led to the development of a successful antitoxin against diphtheria. He demonstrated in 1885, with the Swiss bacteriologist Alexandre YERSIN, that the menace of diphtheria lay not in the bacteria themselves but in a lethal poison produced by them. Roux and Yersin produced large amounts of the diphtheria bacillus in a liquid culture medium. After 42 days they carefully separated the germs from the liquid using fine porcelain filters and injected the germ-free liquid into experimental animals. This liquid contained a toxin so powerful that one ounce (28 grams) of it could kill over half a million guinea pigs.

The next step was to develop a means to neutralize the toxin. Emil von BEHRING had achieved some early success in this by inoculating guinea pigs with diphtheria toxin and collecting the serum, which contained an antitoxin to the poison. Roux himself used horses, which enabled the extraction of much larger quantities of serum. By 1894 he was ready to try his serum on patients in the Enfants Malades Hospital in Paris. Within four months mortality fell from 51% to 24%. In 1903 Roux achieved some success, in collaboration with Elie METCHNIKOFF, in transmitting syphilis to a chimpanzee. Such work greatly facilitated the laboratory investigation of syphilis and the search for a cure.

Rowland, Frank Sherwood (1927–   ) *American chemist*

The work is going well, but it looks like the end of the world.
—On his researches into the destruction of the ozone layer

Born in Delaware, Ohio, Rowland was educated at Wesleyan University, Ohio, and at the University of Chicago, where he gained his PhD in 1952.

After holding teaching posts at Princeton and Kansas, Rowland moved to the University of California, Irvine, in 1964 as professor of chemistry.

Shortly before Christmas 1973, Mario MOLINA took to Rowland, his postdoctoral adviser, some calculations suggesting that CFCs (chlorofluorocarbons), widely used in aerosol propellants, will rise to the upper atmosphere and destroy the ozone layer, located 8 to 30 miles above the Earth. As the layer protects us from harmful ultraviolet rays, its destruction could have disturbing consequences.

Rowland and Molina published their preliminary results in June 1974. They pointed out that in the lower atmosphere CFCs were relatively inert compounds. But at a height of about 15 miles (25 km) in the stratosphere they begin to absorb ultraviolet radiation in the 1,900–2,250 angstrom range and decompose, releasing chlorine atoms which will attack ozone (O_3) atoms in a chain reaction:

$$Cl\cdot + O_3 \rightarrow ClO + O_2$$

$$ClO + O\cdot \rightarrow Cl\cdot + O_2$$

In the first part of the reaction a chlorine atom attacks an ozone molecule and forms chlorine monoxide and normal oxygen; in the second stage of the reaction, involving oxygen atoms, the chlorine is regenerated and is free to enter once more into the first reaction, destroying an ozone molecule in the process. The result is that a relatively small amount of CFC can destroy a large amount of ozone.

Rowland discovered that 400,000 tons of CFCs had been produced in the United States in 1973, and that the bulk of this was being discharged into the atmosphere. He calculated that at the then current production rate there would be a long-term steady-state ozone depletion of 7–13%. The CFC industry responded by pointing out there was no actual proof of Rowland's hypothesis. Further, they argued, even if the hypothesis was true, other atmospheric processes could offset the effects of the reaction. In 1974 it seemed that Rowland had found just such a process with the possible formation of chlorine nitrate ($ClONO_2$) in the atmosphere. Thus it seemed possible that the reaction:

$$ClO + NO_2 \rightarrow ClONO_2$$

would remove chlorine monoxide, leaving less chlorine to react with ozone. More detailed analysis revealed that chlorine nitrate might change the distribution of ozone in the atmos-phere without significantly minimizing its depletion rate. The National Academy of Sciences published a report in September 1976 supporting the work of Rowland and Molina, and in October 1978 CFC use in aerosols was banned in the United States. Final confirmation came when Joe Farman discovered in late 1984 a 40% ozone loss over Antarctica.

For his work on CFCs Rowland shared the 1995 Nobel Prize for chemistry with Mario Molina and Paul CRUTZEN.

Rowland, Henry Augustus (1848–1901) *American physicist*

> He who makes two blades of grass grow where one grew before is the benefactor of mankind, but he who obscurely worked to find the laws of such growth is the intellectual superior as well as the greater benefactor of mankind.
> —Quoted by D. S. Greenburg in *The Politics of Pure Science* (1967)

Born at Honesdale in Pennsylvania, Rowland graduated in engineering from the Rensselaer Polytechnic, New York, in 1870, and worked for a time as a railroad engineer. He then taught at Wooster, Ohio, before returning to Rensselaer to take up a professorial appointment in 1874. Following a year's study in Berlin under Hermann von HELMHOLTZ, Rowland returned to America and joined Johns Hopkins University as their first professor of physics (1876).

While in Berlin Rowland contributed to electromagnetic theory by showing that a magnetic field accompanies electric charge. The strength of the magnetic field equals the velocity of the charge multiplied by its quantity, and the phenomenon is analogous to an electric current flowing in a conductor. At Johns Hopkins, Rowland worked toward achieving more accurate values for units.

Rowland is best remembered for his design, in 1882, of the concave diffraction grating, which enables spectra to be photographed without the need for lenses or prisms. Moreover the grating (with 20,000 grooves to the inch) gave a greater resolving power and dispersion, and it has since been used in many areas of spectroscopy.

Rubbia, Carlo (1934–) *Italian physicist*

Born at Gorizia, Trieste, Rubbia (roo-**bee**-a) was educated at the University of Pisa, where he obtained his PhD in 1958. After spending a year each at Columbia, New York, and Rome, he took up an appointment in 1960 at the European Laboratory for Particle Physics (CERN), Geneva, becoming its director-general in 1989. He has also held since 1972 a professorship of physics at Harvard.

Rubbia is noted for his work in high-energy physics using the considerable accelerator capacity of CERN. He set himself an ambitious target in the mid 1970s, namely, the discovery of the intermediate vector bosons. Forces operate by interchanging particles. Thus the electromagnetic force works by exchange of virtual photons. The weak interaction would, therefore, require a comparable particle; in actual fact three such particles would be needed, W^+, W^-, and Z^0. Further, as the weak force acts at distances below about 10^{-13} centimeters, and as the shorter the distance the larger the particle would have to be, the bosons would have to be massive, some 80 times bigger than a proton. To produce such particles in an accelerator requires enormous energies and it was not expected that CERN would be able to obtain such energies for more than a decade. Rubbia proposed in 1976 that the existing super proton synchroton should be changed from a fixed-target accelerator to one producing collisions between beams of protons and antiprotons traveling in opposite directions. If feasible, and given that a particle's kinetic energy increases as the square of its velocity, much higher energies would be attained. As redesigned by his CERN colleague, Simon VAN DER MEER, the SPS produced energies of 540 billion electronvolts (540 GeV) – the equivalent of the 155,000 GeV achieved by striking a stationary target.

Rubbia faced two further problems: how to produce enough antiprotons, and how to recognize the W and Z particles. Antiprotons were produced by accelerating protons in the SPS and firing them at a metal target. A new detector, designed by CHARPAK, was built. To detect a W particle the experimenters looked for its characteristic interactions. They should see antiprotons collide with protons and emit a W particle, which in 10^{-20} second should decay into an electron and a neutrino. The experiment began in September 1982 and ran until December 6, leaving millions of collisions to analyze. Among them they found six possible W particles, five of which were accepted as genuine. The Z particle was subsequently discovered in May 1983.

Rubbia published the discovery of the W particle in January 1983 in a paper listing 130 coauthors. For their part in the discovery of the W and Z particles Rubbia and van der Meer shared the 1984 Nobel Prize for physics.

Rubik, Erno (1944–) Hungarian architect and designer

Rubik (**roo**-bik) was educated at the Technical University and the Academy of Applied Arts in his native city of Budapest; he became director of postgraduate studies at the Academy in 1983.

In order to stimulate the ability of his students to think in three-dimensional terms, he invented in 1975 a $3 \times 3 \times 3$ multicolored cube having 9 squares on each face. The cube could be twisted so as to adopt any one of 43,252,003,274,489,856,000 possible positions.

The aim was to restore a scrambled cube to its original state in which each face presents a single color. As some 20 million cubes were sold in the two years of its greatest popularity, few alive in 1980 can have been unaware of the cube. Speed competitions were held at so-called "world championships" with restorations being achieved in less than 23 seconds. Mathematicians sought for algorithms and Morwen Thistlethwaite proved that any cube could be restored in at most 50 moves. The search for "God's algorithm," the minimum number needed for a restoration, led to a figure, without proof, of 22. A rigorous mathematical treatment of the subject, David Singmaster's *Notes on Rubik's Magic Cube* (London, 1980), went through five editions.

The cube first became known to the West when Hungarian scientists began to arrive at international conferences offering to trade cubes for dollars. In 1979 Rubik licensed the Ideal Toy Company to sell the cube. Despite vast sales they found their profits were being threatened by cheap Taiwanese imports. In 1986 Rubik left the Academy to set up the Rubik Studio. While he has produced a number of popular items, including Rubik's snake, Rubik's domino, and Rubik's revenge, he has so far failed to rekindle the mania induced by his cube in the early 1980s. Nonetheless, sales of the cube have made Rubik a very wealthy man.

Rubin, Vera Cooper (1928–) American astronomer

Born in Philadelphia, Pennsylvania, Rubin was educated at Vassar, at Cornell, and at Georgetown University, Washington, DC, where she obtained her PhD in 1954. Since 1965 she has worked at the Carnegie Institution, Washington, DC, while also being an adjunct staff member of the Mount Wilson and Las Campanas observatories.

Rubin's main work has long been concerned with galactic rotation measurements and it has led to one of the more persistent problems of modern astronomy. She has concentrated on spiral galaxies and has measured the rotational velocities of the arms of the galaxy as their distance from the center increases. The velocities of the spiral arms are measured by determining their doppler shifts. That is light emitted from a body moving away from an observer will show a red shift, and a blue shift when emitted from a body moving toward the observer. The degree of spectral shift is proportional to the velocity of the source.

The initial assumption, based upon KEPLER's laws, was that rotational velocity would decrease with distance. Thus the theoretical expectation was that: $v^2 = GM/r^2$ where G is the gravitational constant, M the attracting mass, and r the orbital radius. It is clear from the equation that as r increases, v will decrease.

Rubin, however, found that the rotational velocity of spiral galaxies either remains constant with increasing distance from the center or rises slightly. The only possible conclusion, assuming the laws of motion, was that the figure for M was too low. But as all visible matter had been taken into account in assessing the mass of the galaxy, the missing mass must be present in the form of "dark matter." Rubin found similar results as she extended her survey. It seemed to her in 1983 that as much as 90% of the universe is not radiating sufficiently strongly on any wavelength detectable on Earth.

Rubin's work has presented modern astronomy with two major problems. Firstly to calculate the amount of dark matter in the universe and describe its distribution, and secondly to identify particles that make up the dark matter.

Earlier in her career, in collaboration with Kent Ford, Rubin made the extraordinary discovery that the Milky Way had a peculiar velocity of 500 kilometers per second quite independently of the expansion of the universe. When their results were published in 1975 they were met with considerable skepticism and it was assumed they had miscalculated the distances of the measured galaxies. However, later work by John HUCHRA and others in 1982 seems to have confirmed their measurements.

Rubner, Max (1854–1932) *German physiologist and hygienist*

Rubner (**roob**-ner), who was born at Munich in Germany, was a student of Karl von Voit and became professor of physiology at both Marburg and Berlin universities. He is best known for his work on mammalian heat production and for exposition of the surface law, which indicates that the rate of metabolism is proportional to the surface area of the body, not to weight. This principle appears to be related to mammalian temperature regulation and heat loss via the skin. Rubner also showed that recently fed animals lost heat more readily than fasting ones, indicating some sort of cellular regulatory system. In 1894 Rubner demonstrated the direct analogy between human energy production, accompanied by heat, and actual burning (as in a fire). Since the same amount of energy is released in both cases, given that the quantity of food digested/burned is the same, this clearly indicated that the first law of thermodynamics applies to both animate and inanimate objects, thus refuting the theory of vitalism. Rubner also investigated and compared the energy-producing potential of various foodstuffs, showing that carbohydrates, fats, and proteins were broken down with equal readiness, and proved that an animal's energy consumption for growth purposes is a constant proportion of its total energy output. Rubner's

other work included studies of the nutritional requirements of infants and the physiological effects of different climates on man.

Rudbeck, Olof (1630–1702) *Swedish naturalist*

Rudbeck (**rood**-bek), who was born at Westerås in Sweden, studied at the University of Uppsala, where he taught as professor of anatomy, botany, chemistry, and mathematics at the medical school. He became chancellor of Uppsala University at the age of 31, built a botanic garden, and founded a polytechnic institute, of which he became curator. Rudbeck's scientific investigations are a mixture of fact and fancy. In 1651 he discovered the vertebrate lymphatic system, in particular that of the intestine and its connection with the thoracic duct; but he also believed that Sweden was the site of PLATO's Atlantis and the cradle of civilization (described in his *Atlantikan*, 1675–98; Atlantis). The plant genus *Rudbeckia* is named for him.

Ruelle, David (1935–) *Belgian mathematical physicist*

Ruelle (roo-**el**) was born in Ghent, Belgium, and educated at the Mons Polytechnic and the Université Libre de Bruxelles. After holding short appointments at the Federal Institute of Technology, Zurich, and at the Institute for Advanced Studies, Princeton, he was appointed in 1964 to the Institut des Hautes Etudes Scientifiques, Bures-sur-Yvette, Paris. Since 2000, Ruelle has also been visiting professor at Rutgers University.

In 1971, in collaboration with the Dutch mathematician Floris Takens, Ruelle published an important paper entitled *On the Nature of Turbulence*. The prevailing theory of turbulence at the time was developed by Lev LANDAU. He had argued that when a fluid is set into motion a number of modes are excited. A single mode produces a periodic oscillation, and no mode allows a steady flow; an irregular flow is produced by several excited modes, while many modes will set up a turbulent flow.

Ruelle argued that turbulence was not really a superposition of many modes, but resulted from what he termed to be "a strange attractor." The origin of the term derives from the LORENZ attractor used by Edward Lorenz to describe certain chaotic systems. The Lorenz attractor was strange, Ruelle added, because it was fractal in the sense of Benoit MANDELBROT, it had a sensitive dependence on initial conditions, and a continuum of frequencies.

Rumford, Benjamin Thompson, Count (1753–1814) *American–British physicist*

Benjamin Thompson was the son of a farmer from Woburn, Massachusetts. He started his career apprenticed to a merchant but was injured in a fireworks accident and moved to Boston. In 1772 he married a rich widow and moved to live in Rumford (now Concord, New Hampshire). When the American Revolution broke out, he took the English side and spied for them. By 1775 the hostility of his countrymen toward him had grown to such a pitch that he was forced to sail to England, leaving his wife and daughter behind. Once in England, his opportunist nature quickly raised him to the position of colonial undersecretary of state but, with the end of the American Revolution, he moved to Bavaria.

Here he rose rapidly to high government administrative positions and initiated many social reforms, such as the creation of military workhouses for the poor and the introduction of the potato as a staple food. In 1790 he was made a count in recognition of his service to Bavaria, taking the name of his title from Rumford, New Hampshire.

It was in Bavaria that he first became interested in science, when he was commissioned to oversee the boring of cannon at the Munich Arsenal. Rumford was struck by the amount of heat generated and suggested that it resulted from the mechanical work performed.

According to the old theory of heat, heat produced by friction was caloric "squeezed" from the solid, although it was difficult to explain why the heat should be released indefinitely. Rumford, in his paper to the Royal Society *An Experimental Enquiry concerning the Source of Heat excited by Friction* (1798), suggested the direct conversion of work into heat and made quantitative estimates of the amount of heat generated. It was suggested that the heat came from the lower heat capacity of the metal turnings, although Rumford could discount this by using a blunt borer to show that the turnings produced were not important. Another objection – that the heat came from chemical reaction of air with the fresh surface – was disproved by an experiment of Humphry DAVY (1799) in which pieces of ice were rubbed together in a vacuum. The idea that heat was a form of motion replaced LAVOISIER's caloric theory over the first half of the 19th century.

Rumford returned to London in 1798 and there began work on a series of inventions, including a kitchen stove. More lastingly, he established the Royal Institution of Great Britain (1800), introducing Davy as director. He went to France in 1804 and settled in Paris, where he married Lavoisier's widow. The marriage was unhappy and ended after four years (Rumford is said to have suggested that Lavoisier was lucky to have been guillotined). Rumford himself appears to have been a disloyal and unappealing character, although at the end of his life he left most of his estate to the United States.

Runcorn, Stanley Keith (1922–1995) *British geophysicist*

Runcorn was born at Southport in Lancashire, England, and was educated at Cambridge University. After working on radar during World War II he held teaching appointments at the University of Manchester (1946–50) and at Cambridge (1950–55), before being appointed in 1956 to the chair of physics at King's College, Newcastle, which became the University of Newcastle upon Tyne in 1963. He became Senior Research Fellow at Imperial College, London, in 1989.

Under the early influence of Patrick BLACKETT, Runcorn began research on geomagnetism. From detailed field surveys in both Europe and America they were eventually able to reconstruct the movements of the North Magnetic Pole over the past 600 million years. Runcorn found that he obtained different routes for the migration depending on whether he used European or American rocks. Also, the European rocks always pointed to a position to the east of that indicated by the American rocks for the magnetic pole. From this evidence Runcorn argued, in his paper *Paleomagnetic Evidence for Continental Drift* (1962), that if the two continents were brought close to each other they could be so aligned that the magnetic evidence of their rocks pointed to a single path taken by the magnetic pole. This led Runcorn to become an early supporter of the newly emerging theory of continental drift.

Runcorn died in somewhat mysterious circumstances: he was found battered to death in a motel room in Los Angeles, where he had been attending a conference.

Runge, Friedlieb Ferdinand (1795–1867) *German chemist*

Runge (**ruung**-e), who was born at Hamburg in Germany, was apprenticed to an apothecary in 1810. He studied medicine at Jena, graduating in 1819, and later received his PhD from Berlin. He held the chair of chemistry at Breslau before becoming an industrial chemist in 1830.

His main work was concerned with the chemistry of dyes; he produced a massive three-volume work on the subject, *Farbenchemie* (Part 1, 1834; Part 2, 1842; Part 3, 1850; Color Chemistry). In 1834, using coal tar, he isolated quinoline, a compound that was to have considerable value later in the century. He also pioneered the use of paper chromatography as an analytical tool; he produced a work in 1855 that actually had chromatograms incorporated in the text.

Rush, Benjamin (1745–1813) *American physician, chemist, and politician*

Man is said to be a compound of soul and body.

However proper this language may be in religion, it is not so in medicine. He is, in the eye of a physician, a single and indivisible being, for so intimately united are his soul and body, that one cannot be moved without the other.
—*Sixteen Introductory Lectures* (1811)

Rush was born in Philadelphia, Pennsylvania. After studying at Jersey College and being apprenticed to the physician Redman (a pupil of Boerhaave), he came to Europe to study medicine at Paris and at Edinburgh, where he received his MD in 1768. He was appointed professor of chemistry at Philadelphia University in 1769; he was also elected to Congress and was one of the signatories of the Declaration of Independence. In 1777 he served for a year as surgeon-general but resigned for political reasons and returned to his chair. In 1799 he became treasurer of the U.S. mint, a post he held until his death.

As a chemist Rush's main role was as a teacher and supporter of others rather than as a basic researcher. In medicine he pioneered more humane therapy for the insane; his *Medical Inquiries and Observations upon Diseases of the Mind* (1812) was one of the earliest modern works on the subject.

Rushton, William Albert Hugh (1901–1980) *British physiologist*

Rushton, the son of a London dental surgeon, studied medicine at Cambridge University and University College Hospital, London. He worked in Cambridge from 1931 until 1968, being appointed professor of visual physiology in 1966. In 1968 he moved to the Florida State University, Tallahassee, to serve as research professor of psychobiology until his retirement in 1976.

Rushton studied the theory of nerve excitation until the early 1950s but changed to studying visual pigments following some work with Ragnar GRANIT in Stockholm. Rushton's novel technique was to shine light into the eye and measure the amount reflected back with a photocell. The effects of rhodopsin, the "visual purple" of Willy Kuhne, could be discounted by working with the fovea, the retinal region of sharpest vision devoid of rods containing rhodopsin. As the fovea is also deficient in blue cones, Rushton argued that by limiting pigment absorption measurements to this small area the properties of the red and green cones alone should be revealed.

By examining color-blind individuals, Rushton showed that red-blind defectives lack the red-sensitive pigment erythrolabe and that people who cannot distinguish between red and green lack the green-sensitive pigment chlorolabe.

Ruska, Ernst August Friedrich (1906–1988) *German physicist*

Born in Heidelberg, Germany, Ruska (**ruus-ka**) was educated at the Munich Technical University and at Berlin University, where he obtained his PhD in 1934. He worked in industry until 1955 when he became professor of electron microscopy at the Haber Institute, Berlin, a post he held until his retirement in 1972.

It had long been known that optical microscopes are limited by the wavelength of light to a magnifying power of about 2,000, and the ability to resolve images no closer together than 2–3,000 angstroms (1 angstrom = 10^{-10} meter). In 1927, however, G. P. THOMSON first demonstrated that electrons can behave like waves as well as like particles. The wavelength of the electron depends on its momentum according to de BROGLIE's equation $\lambda = h/p$. The higher the momentum of the electron, the shorter the wavelength. It should be possible to focus short-wavelength electrons and obtain better resolving powers.

In 1928 Ruska attempted to focus an electron beam with an electromagnetic lens. He went on to add a second lens and thus produced the first electron microscope; it had a magnifying power of about seventeen. Improvements, however, came quickly and by 1933 the magnifying power had been increased to 7,000. Soon after he joined the firm of Siemens and began to work on the production of commercial models. The first such model appeared on the market in 1939. It had a resolution of about 250–500 angstroms.

For his work in this field Ruska shared the 1986 Nobel Prize for physics with Binning and ROHRER.

Russell, Bertrand, 3rd Earl Russell (1872–1970) *British philosopher and mathematician*

Mathematics may be defined as the subject in which we never know what we are talking about, nor whether what we are saying is true.
—*Mysticism and Logic*

Even if the open windows of science at first make us shiver after the cosy indoor warmth of traditional humanizing myths, in the end the fresh air brings vigor, and the great spaces have a splendor of their own.
—*What I Believe* (1925)

Russell, who was born at Trelleck, England, was orphaned at an early age and brought up in the home of his grandfather, the politician Lord John Russell. He was educated privately before attending Cambridge University (1890), from which he graduated (1893) in mathematics. In 1895 he became a fellow and lecturer at Cambridge. His work after 1920 was mainly devoted to the development of his philosophical and political opinions. He became well known for his popularization of many areas of philosophy and also, in works such as *The ABC of Atoms* (1923) and *The ABC of Relativity* (1925), of the new trends in scientific thought. For his writings he was awarded the 1950 Nobel Prize for literature. He succeeded his brother to become the 3rd Earl Russell in 1931. Throughout much of his life Russell was an intense advocate of pacifism and during World War I he was imprisoned for expressing these views. Later, in the 1950s and 1960s, he became a central figure in the movements criticizing the use of the atomic bomb, leading demonstrations and mass sit-downs and becoming president of the British Campaign for Nuclear Disarmament in 1958.

At Cambridge Russell became interested in the relatively new discipline of mathematical logic in which he was to be a pioneer. With Giuseppe PEANO he was one of the few to recognize the genius of Gottlob FREGE and his new system of logic. In 1902 he wrote to Frege, presenting what is now known as *Russell's paradox*, and asking how Frege's system would deal with it. (Unfortunately, as Frege acknowledged, the system could not accommodate it.) The paradox is one of the paradoxes of set theory and rests on the (then ill-defined) notion of a set. Some sets are members of themselves (the set of all sets is an example because it is itself a set; the set of cats is not an example, as it is not itself a cat). Consider the set of all sets that are not members of themselves: is it a member of itself? If it is, it is not and vice versa. To avoid such paradoxes Russell formulated his logical theory of types. In 1903 he began his collaboration with A. N. WHITEHEAD on their ambitious, if not entirely successful, project of placing mathematics on a sound axiomatic footing by deriving it from logic. This culminated in the publication of *Principia Mathematica* (1910, 1912, 1913), containing major advances in logic and the philosophy of mathematics.

Russell, Sir Edward John (1872–1965)
British agricultural scientist

When Russell, who was born at Frampton in England, left school at 14, he had already decided to pursue a career in chemistry. His first job in a chemist's shop disappointed him, but he attended evening classes and gained entrance to Owens College, Manchester, graduating in chemistry in 1896. While doing social work in Manchester, Russell conceived the idea of a rural settlement where the poor could be trained in agriculture and prepared for a healthier life in the country. To learn more about agriculture he applied, in 1901, for a lectureship at Wye College, London, against the advice of his superiors who could see no future for a chemist in agriculture.

Russell soon saw the impracticality of his settlement scheme but nevertheless quickly realized how greatly agriculture could be advanced by planned scientific research. At Wye he met A. D. Hall, with whom he produced *The Soils and Agriculture of Kent, Surrey and Sussex* (1911), which was recognized on publication as a classic example of a regional agricultural survey.

In 1907 Russell followed Hall to Rothamsted Experimental Station, succeeding Hall as director in 1912. In 1912 he also published his most famous book, *Soil Conditions and Plant Growth*, which has expanded over the years through many editions, the later ones being edited by his son E. W. Russell. He did some important research, particularly on soil sterilization, but his major achievement was to extend the staff and facilities at Rothamsted at a time when the British government seemed quite blind to the benefits of agricultural research.

Russell remained active into his old age, finishing his last book, a history of agricultural research, only a few weeks before his death at the age of 92.

Russell, Sir Frederick Stratten
(1897–1984) *British marine biologist*

Russell was born at Bridport in Dorset, England. Although he had plans to study medicine at Cambridge University, these were delayed by the outbreak of World War I, during which he served with the Royal Naval Air Service. His interest turned from medicine to biology at Cambridge and on graduation he spent a year in Alexandria, Egypt, studying the eggs and larvae of marine fish. He then joined the Marine Biological Association's Laboratory at Plymouth, serving as its director from 1945 to 1965. Russell did extensive research on marine planktonic organisms, their life histories, behavior, and relation to fisheries and water movements. With Sir Maurice Yonge he wrote *The Seas* (first published 1928, and many subsequent editions) as well as monographs on the British medusae (Coelenterata) and the eggs and planktonic stages of British sea fishes. He was knighted in 1965.

Russell, Henry Norris (1877–1957) *American astronomer*

> The pursuit of an idea is as exciting as the pursuit of a whale.
> —Quoted by A. L. Mackay in *A Dictionary of Scientific Quotations* (1991)

Russell, the son of a Presbyterian minister, was born in Oyster Bay, New York. A brilliant scholar at Princeton, he graduated in 1897 and obtained his PhD in 1899. He spent the period 1902–05 as a research student and assistant at Cambridge University, England, returning then to Princeton where he served as professor of astronomy from 1911 to 1927 and director of the university observatory from 1912 to 1947. He was also a research associate at the Mount Wilson Observatory in California (1922–42) and, after his retirement, at the Harvard and Lick observatories.

Russell's great achievement was his publication in 1913 of a major piece of research contained in what is now called the *Hertzsprung–Russell diagram* (H–R diagram). The same results had in fact been published earlier and independently by Ejnar HERTZSPRUNG with little impact. Russell's work was based upon determinations of absolute magnitudes, i.e., intrinsic brightness, of stars by the measurement of stellar parallax. His measurement technique was developed in collaboration with Arthur Hinks while he was at Cambridge and involved photographic plates, then a fairly recent scientific tool. He found that values of absolute magnitude correlated with the spectral types of the stars. Spectral type was derived from the Harvard system of spectral classification as revised by Annie CANNON and indicated surface temperature.

A graph of absolute magnitude versus spectral type produced the H–R diagram and showed that the majority of stars lie on a diagonal band, now called the "main sequence," in which magnitude increases with increasing surface temperature. A separate group of very bright stars lie above the main sequence. This meant that there could be stars of the same spectral type differing enormously in magnitude. To describe such a difference the now familiar terminology of "giant" and "dwarf" stars was introduced into the literature.

The most obvious feature of the diagram for Russell, however, was that it was not completely occupied by stars. This led him to propose a path of stellar evolution, which he put forward in 1913 at the same time as the diagram. He argued that stars evolve from hot giants, pass down the main sequence and end as cold dwarfs. The mechanism driving the change was that of contraction. The bulky giants of spectral type M contract and with the resulting rise of temperature move leftward in the diagram, gradually becoming B-type dwarfs. But at some stage the contraction and density become too great for the gas laws to apply and the star cools, slipping down the main sequence and evolving finally to an M-type dwarf.

By 1926, however, Arthur EDDINGTON could talk confidently of the overthrow of the "giant and dwarf theory"; it was too simple to fit the growing data on the distribution of mass and luminosity among the different spectral types of stars. Although Russell's evolutionary theory quickly fell from favor the H–R diagram has continued to be of enormous importance and the start for any new theory of stellar evolution.

Eclipsing binary stars, such as Algol, the "winking demon," were also of great interest to Russell. He devised methods by which both orbital and stellar size could be determined and which became widely used. He also analyzed the variations in light output of a large number of eclipsing binaries, which again became invaluable to later researchers.

Another major line of research for Russell was his investigation of the solar spectrum, which began as a result of the publication in 1921 of the ionization equation of Meghnad SAHA. The Saha equation was tested and modified by Russell, using the solar spectrum, and was then used by him to calculate the abundance of the chemical elements in the Sun's atmosphere. He realized that the abundances in other stars could also be calculated from their spectra. He showed that the abundance of elements within the Sun itself could be found and in 1929 published the first reliable determination of this, demonstrating surprisingly that 60% of the Sun's volume was hydrogen. Although this was an underestimate, as Donald MENZEL was later able to show that a figure of over 80% was more accurate, it did pose the problem as to why the Sun, and presumably other stars too, should contain so much hydrogen. The answer to this question was given in the version of the big-bang theory proposed by George GAMOW.

Rutherford, Daniel (1749–1819) *British chemist and botanist*

Rutherford, the son of an Edinburgh physician, studied under William Cullen and Joseph BLACK at Edinburgh University and became a doctor of medicine in 1777. In 1786 he was made professor of botany and keeper of the Royal Botanic Garden at Edinburgh.

Rutherford was the first to distinguish between carbon dioxide and nitrogen. A thesis he wrote in 1772, *De aere fixo dicto aut mephitico* (On Air said to be Fixed or Mephitic), contains some of Joseph PRIESTLEY's later discoveries. In his experiment mice were allowed to breathe in a closed container. The fixed air (carbon dioxide) was absorbed by caustic potash. The remaining air, Rutherford pointed out, was not fixed but would not support life or combustion and he called it "mephitic air." He had in fact isolated nitrogen about the same time as Karl SCHEELE.

Rutherford, Ernest, 1st Baron (1871–1937) *New Zealand physicist*

> Don't let me catch anyone talking about the Universe in my department.
> —Quoted by John Kendrew in "J. D. Bernal and the Origin of Life," BBC radio talk, 26 July 1968

> When we have found how the nucleus of atoms is built up we shall have found the greatest secret of all – except life. We shall have found the basis of everything – of the earth we walk on, of the air we breathe, of the sunshine, of our physical body itself, of everything in the

world, however great or however small –
except life.
—Quoted in *Passing Show*

Rutherford, who was born at Nelson in New Zealand, was certainly the greatest scientist to emerge from that country; he can also fairly be claimed to be one of the greatest experimental physicists of all time. His career almost exactly spans the first great period of nuclear physics, a field he did much to advance and which he dominated for so long. This period stretches from the detection of radioactivity by Henri BECQUEREL in 1896 to the discovery of nuclear fission by Otto HAHN in 1938, the year after Rutherford's death. He came from a fairly simple background, the fourth of twelve children, the son of a man variously described as a farmer, wheelwright, and miller. He was educated at Canterbury College, Christchurch, and in 1895 won a scholarship to Cambridge University, England.

In New Zealand Rutherford had done some work on high-frequency magnetic fields. At Cambridge, working under J. J. THOMSON, he first continued this research, and then in 1896 began to work on the conductivity of air ionized by x-rays. In 1898 he moved to become professor at McGill University in Canada. This was a good appointment for Rutherford in two respects. McGill had one of the best-equipped physics laboratories in the world and, in particular, there was a good supply of the then very costly radium bromide. The other main gain for Rutherford in Montreal was the presence of Frederick SODDY, an Oxford-trained chemist with whom he entered into a most rewarding eighteen-month collaboration, from October 1901 to April 1903, during which time they produced nine major papers laying the foundations for the serious study of radioactivity.

When Rutherford began working on radioactivity at the end of the century little was known about it apart from the result of Pierre and Marie CURIE that it was not limited to uranium alone but was also a property of thorium, radium, and polonium. Rutherford's first important advance, in 1899, was his demonstration that there were two quite different kinds of emission, which he referred to as alpha and beta rays. The first kind had little penetrating power but produced considerable ionization while the beta rays (electrons) were as penetrating as x-rays but possessed little ionizing power. To find out exactly what they were took Rutherford the best part of a decade of careful experimentation but, long before he had the final answer, he had used their existence to work out with Soddy a daring theory of atomic transmutation. In 1900 Rutherford showed that a third type of radiation, undeflected by magnetic fields, was high-energy electromagnetic radiation. He called this radiation "gamma rays."

Rutherford also began to investigate the radioactive element thorium, which in addition to alpha, beta, and gamma rays also emits a radioactive gas that he called "emanation." He showed that the emanation decayed in activity at a particular rate, losing half its activity in a fixed period of time (the half-life). Rutherford and Soddy began an intensive investigation of thorium compounds and showed that a more active substance, thorium X, was present. They eventually came to the view that the emanation was produced from the thorium X, which in turn came from the original thorium. In other words, there was a series in which chemical elements were being changed (transmuted) into other elements. Rutherford and Soddy published their theory of a series of transformations in 1905. Rutherford later published a book with the title *The Newer Alchemy* (1937). Soddy went on to continue this work, eventually introducing the idea of isotopes.

Rutherford directed his attention to the alpha radiation emitted in radioactive decay, proving that it consisted of helium atoms that have each lost two electrons. He continued to study alpha radiation, moving to the University of Manchester, England, in 1907. At Manchester Rutherford and Hans GEIGER invented the Geiger counter in 1908. Here too Geiger and E. Marsden in 1910, at Rutherford's suggestion, studied the scattering of alpha particles passing through thin metal foils. The particles were detected by a screen coated with zinc sulfide, which gives brief flashes of light (scintillations) when hit by high-energy particles.

Geiger and Marsden found that most of the particles were deflected only slightly on passing through the foil but that a small proportion (about 1 in 8,000) were widely deflected. Rutherford later described this as "quite the most incredible event that has ever happened to me in my life...It was almost as incredible as if you fired a 15-inch shell at a piece of tissue paper and it came back and hit you." To make sense of the results, Rutherford published in 1911 a model of the atom in which he suggested that almost all the mass was concentrated in a very small region and that most of the atom was "empty space." This was the nuclear atom (although Rutherford did not use the term "nucleus" until 1912). He also produced a theoretical formula giving the numbers of particles that would be scattered by a nucleus at different angles. The idea of the nuclear atom was developed further by Niels BOHR.

After the war, which Rutherford spent working for the Admiralty on sonic methods of detecting submarines, he moved in 1919 to take the Cavendish chair of physics and the directorship of the Cavendish Laboratory at Cambridge University, England. It was there that he announced in 1919 his third major discovery, the artificial disintegration of the nucleus. Following some earlier experiments of Marsden he placed an alpha-particle source in a cylinder into which various gases could be introduced. At one end of the cylinder a small hole was covered with a metal disk through which some atoms could escape and register their presence on a zinc sulfide screen. The introduction of nitrogen produced highly energetic particles, which turned out to be hydrogen nuclei (that is, protons). The implications were not lost on Rutherford who concluded that "the nitrogen atom is disintegrated under the intense forces developed in a close collision with the swift alpha particle, and that the hydrogen atom which is liberated formed a constituent part of the nitrogen nucleus." Occasionally, it was later shown, "the alpha particle actually enters the nitrogen nucleus ... breaks up ... hurling out a proton and leaving behind an oxygen nucleus of mass 17." Rutherford had thus succeeded in bringing about the first transmutation, although when described in nuclear terms it seems a simple enough process. With James CHADWICK he went on to show between 1920 and 1924 that most of the lighter elements emitted protons when bombarded with alpha particles.

By his work in 1911 and 1919 Rutherford had shown that not only does the atom have a nucleus but that the nucleus has a structure from which pieces can be knocked out and by which other particles can be absorbed. It was this work which virtually created a whole new discipline, that of nuclear physics. Rutherford received the Nobel Prize for chemistry in 1908.

Ružička, Leopold (1887–1976) *Croatian–Swiss chemist*

Ružička (**roo**-zheech-ka or **roo**-zich-ka) was born in Vukovar, Croatia, the son of a cooper. He graduated in chemistry from the Karlsruhe Institute of Technology, in Germany, where he became assistant to Hermann STAUDINGER, following him to Zurich in 1912. In 1926 he was appointed professor of organic chemistry at the University of Utrecht but in 1929 he returned to the Federal Institute of Technology at Zurich to take up a similar chair.

Beginning in 1916 Ružička worked on the chemistry of natural odorants. While investigating such compounds as musk and civet he discovered a number of ketone compounds containing large rings of carbon atoms.

From the early 1920s Ružička also worked on terpenes. By dehydrogenating the higher terpenes to give aromatic hydrocarbons he was able to determine the structure of pentacyclic triterpenes. He also corrected the formulas of the bile acids and cholesterol proposed by Adolf WINDAUS and Heinrich WIELAND. Ružička's theory that the carbon skeleton of higher terpenes could be seen as consisting of isoprene units proved a useful hypothesis in further work.

In the 1930s Ružička moved into the field of sex hormones. In 1931 Adolf BUTENANDT, with whom Ružička shared the 1939 Nobel Prize for chemistry, isolated 15 milligrams of the steroid hormone androsterone from 7,000 gallons of urine. Androsterone is a male hormone secreted by the adrenal gland and testis, which when released at puberty causes the development of male sexual characteristics. In 1934 Ružička succeeded in synthesizing it, the first of several such triumphs.

Rydberg, Johannes Robert (1854–1919) *Swedish physicist and spectroscopist*

Johannes Rydberg (**rid**-berg) was born in Halmstad, Sweden, and educated at Lund University, where he received a PhD in 1879. The next year he started teaching mathematics there and stayed at Lund for the rest of his life, taking the chair of physics in 1901.

All of Rydberg's work arose from his interest in the periodic classification of the elements introduced by Dmitri MENDELEEV. Rydberg's great intuition was that the periodicity was a result of the structure of the atom. His first research was into the relationship between the spectral lines of elements. In 1890 he found a general formula giving the frequency of the lines in the spectral series as a simple difference between two terms. His formula for a series of lines is:

$$\nu = R(1/m^2 - 1/n^2)$$

where n and m are integers. The constant R is now known as the *Rydberg constant*.

In the early 1900s Rydberg continued to work on the periodic table, reorganizing it, finding new mathematical patterns, and even casting it into spiral form. In the main his theoretical work was confirmed by Henry MOSELEY's discovery that the positive charge on the nucleus gave a better periodic ordering than the atomic weight.

Ryle, Sir Martin (1918–1984) *British radio astronomer*

Ryle, the son of a physician, was born at Brighton on the south coast of England and studied at Oxford University. He spent the war with the Telecommunications Research Establishment in Dorset working on radar. After the war he received a fellowship to the Cavendish Laboratory of Cambridge University and in 1948 was appointed lecturer in physics. In 1959 he became the first Cambridge professor of radio astronomy, having been made in 1957 the director of the Mullard Radio Astronomy Observatory in Cambridge. Ryle was appointed Astronomer Royal in 1972 and in 1974 was awarded, jointly with Antony HEWISH, the Nobel Prize for physics. He was knighted in 1966.

It was mainly due to Ryle and his colleagues that Cambridge, after the war, became one of the leading centers in the world for astronomical research. He realized that one of the first jobs to be done was simply to map the radio sky. He therefore began in 1950 the important series of Cambridge surveys. The first survey used the principle of interferometry and discovered some 50 radio sources. The second survey in 1955 listed nearly 2,000 sources, many of which turned out to be spurious. The crucial survey was the third one, the results of which were published in 1959 in the *Third Cambridge Catalogue* (3C). This listed the positions and strengths of 500 sources and has since become the definitive catalog used by all radio astronomers. The use of more sensitive receivers in 1965 enabled the 4C survey to detect sources five times fainter than those in the 3C and covered the whole of the northern sky; 5,000 sources were cataloged. Finally, with the opening of two highly sensitive radio telescopes in 1965 and 1971, important areas of the sky are being surveyed in depth: a full survey would take over 2,000 years.

The two new telescopes, the One Mile telescope and then the Five Kilometer telescope, operate by a technique developed by Ryle and called "aperture synthesis." A number of radio dishes are used to give a very large effective aperture, much larger than the aperture of a single dish, and hence produce very considerable resolution of detail in a radio map of an area of the sky. The dishes are mounted along a line, some fixed in position, some movable, and are used in pairs, at different distances apart, to form interferometers.

S

Sabatier, Paul (1854–1941) *French chemist*

Sabatier (sa-ba-**tyay**), who was born at Carcassone in southwest France, was a student at the Ecole Normale, Paris, and gained his PhD from the Collège de France in 1880. He became professor of chemistry at the University of Toulouse (1884–1930).

In 1897 Sabatier showed how various organic compounds could undergo hydrogenation, e.g., ethylene will not normally combine with hydrogen but when a mixture of the gases is passed over finely divided nickel, ethane is produced. Benzene can be converted into cyclohexane in the same way. Sabatier discussed the whole problem in his book *Le catalyse en chimie organique* (1912; Catalysis in Organic Chemistry), published the same year in which he was awarded the Nobel Prize for chemistry for his work on catalytic hydrogenations.

Sabin, Albert Bruce (1906–1993) *Polish–American microbiologist*

Sabin (**say**-bin), who was born at Bialystok in Poland, emigrated with his parents to America in 1921; he was educated at New York University, where he gained his MD in 1931. He later joined the staff of the Rockefeller Institute of Medical Research but in 1939 moved to the University of Cincinnati. Following war service with the U.S. Army Medical Corps, he was appointed (1946) research professor of pediatrics at Cincinnati, a post he held until his retirement in 1971.

It was clear to Sabin from the success of John ENDERS in growing polio virus in tissue culture that a vaccination against the disease was only a matter of time. Sabin already had experience in this field having worked on developing vaccines against dengue fever and Japanese B encephalitis. He therefore began work on the polio vaccine but, unlike Jonas SALK, he was determined to develop a live attenuated vaccine rather than a killed one.

Sabin was not the first to attempt to produce a live vaccine. Herald Cox and Hilary Koprowski working at the Lederle Laboratories of the Cyanamid Company produced such a vaccine, attenuated by repeated passages through mouse brains, in 1952. However they failed to produce convincing tests for their vaccines and, disillusioned, split up in 1956.

This left the field open to Sabin who by then had developed his own live virus, attenuated in monkey kidney tissue. It was impossible to persuade the American public to submit to the testing of another vaccine after the difficulties involved with the Salk campaign of 1954. Sabin therefore hit on the audacious idea of attempting to arouse the interest of the Russians and to persuade them to do his tests for him.

In 1959 Sabin was able to produce the results of 4.5 million vaccinations. They were completely safe, nor was there any reversion to the more virulent form found to develop in the vaccine of Cox and Koprowski. The vaccine also possessed a number of advantages over that of Salk: it gave a stronger longer lasting immunity thus making it unnecessary to give more than a single injection, which was in any case dispensable as the Sabin vaccine could be administered orally. Almost immediately the new vaccine began to take over from that of Salk. Great Britain changed over to the Sabin vaccine as early as 1962 with most other countries following soon afterward.

In 1973 Sabin reported a further major advance, this time in cancer research. With his collaborator G. Tarro he claimed to have evidence linking the herpes virus with a number of cancers. If this were true it would have been the first solid evidence in support of the viral origin of human cancer. In the following year Sabin took the difficult step of completely rejecting his own work when he found that he could no longer repeat his results.

Sabine, Sir Edward (1788–1883) *British geophysicist*

> To many men speculations are far more attractive than facts.
> —Letter to John Tyndall, 24 April 1855

Sabine, a Dubliner by birth, was educated at the Royal Military Academy, Woolwich, near London, and was commissioned in the Royal Artillery in 1803. He took part in a number of expeditions, sailing as astronomer and meteorologist with John Ross in 1818 and William Parry in 1820 in their search for the Northwest Passage. He made other trips to the tropics and Greenland. Using Henry Katers's pendulum, he made observations at different latitudes to investigate the figure (shape) of the Earth but his results overestimated its ellipticity. He also established magnetic observatories in several British colonies.

Sabine's main scientific achievement was in the field of geomagnetism. In 1851 he announced that he had detected a periodicity of about 10–11 years in the occurrence of magnetic perturbations, in which the magnetic needle deviates abnormally from its average position. This was also discovered by Johann von LAMONT at about the same time but Sabine took the further step of correlating the variations in magnetic activity with the sunspot cycle discovered by Heinrich Schwabe in 1843.

Sabine was secretary of the British Association (1838–59) and while in charge of the Royal Observatory at Kew he attempted to organize a number of small observatories throughout the world sending him data to be processed at Kew. He developed a theory in which the Earth's magnetic field was part of the atmosphere, but in 1839 Karl GAUSS succeeded in demonstrating that the magnetic field was restricted to the interior and surface of the Earth.

Sabine was knighted in 1869.

Sabine, Wallace Clement Ware (1868–1919) *American physicist*

Born in Richwood, Ohio, Sabine graduated from Ohio State University in 1886 and went on to do graduate work at Harvard. He was employed at Harvard from 1889, being made professor of physics in 1905.

Sabine is recognized as the founder of scientific architectural acoustics. In 1895 Harvard's Fogg Art Museum was opened and its lecture theater found to be "monumental in its acoustic badness." Sabine was asked if he could correct the fault. He found that a normally spoken word would reverberate for as long as 51/2 seconds, with the result that a speaker could be completing a long sentence while his first words were still reverberating through the hall. Sabine found that if he covered all the seats, the stage, and most of the floor with ordinary cushions he could reduce the reverberation time to a little more than a second.

Sabine went on to make a systematic study of the acoustics of buildings and eventually formulated a general law relating the reverberation time to the volume, surface area, and absorption coefficient of the room. His *Collected Papers on Acoustics* was published in 1922.

Sachs, Julius von (1832–1897) *German botanist*

> More and more I find that physiology achieves its most important results when it goes its own way entirely, without concerning itself very much with physics and chemistry.
> —Letter, 15 May 1879

Sachs (zahks), who was born at Breslau (now Wrocław in Poland), started his career as assistant to Johannes PURKINJE at the University of Prague, where he gained his PhD in 1856. The following year he qualified as a lecturer in plant physiology and taught at Prague until being appointed assistant in physiology at the Agricultural and Forestry College in Tharandt. (At this time plant physiology encompassed the whole of botany except systematics.) In 1861 Sachs obtained a teaching post at the Agricultural College, Poppelsdorf, and in 1867 succeeded Anton DE BARY as professor of botany at Freiburg University. A year later he became professor of botany at Würzburg University, where he remained until his death.

In 1865 Sachs established that the green pigment of plants, chlorophyll, is not distributed throughout the plant but is confined to discrete bodies, later named "chloroplasts." He showed that the starch in chloroplasts is the first visible product of photosynthesis, the process whereby carbon dioxide is taken up by the plant and converted to complex organic compounds. He also demonstrated that plants as well as animals respire, consuming oxygen and producing carbon dioxide. Sachs was interested in water movement in plants and in 1874 announced his erroneous "inhibition theory," in which he stated that absorbed water does not travel within the cell cavities but moves in tubes in the plant walls. He also studied the formation of annual growth rings in trees and the importance of turgidity in plant tissues for the mechanical support of the plant. He investigated plant growth responses (tropisms) to light and gravity and for these experiments invented the clinostat, which measures the effects of external agents on the movements of plants.

Of his many publications the best known is *Lehrbuch der Botanik* (1868; Textbook of Botany), translated into English in 1875.

Sadron, Charles Louis (1902–1990) *French physical chemist and biophysicist*

Born at Cluis in France, Sadron (sa-**dron**) was educated at the universities of Poitiers and Strasbourg. After teaching in a number of lycées, he was appointed professor of physics at the University of Strasbourg in 1937 while also serving as director of the Center for Research on Macromolecules from 1947 onward. In 1967 he moved to Orleans where he set up the Center for Molecular Biophysics until his retirement in 1974.

Sadron made substantial contributions to the study of macromolecules, particularly in solution. With his collaborators he developed a wide variety of techniques for investigating the properties of large molecules and for studying polymerization reactions. His group discovered the block copolymers – materials with a regular open matrix of one polymer with filaments or spheres of another polymer dispersed through it. In the early 1960s he turned his attention to studies of nucleic acids, proteins, and other biological macromolecules.

Sagan, Carl Edward (1934–1996) *American astronomer*

> Our loyalties are to the species and the planet. We speak for Earth. Our obligation to survive is owed not just to ourselves but also to that cosmos, ancient and vast, from which we spring.
> —*Cosmos* (1980)

Born in New York City, Sagan studied at the University of Chicago where he obtained his BS in 1955 and his PhD in 1960. He was a research fellow at the University of California, Berkeley, from 1960 to 1962 when he moved to the Smithsonian Astrophysical Observatory in Cambridge, Massachusetts, working at Harvard as lecturer then assistant professor. In 1968 he moved to Cornell where he was appointed director of the Laboratory for Planetary Studies and in 1970 became professor of astronomy and space science. Sagan's main work was on virtually all aspects of the solar system. One major line of research was on the physics and chemistry of planetary atmospheres and surfaces, especially of Mars. His other primary interest was the origin of life on Earth and the possibility of extraterrestrial life and he did much to interest the general public in the new field of exobiology. In laboratory experiments simulating the primitive atmosphere of Earth, he and his colleagues showed how a variety of organic molecules, such as amino acids, which are the building blocks of proteins, could readily be produced. The energy sources used in these syntheses included ultraviolet radiation, which would have flooded the Earth's primitive atmosphere, and high-pressure shock waves. It was while working with C. Ponnamperuna and Ruth Mariner at NASA's exobiology division in 1963 that he showed how the fundamental molecule adenosine triphosphate, ATP, could have been produced. ATP is the universal energy intermediary of living organisms and without its presence it is difficult to see how life could have ever originated on Earth.

In 1984 Sagan coauthored, with R. Turco, O. Toon, T. Ackerman and J. Pollock, an influential paper, *Nuclear Winter: Global Consequences of Multiple Nuclear Explosions*, referred to since as the TTAPS paper. The authors argued that even a relatively small-scale nuclear bomb of 5,000 megatons would create enough atmospheric smoke (300 million tons) and dust (15 million tons) to produce a temperature drop of 20–40°C, which would persist for many months. This prolonged nuclear winter would destroy much of the world's agriculture and industry. The impact of the paper on politicians and the public was dramatic.

The nuclear-winter argument itself was heavily criticized by scientists. "These guys don't know what they are talking about," commented Richard FEYNMAN, and Freeman DYSON dismissed the paper as an "absolutely atrocious piece of science." The meteorologist S. Schneider pointed out that the TTAPS model was one-dimensional in that it represented only the vertical structure of the atmosphere and ignored the oceans and seasonal changes. It was, he concluded, "a first generation assessment whose conclusions would have to be modified," and claimed that it threatened more a "nuclear fall" than a nuclear winter.

Sagan wrote extensively on the results of planetary science. His *Cosmic Connection* (1973) introduced these results to a wider audience. His subsequent works *The Dragons of Eden* (1977) and *Broca's Brain* (1979) have tried to do the same for recent advances in the theory of evolution and neurophysiology. In a further work, *Cosmos* (1980), based on a major TV series, Sagan charted the history of physics and astronomy. His later books include *Pale Blue Dot: A Vision of the Human Future in Space* (1994) and *The Demon-Haunted World* (1996).

Saha, Meghnad N. (1894–1956) *Indian astrophysicist*

Born the son of a small shopkeeper in Dacca (now in Bangladesh), Saha (**sah**-hah) won a scholarship to the Government School there in 1905 but was expelled for participating in the boycott of a visit by the Governor of Bengal. He completed his education in Calcutta, at the Presidency College, where he obtained his MA in applied mathematics in 1915. After lecturing in mathematics and then physics at Calcutta's University College of Science (1916–19), he visited London and Berlin on a traveling scholarship. He returned to India in 1921 and from 1923 to 1938 taught at the University of Allahabad. In 1938 he moved to the chair of physics at Calcutta where he remained until his death.

Early in his career Saha became interested in both thermodynamics and astrophysics and this led to his work on the thermal ionization that occurs in the very hot atmospheres of stars. In 1920 he published a fundamental paper, *On Ionization in the Solar Chromosphere*, in which he stated his ionization equation. The absorption lines of stellar spectra differ widely, with some stars showing virtually nothing but hydrogen and helium lines while others show vast numbers of lines of different metals. Saha's great insight was to see that all these spectral lines could be represented as the result of ionization. He saw that the degree of ionization, i.e., the number of electrons stripped away from the nucleus, would depend primarily on temperature. As the temperature increases, so does the proportion of ionized atoms. The remaining neutral atoms will thus produce only weak absorption lines that, when the temperature gets high enough, will disappear entirely. But the singly, doubly, and even triply ionized atoms will absorb at different sets of wavelengths, and different sets of lines will appear in stellar spectra, becoming stronger as the proportions of these ions grow.

In later years Saha moved into nuclear physics and worked for the creation of an institute for its study in India, which was later named for him. He also devoted much time to social, economic, and political problems in India.

Sakharov, Andrei Dmitriyevich (1921–1989) *Russian physicist*

> Every day I saw the huge material, intellectual and nervous resources of thousands of people being poured into the creation of a means of total destruction...I noticed that the control levers were in the hands of people who, though talented in their own ways, were cynical ... Beginning in the late fifties, one got an increasingly clearer picture of the collective might of the military-industrial complex and of its vigorous, unprincipled leaders, blind to everything except their "job."
> —*Sakharov Speaks* (1974)

Born the son of a physics teacher in Moscow, Sakharov (**sah**-ka-rof) graduated from the university there in 1942 just before the German invasion. He spent the war working as an engineer. In 1945 he joined the Lebedev Physics Institute in Moscow and began to work on cosmic rays. In 1950, in collaboration with Igor TAMM, Sakharov described a process whereby a deuterium plasma could be confined in a magnetic bottle in such a way as to extract energy produced by nuclear fusion. The design became known later in the West as the tokamak.

Soon after, Sakharov was deployed to secret research on the development of nuclear weapons. It is widely assumed that he played a crucial role in the explosion of the first Soviet hydrogen bomb in 1954. He continued this work until 1968, when he published his famous pamphlet, *Progress, Peaceful Coexistence and Intellectual Freedom*, which argued for a global reduction in nuclear weapons and the granting of civil rights in the Soviet Union.

Sakharov was immediately moved to nonclassified work back at the Lebedev Institute, and further agitation led to increased harassment by the authorities. He was awarded the Nobel Peace prize in 1975 and was finally sent into exile to Gorky in 1980. He was eventually released in 1986 and lived just long enough to see the fruition of Gorbachev's reforms, being elected shortly before his death to the Congress of the USSR.

Sakharov also made major contributions to several areas of theoretical physics. In 1966 he offered important evidence in support for the existence of quarks, the particles first proposed in 1964 by GELL-MANN. Also in 1966 he offered an explanation for the apparent dearth of antimatter in the observed universe.

Sakmann, Bert (1942–) *German biophysicist*

Born in Stuttgart, Germany, Sakmann (**zahk**-man) attended the universities of Tübingen and Munich, and gained his MD from the University of Göttingen. He became a research assistant at the Max Planck Institute of Psychiatry, Munich (1969–70), and subsequently spent two years as a British Council Fellow at the Biophysics Department of University College, London (1971–73). In 1974 he joined the Max Planck Institute for Biophysical Chemistry at the University of Göttingen, becoming head of the membrane physiology unit in 1983 and director in 1985. Two years later he was appointed professor in the department of cell physiology. In 1989 he moved to Heidelberg as director of the cell physiology department of the Max Planck Institute for Medical Research.

While working at Göttingen in the mid-1970s, in collaboration with the biophysicist Erwin NEHER, Sakmann developed the so-called "patch-clamp" technique for studying ion channels in cell membranes. This, together with their descriptions of the biophysical properties of the channels, earned them the 1991 Nobel Prize for physiology or medicine. Sakmann still works at the Max Planck Institute.

Salam, Abdus (1926–1996) *Pakistani physicist*

> One-eighth of the Koran is an exhortation to the believers to study nature and to find the signs of God in the phenomena of nature. So Islam has no conflict with science.
> —Wolpert and Richards, *A Passion for Science* (1988)

> The whole history of particle physics, or of physics, is one of getting down the number of concepts to as few as possible.
> —As above

Salam (sah-**lahm**), who was born at Jhang in Pakistan, attended Punjab University and Cambridge University, where he received his PhD in 1952. From 1951 to 1954 he was a professor of mathematics at the Government College of Lahore, concurrently with a post as head of the mathematics department of Punjab University. From 1954 until 1956 he lectured at Cambridge and from 1957 to 1993 he was a professor of theoretical physics at the Imperial College of Science and Technology, London. He was largely responsible for the establishment in 1964 of the International Center for Theoretical Physics, Trieste, as an institute to assist physicists from developing countries. He was director of the center from its inception until

1994, dividing his time between there and Imperial College.

Salam's work has been concerned with the theories describing the behavior and properties of elementary particles; for this he received the 1979 Nobel Prize for physics, shared with Sheldon GLASHOW and Steven WEINBERG. Although the three men did most of their work independently, they each contributed to the development of a theory that could take account of the "weak" and "electromagnetic" interactions. One of their predictions was the phenomenon of neutral currents and their strengths, which was first confirmed in 1973 at the European Organization for Nuclear Research (CERN) and later by other groups. A further prediction of the theory is that of the existence of "intermediate vector bosons" with high masses. The discovery of a vector boson was reported in 1983 by two teams (comprising 180 scientists) working at the European Laboratory for Particle Physics near Geneva.

Salisbury, Sir Edward James (1886–1978) *British botanist*

Salisbury, who was born at Harpenden in Hertfordshire, England, was educated at University College, London, where he joined the faculty at the end of World War I. He served as professor of botany from 1929 until 1943 when he was appointed director of the Royal Botanic Gardens, Kew, an office he held until his retirement in 1956.

Salisbury's first substantial work was his *Plant Form and Function* (1938) written in collaboration with Felix FRITSCH, a widely used textbook. However Salisbury was primarily a plant ecologist and did much work both on the effects of soil conditions and on the seed-producing capacity of British species. This was presented in his *The Reproductive Capacity of Plants* (1942). He also carried out a long-term study of sand-dune ecology, the results of which were published in his *Downs and Dunes* (1952). He also wrote the popular horticultural work *The Living Garden* (1935).

Salk, Jonas Edward (1914–1995) *American microbiologist*

> We are on our way to a new age of immunization throughout the world. I even look forward to the day when the principle of the killed virus will be used in a single vaccine against virtually all the virus diseases of man.
> —On the future of immunization

> The people – could you patent the sun?
> —On being asked who owned the patent on his polio vaccine. Quoted by S. Bolton in *Famous Men of Science*

Salk was born in New York City, the son of a garment worker. He was educated at the City College of New York and at New York University Medical School, where he obtained his MD

in 1939. In 1942 he went to the University of Michigan where he worked as a research fellow on influenza vaccine under Thomas Francis. In 1947 he moved to the University of Pittsburgh, serving as professor of bacteriology from 1949 onward. Here Salk began the work that eventually led to the discovery of a successful polio vaccine in 1954. After this breakthrough Salk was invited in 1963 to become director of the Institute for Biological Studies at San Diego, California, known more simply as the Salk Institute – soon to emerge as one of the great research centers of the world.

Salk was not the first to develop a vaccine against polio for in 1935 killed and attenuated vaccines were tested on over 10,000 children. It turned out, however, that not only were the vaccines ineffective, but they were also unsafe and probably responsible for some deaths and a few cases of paralysis.

By the time Salk began his work in the early 1950s a number of crucial advances had been made since the 1935 tragedy. In 1949 John ENDERS and his colleagues had shown how to culture the polio virus in embryonic tissue. Another essential step was the demonstration, in 1949, that there were in fact three types of polio virus with the inconvenient consequence that vaccine effective against any one type was likely to be powerless against the other two.

Salk had to develop a vaccine that was safe but potent. To ensure its safety he used virus exposed to formaldehyde for up to 13 days and afterward tested for virulence in monkey brains. This, in theory, allowed a large safety margin, for Salk could detect no live virus after only three days.

To test its potency Salk injected children who had already had polio and noted any increase in their antibody level. When it became clear that high antibody levels were produced by the killed vaccine, Salk moved on to submitting it to the vital test of a mass trial. Two objections were raised to this. One from Albert SABIN that killed vaccine was simply the wrong type to be used and a second, from various workers, who claimed to find live virus in the supposedly killed vaccine.

Despite such qualms Salk continued with the trial, administering in 1954 either a placebo or killed vaccine to 1,829,916 children. The evaluation of the trial was put into the hands of the virologist Thomas Francis, who, in March 1955, reported that the vaccination was 80–90% effective. The vaccine was released for general use in the United States in April 1955.

Salk became a national hero overnight and plans went ahead to vaccinate 9 million children. However, within weeks there were reports from California in which children developed paralytic polio shortly after being vaccinated. After a period of considerable confusion it became clear that all such cases involved vaccine prepared in a single laboratory.

After several days of almost continuous debate, the decision was taken to proceed and, by the end of 1955, 7 million doses had been administered. Further technical improvements were made in the production process. These safeguards would either eliminate any further cases of live vaccine or if a live virus did manage to penetrate all defenses it should make its presence known long before its use in a vaccine.

The courage shown by Salk in persisting with his vaccine was clearly justified by the results for the period 1956–58 – 200 million injections without a single case of vaccine-produced paralysis.

Samuelsson, Bengt Ingemar (1934–   ) *Swedish biochemist*

Born at Halmstad in Sweden, Samuelsson (**sam**-oo-el-son) was educated at the Karolinska Institute, Stockholm, where he gained his MD in 1961. He continued to work there until 1966, when he moved to the Royal Veterinary College, Stockholm, as professor of medical chemistry. In 1972 he returned to the Karolinska Institute as professor of medicine and physiological chemistry.

Samuelsson has worked extensively on the hormonelike prostaglandins first identified by Ulf VON EULER in the 1930s. With Sune BERGSTRÖM in the 1950s he worked out the structure of several prostaglandins and went on to explore some of their physiological properties. Later work by Samuelsson and his colleagues indicated a relationship between prostaglandins and the chemicals involved in transmitting nerve impulses. He discovered the prostaglandin PGA_2, thromboxane, which causes blood vessels to contract and platelets to clump. A second prostaglandin, PGE_2, inhibits the release of norepinephrine, a neurotransmitter of the sympathetic nervous system, and thereby blocks the transmission of nerve impulses.

Samuelsson also worked out the structure of a number of prostaglandins. They were shown to be closely related and synthesized in the body from a number of polyunsaturated fatty acids. All prostaglandins were found to be 20-carbon carboxylic acids with a five-member carbon ring. They are divided into three series – PG_1, PG_2, and PG_3 – depending on whether they have one, two, or three double bonds. The different prostaglandins could be distinguished by the location of an oxygen atom or a hydroxyl group (OH). For his work on prostaglandins Samuelsson shared the 1982 Nobel Prize for physiology or medicine with John VANE and Sune Bergström.

Sanctorius, Sanctorius (1561–1636) *Italian physiologist*

No medical book has attained this perfection.
—Herman Boerhaave on Sanctorius's *De statica medicina* (On Medical Measurement)

Sanctorius (sangk-**tor**-ee-us), who was born at Capodistria (now in Croatia), graduated in medicine from the University of Padua in 1582. It is believed he may then have spent some years in Poland serving the king and his court. On his return he was appointed professor of theoretical medicine at the University of Padua from 1611 until 1624 when he retired to devote himself to private research.

Sanctorius was a highly original and creative scientist whose work was characterized by a determination to introduce measurement and quantification into physiology. Thus in 1612 he was probably the first to describe and construct a clinical thermometer. This was just one of a number of instruments designed by him. He also devised a "pulsilogium," which measured the pulse rate by means of a pendulum.

With his battery of instruments Sanctorius conducted a series of classic experiments that, recorded in his *De statica medicina* (1614; On Medical Measurement), virtually founded the modern study of animal metabolism. Starting from GALEN's proposal that the skin "breathed" an "insensible perspiration," Sanctorius introduced an experimental setup incorporating a chair attached to a large balance in which such changes could be accurately measured. By recording his weight before and after various activities, he was able to demonstrate the existence of "insensible perspiration" and to show that its production "was more extensive than all the visible and palpable excreta taken together," accounting for about 62% of weight loss.

His work, however, failed to have any great impact on contemporary science, one reason being that Sanctorious was somewhat ahead of his time, for there seemed little use to which his insights could be put in the context of 17th-century physiology.

Sandage, Allan Rex (1926–   ) *American astronomer*

Born in Iowa City, Sandage graduated from the University of Illinois in 1948 and obtained his PhD in 1953 from the California Institute of Technology. He was on the staff of the Hale Observatories at Mount Wilson and Mount Palomar from 1952 when he began as an assistant to Edwin HUBBLE. He was professor of physics at Johns Hopkins University (1987–88) and is on the staff of the Carnegie Observatories, Pasadena, where he still works as an emeritus professor.

In 1960, using the 200-inch (5-m) reflecting telescope, Sandage in collaboration with Thomas Matthews succeeded in making the first optical identification of a quasar or quasi-stellar object. Quasars first came to the notice of astronomers when a number of compact, rather than extended, radio sources were detected by the Cambridge surveys of the radio sky carried out in the 1950s. Sandage and Matthews showed that 3C 48, a compact radio source from the third Cambridge survey, was at the same position as a faint apparently starlike object. Sandage and others succeeded in obtaining spectra of 3C 48 that were found to be quite unlike those of any other star. The mystery of these strange new objects was partially cleared up by Maarten SCHMIDT in 1963 when he showed that the spectral lines of a quasar have undergone an immense shift in wavelength.

Sandage continued to work on quasars and in 1965 introduced a method of identifying them by searching at an indicated radio position for objects emitting an excessive amount of ultraviolet or blue radiation. He found that many ultraviolet objects, which he named blue stellar objects or BSOs, were not radio emitters but could still be classed as quasars because they had the characteristic immense red shift first detected by Schmidt. He speculated that these might be older quasars that had passed beyond the radio phase of their life cycle. It is now known that the vast majority of quasars are not radio sources.

Following the work of Hubble on the expanding universe, Sandage has also tackled the difficult question of the age of the universe and whether it will continue to expand forever. He has claimed that the rate of expansion is slowing down so that the universe will eventually stop expanding and start contracting back to the original primal atom, at which point the cycle will begin again. The time scale of this "oscillating universe" he predicted as about 80 billion years: there would thus be about another 25 billion years of expansion to complete before beginning the 40 billion year contraction. Such theories, depending as they do on such difficult measurements as the density of matter in the universe as a whole, are difficult to test and are not surprisingly frequently revised.

Sänger, Eugen (1905–1964) *Austrian rocket scientist*

Sänger (**seng**-er) was born at Pressnitz (now Přisečnice in the Czech Republic) and was educated at the universities of Graz and Vienna, where he graduated in aeroscience in 1929. He taught at the Vienna Technical Institute before moving to Germany in 1936. He remained in Germany until 1945, serving at the Institute for Rocket Flight at Trauen and from 1942 at the Gliding Research Station, Ainring, Bavaria. From 1945 Sänger advised the French ministry of armaments until his appointment as head of the Jet Propulsion Institute, Stuttgart, in 1954. A dispute with the Federal Government led to his resignation and acceptance of a chair of space-travel research in 1963 at the Technical University, West Berlin.

Sänger was one of the leading pioneers of rockets and space travel. He published in 1933 the influential *Technique of Rocket Flight* and did much in the 1930s to work out both the theory and practice of liquid-propelled rockets. During the war he worked on the development of the ram jet and went on to draw up plans for his proposed "antipodal" plane capable of bombing New York from Berlin. Launched from a sled and flying at an altitude of 150 kilometers (93 mi), it had a planned range of 23,000 kilometers (14,300 mi). He also produced, with Irene Bredt, later to be his wife, the secret report *Rocket Propulsion of Long-Range Bombers*.

Sanger, Frederick (1918–) *British biochemist*

Sanger, a physician's son from Rendcombe in England, received both his BA and his PhD from Cambridge University (in 1939 and 1943 respectively). He continued his research at the university and from 1951 until 1983 was a member of the scientific staff of the Medical Research Council. In 1955, after some ten years' work, Sanger established the complete amino-acid sequence of the protein bovine insulin. This was one of the first protein structures identified, and Sanger received the Nobel Prize for chemistry in 1958 in recognition of his achievement. Sanger's work enabled chemists to synthesize insulin artificially and also generally stimulated research in protein structure.

In 1977 Sanger's team at the MRC laboratories, Cambridge, published the complete nucleotide (base) sequence of the genetic material (DNA) of the virus Phi X 174. This involves determining the order of 5,400 nucleotides along the single circular DNA strand. Moreover they found two cases of genes located within genes. Previously it had been thought that genes could not overlap. Sanger's research required the development of new techniques for splitting the DNA into different-sized fragments. These are radioactively labeled and then separated by electrophoresis. The base sequence can then be worked out because it is known which base is located at the end of each fragment due to the specificity of the enzymes (the so-called restriction enzymes) used to split the DNA. Sanger was awarded the Nobel Prize for chemistry a second time (1980) for his work on determining the base sequences of nucleic acids. He retired from the Sanger Institute in 1982.

Sarich, Vincent M. (1934–) *American chemist and anthropologist*

Chicago-born Sarich was educated at the University of California, Berkeley, where he gained his PhD in 1967. He has remained at Berkeley, being appointed professor of biochemistry in 1981.

Sarich was struck by the range of dates, from 4 million to 30 million years ago, within which anthropologists of the early 1960s placed the origin of the split between the hominids and the great apes. He began work, in collaboration with his Berkeley colleague Allan WILSON, to see if there was a more precise method of dating using the genetic relationship between man and apes. They chose to work with proteins, which closely reflect genes, choosing the blood-serum protein, albumin. As man and apes diverged further from their common ancestor, their albumins would also have diverged and would now be recognizably different.

Serum samples from apes, monkeys, and man were purified and then injected into rabbits to produce antiserums. A rabbit immunized against a human sample (antigen) will also react to other anthropoid antigens, only not as strongly. As antigenic differences are genetically based, response differences will therefore measure genetic differences between species. Sarich chose to work with a group of proteins found in blood serum known collectively as the "complement system." Antigens tend to attract and fix some of the complement. The amount of complement fixed could be measured precisely. Thus differences in complement-fixation rates produced by the albumin of man and gorilla when injected into immunized rabbits would measure their immunological distance.

If it could be assumed that protein differences between species have evolved at a constant rate, then immunological distance would also be a measure of evolutionary separation. It remained to calibrate the clock. Sarich and Wilson took as their base line the date 30 million years ago marking the separation between the hominoids and the Old-World monkeys. Thereafter it was a relatively easy matter to turn immunological distance into dates.

The results were clear but surprising. *Homo*, on this scheme, separated from the chimps and gorillas only 5 million years ago. This was a bold claim to make in 1967 as orthodox opinion, argued for example by David PILBEAM, placed the split between hominids and hominoids closer to 15 million years. What is more they had the skull of *Ramapithecus*, dating from this period, to prove their point. Initially, therefore, Sarich's views were rejected out of hand by paleontologists. Slowly, however, Sarich made converts. He argued, "I know my molecules have ancestors, you must prove your fossils had descendants." They found it more and more difficult to do so. Consequently, when it became clear that *Ramapithecus* was the ancestor not of man but of the orang-utan, opposition to Sarich largely disappeared.

Saussure, Horace Bénédict de (1740–1799) *Swiss physicist and geologist*

> In my youth, when I had crossed the Alps only a few times, I believed I had grasped all the facts ... Since then, further explorations of different parts of the mountain chain have shown me that the only constant thing in the Alps is their great variety.
> —*Voyages dans les Alpes* (1779–96; Travels in the Alps)

Saussure (soh-**soor**) was the son of an agriculturist. He was educated at the university in his native city of Geneva, graduating in 1759, and became professor of physics and philosophy at the Geneva Academy (1762–86).

He was the first to make a systematic and prolonged study of the Alps, his work being published in his classic *Voyages dans les Alpes* (1779–96; Travels in the Alps). He began as a disciple of Abraham WERNER, accepting that the mass of the Alps had crystallized from the primitive ocean. However, 17 years of studying the convolutions and folds of the Alps led him to state in 1796 that such folds could only be produced by some force acting from below, or that they must have actually been laid down folded. Both alternatives preclude deposition by water. He was unwilling to decide in favor of the plutonist theories of James HUTTON and thus to introduce fire as an agent, for he could recognize no sign of it.

He also collected considerable meteorological data and developed, in 1783, an improved hygrometer to measure humidity, using a human hair. He made the second ascent of Mont Blanc (1787) and at the summit took many scientific recordings. He also made nocturnal recordings and, in 1788, stayed for 17 days on the 11,000-foot (3,358 m) summit of Col du Géant.

Savery, Thomas (c. 1650–1715) *British engineer and inventor*

> ... vessells or engines for raiseing water or occasioning motion to any sort of millworks by the impellent force of fire.
> —Terms of Savery's patent for his steam pump (1698)

Little is known of Savery's early life except that he was born at Shilstone in England and that he probably trained as a military engineer. During the 1690s he tried to find a solution to the problem of pumping water from mines and built the first steam engine to be used during the Industrial Revolution. In order to pump water from coal mines, he developed a way of creating a vacuum in a closed vessel by filling it with steam and then condensing the steam with a spray of cold water. His water-raising engine was patented in 1698. The engine could not be used with high-pressure steam and could raise water by only 20 feet (6 m), but the design was improved by Thomas NEWCOMEN, with whom he later went into partnership.

Scarpellini, Caterina (1808–1873) *Italian astronomer and meteorologist*

Scarpellini was one of the outstanding women scientists of the 19th century, noted for her many astronomical and meteorological observations, including the discovery of a comet in 1854. She was born in Foligno, Perugia, into a prominent family; her uncle, Abbé Feliciano Scarpellini, was among the first papal scientific appointees and director of a papal observatory. Caterina was interested in science from an early age and developed a special interest in astronomy, living next to the observatory on the Capitoline Hill after settling in Rome. From 1847 she edited the scientific bulletin, *Correspondenza Scientifica*, and in the 1850s reorganized the observatory as the Meteorological Ozonometric Station, acting as its director from 1859. She made observations of weather conditions six times daily, keeping careful records over a number of years. Among her astronomical observations were details of comets, and she compiled the first Italian meteor catalogue. Moreover, she wrote papers on diverse scientific topics, such as the speed of light, magnetism, the electrical phenomena associated with muscle contraction, and the influence of the Moon on earthquakes. Plaudits for her scientific work included a silver medal from the Italian Ministry of Education, awarded in 1872. After her death, her statue was erected in the Campo Verano, Rome.

Schaefer, Vincent Joseph (1906–1993) *American physicist*

Schaefer, who was born at Schenectady in New York, graduated from the Davey Institute of Tree Surgery in 1928. He was appointed as assistant to Irving Langmuir at the research laboratory of the General Electric Company in 1931 and remained there until 1954, becoming a research associate in 1938. In 1954 he was appointed director of research at the Munitalp Foundation, where he remained until 1959 when he was appointed to a chair of physics at the State University of New York, Albany. He retired in 1976.

In 1946 Schaefer was the first to demonstrate that it was possible to induce rainfall. Tor Bergeron had earlier argued that the presence of ice crystals in the atmosphere was a necessary precondition for the formation of rain. During World War II Schaefer had worked on atmospheric research and, more specifically, the problem of airplane wings icing up, and had discovered that he could produce a snow storm in the laboratory by dropping dry ice (solid carbon dioxide) into a container filled with a supercooled mist. In 1946 he seeded clouds over Massachusetts with dry ice pellets and produced the first man-made precipitation.

Following the success of this experiment the atmospheric research program known as Project Cirrus was established during which Bernard Vonnegut discovered the effectiveness of silver iodide as a cloud-seeding material.

Schally, Andrew Victor (1926–) *Polish–American physiologist*

Schally (**shal**-ee), who was born at Wilno in Poland, left his native country for Britain in 1939. After graduating from the University of London, he worked at the National Institute for Medical Research from 1949 until 1952 when he moved to Canada. There he worked at McGill University, Montreal, obtaining his PhD in biochemistry in 1957, before joining the staff of Baylor University Medical School, Houston. In 1962 he became head of the endocrine and polypeptide laboratories at the Veterans Administration Hospital, New Orleans.

Like his great rival Roger Guillemin, Schally spent much of his early career trying to confirm the hypothesis of Geoffrey Harris on the existence and role of hypothalamic hormones. It was not in fact until he had been donated a million pig's hypothalami by a meat packer that, independently of Guillemin, he isolated some 3 milligrams of the thyrotropin-releasing factor in 1966.

He followed this in 1971 by detecting the luteinizing releasing factor, showing it to be a decapeptide and working out its sequence of ten amino acids, thus permitting its synthesis.

For his work on the hypothalamic hormones Schally shared the 1977 Nobel Prize for physiology or medicine with Guillemin and Rosalyn Yalow.

Schank, Roger Carl (1946–) *American computer scientist*

Schank, a New Yorker by birth, was educated at the University of Texas where he obtained his PhD in 1969. After working at Stanford until 1974, Schank moved to Yale, becoming professor of computing science and psychology in 1976. In 1989 he was appointed professor of computer science at Northwestern University, Evanston, Illinois. In 2002 he founded Socratic Arts, with the aim of promoting electronic learning.

Schank's work has concerned the fundamental problem of understanding natural language. One difficulty is the way in which simple verbs like "gave" are highly ambiguous in such uses as:

John gave Mary a book.
John gave Mary a hard time.
John gave Mary a night on the town.
John gave up.

and many more. While the computer can be instructed to infer from "John gave Mary a book" the conclusion: "Mary now possesses a book," it must not be allowed to infer "Mary now possesses a hard time" from "John gave Mary a hard time." Such complications seem endless.

Schank sought to avoid this particular problem by identifying different classes of verbs. Among action verbs he defined the class PTRANS, which stands for the ways objects can be transferred by such processes as carrying or throwing. Another class, MTRANS, describes the ways in which mental information is transferred by such methods as reading or writing. Knowing which class a verb belongs to, and some of the rules governing its use, computers are less likely to be misled by pervasive ambiguities.

In addition to this ambiguity, there are more fundamental differences between the way in which a computer "understands" and the way a person understands. For instance, a computer can be told that John went to a restaurant, ordered a meal, paid the check, and left, without being aware that John actually ate the meal. It can of course be informed of this fact. But if everything implicit in language must be made explicit, then programming a computer to deal with language would become an impossible task. Schank tried to deal with this by educating computers with the aid of various "scripts," which give the computer information about the "real world." In the case of the restaurant, the computer lacks a restaurant script. Schank filled the gap by telling the computer the props met with in restaurants (menus, plates, etc.), the players (cooks, waiters, etc.) and, among other items, some results (customer no longer hungry, has less money, etc.).

Are computers programmed in this way any more intelligent or knowledgeable? One difficulty, Schank has noted, is that it is impossible "to write down scripts for all the things stories and texts can be about." What is needed is a computer that can recognize its own ignorance and know where to find the data it needs. But, Schank notes, although "such a system is possible theoretically, we are still very far from developing one."

Schaudinn, Fritz Richard (1871–1906)
German zoologist

Schaudinn (**show**-din), who was born at Röseningken (now in Poland), took a doctorate in zoology at the University of Berlin (1894), where he became lecturer (1898). In 1904 he was appointed director of the protozoology laboratory in the Imperial Office, Berlin, and was subsequently director of the department of protozoological research in the Hamburg Institute for Tropical Diseases.

Some of Schaudinn's most important work was in his studies of those protozoans (notably trypanosomes) that cause tropical diseases in man. He distinguished between the amoeboid cause of tropical dysentery, *Entamoeba histolytica*, and its innocuous relative *Escherichia*

coli, which lives in the lining of the intestine where it is actually beneficial in engulfing bacteria.

With the dermatologist Erich Hoffmann he isolated (1905) the spirochete cause of syphilis (*Treponema pallidum*, formerly *Spirochaeta pallida*) and, confirming an earlier conjecture, proved that hookworm infection occurs through the skin.

Schaudinn also carried out research into human and animal (bird) malaria, providing the basis for subsequent researchers to discover the causative blood parasite. Schaudinn's other work included the demonstration of an alternation of generations in the Foraminifera (rhizopod protozoans) and the Coccidae (scale insects).

Schawlow, Arthur Leonard (1921–1999)
American physicist

> To do successful research you don't need to know everything. You just need to know of the one thing that isn't known.
> —Springer house magazine

Born in Mount Vernon, New York, Schawlow was educated at the University of Toronto and worked at Columbia (1949–51) and at the Bell Telephone Laboratories (1951–61). He became professor of physics at Stanford University in 1961, retiring in 1991.

Schawlow is noted for his work on the development and use of lasers. He collaborated with Charles TOWNES in early work on maser principles and is generally credited as a coinventor of the laser. Although he did not share in Townes's Nobel award (1964), Schawlow did share the 1981 Nobel Prize for physics with Nicholaas BLOEMBERGEN for their (independent) research in laser spectroscopy. In particular, Schawlow, with Theodor HÄNSCH, worked on the use of tunable dye lasers for high-resolution spectroscopy.

Scheele, Karl Wilhelm (1742–1786)
Swedish chemist

> I realized the necessity to learn about fire ... But I soon realized that it was not possible to form an opinion on the phenomena of fire as long as one did not understand air.
> —On his studies of the composition of air. Letter to Anders Retzius (1768)

Scheele (**shay**-le), who came from a poor background in Straslund (now in Germany), received little schooling and was apprenticed to an apothecary in Göteborg when he was 14 years old. In 1770 he moved to Uppsala to practice as an apothecary. He met and impressed Torbern BERGMAN, the professor of chemistry there, and was elected to the Stockholm Royal Academy of Sciences in 1775. Also in 1775 he moved to Köping where he established his own pharmacy.

In 1777 Scheele published his only book,

Chemical Observations and Experiments on Air and Fire. In this work he stated that the atmosphere is composed of two gases, one supporting combustion, which he named "fire air" (oxygen), and the other preventing it, which he named "vitiated air" (nitrogen). He was successful in obtaining oxygen in about 1772, two years before Joseph PRIESTLEY. He also discovered chlorine, manganese, barium oxide, glycerol, silicon tetrafluoride, and a long list of acids, both organic and inorganic, including citric, prussic, and tartaric acids. One further piece of work that had unexpectedly important consequences was his demonstration of the effects of light on silver salts.

Despite receiving many lucrative offers from Germany and England, Scheele remained at Köping for the rest of his life devoting himself to his chemical researches. Although his work must have suffered from his isolation, and he lost priority in many discoveries owing to delay in publication, he is still frequently referred to as the greatest experimental chemist of the 18th century.

Scheiner, Christoph (1575–1650) *German astronomer*

Scheiner (**shI**-ner), a Jesuit who became professor of mathematics and Hebrew at the university in his native city of Ingolstadt, claimed to be the discoverer of sunspots. In 1611 Scheiner was observing the Sun telescopically through a thick mist when he discovered "several black drops." His observations were published in 1612. GALILEO replied with his *Letters on the Solar Spots* (1613) claiming that he had first observed such spots in 1610. He also argued against Scheiner's interpretation of such spots as small planets circling the Sun. Scheiner, although a Copernican, campaigned against the publication of Galileo's dialogues and did much to turn the Jesuits against Galileo. In 1615 he made the first Keplerian telescope and in 1630 he published his basic treatise on sunspots *Rosa ursina* (The Red Bear).

Schiaparelli, Giovanni Virginio (1835–1910) *Italian astronomer*

> The meteor showers are the product of the dissolution of the comets and consist of very minute particles.
> —On the origin of meteors. Letter, December 1872

Schiaparelli (skyah-pa-**rel**-ee) was born at Savigliano in Italy. After graduating from Turin in 1854, he studied under Johann ENCKE in Berlin and Friedrich STRUVE in St. Petersburg. In 1860 he became director of the Brera Observatory, Milan, where he remained until he retired in 1900.

Schiaparelli worked mainly on the solar sys-tem, discovering the planetoid Hesperia in 1861. He contributed to the theory of meteors when he showed in 1866 that they follow cometary orbits. He also made careful studies of Mars, Venus, and Mercury. In 1877 Mars approached Earth at its nearest point, a mere 35 million miles. He observed what he called "canali." In Italian this means not "canals" but "channels," but the word was mistranslated into English as the former, which led to much controversy. Schiaparelli himself was neutral as to their origin. He would not rule out that they were constructed rather than natural but nor would he conclude from their geometrical precision that they were buildings, for he pointed out that other examples of regularity, such as Saturn's rings, had not been man-made. It was other astronomers, such as Percival LOWELL and Camille FLAMMARION, who made extravagant claims about the "canals," not Schiaparelli.

After detailed observations of Venus and Mercury he announced that their period of axial rotation was the same as their sidereal period (the time taken to orbit the Sun, relative to the stars). Thus they would always keep the same face to the Sun. It was not until the early 1960s that this view was disproved, and then only by the use of sophisticated radar techniques.

Schickard, Wilhelm (1592–1635) *German scientist*

The son of a carpenter from Herrenberg in Germany, Schickard (**shik**-art) was educated at the local monastery school and the University of Tübingen. After several years working as a Lutheran pastor, Schickard returned to Tübingen as professor of oriental languages. His interests also extended to the natural sciences, a combination of interests not uncommon in the 17th century.

Schickard became a friend of the great Johannes KEPLER when he visited Tübingen in 1617. In a 1623 letter to Kepler he spoke of a machine he had constructed "consisting of eleven complete and six incomplete sprocket wheels which can calculate." He promised to send one to Kepler but later apologized that it had been destroyed in a fire. He also referred to his invention as a "calculating clock." Little more is known about his life other than that he succeeded Michael Mastlin as professor of astronomy at Tübingen in 1631 and that he died of bubonic plague in 1635.

At this point Schickard was ignored for three centuries. In 1935 Fritz Hammer came across Schickard's sketch of his calculating clock while going through Kepler's papers. Many years later, in 1956, Hammer found another diagram and some instructions in Schickard's papers in Stuttgart. Hammer published his finds in 1957. Upon hearing of this a modern Tübingen professor, Bruno von Freytag, produced a working model of Schickard's invention.

The device consisted of six numbered dials connected to six axles. Tens were carried or borrowed by installing a single toothed gear on each axle, linked by an intermediate gear to the adjacent axle. The machine is of considerable interest to historians as it preceded the work of PASCAL – long thought to have invented the first mechanical calculator – by twenty years.

Schiff, Moritz (1823–1896) *German physiologist*

Schiff (shif), the son of a merchant from Frankfurt am Main in Germany, obtained his doctorate from the University of Göttingen in 1844. After working in the Frankfurt Zoological Museum, Schiff moved to Bern in 1854 as professor of comparative anatomy where he remained until 1863 when he was appointed to the chair of physiology in Florence. A campaign against vivisection forced Schiff to leave Florence in 1876 when he accepted the professorship of physiology at Geneva.

In 1856 Schiff demonstrated that removal of the thyroid gland in dogs and guinea pigs resulted in their death. He also showed, in 1885, that the effect could be postponed by grafting a piece of the gland elsewhere in the animal before its removal. The relief was however only temporary as the gland was absorbed by the body. Unfortunately Schiff's earlier work was unknown to surgeons like Theodor KOCHER when, in the early 1870s, they began to perform thyroidectomies in humans, operations that often led to tragic ends.

Schiff had also worked in the 1850s on nervous control of the blood supply. By cutting the brainstem he was able to show the existence of special centers in the brain for the control of vasomotor nerves, nerves that narrow or widen blood vessels as the body's demand rises and falls. The same results were independently obtained by Claude Bernard.

Schimper, Andreas Franz Wilhelm (1856–1901) *German plant ecologist*

Schimper (shim-per) was born at Strasbourg, which is now in France. He first became interested in natural history while on excursions with his father, Wilhelm Philipp, who was professor of natural history and geology at the University of Strasbourg. Andreas entered the university in 1874, obtained his doctorate in natural philosophy in 1878, and in 1880 earned a fellowship to Johns Hopkins University. He returned to Germany in 1882 and became lecturer and eventually professor at the University of Bonn (1886), where he remained until in 1898 he was appointed professor of botany at the University of Basel.

While at Strasbourg Schimper made an important study of the nature and growth of starch grains showing that they arise in specific organelles, which he named chloroplasts. However, it is to the study of plant geography and ecology that he made his most significant contributions. During travels to the West Indies in 1881 and 1882–83, Brazil (1886), Ceylon (Sri Lanka) and Java (1889–90), and the Canary Islands, Cameroons, East Africa, Seychelles, and Sumatra (1898–99) with the *Valdivia* deep-sea expedition, he made ecological studies of tropical vegetation. His results led to publication of important papers on the morphology and biology of epiphytes and littoral vegetation, culminating with his masterpiece, *Pflanzengeographie auf physiologischer Grundlage* (1898; Plant Geography Upon a Physiological Basis), which relates the physiological structure of plants to their type of environment.

Schleiden, Matthias Jakob (1804–1881) *German botanist*

> As a popularizer he was a model, as a scientist an initiator.
> —L. Errera, *Revue scientifique de la France et de l'étranger* (1882; Scientific Review of France and Other Countries)

Schleiden (shll-den) was born at Hamburg in Germany and studied law at Heidelberg; he then returned to Hamburg to practice as a lawyer. However, he soon became fully occupied by his interest in botany and graduated in 1831 from the University of Jena, where he became professor of botany in 1839.

Instead of becoming involved in plant classification – the pursuit of most of his botanical contemporaries – Schleiden studied plant growth and structure under the microscope. This led to his *Contributions to Phytogenesis* (1838), which stated that the various structures of the plant are composed of cells or their derivatives. He thus formulated the cell theory for plants, which was layer elaborated and extended to animals by the German physiologist Theodor SCHWANN. Schleiden recognized the significance of the cell nucleus and sensed its importance in cell division, although he thought (wrongly) that new cells were produced by budding from its surface. He was one of the first German biologists to accept Darwinism.

Schmidt, Bernhard Voldemar (1879–1935) *Estonian telescope maker*

Schmidt (shmit), who was born on the island of Naissaar, in Estonia, received little education. After working in Gothenburg, Sweden, he went in 1901 to study engineering at Mittweida in Germany, near Jena. He set up his own workshop in 1904 in Mittweida and manufactured high-grade mirrors to be used in telescopes. He also built some reflecting telescopes, including one for the Potsdam Astrophysical Observatory, and set up his own observatory. In 1926 he moved to the Hamburg Observatory in Bergedorf. As a master craftsman he worked unaided even though he had lost his right arm as a boy. He was also an alcoholic and claimed to have his best ideas after prolonged drinking bouts. He died in a mental hospital.

His name is known to all astronomers as the designer of one of the most basic items of observatory equipment, the Schmidt telescope. This was built to overcome some of the penalties inherent in the design of the large parabolic reflectors like the Mount Wilson 100-inch (2.5-m) telescope. Parabolic mirrors are used rather than spherical ones in telescopes to correct the optical defect known as spherical aberration and thus allow the light from an object to be accurately and sharply focused. This accurate focusing only occurs, however, for light falling on the center of a parabolic mirror. Light falling at some distance from the center is not correctly focused owing to a different optical distortion in the image, known as coma.

This limits the use of parabolic reflectors to a narrow field of view and thus precludes them from survey work and the construction of star maps.

Schmidt replaced the primary parabolic mirror with a spherical mirror, which though coma-free did however suffer from spherical aberration, thus preventing the formation of a sharp image. To overcome this fault Schmidt introduced a "corrector plate" through which the light passed before reaching the spherical mirror. It was so shaped to be thickest in the center and least thick between its edges and the center. In this way a comparatively wide beam of light passing through it is refracted in such a way as to just compensate for the aberration produced by the mirror and produce an overall sharp image on a (curved) photographic plate.

Schmidt's first hybrid reflector/refractor was ready and installed in the Hamburg Observatory in the early 1930s. Observatories have since used the Schmidt telescope to photograph large areas of the sky. The whole sky has now been surveyed with these instruments and the results, which include the very faintest objects down to a magnitude of 21, are published in the Palomar Sky Survey and the Southern Sky Survey.

Schmidt, Brian P. (1967–) *American–Australian astrophysicist*

Schmidt grew up in Montana and Alaska, and received BS degrees in physics and astronomy from the University of Arizona (1989). In 1993 he was awarded his PhD from Harvard University for a thesis on supernovae (exploding stars), under the supervision of Robert Kirshner. At around this time a group of astronomers

in Chile discovered that certain supernovae of similar absolute brightness could be used to measure distances in the universe according to how bright they appeared on earth. By comparing distances of near and distant supernovae, astronomers could gain crucial data about the rate of expansion of the universe. With his Australian wife, Schmidt emigrated to Australia in 1994, having secured a fellowship at the Australian National University's Mount Stromlo Observatory. Later that year he formed the High-Z SN Search team, an international collaboration between 20 astronomers to search for supernovae. By the end of 1997 Schmidt's team had discovered 15 distant supernovae that were accelerating as they moved away. This confounded accepted cosmological theory, which held that after the Big Bang gravity would slow the expansion of the universe, and ultimately reverse it, causing an eventual collapse. In 1998 Schmidt and his team, including the astrophysicist Adam Riess, reported their evidence for an "accelerating universe." Almost at the same time, similar results were reported by another team, the Supernova Cosmology Project, led by Saul Perlmutter. These findings led to a reappraisal of the nature of the universe, and the proposition that a form of matter/energy called dark energy is responsible for the accelerating expansion. Schmidt's work has brought him many accolades, including the 2006 Shaw Prize and the 2007 Gruber Cosmology Prize. He continues to work at the Research School of Astronomy and Astrophysics at the Australian National University, and is leading the SkyMapper project to provide a deep digital map of the southern sky.

Schmidt, Ernst Johannes (1877–1933) *Danish biologist*

Schmidt was born at Jaegerspris in Denmark and became director of the Carlsberg Physiological Laboratory, Copenhagen. He is chiefly known for his discovery of the breeding ground and life history of the European eel in 1904. Schmidt attained this end by a careful compilation of statistics of the length of eel larvae (leptocephali) found at different points and at different times in the Atlantic. From these he was able to link together leptocephali of similar size, radiating from a central area, the smaller and younger nearer the center. The center of radiation, and the breeding ground of all European eels, proved to be the Sargasso Sea, near Bermuda. Schmidt also carried out research and produced publications on bacteria and the flora of the island of Ko Chang, Thailand.

Schmidt, Maarten (1929–) *Dutch–American astronomer*

Born in Groningen in the Netherlands, Schmidt

graduated from the university there in 1949 and obtained his PhD in 1956 from Leiden University. After working at the Leiden Observatory from 1953 to 1959, he moved to America, taking up an appointment at the California Institute of Technology as a staff member of the Hale Observatories. He was made professor of astronomy in 1964 and also served as director of the Hale Observatories from 1978 to 1980.

Schmidt has investigated the structure and dynamics of our Galaxy and the formation of stars but he is best known for his research on quasars, or quasi-stellar objects. In 1960 Allan SANDAGE and Thomas Matthews identified a compact radio source, known as 3C 48, with a 16th-magnitude starlike object that was found to have a most curious spectrum. Soon, other optical identifications were made, including that of 3C 273 with a 13th-magnitude object that had an equally puzzling spectrum. These objects became known as quasars. In 1963 Schmidt was the first to produce a satisfactory interpretation of the spectrum of a quasar.

Schmidt realized that certain broad emission lines in the spectrum of 3C 273 were the familiar hydrogen lines but shifted in wavelength by an unprecedented amount. According to the DOPPLER effect, light emitted from a source that is moving away from an observer increases its wavelength, i.e., its spectral lines shift toward the red end of the spectrum. The faster an object is moving away, the greater the so-called red shift. HUBBLE had assumed that the red shift of the galaxies was explained by the Doppler effect: the galaxies were receding as the universe expanded and that as the velocity and hence the distance of a galaxy increased, its red shift increased accordingly. 3C 273 had an immense red shift. Assuming it to be a Doppler shift resulting from the expansion of the universe, Schmidt was amazed when he found that 3C 273 must be a billion light-years away. In that case, how could such a small source be visible at such an enormous distance? It would need to be as luminous as a hundred galaxies and it was by no means clear what physical mechanism could yield so much energy from such a compact source. Schmidt's work was soon confirmed by the red-shift interpretation of the spectra of other quasars; they all possessed unusually large red shifts. There arose a long debate as to whether the Doppler effect did explain the quasar red shift but it is now generally accepted that this is the case.

By the end of the 1960s many quasars had been discovered and their distribution mapped in the heavens. Schmidt realized that this allowed him to test the cosmological steady-state doctrine of Thomas GOLD and others, which assumes that the universe on a large scale looks the same at all times and all places. He found, however, on examining the distribution of quasars and using the Doppler interpretation of

their red shifts, that their numbers increase with distance and that they are indeed the most distant objects in the universe. Assuming that the big-bang rather than the steady-state theory is correct, they are also the youngest objects in the universe.

The discovery of the quasars with the problems they posed produced an enormous growth in astronomical research that led to the discovery of even stranger objects, such as pulsars, and the continued search for black holes. Huge black holes are indeed thought to be the source of the prodigious energy of quasars.

Schönbein, Christian Friedrich (1799–1868) *German chemist*

Schönbein (**shu(r)n**-bīn) was born at Metzingen in Germany. After studying at the universities of Tübingen and Erlangen, he took up an appointment at the University of Basel in 1828, staying there for the rest of his life.

Many stories relate how he discovered nitrocellulose (guncotton) in 1846. In all of them a bottle in which he had been distilling nitric and sulfuric acids broke on the floor of the kitchen. In some stories, as he was forbidden by his wife to experiment in the kitchen, he is supposed to have panicked and wiped the mess up with his wife's cotton apron. In others he is unable to find a mop and uses the nearest thing to hand, his wife's cotton apron. Put to dry over the stove it flared up without smoke: Schönbein had discovered the first new explosive since gunpowder. (He was nearly anticipated in his discovery of guncotton: Théophile Pelouze had obtained an inflammable material in 1838 by treating cotton with nitric acid, but he failed to follow it up.)

Schönbein saw what a valuable commodity he had and quickly secured the appropriate patents on it. He gave exclusive rights of manufacture to John Hall and Sons in Britain but, unfortunately, their factory at Faversham blew up in July 1847, killing 21 workers. Similar lethal explosions occurred in France, Russia, and Germany. Its properties were too valuable to allow chemists to abandon it altogether: it was smokeless and four times more powerful than gunpowder; properly controlled it would make an ideal propellant. It was finally modified by Frederick ABEL and James DEWAR later in the century in the forms of Poudre B and cordite.

In 1840 Schönbein discovered ozone, the allotropic form of oxygen. Investigating the curious smell that seemed to linger around electrical equipment, he traced it to a gas (O_3) that he named after the Greek word for smell (*ozon*).

Schramm, David Norman (1945–1997) *American physicist*

Born in St. Louis, Missouri, Schramm was educated at the Massachusetts Institute of Technology and the California Institute of Technology, where he took his PhD in 1971. After a four-year spell at the University of Texas, he moved to the University of Chicago, becoming professor of physics in 1977.

Cosmology has long been a subject rich in speculation but poor in experimental control. Schramm sought to bridge the gap through a union between cosmology and particle physics. From one side, Schramm saw, cosmology could offer insights to physics. Thus he noted physicists spoke of what happened at high energies well beyond the reach of any actual accelerator. They were not, however, beyond the capacity of the big bang. Hence the study of the big bang could perhaps throw light on and control the speculations of high-energy physicists.

Conversely, a well-established theory of elementary particles could test speculative cosmological theory. For example, Schramm argued, from the nuclear reactions occurring when the universe was one second old, it follows that the number of fundamental particles must be small. Because of the amount of helium-4 in the universe, a consequence of the big bang, there can be no more than four families of quarks and leptons.

Schrieffer, John Robert (1931–) *American physicist*

Born in Oak Park, Illinois, Schrieffer was educated at the Massachusetts Institute of Technology and the University of Illinois, where he obtained his PhD in 1957. After serving as a postdoctoral fellow in Europe at Birmingham and Copenhagen he worked at the University of Illinois from 1959 until 1962 when he moved to the University of Pennsylvania, Philadelphia, being appointed professor of physics there in 1964. He was professor of physics at the University of California, Santa Barbara (1980–91), moving to Florida State University, Tallahassee, in 1992. He retired in 2006.

Schrieffer worked on superconductivity. In 1972 he was awarded the Nobel Prize for physics with John BARDEEN and Leon N. COOPER for their formulation in 1957 of the first successful theory of superconductivity, now known as the BCS theory.

Schrock, Richard R. (1945–2006) *American chemist*

Schrock was born in Berne, Indiana, and was educated at Harvard University, obtaining his PhD in chemistry in 1971. He subsequently became the Frederick G. Keyes Professor of Chemistry at the Massachusetts Institute of Technology.

Much of Schrock's research was concerned with metathesis reactions in organic synthe-

sis. After Yves CHAUVIN put forward his mechanism for metathesis reactions in 1971, Schrock and others performed experimental investigations which provided strong evidence in favour of this mechanism. Starting in the early 1970s while working at Du Pont, Schrock investigated many metal compounds in the hope that they would act as catalysts in metathesis reactions without the difficulties associated with many organic catalysts, such as sensitivity to air and moisture and loss of activity with time. He found that certain compounds of molybdenum, tantalum, and tungsten were effective, in particular; Schrock and his colleagues constructed a very efficient and active set of catalysts based on molybdenum in 1990.

The discovery of effective catalysts for metathesis reactions stimulated a great deal of subsequent interest in using this technique. Schrock shared the 2005 Nobel Prize for chemistry with Yves Chauvin and Robert GRUBBS for the development of the metathesis method in organic synthesis.

Schrödinger, Erwin (1887–1961) *Austrian physicist*

> Thus the task is not so much to see what no one has yet seen; but to think what nobody has yet thought, about that which everybody sees.
> —Quoted by L. Bertlanffy in *Problems of Life* (1952)

Schrödinger (**shru(r)**-ding-er or **shroh**-ding-er), the son of a prosperous Viennese factory owner, was educated at both the gymnasium and the university in his native city, where he obtained his doctorate in 1910. After serving as an artillery officer in World War I, he taught at various German-speaking universities before he succeeded Max PLANCK as professor of physics at the University of Berlin in 1927.

Before long, however, Schrödinger's bitter opposition to the Nazis drove him, in 1933, into his first period of exile, which he spent in Oxford, England. Homesick, he allowed himself to be tempted by the University of Graz in Austria in 1936 but, after the Anschluss in 1938, he found himself once more under a Nazi government which this time was determined to arrest him. Schrödinger had no alternative but to flee. He was however fortunate in that the prime minister of Eire, Eamonn De Valera, himself a mathematician, was keen to attract him to a newly established Institute of Advanced Studies in Dublin. Working there from 1939 Schrödinger gave seminars that attracted many eminent foreign physicists (as well as the frequent presence of De Valera) until his retirement in 1956 when he returned to Austria.

Starting from the work of Louis de BROGLIE, Schrödinger in the period 1925–26 developed wave mechanics, one of the several varieties of quantum theory that emerged in the mid-1920s. He was deeply dissatisfied with the early

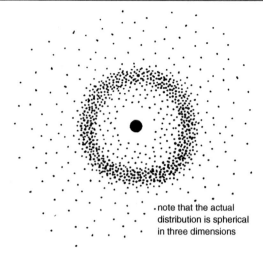

note that the actual
distribution is spherical
in three dimensions

HYDROGEN ATOM   The hydrogen atom according to Erwin Schrödinger.

quantum theory of the atom developed by Niels Bohr, complaining of the apparently arbitrary nature of a good many of the quantum rules. Schrödinger took the radical step of eliminating the particle altogether and substituting for it waves alone. His first step was to derive an equation to describe the behavior of an electron orbiting an atomic nucleus. The de Broglie equation giving the wavelength $\lambda = h/mv$ (where h is the Planck constant and mv the momentum) presented too simple a picture for in reality, particularly with the inner orbits, the attractive force of the nucleus would result in a very complex and variable configuration. He eventually succeeded in establishing his famous wave equation, which when applied to the hydrogen atom yielded all the results of Bohr and de Broglie. It was for this work that he shared the 1933 Nobel Prize for physics with Paul Dirac.

Despite the considerable predictive success of wave mechanics, as the theory became known, there remained two problems for Schrödinger. He still had to attach some physical meaning to ideas of the nature of an electron, which was difficult if it was nothing but a wave; he also had to interpret the ψ function occurring in the wave equation, which described the wave's amplitude. He tried to locate the electron by constructing stable "wave packets" from many small waves, which it was hoped would behave in the same way as a particle in classical mechanics. The packets were later shown to be unstable.

Nor was his interpretation of the ψ function as a measure of the spread of an electron any more acceptable. Instead the probabilistic interpretation of Max Born soon developed into a new orthodoxy. Schrödinger found such a view totally unacceptable, joining those other founders of quantum theory, Einstein and de Broglie, in an unrelenting opposition to the indeterminism entering physics.

In 1944 Schrödinger published his *What Is Life?*, one of the seminal books of the period. Partly due to its timely publication it influenced a good many talented young physicists who, disillusioned by the bombing of Hiroshima, wanted no part of atomic physics. Schrödinger solved their problem by revealing a discipline free from military applications, significant and, perhaps just as important, largely unexplored. He argued that the gene was not built like a crystal but that it was rather what he termed an "aperiodic solid." He went on to talk of the possibility of a "code" and observed that "with the molecular picture of the gene it is no longer inconceivable that the miniature code should precisely correspond with a highly complicated and specified plan of development." It is not surprising that such passages, written with more insight than that contained in most contemporary biochemical works, inspired a generation of scientists to explore and decipher such a code.

Schrötter, Anton (1802–1875) *Austrian chemist*

Schrötter (**shru(r)**-ter) was the son of an apothecary from Olomouc (now in the Czech Republic). He taught chemistry at Graz and the Vienna Polytechnical Institute. From 1845 he was master of the Austrian mint.

Schrötter was the first to recommend hydrogen peroxide for bleaching hair but his greatest achievement was the discovery in 1845 of red phosphorus. Until this time white phosphorus had been used in the production of matches. This was highly flammable and also toxic, producing inflammation and eventual necrosis of the jaw, a condition popularly known as "phossy jaw."

Schrötter found that if ordinary phosphorus was heated in an enclosed vessel the allotrope, red phosphorus, would be produced. This had the advantage of being neither spontaneously flammable nor toxic. He lectured on his discovery in Birmingham, England, in 1849 and Arthur ALBRIGHT, the British phosphorus producer, quickly introduced the new process into his Oldbury factory. Schrötter's discovery was eventually given legal backing in Britain under the White Phosphorus Prohibition Act of 1908.

Schultze, Max Johann Sigismund
(1825–1874) *German zoologist*

Schultze (**shuult**-se) was born at Freiberg in Germany. He was educated at the University of Griefswald, where his father was professor of anatomy, and the University of Berlin. After a brief period on the staff of the University of Halle he moved in 1859 to Bonn where he served as professor of zoology until his sudden death in 1874 from a perforated ulcer.

In 1861 Schultze, who had worked on the cellular structure of a wide variety of animals, published a famous paper in which he emphasized the role of protoplasm in the workings of the cell. Cells, he argued, were "nucleated protoplasm" or "the physical basis of life," the protoplasm and not the cell wall being the important constituent. This he illustrated by pointing out that some cells, for example those of the embryo, do not have bounding membranes.

In 1866 Schultze went on to formulate the so-called duplicity theory of vision. He had noticed that in diurnal birds the retina consisted mainly of cones but nocturnal birds possessed a retina with an abundance of rods. This led him to propose that cones must respond to colored light while rods should be more sensitive to black and white.

Schuster, Sir Arthur (1851–1934) *British physicist and spectroscopist*

Schuster (**shoo**-ster) was the son of a Frankfurt textile merchant and banker who, unwilling to remain in the city after its annexation by Prussia in the wake of the 1866 war, moved with his family to Manchester, England. Schuster became a British citizen in 1875 and studied physics at Owens College, Manchester, and the University of Heidelberg where he obtained his doctorate in 1873. Schuster then spent the period 1875–81 at the Cavendish Laboratory, Cambridge, but returned to Manchester to serve first as professor of applied mathematics and from 1889 to 1907 as professor of physics. His somewhat premature retirement at the age of 56 was spent on his own research and the for-

mation of the International Research Council, which he served as first secretary from 1919 to 1928.

Initially Schuster worked as a spectroscopist. In 1881 he refuted the speculation of George STONEY that spectral lines could be regarded as the harmonics of a fundamental vibration. This was done by a statistical analysis of the spectral lines of five elements in which he showed their random distribution. Somewhat discouraged by this result he turned to the study of the passage of an electric current through a gas.

In the 1880s he was the first to show that an electric current is conducted by ions. He went on to propose how the ratio between the charge and the mass of cathode rays could be calculated and in fact described the technique later used by J. J. THOMSON in his determination of the charge on the electron. He further proposed, in 1896, that the new x-rays of Wilhelm RÖNTGEN were in fact transverse vibrations of the ether of very small wavelength.

One of Schuster's greatest achievements was the physics department of Manchester University. He raised funds to construct a new laboratory in 1897, was bold enough to create new departments, such as that of meteorology in 1905, and made ample provisions for research. He also had the insight to pick as his successor Ernest RUTHERFORD who, despite invitations from the Smithsonian, Yale, and King's College, London, recognized the value of Schuster's achievement by preferring the Manchester invitation.

Schwann, Theodor (1810–1882) *German physiologist*

> The treatise [is] worthy to be ranked among the most important steps by which the science of physiology has ever been advanced.
> —Henry Smith, from the introduction to the first English edition (1847) of Schwann's *Mikroskopische Untersuchungen* (1839; Microscopical Researches)

Born at Neuss in Germany, Schwann (shvahn) was educated at the universities of Bonn, Würzburg, and Berlin where he obtained his MD in 1834. He worked with Johannes MÜLLER in Berlin from 1834 until 1838 when he moved to Belgium, serving as professor of anatomy first at Louvain (1838–47) and thereafter at Liège until his death.

Schwann's first experiments at Berlin were on muscle contraction. He showed that the mechanism of contraction could be explained without invoking any vital principles – a marked departure from the teachings of Müller. This mechanistic philosophy was fruitfully developed by Schwann's successors at Berlin, Emil du Bois-Reymond and Hermann von HELMHOLTZ. Schwann next conducted some experiments to disprove (again) the theory of spontaneous generation, which was enjoying a renaissance in the mid 1830s. One unexpected outcome of his experiments on putrefaction and fermentation was his discovery in 1836, independently of CAGNIARD DE LA TOUR, that yeast is involved in fermentation. The same year Schwann also discovered the digestive enzyme pepsin.

His most memorable achievement however is his *Mikroskopische Untersuchungen* (1839; Microscopical Researches) in which he first formulated, at the same time as Matthias SCHLEIDEN, the most important of all ideas in modern biology, namely that "cellular formation might be a widely extended, perhaps a universal principle for the formation of organic substances."

In 1838 Schleiden had proposed that all plant tissue was composed of nucleated cells. Using the newly introduced achromatic microscope Schwann went on to examine a variety of tissues taken from several different animals. He surmised that fibers, ducts, etc., do not form directly from molecules but rather are built up from cells. The process of cell formation he saw as something like that of crystallization: cells were not formed from other cells but somehow condensed out of intercellular "nutrient liquid." One further radical misconception was that the cellular material, Schwann's cytoblastema, was devoid of structure.

Despite such errors the cell theory met with rapid acceptance. Improvements were soon made. Robert REMAK first described cell division in 1841 and by 1855 Rudolf VIRCHOW could issue the new dogma *omnis cellula e cellula* (all cells come from cells). The cytoblastema also came in for revision; renamed protoplasm, it was shown by Max SCHULTZE in 1861 to have definite properties and a structure.

Despite these successes Schwann's work on fermentation was savagely criticized by leading chemists of the time, notably Justus von LIEBIG and Friedrich WÖHLER. A particularly damaging paper by the pair was published in 1839 after which Schwann found it impossible to continue his career in Germany. In Belgium he conscientiously carried out his professional duties and invented some useful equipment for the mining industry. His brilliant contributions to physiology, however, virtually ceased. Not until PASTEUR's work in the 1850s was Schwann vindicated.

Schwartz, Melvin (1932–   ) *American physicist*

Schwartz was born in New York and educated at Columbia University, where he obtained his PhD in 1958. He remained at Columbia until 1966, when he was appointed to the chair of physics at Stanford, California.

In 1961 Schwartz, in collaboration with his Columbia colleagues Leon LEDERMAN and Jack STEINBERGER, performed what has since come to be known as the two-neutrino experiment. All three shared the 1988 Nobel Prize for physics.

Long before his Nobel award Schwartz had abandoned academic science. Physics, he said, in the earlier days had been "a lot of fun." But after the bureaucrats moved in and set up committees to control finance and direct research, he found his work obstructed and his time spent less and less on physics and more and more on the preparation of reports. Consequently he left Stanford in 1979, moved to "Silicon Valley" in California, and set up his own company, Digital Pathways, specializing in computer security. He returned to academic life in 1991, when he was appointed professor of physics at Columbia, New York.

Schwarz, John Henry (1941–   ) *American mathematical physicist*

Born in North Adams, Massachusetts, Schwarz was educated at Harvard and the University of California, Berkeley, where he completed his PhD in 1966. After working at Princeton until 1972, he moved to the California Institute of Technology and was appointed professor of theoretical physics there in 1985.

In 1970 Yoichiro NAMBU had proposed that elementary particles may not be particles at all but could be vibrating rotating strings. Schwarz saw in this idea a way to explain the behavior of hadrons – i.e., particles such as protons and neutrons, which respond to the strong nuclear force. Hadrons were known to be composed of quarks and, in the new model, the quarks could be seen as joined by stringlike connections. The theory required space to have 26 dimensions, the existence of a massless particle, and the presence of tachyons (particles supposedly able to travel faster than light). More significantly, however, string theory had to compete with the more plausible account of hadrons proposed by Sheldon GLASHOW and others, known as quantum chromodynamics (QCD). Against competition from this source string theory withered away.

Schwarz, however, continued to work on string theory and in collaboration with Michael Green reduced the dimensions demanded by early theories to ten. They also eliminated the need for tachyons. As for the massless particle, it seemed to possess precisely the properties demanded by EINSTEIN's theory of general relativity for the carrier of the gravitational force, a fact that was of interest to cosmologists.

Further, the new theory carried other implications. It exhibited a deep symmetry, known as supersymmetry, which seemed able to unify two fundamentally different categories of particles: fermions and bosons. Bosons are not conserved and have an integral spin; fermions are conserved and have a spin of one half. Schwarz demonstrated how they could be seen as waves moving in a closed loop, with fermions moving in one direction and bosons in the other direction.

The fact that the theory needs space–time to have ten dimensions rather than the four observed is explainable if the six "extra" dimensions are extremely small, curled into six-dimensional balls with a diameter of about 10^{-35} meter.

Schwarz has published a full account of his work, in collaboration with Green and Edward WITTEN, in *Superstring Theory* (1987, 2 vols.).

Schwarzschild, Karl (1873–1916) *German astronomer*

> The greatest German astronomer of the last 100 years.
> —Berlin Academy of Sciences dedication on inaugurating the Karl Schwarzschild telescope (1960)

Schwarzschild (**shvarts**-shilt or **shworts**-chIld) was the son of a prosperous Jewish businessman from Frankfurt am Main in Germany. His interest in astronomy arose while he was at school and he had published two papers on binary orbits by the time he was 16. Following two years at the University of Strasbourg, he went in 1893 to the University of Munich, obtaining his PhD in 1896. He worked at the Kuffner Observatory in Vienna from 1896 to 1899 and after a period of lecturing and writing became in 1901 associate professor, later professor, at the University of Göttingen and director of its observatory. In 1909 he was appointed director of the Astrophysical Observatory in Potsdam. He volunteered for military service in 1914 at the beginning of World War I and was invalided home in 1916 with a rare skin disease from which he died.

Schwarzschild's practical skill was demonstrated by the instruments he designed, the measuring techniques he devised, and the observations he made. In the 1890s, while the use of photography for scientific purposes was still in its infancy, he developed methods whereby the apparent magnitude, i.e., observed brightness, of stars could be accurately measured from a photographic plate. At that time stellar magnitudes were usually determined by eye. He was then able to establish the photographic magnitude of 3,500 stars brighter than magnitude 7.5 and lying between 0° and 20° above the celestial equator. He also determined the magnitude of the same stars visually, demonstrating that the two methods do not yield identical results. This difference between the visual and photographic magnitude of a star, measured at a particular wavelength, is known as its color index.

Schwarzschild also made major contributions to theoretical astronomy, the subjects including orbital mechanics, the curvature of space, and the surface structure of the Sun. In 1906 he published a paper showing that stars could not just be thought of as a gas held together by its own gravity. Questions of thermodynamics arise, concerning the transfer of heat within the star both by radiation and convection, that need a full mathematical treatment.

EINSTEIN's theory of general relativity was published in 1916. While serving in Russia, Schwarzschild wrote two papers on the theory, which were also published in 1916. He gave a solution – the first to be found – of the complex partial differential equations by which the theory is expressed mathematically and introduced the idea of what is now called the *Schwarzschild radius*. When a star, say, is contracting under the effect of gravity, if it attains a particular radius then the gravitational potential will become infinite. An object will have to travel at the velocity of light to escape from the gravitational field of the star. The value of this radius, the Schwarzschild radius, SR, depends on the mass of the body. If a body reaches a radius less than its SR nothing, including light, will be able to escape from it and it will be what is now known as a "black hole." The SR for the Sun is 3 kilometers while its actual radius is 700,000 kilometers. The theoretical study of black holes and the continuing search for them has become an important field in modern astronomy.

Schwarzschild's son, Martin, is also a noted astronomer.

Schwinger, Julian Seymour (1918–1994) *American physicist*

Schwinger, who was born in New York City, developed his prowess for mathematics and physics at an early age. At 14 he entered the City College of New York, but later transferred to Columbia University. He received his BA degree from Columbia at the age of 17 (1936) and his doctorate three years later. Moving to the University of California at Berkeley, he worked as a research associate under J. Robert OPPENHEIMER. In the war years (1943–45) he worked at the Metallurgical Laboratory of the University of Chicago and at the Radiation Laboratory of the Massachusetts Institute of Technology. In 1945 he joined the faculty of Harvard as an associate professor of physics, and the next year was made full professor, one of the youngest in Harvard's history. He became professor of physics at the University of California, Los Angeles, in 1972.

Schwinger's most notable contribution to physics was in the fusion of electromagnetic theory and quantum mechanics into the science of quantum electrodynamics (the foundations of which had been laid by Paul DIRAC, Werner HEISENBERG, and Wolfgang Pauli). Shortly after World War II, Schwinger, and others such as Richard FEYNMAN, Sin-Itiro TOMONAGA, and Frank DYSON, developed the mathematical formulation of quantum electrodynamics in a way that was consistent with EINSTEIN's theory of relativity. The new theory led to a better understanding of the interactions between charged particles and electromagnetic fields and proved useful in measuring and explaining the behavior of atomic and subatomic particles. Schwinger, Feynman, and Tomonaga, who had conducted their work independently – Feynman at the California Institute of Technology and Tomonaga at the Tokyo Education University – were subsequently rewarded with the 1965 Nobel Prize for physics. Schwinger made a number of other important contributions to quantum mechanics and quantum field theory. He was one of the first people to investigate the quantization of non-Abelian gauge theories.

Schwinger also conducted significant research into the properties of synchrotron radiation, produced when a fast-moving charged particle is diverted in a magnetic field. He wrote several books, notably *Quantum Kinematics and Dynamics* (1970) and *Particles, Sources and Fields* (3 vols.; 1970, 1973, 1989). He also wrote a semi-popular book on relativity theory, *Einstein's Legacy* (1986). After his death books were published based on his lectures: *Classical Electrodynamics* and *Quantum Mechanics*.

Scott, Dukinfield Henry (1854–1934)
British paleobotanist

Scott, the son of the architect Gilbert Scott, was born in London and studied classics at Oxford University before studying engineering for three years. He finally decided to pursue a career in botany rather than architecture and studied for his doctorate under Julius von SACHS at the University of Wurzburg. Scott gained his PhD in 1882 and was appointed to an assistantship at University College, London. He held various junior posts before becoming director of research at the Jodrell Laboratory, Kew, in 1892, a post he held until his retirement in 1906 after which he devoted himself to full-time research.

In 1889 Scott met William WILLIAMSON who awoke in him a passion for paleobotany. He worked mainly on the plants of the Carboniferous but also published such general works as his *Extinct Plants and Problems of Evolution* (1924). With Williamson and Albert SEWARD, Scott was one of the founders of the scientific study of fossil plants, a task as crucial to the study of evolution as better-known searches for fossil vertebrates.

Seaborg, Glenn Theodore (1912–1999)
American nuclear chemist

> People must understand that science is inherently neither a potential for good nor for evil. It is a potential to be harnessed by man to do his bidding.

> —Associated Press interview with Alton Blakeslee, 29 September 1964

Born in Ishpeming, Michigan, Seaborg graduated in 1934 and gained his doctorate in 1937 at the University of California. He rose to become full professor in the Berkeley faculty in 1945.

Over the period 1948–58 Seaborg and his collaborators extended the periodic table beyond uranium (element 92) – hence the term "transuranics" applied to artificial elements with atomic numbers higher than 92. Chief among his collaborators in the early days was Edwin McMILLAN (with whom he shared the Nobel Prize for chemistry in 1951), who with Philip ABELSON had discovered the first transuranic element – neptunium (element 93) – in the spring of 1940. Neptunium was found to be a beta-particle emitter and it was thus expected that its decay product should be a transuranic element – the next in the series. In December 1940 Seaborg, McMillan, and their collaborators isolated element 94 – plutonium. It was realized that the transuranics formed a special series – the actinide transition series of elements, similar to the lanthanide series of rare-earth elements. Thus Seaborg and his coworkers were able to predict the chemical properties of further then-unknown transuranics and enabled them to be isolated. Seaborg's name is associated with the discovery or first isolation of a number of transuranics:

Element 93 neptunium 1940
Element 94 plutonium 1940
Element 95 americium 1944
Element 96 curium 1944
Element 97 berkelium 1949
Element 98 californium 1950
Element 101 mendelevium 1955
Element 102 nobelium 1958

It should be noted that all of these discoveries are correctly attributed to groups or teams of researchers and that attribution has been disputed in the past. In particular the first, unconfirmed, report of element 102 was in 1957 by an international group of physicists working at the Nobel Institute in Stockholm. The Berkeley team subsequently confirmed the discovery the next year. Similar work to Seaborg's was done in the former Soviet Union by Georgii FLEROV.

Another discovery with which Seaborg was associated is the isolation of the isotope uranium-233 from thorium (1941). This may be an alternative source of fuel for nuclear fission – a route to nuclear energy that is still relatively unexplored.

The work on elements was directly relevant to the World War II effort to develop an atomic bomb, and from 1942 until 1946 Seaborg was on leave to the University of Chicago metallurgical laboratory, where he was made head of the laboratory's chemical-separation section. It is said that he was influential in determining the choice of plutonium rather than uranium in the first atomic-bomb experiments.

Seaborg went on to become chancellor of the University of California from 1958 to 1962 and was then chair of the U.S. Atomic Energy Commission until 1971. Element 106, synthesized in 1974, was named *seaborgium* in his honor.

Secchi, Pietro Angelo (1818–1878) *Italian astronomer*

> Who knows whether or not an intimate relationship may exist between certain solar phenomena and some terrestrial ones?
> —*Le soleil* (1875–77; The Sun)

Secchi (**sek**-ee), who was born at Reggio in Italy, was a Jesuit who lectured in physics and mathematics. He spent some time abroad when the Jesuits were expelled from Rome, being at one time professor of physics at Georgetown University, Washington. He returned to Italy in 1849, becoming professor of astronomy and director of the observatory of the Roman College, which he rebuilt and reequipped.

He researched in stellar spectroscopy and his main work was done on spectral types. He introduced some order into the mass of new observations that was pouring in from the early spectroscopists. In 1867 he proposed four spectral classes. Class 1 had a strong hydrogen line and included blue and white stars; class 2 had numerous lines and included yellow stars; class 3 had bands rather than lines, which were sharp toward the red and fuzzy toward the violet and included both orange and red lines; finally, class 4 had bands that were sharp toward the violet and fuzzy toward the red and included red lines alone. These spectral types mark an important, and fairly straightforward, temperature sequence. Secchi's classification, as very much extended and modified by Edward PICKERING and Annie CANNON, has become one of the basic tools of astrophysicists.

Sedgwick, Adam (1785–1873) *British geologist and mathematician*

Sedgwick was the son of the vicar of Dent in England. He graduated in mathematics from Cambridge University in 1808, was made a fellow in 1810, and was elected to the Woodwardian Chair of Geology in 1818, a post he retained until his death. He was made president of the Geological Society in 1829.

In 1831 he began a study of the Paleozoic rocks of Wales, choosing an older region than the Silurian recently discovered by Roderick MURCHISON. In 1835 he named the oldest fossiliferous strata the Cambrian (after Cambria, the ancient name for Wales). This immediately caused a problem for there was no reliable way to distinguish the Upper Cambrian from the Lower Silurian.

Sedgwick formed a close friendship with Murchison. The two made their most significant joint investigation with their identification of the Devonian System from studies in southwest England in 1839. The partnership between Sedgwick and Murchison was broken when Murchison annexed what Sedgwick considered to be his Upper Cambrian into the Silurian. The bitter dispute between the two over these Lower Paleozoic strata was not resolved until after Sedgwick's death, when Charles LAPWORTH proposed that the strata should form a new system – the Ordovician.

Sedgwick's works included *A Synopsis of the Classification of the British Paleozoic Rocks* (1855). In 1841, largely due to Sedgwick, a museum now bearing his name was opened to house the growing geological collection. Sedgwick was, throughout his life, a committed opponent of DARWIN's theory of evolution.

Seebeck, Thomas Johann (1770–1831) *Estonian–German physicist*

Seebeck (**zay**-bek or **see**-bek) was born into a wealthy family in Tallinn, the Estonian capital, and moved to Germany at the age of 17. He studied medicine in Berlin and in 1802 received an MD from the University of Göttingen. More interested in science than medical practice, he was wealthy enough to be able to devote his time to scientific research. In the early years of the 19th century he moved to Jena, where he became acquainted with an important intellectual circle of scientists and philosophers. His subsequent researches made him one of the most distinguished experimental physicists of his day.

Seebeck made investigations into photoluminescence (the luminescent emission from certain materials excited by light), the heating and chemical effects of different parts of the solar spectrum, polarization, and the magnetic character of electric currents. His most important work however came in 1822, after he had moved to Berlin. His discovery of thermoelectricity (the *Seebeck effect*) showed that electric currents could be produced by temperature differences. Seebeck joined two wires of different metals to form a closed circuit and applied heat to one of the junctions; a nearby magnetic needle behaved as if an electric current flowed around the circuit. He called this effect "thermomagnetism" (and later objected to the term thermoelectricity).

Sefström, Nils Gabriel (1787–1845) *Swedish chemist*

Born at Ilsbo in Sweden, Sefström (**sev**-stru(r)m) studied under Jöns BERZELIUS in Stockholm, graduating in 1813. He taught chemistry at the School of Mines from 1820. While there he was informed that the local steelmakers were able to predict whether iron ore delivered at the foundries would produce steel that was brittle or not. On investigation he found that they tested the ore by dissolving it in hydrochloric acid and if a black powder resulted the steel was likely to be brittle. Sefström investigated and found that what was important in the test was the presence or absence in the ore of a new element. In 1830 he isolated the new element, which he named vanadium after the Norse goddess Vanadis. This proved to be identical to the metal discovered by Andrès DEL RIO in 1801 and named erythronium. Del Rio had been dissuaded that this was in fact a new element and had abandoned his claim.

Segrè, Emilio Gino (1905–1989) *Italian–American physicist*

Segrè (se-**gray**), who was born at Tivoli in Italy, studied at the University of Rome under Enrico FERMI and obtained his doctorate there in 1928. He worked with Fermi until 1936, when he was appointed director of the physics laboratory at Palermo. He was dismissed for political reasons in 1938 and moved to America where he worked at the University of California, Berkeley, from 1938 to 1972, apart from the years 1943–46, which he spent at Los Alamos working on the development of the atom bomb. He became a professor in 1945.

Segrè made a number of significant discoveries in his career. In 1937 he filled one of the gaps in the periodic table at atomic number 43 when he showed that some molybdenum that had been irradiated with deuterium nuclei by Ernest LAWRENCE contained traces of the new element. As the first completely man-made element they gave it the appropriate name, technetium. Segrè played a part in the detection of element 85, astatine, and also plutonium in 1940.

His main achievement however was the discovery of the antiproton with Owen CHAMBERLAIN in 1955, for which they shared the 1959 Nobel Prize for physics. Segrè calculated that producing an antiproton would require about 6 billion electron-volts (Bev), which could be provided by the recently constructed bevatron at the University of California. He therefore went on to bombard copper with protons that had been accelerated to 6 Bev, thus yielding a large number of particles. As only one antiproton was produced to about 40,000 other particles his next problem was to detect this rare event.

This was done by noting that the antiproton would travel much faster than the other particles and at such speeds it would give off CHERENKOV radiation in certain media and could thus be detected. The few particles that produced this radiation could more easily be screened to see if any of them possessed the necessary properties. Before long Segrè had identified antiprotons at the rate of about four per hour. The work of the California group was soon confirmed by Italian physicists who began to detect the tracks of antiprotons on photographic plates.

Seleucus (about 2nd century BC) *Babylonian astronomer*

According to Strabo, Seleucus (se-**loo**-kus) was a Babylonian, an inhabitant of Seleucia on the Tigris (in modern Iraq). He is unique in that he is the only named astronomer in antiquity who followed ARISTARCHUS of Samos in believing that the Earth moves. What his precise belief was is unclear. He is reported by Plutarch as asserting that the Earth rotates on its axis but whether he also accepted the annual motion around the Sun is still a matter of speculation. He is also reported as writing on the theory of the tides. He attributed the tides to the movement of the Moon, thinking that its motion pushed the air between it and the Earth, which in turn pushed the seas and oceans to produce the tides.

Semenov, Nikolay Nikolaevich (1896–1986) *Russian chemist*

Semenov (sye-**myen**-of) was born in Saratov in Russia and educated at the University of Petrograd (now St. Petersburg). After working in various institutes in St. Petersburg he moved to Moscow University in 1944 as head of the department of chemical kinetics.

It was for work in this field that Semenov was awarded the 1956 Nobel Prize for chemistry, the first Russian to be so honored. He shared the prize with Sir Cyril HINSHELWOOD. His particular contribution was in the study of chemical chain reactions – an idea introduced by Max BODENSTEIN in 1913. Semenov investigated the idea of a chain reaction in the 1920s and was able to show that such reactions can lead to combustion and violent explosions when the chain branches, spreading with explosive rapidity. In 1934 he published a book on the subject, which was translated into English the following year, *Chemical Kinetics and Chain Reactions*.

Semmelweis, Ignaz Philipp (1818–1865) *Hungarian physician*

Semmelweis (**zem**-el-vIs), the son of a storekeeper in Buda (now Budapest in Hungary), graduated in medicine from the University of Vienna in 1844. He specialized in obstetrics, being appointed assistant under J. Klein in the clinic of the Vienna General Hospital. Here he became concerned about the high incidence of puerperal fever, a streptococcal infection then endemic throughout 19th-century maternity wards. Although 5–10% of women in such wards died from the disease, particularly virulent outbreaks could more than treble such appalling figures. Few physicians had any idea of the cause of the complaint; although theories that emphasized an "atmosphere of infection" were known, they were largely discounted by the Viennese authorities.

Semmelweis began searching through the excellent hospital records maintained by the Hapsburg bureaucracy. He soon came across a striking statistic. Obstetric services of the hospital were provided by two distinct clinics, one staffed mainly by medical students and the other staffed by midwives. The mortality rate from puerperal fever in the former was three times higher than in the latter. Moreover Semmelweis pointed out that mortality figures had significantly increased with the appointment of Klein as professor of obstetrics in 1822, an observation that led to Semmelweis's virtual demotion.

Semmelweis tested many theories to account for such figures. The answer came to him early in 1847 when he heard of the death of his colleague Jakob Kolletschka, professor of forensic medicine, from an infection arising from a scalpel wound incurred during a postmortem. Semmelweis realized that the symptoms of Kolletschka's disease had been identical to puerperal fever. He proposed that "cadaveric particles" had fatally infected Kolletschka, the same particles that would be found on the hands of medical students (but not midwives) who routinely moved from assisting in postmortems to aid in the delivery ward. In May 1847 he made all students and doctors wash their hands thoroughly in a solution of chlorinated lime at the entrance of the wards. In 1848 deaths from puerperal fever had dropped from just under 10% to 1.27%.

His impressive results still failed to convince many of his colleagues and, disheartened, Semmelweis returned to his native Buda in 1850. He took up an appointment at the St. Rochus Hospital where his ideas were accepted. However, in later years his mind became unbalanced and he was committed to an asylum where, ironically, he died within a matter of days of an infection variously described as gangrene or puerperal fever. It was left to the more conventional Joseph LISTER to rediscover the results of Semmelweis in the very year of his death.

Serre, Jean-Pierre (1926–   ) *French mathematician*

Serre (sair), one of the outstanding French mathematicians of his generation, was born at Bages in France and studied at the Ecole Normale Supérieure. He was a professor at the Collège de France from 1956 until his retirement in 1994 and has been a member of the French Academy of Sciences since 1976, as well as teaching at both the University of Nancy and Princeton University.

Serre's mathematical work began with a collaboration with Henri CARTAN in which they were able to bring about a decisive reorientation of the theory of a complex variable. What Serre and Cartan did was to reformulate, in a completely original way, the central problems and results of the subject in terms of what is known as cohomology theory. Serre himself also did work of great originality and importance on the homotopy theory of spheres.

Homotopy and homology theory have been the area of his greatest mathematical achievements. Before his revolutionary work, the two subjects had been thought of as quite unconnected and unrelated. But Serre was able to show how to associate the homotopy group of a space with the homology groups of suitably constructed auxiliary spaces. By forging this link he succeeded in bringing to bear purely algebraic methods and techniques on the central problems of homotopy theory in a way that was entirely new and immensely fruitful.

In recognition of this fundamental work in revolutionizing homotopy theory Serre, aged only 28, was awarded a Field's Medal in 1954.

Sertürner, Friedrich Wilhelm Adam Ferdinand (1783–1841) *German apothecary*

Sertürner (zair-**toor**-ner), whose father was Austrian, was a civil engineer in the service of the prince of Paderborn. Sertürner was born at Neuhaus (now in Germany) and on the death of his father was apprenticed to the court apothecary. Eventually, in 1820, he became the town pharmacist of Hameln.

In 1805 Sertürner published an account of a substance he had separated from opium. It turned out to be the sleep-inducing factor he was searching for so he consequently named it morphine after the Roman god of sleep, Morpheus. His paper received little attention and it was not until he republished his results in 1817 that it was picked up by physicians and rapidly entered their pharmacopoeia. The term was first reported in England as early as 1828.

Servetus, Michael (1511–1553) *Spanish physician*

> I shall burn, but this is a mere incident. We shall continue our discussion in eternity.
> —Comment to his judges on being condemned to be burned at the stake as a heretic

Servetus (ser-**vee**-tus), the son of a notary, was born at Tudela in Spain and studied law in Toulouse and medicine at the University of Paris. After practicing medicine in Lyons for some years he considered it prudent to leave France for Italy but, for reasons never explained, he decided to travel via Calvin's

Geneva where, in 1553, he was burned at the stake as a heretic. Servetus appears as the kind of man who once having seen the strength of a particular idea could never let it go. Thus when it struck him sometime in the late 1520s that the word "trinity" appears nowhere in the Bible he was a lost man, and he further invited disaster by sending his heretical works to Calvin in 1546. Calvin swore that if Servetus ever visited Geneva he would not get out alive.

However, before Calvin was allowed to fulfill his threat Servetus had made a noteworthy contribution to science by describing in *Christianismi restitutio* (1553; Restoring Christianity) the lesser circulation. This was the proposal that blood traveled from the right to the left side of the heart via the lungs and not, as had been supposed since GALEN, through some minute perforations in the dividing wall or septum. It is not clear what led Servetus to this conclusion but it was certainly not derived from anatomical dissection.

As it was contained in but a few lines of a bulky theological work that was declared heretical and suffered in most cases the same fate as the author (only three copies have survived), it is not surprising that his views were virtually unknown in his own day and had no effect on the work of William HARVEY.

Seward, Albert Charles (1863–1941)
British paleobotanist

Seward, the son of a hardware dealer from Lancaster in northwest England, studied natural sciences at Cambridge University. After graduating in 1886 he spent a year in Manchester with William WILLIAMSON. Seward then spent his whole career at Cambridge, beginning in 1890 as a lecturer in botany and serving from 1906 until his retirement in 1936 as professor of botany. He had also held, from 1915, the office of master of Downing College.

With Dukinfield SCOTT, Seward did much to establish the foundations of the new discipline of paleobotany. Although he began with the study of Paleozoic plants he switched to the Mesozoic when invited by the British Museum (Natural History) to catalog their collection. The results were his *Jurassic Flora* (1900–1904) and his four volumes of *Fossil Plants for Students of Botany and Geology* (1898–1919). Seward also published a more discursive and popular work, *Plant Life through the Ages* (1931).

Seyfert, Carl Keenan (1911–1960) *American astronomer*

Seyfert was born in Cleveland, Ohio, and studied at Harvard where he graduated in 1933 and obtained his PhD in 1936. He worked at the McDonald Observatory in Texas from 1936 to 1940, the Mount Wilson Observatory in California from 1940 to 1942 and the Case Institute of Technology in Cleveland, Ohio, from 1942 to 1946. Seyfert then moved to Vanderbilt University in Tennessee as associate professor and where from 1951 until his death he was director of the Dyer Observatory.

In 1943, while observing at Mount Wilson, Seyfert discovered an unusual class of spiral galaxies that have since been named for him. Optically they presented a very small intensely bright nucleus; spectroscopically they had very broad emission lines indicating the presence of very hot gas moving at considerable velocities.

Since their discovery Seyfert galaxies have become even more puzzling as they are now recognized to be emitters of prodigious amounts of energy from a very compact area. Energy is released not just in the form of light but also as x-rays and radio and infrared waves. It is felt that they are related in some way to quasars, discovered in the 1960s, although possessing a less powerful source of energy.

Shannon, Claude Elwood (1916–2001)
American mathematician

Born in Gaylord, Michigan, Shannon graduated from the University of Michigan in 1936. He later worked both at the Massachusetts Institute of Technology and the Bell Telephone Laboratories. In 1958 he returned to MIT as Donner Professor of Science, a post he held until his retirement in 1978.

Shannon's greatest contribution to science was in laying the mathematical foundations of communication theory. The central problem of communication theory is to determine the most efficient ways of transmitting messages. What Shannon did was to show a precise way of quantifying the information content of a message, thus making the study of information flow amenable to exact mathematical treatment. He first published his ideas in 1948 in *A Mathematical Theory of Communication*, written in collaboration with Warren Weaver.

The resulting theory found wide application in such wide-ranging fields as circuit design, computer design, communication technology in general, and even in biology, psychology, semantics, and linguistics. Shannon's work made extensive use of the theory of probability; he also extended the concept of entropy from thermodynamics and applied it to lack of information.

Shannon also made important contributions to computer science. In his paper *A Symbolic Analysis of Relay and Switching Circuits* (1938) he drew the analogy between truth values in logic and the binary states of circuits. He also coined the term "bit" for a unit of information.

Shapley, Harlow (1885–1972) *American astronomer*

Shapley came from a farming background in Nashville, Missouri. He began his career as a crime reporter on the *Daily Sun* of a small Kansas town when he was 16. He entered the University of Missouri in 1907 intending to study journalism but took astronomy instead, gaining his MA in 1911. He then went on a fellowship to Princeton where he studied under Henry RUSSELL and gained his PhD in 1913. From 1914 to 1921 he was on the staff of the Mount Wilson Observatory in California. Finally Shapley was appointed in 1921 to the directorship of the Harvard College Observatory where he remained until 1952, also serving for the period 1922–56 as Paine Professor of Astronomy.

Shapley's early work, under Russell, on eclipsing binaries proved that the group of stars, known as Cepheids, were not binary but were single stars that changed their brightness as they changed their size. Cepheids were thus the first "pulsating variables" to be discovered, the theory of the pulsation being supplied subsequently by Arthur EDDINGTON.

Once at Mount Wilson, Shapley began to study Cepheids in globular clusters, huge spherical groups of closely packed stars. From this stemmed his fundamental work on the size and structure of our Galaxy. In 1915 he was able to make a bold speculation about the galactic structure. Using the relation between the period of Cepheids and their observed brightness, discovered in 1912 by Henrietta LEAVITT, he was able to map the relative distances of clusters from us and from each other. To his surprise he found that they were widely and randomly distributed both above and below the plane of the Milky Way and appeared to be concentrated in one smallish area in the direction of the constellation Sagittarius. He argued that such a distribution would make sense if the Galaxy had the shape of a flattened disk with the clusters grouped around the galactic center. This required that the solar system be displaced from its accepted central position by a considerable distance.

Thus Shapley had found the general structure of the Galaxy but not its size. Here the Cepheids were of limited use as they could only provide a relative scale. Absolute distances could at that time only be determined for small distances. In order to calibrate his galactic structure Shapley needed to measure the distance of a few Cepheids. He used a statistical method pioneered by Ejnar HERTZSPRUNG in 1913. Since the intrinsic brightness, or luminosity, of stars can be determined once their distance is known, Shapley's measurements allowed him to produce a quantified form of the relationship between Cepheid period and observed brightness, i.e., a period-luminosity relationship. This P-L relationship meant that a measure of the period of any Cepheid would

reveal its luminosity and hence its distance and the distance of the stars surrounding it.

By 1920 Shapley felt that he had finally cracked the fundamental problem of the scale of the Galaxy. The Sun, he declared, was some 50,000 light-years from the center of the Galaxy while the diameter of the galactic disk could be perhaps 300,000 light-years. Actually Shapley's calculations were too generous as he was unaware of the interstellar matter that absorbs some of the light from stars and thus affects determinations of stellar brightness. Consequently his figures were later revised to 30,000 light-years for the distance to the galactic center and 100,000 light-years for the diameter.

Shapley was however less successful with his work on the scale of the universe. In 1920 he took part with Heber CURTIS in a famous debate organized by the National Academy of Sciences at the Smithsonian in Washington. Using the brightness of novae in the Andromeda nebula, Curtis gave an estimate approaching 500,000 light-years for its distance and maintained that it was an independent star system. Shapley, misled by the measurements of Adriaan van Maanen, argued that this distance was far too great and that the Andromeda nebula and the other spiral nebulae lay within the Galaxy. It was left to Edwin HUBBLE to show, some years later, that Curtis had in fact underestimated rather than overestimated the distance of the Andromeda nebula and that it was in fact a separate star system.

Not the least of Shapley's achievements was his development of the Harvard Observatory into one of the major research institutions of the world. He introduced a graduate program and attracted a distinguished and much increased permanent staff. During his time there his interest turned to "galaxies," as he called them, or "extragalactic nebulae" in Hubble's terminology. Northern and southern skies were surveyed for galaxies and tens of thousands were recorded. In 1932 he produced a catalog, with Adelaide Ames, of 1,249 galaxies, which included over a thousand galaxies brighter than 13th magnitude. In 1937 he published a survey of 36,000 southern galaxies. He also studied the Magellanic Clouds and identified the first two dwarf galaxies, the Fornax and Sculptor systems, which are members of the Local Group of galaxies.

Shapley wrote several books on astronomy and left an account of his scientific life in his informal *Through Rugged Ways to the Stars* (1969).

Sharp, Phillip Allen (1944–) *American molecular biologist*

Born in Falmouth, Kentucky, Sharp was educated at Union College, Kentucky, and the University of Illinois, Urbana, where he obtained his PhD in 1969. After spending short periods as a postdoctoral fellow at Caltech and Cold Spring Harbor Laboratory, New York, Sharp joined the Massachusetts Institute of Technology in 1974 and was appointed professor of biology in 1979.

Much of the early work in molecular genetics had been carried out on prokaryotes, cells which lack a nucleus. It was found that continuous stretches of DNA were converted into various proteins. The DNA was first transcribed into a continuous sequence of messenger RNA (mRNA), triplets of which coded for one of the amino acids from which proteins were assembled:

DNA	CCC	TGA	TCG	AAA	ATA	CAG	...
mRNA	GGG	ACU	AGC	UUU	UAU	GUC	...
amino acid	gly	thr	ser	phe	tyr	val	...

It was automatically assumed that similar mechanisms would be found to operate in eukaryotic cells, cells with a nucleus.

In 1977, however, Sharp demonstrated that this assumption was baseless. Sharp worked with adenoviruses, the viruses responsible for, among other things, the common cold. He explored the process of protein production by forming double stranded hybrids of adenovirus DNA and mRNA. The hybrids were then displayed on an electron micrograph. To Sharp's surprise the mRNA hybridized with only four regions of DNA, and these were separated by long stretches of DNA looping out from the hybrid. The intervening loops, later to be termed "introns" by Walter GILBERT, it was presumed, were later snipped off and the four remaining groups, "exons" in Gilbert's terminology, would be spliced together to form the mature mRNA. This mature mRNA would then leave the cell's nucleus and serve as the template upon which proteins could be assembled.

Sharp's work was confirmed independently by Richard ROBERTS. The "split genes" identified in adenoviruses by Sharp were quickly shown to be fairly standard in eukaryotic cells. The phenomenon has proved highly puzzling. In some organisms as much as 90% of nuclear DNA is snipped away as introns and consequently seems to serve no purpose at all. Why there should be so much "junk" DNA as it has sometimes been described remains a mystery.

For his discovery of split genes Sharp shared the 1993 Nobel Prize for physiology or medicine with Richard Roberts.

Sharpey-Schäfer, Sir Edward Albert
(1850–1935) *British physiologist*

Schäfer (**shay**-fer), the son of a city merchant in London, qualified in medicine at University College there in 1874. He joined the staff of the college and served as Jodrell Professor of Physiology from 1883 until 1899. He then moved to a similar chair at the University of Edinburgh where he remained until his retirement in 1933.

In 1896 Schäfer, working with George Oliver, discovered that an extract from the medulla of the adrenal gland produced an immediate elevation of blood pressure when injected into animals. The substance, adrenaline (epinephrine), was later isolated and crystallized by Jokichi TAKAMINE in 1901. Schäfer also worked on pituitary extracts. He is further remembered in the field of endocrinology for his proposal that the active pancreatic substance in the islets of Langerhans should be called "insuline," some eight years before its discovery by Frederick BANTING and Charles BEST.

Schäfer published two influential works: *Essentials of Histology* (1885) and *Endocrine Organs* (1916). He also founded the important *Quarterly Journal of Experimental Physiology* in 1898.

Schäfer had named one of his two sons Sharpey after his much admired anatomy teacher, William Sharpey, at University College. But after the tragic death of both his sons in World War I Schäfer changed his own name to the hyphenated Sharpey-Schäfer.

Sharpless, K. Barry (1941–) *American chemist*

Sharpless was born in Philadelphia, Pennsylvania, and educated at Dartmouth College, gaining a BA in 1963, and at Stanford University, where he obtained a PhD in 1968. He did postdoctoral research at Stanford University in 1968 and at Harvard University in 1969. Between 1970 and 1977 he was in the chemistry faculty of the Massachusetts Institute of Technology (MIT), joining the chemistry faculty of Stanford University in 1977 and rejoining MIT in 1980. Between 1987 and 1990 he was the Arthur C. Cope Professor at MIT. Since 1990 he has been W. M. Keck Professor of Chemistry at the Scripps Research Institute, La Jolla, California.

Along with William KNOWLES and Ryoji NOYORI, Sharpless was one of the pioneers in the development of catalytic asymmetric synthesis. In contrast to the work of Knowles and Noyori, who were concerned with chirally catalyzed hydrogenation reactions, the work of Sharpless involved chirally catalyzed oxidation reactions. Knowles, Noyori, and Sharpless shared the 2001 Nobel Prize for chemistry for their work on chiral catalysts.

Shaw, Sir William Napier (1854–1945) *British meteorologist*

Born in the English Midlands city of Birmingham, Shaw studied at Cambridge University, England. After graduating in 1876 he was appointed as a lecturer in experimental physics in the Cavendish Laboratory, Cambridge. In 1900 he took the post of secretary of the Meteorological Council, later becoming (1905) director of the Meteorological Office. After his retirement in 1920 he was appointed professor of meteorology at Imperial College, London.

Shaw established the Meteorological Office as an efficient organization and demonstrated the use of meteorology as an applied science. He carried out research into air currents and in his *Life History of Surface Air Currents* (1906) anticipated the "polar front" theory of cyclones later developed by Jacob BJERKNES. He also worked on the upper atmosphere using instrumented balloons. After he retired from teaching in 1924 he produced a major account of the discipline in his four-volume *Manual of Meteorology* (1926–31). He was knighted in 1915.

Shemin, David (1911–1991) *American biochemist*

Shemin was born in New York City and educated at the City College there before obtaining his PhD from Columbia in 1938. He served as professor of biochemistry at Columbia from 1953 until 1968, when he moved to a similar post at Northwestern University in Chicago, which he held until 1979.

In the 1940s Shemin and his colleagues succeeded in working out some of the details of porphyrin synthesis. Several biologically important molecules, including hemoglobin and chlorophyll, are porphyrins, i.e., they contain a porphyrin ring in their structure. This consists of carbon and nitrogen atoms forming four pyrrole rings, which are linked by carbon atoms and joined to a metal atom in the center – iron in the case of hemoglobin and magnesium in chlorophyll.

Using radioactive tracers, Shemin and his coworkers were able to show that all the carbon and nitrogen atoms of the porphyrin ring are derived from glycine and succinyl-CoA. Such work is not merely of academic interest, since errors in the metabolism of porphyrin are responsible for a number of diseases. Porphyria, an excess of porphyrins in the blood affecting the nervous system, was possibly responsible for the madness of the English king, George III.

Shen Kua (about 1031–1095) *Chinese scientist and scholar*

> As for the waxing and waning of the moon, although some phenomena such as pregnancy and the tides are tied to them, they have nothing to do with seasons or changes of climate.
> —*Meng ch'i pi t'an* (c. 1086; Dream Pool Essays)

Shen Kua (shen kwah) spent his life in government service in his native China, working variously as an ambassador, a military commander, and a director of water systems. In around 1086 he completed his *Meng ch'i pi t'an* (Dream Pool Essays) in which he recorded the scientific observations made on his extensive travels, the results of his own researches, and the general scientific activity of his day.

One of the work's most important parts is its description of the discovery of the magnetic compass. This is first reported in Europe in the 12th century but Shen Kua described it a century before. He also discussed fossils that he had been shown, giving an accurate account of their origin. Other passages in the book contain detailed descriptions of meteorological phenomena he had seen, experiments done on mirrors, and mathematical problems.

Sherman, Henry Clapp (1875–1955) *American biochemist*

Sherman, one of ten children, was born on a farm in Ash Grove, Virginia. He attended Maryland Agricultural College, graduating in 1893 in general science and chemistry. He then went to Columbia University, where he received his doctorate in chemistry and physiology in 1897, and he remained working in the chemistry department at Columbia for the rest of his career.

His most important work was on the development of quantitative biological methods for assaying the vitamin content of food. He studied enzyme chemistry, producing experimental evidence for the protein nature of digestive enzymes and contributing to the development of greater precision in measuring amylase activity. He investigated the calcium and phosphorus requirements of the body, showing that both are needed in an appropriate ratio and that rickets can be induced on a low-phosphorus diet even when calcium supplies are more than adequate.

Sherrington, Sir Charles Scott (1857–1952) *British physiologist*

> Like this old Earth that rolls through sun and shade,
> Our part is less to make than to be made.
> —*Man on his Nature* (1955)

Sherrington, a Londoner by birth, was educated at Cambridge University and St. Thomas's Hospital, London, gaining his BA in natural science in 1883 and his MB in 1885. He then traveled to Europe to study under Rudolf VIRCHOW and Robert KOCH in Berlin. After lecturing in physiology at St. Thomas's Hospital, Sherrington was superintendent of the Brown Institute (1891–95), a veterinary hospital of the University of London. He then became professor of physiology, firstly at the University of Liverpool (1895–1913) and then at Oxford University, holding the latter post until his retirement in 1935.

Sherrington's early medical work was in bacteriology. He investigated cholera outbreaks in Spain and Italy and was the first to use diphtheria antitoxin successfully in England, his nephew being the patient. During the war he tested antitetanus serum on the wounded and also worked (incognito) as a laborer in a munitions factory. He then turned his attention to studies of the reflex actions in man, demonstrating their effect in enabling the nervous system to function as a unit and anticipating Ivan PAVLOV in his discovery of the "conditioned reflex." These researches culminated in his most celebrated publication, *The Integrative Action of the Nervous System* (1906), a landmark in modern physiology. Sherrington also did much work on decerebrate rigidity and the renewal of nerve tissue. For their work on the function of the neuron, Sherrington and Edgar ADRIAN were jointly awarded the Nobel Prize for physiology or medicine in 1932. Sherrington was knighted in 1922.

Shibukawa, Harumi (1639–1715) *Japanese astronomer*

> Barbarians who may have theories but cannot prove methods.
> —On Western astronomers

Shibukawa (shee-buu-**kah**-wah) was born at Kyoto in Japan, the son of a professional go player. Initially he followed his father's career, even adopting for a time his father's name, Yasoi Santetsu II. He also studied mathematics and astronomy and became increasingly interested in calendar reform.

In 862 Japan had adopted the Chinese T'ang calendar, the Hsuan-Ming, compiled in 822 and unreformed in Japan until Shibukawa's time. As the Chinese calendar consisted of 12 lunar months (six of 30 days and six of 29 days) and 354 days it rapidly fell out of phase with the tropical year of 365.2422 days. The simplest solution to the problem was to add intercalary months when needed. Thus seven months added in a 19-year cycle would be no more than six hours out of phase. But, six hours per 19 years does mount up and, by the time of Shibukawa, the calendar had become useless for eclipse prediction and could not even establish the date of the winter solstice.

Shibukawa began to make observations with a gnomon to determine the precise solstice date. In 1684 he resigned his position as go master but remained at court as head of a new department, the Tenmongata (or Bureau of Astronomy). In this office he was largely responsible for the annual production of the civil calendar. His new system, the Jujireki, was based upon the Shou-shih calendar of the Yuan dynasty (1279–1368) corrected for the differences in latitude and longitude between Peking and Kyoto. However, as Shibukawa's instruments were inferior to those used by the Yuan astronomers, discrepancies soon arose

and Shibukawa's system remained in force only until 1754.

Shirakawa, Hideki (1936–   ) *Japanese chemist*

Shirakawa (**shee**-ra-kah-wa) was born in Tokyo and educated at the Tokyo Institute of Technology. For more than 20 years he has been a faculty member of the Institute of Materials Science at the University of Tsukuba, Japan, where he is now a professor.

Along with his collaborators Alan HEEGER and Alan MACDIARMID, Shirakawa was one of the pioneers of electrically conducting polymers. It is frequently said that the discovery of these materials was initiated by a mistake by Shirakawa when he was a research associate at the Chemical Resources Laboratory at the Tokyo Institute of Technology. When synthesizing a polymer a large excess of catalyst was accidentally added. This resulted in a silvery film with some metallic properties. When MacDiarmid heard about this he invited Shirakawa to the University of Pennsylvania as a postdoctoral fellow. MacDiarmid, Shirakawa, and Heeger then investigated how electrical conductivity arose in polymers, which were initially insulators. They concluded that it is possible to introduce carriers of electric charge into polymers by doping. The example they studied was the modification of polyacetylene by oxidizing it with halogen vapor. This discovery, which was published in 1977, has led to a vast amount of work on electrically conducting polymers and their technological applications. Heeger, MacDiarmid, and Shirakawa were awarded the 2000 Nobel Prize for chemistry for their discovery.

Shizuki, Tadao (1760–1806) *Japanese astronomer and translator*

> The cause of gravity is quite inscrutable. Even with advanced Western instruments and mathematics, the fundamental cause is indeterminable.
> —*Rekisho Shinsho* (1802; New Treatise on Calendrical Phenomena)

Born in Nagasaki, the son of an interpreter, Shizuki (shee-**zuuk**-ee) studied under another interpreter, Motoki Ryoei. By the end of the 17th century the only foreign ideas allowed to enter Japan were traditional ones from China. Trade was permitted with the Dutch and Portuguese but since 1641 was confined to the island of Dejima. No other intellectual contact was permitted.

Some relaxation, however, was allowed during the reign of Yoshimune (1716–45), and Western texts on such subjects as astronomy and gunnery began to circulate in Japan in Chinese translation. Shizuki took advantage of the more liberal regime and began work in the 1780s on the translation of *Introductio ad veram physicam et veram astronomiam* (1739; Introduction to True Physics and True Astronomy), a commentary by John Keill (1671–1721) on Newtonian physics. Shizuki worked from a 1741 Dutch translation by Johan Lulof. The work was completed by 1802 and published under the title *Rekisho Shinso* (New Treatise on Calendrical Phenomena) and was in fact the first book on modern physics and astronomy to be published in Japanese. It contained an elementary account of universal gravitation, the laws of motion, and the properties of ellipses.

Shockley, William Bradford (1910–1989)
British–American physicist

Shockley, born the son of a mining engineer in London, was educated at the California Institute of Technology and at Harvard, where he obtained his PhD in 1936. He started work at the Bell Telephone Laboratories in 1936. In 1963 he took up an appointment as professor of engineering at Stanford University.

Shockley is noted for his early work in the development of the transistor – an invention that has had a profound effect on modern society. He collaborated with John BARDEEN and Walter BRATTAIN in their work on the point-contact transistor (1947). The following year Shockley developed the junction transistor.

In semiconductors such as germanium and silicon the electrical conductivity is strongly affected by impurities. The germanium and silicon atoms have four outermost electrons and an impurity such as arsenic, with five outer electrons, contributes extra electrons to the solid. In such materials the current is carried by negative electrons and the conductivity is said to be

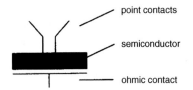

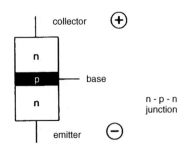

TRANSISTORS   Point-contact and junction transistors.

n-type. Alternatively, impurities such as boron, with three outer electrons, have a different effect in that they introduce "holes" – i.e., "missing" electrons. An electron on an adjacent atom can move to "fill" the hole, leaving another hole. By this mechanism electrical conduction is by movement of positive holes through the solid – the conductivity is said to be *p*-type.

Shockley experimented with junctions of *p*- and *n*-type material, showing how they act as rectifiers. He formed the first junction transistor of a thin layer of *p*-type material sandwiched between two *n*-type regions. This *n–p–n* junction transistor could be used to amplify current. Shockley shared the 1970 Nobel Prize for physics with Bardeen and Brattain.

Shoemaker, Carolyn Jean (1929–   )
American astronomer

Despite starting her career in astronomy relatively late, at the age of 51, Shoemaker identified some 370 named asteroids and 32 comets, a record number for an individual. Born Carolyn Spellman in Gallup, New Mexico, she moved with her family to Chico, California, where she attended Chico State College, receiving a BS and MS in history and political science. After a spell as a schoolteacher, she married the astrogeologist Gene Shoemaker in 1951, and settled down eventually in Flagstaff, Arizona, where her husband established the United States Geological Survey Center for Astrogeology. In the late 1960s, after the Shoemakers' three children had left home, Carolyn decided to help out with a program at the California Institute of Technology to identify asteroids approaching the earth and predict the likelihood of their colliding with our planet. She trained to analyze photographic plates of the night sky using a stereomicroscope, pinpointing asteroids and comets and determining their positions relative to known stars. Carolyn worked closely with her husband, regularly spending nights observing the sky at the Palomar Observatory in California. In 1993 the husband and wife team, working with David Levy, discovered the comet Shoemaker–Levy 9. This achieved international prominence when its remnants collided with Jupiter in 1994, a rare event witnessed by astronomers around the world, and by the Voyager 2 spacecraft. While in Australia in 1997, a car crash killed her husband and left Carolyn Shoemaker seriously injured. However, she recovered and resumed her astronomical work with Levy and his wife Wendee.

Shull, Clifford Glenwood (1915–2001)
American physicist

Shull, who was born in Pittsburgh, Pennsylvania, was educated at the Carnegie Institute of Technology and New York University, where he obtained his PhD in 1941. He began as a research physicist working first for the Texas Company, from 1941 to 1946, and then with the Oak Ridge National Laboratory from 1946 until 1955, when he entered academic life as professor of physics at the Massachusetts Institute of Technology. He remained at the Institute until his retirement in 1986.

Shull's main research interest was in the diffraction of slow neutrons by crystals. Just as x-rays can be diffracted by a crystal lattice, neutron beams of suitable energy also show diffraction effects. In the case of x-rays, diffraction is mainly by the electrons in the atom, whereas in neutron diffraction the nuclei scatter the neutrons. From about 1946 onward Shull applied neutron diffraction to determining crystal structure, showing that the method could indicate the position of light atoms such as hydrogen (which are not detected by x-ray methods).

Shull also showed that an additional effect occurred in neutron diffraction – magnetic scattering by interaction of the neutron's magnetic moment with that of the atom. He demonstrated antiferromagnetism in manganese(II) oxide using this technique in 1949. He subsequently did considerable work in "magnetic diffraction" of neutrons and in other aspects of neutron interaction with matter. For his work on neutron diffraction Shull shared the 1994 Nobel Prize for physics with Bertram BROCKHOUSE.

Sidgwick, Nevil Vincent (1873–1952) *British chemist*

Sidgwick, who was born in Oxford, England, came from a distinguished intellectual family; both his father and an uncle were Oxford classicists and another uncle was a professor of moral philosophy at Cambridge University. He was educated at Oxford University obtaining first class degrees in both chemistry (1895) and classics (1897). After further study in Germany, during which he obtained his PhD in 1901 from Tübingen, he returned to Oxford as a fellow and spent the remainder of his life there.

Sidgwick began his career working on organic compounds and in 1910 he produced *The Organic Chemistry of Nitrogen*, a classic text on the subject. In 1914 Sidgwick attended a meeting of the British Association in Australia and there he met Ernest RUTHERFORD with whom he formed a lasting friendship. The meeting marked a turning point in his career; he became interested in atomic structure and tried to explain chemical reactions through this.

Sidgwick's theory was eventually published in 1927 in his *Electronic Theory of Valency*, which established his international reputation. The significance of his work was that it extended the idea of valency developed by Gilbert LEWIS and Irving LANGMUIR to inorganic compounds, emphasizing the necessity of assuming the Bohr–Rutherford model of the atom. He introduced what he termed a coordinate bond in which, unlike the covalent bond of Lewis, both electrons are donated by one atom and accepted by the other. This explained the coordination compounds of Alfred WERNER.

In his later years Sidgwick worked on his two-volume *The Chemical Elements and their Compounds* (1950), a massive work that attempted to demonstrate the adequacy of valency theory by showing that it applied to all compounds. The work took 25 years of Sidgwick's life and for it he was reported to have examined 10,000 scientific papers.

Siebold, Karl Theodor Ernst von (1804–1885) *German zoologist and parasitologist*

> Instead of being able to relax in my later years, I have to learn just as much – no, even more – than I did during all my younger days. If you reflect that in old age it is much harder to learn than to forget, you will bear with me.
> —Letter to Ernst Haeckel, 16 February 1874

Siebold (**zee**-bohlt), who came from a family of physicians in Würzburg, Germany, began his academic education by studying medicine at Berlin and Göttingen. He later practiced briefly in Danzig. In his spare time he studied marine fauna and it was largely owing to his published research on these animals that he became elected to the chair of anatomy and physiology at Erlangen in 1840. Siebold later held professorships at Freiburg, Breslau, and Munich. Apart from some work on the salamanders and freshwater fish of central Europe, Siebold devoted his attention almost entirely to the invertebrates, especially the vermiform parasites. Of particular importance was his demonstration that the various stages in the life cycles of parasites develop in different host animals, proving, for example, that the sheep disease of the brain known as "gid" or "the staggers" is caused by the adolescent stage of the tapeworm that passes its adult life in the intestine of the dog. Siebold also studied insect parthenogenesis, proving that drone (male) bees develop from unfertilized eggs, and made valuable contributions to invertebrate classification, especially with regard to the Protozoa. Siebold's major publication, which he produced in collaboration with Hermann Stannius, is *Lehrbuch der vergleichenden Anatomie* (1846; Textbook of Comparative Anatomy). Embracing both invertebrates (Siebold) and vertebrates (Stannius), it is regarded as one of the first important texts on comparative anatomy, based as it is on observed facts rather than abstruse philosophy. In 1848 Siebold founded, with Rudolph von KÖLLIKER, the prestigious *Zeitschrift für wissenschaftliche Zoologie* (Journal of Scientific Zoology).

Siegbahn, Kai Manne Börje (1918–)
Swedish physicist

Siegbahn (**seeg**-bahn), who was born at Lund in Sweden, was the professor of physics at the Royal Institute of Technology, Stockholm, from 1951 to 1954. He taught at the University of Uppsala from 1954 to 1984. Here he worked on the emission of electrons from substances irradiated with x-rays. Siegbahn's technique was to subject a specimen to a narrow beam of x-rays with a single wavelength (i.e., energy) and measure the energy spectrum of the ejected electrons by magnetic or electrostatic deflection. The spectrum shows characteristic peaks formed by electrons ejected from different inner energy levels of atoms. Moreover, the positions of these peaks depend to a slight extent on the way in which the atom is linked to other atoms in the molecule. These "chemical shifts" allow the technique to be used as an analytical tool – Siegbahn has named it ESCA (electron spectroscopy for chemical analysis). He has also worked on the related technique of ultraviolet photoelectron spectroscopy developed by David TURNER.

Siegbahn is the son of Karl Manne Siegbahn, who won the Nobel Prize for physics in 1924. Kai Siegbahn was awarded the Nobel Prize in 1981.

Siegbahn, Karl Manne Georg (1886–1978) *Swedish physicist*

Siegbahn, who was born at Örebro in Sweden, was educated at the University of Lund, where he studied astronomy, mathematics, physics, and chemistry, obtaining his doctorate in 1911. In 1914 he turned his attention to the new science of x-ray spectroscopy. It had already been established from x-ray spectra that there were two distinct "shells" of electrons within atoms, each giving rise to groups of spectral lines, labeled "*K*" and "*L*." In 1916 Siegbahn discovered a third, or "*M*," series. (More were to be found later in heavier elements.)

Through successive refinement of his x-ray equipment and technique, Siegbahn was able to achieve a significant increase in the accuracy of his determinations of spectral lines. This allowed him to make corrections to BRAGG's equation for x-ray diffraction to allow for the finer details of crystal diffraction. Besides working with crystals, he performed x-ray spectroscopy at longer wavelengths using gratings. Here again his accurate measurements revealed discrepancies that were later shown to result from inaccuracies in the value assumed for the electronic charge.

In 1920 Siegbahn was made professor and head of the physics department at the University of Lund and in 1923 he moved to the University of Uppsala to become chairman of the physics department. In 1924 he received the Nobel Prize for physics, cited for "his discover-

ies and research in the field of x-ray spectroscopy," and the following year saw publication of his influential book *Spectroscopy of X-rays* (1925). In the same year Siegbahn and his colleagues showed that x-rays are refracted as they pass through prisms, in the same way as light.

When, in 1937, the Swedish Royal Academy of Sciences created the Nobel Institute of Physics at Stockholm, Siegbahn was appointed its first director. In the same year he became professor of physics at the University of Stockholm, retaining this post until his retirement in 1964. He was responsible for the building of accelerators, laboratory spectrometers, and other equipment at the Nobel Institute.

Siemens, Ernst Werner von (1816–1892)
German engineer

> To measure is to know.
> —Attributed

The son of a farmer from Lenthe in Germany, Siemens (**zee**-mens) was the eldest of fourteen highly talented children, who included William Siemens. In 1834 he joined the Prussian army and spent three years in Berlin receiving a thorough training in science and mathematics. Afterward duties were so light as to allow him to pursue his growing interest in chemistry and electricity. From this work he derived his first invention, a new system of electroplating sold by his brother William in London for 1,600 pounds in 1843.

In 1847 Siemens founded, with Johann Halske (1814–90), the firm of Siemens and Halske, later to become one of the major industrial concerns in Europe. Initially Siemens hoped to move into the rapidly growing telegraphy business. In 1847 he built the Berlin–Frankfurt line, insulating the underground cables with the newly introduced gutta-percha. Unfortunately for Siemens the gutta-percha had been vulcanized and the copper wire and sulfur destroyed the insulation. Contracts were canceled and, for a time, Siemens found it necessary to work outside Germany. The period, nonetheless, gave Siemens time to experiment, refine, and improve on the basic principles of telegraphy. He was consequently selected to construct the telegraph line connecting London to Calcutta, a distance of 11,000 kilometers, which was completed in 1870. Siemens's other main interest lay in power generation. The early electric generators were cumbersome machines using large steel permanent magnets and delivering very little power. In 1867 Siemens proposed to replace them with the self-activating dynamo. The permanent magnets were replaced by electromagnets and these were fed by a current obtained from an armature and commutator. Once the dynamo had been perfected it made possible the many manifestations of electric power – lighting, both domestic and public, transport, heating, cooling, and so on. The Siemens companies were well placed to take advantage of the commercial and industrial revolution created by the dynamo. Perhaps the clearest measure of the achievements of Siemens could once have been found in Berlin, where the suburb in which the firm's factories were located and where 120,000 men were employed was named Siemensstadt.

Siemens, Sir William (1823–1883) *British engineer*

Born Carl Wilhelm in Lenthe, Germany, Siemens (**see**-menz or **zee**-mens) was the son of a tenant farmer and a younger brother of Ernst Werner Siemens. He was educated at Göttingen and first visited Britain in 1843 as an agent of his brother Ernst Werner. He settled in England shortly afterwards, becoming a naturalized citizen in 1859.

Siemens worked in two main areas, namely, heat and electricity. In the age of JOULE he was aware of the potential value to be gained by conserving heat. Early attempts to redesign the steam engine proved impractical. More successful was his introduction, aided by his younger brother Friedrich (1826–1904), of the regenerator furnace (1856). In the Siemens furnace the hot combustion gases were not simply discharged into the air but used to heat the air supply to the chamber. The process was first used in the manufacture of steel by an open-hearth process known as the *Siemens–Martin process* (after the French engineer Pierre Blaise Emile Martin, 1824–1915) in the 1860s. It proved to be the first serious challenge to the BESSEMER process and by the century's end had become the favored method of steel production. On the strength of this a steel foundry was opened at Landore, South Wales, in 1869. As it failed to prosper it was abandoned in 1888.

He was more successful with his work in electric telegraphy. Siemens designed the cable-laying ship *Faraday* for laying a new trans-Atlantic cable in 1874. He also worked on electric lighting and on the Portrush electric railway in Northern Ireland. He died suddenly from a heart attack in 1883. When, forty years before, he had first arrived in Britain his English had been so poor that he looked for legal advice from an undertaker. Since then, it was said of him, he made three fortunes: one he lost, one he gave away, and one he bequeathed to his brothers. The electrical unit of conductance, the *siemens*, is named in his honor.

Sierpiński, Waclaw (1882–1969) *Polish mathematician*

The son of a physician, Sierpiński (sher-**pin**-ye-skee or seer-**pin**-skee) was educated at the university in his native city of Warsaw. After first

teaching at Lvov University he moved in 1919 to the University of Warsaw, where he remained for the rest of his career.

Although Sierpiński had written 600 papers on set theory and another 100 on number theory, his name only became widely known when popular books on chaos began to talk about *Sierpiński carpets*. The carpet begins with a square divided into nine equal squares. The central square is removed and the operation is repeated on the eight other squares, and the process is continued indefinitely. The figure is a fractal with a dimension of log 8/log 3 = 1.8928.

Silliman, Benjamin (1779–1864) *American chemist*

Born at North Stratford (now Trumbull) in Connecticut, Silliman graduated in law from Yale and was called to the bar in 1802. Instead of taking up law he was appointed professor of chemistry and natural history at Yale. To prepare himself he spent two years at the University of Pennsylvania studying chemistry before taking up his appointment in 1804.

Silliman did much to stimulate the interest in and growth of science in 19th-century America. He also prepared and founded the organizational structure of American science in anticipation of its later growth. He was a great teacher and popularizer willing to lecture on most aspects of science. In 1813 he introduced lectures on geology at Yale and in 1818 founded the *American Journal of Science*, the only journal at the time which had a nationwide distribution as well as being read in Europe. In 1830 he published his textbook *Elements of Chemistry*. He became (1840) the first president of the Association of American Geologists, an organization which later became the American Association for the Advancement of Science. He was also a founder member of the National Academy of Science in 1863. His son, Benjamin Jr., succeeded him to his chair at Yale.

Simon, Sir Francis Eugen (1893–1956) *German–British physicist*

Simon, the son of a wealthy Berlin merchant, was educated at the universities of Munich, Göttingen, and Berlin, where he worked with Walther NERNST and obtained his doctorate in 1921. He moved to Breslau in 1931 to take the chair of physical chemistry but the rise of Hitler forced him into exile in 1933. He was invited by Frederick LINDEMANN to the Clarendon Laboratory in Oxford and was made reader in thermodynamics there in 1935. He succeeded Lindemann as professor in 1956 but died a month later.

Under Simon, Oxford became one of the leading centers in the world for low-temperature physics. Shortly before his death a temperature of 0.000016 kelvin was achieved in his laboratory – less than 1/200,000 of a degree above absolute zero – using the method of magnetic cooling introduced by Francis GIAUQUE and Peter DEBYE in 1926.

Simon was able to claim one further distinction, that he was the only man to hold the Iron Cross of Imperial Germany and a knighthood of the British Empire.

Simpson, Sir George Clark (1878–1965)
British meteorologist

Simpson, who was born at Derby in England, was educated at Manchester University. After a year at Göttingen (1902) and fieldwork in Lapland investigating atmospheric electricity, he was appointed lecturer in meteorology at Manchester (1905), the first such post to be created in a British university. He was meteorologist on Scott's final expedition to the Antarctic in 1910, publishing his account of his researches in three volumes (1919–23). He succeeded William Napier SHAW as director of the Meteorological Office in 1920 and was knighted in 1935.

His main research was in the field of atmospheric electricity and he attempted to establish the process by which clouds and rain become electrified. He also improved the scale of wind speeds worked out by Francis BEAUFORT in 1806. The Beaufort scale was defined in numerical terms of the behavior of sails in varying winds. Simpson brought this up to date by standardizing Beaufort numbers against wind speeds in miles per hour at a height of 33 feet (10 m) above ground level. Thus force 10 on the Beaufort scale became the equivalent of 55–63 miles per hour (88–101 km per hour).

Simpson, George Gaylord (1902–1984)
American paleontologist

Simpson, born the son of a lawyer in Chicago, Illinois, was educated at the University of Colorado and at Yale where he obtained his PhD in 1926. He joined the staff of the American Museum of Natural History in the following year and after serving as curator from 1942 to 1959 moved to Harvard as Agassiz Professor of Vertebrate Paleontology at the Museum of Comparative Zoology, a post he retained until his retirement in 1970. In 1967 he became professor of geosciences at the University of Arizona, Tucson.

Simpson worked extensively on the taxonomy and paleontology of mammals. His main contribution was his elucidation of the history of the early mammals in the late Mesozoic and the Paleocene and Eocene. To this end, beginning with field trips to Patagonia and Mongolia in the 1930s, he traveled to most areas of the world. In addition to a number of monographs he produced, in 1945, *Principles of Classifica-* *tion and the Classification of Mammals,* a major reference work on the subject.

In such works as *The Meaning of Evolution* (1949) and *The Major Features of Evolution* (1949) he also did much to establish a neo-Darwinian orthodoxy, as did similar works by Theodosius DOBZHANSKY and Ernst MAYR. Such works brought together population genetics, paleontology, and chromosomal studies to establish for the first time a broad consensus concerning the nature and mechanism of evolution.

Simpson, Sir James Young (1811–1870)
British obstetrician

Simpson, a baker's son, was born at Bathgate in Scotland. He obtained his MD in 1832 from the University of Edinburgh, where, in 1839, he was appointed professor of midwifery.

On 21 December 1846 Robert Liston performed the first operation under anesthetic in Britain at University College Hospital, London. Simpson immediately traveled to London to confirm the reports with Liston. On 19 January 1847 he used ether for the first time in a difficult delivery with complete success, establishing that the anesthetic did not, as had been feared, inhibit uterine contraction. He was not, however, completely satisfied with ether and later used chloroform (first tested on lower animals by Pierre Flourens) as an alternative. By the end of November 1847 he had not only used chloroform in three surgical cases but had also written and published an important pamphlet, *Account of a New Anaesthetic Agent.*

This pamphlet initiated a major controversy. There was little objection to performing major surgical operations under ether or chloroform but the extension to childbirth resulted in the publication of dozens of papers arguing both ingeniously and crudely against the innovation on supposed biblical and medical grounds. The practice became respectable when Queen Victoria decided in 1853 to permit John SNOW to use chloroform during her eighth confinement.

Simpson was however less sympathetic to the introduction of antiseptic techniques proposed by Joseph LISTER and saw hospital infection more as a problem of architectural design. He thus proposed replacing the large overcrowded hospitals with smaller units, open to the fresh air. The future however lay with Lister.

In 1866 Simpson was made a baronet, the first physician practicing in Scotland to be so honored.

Simpson, Thomas (1710–1761) *British mathematician*

> The ablest Analyst ... that this country [Britain] can boast of.
> —R. Woodhouse, introduction to the 1823 edition of Simpson's *A New Treatise on Fluxions*

The son of a weaver from Market Bosworth in Leicestershire, England, Simpson was largely self-educated. His mathematical interests were aroused when a peddler gave him a copy of the popular textbook, *Cocker's Arithmetic*. He left home early, for by 1724 he was reported to be in nearby Nuneaton, practicing as an astrologer. By 1735 he had arrived in London where he worked initially as a weaver but also as a part-time teacher of mathematics. He soon became well known through a series of popular textbooks among which were *A New Treatise on Fluxions* (1737) and his *Treatise of Algebra* (1745). In 1743 he was appointed to the Royal Military Academy at Woolwich. He also served (1754–60) as editor of the *Ladies Diary* – a journal that sought to interest the "fair sex" in "Mathematicks and Philosophical Knowledge."

In mathematics he is best known for his formulation in 1743 of what has since been known as *Simpson's rule*, allowing the area under a curve to be approximated by using parabolic arcs.

Sitter, Willem de *See* DE SITTER, WILLEM.

Skoda, Josef (1805–1881) *Austrian physician*

The son of a poor Bohemian locksmith from Pilsen (now in the Czech Republic), Skoda (**shkaw**-dah) studied medicine at the University of Vienna where he obtained his doctorate in 1831. After some years in practice in Pilsen he returned to the University of Vienna and served as professor of internal medicine from 1846 onward.

In 1839 he published an important work on percussion and auscultation, which helped to establish the modern techniques of Leopold AUENBRUGGER and René LAËNNEC. Skoda, with Ferdinand von Hebra, was one of the few members of the Viennese establishment actively to support the work of Ignaz SEMMELWEIS on puerperal fever and it was owing to him that his colleague's claims were investigated, however incompetently, and first published in 1849.

Skolem, Thoralf Albert (1887–1963) *Norwegian mathematician*

The son of a teacher, Skolem (**skoh**-lem) was born at Sandsvaer in Norway and educated at the University of Oslo. He joined the faculty in 1911 and was appointed professor of mathematics in 1938, a post he held until his retirement in 1950.

Skolem is best known for his work in mathematical logic, including his contribution to the proof of the *Lowenheim-Skolem theorem* and the construction of the *Skolem paradox*.

The Lowenheim theorem of 1915, as generalized by Skolem in 1920, simply states that any schema satisfiable in some domain is satisfiable in a denumerably infinite domain. Denunerably infinite domains, also known as countable domains, can be matched in a one–one correspondence with the domain of natural numbers. Clearly the natural numbers are denumerably infinite, as are the even numbers and the rational numbers. There are in fact just as many even numbers as natural numbers, as can be seen from the correspondences set out below:

2	4	6	8	10	12	14	16 ...
1	2	3	4	5	6	7	8 ...

There are, however, domains, such as the domain of the real numbers, which cannot be put into such a correspondence with the natural numbers. Such domains are nondenumerably infinite.

Skolem pointed out in 1922 that this result leads to a new paradox in set theory. There are sets, the set of real numbers for example, which are nondenumerable. Yet, by the Lowenheim-Skolem theorem, they must be satisfiable in a denumerably infinite domain. Skolem proposed to defuse the paradox by claiming that notions such as nondenumerability have no absolute meaning, but can only be understood within the confines a particular axiomatic system. Thus a set may be nondenumerable within a system and denumerable outside. Consequently, the paradox will fail to arise.

Skou, Jens Christian (1918–   ) *Danish biologist*

Skou was educated at the University of Copenhagen. In 1946 he joined the staff of the Institute of Physiology, Aarhus University, where he served as professor of biophysics from 1978 until his retirement in 1988.

Skou sought to understand how the energy-rich adenosine triphosphate (ATP) molecule fuels cellular activities. It had been established by Peter MITCHELL that a greater concentration of ions was normally found on one side of the cell's membrane. How the ions were transported across the membrane, however, remained a mystery. Skou suspected that the process was mediated by an enzyme. In 1957, working with crab nerve cells, Skou found an enzyme in the cell's membranes that appeared to pump sodium out of cells while transporting potassium within, a process since referred to as the Na^+/K^+ pump. The enzyme, known as sodium-potassium-ATPase, proved to be the first of many ATP-based enzymes responsible for transporting molecules through cellular membranes.

For his discovery of the Na^+/K^+ pump, Skou shared the 1997 Nobel Prize for chemistry with Paul BOYER and John WALKER.

Skraup, Zdenko Hans (1850–1910) *Austrian chemist*

Skraup (skrowp) was born in Prague (now in the Czech Republic) and educated at the university there. He worked at the University of Vienna and in the Mint before being appointed professor of chemistry at the Institute of Technology in Graz in 1886. He returned to the University of Vienna in 1906 to take the chemistry chair.

In 1880 Skraup synthesized quinoline, which had been discovered by Friedlieb RUNGE in 1834, by heating a mixture of aniline and glycerol with sulfuric acid and an oxidizing agent, nitrobenzene. Skraup also worked on carbohydrates and proteins.

Slipher, Vesto Melvin (1875–1969) *American astronomer*

> It seems to me that with this discovery, the great question, if the spirals belong to the Milky Way or not, is answered with great certainty, to the end that they do not.
> —On Slipher's research into spiral nebulae.
> Ejnar Hertzsprung, letter to Slipher, 14 March 1914

Born in Mulberry, Indiana, Slipher graduated from the University of Indiana in 1901 and obtained his PhD there in 1909. He spent the whole of his career from 1901 to 1952 at Percival LOWELL's observatory in Flagstaff, Arizona, being made acting director on Lowell's death in 1916 and director in 1926.

Slipher was basically a spectroscopist. One of his major achievements was to determine the rotation periods of some of the planets by spectroscopic means. Thus in 1912, in collaboration with Lowell, he found that the spectral lines at the edge of the disk of Uranus were displaced by an amount corresponding to a speed of 10.5 miles (16.8 km) per second. Knowing the circumference it was easy to work out the rotation period as 10.8 hours. Although still the accepted figure, it is now thought that this rotation period could be considerably longer. Slipher also produced comparable data for Venus, Mars, Jupiter, and Saturn and showed that Venus's period was much longer than expected.

Slipher also studied the matter lying between the stars in our Galaxy. Like Johannes HARTMANN he concluded in 1908 from his spectroscopic research that there must be gaseous material lying between the stars. He also studied the spectra of the luminous nebulae in the Pleiades cluster of stars and proposed in 1912 that they were illuminated by starlight reflected off dust grains. This was an early indication of the presence of solid material in nebulae and other interstellar clouds.

Slipher's most important achievement, however, was his determination of the radial velocities of spiral nebulae by the measurement of the displacement of their spectral lines. Such measurement relies on the DOPPLER effect by which the wavelength of light from an object moving away from an observer will be lengthened, i.e., shifted toward the red end of the spectrum, while light from an object moving toward an observer will have its wavelength shortened, i.e., moved toward the blue end of the spectrum. By measuring the change in wavelength, known as the Doppler shift, the velocity of the moving object can be determined easily.

Slipher's work produced two surprising results. The first was the immense speed of the Andromeda nebula (galaxy), which he first successfully measured in 1912. He found it to be moving toward the Earth with a velocity of nearly 300 kilometers per second, which was then the greatest velocity ever observed. Secondly, by 1917 he had obtained the radial velocities of 15 spiral nebulae of which it would have been thought that roughly half would be receding while the other half would be approaching. But he found that 13 out of the 15 were receding. What was equally significant was their velocity, which in many cases exceeded that of the 300 kilometers per second of the Andromeda nebula. Many astronomers questioned these findings. At that time there was considerable controversy over whether the spiral nebulae were part of our Galaxy or lay far beyond it as independent star systems. Slipher's work was, in retrospect, evidence both for the extragalactic hypothesis, since the velocities of the spiral nebulae were too great for them to be members of the Galaxy, and for the expanding universe, which was first proposed by Aleksandr FRIEDMANN in 1922 and later shown to be correct by Edwin HUBBLE.

Smalley, Richard Errett (1943–2005) *American chemist*

Born in Akron, Ohio, Smalley was educated at Michigan University and at Princeton where he obtained his PhD in 1973. After spending the period from 1973 to 1976 at the James Franck Institute, Chicago, Smalley moved to Rice University, Houston, and was appointed professor of chemistry in 1981.

In 1981 Smalley devised a procedure to produce microclusters of a hundred or so atoms. The technique is to vaporize the metal by a laser. The released atoms are cooled by a jet of helium and condense into variously sized clusters. In 1985 a visiting British chemist, Harry KROTO, persuaded Smalley to direct his laser beam at a graphite target. Smalley knew that an Exxon group had already used graphite and produced carbon molecules with an even number of atoms. At first Smalley was reluctant to repeat this work but he was eventually persuaded. They soon had spectroscopic evidence for the presence of an apparently large, stable molecule of sixty carbon atoms – now known as buckminsterfullerene.

Smellie, William (1697–1763) *Scottish obstetrician*

Smellie was a medical practitioner in his native Lanark until 1739, when he moved to London. Here he studied obstetrics, becoming a close colleague of another Scottish obstetrician, William HUNTER. In 1741 he began teaching midwifery and in 1745 he received a medical degree from Glasgow University. His *Treatise on the Theory and Practice of Midwifery* (1752) provided much-improved descriptions of labor and childbirth. Smellie gave his name to several obstetrical techniques, including a method of breech delivery and to a special type of delivery forceps.

Smith, Hamilton Othanel (1931–) *American molecular biologist*

Born in New York City, Smith studied at the University of Illinois, graduated in mathematics (1952) from the University of California, Berkeley, and took his MD at Johns Hopkins in 1956. He taught at the University of Michigan from 1962 until 1967, when he returned to Johns Hopkins where later, in 1973, he became professor of microbiology.

In 1970 Smith succeeded in confirming the hypothesis of Werner ARBER to explain the phenomenon of "host-controlled variation" in phages. He identified an enzyme extracted from the bacterium *Hemophilus influenzae*, later to be known as Hind II, which cut DNA at a specific site. This was the first of the so-called restriction enzymes to be identified – by 1978 over 80 were known. Such enzymes, by permitting the controlled splitting of genes, opened up the possibility of the genetic engineering so energetically pursued from the 1970s onward.

For his work on restriction enzymes Smith shared the 1978 Nobel Prize for physiology or medicine with Arber and Daniel NATHANS.

Smith, Henry John (1826–1883) *Irish mathematician*

> The bond of union among the physical sciences is the mathematical spirit and the mathematical method which pervades them.
> —*Nature*, Vol. VIII

Dublin-born Smith studied at Oxford and had a great interest in classics – it was only after a good deal of hesitation that he chose mathematics as a profession instead. He remained in Oxford in various capacities for most of the rest of his life. In 1860 he became Savilian Professor of Geometry there.

Smith's main work was in number theory and his greatest contribution was his development of a general theory of *n* indeterminates, which enabled him to establish results about the possibility of expressing positive integers as sums of five and seven squares. This achievement ought to have won Smith the prestigious prize offered in 1882 by the French Academy for their mathematical competition. However Smith, a notably unambitious man, did not enter and the prize was in fact given to Hermann MINKOWSKI. When it was discovered that Minkowski had made use of crucial results published by Smith, the French Academy hastened to transfer the prize to Smith, but as he had unfortunately died in the meantime his fame was only posthumous. Smith also worked on the theory of elliptic functions.

Smith, Michael (1932–2000) *Canadian biochemist*

Born in the Lancashire coastal town of Blackpool, England, Smith was educated at the University of Manchester where he obtained his PhD in 1956. He moved soon after to Canada, working initially as a postdoctoral fellow at the University of British Columbia, Vancouver. From 1961 until 1966 Smith served with the Fisheries Research Board of Canada, Vancouver, but returned to the University of British Columbia in 1966 and was appointed professor of biochemistry in 1970.

In 1978 Smith introduced a basic new technique known as "site-specific mutagenesis" into molecular biology. In order to establish the function of a particular protein or gene, it had long been an established procedure to induce a mutation in the gene and observe the consequences. Thus if changes to a gene prevented an organism from making a particular enzyme, then it was reasonable to conclude that the gene controlled some part of the production of that enzyme. The difficulty with this approach was that the available mutagens, radiation and chemicals, produced random and multiple mutations. The precise effects of a single mutant gene could seldom, therefore, be distinguished from the other consequences of the mutagens.

Smith demonstrated how to introduce specific mutations into genes. He worked with a single strand of viral DNA. A short segment of complementary DNA differing at a single site was assembled and allowed to bind to the original viral DNA. The second strand was then completed in the normal way and the double-stranded DNA inserted into the viral genome. The virus would develop with normal and mutated versions of the gene, which would in turn produce normal and mutated proteins. When the different protein molecules were compared, the role of the initial mutation would become apparent.

The technique has been widely used in protein chemistry and molecular biology. Smith himself used it to investigate the role of cytochrome c in cellular respiration, and of myoglobin in oxygen storage. For his work in this field Smith shared the 1993 Nobel Prize for chemistry with Kary MULLIS.

Smith, Theobald (1859–1934) *American bacteriologist*

> There was about him [Smith] an unobtrusive pride, a reserve tinged with austerity.
> —H. Zinsser, *Biographical Memoirs of the National Academy of Sciences* (1936)

Smith, the son of a German immigrant tailor, was born in Albany, New York, and educated at Cornell and at Albany Medical College, where he obtained his MD in 1883. He began his career at the U.S. Bureau of Animal Husbandry but, dissatisfied with the recognition his work received, moved in 1895 to Harvard, where he served as professor of comparative pathology. In 1914 Smith was appointed director of animal pathology at the Rockefeller Institute, Princeton, a post he held until his retirement in 1929.

For more than a generation Smith was undoubtedly the leading American bacteriologist. He published numerous monographs on a wide variety of infections, beginning with his work on hog cholera in 1889. He went on in 1896 to distinguish between the bacilli responsible for bovine and human tuberculosis.

His name is mainly remembered, however, for his work in 1893 on Texas cattle fever, one of the earliest demonstrations of the transmission of a disease via a tick. He noted that northern cattle seemed to acquire their infection not from direct contact with southern diseased animals but by passing over territory previously occupied by such cattle. In 1889 he discovered a microorganism in the red blood cells of infected cattle and began to suspect the cattle tick, *Boophilus bovis*, of transmitting it. Smith went on to confirm his hypothesis in 1890 by noting that when ticks were placed on cattle there was a sudden fever and extensive loss of red cells that could not be explained by a mere loss of blood. Such work led to immediate and simple methods to control the disease. Smith published a full account of his work in *Nature, Causation and Prevention of Southern Cattle Fever* (1893).

Smith, William (1769–1839) *British surveyor and geologist*

> In consideration of his being a great original discoverer in English Geology; and especially for his having been the first, in this country, to discover and teach the identification of strata, and to determine their succession by means of their imbedded fossils.
> —Citation on the award of the Wollaston Medal of the Geological Society of London to Smith (1831)

Smith was born in Churchill, England, the son of the village blacksmith. He was educated at local schools and in 1787 began work with the surveyor who had been commissioned to make a survey of his parish. This was at the time of the great canal boom and Smith soon found himself fully employed conducting surveys for proposed canals. He was also regularly employed to report on coal deposits in different parts of the country while making his canal surveys. During this work he traveled considerable distances throughout Britain.

From his surveying and observation of rock strata Smith formulated very clearly two basic principles of geology for which he is often known as the father of British geology. As early as 1791 he had noted that certain strata, wherever they occurred, all contained the same invertebrate fossils and that different strata could be distinguished by a difference in their fossil content. Previously geologists had relied upon the nature of the rocks to identify and discriminate between strata. This could work well in some favored cases but, in general, was far from reliable. Smith's other major theory was the law of superposition, which simply states that if one layer of sedimentary rocks overlays another then it was formed later, unless it can be shown that the beds have been inverted.

Smith never managed to produce a major book on the geology of Britain although he did publish two small pamphlets: *Strata Identified by Organized Fossils* (1816) and *Stratigraphical Systems of Organized Fossils* (1819). His major productions were instead in the field of maps. In 1815 he produced the first geological map of England and Wales. This was published in 15 sheets at a scale of 5 miles to the inch. During the period 1819–24 he published a series of geological maps of 21 counties. The plates were still being used to produce maps as late as 1911.

Smith seems to have received little recognition and reward during his early life; he was forced to sell his collection of fossils to the British Museum to overcome his financial difficulties. However, in the 1830s he began to receive the recognition he deserved. In 1831 the Geological Society of London awarded him their first Wollaston medal and in 1835 he received an honorary LLD from Dublin.

Smithies, Oliver (1925–   ) *American geneticist and pathologist*

Smithies was born in Yorkshire, England, and won a scholarship to Oxford University, from where he graduated with a BA in physiology in 1946. He remained at Oxford for postgraduate studies in biochemistry, gaining his doctorate in 1951. Two years later he moved to the Connaught Medical Research Laboratory, University of Toronto, and in 1960 joined the University of Wisconsin, Madison, as an assistant, later becoming professor of genetics and medical genetics. In 1988 he was appointed Excellence Professor of Pathology and Laboratory Medicine at the University of North Carolina.

Smithies grew up with a fascination for inventing and building things, and is credited with devising, in 1955, starch-gel electrophoresis, a technique for separating proteins and other large molecules. However, it is in genetics where his most significant achievements lie. As early as the 1960s he had begun to appreciate the potential significance of recombination in altering human genes, and by the early 1980s was seeking to devise a method of repairing defective human genes using homologous recombination. Essentially the 'foreign' DNA swaps places with its counterpart, or homolog, located on one of the chromosomes in the cell nucleus. However, such events are relatively rare, and special techniques are required to select and isolate recombinant cells.

A major step forward came in 1985 when Smithies reported the successful insertion of a DNA sequence into a specific target gene, the beta-globin gene, in cultured human leukemia cells. Independently and around the same time, Mario Capecchi was also investigating homologous recombination as a strategy for introducing targeted genetic modifications to cultured cells. Both Smithies and Capecchi then adapted their techniques to modify embryonic stem (ES) cells, recently isolated by Martin EVANS. In 1987, Smithies described how his team had repaired a defective mutant for the enzyme hypoxanthine phosphoribosyltransferase (HPRT) in a line of ES cells. This demonstrated the feasibility of the technique for predetermined and targeted manipulation of mammalian cells. The team went on to show that the repaired gene could be inherited in the germline of mice derived from embryos treated with the repaired ES cells.

This work, combined with that of Evans and Capecchi, paved the way for the era of gene targeting, and opened new avenues for research into human diseases and developmental disorders. Smithies made full use of these new techniques, exploring possible methods of gene therapy, and creating mice with genetic modifications that mimic complex human disorders, such as hypertension and atherosclerosis.

In 2001 Smithies was awarded the Albert Lasker Award for basic medical research, jointly with Capecchi and Evans. He also received the 2007 Nobel Prize for physiology or medicine, with the same co-recipients, for discovering principles for introducing specific gene modifications in mice by the use of embryonic stem cells.

Smithson, James (1765–1829) *British geologist*

Smithson was born in France, the illegitimate son of Hugh Smithson Percy, duke of Northumberland. His mother, a cousin of his father's wife, had been twice married to wealthy husbands and was a descendant of Henry VII. Smithson was first known as James Lewis or Louis Macie, only taking his father's name after 1800. He was educated at Oxford University where he received his MA in 1786. He became interested in chemistry and mineralogy and published many papers in these fields, which were collected in 1879. Zinc carbonate was named smithsonite for him.

Smithson inherited a large fortune, chiefly from his mother's family, and in his will, drawn up in 1826, he left the income of this fortune (well over £100,000) to his nephew. If his nephew died without issue then, under the terms of the will, the capital was to be offered "to the United States of America, to found at Washington, under the name of the Smithsonian Institution, an establishment for the increase and diffusion of knowledge among men." Why he did this is not clear; he had never visited America although he was widely traveled in Europe. It was not immediately acceptable to the Americans. President Jackson ignored it and it was not until John Quincy Adams argued Congress into acceptance that the research institution was finally established by act of Congress in 1846 with Joseph HENRY as its first secretary.

In 1904 Smithson's remains were removed to America where they were fittingly interred in the original Smithsonian building.

Smoot, George Fitzgerald III (1945–   ) *American astrophysicist*

> We now have direct evidence of the birth of the Universe and its evolution ... ripples in space–time laid down earlier than the first billionth of a second. If you're religious, it's like seeing God.
> —*Wrinkles in Time* (1993)

Born at Yukon in Florida, Smoot was educated at the Massachusetts Institute of Technology where he took his PhD in physics in 1970. He moved to the University of California, Berkeley, in 1971 as a research physicist and in 1974 was appointed team leader for the differential microwave radiometers on board the Cosmic Background Explorer Satellite (COBE).

In 1965 PENZIAS and WILSON had discovered the cosmic background radiation. Initially it appeared to be perfectly isotropic, exactly the same whatever part of the universe it came from. Theorists found it difficult to account for such uniformity, and experimentalists began to wonder if it really was as uniform as it appeared.

The first disproof of isotropy came in 1977 from observations taken on board a high-flying U2 plane. The dipole anisotropy, as it was called, was small and was connected with the position of the Milky Way. Clearly further work was called for. After a number of delays, COBE was launched in 1989. Three instruments were carried. The differential microwave radiometer would measure differences in radiation from two points in the sky and could pick out differences between them of 1 part in 100,000. Also, a photometer measured the absolute brightness of the sky and searched for diffuse infrared radiation from the early universe. Finally, an interferometer measured the spectrum of the background radiation from 1 centimeter to 100 micrometers.

As the results emerged Smoot saw within the assumed uniformity "islands of structure." A year was spent checking the reliability of the data – prizes were offered to anyone on the team who could identify a significant flaw. Finally the material was checked against a list of all the systematic errors ever noted during the years of preparation. After four papers describing the initial results had been revised more than a hundred times, Smoot was ready to go public.

The results seemed to show that there were bright spots in the universe, 30 millionths of a degree warmer than the average temperature. This was precisely the result predicted by the inflationary model of Alan GUTH. It might also be possible, Smoot considered, to find in the ripples in the radiation the galactic clusters that populate the universe. Smoot has published a valuable popular account of the COBE mission in his *Wrinkles in Time* (1993). He shared the 2006 Nobel Prize for physics with John MATHER for their work on COBE.

Smyth, Charles Piazzi (1819–1900)
British astronomer

The son of William Henry Smyth (1788–1865), a naval officer and astronomer, Smyth was born at Naples in Italy and named after a friend of his father, the astronomer Giussepe PIAZZI (1746–1826). They returned to England in 1825 and settled in Bedford where William Smyth established his own observatory; he also published a pioneering work, *Sidereal Chromatics or Colours of Double Stars* (1864), as well as a catalog of 850 double stars, the *Bedford Catalogue* (1844).

Charles Smyth left school at sixteen and went straight to work at the Cape of Good Hope Observatory in South Africa as assistant to the director, Thomas Maclear, another friend of his father. William further exercised his influence in 1845 to make his son Astronomer Royal of Scotland, a post that carried with it the directorship of the Edinburgh Observatory and the position of Regius Professor of Practical Astronomy at Edinburgh University.

Much of Smyth's work was devoted to spectroscopy. In 1872 on an expedition to Sicily, he succeeded in obtaining the spectrum of the zodiacal light – a luminous triangular patch seen in the eastern sky just before sunrise and in the west just after sunset. The spectrum was continuous, from which he concluded that zodiacal light could not be the same as the Aurora, as was widely believed, which possessed a line spectrum. It is probably caused by the reflection of sunlight from cosmic dust.

Smyth also worked on the solar spectrum. He sought to show that not all the spectral lines were solar in origin; some originated in the Earth's atmosphere. On an expedition to Tenerife in 1856 he had noted that FRAUNHOFER's C band (the hydrogen alpha line) remained unchanged whatever the Sun's altitude; the red A and B bands grew visibly at lower altitudes. Similar results were obtained by Smyth in 1878 on an expedition to Lisbon.

In 1884 working on the spectrum of oxygen under high dispersion Smyth found that the single lines in the spectrum were in fact triplets when viewed at high resolution. He tried unsuccessfully to identify triplets in the solar spectrum. They were recorded photographically by F. McClean in 1897.

Smyth, like many an astronomer both before and after, became obsessed by the metrology of the Great Pyramid. He visited the pyramid in 1864 and made thousands of very accurate measurements, which he recorded in his *Life and Work at the Great Pyramid* (3 vols., Edinburgh, 1867).

Snell, George Davis (1903–1996) *American geneticist*

> With Snell's fundamental discoveries came the birth of transplantation immunology.
> —Nobel Prize citation (1980)

Snell was born in Bradford, Massachusetts, and educated at Dartmouth and Harvard, where he obtained his doctorate in 1930. After brief appointments at Texas, Brown, and Washington University, St. Louis, he joined the staff of the Jackson Laboratory, Bar Harbor, Maine, in 1935 and remained there for his entire career, retiring finally in 1969.

Early in his career, while at the University of Texas, Snell was the first to show that x-rays can cause mutations in mammals, by his demonstration that x-rays induce chromosome translocations in mice. His main work concerned what he called the major histocompatibility complex. It had been known since the 1920s that although skin grafts between mice are generally rapidly rejected they survive best when made between the same inbred line. Snell's coworker Peter GORER showed in 1937 that this was due to the presence of certain histocompatibility antigens found on the surface of mouse cells and since known as the H-2 antigens. In the 1940s Snell began a detailed study of the system.

His first task was to develop inbred strains of mice through backcrossing, genetically identical except at the H-2 locus. After much effort he was able to show that the H-2 antigens were controlled by the genes at the H-2 complex of chromosome 17, described by him as the major histocompatibility complex (MHC).

It was for this work that Snell shared the 1980 Nobel Prize for physiology or medicine with Jean DAUSSET and Baruj BENACERRAF.

Snell, Willebrord van Roijen (1580–1626) *Dutch mathematician and physicist*

Snell, who was born at Leiden in the Netherlands, received his initial training in mathematics from his father, who taught at Leiden University. He traveled widely in Europe, visiting Paris, Würzburg, and Prague, and among the celebrated scientists he met were Johannes KEPLER and Tycho BRAHE. Once he had returned to Leiden, Snell published a number of editions of classical mathematical texts. On the death of his father (1613) Snell succeeded him as professor of mathematics at the university.

He was involved in practical work in geodesy and took part in an attempt to measure the length of the meridian. In this project he was one of the first to see the full usefulness of triangulation and published his method of measuring the Earth in his *Eratosthenes Batavus* (1617; The Dutch Eratosthenes). In 1621 Snell discovered his famous law of refraction, based on a constant known as the refractive index, after much practical experimental work in optics. Snell did not, however, publish his discovery and the law first reached print in DESCARTES's *La Dioptrique* (1637; Dioptrics). However, Descartes had arrived at the law in a totally different way from Snell and made no use of practical observation.

Snow, John (1813–1858) *British physician*

Snow was born at York in England and apprenticed to a surgeon in Newcastle-upon-Tyne from the age of 14. In 1836 he enrolled at the Great Windmill Street School of Medicine in London and also studied at Westminster Hospital, receiving his MD from the University of London in 1844.

Snow had harbored a concern for the causes and spread of cholera since working amid an outbreak of the disease in northeast England. In 1849 he published an essay in which he asserted that cholera was transmitted via a contaminated water supply. He advocated simple hygiene precautions, such as boiling drinking water, to avoid infection. Snow's reputation, however, rested upon his skill as an anesthetist. He introduced a device for the controlled administration of ether to patients and did likewise for chloroform when it superseded ether. Snow personally administered chloroform to Queen Victoria at the birth of two of her children.

Snyder, Solomon Halbert (1938–) *American psychopharmacologist*

Snyder was born in Washington, DC, and educated at Georgetown University. After receiving his MD in 1962 he moved to the National Institutes of Health, Bethesda, Maryland, as a research associate. He later (1965) joined the staff of Johns Hopkins where he has served since 1970 as professor of psychiatry and pharmacology.

Many drugs, hormones, and neurotransmitters are effective at very low concentrations. The synthetic opiate etorphine produces euphoria and relieves pain in doses as low as one ten-thousandth of a gram. It was inferred from this that for such small doses to be effective they must bind to highly selective receptor sites. Snyder, in collaboration with C. Pert, began the search for such receptor sites using radioactively labeled opiates. By 1973, despite many complications, they were able to report the presence of receptors in the mammalian limbic system, a primitive region in the center of the brain associated with the perception and integration of pain and emotional experiences. To identify the receptors Snyder, working with Candice Pert, developed the widely used technique of "reversible ligand binding." Brain tissue was exposed to opiates labeled with radioactive isotopes. These were quickly washed away. The assumption was that, in low concentrations, opiates would first bind tightly to receptors. The opiates which would normally bind loosely with other tissue would be washed away. The binding sites could then be identified by locating the radioactive isotopes.

The implications of such a discovery were far-reaching for it is clear that opiate receptors had not evolved to await the isolation of morphine. The alternative is that there must be natural morphinelike substances in the brain that bind at these sites. Within a few years the first such enkephalins or endorphins, as they were named, were discovered by J. Hughes and Hans KOSTERLITZ.

In more recent work Snyder has claimed to have identified a new kind of neurotransmitter, namely, the unlikely and highly toxic nitric oxide (NO). In 1987 it had been discovered that NO diffused from blood-vessel walls causing adjacent muscles to relax and the vessels to dilate. Snyder set out to find if NO was made in the brain. He found NO present bound with iron in an enzyme, cyclic guanosine monophosphate (cGMP). NO acts in an unusual way by initiating a three-dimensional change in the shape of the enzyme. Snyder has also suggested that nitric oxide in the brain could be involved in changes connected with learning and memory processes. He has proposed that carbon monoxide may also belong to this novel class of neurotransmitters.

In another breakthrough in 1987 Snyder succeeded in cultivating *in vitro* human cerebral cortex tissue. For some unknown reason neurons, which do not normally divide, taken from a child undergoing an operation have continued to divide. "They have all the properties of neurons," Snyder emphasized, "they do everything that neurons do." Snyder has speculated that it might in the future be possible to implant such neurons in badly damaged brains.

Sobrero, Ascanio (1812–1888) *Italian chemist*

Born at Casal in Italy, Sobrero (soh-**brair**-oh) began by studying medicine but changed to chemistry, attending the universities at Turin, Paris, and Giessen. He became professor of chemistry at Turin in 1849, staying there until his retirement in 1882.

In 1846 – the year that Christian SCHÖNBEIN discovered nitrocellulose – Sobrero discovered an even more powerful explosive, nitroglycerin. By slowly stirring drops of glycerin into a cooled mixture of nitric and sulfuric acids he produced a new but unpredictable explosive. Unlike Schönbein, Sobrero showed no desire to exploit the commercial value of his discovery. As it was liable to explode on receiving the slightest vibration there seemed to be no way to develop it, and its liquid nature made it difficult to use as a blaster. It was not utilized until 1866, when Alfred Nobel mixed it with the earth kieselguhr to produce a compound that could be transported and handled without too much difficulty. In this form – dynamite – it was used extensively in the great engineering programs of roads, railroads, and harbors of the late 19th century.

Socrates (*c.* 469 BC–399 BC) *Greek philosopher*

> It seemed to me a superlative thing to know the explanation of everything, why it comes to be, why it perishes, why it is.
> —Quoted by Plato in *Phaedo*

Through Plato's writings Socrates (**sok**-rateez), born the son of a stonemason in Athens, has become one of the best known figures of antiquity. Apart from a public life of service in the Peloponnesian War and membership of the council of 500, in 406 BC, Socrates's life appears to have been spent in discussion and debate with both the youth of Athens and the leading philosophers of his day. From such discussions emerged his reputation, confirmed by the Delphic oracle, as the wisest of all men. Socrates characteristically interpreted this to mean that his wisdom lay in his awareness of his own lack of wisdom. In 399 he was accused of neglecting the gods worshiped by the city and of corrupting the young, tried, found guilty, sentenced to death, and forced to drink hemlock – scenes fully described in some of the most remarkable pages in the whole corpus of Western literature in the *Apology* and *Phaedo* of Plato.

Socrates's innovations were dialectical rather than scientific. His interests lay in the behavior of man rather than the planets. Despite this he introduced a new critical spirit, a deliberate and methodical refusal to accept either established opinion or recognized authority, as important for the development of science as for any other discipline.

Soddy, Frederick (1877–1956) *British chemist*

Soddy, born the son of a corn merchant in the coastal town of Eastbourne, England, was educated at the University College of Aberystwyth and Oxford University. After working with Ernest RUTHERFORD in Canada and William RAMSAY in London, Soddy took up an appointment at Glasgow University in 1904. He moved to take the chair of chemistry at Aberdeen University in 1914 where he remained until 1919 when he accepted the post of Dr. Lee's Professor of Chemistry at Oxford.

Soddy's work was quite revolutionary in that he succeeded in overthrowing two deep assumptions of traditional chemistry. The first arose out of his period of collaboration with Rutherford, from 1901 to 1903. Together they established that radioactive elements could change into other elements through a series of stages.

Soddy's next major achievement was to make some kind of sense of the bewildering variety of new elements that had been found as decay products of radium, thorium, and uranium. The books of the period refer to such strange entities as mesothorium, ionium, radium A, B, C, D, E, and F, and uranium X. Such entities were clearly distinct for they had markedly different half-lives. But what they were and, more significantly, where they fitted in the periodic table were difficult questions. There were gaps in the periodic table but far too few to accommodate so many new elements. One further difficulty soon forced itself on chemists. Attempts to separate thorium from radiothorium by Otto HAHN in 1905 and radium D from lead by Georg von HEVESY a few years later had failed, as had numerous other attempts to separate various radioactive elements by chemical means.

Finally Soddy made the bold claim that the reason such substances could not be separated was because they were in fact identical. Consequently some kind of modification of the periodic table was called for. In his view (1913) "it would not be surprising if the elements … were mixtures of several homogeneous elements of similar but not completely identical atomic weights." He called such chemically identical elements, with slightly differing atomic weights, isotopes (from the Greek words meaning in the same place). He could thus assert that both radium D and thorium C were in fact isotopes of lead. Radium D has a half-life of 24 years and an atomic weight of 210 while thorium C has a half-life of 87 minutes and an atomic weight of 212; but, although they have different half-lives and differing weights, they were both chemically indistinguishable from lead.

Until the discovery of the neutron by James CHADWICK a complete understanding of this enormously fruitful idea was not available to Soddy. All he could propose, somewhat vaguely, as an explanation was different numbers of positive and negative charges in the nucleus. As yet, no one seemed to suspect the existence of a neutral particle without a charge.

He did, however, go on to explain the transformation of atoms by his displacement law. In this the emission of an alpha particle, a helium nucleus of two protons and two neutrons, lowers the atomic weight by four while the emission of a beta particle, an electron, raises the atomic number by one. Given these rules Soddy could show how, for example, uranium and thorium could both decay, by different paths, to different isotopes of lead (Casimir FAJANS independently suggested the same law).

Despite the award of the 1921 Nobel Prize for chemistry for his work on the origins and nature of isotopes Soddy became disillusioned with science and his place in it. After 1919 Soddy took no further part in creative science. He wrote a good deal, mainly in the fields of economic and social questions, which raised little interest or support. On the issue of energy, however, he was remarkably perceptive. As early as 1912 he could comment that "the still unrecognized 'energy problem' ... awaits the future," continuing with the by now familiar refrain of our profligate use of hydrocarbons, "a legacy from the remote past," and concluding with what he saw as our only hope, atomic energy, which "could provide anyone who wanted it with a private sun of his own."

Sokoloff, Louis (1921–   ) *American neurophysiologist*

Born in Philadelphia, Pennsylvania, Sokoloff was educated at the University of Pennsylvania. After gaining his MD in 1946 he joined the staff of the medical school there, leaving in 1956 for the National Institute of Mental Health, Bethesda, Maryland, where he served from 1968 as chief of the Cerebral Metabolism Laboratory.

Sokoloff and his colleagues at the NIMH have recently established a relatively simple technique for determining which brain cells respond to experimental stimuli. The technique, described as opening up "an entirely new realm in brain research," depends on the fact that active neurons in brain cells absorb more glucose than inactive ones. The glucose is normally metabolized rapidly but if a similar sugar is substituted, 2-deoxyglucose, it tends to accumulate in the active neurons. When radioactively labeled, such accumulated deoxyglucose can be readily detected and measured.

The value of this new approach was shown by David HUBEL and Torsten Weisel in 1978 when they succeeded in confirming much of their earlier work by obtaining from neurons of the visual cortex autoradiographs of the regions responsive to vertical lines.

Solvay, Ernest (1838–1922) *Belgian industrial chemist*

Solvay (**sol**-vay) was born at Rebecq-Rognon in Belgium. As the son of a salt refiner and the nephew of a manager of a gas plant, he was introduced at an early age to the techniques and problems of the chemical industry. He devised several methods for purifying gases but is best known for the ammonia–soda process named for him.

For most of the 19th century soda was produced by the LEBLANC process. This had a number of disadvantages: it produced toxic hydrochloric acid fumes and also a number of expensive and irrecoverable waste products. As early as 1811 Augustin FRESNEL had proposed an ammonia–soda process. However, although chemists succeeded in the laboratory, when they tried to translate their results onto an industrial scale they invariably ended up like James MUSPRATT, who lost £8,000 in the period 1840–42. Solvay was the first to solve the engineering problems of the process. He later confessed that he was completely ignorant of all these earlier failures, adding that he would probably never have tried if he had known.

In 1861 Solvay took out his first patent for soda production and in 1863 set up his first factory at Charleroi, in partnership with his brother. The process involved mixing brine with ammonium carbonate, which produced sodium carbonate and ammonium chloride. The sodium carbonate yielded soda on being heated and the ammonium chloride, when mixed with carbon, regenerated the ammonium carbonate the process started from. Solvay's innovation was to introduce pressurized carbonating towers.

The system was soon adopted throughout the world and by 1900 95% of a greatly increased world production of soda came from the Solvay process. The price of soda fell by more than half in the last quarter of the 19th century.

Solvay is also remembered for financing the great series of international conferences of physicists starting in 1911, in which much of the new nuclear and quantum physics was discussed.

Somerville, Mary (1780–1872) *British astronomer and physical geographer*

Mary Fairfax, as she was born in Jedburgh,

Scotland, was the daughter of a naval officer. She received precisely one year of formal education before her marriage to a cousin in 1804 who was a captain in the Russian navy. After his death in 1807 she married another cousin, W. Somerville, an army physician, in 1812.

Somerville was unique in 19th-century British science as she was an independent female. Virtually all other women participated in science as the wife or sister of a husband or brother whom they assisted and sometimes went on to make some small contribution of their own. Her interest in science began when as a young girl she first heard of algebra and EUCLID and satisfied her curiosity as to the nature of these subjects from books she purchased. She certainly received no encouragement from her father nor was her first husband much more sympathetic. She was more fortunate with her second husband, who encouraged and assisted her.

Living with her husband in London from 1816 she soon became a familiar and respected figure in the scientific circles of the capital. Her first significant achievement was her treatise on the *Mécanique céleste* (Celestial Mechanics) of Pierre Simon de LAPLACE. She was persuaded to undertake this difficult task by John HERSCHEL and in 1831 750 copies of *The Mechanism of the Heavens* were published. The work was a great success and was used as a basic text in advanced astronomy for the rest of the century.

She followed this in 1834 with her *On the Connexion of the Physical Sciences*, a more popular but still serious work. In this she suggested that the perturbations of Uranus might reveal the existence of an undiscovered planet. Somerville was of course denied such obvious honors as a fellowship of the Royal Society as a result of her work. She was, however, granted a government pension of £300 a year in 1837.

From 1840, because of the health of her husband, she moved to Europe, living mainly in Italy. It was there that she produced her third and most original work, *Physical Geography* (1848). It was widely used as a university text book to the end of the century, although overshadowed by the *Kosmos* (Cosmos) of Friedrich von HUMBOLDT, which came out in 1845.

She produced her fourth book, *On Molecular and Microscopic Science* (1869), at the age of 89 and was working on a second edition when she died.

When a hall was opened in Oxford in 1879 for the education of women it was appropriately named for her and was to produce sufficient talent to refute her own belief that genius was a gift not granted to the female sex.

Sommerfeld, Arnold Johannes Wilhelm (1868–1951) *German physicist*

The misuse of the word "national" by our rulers has thoroughly broken me of the habit of national feeling that was so pronounced in my case. I would now be willing to see Germany disappear as a power and merge into a pacified Europe.
—On the Nazi regime. Letter to Albert Einstein (1934)

Sommerfeld (**zom**-er-felt), the son of a physician, was born in Königsberg (now Kaliningrad in Russia) and educated at the university in his native city. He later taught at Göttingen, Clausthal, and Aachen before being appointed to the chair of theoretical physics at the University of Munich in 1906.

In 1916 Sommerfeld produced an important modification to the model of the atom proposed by Niels BOHR in 1913. In Bohr's model an atom consists of a central nucleus around which electrons move in definite circular orbits. The orbits are quantized, that is, the electrons occupy only orbits that have specific energies. The electrons can "jump" to higher or lower levels by either absorbing or emitting photons of the appropriate frequency. It was the emission of just those frequencies that produced the familiar lines of the hydrogen spectrum. Increasing knowledge of the spectrum of hydrogen showed that Bohr's model could not account for the fine structure of the spectral lines. What at first had looked like a single line turned out to be in certain cases a number of lines close to each other. Sommerfeld's solution was to suggest that some of the electrons moved in elliptical rather than circular orbits. This required introducing a second quantum number, the azimuthal quantum number, l, in addition to the principal quantum number of Bohr, n. The two are simply related and together permit the fine structure of atomic spectra to be satisfactorily interpreted.

Sommerfeld was the author of an influential work that went through a number of editions in the 1920s, *Atombau und Spektrallinien* (Atomic Structure and Spectral Lines).

Sondheimer, Franz (1926–1981) *British chemist*

Sondheimer (**sond**-hI-mer) was born at Stuttgart in Germany but moved with his family to Britain in 1937. He was educated at Imperial College, London, where he obtained his PhD in 1948. After serving as a research fellow with Robert WOODWARD at Harvard from 1949 to 1952 he spent a brief period as associate director of chemical research with Syntex in Mexico City before being appointed in 1956 professor of organic chemistry at the Weizmann Institute, Rehovoth, Israel. In 1964 Sondheimer returned to Britain where he held a Royal Society research professorship first at Cambridge and from 1967 at University College, London.

In 1952, while with Syntex, Sondheimer collaborated with Carl DJERASSI in the synthesis of an oral estronelike compound, the precursor of the contraceptive pill. In the 1960s Sondheimer deployed his synthetic skills on the annulenes, monocyclic hydrocarbons with alternating double and single bonds like the familiar benzene ring. Such molecules were important in theoretical chemistry following the formulation in 1931 by Erich HÜCKEL of his rule claiming that compounds with monocyclic planar rings containing $(4n + 2)\pi$ electrons should be aromatic. The rule obviously held for benzene (n = 1).

In 1956 Sondheimer and his colleagues discovered a relatively simple way to synthesize large-ring hydrocarbons and by the early 1960s they had produced annulenes with 14 and 18 carbon atoms, n = 2 and n = 4 respectively. 14–annulene is a highly unstable compound, which is not planar because of the positions of the hydrogen atoms on the molecule. 18–annulene has a planar ring obeying the Hückel rule and does have aromatic properties. The group also synthesized 30–annulene – a planar compound (n = 7). In 1981 while a visiting professor at Stanford, Sondheimer was found dead in his laboratory beside an empty cyanide bottle. He had apparently been suffering from depression.

Sonneborn, Tracy Morton (1905–1981) American geneticist

Born in Baltimore, Maryland, Sonneborn received his PhD from Johns Hopkins in 1928 and soon afterward joined the faculty there. In 1939 he moved to Indiana University, later serving as professor of zoology from 1943 until his retirement in 1976.

In 1937 Sonneborn made the important discovery of sexuality in the protozoan *Paramecium aurelia*. He reported the existence of two mating types between which conjugation could occur but within which conjugation is prohibited. This allowed Mendelian breeding experiments to be carried out with a unicellular organism and led to the demonstration that Mendelian principles, which had been established using higher animals and plants, are also applicable to microorganisms.

In 1943 Sonneborn announced that he had found evidence for what he believed to be cytoplasmic inheritance in *Paramecium*. He found two strains one of which was lethal to the other when mixed together, and yet both strains were identical genetically. He devised an experiment that showed the killer factor is inherited via the cytoplasm, and later found many different types of killer paramecia each with cytoplasmic particles containing a different killer gene. However, the killer trait still depended for its expression upon the presence of a certain gene in the nucleus.

From such work Sonneborn began to talk of the gene as consisting of two parts, one localized in the chromosome and responsible for replication while the other part was present in the cytoplasm and served the varied demands of cellular differentiation. Such latter bodies soon became known as plasmagenes, a term introduced by Cyril DARLINGTON in 1944.

However, it was later shown that Sonneborn's cytoplasmic particles are in fact symbiotic bacteria. Nevertheless, Sonneborn's work is still of significance in showing that cellular cytoplasm is something more than a "playground for the genes."

Sorby, Henry Clifton (1826–1908) British geologist

> Not to pass an examination, but to qualify myself for a career of original investigation.
> —The stated aim of his education

Sorby was born in Woodbourne, England, the only son of a cutlery manufacturer. He was educated at schools in Harrogate and Sheffield and under a private tutor, who stimulated his interest in science. On the death of his father (1847) he inherited sufficient money to allow him to set up a laboratory and devote himself to science.

Sorby became particularly interested in geology and he recognized the value of the microscope in the science. Using the NICOL prism introduced by William Nicol and thin slices of rock, about 0.001 inch (0.0025 cm) wide, he was able to study the microscopic structure of minerals and metals. By polishing steel, etching it, and then examining it under a microscope, he could see the crystalline structure of the metal and the presence of impurities, and he concluded that steel is a crystallized igneous rock. This has been referred to as the beginning of the technique of metallography, which is an essential feature of modern metallurgy. He also attempted to model geological processes in the laboratory; he tried to throw light on the process of sedimentation by examining in great detail the rate at which particles of different sizes settle in water.

Sørensen, Søren Peter Lauritz (1868–1939) Danish chemist

Sørensen (**su(r)**-ren-sen), the son of a farmer from Haurebjerg in Denmark, was educated at the University of Copenhagen, obtaining his PhD in 1899. After serving as a consultant chemist to the Royal Danish Laboratory from 1892 he became director of the Carlsberg Laboratory (1901).

Sørensen is noted for introducing the concept of pH (potential of hydrogen) as a measure of the acidity or alkalinity of solutions. The pH value is defined as $\log_{10}(1/[H^+])$, where $[H^+]$ is the concentration of hydrogen ions. On this scale, a neutral solution has a pH of 7; alkaline solutions have values higher than 7 and acidic solutions have values lower than 7.

Sosigenes (about 1st century BC) *Egyptian astronomer*

Nothing seems to be known about Sosigenes (soh-**sij**-e-neez) apart from his design of the Julian calendar. By the time of Julius Caesar the Roman calendar was hopelessly out of alignment with the seasons. The Romans had traditionally had a lunar 12-month calendar of 355 days. To bring it into phase with the solar year an intercalary (inserted) month of 27 days was supposed to be added every other year to a reduced February of 23 or 24 days. In theory this should have produced a year of 366¼ days, which would have proved inaccurate in the long run but should have been controllable by skipping an intercalation whenever the discrepancy became too uncomfortable. For whatever reason the practice had not been followed and to cure the confusion Caesar felt in need of expert foreign advice. He called in Sosigenes, an Alexandrian Greek from Egypt. To restore the situation to normal he introduced two intercalary months between November and December totaling 67 days and one of 27 days after February producing the famous year of "ultimate confusion" of 445 days in 46 BC. To ensure that harmony would continue he introduced what was the basic Egyptian year of 365 days plus a leap year every four years. This is in fact eleven minutes too long but it is a tribute to Sosigenes that it lasted for 1,500 years before being modified.

Spallanzani, Lazzaro (1729–1799) *Italian biologist*

> You have discovered more truths in five years than entire academies in half a century.
> —Charles Bonnet, letter to Spallanzani

Spallanzani (spah-lan-**tsah**-nee) was born at Scandiano in Italy and educated at the Jesuit College, Reggio, before leaving to study jurisprudence at Bologna University. While at Bologna he developed an interest in natural history, which was probably encouraged by his cousin, Laura Bassi, who was professor of physics there. After receiving his doctorate he took minor orders and a few years later became a priest, although he continued to pursue his researches into natural history.

Spallanzani's most important experiments, published in 1767, questioned John NEEDHAM's "proof" 20 years earlier of the spontaneous generation of microorganisms. He took solutions in which microorganisms normally breed and boiled them for 30 to 45 minutes before placing them in sealed flasks. No microorganisms developed, demonstrating that Needham's broth had not been boiled for long enough to sterilize it. Opponents of Spallanzani asserted, however, that he had destroyed a vital principle in the air by prolonged boiling. While conducting these experiments, Spallanzani showed that some organisms can survive for long periods in a vacuum: this was the first demonstration of anaerobiosis (the ability to live and grow without free oxygen).

In 1768 he submitted papers to the Royal Society on his findings concerning the regeneration of amputated parts in lower animals, and on the strength of this was elected a fellow of the Royal Society. In the same year Maria Theresa of Austria appointed Spallanzani to the chair of natural history at Pavia, which at that time was under Austrian dominion, and here he remained until his death. He was also in charge of the museum at Pavia and made many journeys around the Mediterranean collecting natural-history specimens for the museum.

Spallanzani's research interests covered a wide area and during his career he made important contributions to the understanding of digestion, reproduction, respiration, and blood circulation, as well as sensory perception in bats. He also (in 1785) managed to accomplish the artificial insemination of a dog.

Spedding, Frank Harold (1902–1984) *American chemist*

Spedding was born at Hamilton, Ontario, in Canada and educated at the University of Michigan and Berkeley where he obtained his PhD in 1929. After working at Cornell from 1935 to 1937, he moved to Iowa State University, where he remained for the rest of his career. He was appointed professor of chemistry in 1941 and director of the Institute of Atomic Research from 1945 to 1968.

In 1942, at the request of Arthur COMPTON, Spedding and his Iowa colleagues devised new techniques for the purification of the uranium required urgently for the development of the atomic bomb. Their method reduced the price of uranium from $22 per pound to $1 per pound.

After the war Spedding put his new skills to separating the lanthanide elements, an extremely difficult task because of the similarity of their physical and chemical properties. The technique used on uranium, and later successfully applied to the lanthanides, was that of ion-exchange chromatography. A simple example is seen when hard water is allowed to percolate through a column of the mineral zeolite; calcium ions are absorbed by the mineral, which releases its own sodium ions into the water – in effect, an exchange of ions. In the late 1930s more efficient synthetic resins were introduced as ion exchangers.

Spedding passed lanthanide chlorides through an exchange resin that differentially absorbed the compounds present, thus allowing them to be separated. As a result, for the first time, chemists could deal with lanthanoids in substantial quantities.

In 1965 Spedding published, with Adrian Daane, an account of his work in his *Chemistry of Rare Earth Elements*.

Spemann, Hans (1869–1941) *German zoologist, embryologist, and histologist*

Spemann (**shpay**-mahn), who was born at Stuttgart in Germany, worked for a time in his father's bookshop there before graduating in zoology, botany, and physics at the universities of Heidelberg, Munich, and Würzburg. He was first an assistant, then a lecturer, at the Zoological Institute of Würzburg (1894–1908) before becoming professor at Rostock (1909–14). He was then successively associate director of the Kaiser Wilhelm Institute of Biology, Berlin (1914–19), and professor of zoology at Freiburg (1919–35).

Spemann's concept of embryonic induction, based on a lifetime's study of the development of amphibians such as newts, showed that certain parts of the embryo – the organizing centers – direct the development of groups of cells into particular organs and tissues. He further demonstrated an absence of predestined organs or tissues in the earliest stages of embryonic development; tissue excised from one part of the embryo and grafted onto another part will assume the character of the latter, losing its original nature. Spemann's highly original work, for which he received the Nobel Prize in 1935, paved the way for subsequent recognition of similar organizing centers in other animal groups. It is elaborated in *Experimentelle Beiträge zu einer Theorie der Entwicklung* (1936; Embryonic Development and Induction).

Spence, Peter (1806–1883) *British chemical industrialist*

Born in Brechin, Scotland, Spence started his career as a grocer's apprentice. He was later employed in the Dundee gas plant before setting up as a small chemical manufacturer in London.

In 1845 his introduction of a new process revolutionized the production of potash alum (potassium aluminum sulfate), which was widely used as a mordant in the textile industry. It was obtained by burning shales sufficiently rich in alum, found near Guisborough and Whitby. Some of the waste from Scottish coal mines consisted of shales containing alum, which was extracted on a small scale, producing 2,000 tons a year by 1835.

Spence patented his new process in 1845 and established his factory in Manchester. He used sulfuric acid to treat a mixture of shale and iron pyrites (iron sulfide), producing alum and iron sulfate. The spent shale could be used in cement production. Spence became the world's largest producer of alum. In an early pollution case in 1857 he was prosecuted and forced to move his factory from Manchester.

Spencer, Herbert (1820–1903) *British philosopher*

> Progress ... is not an accident, but a necessity ... it is part of nature.
> —*Social Statics*

> Science is organized knowledge.
> —*Education*

Spencer was born in Derby, England, and educated by his father and uncle. Having turned down his uncle's offer to send him to Cambridge University, he worked instead as a railroad civil engineer. He began submitting articles on sociology and psychology to various papers in 1842 and later became a subeditor with *The Economist*.

Some years before the publication of Charles DARWIN's *The Origin of Species*, Spencer had formulated his own theory of evolution and applied it to the development of human societies. A generation earlier, Karl von BAER had demonstrated that heterogenous organs develop from homogenous germ layers in the embryo. Spencer adopted this observation, applying it to the development of animal species, industry, and culture, defining evolution as the progression from "an indefinite incoherent homogeneity to a definite coherent heterogeneity." He believed in the inheritance of acquired characteristics, but adapted his views when *The Origin* was published, integrating the theory of natural selection into his scheme and popularizing the term "survival of the fittest."

Even though it was quite inappropriate to apply Darwin's theory to social development, Spencer's "social Darwinism" was very influential outside scientific circles and was used to justify many industrial and social malpractices. Darwin himself remarked that Spencer's habit was to think very much and observe very little. The same criticism may be seen in T. H. HUXLEY's comment that Spencer's idea of a tragedy was a "deduction killed by a fact."

Spencer Jones, Sir Harold (1890–1960) *British astronomer*

The son of a London accountant, Spencer Jones won a scholarship to Cambridge University, graduating in 1912. He was chief assistant to the Astronomer Royal from 1913 to 1923 and from 1923 to 1933 served as astronomer at the Royal Observatory at the Cape of Good Hope. In 1933 he returned to Greenwich to become Astronomer Royal, retaining this office until his retirement in 1955. He was knighted in 1943.

Spencer Jones made many contributions to astronomy both in terms of research and organizing ability. In 1931 the minor planet Eros, discovered in 1898, was due to come within 16 million miles (26 million km) of the Earth. The accurate determination of its position would allow an improved calculation of the solar parallax, i.e., the angle subtended by the Earth's radius at the center of the Sun. This in turn would give a more accurate value to the astronomical unit, the mean distance of the Earth from the Sun. Under the organization of Spencer Jones, astronomers all over the world cooperated in the collection of the appropriate data. Eventually over 3,000 photographs arrived at Greenwich and Spencer Jones, after nearly ten years unremitting calculation, announced in 1941 a new figure for the solar parallax of 8.7904 seconds of arc. Although a great improvement over previous results, and thus a great achievement, more recent radar measurements give a value of 8".7941.

Spencer Jones's second major contribution to astronomy concerned the Earth's rotation. Using new quartz clocks accurate to a thousandth of a second per year, he found that the Earth did not rotate regularly but kept time like a "cheap watch." In 1939 he announced that the Earth was running slow by about a second per year, which was sufficient to explain a number of anomalies in the orbits of the Moon and the planets. As a result of this work a new system of measuring time, ephemeris time, independent of the Earth's rotation rate, was brought into use in 1958.

Finally, from 1948, Spencer Jones organized the move of the Royal Observatory from Greenwich to Herstmonceux in Sussex. In London the Milky Way could no longer be seen, the silvered mirrors were being corroded by sulfur dioxide, and street lighting precluded the taking of long-exposure photographs. The move was thus long overdue but was only completed in 1958 after his retirement.

Sperry, Elmer Ambrose (1860–1930)
American inventor

> No one American has contributed so much to our naval technical progress.
> —Charles Francis Adams

Born in Cortland, New York, Sperry is best known for the gyroscopic compass and gyroscopic stabilizers, now crucial to marine navigation. From the age of 19 he developed various inventions, from an improved dynamo and an arc light to electric cutting machinery for mines. He founded his first company in 1880.

The gyrocompass, invented by him in 1896, was developed from a toy, the gyroscope. It was not until 1911 that one was installed in a U.S. Navy battleship. The gyro principle was also used in gyropilots (for steering ships) and in gyroscopic stabilizers.

Sperry, Roger Wolcott (1913–1994)
American neurobiologist

Sperry, who was born in Hartford, Connecticut, studied psychology at Oberlin College and zoology at the University of Chicago, where he obtained his PhD in 1941. He worked at Harvard, the Yerkes Primate Center, and at Chicago before he moved to the California Institute of Technology in 1954 as professor of psychobiology where he remained until 1984.

Sperry worked on the hemispheres of the brain. Architecturally the brain consists of two apparently identical halves constructed in such a way that each half controls the opposite side of the body. The language center of the human brain is located in most people in the left side alone. The two cerebral hemispheres are far from distinct anatomically, with a number of bands of nervous tissue (commissures) carrying many fibers from one side to the other. In the early 1950s Sperry set out to find how a creature would behave if all such commissures were severed resulting in a "split brain." To his surprise he found that monkeys and cats with split brains act much the same as normal animals. However, where learning was involved the creatures behaved as if they had two independent brains. Thus if a monkey was trained to discriminate between a square and a circle with one eye, the other being covered with a patch, then, if the situation was reversed the animal would have to relearn how to make the discrimination.

He also studied a 49-year-old man whose brain had been "split" to prevent the spread of severe epileptic convulsions from one side to the other. He found that, though normal in other ways, the patient showed the effect of cerebral disconnection in any situation that required judgment or interpretation based on language. Sperry's work immediately posed the problem of whether there is any comparable specialization inherent in the human right-hand brain. This topic is receiving much attention.

Sperry also performed some equally dramatic experiments on nerve regeneration in amphibians. Although in mammals a severed optic nerve remains permanently severed, in certain amphibians such as the salamander it will regenerate. Sperry wondered if the nerves regenerate along the old pathway or whether a new one is formed. He found that whatever obstacles were placed before the nerve fiber it would invariably, however tortuous the path might be, find its way back to its original synaptic connection in the brain. This was shown most convincingly when, after severing the optic nerve, Sperry removed the eye, rotated it through 180° and replaced it. When food was presented to the right of the animal it would aim to the left, thus clearly showing the fibers had made their old functional connection. Sperry shared the 1981 Nobel Prize for physiology or medicine with David HUBEL and Torsten WIESEL.

Spiegelman, Sol (1914–1983) *American microbiologist*

Spiegelman was educated at the City College in his native New York, at Columbia, and at Washington University where he obtained his PhD in 1944. He initially taught physics and mathematics at Washington before moving to the University of Illinois, Urbana, in 1949 to serve as professor of microbiology. In 1969 he was appointed director of the Institute of Cancer Research and professor of human genetics and development at Columbia.

In 1958 Spiegelman, in collaboration with Masayasu Nomura and Benjamin Hall, introduced the techniques for the construction of hybrid nucleic acids. After considerable effort they finally succeeded in joining single-stranded DNA from the virus bacteriophage T2 with RNA from the bacterium *Escherichia coli* infected with T2 phage. They further demonstrated that the RNA would hybridize with DNA from no other source, not even DNA from the closely related phage, T5. Spiegelman concluded that the base sequence of *E. coli* RNA is complementary to at least one of the two strands in T2 DNA.

The significance of this work was to provide support for the existence of an "informational intermediary," an RNA template, between the DNA of the genes and the synthesis of the proteins. Such an intermediary, referred to as "translatable RNA" by Spiegelman but later known as messenger RNA, was incorporated in the detailed theory of protein synthesis of François Jacob and Jacques MONOD in 1961.

Spitzer, Lyman Jr. (1914–1997) *American astrophysicist*

Born at Toledo in Ohio, Spitzer gained his BA in 1935 from Yale University and then spent a year at Cambridge University, England. Returning to America he gained his PhD in 1938 from Princeton University. From Princeton he went back to Yale as an instructor in physics and astronomy, later (in 1946) to become an associate professor in astrophysics. World War II intervened, however, and in the years 1942–46 he was involved in sonar and undersea-warfare development at Columbia University.

From 1947 to 1979 Spitzer was professor of astronomy and director of the observatory at Princeton. His major research interest was the physical processes that occur in interstellar space, particularly those that lead to matter condensation and the formation of new stars. In considering how ionized matter behaves at very high temperatures inside stars he (and others) saw the possibility of creating suitable conditions for the fusion of hydrogen into helium here on Earth, thus liberating vast amounts of energy. In stars the ionized gases (plasma) are contained by intense gravitational fields but on Earth they must be contained by magnetic fields. Spitzer devised a system of magnetic containment in a figure-eight shaped loop. The machine, known as the "Stellerator," was built at Princeton's plasma physics laboratory and is one of a succession of machines by which physicists are still pursuing the goal of controlled thermonuclear fusion.

Spörer, Gustav Friedrich Wilhelm (1822–1895) *German astronomer*

Spörer (**shpu(r)**-rer) studied astronomy at the university in his native city of Berlin and wrote a thesis on comets. He became professor at the Aachen Gymnasium and in 1875 began work in the astrophysical laboratory at Potsdam. He showed, at the same time as Richard Carrington, that the Sun does not rotate as a rigid body. He also discovered the regular latitude variation of sunspots, known as *Spörer's law*. This showed that sunspots are restricted to latitudes 5°–35°. After the minimum phase, spots appear in high latitudes, gradually moving toward the equator where the old equatorial spots are just dying away.

Stahl, Franklin William (1929–   ) *American molecular biologist*

Born in Boston, Massachusetts, Stahl was educated at Harvard and the University of Rochester where he obtained his PhD in 1956. After spending the period 1955–58 at the California Institute of Technology and a year at the University of Missouri, he moved in 1959 to the University of Oregon where he was later, in 1970, appointed to the chair of biology.

In 1958 Stahl published a joint paper with Matthew MESELSON in which they reported the results of their classic experiment establishing the "semiconservative" nature of DNA replication.

Stahl, Georg Ernst (1660–1734) *German chemist and physician*

> Where there is doubt, whatever the great majority of people hold to be true is wrong.
> —His motto

Stahl (shtahl) was the son of a protestant minister from Ansbach in Germany. He studied medicine at Jena, graduating in 1684, and in 1687 was appointed physician to the duke of Sachsen-Weimar. He moved to Halle in 1694 where he became professor of medicine in the newly founded university. In 1716 he became physician to the king of Prussia.

Stahl developed phlogiston from the vague speculations of Johann BECHER into a coherent theory, which dominated the chemistry of the latter part of the 18th century until replaced by that of Antoine LAVOISIER. Phlogiston was the combustible element in substances. If substances contained phlogiston they would burn,

and the fact that charcoal could be almost totally consumed meant that it was particularly
rich in phlogiston. When a metal was heated it
left a calx (a powdery substance) from which it
was deduced that a metal was really calx plus
phlogiston. The process could be reversed by
heating the calx over charcoal, when the calx
would take the phlogiston driven from the charcoal and return to its metallic form. It seemed
to chemists that for the first time ever they
could begin to understand the normal transformations that went on around them and the
theory was the first rational explanation of
combustion. It is no wonder that Stahl's theory
was eagerly accepted and passionately supported.

As principles in addition to phlogiston Stahl
accepted water, salt, and mercury. He also
adopted the law of affinity that like reacts with
like. However, there were difficulties with the
theory for it seemed that, to explain some interactions, phlogiston must have no weight or
even negative weight for the bodies that gain it,
far from becoming heavier, sometimes become
lighter.

Stanley, Wendell Meredith (1904–1971)
American biochemist

Stanley, who was born in Ridgeville, Indiana,
gained his doctorate in chemistry from the University of Illinois in 1929 and then traveled to
Munich to work on sterols. On his return to
America he joined the Rockefeller Institute for
Medical Research in Princeton, New Jersey,
where he began research with the tobacco mosaic virus (TMV).

Stanley was impressed by John NORTHROP's
success in crystallizing proteins and applied
Northrop's techniques to his extracts of TMV.
By 1935 he had obtained thin rodlike crystals
of the virus and demonstrated that TMV still
retained its infectivity after crystallization. Initially this achievement met with skepticism
from many scientists who had thought viruses
were similar to conventional living organisms
and thus incapable of existence in a crystalline
form. In 1946 Stanley's research was recognized by the award of the Nobel Prize for chemistry, which he shared with Northrop and with
James SUMNER, who had crystallized the first
enzyme.

During World War II Stanley worked on isolating the influenza virus and prepared a vaccine against it. From 1946 until his death he
was director of the virus research laboratory at
the University of California.

Stark, Johannes (1874–1957) *German physicist and spectroscopist*

> The agents of Judaism in German intellectual
> life will have to disappear, just like the Jews
> themselves.

> —*Das Schwarze Korps* (The Black Corps),
> magazine of the SS, 15 July 1937

Born at Schickenhof in Germany, Stark (shtark)
was educated at the University of Munich
where he obtained his doctorate and began his
teaching career in 1897. Between 1906 and
1922 he taught successively at the universities
of Göttingen, Hannover (where he first became
a professor), Aachen, Griefswald, and Würzburg. At this point his academic career came to
an end. He first tried to start a porcelain industry in northern Germany but the years following World War I were not generous to new
businesses. Despite the award of the Nobel
Prize for physics in 1919 his attempt to return
to academic life was not successful and he had
been rejected by six German universities by
1928.

This was due to his general unpopularity
and because he had become somewhat extreme
in his denunciation of quantum theory and the
theories of Albert EINSTEIN as being the product
of "Jewish" science. Thus Stark, like Philipp
LENARD, began to drift into Nazi circles and in
1930 joined the party. Unlike Lenard, who was
content merely to rewrite the history of physics
in the Aryan mode, Stark made a real bid for
the control of German science. In 1933, although he was rejected by the Prussian Academy of Science, he succeeded in obtaining the
presidency of the Imperial Institute of Physics
and Technology, which he tried to use as a
power base in his attempt to gain control of
German physics. His attempt brought him into
conflict with senior politicians and civil ser-
vants at the Reich Education Ministry, who
saw him as too erratic and disruptive a force to
be of much use to them, and consequently
forced his resignation in 1939. Stark's final hu-
miliation came in 1947, when he was sentenced
to four years in a labor camp by a German de-
Nazification court.

Stark first observed (1905) a shift of frequency in the radiation emitted by fast-moving
charged particles (i.e., a DOPPLER effect). His
other main scientific achievement was his discovery in 1913 of the spectral effect now known
as the *Stark effect*, which won him the Nobel
Prize. In this, following Pieter ZEEMAN's demonstration of the splitting of the spectral lines of
a substance in a magnetic field, Stark succeeded in obtaining a similar phenomenon in
an electric field. This is a quantum effect but
Stark, although an early supporter of quantum
theory, began to argue, with typical perversity,
against the new theory as evidence for it
mounted.

Starling, Ernest Henry (1866–1927)
British physiologist

A Londoner by birth, Starling studied medicine at Guy's Hospital, London, where he obtained his MB in 1889 and eventually became

head of the department of physiology. In 1899 he moved to University College, London, to become Jodrell Professor of Physiology, a position he held until his death.

In 1896 Starling introduced the concept of the *Starling equilibrium*, which tried to relate the pressure of the blood to its behavior in the capillary system. He realized that the high pressure of the arterial system is enough to force fluids through the thin-walled capillaries into the tissues. But as the blood is divided through more and more capillaries its pressure falls. By the time it reaches the venous system the pressure of the fluid in the surrounding tissues is higher than that of blood in the venous capillaries, allowing much of the fluid lost from the arterial side to be regained. In theory the two systems should be in a state of equilibrium. In reality the system is complicated by the hydrostatic pressure of the blood and the osmotic pressure arising from the various salts and proteins dissolved in it.

In 1915 Starling formulated the important law (*Starling's law*) stating that the energy of contraction of the heart is a function of the length of the muscle fiber. As the heart fills with blood the muscle is forced to expand and stretch; the force with which the muscle contracts to expel the blood from the heart is simply a function of the extent to which it has been stretched. The curve that relates the two variables of the heart – pressure and volume – is known as *Starling's curve*.

Starling's best-known work was his collaboration with William BAYLISS in the discovery of the hormone secretin in 1902. The normal pancreas releases a number of juices into the duodenum to aid in the process of digestion. By cutting all the pancreatic nerves and noting the continuing secretion of the pancreatic juices Starling and Bayliss showed that the release of the juices was not under nervous control. They concluded that a chemical, rather than a nervous, message must be sent to the pancreas through the blood when food enters the duodenum. They proposed to call the chemical messenger secretin. For the general class of such chemicals Starling proposed, in 1905, the term "hormone," from the Greek root meaning to excite. Thus endocrinology – a major branch of medicine and physiology – had been created.

It was widely known before the outbreak of World War I that Starling and Bayliss had been the strongest of the candidates for the Nobel Prize for physiology and medicine. However, as no awards were made during the war, they missed out completely; the prizes after 1918 were awarded for more recent work. As for honors from his own country, Starling was far too outspoken about the incompetent direction of the war even to be considered.

Stas, Jean Servais (1813–1891) *Belgian chemist*

Stas (stahs), who was born at Louvain in Belgium, trained initially as a physician. He later switched to chemistry, serving as assistant to Jean DUMAS before being appointed to the chair of chemistry at the Royal Military School in Brussels in 1840. He had to retire in 1869 because of trouble with his voice through a throat ailment and became instead commissioner of the mint, but retired from this in 1872.

Stas was well known in his time for his extremely accurate determination of atomic weights. At first he supported William PROUT's hypothesis that the weight of all elements is an exact multiple of that of the hydrogen atom. All his early measurements seemed to agree with this theory, but as his work progressed he seemed to be getting more and more fractional numbers and this turned him into the most articulate and damaging opponent of Prout. His work laid the foundations for the eventual formation of the periodic system.

Stas also carried out chemical analysis on potato blight and nicotine poisoning.

Staudinger, Hermann (1881–1965) *German chemist*

Staudinger (**shtow**-ding-er) was born in Worms, Germany, the son of a philosophy professor; he was educated at the universities of Darmstadt, Munich, and Halle where, in 1903, he obtained his doctorate. He taught at the University of Strasbourg, the Karlsruhe Technical College, and the Federal Institute of Technology in Zurich before taking up an appointment at the University of Freiburg in 1926, where he remained until his retirement in 1951.

In 1922 Staudinger introduced the term "macromolecule" into chemistry and went on to propound the unorthodox view that there was no reason why molecules could not reach any size whatever. He argued that chain molecules could be constructed of almost any length in which the atoms were joined together by the normal valence bonds. Innocuous as such a view may now sound, at the time it was considered very strange and, by some, quite absurd.

The accepted view, the aggregate theory, regarded molecules in excess of a molecular weight of 5,000 as aggregates of much smaller molecules joined together by the secondary valence (nebenvalenzen) of Alfred WERNER. Staudinger argued for his theory at length at a stormy meeting of the Zurich Chemical Society in 1926 in front of his most important critics. Within a few years the issue would be decisively settled in favor of Staudinger by the development of the ultracentrifuge by Theodor SVEDBERG. Consequently Staudinger was awarded the Nobel Prize for chemistry in 1953.

Stebbins, George Ledyard (1906–2000)
American geneticist

Born in Lawrence, New York, Stebbins studied biology at Harvard where he obtained his PhD in 1931. After working at Colgate University he moved to Berkeley in 1935 and to Davis in 1950, where he established the department of genetics, holding the chair until his retirement in 1973.

In his *Variation and Evolution in Plants* (1950) Stebbins was the first to apply the modern synthesis of evolution, as expounded by Julian HUXLEY, Ernst MAYR, and others, to plants. In collaboration with Ernest Babcock, Stebbins also studied polyploidy – the occurrence of three or more times the basic (haploid) number of chromosomes. When an artificial means of inducing polyploidy was developed Stebbins applied it to wild grasses, and in 1944 managed to establish an artificially created species in a natural environment. He also used the technique to double the chromosome number of sterile interspecific hybrids and in so doing created fertile polyploid hybrids. Fertility tends to be restored in polyploid hybrids because the two different sets of chromosomes from the parent species will each have an identical set to pair with at meiosis and so the formation of gametes is not disturbed. Polyploids have proved extremely useful in plant-breeding work. Knowledge of naturally occurring polyploid systems has also helped greatly in understanding the relationships and consequently in classifying difficult genera such as *Taraxacum* (the dandelions).

Stebbins also studied gene action and proposed that mutations that result in a change in morphology act by regulating the rate of cell division in specific areas of the plant.

Steenstrup, Johann Japetus Smith (1813–1897) *Norwegian–Danish zoologist*

> We may consider the bogs as annual reviews in which we can see how the flora and fauna of our country have developed and changed ...
> The further we go back in time, the colder was the climate.
> —*Geognostik-geologisk Undersögelse af Skovmoserne Vidnesdam og Lillemose i det nordlige Sjaeland* (1842; Geological Examination of Forest Bogs and Small Bogs in the North of Zealand)

Born at Vang in Norway, Steenstrup (**steenstruup**) was educated in Copenhagen and obtained his first academic post as lecturer in botany and mineralogy at the Zealand Academy at Soro. After explorative expeditions to Jutland, Norway, Scotland, and Ireland (1836–44) he became professor of zoology at the University of Copenhagen Museum of Natural History. Steenstrup's most important contribution to zoology lies in his discovery and recognition, independently of Adelbert von CHAMISSO, of the

alternation of sexual and asexual generations in certain animals, for example, the coelenterates, which produce both sexual (medusa) and asexual (polyp) forms. He also carried out anatomical and embryological studies of various marine animals, demonstrating hermaphroditism (bisexuality) in cephalopods and other organisms and investigating vision in flounders. His findings are published in such works as *Researches on the Existence of Hermaphrodites in Nature* (1846) and *Propagation and Development of Animals through Alternate Generations* (1892). Steenstrup also carried out pioneering studies of peat mosses as an aid to archeological dating and made investigations of prehistoric Danish kitchen middens.

Stefan, Josef (1835–1893) *Austrian physicist*

Stefan (**shte**-fahn) was educated in his native Klagenfurt and at the University of Vienna. In 1863 he became professor of mathematics and physics at Vienna University and remained there for the rest of his life.

Stefan's wide-ranging work included investigations into electromagnetic induction, thermomagnetic effects, optical interference, thermal conductivity, diffusion, capillarity, and the kinetic theory of gases. However, he is best remembered for his work on heat radiation in 1879. After examining the heat losses from platinum wire he concluded that the rate of loss was proportional to the fourth power of the absolute temperature; i.e., rate of loss = σ^4. In 1884 one of his students, Ludwig BOLTZMANN, showed that this law was exact only for black bodies (ones that radiate all wavelengths) and could be deduced from theoretical principles. The law is now known as the *Stefan–Boltzmann law*; the constant of proportionality, σ, as *Stefan's constant*.

Stefan was a good experimental physicist and a well-liked teacher. During his lifetime he held various important positions, including *Rector Magnificus* of the University (1876) and secretary (1875) then vice-president (1885) of the Vienna Academy of Sciences.

Stein, William Howard (1911–1980) *American biochemist*

Stein, who was born in New York City, graduated in chemistry from Harvard in 1933 and obtained his PhD in biochemistry from Columbia in 1938. He moved to the Rockefeller Institute being appointed professor of biochemistry in 1954.

From 1950 onward Stein, with his colleague Stanford MOORE, worked on the problem of determining the amino-acid sequence of the enzyme ribonuclease – work which took them most of the decade. For their success in being the first to work out the complete sequence of an enzyme they shared the 1972 Nobel Prize for physiology or medicine with Christian ANFINSEN.

Steinberger, Jack (1921–) *American physicist*

Steinberger, who was born at Bad Kissingen in Germany, emigrated with his family to America in 1934 as Jewish refugees. He studied chemistry at the University of Chicago and spent World War II working in the Radiation Laboratory at the Massachusetts Institute of Technology. After the war he switched to physics and moved to Columbia, New York, where he was appointed professor of physics in 1950. In 1968 Steinberger returned to Europe to work as a senior physicist at the European Laboratory for Particle Physics in Geneva.

In the early 1960s Steinberger collaborated with his Columbia colleagues Leon LEDERMAN and Melvin SCHWARTZ on the two-neutrino experiment, which won for them the 1988 Nobel Prize for physics. To win the prize, Steinberger noted, for an experiment performed a quarter of a century earlier, you needed not only significant results but longevity as well.

Steno, Nicolaus (1638–1686) *Danish anatomist and geologist*

The son of a goldsmith, Steno (**stee**-noh) was educated in his native city of Copenhagen before beginning his travels and studies abroad in 1660. While studying anatomy in Amsterdam he discovered the parotid salivary duct, also called *Stensen's duct* after the Danish form of his name. Other important anatomical findings included his realization that muscles are composed of fibrils and his demonstration that the pineal gland exists in animals other than man. (René DESCARTES had considered the pineal gland the location of the soul, believing that both were found only in man.)

Steno obtained his medical degree from Leiden in 1664 and the following year went to Florence, where he became physician to the grand duke Ferdinand II. In the field of geology he made important contributions to the study of crystals and fossils. His observations on quartz crystals showed that, though the crystals differ greatly in physical appearance, they all have the same angles between corresponding faces. This led to the formulation of *Steno's law*, which states that the angles between two corresponding faces on the crystals of any chemical or mineral species are constant and characteristic of the species. It is now known that this is a consequence of the internal regular ordered arrangement of the atoms or molecules.

Steno's geological and mineralogical views were expressed in his *De solido intra solidum naturaliter contento dissertationis prodromus* (1669; An Introductory Discourse on a Solid Body Contained Naturally within a Solid). The curious title refers to the solid bodies we refer to as fossils found in other solid bodies. Steno was particularly concerned with the common Mediterranean fossils known at the time as "glossipetrae" (tongue stones), thought by some to have fallen from the sky and by others to have grown in the earth like plants. They were triangular, flat, hard, and with discernible crenellations along two sides.

In 1666 Steno was presented with the head of a giant shark. He was immediately struck by the close similarity between the glossipetrae and sharks' teeth. In attempting to understand this correlation Steno formulated two important principles to explain how solids form in solids. By the first, an ordering rule, it proved possible to tell which solidified first by noting which solid was impressed on the other. As glossipetrae left their imprint in the surrounding rocks they must have been formed first. Therefore it made no sense to suppose that they grew in the strata.

Steno's second rule proclaimed that if two solids were similar in all observed respects then they were likely to have been produced in the same way. It followed that the similarity between the glossipetrae and sharks' teeth revealed them as fossilized teeth, a revolutionary claim at the time. But *Steno's rules* offered more than an explanation of glossipetrae; they in fact offered a novel way of interpreting the fossil record, one which would be followed increasingly by later geologists.

Steno was brought up a Lutheran but converted to Catholicism in 1667, taking holy orders in 1675. In 1677 he was appointed Titular Bishop of Titopolis (in Turkey), catering for the spiritual needs of the few Catholics surviving in Scandinavia and Northern Germany.

Stephenson, George (1781–1848) *British engineer and inventor*

> He could no more explain to others what he meant to do and how he meant to do it than he could fly; and therefore members of the House of Commons, after saying, "There is rock to be excavated to a depth of more than sixty feet … there is a swamp of five miles in length to be traversed … how will you do this?" and receiving no answer but a broad Northumbrian "I can't tell you how I'll do it, but I can tell you I *will* do it," dismissed Stephenson as a visionary.
> —Fanny Kemble, *Records of a Girlhood* (1878)

Stephenson, who was born at Wylam, Northumberland, in the north of England, started adult life as the operator of a steam pump in a coal mine in Newcastle. He attended night school to learn to read and write. He married and had one son, Robert, from whom he learned mathematics as they went over school homework together.

In 1813, when Stephenson was an enginewright in a coal mine, he had the idea of improving a "steam boiler on wheels" that he had seen on a neighboring mine. By introducing a steam blast, and exhausting steam into a chimney, more air could be drawn in. He built several locomotives for his colliery and in 1821, when he heard that there was to be a horse-drawn railroad to carry coal from Stockton to Darlington, he persuaded the project's sponsors to let him build a locomotive for the line. The first train traveled on the line on 27 September 1825 and carried 450 people at 15 miles (24 km) per hour.

Stephenson was then commissioned to build a 40-mile line from Liverpool to Manchester. On the line's completion, his locomotive design the *Rocket* won the famous Rainhill trials (at Rainhill near Liverpool in 1829) by achieving a speed of 36 miles (58 km) per hour. For the rest of his life Stephenson continued to design and advise on railroads throughout the world.

Stern, Curt (1902–1981) *German–American geneticist*

Born at Hamburg in Germany, Stern received his PhD in zoology from the University of Berlin in 1923. He spent two years as a post-doctoral fellow with T. H. MORGAN at Columbia before being appointed *Privatdozent* at the University of Berlin in 1928. Stern returned to America as a refugee from Nazi Germany in 1933 and settled first at the University of Rochester, serving as professor of zoology from 1941 until 1947. He then moved to the chair of zoology and genetics at the University of California, Berkeley, from which post he retired in 1970.

Stern, in 1931, was the first geneticist actually to demonstrate the phenomenon of crossing over in the chromosomes of *Drosophila*. That crossing over did occur had been assumed and widely used by Morgan and his school since about 1914. It was, however, only when Stern managed to get flies with a pair of homologous chromosomes that were structurally markedly different from each other at both ends that experimental support could be produced. (Normally chromosomes that pair together are structurally identical.) Stern knew that the longer chromosome (long–long) carried the genes AB while the shorter chromosome (short–short) carried the genes A^1B^1. Cytological examination of the offspring revealed that those carrying the genes AB^1 or A^1B had long–short and short–long chromosomes respectively, showing that crossing over had indeed occurred. In the same year comparable evidence was provided by Harriett Creighton and Barbara McCLINTOCK from their work with maize.

Stern later worked on problems concerned with genetic mosaics and demonstrated that crossing over can occur in the somatic (nonre-productive) cells as well as the germ cells. He also produced the widely read textbook *Principles of Human Genetics* (1949).

Stern, Otto (1888–1969) *German–American physicist*

Stern, who was born at Sohrau (now in Poland), was educated at the University of Breslau where he obtained his doctorate in 1912. He joined EINSTEIN at the University of Prague and later followed him to Zurich (1913). After teaching at a number of German universities he was appointed an associate professor of theoretical physics at Rostock in 1921. He later moved (1923) to the University of Hamburg as professor of physical chemistry, but resigned in opposition to Hitler in 1933 and emigrated to America, where he took up an appointment with the Carnegie Institute of Technology at Pittsburgh. He retired in 1945.

Stern's main research came from his work with molecular beams of atoms and molecules (beams of atoms traveling in the same direction at low pressure, with no collisions occurring within the beam). Using such beams it is possible to measure directly the speeds of molecules in a gas. In 1920 Stern used a molecular beam of silver atoms to test an important prediction of quantum theory – namely, that certain atoms have magnetic moments (behave like small magnets) and that in a magnetic field these magnets take only certain orientations to the field direction. The phenomenon is known as space quantization, and it could be predicted theoretically that silver atoms could have only two orientations in an external field. To test this, Stern with Walter Gerlach passed a beam of silver atoms through a nonuniform magnetic field and observed that it split into two separate beams. This, the famous *Stern–Gerlach experiment*, was a striking piece of evidence for the validity of the quantum theory and Stern received the 1943 Nobel Prize for physics for this work.

Stern used molecular beams for other measurements. Thus he was able to measure the magnetic moment of the proton by this technique. He also succeeded in demonstrating that atoms and molecules had wavelike properties by diffracting them in experiments similar to those of Clinton J. DAVISSON on the electron.

Stevens, Nettie Maria (1861–1912) *American biologist*

Stevens, one of the outstanding women in the development of American science, is noted chiefly for demonstrating that sex is determined by the inheritance of specific X and Y chromosomes. She was born in Cavendish, Vermont, attended Westford Academy, and in 1881 enrolled for teacher training at Westfield Normal School. She graduated with flying colors in just two years, then worked as a teacher and librarian. At the age of 35 she enrolled at Stanford University, California, receiving her BA in biology (1899) and MA in physiology (1900). She then went back east to Bryn Mawr College, Pennsylvania. Her graduate work included spells of foreign study, at the Zoological Station, Naples, and the University of Würzburg, Germany, where she assisted the renowned biologist Theodor BOVERI. In 1903 she received her PhD in morphology, and remained at Bryn Mawr, becoming (in 1905) associate in experimental morphology.

One of Stevens' early research areas was the taxonomy and morphology of ciliate protozoa, and later she collaborated with Thomas Hunt MORGAN investigating the regeneration of cells. However, at Bryn Mawr she turned her attention to discovering what determined the sex of organisms. With the help of a Carnegie Institution grant (1903–1905), she examined the eggs and sperm of grain beetles and other insects, and hypothesized that gender is determined at conception: sperm cells carrying an X chromosome give rise to females, and ones carrying a Y chromosome produce males. She confirmed her hypothesis in experimental matings, and produced a landmark paper. However, male colleagues and fellow scientists were skeptical of her findings, including the geneticist Edmund WILSON. Wilson later revised his opinion in favor of Stevens' position, and received much of the credit for establishing the chromosomal determination of sex. In 1905 Stevens received the Ellen Richards Prize, given to outstanding women researchers. Bryn Mawr established a research professorship for her in 1912, shortly before her death from breast cancer, but she was too ill to occupy the position.

Stevin, Simon (1548–1620) *Flemish mathematician and engineer*

> What appears to be a marvel is no marvel at all.
> —Motto, expressing Stevin's view that everything is logically intelligible

Stevin (ste-**vIn**) was also known as Stevinus, the Latinized form of his name. Born in the city of Bruges, he worked for a time as a clerk in Antwerp, eventually working his way up to become quartermaster of the army under Prince Maurice of Nassau. While in this post he devised a system of sluices, which could flood the land as a defense should Holland be attacked.

Stevin was a versatile man who contributed to several areas of science. Mathematics owes to him the introduction of the decimal system of notating fractions. This system was perfected when John NAPIER invented the decimal point. Stevin helped to popularize the practice of writing scientific works in modern languages (in his case Dutch) rather than Latin, which for so long had been the traditional European language of learning. However such was the hold of the old ways that Willebrord SNELL thought it was worthwhile to translate some of Stevin's work into Latin. To hydrostatics he contributed the discovery that the shape of a vessel containing liquid is irrelevant to the pressure that liquid exerts. He also did some important experimental work in statics and in the study of the Earth's magnetism.

Stewart, Balfour (1828–1887) *British physicist*

Stewart was educated at the universities of St. Andrews and Edinburgh (his native city). He spent the years between 1846 and 1856 in the business world, but from 1856, when he joined the staff of Kew Observatory, he devoted himself to science. In 1856 he became assistant to J. D. Forbes at Edinburgh. He was appointed director of Kew Observatory in 1859. In 1870 he was made professor of natural philosophy at Owens College, Manchester, remaining there until his death.

Stewart's most original contributions to science were his researches into radiant heat. He extended Pierre PRÉVOST's work on heat exchange and investigated absorption and radiation. However, Stewart's work did not become widely known and when, two years later, KIRCHHOFF independently made similar investigations his more rigorous treatment and the practical applications suggested by his work had the decisive influence on subsequent developments. After 1859 Stewart devoted himself mainly to meteorology, his investigations including terrestrial magnetism and sunspots. One of his suggestions (1882) was that daily variations in the Earth's magnetic field were caused by electric currents in the upper atmosphere.

Stibitz, George Robert (1904–1995) *American computer pioneer*

The son of a theologian, Stibitz was born in York, Pennsylvania, and educated at Cornell where he gained his PhD in 1930. He worked at the Bell Telephone Laboratories in New York as a mathematical engineer from 1930 to 1945, when he moved to the Vermont firm Burlington and Underhill as consultant in applied mathematics. Finally, from 1964 until his retirement in 1972, he worked as a research associate at the Dartmouth Medical School, New Hampshire.

At Bell Stibitz worked on the design of relay switches and in 1937 he was struck by the thought that relays could be used to represent binary numbers. Thus a lighted bulb could represent 1 and an unlighted bulb 0. Using this simple idea any number could be represented. So, with a "scrap of board, some snips of metal from a tobacco can, 2 relays, 2 flashlight bulbs, and a couple of dry cells," Stibitz built his first calculator.

Stibitz was asked if he could extend his work to deal with complex numbers. The result was his Model I at a cost of $20,000, based on about 450 electromagnetic relays. It was slow, limited by the speed of the relays, not programmable, and very restricted in its use. With the outbreak of war, Stibitz undertook to prepare a Mark II version for the NDRC (National Defense Research Committee) for use by the artillery. Two further models were developed during the war for military use.

Finally, in 1945, Model V appeared, an all-purpose machine containing 9,000 relays. It was the first to incorporate a floating-point decimal, that is, to represent a number by a fraction multiplied by a power of ten. Two copies of the model were built, both for military use.

Stirling, Robert (1790–1878) *Scottish inventor*

Stirling was born at Cloag in Scotland and attended the universities of Glasgow and Edinburgh. Having been ordained in the Church of Scotland in 1816, he spent most of his life from 1837 until his retirement in 1876 as minister at Galston, Ayrshire.

As engineers of today search for a reasonably efficient, nonpolluting, and silent engine, they seem unaware that precisely such an engine had been developed by a Scottish clergyman in 1816. A *Stirling engine* consists of a closed cylinder with a regenerator separating two opposing pistons. The compression space is kept at a low temperature by an external cooling system; the expansion chamber at a high temperature by an external heat supply. At the beginning of the cycle air, the working fluid, is compressed by the piston and moves into the expansion space while absorbing heat stored in the regenerator in the previous cycle. As the air expands and drives the piston work is done. The motion of the pistons and the gas is then reversed as heat is supplied to the expansion space from an external source.

As the source of heat is external, any fuel can be used and sited wherever convenient. Stirling's first engine generated about two horsepower and ran for about two years pumping water out of a quarry. A later model in 1843 was a modified steam engine that produced 37 horsepower more efficiently than a steam engine. Because the cylinders tended to burn out rather quickly, a common fault of Stirling engines, it was converted to steam use. In drawing up the patent, Robert Stirling was assisted by his brother James – a mechanical engineer who managed the foundry in Dundee where the engine was constructed.

In more recent years with improved materials major industrialists have investigated the possibility of developing the Stirling engine. It has, however, found only marginal use to power small generators and certain specialized engines.

Stock, Alfred (1876–1946) *German chemist*

Stock (shtok) was born at Danzig (now Gdańsk in Poland) and educated at the University of Berlin, where he worked for some time as assistant to Emil FISCHER. In 1909 he joined the staff of the Inorganic Chemistry Institute at Breslau, later moving to the Kaiser Wilhelm Institute in Berlin. He then served as director of the Chemical Institute at the Karlsruhe Technical Institute (1926–36).

Stock began studying boron hydrides in 1909. Boron had previously been little studied and was thought only to react with strongly electronegative elements, such as oxygen. However Stock found that magnesium boride and acid produced B_4H_{10}, the first of the several boron hydrides he discovered. It was clear from this and the other hydrides, B_2H_6 and $B_{10}H_{14}$, that the bond between boron and hydrogen could not be the familiar covalent bond which holds between carbon and hydrogen. Its actual form was not solved until the work of William LIPSCOMB in the 1950s. The boranes have since become useful in rocket fuels.

Stock is also remembered as being the first working chemist to be fully aware that he was suffering from mercury poisoning. By 1923 he was complaining of a large number of such distressing symptoms as deafness, vertigo, headaches, and amnesia, which he traced to mercury. Not only did he give his findings maximum publicity, broadcasting his infirmities to the chemical world, but he also devised techniques of laboratory practice to reduce the risk of such poisoning and introduced tests for the presence of mercury.

Stokes, Adrian (1887–1927) *Anglo-Irish bacteriologist*

Stokes, whose father worked in the Indian Civil Service, was born at Lausanne in Switzerland and educated at Trinity College, Dublin, where he obtained his MD in 1911. After serving in the Royal Army Medical Corps during the war, in which he was awarded the DSO, he returned to Dublin in 1919 as professor of bacteriology but soon moved to London, where in 1922 he became professor of pathology at Guy's Hospital.

In 1920 Stokes visited Lagos to study yellow fever. He was anxious to test the suggestion of the Japanese bacteriologist Hideyo NOGUCHI that yellow fever was caused by the bacillus *Leptospira icteroides*, but it was not until his second visit to Lagos in 1927 that he made the vital breakthrough.

Stokes succeeded, for the first time, in infecting an experimental animal (the rhesus monkey) with the disease. He went on to show that while he could pass yellow fever from monkey to monkey there was no evidence that Noguchi's bacillus was also transmitted. But before he could proceed further Stokes, who was daily handling infected monkey blood, contracted the disease and joined the growing list of bacteriologists who had fallen victim to the virus.

Stokes, Sir George Gabriel (1819–1903)
British mathematician and physicist

Stokes was born at Skreen (now in the Republic of Ireland) and studied at Cambridge, remaining there throughout his life. In 1849 he became Lucasian Professor of Mathematics, but he found it necessary to supplement his slender income from this post by teaching at the Government School of Mines in London. He held his Cambridge chair until his death aged 84. He was the member of parliament for the university and among his many honors were a baronetcy conferred on him in 1889.

Stokes was equally interested in the theoretical and experimental sides of physics and did important work in a wide area of fields, including hydrodynamics, elasticity, and the diffraction of light. In hydrodynamics he derived the formula now known as *Stokes's law*, giving the force resisting motion of a spherical body through a viscous fluid. Among Stokes's other fields of study was fluorescence – one of his experimental discoveries was the transparency of quartz to ultraviolet light. He was also much interested in the then influential concept of the ether as an explanation of the propagation of light. Stokes became aware of some inherent difficulties with the concept, but rather than rejecting the whole idea of an ether he tried to explain these problems away by using work he had done on elastic solids, though naturally enough problems arose with his own ideas.

Stokes was perceptive in his views of other physicists' work. For example, he was among the first to appreciate the importance of the work of James JOULE and to see the true meaning of the spectral lines discovered by Joseph von FRAUNHOFER.

Stoney, George Johnstone (1826–1911)
Irish physicist

Stoney, the son of an impoverished landowner, was born at Oakley Park (now in the Republic of Ireland) and educated at Trinity College, Dublin. After graduation in 1848 he worked as an assistant to the astronomer, the earl of ROSSE, at his observatory at Parsonstown until 1853 when he was appointed professor of natural philosophy at Queen's College, Galway. However, from 1857 onwards Stoney worked as an administrator, first as secretary of

Queen's University, Belfast, and finally, from 1882 until 1893, as superintendent of civil service examinations.

Stoney is best known for his introduction of the term "electron" into science. Although he is reported to have spoken of "an absolute unit of electricity" as early as 1874, his first public use of the term in print was in 1891 when he spoke of "these charges, which it will be convenient to call electrons" before the Royal Society of Dublin.

He did however make more substantial contributions to science than this and in early spectroscopy his work was of considerable significance. He began, in 1868, by making a crucial distinction between two types of molecular motion. There was the motion of a molecule in a gas relative to other molecules, which Stoney was able to exclude as the cause of spectra. There was also internal motion of a molecule, which according to Stoney produces the spectral lines. He went on to tackle, with little real success, the difficult problem of establishing an exact formula for the numerical relationship between the lines in the hydrogen spectrum. This problem was solved by the quantum theory of Niels BOHR.

Störmer, Horst L. (1949–   ) *German–American physicist*

Störmer was born in Frankfurt-am-Main, Germany, and gained his PhD in physics in 1977 from Stuttgart University. From 1992 to 1998 he was supervisor of the Physical Research Laboratory, Bell Laboratories, where he studied the fractional quantum Hall effect. In 1998 he moved to Columbia University, New York. In 1998 Störmer shared the Nobel Prize for physics with Robert LAUGHLIN and Daniel TSUI, for their discovery and explanation of a new form of quantum fluid with fractionally charged excitations.

Strabo (*c.* 63 BC–*c.* 23 AD) *Greek geographer and historian*

> The poets were not alone in sponsoring myths. Long before them cities and lawmakers had found them a useful expedient … They needed to control the people by superstitious fears, and these cannot be aroused without myths and marvels.
> —*Geography*, Book I

Strabo (**stray**-boh), who was born at Amaseia (now Amasya in Turkey), traveled to Rome in 44 BC and remained there until about 31 BC. He visited Corinth in 29 BC and in about 24 BC sailed up the Nile.

Although the historical writings of Strabo, including his *Historical Sketches*, in 47 books, have been almost entirely lost, his *Geography*, in 17 books, has survived virtually intact. This major geographical work is an important source of information on the ancient world. In it Strabo accepted the traditional description of the Earth as divided into five zones with the *oikoumene*, or inhabited part, represented as a parallelogram spread over eight lines of latitude and seven meridians of longitude. Where he excelled, however, was in the field of historical and cultural geography and he gave a detailed account of the history and culture of the lands and people of the Roman Empire and of such areas as India, which lay beyond the dominion of Augustus. In this he quoted much from the earlier Greeks, including ERATOSTHENES and Artemidorus.

Strabo, not content merely to describe the lands of the civilized world, also wished to understand its enormous diversity. He rejected the simple climatic determinism that he attributed to the Stoic POSEIDONIUS, arguing in its place for the role of institutions and education. Despite the value of this work Strabo seemed to exercise little influence until Byzantine times.

Strachey, Christopher (1916–1975)
British computer scientist

The son of a cryptographer and a nephew of the writer Lytton Strachey, Christopher Strachey was born at Hampstead in north London and educated at Cambridge. After graduating in 1939 he spent the war working on the development of radar. On demobilization in 1945 he became a schoolmaster. During this time he developed an interest in large computers.

In 1951 he wrote a successful checkers program for ACE (Automatic Computing Engine), one of the three computers then in use in Britain. Soon after he moved to the National Research and Development Corporation where he worked on the design of the new Ferranti computer, Pegasus. He also worked on computer models of the St. Lawrence Seaway, one of the great postwar engineering projects.

After spending some time as a private consultant, Strachey opted for academic life, accepting an appointment at Cambridge in 1962. He moved to Oxford in 1966 where he became professor of computation in 1971, a post he held until his death in 1975.

During this latter academic period Strachey worked on the design of the new high level computer language, CPL, later developed into BCPL (Basic Computer Programming Language).

Strasburger, Eduard Adolf (1844–1912)
German botanist

Strasburger (**shtrahs**-buur-ger) was born in the Polish capital Warsaw and educated at the universities of Paris, Bonn, and Jena, where he received his PhD in 1866. In 1868 he taught at Warsaw but in 1869 returned to Jena, where he became professor of botany in the same year and director of the botanical gardens in 1873.

His early work extended the researches of Wilhelm HOFMEISTER into the regular alternation of generations in the plant kingdom. Strasburger was the first to describe accurately the embryo sac in gymnosperms (conifers and their allies) and angiosperms (flowering plants) and to demonstrate the process of double fertilization in angiosperms. In 1875 he laid down the basic principles of mitosis in his *Cell Formation and Cell Division*. The third edition of this work contained one of the now well-established laws of cytology: that new nuclei can arise only from the division of existing nuclei.

From 1880 until his death Strasburger worked at Bonn, establishing it as one of the world's leading centers for cytological research. In 1891 he demonstrated that physical forces (e.g., capillarity), and not physiological forces, are primarily responsible for the movement of liquids up the plant stem. In 1894, together with three other eminent botanists, he founded the famous *Strasburger's Textbook of Botany*, which was run to 30 editions and is still important as a general course book today.

Strassmann, Fritz (1902–1980) *German chemist*

Strassmann (**shtrahs**-mahn) was born at Boppard in Germany and educated at the Technical University at Hannover. He taught at Hannover and at the Kaiser Wilhelm Institute before being appointed to the chair of inorganic and nuclear chemistry at the University of Mainz in 1946. In 1953 he became director of chemistry at the Max Planck Institute.

In 1938 Strassmann collaborated with Otto HAHN on the experiment that first clearly revealed the phenomenon of nuclear fission.

Strato of Lampsacus (about 287 BC) *Greek philosopher*

Little is known about the life of Strato (or Straton; **stray**-toh or **stray**-ton). Born at Lampsacus in Greece, he probably studied under THEOPHRASTUS at the Lyceum and later became tutor in Alexandria to the son of the first king Ptolemy of Egypt. He succeeded Theophrastus as head of the Peripatetic school of philosophy in 287 BC and remained in Athens until his death in about 269. Although only fragments of his writings remain, Diogenes Laertius names about 40 of his works.

Many of Strato's ideas were developments of Aristotelianism with some resemblances to atomism. He believed that all bodies, even light rays, consisted of tiny particles separated by a void; this void existed as discrete interstices where the particles did not fit together exactly; space is always filled with some kind of matter. Strato gave explanations of the nature of light and sound, and had a considerable reputation as an experimenter.

Strato, held in high esteem by later writers, represents a link between the Lyceum of Athens and its offshoot, the Museum of Alexandria. He was known in antiquity as Strato the physicist.

Strohmeyer, Friedrich (1776–1835) *German chemist*

Strohmeyer (**shtroh**-mI-er) was educated at the university in his native city of Göttingen, where his father was a professor of medicine, and also studied in Paris under Louis VAUQUELIN. He started teaching in 1802 at Göttingen, becoming professor of chemistry in 1810. He was appointed inspector-general of the apothecaries of Hanover in 1817.

In 1817 while working with what he thought was zinc carbonate he found that it turned yellow when heated. At first he thought this indicated the presence of iron, but after failing to discover this he traced the color to a new element, which he named cadmium after the Latin name for zinc ore, cadmia.

Strömgren, Bengt Georg Daniel (1908–1987) *Swedish–Danish astronomer*

Elis Strömgren (**stru(r)m**-grayn), father of Bengt, was an astronomer of distinction who served as director of the Copenhagen Observatory. His son was born at Gothenburg in Sweden and studied at the University of Copenhagen. After obtaining his PhD there in 1929 he joined the staff and was appointed professor of astronomy in 1938. He succeeded his father as director in 1940. He later moved to America, serving from 1951 to 1957 as professor at the University of Chicago and director of the Yerkes and McDonald observatories. He was a member of the Institute for Advanced Study, Princeton, from 1957 until 1967, when he returned to Copenhagen as professor of astrophysics.

In the 1930s and 1940s Strömgren engaged in pioneering work on emission nebulae – huge clouds of interstellar gas and dust shining by their own light. He showed that they consist largely of ionized hydrogen, H II to the spectroscopist. If hot young stars were embedded in uniformly but thinly distributed neutral hydrogen, then the emission by them of ultraviolet radiation would virtually ionize the gas completely. To meet this condition the stars would need a surface temperature of some 25,000 kelvin. At a certain distance from the star, the *Strömgren radius*, the emitted photons of radiation would no longer possess sufficient energy to ionize the hydrogen, leading to a sharp boundary between ionized and cooler nonionized regions. Strömgren showed that this distance would depend on the density of the hydrogen and the stellar temperature.

A typical example of the process described by Strömgren is to be found in the Orion nebula. Later work has however shown that there are three types of emission nebulae, two of which are produced by different mechanisms.

Struve, Friedrich Georg Wilhelm von (1793–1864) *German–Russian astronomer*

> A magnificent work ranking among the greatest performed by astronomical observers in recent times.
> —Friedrich W. Bessel, reviewing Struve's *Stellarum Duplicium Mensurae Micrometricae* (1837; Micrometric Measurements of Double Stars)

Struve (**shtroo**-ve), who was born at Altona in Germany, moved to Dorpat in Latvia in 1808 in order to escape conscription into the Napoleonic army then in control of Germany. He took a degree in philology in 1811 before becoming professor of astronomy and mathematics in Dorpat in 1813. In 1817 he became director of the Dorpat Observatory, which he equipped with a 9.5-inch (24-cm) refractor that he used in a massive survey of binary stars from the north celestial pole to 15°S. He measured 3,112 binaries – discovering well over 2,000 – and cataloged his results in *Stellarum Duplicium Mensurae Micrometricae* (1837; Micrometric Measurements of Double Stars).

In 1835 Czar Nicholas I persuaded Struve to set up a new observatory at Pulkovo, near St. Petersburg. There in 1840 Struve became, with Friedrich Bessel and Thomas HENDERSON, one of the first astronomers to detect parallax. He chose Vega, a bright star with a larger-than-normal proper motion and soon established a parallactic measurement (that was, however, too high).

Struve founded a dynasty of astronomers that is still in existence. He was succeeded by his son Otto at Pulkovo, his grandson Hermann became director of the Berlin Observatory, and his great-grandson, Otto Struve, became director of the Yerkes Observatory in Wisconsin.

Struve, Otto (1897–1963) *Russian–American astronomer*

> It is not an exaggeration to say that almost all our knowledge of the structure of the Milky Way which has developed during the past quarter of a century has come from the Mount Wilson discovery of spectroscopic luminosity criteria.
> —*The Science Counselor* (1948)

Struve (**stroo**-ve), who was born at Kharkov in Russia, came from a long line of distinguished astronomers, being the great grandson of its founder Friedrich Georg von Struve. His father was the professor of astronomy and director of the observatory at the University of Kharkov and he had two uncles who were directors of German observatories. His studies at the university were interrupted by World War I but he finally graduated in 1919. Called up again in 1919 after the revolution, he ended up destitute in Turkey in 1920. Following a journey of some difficulty he finally arrived in America in 1921 where he attended the University of Chicago, obtaining his PhD in 1923. He worked at the Yerkes Observatory, serving as director from 1932 to 1947 as well as professor of astrophysics at Chicago for the same period. He played an important role in the founding of the McDonald Observatory on Mount Locke in Texas and the planning of its 82-inch (2.1-m) reflecting telescope, then the second largest in the world. He served as McDonald's first director from 1939 to 1950. Struve moved to a less demanding position at the University of California at Berkeley in 1950 but agreed in 1959 to become the first director of the National Radio Astronomy Observatory at Green Bank in West Virginia. Forced to resign in 1962 owing to ill health, he died shortly after.

Although Struve spent much of his time in administration and organization, he was able to conduct some major observational work. He made spectroscopic studies of binary and variable stars, stellar rotation, stellar atmospheres, and, possibly most important, of interstellar matter.

One of the problems facing astronomers at the beginning of the century was whether there was any interstellar matter and if so, did it significantly absorb or distort distant starlight. This was no trivial question for the answer could make nonsense of many accounts of the distribution of stars. In 1904 Johannes HARTMANN had argued for the presence of interstellar calcium by pointing out that the calcium spectral lines associated with the binary system Delta Orionis did not oscillate with the other spectral lines as the stars orbited each other. This work was extended by Vesto SLIPHER in 1908 and 1912.

Struve produced evidence on the next crucial point as to whether the interstellar matter was diffuse and pervasive or only local and associated with individual star systems. In 1929, in collaboration with B. P. Gerasimovic, he showed that it exists throughout the Galaxy. This work was also done independently by John PLASKETT. In 1937 Struve discovered the presence of interstellar hydrogen, in ionized form, which though much more prevalent than calcium was initially more difficult to detect.

Sturgeon, William (1783–1850) *British physicist*

Sturgeon's father, a shoemaker of WHITTINGTON, England, has been described as an "idle poacher who neglected his family." Seeing little future as an apprentice cobbler, Sturgeon enlisted in the army in 1802. While serving in Newfoundland his interest in science was aroused while watching a violent thunderstorm. Finding that no one seemed able to explain satisfactorily to him the cause and nature of lightning, he started reading whatever science books were available. This led him to the study of mathematics and Latin. When he left the army in 1820 he had acquired a considerable amount of scientific knowledge and skill. He began to write popular articles, joined the Woolwich Literary Society, and must have so impressed his associates that a move was made to find him a more suitable job than the shoemaking he was being forced back into. Thus in 1824 he was appointed to a lectureship in experimental philosophy at the East India Company's Royal Military College at Addiscombe.

In 1840 he moved to Manchester as the superintendent of the Royal Victoria Gallery of Practical Science. In 1836 he began the publication of the *Annals of Electricity*, the first periodical of its kind to be issued in Britain.

After various further appointments as an itinerant lecturer he was awarded a government pension of £200 a year for his services to science. His collected papers, *Scientific Researches*, were published in 1850.

Sturgeon made several fundamental contributions to the new science of electricity. The cell devised by Alessandro VOLTA had certain inherent weaknesses – any impurity in the zinc plates used caused erosion of the electrode. In 1828 Sturgeon found that amalgamating the plate with mercury made it resistant to the electrolyte. More important was his construction in about 1821 of the first electromagnet. Following the work of François ARAGO he wound 16 turns of copper wire around a one-foot iron bar, which, when bent into the shape of a horseshoe, was powerful enough to lift a weight of 9 pounds when the wire was connected to a single voltaic cell. He demonstrated his magnets in 1825 in London. More powerful ones were soon built by Joseph HENRY and Michael FARADAY.

In later years Sturgeon also made improvements to the design of the galvanometer, inventing the moving-coil galvanometer in 1836. In the same year he introduced the first commutator for a workable electric motor (1836).

Sturtevant, Alfred Henry (1891–1970) *American geneticist*

> I went home, and spent most of the night (to the neglect of my undergraduate homework) in producing the first chromosome map.
> —On his early work on chromosomes, c. 1910.
> *History of Genetics* (1965)

Born in Jacksonville, Illinois, Sturtevant graduated at Columbia University in 1912 and continued there, working for his PhD under the supervision of T. H. MORGAN. His thesis dealt with certain aspects of fruit fly (*Drosophila*) genetics, the research being conducted in the famous "fly room" at Columbia.

During this period Sturtevant developed a method for finding the linear arrangement of genes along the chromosome. This technique, termed "chromosome mapping," relies on the analysis of groups of linked genes. His paper, published in 1913, describes the location of six sex-linked genes as deduced by the way in which they associate with each other: it is one of the classic papers in genetics.

Sturtevant later discovered the so-called "position effect," in which the expression of a gene depends on its position in relation to other genes. He also demonstrated that crossing over between chromosomes is prevented in regions where a part of the chromosome material is inserted the wrong way round. This had important implications for genetic analysis. Although employed by the Carnegie Institution in 1915, Sturtevant continued working at Columbia until 1928. He then moved to the California Institute of Technology, where he was professor of genetics and biology until his death. He wrote many important papers and books and was one of the authors of *The Mechanism of Mendelian Heredity* (1915).

Suess, Eduard (1831–1914) *Austrian geologist*

> The history of the continents results from that of the seas.
> —*The Face of the Earth*, Vol. II

Suess (zoos), born the son of a businessman in London, was educated at the University of Prague. He began work, in 1852, in the Hofmuseum, Vienna, before moving to the University of Vienna in 1856 where he became professor of geology in 1861. Besides being an academic Suess served as a member of the Reichsrat (parliament) from 1872 to 1896. He was responsible for the provision of pure water to Vienna by the construction of an aqueduct in 1873 and the prevention of frequent flooding by the opening of the Danube canal in 1875.

His major work as a geologist was his publication of *Das Antlitz der Erde* (1883–88), translated into English as *The Face of the Earth* (5 vols., 1904–24). This was not a particularly original work but acquired significance as being the great synthesis of the achievements of the later 19th-century geologists, geographers, paleontologists, and so on. He also published, in 1857, a classic work on the origin of the Alps.

Suess was the first to propose the existence of the great early southern continent, Gondwanaland. He was impressed by the distribution of a fern, *Glossopteris*, present during the Carboniferous period. It was found in such widely scattered lands as Australia, India, South Africa, and South America. Suess therefore proposed that these lands had once formed part of one great continent, which he named for the Gonds, the supposed aboriginal Indians.

Sugden, Samuel (1892–1950) *British chemist*

Born in Leeds, England, Sugden was educated at the Royal College of Science, London, where he was awarded a DSc in 1924. He joined the staff of Birkbeck College, London, where he served as professor of chemistry (1932–37) before moving to University College, London. During World War II he served in the Ministry of Supply and although he returned to his chair after the war his ill health virtually ended his research career.

Sugden worked on the physical properties of liquids, particularly surface tension. He introduced the measure that he named the "parachor" (1924) – a formula involving the molecular weight of a substance, its surface tension, and its densities in the liquid and vapor states. The parachor was a standard measure of molecular volume – i.e., a measure of the volume of molecules independent of attractive forces between molecules within the liquid. As such it was hoped that it could be used for determining the structures of organic compounds by adding values for different atoms and types of bond. Although the idea aroused much interest at the time it had little practical success, and the technique is now obsolete. Sugden also worked on radioactive tracers and on the magnetic properties of chemical compounds. In 1930 he published *The Parachor and Valency*.

Sugita, Genpaku (1738–1818) *Japanese physician*

The son of a physician, Sugita (suu-**gee**-ta) learned surgery in Edo. While there he saw a Dutch translation of a German anatomical text, *Anatomische Tabellen* (1722; Anatomical Tables) by Johan Kulmus. He was surprised to see how radically the anatomical illustrations differed from those contained in Chinese texts. Some time soon after he attended a postmortem dissection of a prisoner executed by decapitation. He took with him the illustrations from Kulmus and was amazed to note the agreement between its diagrams and the body. No such concord could be found with Chinese texts. Consequently Sugita decided to translate a Dutch version of the text into Japanese. It was published in 1774 as *Kaito Shinso* (New Book of Anatomy) and was the first translation of a European medical text into Japanese.

The influence of Sugita's work was considerable. It created a demand for further Dutch texts, as well as a new school of "Dutch medicine" called Ranpo, which was openly critical of traditional Chinese-style medicine.

Sulston, John E. (1942–) *British geneticist*

Sulston studied at Cambridge University, gaining his BA in 1963 and PhD in 1966. After three years as a postdoctoral fellow at the Salk Institute for Biological Studies, San Diego, California, he returned to the UK to join the Medical Research Council's Laboratory of Molecular Biology in Cambridge. Between 1992 and 2000 he was director of the Sanger Centre, Cambridge, heading the UK's contribution to the international Human Genome Project.

Throughout his career, Sulston has been interested in the development and genetics of the tiny nematode (roundworm) *Caenorhabditis elegans*. This interest was sparked by the work of Sydney BRENNER, with whom Sulston collaborated at the MRC's Laboratory of Molecular Biology. For some years, Brenner had been convinced that *C. elegans* was a valuable experimental model for unraveling the complexities of the developmental biology and genetics of higher organisms. It is about 1 mm long, is almost completely transparent, and consists of roughly 1000 cells. Moreover, because it is hermaphrodite, all the progeny of any worm are genetically identical, and develop according to the exact same pattern.

Sulston's early breakthrough came when he realized that it was possible to observe, through a microscope, individual cells dividing, migrating, and sometimes dying inside the worm during its larval development. He and his colleagues traced the lineage of every cell of an adult worm back to the point at which it hatched from the egg. This work clearly demonstrated the importance of *programmed cell death (apoptosis)* – the genetically predetermined elimination of certain healthy cells – in shaping the development of body organs and systems, such as the nervous system. It opened up fruitful avenues of research into cellular control mechanisms, some of which have been explored by H. Robert HORVITZ, one of Sulston's colleagues at Cambridge in the mid-1970s.

Sulston then took up the even greater challenge of following the fate of cells inside the egg during embryonic development. With painstaking effort over a period of 18 months, he tracked the origin of each cell, and how it grew, migrated, divided, and in some cases died. His labors resulted in a definitive chart showing the formation, through space and time, of all the constituent cells of the worm's tissues and organs.

During the 1980s Sulston's attention turned to mapping the genome of *C. elegans*. Working with a British biochemist, Alan Coulson, he started to devise computer programs that would speed the process of identifying and ordering the numerous cloned DNA fragments derived from the worm's genetic material. By 1989 Sulston was at the hub of an international but informal effort to sequence the worm's genome. Sulston resisted lucrative offers to move his research to America and into the commercial sector. Instead he was appointed head of a new project, the Sanger Centre, that came

to house the UK's contribution to the Human Genome Project. This was funded by the Wellcome Trust, the UK's leading scientific charitable institution. The ethos of the Sanger Centre embodied Sulston's own principles: discoveries made by the Sanger team were made freely accessible to other scientists without delay, and their DNA sequence data were not subject to patenting or other restrictions.

In 2002, Sulston was awarded the Nobel Prize in physiology or medicine, jointly with Brenner and Horvitz, for their work on the genetic regulation of organ development and programmed cell death.

Sumner, James Batcheller (1877–1955)
American biochemist

> Nothing can take the place of intelligence.
> —TIBS, July 1981

Sumner, a wealthy cotton manufacturer's son from Canton, Massachusetts, was educated at Harvard, where he obtained his PhD in 1914. In the same year he took up an appointment at the Cornell Medical School where, in 1929, he became professor of biochemistry.

Despite having lost an arm in a shooting accident at 17, Sumner persisted in his desire to become an experimental chemist. In 1917 he began his attempt to isolate a pure enzyme. He chose for his attempt urease, which catalyzes the breakdown of urea into ammonia and carbon dioxide and is found in large quantities in the jack bean. After much effort he found, in 1926, that if he dissolved urease in 30% acetone and then chilled it, crystals formed. The crystal had high urease activity. Moreover Sumner's crystals were clearly protein and however hard he tried to separate the protein from them he always failed. He was therefore forced to conclude that urease, an enzyme, was a protein. However, this ran against the authority of Richard WILLSTÄTTER who had earlier isolated enzymes in which no protein was detectable. In fact, protein was in Willstätter's samples, but in such small quantities as to be undetected by his techniques.

Consequently little attention was paid to Sumner's announcement and it was only when John NORTHROP succeeded in crystallizing further protein enzymes in the early 1930s that his work was properly acknowledged. In 1946 for "his discovery that enzymes can be crystallized" he was awarded the Nobel Prize for chemistry jointly with Northrop and Wendell Stanley.

Su Sung (1020–1101) *Chinese inventor*

Su Sung (soo suung), who was born at Nan-an in the Fujian province of China, was a mandarin and, like others of his class, he had considerable scientific knowledge and insight. In 1086 the emperor, Che Tsung, issued instructions for the construction of an astronomical clock that would surpass those of earlier dynasties. By clock the emperor meant a celestial globe or armillary sphere, that is, a set of connected rings corresponding to the main circles of the celestial sphere, which revolved in harmony with the heavens. There is a long tradition of such instruments in China going back as far as CHANG HENG in the second century AD and including major improvements by I-HSING in the eighth century.

By 1090 Su Sung had built his clock, details of which he revealed in his *New Design for an Astronomical Clock* (1094), which were clear enough to permit the construction of a working model in 1961. The clocktower was 30 feet tall and contained a revolutionary mechanism. The problem facing clock makers is how to release the power in the mechanism, whether it be a falling weight or water, gradually, measurably, and under control. The early water clocks avoided the problem by using a continuous flow of water on the surface of which a pointer could float. The great advance beyond this was the escapement, which only appeared in Europe with Giovanni DONDI in the 14th century but which was first introduced by Su Sung. Water poured into scoops on a waterwheel, which when full tripped a lever that allowed the wheel to turn a measurable amount. The next scoop then filled and repeated the process. This turned the main drive shaft that worked the celestial globes, spheres, and chiming bells.

There is no direct evidence supporting the claim that Su Sung's work was known to the European clockmakers. The fate of the clock itself is obscure. It appears to have been damaged in a storm and finally destroyed in the Mongol invasion of the 13th century. The tradition of celestial clockwork survived only until the Ming dynasty in the 14th century, for when Matteo RICCI arrived in China in 1600 he could find no one who could still understand the astronomical instruments of their ancestors.

Sutherland, Earl (1915–1974) *American physiologist*

Born in Burlingame, Kansas, Sutherland was educated at Washington University, St. Louis. After serving in World War II as an army doctor he returned to St. Louis but in 1963 moved to Vanderbilt University, Tennessee, as professor of physiology. In the year before his death Sutherland joined the University of Miami Medical School in Florida.

In 1957 Sutherland discovered a molecule of great biological significance – 3,5–adenosine monophosphate, more familiarly known as cyclic AMP. At that time he was working with T. Rall on the way in which the hormone adrenaline (epinephrine) effects an increase in the amount of glucose in the blood. They found that the hormone stimulated the release of the enzyme adenyl cyclase into liver cells. This in turn converts adenosine triphosphate (ATP) into cyclic AMP, which then initiates the complex chain converting the glycogen stored in the liver into glucose in the blood. The significance of this reaction is that adrenaline does not act directly on the molecules in the liver cell; it apparently needs and "calls for" what soon became described as a "second messenger," cyclic AMP.

Sutherland went on to show that other hormones, such as insulin, also used cyclic AMP as a second messenger and that it was in fact used to control many processes of the cell. For his discovery of cyclic AMP Sutherland was awarded the 1971 Nobel Prize for physiology or medicine.

Svedberg, Theodor (1884–1971) *Swedish chemist*

Svedberg (**sved**-berg or **svayd**-bar-ye), born the son of a civil engineer in Fleräng, Sweden, was educated at the University of Uppsala, where he obtained his doctorate in 1908. He spent his whole career at the university, becoming a lecturer in physical chemistry in 1907, a professor (1912–49), and finally, in 1949, director of the Institute of Nuclear Chemistry.

In 1924 he introduced the ultracentrifuge as a technique for investigating the molecular weights of very large molecules. In a suspension of particles, there is a tendency for the particles to settle (under the influence of gravity); this is opposed by Brownian motion, i.e., by collision with molecules. The rate of sedimentation depends on the size and weight of the particles, and can be used to measure these.

Svedberg applied this to measuring the sedimentation of proteins in solution, using an ultracentrifuge that generated forces much greater than that of the Earth's gravitational field. Using this, he could measure the molecular weights of proteins and was able to show that these were much higher than originally thought (hemoglobin, for instance, has a molecular weight of about 68,000).

Apart from confirming the claim made by Hermann STAUDINGER for the existence of giant molecules, Svedberg's invention also settled one further question. The same protein invariably yielded the same weight, thus implying that they did have a definite size and composition and were not, as Wilhelm OSTWALD had earlier maintained, irregular assemblies of smaller molecules. For his work on the ultracentrifuge Svedberg was awarded the Nobel Prize for chemistry in 1926.

Svedberg was less successful with the inference he drew from his measurements of protein molecular weights. He thought that the molecular weight of egg albumin formed the basic protein unit of which all the other proteins were multiples. Following later research by crystallographers in the 1930s this view was disproved.

Swammerdam, Jan (1637–1680) *Dutch naturalist and microscopist*

> The matter being properly considered, a worm or caterpillar does not change into a pupa, but becomes a pupa by the growing of parts.
> —*Historia insectorum* (1669; Account of Insects)

Swammerdam (**svahm**-er-dahm), an Amsterdam apothecary's son, studied medicine at Leiden University, graduating in 1667. However, he never practiced and instead devoted his life to microscopical studies of a widely varying nature. His most important work, namely the discovery and description of red blood corpuscles in 1658, was completed before he went to university. He later demonstrated experimentally that muscular contraction involves a change in the shape but not volume of the muscle. He also studied movements of the heart and lungs and discovered the valves in the lymph vessels that are named for him.

Swammerdam is also remembered for his pioneering work on insects. He collected some 3,000 different species and illustrated and described the anatomy, reproductive processes, and life histories of many of these. This work, together with his system of insect classification, laid the foundations of modern entomology. Swammerdam's *Biblia naturae* (Book of Nature), published long after his death (1737–38), still stands as one of the finest one-man collections of microscopical observations.

At the theoretical level Swammerdam developed a new argument in support of the preformationist position, the view that organisms are born already formed. His argument, first presented in his *Historia insectorum* (1669; Account of Insects), was based upon the nature of insect metamorphosis. At first sight it might appear that the metamorphic process supported the alternative view of development, epigenesis, the claim that organisms develop gradually and in sequence. Swammerdam, however, revealed a different picture when, with the aid of a microscope, he succeeded in identifying structures belonging to butterflies in pupae and caterpillars. The caterpillar, Swammerdam insisted, was not changed into a butterfly, rather it grew by the expansion of parts already formed. Nor does the tadpole change into a frog; it becomes a frog "by the infolding and increasing of some of its parts." In proof of his position Swammerdam would display a silkworm to his critics, peel off the outer skin, and display the rudiments of the wings within.

In the same work Swammerdam added one more piece of evidence against the claim that organisms can generate spontaneously. Insects found in plant galls, he pointed out, developed from eggs laid therein by visiting flies.

Swan, Sir Joseph Wilson (1828–1914) *British inventor and industrialist*

Swan, the son of a dealer in anchors and chains from Sunderland in the northeast of England, was educated at local schools and apprenticed to an apothecary at the age of 14. After three years he left to work for a chemical supplier, John Mawson, eventually becoming his partner. One part of their business was supplying chemicals to photographers and Swan made an early improvement in the new collodion process when, in 1864, he patented the carbon process for producing permanent photographic prints.

But long before this, in 1848, he had already become interested in the construction of electric lights – the development for which he is best known. Early patent specifications on platinum filaments heated in a vacuum led him to construct similar lamps with carbon filaments. He failed, however, to solve the problem of all the early bulbs – the filament rapidly oxidized at the high temperatures necessary for incandescence. The obvious solution – to exclude oxygen by using a better vacuum tube – was technically beyond him. So, after about 1860, he put the problem to one side.

He was fortunate in that in 1865 Hermann Sprengel invented an efficient mercury vacuum pump. Swan seems not to have heard of this basic tool until about 1877 but then lost little time in fitting a carbon filament into one of the improved vacuum tubes and in December 1878 he demonstrated his new invention to the Newcastle Chemical Society. He patented his discovery in 1880, forming the Swan Electric Company the following year. The House of Commons tried the new bulbs almost immediately, the British Museum in 1882. His work was an instant success and he stood to make a vast fortune.

The one drawback was that Thomas EDISON had made a similar discovery in America and was intent on developing it exclusively for the use of his own company. Both men eventually came to an agreement and merged to set up a joint company, the Edison and Swan United Electric Light Company. Swan's discovery brought him considerable renown and he was knighted in 1904.

Sydenham, Thomas (1624–1689) *English physician*

Born at Wynford Eagle in Dorset, England, Sydenham studied medicine at Oxford University. Although his education was interrupted by the English Civil War, in which he fought for the Parliamentarians, he returned to Oxford to gain his MB in 1648 and became a fellow of All Souls College. He later moved to London where he spent the rest of his life in private practice.

For his insistence on painstaking observation and careful recording, expressed in his

Observationes medica (1676; Medical Observations), Sydenham became known as the English HIPPOCRATES. He rejected speculative theories in medicine, the justification for this resting heavily on a frequently used analogy between diseases and plants. Sydenham saw similarities between the life cycle of a plant and the course of a disease, declaring that, by careful observation, one could predict the course of a disease in the same way as a botanist can predict the time of flowering, leaf fall, etc. He was one of the first to describe scarlet fever and Sydenham's chorea.

By way of therapy Sydenham followed such traditional practices as bleeding and purging while making full use of such newly introduced drugs as quinine for malaria and the mercury and laudanum of PARACELSUS. He is, however, more accurately represented by his own statement, "I have consulted my patient's safety and my own reputation most effectually by doing nothing at all."

Sylvester, James Joseph (1814–1897) *British mathematician*

> As the prerogative of Natural Science is to cultivate a taste for observation, so that of Mathematics is, almost from the starting point, to stimulate the faculty of invention.
> —*The Collected Mathematical Papers of James Joseph Sylvester* (1904–12)

Born in London, Sylvester studied mathematics at Cambridge but was not granted his BA degree since he was a practicing Jew. The relevant statute was later revoked and Sylvester was granted both his BA and MA in 1871. He was widely read in a number of languages and was a keen amateur musician and a prolific poet. Feeling unable to keep an academic post as a mathematician Sylvester worked first in an insurance company and later as a lawyer. In 1876 he went to America to become the first professor of mathematics at Johns Hopkins University. He became the first editor of the *American Journal of Mathematics* and did much to develop mathematics in America. He returned to England in 1883 to become Savilian Professor of Geometry at Oxford University.

Sylvester's best mathematical work was in the theory of invariants and number theory. With his lifelong friend the British mathematician Arthur CAYLEY, he was one of the creators of the theory of algebraic invariants, which proved to be of great importance for mathematical physics.

Sylvius, Franciscus (1614–1672) *Dutch physician*

Sylvius (**sil**-vee-us) was born at Hannau in Prussia (now in Germany), where his parents had fled from the Dutch struggles for independence from Spain. He Latinized his name from Franz de le Boë and attended the University of Basel where he qualified as a doctor in 1637. He later returned to the (independent) Netherlands and practiced medicine in Amsterdam from 1641 until 1658 when he was appointed professor of medicine at the University of Leiden, a post he held until his death.

Much influenced by PARACELSUS and Jan van HELMONT, he attempted to break away from the traditional theories of disease and tried to formulate a straightforward chemical account. His system, no less simplistic than that of the traditional humoral theory of HIPPOCRATES, began with the fermentation of food into blood. From this process arose acids and alkalis, which could balance each other or lead to an acid- or an alkali-dominated disturbance of the body. Although the acid–alkali theory of disease had a short life, it was important in showing that there were alternatives to classical theory.

Sylvius was also an anatomist and described the division between the temporal and frontal lobes of the brain (the *fissure of Sylvius*). He also described the *aqueduct of Sylvius*, connecting the third and fourth ventricles of the brain while the ordinarily named middle cerebral artery is more accurately and impressively known as *arteria cerebri media Sylvii*.

Synge, Richard Laurence Millington (1914–1994) *British chemist*

Synge was born in Liverpool, where his father was a member of the Liverpool Stock Exchange, and educated at Cambridge University, obtaining his doctorate in 1941. He worked in a number of research institutes, including the Wool Industries Research Association at Leeds (1941–43) and the Lister Institute of Preventive Medicine (1943–48). From 1948 to 1967 he worked as a protein chemist at the Rowett Research Institute near Aberdeen and finally, from 1967 to 1976, at the Food Research Institute, Norwich.

Synge was jointly awarded the 1952 Nobel Prize for chemistry with Archer MARTIN for their development in 1941 of partition chromatography, especially paper chromatography, which was to become one of the basic analytical tools of the revolution in biochemistry and molecular biology. Synge used the method to find the structure of the simple protein gramicidin S, working out the sequence of the different amino acids in the molecule. This work soon proved of great use to Frederick SANGER in his elucidation of the sequence of amino acids in the insulin molecule.

Szent-Györgi, Albert von (1893–1986) *Hungarian–American biochemist*

> A substance that makes you ill if you don't eat it.
> —His definition of a vitamin

Discovery consists of seeing what everybody has seen and thinking what nobody has thought.
—Quoted by I.-J. Good in *The Scientist Speculates*

Szent-Györgi (sent-**jur**-jee), who was born in the Hungarian capital Budapest, studied anatomy at the university there, obtaining his MD in 1917. He continued his studies in Hamburg, Groningen, and at Cambridge University where he received his PhD in 1927. He also spent some time at the Mayo Clinic, Minnesota, before returning to Hungary as professor of medical chemistry at the University of Szeged. In 1947, however, he emigrated to America, becoming director of the Institute for Muscle Research at the Marine Biological Station, Woods Hole, Massachusetts.

Szent-Györgi first became widely known in the late 1920s for his work on the adrenal glands. In the usually fatal condition ADDISON's disease, where the adrenal glands cease to function, one symptom is a brown pigmentation of the skin. Szent-Györgi wondered if there was a connection between this and the browning of certain bruised fruits, which is due to the oxidation of phenolics to quinole. Some fruits, notably citrus, do not turn brown because they contain a substance that inhibits this reaction. Szent-Györgi isolated a substance from adrenal glands, which he named hexuronic acid, that was also present in nonbruising citrus fruits with a high vitamin C content. He suspected he had finally succeeded in isolating the elusive vitamin but was anticipated in announcing his discovery by Charles King, who published his own results two weeks earlier. The main reason for Szent-Györgi's delay was the problem of supply. However, when he began work in Szeged, with its paprika milling industry, he found a rich supply of the vitamin in Hungarian paprika and was soon able to confirm his suppositions and further investigate the action of the vitamin in the body.

Szent-Györgi also studied the uptake of oxygen in isolated muscle tissue and found that he could maintain the rate of uptake by adding any one of the four acids – succinic, fumaric, malic, or oxaloacetic. This work was extended by Hans KREBS and led to the elucidation of the Krebs cycle. For his studies into "biological combustion processes" Szent-Györgi was awarded the 1937 Nobel Prize for physiology or medicine.

Szent-Györgi also became widely known for his studies of the biochemistry of muscular contraction. It was known that the contractile part of muscle was made mainly from the two proteins, actin and myosin. In 1942, in collaboration with Ferenc Straub, Szent-Györgi showed that the two proteins can be encouraged to form fibers of actomyosin which, in the presence of ATP, the cell's energy source, will contract spontaneously. Just how the combining of the two proteins can lead to muscular contraction was illuminated by Hugh HUXLEY.

Szilard, Leo (1898–1964) *Hungarian–American physicist*

If you want to succeed in this world you don't have to be much cleverer than other people, you just have to be one day earlier.
—*Leo Szilard: His Version of the Facts* (1978)

Szilard (**sil**-ard), the son of an architect, studied engineering in his native city of Budapest before moving to the University of Berlin where he began the study of physics and obtained his doctorate in 1922. He remained there until 1933 when, after spending a few years in England working at the Clarendon Laboratory, Oxford, and at St. Bartholomew's Hospital, London, he emigrated to America in 1938. After the war Szilard moved into biology and in 1946 was appointed to the chair of biophysics at the University of Chicago, where he remained until his death. He became a naturalized American in 1943.

Szilard was one of the first men in the world to see the significance of nuclear fission and the first to bring it to the attention of Roosevelt. In 1934, after hearing of the dismissal of the possibility of atomic energy by Ernest RUTHERFORD, he worked out that an element that is split by neutrons and that would emit two neutrons when it absorbed one neutron could, if assembled in sufficiently large mass, sustain a nuclear chain reaction. Szilard applied for a patent, which he assigned to the British Admiralty to preserve secrecy.

When in 1938–39 he heard of the work of Otto HAHN and Lise MEITNER on the fission of uranium he was well prepared. After quickly confirming that the necessary neutrons would be present Szilard, fearing the consequences that would ensue from Hitler's possession of such a weapon, decided that the only sound policy was for America to develop such a weapon first. To this end he approached Albert EINSTEIN, with whom he had worked earlier and who commanded sufficient authority to be heard by all, and invited him to write a letter to the President of the United States. This initiated the program that was to culminate in the dropping of the atomic bomb on Hiroshima six years later. During the war Szilard worked on the development of the bomb and, in particular, worked with Enrico FERMI on the development of the uranium–graphite pile.

If Szilard was one of the first to see the possibility and necessity to develop the bomb he was also one of the earliest to question the wisdom and justice of actually using it against the Japanese. He was the dominant spirit behind the report submitted by James FRANCK to the Secretary of War in 1945 forecasting the nuclear stalemate that would follow a failure to ban the bomb.

T

Tanaka, Koichi (1959–) *Japanese chemist*

Tanaka (**ta**-nah-ka) was born in Toyama City, Japan, and educated at Tohoku University, Japan. In 1983 he joined the Shimadzu Corporation, Japan, where he has remained.

Tanaka announced a method for ionizing protein molecules at a symposium in 1987 and in a paper published in 1988. In this method, known as *soft laser desorption* (SLD), a sample, which is in a solid or viscous liquid phase, is struck by a laser pulse. The resulting ionized protein molecules then have their masses determined by having their time of flight over some fixed distance measured. Like the method of *electrospray ionization* (ESI) developed by John FENN, soft laser desorption is a technique of mass spectrometry that is useful in measuring the masses of large protein molecules. Tanaka was the first person to demonstrate that lasers could be used to help to determine the masses of large molecules of biological significance. There are many important applications of SLD in biology including the early diagnosis of cancer and malaria. Tanaka and Fenn won a share in the 2002 Nobel Prize for chemistry for developing methods of mass spectrometry which are applicable to very large molecules such as proteins.

Takamine, Jokichi (1854–1922) *Japanese–American chemist*

Takamine (tah-kah-**mee**-nee) was born at Takaoka in Japan, the son of a physician. Although brought up along traditional lines, he nevertheless received a modern scientific education at the Tokyo College of Science and Engineering, where he graduated in chemical engineering in 1879. After two years' training at Anderson's College, Glasgow, he returned to Japan in 1883 and entered the government department of agriculture and commerce. In 1887 he left to establish the first factory for the manufacture of superphosphates in Japan.

In 1890, having married an American, he settled permanently in America. He set up a private laboratory and in 1894 produced Takadiastase, a starch-digesting enzyme, which had applications in medicine and the brewing industry.

It had been demonstrated in 1896 that an injection of an extract from the center of the suprarenal (adrenal) gland causes blood pressure to rise rapidly. In 1901 Takamine managed to isolate and purify the substance involved – adrenaline (epinephrine). This was the first isolation and purification of a hormone from a natural source.

Tamm, Igor Yevgenyevich (1895–1971) *Russian physicist*

Born the son of an engineer in Vladivostock, Tamm (tahm) was educated at the universities of Edinburgh and Moscow, where he graduated in 1918. After a short period at the Odessa Polytechnic, Tamm taught at Moscow University (1924–34) then moved to the Physics Institute of the Academy.

In 1958, in collaboration with Ilya FRANK and Pavel CHERENKOV, Tamm was awarded the Nobel Prize for physics for his work in explaining Cherenkov radiation.

Tansley, Sir Arthur George (1871–1955) *British plant ecologist*

Tansley was born in London. Having found his school science teaching "farcically inadequate," he attended lectures at University College, London, where he received his first proper tuition in botany from Francis Oliver. In 1890 he went to Cambridge University and on graduation returned to London as Oliver's assistant.

In the following years Tansley's thinking was greatly influenced by two major books, *Ecological Plant Geography* (1895) by E. WARMING and *The Physiological Basis of Plant Geography* (1898) by Andreas SCHIMPER. These – together with his travels in Ceylon, Malaya, and Egypt – stimulated his interest in different vegetation types.

In 1902 Tansley founded *The New Phytologist*, a journal designed to promote botanical communication and debate in Britain. In 1913 he founded and became the first president of the British Ecological Society and four years later founded and edited the *Journal of Ecology*. These activities, and his ecology courses at Cambridge, played a large part in establishing the science of ecology.

After World War I Tansley turned to psychology and resigned from Cambridge in 1923 to spend time studying under Sigmund FREUD in Austria. In 1927 he became professor of botany at Oxford University, a position held until his retirement in 1937. He continued to exert much influence, however, becoming president of the Council for the Promotion of Field Studies in 1947 and chairman of the Nature Conservancy Council in 1949, both bodies that he had helped to create. Probably his most important book, *The British Islands and Their Vegetation* (1939), was also published after his retirement.

Tarski, Alfred (1902–1983) *Polish–American mathematician and logician*

Tarski served as a professor at the university in his native city of Warsaw (1925–39). In 1942 he joined the staff at the University of California and became professor there in 1949 and research professor at the Miller Institute (1959–60).

Tarski worked on set theory and algebra and is noted as one of the pioneers in the study of formalized logical systems as purely algebraic structures. He emphasized the difference between the metalanguage, used to talk about these structures, and the formal language whose syntax formed the system being studied. His famous paper *The Concept of Truth in Formalized Languages* (1935) was one of the foundation stones of model theory and has had a profound influence both in logic and the philosophy of language.

Tartaglia, Niccoló (1500–1557) *Italian mathematician, topographer, and military scientist*

> I ... continued to labor by myself over the works of dead men, accompanied only by the daughter of poverty that is called industry.
> —On teaching himself mathematics. *Quesiti et inventioni diverse* (1546; Various Inquiries and Discoveries)

Tartaglia (tar-**tah**-lya) was born Niccoló Fontana but as a boy he suffered a saber wound to his face during the French sack of Brescia (1512), his native city; this left him with a speech defect and he adopted the nickname Tartaglia (Stammerer) as a result. Tartaglia began his studies as a promising mathematician, but his interests soon became very wide-ranging. He held various posts, including school teacher, before he eventually became a professor of mathematics in Venice where he stayed.

Tartaglia is remembered chiefly for his work on solving the general cubic equation. He discovered a method in 1535 but did not publish it. Incautiously he revealed his new method to his friend the mathematician Girolamo CARDANO, who published it in *Ars magna* (1545; The Great Skill). This, not surprisingly, was the end of their friendship and led to a violent contro-

versy. Tartaglia eventually lost the quarrel and with it his post as lecturer at Brescia in 1548.

Tartaglia's other chief mathematical interests were in arithmetic and geometry. The pattern now known as "Pascal's triangle" appeared in a work of Tartaglia's. His geometrical work centered on problems connected with the tetrahedron and he helped further the diffusion of classical mathematics by making the first translation of EUCLID's *Elements* into a modern European language. His chief published work was the three-volume *Trattato di numeri et misure* (1556–60; Treatise on Numbers and Measures), an encyclopedic work on elementary mathematics. Apart from these mathematical activities Tartaglia made notable innovations in topography and the military uses of science, such as ballistics.

Tatum, Edward Lawrie (1909–1975) *American biochemist*

Tatum, who was born in Boulder, Colorado, studied chemistry at the University of Wisconsin, where his father was professor of pharmacology. He obtained the BA degree in 1931, then undertook research in microbiology for his master's degree, conferred the following year. His PhD was more biochemically oriented and after receiving his doctorate he worked as a research assistant in biochemistry for a year. He studied bacteriological chemistry at Utrecht University from 1936 to 1937 and on returning to America was appointed research associate at Stanford University.

His early experiments at Stanford concentrated on the nutritional requirements of the fruit fly, *Drosophila melanogaster*, but in 1940, in collaboration with George BEADLE, he began working on the pink bread mold, *Neurospora crassa*. They irradiated the mold with x-rays to induce mutations and were then able to isolate a number of lines with different nutritional deficiencies. These lines needed special supplements to the basic growth medium to enable growth to continue as normal. When a mutant mold was crossed with the normal wild-type mold, the dietary deficiency was inherited in accordance with expected Mendelian ratios. Such studies established that genes act by regulating specific chemical processes. During World War II this work was of use in maximizing penicillin production, and it has also made possible the introduction of new methods for assaying vitamins and amino acids in foods and tissues.

In 1945 Tatum moved to Yale University where he extended his techniques to yeast and bacteria. Through studying nutritional mutations of the bacterium *Escherichia coli*, he and Joshua LEDERBERG were able to demonstrate, in 1946, that bacteria can reproduce sexually. Following this work, bacteria have become the primary source of information on the genetic control of biochemical processes in the cell.

Tatum returned to Stanford in 1948 and in 1957 joined the Rockefeller Institute for Medical Research. In 1958, together with Beadle and Lederberg, he received the Nobel Prize for physiology or medicine in recognition of the work that helped create the modern science of biochemical genetics.

Taube, Henry (1915–2005) American inorganic chemist

Taube, who was born in Saskatchewan, Canada, moved to America in 1937 and became naturalized in 1942. He was educated at the University of Saskatchewan and the Berkeley campus of the University of California, where he gained his PhD in 1940. After working at Cornell University (1941–46), Taube moved to the University of Chicago and in 1952 was appointed professor of chemistry, a post he held until 1962 when he accepted a comparable appointment at Stanford University.

As a leading inorganic chemist Taube succeeded in developing a range of experimental techniques for studying the kinetics and mechanism of inorganic reactions, in particular electron-transfer reactions. Transition metals such as iron, copper, cobalt, and molybdenum form coordination compounds of a type first described by Alfred WERNER. In a typical coordination compound a metal ion is attached to a number of ligands, such as water or ammonia. It was thought that the ligands would keep the ions apart and inhibit electron transfer between ions. Taube showed experimentally that ligand bridges form between interacting complexes, thus allowing electrons to be transferred.

For his work in this field Taube was awarded the 1983 Nobel Prize for chemistry. He maintained a lifelong interest in redox reactions in biological processes and also researched photochemistry, being among the first to use isotopes to investigate metabolic pathways.

Taylor, Brook (1685–1731) British mathematician

Born in Edmonton, near London, Taylor studied at Cambridge University, and was secretary to the Royal Society during the period 1714–18. He made important contributions to the development of the differential calculus in his *Methodus incrementorum directa et inversa* (1715; Direct and Indirect Methods of Incrementation). This contained the formula known as *Taylor's theorem*, which was recognized by Joseph LAGRANGE in 1772 as being the principle of differential calculus. The *Methodus* also contributed to the calculus of finite differences, which Taylor applied to the mathematical theory of vibrating strings.

Outside mathematics Taylor was an accomplished artist and this led him to an interest in the theory of perspective, publishing his work on this subject in *Linear Perspective* (1715). *Taylor expansions* are named for him.

Taylor, Sir Geoffrey Ingram (1886–1975) British physicist

> Many of his scientific contributions opened up whole new fields; he had the knack of being first.
> —Obituary notice in *The Times*, 30 June 1975

Taylor, a Londoner by birth, was educated at Cambridge University where, apart from absence on service in two wars, he spent the whole of his career. He served initially in 1911 as reader in dynamic meteorology and from 1923 until his retirement in 1952 as research professor in physics.

Taylor was an original researcher and a prolific author, publishing on a wide variety of topics. Studies by him over many years into fluid turbulence have yielded applications in fields as varied as meteorology, aerodynamics, and Jupiter's Great Red Spot. One of his most important ideas, that of dislocation in crystals, was first formulated in 1934. By introducing the idea of an edge dislocation in which one layer of atoms is slightly displaced relative to a neighboring layer he was able to explain much about the properties of metals. The same proposal was made independently by Michael POLANYI and E. Orowan.

Taylor, Sir Hugh (Stott) (1890–1974) British chemist

Taylor, the son of a glass technologist, was born in St. Helens, in northwest England. He studied chemistry at the University of Liverpool (1906–12), where he obtained a DSc in 1914, and in Stockholm under Svante ARRHENIUS. In 1914 he moved to Princeton University, remaining there for the whole of his career and serving as professor of chemistry from 1922 until his retirement in 1958.

At the beginning of his career Taylor worked on problems of catalysis; during World War I he concentrated on the catalytic synthesis of ammonia from nitrogen and hydrogen (the Haber process). With Eric RIDEAL, he produced the comprehensive work *Catalysis in Theory and Practice* (1919). In 1925, in his paper *A Theory of the Solid Catalytic Surface* (1925), he suggested that catalyst surfaces could not be homogeneous, i.e., that only a fraction of the surface of the catalyst was important and that the catalyst, in effect, had "active centers."

During World War II Taylor worked on the Manhattan Project at the Columbia Nuclear Laboratory to develop the gaseous diffusion process introduced by John DUNNING for the enrichment of uranium.

Taylor, Joseph Hooton (1941–   ) American astrophysicist

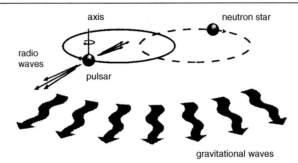

BINARY PULSAR   The pulsar discovered by J. H. Taylor and Russell Hulse has a companion, which it orbits approximately every eight hours.

Born in Philadelphia, Taylor was educated at Haverford College, Pennsylvania, and at Harvard, where he gained his PhD in astronomy in 1968. He moved to the University of Massachusetts, Amherst, in 1969 and was appointed professor of astronomy in 1977, a post he held until 1980 when he was elected professor of physics at Princeton. He retired in 2006.

In 1974 Russell HULSE, a research student of Taylor, while working at the Arecibo Radio Telescope in Puerto Rico, discovered a binary pulsar. The pulsar orbited its invisible companion with a period of 7.75 hours, and rotated about its axis every 0.05903 seconds. Taylor and Hulse continued to observe the pulsar and to establish the details of its orbital behavior as precisely as possible.

Taylor also saw that the pulsar could provide an important observational test of EINSTEIN's theory of general relativity. In 1916 Einstein had argued that an accelerating mass should radiate energy in the form of gravitational waves. Any such energy radiated by Hulse's pulsar, 16,000 light years away, would be so weak by the time it reached Earth as to be undetectable. In fact, so far no direct reproducible evidence has been obtained for the existence of gravitational waves, despite the experiments of Joseph WEBER carried out since the 1960s.

Taylor realized there was another way for the gravitational waves to be detected. Any system radiating gravitational waves will be losing energy. This loss of energy will cause the pulsar and its companion to approach closer to each other and a consequent decrease in the pulsar's orbital period. The orbital shrinkage would amount to only 3.5 meters a year, too small to be detected; a decrease of 75 millionth of a second per year in the orbital period, however, should be detectable. After four years careful observation and analysis Taylor announced in 1978 that he had detected just such a decrease in the orbital period. "Hence 66 years after Einstein predicted the existence of gravitational waves," Taylor concluded, "an experiment has been done that yields clear evidence for their existence." For his discovery of this evidence Taylor shared the 1993 Nobel Prize for physics with Hulse.

Taylor, Richard E. (1929–   ) *Canadian physicist*

Taylor, who was born in Medicine Hat, Alberta, was educated at the University of Alberta and at Stanford University, California, where he obtained his PhD in 1962. After two years at the University of California, Berkeley, he returned to Stanford in 1964 and was appointed to the chair of physics in 1970, becoming an emeritus professor in 2003.

In 1990 Taylor shared the Nobel Prize for physics with Jerome FRIEDMAN and Henry KENDALL for their discovery in 1967 that the proton has an inner structure. Their work presented the first convincing experimental evidence for the existence of quarks.

Taylor, Stuart (1937–   ) *American physiologist*

The son of a physician, Taylor was educated at Cornell, Columbia, and New York University where he gained his PhD in 1966. From 1970 he has worked at the Mayo Medical School, Minnesota. Since the early 1990s he has also served as Distinguished Professor at Hunter College, New York.

During the mid-1950s physiologists began at last to understand in some detail the manner in which muscles contract. One of the leading workers in this field is Hugh HUXLEY, whom Taylor heard lecturing at Columbia in 1964. Taylor became so interested in the subject that he served a two-year postdoctoral fellowship in Huxley's laboratory at University College, London.

Teisserenc de Bort, Léon Philippe (1855–1913) *French meteorologist*

Teisserenc de Bort worked as chief meteorologist for the Central Meteorological Bureau in his native city of Paris from 1892 until 1896, when he opened his own meteorological observatory at Trappes, near Versailles.

In 1902 he put forward his conclusions on the atmosphere. Early balloonists had established that temperature decreased with height by about 6°C per 330 feet (100 m). Using unmanned instrumented balloons Teisserenc de Bort found that above an altitude of 7 miles (11 km) temperature ceased to fall and sometimes increased slightly. He named this upper part of the atmosphere the stratosphere, because he thought that the different gases would lie in distinct strata as, without temperature differentials, there would be no mechanism to disturb them. The lower part of the atmosphere he named the troposphere (Greek: sphere of change) as here, with abundant temperature differentials, constant change and mingling of atmospheric gases occurred.

The phenomenon was explained by E. Gold in 1909 by reference to cooling of rising air in the troposphere and the absence of convection currents in the stratosphere.

Telford, Thomas (1757–1834) *British architect and engineer*

Telford was brought up in rural poverty in Westerkirk, Dumfriesshire, Scotland. When he was 15 years old and working as an apprentice stonemason, he was allowed access to the library of a local lady. He moved to Edinburgh in 1780 then to London in 1782 and in 1786 was appointed surveyor of Shropshire, where his work included the construction of bridges and buildings. Three bridges over the River Severn gained him notice and as a result he was appointed (1793) to plan and construct the Ellesmere Canal (1793–1805). His two great aqueducts, built to carry the canal, brought him national recognition. Many of his projects were noted for their architectural beauty.

In 1803 Telford presented a report on opening up access to the Scottish Highlands. The plan included extensive road and bridge building as well as a continuous channel from the North Sea to the Atlantic for warships and commercial ships – the Caledonian Canal. In 18 years 900 miles (1,450 km) of road and 120 bridges were built. He designed a system of locks for the Caledonian Canal to cope with a 90-foot (27-m) difference in water level.

Telford's other projects included the Göta Canal in Sweden, for which he received a Swedish knighthood, but his greatest achievement was the Menai Bridge between Wales and the Island of Anglesey. This was the first really successful suspension bridge in Britain. It took ten years to build and is still in use. He used wrought-iron chains, which were towed across the Menai Strait and hoisted, and from which the deck of the bridge was suspended.

Telford was the first president of the Institution of Civil Engineers, London.

Teller, Edward (1908–2003) *Hungarian–American physicist*

> Two paradoxes are better than one; they may even suggest a solution.
> —Attributed remark

The son of a lawyer, Teller was born in the Hungarian capital Budapest. Having attended the Institute of Technology there, he continued his education in Germany at the universities of Karlsruhe, Munich, Leipzig (where he obtained his PhD in 1930), and Göttingen. He left Germany when Hitler came to power in 1933. After a short period in Denmark and London, he emigrated to America where in 1935 he took up an appointment as professor of physics at George Washington University. During the war he worked in the theoretical physics division at Los Alamos on the development of the atom bomb, resuming his academic career in 1946 at the University of Chicago. Teller moved in 1953 to the University of California at Berkeley, where he remained until his retirement in 1975.

Teller is often referred to as the "father of the hydrogen bomb." Insofar as he made the initial proposal in 1942, worked longest on the project, and campaigned most vigorously for its completion, the description is accurate. If, however, the phrase is taken to mean that the bomb exploded in 1951 was Teller's own design, the description is misleading.

From the mid-1940s Teller was working with three possible bomb designs, A, B, and C. By 1947 it was clear that B would fail and that C, though viable, would produce too small an explosion to be worthwhile. Research effort was therefore concentrated on design A, which required horrendous amounts of calculation; these were carried out by the mathematician Stanislaw ULAM. In 1950 President Truman, following the Russian atomic bomb explosion, ordered the program to be speeded up. At this point Ulam revealed that option A was hopeless, requiring such massive amounts of tritium as to be virtually unworkable. At this point Teller found himself, after nearly a decade of intensive and costly research, without any viable design or idea in response to the President's urgent call.

By early 1951 Ulam and Teller had worked out a fourth and effective method, the origin of which remains a matter of dispute. The basic problem of the early designs was that the bomb would fly apart under the explosion of the fission device before the fusion material was sufficiently compressed. Ulam's solution was to use x-rays from the fission device to produce compression waves long before any shock waves could strangle the planned thermonuclear reaction. While Hans BETHE attributes the idea to

Ulam, Teller claimed the credit for himself, insisting that "Ulam triggered nothing."

After the successful explosion of the first thermonuclear device in 1951 Teller moved to the Livermore laboratories, which were performing research for the Atomic Energy Commission. Shortly afterward, loyalty hearings were held against OPPENHEIMER, who had been director of the Los Alamos laboratory while the atom bomb was developed. Teller, despite much pressure from the scientific community, decided to testify against him and although he did not accuse Oppenheimer of disloyalty, still less of treason, he did complain that "his actions frankly appeared to me confused and complicated ... I would personally feel more secure if public matters could rest in other hands." In the charged atmosphere of the 1950s this contributed to the withdrawal of Oppenheimer's security status and to the ostracism of Teller by many of his old friends and colleagues. For ten years Teller was neither invited to nor visited Los Alamos.

Teller continued to be a powerful figure in American science after his retirement from Berkeley in 1975. In 1984, for example, he advised Washington that the influential paper of Carl SAGAN and his colleagues on the dangers of a nuclear winter was far from convincing. At Livermore he worked on and campaigned vigorously for the Strategic Defense Initiative (SDI), better known as "Star Wars." The key ingredient of SDI was the x-ray laser, which would supposedly destroy enemy missiles in space. Teller had to face charges against the program from Hans Bethe that it was "unwieldy, costly, easily countered, destabilizing, and uncertain." Despite these and other charges, and despite a tendency for supporters of the SDI to promise more than they could ever hope to deliver, Teller's support for the project never wavered. He published a forthright defense of the program in his book *Better a Shield than a Sword* (1987). However, in 1993 the project was abandoned amid accusations that test results had been distorted. The SDI department was renamed the Ballistic Missile Defense Organization and now researches ways of eliminating enemy missiles nearer the ground. In 2001 he published his *Memoirs*.

Temin, Howard Martin (1934–1994)
American molecular biologist

Born in Philadelphia, Pennsylvania, Temin studied biology at Swarthmore College and at the California Institute of Technology, where he obtained his PhD in animal virology in 1959. He worked at the University of Wisconsin from 1960 onward, serving as professor of oncology there from 1969 until his death from lung cancer.

There are two classes of viruses, those with DNA and those with RNA genes. The former replicate by transforming their DNA into new DNA and transmit information from DNA through RNA into protein. The latter class of viruses replicate RNA into RNA and transmit information directly into protein without the need for DNA. That is, all such reactions fitted into the general sequence DNA to RNA to protein, the so-called Central Dogma of molecular biology. In the early 1960s Temin discovered a curious feature of the RNA ROUS chicken sarcoma virus (RSV): he found that it would not grow in the presence of the antibiotic actinomycin D, a drug known to inhibit DNA synthesis. Temin realized that this might mean the RSV replicated through a DNA intermediate, which he called the provirus. That is, Temin was proposing the sequence RNA (of the RSV) to DNA (provirus) to RNA (replicated RSV) which, while not actually excluded by the Central Dogma, was not implied by it either.

If such a reaction did take place then it would certainly require the presence of an enzyme capable of transcribing RNA into DNA. It was not until 1970 that Temin identified the enzyme (discovered independently by David BALTIMORE) known variously as reverse transcriptase or RNA-directed DNA polymerase. It was for this work that Temin shared the 1975 Nobel Prize for physiology or medicine with Baltimore and Renato DULBECCO.

Tennant, Smithson (1761–1815) *British chemist*

Tennant was born in Richmond, England, the son of a vicar. After studying medicine at Edinburgh University, where he attended the lectures of Joseph BLACK, he went on to study chemistry and botany at Cambridge University (1782). He received his medical degree in 1796 and was professor of chemistry at Cambridge from 1813 until his death.

Tennant was one of the first professional chemists to be seriously interested in chemistry's application to farming. He purchased land in Somerset where he farmed and showed, among other things, that lime from many parts of England contains magnesium and is positively harmful to crops. In pure chemistry he showed that diamond is a form of carbon (1797). While working with platinum minerals dissolved in aqua regia (1803–04) he discovered two new elements. The first he named iridium from the Greek for rainbow to mark the many colors its compounds form and the second he called osmium from the Greek for malodorous.

He died while on a visit to France when a drawbridge he was crossing on horseback broke, resulting in his drowning.

Tesla, Nikola (1856–1943) *Croatian–American physicist*

It has been argued that the perfection of guns

of great destructive power will stop warfare... On the contrary, I think that every new arm that is invented, every new departure that is made in this direction, merely ... gives a fresh impetus to further development.
—Quoted by H. Bruce Franklin in *War Stars: The Superweapon and the American Imagination* (1988)

Tesla (**tes**-la) was born at Smiljan in Croatia, at that time within the Austro-Hungarian empire. He studied mathematics and physics at the University of Graz and philosophy at Prague. In 1884 he emigrated to America where he worked for Thomas EDISON for a while before a bitter quarrel led to his resigning and joining the Westinghouse Company. After his invention of the first alternating current (a.c.) motor in 1887 he left to set up his own research laboratory.

Most commercially generated electricity at that time (including that of Edison) was direct current (d.c.). Tesla saw some fundamental weaknesses in the d.c. system: it required a commutator and needed costly maintenance. The main advantage of a.c. was that, with transformers, it was easier and cheaper to transmit very high voltages over long distances. Tesla's invention was soon taken up by Westinghouse and led to intense competition with Edison and the other d.c. users. Edison was not beyond suggesting that a.c. was inherently dangerous and when in 1889 the first criminal was electrocuted, Edison proposed that being "westinghoused" would be a good term to describe death by the electric chair.

In 1891 the transformer was first demonstrated at the Frankfurt fair when it was shown that 25,000 volts (alternating) could be transmitted for 109 miles (175 km) with an efficiency of 77%. The a.c. system soon replaced d.c. electricity, which was confined to specialized uses.

Thales (*c.* 625 BC–*c.* 547 BC) *Greek philosopher, geometer, and astronomer*

> Water is the principle, or the element, of things. All things are water.
> —Quoted by Plutarch in *Placita Philosophorum*

> To Thales, the primary question was not what do we know, but how do we know it.
> —Aristotle

It is with Thales (**thay**-leez) that physics, geometry, astronomy, and philosophy have long been thought to begin. However, little is known of the first supposed identifiable "scientist" apart from the fact that he was born at Miletus, now in Turkey, and a number of anecdotes that clearly originate in folklore.

Thus he is traditionally supposed to have acquired his learning from Egypt, an implausible claim when the modest mathematical skills of sixth-century Egypt are contrasted with the supposed achievements of Thales. To him is even attributed a proof of the proposition that the circle is divided into two equal parts by its

diameter, a theorem not to be found in EUCLID some 300 years later. It is also reported by the historian Herodotus that Thales gave a successful prediction of a solar eclipse in 585 BC.

The remaining claim for Thales rests on his introduction of naturalistic explanations of physical phenomena in opposition to the customary understanding of nature in terms of the behavior of the gods. Hence the importance of his claim that everything is water, perhaps the first recorded general physical principle in history.

Theiler, Max (1899–1972) *South African–American virologist*

Theiler, the son of a physician from Pretoria in South Africa, was educated at the University of Cape Town; he received his MD in 1922 after attending St. Thomas's Hospital, London, and the London School of Tropical Medicine. The same year he left for America to take up a post at the Harvard Medical School. In 1930 Theiler moved to the Rockefeller Foundation in New York, where he later became director of the Virus Laboratory and where he spent the rest of his career.

When Theiler began at Harvard it was still a matter of controversy whether yellow fever was a viral infection, as Walter REED had claimed in 1901, or whether it was due to *Leptospira icteroides*, the bacillus discovered by Hideyo NOGUCHI in 1919. Theiler's first contribution was to reject the latter claim by showing that *L. icteroides* is responsible for WEIL's disease, an unrelated jaundice.

Little can normally be done in the development of a vaccine without an experimental animal in which the disease can be studied and in which the virus can spread. The breakthrough here came in 1927 when Adrian STOKES found that yellow fever could be induced in Rhesus monkeys from India. Within a year both Stokes and Noguchi had died from yellow fever and did not witness Theiler's next major advance. As monkeys tend to be expensive and difficult to handle, researchers much prefer to work with such animals as mice or guinea pigs. Attempts to infect mice had all failed when Theiler tried injecting the virus directly into their brains. Although the animals failed to develop yellow fever they did die of massive inflammation of the brain (encephalitis). In the course of this work Theiler himself contracted yellow fever but fortunately survived and developed immunity.

Although, he reported in 1930, the virus caused encephalitis when passed from mouse to mouse, if it was once more injected into the monkey it revealed itself still to be functioning and producing yellow fever. Yet there had been one crucial change: the virus had been attenuated and while it did indeed produce yellow fever it did so in a mild form and, equally important, endowed on the monkey immunity from a later attack of the normal lethal variety. All was thus set for Theiler to develop a vaccine against the disease. It was not however until 1937, after the particularly virulent Asibi strain from West Africa had passed through more than a hundred subcultures, that Theiler and his colleague Hugh Smith announced the development of the so-called 17-D vaccine. Between 1940 and 1947 Rockefeller produced more than 28 million doses of the vaccine and finally eliminated yellow fever as a major disease of man. For this work Theiler received the 1951 Nobel Prize for physiology or medicine.

Thénard, Louis-Jacques (1777–1857)
French chemist

Thénard (tay-**nar**) was the son of a peasant from Louptière in France. He studied pharmacy in Paris in poverty until he was befriended by Louis VAUQUELIN. He became assistant professor at the Ecole Polytechnique (1798) and succeeded Vauquelin as professor of chemistry at the Collège de France (1802). He also held a chair at the University of Paris where he became chancellor in 1832. In 1797 he made his fortune by his discovery of Thénard blue, a pigment that consisted of a fusion of cobalt oxide and alumina capable of withstanding the heat of furnaces used in porcelain production.

Thénard collaborated with his lifelong friend Joseph GAY-LUSSAC on studies of the alkali metals sodium and potassium, which they obtained by heating potash and soda with iron. However, it was not immediately clear whether these were indeed elements. Humphry DAVY also discovered them in 1807 and competed with the French chemists to work out their properties. Thénard and Gay-Lussac heated potassium with ammonia and found more hydrogen liberated than could have come from the ammonia. They therefore concluded that potassium was a hydride but Davy showed that they had taken moist ammonia, the water in which could account for the excess hydrogen.

Working on his own Thénard discovered hydrogen peroxide (1818). He also produced a four-volume standard textbook, *Traité elémentaire de chimie* (Elementary Treatise on Chemistry), which remained a standard text for 25 years and by 1834 had gone into its sixth edition.

Theophrastus (c. 372 BC–c. 287 BC) *Greek botanist and philosopher*

> It is manifest that Art imitates Nature and sometimes produces very peculiar things.
> —*History of Stones*

Theophrastus (thee-oh-**fras**-tus), who was born at Eresus on Lesbos (now in Greece), attended the Academy at Athens as a pupil of PLATO.

After Plato's death he joined ARISTOTLE and became his chief assistant when Aristotle founded the Lyceum at Athens. On Aristotle's retirement Theophrastus became head of the school. The school flourished under him and is said to have numbered two thousand pupils at this time.

Of Theophrastus's many works, his nine-volume *Enquiry into Plants* is considered the most important. This is a systematically arranged treatise that discusses the description and classification of plants and contains many personal observations. A second series of six books, the *Etiology of Plants*, covers plant physiology. Theophrastus appreciated the connection between flowers and fruits and, from his description of germination, it is seen that he realized the difference between monocotyledons and dicotyledons.

Many of Theophrastus's pupils lived in distant regions of Greece and he encouraged them to make botanical observations near their homes. This practice probably helped him to conclude that plant distribution depends on soil and climate. Theophrastus was the first to invent and use botanical terms and is often called "the father of scientific botany."

Theorell, Axel Hugo Teodor (1903–1982)
Swedish biochemist

Theorell (**tay**-oh-rel), who was born at Linköping in Sweden, received his MD from the Karolinska Institute in Stockholm in 1930. However, he did not pursue a career in medicine because of a polio attack. Instead, he became assistant professor of biochemistry at Uppsala University (1932–33, 1935–36), spending the intervening years with Otto WARBURG at the Kaiser Wilhelm Institute in Berlin.

Theorell found that the sugar-converting (yellow) enzyme isolated from yeast by Warburg consisted of two parts: a nonprotein enzyme (of vitamin B_2 plus a phosphate group) and the protein apoenzyme. He went further to show that the coenzyme oxidizes glucose by removing a hydrogen atom, which attaches at a specific point on the vitamin molecule. This was the first detailed account of enzyme action.

Theorell studied cytochrome c (important in the electron-transport chain) and was the first to isolate crystalline myoglobin. His research on alcohol dehydrogenase resulted in the development of blood tests that may be used to determine alcohol levels.

In 1937 Theorell became director of the biochemistry department of the Nobel Medical Institute, Stockholm, and in 1955 received the Nobel Prize for physiology or medicine.

Thiele, Friedrich Karl Johannes (1865–1918) *German chemist*

Born at Ratibor (now Racibórz in Poland),

Thiele (**tee**-le) studied mathematics at the University of Breslau but later turned to chemistry, receiving his doctorate from Halle in 1890. He taught at the University of Munich from 1893 to 1902, when he was appointed professor of chemistry at Strasbourg.

In 1899 Thiele proposed the idea of partial valence to deal with a problem that had been troubling theoretical chemists for some time. The structures produced by August KEKULÉ in 1858 and 1865 had revolutionized chemical thought but had also produced major problems. Double bonds in chemistry usually indicate reactivity but Kekulé's proposed structure for benzene, C_6H_6, contained three double bonds in its ring and yet benzene is comparatively unreactive. Thus it tends to undergo substitution reactions rather than addition reactions.

Thiele proposed that, when double and single bonds alternate, a pair of single bonds affects the intervening double bond in such a way as to give it some of the properties of a single bond. Given the ring structure of benzene, this occurs throughout the molecule and so neutralizes the activity of the double bonds. The same argument cannot be used with double bonds in a carbon chain for there the ends of the chain will be open to addition. Thiele's problem could not be completely solved until the development of quantum theory. Thiele's ideas are similar to the later concept of resonance structures – intermediate forms of molecules with bonding part way between conventional forms.

Thom, Alexander (1894–1985) *British engineer and archeoastronomer*

Thom (tom) studied at the University of Glasgow where he obtained his BSc in 1915 and his PhD in 1926. He lectured in engineering at Glasgow from 1922 until 1939 when he joined the Royal Aircraft Establishment at Farnborough. After the war he was appointed professor of engineering science at Oxford, a post he held until his retirement in 1961.

From 1934 Thom was engaged upon an investigation of the stone circles distributed so profusely throughout the British Isles and Brittany. He published two major works in this field after his retirement: *Megalithic Sites in Britain* (1967) and *Megalithic Lunar Observatories* (1971). The investigation of ancient remains for astronomical significance really begins with Norman LOCKYER at the turn of the century. Thom's preeminence in this field was based upon his personal knowledge of hundreds of sites and his mathematical and surveying skills.

The twenty years he had spent making accurate surveys of stone circles allowed him to begin publishing in the 1950s some of his startling conclusions. That the stone circles clearly presupposed a considerable knowledge of the 19-year lunar cycle was not particularly origi-

nal, although Thom produced an unprecedented amount of hard data in its favor. More original was his claim to have discovered certain widespread units of measurement used in the construction of the monuments.

These consist of a "megalithic yard" of 2.72 feet or 0.829 meter and a "megalithic inch," 0.816 of a modern inch, 40 of which made a megalithic yard. The evidence for the existence of such units is provided by a formidable statistical analysis of the rings themselves, which are not actually circles but more complicated geometrical figures with dimensions that are multiples of megalithic yards.

Although Thom's conclusions are not acceptable to all scholars and have been much exaggerated by others, there is no doubt of the importance and rigor of his work, which has raised many still unsolved problems.

Thomas, Edward Donnall (1920–   ) *American physician*

Thomas was born at Mart in Texas and educated at the University of Texas, receiving his BA in 1941 and MA two years later. In 1946 he was awarded his MD by Harvard University. He worked at the Peter Bent Brigham Hospital, Boston, from 1946, eventually specializing in hematology. After two years as a research associate with the Cancer Research Foundation at the Children's Medical Center, Boston (1953–55), he moved to the Mary Imogene Bassett Hospital as hematologist and assistant physician. In 1956 he became associate clinical professor of medicine at the College of Physicians and Surgeons, Columbia University, and from 1963 until 1990 he served with the University of Washington Medical School in Seattle as professor of medicine.

Although primarily a clinician rather than a research scientist, Thomas was instrumental in gaining new insights into how the body's immune system rejects tissue transplants. His expertise in hematology and cancer biology enabled him to develop the technique of bone-marrow transplantation to treat patients suffering from leukemia or other cancers of the blood. This involves the transfer of bone-marrow cells from a healthy donor to the bone marrow of the patient, so that the patient can resume production of healthy white blood cells to replace the cancerous cells.

In experiments using dogs, Thomas demonstrated the importance of matching donor tissue and the recipient's tissue as closely as possible, so as to minimize the risk of rejection of the transplanted tissue. He also showed that by treating the recipient with cell-killing drugs, it was possible to avoid another of the pitfalls associated with tissue transplantation, namely, the graft-versus-host reaction. This occurs when immune cells belonging to the donor and transferred with the transplant recognize the host's tissues as foreign and start to attack them.

Under Thomas's leadership the University of Washington Medical School became the preeminent North American center for bone-marrow transplants, where physicians from all over the world came to learn the pioneering technique. For his work Thomas was awarded the 1990 Nobel Prize for physiology or medicine, jointly with Joseph MURRAY.

Thomas, Sidney Gilchrist (1850–1885) *British chemist*

Thomas came from a nonconformist background in London. Owing to the death of his father in 1866 he was unable to complete his education and instead became a clerk in a police court. In 1870 he started studying in the evenings at Birkbeck College, London, and passed exams in metallurgy and chemistry at the Royal College of Mines.

Thomas heard of the need to eliminate phosphorus from the steelmaking process introduced by Henry BESSEMER in 1856. If iron ores containing phosphorus were used the resulting steel was brittle and useless but large reserves of phosphorus-rich ores existed in France and Germany, and also in Britain, in Wales and Cleveland. Thomas began experimenting with various substances, including magnesium and lime. He made trial tests using a converter at the Blaenafon steel plant with the assistance of his cousin, Percy Carlyle GILCHRIST, who was a chemist there.

In 1875 Thomas developed a lining for the Bessemer converter made of calcined dolomite with a lime binding, which eliminated the phosphorus and also produced a valuable fertilizer – basic slag – as a by-product. Demonstrated in 1879, the *Thomas–Gilchrist basic process* was an immediate success. Thomas resigned his job to manage the issue of licenses from his patents. He was awarded the Bessemer gold medal in 1883 and made a fortune but died young from tuberculosis. His money, inherited by his sister, was spent on the philanthropic works he was keen to promote. The new process made steel cheap and abundant; in 1879 world production of steel was about 4 million tons – by 1929 it was over 104 million tons, 90% of which was produced by the basic process.

Thompson, Sir D'Arcy Wentworth (1860–1948) *British biologist*

> It behoves us always to remember that in physics it has taken great men to discover simple things. They are very great names indeed which we couple with the explanation of the path of a stone, the droop of a chain, the tints of a bubble, the shadows in a cup.
> —*On Growth and Form* (1917)

Thompson studied medicine at the university in his native city of Edinburgh, where he was greatly influenced by Charles THOMSON, who had recently returned from the Challenger Expedition. In 1884 he became professor of biology (subsequently natural history) at University College, Dundee. In 1917, when he became senior professor, Thompson published *On Growth and Form*, in which he developed the notion of evolutionary changes in animal form in terms of physical forces acting upon the individual during its lifetime, rather than as the sum total of modifications made over successive generations – the latter being the traditional credo postulated by Darwinists. In a later edition (1942), however, Thompson admitted the difficulty of explaining away the cumulative effect of physical and mental adaptations, which can scarcely be accounted for in the experience of one generation. In addition to such theoretical work, Thompson was much involved in oceanographic studies, as well as fisheries and fur-seal conservation in northern Europe. He was one of the British representatives on the International Council for the Exploration of the Sea, from its foundation in 1902. He was also interested in classical science, publishing works on the natural history of ancient writers, including an edition of ARISTOTLE's *Historia Animalium* (1910; History of Animals) and accounts of Greek birds and fishes.

Thompson, Sir Harold Warris (1908–1989) *British physical chemist*

Thompson, who was born at Wombwell in Yorkshire, England, was educated at the universities of Oxford and Berlin, where he obtained his PhD in 1930. He spent his entire career at Oxford, serving as professor of chemistry from 1964 until his retirement in 1975.

His early work was on chemical kinetics and photochemistry, later concentrating on absorption spectroscopy with ultraviolet radiation. In the 1930s he began to study absorption spectra taken in the infrared region. Absorption at a particular wavelength in the infrared corresponds to a change in vibrational energy of the molecule, and Thompson was able to use infrared spectra to measure the vibrational frequencies of bonds and the shapes and sizes of molecules.

During World War II he started to develop infrared spectroscopy as an analytical tool. Various bonds and groups in molecules have characteristic vibrational frequencies, and Thompson showed how analysis of the spectra could be used to work out the structures of complex organic molecules. Infrared spectroscopy is now a routine technique in organic chemistry for investigating and "fingerprinting" compounds.

Thompson also made significant advances in both the instrumentation used and in the theory of intensities of infrared absorption lines. He used infrared spectroscopy in studies of intermolecular forces and also investigated rotational spectra in the far infrared (at low frequencies).

Thomsen, Christian Jürgensen (1788–1865) *Danish archeologist*

> Mr. Thomsen is admittedly only a dilettante, but a dilettante with a wide range of knowledge. He has no university degree, but in the present state of scientific knowledge I hardly consider that fact as being a disqualification.
> —A member of the Royal Commission for the Preservation of Danish Antiquities, recommending Thomsen's employment as curator of the National Museum (1816)

Thomsen (**tom**-sen), the son of a Copenhagen merchant, was educated privately. He worked in the family business until 1840 and also, from 1816 until his death, served as curator of the Danish National Museum, first opened in 1819.

In a guide for the museum, *Ledetraad til nordisk Oldkyndighed* (A Guide to Northern Antiquities) published in 1836 but written much earlier, Thomsen introduced his revolutionary proposal on "the different periods to which the heathen antiquities may be referred," namely, the Stone, Bronze, and Iron ages. For Thomsen they represented probable technological stages; they were soon to become the essential basis for prehistory when shown by Jens WORSAAE in the 1840s to correlate with the chronological sequence of artefacts found in the earth.

Although much modified and extended, complicated with subdivisions and insertions, the framework of Thomsen's three-age system, though not without its critics, is still very much part of the classification of prehistory.

Thomsen, Hans Peter Jörgen Julius (1826–1909) *Danish chemist*

Thomsen taught at the Technical University in his native city of Copenhagen (1847–56) and at the Military High School (1856–66) before becoming professor of chemistry at the University of Copenhagen (1866–91). He also served as director of the Technical University (1883–92).

Thomsen's main work was in thermochemistry, published over the years in four volumes of his *Thermochemische Untersuchungen* (1882–86; Thermochemical Analyses). He performed numerous calorimetric experiments and in 1854 stated his principle, namely, that every reaction of a purely chemical character, simple or complex, is accompanied by an evolution of heat. Marcellin BERTHELOT enunciated a similar principle in 1873 without acknowledgment, which led to an acrimonious dispute between the two chemists. Thomsen did not go so far as Berthelot in extending his principle but, by careful measurement of the heat liberated in

different reactions, he hoped to be able to work out the affinity between substances. Unfortunately both were wrong, and the emphasis on heat misled chemists for some time. There are reactions that absorb heat rather than liberate it.

Thomsen made a substantial fortune for himself by discovering in 1853 a method for manufacturing sodium carbonate from cryolite.

Thomson, Benjamin *See* RUMFORD, BENJAMIN THOMSON, COUNT.

Thomson, Sir Charles Wyville (1830–1882) *British marine biologist*

Thomson was born at Bonsyde in Scotland and educated at Edinburgh University; his first academic posts were as lecturer in botany at Aberdeen University (1850–51) and Marischal College (1851–52). He was then appointed to the chairs of natural history at Cork (1853) and Belfast (1854–68). From 1870 he was professor of natural history at Edinburgh.

Thomson is chiefly remembered for his extensive studies of deep-sea life, and particularly of marine invertebrates, in which he came to specialize. He made a number of oceanic expeditions to various parts of the world. In 1868–69 he led two deep-sea biological and depth-sounding expeditions off the north of Scotland, discovering, at a depth of some 650 fathoms, a wide variety of invertebrate forms, many of them previously unknown. To explain the variations in temperature that occurred at great depths he postulated the existence of oceanic circulation. After a further expedition to the Mediterranean (1870), Thomson published *The Depths of the Sea* (1872), in which he described his researches and findings. This culminated in his appointment as scientific head of the Challenger Expedition to the Atlantic, Pacific, and Antarctic oceans (1872–76), during which soundings and observations were made at 362 stations in a circumnavigation of some 70,000 miles. Using temperature variations as indicators, Thomson produced evidence to suggest the presence of a vast mountain range in the depths of the Atlantic – the Mid-Atlantic Ridge. His findings were later confirmed by a German expedition in 1925–27. Knighted on his return from the Challenger voyage, Thomson began preparation of the expedition's scientific reports – a work that eventually ran to 50 volumes – but had to resign in 1881 due to ill health. Thomson also wrote a general account of the expedition in *The Voyage of the Challenger* (1877).

Thomson, Elihu (1853–1937) *American electrical engineer*

Although born at Manchester in England, Thomson moved to Philadelphia as a young child. He graduated from Philadelphia Central High School in 1870 and from 1870 to 1880 taught chemistry and physics there, gaining his MA in 1875.

It was with a fellow teacher, Edwin J. Houston, that he formed an innovative partnership, designing an arc-lighting system that attracted the financial backing to form the American Electric Company in New Britain, Connecticut. This was the precursor of the Thomson–Houston Electric Company of Lynn, Massachusetts, founded in 1883, which in 1892 merged with the Edison General Electric Company to become the General Electric Company – now the world's largest producer of electrical machinery. After the merger Thomson remained with General Electric as a consultant and was director of their Thomson Research Laboratory.

He was a prolific inventor and is associated with some 700 patents in the field of electrical engineering. General Electric operated under his patents, which, besides those for stable and efficient arc lighting, included ones for the three-phase alternating current generator, the high-frequency generator, the three-coil armature for dynamos and motors, electric welding by the incandescent method, the induction coil system of distribution, the induction motor, and meters for direct and alternating current.

Thomson also made his mark in the field of radiology, improving x-ray tubes and making stereoscopic x-ray pictures. He was also the first to suggest the use of a helium–oxygen breathing mixture for workers in underwater pressurized vessels to overcome caisson disease ("the bends"). He wrote many scientific papers and received awards and distinctions from scientific and engineering societies throughout the world.

Thomson, Sir George Paget (1892–1975)
British physicist

George Thomson was the son of J. J. Thomson, the discoverer of the electron. He was born in Cambridge and educated at the university there, where he taught (1914–22). He was then appointed to the chair of physics at Aberdeen University. Thomson moved to take the chair of physics at Imperial College, London, in 1930. He remained there until 1952 when he returned to Cambridge as master of Corpus Christi College, a position he held until his retirement in 1962.

His early work was in investigating isotopic composition by a mass spectrograph method. In 1927 he also performed a classic experiment in which he passed electrons through a thin gold foil onto a photographic plate behind the foil. The plate revealed a diffraction pattern, a series of concentric circles with alternate darker and lighter rings. The experiment provided crucial evidence of the wave–particle duality of the electron. Thomson shared the 1937 Nobel Prize for physics for this work with Clinton J. DAVISSON who had made a similar discovery independently in the same year.

During the war Thomson was chairman of the "Maud committee" to advise the British government on the atom bomb. (The name of this committee arose from a telegram message that Niels BOHR had managed to convey to England shortly after the German invasion of Denmark. To assure his friends of his well-being he instructed: "Please inform Cockroft and Maud Ray, Kent," which was mistakenly interpreted as a secret message to "make uranium day and night"; Maud Ray was Bohr's former governess.) It was this committee that, in 1941, gave the crucial advice to Churchill that it was indeed possible to make an effective uranium bomb and elicited from him the minute: "Although personally I am quite content with existing explosives, I feel that we must not stand in the path of improvement."

Thomson, Sir Joseph John (1856–1940)
British physicist

A research on the lines of applied science would doubtless have led to improvement and development of the older methods – the research in pure science has given us an entirely new and much more powerful method. In fact, research in applied science leads to reforms, research in pure science leads to

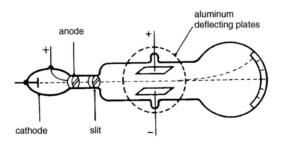

aluminum deflecting plates
anode
cathode slit

THOMSON'S APPARATUS   Diagram of the original apparatus used by J. J. Thomson to investigate cathode rays.

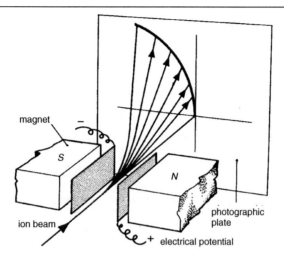

PARABOLIC MASS SPECTROMETER   *The arrangement used by J. J. Thomson to produce a mass spectrum.*

revolutions, and revolutions, whether political or industrial, are exceedingly profitable things if you are on the winning side.
—On his experimental work at the Cavendish Laboratory. Quoted by John Rayleigh in *J. J. Thomson*

Thomson, who was born at Manchester in England, entered Owens College (later Manchester University) at the age of 14. After studying engineering and then the sciences, he won a scholarship to Trinity College, Cambridge University. He graduated in 1876 and remained a member of the college in various capacities for the rest of his life. After graduating he worked at the Cavendish Laboratory under John RAYLEIGH, whom he succeeded as Cavendish Professor of Experimental Physics in 1884. As professor, Thomson built up the Cavendish Laboratory as a great and primarily experimental research school. Thomson was succeeded as professor in 1919 by his student Ernest RUTHERFORD.

Thomson's most brilliant and famous scientific work was his investigations into cathode rays, in which he is considered to have discovered the electron. Using a highly evacuated discharge tube he calculated the velocity of the rays by balancing the opposing deflections caused by magnetic and electric fields. Knowing this velocity, and using a deflection from one of the fields, he was able to determine the ratio of electric charge (e) to mass (m) of the cathode rays. Thomson found that the ratio e/m was the same irrespective of the type of gas in the tube and the metal of the cathode, and was about a thousand times smaller than the value already obtained for hydrogen ions in the electrolysis of liquids. He later measured the charge of electricity carried by various negative ions and found it to be the same in gaseous discharge as in electrolysis. He thus finally established that cathode rays were negatively charged particles, fundamental to matter, and much smaller than the smallest atoms known. This opened up the way for new concepts of the atom and for the study of subatomic particles. Thomson announced his discovery of a body smaller than the hydrogen atom in April 1897.

His later researches included studies of Eugen GOLDSTEIN's canal rays, which he named positive rays. These studies gave a new method (1912) of separating atoms and molecules by deflecting positive rays in magnetic and electric fields. Ions of the same charge-to-mass ratio form a parabola on a photographic plate. Using this arrangement, Thomson first identified the isotope neon–22. This work was taken up by Francis W. ASTON, who later developed the mass spectrograph.

Thomson's treatises were widely used in British universities. His *Conduction of Electricity through Gases* (1903) describes the work of his great days at the Cavendish; his autobiography, *Recollections and Reflections*, was published in 1936. He was awarded the 1906 Nobel Prize for physics for his work on the conduction of electricity through gases and was knighted in 1908.

't Hooft, Gerardus (1946–   ) *Dutch theoretical physicist*

't Hooft (toof) was born in Den Helder in the Netherlands. He was educated at the University of Utrecht, gaining his doctoral degree in 1972. From 1972 to 1974 he was at CERN, Geneva. Since that time he has been based at the University of Utrecht, becoming a full professor in 1977.

In 1971 't Hooft showed that it is possible to

obtain sensible results in calculations for a type of quantum field theory called a non-Abelian gauge theory, in particular one used in models that unify the electromagnetic with the weak nuclear interactions responsible for (among other things) beta decay.

Classical electrodynamics, as described by the MAXWELL equations, is the earliest example of a gauge theory. It is possible to express the electric and magnetic fields in electrodynamics in terms of potentials. Moreover, it is possible to change (gauge transform) these potentials in a certain way without changing the fields. The simplest example of a gauge transformation is the addition of a constant to the electric potential. This is in accord with the result of electrostatics that the zero point of the electric potential is arbitrary, with only the differences in potential being of significance.

The arbitrariness of this change in the zero point of the electric potential is associated with a symmetry, called a *gauge symmetry*. In electrodynamics the order in which two gauge transformations are performed does not matter. It is said that electrodynamics is an *Abelian gauge theory*, named after Niels Henrik ABEL. In general, a group of Abelian transformations is a set of transformations in which the result of applying two of the transformations does not depend on the order in which the two transformations are applied. An example of a group of Abelian transformations is the set of rotations in two dimensions. By contrast, a group of *non-Abelian* transformations is a set of transformations in which the result of applying two of the transformations does, in general, depend on the order in which the two transformations are applied. An example of a group of non-Abelian transformations is the set of rotations in three dimensions.

When formulating quantum electrodynamics (QED) in 1929–30 Werner HEISENBERG and Wolfgang PAULI made use of the fact that QED is an Abelian gauge theory. Heisenberg and Pauli, as well as subsequent investigators such as Robert OPPENHEIMER, found that all but the simplest calculations in QED led to the answer infinity. This problem was solved in the 1940s by Richard FEYNMAN, Julian SCHWINGER, and Sin-Itiro TOMONAGA using the technique of *renormalization*, which they developed.

In 1954 Chen Ning YANG and Robert Mills constructed a quantum field theory with a non-Abelian group of gauge transformations, this type of theory being called a *non-Abelian gauge theory*. At first, it was unclear whether this type of theory had any relevance to nuclear interactions, particularly since both strong and weak nuclear interactions are short range, whereas gauge theories are associated with long-range interactions, such as the COULOMB force for electrically charged particles, and with massless particles, such as the photon in the case of QED. Nevertheless, in spite of these se-

rious difficulties the theory attracted considerable interest from theoretical physicists. In particular, Steven WEINBERG in 1967 and Abdus SALAM in 1968 independently put forward a theory, now known as the *Weinberg–Salam model*, which unifies the weak and electromagnetic interactions in a non-Abelian gauge theory. In the Weinberg–Salam model the gauge symmetry is a symmetry which is broken by the mechanism put forward by Peter HIGGS. This mechanism has the consequence that the vector mesons mediating the weak interactions become massive. This meant that the Weinberg–Salam model could explain why weak interactions have a short range but made it difficult to see how renormalization could be applied to this model.

In 1971 't Hooft made a major breakthrough when he showed how renormalization could be applied to non-Abelian gauge theories, including the case of gauge symmetry broken by the Higgs mechanism. This immediately made the Weinberg–Salam model a viable theory on which calculations could be performed. 't Hooft, together with his supervisor Martinus VELTMAN, continued his investigations on the renormalization of gauge theories in the early 1970s.

The centrality of gauge theories in describing the fundamental forces of Nature was given another major boost in the early 1970s by the emergence of a theory called quantum chromodynamics (QCD) to describe the strong nuclear interactions. Almost as soon as QCD was proposed by Murray GELL-MANN and others it was suggested that the nature of the vacuum state in QCD could explain why isolated quarks had never been observed. Although this hypothesis of quark confinement has not been given a conclusive mathematical proof a very attractive and plausible picture of quark confinement was given by 't Hooft and others in the late 1970s and early 1980s in terms of an analogy between the vacuum state in QCD and superconductivity. Subsequent work by Edward WITTEN and others has given support for this point of view.

Since the mid-1980s the main focus of 't Hooft's research has been to attempt to understand black holes in terms of quantum mechanics. In the mid-1990s he found a feature of the quantum theory of gravitational systems called the *holographic hypothesis*, which postulates that information on the quantum states in the bulk of a gravitational system is encoded in the surface of that system. In the late 1990s evidence emerged that the holographic hypothesis is realized in superstring theory.

't Hooft and Veltman shared the 1999 Nobel Prize for physics for their work on the renormalization of gauge theories.

Thorne, Kip (Stephen) (1940–) *American physicist*

Born at Logan in Utah, Thorne was educated at

the California Institute of Technology (Cal Tech) and at Princeton, where he obtained his PhD in 1965. He has been professor of theoretical physics at Cal Tech since 1970. His main concern has been the astrophysical and cosmological consequences of general relativity theory, including black holes and gravitational waves. He coauthored *Gravitation Theory and Gravitational Collapse* (1965) and, with J. A. WHEELER and C. W. MISNER, wrote an authoritative work, *Gravitation* (1973).

Thorne has also been involved in the search for a black hole. The technique, proposed by Y. B. Ze'ldovich and O. Gusyenov, is to search for binary systems that have an invisible companion with a large mass. It was also realized that a black hole could pull gas off its companion, which should heat up as it becomes attracted to the black hole and emit x-rays. It was only after the Uhuru satellite had identified 125 x-ray sources in 1972 that this proposal could be tested. This produced six x-ray binaries of which the most plausible candidate was the star known as Cygnus X-I. The launching of the Einstein X-ray Observatory in 1978 produced new evidence showing that the mass of the companion of Cygnus X-I is greater than six solar masses. As it is too massive to be either a white dwarf or a neutron star, and it is also very compact, it may be a black hole. In the 1980s Thorne was involved in promoting the *membrane paradigm* of black-hole theory, in which the properties of a black hole are analyzed in terms of an imaginary membrane at its surface. Thorne coauthored a book entitled *The Membrane Paradigm* (1986). Thorne published his views on black holes, along with some more speculative material, in his *Black Holes and Time Warps* (1994).

Thorpe, Sir Thomas Edward (1845–1925) *British chemist*

Thorpe was the son of a cotton merchant from Manchester, England. He was educated at Owens College, Manchester, and at the University of Heidelberg where he gained his doctorate in 1869. Having been appointed to the chemistry chair at Anderson's College, Glasgow, in 1870, he moved to the Yorkshire College of Science at Leeds four years later. He later became professor of chemistry at the Royal College of Science, London (1885–94, 1909–12), and also served as director of the British government laboratories (1894–1909).

Thorpe worked with Henry ROSCOE on the new element vanadium, made some precise atomic weight measurements, and studied phosphorus compounds. In his work as a government chemist he managed to institute controls over such chemical hazards as white phosphorus, arsenic in beer, and lead glazes. He produced a number of works on the history of chemistry, including a biography of Joseph

PRIESTLEY in 1906. He also published a multi-volume *Dictionary of Applied Chemistry* (1890–93). He was knighted in 1909.

Thunberg, Carl Per (1743–1828) *Swedish botanist*

Thunberg, who was born at Jönköping in Sweden, studied medicine at Uppsala University, graduating in 1770. While there he went on botanical collecting trips for his teacher LINNAEUS. After a year's study in Paris he accepted an invitation to make a plant-collecting trip aboard a Dutch merchant ship bound for Japan. By April 1772 Thunberg had completed the first stage of the journey with his arrival in Cape Town, South Africa. He remained there for three years, during which time he made three excursions into the interior on which he collected over 3000 species of plants. About a third of these were new to science. On two of his trips he was accompanied by the British plant collector Francis Masson.

In March 1775 Thunberg sailed on to Japan, arriving at Nagasaki in August. Restrictions imposed on the movements of foreigners at that time prevented him from making long botanical excursions. However, with the aid of some young Japanese physicians whom the traders employed as interpreters, Thunberg was able to obtain species of Japanese plants in exchange for his knowledge of modern European medicine. In December 1776 he left Japan and visited Java, Colombo (Sri Lanka), Cape Town, and London before reaching Sweden in March 1779. On his return Thunberg took up his appointment as botanical demonstrator at Uppsala University and in 1784 he succeeded Carl von Linné (the younger) as professor of botany, a post he held until his death.

Thunberg's time as professor was mainly occupied with writing about his extensive collections. In 1784 he published his *Flora Japonica*, describing 21 new genera and several hundred new species of Japanese plants. His works on floras of the Cape colony include *Prodromus Plantarum Capensis* (1794–1800; Foreword to the Plants of the Cape) and the more important *Flora Capensis* (1807–23; Flowers of the Cape), completed with the help of the German botanist Joseph Schultes. His shorter works include writings on *Protea*, *Oxalis*, *Ixia*, and *Gladiolus*.

Tiemann, Johann Carl Wilhelm Ferdinand (1848–1899) *German chemist*

Tiemann (**tee**-mahn) was born at Rubeland in Germany and started his career as an apprentice to an apothecary; he later studied chemistry at the University of Berlin, where he was appointed to the chair of chemistry in 1882.

Tiemann worked mainly on the chemistry of plant products, particularly glycosides and essential oils, which led to contacts with and commercial interests in the perfume industry. His work on ionone, the perfume of violets, led to protracted litigation.

Tilden, Sir William Augustus (1842–1926) *British chemist*

Tilden, a Londoner by birth, was apprenticed to a pharmacist but allowed to attend the lectures of August Hofmann at the Royal College of Chemistry in his spare time. He worked as a demonstrator for the Pharmaceutical Society from 1863 until 1872. He then taught chemistry at Clifton College until 1880, when he became professor of chemistry at Mason College, Birmingham. He finally moved to the Royal College of Science in London, where he held the chair of chemistry from 1894 until 1909.

Tilden's main work was in organic chemistry. He synthesized the compound nitrosyl chloride (1874) and discovered that it formed a crystalline derivative with turpentine. This led him into the study of terpenes – his main field of research. He made the important discovery in 1884 that, when heated, terpenes – hydrocarbons with the general formula $(C_5H_8)_n$ – decomposed to a mixture containing a closely related hydrocarbon, isoprene. The significance of this is that isoprene molecules can be polymerized (joined together in a long chain) to produce rubber. Tilden observed this when he made a sample of isoprene by passing turpentine through a heated tube and left part of it on a shelf. When he examined it several years later he found "a dense syrup in which was floating several large masses of solid." This was a form of rubber. Unfortunately the polymerization took place very slowly and attempts to speed up the process were unsuccessful at the time. Tilden also wrote on the history of chemistry.

Tinbergen, Nikolaas (1907–1988) *Dutch–British zoologist and ethologist*

Tinbergen (**tin**-ber-gen) was born in The Hague, the Dutch capital, and educated at Leiden University, where he gained his doctorate in 1932 for a thesis on insect behavior. Soon after he joined a Dutch meteorological expedition to East Greenland. The results of his Arctic year observing huskies, buntings, and phalaropes were later described in his *Curious Naturalists* (1958). In 1936 Tinbergen was appointed lecturer in experimental zoology at Leiden. Contact with Konrad LORENZ in 1937 led to an early collaboration. Tinbergen's work, however, was interrupted by the war. He refused to cooperate with plans to Nazify Leiden University and was consequently imprisoned in a concentration camp from 1942 to 1944. Although he was appointed a full professor by Leiden in 1947, Tinbergen chose to move to Oxford in 1947 to escape administrative duties. Here he took up the more junior post of lec-

turer in animal behavior. Tinbergen remained in Oxford until his retirement in 1974, having been appointed professor in animal behavior in 1966. He became a naturalized British subject in 1954.

Tinbergen demonstrated that ethology was basically an observational and experimental science. Unlike Lorenz, who tended to work with a large number of pets, Tinbergen worked with animals in their natural setting. Much of his early work dealt with identifying the mechanisms by which animals found their way around. How, for example, does a digger wasp recognize its burrow? In a few simple experiments with nothing more elaborate than a handful of pine cones, Tinbergen was able to show that the wasps were guided by the spatial arrangement of landmarks at the nest entrance. He also studied the social control of behavior in his work on the mating habits of sticklebacks.

Much of this early work was brought together in his classic text, *The Study of Instinct* (Oxford, 1951). In his other major work, *The Herring Gull's World* (London, 1953), Tinbergen began by recognizing the diversity of behavioral signals found in different species of gulls. Such a diversity had as much an evolutionary origin and history as more obvious anatomical features. Tinbergen set out to recover some of this history.

In his later years Tinbergen attempted to apply some of the principles of ethology to problems in human behavior. In particular, he worked with autistic children, publishing his results in *Autistic Children* (1983), a book he wrote in collaboration with his wife.

For his achievements in the field of animal behavior Tinbergen shared the 1973 Nobel Prize for physiology or medicine with Lorenz and FRISCH. His brother Jan Tinbergen had been awarded the Nobel Prize for economics four years earlier.

Ting, Samuel Chao Chung (1936–) *American physicist*

Although born at Ann Arbor, Michigan, Ting was educated at primary and secondary schools in China; he subsequently moved to Taiwan and returned to America to study at the University of Michigan. At Michigan he gained his bachelor's, master's, and doctoral degrees in the six years 1956–62.

His interest in elementary-particle physics, which was to lead to his sharing the Nobel Prize for the discovery of a significant new particle, took him to the European Organization for Nuclear Research (CERN) in Geneva (1963) and then Columbia University, New York (1964), where he became an associate professor in 1965. In 1966 Ting was given a group leader post at DESY, the German electron synchrotron project at Hamburg, and in 1967 joined the

Massachusetts Institute of Technology; he was appointed professor of physics there in 1969.

Working at Long Island, New York, on the Brookhaven National Laboratory's alternating-gradient synchrotron, Ting and his collaborators performed experiments in which streams of protons were fired at a stationary beryllium target. In such an experiment a particle was observed that had a lifetime almost 1,000 times greater than could be expected from its observed mass. Announcement was made in a 14-author paper in the *Physical Review Letters* in 1974. The discovery was made independently and almost simultaneously by Burton RICHTER and his colleagues some 2,000 miles away at the Stanford Linear Accelerator Center. Ting called the new particle J, Richter named it psi (ψ); it is now known as the J/psi in recognition of the simultaneity of its discovery. Confirmation came quickly from other high-energy physics laboratories and a whole family of similar particles has since been created and detected.

Ting and Richter were very quickly honored for their discovery by the award, jointly, of the 1976 Nobel Prize for physics. By 1976 Ting was directing three research groups, at Brookhaven, CERN, and DESY.

Tiselius, Arne Wilhelm Kaurin (1902–1971) *Swedish chemist*

Tiselius (tee-**say**-lee-uus) was born in Stockholm, Sweden, and educated at the University of Uppsala, where he became the assistant of Theodor SVEDBERG in 1925. He obtained his PhD in 1930 and his whole career was spent at Uppsala where, in 1938, a special research chair in biochemistry was created for him, which he occupied until 1968.

Tiselius's doctoral thesis was on electrophoresis – a method of separating chemically similar charged colloids. An electrical field is applied to the sample, and particles with different sizes migrate at different rates to the pole of opposite charge, enabling them to be detected and identified. The method was not initially very successful but by 1937 Tiselius had made a number of improvements to the apparatus. Using the technique on blood serum Tiselius confirmed the existence of four different groups of proteins – albumins and alpha, beta, and gamma globulins. Tiselius also conducted work on other methods for the separation of proteins and other complex substances in biochemistry including chromatography (from 1940) and partition and gel filtration (from the late 1950s).

In 1948 he was awarded the Nobel Prize for chemistry for his work on electrophoresis and other new methods of separating and detecting colloids and serum proteins. After the war Tiselius played an important role in the development and organization of science in Sweden, serving (1946–50) as chairman of the Swedish Natural Science Research Council.

Tishler, Max (1906–1989) *American pharmacologist*

Born in Boston, Massachusetts, Tishler was educated at Tufts University and Harvard, where he obtained his PhD in 1934. He taught at Harvard until 1937 when he joined Merck and Company, the pharmaceutical firm, rising to the position of head of the research division in 1956. He retired from Merck in 1970, joining Wesleyan University as professor of chemistry.

Tishler worked with Selman WAKSMAN on the important antibiotic actinomycin and headed the research groups investigating the production of streptomycin and penicillin. He also did much to advance the commercial production of cortical steroids, developing a method whereby oxygen is introduced at a certain position in the steroid nucleus, which is vital to the antiinflammatory properties of such compounds. Tishler also contributed to vitamin chemistry, his investigations with Louis Fieser at Harvard leading to the syntheses of many important vitamins, such as vitamin A and riboflavin.

Tizard, Sir Henry (Thomas) (1885–1959) *British chemist and administrator*

> The secret of science is to ask the right question, and it is the choice of problem more than anything else that marks the man of genius in the scientific world.
> —Quoted by C. P. Snow in *A Postscript to Science and Government* (1962)

Tizard, who was born at Gillingham in England, was the son of a naval officer who served as the navigator on the Challenger voyage. Barred from a similar career by an eye accident, Tizard instead went to Oxford University where he studied chemistry under Nevil SIDGWICK. After spending a year in Berlin working under Walther NERNST, he returned to Oxford in 1911 to take a fellowship. It was in Berlin that he first met and became friendly with Frederick LINDEMANN, who was later to become his principal opponent for positions of power in British scientific government circles.

Tizard spent World War I in the Royal Flying Corps working on the development of bomb sights and the testing of new planes. After the war he realized, as he put it in his unpublished autobiography, that he "would never be outstanding as a pure scientist." Having developed a taste for the application of science to military problems, he took the post of assistant secretary at the Department of Scientific and Industrial Research (DSIR) in 1920 with specific responsibility for coordinating research relevant to the needs of the armed forces. In 1929, largely for financial reasons, Tizard accepted the position of rector of Imperial College, London, where he remained until 1942.

Tizard quickly established a reputation for having an expert and practical knowledge of service needs. He had the rare ability to distinguish between a crankish, totally unsound, idea and one that, though strange and new, was basically sound and could find practical military application. Thus it was that Tizard backed the young Frank WHITTLE in the development of jet propulsion of aircraft in 1937 and also Barnes WALLIS in 1940 in his development of the bouncing bomb.

But, above all else, it was Tizard's support for the development of radar that will be remembered. In 1934 the Air Ministry set up the Committee for the Scientific Survey of Air Defence, under the chairmanship of Tizard. This was the famous "Tizard committee," which, in 1935, decided that radar was a workable means of air defense and should receive top priority.

The decision was not taken without dissent. In particular, Lindemann, then Churchill's scientific adviser, while recognizing the potential of radar, did not agree with the overriding priority demanded for it by Tizard and his associates. There was a further disagreement between the two men in that Lindemann advocated mass bombing of Germany while Tizard proposed instead (in 1942) a more balanced bombing policy with adequate aircraft being committed to the Battle of the Atlantic.

As a chemist Tizard's most significant work was on the ignition of gases in the internal combustion engine. He was editor of *Science of Petroleum* (1938), a standard multivolume work on the subject. He was knighted in 1937.

Todd, Alexander Robertus, Baron
(1907–1997) *British biochemist*

Todd graduated from the university in his native city of Glasgow in 1928 and spent a further year there on a Carnegie Scholarship before going to the University of Frankfurt, Germany. He received doctorates from both Frankfurt (1931) and Oxford (1933) universities and in 1934 joined the medical faculty at Edinburgh University where he began work on thiamine (vitamin B_1). He continued this research at the Lister Institute of Preventive Medicine, London, and worked out the structure and synthesis of thiamine.

In 1938 Todd became professor of chemistry at Manchester University and continued vitamin studies on vitamins E and B_{12}. He also isolated the active principle from *Cannabis*, extracting the compound cannabinol from cannabis resin. Todd transferred to Cambridge in 1944 to become professor of organic chemistry, a post he held until his retirement in 1971. At Cambridge he synthesized all the purine and pyrimidine bases that occur in nucleic acids (DNA and RNA) and found their structures. This was an important development in the understanding of the structure of the hereditary material and verified the formulae that Phoebus LEVENE had suggested for the nucleotide bases. Todd also synthesized various coenzymes related to these compounds, e.g., flavin adenine dinucleotide (FAD), and synthesized the energy-rich compounds adenosine di- and triphosphate (ADP and ATP) so important in energy transfer in living cells.

The 1957 Nobel Prize for chemistry was awarded to Todd for these contributions to biochemistry and the understanding of the gene. Todd was knighted in 1954, raised to the British peerage as Baron Todd of Trumpington in 1962, and from 1975 to 1980 was president of the Royal Society. He also published an account of his busy life in his autobiography *A Time to Remember* (1983).

Tombaugh, Clyde William (1906–1997)
American astronomer

Tombaugh (**tom**-baw) came from a poor farming background in Streator, Illinois; although he never went to college he managed to teach himself enough of the basic observational skills to be taken on by the Lowell Observatory in Flagstaff, Arizona, in 1929. He transferred to New Mexico in 1955 and served as professor between 1965 and 1973.

Percival Lowell had begun his search for a planet lying beyond the orbit of Neptune in 1905 and this search continued at the Lowell Observatory after his death. Tombaugh was given the job of systematically photographing the sky along the ecliptic where it was thought any trans-Neptunian planet would be found. He used a specially designed wide-field telescope and for each region took two long-exposure photographs separated by several days. He then examined the pairs by means of a blink comparator, an instrument that allows two plates to be alternately observed in rapid succession. Any object that has moved against the background of the stars in the interval between the two exposures will appear to jump backward and forward. Methods were therefore devised for distinguishing asteroids and other moving bodies from the sought-for planet. After a year's observation Tombaugh was able to announce on 13 March 1930 the detection of the new planet, later named Pluto, at a point agreeing closely with the position predicted by Lowell.

The question naturally arose as to whether there were any further trans-Neptunian planets. Consequently Tombaugh continued the search but, although he examined 90 million star images and discovered 3,000 asteroids, no other new planet was detected. It is likely that Tombaugh has the honor of discovering the final planet of the solar system.

Tomonaga, Sin-Itiro (1906–1979) *Japanese theoretical physicist*

Tomonaga (to-mo-**nah**-ga), who was born at Tokyo in Japan, graduated from Kyoto University in 1929 and then went to work in his native city. He remained there for the rest of his academic career, becoming professor of physics in 1941 and president of the university in 1956.

He was one of the first to develop a consistent theory of relativistic quantum electrodynamics. The problem at the time was that there was no quantum theory applicable to subatomic particles with very high energies. Tomonaga's first step in forming such a theory was his analysis of intermediate coupling – the idea that interactions between two particles take place through the exchange of a third (virtual particle), like one ship affecting another by firing a cannonball. Between 1941 and 1943 he used this concept to develop a quantum field theory that was consistent with the theory of special relativity. However, World War II prevented news of his work from reaching the West until 1947, at about the time that Richard FEYNMAN and Julian SCHWINGER published their own independent solutions to the same problem. All three shared the Nobel Prize for physics for this work in 1965.

Tonegawa, Susumu (1939–   ) *Japanese immunologist*

Tonegawa (ton-e-**gah**-wa) was born at Nagoya in Japan and educated at Kyoto University and the University of California, San Diego. He worked at the Basle Institute for Immunology from 1971 and in 1981 was appointed professor of biology at the Center for Cancer Research and Department of Biology of the Massachusetts Institute of Technology.

Working at Basle in collaboration with Niels JERNE and Nobumichi Hozumi, Tonegawa revealed how the immune system is capable of generating the enormous diversity of antibodies required so that whatever the nature of the invading organism or "foreign" tissue a suitable antibody is available to bind specifically to it. This implies that the immune system's antibody-producing cells – the B-lymphocytes – can potentially manufacture billions of different antibodies; yet, even in humans, these cells carry only about 100,000 genes on their chromosomes.

Working with mouse cells, Tonegawa showed that during the development of the antibody-producing cell, the genes coding for antibody are shuffled at random, so that in the mature cell a cluster of functional genes is formed specific to that cell. Each individual mature cell thus produces its own specific antibody. This potential diversity is amplified by the fact that each antibody molecule comprises four protein chains, all with a highly variable terminal region. Hence the diversity of antibodies is a con-

sequence of the huge numbers of lymphocytes present in the body, each with its own combination of functional antibody-producing genes. In recognition of the significance of his discovery Tonegawa was awarded the 1987 Nobel Prize for physiology or medicine.

Topchiev, Alexsandr Vasil'evich (1907–1962) *Russian chemist*

Topchiev (**top**-chyef) was born in Russia and educated at the Moscow Chemical Technology Institute, graduating in 1930. After working at the Institute of Technology of the Food Industry he joined the Moscow Petroleum Institute in 1940, becoming director in 1943. Although Topchiev served as deputy minister of higher education (1947–49) and from 1949 held high office in the Soviet Academy of Sciences he continued to retain his link with the Petroleum Institute.

His work was mainly in the field of hydrocarbon chemistry and he was an important figure in the development of the Russian petroleum industry.

Torrey, John (1796–1873) *American botanist*

Torrey was the son of a New York alderman, who had responsibility for the city's prisons. This led to Torrey meeting Amos Eaton, a lawyer turned natural historian and imprisoned for forgery, who encouraged Torrey's early interest in botany. He obtained an MD in 1818 from the College of Physicians and Surgeons, to which he returned in 1827 as professor of chemistry after a spell (1824–27) at the Military Academy, West Point, as professor of chemistry, mineralogy, and geology. Torrey also held from 1830 a professorship of chemistry at Cornell but in 1855 he resigned both chairs to become assayer of the New York branch of the Mint, a post he retained until his death.

However, it little mattered what post Torrey held for his life was exclusively devoted to the description and classification of the flora of America. In 1824 he published *Flora of the Northern and Middle Sections of the United States* and continued, in collaboration with his pupil Asa GRAY, with *A Flora of North America* (2 vols., 1838–43) and his own *Flora of the State of New York* (1843).

In later life Torrey devoted himself to the description in numerous monographs of the flora sent him by the explorers of the North American continent. From such a source he built up a collection of some 40,000 species, which was given to the New York Botanical Gardens in 1899, forming the nucleus of their herbarium.

Nor was Torrey simply a collector for, against considerable opposition, he discarded the sexual classification of LINNAEUS and introduced into North America the "natural" system of Antoine de JUSSIEU and Alphonse de CANDOLLE. A genus of ornamental trees and shrubs, *Torreya*, in the yew family is named for him, and his name is also remembered with the Torrey Botanical Club.

Torricelli, Evangelista (1608–1647) *Italian physicist*

> A Galileist by profession and sect.
> —Describing himself in a letter to Galileo Galilei, 11 September 1632

Born at Faenza in Italy, Torricelli (tor-i-**chel**-ee) was educated at the Sapienza College, Rome. His *De motu* (1641; On Movement) attracted the attention of Galileo, who invited him to come to Florence to work and live with him. After the death of Galileo in 1642 Torricelli was appointed professor of mathematics in Florence, where he remained until his death.

He had been introduced by Galileo to the problem of why water, in the duke of Tuscany's well, could not be raised higher than 30 feet (9 m). Dissatisfied with earlier explanations, he used Galileo's earlier demonstration that the atmosphere has weight to offer a more satisfactory account. He argued that as the atmosphere has weight, it must also have pressure that can force water up a pipe but only until the weight of the water produced an equivalent counterpressure. Thirty feet of water was equal to the pressure exerted by the atmosphere.

Torricelli realized that he could test this argument by substituting mercury for water. A tube of mercury inverted over a dish – given that mercury is 14 times heavier than water – should be supported by the atmosphere to only one-fourteenth the height of an equivalent amount of water. This was confirmed by his pupil VIVIANI in 1643. Torricelli noticed that over time there was a variation in the height of the mercury in the tube, and reasoned that this was due to variations in the pressure of the atmosphere. This led to his construction of the barometer in 1644. The vacuum above the mercury in a closed tube is called a *Torricellian vacuum* and the unit of pressure, the *torr*, was named in his honor. Torricelli also made advances in pure mathematics and geometry, in particular in his calculations on the cycloid.

Toscanelli, Paolo (1397–1482) *Italian astronomer and cartographer*

Toscanelli (tos-kah-**nel**-ee), who was born at Florence in Italy, studied medicine at Padua and had a reputation among his contemporaries as being the most learned man of his time. Only a few extracts of his work have survived. He is best known for having convinced Columbus that the East Indies could be reached by sailing a moderate distance across the Atlantic. He mistakenly thought (as did many others) that Asia was a mere 3,000 miles west of Europe and drew up a map showing this. Columbus saw a copy of this and, not satisfied with a distance of 3,000 miles to Japan, convinced himself that it was even less.

Toscanelli also made careful recordings of the comets of 1433, 1449, 1456 (Halley's comet), 1457, and 1472.

Tousey, Richard (1908–1997) *American physicist*

Born in Somerville, Massachusetts, Tousey was educated at Tufts, where he graduated in physics and mathematics (1928), and Harvard, where he gained his PhD (1933). After holding junior appointments at Harvard (1933–36) and Tufts (1936–41), he joined the U.S. Naval Research Laboratory, serving in the Optics Division until 1958 when he was appointed head of the spectroscopy branch, a post he retained until his retirement in 1978.

In 1946 Tousey initiated a new era in astronomy by launching a German V-2 rocket into the upper atmosphere with a spectrograph in its nose cone. With it he succeeded in photographing the solar spectrum from above the ozone layer and observing the extreme short-wavelength ultraviolet region down to a wavelength of 2,200 angstroms. Such a pioneering attempt was followed in the 1960s by more sophisticated and powerful orbiting solar observatories (OSOs) launched by NASA.

Tousey was also involved in NASA's Skylab program, in which he conducted experiments using powerful solar instruments operated by the astronauts. He was awarded the Exceptional Scientific Achievement Medal of NASA in 1974.

Townes, Charles Hard (1915–) *American physicist*

Born in Greenville, South Carolina, Townes was educated at Furman and Duke universities in his home state and at the California Institute of Technology, where he obtained his PhD in 1939. He worked at the Bell Telephone Laboratories (1939–47) before he took up an appointment at Columbia University, New York, where he became a full professor in 1950. He moved to the Massachusetts Institute of Technology in 1961 and then served as professor of physics at the University of California at Berkeley from 1967 until 1986, when he retired.

In 1953 Townes designed the first maser (*mi*crowave *a*mplification by *s*timulated *e*mission of *r*adiation). The maser works on the realization in quantum theory that molecules can only adopt a certain number of discrete characteristic energy states and that in their movement from one energy level to another they emit or absorb precisely determined amounts of radiation. Townes knew that the ammonia molecule

(NH$_3$) could occupy one of two energy levels, the difference between them equaling a particular energy of a photon.

The question Townes went on to ask is, what happens if the ammonia molecule absorbs the photon of the appropriate frequency while it is at its higher energy level? Albert EINSTEIN had answered the question in 1917 and shown that the molecule would fall to its lower state emitting a photon of the same frequency. There would then in fact be two photons of the right frequency to repeat the process and produce four photons. This is a rapid and powerful amplification producing a narrow beam of radiation with a single frequency. In this case the radiation emitted would have a frequency of 1.25 centimeters and thus fall in the microwave band, hence the name maser.

In any normal sample of ammonia only a few of the molecules would be in the higher energy state. Townes's problem was to devise a technique for the separation of molecules of the higher energy level ("population inversion") and he did this using a nonuniform electric field; molecules in the higher state were repelled and focused into the resonator while those in the lower state were attracted to it. By 1953 Townes had a working model of a maser.

Masers quickly found use in atomic clocks, receivers in radio telescopes, and numerous other uses. The maser led to the development of the laser, where light is amplified rather than microwave radiation and which was known in its earlier days as an "optical maser." Townes, in fact, published a paper with A. L. SCHAWLOW in 1958 showing the theoretical possibility of the laser. He was, however, beaten in the race to construct it by Theodore MAIMAN in 1960.

For his work on the maser Townes shared the 1964 Nobel Prize for physics with the Russians Nicolai BASOV and Aleksandr PROKHOROV, who had independently produced a maser in the Soviet Union in 1955.

Townsend, Sir John Sealy Edward (1868–1957) *Irish physicist*

Townsend, the son of a professor of civil engineering from Galway (now in the Republic of Ireland), was educated at Trinity College, Dublin. In 1895 he took advantage of a change in the Cambridge examination statutes and, together with Ernest RUTHERFORD, entered the Cavendish Laboratory as one of the first two non-Cambridge graduates. There they worked as research students of J. J. THOMSON. In 1900 Townsend moved to Oxford as the first Wykeham Professor of Experimental Physics, a post he retained until his retirement in 1941.

In 1898 Townsend achieved his first major scientific success when he measured the fundamental unit of electric charge (the charge of the electron). The previous year Thomson had reported the discovery of the electron, whose mass he estimated at about 1/1,000 that of the hydrogen atom.

Townsend, working with gases released by electrolysis, was able to form charged clouds of water droplets and, by measuring the rate of fall of a water drop in the cloud, he could calculate the charge on each drop. More accurate work of this type was done by Robert MILLIKAN in 1911.

Townsend's main work however was on, to take the title of his important book, *Electricity in Gases* (1915). He formulated a theory of ionization by collision, showing that the motion of electrons in an electric field would release more electrons by collision. These in turn would release even more electrons, and so on. This multiplication of charges, known as an avalanche, allowed him to explain the passage of currents through gases where the electric field was thought to be too weak.

Townsend was knighted in 1941.

Traube, Moritz (1826–1894) *German chemist*

Born at Ratibor (now Racibórz in Poland), Traube (**trow**-be) studied at the universities of Berlin and Giessen, where he worked on fermentation under Justus von LIEBIG. Owing to the deaths of his two older brothers he was forced to give up his chemical career to run the family wine business but he carried on his research privately at the University of Breslau, producing over 50 papers.

Although having worked under Liebig, Traube differed from him in his dispute with Louis PASTEUR over the nature of fermentation. He showed that it was caused by what he called "a nonliving unorganized ferment" produced by a yeast organism. He tried unsuccessfully to isolate this "ferment" (which later became better known as an enzyme). Traube also worked on osmosis, investigated first by Jean NOLLET in 1748, and introduced various semipermeable membranes, which could serve as models for transfer in living cells.

Travers, Morris William (1872–1961) *British chemist*

The son of a London physician, Travers was educated at University College, London, where he obtained a DSc in 1898, and in Nancy, France. He worked with William RAMSAY in London (1894–1904) and then moved to Bristol as professor of chemistry. Travers served in India as the director of the Indian Institute for Science at Bangalore (1906–14). On returning to England he worked in the glass industry for some years before returning to Bristol in 1927 as honorary professor and Nash Lecturer in Chemistry and remained there until his retirement in 1937.

Travers was associated with Ramsay in his

work on the rare gases, discovering in 1898 krypton, neon, and xenon, and later gave a full account of their work in *The Discovery of the Rare Gases* (1928). He also carried out research into low-temperature phenomena and into gas reactions.

Trembley, Abraham (1710–1784) *Swiss zoologist*

Trembley (trahn-**blay**) was born at Geneva in Switzerland. In the uprising of 1733, however, his father Jean, a leading politician and soldier, was driven from office and into exile. The fall in the family's fortune forced Trembley to seek employment. He moved to Holland in 1733 and served as tutor to the children of various noblemen. While working for Count Bentinck he read the *Mémoires* of René RÉAUMUR, which so stimulated him that he began to observe nature himself. During the next few years he made a number of discoveries which were to astound Europe.

At first he worked on parthenogenesis in aphids. Though he achieved some interesting results the work was basically derivative, having already been carried out by Réaumur and Charles BONNET. At the same time he was studying polyps or, in modern terminology, hydra, of the species *Chlorohydra viridissima*. They were assumed to be plants but one day in the summer of 1740 Trembley observed them to move their "arms." At first he assumed that the movement was caused by currents in the water. When he swirled the jar around, expecting to see the hydra sway with the vortex, he noted that they suddenly contracted to a point, their tentacles seeming to disappear in the process. As the water calmed, the polyps stretched and once more revealed their tentacles. He was still unsure, however, whether they were plants or animals. There was, he realized, a simple procedure he could carry out to resolve the dilemma. If they were animals, then they would surely die if cut into two; plants would continue to grow.

In late 1740 he divided a number of polyps transversely. To his surprise he found that within a few days the two parts had not merely grown, but had developed into two perfect polyps: the tail grew a head, and the head a tail. He published his results in his *Mémoires … à l'histoire d'un genre de polypes d'eau douce* (1744; Memoirs on the Natural History of Freshwater Polyps). Polyp samples were also dispatched to scholars and institutions throughout Europe. Some, particularly vitalists, found Trembley's work difficult to accept. What happened to the animal's soul? Was that also divided into two? Others saw it as evidence for the oneness of nature, the "Great Chain of Being," with the polyp being the link between plants and animals. Embryologists saw in the polyp conclusive proof against the preforma-

tionists who claimed that embryos were minute but preformed individuals. These and other issues were endlessly debated throughout the century, and Trembley's polyps became the best-known invertebrates in Europe.

Trembley himself had little more to contribute to science. He left his post with Bentinck in 1747. Thereafter he traveled around Europe, engaged in some kind of diplomatic activity for Britain, for which he received a pension of 300 pounds per annum for life, and wrote a number of books on education, politics, and philosophy.

Trumpler, Robert Julius (1886–1956) *Swiss–American astronomer*

Trumpler was the son of an industrialist. He studied at the university in his native city of Zurich in Switzerland (1906–08) and then at the University of Göttingen in Germany, where he obtained his PhD in 1910. After spending four years with the Swiss Geodetic Survey he emigrated to America in 1915, where he worked first at the Allegheny Observatory near Pittsburgh before moving to the Lick Observatory in California in 1919. He remained there until his retirement in 1951, also holding from 1938 to 1951 a chair of astronomy at Berkeley.

Trumpler's most important work was his discovery in 1930 of conclusive evidence for interstellar absorption. He had examined over 300 open clusters of stars and found that remote clusters seemed to be about twice the size of nearer ones. He could find no observational error nor could he believe that he was witnessing a real phenomenon. He did, however, appreciate that this could be due to the presence of an absorbing medium occurring between the clusters and the observer on Earth. Trumpler assumed correctly that the quantity of absorbing medium increased with distance so that this would cause more distant clusters to appear fainter and would lead to an overestimate of their distance and size. For nearby objects only a small correction would be needed but for distant ones it could be quite considerable. He went on to estimate the effect of the absorbing medium, which was interstellar dust, on the dimming of the received light as 0.2 of a magnitude per thousand light-years. This means that the brightness of a star is decreased by interstellar dust by a factor of 1.208 for every thousand light-years that the starlight travels toward Earth. This had far-reaching implications for the work of such astronomers as Harlow SHAPLEY who had been working on the size and structure of our Galaxy. It forced him to reduce the scale of his model by a factor of three.

Trumpler was also involved in 1922 in a test of EINSTEIN's general theory of relativity. Einstein had predicted the amount by which starlight would be bent when it passed close to the Sun's limb. Trumpler assisted W. W. Campbell of the Lick Observatory to make the relevant measurements at Wallal in Australia during the total solar eclipse of 1922. The value they obtained for the deflection, 1.75±0.09 seconds of arc, was much more accurate than the value Arthur EDDINGTON had found in 1919 and was very close to Einstein's prediction of 1″.745.

Tsiolkovsky, Konstantin Eduardovich
(1857–1935) *Russian physicist*

> The Earth is the cradle of mankind. But one does not live in the cradle forever.
> —Quoted by Alan L. Mackay in *A Dictionary of Scientific Quotations* (1991)

Tsiolkovsky (tsyol-**kof**-skee), who was born at Ryazan in Russia, was the son of a forester whom he described as being an "impractical inventor and philosopher." He was educated at local schools but when nine years old became almost completely deaf. He was to a large extent self-educated, his deafness driving him to a solitary life of reading. He studied science for three years in Moscow (1873–76), returned home and took his teacher's examination in 1879, and in 1880 obtained a post at a school in Borovsk. In 1892 he moved to a school and later to a college in Kaluga where he remained until his retirement in 1920.

Such was Tsiolkovsky's isolation from the world of science that when he submitted some of his work to the St. Petersburg Society of Physics and Chemistry in 1880 he had to be told that his results on the gas laws and the speed of light were by no means new. What was new, however, were the thoughts that he was beginning to have about space travel.

Tsiolkovsky had worked for many years on various problems in aeronautics, including the design and testing, in a wind tunnel, of an all-metal airship. These investigations led him in the late 1890s to consider and write about the possibility of space flight and he is thus considered the pioneer in this field. In 1896 he began to write his famous work, published in 1903, *Exploration of Cosmic Space by Means of Reaction Devices*, in which he clearly stated most of the basic principles of rocketry and space travel. He realized that a reaction engine was required and discussed the problems of using such engines in space and here and in subsequent articles described the basic theory and design of a liquid-fuel rocket engine. Tsiolkovsky had no funds or means to test his theories, unlike the American Robert GODDARD, who actually built and launched the first liquid-fuel rocket. He still continued into late life to put forward his ideas. His main method of giving them wider circulation was to write science-fiction stories.

After his death he was much honored as a Soviet pioneer of the space age. Appropriately enough the launching of the first satellite, Sputnik 1, was planned (although not realized) for the 22nd anniversary of his death and a giant crater on the reverse side of the Moon, unseen by all until recently, has been named for him.

Tsui, Daniel C. (1939–) *Chinese–American physicist*

Tsui was born in Henan in China, and gained his PhD in physics in 1967 from the University of Chicago. In 1998 he became professor at Princeton University, where he studied the fractional quantum Hall effect. In 1998 he shared the Nobel Prize for physics, with Robert LAUGHLIN and Horst STÖRMER, for their discovery and explanation of a new form of quantum fluid with fractionally charged excitations.

Tsui, Lap-Chee (1950–) *Chinese–Canadian molecular biologist*

Tsui (**tsoo**-ee) was born at Shanghai in China and educated at the Chinese University, Hong Kong, and at the University of Pittsburgh, where he obtained his PhD in 1979. He joined the staff of the Hospital for Sick Children, Toronto, in 1980 and has continued to work there while also holding (since 1990) a professorship of medical genetics at the University of Toronto.

In 1989, in collaboration with his Toronto colleague Jack Riordan and Francis Collins of the University of Michigan and against strong competition, Tsui announced that he had located the cystic fibrosis gene. Cystic fibrosis (CF) is caused by a recessive gene widely distributed among Caucasians; about 1,000 children with CF are born in the United States each year. The disease affects secretory epithelia and, until recently, patients tended to die from lung infections and heart failure in their twenties.

When Tsui arrived in Toronto in 1980 new techniques were being proposed which, if effective, would allow defective genes to be identified. The procedure, introduced by Ray White and his colleagues, involved identifying an appropriate RFLP (restriction fragment length polymorphism), i.e., a fragment of DNA for use as a genetic marker to locate the site of the defective gene. The difficulty was that the human genome contains 3 billion base pairs of DNA; it was, therefore, a very remote possibility that any genetic marker would be within a reasonable distance of the CF gene. The chances would, however, be much improved if it was known on which chromosome the gene was sited. This was established as chromosome 7 by a Massachusetts biotechnology firm, Collaborative Research Inc (CRI), in 1985.

Thereafter it was a matter of using the available probes to focus on the area in which the gene could be found. By 1989 it had been narrowed to 300,000 base pairs. Further work demonstrated that the defective gene encodes a membrane protein of 1,480 amino acids, dubbed the CF transmembrane conductance regulator (CFTR). Tsui and his colleagues have shown that a loss of one amino acid one third of the way along the gene is responsible for 68% of cases of CF. He has continued to search for the mutations responsible for the other cases and has also sought to understand at the molecular level variations in the severity of the disease. In 2002 he returned to Hong Kong to become Vice-Chancellor of the University of Hong Kong.

Tsvet, Mikhail Semenovich (1872–1919) *Russian botanist*

Tsvet (tsvyayt), who was born at Asti in Italy, entered Geneva University in 1891 and followed courses in physics, chemistry, and botany. In 1896, having presented his thesis on cell physiology, he moved to the biological laboratory at St. Petersburg, where he began working on plant pigments.

Before Tsvet started applying chemical and physical methods to pigment analysis it was thought that only two pigments, chlorophyll and xanthophyll, existed in plant leaves. Following established procedures, Tsvet soon demonstrated the existence of two forms of chlorophyll. However, the isolation of pigments became a much simpler matter once he had developed, in 1900, the technique of adsorption analysis. By 1911 Tsvet had found eight different pigments.

His technique involved grinding leaves in organic solvent to extract the pigments and then washing the mixture through a vertical glass column packed with a suitable adsorptive material (e.g., powdered sucrose). The various pigments traveled at different rates through the column due to their different adsorptive properties and were therefore separated into colored bands down the column. Tsvet first described this method in 1901 and in a publication of 1906 suggested it should be called "chromatography."

The technique is extremely useful in chemical analysis, being simple, quick, and sensitive, but it was not much used until the 1930s. Tsvet died when only 47 from overwork and the stress of the war, during which he was frequently transferred from one institute to another. He thus did not live to see the fruits of the wider application of chromatography in the hands of such scientists as Richard KUHN.

Tull, Jethro (1674–1741) *British agriculturalist, writer, and inventor*

> I am told that some pretenders to making the hoe-plow have fix'd its bottom to the plank immoveable, which makes it as useless for hoeing betwixt rows, as a violin with but one peg to its four strings would be for playing a sonata.
> —*The New Horse-Hoeing Husbandry* (1731)

Born at Basildon in the eastern English county

of Essex, Tull trained in law at Oxford University and was called to the bar in 1699 but gave this up to farm. He developed farming methods that became the basis of modern British agriculture. In about 1701 he invented a seed drill that sowed in regular straight rows. He also introduced a system of plowing that increased the water supply to plant roots and reduced the need for fallow land.

Tull's techniques were published in *The New Horse-Hoeing Husbandry: Or an Essay on the Principles of Tillage and Vegetation* (1731). At first they encountered opposition in Britain but they were more readily accepted in France, where VOLTAIRE supported them.

Turing, Alan Mathison (1912–1954) *British mathematician*

> We do not need to have an infinity of different machines doing different jobs. A single one will suffice. The engineering problem of producing various machines for various jobs is replaced by the office work of "programming" the universal machine to do these jobs.
> —Quoted by A. Hodges in *Alan Turing: the Enigma of Intelligence*

Turing, who was born in London, saw little of his parents in his early years. His father served in the Indian Civil Service before retiring in 1926; thereafter they lived in France to eke out a small pension. Turing was educated at Cambridge University, where he gained a fellowship at King's College in 1935. He spent the period from 1936 to 1938 at Princeton. During this time he published one of the most significant mathematical papers of the century, *On Computable Numbers* (1937).

He began by describing a hypothetical universal computer, since known as a *Turing machine*. It consists of an infinite length of tape divided into cells, a movable scanner/printer capable of reading the tape, printing, erasing, moving to the left and right, and halting. In each cell the tape has a symbol taken from a finite set of symbols (in a simple case 0 and 1). The control unit of the machine can be in one of a finite set of internal states (states S_1, S_2, S_3, etc.). The machine has a "program," which is a set of groups of five symbols. For example, one set of five symbols might be $S_1 01XS_2$, where X is R, L, or N. This is interpreted as meaning that the machine is in state S_1 (the first symbol of the five) and reading 0 (the second symbol). In this state it replaces the 0 by 1 (third symbol). If X = R it moves to the next cell on the right, if X = L it moves to the next on the left, and if X = N it does not move (it halts). Finally it goes into state S_2. The program, which can be used to do calculations, consists of a set of such quintuples (e.g., $S_1 00RS_1$, $S_1 10RS_2$, $S_2 01RS_3$, etc.).

Turing went on to define a set of integers N as computable if there was a Turing machine which, given any number m as input, will halt on 1 if m is a member of N, and halt on 0 otherwise. Using a variant of CANTOR's diagonal argument Turing proved, echoing earlier work by Kurt Godel and Alonzo Church, that some sets of integers are not computable.

Soon after the outbreak of war in 1939 Turing joined the government Code and Cypher School at Bletchley Park, Buckinghamshire. The Germans were known to be using a coding device called "Enigma." The basic model looked like an electric typewriter; the keyboard (input) was connected to the typed output by three rotors which changed position after each letter. This was equivalent to a polyalphabetic system with a periodicity of $26^3 = 17,576$. The military version of Enigma added a number of complications. The positions of the rotors could be changed, increasing the periodicity to 105,456. Further improvements increased the periodicity to over a trillion. The Germans felt that the military Enigma was "very close to practical insolvability."

Turing, along with a motley collection of mathematicians, linguists, and chess grandmasters, worked initially on the naval version of Enigma. The key innovation was the development of a computer, known as a "Bombe," to handle the vast amounts of traffic. The name derived from the fact that early models designed by Polish analysts ticked very loudly. The Bombe allowed numerous possible solutions to be quickly checked against traffic and eliminated. By March 1940 they were reading some of the traffic and finding that it consisted of nursery rhymes sent as practice transmissions. By June 1941 they were reading operational naval traffic. This, however, was not the end of Turing's work. In early 1942 the German navy adopted a new Enigma system and added a fourth rotor to its design. The sinking of Allied shipping rapidly increased. The breakthrough came in December 1942. The Germans were transmitting weather reports daily by Enigma machines using only the first three rotors. Cribs of the weather reports were provided rapidly from other sources. It was therefore only necessary for the Bombes to work through the 26 possibilities of the fourth rotor to decipher fully encrypted traffic. By the new year Allied shipping could once more be diverted away from known U-boat positions. Shortly after this Turing left Bletchley Park for nearby Hanslope Park where he worked for the rest of the war on speech encipherment. Few individuals can have contributed so much to the war effort, or have saved so many lives, as Turing. In 1946 he was awarded an OBE for his services.

Turing was reluctant to return to Cambridge and a career in pure mathematics. Consequently he accepted a position at the National Physical Laboratory working on the design of a new computer, ACE (Automatic Computing Engine). He moved to Manchester University in 1948 to undertake similar work on the development of MADAM (Manchester Automatic Digital Machine).

During this period Turing produced two influential papers. In his *Computing Machinery and Intelligence* (1950) he challenged his critics to specify how computers could be distinguished from intelligent human beings. In the imitation game the interrogator posed questions to two "individuals" A and B; he was asked to determine from their written replies which answer came from a computer. Both can of course lie. The *Turing test*, as the procedure is now called, has been cited as a way of testing machine intelligence and still causes debate and experiment. Turing himself was in no doubt that one day computers would be able to think.

Turing's other paper, *The Chemical Basis of Morphogenesis* (1952), concerned the generation of form. How can an assemblage of cells develop, as with the case of a starfish, a five-fold symmetry? Or how does a sphere of cells, in the process of gastrulation, form a groove at a specific point? He argued that it was possible for differences in chemical concentration to develop, even though the original situation had a uniform concentration. Turing's original model was mathematical. Chemists have, however, since found that there are systems, namely reaction–diffusion models, which mimic Turing's "morphogens."

Before he could develop his ideas further Turing committed suicide. He was homosexual and, for the times, fairly open about his life. A friend of a casual pick-up had stolen a few items of no great value from Turing's house and Turing reported the theft to the police. The culprit was arrested and in the course of investigations it was revealed that Turing had been sexually involved with a 19-year-old man. He was charged with gross indecency. Reluctantly he allowed himself to be persuaded to plead guilty and the court placed him on probation on the condition that he undergo hormone treatment. Although he seemed to find the process no more than irritating, and although his job remained secure, to the surprise of his family and friends he took his own life in 1954 by eating an apple dosed with cyanide.

Turner, David Warren (1927–) *British physical chemist*

Born at Leigh-on-Sea in Essex, England, Turner was educated at Exeter University and began his research at Imperial College, London. In 1965 he moved to Oxford. He is noted for his development of the technique known as molecular photoelectron spectroscopy. In this a narrow beam of monochromatic ultraviolet radiation is directed into a sample of gas; usually helium radiation (584 Å) is used. The gas is ionized and the energies of the ejected electrons are analyzed by electrostatic deflection. The energy of a particular electron equals the energy of the radiation minus the binding energy (ionization potential) of the electron in the atom or molecule. The technique thus determines the outer energy levels of the molecule, as well as information about the ions formed. It can also be used with solid samples to determine the energy bands. It is similar to the technique developed by Kai Siegbahn (known as *ESCA* – electron spectroscopy for chemical analysis), which uses higher-energy x-rays and measures the binding energies of inner electrons.

Turner subsequently developed ultraviolet photoelectron microscopy – a form of electron microscopy in which the electrons are ejected from the sample by x-rays or ultraviolet radiation. It is particularly suitable for studying processes on solid surfaces.

Turner, Herbert Hall (1861–1930) *British astronomer*

Turner, born the son of an artist in Leeds, England, studied mathematics at Cambridge University from which he graduated in 1882. On the recommendation of George Airy he was appointed chief assistant at the Royal Observatory, Greenwich, in 1884. He moved to Oxford in 1893 as professor of astronomy, a post he held until his death.

At Greenwich Turner did much to organize the Royal Observatory's contribution to the catalog and photographic chart of the sky, the Carte du Ciel, proposed in Paris by E. Mouchez in 1887. Once at Oxford he devoted much time and energy to the completion of Oxford's share in the project and the work was published shortly after Greenwich's contribution in 1909. Following the formation of the International Astronomical Union in 1919 he became president of the committee in charge of the Carte du Ciel and did much to try and get the chart and catalog finished.

Turner also became interested in seismology and after the death of his friend John Milne in 1913 took on the responsibility for the collection and dissemination of basic seismological data. He consequently started in 1918 and continued to edit until 1927 the quarterly *International Seismological Summary*, providing the origins and times of all detected earthquakes.

Tuve, Merle Antony (1901–1982) *American geophysicist*

Born at Canton in South Dakota, Tuve gained his BS degree in electrical engineering in 1922 from the University of Minnesota. He held posts at Princeton (1923–24) and Johns Hopkins (1924–26), receiving his PhD from the latter in 1926. From 1926 he was a staff member of the department of terrestrial magnetism of the Carnegie Institution of Washington.

Tuve is known principally for his techniques of radio-wave exploration of the upper atmosphere. In 1925 Tuve and Gregory Breit at Carnegie conducted some of the first experiments in range-finding using radio-waves in which they measured the height of the ionosphere. They transmitted a train of pulses of waves and determined the time each pulse took to return to Earth. Thereafter, pulse-ranging became the standard procedure for ionospheric research and laid the foundation for much of the later work on the development of radar.

In 1926 Tuve investigated long-range seismic refraction – the effect of different materials in the Earth's crust on the propagation of a seismic disturbance. He went on to construct an "upper-mantle velocities map" of America, which has been found to accord with theories of isostasy – a hydrostatic state of equilibrium in the distribution of materials of varying density in the Earth's interior.

During World War II Tuve worked for the Office of Scientific Research and Development, developing the proximity fuse, which stopped the "buzz bomb" attacks on Britain and Antwerp, among other projects. He returned to Carnegie in 1946 to become director of the department of terrestrial magnetism, a position he held up to 1966.

As well as seismic refraction and range-finding, Tuve also made studies of artificially produced beta and gamma rays, transmutations of atomic nuclei, and artificial radioactivity. He was for nine years editor of the *Journal of Geophysical Research*.

Twort, Frederick William (1877–1950) *British bacteriologist*

Twort, a doctor's son from Camberley in England, qualified in medicine in 1900 and after various appointments in London hospitals became professor of bacteriology at London University. His most important discovery was made during an attempt to grow viruses in artificial media: he noticed that bacteria, which were infecting his plates, became transparent. This phenomenon proved to be contagious and was the first demonstration of the existence of bacteria-infecting viruses. These were later called "bacteriophages" by the Canadian bacteriologist Felix d'Herelle, who discovered them independently.

Twort was also the first to culture the causative organism of Johne's disease, an important intestinal infection of cattle.

Tyndall, John (1820–1893) *British physicist*

> Watch the cloud-banner from the funnel of a running locomotive; you see it growing gradually less dense. It finally melts away

altogether, and if you continue your observations, you will not fail to notice that the speed of its disappearance depends upon the character of the day. In humid weather the cloud hangs long and lazily in the air; in dry weather it is rapidly licked up. What has become of it? It has been converted into true invisible vapor.
—"The Forms of Water" (1872)

Tyndall was born at Carlow (now in the Republic of Ireland) and after leaving school began work as a draftsman and civil engineer in the Irish Ordnance Survey. He later became a railway engineer for a Manchester firm. His drive for knowledge caused him to read widely and attend whatever public lectures he could. In 1847 he became a teacher of mathematics, surveying, and engineering physics at the Quaker school, Queenwood College, Hampshire.

The following year Tyndall entered the University of Marburg, Germany, to study mathematics, physics, and chemistry; after graduating in 1850 he worked in H. G. Magnus's laboratory in Berlin on diamagnetism. He was appointed professor of natural philosophy at the Royal Institution in 1853 and became a colleague and admirer of Michael Faraday; he succeeded FARADAY as director of the Royal Institution in 1867 and held this position until his retirement in 1887.

Tyndall's activities were many-sided. His chief scientific work is considered to be his researches on radiant heat; these included measurements of the transmission of radiant heat through gases and vapors published in a series of papers starting in 1859. But he is perhaps better known for his investigations on the behavior of light beams passing through various substances; he gave his name to the *Tyndall effect* – the scattering of light by particles of matter in its path, thus making the light beam visible – which he discovered in 1859. Tyndall elucidated the blue of the sky following the work of John RAYLEIGH on the scattering of light. He also discovered the precipitation of organic vapors by means of light, examined the opacity of the air for sound in connection with lighthouses and siren work, demonstrated that dust in the atmosphere contained microorganisms, and verified that germ-free air did not initiate putrefaction.

Tyndall was especially noted in his day as a great popularizer of science and advocate of scientific education, rather than as a great scientist. Among his many books for the nonspecialist the famous *Heat Considered as a Mode of Motion* (1863), the first popular exposition of the mechanical theory of heat, went through numerous editions. He was a member of the "X" Club, a group of prominent British scientists formed to ensure that the claims of science and scientific education were kept before the government of the day. He also helped to inaugurate the British scientific journal *Nature*. In 1872 and 1873 he undertook public lecture tours in America, giving the proceeds to a trust set up to benefit American science.

U

Uhlenbeck, George Eugene (1900–1988)
Dutch–American physicist

Uhlenbeck (**oo**-len-bek), who was born at Batavia (now Djakarta) in Indonesia, was educated at the University of Leiden, where he obtained his PhD in 1927. He then emigrated to America where he worked at the University of Michigan (1927–60), being appointed professor of theoretical physics in 1939. In 1960 he moved to the Rockefeller Medical Research Center at the State University of New York (now Rockefeller University) where he remained until his retirement in 1974.

Uhlenbeck's chief contribution to physics was made in 1925 when, in collaboration with Samuel GOUDSMIT, he discovered electron spin.

Ulam, Stanislaw Marcin (1909–1986)
Polish–American mathematician

The son of a lawyer, Ulam (**oo**-lam) was educated at the polytechnic in his native city of Lwow, Poland, where he completed his doctorate in 1933. This was a golden age of Polish mathematics. At coffee houses in Lwow and Warsaw, Ulam reported, BANACH, SIERPIŃSKI, and TARSKI could be found discussing the latest advances in set theory. Problems too difficult for an immediate solution were entered in a large notebook known as "The Scottish Book." It survived the war, and the occupation of Lwow by the Soviet Union and Germany, and was eventually translated and published by Ulam as *The Scottish Book* (Los Alamos, 1957).

Ulam quickly developed a reputation as an original mathematician and in 1936 he was invited to visit the Institute for Advanced Study, Princeton, for a year. He decided to remain in the United States and spent the period from 1937 to 1940 at Harvard before being appointed professor of mathematics at the University of Wisconsin. He became a naturalized American citizen in 1943.

At the same time Ulam was invited to work on the development of the atom bomb at Los Alamos, New Mexico, and he remained associated with Los Alamos until 1967. Here he worked on the theory of nuclear reactions. When neutrons are released in a reactor some scatter, some are absorbed, others escape or collide, etc. The actual process is too complex for analytical calculation. If, however, the fate of a practical number of neutrons is followed, and if at each branch one outcome with a suitable probability is selected, it is possible to derive a reasonably accurate mathematical model of the process. Ulam's technique, which is known as the "Monte Carlo method" (after the casino), is a widely used numerical method in many different fields.

After the war Ulam worked with TELLER on the development of the hydrogen bomb. He served as research adviser to the director of Los Alamos from 1957 to 1967, when he was appointed professor of mathematics at the University of Colorado, Boulder, a post he held until his retirement in 1977. He left an account of his life in his *Adventures of a Mathematician* (New York, 1976).

Ulugh Beg (1394–1449) *Persian astronomer*

> It is the duty of every true Muslim, man and woman, to strive after knowledge.
> —Quoting from the Hadith, the body of received tradition about the teachings and lives of Mohammed and his followers

Ulugh Beg (**oo**-luug beg) was a grandson of Tamerlane. He was born at Soltaniyeh (now in Iran) and succeeded to the throne in 1447 but was killed by his rebellious son two years later. He began building an observatory at Samarkand in 1428. He was himself an observer and published planetary tables and a catalog of stars in 1437. They were published in Europe in 1665. The observatory was reported to have had a gnomon for measuring the elevation of the Sun as high as the dome of St. Sophia in Byzantium (180 feet).

Unverdorben, Otto (1806–1873) *German chemist*

Born at Potsdam in Germany, Unverdorben (uun-fer-**dor**-ben) was a self-taught chemist who worked on the distillation of various organic compounds, publishing his results between 1824 and 1829. In 1826 he isolated a new substance from indigo, which he named "crystalline." Some years later August Hofmann isolated a substance from coal tar that he recognized as being the same as that obtained by Unverdorben from indigo. He called it aniline, from *al nīl*, the Arabic name for the indigo plant.

Urbain, Georges (1872–1938) *French chemist*

Urbain (oor-**ban**) studied chemistry at the Sorbonne, in his native city of Paris, receiving his PhD in 1899. After working in industry for some years he returned to the Sorbonne in 1908 as professor of mineral chemistry and became, in 1928, professor of general chemistry.

He is noted for his work on the lanthanoid elements and discovered the element lutetium in a sample of ytterbium in 1907.

Ure, Andrew (1778–1857) *British chemist*

Born in Glasgow, Scotland, Ure was educated at Edinburgh and Glasgow universities where he studied medicine. He succeeded George Birkbeck in 1804 as professor of chemistry and natural philosophy at Anderson's College, Glasgow. In 1830 he set himself up in London as an analytical and consultant chemist and in 1834 he became analytical chemist to the Board of Customs.

Ure's chief significance was as an educationalist and an apologist for the industrial revolution. In 1809 he started popular science lectures in Glasgow following the tradition established by George Birkbeck. He published a two-volume *Dictionary of Chemistry* (1821) and a *Dictionary of Arts, Manufactures, and Mines* (1839), which are in this popular tradition. Ure also wrote a more original and unusual work; in his *Philosophy of Manufactures* (1835) he portrayed a very different picture of the industrial revolution from that which is usually represented. In it he welcomed and praised the benefits of industrialization and the actual mode of life of the workers in the mines and factories. He argued, for example, that gaslight was an adequate replacement for sunlight and that machinery reduced the laborious work.

Ure's chemical studies included work on the specific gravity of solutions of sulfuric acid. He also invented a method of mercury extraction.

Urey, Harold Clayton (1893–1981) *American physical chemist*

Urey, born the son of a teacher and lay minister in Walkerton, Indiana, was educated at the universities of Montana, where he studied zoology, and California, where he obtained a PhD in chemistry (1923). After a year at the Institute of Theoretical Physics in Copenhagen, he began his teaching career at Johns Hopkins in 1924. In 1929 he moved to Columbia University, remaining there until 1958 when he became a professor at the University of California.

He is best known as the discoverer of deuterium – the isotope of hydrogen containing one proton and one neutron in its nucleus. This followed the accurate measurement of the atomic weights of hydrogen and oxygen by Francis W. Aston and the discovery of oxygen isotopes by William Giauque.

To obtain deuterium Urey used the fact that it would evaporate at a slightly slower rate than normal hydrogen. He took some four liters of liquid hydrogen, which he distilled down to a volume of one cubic centimeter. The presence of deuterium was then proved spectroscopically. Urey went on to investigate differences in chemical-reaction rate between isotopes. During World War II he was in charge of the separation of isotopes in the atomic-bomb project. Urey's research also led to a large-scale method of obtaining deuterium oxide (heavy water) for use as a neutron moderator in reactors.

His interest in isotope effects in chemical reactions gave him the idea for a method of measuring temperatures in the oceans in the past. It depended on the fact that the calcium carbonate in shells contains slightly more oxygen–18 than oxygen–16, and the ratio depends on the temperature at which the shell formed.

V

Vallisneri, Antonio (1661–1730) *Italian physician and biologist*

Born at Trassilico, near Modena in Italy, Vallisneri (or Vallisnieri; val-eez-**nyair**-ee) studied medicine at Bologna under Marcello MALPIGHI and at Reggio where he obtained his MD in 1684. After practicing medicine in Reggio, Vallisneri was appointed to the chair of medicine at the University of Padua in 1700 where he remained until his death.

Francesco REDI in 1668 had performed a famous experiment proving that the maggots in rotten meat were not spontaneously generated but arose, in the normal manner, from eggs laid by flies. He did however spoil the force of his argument by conceding that the larvae found in galls, for which he could find no eggs, were spontaneously generated. In 1700 Vallisneri plugged this gap in his *Sopra la curiosa origine di molti insetti* (On the Strange Origin of Many Insects) in which he reported success in detecting eggs of the insects in plant galls. In 1715 Vallisneri published *Origine delle fontane* (Origin of Fountains), which threw much light on another longstanding problem. Many ancient and medieval authorities were convinced that springs and rivers originated in the sea, and consequently the source of artesian wells, such as those at Modena, presented a problem. By exploring the local mountains, Vallisneri found that the rain and melting snow ran into fissures and formed subterranean rivers. Such rivers, passing under Modena at high pressure, would readily produce "fontane" if deep enough shafts were sunk.

As a biologist and anatomist Vallisneri also produced a number of treatises on such unfamiliar animals as the ostrich (1712) and the chameleon (1715). His studies of a group of aquatic plants led to the genus *Vallisneria* being named for him.

Van Allen, James Alfred (1914–2006) *American physicist*

Van Allen was born at Mount Pleasant, Iowa. After graduating from the Iowa Wesleyan College in 1935 he went on to the University of Iowa, where he gained his PhD in 1939. His subsequent career took him to the Carnegie Institution of Washington (1939–42), as a physics research fellow in the department of terrestrial magnetism, and to the applied physics laboratory at the Johns Hopkins University (1942 and 1946–50). In 1951 Van Allen returned to the University of Iowa as a professor of physics and head of the department of physics and astronomy, retiring in 1990.

During the war years (1942–46) he served as an ordnance and gunnery specialist and combat observer with the U.S. Navy and developed the radio proximity fuse, a device that guided explosive weapons, such as antiaircraft shells, close to their targets and then detonated them. He also gained considerable expertise in the miniaturization of electronics and rocket controls, which he later put to use in the scientific exploration of the Earth's upper atmosphere.

After the war Van Allen was able to use German V-2 rockets for atmospheric studies and was associated with the development of the Aerobee sounding rocket. He also used rocket–balloon combinations that could carry small rockets to higher altitudes. In the years 1949–57 he organized and led several scientific expeditions (to Peru, the Gulf of Alaska, Greenland, and Antarctica) to study cosmic rays – highly energetic particles arriving from space. The direction of all this work led to the launching in January 1958 of America's first satellite, Explorer I (as part of a major International Geophysical Year series of experiments), which carried experiments designed to measure cosmic rays and other energetic particles. Unexpectedly high radiation levels were found in certain regions of the Earth's atmosphere – so high that the satellite's Geiger counters jammed. This observation was contrary to the observation of the first Russian satellite, Sputnik I, launched five months earlier, and gave impetus to further satellite exploration. Subsequent observations by a succession of satellites (Explorer, Pioneer, Sputnik, Mechta, Lunik) have shown that the Earth's magnetic field traps high-speed charged particles in two zones girdling the Earth, with the greatest particle concentration above the equator. One zone lies roughly 600–3,000 miles (1,000–5,000 km) above the Earth's surface; the other is 9,000–15,000 miles (15,000–25,000 km) above the equator, curving down toward the magnetic poles. These regions were later to be named the *Van Allen belts*. The particles in the belts are electrons and protons (as suggested by

F. Singer early in 1957) originating in cosmic-ray collisions or captured from the "solar wind" of particles that streams out from the Sun. In 1958 a controversial experiment, known as "Project Argus," was carried out – also as part of the International Geophysical Year. This involved the detonation of three small nuclear bombs, at altitudes over 300 miles (480 km) over the South Atlantic Ocean, to inject very energetic particles into the upper atmosphere. These were subsequently found to have been captured in the Van Allen belts.

Van Allen produced over 200 scientific papers, received a great number of scientific awards, and was a member of several U.S. governmental committees concerned with space exploration.

Van de Graaff, Robert Jemison (1901–1967) *American physicist*

Born in Tuscaloosa, Alabama, Van de Graaff studied engineering at the University of Alabama, gaining his BS in 1922 and his MS in 1923. He enrolled in 1924 at the Sorbonne in Paris where he was inspired by the lectures of Marie CURIE to study physics. In 1928 he obtained a PhD from Oxford University for research into the motion of ions in gases. It was during these studies that he conceived of an electrostatic generator that could radically improve on existing types, such as the Wimshurst machine, by building up electric charge on a hollow insulated metal sphere. A year later he returned to America and started working as a research fellow at Princeton. In 1931 he moved to the Massachusetts Institute of Technology as a research associate, serving as associate professor of physics from 1934 until he resigned in 1960.

While at Princeton he constructed, in 1931, the first model of his generator, now known as the *Van de Graaff generator*. The charge was carried to the hollow sphere by means of an insulated fabric belt and once transferred could accumulate on the outer surface of the sphere, leading ultimately to potentials of 80,000 volts. This was eventually increased to over a million volts.

At MIT Van de Graaff developed the generator for use as a particle accelerator. This *Van de Graaff accelerator* used the generator as a source of high voltage that could accelerate charged particles, such as electrons, to high velocities and hence high energies. It was thus to be a major tool in the developing fields of atomic and nuclear physics. One of Van de Graaff's aims was to explore the possibility of uranium fission and to try to create elements with larger atoms than uranium.

In collaboration with John Trump, an electrical engineer, he adapted the generator to produce high-energy x-rays, which could be used in the treatment of cancer. The first x-ray

generator began operation in a Boston hospital in 1937. During World War II Van de Graaff was director of the radiographic project of the Office of Scientific Research and Development in which the generator was developed for another use: the examination of the interior structure of heavy ordnance by means of x-rays.

In 1946 Trump and Van de Graaff formed the High Voltage Engineering Corporation to market Van de Graaff accelerators and x-ray generators to hospitals, industry, and scientific research establishments. Van de Graaff was director and chief physicist and in 1960 left MIT to work there full time as chief scientist.

van de Hulst, Hendrik Christoffel (1918–2000) *Dutch astronomer*

Van de Hulst (van de hoolst) studied at the university in his native city of Utrecht where he obtained his PhD in 1946. He spent two years at the University of Chicago as a postdoctoral fellow (1946–48) then took up an appointment at the University of Leiden. Following several years at different universities in America he became professor of astronomy at Leiden and director of the Leiden Observatory.

In the German-occupied Netherlands a group of astronomers began to think about the implications of Grote REBER's discovery of radio emission from the Milky Way. Jan OORT, then director of the Leiden Observatory, wondered whether the emission was merely a uniform noise or could the equivalent of the absorption and emission lines of light be detected. In 1944 van de Hulst came up with the answer.

He proposed that hydrogen atoms, which occur in diffuse but widespread regions in interstellar space, can exist in two forms. In ordinary hydrogen the proton and its orbiting electron spin in the same direction. There is a very small chance that the electron will spontaneously flip over and spin in the opposite direction to the proton and in the process emit radiation with a wavelength of 21 centimeters, i.e., at a frequency of 1,420.4 megahertz. Although emission from a single atom is a very rare event, there is such an abundance of hydrogen in the universe that the process should be taking place with sufficient frequency to be detectable.

It was not, however, until 1951 that van de Hulst's proposal was confirmed: the 21-centimeter hydrogen line emission was detected at Harvard by Edward M. PURCELL and Harold Ewen, who used equipment specially built for the purpose. It has since proved a crucial tool for the investigation of the distribution and movement of neutral hydrogen in our own galaxy and in other spiral galaxies. Since the hydrogen lies in the spiral arms and since the 21-centimeter emission passes unimpeded through the dust that prevents optical observation, this has led to a greatly increased knowledge of galactic structure.

It was also felt that if intelligent life did exist outside the solar system and wished to communicate with other civilizations, then it would be highly rational to transmit signals at the hydrogen-emission wavelength of 21 centimeters. Although much expensive telescope time has been spent listening on this wavelength, nothing has yet been heard that does not come from hydrogen itself.

van de Kamp, Peter (1901–1995) *Dutch–American astronomer*

Van de Kamp, who was born at Kampen in the Netherlands, studied at the University of Utrecht, obtaining his PhD in 1922. He emigrated to America in 1923 and, after appointments at the Lick Observatory (1924–25) and the University of Virginia (1923–24, 1925–37), became director of the Sproul Observatory and professor of astronomy at Swarthmore College, Pennsylvania. Following his retirement in 1972 he became research astronomer there.

In the 1960s van de Kamp found strong evidence for a new celestial phenomenon: planets orbiting stars other than the Sun. Since 1937 he had been studying the motion of BARNARD's star, a nearby red star with a very large proper motion of 10.3 seconds of arc per year. By 1969 he was able to state that the star was oscillating very slightly in position about a straight line and that this wobbling motion was caused by an unseen companion. This companion was orbiting Barnard's star in about 25 years and was only about 1.5 times the mass of Jupiter. As this is too small a mass for a star, van de Kamp concluded that it was a planet. After further calculations he said that it was more likely that there were two planets, both of a similar mass to Jupiter, one orbiting in about 12 years and the other in 26 years.

This was not the only search for planets. In 1943 K. A. Strand claimed that he had detected a planetary body in the binary star 61 Cygni. His evidence was not as good as van de Kamp's however, although more recent observations indicate a planet about eight times the mass of Jupiter orbiting the brighter component of 61 Cygni in 4.8 years. There are other stars that have also shown perturbed motions from planet-sized companions.

The planets belong to stars that are relatively close. It is unlikely that this is merely coincidental and consequently van de Kamp and others concluded that planetary systems are likely to be widespread. Such a conclusion implies that some form of life may well be present outside the solar system.

van der Meer, Simon (1925–) *Dutch engineer and physicist*

Van der Meer (van der mayr) was educated at the Gymnasium in his native city of The Hague and at the Technical University, Delft, where he gained his PhD in 1956. He immediately joined the staff at the European Organization for Nuclear Research (CERN) and remained there until his retirement in 1990.

In 1979 the Nobel Prize for physics was awarded to Sheldon GLASHOW and two colleagues for their unification of the electromagnetic and weak forces. Although the neutral currents predicted by the theory were detected in 1973, it still remained to discover the charged W^+ and W^- and the neutral Z^0 bosons whose existence was a consequence of the theory. As the masses of the particles were about 80 times that of the proton, the energy required for their production outstripped the capacity of any existing accelerator. In 1978 Carlo RUBBIA, a colleague at CERN, asked van der Meer if there was any way to conjure such high energies from the existing accelerators.

CERN's SPS (Super Proton Synchroton) could deliver about 450 billion electronvolts (450 GeV). One possible solution would be to convert the SPS into a colliding-beam machine, that is, protons and antiprotons would be accelerated, stored separately, and then induced to collide with each other head-on. Proton–antiproton collisions, it was calculated, with an energy of 270 GeV per beam were equivalent to a beam of 155,000 GeV colliding with a stationary target.

The problem facing van der Meer was how to concentrate the beams. Protons normally repel each other, as do antiprotons, and, consequently, charged particle beams tend to spread out in space. To maximize the colliding power of the beams van der Meer somehow needed to focus them. He proposed to use the technique of "stochastic cooling," first described by him in 1972 as a way of reducing random motion in the beam. To achieve this the exact center of the beam was calculated and correcting magnetic fields were applied by a system of "kickers" placed around the ring. By this means particles out of line were nudged back into position. The system was successfully tested in May 1979 and was used in 1983 to create the W and the Z particles. Van der Meer shared the 1984 Nobel Prize for physics with Rubbia for this work.

van der Waals, Johannes Diderik (1837–1923) *Dutch physicist*

> This at once put his name among the foremost in science.
> —James Clerk Maxwell, on van der Waals's thesis *On the Continuity of the Liquid and Gaseous States* (1873)

Van der Waals (van der vahls or van der wahlz) was born at Leiden in the Netherlands. He was largely self-taught in science and originally worked as a school teacher. He later managed to study at the University of Leiden, having been exempted from the Latin and Greek entrance requirements. In 1877 he became professor of physics at the University of Amsterdam.

Van der Waals studied the kinetic theory of gases and fluids and in 1873 presented his influential doctoral thesis, *On the Continuity of the Liquid and Gaseous States*. His main work was to develop an equation (the *van der Waals equation*) that – unlike the gas laws of Robert BOYLE and Jacques CHARLES – applied to real gases. The Boyle–Charles law, strictly speaking, applies only to "ideal" gases but can be derived from the kinetic theory given the assumptions that there are no attractive forces between gas molecules and that the molecules have zero volume.

Since the molecules do have attractive forces and volume (however small), van der Waals introduced into the theory two further constants to take these properties into account. Initially these constants had to be specific to each gas since the size of the molecules and the attractive force between them is different for each gas. Further work by van der Waals yielded the law of corresponding states – an equation that is the same for all substances. His valuable results enabled James DEWAR and Heike KAMERLINGH-ONNES to work out methods of liquefying the permanent gases.

In 1910 van der Waals was awarded the Nobel Prize for physics for his work on the equation of state of gases and liquids. The weak electrostatic attractive forces between molecules and between atoms are called *van der Waals forces* in his honor.

Vane, Sir John Robert (1927–2004)
British pharmacologist

Vane studied chemistry at the University of Birmingham and pharmacology at Oxford, where he obtained his DPhil in 1953. He then worked at the Royal College of Surgeons, serving as professor of experimental pharmacology from 1966 until 1973, when he moved to the Wellcome Foundation as director of research and development. In 1985 Vane left Wellcome to serve as director of the William Harvey Research Institute, St. Bartholomew's Hospital, London. He was knighted in 1984.

Vane worked on hormonelike substances, the prostaglandins, first observed by Ulf VON EULER in the 1930s. In the 1960s he began to explore their physiological roles. He extracted in 1969 a substance from the lung tissue of rats sensitive to an allergen. As it caused rabbit aortas to contract it was named "rabbit aorta contracting substance" (RCS). He also found that RCS caused blood platelets to clot. It was later shown by Bengt SAMUELSSON that RCS contained the prostaglandin PGH_2 as an active in-

gredient. But Vane had earlier shown that the effects of RCS could be inhibited by aspirin and other antiinflammatory drugs. This allowed him to propose a mechanism for both the effects of aspirin and prostaglandins. Aspirin, he argued, reduced pain, inflammation, and fever by blocking the action of prostaglandins which, at least in some cases, seemed to produce precisely these effects. For his work in this field Vane shared the 1982 Nobel Prize for physiology or medicine with Bengt Samuelsson and Sune BERGSTRÖM.

He also worked on the pharmacological effects of adrenaline (epinephrine) and edited the CIBA Foundation symposium on the subject, *Adrenergic Mechanisms* (1960).

van Maanen, Adriaan (1884–1946)
Dutch–American astronomer

Van Maanen (van **mah**-nen), who was born at Sneek in the Netherlands, studied at the University of Utrecht, where he obtained his doctorate in 1911. He worked at the University of Groningen from 1909 to 1911 when he moved to America. After working briefly at the Yerkes Observatory he took up an appointment in 1912 at the Mount Wilson Observatory in California, where he remained for the rest of his career.

Van Maanen specialized in measuring the minute changes in position of astronomical objects over a period of time, from which he could determine their proper motion and parallax. These objects included stars, clusters of stars, and nebulae. Between 1916 and 1923 he produced a number of measurements of spiral nebulae from which he calculated their rotation rate. This was at the time of the controversy between astronomers as to whether such nebulae were island universes in their own right or part of our own Galaxy. Van Maanen's results were therefore of considerable significance, for whether one can detect rotation in a distant body and measure its rate of rotation partly depend on the distance of the object from the observer. The rotation of about 0.02 seconds of arc per year that he obtained from a number of nebulae over a period of seven years seemed to be indisputable evidence against the emerging view that such objects as the Andromeda nebula were really separate remote star systems. This view was certainly the view that Harlow SHAPLEY took in 1920 as he could see no reason to doubt the work of van Maanen, his close friend, who was known to be a careful and competent observer.

Van Maanen's work was, however, incompatible with the growing body of measurements produced by Edwin HUBBLE, also at Mount Wilson. These suggested that nebulae like Andromeda were as much as 800,000 light-years away, at which distance it was inconceivable that any internal motion should be detectable. As no one, including van Maanen himself, had any idea where he had gone wrong, his work became something of a curiosity and tended to be ignored. Other astronomers, like Knut LUND-MARK in 1923, failed to reproduce his results while in 1935 van Maanen was able to detect a displacement only about half that found in the 1920s. When in the same year Hubble also reported failure to detect the rotation, it became widely accepted that there was some unknown instrumental or personal error in van Maanen's work.

It is unlikely to have been an instrumental error as Hubble used the same instruments while a later computer analysis of his work has revealed no major computational errors. This only leaves unconscious personal error and recent work has shown that a systematic error of only 0.002 millimeter in the measurements of points on photographic plates would be sufficient to produce his results. The same systematic error also occurred in his work on the strength of the solar magnetic field, which he considerably overestimated.

He did however discover the second white dwarf, since named *van Maanen's star*, with a density some 400,000 times that of the Sun.

van't Hoff, Jacobus Henricus (1852–1911) *Dutch theoretical chemist*

> A Dr. J. H. van't Hoff, of the veterinary school at Utrecht, has as it seems, no taste for exact chemical investigation.
> —Hermann Kolbe attacking van't Hoff's theory of molecular structure in *Journal für praktische Chemie* (1877; Journal of Practical Chemistry)

Van't Hoff (vahnt hoff) was born at Rotterdam in the Netherlands, the son of a physician. He studied at Delft Polytechnic and the University of Leiden before going abroad to work with August KEKULÉ in Bonn (1872) and with Charles Adolphe WURTZ in Paris (1874), where he met Joseph-Achille LE BEL. In 1878 he was appointed to the Amsterdam chair of chemistry where he remained until moving to the University of Berlin in 1896.

In 1874 van't Hoff published a paper entitled *A Suggestion Looking to the Extension into Space of the Structural Formulas at Present Used in Chemistry*, which effectively created a new branch of science – stereochemistry. The problem began with the discovery of optically active compounds. Louis PASTEUR later established the asymmetry of crystals of tartaric acid: some would rotate polarized light to the right and others to the left. This was explained by the actual asymmetry of the crystal: the crystals were mirror images of each other. Pasteur thought that the molecules themselves were asymmetric but could offer no proof. This would explain the further problem of the optical activity of noncrystalline solutions. Van't Hoff solved these problems and offered an account of molecular asymmetry by concentrating on the structure of the carbon atom, newly established by Kekulé. He announced (1874) that the four chemical bonds that carbon can form are directed to the corners of a tetrahedron. With this structure, certain molecules can have left-and right-handed isomers, which have opposite effects on polarized light. It also explained why certain isomers do not occur.

Van't Hoff's account of molecular structure was attacked by Hermann Kolbe but a similar theory was put forward simultaneously by Joseph-Achille Le Bel, independently of van't Hoff. Despite the hostility, his ideas were soon vindicated by Emil FISCHER's researches into sugars in the 1880s.

Major contributions were also made by van't Hoff to the thermodynamics and kinetics of solutions. Many of these results are reported in his book *Etudes de dynamique chimie* (1884; Studies in Chemical Kinetics). He had the central insight in 1886 that there is a similarity between solutions and gases provided that osmotic pressure is substituted for the ordinary pressure of gases, and derived laws for dilute solutions similar to those of Robert BOYLE and Joseph GAY-LUSSAC for gases. This fundamental result could be used to determine the molecular weight of a substance in solution.

In 1901 van't Hoff was awarded the first Nobel Prize for chemistry.

Van Vleck, John Hasbrouck (1899–1980) *American physicist*

Van Vleck was born at Middletown, Connecticut, and educated at the University of Wisconsin, where he graduated in 1920. Moving to Harvard University he gained his master's degree (1921) and his doctorate (1922) and stayed for a further year as an instructor. From Harvard he went to the University of Minnesota, where he became a full professor in 1927, returned to Wisconsin in 1928, and then went back to Harvard in 1934.

Van Vleck is regarded as the founder of the modern quantum mechanical theory of magnetism. His earliest papers were on the old quantum theory, but with the advent of wave mechanics pioneered by Paul DIRAC he began to look at the implications for magnetism in particular. In the field of paramagnetism he introduced the concept of temperature-independent susceptibility, now known as *Van Vleck paramagnetism*. He also made calculations of molecular structure that shed new light on chemical bonding and he developed ways of describing the behavior of an atom or an ion in a crystal. Another important contribution of Van Vleck was to point out the importance of electron correlation – the interaction between the motion of electrons – for the appearance of local magnetic moments in metals.

During World War II Van Vleck worked on radar, showing that at about 1.25-centimeter wavelength water molecules in the atmosphere would lead to troublesome absorption and that at 0.5-centimeter wavelength there would be a similar absorption by oxygen molecules. This was to have important consequences not just for military (and civil) radar systems but later for the new science of radioastronomy.

In 1977, together with Nevill MOTT and Philip ANDERSON, he shared the Nobel Prize for physics for "fundamental theoretical investigations of the electronic structure of magnetic and disordered systems." (Anderson was once a student of Van Vleck's at Harvard.) Van Vleck's work on electron correlation was mentioned specifically for the central role it played in the later development of the laser.

Varenius, Bernhard (1622–1650) *German physical geographer*

Varenius (far-**ay**-nee-uus) was born at Hitzacker in Germany and educated at the Hamburg gymnasium and the universities of Königsberg and Leiden, where he studied medicine. In 1649 he published his first geographical work, *Descriptio Regni Japoniae* (Description of the Kingdom of Japan).

His main work was his *Geographia generalis* (1650), published only shortly before his death. Some idea of the importance the 17th century attached to the work can be discerned by noting that none less than Isaac NEWTON issued a revised and enlarged edition in 1672. It was first translated into English in 1693 and went into its last English edition in 1765. In this work Varenius laid down a framework for physical geography. He defined his universal geography as that which "considers the whole Earth in general, and explains its properties without regard to particular countries." He divided it into three parts, the first part consisting of facts about the Earth, such as its dimensions. The second part consisted of facts related to celestial events, such as the length of days and nights or the seasons, and the third part was comparative.

Varmus, Harold Eliot (1939–   ) *American microbiologist*

Born at Oceanside, New York, Varmus was educated at Amherst College, Harvard, and at Columbia University, where he studied medicine. After working at the Presbyterian Hospital, New York (1966–68), he joined the National Institutes of Health, Bethesda, as clinical associate (1968–70) before moving to the Department of Microbiology at the University of California at San Francisco. He was appointed full professor in 1979. Varmus was director of the National Institutes of Health (1993–2000).

In the 1970s, working in collaboration with Michael BISHOP, Dominique Stehelin, and Peter Vogt, Varmus made a crucial breakthrough in our understanding of cancer. The ROUS sarcoma virus, which causes cancer in chickens, was known to have a particular gene (called an "oncogene") associated with its cancer-causing capability. Varmus and his colleagues prepared a molecular probe for this gene – a fragment of DNA with a base sequence complementary to the gene and capable of pairing with it – and demonstrated that the gene was present in the cells of normal chickens.

This showed for the first time that viral oncogenes are derived from genes of the virus's host, incorporated into the genetic material of the virus in a modified form. This breakthrough led to the discovery of a large number of similar cellular genes, subsequently termed "proto-oncogenes," that acted as a source of oncogenes. The key to understanding how viral oncogenes transform a normal cell into a cancerous one thus lies in determining how their equivalent proto-oncogenes function in normal cells. Several roles have been elucidated for these genes, principally the regulation of cell growth, division, and differentiation. Interference or disturbance in these processes, as may occur in the presence of an oncogene, could lead to uncontrolled cell proliferation, as in cancer.

The results of Varmus's work were published in 1976 and opened the door to a major new field in cancer research. For this work Varmus and Bishop were jointly awarded the 1989 Nobel Prize for physiology or medicine.

Varolio, Constanzo (1543–1575) *Italian anatomist*

Varolio (vah-**roh**-lee-oh) was educated at the university in his native city of Bologna where he qualified as a physician in 1567; he later served as professor of anatomy and surgery there from 1569 until 1572. He then moved to Rome and, although he is thought to have served as physician to Pope Gregory XIII and taught at the Sapienza or Papal University, there is no real documentary evidence for either claim.

He is mainly remembered for his *De nervis opticis* (1573; On the Optic Nerves) in which he described the part of the brain situated between the midbrain and the medulla oblongata, since known as the pons varolii. His work contains what has been called a "crude illustration" of the base of the brain clearly showing the pons. A far superior drawing had been made by Bartolommeo EUSTACHIO 20 years previously but his work was lost to science for about 150 years.

Vauquelin, Louis Nicolas (1763–1829) *French chemist*

The son of a farm laborer from Saint-André

d'Hebertot in France, Vauquelin (voh-**klan**) began work as an apprentice to a Rouen apothecary. He became a laboratory assistant to Antoine-François FOURCROY (1783–91), with whom he later collaborated. Vauquelin became a member of the French Academy of Sciences in 1791 and professor of chemistry in the School of Mines in 1795. In 1799 he wrote *Manuel de l'essayeur* (An Assayer's Manual), which led to his being appointed assayer to the mint in 1802 and professor of chemistry at the University of Paris in 1809.

Vauquelin is best known for his discovery of the elements chromium and beryllium. In 1798, while working with a red lead mineral from Siberia known as crocolite, he isolated the new element chromium – so called because its compounds are very highly colored. Martin KLAPROTH made a similar discovery shortly afterward. In the same year Vauquelin also isolated a new element in the mineral beryl. It was initially called glucinum because of the sweetness of its compounds, but later given its modern name of beryllium. He was the first to isolate an amino acid: asparagine from asparagus.

Vavilov, Nikolai Ivanovich (1887–1943)
Russian plant geneticist

> We shall go to the pyre, we shall burn, but we shall not renounce our convictions.
> —Quoted by Z. A. Medvedev in *The Rise and Fall of T. D. Lysenko* (1969)

Having graduated from the Agricultural Institute in his native city of Moscow, Vavilov (**vav**-i-lof) continued his studies firstly in England under William BATESON and then in France at the Vilmoren Institution. Back in Russia he was appointed, in 1917, both professor of genetics and selection at the Agricultural Institute, Voronezh, and professor of agriculture at Saratov University. Three years later he took over the directorship of the Bureau of Applied Botany, Petrograd (now St. Petersburg), which later became the All Union Institute of Plant Industry. The institute flourished under Vavilov's leadership, becoming the center for over 400 research institutes throughout the Soviet Union. In 1929 he became the first president of the Academy of Agricultural Sciences.

During the years 1916–1933 Vavilov led several plant-collecting expeditions to countries all around the globe. The purpose was to gather material of potential use in crop-breeding programs, particularly the wild relatives and ancestors of cultivated plants. He was highly successful in this, his collection numbering some 250,000 accessions by 1940. This was the first large-scale attempt to conserve and utilize the immensely valuable genetic resources upon which crop improvement relies.

A second important consequence of these travels was Vavilov's observation that the genetic diversity of crop relatives is concentrated in certain areas that he termed "gene centers," postulating that these correspond to regions where agriculture originated. The theory and the exact number of centers have since been modified but the recognition of such areas is an invaluable aid to other plant hunters. He also found certain regularities between unrelated genera in such centers, described in *The Law of Homologous Series in Variation* (1922).

Vavilov's excellent work was gradually stifled by the intrusion of politics into Soviet biology in the 1930s. His belief in the advances in genetics made by MENDEL and T. H. MORGAN brought him into conflict with the government-backed Trofim Lysenko, who was returning to a Lamarckian view of inheritance. The 1937 International Congress of Genetics, due to be held in Moscow in view of the strides made in Soviet genetics under Vavilov, was canceled by the Lysenkoists. Vavilov was arrested in 1940 while plant collecting and died three years later in a Siberian labor camp.

Today Vavilov is recognized in his own country as an outstanding scientist, the Vavilov Institute being named in his honor.

Veksler, Vladimir Iosofich (1907–1966)
Ukrainian physicist

Veksler (**vyayks**-ler), who was born at Zhitomar in the Ukraine, graduated from the Moscow Energetics Institute in 1931. He first worked for the Electrochemical Institute (1930–36), then moved to the Lebedev Physics Institute of the Academy. In 1956 he was appointed director of the High Energy Laboratory of the Institute of Nuclear Physics. In 1944 Veksler invented the synchrocyclotron independently of Edwin MCMILLAN.

Veltman, Martinus J(ustinus) G(odefridus) (1931–) *Dutch theoretical physicist*

Veltman (**velt**-man) was born in Waalwijk in the Netherlands and educated at the University of Utrecht, gaining his doctoral degree in physics in 1963. After obtaining his doctoral degree he did research at CERN, SLAC, and Brookhaven. In 1966 he returned to the University of Utrecht as professor of physics. In 1981 he left Utrecht to become a professor of physics at the University of Michigan. He retired in 1996 and has lived in Bilthoven in the Netherlands since then.

In the late 1960s Veltman was one of the few people working on the technically difficult problem of the renormalization of non-Abelian gauge theories. The major breakthrough which solved this problem came in 1971 with the work of Gerardus 't Hooft, a graduate student of Veltman. 't Hooft and Veltman then studied the renormalization of gauge theories in great detail. This work was significant because it meant that finite answers could be obtained in calculations in the gauge theories, which describe the strong nuclear interactions and which unify the weak nuclear and electromagnetic interactions, thereby setting the stage for what has become known as the *standard model* of the elementary particles and the interactions which govern their behavior.

In the mid-1970s Veltman initiated the study of high-order corrections in the non-Abelian gauge theory that unifies the weak and electromagnetic interactions. These calculations go beyond the simplest type of calculations and are very important because comparison of the results with precise experimental observations enables the standard model to be tested stringently and may provide evidence for physics beyond the standard model.

Veltman shared the 1999 Nobel Prize for physics with 't Hooft for their fundamental work in showing that the process of renormalization could be applied to non-Abelian gauge theories, including the type of non-Abelian gauge theory that unifies the weak and electromagnetic interactions.

Vening Meinesz, Felix Andries (1887–1966) *Dutch geologist*

The son of the burgomaster of Rotterdam, Vening Meinesz (**vay**-ning **mI**-nes) was born in the Dutch capital The Hague. He graduated from the University of Delft in 1910 and, after working on a gravimetric survey of the Netherlands, was appointed in 1927 to the chair of geodesy, cartography, and geophysics at Delft.

Vening Meinesz developed, in 1923, a method for measuring gravity below the oceans using instruments carried by submarine. Measurements of the Earth's gravitational field had been made since Jean RICHER's pendulum measurements in Cayenne in 1671. Despite the sophistication of later expeditions the measurements all suffered from the limitation of being continental; it was not possible to obtain accurate pendulum recordings at sea. Vening Meinesz first used his new method in a submarine voyage to Java (1923) and collected significant new data. During the period 1923–39 he made 11 voyages throughout the oceans of the world making 843 observations.

One of his most significant findings was that there are areas of weak gravity extending in long arcs through both the East and West Indies and along the oceanic side of Japan. These areas appear to coincide with the deep ocean trenches. Vening Meinesz further developed a theory of convection currents within the Earth similar to that formulated by Arthur HOLMES in 1929.

Venter, J(ohn) Craig (1946–   ) *American geneticist*

For over two decades Venter has been a leading and controversial figure in the rapidly developing field of genomics. He was born in Salt Lake City, and served in the US Navy in Vietnam (1967–68) before enrolling at the University of California, San Diego, where he received his bachelor's degree in biochemistry and a PhD in physiology and pharmacology. After serving as professor at the State University of New York and Roswell Park Cancer Institute, he moved in 1984 to the National Institutes of Health (NIH). He founded The Institute for Genomic Research (TIGR) in 1992, and in 1998 Celera Genomics, serving as its first president. In 2006, TIGR and several other affiliated organizations were merged to form the J. Craig Venter Institute, based in Rockville, Maryland, and La Jolla, California.

Throughout his career, Venter has been a pioneer of new techniques for investigating genetic material of living organisms, the field known as genomics. Central to this are automated methods for reading the base sequence of DNA, and computerized methods for storing, retrieving, and analyzing the data. When working for the NIH, he devised the technique of characterizing genes using short partial sequences called expressed sequence tags (ESTs). These serve as unique "labels" for genes, and are determined from a single read of a cloned DNA fragment, much more rapidly than the multiple reads of numerous clones required to determine a complete gene sequence. After leaving the NIH to set up TIGR, Venter proposed a novel approach to sequencing entire genomes called the whole-genome shotgun technique. This relied on computational power to assemble the sequence reads from thousands of clones into the correct order, without the time-consuming process of constructing a map indicating the location of sequences relative to each other. Venter's approach was rather like fitting the pieces of a jigsaw together without the full image to refer to. Venter and his colleague Hamilton SMITH applied the technique to the bacterium *Haemophilus influenzae*, and in 1995 announced its complete DNA sequence – the first genome of a free-living organism to be sequenced.

In 1998 Venter announced that his new company, Celera, intended to use high-tech automated DNA sequencing machines to sequence the entire human genome in just 2–3 years. This meant that Celera would potentially be able to hold patent rights over fundamental biological information that most scientists thought should be freely available to all. A race then ensued between Celera and the publicly financed collaborative project, the International Human Genome Sequencing Consortium (IHGSC). Venter lobbied the US Congress to withdraw funding from the IHGSP, but to no avail. Already Venter's team had used the whole-genome shotgun approach to sequence the more complex genomes of animals such as the fruit fly *Drosophila*. The IHGSC was using a more systematic approach, including the initial mapping step. By 2000 both groups were able to announce draft sequences, which were incomplete and error-strewn. Nonetheless, Celera imposed restrictions on the use of its data. However, it was left to the IHGSC to produce the finished sequence, in April 2003. Meanwhile, in 2002, after leaving Celera, Venter announced that he had secretly bypassed company procedures ensuring anonymity of DNA donors and used his own DNA for much of the private sequencing project.

He was censured by Celera's own ethics board. A project to collect DNA samples from microbes in the world's oceans was launched by Venter in 2003. The Global Ocean Sampling program collects bacteria and analyzes their DNA en masse. This metagenomics approach reveals the diversity of organisms found at various sites, and provides insights into their evolutionary and functional diversity. Venter hit the headlines again in 2008 when he announced the construction of a synthetic chromosome, based on sequence data derived from the bacterium *Mycobacterium genitalum*. The goal is to transfer the artificial genome to a bacterium to direct the functions of the recipient cell, thereby effectively creating a new, patented, life form. The news was greeted with alarm by some bioethics groups.

Vernier, Pierre (*c.* 1580–1637) *French mathematician*

Born at Ornans in France, Vernier (vair-**nyay** or **ver**-nee-er) was educated by his father, a scientist, and became interested in scientific instruments. He was employed as an official with the government of Spain and then held various offices under the French government.

In 1631 Vernier invented the caliper named for him, an instrument for taking very precise measurements. The principle of the vernier scale is described in his book *La Construction, l'usage, et les propriétés du quadrant nouveau de mathématique* (1631; The Construction, Uses, and Properties of a New Mathematical Quadrant), which also contained some of the earliest tables of trigonometric functions and formulas for deriving the angles of a triangle from the lengths of its sides.

Vesalius, Andreas (1514–1564) *Belgian anatomist*

> It was when the more fashionable doctors in Italy, in imitation of the old Romans, despising the work of the hand, began to delegate to slaves the manual attentions they deemed necessary for their patients ... that the art of medicine went to ruin.
> —Quoted by B. Farrington in *Science in Antiquity* (1936)

Vesalius (ve-**say**-lee-us), born the son of a pharmacist in Brussels, Belgium, was educated at the universities of Louvain, Paris, and Padua, receiving his MD from the last in 1537. He was immediately appointed professor of anatomy and surgery at Padua where he remained until 1543 when, at the age of 28, he joined the Hapsburg court. Here Vesalius successively served as physician to the Emperor Charles V and King Philip II of Spain. For reasons unknown, he left their service sometime after 1562 and died while on a pilgrimage to the Holy Land.

Vesalius thus completed his anatomical researches in the short period between 1538, when he produced his six anatomical plates, the *Tabulae sex* (Six Tables), and 1543, when his masterpiece *De humani corporis fabrica* (On the Structure of the Human Body) was printed in Basel. With this work he gained the reputation of being the greatest of Renaissance anatomists.

The work generally followed the physiological system of GALEN and repeated some traditional errors; for example, he described the supposed pores in the septum of the heart despite confessing his inability to detect them. Other parts, such as the female generative organs, were treated inadequately because of a lack of the appropriate cadavers. However Vesalius's main innovation was to insist on conducting, personally, dissections on human cadavers, which taught him that Galenic anatomy was not to be treated unquestioningly.

The work of Vesalius was of considerable significance in marking the departure from ancient concepts. The *Fabrica* presented in a single, detailed, comprehensive, and accessible work (superbly illustrated, probably at the Titian school in Venice), a basis for following generations of anatomists to compare with their own dissections. It has been said that after Vesalius medicine became a science.

Vidal de la Blache, Paul (1845–1918) *French geographer*

Vidal (vee-**dal**) was born at Pézenas in France and studied history and geography at the Ecole Normale Supérieure, Paris, graduating in 1865. After some time at the French School in Athens and teaching at Nancy he returned, in 1877, to the Ecole Normale Supérieure, remaining there until he became professor of geography at the Sorbonne (1898–1918).

He is recognized as being the founder of French human geography and virtually all chairs of geography in France were occupied by his students for the first half of the 20th century. His approach was to study the interaction of people's activities with the environment in which they lived, examining in particular small areas. He argued against physical determinism, pointing out that man is able to modify his environment to a certain extent.

Vidal founded, in 1891, the important journal *Annales de géographie* (Annals of Geography), which he edited until his death, and began the publication of an annual *Bibliographie géographique* (Geographical Bibliography). Some of his more important papers were collected in *Principes de géographie humaine* (1922), published in English in 1950 as *Principles of Human Geography*.

Viète, François (1540–1603) *French mathematician*

> I ... do not profess to be a mathematician, but ... whenever there is leisure, delight in mathematical studies.
> —Introduction to *Reply ... to a Problem Posed by Adrian Romanus* (1595)

Viète (vyet), who was born at Fontenay-le-Comte in France, is also known by the Latinized form of his name, Franciscus Vieta (vI-**ee**-ta). He was educated at Poitiers where he studied law and for a time he practiced as a lawyer. He was a member of the *parlement* of Brittany but because of his Huguenot sympathies he was forced to flee during the persecution of the Huguenots. On Henry IV's accession, however, he was able to hold further offices and became a privy councillor to the king. He put his mathematical abilities to practical use in deciphering the code used by Spanish diplomats.

Viète's chief work was in algebra. He made a number of innovations in the use of symbolism and several technical terms still in use (e.g., coefficient) were introduced by him. His work is important because of his tendency to solve problems by algebraic rather than geometric methods. By bringing algebraic techniques to bear on them Viète was able to solve a number of geometrical problems. A particularly long-standing problem – formulated by the Greek geometer APOLLONIUS of Perga – namely, how to construct a circle that touches three given circles, was solved in this way by Viète.

Viète's major work is contained in his treatise *In artem analyticem isagoge* (1591; Introduction to the Analytical Arts) and among other advances in algebra that it contains are new and improved methods for solving cubic equations. Among these are techniques that make use of trigonometric methods. Viète also developed methods of approximating the solutions to equations.

Villemin, Jean Antoine (1827–1892) *French physician*

Villemin (veel-**man**), a farmer's son from Prey in France, was educated at the military medical schools of Strasbourg and Val-de-Grâce, Paris. He later taught at the latter school holding the chair of hygiene from 1869 to 1873, when he succeeded to the chair of clinical medicine.

In 1865 Villemin made a major breakthrough in our understanding of tuberculosis. Before Villemin it was widely thought that tuberculosis arose from an inherited predisposition to develop the disease. Villemin, however, pointed out that, while the disease was rare in the sparsely populated areas of the countryside, in situations where people were living at close quarters, such as army barracks, the incidence was high. Concluding then that the disease must be contagious, he took fluid from the lungs of a dead patient and injected it into some rabbits. When he killed the rabbits three months later he found their bodies riddled with tuberculous lesions. He also showed that bovine tuberculosis could be transmitted to rabbits, in this case with fatal results.

Villemin published his views fully in his *Etudes sur la tuberculose* (1868; Studies on Tuberculosis). They seemed, however, to have little effect on the work of his contemporaries. Robert KOCH in particular, who later actually discovered the causative bacillus, went out of his way to belittle the work of Villemin. Julius COHNHEIM, however, confirmed Villemin's results using an elegant experimental technique on rabbits.

Vine, Frederick John (1939–1988) *British geologist*

> A sudden change in attitude towards the matter is generally attributed to the publication in 1963 of a scientific article by Fred Vine and Drum Matthews ... in spite of a few serious contrarians and the waggish carping of a group that called itself the Stop Continental Drift Society, continental drift is now accepted as the explanation for the present configuration of the continents.
> —Charles Officer and Jake Page, *Tales of the Earth: Paroxysms and Perturbations*

Vine was educated at Cambridge University. After a period in America teaching at Princeton University (1965–70) he returned to Britain, becoming reader (1970) and professor of environmental science (1974) at the University of East Anglia.

In 1963, in collaboration with his supervisor Drummond Hoyle Matthews, Vine produced a paper, *Magnetic Anomalies over Ocean Ridges*, which provided additional evidence for, and modified, the sea-floor spreading hypothesis of Harry H. HESS, published in 1962. The fact that magnetic reversals had occurred during the Earth's history had been known since the work of Motonori MATUYAMA in the 1920s and B. Brunhes earlier in the century. Vine and Matthews realized that if Hess was correct, the new rock emerging from the oceanic ridges would, on cooling, adopt the prevailing magnetic polarity. Newer rock emerging would push it further away from the ridge and, intermittently, as magnetic reversals occurred, belts of material of opposing magnetic polarity would be pushed out.

From examining several ocean ridges in the North Atlantic, Vine and Matthews established that the parallel belts of different magnetic polarities were symmetrical on either side of the ridge crests. This provided crucial evidence for the sea-floor spreading hypothesis. Correlation between the magnetic anomalies of ocean ridges in other oceans was also established.

Vinogradov, Ivan Matveïevitch (1891–1983) *Soviet mathematician*

Vinogradov (vye-no-**grah**-dof), who was born at Milolyub (now Velikiye Luki) in Russia, held a number of posts at various institutions in the Soviet Union. Initially he taught at the University of Perm (1918–20) until appointed professor of mathematics at the Leningrad Polytechnic Institute. In 1925 he became a professor at Leningrad State University and in 1932 was appointed chairman of the National Committee of Soviet Mathematicians of the Soviet Academy of Sciences. From 1934 he was professor of mathematics at Moscow State University.

Vinogradov was preeminent in the field of analytical number theory, i.e., the study of problems posed in purely number-theoretic terms by means of the techniques of analysis. He published several books, chiefly on various aspects of number theory, which include *A New Method in the Analytical Theory of Numbers* (1937) and *The Method of Trigonometric Sums in the Theory of Numbers* (1947).

One of his most outstanding achievements was his solution in 1937 of a problem of Christian Goldbach's (not to be confused with Goldbach's famous conjecture about even numbers and primes, which is still undecided). Vinogradov showed that every sufficiently large odd natural number is a sum of three primes.

Virchow, Rudolf Carl (1821–1902) *German pathologist*

> Pathology is the science of disease, from cells to societies.
> —*Archiv für pathologische Anatomie* (Archive of Pathological Anatomy)

> There can be no scientific dispute with respect to faith, for science and faith exclude one another.
> —*Disease, Life, and Man*

The son of a merchant from Schivelbein (now Świdwin in Poland), Virchow (**feer**-koh) graduated in medicine from the Army Medical School, Berlin, in 1843. He then worked at the Charité Hospital in Berlin where he wrote a classic paper on one of the first known cases of leukemia. In 1849 he moved to Würzburg as professor of pathological anatomy. He returned to Berlin in 1856 as director of the university's Institute of Pathology, where he remained until his death.

In 1858 Virchow published *Die Cellularpathologie* (Cellular Pathology), in which he formulated two propositions of fundamental importance. The first, consciously echoing the words of William HARVEY: "Omne vivum ex ovo" (All life is derived from an egg), declared "Omnis cellula e cellula" (Every cell is derived from a preexisting cell). Others, such as John Goodsir, had already advanced such ideas but Virchow differed in applying them to pathology, his second major thesis being that disease was a pathological cellular state. The cells are the "seat" of disease, or, disease is simply the response of a cell to abnormal conditions. This by itself immediately generated the immense research program of collecting, examining, and classifying different types of cells and noting their variety and development, both normal and abnormal.

Virchow consequently had little time for the emerging germ theory of disease, which later in the century would sweep all other theories out of the way. In fact after 1870 Virchow tended to pursue interests other than pure science. Dissatisfied not only with the new germ theory but also with the theory of evolution, which he tried to have banned from school curricula, Virchow seemed more interested in archeology and politics than science.

Thus Virchow encouraged his friend Heinrich Schliemann in his determination to discover the site of Homer's Troy and actually worked on the dig at Hissarlik in 1879. In politics he was a member of the Reichstag from 1880 to 1893 and, as a leading liberal, was a bitter opponent of Bismarck who went so far as to challenge him to a duel in 1865.

Virchow was also widely known for founding, in 1847, the journal *Archiv für pathologische Anatomie* (Archive of Pathological Anatomy), which he continued to edit for 50 years.

Virtanen, Artturi Ilmari (1895–1973) *Finnish chemist*

Virtanen (**veer**-ta-nen) was educated at the university in his native city of Helsinki, where he obtained his PhD in 1919, and at the universities of Zurich and Stockholm. He worked from 1921 to 1931 as director of the Finnish Cooperative Dairies Association Laboratory and from 1924 at the University of Helsinki where, in 1931, he became director of the Biochemical Institute.

In 1945 Virtanen was awarded the Nobel Prize for chemistry for his method of fodder preservation. This AIV method, as it became known, named for his initials, was designed to stop the loss of nitrogenous material in storage. By storing green fodder in an acid medium he hoped to prevent spoilage and still retain nutritious fodder. After much experimentation he finally found that a mixture of hydrochloric and sulfuric acid was adequate as long as its strength was kept within certain precise limits. Specifically, this demanded a pH of about four. In 1929 Virtanen found that cows fed on silage produced by his method gave milk indistinguishable in taste from that of cows fed on normal fodder. Further, it was just as rich in both vitamin A and C.

Viviani, Vincenzo (1622–1703) *Italian mathematician*

Viviani (vee-**vyah**-nee), who was born at Florence in Italy, was an associate and pupil of GALILEO, although his chief interest was in mathematics rather than in physics. After the condemnation of Galileo's ideas by the Catholic Church it was unsafe for Viviani to pursue his work on Galileo's mathematics. Accordingly Viviani devoted himself to the thorough study of the mathematics of the Greeks, in particular their geometry, and in this field of work he achieved wide fame. In 1696 he was elected a fellow of the Royal Society of London. Viviani was particularly interested in trying to reconstruct lost sections of works by ancient Greek mathematicians, such as the missing fifth book of APOLLONIUS's *Conics*. He also published Italian translations of the works of classical mathematicians including EUCLID and ARCHIMEDES. He was an associate of the physicist Evangelista TORRICELLI and collaborated with him in his work on atmospheric pressure and in the invention of the mercury barometer.

Vogel, Hermann Karl (1842–1907) *German astronomer*

Born at Leipzig in Germany, Vogel (**foh**-gel) began as an assistant at the observatory there. He later directed a private observatory and finally moved to Potsdam to work in the new astrophysical observatory, of which he became director in 1882. He was one of the earliest astronomers to devote himself almost exclusively to spectroscopy. His first discovery came in 1871 when he showed that the solar rotation could be measured using spectroscopic DOPPLER effects, obtaining identical results to those achieved using sunspots as markers.

In 1890 he came across some unusual stellar spectra – in particular that of the variable star Algol. He found that some stars seemed to be both advancing and receding, for their spectral lines periodically doubled showing both a red and a blue shift. He correctly interpreted this as indicating a binary system with two stars so close together that they could not be separated optically, with one star advancing and one receding. When one star is eclipsed by its companion just one spectra will be visible, but as the other emerges the spectral line will appear to double only to disappear again in the next

eclipse. Such systems are known as eclipsing binaries.

Volta, Count Alessandro (1745–1827) *Italian physicist*

> Nothing good can be done in physics unless things are reduced to degrees and numbers.
> —*Works*, Vol. I

> [Volta] understood a lot about the electricity of women.
> —Georg Christoph Lichtenberg, quoted in Volta's *Letters*, Vol. II

Volta (**vohl**-ta), who was born at Como in Italy, grew up in an atmosphere of aristocratic religiosity with almost all his male relations becoming priests. However, Volta decided early that his life's work lay in the study of electricity and, by the age of 24, had developed his own version of Benjamin FRANKLIN's electrical fluid theory. In 1774 he started teaching physics at the gymnasium in Como, where, a few months later, he invented the electrophorus – a device for producing electric charge by friction and, at the time, the most efficient way of storing electric charge. On the strength of this invention he was promoted, in 1775, to the position of professor of physics at Como and, three years later, took up a similar appointment at Pavia University. Here, stimulated by the experiments of his friend Luigi GALVANI, he started investigating the production of electric current. In 1795 he was appointed rector of Pavia but his work was disturbed by the political upheavals in Lombardy at the time. The state was oscillating between French and Austrian control in the Napoleonic campaigns and in 1799–1800 the Austrians closed the university.

Volta chose this time to make public his great discovery that the production of electric current did not need the presence of animal tissue, as Galvani and others had supposed. Volta produced the famous *voltaic pile*, consisting of an alternating column of zinc and silver disks separated by porous cardboard soaked in brine. This instrument revolutionized the study of electricity by producing a readily available source of current, leading almost immediately to William NICHOLSON's decomposition of water by electrolysis and later to Humphry DAVY's discovery of potassium and other metals by the same process.

In 1800 Napoleon returned in victory to Pavia, reopened the university, and invited Volta to Paris to demonstrate his pile. He awarded Volta the medal of the Legion of Honor and made him a count. In his honor the unit of electric potential (or potential difference or electromotive force) was called the volt.

Voltaire, François Marie Arouet (1694–1778) *French writer*

Facts must prevail.
—Motto

You have confirmed in these tedious places what Newton found out without leaving his room.
—On Maupertius's expedition to the Arctic circle to verify the flattening of the Earth at the poles

The son of a Parisian lawyer, Voltaire (vol-**tair**) was educated in a Jesuit seminary. He began to train as a lawyer but his progress was impeded by political and literary scandals, as well as the occasional indiscreet love affair. In 1726 he began a famous quarrel with the Chevalier de Rohan who, unable to compete against Voltaire's wit, had Voltaire imprisoned in the Bastille. Released on the condition that he keep out of trouble, he thought it prudent to leave the country.

Voltaire arrived in London during the last few months of Newton's life and, although they never met, he attended Newton's funeral and spoke to many of Newton's friends and disciples. The result was that his *Letters Concerning the English Nation* (London, 1733) contained a comparison between the science of DESCARTES and Newton. Although Voltaire did his best to present in a brief compass some of the main features of Newton's thought, the gossip in Voltaire frequently got the better of him and many now-familiar anecdotes made their first appearance in the *Letters*. It was here that the story of the falling apple first received wide publicity, as did the news of Newton's unitarianism and chastity.

Voltaire went on to publish a much longer and more comprehensive study with his *Les éléments de la philosophie de Newton* (1738; Rudiments of Newton's Philosophy). At the same time he was helping Madame du CHÂTELET complete her translation of Newton's *Principia* and actually contributed a preface and a poem. Also, in the late 1730s, with Madame du Châtelet at her chateau at Cirey, Voltaire engaged upon some genuine scientific research. The issue was the nature of fire, set by the Académie as a prize competition. He tackled the old problem of whether or not fire has weight. His results were variable, he complained, and "To discover the least scrap of truth entails endless labor." The prize went to the mathematician Leonhard EULER. Following the death of Madame du Châtelet in 1749, Voltaire's interests were directed elsewhere.

von Braun, Wernher Magnus Maximilian (1912–1977) *German–American rocket engineer*

It [the rocket] will free man from the remaining chains, the chains of gravity which still tie him to this planet.
—*Time*, 10 February 1958

Basic research is when I'm doing what I don't know I'm doing.
—Attributed remark

Von Braun (fon brown) came from an affluent background in Wirsitz (now Wyrzysk in Poland). His father, a high government official, had served as minister of education and agriculture in the Weimar Republic. He studied at the Federal Institute of Technology, Zurich, received his BS in 1932 from the Institute of Technology in Berlin, and obtained his PhD in 1934 from the University of Berlin. He had already shown his interest in rockets, joining the VfR (*Verein für Raumschiffahrt* or Society for Space Travel) in 1930. The society consisted of a group of enthusiasts, numbering 870 in 1929, and included such talented engineers as Hermann OBERTH, Willie Ley, and Rudolph Nebel, who took the problem of building rockets seriously.

By the early 1930s the VfR and von Braun had come to the attention of the German military. Barred from openly developing conventional weapons by the Versailles Treaty, military interest turned to unconventional forms of weaponry such as rockets. Von Braun was recruited by the German Ordnance Department in 1932 and by 1934 had already developed the A–2 rocket which, using liquid fuel, reached an altitude of 1.6 miles (2.6 km).

In 1937 von Braun moved to Peenemünde, on the Baltic Sea, as civilian head of the technical department of the German rocket development center. There in October 1942 he successfully tested his A–4 rocket by delivering the missile directly on its target some 120 miles away. It was this rocket that became known as "Vengeance weapon 2" – the supersonic ballistic V–2 missile used against Britain in 1944. Von Braun also developed the supersonic anti-aircraft missile Wasserfall but this never became operational. Not all his efforts, however, were directed to military ends. Even in the limited atmosphere of Peenemünde, von Braun was drawing up plans for his A–9, A–10, and A–11 rockets with which he could launch a payload of some 30 tons into space "and, maintaining a regular shuttle service to the orbit … permit the building of a space station there." His commitments to the dreams of his old VfR colleagues never lapsed.

With the collapse of Germany, von Braun delivered his rocket team en masse to the American army. He was still only 33. Initially they worked with the U.S. Army Ordnance Corps at Fort Bliss in Texas and at the White Sands testing facility in New Mexico, studying potential ramjet and rocket missiles and developing the V–2 for high-altitude research. In 1950 the team moved to the Redstone Arsenal in Huntsville, Alabama, to form what was later to become the Marshall Space Flight Center of NASA and which von Braun was to direct from 1960 to 1970. It was here in the 1950s that von

Braun headed the American ballistic weapons program and he and his team designed the Jupiter–C, Juno, and Redstone rockets used in the American space program: Jupiter–C launched Explorer I, the first American satellite, into orbit on 31 January 1958. This was followed by the successful Saturn I, IB, and V rockets. Saturn V was used to launch the Apollo craft that landed Americans on the Moon in 1969.

In 1970 von Braun left Huntsville for NASA headquarters in Washington to serve as deputy associate administrator for planning. It soon became clear to him that there no longer existed any deep commitment to space exploration at the highest levels of American government. He consequently resigned from NASA in 1972 and moved into private industry. He worked for Fairchild Industries in Germantown, Maryland, as vice-president for engineering and research. After unsuccessful surgery for cancer he resigned in December 1976, shortly before his death.

Von Braun wrote several books, including *History of Rocketry and Space Travel* (1967) and *Moon* (1970).

von Buch, Christian Leopold (1774–1853) *German geographer and geologist*

Von Buch (fon book) was the son of a wealthy nobleman from Angermünde in Germany. He was educated in Berlin and at the Mining Academy, Freiberg (1790–93), where he studied under Abraham WERNER and formed his lifelong friendship with Alexander von HUMBOLDT. He worked briefly in the mining service of Silesia but resigned in 1797 to devote himself exclusively to geological investigation.

He became one of the great geological travelers of his day. Apart from working on the Alps and the other popular geological parts of western Europe he also spent two years in Scandinavia, worked in North Africa and the Canaries, and gave one of the first clear descriptions of the geology of central Europe.

Von Buch left Freiberg committed to Werner's neptunism. However, unlike Werner who rarely ventured beyond Saxony, von Buch traveled extensively and had the problem of fitting his numerous observations into the rigid system of his master. He realized after his travels through Italy and the Auvergne district of France that volcanoes were more numerous and much more significant than Werner allowed. Nor could he locate the coal beds that supposedly fueled the volcanoes. Thus he was gradually converted to the vulcanist school of thought, believing that many rocks, such as basalt, were of igneous origin.

von Euler, Ulf Svante (1905–1983) *Swedish physiologist*

Von Euler (fon **oi**-ler) was the son of Karl von EULER-CHELPIN, Nobel Prize winner in 1929. He was born in the Swedish capital of Stockholm and educated at the Karolinska Institute, where he obtained his MD in 1930. He taught there from 1930 onward becoming, in 1939, professor of physiology. In 1966 von Euler was elected to the powerful position of president of the Nobel Foundation, which he held until 1975.

In 1906 the idea that nerve cells communicate with each other and the muscles they control by the release of chemicals was first proposed by Thomas ELLIOTT. Since then there had been much searching for the elusive neurotransmitters and it was not until 1946 that von Euler succeeded in isolating that of the sympathetic system and showed it to be noradrenaline (norepinephrine). For this work von Euler shared the 1970 Nobel Prize for physiology or medicine with Julius AXELROD and Bernard KATZ.

Von Euler had earlier, in 1935, discovered a substance in human semen showing great physiological potency. As he assumed it came from the prostate gland he named it "prostaglandin." It later turned out that prostaglandins could be found in many other human tissues; however, his deduction that they were fatty acids has since been confirmed.

von Kármán, Theodore (1881–1963) *Hungarian–American aerodynamicist*

> Scientists as a group should not try to force or even persuade the government to follow their decisions.
> —*Collected Works of Dr. Theodore von Kármán* (1956)

The son of a distinguished educationalist, von Kármán (von **kar**-mahn) studied engineering at the Polytechnic in his native city of Budapest. After graduating in 1902 he taught at the Polytechnic until 1906 when he moved to Göttingen, where he completed his PhD in 1908. At about this time his interest in aeronautics was aroused when he saw Henri FARMAN fly a biplane in Paris. He pursued his new interests further at Göttingen when he was asked to help Ludwig Prandtl design a wind tunnel for research on airships. Von Kármán continued to work in aeronautics and in 1912 he was invited to establish and direct a new institute of aerodynamics at the University of Aachen. Here he remained until 1930, apart from the war years spent in Austria working at the Military Aircraft Factory, Fischamend. In 1930, unhappy with political conditions in Germany, he moved to the California Institute of Technology to set up and direct another new institute, the Guggenheim Aeronautic Laboratory at Pasadena, California. He became a naturalized American in 1936.

Von Kármán remained director of the

Guggenheim Laboratory until 1949. During this time he contributed to many branches of aeronautics and encouraged work on jet propulsion, rockets, and supersonic flight. At von Kármán's insistence the world-famous Jet Propulsion Laboratory was set up in 1938 and he served as its director until 1945. He also served as a consultant to the U.S. Army Air Corps from 1939 onward. After his retirement from Pasadena he organized the Advisory Group for Aeronautical Research and Development to provide NATO with technical advice.

Among his many contributions to aerodynamics, von Kármán is probably best known for his discovery in 1911 of what have since been called *Kármán vortices* – the alternating vortices found behind obstacles placed in moving fluids. The basic idea was drawn to his attention by a graduate student in Prandtl's laboratory. He had been asked to measure the pressure distribution around a cylinder placed in a steady flow. But, the student found, the flow refused to move steadily and invariably oscillated violently. Prandtl insisted the fault lay with the student who had not bothered to machine circular cylinders. Von Kármán would enquire of the student daily how the flow was behaving and was daily given the sad reply that the flow still oscillated. Eventually von Kármán came to see that the student had stumbled upon a genuine effect. Over a weekend he calculated that the wake should indeed separate into two periodic vortices. Further, there is a symmetric arrangement of vortices which is unstable; only an asymmetric arrangement of vortices persists when the conditions are changed.

Von Kármán went on to demonstrate that above a certain velocity v, where d is the cylinder's diameter, vibrations will be induced with a frequency v/d cycles per second. It was precisely these vibrations which were induced in 1940 in the Tacoma Narrows suspension bridge when v exceeded its critical velocity of 42 mph.

von Laue, Max Theodor Felix (1879–1960) *German physicist*

> No matter how great the repression, the representative of science can stand erect in the triumphant certainty that is expressed in the simple phrase: And yet it moves!
> —Alluding to Galileo's supposed defiance of the Inquisition. Address to the Physics Congress, Würzburg, 18 September 1933

The son of a civil servant, von Laue (fon **low**-e) was born in Pfaffendorf, Koblenz, and educated at the universities of Strasbourg, Göttingen, Munich, and Berlin, where he obtained his doctorate in 1903. He worked in various universities before his appointment as professor of theoretical physics at Berlin University in 1919. He remained there until 1943 when he moved to Göttingen as director of the Max Planck Institute.

Although von Laue began his research career working on relativity theory, his most important work was the discovery of x-ray diffraction in 1912 for which he was awarded the 1914 Nobel Prize for physics. From this discovery much of modern physics was to develop and, some forty years later, the new discipline of molecular biology was to emerge.

Von Laue put together two simple and well known ideas. He knew that x-rays had wavelengths shorter than visible light; he also knew that crystals were regular structures with their atoms probably lined up neatly in rows. Thus, he concluded, if the wavelength of x-rays was similar to the interatomic distance of the atoms in the crystal, then x-rays directed onto a crystal could be diffracted and form a characteristic and decipherable pattern on a photographic plate.

He passed the actual experimental work to two of his students, Walter Friedrich and Paul Kipping, who first tried copper sulfate (1912), which yielded a somewhat unclear pattern. When they changed to zinc sulfide they almost immediately obtained a clear photograph marking out the regular and symmetric arrangement of the atoms in the crystal.

Vonnegut, Bernard (1914–1997) *American physicist*

Born in Indianapolis, Indiana, Vonnegut was educated at the Massachusetts Institute of Technology, where he obtained his PhD in 1939. After working in the Research Laboratory of the General Electric Company under Vincent SCHAEFER (1945–52), he moved to the Arthur D. Little Company and remained there until 1967, when he was appointed professor of atmospheric science at the New York State University, Albany.

In 1947, while with the General Electric Research Laboratory, Vonnegut made a major advance in the rain-making techniques developed by Schaefer, when he found that he obtained much better results with silver iodide crystals for cloud seeding than the dry ice used by Schaefer.

von Neumann, John (1903–1957) *Hungarian–American mathematician*

> He [von Neumann] was a genius, in the sense that a genius is a man who has *two* great ideas.
> —Jacob Bronowski, *The Ascent of Man* (1973)

> In mathematics you don't understand things. You just get used to them.
> —Quoted by G. Zukav in *The Dancing Wu Li Masters*

John (originally Johann) von Neumann (von **noi**-man) was born in Budapest, Hungary, and studied at the University of Berlin, the Berlin Institute of Technology, and the University of Budapest, where he obtained his doctorate in 1926. He was *Privatdozent* (nonstipendiary lecturer) at Berlin (1927–29) and taught at Hamburg (1929–30). He left Europe in 1930 to work in Princeton, first at the university and later at the Institute for Advanced Study. From 1943 he was a consultant on the atomic-bomb project.

Von Neumann may have been one of the last people able to span the fields of pure and applied mathematics. His first work was in set theory (the subject of his doctoral thesis). Here he improved the axiomatization given by Ernst Zermelo and Abraham Fraenkel. In 1928 he published his first paper in the field for which he is best known, the mathematical theory of games. This work culminated in 1944 with the publication of *The Theory of Games and Economic Behavior*, which von Neumann had coauthored with Oskar Morgenstern. Not all the results in this work were novel, but it was the first time the field had been treated in such a large-scale and systematic way.

Apart from the theory of games von Neumann did important work in the theory of operators. Dissatisfied with the resources then available for solving the complex computational problems that arose in hydrodynamics, von Neumann developed a broad knowledge of the design of computers and with his interest in the general theory of automata became one of the founders of a whole new discipline. He was much interested in the general role of science and technology in society and this led to his becoming increasingly involved in high-level government scientific committees. Von Neumann died at the early age of 54 from cancer.

von Ohain, Hans Joachim Pabst (1911–1998) *German–American aeronautical engineer*

Born at Dessau in Germany, von Ohain (von **oh**-In) took his PhD in aerodynamics at Göttingen in 1935. He immediately joined the Heinkel Aircraft Company at Rostock. It had long been apparent to engineers that if planes were to fly faster they would have to fly higher and so benefit from the lower air resistance. But in a thinner atmosphere propellers and piston engines worked badly. The dilemma, von Ohain realized, could be resolved if turbojets were used. Thus in 1935, four years after Frank WHITTLE, von Ohain took out his first patent on the gas-turbine jet engine.

Backed by Ernst Heinkel (1885–1958) he began to work on the He 178. In September 1937 a hydrogen-fueled bench model produced a 250-kilogram thrust. The plane was ready for its test flight, the first jet flight ever, in August 1939, when it reached a top speed of about 350 miles per hour. Whittle's first jet, the Gloster E28/39 prototype, had its maiden flight in 1941.

Heinkel went on to develop the He 280, powered by two von Ohain engines. By this time, however, Heinkel had lost the confidence of the Nazis and the contract to develop a jet fighter was awarded to Messerschmitt. The Me 262, powered by Junkers-built jet engines, entered service in late 1944 with a top speed of 550 miles per hour. Although 1,430 were built, only about 400 actually saw combat and they arrived too late to influence the war's outcome.

Despite this von Ohain found himself in great demand when peace came and in 1947 he began work for the U.S. Airforce on the design of a new generation of military jets at the Wright-Patterson base, where he remained until 1975. After a further spell as chief scientist at the Aero-Propulsion Laboratory, von Ohain retired in 1979.

In 1991 he shared with Whittle the Draper Prize – the engineering equivalent of the Nobel Prize – for their independent invention of the jet engine.

von Richthofen, Baron Ferdinand (1833–1905) *German geographer*

Von Richthofen (fon **rikt**-hoh-fen) was born at Karlsruhe (now in Poland) and educated at the University of Berlin. After extensive travels in southeast Asia, China, and America, he occupied chairs of geography at the universities of Bonn (1877–83), Leipzig (1883–86), and Berlin.

He was the author of a classic five-volume work on the geography of China, *China, Ergebnisse eigener Reisen und darauf gegründeter Studien* (1877–1912; China, the Results of My Travels and the Studies Based Thereon), based on his own intensive fieldwork.

W

Waage, Peter (1833–1900) *Norwegian chemist*

Waage (**vaw**-ge), who was born at Flekkefjord in Norway, became professor of chemistry at the University of Christiania (now Oslo) in 1862, remaining there until 1900. He is remembered for his collaboration with his brother-in-law, Cato GULDBERG, for their discovery of the law of mass action in 1864.

Waals, Johannes Diderik van der *See* VAN DER WAALS, JOHANNES DIDERIK.

Waddington, Conrad Hal (1905–1975) *British embryologist and geneticist*

> DNA plays a role in life rather like that played by the telephone directory in the social life of London: you can't do anything much without it, but, having it, you need a lot of other things – telephones, wires, and so on – as well.
> —Review of James Dewey Watson's *The Double Helix, The Sunday Times*, 25 May 1968

Waddington, the son of a tea planter in southern India, was born at Evesham in England and graduated in geology from Cambridge University. In 1933 he was appointed embryologist at the Strangeways Research Laboratory, Cambridge, and in 1947 he moved to the University of Edinburgh where he served as professor of animal genetics until his retirement in 1970.

As a geneticist and a Darwinian, Waddington introduced two important concepts into the discussion of evolutionary theory. The first dealt with developmental reactions that occur in organisms exposed to natural selection and proposed that such reactions are generally canalized. In other words, they adjust to bring about one definite end result notwithstanding small changes in conditions over the course of the reaction.

The second idea was introduced in his 1953 paper *Genetic Assimilation of an Acquired Character*, in which he tried to show that the inheritance of acquired characteristics, the "heresy" of Jean LAMARCK, could in fact be incorporated into orthodox genetics and evolutionary theory. As an example Waddington quoted the calluses formed on the embryonic rump of an ostrich. If the Lamarckian explanation of the inheritance of an earlier acquired characteristic is rejected then what remains is the convenient but implausible appearance of a random mutation.

Waddington claimed to have demonstrated experimentally the process of genetic assimilation in normal fruit flies (*Drosophila*). He subjected the pupae of the flies to heat shock and noted that a small proportion developed lacking the posterior cross-vein in their wings. Careful breeding increased the proportion of such flies and eventually Waddington built up a stock of flies without cross-veins that had never been subjected to heat shock. The experiment has been criticized as differing from the calluses of the ostrich in dealing with nonadaptive traits. It also appears to be the case that other genetic mechanisms are available to explain the data without assuming the reality of genetic assimilation.

As an embryologist Waddington was the author of the standard textbook *Principles of Embryology* (1956). He had earlier worked on the powers of the "organizer" of Hans SPEMANN, showing that Spemann's results can be extended to warm-blooded animals. Waddington also showed that the action of certain embryonic tissues in inducing organ formation is retained even when the tissue is dead.

Waddington was a well-known popularizer of science and as well as his important works in embryology and genetics he also wrote more general texts, such as *The Ethical Animal* (1960) and *Biology for the Modern World* (1962).

Wagner-Jauregg, Julius (1857–1940) *Austrian psychiatrist*

Wagner-Jauregg (vahg-ner-**yow**-rek) was born at Wels in Austria and educated at the University of Vienna, where he gained his MD in 1880. Finding it difficult to obtain an academic post in orthodox medicine, he turned to psychiatry in 1883 and in 1889 succeeded Krafft-Ebbing as professor of psychiatry at the University of Graz. In 1893 he returned to Vienna as director of the Psychiatric and Neurological Clinic, where he remained until his retirement in 1928.

In 1917 he proposed a new treatment for general paralysis of the insane (GPI), then a relatively common complication of late syphilis. As early as 1887 he had noticed that rare cases of remission were often preceded by a feverish infection, suggesting that the deliberate production of a fever could have a similar effect. Consequently, in 1917 he inoculated nine GPI patients with tertian malaria – a form of malaria that gives a two-day interval between fever attacks. He later reported that in six of these patients extensive remissions had taken place. It was for this work that Wagner-Jauregg received the Nobel Prize for physiology or medicine in 1927. Although therapeutic malaria inoculations were used in the treatment of GPI for some time, demand for them ceased with the discovery of penicillin.

Wagner-Jauregg also proposed in 1894 that cretinism, a thyroid deficiency disease, could be successfully controlled by iodide tablets.

Waksman, Selman Abraham (1888–1973) *Russian–American biochemist*

Waksman (**waks**-man), who was born at Priluki in Russia, emigrated to America in 1910; he graduated from Rutgers University in 1915 and obtained his American citizenship the following year. He studied for his doctorate at the University of California, receiving his PhD in 1918, and then returned to Rutgers, becoming professor of soil microbiology in 1930.

A new area in the science of soil microbiology was opened up with the discovery by René DUBOS, in 1939, of a bacteria-killing agent in a soil microorganism. This stimulated renewed interest in FLEMING's penicillin and, with the value of penicillin at last established, Waksman began a systematic search for antibiotics among microorganisms. In 1943 he isolated streptomycin from the mold *Streptomyces griseus* and found that it was effective in treating tuberculosis, caused by Gram-negative bacteria. This was a breakthrough as previously discovered antibiotics had proved useful only against Gram-positive bacteria. This work gained Waksman the 1953 Nobel Prize for physiology or medicine; he donated the prize money to a research foundation at Rutgers.

Waksman isolated and developed many other antibiotics, including neomycin. From 1940 until his retirement in 1958 he was professor of microbiology and chairman of the department at Rutgers; from 1949 he also held the post of director of the Rutgers Institute of Microbiology.

Walcott, Charles Doolittle (1850–1927) *American paleontologist*

Walcott was born into a poor family in Utica, New York, and educated in the public schools there. He began work as a farm laborer and took to collecting the trilobites he found scattered around the farm, some of which he sold to Louis AGASSIZ. In 1876 he was taken as assistant to the New York state geologist. He moved to the U.S. Geological Survey in 1879 as a field geologist and by 1894 had risen to be its direc-

tor. In 1907 he accepted the important post of secretary of the Smithsonian Institution, a position he held, along with a number of other offices in scientific administration, until his death in 1927.

Walcott specialized in the Cambrian, the period 550 million years ago when multicellular organisms first appeared. In this field he is best known for his discovery in 1909 of the much discussed Burgess Shale fossils. The shale lies 8,000 feet high in the Rockies on the eastern border of British Columbia. Within two strata he found thousands of fossils representing 120 species of marine invertebrates. Further, while most fossils preserve only such hard parts as shells, bones, and teeth, the Burgess specimens by some geological fluke had preserved their soft tissues.

Walcott shipped his material back to Washington. Between 1910 and 1912 he published a few preliminary reports on the "abrupt appearance of the Cambrian fauna." His initial view that his specimens were early forms of modern groups remained unchallenged. Walcott himself was too concerned with administering American science to have time to reconsider his early ideas. It was not until the 1970s when Harry WHITTINGTON began to review Walcott's specimens that it was appreciated that another, more radical, view was possible. Walcott's story is vividly told in S. J. GOULD's popular work *Wonderful Life* (1989).

Wald, George (1906–1997) *American biochemist*

> We already know enough to begin to cope with all the major problems that are now threatening human life and much of the rest of life on earth. Our crisis is not a crisis of information; it is a crisis of decision of policy and action.
> —*Philosophy and Social Action* (1979)

Born in New York City, Wald was educated at New York University and at Columbia where he obtained his PhD in 1932. After spending the period 1932–34 in Europe, where he worked under Otto WARBURG in Berlin and Paul KARRER in Zurich, he returned to America where he took up an appointment at Harvard. Wald remained at Harvard for the whole of his career, becoming professor of biology in 1948 and emeritus professor in 1977.

Wald did fundamental work on the chemistry of vision. In 1933 he discovered that vitamin A is present in the retina of the eye, and thereafter tried to find the relationship between this vitamin and the visual pigment rhodopsin. The first clue came from the constitution of rhodopsin. It was found to consist of two parts: a colorless protein, opsin, and a yellow carotenoid, retinal, which is the aldehyde of vitamin A. Wald was now in a position to work out the main outlines of the story.

Rhodopsin is light sensitive and splits into its two parts when illuminated, with the retinal being reduced further to vitamin A by the enzyme alcohol dehydrogenase. In the dark the procedure is reversed. What was further needed was some indication of how the splitting of the rhodopsin molecule could somehow generate electrical activity in the optic nerve and visual cortex. Part of the answer came from Haldan HARTLINE and Ragnar GRANIT who shared the 1967 Nobel Prize for physiology or medicine with Wald.

Wald speculated that since retinal is a carotenoid pigment, and such pigments are also found in plants, then it is possible that the phototropic responses of plants may rely on a similar mechanism. Wald later became widely known for his opposition to the Vietnam War.

Walden, Paul (1863–1957) *Russian–German chemist*

The son of a farmer from Cēsis (now in Latvia), Walden (**vahl**-den) was educated at Riga Polytechnic, where he studied under Wilhelm OSTWALD. Having become professor of chemistry in 1894, he remained at the polytechnic until the Russian Revolution, when he moved to Germany. From 1919 to 1934 he served as professor of chemistry at the University of Rostock.

In 1896 Walden found that if he took a sample of malic acid that rotated polarized light in a clockwise direction and allowed it to react in a certain way, then on recovery it would be found to rotate polarized light in a counterclockwise direction. The actual reaction involved first combining the malic acid with phosphorus pentachloride to give chlorosuccinic acid. This converts back into malic acid under the influence of silver oxide and water but the malic acid has an inverted configuration. Such inversions later became a useful tool for studying the detail of organic reactions. *Walden inversions*, as they are called, occur when an atom or group approaches a molecule from one direction and displaces an atom or group from the other side of the molecule.

Walden also worked on the electrochemistry of nonaqueous solutions and formulated *Walden's rule*, which relates conductivity and viscosity in such solutions. In later life he turned to the history of chemistry on which topic he is notable for having regularly lectured at the University of Tübingen while well into his nineties.

Waldeyer-Hartz, Wilhelm von (1836–1921) *German anatomist and physiologist*

Waldeyer-Hartz (vahl-dI-er-**harts**), who is better known by his original surname, Waldeyer, was born in Hehlen, Germany, the son of an estate manager. He studied at the universities of Göttingen and Griefswald and at Berlin,

where he graduated in medicine, and later taught at the University of Breslau, where he was professor of pathology from 1864 until 1872; he then moved to the chair of anatomy at Strasbourg. In 1883 he returned to Berlin where he remained until his retirement in 1917.

Waldeyer is mainly remembered for the introduction of two basic scientific terms. In 1888 he proposed the term "chromosome" to refer to the rods that appear in the cell nucleus before division occurs and which readily take up stain. He followed this in 1891 by coining the term "neuron" to explain the work of Santiago Ramón y Cajal, although he did not contribute any original work on nerve cells.

Waldeyer is known to anatomists for his description of the lymphoid tissue encircling the throat, known since as *Waldeyer's ring*. The tissue is presumed to be part of the immune system.

In 1863 Waldeyer established that cancer begins as a single cell and spreads to the other parts of the body by cells migrating from the original site via the blood or lymphatic system. Such an observation carried with it the implication that if the original cells could somehow be removed the cancer would be completely cured, a more congenial view than the bleak alternative that cancer was such a generalized attack on the body that removal of a particular focus was pointless.

Walker, Sir James (1863–1935) *British chemist*

Walker, the son of a flax merchant from Dundee in Scotland, was apprenticed to a flax spinner before attending Edinburgh University (1882–85). After a year in Germany under Johann von BAEYER and Wilhelm OSTWALD he returned to Edinburgh as assistant to Alexander CRUM BROWN before taking a post at University College, London, in 1892. In 1894 he was appointed to the chemistry chair at Dundee, where he remained until he moved to a similar chair at Edinburgh in 1908.

Walker was a physical chemist and did much to popularize the new subject. He himself did valuable work on ionization constants and osmotic pressure but his main importance was as a channel for the ideas of Ostwald; his textbook *Introduction to Physical Chemistry* (1899) was significant in this respect. Walker was knighted in 1921 for his work on TNT during World War I.

Walker, Sir John Ernest (1941–) *British molecular biologist*

Walker was educated at Oxford University, gaining his DPhil in 1969. In 1974 he joined the staff of the Medical Research Council at the Molecular Biology Laboratory, Cambridge.

Since 1999 he has been director of the MRC Dunn Nutrition Unit, Cambridge.

In the 1970s Paul BOYER had proposed a theoretical model by which the enzyme ATP-synthase operating in mitochondria could catalyze the production of the adenosine triphosphate molecule (ATP), the main source of cellular energy. The model was partially verified by Walker in 1994. He determined the structure of the enzyme, first by low-resolution electron microscopy and eventually, using x-ray crystallography, constructed a three-dimensional model of the enzyme.

For his contribution to this field Walker shared the 1997 Nobel Prize for chemistry with Paul Boyer and Jens SKOU.

Wallace, Alfred Russel (1823–1913)
British naturalist

> These checks – war, disease, famine and the like – must, it occurred to me, act on animals as well as man … and while pondering vaguely on this fact there suddenly flashed upon me the idea of the survival of the fittest – that the individuals removed by these checks must be on the whole inferior to those that survived. In the two hours that elapsed before my ague fit was over, I had thought out almost the whole of the theory: and the same evening I sketched the draft of paper, and in the two succeeding evenings wrote it out in full, and sent it by the next post to Mr. Darwin.
> —Quoted by Basil Willey in *Darwin and Butler* (1960)

Wallace, who was born at Usk in Wales, received only an elementary schooling before joining an elder brother in the surveying business. In 1844 he became a master at the Collegiate School, Leicester, where he met the entomologist Henry BATES. Wallace persuaded Bates to accompany him on a trip to the Amazon, and they joined a scientific expedition as naturalists in 1848.

Wallace published an account of his expedition in his *A Narrative of Travels on the Amazon and River Negro* (1853). In 1854 he traveled to the Malay Archipelago, where he spent eight years and collected over 125,000 specimens, a journey described in his *Malay Archipelago* (1869). In this region he noted the marked differences between the Asian and Australian faunas, the former being more advanced than the latter, and proposed a line, still referred to as *Wallace's line*, separating the two distinct ecological regions. He suggested that Australian animals are more primitive because the Australian continent broke away from Asia before the more advanced Asian animals evolved and thus the marsupials were not overrun and driven to extinction. This observation, together with a reconsideration of Thomas MALTHUS's essay on population, led him to propose the theory of evolution by natural selection. He wrote an essay entitled *On the Tendency of Va-rieties to Depart Indefinitely from the Original Type*, which he sent to Darwin for his opinion. On receipt, Darwin realized this was a summary of his own views and the two papers were jointly presented at a meeting of the Linnaean Society in July 1858.

Wallace continued to collect evidence for this evolutionary theory, making an important study on mimicry in the swallowtail butterfly and writing pioneering works on the geographical distribution of animals, including his *Geographical Distribution of Animals* (2 vols., 1876) and *Island Life* (1880). He was also an active socialist, having been introduced to the ideas of the reformer Robert Owen at an early age, and he campaigned for land nationalization and women's suffrage.

In addition to his scientific and political pursuits, Wallace also participated in many of the more dubious intellectual movements of the 19th century. He supported spiritualism, phrenology, and mesmerism. He testified in 1876 on behalf of Henry Slade, a professional medium, charged on evidence submitted by Ray LANKESTER with being a "common rogue." His views on these matters led Wallace to disagree with Darwin on the evolution of man. Man's spiritual essence, Wallace insisted, could not have been produced by natural selection. "I hope you have not murdered our child," Darwin commented. Wallace also campaigned persistently against the practice of vaccination. He published a pamphlet in 1885 claiming British and U.S. statistics showed it to be "both useless and dangerous." He testified in a similar manner before a Royal Commission in 1890 and published his evidence in a pamphlet, *Vaccination, a Delusion* (1895).

Throughout his career Wallace never held an academic appointment and after 1848 no appointment of any kind. He hoped to live on the sale of specimens collected during his Amazon and Malay expeditions. Unfortunately, however, the bulk of the Amazonian material was lost at sea, while funds gathered from the sale of his Malay collection were squandered in unwise investments and expensive disputes with builders. Wallace was therefore forced to earn his living by writing and lecturing. The award of a civil list pension of £200 a year from 1880 greatly eased Wallace's financial burdens.

Wallace also published a spirited account of his life in *My Life* (London, 1905).

Wallach, Otto (1847–1931) *German chemist*

Born at Königsberg (now Kaliningrad in Russia), Wallach (**vahl**-ahk or **wol**-ak) studied at Berlin and at Göttingen, where he obtained his PhD in 1869. After a period in industry in Berlin he moved to Bonn (1870), becoming August KEKULÉ's assistant and later (1876) professor of chemistry. He remained at Bonn until

1889, when he moved to a similar chair at Göttingen.

When Wallach began to give regular classes in pharmacy he became interested in essential oils – oils removed from plants by steam distillation with wide uses in medicine and the perfume industry – and started research into determining their molecular structure. This study led to what was to become his major field of research, the chemistry of the terpenes.

These had hitherto presented considerable difficulties to the analytic chemist. Wallach succeeded in determining the structure of several terpenes, including limonene, in 1894. His greatest achievement, however, was his formulation of the isoprene rule in 1887. Isoprene, with the formula C_5H_8, had been isolated from rubber in the 1860s by C. WILLIAMS. Wallach showed that terpenes were derived from isoprene and therefore had the general formula $(C_5H_8)_n$; limonene is thus $C_{10}H_{16}$. Terpenes were of importance not only in the perfume industry but also as a source of camphors. It was also later established that vitamins A and D are related to the terpenes.

Wallach published 126 papers on the terpenes – work for which he was awarded the Nobel Prize for chemistry in 1910.

Wallis, Barnes Neville (1887–1979) *British engineer*

Wallis was born at Ripley in England, the son of a doctor. Having served an apprenticeship with the Thames Engineering Company (1904–08), he moved to a marine engineering company at Cowes on the Isle of Wight. In 1913 Wallis joined the aeronautical company Vickers, initially to work on the development of the airship, and spent most of his professional career with this company working on a variety of projects. He served with them as chief designer of structures from 1930 and as assistant chief designer of aviation from 1937. After World War II until his retirement in 1971 he worked as head of the department of aeronautical research and development.

His first major success came with the R100 airship in which he displayed his originality of structural design. In 1930 he invented the geodetic construction (a lattice structure in which compression loads on any member are balanced by tension loads in a crossing member), which he applied to aircraft design during the 1930s. As a result he designed the Wellesley, the first geodetic aircraft to enter service in the Royal Air Force, and the Wellington bomber. These aircraft, with the new construction, were lighter and stronger than earlier designs.

Wallis's main war work has become widely known with the massive success of the book and the film about 617 Squadron, better known as the "Dam Busters." Wallis's original idea for a bouncing bomb was contained in his 1940 paper *A Note on a Method of Attacking the Axis Powers*. Using this weapon the successful raid by 617 Squadron on the Möhne and Eder dams took place on 16–17 May 1943. Wallis's originality persisted throughout the postwar years, which were devoted to the design of an entirely new type of aircraft. He first described in 1945 a design so different from traditional aircraft that he gave it a new name, the "aerodyne." This was based on his realization that no conventional airplane could be efficient under the great variety of conditions encountered in subsonic and supersonic flight. His solution, the "wing controlled aerodyne," more popularly known as the swing-wing airplane or the plane with variable geometry, received some backing from Vickers and an experimental version, the Swallow, was built and tested in the 1950s. The project was, however, abandoned when the British government decided to withdraw its backing, but some of Wallis's ideas were nonetheless incorporated in the American fighter airplane, the F1-11.

Although Wallis was made a member of the Royal Society in 1945, then a rare honor for an engineer, it was not until 1968 when, over 80, he was belatedly knighted.

Wallis, John (1616–1703) *English mathematician and theologian*

Born at Ashford in England, Wallis was educated at Cambridge University (1632–40), obtaining his MA in 1640. His early training was in theology and it was as a theologian that he first made his name. He took holy orders and eventually became bishop of Winchester. He moved to London in 1645 where he became seriously interested in mathematics and in 1649 he was appointed to the Savilian Chair in Geometry at Oxford University.

Wallis's most celebrated mathematical work is contained in his treatise the *Arithmetica infinitorum* (1655; The Arithmetic of Infinitesimals). In this work he gave his famous infinite series expression for π:

$$2/\pi = (1.3.3.5.5.7....)/(2.2.4.4.6.6....)$$

Generally the treatise took the development of 17th-century mathematics a significant step nearer NEWTON's creation of the infinitesimal calculus. Wallis was one of the first mathematicians to introduce the functional mode of thinking, which was to be of such importance in Newton's work. He also did notable work on conic sections and published a treatise on them, *Tractatus de sectionibus conicis* (1659; Tract on Conic Sections), which developed the subject in an ingeniously novel fashion. His writings were certainly read by Newton and are known to have made a considerable impact on him. Before Newton, Wallis was probably one of the most influential of English mathematicians.

Wallis wrote a substantial history of mathematics. His other interests included music and the study of language. He was active in the weekly scientific meetings that eventually led to the foundation of the Royal Society in 1662. During the English Civil War he was a Parliamentarian and put his mathematical talents to use in decoding enciphered letters.

Walton, Ernest Thomas Sinton (1903–1995) *Irish physicist*

Walton, who was born at Dungarvan in Ireland, studied at the Methodist College, Belfast, where he excelled at mathematics and science. In 1922 he entered Trinity College, Dublin, graduating in mathematics and experimental science in 1926. In 1927 he went to Cambridge University on a research scholarship and worked in the Cavendish Laboratory under Ernest RUTHERFORD. It was here that he performed experiments, together with John COCKCROFT, with accelerated particles. The experiments were to lead to the two men sharing the 1951 Nobel Prize for physics for "their pioneer work on the transmutation of atomic nuclei by artificially accelerated atomic particles," more commonly known as "splitting the atom."

In 1934 Walton gained his PhD from Cambridge and returned to Dublin as a fellow of Trinity College. He was appointed Erasmus Smith Professor of Natural and Experimental Philosophy in 1946 and was elected a senior fellow in 1960. In 1952 he became chairman of the School of Cosmic Physics of the Dublin Institute for Advanced Studies, where he remained until his retirement in 1974.

Wambacher, Hertha (1903–1950) *Austrian physicist*

Born in Vienna, Wambacher studied at Vienna University, initially chemistry and then physics, receiving her PhD in 1932. She joined her erstwhile university supervisor, Marietta Blau, at Vienna's Radiuminstitut, and collaborated with her in developing photographic emulsions capable of detecting the spectrum of neutrons arising from reactions in the emulsion. For this they received the Lieben Prize, a prestigious Austrian scientific award. The pair then showed how the photographic technique could be used to detect cosmic rays. They exposed photographic plates over several months at Hafelekar high-altitude observatory and discovered the phenomenon of disintegration stars – starlike tracks indicating nuclear reactions induced by cosmic rays, now also known as Blau–Wambacher stars. However, the fruitful collaboration between the two women was severed by growing political turmoil. Wambacher had for some years been a member of the illegal Nazi party, and came increasingly under the influence of Georg Stetter, a colleague and fellow Nazi. Blau left for Norway on the day before the German annexation of Austria, whereas Wambacher continued research on cosmic rays, and working as a teacher at Vienna University during World War II. In 1945 she was removed from her post and briefly exiled to the USSR, returning to Vienna in 1946. In 1962 she was posthumously awarded the Erwin Schrödinger Prize, shared with Blau.

Wankel, Felix (1902–1988) *German mechanical engineer and inventor*

Wankel (**vahng**-kel or **wang**-kel) was born at Lahr in Germany. The son of a civil servant who was killed in World War I, he left school early when the family savings were rendered valueless by the hyperinflation of the 1920s. He began work in 1921 in a Heidelberg bookshop while pursuing his engineering interests by attending night classes and by taking correspondence courses. In 1924 he began to consider the possibility of constructing a rotary engine and would spend the rest of his life trying to perfect one.

There was nothing new in the idea of a rotary engine; over 2,000 patents had been taken out in Britain before 1910. None, however, worked adequately, for none had solved the problem of designing a gas-tight seal. In a piston engine it is a simple matter to maintain a good seal with piston rings. In Wankel's rotary engine the whole rotor circumference has to be sealed. If the seal is too loose, power is lost as gas escapes; if too tight, friction increases, power falls, and rotors wear out.

While working on his own engine Wankel also designed a rotary disk valve consisting of a flat disk with a hole that revolves to allow air or gas entry into the cylinder at the appropriate time. The success of his design enabled Wankel to leave the bookshop in 1926 and to set up his own workshop. He also found a patron in Wilhelm Keppler, a prosperous businessman and economic adviser to Hitler. With Keppler's aid Wankel received commissions from Daimler in 1933 and BMW in 1934. In 1936, again at Keppler's behest, Wankel was installed by the German Air Ministry in a workshop in Lindau to work on rotary valves for aircraft engines.

Wankel had in fact been a member of the Hitler Youth and the Nazi party but, he later claimed, he had left the party in 1932, and after exposing a corrupt party official was actually imprisoned in 1933 for several months. Despite this, his workshop was occupied and destroyed by the French in 1945.

It was not until 1951 that Wankel was allowed to reopen his workshop. During the interval he had worked intensively on rotary-engine design and by 1953 he had a workable model – a triangular rotor housed in a trochoid-shaped chamber fitted with both apex and side seals. As the rotor revolves in its housing three chambers of variable size are produced, in each of which the four-stroke cycle of intake, compression, ignition, and exhaust takes place. Compared with reciprocating piston engines, Wankel's design was much smaller, required half the number of moving parts, and was virtually free of vibrations.

Over the years rights to manufacture Wankel's engine were sold to NSU in Germany, Curtis-Wright in America, and Toyo Kogyo in Japan. While most major motor companies, including Ford, General Motors, and Rolls Royce, showed some interest in the rotary engine, none, with the exception of Mazda, pursued the matter further. In 1971 Mazda tried to break into the American market with their R-100 and RX-2. Initially sales were good and 42,000 models were sold in 1972, but this was to prove the high point. The oil crisis of 1973 was no time to introduce a revolutionary new model. Consequently, Mazda's rotary engine found a more durable niche in one of their low-volume high-priced sports cars.

Wankel subsequently became director of his own research establishment at Lindau, where he did further work on rotary engines.

While few seem to have profited from Wankel's design, he himself accumulated a fortune of $20 million from the sale of manufacturing rights, the bulk of which was spent building homes for cats and dogs.

Warburg, Otto Heinrich (1883–1970)
German physiologist

> A scientist must have the courage to attack the great unsolved problems of his time.
> —Attributed remark

Warburg (**var**-buurk or **wor**-burg), who was born at Freiburg im Breisgau in Germany, was the son of Emil Warburg, a distinguished professor of physics at Berlin. Otto was educated at the University of Berlin, where he obtained his PhD in 1906, and at Heidelberg, where he gained his MD in 1911. He joined the Kaiser Wilhelm Institute for Biology in 1913, attaining professorial status in 1918, and in 1931 became director of the Kaiser Wilhelm Institute for Cell Physiology, renamed after Max PLANCK following World War II. Here Warburg remained in charge until his death at the age of 86.

When the human body converts lactic acid into carbon dioxide and water it consumes oxygen. In the early 1920s Warburg began to investigate just how such aerobic metabolism works. To do this he designed, in 1923, the *Warburg manometer*, which is used to measure the rate of oxygen uptake by human tissue. It was clear to Warburg that such a reaction could only take place at normal temperatures with the aid of enzymes but, because of the tiny amounts involved, such enzymes would be impossible to isolate by orthodox analytical techniques. He suspected the respiratory enzymes to be the cytochromes discovered a decade ear-

lier and consequently set out to explore their nature by noting which substances affected the rate of oxygen uptake. He first noted that intercellular respiration was blocked by hydrogen cyanide and by carbon monoxide. This suggested to Warburg that the respiratory enzymes contained iron on the analogy that carbon monoxide acts on hemoglobin by breaking the oxygen–iron bonds. Support for such a supposition was derived from the similarity between the spectrum of the carbon monoxide–hemoglobin complex and that of the carbon monoxide–respiratory enzyme complex.

Warburg also studied the metabolism of cancerous cells and in 1923 discovered that malignant cells use far less oxygen than normal cells and can in fact live anaerobically. This extremely interesting observation led him to speculate that cancer is caused by a malfunction of the cellular respiratory system. He advocated that cancer might be prevented by avoiding foods and additives that impair cellular activity and by ensuring a high level of respiratory enzyme in the body by taking plenty of iron and vitamin B.

Warburg also worked on other enzyme systems, particularly the flavoproteins, or yellow enzymes, active in cellular dehydrogenation. He found that the coenzyme flavin adenine dinucleotide (FAD) is the active part of flavoproteins and later demonstrated that nicotinamide is similarly the active part of nicotinamide adenine dinucleotide (NAD^+). Following these discoveries he showed that in alcohol fermentation a hydrogenated form of NAD^+ ($NADH_2$) reacts with acetaldehyde to give NAD^+ and ethyl alcohol.

For his contributions to biochemistry, Warburg was nominated three times for the Nobel Prize for physiology or medicine, in 1926, 1931, and 1944, although he only actually received the award in 1931.

Ward, Joshua (1685–1761) *English physician and chemist*

Ward, who was born in the eastern English county of Suffolk, was elected to parliament as member for Marlborough in 1717. However, he was prevented from taking his seat because of electoral irregularities and was forced to flee to France. He returned on being pardoned in 1733. Ward became the most famous (or notorious) of the pill peddlers. He won the protection of George II when he cured him of a dislocated thumb and treated many of the 18th-century notables, including Edward Gibbon and Henry Fielding. The contents of his pills were revealed in an unsuccessful legal suit brought by him in 1734. Most of his pills, including the "white drops," contained antimony and some also contained arsenic. His popularity was such, however, that he was specifically excluded from the

Apothecaries Act of 1748 designed to prevent unlicensed prescribing.

One of the great problems facing 18th-century industry was the production of sulfuric acid in quantity. In 1736 Ward opened a factory in Surrey where he burned sulfur in 50-gallon glass globes together with a sodium nitrate catalyst to produce sulfuric acid. He was able to produce enough sulfuric acid in this way both to make his own fortune and to reduce the price of the acid to an eighth of its original cost. This increase in quantity and decrease in price contributed to the growth of such industries as dyeing and bleaching later in the century.

Warming, (Johannes) Eugenius Bülow
(1841–1924) *Danish botanist*

Warming was born on the island of Manö in the Frisian islands. He studied botany at Munich and later became professor of botany at Stockholm (1882–85) and at Copenhagen (1885–1911).

In 1895 he published his book *Plantesamfund* on plant ecology and is regarded as one of the founders of the subject.

Warren, J. Robin (1937–   ) *Australian pathologist*

Warren was born in Adelaide, South Australia, and gained his MB BS from the University of Adelaide in 1961. After a brief stint as Junior Resident Medical Officer in the Queen Elizabeth Hospital in Woodville, South Australia, he specialized in pathology, in 1962 becoming registrar in haematology and clinical pathology at Adelaide's Institute of Medical and Veterinary Science. His training continued at the Royal Melbourne Hospital (1964–68), and in 1967 he was made a fellow of the Royal College of Pathologists of Australasia. In 1968 he moved to the Royal Perth Hospital as senior pathologist, where he remained until 1999.

Warren's greatest contribution to medicine has been in discovering the role played by certain bacteria in ulcers of the duodenum and stomach (peptic ulcers). In 1979 he first reported seeing a novel type of spiral-shaped bacterium in pathology specimens (biopsies) obtained from patients with stomach inflammation (gastritis). This contradicted medical orthodoxy, which held that no bacteria could survive the highly acidic environment of the stomach, and scientific colleagues were disdainful of Warren's findings. The conventional wisdom was that gastritis and peptic ulcers were physiological diseases caused by overproduction of acid; treatment entailed often long-term treatment with acid-suppressing drugs. However, in 1981 Warren began what was to prove a highly fruitful collaboration with Barry MARSHALL, who was training to be a physician at the Royal Perth Hospital. With Marshall's as-

sistance, Warren undertook a systematic two-year study of biopsy specimens from 100 patients with gastric disorders. This showed a strong association between presence of the bacteria and the occurrence of peptic ulcers; the bacteria were also found in about half the cases of stomach cancer. In 1982 they successfully cultured the bacterium, and showed that it was indeed new to science. They tentatively identified it as belonging to the genus *Campylobacter*, but it was subsequently placed in its own genus as *Helicobacter pylori*.

Warren found that the bacteria evade the acidic stomach contents by living in the thick layer of mucus that coats the stomach wall, where they trigger an inflammatory response by the body's immune cells. Marshall went on to show that the bacteria use an enzyme to break down urea into ammonia and carbon dioxide, thereby creating an alkaline microenvironment for themselves. This further safeguards them from attack by the stomach acid.

Warren and Marshall argued that the bacterium causes gastritis, which in turn can lead in some cases to the development of ulcers. They published their results in the medical journal *The Lancet* in 1983, but met with a largely hostile reception from their peers. Their frustration with the scepticism of the medical establishment was immense, and in 1985 this led Marshall to deliberately infect himself with a culture of *H. pylori*. As predicted, he duly developed gastritis. The pair went on to investigate the most effective drug combinations against *H. pylori*, and showed that by administering combinations of bismuth and antibiotics, the bacteria could be completely eliminated and patients cured of their peptic ulcers.

Eventually, the medical community began to accept that this radical new approach offered the prospect of cure for many patients who had long suffered from chronic gastric disease, and could help to identify people at risk of developing not only gastric and duodenal ulcers, but also certain forms of stomach cancer. Since the 1990s, in recognition of their achievement, Warren and Marshall have received numerous honours and awards, both individually and jointly, the latter including the 1994 Warren Alpert Prize, the 1997 Paul Ehrlich Prize, and the 2005 Nobel Prize for physiology or medicine, "for their discovery of the bacterium *Helicobacter pylori* and its role in gastritis and peptic ulcer disease."

Wassermann, August von (1866–1925)
German bacteriologist

Wassermann (**vah**-ser-mahn or **wah**-ser-man), who was born at Bamberg in Germany, was educated at the universities of Erlangen, Vienna, Munich, and Strasbourg, where he graduated in 1888. From 1890 he worked under Robert KOCH at the Institute for Infectious Diseases in

Berlin, becoming head of the department of therapeutics and serum research in 1907. In 1913 he moved to the Kaiser Wilhelm Institute, where he served as director of experimental therapeutics until his death.

Wassermann is best remembered for the *Wassermann test* (or *reaction*), which he introduced in 1906 for the diagnosis of syphilis. The test depends upon an infected person producing in his or her blood the antibody to syphilis, which will combine with known antigens, such as beef liver or heart, to form a complex. The test is regarded as positive by the ability of the complex to fix complement, the serum protein discovered by Jules BORDET in the 1880s. The test is still widely used as a diagnostic tool.

Waterston, John James (1811–1883) *British physicist*

Waterston graduated from the university in his native city of Edinburgh, where he studied medicine and science. He worked as an engineer in London and the Far East before returning to Edinburgh.

Apparently he was stimulated by the works of Julius MAYER and James JOULE to work out for himself the kinetic theory of gases, which later won great fame for James Clerk MAXWELL and Rudolf CLAUSIUS. Waterston submitted a paper to the Royal Society in 1845 describing his great discovery but it was rejected by their referee as being "nothing but nonsense." Lord RAYLEIGH found the paper in the Royal Society archives and arranged for it to be published in 1892, nine years after Waterston's death.

Watson, David Meredith Seares (1886–1973) *British paleontologist*

Watson was born in Manchester, England, the son of a prosperous industrialist. He studied chemistry and geology at the university in his native city, obtaining his MSc in 1909. For some years he worked in unofficial posts at the British Museum and University College, London; he also traveled in Australia and South Africa studying both their geology and fossils. In 1921 he succeeded Peter Hill as Jodrell Professor of Zoology and Comparative Anatomy at University College, London, a post he held until his retirement in 1951.

Watson, originally trained as a geologist, began his career working with Marie Stopes, then famous as the first female science lecturer at Manchester, on fossil plants in coal mines. He soon turned, however, to the main work of his life, vertebrate paleontology, on which topic he published over a hundred papers. His views were published in less specialized form in his *Palaeontology and Modern Biology* (1951).

Watson, James Dewey (1928–) *American biochemist*

It is necessary to be slightly underemployed if you want to do something significant.
—Quoted by H. Judson in *The Eighth Day of Creation*

Watson entered the university in his native city of Chicago at the early age of 15, graduating in 1947. He obtained his PhD (1950) for studies of viruses at the University of Indiana and continued this work at the University of Copenhagen. In Copenhagen he realized that one of the major unsolved problems of biology lay in identifying the structure of the nucleic-acid molecules making up chromosomes. In 1951 he moved to the Cavendish Laboratory in Cambridge, England, to study the structure of DNA.

Early in 1953 Watson and Francis CRICK published a molecular structure of DNA having two cross-linked helical chains (*General Implications of the Structure of Deoxyribonucleic Acid*). They arrived at this by considering possible geometric models, which they based on two independent sets of experimental work: the x-ray crystallography of Maurice WILKINS and Rosalind FRANKLIN at King's College, London, and the earlier work of Erwin CHARGAFF, which had established the relative quantities of the organic bases present in the nucleic acids. Watson and Crick were able to show that certain organic bases linked the chains together by hydrogen bonds.

The model explains the three basic characteristics of heredity. It shows how genetic information can be expressed in the form of a chemical code; it demonstrates the way in which genes replicate themselves – when the two chains separate each can serve as a template for the synthesis of a new chain; and finally it provides an explanation of how mutations occur in genes, in terms of changes in the chemical structure of DNA. Watson, Crick, and Wilkins shared the Nobel Prize for physiology or medicine for this work in 1962.

Watson left Cambridge in 1953 for the California Institute of Technology. From 1955 to 1968 he worked at Harvard, becoming professor of biology in 1961. Here he continued to study the genetic code. In 1968 he became director of the Cold Spring Harbor Laboratory, New York, where he concentrated effort on cancer research. The same year he published *The Double Helix*, an informal, highly personal, and somewhat controversial account of the discovery of the structure of DNA. In 2007 he brought out an autobiography, *Avoid Boring People: Lessons from a Life in Science*. In October 2007 a biographical article by one of his former assistants quoted him as making remarks that were construed as implying that there is a link between race and intelligence. The publicity caused a considerable amount of controversy, with Watson attracting much criticism. The London Science Museum canceled a talk that Watson was to give, and he was suspended from Cold Spring

Harbor. Despite an apology and explanation Watson resigned from Cold Spring Harbor.

Watson, Sir William (1715–1787) *British physicist, physician, and botanist*

> Never indolent in the slightest degree ... an exact economist of his time.
> —R. Pulteney, *Sketches of the Progress of Botany in England* (1790)

Watson, the son of a London tradesman, was apprenticed to an apothecary from 1731 to 1738. After working for many years at that trade, Watson was made a licentiate of the Royal College of Physicians. This was later followed in 1762 with an appointment as physician to the Foundling Hospital.

Watson had a great interest in natural history and was instrumental in introducing the Linnaean system of botanical classification into Britain. He is, however, mainly remembered for his account of the nature of electricity. This was based on a series of experiments with the Leyden jar, discovered by Pieter van MUSS-CHENBROEK in 1746. Watson not only improved the device by coating the inside of it with metal foil but also realized that the pattern of discharge of the jar suggested that electricity was simply a single fluid or, as he termed it, an "electrical ether." Normally bodies have an equal density of this fluid so that when two such bodies meet there will be no electrical activity. If, however, their densities are unequal the fluid will flow and there will be an electric discharge. That is, electricity can only be transferred from one body to another; it cannot be created or destroyed. Such a theory was also developed with greater depth at about the same time by Benjamin FRANKLIN and was to emerge as the orthodox position by the end of the century.

Watson also made an early and unsuccessful attempt in 1747 to measure the velocity of electricity over a four-mile (6.4-km) circuit. Although it appeared to complete its journey in no time at all Watson sensibly concluded that it probably traveled too fast to be measured.

Watson-Watt, Sir Robert Alexander (1892–1973) *British physicist*

> The wings of an aeroplane act like a kind of horizontal wire in the air. When you aim a powerful wireless beam at them, they turn into a "secondary transmitter" and send the waves back at the angle of incidence, just as a mirror reflects light rays.
> —On the principle of radar. Quoted by Egon Larsen in *A History of Invention* (1969)

Watson-Watt was born Robert Watt at Brechin in Angus, Scotland. The Watson part of his name came from his mother's family and the hybrid Watson-Watt was adopted in 1942 on receipt of his knighthood. He was the son of a carpenter and was educated at the University of St. Andrews. After graduating in 1912 he immediately joined the faculty but found his academic career disrupted by World War I. He spent much of the war working as a meteorologist at the Royal Aircraft Establishment, Farnborough, attempting to locate thunderstorms with radio waves.

He remained in the scientific civil service after the war and in 1921 was appointed superintendent of the Radio Research Station at Slough. In 1935 he was asked by the Air Ministry if a "death ray" could be built – one capable of eliminating an approaching enemy pilot. Watson-Watt asked a colleague to calculate how much energy would be needed to raise a gallon of water from 98°F to 105°F at a distance of a mile, i.e., a significant rise in body temperature. He advised the Ministry that the energy needed outstripped the available technology.

Watson-Watt also pointed out that Post Office engineers had noted interference in radio reception as aircraft flew close to their receivers. Interference of this kind, he suggested, could perhaps be used to detect the approach of enemy aircraft. In 1935 he submitted an important paper, *The Detection of Aircraft by Radio Methods*, to TIZARD at the Ministry. Watson-Watt was normally a man, it was said, who could never say in one word what could be said in a thousand. This time, however, the report was terse and to the point. Tizard asked for a demonstration. In February 1935 the BBC short-wave transmitter at Daventry was successfully used to identify the approach of a Heyford bomber eight miles away.

Tizard moved quickly. Watson-Watt was invited to set up a research station at Bawdsley in Suffolk to develop radio detection and ranging; the acronym "radar" was first recorded in use in the *New York Times* in 1941.

The principles behind radar are relatively simple. Radio waves are reflected strongly off large objects such as airplanes. The difficulty was that very little, something of the order of 10^{-12}, of the transmitted signal would be picked up by the receiving antennae. Both high transmitting power and high amplification would therefore be needed. Watson-Watt assembled a talented team at Bawdsley and by the outbreak of World War II an operational chain of eight stations, known as "Chain Home," defended Britain's eastern and southern coasts. They operated in the high-frequency bands and required very visible 360-foot-high transmitters and 240-foot-high receivers.

Watson-Watt left Bawdsley in 1938 for the Air Ministry and the post of director of communication development. His main task was to make radar workable, to ensure that it was acceptable to the RAF and that they could actually operate the new equipment. He also had to arrange for the manufacture of the relevant transmitters, receivers, and electron tubes.

He finally left the civil service in 1945 to set

up as a consultant. He was also invited to give evidence before the Royal Commission on Inventors on behalf of his colleagues and himself. After speaking for six days Watson-Watt was awarded £52,000 for his work on radar.

Watt, James (1736–1819) *British instrument maker and inventor*

> I have now got an engine that shall not waste a particle of steam. It shall be all boiling hot; – aye, and hot water injected if I please.
> —Quoted by John Robison in *A Narrative of Mr. Watt's Invention of the Improved Engine* (1796)

> James Watt, who, directing the force of an original genius, Early exercised in philosophic research to the improvement of THE STEAM ENGINE, enlarged the resources of his country, increased the power of men, and rose to an eminent place among the most illustrious followers of science and the real benefactors of the World.
> —Epitaph in Westminster Abbey, London

The son of a Clydeside shipbuilder and house builder, Watt was born in Greenock, Scotland. At the age of 17 he started a career in Glasgow as a mathematical-instrument maker. Through his shop, opened in 1757, he met many of the scientists at Glasgow University.

In 1764 it occurred to Watt that the NEW-COMEN steam engine, a model of which he had been repairing, wasted a great deal of energy by dissipating the latent heat given up by steam condensing to water. The solution was to build an engine with a separate condenser, so that there was no need to heat and cool the cylinder at each stroke. In 1768 Watt entered into partnership with John Roebuck, who had established an iron foundry, to produce the steam engine but his duties as a land surveyor, taken up in 1766, left him little time to develop this and Roebuck went bankrupt in 1772. A second partnership (1775) with Matthew BOULTON proved more productive although it took Watt until 1790 to perfect what became known as the *Watt engine*.

This engine, throughout its various stages of improvement, was one of the main contributors to the Industrial Revolution. Early reciprocating versions were used for pumping water out of Cornish copper and tin mines. A rotating engine with the sun-and-planet gearing system invented by Watt in 1781 was used in flour mills, cotton mills, and paper mills. An automatic speed control mechanism, the centrifugal governor invented in 1788, was another improvement.

Watt made a great deal of money from the sale of his engines and became accepted into the scientific establishment. He retired from the business of steam-engine manufacture in 1800 and spent his time traveling, working as a consultant, and working on minor inventions in his workshop at home. Watt was the first to use the term horsepower as a unit of power and the *watt*, a unit of power, was named for him.

Weber, Ernst Heinrich (1795–1878) *German physiologist and psychologist*

Weber was the eldest of three brothers who all made important contributions to science. He was born at Wittenberg in Germany and became a professor at the University of Leipzig in 1818, a position he held until his death.

Weber is best known for his work on sensory response to weight, temperature, and pressure. In 1834 he conducted research on the lifting of weights. From his researches he discovered that the experience of differences in the intensity of sensations depends on percentage differences in the stimuli rather than absolute differences. This is known as the just-noticeable difference (j.n.d.), difference threshold, or limen. The work was published in *Der Tastsinn und das Gemeingefühl* (1851; The Sense of Touch and the Common Sensibility) and was given mathematical expression by Weber's student Gustav Theodor FECHNER as the *Weber–Fechner* law.

Weber is regarded as a founder of experimental psychology and psychophysics. He also conducted important anatomical work.

Weber, Joseph (1919–2000) *American physicist*

Born in Paterson, New Jersey, Weber graduated from the U.S. Naval Academy in 1940 and served in the Navy until 1948, when he joined the faculty of the University of Maryland, College Park. He completed his doctorate at the Catholic University of America, Washington, DC, in 1951 and was appointed professor of physics at Maryland in 1959, a post which he held for the remainder of his career.

EINSTEIN's general theory of relativity predicts that accelerated masses should radiate gravitational waves. Like electromagnetic waves, these should carry energy and momentum and should be identifiable with a suitable detector. For gravitational waves, this would be an object of large mass with a method of detecting any disturbance of it. In 1958 Weber began the design and construction of just such a device. By 1965 he had built a solid aluminum cylinder detector, 3 feet in diameter and weighing 3.5 tons. Bonded around the cylinder were a number of piezoelectric crystals, which generate a voltage when the bar is compressed or extended. Weber claimed that his instruments could detect deformations corresponding to 1 part in 10^{16}, a difference of about 1/100th the diameter of an atomic nucleus. Weber was aware of the problems involved in this kind of design. To rule out causes other than gravitational – acoustic, thermal, seismic, etc. – he suspended the cylinder in a vacuum. More significantly he built a second detector 600 miles away from Maryland at the Argonne National Laboratory in Chicago, and only recognized coincident readings as evidence for gravitational waves. He reported the first coincident readings in 1968. He also noted, in 1970, that such readings reached a peak when the cylinders were both oriented in the direction of the galactic center.

Unfortunately, although there were several attempts to replicate Weber's work in the 1970s, none proved successful. Work has, however, continued with more sensitive antennas. Supercooled niobium rods have been installed at the European Laboratory for Particle Physics in Geneva, at Stanford, and at the Louisiana State University for a three-way coincidence experiment. Despite recognizing 60–70 events a day, none have yet conclusively proved to be coincidental.

Attempts to resolve Weber's conjecture were also begun in 1991 with the announcement of the LIGO project (Laser Interferometer Gravitational-Wave Observatory). Two detectors are in use as part of the project, one at Livingston, Louisiana, and the other at Richland, Washington. Both use laser interferometers to detect events. By 2005 the detectors had reached the required sensitivity, although no positive events have so far been detected.

Weber, Wilhelm Eduard (1804–1891) *German physicist*

Weber (**vay**-ber) was the son of a professor of divinity and brother of the noted scientists Ernst Heinrich Weber and Eduard Friedrich Weber, both of whom worked in anatomy and physiology. He was born in Wittenberg in Germany and studied physics at Halle, where his early research concerned acoustics. He obtained his PhD in 1826 for a thesis on reed organ pipes. He remained teaching at Halle until 1831 when he was made professor of physics at Göttingen on the recommendation of the mathematician Karl Friedrich GAUSS.

Some of Weber's research was done in collaboration with his brothers. Thus in 1824 he published work on wave motion with Ernst, and in 1833 he and Eduard investigated the mechanism of walking. However, most of his academic life was spent working with Gauss. In 1833 they built the first practical telegraph between their laboratories to coordinate their experiments on geomagnetism. In 1837 Weber lost his post for opposing the new king of Hannover's interference with the State constitution. Nevertheless, he stayed in Göttingen for a further six years until he was appointed professor at Leipzig. Here, he improved the tangent galvanometer invented by Hermann von HELMHOLTZ and built an electrodynamometer suitable for studying the force produced by one electric current on another.

Weber's main work was the development of a system of units that expressed electrical concepts in terms of mass, length, and time. Gauss had previously done this for magnetism. Since force was expressed in these dimensions, he was then able to find his law of electric force. The principle was not very satisfactory because it did not conserve energy, but with it Weber publicized the view that matter was made up of charged particles held together by the force. This inspired the direction that physics took in the latter half of the 19th century. The units of Gauss and Weber were adopted at an international conference in Paris in 1881. The unit of magnetic flux (the *weber*) is named in his honor.

In 1849 he returned to his post in Göttingen and collaborated with R. H. A. Kohlrausch in measuring the ratio between static and dynamic units of electric charge. This turned out to be the speed of light; this unexpected link between electricity and optics became central to James Clerk MAXWELL's great development of electromagnetic field theory.

Wegener, Alfred Lothar (1880–1930) *German meteorologist and geologist*

Wegener (**vay**-ge-ner), who was born in Berlin, was educated at the universities of Heidelberg, Innsbruck, and Berlin, where he obtained his doctorate in astronomy in 1905. In 1906 he went on his first meteorological research trip to Greenland and, on his return (1908), was appointed to a lectureship in astronomy and meteorology at the University of Marburg. After World War I he moved to a special chair of meteorology and geophysics at the University of Graz, Austria, in 1924. He made further expeditions to Greenland, where he died on his fourth visit.

In 1915 Wegener produced his famous work *Die Enstehung der Kontinente und Ozeane* (translated as *Origin of Continents and Oceans*, 1924), in which he formulated his hypothesis of continental drift. In this he proposed that the continents were once contiguous, forming one supercontinent, Pangaea, which began to break up during the Mesozoic Era and drifted apart to form the continents we know today. To support his theory Wegener produced four main arguments. He first pointed to the obvious correspondence between such opposite shores as those of Atlantic Africa and Latin America. An even better fit was evident if the edges of the continental shelves were matched instead of the coastlines. Secondly he argued that geodetic measurements indicated that Greenland was moving away from Europe. This supported his third argument that a large proportion of the Earth's crust is at two separate levels, the continental and the ocean floor, and that the crust is made of a lighter granite floating on a heavier basalt. His final argument was that there were patterns of similarities between species of the flora and fauna of the continents.

Wegener's theory at first met with considerable hostility. However, in 1929 Arthur HOLMES

was able to suggest a plausible mechanism to account for continental movement and this, together with advances in geomagnetism and oceanography, was to lead to the full acceptance of Wegener's theory and the creation of the

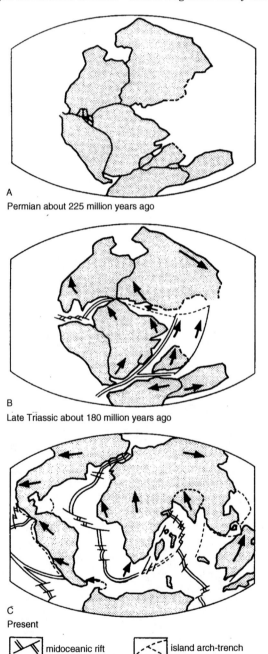

A

Permian about 225 million years ago

B

Late Triassic about 180 million years ago

C

Present

| ⬡ | midoceanic rift | ⬡ | island arch-trench |

CONTINENTAL DRIFT   Originally the present continents formed a single landmass – a supercontinent called "Pangaea." This probably began to break up during the Mesozoic era. The stages are shown above.

new geophysical discipline of plate tectonics after World War II.

Wegener's meteorological works include *Die Klimate der Geologischen Vorzeit* (1924; Climates in Geological Antiquity) published in association with his father-in-law, Wladimir KÖPPEN.

Weierstrass, Karl Wilhelm Theodor
(1815–1897) *German mathematician*

> A mathematician who is not somewhat of a poet will never be a true mathematician.
> —Attributed remark

> The general theories [of Weierstrass] answer all our possible questions; unfortunately they answer them far too readily, without requiring any effort on our part.
> —H. Lebesgue, *Notes on the History of Mathematics* (1958)

Weierstrass (**vI**-er-shtrahs or **vI**-er-strahs), who was born at Ostenfelde in Germany, spent four years at the University of Bonn studying law to please his father. After abandoning law he trained as a school teacher and spent nearly 15 years teaching at elementary schools in obscure German villages. However, he found time to combine his mathematical researches with his school teaching and in 1854 he attracted considerable favorable attention with a memoir on Abelian functions, which he published in Crelle's journal. The fame this work brought him resulted in his obtaining a post as professor of mathematics at the Royal Polytechnic School in Berlin and he soon moved on to the University of Berlin.

Weierstrass's work on Abelian functions is generally considered to be his finest, but he made numerous other contributions to many other areas of mathematics. He was one of the first to make systematic use in analysis of representations of functions by power series. He was a superb and very influential teacher, an excellent fencer, and, unlike many mathematicians, he intensely disliked music. His work in "arithmetizing" analysis led him into a fierce controversy with the constructivist Leopold KRONECKER, who thought that Weierstrass's widespread use of nonconstructive proofs and definitions was unsound.

It is to Weierstrass together with Augustin CAUCHY that modern analysis is indebted for its high standards of rigor. Weierstrass gave the first truly rigorous definitions of such fundamental analytical concepts as limit, continuity, differentiability, and convergence. He also did very important work in investigating the precise conditions under which infinite series converged. Tests for convergence that he devised are still in use.

Weil, André (1906–1998) *French mathematician*

Weil (**vIl**) studied at the Ecole Normale Supérieure in his native city of Paris and at the universities of Rome and Göttingen. He held teaching posts in many countries, including posts at the Aligarh Muslim University in India and at the universities of Strasbourg, São Paulo in Brazil, and Chicago. In 1958 he moved to the Institute for Advanced Study at Princeton, where he was made professor emeritus in 1976.

Weil's mathematical work centered on number theory, algebraic geometry, and group theory. He proved one of the central results in the theory of algebraic fields. His publications include *Foundations of Algebraic Geometry* (1946).

The religious philosopher and mystic Simone Weil, who died in 1943, was his sister.

Weinberg, Steven (1933–) *American physicist*

Born in New York, Weinberg was educated (as was his future colloborator, Sheldon Glashow) at the Bronx High School of Science, at Cornell, and at Princeton, where he gained his PhD in 1957. Following appointments at Columbia (1957–59), Berkeley (1959–69), the Massachusetts Institute of Technology (1969–73), and Harvard (1973–83), he was appointed professor of physics at the University of Texas, Austin.

In 1967 Weinberg published a paper, *A Model of Leptons*, which proposed a unification of the weak and electromagnetic interactions, since known as the "electroweak theory." In modern particle physics forces operate through the interchange of particles: the electromagnetic force by interchanging photons, and the weak force by the interchange of the W and Z bosons. The claim that the forces had been united into a single force would imply that photons and bosons belonged to the same family of particles. But it is only too clear that this could not be the case; the photon was virtually massless, while the bosons were even more massive than the proton.

The difference was explained by Weinberg in terms of spontaneous symmetry breaking (SSB). At the extremely high temperatures present shortly after the big bang, photons and bosons would have been indistinguishable. At some point during the cooling the initial symmetry was spontaneously broken, and during this breakage some particles acquired different properties. Weinberg likens the process to what happens when a piece of iron is cooled below a temperature of 770°C. Below this point the material becomes ferromagnetic and a magnetic field pointing in some unpredictable direction can appear, spontaneously breaking the symmetry between different directions.

The question of the origin of the mass of the bosons remained. Weinberg proposed that the Higgs mechanism, described by Peter HIGGS in 1964, though hypothetical, would suffice. As a consequence of Weinberg's theory the existence of "neutral weak currents" was predicted. It had previously been supposed that weak interactions invariably involved a transfer of electric charge carried by the bosons W^+ and W^-. In electromagnetic interactions the photon is exchanged setting up a neutral current. The weak interaction should be able to proceed in the same way with the transfer of the neutral boson Z^0. Neutral weak currents were first observed in 1973, and the bosons of the electroweak theory were detected by Carlo RUBBIA at the European Laboratory for Particle Physics in 1983. It was for this work that Weinberg shared the 1979 Nobel Prize for physics with his schoolmate from the Bronx, Sheldon GLASHOW, and with Abdus SALAM.

Weinberg has also worked in the field of cosmology, publishing in 1972 a substantial treatise on the subject, *Gravitation and Cosmology*. This was followed by *The First Three Minutes* (1977), an extremely popular account of the three minutes following "about one hundredth of a second after the beginning when the temperature had cooled to a mere hundred thousand million degrees Kelvin." In 1982 he published another popular book, *The Discovery of Subatomic Particles*.

In a later work, *Dreams of a Final Theory* (1993), Weinberg argued that in today's theories we are already beginning to catch glimpses of the outlines of a final theory. His three-volume treatise *The Quantum Theory of Fields* (1995, 1996, 2000) is the standard text on the subject.

Weismann, August Friedrich Leopold
(1834–1914) *German biologist*

> He has done more than any other man to focus attention on the mechanism of inheritance.
> —Citation on the award of the Royal Society's Darwin Medal to Weismann in 1908

Born at Frankfurt am Main in Germany, Weismann (**vIs**-mahn) studied medicine at Göttingen, graduating in 1856. He took several temporary jobs before joining the medical faculty of the University of Freiburg in 1863.

In his early work Weismann made much use of the microscope, but failing eyesight forced him to abandon microscopy for theoretical biology. His microscopic observations, especially those on the origin of the germ cells of hydrozoans, were nevertheless put to good use in the formulation of his theory of the continuity of the germ plasm, which he published in 1886 (English translation, 1893; *The Germ-Plasm: A Theory of Heredity*). Weismann had noted that germ cells can be distinguished from somatic cells early in embryonic development, and from this he visualized the protoplasm of the germ cell (germ plasm) as being passed on unchanged through the generations and therefore responsible for inheritance. Although the body might be modified by environmental effects, the germ plasm – well protected within it – could not be. This insulation of the germ plasm from environmental influences – the so-called *Weismann barrier* – is one of the fundamental tenets of modern Darwinian theory. Weismann himself argued strongly against the Lamarckian theory of the inheritance of acquired characteristics. His publication *Studies in the Theory of Descent* (1882) contained a preface by Darwin.

Weismann closely followed Edouard van BENEDEN's work on meiosis (reduction division of cells) and arrived at the correct explanation for this process – that a reduction division is necessary to prevent chromosome numbers doubling at fertilization. Weismann became director of the new museum and zoological institute built at Freiburg and remained at the university until his retirement in 1912.

Weizmann, Chaim Azriel (1874–1952)
Russian–British–Israeli chemist

Born in Motol (now in Belarus), the son of a timber merchant, Weizmann (**vIts**-mahn or **wIts**-man) was brought up in an orthodox segregated Jewish community. His early promise was recognized but, owing to difficulties placed in the way of Jews seeking higher education in Russia, Weizmann was sent to Germany instead. In Germany he received a PhD from the University of Freiburg in 1899 and was also converted to the Zionist doctrines of Theodor Herzl. After three years working at the University of Geneva, Weizmann decided to settle in England in 1904, eventually becoming naturalized in 1910.

Weizmann entered Manchester University, first as a student, but in 1907 he was appointed lecturer in biochemistry. Now began his most creative period as a scientist. He started by attempting to develop a synthetic rubber, which produced a need for the alcohol butanol. Butanol was not available commercially so Weizmann began searching for a ferment that would produce as much butanol as he needed from a cheap source. Eventually he came up with the bacterium *Clostridium acetobutylicum*, which turns cooked corn into butanol and acetone.

It happened to be the acetone which made Weizmann's name rather than the butanol as acetone is an important ingredient in the manufacture of cordite. With the start of World War I in 1914, cordite became a valuable commodity and consequently factories were set up in several countries to manufacture acetone by Weizmann's method.

With the commitment to a national home in Palestine clearly stated in the Balfour Declaration of 1917, Weizmann became almost exclusively concerned with Zionist politics. He was apppointed head of the World Zionist Movement in 1920 and of the Jewish Agency for Palestine in 1929. Finally, when Israel was created in 1948, Weizmann was elected its first president.

Weizsäcker, Baron Carl Friedrich von
(1912–2007) *German physicist*

> Those reductionists who try to reduce life to physics usually try to reduce it to primitive physics – not to good physics.
> —*Theoria to Theory* (1968)

> Classical physics has been superseded by quantum theory: quantum theory is verified by experiments. Experiments must be described in terms of classical physics.
> —Attributed comment

Weizsäcker (**vIts**-zek-er), who was born at Kiel in Germany, studied at the universities of Berlin, Göttingen, and Leipzig, obtaining his PhD from Leipzig in 1933. Between 1933 and 1945 he taught successively at the universities of Leipzig, Berlin, and Strasbourg. In 1946 he returned to Göttingen as director of physics at the Max Planck Institute where he remained until 1957, when he was appointed professor of philosophy at Hamburg. In 1970 he moved to Starnberg as director of the Max Planck Institute on the Preconditions of Human Life in the Modern World, a post he occupied until his retirement in 1980.

Weizsäcker proposed solutions to two fundamental problems of astrophysics. In 1938 he tackled the problem of how stars like the Sun can continue to radiate colossal amounts of energy for billions of years. Independently of Hans BETHE, he proposed a chain of nuclear-fusion reactions that could proceed at the high temperatures occurring in the dense central cores of stars. In this sequence, called the "carbon cycle," one carbon nucleus and four hydrogen nuclei, or protons, undergo various transformations before ending the cycle as one carbon nucleus and one helium nucleus. The process involves the release of an immense amount of energy that is eventually radiated from the star's surface mainly as heat, light, and ultraviolet radiation. As the stars are rich in hydrogen, it was now clear that they could continue radiating until their core hydrogen was consumed.

In 1944 Weizsäcker proposed a variation of the nebular hypothesis of Pierre Simon de LAPLACE to account for the origin of the planets. Beginning with the Sun surrounded by a disk of rotating gas he argued that such a mass would experience turbulence and break up into a number of smaller vortices and eddies. Where the eddies met, conditions were supposed to be suitable for planets to form from the continuous aggregation of progressively larger bodies. The system did not, however, explain the crucial point of how the planets managed to acquire so much angular momentum, a property that is conserved and cannot just be created out of nothing. Modifications and additions later proposed by Hannes ALFVÉN and Fred HOYLE on this issue used forces generated by the Sun's magnetic field as the means of transmitting momentum and won a fair amount of support for the theory.

Welch, William Henry (1850–1934) *American pathologist and bacteriologist*

> Medical education is not completed at the medical school: it is only begun.
> —Attributed comment

Welch was born in Norfolk, Connecticut. With a father, grandfather, and great-grandfather all physicians, it is not surprising that he too studied medicine at the College of Physicians and Surgeons, New York. After obtaining his MD in 1875, he spent some time in Germany working with Carl Ludwig and Julius COHNHEIM, learning techniques in pathology and microscopic anatomy. On his return to America in 1879 he became professor of pathology and anatomy at the Bellevue Hospital Medical College, New York. He was then appointed professor of pathology at Johns Hopkins University, Baltimore, in 1884 even though the medical school there was not actually opened until 1893. Following his retirement in 1916, Welch founded and became director of the Baltimore School of Hygiene and Public Health attached to Johns Hopkins University. From his second retirement in 1926, he served as professor of the history of medicine at the University until 1931.

Welch did much to establish Johns Hopkins as a major center for medical research. To it, and thus to America also, he introduced the new techniques for the culture and investigation of microbes that he had learned in Germany.

He achieved his own personal triumph in 1892 with the discovery and identification of the organism responsible for gas gangrene, *Clostridium perfringens*, later named *Clostridium welchii*.

Weldon, Walter (1832–1885) *British industrial chemist*

Weldon, born the son of a manufacturer at Loughborough in Leicestershire, England, traveled to London at the age of 22 to become a journalist. There he founded the short-lived journal *Weldon's Register of Facts and Occurrences in Literature, Science, and Art* (1860–64). Weldon had no formal training in chemistry although he acquired a working knowledge of it. He became interested in how manganese used in the production of chlorine could be recovered. One of the major problems facing heavy chemical industries during the 19th century was the wastage of expensive ingredients, which were converted to unusable waste products.

During the period 1866–69 he took out six

patents covering the *Weldon process*. In the Scheele process of chlorine production, hydrochloric acid and manganese dioxide yield chlorine and the waste product, manganese chloride. Weldon's method was to regenerate manganese dioxide by treating the manganese salt with milk of lime (calcium hydroxide in water) and blowing air through it. The liberated manganese dioxide could be pumped straight back into the chlorine stills. This considerably reduced the price of chlorine and quadrupled its production. His achievement was described by the great French chemist Jean DUMAS: "By Mr. Weldon's invention, every sheet of paper and every yard of calico has been cheapened throughout the world."

Weldon became a fellow of the Royal Society (1882) and was active in founding the Society of Chemical Industry, becoming its president during the period 1883–84.

Weller, Thomas Huckle (1915–) *American microbiologist*

Born in Ann Arbor, Michigan, Weller was educated at the University of Michigan, where his father was professor of pathology, and at Harvard, where he gained his MD in 1940. After serving in the U.S. Army Medical Corps from 1942 until 1945 Weller worked with John EN-DERS at the Children's Medical Center, Boston. In 1954 he returned to Harvard as professor of tropical public health, becoming professor emeritus in 1985.

In 1948 Weller, in collaboration with Franklin Neva, succeeded in growing the German measles (rubella) virus in tissue culture. They later went on to grow and isolate the chickenpox virus in a culture of human embryonic muscle and skin. With Enders and Frederick ROBBINS, Weller successfully applied the same method to the culture of poliomyelitis virus. By making adequate supplies of polio virus available to laboratory workers, this opened the way for the development of a successful polio vaccine.

For this work Weller shared the 1954 Nobel Prize for physiology or medicine with Enders and Robbins.

Wells, Horace (1815–1848) *American dentist*

Wells was born in Hartford, Vermont, and studied dentistry in Boston. Later, while in practice, he formed a brief partnership in Boston with William MORTON.

In 1844 he attended a demonstration of the effects of nitrous oxide (laughing gas) staged by a visiting showman. Like Crawford LONG before him, Wells noted that although the subjects fell about frequently they did not seem to feel their bumps and bangs. Wells quickly saw the significance of this and persuaded the showman to administer the gas to him while a colleague, John Riggs, removed Wells's troublesome wisdom tooth.

In January 1845, considerably excited by his discovery, Wells informed Morton and the well-known Boston chemist Charles Jackson of his success, asking them to use their influence to arrange a demonstration for the surgeons of Boston. In early 1845 Wells was invited to demonstrate his invention before the leading Boston surgeon, John Collins Warren. He presumably, however, got his dosage wrong for the student who volunteered for an extraction seemed to scream in pain as soon as his tooth was touched.

Wells gave up his practice and took to selling first canaries, later shower baths, to the citizens of Connecticut. However, when, in 1846, Morton achieved success with ether Wells made extravagant claims for his priority, which he could never hope to establish.

Few of the early workers in anesthesia gained from their labor and many were to lose much. Of these Wells lost the most. In 1848, addicted to chloroform and deranged, he threw vitriol in the face of a New York prostitute. Arrested and imprisoned in the Tombs of New York, he inhaled some chloroform and committed suicide.

Wendelin, Gottfried (1580–1667) *Belgian astronomer*

Wendelin (**vent**-e-lin), who was born at Herckla-Ville in Belgium and educated at Tournai and Louvain, became an official at Tournai cathedral. He was much respected in his time as an astronomer, being referred to as the Ptolemy of his age. Wendelin was an early Copernican who has been credited with attempting to determine the solar parallax, in 1626. He is also remembered as the teacher of Marin MERSENNE, Pierre GASSENDI, and Christiaan HUYGENS.

Wenzel, Carl Friedrich (c. 1740–1793) *German chemist*

Wenzel (**vent**-sel) was born in Dresden, Germany, the son of a bookbinder. He studied surgery and medicine at Amsterdam and served as a ship's surgeon in the Dutch naval service. He was made chemist to the Freiberg mines in 1780, becoming chief assessor in 1784. In 1786 he became metallurgical chemist at the Meissen porcelain factory in Saxony.

Wenzel published, in 1777–82, his *Lehre von der Verwandtschaft der Körper* (Principles of the Affinity of Solids), which went through three editions. In it he tackled, among other problems, that of the affinity of substances for each other. He measured the varying rates at which an acid will dissolve cylinders of different metals and tried to use these figures to calculate their relative attractive forces.

Werner, Abraham Gottlob (1750–1817)
German mineralogist and geologist

> It is Werner's great merit, to have drawn attention to this sequence [of rock formations] and to have looked at it throughout with correct eyes ... To determine how things were millions of years ago ... is not interesting, the real interest being confined to that which is now in existence – the system of different formations.
> —G. W. F. Hegel, *Encyclopädie der philosophischen Wissenschaften* (1830; Encyclopedia of Philosophical Sciences)

Werner (**vair**-ner) was born in the traditional mining town of Wehrau, which is now in Poland. Most of his ancestors had worked in some position or other in the industry and his father was inspector of the iron foundry at the town. He began work as an assistant to his father before entering the new Freiberg Mining Academy in 1769. He studied at the University of Leipzig (1771–75) before returning to teach at the Freiberg Mining Academy. There he established his neptunist views on the aqueous origin of rocks and attracted a considerable following.

Werner's neptunian theory explained the surface of the Earth and the distribution and sequence of rocks in terms of a deluge, which had covered the entire Earth, including the highest mountains. The rock formations were laid down when the flood subsided in a universal and specific sequence. The first layer consisted of primitive rocks, such as granite, gneiss, and slates, and contained no fossils. The next strata (the transitional) consisted of shales and graywacke and contained fossilized fish. Above this were the limestones, sandstones, and chalks of the secondary rocks and then the gravels and sands of the alluvial strata. Finally, after the waters had completely disappeared, local volcanic activity produced lavas and other deposits.

However, this fivefold scheme, while no doubt applicable in Werner's region of Saxony, presented great difficulties outside the area. There was much that Werner could not explain, such as where the enormous flood had gone to and the presence of large basalt tracts in Europe, which were found in areas free of volcanoes. For many years Werner's theories eclipsed those of the plutonists, led by James HUTTON, who emphasized the origin of igneous rocks from molten material. But as knowledge of the strata of Europe increased it became clear that there were too many regions in which Werner's sequence bore no relation to reality.

Yet neptunism certainly had its attractions, with Werner's disciples distributed throughout Europe. The advantages of the theory were that it was theologically acceptable, it was simple,

and it showed how the Earth could be formed in the short time available.

Werner was also a mineralogist and he constructed a new classification of minerals. There was a major split among 18th-century mineralogists as to whether minerals should be classified according to their external form (the natural method) or by their chemical composition (the chemical method). Werner finally adopted, in 1817, a mixed set of criteria by which he divided minerals into four main classes – earthy, saline, combustible, and metallic.

Werner, Alfred (1866–1919) *French chemist*

> He has thrown fresh light on old problems and opened new fields of research, particularly in inorganic chemistry.
> —Citation on the award of the Nobel Prize for chemistry to Werner in 1913

Werner (**vair**-ner or **wer**-ner) was born the son of an ironworker at Mulhouse in France. He was educated at the University of Zurich, where he gained his PhD in 1890. After a year in Paris working with Marcellin BERTHELOT he returned to Zurich, where he was appointed professor of chemistry in 1895.

In 1905 Werner produced a work, later translated into English as *New Ideas on Inorganic Chemistry* (1911), which was to revolutionize inorganic chemistry and earn him the Nobel Prize for chemistry in 1913. Although the ideas introduced by August KEKULÉ had contributed greatly to organic chemistry, attempts to apply his valence theory to inorganic molecules were much less successful. Many metals appeared to show variable valence and form complex compounds.

Werner proposed distinguishing between a primary and a secondary valence of a metal. The primary was concerned with binding ions, while the secondary valence applied not only to atoms but also to molecules, which can have an independent existence. Certain metals, such as cobalt and platinum, were capable through their secondary valences of joining to themselves a certain number of atoms or molecules. These were termed by Werner "coordination compounds" and the maximum number of atoms (or "ligands" as he called them) that can be joined to the central metal is its coordination number. This led Werner to make very detailed predictions about the existence of certain hitherto unsuspected isomers. He managed to resolve optical isomers of an inorganic compound in 1911.

Wernicke, Carl (1848–1905) *German neurologist and psychiatrist*

Wernicke (**vair**-ni-ke) born at Tarnowitz (now Tarnowskie Gory in Poland) and studied medicine at the University of Breslau, qualifying in psychiatry in 1875. He later taught neurology and psychiatry in various institutions in Berlin, Breslau, and Halle.

In 1874 he published his *Der aphasische Symptomencomplex* (The Complex of Aphasic Symptoms) in which he described a new type of aphasia. In 1861 Paul BROCA had shown that damage to a particular part of the frontal cerebral cortex produced an aphasia characterized by abnormally slow and labored speech but, he went on to show in 1865, only if the damage was located in the left-hand side of the brain.

Wernicke identified a second area, also in the left part of the brain, but this time in the temporal lobe, damage to which had no effect on the mode of articulation but did seem to effect comprehension. This allowed him to define, rather neatly, three types of aphasia. Damage to Wernicke's area would produce sensory aphasia characterized by difficulty in comprehending, while damage to Broca's area would produce motor aphasia, i.e., difficulty in articulating. As he assumed the two areas to be connected, damage to the connecting area should produce a third type of disturbance later known as conduction aphasia. It was thus the work of Wernicke that opened up the modern study of the various aphasias.

Wernicke also described, in 1881, a condition involving disorders of gait and consciousness and paralysis of the eye muscles, later to be known as *Wernicke's encephalopathy* and shown to be due to thiamine deficiency.

West, Harold Dadford (1904–1974) *American biochemist*

West was educated at the University of Illinois, Urbana, graduating in 1925 and gaining his PhD in 1937. After working initially at Morris Brown College, Atlanta, he moved to Meharry Medical College, Nashville, in 1927. West spent his entire career at Meharry serving as professor of biochemistry from 1938 until his retirement in 1973, apart from a break between 1952 and 1963, when he held the position of college president.

Although West worked in a number of fields and conducted research on the biochemistry of various bacilli, the B vitamins, and antibiotics, he is best known for his studies of the amino acids. In particular it was West who first synthesized the essential amino acid threonine.

Westinghouse, George (1846–1914) *American inventor*

Westinghouse, who was born at Central Bridge in New York, served in both the army and the navy during the American Civil War. After the war he began to design and patent several devices, the most important of which was an air brake for trains, patented in 1869. This was widely used throughout Europe and America and later formed the basis of a standardized air-brake system, also devised by Westinghouse, that could be used on different types of train. Later he applied his knowledge of hydraulics to the development of an electrical and compressed-air railroad signaling system.

In the 1880s Westinghouse played a central role in the development of energy-supply systems. Not only did he design equipment for the safe piping of natural gas, but he was also instrumental in the introduction of alternating-current (a.c.) electricity supply to America. At this time a.c. electricity, which permits voltage changes by transformers, was being developed in Europe, but the American electrical systems were direct current (d.c.). Westinghouse imported and improved European a.c. generators and transformers and set up an a.c. electrical supply in Pittsburgh.

Westinghouse founded, in 1886, the Westinghouse Electric Company, which successfully used the patents that it received or bought to develop a profitable business. Although he gave up all connections with his companies a few years before his death, Westinghouse's name has continued to be associated with developments in energy technology, particularly in the nuclear-power industry.

Wexler, Nancy (1946–) *American clinical psychologist*

Wexler was born in New York, the daughter of the well-known psychoanalyst Milton Wexler. She was educated at Harvard and the University of Michigan where she completed her doctorate. In 1968 her mother developed Huntington's disease (HD), an untreatable and incurable condition that leads inevitably to the destruction of the mind. The disease usually appears between the ages of 35 and 50 and, as it is caused by a dominant gene, Nancy and her sister had a 50% chance of inheriting the condition. Milton Wexler's response was to set up the Hereditary Disease Foundation to stimulate and organize research into Huntington's disease and other hereditary complaints.

After completing her doctorate, Nancy Wexler moved to Columbia University as professor of clinical psychology. Following the death of her mother in 1978 she began to devote more of her time to work on HD. In 1981 she heard from a Venezuelan biochemist, Americo Negrette, of an extended family on the shores of Lake Maracaibo in which HD was rife. Wexler sought to trace the gene through the family tree, which began with Maria Concepcion and a presumed encounter with a European seaman in about 1800. Of her 9,000 living descendants Wexler traced 371 with HD and found 1,200 with a 50% chance of contracting the disease and a further 2,400 with a 25% chance. Wexler realized that, given the current state of molecular biology, the material she had gathered could be used to identify the gene responsible for HD. The key step in this process had been the discovery by Ray White and his colleagues of RFLPs, DNA sequences that could be used as genetic markers. Blood samples were taken and sent immediately to James Gusella at Massachusetts General Hospital and he began what he thought would be a lengthy search for the appropriate RFLP. But Gusella was extremely lucky and the twelfth probe he tried seemed to be linked with HD. Further work established that the gene was on the short arm of chromosome 4.

Wexler and Gusella announced their results in 1983. They continued to home in on the gene and by 1992 had restricted it to a segment of 500,000 bases. More precisely, a stretch of DNA had been identified in which the nucleotide triplet C–A–G is repeated. In people without HD there seem to be 11–34 copies of the triplet, while those affected by HD have 42–86 triplets. Wexler is currently Higgins Professor of Neuropsychology at Columbia University.

Weyl, Hermann (1885–1955) *German mathematician*

> Symmetry, as wide or as narrow as you may define its meaning, is one idea by which man through the ages has tried to comprehend and create order, beauty, and perfection.
> —*Symmetry*

Born at Elmshorn in Germany, Weyl (vIl) studied at Göttingen, where he was one of David HILBERT's most oustanding students. He became a coworker of Hilbert, who influenced his particular interests and his general outlook on mathematics. Weyl taught at the Federal Institute of Technology in Zurich from 1913, and this too had a decisive influence in directing his mathematical interests through the presence there of Albert EINSTEIN. In 1930 he returned to Göttingen to take up the chair vacated by Hilbert. With the Nazis' rise to power in 1933, Weyl, with many other members of the Göttingen scientific community, went into exile in America. Weyl found a post at the Institute for Advanced Study in Princeton along with other exiles, such as Einstein and Kurt GÖDEL.

Weyl's mathematical interests, like those of Hilbert, were exceptionally wide, ranging from mathematical physics to the foundations of mathematics. He worked on two areas of pure mathematics: group theory and the theory of Hilbert space and operators, which, although developed for purely mathematical purposes, later turned out to be precisely the mathematical framework needed for the revolutionary physical ideas of quantum mechanics. Weyl also wrote a number of books on the theory of groups and he was particularly interested in symmetry and its relation to group theory. One of his most important results in group theory was a key theorem about the application of representations to LIE algebras. Weyl's work on Hilbert space had grown out of his interest in Hilbert's work on integral equations and operators. The theory of Hilbert space (infinite-dimensional space) was recognized by Erwin SCHRÖDINGER and Werner HEISENBERG in the mid-1920s as the necessary unifying systematization of their theories of quantum mechanics.

Weyl's contact with Einstein at Zurich was responsible for an interest in the mathematics of relativity, and especially Riemannian geometry, which plays a central role. Weyl initiated the whole project of trying to generalize Riemannian geometry. He himself worked chiefly on the geometry of affinely connected spaces, but this was only one of many generalizations that resulted from his work. Weyl also did similar work on generalizing and refining the basic concepts of differential geometry. All this work was to be of importance for relativity. Weyl's views on relativity were expounded in his book *Raum-Zeit-Materie* (1919; Space-Time-Matter).

Weyl, like his teacher Hilbert, was always interested in the philosophical aspects of mathematics. However, in contrast to Hilbert, his general attitude was similar to that of L. E. J. BROUWER with whom he shared constructivist leanings developed from work in analysis. Weyl expounded his philosophical ideas in another book, *Philosophy of Mathematics and Natural Sciences* (1949). Unlike Brouwer, however, Weyl was less rigorous in avoiding nonconstructive mathematics, and doubtless his interest in physics contributed to this.

Wharton, Thomas (1614–1673) *English physician*

> A man of great learning and experience, and of equal freedom to communicate it.
> —Izaak Walton, *The Compleat Angler* (1653)

Wharton was born at Winson-on-Tees in England and educated at the universities of Cambridge and Oxford, obtaining his MD from Oxford in 1647. He later spent much of his life in private practice in London, being appointed in 1659 physician at St. Thomas's Hospital. During the Great Plague (1665) he remained in London, unlike most physicians, to treat his patients.

In 1656 he published *Adenographia* (Textbook on Glands), the first comprehensive survey of the glands of the body. In it he named the thyroid, from the Greek "thyreos," a shield. He also described, for the first time, the duct of the submaxillary salivary gland, since known as *Wharton's duct*.

Wheatstone, Sir Charles (1802–1875) *British physicist*

After a private education in his native city of

Gloucester, Wheatstone began business in London as a musical-instrument maker (1823). His early scientific researches were in acoustics and optics and his contributions were numerous. Thus he devised a "kaleidophone" to illustrate harmonic motions of different periods; he suggested a stereoscope (1838) that, using two pictures in dissimilar perspective, could give the appearance of solidity; he showed that every CHLADNI figure was the resultant of two or more sets of isochronous parallel vibrations; and he demonstrated how minute quantities of metals could be detected from the spectral lines produced by electric sparks.

Perhaps his most important work, however, was to produce, with William COOKE, the first practical electric telegraph system. In 1837, in conjunction with the new London and Birmingham Railway Company, Cooke and Wheatstone installed a demonstration line about one mile long. Improvements rapidly followed and, with the needs of the railroads providing the impetus and finance, by 1852 more than 4,000 miles of telegraph lines were in operation throughout Britain. Wheatstone constructed the first printing telegraph (1841) and a single-needle telegraph (1845). He made contributions to the development of submarine telegraphy and to dynamos. The *Wheatstone bridge*, a device for comparing electrical resistances, was not invented by Wheatstone but brought to notice by him.

Wheatstone was appointed professor of experimental philosophy at King's College, London, in 1834 and was knighted in 1868. At his death he held about 40 awards and distinctions. He was prolific in his inventions and had an extraordinary ability to turn his theoretical knowledge to practical account.

Wheeler, John Archibald (1911–)
American theoretical physicist

> Time is what prevents everything from happening at once.
> —*American Journal of Physics* (1978)

> We shall call this completely collapsed gravitational object a "black hole."
> —Address given at a conference on relativity, 1967

Born in Jacksonville, Florida, Wheeler was educated at Johns Hopkins University, where he obtained his PhD in physics in 1933. After spending the period 1933–35 in Copenhagen working with Niels BOHR, he returned to America to take up a teaching position at the University of North Carolina. In 1938 he went to Princeton, where he served as professor of physics from 1947 until his move to the University of Texas in 1976 to become professor of physics. He retired in 1986.

Wheeler has been active in theoretical physics. One of the problems tackled by him has been the search for a unified field theory.

His earlier papers on the subject were collected in 1962 in his *Geometrodynamics*. It was here that he introduced the concept of a *geon* – a bunch of gravitational waves (or a mixture of gravitational and electromagnetic waves) held together by their own attraction – with which he aimed to achieve the unification of the two fields. He also collaborated with Richard FEYNMAN in two papers in 1945 and 1949 on the important concept of action at a distance. They formulated a problem that arises when it is accepted that such action cannot take place instantaneously. If X and Y are at rest and one light-minute apart, then any electromagnetic signal emitted by X will reach Y one minute later. This is described by saying X acts on Y by a retarded effect. But by NEWTON's third law, to each action there corresponds an opposite and equal reaction. This must mean that from Y to X there should also be an advanced effect acting backward in time. Feynman and Wheeler demonstrated how the advanced wave could be eliminated from the model to account for the fact that the Universe displays only retarded effects.

Wheeler also made important contributions to nuclear physics. With Niels Bohr he put forward an explanation of the mechanism of nuclear fission in 1939. He joined the Los Alamos group exploring the possibility of producing an explosive device using heavy hydrogen in 1949–50. Wheeler has provided a popular account of his work in his *Journey into Gravity and Spacetime* (1990); he has also published two autobiographical volumes: *At Home in the Universe* (1993) and *Geons, Black Holes and Quantum Foam* (1998).

Wheeler was one of the main people responsible for the revival of interest in general relativity theory in the 1950s and 1960s. He also coined the expression "black hole." Wheeler has also coauthored several influential works on general relativity theory, including the large book *Gravitation* (1973) with Charles MISNER and Kip THORNE, which has remained a standard work on this topic.

Wheeler has been a visionary thinker about quantum gravity, putting forward the idea of *space-time foam*, i.e., the seething foamlike structure which may replace Riemannian geometry at the extremely short-length scales characteristic of quantum gravity. He has also emphasized the possible importance of information theory in contemporary physics.

Whewell, William (1794–1866) *British scientist, historian, and philosopher*

> We need very much a name to describe a cultivator of science in general. I should incline to call him a scientist.
> —The first recorded use of the word. *The Philosophy of the Inductive Sciences*, Vol. 1 (1840)

Whewell was the son of a master carpenter from Lancaster in northwest England; his talents were recognized by a local schoolmaster. In 1812 he won a £50 exhibition to Trinity College, Cambridge, where he remained for the rest of his life – as a fellow from 1817 and as master from 1841 until his death in 1866 from a horse-riding accident. He also served as professor of mineralogy (1828–32) and as professor of moral philosophy (from 1838).

Whewell was one of the great Victorian poly-maths, the subject of Sydney Smith's aphorism that "Science was his forte, omniscience his foible." He published works on dynamics, mineralogy, chemistry, architecture, moral philosophy, history, political economy, geometry, and the theory of the tides, as well as translating German verse and Greek philosophy into English.

His most original scientific work can be found in the 14 memoirs on tides published between 1833 and 1850. Here he collected a vast amount of data in an attempt to establish cotidal lines (points of simultaneous high tide). He also investigated the diurnal inequality, that is, the differences found between the heights and times of tides at the same place on the same day. Whewell found the whole system so complex that he despaired of ever constructing a general theory of the tides.

Two other works of Whewell are still read today, namely, his *History of the Inductive Sciences* (3 vols., 1837) and his *Philosophy of the Inductive Sciences* (3 vols., 1840). In the former he presented the growth of science as the development of theories of ever-increasing generality. He spoke of development in terms of a prelude in which old theories began to face difficulties, an inductive epoch where new theories were established, and a final sequel in which the new theory was refined and applied to nature.

As a philosopher of science Whewell dismissed the view that induction was merely reasoning from the particular to the general. It also involves a process of "colligation" in which ideas are best understood by being placed in a common framework. In addition he spoke of the highest form of induction, "consilience," in which inductions from different fields are united, as when NEWTON demonstrated that the same law could describe both the vertical fall of an apple and the circular orbit of the moon. Whewell has also left a permanent mark on the language of science. In 1834 he proposed to Michael FARADAY the terms "anode" and "cathode" to distinguish between negative and positive electrodes. Other words proposed to Faraday include "anion," "cation," "ion," and "dielectric." To Charles LYELL he offered the terms "Eocene," "Miocene," and "Pliocene" to describe Tertiary epochs. As a final contribution Whewell in 1840 introduced into English the two terms "physicist" and "scientist."

Whipple, Fred Lawrence (1906–2004)
American astronomer

Born in Red Oak, Iowa, Whipple graduated from the University of California in Los Angeles in 1927 and obtained his PhD from Berkeley in 1931. He then moved to Harvard where he became professor of astronomy in 1945, Philips Professor of Astronomy in 1950, and director of the Smithsonian Astrophysical Observatory from 1955 until his retirement in 1973.

He described his research as centering on "physical processes in the evolution of the solar system" and produced in this field a much admired work, *Earth, Moon and Planets* (1941 and many subsequent editions).

Whipple is also well known for his work on comets. In 1950 he proposed an icy-nucleus model in which he described the nucleus of a comet as a "dirty snowball," made from a mixture of water ice and dust, plus carbon dioxide, carbon monoxide, methane, and ammonia ices, and only becoming active when passing close to the Sun. The main advantage of this model is that it can account for such distinctive features of comets as their orbital motion. It had been long known that some comets, such as ENCKE's, persist in returning earlier than Newtonian theory would predict while others, such as HALLEY's, arrive over four days later than expected. Whipple proposed that solar radiation would cause the ices on the outside of the cometary nucleus to evaporate, leaving a thin insulating layer of dust particles, and that this would set up a delayed jet reaction. The radiation has the effect of pushing further out those comets that are rotating in the same direction as their orbit. This will increase their orbit and delay their return. The radiation will produce a drag force on those comets rotating counter to their orbit, causing them to drift in toward the Sun, reducing their period and thus hastening their return.

As there should be no preferred direction of rotation, Whipple predicted that about half the comets should appear to be retarded and half accelerated in their orbit, an effect since confirmed.

Whipple, George Hoyt (1878–1976)
American physician and physiologist

Whipple, born the son of a physician in Ashland, New Hampshire, was educated at Yale and Johns Hopkins University, where he obtained his MD in 1905. After working at the University of California he moved in 1921 to the University of Rochester, where he served as professor of pathology until his retirement in 1955.

Whipple began his research career by working on bile pigments but went on to study the formation and breakdown of the blood pigment, hemoglobin, of which bile pigments are the breakdown products. To do this he bled dogs until he had reduced their hemoglobin level to a third, then measured the rate of hemoglobin regeneration. He soon noted that this rate varied with the diet of the dogs and by 1923 reported that liver in the diet produced a significant increase in hemoglobin production.

It was this work that led George MINOT and William MURPHY to develop a successful treatment for pernicious anemia and earned all three men the Nobel Prize for physiology or medicine in 1934.

Whiston, William (1667–1752) *British mathematician and geologist*

Whiston, the son of a parish priest from Norton in England, was educated at Cambridge University, where he came to the attention of Isaac NEWTON. He was selected, on Newton's recommendation, to succeed him as Lucasian Professor of Mathematics (1703) and to edit his *Arithmetica universalis* (1707; Universal Arithmetic). In 1710 he was dismissed from the university for his unorthodox religious belief in the Arian heresy, after which time he supported himself by giving public lectures on popular science.

Whiston's chief scientific work, *A New Theory of the Earth* (1696), was praised by Newton and by John LOCKE. In this Whiston followed the tradition recently established by Thomas BURNET in attempting to explain biblical events, such as the Creation, scientifically. The Flood, he believed, was caused by a comet passing close to the Earth on 28 November 2349 BC. This put stress on the Earth's crust, causing it to crack and allow the water to escape and flood the Earth.

White, Gilbert (1720–1793) *British naturalist*

> A more delightful, or more original work than Mr. White's *History of Selborne* has seldom been published ... The book is not a compilation from former publications, but the result of many years' attentive observations to nature itself, which are told not only with the precision of a philosopher, but with that happy selection of circumstances, which mark the poet.
> —*The Topographer* (1789)

White was educated at Oxford University but returned to his native village of Selborne to become a curate. He soon began making detailed observations on the flora and fauna in his garden and the surrounding countryside, concentrating particularly on the behavior and habitats of birds. His letters to friends on these matters provided the material for his book *The Natural History and Antiquities of Selborne* (1789), which collected together more than 20 years' work.

The book won immediate recognition from contemporary naturalists who praised White's meticulous observations and scientific procedure. Many important discoveries were recorded, such as the addition of the noctule bat and the harvest mouse to the list of British mammals. He also recognized the three different kinds of British leaf warblers by the differences in their song and recorded the migration of swallows.

The book has become a natural-history classic, going through several editions and still selling well today.

White, Raymond L. (1943–   ) *American molecular biologist*

Born in Eugene, Oregon, the son of a dentist, White originally intended to follow a medical career. However, while a student at the University of Oregon he became more interested in molecular biology and genetics. Consequently he moved to the Massachusetts Institute of Technology, where he completed his PhD in molecular biology in 1971. After a postdoctoral period at Stanford, he joined the staff of Massachusetts University Medical School, Worcester, in 1975. In 1980 he was appointed to the faculty at the Hughes Medical Institute, Utah. He is now director of the Ernest Gallo Clinic and Research Center at the University of California, San Francisco.

During the 1970s scientists had begun to ask whether ways could be found to determine which of the 100,000 human genes scattered along 23 chromosomes are responsible for a large number of inherited diseases. A possible answer was proposed to White in 1978 by David Botstein. Biologists knew that while most of the sequences of nucleotides found in human DNA were identical, there were also stretches which varied from person to person. If "polymorphisms" of this kind were widely scattered throughout the genome, then it was possible that some of them could be reasonably close to a defective gene and that the presence or absence of the polymorphism could be used to indicate the presence or absence of a defective gene. As the polymorphisms varied in length, and as they were normally isolated by the use of a restriction enzyme, they soon became known as restriction fragment length polymorphisms, or RFLPs, pronounced "**rif**-lips."

The first test of the theory would be to see if RFLPs were actually collectable. White began the search in collaboration with Arlene Wyman in 1979. They began by eliminating sequences of repetitious DNA widely scattered throughout the genome. After much effort they were left with five unique sequences of about 15–20,000 bases. Complementary copies of unique sequences were made (cDNA) and mixed with a donor's DNA. If the donor also carried the unique RFLP, the cDNA, radioactively labeled, would coordinate with the RFLP and be easily detectable. White and Wyman worked with blood cells from 56 subjects and identified eight distinct RFLPs.

Having shown that RFLPs could be found, it remained for White to show that they could be used as genetic markers. If the RFLP was close enough to a gene it would be inherited along with the gene. This was shown in 1981 when Davies and MURRAY at St. Mary's Hospital, London, discovered a RFLP linked with Duchenne muscular dystrophy on the X chromosome. Since then several other links have been identified.

The situation was sufficiently promising, White argued, to make it worthwhile constructing a complete RFLP map of the human genome. White and his colleagues in Utah spent much of the 1980s on this project. By 1988 White could announce that "preliminary maps for most of the human chromosome" had been completed. He then began work upon a higher resolution map, which would indicate the location of RFLPs and genes with greater precision.

Whitehead, Alfred North (1861–1947)
British mathematician and philosopher

> A science which hesitates to forget its founders, is lost.
> —Attributed remark. Quoted by Alan L. Mackay in *A Dictionary of Scientific Quotations* (1991)

Whitehead, who was born at Ramsgate on the south coast of England, obtained his PhD from Cambridge University in 1884. For the next few years he taught there and met Bertrand RUSSELL, who was one of his students and with whom he was later to collaborate. Whitehead was one of a growing section of philosophers of science who criticized the deterministic and materialistic views prevalent in 19th-century science. The main theme of this critique was that scientific theories were patterns derived from our way of measuring and perceiving the world and not innate properties of the underlying reality. While in Cambridge his mathematical work reflected this viewpoint, developed in his book *Treatise of Universal Algebra* (1898), which treated algebraic structures as objects worthy of study in their own right, independent of their relationship to real quantities. In 1910 he published, with Russell, the first volume of the vast *Principia Mathematica* (Mathematical Principles) which was an attempt, inspired by the work of Gottlob FREGE and Giuseppe PEANO, to clarify the conceptual foundations of mathematics using the formal methods of symbolic logic.

Whitehead then moved to London, where he taught at University College and later became professor at Imperial College. Here he developed his action-at-a-distance theory of relativity, which challenged EINSTEIN's field-theoretic viewpoint but never gained wide acceptance. In 1924 he emigrated to America and worked at Harvard, developing his antimechanistic philosophy of science and a system of metaphysics, until he retired in 1937.

Whittaker, Sir Edmund Taylor (1873–1956) *British mathematician and physicist*

Born at Birkdale in northwest England, Whittaker studied at Cambridge and taught there from 1896. In 1906 he became Astronomer Royal for Ireland. From 1912 until 1946 he was professor of mathematics at Edinburgh.

Whittaker was primarily a mathematical physicist and was appropriately much interested in the theory of differential equations. In this field he obtained the general solution of LAPLACE's equation in three dimensions. This led Whittaker to finding the general integral representation of any harmonic function, which greatly simplified potential theory as well as opening up new areas of research. He also did notable work on the theory of automorphic functions. One of Whittaker's other great interests was in relativity theory. He wrote a number of influential textbooks on analysis, on quantum mechanics, and on dynamics, including his *History of the Theories of Aether and Electricity* (1910; revised 1941 and 1951), a comprehensive history of theories of electromagnetism and atomism. Whittaker was known as a superb and highly effective teacher. He was also a deeply religious man with an interest in the philosophical and religious importance of contemporary physics and wrote a number of books on these questions.

Whittington, Harry Blackmore (1916–　) *British geologist*

Whittington was born at Handsworth in Yorkshire, England. After gaining his PhD from the University of Birmingham, he spent the years of World War II teaching in the Far East, first at the University of Rangoon, Burma, and for the rest of the war at Ginling College, West China. He returned to Birmingham in 1945 but moved to Harvard as curator of vertebrate paleontology at the Museum of Comparative Zoology. In 1966 Whittington left America for the post of Woodward Professor of Geology at Cambridge, a position he held until his retirement in 1983.

Since the early 1960s Whittington has devoted the bulk of his time to the study of the Burgess Shale fossils, discovered and described by C. D. WALCOTT earlier in the century. He made two expeditions to the site in 1966 and 1967 and recruited two assistants, Derek Briggs and Simon Conway Morris, to help him reexamine the entire collection.

Whittington's first report, published in 1971,

was devoted to *Marella splendens*, identified by Walcott as a trilobite, a primitive and long-extinct arthropod. Whittington, after four years' work on several thousand specimens, found too many uncharacteristic trilobite features to be happy with Walcott's classification. He compromised by calling it *Trilobitoidea* (trilobite-like). His suspicion that many of Walcott's arthropods had been wrongly classified was increased when Whittington next looked at the subject of his 1975 monograph, *Opabinia*. As he could find no jointed appendages it was clear to Whittington that *Opabinia* was not an arthropod. What its affinities were, however, remained uncertain.

By the time he came to deal in 1985 with *Anamalocaris* he could state confidently that it was no arthropod but "the representative of a hitherto unknown phylum." Whittington and his colleagues went on to identify ten invertebrate genera "that have so far defied all attempts to link them with known phyla."

Whittington's labors thus presented a dramatic new picture. The Cambrian is now seen as a period in which many new complex species suddenly appear. Further, relatively few of these seemingly advanced groups lasted beyond the Cambrian. The full significance of Whittington's work has yet to be worked out.

Whittle, Sir Frank (1907–1996) *British aeronautical engineer*

> Jet-propelled fighter aircraft have successfully passed experimental tests and will soon be in production ... The greatest credit should be given to Group Captain Whittle for this fine performance, for it was his genius and energy that made this possible.
> —*The Engineer*, 14 January 1944

Whittle, the son of a mechanic from Coventry in the English Midlands, joined the Royal Air Force as an apprentice in 1923. He was trained at the RAF College, Cranwell, and Cambridge University, where he studied mechanical sciences (1934–37).

While still a student at Cranwell, Whittle had expressed his prediction that there would soon emerge a demand for high-speed high-altitude aircraft. He recognized the inadequacies of the conventional airscrew to meet these needs and took out his first patent for the turbojet engine in 1930. He gained little government backing but with the assistance of friends he formed, in 1936, the company Power Jets. By the following year, his first engine, the W1, was ready for testing. With the advent of World War II government funds were rapidly awarded to develop this and the jet engine was fitted to the specially built Gloster E28/39 aircraft. It made its first flight on 15 May 1941 and by 1944 was in service with the RAF.

For his work Whittle was made a fellow of the Royal Society in 1947, knighted in 1948, and awarded a tax-free gift of £100,000 by the British government. He left the RAF in 1948 and served as a consultant with the British Overseas Airways Corporation (1948–52), the Shell Group (1952–57), and Bristol Siddeley Engines (1961–70). In 1977 Whittle accepted the post of research professor at the U.S. Naval Academy, Annapolis.

Whitworth, Sir Joseph (1803–1887) *British engineer*

The son of a schoolmaster and Congregational minister from Stockport in northwest England, Whitworth left school at fourteen to work with his uncle, a cotton spinner, in Derbyshire. He soon gained an intimate knowledge of the available machinery and was immediately struck by its crudity. Four years later he moved to London to gain further experience with machine tools working for Henry MAUDSLAY (1771–1831), one of the great engineers of the day. During this period Whitworth also worked on the mechanical calculating machine being constructed by Charles BABBAGE.

In 1833 he returned to Manchester to set up his own machine-tool factory and at once sought to introduce new standards of accuracy into tool manufacture. He introduced sets of standard gauges and measures and these soon gained wide use. By 1859 he had built a measuring machine capable of detecting a difference of one millionth of an inch.

Earlier, in 1841, he had argued for a uniform system of screw threads. Before Whitworth, nuts and bolts were so individually matched that once separated they might as well be thrown away. Whitworth proposed that bolts of a given size would have threads of identical pitch and depth and a common thread angle of 55°. The Whitworth system rapidly became the standard for British engineering. Unfortunately the United States adopted in 1864 a different system based on a thread angle of 60° as proposed by William Sellers. Despite the difficulties imposed on the Allied forces in two world wars the two standards persisted until 1948 when the Unified Thread System was adopted by the United States, Canada, and Britain. It incorporated features of both the Whitworth and Sellers systems.

In the wake of the Crimean War Whitworth was invited by the War Office to redesign the British army's ordnance. After much research and testing he produced a rifle with a smaller bore, a hexagonal barrel, an elongated projectile, and a more rapid twist. The War Office rejected Whitworth's work, although many features of Whitworth's designs subsequently found their way into the army's ordnance.

Whytt, Robert (1714–1766) *British physician and anatomist*

Whytt graduated in arts at St. Andrews University and then studied medicine at the university in his native city of Edinburgh. He also spent some time pursuing medical studies in London, Paris, Rheims – where he obtained his MD in 1736 – and Leiden. He later returned to the University of Edinburgh where he served as professor of medicine from 1747 until his death.

In 1743 Whytt published his *Virtues of Lime Water in the Cure of the Stone*, a work that basically proposed the treatment of stones (calculi) with alkaline solutions and was probably the starting point for Joseph BLACK's discovery of carbon dioxide in 1756.

His most important work, however, was his *On the Vital and Other Involuntary Motions of Animals* (1751) in which, though not the first to talk of reflex actions (both René DESCARTES and Stephen HALES had preceded him), he was the first to accord them a proper prominence. Whytt argued that such "vital motions" as he called them were not dependent on the brain, for a decapitated frog continued for some time to move in a coordinated way when receiving the appropriate stimulation. The cause of such motions must then be in the spinal cord, for when particular parts were removed the reflexes could no longer be elicited. Individual reflexes described by Whytt included the pupillary response to light, sometimes known as *Whytt's reflex*.

The idea that actions could be mediated by parts of the body other than the brain or rational soul was a difficult one for the vitalistic and theologically based physiology of the 18th century. Incapable of seeing the "vital motion" in a mechanical way, Whytt actually went so far as to introduce the idea of an unconscious sentient principle in the spinal cord as the vital cause.

Wieland, Heinrich Otto (1877–1957) *German chemist*

Wieland (**vee**-lahnt) was born in Pforzheim, Germany, the son of a chemist in a gold and silver refinery. He was educated at the University of Munich where he obtained his PhD in 1901. After teaching at the Munich Technical Institute and the University of Freiburg, Wieland succeeded Richard WILLSTÄTTER in 1925 as professor of chemistry at the University of Munich, a post he retained until his retirement in 1950.

In 1912 Wieland began work on the bile acids. These secretions of the liver had been known for the best part of a century to consist of a large number of substances. He began by investigating three of them: cholic acid, deoxycholic acid, and lithocholic acid, finding that they were all steroids, very similar to each other, and all convertible into cholanic acid.

As Adolf WINDAUS had derived cholanic acid from cholesterol, an important biological sterol,

this led Wieland to propose a structure for cholesterol. For his contributions to steroid chemistry Wieland was awarded the 1927 Nobel Prize for chemistry.

After 1921 Wieland worked on a number of curious alkaloids including toxiferin, the active ingredient in curare, bufotalin, the venom from toads, and phalloidine and amatine, the poisonous ingredients in the deadly amanita mushroom.

Wieman, Carl E. (1951–) *American physicist*

Wieman was born in Corvallis, Oregon. He was educated at MIT, gaining a BS in 1973, and at Stanford University, gaining a PhD in 1977. From 1977 to 1979 he was an assistant research scientist at the department of physics, University of Michigan. From 1979 to 1984 he was an assistant professor of physics at the University of Michigan. He joined the University of Colorado in 1984 as an associate professor of physics, becoming a professor of physics in 1987, a position he still holds. Also since 1985 he has been a fellow at the Joint Institute for Laboratory Astrophysics (JILA), Boulder.

In 1995 together with Eric CORNELL and independently of Wolfgang KETTERLE, he obtained a type of matter known as a Bose–Einstein condensate. Cornell, Ketterle, and Wieman shared the 2001 Nobel Prize for physics for their discovery.

Wien, Wilhelm Carl Werner Otto Fritz Franz (1864–1928) *German physicist*

The son of a farmer from Gaffken in Eastern Europe, Wien (veen) studied mathematics and physics for a brief period in 1882 at the University of Göttingen. Having recommenced his studies in 1884 at the University of Berlin, he received a doctorate in 1886 for a thesis on the diffraction of light. At various times he considered becoming a farmer but after his parents were forced to sell their land he decided on an academic career in physics. In 1890 he joined the new Imperial (now Federal) Institute for Science and Technology in Charlottenburg, Berlin, as assistant to Hermann von HELMHOLTZ, under whom he had studied. From 1896 to 1899 he worked at the technical college in Aachen and in 1900 was appointed professor of physics at the University of Würzburg. In 1920 he became professor at the University of Munich.

Wien was highly competent in both theoretical and experimental physics. His major research was into thermal or black-body radiation. In 1893 he showed that the wavelength at which the maximum energy is radiated from a source is inversely proportional to the absolute temperature of the source. Thus in heating an object it first glows red hot, emitting most of its energy at the wavelengths of red light; as the temperature is increased, the wavelength at which maximum energy is emitted becomes shorter, and the body becomes white hot. This behavior is known as *Wien's displacement law*. In 1896 Wien derived a formula, now known as *Wien's formula*, for the distribution of energy in black-body radiation for a whole range of wavelengths. Its importance for future research lay in the fact that although successful at short wavelengths it disagreed with experiments at longer wavelengths. The discrepancy, which is sometimes known as the "ultraviolet catastrophe," highlighted the inadequacies of classical mechanics and inspired Max PLANCK to develop the quantum theory. Wien was awarded the Nobel Prize for physics in 1911 for his discoveries regarding the laws governing the radiation of heat.

Wien also studied the conduction of electricity in gases and, while teaching in Aachen, confirmed that cathode rays consisted of high-velocity particles (1897) and were negatively charged (1898). In addition he showed that canal rays were positively charged particles. He later conducted research into x-rays.

Wiener, Norbert (1894–1964) *American mathematician*

Born in Columbia, Missouri, Wiener was a child prodigy in mathematics who sustained his early promise to become a mathematician of great originality and creativity. He is probably one of the most outstanding mathematicians to have been born in the United States. Such was Wiener's precocity that he took his degree in mathematics, from Tufts University, at the age of 14 in 1909.

Throughout his life Wiener had many extramathematical interests, especially in biology and philosophy. At Harvard his studies in philosophy led him to an interest in mathematical logic and this was the subject of his doctoral thesis, which he completed at the age of 18. Wiener went from Harvard to Europe to pursue his interest in mathematical logic with Bertrand RUSSELL in Cambridge and with David HILBERT in Göttingen. After he returned from Europe, Wiener's mathematical interests broadened but, surprisingly, he was unable to get a suitable post as a professional mathematician and for a time tried such unlikely occupations as journalism and even writing entries for an encyclopedia. In 1919 Wiener finally obtained a post in the mathematics department of the Massachusetts Institute of Technology, where he remained for the rest of his career.

After his arrival at MIT Wiener began his extremely important work on the theory of stochastic (random) processes and Brownian motion. Among his other very wide mathematical interests at this time was the generaliza-

tion of FOURIER's work on resolving functions into series of periodic functions (this is known as harmonic analysis). He also worked on the theory of Fourier transforms. During World War II Wiener devoted his mathematical talents to working for the military – in particular to the problem of giving a mathematical solution to the problem of aiming a gun at a moving target. In the course of this work Wiener discovered the theory of the prediction of stationary time series and brought essentially statistical methods to bear on the mathematical analysis of control and communication engineering.

From here it was a short step to his important work in the mathematical analysis of mechanical and biological systems, their information flow, and the analogies between them – the subject he named "cybernetics." It allowed full rein to his wide interests in the sciences and philosophy and Wiener spent much time popularizing the subject and explaining its possible social and philosophical applications. Wiener also worked on a wide range of other mathematical topics, particularly important being his work on quantum mechanics.

Wieschaus, Eric (1947–) *American biologist*

While working in the late 1970s in the European Molecular Biology Laboratory, Heidelberg, Wieschaus collaborated with Christiane NÜSSLEIN-VOLHARD on a study of the genetic factors producing segmentation in the fruit fly *Drosophila melanogaster*. Flies were exposed to mutagenic chemicals and thousands of their larval descendants examined to see if particular types of mutants emerged. Their work, published in 1980, established 15 mutant loci that radically altered the segmental pattern of the *Drosophila* larvae. It allowed the main development sequences to be marked out – to show, in effect, how the embryo becomes increasingly segmented. At first the gap genes divide the embryo into its main regions. Pair-rule genes then subdivide these regions into segments and, finally, polarity genes mark out repeating patterns in each segment. This has proved to be especially significant as it is suspected that the same development pattern may be found in other organisms. It has even been suggested that Waardenburg's syndrome, a rare disease in humans that leads to deafness and albinism, is caused by mutations in the human version of the pair-rule gene.

Wieschaus shared the 1995 Nobel Prize for physiology or medicine with his collaborators Nüsslein-Volhard and Edward LEWIS.

Wiesel, Torsten Nils (1924–) *Swedish neurophysiologist*

Wiesel (**vee**-sel), who was born at Uppsala in Sweden, obtained his MD from the Karolinska

Institute, Stockholm. He moved to America shortly afterward, working first at Johns Hopkins before moving to Harvard (1959), where he was appointed professor of neurophysiology in 1974. In 1983 he moved to Rockefeller University, New York, where he served as head of the neurobiology laboratory.

Since his arrival in America Wiesel has been engaged upon a most productive investigation with David HUBEL, into the mammalian visual system. Their 20-year collaboration led to the formulation of the influential hypercolumn theory. Wiesel and Hubel received the 1981 Nobel Prize for physiology or medicine for their work, sharing the prize with Roger SPERRY. Wiesel has gone on to investigate the chemical transmitters involved in the nerve cells of the visual system.

Wigglesworth, Sir Vincent Brian (1899–1994) *British entomologist*

Born at Kirkham in England, Wigglesworth was educated at Cambridge University and at St. Thomas's Hospital, London. He subsequently held posts as lecturer in medical entomology at the London School of Hygiene and Tropical Medicine and as reader in entomology at the universities of London and Cambridge. He was director of the Agricultural Research Council Unit of Insect Physiology at Cambridge (1943–1967) and from 1952 was Quick Professor of Biology. Wigglesworth's main line of research was in insect physiology, much of his work being done using the bloodsucking bug *Rhodnius prolixus*. He carried out research on hormonal stimulation in insect ecdysis (molting of the cuticle), glandular growth and reproductive secretions, external stimuli perception (e.g., heat receptors on antennae, and body hairs), and insects' perception of time, due to metabolic rate and daily rhythm. His most important publications are *The Physiology of Insect Metamorphosis* (1954), *The Control of Growth and Form* (1959), and *The Principles of Insect Physiology* (1939; 6th edition, 1965).

Wigner, Eugene Paul (1902–1995) *Hungarian–American physicist*

> The simplicities of natural laws arise through the complexities of the languages we use for their expression.
> —*Communications on Pure and Applied Mathematics* (1959)

Born the son of a businessman in Budapest, Hungary, Wigner (**wig**-ner) was educated at the Berlin Institute of Technology, where he obtained a doctorate in engineering in 1925. After a period at Göttingen he moved to America in 1930 and took a part-time post at Princeton. He became a naturalized American citizen in 1937 and in 1938 was appointed to the chair of theoretical physics. Wigner remained at Prince-

ton until his retirement in 1971 apart from leave of absence when he served at the Metallurgical Laboratory, Chicago (1942–45), and at Oak Ridge as director of the Clinton Laboratories (1946–47).

Wigner made many fundamental contributions to quantum and nuclear physics. He did some early work on chemical reactions and on the spectra of compounds. In 1927 he introduced the idea of parity as a conserved property of nuclear reactions. The basic insight was mathematical and arose from certain formal features Wigner had identified in transformations of the wave function of Erwin SCHRÖDINGER. The function $\psi(x,y,z)$ describes particles in space, and parity refers to the effect of changes in the sign of the variables on the function: if it remains unchanged the function has even parity while if its sign changes it has odd parity. It was proposed by Wigner that a reaction in which parity is not conserved is forbidden.

In physical terms this meant that a nuclear process should be indistinguishable from its mirror image; for example, an electron emitted by a nucleus should be indifferent as to whether it is ejected to the left or the right. Such a consequence seemed natural and remained unquestioned until 1956 when Tsung Dao LEE and Chen Ning YANG shocked the world of physics by showing that parity was not conserved in the weak interaction.

In the 1930s Wigner made major contributions to nuclear physics. Working particularly on neutrons he established early on that the nuclear force binding the neutrons and protons together must be short-range and independent of any electric charge. He also with Gregory BREIT in 1936 worked out the *Breit–Wigner formula*, which did much to explain neutron absorption by a compound nucleus. Wigner was involved in much of the early work on nuclear reactors leading to the first controlled nuclear chain reaction.

Wigner shared the 1963 Nobel Prize for physics with Maria GOEPPERT MAYER and J. Hans JENSEN for "systematically improving and extending the methods of quantum mechanics and applying them widely."

Wilcke, Johan Carl (1732–1796) *German–Swedish physicist*

Wilcke (**vil**-ke) was born at Wisman in Germany but moved to Stockholm with his parents in 1739. His father was a priest. After studying at Uppsala he spent the period 1751–57 traveling on the Continent. On his return to Sweden he was appointed lecturer in physics at the Military Academy in Stockholm in 1759. He later became professor and in 1784 was made the secretary of the Stockholm Academy of Sciences.

Although he was a prolific and inventive

physicist who worked in many fields he is best known for his work on latent and specific heat – ideas that he developed independently of Joseph BLACK. Wilcke started with a formula given by G. W. Richmann for calculating the final temperature of a mixture obtained by mixing liquids of different temperatures. In 1772 he found that the formula worked reasonably well with most mixtures but that if he took snow and water the mixture would not have the temperature predicted.

Wilcke calculated the amount of heat that was needed to melt the snow before any heat could be used to raise the temperature (i.e., the latent heat) and produced a new formula for such conditions. He followed this by producing tables of specific heats in 1781, which were calculated by mixing various hot substances with snow.

He had earlier, in 1757, published an important work on electricity, *De Electricitatibus Contrariis* (On Opposing Types of Electricity), in which he demonstrated that it was too simplistic to suppose that there were two absolutely distinct types of electricity – the vitreous and resinous types of Charles Dufay – which repel their own kind and attract the other. It was certainly true that two pieces of material rubbed on glass would repel each other and attract a piece rubbed on silk or amber, but Wilcke showed that these classifications were not absolute, for friction could produce both kinds of charge.

He also demonstrated other electrical effects, collaborating with Franz Aepinus. Together they established the phenomenon of pyroelectricity, finding that certain crystals would take opposite electric charges on opposite faces on being heated.

Wilczek, Frank (1951–) *American theoretical physicist*

Wilczek was born in Queens, New York, and educated at Princeton University, where he obtained his PhD in physics under the supervision of David GROSS. He has subsequently held appointments at the Department of Physics, Princeton University, the Institute for Theoretical Physics at the University of California, Santa Barbara, the Institute for Advanced Study, Princeton, and the Department of Physics, Massachusetts Institute of Technology (MIT).

In 1973 Wilczek and Gross, and independently David POLITZER, discovered asymptotic freedom. Gross, Wilczek, and Politzer shared the 2004 Nobel Prize for physics for this discovery. Wilczek has also made many other contributions to quantum field theory.

Wild, John Paul (1923–) *British–Australian radio astronomer*

Wild, who was born at Sheffield in England, studied at Cambridge University. After service with the Royal Navy as a radar officer from 1943 to 1947, he joined the Commonwealth Scientific and Industrial Research Organization (CSIRO) in New South Wales, Australia, to work in the new field of radio astronomy. He later served as director of the CSIRO Solar Observatory at Culgoora from 1966, as chief of the radiophysics division from 1971 until 1977, and as chairman of the CSIRO from 1979 until 1985.

Wild first worked on solar bursts – intense outbursts of radio waves that frequently accompany solar flares. Normal radio telescopes measure radio noise over a narrow band of frequencies, making them of little use in the study of solar bursts with their rapid frequency drifts of in some cases 20 megahertz. Wild overcame this difficulty, in collaboration with L. McCready, with the design (1950) of a radio spectrometer capable of sweeping rapidly across a wide frequency band. With this "panoramic receiver" he investigated and classified the complicated pattern of solar outbursts.

He went on to construct at Culgoora an enormous radio heliograph, consisting of 96 dish antennas, equally spaced around the circumference of a 1.8-mile (3-km) diameter circle. When completed in 1967 it was capable of providing a practically instantaneous moving display of radio activity in the Sun's atmosphere.

Wildt, Rupert (1905–1976) *German–American astronomer*

Wildt (vilt) was born in Munich, Germany, and studied at the University of Berlin, obtaining his PhD there in 1927. After a period of teaching at Göttingen he emigrated to America in 1934 and worked at various institutions including the Mount Wilson Observatory (1935–36), Princeton University (1937–42), and the University of Virginia (1942–46). In 1948 he took up an appointment at Yale University where he served as professor of astrophysics from 1957 until his retirement in 1973.

Wildt worked in stellar spectroscopy, theoretical astrophysics, and geochemistry, but his main interest lay in planetary studies. Since much of his work was done before the start of the space program it is not surprising that many of his speculations have since been rejected. Thus his claim that the Venusian clouds contained formaldehyde (CH_2O) formed under the influence of ultraviolet rays has not been confirmed by space probes.

He was, however, successful in his identification in 1932 of certain absorption bands observed by Vesto SLIPHER in the spectrum of Jupiter as being due to ammonia and methane. He also, in 1943, proposed a model of the constitution of Jupiter based on its density of 1.3 arguing that it consisted of a large metallic rocky core surrounded by ice and compressed hydrogen above which lies a thick layer of atmosphere. Saturn, Uranus, and Neptune were thought to have a similar constitution, but more recent models, together with the Pioneer and Voyager space flights, have suggested that the giant planets are made mainly of hydrogen in various forms with much smaller cores than Wildt proposed.

Wiles, Andrew John (1953–) *British mathematician*

The son of a theology professor, Wiles was educated at Cambridge University, where he gained his PhD in 1980. He immediately took up an appointment as professor of mathematics at Princeton.

Wiles worked on the most famous of all mathematical problems, namely, "Fermat's last theorem" (FLT). Pierre FERMAT claimed in 1637 that he had proved that there are no numbers $n > 2$ such that:

$$a^n + b^n = c^n$$

It was, of course, well known that where $n = 2$ there were many solutions to the equation as when $a = 3$, $b = 4$, and $c = 5$. Fermat had merely stated the theorem, as was his custom, in the margins of a book, in this case the *Arithmetica* of DIOPHANTUS. He simply added that there was insufficient room in the margin to record the details of the proof.

As over the years all Fermat's other marginalia have turned out to be accurate and proofs found, mathematicians were optimistic that the one unproved proposition – the last theorem – would also succumb. Yet 300 years later not only had no proof been found, but mathematicians seemed unaware in which direction a proof could be found. They could, of course, simply show the proposition to be false by finding an $n > 3$ which does satisfy the equation. But that was going to be no easy matter either, for it had been shown that any such number n must be very large – by 1992 all exponents up to 4 million had been tested and failed.

An alternative approach was suggested by some work in 1954 on elliptic curves by the Japanese mathematician Yutaka Taniyama. An elliptic curve is a set of solutions to an equation relating a quadratic in one variable to a cubic in another as in:

$$y^2 = ax^3 + bx^2 + cx + d$$

The Taniyama conjecture asserts that associated with every elliptical curve was a function with certain very precise specific properties. The German mathematician Gerhard Frey argued in 1985 that the Taniyama conjecture had important implications for Fermat's last theorem. He demonstrated that any possible solution to the theorem would give rise to a class of elliptical curves, referred to as Frey curves, which could not satisfy the conditions of the Taniyama conjecture. Thus a proof of the Taniyama conjecture would show that there could be no solutions to Fermat's last theorem.

In 1986 Wiles set out to show that Frey curves could not exist. Seven years later he had established a 200-page proof, which he revealed to the public for the first time at a mathematical conference in Cambridge, England, in 1993. Although Wiles's proof made headline news around the world it soon became evident that gaps still existed. After a further year the gaps had been eliminated and the 200-page paper had been accepted for publication.

Wilkes, Sir Maurice Vincent (1913–) *British computer scientist*

Wilkes was born at Dudley in Worcestershire, England, and educated at Cambridge University. After working on operational research during World War II, he returned to Cambridge where he was appointed professor of computing technology in 1965, a post he held until 1980 when he joined Digital Equipment Corporation in Massachusetts. From 1986 he was on the board of Olivetti. Wilkes was knighted in 2000. In 2002 he moved back to Cambridge.

In 1946 Wilkes attended a course on the design of electronic computers at the Moore School of Electrical Engineering at the University of Pennsylvania. Here Wilkes learned of the direction modern computers would have to follow. Earlier models, such as the Moore School's ENIAC, were really designed to deal with one particular type of problem. To solve a different kind of problem thousands of switches would have to be reset and miles of cable rerouted. The future of computing lay with the idea of the "stored program," as preached by John VON NEUMANN at the Moore School.

Consequently Wilkes returned to Cambridge to begin work on EDSAC (Electronic Delay Storage Automatic Computer). In order to store a program, the computer must first have a memory, something lacking from ENIAC and earlier devices. Wilkes chose to adopt the mercury delay lines suggested by J. P. ECKERT to serve as an internal memory store. In a delay tube an electrical signal is converted into a sound wave traveling through a long tube of mercury with a speed of 1,450 meters per second. It can be reflected back and forth along the tube for as long as necessary. Thus assigning the number 1 to be represented by a pulse of 0.5 microsecond, and 0 by no pulse, a 1.45-meter-long tube could retain 1,000 binary digits.

EDSAC came into operation in May 1949, gaining for Wilkes the honor of building the first working computer with a stored program; it remained in operation until 1958. The future, however, lay not in delay lines but in magnetic storage, and EDSAC soon became as obsolete as ENIAC. Wilkes provided a lively account of his work in his *Memoirs of a Computer Pioneer* (1985).

Wilkins, John (1614–1672) *English mathematician and scientist*

> Perhaps there may be some other means invented for a conveyance to the Moon … We have not now any Drake or Columbus to undertake this voyage, or any Daedalus to invent a conveyance through the aire.
> However, I doubt not but that time … will also manifest to our posterity that which we now desire but cannot know.
> —*The Discovery of a World in the Moon* (1638)

Born at Fawsley in Northamptonshire, England, Wilkins was educated at Oxford University, graduating in 1631. He was a parliamentarian during the English Civil War and became warden of Wadham College, Oxford University. In 1659 he was appointed master of Trinity College, Cambridge University. After the Restoration he lost his post but regained favor to become bishop of Chester.

Wilkins's chief contribution to the development of science was his part in founding the Royal Society. His influence can be traced back to his student days at Oxford when he collected around him a lively group of philosophers and scientists who later became founder members of the society in 1662. His own writings covered a wide range of fields and although he had a certain amount of mathematical knowledge he was more a practical scientist. His *Discovery of a World in the Moon* (1638) is a fantasy in which he speculated about the structure of the Moon. A later semimathematical work, *Mathematical Magick*, deals with the principles of machine design and in it Wilkins argued that perpetual motion is a theoretical possibility. One nonscientific interest to which Wilkins devoted much time was his project of devising a universal language.

Wilkins, Maurice Hugh Frederick (1916–2004) *New Zealand–British biophysicist*

Wilkins was born at Pongaroa in New Zealand. After graduating in physics from Cambridge University in England in 1938, he joined John RANDALL at Birmingham University to work on the improvement of radar screens. He received his PhD in 1940 for an electron-trap theory of phosphorescence and soon after went to the University of California, Berkeley, as one of the British team assigned to the Manhattan project and development of the atomic bomb. The results and implications of this work caused him to turn away from nuclear physics and in 1945 he began a career in biophysics, firstly at St. Andrews University, Scotland, and from 1946 at the Biophysics Research Unit, King's College, London.

The same year that Wilkins joined King's College, scientists at the Rockefeller Institute announced that genes consist of deoxyribonucleic acid (DNA). Wilkins began studying DNA molecules by optical measurements and chanced to observe that the DNA fibers would be ideal material for x-ray diffraction studies. The diffraction patterns showed the DNA molecule to be very regular and have a double-helical structure. The contributions of Wilkins's colleague, Rosalind FRANKLIN, were especially important in showing that the phosphate groups are to the outside of the helix, so disproving Linus PAULING's theory of DNA structure.

Wilkins passed on his data to James WATSON and Francis CRICK in Cambridge who used it to help construct their famous molecular model of DNA. For their work in elucidating the structure of the hereditary material, Wilkins, Watson, and Crick were awarded the 1962 Nobel Prize for physiology or medicine. Wilkins went on to apply his techniques to finding the structure of ribonucleic acid (RNA). From 1955 he was deputy director of the Biophysics Research Unit and from 1963 he was professor at King's College, firstly of molecular biology and from 1970 of biophysics. He retired in 1981.

Wilkins, Robert Wallace (1906–2003) *American physician*

Born at Chattanooga in Tennessee, Wilkins was educated at the University of North Carolina and at Harvard Medical School, where he obtained his MD in 1933. He then held various posts at Harvard and Johns Hopkins University before moving to Boston University in 1940, where he served as professor of medicine from 1955 until his retirement in 1972.

Wilkins investigated means of relieving the condition of hypertension (high blood pressure). He demonstrated that the disease is reversible, at least in its early stages, and in searching for suitable hypotensive drugs he helped to introduce the drug reserpine into the modern pharmacopoeia. The drug, an alkaloid derived from the tropical plant *Rauwolfia serpentina*, had been used extensively in India for a wide variety of complaints for many centuries. It was originally tested for its ability to reduce high blood pressure and has since been used as a hypotensive.

It was, however, soon noted that on some patients it had a marked calming effect without actually sending them to sleep. It was in fact the first of the "tranquilizers." Once widely used in the treatment of schizophrenia, it has been largely replaced by the phenothiazine drugs, such as chlorpromazine, because of its serious side effects.

Wilkinson, Sir Denys Haigh (1922–)
British physicist

Wilkinson, who was born at Leeds in England, was educated at the University of Cambridge, where he obtained his PhD in 1947. He remained at Cambridge, serving as professor of experimental physics and head of the Nuclear Physics Laboratory from 1957 until 1976, when he was appointed vice-chancellor of Sussex University. He retired in 1987.

Wilkinson has worked mainly in the field of elementary-particle physics. In particular he has investigated the phenomenon of charge independence found in the so-called strong force, which holds together the particles of the atomic nucleus. Following the collapse of parity in weak interactions as reported by Chen Ning YANG and Tsung Dao LEE he has also investigated the validity of parity conservation in strong nuclear interactions. He was knighted in 1974.

Wilkinson, Sir Geoffrey (1921–1996)
British inorganic chemist

Wilkinson was born at Todmorden in Yorkshire, England, and educated at Imperial College, London; after spending World War II working in North America on the development of the atomic bomb, he finally obtained his PhD in 1946. He later worked at the Massachusetts Institute of Technology and at Harvard, before being elected to the chair of inorganic chemistry at Imperial College, a post he held from 1956 until 1988. He was knighted in 1976.

Wilkinson is noted for his studies of inorganic complexes. He shared the Nobel Prize for chemistry in 1973 with Ernst FISCHER for work on "sandwich compounds." A theme of Wilkinson's work in the 1960s was the study and use of complexes containing a metal–hydrogen bond. Thus complexes of rhodium with triphenyl phosphine $((C_6H_5)_3P)$ can react with molecular hydrogen. The compound $RhCl(P(C_6H_5)_3)$, known as *Wilkinson's catalyst*, was the first such complex to be used as a homogeneous catalyst for adding hydrogen to the double bonds of alkenes (hydrogenation). This type of compound can also be used as a catalyst for the reaction of hydrogen and carbon monoxide with alkenes (hydroformylation). It is the basis of industrial low-pressure processes for making aldehydes from ethene and propene.

Willadsen, Steen (1944–) *Danish embryologist*

Born in Copenhagen, Willadsen attended the Royal Veterinary College in his home city, graduating in veterinary science. After spending some time in the veterinary field, he returned to the Royal Veterinary College to study reproductive physiology, gaining a PhD. In the early 1970s he moved to the UK to join the Agricultural Research Council's Institute of Animal Physiology at Babraham near Cambridge. In

1985 he joined the Texas-based biotech company, Grenada Genetics. Subsequently, Willadsen worked as a senior scientist at the Institute for Reproductive Medicine and Science of Saint Barnabas in New Jersey.

At Cambridge in the 1970s, Willadsen developed his expertise in handling mammalian embryos. In an early project, he devised a technique for successfully freezing livestock embryos so that they remained viable on thawing. Then he discovered that coating embryos in agar, a gelatinous substance derived from seaweed, protected them from damage during manipulation.

Willadsen quickly began to demonstrate the potential of embryo manipulation. He created *chimeras* – embryos consisting of cells from different animals. Using an ultrafine pipette, he would suck out single cells from a developing sheep's embryo and replace them with cells from another sheep's embryo. The chimeric embryo would then be implanted in the uterus of a surrogate mother to develop. Willadsen also used cells from different species: he created sheep–goat chimeras, which exhibited characteristics of both species, and even sheep–cow chimeras.

He soon went on to show that a technique known as *nuclear transfer* could be used to clone sheep embryos. The technique involved removing the nucleus from an unfertilized sheep's egg, and then inserting the nucleus from an early sheep's embryo cell. The egg could be stimulated to continue its development with the new nucleus, and then implanted into a surrogate mother. In 1984 Willadsen produced two lambs with this method, the first farm animals to be cloned.

At Grenada Genetics in Texas, Willadsen focused on cloning cattle embryos. In an unpublished experiment, he managed to clone a calf using nuclei obtained from a week-old differentiated embryo – a feat hitherto regarded as impossible because it was thought that such nuclei could not be reprogrammed to direct embryological development from scratch. Despite the lack of publicity surrounding Willadsen's achievements, they inspired other cloning scientists, notably the creator of Dolly the sheep, Ian WILMUT.

During the 1990s Willadsen turned to human fertility problems, and devised a technique called *cytoplasmic transfer*. This is intended to assist the development of human embryos in cases in which the patient has a history of abnormal egg development due to defective cytoplasm. A controversial procedure, banned in the United States in 2001, it involves injecting cytoplasm from a healthy donor egg into a recipient patient's egg produced by in vitro fertilization.

Williams, Robert R. (1886–1965) *American chemist*

The son of a Baptist missionary, Williams was born at Nellore in India and educated at the universities of Ottawa, Kansas, and Chicago. He began his career in government service, serving as chemist to the Bureau of Science in Manila before returning to America, where he worked at the Bureau of Chemistry in the agriculture department until 1918. He then moved into industry, working first for Western Electric before joining the Bell Telephone Laboratories in 1924, where he directed the chemistry laboratory until 1945.

With considerable single-mindedness Williams, early in his career, set himself the task of isolating the cause of beriberi. As early as 1896 Christiaan EIJKMAN had shown that it was a deficiency disease while Casimir FUNK had demonstrated that the vitamin whose absence caused the disease was an amine. Beyond that nothing was known when Williams began his work in the Philippines.

Working mainly in his spare time, in 1934 he managed to isolate, from several tons of rice husks, enough of the vitamin, B_1, to work out its formula. In 1937 he succeeded in synthesizing it.

His brother Roger, also a chemist, discovered pantothenic acid, another important vitamin in the B complex.

Williams, Robley Cook (1908–1995) *American biophysicist*

Williams, who was born in Santa Rosa, California, received a PhD in physics from Cornell University in 1935 and then went on to teach astronomy at the University of Michigan, transferring to the physics department in 1945. He remained there until 1950, when he transferred to a lecturing post in the biochemistry department at the University of California, Berkeley. In 1964, when a department of molecular biology was established at Berkeley, Williams became its chairman.

As an astronomer Williams worked on the estimation of stellar surface temperatures. Military research during World War II turned his attention to electron microscopy, and an insight drawn from his knowledge of astronomical techniques led to a fruitful collaboration with the crystallographer Ralph WYCKOFF. The early electron microscopes were transmission microscopes, i.e., the beam of electrons passes through the sample, giving a two-dimensional image. Working with Ralph Wyckoff at Michigan, Williams developed a technique of preparing specimens so that they could be observed with reflected beams of electrons. The technique involves depositing metal obliquely on the specimen. This effectively "casts shadows" and creates a vivid three-dimensional effect in the image.

Williams turned to the study of viruses, using his shadowing technique, and made important contributions to an understanding of viral structure. In 1955 (in collaboration with Heinz Fraenkel-Conrat) he achieved, with the tobacco mosaic virus, the first reconstitution of a biologically active virus from its constituent proteins and nucleic acids.

Williamson, Alexander William (1824–1904) *British chemist*

Williamson's father was a clerk in the East India Company in London. After his retirement in 1840 the family lived on the Continent, where Williamson was educated. He studied at Heidelberg and at Giessen (under Justus von Liebig), where he received his PhD in 1846. He also studied mathematics in Paris. In 1849 he took up the chair of chemistry at London University, a post he occupied until 1887.

Between 1850 and 1856 Williamson showed that alcohol and ether both belong to the water type. Type theory, developed by Charles GERHARDT and Auguste LAURENT, was based on the idea that organic compounds are produced by replacing one or more hydrogen atoms of inorganic compounds (which form the types) by radicals. Using the correct formula for alcohol (which he had recently established) Williamson represented the water type as: H_2O (water); C_2H_5OH (alcohol); $C_2H_5OC_2H_5$ (ether), where the H of water is progressively replaced by C_2H_5.

A further contribution to chemical theory was his demonstration (in 1850) of reversible reactions: two substances, A and B, react to form the products X and Y, which in turn react to produce the original A and B. Under certain conditions the system could be in dynamic equilibrium, when the amount of A and B reacting to form X and Y is equal to the amount of A and B produced by X and Y. He is remembered for what is now known as *Williamson's synthesis*, a method of making ethers by reacting a sodium alcoholate with a haloalkane.

Williamson, William Crawford (1816–1895) *British paleobotanist*

The son of a gardener and naturalist, Williamson was born in Scarborough in northeast England and was apprenticed in 1832 to an apothecary. In 1835, however, Williamson left for Manchester to take up the appointment of curator of the newly formed Museum of the Natural History Society. He later completed his medical training at University College, London, and after some years in private practice in Manchester entered academic life in 1851 as professor of natural history at Owens College, in which post he remained until his retirement in 1890.

In the 1850s Williamson's interest in the

plants of the Carboniferous was aroused when he was presented with a number of coal balls – aggregates of petrified plants found in the coal itself. To their study he devoted the rest of his career, publishing his results in his *On the Organization of Fossil Plants of the Coal Measures* (19 parts; 1871–93). The work contains much information on the early pteridophytes and seed plants.

Willis, Thomas (1621–1675) *English anatomist and physician*

Willis, who was born at Great Bedwin in England, entered Oxford University in 1636 and graduated firstly in arts before returning to take his MB in 1646. He fought for the royalists in the siege of Oxford during the English Civil War and afterwards he was a member of the "Invisible College," the informal group of scientists from whose meetings the Royal Society emerged in 1660. With the Restoration of Charles II the loyalty of Willis was rewarded with his appointment to the chair of natural philosophy at Oxford in 1660. Willis resigned in 1666 and moved to London where he set up in private practice.

It was as an anatomist that Willis made his most significant contribution to medicine. In 1664 he published his *Cerebri anatome* (Anatomy of the Brain), a work still in use a century later. It contains a much improved treatment of the cranial nerves, being in complete agreement with modern neurology on the first six and giving the first description of the eleventh cranial nerve. It also contains the first account of the so-called *circle of Willis* at the base of the brain.

This arose out of an observation at an autopsy of a body with a completely occluded right carotid artery in the neck. Surprisingly this appeared not to have obstructed the blood supply to the brain nor to have produced any obvious symptoms during the life of the deceased. Careful dissection revealed an interconnected ring of arteries at the base of the brain. The effect of this, Willis saw, would be to preserve a continuous supply of blood to both sides of the brain even if there had been a failure on one side.

Willis also made an attempt to apply the advances in chemistry and atomic theory to the problems of disease. His attempt was premature but nevertheless he was the first to describe a number of diseases, such as diabetes mellitus, puerperal fever, and general paralysis of the insane (a late effect of syphilis).

Willstätter, Richard (1872–1942) *German chemist*

Willstätter (**vil**-shtet-er) was the son of a textile merchant from Karlsruhe in Germany. He was educated at the University of Munich, receiving his PhD in 1894 for work on cocaine. He was professor of chemistry in Zurich from 1905 to 1912, when he left to work in the Kaiser Wilhelm Institute in Berlin. In 1916 he succeeded Adolf von BAEYER to the chemistry chair at Munich. He resigned in 1924 in protest at the growing anti-Semitism in Germany but remained in his homeland until he felt his own life was no longer safe, going into exile in Switzerland in 1939.

His early work was mainly on the structure of alkaloids – he managed to throw light on such important compounds as cocaine, which he synthesized in 1923, and atropine. In 1905 he began work on the chemistry of chlorophyll. By using the chromatographic techniques developed by Mikhail TSVET, he was soon able to show that it consists of two compounds, chlorophyll a and b, and to work out their formulas. One of the significant features he noted was that chlorophyll contains a single atom of magnesium in its molecule, just as hemoglobin contains a single iron atom. He also investigated other plant pigments, including the yellow pigment carotene and the blue pigment anthocyanin. His work on chlorophyll was justified in 1960 when Robert WOODWARD succeeded in synthesizing the compounds described by his formulas and came up with chlorophyll.

Willstätter was less successful with his enzyme theory. In the 1920s he claimed to have isolated active enzymes with no trace of protein. His views were widely accepted until protein was restored to its rightful place in enzyme activity by the work of John NORTHROP in 1930.

For his work on plant pigments Willstätter was awarded the Nobel Prize for chemistry in 1915.

Wilmut, Ian (1944–) *British embryologist*

Born in Hampton Lucy, England, Wilmut studied agricultural science at Nottingham University, obtaining a BSc, and went on to Cambridge University, receiving his PhD in 1971 for work on the freezing of boar semen. He stayed on for further research into freezing cattle embryos, and in 1974 he joined the Animal Breeding Research Station – now the Roslin Institute – near Edinburgh.

In 1973, while still at Cambridge University, Wilmut became the first person to obtain a live calf from a frozen embryo – he named the animal "Frosty." During the 1970s and early 80s Wilmut's work concerned prenatal mortality of sheep and pigs, and later the development of techniques for egg retrieval and embryo transfer in sheep and cattle.

By the mid-1980s his attentions were focused on the challenge of manually inserting genes into sheep embryos. After hearing of the success of Steen WILLADSEN in cloning sheep and cattle from embryo cells, Wilmut switched to this more promising line of research. In 1990 the cell biologist Keith CAMPBELL joined Wilmut at the Roslin Institute, and they set about refining the technique of nuclear transfer. They found that by starving embryo cells before allowing them to fuse with "empty" unfertilized egg cells (i.e., egg cells with the nucleus removed), the cell cycle of both nucleus and recipient cell could be synchronized, producing much better results.

Initial success came in 1985 with the birth of Megan and Morag, Welsh Mountain sheep cloned from embryo cells. In 1996, this was eclipsed by the birth of Dolly, a Finn Dorset lamb named after the US singer Dolly Parton. Dolly was the first mammal to be cloned from fully differentiated adult body cells – in this case mammary cells. The event received wide press coverage and Wilmut became the focus of public debate about the ethics of cloning, especially the possibility of human cloning. The following year saw the birth of another lamb, called Polly, cloned from fetal skin cells containing a human gene.

Wilmut sees the usefulness of his work primarily in creating animals that are genetically altered so that they will, for example, secrete drugs or other substances in their milk, or provide a source of tissues and organs for human transplantation. He reportedly views human cloning as broadly unacceptable, while not ruling out the use of nuclear transfer techniques to avoid certain congenital disorders.

Wilson, Alexander (1766–1813) *Scottish–American ornithologst*

> The achievement of this man is little short of marvelous.
> —Frances Herrick on Wilson's publication of *The American Ornithology*

Wilson was born at Paisley in Scotland and spent his early life as a weaver and peddler. He was also a poet, but the nature of his poems, championing the weaver's cause in Scotland, led to fines and imprisonment. In 1894 he emigrated to America, where he spent most of the remainder of his life compiling his monumental *American Ornithology*, seven volumes of which he completed from 1808 to 1813, others being added after his death. He made extensive and often arduous travels throughout America and listed, described, and illustrated a total of 264 species, many of them new to science. Wilson also made many contributions to knowledge of bird calls and songs, egg numbers, plumage and sexual differences, behavior, migration numbers, and so on. His analyses of bird stomach contents produced new information about food preferences and requirements, while he also demonstrated, by means of dissection, the stereoscopic vision of owls. Characterized by accuracy and great attention to detail, Wilson's *American Ornithology* has been described as probably the most important pioneering contribution to ornithological science in America.

Wilson, Allan Charles (1934–1991) *New Zealand biochemist*

Born at Ngaruawakia in New Zealand, Wilson was educated at the University of Otago, Dunedin, and at Washington State University before completing his PhD in 1961 at the University of California, Berkeley. After working at the Weizmann Institute in Israel and at universities in Nairobi, Kenya, and Harvard, he returned to Berkeley serving as professor of biochemistry until his death from leukemia in 1991.

In 1967, in collaboration with Vincent SARICH, Wilson, following the work of Emile Zuckerandl, argued that molecular clocks could reveal much about the early history of man. Against the opposition of paleontologists they claimed that the divergence between man and the great apes began only 5 million years ago. Their view seems to have prevailed.

In the 1980s Wilson sought to challenge the paleontologists once more, this time on the issue of the emergence of modern man. While anthropologists favored a date of 1 million years, Wilson's work suggested a time no later than 200,000 years ago.

He chose to work with mitochondria, the cellular organelles which convert food into energy. Like a cell nucleus, mitochondria also contain DNA. It encodes, however, only 37 genes as opposed to the 100,000 of nuclear DNA. Further, mitochondrial DNA evolved rapidly and regularly and, surprisingly, it is inherited from the mother alone. It follows, Wilson pointed out, that "all human mitochondrial DNA must have had an ultimate common female ancestor." Where and when, he went on to ask, could she be found?

Wilson adopted the parsimony principle that subjects are connected in the simplest possible way. That is, the fewer differences found in mitochondrial DNA, the closer they were connected. Mitochondria from 241 individuals from all continents and races were collected and analyzed. The tree constructed had two branches, both of which led back to Africa.

What was the date of this "African Eve" as she was quickly dubbed by the press? Wilson measured the ratio of mitochondrial DNA divergence between humans to the divergence between humans and chimpanzees. The ratio was found to be 1:25 and, as human and chimpanzee lineages diverged 5 million years ago, human maternal lineages must have separated by 1/25 of this time, namely, 200,000 years ago.

Wilson's hypothesis, first presented in 1987, has provoked considerable opposition. Those who prefer a multiregional explanation of human evolution have questioned most of Wilson's assumptions and have argued that until it is backed up by unequivocal fossil evidence it must remain speculative.

Wilson, Charles Thomson Rees
(1869–1959) *British physicist*

> In September 1894 I spent a few weeks ... on the summit of Ben Nevis. The wonderful optical phenomena shown when the sun shone on the clouds surrounding the hill-top, and especially the coloured rings surrounding the sun ... made me wish to imitate them in the laboratory.
> —*Weather* (1954)

Charles Wilson was the son of a sheep farmer from Glencorse in Scotland but his father died when he was four and Charles and his mother moved to Manchester. He was educated there and started to specialize in biology but moved to Cambridge University to study physics. There he started work with J. J. THOMSON.

Wilson began experiments to duplicate cloud formation in the laboratory by letting saturated air expand, thus cooling it. He found that clouds seemed to need dust particles to start the formation of water droplets and that x-rays, which charged the dust, greatly speeded up the process. Inspired by this, he showed that charged subatomic particles traveling through supersaturated air also formed water droplets. This was the basis of the cloud chamber, which Wilson perfected in 1911 and for which he received the Nobel Prize for physics in 1927. The cloud chamber became an indispensable aid to research into subatomic particles and, with the addition of a magnetic field, made different particles distinguishable by the curvature of their tracks.

Returning to the study of real clouds, Wilson also investigated atmospheric electricity and developed a sensitive electrometer to measure it. A result of this work was his determination of the electric structure of thunder clouds.

Wilson, Edmund Beecher (1856–1939)
American biologist

Wilson, born the son of a judge in Geneva, Illinois, was educated at Yale and Johns Hopkins where, in 1881, he gained his PhD. After further training abroad at Cambridge, Leipzig, and Naples he returned to America and taught for some years at the Massachusetts Institute of Technology and Bryn Mawr College. In 1891 Wilson moved to Columbia where he spent the rest of his career, serving as professor of zoology from 1894 until his retirement in 1928.

In his *Cell Development and Inheritance* (1896) Wilson stressed the importance of the cell theory, hoping that its application to the problems of development and heredity would lead to advances in these areas. He did important work on the concept of cell lineage, studied internal cellular organization, and investigated the part played by the chromosomes in the determination of sex. It was in Wilson's department that the science of genetics really became established through the work of T. H. MORGAN and Hermann MULLER.

Wilson, Edward Osborne (1929–)
American entomologist, ecologist, and sociobiologist

> Marxism is sociobiology without biology ... Although Marxism was formulated as the enemy of ignorance and superstition, to the extent that it has become dogmatic it ... is now mortally threatened by the discoveries of human sociobiology.
> —*On Human Nature* (1978)

Wilson, who was born in Birmingham, Alabama, graduated in biology from the University of Alabama in 1949 and obtained his PhD from Harvard in 1955. He joined the Harvard faculty the following year, becoming professor in 1964 and curator of entomology at the Museum of Comparative Zoology in 1971.

Much of Wilson's entomological work was with ants and other social insects and was comprehensively surveyed in his massive *Insect Societies* (1971). He has also worked on speciation and with William Brown introduced the term "character displacement" to describe the process that frequently takes place when closely related species that have previously been isolated begin to overlap in distribution. The differences that do exist between the species become exaggerated to avoid competition and hybridization.

Wilson collaborated with Robert MACARTHUR in developing a theory on the equilibrium of island populations from which emerged their *Theory of Island Biogeography* (1967). To test such ideas Wilson conducted a number of remarkable experiments with Daniel Simberloff in the Florida Keys. They selected six small mangrove clumps and made a survey of the number of insect species present. They then fumigated the islands to eliminate all the 75 insect species found. Careful monitoring over the succeeding months revealed that the islands had been recolonized by the same number of species, thus confirming the prediction that "a dynamic equilibrium number of species exists for any island."

It was, however, with his *Sociobiology* (1975) that Wilson emerged as a controversial and household name. He argued that "a single strong thread does indeed run from the conduct of termite colonies and turkey brotherhoods to the social behavior of man." Using the arguments of William Hamilton and Robert Trivers, Wilson had little difficulty in showing the deep biological and genetic control exercised over many apparently altruistic acts in insects, birds, and mammals. He also proposed plausible mechanisms to explain much of the social behavior and organization of many species. Many believe, however, that he is on very shaky ground when he extends such arguments to human social evolution. Wilson has continued to produce a large number of popular, personal, and technical works. Among these are a work on ecology, *The Diversity of Life* (1992); his Pulitzer prize-winning *The Ants* (1988, written in collaboration with B. Hölldobler); and his revealing autobiography, *Naturalist* (1994).

Wilson, John Tuzo (1908–1993) *Canadian geophysicist*

Born in Ottawa, Canada, Wilson was educated at the University of Toronto and at Princeton, where he obtained his PhD in 1936. After working for the Canadian Geological Survey (1936–39) and war service, he was appointed professor of geophysics at the University of Toronto (1946) where he remained until his retirement in 1974.

Wilson did much to establish the new discipline of plate tectonics during the early 1960s and was the first to use the term "plate" to refer to the rigid portions (oceanic, continental, or a combination of both) into which the Earth's crust is divided. In 1963 he produced some of the earliest evidence in favor of the sea-floor spreading hypothesis of Harry H. HESS when he pointed out that the further away an island lay from the midocean ridge the older it proved to be.

His most significant work, however, was contained in his important paper of 1965, *A New Class of Faults and Their Bearing on Continental Drift*, in which he introduced the idea of a transform fault. Plate movement had been identified as divergent, where plates are being separated by the production of new oceanic crust from the midocean ridges, and convergent, where plates move toward each other with one plate sliding under the other. Wilson realized a third kind of movement was needed to explain the distribution of seismic activity and the way in which the ocean ridges do not run in continuous lines but in a series of offsets joined by the transform faults. Here the plates slide past each other without any creation or destruction of material.

Wilson replied to critics of the plate tectonics theory, such as Vladimir BELOUSOV, in his *A Revolution in Earth Science* (1967).

Wilson, Kenneth G. (1936–) *American theoretical physicist*

Wilson was educated at Harvard and Cal Tech where he gained his PhD in 1969. He taught at Cornell University from 1963, serving as professor of physics there from 1970 to 1988, when he moved to a similar post at Ohio State University, Columbus.

Wilson received the 1982 Nobel Prize for physics for theoretical work on critical phenomena in connection with phase transitions. He first applied his methods to the problem of ferromagnetic materials. Above a certain temperature, known as the CURIE point, such materials become paramagnetic. This behavior results from the individual magnetic moments of the atoms. In the ferromagnetic state numbers of individual atoms "couple" together so that their spins are aligned, and there is a resulting long-range interaction over a region of the solid. Above the critical point (the Curie point) this long-range order breaks down. Wilson's achievement was to develop a theory that could apply to the system near the critical point.

He did this using an idea first suggested by Leo Kadanoff. He took a block of atoms and calculated the effective spin of the block, then took a number of blocks and calculated the value for the larger block, and so on. The method involves a mathematical technique known as renormalization.

Using such methods Wilson could go from the properties of individual atoms to properties characteristic of many atoms acting together, and the resulting theory could be applied to properties other than magnetism. Thus it can be used for the critical state observed in the change between liquid and gas and to changes in alloy structure. Wilson is now applying his methods to the strong forces between nucleons.

Wilson, Robert Woodrow (1936–) *American astrophysicist*

Wilson studied initially at Rice University in his native city of Houston, where he gained his BA in physics in 1957; he went on to obtain his PhD from the California Institute of Technology in 1962. He joined the Bell Laboratories, Holmdel, New Jersey, in 1963 and served as head of the radiophysics research department from 1976 to 1990.

It was at the Bell Laboratories that he and his coworker Arno PENZIAS found the first evidence in 1964 of the cosmic microwave background radiation, which is now widely interpreted as being the remnant radiation from the "big bang" creation of the universe several billion years ago. The two men were jointly honored with the 1978 Nobel Prize for physics, which they shared with Pyotr L. KAPITZA for his (unrelated) discoveries in low-temperature physics.

Wilson is continuing his astrophysics work with Penzias, looking for interstellar molecules and determining the relative abundances of interstellar isotopes.

Windaus, Adolf Otto Reinhold (1876–1959) *German chemist*

Windaus (**vin**-dows) studied medicine at the university in his native city of Berlin and at Freiburg University, where he changed to chemistry under the influence of Emil FISCHER. After holding chairs in Freiburg and Innsbruck he became, in 1915, professor of chemistry at Göttingen, where he remained until his retirement in 1944.

In 1901 Windaus began his study of the steroid cholesterol, a compound of considerable biological significance. Over the years he threw considerable light on its structure and in 1928 was awarded the Nobel Prize for chemistry for this work and for showing the connection between steroids and vitamins.

It was known that cod-liver oil prevents rickets because it contains vitamin D. It was also known that sunlight possesses antirachitic properties and, further, that mere exposure of certain foods to sunlight could make them active in preventing rickets. Clearly something in the food is converted photochemically into vitamin D but nobody knew what.

As vitamin D is fat soluble, the precursor of vitamin D (the provitamin) was not surprisingly found to be a steroid. In 1926 Windaus succeeded in showing that the provitamin is present as an impurity of cholesterol, ergosterol, which is converted into vitamin D by the action of sunlight.

Winkler, Clemens Alexander (1838– 1904) *German chemist*

Winkler (**vingk**-ler) was born at Freiberg in Germany and studied at the School of Mines there. He was later appointed to the chair of chemical technology and analytical chemistry at Freiberg in 1871.

In 1885 a new ore – argyrodite – was discovered in the local mines. Winkler, who had a considerable reputation as an analyst, was asked to examine it and to his surprise the results of his analysis consistently came out too low. He discovered that this was due to the presence of a new element, which, after several months' search, he isolated and named germanium after his fatherland. The properties of germanium matched those of the eka-silicon whose existence had been predicted in 1871 by Dmitri MENDELEEV. Winkler's discovery completed the detection of the three new elements predicted by Mendeleev nearly 20 years before.

Winograd, Terry Allen (1946–) *American computer scientist*

Born at Tacoma Park in Maryland, Winograd was educated at the University of Colorado and at the Massachusetts Institute of Technology, where he obtained his PhD in 1970. He remained at MIT until 1973 when he moved to Stanford, California, as professor of computer science and linguistics.

Much of Winograd's work has been concerned with artificial intelligence (AI). In this field he developed in 1970 the successful program SHRDLU; the name is taken from the letters SHRDLU ETAOIN, which appear on the keyboard of a typesetting machine and are used by printers to mark something unintelligible. One pervasive problem facing AI programs is that of background knowledge. For example, in translating technical texts between Russian and English, readers were puzzled to find frequent references to a "water sheep" – the computer was doing its best to cope with "hydraulic ram." To avoid semantic problems of this kind Winograd created a simple closed world that could be completely described. It consisted of a small number of differently shaped colored bricks. The computer was then instructed to perform such simple tasks as "put a small cube onto the green cube which supports a pyramid," and to answer correctly questions of the form: "How many things are on top of green cubes?" The actions and answers took place only within the confines of a computer program, thus avoiding the problem of designing a robotic arm. Despite this SHRDLU worked effectively. It was described by Winograd in his *Understanding Natural Languages* (1972).

He noted, further, in 1984 that there is too much deep ambiguity in a natural language to permit unrestricted machine translation. Such sentences as "the chickens are ready to eat," and "he saw that gasoline can explode," with their deep structural ambiguity, are common in a natural language and still tend to defeat any computer. The solution, to preedit and postedit the text, is expensive and time-consuming. But, Winograd notes, such a system was used by the Pan American Health Organization to translate a million words from Spanish into English between 1980 and 1984. Winograd has concluded, however, that there is no software currently capable of dealing with meaning over a significant subset of English.

Wislicenus, Johannes (1835–1902) *German chemist*

Wislicenus (vis-lee-**tsay**-nuus) was born in Klein-Eichstadt, Germany, the son of a Lutheran pastor who was forced to flee Europe in 1853 because of his political views. Wislicenus accompanied his father to America, where he attended Harvard until returning to Europe in 1856. He continued his education at Halle and at Zurich, where he was appointed professor of chemistry in 1870. In 1872 he moved to Würzburg, where he stayed until he succeeded Adolph KOLBE as professor of chemistry at the University of Leipzig (1885).

In 1872 Wislicenus showed that there were two forms of lactic acid having the formula $CH_3CH(OH)COOH$. One, derived from sour milk by Karl SCHEELE in 1870, was optically inactive; the other, discovered by Jöns Jacob BERZELIUS in 1808, was active. Wislicenus suggested that this was caused by different arrangements of the same atoms in space producing different properties in the compounds. Wislicenus's findings and similar work led Jacobus VAN'T HOFF and Joseph LE BEL to establish the new discipline of stereochemistry a few years later. Wislicenus went on to study "geometrical isomerism" – the existence of isomers because of different arrangements of groups or atoms about a double bond in the molecule.

Withering, William (1741–1799) *British physician*

> Time will fix the real value upon this discovery.
> —On his discovery of the medicinal properties of the foxglove

Withering was born at Wellington in England and educated at the University of Edinburgh, where he obtained his MD in 1766. He then practiced in Stafford but in 1775 moved to a more prosperous practice in Birmingham where, as a member of the Lunar Society, he met such eminent scientists as Joseph PRIESTLEY and Erasmus DARWIN. Withering retired in 1783 as he was then suffering from tuberculosis. His house, like that of his friend Priestley, was sacked in 1791 by the Birmingham mob for his sympathy with the French Revolution.

Withering was an expert botanist and in 1776 began the publication of his *Botanical Arrangement*. Because of his reputation his advice was sought on a mixture of herbs an old Shropshire woman was using to cure dropsy. He realized that the active herb was the foxglove and began a more careful study of its properties in 1775. He decided that the leaves were preferable to the root of the plant and set about establishing a standardized dose. He recommended its use as a diuretic but was in error in supposing that the foxglove acted directly as a diuretic. In fact it acted directly on the heart, whose beat it both slowed and strengthened. He published a full report of his work in his *An Account of the Foxglove* (1785). Since then drugs derived from digitalis, an extract of foxglove leaves, have come to have an important role in the treatment of heart failure. They are, however, no longer used for "dropsy" as such because more powerful and specialized diuretics are now available.

Withering was also an authority in mineralogy and the ore witherite (barium carbonate) is named for him.

Witten, Ed(ward) (1951–) *American mathematical physicist*

Witten was born in Baltimore, Maryland. Having graduated from Brandeis College, Massachusetts, in 1971 with a degree in history, he intended to pursue a career in journalism. However, after working on George McGovern's 1972 presidential campaign, he realized that he was ill-suited to the world of political journalism and returned to university to study physics at Princeton. After completing his PhD in 1976 Witten moved to Harvard for several years before returning to Princeton, where he was appointed professor of physics – a post he occupied until 1987, when he moved to the Institute for Advanced Study.

Witten has worked mainly in the development of string theory. In the 1960s Yoichiro NAMBU and others had shown that elementary particles could be treated as strings of a certain kind, but it was soon shown that the theory only worked satisfactorily in 26 dimensions. A more ambitious theory of superstrings was promoted by Mike Green and others in the 1970s. When Witten came across the theory in 1975 he saw that it could throw light on what he termed "the single biggest puzzle in physics," namely, how to unify general relativity, which deals with gravity and space, with quantum mechanics, which explains events at the nuclear level. The realization in 1982 that superstring theory demanded the presence of gravity in its working was, for Witten, "the greatest intellectual thrill of my life." In the early 1980s Witten ruled out field theories in 11 dimensions derived from models based on the approach of Theodor KALUZA and Oskar KLEIN. Further, he argued that a number of mathematical anomalies would emerge in spaces with two, six, or ten dimensions. In 1984, however, Green and John SCHWARZ were able to show that, under certain special assumptions, a theory of ten dimensions could be developed that avoided the anomalies and explained the existence of particles with a built-in handedness (chirality).

Witten then began work showing how, in a ten-dimensional universe, the hidden extra six dimensions could be compacted and how they could interact with particles in detectable ways. He has sought for an analysis, based on geometric foundations, and has attempted to develop a topological quantum-field theory that gives due regard to the fundamental geometrical properties of matter. Witten's work is important in pure mathematics as well as in physics. In 1990 he was awarded the Fields Medal (regarded as the mathematics equivalent of a Nobel prize), mostly for his work on knot theory.

Witten has also made important contributions to quantum field theory, particularly the theory of anomalies, i.e., quantum mechanical symmetry breaking, quark confinement, and supersymmetric field theories. In 1981 Witten made a significant contribution to general relativity theory when he found a relatively simple proof of the *positive energy (mass) theorem*, stating that the energy (mass) of an isolated system described by general relativity theory cannot be negative.

Perhaps the most significant of Witten's many contributions to superstring theory was his demonstration in 1995 that the five superstring theories that are mathematically consistent and 11 space-time dimensional supergravity theory are all related to each other mathematically by transformations called *duality transformations*. This has led to superstring theory being superseded by *M-theory*, an 11-dimensional theory which has yet to receive its definitive formulation. In spite of that, this breakthrough has led to further major developments such as an understanding of black holes in terms of superstring theory/M-theory and the possibility that the discrete structures predicted by M-theory will be observed in future detailed observations of the cosmic microwave background radiation.

Wittig, Georg (1897–1987) *German organic chemist*

Born in Berlin, Germany, Wittig (**vit**-ik) was educated at the university of Marburg. He worked at Braunschweig (1932–37), at Freiburg im Breisgau (1937–44), and at Tubingen (1944–65). In 1965 he became director of the Chemical Institute at Heidelberg, a post he held until his retirement in 1967.

Wittig worked extensively in organic chemistry, in particular on the chemistry of carbanions – negatively charged organic ions such as $C_6H_5^-$. In this work he discovered a class of reactive phosphorus compounds of the type $(C_6H_5)_3P{:}CH_2$. Such compounds (known as ylides) are able to replace the oxygen of a carbonyl group $C{=}O$ by a CH_2 group, to give $C{=}CH_2$. This reaction, known as the *Wittig reaction*, is of immense importance in the synthesis of certain natural compounds, such as prostaglandin and vitamins A and D_2. Wittig also discovered a useful directed form of the aldol condensation. He was awarded the Nobel Prize for chemistry in 1979.

Wöhler, Friedrich (1800–1882) *German chemist*

> Organic chemistry just now is enough to drive one mad. It gives one the impression of a primeval, tropical forest full of the most remarkable things, a monstrous and boundless thicket, with no way of escape, into which one may well dread to enter.
> —Letter to J. J. Berzelius, 28 January 1835

Wöhler (**vu(r)**-ler), who was born at Eschersheim near Frankfurt, acquired from his father, the master of horse of the crown prince of Hesse-Cassel, an interest in mineralogy – he actually met the aged poet Johann Wolfgang von GOETHE, another devotee, in the shop of a Frankfurt mineral dealer. He began training as a physician at Marburg and Heidelberg but was persuaded by Leopold GMELIN to change to chemistry. After a year in Sweden with Jöns Jacob Berzelius he taught chemistry in Berlin and Cassel before his appointment to the chair of chemistry at Göttingen (1836), where he remained for the rest of his life.

Wöhler's most famous discovery occurred in 1828, when he synthesized crystals of urea while evaporating a solution of ammonium cyanate. He wrote excitedly to Berzelius, "I must tell you I can prepare urea without requiring a kidney of an animal, either man or dog." The significance of his achievement was that urea is an organic substance, which it was hitherto thought could be synthesized only by a living organism. If the constituents of a living body can be put together in the laboratory like common salt or sulfuric acid then there is apparently nothing left to distinguish the living from the nonliving. For this reason Wöhler's work is frequently cited as marking the death of vitalism, although at the time Wöhler was probably more concerned with the chemical reactions involved.

However, it was not seen by Wöhler's contemporaries as having that significance. Just because one substance had been synthesized, no one was prepared to claim that all organic substances could be so created. Justus von LIEBIG, who knew Wöhler's work well and collaborated with him over a long period of time, was a vitalist and vitalism was too complex and deep-seated an idea to disappear as a result of one experiment. Wöhler's work was more important in opening up whole new dimensions of biochemistry, stimulating work on the chemistry of digestion, respiration, growth, and reproduction.

Wöhler made other contributions to organic chemistry. In 1832, in collaboration with Liebig, he showed that the benzoyl radical (C_6H_5CO) could enter unchanged in a series of compounds: the hydride, chloride, cyanide, and oxide. Thus organic chemistry became, for a time, the chemistry of compound radicals. With the theories of Berzelius, this approach led to a great increase in the knowledge of organic compounds without a corresponding understanding of their chemistry.

In fact organic chemistry became so confusing that Wöhler returned to inorganic chemistry. In later years he tended to concentrate on the chemistry of metals, in particular the production of pure samples of some of the less common metals. He succeeded, at great expense, in obtaining pure aluminum (1827) and beryllium (1828).

With Liebig he was partly responsible for the discovery of isomerism. In 1823, while working in the laboratory of Berzelius, he prepared silver cyanate; at the same time Liebig produced silver fulminate, a compound with very different properties. To their surprise they found both compounds had identical formulas. Berzelius named the phenomenon "isomerism."

One final achievement of Wöhler was the creation of one of the first great teaching laboratories of Europe at Göttingen. From the 1830s nearly all creative chemists of the 19th century spent some time at Göttingen; students came not just from the Continent but also from America and Britain. The tradition persisted until the time of Hitler.

Wolf, Johann Rudolf (1816–1893) *Swiss astronomer*

Wolf (volf), the son of a minister from Fällenden in Switzerland, studied at the universities of Zurich, Vienna (1836–38), and Berlin (1838). He began his career in 1839 at the University of Bern where he served as director of the observatory from 1847 and professor of astronomy from 1844 until 1855, when he moved to Zurich. He there held chairs of astronomy at the university and the Institute of Technology as well as serving as director of the newly opened Federal Observatory.

It was not until 1851 that the work of Heinrich Schwabe on sunspot cycles became widely known. Wolf immediately began a study of solar observations dating back to the 17th century from which he was able to demonstrate that the cycle had a mean period of 11.1 years. He further recorded the maxima and minima from 1610 onward. With the announcement in 1851 by Johann von LAMONT of an approximately 10-year variation in the terrestrial magnetic field, it occurred to Wolf, as it did independently to Edward SABINE and Alfred Gautier, that the two cycles were in fact connected.

Under Wolf the Federal Observatory became the world center for sunspot information and from 1855 daily counts of sunspots were made. To establish some kind of common statistical basis, Wolf proposed the "Zurich relative sunspot number" that is still used to indicate the level of sunspot activity. It is calculated from a formula that takes into account the number of sunspot groups, the total number of component spots in the groups, and in addition the competence of the observer and the quality of the instrument used.

Wolf, Maximilian Franz Joseph Cornelius (1863–1932) *German astronomer*

The son of a wealthy physician, Wolf studied at the university in his native city of Heidelberg, where he obtained his PhD in 1888. After spending two years in Stockholm he returned to the University of Heidelberg, where he was appointed to the chair of astronomy and astrophysics in 1901 and where he remained until his death in 1932.

Wolf is best known for his discovery of hundreds of new asteroids or minor planets. These are small rocky bodies that orbit the Sun, mainly in a belt between the orbits of Mars and Jupiter. The first to be discovered was Ceres by Giuseppe PIAZZI in 1801 and by 1891 a further 300 had been identified. It was then that Wolf introduced his "labor-saving photographic method." A camera was attached to a telescope that moved at the same speed as the stars and thus gave point images on the photographic plate; asteroids, moving at a different speed relative to the stars, showed up on the plates as streaks. Wolf worked first in his private observatory in the center of Heidelberg and then at a new observatory, the Baden Observatory, built for him on the Königstuhl. He managed to discover over 500 new asteroids including, in 1906, Achilles, the first of the Trojan asteroids, which lie in two groups in Jupiter's orbit and form an equilateral triangle with Jupiter and the Sun.

Wolf also drew attention to dark regions of the Milky Way that appeared to be devoid of stars. He argued that they were caused by the presence of obscuring clouds of dust and gas and in 1923 devised the *Wolf diagram* – a method for determining the distance and absorption characteristics of the dark nebulae. By 1911 he had established the means of differentiating between spiral nebulae, later shown to be star systems lying far beyond our Galaxy, and the dark and the luminous nebulae within the Galaxy. These methods were soon widely adopted.

Wolff, Kaspar Friedrich (1733–1794) *German anatomist and physiologist*

Wolff, who served as a surgeon during the Seven Years' War, studied in his native city of Berlin and at Halle, presenting in 1759 his thesis *Theoria generationis* (Theory of Reproduction). In this he destroyed the preformation or homunculus theory of development, which had postulated that the embryo contained all the adult organs preformed in miniature. In its place Wolff advanced the now accepted idea that organs develop from undifferentiated tissue. His theories were initially based on philosophical grounds but he later carried out research at St. Petersburg that substantiated his claims. While examining chick embryos Wolff also discovered an embryonic form of the kidney – now called the *Wolffian body* – that precedes the true organ in developing animals.

Wolfram, Stephen (1959–   ) *British theoretical physicist*

The son of a novelist father and philosopher mother, Wolfram showed an early passion for and understanding of science. Born in London, he entered Oxford University when he was sixteen, having already written several papers in particle physics. As he could find no one at Oxford capable of teaching him anything interesting, he attended no lectures and pursued his own interests. At the suggestion of Murray GELL-MANN he moved to the California Institute of Technology in 1978 and completed his PhD within a year. He immediately joined the Institute staff and began work on the development of a new computer language, SMP, capable of manipulating algebraic formulas as well as numbers. Wolfram subsequently developed another computer system called *Mathematica*, which has been very widely used. In 1983 Wolfram moved to the Institute for Advanced Studies, Princeton, where he remained until 1986, when he was appointed professor of physics at the University of Illinois where he set up the Center for Complex Systems Research. He is also the founder and president of the Wolfram Research Company, Champaign, Illinois.

In 1982 Wolfram became interested in cellular automata (CA), a subject devised by John VON NEUMANN in the 1940s, and has worked almost entirely in this field ever since. One-dimensional cellular automata begin with a horizontal line of cells with each cell in one of two states, blank or filled. Further states displayed below each other are generated by a rule; for example, the rule that a cell is filled if and only if one and only one of the three cells immediately above is also filled.

With the aid of a computer the outcome of various rules can be examined over hundreds of generations. Wolfram has claimed to be able to distinguish between four classes of one-dimensional automata:

1. The starting pattern dissolves to a uniform, homogenous state, e.g., all cells filled or all cells blank.
2. The pattern evolves to a fixed finite size or oscillates endlessly between two fixed patterns.
3. Patterns continue to grow at a fixed rate and often produce self-similar fractal patterns.
4. Patterns show apparently chaotic behavior; they grow and contract unpredictably and irregularly.

Behind Wolfram's manipulation of CA on computer screens is the hope that it will lead to a greater insight into the development of complex systems. In 2002 he published *A New Kind of Science*, summarizing his ideas on complexity theory.

Wollaston, William Hyde (1766–1828)
British chemist and physicist

[At a] distance from the misrepresentations of narrow-minded bigots.
—Referring to his choice of astronomy as a specialty. *The Secret History of a Private Man* (1795)

Wollaston, the son of a clergyman from East Dereham in Norfolk, England, was educated at Cambridge University, England, where he grad-
uated in 1788. He practiced as a physician before moving to London (1801) to devote himself to science, working in a variety of fields, including chemistry, physics, and astronomy, and making several important discoveries, both theoretical and practical.

Wollaston made himself financially independent by inventing, in 1804, a process to produce pure malleable platinum, which could be welded and made into vessels. He is reported to have made about £30,000 from his discovery, as he kept the process secret until shortly before his death, allowing no one to enter his laboratory. Working with platinum ore, he also isolated two new elements: palladium (1804), named for the recently discovered asteroid Pallas, and rhodium (1805), named for the rose color of its compounds. In 1810 he discovered the second amino acid, cystine, in a bladder stone.

In optics Wollaston developed the reflecting goniometer (1809), an instrument for the measurement of angles between the faces of a crystal. He also patented the camera lucida in 1807. In this device an adjustable prism reflects light from the object to be drawn and light from the paper into the draftsman's eye. This produces the illusion of the image on the paper, allowing him to trace it. Wollaston was a friend of Thomas YOUNG and a supporter of the wave theory of light. One opportunity he missed occurred when, in 1802, he observed the dark lines in the solar spectrum but failed to grasp their importance, taking them simply to be the natural boundaries of colors. He missed a similar chance in 1820 when he failed to pursue the full implications of Hans OERSTED's 1820 demonstration that an electric current could cause a deflection in a compass needle. Although he performed some experiments it was left to Michael FARADAY in 1821 to discover and analyze electromagnetic rotation. Wollaston was successful in showing that frictional and galvanic electricity were identical in 1801. In 1814 he proposed the term "chemical equivalents."

Wolpert, Lewis (1929–) *British embryologist*

Wolpert, who was born at Johannesburg in South Africa, trained initially as an engineer at the University of Witwatersrand. After working as an engineer in Britain and Israel, Wolpert's interests turned to biology and he began to study for a PhD in embryology at King's College, London. He taught there from 1958 until 1966, when he was appointed professor of biology at the Middlesex Hospital Medical School. From 1987 he has worked at University College, London.

Wolpert has worked mainly on the problem of pattern formation in biological development. How is it, for example, that the same differentiated cells – muscle, cartilage, skin, connective tissue – arrange themselves as legs in one place and arms in another. Francis CRICK proposed in 1970 that patterns could be produced through the action of a "morphogen," a substance whose concentration throughout the field could be sensed by individual cells. Wolpert illustrated the mechanism with his flag analogy.

Imagine a line of cells capable of turning blue, red, or white. What simple mechanism could generate the pattern of the French tricolor? One way would be to have a chemical whose concentration, while fixed at one end of the line, decreased along the line. Cells could respond to a certain concentration of the morphogen and turn blue, red, or white accordingly.

Wolpert found that in such a system it should be possible to specify some 30 different cell states along a line of about 100 cells. The limiting factor was the accuracy with which cells can identify thresholds of concentration.

Wolpert identified a second positional system, one dependent upon time. Wing growth in chicks, for example, is mainly due to cell multiplication at the tip of the limb in a region known as the "progress zone." The cells learn their position by responding to the length of time they remain in the progress zone. Thus the cells which stay in the progress zone the shortest time form the humerus, and those that stay in the longest develop into digits.

Much recent work in developmental biology was described by Wolpert in his *The Triumph of the Embryo* (Oxford, 1991), material that was originally presented in his 1986 Royal Institution Christmas Lectures. Wolpert has also taken it upon himself to speak for science in such works as his *The Unnatural Nature of Science* (London, 1993), against what he saw as increasingly philistine attacks from journalists and politicians against the aims and methods of modern science.

Wood, Robert Williams (1868–1955) *American physicist*

Wood was born in Concord, Massachusetts, and majored in chemistry at Harvard, graduating in 1891. However, during his postgraduate research (1891–96) at Johns Hopkins University, Baltimore, and the universities of Chicago and Berlin, he turned toward physics. In 1897 he started teaching at the University of Wisconsin and four years later was appointed professor of experimental physics at Johns Hopkins. On his retirement in 1938 he was reappointed research professor.

Wood had wide-ranging interests in science and technology but his major contributions were in optics and spectroscopy. At Johns Hopkins he started his lifelong study of the optical properties of fluorescent gases and his spectroscopic data formed much of the experimental foundation of the model of the atom put forward by Niels BOHR in 1913.

Wood liked experimenting with slightly mysterious things that would grip the public imagination. This predilection led to his work on infrared and ultraviolet photography and he invented a filter, since named *Wood's glass*, that is almost opaque to visible light but lets through the ultraviolet. As well as investigating "invisible light" during World War I, he became interested in "inaudible sounds," especially in the biological effects of these ultrasonic vibrations. His work on ultrasonics aroused considerable interest in the subject. His book *Physical Optics* (1905) became a standard work on the subject and he was also well known as a popular writer and lecturer.

Woodward, Sir Arthur Smith (1864–1944) *British paleontologist*

Woodward, the son of a silk dyer from Macclesfield in northwest England, was educated at Owens College, Manchester. In 1882 he was appointed to the staff of the department of geology at the British Museum (Natural History), where he remained until 1924, being made keeper in 1901.

Woodward was a most conscientious worker with over six hundred publications to his credit. His most substantial work was on the topic of fossil fish, with his catalog of the Natural History Museum collection (4 vols., 1889–1901) his most solid achievement. However, despite his devotion to such an esoteric subject, he became known to a wide public for a discovery made by him and some friends in 1912.

While exploring the gravels of the Sussex Ouse with Charles Dawson they discovered the remains of Piltdown man. By the end of the year Woodward was ready to describe the skull to a meeting of the Geological Society. It possessed, he declared, a cranium like that of modern man with a jaw similar to an ape's, in which were found two molar teeth with a marked regular flattening that had not been observed among apes. Without the molars, Woodward stated, it would be impossible to tell the jaw was human. The evidence of geology made Piltdown older than Neanderthal man, thus leading Woodward to his main conclusion that *Eoanthropus dawsoni*, as he named it, was the true ancestor of modern man.

Woodward never wavered in his view. He in fact died some years before the exposure of the Piltdown fraud by Kenneth OAKLEY in 1953.

Woodward, John (1665–1728) *British geologist*

Little is known of Woodward's early life except that he was born in England and apprenticed to a draper; he later came to the attention of a physician to Charles II who organized his education. In 1692 he was appointed professor of physics at Gresham College, London.

In 1695 Woodward published his *Essays Toward a Natural History of the Earth*. Following Thomas BURNET, he attempted to give a clear and naturalistic account of the Earth's history in keeping with the Creation as told in Genesis. Unlike the universal decay that Burnet proclaimed, Woodward saw the deluge as creating the Earth very much as we now know it. Woodward thus had the problem of explaining the facts of geological change. He denied that earthquakes and volcanoes altered the topography and would not accept that valleys and mountains could be worn away. His answer to the obvious signs of denudation was to propose a compensating mechanism by which the materials eroded would be washed into the rivers, picked up by the winds, and fall back onto the mountains.

In other fields Woodward conducted a series of experiments on plant nutrition in 1691 and was the first to show that much of the moisture absorbed by plants is transpired. He also, from 1704 until his death, worked on systems of mineral classification. His final achievement was the creation of a chair of geology at Cambridge University named for him.

Woodward, Robert Burns (1917–1979) *American chemist*

Born in Boston, Massachusetts, Woodward was educated at the Massachusetts Institute of Technology, obtaining his PhD in 1937. His whole career was spent at Harvard where, starting as a postdoctoral fellow in 1937, he became Morris Loeb Professor of Chemistry in 1953.

In 1944 Woodward, with William von Eggers Doering, synthesized quinine from the basic elements. This was an historic moment for it was the quinine molecule that William PERKIN had first, somewhat prematurely, attempted to synthesize in 1855.

Woodward and his school later succeeded in synthesizing an impressive number of molecules, many of which are important far beyond the field of chemistry. Thus among the most important were cholesterol and cortisone in 1951, strychnine and LSD in 1954, reserpine in 1956, chlorophyll in 1960, a tetracycline antibiotic in 1962, and vitamin B_{12} in 1971. The work on the synthesis of B_{12} led Woodward and Roald Hoffman to introduce the principle of conservation of orbital symmetry. This major theoretical advance has provided a deep understanding of a wide group of chemical reactions.

He received the Nobel Prize for chemistry in 1965. Woodward's death in 1979 deprived him of a second Nobel award, namely, the chemistry prize awarded to his colleague HOFFMANN in 1981 for their work on orbital theory.

Woolley, Sir Richard van der Riet (1906–1986) *British astronomer*

Woolley was born in Weymouth on the south coast of England. The son of a rear-admiral, he was educated at the University of Cape Town, where his father had moved on retirement, and at Cambridge. After spending two years at Mount Wilson Observatory, California, he returned to Cambridge in 1931. Shortly afterward he was appointed for the first time to the Royal Observatory, Greenwich, as chief assistant.

From 1933 to 1937 Woolley worked mainly in the solar department and published jointly with F. DYSON *Eclipses of the Sun and Moon* (London, 1937). After a second spell at Cambridge he moved to Australia in 1939 as director of the Commonwealth Solar Observatory, Mt. Stromlo, Canberra. During his long stay he sought to move the observatory into astrophysics and to concentrate its observational work on the southern stars. To this end he installed a 74-inch reflector in 1950.

In 1956 Woolley returned to Greenwich as the 11th Astronomer Royal, the last one to be in charge of the Royal Observatory, which, by 1958, had completed its move from Greenwich to Herstmonceaux, Sussex. One of Woolley's first tasks was to revive plans to build the 100-inch Isaac Newton telescope first proposed in 1946. Under Woolley the telescope was eventually opened in 1967. Woolley retired as Astronomer Royal in 1971 but served a further period in South Africa as Director of the Cape Observatory (1972–76).

For many Woolley will always be remembered for his 1956 judgment that space travel was "utter bilge," just one year before Sputnik 1 was launched.

Worsaae, Jens Jacob Asmussen (1821–1885) *Danish archeologist*

> The actual founder of antiquarian research as an independent science.
> —Johannes Brønsted, describing Worsaae

Worsaae (**vor**-saw), the son of a government official, was born at Vejle in Denmark. Although he was initially trained as a law student his real interest, going back to his schooldays, lay in the collection of antiquities. He consequently joined Christian THOMSEN as an unpaid assistant at the Danish National Museum in 1836 and in 1847 he was appointed inspector of Danish historic and prehistoric monuments. In 1854, as professor of archeology at the University of Copenhagen, he became the first ever paid full-time archeologist. In 1865 he succeeded Thomsen as director of the National Museum, a post he occupied until his death in 1885.

With his *Danmark's Oldtid oplyst ved Oldsager og Gravhøie* (1843), translated into English in 1849 as *Primeval Antiquities of Den-*

mark, Worsaae made the first application of Thomsen's three-age system to actual excavations in the field as opposed to exhibits in museums. He was thus able to show that the Danish grave-hills are usually divisible into three classes, corresponding to the Stone, Bronze, and Iron age periods.

He further argued for the importance of comparative studies, emphasizing the necessity of studying similar monuments in other countries to gain a satisfactory knowledge of Danish memorials. In general, Worsaae is regarded as one of the people responsible for establishing scientific standards in the field of archeology.

Wright, Sir Almroth Edward (1861–1947) *British physician*

Wright was born in Richmond, England, and educated at Trinity College, Dublin, where he received a BA in modern literature (1882) and his bachelor of medicine (1883). He completed his medical studies in London, Europe, and at the University of Sydney, Australia, before being appointed to an army medical school. Here he developed a vaccine against typhoid fever. This later proved to be of considerable strategic importance in combat.

In 1902 Wright began his long association with St. Mary's Hospital, London, as professor of pathology. In 1911 he went to South Africa to develop an antipneumonia inoculation. During World War I he served in France, investigating techniques for treating infected wounds. Several of Wright's students at St. Mary's subsequently achieved fame, including the discoverer of penicillin, Alexander FLEMING. Wright was also noted as a vehement antifeminist and received a public rebuke for this failing from the writer George Bernard Shaw. He was knighted in 1906.

Wright, Orville *See* WRIGHT, WILBUR.

Wright, Wilbur (1867–1912) *American aeronautical engineer*

> For some years I have been afflicted with the belief that flight is possible to man. My disease has increased in severity and I feel that it will soon cost me an increased amount of money if not my life. I have been trying to arrange my affairs in such a way that I can devote my entire time for a few months to experiment in this field.
> —Wilbur Wright, letter to Octave Chanute, 13 May 1900

The sons of a bishop in the United Brethren Church, Wilbur Wright was born in Millville, Indiana, and Orville (1871–1948) in Dayton, Ohio. Neither brother received more than a high-school education. They had, however, shown a certain inventiveness and an interest in things mechanical. They were the kind of boys who having seen a woodcut in a magazine would immediately make their own woodcut. Thus on leaving school they first experimented with printing, publishing the weekly *West Side News* for over a year. By 1892 they had lost their interest in printing and decided instead to open a bike shop in which they not only sold and repaired bikes but made them themselves.

They later reported that their interest in flight had been stimulated by reading in 1896 of the death of the German engineer LILIENTHAL in a gliding accident. They first devoured the available literature describing the machines and flights of Lilienthal, Samuel LANGLEY, and others. Above all they were struck by the lack of control mechanisms in the early machines. The early designers had merely sought to maintain equilibrium, but the Wrights saw that flying meant directing and upsetting equilibrium in a carefully controlled manner. A specific control mechanism was suggested to them by observing the flight of pigeons, and how they maintained their balance by twisting their wing tips. A comparable effect could be achieved in a plane by warping the wings' ends. But how could this be produced? The answer came to Wilbur when, while fiddling with a narrow rectangular box, he noted how easily the ends could be twisted in opposite directions.

The principle was incorporated in a biplane kite which they tested in 1900 on the sandhills at Kitty Hawk, North Carolina. A larger kite was tested in 1901 and in 1902 further data was collected from trials in a wind tunnel constructed by the brothers. One result of this work was the installation of a vertical tail on the 1902 glider. At this point they considered converting their kite-glider to a powered aircraft. Characteristically they designed and built their own 12 horsepower model and fitted two propellers with a diameter of 8.5 feet. Wilbur piloted the first flight on 14 December 1903 but induced a stall; during the second flight, on 17 December, Orville covered 120 feet at an average speed of 7 miles per hour. The plane, known as "the Flyer," was damaged in a later flight that day and was never flown again; it was later placed as a permanent exhibit at the Smithsonian, Washington, DC.

The brothers continued to work on their design and only when completely satisfied with a new version of the Flyer were they prepared to demonstrate powered flight to the public. Wilbur first flew publicly near Le Mans in France in August 1908, and Orville a few days later at Fort Meyer, Virginia.

In 1909 they set up the Wright Company, with considerable financial backing, to build versions of the Flyer. They also received license fees from European manufacturers. Much of their time, however, must have been spent in patent disputes which dragged on in one form or another until 1928. Wilbur died from typhoid fever in 1912. Orville sold the business in 1915 for a sum said to be $1.5 million, while remaining as a consultant for $25,000 a year. Much of his later life was spent ensuring the contribution of the Wright brothers to the early history of aviation was properly recognized. He died of a heart attack in 1948.

Wright, Sewall (1889–1988) *American statistician and geneticist*

Born in Melrose, Massachusetts, Wright graduated from Lombard College in 1911; he gained his master's degree from the University of Illinois the following year and his doctorate from Harvard in 1916. He then worked as senior animal husbandman for the U.S. Department of Agriculture and began his researches into the population genetics of guinea pigs. His first work aimed to find the best combination of inbreeding and crossbreeding to improve stock, this having practical application in livestock breeding. From this he also developed a mathematical theory of evolution.

His name is best known, however, in connection with the process of genetic drift, which is also termed the *Sewall Wright effect*. He demonstrated that in small isolated populations certain forms of genes may be lost quite randomly, simply because the few individuals possessing them happen not to pass them on. The loss of such characters may lead to the formation of new species without natural selection coming into operation. Wright held professorial positions at the University of Chicago and Edinburgh University and was emeritus professor at the University of Wisconsin.

Wrinch, Dorothy (1894–1976) *British–American mathematician and biochemist*

Wrinch was born at Rosario in Argentina and educated at Cambridge University, where she held a research fellowship from 1920 to 1924. She then taught physics at Oxford until 1939, when she moved to America to take up an appointment as lecturer in chemistry at Johns Hopkins University. In 1942 she moved to Smith College, remaining there until her retirement in 1959.

In 1934 Wrinch tackled the important problem of identifying the chemical carriers of genetic information. In common with other scientists at that time, she argued that chromosomes consisted of sequences of amino acids; these were the only molecules thought to possess sufficient variety to permit the construction of complex molecules. She proposed a model of the gene in the form of a T-like structure with a nucleic-acid stem and a sequence of amino acids as the cross bar.

In actual fact there were many such models in the 1930s. If it was not accepted that genes were made from specific sequences of amino acids then it became very difficult to see what

they could come from. The trouble with all these models was that the experimentalists quickly found serious defects in them. Thus W. Schmidt in 1936 was able to show that Wrinch's model was incompatible with the known optical properties of nucleic acid and the chromosomes. The first suggestion that there might be an alternative to the protein structure of the gene came with the famous experiment of Oswald AVERY in 1944.

Wróblewski, Zygmunt Florenty von (1845–1888) *Polish physicist*

Wróblewski (vroo-**blef**-skee), who was born at Grodno (now in Belarus), was a student at Kiev. However, his education was interrupted when he was exiled to Silesia for his part in the 1863 uprising against Russia. On his release he studied abroad with Hermann von HELMHOLTZ in Berlin and at Strasbourg before being appointed to the chair of physics at the University of Cracow in 1882. He died in the flames of his laboratory, apparently having knocked over a kerosene lamp while working late.

Wróblewski was one of the 19th-century physicists who worked on the liquefaction of gases following the success of Raoul PICTET and Louis CAILLETET. In 1883 Wróblewski and K. Olszewski developed a method of making liquid oxygen in usable quantities.

Wu, Chien-Shiung (1912–1997) *Chinese–American physicist*

One of the world's leading experimental physicists, Wu (woo), who was born in Shanghai, China, gained her BS from the National Central University of China before moving to America in 1936. Here she studied under Ernest O. LAWRENCE at the University of California, Berkeley. She gained her PhD in 1940, then went on to teach at Smith College, Northampton, Massachusetts, and later at Princeton University. In 1946 she became a staff member at Columbia University, advancing to become professor of physics in 1957.

Her first significant research work was on the mechanism of beta disintegration (in radioactive decay). In particular, she demonstrated in 1956 that the direction of emission of beta rays is strongly correlated with the direction of spin of the emitting nucleus, showing that parity is not conserved in beta disintegration. This experiment confirmed the theories advanced by Tsung Dao LEE of Columbia and Chen Ning YANG of Princeton that in the so-called "weak" nuclear interactions the previously held "law of symmetry" was violated. Yang and Lee later received the Nobel Prize for physics for their theory, and the discovery overturned many central ideas in physics.

In 1958 Richard FEYNMAN and Murray GELL-MANN proposed the theory of conservation of vector current in beta decay. This theory was experimentally confirmed in 1963 by Wu, in collaboration with two other Columbia University physicists.

Wu's other contributions to elementary-particle physics include her demonstration that the electromagnetic radiation from the annihilation of positrons and electrons is polarized – a finding in accordance with DIRAC's theory, proving that the electron and positron have opposite parity. She also undertook a study of the x-ray spectra of muonic atoms. Later in her career she became interested in biological problems, especially the structure of hemoglobin.

Wu, Hsien (1893–1959) *Chinese biochemist*

Born into a scholarly family in Fouchow, China, Wu won a scholarship to the Massachusetts Institute of Technology in 1911 to study naval architecture. A reading of T. H. HUXLEY's "On the physical basis of life" diverted him to the study of biology and chemistry at Harvard, where he completed his PhD in 1919.

While at Harvard, working under Otto Folin, Wu developed important analytical techniques that allowed most clinical tests to be carried out with blood samples as small as 10 milliliters. In 1920 he returned to China to the Peking Union Medical College, being appointed in 1928 professor of biochemistry, a post he continued to hold until the Japanese occupation in 1942. Much of Wu's work was carried out in the field of nutrition. He produced a textbook on the subject, *The Principles of Nutrition* (1929). After the war Wu was appointed director of the Nutrition Institute in Peking. But, soon after, with the rise of communism, Wu decided to settle in America and spent the latter part of his life as professor of biochemistry at the University of Alabama.

Wunderlich, Carl Reinhold August (1815–1877) *German physician*

The son of a physician, Wunderlich (**vuun**-der-lik) was born in Sulz, Germany. He studied medicine at the University of Tübingen, qualifying as a doctor in 1837. After a few years' further study in Paris and Vienna he took up an appointment at the Tübingen clinic in 1839. Wunderlich moved in 1850 to Leipzig University where he served as professor of anatomy until his death.

In 1868 Wunderlich published one of the classic works of modern medicine, *Das Verhalten der Eigenwärme in Krankheiten*, translated into English in 1871 as *On the Temperature in Disease*. He had begun in 1848 to record systematically the temperature of his patients, a most unusual event at the time if only because the thermometers of his day were bulky, uncomfortable, and required a good deal of time to register accurately.

Wunderlich demonstrated two basic principles: constant temperature in health but variation in disease, and the normal range of temperature. More importantly, however, he noted that certain diseases are characterized by a given pattern of changes in temperature. He had thus shown that clinical thermometry could be used as an important additional diagnostic technique.

Wurtz, Charles Adolphe (1817–1884) *French chemist*

Wurtz (voorts) was educated at the university in his native city of Strasbourg. He worked under Justus von LIEBIG in Giessen and under Jean DUMAS in Paris. In 1853 he was appointed professor of chemistry at the Ecole de Médicine until he moved to the chair of organic chemistry at the Sorbonne in 1874.

Wurtz contributed to the development of the type theory of Charles GERHARDT and Auguste Laurente by introducing the ammonia type in 1849. He synthesized ethylamine from ammonia and constructed his ammonia type by substituting the carbon radical C_2H_5 for one or more of the hydrogen atoms in ammonia (NH_3). He thus produced the series ammonia (NH_3); ethylamine ($C_2H_5NH_2$); diethylamine (($C_2H_5)_2NH$); triethylamine (($C_2H_5)_3N$). Other types were added by Gerhardt.

In 1855 Wurtz developed a method of synthesizing hydrocarbons by reacting alkyl halides with sodium (still known as the *Wurtz reaction*). With Rudolph FITTIG he developed a similar reaction for synthesizing aromatic hydrocarbons. In 1860 Wurtz was involved, with August KEKULÉ, in initiating the first conference of the International Chemical Congress at Karlsruhe. He was also involved in the COUPER tragedy. In 1858 Archibald Couper had apparently anticipated Kekulé in working out the structure of the carbon atom and asked Wurtz to present his paper to the Académie des Sciences. Wurtz delayed and Kekulé published. When Couper remonstrated with Wurtz he was expelled from Wurtz's laboratory. Couper had a breakdown on his return to Scotland and never did any serious chemistry again.

Wurtz was a prolific author, his *La Théorie atomique* (1879; Atomic Theory) being his best-known work.

Wüthrich, Kurt (1938–) *Swiss chemist*

Wüthrich (**wu(r)**-thrich) was born in Aarberg, Switzerland, and educated at the University of Bern between 1957 and 1962 and the University of Basel between 1962 and 1964, gaining a PhD in inorganic chemistry in 1964. He did postdoctoral research at the University of Basel (1964–65) and the University of California at Berkeley (1965–67). Between 1967 and 1969 he was a member of the technical staff at Bell

Telephone Laboratories in Murray Hill, New Jersey. He joined the Eidgenössische Technische Hochschule (ETH) in Zürich, Switzerland, in 1969, becoming a professor of molecular biophysics in 1980. Since 2001 he has divided his time between the ETH and the Scripps Research Institute, La Jolla, California, where he is the Cecil H. and Ida M. Green Visiting Professor of Structural Biology.

In the early 1980s Wüthrich found a way of making nuclear magnetic resonance (NMR) a useful technique for the determination of the structures of large molecules such as proteins. This method, called *sequential assignment*, makes it possible to determine the distances between many pairs of hydrogen nuclei, and hence to determine the three-dimensional structure of large molecules. Many protein structures have been determined in this way. Wüthrich has written three monographs on this work. In 2002 he shared the Nobel Prize for chemistry with John FENN and Koichi TANAKA for pioneering work on determining the structures of large molecules.

Wyckoff, Ralph Walter Graystone (1897–1994) *American crystallographer and electron microscopist*

Wyckoff was born in Geneva, New York, and graduated from Hobart College. He obtained his PhD from Cornell in 1919. Between 1919 and 1938 Wyckoff worked first in the Geophysical Laboratory, New York, and then at the Rockefeller Institute before transferring to the Lederle Laboratories in 1938. He then worked at the University of Michigan (1943–45) and the National Institutes of Health (1945–59). He was appointed to the chair of physics and microbiology at the University of Arizona in 1959.

While at the Rockefeller Institute Wyckoff managed to purify various viruses, including that causing equine encephalomyelitis, using an ultracentrifuge. The pure preparations of encephalomyelitis virus were used to develop a killed-virus vaccine which proved effective against the epidemic that was affecting horses in America. This success led to a program for producing typhus vaccine.

In 1944 Wyckoff entered into an unusual and profitable collaboration with the astronomer-turned-biophysicist Robley WILLIAMS. Wyckoff was using the electron microscope to photograph viruses but found, as did other virolo-gists of the time, that the amount of information conveyed about the size and shape of the virus was strictly limited.

Wyckoff discussed with Williams the problem of determining the size of a speck of dust that had fallen onto a specimen and been photographed with it. To an astronomer the solution was obvious, for it is a standard procedure to measure the heights of lunar mountains from the length of the shadow cast by them and knowledge of the angle of the incident light source. The problem was to make viruses cast shadows. They placed the specimen in a vacuum together with a heated tungsten filament covered with gold. This vaporized and coated the side of the specimen nearest the filament, leaving a "shadow" on the far side.

This technique of "metal shadowing" opened a new phase in the study of viruses allowing better estimates to be made of their size and shape, as well as revealing details of their structure.

Wynne-Edwards, Vero Copner (1906–1997) *British zoologist*

Wynne-Edwards, who was born at Leeds in England, graduated in natural science from Oxford University in 1927. After leaving Oxford in 1929 he taught zoology at Bristol University (1929–30) and at McGill University, Montreal (from 1930). He returned to Britain in 1946 and served as professor of natural history at Aberdeen University until his retirement in 1974.

In 1962 Wynne-Edwards published his *Animal Dispersion in Relation to Social Behaviour*, one of the most influential zoological works of the postwar years. Much of it became known to a wider public through the popular writings of Robert Ardrey. In it he put the strongest possible case for group selection, the view that animals sacrifice personal survival and fertility to control population growth, that is, for the good of the group as a whole. They behave, in fact, altruistically.

Thus for Wynne-Edwards all such animal behavior as territoriality, dominance hierarchies, and grouping in large flocks (epideictic behavior) were simply devices for the control of population size. Such views stimulated a strong reaction, forcing his opponents to develop alternative accounts of altruism and population control in as much depth as his own.

X

Xenophanes of Colophon (*c.* 560 BC–*c.* 478 BC) *Greek philosopher*

XENOPHANES (ze-**nof**-a-neez) appears to have moved from his birthplace, Colophon (which is now in Turkey), when the Persians invaded Asia Minor in about 546 BC. First he went to Sicily and then to Elea in southern Italy. Although PARMENIDES is usually considered the founder of the Eleatic school, Xenophanes anticipated his views.

It is Xenophanes who made the first explicit and comprehensive attack on the anthropomorphic view of nature so prevalent in the ancient world. Drawing attention to the relativity of belief he commented that: "The Ethiopians say their gods are snub-nosed and black, the Thracians that theirs have light-blue eyes and red hair." More radical, however, was his claim that if horses and cattle had hands, "they would draw the forms of the gods like horses, and ... like cattle." In place of the anthropomorphic polytheism of Homer he substituted the cryptic "the One is god."

Y

Yalow, Rosalyn Sussman (1921–)
American physicist

Yalow was born in New York City and educated at Hunter College and at the University of Illinois, where she obtained her PhD in nuclear physics in 1945. Since 1947 she has worked at the Veterans Administration Hospital in the Bronx as a physicist and, since 1968, she has also held the post of research professor at the Mount Sinai School of Medicine.

In the 1950s, working with Solomon Berson, Yalow developed the technique of radioimmunoassay (RIA), which permits the detection of extremely small amounts of hormone. The technique involves taking a known amount of radioactively labeled hormone, together with a known amount of antibody against it, and mixing it with human serum containing an unknown amount of unlabeled hormone. The antibodies bind to both the radioactive and normal hormone in the proportions in which they are present in the mixture. It is then possible to calculate with great accuracy the amount of unlabeled hormone present in the original sample; using this technique, amounts as small as one picogram (10^{-12} g) can be detected.

This technique enabled Roger GUILLEMIN and Andrew SCHALLY to detect the hypothalamic hormones; Yalow, Guillemin, and Schally shared the Nobel Prize for physiology or medicine in 1977.

Yang, Chen Ning (1922–) *Chinese–American physicist*

Yang, who was born the son of a mathematics professor at Hefei in China, graduated from the National Southwest Associated University in Kunming and received an MSc from Tsinghua. A fellowship enabled him to travel to America, where he studied for his PhD at the University of Chicago, under Enrico FERMI. After teaching at Chicago he joined the Institute for Advanced Study, Princeton, becoming professor of physics in 1955. In 1965 he was appointed Einstein Professor of Physics and director of the Institute of Theoretical Physics at the State University of New York, Stony Brook.

Yang collaborated with Tsung Dao LEE, and in 1956 they made a fundamental theoretical breakthrough in predicting that the law of conservation of parity would break down in the so-called weak interactions. Their startling prediction was quickly confirmed experimentally, by Chien-Shiung WU, and in 1957 Yang and Lee were awarded the Nobel Prize for physics.

Yang has also made other advances in theoretical physics. In collaboration with R. L. Mills he proposed a non-Abelian gauge theory – also known as the *Yang–Mills theory* – a mathematical principle describing fundamental interactions for elementary particles and fields. Yang has also made contributions to statistical mechanics.

Yanofsky, Charles (1925–) *American geneticist*

Born in New York City, Yanofsky graduated in chemistry from the City College of New York in 1948 and went on to gain his PhD in microbiology from Yale University in 1951. During three years' postdoctoral work at Yale on gene mutations, he demonstrated that the effect of one mutation may be compensated for by another "suppressor" mutation, which supplies the enzyme lacking, or rendered ineffective, in the original mutant. In 1958 Yanofsky joined Stanford University as associate professor in microbiology. His research at Stanford provided evidence that the linear sequence of amino acid molecules in proteins is determined by the order of nucleotide molecules in the hereditary material (DNA). This concept had been central to genetics since James WATSON and Francis CRICK proposed their molecular structure of DNA, but the theory still remained to be proved.

Yanofsky demonstrated the colinearity of DNA and proteins by using Seymour BENZER's strain of the bacteriophage T4 with the gene rII, rendering it incapable of multiplying on a certain type of the bacterium *Escherichia coli*. Many different mutants of the rII gene were isolated, and by recombination studies the positions of these mutations within the gene were mapped. The amino acid sequence of the enzyme produced by the rII gene was established, as were the sequences of the various mutant forms of the enzyme. It was then seen that the positions of amino acid changes corresponded to the mutant sites on the genetic map, indicating that proteins are indeed colinear with DNA.

Yanofsky has been professor of biology at

Stanford University since 1961 and has received many awards for his research in molecular biology.

Yersin, Alexandre Emile John (1863–1943) *Swiss bacteriologist*
Yersin (yair-**san**) was born at Aubonne, near Lausanne, in Switzerland. He studied medicine at the universities of Marburg and Paris and, in 1888, became an assistant to the bacteriologist Emile ROUX, at the Pasteur Institute, Paris. They collaborated in investigating the toxins produced by the diphtheria bacterium. Yersin also briefly worked under Robert KOCH in Berlin before making the first of several trips to Southeast Asia. In 1894, while based in Hong Kong, he discovered the bacterium responsible for causing bubonic plague, *Pasteurella pestis* (now renamed *Yersinia pestis*). Simultaneously and independently, the same discovery was made by the Japanese bacteriologist, Shibasaburo KITASATO. In the following year Yersin prepared the first antiplague serum.

Yersin made a great contribution to fighting disease in Indochina and helped to found a medical school in Hanoi. He also had a keen interest in the fauna and flora of the region, especially its agricultural crops.

Yoshimasu, Todu (1702–1773) *Japanese physician*
Yoshimasu, also known as Yoshimasu Shusuke (yoh-shee-**mah**-soo shuu-**soo**-ke); was actually named Tamenori; he was born at Hiroshima in Japan and studied medicine in Kyoto where he also practiced. As the 18th century began, Japanese medicine was derived almost entirely from traditional Chinese sources. The particular form prevalent in Japan has become known as "phase energetics" and involved correlating bodily states with the presence in the body of yin–yang and the five elements (earth, fire, wood, metal, and water), all against a background of calendrical and meteorological conditions.

Yoshimasu made clear his opposition to Chinese medicine. The yin and the yang, he insisted, were the "ch'i" (energy) of the universe and had nothing to do with medicine. In the place of such speculative theories, he insisted, observation should rule. Only then could we establish the important correspondence – that between specific symptoms and drugs capable of curing them. To this effect Yoshimasu published his *Ruijuho* (Classified Prescription; 1764); it contained 220 prescriptions, all derived from Chinese sources, reportedly based upon observation.

Yet despite his insistence upon observation in therapeutics, Yoshimasu could be as dogmatic as any traditionalist in his diagnostics. Thus in an earlier work, *Idan* (New Perspective on Medicine; 1759), Yoshimasu claimed that all disease originated in the abdomen and could be attributed to a single toxic principle.

Young, James (1811–1883) *British chemist*
Young was the son of a carpenter from Glasgow in Scotland. He worked with his father by day and attended the classes of Thomas GRAHAM at Anderson's College, Glasgow, by night. There he made friends with Lyon PLAYFAIR and the explorer David Livingstone, two fellow students. He became Graham's assistant in 1832 and moved with him when he took up a post at University College, London (1837).

Young started his industrial career as a manager with James MUSPRATT at his alkali factory in 1839, moving to Charles TENNANT's factory in Manchester in 1844. While there he heard from Lyon Playfair of oil seepage at Alfreton, Derbyshire. In 1847 Young acquired the mineral rights and was soon producing 300 gallons daily for use in lubrication and lighting. The source was exhausted in 1851 when he moved to Lothian, where once more his source soon ran out.

Rather than rely on oil springs Young pioneered the low-temperature distillation of oil-rich shales and coals to produce yields of paraffin oil. He moved to Glasgow where he produced oils for use in lighting, heating, and lubrication. He sold his business in 1866.

Young's interest in science persisted and in 1881 he tried to determine the speed of light using FIZEAU's technique, obtaining an answer in excess of that of Albert MICHELSON at a little over 300,000 kilometers per second. He also established that blue light travels faster than red.

He gave substantially to Livingstone to finance his travels and also endowed the Young Chair of Technical Chemistry at Anderson's College, becoming its president during the period 1868–77.

Young, Thomas (1773–1829) *British physicist, physician, and Egyptologist*

> Acute suggestion was … always more in the line of my ambition than experimental illustration.
> —Quoted by George Peacock in his *Life of Thomas Young* (1855)

Young, who was born at Milverton in southwest England, was a child prodigy. He could read with considerable fluency at the age of 2 and by 13 he had a good knowledge of Latin, Greek, French, and Italian. He had also begun to study natural history and natural philosophy and could make various optical instruments. A year later he began an independent study of the Hebrew, Chaldean, Syriac, Samaritan, Arabic,

Persian, Turkish, and Ethiopic languages. When he was 19 he was a highly proficient Latin and Greek scholar, having mastered many literary and scientific works including NEWTON's *Principia* and *Opticks* and Antoine LAVOISIER's *Traité élémentaire de chimie* (Elementary Treatise on Chemistry).

In 1793 Young began a medical education, studying first at St. Bartholomew's Hospital, London, and then at the universities of Edinburgh (1794), Göttingen (1795), and Cambridge (1797). In 1800, after receiving a considerable inheritance, he set up a medical practice in London; this practice, however, never really flourished. In 1801 he was appointed professor of natural philosophy at the Royal Institution. Although his lectures were erudite, and remarkable for their scope and originality, they were not successful. They were too technical and detailed for popular audiences and compared unfavorably with those of his colleague, Humphry DAVY. In 1803 Young resigned his post. From then until his death he held various medical appointments and several offices related to science.

Young's early scientific researches were concerned with the physiology of the eye. He was elected a fellow of the Royal Society in 1794 for his explanation of how the ciliary muscles change the shape of the lens to focus on objects at differing distances (accommodation); in 1801 he gave the first descriptions of the defect astigmatism and of color sensation.

Young's most lasting contribution to science was his work in helping to establish the wave theory of light. Between 1800 and 1804 he revived an interest in this theory and gave it strong support. He compared the ideas of Newton and Christiaan HUYGENS on the nature of light, criticizing the corpuscular theory for its inadequacy in explaining such phenomena as simultaneous reflection and refraction. He introduced the idea of interference of light, which he explained by the superposition of waves – a principle that he applied to a range of optical phenomena including Newton's rings, diffraction patterns, and the color of the supernumerary bows of the rainbow. In his best-known demonstration of interference he passed light first through a single pinhole, then through two further pinholes close together; the light then fell upon a screen and gave a series of light and dark bands. The apparatus is known as *Young's slits*. Young's views were very badly received in England, where opposition to Newton's corpuscular theory was unthinkable. During this period Young was savagely and maliciously attacked by the writer and politician (and amateur scientist) Henry Brougham in the fashionable *Edinburgh Review*.

From about 1804 Young devoted himself more to medical practice and the study of philology, especially the decipherment of hieroglyphic writing. He made very important contributions to the latter field and, independently of Jean François Champollion, helped in translating the text of the Rosetta Stone. His interest in optics was revived in about 1816 by the work of François ARAGO and Augustin FRESNEL. In a letter to Arago he suggested that light might be propagated as a transverse wave (in which the vibrations of the medium are perpendicular to the direction of propagation). This allowed polarization to be explained on the wave theory and gave a satisfactory explanation of the known optical phenomena. The decisive test on the nature of light came later when its speed in the air and water could be accurately measured. Young's other scientific contributions included researches into sound, capillarity, and the cohesion of fluids. Because he gave a physical meaning to the constant of proportionality (E) in Hooke's law, E is called *Young's modulus*.

Yukawa, Hideki (1907–1981) *Japanese physicist*

Yukawa (yoo-**kah**-wah) was born Hideki Ogawa at Kyoto in Japan, the son of the professor of geography at the university there; he assumed the name of his wife, Sumi Yukawa, on their marriage in 1932. He was educated at the university of Kyoto and at Osaka, where he joined the faculty in 1933 and where he completed his doctorate in 1938. In the following year Yukawa was appointed professor of physics at Kyoto University, a position he continued to hold until his retirement in 1970.

Yukawa was concerned with the force that binds the neutrons and protons together in the nucleus. At first sight, any nucleus containing more than one proton should be unstable since positively charged particles repel each other; squeezing a number of positively charged protons into the nucleus of an atom should generate powerful repulsive forces. The obvious answer is that there must be another, attractive, force that operates only at short range and holds the nucleons together. Such a force became known to physicists as the "strong interaction."

Yukawa sought to find the mechanism of the strong force and used the electromagnetic force as an analogy. Here the interaction between charged particles is seen as the result of the continuous exchange of a quantum or unit of energy carried by a "virtual particle" – in this case the photon. So, just as electrons and protons interact by exchanging photons, the nucleons interact by exchanging the appropriate particle. Yukawa could predict its mass from quantum theory as the range over which a particle operates is inversely proportional to its mass. The massless photon is thus thought to operate over an infinite distance; as the strong force operates over a distance of less than 10^{-12} cm it must be mediated by a particle,

Yukawa predicted, with a mass of about 200 times that of the electron.

Yukawa made his prediction in 1935 and when two years later Carl ANDERSON found signs of such a particle in cosmic-ray tracks physicists took this as supporting Yukawa's hypothesis and named the particle a mu-meson (now called a muon). However, although the muon had the appropriate mass it interacted with nucleons so infrequently that it could not possibly be the nuclear "glue." Yukawa's theory was saved, however, by the discovery in 1947 by Cecil POWELL, once more in cosmic-ray tracks, of a particle with a mass of 264 times that of the electron and of which the muons were the decay product. The pi-meson, or pion as it became known, interacted very strongly with nucleons and thus filled precisely Yukawa's predicted role.

Z

Zeeman, Pieter (1865–1943) *Dutch physicist*

Born at Zonnemair in the Netherlands, Zeeman (**zay**-mahn) studied at Leiden University and received a doctorate in 1893. This was for his work on the Kerr effect, which concerns the effect of a magnetic field on light. In 1896 he discovered another magnetooptical effect, which now bears his name – he observed that the spectral lines of certain elements are split into three lines when the sample is in a strong magnetic field perpendicular to the light path; if the field is parallel to the light path the lines split into two. This work was done before the development of quantum mechanics, and the effect was explained at the time using classical theory by Hendrik Antoon LORENTZ, who assumed that the light was emitted by oscillating electrons.

This effect (splitting into three or two lines) is called the *normal Zeeman effect* and it can be explained using Niels BOHR's theory of the atom. In general, most substances show an *anomalous Zeeman effect*, in which the splitting is into several closely spaced lines – a phenomenon that can be explained using quantum mechanics and the concept of electron spin.

Zeeman was a meticulous experimenter and he applied his precision in measurement to the determination of the speed of light in dense media, confirming Lorentz's prediction that this was related to wavelength. Also, in 1918, he established the equality of gravitational and inertial mass thus reconfirming EINSTEIN's equivalence principle, which lies at the core of general relativity theory.

Zeeman and Lorentz shared the 1902 Nobel Prize for physics for their work on magnetooptical effects.

Zel'dovich, Yakov Borisovitch (1914–1987) *Russian physicist*

Zel'dovich was born in Minsk, Russia, and educated at the University of Leningrad, graduating in 1931. He then worked at the Institute of Chemical Physics of the Soviet Academy of Sciences, Moscow. He was involved in the Soviet war effort during World War II and subsequently in work on the Soviet hydrogen bomb. After the war, Zel'dovich worked at the Institute of Cosmic Research, which was part of the Space Research Institute, Moscow.

In the 1930s and 1940s Zel'dovich worked on chemical explosives and nuclear explosions. He also made many important contributions to astrophysics and cosmology. For example, he was one of the first people to suggest that quasars are powered by supermassive black holes and that black holes in binary systems could be a source of x-rays because of the matter falling on to them. He was also one of the pioneers who investigated the consequences of the big-bang model of the Universe, such as the cosmic microwave background radiation and the production of helium in the early Universe. He wrote or co-authored a number of influential books and reviews on cosmology and astrophysics and was also influential as a mentor to many astrophysicists and cosmologists in the former Soviet Union.

Zeno of Elea (*c.* 490 BC–*c.* 430 BC) *Greek philosopher*

Zeno (**zee**-noh) was born at Elea (now Velia in Italy) and in about 450 BC accompanied his teacher, PARMENIDES, to Athens. There he propounded the theories of the Eleatic school and became famous for his series of paradoxes and his invention of dialectic.

Little survives of Zeno's written work and this only in other authors' writings. He proposed that motion and multiplicity are unreal (thus supporting Parmenides's theories) since assumption of their existence gave rise to contradictory propositions. One of the most famous arguments against plurality and motion is that of Achilles and the tortoise: if the tortoise is given a start in a race against Achilles, when Achilles reaches the tortoise's starting position, the tortoise will have advanced a small way to a new position. Endless repetition of this argument means that Achilles can never overtake the tortoise.

Zeno's paradoxes remained unresolved for about 20 centuries, in fact until the advances in rigor of mathematical analysis (to the development of which these paradoxes may be said to have contributed). These advances included the study of convergent series (infinite series with a finite sum), the invention by Gottfried LEIBNIZ and Isaac NEWTON of calculus, and Georg

CANTOR's theory of the infinite in the 19th century.

Following his return to Elea Zeno died while joining a coup against the tyrant Nearchus.

Zernike, Frits (1888–1966) *Dutch physicist*

Zernike (**zair**-ni-ke), who was the son of mathematics teachers at Amsterdam in the Netherlands, studied at the university there, obtaining a doctorate in 1915. In 1913 he became assistant to the astronomer Jacobus KAPTEYN at the University of Groningen, where he remained until his retirement in 1958, becoming professor of theoretical physics in 1920 and later of mathematical physics and theoretical mechanics.

Zernike's interest centered around optics and, more particularly, diffraction and in 1935 he developed the phase-contrast microscope. This uses the fact that light passing through bodies with a different refractive index from the surrounding medium has a different phase. The microscope contains a plate in the focal plane, which causes interference patterns and thus increases the contrast. For instance, it can make living cells observable without killing them by staining and fixing. The method of phase contrast also allows the detail in transparent objects or on metal surfaces to be observed.

Zewail, Ahmed H. (1946–   ) *Egyptian-- American chemist*

Zewail (ze-**wil**) was born in Damanhur, Egypt, and educated at the University of Alexandria and the University of Pennsylvania, from which he graduated with a PhD in 1974. From 1974 to 1976 he did postdoctoral research at the University of California at Berkeley. Since 1976 he has been based at the California Institute of Technology, where he has occupied the Linus Pauling Chair of Chemical Physics since 1990.

Zewail has been able to study the dynamics of atoms and molecules taking part in chemical reactions in great detail by using ultra-short flashes of laser light. The laser flashes are of such a short duration that they have the same time-scale as chemical reactions. This time-scale is the femtosecond, equal to 10^{-15} seconds. The study of chemical reactions using Zewail's method is often known as *femtochemistry*. Femtochemistry has given information about why certain chemical reactions occur and it also enables the temperature dependence of the speed of chemical reactions to be understood.

In 1889 Svante ARRHENIUS postulated that a molecule will only take part in a chemical reaction if it has sufficient energy to overcome an energy barrier. He was able to explain the temperature dependence of chemical reactions in this way, with an increase in temperature giving molecules more energy, and hence increasing the number of molecules with sufficient energy to overcome the barrier.

It was not possible to give a description of chemical reactions from first principles until the discovery of quantum mechanics. In 1928 Fritz London showed that the energy barrier for molecules could be calculated by the construction of potential energy surfaces, i.e., the potential energy of a set of atoms plotted as a function of the distances between the atoms. The theory of chemical reactions in terms of potential energy surfaces was developed extensively by Henry EYRING and Michael POLANYI in the late 1920s and 1930s. In this theory it was envisaged that the time-scale for processes to occur when the molecule is at the top of the energy barrier is similar to that for molecular vibrations. At that time it was inconceivable that experiments would be able to investigate such short time-scales, i.e., between 10 and 100 femtoseconds.

In the original femtochemistry experiments in the late 1980s the starting substances in a reaction are mixed as beams of molecules in a vacuum chamber. Two pulses are then injected from an ultrafast laser. The first pulse, known as the pump pulse, initiates the chemical reaction by exciting a molecule to a higher excited state, while the second pulse, known as the probe pulse, is used to examine what is happening. It is possible to see how quickly the original molecule is changing by changing the time interval between the two pulses using mirrors. The spectra of the molecules that are obtained in this way are compared with theoretical quantum-mechanical calculations of the processes involved.

Zewail and his colleagues studied many chemical reactions in this way. As well as providing solid experimental evidence for the ideas about the general mechanisms of chemical reactions put forward by Arrhenius, London, Eyring, and Polanyi, femtochemistry provided detailed descriptions of the processes involved, with many interesting surprises. The technique has been extended to investigate surfaces, liquids, and polymers, including many biological systems.

Zewail has won many awards for his pioneering work in femtochemistry, including the Nobel Prize for chemistry in 1999.

Ziegler, Karl (1898–1973) *German chemist*

Ziegler (**tsee**-gler or **zee**-gler) was born at Helsa in Germany, the son of a minister. He received his doctorate from the University of Marburg in 1923 and then taught at Frankfurt, Heidelberg, and Halle before becoming director of the Max Planck Institute for Coal Research in 1943. In 1963 he was awarded the Nobel Prize for chemistry with Giulio NATTA for their discovery of *Ziegler–Natta catalysts*.

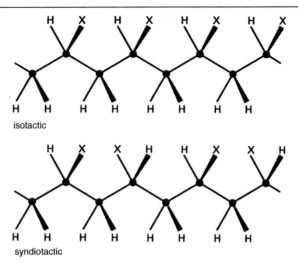

isotactic

syndiotactic

STEREOSPECIFIC POLYMERS   Examples of stereospecific polymers produced by the Ziegler process.

One of the earliest plastics, polyethylene, was simply made by polymerization of the ethylene molecule into long chains containing over a thousand ethylene units. In practice, however, the integrity of the chain tended to be ruined by the development of branches weakening the plastic and endowing it with a melting point only slightly above the boiling point of water.

In 1953 Ziegler introduced a family of catalysts that prevented such branching and produced a much stronger plastic, one which could be soaked in hot water without softening. The catalysts are mixtures of organometallic compounds containing such metallic ions as titanium and aluminum. The new process had the additional advantage that it requires much lower temperatures and pressures than the old method.

Zinder, Norton David (1928–   ) *American geneticist*

Born in New York City, Zinder graduated from Columbia University and did graduate work with Joshua LEDERBERG at the University of Wisconsin. After obtaining his PhD in medical microbiology in 1952 he moved to the Rockefeller Institute (now Rockefeller University), where he has served as professor of genetics since 1964.

While working under Lederberg, Zinder attempted to extend Lederberg's observation of mating (conjugation) in the bacterium *Escherichia coli* to the closely related species *Salmonella*. He obtained large numbers of nutritional mutants of *Salmonella* by developing an effective new and much less laborious screening method involving the antibiotic penicillin. This method utilizes the fact that penicillin only kills growing bacteria and thus any mutant unable to grow on a certain medium would survive the application of penicillin while all the normal cells would succumb.

Having thus obtained his mutants, and maintained them on selectively enriched media, Zinder began looking for conjugation, but instead found a completely new means of genetic transfer in bacteria – bacterial transduction. Investigations revealed that small portions of genetic material are transferred from one bacterium to another via infective bacterial viruses or phage particles. This discovery allowed small portions of the bacterial genome to be investigated and enabled Milislav DEMEREC to show that the genes controlling sequential steps in a given biosynthetic pathway are clustered together in what are now termed "operons."

Zinder's group also discovered the F2 phage, a small virus that has RNA for its genetic material. Studies of the messenger RNA produced by this virus showed that, in addition to the codons for specific amino acids, there are "punctuation" points to signal the initiation and termination of protein chains.

Zinkernagel, Rolf (1944–   ) *Swiss immunologist*

Zinkernagel qualified as an MD at the University of Basle in 1970. He moved to the Australian National University, Canberra, where he gained his PhD in 1975. Zinkernagel returned to Switzerland in 1979 to work at the University of Zurich. In 1992 he was appointed head of the Department of Experimental Immunology.

While in Canberra, when still a graduate student, Zinkernagel worked with Peter DO-HERTY on the role of T lymphocytes and major histocompatibility complexes in fighting infection. For their work in this field they shared the 1997 Nobel Prize for physiology or medicine.

Zinn, Walter Henry (1906–2000) *Canadian–American physicist*

Born at Kitchener in Ontario, Canada, Walter Zinn moved to America in 1930 and received a PhD from Columbia University four years later. He continued research there in collaboration with Leo SZILARD, investigating atomic fission. In 1938 he became a naturalized American citizen.

A year later, Zinn and Szilard demonstrated that uranium underwent fission when bombarded with neutrons and that part of the mass was converted into energy according to EINSTEIN's famous formula, $E = mc^2$. This work led him, during World War II, into research into the construction of the atomic bomb. After the war Zinn started the design of an atomic reactor and, in 1951, he built the first breeder reactor. In a breeder reactor, the core is surrounded by a "blanket" of uranium-238 and neutrons from the core convert this into plutonium-239, which can also be used as a fission fuel.

Zozimus of Panopolis (born *c.* 250) *Greek–Egyptian alchemist*

Little seems to be known about the life of Zozimus (or Zosimos; **zoh**-si-mus or **zoh**-si-mos) except that he was born at Panopolis (now Alchmon in Egypt). He is best known for his writings – a 28-volume encyclopedia of chemical arts. Zozimus showed how alchemy had progressed since the time of Bolos of Mende, a hellenized Egyptian who lived in the Nile Delta. Sulfur, mercury, and arsenic were essential ingredients in this alchemy, mercury being alluded to variously as "divine dew" or "Scythian water." The main aim was the preparation of gold from base metals and its success depended on the production of a series of colors, usually from black to red, yellow, white, black, green, and finally purple. Although the search for the philosopher's stone had not yet begun Zozimus refers to "the tincture," a substance the alchemists believed to exist, that could instantly transform base metals to gold.

Zsigmondy, Richard Adolf (1865–1929) *Austrian chemist*

> A swarm of dancing gnats in a sunbeam will give one an idea of the motion of the gold particles in the hydrosol of gold.
> —*Kolloidchemie* (1912; Colloidal Chemistry)

The son of a Viennese doctor, Zsigmondy (**zhig**-mon-dee) was educated at the universities of Vienna and Munich, where he acquired his PhD in 1890. After periods at the University of Graz and in a glass factory in Jena, in 1908 he became professor of inorganic chemistry at the University of Göttingen, where he remained until his death.

Zsigmondy's first interest was in the chemistry of glazes applied to glass and ceramics. Studies on colored glasses led him into the field of colloids, first distinguished and named by Thomas GRAHAM. Little advance had been made since Graham's time as it was not clear how to study them; the conventional microscope was not powerful enough to detect the particles. In 1903 Zsigmondy remedied this when, in collaboration with Henry Siedentopf, he invented the ultramicroscope in which the particles were illuminated with a cone of light at right angles to the microscope. Although still too small to be seen the particles would diffract light shone on them and therefore appeared as disks of light against a dark background. The particles could be counted, measured, and have their velocity and path determined. Zsigmondy published his work in this field in his book *Kolloidchemie* (1912; Colloidal Chemistry). In 1925 he was awarded the Nobel Prize for chemistry for his work on colloids.

Zuckerkandl, Emile (1922–   ) *Austrian molecular evolutionary biologist*

Zuckerkandl (**tsuuk**-er-an-dl) was born in the Austrian capital of Vienna and educated at the University of Illinois and at the Sorbonne, where he gained his PhD in 1959. He worked with Linus PAULING at the California Institute of Technology from 1959 until 1964, when he moved to the CNRS at Montpellier. In 1980 Zuckerkandl returned to California to serve as head of the Linus Pauling Institute, Palo Alto, California.

Zuckerkandl has examined the hemoglobin from a number of animal species. Human hemoglobin is composed of four chains. The beta-chain consists of 146 amino acids. When compared with the beta-chain of a gorilla it differs at just one point, containing arginine where the gorilla has lysine. In contrast horse beta-chain differs at 26 sites and fish hemoglobin has a total lack of overlap.

Zuckerkandl went on to argue that comparison of hemoglobin chains offered a way to measure the rate at which evolution works. Thus comparison of the alpha and beta hemoglobin chains of humans with those of horses, pigs, cattle, and rabbits produced a mean number of differences of 22. If the estimated age of their common ancestor is 80 million years then it can be estimated that there should be a change of one amino acid per seven million years.

Zuckerkandl's approach has been adapted by other workers. Vincent SARICH in 1967 tried to use the protein albumin to establish a molecular clock, as did FINCH and Margoliash with cytochrome c. More recently Allan WILSON has used mitochondrial DNA.

Zuckerman, Solly, Baron Zuckerman
(1904–1993) *British zoologist and educationalist*

> The nuclear world, with all its perils is the scientists' creation; it is certainly not a world that came about in response to any external demand.
> —*Apocalypse Now?* (1980)

Zuckerman, who was born at Cape Town in South Africa, received his education at the university there and at University College Hospital, London. He then spent a year in America at Yale University before joining the faculty of Oxford University (1934) where he remained until the outbreak of World War II. During the war he investigated the biological effects of bomb blasts and later, as adviser to the RAF chief of combined operations, developed the plan for selective bombing of coastal defenses in Europe in preparation for D-Day. He then became Sands Cox Professor of Anatomy at Birmingham (1946–68) and was appointed secretary of the Zoological Society of London in 1955. Zuckerman served as scientific adviser to the secretary of state for defence (1960–66), working in collaboration with Lord Mountbatten, and was chairman of the United Kingdom Natural Resources Committee (1951–64).

Zuckerman did extensive research on vertebrate anatomy, zoology, and endocrinology. As anatomical research fellow of the Zoological Society he carried out important studies of the menstrual cycle of primates (1930), while studies of captive apes, such as hamadryad baboons, resulted in such classic works as *The Social Life of Monkeys and Apes* (1932) and *Functional Affinities of Man, Monkeys, and Apes* (1933). His interest in ethology led him to oppose the Konrad Lorenz–Robert Ardrey view that man's aggressiveness is instinctive, his critique being published as a collection of essays entitled *Man and Aggression* (1968). From 1969 Zuckerman was professor-at-large of the University of East Anglia.

Zumino, Bruno (1923–   ) *Italian-born American physicist*

Zumino (**zoo**-mee-no) was educated at the University of Rome, where he was awarded a PhD in 1945. Between 1953 and 1968 he was at New York University and subsequently held a position at CERN (1968–81). From 1981 he was based at the University of California at Berkeley, retiring in 1984.

Zumino is best known for playing a leading role in the development of *supersymmetry*, a symmetry which combines fermions and bosons into one mathematical structure. In 1974 Zumino and Julius Wess showed that it is possible to construct quantum field theories that incorporate supersymmetry. The concept is a key ingredient in attempts to find a unified theory of all particles and forces. Supersymmetry predicts that all the known elementary particles – quarks, electrons, photons, gluons, etc. – should have supersymmetric partners, with the supersymmetric partners of bosons being fermions and vice versa. It is hoped that these supersymmetry particles will be discovered in future accelerators. In 1976 Zumino and Stanley Deser discovered *supergravity*, a supersymmetric version of general relativity theory. Zumino has also made a number of other important contributions to quantum field theory.

Zuse, Konrad (1910–1995) *German computer engineer*

Zuse (**tsoo**-ze) graduated in engineering from the Technical College in his native city of Berlin in 1935. He immediately started work with the Henschel Aircraft Company in Berlin as a stress analyst. Part of his duties involved solving large numbers of linear equations, a tedious and time-consuming task. Consequently, as he later put it, he sat down one day in 1936 and invented the computer, "out of laziness."

Like George STIBITZ, Zuse worked with binary rather than decimal numbers and abandoned wheels and gears in favor of electromagnetic relays. In this way he avoided the problem of how to carry and borrow numbers. His first model in 1938, the ZI, worked badly. The defects were eliminated in Zuse's second model, the ZII, which was completed in 1939. The models were originally named V1, V2, etc., where V stood for *Versuchsmodall*, or experimental model. He later renamed them ZI, ZII, etc., so that they would not become confused with the V1 and V2 rockets.

Before he could develop his work, war broke out and Zuse found himself drafted. When he offered his computing skills to the Luftwaffe to help design aircraft, he was told that German aircraft were already the best in the world and his services would not be needed. A few months later, however, he was released from military service to work as an engineer in the aircraft industry. He also worked on a new version of his machine, ZIII, completed in 1941, which he used to handle equations describing wing flutter. This was in fact the first "fully functional, program-controlled, general purpose, digital computer." It was also built two years before ENIAC, designed by J. V. MAUCHLY and J. P. ECKERT, was even started. ZIII was destroyed in the bombing of 1944.

By this time Zuse was working on a larger version, the ZIV, completed in 1945. It was removed from Berlin in the final months of the war and hidden in the village of Hopferau in the Bavarian Alps. In 1950 it was installed at the Zurich Technical Institute where at the time it was the only modern computer in continental Europe.

Zuse had earlier attempted to develop an electronic computer, one operating with vacuum tubes. Although the idea was a clear advance on working with relays it had to be abandoned because of the shortage of electron tubes in wartime Germany.

After 1945 Zuse set up his own business, Zuse Apparatebau, a small computer company which was taken over by Siemens in 1956. He also sought from 1945 onward to devise Plankalkul – a universal algorithmic language. He left a full account of his work in his *Der Computer, mein Lebenswerk* (1970; The Computer, My Life's Work).

Zwicky, Fritz (1898–1974) *Swiss–American astronomer and physicist*

Zwicky (**tsvik**-ee), who was born at Varna in Bulgaria, studied at the Federal Institute of Technology, Zurich, where he obtained his BS in 1920 and his PhD in 1922. He moved to America in 1925, working at the California Institute of Technology and the Mount Wilson and Palomar Observatories until his retirement in 1968. He was associate professor of theoretical physics from 1929 to 1942 and professor of astrophysics from 1942 to 1968.

Zwicky worked in various fields of physics, including jet propulsion and the physics of crystals, liquids, and gases. He is, however, better known for his astronomical research. In 1936 he began an important search for supernovas. These are celestial bodies whose brightness suddenly increases by an immense amount as a result of a catastrophic explosion. They had been observed over several centuries in our Galaxy and one had been detected in the Andromeda galaxy as long ago as 1885. But when Edwin HUBBLE showed in 1923 that the Andromeda galaxy was about 900,000 light-years away, the question arose as to how anything could appear so bright over such a vast distance.

Zwicky worked out their frequency as about three per millennium per galaxy. Although many have passed unobserved in our Galaxy, five supernovas have been reported since AD 1000, including one in 1054 that produced the Crab nebula, Tycho's star in 1572, and KEPLER's star in 1604. Zwicky also showed that supernovas characteristically have an absolute magnitude of −13 to −15, which makes them up to 100 million times brighter than the Sun.

In 1932 Lev LANDAU introduced the concept of a neutron star into astronomy and in 1934 Zwicky and Walter BAADE suggested that these

compact superdense objects might be produced in the cores of supernovas. This was later developed by Robert OPPENHEIMER, G. M. Volkoff, and others in 1939 into an important theory of stellar evolution.

In more recent years Zwicky and his colleagues carefully studied both galaxies and clusters of galaxies. One result of this work is the so-called *Zwicky catalog*, which gives the positions and magnitudes of over 30,000 galaxies and almost 10,000 clusters lying mainly in the northern-hemisphere sky.

Zworykin, Vladimir Kosma (1889–1982) *Russian–American physicist*

Born at Mouron in Russia, Zworykin (**zwor**-i-kin) studied electrical engineering at Petrograd (now St. Petersburg), graduating in 1912. During World War I he served as a radio officer in the Russian army. He moved to America in 1919 and joined the Westinghouse Electric Corporation in 1920. He did graduate research at Pittsburgh University, receiving a PhD in 1926. In 1929 he joined the Radio Corporation of America. Zworykin made a number of contributions to electron optics and was the inventor of the first electronic-scanning television camera – the iconoscope.

The first such device was constructed at Westinghouse in 1923. The principle was to focus an image on a screen made up of many small photoelectric cells, each insulated, which developed a charge that depended on the intensity of the light at that point. An electron beam directed onto the screen was scanned in parallel lines over the screen, discharging the photoelectric cells and producing an electrical signal.

Zworykin also used the cathode-ray tube invented in 1897 by Karl Ferdinand BRAUN to produce the image in a receiver. The tube (which he called a "kinescope") had an electron beam focused by magnetic and electric fields to form a spot on a fluorescent screen. The beam was deflected by the fields in parallel lines across the screen, and the intensity of the beam varied according to the intensity of the signal. In this way it was possible to reconstruct the electrical signals into an image. In 1923 an early version of the system was made and Zworykin managed to transmit a simple picture (a cross). By 1929 he was able to demonstrate a better version suitable for practical use.

Zworykin also developed other electron devices, including an electron-image tube and electron multipliers. In 1940 he invited James HILLIER to join his research group at RCA, and it was here that Hillier constructed his electron microscope.

Chronology

***c.* 590 BC**
Physics
Thales proposes that water is fundamental principle of all matter

***c.* 546 BC**
Physics
Anaximenes proposes air as primal substance

***c.* 540 BC**
Mathematics
Pythagorean mathematics

***c.* 500 BC**
Physics
On Nature (Heraclitus) – doctrine of change

***c.* 480 BC**
Physics
On Nature (Anaxagoras)

***c.* 465 BC**
Mathematics
Zeno proposes his paradoxes

***c.* 450 BC**
Physics
Empedocles proposes four-element theory (earth, air, fire, water)

***c.* 400 BC**
Medicine
Corpus Hippocraticum
Physics
Democritus expounds atomism

***c.* 387 BC**
Philosophy
Plato founds Academy in Athens

***c.* 300 BC**
Physics
Epicurus revives atomism

***c.* 3rd century BC**
Mathematics
Elements (Euclid)

***c.* 200 BC**
Mathematics
Conics (Apollonius of Perga)

***c.* 100 BC**
Earth Science
Poseidonius measures circumference of Earth

56 BC
General
De rerum natura (Lucretius)

***c.* AD 62**
Technology
Hero describes a simple steam engine

***c.* AD 77**
Medicine
De materia medica (Dioscorides)

***c.* 150**
Astronomy
Almagest (Ptolemy) – describes Ptolemaic system with Earth at center of universe

***c.* 175**
Biology
Galen shows that the arteries carry blood not air

***c.* 250**
Mathematics
Arithmetica (Diophantus) – contains Diophantine equations

***c.* 464**
Mathematics
Ch'ung-Chih Tsu calculates the value of pi

628
Astronomy
Brahmasiddhanta (The Opening of the Universe; Brahmagupta)

***c.* 800**
Chemistry
Geber introduces idea that all metals are formed from mercury and sulfur

***c.* 1020**
Medicine
Al Qanun (The Canon; Avicenna)

***c.* 1086**
Physics
Shen Kua describes discovery of magnetic compass

1459
Astronomy
Purbach publishes his table of lunar eclipses

1492
General
Columbus lands in Bahamas (Oct. 12)

1535
Mathematics
Tartaglia discovers a method to solve the general cubic equation

1536
Medicine
Paré shows that the practice of cauterizing wounds is harmful

1543
Astronomy
De revolutionibus orbium coelestium (Copernicus) – heliocentric system of universe
Medicine
De humani corporis fabrica (Vesalius)

1545
Mathematics
Ars magna (Cardano) – introduces the solutions for the general cubic and general quartic equations

1546
Medicine
Fracastoro publishes a work in which he anticipates the germ theory of disease

1553
Biology
Servetus describes the lesser circulation

1555
Biology
Belon publishes a work on birds in which he draws attention to homologies between bird and human skeletons

1556
Chemistry
De re metallica (Agricola)

1561
Biology
Observationes anatomicae

(Fallopius) – describes the fallopian tubes

1564
Biology
Opuscula anatomica (Eustachio) – describes the eustachian tubes of the ear and also the first description of the adrenal glands

1569
Chemistry
De mineralibus (Albertus Magnus)
Earth Science
Mercator introduces a cartographic projection (Mercator's projection) that allows navigators to chart courses on a straight line

1570
Earth Science
Theatrum orbis terrarum (Ortelius) – first modern atlas of world

1572
Physics
Publication of Latin translation of Alhazen's work on optics as *Opticae thesaurus*

1573
Astronomy
De nova stella (Tycho Brahe)

1581
Physics
Norman publishes discovery of magnetic dip

1596
Astronomy
D. Fabricius discovers first variable star (Mira Ceti)
Medicine
Pen Tshao Kang Mu (The Great Pharmacopoeia; Shih-Chen Li) – published posthumously

1597
Biology
Herball (Gerard)

1600
Physics
De magnete (Gilbert) – introduces idea that Earth acts as a large bar magnet

1603
Biology
De venarum ostiolis (Fabricius ab Aquapendente) – describes the venous system
General
Accademia dei Lincei founded in Rome by Prince Federico Cesi

1604
Physics
Galileo proves falling bodies move with uniform acceleration

1605
General
Advancement of Learning (Bacon)

1606
Chemistry
Alchymia (Libavius)

1608
Physics
Lippershey makes first telescope

1609
Astronomy
Astronomia Nova (Kepler) – first two laws of planetary motion
(to 1610) Galileo makes astronomical observations using telescope – studies Moon and discovers satellites of Jupiter

1612
Medicine
Sanctorius constructs and describes a clinical thermometer

1613
Astronomy
Letters on Sunspots (Galileo)

1614
Biology
Sanctorius publishes results revealing the extent of water loss by perspiration
Mathematics
Mirifici logarithmorum (Napier) – introduction of logarithms

1619
Astronomy
De harmonices mundi (Kepler) – includes his 3rd law of planetary motion

1621
Physics
Snell discovers his law of refraction (not published until 1638 in a work by Descartes)

1622
Biology
Aselli discovers the lacteals
Mathematics
Oughtred invents the slide rule

1623
Astronomy
The Assayer (Galileo)

1627
Astronomy
Kepler publishes the Rudolphine Tables

1628
Biology
De motu cordis (Harvey) – describes the circulation of the blood

1631
Technology
Vernier invents the caliper and scale named for him

1632
Astronomy
Dialogo di Galileo Galilei (Galileo) – exposition of Copernican system of universe

1635
Mathematics
Cavalieri introduces his method of indivisibles

1637
Mathematics
La Géométrie (Descartes) – introduces coordinate geometry

1638
Physics
Discorsi e dimostrazioni... (Galileo) – describes work on motion

1644
Physics
Torricelli constructs first mercury barometer

1650
Earth Science
Geographia generalis (Toscanelli)
Physics
Guericke constructs first vacuum pump

1651
Biology
Rudbeck discovers the vertebrate lymphatic system
Pecquet publishes the first description of the thoracic duct

1654
Physics
Guericke demonstrates pressure of atmosphere using Magdeburg hemispheres

1655
Astronomy
Huygens discovers Titan (satellite of Saturn)
Mathematics
Arithmetica infinitorum (Wallis)
General
Philosophical and Physical Opinions (Margaret Cavendish)

1657
Physics
Huygens designs first
pendulum clock
General
Accademia del Cimento
founded by Ferdinand and
Leopold de Medici
1658
Biology
Swammerdam discovers red
blood corpuscles
1660
Biology
Malpighi describes blood
capillaries in the lungs
Physics
*New Experiments Physico-
Mechanicall, Touching the
Spring of the Air and its
Effects* (Boyle)
General
The Royal Society founded on
November 29. Its first patron
was Charles II and its first
secretary was Henry
Oldenburg
1661
Chemistry
The Sceptical Chymist (Boyle)
1662
Astronomy
Venus in Sole Vista (Venus in
the Face of the Sun; Jeremiah
Horrocks – published
posthumously)
Physics
Explicit statement of Boyle's
law concerning the effect of
pressure on the volume of a
gas
Incorporation of Royal Society
1665
Physics
Newton starts work on
gravitation, calculus, and
optics
Grimaldi publishes discovery
of diffraction of light
(posthumously)
Micrographica (Hooke)
General
*The Philosophical
Transactions*, the first genuine
scientific periodical, is
published by the Royal Society
1668
Astronomy
Newton makes first reflecting
telescope
Biology
Graaf describes the structure
of the testicles

1669
Biology
Historia insectorum (Jan
Swammerdam)
Physics
Bartholin announces his
discovery of double refraction
c. **1669**
Chemistry
Brand discovers phosphorus
1670
Mathematics
Lectiones geometricae (Isaac
Barrow)
Methodis fluxionum (Isaac
Newton) – published
posthumously in 1736
1671
Astronomy
Cassini discovers Iapetus
(satellite of Saturn)
1672
Astronomy
Cassini discovers Rhea
(satellite of Saturn)
Cassegrain invents a
reflecting telescope
Construction of Paris
Observatory completed by the
Académie des Sciences
Physics
*New Theory about Light and
Colours* (Isaac Newton)
1673
Biology
Graaf describes the ovarian
follicles
Physics
Horologium oscillatorium
(Huygens)
1675
Astronomy
Cassini discovers Cassini's
division of Saturn's rings
John Flamsteed appointed
Astronomer Royal by Charles
II and provided with an
observatory at Greenwich
1676
Astronomy
Rømer discovers a procedure
for measuring the speed of
light
Royal Greenwich Observatory
opened
Physics
Mariotte independently
formulates, and refines,
Boyle's law relating pressure
and volume of gas
1678
Physics
Huygens advances his wave
theory of light

1679
Biology
Antony van Leeuwenhoek
announces his discovery of
human spermatozoa
Physics
Hooke announces his law of
elasticity (Hooke's law)
Papin invents pressure cooker
(steam digester)
1682
Astronomy
Halley observes comet now
named for him and calculates
the date of return
1683
Mathematics
Lectiones mathematicae (Isaac
Barrow) – published
posthumously
1684
Astronomy
Cassini discovers Dione and
Tethys (satellites of Saturn)
Biology
Antony van Leeuwenhoek
describes red blood cells
Mathematics
*Nova methodus pro maximis
et minimis* (Leibniz) –
publication of invention of
calculus
1687
Physics
Principia (Newton) – theory of
gravitation and laws of motion
Amontons invents hygrometer
1688
Earth Science
Halley produces
meteorological map
1690
Physics
Traité de la lumière (Huygens)
1698
Technology
Savery's steam engine is
patented
1699
Physics
Amontons does early work on
variation of gas volume with
temperature
c. **1701**
Agriculture
Tull introduces a seed drill
that sows in straight rows
Earth Science
Halley produces magnetic
map
1702
Physics
Amontons invents constant-
volume air thermometer

1704
Physics
Opticks (Newton)
1707
Physics
Papin publishes design for steam engine
1712
Technology
The first of Newcomen's piston-operated steam engines is installed
1714
Physics
Fahrenheit invents mercury thermometer
1715
Mathematics
Taylor publishes his work containing Taylor's theorem
1717
Medicine
Lancisi draws attention to the connection between swamps, mosquitoes, and malaria
1718
Medicine
Petit invents the screw tourniquet
1721
Astronomy
John Hadley produces his first reflecting telescope
1725
Astronomy
Historia coelestis Britannica (Flamsteed)
1727
Biology
Vegetable staticks (Hales) – observations on plant physiology
1729
Astronomy
Bradley discovers aberration of light
1733
Medicine
Haemastaticks (Hales) – observations on circulation of the blood, including measurements of blood pressure
Physics
Du Fay introduces two-fluid theory of electricity (vitreous and resinous)
1735
Biology
Systema Naturae (Linnaeus) – systematic classification of animal, plant, and mineral kingdoms

Earth Science
Concerning the Cause of the General Trade Winds (Hadley) – introduces idea of Hadley cell
c. 1740
Biology
Bonnet discovers parthenogenesis in aphids
1742
Chemistry
Brandt isolates cobalt
1743
Mathematics
Traité de dynamique (D'Alembert)
Simpson's rule formulated by Thomas Simpson
Physics
Théorie de la figure de la terra (Theory of the Shape of the Earth; Alexis Claude Clairaut) – confirms that the Earth is shaped like an oblate spheroid
1744
Physics
Maupertuis formulates the principle of least action
1746
Physics
Musschenbroek reports his discovery of the Leyden jar
1747
Chemistry
Marggraf discovers sucrose in beet
1749
Biology
Linnaeus introduces binomial nomenclature in taxonomy
1751
Astronomy
Lacaille and Lalande observe lunar parallax
Biology
Whytt describes and emphasizes the importance of reflex reactions
Chemistry
Cronstedt discovers nickel
Earth Science
Guettard observes evidence for ancient volcanic activity
Technology
La Condamine first reports existence of rubber in Europe
1752
Physics
Franklin performs his electrical experiment with a kite in a thunderstorm
1753
Biology
Species Plantarum (Linnaeus)

1754
Medicine
Lind publishes a cure for scurvy
1756
Chemistry
Experiments upon Magnesia Alba, Quicklime, and some other Alcaline Substances (Black)
1757
Biology
Monro publishes a work distinguishing clearly between the lymphatic and circulatory systems
Physics
Black introduces concept of latent heat
Carton observes fluctuations in Earth's magnetic field
1758
Physics
Dollond invents achromatic lens system
Alexis Claude Clairaut informs the Académie des Sciences in November that Halley's Comet would be at perihelion on April 13, 1759 (the actual date was March 13)
1759
Biology
Wolff presents the thesis that destroyed the preformation theory of development and held instead that organs develop from undifferentiated tissue
c. 1760
General
The Lunar Society formed by Erasmus Darwin, Mathew Boulton, and William Small in Birmingham, England
1761
Astronomy
Nathaniel Bliss observes a transit of Venus enabling him to calculate the horizontal parallax of the Sun as 10.3″
Medicine
Auenbrugger publishes his work on chest percussion
De sedibus et causis morborum (Morgagni)
Physics
Black measures latent heat of fusion of ice
1763
General
The Royal Society of Edinburgh is founded

1764
Technology
Watt repairs a model of a Newcomen steam engine and begins to develop his own improved engine
Hargreaves invents the spinning jenny
1765
Astronomy
Nevil Maskelyne, fifth Astronomer Royal, first publishes the *Nautical Almanac*, which uses Greenwich as its prime meridian
Physics
Cavendish formulates concept of specific heat (unpublished)
1766
Physics
Three Papers containing Experiments of Factitious Airs (Cavendish) – distinguishing hydrogen (inflammable air) and carbon dioxide (fixed air) from common air
1767
Biology
Spallanzani disproves theory of spontaneous generation in microorganisms
1768
Earth Science
Desmarest shows that not all rocks are sedimentary
James Cook sets sail in the *Endeavour* for the South Pacific to observe the planet Venus and chart the coast of New Zealand and E. Australia
Mathematics
Lambert publishes his proof that π and e are irrational
1769
Technology
Arkwright patents his spinning machine
1770
Technology
Cugnot builds a two-piston steam boiler for military use, probably the first fuel-driven vehicle
1771
Earth Science
(to 1775) Werner develops his theory of neptunism
1772
Astronomy
Bode publishes Bode's law
Chemistry
Lavoisier experiments on combustion

1774
Chemistry
Gahn discovers manganese
Priestley discovers oxygen
1776
Anthropology
De generis humani variatate nativa (Blumenbach) – classifies humans into five races
Chemistry
(to 1777) *Éléments de chimie théorique et pratique* (Guyton de Morveau) – an attempt to quantify chemical affinity
1777
Chemistry
Chemical Observations and Experiments on Air and Fire (Scheele)
1779
Botany
Ingenhousz shows that plants absorb carbon dioxide and give off oxygen (photosynthesize) in the presence of light
1781
Astronomy
Herschel discovers Uranus
Chemistry
Experiments on Air (Cavendish) – showing that air is oxygen and nitrogen in ratio 1:4 and water is formed of hydrogen and oxygen in ratio 2:1
Hauy discovers the geometrical law of crystallization
1782
Astronomy
Goodricke proposes that Algol is an eclipsing binary
Chemistry
Hjelm discovers molybdenum
Müller first isolates the element tellurium
1783
Chemistry
Méthode de nomenclature chimique (Lavoisier and others) – proposes new names for elements
Lavoisier experiments on formation of water from hydrogen and oxygen
D. F. and J. J. D'Elhuyar discover tungsten
Physics
Saussure invents hair hygrometer
Technology
Cort patents method of rolling iron into bars using grooved rollers
The Montgolfier brothers

demonstrate the first hot-air balloon on June 4 at Annonay
In October the first manned flight in a hot-air balloon is made by François de Rozier in Paris
1784
Astronomy
Messier publishes first nebula catalog
Earth Science
Elements of Mineralogy (Kirwan)
Technology
Cort patents dry-puddling process for converting pig iron into wrought iron
1785
Medicine
Withering publishes a report on the use of the foxglove as a source of the drug digitalis
Physics
Coulomb publishes his inverse square law of electrical attraction and repulsion (Coulomb's law)
Technology
Cartwright's power loom is patented
1786
Astronomy
Discovery of Encke's Comet (orbit computed by Encke in 1819)
1787
Astronomy
Herschel discovers Titania and Oberon (satellites of Uranus)
Chemistry
Hope isolates strontium
Physics
Charles discovers law governing change in volume of a gas with temperature (Charles's law)
1788
Biology
Linnean Society founded
Chemistry
Blagden discovers that lowering of freezing point of a solution is proportional to amount of solute (Blagden's law)
Earth Science
Hutton first publishes his uniformitarian theory
Physics
Mécanique analytique (Lagrange)
1789
Astronomy
(to 1790) Herschel discovers

Mimas and Enceladus (satellites of Saturn)
Biology
Genera Plantarum (Jussieu) – classification of plants
Chemistry
Klaproth discovers zirconium and uranium
Traité élémentaire de chimie (Lavoisier) – states law of conservation of mass

1791
Biology
Galvani publishes his findings in electrophysiology
Chemistry
Nicolas Leblanc patents his process for producing soda ash
Gregor discovers titanium (not identified as new element until 1795 when isolated by Klaproth)
Physics
Prévost puts forward his theory of exchanges for heat

1794
Biology
Young explains accommodation of the eye
Medicine
Dalton describes the condition of color blindness

1795
Chemistry
Klaproth rediscovers titanium
Earth Science
Theory of the Earth (Hutton)

1796
Medicine
Jenner first uses cowpox vaccine

1798
Biology
An Essay on the Principle of Population (Malthus)
Chemistry
Klaproth discovers chromium
Vauquelin discovers chromium and beryllium
Physics
Cavendish measures density of Earth by torsion-balance experiment
Enquiry concerning the Source of Heat excited by Friction (Rumford)

1799
Astronomy
(to 1825) *Traité de mécanique céleste* (Laplace)
Chemistry
Proust formulates his law of definite proportions

Mathematics
Gauss's proof of fundamental theorem of algebra

1799
General
The Royal Institution is founded by Count Rumford at 21 Albemarle St., London

1800
Chemistry
Nicholson discovers electrolysis
Physics
Volta invents voltaic pile – first battery

1801
Astronomy
Piazzi discovers Ceres (first minor planet)
Chemistry
Del Rio discovers vanadium (rediscovered by Sefström in 1830)
Dalton reads his paper on the constitution of gas mixtures, which contains the law of partial pressures
Clément and Désormes discover carbon monoxide
Hatchett discovers the element now known as niobium (originally named columbium)
Henry formulates his law for the solubility of gases (Henry's law)
Earth Science
Matthew Flinders sets sail in the *Investigator* to chart the coast of Australia
Mathematics
Disquisitiones arithmeticae (Gauss)
Physics
Ritter identifies ultraviolet radiation
Wollaston shows that frictional and galvanic electricity are the same

1802
Astronomy
Olbers discovers asteroid Pallas
Chemistry
Ekeberg discovers tantalum
Dalton compiles his first table of atomic weights
Gay-Lussac establishes Gay-Lussac's law
Earth Science
Illustrations of the Huttonian Theory (J. Playfair) – summarizes Hutton's uniformitarian theory of the Earth

1803
Chemistry
Klaproth discovers cerium oxide
Hisinger and Berzelius discover cerium
Dalton expounds his atomic theory
(to 1804) S. Tennant discovers iridium and osmium
Physics
Young provides evidence for the wave nature of light by demonstrating interference using Young's slits

1804
Chemistry
Wollaston discovers palladium

1805
Chemistry
Wollaston discovers rhodium
Medicine
Sertürner isolates morphine from opium

1806
Biochemistry
Vauquelin isolates the first amino acid – asparagine
Earth Science
Beaufort proposes scale for wind speed (Beaufort scale)

1807
Astronomy
Olbers discovers asteroid Vesta

1808
Biology
(to 1813) *American Ornithology* (Wilson)
Chemistry
A New System of Chemical Philosophy (Dalton)
Gay-Lussac formulates his law of combining volumes
Physics
Malus describes the polarization of light

1809
Biology
Philosophie zoologique (Lamarck) – includes theory of evolution by inheritance of acquired characteristics
Physics
Chladni demonstrates the patterns (Chladni's figures) formed by vibrations of a plate

1810
Chemistry
Davy establishes that chlorine does not contain oxygen and is an element in its own right
Physics
Zur Farbenlehre (On the Science of Colors; Goethe)

1811
Biology
New Idea of Anatomy of the Brain (Charles Bell)
Chemistry
Avogadro publishes the paper containing Avogadro's hypothesis
Courtois discovers iodine (published 1813)
Dulong discovers nitrogen trichloride
Earth Science
Brongniart and Cuvier first identify rock strata by their fossil content
Medicine
Bell partly anticipates Magendie in showing functional differentiation of spinal nerve roots (Bell–Magendie law) but does not publish
Physics
Arago discovers chromatic polarization
1812
Earth Science
Mohs introduces a scale of hardness for minerals (Mohs scale)
Medicine
Parkinson establishes perforation as the cause of death in appendicitis
1813
Biology
Théorie élémentaire de la botanique (Candolle) – system of classifying plants
Chemistry
Clément and Désormes confirm Courtois's discovery of iodine
Physics
Brewster formulates a law for polarization by reflection (Brewster's law)
1814
Physics
Fraunhofer observes absorption lines (Fraunhofer lines) in the solar spectrum
1815
Biology
(to 1822) *Natural History of Invertebrates* (Lamarck) – reorganizes the classification of invertebrate animals
Chemistry
Biot shows that solutions (as well as solids) can show optical activity
Prout suggests that the atomic weight of any atom is an exact multiple of the atomic weight

of hydrogen (Prout's hypothesis)
1816
Chemistry
The Davy lamp is produced
Medicine
Laënnec invents the stethoscope
Technology
Stirling invents the Stirling engine
1817
Biology
Pelletier and Caventou discover chlorophyll
Chemistry
Strohmeyer discovers cadmium
Berzelius discovers selenium (to 1820)
Caventou and Pelletier isolate the alkaloids strychnine, brucine, cinchonine, quinine, veratrine, and colchicine
Medicine
Parkinson publishes his description of Parkinson's disease
1818
Chemistry
Discovery of lithium by J.A. Arfvedson
1819
Biology
Chamisso discovers the alternation of generations in mollusks and tunicates
Chemistry
Mitscherlich formulates his law of isomorphism
Dulong and Petit discover the law of atomic heats (Dulong and Petit's law)
Kidd and Garden obtain naphthalene from coal tar
Physics
Clément and Désormes publish paper on principal specific heats of gases
1820
Physics
Arago demonstrates electromagnetic effect to French Academy of Sciences
Oersted publishes his discovery of electromagnetism
Schweigger invents multiplier (galvanometer)
Ampère begins his work on electromagnetism
Faraday begins work on electromagnetism
1821
Physics
Faraday constructs simple electric motor

1822
Biology
Magendie (and, earlier, Charles Bell) discovers that the roots of the anterior nerves of the spinal cord control motion, while the roots of the posterior nerves control sensation
Mathematics
Théorie analytique de chaleur (Fourier) – introduces the use of Fourier series
Medicine
Magendie publishes his work on alkaloids, introducing such compounds as morphine, strychnine, and quinine into medical practice
Physics
Latour discovers the critical state
Seebeck discovers thermoelectric effect (Seebeck effect)
1823
Mathematics
Babbage designs his analytical engine (computer)
Technology
Macintosh produces a flexible waterproof material
1824
Biology
Flourens publishes his work demonstrating the major roles of different parts of the central nervous system
Physics
Réflexions sur la puissance motrice du feu (Carnot) – introduces Carnot cycle
1825
Chemistry
Faraday discovers benzene
Balard discovers bromine
Medicine
Bretonneau performs the first successful tracheotomy to treat diphtheria
Physics
Nobili invents astatic galvanometer
Ampère deduces law for force between current-carrying conductors (Ampère's law)
Technology
Stockton–Darlington railway opened (September 27)
1826
Astronomy
Olbers proposes his paradox
Discovery of Biela's comet
Biology
Müller shows that sensory

nerves can only interpret stimuli in one way

Chemistry
Unverdorben synthesizes aniline (which he names crystalline)

Mathematics
Lobachevsky announces his non-Euclidean geometry

Technology
Niepce produces the first permanent camera photograph

General
Michael Faraday introduces the Christmas lectures for children at the Royal Institution, London. He also begins the Friday evening discourses, in which leading scientists speak about their work to Royal Institution members

1827

Biology
Baer publishes his discovery of the mammalian ovum within the Graafian follicle (to 1838) *Birds of America* (Audubon)

Chemistry
Brownian motion first described, by Robert Brown
Wöhler synthesizes aluminum

Physics
Ohm publishes his law relating current and voltage (Ohm's law)
Mémoires sur la théorie mathématique des phénomènes electrodynamiques uniquement déduite de l'expérience (Ampère)

Technology
Fourneyron constructs his first outward-flow turbine, a six-horsepower unit

1828

Chemistry
Wöhler isolates beryllium
Berzelius discovers thorium
Wöhler synthesizes urea, the first organic compound to be synthesized from inorganic materials

Physics
Sturgeon constructs first electromagnet
Invention of Nicol prism

1829

Chemistry
Döbereiner formulates his law of triads
Graham publishes a paper on the diffusion of gases, which

contains his law of diffusion (Graham's law)

Physics
Nobili makes first thermopile
Babinet first suggests that wavelengths of spectral lines could be used as standards of length
Henry invents first practical electric motor

1830

Chemistry
Sefström discovers vanadium

Earth Science
The Principles of Geology (Lyell) – establishes Hutton and J. Playfair's uniformitarian theories

Physics
On the Improvement of Compound Microscopes (J. J. Lister) – showing how distortions can be reduced by use of achromatic lenses

Technology
Stephenson's *Rocket* wins the Rainhill trials in England

1831

Biology
Müller confirms the Bell–Magendie law distinguishing between motor and sensory nerves
(to 1836) Expedition of HMS *Beagle* to South America, with Darwin aboard as naturalist

Chemistry
Guthrie and Soubeiran independently discover chloroform

Physics
Mellini and Nobili publish a description of a sensitive thermopile
Experimental Researches in Electricity, 1st series (Faraday)

Technology
Philips patents the contact process for making sulfuric acid

General
The British Association for the Advancement of Science established in York, with Vernon Harcourt as its first chairman

1832

Astronomy
Henderson measures parallax of Alpha Centauri (but does not publish until 1839)

Biology
Dutrochet demonstrates that gas exchange in plants is via the stomata

Gideon Algernon Mantell discovers the first armored dinosaur, *Hylaeosaurus*

Mathematics
Bolyai publishes his version of non-Euclidean geometry
Galois outlines his ideas on group theory

Medicine
T. Hodgkin describes lymphadenoma (Hodgkin's disease)

Physics
Faraday performs experiments on electromagnetic induction
Henry discovers self-induction

1833

Medicine
Observations on the Gastric Juice and the Physiology of Digestion (W. Beaumont)

1834

Chemistry
Runge discovers quinoline
Dumas and Peligot discover methyl alcohol (methanol)

Physics
Peltier demonstrates thermoelectric effect (Peltier effect)
Lenz discovers law for magnetic induction (Lenz's law)

1835

Biology
Dujardin is able to refute the claim that microorganisms have the same organs as higher animals

Earth Science
Charpentier first presents his glaciation theory

Physics
Coriolis describes the inertial Coriolis force
Henry develops the electric relay

Technology
Morse produces a working model of the electric telegraph

1836

Anthropology
Lartet discovers fossil ancestors of the gibbon (Pliopithecus)

Astronomy
Baily observes Baily's beads during total eclipse

Biology
Schwann and Cagniard de la Tour independently discover the role of yeast in alcoholic fermentation
Schwann prepares the first precipitate of an enzyme (pepsin) from animal tissue

Chemistry
The Daniell cell is introduced
Gossage improves process for
manufacture of alkali
Physics
Sturgeon designs moving-cell
galvanometer and introduces
commutator for electric motor

1837
Biology
Purkinje locates the Purkinje
cells in the brain
Chemistry
Boussingault begins his
experiments proving that
leguminous plants are capable
of using atmospheric nitrogen
Earth Science
A System of Mineralogy
(Dana)
Mathematics
*Recherches sur la probabilité
des jugements* (Poisson) –
introduces law of large
numbers and Poisson
distribution
Technology
Wheatstone and Cooke
produce first practical
telegraph system

1838
Astronomy
Bessel announces detection of
parallax of 61 Cygni – first
determination of stellar
distance
Biochemistry
Peligot and Bouchardat
demonstrate glucose in the
urine of diabetics
Biology
Beiträge zur Phytogenesis
(Schleiden) – states that all
structures of plants composed
of cells or their derivatives
Remak shows that nerves are
not hollow tubes
(to 1843) *Flora of North
America* (A. Gray)

1839
Biology
*A Naturalist's Voyage on the
Beagle* (Darwin)
Purkinje discovers the
Purkinje fibers in the heart
Chemistry
Mosander discovers
lanthanum
Technology
Nasmyth designs the steam
hammer
Daguerre perfects the
daguerreotype

1840
Chemistry
Hess formulates the law of
constant heat summation
Schönbein discovers ozone
Earth Science
Etudes sur les glaciers
(Agassiz) – postulates
existence of ice ages
General
William Whewell introduces
the terms 'physicist' and
'scientist' into the English
language

1841
Earth Science
Essai sur les glaciers
(Charpentier)
Technology
Sir Joseph Whitworth first
argues for a unified system of
screw threads – the
Whitworth system, which
becomes the standard for
British engineering

1842
Biology
Naegeli publishes his
observations on cell division
during pollen formation, so
disproving Schleiden's theory
of cell formation by budding
off the nuclear surface
Steenstrup publishes his
observations of the alternation
of generations in certain
animals
Bowman publishes his work
on the Malpighian bodies of
the kidneys, which includes a
description of the structure
and role of the Bowman's
capsules
Medicine
Braid suggests the use of
hypnosis to treat pain, anxiety,
and certain nervous disorders
Physics
Doppler discovers the Doppler
effect for wave motions
Mayer's paper containing the
law of conservation of energy
is published

1843
Astronomy
Schwabe announces discovery
of sunspot cycle
Biology
Du Bois-Reymond shows that
applying a stimulus to a nerve
causes a drop in electrical
potential at the point of
stimulus
Chemistry
Mosander discovers erbium
and terbium

Medicine
Holmes publishes a paper on
the contagious nature of
puerperal fever
General
Women first admitted to
membership of the British
Association for the
Advancement of Science

1844
Astronomy
Bessel announces that Sirius
is a double star system
Biology
Remak discovers ganglion
cells in the heart
Chemistry
Gerhardt develops his type
theory of organic compounds

1845
Astronomy
J. C. Adams works out position
of Neptune
Biology
(to 1848) *Viviparous
Quadrupeds of North America*
(Audubon)
Chemistry
Schröder discovers red
phosphorus
Physics
Waterson's paper on the
kinetic theory of gases is
rejected by the Royal Society
Faraday discovers effect of
magnetic field on polarized
light (Faraday effect)
Kirchhoff formulates his laws
for electric circuits
(Kirchhoff's laws)
*An Essay on the Application of
Mathematics to Electricity and
Magnetism* (George Green),
reissued by Lord Kelvin in
Crelle's Journal thus
introducing *Green's theorem*,
Green's function, and the
notion of electric potential into
science

1846
Astronomy
Leverrier works out position of
Neptune – discovered by Galle
Lassell discovers Triton
(satellite of Neptune)
Harvard College Observatory
founded with W. C. Bond as
director at Cambridge,
Massachusetts
Biology
Siebold and Stannius publish
their textbook of comparative
anatomy
Hugo von Mohl coins the term
protoplasm for the substance
surrounding the cell nucleus

Chemistry
Sobrero discovers
nitroglycerine
Schönbein discovers guncotton
Earth Science
Loomis publishes first
synoptic weather map
Medicine
Morton demonstrates the use
of ether inhalation to produce
general anesthesia
1847
Biology
The first volume of *Antiquités
celtiques et antédiluviennes*
(Boucher de Perthes)
Ludwig invents the
kymograph
Chemistry
Babo formulates his law for
the depression of vapor
pressure by a solute (Babo's
law)
Medicine
J. Y. Simpson introduces
chloroform as an anesthetic in
hospital operations
Semmelweis discovers the
nature of puerperal fever and
implements a successful
preventative policy
Physics
Über die Erhaltung der Kraft
(Helmholtz) – develops the
law of conservation of energy
Kelvin draws an analogy
between an electrostatic field
and an incompressible elastic
solid
1848
Astronomy
W. C. Bond and G. Bond
discover Hyperion (satellite of
Saturn)
Biology
Kölliker isolates cells of
smooth muscle
(to 1854) *Untersuchungen über
tierische Elektricität* (Du Bois-
Reymond) – reports his work
on nerve and muscle activity
Chemistry
Pasteur discovers molecular
asymmetry (as cause of optical
activity)
Physics
Joule publishes a paper on the
kinetic theory of gases in
which he makes the first
estimate of the speed of gas
molecules
Kelvin introduces the absolute
temperature scale
General
American Association for the
Advancement of Science

founded with W. C. Redfield as
its first president
1849
Chemistry
Wurtz synthesizes ethylamine
from ammonia
Deville discovers nitrogen
pentoxide
Medicine
Addison describes Addison's
disease
Physics
Fizeau measures the speed of
light by a toothed-wheel
method
*On the Mechanical Equivalent
of Heat* (Joule)
1850
Astronomy
W. C. Bond makes first
photograph of a star (Vega)
and discovers third ring of
Saturn (crepe ring)
Roche proposes and calculates
the Roche limit
Lamont announces discovery
of magnetic cycle of Earth
Earth Science
Mallet begins precise
experiments to determine
speed of earthquake waves
Physics
Fizeau and Foucault provide
experimental support for the
wave theory of light by
measuring the speed in air
and in water
*Über die bewegende Kraft der
Wärme* (Clausius) – first
formulates the second law of
thermodynamics
1851
Astronomy
Lassell discovers Ariel and
Umbriel (satellites of Uranus)
Biology
Helmholtz invents the
ophthalmoscope
Vergleichende Untersuchungen
(Hofmeister) – describes
discovery of alternation of
generations in lower plants
and similarities between seed-
bearing and non-seed-bearing
plants
Chemistry
Thomas Anderson extracts
pyridine from bone oil
Physics
Foucault uses pendulum to
demonstrate rotation of Earth
1852
Chemistry
Frankland originates the
theory of valence

Mathematics
Francis Guthrie first notices
the map-coloring problem
Physics
Kelvin sets out the law of
conservation of energy
(to 1862) Joule and Kelvin
investigate the Joule–Kelvin
effect
1853
Aeronautics
Cayley constructs the first
successful man-carrying glider
Chemistry
Hittorf introduces idea of ionic
mobility in electrolysis
Cannizzaro discovers the
reaction known as
Cannizzaro's reaction
Mathematics
Hamilton introduces
quaternions
1854
Chemistry
Berthelot synthesizes fats
from glycerin and fatty acids
Mathematics
The Laws of Thought (Boole)
Reimann puts forward his
general formulation of
Reimannian geometry
1855
Biology
Pringsheim confirms the
occurrence of sexuality in
algae
Chemistry
Wurtz develops the Wurtz
reaction for synthesizing
hydrocarbons
Deville develops a process for
the large-scale production of
aluminum
1856
Anthropology
Lartet discovers fossil
ancestor of apes
(Dryopithecus)
Biology
Pasteur declares that
fermentation is caused by
living organisms
Chemistry
Perkin discovers first
synthetic dye (mauveine)
Medicine
Brown-Séquard shows there is
a connection between excising
the adrenal glands and
Addison's disease
Technology
Bessemer announces his
process for converting pig iron
into steel
Charles William Siemens and

Friedrich Siemens introduce the regenerator furnace

1857

Astronomy
G. Bond shows how stellar magnitude may be calculated from photographs

Earth Science
Buys Ballot formulates law concerning wind direction

1858

Biology
On the Tendency of Varieties to Depart Indefinitely from the Original Type (Wallace)
Joint paper by Darwin and Wallace, outlining their theory of evolution by natural selection, read at meeting of Linnean Society
Die Cellularpathologie (Virchow)

Chemistry
Kekulé publishes his paper on bonding in carbon compounds

1859

Astronomy
(to 1862) *Bonner Durchmusterung* (Argelander) – survey of northern stars

Biology
On the Origin of Species by Means of Natural Selection (Darwin)

Physics
Tyndall discovers scattering by colloidal suspension (Tyndall effect)
Planté invents Planté battery – the first storage battery

1860

Biology
Oxford meeting of the British Association for the Advancement of Science, in which T. H. Huxley and others defend Darwin's theory of evolution

Chemistry
Bunsen and Kirchhoff publish their paper on spectroscopy – spectroscopic methods enabled Bunsen to discover rubidium and cesium
First meeting of the International Chemical Congress at Karlsruhe
Graham distinguishes colloids

***c.* 1860**

Physics
Illustrations of Dynamical Theory of Gases (Maxwell)

Technology
Lenoir takes out a patent for an internal combustion engine

1861

Astronomy
A. G. Clark discovers Sirius B
Schiaparelli discovers the asteroid Hesperia

Biology
H. W. Bates publishes his paper on Batesian mimicry in butterflies
Gegenbaur shows that all vertebrate eggs and sperm are unicellular

Chemistry
Bunsen and Kirchhoff discover rubidium
Crookes discovers thallium

Technology
Solvay patents his process for soda production
Massachusetts Institute of Technology (MIT) founded in Boston with William Barton Rogers as its first president

1862

Biology
(to 1883) *Genera Plantarum* (Bentham and Hooker)

Chemistry
Béguyer de Chancourtois proposes his periodic classification of the elements (telluric screw)

1863

Astronomy
Huggins shows the stars to be composed of known elements

Biology
Helmholtz publishes his theory of hearing

Chemistry
Reich and Richter discover indium

Earth Science
Galton publishes his work on weather systems, outlining the modern technique of weather mapping

Medicine
Leuckart publishes the first volume of his work on the parasites of man
Davaine shows that anthrax can be transmitted to healthy cattle by injecting them with the blood of diseased cattle
Waldeyer establishes that cancer begins as a single cell

General
National Academy of Sciences created

1864

Astronomy
Donati first observes spectrum of comet – discovers gaseous composition of tails

W. Huggins discovers some nebulae to be gaseous

Chemistry
Guldberg and Waage discover the law of mass action
Newlands publishes his law of octaves

Physics
Dynamical Theory of the Electric Field (Maxwell) – contains Maxwell's differential equations describing the propagation of electromagnetic waves

1865

Biology
Sachs shows that chlorophyll is confined to discrete bodies – chloroplasts

Chemistry
Kekulé proposes a ring structure for benzene
(to 1885) Baeyer researches indigo
Zur Gross der Loftmolecule (On the Magnitude of Air Molecules; Loschmidt)

Medicine
Villemin shows that tuberculosis is contagious

Physics
Clausius introduces the term entropy as a measure of the availability of heat

1866

Astronomy
Schiaparelli shows meteors follow cometary orbits

Biology
De Bary shows that a lichen is a close association between an alga and a fungus
Mendel publishes the results of his plant-breeding experiments, summarized later as the law of segregation and the law of independent assortment
His introduces the microtome

Medicine
Allbutt invents the short clinical thermometer

Physics
Kundt develops the Kundt tube to measure the speed of sound

Technology
Leclanché invents the electrical battery (dry cell)

1867

Biology
Handbuch der physiologische Optik (Helmholtz)

Medicine
Cohnheim publishes his work on inflammation, disproving

Virchow's theory that pus corpuscles originate at the point of wounding, and demonstrating that swelling is due to the passage of leukocytes from the veins into the wound

Joseph Lister introduces the principles of antiseptic surgery into medicine

Technology
Nobel patents dynamite and the detonating cap in the UK

1868
Anthropology
Lartet discovers remains of Cro-Magnon man

Astronomy
W. Huggins discovers red shift in spectrum of Sirius
Lockyer identifies lines of helium in solar spectrum

Medicine
Wunderlich publishes his investigations into body temperature and disease, so introducing clinical thermometry as an important diagnostic technique

1869
Chemistry
Mendeleev proposes his periodic table of the elements
T. Andrews publishes his work on the liquefaction of gases, which includes the concept of critical temperature
Caro, Graebe, and Liebermann synthesize alizarin, the first natural dye to be synthesized

Medicine
Joseph Lister uses catgut ligatures instead of silk

1870
Biology
Hitzig and Fritsch publish a paper containing the first experimental evidence for cerebral localization

Chemistry
J. L. Meyer publishes his periodic classification of elements

1871
Biochemistry
Miescher announces the discovery of nuclein (later renamed nucleic acid) in cell nuclei

Biology
Hoppe-Seyler discovers the enzyme invertase
Ranvier describes the nodes of Ranvier
The Descent of Man (Darwin)

1872
Astronomy
H. Draper photographs first stellar spectrum (of Vega)

Biology
Untersuchungen über Bacterien (Cohn) – provides basis for modern bacterial nomenclature

Chemistry
Dewar devises the Dewar vacuum flask
Wislicenus demonstrates geometrical isomerism in lactic acid

Earth Science
(to 1876) C. W. Thomson collects evidence for the presence of the Mid-Atlantic Ridge

Medicine
George Huntington describes Huntington's chorea

Physics
E. Abbe invents Abbe condenser lens for microscopes
Amagat experiments on behavior of gases at high pressures

1873
Biology
Osler discovers the blood platelets

Chemistry
(to 1876) Gibbs develops the theory of chemical thermodynamics
On the Continuity of the Liquid and Gaseous States (van der Waals) – includes the van der Waals equation

Medicine
Billroth performs first laryngectomy
Hansen observes rod-shaped bacilli in the tissues of leprosy patients

1874
Chemistry
A Suggestion Looking to the Extension into Space of the Structural Formulae at Present Used in Chemistry (van't Hoff) – introduces idea of tetrahedral arrangement of bonds to carbon atom
Lecoq de Boisbaudran discovers gallium

Physics
Braun discovers use of semiconducting crystals as rectifiers

Technology
A. G. Bell's multiple telegraph is patented

General
The Cavendish laboratory opens in Cambridge, England, with Clerk Maxwell as its first director

1876
Biology
Geographical Distribution of Animals (Alfred Russell Wallace)

Chemistry
On the Equilibrium of Heterogeneous Substances (J. W. Gibbs) – work on chemical thermodynamics introducing phase rule

Medicine
Koch publishes the life cycle of the anthrax bacillus

Technology
A. G. Bell's invention of the telephone is patented

1877
Astronomy
A. Hall discovers Phobos and Deimos (satellites of Mars)

Chemistry
Friedel and Crafts discover the Friedel–Crafts reaction for the alkylation or acylation of aromatic hydrocarbons

Medicine
Manson demonstrates, from his investigations of elephantiasis, that certain diseases are transmitted by insects

Physics
Cailletet succeeds in producing liquid oxygen

Technology
Edison invents the phonograph

1878
Chemistry
Marignac isolates ytterbium

1879
Biology
De Bary coins the term symbiosis to describe the mutually beneficial association between two unrelated organisms

Chemistry
Cleve discovers thulium
Lecoq de Boisbaudran discovers samarium
Nilson discovers scandium
Fahlberg discovers saccharin

Medicine
Neisser discovers the gonococcus responsible for gonorrhea

Physics
Stefan formulates law for heat

radiation (Stefan–Boltzmann law)

Swan invents carbon-filament incandescent electric lamp

Edwin Hall discovers the Hall effect

Technology

Edison demonstrates his electric light bulb

The Thomas–Gilchrist steel-making process is demonstrated

1880

Chemistry

Skraup synthesizes quinoline

Marignac discovers gadolinium

Earth Science

Milne invents seismograph

Medicine

Laveran discovers the causative agent of malaria – the *Plasmodium* parasite

Physics

P. Curie discovers piezoelectricity

Technology

Eadweard James Muybridge demonstrates the first moving picture in public using the 'zoopraxiscope'

1881

Biology

T. W. Engelmann demonstrates chemotactic response of bacteria

Medicine

Finlay publishes a paper naming the mosquito as the vector of yellow fever

Physics

Ewing discovers hysteresis

1882

Biology

Ants, Bees and Wasps (Lubbock)

Zell-substanz, Kern und Zelltheilung (W. Flemming)

Chemistry

Raoult shows that the depression in freezing point of a solvent is proportional to the mass of substance dissolved divided by the substance's molecular weight (Raoult's rule)

Earth Science

Anthropogeographie (Ratzel, 2nd vol. 1891) – lays foundation for modern study of human geography

Medicine

Metchnikoff describes phagocytosis

Pasteur produces an

attenuated vaccine for anthrax

Koch discovers the bacillus responsible for tuberculosis

Physics

Ritter builds first dry voltaic pile

Rowland designs diffraction grating for spectral photography

1883

Biology

Galton coins the term eugenics

Medicine

Kocher demonstrates that operative myxedema (cretinism) following removal of the thyroid can be avoided by leaving a portion of the thyroid in the body

Physics

Wroblewski and Olszewski achieve liquefaction of large amounts of oxygen

1884

Biology

Gram discovers the Gram stain

Chemistry

Chardonnet's rayon, the world's first artificial fiber, is patented

Le Chatelier discovers Le Chatelier's principle

Tilden synthesizes isoprene, leading to the production of synthetic rubber

Mathematics

Die Grundlagen der Arithmetik (Frege) – defines cardinal number and derives properties of numbers

Medicine

Löffler isolates and cultivates the bacillus responsible for diphtheria

Gaffky obtains a pure culture of the typhoid bacillus

Koller publishes his experiments on the use of cocaine as a local anesthetic

Physics

Poynting introduces Poynting's vector

Technology

Parsons invents the multistage steam turbine

General

The International Meridian Conference in Washington adopts a single prime meridian passing through the center of the transit instrument at the Royal Greenwich Observatory

1885

Biology

Kossel isolates adenine from his newly discovered nucleic acids

Chemistry

Winkler discovers germanium

Medicine

Roux and Yersin show that diphtheria is caused by a toxin released from diphtheria bacteria

Pasteur successfully uses a rabies vaccine to cure a rabies victim

Physics

Balmer discovers formula for series of lines in hydrogen spectrum (Balmer series)

Technology

Daimler patents the first internal-combustion engine

1886

Biology

Eichler publishes his plant classification system

Weismann publishes his theory of the continuity of the germ plasm

Chemistry

Beckmann discovers the Beckmann rearrangement in organic chemistry

Moissan isolates fluorine gas

C. M. Hall and Héroult independently discover a cheap process for extracting aluminum

Lecoq de Boisbaudran discovers dysprosium

Physics

Boltzmann derives distribution law for energies of gas molecules (Maxwell–Boltzmann distribution)

E. Abbe invents apochromatic lens system

Goldstein discovers canal rays

Technology

Benz patents his three-wheeled automobile, the first practical automobile powered by an internal-combustion engine

1887

Astronomy

Paris Observatory organizes a congress that decides to construct the *Carte du Ciel*, a photographic map of the whole sky

Biology

Henson introduces the term plankton

Petri introduces the Petri dish for cultures

Chemistry
Arrhenius publishes his theory on ionic dissociation in solution
Wallach characterizes the terpenes

Physics
Hertz discovers photoelectric effect
Michelson and Morley attempt to detect the ether and obtain negative result (the Michelson–Morley experiment)
(to 1888) Lodge discovers propagation of electromagnetic waves along conducting wires

Technology
Tesla invents the first alternating-current motor

1888
Astronomy
New General Catalogue of Nebulae and Clusters of Stars (NGC) (Dreyer)
Lick Observatory completed at a site on Mount Hamilton overlooking San José, California

Chemistry
Baeyer makes the first synthesis of a terpene

Earth Science
(to 1889) Nansen explores Greenland icecap

Mathematics
Was sind und was sollen die Zahlen (Dedekind) – the first formulation of the axioms of arithmetic

Physics
Hertz demonstrates propagation of electromagnetic waves (radio waves)

Technology
Eastman markets his hand-held box camera

1889
Biology
Pavlov shows the secretion of gastric juices is prompted by the sight of food
Intracellular Pangenesis (de Vries) – the theory that hereditary traits are determined by 'pangenes' in the cell nucleus

Chemistry
F. A. Abel and J. Dewar introduce cordite

Mathematics
Peano presents axioms of arithmetic

Physics
Sur les trois corps et les equations de la dynamique (Henri Poincaré) – discusses the three-body problem

Technology
Otto Lilienthal publishes *Der Vogelflug als Grundlage der Fliegerkunst* (Bird Flight as the Basis of Aviation) and builds his first model glider

1890
Astronomy
Vogel discovers eclipsing binaries spectroscopically

Biology
Kitasato and von Behring announce the discovery of diphtheria and tetanus antitoxins
Koch formulates Koch's postulates for establishing that an organism is the cause of a disease

Chemistry
Guillaume begins work on nickel alloys (invar, elinvar)
Beilby patents a process for the synthesis of potassium cyanide
Commercial synthetic indigo is produced

Computing
Herman Hollerith invents a tabulating system using punched cards for the 1890 US census

Medicine
Grassi, Marchiafava, and Celli show there are a number of protozoan species that can produce malaria

Physics
Rydberg discovers formula for spectral-line frequencies, involving Rydberg constant

1891
Astronomy
Chandler announces discovery of Earth's Chandler wobble

Biology
Strasburger demonstrates that physiological rather than physical forces are responsible for the movement of liquids up the plant stem

Chemistry
E. H. Fischer deduces the configurations of the 16 possible aldohexose sugars

Physics
Poynting determines the mean density of the Earth

Stoney coins term electron
Lippmann produces the first color photograph

General
The Throop Polytechnic Institute, the forerunner of the California Institute of Technology, is founded in Pasadena, California

1892
Astronomy
E. E. Barnard discovers Amalthea (satellite of Jupiter)

Biology
Ivanovsky, working on tobacco mosaic disease, finds the first evidence of viruses

Chemistry
Bevan and Cross patent the viscose process of rayon manufacture
Crum Brown and Gibson propose Crum Brown's rule for substitution reactions of benzene compounds

Medicine
Welch discovers the bacterium that causes gas gangrene

Physics
(to 1904) Lorentz develops his electron theory

1893
Astronomy
Maunder discovers Maunder minimum for sunspot activity

Chemistry
Kjeldahl develops the Kjeldahl method for determining the amount of nitrogen in organic compounds

Mathematics
(and 1903) *Grundgesetze der Arithmetik* (Frege)

Physics
Wien derives his displacement law for radiation from hot bodies
Poynting determines the gravitational constant

Technology
Diesel demonstrates his first engine

1894
Biology
Pfeiffer discovers bacteriolysis
Rubner demonstrates that human energy production can be explained by thermodynamics

Chemistry
Ramsay and Rayleigh discover argon
Sulfur is raised from deep underground by means of superheated water – the Frasch process

Medicine
Yersin and Kitasato discover the bacterium responsible for bubonic plague
Physics
Lodge develops his coherer for detecting radio waves
Technology
Marconi begins experiments on communicating using radio waves
1895
Astronomy
Keeler shows the rings of Saturn do not rotate uniformly
Chemistry
Ramsay and Crookes identify helium
Medicine
Bruce discovers the trypanosome parasite that causes sleeping sickness
Physics
Lorentz describes force (Lorentz force) on a moving charged particle in electromagnetic field
P. Curie discovers transition temperature for change from ferromagnetic to paramagnetic behavior (Curie point)
Röntgen discovers x-rays
Psychology
Studien über Hysterie (Freud) – beginning of psychoanalysis
1896
Biology
The first observation of the fermentation of sugar by cell-free extracts of yeast, proving the process is purely chemical and does not require intact cells
Schäfer and Oliver discover that injecting an extract from the adrenal gland increases blood pressure
Chemistry
Walden discovers the Walden inversion in organic substitution reactions
Mathematics
Jacques Salomon Hadamard sets out his proof of the prime number theorem (independently of Charles de la Vallée-Poussin)
Physics
Becquerel discovers radioactivity (in uranium salts)
Wien derives a formula for the distribution of energy in

black-body radiation (Wien's formula)
Boltzmann introduces the equation relating entropy to probability
Zeeman discovers splitting of spectral lines in magnetic field (Zeeman effect)
Technology
Marconi patent on radio telegraphy
Popov in Russia experiments with radio communication
Sperry invents the gyrocompass
A plant using the Castner–Kellner process is opened at Niagara Falls
General
Alfred Nobel dies leaving a fortune and instructions in his will to set up the Nobel Foundation
1897
Astronomy
Yerkes Observatory opened under the directorship of G. Hale at Williams Bay, Wisconsin
Biology
J. J. Abel isolates a physiologically active substance from the adrenal gland – adrenaline (epinephrine)
Chemistry
Sabatier demonstrates hydrogenation of organic compounds
Medicine
Ross identifies malaria parasites (plasmodia) in the bodies of *Anopheles* mosquitoes fed on blood from infected patients
Eijkman discovers a cure for beriberi
Physics
Larmor predicts precession of electron orbits in atoms (Larmor precession)
J. J. Thomson discovers the electron
Boys determines the gravitational constant
1898
Biology
Bordet discovers alexin (complement)
Pflanzengeographie auf physiologischer Grundlage (Plant Geography Upon a Physiological Basis; Schimper)
Löffler and Frosch demonstrate that viruses can cause diseases in animals

Golgi describes the Golgi apparatus
Beijerinck coins the term filterable virus for the causative agent of tobacco mosaic disease
Chemistry
Lowry describes mutarotation of optically active compounds
E. H. Fischer synthesizes purine
Ramsay and Travers discover neon, krypton, and xenon
Caro discovers Caro's acid, a powerful oxidizing agent
J. Dewar produces liquid hydrogen
Medicine
Grassi demonstrates that only the *Anopheles* species of mosquito can transmit malaria to humans
Physics
M. Curie discovers radium and polonium
Townsend measures the charge of the electron
1899
Astronomy
W. H. Pickering discovers Phoebe (satellite of Saturn)
Chemistry
Thiele proposes the idea of partial valence in chemical compounds
J. Dewar produces solid hydrogen
Physics
Rutherford distinguishes alpha and beta rays in radioactivity
Technology
Marconi achieves first international radio transmission (England–France)
1900
Biology
Correns rediscovers Mendel's work and publishes his own work on crossing experiments confirming Mendel's results
Chemistry
The discovery of the Grignard reagents
Gomberg obtains triphenylmethyl – a stable free radical
Tsvet develops the technique of adsorption analysis, which he later named chromatography
Earth Science
Oldham clearly identifies primary (P) and secondary (S)

seismic waves for the first time

Köppen first formulates his climatic classification

Mathematics

Hilbert presents list of 23 research problems at International Congress of Mathematicians in Paris

Medicine

Leishman discovers the protozoan parasite causing kala-azar (now known as leishmaniasis)

Grassi and Manson provide evidence that malaria is transmitted by mosquitoes

Physics

Debierne discovers actinium

Rutherford discovers gamma radiation

Dorn discovers that radium evolves radioactive gas (radon)

Planck publishes paper giving theory of energy distribution in black-body radiation, introducing the idea of quantized energy transfer

1901

Biology

Takamine achieves the first isolation and purification of a hormone (adrenaline) from a natural source

Hopkins discovers the amino acid tryptophan

J. N. Langley finds that extract from adrenal gland stimulates effect of sympathetic nervous system

(to 1903) *The Mutation Theory* (de Vries) – proposes that new species evolve by genetic mutations

Chemistry

Demarçay discovers europium

Earth Science

(to 1909) *Die Alpen im Eiszeitalter* (Penck) – identification of four ice ages

Medicine

Einthoven develops the electrocardiogram

Reed and Carroll establish that yellow fever is caused by a virus

Physics

Richardson formulates a law to explain the emission of electrons from hot surfaces

Technology

Marconi makes first transatlantic radio transmission

1902

Biology

Landsteiner announces his discovery of the ABO blood-group system

Bayliss and Starling discover the role of the hormone secretin in digestion

Medicine

Richet coins the term anaphylaxis to describe the reaction whereby a second injection of an antigen proves fatal

Physics

Heaviside (and Kennelly) propose the existence of the Kennelly–Heaviside layer in the upper atmosphere

1903

Astronomy

Exploration of Cosmic Space by means of Reaction Devices (Tsiolkovsky)

Chemistry

Zsigmondy and Siedentopf invent the ultramicroscope

Birkeland and Eyde develop their process of nitrogen fixation

Physics

Rutherford and Soddy publish account of radioactive series involving transmutation of elements

Technology

Wilbur Wright pilots the first powered flight on December 14; Orville Wright pilots the second flight on December 17

1904

Astronomy

Moulton and Chamberlin formulate planetismal theory

Hartmann finds first evidence of interstellar gas

Solar observatory set up on Mount Wilson at the south end of the Sierra Madre range by G. E. Hale

Chemistry

Ramsay and Whytlaw-Gray codiscover niton (now known as radon)

Medicine

Aschoff describes inflammatory nodules in heart muscle (Aschoff's bodies)

Physics

Lorentz and Fitzgerald independently propose Lorentz–Fitzgerald contraction

Lorentz publishes transformations for space and time coordinates between

frames of reference (Lorentz transformations)

J. A. Fleming invents thermionic vacuum tube

Barkla observes the polarization of x-rays

1905

Biology

Optima and Limiting Factors (Blackman)

Elements of Heredity (Johannsen)

J. S. Haldane and J. G. Priestley demonstrate the role of carbon dioxide in the regulation of breathing

Johannsen introduces the terms genotype and phenotype

(to 1909) Lucas formulates all-or-none law of nervous stimulation

(to 1915) Willstätter works out formulae of chlorophylls a and b

Chemistry

Goldschmidt introduces his method for the reduction of metallic oxides to metals – the Thermit process

Earth Science

On the Influence of the Earth's Rotation on Ocean Currents (Ekman)

Medicine

Schaudinn and Hoffmann isolate the spirochete that causes syphilis

Physics

Boltwood demonstrates that lead is final product of uranium decay

Rutherford and Soddy publish theory of nuclear transmutation

Einstein publishes papers on Brownian motion, the photoelectric effect, and special relativity; relates mass and energy ($E = mc^2$)

1906

Biology

The Integrative Action of the Nervous System (Sherrington)

(to 1907) Hopkins shows that a diet of proteins, carbohydrates, fats, and salts will not support growth in mice and postulates other essential accessory substances (later to be called vitamins) in the normal diet

Chemistry

Tsvet coins the term chromatograph for the analytic procedure he developed to separate plant pigments

Earth Science
Oldham provides evidence that the Earth has a central core
Medicine
Wasserman develops a test to diagnose syphilis (the Wasserman test)
Physics
Nernst formulates the Nernst heat theorem (third law of thermodynamics)
Lyman series identified in hydrogen spectrum
Technology
First music and speech broadcast in America (transmitted on Christmas Eve by Fessenden)
1907
Biology
Hopkins and Fletcher demonstrate that lactic acid accumulates in working muscle
Harrison develops the technique of tissue culture
Chemistry
Urbain discovers lutetium
(to 1909) Haber demonstrates his process for synthesizing ammonia from nitrogen and hydrogen
Medicine
Alois Alzheimer diagnoses Alzheimer's disease
Physics
Minkowski puts forward concept of 4-dimensional space–time
Technology
De Forest patents the Audion tube
1908
Astronomy
Hale finds first extraterrestrial magnetic field in sunspots
Henrietta Leavitt shows that cepheid variables can be used to estimate stellar distances
Biology
Inborn Errors of Metabolism (Garrod) – introduces the idea of genetic diseases
Hardy and Weinberg discover Hardy–Weinberg law
Medicine
Landsteiner manages to transmit polio from humans to monkeys
Physics
Kamerlingh-Onnes succeeds in liquefying helium
Perrin confirms experimentally Einstein's

predictions concerning Brownian motion
Technology
Wilbur Wright first demonstrates powered flight publicly at Le Mans in August; Orville Wright gives a demonstration a few days later at Fort Meyer, Virginia
1909
Biology
Ricketts describes the microorganism causing Rocky Mountain spotted fever – an organism intermediate between a bacterium and a virus
Correns obtains the first evidence for cytoplasmic inheritance
Johannsen introduces the term genes to describe Mendel's factors of inheritance
Chemistry
Levene shows that the carbohydrate in yeast nucleic acid is the pentose sugar ribose
Bakelite is invented by Baekeland
Earth Science
Mohorovičić discovers discontinuity between Earth's mantle and crust (Mohorovičić discontinuity)
Charles Dolittle Walcott discovers the Burgess Shale fossils
Medicine
Nicolle discovers that typhus is carried by the louse
Physics
Millikan starts series of experiments to determine the charge on the electron
1910
Geology
The California Earthquake of April 19, 1906 (Harry Fielding Reid) – formulates the elastic rebound theory
Mathematics
(to 1913) *Principia Mathematica* (B. Russell and Whitehead)
Medicine
Ehrlich finds a chemical – Salvarsan – effective in treating syphilis
Psychology
Formation of the International Psycho-Analytical Association
Technology
Claude introduces neon lighting

1911
Biology
Krogh shows that movement of oxygen from the lungs is by simple diffusion (no active secretion is involved)
T. H. Morgan and colleagues produce the first chromosome maps
Medicine
Rous shows that cancer in chickens can be transmitted by a virus
Physics
C. T. R. Wilson invents cloud chamber
Rutherford proposes nuclear model of the atom
Kamerlingh-Onnes discovers superconductivity
(to 1912) V. F. Hess discovers cosmic rays
Einstein predicts gravitational field should deflect light
General
The Kaiser Wilhelm Gesselschaft founded in the Dahlem suburb of Berlin under the directorship of the historian Adolf von Harnack – the forerunner of the Max Planck Institute
1912
Anthropology
A. S. Woodward and Dawson find Piltdown man
Astronomy
Slipher measures velocity of Andromeda
Leavitt discovers period–luminosity relation of Cepheid variables
Biology
Funk suggests that certain diseases are caused by deficiencies of certain ingredients (vitamins) from the diet
Physics
Von Laue discovers x-ray diffraction by crystals
W. L. Bragg formulates *Bragg's law* for x-ray diffraction
1913
Astronomy
H. N. Russell publishes Hertzsprung–Russell diagram
Biology
Michaelis and Menten propose the Michaelis–Menten equation to describe the variation in rate found when an enzyme acts on a substrate
Sturtevant publishes a linkage map constructed from

data on recombinant frequencies in *Drosophila*

Earth Science
Gutenberg identifies the Earth's core as being liquid from seismological evidence
A. Holmes proposes the first quantitative geological time scale

Physics
Soddy proposes existence of isotopes
Stark discovers splitting of spectral lines in strong electric field (Stark effect)
On the Constitution of Atoms and Molecules (N. Bohr) – introduces his quantum theory of the atom and atomic spectra
Moseley demonstrates characteristic x-ray wavelengths of elements depend on squares of atomic number plus a constant (Moseley's law)
Fajans proposes laws governing products of radioactive decay (independently of Soddy)

1914
Astronomy
Nicholson discovers Sinope (satellite of Jupiter)
W. S. Adams investigates spectra of stars; leading to stellar parallax method of finding distances

Biology
Ewins and Dale isolate acetylcholine from ergot
Kendall isolates the active ingredient of the thyroid gland

Medicine
Respiratory Function of the Blood (Barcroft)

Physics
Frank and G. Hertz demonstrate quantized nature of energy transfer

1915
Astronomy
Shapley proposes disk-shaped structure for Galaxy
W. S. Adams identifies first white dwarf (Sirius B)
Einstein and Grossmann explain advance in the perihelion of Mercury

Biology
D'Herelle and Twort independently discover the bacteriophage

Chemistry
Debye gives theory of electron diffraction by gases

Earth Science
Die Enstehung der Kontinente und Ozeane (Wegener) – postulates original super-continent of Pangaea and idea of continental drift

Mathematics
Fisher describes a solution for the exact distribution of the correlation coefficient in statistics

Physics
W. H. Bragg constructs first x-ray spectrometer

1916
Astronomy
E. E. Barnard discovers Barnard's star
Schwarzschild introduces the idea of the Schwarzschild radius

Biology
The Involuntary Nervous System (Gaskell)

Physics
Sommerfeld proposes elliptical orbits for electrons in atoms
N. Bohr formulates correspondence principle
Einstein publishes general theory of relativity

1917
Astronomy
De Sitter proposes a mathematical model of the universe (de Sitter universe)
The 100-inch reflector first used at the Mount Wilson Observatory

Chemistry
Meitner and Hahn discover protactinium

Physics
Chapman predicts phenomenon of thermal diffusion of gases

1918
Astronomy
100-inch Hooker telescope in operation at Mount Wilson (largest until 1948)
(to 1924) *Henry Draper Catalogue* (E. C. Pickering and A. Cannon) – records spectra of stars

Biology
The Correlation between Relatives on the Supposition of Mendelian Inheritance (Fisher) – demonstrates that continuous variations are inherited in Mendelian fashion
Meyerhof investigates cellular metabolism – leads to

discovery of Embden–Meyerhof pathway

1919
Astronomy
Eddington reports results of observations during solar eclipse, verifying bending of light passing close to Sun, as predicted by general theory of relativity

Physics
Aston produces his first mass spectrograph
Barkhausen discovers that magnetization tends to occur in discrete steps
Rutherford identifies first artificial transmutation of a nucleus

1920
Astronomy
Baade discovers the asteroid Hidalgo
Saha publishes the equation for degree of ionization of atoms in stars

Mathematics
Thoralf Albert Skolem contributes to the Lowenheim–Skolem theorem and constructs the Skolem paradox inset theory

Medicine
Goldberger finds a cure for pellagra

Physics
O. Stern confirms experimentally the phenomenon of space quantization

General
The Throop Polytechnic Institute in Pasadena, California, is renamed the California Institute of Technology

c. 1920
Biology
(to c. 1925) Keilin discovers cytochrome

1921
Biology
Banting and Best isolate insulin and show its effectiveness in treating diabetes
Loewi discovers that chemicals are released from nerve endings when the nerve is stimulated

Chemistry
Midgley discovers tetraethyl lead as an antiknock additive for fuel

Earth Science
On the Dynamics of the

Circular Vortex with Applications to the Atmospheric Vortex and Wave Motion (V. Bjerknes)
Physics
Hahn defines nuclear isomerism
Theodor Franz Eduard Kaluza supplements Einstein's 4-dimensional space–time model with a fifth dimension
1922
Biology
A. Fleming publishes his discovery of lysozyme
Hopkins isolates glutathione
Chemistry
Kraus develops processes for the commercial production of tetraethyl lead
Heyrovský describes polarography
Earth Science
Weather Prediction by Numerical Process (Richardson) – the first attempt to apply mathematical techniques to weather prediction
Physics
Friedmann publishes his paper on the expanding universe
Technology
Sabine publishes his papers on architectural acoustics
1923
Astronomy
Wolf devises Wolf diagram for dark nebulae
Biology
Conant shows that oxyhemoglobin contains ferrous iron
Euler-Chelpin works out the structure of the yeast coenzyme
Heidelberger and Avery show that certain polysaccharides in the capsules of pneumococci are responsible for specific antigenic properties
Hevesy makes first application of a radioactive tracer – ^{212}Pb – to a biological system
Whipple reports that liver in the diet increases hemoglobin production
Warburg designs instrument (Warburg manometer) to measure uptake of oxygen by human tissue
Chemistry
Lowry and Brønsted independently formulate a

theory of acids and bases (the Lowry–Brønsted theory)
Valence and the Structure of Atoms and Molecules (G. N. Lewis) – proposes octet theory of valence
Hevesy and Coster discover hafnium
Debye and Huckel publish their theory of electrolytes
Physics
Zworykin constructs the first iconoscope
Compton discovers scattering effect of radiation by electrons (Compton effect)
1924
Archeology
Dart describes the Taung skull
Astronomy
Eddington introduces mass–luminosity relationship for stars
Biology
Berger makes the first human encephalogram
Chemistry
Svedberg introduces the ultracentrifuge as a technique for investigating the molecular weights of very large molecules
Earth Science
The Earth: its Origin, History, and Physical Constitution (Jeffreys)
Mathematics
Banach and Tarski prove the Banach–Tarski paradox
Physics
Pauli proposes his exclusion principle
De Broglie proposes that particles can behave as waves
Appleton establishes experimentally the presence of the Heaviside-Kennelly layer
1925
Astronomy
The Draper Extension first published by Harvard College Observatory, Massachusetts
Biology
Adrian finds that nerve messages are relayed by changes in the frequency of the discharge
Chemistry
Nieuwland polymerizes acetylene to give divinylacetylene
Physics
Giauque introduces the method of adiabatic demagnetization for low temperatures

Goudsmit and Uhlenbeck propose electron spin
Tuve and Breit use pulse-ranging to determine the height of the ionosphere
Heisenberg formulates matrix mechanics
Born and Jordan develop matrix mechanics
Auger discovers that an excited atom can emit an electron in reverting to a lower energy state (Auger effect)
(to 1926) Schrödinger develops wave mechanics
c. **1925**
Biology
Hecht formulates photochemical theory of visual adaptation
1926
Astronomy
The Internal Constitution of the Stars (Eddington)
Hubble publishes Hubble's classification of galaxies
Lindblad proposes rotation of Galaxy
Biology
Introduction to Experimental Biology (de Beer) – disproves germ-layer theory
Windaus discovers that ergosterol is a precursor of vitamin D
Sumner achieves the first isolation of a pure enzyme (urease) and shows it to be a protein
J. J. Abel announces the crystallization of insulin
H. J. Muller discovers that x-rays can induce mutations
Chemistry
I. Noddack, W. Noddack, and O. Berg discover rhenium
Ingold introduces the concept of mesomerism
Medicine
Minot and Murphy discover a cure for pernicious anemia
Physics
Oskar Klein revises Kaluza's theory of a fifth dimension by formulating the Kaluza–Klein theory
Technology
Goddard launches first successful liquid-fuel rocket flight
Baird demonstrates his apparatus for transmitting images

1927
Astronomy
Lemaître provides cosmological model of expanding universe
Oort confirms galactic rotation
Biology
Landsteiner discovers the MN and the P blood-group systems
Chemistry
Electronic Theory of Valency (Sidgwick)
Medicine
Stokes manages to transmit yellow fever to the rhesus monkey
Physics
Wigner introduces parity as a conserved property of nuclear reactions
G. Thomson and (independently) Davisson and Germer demonstrate wave property (diffraction) of electrons
N. Bohr formulates complementarity principle
Heisenberg formulates his uncertainty principle
(to 1932) Fock and Hartree show how Schrödinger's wave equation can be applied to atoms with more than one electron
Psychology
Practice and Theory of Individual Psychology (Adler)
1928
Astronomy
Ira Bowen explains forbidden emission lines in spectra of nebulae
Biology
Riddle isolates prolactin
Griffith discovers bacterial transformation in pneumococci
Medicine
A. Fleming discovers penicillin
Physics
Geiger–Müller counter invented
Dirac introduces relativity into the Schrödinger wave equations to account for electron spin
Raman discovers Raman scattering effect
Ernst August Friedrich Ruska produces the first electron microscope
1929
Astronomy
Struve shows interstellar matter exists throughout Galaxy

Hubble announces his law that the recessional velocity of galaxies is proportional to their distance (Hubble's law)
Biology
Butenandt isolates the first pure sex hormone, estrone
Edward Adelbert Doisy isolates the hormone oestrone and soon afterwards, oestradiol
(to 1931) Barbara McClintock establishes linkage groups in maize
Chemistry
Paneth demonstrates the existence of the methyl radical
Levene identifies the carbohydrate in thymus nucleic acid as deoxyribose
The Constitution of the Sugars (Haworth) – introduces the idea that sugar molecules can exist in ring form
H. Fischer synthesizes hemin
Eyring and Polyani calculate potential-energy surface for a chemical reaction – Eyring later develops absolute-rate theory of reaction rates
Earth Science
A. Holmes proposes his convection-current theory to explain continental drift
On the Direction of Magnetization of Basalt (Matuyama) – discovery of remnant magnetization
Medicine
Forssman introduces the technique of cardiac catheterization
Physics
Bothe uses coincidence method to show cosmic rays are particles
(to 1933) Rabi invents the atomic- and molecular-beam magnetic-resonance method
1930
Aeronautics
B. N. Wallis invents the geodetic construction
Astronomy
Tombaugh discovers Pluto
Trumpler discovers interstellar absorption
Lyot invents the coronagraph
Biology
Karrer determines the structure of carotene
Physics
Lawrence begins to construct cyclic particle accelerator (cyclotron)
Pauli suggests the existence of the neutrino

Dirac proposes negative energy states for the electron, which led subsequently to the appreciation of antimatter
Néel proposes the existence of antiferromagnetism
1931
Biology
Goodpasture introduces the method of culturing viruses in fertile eggs
Karrer determines the structure of, and synthesizes, vitamin A
C. Stern provides experimental evidence for crossing-over between homologous chromosomes
Chemistry
Carothers produces the first synthetic rubber (neoprene)
Urey discovers heavy water (deuterium oxide)
Mathematics
Gödel proves incompleteness of arithmetic
Physics
Van de Graaf builds his high-voltage electrostatic generator
1932
Astronomy
Jansky publishes his discovery of radio interference from the stars, so founding the new field of radio astronomy
Biology
J. B. S. Haldane makes the first estimate of mutation rates in humans
Krebs and Henseleit introduce the urea cycle
King and Szent-Györgi announce their (independent) isolations of vitamin C
Chemistry
Pauling introduces the idea of resonance hybrids
Bergmann discovers the carbobenzoxy method of peptide synthesis
Physics
Chadwick discovers the neutron
Anderson discovers the positron
Cockcroft and Walton achieve first artificial nuclear transformation
1933
Biology
King determines the formula of vitamin C
The first synthesis of a vitamin, vitamin C, by Haworth and Hirst
Brachet demonstrates that

both DNA and RNA occur in animal and in plant cells

Bernal obtains the first x-ray photograph of a single-crystal protein

Mathematics

The Foundations of the Theory of Probability (Kolmogorov)

Physics

Anderson discovers the muon

Occhialini and Blackett obtain cloud-chamber tracks of the positron

Fermi proposes weak interaction as one of fundamental interactions in physics

Walther Meissner and R. Oschenfeld discover the Meissner effect, now used as a routine test for superconductivity

General

Institute for Advanced Study opens in Princeton, New Jersey

1934

Biology

R. R. Williams works out the formula of vitamin B_1

Ružička is the first to synthesize a sex hormone, androsterone

(to 1935) Theorell gives the first detailed account of enzyme action

Chemistry

Oliphant and Harteck produce the hydrogen isotope tritium

Earth Science

(to 1939) *On Seismic Waves* (Gutenberg and Richter)

Physics

Casimir and Gorter advance a two-fluid model of superconductivity

Cherenkov discovers radiation from fast particles (Cherenkov radiation)

Fermi discovers that slow neutrons more readily interact with nuclei

Noddack anticipates Frisch in suggesting the hypothesis of nuclear fission

I. and P. Joliot-Curie discover artificial radioactivity

Goldhaber and Chadwick discover the nuclear photoelectric effect

1935

Biology

Dam discovers vitamin K

Stanley crystallizes the tobacco mosaic virus and demonstrates the retention of infectivity after crystallization

Von Euler isolates the first of the prostaglandins

Chemistry

Carothers produces nylon

Earth Science

Seismological Tables (Jeffreys and Bullen)

Richter develops scale for earthquake magnitudes (Richter scale)

Mathematics

The Concept of Truth in Formalized Languages (Tarski)

Medicine

Egas Moniz introduces the operation of prefrontal leukotomy

Domagk reports on the effectiveness of prontosil against streptococcal infection

Physics

Yukawa predicts the existence of the pion

Can Quantum Mechanical Description of Physical Reality be Considered Complete? (Albert Einstein, Boris Podolsky, and Nathan Rosen)

Technology

The Detection of Aircraft by Radio Methods (Watson-Watt)

1936

Biology

Kuhn synthesizes vitamin B_6

Rose discovers the first essential amino acid – threonine

Experimentelle Beiträge zu einer Theorie der Entwicklung (Spemann)

Dale demonstrates that acetylcholine is released at motor-nerve endings of voluntary muscle

Computers

Konrad Zuse invents the computer 'out of laziness'

Earth Science

Daly suggests existence of turbidity currents in oceans

Physics

Wigner and Breit work out the Breit–Wigner formula

Mueller invents the field-emission microscope

The Structure of Metals and Alloys (William Hume Rothery)

1937

Biology

Gorer discovers the histocompatibility antigens and establishes their control at the genetic level

Williams synthesizes vitamin B_1

Electrical Signs of Nervous Activity (Erlander and Gasser)

Bawden and Pirie demonstrate that tobacco mosaic virus contains RNA

Blakeslee discovers that colchicine induces multiple sets of chromosomes in plants

Krebs discovers the tricarboxylic acid cycle (Krebs cycle)

Genetics and the Origin of Species (Dobzhansky)

Sonneborn discovers sexuality in the protozoa

Chemistry

Segrè discovers technetium

Computers

George Robert Stibitz builds his first calculator

Earth Science

Our Wandering Continents (Du Toit) – suggests separation of Pangaea into Laurasia and Gondwanaland

Mathematics

On Computable Numbers (Alan Turing)

Medicine

Bovet develops the antihistamine drug 933F

Theiler and Smith announce the development of a vaccine against yellow fever

Elvehjem shows active ingredient in pellagra-preventive factor is nicotinic acid

Physics

Frank and Tamm explain Cherenkov radiation

Kapitza discovers the superfluidity of helium

Technology

Chester Floyd Carlson and Otto Kornei make the first electrostatic (xerographic) copier

1938

Astronomy

Nicholson discovers Lysithea and Carme (satellites of Jupiter)

Hahn discovers nuclear fission (interpreted by Meitner and Frisch)

Physics

Alvarez describes K-capture

Technology

Ladislao and Georg Biró patent the ballpoint pen

1939

Astronomy

Introduction to the Study of

Stellar Structure
(Chandrasekhar) – introduces idea of Chandrasekhar limit

Biology
Dam and Karrer isolate vitamin K
Doisy synthesizes and characterizes vitamin K

Chemistry
The Nature of the Chemical Bond (Pauling)
Perey discovers francium

Earth Science
Bergeron–Findeisen theory of precipitation

Medicine
Ewin's research team develops the drug sulfapyridine
Dubos discovers the antibiotic tyrothricin
Florey and Chain extract penicillin

Physics
Energy Production in Stars (Bethe)
Meitner and O. Frisch interpret Hahn's experiments as evidence of nuclear fission
Meitner announces the discovery of light nuclei (fission products) following neutron bombardment of uranium nuclei
Luis Walter Alvarez and Felix Bloch make the first measurement of the magnetic moment of the neutron
Alvarez demonstrates that tritium is radioactive
Boot and Randall develop the cavity magnetron

Technology
The first jet flight is powered by an engine designed by Hans Joachim Pabst von Ohain

1940

Astronomy
Reber publishes first radio map of sky

Biology
Landsteiner announces his discovery of the rhesus factor

Chemistry
McMillan and Abelson discover the first transuranic element – neptunium
Patent taken out on the insecticide DDT (P. H. Müller)
Kamen and Ruben discover the radioisotope carbon–14

Earth Science
Rossby discovers long sinusoidal atmospheric waves of large amplitude (Rossby waves)

Physics
Hillier builds the first successful American high-resolution electron microscope
(to 1950) Development of modern theory of quantum electrodynamics by Feynman, Schwinger, and Tomonaga (independently)

1941

Aeronautics
The jet engine, designed by Whittle and fitted to a Gloster E28/39 aircraft, makes its first flight (May 15)

Astronomy
Edlen shows 'coronium' to be highly ionized heavy atoms

Chemistry
Martin and Synge develop partition chromatography

Medicine
Cournand and Ranges develop cardiac catheterization
C. B. Huggins pioneers the use of female sex hormones in treating cancer of the prostate gland

Physics
(to 1943) Tomonaga develops a quantum field theory consistent with the theory of special relativity

1942

Biology
Brachet suggests RNA granules might be the agents of protein synthesis
(to 1953) K. E. Bloch and others determine biosynthesis of cholesterol

Computers
John Vincent Atanasoff and Clifford Berry complete the Atanasoff–Berry Computer (ABC)

Earth Science
Belousov suggests that Earth movements caused by density of crust

Physics
First self-supporting nuclear-fission chain reaction (Stagg Field, Chicago)
Alfvén postulates hydromagnetic waves (Alfvén waves) in plasmas

1943

Astronomy
MKK system of stellar classification introduced (by W. W. Morgan, Keenan, and Kellman)
Seyfert discovers Seyfert galaxies

Baade introduces Populations I and II classification of stars

Biology
Luria and Delbrück demonstrate spontaneous mutation in bacteria
B. Chance provides experimental evidence for the formation of an enzyme–substrate complex, as proposed by Michaelis
Waksman isolates streptomycin from the mold *Streptomyces griseus*

Chemistry
Longuett-Higgins suggests that borane contains hydrogen bridges

Computers
Howard Hathaway Aiken completes the Harvard Mark I or ASCC (Automatic Sequence Controlled Calculator)

Technology
Luis Walter Alvarez guides a distant plane to land using radar

1944

Biology
An artificially produced plant species *Ehrharta erecta* is established in a natural environment
MacLeod, Avery, and McCarty provide evidence that DNA is responsible for bacterial transformation
Wyckoff and R. C. Williams develop the metal-shadowing technique to give three-dimensional photographs with the electron microscope
What is Life? (Schrödinger)

Earth Science
Principles of Physical Geography (A. Holmes)

Physics
Veksler invents synchrocyclotron

1945

Biology
Beadle and Tatum formulate the one gene–one enzyme hypothesis
Alfred Day Hershey demonstrates (independently of Salvador Edward Luria) spontaneous mutations in bacteriophages

Computers
Mauchly and Eckert complete ENIAC (Electronic Numerator Integrator and Calculator)

Physics
Atomic bomb first tested (July 16)

General
Studies in the Logic of Confirmation (Carl Gustav Hempel) – challenges the foundations of inductive logic
1946
Astronomy
Tousey photographs the solar spectrum from above the ozone layer
Biology
Delbrück and Hershey demonstrate genetic recombination in viruses
Von Euler isolates noradrenaline (norepinephrine)
Tatum and Lederberg demonstrate occurrence of sexual reproduction in bacteria
Cohen introduces radioactive labeling of microorganisms
Earth Science
Schaefer seeds clouds and produces the first artificial rainfall
Physics
Bloch and Purcell independently introduce the technique of nuclear magnetic resonance
General
The Kaiser Wilhelm Gesselschaft in Berlin is renamed the Max Planck Institute
1947
Astronomy
Bok discovers Bok globules
Ambartsumian introduces idea of stellar association
Biology
Lipmann discovers coenzyme A
Todd synthesizes ADP and ATP
Chemistry
Libby develops the radiocarbon dating technique
Physics
Kusch measures the magnetic moment of the electron and finds a discrepancy between the experimental and theoretical values
Brattain, Bardeen, and Shockley invent the point-contact transistor
Lamb announces the *Lamb shift*
Occhialini and Powell observe tracks of the pion

1948
Anthropology
Sexual Behavior in the Human Male (Kinsey)
Astronomy
Burnright detects solar x-rays
H. D. and H. W. Babcock develop magnetograph and detect solar and stellar magnetic fields
Alpher, Bethe, and Gamow publish theory on origin of elements
Alpher and Herman first predict cosmic background radiation from big bang
The Steady-State Theory of the Expanding Universe (Gold and Bondi)
Kniper discovers Miranda (satellite of Uranus)
200-inch Hale telescope in operation at Mount Palomar (largest until 1977)
Biology
Folkers isolates vitamin B_{12}
The Functional Organization of the Diencephalon (W. H. Hess)
Medicine
Weller and Neva succeed in growing the rubella virus in tissue culture
Duggar discovers chlortetracycline (Aureomycin), the first tetracycline antibiotic
Hench introduces cortisone to treat rheumatoid arthritis
Physics
Shockley develops the junction transistor
Gabor invents the technique of holography
Richard Feynman's paper *Space–Time Approach to Non-Relativistic Quantum Mechanics*, in which he introduces path integrals, is turned down by *Physical Review*
General
Studies in the Logic of Explanation (Carl Gustav Hempel and Paul Openheim) – proposes the Hypothetico-Deductive method
1949
Astronomy
Kniper discovers Nereid (satellite of Neptune)
Baade discovers asteroid Icarus
Biology
Enders, Weller, and Robbins cultivate polio virus *in vitro* on human embryonic tissue

Sir Frank Macfarlane Burnet postulates that immunological tolerance has not yet developed in the embryo stage
Murray Llewellyn Barr with Ewart Bertram discover Barr bodies in female somatic cells
Chemistry
G. Porter and Norrish begin developing methods of flash photolysis
Computers
Maurice Wilkes builds EDSAC, the first working computer with a stored program
Mathematics
A Mathematical Theory of Communication (Shannon and Weaver)
Medicine
Burnet discovers the phenomenon of acquired immunologic tolerance
Physics
Shull demonstrates antiferromagnetism using neutron diffraction
Space–Time Approach to Quantum Electrodynamics (Richard Feynman) – introduces use of Feynman diagrams
1950
Astronomy
The Nature of the Universe (Hoyle)
Oort proposes comets originate from reservoir far out in solar system (Oort's cloud)
Whipple proposes icy-nucleus theory of comets
R. H. Brown plots the first radio map of an external galaxy
Biology
Pontecorvo and Roper discover the parasexual cycle in fungi
Chargaff announces base ratios of DNA, important in the construction of Watson–Crick model of DNA (to 1952) Harris shows that release of pituitary hormones is controlled by the hypothalamus
Computers
Computing Machinery and Intelligence (Alan Turing)
Physics
The betatron at the University of Illinois is completed
Rainwater suggests his modification to the shell model of nuclear structure
(to 1953) Mottelson and A.

Bohr advance a theory of nuclear structure

Andrei Dmitriyevich Sakharov and Igor Tamm describe a process whereby a deuterium plasma could be confined in a magnetic bottle so as to extract energy produced by nuclear fusion – the tokamak

1951

Astronomy

Brouwer, Clemence, and Eckert publish table for accurate orbits of outer planets

Purcell and Ewen detect 21-centimeter hydrogen emission line (predicted 1944, van de Hulst)

Nicholson discovers Ananke (satellite of Jupiter)

Biology

The Study of Instinct (Tinbergen)

Chromosome Organization and Genetic Expression (Barbara McClintock) – description of transposition of genes

Chemistry

Pauling suggests protein molecules are arranged in helices

Physics

The first breeder reactor is built following Zinn's design

1952

Astronomy

Merrill detects technetium in S-type stars

Biology

Hershey and Chase prove that DNA is the genetic material of bacteriophages

Wilkins and Franklin obtain evidence for the double helical structure of DNA

Lederberg and Zinder discover bacterial transduction

Cohen and Levi-Montalcini determine the nature of a nerve-growth factor from mouse tumor cells

Chemistry

E. O. Fischer confirms sandwich structure for ferrocene by x-ray crystallography

Computers

EDVA (Electronic Digital Variable Computer) is completed

Earth Science

Heezen provides evidence of turbidity currents in oceans

Medicine

Pathology of the Cell (Cameron)

Physics

Glaser produces the first bubble chamber

Bohr and Mottelson propose their collective model of nuclear structure

CERN (Conseil Européen pour la Recherche Nucléaire) set up by eleven European governments in Geneva under its first director, Felix Bloch

1953

Anthropology

Sexual Behavior in the Human Female (Kinsey)

Biology

H. E. Huxley and Hanson propose the sliding-filament theory of muscle contraction

Medawar shows that adult mammals injected with foreign tissues at the embryonic stage or at birth will later accept grafts from the original tissue donor

Watson and Crick construct a three-dimensional model of the DNA molecule

Kettlewell demonstrates industrial melanism in moths

Chemistry

Ziegler introduces the Ziegler catalysts for polymerization of ethane

Karl and Hauptman publish *The Phases and Magnitudes of the Structure Factors*, in which they show how phase structures could be inferred directly from diffraction patterns

Physics

Lax detects cyclotron resonance in germanium

Gell-Mann introduces the concept of strangeness

Townes makes working model of maser

1954

Biology

Du Vigneaud synthesizes the first protein – oxytocin

Chemistry

Natta uses improved Ziegler catalysts to produce stereospecific polymers

Eigen introduces relaxation techniques to study extremely fast chemical reactions

Earth Science

Seismicity of the Earth (Richter and Gutenberg)

Medicine

First clinical trials of oral contraceptives are conducted

Joseph Edward Murray carries out the first kidney transplant into the pelvis of the recipient

Physics

Construction of the 25 GeV Synchrotron begins at CERN in Geneva

1955

Astronomy

US Navy begins Vanguard project to launch US satellite

Biology

Ochoa discovers the enzyme polynucleotide phosphorylase, later used to synthesize artificial RNA

The complete amino-acid sequence of a protein (insulin) is established by Sanger

Chemistry

Thermodynamics of Irreversible Processes (Prigogine)

Medicine

Jerne proposes the clonal selection theory of antibody selection to account for white blood cells being able to produce a large range of antibodies

Physics

Chamberlain, Segrè, Weigand, and Ypsilantis discover the antiproton

Basov and Prokhorov develop the maser

Bridgman transforms graphite into synthetic diamond using extremely high pressures

Dehmelt begins work on the Penning trap, to isolate a single electron

1956

Astronomy

Oort and Walraven show that light from the Crab nebula is synchrotron radiation

Sir Richard van der Riet declares that space travel is 'utter bilge'

Biology

Kornberg discovers DNA polymerase

Li determines the sequence of amino acids in adrenocorticotropic hormone

Berg identifies the molecule later known as transfer RNA

Hodgkin establishes the structure of vitamin B_{12}

Earth Science
Ewing plots Mid-Atlantic Ridge
Physics
Lee and Yang argue that parity is not conserved in weak interactions
Reines and Cowan confirm the existence of the neutrino
Wu experimentally confirms that parity is not conserved in beta disintegration
Cooper shows that, at low temperatures, electrons in a conductor can act as bound pairs (Cooper pairs)
1957
Astronomy
Hoyle, Fowler, and E. M. and G. Burbidge publish theory of formation of elements in stars
250-foot radio telescope in use at Jodrell Bank
Launch (Oct. 4) of Soviet satellite Sputnik 1 – first artificial satellite
Launch of Soviet satellite Sputnik 2 with dog Laika – proves living organisms can survive in space
Biology
The Physiology of Nerve Cells (Eccles)
Sutherland discovers cyclic AMP
Isaacs and Lindenmann report discovery of interferon
Kendrew determines the first three-dimensional model of a protein – myoglobin
Leloir announces his discovery of the mechanism whereby glycogen is synthesized in the body
The Path of Carbon in Photosynthesis (Calvin)
Antibody Production Using the Concept of Clonal Selection (Burnett)
Computers
John McCarthy sets up the Artificial Intelligence Laboratory at the Massachusetts Institute of Technology
Physics
Bardeen, Cooper, and Schrieffer formulate their theory of superconductivity
Mössbauer discovers recoilless nuclear resonance absorption (Mössbauer effect)
1958
Astronomy
Launch of Explorer 1, which discovers Van Allen belts –

first in major series of US research satellites
Launch of Pioneer 1 – start of series of US probes to study Moon and solar system
Ephemeris time adopted by astronomical scientists
The Royal Greenwich Observatory moved to Herstmonceux Castle in Sussex
Biology
Evolution of Genetic Systems (Darlington)
Meselson and Stahl demonstrate the semiconservative nature of DNA replication
Dausset discovers the human histocompatibility system
Medicine
Denis Burkitt publishes his first account of Burkitt's lymphoma
Physics
Esaki discovers tunneling effect in semiconductor junctions
Technology
Jack St. Clair Kilby makes the first integrated circuit using silicon
1959
Astronomy
Third Cambridge Catalogue (3C) – catalog of known radio sources
Launch of Lunar 1 – first in series of Soviet lunar space probes
Launch of Discoverer 1 – first in series of US unmanned probes in orbit over poles; later Discoverer probes used for military surveillance
Biology
Dulbecco introduces the idea of cell transformation
Moore and Stein establish the amino-acid sequence of the enzyme ribonuclease
Medicine
Sabin announces successful results from testing live attenuated polio vaccine
Kouwenhoven introduces closed-chest cardiac massage
Physics
Luis Walter Alvarez builds a 72″ bubble chamber
Lederman, Lee, Schwartz, and Steinberger carry out the two-neutrino experiment
1960
Anthropology
Louis Leakey discovers the

skull of *Homo erectus* at Olduvai
Astronomy
Sandage and Matthews discover quasars
Launch of Tiros 1 – first in series of US weather satellites
Biology
R. B. Woodward synthesizes chlorophyll
Jacob and Monod propose the existence of the operon
Physics
Giaever investigates electron tunneling in superconducting junctions
Maiman designs the laser
Technology
Echo I, simple radio reflector balloon, launched
1961
Astronomy
Launch of Venera 1 (also called Venus 1) – first in series of Soviet Venus space probes (first mission failed)
Launch of Vostok 1 – first in series of Soviet manned space flights putting first man in space (Yuri Gagarin, Apr. 12)
Launch (May 5) of US Mercury-Redstone 3 mission – first American in space (Alan B. Shepard in suborbital flight)
Launch of Ranger 1 – first in series of US probes to investigate Moon (detailed photographs by Ranger 7, 1964)
Biology
Brenner and Crick show that the genetic code consists of a continuous string of nonoverlapping base triplets
Good and J. Miller independently discover the role of the thymus in the development of immunity
Nirenberg makes the first step in breaking the genetic code by finding that UUU codes for phenylalanine
Physics
Brans and Dicke propose Brans–Dicke theory of gravitation
Ne'eman and Gell-Mann independently develop a mathematical representation for the classification of elementary particles
1962
Astronomy
Rossi detects cosmic x-ray source in Scorpio (Sco X–1)
Launch (Feb. 20) of US

Mercury–Atlas 6 mission – first American in orbit (John H. Glenn)

Launch of OSO1 (Orbiting Solar Observatory) – first in series of US probes for studies of Sun

Launch of Ariel 1 – first in series of US-launched British satellites for studies of ionosphere and x-ray sources

Launch of Mariner 1 and Mariner 2 – first in series of US planetary space probes: Mariner 2 flies past Venus

Launch of Mars 1 – first in series of Soviet probes to study Mars

Chemistry
Bartlett demonstrates that not all noble gases are inert by forming the compound xenon fluoroplatinate

Earth Science
History of Ocean Basins (H. H. Hess) – introduction of sea-floor spreading hypothesis

Physics
Josephson predicts effects occurring in superconducting junctions (Josephson effects)

General
The Structure of Scientific Revolutions (Thomas Kuhn)

1963

Astronomy
Schmidt investigates red shifts of quasars

Launch of Soviet spacecraft Vostok 6 – first woman in space (Valentina Tereshkova, June 16)

Biology
Blumberg discovers the Australian antigen

On Aggression (Lorenz)

Medicine
Gajdusek describes the first slow-virus infection to be identified in humans

Physics
(to 1964) Cormack publishes papers on the mathematical basis of x-ray tomography

1964

Astronomy
Dicke and Peebles predict cosmic background radiation and start searching for it

Penzias and Wilson discover cosmic microwave background radiation

Launch of unmanned test flight Gemini 1 – first in a US series of manned space missions

Ranger 7 takes first detailed photographs of Moon's surface

Launch of US planetary probe Mariner 4 to investigate Mars (1965)

Biology
Merrifield synthesizes bradykinin

Conduction of the Nervous Impulse (A. L. Hodgkin)

Genetical Theory of Social Behaviour (W. D. Hamilton) – beginning of sociobiology

Medicine
Clarke announces injection of Rh-negative mothers with Rh-antibodies prevents subsequent Rhesus babies

Physics
The Eightfold Way (Gell-Mann and Ne'eman)

Cronin, Fitch, Christenson, and Turley find that charge-conjugation-parity (CP) conservation is violated in the decay of neutral kaons

On the Einstein–Podolsky–Rosen Paradox (John Stuart Bell) – includes Bell's theorem on quantum mechanics

Peter Higgs first postulates the Higgs field and the Higgs boson

1965

Astronomy
Nucleosynthesis in Massive Stars and Supernovae (Hoyle and Fowler)

Friedman first accurately locates nonsolar x-ray source in Crab nebula (Tau X–1)

Mariner 4 flies past Mars and relays photographs of surface

Biology
Holley announces the nucleotide sequence of alanine tRNA

R. B. Woodward and Eschenmoser synthesize vitamin B_{12}

Earth Science
Launch of US probe Explorer 29; Geodetic Earth-Orbiting Satellite (GEOS), used for geodetic measurements

Physics
Joseph Weber completes building his detector for gravitational waves

Technology
Launch of Intelsat 1 (called Early Bird) – International Communications Satellite stationed over Atlantic

1966

Astronomy
Launch of Soviet Moon probe Luna 9 – first probe to make soft landing on Moon and return photographs

Launch of Apollo 1 – the first in a series of US manned space probes in program to land men on the Moon (Apollo 11, 1969)

Launch of Soviet Moon probe Luna 10 – first probe to go into lunar orbit

Launch of OAO1 (Orbiting Astronomical Observatory) – first in series of US probes containing telescopes, spectrometers, and detectors for ultraviolet and x-ray studies of universe

Launch of Surveyor 1 – first in series of US Moon probes to make soft landing and relay photographs and data on Moon's surface

Launch of Lunar Orbiter 1 – first in series of US Moon probes to photograph lunar surface

Biology
Gilbert and Muller-Hill isolate the lac repressor

1967

Astronomy
Bell and Hewish discover first pulsar

Launch of Soyuz 1 – first in major series of Soviet manned spacecraft (Vladimir Komarov killed during re-entry on first mission)

Chemistry
Pederson publishes work on crown ethers

Earth Science
A Revolution in Earth Science (J. T. Wilson)

Medicine
The first heart transplant performed on a human patient (by C. N. Barnard)

Physics
Friedman, Kendall, and Taylor demonstrate that the proton has an inner structure

A Model of Leptons (Weinberg) – introduces electroweak theory

1968

Astronomy
Discovery of first mascons on Moon's surface – local areas of increased gravity discovered by tracking spacecraft Lunar Orbiter 5

Physics
Salam and Weinberg
independently formulate
gauge theory to give unified
description of weak and
electromagnetic interactions –
predict existence of neutral
currents
Georges Charpak describes
his drift chamber
Feynman introduces the idea
of 'partons'
Sakharov publishes *Progress,
Peaceful Coexistence and
Intellectual Freedom*, arguing
for a global reduction in
nuclear weapons

1969
Astronomy
N. A. Armstrong and E. E.
Aldrin make first manned
lunar landing (Apollo 11, July
20 in Sea of Tranquillity) –
bring back samples of lunar
material
Van de Kamp detects planet
orbiting Barnard's star
Mount Wilson and Palomar
Observatories renamed the
Hale Observatories
Biology
Edelman announces the
amino-acid sequence of
immunoglobulin G
Guillemin and Schally
independently determine
structure of thyrotropin
releasing factor

1970
Astronomy
Launch of Soviet Moon probe
Luna 16 – first successful
automatic return of lunar
sample
Launch of Explorer 42 (called
Uhuru) – US probe for
research on cosmic x-rays
Biology
Baltimore and Temin
independently discover the
enzyme reverse transcriptase
Khorana announces the
synthesis of the first artificial
gene
H. O. Smith identifies the first
restriction enzyme (Hind II)
Neher and Sakmann develop
the patch-clamp technique for
studying ion channels in cell
membrane
Duesberg discovers oncogenes
Physics
Glashow extends the
Weinberg–Salam theory by
introducing the property of
charm

Yoichipo Nambu introduces
string theory

1971
Astronomy
Launch of Salyut 1 – first in a
series of Soviet space stations
for scientific research and
military surveillance
Launch of Mariner 9 – goes
into orbit around Mars and
relays data and photographs
of surface

1972
Anthropology
Richard Erskine discovers
Homo habilis
Astronomy
Launch of US planetary probe
Pioneer 10 to investigate
Jupiter (1973)
Biology
Nathans determines the order
of the 11 cleavage fragments
of the SV40 virus
Gould and Eldredge propose
the punctuated equilibrium
hypothesis, that views
evolution as episodic rather
than continuous
Medicine
Gallo and his team identify
interleukin-2 – a factor that
stimulates growth in human
T-cells
Technology
Land launches the first
polaroid camera

1973
Astronomy
Launch of US planetary probe
Pioneer 11 to investigate
asteroid belt, Jupiter (1974),
and Saturn (1979)
Launch of Skylab 1 – first in
series of US space stations for
scientific research
Launch of Skylab 3 –
observations made on solar
flares
Pioneer 10 flies past Jupiter –
relays photographs of planet
and satellites
Launch of US planetary probe
Mariner 10 – to investigate
Venus and Mercury (1974)
*Large Scale Structure of Space
and Time* (Hawking and Ellis)
Biology
Snyder and Pert find receptor
sites for opiates in the limbic
system of the brain
Boyer and Helling construct
functional DNA by joining
together DNA from two
different sources

Medicine
CAT body scanners introduced
Physics
Dehmelt isolates and traps a
single electron

1974
Anthropology
Johanson discovers the oldest-
known human fossil
Australopithecus afarensis
(known as 'Lucy')
Astronomy
Mariner 10 flies past Venus
and Mercury and relays
photographs of surface
Pioneer 11 flies past Jupiter
Hawking proposes that black
holes can emit particles
Biology
The Berg letter warns of the
dangers of uncontrolled
experiments on DNA
recombinants
Chemistry
Rowland and Molina point out
that CFCs in the stratosphere
attack ozone
Physics
Richter and Ting
independently discover the
J/psi particle
Perl identifies a heavy lepton
Penrose shows that a non-
periodic tiling can have an
almost five-fold symmetry
The Physics of Liquid Crystals
(de Gennes)

1975
Anthropology
Richard Leakey discovers
Homo erectus
Astronomy
Launch of Venera 9 and
Venera 10 – Soviet probes that
land and relay data and
photographs of surface of
Venus
Apollo–Soyuz Test Project –
joint US and Soviet space
project with docking of two
spacecraft in orbit
Biology
Milstein produces monoclonal
antibodies
Kosterlitz and Hughes
discover endogenous opiates
(enkephalins) in brain tissue

1976
Chemistry
The National Academy of
Sciences publishes a report
supporting the work of
Rowland and Molina on the
effect of CFCs on the ozone
layer

Mathematics
Haker and Appel announce the solution of the four-color map problem

Medicine
Varmus, Bishop, Stehelin, and Vogt show that viral oncogenes are derived from the genes of the host, incorporated into the genehi material of the virus in a modified form

Physics
The 450 GeV Super Proton Synchrotron is installed at CERN

1977

Astronomy
Soviet 6-meter reflector in operation at Zelenchukskaya
Launch of Voyager 1 and Voyager 2 (originally Mariner 11 and Mariner 12) – US planetary probes to study Jupiter (1979) and Saturn (1980/81)

Biology
The first complete nucleotide sequence of the genetic material of an organism, the bacteriophage OX174, is published

Physics
Lederman and co-workers discover the bottom quark
Fairbank announces the first experimental evidence for an isolated quark

1978

Medicine
Edwards and Steptoe achieve successful transplantation of human ova fertilized *in vitro*

1979

Astronomy
Voyager 1 and Voyager 2 fly by Jupiter – study and photograph atmosphere and surfaces of main satellites
Pioneer 11 flies past Saturn

1980

Anthropology
Pilbeam discovers a partial skull of a hominid fossil, *Sivapithecus*

Astronomy
Voyager 2 flies past Saturn – relays pictures of rings

Physics
Alan Guth proposes his inflationary universe model
Fred Hoyle suggests that cosmic dust contains bacterial cells

Geology
Luis Alvarez publishes his

impact theory of the extinction of dinosaurs

Physics
von Klitzing discovers the quantum Hall effect

1981

Astronomy
Successful launch and recovery of space shuttle – first reusable spacecraft

Biology
K. E. Davies and J. M. Murray discover a restriction fragment length polymorphism (RFLP) on the X chromosome linked with Duchenne muscular dystrophy
Bining and Roher use their scanning-tunneling microscope (STM) to resolve surface details of crystals

1982

Astronomy
Venera 13 and 14 land on Venus and relay color pictures of surface
John Huchra and colleagues put forward evidence for *The Great Attractor* lying beyond the Hydra–Centaurus supercluster

Biology
Hartmut Michael crystallizes the membrane proteins of the bacterium *Rhodopseudomonas viridis*
Thomas Cech concludes that RNA is self-splicing and proposes the term 'ribozyme'

Chemistry
Huffman and Kratschmer find spectral evidence for what is later recognized to be a new form of carbon (fullerene)

Mathematics
Benoit Mandelbrot introduces the term 'fractal'

Medicine
Robert Gallo discovers the viruses HTLV-1 and HTLV-2
Discovery of cancer-causing genes (oncogenes)

1983

Astronomy
First American woman in space (Sally Ride on space-shuttle flight)

Biology
Mullis invents his technique for producing copies of DNA – the polymerase chain reaction

Medicine
Nancy Wexler and James Gusella discover the position of the gene responsible for Huntington's disease

Luc Montagnier becomes convinced that LAV (lymphadenopathy associated virus) is the cause of AIDS
Robert Gallo isolates the virus HTLV-3

Physics
Rubbia and Charpak show evidence for the W particle at CERN

1984

Astronomy
The 98-inch Isaac Newton telescope is moved from the Royal Greenwich Observatory at Herstmonceux Castle, Sussex, to a new site at La Palma in the Canary Islands

Biology
Alec Jeffreys discovers the technique of genetic fingerprinting

Chemistry
Dany Schectman and colleagues find crystals with a five-fold symmetry (quasicrystals)

Physics
Carl Sagan and colleagues publish the controversial *Nuclear Winter: Global Consequences of Multiple Nuclear Explosions*

1985

Chemistry
Richard Smalley, Harold Kroto, and others discover a new form of carbon C_{60} and propose the name 'buckminsterfullerene'
Joe Farman publishes his results on the connection between ozone depletion and atmospheric CFCs

Biology
Leroy Hood and Lloyd Smith develop the automatic sequencer for determining the sequence of DNA base pairs – a useful tool for mapping the human genome

Medicine
Gallo patents a test for HIV
Luc Montagnier isolates the virus HIV-2

1986

Astronomy
The space shuttle Challenger blows up shortly after launch on Jan. 28 with the death of six astronauts and an observer
Voyager 2 passes close to Uranus – 10 more satellites discovered
Mir (meaning 'Peace')

launched by the Soviet Union – the first permanently manned space station

Biology
Charles Delisi organizes a meeting in Santa Fe, New Mexico, to initiate the setting up of the Human Genome Project

Medicine
The International Committee on the Taxonomy of Viruses officially names the AIDS virus as HIV (human immunodeficiency virus)
Gallo and Montagnier agree to share the patent for the AIDS test

Physics
Georg Bednorz and Alex Muller discover a new type of superconductor with a high critical temperature (35K)

1987

Physics
Paul Ching-Wu Chu makes a stable mixed yttrium–barium–copper oxide superconductor with the high critical temperature of 93K
Superstring Theory (John Henry Schwarz, Michael Green, and Ed Witten)
John Stuart Bell publishes *Speakable and Unspeakable in Quantum Mechanics*, giving his own views on 'Bell's paradox'

1988

Biology
Ray White announces that preliminary maps have been completed for most of the human chromosome

Physics
Martin Fleischmann and Stanley Pons claim to have discovered an electrolytic method of achieving cold nuclear fusion – other research groups fail to reproduce the effects

1989

Astronomy
John Huchra and Margaret Geller find the cluster of galaxies 'The Great Wall'
Voyager 2 passes Neptune – discovers a 'Great Dark Spot' and 4 complete rings

Biology
Lap-Chee Tsui, Jack Riordan, and Francis Collins announce the location of the cystic fibrosis gene

Earth Science
Global Warming (Stephen Henry Schneider) – discussing the greenhouse effect

Physics
The Large Electron–Positron Collider is installed at CERN, Geneva
Richard Muller publishes *Nemesis*, claiming that the Sun has a companion star

1990

Astronomy
The Hubble Space Telescope is launched – a fault in the primary mirror means that it has a lower angular resolution than expected
The Cosmic Background Explorer (COBE) starts mapping the sky
An 18th moon of Saturn is discovered (from Voyager 2 results)

Medicine
Start of The Human Genome Project with the selection of six institutions to do the work

1991

Astronomy
First British woman in space (Helen Sharman on a Russian Mir mission)

Chemistry
Doped fullerenes discovered to have superconductivity

1993

Astronomy
Results from COBE show ripples in the cosmic background radiation (supporting the inflationary theory of the universe)
NASA sends space shuttle mission to repair Hubble Space Telescope

Mathematics
Andrew Wiles produces a proof of Fermat's last theorem

1994

Chemistry
Element 110 (darmstadtium) and element 111 (roentgenium) discovered by workers at Darmstadt in Germany

1995

Astronomy
Galileo spacecraft sends back pictures of Jupiter and its moons
Comet Hale–Bopp first observed

Biology
Publication of the first complete genome of a free-

living organism (the bacterium *Haemophilus influenzae*)

Physics
E. Cornell and C. Wieman produce a Bose–Einstein condensate
Researchers using the Tevatron at Fermilab near Chicago discover the top quark

1996

Astronomy
Launch (Feb. 17) of Near Earth Asteroid Rendevous (NEAR) to investigate the asteroid belt

Biology
Dolly the sheep born – the first mammal to be cloned from adult body cells – as a result of work by scientists at the Roslin Institute near Edinburgh

Chemistry
Element 112 (ununbium) discovered by workers at Darmstadt in Germany

1997

Earth Science
Kyoto Protocol on reduction of carbon dioxide emissions agreed

1998

Biology
A. Fire and C. Mello first report RNA interference in cells
R. MacKinnin obtains first high-resolution image of a protein channel

Physics
Workers at the neutrino detector Super Kamiokande in Japan observe neutrino oscillations

1999

Astronomy
Chandra X-ray Observatory launched (July 23)
Launch of the Stardust probe (Feb. 7) to collect cometary dust

Chemistry
Element 114 (ununquadium) discovered by workers at Dubna in Russia

2000

Astronomy
Near Earth Asteroid Rendezvous (NEAR) spacecraft investigates the asteroid Eros

Physics
Direct evidence obtained for the tau neutrino by workers at

the Fermi National Accelerator Laboratory near Chicago

2001

Astronomy
Russian Mir space station is de-orbited into the Earth's atmosphere (Mar. 23)

Biology
R. Kornberg et al. publish crystal structure of a yeast RNA polymerase in action

Chemistry
Element 116 (ununhexium) discovered at Dubna in Russia

Earth Science
President Bush rejects the Kyoto Protocol on global warming

Technology
IBM demonstrates a quantum computer

2002

Astronomy
Trans-neptunian object Quaoar, about half the size of Pluto, discovered by C. Trajillo and M. Brown

Earth Science
Earth Summit (World Summit on Sustainable Development) held in Johannesburg

Biology
Publication of the genome sequence of the plant *Arabidopsis thaliana*

Medicine
First outbreak of SARS (severe acute respiratory syndrome) in China

2003

Astronomy
First manned Chinese space flight, headed by Yang Liwei
Galileo spacecraft intentionally crashed into Jupiter
Space shuttle Columbia breaks up during re-entry, killing 7 crew members (Feb. 1)

NASA satellite WMAP produces detailed map of the microwave background, supporting inflationary theory

Biology
The first cloned pet, 'Copy Cat,' born at Texas A&M University

2004

Astronomy
Cassine–Huygens probe investigates Jupiter and Saturn
A new dwarf planet, Sedna, discovered at the Palomar Observatory in California
Transit of Venus across the Sun occurs

Chemistry
Element 113 (ununtrium) and element 115 (ununpentium) reported jointly by workers at Dubna in Russia and at the Lawrence Livermore Laboratory in the U.S.

2005

Astronomy
Deep Impact mission successfully crashes a probe into the comet Temple-1
Space Shuttle Discovery resumes the program following the 2003 Columbia accident

Earth Science
Kyoto Protocol comes into force

Medicine
French surgeons carry out the first human face transplant in Amiens

Paleontology
Two fossil skulls discovered by Richard Leakey in 1967 redated at 195,000 years old, making them the oldest evidence of *Homo sapiens* remains

2006

Astronomy
International Astronomical Union introduces new

definition of a planet and creates the new classification of dwarf planet
Pluto is reassigned as a dwarf planet
A new dwarf planet, Eris, is found, larger then Pluto
NASA spacecraft New Horizons launched (Jan. 19) in mission to Pluto
Stardust mission brings cometary dust samples back to Earth

Chemistry
Element 118 (ununoctium) reported jointly by workers at Dubna in Russia and at the Lawrence Livermore Laboratory in the U.S.

2007

Astronomy
The Nasa space probe Dawn launched (Sept. 27) to investigate Vesta and Ceres in the asteroid belt
Gravity Probe B finds evidence that the Earth affects the curvature of space–time (the geodetic effect)

Biology
US workers report the cloned embryo of a monkey from a skin cell
Remains of a new large birdlike dinosaur, *Gigantoraptor erianesis*, found in the Gobi desert

2008

Astronomy
NASA Messenger spacecraft makes its first flyby of Mercury (Jan. 15)

Biology
C. Venter reports the synthesis from scratch of a complete genome (the bacterium *Mycoplasma genitalium*)

Physics
The Large Hadron Collider scheduled to begin work at CERN

Useful Web Sites

The Alchemy Web Site
http://www.levity.com/alchemy/index.html
A very extensive web site containing over 240 complete alchemical texts and a large amount of introductory and general reference material on alchemy.

Biographies of Astronomers
http://www.astro.uni-bonn.de/~pbrosche/hist_astr/ha_pers.html
A web site at the University of Bonn giving short biographies for about 1,750 astronomers and people connected to astronomy (scholars from other fields who did investigations in astronomy, makers of astronomical instruments, etc.). It includes over 4,000 links to extra information.

Classic Chemistry
http://web.lemoyne.edu/~GIUNTA/papers.html
A web site at Le Moyne College containing the text of a large number of papers from the history of chemistry. It includes papers from the historical papers section of the ChemTeam site: http://dbhs.wvusd.k12.ca.us/webdocs/Chem-History/Classic-Papers-Menu.html.

Discoveries in Chemistry
http://pubs.acs.org/archives/promo/timeline01.html
A timeline of discoveries in chemistry (1870 to 1995) from the American Chemical Society. There is also a timeline of industrial chemistry milestones (1900 to 1999): http://pubs.acs.org/journals/iecred/promo/100th/timeline.html.

ECHO
http://echo.gmu.edu/
A directory to over 5,000 web sites concerning the history of science, technology, and industry. The ECHO (Exploring and Collecting History Online) project is based at George Mason University's Center for History and New Media.

Eric Weisstein's World of Science
http://scienceworld.wolfram.com/biography/
The biography section of the World of Science collection of online science encylopedias, assembled by Eric W. Weisstein with assistance from the internet community.

Fields Medal
http://www.mathunion.org/Prizes/Fields/Prizewinners.html
The International Mathematical Union web site, giving a list of winners of the Fields Medal for mathematics. The list links to biographical information about the people awarded the medal.

History of Biology
http://www.pasteur.fr/recherche/unites/REG/causeries/Antiquity.html
A timeline from the Pasteur Institute.

History of Chemistry
http://www.chemheritage.org/exhibits/exhibits.html
The web site of the Chemical Heritage Foundation in Philadelphia, formed by the American Chemical Society and the American Institute of Chemical Engineers. The site includes selected online exhibits.

History of Physics
http://www.aip.org/history/
The American Institute of Physics Center for History of Physics web site offering archival collections and educational initiatives.

History of Mathematics
http://www.maths.tcd.ie/pub/HistMath/
A web site maintained at the School of Mathematics, Trinity College, Dublin. Among other online resources it includes accounts of the lives and works of seventeenth and eighteenth century mathematicians (and some other scientists), adapted from *A Short Account of the History of Mathematics* by W. W. Rouse Ball (4th Edition, 1908).

History of Medicine
http://www.nlm.nih.gov/onlineexhibitions.html
Online exhibitions and digital projects from the web site of the US National Library of Medicine.

History of Space Exploration
http://www.hq.nasa.gov/office/pao/History/on-line.html
A comprehensive resource on the history of space exploration run by the US National Aeronautics and Space Administration (NASA). It also includes a chronology: http://history.nasa.gov/timeline.html.

Institute and Museum of the History of Science
http://biblioteca.imss.fi.it/index.html
The web site of the Institute and Museum of the History of Science in Florence. It has an online catalogue giving descriptions and photographs of more than 1,200 items on permanent exhibition. There is also a section giving biographical entries for about 400 scientists, inventors, and instrument makers, with links to background information.

National Maritime Museum
http://www.nmm.ac.uk/collections/index.cfm
The National Maritime Museum and Royal Observatory, Greenwich, have a collection of over 2 million objects about the sea, ships, astronomy and time. Collections Online gives access to over 10,000 of these and is frequently updated.

Nobel Prizes
http://nobelprize.org/
The official web site of the Nobel Foundation, listing every Nobel Prize since 1901. It includes biographies, photographs, background information, Nobel Lectures, and other material.

The Oxford Museum of the History of Science
http://www.mhs.ox.ac.uk/
The web site has a number of selected online exhibits. There is also a database (in development) containing information about and photos of almost 15,000 objects in the collections.

Science Timeline
http://www.rsc.org/Chemsoc/Timeline/index.asp
A visual exploration of key events in the history of science on the Royal Society of Chemistry web site. Individual items are illustrated and linked to further information on each subject.

Whipple Museum of the History of Science
http://www.hps.cam.ac.uk/whipple/index.html
The web site of the Whipple Museum of the History of Science in Cambridge. It contains a number of selected online exhibits.

Women in Mathematics
http://www.scottlan.edu/lriddle/women/women.htm
Part of an on-going project at Agnes Scott College in Atlanta, Georgia, to illustrate the numerous achievements of women in the field of mathematics. It includes biographical essays on a selection of women mathematicians, as well as additional resources about women in mathematics.

Women in Science
http://library.thinkquest.org/20117/mainbio.html
A selection of biographies featuring women scientists. It is sponsored by the Oracle Education Foundation.

Index